List of the Elements with their Symbols and Atomic Masses*

Element	Symbol	Atomic number	Atomic mass†	Element	Symbol	Atomic number	Atomic mass†
Actinium	Ac	89	(227)	Neon	Ne	10	20.18
Aluminum	Al	13	26.98	Neptunium	Np	93	(237)
Americium	Am	95	(243)	Nickel	Ni	28	58.69
Antimony	Sb	51	121.8	Niobium	Nb	41	92.91
Argon	Ar	18	39.95	Nitrogen	N	7	14.01
Arsenic	As	33	74.92	Nobelium	No	102	(253)
Astatine	At	85	(210)	Osmium	Os	76	190.2
Barium	Ba	56	137.3	Oxygen	O	8	16.00
Berkelium	Bk	97	(247)	Palladium	Pd	46	106.4
Beryllium	Be	4	9.012	Phosphorus	P	15	30.97
Bismuth	Bi	83	209.0	Platinum	Pt	78	195.1
Boron	B	5	10.81	Plutonium	Pu	94	(242)
Bromine	Br	35	79.90	Polonium	Po	84	(210)
Cadmium	Cd	48	112.4	Potassium	K	19	39.10
Calcium	Ca	20	40.08	Praseodymium	Pr	59	140.9
Californium	Cf	98	(249)	Promethium	Pm	61	(147)
Carbon	C	6	12.01	Protactinium	Pa	91	(231)
Cerium	Ce	58	140.1	Radium	Ra	88	(226)
Cesium	Cs	55	132.9	Radon	Rn	86	(222)
Chlorine	Cl	17	35.45	Rhenium	Re	75	186.2
Chromium	Cr	24	52.00	Rhodium	Rh	45	102.9
Cobalt	Co	27	58.93	Rubidium	Rb	37	85.47
Copper	Cu	29	63.55	Ruthenium	Ru	44	101.1
Curium	Cm	96	(247)	Samarium	Sm	62	150.4
Dysprosium	Dy	66	162.5	Scandium	Sc	21	44.96
Einsteinium	Es	99	(254)	Selenium	Se	34	78.96
Erbium	Er	68	167.3	Silicon	Si	14	28.09
Europium	Eu	63	152.0	Silver	Ag	47	107.9
Fermium	Fm	100	(253)	Sodium	Na	11	22.99
Fluorine	F	9	19.00	Strontium	Sr	38	87.62
Francium	Fr	87	(223)	Sulfur	S	16	32.07
Gadolinium	Gd	64	157.3	Tantalum	Ta	73	180.9
Gallium	Ga	31	69.72	Technetium	Tc	43	(99)
Germanium	Ge	32	72.59	Tellurium	Te	52	127.6
Gold	Au	79	197.0	Terbium	Tb	65	158.9
Hafnium	Hf	72	178.5	Thallium	Tl	81	204.4
Helium	He	2	4.003	Thorium	Th	90	232.0
Holmium	Ho	67	164.9	Thulium	Tm	69	168.9
Hydrogen	H	1	1.008	Tin	Sn	50	118.7
Indium	In	49	114.8	Titanium	Ti	22	47.88
Iodine	I	53	126.9	Tungsten	W	74	183.9
Iridium	Ir	77	192.2	Unnilennium	Une	109	(266)
Iron	Fe	26	55.85	Unnilhexium	Unh	106	(263)
Krypton	Kr	36	83.80	Unniloctium	Uno	108	(265)
Lanthanum	La	57	138.9	Unnilpentium	Unp	105	(260)
Lawrencium	Lr	103	(257)	Unnilquadium	Unq	104	(257)
Lead	Pb	82	207.2	Unnilseptium	Uns	107	(262)
Lithium	Li	3	6.941	Uranium	U	92	238.0
Lutetium	Lu	71	175.0	Vanadium	V	23	50.94
Magnesium	Mg	12	24.31	Xenon	Xe	54	131.3
Manganese	Mn	25	54.94	Ytterbium	Yb	70	173.0
Mendelevium	Md	101	(256)	Yttrium	Y	39	88.91
Mercury	Hg	80	200.6	Zinc	Zn	30	65.39
Molybdenum	Mo	42	95.94	Zirconium	Zr	40	91.22
Neodymium	Nd	60	144.2				

* All atomic masses have four significant figures. These values are recommended by the Committee on Teaching of Chemistry, International Union of Pure and Applied Chemistry.

† Approximate values of atomic masses for radioactive elements are given in parentheses.

PRATEEK LALA (IF FOUND, PLEASE CALL 738-0466)

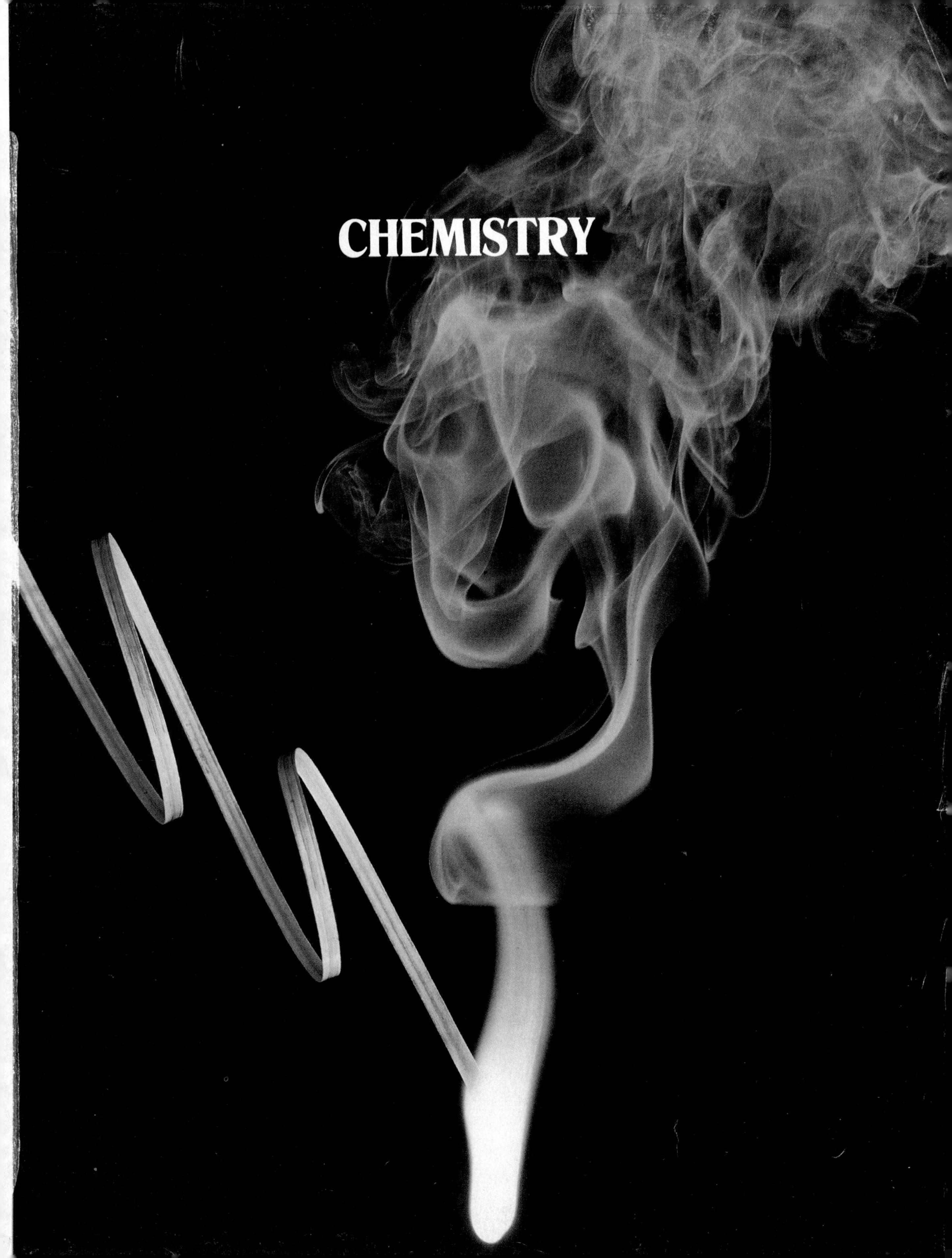
CHEMISTRY

CHEMISTRY

FOURTH EDITION

Raymond Chang

WILLIAMS COLLEGE

McGRAW-HILL, INC.

NEW YORK ST. LOUIS SAN FRANCISCO AUCKLAND BOGOTÁ
CARACAS LISBON LONDON MADRID MEXICO
MILAN MONTREAL NEW DELHI PARIS SAN JUAN
SINGAPORE SYDNEY TOKYO TORONTO

CHEMISTRY

4 5 6 7 8 9 0 VNH VNH 9 5 4 3 2

ISBN 0-07-010518-9

Acknowledgments appear on pages I 14–I 16, and on this page by reference.

This book was set in Times Roman by York Graphic Services, Inc.
The editors were Randi B. Rossignol, Denise L. Schanck, and Jack Maisel;
the designer was Hermann Strohbach;
the production supervisor was Janelle S. Travers.
Von Hoffmann Press, Inc., was printer and binder.

Cover photograph: Ken Karp. The cover photo shows magnesium burning in air. When magnesium is ignited, it burns instantly and dramatically.

Library of Congress Cataloging-in-Publication Data

Chang, Raymond.
Chemistry/Raymond Chang.—4th ed.
p. cm.
ISBN 0-07-010518-9
1. Chemistry I. Title.
QD31.2.C37 1991
540—dc20 90-5718

ABOUT THE AUTHOR

Raymond Chang was born in Hong Kong and grew up in Shanghai, China, and Hong Kong. He received his B.Sc. degree in chemistry from London University, England, and his Ph.D. in chemistry from Yale University. After doing postdoctoral research at Washington University and teaching for a year at Hunter College, he joined the chemistry department at Williams College, where he has taught since 1968. Professor Chang has written books on spectroscopy, physical chemistry, and industrial chemistry and has coauthored books on the Chinese language and a novel for juvenile readers.

For relaxation, Professor Chang maintains a forest garden, plays tennis, and practices the violin.

CONTENTS

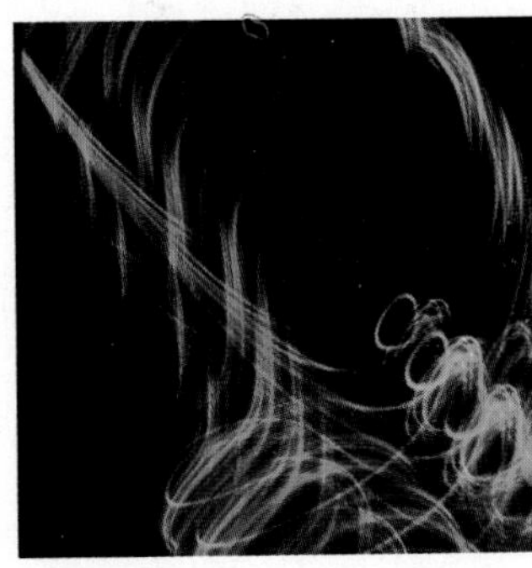

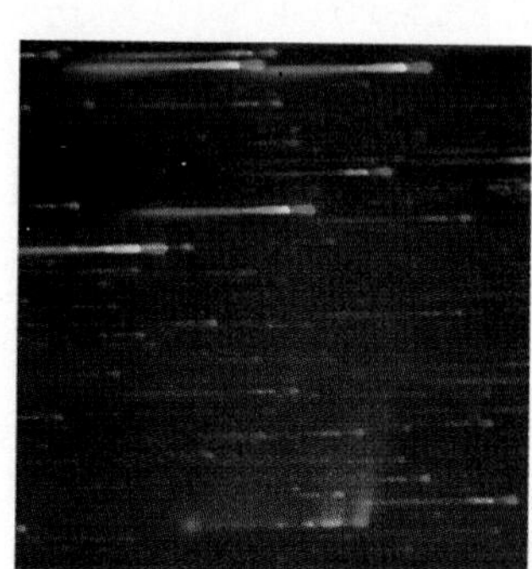

PHILLIPS'
MILK OF MAGNESIA
PARSONS'
AMMONIA
HEINZ
VINEGAR
Alka-Seltzer

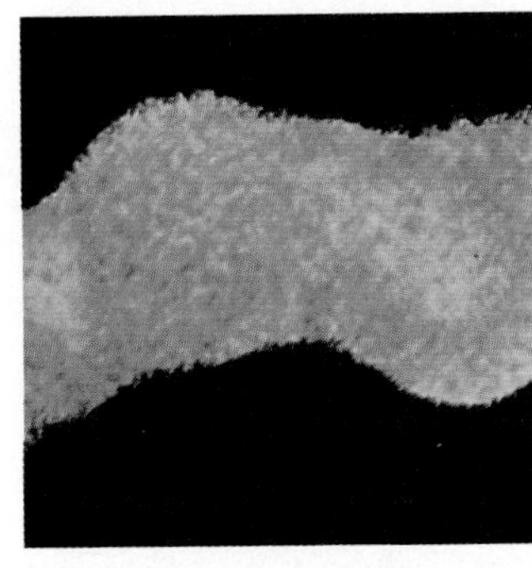

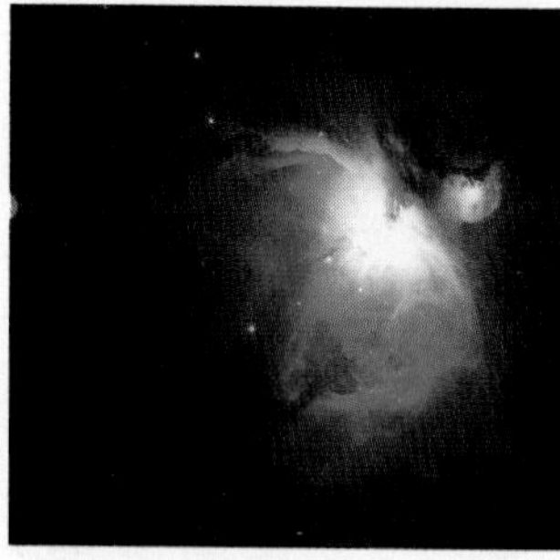

PREFACE

This text is for students taking a full-year general chemistry course. As with previous editions, my aim in preparing this edition has been to present chemistry in a manner that is understandable, interesting, and meaningful to the beginning student. I have tried to strike a balance between theory and application, and whenever possible, to illustrate the basic principles with everyday examples. The encouraging feedback from numerous users have convinced me that I am on the right track. Therefore, in this edition I have retained the basic format of the text. Some of the important changes are as follows:

- Chemical reactions are now discussed in two chapters (Chapters 3 and 4). The discussion is focused on three major types of reactions: precipitation, oxidation-reduction, and acid-base.
- Gases (Chapter 5) now precedes Thermochemistry (Chapter 6).
- Nuclear Chemistry (Chapter 23) has been moved just before Organic Chemistry.
- A new chapter, Industrial Chemistry (Chapter 26), has been added. This chapter surveys the source of raw materials used in the chemical industry, important inorganic and organic chemicals, and the impact of the chemical industry on the environment.
- A new type of worked example, Applied Chemistry Example, now appears in many chapters. These multiconcept problems give students additional problem-solving practice while exposing them to interesting and important industrial processes.
- Many new end-of-chapter problems have been added to each chapter.

The main features of the text are described below.

Organization. The chapters of this edition follow a generally accepted sequence that users of the text have found to correspond well with their individual syllabuses while still allowing for flexibility in class assignments. After an introductory chapter, Chapter 2 studies the basic concepts of atoms, molecules, and ions. It also has a useful section on inorganic nomenclature to coordinate with laboratory work. Chapters 3 and 4 deal with the nature of precipitation, oxidation-reduction, and acid-base reactions and their mass relationships. The placement of gases (Chapter 5) before thermochemistry (Chapter 6) makes it easier to discuss thermochemical processes involving gases and the first law of thermodynamics. Chapter 7 takes up the electronic structure of atoms and provides a logical introduction to the discussion of periodic relationships among the elements in the following chapter. Chapter 9 presents an elementary treatment of ionic and covalent bonding, while Chapter 10 concentrates on molecular structure and modern theories of covalent bonding. Chapter 11 emphasizes the effects of intermolecular forces on the properties of liquids and solids. The following chapter discusses the physical properties of solutions.

As a group, Chapters 13–19 deal with the quantitative aspects of chemistry. Even the most abstract topics, such as thermodynamics, are presented at a level that a beginning student can understand. Chapters 20–22 cover inorganic descriptive chemistry while Chapter 23 deals with nuclear processes. Chapters 24–25 introduce organic chemistry and synthetic and natural polymers. The theme of the final chapter is on industrial chemistry, a very important topic that is often missing in the chemistry curriculum.

Problem-Solving Pedagogy. Problem solving is the nemesis of both students and professors. The pedagogy in this book incorporates the ideas of reviewers who see a need for additional support in this quantitative area of chemistry. Students are encouraged to use a thoughtful and logical approach to solving problems. Frequently, they are asked to examine the reasonableness of the answer which is an important but often neglected procedure.

In this edition, about 250 worked examples are integrated into the text discussion by a statement describing the particular concept, technique, or calculation that is to be illustrated. This procedure gives students some idea of the type of problem to be solved *before* they have to confront the logic necessary to solve it. Each example ends with a reference to similar problems at the end of the chapter so that students have the opportunity to practice a particular problem-solving technique.

The end-of-chapter exercises consist of about 2,300 review questions, problems, and miscellaneous problems. Prominent headings in the exercise section indicate the major topics and concepts covered in the chapter. Under each heading are review questions and problems.

The review questions cover the concepts that explain the "why" of chemistry. To answer them, the student must apply the conceptual logic needed to perform the calculations. The problems cover the quantitative, experimental "how" of chemistry that yields numerical results. They test the students' ability to apply the logic they have just reviewed as they perform specific calculations. The miscellaneous problems include more challenging and multi-concept problems. They give students practice in identifying the concept, topic, or technique to be applied—just as they would have to do on a test or exam.

Readability. I have benefited from users' comments regarding my explanations and descriptions of each and every concept, term, process, and reaction. The narrative has been revised wherever appropriate to provide smooth transitions from topic to topic, to define complex terms in a clear manner, and to explain difficult concepts carefully. As an example of how an analogy is used to introduce a chemical concept, see the comparison between downhill skiing and dynamic chemical equilibrium on p. 596.

Real-World Applications. One of the joys of learning chemistry is seeing how chemical principles can be applied to everyday experience. The Chemistry in Action sections show the relevance of chemistry to biological, medical, technological, and engineering fields, as well as current news topics. Several new essays have been added and a number of others have been revised and updated in this edition.

Balance of Theory and Descriptive Chemistry. There is a generous amount of information in this text about the reactions that take place among elements and compounds. This descriptive material provides sound support for the principles on which chemistry is based and clearly substantiates the important interrelationship between experimentation and fact. I have incorporated descriptive chemistry meaningfully into the discussion of principles, as I hope will be evident, especially in Chapters 3, 4, 8, and 20–22.

Attractive and Functional Format. The switch to 5-color design in the last edition has helped students to visualize the appearance of compounds and various chemical processes. Many new full-color photographs and line drawings have been added in this edition. A number of marginal arts have been introduced to enhance discussion and to accompany worked examples. Wherever appropriate, there is consistent use of color to illustrate similar concepts.

The supplements available for use with this text are:

Student Solutions Manual by Jerry L. Mills and Raymond Chang. This supplement contains complete solutions for about two-third of the problems in the text. Directed toward a student audience, this innovative manual includes a three-pronged approach to problem analysis. The first part is an introductory discussion with text references, comments, and hints. The second part includes questions that test the student's understanding of the logic behind solving the problem. The third part contains afterthoughts on the problem and additional questions to help students deal with similar type of problems.

Microscale and Macroscale Experiments for General Chemistry by Jerry L. Mills and Michael D. Hampton. This manual contains over 40 general chemistry experiments based on both macroscale and microscale laboratory techniques. Special attention is given to the cost and safety of experiments and the safe disposal of wastes.

Study Guide by Kenneth W. Watkins. This valuable ancillary includes study objectives, a review of important concepts, detailed methods for solving problems where appropriate, true-false questions, and self-test questions for each chapter.

Instructor's Manual by Raymond Chang and Jerry L. Mills. This manual includes for each chapter a brief explanation of the chapter contents, learning goals, a list of more challenging problems, and solutions to all the problems in the text.

Test Bank. This print test file contains over 2000 exam questions in both multiple-choice and short-answer formats.

R-H Test. A computerized version of the print Test Bank, available for IBM PC and Macintosh computers.

Overhead Transparencies. About 200 full-color acetates of important illustrations in the text.

Chemistry at Work Videodisc. This valuable teaching tool provides instant access to all the tables, illustrations, and photos from the text in addition to 50 live laboratory demonstrations.

Micro Guide. Tutorial software specifically correlated to the fourth edition. Available in IBM and Macintosh formats.

Acknowledgments

I would like to thank the following individuals, whose comments on the third edition and/or the manuscript for the fourth edition have been so important in shaping this revision:

RICHARD C. ARMSTRONG, Madison Area Technical College
DAVID B. ARNOLD, Widener University
PETER BAINE, California State University—Long Beach
JOHN E. BAUMAN, JR., University of Missouri—Columbia
RAYMOND F. BOGUCKI, University of Hartford
WILLIAM H. BREAZEALE, Francis Marion College
ALAN CAMPION, University of Texas—Austin
ROBERT C. COSTELLO, University of South Carolina—Sumter
JAMES E. DAVIS, Harvard University
MICHAEL I. DAVIS, The University of Texas at El Paso
CLYDE K. EDMISTON, University of Wyoming
SIDNEY EMERMAN, Kingsborough Community College of CUNY
GORDON J. EWING, New Mexico State University—Las Cruces
ROBERT D. FLANAGAN, Diablo Valley College
RAYMOND C. FORT, University of Maine—Orono
DONNA FREDEEN, South Connecticut State University
H. L. FRISCH, State University of New York at Albany
HENRY GEHRKE, JR., South Dakota State University
PAULA M. GETZIN, Kean College of New Jersey
KIM E. GILMORE, University of Idaho
L. PETER GOLD, Pennsylvania State University
FRANK J. GOMBA, United States Naval Academy
DANIEL T. HAWORTH, Marquette University
JOHN F. HELLING, University of Florida—Gainesville
DAVID C. HILDEBRAND, South Dakota State University
JEFFREY A. HURLBUT, Metropolitan State College
RONALD C. JOHNSON, Emory University
STANLEY N. JOHNSON, Orange Coast College
MURRAY V. JOHNSTON, University of Colorado at Boulder
NADYA KELLER, Northwestern State University of Louisiana
LESLIE N. KINSLAND, University of Southwestern Louisiana
DAVID F. KOSTER, Southern Illinois University at Carbondale
PHILIP S. LAMPREY, University of Lowell
LARRY D. MARTIN, Morningside College
GEORGE E. MILLER, University of California—Irvine
JERRY L. MILLS, Texas Tech University
HUGGINS Z. MSIMANGA, Kennesaw College
M. ALI NABI RAHNI, Pace University—Pleasantville
JEROME D. ODOM, University of South Carolina
WILLIAM H. ORTTUNG, University of California—Riverside
WAYNE H. PEARSON, United States Naval Academy
BARBARA PEREZ, El Camino Community College
JACK E. POWELL, Iowa State University of Science and Technology
THOMAS PYNADATH, Kent State University
CHARLES R. RUSS, University of Maine—Orono
MARTHA E. RUSSELL, Iowa State University of Science and Technology
SAUL I. SHUPACK, Villanova University
JOYCE E. SHADE, United States Naval Academy
CHARLES J. THOMAN, Stephen F. Austin University
BOYD WAITE, United States Naval Academy

DARRELL WATSON, GMI Engineering and Management Institute
JOHN W. WAY, Widener University
DON B. WESER, Frostburg State College
STANLEY M. WILLIAMSON, University of California—Santa Cruz

I would also like to thank George Fleck, Smith College; Michael D. Hampton, University of Central Florida; John S. Hutchinson, Rice University; Stephen McMillan, Clarkson University; James L. Stewart, Cypress College; and Naola VanOrden, Sacramento City College for their suggestions for end-of-chapter problems, Wayne R. Tikkanen, California State University at Los Angeles for his contribution to applied chemistry examples, and Cynthia Friend, Harvard University for useful source of photographs. Thanks are due Kerry Cho and Jessica Walker for checking the end-of-chapter problems.

In addition, it is a pleasure to acknowledge the enthusiastic support and assistance given to me by the following members of McGraw-Hill's College Division: Seibert Adams, Kirk Emry, Barry Fetterolf, Mel Haber, Joan McNamara, June Smith, and Jane Vaicunas. In particular, I would like to mention Kathy Bendo, Jack Maisel, and Janelle Travers who are responsible for the smooth production of the book, Judith Duguid for copyediting, and Hermann Strohbach for design. My deep appreciation goes to Denise Schanck for guidance and encouragement and Randi Rossignol for her inputs and suggestions that have improved the text in so many ways.

I would greatly appreciate your comments and suggestions.

Raymond Chang
Department of Chemistry
Williams College
Williamstown, MA 01267

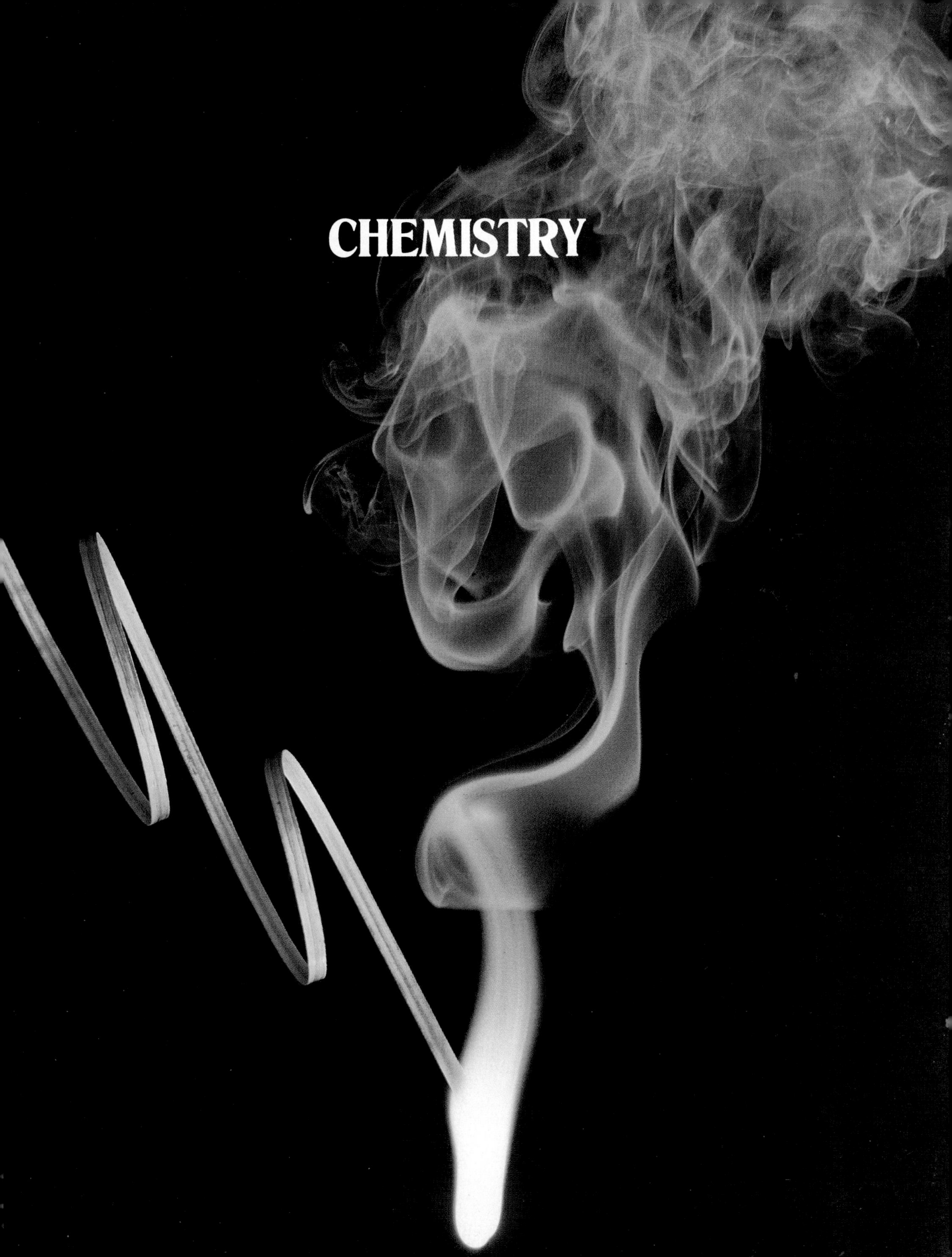

CHEMISTRY

Opposite page: The photo shows a student performing a simulated experiment at a work station equipped with a touch-sensitive computer screen. This is one of many creative uses of technology in the chemistry curriculum.

C H A P T E R 1

TOOLS OF CHEMISTRY

Chemistry is an active and continually growing science that has vital importance to our world, in both the realm of nature and the realm of society. Its roots are ancient, but as we will soon see, chemistry is every bit a modern science, an experimental science that rests on a foundation of precise vocabulary and established methods.

1.1 Chemistry Today

As recently as thirty years ago, the word "chemist" conjured up visions of someone in a white coat, working busily in the laboratory, with test tubes and beakers in hand and Bunsen burners aglow. This image, dear to the hearts of film makers, is outdated. Great advances made during the last few decades have transformed chemistry. The science of ***chemistry*** can still be defined as *the study of the materials that make up the universe and the changes these materials undergo*. But not all chemists today work in laboratories full of bubbling solutions, and we can no longer describe what a chemist does in just a few words. For example, a chemist may develop new drugs or agricultural chemicals, measure the speed of chemical and biological reactions, or probe the structure and function of protein molecules. Some chemists don't even wear white coats or spend much time in the laboratory making solids appear miraculously from solutions. Today's chemist may spend at least part of his or her time sitting before a computer studying molecular structure and properties, or using sophisticated electronic equipment to analyze pollutants from auto emissions or toxic substances in the environment. In fact, chemistry is so diversified that a person would no longer say simply, "I am a chemist." He or she would be identified as, for example, an analytical chemist, a physical chemist, an inorganic chemist, a biochemist, an organic chemist, an agricultural chemist, or an environmental chemist.

Some science historians believe that the word "chemistry" derives from the Greek word *chemeia,* meaning "the art of metalworking." Obviously, modern chemistry involves a great deal more than this. Chemistry has become an interdisciplinary science, and today no scientific work can escape chemistry. Many of the frontiers in medicine and biology are being explored at the level of atoms and molecules, which are the tiny bits of matter on which the study of chemistry is based. Chemists are now involved in the design and synthesis of drugs to treat a variety of diseases and to combat cancer. In addition, there is great public and governmental concern with keeping our environment clean and with finding new sources of energy. Such problems can be solved only by the inventive application of what we know about the chemistry of the systems involved. And most industries, whatever their products, are dependent on the work of chemists. For example, through years of research, chemists have learned to manufacture polymers (molecules that contain thousands of atoms) of various sizes and shapes that are used in the clothing, cooking utensils, and toys in our households. Chemists devise new products and better ways to make old ones. They monitor the composition of raw materials that enter a manufacturing plant and check the quality of the finished product that goes out.

As with any discipline, we must learn the necessary vocabulary before we can begin to understand and appreciate how the principles of chemistry are applied to practical systems. The first 12 chapters of this book provide the basic definitions in chemistry, as well as the tools we will need to study the quantitative relationships in chemical reactions, the structure and properties of atoms and molecules, and the forces responsible for the existence of gases, liquids, and solids. Once we have acquired a basic knowledge of chemistry, we can take a closer look at many different topics dealing with the physical and chemical properties of matter. Throughout this book we will see that chemistry, by virtue of its interdisciplinary nature, is concerned with subjects ranging from the synthesis of ammonia, the cracking of petroleum, the depletion of the ozone in the stratosphere, high-temperature superconductors, and the understanding of biological processes to dental filling, lasers, photography, and even ice skating.

1.2 Science and Its Methods

All sciences, including the social sciences, employ variations of what is called the ***scientific method***—*a systematic approach to research*. For example, a psychologist who wants to know how noise affects people's ability to learn chemistry and a chemist interested in measuring the heat given off when hydrogen gas burns in air would follow roughly the same procedure in carrying out their investigations. In this procedure, the first step involves carefully defining the problem. The next step includes performing experiments, making careful observations, and collecting bits of information about the system. The bits of information are called "data." The word "system" here means that part of the universe that is under investigation. Referring to the two examples above, the system may be a group of college students or a mixture of hydrogen and air. The information obtained may be both ***qualitative***, *consisting of general observations about the system*, and ***quantitative***, *comprising numbers obtained by various measurements of the system*.

When enough information has been gathered, a ***hypothesis***—*a tentative explanation for a set of observations*—can be formulated. Further experiments are devised to test the validity of the hypothesis in as many ways as possible.

After a large amount of data has been collected, it is often desirable to summarize the information in a concise way. A ***law*** is *a concise verbal or mathematical statement of a relationship between phenomena that is always the same under the same conditions*. As mentioned earlier, hypotheses provide only tentative explanations that must be tested by many experiments. If they survive such tests, hypotheses may develop into theories. A ***theory*** is *a unifying principle that explains a body of facts and those laws that are based on them*. Theories too are constantly being tested. If a theory is proved incorrect by experiment, then it must be discarded or modified so that it becomes consistent with experimental observations.

Perhaps the best way to appreciate the scientific method is to follow the steps of scientists in solving a particular problem. One example is the question of what happened to the dinosaurs, as described in the Chemistry in Action section on p. 30.

Scientific progress is seldom, if ever, made in a rigid, step-by-step fashion. Sometimes a law precedes a theory; sometimes it is the other way around. Two scientists may start working on a project with exactly the same objective, but will take drastically different approaches. They may be led in vastly different directions. Scientists are, after all, human beings, and their modes of thinking and working habits are very much influenced by their background, training, and personalities.

When we study the development of science over the last two to three hundred years, we find that it is often irregular and, sometimes, even illogical. Great discoveries are usually the result of the cumulative contributions and experience of many workers, even though the credit for formulating a theory or a law is usually given to only one individual. There is also, of course, an element of luck involved in scientific discoveries. But it has been said that "chance favors the prepared mind." It takes an alert and well-trained person to recognize the significance of an accidental discovery and to take full advantage of the fortunate finding. More often than not, the public learns only of spectacular scientific breakthroughs. For every success story, however, hundreds of scientists may have spent years working on projects that seemed promising at first but turned out to be dead ends, or in which positive achievements were made only after many wrong turns and at such a slow pace that they went unheralded. It is the love of the search that keeps many scientists in the laboratory.

The Study of Chemistry

Chemistry is largely an experimental science. Most chemists work in a laboratory of one kind or another. In a broad sense we can view chemistry on three levels. The first level is *observation*. At this level the chemist observes what actually takes place in an experiment—a rise in temperature, a change in color, evolution of a gas, and so on. The second level is *representation*. The chemist records and describes the experiment in scientific language using shorthand symbols and equations. This shorthand helps to simplify the description and establishes a common base with which chemists can communicate with one another. The third level is *interpretation,* meaning that the chemist attempts to explain the phenomenon (Figure 1.1).

All of us have witnessed the rusting of iron at one time or another (Figure 1.2). This is a process that takes place in the *macroscopic world,* where we deal with things that can be seen, touched, weighed, and so on. If we were studying the rusting of iron as a chemistry project, our next step would be to describe this process with a "chemical equation" that tells us how rust is formed from iron, oxygen gas, and water, under a given set of conditions. Lastly, we would address questions like "What actually happens when iron rusts?" and "Why does iron rust but gold does not, under similar

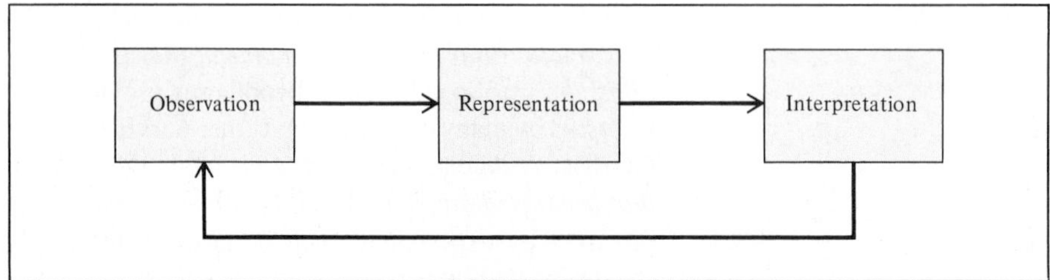

Figure 1.1 *The three levels of studying chemistry and their relationships. Observation deals with events in the macroscopic world; atoms and molecules constitute the microscopic world. Representation is a scientific shorthand for describing an experiment in symbols and chemical equations. Chemists use their knowledge of atoms and molecules to explain an observed phenomenon.*

Figure 1.2 *A badly rusted car. Corrosion of iron costs the U.S. economy tens of billions of dollars every year.*

conditions?'' To answer these and other related questions, we have to know the behavior of the fundamental units of substances, which are atoms and molecules. Because atoms and molecules are extremely small compared with macroscopic objects, the interpretation of an observed phenomenon takes us into the *microscopic world.*

The study of chemistry requires that we consider both the macroscopic and the microscopic worlds. The data for chemical investigations most often come from large-scale phenomena and observations. But the testable hypotheses, theories, and explanations, which make chemistry an experimental science, are frequently expressed in terms of the unseen and partially imagined microscopic world of atoms and molecules.

Some students find it confusing at first that the instructor or the textbook seems to be continually shifting back and forth between these two distinctly different ways of considering our physical universe. It has been said that often a chemist *sees* one thing (in the macroscopic world) and *thinks* another (in the microscopic world). For example, a chemist will look at the rusted car in Figure 1.2, and think about the basic properties of individual units of iron and how these units interact to produce the observed change. Welcome to this fascinating ''dual world'' of chemistry!

1.3 Some Basic Definitions

Words that we use in everyday life often take on a new meaning in a scientific context. In this course you will soon learn how a number of terms, some familiar and some unfamiliar, are used in chemistry. These terms are usually best introduced and explained as they are needed, in their proper context. However, a few of them are so important and so fundamental that they are necessary to understanding nearly all the subjects discussed in this book. These terms are briefly introduced and defined here.

Matter

Anything that occupies space and has mass is called ***matter.*** Matter is all around us. Matter includes things we can see and touch (such as water, earth, and trees) as well as things we cannot (such as air).

Mass and Weight

Mass is *a measure of the quantity of matter in an object.* The terms ''mass'' and ''weight'' are often used interchangeably, although, strictly speaking, they refer to different quantities. In scientific language, ***weight*** refers to *the force that gravity exerts on an object.* An apple that falls from a tree is pulled downward by Earth's gravity. The mass of the apple is constant and does not depend on its location, but its weight does. For example, on the surface of the moon the apple would weigh only one-sixth what it does on Earth, since the moon's gravity is only one-sixth that of Earth. This is why astronauts are able to jump about rather freely on the moon's surface despite their bulky suits and equipment. The mass of an object can be determined readily with a balance, and this process, oddly, is called weighing.

Substances and Mixtures

A ***substance*** is *a form of matter that has a definite or constant composition* (the number and type of basic units present) *and distinct properties.* Examples include water, am-

(a)

(b)

Figure 1.3 *Separating iron filings from a heterogeneous mixture. The same technique is used on a larger scale to separate iron and steel from nonmagnetic objects such as aluminum, glass, and plastics.*

monia, table sugar (sucrose), gold, oxygen, and so on. Substances differ from one another in composition and can be identified by their appearance, smell, taste, and other properties. At present, the number of known substances exceeds 5 million, and the list of new substances is growing rapidly.

A ***mixture*** is *a combination of two or more substances in which the substances retain their identity*. Some familiar examples are air, soft drinks, milk, and cement. Mixtures do not have constant compositions; samples of air collected over two different cities will probably have different compositions, as a result of their differences in altitude, pollution, and so on.

Mixtures are either homogeneous or heterogeneous. When a spoonful of sugar dissolves in water, *the composition of the mixture,* after sufficient stirring, *is the same throughout the solution*. This solution is a ***homogeneous mixture.*** If sand is mixed with iron filings, however, the result is a heterogeneous mixture (Figure 1.3). In a ***heterogeneous mixture,*** *the individual components remain physically separated and can be seen as separate components*. Any mixture, whether homogeneous or heterogeneous, can be put together and then separated by physical means into pure components without changing the identity of the components. Thus, sugar can be recovered from the homogeneous mixture we described above by evaporating the solution to dryness. Condensing the water vapor that comes off will give us back the water component. We can use a magnet to remove the iron filings from the sand, since sand is not attracted to the magnet [see Figure 1.3(b)]. After separation, there will have been no change in the composition of the substances making up the mixture.

Physical and Chemical Properties

Substances are characterized by their individual and sometimes unique properties. The color, melting point, boiling point, and density of a substance are examples of its physical properties. A ***physical property*** *can be measured and observed without changing the composition or identity of a substance*. For example, we can measure the melting point of ice by heating a block of ice and recording the temperature at which the ice is converted to water. But since water differs from ice only in appearance and not in composition, this is a physical change; we can freeze the water to recover the original ice. Therefore, the melting point of a substance is a physical property. Similarly, when we say that helium gas is lighter than air, we are referring to a physical property. On the other hand, the statement "Hydrogen gas burns in oxygen gas to form water" describes a ***chemical property*** of hydrogen because *in order to observe this property we must carry out a chemical change,* in this case burning. After the change, the original chemical substance, the hydrogen gas, will have vanished, and all that will be left is a different chemical substance, water. We *cannot* recover the hydrogen from the water by a physical change such as boiling or freezing the water.

Every time we hard-boil an egg for breakfast, we are causing chemical changes. When subjected to a temperature of about 100°C, the yolk and the egg white undergo reactions that alter not only their physical appearance but their chemical makeup as well. When eaten, the egg is altered again by substances in the stomach called *enzymes*. This digestive action is another example of a chemical change. The specific way such a process is carried out depends on the chemical properties of the specific enzymes and of the food involved.

All measurable properties of matter fall into two categories: extensive properties and intensive properties. The measured value of an ***extensive property*** *depends on how much matter is being considered.* Mass, length, and volume are extensive properties. More matter means more mass. Values of the same extensive property can be added together. For example, two pieces of copper wire will have a combined mass that is the sum of the two separate wires, and the volume occupied by the water in two beakers is the sum of the volumes of the water in each of the individual beakers. The value of an extensive quantity depends on the amount of matter.

The measured value of an ***intensive property*** *does not depend on how much matter is being considered.* Temperature is an intensive property. Suppose that we have two beakers of water at the same temperature. If we combine them to make a single quantity of water in a larger beaker, the temperature of the larger quantity of water will be the same. Unlike mass and volume, temperature is not additive.

Atoms and Molecules

Atoms and molecules will be defined formally in Chapter 2. Here you simply need to know that *all* matter is composed of atoms of different kinds, combined in many different ways. *Atoms* are the smallest possible units of a substance. *Molecules* are made up of atoms held together by special forces.

Elements and Compounds

Substances can be either elements or compounds. An ***element*** is *a substance that cannot be separated into simpler substances by chemical means.* At present, 109 elements have been positively identified. Eighty-three of them occur naturally on Earth. The others have been created by scientists in nuclear reactions (about which you will learn more later).

Elements are indicated by symbols that are combinations of letters. The first letter of the symbol for an element is *always* capitalized, but the second and third letters are *never* capitalized. For example, Co is the symbol for the element cobalt, whereas CO is the formula for the molecule carbon monoxide. Table 1.1 shows some common elements and their symbols. The symbols of some elements are derived from their Latin

Table 1.1 Some common elements and their symbols

Name	Symbol	Name	Symbol	Name	Symbol
Aluminum	Al	Fluorine	F	Oxygen	O
Arsenic	As	Gold	Au	Phosphorus	P
Barium	Ba	Hydrogen	H	Platinum	Pt
Bismuth	Bi	Iodine	I	Potassium	K
Bromine	Br	Iron	Fe	Silicon	Si
Calcium	Ca	Lead	Pb	Silver	Ag
Carbon	C	Magnesium	Mg	Sodium	Na
Chlorine	Cl	Manganese	Mn	Sulfur	S
Chromium	Cr	Mercury	Hg	Tin	Sn
Cobalt	Co	Nickel	Ni	Tungsten	W
Copper	Cu	Nitrogen	N	Zinc	Zn

names—for example, Au from *aurum* (gold), Fe from *ferrum* (iron), and Na from *natrium* (sodium)—while most of them can be derived from their English names. Special three-letter element symbols have been proposed for the most recently synthesized elements.

As we mentioned earlier, water can be formed by burning hydrogen in oxygen. The atoms of hydrogen and oxygen combine to form water, which has properties that are distinctly different from those of the starting materials. Thus, water is an example of a ***compound,*** *a substance composed of atoms of two or more elements chemically united in fixed proportions.* In every water unit, there are two H atoms and one O atom. This composition does not change, regardless of whether the water is in the United States, in Outer Mongolia, or on Mars. Unlike mixtures, compounds cannot be separated into their pure components, which are the atoms of the elements present, except by chemical methods.

Elements, compounds, and other categories of matter are summarized in Figure 1.4.

1.4 Chemical Elements and the Periodic Table

The beginning of a systematic arrangement of the elements occurred in the nineteenth century, when chemists began to realize that many elements show very strong similarities to one another and demonstrate regularities in their physical and chemical behavior. This knowledge led to the development of the ***periodic table****—a tabular arrangement of the elements* (Figure 1.5). *The elements in a column of the periodic table* are known as a ***group*** or ***family.*** *Each horizontal row of the periodic table* is called a ***period.***

The elements can be divided into three categories—metals, nonmetals, and metalloids. A ***metal*** is *a good conductor of heat and electricity.* With the exception of mercury (which is a liquid), all metals are solids at room temperature (usually designated as 25°C). A ***nonmetal*** is *usually a poor conductor of heat and electricity, and it has more varied physical properties than a metal.* A **metalloid** *has properties that fall between those of metals and nonmetals.*

For convenience, some element groups have special names. *The Group 1A elements (Li, Na, K, Rb, Cs, and Fr)* are called ***alkali metals,*** and *the Group 2A elements (Be, Mg, Ca, Sr, Ba, and Ra)* are called ***alkaline earth metals.*** *Elements in Group 7A (F,*

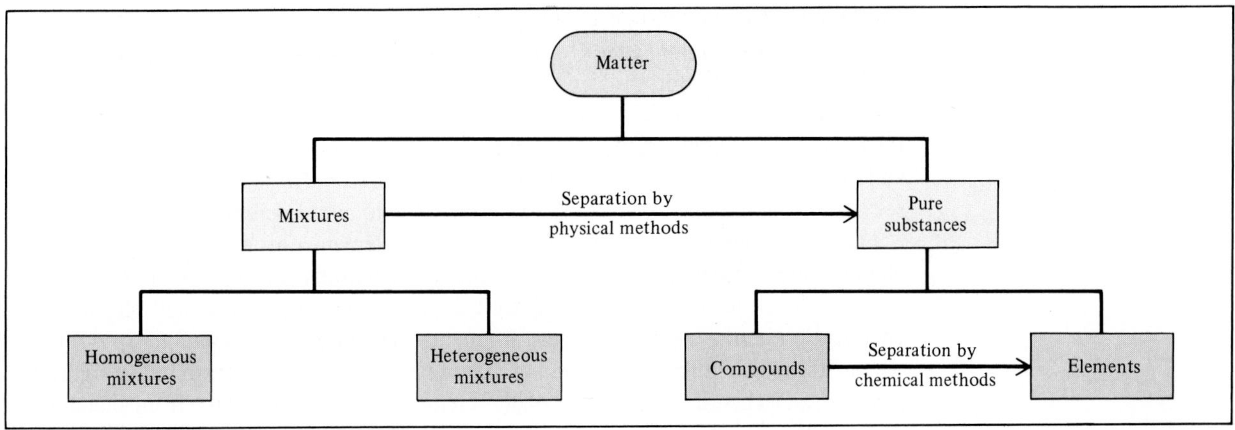

Figure 1.4 *Classification of matter and relationships between mixtures and pure substances and between compounds and elements.*

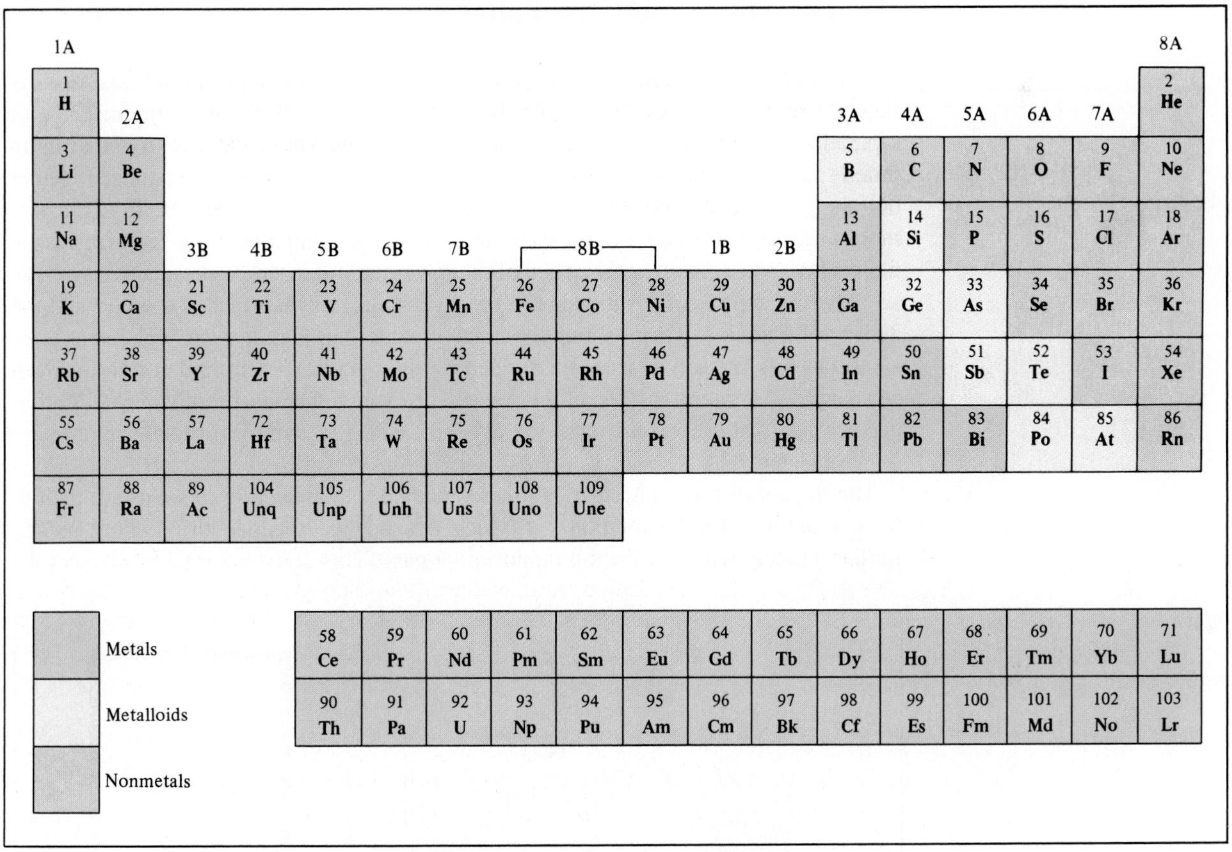

1A	2A	3B	4B	5B	6B	7B	8B	8B	8B	1B	2B	3A	4A	5A	6A	7A	8A
1 **H**																	2 **He**
3 **Li**	4 **Be**											5 **B**	6 **C**	7 **N**	8 **O**	9 **F**	10 **Ne**
11 **Na**	12 **Mg**											13 **Al**	14 **Si**	15 **P**	16 **S**	17 **Cl**	18 **Ar**
19 **K**	20 **Ca**	21 **Sc**	22 **Ti**	23 **V**	24 **Cr**	25 **Mn**	26 **Fe**	27 **Co**	28 **Ni**	29 **Cu**	30 **Zn**	31 **Ga**	32 **Ge**	33 **As**	34 **Se**	35 **Br**	36 **Kr**
37 **Rb**	38 **Sr**	39 **Y**	40 **Zr**	41 **Nb**	42 **Mo**	43 **Tc**	44 **Ru**	45 **Rh**	46 **Pd**	47 **Ag**	48 **Cd**	49 **In**	50 **Sn**	51 **Sb**	52 **Te**	53 **I**	54 **Xe**
55 **Cs**	56 **Ba**	57 **La**	72 **Hf**	73 **Ta**	74 **W**	75 **Re**	76 **Os**	77 **Ir**	78 **Pt**	79 **Au**	80 **Hg**	81 **Tl**	82 **Pb**	83 **Bi**	84 **Po**	85 **At**	86 **Rn**
87 **Fr**	88 **Ra**	89 **Ac**	104 **Unq**	105 **Unp**	106 **Unh**	107 **Uns**	108 **Uno**	109 **Une**									

58 **Ce**	59 **Pr**	60 **Nd**	61 **Pm**	62 **Sm**	63 **Eu**	64 **Gd**	65 **Tb**	66 **Dy**	67 **Ho**	68 **Er**	69 **Tm**	70 **Yb**	71 **Lu**
90 **Th**	91 **Pa**	92 **U**	93 **Np**	94 **Pu**	95 **Am**	96 **Cm**	97 **Bk**	98 **Cf**	99 **Es**	100 **Fm**	101 **Md**	102 **No**	103 **Lr**

Metals
Metalloids
Nonmetals

Figure 1.5 *The positions of metals, nonmetals, and metalloids in a modern version of the periodic table. With the exception of hydrogen (H), nonmetals appear at the far right of the table. The elements are arranged according to the numbers above their symbols. These numbers are based on a fundamental property of the elements that will be discussed in Chapter 8.*

Cl, Br, I, and At) are known as ***halogens,*** and *those in Group 8A (He, Ne, Ar, Kr, Xe, and Rn)* are called ***noble gases*** (or ***rare gases***). The names of other groups or families will be introduced later.

Looking at Figure 1.5, we see that two horizontal rows of elements are set aside at the bottom of the chart. Actually, cerium (Ce) should come right after lanthanum (La), and thorium (Th) should come right after actinium (Ac). These two rows of metals are set aside from the main body of the periodic table to avoid making the table too wide.

Figure 1.5 shows that the majority of known elements are metals, only seventeen elements are nonmetals, and eight elements are metalloids. The characteristic shape of the periodic table was originally devised to place elements with related physical and chemical properties near each other. Elements belonging to the same group resemble one another in chemical properties. In moving across any period from left to right, the physical and chemical properties of the elements change gradually from metallic to nonmetallic. The periodic table correlates the chemical behavior of the elements in a systematic way and helps us remember and understand many facts.

1.5 The Three States of Matter

All substances, at least in principle, can exist in three states: solid, liquid, and gas. As Figure 1.6 shows, gases differ from the other two states of matter in the distance of separation between the molecules. In a solid, molecules are held close together in an orderly fashion with little freedom of motion. Molecules in a liquid are close together but are not held so rigidly in position and can move past one another. In a gas, the molecules are separated by distances that are large compared with the size of the molecules. When viewing the states of matter in the microscopic perspective, we can see why the density of a solid should be slightly greater than that of a liquid and the density of a liquid, in turn, should be much greater than that of a gas. (The density of a substance is defined as its mass divided by its volume.) This trend is true for most substances. An important exception is water, for which the density of its liquid state is *greater* than that of its solid state, ice. We will see the reason for this unusual property of water later.

The three states of matter can be interconverted. Upon heating, a solid will melt to form a liquid. (The temperature at which this occurs is called the *melting point.*) Further heating will convert the liquid into a gas. (This conversion takes place at the

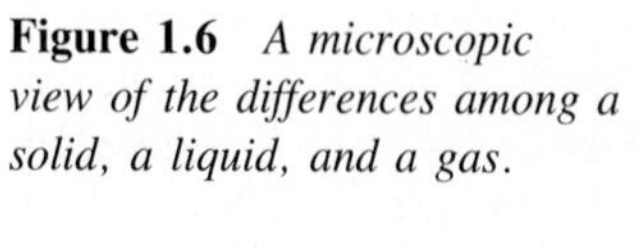

Figure 1.6 *A microscopic view of the differences among a solid, a liquid, and a gas.*

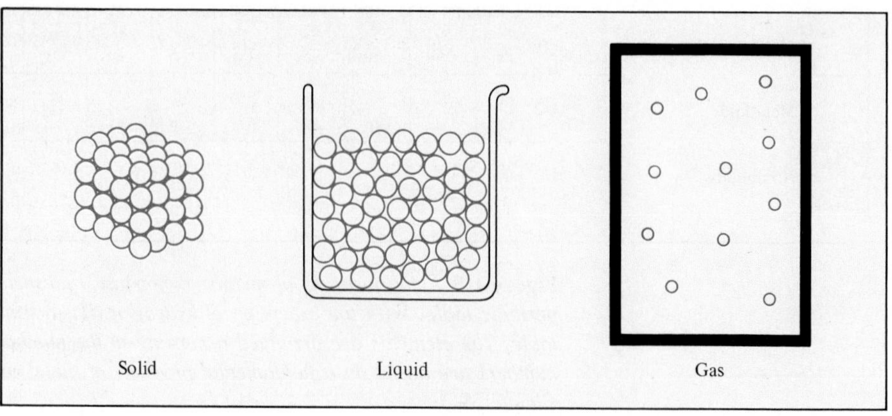

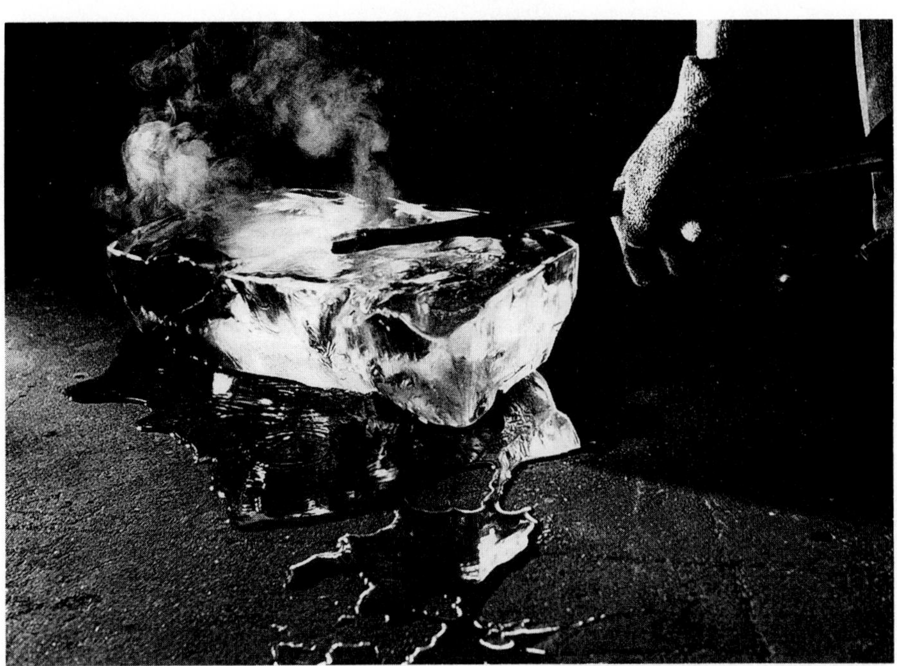

Figure 1.7 *The three states of matter. A hot poker changes ice into water and steam.*

boiling point of the liquid.) On the other hand, cooling a gas will condense it into a liquid. When the liquid is cooled further, it will freeze into the solid form. Figure 1.7 shows the three states of water.

1.6 Measurement

As mentioned earlier, chemistry is largely an experimental science and deals with things that can be measured. The measurements we make are often used in calculations to obtain other related quantities. Our ability to measure properties depends to a large degree on the current state of technology. For instance, radiant energy from the sun could not be measured before the invention of devices that could detect it. The scope of chemistry continually expands as new instruments increase the range and precision of possible measurements.

A number of common devices that you will be using in the laboratory enable you to make simple measurements: The meter stick measures length or scale; the buret, the pipet, the graduated cylinder, and the volumetric flask measure volume (Figure 1.8); the balance measures mass; the thermometer measures temperature.

But suppose we want to know the mass of a single iron atom. Atoms are so small that we cannot handle individual atoms, nor do we have balances sensitive enough to weigh lone atoms. We must therefore take an indirect approach. We can, for example, measure the mass of a piece of iron metal. If we know the total number of iron atoms present in the metal, we can calculate the mass of one iron atom as follows:

$$\text{mass of one iron atom} = \frac{\text{mass of iron metal sample}}{\text{total number of iron atoms present}}$$

This approach is indeed plausible, as we will see in the next chapter.

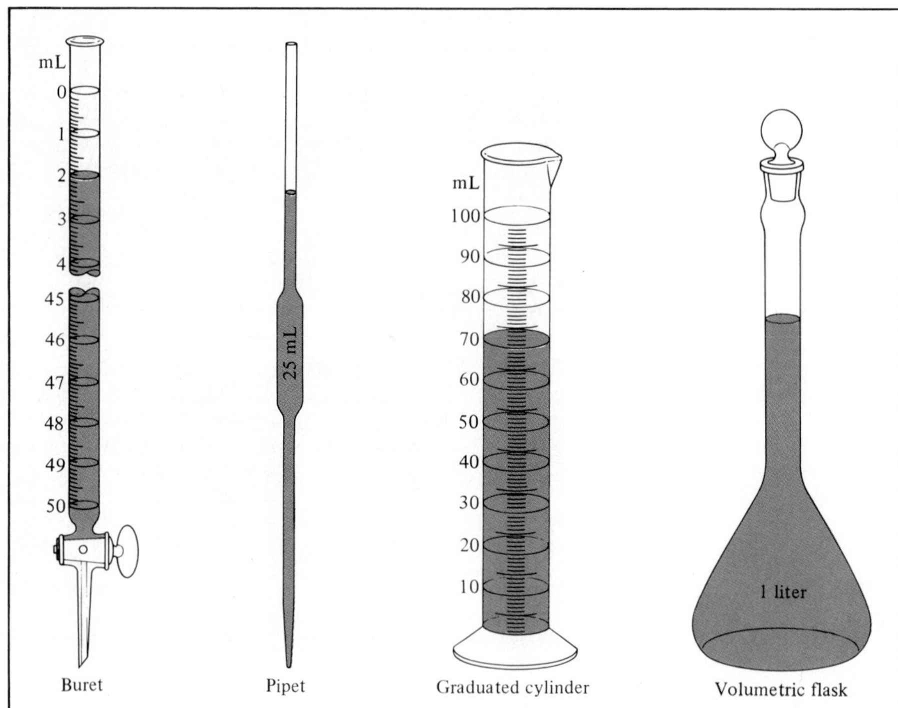

Figure 1.8 *Some common measuring devices found in a chemistry laboratory. These devices are not drawn to scale relative to one another. We will discuss the uses of these measuring devices in Chapter 4.*

When we speak of measurements, we may be referring either to a ***macroscopic property,*** which is *determined directly,* or to a ***microscopic property*** on the atomic or molecular scale, which must be *determined by an indirect method.* A measured quantity is usually written as a number with an appropriate unit. To say that the distance between New York and San Francisco by car along a certain route is 5166 is meaningless. We must say that the distance is 5166 kilometers. The same is true in chemistry; units are essential to stating a measurement correctly.

1.7 Units of Measurement

For many years, the units used in science (including chemistry) were, in general, *metric units,* which were developed in France in the eighteenth century. Metric units are related decimally, that is, by powers of 10. This relationship is usually indicated by a prefix attached to the unit.

SI Units

In 1960 the General Conference of Weights and Measures, the international authority on units, proposed a revised and modernized metric system called the *International System of Units* (abbreviated *SI*, from the French Le *S*ystem *I*nternational d'Unites). Table 1.2 shows the seven SI base units. As we will see shortly, all other units needed for measurement can be derived from these base units. Like metric units, SI units are modified in decimal fashion by a series of prefixes like those shown in Table 1.3. Most of these prefixes will be used often in this book, and you should become familiar with their meanings.

Table 1.2 SI base units

Base quantity	Name of unit	Symbol
Length	Meter	m
Mass	Kilogram	kg
Time	Second	s
Electrical current	Ampere	A
Temperature	Kelvin	K
Amount of substance	Mole	mol
Luminous intensity	Candela	cd

Table 1.3 Some prefixes used with SI units

Prefix	Symbol	Meaning	Example
Tera-	T	1,000,000,000,000, or 10^{12}	1 terameter (Tm) = 1×10^{12} m
Giga-	G	1,000,000,000, or 10^9	1 gigameter (Gm) = 1×10^9 m
Mega-	M	1,000,000, or 10^6	1 megameter (Mm) = 1×10^6 m
Kilo-	k	1,000, or 10^3	1 kilometer (km) = 1×10^3 m
Deci-	d	1/10, or 10^{-1}	1 decimeter (dm) = 0.1 m
Centi-	c	1/100, or 10^{-2}	1 centimeter (cm) = 0.01 m
Milli-	m	1/1,000, or 10^{-3}	1 millimeter (mm) = 0.001 m
Micro-	μ	1/1,000,000, or 10^{-6}	1 micrometer (μm) = 1×10^{-6} m
Nano-	n	1/1,000,000,000, or 10^{-9}	1 nanometer (nm) = 1×10^{-9} m
Pico-	p	1/1,000,000,000,000, or 10^{-12}	1 picometer (pm) = 1×10^{-12} m

Length. The SI base unit of length is the *meter* (m). Other common units used are the centimeter (cm), decimeter (dm), and kilometer (km):

$$1 \text{ cm} = 0.01 \text{ m} = 1 \times 10^{-2} \text{ m}$$

$$1 \text{ dm} = 0.1 \text{ m} = 1 \times 10^{-1} \text{ m}$$

$$1 \text{ km} = 1000 \text{ m} = 1 \times 10^{3} \text{ m}$$

Mass. The SI base unit of mass is the *kilogram* (kg). In chemistry, the most frequently used unit for mass is the gram (g):

$$1 \text{ kg} = 1000 \text{ g} = 1 \times 10^{3} \text{ g}$$

Recall that length and mass are extensive properties.

Temperature. The SI base unit of temperature is the *kelvin* (K). (Note the absence of "degree" in the name and in the symbol.) The degree Celsius (°C), formerly termed the degree centigrade, is also allowed with SI. We will consider the relationship of the Celsius temperature scale to the nonmetric Fahrenheit scale later in this chapter. The Kelvin scale will play a major role when we study gas behavior (Chapter 5) and thermodynamics (Chapter 18).

Temperature is an intensive property: It does not depend on the amount of matter being considered.

Time. The SI base unit of time is the *second* (s). Greater time intervals can be expressed either with suitable prefixes, such as in kiloseconds (ks), or with the familiar units called minutes (min) and hours (h).

Of the remaining base units, the mole (mol) and the ampere (A) are important to the practice of chemistry. They will be defined and explained in context later in this book. The seventh unit, the candela (cd), is not involved in the development of general chemistry.

Derived SI Units

The units for a number of other quantities can be derived from the SI base units. For example, from the base unit of length, we can define volume, and from the base units of length, mass, and time, we can define energy. The following quantities are frequently encountered in the study of chemistry.

Volume. Since *volume* is length cubed, its SI derived unit is m^3. Related units are the cubic centimeter (cm^3) and the cubic decimeter (dm^3):

$$1\ cm^3 = (1 \times 10^{-2}\ m)^3 = 1 \times 10^{-6}\ m^3$$

$$1\ dm^3 = (1 \times 10^{-1}\ m)^3 = 1 \times 10^{-3}\ m^3$$

Another common (but non-SI) unit of volume is the liter (L). A ***liter*** is *the volume occupied by one cubic decimeter*. One liter of volume is equal to 1000 milliliters (mL), and one milliliter of volume is equal to one cubic centimeter:

$$1\ L = 1000\ mL$$

$$1\ mL = 1 \times 10^{-3}\ L$$

$$1\ L = 1\ dm^3$$

$$1\ dm^3 = 1 \times 10^3\ cm^3$$

$$1\ mL = 1\ cm^3$$

Figure 1.9 compares the relative sizes of two volumes. This book mainly uses the units L and mL for volume.

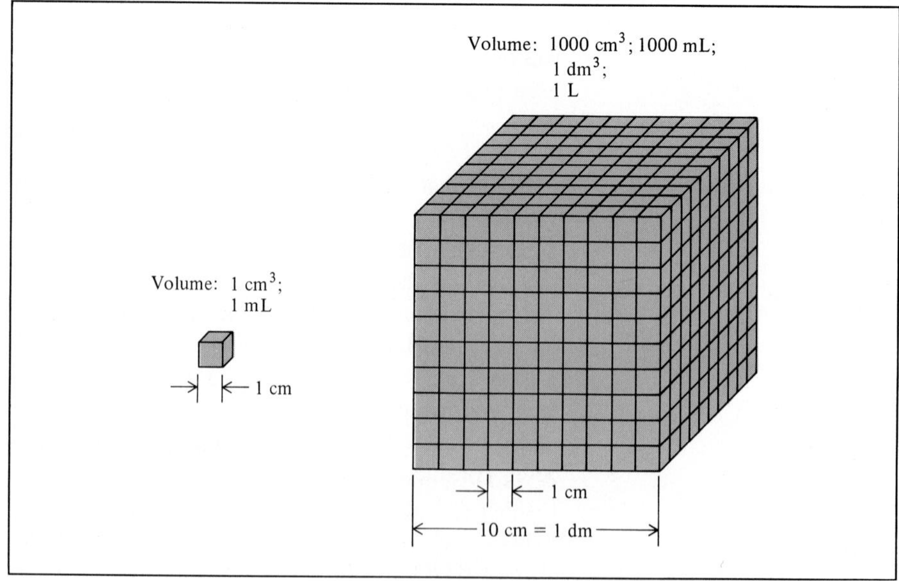

Figure 1.9. *Comparison of two volumes, 1 mL and 1000 mL.*

Velocity and Acceleration. By definition, *velocity* is the change in distance with elapsed time; that is,

$$\text{velocity} = \frac{\text{distance}}{\text{time}}$$

Acceleration is the change in velocity with time; that is,

$$\text{acceleration} = \frac{\text{change in velocity}}{\text{time}}$$

Therefore, velocity has the units of m/s (or cm/s), and acceleration has the units of m/s^2 (or cm/s^2). Velocity is needed to define acceleration, which in turn is required to define force and hence energy. Both force and energy play important roles in many areas in chemistry.

Force. According to Newton's second law of motion,†

$$\text{force} = \text{mass} \times \text{acceleration}$$

In common language, a force is often regarded as a push or a pull. Chemistry is concerned mainly with the electrical forces that exist among atoms and molecules. The nature of these forces will be explored in later chapters. *The derived SI unit for force* is the ***newton (N),*** where

$$1 \text{ N} = 1 \text{ kg m/s}^2$$

Pressure. ***Pressure*** is defined as *force applied per unit area;* that is,

$$\text{pressure} = \frac{\text{force}}{\text{area}}$$

The force experienced by any area exposed to Earth's atmosphere is equal to *the weight of the column of air above it. The pressure exerted by this column of air* is called ***atmospheric pressure.*** The actual value of atmospheric pressure depends on location, temperature, and weather conditions. A common reference pressure of *one atmosphere* (1 atm) represents the atmospheric pressure exerted by a column of dry air at sea level and 0°C. The derived SI unit of pressure is obtained by applying the derived force unit of one newton over one square meter, which is the derived unit of area. *A pressure of one newton per square meter (1 N/m²)* is called a ***pascal (Pa).***§ One atmosphere is now defined by the exact relation

$$\begin{aligned} 1 \text{ atm} &= 101{,}325 \text{ Pa} \\ &= 101.325 \text{ kPa} \end{aligned}$$

Energy. ***Energy*** is *the capacity to do work or to produce change.* In chemistry, we are interested mainly in energy effects that result from chemical or physical changes. In

† Sir Isaac Newton (1642–1726). English mathematician, physicist, and astronomer. Newton is regarded by many as one of the two greatest physicists the world has known (the other is Albert Einstein). There was hardly a branch of physics to which Newton did not make a significant contribution. His book *Principia,* published in 1687, marks a milestone in the history of science.

§ Blaise Pascal (1623–1662). French mathematician and physicist. Pascal's work ranged widely in mathematics and physics, particularly in the area of hydrodynamics (the study of the motion of fluids). He also invented a calculating machine.

mechanics, work is defined as "force × distance." Since energy can be measured as work, we can write

$$\begin{aligned} \text{energy} &= \text{work done} \\ &= \text{force} \times \text{distance} \end{aligned}$$

Thus *the SI derived unit of energy has the units of newtons × meter (N m)* or kg m^2/s^2. This SI derived energy unit is more commonly called a ***joule (J)***.†

$$\begin{aligned} 1\ \text{J} &= 1\ \text{kg m}^2/\text{s}^2 \\ &= 1\ \text{N m} \end{aligned}$$

Sometimes energy is expressed in kilojoules (kJ):

$$1\ \text{kJ} = 1000\ \text{J}$$

Traditionally, chemists have expressed energy in *calories* (cal). The calorie is defined by the relationship

$$1\ \text{cal} = 4.184\ \text{J}$$

In this book we will mainly use joules and kilojoules as the units for energy.

Density. The ***density*** of an object is *the mass of the object divided by its volume:*

$$\text{density} = \frac{\text{mass}}{\text{volume}}$$

or

$$d = \frac{m}{V}$$

where d and V denote density and volume, respectively. Note that the density of a given material does not depend on the quantity of mass present. This is because V increases as m does, so that the ratio of the two quantities always remains the same for the given material. Thus density is an intensive quantity.

The SI derived unit for density is the kilogram per cubic meter (kg/m^3). This unit is awkwardly large for most chemical applications. The unit g/cm^3 and its equivalent, g/mL, are more commonly used for solid and liquid densities. Because gas densities are often very low, we use the units of g/L:

$$\begin{aligned} 1\ \text{g/cm}^3 &= 1\ \text{g/mL} = 1000\ \text{kg/m}^3 \\ 1\ \text{g/L} &= 0.001\ \text{g/mL} \end{aligned}$$

The following example shows a density calculation.

† James Prescott Joule (1818–1889). English physicist. Joule's work was mainly in the kinetic molecular theory of gases and thermodynamics.

EXAMPLE 1.1

A piece of platinum metal of mass 96.4 g has a volume of 4.49 cm^3. Calculate the density of the element platinum (Pt).

A platinum crucible.

Answer

The density of the platinum metal is given by

$$d = \frac{m}{V}$$
$$= \frac{96.4\ \text{g}}{4.49\ \text{cm}^3}$$
$$= 21.5\ \text{g/cm}^3$$

Similar problems: 1.29, 1.31.

The density can be rearranged to solve for mass or volume, if the other two values in the equation are known. The following example shows the calculation of mass from density and volume data.

EXAMPLE 1.2

The density of ethanol is 0.798 g/mL. Calculate the mass of 17.4 mL of the liquid.

Answer

The mass of ethanol is found by rearranging the density equation $d = m/V$ as follows:

$$m = d \times V$$
$$= 0.798 \frac{\text{g}}{\text{mL}} \times 17.4\ \text{mL}$$
$$= 13.9\ \text{g}$$

Similar problem:1.30.

Temperature Scales

Three temperature scales are currently in use. Their units are K (kelvin), °C (degree Celsius), and °F (degree Fahrenheit). The Fahrenheit scale defines the normal freezing and boiling points of water to be exactly 32°F and 212°F, respectively. It is the most commonly used temperature scale in the United States. The Celsius scale divides the range between the freezing point (0°C) and boiling point (100°C) of water into 100 degrees (Figure 1.10). The Celsius scale is the most generally used scale in science (including this course).

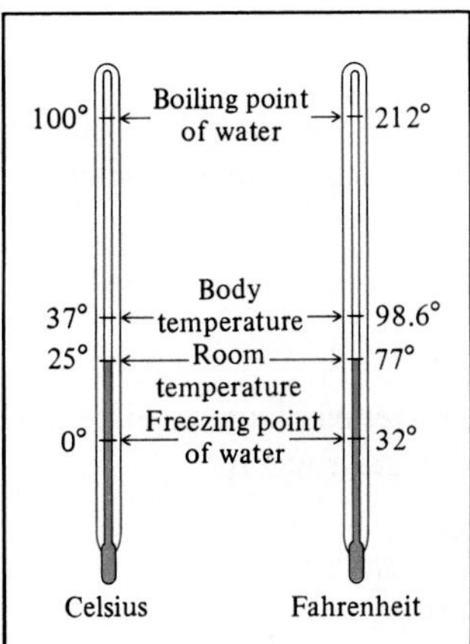

Figure 1.10 *Comparison of the Celsius and Fahrenheit temperature scales. Note that there are 100 divisions, or 100 degrees, between the freezing point and the boiling point of water on the Celsius scale, and there are 180 divisions, or 180 degrees, between the same two temperature limits on the Fahrenheit scale. The Celsius scale was formerly called the centigrade scale.*

The size of a degree on the Fahrenheit scale is only 100/180, or 5/9 times that on the Celsius scale. To convert degrees Fahrenheit to degrees Celsius, we write

$$?^\circ C = (^\circ F - 32^\circ F) \times \frac{5^\circ C}{9^\circ F}$$

To convert degrees Celsius to degrees Fahrenheit, we write

$$?^\circ F = \frac{9^\circ F}{5^\circ C} \times (^\circ C) + 32^\circ F$$

The Kelvin temperature scale is discussed in Chapter 5.

The following example shows conversions between degrees Celsius and degrees Fahrenheit.

EXAMPLE 1.3

(a) The normal body temperature is 98.6°F. What is the temperature in degrees Celsius? (b) The temperature on a winter day in Boston is −10.3°C. Convert this temperature to degrees Fahrenheit.

Answer

(a) This conversion is carried out by writing

$$(98.6^\circ F - 32^\circ F) \times \frac{5^\circ C}{9^\circ F} = 37.0^\circ C$$

(b) Here we have

$$\frac{9^\circ F}{5^\circ C} \times (-10.3^\circ C) + 32^\circ F = 13.5^\circ F$$

Similar problems: 1.33, 1.34.

1.8 Handling Numbers

Having surveyed some of the units used in chemistry, we are now ready to take a closer look at two techniques for handling numbers associated with measurements: scientific notation and significant figures.

Scientific Notation

In chemistry, we often deal with numbers that are either extremely large or extremely small. For example, in 1 g of the element hydrogen there are roughly

602,200,000,000,000,000,000,000

hydrogen atoms. Each hydrogen atom has a mass of only

0.00000000000000000000000166 g

These numbers are cumbersome to handle, and it is easy to make mistakes when using them in arithmetic computations. Consider the following multiplication:

$$0.0000000056 \times 0.00000000048 = 0.000000000000000002688$$

It would be easy for us to miss one zero or add one more zero after the decimal point. To handle these very large and very small numbers, we use a system called *scientific notation*. Regardless of their magnitude, all numbers can be expressed in the form

$$N \times 10^n$$

where N is a number between 1 and 10 and n is an *exponent* that can be a positive or negative *integer* (whole number). Any number expressed in this way is said to be written in scientific notation.

Suppose that we are given a certain number and asked to express it in scientific notation. Basically, this amounts to finding n. We count the number of places that the decimal point must be moved to give the number N (which is between 1 and 10). If the decimal point has to be moved to the left, then n is a positive integer; if it has to be moved to the right, n is a negative integer. The following examples illustrate the use of scientific notation:

(a) Express 568.762 in scientific notation:

$$568.762 = 5.68762 \times 10^2$$

Note that the decimal point is moved to the left by two places and $n = 2$.

(b) Express 0.00000772 in scientific notation:

$$0.00000772 = 7.72 \times 10^{-6}$$

Note that the decimal point is moved to the right by six places and $n = -6$.

Next, we consider how scientific notation is handled in arithmetic operations.

Addition and Subtraction. To add or subtract using scientific notation, we first write each quantity with the same exponent n. Then we add or subtract the N parts of the numbers; the exponent parts remain the same. Consider the following examples:

$$(7.4 \times 10^3) + (2.1 \times 10^3) = 9.5 \times 10^3$$

$$\begin{aligned}(4.31 \times 10^4) + (3.9 \times 10^3) &= (4.31 \times 10^4 + 0.39 \times 10^4)\\ &= 4.70 \times 10^4\end{aligned}$$

$$\begin{aligned}(2.22 \times 10^{-2}) - (4.10 \times 10^{-3}) &= (2.22 \times 10^{-2}) - (0.41 \times 10^{-2})\\ &= 1.81 \times 10^{-2}\end{aligned}$$

Multiplication and Division. To multiply numbers expressed in scientific notation, we multiply the N parts of the numbers in the usual way, but *add* the exponents n together. To divide using scientific notation, we divide the N parts of the numbers as usual and *subtract* the exponents n. The following examples show how these operations are performed:

$$\begin{aligned}(8.0 \times 10^4) \times (5.0 \times 10^2) &= (8.0 \times 5.0)(10^{4+2})\\ &= 40 \times 10^6\\ &= 4.0 \times 10^7\end{aligned}$$

$$
\begin{aligned}
(4.0 \times 10^{-5}) \times (7.0 \times 10^{3}) &= (4.0 \times 7.0)(10^{-5+3}) \\
&= 28 \times 10^{-2} \\
&= 2.8 \times 10^{-1}
\end{aligned}
$$

$$
\begin{aligned}
\frac{8.5 \times 10^{4}}{5.0 \times 10^{9}} &= \frac{8.5}{5.0} \times 10^{4-9} \\
&= 1.7 \times 10^{-5}
\end{aligned}
$$

$$
\begin{aligned}
\frac{6.9 \times 10^{7}}{3.0 \times 10^{-5}} &= \frac{6.9}{3.0} \times 10^{7-(-5)} \\
&= 2.3 \times 10^{12}
\end{aligned}
$$

Significant Figures

Except when all the numbers involved are integers (for example, in counting the number of students in a class), it is often impossible to obtain the exact value of the quantity under investigation. For this reason, it is important to indicate the margin of error in a measurement by clearly indicating the number of ***significant figures,*** which are *the meaningful digits in a measured or calculated quantity*. When significant figures are counted, the last digit is understood to be uncertain. For example, we might measure the volume of a given amount of liquid using a graduated cylinder (see Figure 1.8) with a scale that gives an uncertainty of 1 mL in the measurement. If the volume is found to be 6 mL, then the actual volume is in the range of 5 mL to 7 mL. We represent the volume of the liquid as (6 ± 1) mL. In this case, there is only one significant figure (the digit 6) that is uncertain by either plus or minus 1 mL. As an improvement, we might use a graduated cylinder that has finer divisions, so that the volume we measure is now uncertain by only 0.1 mL. If the volume of the liquid is now found to be 6.0 mL, we may express the quantity as (6.0 ± 0.1) mL, and the actual value is somewhere between 5.9 mL and 6.1 mL. We can further improve the measuring device and obtain more significant figures. In each case, the last digit is always uncertain; the amount of this uncertainty depends on the particular measuring device we use.

Figure 1.11 shows a modern balance. Balances such as this are available in many general chemistry laboratories; they readily measure the mass of objects to four decimal places. This means that the measured mass will in general have four significant figures (for example, 0.8642 g) or more (for example, 3.9745 g). Keeping track of the

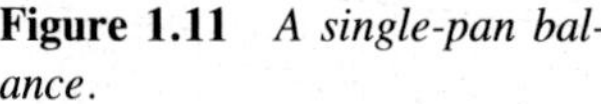
Figure 1.11 *A single-pan balance.*

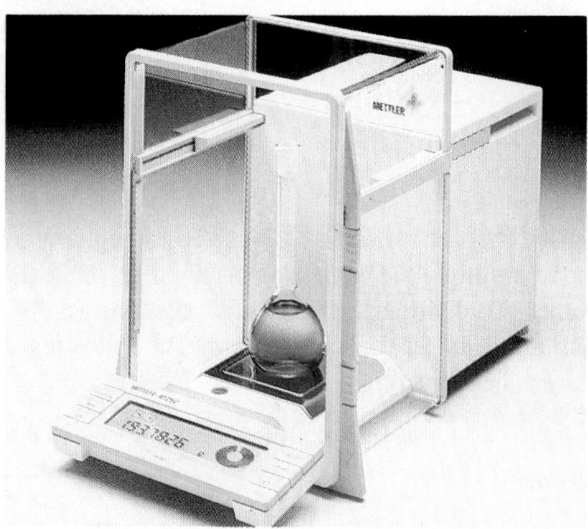

number of significant figures in a measurement such as mass ensures that calculations involving the data will correctly represent the precision of the measurement.

Guidelines for Using Significant Figures. The previous discussion shows that we must always be careful in scientific work to write the proper number of significant figures. In general, it is fairly easy to determine how many significant figures are present in a number by following these rules:

- Any digit that is not zero is significant. Thus 845 cm has three significant figures, 1.234 kg has four significant figures, and so on.
- Zeros between nonzero digits are significant. Thus 606 m contains three significant figures, 40,501 J contains five significant figures, and so on.
- Zeros to the left of the first nonzero digit are not significant. These zeros are used to indicate the placement of the decimal point. Thus 0.08 L contains one significant figure, 0.0000349 g contains three significant figures, and so on.
- If a number is greater than 1, then all the zeros written to the right of the decimal point count as significant figures. Thus 2.0 mg has two significant figures, 40.062 mL has five significant figures, and 3.040 dm has four significant figures. If a number is less than 1, then only the zeros that are at the end of the number and the zeros that are between nonzero digits are significant. Thus 0.090 kg has two significant figures, 0.3005 J has four significant figures, 0.00420 min has three significant figures, and so on.
- For numbers that do not contain decimal points, the trailing zeros (that is, zeros after the last nonzero digit) may or may not be significant. Thus 400 cm may have one significant figure (the digit 4), two significant figures (40), or three significant figures (400). We cannot know which is correct without more information. By using scientific notation, however, we avoid this ambiguity. In this particular case, we can express the number 400 as 4×10^2 for one significant figure, 4.0×10^2 for two significant figures, or 4.00×10^2 for three significant figures.

The following example shows the determination of significant figures.

EXAMPLE 1.4

Determine the number of significant figures in the following measured quantities: (a) 478 cm, (b) 6.01 g, (c) 0.825 m, (d) 0.043 kg, (e) 1.310×10^{22} atoms, (f) 7000 mL.

Answer

(a) Three.
(b) Three.
(c) Three.
(d) Two.
(e) Four.
(f) This is an ambiguous case. The number of significant figures may be four (7.000×10^3), three (7.00×10^3), two (7.0×10^3), or one (7×10^3).

Similar problems: 1.41, 1.42.

Our next step is to see how significant figures are handled in calculations. We can adhere to the following rules:

- In addition and subtraction, the number of significant figures to the right of the decimal point in the final sum or difference is determined by the lowest number of significant figures to the right of the decimal point in any of the original numbers. Consider the examples:

$$\begin{array}{r} 89.332 \\ +\ 1.1 \\ \hline 90.432 \end{array} \quad \begin{array}{l} \longleftarrow \text{ one significant figure after the decimal point} \\ \nwarrow \text{ round off to } 90.4 \end{array}$$

$$\begin{array}{r} 2.097 \\ -0.12 \\ \hline 1.977 \end{array} \quad \begin{array}{l} \longleftarrow \text{ two significant figures after the decimal point} \\ \nwarrow \text{ round off to } 1.98 \end{array}$$

  The rounding-off procedure is as follows. If we wish to round off a number at a certain point, we simply drop the digits that follow if the first of them is less than 5. Thus 8.724 rounds off to 8.72 if we want only two figures after the decimal point. If the first digit following the point of rounding off is equal to or greater than 5, we add 1 to the preceding digit. Thus 8.727 rounds off to 8.73, and 0.425 rounds off to 0.43.
- In multiplication and division, the number of significant figures in the final product or quotient is determined by the original number that has the smallest number of significant figures. The following examples illustrate this rule:

$$2.8 \times 4.5039 = 12.61092 \longleftarrow \text{round off to } 13$$

$$\frac{6.85}{112.04} = 0.0611388789 \longleftarrow \text{round off to } 0.0611$$

- Keep in mind that *exact numbers* obtained from definitions or by counting numbers of objects can be considered to have an infinite number of significant figures. If an object has the mass 0.2786 g, then the mass of eight such objects is

$$0.2786 \text{ g} \times 8 = 2.229 \text{ g}$$

  We do *not* round off this product to one significant figure, because the number 8 is actually 8.00000 . . . , by definition. Similarly, to take the average of the two measured lengths 6.64 cm and 6.68 cm, we write

$$\frac{6.64 \text{ cm} + 6.68 \text{ cm}}{2} = 6.66 \text{ cm}$$

  because the number 2 is actually 2.00000 . . . , by definition.

The following example shows how significant figures are handled in arithmetic operations.

EXAMPLE 1.5

Carry out the following arithmetic operations: (a) 11,254.1 + 0.1983, (b) 66.59 − 3.113, (c) 8.16 × 5.1355, (d) 0.0154 ÷ 883, (e) $2.64 \times 10^3 + 3.27 \times 10^2$.

Answer

(a)

$$\begin{array}{r} 11{,}254.1 \\ +\quad 0.1983 \\ \hline 11{,}254.2983 \end{array} \longleftarrow \text{round off to } 11{,}254.3$$

(b)

$$\begin{array}{r} 66.59 \\ -\ 3.113 \\ \hline 63.477 \end{array} \longleftarrow \text{round off to } 63.48$$

(c) $8.16 \times 5.1355 = 41.90568 \longleftarrow$ round off to 41.9

(d) $\dfrac{0.0154}{883} = 0.0000174405436 \longleftarrow$ round off to 0.0000174, or 1.74×10^{-5}

(e) First we change 3.27×10^2 to 0.327×10^3 and then carry out the addition $(2.64 + 0.327) \times 10^3$. Following the procedure in (a), we find the answer is 2.97×10^3.

Similar problem: 1.43.

The above rounding-off procedure applies to one-step calculations. In *chain calculations,* that is, calculations involving more than one step, we use a modified procedure. Consider the following two-step calculation:

First step: $A \times B = C$
Second step: $C \times D = E$

Let us suppose that $A = 3.66$, $B = 8.45$, and $D = 2.11$. Depending on whether we round off C to three or four significant figures, we obtain a different number for E:

Method 1	Method 2
3.66 × 8.45 = 30.9	3.66 × 8.45 = 30.93
30.9 × 2.11 = 65.2	30.93 × 2.11 = 65.3

However, if we had carried out the calculation as 3.66 × 8.45 × 2.11 on a calculator without rounding off the intermediate result, we would have obtained 65.3 as the answer for E. The procedure of carrying the answers for all the intermediate calculations to *one more significant figure* and only rounding off the final answer to the correct number of significant figures will be used in some of the worked examples in this book. In general, we will show the correct number of significant figures in each step of the calculation.

Accuracy and Precision. In discussing measurements and significant figures it is useful to distinguish two terms: *accuracy* and *precision*. ***Accuracy** tells us how close a measurement is to the true value of the quantity that was measured.* ***Precision** refers to how closely two or more measurements of the same quantity agree with one another.* Suppose that three students are asked to determine the mass of a piece of copper wire of mass 2.000 g. The results of two successive weighings by each student are

	Student A	**Student B**	**Student C**
	1.964 g	1.972 g	2.000 g
	1.978 g	1.968 g	2.002 g
Average value	1.971 g	1.970 g	2.001 g

Student B's results are more *precise* than those of Student A (1.972 g and 1.968 g deviate less from 1.970 g than 1.964 g and 1.978 g from 1.971 g). Neither set of results is very *accurate,* however. Student C's results are not only *precise* but also the most *accurate,* since the average value is closest to the true value. Highly accurate measurements are usually precise too. On the other hand, highly precise measurements do not necessarily guarantee accurate results. For example, an improperly calibrated meter stick or a faulty balance may give precise readings that are in error.

1.9 The Factor-Label Method of Solving Problems

The procedure we will use in solving problems is called *the factor-label method* (also called *dimensional analysis*). A simple and powerful technique requiring little memorization, the factor-label method is based on the relationship between different units that express the same physical quantity.

We know, for example, that the unit "dollar" for money is different from the unit "penny." However, we say that 1 dollar is *equivalent* to 100 pennies because they both represent the same amount of money. This equivalence allows us to write

$$1 \text{ dollar} = 100 \text{ pennies}$$

Because 1 dollar is equal to 100 pennies, it follows that their ratio has a value of 1; that is,

$$\frac{1 \text{ dollar}}{100 \text{ pennies}} = 1$$

This ratio can be read as 1 dollar per 100 pennies. This fraction is called a *unit factor* (equal to 1) because the numerator and denominator describe the same amount of money.

We could also have written the ratio as

$$\frac{100 \text{ pennies}}{1 \text{ dollar}} = 1$$

This ratio reads as 100 pennies per dollar. The fraction 100 pennies/1 dollar is also a unit factor. We see that the reciprocal of any unit factor is also a unit factor. The usefulness of unit factors is that they allow us to carry out conversions between different units that measure the same quantity. Suppose that we wish to convert 2.46 dollars into pennies. This problem may be expressed as

$$? \text{ pennies} = 2.46 \text{ dollars}$$

Since this is a dollar-to-penny conversion, we choose the unit factor that has the unit ''dollar'' in the denominator (to cancel the ''dollars'' in 2.46 dollars) and write

$$2.46 \cancel{\text{dollars}} \times \frac{100 \text{ pennies}}{1 \cancel{\text{dollar}}} = 246 \text{ pennies}$$

Note that the unit factor 100 pennies/1 dollar contains exact numbers, so it does not affect the number of significant figures in the final answer.

Next let us consider the conversion of 57.8 meters to centimeters. This problem may be expressed as

$$? \text{ cm} = 57.8 \text{ m}$$

By definition,

$$1 \text{ cm} = 1 \times 10^{-2} \text{ m}$$

Since we are converting ''m'' to ''cm,'' we choose the unit factor that has meters in the denominator,

$$\frac{1 \text{ cm}}{1 \times 10^{-2} \text{ m}} = 1$$

and write the conversion as

$$\begin{aligned} ? \text{ cm} &= 57.8 \cancel{\text{m}} \times \frac{1 \text{ cm}}{1 \times 10^{-2} \cancel{\text{m}}} \\ &= 5780 \text{ cm} \\ &= 5.78 \times 10^3 \text{ cm} \end{aligned}$$

Note that scientific notation is used to indicate that the answer has three significant figures. The unit factor 1 cm/1 × 10^{-2} m contains exact numbers; therefore, it does not affect the number of significant figures.

The advantage of the factor-label method is that if the equation is set up correctly, then all the units will cancel except the desired one. If this is not the case, then an error must have been made somewhere, and it can usually be spotted by inspection.

The following examples illustrate the factor-label method.

EXAMPLE 1.6

Convert 5.6 dm to meters.

Answer

The problem is

$$? \text{ m} = 5.6 \text{ dm}$$

By definition,

$$1 \text{ dm} = 1 \times 10^{-1} \text{ m}$$

The unit factor is

$$\frac{1 \times 10^{-1} \text{ m}}{1 \text{ dm}} = 1$$

Therefore we write

$$? \text{ m} = 5.6 \cancel{\text{dm}} \times \frac{1 \times 10^{-1} \text{ m}}{1 \cancel{\text{dm}}} = 0.56 \text{ m}$$

Similar problem: 1.44.

EXAMPLE 1.7

A man weighs 162 pounds (lb). What is his mass in milligrams (mg)?

Answer

The problem can be expressed as

$$? \text{ mg} = 162 \text{ lb}$$

The conversion factors are

$$1 \text{ lb} = 453.6 \text{ g}$$

so the unit factor is

$$\frac{453.6 \text{ g}}{1 \text{ lb}} = 1$$

and

$$1 \text{ mg} = 1 \times 10^{-3} \text{ g}$$

so the unit factor is

$$\frac{1 \text{ mg}}{1 \times 10^{-3} \text{ g}} = 1$$

Thus

$$? \text{ mg} = 162 \cancel{\text{lb}} \times \frac{453.6 \cancel{\text{g}}}{1 \cancel{\text{lb}}} \times \frac{1 \text{ mg}}{1 \times 10^{-3} \cancel{\text{g}}} = 7.35 \times 10^{7} \text{ mg}$$

Similar problem: 1.46.

Note that unit factors may be squared or cubed, because $1^2 = 1^3 = 1$. The use of such factors is illustrated with Examples 1.8 and 1.9.

EXAMPLE 1.8

Calculate the number of cubic centimeters in 6.2 m^3.

Answer

The problem can be stated as

$$? \text{ cm}^3 = 6.2 \text{ m}^3$$

By definition,

$$1 \text{ cm} = 1 \times 10^{-2} \text{ m}$$

The unit factor is

$$\frac{1 \text{ cm}}{1 \times 10^{-2} \text{ m}} = 1$$

It follows that

$$\left(\frac{1 \text{ cm}}{1 \times 10^{-2} \text{ m}}\right)^3 = 1^3 = 1$$

Therefore we write

$$? \text{ cm}^3 = 6.2 \text{ m}^3 \times \left(\frac{1 \text{ cm}}{1 \times 10^{-2} \text{ m}}\right)^3 = 6{,}200{,}000 \text{ cm}^3$$
$$= 6.2 \times 10^6 \text{ cm}^3$$

Similar problem: 1.47.

EXAMPLE 1.9

The density of gold is 19.3 g/cm^3. Convert the density to units of kg/m^3.

Answer

The problem can be stated as

$$? \text{ kg/m}^3 = 19.3 \text{ g/cm}^3$$

We need two unit factors—one to convert g to kg and the other to convert cm^3 to m^3. We know that

$$1 \text{ kg} = 1000 \text{ g}$$

so

$$\frac{1 \text{ kg}}{1000 \text{ g}} = 1$$

Second, from Example 1.8,

$$\left(\frac{1 \text{ cm}}{1 \times 10^{-2} \text{ m}}\right)^3 = 1$$

Thus we write

$$? \text{ kg/m}^3 = \frac{19.3 \text{ g}}{1 \text{ cm}^3} \times \frac{1 \text{ kg}}{1000 \text{ g}} \times \left(\frac{1 \text{ cm}}{1 \times 10^{-2} \text{ m}}\right)^3 = 19{,}300 \text{ kg/m}^3$$
$$= 1.93 \times 10^4 \text{ kg/m}^3$$

Similar problem: 1.55.

CHEMISTRY IN ACTION

The Scientific Method and the Extinction of the Dinosaurs

Dinosaurs, which dominated life on Earth for millions of years, disappeared very suddenly (Figure 1.12), and the puzzle of their disappearance has apparently been solved by applying the scientific method. In the experimentation and data-collection stage, paleontologists studied fossils and skeletons found in rocks in various layers of Earth's crust. Their findings enabled them to map out which species existed on Earth during specific geologic periods. Their studies also showed that in the rocks that were formed immediately after the Cretaceous period, which dates back some 65 million years, not one dinosaur skeleton has been found. It is therefore assumed that the dinosaurs became extinct about 65 million years ago. What happened?

Figure 1.12 *Where have all the dinosaurs gone, long time passing? The study of dinosaur extinction illustrates what we mean by the scientific method.*

Among the many hypotheses put forward to account for their disappearance were disruptions of the food chain and a dramatic change in climate caused by violent volcanic eruptions. However, the evidence was not convincing for any one hypothesis until 1977. It was then that a group of paleontologists working in Italy obtained some very puzzling data at a site near Gubbio. Through the chemical analysis of a layer of clay located above the sediments formed during the Cretaceous period (and therefore a layer that records events occurring *after* the Cretaceous period), they found a surprisingly high content of the element iridium. Iridium is very rare in Earth's crust but is comparatively abundant in asteroids.

This investigation led scientists to hypothesize the extinction of dinosaurs as follows. To account for the quantity of iridium found, they suggested that a large asteroid several miles in diameter must have hit Earth at that time. Presumably the impact of the asteroid on Earth's surface was so tremendous that it literally vaporized a large quantity of surrounding rocks, soils, and other objects. The resulting dust and debris floated through the air and blocked the sunlight for months or perhaps even years. Without ample sunlight most plants could not grow, and the fossil record shows that many types of plants did indeed die out at this time. Consequently, of course, many plant-eating animals gradually perished, and then, in turn, meat-eating animals began to starve. Limitation of food sources would obviously affect large animals needing great amounts of food more quickly and more severely than small animals.

The hypothesis about dinosaur extinction is both interesting and provocative. It can, of course, be tested further. If the hypothesis is correct, then we should find similar high iridium content in corresponding layers at different locations on Earth. Moreover, we would expect the simultaneous extinction of other large species in addition to dinosaurs. Both of these predictions are strongly supported by recently collected data. In fact, the evidence has become so convincing that some scientists now refer to the explanation as the *theory* of dinosaur extinction.

Summary

1. The scientific method involves gathering information by making observations and measurements. In the process, hypotheses, laws, and theories are devised and tested.
2. The study of chemistry involves three basic steps: observation, representation, and interpretation. Observations refer to measurements in the macroscopic world; representations involve the use of shorthand symbols and equations for communication; interpretations are based on atoms and molecules, which belong to the microscopic world.
3. Substances have unique physical properties that can be observed without changing the identity of the substances, and unique chemical properties that, when they are demonstrated, change the identity of the substances.
4. The simplest substances in chemistry are elements. Compounds are formed by the combination of atoms of different elements.
5. Elements can be grouped together according to their related properties in a pattern called the periodic table. The periodic table is a central source of chemical information.
6. All substances can in principle exist in three states: solid, liquid, and gas. The interconversion between these states can be effected by a change in temperature.
7. SI units are used to express physical quantities in all sciences, including chemistry.
8. Numbers expressed in scientific notation in the form $N \times 10^n$, where N is between 1 and 10 and n is a positive or negative integer, help us handle very large and very small quantities.
9. The degree of certainty in a measured number is indicated by expressing only the significant figures and by rounding off answers to calculations involving that value to correct numbers of significant figures.
10. Chemical problems can be conveniently solved by the factor-label method.

Key Words

Accuracy, p. 26
Alkali metals, p. 10
Alkaline earth metals, p. 10
Atmospheric pressure, p. 17
Chemical property, p. 9
Chemistry, p. 4
Compound, p. 10
Density, p. 18
Element, p. 9
Energy, p. 17
Extensive property, p. 9
Family (in periodic table), p. 10
Group (in periodic table), p. 10
Halogens, p. 12
Heterogeneous mixture, p. 8
Homogeneous mixture, p. 8
Hypothesis, p. 5
Intensive property, p. 9
Joule (J), p. 18
Law, p. 5
Liter, p. 16
Macroscopic property, p. 14
Mass, p. 7
Matter, p. 7
Metal, p. 10
Metalloid, p. 10
Microscopic property, p. 14
Mixture, p. 8
Newton (N), p. 17
Noble gases, p. 12
Nonmetal, p. 10
Pascal (Pa), p. 17
Period (in periodic table), p. 10
Periodic table, p. 10
Physical property, p. 8
Precision, p. 26
Pressure, p. 17
Qualitative, p. 5
Quantitative, p. 5
Rare gases, p. 12
Scientific method, p. 5
Significant figures, p. 22
Substance, p. 7
Theory, p. 5
Weight, p. 7

Exercises

The Scientific Method

REVIEW QUESTIONS

1.1 Explain what is meant by the scientific method.
1.2 What is the difference between qualitative data and quantitative data?
1.3 Define the following terms: (a) hypothesis, (b) law, (c) theory.

PROBLEMS

1.4 Classify the following as qualitative or quantitative statements, giving your reasons. (a) The sun is approximately 93 million miles from Earth. (b) Leonardo da Vinci was a better painter than Michelangelo. (c) Ice is less dense than water. (d) Butter tastes better than margarine. (e) A stitch in time saves nine.
1.5 Classify each of the following statements as a hypothesis, a law, or a theory. (a) Beethoven's contribution to music would have been much greater if he had married. (b) An autumn leaf gravitates toward the ground because there is an attractive force between the leaf and Earth. (c) All matter is composed of very small particles called atoms.
1.6 Discuss whether the methods of investigation used by your favorite detective are scientific or not.

Basic Definitions

REVIEW QUESTIONS

1.7 Define the following terms: (a) matter, (b) mass, (c) weight, (d) substance, (e) mixture.
1.8 Which of the following statements is scientifically correct? "The mass of the student is 56 kg." "The weight of the student is 56 kg."
1.9 Give an example of a homogeneous mixture and an example of a heterogeneous mixture.
1.10 What is the difference between a physical property and a chemical property?
1.11 Give an example of an intensive property and an example of an extensive property.
1.12 Define the following terms: (a) element, (b) compound.

PROBLEMS

1.13 Do the following statements describe chemical or physical properties? (a) Oxygen gas supports combustion. (b) Fertilizers help to increase agricultural production. (c) Water boils below 100°C on top of a mountain. (d) Lead is more dense than aluminum. (e) Sugar tastes sweet.
1.14 Does each of the following describe a physical change or a chemical change? (a) The helium gas inside a balloon tends to leak out after a few hours. (b) A flashlight beam slowly gets dimmer and finally goes out. (c) Frozen orange juice is reconstituted by adding water to it. (d) The growth of plants depends on the sun's energy in a process called photosynthesis. (e) A spoonful of table salt dissolves in a bowl of soup.
1.15 Which of the following properties are intensive and which are extensive? (a) length, (b) area, (c) volume, (d) temperature, (e) mass.
1.16 Give the names of the elements represented by the chemical symbols Li, F, P, Cu, As, Zn, Cl, Pt, Mg, U, Al, Si, Ne. (See Table 1.1 and the inside front cover.)
1.17 Give the chemical symbols for the following elements: (a) potassium, (b) tin, (c) chromium, (d) boron, (e) barium, (f) plutonium, (g) sulfur, (h) argon, (i) mercury. (See Table 1.1 and the inside front cover.)
1.18 Classify each of the following substances as an element or a compound: (a) hydrogen, (b) water, (c) gold, (d) sugar.
1.19 Classify each of the following, with a brief explanation, as an element, a compound, a homogeneous mixture, or a heterogeneous mixture: (a) seawater, (b) helium gas, (c) sodium chloride (table salt), (d) a bottle of soft drink, (e) a milkshake, (f) air, (g) concrete.

The Periodic Table

REVIEW QUESTIONS

1.20 What is the periodic table, and what is its significance in the study of chemistry? What are groups and periods in the periodic table?
1.21 Give two differences between a metal and a nonmetal.
1.22 Write the names and symbols for four elements in each of the following categories: (a) nonmetal, (b) metal, (c) metalloid.
1.23 Define, with two examples, the following terms: (a) alkali metals, (b) alkaline earth metals, (c) halogens, (d) noble gases.

Units

REVIEW QUESTIONS

1.24 What are the SI base units of importance to chemistry?
1.25 Give the SI units for expressing the following: (a) length, (b) area, (c) volume, (d) mass, (e) time, (f)) force, (g) energy, (h) temperature.
1.26 Write the numbers for the following prefixes: (a) mega-, (b) kilo-, (c) deci-, (d) centi-, (e) milli-, (f) micro-, (g) nano-, (h) pico-.

1.27 Define density. What are the units normally used for density? Is density an intensive or extensive property?

1.28 Write equations that would enable you to convert degrees Celsius to degrees Fahrenheit and degrees Fahrenheit to degrees Celsius.

PROBLEMS

1.29 A lead sphere has a mass of 1.20×10^4 g, and its volume is 1.05×10^3 cm^3. Calculate the density of lead.

1.30 Mercury is the only metal that is a liquid at room temperature. Its density is 13.6 g/mL. How many grams of mercury will occupy a volume of 95.8 mL?

1.31 Bromine is a reddish-brown color. Calculate the density of the liquid (in g/mL) if 586 g of the substance occupies 188 mL.

1.32 Lithium is the least dense metal known (density: 0.53 g/cm^3). What is the volume occupied by 1.20×10^3 g of lithium?

1.33 Calculate the temperature in degrees Celsius of the following: (a) a hot summer day of 95°F, (b) a cold winter day of 12°F, (c) a fever of 102°F, (d) a furnace operating at 1852°F.

1.34 (a) Normally our bodies can endure a temperature of 105°F for only short periods of time without permanent damage to the brain and other vital organs. What is the temperature in degrees Celsius? (b) Ethylene glycol is a liquid organic compound that is used as an antifreeze in car radiators. It freezes at −11.5°C. Calculate its freezing temperature in degrees Fahrenheit. (c) The temperature on the surface of our sun is about 6.3×10^3°C. What is this temperature in degrees Fahrenheit?

1.35 At what temperature does the numerical reading on a Celsius thermometer equal that on a Fahrenheit thermometer?

Scientific Notation

REVIEW QUESTION

1.36 What is the advantage of using scientific notation over decimal notation?

PROBLEMS

1.37 Express the following numbers in scientific notation: (a) 0.000000027, (b) 356, (c) 47,764, (d) 0.096.

1.38 Express the following numbers as decimals: (a) 1.52×10^{-2}, (b) 7.78×10^{-8}.

1.39 Express the answers to the following in scientific notation:
(a) $145.75 + (2.3 \times 10^{-1})$
(b) $79{,}500 \div (2.5 \times 10^2)$
(c) $(7.0 \times 10^{-3}) - (8.0 \times 10^{-4})$
(d) $(1.0 \times 10^4) \times (9.9 \times 10^6)$
(e) $0.0095 + (8.5 \times 10^{-3})$
(f) $653 \div (5.75 \times 10^{-8})$
(g) $850{,}000 - (9.0 \times 10^5)$
(h) $(3.6 \times 10^{-4}) \times (3.6 \times 10^6)$

Significant Figures

REVIEW QUESTION

1.40 Define significant figure. Discuss the importance of using the proper number of significant figures in measurements and calculations.

PROBLEMS

1.41 What is the number of significant figures in each of the following measured quantities?
(a) 4867 mi
(b) 56 mL
(c) 60,104 ton
(d) 2900 g
(e) 40.2 g/cm^3
(f) 0.0000003 cm
(g) 0.7 min
(h) 4.6×10^{19} atoms

1.42 How many significant figures are there in each of the following? (a) 0.006, (b) 0.0605, (c) 60.5, (d) 605.5, (e) 960×10^{-3}, (f) 6, (g) 60.

1.43 Carry out the following operations as if they were calculations of experimental results, and express each answer in the correct units and with the correct number of significant figures:
(a) 5.6792 m + 0.6 m + 4.33 m
(b) 3.70 g − 2.9133 g
(c) 4.51 cm × 3.6666 cm
(d) 7.310 km ÷ 5.70 km
(e) 3.26×10^{-3} mg − 7.88×10^{-5} mg
(f) 4.02×10^6 dm + 7.74×10^7 dm

The Factor-Label Method

PROBLEMS

1.44 Convert 22.6 m to decimeters.

1.45 Convert 25.4 mg to kilograms.

1.46 Convert 242 lb to milligrams.

1.47 Convert 68.3 cm^3 to m^3.

1.48 The price of gold on a certain day in 1989 was $327 per ounce. How much did 1.00 g of gold cost that day? (1 ounce = 28.4 g.)

1.49 How many seconds are in a solar year (365.24 days)?

1.50 How many minutes does it take light from the sun to reach Earth? (The distance from the sun to Earth is 93 million mi; the speed of light = 3.00×10^8 m/s.)

1.51 A slow jogger runs a mile in 13 min. Calculate the speed in (a) in/s, (b) m/min, (c) km/h. (1 mi = 1609 m; 1 in = 2.54 cm.)

1.52 Carry out the following conversions: (a) A 5.9-ft person weighs 158 lb. Express this person's height in meters and weight in kilograms. (1 lb = 453.6 g; 1 m = 3.28 ft.) (b) The current speed limit in some states in the U.S. is 55 miles per hour. What is the speed limit in kilometers per hour? (c) The speed of light is 3.0×10^{10} cm/s. How many miles does light travel in 1 hour? (d) Lead is a toxic substance. The "normal" lead content in human blood is about 0.40 parts per million (that is, 0.40 g of lead per million grams of blood). A value of 0.80 parts per million (ppm) is considered to be dangerous. How many grams of lead are contained in 24.5 lb of blood if the lead content is 0.62 ppm?

1.53 In a city with heavy automobile traffic such as Los Angeles or New York, it is estimated that about 9.0 tons of lead from exhausts are deposited on or near the highway every day. What is this amount in kilograms per month? (1 ton = exactly 2000 lb; 1 month = 31 days.)

1.54 Carry out the following conversions: (a) 1.42 light-years to miles (a light-year is an astronomical measure of distance—the distance traveled by light in a year, or 365 days), (b) 32.4 yd to centimeters, (c) 3.0×10^{10} cm/s to ft/s, (d) 47.4°F to degrees Celsius, (e) −273.15°C to degrees Fahrenheit, (f) 71.2 cm^3 to m^3, (g) 7.2 m^3 to liters.

1.55 The density of aluminum is 2.70 g/cm^3. What is its density in kg/m^3?

1.56 The density of ammonia gas under certain conditions is 0.625 g/L. Calculate its density in g/cm^3.

Miscellaneous Problems

1.57 Give one qualitative and one quantitative statement about each of the following: (a) water, (b) carbon, (c) iron, (d) hydrogen gas, (e) sucrose (cane sugar), (f) table salt (sodium chloride), (g) mercury, (h) gold, (i) air.

1.58 How is it that an astronaut of mass 75 kg achieves a weightless condition when aboard a Skylab orbital space laboratory?

1.59 The elements in Group 8A of the periodic table are called noble gases. Can you guess the meaning of "noble" in this context?

1.60 Which of the following describe physical and which describe chemical properties? (a) Iron has a tendency to rust. (b) Rainwater in industrialized regions tends to be acidic. (c) Hemoglobin molecules have a red color. (d) When a glass of water is left out in the sun, the water gradually disappears. (e) Carbon dioxide in air is converted to more complex molecules by plants during photosynthesis.

1.61 Describe a chemical property exhibited by each of the following: (a) air, (b) water, (c) ethanol, (d) wax, (e) bread, (f) sodium metal.

1.62 Consult a handbook of chemical and physical data (ask your instructor where you can locate a copy of the handbook) to find (a) two metals less dense than water, (b) two metals more dense than mercury, (c) the densest known solid metallic element, (d) the densest known nonmetallic element.

1.63 Suppose that a new temperature scale has been established on which the melting point of ethanol (−117.3°C) and the boiling point of ethanol (78.3°C) are taken as 0°S and 100°S, respectively, where S is the symbol for the new temperature scale. Derive an equation relating a reading on this scale to a reading on the Celsius scale. What would be the reading of this thermometer at 25°C?

1.64 Group the following elements in pairs that you would expect to show similar physical and chemical properties: F, K, P, Na, Cl, and As. (*Hint:* See Figure 1.5.)

1.65 In the determination of the density of the material of a rectangular metal bar, a student made the following measurements: length, 8.53 cm; width, 2.4 cm; height, 1.0 cm; mass, 52.7064 g. Calculate the density of the material, giving your answer with the correct number of significant figures.

1.66 Calculate the mass of each of the following: (a) a sphere of gold of radius 10.0 cm (the volume of a sphere of radius r is $V = (4/3)\pi r^3$; the density of gold = 19.3 g/cm^3), (b) a cube of platinum of edge length 0.040 mm (the density of platinum = 21.4 g/cm^3), (c) 50.0 mL of ethanol (the density of ethanol = 0.798 g/mL).

1.67 A cylindrical glass tube 12.7 cm in length is filled with mercury. The mass of mercury needed to fill the tube is found to be 105.5 g. Calculate the inner diameter of the tube. (The density of mercury = 13.6 g/mL.)

1.68 The following procedure was carried out to determine the volume of a flask. The flask was weighed dry and then filled with water. If the masses of the empty flask and filled flask were 56.12 g and 87.39 g, respectively, and the density of water is 0.9976 g/cm^3, calculate the volume of the flask in cm^3.

1.69 A piece of silver (Ag) metal weighing 194.3 g is placed in a graduated cylinder containing 242.0 mL of water. The volume of water now reads 260.5 mL. From these data calculate the density of silver.

1.70 The experiment described in Problem 1.69 is a crude but convenient way to determine the density of some solids. Describe a similar experiment that would allow you to measure the density of ice. Specifically, what would be the requirements for the liquid used in your experiment?

1.71 The speed of sound in air at room temperature is about 343 m/s. Calculate this speed in miles per hour (mph).

1.72 Of the 109 elements known, only two are liquids at room temperature (25°C). What are they?

1.73 List the elements that exist as gases at room temperature.

1.74 Referring to Appendix 1, list five elements each that are (a) named after places, (b) named after people, (c) named after color.

1.75 Describe the changes in properties (from metals to nonmetals or from nonmetals to metals) as we move (a) down a periodic group and (b) across the periodic table.

1.76 Elements whose names end with ium are usually metals. Name a nonmetal whose name also ends with ium.

1.77 The medicinal thermometer used in homes can be read to ±0.1°F, while those in the doctor's office may be accurate to ±0.1°C. In degrees Celsius, express the percent error expected from each of these thermometers in measuring a temperature of 38.9°C.

Opposite page: Illustration depicting Marie and Pierre Curie at work in their laboratory. The Curies studied and identified many radioactive elements.

C H A P T E R 2

ATOMS, MOLECULES, AND IONS

Since ancient times humans have pondered the nature of matter. Our modern ideas of the structure of matter began to take shape in the early nineteenth century with Dalton's atomic theory. We now know that all matter is made of atoms, molecules, and ions. All of chemistry is concerned in one way or another with these species.

2.1 The Atomic Theory—From Early Ideas to John Dalton

In the fifth century B.C. the Greek philosopher Democritus expressed the belief that all matter is composed of very small, indivisible particles, which he named *atomos* (meaning uncuttable or indivisible). Although Democritus' idea was not accepted by many philosophers of his day (notably Plato and Aristotle), his suggestion persisted through the centuries. Experimental evidence from early scientific investigations provided support for the notion of "atomism" and gradually gave rise to our modern definitions of elements and compounds. However, it was not until 1808 that an English scientist and school teacher, John Dalton,† formulated a precise definition of the indivisible building blocks of matter that we call atoms.

Dalton's atomic theory marks the beginning of the modern era of chemistry. The hypotheses about the nature of matter on which Dalton based his theory can be summarized as follows:

- Elements are composed of extremely small particles, called atoms. All atoms of a given element are identical, having the same size, mass, and chemical properties. The atoms of one element are different from the atoms of all other elements.
- Compounds are composed of atoms of more than one element. In any compound, the ratio of the numbers of atoms of any two of the elements present is either an integer or a simple fraction.
- A chemical reaction involves only the separation, combination, or rearrangement of atoms; it does not result in their creation or destruction.

Figure 2.1 is a schematic representation of the first two hypotheses.

As you can see, Dalton's concept of an atom was far more detailed and specific than the description of Democritus. The first hypothesis states that atoms are different for different elements. Dalton made no attempt to describe the structure or composition of atoms—he had no idea what an atom is really like. But he did realize that the different properties shown by elements such as hydrogen and oxygen, for example, can be explained by assuming that hydrogen atoms are not the same as oxygen atoms.

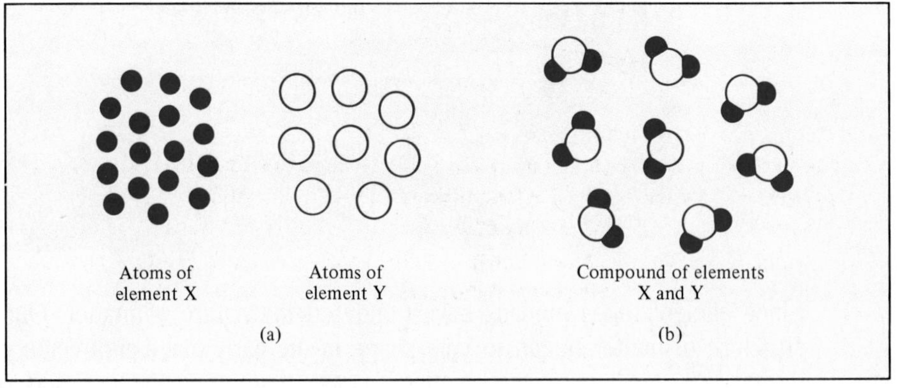

Figure 2.1 *(a) According to Dalton's atomic theory, atoms of the same element are identical, but atoms of one element are different from atoms of other elements. (b) Compound formed from atoms of elements X and Y. In this case, the ratio of the atoms of element X to the atoms of element Y is 2:1.*

†John Dalton (1766–1844). English chemist, mathematician, and philosopher. In addition to the atomic theory, he also formulated several gas laws and gave the first detailed description of color blindness, from which he suffered. Dalton was described as an indifferent experimenter, and singularly wanting in the language and power of illustration. His only recreation was lawn bowling on Thursday afternoons. Perhaps it was the sight of those wooden balls that provided him with the idea of the atomic theory.

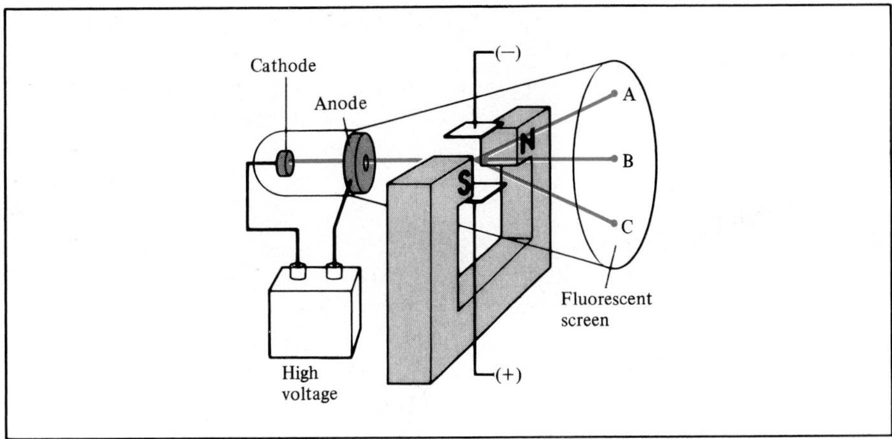

Figure 2.2 *A cathode ray tube with an electric field perpendicular to the direction of the cathode rays and an external magnetic field. The symbols N and S denote the north and south poles of the magnet. The cathode rays will strike the end of the tube at A in the presence of a magnetic field; at C in the presence of an electric field; at B when there are no external fields present or when the effects of the electric field and magnetic field cancel each other.*

The second hypothesis suggests that, in order to form a certain compound, we need not only atoms of the right kinds of elements, but the correct numbers of these atoms as well. The last hypothesis is another way of stating the ***law of conservation of mass,*** which says that *matter can be neither created nor destroyed.* Since matter is made of atoms that are unchanged in a chemical reaction, it follows that mass must be conserved as well. Dalton's brilliant insight into the nature of matter was the main cause of the rapid progress of chemistry in the nineteenth century.

2.2 The Structure of the Atom

On the basis of Dalton's atomic theory, we can define an ***atom*** as *the basic unit of an element that can enter into chemical combination.* Dalton imagined an atom that was both extremely small and indivisible. However, a series of investigations that began in the 1850s and extended into the twentieth century clearly demonstrated that atoms actually possess an internal structure; that is, they are made up of even smaller particles, which are called *subatomic particles.* Research led to the discovery of three such particles—electrons, protons, and neutrons.

The Electron

Electrons are normally associated with atoms. However, they can also be studied individually.

The discovery of electrons and the first detailed study of their behavior came about with the invention of the cathode ray tube, which was the forerunner of today's television tube. Figure 2.2 shows a schematic diagram of a cathode ray tube. *Negatively charged particles,* or ***electrons,*** emitted from the cathode are drawn to a positively charged plate, the *anode.* A hole in the anode allows electrons to pass through. The stream of electrons forms what early investigators named a *cathode ray.* The cathode ray goes on to strike the inside surface of the end of the tube. The surface is coated with a fluorescent material, such as zinc sulfide, so that a strong fluorescence, or emission of light, is observed when the surface is bombarded by the electrons.

In some experiments two electrically charged plates and a magnet were added to the cathode ray tube, as shown in Figure 2.2. When the magnetic field is on and the electric field is off, the cathode ray strikes point A. When only the electric field is on, the ray strikes point C. When both the magnetic and the electric fields are off or when

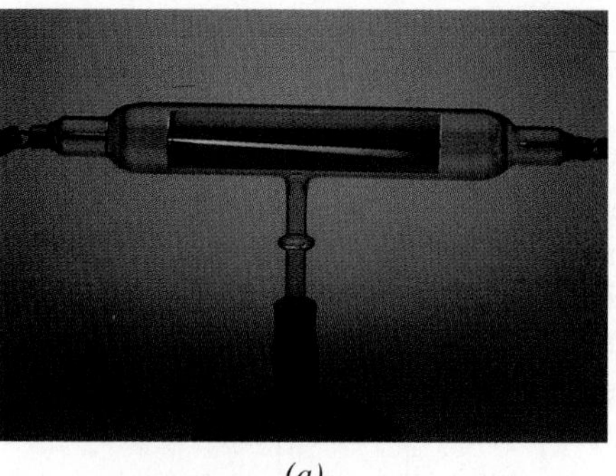

(a)

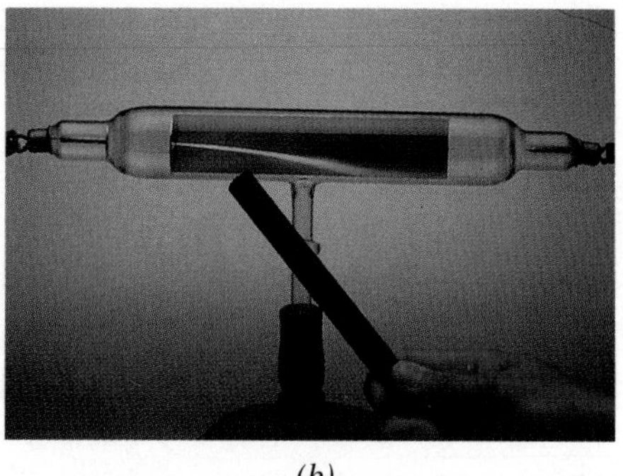

(b)

Figure 2.3 *(a) A cathode ray produced in a discharge tube. The ray itself has no color; the green color is due to the fluorescence of zinc sulfide coated on the screen in contact with the ray. (b) The cathode ray is bent in the presence of a magnet.*

they are both on but balanced so that they cancel each other's influence, the ray strikes point B. Such behavior is consistent with the fact that electrons possess a negative charge. Electromagnetic theory tells us that a moving charged body behaves like a magnet and can interact with electric and magnetic fields through which it passes. Since the cathode ray is attracted by the plate bearing positive charges and repelled by the plate bearing negative charges, it is clear that it must consist of negatively charged particles. Figure 2.3 shows an actual cathode ray tube and the effect of a bar magnet on the cathode ray.

Around the turn of the twentieth century, J. J. Thomson† used a cathode ray tube and his knowledge of the effects of electrical and magnetic forces on a negatively charged particle to obtain the ratio of electric charge to mass for an electron. Thomson found the ratio to be -1.76×10^{8} C/g, where C stands for coulomb, which is the unit of electric charge. Thereafter, in a series of experiments carried out between 1908 and 1917, R. A. Millikan‡ found the charge of an electron to be -1.60×10^{-19} C. From these data we can calculate the mass of an electron:

$$
\begin{aligned}
\text{mass of an electron} &= \frac{\text{charge}}{\text{charge/mass}} \\
&= \frac{-1.60 \times 10^{-19}\ \text{C}}{-1.76 \times 10^{8}\ \text{C/g}} \\
&= 9.09 \times 10^{-28}\ \text{g}
\end{aligned}
$$

which is an exceedingly small mass.

X Rays and Radioactivity

In the 1890s many scientists became caught up in the study of cathode rays and other kinds of rays. Some of these rays were associated with the newly discovered phenomenon called ***radioactivity,*** which is *the spontaneous emission of particles and/or radiation.* ***Radiation*** is the term used to describe *the emission and transmission of energy through space in the form of waves.* A radioactive substance *decays,* or breaks down, spontaneously. By the early twentieth century scientists had discovered several types

†Joseph John Thomson (1856–1940). British physicist who received the Nobel Prize in physics in 1906 for discovering the electron.

‡Robert Andrews Millikan (1868–1953). American physicist who was awarded the Nobel Prize in physics in 1923 for his experiments in determining the charge of the electron.

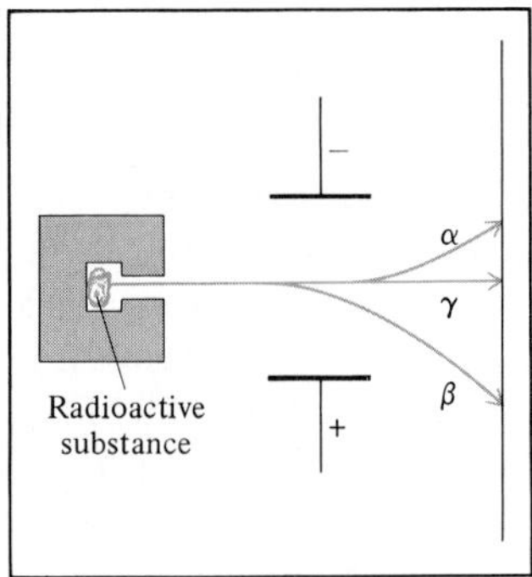

Figure 2.4 *Three types of rays emitted by radioactive elements. β Rays consist of negatively charged particles (electrons) and are therefore attracted by the positively charged plate. The opposite holds true for α rays—they are positively charged and are drawn to the negatively charged plate. Because γ rays do not consist of charged particles, their movement is unaffected by an external electric field.*

of radioactive "rays." Information gained by studying these rays and their effects on other materials contributed greatly to the growing understanding of the structure of the atom.

In 1895 Wilhelm Röntgen† noticed that when cathode rays struck glass and metals, new and very unusual rays were emitted. These rays were highly energetic and could penetrate matter. They also darkened covered photographic plates and could produce fluorescence in various substances. Since these rays could not be deflected by a magnet, they did not consist of charged particles as did cathode rays. Röntgen called them *X rays*. They were later identified as a type of high-energy radiation.

Not long after Röntgen's discovery, Antoine Becquerel,‡ a professor of physics in Paris, began to study fluorescent properties of substances. Purely by accident, he noticed that a certain compound containing uranium was able to darken photographic plates that were wrapped in thick papers or even in thin metal sheets, without the stimulation of cathode rays. The nature of the radiation that was doing this was not known, although it seemed to resemble X rays in being highly energetic and not consisting of charged particles. One of Becquerel's students, Marie Curie,§ suggested the name "radioactivity" for this phenomenon. Any element, such as uranium, that exhibits radioactivity is said to be radioactive. Marie Curie and her husband, Pierre, later studied and identified many radioactive elements.

Further investigation showed that three types of rays can be emitted by radioactive elements. These rays were studied by using an arrangement similar to that shown in Figure 2.4. It was found that two of the three types of rays could be deflected when they passed between two oppositely charged metal plates.

Depending on the deflection, these two rays are called alpha (α) rays and beta (β) rays. The third type of ray, which is unaffected by charged plates, is called a gamma (γ) ray. ***α Rays*** or ***α particles*** were found to be *particles with a positive charge of +2*. They are therefore attracted by the negatively charged plate. The opposite holds true for ***β rays*** or ***β particles***—they consist of *negatively charged electrons,* which are drawn to the positively charged plate. Because ***γ rays*** *do not consist of charged particles,* their movement is unaffected by an external electric field. *They are high-energy radiation.*

The Proton and the Nucleus

By the early 1900s, two features of atoms had become clear: They contain electrons, and they are electrically neutral. Since it is neutral, every atom must contain an equal number of positive and negative charges, to maintain the electrical neutrality. Around the turn of the century, the accepted model for atoms was the one that was proposed by

† Wilhelm Konrad Röntgen (1845–1923). German physicist who received the Nobel Prize in physics in 1901 for the discovery of X rays.

‡ Antoine Henri Becquerel (1852–1908). French physicist who was awarded the Nobel Prize in physics in 1903 for discovering radioactivity in uranium.

§ Marie (Marya Sklodowska) Curie (1867–1934). Polish-born chemist and physicist. In 1903 she and her French husband, Pierre Curie, were awarded the Nobel Prize in physics for their work on radioactivity. In 1911, she again received the Nobel Prize, this time in chemistry, for her work on the radioactive elements radium and polonium. She is one of only three people to have received two Nobel Prizes in science. Despite her great contribution to science, her nomination to the French Academy of Sciences in 1911 was rejected by one vote because she was a woman! Her daughter and son-in-law, Irene and Frederic Joliot-Curie, shared the Nobel Prize in chemistry in 1935.

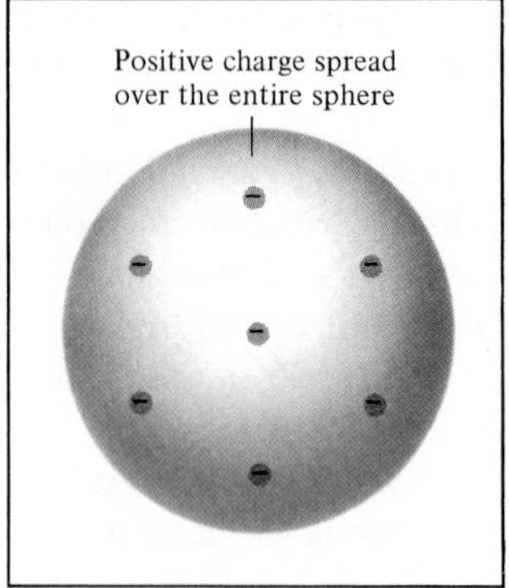

Figure 2.5 *Thomson's model of the atom, sometimes described as the "plum pudding" model, from a traditional English dessert containing raisins. The electrons are embedded in a uniform, positively charged sphere.*

J. J. Thomson. According to his description, an atom could be thought of as a uniform, positive sphere of matter in which electrons are embedded (Figure 2.5).

In 1910 Ernest Rutherford,† who had earlier studied under Thomson at Cambridge University, decided to use α particles to probe the structure of atoms. Together with his associate Hans Geiger‡ and an undergraduate named Ernest Marsden,§ Rutherford carried out a series of experiments in which very thin foils of gold and other metals were used as targets for α particles emitted from a radioactive source (Figure 2.6). They observed that the majority of the particles penetrated the foil either undeflected or with only a slight deflection. They also noticed that every now and then an α particle would be scattered (or deflected) at a large angle. In some instances, an α particle would even be turned back in the direction from which it had come! This was a most surprising finding, for in Thomson's model the positive charge of the atom was so diffuse that the positive α particles were expected to pass through with very little deflection. To quote Rutherford's initial reaction when told of this discovery: "It was as incredible as if you had fired a 15-inch shell at a piece of tissue paper and it came back and hit you."

Rutherford was later able to explain the results of the α-scattering experiment, but he had to abandon Thomson's model and propose a new model for the atom. According to Rutherford, most of the atom must be empty space. This explains why the majority of α particles passed through the gold foil with little or no deflection. The atom's positive charges, Rutherford proposed, are all concentrated in *a central core within the atom,* which he called the ***nucleus.*** Whenever an α particle came close to a nucleus in the scattering experiment, it experienced a large repulsive force and therefore a large deflection. If an α particle traveled directly toward a nucleus, it would experience an enormous repulsion that could completely reverse the direction of the moving particle.

The positively charged particles in the nucleus are called ***protons,*** and each has a

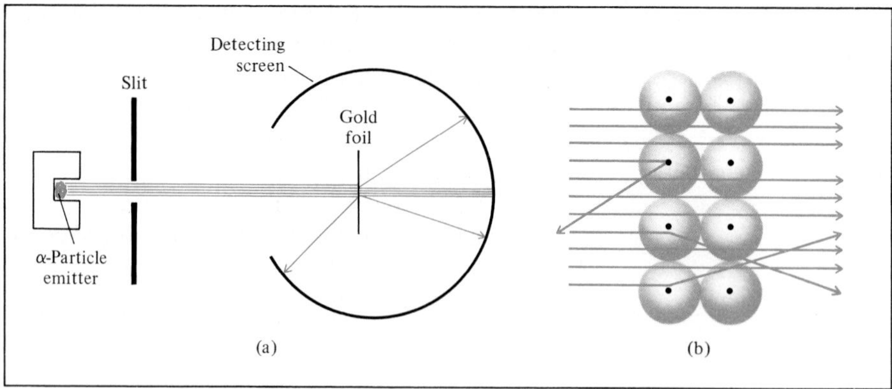

Figure 2.6 *(a) Rutherford's experimental design for measuring the scattering of α particles by a piece of gold foil. Most of the α particles passed through the gold foil with little or no deflection. A few were deflected at wide angles. Occasionally an α particle was turned back. (b) Magnified view of α particles passing through and being deflected by nuclei.*

†Ernest Rutherford (1871–1937). New Zealand physicist. Rutherford did most of his work in England (Manchester and Cambridge universities). He received the Nobel Prize in chemistry in 1908 for his investigations into the structure of the atomic nucleus. His often-quoted comment to his students was that "all science is either physics or stamp-collecting."

‡Johannes Hans Wilhelm Geiger (1882–1945). German physicist. Geiger's work was mainly in solving the structure of the atomic nucleus and in radioactivity. He invented a device for measuring radiation that is now commonly called the Geiger counter.

§Ernest Marsden (1889–1970). English physicist. It is gratifying to know that at times an undergraduate can assist in winning a Nobel Prize. Marsden went on to contribute significantly to the development of science in New Zealand.

mass of 1.67252×10^{-24} g. In separate experiments, it was found that each proton carries the same *quantity* of charge as an electron and is about 1840 times heavier than the oppositely charged electron.

At this stage of investigation, scientists perceived the atom as follows. The mass of a nucleus comprises most of the mass of the entire atom, but the nucleus occupies only about $1/10^{13}$ of the volume of the atom. For atomic (and molecular) dimensions, we will express lengths in terms of the SI unit called the *picometer (pm),* where

A common non-SI unit for atomic length is the angstrom (Å: 1 Å = 100 pm.)

$$1 \text{ pm} = 1 \times 10^{-12} \text{ m}$$

A typical atomic radius is about 100 pm, whereas the radius of an atomic nucleus is only about 5×10^{-3} pm. You can appreciate the relative sizes of an atom and its nucleus by imagining that if an atom were the size of the Houston Astrodome, the volume of its nucleus would be comparable to that of a small marble. While the protons are confined to the nucleus of the atom, the electrons are conceived of as being spread out about the nucleus at some distance from it.

The concept of atomic radius is useful experimentally, but it should not be inferred that atoms have well-defined boundaries or surfaces. We will learn later that the outer regions of atoms are relatively "fuzzy."

The Neutron

In spite of Rutherford's success in explaining atomic structure, one major problem remained unsolved. It was known that hydrogen, the simplest atom, contains only one proton, and that the helium atom contains two protons. Therefore, the ratio of the mass of a helium atom to that of a hydrogen atom should be 2:1. (Because electrons are much lighter than protons, their contribution can be ignored.) In reality, however, the ratio is 4:1. Earlier Rutherford and others had postulated that there must be another type of subatomic particle in the atomic nucleus; the proof was provided by James Chadwick† in 1932. When Chadwick bombarded a thin sheet of beryllium with α particles, a very high-energy radiation that somewhat resembled γ rays was emitted by the metal. Later experiments showed that the rays actually consisted of *electrically neutral particles having a mass slightly greater than that of protons*. Chadwick named these particles ***neutrons.***

Physicists have discovered that atoms release many different kinds of subatomic particles when bombarded by extremely energetic particles under special conditions in "atom smashers." Chemists deal only with electrons, protons, and neutrons, however, because most chemical reactions are carried out under normal conditions.

The mystery of the mass ratio could now be explained. In the helium nucleus there are two protons and two neutrons, and in the hydrogen nucleus there is only one proton and no neutrons; therefore, the ratio is 4:1.

Table 2.1 summarizes the mass and charge of the three subatomic particles that are important in chemistry—the electron, the proton, and the neutron.

Table 2.1 Mass and charge of subatomic particles

		Charge	
Particle	**Mass (g)**	**Coulomb**	**Charge unit**
Electron*	9.1095×10^{-28}	-1.6022×10^{-19}	-1
Proton	1.67252×10^{-24}	$+1.6022 \times 10^{-19}$	$+1$
Neutron	1.67495×10^{-24}	0	0

*More refined experiments have given us a more accurate value of an electron's mass than Millikan's.

†James Chadwick (1891–1972). British physicist. He received the Nobel Prize in physics in 1935 for proving the existence of neutrons.

2.3 Mass Relationships of Atoms

Chemistry, as we have noted, is a quantitative science. In this and the following sections, we will learn how to identify and describe atoms quantitatively.

Atomic Number, Mass Number, and Isotopes

Subatomic particles can help us better understand the properties of individual atoms. All atoms can be identified by the number of protons and neutrons they contain.

Atomic number is actually proton number.

The ***atomic number (Z)*** is *the number of protons in the nucleus of each atom of an element*. In a neutral atom the number of protons is equal to the number of electrons, so that the atomic number also indicates the number of electrons present in the atom. The chemical identity of an atom can be determined solely by its atomic number. For example, the atomic number of nitrogen is 7; this means that each neutral nitrogen atom has 7 protons and 7 electrons. Or, viewing it another way, every atom in the universe that contains 7 protons is correctly named "nitrogen."

The ***mass number (A)*** is *the total number of neutrons and protons present in the nucleus of an atom of an element*. Except for the atom of the most common form of hydrogen, which has one proton and no neutrons, all atomic nuclei contain both protons and neutrons. In general the mass number is given by

$$\begin{aligned}\text{mass number} &= \text{number of protons} + \text{number of neutrons}\\ &= \text{atomic number} + \text{number of neutrons}\end{aligned}$$

The number of neutrons in an atom is equal to the difference between the mass number and the atomic number, or $(A - Z)$. For example, the mass number of fluorine is 19 and the atomic number is 9 (indicating 9 protons in the nucleus). Thus the number of neutrons in an atom of fluorine is $19 - 9 = 10$. Note that all three quantities (atomic number, number of neutrons, and mass number) must be positive integers, or whole numbers.

In most cases atoms of a given element do not all have the same mass. For example, there are three types of hydrogen atoms, which differ only in their number of neutrons. They are hydrogen, with one proton and no neutrons; deuterium, with one proton and one neutron; and tritium, with one proton and two neutrons. *Atoms that have the same atomic number but different mass numbers* are called ***isotopes***.

The name "protium" has been proposed for the isotope of hydrogen containing one proton and no neutrons, but it is not widely used by chemists.

The accepted way to denote the atomic number and mass number of an atom of element X is as follows:

$$\begin{matrix}\text{mass number} \searrow \\ \text{atomic number} \nearrow\end{matrix} {}^{A}_{Z}\text{X}$$

Thus, for the isotopes of hydrogen, we write

$$\underset{\text{hydrogen}}{{}^{1}_{1}\text{H}} \qquad \underset{\text{deuterium}}{{}^{2}_{1}\text{H}} \qquad \underset{\text{tritium}}{{}^{3}_{1}\text{H}}$$

As another example, consider two common isotopes of uranium with mass numbers of 235 and 238, respectively:

$${}^{235}_{92}\text{U} \qquad {}^{238}_{92}\text{U}$$

If we know the name of the element, we always know its atomic number, but we may not know its mass number.

The first isotope is used in nuclear reactors and atomic bombs, whereas the second isotope lacks the properties to be utilized in these respects. With the exception of hydrogen, isotopes of elements are identified by their mass numbers. Thus the above two isotopes are called uranium-235 (pronounced "uranium two thirty-five") and uranium-238 (pronounced "uranium two thirty-eight").

The chemical properties of an element are determined primarily by the protons and electrons in its atoms; neutrons do not take part in chemical changes under normal conditions. Therefore, isotopes of the same element have similar chemistries, forming the same types of compounds and displaying similar reactivities.

The following example illustrates how to determine the number of elementary particles in atoms from the atomic and mass numbers.

EXAMPLE 2.1

Give the number of protons, neutrons, and electrons in each of the following species: (a) $^{11}_{5}B$, (b) $^{199}_{80}Hg$, (c) $^{200}_{80}Hg$.

Answer

(a) The atomic number is 5, so there are 5 protons. The mass number is 11, so the number of neutrons is $11 - 5 = 6$. The number of electrons is the same as the number of protons, that is, 5.
(b) The atomic number is 80, so there are 80 protons. The mass number is 199, so the number of neutrons is $199 - 80 = 119$. The number of electrons is 80.
(c) Here the number of protons is the same as in (b), or 80. The number of neutrons is $200 - 80 = 120$. The number of electrons is also the same as in (b), 80. The species in (b) and (c) are two isotopes of mercury that are chemically similar.

Similar problems: 2.16, 2.17.

Atomic Masses

One of the fundamental properties of an atom is its mass. The mass of an atom is related to the number of electrons, protons, and neutrons in the atom. Knowledge of an atom's mass is also important in laboratory work. But atoms are extremely small particles—even the smallest speck of dust that our unaided eyes can detect contains as many as 1×10^{16} atoms! If atoms are so tiny, how can we ever hope to determine their mass? We cannot weigh a single atom, but there are experimental methods of determining the mass of one atom *relative* to another. The first step is to assign a value to the mass of one atom of a given element so that it can be used as a standard.

Section 2.8 describes a method for determining atomic mass.

By international agreement, an atom of the isotope of carbon (called carbon-12) that has six protons and six neutrons has a mass of exactly 12 ***atomic mass units (amu).*** This carbon-12 atom serves as the standard, so one atomic mass unit (amu) is defined as *a mass exactly equal to one-twelfth the mass of one carbon-12 atom.*

One atomic mass unit is also called 1 dalton.

$$\text{mass of one carbon-12 atom} = 12 \text{ amu}$$

$$1 \text{ amu} = \frac{\text{mass of one carbon-12 atom}}{12}$$

The term "atomic weight" has also been used to mean atomic mass.

Experiments have shown that, on average, a hydrogen atom is only 8.400 percent as massive as the standard carbon-12 atom. Thus if we accept the mass of one carbon-12 atom to be exactly 12 amu, then the ***atomic mass*** (that is, *the mass of the atom in atomic mass units*) of hydrogen must be $0.08400 \times 12 = 1.008$ amu. Similar calculations show that the atomic mass of oxygen is 16.00 amu and that of iron is 55.85 amu. Note that although we do not know just how much an average iron atom's mass is, we know that it is approximately fifty-six times as massive as a hydrogen atom.

Average Atomic Mass

When you look up the atomic mass of carbon in a table such as the one on the inside front cover of this book, you will find that its value is not 12.00 amu but 12.01 amu. The reason for the difference is that most naturally occurring elements (including carbon) have more than one isotope. This means that when we measure the atomic mass of an element, we must generally settle for the *average* mass of the naturally occurring mixture of isotopes. For example, the natural abundances of carbon-12 and carbon-13 are 98.89 percent and 1.11 percent, respectively. The atomic mass of carbon-13 has been determined to be 13.00335 amu. Thus the average atomic mass of carbon can be calculated as follows:

$$\text{average atomic mass of natural carbon} = (0.9889)(12.00000\text{ amu}) + (0.0111)(13.00335\text{ amu}) = 12.0\text{ amu}$$

A more accurate determination gives the atomic mass of carbon as 12.01 amu. Note that in calculations involving percentages, we need to convert percentages to fractions. For example, 98.89 percent becomes 98.89/100, or 0.9889. Because there are many more carbon-12 than carbon-13 atoms in naturally occurring carbon, the average atomic mass is much closer to 12 amu than 13 amu.

It is important to understand that when we say that the atomic mass of carbon is 12.01 amu, we are referring to the *average* value. If carbon atoms could be examined individually, we would find either an atom of atomic mass 12.00000 amu or one of 13.00335 amu, but never one of 12.01 amu.

The following example shows how the average atomic mass of an element is calculated.

Copper.

EXAMPLE 2.2

Copper, a metal known since ancient times, is used in electrical cables and pennies, among other things. The atomic masses of its two stable isotopes, $^{63}_{29}Cu$ (69.09%) and $^{65}_{29}Cu$ (30.91%), are 62.93 amu and 64.9278 amu, respectively. Calculate the average atomic mass of copper. The percentages in parentheses denote the relative abundances.

Answer

Converting the percentages to fractions, we calculate the average atomic mass as follows:

$$(0.6909)(62.93\text{ amu}) + (0.3091)(64.9278\text{ amu}) = 63.55\text{ amu}$$

Similar problem: 2.30.

The atomic masses of many elements have been accurately determined to five or six significant figures. However, for purposes of calculation in this text, we will normally use atomic masses accurate only to four significant figures (see the table of atomic masses inside the front cover).

Molar Mass of an Element and Avogadro's Number

We have seen that atomic mass units provide a relative scale for the masses of the elements. But since atoms have such small masses, no usable scale can be devised to weigh them in calibrated units of atomic mass units. In any real situation (for example, in the laboratory) we deal with samples of substances containing enormous numbers of atoms. Therefore it would be convenient to have a special unit to describe a very large number of atoms. The idea of a unit to describe a particular number of objects is not new. For example, the pair (2 items), the dozen (12 items), and the gross (144 items) are all familiar units.

The adjective formed from the noun "mole" is "molar."

The unit defined by the SI system is the ***mole (mol),*** which is *the amount of substance that contains as many elementary entities (atoms, molecules, or other particles) as there are atoms in exactly 12 grams (or 0.012 kilogram) of the carbon-12 isotope.* Notice that this definition specifies only the *method* by which the number of elementary entities in a mole may be found. The actual *number* is determined experimentally. The currently accepted value is

$$1 \text{ mole} = 6.022045 \times 10^{23} \text{ particles}$$

This number is called ***Avogadro's number,*** in honor of the Italian scientist Amedeo Avogadro.† In most of our calculations, we will round this number to 6.022×10^{23}. The term "mole" and its symbol "mol" are used to represent Avogadro's number. Just as one dozen oranges contains twelve oranges, 1 mole of hydrogen atoms contains 6.022×10^{23} H atoms. Figure 2.7 shows 1 mole each of several common elements.

Figure 2.7 *One mole each of several common elements: copper (as pennies), iron (as nails), carbon (black charcoal powder), sulfur (yellow powder), and mercury (shiny liquid metal).*

† Lorenzo Romano Amedeo Carlo Avogadro di Quaregua e di Cerreto (1776–1856). Italian mathematical physicist. He practiced law for many years before he became interested in science. His most famous work, now known as Avogadro's law (see Chapter 5), was largely ignored during his lifetime, although it became the basis for determining atomic masses in the late nineteenth century.

The term "gram molecular weight" has also been used to mean molar mass.

The units of molar mass are g/mol or kg/mol. However, it is also acceptable to say that the molar mass of Na is 22.99 g rather than 22.99 g/mol.

We have seen that 1 mole of carbon-12 atoms has a mass of exactly 12 grams and contains 6.022×10^{23} atoms. This quantity is called the ***molar mass*** of carbon-12, *the mass (in grams or kilograms) of 1 mole of units* (such as atoms or molecules) of the substance. Since each carbon-12 atom has an atomic mass of exactly 12 amu, it is useful to observe that the molar mass of an element (in grams) is numerically equal to its atomic mass in amu. Thus the atomic mass of sodium (Na) is 22.99 amu and its molar mass is 22.99 grams; the atomic mass of copper (Cu) is 63.55 amu and its molar mass is 63.55 grams; and so on. If we know the atomic mass of an element, we also know its molar mass.

We can now calculate the mass (in grams) of a single carbon-12 atom. From our discussion we know that 1 mole of carbon-12 atoms weighs exactly 12 grams. This means that

$$12.00 \text{ g carbon-12} = 1 \text{ mol carbon-12 atoms}$$

Therefore, we can write the unit factor as

$$\frac{12.00 \text{ g carbon-12}}{1 \text{ mol carbon-12 atoms}} = 1$$

Similarly, since there are 6.022×10^{23} atoms in 1 mole of carbon-12 atoms, we have

$$1 \text{ mol carbon-12 atoms} = 6.022 \times 10^{23} \text{ carbon-12 atoms}$$

and the unit factor is

$$\frac{1 \text{ mol carbon-12 atoms}}{6.022 \times 10^{23} \text{ carbon-12 atoms}} = 1$$

We can now write

$$\begin{aligned} &\text{mass (in grams) of} \\ &1 \text{ carbon-12 atom} = 1 \cancel{\text{carbon-12 atom}} \times \frac{1 \cancel{\text{mol carbon-12 atoms}}}{6.022 \times 10^{23} \cancel{\text{carbon-12 atoms}}} \times \frac{12.00 \text{ g carbon-12}}{1 \cancel{\text{mol carbon-12 atoms}}} \\ &= 1.993 \times 10^{-23} \text{ g} \end{aligned}$$

Furthermore, we can find the relationship between atomic mass units and grams by noting that since the mass of every carbon-12 atom is exactly 12 amu, the number of grams equivalent to 1 amu is

$$\begin{aligned} \frac{\text{gram}}{\text{amu}} &= \frac{1.993 \times 10^{-23} \text{ g}}{1 \cancel{\text{carbon-12 atom}}} \times \frac{1 \cancel{\text{carbon-12 atom}}}{12 \text{ amu}} \\ &= 1.661 \times 10^{-24} \text{ g/amu} \end{aligned}$$

Thus

$$1 \text{ amu} = 1.661 \times 10^{-24} \text{ g}$$

and

$$1 \text{ g} = 6.022 \times 10^{23} \text{ amu}$$

This example shows that Avogadro's number can be used to convert from the atomic mass unit to the mass in grams and vice versa.

The notions of Avogadro's number and molar mass enable us to carry out conversions between mass of atoms and moles of atoms, number of atoms and mass of atoms, and to calculate the mass of a single atom. Examples 2.3–2.6 show how these conversions are carried out. We will employ the following unit factors in the calculations:

$$\frac{1 \text{ mol X}}{\text{molar mass of X}} = 1 \qquad \frac{1 \text{ mol X}}{6.022 \times 10^{23} \text{ X atoms}} = 1$$

where X represents the symbol of an element.

EXAMPLE 2.3

Helium (He) is a valuable gas used in industry, low-temperature research, deep-sea diving, and balloons. How many moles of He are in 6.46 g of He?

Answer

The atomic masses of the elements are given in the inside front cover of the book.

First we find that the molar mass of He is 4.003 g. This can be expressed by the equality 1 mole of He = 4.003 g of He. To convert the amount of He in grams to He in moles, we write

$$6.46 \cancel{\text{g He}} \times \frac{1 \text{ mol He}}{4.003 \cancel{\text{g He}}} = 1.61 \text{ mol He}$$

Thus there are 1.61 moles of He atoms in 6.46 g of He.

Similar problem: 2.37.

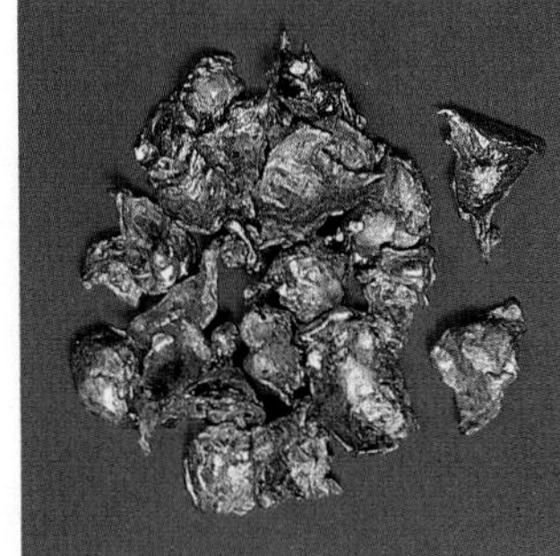

Zinc.

EXAMPLE 2.4

Zinc (Zn) is a silvery metal that is used to form brass (with copper) and plate iron to prevent corrosion. How many grams of Zn are there in 0.356 mol of Zn?

Answer

Since the molar mass of Zn is 65.39 g, the mass of Zn in grams is given by

$$0.356 \cancel{\text{mol Zn}} \times \frac{65.39 \text{ g Zn}}{1 \cancel{\text{mol Zn}}} = 23.3 \text{ g Zn}$$

Thus there are 23.3 g of Zn in 0.356 mole of Zn.

Similar problem: 2.38.

Silver.

EXAMPLE 2.5

Silver (Ag) is a precious metal used mainly in jewelry. What is the mass (in grams) of 1 Ag atom?

Answer

The molar mass of silver is 107.9 g. Since there are 6.022×10^{23} Ag atoms in 1 mole of Ag, the mass of 1 Ag atom is

$$1 \cancel{\text{Ag atom}} \times \frac{1 \cancel{\text{mol Ag atoms}}}{6.022 \times 10^{23} \cancel{\text{Ag atoms}}} \times \frac{107.9 \text{ g Ag}}{1 \cancel{\text{mol Ag atoms}}} = 1.792 \times 10^{-22} \text{ g}$$

Similar problem: 2.39.

Sulfur.

EXAMPLE 2.6

Sulfur (S) is a nonmetallic element. Its presence in coal gives rise to the acid rain phenomenon. How many atoms are in 16.3 g of S?

Answer

Solving this problem requires two steps. First, we need to find the number of moles of S in 16.3 g of S (as in Example 2.3). Next, we need to calculate the number of S atoms from the known number of moles of S. We can combine the two steps as follows:

$$16.3 \cancel{\text{g S}} \times \frac{1 \cancel{\text{mol S}}}{32.07 \cancel{\text{g S}}} \times \frac{6.022 \times 10^{23} \text{ S atoms}}{1 \cancel{\text{mol S}}} = 3.06 \times 10^{23} \text{ S atoms}$$

Similar problem: 2.41.

2.4 Molecules: Atoms in Combination

Molecules and Chemical Formulas

We will discuss the nature of these chemical forces (also called chemical bonds) in later chapters.

A ***molecule*** *is an aggregate of at least two atoms in a definite arrangement held together by chemical forces.* In Section 1.3 we discussed the symbols used to represent individual elements. To denote molecules we combine these symbols into chemical formulas. A ***chemical formula*** *expresses the composition of a compound in terms of the symbols of the atoms of the elements involved.* By composition we mean not only the elements present in the compound but also the ratios in which atoms occur in the compound. The two types of chemical formulas we need to become familiar with are molecular formulas and empirical formulas.

Molecular Formula

A ***molecular formula*** *shows the exact number of atoms of each element in a molecule. The simplest type of molecule contains only two atoms* and is called a ***diatomic molecule.*** Elements that exist as diatomic molecules under atmospheric conditions include hydrogen (H_2), nitrogen (N_2), and oxygen (O_2), as well as the Group 7A halogens—

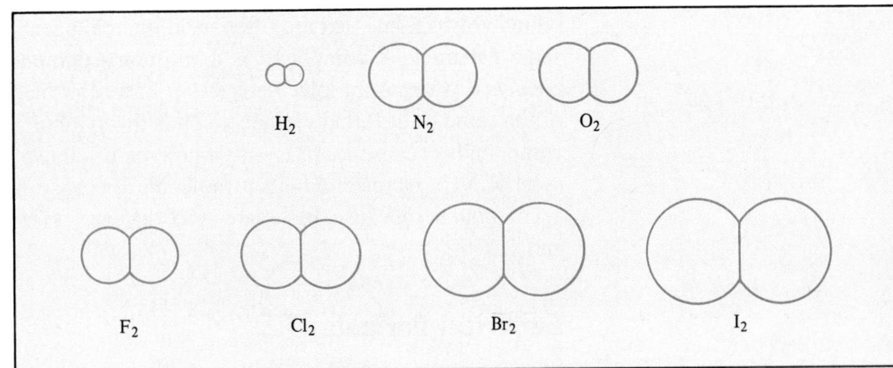

Figure 2.8 *Relative sizes of the diatomic molecules H_2, N_2, and O_2, and the halogens (F_2, Cl_2, Br_2, and I_2).*

fluorine (F_2), chlorine (Cl_2), bromine (Br_2), and iodine (I_2). Figure 2.8 shows models for some of these diatomic molecules. In each case the subscript (2) in the formula indicates the number of atoms in the molecule. Of course, a diatomic molecule can contain atoms of two different elements, as in hydrogen chloride (HCl) and carbon monoxide (CO). These formulas have no subscripts because when the number of a particular type of atom present is one, the number is not shown as a subscript.

Sometimes chemists are rather sloppy in terminology. For example, when a chemist says "hydrogen," it is not always clear whether he or she means atomic hydrogen (H) or a hydrogen molecule (H_2). To avoid such confusion, we will adhere to the practice, whenever appropriate, of using the term "atomic hydrogen" for hydrogen atoms, "molecular hydrogen" for hydrogen molecules, and "the hydrogen element" when we are discussing the properties of the element hydrogen. The same practice also applies to other substances.

A molecule may contain more than two atoms either of the same type, as in ozone (O_3), or of different types, as in water (H_2O) and ammonia (NH_3). Figure 2.9 shows models for these three molecules. *Molecules containing more than two atoms* are called ***polyatomic molecules.*** Note that both oxygen (O_2) and ozone (O_3) are elemental forms of the same element, oxygen. *Different forms of the same element* are called ***allotropes.*** Two allotropic forms of the element carbon—diamond and graphite—present dramatic differences not only in properties but also in their relative cost (see Chemistry in Action, p. 76).

In their simplest forms, molecular formulas are written by grouping the atoms of the same elements together. For example, the molecular formula of ethanol is C_2H_6O. However, chemists often write its molecular formula as CH_3CH_2OH (or in a more abbreviated form as C_2H_5OH). This kind of representation tells us how the atoms in the molecule are linked together, and as we will see later, it also reveals some of the chemical properties of the molecule.

You should be aware of the important differences between "molecule" and "com-

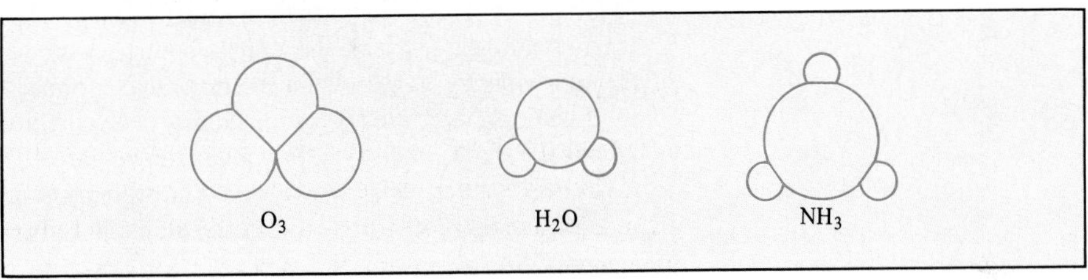

Figure 2.9 *Relative sizes and approximate shapes of the polyatomic molecules O_3, H_2O, and NH_3.*

pound,'' two terms that are often used interchangeably but do not necessarily have the same meaning. A compound is a *substance* composed of the atoms of two or more *elements,* whereas a molecule is a *unit* of a substance composed of two or more *atoms* of the same or different elements. Thus the symbol F_2 represents a molecule but not a compound, because there is only one type of element present. On the other hand, the symbol NH_3 represents both a molecule (because there are four atoms present) and a compound (because there are two different elements present). (See Figures 2.8 and 2.9.)

Empirical Formula

The word ''empirical'' means ''derived from experiment.'' As we will see shortly, empirical formulas are determined experimentally.

The molecular formula of hydrogen peroxide, a substance used as an antiseptic and as a bleaching agent for textiles and hair, is H_2O_2. This formula indicates that each hydrogen peroxide molecule consists of two hydrogen atoms and two oxygen atoms. The ratio of hydrogen to oxygen atoms in this molecule is 2:2 or 1:1. The empirical formula of hydrogen peroxide is written as HO. Thus the ***empirical formula*** *tells us which elements are present and the simplest whole-number ratio of their atoms,* but not necessarily the actual number of atoms present in the molecule. As another example, consider the compound hydrazine (N_2H_4), which is used as a rocket fuel. The empirical formula of hydrazine is NH_2. Although the ratio of nitrogen to hydrogen is 1:2 in both the molecular formula (N_2H_4) and the empirical formula (NH_2), only the molecular formula tells us the actual number of N atoms (two) and H atoms (four) present in a hydrazine molecule. Empirical formulas are therefore the simplest chemical formulas; they are always written so that the subscripts in the molecular formulas are converted to the *smallest* possible whole numbers. Molecular formulas are the *true* formulas of molecules.

For many molecules, the molecular formula and the empirical formula are one and the same. Some examples are water (H_2O), ammonia (NH_3), carbon dioxide (CO_2), and methane (CH_4).

The relationship between the empirical formula and the molecular formula for some molecules is shown in the following example.

EXAMPLE 2.7

Write the empirical formulas for the following molecules: (a) acetylene (C_2H_2), (b) dinitrogen tetroxide (N_2O_4), (c) glucose ($C_6H_{12}O_6$), (d) diiodine pentoxide (I_2O_5).

Answer

(a) There are two carbon atoms and two hydrogen atoms in acetylene. Dividing the subscripts by 2, we obtain the empirical formula CH.

(b) In dinitrogen tetroxide there are two nitrogen atoms and four oxygen atoms. Dividing the subscripts by 2, we obtain the empirical formula NO_2.

(c) In glucose there are six carbon atoms, twelve hydrogen atoms, and six oxygen atoms. Dividing the subscripts by 6, we obtain the empirical formula CH_2O. Note that if we had divided the subscripts by 3, we would have obtained the formula $C_2H_4O_2$. Although the ratio of carbon to hydrogen to oxygen atoms in $C_2H_4O_2$ is the same as that in $C_6H_{12}O_6$

(1:2:1), $C_2H_4O_2$ is not the simplest formula because its subscipts have not been converted to their smallest whole numbers.
(d) Since the subscripts in I_2O_5 are already the smallest whole numbers (there is no number by which both 2 and 5 are divisible), the empirical formula for diiodine pentoxide is the same as its molecular formula.

Similar problem: 2.53.

Molecular Mass

The term "molecular weight" has also been used to mean molecular mass.

Once we know the atomic masses, we can proceed to calculate the masses of molecules. The ***molecular mass*** is *the sum of the atomic masses (in amu) in the molecule*. For example, the molecular mass of H_2O is

$$2(\text{atomic mass of H}) + \text{atomic mass of O}$$

or

$$2(1.008 \text{ amu}) + 16.00 \text{ amu} = 18.02 \text{ amu}$$

EXAMPLE 2.8

Calculate the molecular masses of the following compounds: (a) sulfur dioxide (SO_2); (b) ascorbic acid, or vitamin C ($C_6H_8O_6$).

Answer

(a) From the atomic masses of S and O we get

$$\begin{aligned}\text{molecular mass of } SO_2 &= 32.07 \text{ amu} + 2(16.00 \text{ amu})\\ &= 64.07 \text{ amu}\end{aligned}$$

(b) From the atomic masses of C, H, and O we get

$$\begin{aligned}\text{molecular mass of } C_6H_8O_6 &= 6(12.01 \text{ amu}) + 8(1.008 \text{ amu}) + 6(16.00 \text{ amu})\\ &= 176.12 \text{ amu}\end{aligned}$$

Similar problem: 2.56.

The ***molar mass of a compound*** is *the mass (in grams or kilograms) of 1 mole of the compound*. It is useful to remember that the molar mass of a compound (in grams) is numerically equal to its molecular mass (in amu). For example, the molecular mass of water is 18.02 amu, so its molar mass is 18.02 g. Both the molecular mass and the molar mass of a compound have the same number, but they differ from each other in their units (amu versus g). The principle here is similar to that regarding atomic mass and molar mass of an atom discussed in Section 2.3. One mole of water weighs 18.02 g and contains 6.022×10^{23} H_2O molecules.

As the following two examples show, a knowledge of the molar mass enables us to calculate the number of moles and amounts of individual atoms in a given quantity of a compound.

EXAMPLE 2.9

Methane (CH_4) is the principal component of natural gas. How many moles of CH_4 are present in 6.07 g of CH_4?

Answer

First we calculate the molar mass of CH_4:

$$\text{molar mass of } CH_4 = 12.01 \text{ g} + 4(1.008 \text{ g}) = 16.04 \text{ g}$$

We then follow the procedure in Example 2.3:

$$6.07 \cancel{\text{g } CH_4} \times \frac{1 \text{ mol } CH_4}{16.04 \cancel{\text{g } CH_4}} = 0.378 \text{ mol } CH_4$$

Similar problem: 2.61.

EXAMPLE 2.10

How many hydrogen atoms are present in 25.6 g of sucrose, or table sugar ($C_{12}H_{22}O_{11}$)? The molar mass of sucrose is 342.3 g.

Answer

There are 22 hydrogen atoms in every sucrose molecule; therefore, the total number of hydrogen atoms is

$$25.6 \cancel{\text{g } C_{12}H_{22}O_{11}} \times \frac{1 \cancel{\text{mol } C_{12}H_{22}O_{11}}}{342.3 \cancel{\text{g } C_{12}H_{22}O_{11}}} \times \frac{6.022 \times 10^{23} \cancel{\text{molecules } C_{12}H_{22}O_{11}}}{1 \cancel{\text{mol } C_{12}H_{22}O_{11}}} \times \frac{22 \text{ H atoms}}{1 \cancel{\text{molecule } C_{12}H_{22}O_{11}}} = 9.91 \times 10^{23} \text{ H atoms}$$

We could calculate the number of carbon and oxygen atoms by the same procedure. However, there is a shortcut. Note that the ratio of carbon to hydrogen atoms in sucrose is 12/22, or 6/11, and that of oxygen to hydrogen atoms is 11/22, or 1/2. Therefore the number of carbon atoms in 25.6 g of sucrose is $(6/11)(9.91 \times 10^{23})$, or 5.41×10^{23} atoms. The number of oxygen atoms is $(1/2)(9.91 \times 10^{23})$, or 4.96×10^{23} atoms.

Similar problems: 2.64, 2.65.

2.5 Ions and Ionic Compounds

In a neutral atom, the number of electrons is equal to the number of protons.

The positively charged protons in the nucleus of an atom remain there during ordinary chemical changes (also called chemical reactions), but the negatively charged electrons in atoms are readily gained or lost. *When electrons are removed from or added to a neutral atom (or molecule), a charged particle called an* ***ion*** *is formed. An ion that bears a net positive charge* is called a ***cation;*** *an ion whose net charge is negative* is called an ***anion.*** For example, a sodium atom (Na) can readily lose an electron to become the cation represented by Na^+ (called the sodium cation):

Na atom	*Na^+ ion*
11 protons	11 protons
11 electrons	10 electrons

A chlorine atom (Cl) can gain an electron to become the anion represented by Cl^- (called the chloride ion):

Cl atom	*Cl^- ion*
17 protons	17 protons
17 electrons	18 electrons

Sodium chloride (NaCl), ordinary table salt, is a compound formed from Na^+ cations and Cl^- anions.

Of course, an atom can lose or gain more than one electron, as in Mg^{2+}, Fe^{3+}, S^{2-}, and N^{3-}. Furthermore, a group of atoms may join together as in a molecule, but may also form an ion that has a net positive or a net negative charge, as in OH^- (hydroxide ion), CN^- (cyanide ion), NH_4^+ (ammonium ion), NO_3^- (nitrate ion), SO_4^{2-} (sulfate ion), and PO_4^{3-} (phosphate ion). *Ions that contain only one atom* are called ***monatomic ions,*** and *ions that contain more than one atom* are called ***polyatomic ions.***

A solid sample of sodium chloride (NaCl) consists of equal numbers of Na^+ and Cl^- ions arranged in a three-dimensional network (Figure 2.10). In such a compound there is 1:1 ratio of cations to anions, so the compound is electrically neutral. *Compounds containing cations and anions* are called ***ionic compounds.***

In most cases, ionic compounds contain a metallic element as the cation and a nonmetallic element as the anion. Metals tend to form cations in compounds. On the other hand, nonmetallic elements form anions when combined with metals. There is one important exception to this description of ionic compounds, however. The ammonium ion (NH_4^+), which is made up of two nonmetallic elements (N and H), is the cation in a number of ionic compounds.

As you can see from Figure 2.10, a particular Na^+ ion in NaCl is not associated with just one particular Cl^- ion. In fact, each Na^+ ion is equally held by six Cl^- ions surrounding it, and vice versa. The NaCl ionic network does *not* contain discrete

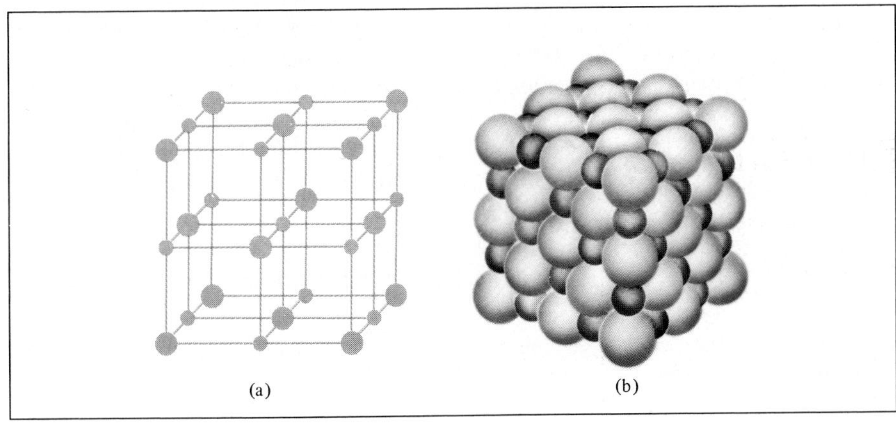

Figure 2.10 *(a) Structure of solid NaCl. (b) In reality, the cations are in contact with the anions. In both (a) and (b), the smaller spheres represent Na^+ ions and the larger spheres Cl^- ions.*

molecular NaCl units. For other ionic compounds the actual structure may be different, but the arrangement between cations and anions is similar to that shown for NaCl. Thus the formula NaCl represents the empirical formula for sodium chloride. One mole of NaCl, then, refers to 6.022×10^{23} NaCl *formula units*. The molar mass of an ionic compound is the sum of the molar masses of the cations and anions present. Thus the molar mass of NaCl is given by

$$\begin{aligned}\text{molar mass of NaCl} &= \text{molar mass of Na}^+ \text{ ion} + \text{molar mass of Cl}^- \text{ ion} \\ &= 22.99 \text{ g} + 35.45 \text{ g} \\ &= 58.44 \text{ g}\end{aligned}$$

Note that the total number of electrons in Na^+ and Cl^- is the same as that in Na and Cl.

Because the mass of an electron is very small compared with that of the entire atom, the molar mass of an ion is virtually the same as that of the uncharged atom from which it is derived. Thus Na and the Na^+ ion are regarded as having the same mass; and Cl and the Cl^- ion also have essentially the same mass.

EXAMPLE 2.11

Sodium fluoride (NaF) is the substance used in some toothpastes to fight tooth decay. How many formula units of NaF are present in 253.6 g of NaF? (The molar mass of NaF is 41.99 g.)

Answer

The number of NaF formula units in 253.6 g of NaF is

$$253.6 \text{ g NaF} \times \frac{1 \text{ mol NaF}}{41.99 \text{ g NaF}} \times \frac{6.022 \times 10^{23} \text{ formula units NaF}}{1 \text{ mol NaF}} = 3.637 \times 10^{24} \text{ NaF formula units}$$

Similar problem: 2.72.

2.6 Percent Composition by Mass of Compounds

One reason for wanting to know the percent composition by mass of a compound is that by comparing the calculated value with that found experimentally, we can verify the purity of the compound.

As we have seen, the formula of a compound tells us the composition of the compound. The composition of a compound is conveniently expressed as the ***percent composition by mass*** (also known as the ***percent composition by weight***), which is *the percent by mass of each element in a compound*. It is obtained by dividing the mass of each element in 1 mole of the compound by the molar mass of the compound and multiplying by 100 percent. In 1 mole of hydrogen peroxide (H_2O_2), for example, there are 2 moles of H atoms and 2 moles of O atoms. The molar masses of H_2O_2, H, and O are 34.02 g, 1.008 g, and 16.00 g, respectively. Therefore, the percent composition of H_2O_2 is calculated as follows:

$$\%\text{H} = \frac{2.016 \text{ g}}{34.02 \text{ g}} \times 100\% = 5.926\%$$

$$\%O = \frac{32.00 \text{ g}}{34.02 \text{ g}} \times 100\% = 94.06\%$$

The sum of the percentages is 5.926 percent + 94.06 percent = 99.99 percent. The small discrepancy from 100 percent is due to the way we rounded off the molar masses of the elements. If we had used the empirical formula HO for the calculation, we would have written

$$\%H = \frac{1.008 \text{ g}}{17.01 \text{ g}} \times 100\% = 5.926\%$$

$$\%O = \frac{16.00 \text{ g}}{17.01 \text{ g}} \times 100\% = 94.06\%$$

Since both the molecular formula and the empirical formula tell us the composition of the compound, it is not surprising that they give us the same percent composition by mass.

The following example illustrates how to calculate the percent composition of a compound.

EXAMPLE 2.12

Phosphoric acid (H_3PO_4) is used in detergents, fertilizers, toothpastes, and carbonated beverages. Calculate the percent composition by mass of H, P, and O in this compound.

Answer

The molar mass of H_3PO_4 is given by

$$3(1.008 \text{ g}) + 30.97 \text{ g} + 4(16.00 \text{ g}) = 97.99 \text{ g}$$

Therefore, the percent by mass of the elements in H_3PO_4 is

$$\%H = \frac{3(1.008 \text{ g})}{97.99 \text{ g}} \times 100\% = 3.086\%$$

$$\%P = \frac{30.97 \text{ g}}{97.99 \text{ g}} \times 100\% = 31.61\%$$

$$\%O = \frac{4(16.00 \text{ g})}{97.99 \text{ g}} \times 100\% = 65.31\%$$

The sum of the percentages is (3.086% + 31.61% + 65.31%) = 100.01%. Again, the discrepancy from 100 percent is due to the way we rounded off.

Similar problems: 2.79, 2.80.

The procedure in the example can be reversed if necessary. Suppose we are given the percent composition by mass of a compound. We can determine the empirical formula of the compound, as the following examples show.

EXAMPLE 2.13

Ascorbic acid (vitamin C) cures scurvy and may help prevent the common cold. It is composed of 40.92 percent carbon (C), 4.58 percent hydrogen (H), and 54.50 percent oxygen (O) by mass. Determine its empirical formula.

Answer

Because the sum of all the percentages is 100 percent, it is convenient to solve this type of problem by considering exactly 100 g of the substance. In 100 g of ascorbic acid there will be 40.92 g of C, 4.58 g of H, and 54.50 g of O. Next, we need to calculate the number of moles of each element in the compound. Let n_C, n_H, and n_O be the number of moles of elements present. Using the molar masses of these elements, we write

$$n_C = 40.92 \cancel{\text{g C}} \times \frac{1 \text{ mol C}}{12.01 \cancel{\text{g C}}} = 3.407 \text{ mol C}$$

$$n_H = 4.58 \cancel{\text{g H}} \times \frac{1 \text{ mol H}}{1.008 \cancel{\text{g H}}} = 4.54 \text{ mol H}$$

$$n_O = 54.50 \cancel{\text{g O}} \times \frac{1 \text{ mol O}}{16.00 \cancel{\text{g O}}} = 3.406 \text{ mol O}$$

Thus, we arrive at the formula $C_{3.407}H_{4.54}O_{3.406}$, which gives the identity and the ratios of atoms present. However, since chemical formulas are written with whole numbers, we cannot have 3.407 C atoms, 4.54 H atoms, and 3.406 O atoms. We can make some of the subscripts whole numbers by dividing all the subscripts by the smallest subscript (3.406):

$$\text{C: } \frac{3.407}{3.406} = 1 \qquad \text{H: } \frac{4.54}{3.406} = 1.33 \qquad \text{O: } \frac{3.406}{3.406} = 1$$

This gives us $CH_{1.33}O$ as the formula for ascorbic acid. Next, we need to convert 1.33, the subscript for H, into an integer. This can be done by a trial-and-error procedure:

$$1.33 \times 1 = 1.33$$

$$1.33 \times 2 = 2.66$$

$$1.33 \times 3 = 3.99 \simeq 4$$

Because 1.33×3 gives us an integer (4), we can multiply all the subscripts by 3 and obtain $C_3H_4O_3$ as the empirical formula for ascorbic acid.

Similar problems: 2.90, 2.91.

EXAMPLE 2.14

The major air pollutant in coal-burning countries is a colorless, pungent gaseous compound containing only sulfur and oxygen. Chemical analysis of a 1.078-g sample of this gas showed that it contained 0.540 g of S and 0.538 g of O. What is the empirical formula of this compound?

Answer

Our first step is to calculate the number of moles of each element present in the sample of the compound. Let these numbers be n_S and n_O. The molar masses of S and O are 32.07 g and 16.00 g, respectively, so we proceed as follows:

$$n_S = 0.540 \not{g}\,\not{S} \times \frac{1 \text{ mol S}}{32.07 \not{g}\,\not{S}} = 0.0168 \text{ mol S}$$

$$n_O = 0.538 \not{g}\,\not{O} \times \frac{1 \text{ mol O}}{16.00 \not{g}\,\not{O}} = 0.0336 \text{ mol O}$$

From these results we could say that the formula is $S_{0.0168}O_{0.0336}$, but we know that we must convert the subscripts to whole numbers. Following the procedure in Example 2.13, we divide each number by the smaller of the two subscripts, that is, 0.0168, so that the subscript of S becomes 1. This procedure gives us $SO_{2.0}$. We know immediately that the empirical formula is SO_2, since the subscript of O must be an integer.

Chemists often want to know the actual mass of an element in a certain mass of a compound. Since the percent composition by mass of the element in the substance can be readily calculated, such a problem can be solved in a rather direct way, as the next example shows.

Chalcopyrite.

EXAMPLE 2.15

Chalcopyrite ($CuFeS_2$) is a principal ore of copper. Calculate the number of kilograms of Cu contained in 3.71×10^3 kg of chalcopyrite.

Answer

The molar masses of Cu and $CuFeS_2$ are 63.55 g and 183.5 g, respectively, so the percent composition by mass of Cu is

$$\%\text{Cu} = \frac{63.55 \text{ g}}{183.5 \text{ g}} \times 100\% = 34.63\%$$

To calculate the mass of Cu in a 3.71×10^3-kg sample of $CuFeS_2$, we need to convert the percentage to a fraction (that is, convert 34.63% to 34.63/100, or 0.3463) and write

$$\text{mass of Cu in CuFeS}_2 = 0.3463 \times 3.71 \times 10^3 \text{ kg} = 1.28 \times 10^3 \text{ kg}$$

This calculation can be simplified by combining the above two steps as follows:

$$\text{mass of Cu in CuFeS}_2 = 3.71 \times 10^3 \text{ kg CuFeS}_2 \times \frac{63.55 \text{ g Cu}}{183.5 \text{ g CuFeS}_2}$$
$$= 1.28 \times 10^3 \text{ kg Cu}$$

Similar problems: 2.85, 2.88.

Experimental Determination of Empirical Formulas

We have seen that if we know the percent composition of a compound we can determine its empirical formula. This is in fact one procedure used to identify compounds experimentally. The basic steps are as follows. First, chemical analysis determines the number of grams of each element present in a given amount of a compound. Then the quantities in grams are converted to numbers of moles of the elements. Finally the empirical formula of the compound is determined by the method of Example 2.14.

As a specific example let us consider the compound ethanol. When ethanol is burned in an apparatus such as that shown in Figure 2.11, carbon dioxide (CO_2) and water (H_2O) are given off. Since neither carbon nor hydrogen was in the inlet gas, we can conclude that both carbon (C) and hydrogen (H) were present in ethanol and that oxygen (O) may also be present. (Molecular oxygen was added in the combustion process, but some of the oxygen may also have come from the original ethanol sample.)

The mass of CO_2 and of H_2O produced can be determined by measuring the increase in mass of the CO_2 and H_2O absorbers, respectively. Suppose that in one experiment the combustion of 11.5 g of ethanol produced 22.0 g of CO_2 and 13.5 g of H_2O. We can calculate the masses of carbon and hydrogen in the original 11.5-g sample of ethanol as follows:

$$\begin{aligned}\text{mass of C} &= 22.0\ \cancel{\text{g CO}_2} \times \frac{1\ \cancel{\text{mol CO}_2}}{44.01\ \cancel{\text{g CO}_2}} \times \frac{1\ \cancel{\text{mol C}}}{1\ \cancel{\text{mol CO}_2}} \times \frac{12.01\ \text{g C}}{1\ \cancel{\text{mol C}}} \\ &= 6.00\ \text{g C}\end{aligned}$$

$$\begin{aligned}\text{mass of H} &= 13.5\ \cancel{\text{g H}_2\text{O}} \times \frac{1\ \cancel{\text{mol H}_2\text{O}}}{18.02\ \cancel{\text{g H}_2\text{O}}} \times \frac{2\ \cancel{\text{mol H}}}{1\ \cancel{\text{mol H}_2\text{O}}} \times \frac{1.008\ \text{g H}}{1\ \cancel{\text{mol H}}} \\ &= 1.51\ \text{g H}\end{aligned}$$

Thus, in 11.5 g of ethanol, 6.00 g are carbon and 1.51 g are hydrogen. The remainder must be oxygen, whose mass is

$$\begin{aligned}\text{mass of O} &= \text{mass of sample} - (\text{mass of C} + \text{mass of H}) \\ &= 11.5\ \text{g} - (6.00\ \text{g} + 1.51\ \text{g}) \\ &= 4.0\ \text{g}\end{aligned}$$

The number of moles of each element present in 11.5 g of ethanol is

$$\text{moles of C} = 6.00\ \cancel{\text{g C}} \times \frac{1\ \text{mol C}}{12.01\ \cancel{\text{g C}}} = 0.500\ \text{mol C}$$

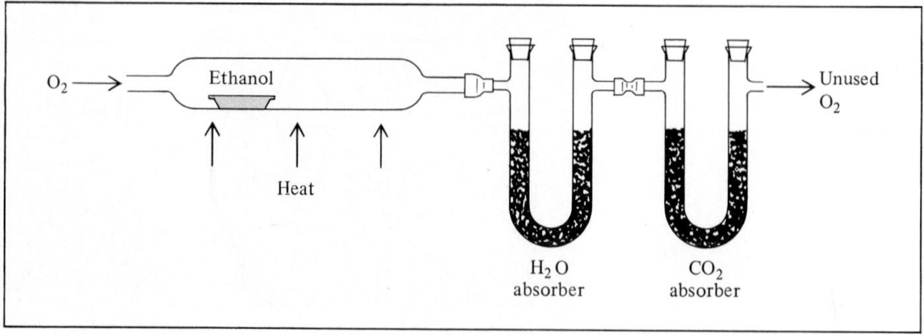

Figure 2.11 *Apparatus for determining the empirical formula of ethanol.*

$$\text{moles of H} = 1.51\ \cancel{\text{g H}} \times \frac{1\ \text{mol H}}{1.008\ \cancel{\text{g H}}} = 1.50\ \text{mol H}$$

$$\text{moles of O} = 4.0\ \cancel{\text{g O}} \times \frac{1\ \text{mol O}}{16.00\ \cancel{\text{g O}}} = 0.25\ \text{mol O}$$

The formula of ethanol is therefore $C_{0.50}H_{1.5}O_{0.25}$ (we round off the number of moles to two significant figures). Since the number of atoms must be an integer, we divide the subscripts by 0.25, the smallest subscript, and obtain for the empirical formula C_2H_6O.

Now we can better understand the word "empirical," which literally means "based only on observation and measurement." The empirical formula of ethanol is determined from analysis of the compound in terms of its component elements. No knowledge of how the atoms are linked together in the compound is required.

Determination of Molecular Formulas

The formula calculated from data about the percent composition by mass is always the empirical formula. To calculate the actual or molecular formula we must know the approximate molar mass of the compound in addition to its empirical formula. The following examples illustrate the point.

EXAMPLE 2.16

The empirical formula of acetic acid (the important ingredient of vinegar) is CH_2O. What is the molecular formula of the compound, given that its molar mass is approximately 60 g?

Answer

First we calculate the molar mass that corresponds to the empirical formula CH_2O:

$$\begin{array}{ccccc} C & +2 & H & + & O \end{array}$$

$$12.01\ \text{g} + 2(1.008\ \text{g}) + 16.00\ \text{g} = 30.03\ \text{g}$$

Then we divide the estimated molecular molar mass by the empirical molar mass:

The answer is rounded off to 2 because molecular molar mass is an integral multiple of empirical molar mass.

$$\frac{\text{estimated molar mass}}{\text{empirical molar mass}} = \frac{60\ \text{g}}{30.03\ \text{g}} = 2$$

Thus there is twice the mass and therefore twice as many atoms in the molecular formula as in the empirical formula. The molecular formula must be $(CH_2O)_2$, or $C_2H_4O_2$, a more conventional way to write it.

Similar problem: 2.92.

EXAMPLE 2.17

A compound of oxygen (O) and nitrogen (N) has the composition 1.52 g of N and 3.47 g of O. The molar mass of this compound is known to be between 90 g and 95 g. Determine its molecular formula and the molecular molar mass of the compound to four significant figures.

Answer

We first determine the empirical formula as outlined in Example 2.14. Let n_N and n_O be the number of moles of nitrogen and oxygen. Then

$$n_N = 1.52 \text{ g N} \times \frac{1 \text{ mol N}}{14.01 \text{ g N}} = 0.108 \text{ mol N}$$

$$n_O = 3.47 \text{ g O} \times \frac{1 \text{ mol O}}{16.00 \text{ g O}} = 0.217 \text{ mol O}$$

Thus the formula of the compound is $N_{0.108}O_{0.217}$. As in Example 2.14, we divide the subscripts by the smaller subscript, 0.108. After rounding off, we obtain NO_2 as the empirical formula. The molecular formula will be equal to the empirical formula or to some integral multiple of it (for example, two, three, four, or more times the empirical formula). The molar mass of the empirical formula NO_2 is

$$\text{empirical molar mass} = (14.01 \text{ g}) + 2(16.00 \text{ g}) = 46.02 \text{ g}$$

Next we determine the number of (NO_2) units present in the molecular formula. This number is found by taking the ratio

$$\frac{\text{molar mass}}{\text{empirical molar mass}} = \frac{95 \text{ g}}{46.02 \text{ g}} = 2.1 \approx 2$$

Thus there are two NO_2 units in each molecule of the compound, so the molecular formula is $(NO_2)_2$, or N_2O_4. The molar mass of the compound is 2(46.02 g), or 92.04 g.

Similar problem: 2.93.

2.7 Laws of Chemical Combination

Having discussed chemical formulas of molecules and compounds, we will now consider two important laws that played a major role in the early steps toward understanding chemical compounds.

The ***law of definite proportions*** states that *different samples of the same compound always contain its constituent elements in the same proportions by mass*. This law is generally attributed to Joseph Proust,† a French chemist who published it in 1799, eight years before Dalton's atomic theory was advanced. The law says that if, for example, we analyze samples of carbon dioxide (CO_2) gas obtained from different

† Joseph Louis Proust (1754–1826). French chemist. Proust was the first person to isolate sugar from grapes.

sources, we will find in each sample the same ratio by mass of carbon to oxygen. This statement seems obvious today, for we normally expect all molecules of a given compound to have the same composition, that is, to contain the same numbers of atoms of its constituent elements. If the ratios of different types of atoms are fixed, then the ratios of the masses of these atoms must also be fixed.

The other fundamental law is the ***law of multiple proportions,*** which states that *if two elements can combine to form more than one compound, the masses of one element that combine with a fixed mass of the other element are in the ratios of small whole numbers*. For example, carbon forms two stable compounds with oxygen, namely, CO (carbon monoxide) and CO_2 (carbon dioxide). Chemical analysis of the compounds shows the following data:

First compound (CO)

The mass of O that combines with 12 g of C is 16 g, so the ratio is

$$\frac{\text{mass of C}}{\text{mass of O}} = \frac{12\text{ g}}{16\text{ g}}$$

Second compound (CO_2)

The mass of O that combines with 12 g of C is 32 g, so the ratio is

$$\frac{\text{mass of C}}{\text{mass of O}} = \frac{12\text{ g}}{32\text{ g}}$$

The ratio of masses of O that combine with 12 g of C in these two compounds is given by

$$\frac{\text{mass of O in CO}}{\text{mass of O in CO}_2} = \frac{16\text{ g}}{32\text{ g}} = \frac{1}{2}$$

The ratio 1:2 supports the law of multiple proportions.

Dalton's atomic theory explains the law of multiple proportions quite simply. Compounds differ in the number of atoms of each kind that combine. For the two compounds formed between carbon and oxygen, measurements suggest that one atom of carbon combines with one atom of oxygen in one compound (that is, in CO) and that one carbon atom combines with two oxygen atoms in the other compound (that is, CO_2).

The following example deals with the law of multiple proportions of two xenon compounds.

EXAMPLE 2.18

For many years the Group 8A noble gas elements (He, Ne, Ar, Kr, and Xe) were called "inert gases" because no one had succeeded in synthesizing compounds containing them. Since 1962, however, a number of stable xenon (Xe) compounds have been prepared. Fluorine (F) forms two compounds with xenon. In one of them, 0.312 g of F is combined with 1.08 g of Xe; in the other compound, 0.426 g of F is combined with 0.736 g of Xe. Prove that these data are consistent with the law of multiple proportions.

You can arrive at the same conclusion by calculating the number of grams of Xe that combines with 1.00 g of F in each of these two compounds.

Answer

We need to calculate the masses of F that combine with a fixed mass of Xe in each of the two compounds. Since we are only interested in the ratio of the masses of F, any quantity of Xe will be acceptable. For convenience we choose 1.00 g of Xe for our calculation.

First compound
First we set up the ratios

$$\frac{x \text{ g F}}{1.00 \text{ g Xe}} = \frac{0.312 \text{ g F}}{1.08 \text{ g Xe}}$$

Thus
$$x \text{ g F} = \frac{(0.312 \text{ g F})(1.00 \text{ g Xe})}{1.08 \text{ g Xe}} = 0.289 \text{ g F}$$

Second compound
Similarly,

$$\frac{x \text{ g F}}{1.00 \text{ g Xe}} = \frac{0.426 \text{ g F}}{0.736 \text{ g Xe}}$$

Thus
$$x \text{ g F} = \frac{(0.426 \text{ g F})(1.00 \text{ g Xe})}{0.736 \text{ g Xe}} = 0.579 \text{ g F}$$

The ratios of F to Xe in these two compounds are compared as follows:

$$\frac{0.579 \text{ g}}{0.289 \text{ g}} = \frac{2}{1}$$

A ratio of 2:1 means that there are twice as many F atoms in one compound than the other. This finding is consistent with the law of multiple proportions.

Similar problems: 2.95, 2.96, 2.97.

2.8 Experimental Determination of Atomic and Molecular Masses

The most direct and accurate method for determining atomic and molecular masses uses mass spectrometry. The operation of a *mass spectrometer* is shown in Figure 2.12. A gaseous sample is first bombarded by a stream of high-energy electrons.

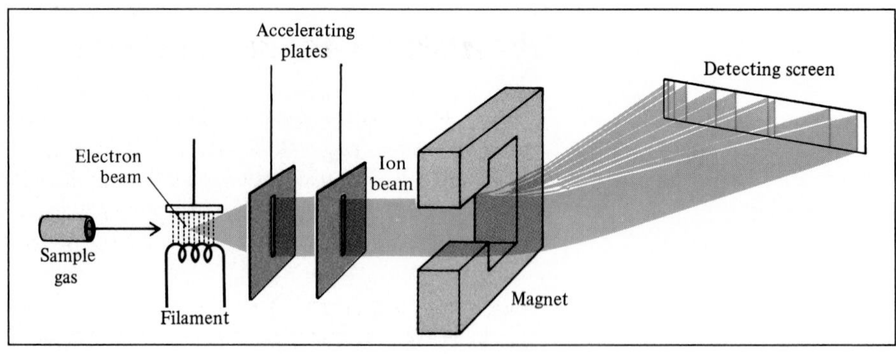

Figure 2.12 *Schematic diagram of one type of mass spectrometer. The gas molecules are first ionized by an electron beam and then accelerated by an electric field. The ion beam is then passed through a magnetic field to be separated into different components according to the masses of the ions.*

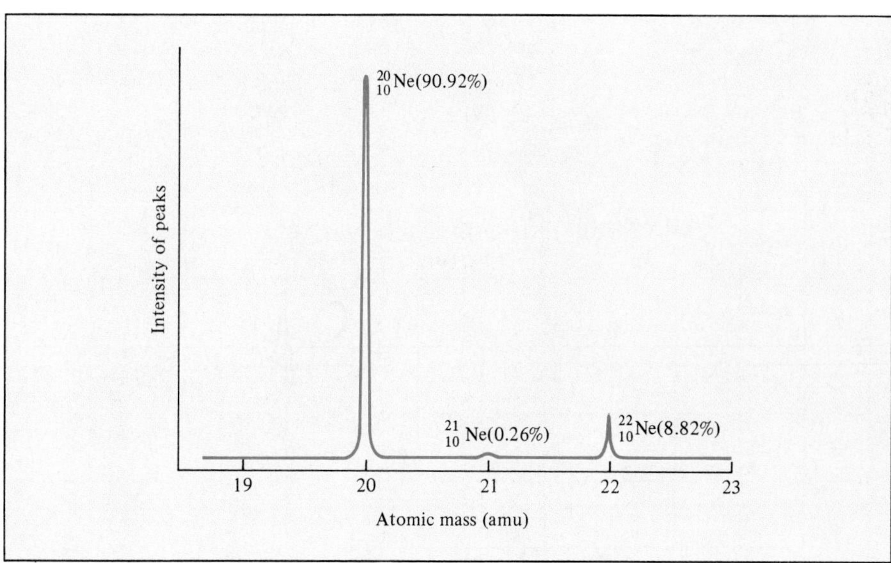

Figure 2.13 *The mass spectrum of the three Ne isotopes.*

Collisions between the electrons and the atoms produce positive ions by removing an electron from each atom or molecule. These positive ions (of mass *m* and charge *e*) are accelerated by passing them through two oppositely charged plates. The emerging ions are then passed between the poles of a magnet, which forces them into a circular path. The radius of the path depends on the charge-to-mass ratio (that is, e/m). Ions of smaller e/m ratio follow a curve of larger radius than those having a larger e/m ratio. In this manner ions with equal charges but different masses can be separated from one another.

One of the first applications of the mass spectrometer was the demonstration by F. W. Aston† that naturally occurring neon consists of the isotopes $^{20}_{10}Ne$, $^{21}_{10}Ne$, and $^{22}_{10}Ne$. Figure 2.13 shows a *mass spectrum,* or a graph of the intensity (that is, the amount of each isotope present) versus atomic mass for the Ne isotopes. By comparing the positions of the peaks with the calibrated horizontal scale, we can determine the atomic masses of these isotopes. In addition, the peak intensities tell us the natural abundances of the isotopes. A knowledge of both the natural abundance and atomic mass of the isotopes allows us to calculate the *average* atomic mass of an element, as discussed in Section 2.3. The masses of molecules are determined in a similar manner.

Another early important application of the mass spectrometer was in the separation of $^{235}_{92}U$ and $^{238}_{92}U$ isotopes, which helped scientists select the isotope to use in atomic bombs and nuclear reactors ($^{235}_{92}U$).

2.9 Naming Inorganic Compounds

At this stage in your chemistry course, you have been introduced to experimental work in the laboratory. You will be faced with learning the names of the compounds you encounter. When chemistry was a young science and the number of known compounds was small, it was possible to memorize their names. Many of the names were derived from their physical appearance, properties, origin, or application; for example, milk of

† Francis William Aston (1877–1945). English chemist and physicist. He was awarded the Nobel Prize in chemistry in 1922 for developing the mass spectrometer.

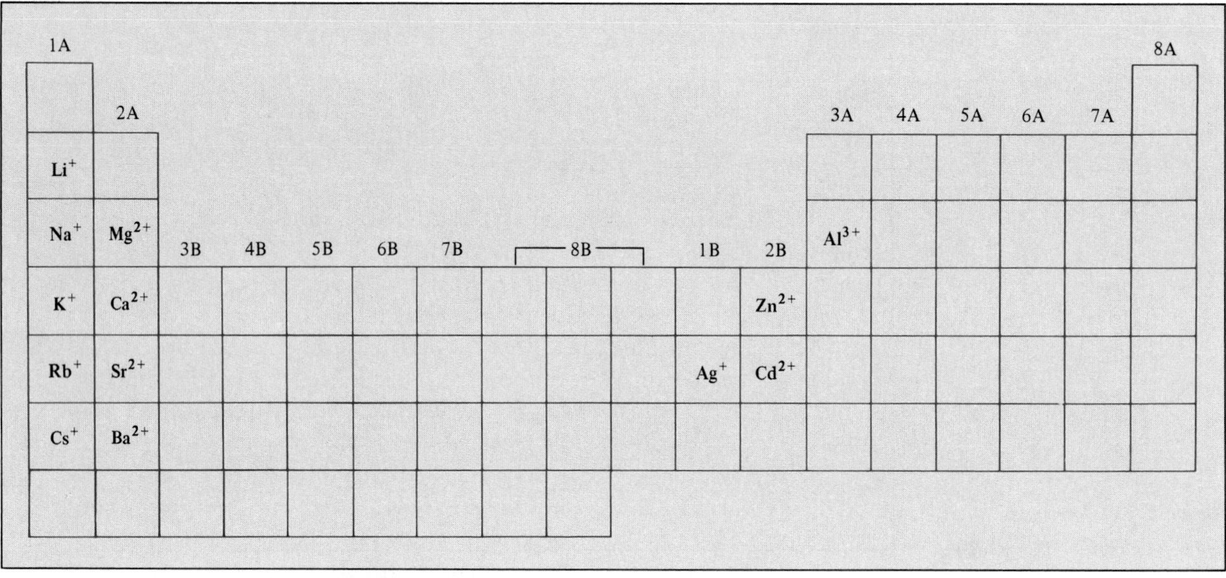

Figure 2.14 *Metal cations that can bear only one type of charge. The cations are arranged according to the positions of the elements in the periodic table.*

magnesia, laughing gas, limestone, caustic soda, lye, washing soda, and baking soda.

Today the number of known compounds is well over 6 million. Fortunately, it is not necessary to memorize their names, even if it were possible to do so. Over the years chemists have devised clear, systematic ways of naming chemical substances. The naming schemes are accepted worldwide, facilitating communication among chemists and providing useful ways of dealing with the overwhelming variety of substances currently identified.

To begin our study of nomenclature, the naming of chemical compounds, we must first distinguish between inorganic and organic compounds. ***Organic compounds*** *contain carbon, usually in combination with elements such as hydrogen, oxygen, nitrogen, and sulfur. All other compounds are classified as* ***inorganic compounds.*** Some carbon-containing compounds such as carbon monoxide (CO), carbon dioxide (CO_2), carbon disulfide (CS_2), compounds containing the cyanide group (CN^-), and carbonate (CO_3^{2-}) and bicarbonate (HCO_3^-) groups are considered for convenience to be inorganic compounds. Although the nomenclature of organic compounds will not be discussed until Chapter 24, we will use some organic compounds to illustrate chemical principles throughout this book.

To organize and simplify our venture into naming compounds, we can divide inorganic compounds into four categories: ionic compounds, molecular compounds, acids and bases, and hydrates.

Ionic Compounds

In Section 2.5 we learned that ionic compounds are made up of cations (positive ions) and anions (negative ions). With the important exception of the ammonium ion, NH_4^+, all cations of interest to us are derived from metal atoms. Metal cations take their names from the elements. For example:

	Element		*Name of cation*
Na	sodium	Na^+	sodium ion (or sodium cation)
K	potassium	K^+	potassium ion (or potassium cation)
Mg	magnesium	Mg^{2+}	magnesium ion (or magnesium cation)
Al	aluminum	Al^{3+}	aluminum ion (or aluminum cation)

Figure 2.14 shows the ionic charges of some metals according to their positions in the periodic table. These are the only charges that the metal cations can bear.

Many ionic compounds are ***binary compounds,*** or *compounds formed from just two elements*. For binary compounds the first element we write is the metal cation, followed by the nonmetallic anion. Thus NaCl is sodium chloride, where the anion is named by taking the first part of the element name (chlorine) and adding *-ide*. The charges on the cation and anion are not shown in the formula. Table 2.2 shows the "-ide" nomenclature of some common monatomic anions according to their positions in the periodic table.

The "-ide" ending is also used for certain anion groups containing two different elements, such as hydroxide (OH^-) and cyanide (CN^-). Thus the compounds LiOH and KCN are named lithium hydroxide and potassium cyanide. These and a number of other such ionic substances are called ***ternary compounds,*** meaning *compounds consisting of three elements*. Another example of a ternary ionic compound is ammonium chloride (NH_4Cl). In this case the cation (NH_4^+) is made up of two different elements. Table 2.3 lists alphabetically the names of a number of common inorganic cations and anions.

An important guideline for writing the correct formulas of ionic compounds is that each compound must be electrically neutral. This means that the sum of charges on the cation and the anion in each formula unit must add up to zero. Electrical neutrality can be maintained by following the useful rule: *The subscript of the cation is numerically equal to the charge on the anion, and the subscript of the anion is numerically equal to the charge on the cation*. If the charges are numerically equal, then the subscripts for both the cation and the anion will be 1. This follows from the fact that because the formulas of ionic compounds are empirical formulas, the subscripts must always be reduced to the smallest ratios. Let us now consider some examples.

- *Potassium bromide*. Figure 2.14 shows that the cation is K^+, and Table 2.2 shows that the anion is Br^-. Therefore, the formula is KBr. The sum of the charges is $+1 + (-1) = 0$.
- *Zinc iodide*. Figure 2.14 shows that the cation is Zn^{2+}, and Table 2.2 shows that the anion is I^-. Therefore, the formula is ZnI_2. The sum of the charges is

Table 2.2 The "-ide" nomenclature of some common monatomic anions according to their positions in the periodic table

Group 4A	Group 5A	Group 6A	Group 7A
C Carbide (C^{4-})*	N Nitride (N^{3-})	O Oxide (O^{2-})	F Fluoride (F^-)
Si Silicide (Si^{4-})	P Phosphide (P^{3-})	S Sulfide (S^{2-})	Cl Chloride (Cl^-)
		Se Selenide (Se^{2-})	Br Bromide (Br^-)
		Te Telluride (Te^{2-})	I Iodide (I^-)

*The word "carbide" is also used for the anion C_2^{2-}.

Table 2.3 Names and formulas of some common inorganic cations and anions

Cation	Anion
Aluminum (Al^{3+})	Bromide (Br^-)
Ammonium (NH_4^+)	Carbonate (CO_3^{2-})
Barium (Ba^{2+})	Chlorate (ClO_3^-)
Cadmium (Cd^{2+})	Chloride (Cl^-)
Calcium (Ca^{2+})	Chromate (CrO_4^{2-})
Cesium (Cs^+)	Cyanide (CN^-)
Chromium(III) or chromic (Cr^{3+})	Dichromate ($Cr_2O_7^{2-}$)
Cobalt(II) or cobaltous (Co^{2+})	Dihydrogen phosphate ($H_2PO_4^-$)
Copper(I) or cuprous (Cu^+)	Fluoride (F^-)
Copper(II) or cupric (Cu^{2+})	Hydride (H^-)
Hydrogen (H^+)	Hydrogen carbonate or bicarbonate (HCO_3^-)
Iron(II) or ferrous (Fe^{2+})	Hydrogen phosphate (HPO_4^{2-})
Iron(III) or ferric (Fe^{3+})	Hydrogen sulfate or bisulfate (HSO_4^-)
Lead(II) or plumbous (Pb^{2+})	Hydroxide (OH^-)
Lithium (Li^+)	Iodide (I^-)
Magnesium (Mg^{2+})	Nitrate (NO_3^-)
Manganese(II) or manganous (Mn^{2+})	Nitride (N^{3-})
Mercury(I) or mercurous (Hg_2^{2+})*	Nitrite (NO_2^-)
Mercury(II) or mercuric (Hg^{2+})	Oxide (O^{2-})
Potassium (K^+)	Permanganate (MnO_4^-)
Silver (Ag^+)	Peroxide (O_2^{2-})
Sodium (Na^+)	Phosphate (PO_4^{3-})
Strontium (Sr^{2+})	Sulfate (SO_4^{2-})
Tin(II) or stannous (Sn^{2+})	Sulfide (S^{2-})
Zinc (Zn^{2+})	Sulfite (SO_3^{2-})
	Thiocyanate (SCN^-)

*Mercury(I) exists as a pair as shown.

The use of Roman numerals in parentheses such as (III) will be discussed shortly.

$+2 + 2(-1) = 0$. Note that because the zinc cation always carries a $+2$ charge and the iodide anion always carries a -1 charge, there is no need to indicate the presence of two I^- ions in the name. To maintain electrical neutrality, the subscript for I must be 2 in zinc iodide.

- *Sodium nitride*. From Figure 2.14 and Table 2.2 we see that the cation is Na^+ and the anion is N^{3-}. Therefore, the formula is Na_3N. The sum of the charges is $3(+1) + (-3) = 0$. Again, there is no need to indicate the presence of three Na^+ ions in the name, because a sodium cation is always Na^+ and a nitride anion is always N^{3-}.
- *Barium sulfide*. The cation is Ba^{2+} and the anion is S^{2-}. Applying our rule for writing subscripts, we obtain Ba_2S_2. However, since the subscripts must be reduced to the smallest ratios, the formula is BaS.
- *Aluminum oxide*. The cation is Al^{3+} and the anion is O^{2-}. The following diagram helps us determine the subscripts for the cation and the anion:

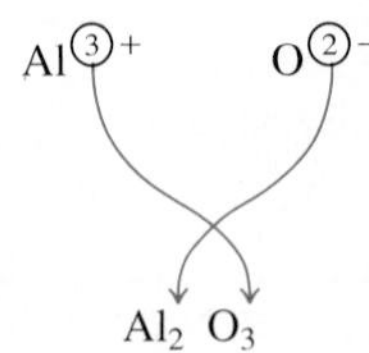

The sum of the charges is 2(+3) + 3(−2) = 0. As before, we do not need to indicate the presence of two Al^{3+} cations and three O^{2-} anions in the name.

The transition metals are the elements in Groups 1B and 3B–8B (see Figure 1.5).

Certain metals, especially the *transition metals,* can form more than one type of cation. Take iron as an example. Iron can form two cations: Fe^{2+} and Fe^{3+}. An older method that still finds limited use assigns the ending "-ous" to the cation with fewer positive charges and the ending "-ic" to the cation with more positive charges:

Fe^{2+}	ferrous ion
Fe^{3+}	ferric ion

The names of the compounds that these iron ions form with chlorine would thus be

$FeCl_2$	ferrous chloride
$FeCl_3$	ferric chloride

This method of naming has some distinct limitations. First, the "-ous" and "-ic" suffixes do not provide information regarding the actual charges of the two cations involved. Thus ferric ion is Fe^{3+}, but the cation of copper named cupric has the formula Cu^{2+}. In addition, the "-ous" and "-ic" designations provide element cation names for only two different positive charges. Some metallic elements can assume three or more different positive charges in compounds. Therefore, it has become increasingly common to designate different cations with Roman numerals. This is called the Stock† system. In this system, the Roman numeral I is used for one positive charge, II for two positive charges, and so on. For example, manganese (Mn) atoms can assume several different positive charges:

Mn^{2+}: MnO	manganese(II) oxide
Mn^{3+}: Mn_2O_3	manganese(III) oxide
Mn^{4+}: MnO_2	manganese(IV) oxide

These compound names are pronounced manganese-two oxide, manganese-three oxide, and manganese-four oxide. Using the Stock system, we can express the ferrous ion and the ferric ion as iron(II) and iron(III), respectively; ferrous chloride becomes iron(II) chloride; and ferric chloride is called iron(III) chloride. In keeping with modern practice, we will favor the use of the Stock system of naming compounds in this textbook.

The following two examples deal with naming ionic compounds and writing formulas from names based on the information given in Figure 2.14 and Tables 2.2 and 2.3.

†Alfred E. Stock (1876–1946). German chemist. Stock did most of his research in the synthesis and characterization of boron, beryllium, and silicon compounds. He was the first scientist to explore the dangers of mercury poisoning.

EXAMPLE 2.19

Name the following ionic compounds: (a) $CuBr_2$, (b) KH_2PO_4, (c) NH_4ClO_3, (d) Na_2O_2.

Since potassium only forms one type of ion (K^+), there is no need to use potassium(I) in the name.

Answer

(a) Since the bromide ion (Br^-) bears one negative charge, the copper ion must have two positive charges. Therefore, the compound is copper(II) bromide.
(b) The cation is K^+ and the anion is $H_2PO_4^-$ (dihydrogen phosphate). The compound is potassium dihydrogen phosphate.
(c) The cation is NH_4^+ (ammonium ion) and the anion is ClO_3^- (chlorate ion). The compound is ammonium chlorate.
(d) Since each sodium ion bears a positive charge and there are two Na^+ ions present, the total negative charge must be -2. Thus the anion is the peroxide ion O_2^{2-} and the compound is sodium peroxide.

Similar problem: 2.110.

EXAMPLE 2.20

Write chemical formulas for the following compounds: (a) mercury(I) nitrite, (b) cesium sulfide, (c) calcium phosphate, (d) potassium dichromate.

Answer

(a) The mercury(I) ion is diatomic, namely, Hg_2^{2+}, and the nitrite ion is NO_2^-. Therefore, the formula is $Hg_2(NO_2)_2$.
(b) Each sulfide ion bears two negative charges, and each cesium ion bears one positive charge (cesium is in Group 1A, as is sodium). Therefore, the formula is Cs_2S.
(c) Each calcium ion (Ca^{2+}) bears two positive charges, and each phosphate ion (PO_4^{3-}) bears three negative charges. To make the sum of charges equal to zero, the numbers of cations and anions must be adjusted:

$$3(+2) + 2(-3) = 0$$

Thus the formula is $Ca_3(PO_4)_2$.
(d) Each dichromate ion ($Cr_2O_7^{2-}$) has two negative charges and each potassium ion carries one positive charge, so the formula is $K_2Cr_2O_7$.

Similar problem: 2.111.

Molecular Compounds

Unlike ionic compounds, molecular compounds contain discrete molecular units. They are usually composed of nonmetallic elements (see Figure 1.5). Let us focus mainly on binary compounds, since many inorganic molecular compounds are made up of only two elements. Naming binary molecular compounds is similar to naming binary ionic compounds; that is, we name the first element first, and the second element is named by taking the first part of the element name and adding *-ide*. Some examples are

HCl	hydrogen chloride
HBr	hydrogen bromide

Table 2.4 Greek prefixes used in naming molecular compounds

Prefix	Meaning
Mono-	1
Di-	2
Tri-	3
Tetra-	4
Penta-	5
Hexa-	6
Hepta-	7
Octa-	8
Nona-	9
Deca-	10

SiC silicon carbide

Frequently we find that one pair of elements can form several different compounds. In these cases, confusion in naming the compounds is avoided by using Greek prefixes to denote the number of atoms of each element present (see Table 2.4). Consider the following examples:

CO	carbon monoxide
CO_2	carbon dioxide
SO_2	sulfur dioxide
SO_3	sulfur trioxide
PCl_3	phosphorus trichloride
PCl_5	phosphorus pentachloride
NO_2	nitrogen dioxide
N_2O_4	dinitrogen tetroxide
Cl_2O_7	dichlorine heptoxide

The following guidelines are helpful when you are naming compounds with prefixes:

- The prefix ''mono-'' may be omitted for the first element. For example, SO_2 is named sulfur dioxide, not monosulfur dioxide. Thus the absence of a prefix for the first element usually implies there is only one atom of that element present in the molecule.
- For oxides, the ending ''a'' in the prefix is sometimes omitted. For example, N_2O_4 may be called dinitrogen tetroxide rather than dinitrogen tetraoxide.

An exception to the use of Greek prefixes involves molecular compounds containing hydrogen. Traditionally, many of these compounds are called either by their common, nonsystematic names or by names that do not specifically indicate the number of H atoms present:

B_2H_6	diborane
CH_4	methane
SiH_4	silane
NH_3	ammonia
PH_3	phosphine
H_2O	water
H_2S	hydrogen sulfide

Note that even the order of writing the elements in the formulas is irregular. The above examples show that H is written first in water and hydrogen sulfide, whereas H is written last in the other compounds.

Writing formulas for molecular compounds is usually straightforward. Thus the name arsenic trifluoride means that there are one As atom and three F atoms in each molecule and the molecular formula is AsF_3. Note that the order of elements in the formula is the same as that in its name.

The following two examples deal with naming molecular compounds and writing formulas from names.

EXAMPLE 2.21

Name the following molecular compounds: (a) BF_3, (b) $SiCl_4$, (c) IF_7, (d) P_4O_{10}.

Answer

(a) Since there are three fluorine atoms present, the compound is boron trifluoride. (b) There are four chlorine atoms present, so the compound is silicon tetrachloride. (c) Since there are seven fluorine atoms present, the compound is iodine heptafluoride. (d) There are four phosphorus atoms and ten oxygen atoms present, so the compound is tetraphosphorus decoxide.

Similar problem: 2.110.

EXAMPLE 2.22

Write chemical formulas for the following molecular compounds: (a) nitrogen trichloride, (b) carbon disulfide, (c) disilicon hexabromide, (d) tetranitrogen tetrasulfide.

Answer

(a) Since there are one nitrogen atom and three chlorine atoms present, the formula is NCl_3.
(b) There are one carbon atom and two sulfur atoms present, so the formula is CS_2.
(c) Since there are two silicon atoms and six bromine atoms present, the formula is Si_2Br_6.
(d) There are four nitrogen atoms and four sulfur atoms present, so the formula is S_4N_4.

Similar problem: 2.111.

Acids and Bases

The H^+ ion is also called the *proton*.

Naming Acids. For present purposes an ***acid*** is defined as *a substance that yields hydrogen ions (H^+) when dissolved in water*. Formulas for inorganic acids contain one or more hydrogen atoms as well as an anionic group. Anions whose names end in "-ide" have associated acids with a "hydro-" prefix and an "-ic" ending, as shown in Table 2.5. You may have noticed that in some instances two different names seem to be assigned to the same chemical formula, for example,

HCl	hydrogen chloride
HCl	hydrochloric acid

Table 2.5 Some simple acids

Anion	Corresponding acid
F^- (fluoride)	HF (hydrofluoric acid)
Cl^- (chloride)	HCl (hydrochloric acid)
Br^- (bromide)	HBr (hydrobromic acid)
I^- (iodide)	HI (hydroiodic acid)
CN^- (cyanide)	HCN (hydrocyanic acid)
S^{2-} (sulfide)	H_2S (hydrosulfuric acid)

The name assigned to the compound depends on its physical state. When HCl exists in the gaseous or pure liquid state, it is a molecular compound and we call it hydrogen chloride. When it is dissolved in water, the molecules break up into H^+ and Cl^- ions; in this state, the substance is called hydrochloric acid.

Acids that contain hydrogen, oxygen, and another element (the central element) are called ***oxoacids.*** The formulas of oxoacids are usually written with the H first, followed by the central element and then O, as illustrated by the following series of oxoacids:

H_2CO_3	carbonic acid
HNO_3	nitric acid
H_2SO_4	sulfuric acid
$HClO_3$	chloric acid

Often two or more oxoacids have the same central atom but a different number of O atoms. Starting with the oxoacids whose names end with "-ic," we use the following rules to name these compounds.

- Addition of one O atom to the "-ic" acid: The acid is called "per. . .-ic" acid. Thus when we convert $HClO_3$ to $HClO_4$, the acid is called perchloric acid.
- Removal of one O atom from the "-ic" acid: The acid is called "-ous" acid. Thus when HNO_3 is converted to HNO_2, it is called nitrous acid.
- Removal of two O atoms from the "-ic" acid: The acid is called "hypo. . .-ous" acid. Thus when $HBrO_3$ is converted to HOBr, the acid is called hypobromous acid.

The rules for naming *anions of oxoacids,* called ***oxoanions,*** are as follows.

- When all the H ions are removed from the "-ic" acid, the anion's name ends with "-ate." For example, the anion CO_3^{2-} derived from H_2CO_3 is called carbonate.
- When all the H ions are removed from the "-ous" acid, the anion's name ends with "-ite." Thus the anion ClO_2^- derived from $HClO_2$ is called chlorite.
- The names of anions in which one or more but not all the hydrogen ions have been removed must indicate the number of H ions present. For example, consider the anions derived from phosphoric acid:

H_3PO_4	phosphoric acid
$H_2PO_4^-$	dihydrogen phosphate
HPO_4^{2-}	hydrogen phosphate
PO_4^{3-}	phosphate

Note that we usually omit the prefix "mono-" when there is only one H in the anion. Figure 2.15 summarizes the nomenclature for the oxoacids and oxoanions, and Table 2.6 gives the names of the oxoacids and oxoanions that contain chlorine.

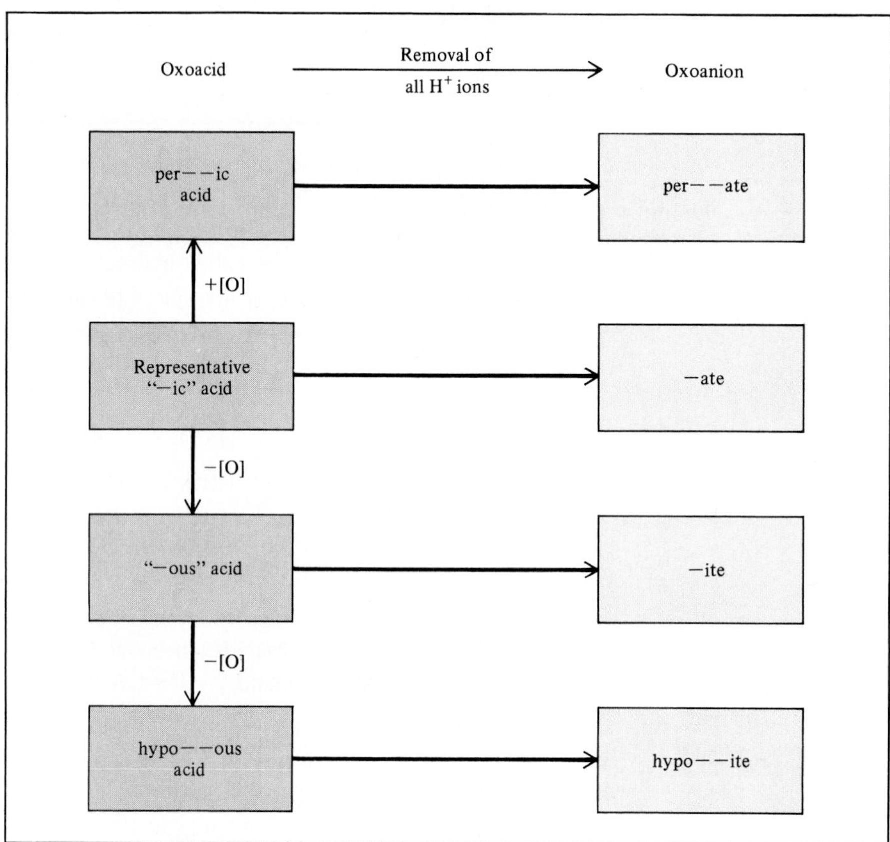

Figure 2.15 *Naming oxoacids and oxoanions.*

Table 2.6 Names of oxoacids and oxoanions that contain chlorine

Acid	Anion
$HClO_4$ (perchloric acid)	ClO_4^- (perchlorate)
$HClO_3$ (chloric acid)	ClO_3^- (chlorate)
$HClO_2$ (chlorous acid)	ClO_2^- (chlorite)
HOCl (hypochlorous acid)	OCl^- (hypochlorite)

Example 2.23 deals with naming an oxoacid and oxoanions.

EXAMPLE 2.23

Name the following oxoacid and oxoanions: (a) H_3PO_3, (b) IO_4^-, (c) HSO_3^-.

Answer

(a) We start with our reference acid, phosphoric acid (H_3PO_4). Since H_3PO_3 has one fewer O atom, it is called phosphorous acid.
(b) The parent acid is HIO_4. Since the acid has one more O atom than our reference iodic acid (HIO_3), it is called periodic acid. Therefore, the anion derived from HIO_4 is called periodate.
(c) The parent acid is H_2SO_3, which is called sulfurous acid because it has one fewer O atom than sulfuric acid. The anion SO_3^{2-} is called sulfite; therefore, HSO_3^- is hydrogen sulfite.

Similar problems: 2.110, 2.111.

Naming Bases. For present purposes, a ***base*** is defined as *a substance that yields hydroxide ions (OH^-) when dissolved in water*. Some examples are

NaOH	sodium hydroxide
KOH	potassium hydroxide
$Ba(OH)_2$	barium hydroxide

Ammonia (NH_3), a molecular compound in the gaseous or pure liquid state, is also classified as a common base. At first glance this may seem to be an exception to the definition for base that we have given. But note that all that is required is that a base *yields* hydroxide ions when dissolved in water. It is not necessary for the original base to contain hydroxide ions in its structure. In fact, when ammonia dissolves in water, NH_3 reacts at least partially with water to yield NH_4^+ and OH^- ions. Thus it is properly classified as a base.

Hydrates

Compounds that have a specific number of water molecules attached to them are called ***hydrates***. For example, in its normal state (say, from a reagent bottle), each unit of copper(II) sulfate has five water molecules associated with it. The formula of copper(II) sulfate pentahydrate is $CuSO_4 \cdot 5H_2O$. These water molecules can be driven off by heating. When this occurs, the resulting compound is $CuSO_4$, which is sometimes called *anhydrous* copper(II) sulfate, where "anhydrous" means that the compound no longer has water molecules associated with it (Figure 2.16). Some examples of hydrates are

Pentahydrate means five water molecules associated with each unit of $CuSO_4$.

Figure 2.16 *$CuSO_4 \cdot 5H_2O$ (left) is blue; $CuSO_4$ (right) is white.*

$BaCl_2 \cdot 2H_2O$	barium chloride dihydrate
$LiCl \cdot H_2O$	lithium chloride monohydrate
$MgSO_4 \cdot 7H_2O$	magnesium sulfate heptahydrate
$Sr(NO_3)_2 \cdot 4H_2O$	strontium nitrate tetrahydrate

Familiar Inorganic Compounds

Some compounds are better known by their common names rather than by systematic chemical names. The names of some familiar inorganic compounds are given in Table 2.7.

Table 2.7 Common and systematic names of some compounds

Formula	Common name	Systematic name
H_2O	Water	Dihydrogen oxide
NH_3	Ammonia	Trihydrogen nitride
CO_2	Dry ice	Solid carbon dioxide
NaCl	Table salt	Sodium chloride
N_2O	Laughing gas	Dinitrogen oxide (nitrous oxide)
$CaCO_3$	Marble, chalk, limestone	Calcium carbonate
CaO	Quicklime	Calcium oxide
$Ca(OH)_2$	Slaked lime	Calcium hydroxide
$NaHCO_3$	Baking soda	Sodium hydrogen carbonate
$Na_2CO_3 \cdot 10H_2O$	Washing soda	Sodium carbonate decahydrate
$MgSO_4 \cdot 7H_2O$	Epsom salt	Magnesium sulfate heptahydrate
$Mg(OH)_2$	Milk of magnesia	Magnesium hydroxide
$CaSO_4 \cdot 2H_2O$	Gypsum	Calcium sulfate dihydrate

CHEMISTRY IN ACTION

Allotropes

It is an interesting chemical phenomenon that certain elements can exist in more than one stable form. A chemical element is said to exhibit *allotropy* when it occurs in two or more forms; the forms are called *allotropes*. Allotropes generally differ in structure and in both physical and chemical properties. The familiar elements that exhibit allotropy are carbon, oxygen, sulfur, phosphorus, and tin. Here we will briefly describe the allotropes of carbon and oxygen.

Carbon

Figure 2.17 shows the two allotropes of carbon—graphite and diamond. Looking at the figure, you may find it hard to believe that both substances are made of the same carbon atoms. Yet their different physical appearance and properties are determined only by the manner in which the carbon atoms are linked together (Figure 2.18). Graphite is a soft, dark black solid with a metallic luster. It is a good conductor of electricity and is used as an electrode (electrical connection) in batteries. The so-called lead in ordinary pencils is in reality a mixture of graphite and clay. Graphite is also used in stove polish, in typewriter ribbons, and as a lubricant. Diamond is formed over long periods on the geologic scale when graphite is subjected to tremendous pressure underground. In pure form, diamond is a transparent solid. Diamond is the less stable of the two allotropes, and in time it will turn back into graphite. Fortunately for diamond jewelry owners, this process takes millions of years. The hardest natural substance known, diamond is used in industry as an abrasive and to cut concrete and other hard substances.

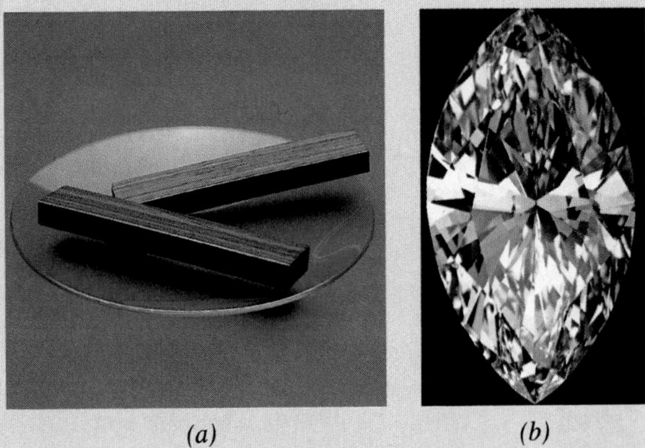

(a) *(b)*

Figure 2.17 *The two allotropes of carbon: (a) graphite and (b) diamond.*

Oxygen

Molecular oxygen is a diatomic molecule, whereas ozone, the less stable allotrope of oxygen, is triatomic (Figure 2.19). Molecular oxygen is a colorless and odorless gas; it is essential for our survival. Metabolism, the process by which energy stored in the food we eat is extracted for growth and function, cannot take place without oxygen. Combustion also requires oxygen. Air is about 20 percent oxygen gas by volume. Oxygen is used in steelmaking and also in medicine.

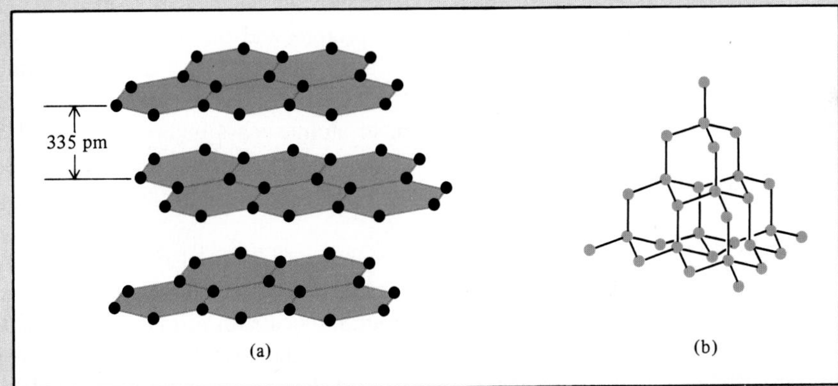

Figure 2.18 *Models showing how carbon atoms are joined together in (a) graphite and (b) diamond.*

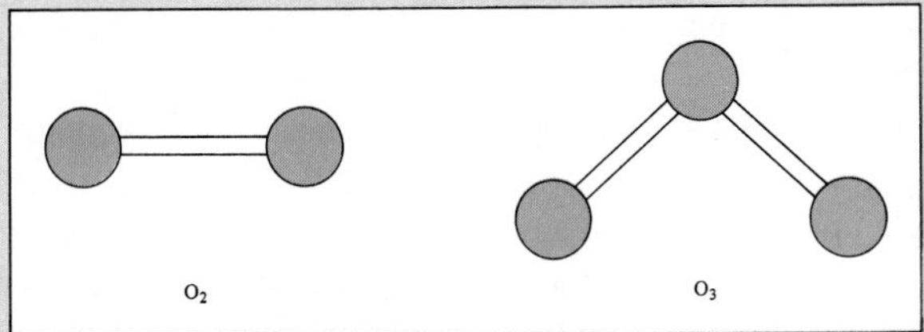

Figure 2.19 *An oxygen (O_2) molecule and an ozone (O_3) molecule.*

Ozone can be prepared from molecular oxygen by subjecting the latter to an electrical discharge. In fact, the pungent odor of ozone is often noticeable near a subway train (where there are frequently electrical discharges). Ozone is a toxic, light blue gas. It is used mainly to purify drinking water, to deodorize air and sewage gases, and to bleach waxes, oils, and textiles. Although ozone is present in the atmosphere only in trace amounts, it plays a central role in two processes that affect our lives. Near the surface of Earth, ozone promotes the formation of smog, which is detrimental to all living things. Ozone is also present in the stratosphere, a region about 40 km (25 miles) above the surface of Earth. There, the ozone molecules absorb much of the harmful high-energy radiation from the sun and thus protect the life beneath.

Summary

1. Modern chemistry began with Dalton's atomic theory, which states that all matter is composed of tiny, indivisible particles called atoms; that all atoms of the same element are identical; that compounds contain atoms of different elements combined in whole-number ratios; and that atoms are neither created nor destroyed in chemical reactions (the law of conservation of mass).
2. An atom consists of a very dense central nucleus containing protons and neutrons, with electrons moving about the nucleus at a relatively large distance from it.
3. Protons are positively charged, neutrons have no charge, and electrons are negatively charged. Protons and neutrons have roughly the same mass, which is about 1840 times greater than the mass of an electron.
4. Electrons were discovered in experiments with cathode ray tubes. Protons and the nucleus were discovered in scattering experiments using α particles from radioactive elements to bombard gold foil. Neutrons were discovered in the rays produced by α-particle bombardment of beryllium.
5. The atomic number of an element is the number of protons in the nucleus of an atom of the element; it determines the identity of an element. The mass number is the sum of the number of protons and the number of neutrons in the nucleus.
6. Isotopes are atoms of the same element, with the same number of protons but different numbers of neutrons.
7. Atomic masses are given in atomic mass units (amu), a relative unit based on exactly 12 for the ${}^{12}_{6}C$ isotope. The atomic mass given for the atoms of a particular element is usually the average of the naturally occurring isotope distribution of that element.
8. A mole is an Avogadro's number (6.022×10^{23}) of atoms, molecules, or other particles. The molar mass (in grams) of an element or a compound is numerically equal to the mass of the atom, molecule, or formula unit (in amu), and contains an Avogadro's number of atoms (in the case of elements), molecules, or simplest units of cations and anions (in the case of ionic compounds).
9. Chemical formulas combine the symbols for the constituent elements with whole-number subscripts to show the type and number of atoms contained in the smallest unit of a compound.
10. The molecular formula shows the specific number and type of atoms combined in each molecule of a compound. The empirical formula shows the simplest ratios of the atoms combined in a molecule.
11. The molecular mass of a molecule is the sum of the atomic masses of the atoms in the molecule.
12. Ionic compounds are made up of cations and anions, formed when atoms lose and gain electrons, respectively. Chemical compounds are either molecular compounds (in which the smallest units are discrete, individual molecules) or ionic compounds (in which positive and negative ions are held together by mutual attraction).
13. The percent composition by mass of a compound is the percent by mass of each element present. If we know the percent composition by mass of a compound, we can deduce the empirical formula of the compound and also the molecular formula of the compound if the approximate molar mass is known.
14. Atoms of constituent elements in a particular compound are always combined in

the same proportions by mass (law of definite proportions). When two elements can combine to form more than one type of compound, the masses of one element that combine with a fixed mass of the other element are in the ratio of small whole numbers (law of multiple proportions).

15. Atomic and molecular masses are determined accurately in a mass spectrometer. Positive ions are created by bombarding a test sample with high-energy electrons. These positive ions then pass through electric and magnetic fields that separate them according to their masses.
16. The names of many inorganic compounds can be deduced from a set of simple rules. The formulas can be written from the names of the compounds.

Applied Chemistry Example: Potash

Potash is any potassium mineral which is used for its potassium content. It ranked 34th among the top 50 industrial chemicals produced in the United States in 1988 (3.4 billion pounds). About 95 percent of potash is used for fertilizer and the remaining 5 percent is used to make potassium hydroxide, potassium permanganate, and other chemicals which require the potassium atom. Relatively small amounts of potassium chloride are used as a salt substitute in low sodium diets.

The major minerals for potash are potassium chloride, potassium sulfate, and the double salt, $K_2SO_4 \cdot 2MgSO_4$. The three primary methods currently used for the production of potash all involve isolating it from natural deposits. One method extracts potash from surface deposits in salt lakes and the other two work on underground deposits either by mining the solid in a fashion similar to coal mining or by dissolving the potassium salts and transporting the solution to the surface for processing.

In fertilizer manufacture, the main concern is potassium content, so a company might switch from one mineral to another depending on cost. In fact, the potash production is often reported as the potassium oxide (K_2O) equivalent or the amount of K_2O which could be made from the mineral being considered.

1. Give the atomic number, number of neutrons, and mass number for the most abundant isotope of potassium, given that the three stable isotopes of potassium are K-39, K-40, and K-41.

   Answer: From the periodic table (see the inside front cover of the book) we find that potassium has atomic number 19 and an atomic mass 39.10 amu. Since the atomic mass is so close to 39, the mass number of the most abundant isotope must be 39. Then, the number of neutrons is $39 - 19 = 20$.

2. If potassium chloride costs \$0.055 per kg, at what price (dollar per kg) must potassium sulfate be sold at in order to supply the same amount of potassium on a per dollar basis?

   Answer: From Section 2.9 we find that potassium chloride is KCl and potassium sulfate is K_2SO_4. To solve this problem, we need to first calculate the percent composition of K in KCl and in K_2SO_4. Following the procedure in Example 2.15, we write

$$\%\text{K in KCl} = \frac{39.10 \text{ amu}}{39.10 \text{ amu} + 35.45 \text{ amu}} \times 100\% = 52.45\%$$

Similarly, we have

$$\%K \text{ in } K_2SO_4 = \frac{2 \times 39.10 \text{ amu}}{2 \times 39.10 \text{ amu} + 32.07 \text{ amu} + 4 \times 16.00 \text{ amu}} \times 100\% = 44.87\%$$

The price at which K_2SO_4 supplies the same amount of potassium per dollar is found by the following ratio:

$$\frac{\text{price of } K_2SO_4}{\text{price of KCl}} = \frac{\%K \text{ in } K_2SO_4}{\%K \text{ in KCl}}$$

Rearranging, we obtain

$$\begin{aligned} \text{price of } K_2SO_4 &= \frac{\%K \text{ in } K_2SO_4}{\%K \text{ in KCl}} \times \text{price of KCl} \\ &= \frac{44.87\%}{52.45\%} \times \$0.055/\text{kg} \\ &= \$0.047/\text{kg} \end{aligned}$$

Thus the price of K_2SO_4 must be \$0.047/kg to give us the same amount of potassium on a per dollar basis as KCl would at \$0.055/kg.

To check the reasonableness of the answer, we see that because the percent composition by mass of K in K_2SO_4 is less than that in KCl, the cost of K_2SO_4 must be *lower* to supply the same amount of K per dollar.

3. What mass (in kg) of K_2O contains the same number of moles of K atoms as 1.00 kg of KCl?

Answer: The first step is to find the number of moles of K atoms in 1.00 kg of KCl. From the percent by mass of K in KCl in 2, we write

$$\begin{aligned} \text{moles K} &= \frac{52.45 \text{ g K}}{100 \text{ g KCl}} \times 1.00 \times 10^3 \text{ g KCl} \times \frac{1 \text{ mol K}}{39.10 \text{ g K}} \\ &= 13.4 \text{ mol K} \end{aligned}$$

Next we calculate the number of moles of K_2O that contain 13.4 moles of K. Since every K_2O unit contains two K atoms, this number turns out to be 13.4×0.5 or 6.70 moles. Finally, the mass of K_2O that supplies the same number of moles of K as 1.00 kg of KCl is

$$\begin{aligned} \text{mass of } K_2O &= 6.70 \text{ mol } K_2O \times \frac{94.20 \text{ g } K_2O}{1 \text{ mol } K_2O} \\ &= 631 \text{ g } K_2O \\ &= 0.631 \text{ kg } K_2O \end{aligned}$$

Since the percent composition by mass of K in K_2O is higher than that in KCl, we expect the mass of K_2O to be less than 1 kg. Therefore, the answer is reasonable.

Key Words

α Particles, p. 41
α Rays, p. 41
Acid, p. 72
Allotrope, p. 51
Anion, p. 55
Atom, p. 39
Atomic mass, p. 46
Atomic mass unit (amu), p. 45
Atomic number (Z), p. 44
Avogadro's number, p. 47
β Particles, p. 41
β Rays, p. 41
Base, p. 75
Binary compound, p. 67
Cation, p. 55
Chemical formula, p. 50

Diatomic molecule, p. 50
Electron, p. 39
Empirical formula, p. 52
γ Rays, p. 41
Hydrate, p. 75
Inorganic compound, p. 66
Ion, p. 55
Ionic compound, p. 55
Isotope, p. 44
Law of conservation of mass, p. 39
Law of definite proportions, p. 62
Law of multiple proportions, p. 63
Mass number (*A*), p. 44
Molar mass, p. 48
Molar mass of a compound, p. 53
Mole (mol), p. 47
Molecular formula, p. 50
Molecular mass, p. 53
Molecule, p. 50
Monatomic ion, p. 55
Neutron, p. 43
Nucleus, p. 42
Organic compound, p. 66
Oxoacid, p. 73
Oxoanion, p. 73
Percent composition by mass, p. 56
Percent composition by weight, p. 56
Polyatomic ion, p. 55
Polyatomic molecule, p. 51
Proton, p. 42
Radiation, p. 40
Radioactivity, p. 40
Ternary compound, p. 67

Exercises

Structure of the Atom

REVIEW QUESTIONS

2.1 Define the following terms: (a) α particle, (b) β particle, (c) γ ray, (d) X ray.
2.2 List the types of radiation that are known to be emitted by radioactive elements.
2.3 Compare the properties of the following: α particles, cathode rays, protons, neutrons, electrons.
2.4 What is meant by the term "fundamental particle"?
2.5 Which of the following are subatomic particles? α particle, electron, hydrogen atom, proton, γ ray, neutron.
2.6 Identify the following scientists with their contributions to our knowledge of atomic structure: J. J. Thomson, R. A. Millikan, Ernest Rutherford, James Chadwick.

PROBLEMS

2.7 A sample of a radioactive element is found to be losing mass gradually. Explain what is happening to the sample.
2.8 Describe the experimental basis for believing that the nucleus occupies a very small fraction of the volume of the atom.
2.9 The diameter of a neutral helium atom is about 1×10^2 pm. Suppose that we could line up helium atoms side by side in contact with one another. Approximately how many atoms would it take to make the distance between the first and last atoms 1 cm?
2.10 Roughly speaking, the radius of an atom is about 10,000 times greater than that of its nucleus. If an atom were magnified so that the radius of its nucleus became 10 cm, what would be the radius of the atom in miles? (1 mi = 1609 m)
2.11 Describe the observations that would be made in an α-particle scattering experiment if (a) the nucleus of an atom were negatively charged and the protons occupied the empty space outside the nucleus and (b) the electrons were embedded in a positively charge sphere (Thomson's model).

Atomic Number, Mass Number, and Isotopes

REVIEW QUESTIONS

2.12 Define the following terms: (a) atomic number, (b) mass number. Why does a knowledge of atomic number enable us to deduce the number of electrons present in an atom?
2.13 Why do all atoms of an element have the same atomic number, although they may have different mass numbers? What do we call atoms of the same elements with different mass numbers?
2.14 Explain the meaning of each term in the symbol ${}^{A}_{Z}X$.

PROBLEMS

2.15 What is the mass number of an iron atom if that atom has 28 neutrons?
2.16 For each of the following species, determine the number of protons and the number of neutrons in the nucleus:

$${}^{3}_{2}He,\ {}^{4}_{2}He,\ {}^{24}_{12}Mg,\ {}^{25}_{12}Mg,\ {}^{48}_{22}Ti,\ {}^{79}_{35}Br,\ {}^{195}_{78}Pt$$

2.17 Indicate the number of protons, neutrons, and electrons in each of the following species:

$${}^{15}_{7}N,\ {}^{33}_{16}S,\ {}^{63}_{29}Cu,\ {}^{84}_{38}Sr,\ {}^{130}_{56}Ba,\ {}^{186}_{74}W,\ {}^{202}_{80}Hg$$

2.18 For the noble gases (the Group 8A elements),

$${}^{4}_{2}He \quad {}^{20}_{10}Ne \quad {}^{40}_{18}Ar \quad {}^{84}_{36}Kr \quad {}^{132}_{54}Xe$$

(a) determine the number of protons and neutrons in the nucleus of each atom and (b) determine the ratio of neutrons to protons in the nucleus of each atom. Describe any general trend you discover in the way this ratio changes with increasing atomic number.

2.19 Write the appropriate symbol for each of the following isotopes: (a) $Z = 11$, $A = 23$; (b) $Z = 28$, $A = 64$; (c) $Z = 74$, $A = 186$; (d) $Z = 80$, $A = 201$.
2.20 Which of the following symbols provides us with more information about the atom: ${}^{23}Na$ or ${}_{11}Na$? Explain.

Atomic Mass and Avogadro's Number

REVIEW QUESTIONS

2.21 What is an atomic mass unit?

2.22 What is the mass (in amu) of a carbon-12 atom?

2.23 When we look up the atomic mass of carbon, we find that its value is 12.01 amu rather than 12.00 amu as defined. Why?

2.24 Define the term "mole." What is the unit for mole in calculations? What does the mole have in common with the pair, the dozen, and the gross?

2.25 What does an Avogadro's number represent?

2.26 What does 1 mole of oxygen atoms mean?

2.27 Define "molar mass." What are the commonly used units for molar mass?

2.28 Calculate the charge (in coulombs) and mass (in grams) of 1 mole of electrons.

2.29 Explain clearly what is meant by the statement "The atomic mass of gold is 197.0 amu."

PROBLEMS

2.30 The atomic masses of $^{35}_{17}Cl$ (75.53%) and $^{37}_{17}Cl$ (24.47%) are 34.968 amu and 36.956 amu, respectively. Calculate the average atomic mass of chlorine. The percentages in parentheses denote the relative abundances.

2.31 Earth's population is about 5.0 billion. Suppose that every person on Earth participates in a process of counting identical particles at the rate of two particles per second. How many years would it take to count 6.0×10^{23} particles? Assume that there are 365 days in a year.

2.32 Avogadro's number has sometimes been described as a conversion factor between the units of amu and grams. Use the fluorine atom (19.00 amu) as an example to show the relation between the atomic mass unit and the gram.

2.33 What is the mass in grams of 13.2 amu?

2.34 How many amu are there in 8.4 g?

2.35 How many atoms are there in 5.10 moles of sulfur (S)?

2.36 How many moles of cobalt atoms are there in 6.00×10^9 (6 billion) Co atoms?

2.37 How many moles of calcium (Ca) atoms are in 77.4 g of Ca?

2.38 How many grams of gold (Au) are there in 15.3 moles of Au?

2.39 What is the mass in grams of a single atom of each of the following elements? (a) Hg, (b) Ne, (c) As, (d) Pb.

2.40 What is the mass in grams of 1.00×10^{12} lead (Pb) atoms?

2.41 How many atoms are present in 3.14 g of copper (Cu)?

2.42 Which of the following has more atoms: 1.10 g of hydrogen atoms or 14.7 g of chromium atoms?

2.43 In the formation of carbon monoxide, CO, it is found that 2.445 g of carbon combines with 3.257 g of oxygen. What is the atomic mass of oxygen if the atomic mass of carbon is 12.01 amu?

2.44 A certain metal oxide has the formula MO. A 39.46-g sample of the compound is strongly heated in an atmosphere of hydrogen to remove oxygen as water molecules. At the end, 31.70 g of the metal M is left over. If O has an atomic mass of 16.00 amu, calculate the atomic mass of M and identify the element.

Molecules and Chemical Formulas

REVIEW QUESTIONS

2.45 What is the difference between an atom and a molecule?

2.46 What does a chemical formula represent?

2.47 What is the ratio (in number of atoms and in number of moles) of the atoms in the following molecular formulas? (a) NO, (b) NCl_3, (c) N_2O_4, (d) P_4O_6.

2.48 Define molecular formula and empirical formula. What are the similarities and differences between the empirical formula and molecular formula of a compound?

2.49 Give an example of a case in which two molecules have different molecular formulas but the same empirical formula.

2.50 What does the expression P_4 signify? How does this differ from 4P?

2.51 What is the difference between a molecule and a compound? Show a molecule that is also a compound and a molecule that is not a compound.

2.52 Give two examples for each of the following: (a) a diatomic molecule containing atoms of the same element, (b) a diatomic molecule containing atoms of different elements, (c) a polyatomic molecule containing atoms of the same element, (d) a polyatomic molecule containing atoms of different elements.

PROBLEM

2.53 What are the empirical formulas of the following compounds? (a) C_2N_2, (b) C_6H_6, (c) C_9H_{20}, (d) P_4O_{10}, (e) B_2H_6, (f) Al_2Br_6, (g) $Na_2S_2O_4$, (h) N_2O_5, (i) $K_2Cr_2O_7$.

Molecular Mass

REVIEW QUESTIONS

2.54 What is the difference between atomic mass and molecular mass?

2.55 Why do chemists prefer not to express the masses of individual atoms and molecules in grams in practical work?

PROBLEMS

2.56 Calculate the molecular mass (in amu) of each of the following substances: (a) CH_4, (b) H_2O, (c) H_2O_2, (d) C_6H_6, (e) PCl_5.

2.57 Calculate the molar mass of the following substances: (a) S_8, (b) CS_2, (c) $CHCl_3$ (chloroform), (d) $C_6H_8O_6$ (ascorbic acid, or vitamin C).

2.58 Calculate the molar mass of $C_{55}H_{72}MgN_4O_5$ (chlorophyll), the green plant pigment that plays the central role in photosynthesis.

2.59 Calculate the molar mass of a compound if 0.372 mole of it has a mass of 152 g.

2.60 Consider the B_2H_6 molecule. What is the relationship between the molar mass of the molecule and the molar mass calculated from the empirical formula of the molecule?

2.61 How many moles of ethane (C_2H_6) are present in 0.334 g of C_2H_6?

2.62 The density of water is 1.00 g/mL at 4°C. How many water molecules are present in 2.56 mL of water at this temperature?

2.63 What mole ratio of molecular chlorine (Cl_2) to molecular oxygen (O_2) would result from the breakup of the compound Cl_2O_7 into its constituent elements?

2.64 Calculate the numbers of C, H, and O atoms in 1.50 g of glucose ($C_6H_{12}O_6$), a sugar.

2.65 Urea [$(NH_2)_2CO$] is a compound used for fertilizer and many other things. Calculate the number of N, C, O, and H atoms in 1.68×10^4 g of urea.

2.66 Pheromones are a special type of compound secreted by the females of many insect species to attract the males for mating. One pheromone has the molecular formula $C_{19}H_{38}O$. Normally, the amount of this pheromone secreted by a female insect is about 1.0×10^{-12} g. How many molecules are there in this quantity?

Ionic Compounds

REVIEW QUESTIONS

2.67 What is an ionic compound? How is electrical neutrality maintained in an ionic compound?

2.68 Which elements are most likely to form ionic compounds?

2.69 Why are ionic compounds normally not called molecules?

2.70 Compare and contrast the properties of molecular compounds and ionic compounds.

PROBLEMS

2.71 Give the number of protons and electrons in each of the following ions present in ionic compounds: Na^+, Ca^{2+}, Al^{3+}, Fe^{2+}, I^-, F^-, S^{2-}, O^{2-}, and N^{3-}.

2.72 Cyanides are compounds that contain the CN^- anion. Most cyanides are deadly poisonous compounds. For example, ingestion of a quantity as small as 1.00×10^{-3} g of potassium cyanide (KCN) may prove fatal. How many KCN units does this quantity of KCN contain?

2.73 Which of the following compounds are likely to be ionic? Which are likely to be molecular? $SiCl_4$, LiF, $BaCl_2$, B_2H_6, KF, C_2H_4.

Percent Composition and Chemical Formulas

REVIEW QUESTIONS

2.74 Define percent composition by mass of a compound.

2.75 Describe how the knowledge of the percent composition by mass of an unknown compound of high purity can help us identify the compound.

2.76 What does the word ''empirical'' mean?

2.77 If we know the empirical formula of a compound, what additional information do we need in order to determine its molecular formula?

PROBLEMS

2.78 Tin (Sn) exists in Earth's crust as SnO_2. Calculate the percent composition by mass of Sn and O in SnO_2.

2.79 Sodium, a member of the alkali metal family (Group 1A), is a very reactive metal and thus is never found in nature in its elemental state. It forms ionic compounds with members of the halogens (Group 7A). Calculate the percent composition by mass of all the elements in each of the following compounds: (a) NaF, (b) NaCl, (c) NaBr, (d) NaI.

2.80 For many years chloroform ($CHCl_3$) was used as an inhalation anesthetic in spite of the fact that it is also a toxic substance that may cause severe liver, kidney, and heart damage. Calculate the percent composition by mass of this compound.

2.81 Allicin is the compound responsible for the characteristic smell of garlic. An analysis of the compound gives the following percent composition by mass: C: 44.4%; H: 6.21%; S: 39.5%; O: 9.86%. Calculate its empirical formula. What is its molecular formula given that its molar mass is about 162 g?

2.82 Cinnamic alcohol is used mainly in perfumery, particularly soaps and cosmetics. Its molecular formula is $C_9H_{10}O$. (a) Calculate the percent composition by mass of C, H, and O in cinnamic alcohol. (b) How many molecules of cinnamic alcohol are contained in a sample of mass 0.469 g?

2.83 All of the substances listed below are fertilizers that contribute nitrogen to the soil. Which of these is the richest source of nitrogen on a mass percentage basis?
(a) Urea, $(NH_2)_2CO$
(b) Ammonium nitrate, NH_4NO_3
(c) Guanidine, $HNC(NH_2)_2$
(d) Ammonia, NH_3

2.84 Which of the following substances contains the greatest mass of chlorine? (a) 5.0 g Cl_2, (b) 60.0 g $NaClO_3$, (c) 0.10 mol KCl, (d) 30.0 g $MgCl_2$, (e) 0.50 mol Cl_2.

2.85 The formula for rust can be represented by Fe_2O_3. How many moles of Fe are present in 24.6 g of the compound?

2.86 How many grams of sulfur (S) are needed to combine with 246 g of mercury (Hg) to form HgS?

2.87 Calculate the mass in grams of iodine (I_2) that will react completely with 20.4 g of aluminum (Al) to form aluminum iodide (AlI_3).

2.88 Tin(II) fluoride (SnF_2) is often added to toothpaste as an ingredient to prevent tooth decay. What is the mass of F in grams in 24.6 g of the compound?

2.89 Platinum forms two different compounds with chlorine. One contains 26.7 percent Cl by mass, and the other contains 42.1 percent Cl by mass. Determine the empirical formulas of the two compounds.

2.90 What are the empirical formulas of the compounds with the following compositions? (a) 2.1 percent H, 65.3 percent O, 32.6 percent S; (b) 20.2 percent Al, 79.8 percent Cl; (c) 40.1 percent C, 6.6 percent H, 53.3 percent O; (d) 18.4 percent C, 21.5 percent N, 60.1 percent K.

2.91 Peroxyacylnitrate (PAN) is one of the components of smog. It is a compound of C, H, N, and O. Determine the percent composition of oxygen and the empirical formula from the following percent composition by mass: 19.8 percent C, 2.50 percent H, 11.6 percent N.

2.92 The molar mass of caffeine is 194.19 g. Is the molecular formula of caffeine $C_4H_5N_2O$ or $C_8H_{10}N_4O_2$?

2.93 Monosodium glutamate (MSG) has sometimes been suspected as the cause of "Chinese restaurant syndrome" because this food-flavor enhancer can induce headaches and chest pains. MSG has the following composition by mass: 35.51 percent C, 4.77 percent H, 37.85 percent O, 8.29 percent N, and 13.60 percent Na. What is its molecular formula if its molar mass is 169 g?

Law of Multiple Proportions

REVIEW QUESTION

2.94 Define the law of definite proportions and the law of multiple proportions.

PROBLEMS

2.95 Nitrogen forms a number of compounds with oxygen. It is found that in one compound, 0.681 g of N is combined with 0.778 g of O; in another compound, 0.560 g of N is combined with 1.28 g of O. Prove that these data support the law of multiple proportions.

2.96 Chemical analysis of two iron chlorides shows their composition to be 44.06 g of Fe and 55.94 g of Cl for one, and 34.43 g of Fe and 65.57 g of Cl for the other. Are these data consistent with the law of multiple proportions?

2.97 Phosphorus forms two compounds with chlorine. In one compound, 1.94 g of P combines with 6.64 g of Cl; in another, 0.660 g of P combines with 3.78 g of Cl. Are these data consistent with the law of multiple proportions?

Mass Spectrometry

REVIEW QUESTIONS

2.98 Describe the operation of a mass spectrometer.

2.99 Describe how you would determine the isotopic abundance of an element from its mass spectrum.

PROBLEMS

2.100 Carbon has two stable isotopes, ${}^{12}_{6}C$ and ${}^{13}_{6}C$, and fluorine has only one stable isotope, ${}^{19}_{9}F$. How many peaks would you observe in the mass spectrum of the positive ion of CF_4^+? Assume no decomposition of the ion into smaller fragments.

2.101 Hydrogen has two stable isotopes, ${}^{1}_{1}H$ and ${}^{2}_{1}H$, and sulfur has four stable isotopes, ${}^{32}_{16}S$, ${}^{33}_{16}S$, ${}^{34}_{16}S$, ${}^{36}_{16}S$. How many peaks would you observe in the mass spectrum of the positive ion of hydrogen sulfide, H_2S^+? Assume no decomposition of the ion into smaller fragments.

2.102 A compound containing only C, H, and Cl was examined in a mass spectrometer. The highest mass peak seen corresponds to an ion mass of 52 amu. The most abundant mass peak seen corresponds to an ion mass of 50 amu and is about three times as intense as the peak at 52 amu. Deduce a reasonable molecular formula for the compound and explain the positions and intensities of the mass peaks mentioned. (*Hint:* Chlorine is the only element that has isotopes in comparable abundances: ${}^{35}_{17}Cl$: 75.5%; ${}^{37}_{17}Cl$: 24.5%. For H, use ${}^{1}_{1}H$; for C, use ${}^{12}_{6}C$.)

2.103 A mixture of D_2 and Br_2 was allowed to react, and the products are analyzed using a mass spectrometer and the following mass spectrum obtained:

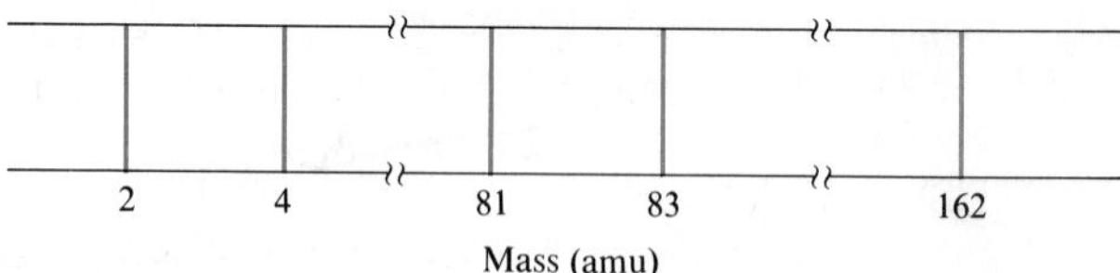

Assume all the species are singly charged. Identify each of the lines. (Use ${}^{2}_{1}H$ for hydrogen and ${}^{81}_{35}Br$ for bromine.)

Naming Inorganic Compounds

REVIEW QUESTIONS

2.104 What is the difference between an inorganic compound and an organic compound?

2.105 Label each of the following compounds as inorganic or organic: (a) NH_4Br, (b) CCl_4, (c) $KHCO_3$, (d) NaCN, (e) CH_4, (f) $CaCO_3$, (g) CO, (h) CH_3COOH.

2.106 Define the following terms: binary compound, ternary compound, acid, oxoacid, oxoanion, base, hydrate.

2.107 Describe the use and advantages of the Stock system in naming inorganic compounds.

2.108 Give two examples each of (a) an acid that contains one or more oxygen atoms and (b) an acid that does not.

2.109 Give (a) two examples of a base that contains a hydroxide (OH^-) group and (b) one example of a base that does not.

PROBLEMS

2.110 Name the following compounds: (a) KH_2PO_4, (b) K_2HPO_4, (c) HBr (gas), (d) HBr (in water), (e) Li_2CO_3, (f) $K_2Cr_2O_7$, (g) NH_4NO_2, (h) PF_3, (i) PF_5, (j) P_4O_6, (k) CdI_2, (l) $SrSO_4$, (m) $Al(OH)_3$, (n) KClO, (o) Ag_2CO_3, (p) $FeCl_2$, (q) $KMnO_4$, (r) $CsClO_3$, (s) KNH_4SO_4, (t) FeO, (u) Fe_2O_3, (v) $TiCl_4$, (w) NaH, (x) Li_3N, (y) Na_2O, (z) Na_2O_2.

2.111 Write the formulas for the following compounds: (a) rubidium nitrite, (b) potassium sulfide, (c) sodium hydrogen sulfide, (d) magnesium phosphate, (e) calcium hydrogen phosphate, (f) potassium dihydrogen phosphate, (g) iodine heptafluoride, (h) ammonium sulfate, (i) silver perchlorate, (j) iron(III) chromate, (k) copper(I) cyanide, (l) strontium chlorite, (m) perbromic acid, (n) hydroiodic acid, (o) disodium ammonium phosphate, (p) lead(II) carbonate, (q) tin(II) fluoride, (r) tetraphosphorus decasulfide, (s) mercury(II) oxide, (t) mercury(I) iodide, (u) copper(II) sulfate pentahydrate.

2.112 Write the formulas for the compounds with the following common names: (a) dry ice, (b) baking soda, (c) milk of magnesia, (d) laughing gas, (e) table salt, (f) marble, (g) chalk, (h) limestone, (i) quicklime, (j) slaked lime.

Miscellaneous Problems

2.113 The atomic masses of ${}^{6}_{3}Li$ and ${}^{7}_{3}Li$ are 6.0151 amu and 7.0160 amu, respectively. Calculate the natural abundances of these two isotopes. The average atomic mass of Li is 6.941 amu.

2.114 One isotope of a metallic element has mass number 65 and 35 neutrons in the nucleus. The cation derived from the isotope has 28 electrons. Write the symbol for this cation.

2.115 In which one of the following pairs do the two species resemble each other most closely in chemical properties? (a) ${}^{1}_{1}H$ and ${}^{1}_{1}H^+$, (b) ${}^{14}_{7}N$ and ${}^{14}_{7}N^{3-}$, (c) ${}^{12}_{6}C$ and ${}^{13}_{6}C$.

2.116 The table gives numbers of electrons, protons, and neutrons in atoms or ions of a number of elements. Answer the following: (a) Which of the species are neutral? (b) Which are negatively charged? (c) Which are positively charged? (d) What are the conventional symbols for all the species?

Atom or ion of element	A	B	C	D	E	F	G
Number of electrons	5	10	18	28	36	5	9
Number of protons	5	7	19	30	35	5	9
Number of neutrons	5	7	20	36	46	6	10

2.117 What is wrong or ambiguous about each of the following statements? (a) 1 mole of hydrogen. (b) The molecular mass of NaCl is 58.5 amu.

2.118 Which of the following are elements, which are molecules but not compounds, which are compounds but not molecules, and which are both compounds and molecules? (a) SO_2, (b) S_8, (c) Cs, (d) N_2O_5, (e) O, (f) O_2, (g) O_3, (h) CH_4, (i) KBr, (j) S, (k) P_4, (l) LiF.

2.119 A hydrate compound contains 4.00 g of calcium, 7.10 g of chlorine, and 7.20 g of water. What is its formula?

2.120 Which one of the following samples contains the largest number of atoms? (a) 2.5 mol CH_4, (b) 10.0 mol He, (c) 4.0 mol SO_2, (d) 1.8 mol S_8, (e) 3.0 mol NH_3.

2.121 Which of the following has the greater mass: 0.72 g of O_2 or 0.0011 mol of chlorophyll ($C_{55}H_{72}MgN_4O_5$)? (For molar mass of chlorophyll, see Problem 2.58.)

2.122 Arrange the following in order of increasing mass: (a) 16 water molecules, (b) 2 atoms of lead, (c) 5.1×10^{-23} mole of helium.

2.123 The atomic mass of element X is 33.42 amu. A 27.22-g sample of X combines with 84.10 g of another element Y to form a compound XY. Calculate the atomic mass of Y in amu.

2.124 Analysis of a metal chloride XCl_3 shows that it contains 67.2% Cl by mass. Calculate the atomic mass of X in amu and identify the element.

2.125 Ionic compounds are most likely to be formed as a result of the combination between the metals in Group 1A, Group 2A, and aluminum (a Group 3A metal) and the nonmetals oxygen, nitrogen, and the halogens (the Group 7A elements). Write chemical formulas and names of all the binary compounds that can result from such combinations.

2.126 The following phosphorus sulfides are known: P_4S_3, P_4S_7, and P_4S_{10}. Do these compounds obey the law of multiple proportions? (*Hint:* Since the number of P atoms is the same in all three compounds, you need only look at the ratio of the S atoms in these compounds.)

2.127 Hemoglobin ($C_{2952}H_{4664}N_{812}O_{832}S_8Fe_4$) is the oxygen carrier in blood. (a) Calculate its molar mass. (b) How many molecules are present in 74.3 g of hemoglobin?

2.128 Myoglobin stores oxygen for metabolic processes in muscle. Chemical analysis shows that it contains 0.34 percent Fe by mass. What is the molar mass of myoglobin?

2.129 How many moles of O are needed to combine with 0.212 mole of C to form (a) CO and (b) CO_2?

2.130 Calculate the number of cations and anions in each of the following cases: (a) 8.38 g of KBr, (b) 5.40 g of Na_2SO_4, (c) 7.45 g of $Ca_3(PO_4)_2$.

2.131 The aluminum sulfate hydrate $[Al_2(SO_4)_3 \cdot xH_2O]$ contains 8.20% of Al by mass. Calculate x.

2.132 (a) Derive an expression relating the number of moles of a molecule (n) to its mass in grams (m) and its molar mass in g/mol ($\mathcal{M}$). (b) Derive an expression relating the number of molecules (N) to the number of moles of the molecule (n) and Avogadro's number (N_A). Make sure that your equations are dimensionally correct; that is, the same units must appear on both sides of the equation.

2.133 A sample containing NaCl, Na_2SO_4, and $NaNO_3$ gives the following elemental analysis: Na: 32.08%; O: 36.01%; Cl: 19.51%. Calculate the mass percent of each compound in the sample.

2.134 Lysine, an essential amino acid in the human body, contains C, H, O, and N. In one experiment, the complete combustion of 2.175 g of lysine gave 3.94 g CO_2 and 1.89 g H_2O. In a separate experiment, 1.873 g of lysine gave 0.436 g NH_3. (a) Calculate the empirical formula of lysine. (b) The approximate molar mass of lysine is 150 g. What is the molecular formula of the compound?

2.135 The thickness of a piece of paper is 0.0036 in. Suppose a certain book has an Avogadro's number of pages; calculate the thickness of the book in light-years. (*Hint:* See Problem 1.54 for the definition of light-year.)

2.136 (a) For molecules having small molecular masses, mass spectrometry can be used to identify their formulas. To illustrate this point, identify the molecule which most likely accounts for the observation of a peak in a mass spectrum at: 16 amu, 17 amu, 18 amu, and 64 amu. (b) Note that there are (among others) two likely molecules which would give rise to a peak at 44 amu, namely, C_3H_8 and CO_2. In such cases, a chemist might try to look for other peaks generated when some of the molecules break apart in the spectrometer. For example, if a chemist sees a peak at 44 amu and also one at 15 amu, which molecule is producing the 44-amu peak? Why? (c) Using the following precise atomic masses: ^1H(1.00797 amu), ^{12}C(12.00000 amu), and ^{16}O(15.99491 amu), how precisely must the masses of C_3H_8 and CO_2 be measured to distinguish between them?

2.137 An unidentified metal is known to form an oxide which is 77.73 percent metal by mass. Assume that the atomic mass of oxygen is 16.00 amu. (a) Provide at least three possible atomic masses for the metal that are consistent with this percent composition. (b) The metal forms a second oxide which is 69.94 percent metal by mass. Show that these two oxides obey the law of multiple proportions. (c) Show that we still cannot unambiguously assign the atomic mass for the metal by using two possible atomic masses derived in part (a) along with the corresponding empirical formulas which are consistent with the percent composition of both oxides. (d) The unidentified metal also forms two chlorides whose compositions are 43.90 percent and 34.45 percent metal by mass. Assume that the atomic mass of chlorine is 35.45 amu; show that these two chlorides obey the law of multiple proportions. (e) Based on all the data presented, deduce the most likely atomic mass of the metal.

Opposite page: The reaction between aluminum metal and iron(III) oxide (called the thermite reaction) generates much heat and light.

C H A P T E R 3

CHEMICAL REACTIONS I: CHEMICAL EQUATIONS AND REACTIONS IN AQUEOUS SOLUTION

Having discussed the masses of atoms and molecules, our next concern is what happens to atoms and molecules when chemical changes occur. This is the first of two chapters that deal with the study of chemical reactions.

3.1 Chemical Equations

A *chemical change* is called a ***chemical reaction.*** In order to communicate with one another about chemical reactions, chemists have devised a standard way to represent them. This section deals with two related aspects of describing chemical reactions by the use of chemical symbols—writing chemical equations and balancing chemical equations.

Writing Chemical Equations

Hydrogen gas burning in air.

Consider what happens when hydrogen gas (H_2) burns in air (which contains oxygen, O_2) to form water (H_2O). This reaction can be represented by the chemical equation

$$H_2 + O_2 \longrightarrow H_2O \tag{3.1}$$

where the + sign means "reacts with" and the $\longrightarrow$ sign means "to yield." Thus, this symbolic expression can be read: "Molecular hydrogen reacts with molecular oxygen to yield water." The reaction is assumed to proceed from left to right as the arrow indicates.

The expression (3.1) is not complete, however, because there are twice as many oxygen atoms on the left side of the arrow (two) as on the right side (one). To conform with the law of conservation of mass, there must be the same number of each type of atom on both sides of the arrow; that is, we must have as many atoms after the reaction ends as we did before it started. We can *balance* this expression by placing an appropriate coefficient (2 in this case) in front of H_2 and H_2O:

When the coefficient is 1, as in the case of O_2, it is not shown.

$$2H_2 + O_2 \longrightarrow 2H_2O$$

This *balanced chemical equation* shows that "two hydrogen molecules can combine or react with one oxygen molecule to form two water molecules" (Figure 3.1). Since the ratio of the number of molecules is equal to the ratio of the number of moles, the equation can also be read as "2 moles of hydrogen molecules react with 1 mole of oxygen molecules to produce 2 moles of water molecules." We know the mass of a mole of each of these substances, so we can also interpret the equation as "4.04 g of H_2 react with 32.00 g of O_2 to give 36.04 g of H_2O." These three ways of reading the equation are summarized in Table 3.1.

We refer to H_2 and O_2 in Equation (3.1) as ***reactants,*** which are *the starting materials in a chemical reaction.* Water is the ***product,*** which is *the substance formed as a*

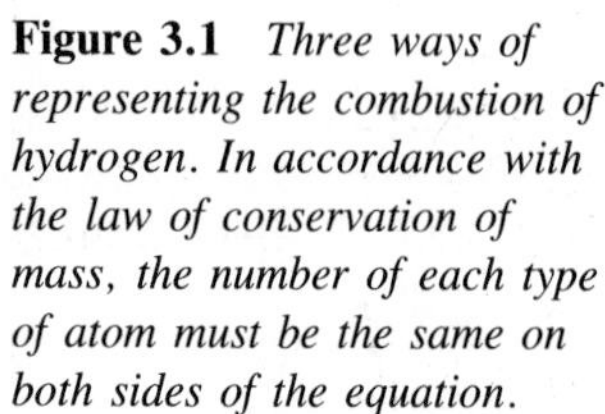

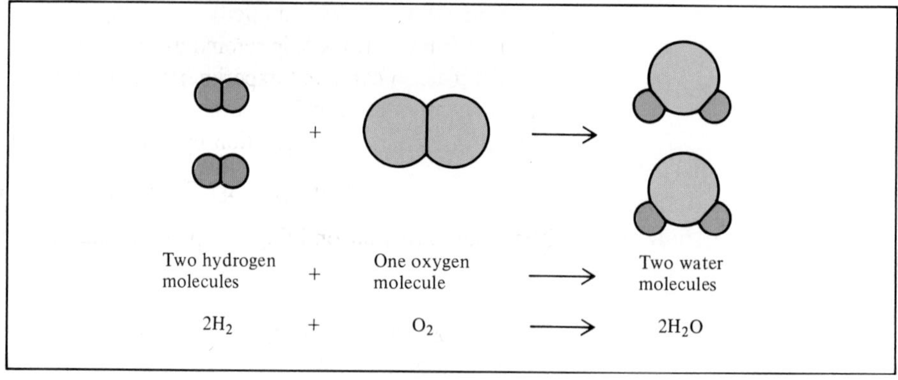

Figure 3.1 *Three ways of representing the combustion of hydrogen. In accordance with the law of conservation of mass, the number of each type of atom must be the same on both sides of the equation.*

Table 3.1 Interpretation of a chemical equation

$2H_2$	$+ O_2$	$\longrightarrow$	$2H_2O$
Two molecules	+ one molecule	$\longrightarrow$	two molecules
2 moles	+ 1 mole	$\longrightarrow$	2 moles
2(2.02 g) = 4.04 g	+ 32.00 g	$\longrightarrow$	2(18.02 g) = 36.04 g
36.04 g reactants			36.04 g products

result of a chemical reaction. A chemical equation, then, can be thought of as the chemist's shorthand description of a reaction. In a chemical equation the reactants are conventionally written on the left and the products on the right of the arrow:

$$\text{reactants} \longrightarrow \text{products}$$

To provide additional, and often very useful, information when writing chemical equations, chemists often indicate the physical states of the reactants and products by using the abbreviations *g*, *l*, and *s* in parentheses to denote the gas, liquid, and solid state, respectively. For example,

A detailed discussion of balancing chemical equations begins on p. 92.

$$2CO(g) + O_2(g) \longrightarrow 2CO_2(g)$$

$$2P(s) + 3Cl_2(g) \longrightarrow 2PCl_3(l)$$

$$2HgO(s) \longrightarrow 2Hg(l) + O_2(g)$$

Chemists also write equations to represent physical processes. Freezing (liquid to solid) and evaporating (liquid to vapor) water, for example, can be represented as follows:

$$H_2O(l) \longrightarrow H_2O(s)$$

$$H_2O(l) \longrightarrow H_2O(g)$$

To describe the processes of sodium chloride and hydrogen chloride gas dissolving in water, we write

$$NaCl(s) \xrightarrow{H_2O} NaCl(aq)$$

$$HCl(g) \xrightarrow{H_2O} HCl(aq)$$

where *aq* denotes the aqueous (that is, water) environment. Writing H_2O above the arrow indicates the physical process of dissolving a substance in water, although it is sometimes left out for simplicity.

These abbreviations help remind us of the state of the reactants and products and are useful for us to carry out experiments. For example, when potassium bromide (KBr) and silver nitrate ($AgNO_3$) react in an aqueous environment, a solid, silver bromide (AgBr) is formed. This reaction can be represented by the equation:

$$KBr(aq) + AgNO_3(aq) \longrightarrow KNO_3(aq) + AgBr(s)$$

The same equation omitting the physical states of reactants and products would be

$$KBr + AgNO_3 \longrightarrow KNO_3 + AgBr$$

Since no physical states are given, an uninformed person might try to carry out this reaction by mixing solid potassium bromide with solid silver nitrate. Solid potassium bromide and silver nitrate, if mixed together, would react very slowly or not at all. Viewing the process on the microscopic level, we see that for a product like silver bromide to form, the Ag^+ and Br^- ions would have to come into contact. However, these ions are locked in place in the solid state and have little mobility. (Here is an example of how we explain a phenomenon by thinking at the molecular level, as discussed in Section 1.2).

The following example shows the interpretation of a chemical equation.

EXAMPLE 3.1

When iron (Fe) is exposed to moist air (which contains oxygen), it slowly forms rust, which is iron(III) oxide (Fe_2O_3). Iron(III) oxide can be formed more rapidly by burning iron in an atmosphere of oxygen gas. Describe in words what can be deduced from the following chemical equation:

$$4Fe(s) + 3O_2(g) \longrightarrow 2Fe_2O_3(s)$$

Answer

1. Solid iron reacts with O_2 (oxygen gas) to form solid iron(III) oxide.
2. Four iron atoms react with three oxygen molecules to produce two iron(III) oxide formula units (Fe_2O_3 is an ionic compound, consisting of Fe^{3+} and O^{2-} ions).
3. Four moles of iron react with three moles of oxygen molecules to produce two moles of iron(III) oxide.
4. 223.4 g of iron react with 96.00 g of oxygen to produce 319.4 g of iron(III) oxide.

Similar problem: 3.5.

Some reactions never go to completion.

A chemical equation is not a complete description of what actually happens during a chemical reaction. It generally describes only the *overall* change, that is, the number of atoms, molecules, or ions before and after a reaction; it says nothing about *how* products are formed from reactants. Furthermore, a chemical equation does not tell us how long it will take for the change to occur. The reaction discussed in Example 3.1 is a nice illustration of the limitation of chemical equations. Normally, the rusting of an iron nail would take weeks or months. On the other hand, iron(III) oxide can be formed within seconds when iron is burned in oxygen. Yet both of these processes are represented by the same equation.

Balancing Chemical Equations

Suppose we want to write an equation to describe a chemical reaction that we have just carried out in the laboratory. How should we go about doing this? Because we know the identities of the reactants, we can write the chemical formulas for them. The identities of products are more difficult to establish. For simple reactions it is often possible to guess the product(s). For more complicated reactions, where there may be three or more products, chemists may need to carry out further tests to establish the presence of specific compounds. There are additional aids that help us in this respect. As we will see later in this chapter, knowing the type of reactants will often help us guess intelligently what products may be formed. Furthermore, we can learn a great deal about a reaction by *observing* its progress. Thus we can conclude that a gaseous product is formed if we see bubbles appearing in water during a reaction carried out in the aqueous environment. Color change is another indication that a chemical reaction has occurred. Once we have identified all the reactants and products and have written the correct formulas for them, we assemble them in the conventional sequence—reactants on the left separated from products on the right by an arrow. The equation written at this point is most likely to be *unbalanced,* that is, the number of each type of atom on both sides of the equation is not equal. In general, we can balance a chemical equation by the following steps:

- Identify all reactants and products and write their correct formulas on the left and right side of the equation, respectively.
- Begin balancing the equation by trying suitable coefficients that will make the number of atoms of each element the same on both sides of the equation. You can change only the coefficients (the numbers before the formulas) but not the subscripts (the numbers within formulas). For example, $2NO_2$ means "two molecules of nitrogen dioxide," but if we double the subscripts, we have N_2O_4, which is the formula of dinitrogen tetroxide, a completely different compound.
- Look for elements that appear only once on each side of the equation and with equal numbers of atoms on each side—the formulas containing these elements must have the same coefficient. Next, look for elements that appear only once on each side of the equation but in unequal numbers of atoms. Balance these elements. Finally, balance elements that appear in two or more formulas on the same side of the equation.
- Check your balanced equation to be sure that you have the same total number of each type of atoms on both sides of the equation arrow.

Let's consider a specific example. In the laboratory, small amounts of oxygen gas can be conveniently prepared by heating potassium chlorate ($KClO_3$). The products are oxygen gas (O_2) and potassium chloride (KCl). From this information, we write

For simplicity, the physical states of reactants and products are omitted.

$$KClO_3 \longrightarrow KCl + O_2$$

We see that all three elements (K, Cl, and O) appear only once on each side of the equation, but only K and Cl appear in equal numbers of atoms on both sides. Thus $KClO_3$ and KCl must have the same coefficient. The next step is to make the number of O atoms the same on both sides of the equation. Since there are three O atoms on the left and two O atoms on the right of the equation, we can balance the O atoms by placing a 2 in front of $KClO_3$ and a 3 in front of O_2.

$$2KClO_3 \longrightarrow KCl + 3O_2$$

Finally, we balance the K and Cl atoms by placing a 2 in front of KCl:

$$2KClO_3 \longrightarrow 2KCl + 3O_2 \qquad (3.2)$$

As a final check, we can draw up a balance sheet for the number of atoms of each element in both the reactants and products:

Reactants	*Products*
K (2)	K (2)
Cl (2)	Cl (2)
O (6)	O (6)

where the number in parentheses indicates the number of atoms.

Note that this equation could also be balanced with coefficients which are multiples of 2 (for $KClO_3$), 2 (for KCl), and 3 (for O_2); for example,

$$4KClO_3 \longrightarrow 4KCl + 6O_2$$

$$20KClO_3 \longrightarrow 20KCl + 30O_2$$

However, it is common practice to use the *simplest* possible set of whole-number coefficients to balance the equation. Equation (3.2) conforms with this convention.

Ethane is a component of natural gas.

Now let us consider the combustion (that is, burning) of ethane (C_2H_6) in oxygen or air, which yields carbon dioxide (CO_2) and water. We write

$$C_2H_6 + O_2 \longrightarrow CO_2 + H_2O$$

We see that the number of atoms is not the same on both sides of the equation for any of the elements (C, H, and O). We see that C and H appear only once on each side of the equation; O appears in two compounds on the right side (CO_2 and H_2O). To balance the C atoms, we place a 2 in front of CO_2:

$$C_2H_6 + O_2 \longrightarrow 2CO_2 + H_2O$$

To balance the H atoms, we place a 3 in front of H_2O:

$$C_2H_6 + O_2 \longrightarrow 2CO_2 + 3H_2O$$

At this stage, the C and H atoms are balanced, but the O atoms are not balanced, because there are seven O atoms on the right-hand side and only two O atoms on the left-hand side of the equation. This inequality of O atoms can be eliminated by writing $\frac{7}{2}$ in front of the O_2 on the left-hand side:

$$C_2H_6 + \tfrac{7}{2}O_2 \longrightarrow 2CO_2 + 3H_2O$$

The "logic" of the $\frac{7}{2}$ factor can be clarified as follows. There were seven oxygen atoms on the right-hand side of the equation, but only a pair of oxygen atoms (O_2) on the left. To balance them we ask how many *pairs* of oxygen atoms are needed to equal seven oxygen atoms. Just as 3.5 pairs of shoes equal seven shoes, $\frac{7}{2}O_2$ molecules equal seven O atoms. As the following tally shows, the equation is now completely balanced:

Reactants	*Products*
C (2)	C (2)
H (6)	H (6)
O (7)	O (7)

However, we normally prefer to express the coefficients as whole numbers rather than as fractions. Therefore, we multiply the entire equation by 2 to convert $\frac{7}{2}$ to 7:

$$2C_2H_6 + 7O_2 \longrightarrow 4CO_2 + 6H_2O$$

The final tally is

Reactants	*Products*
C (4)	C (4)
H (12)	H (12)
O (14)	O (14)

Note that the coefficients used in balancing the equation are the smallest possible set of whole numbers.

We now apply the procedure for balancing chemical equations to the following examples.

EXAMPLE 3.2

(a) When aluminum metal is exposed to air, a protective layer of aluminum oxide (Al_2O_3) forms on its surface. This layer prevents further reaction between aluminum and oxygen, and this is the reason that aluminum beverage cans do not corrode. (In the case of iron, the rust, or iron(III) oxide, that forms is too porous to protect the iron metal underneath, so rusting continues.) Write a balanced equation for this process. (b) The first step in the industrial preparation of nitric acid (HNO_3), an important industrial chemical used in the manufacture of fertilizer, drugs, and other substances, involves the reaction between ammonia and oxygen gas to form nitric oxide (NO) and water. Write a balanced equation for this reaction.

Answer

(a) The unbalanced equation is

$$Al + O_2 \longrightarrow Al_2O_3$$

We see that both Al and O appear only once on each side of the equation but in unequal numbers. To balance the Al atoms, we place a 2 in front of Al:

$$2Al + O_2 \longrightarrow Al_2O_3$$

There are two O atoms on the left-hand side and three O atoms on the right-hand side of the equation. This inequality of O atoms can be eliminated by writing $\frac{3}{2}$ in front of the O_2 on the left-hand side of the equation.

$$2Al + \tfrac{3}{2}O_2 \longrightarrow Al_2O_3$$

As in the case for ethane shown above, we multiply the entire equation by 2 to convert $\frac{3}{2}$ to 3:

$$4Al + 3O_2 \longrightarrow 2Al_2O_3$$

The final tally is

Reactants	*Products*
Al (4)	Al (4)
O (6)	O (6)

(b) The unbalanced equation is

$$NH_3 + O_2 \longrightarrow NO + H_2O$$

Of these three elements (N, H, and O), only N appears once on each side of the equation and in equal numbers. Therefore, the coefficient for NH_3 must be the same as that for NO. We note that H appears once on each side of the equation but in unequal numbers of atoms (three on the left and two on the right). To balance H, we need to place coefficients in front of NH_3 and H_2O such that the number of H atoms on each side is the same. In this example, $2NH_3$ and $3H_2O$ give 6H's on both sides of the equation. We now write

$$2NH_3 + O_2 \longrightarrow NO + 3H_2O$$

Next we balance the N atoms by placing a 2 in front of NO:

$$2NH_3 + O_2 \longrightarrow 2NO + 3H_2O$$

There are two O atoms on the left and five O atoms on the right, so we place the coefficient $\frac{5}{2}$ in front of O_2:

$$2NH_3 + \tfrac{5}{2}O_2 \longrightarrow 2NO + 3H_2O$$

To avoid using fractions, we multiply the equation throughout by 2 to obtain

$$4NH_3 + 5O_2 \longrightarrow 4NO + 6H_2O$$

Similar problems: 3.7, 3.8.

3.2 Properties of Aqueous Solutions

Now that we have learned the fundamentals for writing and balancing chemical equations, we are ready to study chemical reactions in more detail. However, before we begin, it will be useful to have some knowledge about the medium in which reactions take place. Because many important chemical reactions and virtually all biological processes take place in an aqueous environment, it is important to consider the properties of water solutions. We begin by defining some useful terms.

A ***solution*** is *a homogeneous mixture of two or more substances. The substance present in smaller amount* is called the ***solute,*** while *that present in larger amount* is called the ***solvent.*** A solution may be gaseous (such as air), solid (such as an alloy), or a liquid (seawater, for example). In this section we will discuss only cases in which the solute is a liquid or a solid and the solvent is water—that is, aqueous solutions.

Electrolytes versus Nonelectrolytes

All solutes in aqueous solution can be divided into two categories: electrolytes and nonelectrolytes. An ***electrolyte*** is *a substance that, when dissolved in water, results in a solution that can conduct electricity.* A ***nonelectrolyte*** *does not conduct electricity when dissolved in water.* Figure 3.2 shows an easy and straightforward method of distinguishing between electrolytes and nonelectrolytes. A pair of platinum electrodes is immersed in a beaker containing water. To light the bulb, electric current must flow from one electrode to the other, thus completing the circuit. Pure water is a very poor conductor of electricity. However, if we add a small amount of an ionic compound such as sodium chloride (NaCl), the bulb will glow as soon as the salt dissolves in the water. When solid NaCl dissolves in water, it breaks up into Na^+ and Cl^- ions. The movement of Na^+ ions toward the negative electrode and Cl^- ions toward the positive electrode is equivalent to the flow of electrons along a metal wire. In this way a solution containing an electrolyte is able to conduct electricity. Thus we say that NaCl is an electrolyte. There are very few ions present in pure water, so it cannot conduct electricity.

Comparison of the light bulb's brightness for the *same molar amounts* of dissolved substances helps us distinguish between strong and weak electrolytes. Ionic compounds such as sodium chloride (NaCl), potassium iodide (KI), and calcium chloride ($CaCl_2$) are strong electrolytes. A characteristic of strong electrolytes is that the solute

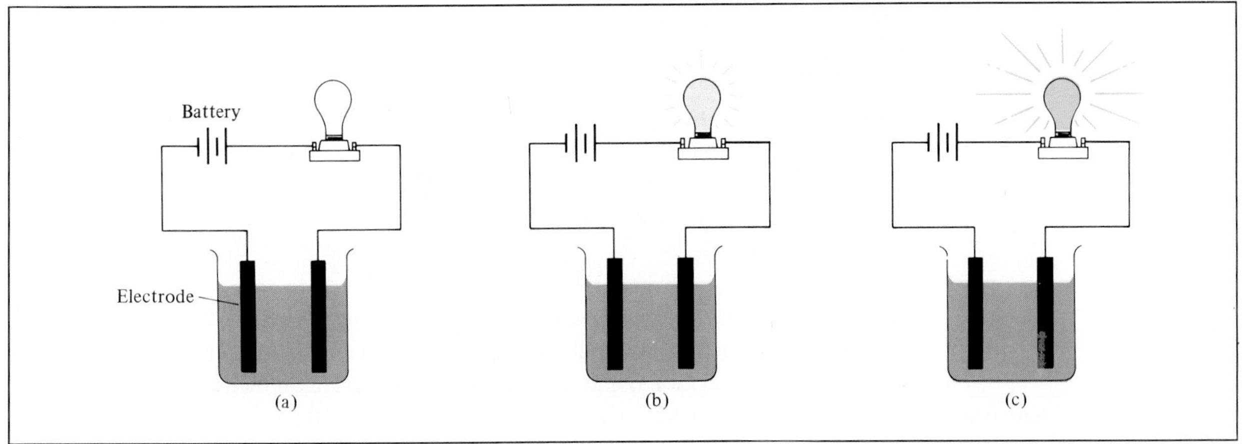

Figure 3.2 *An arrangement for distinguishing between electrolytes and nonelectrolytes. A solution's ability to conduct electricity depends on the number of ions it contains. (a) A nonelectrolyte solution does not contain ions, and the light bulb is not lit. (b) A weak electrolyte solution contains a small number of ions, and the light bulb is dimly lit. (c) A strong electrolyte solution contains a large number of ions, and the light bulb is brightly lit. The molar amounts of the dissolved solutes are equal in all three cases.*

is assumed to be 100 percent dissociated into ions in solution. (By dissociation we mean the splitting of the compound into cations and anions.) Thus we can represent the dissolving of sodium chloride in water as

$$\mathrm{NaCl}(s) \xrightarrow{\mathrm{H_2O}} \mathrm{Na^+}(aq) + \mathrm{Cl^-}(aq)$$

What this equation says is that all the sodium chloride that enters the solution ends up as Na^+ and Cl^- ions; there are no undissociated NaCl units in solution.

Water is a very effective solvent for ionic compounds. It is often referred to as a *polar* solvent. Although water is an electrically neutral molecule, it has a positive end (the H atoms) and a negative end (the O atom), or positive and negative "poles." When a substance such as sodium chloride dissolves in water, the three-dimensional network of the ions in the solid is destroyed, and the Na^+ and Cl^- ions are separated from each other. In solution, each Na^+ ion is surrounded by a number of water molecules orienting their negative ends toward the cation. Similarly, each Cl^- ion is surrounded by a number of water molecules with their positive ends oriented toward the anion (Figure 3.3). *The process in which an ion is surrounded by water molecules*

Figure 3.3 *Hydration process of Na^+ and Cl^- ions.*

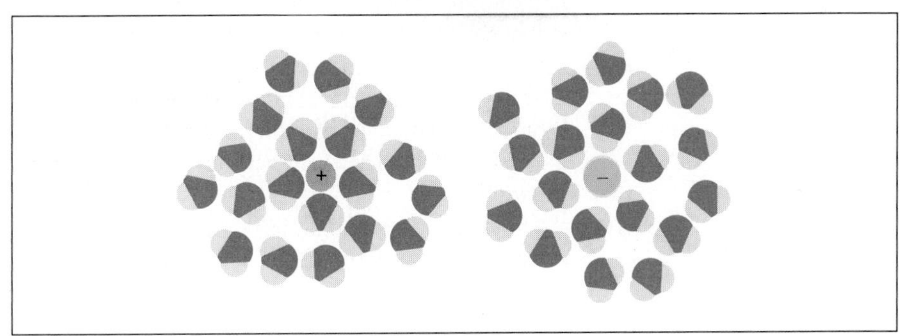

arranged in a specific manner is called ***hydration.*** Hydration helps to stabilize ions in solution and prevent cations from combining with anions.

Two other classes of compounds, acids and bases, are also electrolytes. Certain acids such as hydrochloric acid (HCl) and nitric acid (HNO_3) are strong electrolytes. They share the common characteristic of ionizing completely when dissolved in water; for example, when hydrogen chloride gas dissolves in water, it forms hydrated H^+ and Cl^- ions:

$$HCl(g) \xrightarrow{H_2O} H^+(aq) + Cl^-(aq)$$

In other words, *all* the dissolved HCl molecules give hydrated H^+ and Cl^- ions in solution. Thus when we write HCl(*aq*), it is to be understood that it is a solution of only $H^+(aq)$ and $Cl^-(aq)$ ions, and there are no hydrated HCl molecules present. On the other hand, certain acids such as acetic acid (CH_3COOH), which is found in vinegar, ionize to a much less extent. We represent the ionization of acetic acid as

The formula CH_3COOH tells us that the ionizable proton is in the COOH group.

$$CH_3COOH(aq) \rightleftharpoons CH_3COO^-(aq) + H^+(aq)$$

where CH_3COO^- is called the acetate ion. (In this book we will use the term "dissociation" for ionic compounds and "ionization" for acids and bases.)

The double arrow $\rightleftharpoons$ means that the reaction is ***reversible;*** that is, *the reaction can occur in both directions.* Initially, a number of CH_3COOH molecules break up to yield CH_3COO^- and H^+ ions. As time goes on, some of the CH_3COO^- and H^+ ions recombine to give CH_3COOH molecules. Eventually, a state is reached in which the acid molecules break up as fast as its ions recombine. Such *a chemical state, in which no net change can be observed* (although continuous activity is taking place on the molecular level), is called ***chemical equilibrium.*** Acetic acid, then, is a weak electrolyte because its ionization in water is incomplete. By contrast, in a hydrochloric acid solution the H^+ and Cl^- ions have no tendency to recombine to form molecular HCl. Therefore, we use the single arrow to represent the fact that ionization is complete.

There are many different types of chemical equilibrium. We will return to this very important topic in later chapters.

Substances that dissolve in water as neutral molecules rather than ions are nonelectrolytes because their solutions do not conduct electricity. Various sugars and alcohols are examples of nonelectrolytes. Table 3.2 lists some examples of strong electrolytes, weak electrolytes, and nonelectrolytes. It is interesting to note that the human body fluid consists of many strong and weak electrolytes.

3.3 Precipitation Reactions

Having discussed the general properties of aqueous solutions, we are now ready to study some common and important reactions in this medium in this and the next two sections.

A precipitation reaction is characterized by the formation of an insoluble product, or precipitate. A ***precipitate*** is *an insoluble solid that separates from the solution.* Precipitation reactions usually involve ionic compounds. For example, when an aqueous solution of lead nitrate [$Pb(NO_3)_2$] is added to sodium iodide (NaI), a yellow precipitate, lead iodide (PbI_2), is formed (Figure 3.4):

$$Pb(NO_3)_2(aq) + 2NaI(aq) \longrightarrow PbI_2(s) + 2NaNO_3(aq)$$

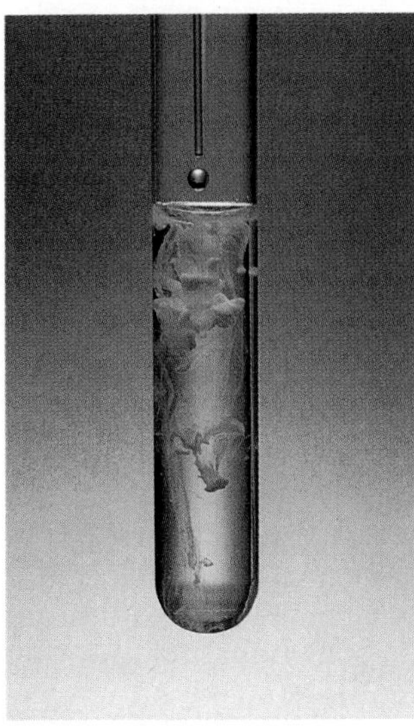

Figure 3.4 *Formation of yellow PbI_2 precipitate as a solution of $Pb(NO_3)_2$ is added to a solution of NaI.*

Table 3.2 Classification of solutes in aqueous solution

Strong electrolyte	Weak electrolyte	Nonelectrolyte
HCl	CH_3COOH	$(NH_2)_2CO$ (urea)
HNO_3	HF	CH_3OH (methanol)
$HClO_4$	HNO_2	C_2H_5OH (ethanol)
H_2SO_4*	NH_3	$C_6H_{12}O_6$ (glucose)
NaOH	H_2O†	$C_{12}H_{22}O_{11}$ (sucrose)
$Ba(OH)_2$		
Ionic compounds		

* H_2SO_4 has two ionizable H^+ ions.
† Pure water is an extremely weak electrolyte.

The above equation is called a ***molecular equation*** because *the formulas of the compounds are written as though all species existed as molecules or whole units.* A molecular equation is useful because it tells us the identity of the reagents (that is, lead nitrate and sodium iodide). If we were to carry out this reaction, this would be the equation for us to follow. However, a molecular equation does not accurately describe what actually is happening from a microscopic perspective. As pointed out earlier, when ionic compounds dissolve in water, they break apart completely into their component cations and anions. Therefore these equations, to be more faithful to reality, should be written to indicate the dissociation of dissolved ionic compounds into ions. Returning to the reaction between sodium iodide and lead nitrate, we would thus write

$$Pb^{2+}(aq) + 2NO_3^-(aq) + 2Na^+(aq) + 2I^-(aq) \longrightarrow PbI_2(s) + 2Na^+(aq) + 2NO_3^-(aq)$$

Such an equation, which *shows dissolved ionic compounds in terms of their free ions,* is called an ***ionic equation.*** *Ions that are not involved in the overall reaction,* in this case the Na^+ and NO_3^- ions, are called ***spectator ions.*** Since the spectator ions appear on both sides of the equation and are unchanged in the chemical reaction, they can be canceled. To focus on the change that actually occurs, we write the ***net ionic equation***—that is, *the equation that indicates only the species that actually take part in the reaction,* as follows:

$$Pb^{2+}(aq) + 2I^-(aq) \longrightarrow PbI_2(s)$$

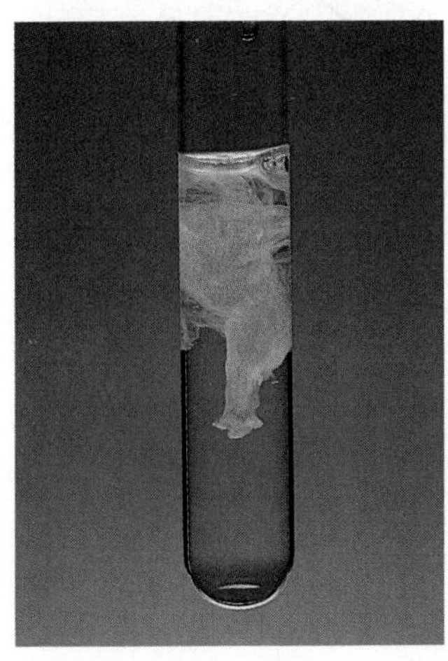

Formation of $BaSO_4$ precipitate.

Similarly, when an aqueous solution of barium chloride ($BaCl_2$) is added to an aqueous solution of sodium sulfate (Na_2SO_4), a white precipitate of barium sulfate ($BaSO_4$) is formed:

$$BaCl_2(aq) + Na_2SO_4(aq) \longrightarrow BaSO_4(s) + 2NaCl(aq)$$

The ionic equation for the reaction is

$$Ba^{2+}(aq) + 2Cl^-(aq) + 2Na^+(aq) + SO_4^{2-}(aq) \longrightarrow BaSO_4(s) + 2Na^+(aq) + 2Cl^-(aq)$$

Canceling the spectator ions (Na^+ and Cl^-) on both sides of the equation gives us the net ionic equation

$$Ba^{2+}(aq) + SO_4^{2-}(aq) \longrightarrow BaSO_4(s)$$

The following steps are useful for writing ionic and net ionic equations.

- Write a balanced molecular equation for the reaction.
- Rewrite the equation to indicate which substances are in ionic form in solution. Remember that all strong electrolytes, when dissolved in solution, are completely dissociated into cations and anions. This procedure gives us the ionic equation.
- Lastly, identify and cancel spectator ions on both sides of the equation to arrive at the net ionic equation.

In order to be able to predict whether a precipitate will be formed when two solutions are mixed or when a compound is added to a solution, we need to know the ***solubility,*** that is, *the maximum amount of solute that can be dissolved in a given quantity of solvent at a specific temperature*. Table 3.3 lists the solubilities of many common ionic compounds. We divide the compounds into three broad categories called "soluble," "slightly soluble," and "insoluble."

Table 3.3 Solubility of ionic compounds in water at 25°C

1. All alkali metal (Group 1A) compounds are soluble.
2. All ammonium (NH_4^+) compounds are soluble.
3. All compounds containing nitrate (NO_3^-), chlorate (ClO_3^-), and perchlorate (ClO_4^-) are soluble.
4. Most hydroxides (OH^-) are insoluble. The exceptions are the alkali metal hydroxides and barium hydroxide [$Ba(OH)_2$]. Calcium hydroxide [$Ca(OH)_2$] is slightly soluble.
5. Most compounds containing chlorides (Cl^-), bromides (Br^-), or iodides (I^-) are soluble. The exceptions are those containing Ag^+, Hg_2^{2+}, and Pb^{2+}.
6. All carbonates (CO_3^{2-}), phosphates (PO_4^{3-}), and sulfides (S^{2-}) are insoluble; the exceptions are those of alkali metals and the ammonium ion.
7. Most sulfates (SO_4^{2-}) are soluble. Calcium sulfate ($CaSO_4$) and silver sulfate (Ag_2SO_4) are slightly soluble. Barium sulfate ($BaSO_4$), mercury(II) sulfate ($HgSO_4$), and lead sulfate ($PbSO_4$) are insoluble.

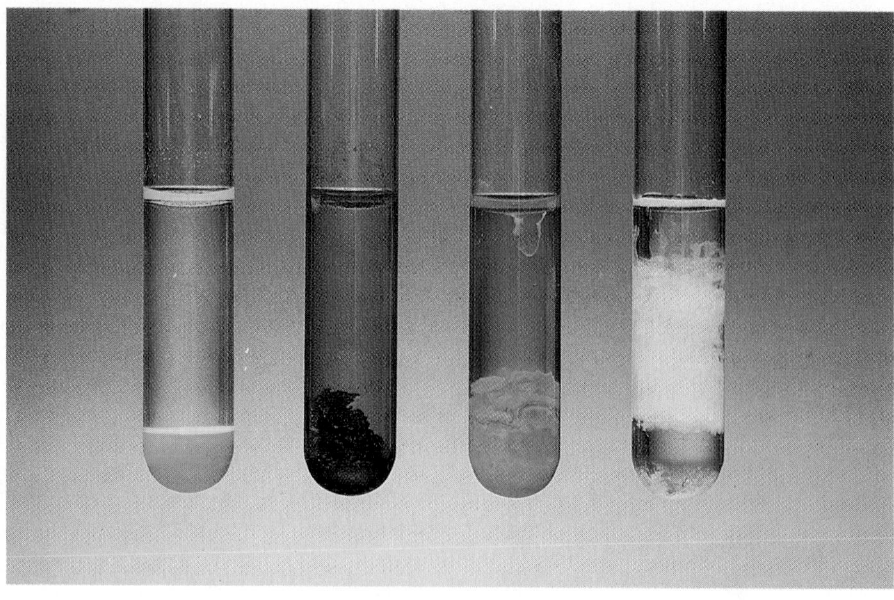

Figure 3.5 *Appearance of several precipitates. From left to right: CdS, PbS, $Ni(OH)_2$, $Al(OH)_3$.*

Figure 3.5 shows the appearance of several insoluble substances.

The following examples illustrate the application of the solubility rules.

EXAMPLE 3.3

Classify the following ionic compounds as soluble, slightly soluble, or insoluble: (a) silver sulfate (Ag_2SO_4), (b) calcium carbonate ($CaCO_3$), (c) sodium phosphate (Na_3PO_4).

Answer

(a) According to rule 7 in Table 3.3, we see that Ag_2SO_4 is slightly soluble.
(b) According to rule 6, $CaCO_3$ is insoluble.
(c) Sodium is an alkali metal (a member of the Group 1A metals). According to rule 1, Na_3PO_4 is soluble.

Similar problem: 3.19.

Formation of $Ca_3(PO_4)_2$ precipitate.

EXAMPLE 3.4

Predict the products of the following reaction and write a net ionic equation for the reaction

$$K_3PO_4(aq) + Ca(NO_3)_2(aq) \longrightarrow ?$$

Answer

Both reactants are soluble salts, but according to solubility rule 6, calcium ions (Ca^{2+}) and phosphate ions (PO_4^{3-}) can form an insoluble compound, calcium phosphate [$Ca_3(PO_4)_2$]. Therefore this is a precipitation reaction. The other product, potassium nitrate (KNO_3), is soluble and therefore remains in solution. The molecular equation is

$$2K_3PO_4(aq) + 3Ca(NO_3)_2(aq) \longrightarrow 6KNO_3(aq) + Ca_3(PO_4)_2(s)$$

and the ionic equation is

$$3Ca^{2+}(aq) + 6NO_3^-(aq) + 6K^+(aq) + 2PO_4^{3-}(aq) \longrightarrow Ca_3(PO_4)_2(s) + 6K^+(aq) + 6NO_3^-(aq)$$

Canceling the spectator K^+ and NO_3^- ions, we obtain the net ionic equation

$$3Ca^{2+}(aq) + 2PO_4^{3-}(aq) \longrightarrow Ca_3(PO_4)_2(s)$$

Similar problem: 3.21.

CHEMISTRY IN ACTION

An Undesirable Precipitation Reaction

Limestone ($CaCO_3$) and dolomite ($CaCO_3 \cdot MgCO_3$), which are widespread on Earth's surface, often enter the water supply. According to solubility rule 6, calcium carbonate is insoluble in water. However, in the presence of dissolved carbon dioxide (from the atmosphere), the following reaction takes place:

$$CaCO_3(s) + CO_2(aq) + H_2O(l) \longrightarrow Ca^{2+}(aq) + 2HCO_3^-(aq)$$

where HCO_3^- is the bicarbonate ion.

Water containing Ca^{2+} and/or Mg^{2+} ions is called ***hard water,*** and *water that is mostly free of these ions* is called ***soft water.*** The presence of these ions makes water unsuitable for some household and industrial uses.

When water containing Ca^{2+} and HCO_3^- ions is heated or boiled, the solution reaction is reversed to produce the $CaCO_3$ precipitate:

$$Ca^{2+}(aq) + 2HCO_3^-(aq) \longrightarrow CaCO_3(s) + CO_2(aq) + H_2O(l)$$

and the carbon dioxide gas is driven off from the solution:

$$CO_2(aq) \longrightarrow CO_2(g)$$

Solid calcium carbonate formed in this way is the main component of the scale that accumulates in boilers, water heaters, pipes, and tea kettles. The thick layer formed reduces heat transfer and decreases the efficiency and durability of boilers, pipes, and appliances. In household hot-water pipes it can restrict or totally block the flow of water (Figure 3.6). A simple way used by plumbers to remove the deposit from these pipes is to introduce a small amount of hydrochloric acid:

$$CaCO_3(s) + 2HCl(aq) \longrightarrow CaCl_2(aq) + H_2O(l) + CO_2(g)$$

In this way, $CaCO_3$ is converted to the soluble $CaCl_2$.

Figure 3.6 *Boiler scale almost fills this hot-water pipe. The deposit consists mostly of $CaCO_3$ with some $MgCO_3$.*

3.4 Acid-Base Reactions

Acid-base reactions are among the most common and important reactions in chemical and biological systems. Before we discuss reactions involving these compounds, it is useful first to examine their general properties and definitions.

General Properties of Acids and Bases

Acids

- Acids have a sour taste; for example, vinegar owes its taste to acetic acid, and lemons and other citrus fruits contain citric acid.

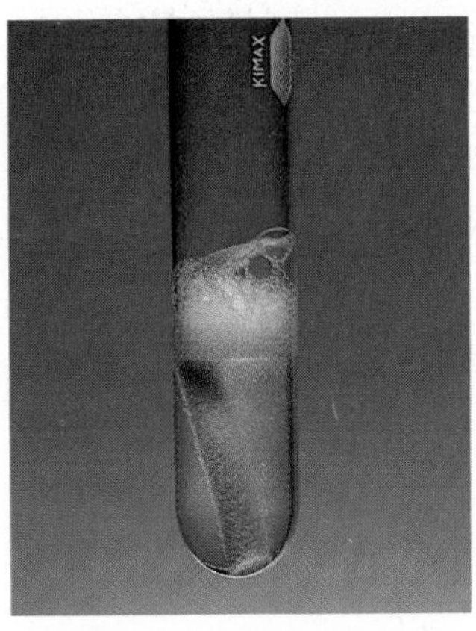

Figure 3.7 *A piece of blackboard chalk, which is mostly $CaCO_3$, reacts with hydrochloric acid.*

- Acids cause color changes in plant dyes; for example, they change the color of litmus from blue to red.
- Acids react with certain metals such as zinc, magnesium, and iron to produce hydrogen gas. A typical reaction is that between hydrochloric acid and magnesium:

$$2HCl(aq) + Mg(s) \longrightarrow MgCl_2(aq) + H_2(g)$$

- Acids react with carbonates and bicarbonates (for example, Na_2CO_3, $CaCO_3$, $NaHCO_3$) to produce carbon dioxide gas (Figure 3.7). For example,

$$2HCl(aq) + CaCO_3(s) \longrightarrow CaCl_2(aq) + H_2O(l) + CO_2(g)$$

$$HCl(aq) + NaHCO_3(s) \longrightarrow NaCl(aq) + H_2O(l) + CO_2(g)$$

- Aqueous acid solutions conduct electricity.

Bases

- Bases have a bitter taste.
- Bases feel slippery; for example, soaps, which contain bases, exhibit this property.
- Bases cause color changes in plant dyes; for example, they change the color of litmus from red to blue.
- Aqueous base solutions conduct electricity.

Definitions of Acids and Bases

Note that the Brønsted definitions, unlike the Arrhenius definitions, do not restrict acid and base behavior to water solutions.

In the late nineteenth century Svante Arrhenius† formulated a theory of acids and bases that defines an acid as a substance that ionizes in water to produce H^+ ions and a base as a substance that ionizes in water to produce OH^- ions. These are the definitions used in Section 2.9 for acids and bases. A general definition of acids and bases, which was originally proposed by the Danish chemist Johannes Brønsted‡ in 1932, describes an acid as *a proton donor* and a base as *a proton acceptor*. Substances that behave according to this definition are called ***Brønsted acids*** and ***Brønsted bases.*** Nitric acid is a Brønsted acid since it donates a proton in water:

$$HNO_3(aq) \longrightarrow H^+(aq) + NO_3^-(aq)$$

Because hydrogen ions (or protons) are hydrated in solution, they are often represented as H_3O^+, which is called a ***hydronium ion.*** Thus the ionization of nitric acid is better represented as

$$HNO_3(aq) + H_2O(l) \longrightarrow H_3O^+(aq) + NO_3^-(aq)$$

In this text we will use both $H^+(aq)$ and H_3O^+ to represent a hydrated proton—the $H^+(aq)$ notation is for convenience and the H_3O^+ notation is closer to reality. Keep in mind that they represent the same species in aqueous solution.

Among the acids commonly used in the laboratory are hydrochloric acid (HCl), nitric acid (HNO_3), acetic acid (CH_3COOH), sulfuric acid (H_2SO_4), and phosphoric

† Svante August Arrhenius (1859–1927). Swedish chemist. Arrhenius made important contributions in the study of chemical kinetics and electrolyte solutions. He also speculated that life on Earth had come from other planets, a theory now known as *panspermia*. Arrhenius was awarded the Nobel Prize in chemistry in 1903.

‡ Johannes Nicolaus Brønsted (1879–1947). Danish chemist. In addition to his theory of acids and bases, Brønsted worked on thermodynamics and the separation of mercury isotopes.

acid (H_3PO_4). The first three acids are ***monoprotic acids;*** that is, *each unit of the acid yields one hydrogen ion:*

$$HCl(aq) \longrightarrow H^+(aq) + Cl^-(aq)$$

$$HNO_3(aq) \longrightarrow H^+(aq) + NO_3^-(aq)$$

$$CH_3COOH(aq) \rightleftharpoons CH_3COO^-(aq) + H^+(aq)$$

As mentioned earlier, because the ionization of acetic acid is incomplete (note the use of double arrows), it is a weak electrolyte. For this reason it is called a weak acid. On the other hand, both HCl and HNO_3 are strong acids because they are strong electrolytes, so they are completely ionized in solution (note the use of single arrows).

Sulfuric acid (H_2SO_4) is an example of a ***diprotic acid*** because *each unit of the acid yields two H^+ ions,* in two separate steps:

$$H_2SO_4(aq) \longrightarrow H^+(aq) + HSO_4^-(aq)$$

$$HSO_4^-(aq) \rightleftharpoons H^+(aq) + SO_4^{2-}(aq)$$

H_2SO_4 is a strong electrolyte or strong acid (note that the first step of ionization is complete), but HSO_4^- is a weak acid and we need a double arrow to represent its incomplete ionization. There are relatively few ***triprotic acids,*** or *acids that yield three H^+ ions*. The best known example of a triprotic acid is phosphoric acid, whose ionizations are

$$H_3PO_4(aq) \rightleftharpoons H^+(aq) + H_2PO_4^-(aq)$$

$$H_2PO_4^-(aq) \rightleftharpoons H^+(aq) + HPO_4^{2-}(aq)$$

$$HPO_4^{2-}(aq) \rightleftharpoons H^+(aq) + PO_4^{3-}(aq)$$

Note that all three species (H_3PO_4, $H_2PO_4^-$, and HPO_4^{2-}) are weak acids, and we use the double arrows to represent each ionization step. Anions such as $H_2PO_4^-$ and HPO_4^{2-} are generated when compounds such as NaH_2PO_4 and Na_2HPO_4 dissolve in water.

Table 3.2 shows that sodium hydroxide (NaOH) and barium hydroxide [$Ba(OH)_2$] are strong electrolytes. This means that they are completely ionized into the metal ions and hydroxide (OH^-) ions in solution:

$$NaOH(s) \xrightarrow{H_2O} Na^+(aq) + OH^-(aq)$$

$$Ba(OH)_2(s) \xrightarrow{H_2O} Ba^{2+}(aq) + 2OH^-(aq)$$

The OH^- ion can accept a proton as follows:

$$H^+(aq) + OH^-(aq) \longrightarrow H_2O(l)$$

Thus OH^- is a Brønsted base.

Ammonia (NH_3) is classified as a Brønsted base because it can accept a H^+ ion as shown below:

$$NH_3(aq) + H_2O(l) \rightleftharpoons NH_4^+(aq) + OH^-(aq)$$

Aqueous ammonia is sometimes erroneously called ammonia hydroxide. There is no evidence that the species NH_4OH actually exists.

Ammonia is a weak electrolyte (and therefore a weak base) because only a small fraction of dissolved NH_3 molecules react with water to form NH_4^+ and OH^- ions.

The most commonly used strong base in the laboratory is sodium hydroxide. It is cheap and soluble. (In fact, all of the alkali metal hydroxides are soluble.) The most commonly used weak base is aqueous ammonia solution. All of the Group 2A elements form hydroxides of the type $M(OH)_2$, where M denotes an alkaline earth metal. Of these hydroxides, only $Ba(OH)_2$ is soluble. Magnesium and calcium hydroxides are used in medicine and industry. Hydroxides of other metals such as $Al(OH)_3$ and $Zn(OH)_2$ are insoluble and are less commonly used.

The following example classifies substances as Brønsted acid or base.

EXAMPLE 3.5

Classify each of the following species as a Brønsted acid or base: (a) HBr, (b) NO_2^-, (c) HCO_3^-.

Answer

(a) HBr dissolves in water to yield H^+ and Br^- ions:

$$HBr(aq) \longrightarrow H^+(aq) + Br^-(aq)$$

Therefore HBr is a Brønsted acid.

(b) In solution the nitrite ion can accept a proton to form nitrous acid:

$$NO_2^-(aq) + H^+(aq) \longrightarrow HNO_2(aq)$$

This property makes NO_2^- a Brønsted base.

(c) The bicarbonate ion is a Brønsted acid because it ionizes in solution as follows:

$$HCO_3^-(aq) \rightleftharpoons H^+(aq) + CO_3^{2-}(aq)$$

It is also a Brønsted base because it can accept a proton:

$$HCO_3^-(aq) + H^+(aq) \longrightarrow H_2CO_3(aq)$$

Similar problem: 3.30.

Acid-Base Neutralization

An acid-base reaction, also called a ***neutralization reaction,*** is *a reaction between an acid and a base*. Aqueous acid-base reactions are generally characterized by the following equation:

Acid-base reactions generally go to completion.

$$\text{acid} + \text{base} \longrightarrow \text{salt} + \text{water}$$

A ***salt*** is *an ionic compound made up of a cation other than H^+ and an anion other than OH^- or O^{2-}*. Acid-base reactions can be classified into four categories, discussed below.

Reaction between a Strong Acid and a Strong Base. An example of a strong acid reacting with a strong base is

$$HCl(aq) + NaOH(aq) \longrightarrow NaCl(aq) + H_2O(l)$$

However, since both the acid and the base are strong electrolytes, they are completely ionized in solution. The ionic equation is

$$H^+(aq) + Cl^-(aq) + Na^+(aq) + OH^-(aq) \longrightarrow Na^+(aq) + Cl^-(aq) + H_2O(l)$$

Therefore, the reactions can be represented by the net ionic equation

$$H^+(aq) + OH^-(aq) \longrightarrow H_2O(l)$$

Both Na^+ and Cl^- are spectator ions.

Reaction between a Strong Acid and a Weak Base. A typical reaction is

$$HNO_3(aq) + NH_3(aq) \longrightarrow NH_4NO_3(aq)$$

Since HNO_3 is completely ionized but NH_3 is largely un-ionized in solution, we write the ionic equation for the reaction as

$$H^+(aq) + NH_3(aq) \longrightarrow NH_4^+(aq)$$

The spectator ion in this reaction is NO_3^-.

Reaction between a Weak Acid and a Strong Base. Consider the reaction

$$HF(aq) + NaOH(aq) \longrightarrow NaF(aq) + H_2O(l)$$

where HF is a weak acid. Because HF is largely un-ionized, we write the ionic equation as

$$HF(aq) + OH^-(aq) \longrightarrow F^-(aq) + H_2O(l)$$

The spectator ion here is Na^+.

Reaction between a Weak Acid and a Weak Base. Since neither weak acids nor weak bases are appreciably ionized in solution, the equation representing this type of reaction *must be* written in molecular form:

$$CH_3COOH(aq) + NH_3(aq) \longrightarrow CH_3COO^-(aq) + NH_4^+(aq)$$

In the above four types of reactions we see that if we started with equal molar amounts of an acid and a base, then at the end of the reaction only a salt is formed and no acid or base is left. This is a characteristic of acid-base neutralization reaction.

Finally we note that in Brønsted's framework acid-base reactions can be viewed as proton transfer reactions. In such a reaction, one or more protons are transferred from an acid to a base in solution. This generalization is consistent with the Brønsted definition that acids are proton donors and bases are proton acceptors.

3.5 Oxidation-Reduction Reactions

Oxidation-reduction reactions are very much a part of the world around us. They range from combustion of fossil fuels to the action of household bleaching agents. Most metallic and nonmetallic elements are obtained from their ores by the process of oxidation or reduction. This section will deal with some basic definitions, types of oxidation-reduction reactions, and a method for balancing the equations representing these reactions.

Consider the reactions of metals and nonmetals with oxygen to form oxides:

$$2Ca(s) + O_2(g) \longrightarrow 2CaO(s)$$

$$P_4(s) + 5O_2(g) \longrightarrow P_4O_{10}(s)$$

CHEMISTRY IN ACTION

Salvaging the Recorder Tape from the *Challenger*

When the space shuttle *Challenger* exploded in flight on January 28, 1986, the crew cabin separated from the rest of the orbiter and broke up when it hit the water. The cabin was equipped with tape recorders to collect shuttle data and record conversations among the crew. However, there was no "black box" to protect the tapes as is used in airplanes. Thus, when the tapes were found six weeks later in 90 feet of water, they were considerably damaged by exposure to seawater and resultant chemical reactions. The tapes were described as "a foaming, concretelike mess, all glued together."

The major problem was the formation of magnesium hydroxide [$Mg(OH)_2$] by reaction of seawater with magnesium used in the tape reel:

$$Mg^{2+}(aq) + 2OH^-(aq) \longrightarrow Mg(OH)_2(s)$$

(Seawater is somewhat basic and therefore contains enough hydroxide ions to react with the Mg^{2+} ions formed when Mg metal comes in contact with ions of less active metals.) The magnesium hydroxide gradually covered the tape layers and glued them together. In addition, binders holding the iron(II) oxide (the magnetic material used in tapes) to the plastic backing were weakened, exposing bare tape in some places. After experimenting with the recovery process using less important retrieved tapes, a team of scientists prepared to salvage the central tape—the one that recorded the crew conversations. In a very tedious and lengthy process, they carefully neutralized the magnesium hydroxide, removing it from the tape, and stabilized the iron oxide layer. All the work had to be done with the tape still coiled. The tape was alternately treated with nitric acid and distilled water (Figure 3.8). The acid-base neutralization reaction is

$$Mg(OH)_2(s) + 2HNO_3(aq) \longrightarrow Mg(NO_3)_2(aq) + 2H_2O(l)$$

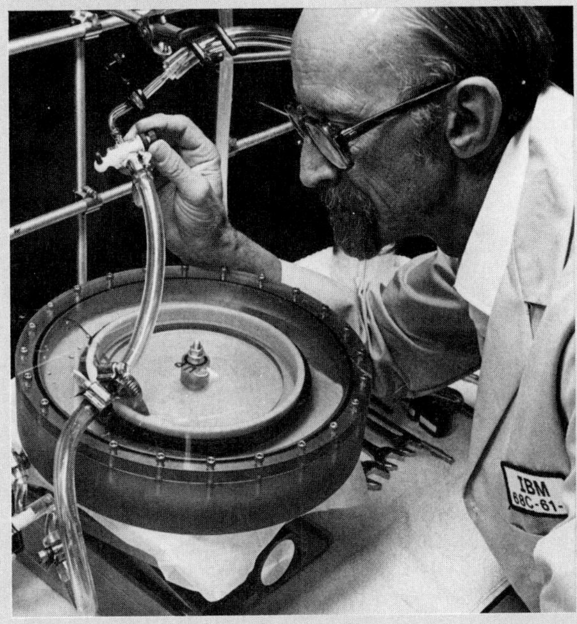

Figure 3.8 *A scientist examines the magnetic tape from the space shuttle* Challenger *as the tape soaks in a solution.*

The purpose of the distilled water was to slowly rinse the tape as the magnesium hydroxide was removed. The tape was then rinsed with methanol to remove the water and treated with the lubricant methyl silicone to protect the tape layers. Finally, the 350-foot-long tape was unwound, transferred to a new reel, and rerecorded on a fresh tape.

The recording showed that at least some of the crew members were aware in the final seconds that the shuttle was in trouble. The impressive fact about this tape-salvaging project is that the principle involved is no more complex than what you would encounter in an introductory chemistry experiment!

These processes fit into a broad class of reactions called ***oxidation-reduction*** (or ***redox***) ***reactions.*** (The term "oxidation" was originally used by chemists to denote combinations of elements with oxygen. However, it now has a broader meaning that includes reactions not involving oxygen.) Let us focus for the moment on the formation of calcium oxide (CaO), an ionic compound made up of Ca^{2+} and O^{2-} ions. In this reaction, two Ca atoms give up or transfer four electrons to two O atoms (in O_2). For convenience, this process can be considered as two separate steps, one involving the loss of four electrons by the two Ca atoms and the other the gain of four electrons by an O_2 molecule:

$$2Ca \longrightarrow 2Ca^{2+} + 4e^-$$

$$O_2 + 4e^- \longrightarrow 2O^{2-}$$

Each of these steps is called a ***half-reaction,*** which *explicitly shows electrons involved.* The sum of the half-reactions gives the overall reaction:

$$2Ca + O_2 + 4e^- \longrightarrow 2Ca^{2+} + 2O^{2-} + 4e^-$$

or, if we cancel the electrons that appear on both sides of the equation,

$$2Ca + O_2 \longrightarrow 2Ca^{2+} + 2O^{2-}$$

Finally, the Ca^{2+} and O^{2-} ions combine to form CaO:

$$2Ca^{2+} + 2O^{2-} \longrightarrow 2CaO$$

By convention, we do not show the charges in the formula of an ionic compound, so that calcium oxide is normally represented as CaO rather than $Ca^{2+}O^{2-}$.

Oxidizing agents are always reduced, and reducing agents are always oxidized. This statement, which sounds somewhat confusing, is simply a consequence of the definitions of the two processes.

The *half-reaction that involves loss of electrons* is called an ***oxidation reaction;*** the *half-reaction that involves gain of electrons* is called a ***reduction reaction.*** In this example, calcium is oxidized. It is said to act as a ***reducing agent*** because it *donates two electrons* to oxygen and causes oxygen to be reduced. Oxygen is reduced and acts as an ***oxidizing agent*** because it *accepts electrons* from calcium, causing calcium to be oxidized. Note that the extent of oxidation in a redox reaction must be equal to the extent of reduction; that is, the number of electrons lost by a reducing agent must be equal to the number of electrons gained by an oxidizing agent.

Oxidation Number

The definitions of oxidation and reduction in terms of loss and gain of electrons apply to the formation of ionic compounds such as CaO. However, these definitions do not apply to the formation of tetraphosphorus decoxide (P_4O_{10}, shown above) and hydrogen chloride (HCl):

$$H_2(g) + Cl_2(g) \longrightarrow 2HCl(g)$$

Compounds such as HCl and P_4O_{10} are not ionic; they are made up of molecules of the indicated formula. Thus it is not possible to view the formation of these compounds from the elements in terms of electron transfer as in the case of CaO. However, there is good reason to treat these reactions (that is, the formation of HCl and P_4O_{10}) as redox reactions. For one thing, the combination of an element (phosphorus in our case) with oxygen is classified as a redox reaction according to the older definition mentioned above. To deal with the range of redox processes, chemists find it convenient to introduce the concept of oxidation number to help keep track of electrons in these reactions. ***Oxidation number*** (also called ***oxidation state***) refers to the *number of charges an atom would have in a molecule (or an ionic compound) if electrons were* *transferred completely*. We can now redefine redox reactions more generally in terms of oxidation number, as follows: *An element is said to be oxidized if its oxidation number increases in a reaction; if the oxidation number of the element decreases in a reaction, it is said to be reduced.*

The following rules help us assign the oxidation number of elements.

1. In free elements (that is, in uncombined state), each atom has an oxidation number of zero. Thus each atom in H_2, Br_2, Na, Be, K, O_2, P_4, and S_8 has the same oxidation number: zero.
2. For ions composed of only one atom, the oxidation number is equal to the charge on the ion. Thus Li^+ has an oxidation number of +1; Ba^{2+} ion, +2; Fe^{3+} ion, +3; I^- ion, −1; O^{2-} ion, −2; and so on. All alkali metals have an oxidation number of +1, and all alkaline earth metals have an oxidation number of +2 in their compounds. Aluminum always has an oxidation number of +3 in all its compounds.
3. The oxidation number of oxygen in most compounds (for example, MgO and H_2O) is −2, but in hydrogen peroxide (H_2O_2) and peroxide ion (O_2^{2-}), its oxidation number is −1.
4. The oxidation number of hydrogen is +1, except when it is bonded to metals in binary compounds (that is, compounds containing two elements). For example, in LiH, NaH, and CaH_2, its oxidation number is −1.
5. Fluorine has an oxidation number of −1 in *all* its compounds. Other halogens (Cl, Br, and I) have negative oxidation numbers when they occur as halide ions in their compounds. When combined with oxygen, for example in oxoacids and oxoanions (see Section 2.9), they have positive oxidation numbers.
6. In a neutral molecule, the sum of the oxidation numbers of all the atoms must be zero. In a polyatomic ion, the sum of oxidation numbers of all the elements in the ion must be equal to the net charge of the ion. For example, in the ammonium ion, NH_4^+, the oxidation number of N is −3 and that of H is +1. Thus the sum of the oxidation numbers is $-3 + 4(+1) = +1$, which is equal to the net charge of the ion.

EXAMPLE 3.6

Assign oxidation numbers to all the elements in the following compounds and ion: (a) Li_2O, (b) PF_3, (c) HNO_3, (d) $Cr_2O_7^{2-}$.

Answer

(a) By rule 2 we see that lithium has an oxidation number of +1 (Li^+) and oxygen an oxidation number of −2 (O^{2-}).

(b) By rule 5 we see that fluorine has an oxidation number of −1, so that P must have an oxidation number of +3.

(c) This is the formula for nitric acid, which yields a H^+ ion and a NO_3^- ion in solution. From rule 4 we see that H has an oxidation number of +1. Thus the other group (the nitrate ion) must have a net oxidation number of −1. Since oxygen has an oxidation number of −2, the oxidation number of nitrogen (labeled x) is given by

$$-1 = x + 3(-2)$$

or

$$x = +5$$

(d) From rule 6 we see that the sum of the oxidation numbers in $Cr_2O_7^{2-}$ must be -2. We know that the oxidation number of O is -2, so all that remains is to determine the oxidation number of Cr, which we call y. The sum of the oxidation numbers in the ion is

$$-2 = 2(y) + 7(-2)$$

or

$$y = +6$$

Similar problems: 3.40, 3.41, 3.42.

Figure 3.9 *The oxidation numbers of elements in their compounds. The more common oxidation numbers are in color.*

1A	2A	3B	4B	5B	6B	7B	8B	8B	8B	1B	2B	3A	4A	5A	6A	7A	8A
1 **H** +1 −1																	2 **He**
3 **Li** +1	4 **Be** +2											5 **B** +3	6 **C** +4 +2 −4	7 **N** +5 +4 +3 +2 +1 −3	8 **O** +2 $-\frac{1}{2}$ −1 −2	9 **F** −1	10 **Ne**
11 **Na** +1	12 **Mg** +2											13 **Al** +3	14 **Si** +4 −4	15 **P** +5 +3 −3	16 **S** +6 +4 +2 −2	17 **Cl** +7 +6 +5 +4 +3 +1 −1	18 **Ar**
19 **K** +1	20 **Ca** +2	21 **Sc** +3	22 **Ti** +4 +3 +2	23 **V** +5 +4 +3 +2	24 **Cr** +6 +5 +4 +3 +2	25 **Mn** +7 +6 +4 +3 +2	26 **Fe** +3 +2	27 **Co** +3 +2	28 **Ni** +2	29 **Cu** +2 +1	30 **Zn** +2	31 **Ga** +3	32 **Ge** +4 −4	33 **As** +5 +3 −3	34 **Se** +6 +4 −2	35 **Br** +5 +3 +1 −1	36 **Kr** +4 +2
37 **Rb** +1	38 **Sr** +2	39 **Y** +3	40 **Zr** +4	41 **Nb** +5 +4	42 **Mo** +6 +4 +3	43 **Tc** +7 +6 +4	44 **Ru** +8 +6 +4 +3	45 **Rh** +4 +3 +2	46 **Pd** +4 +2	47 **Ag** +1	48 **Cd** +2	49 **In** +3	50 **Sn** +4 +2	51 **Sb** +5 +3 −3	52 **Te** +6 +4 −2	53 **I** +7 +5 +1 −1	54 **Xe** +6 +4 +2
55 **Cs** +1	56 **Ba** +2	57 **La** +3	72 **Hf** +4	73 **Ta** +5	74 **W** +6 +4	75 **Re** +7 +6 +4	76 **Os** +8 +4	77 **Ir** +4 +3	78 **Pt** +4 +2	79 **Au** +3 +1	80 **Hg** +2 +1	81 **Tl** +3 +1	82 **Pb** +4 +2	83 **Bi** +5 +3	84 **Po** +2	85 **At** −1	86 **Rn**

Figure 3.9 shows the known oxidation numbers of the more familiar elements, arranged according to their positions in the periodic table. This arrangement is useful because it shows up the following general features of oxidation numbers:

- Metallic elements mostly have positive oxidation numbers, whereas nonmetallic elements may have either positive or negative oxidation numbers.
- The highest oxidation number a representative element can have is its group number in the periodic table. For example, the halogens are in Group 7A, so their highest possible oxidation number is +7.
- The transition metals (Groups 1B, 3B–8B) usually have several oxidation numbers.

Types of Redox Reactions

Having discussed the concept of oxidation number, we are now ready to look at several types of redox reactions. To begin with, we should realize that in the vast majority of cases, reactions involving elements as reactants are redox reactions. Thus *all combustion reactions* (which involve elemental oxygen) as well as many reactions involving elements are redox in nature (Figure 3.10):

Figure 3.10 *Some simple redox reactions. (a) Hydrogen gas burning in air to form water. (b) Sulfur burning in air to form sulfur dioxide. (c) Magnesium burning in air to form magnesium oxide and magnesium nitride. (d) Sodium burning in chlorine to form sodium chloride. (e) Aluminum reacting with bromine to form aluminum bromide.*

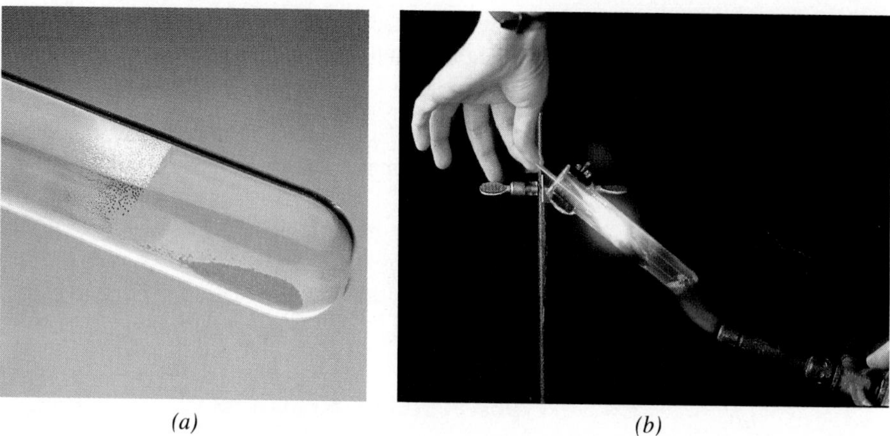

Figure 3.11 *(a) On heating, mercury(II) oxide (HgO) decomposes to form mercury and oxygen. (b) On heating, potassium chlorate ($KClO_3$) produces oxygen which supports the combustion of the wood splint.*

The number 0 over O_2 means each O atom in O_2 has an oxidation number of 0. The same interpretation applies to other assignments.

$$\overset{0}{S}(s) + \overset{0}{O_2}(g) \longrightarrow \overset{+4-2}{SO_2}(g)$$

$$3\overset{0}{Mg}(s) + \overset{0}{N_2}(g) \longrightarrow \overset{+2\ -3}{Mg_3N_2}(s)$$

where the number above each element denotes the oxidation number. This format allows us to identify elements that are oxidized (an increase in oxidation number) and reduced (a decrease in oxidation number) at a glance. *Decomposition reactions* that produce one or more free elements are mostly redox reactions (Figure 3.11). For example,

$$2\overset{+2\ -2}{HgO}(s) \longrightarrow 2\overset{0}{Hg}(l) + \overset{0}{O_2}(g)$$

$$2\overset{+5-2}{KClO_3}(s) \longrightarrow 2\overset{-1}{KCl}(s) + 3\overset{0}{O_2}(g)$$

$$2\overset{+1\ -1}{NaH}(s) \longrightarrow 2\overset{0}{Na}(s) + \overset{0}{H_2}(g)$$

A particularly important type of redox reaction is called a ***displacement reaction,*** in which *an ion (or atom) in a compound is replaced by another ion (or atom) of another element*. It is represented by

$$A + BC \longrightarrow AC + B$$

Most displacement reactions can be conveniently divided into three major categories: hydrogen displacement, metal displacement, and halogen displacement.

Hydrogen Displacement. All alkali metals and some alkaline earth metals (Ca, Sr, and Ba), which are the most reactive of the metallic elements, will displace hydrogen from cold water (Figure 3.12):

$$2\overset{0}{Na}(s) + 2\overset{+1}{H_2}O(l) \longrightarrow 2\overset{+1}{Na}O\overset{+1}{H}(aq) + \overset{0}{H_2}(g)$$

$$\overset{0}{Ca}(s) + 2\overset{+1}{H_2}O(l) \longrightarrow \overset{+2}{Ca}(O\overset{+1}{H})_2(s) + \overset{0}{H_2}(g)$$

(a)

(b)

Figure 3.12 *Reactions of (a) sodium (Na) and (b) calcium (Ca) with cold water. Note that the reaction is more violent with Na than with Ca.*

Note that we only show oxidation numbers of elements that are oxidized or reduced. Less reactive metals, such as aluminum and iron, react with steam to give hydrogen gas:

$$2\overset{0}{\text{Al}}(s) + 3\overset{+1}{\text{H}}_2\text{O}(g) \longrightarrow \overset{+3}{\text{Al}}_2\text{O}_3(s) + 3\overset{0}{\text{H}}_2(g)$$

$$2\overset{0}{\text{Fe}}(s) + 3\overset{+1}{\text{H}}_2\text{O}(g) \longrightarrow \overset{+3}{\text{Fe}}_2\text{O}_3(s) + 3\overset{0}{\text{H}}_2(g)$$

Many metals, including those that do not react with water, are capable of displacing hydrogen from acids. For example, cadmium (Cd) and tin (Sn) do not react with water but do react with hydrochloric acid, as follows:

$$\overset{0}{\text{Cd}}(s) + 2\overset{+1}{\text{H}}\text{Cl}(aq) \longrightarrow \overset{+2}{\text{Cd}}\text{Cl}_2(aq) + \overset{0}{\text{H}}_2(g)$$

$$\overset{0}{\text{Sn}}(s) + 2\overset{+1}{\text{H}}\text{Cl}(aq) \longrightarrow \overset{+2}{\text{Sn}}\text{Cl}_2(aq) + \overset{0}{\text{H}}_2(g)$$

The ionic equations are

$$\overset{0}{\text{Cd}}(s) + 2\overset{+1}{\text{H}^+}(aq) \longrightarrow \overset{+2}{\text{Cd}^{2+}}(aq) + \overset{0}{\text{H}}_2(g)$$

$$\overset{0}{\text{Sn}}(s) + 2\overset{+1}{\text{H}^+}(aq) \longrightarrow \overset{+2}{\text{Sn}^{2+}}(aq) + \overset{0}{\text{H}}_2(g)$$

Some metals such as copper (Cu), silver (Ag), and gold (Au) do not react even with hydrochloric acid. Figure 3.13 shows the reactions between hydrochloric acid (HCl) and iron (Fe), zinc (Zn), and magnesium (Mg). These reactions are used to prepare hydrogen gas in the laboratory.

Metal Displacement. A metal in a compound can also be displaced by another metal in the uncombined state. For example, when metallic zinc is added to a solution

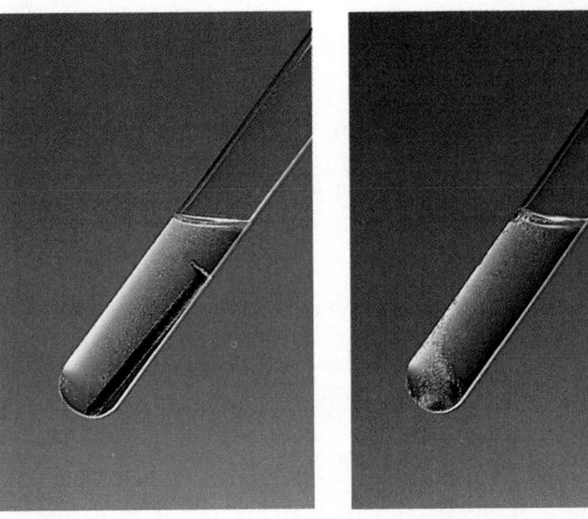

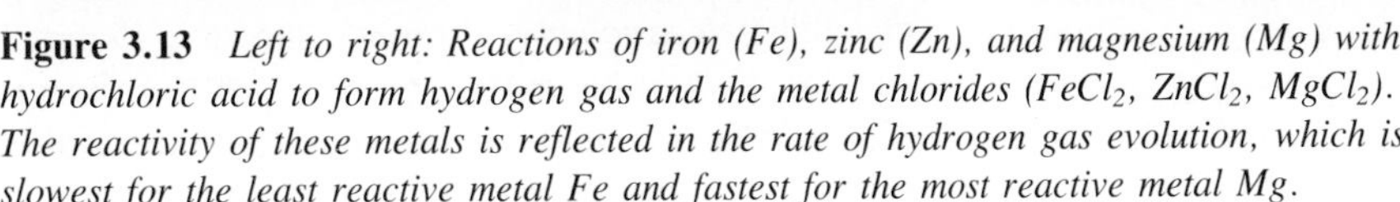

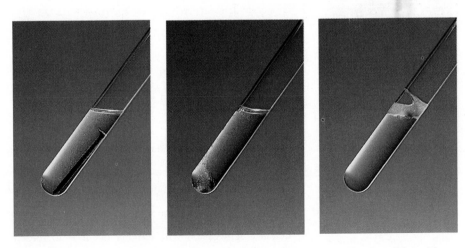

Figure 3.13 *Left to right: Reactions of iron (Fe), zinc (Zn), and magnesium (Mg) with hydrochloric acid to form hydrogen gas and the metal chlorides ($FeCl_2$, $ZnCl_2$, $MgCl_2$). The reactivity of these metals is reflected in the rate of hydrogen gas evolution, which is slowest for the least reactive metal Fe and fastest for the most reactive metal Mg.*

Figure 3.14 *Left to right: When a piece of Zn metal is placed in an aqueous $CuSO_4$ solution, Zn atoms enter the solution as Zn^{2+} ions and Cu^{2+} ions are converted to Cu (first beaker). In time, the blue color of the $CuSO_4$ solution disappears (second beaker). When a piece of Cu wire is placed in an aqueous $AgNO_3$ solution, Cu atoms enter the solution as Cu^{2+} ions and Ag^+ ions are converted to Ag (third beaker). Gradually, the solution acquires the characteristic blue color due to the hydrated Cu^{2+} ions (fourth beaker).*

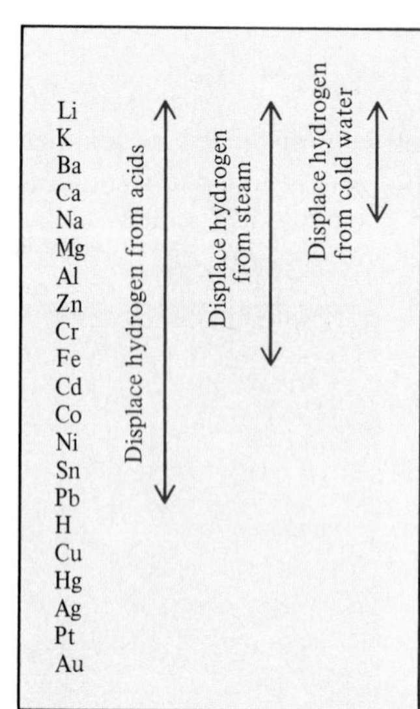

Figure 3.15 *The activity series for metals. The metals are arranged according to their ability to displace hydrogen from an acid or water. Li (lithium) is the most reactive metal, and Au (gold) is the least reactive.*

containing copper sulfate ($CuSO_4$), the zinc metal displaces Cu^{2+} ions from the solution (Figure 3.14):

$$\overset{0}{Zn}(s) + \overset{+2}{Cu}SO_4(aq) \longrightarrow \overset{+2}{Zn}SO_4(aq) + \overset{0}{Cu}(s)$$

or

$$\overset{0}{Zn}(s) + \overset{+2}{Cu^{2+}}(aq) \longrightarrow \overset{+2}{Zn^{2+}}(aq) + \overset{0}{Cu}(s)$$

Similarly, metallic copper displaces silver ions from a solution containing silver nitrate ($AgNO_3$):

$$\overset{0}{Cu}(s) + 2\overset{+1}{Ag}NO_3(aq) \longrightarrow \overset{+2}{Cu}(NO_3)_2(aq) + 2\overset{0}{Ag}(s)$$

or

$$\overset{0}{Cu}(s) + 2\overset{+1}{Ag^+}(aq) \longrightarrow \overset{+2}{Cu^{2+}}aq) + 2\overset{0}{Ag}(s)$$

Reversing the roles of the metals would result in no reaction. Thus copper metal will not displace zinc ions from zinc sulfate, and silver metal will not displace copper ions from copper nitrate.

An easy way to predict whether a displacement reaction (in both hydrogen and metal displacement cases) will actually occur is to refer to an ***activity series*** (sometimes called the ***electrochemical series***), shown in Figure 3.15. Basically, an activity series such as this can be considered simply *a convenient summary of the results of many possible displacement reactions* similar to those already discussed. According to this series, any metal above hydrogen will displace it from water or from an acid, but metals below hydrogen will not react with either water or an acid. In fact, any species listed in the series will react with any species (in a compound) below it. For example, Zn is above Cu, so zinc metal will displace copper ions from copper sulfate.

Metal displacement reactions find many applications in metallurgical processes in which pure metals are obtained from their compounds in ores. For example, vanadium is obtained by treating vanadium(V) oxide with metallic calcium:

$$\mathrm{V_2O_5}(s) + 5\mathrm{Ca}(l) \longrightarrow 2\mathrm{V}(l) + 5\mathrm{CaO}(s)$$

Similarly, titanium is obtained from titanium(IV) chloride according to the reaction

$$\mathrm{TiCl_4}(g) + 2\mathrm{Mg}(l) \longrightarrow \mathrm{Ti}(s) + 2\mathrm{MgCl_2}(l)$$

In each case the metal that acts as the reducing agent lies above the metal that is reduced (that is, Ca is above V and Mg is above Ti in the activity series). We will see more examples of this type of reaction in Chapter 20.

Halogen Displacement. An analogous activity series also applies to the halogens. The power of these elements as oxidizing agents decreases as we move down from fluorine to iodine in Group 7A, so molecular fluorine can replace chloride, bromide, and iodide ions in solution. In fact, molecular fluorine is so reactive that it also attacks water; thus these reactions cannot be carried out in aqueous solutions. On the other hand, molecular chlorine can displace bromide and iodide ions in solution (Figure 3.16):

$$\overset{0}{\mathrm{Cl}}_2(g) + 2\mathrm{K}\overset{-1}{\mathrm{Br}}(aq) \longrightarrow 2\mathrm{K}\overset{-1}{\mathrm{Cl}}(aq) + \overset{0}{\mathrm{Br}}_2(l)$$

$$\overset{0}{\mathrm{Cl}}_2(g) + 2\mathrm{Na}\overset{-1}{\mathrm{I}}(aq) \longrightarrow 2\mathrm{Na}\overset{-1}{\mathrm{Cl}}(aq) + \overset{0}{\mathrm{I}}_2(s)$$

Written as ionic equations, we have

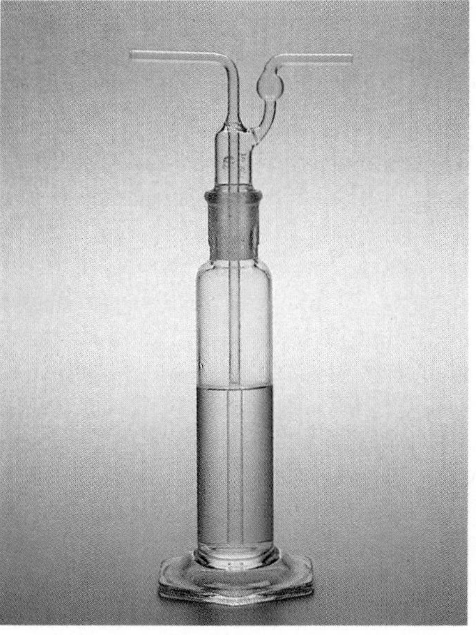

(a)

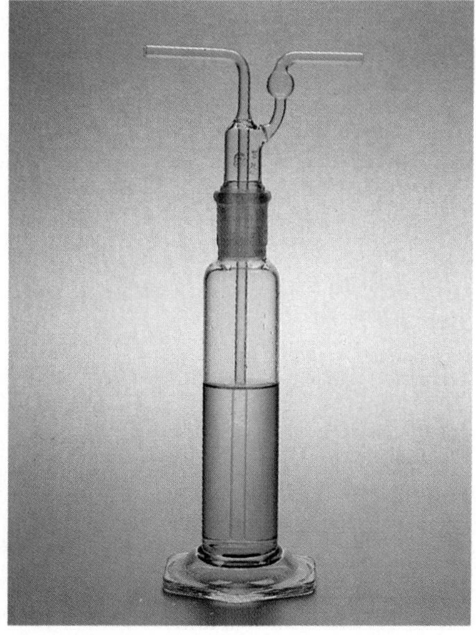

(b)

Figure 3.16 *(a) An aqueous KBr solution. (b) After bubbling chlorine gas through the solution, most of the bromide ions are converted to liquid bromine.*

$$\overset{0}{Cl_2}(g) + 2\overset{-1}{Br^-}(aq) \longrightarrow 2\overset{-1}{Cl^-}(aq) + \overset{0}{Br_2}(l)$$

$$\overset{0}{Cl_2}(g) + 2\overset{-1}{I^-}(aq) \longrightarrow 2\overset{-1}{Cl^-}(aq) + \overset{0}{I_2}(s)$$

Molecular bromine, in turn, can displace iodide ion in solution:

$$\overset{0}{Br_2}(l) + 2\overset{-1}{I^-}(aq) \longrightarrow 2\overset{-1}{Br^-}(aq) + \overset{0}{I_2}(aq)$$

Reversing the roles of the halogens leads to no reaction. Thus bromine cannot displace chloride ions, and iodine cannot displace bromide and chloride ions.

The halogen displacement reactions have a direct industrial application. The halogens (the Group 7A elements) as a group are the most reactive of the nonmetallic elements. They are all strong oxidizing agents. As a result, they are found in nature in the combined state (with metals) as halides and never as free elements. Of these four elements, chlorine is by far the most important industrial chemical. In 1988 its annual production was 23 billion pounds, which ranked it ninth among the major industrial chemicals. (See Chapter 26 for a discussion of the top industrial chemicals produced in the United States.) The annual production of bromine is only one-hundredth that of chlorine, while those of fluorine and iodine are much less.

The recovery of the halogens from their halides requires an oxidation process, which is represented by

$$2X^- \longrightarrow X_2 + 2e^-$$

where X denotes a halogen element. Seawater and natural brine (for example, underground water in contact with salt deposits) are rich sources of Cl^-, Br^-, and I^- ions, while minerals such as fluorite (CaF_2) and cryolite (Na_3AlF_6) are used to prepare fluorine. Because fluorine is the strongest oxidizing agent known, there is no way to convert F^- ions to F_2 by chemical means. The only way to carry out the oxidation is by electrolytic means, the details of which will be discussed in Chapter 19. Industrially, chlorine is produced electrolytically as in the case of fluorine.

Bromine is prepared industrially by the oxidation of Br^- ions by chlorine, which is a strong enough oxidizing agent to oxidize Br^- ions but not water:

$$2Br^-(aq) \longrightarrow Br_2(l) + 2e^-$$

One of the richest sources of Br^- ions is the Dead Sea (about 4000 parts per million, or 4000 ppm, by mass of all dissolved substances is Br). Bromine is a red fuming liquid. Following the oxidation of Br^- ions, bromine is removed from the solution by blowing air over the solution, and the air-bromine mixture is then cooled to condense the bromine (Figure 3.17). Iodine is also prepared from seawater and natural brine by oxidation of I^- ions with chlorine. Because Br^- and I^- ions are invariably present in the same source, they are both oxidized by chlorine. However, it is relatively easy to separate Br_2 from I_2 because iodine is a solid that is sparingly soluble in water. The air-blowing procedure will remove most of the bromine formed but will not affect the iodine present.

Among the elements that are most likely to undergo disproportionation are N, P, O, S, Cl, Br, I, Mn, Cu, Au, and Hg.

Disproportionation Reaction. A special type of redox reactions is called disproportionation reactions. In a ***disproportionation reaction*** *an element in one oxidation state is both oxidized and reduced.* One reactant in a disproportionation reaction *always* contains an element that can have at least three oxidation states. The reactant

Figure 3.17 *The industrial manufacture of liquid bromine by oxidizing an aqueous solution containing Br^- ions with chlorine gas.*

itself is in one of these states, and there are both a higher and a lower possible oxidation state for the same element. The decomposition of hydrogen peroxide is an example of a disproportionation reaction:

$$2H_2\overset{-1}{O_2}(aq) \longrightarrow 2H_2\overset{-2}{O}(l) + \overset{0}{O_2}(g)$$

Here the oxidation number of oxygen (-1) in the reactant both increases to zero in O_2 and decreases to -2 in H_2O. Another example is the reaction between molecular chlorine and a NaOH solution:

$$\overset{0}{Cl_2}(g) + 2OH^-(aq) \longrightarrow O\overset{+1}{Cl}{}^-(aq) + \overset{-1}{Cl}{}^-(aq) + H_2O(l)$$

This reaction describes the formation of household bleaching agents, for it is the hypochlorite ion (OCl^-) that oxidizes the color-bearing substances in stains, converting them to colorless compounds.

Finally it is interesting to compare redox reactions with acid-base reactions. In some ways they are analogous to each other in that acid-base reactions involve the transfer of protons while redox reactions involve the transfer of electrons. However, while acid-base reactions are quite easy to recognize (since they always involve an acid and a base), there is no simple procedure in general to identify a redox process. The only sure way is to compare the oxidation numbers of all the elements in the reactants and products. Any change in oxidation number *guarantees* that the reaction must be redox in nature.

3.6 Balancing Oxidation-Reduction Equations

So far we have discussed several fairly simple types of redox reactions. Equations representing these reactions are relatively easy to balance. However, in the laboratory we often encounter more complex redox reactions, reactions that invariably involve oxoanions such as chromate (CrO_4^{2-}), dichromate ($Cr_2O_7^{2-}$), permanganate (MnO_4^-), nitrate (NO_3^-), and sulfate (SO_4^{2-}). In principle we can balance any redox equation using the procedure outlined in Section 3.1. However, there are some special techniques for handling redox reactions which also give us insight into electron transfer processes. Here we will discuss one such method, called the *ion-electron method*. In this approach, the overall reaction is divided into two half-reactions, one for oxidation and one for reduction (as in the case of the formation of CaO discussed earlier). The equations for the two half-reactions are balanced separately and then added to give the overall balanced equation.

Suppose we are asked to balance the equation showing the oxidation of Fe^{2+} ions to Fe^{3+} ions by dichromate ions ($Cr_2O_7^{2-}$) in an acidic medium. The $Cr_2O_7^{2-}$ ions are reduced to Cr^{3+} ions. The following steps will help us accomplish this task.

Step 1. Write the skeletal equation for the reaction in ionic form. Separate the equation into two half-reactions:

$$Fe^{2+} + Cr_2O_7^{2-} \longrightarrow Fe^{3+} + Cr^{3+}$$

The equations for the two half-reactions are

$$\text{Oxidation: } \overset{+2}{Fe^{2+}} \longrightarrow \overset{+3}{Fe^{3+}}$$

$$\text{Reduction: } \overset{+6}{Cr_2O_7^{2-}} \longrightarrow \overset{+3}{Cr^{3+}}$$

Step 2. Balance the atoms in each half-reaction separately. For reactions in an acidic medium, add H_2O to balance the O atoms and H^+ to balance the H atoms. For reactions in a basic medium, first balance the atoms as you would in an acidic medium. Then, for every H^+ ion, add an equal number of OH^- ions to both sides of the equation. Where H^+ and OH^- appear on the same side of the equation, combine the ions to give H_2O.

This is merely a convenient procedure, and it does not alter the nature of the reaction.

We begin by balancing the atoms in each half-reaction. The oxidation reaction is already balanced for Fe atoms. For the reduction step, we multiply the Cr^{3+} by 2 to balance the Cr atoms. Since the reaction takes place in an acidic medium, we add seven H_2O molecules to the right-hand side of the equation to balance the O atoms:

$$Cr_2O_7^{2-} \longrightarrow 2Cr^{3+} + 7H_2O$$

To balance the H atoms, we add fourteen H^+ ions on the left side:

$$14H^+ + Cr_2O_7^{2-} \longrightarrow 2Cr^{3+} + 7H_2O$$

Step 3. Add electrons to one side of each half-reaction to balance the charges. If necessary, equalize the number of electrons in the two half-reactions by multiplying one or both half-reactions by appropriate coefficients.

For the oxidation half-reaction we write

$$Fe^{2+} \longrightarrow Fe^{3+} + e^-$$

We added an electron to the right-hand side so that there is a 2+ charge on each side of the half-reaction.

In the reduction half-reaction there are net twelve positive charges on the left-hand side and only six positive charges on the right-hand side. Therefore, we add six electrons on the left.

$$14H^+ + Cr_2O_7^{2-} + 6e^- \longrightarrow 2Cr^{3+} + 7H_2O$$

To equalize the number of electrons in both half-reactions, we multiply the oxidation half-reaction by 6:

$$6Fe^{2+} \longrightarrow 6Fe^{3+} + 6e^-$$

Step 4. Add the two half-reactions together and balance the final equation by inspection. The electrons on both sides must cancel.

The two half-reactions are added to give

$$14H^+ + Cr_2O_7^{2-} + 6Fe^{2+} + 6e^- \longrightarrow 2Cr^{3+} + 6Fe^{3+} + 7H_2O + 6e^-$$

The electrons on both sides cancel, and we are left with the balanced net ionic equation:

$$14H^+ + Cr_2O_7^{2-} + 6Fe^{2+} \longrightarrow 2Cr^{3+} + 6Fe^{3+} + 7H_2O$$

Step 5. Verify that the equation contains the same types and numbers of atoms and the same charges on both sides of the equation.

A final check shows that the resulting equation is ''atomically'' and ''electrically'' balanced.

We now apply the ion-electron method to balance the equation of a redox reaction in basic medium.

EXAMPLE 3.7

Write a balanced ionic equation to represent the oxidation of iodide ion (I^-) by permanganate ion (MnO_4^-) in basic solution to yield molecular iodine (I_2) and manganese(IV) oxide (MnO_2).

Answer

Step 1: The skeletal equation is

$$MnO_4^- + I^- \longrightarrow MnO_2 + I_2$$

The two half-reactions are

$$\text{oxidation: } \overset{-1}{I^-} \longrightarrow \overset{0}{I_2}$$

$$\text{reduction: } \overset{+7}{MnO_4^-} \longrightarrow \overset{+4}{MnO_2}$$

Step 2: To balance the I atoms in the oxidation half-reaction, we write

$$2I^- \longrightarrow I_2$$

In the reduction half-reaction, to balance the O atoms we add two H_2O molecules on the right:

$$MnO_4^- \longrightarrow MnO_2 + 2H_2O$$

To balance the H atoms, we add four H^+ ions on the left:

$$MnO_4^- + 4H^+ \longrightarrow MnO_2 + 2H_2O$$

Since the reaction occurs in a basic medium and there are four H^+ ions, we add four OH^- ions to both sides of the equation:

$$MnO_4^- + 4H^+ + 4OH^- \longrightarrow MnO_2 + 2H_2O + 4OH^-$$

Combining the H^+ and OH^- ions to form H_2O, we write

$$MnO_4^- + 2H_2O \longrightarrow MnO_2 + 4OH^-$$

Step 3: Next, we balance the charges of the two half-reactions as follows:

$$2I^- \longrightarrow I_2 + 2e^-$$

$$MnO_4^- + 2H_2O + 3e^- \longrightarrow MnO_2 + 4OH^-$$

To equalize the number of electrons, we multiply the oxidation half-reaction by 3 and the reduction half-reaction by 2:

$$6I^- \longrightarrow 3I_2 + 6e^-$$

$$2MnO_4^- + 4H_2O + 6e^- \longrightarrow 2MnO_2 + 8OH^-$$

Step 4: The two half-reactions are added together to give

$$6I^- + 2MnO_4^- + 4H_2O + 6e^- \longrightarrow 3I_2 + 2MnO_2 + 8OH^- + 6e^-$$

After canceling the electrons on both sides, we obtain

$$6I^- + 2MnO_4^- + 4H_2O \longrightarrow 3I_2 + 2MnO_2 + 8OH^-$$

Step 5: A final check shows that the equation is balanced in terms of both atoms and charges.

Similar problems: 3.55, 3.56.

CHEMISTRY IN ACTION

Black and White Photography

Photography has long been a popular hobby for the young and old. Many amateur photographers send their films away to be developed, although an increasing number prefer to spend long hours in the darkroom developing their own film. The process of film development involves a redox reaction.

Black and white photographic film contains small grains of silver bromide, evenly spread over a thin gelatin coating on paper. Exposure of the film to light activates the silver bromide as follows:

$$AgBr + \text{light} \longrightarrow AgBr^*$$

where the asterisk denotes AgBr excited by light. Next, the exposed film is treated with a developer, a solution containing a mild reducing agent such as hydroquinone.

$$\underset{}{2AgBr^*(s)} + \underset{\text{hydroquinone}}{C_6H_6O_2(aq)} \longrightarrow 2Ag(s) + 2HBr(aq) + \underset{\text{quinone}}{C_6H_4O_2(aq)}$$

In this redox process, the Ag^+ ions in the excited silver bromide, AgBr*, are preferentially reduced to metallic silver, and hydroquinone is oxidized to quinone. The oxidation step, which at first is not obvious, can be clarified by writing the above reaction as two half-reactions:

Oxidation

$$C_6H_6O_2(aq) \longrightarrow C_6H_4O_2(aq) + 2H^+(aq) + 2e^-$$

Reduction

$$2Ag^+(aq) + 2e^- \longrightarrow 2Ag(s)$$

The sum of these half-reactions is

$$2Ag^+(aq) + C_6H_6O_2(aq) \longrightarrow 2Ag(s) + C_6H_4O_2(aq) + 2H^+(aq)$$

which is the net ionic equation for the redox process. The amount of black metallic silver particles formed on the film is directly proportional to the amount or intensity of light that originally fell on the film. The unreacted (that is, the unexcited) AgBr must be removed from the film at this stage; otherwise, it also would slowly be reduced by hydroquinone and the *entire* film would eventually turn black. To prevent this undesired reaction, the film is quickly treated with a "fixer," a sodium thiosulfate ($Na_2S_2O_3$) solution, to remove the silver ions:

$$AgBr(s) + 2S_2O_3^{2-}(aq) \longrightarrow Ag(S_2O_3)_2^{3-}(aq) + Br^-(aq)$$

What has been described is the preparation of a black and white negative. A positive print can be obtained by shining light through the negative onto another piece of photographic paper and repeating the developing procedure. Because white regions of the subject appear black in a negative, they are opaque and leave unexcited (white) areas in the positive print. This process therefore inverts the light and dark areas of the negative to produce the desired picture (Figure 3.18).

(a)

(b)

Figure 3.18 *The negative (a) and positive (b) of a black and white photograph.*

Summary

1. Chemical changes, called chemical reactions, are represented by chemical equations. Substances that undergo change—the reactants—are written on the left and substances formed—the products—appear on the right of the arrow.
2. Chemical equations must be balanced, in accordance with the law of conservation of mass. The number of atoms of each type of element in the reactants and products must be equal.
3. Aqueous solutions are electrically conducting if the solutes are electrolytes. If the solutes are nonelectrolytes, the solutions do not conduct electricity.
4. Three major types of chemical reactions are precipitation reactions, acid-base reactions, and oxidation-reduction reactions.
5. From general rules about solubilities of ionic compounds, we can predict whether a precipitate will form in a reaction.
6. Arrhenius acids ionize in water to give H^+ ions, and Arrhenius bases ionize in water to give OH^- ions. Brønsted acids donate protons, and Brønsted bases accept protons.
7. The reaction of an acid with a base is called neutralization.
8. In oxidation-reduction, or redox, reactions, oxidation and reduction always occur simultaneously. Oxidation is characterized by the loss of electrons; reduction is characterized by the gain of electrons.
9. Oxidation numbers help us keep track of charge distribution and are assigned to all atoms in a compound or ion according to a specific set of rules. Oxidation can be defined as an increase in oxidation number; reduction can be defined as a decrease in oxidation number.
10. Many redox reactions can be classified as combination, decomposition, displacement, or disproportionation reactions.
11. Redox equations can be balanced by the ion-electron method, in which the oxidation and reduction half-reaction equations are balanced separately and then added together.

Applied Chemistry Example: Phosphoric Acid

Phosphoric acid, H_3PO_4, is one of the major industrial chemicals produced in the world. It ranked 8th among the top 50 industrial chemicals produced in the United States in 1988 (23.4 billion pounds).

The major use of phosphoric acid is in the production of fertilizers since phosphorus is essential for plant growth. Phosphoric acid is also used in detergents and food industries. Polyphosphates made from phosphoric acid are used as "builders" in detergents to bind calcium and magnesium ions. These metal ions react with soap to form insoluble salts or curds, rendering soap ineffective. However, because of the damaging effects of phosphates to ponds and lakes, the use of phosphate detergent is waning. Phosphoric acid is used in soft drinks to give them the "tangy" flavor and phosphates are found in products like baking powder.

Phosphoric acid is produced principally by two methods: the *electric furnace method* and the *wet process*. The acid produced by the furnace method has high purity. The bulk of phosphoric acid (about 90 percent) used today is produced by the wet process because the cost is lower.

In the electric furnace method, elemental phosphorus (P_4) burns in air to give tetraphosphorus decoxide (P_4O_{10}). This is followed by reaction of P_4O_{10} with water to give phosphoric acid. The wet process involves the reaction of phosphate rock [$Ca_5(PO_4)_3F$] with sulfuric acid to give phosphoric acid, hydrofluoric acid (HF), and calcium sulfate ($CaSO_4$). In the pure form phosphoric acid is a colorless solid (melting point = 42.2°C)

1. Write balanced equations for the electric furnace method and classify them as precipitation, acid-base, or redox reactions.

   Answer: The combustion of phosphorus is represented as

$$P_4 + O_2 \longrightarrow P_4O_{10}$$

   We see that P and O appear once on each side of the equation, but only P is in equal numbers of atoms on both sides. To balance O, we place a 5 in front of O_2:

$$P_4(s) + 5O_2(g) \longrightarrow P_4O_{10}(s)$$

   The equation is now balanced for both P and O. According to the rules for assigning oxidation numbers (see p. 109) and Example 3.6, we see that both phosphorus and oxygen have an oxidation number of zero as reactants. Oxygen has an oxidation number of -2 in P_4O_{10} and assigning an oxidation number x for P, we write

$$4x + 10(-2) = 0$$

   or

$$x = +5$$

   Since the oxidation number of phosphorus has increased and that of oxygen has decreased in the process, this is a redox reaction.

   The unbalanced equation for the reaction of solid P_4O_{10} with water is

$$P_4O_{10} + H_2O \longrightarrow H_3PO_4$$

   Both P and H appear once on each side of the equation but in different numbers. To balance P, we place a 4 in front of H_3PO_4

$$P_4O_{10} + H_2O \longrightarrow 4H_3PO_4$$

   To balance H, we place a 6 in front of H_2O

$$P_4O_{10}(s) + 6H_2O(l) \longrightarrow 4H_3PO_4(aq)$$

   The equation is now balanced for P, H, and O atoms.

   The reaction does not involve formation of a precipitate (see Table 3.3) nor are there changes in oxidation numbers. There are, however, proton transfers from H_2O to P_4O_{10}. Therefore, it is an acid-base reaction.

2. Write a balanced equation for the wet process and classify it as a precipitation, an acid-base, or a redox reaction.

   Answer: The unbalanced equation is

$$Ca_5(PO_4)_3F + H_2SO_4 \longrightarrow H_3PO_4 + HF + CaSO_4$$

   Of the six elements, only Ca appears once on each side of the equation and in unequal numbers. To balance Ca, we place a 5 in front of $CaSO_4$

$$Ca_5(PO_4)_3F + H_2SO_4 \longrightarrow H_3PO_4 + HF + 5CaSO_4$$

Next we balance S and P by placing a 5 in front of H_2SO_4 and a 3 in front of H_3PO_4 to obtain the balanced equation

$$Ca_5(PO_4)_3F(s) + 5H_2SO_4(aq) \longrightarrow 3H_3PO_4(aq) + HF(aq) + 5CaSO_4(s)$$

From solubility rule 7 (see Table 3.3) we see that $CaSO_4$ is only slightly soluble and will therefore precipitate. Furthermore, the reaction involves proton transfer from H_2SO_4 to H_3PO_4. Therefore, the wet process is classified both as a precipitation reaction and an acid-base reaction. Note that there are no changes in oxidation number. Therefore, it is not a redox reaction.

Key Words

Activity series, p. 114
Brønsted acid, p. 103
Brønsted base, p. 103
Chemical equilibrium, p. 98
Chemical reaction, p. 90
Diprotic acid, p. 104
Displacement reaction, p. 112
Disproportionation reaction, p. 116
Electrochemical series, p. 114
Electrolyte, p. 96
Half-reaction, p. 108
Hard water, p. 102
Hydration, p. 98
Hydronium ion, p. 103
Ionic equation, p. 99
Molecular equation, p. 99
Monoprotic acid, p. 104
Net ionic equation, p. 99
Neutralization reaction, p. 105
Nonelectrolyte, p. 96
Oxidation number, p. 108
Oxidation reaction, p. 108
Oxidation-reduction reaction, p. 107
Oxidation state, p. 108
Oxidizing agent, p. 108
Precipitate, p. 98
Product, p. 90
Reactant, p. 90
Redox reaction, p. 107
Reducing agent, p. 108
Reduction reaction, p. 108
Reversible reaction, p. 98
Salt, p. 105
Soft water, p. 102
Solubility, p. 100
Solute, p. 96
Solution, p. 96
Solvent, p. 96
Spectator ion, p. 99
Triprotic acid, p. 104

Exercises

Writing and Balancing Chemical Equations

REVIEW QUESTIONS

3.1 Define the following terms: chemical reaction, reactant, product.

3.2 What is the difference between a chemical reaction and a chemical equation?

3.3 Why must a chemical equation be balanced? What law is obeyed by a balanced chemical equation?

3.4 Write the symbols used to represent gas, liquid, solid, and the aqueous phase in chemical equations.

PROBLEMS

3.5 Describe in words the meaning of the following equation:

$FeS(s)$ +	$2HCl(aq)$	$\longrightarrow$	$FeCl_2(aq)$ +	$H_2S(g)$
iron(II) sulfide	hydrochloric acid		iron(II) chloride	hydrogen sulfide

3.6 From the chemical equation

$$2NO(g) + O_2(g) \longrightarrow 2NO_2(g)$$

which of the following characteristics or quantities can be deduced? (a) The reaction is started by heating. (b) Two moles of nitric oxide react with one mole of molecular oxygen to form two moles of nitrogen dioxide. (c) The reaction is essentially completed in a few minutes after mixing the reactants. (d) All substances are gases in this reaction. (e) This reaction takes place by two NO molecules and one O_2 molecule colliding with one another. (f) Eight hundred molecules of NO would react with four hundred O_2 molecules.

3.7 Balance the following equations using the method outlined in Section 3.1:

(a) $C + O_2 \longrightarrow CO_2$
(b) $CO + O_2 \longrightarrow CO_2$
(c) $H_2 + Br_2 \longrightarrow HBr$
(d) $K + H_2O \longrightarrow KOH + H_2$
(e) $Mg + O_2 \longrightarrow MgO$

(f) $O_3 \longrightarrow O_2$
(g) $H_2O_2 \longrightarrow H_2O + O_2$
(h) $N_2 + H_2 \longrightarrow NH_3$
(i) $Zn + AgCl \longrightarrow ZnCl_2 + Ag$
(j) $S_8 + O_2 \longrightarrow SO_2$
(k) $NaOH + H_2SO_4 \longrightarrow Na_2SO_4 + H_2O$
(l) $Cl_2 + NaI \longrightarrow NaCl + I_2$
(m) $KOH + H_3PO_4 \longrightarrow K_3PO_4 + H_2O$
(n) $CH_4 + Br_2 \longrightarrow CBr_4 + HBr$
(o) $Fe_2O_3 + CO \longrightarrow Fe + CO_2$

3.8 Balance the following equations using the method outlined in Section 3.1:
(a) $KClO_3 \longrightarrow KCl + O_2$
(b) $KNO_3 \longrightarrow KNO_2 + O_2$
(c) $NH_4NO_3 \longrightarrow N_2O + H_2O$
(d) $NH_4NO_2 \longrightarrow N_2 + H_2O$
(e) $NaHCO_3 \longrightarrow Na_2CO_3 + H_2O + CO_2$
(f) $P_4O_{10} + H_2O \longrightarrow H_3PO_4$
(g) $HCl + CaCO_3 \longrightarrow CaCl_2 + H_2O + CO_2$
(h) $Al + H_2SO_4 \longrightarrow Al_2(SO_4)_3 + H_2$
(i) $CO_2 + KOH \longrightarrow K_2CO_3 + H_2O$
(j) $CH_4 + O_2 \longrightarrow CO_2 + H_2O$
(k) $Be_2C + H_2O \longrightarrow Be(OH)_2 + CH_4$
(l) $Cu + HNO_3 \longrightarrow Cu(NO_3)_2 + NO + H_2O$
(m) $S + HNO_3 \longrightarrow H_2SO_4 + NO_2 + H_2O$
(n) $NH_3 + CuO \longrightarrow Cu + N_2 + H_2O$

Aqueous Solutions

REVIEW QUESTIONS

3.9 Define the following terms: solute, solvent, solution, electrolyte, nonelectrolyte, hydration, reversible reaction, chemical equilibrium.

3.10 What is the difference between the following two symbols used in chemical equations: $\rightarrow$ and $\rightleftharpoons$.

3.11 Water, as we know, is an extremely weak electrolyte and therefore cannot conduct electricity. Yet we are often cautioned not to operate electrical appliances when our hands are wet. Why?

3.12 LiF is a strong electrolyte. When we write LiF(*aq*), what species are actually present in the solution?

PROBLEMS

3.13 Identify each of the following substances as a strong electrolyte, a weak electrolyte, or a nonelectrolyte: (a) H_2O, (b) KCl, (c) HNO_3, (d) CH_3COOH, (e) $C_{12}H_{22}O_{11}$, (f) $Ba(NO_3)_2$, (g) Ne, (h) NH_3, (i) NaOH.

3.14 The passage of electricity through an electrolyte solution is caused by the movement of (a) electrons only, (b) cations only, (c) anions only, (d) both cations and anions.

3.15 Predict and explain which of the following systems are electrically conducting: (a) solid NaCl, (b) molten NaCl, (c) an aqueous solution of NaCl.

3.16 Explain why a solution of HCl in benzene does not conduct electricity but in water it does.

3.17 You are given a water-soluble compound X. Describe how you would determine whether it is an electrolyte or a nonelectrolyte. If it is an electrolyte, how would you determine whether it is strong or weak?

Precipitation Reactions

REVIEW QUESTION

3.18 What is the difference between an ionic equation and a molecular equation? What is the advantage of writing net ionic equations for precipitation reactions?

PROBLEMS

3.19 Characterize the following compounds as soluble or insoluble in water: (a) $Ca_3(PO_4)_2$, (b) $Mn(OH)_2$, (c) $AgClO_3$, (d) K_2S, (e) $CaCO_3$, (f) $ZnSO_4$, (g) $Hg(NO_3)_2$, (h) $HgSO_4$, (i) NH_4ClO_4, (j) $BaSO_3$.

3.20 Give the name and formula of two soluble metal hydroxides and two insoluble hydroxides.

3.21 Which of the following processes will likely result in a precipitation reaction? (a) Mixing a $NaNO_3$ solution with a $CuSO_4$ solution. (b) Mixing a $BaCl_2$ solution with a K_2SO_4 solution.

3.22 Write ionic and net ionic equations for the following reactions:
(a) $2AgNO_3(aq) + Na_2SO_4(aq) \longrightarrow$
(b) $BaCl_2(aq) + ZnSO_4(aq) \longrightarrow$
(c) $(NH_4)_2CO_3(aq) + CaCl_2(aq) \longrightarrow$
(d) $Na_2S(aq) + ZnCl_2(aq) \longrightarrow$
(e) $2K_3PO_4(aq) + 3Sr(NO_3)_2(aq) \longrightarrow$

3.23 On the basis of the solubility rules given in this chapter, suggest one method by which you might separate (a) K^+ from Ag^+, (b) Ag^+ from Pb^{2+}, (c) NH_4^+ from Ca^{2+}, (d) Ba^{2+} from Cu^{2+}. All cations are assumed to be in aqueous solution, and the common anion is the nitrate ion.

Acid-Base Reactions

REVIEW QUESTIONS

3.24 Describe the general properties of acids and bases.

3.25 Give Arrhenius's and Brønsted's definitions of an acid and a base.

3.26 Give an example each of a monoprotic acid, a diprotic acid, and a triprotic acid.

3.27 What are the characteristics of an acid-base neutralization reaction?

3.28 Give four examples of a salt.

3.29 Identify the following as a weak or strong acid or base: (a) NH_3, (b) H_3PO_4, (c) LiOH, (d) HCOOH (formic acid), (e) H_2SO_4, (f) HF, (g) $Ba(OH)_2$.

PROBLEMS

3.30 Identify each of the following species as a Brønsted acid or base, or both: (a) HI, (b) CH_3COO^-, (c) $H_2PO_4^-$, (d) PO_4^{3-}, (e) ClO_2^-, (f) NH_4^+.

3.31 Balance the following equations and write the corresponding ionic and net ionic equations (if appropriate):
(a) $HBr(aq) + NH_3(aq) \longrightarrow$ (HBr is a strong acid)
(b) $CH_3COOH(aq) + KOH(aq) \longrightarrow$
(c) $Ba(OH)_2(aq) + H_3PO_4(aq) \longrightarrow$
(d) $HClO_4(aq) + Mg(OH)_2(s) \longrightarrow$

3.32 A compound has just been synthesized in a chemical laboratory. What tests would you employ to see whether it is an acid? If it is an acid, what further tests are needed to determine whether it is a strong or weak acid? Assume that the compound is soluble in water.

Oxidation-Reduction Reactions

REVIEW QUESTIONS

3.33 Define the following terms: half-reaction, oxidation reaction, reduction reaction, reducing agent, oxidizing agent, redox reaction.

3.34 Is it possible to have a reaction in which oxidation occurs and reduction does not? Explain.

PROBLEMS

3.35 For the complete redox reactions given below, (i) break down each reaction into its half-reactions; (ii) identify the oxidizing agent; (iii) identify the reducing agent.
(a) $2Sr + O_2 \longrightarrow 2SrO$
(b) $2Na + S \longrightarrow Na_2S$
(c) $2Li + H_2 \longrightarrow 2LiH$
(d) $2Cs + Br_2 \longrightarrow 2CsBr$
(e) $3Mg + N_2 \longrightarrow Mg_3N_2$
(f) $Zn + I_2 \longrightarrow ZnI_2$
(g) $2C + O_2 \longrightarrow 2CO$

3.36 For the complete redox reactions given below, write the half-reactions and identify the oxidizing and reducing agents:
(a) $4Fe + 3O_2 \longrightarrow 2Fe_2O_3$
(b) $Cl_2 + 2NaBr \longrightarrow 2NaCl + Br_2$
(c) $Si + 2F_2 \longrightarrow SiF_4$
(d) $H_2 + Cl_2 \longrightarrow 2HCl$

Oxidation Numbers

REVIEW QUESTIONS

3.37 Define oxidation number. Explain why, except for ionic compounds, oxidation number does not have any physical significance.

3.38 Without referring to Figure 3.9, give the oxidation numbers of the alkali and alkaline earth metals in their compounds. Also give the highest oxidation numbers that the Groups 3A–7A elements can have.

3.39 Give the oxidation number for the following species: H_2, Se_8, P_4, O, U, As_4, B_{12}.

PROBLEMS

3.40 Arrange the following species in order of increasing oxidation number of the sulfur atom: (a) H_2S, (b) S_8, (c) H_2SO_4, (d) S^{2-}, (e) HS^-, (f) SO_2, (g) SO_3.

3.41 Indicate the oxidation number of phosphorus in each of the following compounds: (a) HPO_3, (b) H_3PO_2, (c) H_3PO_3, (d) H_3PO_4, (e) $H_4P_2O_7$, (f) $H_5P_3O_{10}$.

3.42 Give the oxidation number of each atom in the following molecules and ions: (a) NaCN, (b) ClF, (c) IF_7, (d) NaSCN, (e) CH_4, (f) C_2H_2, (g) C_2H_4, (h) K_2CrO_4, (i) $K_2Cr_2O_7$, (j) $KMnO_4$, (k) $NaHCO_3$, (l) Li_2, (m) $NaIO_3$, (n) KO_2, (o) PF_6^-, (p) $K_4Fe(CN)_6$, (q) $KAuCl_4$, (r) $K_3Fe(CN)_6$. (*Hint:* In CN^- and SCN^-, the oxidation numbers of C and S are +2 and −2, respectively.)

3.43 Give oxidation numbers for the underlined atoms in the following molecules and ions: (a) $\underline{Cs}_2O$, (b) $Ca\underline{I}_2$, (c) $\underline{Al}_2O_3$, (d) $H_3\underline{As}O_3$, (e) $\underline{Ti}O_2$, (f) $\underline{Mo}O_4^{2-}$, (g) $\underline{Pt}Cl_4^{2-}$, (h) $\underline{Pt}Cl_6^{2-}$, (i) $\underline{Sn}F_2$, (j) $\underline{Cl}F_3$, (k) $\underline{Sb}F_6^-$.

3.44 Give the oxidation numbers of all the elements in the following molecules and ions: (a) Mg_3N_2, (b) CsO_2, (c) CaC_2, (d) CO_3^{2-}, (e) $C_2O_4^{2-}$, (f) ZnO_2^{2-}, (g) $NaBH_4$, (h) WO_4^{2-}.

3.45 Nitric acid is a strong oxidizing agent. State which of the following species is *least* likely to be produced when nitric acid reacts with a strong reducing agent such as zinc metal, and explain why: N_2O, NO, NO_2, N_2O_4, N_2O_5, NH_4^+.

3.46 On the basis of oxidation number considerations, one of the following oxides would not react with molecular oxygen: NO, N_2O, SO_2, SO_3, P_4O_6. Which one is it? Why?

Types of Redox Reactions

REVIEW QUESTIONS

3.47 Give an example of a combination redox reaction, a decomposition redox reaction, and a displacement redox reaction.

3.48 Explain the use of the activity series in displacement reactions.

3.49 All combustion reactions are redox reactions. True or false? Explain.

3.50 What is a disproportionation reaction? What is the criterion for a substance to be able to undergo a disproportionation reaction? Name six elements capable of undergoing disproportionation.

PROBLEMS

3.51 Is the following reaction a redox reaction? Explain.

$$3O_2(g) \longrightarrow 2O_3(g)$$

3.52 Ammonium nitrate (NH_4NO_3) is thermally unstable. Write a balanced equation for its decomposition, and classify the reaction. (*Hint:* One of the products is nitrous oxide.)

3.53 Which of the following metals can react with water? (a) Au, (b) Li, (c) Hg, (d) Ca, (e) Pt.

3.54 Predict the outcome of the reactions represented by the following equations by using the activity series, and balance the equations:
(a) $Cu(s) + HCl(aq) \longrightarrow$
(b) $I_2(s) + NaBr(aq) \longrightarrow$
(c) $Mg(s) + CuSO_4(aq) \longrightarrow$
(d) $Cl_2(g) + KBr(aq) \longrightarrow$
Also write the net ionic equation for each reaction.

Balancing Redox Equations

PROBLEMS

3.55 Balance the following redox equation by the ion-electron method. The reaction occurs in acidic solution.

$$MnO_4^- + SO_2 \longrightarrow Mn^{2+} + HSO_4^-$$

3.56 Balance the following redox equations by the ion-electron method:
(a) $Mn^{2+} + H_2O_2 \longrightarrow MnO_2$ in basic solution
(b) $Cr_2O_7^{2-} + C_2O_4^{2-} \longrightarrow Cr^{3+} + CO_2$ in acidic solution
(c) $ClO_3^- + Cl^- \longrightarrow Cl_2 + ClO_2$ in acidic solution
(d) $S_2O_3^{2-} + I_2 \longrightarrow I^- + S_4O_6^{2-}$ in acidic solution
(e) $H_2O_2 + Fe^{2+} \longrightarrow Fe^{3+}$ in acidic solution
(f) $Cu + HNO_3 \longrightarrow Cu^{2+} + NO$ in acidic solution
(g) $Bi(OH)_3 + SnO_2^{2-} \longrightarrow SnO_3^{2-} + Bi$ in basic solution
(h) $CN^- + MnO_4^- \longrightarrow CNO^- + MnO_2$ in basic solution
(i) $Fe^{2+} + MnO_4^- \longrightarrow Fe^{3+} + Mn^{2+}$ in acidic solution
(j) $Br_2 \longrightarrow BrO_3^- + Br^-$ in basic solution

3.57 Steel is a mixture of iron with small amounts of carbon, manganese, and other elements. To determine the quantity of manganese present, a sample of steel is first dissolved in concentrated nitric acid; the reaction between Mn and HNO_3 gives Mn^{2+} and NO_2. Next, the Mn^{2+} ion is treated with an acidic solution containing the periodate ion (IO_4^-) to give the MnO_4^- and IO_3^- ion. Finally, the concentration of the permanganate solution is determined by reacting with a standard $FeSO_4$ solution. Write a balanced equation for each step.

Miscellaneous Problems

3.58 A quantitative definition of solubility is the number of grams of a solute dissolved in a given volume of water at a particular temperature. Describe an experiment which would enable you to determine the solubility of a soluble compound.

3.59 An ionic compound X is only slightly soluble in water. What test would you employ to show that the compound does indeed dissolve in water to a certain extent?

3.60 Classify the following reactions according to the types discussed in the chapter:
(a) $Cl_2 + 2OH^- \longrightarrow Cl^- + ClO^- + H_2O$
(b) $Ca^{2+} + CO_3^{2-} \longrightarrow CaCO_3$
(c) $NH_3 + H^+ \longrightarrow NH_4^+$
(d) $2CCl_4 + CrO_4^{2-} \longrightarrow 2COCl_2 + CrO_2Cl_2 + 2Cl^-$
(e) $Ca + F_2 \longrightarrow CaF_2$
(f) $2Li + H_2 \longrightarrow 2LiH$
(g) $Ba(NO_3)_2 + Na_2SO_4 \longrightarrow 2NaNO_3 + BaSO_4$
(h) $CuO + H_2 \longrightarrow Cu + H_2O$
(i) $Zn + 2HCl \longrightarrow ZnCl_2 + H_2$
(j) $2FeCl_2 + Cl_2 \longrightarrow 2FeCl_3$

3.61 Gold will not dissolve in either concentrated nitric acid or concentrated hydrochloric acid. However, the metal does dissolve in a mixture of the acids (one part HNO_3 and three parts HCl by volume), called *aqua regia*. Write a balanced equation for this reaction. (*Hint:* Among the products are $HAuCl_4$ and NO_2.)

3.62 A student is given an unknown that is either iron(II) sulfate or iron(III) sulfate. Suggest a chemical procedure for determining its identity.

3.63 Hydrochloric acid is not an oxidizing agent in the sense that sulfuric acid and nitric acid are. Explain why the chloride ion is not a strong oxidizing agent like SO_4^{2-} and NO_3^-.

3.64 Write balanced equations for the following redox processes: (a) When potassium iodide is added to acidified hydrogen peroxide, a brown color appears. (b) When sodium sulfite is added to an acidic solution of potassium dichromate, the solution changes from orange to green. (c) Manganese(IV) oxide can oxidize hydrochloric acid to molecular chlorine.

3.65 The Mn^{3+} ion is unstable in solution and undergoes disproportionation to give Mn^{2+}, Mn(IV) oxide, and H^+ ion. Write a balanced net ionic equation for the reaction.

3.66 On standing a concentrated nitric acid solution turns slightly yellow. Explain. (*Hint:* Nitric acid undergoes decomposition. What are the products?)

3.67 Concentrated sulfuric acid can oxidize many metals and nonmetallic elements. Balance the following equations:

(a) $C + H_2SO_4 \longrightarrow CO_2 + SO_2$
(b) $S + H_2SO_4 \longrightarrow SO_2 + H_2O$

3.68 Fluorides, that is, salts containing F^- ions, are not mentioned in the solubility rules in Section 3.3. Look up the solubilities of some metal fluorides in the *Handbook of Chemistry and Physics,* including LiF, CaF_2, and AgF. Does F^- fit with the other halides (that is, salts of Cl^-, Br^-, and I^-) in the solubility rules? How would you place fluorides in the solubility rules?

3.69 Someone gave you a colorless liquid. Describe three physical and three chemical tests you would perform on the liquid to show that it is water.

3.70 Balance the following equation representing the reaction between aqueous bromine and sodium thiosulfate ($Na_2S_2O_3$) in acid medium:

$$Br_2(aq) + S_2O_3^{2-}(aq) \longrightarrow Br^-(aq) + SO_4^{2-}(aq)$$

Preparation of Inorganic Compounds

The materials covered in this chapter are helpful for preparing a number of common and important inorganic compounds either in the laboratory or in industry. The following problems are concerned with some of their preparations.

3.71 Sodium reacts with water to yield hydrogen gas. Why is this reaction not used in the laboratory preparation of hydrogen?

3.72 Write a balanced equation for the preparation of (a) molecular oxygen, (b) ammonia, (c) carbon dioxide, (d) molecular hydrogen, (e) calcium oxide. Indicate the physical state of the reactants and products in each equation.

3.73 Explain how you would prepare potassium iodide by means of (a) an acid-base reaction and (b) a reaction between an acid and a carbonate compound.

3.74 Describe how you would prepare the following compounds: (a) $Mg(OH)_2$, (b) AgI, (c) $Ba_3(PO_4)_2$.

3.75 How would you prepare anhydrous copper sulfate ($CuSO_4$) in the laboratory?

Opposite page: The presence of ice in clouds makes them glow at night at high altitudes. The amount of ice is related to the amount of methane (CH_4) present in the atmosphere because water is a by-product of methane oxidation: $CH_4 + 2O_2 \rightarrow CO_2 + 2H_2O$. Stoichiometry suggests that an observed increase in brightness of these clouds is related to increased methane concentrations in the atmosphere.

C H A P T E R 4

CHEMICAL REACTIONS II: MASS RELATIONSHIPS

In Chapter 3 we discussed the meaning and use of chemical equations and three types of chemical reactions in aqueous solution. We now focus on the mass relationships in chemical reactions. One important law will be our guide—the law of conservation of mass.

4.1 Amounts of Reactants and Products

Having surveyed several types of reactions, we are now ready to study the quantitative aspects of chemical reactions. *The mass relationships among reactants and products in a chemical reaction* represent the ***stoichiometry*** of the reaction. To interpret a reaction quantitatively, we need to apply our knowledge of molar masses and the mole concept. The basic question posed in many stoichiometric calculations is, "If we know the quantities of the starting materials (that is, the reactants) in a reaction, how much product will be formed?" Or in some cases we may have to ask the reverse question: "How much starting material must be used to obtain a specific amount of product?" In practice, the units used for reactants (or products) may be in moles, grams, liters (for gases), or other units. Regardless of which units are used, *the approach to determine the amount of product formed in a reaction* is called the ***mole method.*** It is based on the fact that *the stoichiometric coefficients in a chemical equation can be interpreted as the number of moles of each substance*. Consider the combustion of carbon monoxide in air to form carbon dioxide:

$$2CO(g) + O_2(g) \longrightarrow 2CO_2(g)$$

The equation and the stoichiometric coefficients can be read as "2 moles of carbon monoxide gas combining with 1 mole of oxygen gas to form 2 moles of carbon dioxide." The mole method as commonly practiced consists of the following steps:

1. Write correct formulas for all reactants and products, and balance the resulting equation.
2. Convert the quantities of some or all given or known substances (usually reactants) into moles.
3. Use the coefficients in the balanced equation to calculate the number of moles of the sought or unknown quantities (usually products) in the problem.
4. Using the calculated numbers of moles and the molar masses, convert the unknown quantities to whatever units are required (typically grams).
5. Check that your answer is reasonable in physical terms.

Step 1 is obviously a prerequisite to any stoichiometric calculation. We must know the identities of the reactants and products, and the mass relationships among them must not violate the law of conservation of mass (that is, we must have a balanced equation). Step 2 is the critical step of converting grams (or other units) of substances to number of moles. This conversion then allows us to analyze the actual reaction in terms of moles only.

To complete step 3, we need the balanced equation, already furnished by step 1. The key point here is that the coefficients in a balanced equation provide us with the ratio in which moles of one substance react with or form moles of another substance. Step 4 is similar to step 2, except that now we are dealing with the quantities sought in the problem. Step 5 is often underestimated but is very important: Chemistry is an experimental science, and your answer must make sense in terms of real species in the real world. If you have set up the problem incorrectly or made a computational error, it will often become obvious when your answer turns out to be much too large or much too small for the amounts of materials you started with. Figure 4.1 shows three common types of stoichiometric calculations.

In discussing quantitative relationships it is useful to introduce the symbol $\approx$, which

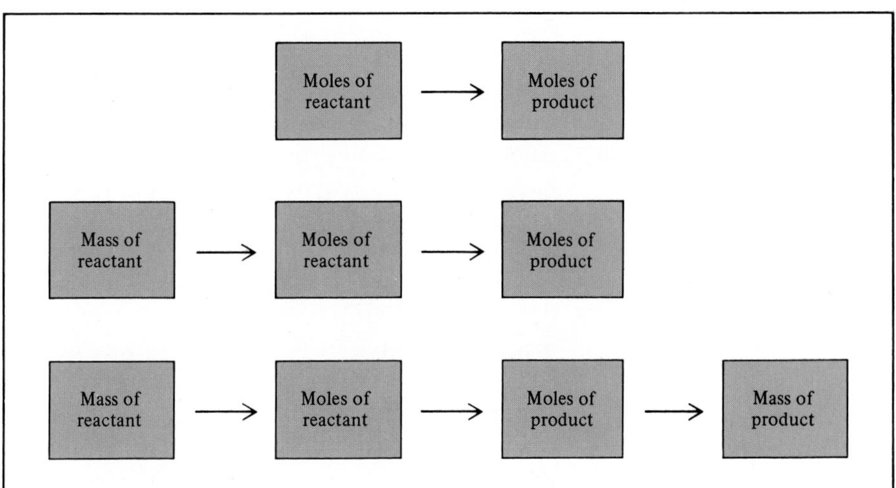

Figure 4.1 *Three types of stoichiometric calculations based on the mole method.*

means "stoichiometrically equivalent to" or simply "equivalent to." In the balanced equation for the formation of carbon dioxide, we see that 2 moles of CO react with 1 mole of O_2, so 2 moles of CO are equivalent to 1 mole of O_2:

$$2 \text{ mol CO} \simeq 1 \text{ mol } O_2$$

In terms of the factor-label method, we can write the unit factor as

$$\frac{2 \text{ mol CO}}{1 \text{ mol } O_2} = 1 \quad \text{or} \quad \frac{1 \text{ mol } O_2}{2 \text{ mol CO}} = 1$$

Similarly, since 2 moles of CO (or 1 mole of O_2) produce 2 moles of CO_2, we can say that 2 moles of CO (or 1 mole of O_2) are equivalent to 2 moles of CO_2:

$$2 \text{ mol CO} \simeq 2 \text{ mol } CO_2$$

$$1 \text{ mol } O_2 \simeq 2 \text{ mol } CO_2$$

Note that in stoichiometric calculations we generally use molecular equations because we are usually interested in the masses of whole units and not just the masses of separate cations and anions.

The following examples illustrate the use of the five-step method in solving some typical stoichiometry problems.

Lithium reacting with water.

EXAMPLE 4.1

All alkali metals react with water to produce hydrogen gas and the corresponding alkali metal hydroxide. A typical reaction is that between lithium and water:

$$2Li(s) + 2H_2O(l) \longrightarrow 2LiOH(aq) + H_2(g)$$

(a) How many moles of H_2 can be formed by the complete reaction of 6.23 moles of Li with water? (b) How many grams of H_2 can be formed by the complete reaction of 80.57 g of Li with water?

Refer to Figure 4.1 for the conversions

Answer

(a)

Step 1: The balanced equation is given in the problem.

Step 2: No conversion is needed because the amount of the starting material, Li, is given in moles.

Step 3: Since 2 moles of Li produce 1 mole of H_2, or 2 mol Li ⇔ 1 mol H_2, we calculate moles of H_2 produced as follows:

$$\text{moles of } H_2 \text{ produced} = 6.23 \cancel{\text{mol Li}} \times \frac{1 \text{ mol } H_2}{2 \cancel{\text{mol Li}}}$$
$$= 3.12 \text{ mol } H_2$$

Step 4: This step is not required.

Step 5: We began with 6.23 moles of Li and produced 3.12 moles of H_2. Since 2 moles of Li produce 1 mole of H_2, 3.12 moles is a reasonable quantity.

(b)

Step 1: The reaction is the same as in (a).

Step 2: The number of moles of Li is given by

$$\text{moles of Li} = 80.57 \cancel{\text{g Li}} \times \frac{1 \text{ mol Li}}{6.941 \cancel{\text{g Li}}} = 11.61 \text{ mol Li}$$

Step 3: Since 2 moles of Li produce 1 mole of H_2, or 2 mol Li ⇔ 1 mol H_2, we calculate the number of moles of H_2 as follows:

$$\text{moles of } H_2 \text{ produced} = 11.61 \cancel{\text{mol Li}} \times \frac{1 \text{ mol } H_2}{2 \cancel{\text{mol Li}}} = 5.805 \text{ mol } H_2$$

Step 4: From the molar mass of H_2 (2.016 g), we calculate the mass of H_2 produced:

$$\text{mass of } H_2 \text{ produced} = 5.805 \cancel{\text{mol } H_2} \times \frac{2.016 \text{ g } H_2}{1 \cancel{\text{mol } H_2}} = 11.70 \text{ g } H_2$$

Step 5: The amount 11.70 g H_2 is a reasonable quantity.

Similar problems: 4.5, 4.6.

EXAMPLE 4.2

The food we eat is degraded, or broken down, in our bodies to provide energy for growth and function. A general overall equation for this very complex process represents the degradation of glucose ($C_6H_{12}O_6$) to carbon dioxide (CO_2) and water (H_2O):

$$C_6H_{12}O_6 + 6O_2 \longrightarrow 6CO_2 + 6H_2O$$

If 856 g of $C_6H_{12}O_6$ is consumed by the body over a certain period, what is the mass of CO_2 produced?

Answer

Step 1: The balanced equation is given.

Step 2: The molar mass of $C_6H_{12}O_6$ is 180.2 g, so the number of moles of $C_6H_{12}O_6$ in 856 g of $C_6H_{12}O_6$ is

$$\text{moles of } C_6H_{12}O_6 = 856\ \text{g } C_6H_{12}O_6 \times \frac{1\ \text{mol } C_6H_{12}O_6}{180.2\ \text{g } C_6H_{12}O_6}$$
$$= 4.75\ \text{mol } C_6H_{12}O_6$$

Step 3: From the balanced equation we see that 1 mol $C_6H_{12}O_6 \Leftrightarrow$ 6 mol CO_2; thus the number of moles of CO_2 produced is given by

$$\text{moles of } CO_2 \text{ produced} = 4.75\ \text{mol } C_6H_{12}O_6 \times \frac{6\ \text{mol } CO_2}{1\ \text{mol } C_6H_{12}O_6}$$
$$= 28.5\ \text{mol } CO_2$$

Step 4: The molar mass of CO_2 is 44.01 g. To convert moles of CO_2 to grams of CO_2 we write

$$\text{mass of } CO_2 \text{ produced} = 28.5\ \text{mol } CO_2 \times \frac{44.01\ \text{g } CO_2}{1\ \text{mol } CO_2}$$
$$= 1.25 \times 10^3\ \text{g } CO_2$$

Step 5: The mass of CO_2 produced is 1.25×10^3 g = 1250 g CO_2, a reasonable amount of carbon dioxide to result from consuming 856 g of $C_6H_{12}O_6$.

Similar problem: 4.10.

After some practice, you may find it convenient to combine steps 2, 3, and 4 in a single factor-label equation, as the following example shows.

EXAMPLE 4.3

The compound cisplatin [$Pt(NH_3)_2Cl_2$] has been used as an antitumor agent. It is prepared by the reaction between potassium tetrachloroplatinate (K_2PtCl_4) and ammonia (NH_3):

$$K_2PtCl_4(aq) + 2NH_3(aq) \longrightarrow Pt(NH_3)_2Cl_2(s) + 2KCl(aq)$$

How many grams of cisplatin can be obtained starting with 0.8862 g of K_2PtCl_4? You may assume that there is enough of NH_3 to react with all of the K_2PtCl_4.

Answer

Step 1: The balanced equation is given.

Steps 2, 3, and 4: From the balanced equation we see that 1 mol $K_2PtCl_4 \Leftrightarrow$ 1 mol $Pt(NH_3)_2Cl_2$. The molar masses of K_2PtCl_4 and $Pt(NH_3)_2Cl_2$ are 415.1 g and 300.1 g, respectively. We combine all of these data into one equation.

$$\text{mass of } Pt(NH_3)_2Cl_2 \text{ produced} = 0.8862 \text{ g } K_2PtCl_4 \times \frac{1 \text{ mol } K_2PtCl_4}{415.1 \text{ g } K_2PtCl_4} \times \frac{1 \text{ mol } Pt(NH_3)_2Cl_2}{1 \text{ mol } K_2PtCl_4} \times \frac{300.1 \text{ g } Pt(NH_3)_2Cl_2}{1 \text{ mol } Pt(NH_3)_2Cl_2} = 0.6407 \text{ g } Pt(NH_3)_2Cl_2$$

Step 5: Since the molar masses of K_2PtCl_4 and $Pt(NH_3)_2Cl_2$ are comparable, and we started with 0.8862 g of K_2PtCl_4, 0.6407 g of cisplatin is a reasonable quantity.

Similar problem: 4.12.

4.2 Limiting Reagents

When a chemist carries out a reaction, the reactants are usually not present in exact ***stoichiometric amounts,*** that is, *in the proportions indicated by the balanced equation. The reactant used up first in a reaction* is called the ***limiting reagent,*** since the maximum amount of product formed depends on how much of this reactant was originally present (Figure 4.2). When this reactant is used up, no more product can be formed. *Other reactants, present in quantities greater than those needed to react with the quantity of the limiting reagent present,* are called ***excess reagents.***

The concept of the limiting reagent is analogous to the relationship between men and women in a dance contest at a club. If there are fifteen men and only nine women, then only nine female/male pairs can compete. Six men will be left without partners. The number of women thus limits the number of men that can dance in the contest, and there is an excess of men.

Sulfur hexafluoride is a colorless, odorless, and extremely stable compound. It is formed by burning sulfur in an atmosphere of fluorine. The equation is

$$S(l) + 3F_2(g) \longrightarrow SF_6(g)$$

This equation tells us that 1 mole of S reacts with 3 moles of F_2 to produce 1 mole of SF_6. Suppose that in a certain preparation 4 moles of S are added to 20 moles of F_2. Since 1 mol S $\Leftrightarrow$ 3 mol F_2, the number of moles of F_2 needed to react with 4 moles of S is

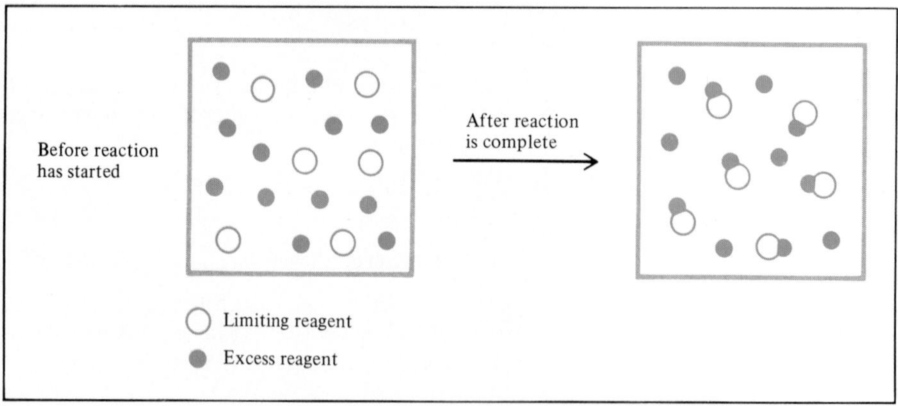

Figure 4.2 *A limiting reagent is completely used up in a reaction.*

$$4 \cancel{\text{mol S}} \times \frac{3 \text{ mol } F_2}{1 \cancel{\text{mol S}}} = 12 \text{ mol } F_2$$

But there are 20 moles of F_2 available, more than needed to completely react with S. Thus S must be the limiting reagent and F_2 the excess reagent. The amount of SF_6 produced depends only on how much S was originally present.

We could, of course, calculate the number of moles of S needed to react with 20 moles of F_2. In this case we write

$$20 \cancel{\text{mol } F_2} \times \frac{1 \text{ mol S}}{3 \cancel{\text{mol } F_2}} = 6.7 \text{ mol S}$$

Since there are only 4 moles of S present, we arrive at the same conclusion that S is the limiting reagent and F_2 is the excess reagent.

In stoichiometric calculations involving limiting reagents, the first step is to decide which reactant is the limiting reagent. After the limiting reagent has been identified, the rest of the problem can be solved as outlined in Section 4.1. The following two examples illustrate this approach. We will not include step 5 in the calculations, but you should always examine the reasonableness of *any* chemical calculation.

Iron(II) sulfide.

EXAMPLE 4.4

At high temperatures, sulfur combines with iron to form the brown-black iron(II) sulfide:

$$Fe(s) + S(l) \longrightarrow FeS(s)$$

In one experiment 7.62 g of Fe are allowed to react with 8.67 g of S. (a) Which of the two reactants is the limiting reagent? (b) Calculate the mass of FeS formed. (c) How much of the excess reagent (in grams) is left at the end of the reaction?

Answer

(a) Since we cannot tell by inspection which of the two reactants is the limiting reagent, we have to proceed by first converting their masses into numbers of moles. The molar masses of Fe and S are 55.85 and 32.07 g, respectively. Thus

$$\text{moles of Fe} = 7.62 \cancel{\text{g Fe}} \times \frac{1 \text{ mol Fe}}{55.85 \cancel{\text{g Fe}}}$$
$$= 0.136 \text{ mol Fe}$$

$$\text{moles of S} = 8.67 \cancel{\text{g S}} \times \frac{1 \text{ mol S}}{32.07 \cancel{\text{g S}}}$$
$$= 0.270 \text{ mol S}$$

From the balanced equation we see that 1 mol Fe $\simeq$ 1 mol S; therefore, the number of moles of S needed to react with 0.136 mole of Fe is given by

$$0.136 \cancel{\text{mol Fe}} \times \frac{1 \text{ mol S}}{1 \cancel{\text{mol Fe}}} = 0.136 \text{ mol S}$$

Since there is 0.270 mole of S present, more than what is needed to react with Fe, S must be the excess reagent and Fe the limiting reagent.

(b) The number of moles of FeS produced is

$$\text{moles of FeS} = 0.136\ \cancel{\text{mol Fe}} \times \frac{1\ \text{mol FeS}}{1\ \cancel{\text{mol Fe}}}$$
$$= 0.136\ \text{mol FeS}$$

The mass of FeS produced is given by

$$\text{mass of FeS} = 0.136\ \cancel{\text{mol FeS}} \times \frac{87.92\ \text{g FeS}}{1\ \cancel{\text{mol FeS}}}$$
$$= 12.0\ \text{g FeS}$$

(c) The number of moles of the excess reagent (S) is 0.270 − 0.136, or 0.134 mole, and therefore

$$\text{mass of S left over} = 0.134\ \cancel{\text{mol S}} \times \frac{32.07\ \text{g S}}{1\ \cancel{\text{mol S}}}$$
$$= 4.30\ \text{g S}$$

Similar problems: 4.21, 4.22.

Urea.

EXAMPLE 4.5

Urea [$(NH_2)_2CO$] is used as fertilizer and as animal feed and is used in the polymer industry. It is prepared by the reaction between ammonia and carbon dioxide:

$$2NH_3(g) + CO_2(g) \longrightarrow (NH_2)_2CO(aq) + H_2O(l)$$

In one process, 637.2 g of NH_3 are allowed to react with 1142 g of CO_2. (a) Which of the two reactants is the limiting reagent? (b) Calculate the mass of $(NH_2)_2CO$ formed. (c) How much of the excess reagent (in grams) is left at the end of the reaction?

Answer

(a) Since we cannot tell by inspection which of the two reactants is the limiting reagent, we have to proceed by first converting their masses into numbers of moles. The molar masses of NH_3 and CO_2 are 17.03 g and 44.01 g, respectively. Thus

$$\text{moles of NH}_3 = 637.2\ \cancel{\text{g NH}_3} \times \frac{1\ \text{mol NH}_3}{17.03\ \cancel{\text{g NH}_3}}$$
$$= 37.42\ \text{mol NH}_3$$

$$\text{moles of CO}_2 = 1142\ \cancel{\text{g CO}_2} \times \frac{1\ \text{mol CO}_2}{44.01\ \cancel{\text{g CO}_2}}$$
$$= 25.95\ \text{mol CO}_2$$

From the balanced equation we see that 2 mol NH_3 ⇔ 1 mol CO_2; therefore, the number of moles of NH_3 needed to react with 25.95 moles of CO_2 is given by

$$25.95\ \cancel{\text{mol CO}_2} \times \frac{2\ \text{mol NH}_3}{1\ \cancel{\text{mol CO}_2}} = 51.90\ \text{mol NH}_3$$

Since there are only 37.42 moles of NH_3 present, not enough to react completely with the CO_2, NH_3 must be the limiting reagent and CO_2 the excess reagent.

(b) The amount of $(NH_2)_2CO$ produced is determined by the amount of limiting reagent present. Thus we write

$$\text{mass of } (NH_2)_2CO = 37.42 \cancel{\text{mol NH}_3} \times \frac{1 \cancel{\text{mol (NH}_2)_2\text{CO}}}{2 \cancel{\text{mol NH}_3}} \times \frac{60.06 \text{ g } (NH_2)_2CO}{1 \cancel{\text{mol (NH}_2)_2\text{CO}}}$$

$$= 1124 \text{ g } (NH_2)_2CO$$

(c) The number of moles of the excess reagent (CO_2) left is

$$25.95 \text{ mol } CO_2 - 37.42 \cancel{\text{mol NH}_3} \times \frac{1 \text{ mol } CO_2}{2 \cancel{\text{mol NH}_3}} = 7.24 \text{ mol } CO_2$$

and

$$\text{mass of } CO_2 \text{ left over} = 7.24 \cancel{\text{mol CO}_2} \times \frac{44.01 \text{ g } CO_2}{1 \cancel{\text{mol CO}_2}}$$

$$= 319 \text{ g } CO_2$$

Similar problems: 4.21, 4.22.

In practice, chemists usually choose the more expensive chemical as the limiting reagent because they want to be sure that all or most of it will be consumed in the reaction. It is often difficult and costly to recover the excess reagent(s), although in many industrial processes the excess reagents are recycled for many runs. Thus for the reaction shown in Example 4.5, NH_3 is invariably the limiting reagent because it is much more expensive than CO_2. In a typical industrial plant, the molar amounts of CO_2 may be as high as 10 to 100 times greater than that for NH_3.

4.3 Yields of Reactions

The amount of limiting reagent present at the start of a reaction is related to *the quantity of product we can obtain from the reaction*. This quantity is called the ***yield of the reaction.*** There are three types of yields that concern us in the quantitative study of chemical reactions.

The ***theoretical yield*** of a reaction is *the amount of product predicted by the balanced equation when all of the limiting reagent has reacted*. The theoretical yield, then, is the *maximum* obtainable yield. In practice, *the amount of product obtained,* called the ***actual yield,*** is almost always less than the theoretical yield. There are many reasons for this. For instance, many reactions are reversible, and so they do not proceed 100 percent from left to right. Even when a reaction is 100 percent complete, it may be difficult to recover all of the product from the reaction medium (say, from an aqueous solution). Some reactions are complex in the sense that the products formed may further react among themselves or with the reactants to form still other products. These further reactions will reduce the yield of the first reaction. Therefore, chemists often use the term ***percent yield (% yield),*** which describes *the proportion of the actual yield to the theoretical yield*. It is defined as follows:

$$\% \text{ yield} = \frac{\text{actual yield}}{\text{theoretical yield}} \times 100\%$$

Percent yields may range from a fraction of 1 percent to 100 percent. An important goal of a chemist involved in synthesis is to maximize the percent yield of product in a reaction. Later we will study factors such as temperature and pressure that can affect the percent yield of a reaction.

The following example shows the calculation of percent yield of an industrial process.

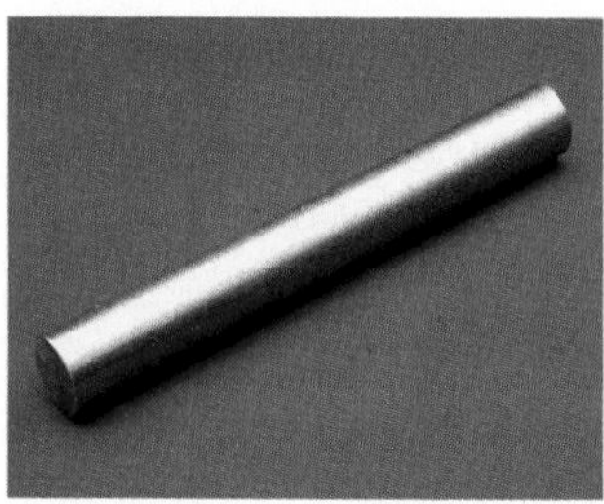

Titanium.

EXAMPLE 4.6

Titanium is a strong, lightweight, corrosion-resistant metal that is used in the construction of rockets, aircrafts, and jet engines. It is prepared by the reduction of titanium(IV) chloride with molten magnesium between 950°C and 1150°C:

$$TiCl_4(g) + 2Mg(l) \longrightarrow Ti(s) + 2MgCl_2(l)$$

In a certain operation 3.54×10^4 kg of $TiCl_4$ is reacted with 1.13×10^4 kg of Mg. (a) Calculate the theoretical yield of Ti in kilograms. (b) Calculate the percent yield if 7.91×10^3 kg of Ti is actually obtained.

Answer

(a) The molar masses of $TiCl_4$ and Mg are 189.7 g and 24.31 g, respectively. Using the conversion factor 1 kg = 1000 g, we write

$$\text{moles of } TiCl_4 = 3.54 \times 10^7 \cancel{\text{g } TiCl_4} \times \frac{1 \text{ mol } TiCl_4}{189.7 \cancel{\text{g } TiCl_4}} = 1.87 \times 10^5 \text{ mol } TiCl_4$$

$$\text{moles of Mg} = 1.13 \times 10^7 \cancel{\text{g Mg}} \times \frac{1 \text{ mol Mg}}{24.31 \cancel{\text{g Mg}}} = 4.65 \times 10^5 \text{ mol Mg}$$

Next, we must determine which of the two substances is the limiting reagent. From the balanced equation we see that 1 mol $TiCl_4 \simeq$ 2 mol Mg; therefore, the number of moles of Mg needed to react with 1.87×10^5 moles of $TiCl_4$ is

$$1.87 \times 10^5 \cancel{\text{mol } TiCl_4} \times \frac{2 \text{ mol Mg}}{1 \cancel{\text{mol } TiCl_4}} = 3.74 \times 10^5 \text{ mol Mg}$$

Since there are 4.65×10^5 moles of Mg present, more than is needed to react with the amount of $TiCl_4$ we have, Mg must be the excess reagent and $TiCl_4$ the limiting reagent.

Since 1 mol $TiCl_4 \simeq$ 1 mol Ti, the theoretical amount of Ti formed is

mass of Ti formed

$$= 3.54 \times 10^4 \text{ kg } \cancel{TiCl_4} \times \frac{1 \cancel{\text{mol } TiCl_4}}{189.7 \cancel{\text{g } TiCl_4}} \times \frac{1 \cancel{\text{mol Ti}}}{1 \cancel{\text{mol } TiCl_4}} \times \frac{47.88 \cancel{\text{g}} \text{ Ti}}{1 \cancel{\text{mol Ti}}}$$

$$= 8.93 \times 10^3 \text{ kg Ti}$$

(b) To find the percent yield, we write

$$\% \text{ yield} = \frac{\text{actual yield}}{\text{theoretical yield}} \times 100\%$$

$$= \frac{7.91 \times 10^3 \text{ kg}}{8.93 \times 10^3 \text{ kg}} \times 100\%$$

$$= 88.6\%$$

Similar problems: 4.28, 4.29.

Industrial processes usually involve products in huge quantities (thousands to millions of tons). Thus even a slight improvement in the yield can significantly reduce the cost of production. The following Chemistry in Action on chemical fertilizers is a good example of this approach.

CHEMISTRY IN ACTION

Chemical Fertilizers

Feeding the world's rapidly increasing population requires that farmers produce ever-larger and healthier crops. Every year they add hundreds of millions of tons of chemical fertilizers to the soil to increase crop quality and yield. In addition to carbon dioxide and water, plants need at least six elements for satisfactory growth. They are N, P, K, Ca, S, and Mg. The preparation and properties of several nitrogen- and phosphorus-containing fertilizers illustrate some of the principles introduced in this chapter.

Nitrogen fertilizers contain nitrate (NO_3^-) salts and ammonium (NH_4^+) salts and other compounds. Plants can absorb nitrogen in the form of nitrate directly, but ammonium salts and ammonia (NH_3) must first be converted to nitrates by the action of soil bacteria. The principal raw material of nitrogen fertilizers is ammonia, prepared by the reaction between hydrogen and nitrogen:

$$3H_2(g) + N_2(g) \longrightarrow 2NH_3(g)$$

(This reaction will be discussed in detail in Chapters 13 and 14.) In its liquid form, ammonia can be injected directly into the soil (Figure 4.3). Alternatively, ammonia can be converted to ammonium nitrate, NH_4NO_3, ammonium sulfate, $(NH_4)_2SO_4$, or ammonium dihydrogen phosphate, $(NH_4)H_2PO_4$, in the following acid-base reactions:

$$NH_3(aq) + HNO_3(aq) \longrightarrow NH_4NO_3(aq)$$

$$2NH_3(aq) + H_2SO_4(aq) \longrightarrow (NH_4)_2SO_4(aq)$$

$$NH_3(aq) + H_3PO_4(aq) \longrightarrow (NH_4)H_2PO_4(aq)$$

Another method of preparing ammonium sulfate requires two steps:

$$2NH_3(aq) + CO_2(aq) + H_2O(l) \longrightarrow (NH_4)_2CO_3(aq) \quad (4.1)$$

Figure 4.3 *Liquid ammonia being applied to the soil before planting.*

$$(NH_4)_2CO_3(aq) + CaSO_4(aq) \longrightarrow (NH_4)_2SO_4(aq) + CaCO_3(s) \quad (4.2)$$

This approach is desirable because the starting materials—carbon dioxide and calcium sulfate—are less costly than sulfuric acid. To increase the yield, ammonia is made the limiting reagent in Reaction (4.1) and ammonium carbonate is made the limiting reagent in Reaction (4.2).

Table 4.1 lists the percent composition by mass of nitrogen in some common fertilizers. The fertilizer $(NH_2)_2CO$, called *urea,* is an organic compound that is prepared by the reaction between ammonia and carbon dioxide:

Table 4.1 Percent composition by mass of nitrogen in five common fertilizers

Fertilizer	%N by mass
NH_3	82.4
NH_4NO_3	35.0
$(NH_4)_2SO_4$	21.2
$(NH_4)H_2PO_4$	21.2
$(NH_2)_2CO$	46.7

$$2NH_3(g) + CO_2(g) \longrightarrow (NH_2)_2CO(aq) + H_2O(l)$$

Several factors influence the choice of one fertilizer over another: (1) cost of raw materials needed to prepare the fertilizer; (2) ease of storage, transportation, and utilization; (3) percent composition by mass of the desired element; and (4) suitability of the compound, that is, whether the compound is soluble in water and whether it can be readily taken up by plants. Considering all these factors together, we find that NH_4NO_3 is the most important nitrogen-containing fertilizer compound in the world, even though ammonia has the highest percentage by mass of nitrogen.

Phosphorus fertilizers are derived from phosphate rock called *fluorapatite,* $Ca_5(PO_4)_3F$. Fluorapatite is insoluble in water, so it must first be converted to the water-soluble calcium dihydrogen phosphate $[Ca(H_2PO_4)_2]$:

$$2Ca_5(PO_4)_3F(s) + 7H_2SO_4(aq) \longrightarrow 3Ca(H_2PO_4)_2(aq) + 7CaSO_4(aq) + 2HF(g)$$

For maximum yield, fluorapatite is made the limiting reagent in this reaction.

The reactions we have discussed for the preparation of fertilizers all appear relatively simple; yet much effort has been expended to improve the yields by changing conditions such as temperature, pressure, and so on. Because huge quantities of fertilizers are produced annually, even a slight improvement in the yield can significantly reduce the cost of production. Industrial chemists usually run promising reactions first in the laboratory, and then test them in a pilot facility before putting them into mass production.

4.4 Concentration and Dilution of Solutions

As mentioned earlier, many reactions occur in the aqueous media. Before we study solution stoichiometry in the next three sections, let us first become familiar with some quantitative aspects of solutions, that is, concentration and dilution of solutions.

Concentration of Solutions

See Section 3.2 for definitions of solute, solvent, and solution.

The ***concentration of a solution*** is *the amount of solute present in a given quantity of solution*. (In our present discussion we will assume the solute to be a liquid or a solid and the solvent to be a liquid). One of the most common units of concentration in chemistry is ***molarity,*** symbolized by *M* and also called ***molar concentration.*** Molarity is *the number of moles of solute in 1 liter of solution*. Molarity is defined by the equation

$$M = \text{molarity} = \frac{\text{moles of solute}}{\text{liters of soln}} \tag{4.3}$$

where soln denotes solution. Thus, a 1.46 molar glucose ($C_6H_{12}O_6$) solution, expressed as 1.46 *M* $C_6H_{12}O_6$, contains 1.46 moles of the solute ($C_6H_{12}O_6$) in 1 liter of the solution; a 0.52 molar urea $[(NH_2)_2CO]$ solution, expressed as 0.52 *M* $(NH_2)_2CO$, contains 0.52 mole of $(NH_2)_2CO$ (the solute) in 1 liter of solution; and so on.

Of course, we do not always need (or desire) to work with solutions of exactly 1 liter (L). Thus, a 500-mL solution containing 0.730 mole of $C_6H_{12}O_6$ still has the same concentration of 1.46 *M*:

$$M = \text{molarity} = \frac{0.730 \text{ mol}}{0.500 \text{ L}}$$
$$= 1.46 \text{ mol/L} = 1.46\ M$$

As you can see, molarity has the units of moles per liter, so 1.46 M is equivalent to 1.46 mol/L. Remember that because concentration, like density, is an intensive property, its value does not depend on how much of the solution is present. The amount of solute in a solution, however, depends on both the concentration and the volume of solution.

It is important to keep in mind that molarity refers only to the amount of solute originally dissolved in water and does not attempt to reflect any subsequent processes such as the dissociation of a salt or the ionization of an acid. Both glucose and urea are nonelectrolytes, so a 1.00 M urea solution means there is 1.00 mole of the urea molecules in 1 liter of solution. The situation is different for electrolytes. Consider what happens when a sample of potassium chloride (KCl) is dissolved in enough water to make up a 1 M solution:

$$\text{KCl}(s) \xrightarrow{\text{H}_2\text{O}} \text{K}^+(aq) + \text{Cl}^-(aq)$$

Because KCl is a strong electrolyte, it undergoes complete dissociation in solution. Thus, in a 1 M KCl solution there are 1 mole of K^+ ions and 1 mole of Cl^- ions, and no KCl units are present. The concentrations of the ions can be expressed as $[K^+] = 1\ M$ and $[Cl^-] = 1\ M$, where the square brackets [ ] indicate that the concentration is expressed in molarity. Similarly, in a 1 M barium nitrate [$Ba(NO_3)_2$] solution

$$\text{Ba(NO}_3)_2(s) \xrightarrow{\text{H}_2\text{O}} \text{Ba}^{2+}(aq) + 2\text{NO}_3^-(aq)$$

we have $[Ba^{2+}] = 1\ M$ and $[NO_3^-] = 2\ M$ and no $Ba(NO_3)_2$ units at all.

In general, it is easier to measure the volume of a liquid than to determine its mass.

The procedure for preparing a solution of known molarity is as follows. A compound (the solute) is first accurately weighed and transferred to a volumetric flask through a funnel (Figure 4.4). Next, water is added to the flask, which is carefully

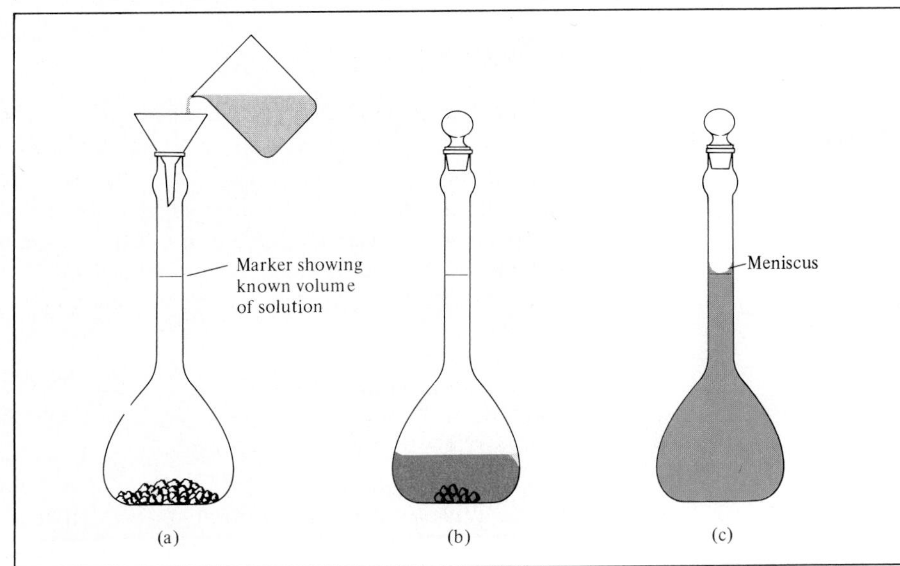

Figure 4.4 *Procedure for preparing a solution of known molarity. (a) A known amount of the substance (the solute) is put into the volumetric flask; then water is added through a funnel. (b) The solid is slowly dissolved by gently shaking the flask. (c) After the solid has completely dissolved, more water is added to bring the level of solution to the mark. Knowing the volume of the solution and the amount of solute dissolved in it, we can calculate the molarity of the prepared solution.*

swirled to dissolve the solid. After *all* the solid has dissolved, more water is added to bring the level of solution to the mark. Knowing the volume of solution (which is the volume of the flask) and the quantity of compound (the number of moles) dissolved, we can calculate the molarity of the solution using Equation (4.3). Note that this procedure does not require us to know the amount of water added, as long as the volume of the final solution is known.

The following two examples are involved with concentration of solutions.

EXAMPLE 4.7

How many grams of sodium sulfate (Na_2SO_4) are required to prepare a 250-mL solution whose concentration is 0.683 *M*?

Answer

The first step is to determine the number of moles of Na_2SO_4 in 250 mL of solution:

$$\text{moles of } Na_2SO_4 = 250 \cancel{\text{mL soln}} \times \frac{0.683 \text{ mol } Na_2SO_4}{1000 \cancel{\text{mL soln}}}$$
$$= 0.171 \text{ mol } Na_2SO_4$$

The molar mass of Na_2SO_4 is 142.1 g, so we write

$$\text{grams of } Na_2SO_4 \text{ needed} = 0.171 \cancel{\text{mol } Na_2SO_4} \times \frac{142.1 \text{ g } Na_2SO_4}{1 \cancel{\text{mol } Na_2SO_4}}$$
$$= 24.3 \text{ g } Na_2SO_4$$

Similar problem: 4.42.

EXAMPLE 4.8

How many grams of ethanol (C_2H_5OH) are present in 85.0 mL of a 0.451 *M* ethanol solution?

Answer

First we find out the number of moles of ethanol present in 85.0 mL of the solution. Next, we calculate the mass of the solute from its molar mass. Combining these two steps, we write

$$\text{mass of ethanol} = 85.0 \cancel{\text{mL soln}} \times \frac{0.451 \cancel{\text{mol } C_2H_5OH}}{1000 \cancel{\text{mL soln}}} \times \frac{46.07 \text{ g } C_2H_5OH}{1 \cancel{\text{mol } C_2H_5OH}}$$
$$= 1.77 \text{ g}$$

Similar problem: 4.37.

Dilution of Solutions

We frequently find it convenient *to prepare a less concentrated solution from a more concentrated solution*. The procedure for this preparation is called ***dilution.*** Suppose that we want to prepare 1 liter of a 0.400 *M* $KMnO_4$ solution from a solution of 1.00 *M* $KMnO_4$. This requires the use of 0.400 mole of $KMnO_4$ from the 1.00 *M* $KMnO_4$ solution. Since there is 1.00 mole of $KMnO_4$ in 1 liter, or 1000 mL, of a 1.00 *M* $KMnO_4$ solution, there is 0.400 mole of $KMnO_4$ in 0.400 × 1000 mL, or 400 mL, of the same solution:

$$\frac{1.00 \text{ mol}}{1000 \text{ mL soln}} = \frac{0.400 \text{ mol}}{400 \text{ mL soln}}$$

Therefore, we must withdraw 400 mL from the 1.00 *M* $KMnO_4$ solution and dilute it to 1000 mL by adding water (in a 1-liter volumetric flask). This method gives us 1 liter of the desired solution of 0.400 *M* $KMnO_4$. Such concentrated solutions (that is, the 1 *M* $KMnO_4$ solution) are often stored and used repeatedly over a prolonged period of time. They are stored as the "stock" of a chemical stockroom. For this reason, they are called *stock* solutions.

In carrying out a dilution process, it is useful to remember that adding more solvent to a given amount of the stock solution changes (decreases) the concentration of the solution without changing the number of moles of solute present in the solution (Figure 4.5); that is,

$$\text{moles of solute before dilution} = \text{moles of solute after dilution}$$

Since molarity is defined as moles of solute per liter of solution, we see that the number of moles of solute is given by

$$\underbrace{\frac{\text{moles of solute}}{\text{liters of soln}}}_{M} \times \underbrace{\text{volume of soln (in liters)}}_{V} = \text{moles of solute}$$

or

$$MV = \text{moles of solute}$$

Because all the solute comes from the original stock solution, we can conclude that

$$\underset{\text{before dilution}}{\underset{\text{moles of solute}}{M_{\text{initial}}V_{\text{initial}}}} = \underset{\text{after dilution}}{\underset{\text{moles of solute}}{M_{\text{final}}V_{\text{final}}}} \tag{4.4}$$

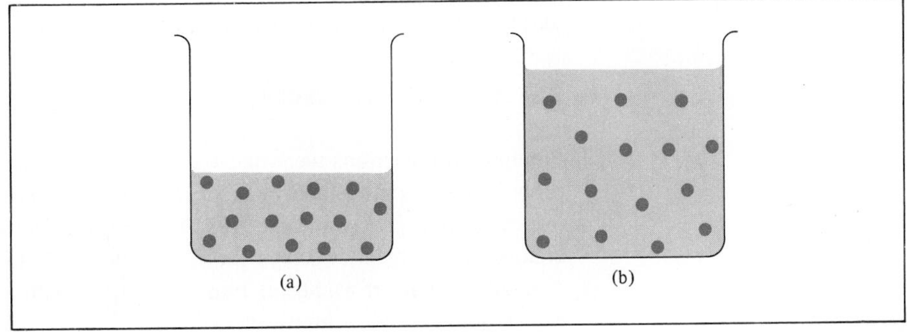

Figure 4.5 *The dilution of a more concentrated solution to a less concentrated one does not change the total number of moles of solute.*

where M_{initial} and M_{final} are the initial and final concentrations of the solution in molarity and V_{initial} and V_{final} are the initial and final volumes of the solution, respectively. Of course, the units of V_{initial} and V_{final} must be the same (mL or L) for the calculation to work. To check the reasonableness of your results, you must always have $M_{\text{initial}} > M_{\text{final}}$ and $V_{\text{final}} > V_{\text{initial}}$.

The following example illustrates the use of Equation (4.4).

EXAMPLE 4.9

Describe how you would prepare 5.00×10^2 mL of 1.75 *M* HNO_3 solution, starting with an 8.61 *M* stock solution of HNO_3.

Answer

Since the concentration of the final solution is less than that of the original one, this is a dilution process. We prepare for the calculation by tabulating our data:

$$M_{\text{initial}} = 8.61\ M \qquad M_{\text{final}} = 1.75\ M$$
$$V_{\text{initial}} = ? \qquad V_{\text{final}} = 5.00 \times 10^2\ \text{mL}$$

Substituting in Equation (4.4)

$$(8.61\ M)(V_{\text{initial}}) = (1.75\ M)(5.00 \times 10^2\ \text{mL})$$

$$V_{\text{initial}} = \frac{(1.75\ M)(5.00 \times 10^2\ \text{mL})}{8.61\ M}$$
$$= 102\ \text{mL}$$

Thus we must dilute 102 mL of the 8.61 *M* HNO_3 solution with sufficient water to give a final volume of 5.00×10^2 mL in a 500-mL volumetric flask to obtain the desired concentration.

Similar problems: 4.47, 4.49.

Now that we have discussed the concentration and dilution of solutions, we can begin to study the quantitative aspects of reactions in aqueous solution. The next three sections will focus on the three types of reactions discussed in Chapter 3, namely, precipitation reactions, acid-base reactions, and redox reactions. These reactions are examples of ***quantitative analysis,*** which is concerned with *the determination of the amount or concentration of a substance in a sample*.

4.5 Gravimetric Analysis

Precipitation reactions were discussed in Section 3.3. They form the basis of a useful procedure in chemical analysis called gravimetric analysis. ***Gravimetric analysis*** is *an analytical procedure that involves the measurement of mass*. One type of gravimetric analysis experiment involves the formation, isolation, and mass determination of a precipitate. This procedure is most often used with ionic compounds. A sample substance of unknown composition is dissolved in water and then converted into a precipitate by allowing it to react with another substance. The precipitate formed is filtered off, dried, and weighed. Knowing the mass and chemical formula of the precipitate formed, we can calculate the mass of a particular chemical component (that is, the anion or cation) of the original sample. From the mass of the component and the mass of the original sample, we can determine the percent composition by mass of the component in the original compound.

A reaction that is often studied in gravimetric analysis is

$$AgNO_3(aq) + NaCl(aq) \longrightarrow NaNO_3(aq) + AgCl(s)$$

or, expressed as a net ionic equation,

$$Ag^+(aq) + Cl^-(aq) \longrightarrow AgCl(s)$$

The precipitate formed is silver chloride (see solubility rule 5 in Section 3.3). As an example, let us say that we wanted to determine *by means of an experiment* the percent by mass of Cl in NaCl. We would first accurately weigh out a sample of NaCl and dissolve it in water. Next, we would add enough $AgNO_3$ solution to the NaCl solution to cause precipitation of all the Cl^- ions present in solution as AgCl. In this procedure NaCl is the limiting reagent and $AgNO_3$ the excess reagent. Finally, the AgCl precipitate is separated from the solution by filtration, dried, and weighed. From the measured mass of AgCl, we can calculate the mass of Cl (using the percent by mass of Cl in AgCl). Since this same amount of Cl was present in the original NaCl sample, we can calculate the percent by mass of Cl in NaCl. Figure 4.6 shows the basic steps in a gravimetric analysis experiment involving the precipitation of AgCl.

Two points regarding gravimetric analysis should be mentioned. First, this is a highly accurate technique, since the mass of samples can be measured accurately.

(a)

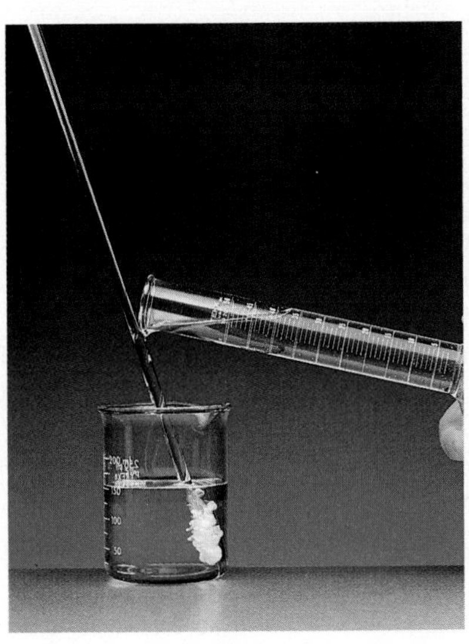

(b)

(c)

Figure 4.6 *Basic steps for gravimetric analysis. (a) A solution containing a known amount of NaCl in a beaker. (b) The precipitation of AgCl upon the addition of $AgNO_3$ solution from a measuring cylinder. In this reaction, $AgNO_3$ is the excess reagent and NaCl the limiting reagent. (c) The solution containing the AgCl precipitate is filtered through a preweighed sintered-disc crucible, which allows the liquid (but not the precipitate) to pass through. The crucible is then removed from the apparatus, dried in an oven, and weighed again. The difference between this mass and that of the empty crucible gives the mass of the AgCl precipitate.*

Second, this procedure is applicable only for reactions that go to completion, or have nearly 100 percent yield. Referring to the above example, if AgCl were slightly soluble instead of being insoluble it would not be possible to remove all of the Cl^- ions from the NaCl solution and the subsequent calculation will be in error.

The following is an example of gravimetric analysis.

EXAMPLE 4.10

A sample of 0.5662 g of an ionic compound containing chloride ions but unknown metal identity is dissolved in water and treated with an excess of $AgNO_3$. If the mass of the AgCl precipitate that forms is 1.0882 g, what is the percent by mass of Cl in the original compound?

Answer

The mass of Cl in the AgCl precipitate is given by

$$\text{mass of Cl} = 1.0882\ \cancel{\text{g AgCl}} \times \frac{1\ \cancel{\text{mol AgCl}}}{143.4\ \cancel{\text{g AgCl}}} \times \frac{1\ \cancel{\text{mol Cl}}}{1\ \cancel{\text{mol AgCl}}} \times \frac{35.45\text{ g Cl}}{1\ \cancel{\text{mol Cl}}}$$
$$= 0.2690\text{ g Cl}$$

We calculate the percent by mass of Cl in the unknown sample as

$$\%\text{Cl by mass} = \frac{\text{mass of Cl}}{\text{mass of sample}} \times 100\%$$
$$= \frac{0.2690\text{ g}}{0.5662\text{ g}} \times 100\%$$
$$= 47.51\%$$

As an exercise, you should compare this value with that of Cl in KCl.

Similar problem: 4.56.

Note that gravimetric analysis does not establish the whole identity of the unknown. Thus in the above example we still do not know what the cation is. However, knowing the percent by mass of Cl greatly helps us in narrowing the possibilities. Because no two compounds containing the same anion (or cation) have the same percent composition by mass, comparison of the percent by mass obtained from gravimetric analysis with that calculated from a series of known compounds would reveal the identity of the unknown.

In cases where the identity of the original compound is known, the gravimetric analysis technique allows us to determine the concentration of the solution containing the compound, as the next example shows.

EXAMPLE 4.11

The concentration of lead ions (Pb^{2+}) in a sample of polluted water that also contains nitrate ions (NO_3^-) is determined by adding solid sodium sulfate (Na_2SO_4) to exactly

Formation of $PbSO_4$ precipitate.

500 mL of the water. (a) Write the molecular and net ionic equations for the reaction. (b) Calculate the molar concentration of Pb^{2+} if 0.450 g of Na_2SO_4 was needed for the complete precipitation of Pb^{2+} ions as $PbSO_4$.

Answer

(a) The molecular equation for the reaction is

$$Pb(NO_3)_2(aq) + Na_2SO_4(aq) \longrightarrow 2NaNO_3(aq) + PbSO_4(s)$$

The net ionic equation is

$$Pb^{2+}(aq) + SO_4^{2-}(aq) \longrightarrow PbSO_4(s)$$

(b) From the molecular equation we see that 1 mol $Na_2SO_4 \simeq$ 1 mol $Pb(NO_3)_2$, or 1 mol $Na_2SO_4 \simeq$ 1 mol Pb^{2+}. The number of moles of Na_2SO_4 (molar mass = 142.1 g) in 0.450 g of the compound is

$$0.450 \text{ g } \cancel{Na_2SO_4} \times \frac{1 \text{ mol } Na_2SO_4}{142.1 \text{ g } \cancel{Na_2SO_4}} = 3.17 \times 10^{-3} \text{ mol } Na_2SO_4$$

This quantity of Na_2SO_4 is needed to precipitate Pb^{2+} ions in a 500-mL solution. Thus, the concentration of Pb^{2+} ions in 1 liter of the solution is given by

$$[Pb^{2+}] = \frac{3.17 \times 10^{-3} \text{ mol}}{500 \text{ mL}} \times \frac{1000 \text{ mL}}{1 \text{ L}}$$
$$= 6.34 \times 10^{-3} \text{ mol/L}$$
$$= 6.34 \times 10^{-3}\ M$$

Similar problem: 4.58.

4.6 Acid-Base Titrations

Quantitative studies of acid-base neutralization reactions (see Section 3.4) are most conveniently carried out using a procedure known as ***titration.*** In a titration experiment, *a solution of accurately known concentration,* called a ***standard solution,*** *is added gradually to another solution of unknown concentration, until the chemical reaction between the two solutions is complete*. If we know the volumes of the standard and unknown solutions used in the titration, and the concentration of the standard solution, we can calculate the concentration of the unknown solution.

Sodium hydroxide is one of the common bases used in the laboratory. However, because it is difficult to obtain solid sodium hydroxide in a pure form, a solution of sodium hydroxide must be *standardized* before it can be used in accurate analytical work. We can standardize the sodium hydroxide solution by titrating it against an acid solution of accurately known concentration. The acid often chosen for this task is a monoprotic acid called potassium hydrogen phthalate (KHP), whose molecular formula is $KHC_8H_4O_4$. KHP is a white, soluble solid that is commerically available in highly pure form. The reaction between KHP and sodium hydroxide is

$$KHC_8H_4O_4(aq) + NaOH(aq) \longrightarrow KNaC_8H_4O_4(aq) + H_2O(l)$$

or, expressed as a net ionic equation,

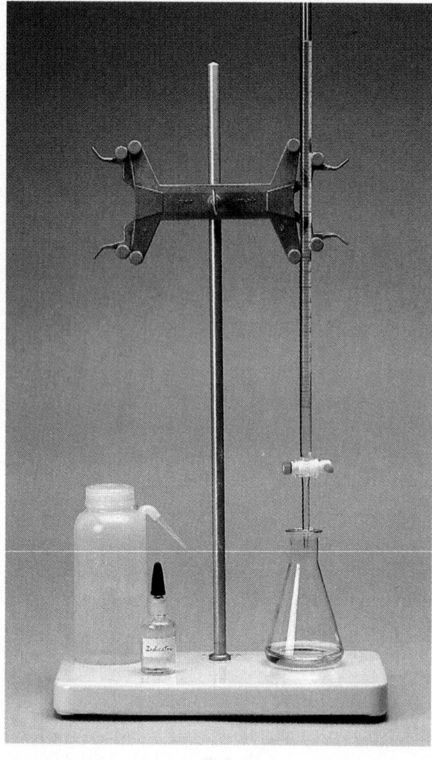

(a)

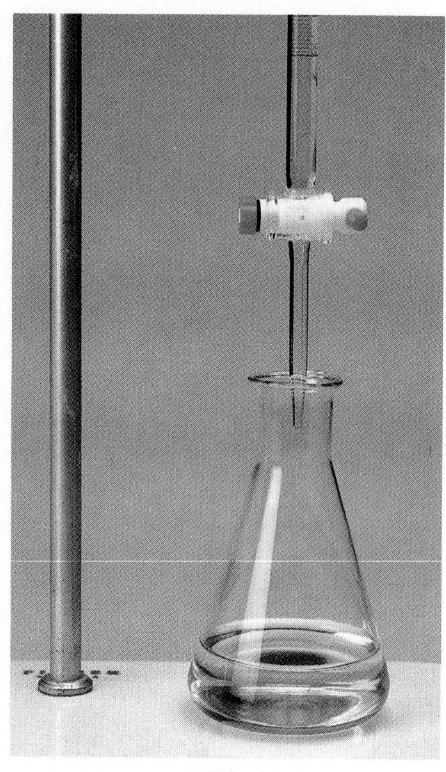

(b)

Figure 4.7 *(a) Apparatus for acid-base titration. A NaOH solution in the buret is added to a KHP solution in an Erlenmeyer flask. (b) A reddish-pink color is observed when the equivalence point is reached. For an accurate titration, the pink color should be very faint. The color here has been intensified for visual display.*

$$HC_8H_4O_4^-(aq) + OH^-(aq) \longrightarrow C_8H_4O_4^{2-}(aq) + H_2O(l)$$

The procedure for the titration is shown in Figure 4.7. First, a weighed amount of KHP is transferred to an Erlenmeyer flask and some distilled water is added to make up a solution. Next, a buret is filled with the NaOH solution. The NaOH solution is carefully added to the KHP solution until we reach the ***equivalence point,*** that is, *the point at which the acid is completely reacted with or neutralized by the base*. The point is usually signaled by a sharp color change of an indicator that has been added to the acid solution. In acid-base titrations, ***indicators*** are usually *substances that have distinctly different colors in acidic and basic media*. One commonly used indicator is phenolphthalein, which is colorless in acidic and neutral solutions but reddish pink in basic solutions. At the equivalence point, all the KHP present has been neutralized by the added NaOH and the solution is still colorless. However, if we add just one more drop of NaOH solution from the buret, the solution will immediately turn reddish pink because the solution is now basic.

EXAMPLE 4.12

In a titration experiment, a student finds that 0.5468 g of KHP is needed to completely neutralize 23.48 mL of a NaOH solution. What is the concentration (in molarity) of the NaOH solution?

Answer

The balanced equation for this reaction is shown above. First we calculate the number of moles of KHP used in the titration:

$$\begin{aligned}\text{moles of KHP} &= 0.5468\ \cancel{\text{g KHP}} \times \frac{1\ \text{mol KHP}}{204.2\ \cancel{\text{g KHP}}}\\ &= 2.678 \times 10^{-3}\ \text{mol KHP}\end{aligned}$$

Since 1 mol KHP ⇔ 1 mol NaOH, there must be 2.678×10^{-3} mole of NaOH in 23.48 mL of NaOH solution. Finally, we calculate the molarity of the NaOH as follows:

$$\begin{aligned}\text{molarity of NaOH soln} &= \frac{2.678 \times 10^{-3}\ \text{mol NaOH}}{23.48\ \cancel{\text{mL soln}}} \times \frac{1000\ \cancel{\text{mL soln}}}{1\ \text{L soln}}\\ &= 0.1141\ M\end{aligned}$$

Similar problem: 4.63.

The acid-base neutralization reaction between NaOH and KHP is one of the simplest types of neutralizations known. Suppose, though, that instead of KHP, a diprotic acid such as H_2SO_4 is used. The reaction is represented by

$$2\text{NaOH}(aq) + \text{H}_2\text{SO}_4(aq) \longrightarrow \text{Na}_2\text{SO}_4(aq) + 2\text{H}_2\text{O}(l)$$

Since 2 mol NaOH ⇔ 1 mol H_2SO_4, we need twice the amount of NaOH to react completely with a H_2SO_4 solution of the *same* concentration as a KHP solution. On the other hand, we would need twice the amount of HCl to completely neutralize a $Ba(OH)_2$ solution having the same concentration as a solution of NaOH:

One mole of $Ba(OH)_2$ yields two moles of OH^- ions.

$$2\text{HCl}(aq) + \text{Ba(OH)}_2(aq) \longrightarrow \text{BaCl}_2(aq) + 2\text{H}_2\text{O}(l)$$

In calculations involving acid-base titrations, regardless of the acid or base involved, keep in mind that the total number of moles of H^+ ions that have reacted at the equivalence point must be equal to the total number of moles of OH^- ions that have reacted. The number of moles of a base in a certain volume is given by

$$\begin{aligned}\text{moles of base} &= \text{molarity (mol/L)} \times \text{volume (L)}\\ &= MV\end{aligned}$$

where M is the molarity and V the volume in liters. A similar expression can be written for an acid.

The following example shows the titration of a NaOH solution with a diprotic acid.

EXAMPLE 4.13

How many milliliters (mL) of a 0.610 M NaOH solution are needed to completely neutralize 20.0 mL of a 0.245 M H_2SO_4 solution?

Answer

The equation for the neutralization reaction is

$$2NaOH(aq) + H_2SO_4(aq) \longrightarrow Na_2SO_4(aq) + H_2O(l)$$

First we calculate the number of moles of H_2SO_4 in 20.0 mL solution:

$$\text{moles } H_2SO_4 = \frac{0.245 \text{ mol } H_2SO_4}{1 \not{\text{L soln}}} \times \frac{1 \not{\text{L soln}}}{1000 \not{\text{mL soln}}} \times 20.0 \not{\text{mL soln}}$$
$$= 4.90 \times 10^{-3} \text{ mol } H_2SO_4$$

From the stoichiometry we see that 1 mol $H_2SO_4 \simeq$ 2 mol NaOH. Therefore, the number of moles of NaOH reacted must be $2 \times 4.90 \times 10^{-3}$ mole, or 9.80×10^{-3} mole. From the definition of molarity [see Equation (4.3)] we have

$$\text{molarity} = \frac{\text{mole of solute}}{\text{liters of soln}}$$

or

$$\text{volume of NaOH (L)} = \frac{\text{mol NaOH}}{\text{molarity of NaOH}}$$

Thus the volume of 0.610 *M* NaOH solution needed is

$$\text{volume of NaOH} = \frac{9.80 \times 10^{-3} \text{ mol NaOH}}{0.610 \text{ mol/L soln}}$$
$$= 0.0161 \text{ L, or } 16.1 \text{ mL}$$

Similar problems: 4.62, 4.64.

4.7 Redox Titrations

In some respects, redox reactions (see Sections 3.5 and 3.6) are similar to acid-base reactions. As mentioned earlier, redox reactions involve the transfer of electrons, and acid-base reactions involve the transfer of protons. Just as an acid can be titrated against a base, we can titrate an oxidizing agent against a reducing agent, using a procedure similar to that described in Section 4.6. We can, for example, carefully add a solution containing an oxidizing agent to a solution containing a reducing agent. The *equivalence point* is reached when the reducing agent is completely oxidized by the oxidizing agent.

Redox indicators are not as common as acid-base indicators.

Like acid-base titrations, redox titrations normally require an indicator, in this case one that has distinctly different colors in its oxidized and reduced forms. In the presence of large amounts of reducing agent, the color of the indicator is characteristic of its reduced form. The indicator assumes the color of its oxidized form when it is present in an oxidizing medium. At or near the equivalence point, a sharp change in the indicator's color will occur as it changes from one form to the other, so the equivalence point can be readily identified.

Two common oxidizing agents you may encounter in the laboratory are potassium dichromate ($K_2Cr_2O_7$) and potassium permanganate ($KMnO_4$). As Figure 4.8 shows, the colors of the dichromate and permanganate anions are distinctly different from those of the reduced species:

Figure 4.8 *Left to right: Colors of solutions containing the MnO_4^-, Mn^{2+}, $Cr_2O_7^{2-}$, and Cr^{3+} ions.*

$Cr_2O_7^{2-}$	$\longrightarrow$	Cr^{3+}
orange yellow		green
MnO_4^-	$\longrightarrow$	Mn^{2+}
purple		light pink

Thus these oxidizing agents can themselves be used as an *internal* indicator in a redox titration.

Redox titrations require the same type of calculations (based on the mole method) as acid-base neutralizations. The difference is that the equations and the stoichiometry tend to be more complex for redox reactions. The following is an example of a redox titration.

Addition of a $KMnO_4$ solution from a buret to a $FeSO_4$ solution.

EXAMPLE 4.14

A 16.42-mL volume of 0.1327 *M* $KMnO_4$ solution is needed to oxidize 20.00 mL of a $FeSO_4$ solution in an acidic medium. What is the concentration of the $FeSO_4$ solution? The net ionic equation is

$$5Fe^{2+} + MnO_4^- + 8H^+ \longrightarrow Mn^{2+} + 5Fe^{3+} + 4H_2O$$

Answer

The number of moles of $KMnO_4$ in 16.42 mL of the solution is

$$\text{moles of KMnO}_4 = 16.42\ \cancel{\text{mL}} \times \frac{0.1327\ \text{mol KMnO}_4}{1\ \cancel{\text{L soln}}} \times \frac{1\ \cancel{\text{L soln}}}{1000\ \cancel{\text{mL soln}}}$$
$$= 2.179 \times 10^{-3}\ \text{mol KMnO}_4$$

and the number of moles of $FeSO_4$ oxidized is

$$\text{moles FeSO}_4 = 2.179 \times 10^{-3}\ \cancel{\text{mol KMnO}_4} \times \frac{5\ \text{mol FeSO}_4}{1\ \cancel{\text{mol KMnO}_4}}$$
$$= 1.090 \times 10^{-2}\ \text{mol FeSO}_4$$

The concentration of the $FeSO_4$ solution is equal to moles of $FeSO_4$ per liter of solution:

$$[\text{FeSO}_4] = \frac{1.090 \times 10^{-2} \text{ mol FeSO}_4}{20.00 \text{ mL soln}} \times \frac{1000 \text{ mL soln}}{1 \text{ L soln}} = 0.5450\ M$$

Similar problems: 4.66, 4.69.

A redox reaction that is useful in chemical analysis is one involving molecular iodine (I_2) and thiosulfate ion ($S_2O_3^{2-}$). The reaction that occurs between them is

$$I_2(s) + 2S_2O_3^{2-}(aq) \longrightarrow 2I^-(aq) + S_4O_6^{2-}(aq)$$

CHEMISTRY IN ACTION

Breath Analyzer

Every year in the United States about 25,000 people are killed and 500,000 more injured as a result of drunk driving. Various organizations have stepped up efforts to educate the public about the dangers of intoxication on America's roads, and stiffer penalties have been imposed for such offenders.

The police often use a device called a breath analyzer to test drivers suspected of being drunk. The chemical basis of this device is a redox reaction. A sample of the driver's breath is drawn into the breath analyzer (Figure 4.9), where it is treated with an acidic solution of potassium dichromate. The alcohol (ethanol) in the breath is converted to acetic acid as shown in the following equation:

$$\underset{\text{ethanol}}{3CH_3CH_2OH} + \underset{\substack{\text{potassium}\\\text{dichromate}\\\text{(orange yellow)}}}{2K_2Cr_2O_7} + \underset{\substack{\text{sulfuric}\\\text{acid}}}{8H_2SO_4} \longrightarrow$$

$$\underset{\substack{\text{acetic}\\\text{acid}}}{3CH_3COOH} + \underset{\substack{\text{chromic}\\\text{sulfate}\\\text{(green)}}}{2Cr_2(SO_4)_3} + \underset{\substack{\text{potassium}\\\text{sulfate}}}{2K_2SO_4} + 11H_2O$$

In this reaction the ethanol is oxidized to acetic acid and the chromium in the orange yellow dichromate ion is reduced to the green chromic ion (see Figure 4.8). The driver's blood alcohol level can be determined readily by measuring the degree of this color change (read from a calibrated meter on the instrument). The current legal limit of blood alcohol content in the United States is 0.1 percent by mass. Any value above this limit constitutes intoxication.

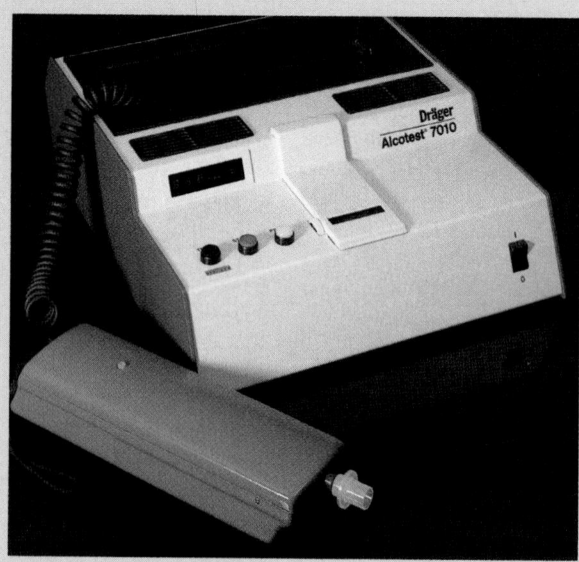

Figure 4.9 *A breath analyzer of the kind used in police stations to test drunk drivers.*

In this reaction, molecular iodine oxidizes the thiosulfate ion to the tetrathionate ion ($S_4O_6^{2-}$) while it itself is reduced to the iodide ion. In practice it is difficult to carry out the reaction as written because molecular iodine is only slightly soluble in water. However, molecular iodine dissolves readily in a solution of iodide ion to form the very soluble triiodide ion (I_3^-):

$$I_2(s) + I^-(aq) \rightleftharpoons I_3^-(aq)$$

An equilibrium is established between I_2 and I_3^- ions. If iodine molecules are removed from solution by a reaction, I_3^- ions dissociate to form more iodine molecules. Thus molecular iodine in a potassium iodide solution can be titrated as though it were a solution of molecular iodine in water. The usefulness of the iodine-thiosulfate couple lies in the fact that the iodide ion can be readily oxidized by a variety of oxidizing agents such as $Cr_2O_7^{2-}$, ClO^-, Br_2, and Cu^{2+}. The half-reaction for the oxidation is

$$2I^-(aq) \longrightarrow I_2(s) + 2e^-$$

By first reacting iodide ions with an oxidizing agent and then titrating the molecular iodine generated with a standard thiosulfate solution, we can determine the concentration of the oxidizing agent. The following example shows the application of this procedure to determine the concentration of ozone in polluted air.

EXAMPLE 4.15

A sample of polluted air containing ozone (O_3) is treated with a solution containing iodide (I^-) ions in acidic medium. The reaction is

$$O_3(g) + 2I^-(aq) + 2H^+(aq) \longrightarrow O_2(g) + I_2(s) + H_2O(l)$$

The iodine formed can be titrated against a standard sodium thiosulfate (NaS_2O_3) solution according to the net ionic equation

$$2S_2O_3^{2-}(aq) + I_2(s) \longrightarrow S_4O_6^{2-}(aq) + 2I^-(aq)$$

If 17.84 mL of a 2.000×10^{-3} *M* $Na_2S_2O_3$ solution are needed to reduce the I_2 generated from a 27.84-g sample of air, calculate the percent by mass of O_3 in the sample.

Answer

The number of moles of $S_2O_3^{2-}$ used in the titration is

$$\text{moles of } S_2O_3^{2-} = 17.84 \, \cancel{\text{mL}} \times \frac{2.000 \times 10^{-3} \text{ mol } S_2O_3^{2-}}{1 \, \cancel{\text{L soln}}} \times \frac{1 \, \cancel{\text{L soln}}}{1000 \, \cancel{\text{mL soln}}}$$
$$= 3.568 \times 10^{-5} \text{ mol } S_2O_3^{2-}$$

The first equation shows that 1 mol $O_3 \simeq$ 1 mol I_2, and the second equation gives 2 mol $S_2O_3^{2-} \simeq$ 1 mol I_2. Combining these two equations, we calculate the mass of O_3 as follows:

$$\text{mass of } O_3 = 3.568 \times 10^{-5} \, \cancel{\text{mol } S_2O_3^{2-}} \times \frac{1 \, \cancel{\text{mol } I_2}}{2 \, \cancel{\text{mol } S_2O_3^{2-}}} \times \frac{1 \, \cancel{\text{mol } O_3}}{1 \, \cancel{\text{mol } I_2}} \times \frac{48.00 \text{ g } O_3}{1 \, \cancel{\text{mol } O_3}}$$
$$= 8.563 \times 10^{-4} \text{ g } O_3$$

Finally, we have

$$\% \text{ by mass of } O_3 = \frac{8.563 \times 10^{-4}\text{ g}}{27.84\text{ g}} \times 100\% = 3.076 \times 10^{-3}\%$$

Similar problems: 4.74, 4.75.

Color of iodine-starch compound.

You may wonder why the above analysis was carried out in an indirect way, that is, why ozone was not directly reacted with the thiosulfate solution. The reason is that there is no suitable indicator for such a titration. On the other hand, starch, which forms a dark blue compound with molecular iodine, can be used as an indicator in the titration of thiosulfate with molecular iodine. At the equivalence point, when all the molecular iodine is used up, the solution changes abruptly from blue to colorless.

Summary

1. Stoichiometric calculations involve the amounts of products and reactants in chemical reactions. The calculations are best done by expressing both the known and unknown quantities in terms of moles and then converting to other units if necessary.
2. A limiting reagent is the reactant that is present in the smallest stoichiometric amount. It limits the amount of product that can be formed.
3. The amount of product obtained in a reaction (the actual yield) may be less than the maximum possible amount (the theoretical yield). The ratio of the two is expressed as the percent yield.
4. The concentration of a solution is the amount of solute present in a given amount of solution. Molarity expresses concentration as the number of moles of solute in 1 liter of solution.
5. Adding a solvent to a solution, a process known as dilution, decreases the concentration (molarity) of the solution without changing the total number of moles of solute present in the solution.
6. Gravimetric analysis is a technique for determining the identity of a compound and/or the concentration of a solution by measuring mass. Gravimetric experiments often involve precipitation reactions.
7. In acid-base titration, a solution of known concentration (say, a base) is added gradually to a solution of unknown concentration (say, an acid) with the goal of determining the unknown concentration. The point at which the reaction in the titration is complete is called the equivalence point.
8. Redox titrations can be carried out like acid-base titrations. The point at which the oxidation-reduction reaction is complete is called the equivalence point.

Applied Chemistry Example: Nitric Acid

Nitric acid (HNO_3) is one of the most important inorganic industrial chemicals. It ranked 12th among the top 50 industrial chemicals produced in the United States in 1988 (15.8 billion pounds). The principal product of nitric acid is ammonium nitrate (NH_4NO_3), which is used in fertilizer. Nitric acid is also used in the manufacture of explosives and preparation of many compounds.

Nitric acid is prepared by the Ostwald process. The first step involves the oxidation of ammonia by air in the presence of a platinum-rhodium catalyst (the substance which promotes the speed of a reaction without itself being used up) at about 800°C:

$$4NH_3(g) + 5O_2(g) \longrightarrow 4NO(g) + 6H_2O(g)$$

The nitric oxide (NO) produced is further oxidized to nitrogen dioxide (NO_2) with air

$$2NO(g) + O_2(g) \longrightarrow 2NO_2(g)$$

Finally, reacting NO_2 with water gives nitric acid

$$2NO_2(g) + H_2O(l) \longrightarrow HNO_3(aq) + HNO_2(aq)$$

On heating, nitrous acid (HNO_2) is converted to nitric acid

$$3HNO_2(aq) \longrightarrow HNO_3(aq) + H_2O(l) + 2NO(g)$$

The NO generated can be recycled to produce NO_2 in the second step.

Pure nitric acid is a colorless liquid with a choking odor. It melts at −41.6°C and boils at 82.6°C. The concentration of ''concentrated'' nitric acid is 15.3 *M*.

1. Describe how you would prepare a 5.00-L solution of 6.00 *M* HNO_3 from concentrated nitric acid.

   Answer: This is a dilution problem. We write

$$M_{\text{initial}} = 15.3\ M \qquad M_{\text{final}} = 6.00\ M$$

$$V_{\text{initial}} = ?\ L \qquad V_{\text{final}} = 5.00\ L$$

   Substituting in Equation (4.4)

$$(15.3\ M)(V_{\text{initial}}) = (6.00\ M)(5.00\ L)$$

$$V_{\text{initial}} = \frac{(6.00\ M)(5.00\ L)}{15.3\ M} = 1.96\ L$$

   Thus the preparation is carried out by diluting 1.96 L of the concentrated acid to 5.00 L with water.

2. In one process 5.00×10^2 kg of ammonia react with 5.60×10^2 kg of HNO_3 to give 6.98×10^2 kg of the fertilizer NH_4NO_3. Calculate the percent yield of NH_4NO_3.

   Answer: Our first step is to determine which of the reactants is the limiting reagent. The reaction is

$$NH_3(g) + HNO_3(aq) \longrightarrow NH_4NO_3(aq)$$

   The stoichiometry shows that 1 mol $NH_3 \Leftrightarrow$ 1 mol HNO_3. To determine the number of moles of reactants present, we write

$$\text{moles of } NH_3 = 5.00 \times 10^5\ g\ NH_3 \times \frac{1\ mol\ NH_3}{17.03\ g\ NH_3} = 2.94 \times 10^4\ mol\ NH_3$$

$$\text{moles } HNO_3 = 5.60 \times 10^5\ g\ HNO_3 \times \frac{1\ mol\ HNO_3}{63.02\ g\ HNO_3} = 8.89 \times 10^3\ mol\ HNO_3$$

Since the number of moles of NH_3 is greater, HNO_3 is the limiting reagent.

Since 1 mol $HNO_3 \approx$ 1 mol NH_4NO_3, the theoretical yield of NH_4NO_3 is

$$\text{theoretical yield of } NH_4NO_3 = 8.89 \times 10^3 \text{ mol } HNO_3 \times \frac{1 \text{ mol } NH_4NO_3}{1 \text{ mol } HNO_3} \times \frac{80.05 \text{ g } NH_4NO_3}{1 \text{ mol } NH_4NO_3}$$
$$= 7.12 \times 10^5 \text{ g } NH_4NO_3$$
$$= 7.12 \times 10^2 \text{ kg } NH_4NO_3$$

Finally, we calculate the percent yield of NH_4NO_3

$$\% \text{ yield} = \frac{\text{actual yield}}{\text{theoretical yield}} \times 100\%$$
$$= \frac{6.98 \times 10^2 \text{ kg}}{7.12 \times 10^2 \text{ kg}} \times 100\%$$
$$= 98.0\%$$

3. The purity of NH_4NO_3 can be analyzed by titrating a solution of NH_4NO_3 with a NaOH solution. (Recall that the NH_4^+ ion is a Brønsted acid.) In one experiment a 0.2041-g sample of industrially prepared NH_4NO_3 required 24.42 mL of 0.1023 M NaOH for neutralization. What is the percent purity of the sample?

Answer: The neutralization reaction is

$$NH_4^+(aq) + OH^-(aq) \longrightarrow NH_3(aq) + H_2O(l)$$

Number of moles of NaOH consumed is given by

$$\text{moles of NaOH} = 24.42 \text{ mL soln} \times \frac{1 \text{ L}}{1000 \text{ mL}} \times \frac{0.1023 \text{ mol NaOH}}{1 \text{ L soln}}$$
$$= 2.498 \times 10^{-3} \text{ mol NaOH}$$

Since 1 mol $NH_4NO_3 \approx$ 1 mol NaOH, the mass of NH_4NO_3 consumed is

$$\text{g } NH_4NO_3 = 2.498 \times 10^{-3} \text{ mol } NH_4NO_3 \times \frac{80.05 \text{ g } NH_4NO_3}{1 \text{ mol } NH_4NO_3}$$
$$= 0.2000 \text{ g } NH_4NO_3$$

Finally, the percent purity of NH_4NO_3 is given by

$$\% NH_4NO_3 = \frac{0.2000 \text{ g}}{0.2041 \text{ g}} \times 100\%$$
$$= 97.99\%$$

Key Words

Actual yield, p. 139
Concentration of a solution, p. 142
Dilution, p. 145
Equivalence point, p. 150
Excess reagent, p. 136
Gravimetric analysis, p. 146
Indicator, p. 150
Limiting reagent, p. 136
Molar concentration, p. 142
Molarity, p. 142
Mole method, p. 132
Percent yield (% yield), p. 139
Quantitative analysis, p. 146
Standard solution, p. 149
Stoichiometric amount, p. 136
Stoichiometry, p. 132
Theoretical yield, p. 139
Titration, p. 149
Yield of the reaction, p. 139

Exercises

Amounts of Reactants and Products

REVIEW QUESTIONS

4.1 Define the following terms: stoichiometry, stoichiometric amounts, limiting reagent, excess reagent, yield of a reaction, theoretical yield, actual yield, percent yield.

4.2 On what law is stoichiometry based?

4.3 Describe the basic steps involved in the mole method.

4.4 Why is it essential to use balanced equations in solving stoichiometric problems?

PROBLEMS

4.5 Consider the combustion of carbon monoxide (CO) in oxygen gas:

$$2CO(g) + O_2(g) \longrightarrow 2CO_2(g)$$

Starting with 3.60 moles of CO, calculate the number of moles of CO_2 produced if there is enough oxygen gas to react with all of the CO.

4.6 Silicon tetrachloride ($SiCl_4$) can be prepared by heating Si in chlorine gas:

$$Si(s) + 2Cl_2(g) \longrightarrow SiCl_4(l)$$

In one reaction, 0.507 mole of $SiCl_4$ is produced. How many moles of molecular chlorine were used in the reaction?

4.7 The annual production of sulfur dioxide from burning coal and fossil fuels, auto exhaust, and other sources is about 26 million tons. The equation for the reaction is

$$S(s) + O_2(g) \longrightarrow SO_2(g)$$

How much sulfur, present in the original materials, would result in that quantity of SO_2?

4.8 When baking soda (sodium bicarbonate or sodium hydrogen carbonate, $NaHCO_3$) is heated, it releases carbon dioxide gas, which is responsible for the rising of cookies, donuts, and bread. (a) Write a balanced equation for the decomposition of the compound (one of the products is Na_2CO_3). (b) Calculate the mass of $NaHCO_3$ required to produce 20.5 g of CO_2.

4.9 When potassium cyanide (KCN) reacts with acids, a deadly poisonous gas, hydrogen cyanide (HCN), is given off. Here is the equation:

$$KCN(aq) + HCl(aq) \longrightarrow KCl(aq) + HCN(g)$$

If a sample of 0.140 g of KCN is treated with an excess of HCl, calculate the amount of HCN formed, in grams.

4.10 Fermentation is a complex chemical process of wine making in which glucose is converted into ethanol and carbon dioxide:

$$\underset{\text{glucose}}{C_6H_{12}O_6} \longrightarrow \underset{\text{ethanol}}{2C_2H_5OH} + 2CO_2$$

Starting with 500.4 g of glucose, what is the maximum amount of ethanol in grams and in liters that can be obtained by this process? (Density of ethanol = 0.789 g/mL.)

4.11 Each copper(II) sulfate unit is associated with five water molecules in crystalline copper(II) sulfate pentahydrate ($CuSO_4 \cdot 5H_2O$). When this compound is heated in air above 100°C, it loses the water molecules and also its blue color:

$$CuSO_4 \cdot 5H_2O \longrightarrow CuSO_4 + 5H_2O$$

If 9.60 g of $CuSO_4$ is left after heating 15.01 g of the blue compound, calculate the number of moles of H_2O originally present in the compound.

4.12 For many years the recovery of gold (that is, the separation of gold from other materials) involved the treatment of gold by isolation from other substances, using potassium cyanide:

$$4Au + 8KCN + O_2 + 2H_2O \longrightarrow 4KAu(CN)_2 + 4KOH$$

What is the minimum amount of KCN in moles needed in order to extract 29.0 g (about an ounce) of gold?

4.13 How many grams of potassium are needed to react completely with 19.2 g of molecular bromine (Br_2) to produce KBr?

4.14 How many moles of sulfuric acid would be needed to produce 4.80 moles of molecular iodine (I_2), according to the following balanced equation:

$$10HI + 2KMnO_4 + 3H_2SO_4 \longrightarrow 5I_2 + 2MnSO_4 + K_2SO_4 + 8H_2O$$

4.15 Limestone ($CaCO_3$) is decomposed by heating to quicklime (CaO) and carbon dioxide. Calculate how many grams of quicklime can be produced from 1.0 kg of limestone.

4.16 Nitrous oxide (N_2O) is also called "laughing gas." It can be prepared by the thermal decomposition of ammonium nitrate (NH_4NO_3). The other product is H_2O. (a) Write a balanced equation for this reaction. (b) How many grams of N_2O are formed if 0.46 mole of NH_4NO_3 is used in the reaction?

4.17 The fertilizer ammonium sulfate [$(NH_4)_2SO_4$] is prepared by the reaction between ammonia (NH_3) and sulfuric acid:

$$2NH_3(g) + H_2SO_4(aq) \longrightarrow (NH_4)_2SO_4(aq)$$

How many kilograms of NH_3 are needed to produce 1.00×10^5 kg of $(NH_4)_2SO_4$?

Limiting Reagents

REVIEW QUESTIONS

4.18 Define limiting reagent and excess reagent. What is the significance of the limiting reagent in predicting the amount of the product obtained in a reaction?

4.19 Give an everyday example that illustrates the limiting reagent concept.

PROBLEMS

4.20 Nitric oxide (NO) reacts instantly with oxygen gas to give nitrogen dioxide (NO_2), a dark brown gas:

$$2NO(g) + O_2(g) \longrightarrow 2NO_2(g)$$

In one experiment 0.886 mole of NO is mixed with 0.503 mole of O_2. Calculate which of the two reactants is the limiting reagent. Calculate also the number of moles of NO_2 produced.

4.21 Balance the equation

$$SF_4 + I_2O_5 \longrightarrow IF_5 + SO_2$$

and calculate the maximum number of grams of IF_5 that can be obtained from 10.0 g of SF_4 and 10.0 g of I_2O_5.

4.22 The depletion of ozone (O_3) in the stratosphere has been a matter of great concern among scientists in recent years. It is believed that ozone can react with nitric oxide (NO) that is discharged from the high-altitude jet plane, the SST. The reaction is

$$O_3 + NO \longrightarrow O_2 + NO_2$$

If 0.740 g of O_3 reacts with 0.670 g of NO, how many grams of NO_2 would be produced? Which compound is the limiting reagent? Calculate the number of moles of the excess reagent remaining at the end of the reaction.

4.23 Propane (C_3H_8) is a component of natural gas and is used in domestic cooking and heating.
(a) Balance the following equation representing the combustion of propane in air:

$$C_3H_8 + O_2 \longrightarrow CO_2 + H_2O$$

(b) How many grams of carbon dioxide can be produced by burning 3.65 moles of propane? Assume that oxygen is the excess reagent in this reaction.

4.24 Consider the reaction

$$MnO_2 + 4HCl \longrightarrow MnCl_2 + Cl_2 + 2H_2O$$

If 0.86 mole of MnO_2 and 48.2 g of HCl react, which reagent will be used up first? How many grams of Cl_2 will be produced?

Yield of the Reaction

REVIEW QUESTIONS

4.25 Define the following terms: yield of a reaction, theoretical yield, actual yield, percent yield.

4.26 Why is the yield of a reaction determined only by the amount of the limiting reagent?

4.27 Why is the actual yield of a reaction almost always smaller than the theoretical yield?

PROBLEMS

4.28 Hydrogen fluoride is used in the manufacture of Freons (which destroy ozone in the stratosphere) and in the production of aluminum metal. It is prepared by the reaction

$$CaF_2 + H_2SO_4 \longrightarrow CaSO_4 + 2HF$$

In one process 6.00 kg of CaF_2 are treated with an excess of H_2SO_4 and yield 2.86 kg of HF. Calculate the percent yield of HF.

4.29 Nitroglycerin ($C_3H_5N_3O_9$) is a powerful explosive. Its decomposition may be represented by

$$4C_3H_5N_3O_9 \longrightarrow 6N_2 + 12CO_2 + 10H_2O + O_2$$

This reaction generates a large amount of heat and many gaseous products. It is the sudden formation of these gases, together with their rapid expansion, that produces the explosion. (a) What is the maximum amount of O_2 in grams that can be obtained from 2.00×10^2 g of nitroglycerin? (b) Calculate the percent yield in this reaction if the amount of O_2 generated is found to be 6.55 g.

4.30 Titanium(IV) oxide (TiO_2) is a white substance produced by the action of sulfuric acid on the mineral ilmenite ($FeTiO_3$):

$$FeTiO_3 + H_2SO_4 \longrightarrow TiO_2 + FeSO_4 + H_2O$$

Its opaque and nontoxic properties make it suitable as a pigment in plastics and paints. In one process 8.00×10^3 kg of $FeTiO_3$ yielded 3.67×10^3 kg of TiO_2. What is the percent yield of the reaction?

Concentration of Solutions

REVIEW QUESTIONS

4.31 Define molarity. Why is molarity a convenient concentration unit?

4.32 Describe steps involved in preparing a solution of known molar concentration using a volumetric flask.

PROBLEMS

4.33 Calculate the mass of NaOH in grams required to prepare a 5.00×10^2-mL solution of concentration 2.80 *M*.

4.34 A quantity of 5.25 g of NaOH is dissolved in sufficient amount of water to make up exactly 1 liter of solution. What is the molarity of the solution?

4.35 How many moles of $MgCl_2$ are present in 60.0 mL of 0.100 *M* $MgCl_2$ solution?

4.36 Calculate the molarity of a phosphoric acid (H_3PO_4) solution containing 1.50×10^2 g of the acid in 7.50×10^2 mL of solution.

4.37 How many grams of KOH are present in 35.0 mL of a 5.50 *M* solution?

4.38 Calculate the molarity of each of the following solutions: (a) 29.0 g of ethanol (C_2H_5OH) in 545 mL of solution, (b) 15.4 g of sucrose ($C_{12}H_{22}O_{11}$) in 74.0 mL of solution, (c) 9.00 g of sodium chloride (NaCl) in 86.4 mL of solution.

4.39 Calculate the number of moles of solute present in (a) 75.0 mL of 1.25 *M* HCl, (b) 100.0 mL of 0.35 *M* H_2SO_4.

4.40 Calculate the molarity of each of the following solutions: (a) 6.57 g of methanol (CH_3OH) in 1.50×10^2 mL of solution, (b) 10.4 g of calcium chloride ($CaCl_2$) in 2.20×10^2 mL of solution, (c) 7.82 g of naphthalene ($C_{10}H_8$) in 85.2 mL of benzene solution.

4.41 Calculate the volume in mL of a solution required to provide the following: (a) 2.14 g of sodium chloride from a 0.270 *M* solution, (b) 4.30 g of ethanol from a 1.50 *M* solution, (c) 0.85 g of acetic acid (CH_3COOH) from a 0.30 *M* solution.

4.42 Determine how many grams of each of the following solutes would be needed to make 2.50×10^2 mL of a 0.100 *M* solution: (a) cesium iodide (CsI), (b) sulfuric acid (H_2SO_4), (c) sodium carbonate (Na_2CO_3), (d) potassium dichromate ($K_2Cr_2O_7$), (e) potassium permanganate ($KMnO_4$).

4.43 A 325-mL sample of solution contains 25.3 g of $CaCl_2$. (a) Calculate the molar concentration of Cl^- in this solution. (b) How many grams of Cl^- are there in 0.100 liter of this solution?

Dilution of Solutions

REVIEW QUESTIONS

4.44 Describe the basic steps involved in the dilution of a solution of known concentration.

4.45 Write the equation that enables you to calculate the concentration of a diluted solution. Define all the terms.

4.46 Using $BaCl_2$ as an example, show that the number of ions will remain the same when a 2 *M* $BaCl_2$ solution is diluted to a 1 *M* solution.

PROBLEMS

4.47 Describe how to prepare 1.00 L of 0.646 *M* HCl solution, starting with a 2.00 *M* HCl solution.

4.48 A quantity of 25.0 mL of a 0.866 *M* KNO_3 solution is poured into a 500-mL volumetric flask, and water is added until the volume of the solution is exactly 500 mL. What is the concentration of the final solution?

4.49 How would you prepare 60.0 mL of 0.200 *M* HNO_3 from a stock solution of 4.00 *M* HNO_3?

4.50 You have 505 mL of a 0.125 *M* HCl solution and you want to dilute it to exactly 0.100 *M*. How much water should you add?

4.51 Describe how you would prepare 2.50×10^2 mL of a 0.450 *M* KOH solution, starting with an 8.40 *M* KOH solution.

4.52 A 46.2-mL, 0.568 *M* calcium nitrate [$Ca(NO_3)_2$] solution is mixed with an 80.5-mL, 1.396 *M* calcium nitrate solution. Calculate the concentration of the final solution.

Gravimetric Analysis

REVIEW QUESTIONS

4.53 Define gravimetric analysis. Describe the basic steps involved in gravimetric analysis. How does such an experiment help us determine the identity of a compound or the purity of a compound if its formula is known?

4.54 Distilled water must be used in the gravimetric analysis of chlorides. Why?

PROBLEMS

4.55 The volume 30.0 mL of 0.150 *M* $CaCl_2$ is added to 15.0 mL of 0.100 *M* $AgNO_3$. What is the mass in grams of AgCl precipitate formed?

4.56 A sample of 0.6760 g of an unknown compound containing barium ions (Ba^{2+}) is dissolved in water and treated with an excess of Na_2SO_4. If the mass of the $BaSO_4$ precipitate formed is 0.4105 g, what is the percent by mass of Ba in the original unknown compound?

4.57 How many grams of NaCl are required to precipitate practically all the Ag^+ ions from 2.50×10^2 mL of 0.0113 *M* $AgNO_3$ solution? Write the net ionic equation for the reaction.

4.58 The concentration of Cu^{2+} ions in the water (which also contains sulfate ions) discharged from a certain industrial plant is determined by adding excess sodium sulfide (Na_2S) solution to 0.800 liter of the water. The molecular equation is

$$Na_2S(aq) + CuSO_4(aq) \longrightarrow Na_2SO_4(aq) + CuS(s)$$

Write the net ionic equation and calculate the molar concentration of Cu^{2+} in the water sample if 0.0177 g of solid CuS is formed.

Acid-Base Titrations

REVIEW QUESTIONS

4.59 Define acid-base titration, standard solution, equivalence point.

4.60 Describe the basic steps involved in acid-base titration. Why is this technique of great practical value?

4.61 How does an acid-base indicator work?

PROBLEMS

4.62 Calculate the volume in mL of a 1.420 *M* NaOH solution required to titrate the following solutions:

(a) 25.00 mL of a 2.430 M HCl solution
(b) 25.00 mL of a 4.500 M H_2SO_4 solution
(c) 25.00 mL of a 1.500 M H_3PO_4 solution

4.63 Acetic acid (CH_3COOH) is an important ingredient of vinegar. A sample of 50.0 mL of a commercial vinegar is titrated against a 1.00 M NaOH solution. What is the concentration (in M) of acetic acid present in the vinegar if 5.75 mL of the base were required for the titration?

4.64 What volume of a 0.50 M KOH solution is needed to neutralize completely each of the following?
(a) 10.0 mL of a 0.30 M HCl solution
(b) 10.0 mL of a 0.20 M H_2SO_4 solution
(c) 15.0 mL of a 0.25 M H_3PO_4 solution

Redox Titrations

REVIEW QUESTION

4.65 What are the similarities and differences between acid-base titrations and redox titrations?

PROBLEMS

4.66 Oxidation of 25.0 mL of a solution containing Fe^{2+} requires 26.0 mL of 0.0250 M $K_2Cr_2O_7$ in acidic solution. Balance the following equation and calculate the molar concentration of Fe^{2+}:

$$Cr_2O_7^{2-} + Fe^{2+} + H^+ \longrightarrow Cr^{3+} + Fe^{3+}$$

4.67 The SO_2 present in air is mainly responsible for the acid rain phenomenon. Its concentration can be determined by titrating against a standard permanganate solution as follows:

$$5SO_2 + 2MnO_4^- + 2H_2O \longrightarrow 5SO_4^{2-} + 2Mn^{2+} + 4H^+$$

Calculate the number of grams of SO_2 in a sample of air if 7.37 mL of 0.00800 M $KMnO_4$ solution is required for the titration.

4.68 A sample of iron ore weighing 0.2792 g was dissolved in dilute acid solution, and all the Fe(II) was converted to Fe(III) ions. The solution required 23.30 mL of 0.0194 M $KMnO_4$ for titration. Calculate the percent by mass of iron in the ore.

4.69 The concentration of a hydrogen peroxide solution can be conveniently determined by titration against a standardized potassium permanganate solution in acidic medium according to the following unbalanced equation:

$$MnO_4^- + H_2O_2 \longrightarrow O_2 + Mn^{2+}$$

(a) Balance the above equation. (b) If 36.44 mL of a 0.01652 M $KMnO_4$ solution is required to completely oxidize 25.00 mL of a H_2O_2 solution, calculate the molarity of the H_2O_2 solution.

4.70 Iodate ion, IO_3^-, oxidizes SO_3^{2-} to SO_4^{2-} in acidic solution. A 100.0-mL sample of solution containing 1.390 g of KIO_3 reacts with 32.5 mL of 0.500 M Na_2SO_3. What is the final oxidation state of the iodine after the reaction has occurred?

4.71 Oxalic acid ($H_2C_2O_4$) is present in many plants and vegetables. (a) Balancing the following equation in acid solution:

$$MnO_4^- + C_2O_4^{2-} \longrightarrow Mn^{2+} + CO_2$$

(b) If a 1.00-g sample of $H_2C_2O_4$ requires 24.0 mL of 0.0100 M $KMnO_4$ solution to reach an equivalence point, what is the percent by mass of $H_2C_2O_4$ in the sample?

4.72 Sulfite ions can be oxidized to sulfate ions by dichromate ions in acid solution:

$$8H^+ + Cr_2O_7^{2-} + 3SO_3^{2-} \longrightarrow 2Cr^{3+} + 3SO_4^{2-} + 4H_2O$$

Calculate the molarity of the $K_2Cr_2O_7$ solution if 28.42 mL of the solution reacts completely with a 25.00-mL solution of 0.3143 M Na_2SO_3.

4.73 A quantity of 25.0 mL of a solution containing both Fe^{2+} and Fe^{3+} ions is titrated with 23.0 mL of 0.0200 M $KMnO_4$ (in dilute sulfuric acid). As a result, all of the Fe^{2+} ions are oxidized to Fe^{3+} ions. Next, the solution is treated with Zn metal to convert all of the Fe^{3+} ions to Fe^{2+} ions. Finally, the solution containing only the Fe^{2+} ions required 40.0 mL of the same $KMnO_4$ solution for oxidation to Fe^{3+}. Calculate the molar concentrations of Fe^{2+} and Fe^{3+} in the original solution.

4.74 Household bleach solutions contain the hypochlorite ion (ClO^-) which is formed when chlorine dissolves in water:

$$Cl_2(g) + H_2O(l) \longrightarrow ClO^-(aq) + Cl^-(aq) + 2H^+(aq)$$

To determine the concentration of hypochlorite in the bleach the solution is first treated with a KI solution:

$$ClO^-(aq) + 2I^-(aq) + 2H^+(aq) \longrightarrow I_2(s) + Cl^-(aq) + H_2O(l)$$

The iodine liberated can then be determined by titration with a standardized thiosulfate solution:

$$I_2(s) + 2S_2O_3^{2-}(aq) \longrightarrow 2I^-(aq) + S_4O_6^{2-}(aq)$$

A 25.00-mL sample of a certain household bleach requires 17.4 mL of a 0.0200 M $Na_2S_2O_3$ solution for titration. Calculate the mass of Cl_2 dissolved in 1 liter of the bleach solution.

4.75 Copper(II) ions oxidize iodide ions to iodine as follows:

$$2Cu^{2+}(aq) + 4I^-(aq) \longrightarrow 2CuI(s) + I_2(s)$$

The amount of iodine liberated can be determined according to the procedure described in Problem 4.74. A 6.00-g "copper coin" is first dissolved in nitric acid to convert all the copper to Cu^{2+} ions. The Cu^{2+} ions are then completely reduced by a KI solution. If the liberated molecular iodine requires 124.2 mL of a 0.624 M $Na_2S_2O_3$ solution to reach the equivalence point, what is the percent by mass of copper in the coin?

4.76 Calcium oxalate (CaC_2O_4) is insoluble in water. This property has been used to determine the amount of Ca^{2+} ions in fluids such as blood. The calcium oxalate isolated from blood is dissolved in acid and titrated against a standardized $KMnO_4$ solution as shown in Problem 4.71. In one test it is found that the calcium oxalate isolated from a 10.0-mL sample of blood requires 24.2 mL of 9.56×10^{-4} *M* $KMnO_4$ for titration. Calculate the number of milligrams of calcium per milliliter of blood.

Miscellaneous Problems

4.77 Which of the following aqueous solutions would you expect to be the best conductor of electricity at 25°C? Explain your answer.
(a) 0.20 *M* NaCl
(b) 0.60 *M* CH_3COOH
(c) 0.25 *M* HCl
(d) 0.20 *M* $Mg(NO_3)_2$

4.78 A quantity of 2.40 g of the oxide of the metal X (molar mass of X = 55.9 g/mol) was heated in carbon monoxide (CO). The products are the pure metal and carbon dioxide. The mass of the metal formed is 1.68 g. From the data given, show that the simplest formula of the oxide is X_2O_3 and write a balanced equation for the reaction.

4.79 An impure sample of zinc (Zn) is treated with an excess of sulfuric acid (H_2SO_4) to form zinc sulfate ($ZnSO_4$) and molecular hydrogen (H_2). (a) Write a balanced equation for the reaction. (b) If 0.0764 g of H_2 is obtained from 3.86 g of the sample, calculate the percent purity of the sample. (c) What assumptions must you make in (b)?

4.80 One of the reactions that occurs in a blast furnace, where iron ore is converted to cast iron, is

$$Fe_2O_3 + 3CO \longrightarrow 2Fe + 3CO_2$$

Suppose that 1.64×10^3 kg of Fe are obtained from a 2.62×10^3-kg sample of Fe_2O_3. Assuming that the reaction goes to completion, what is the percent purity of Fe_2O_3 in the original sample?

4.81 A student took an unknown sample mass of a compound of Ti and Cl and put it into some water. The Ti formed TiO_2, which was removed, dried, and found to weigh 0.777 g. $AgNO_3$ was then added to the solution until all the Cl^- ions were converted to 5.575 g of AgCl. Determine the empirical formula of the unknown compound of Ti and Cl.

4.82 When 0.273 g of Mg is heated strongly in a nitrogen (N_2) atmosphere, a chemical reaction occurs. The product of the reaction weighs 0.378 g. Calculate the empirical formula of the compound containing Mg and N.

4.83 A sample of a compound of Cl and O reacts with excess H_2 to give 0.233 g of HCl and 0.403 g of H_2O. Determine the empirical formula of the compound.

4.84 A certain sample of coal contains 1.6 percent sulfur by mass. When the coal is burned, the sulfur is converted to sulfur dioxide. To prevent air pollution, this sulfur dioxide is treated with calcium oxide (CaO) to form calcium sulfite ($CaSO_3$). Calculate the daily mass (in kilograms) of CaO needed if a power plant uses 6.60×10^6 kg of coal per day.

4.85 A 5.00×10^2-mL sample of 2.00 *M* HCl solution is treated with 4.47 g of magnesium. Calculate the concentration of the acid solution after all the metal has reacted. Assume that the volume remains unchanged.

4.86 A compound X contains 63.3 percent manganese (Mn) and 36.7 percent oxygen by mass. When X is heated, oxygen gas is evolved and a new compound Y containing 72.0 percent Mn and 28.0 percent O is formed. (a) Determine the empirical formulas of X and Y. (b) Write a balanced equation for the conversion of X to Y.

4.87 A common laboratory preparation of oxygen gas is the thermal decomposition of potassium chlorate ($KClO_3$). Assuming complete decomposition, calculate the number of grams of O_2 gas that can be obtained starting with 46.0 g of $KClO_3$. (The products are KCl and O_2.)

4.88 Industrially, hydrogen gas can be prepared by reacting propane gas (C_3H_8) with steam at about 400°C. The products are carbon monoxide (CO) and hydrogen gas (H_2). (a) Write a balanced equation for the reaction. (b) How many kilograms of H_2 can be obtained starting with 2.84×10^3 kg of propane?

4.89 The formula of a hydrate of barium chloride is $BaCl_2 \cdot xH_2O$. If 1.936 g of the compound gives 1.864 g of anhydrous $BaSO_4$ upon treatment with sulfuric acid, calculate the value of *x*.

4.90 A 0.8870-g sample of a mixture of NaCl and KCl is dissolved in water, and the solution is then treated with $AgNO_3$ to yield 1.913 g of AgCl. Calculate the percent by mass of each compound in the mixture.

4.91 A 1.00-g sample of a metal X (that is known to form X^{2+} ions) was added to 0.100 liter of 0.500 *M* H_2SO_4. After all the metal had reacted, the remaining acid required 0.0334 liter of 0.500 *M* NaOH solution for neutralization. Calculate the molar mass of the metal.

4.92 Calculate the volume of a 0.156 *M* $CuSO_4$ solution that would react with 7.89 g of zinc.

4.93 Sodium carbonate (Na_2CO_3) can be obtained in very pure form and can be used to standardize acid solutions. What is the molarity of a HCl solution if 28.3 mL of the solution is required to react with 0.256 g of Na_2CO_3?

4.94 A 3.664-g sample of a monoprotic acid was dissolved in water and required 20.27 mL of a 0.1578 *M* NaOH solution for neutralization. Calculate the molar mass of the acid.

4.95 A mixture of $CuSO_4 \cdot 5H_2O$ and $MgSO_4 \cdot 7H_2O$ is heated until all the water is lost. If 5.020 g of the mixture gives 2.988 g of the anhydrous salts, what is the percent by mass of $CuSO_4 \cdot 5H_2O$ in the mixture?

4.96 A useful application of oxalic acid is the removal of rust (Fe_2O_3) according to the reaction

$$Fe_2O_3 + 6H_2C_2O_4 \longrightarrow 2Fe(C_2O_4)_3^{3-} + 3H_2O + 6H^+$$

Calculate the rust in grams that can be removed by 5.00×10^2 mL of a 0.100 *M* solution of oxalic acid.

4.97 A 60.0-mL 0.513 *M* glucose ($C_6H_{12}O_6$) solution is mixed with a 120.0-mL 2.33 *M* glucose solution. What is the concentration of the final solution? Assume the volumes are additive.

4.98 The alcohol content in a 60.00-g sample of blood from a driver required 28.64 mL of 0.07654 *M* $K_2Cr_2O_7$ for titration. Should the police prosecute the individual for drunken driving? (*Hint:* See Chemistry in Action: Breath Analyzer for the balanced equation and other relevant information.)

4.99 Calculate the mass of Cu^+ in grams in a sample that requires 36.0 mL of 0.146 *M* $KMnO_4$ solution to reach the equivalence point. The products are Mn^{2+} and Cu^{2+} ions. The reaction occurs in an acid solution.

4.100 A mixture of 60.42 g of sodium carbonate (Na_2CO_3) and sodium hydrogen carbonate ($NaHCO_3$) is heated. The loss in mass is 8.415 g. Calculate the percentage by mass of sodium hydrogen carbonate in the mixture. (*Hint:* Only $NaHCO_3$ is decomposed by heat.)

4.101 A sample of methane (CH_4) and ethane (C_2H_6) of mass 13.43 g is completely burned in oxygen. If the total mass of CO_2 and H_2O produced is 64.84 g, calculate the fraction of CH_4 in the mixture.

4.102 Acetylsalicylic acid ($C_9H_8O_4$) is a monoprotic acid commonly known as "aspirin." A typical aspirin tablet, however, contains only a small amount of the acid. In an experiment to determine its composition, an aspirin tablet was crushed and dissolved in water. The solution needed 12.25 mL of 0.1466 *M* NaOH for neutralization. Calculate the number of grains of aspirin in the tablet. (One grain = 0.0648 g.)

4.103 A 0.9157-g mixture of $CaBr_2$ and NaBr is dissolved in water, and $AgNO_3$ is added to the solution to form AgBr precipitate. If the mass of the precipitate is 1.6930 g, what is the percent by mass of NaBr in the original mixture?

4.104 Octane (C_8H_{18}) is a component of gasoline. Complete combustion of octane leads to CO_2 and H_2O. Incomplete combustion produces CO and H_2O, which not only reduces the efficiency of the engine using the fuel but is also toxic. In a certain test run, 1.000 gallon of octane is burned in an engine. The total mass of CO, CO_2, and H_2O produced is 11.53 kg. Calculate the efficiency of the process; that is, calculate the fraction of octane converted to CO_2. The density of octane is 2.650 kg/gallon.

4.105 A 15.00-mL solution of potassium nitrate (KNO_3) was diluted to 125.0 mL, and 25.00 mL of this solution was then diluted to 1.000×10^3 mL. If the concentration of the final solution is 0.00383 *M*, calculate the concentration of the original solution.

4.106 Ethylene (C_2H_4), an important industrial organic chemical, can be prepared by heating hexane (C_6H_{14}) at 800°C:

$$C_6H_{14} \longrightarrow C_2H_4 + \text{other products}$$

If the yield of ethylene production is 42.5 percent, what mass of hexane must be reacted to produce 481 g of ethylene?

4.107 A zinc strip of mass 2.50 g was placed in a $AgNO_3$ solution, and silver metal was observed to form on the surface of the strip. After a certain time the strip was removed from the solution, dried, and weighed. If the mass of the strip was 3.37 g, calculate the mass of Ag and Zn metals present.

Opposite page: The turbulence in air created by heat from passing cars and cross winds. This motion demonstrates the effect of heat on the velocity of gases.

CHAPTER 5

THE GASEOUS STATE

Under certain conditions of pressure and temperature, most substances can exist in any one of the three states of matter: solid, liquid, or gas. Water, for example, exists in the solid state as ice, in the liquid state as water, and in the gaseous state as steam or water vapor. The physical properties of a substance often depend on the state of the substance. In this chapter we will look at the behavior of gases. In many ways, gases are much simpler than liquids or solids. Molecular motion in gases is totally random, and the forces of attraction between molecules are so small that each molecule moves freely and essentially independently of other molecules. A gas subjected to changes in temperature and pressure behaves according to much simpler laws than do solids and liquids. The laws that govern this behavior have played an important role in the development of the atomic theory of matter and the kinetic molecular theory of gases.

5.1 Substances That Exist as Gases

Before we explore the physical behavior of gases, we will briefly survey substances that might be expected to occur as gases (rather than liquids or solids) under normal conditions of pressure and temperature (that is, 1 atm and 25°C). Once we are able to recognize substances that usually exist as gases, we can infer their characteristic physical behavior from our understanding of this distinct state of matter. You should note that ''normal conditions'' really represent a rather narrow range of temperature and pressure. In the laboratory we can create unusual conditions of temperature or pressure. Unusually high or low temperatures or pressures may make the physical state of a substance quite different from what it would be under ''normal conditions.''

In Chapter 2 we learned that compounds can be classified into two categories: ionic and molecular. Ionic compounds do not exist as gases at 25°C and 1 atm, because cations and anions in an ionic solid are held together by very strong electrostatic forces. To overcome these attractions we must apply a large amount of energy, which in practice means strongly heating the solid. Under normal conditions, all we can do is melt the solid; for example, NaCl melts at the rather high temperature of 801°C. In order to boil NaCl, we would have to raise the temperature to well above 1000°C.

The behavior of molecular compounds is more varied. Some—for example, CO, CO_2, HCl, NH_3, and CH_4 (methane)—are gases, but the majority of molecular compounds are liquids or solids at room temperature. However, by heating they can be converted to gases much more easily than ionic compounds can. In other words, molecular compounds usually boil at much lower temperatures than ionic compounds. There is no simple rule to help us determine if a certain molecular compound is a gas under normal atmospheric conditions. To make such a determination we would need to understand the nature and magnitude of the attractive forces among the molecules. In general, the stronger these attractions, the less likely that the compound can exist as a gas at ordinary temperatures.

Figure 5.1 shows the elements that are gases under normal atmospheric conditions. Note that the elements hydrogen, nitrogen, oxygen, fluorine, and chlorine exist as

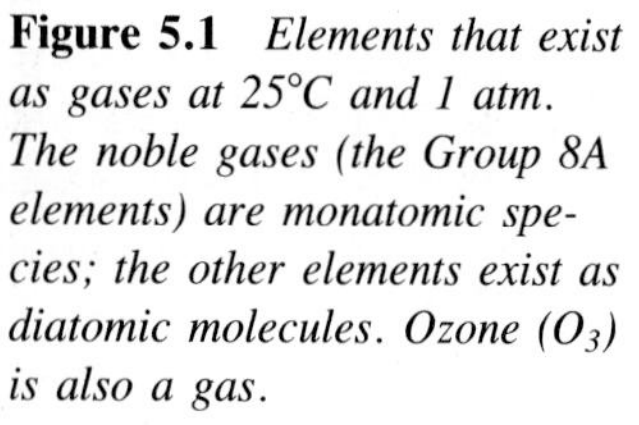

Figure 5.1 *Elements that exist as gases at 25°C and 1 atm. The noble gases (the Group 8A elements) are monatomic species; the other elements exist as diatomic molecules. Ozone (O_3) is also a gas.*

1A	2A	3B	4B	5B	6B	7B	8B	8B	8B	1B	2B	3A	4A	5A	6A	7A	8A
H																	He
Li	Be											B	C	N	O	F	Ne
Na	Mg											Al	Si	P	S	Cl	Ar
K	Ca	Sc	Ti	V	Cr	Mn	Fe	Co	Ni	Cu	Zn	Ga	Ge	As	Se	Br	Kr
Rb	Sr	Y	Zr	Nb	Mo	Tc	Ru	Rh	Pd	Ag	Cd	In	Sn	Sb	Te	I	Xe
Cs	Ba	La	Hf	Ta	W	Re	Os	Ir	Pt	Au	Hg	Tl	Pb	Bi	Po	At	Rn

Table 5.1 Some substances found as gases at 1 atm and 25°C

Elements	Compounds
H_2 (molecular hydrogen)	HF (hydrogen fluoride)
N_2 (molecular nitrogen)	HCl (hydrogen chloride)
O_2 (molecular oxygen)	HBr (hydrogen bromide)
O_3 (ozone)	HI (hydrogen iodide)
F_2 (molecular fluorine)	CO (carbon monoxide)
Cl_2 (molecular chlorine)	CO_2 (carbon dioxide)
He (helium)	NH_3 (ammonia)
Ne (neon)	NO (nitric oxide)
Ar (argon)	NO_2 (nitrogen dioxide)
Kr (krypton)	N_2O (nitrous oxide)
Xe (xenon)	SO_2 (sulfur dioxide)
Rn (radon)	H_2S (hydrogen sulfide)
	HCN (hydrogen cyanide)*
	CH_4 (methane)

*The boiling point of HCN is 26°C, but it is close enough to qualify as a gas at ordinary atmospheric conditions.

Color of NO_2 gas.

gaseous diatomic molecules: H_2, N_2, O_2, F_2, and Cl_2. An allotrope of oxygen, ozone (O_3), is also a gas at room temperature. All the elements in Group 8A, the noble gases, are monatomic gases: He, Ne, Ar, Kr, Xe, and Rn.

Table 5.1 lists the common gases we will use in our study of the physical behavior of gases. Of these, only O_2 is essential for our survival. Hydrogen sulfide (H_2S) and hydrogen cyanide (HCN) are deadly poisons. Several others, such as CO, NO_2, O_3, and SO_2, are somewhat less toxic. The gases He, Ne, and Ar are chemically inert; that is, they do not react with any other substance. Most gases are colorless. Of the gases listed in Table 5.1, only F_2, Cl_2, and NO_2 are colored. The dark brown color of NO_2 is sometimes visible in polluted air.

At this point, it is useful to distinguish between a "gas" and a "vapor," two terms that are often used interchangeably but do not have exactly the same meaning. A *gas* is a substance that is *normally* in the gaseous state at ordinary temperatures and pressures; a *vapor* is the gaseous form of any substance that is a liquid or a solid at normal temperatures and pressures. Thus, at 25°C and 1 atm pressure, we speak of water *vapor* and oxygen *gas*.

5.2 Pressure of a Gas

We live at the bottom of an ocean of air whose composition by volume is roughly 78 percent N_2, 21 percent O_2, and 1 percent other gases, such as Ar and CO_2. The molecules (and atoms) of these gases, like those of all other matter, are subject to Earth's gravitational pull. If we could see air, we would observe that the atmosphere is very dense near the surface of Earth and much less dense at higher altitudes. (The air outside the pressurized cabin of an airplane at 30,000 feet is too thin to breathe.) In fact, the density of air decreases very rapidly with increasing distance from Earth (Figure 5.2). Measurements show that about 50 percent of the atmosphere lies within 4

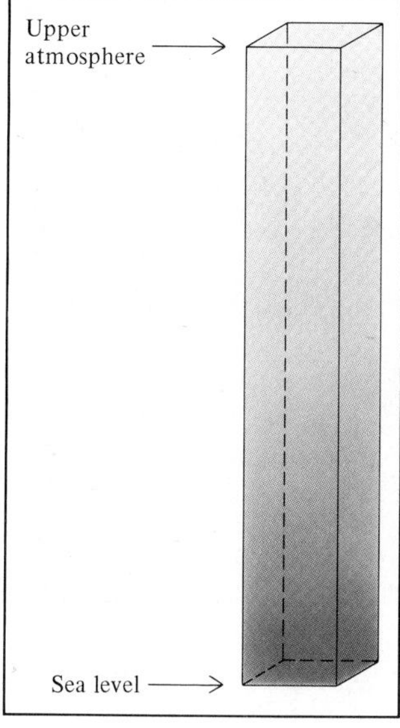

Figure 5.2 *A column of air extending from sea level to the upper atmosphere.*

miles (6.4 km) of Earth's surface, 90 percent within 10 miles (16 km), and 99 percent within 20 miles (32 km).

One of the most readily measurable properties of a gas is its pressure. Gases exert pressure on any surface with which they come into contact, because the gas molecules are constantly in motion and collide with the surface. We humans have adapted so well physiologically to the pressure of the air around us that we are usually unaware of its existence, perhaps as fish are unconscious of the water's pressure on them. There are many ways to demonstrate the existence of atmospheric pressure. One example is our ability to drink a liquid through a straw. Sucking air out of the straw creates a reduced pressure which is filled quickly as the fluid in the container is pushed up into the straw by atmospheric pressure.

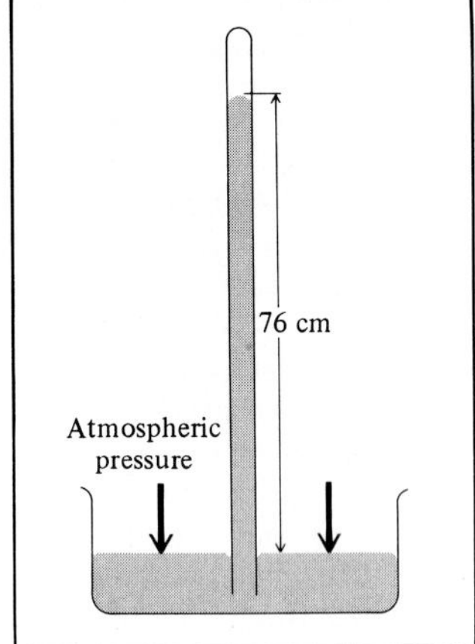

Figure 5.3 *A barometer for measuring atmospheric pressure. Above the mercury in the tube is a vacuum. The column of mercury is supported by the atmospheric pressure.*

A ***barometer*** is *an instrument that measures atmospheric pressure*. A simple barometer can be constructed by filling a long glass tube, closed at one end, with mercury, and then carefully inverting the tube in a dish of mercury, making sure that no air enters the tube. Some mercury in the tube will flow down into the dish, creating a vacuum at the top (Figure 5.3). The weight of the mercury column remaining in the tube is supported by atmospheric pressure acting on the surface of the mercury in the dish. The *standard* atmospheric pressure (1 atm) is equal to the pressure that supports a column of mercury exactly 760 mm (or 76 cm) high at 0°C at sea level. The standard atmosphere thus equals a pressure of 760 mmHg, where mmHg represents the pressure exerted by a column of mercury 1 mm high. The mmHg unit is also called the *torr*, after the Italian scientist Evangelista Torricelli,† who invented the barometer. Thus

$$1 \text{ torr} = 1 \text{ mmHg}$$

and

$$\begin{aligned} 1 \text{ atm} &= 760 \text{ mmHg} \\ &= 760 \text{ torr} \end{aligned}$$

The following example shows the conversion between mmHg and atm.

EXAMPLE 5.1

The pressure outside a jet plane flying at high altitude falls considerably below atmospheric pressure at sea level. The air inside the cabin must therefore be pressurized to protect the passengers. What is the pressure in atmospheres in the cabin if the barometer reading is 688 mmHg?

Answer

The pressure in atmospheres is calculated as follows:

$$\begin{aligned} \text{pressure} &= 688 \text{ mmHg} \times \frac{1 \text{ atm}}{760 \text{ mmHg}} \\ &= 0.905 \text{ atm} \end{aligned}$$

Similar problem: 5.13.

† Evangelista Torricelli (1608–1647). Italian mathematician. Torricelli was supposedly the first person to recognize the existence of atmospheric pressure.

See Section 1.7 to review the definition of a newton.

In SI units, pressure is measured in *pascals (Pa),* defined as one newton per square meter:

$$\text{pressure} = \frac{\text{force}}{\text{area}}$$

$$1 \text{ Pa} = 1 \frac{\text{N}}{\text{m}^2}$$

The relation between atmospheres and pascals (see Appendix 2) is

$$\begin{aligned} 1 \text{ atm} &= 101{,}325 \text{ Pa} \\ &= 1.01325 \times 10^5 \text{ Pa} \end{aligned}$$

and since 1000 Pa = 1 kPa (kilopascal)

This equation gives the SI definition of 1 atm.

$$1 \text{ atm} = 1.01325 \times 10^2 \text{ kPa}$$

Example 5.2 shows the conversion between kPa and mmHg.

EXAMPLE 5.2

The atmospheric pressure in San Francisco on a certain day was 732 mmHg. What was the pressure in kPa?

Answer

The conversion equations show that

$$1 \text{ atm} = 1.01325 \times 10^5 \text{ Pa} = 760 \text{ mmHg}$$

which allows us to calculate the atmospheric pressure in kPa:

$$\begin{aligned} \text{pressure} &= 732 \text{ mmHg} \times \frac{1.01325 \times 10^5 \text{ Pa}}{760 \text{ mmHg}} \\ &= 9.76 \times 10^4 \text{ Pa} \\ &= 97.6 \text{ kPa} \end{aligned}$$

Similar problem: 5.12.

The advantage of using mercury in a barometer is its very high density compared with that of other liquids. As Example 5.3 shows, a water barometer is impractical because the column of the liquid would be too high.

EXAMPLE 5.3

Calculate the height in meters of the column in a barometer for 1 atm pressure if water were used instead of mercury. The density of water is 1.00 g/cm^3 and that of mercury is 13.6 g/cm^3.

Answer

The difference in heights here is the result of the different densities. Therefore

$$\text{height of } H_2O = \text{height of Hg} \times \frac{\text{density of Hg}}{\text{density of } H_2O}$$
$$= 760 \text{ mm} \times \frac{13.6 \text{ g/cm}^3}{1.00 \text{ g/cm}^3}$$
$$\approx 1.03 \times 10^4 \text{ mm}$$
$$= 10.3 \text{ m}$$

The calculation in Example 5.3 shows that there is a limit to how deep an old-fashioned farmyard well can be. Drawing water from such a well by using a suction pump is analogous to drinking soda with a straw. Strictly speaking, only the air in the pipe is pumped out. Water is then pushed up by atmospheric pressure. However, since this operation depends entirely on atmospheric pressure, we cannot get water from more than 10.3 m (or 34 ft) below the surface. Motorized pumps must be used to get water from greater depths.

5.3 The Gas Laws

The gas laws we will study in this chapter can be regarded as useful summaries of the results of countless experiments on the physical properties of gases that were carried out over several centuries. The gas laws are important generalizations regarding the macroscopic behavior of gaseous substances. They have played a major role in the development of many ideas in chemistry.

The Pressure-Volume Relationship: Boyle's Law

The pressure applied on the gas is equal to the gas pressure.

In the seventeenth century, Robert Boyle† studied the behavior of gases systematically and quantitatively. In one series of studies, Boyle investigated the pressure-volume relationship of a gas sample using an apparatus like that shown in Figure 5.4. In Figure 5.4(a) the pressure exerted on the gas is equal to atmospheric pressure. In Figure 5.4(b) there is an increase in pressure due to the added mercury and the resulting unequal levels in the two columns; consequently, the volume of the gas decreases. Boyle noticed that when temperature is held constant, the volume (V) of a given amount of a gas decreases as the total applied pressure (P)—atmospheric pressure plus the pressure due to the added mercury—is increased. This relationship between pressure and volume is shown pictorially in Figure 5.4(b), (c), and (d). Conversely, if the applied pressure is decreased, the gas volume becomes larger. The results of several measurements are given in Table 5.2.

The P-V data actually recorded for such an experiment are consistent with these

†Robert Boyle (1627–1691). British chemist and natural philosopher. Although Boyle is commonly associated with the gas law that bears his name, he made many other significant contributions in chemistry and physics. He was often at odds with contemporary scientists. His book *The Skeptical Chymist* (1661) influenced generations of chemists.

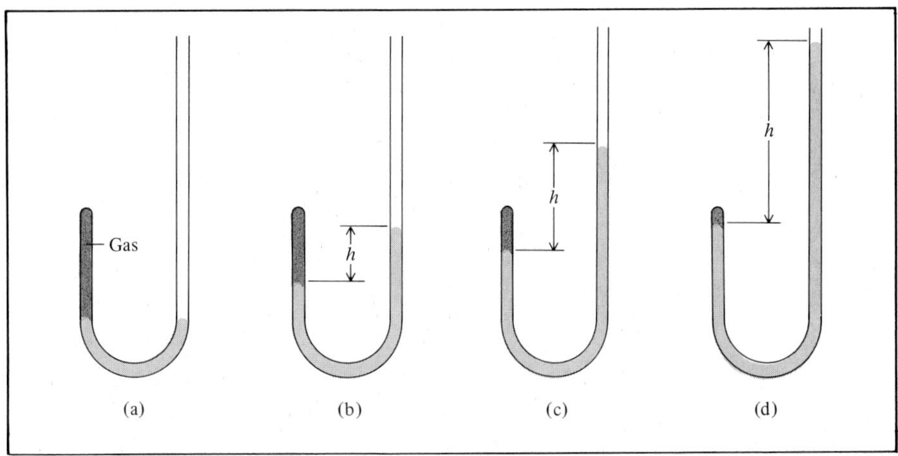

Figure 5.4 *Apparatus for studying the relationship between pressure and volume of a gas. In (a) the pressure of the gas is equal to the atmospheric pressure. The pressure exerted on the gas increases from (a) to (d) as mercury is added, and the volume of the gas decreases, as predicted by Boyle's law. The extra pressure exerted on the gas is shown by the difference in the mercury levels (h mmHg). The temperature of the gas is kept constant.*

mathematical expressions showing an inverse relationship:

$$V \propto \frac{1}{P}$$

where the symbol $\propto$ means *proportional to*. To change $\propto$ to an equals sign, we must write

$$V = k_1 \times \frac{1}{P} \tag{5.1a}$$

where k_1 is a constant called the proportionality constant. Equation (5.1a) is an expression of ***Boyle's law,*** which states that *the volume of a fixed amount of gas maintained at constant temperature is inversely proportional to the gas pressure.* Rearranging Equation (5.1a), we obtain

$$PV = k_1 \tag{5.1b}$$

Equation (5.1b) is another form of Boyle's law, which says that the product of the pressure and volume of a gas at constant temperature is a constant.

The concept of one quantity being proportional to another, and the use of a proportionality constant, can be clarified through the following analogy. The daily income of a movie theater depends on both the price of the tickets (in dollars per ticket) and the number of tickets sold. Thus we write

$$\text{income} = (\text{dollar/ticket}) \times \text{number of tickets sold}$$

Since the number of tickets sold varies from day to day, the income on a given day is said to be proportional to the number of tickets sold:

$$\begin{aligned} \text{income} &\propto \text{number of tickets sold} \\ &= C \times \text{number of tickets sold} \end{aligned}$$

Table 5.2 Typical pressure-volume relationships obtained by Boyle

P (mmHg)	724	869	951	998	1230	1893	2250
V (arbitrary units)	1.50	1.33	1.22	1.16	0.94	0.61	0.51
PV	1.09×10^3	1.16×10^3	1.16×10^3	1.16×10^3	1.2×10^3	1.2×10^3	1.1×10^3

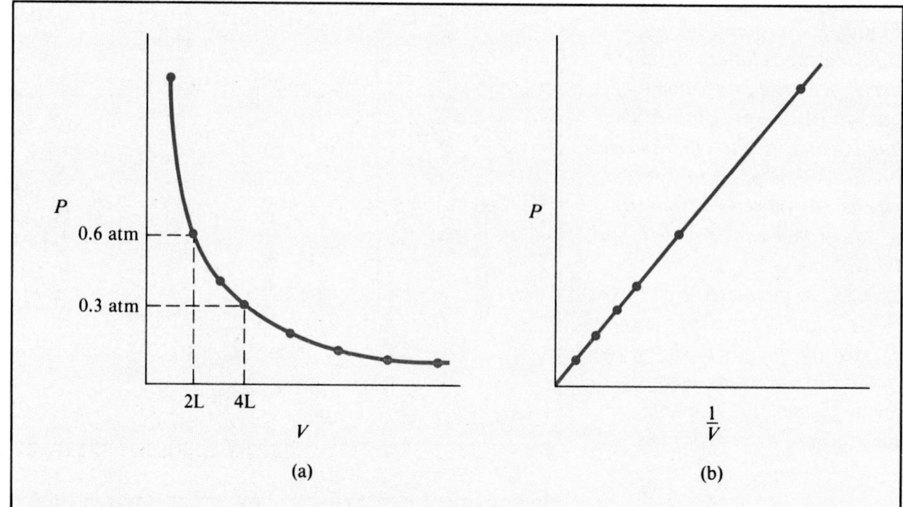

Figure 5.5 *Graphs showing variation of the volume of a gas sample with the pressure exerted on the gas, at constant temperature. (a) P versus V. Note that the volume of the gas doubles as the pressure is halved; (b) P versus 1/V.*

where C, the proportionality constant, is the price per ticket. (We assume that the theater has only one price for all tickets.)

Figure 5.5 shows two conventional ways of expressing Boyle's findings graphically. Figure 5.5(a) is a graph of the equation $PV = k_1$; Figure 5.5(b) is a graph of the equivalent equation $P = k_1 \times 1/V$. Note that the latter equation takes the form of a linear equation $y = mx + b$, where $b = 0$.

Although the individual values of pressure and volume can vary greatly for a given sample of gas, as long as the temperature is held constant and the amount of the gas does not change, P times V is always equal to the same constant. Therefore, for a given sample of gas under two different sets of conditions at constant temperature, we can write

$$P_1V_1 = k_1 = P_2V_2$$

or

$$P_1V_1 = P_2V_2 \tag{5.2}$$

where V_1 and V_2 are the volumes at pressures P_1 and P_2, respectively.

One common application of Boyle's law is to use Equation (5.2) to predict how the volume of a gas will be affected by a change in pressure, or how the pressure exerted by a gas will be affected by a change in volume.

A balloon for scientific research.

EXAMPLE 5.4

An inflated baloon has a volume of 0.55 L at sea level (1.0 atm) and is allowed to rise to a height of 6.5 km, where the pressure is about 0.40 atm. Assuming that the temperature remains constant, what is the final volume of the balloon?

Answer

We use Equation (5.2): $P_1V_1 = P_2V_2$

where

Initial conditions	*Final conditions*
$P_1 = 1.0$ atm	$P_2 = 0.40$ atm
$V_1 = 0.55$ L	$V_2 = ?$

Therefore

$$\begin{aligned} V_2 &= V_1 \times \frac{P_1}{P_2} \\ &= 0.55 \text{ L} \times \frac{1.0 \text{ atm}}{0.40 \text{ atm}} \\ &= 1.4 \text{ L} \end{aligned}$$

Similar problem: 5.20.

It always makes sense to think about the reasonableness of your answer. Since the balloon is rising from a high-pressure region to a low-pressure one, we expect it to expand.

EXAMPLE 5.5

A sample of chlorine gas occupies a volume of 946 mL at a pressure of 726 mmHg. Calculate the pressure of the gas if the volume is reduced to 154 mL. Assume that the temperature remains constant.

Answer

Since

$$P_1V_1 = P_2V_2$$

where

Initial conditions	*Final conditions*
$P_1 = 726$ mmHg	$P_2 = ?$
$V_1 = 946$ mL	$V_2 = 154$ mL

we continue with

$$\begin{aligned} P_2 &= P_1 \times \frac{V_1}{V_2} \\ &= 726 \text{ mmHg} \times \frac{946 \text{ mL}}{154 \text{ mL}} \\ &= 4.46 \times 10^3 \text{ mmHg} \end{aligned}$$

Similar problems: 5.21, 5.22.

Boyle's law can also be written as $P \propto 1/V$. Thus as the volume of the gas decreases, its pressure will increase.

The Temperature-Volume Relationship: Charles and Gay-Lussac's Law

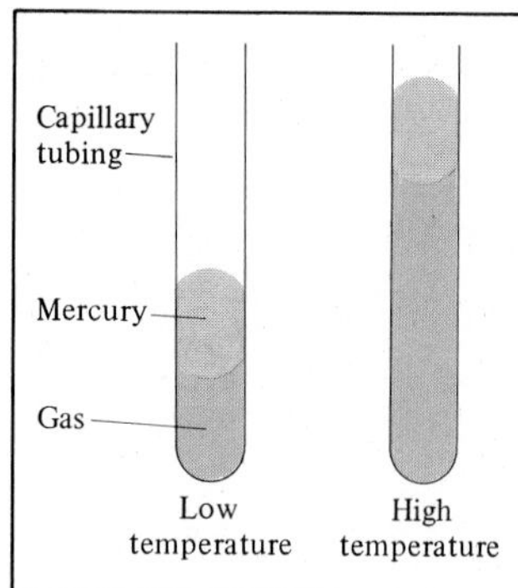

Figure 5.6 *Variation of the volume of a gas sample with temperature, at constant pressure. The pressure exerted on the gas is the sum of the atmospheric pressure and the pressure due to the weight of the mercury.*

Boyle's law depends on the temperature of the system remaining constant. But suppose the temperature changes: How does a change in temperature affect the volume and pressure of a gas? Let us begin with the effect of temperature on the volume of a gas. The earliest investigators of this relationship were French scientists, Jacques Charles† and Joseph Gay-Lussac.‡ Their studies showed that, at constant pressure, the volume of a gas sample expands when heated and contracts when cooled (Figure 5.6). This conclusion may seem almost self-evident, and the quantitative relations involved in these changes in temperature and volume turn out to be remarkably consistent. For example, we observe an interesting phenomenon when we study the temperature-volume relationship at various pressures. At any given pressure, the plot of volume versus temperature yields a straight line. By extending the line to zero volume, we find the intercept on the temperature axis to be −273.15°C. At any other pressure, we obtain a different straight line for the volume-temperature plot, but we get the *same* zero-volume temperature intercept at −273.15°C (Figure 5.7). (In practice, we can measure the volume of a gas over only a limited temperature range, because all gases condense at low temperatures to form liquids.)

In 1848 Lord Kelvin§ realized the significance of this behavior. He identified the temperature −273.15°C as *theoretically the lowest attainable temperature,* called ***absolute zero.*** With *absolute zero as a starting point,* he set up an ***absolute temperature***

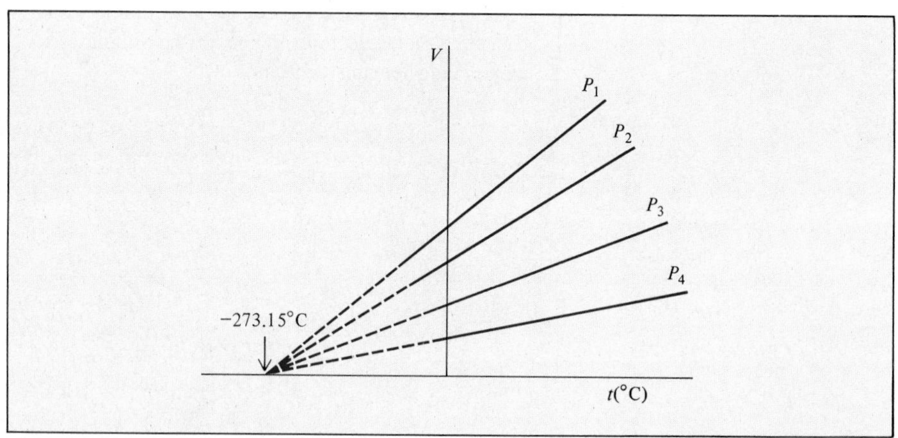

Figure 5.7 *Variation of the volume of a gas sample with temperature, at constant pressure. Each line represents the variation at a certain pressure. The pressures increase from P_1 to P_4. All gases ultimately condense (become liquids) if they are cooled to sufficiently low temperatures; the solid portions of the lines represent the temperature region above the condensation point. When these lines are extrapolated, or extended (the dashed portions), they all intersect at the point representing zero volume and a temperature of −273.15°C.*

† Jacques Alexandre Cesar Charles (1746–1823). French physicist. He was a gifted lecturer, an inventor of scientific apparatus, and the first person to use hydrogen to inflate balloons.

‡ Joseph Louis Gay-Lussac (1778–1850). French chemist and physicist. Like Charles, Gay-Lussac was also a balloon enthusiast. Once he ascended to an altitude of 20,000 feet to collect air samples for analysis.

§ William Thomson, Lord Kelvin (1824–1907). Scottish mathematician and physicist. Kelvin did important work in almost every branch of physics.

scale, now called the ***Kelvin temperature scale.*** One degree Celsius is equal *in magnitude* to one kelvin (K). (Note that the absolute temperature scale has no degree sign, so 25 K is called twenty-five kelvins.) The only difference between the absolute temperature scale and the Celsius scale is that the zero position is shifted. Important points on the scales match up as follows:

Under special experimental conditions, scientists have succeeded in approaching absolute zero to within a small fraction of a kelvin.

Absolute zero:	0 K = −273.15°C
Freezing point of water:	273.15 K = 0°C
Boiling point of water:	373.15 K = 100°C

The relationship between °C and K is

$$T(\mathrm{K}) = t(^\circ\mathrm{C}) + 273.15^\circ\mathrm{C} \tag{5.3}$$

Equation (5.3) shows that to convert a temperature reading in degrees Celsius to kelvins we have to add 273.15 to the reading. This equation can be rewritten as

$$\mathrm{K} = (^\circ\mathrm{C} + 273.15^\circ\mathrm{C})\left(\frac{1\ \mathrm{K}}{1^\circ\mathrm{C}}\right)$$

The unit factor (1 K/1°C) is there to make the units consistent on both sides of the equation. In most calculations we will use 273 instead of 273.15 as the term relating K and °C. By convention we use T to denote absolute (kelvin) temperature and t to indicate temperature on the Celsius scale. Figure 5.8 shows corresponding temperatures on the Celsius and Kelvin scales for a number of systems.

The dependence of volume on temperature is given by

$$V \propto T$$

$$V = k_2 T$$

or

$$\frac{V}{T} = k_2 \tag{5.4}$$

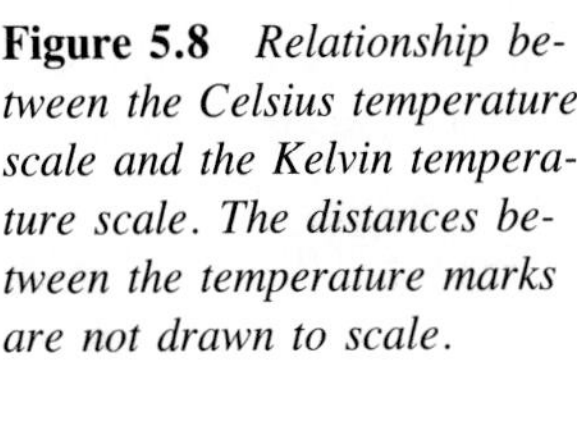

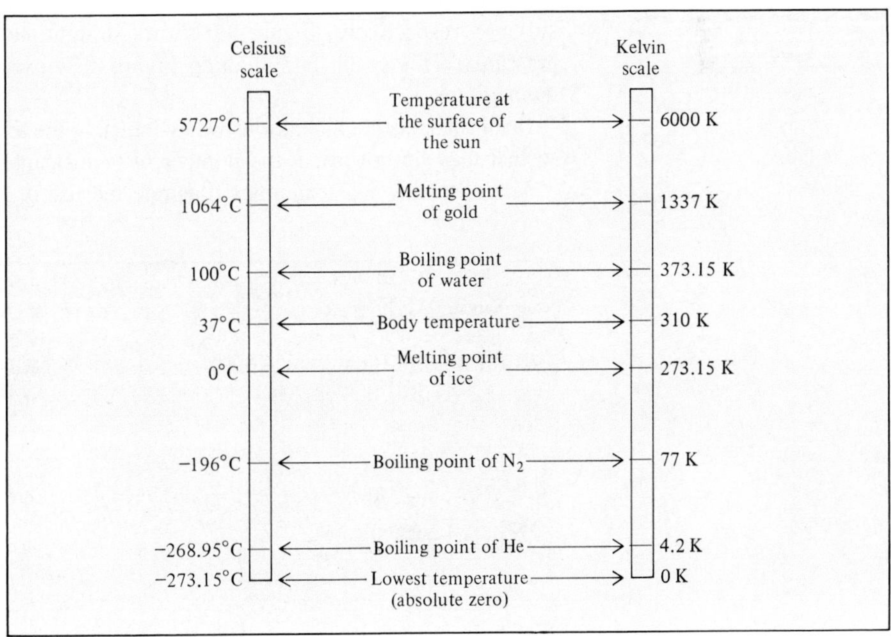

Figure 5.8 *Relationship between the Celsius temperature scale and the Kelvin temperature scale. The distances between the temperature marks are not drawn to scale.*

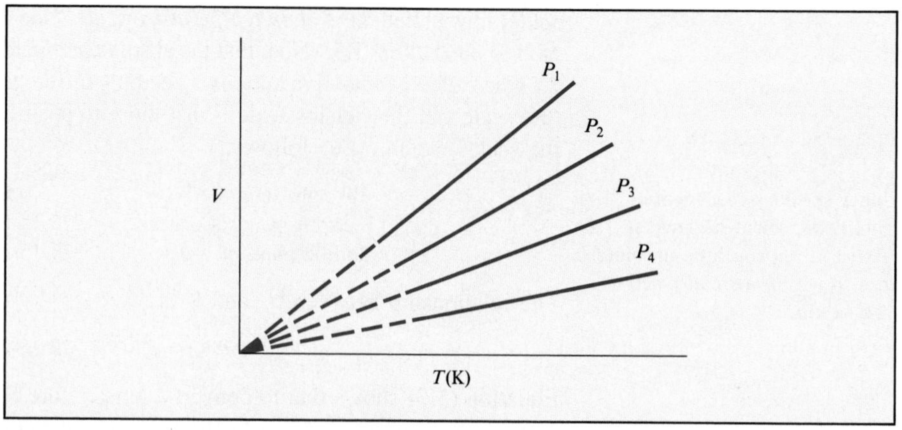

Figure 5.9 *Volume of a gas sample versus temperature at constant pressure. The graph is similar to Figure 5.7, except that the temperature is in kelvins and the line goes through the origin.*

The constant k_2 is not the same as the constant k_1 in Equations (5.1a) and (5.1b).

where k_2 is the proportionality constant. Equation (5.4) is known as ***Charles and Gay-Lussac's law,*** or simply ***Charles' law.*** Charles' law states that *the volume of a fixed amount of gas maintained at constant pressure is directly proportional to the absolute temperature of the gas.*

Just as we did for pressure-volume relationships at constant temperature, we can compare two sets of conditions for a given sample of gas at constant pressure. From Equation (5.4) we can write

$$\frac{V_1}{T_1} = k_2 = \frac{V_2}{T_2}$$

or

$$\frac{V_1}{T_1} = \frac{V_2}{T_2} \tag{5.5}$$

where V_1 and V_2 are the volumes of the gases at temperatures T_1 and T_2 (both in kelvins), respectively. Figure 5.9 shows straightline plots of V versus T at different pressures. This graph is similar to Figure 5.7, except the Kelvin scale is used for temperature.

In all subsequent calculations we assume that the temperatures given in °C are exact, so that they do not affect the number of significant figures.

The following two examples illustrate the use of Charles' law.

EXAMPLE 5.6

Argon is an inert gas used in light bulbs. In one experiment, 452 mL of the gas is heated from 22°C to 187°C at constant pressure. What is its final volume?

Answer

We use Equation (5.5),

$$\frac{V_1}{T_1} = \frac{V_2}{T_2}$$

Remember to convert °C to K when solving gas law problems.

where

Initial conditions	*Final conditions*
$V_1 = 452\ \text{mL}$	$V_2 = ?$
$T_1 = (22 + 273)\ \text{K} = 295\ \text{K}$	$T_2 = (187 + 273)\ \text{K} = 460\ \text{K}$

Hence

$$V_2 = V_1 \times \frac{T_2}{T_1}$$

$$= 452\ \text{mL} \times \frac{460\ \text{K}}{295\ \text{K}}$$

$$= 705\ \text{mL}$$

Similar problem: 5.26.

As you can see, when the gas is heated at constant pressure, it expands.

EXAMPLE 5.7

A sample of carbon monoxide, a poisonous gas, occupies 3.20 L at 125°C. Calculate the temperature at which the gas will occupy 1.54 L if the pressure remains constant.

Answer

Since

$$\frac{V_1}{T_1} = \frac{V_2}{T_2}$$

where

Initial conditions	*Final conditions*
$V_1 = 3.20\ \text{L}$	$V_2 = 1.54\ \text{L}$
$T_1 = (125 + 273)\ \text{K} = 398\ \text{K}$	$T_2 = ?$

then

$$T_2 = T_1 \times \frac{V_2}{V_1}$$

$$= 398\ \text{K} \times \frac{1.54\ \text{L}}{3.20\ \text{L}}$$

$$= 192\ \text{K}$$

or, from Equation (5.3),

$$°\text{C} = T_2 - 273 = 192 - 273$$

$$= -81°\text{C}$$

Similar problem: 5.27.

The above example shows that when a gas is cooled at constant pressure, it contracts.

The Volume-Amount Relationship: Avogadro's Law

The name Avogadro first appeared in Section 2.3.

The work of the Italian scientist Amedeo Avogadro complemented the studies of Boyle, Charles, and Gay-Lussac. In 1811 he published a hypothesis that stated that at the same temperature and pressure, equal volumes of different gases contain the same number of molecules (or atoms if the gas is monatomic). It follows that the volume of any given gas must be proportional to the number of molecules present; that is,

$$V \propto n$$
$$V = k_3 n \qquad (5.6)$$

where n represents the number of moles and k_3 is the proportionality constant.

Equation (5.6) is the mathematical expression of ***Avogadro's law,*** which states that *at constant pressure and temperature, the volume of a gas is directly proportional to the number of moles of the gas present.* From Avogadro's law we learn that when two gases react with each other, their reacting volumes have a simple ratio to each other. If the product is a gas, its volume is related to the volume of the reactants by a simple ratio (a fact demonstrated earlier by Gay-Lussac). For example, consider the synthesis of ammonia from molecular hydrogen and molecular nitrogen:

$$\underset{3\text{ mol}}{3H_2(g)} + \underset{1\text{ mol}}{N_2(g)} \longrightarrow \underset{2\text{ mol}}{2NH_3(g)}$$

Since, at the same temperature and pressure, the volumes of gases are directly proportional to the number of moles of the gases present, we can now write

$$\underset{3\text{ volumes}}{3H_2(g)} + \underset{1\text{ volume}}{N_2(g)} \longrightarrow \underset{2\text{ volumes}}{2NH_3(g)}$$

The ratio of volumes for molecular hydrogen and molecular nitrogen is 3:1, and that for ammonia (the product) and for molecular hydrogen and molecular nitrogen (the reactants) is 2:4, or 1:2 (Figure 5.10).

5.4 The Ideal Gas Equation

Let us summarize the various gas laws we have discussed so far:

$$\text{Boyle's law:} \quad V \propto \frac{1}{P} \quad \text{(at constant } n \text{ and } T)$$

$$\text{Charles' law:} \quad V \propto T \quad \text{(at constant } n \text{ and } P)$$

$$\text{Avogadro's law:} \quad V \propto n \quad \text{(at constant } P \text{ and } T)$$

We can combine all three expressions to form a single master equation for the behavior of gases:

$$V \propto \frac{nT}{P}$$
$$= R\frac{nT}{P}$$

or

$$PV = nRT \qquad (5.7)$$

where R, *the proportionality constant,* is called the ***gas constant.*** Equation (5.7), which *describes the relationship among* the four experimental variables P, V, T, and n,

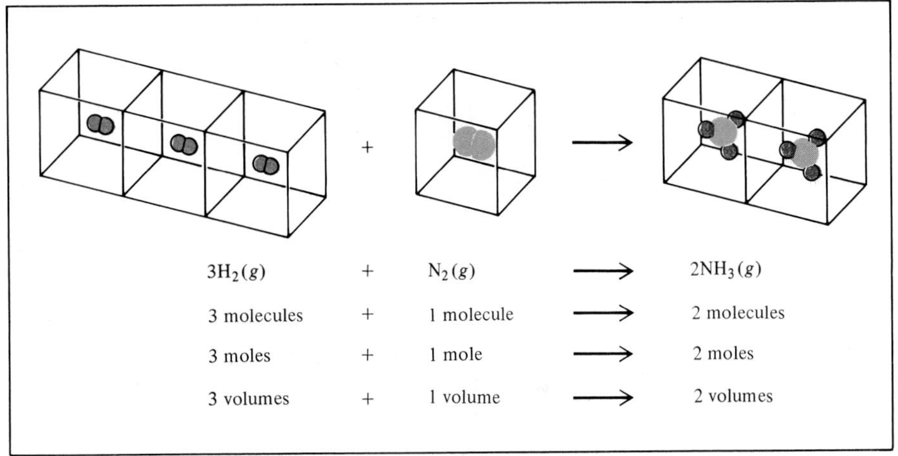

Figure 5.10 *Volume relationship of gases in a chemical reaction. The ratio of volumes for molecular hydrogen and molecular nitrogen is 3:1, and that for ammonia (the product) and molecular hydrogen and molecular nitrogen combined (the reactants) is 2:4, or 1:2.*

is called the ***ideal gas equation.*** An ***ideal gas*** is *a hypothetical gas whose pressure-volume-temperature behavior can be completely accounted for by the ideal gas equation.* The molecules of an ideal gas do not attract or repel one another, and their volume is negligible compared with the volume of the container. Although there is no such thing in nature as an ideal gas, discrepancies in the behavior of real gases over reasonable temperature and pressure ranges do not significantly affect calculations. Thus we can safely use the ideal gas equation to solve many gas problems.

Before we can apply Equation (5.7) to a real system, we must evaluate the gas constant R. At 0°C (273.15 K) and 1 atm pressure, many real gases behave like an ideal gas. Experiments show that under these conditions, 1 mole of an ideal gas occupies 22.414 L (Figure 5.11). Figure 5.12 compares this molar volume with that of a basketball. *The conditions 0°C and 1 atm* are called ***standard temperature and pressure,*** often abbreviated ***STP.*** From Equation (5.7) we can write

The gas constant can also be expressed in other units (see Appendix 2).

$$
\begin{aligned}
R &= \frac{PV}{nT} \\
&= \frac{(1\ \text{atm})(22.414\ \text{L})}{(1\ \text{mol})(273.15\ \text{K})} = 0.082057\frac{\text{L}\cdot\text{atm}}{\text{K}\cdot\text{mol}} \\
&= 0.082057\ \text{L}\cdot\text{atm/K}\cdot\text{mol}
\end{aligned}
$$

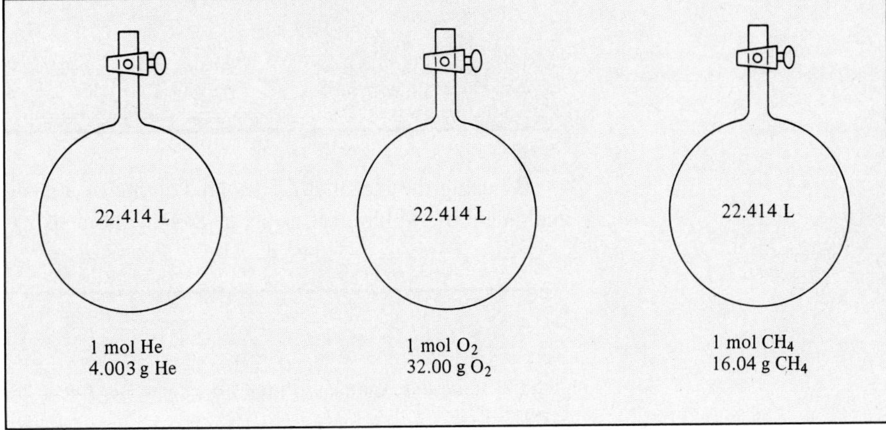

Figure 5.11 *At STP, 1 mole of an ideal gas, regardless of whether it is atomic or molecular, occupies 22.414 L.*

Figure 5.12 *A comparison of the molar volume at STP (22.414 L) with a basketball.*

The dots (between L and atm and between K and mol) remind us that both L and atm are in the numerator and both K and mol are in the denominator. For most calculations, we will round off the value of R to three significant figures (0.0821 L · atm/K · mol) and use 22.41 L for the molar volume of a gas at STP.

The following example shows that if we know the quantity, volume, and temperature of a gas, we can calculate its pressure using the ideal gas equation.

EXAMPLE 5.8

Sulfur hexafluoride (SF_6) is a colorless, odorless, very unreactive gas. Calculate the pressure (in atm) exerted by 1.82 moles of the gas in a steel vessel of volume 5.43 L at 69.5°C.

Answer

From Equation (5.7) we write

$$\begin{aligned} P &= \frac{nRT}{V} \\ &= \frac{(1.82\ \cancel{\text{mol}})(0.0821\ \cancel{\text{L}} \cdot \text{atm}/\cancel{\text{K}} \cdot \cancel{\text{mol}})(69.5 + 273)\ \cancel{\text{K}}}{5.43\ \cancel{\text{L}}} \\ &= 9.42\ \text{atm} \end{aligned}$$

Similar problem: 5.38.

By using the fact that the molar volume of a gas occupies 22.41 L at STP, we can calculate the volume of a gas at STP without using the ideal gas equation.

EXAMPLE 5.9

Calculate the volume (in liters) occupied by 7.40 g of CO_2 at STP.

Answer

Recognizing that 1 mole of an ideal gas occupies 22.41 L at STP, we write

$$V = 7.40 \cancel{g\ CO_2} \times \frac{1 \cancel{mol\ CO_2}}{44.01 \cancel{g\ CO_2}} \times \frac{22.41\ L}{1 \cancel{mol\ CO_2}}$$
$$= 3.77\ L$$

Similar problem: 5.55.

Equation (5.7) is useful for problems that do not involve changes among the variables P, V, T, and n for a gas sample. At times, however, we need to deal with changes in pressure, volume, and temperature, or even the amount of a gas. When pressure, volume, temperature, and amount change, we must employ a modified form of Equation (5.7) that involves initial and final conditions. We derive it as follows. From Equation (5.7),

The subscripts 1 and 2 denote the initial and final states of the gas.

$$R = \frac{P_1V_1}{n_1T_1} \quad \text{(before change)}$$

and

$$R = \frac{P_2V_2}{n_2T_2} \quad \text{(after change)}$$

so that

$$\frac{P_1V_1}{n_1T_1} = \frac{P_2V_2}{n_2T_2}$$

If $n_1 = n_2$, as is usually the case because the amount of gas normally does not change, the equation then becomes

$$\frac{P_1V_1}{T_1} = \frac{P_2V_2}{T_2} \quad (5.8)$$

Applications of Equation (5.8) are the subject of the following two examples.

EXAMPLE 5.10

A small bubble rises from the bottom of a lake, where the temperature and pressure are 8°C and 6.4 atm, to the water's surface, where the temperature is 25°C and pressure is 1.0 atm. Calculate the final volume (in mL) of the bubble if its initial volume was 2.1 mL.

Answer

We start by writing

We can use any appropriate units for volume (or pressure) as long as we use the same units on both sides of the equation.

Initial conditions	*Final conditions*
$P_1 = 6.4$ atm	$P_2 = 1.0$ atm
$V_1 = 2.1$ mL	$V_2 = ?$
$T_1 = (8 + 273)$ K $= 281$ K	$T_2 = (25 + 273)$ K $= 298$ K

The amount of the gas in the bubble remains constant, so that $n_1 = n_2$. To calculate the final volume, V_2, we rearrange Equation (5.8) as follows:

$$\begin{aligned} V_2 &= V_1 \times \frac{P_1}{P_2} \times \frac{T_2}{T_1} \\ &= 2.1 \text{ mL} \times \frac{6.4 \text{ atm}}{1.0 \text{ atm}} \times \frac{298 \text{ K}}{281 \text{ K}} \\ &= 14 \text{ mL} \end{aligned}$$

Thus, the bubble's volume increases from 2.1 mL to 14 mL because of the decrease in water pressure and the increase in temperature.

Similar problems: 5.42, 5.45.

EXAMPLE 5.11

A steel bulb contains xenon (Xe) gas at 2.64 atm and 25°C. Calculate the pressure of the gas when the temperature is raised to 436°C. Assume that the volume of the bulb remains constant.

Answer

The volume and amount of the gas are unchanged, so $V_1 = V_2$ and $n_1 = n_2$. Equation (5.8) can therefore be written as

$$\frac{P_1}{T_1} = \frac{P_2}{T_2}$$

Next we write

Initial conditions	*Final conditions*
$P_1 = 2.64$ atm	$P_2 = ?$
$T_1 = (25 + 273)$ K = 298 K	$T_2 = (436 + 273)$ K = 709 K

The final pressure is given by

$$\begin{aligned} P_2 &= P_1 \times \frac{T_2}{T_1} \\ &= 2.64 \text{ atm} \times \frac{709 \text{ K}}{298 \text{ K}} \\ &= 6.28 \text{ atm} \end{aligned}$$

We see that at constant volume, the pressure of a given amount of gas is directly proportional to its absolute temperature.

Similar problem: 5.43.

One practical consequence of this relationship is that automobile tire pressures should be checked only when the tires are at normal temperatures. After a long drive (especially in the summer), tires become quite hot, leading to a higher air pressure.

Density Calculations

If we rearrange the ideal gas equation, we can use it to calculate the density of a gas:

$$\frac{n}{V} = \frac{P}{RT}$$

The number of moles of the gas, n, is given by

$$n = \frac{m}{\mathcal{M}}$$

where m is the mass of the gas in grams and $\mathcal{M}$ is its molar mass. Therefore

$$\frac{m}{\mathcal{M}V} = \frac{P}{RT}$$

Since density, d, is mass per unit volume, we can write

$$d = \frac{m}{V} = \frac{P\mathcal{M}}{RT} \tag{5.9}$$

Unlike molecules in condensed matter (that is, in liquids and solids), gaseous molecules are separated by distances that are large compared with their size. Consequently, the density of gases is very low under atmospheric conditions. For this reason, the units for gas density are usually expressed as grams per liter (g/L) rather than grams per milliliter (g/mL), as the following example shows.

EXAMPLE 5.12

Calculate the density of ammonia (NH_3) in grams per liter (g/L) at 752 mmHg and 55°C.

Answer

To convert the pressure to atmospheres, we write

$$P = 752 \text{ mmHg} \times \frac{1 \text{ atm}}{760 \text{ mmHg}}$$
$$= \left(\frac{752}{760}\right) \text{atm}$$

Using Equation (5.9) and $T = 273 + 55 = 328$ K, we have

$$d = \frac{P\mathcal{M}}{RT}$$
$$= \frac{(752/760) \text{ atm } (17.03 \text{ g/mol})}{(0.0821 \text{ L} \cdot \text{atm/K} \cdot \text{mol})(328 \text{ K})}$$
$$= 0.626 \text{ g/L}$$

Similar problem: 5.56.

In units of grams per milliliter, the gas density would be 6.26×10^{-4} g/mL.

The Molar Mass of a Gaseous Substance

From what we have said so far, you may have gathered the impression that the molar mass of a substance is found by examining its formula and summing the molar masses

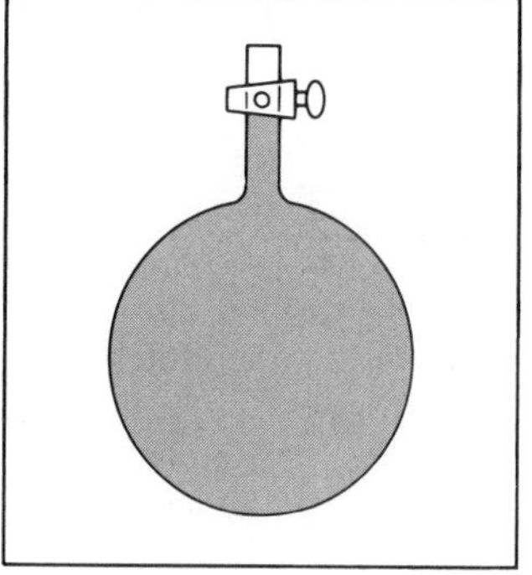

Figure 5.13 *An apparatus for measuring the density of a gas. The bulb of known volume is filled with the gas under study at a certain temperature and pressure. First the bulb is weighed and then it is emptied (evacuated) and weighed again. The difference in masses gives the mass of the gas. Knowing the volume of the bulb, we can calculate the density of the gas.*

of its component atoms. However, this procedure works only if the actual formula of the substance is known. In practice, chemists often deal with substances of unknown or only partially defined composition. If the unknown substance is gaseous, its molar mass can nevertheless be found, thanks to the ideal gas equation. All that is needed is an experimentally determined density value (or mass and volume data) for the gas at a known temperature and pressure. By rearranging Equation (5.9) we get

$$\mathcal{M} = \frac{dRT}{P} \tag{5.10}$$

Under atmospheric conditions, 100 mL of air weigh about 0.12 g, an easily measured quantity.

In a typical experiment, a bulb of known volume is filled with the gaseous substance under study. The temperature and pressure of the contained gas sample are recorded, and the total mass of the bulb plus gas sample is determined (Figure 5.13). The bulb is then evacuated (emptied) and weighed again. The difference in mass is the mass of the gas. The density of the gas is equal to its mass divided by the volume of the bulb. Then we can calculate the molar mass of the substance using Equation (5.10). Example 5.13 shows this calculation.

EXAMPLE 5.13

A chemist has synthesized a greenish-yellow gaseous compound of chlorine and oxygen and finds that its density is 7.71 g/L at 36°C and 2.88 atm. Calculate the molar mass of the compound and determine its molecular formula.

Answer

We substitute in Equation (5.10)

$$\begin{aligned}\mathcal{M} &= \frac{dRT}{P}\\ &= \frac{(7.71 \text{ g/L})(0.0821 \text{ L} \cdot \text{atm/K} \cdot \text{mol})(36 + 273) \text{ K}}{2.88 \text{ atm}}\\ &= 67.9 \text{ g/mol}\end{aligned}$$

We can determine the molecular formula of the compound by trial and error, using only the knowledge of the molar masses of chlorine (35.45 g) and oxygen (16.00 g). We know that a compound containing one Cl atom and one O atom would have a molar mass of 51.45 g, which is too low, while the molar mass of a compound made up of two Cl atoms and one O atom is 86.90 g, which is too high. Thus the compound must contain one Cl atom and two O atoms, or ClO_2, which has a molar mass of 67.45 g.

Similar problems: 5.57, 5.60.

Since Equation (5.10) is derived from Equation (5.7), we can also calculate the molar mass of a gaseous substance using the ideal gas equation. The following example illustrates this approach.

EXAMPLE 5.14

Chemical analysis of a gaseous compound showed that it contained 33.0 percent silicon and 67.0 percent fluorine by mass. At 35°C, 0.210 L of the compound exerted a pressure of 1.70 atm. If the mass of 0.210 L of the compound was 2.38 g, calculate the molecular formula of the compound.

Answer

First we determine the empirical formula of the compound.

$$n_{\text{Si}} = 33.0 \text{ g Si} \times \frac{1 \text{ mol Si}}{28.09 \text{ g Si}} = 1.17 \text{ mol Si}$$

$$n_{\text{F}} = 67.0 \text{ g F} \times \frac{1 \text{ mol F}}{19.00 \text{ g F}} = 3.53 \text{ mol F}$$

Therefore, the formula is $Si_{1.17}F_{3.53}$, or SiF_3. Next, we calculate the number of moles contained in 2.38 g of the compound. From Equation (5.7),

$$n = \frac{PV}{RT}$$

$$= \frac{(1.70 \text{ atm})(0.210 \text{ L})}{(0.0821 \text{ L} \cdot \text{atm/K} \cdot \text{mol})(308 \text{ K})} = 0.0141 \text{ mol}$$

Therefore, the molar mass of the compound is

$$\mathcal{M} = \frac{2.38 \text{ g}}{0.0141 \text{ mol}}$$

$$= 169 \text{ g/mol}$$

The empirical molar mass of SiF_3 (empirical formula) is 85.09 g. Therefore the molecular formula of the compound must be Si_2F_6, because 2 × 85.09, or 170.2 g, is very close to 169 g.

Similar problem: 5.57.

5.5 Stoichiometry Involving Gases

In Chapter 4 we used relationships between amounts (in moles) and masses (in grams) of reactants and products to solve stoichiometry problems. When the reactants and/or products are gases, we can also use the relationships between amounts (moles, n) and volume (V) to solve such problems. The following examples show how the gas laws are used in these calculations.

EXAMPLE 5.15

Calculate the volume of O_2 (in liters) at STP required for the complete combustion of 2.64 L of acetylene (C_2H_2) at STP:

$$2C_2H_2(g) + 5O_2(g) \longrightarrow 4CO_2(g) + 2H_2O(l)$$

Answer

Since both C_2H_2 and O_2 are gases measured at the same temperature and pressure, according to Avogadro's law their reacting volumes are related to their coefficients in the balanced equation; that is, 2 L of C_2H_2 react with 5 L of O_2. Knowing this ratio we can calculate the volume of O_2 that will react with 2.64 L of C_2H_2.

$$\text{volume of } O_2 = 2.64 \text{ L } C_2H_2 \times \frac{5 \text{ L } O_2}{2 \text{ L } C_2H_2}$$
$$= 6.60 \text{ L } O_2$$

Similar problem: 5.61.

EXAMPLE 5.16

An air bag can protect the driver in a collision.

Sodium azide (NaN_3) is used in some air bags in automobiles. The impact of a collision triggers the decomposition of NaN_3 as follows:

$$2NaN_3(s) \longrightarrow 2Na(s) + 3N_2(g)$$

The nitrogen gas produced quickly inflates the bag between the driver and the windshield. Calculate the volume of N_2 generated at 21°C and 823 mmHg upon the decomposition of 60.0 g of NaN_3.

Answer

From the balanced equation we see that 2 mol NaN_3 ≎ 3 mol N_2. The number of moles of N_2 produced by 60.0 g NaN_3 is

$$\text{moles of } N_2 = 60.0 \text{ g } NaN_3 \times \frac{1 \text{ mol } NaN_3}{65.02 \text{ g } NaN_3} \times \frac{3 \text{ mol } N_2}{2 \text{ mol } NaN_3}$$
$$= 1.38 \text{ mol } N_2$$

The volume of N_2 produced can be obtained by using Equation (5.7):

$$V = \frac{nRT}{P} = \frac{(1.38 \text{ mol})(0.0821 \text{ L} \cdot \text{atm/K} \cdot \text{mol})(294 \text{ K})}{(823/760) \text{ atm}}$$
$$= 30.8 \text{ L}$$

Similar problems: 5.62, 5.63.

EXAMPLE 5.17

The first step in the industrial preparation of nitric acid (HNO_3) is the production of nitric oxide (NO) from ammonia (NH_3) and oxygen (O_2):

$$4NH_3(g) + 5O_2(g) \longrightarrow 4NO(g) + 6H_2O(g)$$

If 3.00 L of NH_3 at 802°C and 1.30 atm completely react with oxygen, how many liters of steam measured at 125°C and 1.00 atm are formed?

Answer

Our first step is to find the volume that would be occupied by the NH_3 at 125°C and 1 atm. Since $n_1 = n_2$, we can use Equation (5.8):

$$\frac{P_1V_1}{T_1} = \frac{P_2V_2}{T_2}$$

where

Initial conditions	*Final conditions*
$P_1 = 1.30$ atm	$P_2 = 1.00$ atm
$V_1 = 3.00$ L	$V_2 = ?$
$T_1 = (802 + 273)$ K $= 1075$ K	$T_2 = (125 + 273)$ K $= 398$ K

Rearranging Equation (5.8) gives

$$V_2 = V_1 \times \frac{P_1}{P_2} \times \frac{T_2}{T_1}$$

$$= 3.00\text{ L} \times \frac{1.30\text{ atm}}{1.00\text{ atm}} \times \frac{398\text{ K}}{1075\text{ K}} = 1.44\text{ L}$$

Following the procedure in Example 5.15, the volume of gaseous H_2O (steam) is given by

$$\text{volume of } H_2O = 1.44\text{ L } NH_3 \times \frac{6\text{ L } H_2O}{4\text{ L } NH_3}$$

$$= 2.16\text{ L } H_2O$$

An alternative approach to this problem is first to calculate the number of moles of NH_3 that react, using Equation (5.7). Next calculate the number of moles of H_2O produced from the relationship 4 mol $NH_3 \simeq$ 6 mol H_2O. Finally, calculate the volume of H_2O at 125°C and 1.00 atm, again using Equation (5.7).

EXAMPLE 5.18

A flask of volume 0.85 L is filled with carbon dioxide (CO_2) gas at a pressure of 1.44 atm and a temperature of 312 K. A solution of lithium hydroxide (LiOH) of negligible volume is introduced into the flask. Eventually the pressure of CO_2 is reduced to 0.56 atm because some CO_2 is consumed in the reaction

$$CO_2(g) + 2LiOH(aq) \longrightarrow Li_2CO_3(aq) + H_2O(l)$$

The ability of LiOH to react with CO_2 makes it useful as an air purifier in space vehicles and submarines.

How many grams of lithium carbonate are formed by this process? Assume that the temperature remains constant.

Answer

First we calculate the number of moles of CO_2 consumed in the reaction. The drop in pressure, which is 1.44 atm − 0.56 atm, or 0.88 atm, corresponds to the consumption of CO_2. From Equation (5.7) we write

$$n = \frac{PV}{RT}$$
$$= \frac{(0.88 \text{ atm})(0.85 \text{ L})}{(0.0821 \text{ L} \cdot \text{atm/K} \cdot \text{mol})(312 \text{ K})} = 0.0292 \text{ mol}$$

From the equation we see that 1 mol CO_2 ⇔ 1 mol Li_2CO_3, so the amount of Li_2CO_3 formed is also 0.0292 mole. Then, with the molar mass of Li_2CO_3 (73.89 g) we calculate its mass:

$$\text{mass of } Li_2CO_3 \text{ formed} = 0.0292 \text{ mol } Li_2CO_3 \times \frac{73.89 \text{ g } Li_2CO_3}{1 \text{ mol } Li_2CO_3}$$
$$= 2.2 \text{ g } Li_2CO_3$$

5.6 Dalton's Law of Partial Pressures

Thus far we have concentrated on the behavior of pure gaseous substances. However, experimental studies are very often based on mixtures of gases. For example, we may be interested in the pressure-volume-temperature relationship of a sample of air, which contains several gases. In considering a gaseous mixture, we need to understand how the total gas pressure is related to the *pressures of individual gas components in the mixture,* called ***partial pressures.*** In 1801 Dalton formulated a law, now known as ***Dalton's law of partial pressures,*** which states that *the total pressure of a mixture of gases is just the sum of the pressures that each gas would exert if it were present alone.*

As mentioned earlier, gas pressure results from the impact of gas molecules against the walls of the container.

Consider a case in which two gaseous substances, A and B, are in a container of volume V. The pressure exerted by gas A, according to Equation (5.7), is

$$P_A = \frac{n_A RT}{V}$$

where n_A is the number of moles of A present. Similarly, the pressure exerted by gas B is

$$P_B = \frac{n_B RT}{V}$$

Now, in a mixture of gases A and B, the total pressure P_T is the result of the collisions of both types of molecules, A and B, with the walls. Thus, according to Dalton's law,

$$P_T = P_A + P_B$$
$$= \frac{n_A RT}{V} + \frac{n_B RT}{V}$$

The result shows that P_T depends only on the total number of moles of gas present, not on the nature of the gas molecules.

$$= \frac{RT}{V}(n_A + n_B)$$
$$= \frac{nRT}{V}$$

where n, the total number of moles of gases present, is given by $n = n_A + n_B$, and P_A and P_B are the partial pressures of gases A and B, respectively. In general, the total pressure of a mixture of gases is given by

$$P_T = P_1 + P_2 + P_3 + \ldots \quad (5.11)$$

where P_1, P_2, P_3, . . . are the partial pressures of components 1, 2, 3, ...

To see how each partial pressure is related to the total pressure, consider again the case of a mixture of two gases A and B. Dividing P_A by P_T, we obtain

$$\frac{P_A}{P_T} = \frac{n_A RT/V}{(n_A + n_B)RT/V}$$
$$= \frac{n_A}{n_A + n_B}$$
$$= X_A$$

where X_A is called the mole fraction of gas A. The ***mole fraction*** is *a dimensionless quantity that expresses the ratio of the number of moles of one component to the number of moles of all components present*. It is always smaller than 1, except when A is the only component present. In that case, $n_B = 0$ and $X_A = n_A/n_A = 1$. We can now express the partial pressure of A as

$$P_A = X_A P_T$$

Similarly,

$$P_B = X_B P_T$$

Note that the sum of mole fractions must be unity. If only two components are present, then

$$X_A + X_B = \frac{n_A}{n_A + n_B} + \frac{n_B}{n_A + n_B} = 1$$

If a system contains more than two gases, then the partial pressure of the ith component is related to the total pressure by

$$P_i = X_i P_T \quad (5.12)$$

where X_i is the mole fraction of substance i.

From mole fractions and total pressure, we can calculate the partial pressures of individual components, as Example 5.19 shows.

EXAMPLE 5.19

A mixture of gases contains 4.46 moles of neon (Ne), 0.74 mole of argon (Ar), and 2.15 moles of xenon (Xe). Calculate the partial pressures of the gases if the total pressure is 2.00 atm at a certain temperature.

Answer

The total pressure of the mixture is given by

$$P_T = \frac{nRT}{V} = (n_{Ne} + n_{Ar} + n_{Xe})\frac{RT}{V}$$

where n_{Ne}, n_{Ar}, and n_{Xe} are the numbers of moles of the gases present. The partial pressure of Ne is

$$P_{Ne} = n_{Ne}\frac{RT}{V}$$

Taking the ratio of P_{Ne}/P_T, we get

$$\frac{P_{Ne}}{P_T} = \frac{n_{Ne}}{n_{Ne} + n_{Ar} + n_{Xe}} = X_{Ne} = \frac{4.46 \text{ mol}}{4.46 \text{ mol} + 0.74 \text{ mol} + 2.15 \text{ mol}} = 0.607$$

where X_{Ne} is the mole fraction of Ne. Thus

$$P_{Ne} = 0.607 \times 2.00 \text{ atm} = 1.21 \text{ atm}$$

Similarly,

$$P_{Ar} = 0.10 \times 2.00 \text{ atm} = 0.20 \text{ atm}$$

and

$$P_{Xe} = 0.293 \times 2.00 \text{ atm} = 0.586 \text{ atm}$$

Similar problem: 5.75.

Table 5.3 Pressure of water vapor at various temperatures

Temperature (°C)	Pressure (mmHg) of water vapor
0	4.58
5	6.54
10	9.21
15	12.79
20	17.54
25	23.76
30	31.82
35	42.18
40	55.32
45	71.88
50	92.51
55	118.04
60	149.38
65	187.54
70	233.7
75	289.1
80	355.1
85	433.6
90	525.76
95	633.90
100	760.00

Dalton's law of partial pressures has a practical application in calculating volumes of gases collected over water, for example, oxygen. When potassium chlorate ($KClO_3$) is heated, the products are KCl and O_2:

$$2KClO_3(s) \longrightarrow 2KCl(s) + 3O_2(g)$$

The evolved oxygen gas can be collected over water, as shown in Figure 5.14. Initially, the inverted bottle is completely filled with water. As oxygen gas is generated, the gas bubbles rise to the top and water is pushed out of the bottle. Note that this water displacement method for collecting a gas is based on the assumptions that the gas does not react with water and that it is not appreciably soluble in it. These assumptions are valid for oxygen gas, but the procedure will not work for gases such as NH_3, which dissolves readily in water. The oxygen gas collected in this way is not pure, however, because water vapor is also present in the same space. The total gas pressure is equal to the sum of the pressures exerted by the oxygen gas and the water vapor:

$$P_T = P_{O_2} + P_{H_2O}$$

Consequently, we must allow for the pressure caused by the presence of water vapor when we calculate the amount of O_2 generated. Table 5.3 shows the pressure of water vapor at various temperatures. These data are plotted in Figure 5.15.

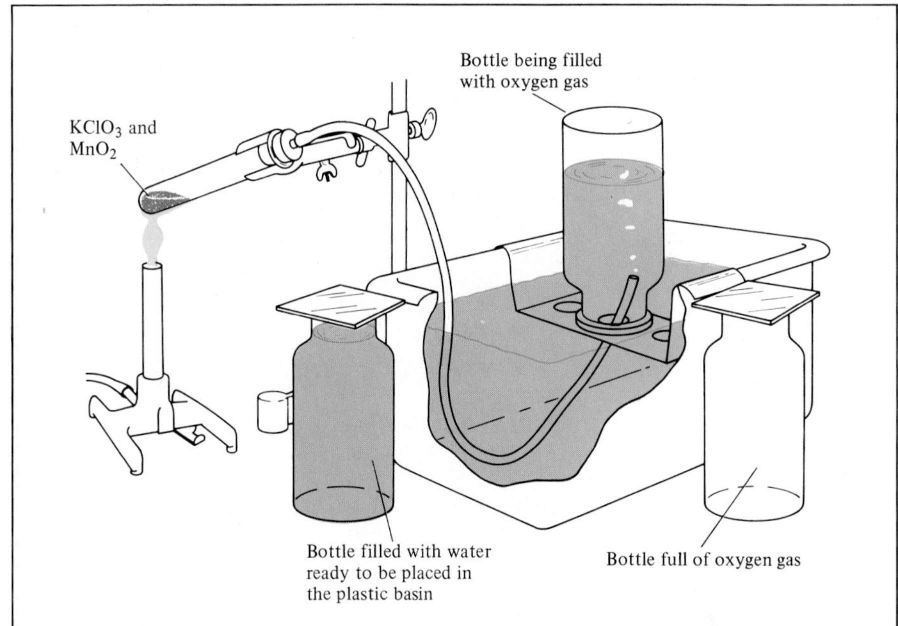

Figure 5.14 *An apparatus for collecting gas over water. The oxygen generated by heating potassium chlorate ($KClO_3$) in the presence of a small amount of manganese dioxide (MnO_2, to speed up the reaction) is bubbled through water and collected in a bottle as shown. Water originally present in the bottle is pushed into the trough by the oxygen gas.*

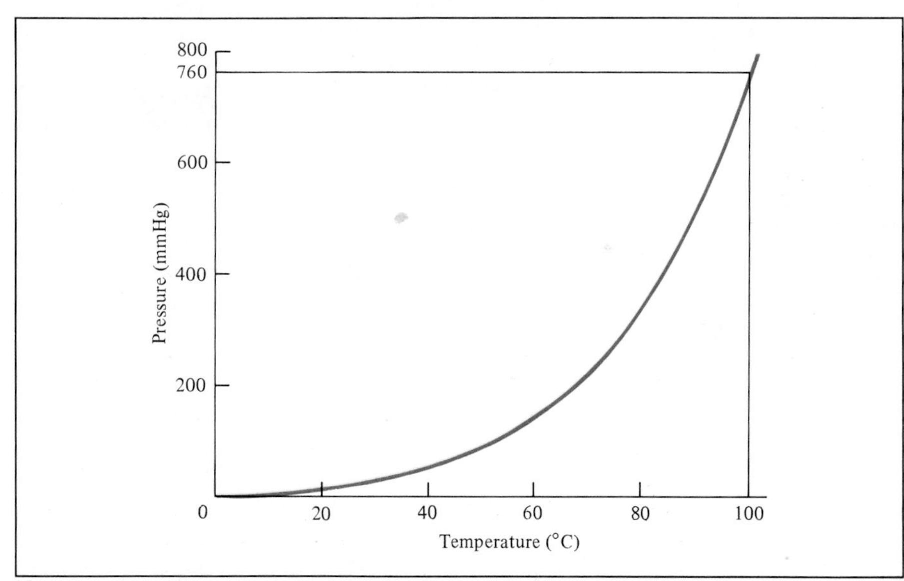

Figure 5.15 *The pressure of water vapor as a function of temperature. Note that at the boiling point of water (100°C) the pressure is 760 mmHg, which is exactly equal to 1 atm of pressure.*

Example 5.20 shows that we can use Dalton's law to calculate the amount of a gas collected over water.

EXAMPLE 5.20

Oxygen gas generated in the decomposition of potassium chlorate is collected as shown in Figure 5.14. The volume of the gas collected at 24°C and atmospheric pressure of 762 mmHg is 128 mL. Calculate the mass (in grams) of oxygen gas obtained. The pressure of the water vapor at 24°C is 22.4 mmHg.

Answer

Our first step is to calculate the partial pressure of O_2. We know that

$$P_T = P_{O_2} + P_{H_2O}$$

therefore

$$\begin{aligned} P_{O_2} &= P_T - P_{H_2O} \\ &= 762 \text{ mmHg} - 22.4 \text{ mmHg} \\ &= 740 \text{ mmHg} \\ &= 740 \text{ mmHg} \times \frac{1 \text{ atm}}{760 \text{ mmHg}} \\ &= 0.974 \text{ atm} \end{aligned}$$

From the ideal gas equation we write

$$PV = nRT = \frac{m}{\mathcal{M}}RT$$

where m and $\mathcal{M}$ are the mass of O_2 collected and the molar mass of O_2, respectively. Rearranging the equation we get

$$\begin{aligned} m &= \frac{PV\mathcal{M}}{RT} = \frac{(0.974 \text{ atm})(0.128 \text{ L})(32.00 \text{ g/mol})}{(0.0821 \text{ L} \cdot \text{atm/K} \cdot \text{mol})(273 + 24) \text{ K}} \\ &= 0.164 \text{ g} \end{aligned}$$

Similar problem: 5.80.

The Chemistry in Action on p. 195 describes some interesting applications of the gas laws.

5.7 The Kinetic Molecular Theory of Gases

The gas laws help us to predict the behavior of gases, but they do not explain, at the molecular level, the changes in volume, pressure, or temperature when conditions are altered. For example, why does the volume of a gas expand upon heating? Starting in the 1850s, a number of physicists, notably Ludwig Boltzmann† in Germany and James Clerk Maxwell‡ in England, found that the physical properties of gases could be satisfactorily explained in terms of the motions of individual molecules. This approach serves to illustrate what we mean by observing phenomena in the macroscopic world (pressure, volume, temperature changes) and interpreting the behavior in the microscopic world (properties of molecules), as discussed in Section 1.2. Boltzmann and Maxwell's work subsequently laid the foundation for the kinetic molecular theory of gases.

†Ludwig Eduard Boltzmann (1844–1906). Austrian physicist. Although Boltzmann was one of the greatest theoretical physicists of all time, his work was not recognized by other scientists in his own lifetime. Suffering from poor health and great depression, he committed suicide in 1906.

‡James Clerk Maxwell (1831–1879). Scottish physicist. Maxwell was one of the great theoretical physicists of the nineteenth century; his work covered many areas in physics, including kinetic theory of gases, thermodynamics, and electricity and magnetism.

CHEMISTRY IN ACTION

Scuba Diving and the Gas Laws

The gas laws discussed in this chapter are vitally important to scuba divers like Jacques Cousteau (Figure 5.16). (''Scuba'' is an acronym for self-contained underwater breathing apparatus.) Two examples of how these laws are applied in scuba diving will be given here.

Seawater has a slightly higher density than fresh water—about 1.03 g/cm^3, compared with 1.00 g/cm^3—therefore, the pressure exerted by a column of 33 ft of seawater is equivalent to 1 atm pressure. Pressure increases with increasing depth, so at a depth of 66 ft the pressure due to water will be 2 atm, and so on.

What would happen if a diver rose to the surface rather quickly without breathing? If the ascent started at 20 ft under water, the total decrease in pressure for this change in depth would be (20 ft/33 ft) × 1 atm, or 0.6 atm. When the diver reached the surface, the volume of air trapped in the lungs would have increased by a factor of (1 + 0.6) atm/1 atm, or 1.6 times. This sudden expansion of air can fatally rupture the membranes of the lungs. Another serious possibility is that an *air embolism* might develop. As air expands in the lungs, it is forced into blood vessels and capillaries. Air bubbles formed in this manner can prevent normal blood flow to the brain. When this happens, a diver may lose consciousness before reaching the surface. The only cure for air embolism is recompression. The victim is placed in a chamber filled with compressed air. Here bubbles in the blood can be slowly squeezed down to harmless size. This painful process may take as long as a day to complete.

Figure 5.16 *A scuba diver.*

Dalton's law also has a direct application to scuba diving. The partial pressure of oxygen gas in air is about 0.20 atm. Because oxygen is so essential for our survival, it is hard to believe that it could be harmful if we breathe more than our normal share. Nevertheless, the toxicity of excess oxygen is well established, although the mechanisms involved are not yet fully understood. For example, newborn infants placed in oxygen tents often develop *retrolental fibroplasia* (damage of the retinal tissue), which can cause partial or total blindness.

Our bodies function best when oxygen gas has a partial pressure of about 0.20 atm. The oxygen partial pressure is given by

$$P_{O_2} = X_{O_2}P_T = \frac{n_{O_2}}{n_{O_2} + n_{N_2}}P_T$$

where P_T is the total pressure. However, since volume is directly proportional to the number of moles of gas present (at constant temperature and pressure), we can now write

$$P_{O_2} = \frac{V_{O_2}}{V_{O_2} + V_{N_2}}P_T$$

Thus the composition of air is 20 percent oxygen gas and 80 percent nitrogen gas by volume. When a diver is submerged, the composition of the air he or she breathes must be changed. For example, at a depth where the total pressure is 2.0 atm, the oxygen content in air should be reduced to 10 percent by volume to maintain the same partial pressure of 0.20 atm; that is,

$$P_{O_2} = 0.20 \text{ atm} = \frac{V_{O_2}}{V_{O_2} + V_{N_2}} \times 2.0 \text{ atm}$$

or
$$\frac{V_{O_2}}{V_{O_2} + V_{N_2}} = \frac{0.20 \text{ atm}}{2.0 \text{ atm}} = 0.10$$

or 10 percent.

Although nitrogen gas may seem to be the obvious choice to mix with oxygen gas, there is a serious problem with it. When the partial pressure of nitrogen gas exceeds 1 atm, a sufficient amount of the gas is dissolved in the blood to cause a condition known as *nitrogen narcosis*. The effects on the diver resemble those associated with alcohol intoxication. Divers suffering from nitrogen narcosis have been known to do strange things, such as dancing on the sea floor and chasing sharks. For this reason, helium is often used to dilute oxygen gas. Helium is an inert gas, much less soluble in blood than nitrogen, and does not produce any narcotic effects.

The kinetic molecular theory of gases (sometimes called simply the kinetic theory of gases) is based on the following assumptions:

1. A gas is composed of molecules that are separated from each other by distances far greater than their own dimensions. The molecules can be considered to be "points"; that is, they possess mass but have negligible volume.
2. Gas molecules are in constant motion in random directions, and they frequently collide with one another. Collisions among molecules are perfectly elastic. Although energy may be transferred from one molecule to another as a result of a collision, the total energy of all the molecules in a system remains the same.
3. Gas molecules exert neither attractive nor repulsive forces on one another.
4. The average kinetic energy of the molecules is proportional to the temperature of the gas in kelvins. Any two gases at the same temperature will have the same average kinetic energy. The average kinetic energy of a molecule is given by

$$\text{KE} = \tfrac{1}{2}m\overline{u^2}$$

where m is the mass of the molecule and u is its speed. The horizontal bar denotes an average value. The quantity $\overline{u^2}$ is called mean square speed; it is the average of the square of the speeds of all the molecules:

$$\overline{u^2} = \frac{u_1^2 + u_2^2 + \cdots + u_N^2}{N}$$

where N is the total number of molecules present.

From assumption 4 we write

$$\text{KE} \propto T$$
$$\tfrac{1}{2}m\overline{u^2} \propto T$$
$$\tfrac{1}{2}m\overline{u^2} = CT \qquad (5.13)$$

where C is the proportionality constant and T the absolute temperature.

According to the kinetic molecular theory, gas pressure is the result of collisions between molecules and the walls of the container. It depends on the frequency of collision per unit area and how "hard" the molecules strike the wall. The theory also provides a molecular interpretation of temperature. According to Equation (5.13), the absolute temperature of a gas is a measure of the average kinetic energy of the molecules. In other words, the absolute temperature is an index of the random motion of the molecules—the higher the temperature, the more energetic the motion. For this reason, random molecular motion is sometimes referred to as thermal motion, because it is related to the temperature of the gas sample.

Application to the Gas Laws

Although the kinetic theory of gases is based on a rather simple model, the mathematical details involved are very complex. However, on a qualitative basis, it is possible to apply the theory to account for the general properties of substances in the gaseous state. The following examples illustrate the range of its usefulness and applicability:

- *Compressibility of gases.* Since molecules in the gas phase are separated by large distances (assumption 1), gases can be compressed easily to occupy smaller volumes.
- *Boyle's law.* The pressure exerted by a gas results from the impact of its molecules on the walls of the container. The rate, or the number of molecular collisions with the walls per second, is proportional to the number density (that is, number of molecules per unit volume) of the gas. Decreasing the volume of a given amount of gas increases its number density and hence its collision rate. For this reason, the pressure of a gas is inversely proportional to the volume it occupies.
- *Charles' law.* Since the average kinetic energy of gas molecules is proportional to the sample's absolute temperature (assumption 4), raising the temperature increases the average kinetic energy. Consequently, molecules will collide with the walls of the container more frequently and with greater impact if the gas is heated, and thus the pressure increases. The volume of gas will expand until the gas pressure is balanced by the constant external pressure (see Figure 5.6).
- *Avogadro's law.* We have shown that the pressure of a gas is directly proportional to both the density and the temperature of the gas. Since the mass of the gas is directly proportional to the number of moles (n) of the gas, we can represent density by n/V. Therefore

$$P \propto \frac{n}{V}T$$

For two gases, 1 and 2, we write

$$P_1 \propto \frac{n_1T_1}{V_1} = C\frac{n_1T_1}{V_1}$$

$$P_2 \propto \frac{n_2T_2}{V_2} = C\frac{n_2T_2}{V_2}$$

Another way of stating Avogadro's law is that at the same pressure and temperature, equal volumes of gases, whether they are the same or different gases, contain equal numbers of molecules.

where C is the proportionality constant. Thus, for two gases under the same conditions of pressure, volume, and temperature (that is, when $P_1 = P_2$, $T_1 = T_2$, and $V_1 = V_2$), it follows that $n_1 = n_2$, which is a mathematical expression of Avogadro's law.

- *Dalton's law of partial pressures.* If molecules do not attract or repel one another (assumption 3), then the pressure exerted by one type of molecule is unaffected by the presence of another gas. Consequently, the total pressure is given by the sum of individual gas pressures.

Distribution of Molecular Speeds

The kinetic theory of gases allows us to investigate molecular motion in more detail. Suppose we have a large number of molecules of a gas, say, 1 mole, in a container. As you might expect, the motion of the molecules is totally random and unpredictable. As long as we hold the temperature constant, however, the average kinetic energy and the mean square speed will remain unchanged as time passes. The characteristic that is of interest to us here is the spread, or distribution, of molecular speeds. At a given instant, how many molecules are moving at a particular speed? An equation to solve this problem was formulated by Maxwell in 1860. The equation, which incorporates the assumptions we have mentioned (p. 196), is based on a statistical analysis of the behavior of the molecules.

Figure 5.17 shows typical *Maxwell speed distribution curves* for an ideal gas at two different temperatures. At a given temperature, the distribution curve tells us the number of molecules moving at a certain speed. The peak of each curve gives the *most probable speed,* that is, the speed of the largest number of molecules. Note that at the higher temperature the most probable speed is greater than that at the lower temperature. Comparing parts (a) and (b) in Figure 5.17 we see that as temperature increases, not only does the peak shift toward the right, but the curve also flattens out, indicating that larger numbers of molecules are moving at greater speeds.

One of the useful results of the kinetic molecular theory is that it enables us to relate macroscopic quantities P and V to molecular parameters such as molar mass ($\mathcal{M}$) and mean-square speed according to the following equation:

This equation can be derived using the assumptions of the kinetic molecular theory.

$$PV = \tfrac{1}{3}n\mathcal{M}\overline{u^2}$$

From the ideal gas equation we have

$$PV = nRT$$

Combining the above two equations, we obtain

$$\tfrac{1}{3}n\mathcal{M}\overline{u^2} = nRT$$

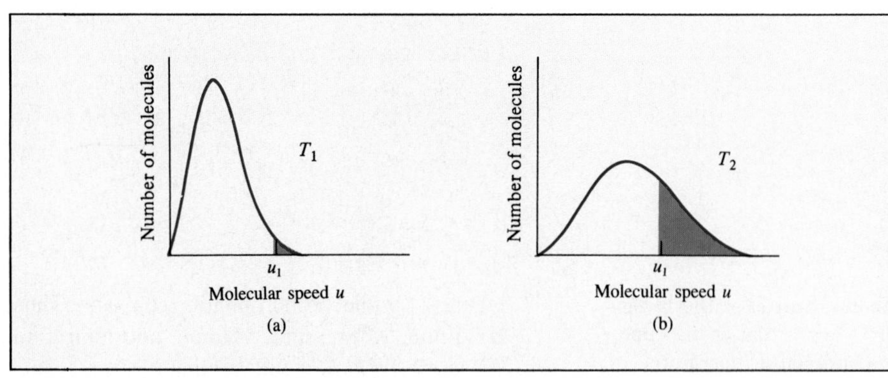

Figure 5.17 *Maxwell's speed distribution for a gas at (a) temperature T_1 and (b) a higher temperature T_2. Note that the curve flattens out at the higher temperature. The shaded areas represent the number of molecules traveling at a speed equal to or greater than a certain speed u_1. The higher the temperature, the greater the number of molecules moving at high speed.*

or
$$\overline{u^2} = \frac{3RT}{\mathcal{M}}$$

Taking the square root of both sides gives

$$\sqrt{\overline{u^2}} = u_{\text{rms}} = \sqrt{\frac{3RT}{\mathcal{M}}} \quad (5.14)$$

where u_{rms} is the root-mean-square speed; that is, it is the square root of the mean-square speed. Equation (5.14) shows that the root-mean-square speed of a gas increases with the square root of its temperature (in kelvins). Because $\mathcal{M}$ appears in the denominator, it follows that the heavier the gas, the more slowly its molecules move. If we use 8.314 J/K · mol for R (see Appendix 2) and convert the molar mass to kg/mol, then u_{rms} will be calculated in meters per second (m/s). This procedure is used in Example 5.21.

The unit joule (J) is defined in Section 1.7.

EXAMPLE 5.21

Calculate the root-mean-square speeds of helium atoms and nitrogen molecules in m/s at 25°C.

Answer

We need Equation (5.14) for this calculation. For He the molar mass is 4.003 g/mol, or 4.003×10^{-3} kg/mol:

$$\begin{aligned} u_{\text{rms}} &= \sqrt{\frac{3RT}{\mathcal{M}}} \\ &= \sqrt{\frac{3(8.314 \text{ J/K} \cdot \text{mol})(298 \text{ K})}{4.003 \times 10^{-3} \text{ kg/mol}}} \\ &= \sqrt{1.86 \times 10^6 \text{ J/kg}} \end{aligned}$$

Using the conversion factor (see Section 1.7)

$$1 \text{ J} = 1 \text{ kg m}^2/\text{s}^2$$

we get

$$\begin{aligned} u_{\text{rms}} &= \sqrt{1.86 \times 10^6 \text{ kg m}^2/\text{kg} \cdot \text{s}^2} \\ &= \sqrt{1.86 \times 10^6 \text{ m}^2/\text{s}^2} \\ &= 1.36 \times 10^3 \text{ m/s} \end{aligned}$$

Similarly for N_2,

$$\begin{aligned} u_{\text{rms}} &= \sqrt{\frac{3(8.314 \text{ J/K} \cdot \text{mol})(298 \text{ K})}{2.802 \times 10^{-2} \text{ kg/mol}}} \\ &= \sqrt{2.65 \times 10^5 \text{ m}^2/\text{s}^2} \\ &= 515 \text{ m/s} \end{aligned}$$

Because of its smaller mass, a helium atom, on the average, moves about 2.6 times faster than a nitrogen molecule at the same temperature (1360 ÷ 515 = 2.64).

Similar problems: 5.91, 5.92.

Figure 5.18 *The distances traveled by a molecule between successive collisions.*

The calculation in Example 5.21 has an interesting relationship to the composition of Earth's atmosphere. Earth, unlike, say, Jupiter, does not have appreciable amounts of gases such as hydrogen or helium in its atmosphere. Why is this the case? A smaller planet, Earth has a weaker gravitational attraction for these lighter molecules. A fairly straightforward calculation shows that to escape Earth's gravitational field, a molecule must possess a speed equal to or greater than 1.1×10^3 m/s. This is usually called the *escape velocity*. Because the average speed of helium is considerably greater than that of molecular nitrogen or molecular oxygen, more helium atoms escape from Earth's atmosphere into outer space. Consequently, only a trace amount of helium is present in our atmosphere. Jupiter, on the other hand, with a mass about 320 times greater than that of Earth, is able to retain both heavy and light gases in its atmosphere.

Mean Free Path

We have seen that the molecules in a gas sample are continually colliding with one another or with the walls of the container. If we could follow a particular molecule from one side of a room to the other, we would find that its motion is entirely random—its path constantly changing as a result of collisions. *The average distance traveled by a molecule between successive collisions* is called the ***mean free path*** (Figure 5.18). The mean free path of a gas is inversely proportional to its density; the higher the density, the smaller the mean free path. Calculations based on the kinetic molecular theory show that the mean free path of air molecules at sea level is about 6×10^{-6} cm. In the upper atmosphere, say, 100 km above sea level, where the air density is much lower, the mean free path is on the order of 10 cm. In intergalactic space, where there are only about one hundred molecules per cubic centimeter, the mean free path is 3×10^{12} cm, or 2×10^7 miles!

5.8 Graham's Laws of Diffusion and Effusion

Diffusion

Gas ***diffusion,*** *the gradual mixing of molecules of one gas with molecules of another by virtue of their kinetic properties,* provides a direct demonstration of random motion. Despite the fact that molecular speeds are very great, the diffusion process itself takes a relatively long time to complete. For example, when a bottle of concentrated ammonia solution is opened at one end of a lab bench, it takes some time before a person at the other end of the bench will smell it. The reason is that in moving from one end of the bench to the other, a molecule experiences numerous collisions, as shown in Figure 5.18. Thus, diffusion of gases always happens gradually, and not instantly as molecular speeds seem to suggest. Furthermore, because the root-mean-square speed of a light gas is greater than that of a heavier gas (see Example 5.21), a lighter gas will diffuse through a certain space more quickly than will a heavier gas.

In 1832 the Scottish chemist Thomas Graham† found that *under the same conditions of temperature and pressure, rates of diffusion for gaseous substances are inversely proportional to the square roots of their molar masses*. This statement, now known as ***Graham's law of diffusion***, is expressed mathematically as

$$\frac{r_1}{r_2} = \sqrt{\frac{\mathcal{M}_2}{\mathcal{M}_1}} \tag{5.15}$$

where r_1 and r_2 are the diffusion rates of gases 1 and 2, and $\mathcal{M}_1$ and $\mathcal{M}_2$ are the molar masses, respectively. Equation (5.15) is based on experimental observation, but it can also be deduced from the kinetic theory of gases as follows. We saw earlier that any two gases at the same temperature will have the same kinetic energy (assumption 4), that is,

$$\tfrac{1}{2}m_1\overline{u_1^2} = \tfrac{1}{2}m_2\overline{u_2^2}$$

Multiplying both sides of the equation by 2 and rearranging, we get

$$\frac{\overline{u_1^2}}{\overline{u_2^2}} = \frac{m_2}{m_1}$$

Taking the square root of both sides of this equation gives us the ratio of the root-mean-square speeds of the gases in terms of the square root of the ratio of their masses:

$$\frac{(u_{\text{rms}})_1}{(u_{\text{rms}})_2} = \sqrt{\frac{m_2}{m_1}}$$

Since the rate of diffusion is proportional to how fast molecules move, or to the root-mean-square speed, and the molar mass of a molecule is proportional to its mass, this equation is equivalent to Equation (5.15).

Figure 5.19 is a demonstration of gaseous diffusion.

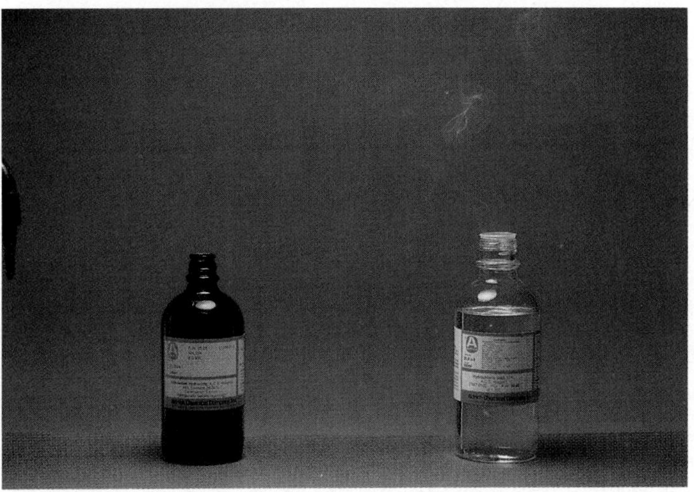

Figure 5.19 *A demonstration of diffusion. NH_3 gas (from a bottle containing aqueous ammonia) combines with HCl gas (from a bottle containing hydrochloric acid) to form solid NH_4Cl. Because NH_3 is lighter and therefore diffuses faster, solid NH_4Cl first appears nearer the HCl bottle (on the right).*

†Thomas Graham (1805–1869). Scottish chemist. Graham did important work on osmosis and characterized a number of phosphoric acids.

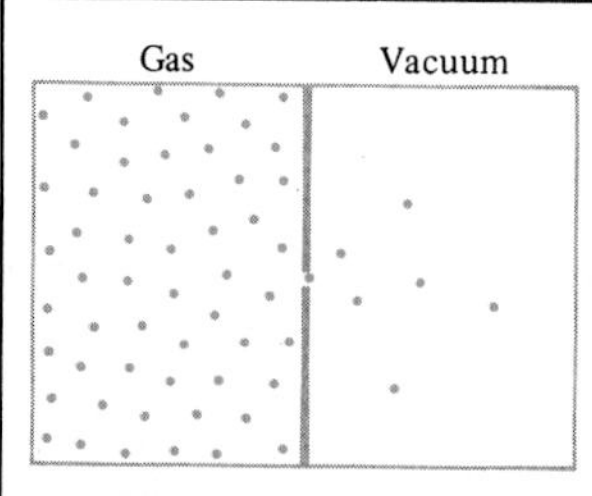

Figure 5.20 *Gaseous effusion. Gas molecules move from a high-pressure region (left) to a low-pressure region through a pinhole.*

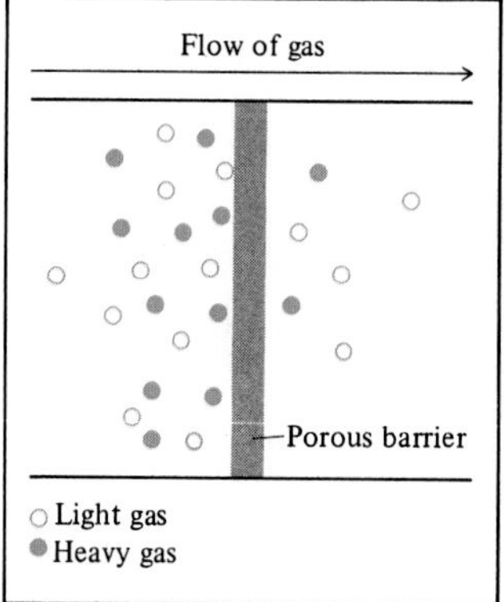

Figure 5.21 *Rates of effusion of a heavy gas (solid circle) and a light gas (open circle) through a porous barrier from a high-pressure region to a low-pressure region. A porous barrier contains many tiny holes suitable for effusion. Light gases effuse more rapidly than heavy gases.*

Figure 5.22 *Left: A porous cup fitted with glass tubing and stopper is attached to a wash bottle filled with colored water. Right: An inverted beaker filled with hydrogen gas is placed over the porous cup. Because molecular hydrogen is lighter than air, it effuses through the cup into the bottle faster than the air in the bottle can effuse out. The difference in the effusion rates results in a greater gas pressure inside the bottle. Consequently, water is squirted out of the bottle.*

Effusion

Whereas diffusion is a process by which one gas gradually mixes with another, ***effusion*** *is the process by which a gas under pressure escapes from one compartment of a container to another by passing through a small opening*. Figure 5.20 shows the effusion of a gas into a vacuum. Although effusion differs from diffusion in nature, the rate of effusion of a gas is also given by Graham's law of diffusion [Equation (5.15)]. As in the case of diffusion, we see that at a given temperature lighter gases effuse faster than heavier gases (Figure 5.21). A practical demonstration of gaseous effusion is shown in Figure 5.22.

The following example compares the effusion rates of two gases.

EXAMPLE 5.22

Compare the effusion rates of helium and molecular oxygen at the same temperature and pressure.

Answer

From Equation (5.15) we write

$$\frac{r_{He}}{r_{O_2}} = \sqrt{\frac{32.00 \text{ g/mol}}{4.003 \text{ g/mol}}} = 2.827$$

Thus helium effuses 2.827 times faster than oxygen.

Similar problem: 5.103.

In practice, the rate of effusion of a gas is inversely proportional to the time it takes for the gas to effuse through a barrier—the longer the time, the slower the effusion rate. From Equation (5.15) we can write

$$\frac{t_1}{t_2} = \frac{r_2}{r_1} = \sqrt{\frac{\mathcal{M}_1}{\mathcal{M}_2}} \tag{5.16}$$

where t_1 and t_2 are the times for effusion for gases 1 and 2, respectively.

Example 5.23 shows that measuring effusion rates can help us identify gases.

EXAMPLE 5.23

A flammable gas made up only of carbon and hydrogen is generated by certain anaerobic bacterium cultures in marshlands and areas where sewage drains. A pure sample of this gas was found to effuse through a certain porous barrier in 1.50 min. Under identical conditions of temperature and pressure, it takes an equal volume of bromine gas (the molar mass of Br_2 is 159.8 g) 4.73 min to effuse through the same barrier. Calculate the molar mass of the unknown gas, and suggest what this gas might be.

Answer

From Equation (5.16) we write

$$\frac{1.50 \text{ min}}{4.73 \text{ min}} = \sqrt{\frac{\mathcal{M}}{159.8 \text{ g/mol}}}$$

where $\mathcal{M}$ is the molar mass of the unknown gas. Solving for $\mathcal{M}$, we obtain

$$\begin{aligned}\mathcal{M} &= \left(\frac{1.50 \text{ min}}{4.73 \text{ min}}\right)^2 \times 159.8 \text{ g/mol} \\ &= 16.1 \text{ g/mol}\end{aligned}$$

Since the molar mass of carbon is 12.01 g and that of hydrogen is 1.008 g, the gas is methane (CH_4).

Similar problems: 5.100, 5.101.

The following Chemistry in Action shows that gaseous effusion played a very important role in the development of the atomic bomb.

CHEMISTRY IN ACTION

Separation of Isotopes by Gaseous Effusion

Graham's law finds an important application in the separation of isotopes. Based on Equation (5.15), we define a quantity called the *separation factor (s)* such that

$$s = \frac{r_1}{r_2} = \sqrt{\frac{\mathcal{M}_2}{\mathcal{M}_1}}$$

Thus the value of s indicates how well gases 1 and 2 can be separated from each other in a one-stage effusion process; the larger the separation factor, the more efficient the separation process. The minimum value of s is 1, which indicates the gases are inseparable. (We assume that $\mathcal{M}_2 > \mathcal{M}_1$.)

Let us consider the separation of two gases: the light and heavy isotopes of hydrogen, H_2 and D_2. We have

$$s = \sqrt{\frac{4.0}{2.0}} = 1.4$$

On the other hand, if we want to separate the two isotopes of neon, ^{20}Ne and ^{22}Ne, the separation factor becomes

$$s = \sqrt{\frac{22.0}{20.0}} = 1.05$$

Because ^{20}Ne and ^{22}Ne differ little in mass, this separation process is less efficient than that of H_2 and D_2.

During the Second World War, scientists discovered that a particular isotope of uranium, ^{235}U, undergoes nuclear breakdown when bombarded with neutrons. The amount of energy released in the process is so enormous that even a relatively small amount of the substance, about 500 g, could be used to build a bomb with a capacity for destruction unparalleled in human history. An intensive effort was launched toward enriching uranium with this isotope. The natural abundance of the two uranium isotopes ^{235}U and ^{238}U are 0.72 percent and 99.28 percent, respectively. Since isotopes of the same element do not differ from each other in chemical properties, the problem of separating ^{235}U from the more abundant ^{238}U isotope was formidable indeed. The separation method finally chosen was gaseous effusion. Uranium can be converted into uranium hexafluoride, UF_6, which is easily vaporized above room temperature. The advantage of this compound is that fluorine consists of a single stable isotope, so the separation involves only two species:

Figure 5.23 *The uranium enrichment process involves thousands of stages. This photo shows what each individual unit in the separation apparatus looks like. Uranium hexafluoride is cycled through the apparatus—consisting of large motors, compressors, valves, diffuser tanks, and pipes—to achieve separation of ^{235}U from ^{238}U. Note that although the separation is brought about by gaseous effusion, both the scientific community and the public call the plant a gaseous diffusion plant.*

$^{235}UF_6$ and $^{238}UF_6$. Note that $^{235}UF_6$ differs from $^{238}UF_6$ by less than 1 percent by mass! The separation factor for these two substances is given by

$$s = \sqrt{\frac{238.05 + (6 \times 18.998)}{235.04 + (6 \times 18.998)}} = 1.0043$$

This is a very small separation factor, but the situation was not entirely hopeless. After a second effusion process, the overall separation factor becomes 1.0043 × 1.0043, or 1.0086, a slight improvement. In the particular case of uranium hexafluoride, then, the separation factor for an n-stage process is 1.0043^n. If n is a large number, say 2000, then it is indeed possible to obtain uranium with about 99 percent enrichment of the ^{235}U isotope. This is essentially what happened. A large-scale effusion plant was built at Oak Ridge, Tennessee. A sample of the gas was passed through a long series of porous barriers (Figure 5.23), and at last $^{235}UF_6$ was obtained with the desired isotopic purity. With this fuel scientists created the atomic bomb.

5.9 Deviation from Ideal Behavior

So far our discussion has assumed that molecules in the gaseous state do not exert any force, either attractive or repulsive, on one another. Further, we have assumed that the volume of the molecules is negligibly small compared with that of the container. A gas that satisfies these two conditions is said to exhibit *ideal behavior*.

These would seem to be fair assumptions, although we cannot expect them to hold under all conditions. For example, without intermolecular forces, gases could not condense to form liquids. The important question is: Under what conditions will gases most likely exhibit nonideal behavior?

When $n = 1$, $PV = nRT$ becomes $PV = RT$, or $PV/RT = 1$.

Figure 5.24 shows the PV/RT versus P plots for three real gases at a given temperature. These plots offer a test of ideal gas behavior. According to the ideal gas equation (for 1 mole of the gas), PV/RT equals 1, regardless of the actual gas pressure. For real gases, this is true only at moderately low pressures ($\lesssim$ 5 atm); significant deviations are observed as pressure increases. The attractive forces among molecules operate at relatively short distances. For a gas at atmospheric pressure, the molecules are relatively far apart and these attractive forces are negligible. At high pressures, the density of the gas increases; the molecules are much closer to one another. Then intermolecular

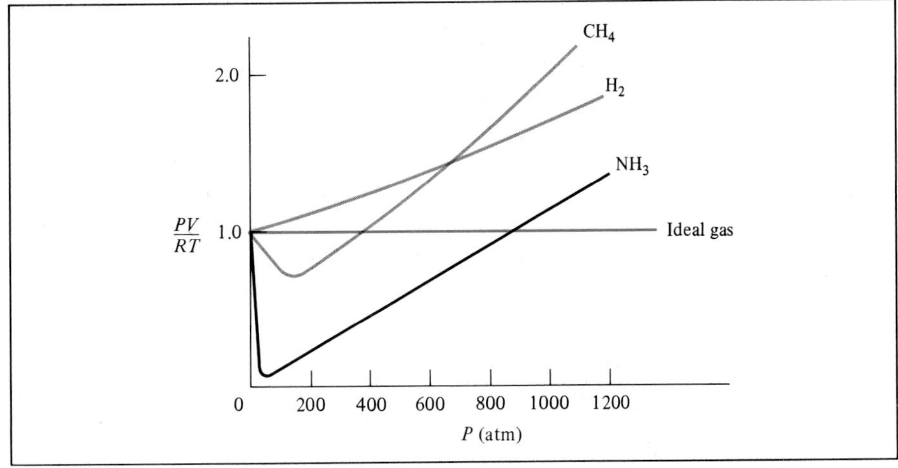

Figure 5.24 *Plot of PV/RT versus P of 1 mole of a gas at 0°C. For 1 mole of an ideal gas, PV/RT is equal to 1, no matter what the pressure of the gas is. For real gases, we observe various deviations from ideality at high pressures. At very low pressures, all gases exhibit ideal behavior; that is, their PV/RT values all converge to 1 as P approaches zero.*

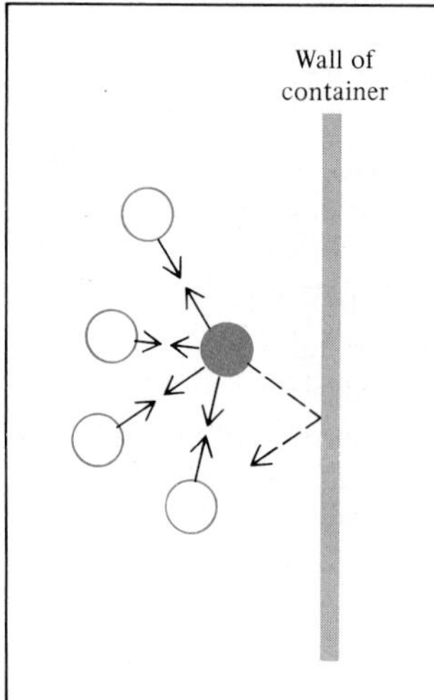

Figure 5.25 *Effect of intermolecular forces on the pressure exerted by a gas. The speed of a molecule that is moving toward the container wall (solid circle) is reduced by the attractive forces exerted by its neighbors (open circles). Consequently, the impact this molecule makes with the wall is less than it would be if no intermolecular forces were present. In general, the measured gas pressure is always lower than the pressure the gas would exert if it behaved ideally.*

forces can become significant enough to affect the motion of the molecules, and the gas will no longer behave ideally.

Another way to observe the nonideality of gases is to lower the temperature. Cooling a gas decreases the molecules' average kinetic energy, which in a sense deprives molecules of the drive they need to break away from their mutual attractive influences.

The nonideality of gases can be dealt with mathematically by modifying Equation (5.7), taking into account intermolecular forces and finite molecular volumes. Such an analysis was first made by the Dutch physicist J. D. van der Waals† in 1873. Besides being mathematically simple, van der Waals' treatment provides us with an interpretation of real gas behavior at the molecular level.

Consider the approach of a particular molecule toward the wall of a container (Figure 5.25). The intermolecular attractions exerted by its neighbors tend to soften the impact made by this molecule against the wall. The overall effect is a lowered gas pressure, as compared with a gas with no such attractive forces present. Van der Waals suggested that the pressure exerted by an ideal gas, P_{ideal}, is related to the experimentally measured pressure, P_{real}, by

$$P_{\text{ideal}} = \underset{\substack{\uparrow \\ \text{observed} \\ \text{pressure}}}{P_{\text{real}}} + \underset{\substack{\uparrow \\ \text{correction} \\ \text{term}}}{\frac{an^2}{V^2}}$$

where a is a constant and n and V are the number of moles and volume of the gas, respectively. The correction term for pressure (an^2/V^2) can be understood as follows. The interaction between molecules that gives rise to nonideal behavior depends on how frequently any two molecules approach each other closely. The number of such "encounters" increases as the square of the number of molecules per unit volume, $(n/V)^2$, because the presence of each of the two molecules in a particular region is proportional to n/V. The quantity P_{ideal} is the pressure we would measure if there were no intermolecular attractions. The quantity a, then, is just a proportionality constant in the correction term for pressure.

Another correction concerns the volume occupied by the gas molecules. The quantity V in Equation (5.7) represents the volume of the container. However, each molecule does occupy a finite, although small, intrinsic volume, so the effective volume of the gas becomes $(V - nb)$, where n is the number of moles of the gas and b is a constant. The term nb represents the volume occupied by n moles of the gas.

Having taken into account the corrections for pressure and volume, we can rewrite Equation (5.7) as

†Johannes Diderick van der Waals (1837–1923). Dutch physicist. Van der Waals received the Nobel Prize in physics in 1910 for his work on the properties of gases and liquids.

Table 5.4 Van der Waals constants of some common gases

Gas	a(atm · L²/mol²)	b(L/mol)
He	0.034	0.0237
Ne	0.211	0.0171
Ar	1.34	0.0322
Kr	2.32	0.0398
Xe	4.19	0.0266
H_2	0.244	0.0266
N_2	1.39	0.0391
O_2	1.36	0.0318
Cl_2	6.49	0.0562
CO_2	3.59	0.0427
CH_4	2.25	0.0428
CCl_4	20.4	0.138
NH_3	4.17	0.0371
H_2O	5.46	0.0305

$$\underbrace{\left(P + \frac{an^2}{V^2}\right)}_{\text{corrected pressure}} \underbrace{(V - nb)}_{\text{corrected volume}} = nRT \tag{5.17}$$

Equation (5.17) is known as the *van der Waals equation*. The van der Waals constants a and b are selected for each gas to give the best possible agreement between the equation and actually observed behavior.

Table 5.4 lists the values of a and b for a number of gases. The value of a is an expression of how strongly molecules of a given type of gas attract one another. We see that the helium atoms have the weakest attraction for one another, because helium has the smallest a value. There is also a rough correlation between molecular size and b. Generally, the larger the molecule (or atom), the greater b is, but the relationship between b and molecular (or atomic) size is not a simple one.

The following example compares the pressure of a gas calculated using the ideal gas equation and the van der Waals equation.

EXAMPLE 5.24

A quantity of 3.50 moles of NH_3 occupies 5.20 L at 47°C. Calculate the pressure of the gas (in atm) using (a) the ideal gas equation and (b) the van der Waals equation.

Answer

(a) We have the following data:

$$V = 5.20 \text{ L}$$
$$T = (47 + 273) \text{ K} = 320 \text{ K}$$
$$n = 3.50 \text{ mol}$$
$$R = 0.0821 \text{ L} \cdot \text{atm/K} \cdot \text{mol}$$

which we substitute in the ideal gas equation:

$$\begin{aligned} P &= \frac{nRT}{V} \\ &= \frac{(3.50 \text{ mol})(0.0821 \text{ L} \cdot \text{atm/K} \cdot \text{mol})(320 \text{ K})}{5.20 \text{ L}} \\ &= 17.7 \text{ atm} \end{aligned}$$

(b) From Table 5.4, we have

$$a = 4.17 \text{ atm} \cdot \text{L}^2/\text{mol}^2$$
$$b = 0.0371 \text{ L/mol}$$

It is convenient to calculate the correction terms first, for Equation (5.17). They are

$$\frac{an^2}{V^2} = \frac{(4.17 \text{ atm} \cdot \text{L}^2/\text{mol}^2)(3.50 \text{ mol})^2}{(5.20 \text{ L})^2} = 1.89 \text{ atm}$$

$$nb = (3.50 \text{ mol})(0.0371 \text{ L/mol}) = 0.130 \text{ L}$$

Finally, substituting in the van der Waals equation, we write

$$(P + 1.89 \text{ atm})(5.20 \text{ L} - 0.130 \text{ L}) = (3.50 \text{ mol})(0.0821 \text{ L} \cdot \text{atm/K} \cdot \text{mol})(320 \text{ K})$$
$$P = 16.2 \text{ atm}$$

Note that the actual pressure measured under these conditions is 16.0 atm. Thus, the pressure calculated by the van der Waals equation (16.2 atm) is closer to the actual value than that calculated by the ideal gas equation (17.7 atm).

Similar problem: 5.110.

Summary

1. Gases exert pressure because their molecules move freely and collide with any surface with which they make contact. Gas pressure units include millimeters of mercury (mmHg), torr, pascals, and atmospheres. One atmosphere equals 760 mmHg, or 760 torr.
2. Under atmospheric conditions, ionic compounds exist as solids rather than as gases. The behavior of molecular compounds is more varied. A number of elemental substances occur as gases: H_2, N_2, O_2, O_3, F_2, Cl_2, and the Group 8A elements (the noble gases).
3. The pressure-volume relationships of ideal gases are governed by Boyle's law: Volume is inversely proportional to pressure (at constant T and n).
4. The temperature-volume relationships of ideal gases are described by Charles and Gay-Lussac's law: Volume is directly proportional to temperature (at constant P and n).
5. Absolute zero (−273.15°C) is the lowest theoretically attainable temperature. The Kelvin temperature scale takes 0 K as absolute zero. In all gas law calculations, temperature must be expressed in kelvins.
6. The amount-volume relationships of ideal gases are described by Avogadro's law: Equal volumes of gases contain equal numbers of molecules (at the same T and P).
7. The ideal gas equation, $PV = nRT$, combines the laws of Boyle, Charles, and Avogadro. This equation describes the behavior of an ideal gas.
8. Dalton's law of partial pressures states that in a mixture of gases each gas exerts the same pressure as it would if it were alone and occupied the same volume.
9. The kinetic molecular theory, a mathematical way of describing the behavior of gas molecules, is based on the following assumptions: Gas molecules are separated by distances far greater than their own dimensions, they possess mass but have negligible volume, they are in constant motion, and they frequently collide with one another. The molecules neither attract nor repel one another.
10. A Maxwell speed distribution curve shows how many gas molecules are moving at various speeds at a given temperature. As temperature increases, more molecules move at greater speeds.
11. The mean free path is the average distance traveled by a molecule between successive collisions. It is inversely proportional to the density of the gas.
12. In diffusion, two gases gradually mix with each other. In effusion, gas molecules move through a small opening under pressure. Both processes demonstrate ran-

dom molecular motion and are governed by the same mathematical laws (Graham's laws of diffusion and effusion).

13. The van der Waals equation is a modification of the ideal gas equation that takes into account the nonideal behavior of real gases. It corrects for the fact that real gas molecules do exert forces on each other and that they do have volume. The van der Waals constants are determined experimentally for each gas.

Applied Chemistry Example: Vinyl Chloride

Vinyl chloride (C_2H_3Cl) is one of the most important commodity chemicals manufactured today. About 9.1 billion pounds of it were produced in the United States in 1988, giving it a rank of 20th among the top 50 industrial chemicals. Vinyl chloride is used chiefly to make poly(vinyl chloride) (PVC), a plastic that is used in piping, siding, floor tile, clothing, and toys. Vinyl chloride is made from ethylene (C_2H_4) in a process called *oxychlorination*. First ethylene is reacted with hydrogen chloride (HCl) and oxygen at 300°C in the presence of a catalyst to give ethylene dichloride ($C_2H_4Cl_2$):

$$C_2H_4(g) + 2HCl(g) + \tfrac{1}{2}O_2(g) \longrightarrow C_2H_4Cl_2(g) + H_2O(g)$$

Next ethylene dichloride is heated at 500°C to yield vinyl chloride:

$$C_2H_4Cl_2(g) \longrightarrow C_2H_3Cl(g) + HCl(g)$$

Vinyl chloride (melting point = −153.8°C and boiling point = −13.4°C) is a colorless gas. It is a recognized carcinogen. The recommended limit for vinyl chloride in the air of working places is 5 ppm (parts per million) by volume.

1. What is the partial pressure of vinyl chloride (in atm) at 5.00 ppm by volume if the total pressure is 743 mmHg and the temperature is 20°C?

   Answer: The question can be solved by applying Avogadro's law. If the concentration of vinyl chloride is 5.00 parts per million by volume, then of every million molecules in a given volume, 5 must be vinyl chloride. Thus the mole fraction of vinyl chloride, $X_{\text{vinyl chloride}}$, must be $5.00/1 \times 10^6$. Next we use Dalton's law to calculate the partial pressure of vinyl chloride:

$$\begin{aligned} P_{\text{vinyl chloride}} &= X_{\text{vinyl chloride}} \times P_T \\ &= \frac{5}{1 \times 10^6} \times 743 \text{ mmHg} \times \frac{1 \text{ atm}}{760 \text{ mmHg}} \\ &= 4.89 \times 10^{-6} \text{ atm} \end{aligned}$$

2. If an average breath is about 0.30 L, how many molecules of vinyl chloride are inhaled in one breath under the conditions in 1?

   Answer: We can find the number of moles of vinyl chloride by using the ideal gas equation [Equation (5.7)] and the answer in 1. From Equation (5.7),

$$\begin{aligned} n &= \frac{PV}{RT} \\ &= \frac{(4.89 \times 10^{-6} \text{ atm})(0.30 \text{ L})}{(0.0821 \text{ L atm/K} \cdot \text{mol})(293 \text{ K})} \\ &= 6.1 \times 10^{-8} \text{ mol} \end{aligned}$$

Next we calculate the number of molecules inhaled per breath as follows:

$$\text{Number of molecules} = 6.1 \times 10^{-8} \text{ mol} \times \frac{6.022 \times 10^{23} \text{ molecules}}{1 \text{ mol}}$$
$$= 3.7 \times 10^{16} \text{ molecules}$$

3. Formaldehyde (CH_2O) is used as a starting material in the polymer industry and in the laboratory as a preservative for dead animals. It is a suspected carcinogen. If a storage vessel leaks 0.10 mL per hour of vinyl chloride by effusion when its pressure is 2.0 atm, would the vessel be suitable for handling formaldehyde (under the same conditions of pressures and temperature) where a leak rate of 0.20 mL per hour can be tolerated?

Answer: The rate at which vinyl chloride (or formaldehyde) leaks out of the vessel can be taken as the rate of effusion. We are given the rate for vinyl chloride so we can calculate the rate for formaldehyde as follows:

$$\frac{\text{rate}_{\text{formaldehyde}}}{\text{rate}_{\text{vinyl chloride}}} = \sqrt{\frac{\mathcal{M}(\text{vinyl chloride})}{\mathcal{M}(\text{formaldehyde})}}$$

$$\text{rate}_{\text{formaldehyde}} = (0.10 \text{ mL/h}) \times \sqrt{\frac{62.49 \text{ g/mol}}{30.03 \text{ g/mol}}}$$
$$= 0.14 \text{ mL/h}$$

Since the rate is less than 0.20 mL/h, the vessel is satisfactory.

Key Words

Absolute temperature scale, p. 176
Absolute zero, p. 176
Avogadro's law, p. 180
Barometer, p. 170
Boyle's law, p. 173
Charles and Gay-Lussac's law, p. 178
Charles' law, p. 178
Dalton's law of partial pressures, p. 190
Diffusion, p. 200
Effusion, p. 202
Gas constant, p. 180
Graham's law of diffusion, p. 201
Ideal gas, p. 181
Ideal gas equation, p. 181
Kelvin temperature scale, p. 177
Mean free path, p. 200
Mole fraction, p. 191
Partial pressures, p. 190
Standard temperature and pressure (STP), p. 181

Exercises

Pressure

REVIEW QUESTIONS

5.1 Define pressure and give the common units for pressure.
5.2 Describe how a barometer is used to measure atmospheric pressure.
5.3 Why is mercury a more suitable substance to use in a barometer than water?
5.4 Explain why the height of mercury in a barometer is independent of the cross-sectional area of the tube.
5.5 Would it be easier to drink water with a straw on the top or at the foot of Mt. Everest? Explain.
5.6 Is the atmospheric pressure in a mine that is 500 m below sea level greater or less than 1 atm?
5.7 What is the difference between a gas and a vapor? At 25°C, which of the following substances in the gas phase should be properly called a gas and which should be called a vapor? Molecular nitrogen (N_2), mercury.
5.8 If the maximum distance that water may be brought up a well by using a suction pump is 34 ft, explain how it is possible to obtain water and oil from hundreds of feet below the surface of Earth.
5.9 How does a vacuum cleaner work?
5.10 Why do astronauts have to wear protective suits when they are on the surface of the moon?

PROBLEMS

5.11 Calculate the pressure (in mmHg) of the gas in the arrangements shown in the illustration below. The liquid is mercury.

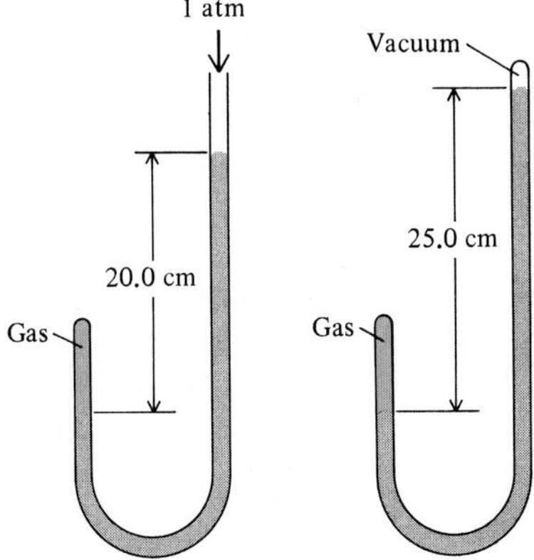

5.12 Convert 562 mmHg to kPa and 2.0 kPa to mmHg.

5.13 The atmospheric pressure at the summit of Mt. McKinley is 606 mmHg on a certain day. What is the pressure in atm?

Substances That Exist as Gases

REVIEW QUESTIONS

5.14 Name five elements and five compounds that exist as gases at room temperature.

5.15 In terms of energy considerations, explain why, in principle at least, all substances can be converted to the gaseous state by heating.

5.16 Under normal conditions, ionic compounds such as NaCl do not exist as gases at room temperature. Explain.

The Gas Laws

REVIEW QUESTIONS

5.17 State the following gas laws in words and also in the form of an equation: Boyle's law, Charles' law, Avogadro's law. In each case, indicate the conditions under which the law is applicable and also give the units for each quantity in the equation.

5.18 Explain why a helium weather balloon expands as it rises in the air. Assume that the temperature remains constant.

5.19 Define absolute zero and absolute zero scale. Write the relationship between °C and K.

PROBLEMS

5.20 A gas occupying a volume of 725 mL at a pressure of 0.970 atm is allowed to expand at constant temperature until its pressure becomes 0.541 atm. What is its final volume?

5.21 At 46°C a sample of ammonia gas exerts a pressure of 5.3 atm. What is the pressure when the volume of the gas is reduced to one-tenth (0.10) of the original value at the same temperature?

5.22 The volume of a gas is 5.80 L, measured at 1.00 atm. What is the pressure of the gas in mmHg if the volume is changed to 9.65 L? (The temperature remains constant.)

5.23 A sample of air occupies 3.8 L when the pressure is 1.2 atm. (a) What volume does it occupy at 6.6 atm? (b) What pressure is required in order to compress it to 0.075 L? (The temperature is kept constant.)

5.24 A diver ascends quickly to the surface of the water from a depth of 4.08 m without exhaling gas from his lungs. By what factor would the volume of his lungs increase by the time he reaches the surface? Assume the density of seawater to be 1.03 g/cm^3 and the temperature to remain constant. (1 atm = 1.01325×10^5 N/m^2 and acceleration due to gravity is 9.8067 m/s^2.) (*Hint:* See Appendix 2.)

5.25 (a) Convert the following temperatures to Kelvin: 0°C, 37°C, 100°C, −225°C. (b) Convert the following temperatures to degrees Celsius: 77 K, 4.2 K, 6.0×10^3 K.

5.26 A quantity of 36.4 L of methane gas is heated from 25°C to 88°C at constant pressure. What is its final volume?

5.27 Under constant-pressure conditions a sample of hydrogen gas initially at 88°C and 9.6 L is cooled until its final volume is 3.4 L. What is its final temperature?

5.28 A dented (but not punctured) Ping-Pong ball can often be restored to its original shape by immersing it in very hot water. Why?

5.29 Ammonia burns in oxygen gas to form nitric oxide (NO) and water vapor. How many volumes of NO are obtained from one volume of ammonia at the same temperature and pressure?

5.30 Molecular chlorine and molecular fluorine combine to form a gaseous product. Under the same conditions of temperature and pressure it is found that one volume of Cl_2 reacts with three volumes of F_2 to yield two volumes of the product. What is the formula of the product?

5.31 Cyanogen is a compound of carbon and nitrogen. On combustion in excess oxygen, 500 mL of cyanogen gives 500 mL of N_2 and 1000 mL of CO_2, measured at the same temperature and pressure. What is the formula of cyanogen?

The Ideal Gas Equation

REVIEW QUESTIONS

5.32 What is an ideal gas?

5.33 Why must we use the absolute temperature scale (instead of the Celsius scale) in gas law calculations?

5.34 Write the ideal gas equation and also state it in words. Give the units for each term in the equation.

5.35 What are standard temperature and pressure (STP)? What is the significance of STP in relation to the volume of 1 mole of an ideal gas?

5.36 Why is the density of a gas much lower than that of a liquid or solid under atmospheric conditions? What are the units normally used to express the density of gases?

PROBLEMS

5.37 A sample of nitrogen gas kept in a container of volume 2.3 L and at a temperature of 32°C exerts a pressure of 4.7 atm. Calculate the number of moles of gas present.

5.38 A sample of 6.9 moles of carbon monoxide gas is present in a container of volume 30.4 L. What is the pressure of the gas (in atm) if the temperature is 62°C?

5.39 What volume would 5.6 moles of sulfur hexafluoride (SF_6) gas occupy if the temperature and pressure of the gas are 128°C and 9.4 atm?

5.40 Nitrous oxide (N_2O) can be obtained by the thermal decomposition of ammonium nitrate (NH_4NO_3). (a) Write a balanced equation for the reaction. (b) In a certain experiment, a student obtains 0.340 L of the gas at 718 mmHg and 24°C. If the gas weighs 0.580 g, calculate the value of the gas constant.

5.41 A certain quantity of gas at 25°C and at a pressure of 0.800 atm is contained in a glass vessel. Suppose that the vessel can withstand a pressure of 2.00 atm. How high can you increase the temperature of the gas without bursting the vessel?

5.42 A gas-filled balloon having a volume of 2.50 L at 1.2 atm and 25°C is allowed to rise to the stratosphere (about 30 km above the surface of Earth), where the temperature and pressure are −23°C and 3.00×10^{-3} atm, respectively. Calculate the final volume of the balloon.

5.43 The temperature of 2.5 L of a gas initially at STP is increased to 250°C at constant volume. Calculate the final pressure of the gas in atm.

5.44 The pressure of 6.0 L of an ideal gas in a flexible container is decreased to one-third of its original pressure, and its absolute temperature is decreased by one-half. What is the final volume of the gas?

5.45 A gas evolved during the fermentation of glucose (wine making) has a volume of 0.78 L when measured at 20.1°C and 1.00 atm. What was the volume of this gas at the fermentation temperature of 36.5°C and 1.00 atm pressure?

5.46 An ideal gas originally at 0.85 atm and 66°C was allowed to expand until its final volume, pressure, and temperature were 94 mL, 0.60 atm, and 45°C, respectively. What was its initial volume?

5.47 The volume of a gas at STP is 488 mL. Calculate its volume at 22.5 atm and 150°C.

5.48 A gas at 772 mmHg and 35.0°C occupies a volume of 6.85 L. Calculate its volume at STP.

5.49 Dry ice is solid carbon dioxide. A 0.050-g sample of dry ice is placed in an evacuated vessel of volume 4.6 L at 30°C. Calculate the pressure inside the vessel after all the dry ice has been converted to CO_2 gas.

5.50 A sample of neon gas occupies 566 cm^3 at 10°C and 772 mmHg pressure. Calculate its volume at 15°C and 754 mmHg.

5.51 A quantity of gas weighing 7.10 g at 741 torr and 44°C occupies a volume of 5.40 L. What is its molar mass?

5.52 The ozone molecules present in the stratosphere absorb much of the harmful radiation from the sun. Typical temperature and pressure of ozone in the stratosphere are 250 K and 1.0×10^{-3} atm, respectively. How many ozone molecules are present in 1.0 L of air under these conditions?

5.53 A 2.10-L vessel contains 4.65 g of gas at 1.00 atm and 27.0°C. (a) Calculate the density of the gas in g/L. (b) What is the molar mass of the gas?

5.54 It is quite easy to achieve a pressure as low as 1.0×10^{-6} mmHg using a diffusion pump. How many molecules of an ideal gas are present in 1.0 L at 25°C under this condition?

5.55 Calculate the volume in liters of the following at STP: (a) 0.681 g of N_2, (b) 7.4 moles of SO_2, (c) 0.58 kg of CH_4.

5.56 Calculate the density of hydrogen bromide (HBr) gas in g/L at 733 mmHg and 46°C.

5.57 A certain anesthetic contains 64.9 percent C, 13.5 percent H, and 21.6 percent O by mass. 1.00 L of the gaseous compound measured at 120°C and 750 mmHg weighs 2.30 g. What is the molecular formula of the compound?

5.58 A volume of 0.280 L of a gas at STP weighs 0.400 g. Calculate the molar mass of the gas.

5.59 Assuming that air contains 78 percent N_2, 21 percent O_2, and 1 percent Ar, all by volume, how many molecules of each type of gas are present in 1.0 L of air at STP?

5.60 A compound has the empirical formula SF_4. 0.100 g of the gaseous compound at 20°C occupies a volume of 22.1 mL and exerts a pressure of 1.02 atm. What is its molecular formula?

Gas Stoichiometry

PROBLEMS

5.61 Propane (C_3H_8) burns in oxygen to produce carbon dioxide gas and water vapor. (a) Write a balanced equation for this reaction. (b) Calculate the number of liters of carbon dioxide measured at STP that could be produced from 7.45 g of propane.

5.62 Some commercial drain cleaners contain two components: sodium hydroxide and aluminum powder. When the mixture is poured down a clogged drain, the following reaction occurs:

$$2NaOH(aq) + 2Al(s) + 6H_2O(l) \longrightarrow 2NaAl(OH)_4(aq) + 3H_2(g)$$

The heat generated in this reaction helps melt away obstructions such as grease, and the hydrogen gas released stirs up the solids clogging the drain. Calculate the volume of H_2 formed at STP if 3.12 g of Al is treated with excess NaOH.

5.63 The equation for the metabolic breakdown of glucose ($C_6H_{12}O_6$) is the same as that for the combustion of glucose in air:

$$C_6H_{12}O_6(s) + 6O_2(g) \longrightarrow 6CO_2(g) + 6H_2O(l)$$

Calculate the volume of CO_2 produced at 37.0°C and 1.00 atm when 5.60 g of glucose is used up in the reaction.

5.64 A compound of P and F was analyzed as follows: Heating 0.2324 g of the compound in a 378-cm^3 container turned all of it to gas, which had a pressure of 97.3 mmHg at 77°C. Then the gas was mixed with calcium chloride solution, which turned all of the F to 0.2631 g of CaF_2. Determine the molecular formula of the compound.

5.65 The volume of a sample of pure HCl gas was 189 mL at 25°C and 108 mmHg. It was completely dissolved in about 60 mL of water and titrated with a NaOH solution; 15.7 mL of the NaOH solution was required to neutralize the HCl. Calculate the molarity of the NaOH solution.

5.66 A quantity of 0.225 g of a metal M (molar mass = 27.0 g/mol) liberated 0.303 L of molecular hydrogen (measured at 17°C and 741 mmHg) from an excess of hydrochloric acid. Deduce from these data the corresponding equation and write formulas for the oxide and sulfate of M.

5.67 A quantity of 73.0 g of NH_3 is mixed with an equal mass of HCl. What is the mass of the solid NH_4Cl formed? What is the volume of the gas remaining, measured at 14.0°C and 752 mmHg? What gas is it?

5.68 Dissolving 3.00 g of an impure sample of calcium carbonate in hydrochloric acid produced 0.656 L of carbon dioxide (measured at 20.0°C and 792 mmHg). Calculate the percent by mass of calcium carbonate in the sample.

5.69 A volume of 5.6 L of molecular hydrogen measured at STP is reacted with an excess of molecular chlorine gas. Calculate the mass in grams of hydrogen chloride produced.

5.70 Ethanol (C_2H_5OH) burns in air:

$$C_2H_5OH(l) + O_2(g) \longrightarrow CO_2(g) + H_2O(l)$$

Balance the equation and determine the volume of air in liters at 35.0°C and 790 mmHg required to burn 227 g of ethanol. Assume air to be 21.0 percent O_2 by volume.

5.71 Nitric oxide (NO) reacts with molecular oxygen as follows:

$$2NO(g) + O_2(g) \longrightarrow 2NO_2(g)$$

Initially NO and O_2 are separated as shown below. When the valve is opened, the reaction quickly goes to completion. Determine what gases remain at the end and calculate their partial pressures. Assume that the temperature remains constant at 25°C.

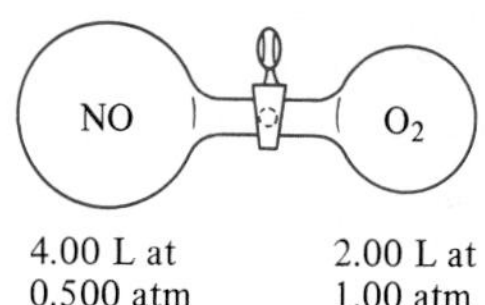

Dalton's Law of Partial Pressures

REVIEW QUESTIONS

5.72 Define Dalton's law of partial pressures and mole fraction. Does mole fraction have units?

5.73 Can partial pressures be measured directly with a barometer? If not, how are the partial pressures of a mixture of gases determined experimentally?

5.74 A sample of air contains only nitrogen and oxygen gases whose partial pressures are 0.80 atm and 0.20 atm, respectively. Calculate the total pressure and the mole fractions of the gases.

PROBLEMS

5.75 A mixture of gases contains CH_4, C_2H_6, and C_3H_8. If the total pressure is 1.50 atm and the numbers of moles of the gases present are 0.31 mole for CH_4, 0.25 mole for C_2H_6, and 0.29 mole for C_3H_8, calculate the partial pressures of the gases.

5.76 A 2.5-L flask at 15°C contains a mixture of three gases, N_2, He, and Ne, at partial pressures of 0.32 atm for N_2, 0.15 atm for He, and 0.42 atm for Ne. (a) Calculate the total pressure of the mixture. (b) Calculate the volume in liters at STP occupied by He and Ne if the N_2 is removed selectively.

5.77 Dry air near sea level has the following composition by volume: N_2, 78.08 percent; O_2, 20.94 percent; Ar, 0.93 percent; CO_2, 0.05 percent. The atmospheric pressure is 1.00 atm. Calculate (a) the partial pressure of each gas in atm and (b) the concentration of each gas in mol/L at 0°C. (*Hint:* Since volume is proportional to the number of moles present, mole fractions of gases can be expressed as ratios of volumes at the same temperature and pressure.)

5.78 A mixture of helium and neon gases is collected over water at 28.0°C and 745 mmHg. If the partial pressure of helium is 368 mmHg, what is the partial pressure of neon? (Vapor pressure of water at 28°C = 28.3 mmHg.)

5.79 A piece of sodium metal undergoes complete reaction with water as follows:

$$2Na(s) + 2H_2O(l) \longrightarrow 2NaOH(aq) + H_2(g)$$

The hydrogen gas generated is collected over water at 25.0°C. The volume of the gas is 246 mL measured at 1.00 atm. Calculate the number of grams of sodium used in the reaction. (Vapor pressure of water at 25°C = 0.0313 atm.)

5.80 A sample of zinc metal is allowed to react completely with an excess of hydrochloric acid:

$$Zn(s) + 2HCl(aq) \longrightarrow ZnCl_2(aq) + H_2(g)$$

The hydrogen gas produced is collected over water at 25.0°C using an arrangement similar to that shown in Figure 5.14. The volume of the gas is 7.80 L and the pressure is 0.980 atm. Calculate the amount of zinc metal in grams consumed in the reaction. (Vapor pressure of water at 25°C = 23.8 mmHg.)

5.81 Helium is mixed with oxygen gas for deep sea divers. Calculate the percent by volume of oxygen gas in the mixture if the diver has to submerge to a depth where the total pressure is 4.2 atm. The partial pressure of oxygen is maintained at 0.20 atm at this depth.

5.82 A sample of ammonia (NH_3) gas is completely decomposed to nitrogen and hydrogen gases over heated iron wool. If the total pressure is 866 mmHg, calculate the partial pressures of N_2 and H_2.

Kinetic Molecular Theory of Gases

REVIEW QUESTIONS

5.83 What are the basic assumptions of the kinetic molecular theory of gases?

5.84 How would you use the kinetic molecular theory to explain the following gas laws? Boyle's law, Charles' law, Avogadro's law, Dalton's law of partial pressures.

5.85 What does the Maxwell speed distribution curve tell us?

5.86 Does Maxwell's theory work for a sample of 200 molecules? Explain.

5.87 Write the expression for the root-mean-square speed for a gas at temperature T. Define each term in the equation and show the units that are used in the calculations.

5.88 What prevents the molecules in the atmosphere from escaping into outer space?

5.89 Which of the following two statements is correct? (a) Heat is produced by the collision of gas molecules against one another. (b) When a gas is heated, the molecules collide with one another more often.

5.90 Define mean free path. Why is the mean free path of a gas inversely proportional to its density?

PROBLEMS

5.91 Compare the root-mean-square speeds of O_2 and UF_6 at 25°C.

5.92 The temperature and pressure in the stratosphere are −23°C and 3.00×10^{-3} atm, respectively. Calculate the root-mean-square speeds of N_2, O_2, and O_3 molecules in this region.

5.93 As we know, UF_6 is a much heavier gas than helium. Yet at a given temperature, the average kinetic energies of the samples of the two gases are the same. Explain.

5.94 Two vessels are labeled A and B. Vessel A contains NH_3 gas at 70°C, and vessel B contains Ne gas at the same temperature. If the average kinetic energy of NH_3 is 7.1×10^{-21} J/molecule, calculate the mean-square speed of Ne atoms in m^2/s^2.

Diffusion and Effusion

REVIEW QUESTIONS

5.95 Define the following terms: diffusion, effusion. Use diagrams to show the difference between a diffusion process and an effusion process.

5.96 Write Graham's law of diffusion and state the law in words.

5.97 How does the rate of effusion of a gas depend on the time it takes for the gas to effuse through a porous barrier?

5.98 Give a practical demonstration of diffusion and an application of the effusion process.

5.99 A helium-filled balloon deflates faster than an air-filled balloon of comparable initial size. Explain.

PROBLEMS

5.100 It requires 57 seconds for 1.4 L of an unknown gas to effuse through a porous wall, and it takes 84 seconds for the same volume of N_2 gas to effuse at the same temperature and pressure. What is the molar mass of the unknown gas?

5.101 A sample of the gas discussed in Problem 5.45 is found to effuse through a porous barrier in 15.0 min. Under the same conditions of temperature and pressure, it takes N_2 12.0 min to effuse through the same barrier. Calculate the molar mass of the gas and suggest what gas it might be.

5.102 List the following gases in order of increasing diffusion rates: PH_3, ClO_2, Kr, NH_3, HI. Calculate the ratio of diffusion rates of fastest to slowest.

5.103 Argon (Ar) effuses through a pinhole at a rate of 3.56 mL/min. At what rate will hydrogen gas (H_2) effuse through the same hole under similar conditions?

5.104 Nickel forms a gaseous compound of the formula $Ni(CO)_x$. What is the value of x given the fact that under the same conditions of temperature and pressure methane (CH_4) effuses 3.3 times faster than the compound.

5.105 Consider the arrangement shown in the illustration. The ammonium chloride sample is being heated. (a) Write a balanced equation for the thermal decomposition of ammonium chloride. (b) Predict the subsequent changes in the color of the blue and the red litmus papers.

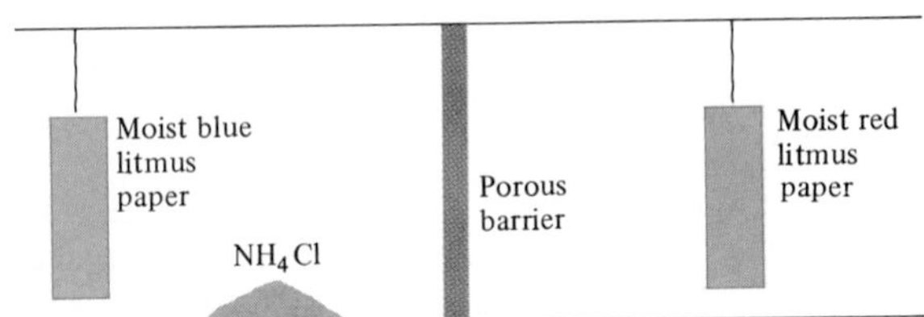

Nonideal Gas Behavior

REVIEW QUESTIONS

5.106 Give two pieces of evidence to show that gases do not behave ideally under all conditions.

5.107 Under what set of conditions would a gas be expected to behave most ideally? (a) High temperature and low pressure, (b) high temperature and high pressure, (c) low temperature and high pressure, (d) low temperature and low pressure.

5.108 Write the van der Waals equation for a real gas. Explain clearly the meaning of the corrective terms for pressure and volume.

PROBLEMS

5.109 The temperature of a real gas that is allowed to expand into a vacuum usually drops. Explain.

5.110 Using the data shown in Table 5.4, calculate the pressure exerted by 2.50 moles of CO_2 confined in a volume of 5.00 L at 450 K. Compare the pressure with that calculated using the ideal gas equation.

5.111 Under the same conditions of temperature and pressure, which of the following gases would behave most ideally: Ne, N_2, or CH_4? Explain.

5.112 10.0 mole of a gas in a 1.50-L container exerts a pressure of 130 atm at 27°C. Is this an ideal gas?

5.113 From the following data collected at 0°C, comment on whether carbon dioxide behaves as an ideal gas. (*Hint:* Determine *PV*. Is it constant?)

P (atm)	0.0500	0.100	0.151	0.202	0.252
V (L)	448.2	223.8	148.8	110.8	89.0

Miscellaneous Problems

5.114 Discuss the following phenomena in terms of the gas laws: (a) the pressure in an automobile tire increasing on a hot day, (b) the "popping" of a paper bag, (c) the expansion of a weather balloon as it rises in the air, (d) the loud noise heard when a light bulb shatters.

5.115 Nitroglycerin, an explosive, decomposes according to the equation

$$4C_3H_5(NO_3)_3(s) \longrightarrow 12CO_2(g) + 10H_2O(g) + 6N_2(g) + O_2(g)$$

Calculate the total volume of gases produced when collected at 1.2 atm and 25°C from 2.6×10^2 g of nitroglycerin. What are the partial pressures of the gases under these conditions?

5.116 Consider the apparatus shown below. When a small amount of water is introduced into the flask by squeezing the bulb of the medicinal dropper, water is squirted upward out of the long glass tubing. Explain this observation. (*Hint:* Hydrogen chloride gas is soluble in water.)

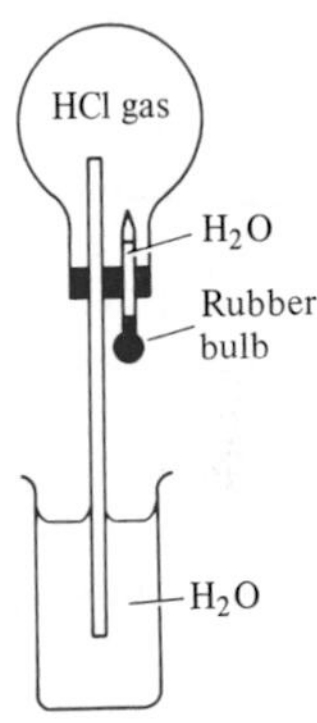

5.117 The empirical formula of a compound is CH. At 200°C, 0.145 g of this compound occupies 97.2 mL at a pressure of 0.74 atm. What is the molecular formula of the compound?

5.118 When ammonium nitrite (NH_4NO_2) is heated, it decomposes to give nitrogen gas. This property is used to inflate some tennis balls. (a) Write a balanced equation for the reaction. (b) Calcuiate the quantity (in grams) of NH_4NO_2 needed to inflate a tennis ball to a volume of 86.2 mL at 1.20 atm and 22°C.

5.119 The percent by mass of bicarbonate (HCO_3^-) in a certain Alka-Seltzer brand is 32.5 percent. Calculate the volume of CO_2 generated (in mL) at 37°C and 1.00 atm when a person ingests a 3.29-g tablet. (*Hint:* The reaction is between HCO_3^- and HCl acid in the stomach.)

5.120 502 mL of Ne at 18°C and 1.04 atm is mixed with 364 mL of SF_6 at 18°C and 0.85 atm in a 250-mL flask. Calculate the partial pressure of each gas.

5.121 The boiling point of liquid nitrogen is −196°C. On the basis of this information alone, do you think nitrogen is an ideal gas?

5.122 In 2.00 min, 29.7 mL of He effuses through a small hole. Under the same conditions of temperature and pressure, it takes 10.0 mL of a mixture of CO and CO_2 to effuse through the hole in the same amount of time. Calculate the percent composition by volume of the mixture.

5.123 A 10.0-mL sample of a gaseous hydrocarbon (that is, containing only C and H) at room temperature was mixed with 60.0 mL of oxygen gas. The mixture is reacted by applying an electric spark. The total volume of the gases left over is 55.0 mL. After shaking the gases with a concentrated NaOH solution, the volume is reduced to 35.0 mL. Assume that all the gas volumes are measured at the same temperature and pressure and that oxygen was the excess reagent; calculate the molecular formula of the compound.

5.124 Describe how you would measure, by either chemical or physical means, the partial pressures of a mixture of gases of the following composition: (a) CO_2 and H_2, (b) He and N_2.

5.125 A certain hydrate has the formula $MgSO_4 \cdot xH_2O$. A quantity of 54.2 g of the compound is heated in an oven to drive off the water. If the steam generated exerts a pressure of 3.55 atm in a 2.00-L container at 120°C, calculate x.

5.126 A mixture of Na_2CO_3 and $MgCO_3$ of mass 7.63 g is reacted with an excess of hydrochloric acid. The CO_2 gas generated occupies a volume of 1.67 L at 1.24 atm and 26°C. From these data, calculate the percent composition by mass of Na_2CO_3 in the mixture.

5.127 In the metallurgical process of refining nickel, the metal is purified by reacting it with carbon monoxide to form tetracarbonylnickel, which is a gas at 43°C:

$$Ni(s) + 4CO(g) \longrightarrow Ni(CO)_4(g)$$

In this way nickel can be separated from other solid impurities. (a) Starting with 86.4 g of Ni, calculate the pressure of $Ni(CO)_4$ in a container of volume 4.00 L. (Assume the above reaction to go to completion.) (b) On further heating the sample above 43°C, it is observed that the pressure of the gas increases much more rapidly than that predicted based on the ideal gas equation. Explain.

5.128 One of the earliest demonstrations that gases participate in chemical reactions, in the same sense as liquids and solids, was performed by Joseph Black in 1756. Black showed that marble ($CaCO_3$) could be converted by heating to quicklime (CaO) and carbon dioxide (CO_2):

$$CaCO_3(s) \longrightarrow CaO(s) + CO_2(g)$$

Calculate the ratio of the volume of CO_2 produced at 25°C and 1.00 atm to the volume of $CaCO_3$ consumed. The density of $CaCO_3$ is 2.930 g/cm^3.

5.129 The lifting ability of a hot-air balloon is based on the fact that the mass of the air inside the balloon is lighter than an equal volume of the cooler air outside the balloon. For a balloon 15.0 m in diameter, calculate the difference in mass (in grams) of the air in the balloon at 80°C and 1.00 atm and at 23°C and 1.00 atm (assume the volume of air in the balloon is the same in each case). The density of dry air at 23°C is 1.12 g/L, and the volume of a sphere is $(\frac{4}{3})\pi r^3$.

5.130 The following apparatus can be used to measure atomic and molecular speed. Suppose that a beam of metal atoms is directed at a rotating cylinder in a vacuum. A small opening in the cylinder allows the atoms to strike a target area. Because the cylinder is rotating, atoms traveling at different speeds will strike the target at different positions. In time, a layer of the metal will deposit on the target area, and the variation in its thickness is found to correspond to Maxwell's speed distribution. In one experiment it is found that at 850°C some bismuth (Bi) atoms struck the target at a point 2.80 cm from the spot directly opposite the slit. If the diameter of the cylinder is 15.0 cm and it is rotating at 130 revolutions per second, calculate (a) the speed (m/s) at which the target is moving (*Hint:* The circumference of a circle is given by $2\pi r$, where r is the radius), (b) the time (in seconds) it takes for the target to travel 2.80 cm, (c) the speed of the Bi atoms. Compare your result in (c) with the u_{rms} of Bi at 850°C. Comment on the difference.

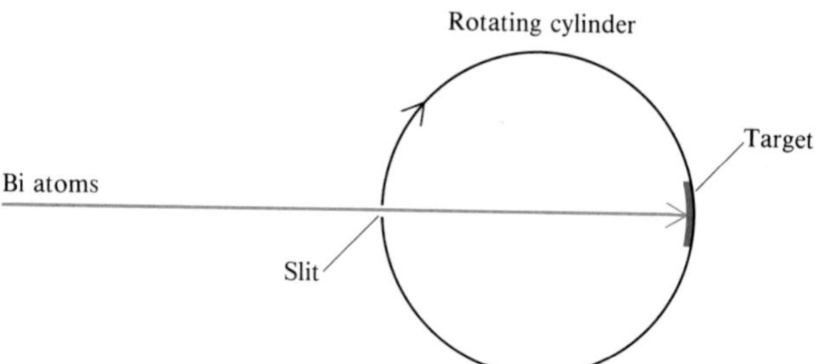

5.131 Recent controversial efforts to generate energy via "cold fusion" of deuterium atoms have centered on the remarkable ability of palladium (Pd) metal to absorb as much as 900 times its own volume in hydrogen gas (or deuterium gas) at 1 atm and 25°C. Calculate the ratio of deuterium atoms to Pd atoms in a piece of fully saturated Pd metal. The density of Pd is 12.02 g/cm^3.

Opposite page: The analysis of particles formed from burning methane (CH_4) in a flame is performed with a visible laser.

CHAPTER 6

THERMOCHEMISTRY

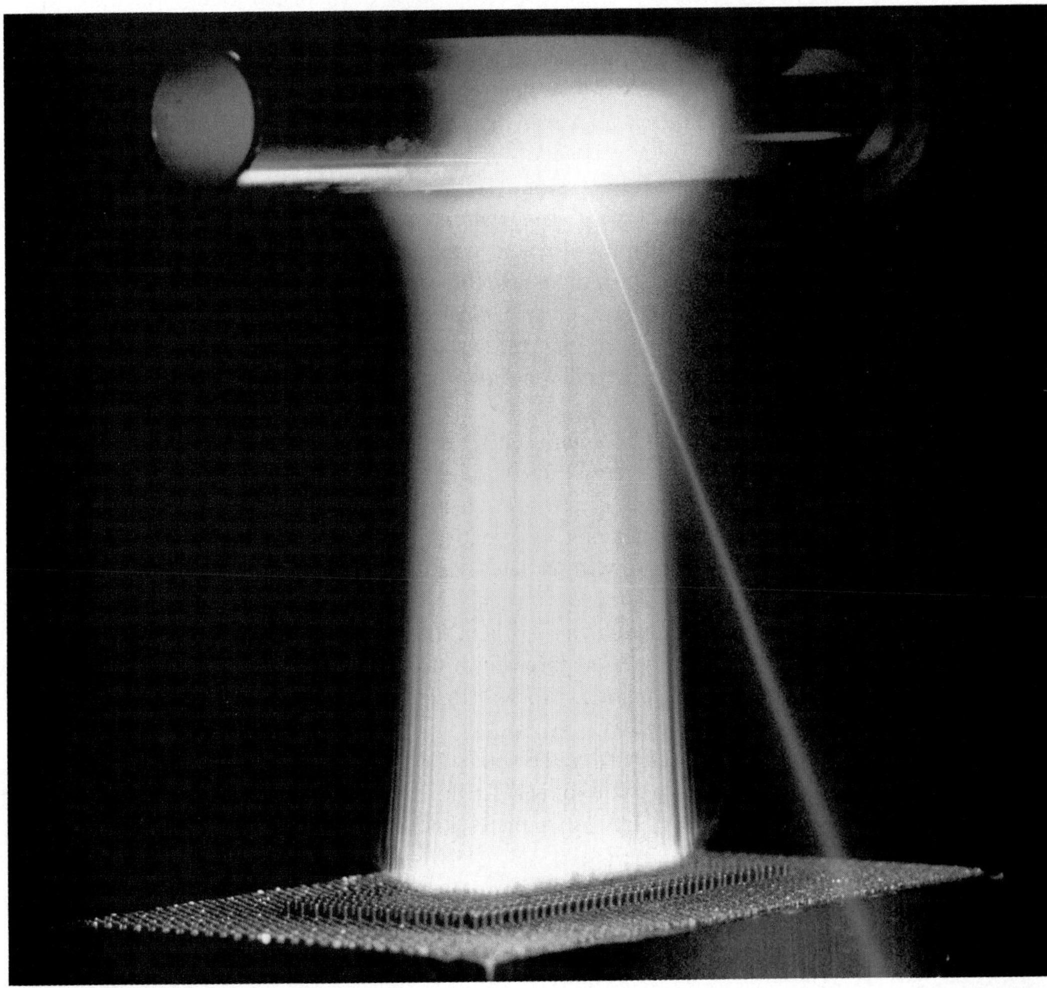

Every chemical reaction obeys two fundamental laws: the law of conservation of mass and the law of conservation of energy. We discussed the mass relationships between reactants and products in Chapter 4; here we will look at the energy changes that accompany chemical reactions.

6.1 Some Definitions

In Chapter 4 we focused on the mass relationships associated with chemical changes. However, the energy changes during chemical reactions are often of equal or greater practical interest. For example, combustion reactions involving fuels such as natural gas and coal are carried on in everyday life more for the thermal energy they release than for the particular quantities of combustion products (whether expressed in grams or in moles) they produce. In order to examine energy changes in chemical reactions more closely, we must first define some fundamental terms associated with this topic.

A ***system*** is *any specific part of the universe that is of interest to us*. For chemists, systems usually include substances involved in chemical and physical changes. For example, in an acid-base neutralization experiment, the system may be a beaker containing 50 mL of HCl to which 50 mL of NaOH is added. *The rest of the universe outside the system* is called the ***surroundings.***

There are three types of systems. An ***open system*** *can exchange mass and energy (usually in the form of heat) with its surroundings*. For example, an open system may consist of a quantity of water in an open container, as shown in Figure 6.1(a). If we close the flask, as in Figure 6.1(b), so that no water vapor can escape from or condense into the container, we create a ***closed system,*** which *allows the transfer of energy (heat) but not mass*. By placing the water in a totally insulated container, we construct an ***isolated system,*** which *does not allow the transfer of either mass or energy,* as shown in Figure 6.1(c).

The unit of energy is the joule (J). We will frequently use kilojoule (kJ) in our calculations. 1 kJ = 1000 J.

"Energy" is a much-used term, although it represents a rather abstract concept. Unlike matter, energy cannot be seen, touched, smelled, or weighed. Energy is known and recognized by its effects. It is usually defined as *the capacity to do work*. "Work" has a special meaning to scientists. In the study of mechanics, work is "force × distance," but we will see later that there are other kinds of work. All forms of energy are capable of doing work (that is, of exerting a force over a distance), but not all of them are equally relevant to chemistry. The energy contained in tidal waves, for example, can be harnessed to perform useful work, but the importance of tidal waves to chemistry is minimal. We will discuss some forms of energy that are of particular interest to chemists.

Radiant energy from the sun (solar energy) is Earth's primary energy source. Solar energy is responsible for heating the atmosphere and Earth's surface, for the growth of vegetation through the process known as photosynthesis, and for global climate patterns.

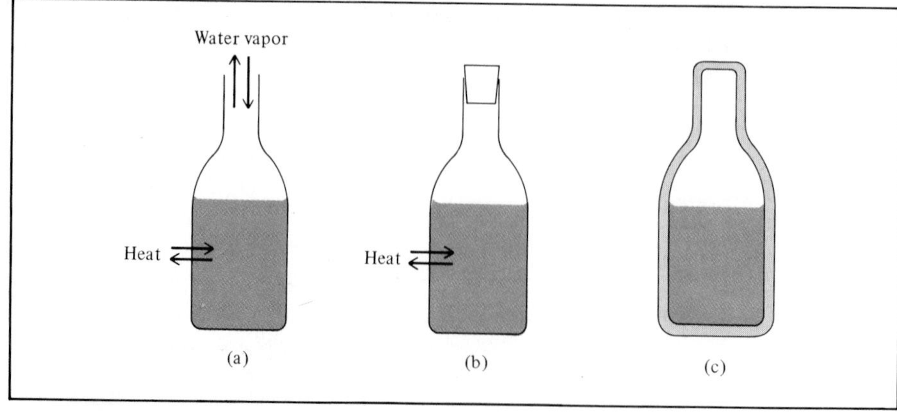

Figure 6.1 *Three systems represented by water in a flask: (a) an open system, which allows both energy and mass transfer; (b) a closed system, which allows energy but not mass transfer; and (c) an isolated system, which allows neither energy nor mass transfer (here the flask is enclosed by a vacuum jacket).*

Thermal energy is *the energy associated with the random motion of atoms and molecules*. In general, thermal energy can be calculated from temperature measurements—the more vigorous the motion of the atoms and molecules in a sample of matter, the hotter the sample is and the greater its thermal energy. However, we need to distinguish carefully between thermal energy and temperature. A cup of coffee at 70°C has a higher temperature than a bathtub filled with warm water at 40°C, but much more thermal energy is stored in the bathtub water because it has a much larger volume and greater mass than the coffee and therefore more water molecules and more molecular motion.

It is important to understand the distinction between thermal energy and heat. ***Heat*** is *the transfer of thermal energy between two bodies that are at different temperatures*. Thus we often speak of the "heat flow" from a hot object to a cold one. Although "heat" itself already implies the transfer of energy, we customarily use terms such as "heat absorbed" or "heat released" to describe the energy changes occurring during a process.

Chemical energy is *a form of energy stored within the structual units of chemical substances;* its quantity is determined by the type and arrangement of atoms in the substance being considered. When substances participate in chemical reactions, chemical energy is released, stored, or converted to other forms of energy.

Energy is also available by virtue of an object's position. This form of energy is called ***potential energy.*** For instance, because of its altitude, a rock at the top of a cliff has more potential energy and will make a bigger splash in the water below than a similar rock located part way down. Chemical energy can be considered a form of potential energy; it is associated with the relative positions and arrangements of atoms within the substances of interest.

Energy available because of the motion of an object is called ***kinetic energy.*** The kinetic energy of a moving object depends on both the mass and the velocity of the object.

All forms of energy can be changed (at least in principle) from one form to another. We feel warm when we stand in sunlight because radiant energy from the sun is converted to thermal energy on our skin. When we exercise, stored chemical energy in our bodies is used to produce kinetic energy of motion. When a ball starts to roll downhill, its potential energy is converted to kinetic energy. You can undoubtedly think of many other examples. Scientists have reached the conclusion that although energy can assume many different forms that are interconvertible, energy can be neither destroyed nor created. When one form of energy disappears, some other form of energy (of equal magnitude) must appear, and vice versa. *The total quantity of energy in the universe is thus assumed to remain constant*. This statement is generally known as the ***law of conservation of energy.***

6.2 Energy Changes in Chemical Reactions

Almost all chemical reactions absorb or produce (release) energy. Heat is the form of energy most commonly absorbed or released in chemical reactions. *The study of heat changes in chemical reactions* is called ***thermochemistry.***

The combustion of hydrogen gas in oxygen is one of many familiar chemical reactions that release considerable quantities of energy (Figure 6.2):

$$2H_2(g) + O_2(g) \longrightarrow 2H_2O(l) + \text{energy}$$

Figure 6.2 *The* Hindenburg *disaster. The* Hindenburg, *a German airship filled with hydrogen gas, was destroyed in a spectacular fire at Lakehurst, New Jersey, in 1937.*

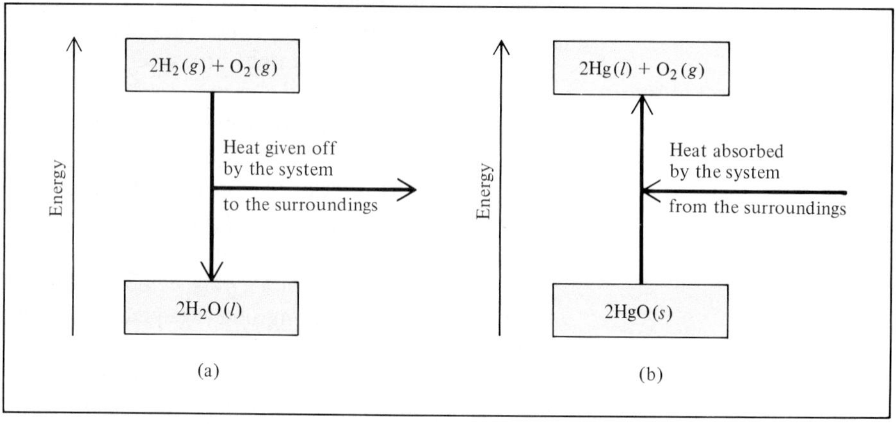

Figure 6.3 *(a) An exothermic process. (b) An endothermic process. The scales in (a) and (b) are not the same. Therefore, the heat released in the formation of H_2O from H_2 and O_2 is not equal to the heat absorbed in the decomposition of HgO.*

In discussing energy changes of this type, we can label the reacting mixture (hydrogen, oxygen, and water molecules) the *system* and the rest of the universe the *surroundings*. Since energy cannot be created or destroyed, any energy lost by the system must be gained by the surroundings. Thus the heat generated by the combustion process is transferred from the system to its surroundings. *Any process that gives off heat* (that is, *transfers thermal energy to the surroundings*) is called an ***exothermic process.*** Figure 6.3(a) shows the energy change for the combustion of hydrogen gas.

Now consider another reaction, the decomposition of mercury(II) oxide (HgO) at high temperatures:

$$\text{energy} + 2\text{HgO}(s) \longrightarrow 2\text{Hg}(l) + \text{O}_2(g)$$

This is an example of an ***endothermic process,*** *in which heat has to be supplied to the system* (that is, to HgO) *by the surroundings* [Figure 6.3(b)].

From Figure 6.3 you can see that in exothermic reactions, the total energy of the products is less than the total energy of the reactants. The difference in the energies is the heat supplied by the system to the surroundings. Just the opposite happens in endothermic reactions. Here, the difference in the energies of the products and reactants is equal to the heat supplied to the system by the surroundings.

6.3 Enthalpy

Most physical and chemical changes, including those that take place in living systems, occur in the constant-pressure conditions of our atmosphere. In the laboratory, for example, reactions are generally carried out in beakers, flasks, or test tubes that remain open to their surroundings and hence to a pressure of approximately one atmosphere (1 atm). To express the heat released or absorbed in a constant-pressure process, chemists use a quantity called ***enthalpy,*** represented by the symbol H. The change in enthalpy of a system during a process at constant pressure, represented by ΔH (''delta H,'' where the symbol Δ denotes change), is equal to the heat given off or absorbed by the system during the process. The ***enthalpy of reaction*** is *the difference between the enthalpies of the products and the enthalpies of the reactants:*

$$\Delta H = H(\text{products}) - H(\text{reactants}) \tag{6.1}$$

Enthalpy of reaction can be positive or negative, depending on the process. For an endothermic process (heat absorbed by the system from the surroundings), ΔH is positive (that is, $\Delta H > 0$). For an exothermic process (heat released by the system to the surroundings), ΔH is negative (that is, $\Delta H < 0$). Now let us apply the idea of enthalpy changes to two common processes—the first involving a physical change, the second a chemical change.

A 0°C and a constant pressure of 1 atm, ice can melt to form liquid water. Measurements show that for every mole of ice converted to liquid water under these conditions, 6.01 kilojoules (kJ) of energy are absorbed by the system (ice). Thus we can write the equation for this physical change as

$$H_2O(s) \longrightarrow H_2O(l) \qquad \Delta H = 6.01\ \text{kJ}$$

where

$$\begin{aligned}\Delta H &= H(\text{products}) - H(\text{reactants})\\ &= H(\text{liquid water}) - H(\text{ice})\\ &= 6.01\ \text{kJ}\end{aligned}$$

Since ΔH is a positive value, this is an endothermic process, which is what we would expect for the energy-absorbing change of melting ice (Figure 6.4).

Methane is the principal component of natural gas.

As another example, consider the combustion of methane (CH_4):

$$CH_4(g) + 2O_2(g) \longrightarrow CO_2(g) + 2H_2O(l) \qquad \Delta H = -890.4\ \text{kJ}$$

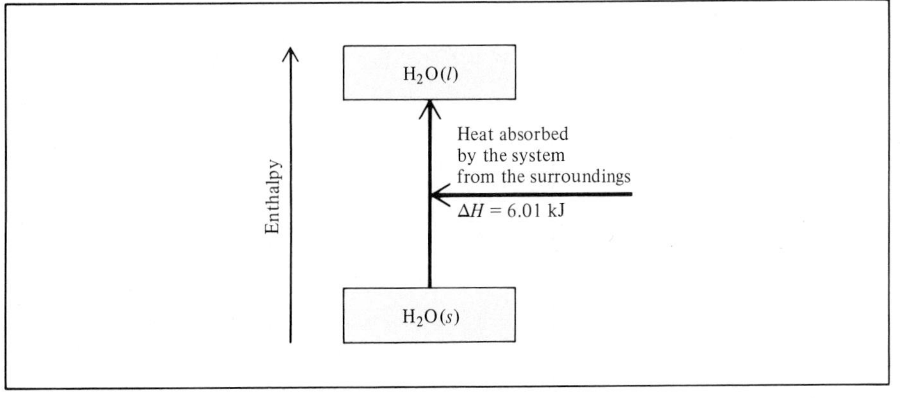

Figure 6.4 *Melting 1 mole of ice at 0°C (an endothermic process) results in an enthalpy increase in the system of 6.01 kJ.*

where $\Delta H = H(\text{products}) - H(\text{reactants})$
$= [H(CO_2, g) + 2H(H_2O, l)] - [H(CH_4, g) + 2H(O_2, g)]$
$= -890.4$ kJ

From experience we know that burning natural gas releases heat to its surroundings, so it is an exothermic process and ΔH must have a negative value (Figure 6.5).

The equations representing the melting of ice and the combustion of methane not only represent the mass relationships involved but also show the enthalpy changes. *Equations showing both the mass and enthalpy relations* are called ***thermochemical equations.*** The following guidelines are helpful in writing and interpreting thermochemical equations:

- The stoichiometric coefficients always refer to the number of moles of each substance. Thus, the equation representing the melting of ice may be "read" as follows: When 1 mole of liquid water is formed from 1 mole of ice at 0°C, the enthalpy change is 6.01 kJ. For the combustion of methane, we interpret the equation this way: When 1 mole of gaseous methane reacts with 2 moles of gaseous oxygen to form 1 mole of gaseous carbon dioxide and 2 moles of liquid water, the enthalpy change is −890.4 kJ.
- When we reverse an equation, we are changing the roles of reactants and products. Consequently, the magnitude of ΔH for the equation remains the same but its sign changes. This will seem reasonable if you consider the processes involved. For example, if a reaction consumes thermal energy from its surroundings (that is, if it is endothermic), then the reverse of that reaction must release thermal energy back to its surroundings (that is, it must be exothermic). This means that the enthalpy change expression must also change its sign. Thus, reversing the melting of ice and the combustion of methane, the thermochemical equations are

$$H_2O(l) \longrightarrow H_2O(s) \qquad \Delta H = -6.01 \text{ kJ}$$

$$CO_2(g) + 2H_2O(l) \longrightarrow CH_4(g) + 2O_2(g) \qquad \Delta H = 890.4 \text{ kJ}$$

and what was an endothermic process now becomes exothermic, and vice versa.

Enthalpy is an extensive property; that is, its magnitude depends on the amount of substance.

- If we multiply both sides of a thermochemical equation by a factor n, then ΔH must also change by the same factor. Thus, for the melting of ice, if $n = 2$, then

$$2H_2O(s) \longrightarrow 2H_2O(l) \qquad \Delta H = 2(6.01 \text{ kJ}) = 12.0 \text{ kJ}$$

and if $n = \frac{1}{2}$, then

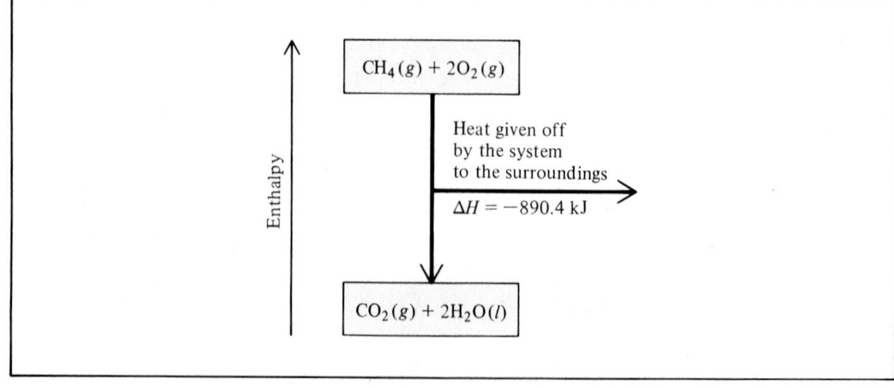

Figure 6.5 *Burning 1 mole of methane in oxygen gas (an exothermic process) results in an enthalpy decrease in the system of 890.4 kJ.*

$$\tfrac{1}{2}H_2O(s) \longrightarrow \tfrac{1}{2}H_2O(l) \qquad \Delta H = \tfrac{1}{2}(6.01 \text{ kJ}) = 3.01 \text{ kJ}$$

- When writing thermochemical equations, we must always specify the physical states of all reactants and products, because they help determine the actual enthalpy changes. For example, in the equation for the combustion of methane, if we show water vapor rather than liquid water as a product,

$$CH_4(g) + 2O_2(g) \longrightarrow CO_2(g) + 2H_2O(g)$$

the enthalpy change is −802.4 kJ rather than −890.4 kJ because 88.0 kJ of energy are needed to convert 2 moles of liquid water to water vapor; that is,

$$2H_2O(l) \longrightarrow 2H_2O(g) \qquad \Delta H = 88.0 \text{ kJ}$$

- The enthalpy of a substance increases with temperature, and the enthalpy change for a reaction also depends on the specified temperature. We will see shortly that enthalpy changes are usually expressed for a reference temperature of 25°C. This convention may seem to present a problem for reactions such as combustion. During the burning of methane, for example, the temperature of the reacting system is considerably higher than 25°C. However, temperature changes during such a reaction should present no problem. We are concerned only with the enthalpy change when we convert 1 mole of methane and 2 moles of oxygen gas at 25°C to 1 mole of carbon dioxide and 2 moles of water at the same temperature. If the products form at a temperature higher than 25°C, they eventually cool down to 25°C, and the heat evolved upon cooling simply becomes part of the overall enthalpy change.

Reactions with large heat changes find many applications in homes and industry, as the following Chemistry in Action shows.

CHEMISTRY IN ACTION

Candles, Burners, and Torches

Combustion reactions are all highly exothermic. Here we will discuss three types of combustions that are commonly seen in homes, laboratories, and industry: candles, burners, and torches.

We burn candles today mostly for aesthetic reasons. The candle's incandescence comes from the presence in the luminous parts of the flame of hot, solid particles [Figure 6.6(a)]. These particles, which are mostly elemental carbon, can be collected by placing a spatula's blade in the flame. The wax of a candle is made of high-molar-mass hydrocarbons. (A *hydrocarbon* is a compound that contains only carbon and hydrogen atoms.) The heat of the candle's flame melts the wax, which is drawn up the wick. The additional heat then vaporizes the wax from the wick. Some of the wax vapor burns to form carbon dioxide and water, and some of it is broken down to form smaller hydrocarbon molecules, fragments of molecules, and carbon. Eventually many of these intermediates are also converted to carbon dioxide and water in the combustion process.

Figure 6.7 shows a Bunsen† burner, which is commonly used in the chemical laboratory. The gas used in the combustion is methane or propane. For a constant

†Robert Wilhelm Bunsen (1811–1899). German chemist. Bunsen did important work in chemical analysis, mostly by spectroscopic methods. He discovered cesium and rubidium and invented the gas burner that now bears his name.

Figure 6.6 *(a) A burning candle. (b) A Bunsen burner with a yellow flame. (c) A Bunsen burner with a blue flame. (d) An oxygen/methane flame.*

gas flow, the temperature of the flame depends on how much air is premixed with the methane gas before combustion. When the air inlet valve at the bottom is closed, the flame has a yellow color, which indicates that the combustion process is incomplete [Figure 6.6(b)]. (Incomplete combustion means that not all the methane gas is converted to carbon dioxide; some of it is converted to elemental carbon as in the case of a burning candle.) With the air inlet valve fully open, the methane gas is largely converted to carbon dioxide and water:

$$CH_4(g) + 2O_2(g) \longrightarrow CO_2(g) + 2H_2O(l)$$

More heat is released in this process, so the temperature of the flame is higher and the color of the flame changes from yellow to blue [Figure 6.6(c)]. The gas used in domestic cooking (methane or propane) is usually premixed with air to make the flame blue.

We can obtain even more complete combustion of methane by premixing it with pure oxygen gas instead of air. An oxygen/methane torch is used in scientific glassblowing; its flame is hot enough to melt quartz (the melting point of quartz is about 1600°C) [Figure 6.6(d)].

Industrially, higher-temperature flames are needed to cut and weld metals. The oxyhydrogen torch premixes hydrogen and oxygen gases prior to combustion:

$$2H_2(g) + O_2(g) \longrightarrow 2H_2O(l)$$

Typically a flame temperature of 2500°C can be obtained from this process. The reaction between acetylene (C_2H_2) and oxygen is even more exothermic:

$$2C_2H_2(g) + 5O_2(g) \longrightarrow 4CO_2(g) + 2H_2O(l) \qquad \Delta H = -2598.8 \text{ kJ}$$

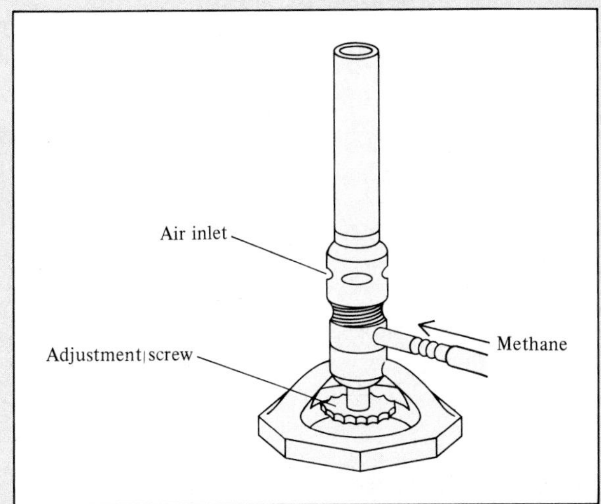

Figure 6.7 *Schematic diagram of a Bunsen burner. The amount of air premixed with methane gas can be controlled by turning the top part of the burner. The lower screw adjusts the amount of methane used in the combustion.*

An oxyacetylene torch has a flame temperature as high as 3000°C; it is used extensively in construction work (Figure 6.8).

Figure 6.8 *An oxyacetylene torch used to weld metals.*

6.4 Calorimetry

In this section we will consider experimental methods for measuring the heat of reaction. Our discussion of ***calorimetry***—that is, *the measurement of heat changes*—will depend on an understanding of specific heat and heat capacity, so let us consider them first.

Specific Heat and Heat Capacity

The ***specific heat(s)*** of a substance is *the amount of heat required to raise the temperature of one gram of the substance by one degree Celsius.* The **heat capacity (*C*)** of a substance is *the amount of heat required to raise the temperature of a given quantity of the substance by one degree Celsius.* The relationship between the heat capacity and specific heat of a substance is

$$C = ms \tag{6.2}$$

where m is the mass of the substance in grams. For example, the specific heat of water is 4.184 J/g · °C, and the heat capacity of 60.0 g of water is

The dot between g and °C reminds us that both g and °C are in the denominator.

$$(60.0\ \text{g})(4.184\ \text{J/g} \cdot {}^\circ\text{C}) = 251\ \text{J/}^\circ\text{C}$$

Note that specific heat has the units J/g · °C and heat capacity has the units J/°C. Table 6.1 shows the specific heats of some common substances.

Table 6.1 The specific heats of some common substances

Substance	Specific heat (J/g · °C)
Al	0.900
Au	0.129
C(graphite)	0.720
C(diamond)	0.502
Cu	0.385
Fe	0.444
Hg	0.139
H_2O	4.184
C_2H_5OH (ethanol)	2.46

If we know the specific heat and the amount of a substance, then the change in the sample's temperature (Δt) will tell us the amount of heat (q) that has been absorbed or released in a particular process. The equation for calculating the heat change is given by

$$q = ms\ \Delta t \tag{6.3}$$

$$q = C\ \Delta t \tag{6.4}$$

where m is the mass of the sample and Δt is the temperature change:

$$\Delta t = t_{\text{final}} - t_{\text{initial}}$$

The sign convention is the same as that for enthalpy change; q is positive for endothermic processes and negative for exothermic processes.

An application of Equation (6.3) is shown in Example 6.1.

EXAMPLE 6.1

A 466-g sample of water is heated from 8.50°C to 74.60°C. Calculate the amount of heat absorbed by the water.

Answer

Using Equation (6.3), we write

$$\begin{aligned} q &= ms\ \Delta t \\ &= (466\ \text{g})(4.184\ \text{J/g}\cdot{}^\circ\text{C})(74.60^\circ\text{C} - 8.50^\circ\text{C}) \\ &= 1.29 \times 10^5\ \text{J} \\ &= 129\ \text{kJ} \end{aligned}$$

Similar problems: 6.21, 6.22.

Constant-Volume Calorimetry

"Constant volume" refers to the volume of the container, which does not change during the reaction. Note that the container remains intact after the measurement. The term "bomb calorimeter" connotes the explosive nature of the reaction (on a small scale) in the presence of excess oxygen gas.

Heats of combustion are usually measured by placing a known mass of the compound under study in a steel container, called a *constant-volume bomb calorimeter,* which is filled with oxygen at about 30 atm of pressure. The closed bomb is immersed in a known amount of water, as shown in Figure 6.9. The sample is ignited electrically, and the heat produced by the combustion can be calculated accurately by recording the rise in temperature of the water. The heat given off by the sample is absorbed by the water and the calorimeter. The special design of the bomb calorimeter allows us to safely assume that no heat (or mass) is lost to the surroundings during the time it takes to make measurements. Therefore we can call the bomb calorimeter and the water in which it is submerged an isolated system. Because there is no heat entering or leaving the system throughout the process, we can write

$$\begin{aligned} q_{\text{system}} &= q_{\text{water}} + q_{\text{bomb}} + q_{\text{rxn}} \\ &= 0 \end{aligned} \tag{6.5}$$

where q_{water}, q_{bomb}, and q_{rxn} are the heat changes for the water, the bomb, and the

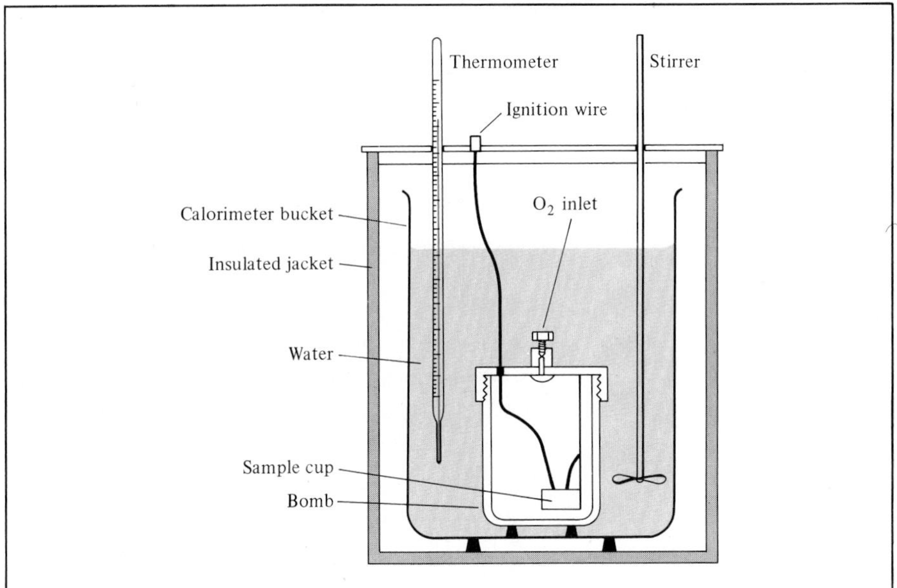

Figure 6.9 *A constant-volume bomb calorimeter. The calorimeter is filled with oxygen gas before it is placed in the bucket. The sample is ignited electrically, and the heat produced by the reaction can be accurately determined by measuring the temperature increase in the known amount of surrounding water.*

reaction, respectively. Thus

$$q_{\text{rxn}} = -(q_{\text{water}} + q_{\text{bomb}}) \tag{6.6}$$

The quantity q_{water} is obtained by

$$q = ms\ \Delta t$$

$$q_{\text{water}} = (m_{\text{water}})(4.184\ \text{J/g} \cdot {}^\circ\text{C})\ \Delta t$$

The product of the mass of the bomb and its specific heat is the heat capacity of the bomb, which remains constant for all experiments carried out in the bomb calorimeter:

$$C_{\text{bomb}} = m_{\text{bomb}} \times s_{\text{bomb}}$$

Hence

$$q_{\text{bomb}} = C_{\text{bomb}}\ \Delta t$$

Note that because reactions in a bomb calorimeter occur under constant-volume rather than constant-pressure conditions, the heat changes do not correspond to the enthalpy change ΔH (see Section 6.3). It is possible to correct the measured heat changes so that they correspond to ΔH values, but the corrections usually are quite small, so we will not concern ourselves with the details of the correction procedure.

A calculation based on data obtained from a constant-volume bomb calorimeter experiment is shown in the following example.

EXAMPLE 6.2

A quantity of 1.435 g of naphthalene ($C_{10}H_8$) was burned in a constant-volume bomb calorimeter. Consequently, the temperature of the water rose from 20.17°C to 25.84°C. If the quantity of water surrounding the calorimeter was exactly 2000 g and the heat capacity of the bomb calorimeter was 1.80 kJ/°C, calculate the heat of combustion of naphthalene on a molar basis; that is, find the molar heat of combustion.

Answer

First we calculate the heat changes for the water and the bomb calorimeter.

$$q = ms\,\Delta t$$

$$\begin{aligned} q_{\text{water}} &= (2000\text{ g})(4.184\text{ J/g}\cdot{}^\circ\text{C})(25.84^\circ\text{C} - 20.17^\circ\text{C}) \\ &= 4.74 \times 10^4\text{ J} \end{aligned}$$

$$\begin{aligned} q_{\text{bomb}} &= (1.80 \times 1000\text{ J/}^\circ\text{C})(25.84^\circ\text{C} - 20.17^\circ\text{C}) \\ &= 1.02 \times 10^4\text{ J} \end{aligned}$$

(Note that we changed 1.80 kJ/°C to 1.80 × 1000 J/°C.) Next, from Equation (6.6) we write

$$\begin{aligned} q_{\text{rxn}} &= -(4.74 \times 10^4\text{ J} + 1.02 \times 10^4\text{ J}) \\ &= -5.76 \times 10^4\text{ J} \end{aligned}$$

The molar mass of naphthalene is 128.2 g, so the heat of combustion of 1 mole of naphthalene is

$$\begin{aligned} \text{molar heat of combustion} &= \frac{-5.76 \times 10^4\text{ J}}{1.435\text{ g }C_{10}H_8} \times \frac{128.2\text{ g }C_{10}H_8}{1\text{ mol }C_{10}H_8} \\ &= -5.15 \times 10^6\text{ J/mol} \\ &= -5.15 \times 10^3\text{ kJ/mol} \end{aligned}$$

Similar problem: 6.26.

Finally, we note that the heat capacity of a bomb calorimeter is usually determined by burning in it a compound with an accurately known heat of combustion value. From the mass of the compound and the temperature rise, we can calculate the heat capacity of the calorimeter (see Problem 6.27).

The bomb calorimeter is ideally suited for measuring the energy content in foods (see Chemistry in Action on p. 231).

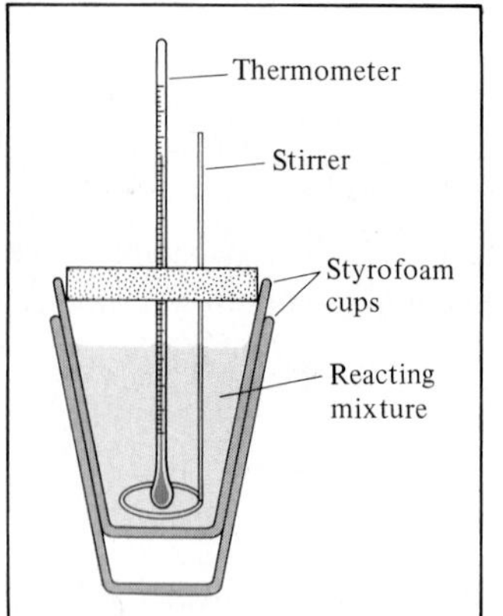

Figure 6.10 *A constant-pressure calorimeter made of two Styrofoam coffee cups. The outer cup helps to insulate the reacting mixture from the surroundings. Two solutions of known volume containing the reactants at the same temperature are carefully mixed in the calorimeter. The heat produced or absorbed by the reaction can be determined by measuring the temperature change.*

Constant-Pressure Calorimetry

A simpler device for determining heats of reactions for other than combustion reactions is a *constant-pressure calorimeter*. In its most basic form, it can be constructed from two Styrofoam coffee cups, as shown in Figure 6.10. Such a calorimeter can be used to measure the heat effect for a variety of reactions, such as acid-base neutralization reactions, as well as heats of solution and heats of dilution. Because the measurements are carried out under constant atmospheric pressure conditions, the heat change for the process (q_{rxn}) is equal to the enthalpy change (ΔH). The measurements are similar to those of a constant-volume calorimeter—we need to know the heat capacity of the calorimeter, as well as the temperature change of the solution. Table 6.3 lists some reactions that have been studied with the constant-pressure calorimeter.

CHEMISTRY IN ACTION

Fuel Values of Foods and Other Substances

The food we eat is broken down in a series of steps by a group of complex biological molecules called enzymes. Most of the energy released at each step is captured for function and growth. This process is called metabolism. One interesting aspect of metabolism is that the overall change in energy is the same as it is in combustion. For example, the total enthalpy change for the conversion of glucose ($C_6H_{12}O_6$) to carbon dioxide and water is the same whether we burn the substance in air or digest it in our bodies:

$$C_6H_{12}O_6(s) + 6O_2(g) \longrightarrow 6CO_2(g) + 6H_2O(l) \qquad \Delta H = -2801 \text{ kJ}$$

The important difference between metabolism and combustion, however, is that the latter is usually a one-step, high-temperature process. Consequently, much of the energy released is lost to the surroundings.

Various foods have different compositions and hence different energy contents. We often speak of the calorie content of the foods we eat (Figure 6.11). The calorie (cal) is a non-SI unit of energy where

$$1 \text{ cal} = 4.184 \text{ J}$$

In the context of nutrition, however, the calorie we speak of (sometimes called a "big calorie") is actually equal to a kilocalorie; that is,

$$1 \text{ Cal} = 1000 \text{ cal} = 4184 \text{ J}$$

Note that we have used capital C to indicate the "big calorie" unit used in food. The bomb calorimeter described in Section 6.4 is ideally suited for measuring the energy content, or "fuel value," of foods. Fuel values are just the enthalpies of combustion, listed in Table 6.2. The foods analyzed this way must be dried, because most of them contain considerable amounts of water. Since the composition of particular foods is often not known, fuel values are expressed in terms of kJ/g rather than kJ/mol.

Table 6.2 Fuel values of foods and some common fuels

Substance	$\Delta H_{\text{combustion}}$ (kJ/g)
Apple	−2
Beef	−8
Beer	−1.5
Bread	−11
Butter	−34
Cheese	−18
Eggs	−6
Milk	−3
Potatoes	−3
Charcoal	−35
Coal	−30
Gasoline	−34
Kerosene	−37
Natural gas	−50
Wood	−20

Kellogg's® Special K®

NUTRITION INFORMATION

SERVING SIZE: 1 OZ. (28.4 g, ABOUT 1 CUP) SPECIAL K ALONE, AND WITH ½ CUP VITAMINS A AND D SKIM MILK.

SERVINGS PER PACKAGE: 7

	CEREAL	WITH SKIM MILK
CALORIES	110	150*
PROTEIN	6 g	10 g
CARBOHYDRATE	20 g	26 g
FAT	0 g	0 g*
CHOLESTEROL	0 mg	0 mg*
SODIUM	230 mg	290 mg
POTASSIUM	50 mg	250 mg

INGREDIENTS: RICE, WHEAT GLUTEN, SUGAR, DEFATTED WHEAT GERM, SALT, CORN SYRUP, WHEY, MALT FLAVORING, CALCIUM CASEINATE,

Figure 6.11 *Many labels on food packages show the calorie content of the food.*

Table 6.3 Heats of some typical reactions measured at constant pressure

Type of reaction	Example	ΔH (kJ)
Heat of neutralization	$HCl(aq) + NaOH(aq) \longrightarrow NaCl(aq) + H_2O(l)$	−56.2
Heat of ionization	$H_2O(l) \longrightarrow H^+(aq) + OH^-(aq)$	56.2
Heat of fusion	$H_2O(s) \longrightarrow H_2O(l)$	6.01
Heat of vaporization	$H_2O(l) \longrightarrow H_2O(g)$	44.0*
Heat of reaction	$MgCl_2(s) + 2Na(l) \longrightarrow 2NaCl(s) + Mg(s)$	−180.2

*Measured at 25°C. At 100°C, the value is 40.79 kJ.

The following example deals with heat of neutralization.

EXAMPLE 6.3

A quantity of 1.00×10^2 mL of 0.500 *M* HCl is mixed with 1.00×10^2 mL of 0.500 *M* NaOH in a constant-pressure calorimeter having a heat capacity of 335 J/°C. The initial temperature of the HCl and NaOH solutions is the same, 22.50°C, and the final temperature of the mixed solution is 24.90°C. Calculate the heat change for the neutralization reaction

$$NaOH(aq) + HCl(aq) \longrightarrow NaCl(aq) + H_2O(l)$$

Assume that the densities and specific heats of the solutions are the same as for water (1.00 g/mL and 4.184 J/g · °C, respectively).

Answer

Assuming that no heat is lost to the surroundings, we write

$$q_{system} = q_{soln} + q_{calorimeter} + q_{rxn} = 0$$

or

$$q_{rxn} = -(q_{soln} + q_{calorimeter})$$

where

Because the density of the solution is 1.00 g/mL, the mass of a 100-mL solution is 100 g.

$$q_{soln} = (1.00 \times 10^2\ g + 1.00 \times 10^2\ g)(4.184\ J/g \cdot °C)(24.90°C - 22.50°C) = 2.01 \times 10^3\ J$$

$$q_{calorimeter} = (335\ J/°C)(24.90°C - 22.50°C) = 804\ J$$

Hence

$$q_{rxn} = -(2.01 \times 10^3\ J + 804\ J) = -2.81 \times 10^3\ J = -2.81\ kJ$$

From the molarities given, we know there is 0.0500 mole of HCl in 1.00×10^2 g of the HCl solution and 0.0500 mole of NaOH in 1.00×10^2 g of the NaOH solution. Therefore the heat of neutralization when 1.00 mole of HCl reacts with 1.00 mole of NaOH is

$$\text{heat of neutralization} = \frac{-2.81\ kJ}{0.0500\ mol} = -56.2\ kJ/mol$$

Note that because the reaction takes place at constant pressure, the heat given off is equal to the enthalpy change.

Similar problem: 6.28.

6.5 Standard Enthalpy of Formation and Reaction

So far we have learned that the enthalpy change of a reaction can be obtained from the heat absorbed or released (at constant pressure). From Equation (6.1) we see that ΔH can also be calculated if we know the actual enthalpies of all reactants and products. Unfortunately there is no way to determine the *absolute* value of the enthalpy of a substance. Only values *relative* to an arbitrary reference can be given. This is similar to the problem geographers faced in expressing the elevations of specific mountains or valleys. Rather than trying to devise some type of "absolute" elevation scale (perhaps based on distance from the center of Earth?), by common agreement all geographic heights and depths are expressed relative to sea level, an arbitrary reference with a defined elevation of "zero" meters or feet. Similarly, chemists have agreed on an arbitrary reference point with which all enthalpies are compared.

The "sea level" reference point for all enthalpy expressions is based on implications of the term "enthalpy of formation" (also called the "heat of formation"). The enthalpy of formation of a compound is the heat change (in kJ) when one mole of the compound is synthesized from its elements under constant-pressure conditions. This quantity can vary with experimental conditions (for example, the temperature and pressure at which the process is carried out). Therefore, we define the ***standard enthalpy of formation*** of a compound **(ΔH_f°)** to be *the heat change that results when one mole of the compound is formed from its elements in their standard states*. ***Standard states*** refers to *the condition of 1 atm*. The superscript $^\circ$ denotes that the measurement was carried out under standard-state conditions (1 atm), and the subscript f denotes formation. Although the standard state does not specify what the temperature should be, we will always use ΔH_f° values measured at 25°C.

The importance of the standard enthalpies of formation is that once we know their values, we can *calculate* the enthalpy of reactions. For example, for the hypothetical reaction

$$aA + bB \longrightarrow cC + dD$$

where a, b, c, and d are stoichiometric coefficients. *The enthalpy of reaction carried out under standard-state conditions*, called the ***standard enthalpy of reaction, ΔH_{rxn}°,*** is given by

$$\Delta H_{\text{rxn}}^\circ = [c\ \Delta H_f^\circ(C) + d\ \Delta H_f^\circ(D)] - [a\ \Delta H_f^\circ(A) + b\ \Delta H_f^\circ(B)] \tag{6.7}$$

where a, b, c, and d all have the unit mol. We can generalize Equation (6.7) as follows:

$$\Delta H_{\text{rxn}}^\circ = \Sigma n\ \Delta H_f^\circ(\text{products}) - \Sigma m\ \Delta H_f^\circ(\text{reactants}) \tag{6.8}$$

where m and n denote the stoichiometric coefficients for the reactants and products, and Σ (sigma) means "the sum of."

Equation (6.7) shows how we can calculate ΔH°_{rxn} from the ΔH°_f values of compounds. The huge amount of calorimetric data that chemists have been accumulating over the years is stored in these ΔH°_f values. By convention, *the standard enthalpy of formation of any element in its most stable form is zero*. Take oxygen as an example. Molecular oxygen (O_2) is a more stable allotropic form of oxygen than ozone (O_3) at 1 atm and 25°C. Thus we can write $\Delta H^\circ_f(O_2) = 0$, but $\Delta H^\circ_f(O_3) \neq 0$. Similarly, graphite is a more stable allotropic form of carbon than diamond at 1 atm and 25°C, so $\Delta H^\circ_f(\text{C, graphite}) = 0$ and $\Delta H^\circ_f(\text{C, diamond}) \neq 0$.

Recall that different forms of the same element are called allotropes.

There are two ways to measure the ΔH°_f values of compounds—the direct method and the indirect method, described below.

The Direct Method. This method applies to compounds which can be readily synthesized from their elements. Now suppose we wish to know the enthalpy of formation of carbon dioxide. This requires that we measure the enthalpy of the reaction when carbon (graphite) and oxygen molecule in their standard states are converted to carbon dioxide in its standard state

$$\text{C(graphite)} + O_2(g) \longrightarrow CO_2(g) \qquad \Delta H^\circ_{rxn} = -393.5 \text{ kJ}$$

As we know from experience, this reaction can be easily carried to completion. From Equation (6.7) we can write

$$\begin{aligned}\Delta H^\circ_{rxn} &= (1 \text{ mol})\, \Delta H^\circ_f(CO_2, g) - [(1 \text{ mol})\, \Delta H^\circ_f(\text{C, graphite}) + (1 \text{ mol})\, \Delta H^\circ_f(O_2, g)] \\ &= -393.5 \text{ kJ}\end{aligned}$$

Since both graphite and O_2 are stable allotropic forms, it follows that $\Delta H^\circ_f(\text{C, graphite})$ and $\Delta H^\circ_f(O_2, g)$ are zero. Therefore

$$\Delta H^\circ_{rxn} = (1 \text{ mol})\, \Delta H^\circ_f(CO_2, g) = -393.5 \text{ kJ}$$

or

$$\Delta H^\circ_f(CO_2, g) = -393.5 \text{ kJ/mol}$$

Note that our arbitrarily assigning zero ΔH°_f for each element in its most stable form at the standard state does not affect our calculations in any way. In thermochemistry we are interested only in the *changes* of enthalpies because they can be determined experimentally but their absolute values cannot. The choice of a zero "reference level" for enthalpy makes calculation easier to handle. Again referring to the terrestrial altitude analogy, we find that Mt. Everest is 8708 ft higher than Mt. McKinley. This difference in altitude is unaffected whether we set the sea level to be 0 ft or 1000 ft.

Other compounds that can be similarly studied are SF_6, P_4O_{10}, and CS_2. The equations representing their syntheses are

$$\text{S(rhombic)} + 3F_2(g) \longrightarrow SF_6(g)$$

$$4\text{P(white)} + 5O_2(g) \longrightarrow P_4O_{10}(s)$$

$$\text{C(graphite)} + 2\text{S(rhombic)} \longrightarrow CS_2(l)$$

Note that S(rhombic) and P(white) are the most stable allotropic forms of the elements at 1 atm and 25°C, so their ΔH°_f values are zero.

The Indirect Method. There are many compounds that cannot be directly synthesized from their elements. For example, the reaction of interest may proceed too slowly, or undesired side reactions may produce substances other than the compound

(a)

(b)

(a) Graphite; (b) diamond.

of interest. In these cases their ΔH_f° values can still be determined by an indirect approach, which is based on Hess's law of heat summation† (or simply Hess's law). ***Hess's law*** can be stated as follows: *When reactants are converted to products, the change in enthalpy is the same whether the reaction takes place in one step or in a series of steps.*

Consider graphite and diamond, the two allotropes of carbon. Under atmospheric conditions graphite is the more stable form so that diamond is slowly converted to graphite:

$$\text{C(diamond)} \longrightarrow \text{C(graphite)}$$

Suppose we are interested in the standard enthalpy of formation of diamond. From the above equation we write

$$\Delta H^\circ_{\text{rxn}} = (1 \text{ mol})\ \Delta H_f^\circ(\text{C, graphite}) - (1 \text{ mol})\ \Delta H_f^\circ(\text{C, diamond})$$

Since $\Delta H_f^\circ(\text{C, graphite})$ is zero by definition, it follows that

$$\Delta H^\circ_{\text{rxn}} = -(1 \text{ mol})\ \Delta H_f^\circ(\text{C, diamond})$$

This equation shows that if we can experimentally determine the standard enthalpy of reaction of the diamond-to-graphite conversion, we can calculate the standard enthalpy of formation of diamond. However, because the reaction is so slow (it may take millions of years to complete), we cannot place diamond in a constant-pressure calorimeter and measure the heat change. Instead, we must take an indirect approach. We know that the enthalpy changes for the following two reactions can be measured:

(a) $\text{C(diamond)} + O_2(g) \longrightarrow CO_2(g) \qquad \Delta H^\circ_{\text{rxn}} = -395.4 \text{ kJ}$

(b) $\text{C(graphite)} + O_2(g) \longrightarrow CO_2(g) \qquad \Delta H^\circ_{\text{rxn}} = -393.5 \text{ kJ}$

Remember to reverse the sign of ΔH when you reverse an equation.

Reversing Equation (b), we get

(c) $CO_2(g) \longrightarrow \text{C(graphite)} + O_2(g) \qquad \Delta H^\circ_{\text{rxn}} = +393.5 \text{ kJ}$

Next, we add Equations (a) and (c) to get the desired equation:

(a) $\text{C(diamond)} + \cancel{O_2(g)} \longrightarrow \cancel{CO_2(g)} \qquad \Delta H^\circ_{\text{rxn}} = -395.4 \text{ kJ}$

(c) $\cancel{CO_2(g)} \longrightarrow \text{C(graphite)} + \cancel{O_2(g)} \qquad \Delta H^\circ_{\text{rxn}} = +393.5 \text{ kJ}$

(d) $\text{C(diamond)} \longrightarrow \text{C(graphite)} \qquad \Delta H^\circ_{\text{rxn}} = -1.9 \text{ kJ}$

Thus we have $\Delta H_f^\circ(\text{C, diamond}) = -\Delta H^\circ_{\text{rxn}}/\text{mol}$, or $+1.9$ kJ/mol. Notice that in adding the two thermochemical equations, we treated the formulas like algebraic expressions. Since 1 mole of O_2 and 1 mole of CO_2 appear on both the left and right sides of Equations (a) and (c), they can be canceled out in the addition.

Recalling Hess's law, we see that the enthalpy change for the overall process (d) is equal to the sum of the enthalpy changes for the two individual steps leading to the overall reaction [(a) + (c)].

Let's consider another example. Suppose we are interested in the standard enthalpy of formation of methane (CH_4). We might represent the synthesis of CH_4 from its elements as

$$\text{C(graphite)} + 2H_2(g) \longrightarrow CH_4(g)$$

†Germain Henri Hess (1802–1850). Swiss chemist. Hess was born in Switzerland, but spent most of his life in Russia. For formulating Hess's law, he is called the founder of thermochemistry.

However, methane cannot be prepared directly from its elements C and H because this reaction does not take place. Therefore we must employ an indirect route, using Hess's law. To begin with, we know that the following reactions involving C, H_2, and CH_4 with O_2 have all been studied and the ΔH°_{rxn} values accurately determined:

$$\text{(a)} \quad C(\text{graphite}) + O_2(g) \longrightarrow CO_2(g) \qquad \Delta H^\circ_{rxn} = -393.5 \text{ kJ}$$

$$\text{(b)} \quad 2H_2(g) + O_2(g) \longrightarrow 2H_2O(l) \qquad \Delta H^\circ_{rxn} = -571.6 \text{ kJ}$$

$$\text{(c)} \quad CH_4(g) + 2O_2(g) \longrightarrow CO_2(g) + 2H_2O(l) \qquad \Delta H^\circ_{rxn} = -890.4 \text{ kJ}$$

Since we want to obtain one equation containing only C and H_2 as reactants and CH_4 as product, we must reverse (c) to get

$$\text{(d)} \quad CO_2(g) + 2H_2O(l) \longrightarrow CH_4(g) + 2O_2(g) \qquad \Delta H^\circ_{rxn} = 890.4 \text{ kJ}$$

The next step is to carry out the operation (a) + (b) + (d):

$$\begin{array}{llcl}
\text{(a)} & C(\text{graphite}) + \cancel{O_2(g)} \longrightarrow \cancel{CO_2(g)} & \Delta H^\circ_{rxn} = & -393.5 \text{ kJ} \\
\text{(b)} & 2H_2(g) + \cancel{O_2(g)} \longrightarrow \cancel{2H_2O(l)} & \Delta H^\circ_{rxn} = & -571.6 \text{ kJ} \\
\text{(d)} & \cancel{CO_2(g)} + \cancel{2H_2O(l)} \longrightarrow CH_4(g) + \cancel{2O_2(g)} & \Delta H^\circ_{rxn} = & 890.4 \text{ kJ} \\
\hline
\text{(e)} & C(\text{graphite}) + 2H_2(g) \longrightarrow CH_4(g) & \Delta H^\circ_{rxn} = & -74.7 \text{ kJ}
\end{array}$$

As you can see, all the nonessential species (O_2, CO_2, and H_2O) cancel in this operation. Because the above equation represents the synthesis of 1 mole of CH_4 from its elements, we have $\Delta H^\circ_f(CH_4) = -74.7$ kJ/mol.

Table 6.4 Standard enthalpies of formation of some inorganic substances at 25°C

Substance	ΔH°_f (kJ/mol)	Substance	ΔH°_f (kJ/mol)
$Ag(s)$	0	$H_2O_2(l)$	−187.6
$AgCl(s)$	−127.04	$Hg(l)$	0
$Al(s)$	0	$I_2(s)$	0
$Al_2O_3(s)$	−1669.8	$HI(g)$	25.94
$Br_2(l)$	0	$Mg(s)$	0
$HBr(g)$	−36.2	$MgO(s)$	−601.8
C(graphite)	0	$MgCO_3(s)$	−1112.9
C(diamond)	1.90	$N_2(g)$	0
$CO(g)$	−110.5	$NH_3(g)$	−46.3
$CO_2(g)$	−393.5	$NO(g)$	90.4
$Ca(s)$	0	$NO_2(g)$	33.85
$CaO(s)$	−635.6	$N_2O_4(g)$	9.66
$CaCO_3(s)$	−1206.9	$N_2O(g)$	81.56
$Cl_2(g)$	0	$O(g)$	249.4
$HCl(g)$	−92.3	$O_2(g)$	0
$Cu(s)$	0	$O_3(g)$	142.2
$CuO(s)$	−155.2	S(rhombic)	0
$F_2(g)$	0	S(monoclinic)	0.30
$HF(g)$	−268.61	$SO_2(g)$	−296.1
$H(g)$	218.2	$SO_3(g)$	−395.2
$H_2(g)$	0	$H_2S(g)$	−20.15
$H_2O(g)$	−241.8	$ZnO(s)$	−347.98
$H_2O(l)$	−285.8		

Looking back, we see that the enthalpy change for the formation of methane is the same whether it takes place in one step [from C and H_2 as shown in (e)] or in a series of steps [(a) + (b) + (d)]. This example should convince you of the usefulness of Hess's law. Since it is not possible to form methane by reacting hydrogen gas with graphite, there is no way of directly measuring the enthalpy change. However, by taking an indirect route, that is, using data on reactions that can be readily studied in the laboratory, we are able to calculate the enthalpy of formation of methane.

A final word on Hess's law. The general rule in applying the law is that we should arrange a series of chemical equations (corresponding to a series of steps) in such a way that, when added together, all species will cancel except for the reactants and products that appear in the overall reaction. To achieve this, we often need to multiply some or all of the equations representing the individual steps with the appropriate coefficients.

Table 6.4 lists the standard enthalpies of formation of a number of elements and compounds. (A more complete list of ΔH_f° values is given in Appendix 3.) In the following examples we show how to calculate $\Delta H_{\text{rxn}}^\circ$ from the known ΔH_f° values (Example 6.4) and the application of Hess's law to determine the ΔH_f° of a compound (Example 6.5).

EXAMPLE 6.4

Pentaborane-9, B_5H_9, is a colorless liquid. It is a highly reactive substance which will burst into flame or even explode when exposed to oxygen. The reaction is

$$2B_5H_9(l) + 12O_2(g) \longrightarrow 5B_2O_3(s) + 9H_2O(l)$$

Pentaborane-9 was once considered as a rocket fuel in the 1950s since it produces a large amount of heat per gram. However, because the solid B_2O_3 formed is an abrasive which quickly destroys the nozzle of the rocket, the idea was abandoned. Calculate the kilojoules of heat released per gram of the compound reacted with oxygen. The standard enthalpy of formation of B_5H_9 is 73.2 kJ/mol.

Answer

Using the ΔH_f° values in Appendix 3 and Equation (6.7), we write

$$\begin{aligned}\Delta H_{\text{rxn}}^\circ &= [(5 \text{ mol})\ \Delta H_f^\circ(B_2O_3) + (9 \text{ mol})\ \Delta H_f^\circ(H_2O) \\ &\qquad - [(2 \text{ mol})\ \Delta H_f^\circ(B_5H_9) + (12 \text{ mol})\ \Delta H_f^\circ(O_2)] \\ &= [(5 \text{ mol})(-1263.6 \text{ kJ/mol}) + (9 \text{ mol})(-285.8 \text{ kJ/mol})] \\ &\qquad - [(2 \text{ mol})(73.2 \text{ kJ/mol}) + (12 \text{ mol})(0)] \\ &= -9036.6 \text{ kJ}\end{aligned}$$

This is the heat released for every 2 moles of B_5H_9 reacted. The heat released per gram of B_5H_9 reacted is

$$\begin{aligned}\text{heat released per gram of } B_5H_9 &= \frac{1 \text{ mol } B_5H_9}{63.12 \text{ g } B_5H_9} \times \frac{-9036.6 \text{ kJ}}{2 \text{ mol } B_5H_9} \\ &= -71.58 \text{ kJ/g } B_5H_9\end{aligned}$$

Similar problems: 6.50, 6.53.

B_5H_9 is an example of the so-called high-energy compounds because of the large amount of energy it releases on combustion. In general, compounds that have positive ΔH_f° values tend to release more heat on combustion and be less stable than those with negative ΔH_f° values.

EXAMPLE 6.5

From the following equations and the enthalpy changes,

(a) $C(\text{graphite}) + O_2(g) \longrightarrow CO_2(g) \qquad \Delta H^\circ_{rxn} = -393.5\ \text{kJ}$

(b) $H_2(g) + \frac{1}{2}O_2(g) \longrightarrow H_2O(l) \qquad \Delta H^\circ_{rxn} = -285.8\ \text{kJ}$

(c) $2C_2H_2(g) + 5O_2(g) \longrightarrow 4CO_2(g) + 2H_2O(l) \qquad \Delta H^\circ_{rxn} = -2598.8\ \text{kJ}$

calculate the standard enthalpy of formation of acetylene (C_2H_2) from its elements:

$$2C(\text{graphite}) + H_2(g) \longrightarrow C_2H_2(g)$$

Answer

Since we want to obtain one equation containing only C, H_2, and C_2H_2, we need to eliminate O_2, CO_2, and H_2O from the first three equations. We note that (c) contains 5 moles of O_2, 4 moles of CO_2, and 2 moles of H_2O. First we reverse (c) to get C_2H_2 on the product side:

(d) $4CO_2(g) + 2H_2O(l) \longrightarrow 2C_2H_2(g) + 5O_2(g) \qquad \Delta H^\circ_{rxn} = +2598.8\ \text{kJ}$

Next, we multiply (a) by 4 and (b) by 2 and carry out the addition of 4(a) + 2(b) + (d):

$$\begin{array}{lll} & 4C(\text{graphite}) + \cancel{4O_2(g)} \longrightarrow \cancel{4CO_2(g)} & \Delta H^0_{rxn} = -1574.0\ \text{kJ} \\ + & 2H_2(g) + \cancel{O_2(g)} \longrightarrow \cancel{2H_2O(l)} & \Delta H^\circ_{rxn} = -571.6\ \text{kJ} \\ + & \cancel{4CO_2(g)} + \cancel{2H_2O(l)} \longrightarrow 2C_2H_2(g) + \cancel{5O_2(g)} & \Delta H^\circ_{rxn} = +2598.8\ \text{kJ} \\ \hline & 4C(\text{graphite}) + 2H_2(g) \longrightarrow 2C_2H_2(g) & \Delta H^\circ_{rxn} = +453.2\ \text{kJ} \end{array}$$

or

$$2C(\text{graphite}) + H_2(g) \longrightarrow C_2H_2(g) \qquad \Delta H^\circ_{rxn} = +226.6\ \text{kJ}$$

When the equation is divided by 2, the ΔH value is halved.

Since the above equation represents the synthesis of C_2H_2 from its elements, we have $\Delta H_f^\circ(C_2H_2) = \Delta H^\circ_{rxn}/\text{mol} = +226.6\ \text{kJ/mol}$.

Similar problems: 6.39, 6.40.

In the Chemistry in Action on page 239 we see the application of Hess's law to a fascinating biological process.

6.6 Heat of Solution and Dilution

So far we have focused mainly on thermal energy effects resulting from chemical reactions. Many physical processes (for example, the melting of ice or the condensation of a vapor) also involve the absorption or release of heat. Enthalpy changes are also involved when a solute dissolves in a solvent or when a solution is diluted. Let us

CHEMISTRY IN ACTION

How a Bombardier Beetle Defends Itself

Survival techniques of insects and small animals in a fiercely competitive environment take many forms. For example, chameleons have developed the ability to change color to match their surroundings; other creatures, like the butterfly *Limenitis,* have evolved into a form that mimics the poisonous and unpleasant-tasting monarch butterfly *Danaus*. A less passive defense mechanism is employed by the bombardier beetles (*Brachinus*). They fight off predators with a "chemical spray."

The bombardier beetle has a pair of glands that open at the tip of its abdomen. Each gland consists basically of two compartments. The inner compartment contains an aqueous solution of hydroquinone and hydrogen peroxide, and the outer compartment contains a mixture of enzymes. (Enzymes are biological molecules that can speed up a reaction.) When threatened, the beetle squeezes some fluid from the inner compartment into the outer compartment, where, in the presence of the enzymes, an exothermic reaction takes place:

(a)

$$\underset{\text{hydroquinone}}{C_6H_4(OH)_2(aq)} + H_2O_2(aq) \longrightarrow \underset{\text{quinone}}{C_6H_4O_2(aq)} + 2H_2O(l)$$

To estimate the heat of reaction, let us consider the following steps:

(b) $C_6H_4(OH)_2(aq) \longrightarrow C_6H_4O_2(aq) + H_2(g)$ $\quad \Delta H° = 177\text{ kJ}$

(c) $H_2O_2(aq) \longrightarrow H_2O(l) + \frac{1}{2}O_2(g)$ $\quad \Delta H° = -94.6\text{ kJ}$

(d) $H_2(g) + \frac{1}{2}O_2(g) \longrightarrow H_2O(l)$ $\quad \Delta H° = -286\text{ kJ}$

Recalling Hess's law, we find that the heat of reaction for (a) is simply the *sum* of those for (b), (c), and (d). Therefore, we write

$$\begin{aligned}\Delta H_a° &= \Delta H_b° + \Delta H_c° + \Delta H_d° \\ &= (177 - 94.6 - 286)\text{ kJ} \\ &= -204\text{ kJ}\end{aligned}$$

The large amount of heat generated is sufficient to heat the mixture to its boiling point. By rotating the tip of its abdomen, the beetle can quickly discharge the vapor in the form of a fine mist toward an unsuspecting predator (Figure 6.12). In addition to the thermal effect, the quinones also act as a repellent to other insects and animals. An average bombardier beetle carries enough reagents in its body to produce about 20 to 30 discharges in quick succession, each with an audible detonation.

Figure 6.12 *A bombardier beetle discharging a chemical spray.*

look more closely at these two related physical processes, involving heat of solution and heat of dilution.

Heat of Solution

In the vast majority of cases, dissolving a solute in a solvent produces measurable heat effects. At constant pressure, the quantity of heat change is equal to the enthalpy change. The ***heat of solution*** (also called ***enthalpy of solution***), ΔH_{soln}, *is the heat generated or absorbed when a certain amount of solute dissolves in a certain amount of solvent*. The quantity ΔH_{soln} represents the difference between the enthalpy of the final solution and the enthalpies of its original components (that is, solute and solvent) before they are mixed. Thus

$$\Delta H_{\text{soln}} = H_{\text{soln}} - H_{\text{components}} \tag{6.9}$$

Neither H_{soln} nor $H_{\text{components}}$ can be measured, but their difference, ΔH_{soln}, can be readily determined. Like other enthalpy changes, ΔH_{soln} is positive for endothermic (heat-absorbing) processes and negative for exothermic (heat-generating) processes.

Consider the heat of solution involving ionic compounds as solute and water as solvent. For example, what happens when solid NaCl dissolves in water? While NaCl is a solid, the Na^+ and Cl^- ions are held together by strong positive-negative (electrostatic) forces, but when a small crystal of NaCl dissolves in water, the three-dimensional network of ions breaks into its individual units. (The structure of solid NaCl is shown in Figure 2.10.) The separated Na^+ and Cl^- ions are stabilized in solution by their interaction with water molecules (see Figure 3.3). These ions are said to be *hydrated*. Here water plays a role similar to that of a good electrical insulator. Water molecules shield the ions (Na^+ and Cl^-) from each other and effectively reduce the electrostatic attraction that held them together in the solid state. The heat of solution is defined by the following process:

$$\text{NaCl}(s) \xrightarrow{H_2O} \text{Na}^+(aq) + \text{Cl}^-(aq) \qquad \Delta H_{\text{soln}} = ?$$

Dissolving an ionic compound such as NaCl in water involves complex interactions among the solute and solvent species. However, for the sake of energy analysis we can imagine that the solution process takes place in only two separate steps. First, the Na^+ and Cl^- ions in the solid crystal are separated from each other in a gas phase (Figure 6.13):

$$\text{energy} + \text{NaCl}(s) \longrightarrow \text{Na}^+(g) + \text{Cl}^-(g)$$

Lattice energy is a positive quantity.

The energy required for this process is called ***lattice energy (U)***, *the energy required to completely separate one mole of a solid ionic compound into gaseous ions*. The lattice energy of NaCl is 788 kJ/mol. In other words, we would need to supply 788 kJ of energy to break 1 mole of solid NaCl into 1 mole of Na^+ ions and 1 mole of Cl^- ions.

Next, the "gaseous" Na^+ and Cl^- ions enter the water and become hydrated:

$$\text{Na}^+(g) + \text{Cl}^-(g) \xrightarrow{H_2O} \text{Na}^+(aq) + \text{Cl}^-(aq) + \text{energy}$$

The enthalpy change associated with the hydration process is called the ***heat of hydration*** (ΔH_{hydr}). (Heat of hydration is a negative quantity for cations and anions.) Applying Hess's law, it is possible to consider ΔH_{soln} as the sum of two related quantities, lattice energy (U) and heat of hydration (ΔH_{hydr}):

$$\begin{array}{rlr} \text{NaCl}(s) \longrightarrow \text{Na}^+(g) + \text{Cl}^-(g) & U = & 788 \text{ kJ} \\ + \text{Na}^+(g) + \text{Cl}^-(g) \xrightarrow{H_2O} \text{Na}^+(aq) + \text{Cl}^-(aq) & \Delta H_{\text{hydr}} = & -784 \text{ kJ} \\ \hline \text{NaCl}(s) \xrightarrow{H_2O} \text{Na}^+(aq) + \text{Cl}^-(aq) & \Delta H_{\text{soln}} = & 4 \text{ kJ} \end{array}$$

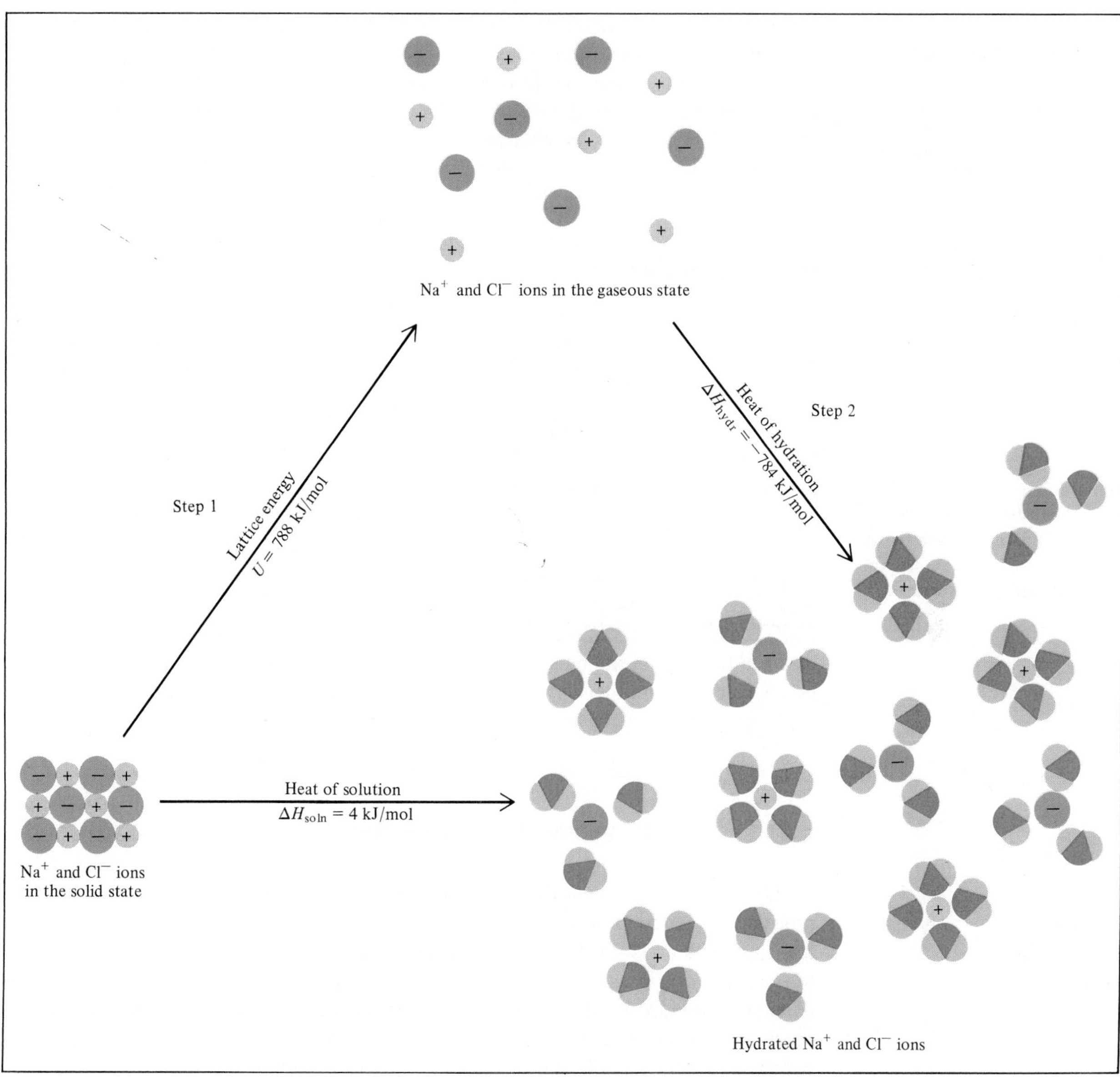

Figure 6.13 *The solution process of NaCl. The process may be imagined to occur in two separate steps: (1) separation of ions from the crystal state to the gaseous state and (2) hydration of the gaseous ions. The heat of solution is equal to the energy changes for these two steps,* $\Delta H_{soln} = U + \Delta H_{hydr}$.

or, summing up the enthalpy changes,

$$\begin{aligned}\Delta H_{\text{soln}} &= U + \Delta H_{\text{hydr}} \\ &= 788 \text{ kJ/mol} - 784 \text{ kJ/mol} \\ &= 4 \text{ kJ/mol}\end{aligned}$$

We see that the solution process for NaCl is slightly endothermic. When 1 mole of

Table 6.5 Heats of solution of some ionic compounds

Compound	ΔH_{soln} (kJ/mol)	
LiCl	−37.1	exothermic
$CaCl_2$	−82.8	
NaCl	4.0	endothermic
KCl	17.2	
NH_4Cl	15.2	
NH_4NO_3	26.2	

NaCl dissolves in water, 4 kJ of heat will be absorbed from the immediate surroundings. We would observe this effect by noting that the beaker containing the solution becomes slightly colder. Table 6.5 lists the ΔH_{soln} of several ionic compounds. Depending on the nature of the cation and anion involved, ΔH_{soln} for an ionic compound may be either negative (exothermic) or positive (endothermic).

Heat of solution has some interesting applications in medicine, discussed in the Chemistry in Action on p. 243.

Heat of Dilution

When a previously prepared solution is *diluted,* that is, when more solvent is added to lower the overall concentration of the solute, additional heat is usually given off or absorbed. *The heat change associated with the dilution process* is called ***heat of dilution.*** If a certain solution process was endothermic, and the solution is diluted again, *more* heat will be absorbed by the same solution from the surroundings. The converse holds true for an exothermic solution process—more heat will be liberated if additional solvent is added to dilute the solution. Therefore, always be cautious when working on a dilution procedure in the laboratory. Because of its highly exothermic heat of dilution, concentrated sulfuric acid (H_2SO_4) poses a particularly hazardous problem if its concentration must be reduced by mixing it with additional water. Concentrated H_2SO_4 is composed of 98 percent acid and 2 percent water by mass. When it is diluted with water, considerable heat is released to the surroundings. This process is so exothermic that you must *never* attempt to dilute the concentrated acid by adding water to it. The heat generated could easily cause the acid solution to boil and splatter. The recommended procedure is to add the concentrated acid slowly into the water (while constantly stirring).

Generations of chemistry students have been urged to remember this procedure by the venerable saying, "Do as you oughter, add acid to water."

6.7 The First Law of Thermodynamics

So far we have discussed the heat changes for a number of physical and chemical processes under conditions of constant pressure and constant volume. Thermochemistry is part of a much more general subject called ***thermodynamics,*** which is *the scientific study of the interconversion of heat and other kinds of energy*. The laws of thermodynamics provide useful guidelines for understanding the energetics and directions of processes. In this section we will concentrate on the first law of thermodynamics, which is particularly relevant to the study of thermochemistry. We will continue our discussion of other aspects of thermodynamics in a later chapter.

Before we introduce the first law of thermodynamics, it is necessary to introduce a few terms that will be used later. In thermodynamics, we often speak of the ***state of a system,*** which means *the values of all pertinent macroscopic properties—for example, composition, energy, temperature, pressure, and volume*. Energy, pressure, volume, and temperature are said to be ***state functions***—*properties that are determined by the state the system is in*. A characteristic of state functions is that when the state of a system changes, the magnitude of change in any state function depends only on the initial and final states of the system and not on how the change is accomplished. Consider a gas at 1 atm, 300 K, and 2 L (the initial state). Let us suppose that after heating, the pressure, temperature, and volume of the gas are converted to 2 atm, 900 K, and 3 L (the final state). The change in volume (ΔV) is

The state of a given amount of a gas is specified by its volume, pressure, and temperature.

CHEMISTRY IN ACTION

Instant Cold and Hot Packs

Athletes often use instant cold packs and hot packs as first-aid devices to treat injuries. These devices operate by utilizing the heat of solution concept discussed in this section. A typical pack consists of a plastic bag containing a pouch of water and a dry chemical (Figure 6.14). Striking the pack causes the pouch to break and the temperature of the pack will be either raised or lowered, depending on whether the heat of solution of the chemical is exothermic or endothermic.

Generally, calcium chloride or magnesium sulfate is used in hot packs, and ammonium nitrate in cold packs. The reactions are

$$CaCl_2(s) \xrightarrow{H_2O} Ca^{2+}(aq) + 2Cl^-(aq) \qquad \Delta H_{soln} = -82.8 \text{ kJ}$$

$$NH_4NO_3(s) \xrightarrow{H_2O} NH_4^+(aq) + NO_3^-(aq) \qquad \Delta H_{soln} = 26.2 \text{ kJ}$$

Experiments show that the addition of 40 g of $CaCl_2$ (0.36 mole) to 100 mL of water raises the temperature from 20°C to 90°C. Similarly, when 30 g of NH_4NO_3 (0.38 mole) is dissolved in 100 mL of water at 20°C, the temperature will be lowered to 0°C. A typical hot or cold pack works for about 20 min.

Figure 6.14 *Commercially available cold pack and hot pack. The cold pack contains ammonium nitrate and water, and the hot pack contains calcium chloride and water.*

$$\begin{aligned}\Delta V &= V_f - V_i \\ &= 3\text{ L} - 2\text{ L} \\ &= 1\text{ L}\end{aligned}$$

where V_i and V_f denote the initial and final volume, respectively. Similarly, the changes in pressure and temperature are (2 atm − 1 atm) or 1 atm and (900 K − 300 K) or 600 K. No matter how the initial state is converted to the final state, the changes in pressure, temperature, and volume will always have the same values because they are state functions.

Recall that an object possesses potential energy by virtue of its position or chemical composition.

Energy is another state function. Using potential energy as an example, we find that the net increase in gravitational potential energy when we go from the same starting point to the top of a mountain is always the same, no matter how we get there (Figure 6.15). As we will see shortly, two important thermodynamic quantities, work and heat, are not state functions.

The First Law of Thermodynamics

The ***first law of thermodynamics*** describes the conservation of energy; it states that *energy can be converted from one form to another, but cannot be created or destroyed.*

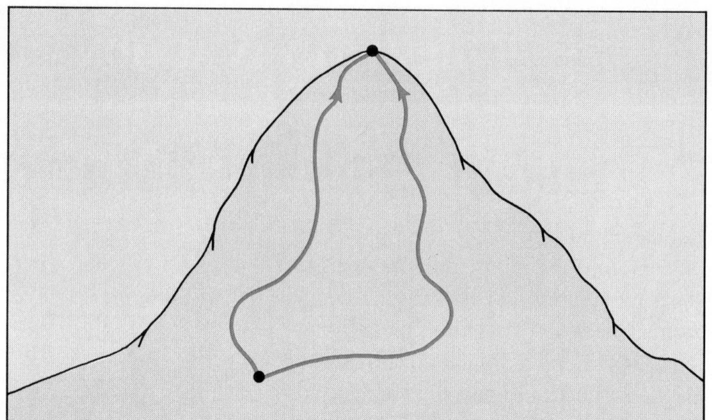

Figure 6.15 *The gain in potential energy in climbing from the base to the top of a mountain is independent of the path taken.*

In other words, the total energy in the universe is a constant. It would be impossible to prove the validity of the first law of thermodynamics if we had to determine the total energy content of the universe. Even determining the total energy content of 1 g of iron, say, would be extremely difficult. Fortunately, in testing the validity of the first law, we need be concerned only with the *change* in the internal energy of a system that results from the difference between its *initial state* and its *final state*. The change in internal energy ΔE is given by

$$\Delta E = E_f - E_i$$

where E_i and E_f are the internal energies of the system in the initial and final states, respectively.

The internal energy of a system has two components: kinetic energy and potential energy. The kinetic energy component consists of various types of molecular motion and the movement of electrons within molecules. Potential energy is determined by the attractive interactions between electrons and nuclei and repulsive interactions between electrons and between nuclei in individual molecules, as well as interaction between molecules. It is impossible to measure all these contributions accurately, so we cannot calculate the total energy of a system with any certainty. Fortunately we do not have to do this and can instead concentrate on the *changes* in energy, which can be determined experimentally.

Consider the reaction between 1 mole of sulfur and 1 mole of oxygen gas to produce 1 mole of sulfur dioxide:

$$S(s) + O_2(g) \longrightarrow SO_2(g)$$

In this case our system is composed of the reactant molecules S and O_2 and the product molecules SO_2. We do not know the internal energy content of either the reactant molecules or the product molecules, but we can accurately measure the *change* in energy content, ΔE, given by

$$\begin{aligned}\Delta E &= \text{energy content of 1 mol } SO_2(g) - [\text{energy content of 1 mol } S(s) + 1 \text{ mol } O_2(g)]\\ &= E(\text{product}) - E(\text{reactants})\end{aligned}$$

Since heat is given off, this is an exothermic reaction. Thus ΔE is negative, which means that the energy of the product is less than that of the reactants. Since the total energy of the universe is unchanged by this reaction, it follows that the sum of energy changes must be zero; that is

$$\Delta E_{sys} + \Delta E_{surr} = 0$$

or

$$\Delta E_{sys} = -\Delta E_{surr}$$

where the subscripts "sys" and "surr" denote system and surroundings, respectively. Thus, if one system undergoes an energy change ΔE_{sys}, the rest of the universe, or the surroundings, must undergo a change in energy that is equal in magnitude but opposite in sign. It follows that energy gained in one place must have been lost somewhere else.

It is just as true that energy lost by one system must show up elsewhere in the universe as an energy gain. Furthermore, because energy can be changed from one form to another, the energy lost by one system can be gained by another system in a different form. For example, the energy lost by burning oil in a power plant may ultimately turn up in our homes as electrical energy, heat, light, and so on.

In chemistry, we are normally interested in the changes associated with the system (which may be a flask containing reactants and products), not with its surroundings. Therefore, a more useful form of the first law is

$$\Delta E = q + w \tag{6.10}$$

For convenience, we sometimes omit the word "internal" when discussing the energy of a system.

(We drop the subscript "sys" for simplicity.) Equation (6.10) says that the change in the internal energy ΔE of a system is the sum of the heat exchange q between the system and the surroundings and the work done w on (or by) the system. Using the sign convention for thermochemical processes (see Section 6.2), we find that q is positive for an endothermic process and negative for an exothermic process. The sign convention for work is that w is positive for work done on the system by the surroundings and negative for work done by the system on the surroundings. Equation (6.10) may appear abstract but is actually quite logical. If a system loses heat to the surroundings or does work on the surroundings, we would expect its internal energy to decrease since both are energy-depleting processes. Conversely, if heat is added to the system or if work is done on the system, then the internal energy of the system would increase, as predicted by Equation (6.10). Table 6.6 summarizes the sign conventions for q and w.

Work and Heat

In Section 6.1 we defined work as force F multiplied by distance d:

$$w = Fd$$

In thermodynamics, work has a broader meaning that includes mechanical work (for example, a crane lifting a steel beam), electrical work (a battery supplying electrons to light the bulb of a flashlight), and so on. In this section we will concentrate on mechanical work; in Chapter 19 we will discuss the nature of electrical work.

Table 6.6 Sign conventions for work and heat

Process	Sign
Work done by the system on the surroundings	−
Work done on the system by the surroundings	+
Heat absorbed by the system from the surroundings (endothermic process)	+
Heat absorbed by the surroundings from the system (exothermic process)	−

A useful example of mechanical work is the expansion of a gas (Figure 6.16). Suppose that a gas is in a cylinder fitted with a weightless, frictionless movable piston, at a certain temperature, pressure, and volume. As it expands, the gas pushes the piston upward against a constant opposing external atmospheric pressure P. The work done by the gas on the surroundings is

For gas expansion, $\Delta V > 0$, so $-P\,\Delta V$ is a negative quantity. For gas compression, $\Delta V < 0$, and $-P\,\Delta V$ is a positive quantity. Thus the signs for w are in accordance with our convention.

$$w = -P\,\Delta V \tag{6.11}$$

where ΔV, the change in volume, is given by $V_f - V_i$.

Equation (6.11) was arrived at by considering the fact that pressure $\times$ volume can be expressed as (force/area) $\times$ volume; that is,

$$P \times V = \underbrace{\frac{F}{d^2}}_{\text{pressure}} \times \underbrace{d^3}_{\text{volume}} = Fd = w$$

where F is the opposing force and d has the dimension of length, d^2 has the dimensions of area, and d^3 has the dimensions of volume. Thus the product of pressure and volume is equal to force times distance, or work. You can see that for a given increase in volume (that is, for a certain value of ΔV), the work done depends on the magnitude of the external, opposing pressure P. If P is zero (that is, if the gas is expanding against a vacuum), the work done must also be zero. If P is some positive, nonzero value, then the work done is given by $-P\,\Delta V$.

The following example shows calculation of work done under different conditions.

EXAMPLE 6.6

A certain gas initially at room temperature undergoes an expansion in volume from 2.0 L to 6.0 L at constant temperature. Calculate the work done by the gas if it expands (a) against a vacuum and (b) against a constant pressure of 1.2 atm.

Answer

(a) Since the external pressure is zero, the work w done is

$$\begin{aligned} w &= -P\,\Delta V \\ &= -0(6.0 - 2.0)\text{L} \\ &= 0 \end{aligned}$$

(b) Here the external pressure is 1.2 atm, so

$$\begin{aligned} w &= -P\,\Delta V \\ &= -(1.2\ \text{atm})(6.0 - 2.0)\text{L} \\ &= -4.8\ \text{L} \cdot \text{atm} \end{aligned}$$

It would be more convenient to express w in units of joules. From Appendix 2 we obtain the conversion factor

$$1\ \text{L} \cdot \text{atm} = 101.3\ \text{J}$$

Thus we can write

$$\begin{aligned} w &= -4.8\ \text{L atm} \times \frac{101.3\ \text{J}}{1\ \text{L} \cdot \text{atm}} \\ &= -4.9 \times 10^2\ \text{J} \end{aligned}$$

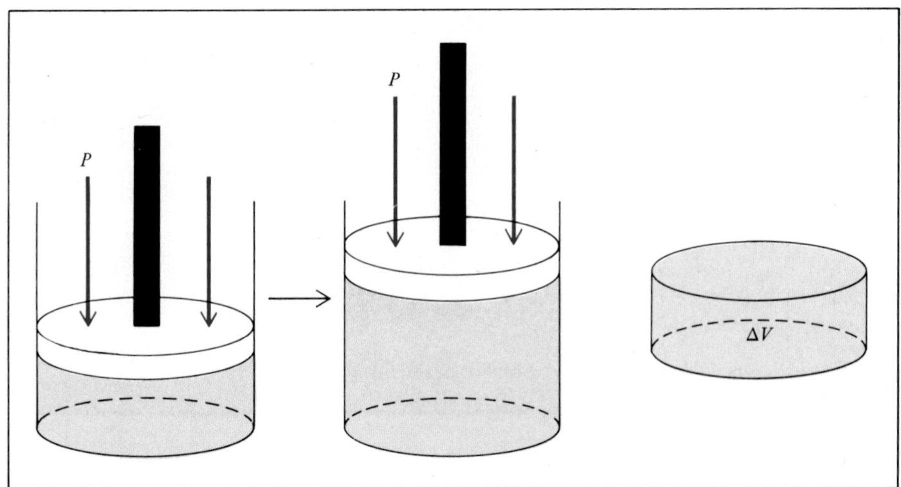

Figure 6.16 *The expansion of a gas against a constant external pressure (such as atmospheric pressure). The gas is in a cylinder fitted with a weightless movable piston. The work done is given by* $-P\ \Delta V$.

Because the temperature is kept constant, you can use Boyle's law to show that the final pressure is the same in (a) and (b).

From Example 6.6 it is clear that work is not a state function, because while the initial and final states are the same in (a) and (b), the amount of work done is different since the external, opposing pressures are different. We *cannot* write $\Delta w = w_f - w_i$ for a change. Work done does not depend only on the initial state and final state; it also depends on how we carry out the process.

The other component of internal energy is q, heat. Heat is the energy transferred from a hot object to a cold one. Like work, heat is also not a function of state. Let us consider a situation in which a system changes from some initial state to some final state. Suppose that the change can be brought about in two different ways. In one case the work done is zero, so that we have

$$\begin{aligned}\Delta E &= q_1 + w_1 \\ &= q_1\end{aligned}$$

In the second case there is both work done and heat transfer, so that

$$\Delta E = q_2 + w_2$$

We use lowercase letters (such as *w* and *q*) to represent thermodynamic quantities that are not state functions.

Since ΔE is the same for both cases (because the initial and final states are the same in both cases), it follows that $q_1 \neq q_2$. This simple example shows that heat is not a state function because the heat associated with a given process, like work, depends on how the process is carried out; that is, we cannot write $\Delta q = q_f - q_i$. It is important to note that although neither heat nor work is a state function, their sum $(q + w)$ is equal to ΔE, and E, as we saw earlier, is a state function.

The following example illustrates the use of equation (6.10).

EXAMPLE 6.7

The work done when a gas is compressed in a cylinder like that shown in Figure 6.16 is 462 J. During this process, there is a heat transfer of 128 J from the gas to the surroundings. Calculate the energy change for this process.

Answer

Compression is work done on the gas, so the sign for w is positive: $w = 462$ J. From the direction of heat transfer (system to surroundings), we know that q is negative: $q = -128$ J. From Equation (6.10) we write

$$\begin{aligned}\Delta E &= q + w \\ &= -128 \text{ J} + 462 \text{ J} \\ &= 334 \text{ J}\end{aligned}$$

As a result of the compression and heat transfer, the energy of the gas increases by 334 J.

Similar problem: 6.67.

Enthalpy

We have so far discussed processes carried out under constant-pressure and constant-volume conditions. It is helpful to apply the first law of thermodynamics to them. If a reaction is run at constant volume, then $\Delta V = 0$, and no work will result from this change. From Equation (6.10) it follows that

$$\Delta E = q + w = q_v$$

that is, the change in energy is equal to the heat change. We have added the subscript v to remind us that this is a constant-volume process. Bomb calorimetry, discussed in Section 6.4, is such a constant-volume process.

Constant-volume conditions are often inconvenient and sometimes impossible to achieve. Most reactions occur under conditions of constant pressure (usually atmospheric pressure). If a reaction conducted at constant pressure results in a net increase in the number of moles of a gas, then the system will do work on the surroundings (expansion). This follows from the fact that the gas formed must push the atmosphere back. Conversely, if more gas molecules are consumed than are produced, work will be done on the system by the surroundings (compression). For reactions that do not involve the evolution or consumption of gases, the changes in volume will be small and we can assume that $\Delta V = 0$.

Sodium reacting with water.

Let us consider a reaction carried out under constant-pressure conditions. When a small piece of sodium metal is added to a beaker of water, the reaction that takes place is

$$2\text{Na}(s) + 2\text{H}_2\text{O}(l) \longrightarrow 2\text{NaOH}(aq) + \text{H}_2(g)$$

The energy released during the reaction is ΔE. One of the products is gaseous hydrogen, which must force its way into the atmosphere by pushing back the air. Consequently, some of the energy produced by the reaction is used to do work of pushing back a volume of air (ΔV) against atmospheric pressure (P). The heat of the reaction under the constant-pressure condition is ΔH, which is the enthalpy change; that is, $q_p = \Delta H$ (the subscript p reminds us that this is a constant-pressure process). We sum up the situation at constant pressure by writing

$$\begin{aligned}\Delta E &= q_p + w \\ &= \Delta H - P\,\Delta V\end{aligned}$$

or

$$\Delta H = \Delta E + P\,\Delta V \tag{6.12}$$

Equation (6.12) says that the enthalpy change of a process is the sum of two parts: the change in energy of the system and the work done (either on the system by the surroundings or on the surroundings by the system).

We can now define the enthalpy of a system as

$$H = E + PV \tag{6.13}$$

where E is the internal energy of the system and P and V are the pressure and volume of the system, respectively. Since E and the product PV have the units of energy, enthalpy also has the units of energy. Furthermore, because E, P, and V are all state functions, the changes in $(E + PV)$ depend only on the initial and final states. It follows that the change in H also depends only on the initial and final states; that is, H is a state function.

The fact that H is a state function allows us to explain Hess's law, discussed in Section 6.5. Because ΔH depends only on the initial and final state (that is, only on the nature of reactants and products), the enthalpy change for a reaction is the same whether it takes one step or many steps.

EXAMPLE 6.8

The oxidation of nitric oxide to nitrogen dioxide is a key step in the formation of smog:

$$2NO(g) + O_2(g) \longrightarrow 2NO_2(g) \qquad \Delta H^\circ = -113.1 \text{ kJ}$$

If 6.00 moles of NO react with 3.00 moles of O_2 at 1.00 atm and 25°C to form NO_2, calculate the work done (in kilojoules) against a pressure of 1.00 atm. What is the ΔE for the reaction? Assume the reaction to go to completion.

Answer

First we calculate the volumes of the reactants and product. From Equation (5.7)

$$V_{\text{reactants}} = \frac{nRT}{P} = \frac{(9.00 \text{ mol})(0.0821 \text{ L}\cdot\text{atm/K}\cdot\text{mol})(298 \text{ K})}{1.00 \text{ atm}}$$

$$= 220 \text{ L}$$

$$V_{\text{product}} = \frac{nRT}{P} = \frac{(6.00 \text{ mol})(0.0821 \text{ L}\cdot\text{atm/K}\cdot\text{mol})(298 \text{ K})}{1.00 \text{ atm}}$$

$$= 147 \text{ L}$$

The pressure-volume work is then

$$w = -P\,\Delta V = -1.00 \text{ atm } (147 \text{ L} - 220 \text{ L})\left(\frac{101.3 \text{ J}}{1 \text{ L}\cdot\text{atm}}\right)$$

$$= 7.39 \times 10^3 \text{ J} = 7.39 \text{ kJ}$$

Note that because the volume of the product is smaller than that of the reactants, work is done on the system and w is a positive quantity.

The change in internal energy can be calculated using Equation (6.12):

$$\Delta E = \Delta H - P\,\Delta V$$

The enthalpy change shown above is for the formation of 2 moles of NO_2. Since 6 moles of NO_2 are formed, the enthalpy change is three times -113.1 kJ, or $\Delta H = 3(-113.1 \text{ kJ})$. Thus

$$\begin{aligned}\Delta E &= 3(-113.1 \text{ kJ}) + 7.39 \text{ kJ} \\ &= -331.9 \text{ kJ}\end{aligned}$$

Similar problem: 6.79.

The Chemistry in Action on p. 251 discusses some interesting applications of the first law of thermodynamics.

Summary

1. Energy is the capacity to do work. There are many forms of energy that are interconvertible. The law of conservation of energy states that although energy can be converted from one form to another, it cannot be created or destroyed.
2. Any process that gives off heat to the surroundings is called an exothermic process; any process that absorbs heat from the surroundings is an endothermic process.
3. The change in enthalpy (ΔH, usually given in kilojoules) is a measure of the heat of reaction (or any other process) at constant pressure.
4. Constant-volume and constant-pressure calorimeters are used to measure heat changes of physical and chemical processes.
5. Hess's law states that the overall enthalpy change in a reaction is equal to the sum of enthalpy changes for the individual steps that make up the overall reaction.
6. The standard enthalpy of reaction can be calculated from the standard enthalpies of formation of reactants and products.
7. The heat of solution of an ionic compound in water is the sum of the lattice energy and the heat of hydration. The relative magnitudes of these two quantities determine whether the solution process is endothermic or exothermic. The heat of dilution is the heat absorbed or evolved when a solution is diluted.
8. The state of a system is defined by variables such as composition, volume, temperature, and pressure.
9. The change in a state function for a system depends only on the initial and final states of the system, and not on the path by which the change is accomplished. Energy is a state function; work and heat are not.
10. Energy can be converted from one form to another, but it cannot be created or destroyed (first law of thermodynamics). In chemistry we are concerned mainly with thermal energy, electrical energy, and mechanical energy, which is usually associated with pressure-volume work.
11. Enthalpy is a state function. A change in enthalpy ΔH is equal to $\Delta E + P\,\Delta V$ for a constant-pressure process.

CHEMISTRY IN ACTION

Making Snow and Inflating a Bicycle Tire

Many phenomena in everyday life can be explained by the first law of thermodynamics. Here we will discuss two examples of interest to lovers of the outdoors.

Making Snow

If you are an avid downhill skier, you have probably skied on artificial snow. How is this stuff made in quantities large enough to meet the needs of skiers on snowless days? The secret of snowmaking is in the equation $\Delta E = q + w$. A snowmaking machine contains a mixture of compressed air and water vapor at about 20 atm. At the appropriate moment, the mixture is sprayed into the atmosphere (Figure 6.17). Because of the large difference in pressure between the tank and the outside atmosphere, the expelled air and water vapor expand so rapidly that, as a good approximation, *no heat exchange occurs between the system* (air and water) *and its surroundings;* that is, $q = 0$. (In thermodynamics, such a process is called an ***adiabatic process.***) Thus we write

$$\Delta E = q + w = w$$

Because the system does work on the surroundings, w is a negative quantity, and there is a decrease in the system's energy. Kinetic energy is part of the total energy of the system. In Section 5.7 we saw that the average kinetic energy of a gas is directly proportional to the absolute temperature [Equation (5.13)]. It follows, therefore, that the change in energy ΔE is proportional to the change in temperature, so that

$$\begin{aligned}\Delta E &\propto \Delta T \\ &= C\,\Delta T\end{aligned}$$

where C is the proportionality constant. Because ΔE is negative, ΔT must also be negative, and it is this cooling effect (or the decrease in the kinetic energy of the water molecules) that is responsible for the formation of snow. Although we need only water to form snow, the presence of air, which also cools on expansion, helps to lower the temperature of the water vapor.

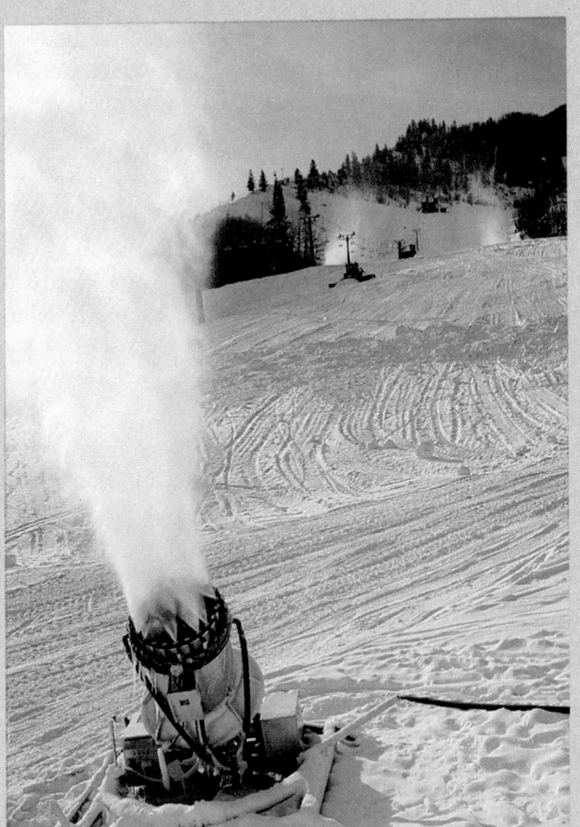

Figure 6.17 *A snowmaking machine in operation.*

Inflating a Bicycle Tire

If you have ever pumped air into a bicycle tire, you probably noticed a warming effect at the valve stem. This phenomenon, too, can be explained by the first law of thermodynamics. The action of the pump compresses the air inside the pump and the tire. The process is rapid enough to be treated as approximately adiabatic, so that $q = 0$ and $\Delta E = w$. Because work is done on the gas in this case (it is being compressed), w is positive, and there is an increase in energy. Hence, the temperature of the system increases also, according to the equation

$$\Delta E = C\,\Delta T$$

Applied Chemistry Example: Lime

Lime is a term that includes both calcium oxide (CaO, also called quicklime) and calcium hydoxide [$Ca(OH)_2$, also called slaked lime]. It ranked 6th among the top 50 industrial chemicals produced in the United States in 1988 (32.3 billion pounds). Lime is one of the oldest chemicals known. It has been used as a building material since 1500 B.C. The largest consumer of lime is the steel industry, which uses the basic properties of lime to remove acidic impurities in iron ores. Lime is also used in air-pollution control and water treatment (to neutralize acids) and to a lesser extent in the food industry.

Industrially quicklime is prepared by heating limestone ($CaCO_3$) above 2000°C:

$$CaCO_3(s) \longrightarrow CaO(s) + CO_2(g) \qquad \Delta H° = 177.8 \text{ kJ}$$

Slaked lime is produced by treating quicklime with water:

$$CaO(s) + H_2O(l) \longrightarrow Ca(OH)_2(s) \qquad \Delta H° = -65.2 \text{ kJ}$$

Quicklime is a white solid that melts at 2570°C. The exothermic reaction of quicklime with water and the rather small specific heats of both quicklime (0.946 J/g · °C) and slakedlime (1.20 J/g · °C) made the storage and transport of the substance hazardous in the old days. Wooden sailing ships would occasionally catch fire when water leaked into the hold containing lime.

1. A 500-g sample of water is reacted with an equimolar amount of CaO (both at an initial temperature of 25°C). What is the final temperature of the product, $Ca(OH)_2$? Assume that the product absorbs all of the heat released in the reaction.

Answer: The number of moles of water present in 500 g of water is

$$\text{moles of } H_2O = 500 \text{ g } H_2O \times \frac{1 \text{ mol } H_2O}{18.02 \text{ g } H_2O} = 27.7 \text{ mol } H_2O$$

From the above equation for the production of $Ca(OH)_2$, we have 1 mol H_2O ≎ 1 mol CaO ≎ 1 mol $Ca(OH)_2$. Therefore, the heat generated by the reaction is

$$27.7 \text{ mol } Ca(OH)_2 \times \frac{65.2 \text{ kJ}}{1 \text{ mol } Ca(OH)_2} = 1.81 \times 10^3 \text{ kJ}$$

Knowing the specific heat and the number of moles of $Ca(OH)_2$ produced, we can calculate the temperature rise using Equation (6.3). First we need to find the mass of $Ca(OH)_2$ in 27.7 mole:

$$27.7 \text{ mol } Ca(OH)_2 \times \frac{74.10 \text{ g } Ca(OH)_2}{1 \text{ mol } Ca(OH)_2} = 2.05 \times 10^3 \text{ g } Ca(OH)_2$$

From Equation (6.3) we write $q = ms\,\Delta t$

Rearranging, we get

$$\Delta t = \frac{q}{ms} = \frac{1.81 \times 10^6 \text{ J}}{(2.05 \times 10^3 \text{ g})(1.20 \text{ J/g} \cdot °\text{C})} = 736°\text{C}$$

and the final temperature is

$$\Delta t = t_{\text{final}} - t_{\text{initial}}$$

$$t_{\text{final}} = 736°C + 25°C = 761°C$$

A temperature of 761°C is high enough to ignite wood.

2. Given that the standard enthalpies of formation of CaO and H_2O are −635.6 kJ/mol and −285.8 kJ/mol, respectively, calculate the standard enthalpy of formation of $Ca(OH)_2$.

Answer: The formation of $Ca(OH)_2$ from its elements is

$$Ca(s) + H_2(g) + O_2(g) \longrightarrow Ca(OH)_2(s)$$

From the standard enthalpies of formation of CaO and H_2O we write

$$H_2(g) + \tfrac{1}{2}O_2(g) \longrightarrow H_2O(l) \qquad \Delta H° = -285.8 \text{ kJ}$$

$$Ca(s) + \tfrac{1}{2}O_2(g) \longrightarrow CaO(s) \qquad \Delta H° = -635.6 \text{ kJ}$$

Applying Hess's law, we find that the sum of the following equations gives the desired overall equation for the formation of $Ca(OH)_2$ and the ΔH_f° of $Ca(OH)_2$:

$$CaO(s) + H_2O(l) \longrightarrow Ca(OH)_2(s) \qquad \Delta H° = -65.2 \text{ kJ}$$

$$H_2(g) + \tfrac{1}{2}O_2(g) \longrightarrow H_2O(l) \qquad \Delta H° = -285.8 \text{ kJ}$$

$$Ca(s) + \tfrac{1}{2}O_2(g) \longrightarrow CaO(s) \qquad \Delta H° = -635.6 \text{ kJ}$$

$$Ca(s) + H_2(g) + O_2(g) \longrightarrow Ca(OH)_2(s) \qquad \Delta H° = -986.6 \text{ kJ}$$

Thus the ΔH_f° of $Ca(OH)_2$ is −986.6 kJ/mol.

Key Words

Exercises

Definitions

REVIEW QUESTIONS

6.1 Define the following terms: system, surroundings, open system, closed system, isolated system, thermal energy, chemical energy, potential energy, kinetic energy, law of conservation of energy.

6.2 What is heat? How does heat differ from thermal energy? Under what condition is heat transferred from one system to another?

6.3 What are the units for energy commonly employed in chemistry?

6.4 Give a few examples of mechanical work.

6.5 A truck initially traveling at 60 kilometers per hour is brought to a complete stop at a traffic light. Does this change violate the law of conservation of energy? Explain.

6.6 The following are various forms of energy: chemical, heat, light, mechanical, and electrical. Suggest ways of interconverting these forms of energy.

6.7 Describe the interconversions of forms of energy occurring in the following processes: (a) You throw a softball up into the air and catch it. (b) You switch on a flashlight. (c) You ride the ski lift to the top of the hill and then ski down. (d) You strike a match and let it burn down.

Energy Changes and Chemical Reactions

REVIEW QUESTIONS

6.8 Define the following terms: thermochemistry, exothermic process, endothermic process.

6.9 Stoichiometry is based on the law of conservation of mass. On what law is thermochemistry based?

6.10 Describe two exothermic processes and two endothermic processes.

6.11 Decomposition reactions are usually endothermic, whereas combination reactions are usually exothermic. Give a qualitative explanation for these trends.

Enthalpy and Enthalpy Changes

REVIEW QUESTIONS

6.12 Define the following terms: enthalpy, enthalpy of reaction.

6.13 Under what condition is the heat of a reaction equal to the enthalpy change of the same reaction?

6.14 In writing thermochemical equations, why is it important to indicate the physical state (that is, gaseous, liquid, solid, or aqueous) of each substance?

6.15 Explain the meaning of the following thermochemical equation:

$$4NH_3(g) + 5O_2(g) \longrightarrow 4NO(g) + 6H_2O(g) \qquad \Delta H = -905 \text{ J}$$

PROBLEMS

6.16 Consider the following reaction:

$$2CH_3OH(l) + 3O_2(g) \longrightarrow 4H_2O(l) + 2CO_2(g) \qquad \Delta H = -1452.8 \text{ kJ}$$

What is the value of ΔH if (a) the equation is multiplied throughout by 2, (b) the direction of the reaction is reversed so that the products become the reactants and vice versa, (c) water vapor instead of liquid water is formed as the product?

Specific Heat and Heat Capacity

REVIEW QUESTIONS

6.17 Define the following terms: specific heat, heat capacity. What are the units for these two quantities?

6.18 Consider two metals A and B, each of mass 100 g and both at an initial temperature of 20°C. A has a larger specific heat than B. Under the same heating conditions, which metal would take longer to increase its temperature to 21°C?

PROBLEMS

6.19 A piece of silver of mass 362 g has a heat capacity of 85.7 J/°C. What is the specific heat of silver?

6.20 Explain the cooling effect experienced when ethanol is rubbed on your skin, given that

$$C_2H_5OH(l) \longrightarrow C_2H_5OH(g) \qquad \Delta H = 42.2 \text{ kJ}$$

6.21 A piece of copper metal of mass 6.22 kg is heated from 20.5°C to 324.3°C. Calculate the heat absorbed (in kJ) by the metal.

6.22 Calculate the amount of heat liberated (in kJ) from 366 g of mercury when it cools from 77.0°C to 12.0°C.

6.23 A sheet of gold weighing 10.0 g and at a temperature of 18.0°C is placed flat on a sheet of iron weighing 20.0 g and at a temperature of 55.6°C. What is the final temperature of the combined metals? Assume that no heat is lost to the surroundings. (*Hint:* The heat gained by gold must be equal to the heat lost by iron.)

Calorimetry

REVIEW QUESTIONS

6.24 Define calorimetry and describe two commonly used calorimeters.

6.25 In a calorimetric measurement, why is it important that we know the heat capacity of the calorimeter?

PROBLEMS

6.26 A 0.1375-g sample of solid magnesium is burned in a constant-volume bomb calorimeter that has a heat capacity of 1769 J/°C. The calorimeter contains exactly 300 g of water, and the temperature increases by 1.126°C. Calculate the heat given off by the burning Mg, in kJ/g and in kJ/mol.

6.27 The enthalpy of combustion of benzoic acid (C_6H_5COOH) is commonly used as the standard for calibrating constant-volume bomb calorimeters; its value has been accurately determined to be −3226.7 kJ/mol. When 1.9862 g of benzoic acid is burned, the temperature rises from 21.84°C to 25.67°C. What is the heat capacity of the calorimeter? (Assume that the quantity of water surrounding the calorimeter is exactly 2000 g.)

6.28 A quantity of 2.00×10^2 mL of 0.862 M HCl is mixed with 2.00×10^2 mL of 0.431 M $Ba(OH)_2$ in a constant-pressure calorimeter having a heat capacity of 453 J/°C. The initial temperature of the HCl and $Ba(OH)_2$ solutions is the same at 20.48°C. Given that the heat of neutralization for the process

$$H^+(aq) + OH^-(aq) \longrightarrow H_2O(l)$$

is −56.2 kJ, what is the final temperature of the mixed solution?

Standard Enthalpy of Formation

REVIEW QUESTIONS

6.29 What is meant by the standard-state condition?

6.30 Define standard enthalpy of formation.

6.31 Which of the following standard enthalpy of formation values is not zero at 25°C? $Na(s)$, $Ne(g)$, $CH_4(g)$, $S_8(s)$, $Hg(l)$, $H(g)$.

6.32 The ΔH_f° values of the two allotropes of oxygen, O_2 and O_3, are 0 and 142.2 kJ/mol at 25°C. Which is the more stable form at this temperature?

6.33 Which is a greater negative quantity at 25°C: ΔH_f° for $H_2O(l)$ or ΔH_f° for $H_2O(g)$? Explain.

6.34 Consider the halogens chlorine, bromine, and iodine. Predict the value of ΔH_f° (greater than, less than, or equal to zero) for these elements present in the following physical states at 25°C: (a) $Cl_2(g)$, (b) $Cl_2(l)$, (c) $Br_2(g)$, (d) $Br_2(l)$, (e) $I_2(g)$, (f) $I_2(s)$.

6.35 In general, compounds with negative ΔH_f° values are more stable than those with positive ΔH_f° values. From Table 6.4 we see that $H_2O_2(l)$ has a negative ΔH_f°. Why, then, does $H_2O_2(l)$ have a tendency to decompose to $H_2O(l)$ and $O_2(g)$?

6.36 Suggest ways (with appropriate equations) that would allow you to measure the ΔH_f° values of $Ag_2O(s)$ and $CaCl_2(s)$ from their elements. No calculations are necessary.

Hess's Law

REVIEW QUESTION

6.37 Define Hess's law. Explain, with one example, the usefulness of Hess's law in thermochemistry.

PROBLEMS

6.38 From these data,

$$S(\text{rhombic}) + O_2(g) \longrightarrow SO_2(g) \qquad \Delta H^\circ_{rxn} = -296.06 \text{ kJ}$$

$$S(\text{monoclinic}) + O_2(g) \longrightarrow SO_2(g) \qquad \Delta H^\circ_{rxn} = -296.36 \text{ kJ}$$

calculate the enthalpy change for transformation

$$S(\text{rhombic}) \longrightarrow S(\text{monoclinic})$$

(Monoclinic and rhombic are different allotropic forms of elemental sulfur.)

6.39 From the following data,

$$C(\text{graphite}) + O_2(g) \longrightarrow CO_2(g) \qquad \Delta H^\circ_{rxn} = -393.5 \text{ kJ}$$

$$H_2(g) + \tfrac{1}{2}O_2(g) \longrightarrow H_2O(l) \qquad \Delta H^\circ_{rxn} = -285.8 \text{ kJ}$$

$$2C_2H_6(g) + 7O_2(g) \longrightarrow 4CO_2(g) + 6H_2O(l) \qquad \Delta H^\circ_{rxn} = -3119.6 \text{ kJ}$$

calculate the enthalpy change for the reaction

$$2C(\text{graphite}) + 3H_2(g) \longrightarrow C_2H_6(g)$$

6.40 From the following heats of combustion,

$$CH_3OH(l) + \tfrac{3}{2}O_2(g) \longrightarrow CO_2(g) + 2H_2O(l) \qquad \Delta H^\circ_{rxn} = -726.4 \text{ kJ}$$

$$C(\text{graphite}) + O_2(g) \longrightarrow CO_2(g) \qquad \Delta H^\circ_{rxn} = -393.5 \text{ kJ}$$

$$H_2(g) + \tfrac{1}{2}O_2(g) \longrightarrow H_2O(l) \qquad \Delta H^\circ_{rxn} = -285.8 \text{ kJ}$$

calculate the enthalpy of formation of methanol (CH_3OH) from its elements:

$$C(\text{graphite}) + 2H_2(g) + \tfrac{1}{2}O_2(g) \longrightarrow CH_3OH(l)$$

6.41 Calculate the standard enthalpy change for the reaction

$$2Al(s) + Fe_2O_3(s) \longrightarrow 2Fe(s) + Al_2O_3(s)$$

given that

$$2Al(s) + \tfrac{3}{2}O_2(g) \longrightarrow Al_2O_3(s) \qquad \Delta H^\circ_{rxn} = -1601 \text{ kJ}$$

$$2Fe(s) + \tfrac{3}{2}O_2(g) \longrightarrow Fe_2O_3(s) \qquad \Delta H^\circ_{rxn} = -821 \text{ kJ}$$

6.42 Calculate the standard enthalpy of formation of carbon disulfide (CS_2) from its elements, given that

$$C(\text{graphite}) + O_2(g) \longrightarrow CO_2(g) \qquad \Delta H^\circ_{rxn} = -393.5 \text{ kJ}$$

$$S(s) + O_2(g) \longrightarrow SO_2(g) \qquad \Delta H^\circ_{rxn} = -296.1 \text{ kJ}$$

$$CS_2(l) + 3O_2(g) \longrightarrow CO_2(g) + 2SO_2(g) \qquad \Delta H^\circ_{rxn} = -1072 \text{ kJ}$$

6.43 The standard enthalpy of formation at 25°C of $HF(aq)$ is −320.1 kJ/mol; of $OH^-(aq)$, it is −229.6 kJ/mol; of $F^-(aq)$, it is −329.1 kJ/mol; and of $H_2O(l)$, it is −285.8 kJ/mol.

(a) Calculate the standard enthalpy of neutralization of $HF(aq)$:

$$HF(aq) + OH^-(aq) \longrightarrow F^-(aq) + H_2O(l)$$

(b) Using the value of −56.2 kJ as the standard enthalpy change for the reaction

$$H^+(aq) + OH^-(aq) \longrightarrow H_2O(l)$$

calculate the standard enthalpy change for the reaction

$$HF(aq) \longrightarrow H^+(aq) + F^-(aq)$$

Enthalpies of Reaction

REVIEW QUESTIONS

6.44 Define standard enthalpy of reaction.

6.45 Write the equation for calculating the enthalpy of a reaction. Define all the terms.

PROBLEMS

6.46 Calculate the heat of decomposition for this process at constant pressure and 25°C:

$$CaCO_3(s) \longrightarrow CaO(s) + CO_2(g)$$

(Look up the standard enthalpy of formation of the reactant and products in Table 6.4.)

6.47 The standard enthalpies of formation of H^+ and OH^- ions in water are zero and −229.6 kJ/mol, respectively, at 25°C. Calculate the enthalpy of neutralization when 1 mole of a strong monoprotic acid (such as HCl) is titrated by 1 mole of a strong base (such as KOH) at 25°C.

6.48 Calculate the heats of combustion of the following from the standard enthalpies of formation listed in Appendix 3:
(a) $2H_2(g) + O_2(g) \longrightarrow 2H_2O(l)$
(b) $2C_2H_2(g) + 5O_2(g) \longrightarrow 4CO_2(g) + 2H_2O(l)$
(c) $C_2H_4(g) + 3O_2(g) \longrightarrow 2CO_2(g) + 2H_2O(l)$
(d) $2H_2S(g) + 3O_2(g) \longrightarrow 2H_2O(l) + 2SO_2(g)$

6.49 Methanol, ethanol, and *n*-propanol are three common alcohols. When 1.00 g of each of three alcohols is burned in air, heat is liberated as follows:
(a) Methanol (CH_3OH): −22.6 kJ
(b) Ethanol (C_2H_5OH): −29.7 kJ
(c) *n*-Propanol (C_3H_7OH): −33.4 kJ
Calculate the heat of combustion of these alcohols in kJ/mol.

6.50 How much heat (in kJ) is given off when 1.26×10^4 g of ammonia is produced according to the equation

$$N_2(g) + 3H_2(g) \longrightarrow 2NH_3(g) \qquad \Delta H^\circ_{rxn} = -92.6 \text{ kJ}$$

Assume the reaction to take place under standard-state conditions and 25°C.

6.51 The standard enthalpy change for the reaction

$$H_2(g) \longrightarrow H(g) + H(g)$$

is 436.4 kJ. Calculate the standard enthalpy of formation of atomic hydrogen (H).

6.52 From the standard enthalpies of formation, calculate ΔH°_{rxn} for the reaction

$$C_6H_{12}(l) + 9O_2(g) \longrightarrow 6CO_2(g) + 6H_2O(l)$$
$$\Delta H^\circ_f \text{ for } C_6H_{12}(l) = -151.9 \text{ kJ/mol}$$

6.53 Industrially, the first step in the recovery of zinc from the zinc sulfide ore is roasting, that is, the conversion of ZnS to ZnO by heating:

$$2ZnS(s) + 3O_2(g) \longrightarrow 2ZnO(s) + 2SO_2(g) \qquad \Delta H^\circ_{rxn} = -879 \text{ kJ}$$

Calculate the heat evolved (in kJ) per gram of ZnS roasted.

6.54 At 850°C, $CaCO_3$ undergoes substantial decomposition to yield CaO and CO_2. Assuming that the ΔH°_f values of the reactant and products at 850°C are the same as those at 25°C, calculate the enthalpy change (in kJ) if 66.8 g of CO_2 is produced in one reaction.

Heat of Solution

REVIEW QUESTIONS

6.55 Define the following terms: enthalpy of solution, hydration, heat of hydration, lattice energy, heat of dilution.

6.56 Why is the lattice energy of a solid always a positive quantity? Why is the hydration of ions always a negative quantity?

6.57 Consider two ionic compounds A and B. A has a larger lattice energy than B. Which of the two compounds is more stable?

6.58 Mg^{2+} is a smaller cation than Na^+ and also carries more positive charge. Which of the two species has a larger hydration energy (in kJ/mol)? Explain.

6.59 Consider the dissolution of an ionic compound such as potassium fluoride in water. Break the process into the following steps: separation of the cations and anions in the vapor phase and the hydration of the ions in the aqueous medium. Discuss the energy changes associated with each step. How does the heat of solution of KF depend on the relative magnitudes of these two quantities? On what law is the relationship based?

6.60 Why is it dangerous to add water to a concentrated acid such as sulfuric acid in a dilution process?

First Law of Thermodynamics

REVIEW QUESTIONS

6.61 On what law is the first law of thermodynamics based?

6.62 Explain what is meant by a state function. Give two examples of quantities that are state functions and two that are not.

6.63 What is an equation of state?

6.64 Explain the sign conventions in the equation $\Delta E = q + w$.

6.65 The internal energy of an ideal gas depends only on its temperature. Do a first-law analysis of the following process. A sample of an ideal gas is allowed to expand at constant temperature against atmospheric pressure. (a) Does the gas do work on its surroundings? (b) Is there heat exchange between the system and the surroundings? If so, in which direction? (c) What is ΔE for the gas for this process?

6.66 Consider the following changes.
(a) $Hg(l) \longrightarrow Hg(g)$
(b) $3O_2(g) \longrightarrow 2O_3(g)$
(c) $CuSO_4 \cdot 5H_2O(s) \longrightarrow CuSO_4(s) + 5H_2O(g)$
(d) $H_2(g) + F_2(g) \longrightarrow 2HF(g)$
At constant pressure, in which of the reactions is work done by the system on the surroundings? By the surroundings on the system? In which of them is no work done?

PROBLEMS

6.67 A gas expands and does P-V work on the surroundings equal to 325 J. At the same time, it absorbs 127 J of heat from the surroundings. Calculate the change in energy of the gas.

6.68 Define the term "enthalpy change" and outline a procedure for measuring the enthalpy change occurring (a) when a sample of metallic magnesium dissolves in sulfuric acid and (b) when a sample of molten naphthalene freezes.

6.69 Calculate the work done when 50.0 g of tin dissolves in excess acid at 1.00 atm and 25°C:

$$Sn(s) + 2H^+(aq) \longrightarrow Sn^{2+}(aq) + H_2(g)$$

Assume ideal gas behavior.

6.70 Calculate the work done in joules when 1.0 mole of water vaporizes at 1.0 atm and 100°C. Assume that the volume of liquid water is negligible compared with that of steam at 100°C and ideal gas behavior.

Miscellaneous Problems

6.71 The convention of arbitrarily assigning a zero enthalpy value for the most stable form of each element in the standard state at 25°C is a convenient way of dealing with enthalpies of reactions. Explain why this convention cannot be applied to nuclear reactions.

6.72 The standard enthalpy change $\Delta H°$ for the thermal decomposition of silver nitrate according to the following equation is +78.67 kJ:

$$AgNO_3(s) \longrightarrow AgNO_2(s) + \tfrac{1}{2}O_2(g)$$

The standard enthalpy of formation of $AgNO_3(s)$ is −123.02 kJ/mol. Calculate the standard enthalpy of formation of $AgNO_2(s)$.

6.73 (a) Calculate $\Delta H°$ for the following reaction:

$$\underset{\text{hydrazine}}{3N_2H_4(l)} \longrightarrow 4NH_3(g) + N_2(g)$$

The standard enthalpy of formation of hydrazine, $N_2H_4(l)$, is 50.42 kJ/mol. (b) Both hydrazine and ammonia burn in oxygen to produce $H_2O(l)$ and $N_2(g)$. (i) Write balanced equations for each of these processes and calculate $\Delta H°$ for each of them. (ii) On a mass basis (per kg) would hydrazine or ammonia be the better fuel?

6.74 Consider the reaction

$$N_2(g) + 3H_2(g) \longrightarrow 2NH_3(g) \qquad \Delta H°_{rxn} = -92.6 \text{ kJ}$$

If 2.0 moles of N_2 react with 6.0 moles of H_2 to form NH_3, calculate the work done (in joules) against a pressure of 1.0 atm at 25°C. What is ΔE for this reaction? Assume the reaction goes to completion.

6.75 Photosynthesis can be represented by the reaction

$$6CO_2 + 6H_2O \longrightarrow C_6H_{12}O_6 + 6O_2$$

where $C_6H_{12}O_6$ is glucose. (a) How would you determine experimentally the $\Delta H°$ value for this reaction? (b) Solar radiation produces about 7.0×10^{14} kg glucose a year on Earth. What is the corresponding $\Delta H°$ change?

6.76 What are the advantages and disadvantages of using H_2 instead of CH_4 as fuel?

6.77 The enthalpy of combustion for the first four hydrocarbons is as follows: CH_4, −890.4 kJ/mol; C_2H_6, −1560 kJ/mol; C_3H_8, −2220 kJ/mol; C_4H_{10}, −2880 kJ/mol. (a) Plot $\Delta H°$ versus the number of C atoms. Estimate $\Delta H°$ for C_5H_{12} from the graph. (b) Determine the value of $\Delta H°$ when the number of C atoms is zero. What does this value represent?

6.78 A 2.10-mole sample of crystalline acetic acid, initially at 17.0°C, is allowed to melt at 17.0°C and is then heated to 118.1°C (its normal boiling point) at 1.00 atm. The sample is allowed to vaporize at 118.1°C and is then rapidly quenched to 17.0°C, re-forming the crystalline solid. Calculate ΔH for the total process as described.

6.79 Consider the reaction

$$H_2(g) + Cl_2(g) \longrightarrow 2HCl(g) \qquad \Delta H° = -184.6 \text{ kJ}$$

If 3 moles of H_2 react with 3 moles of Cl_2 to form HCl, calculate the work done (in joules) against a pressure of 1.0 atm at 25°C. What is ΔE for this reaction? Assume the reaction goes to completion.

6.80 Calculate the work done in joules by the reaction

$$2Na(s) + 2H_2O(l) \longrightarrow 2NaOH(aq) + H_2(g)$$

when 0.34 g of Na reacts with water to form hydrogen gas at 0°C and 1.0 atm.

6.81 Dissolving 2.00 g of LiCl in 1.00×10^2 mL of water results in a temperature change from 18.4°C to 22.6°C. Calculate the enthalpy of solution (kJ/mol) of LiCl. Assume the specific heat of solution to be the same as that for water.

6.82 You are given the following data:

$$H_2(g) \longrightarrow 2H(g) \qquad \Delta H° = 436.4 \text{ kJ}$$

$$Br_2(g) \longrightarrow 2Br(g) \qquad \Delta H^\circ = 192.5 \text{ kJ}$$

$$H_2(g) + Br_2(g) \longrightarrow 2HBr(g) \qquad \Delta H^\circ = -104 \text{ kJ}$$

Calculate ΔH° for the reaction

$$H(g) + Br(g) \longrightarrow HBr(g)$$

6.83 We learned in Chapter 4 that acid-base titrations are carried out using an indicator. Because the reactions are exothermic, they can also be monitored by measuring the temperature change. In a certain experiment 50.0 mL of a 1.00 *M* NaOH solution was placed in a constant-pressure calorimeter and HCl solution was added to it from a buret. The initial temperature of both the acid and base solution was 20.2°C. The data shown at the bottom of the page were recorded. (a) Plot temperature (*y*-axis) versus volume of HCl added (*x*-axis). From your plot, determine the molarity of the HCl solution. (b) From the maximum temperature rise, determine the heat of neutralization. Assume that the specific heat is the same as that of water.

6.84 Methanol (CH_3OH) is an organic solvent and is also used as a fuel in some automobile engines. From the following data, calculate the standard enthalpy of formation of methanol:

$$2CH_3OH(l) + 3O_2(g) \longrightarrow 2CO_2(g) + 4H_2O(l) \qquad \Delta H^\circ_{rxn} = -1452.8 \text{ kJ}$$

6.85 A 44.0-g sample of an unknown metal at 99.0°C was placed in a constant-pressure calorimeter containing 80.0 g of water at 24.0°C. The final temperature of the system was found to be 28.4°C. Calculate the specific heat of the metal. (The heat capacity of the calorimeter is 12.4 J/°C.)

6.86 One of the most important sources of energy in living cells is the transformation of glucose ($C_6H_{12}O_6$) into lactic acid ($C_3H_6O_3$):

$$C_6H_{12}O_6 \longrightarrow 2C_3H_6O_3$$

To determine the standard enthalpy of this reaction, the following calorimetric experiments were performed: (a) A 0.7521-g sample of benzoic acid (C_6H_5COOH) was burned in a bomb calorimeter containing 1.000 L of water. A temperature rise of 3.60°C was observed. Calculate the heat capacity of the calorimeter. The heat of combustion of benzoic acid is −26.43 kJ/g. (b) A 1.251-g sample of glucose was burned in the same calorimeter, and a temperature rise of 3.52°C was observed. Calculate the heat of combustion and the standard enthalpy of formation of glucose. (c) A 0.9485-g sample of lactic acid was burned in the same calorimeter, and a temperature rise of 2.57°C was observed. Calculate the heat of combustion and standard enthalpy of formation of lactic acid. (d) Calculate the enthalpy change for the conversion of glucose to lactic acid.

6.87 A 1.00-mole sample of ammonia at 14.0 atm and 25°C in a cylinder fitted with a movable piston expands against a constant external pressure of 1.00 atm. At equilibrium, the pressure and volume of the gas are 1.00 atm and 23.5 L, respectively. (a) Calculate the final temperature of the sample. (b) Calculate q, w, and ΔE for the process. The specific heat of ammonia is 0.0258 J/g · °C.

6.88 Producer gas (carbon monoxide) is prepared by passing air over red-hot coke:

$$C(s) + \tfrac{1}{2}O_2(g) \longrightarrow CO(g)$$

and water gas (mixture of carbon monoxide and hydrogen) is prepared by passing steam over red-hot coke:

$$C(s) + H_2O(g) \longrightarrow CO(g) + H_2(g)$$

For many years, both producer gas and water gas were used as fuels in industry and for domestic cooking. The large-scale preparation of these gases was made alternately, that is, first producer gas, then water gas, and so on. Using thermochemical reasoning, explain why this procedure was chosen.

6.89 Compare the heat produced by the complete combustion of 1 mole of methane (CH_4) with a mole of water gas (0.50 mole H_2 and 0.50 mole CO) under the same conditions. On the basis of your answer, would you prefer methane over water gas as a fuel? Can you suggest two other reasons why methane is more preferable than water gas?

6.90 The so-called hydrogen economy is based on hydrogen produced from water using solar energy. The gas is then burned as a fuel:

$$2H_2(g) + O_2(g) \longrightarrow 2H_2O(l)$$

A primary advantage of hydrogen as a fuel is that it is nonpolluting. A major disadvantage is that it is a gas and therefore is harder to store than liquids or solids. Calculate the volume of hydrogen gas at 25°C and 1.00 atm required

Volume HCl added (mL)										
0.00	5.00	10.0	15.0	20.0	25.0	30.0	35.0	40.0	45.0	50.0
Temperature (°C)										
20.2	22.4	24.4	26.4	28.1	29.1	28.4	27.9	27.3	26.8	26.2

to produce an amount of energy equivalent to that produced by the combustion of a gallon of octane (C_8H_{18}). The density of octane is 2.66 kg/gal, and its standard enthalpy of formation is −249.9 kJ/mol.

6.91 Ethanol (C_2H_5OH) and gasoline (assumed to be all octane, C_8H_{18}) are both used as automobile fuel. If gasoline is selling for $1.20/gal, what would the price of ethanol have to be in order to provide the same amount of heat per dollar? The density and ΔH_f° of octane are 0.7025 g/mL and −249.9 kJ/mol and of ethanol are 0.7894 g/mL and −277.0 kJ/mol, respectively. 1 gal = 3.785 L.

6.92 What volume of ethane (C_2H_6), measured at 23.0°C and 752 mmHg, would be required to heat 855 g of water from 25.0°C to 98.0°C?

Opposite page: A photograph of Comet West taken with the aid of a prism shows the emission of different colors of light from different atoms and molecules in the comet. The wavelengths of light are emitted when an electron in an excited state falls to a lower energy level. For example, sodium atoms emit primarily yellow light.

C H A P T E R 7

QUANTUM THEORY AND THE ELECTRONIC STRUCTURE OF ATOMS

Individual atoms do not behave like anything we know in the macroscopic world. Scientists' recognition of this fact led to the development of a new branch of physics called quantum theory to explain the behavior of these submicroscopic particles.

Quantum theory allows us to predict and understand the critical role that electrons play in chemistry. In one sense, studying atoms amounts to asking the following questions:

1. How many electrons are present in a particular atom?
2. What energies do individual electrons possess?
3. Where in the atom can electrons be found?

The answers to these questions have a direct relationship to the behavior of all substances in chemical reactions, and the story of the search for answers provides a fascinating backdrop for our discussion.

7.1 From Classical Physics to Quantum Theory

Attempts by nineteenth-century physicists to understand atoms and molecules met with only limited success. By assuming that molecules behave like little rebounding balls, those early physicists were able to predict and explain some macroscopic phenomena, such as the pressure exerted by a gas. However, the same model could not account for the stability of molecules; that is, it could not explain the forces that hold atoms together. It took a long time to realize—and an even longer time to accept—that the properties of atoms and molecules are *not* governed by the same laws that work so well for larger objects.

It all started in 1900 with a young German physicist named Max Planck.† While analyzing the data on the radiation emitted by solids heated to various temperatures, Planck discovered that atoms and molecules emit energy only in whole-number multiples of certain well-defined quantities. Physicists had always assumed that energy is continuous, which meant that any amount of energy could be released in a radiation process. Planck's work, however, showed that energy can be released only in certain definite amounts, called *quanta*. The resulting *quantum theory* turned physics upside down.

Initially the scientific community greeted Planck's work with skepticism. The idea was so revolutionary that Planck himself was not entirely convinced of its validity—he spent years looking for alternative ways to explain the experimental findings. Eventually, however, the scientific community came to accept the quantum theory, and physics was never the same.

Physics before the advent of the quantum theory is generally referred to as classical physics.

In the development of science, a single major experimental discovery or the formulation of one important theory often sets off an avalanche of activity. Thus, in the thirty years that followed Planck's introduction of the quantum theory, a flurry of investigations not only transformed physics but also altered our concept of nature.

To understand Planck's quantum theory, we must first know something about the nature of ***radiation,*** which is *the emission and transmission through space of energy in the form of waves*. We thus start with a discussion of the properties of waves.

Properties of Waves

A ***wave*** can be thought of as *a vibrating disturbance by which energy is transmitted*. The speed of a wave depends on the type of wave and the nature of the medium through which the wave is traveling. The fundamental properties of a wave can be illustrated by considering a familiar example—water waves. Figure 7.1 shows a seagull floating on the ocean. Water waves are generated by pressure differences in various regions in the surface of water. If we carefully observe the motion of a water wave as it affects the motion of the seagull, we find it is periodic in character; that is, the wave form repeats itself at regular intervals.

The distance between identical points on successive waves is called the ***wavelength*** (λ, lambda). The ***frequency*** (ν, nu) of the wave is *the number of waves that pass through a particular point in one second*. In our case here, the frequency corresponds to the number of times per second that the seagull moves through a complete cycle of

†Max Karl Ernst Ludwig Planck (1858–1947). German physicist. Planck received the Nobel Prize in physics in 1918 for his quantum theory. He also made significant contributions in thermodynamics and other areas of physics.

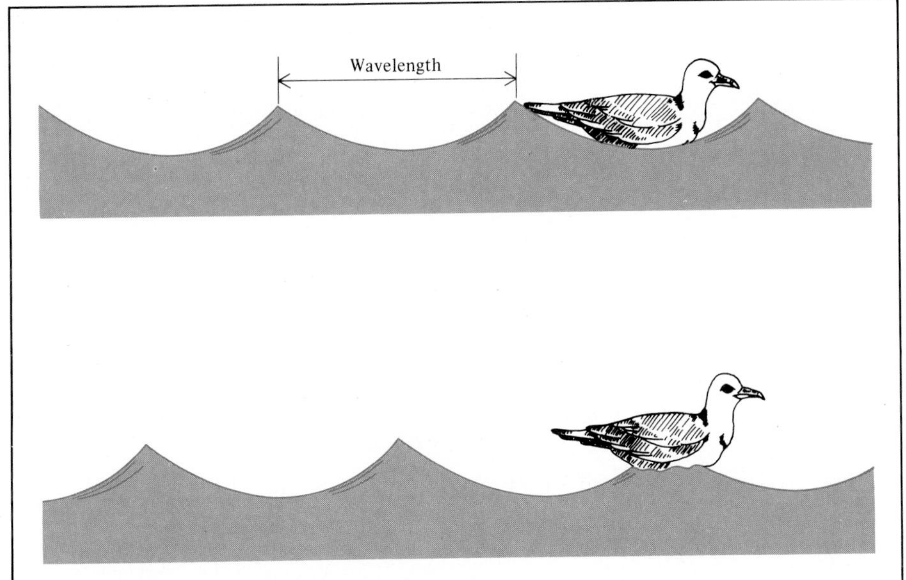

Figure 7.1 *Properties of water waves. The distance between corresponding points on successive waves is called the wavelength, and the number of times the seagull rises up per unit of time is called the frequency. (It is assumed that the seagull does not move horizontally.)*

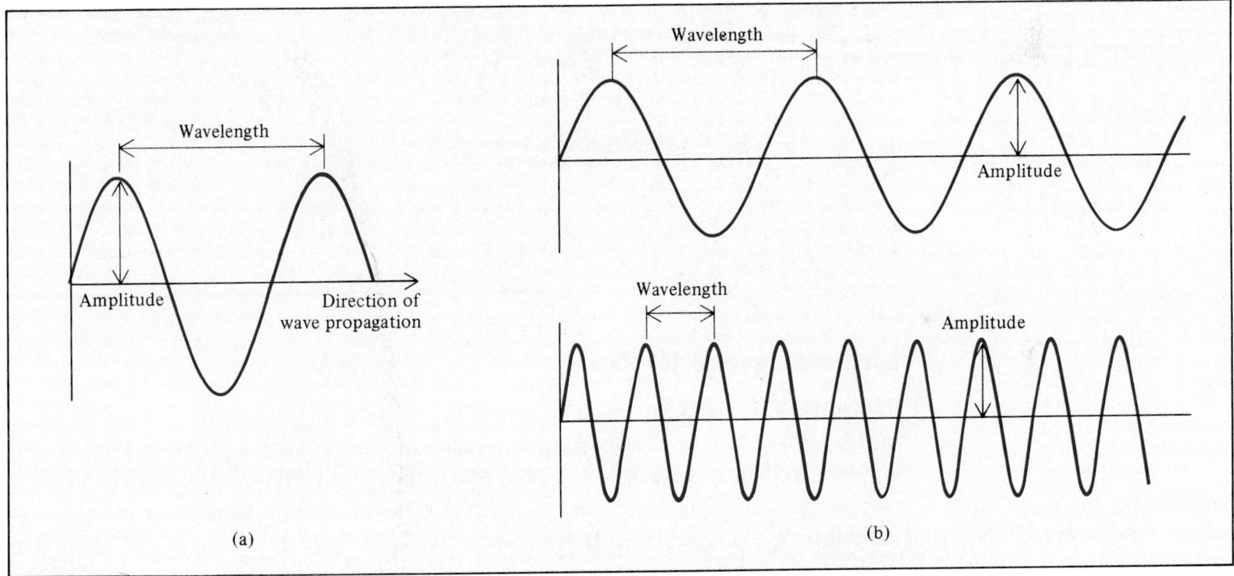

Figure 7.2 *(a) The wavelength and amplitude of an ordinary wave. (b) Two waves having different wavelengths and frequencies. The wavelength of the top wave is three times that of the lower wave, but its frequency is only one-third of the lower wave. Both waves have the same amplitude.*

upward and downward motion. The ***amplitude*** is *the vertical distance from the midline of a wave to the peak or trough* [Figure 7.2(a)]. Figure 7.2(b) shows two waves that have the same amplitude but different wavelengths and frequencies.

An important property of a wave traveling through space is its speed. The speed of a wave depends on the number of cycles of the wave passing through a given point per second (that is, on the frequency) and on the wavelength. In fact, the speed of a wave (u) is given by the product of its wavelength and its frequency:

$$u = \lambda\nu \tag{7.1}$$

The inherent "sensibility" of Equation (7.1) can be seen by analyzing the physical dimensions involved in the three terms. The wavelength (λ) expresses the length of a wave, or distance/wave. The frequency (ν) indicates the number of these waves that pass any reference point per unit of time, or waves/time. Thus the product of these terms results in dimensions of distance/time, which is speed:

$$\frac{\text{distance}}{\text{wave}} \times \frac{\text{waves}}{\text{time}} = \frac{\text{distance}}{\text{time}}$$

$$\lambda \quad \times \quad \nu \quad = \quad u$$

Wavelength is usually expressed in units of meters, centimeters, or nanometers, and frequency is measured in units of hertz (Hz), where

$$1 \text{ Hz} = 1 \text{ cycle/s}$$

The word "cycle" may be left out and the frequency expressed as, for example, 25/s (read as "25 per second").

EXAMPLE 7.1

Calculate the speed of a wave whose wavelength and frequency are 17.4 cm and 87.4 Hz, respectively.

Answer

From Equation (7.1),

$$\begin{aligned} u &= 17.4 \text{ cm} \times 87.4 \text{ Hz} \\ &= 17.4 \text{ cm} \times 87.4/\text{s} \\ &= 1.52 \times 10^3 \text{ cm/s} \end{aligned}$$

Electromagnetic Radiation

Sound waves and water waves are not electromagnetic waves.

Radiation, as we said earlier, is the emission and transmission of energy through space in the form of waves. There are many kinds of waves, such as water waves, sound waves, and light waves. In 1873 James Maxwell showed theoretically that visible light consists of electromagnetic waves. According to Maxwell's theory, an ***electromagnetic wave*** *has an electric field component and a magnetic field component*. These two components have the same wavelength and frequency, and hence the same speed, but they travel in mutually perpendicular planes (Figure 7.3). The significance of Maxwell's theory is that it provides a mathematical description of the general behavior of light. In particular, his model accurately describes how energy in the form of radiation can be propagated through space as vibrating electric and magnetic fields. We now know that light behaves like *electromagnetic radiation,* which is the emission of energy in the form of electromagnetic waves.

A more accurate value for the speed of light is given on the inside back cover of the book.

A feature common to all electromagnetic waves is the speed with which they travel: 3.00×10^8 meters per second, or 186,000 miles per second, which is the speed of light in a vacuum. Although the speed differs from one medium to another, the variations are small enough to allow us to use 3.00×10^8 m/s as the speed of light in our calculations. By convention, we use the symbol c for the speed of light.

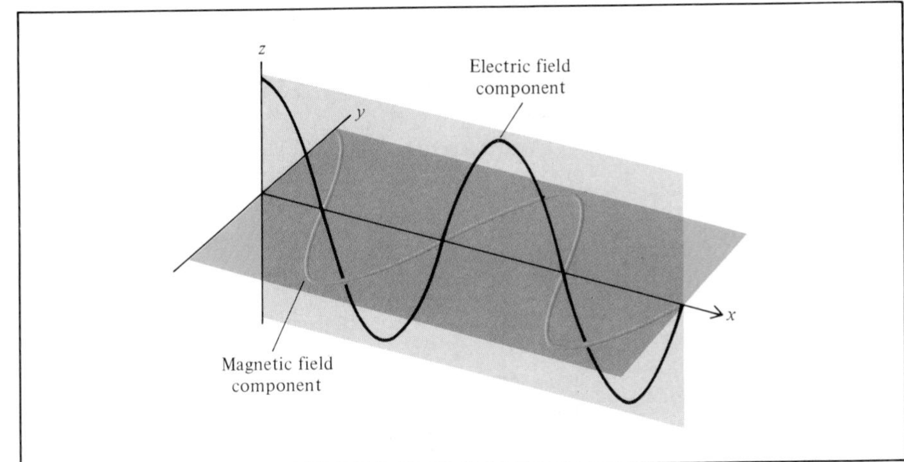

Figure 7.3 *The electric field and magnetic field components of an electromagnetic wave. These two components have the same wavelength, frequency, and amplitude, but they vibrate in two mutually perpendicular planes.*

The wavelength of electromagnetic waves is given in nanometers (nm).

EXAMPLE 7.2

The wavelength of the green light from a traffic signal is centered at 522 nm. What is the frequency of this radiation?

Answer

Rearranging Equation (7.1) we get $\nu = \frac{c}{\lambda}$

Because we are dealing with electromagnetic waves, we replace u in Equation (7.1) with c, which has a value of 3.00×10^8 m/s. Recalling that 1 nm = 1×10^{-9} m (see Table 1.3), we write

$$\nu = \frac{3.00 \times 10^8 \text{ m/s}}{522 \text{ nm} \times \frac{1 \times 10^{-9} \text{ m}}{1 \text{ nm}}}$$

$$= 5.75 \times 10^{14}/\text{s}$$

This odd unit (1/s or s^{-1}) for frequency means that 5.75×10^{14} waves pass a fixed point every second. The very high frequency is in accordance with the very high speed of light.

Similar problems: 7.7, 7.10.

Figure 7.4 shows various types of electromagnetic radiation, which differ from one another in wavelength and frequency. The long radio waves are emitted by large antennas, such as those used by broadcasting stations. The shorter, visible light waves are produced by the motions of electrons within atoms and molecules. The shortest waves, which also have the highest frequency, are those associated with γ (gamma) rays (see Chapter 2), which result from changes within the nucleus of the atom. As we will see shortly, the higher the frequency, the more energetic the radiation. Thus, ultraviolet radiation, X rays, and γ rays are high-energy radiation.

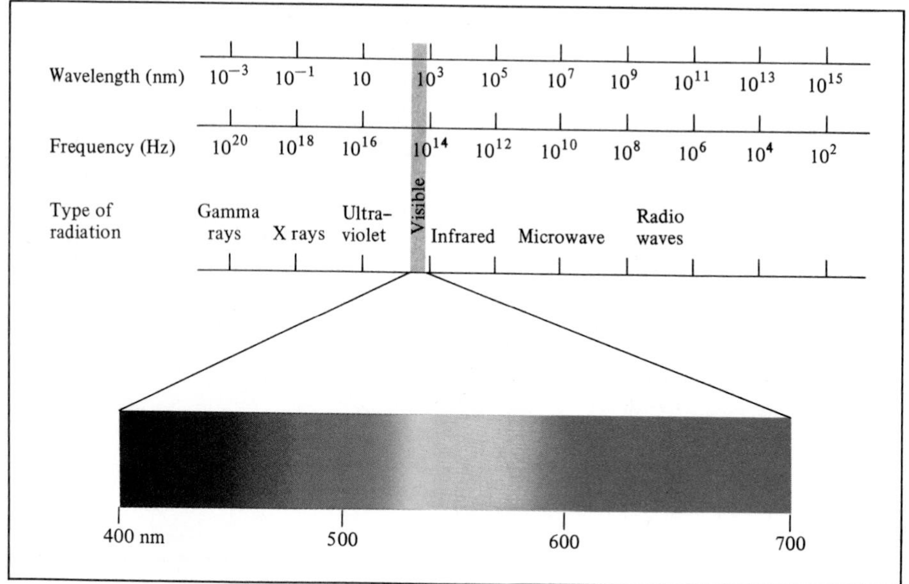

Figure 7.4 *Types of electromagnetic radiation. Gamma rays have the shortest wavelength and highest frequency; radio waves have the longest wavelength and the lowest frequency. Each type of radiation is spread over a specific range of wavelengths (and frequencies). The visible region ranges from 400 nm (violet) to 700 nm (red).*

Planck's Quantum Theory

When solids are heated, they emit radiation over a wide range of wavelengths. The dull red glow of an electric heater and the bright white light of a tungsten light bulb are examples of radiation from solids heated to different temperatures.

Measurements taken in the latter part of the nineteenth century showed that the amount of radiation energy emitted depends on its wavelength. Attempts to account for this dependence in terms of established wave theory and thermodynamic laws were only partially successful. One theory explained short-wavelength dependence but failed to account for the longer wavelengths. Another theory accounted for the longer wavelengths but failed for short wavelengths. It seemed that something fundamental was missing from the laws of classical physics.

In 1900, Planck solved the problem with an assumption that departed drastically from accepted concepts. Classical physics had assumed that atoms and molecules could emit (or absorb) any arbitrary amount of radiant energy. Planck said that atoms and molecules could emit (or absorb) energy only in discrete quantities, like small packages or bundles. Planck gave the name ***quantum*** to *the smallest quantity of energy that can be emitted (or absorbed) in the form of electromagnetic radiation*. The energy E of an emitted single quantum of energy is proportional to the frequency of the radiation:

$$E \propto \nu$$

The proportionality constant for this relationship, symbolized as h, is now called *Planck's constant:*

$$E = h\nu \tag{7.2}$$

where h has the value of 6.63×10^{-34} J s.

According to Planck's quantum theory, energy is always emitted in multiples of $h\nu$; for example, $h\nu$, $2h\nu$, $3h\nu$, . . . , but never, for example, 1.67 $h\nu$ or 4.98 $h\nu$. At the time Planck presented his theory, he could not explain why energies should be fixed or quantized in this manner. Starting with this hypothesis, however, he had no trouble correlating the experimental data for emission by solids over the *entire* range of wavelengths; they all supported the quantum theory.

The idea that energy should be quantized or "bundled" in this manner may seem strange at first, but the concept of quantization has many analogies. For example, an electric charge is also quantized; there can be only whole-number multiples of e, the charge of one electron. Matter itself is quantized, for the numbers of electrons, protons, and neutrons and the numbers of atoms in a sample of matter must also be integers. Our money system is based on a "quantum" of value called a penny. Even processes in living systems involve quantized phenomena. The eggs laid by hens are quantized, and pregnant cats give birth to an integral number of kittens, not to one-half or three-quarters of a kitten.

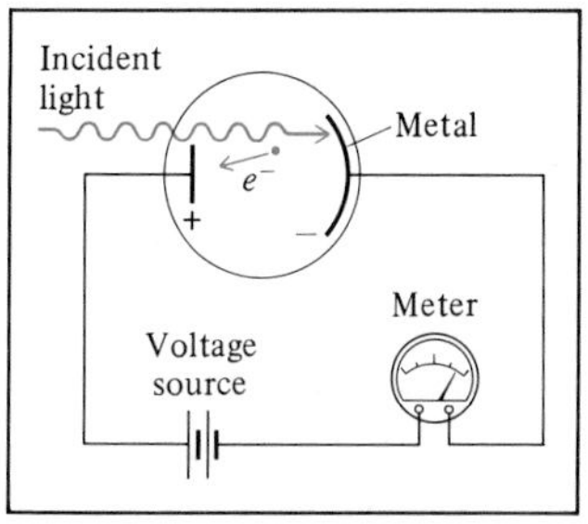

Figure 7.5 *An apparatus for studying the photoelectric effect. Light of a certain frequency falls on a clean metal surface. Ejected electrons are attracted toward the positive electrode. The flow of electrons is indicated by a detecting meter.*

7.2 The Photoelectric Effect

In 1905, only five years after Planck presented the quantum theory, Albert Einstein† used the theory to solve another mystery in physics, the *photoelectric effect*. Experiments had already demonstrated that electrons were ejected from the surface of certain metals exposed to light of at least a certain minimum frequency, called threshold frequency (Figure 7.5). The number of electrons ejected was proportional to the intensity (or brightness) of the light, but the energies of the ejected electrons were not. Below the threshold frequency no electrons were ejected no matter how intense the light.

The observed phenomenon could not be explained by the wave theory of light. However, Einstein was able to account for the photoelectric effect by making an extraordinary assumption. He suggested that we should not think of a beam of light as wavelike but rather as *a stream of particles,* called ***photons.*** Using Planck's quantum theory of radiation as a starting point, Einstein deduced that each photon must possess energy E, given by the equation

This equation has the same form as Equation (7.2) because, as we will see shortly, electromagnetic radiation is emitted as well as absorbed in the form of photons.

$$E = h\nu$$

where ν is the frequency of light. Electrons are held in a metal by attractive forces, and so to remove them from the metal we must employ light of a sufficiently high frequency (which corresponds to sufficiently high energy) to break them free. Shining a beam of light onto a metal surface can be thought of as shooting a beam of particles—photons—at the metal atoms. If the frequency of photons is such that $h\nu$ is exactly equal to the binding energy of the electrons in the metal, then the light will have just enough energy to knock the electrons loose. If we use light of a higher frequency, then the electrons will not only be knocked loose, but they will also acquire some kinetic energy. This situation is summarized by the equation

$$h\nu = \text{KE} + \text{BE} \tag{7.3}$$

† Albert Einstein (1879–1955). German-born American physicist. Regarded by many as one of the two greatest physicists the world has known (the other is Isaac Newton). The three papers (on special relativity, Brownian motion, and the photoelectric effect) that he published in 1905 while employed as a technical assistant in the Swiss patent office in Berne have profoundly influenced the development of physics. He received the Nobel Prize in physics in 1921 for his explanation of the photoelectric effect.

where KE is the kinetic energy of the ejected electron and BE is the binding energy of the electron in the metal. Rewriting Equation (7.3) as

$$\text{KE} = h\nu - \text{BE}$$

shows that the more energetic the photon is (that is, the higher its frequency), the greater is the kinetic energy of the ejected electron.

Now consider two beams of light having the same frequency (which is greater than the threshold frequency) but different intensities. The more intense beam of light consists of a larger number of photons; consequently, the number of electrons ejected from the metal's surface is greater than the number of electrons produced by the weaker beam of light. Thus the more intense the light, the greater the number of electrons emitted by the target metal; the higher the frequency of the light, the greater the kinetic energy of the emitted electrons.

EXAMPLE 7.3

Calculate the energy (in joules) of (a) a photon of wavelength 5.00×10^4 nm (infrared region) and (b) a photon of wavelength 5.00×10^{-2} nm (X-ray region).

Answer

(a) We use Equation (7.2): $E = h\nu$

From Equation (7.1), $\nu = c/\lambda$ (see Example 7.2); therefore

$$\begin{aligned} E &= \frac{hc}{\lambda} \\ &= \frac{(6.63 \times 10^{-34}\ \text{J}\cdot\text{s})(3.00 \times 10^8\ \text{m/s})}{(5.00 \times 10^4\ \text{nm})\left(\dfrac{1 \times 10^{-9}\ \text{m}}{1\ \text{nm}}\right)} \\ &= 3.98 \times 10^{-21}\ \text{J} \end{aligned}$$

This is the energy possessed by a single photon of wavelength 5.00×10^4 nm.

(b) Following the same procedure as in (a), we can show that the energy of the photon of wavelength 5.00×10^{-2} nm is 3.98×10^{-15} J. Thus an "X-ray" photon is 1×10^6, or a million times, more energetic than an "infrared" photon in our case.

Similar problem: 7.16.

Einstein's theory of light has posed a dilemma for scientists. On the one hand, it explains the photoelectric effect satisfactorily. On the other hand, the particle theory of light is not consistent with the known wave behavior of light. The only way to resolve the dilemma is to accept the idea that light possesses *both* particlelike and wavelike properties. Depending on the experiment, we find that light behaves either as a wave or as a stream of particles. This concept was totally alien to the way physicists had thought about matter and radiation, and it took a long time for them to accept it. It turns out that the property of dual nature (particles and waves) is not unique to light but is characteristic of all matter, including submicroscopic particles like electrons, as we will see in Section 7.4.

7.3 Bohr's Theory of the Hydrogen Atom

Emission Spectra

Einstein's work paved the way for the solution of yet another nineteenth-century "mystery" in physics: the emission spectra of atoms. Ever since the seventeenth century, when Newton showed that sunlight is composed of various color components which can be recombined to produce white light, chemists and physicists have studied the characteristics of ***emission spectra*** of various substances, that is, *either continuous or line spectra of radiation emitted by the substances*. The emission of a substance is obtained by energizing a sample of material either with thermal energy or with some other energy form (such as a high-voltage electrical discharge if the substance is gaseous). A "red-hot" or "white-hot" iron bar freshly removed from a high-temperature source glows in a characteristic way. This visible glow is the portion of its emission spectrum that is sensed by eye. The warmth felt at a distance from the same iron bar is another portion of its emission spectrum—this portion in the infrared region. A feature common to the emission spectra of the sun and of a heated solid is that both are continuous; that is, all wavelengths of light are represented in the spectra (see the visible region in Figure 7.4).

The emission spectra of atoms in the gas phase, on the other hand, do not show a continuous spread of wavelengths from red to violet; rather, the atoms *emit light only at specific wavelengths*. Such spectra are called ***line spectra*** because the radiation is identified by the appearance of bright lines in the spectra. Figure 7.6 is a schematic diagram of a discharge tube that is used to study emission spectra. Figure 7.7 shows the color emitted by hydrogen atoms in a discharge tube, and Figure 7.8 shows the portion of the line spectrum of the hydrogen atoms that lies in the visible region.

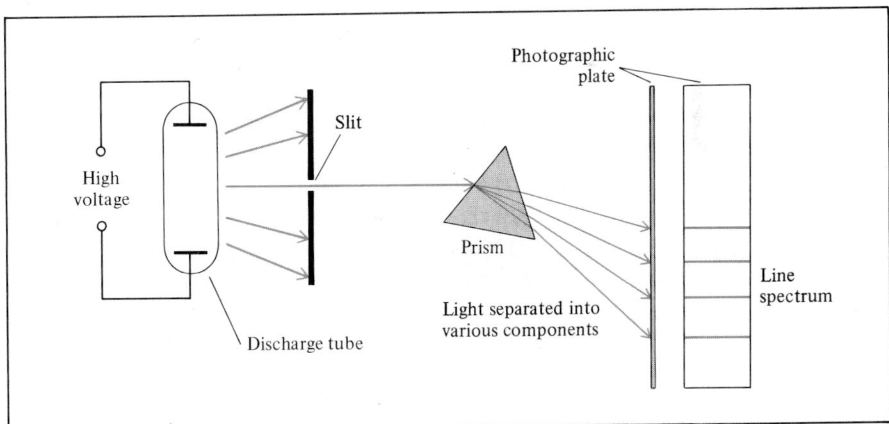

Figure 7.6 *An experimental arrangement for studying the emission spectra of atoms and molecules. The gas under study is in a discharge tube containing two electrodes. As electrons flow from the negative electrode to the positive electrode, they collide with the gas. This collision process eventually leads to the emission of light by the atoms (and molecules). The emitted light is separated into its components by a prism. Each component color is focused at a definite position, according to its wavelength, and forms a colored image of the slit on the photographic plate. The colored images are called spectral lines.*

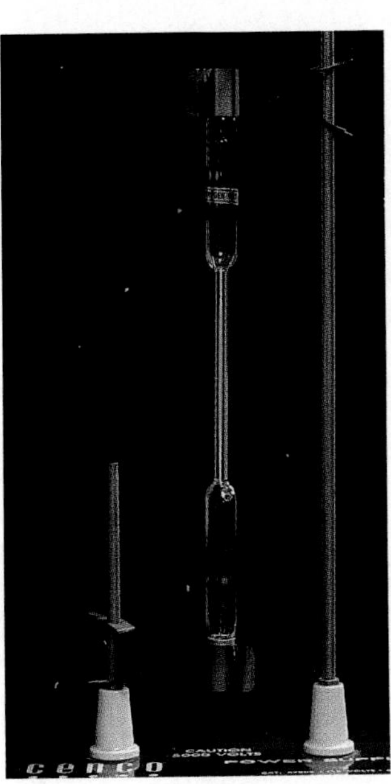

Figure 7.7 *Color emitted by hydrogen atoms in a discharge tube.*

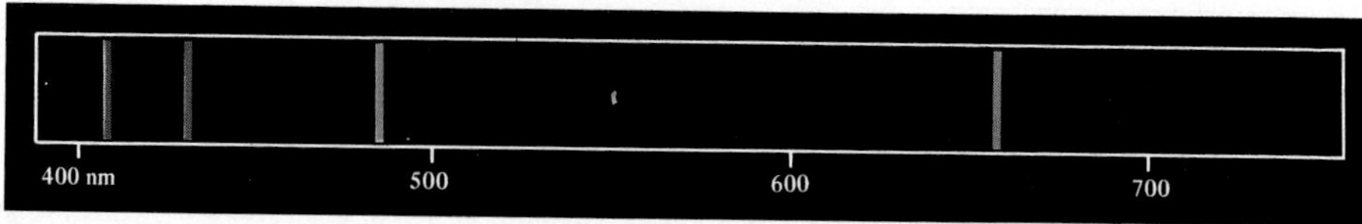

Figure 7.8 *A line emission spectrum of hydrogen atoms.*

Every element has a unique emission spectrum. The characteristic lines in atomic spectra can be used in chemical analysis to identify unknown atoms, much as fingerprints are used to identify people. When the lines of the emission spectrum of a known element exactly match the lines of the emission spectrum of an unknown sample, the identity of the latter is quickly established. Although it was immediately recognized that this procedure is useful, the origin of these lines was unknown until early in this century.

Emission Spectrum of the Hydrogen Atom

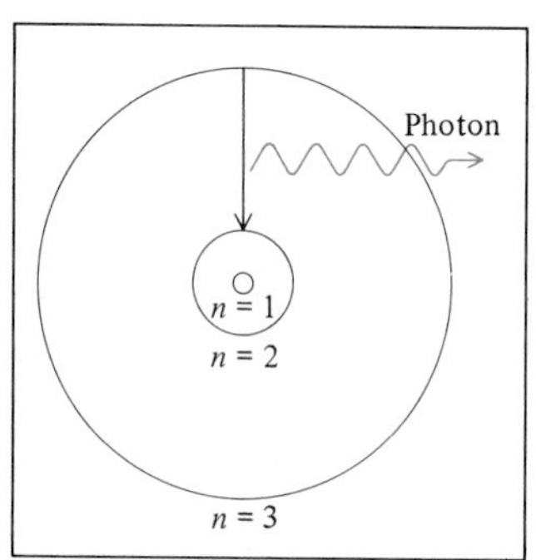

Figure 7.9 *The emission process in an excited hydrogen atom, according to Bohr's theory. An electron originally in a higher-energy orbit ($n = 3$) falls back to a lower-energy orbit ($n = 2$). As a result, a photon with energy $h\nu$ is given off. The quantity $h\nu$ is equal to the difference in energies of the two orbits occupied by the electron in the emission process. For simplicity, only three orbits are shown.*

In 1913, not too long after Planck's and Einstein's discoveries, Niels Bohr† offered a theoretical explanation of the emission spectrum of the hydrogen atom. Bohr's treatment is very complex and is no longer considered to be correct in all its details. Thus, we will concentrate only on his important assumptions and final results, which do account for the positions of the spectral lines.

When Bohr first tackled this problem, physicists already knew that the atom contains electrons and protons. They thought of an atom as an entity in which electrons whirled around the nucleus in circular orbits at high velocities. This was an appealing model because it resembled the well-understood motions of the planets around the sun. In the hydrogen atom, it was believed that the electrostatic attraction between the positive "solar" proton and the negative "planetary" electron pulls the electron inward and that this force is balanced exactly by the acceleration due to the circular motion of the electron.

Bohr's model for the atom included the idea of electrons moving in circular orbits, but he imposed a rather severe restriction: The single electron in the hydrogen atom could be located only in certain orbits. Since each orbit has a particular energy associated with it, Bohr's restriction meant that energies associated with electron motion in the permitted orbits are fixed in value; that is, they are quantized.

The emission of radiation by an energized hydrogen atom could then be explained in terms of the electron dropping from a higher-energy orbit to a lower one and giving up a quantum of energy (a photon) in the form of light (Figure 7.9). Using arguments based on electrostatic interaction and Newton's laws of motion, Bohr showed that the energies that the electron in the hydrogen atom can possess are given by

$$E_n = -R_H\left(\frac{1}{n^2}\right) \tag{7.4}$$

The value of n characterizes a particular orbit.

where R_H, the *Rydberg‡ constant,* has the value 2.18×10^{-18} J. The number n is an integer called the *principal quantum number;* it has the values $n = 1, 2, 3, \ldots$.

The negative sign in Equation (7.4) may seem strange, for it implies that all the

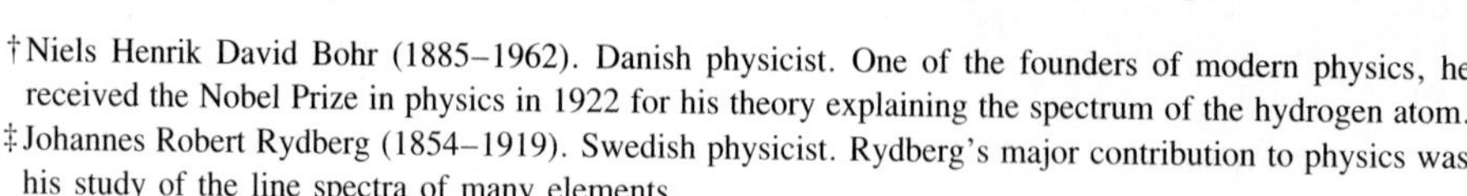

†Niels Henrik David Bohr (1885–1962). Danish physicist. One of the founders of modern physics, he received the Nobel Prize in physics in 1922 for his theory explaining the spectrum of the hydrogen atom.

‡Johannes Robert Rydberg (1854–1919). Swedish physicist. Rydberg's major contribution to physics was his study of the line spectra of many elements.

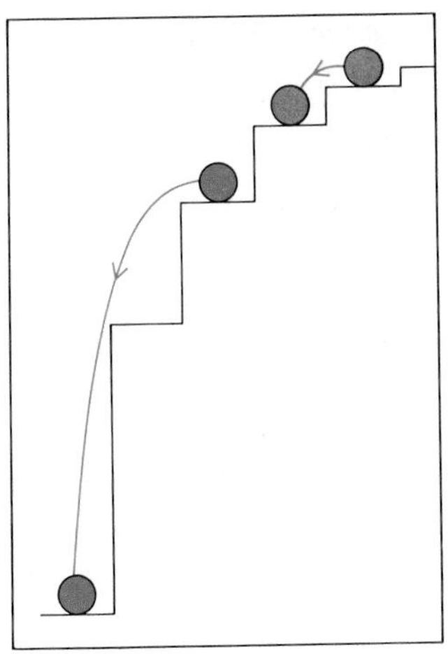

Figure 7.10 *A mechanical analogy for the emission processes.*

allowable energies of the electron are negative. Actually, this sign is nothing more than an arbitrary convention; it says that the energy of the electron in the atom is *lower* than the energy of a *free electron,* or an electron that is infinitely far from the nucleus. The energy of a free electron is arbitrarily assigned a value of zero. Mathematically, this corresponds to setting n equal to infinity in Equation (7.4), so that $E_\infty = 0$. As the electron gets closer to the nucleus (as n decreases), E_n becomes larger in absolute value, but also more negative. The most negative value, then, is reached when $n = 1$, which corresponds to the most stable orbit. We call this the ***ground state,*** or the ***ground level,*** which refers to *the lowest energy state of a system* (which is an atom in our discussion). The stability of the electron diminishes for $n = 2, 3, \ldots$, and each of these is called an ***excited state,*** or ***excited level,*** which is *higher in energy than the ground state*. An electron in a hydrogen atom that occupies an orbit with n greater than 1 is said to be in an excited state. The radius of each circular orbit depends on n^2. Thus, as n increases from 1 to 2 to 3, the orbit radius increases in size very rapidly. The higher the excited state, the farther away the electron is from the nucleus (and the less tightly it is held by the nucleus).

Bohr's theory of the hydrogen atom enables us to explain the line spectra of that atom. Radiant energy absorbed by the atom causes the electron to move from a lower-energy orbit (characterized by a smaller n value) to a higher-energy orbit (characterized by a larger n value). Conversely, radiant energy (in the form of a photon) is emitted when the electron moves from a higher-energy orbit to a lower-energy orbit. The quantized movement of the electron from one orbit to another is analogous to the movement of a tennis ball either up or down a set of stairs (Figure 7.10). The ball can reside at a variety of different stair levels but never between stairs. Its journey from a lower stair to a higher one is an energy-requiring process, whereas that from a higher stair to a lower stair is an energy-releasing process. The quantity of energy involved in either type of change is determined by the distance between the first and last stairs. Similarly, the amount of energy needed to move an electron in the Bohr atom depends on the difference in the levels of the initial and final states.

Let us now apply Equation (7.4) to the emission process in a hydrogen atom. Suppose that the electron is initially in an excited state characterized by the principal quantum number n_i. During emission, the electron drops to a lower energy state characterized by the principal quantum number n_f (the subscripts i and f denote the initial and final states, respectively). This lower energy state may be either another excited state or the ground state. The difference between the energies of the initial and final states is ΔE (delta E), where

$$\Delta E = E_f - E_i$$

From Equation (7.4),

$$E_f = -R_H\left(\frac{1}{n_f^2}\right)$$

and

$$E_i = -R_H\left(\frac{1}{n_i^2}\right)$$

Therefore

$$\begin{aligned}\Delta E &= \left(\frac{-R_H}{n_f^2}\right) - \left(\frac{-R_H}{n_i^2}\right)\\ &= R_H\left(\frac{1}{n_i^2} - \frac{1}{n_f^2}\right)\end{aligned}$$

Because this transition results in the emission of a photon of frequency ν and energy $h\nu$ (Figure 7.9), we can write

$$\Delta E = h\nu = R_H\left(\frac{1}{n_i^2} - \frac{1}{n_f^2}\right) \tag{7.5}$$

When a photon is emitted, $n_i > n_f$. Consequently the term in parentheses is negative and ΔE is negative (energy is lost to the surroundings). When energy is absorbed, $n_i < n_f$ and the term in parentheses is positive, so ΔE is positive. Each spectral line in the emission spectrum corresponds to a particular transition in a hydrogen atom. When we study a large number of hydrogen atoms, we observe all possible transitions and hence the corresponding spectral lines. The brightness of a spectral line depends on how many photons of the same wavelength are emitted.

The emission spectrum of hydrogen covers a wide range of wavelengths from the infrared to the ultraviolet. Table 7.1 shows the series of transitions in the hydrogen spectrum; they are named after their discoverers. The Balmer series was particularly easy to study because a number of its lines fall in the visible range.

Table 7.1 The various series in atomic hydrogen emission spectrum

Series	n_f	n_i	Spectrum region
Lyman	1	2, 3, 4, . . .	Ultraviolet
Balmer	2	3, 4, 5, . . .	Visible and ultraviolet
Paschen	3	4, 5, 6, . . .	Infrared
Brackett	4	5, 6, 7, . . .	Infrared

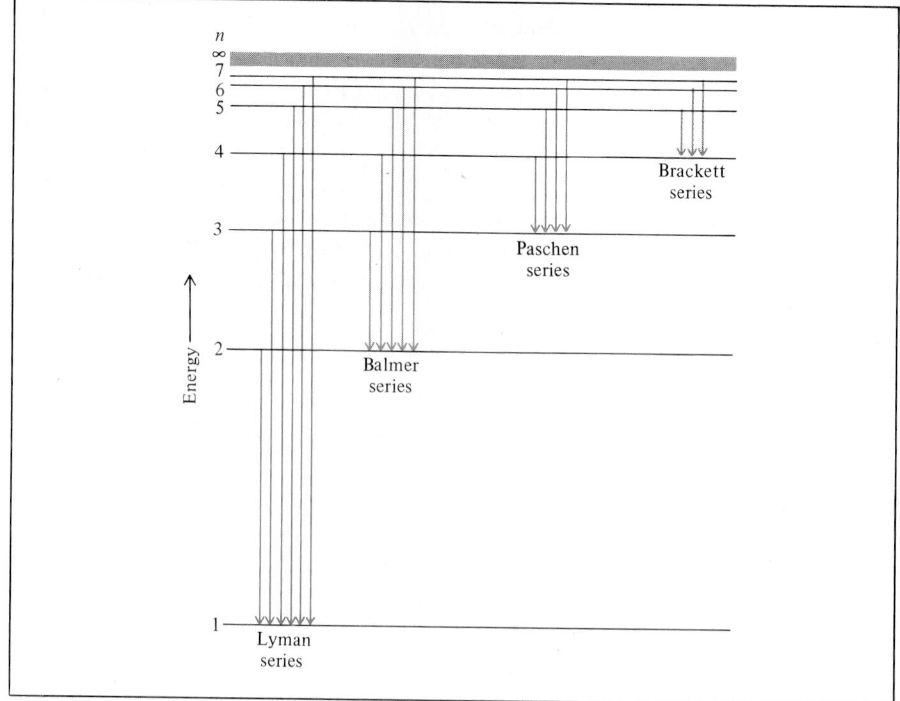

Figure 7.11 *The energy levels in the hydrogen atom and the various emission series. Each energy level corresponds to the energy associated with the motion of an electron in an orbit, as postulated by Bohr and shown in Figure 7.9. The emission lines are labeled according to the scheme in Table 7.1.*

Figure 7.9 shows a single transition. However, it is more informative to express transitions as shown in Figure 7.11. Each horizontal line is called an *energy level*. The position of the energy level, as measured on the energy scale, shows the energy associated with the particular orbit. The orbits are labeled with their principal quantum numbers.

Example 7.4 illustrates the use of Equation (7.5).

EXAMPLE 7.4

What is the wavelength of a photon emitted during a transition from the $n_i = 5$ state to the $n_f = 2$ state in the hydrogen atom?

Answer

Since $n_f = 2$, this transition gives rise to a spectral line in the Balmer series (see Figure 7.11). From Equation (7.5) we write

$$\begin{aligned}\Delta E &= R_H\left(\frac{1}{n_i^2} - \frac{1}{n_f^2}\right)\\ &= 2.18 \times 10^{-18}\ \text{J}\left(\frac{1}{5^2} - \frac{1}{2^2}\right)\\ &= -4.58 \times 10^{-19}\ \text{J}\end{aligned}$$

The negative sign is in accord with our convention that energy is given off to the surroundings.

The negative sign indicates that this is energy associated with an emission process. To calculate the wavelength we will omit the minus sign for ΔE because the wavelength of the photon must be positive. Since $\Delta E = h\nu$ or $\nu = \Delta E/h$, we can calculate the wavelength of the photon by writing

$$\begin{aligned}\lambda &= \frac{c}{\nu}\\ &= \frac{ch}{\Delta E}\\ &= \frac{(3.00 \times 10^8\ \text{m/s})(6.63 \times 10^{-34}\ \text{J s})}{4.58 \times 10^{-19}\ \text{J}}\\ &= 4.34 \times 10^{-7}\ \text{m}\\ &= 4.34 \times 10^{-7}\ \text{m} \times \left(\frac{1 \times 10^9\ \text{nm}}{1\ \text{m}}\right) = 434\ \text{nm}\end{aligned}$$

Similar problems: 7.35, 7.37.

The next two Chemistry in Actions discuss some interesting applications of atomic emission and laser.

7.4 The Dual Nature of the Electron

Physicists were mystified but intrigued by Bohr's theory. The question they asked about it was: Why are the energies of the hydrogen electron quantized? Or, phrasing the question in a more concrete way, Why is the electron in a Bohr atom restricted to

CHEMISTRY IN ACTION

Atomic Emission—Street Lamps, Fluorescent Lights, and Neon Signs

Many substances other than hydrogen (helium, neon, sodium, and mercury, for example) give off visible light when excited in an electric discharge tube (Figure 7.12). The phenomenon of atomic emission turns up frequently in daily living.

Two types of lamps used in street lights rely on atomic emission: the mercury lamp and the sodium lamp. In both cases the light is generated by electron bombardment of either mercury or sodium vapor:

$$\text{Hg} + \text{energy} \longrightarrow \text{Hg}^*$$
$$\text{Hg}^* \longrightarrow \text{Hg} + \text{bluish-green light}$$

$$\text{Na} + \text{energy} \longrightarrow \text{Na}^*$$
$$\text{Na}^* \longrightarrow \text{Na} + \text{yellow light}$$

where the asterisk denotes an atom in an excited electronic state. The visible portion of mercury emission falls mainly in the yellow, green, and purple regions, and sodium emission is centered mainly in the yellow region. The sodium lamp has some advantages over the mercury lamp. For a given input of electricity, the sodium lamp gives off greater light intensity than does the mercury lamp. Moreover, yellow light has a longer wavelength than bluish-green light, so it is not scattered as much by small particles in the air—fog, for example—and penetrates farther than white light. In addition, the human eye exhibits its greatest response to yellow color. (You may have noticed that some automobile fog lights have yellow-tinted glass covers.)

A fluorescent lamp is a discharge tube with the inner surface coated with a fluorescent material such as zinc sulfide (which is also used on TV screens). The tube is filled with mercury vapor at low pressure. Excitation of the mercury atoms by electron bombardment causes the emission of light in the green, blue, and ultraviolet (UV) regions. When the light strikes the inner glass wall, most of the UV light is absorbed by the fluorescent material, which then emits a multitude of longer wavelengths that combine to produce white light. Fluo-

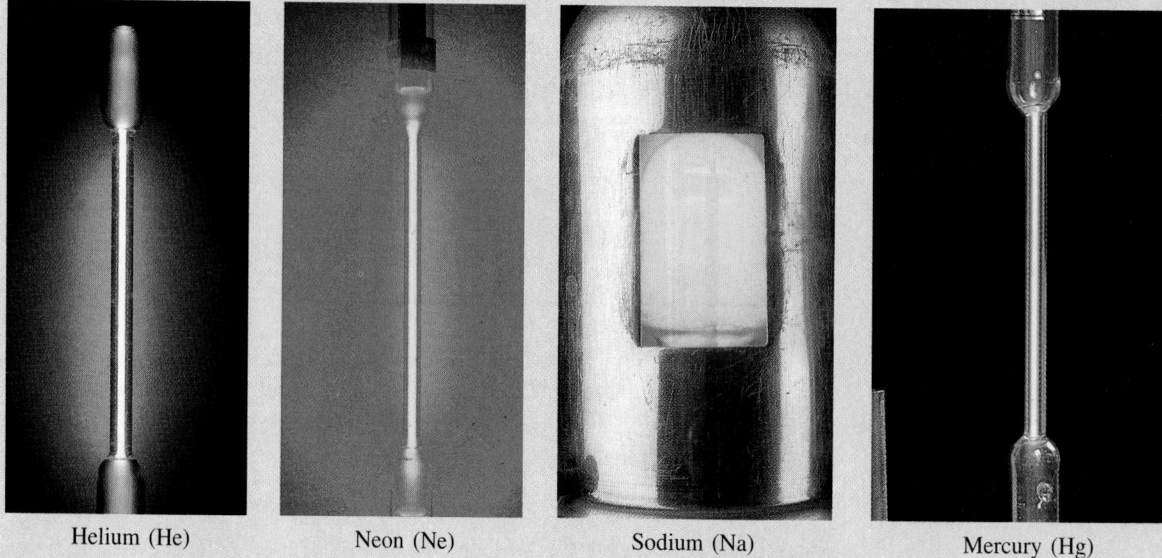

Helium (He) Neon (Ne) Sodium (Na) Mercury (Hg)

Figure 7.12 *From left to right: colors emitted by helium, neon, sodium, and mercury atoms.*

rescent lamps are more energy-efficient (and hence cheaper to operate) than tungsten lamps (ordinary light bulbs) because the generation of fluorescent light involves only the transfer of radiant energy—very little heat is produced. For this reason, a fluorescent lamp is cool to the touch. On the other hand, heating the filament of a tungsten lamp to 3000°C produces visible light and much infrared radiation (heat). Thus a tungsten light bulb is usually too hot to touch after it has been turned on for a few minutes because a substantial fraction of the input energy is reradiated as heat.

The familiar neon signs we see on storefronts are colored by the same atomic emission phenomenon. The procedure for producing neon emission is the same as that for hydrogen, mercury, and sodium. As a result of electron bombardment,

$$\text{Ne} + \text{energy} \longrightarrow \text{Ne}^*$$
$$\text{Ne}^* \longrightarrow \text{Ne} + \text{orange-red light}$$

where the asterisk denotes a neon atom in its excited electronic state. In fact, other gases may be used to make "neon" signs. For example, argon emits a bluish-purple light, and krypton gives off a white light. Special fluorescent materials on the inner glass walls of signs produce greens and other colors as well.

CHEMISTRY IN ACTION

Laser—The Splendid Light

Laser is an acronym for light amplification by stimulated emission of radiation. It is a special type of emission that involves either atoms or molecules. Since the discovery of laser in 1960, it has been used in numerous systems designed to operate in the gas, liquid, and solid states. These systems emit radiation with wavelengths ranging from infrared through visible, UV, to the X ray. The advent of laser has truly revolutionized science, medicine, and technology.

Figure 7.13 shows how a ruby laser (the first known laser) works. (Ruby is a deep-red mineral called corundum, chemical makeup Al_2O_3. In this arrangement, some of the Al^{3+} ions are replaced by Cr^{3+} ions.) First a flashlamp is used to excite the chromium atoms to a higher energy level. The excited atoms are unstable, so that at a given instant some of them will return to the ground state by emitting a photon in the red region of the spectrum. When a photon is emitted, it bounces

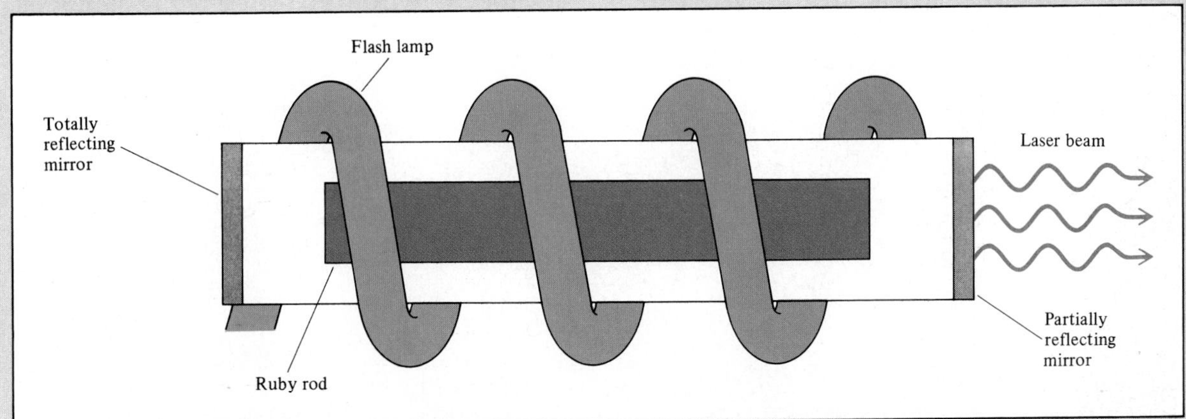

Figure 7.13 *The emission of laser light from a ruby laser.*

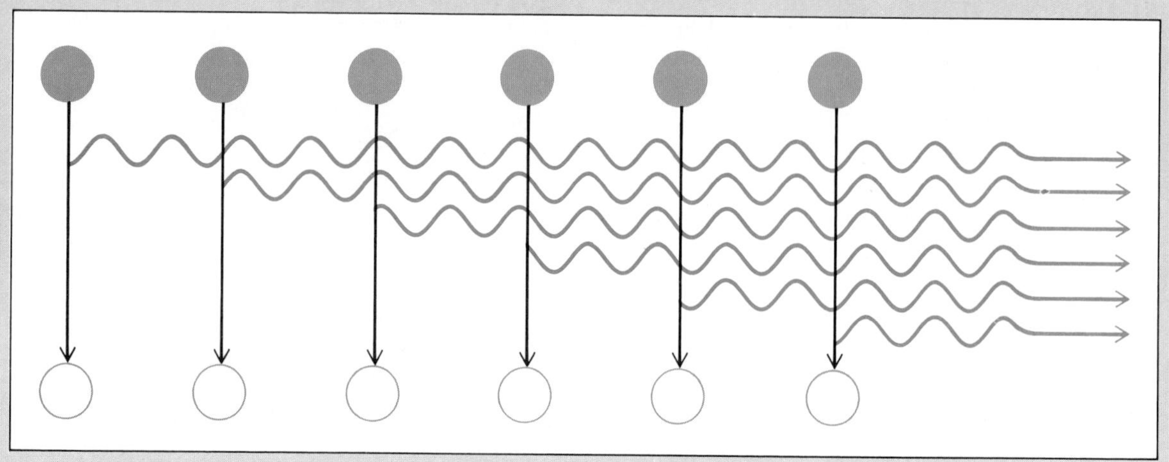

Figure 7.14 *The stimulated emission of one photon by another photon in a cascade event that leads to the emission of laser light.*

back and forth many times between the mirrors. This photon can stimulate the emission of photons of exactly the same wavelength from other excited chromium atoms; these photons in turn can stimulate the emission of more photons, and so on. Because the light waves are *in phase*—that is, their maxima and minima coincide—the photons enhance one another, increasing their power with each passage between the mirrors. One of the end mirrors is made to be only partially reflecting, so that when the light reaches a certain intensity it emerges from the mirror and we have a laser beam (Figure 7.14). Depending on the mode of operation, the laser light may be emitted in pulses (as in the ruby laser case) or it may be emitted in continuous waves.

Laser light is characterized by three properties: It is intense, it has precisely known wavelength and hence energy, and it is coherent. By coherent we mean that the light waves are all in phase. The applications of lasers are quite numerous. Their high intensity and ease of focus make them suitable for doing eye surgery, for drilling holes in metals and welding, and for carrying out nuclear fusion. The fact that they are highly directional and have precisely known wavelengths makes them very useful for telecommunications. Lasers are also used in isotope separation, in holography (three-dimensional photography), in compact discs, and even in supermarkets to read Universal Product Codes. Lasers have played an important role in the spectroscopic investigation of molecular properties and of many chemical and biological processes. Figure 7.15 shows laser lights being used to probe the details of chemical reactions.

Figure 7.15 *State-of-the-art lasers used in the research laboratory of Dr. A. H. Zewail at the California Institute of Technology.*

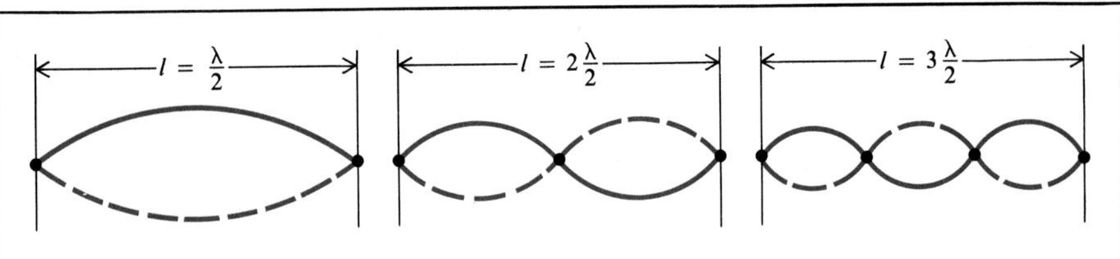

Figure 7.16 *The standing waves generated by plucking a guitar string. Each dot represents a node. The length of the string (ℓ) must be equal to a whole number times one-half the wavelength (λ/2).*

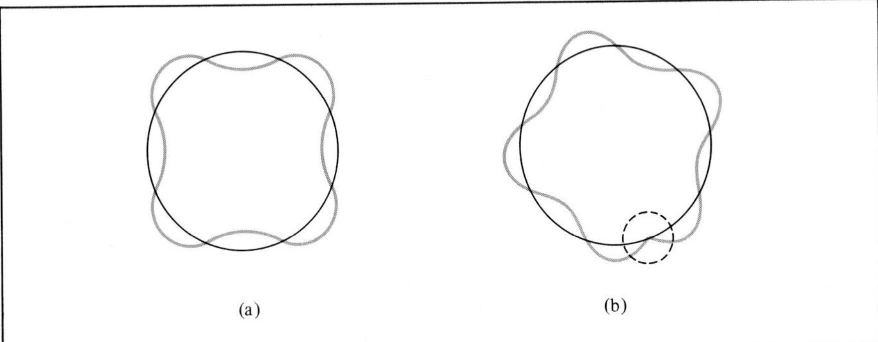

Figure 7.17 *(a) The circumference of the orbit is equal to an integral number of wavelengths. This is an allowed orbit. (b) The circumference of the orbit is not equal to an integral number of wavelengths. As a result, the electron wave does not close in on itself. This is a nonallowed orbit.*

orbiting the nucleus at certain fixed distances? For a decade no one, not even Bohr himself, had a logical explanation. In 1924 Louis de Broglie† provided a solution to this puzzle. De Broglie reasoned as follows: If light waves can behave like a stream of particles (photons), then perhaps particles such as electrons can possess wave properties. According to de Broglie, an electron bound to the nucleus behaves like *standing waves*. Standing waves can be generated by plucking, say, a guitar string (Figure 7.16). The waves are described as standing, or stationary, because they do not travel along the string. Some points on the string, called ***nodes,*** do not move at all, that is, *the amplitude of the wave at these points is zero*. There is a node at each end, and there may be nodes between the ends. The greater the frequency of vibration, the shorter the wavelength of the standing wave and the greater the number of nodes. As Figure 7.16 shows, there can be only certain wavelengths in any of the allowed motions of the string.

De Broglie argued that if an electron does behave like a standing wave in the hydrogen atom, the length of the wave must fit the circumference of the orbit exactly (Figure 7.17). Otherwise the wave would partially cancel itself on each successive orbit; eventually the amplitude of the wave would be reduced to zero, and the wave would not exist.

The relation between the circumference of an allowed orbit ($2\pi r$) and the wavelength (λ) of the electron is given by

$$2\pi r = n\lambda \tag{7.6}$$

where r is the radius of the orbit, λ is the wavelength of the electron wave, and $n = 1, 2, 3, \ldots$. Because n is an integer, it follows that r can have only certain values as n increases from 1 to 2 to 3 and so on. And since the energy of the electron depends on the size of the orbit (or the value of r), its value must be quantized.

†Louis Victor Pierre Raymond Duc de Broglie (1892–1977). French physicist. Member of an old and noble family in France, he held the title of a prince. In his doctoral dissertation, he proposed that matter and radiation have both wave and particle properties. For this work, de Broglie was awarded the Nobel Prize in physics in 1929.

De Broglie's reasoning led to the conclusion that waves can behave like particles and particles can exhibit wave properties. De Broglie deduced that the particle and wave properties are related by the expression

Note that the left side of Equation (7.7) involves the wavelike property of wavelength, whereas the right side makes reference to mass, a distinctly particlelike property.

$$\lambda = \frac{h}{mu} \tag{7.7}$$

where λ, m, and u are the wavelengths associated with a moving particle, its mass, and its velocity, respectively. Equation (7.7) implies that a particle in motion can be treated as a wave, and a wave can exhibit the properties of a particle.

Example 7.5 compares the wavelength of a moving macroscopic object, a tennis ball, with that of an electron in motion.

EXAMPLE 7.5

Calculate the wavelength of the "particle" in the following two cases: (a) The fastest serve in tennis is about 140 miles per hour, or 62 m/s. Calculate the wavelength associated with a 6.0×10^{-2}-kg tennis ball traveling at this speed. (b) Calculate the wavelength associated with an electron moving at 62 m/s.

Answer

(a) Using Equation (7.7) we write

$$\begin{aligned}\lambda &= \frac{h}{mu}\\ &= \frac{6.63 \times 10^{-34}\ \text{J s}}{6.0 \times 10^{-2}\ \text{kg} \times 62\ \text{m/s}}\end{aligned}$$

The conversion factor is $1\ \text{J} = 1\ \text{kg m}^2/\text{s}^2$ (see Section 1.7). Therefore

$$\lambda = 1.8 \times 10^{-34}\ \text{m}$$

This is an exceedingly small wavelength, since the size of an atom itself is on the order of 1×10^{-10} m. For this reason, the wave properties of such a tennis ball cannot be detected by any existing measuring device.

(b) In this case

$$\begin{aligned}\lambda &= \frac{h}{mu}\\ &= \frac{6.63 \times 10^{-34}\ \text{J s}}{9.1095 \times 10^{-31}\ \text{kg} \times 62\ \text{m/s}}\end{aligned}$$

where 9.1095×10^{-31} kg is the mass of an electron. Proceeding as in (a) we obtain

$$\begin{aligned}\lambda &= 1.2 \times 10^{-5}\ \text{m}\\ &= 1.2 \times 10^{-5}\ \text{m} \times \frac{1 \times 10^{9}\ \text{nm}}{1\ \text{m}}\\ &= 1.2 \times 10^{4}\ \text{nm}\end{aligned}$$

A wavelength of 1.2×10^4 nm falls in the infrared region.

Similar problems: 7.44, 7.45, 7.46.

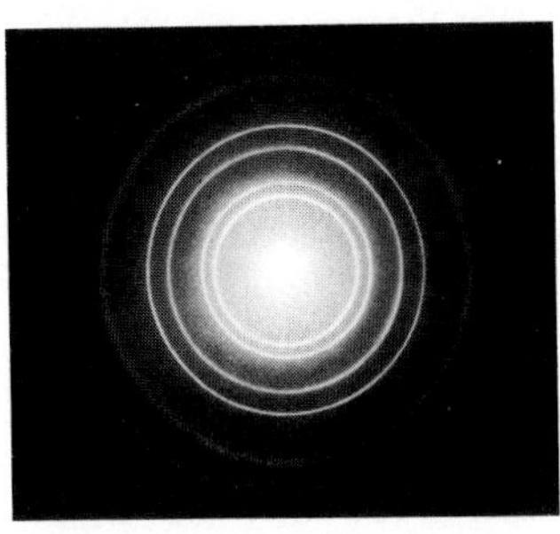

Figure 7.18 *Patterns made by a beam of electrons passing through aluminum. The patterns are similar to those obtained with X rays, which are known to be waves.*

Example 7.5 shows that although de Broglie's equation can be applied to diverse systems, the wave properties become observable only for submicroscopic objects. This distinction is brought about by the smallness of Planck's constant, h, which appears in the numerator in Equation (7.7).

Shortly after de Broglie advanced his equation, Clinton Davisson† and Lester Germer‡ in the United States and G. P. Thomson§ in England demonstrated that electrons do indeed possess wavelike properties. By passing a beam of electrons through a thin piece of gold foil, Thomson obtained a set of concentric rings on a screen, similar to the pattern observed when X rays (which are waves) were used. Figure 7.18 shows such patterns for aluminum.

The Chemistry in Action section below describes how the wave properties of electrons increase the magnification power of microscopes.

† Clinton Joseph Davisson (1881–1958). American physicist. He was awarded the Nobel Prize in physics in 1937 for demonstrating wave properties of electrons.

‡ Lester Halbert Germer (1896–1972). American physicist. Discoverer (with Davisson) of the wave properties of electrons.

§ George Paget Thomson (1892–1975). English physicist. Son of J. J. Thomson, he received the Nobel Prize in physics in 1937 for demonstrating wave properties of electrons.

CHEMISTRY IN ACTION

The Electron Microscope

The electron microscope is an extremely valuable application of the wavelike properties of electrons. According to the laws of optics, it is impossible to form an image of an object that is smaller than half the wavelength of the light used for the observation. Since the range of visible light wavelengths starts at around 400 nm, or 4×10^{-5} cm, we cannot see anything smaller than 2×10^{-5} cm. In order to see objects on the atomic and molecular scale we can in principle use X rays, whose wavelengths range from about 0.01 nm to 10 nm. However, X rays cannot be focused, so they do not produce well-formed images. Because electrons are charged particles, they can be focused easily by applying an electric field or a magnetic field in the same way the image on a TV screen is focused. According to Equation (7.7), the wavelength of an electron is inversely proportional to its velocity. By accelerating electrons to very high velocities, it is possible to obtain wavelengths as short as 0.004 nm. The range of wavelengths available and the relative ease of focusing and operation have made electron microscopes one of the most powerful tools in chemical and biological research. Figure 7.19 shows the electron micrograph of silicon atoms.

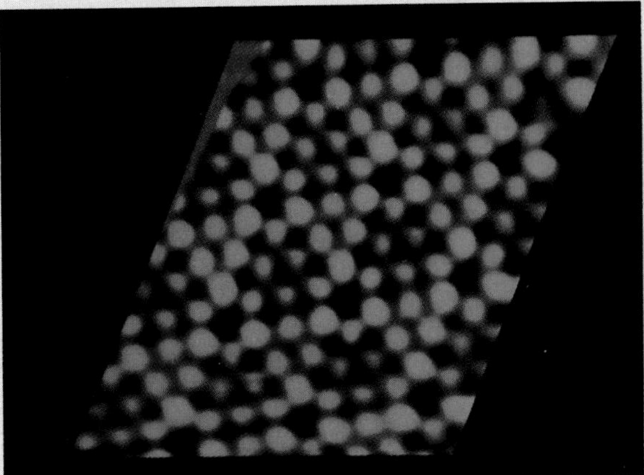

Figure 7.19 *The image of the surface of silicon, magnified about 10 million times, obtained using a scanning tunneling electron microscope.*

7.5 Quantum Mechanics

In reality, Bohr's theory was able to account for the observed emission spectra of He^+ and Li^{2+} ions, as well as that of hydrogen. However, all three systems have one feature in common—each contains a single electron. Thus the Bohr model worked successfully only for the hydrogen atom and for "hydrogenlike ions."

The spectacular success of Bohr's theory was followed by a series of disappointments. For example, Bohr's approach could not account for the emission spectra of atoms containing more than one electron, such as atoms of helium and lithium. The theory also could not explain the appearance of the extra lines in the hydrogen emission spectrum that are observed when a magnetic field is applied. Another problem arose with the discovery that electrons are wavelike: How can the "position" of a wave be specified? We can speak of the amplitude at a certain point in the wave (see Figure 7.2), but we cannot define its precise location because a wave extends in space.

One of the most important consequences of the dual nature of matter is the uncertainty principle, which was formulated by Werner Heisenberg.† The ***Heisenberg uncertainty principle*** states that *it is impossible to know simultaneously both the momentum* (p, defined as mass times velocity) *and the position of a particle* (x) *with certainty.* Stated mathematically, we write

The ≥ sign means that the product $\Delta x \Delta p$ can be greater than or equal to $h/4\pi$, but it can never be smaller than $h/4\pi$.

$$\Delta x \Delta p \geq \frac{h}{4\pi} \tag{7.8}$$

where Δx and Δp are the uncertainties in measuring the position and the momentum, respectively. Equation (7.8) says that if we make the measurement of the momentum of a particle more precise (that is, if we make Δp a small quantity), our knowledge of the position will become correspondingly less precise (that is, Δx will become larger). Similarly, if the position of the particle is known more precisely, then its momentum measurement must be less precise. Applying the Heisenberg uncertainty principle to the hydrogen atom, we see that in reality the electron does not orbit the nucleus in a well-defined path, as Bohr thought. If it did, we could determine precisely both the position of the electron (from the radius of the orbit) and its momentum (from its kinetic energy) at the same time, in violation of the uncertainty principle.

To be sure, Bohr made a significant contribution to our understanding of atoms, and his suggestion that the energy of an electron in an atom is quantized remains unchallenged. But his theory does not provide a complete description of electronic behavior in atoms. When scientists realized this, they began to search for a fundamental equation that would describe the behavior and energies of submicroscopic particles in general, an equation analogous to Newton's laws of motion for macroscopic objects. In 1926 Erwin Schrödinger,‡ using a complicated mathematical technique, formulated the sought-after equation. The *Schrödinger equation* requires advanced calculus to solve, and we will not discuss it here. It is important for us to know, however, that the equation incorporates both particle behavior, in terms of mass m, and wave behavior, in terms of a *wave function* ψ (psi), which depends on the location in space of the system (such as an electron in an atom).

The wave function itself has no direct real physical meaning. However, the square of the wave function, ψ^2, is related to the probability of finding the electron in a certain region in space. We can think of ψ^2 as the probability per unit volume such that the product of ψ^2 and a small volume (called a *volume element*) yields the probability of

† Werner Karl Heisenberg (1901–1976). German physicist. One of the founders of modern quantum theory, Heisenberg received the Nobel Prize in physics in 1932.

‡ Erwin Schrödinger (1887–1961). Austrian physicist. He formulated wave mechanics, which laid the foundation for modern quantum theory. He received the Nobel Prize in physics in 1933.

finding the electron within that volume. (The reason for specifying a small volume is that ψ^2 varies from one region of space to another, but its value can be assumed constant within a small volume.) The total probability of locating the electron in a given volume (for example, around the nucleus of an atom) is then given by the sum of all the products of ψ^2 and the corresponding volume elements.

The idea of relating ψ^2 to the notion of probability stemmed from a wave theory analogy. According to wave theory, the intensity of light is proportional to the square of the amplitude of the wave, or ψ^2. The most likely place to find a photon is where the intensity is greatest, that is, where the value of ψ^2 is greatest. A similar argument thus followed for associating ψ^2 with the likelihood of finding an electron in regions surrounding the nucleus.

Schrödinger's equation began a new era in physics and chemistry, for it launched a new field, *quantum mechanics* (also called *wave mechanics*). We now refer to the developments in quantum theory from 1913—the time Bohr presented his analysis for the hydrogen atom—to 1926 as "old quantum theory."

7.6 Applying the Schrödinger Equation to the Hydrogen Atom

To see how quantum mechanics changed our view of the atom, let us look at the very simplest atom, the hydrogen atom, with one proton and one electron. The Schrödinger equation, when solved for the hydrogen atom, yields two valuable pieces of information: It specifies the possible energy states the electron can occupy, and it identifies the corresponding wave functions (ψ) of the electron associated with each energy state. These energy states and wave functions are characterized by a set of quantum numbers (to be discussed shortly). Recall that the probability of finding an electron in a certain region is given by the square of the wave function, ψ^2. Thus once we know the values of ψ and the energies, we can calculate ψ^2 and construct a comprehensive view of the hydrogen atom.

This information about the hydrogen atom is useful but not quite enough. The world of chemical substances and reactions involves systems considerably more complex than simple atomic species like hydrogen. However, it turns out that the Schrödinger equation cannot be solved exactly for any atom containing more than one electron! Thus even in the case of helium, which contains two electrons, the mathematics becomes too complex to handle. It would appear, therefore, that the Schrödinger equation suffers from the same limitation as Bohr's original theory—it can be applied in practice only to the hydrogen atom. However, the situation is not hopeless. Chemists and physicists have learned to get around this kind of difficulty by using methods of approximation. For example, although the behavior of electrons in ***many-electron atoms*** (that is, *atoms containing two or more electrons*) is not the same as that in the hydrogen atom, we assume that the difference is probably not too great. On the basis of this assumption, we can use the energies and the wave functions obtained for the hydrogen atom as good approximations of the behavior of electrons in more complex atoms. In fact, we find that this approach provides fairly good descriptions of electron behavior in complex atoms.

Although the helium atom has only two electrons, in quantum mechanics it is regarded as a many-electron atom.

Because the hydrogen atom serves as a starting point or model for all other atoms, we need a clear idea of the quantum mechanical description of this system. The solution of the Schrödinger equation shows that the energies an electron can possess in the hydrogen atom are given by the same expression as that obtained by Bohr [see Equation (7.4)]. Both Bohr's theory and quantum mechanics, therefore, show that the energy of an electron in the hydrogen atom is quantized. They differ, however, in their description of the electron's behavior with respect to the nucleus.

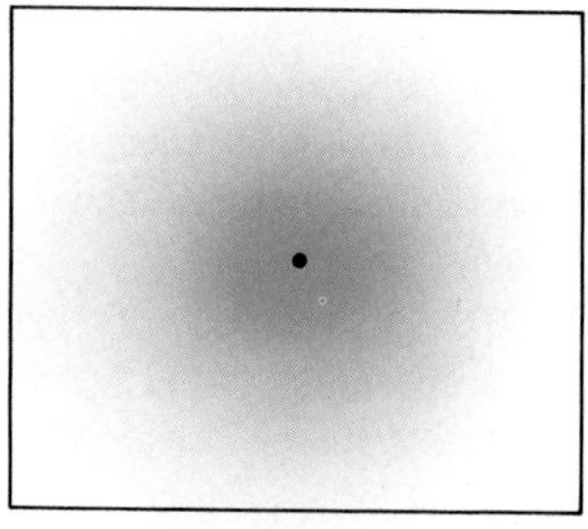

Figure 7.20 *A representation of the electron density distribution surrounding the nucleus in the hydrogen atom.*

Since an electron has no well-defined position in the atom, we find it convenient to use terms like *electron density, electron charge cloud,* or simply *charge cloud* to represent the probability concept (these terms have essentially the same meaning). Basically, ***electron density*** *gives the probability that an electron will be found at a particular region in an atom.* Regions of high electron density represent a high probability of locating the electron, whereas the opposite holds for regions of low electron density (Figure 7.20).

To distinguish the quantum mechanical description from Bohr's model, we replace "orbit" with the term ***orbital*** or ***atomic orbital.*** An orbital can be thought of as *the wave function* (ψ) *of an electron in an atom.* The square of the wave function, ψ^2, defines the distribution of electron density in space around the nucleus. When we say that an electron is in a certain orbital, we mean that the distribution of the electron density or the probability of locating the electron in space is described by the square of the wave function associated with that orbital. An atomic orbital therefore has a characteristic energy, as well as a characteristic distribution of electron density.

7.7 Quantum Numbers

Quantum mechanics tells us that three quantum numbers are required to describe the distribution of electrons in hydrogen and other atoms. These numbers are derived from the mathematical solution of the Schrödinger equation for the hydrogen atom. They are called the principal quantum number, the angular momentum quantum number, and the magnetic quantum number. These quantum numbers will be used to describe atomic orbitals and to label electrons that reside in them. A fourth quantum number that describes the behavior of a specific electron—the spin quantum number—completes the description of electrons in atoms.

The Principal Quantum Number (n)

Equation (7.4) holds only for the hydrogen atom.

The principal quantum number (n) can have integral values 1, 2, 3, and so forth; it corresponds to the quantum number in Equation (7.4). In a hydrogen atom, the value of n determines the energy of an orbital. As we will see shortly, this is not the case for a many-electron atom. The principal quantum number also relates to the average distance of the electron from the nucleus in a particular orbital. The larger the n, the greater the average distance of an electron in the orbital from the nucleus and therefore the larger the orbital (and the less stable the orbital).

The Angular Momentum Quantum Number (ℓ)

The shapes of orbitals are discussed in Section 7.8.

The angular momentum quantum number, ℓ, tells us the "shape" of the orbitals. The values of ℓ depend on the value of the principal quantum number, n. For a given value of n, ℓ has possible integral values from 0 to $(n - 1)$. If $n = 1$, there is only one possible value of ℓ; that is, $\ell = n - 1 = 1 - 1 = 0$. If $n = 2$, there are two values of ℓ, given by 0 and 1. If $n = 3$, there are three values of ℓ, given by 0, 1, and 2. The value of ℓ is generally designated by the letters $s, p, d, \ldots$ as follows:

ℓ	0	1	2	3	4	5
Name of orbital	*s*	*p*	*d*	*f*	*g*	*h*

Thus if $\ell=0$, we have an *s* orbital; if $\ell=1$, we have a *p* orbital; and so on.

The unusual sequence of letters (*s*, *p*, and *d*) has a historical origin. Physicists who studied atomic emission spectra tried to correlate the observed spectral lines with the particular energy states involved in the transitions. They noted that some of the lines were *s*harp; some were rather spread out, or ***d***iffuse; and some were very strong and hence referred to as ***p***rincipal lines. Subsequently, the initial letters of each adjective were assigned to those energy states. However, after the letter *d* and starting with the letter *f*, the orbital designations follow alphabetical order.

Remember that the "2" in 2*s* refers to the value of *n* and the "*s*" refers to the value of ℓ.

A collection of orbitals with the same value of *n* is frequently called a *shell*. One or more orbitals with the same *n* and ℓ values are referred to as *subshells*. For example, the shell with $n = 2$ is composed of two subshells, $\ell=0$ and 1 (the allowed values for $n = 2$). These subshells are called the 2*s* and 2*p* subshells.

The Magnetic Quantum Number (m_ℓ)

The magnetic quantum number, m_ℓ, describes the orientation of the orbital in space (to be discussed in Section 7.8). Within a subshell, the value of m_ℓ depends on the value of the angular momentum quantum number, ℓ. For a certain value of ℓ, there are $(2\ell + 1)$ integral values of m_ℓ, as follows:

$$-\ell, (-\ell + 1), \ldots 0, \ldots (+\ell - 1), +\ell$$

If $\ell = 0$, then $m_\ell = 0$. If $\ell = 1$, then there are $[(2 \times 1) + 1]$, or three values of m_ℓ, namely, -1, 0, and 1. If $\ell = 2$, there are $[(2 \times 2) + 1]$, or five values of m_ℓ, namely, -2, -1, 0, 1, and 2. The number of m_ℓ values indicates the number of orbitals in a subshell with a particular ℓ value.

To summarize our discussion of these three quantum numbers, let us consider the situation in which we have $n = 2$ and $\ell = 1$. The values of *n* and ℓ indicate that we have a 2*p* subshell, and in this subshell we have *three* 2*p* orbitals (because there are three values of m_ℓ, given by -1, 0, and 1).

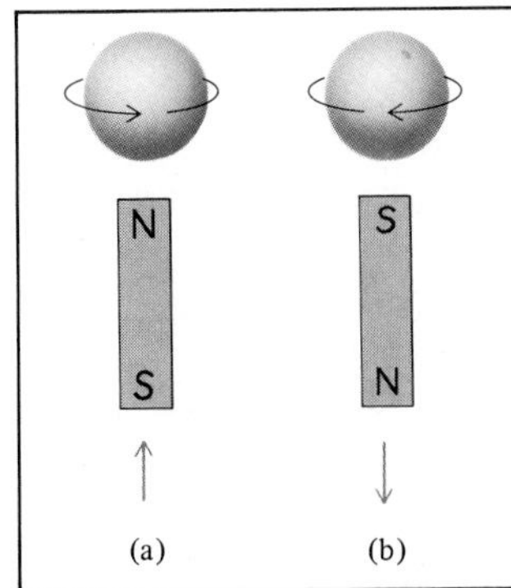

Figure 7.21 *The (a) counterclockwise and (b) clockwise spins of an electron. The magnetic fields generated by these two spinning motions are analogous to those from the two magnets. The upward and downward arrows are used to denote the two directions of spin.*

The Electron Spin Quantum Number (m_s)

Experiments on the emission spectra of hydrogen and sodium atoms indicated the need for a fourth quantum number to describe the electron in an atom. The experiments revealed that lines in the emission spectra could be split by the application of an external magnetic field. The only way physicists could explain these results was to assume that electrons act like tiny magnets. If electrons are thought of as spinning on their own axes, as Earth does, their magnetic properties can be accounted for. According to electromagnetic theory, a spinning charge generates a magnetic field, and it is this motion that causes the electron to behave like a magnet. Figure 7.21 shows the two possible spinning motions of an electron, one clockwise and the other counterclockwise. To take the electron spin into account, it is necessary to introduce a fourth quantum number, called the *electron spin quantum number* (m_s), which has the value of $+\frac{1}{2}$ or $-\frac{1}{2}$. These values correspond to the two possible spinning motions of the electron, shown in Figure 7.21.

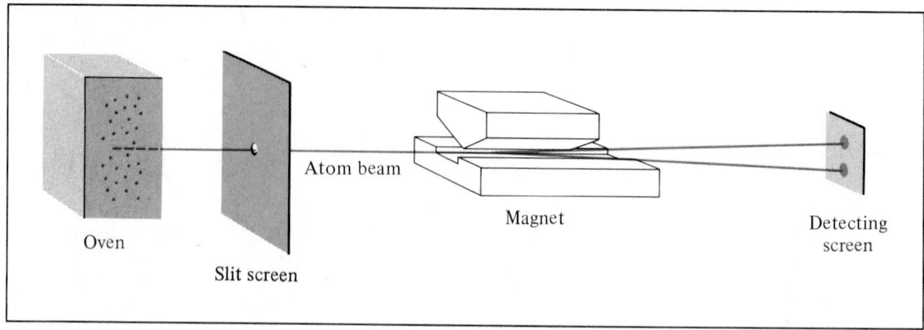

Figure 7.22 *Experimental arrangement for demonstrating the spinning motion of electrons. A beam of atoms is directed through a magnetic field. For example, when a hydrogen atom with a single electron passes through the field, it is deflected in one direction or the other, depending on the direction of the spin. In a stream consisting of many atoms, there will be equal distributions of the two kinds of spins, so that two spots of equal intensity are detected on the screen.*

Conclusive proof of electron spin was provided by Otto Stern† and Walther Gerlach‡ in 1924. Figure 7.22 shows the basic experimental arrangement. A beam of atoms generated in a hot furnace is passed through a nonhomogeneous magnetic field. The interaction between an electron and the magnetic field causes the atom to be deflected from its straight-line path. Because the spinning motion is completely random, half of the atoms will be deflected one way and half the other, and two spots of equal intensity are observed on the detecting screen.

7.8 Atomic Orbitals

Quantum numbers enable us to discuss atomic orbitals in hydrogen and many-electron atoms more fully. Table 7.2 shows the relation between quantum numbers and atomic orbitals. We see that when $\ell = 0$, $(2\ell + 1) = 1$ and there is only one value of m_ℓ, thus we have an *s* orbital. When $\ell = 1$, $(2\ell + 1) = 3$ and hence there are three values of m_ℓ

An *s* subshell has one orbital, a *p* subshell has three orbitals, and a *d* subshell has five orbitals.

Table 7.2 Relation between quantum numbers and atomic orbitals

n	ℓ	m_ℓ	Number of orbitals	Atomic orbital designations
1	0	0	1	$1s$
2	0	0	1	$2s$
	1	$-1, 0, 1$	3	$2p_x, 2p_y, 2p_z$
3	0	0	1	$3s$
	1	$-1, 0, 1$	3	$3p_x, 3p_y, 3p_z$
	2	$-2, -1, 0, 1, 2$	5	$3d_{xy}, 3d_{yz}, 3d_{xz}, 3d_{x^2-y^2}, 3d_{z^2}$
⋮	⋮	⋮	⋮	⋮

†Otto Stern (1888–1969). German physicist. He made important contributions in the study of magnetic properties of atoms and the kinetic theory of gases. Stern was awarded the Nobel Prize in physics in 1943.
‡Walther Gerlach (1889–1979). German physicist. His main area of research was in quantum theory.

or three p orbitals, labeled p_x, p_y, and p_z. When $\ell = 2$, $(2\ell + 1) = 5$ and there are five values of m_ℓ, and the corresponding five d orbitals are labeled with more elaborate subscripts. The following sections examine s, p, and d orbitals separately.

Since the wave function for an orbital has no theoretical outer limit as one moves outward from the nucleus, this raises interesting philosophical questions regarding the sizes of atoms. Rather than pondering these issues, chemists have agreed to define an atom's size operationally, as we will see in later chapters.

***s* Orbitals.** One of the important questions we ask when studying the properties of atomic orbitals is, What are the shapes of the orbitals? Strictly speaking, an orbital does not have a well-defined shape because the wave function characterizing the orbital extends from the nucleus to infinity. In that sense it is difficult to say what an orbital looks like. On the other hand, it is certainly convenient to think of orbitals in terms of specific shapes, particularly in discussing the formation of chemical bonds between atoms, as we will do in Chapters 9 and 10.

Although in principle an electron can be found anywhere, we know that most of the time it is quite close to the nucleus. Figure 7.23(a) shows a plot of the electron density in a hydrogen atomic 1s orbital versus the distance from the nucleus. As you can see, the electron density falls off rapidly as the distance from the nucleus increases. Roughly speaking, there is about a 90 percent probability of finding the electron within a sphere of radius 100 pm (1 pm = 1×10^{-12} m) surrounding the nucleus. Thus, we can represent the 1s orbital by drawing a ***boundary surface diagram*** that *encloses about 90 percent of the total electron density in an orbital,* as shown in Figure 7.23(b). A 1s orbital represented in this manner is merely a sphere.

Figure 7.24 shows boundary surface diagrams for the 1s, 2s, and 3s hydrogen

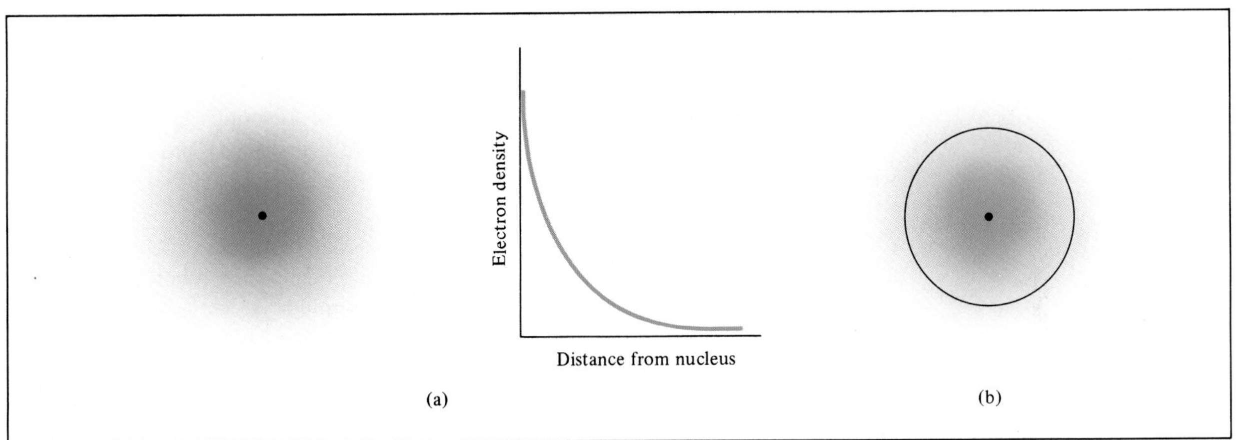

Figure 7.23 *(a) Plot of electron density in the hydrogen 1s orbital as a function of the distance from the nucleus. The electron density falls off rapidly as the distance from the nucleus increases. (b) Boundary surface diagram of the hydrogen 1s orbital.*

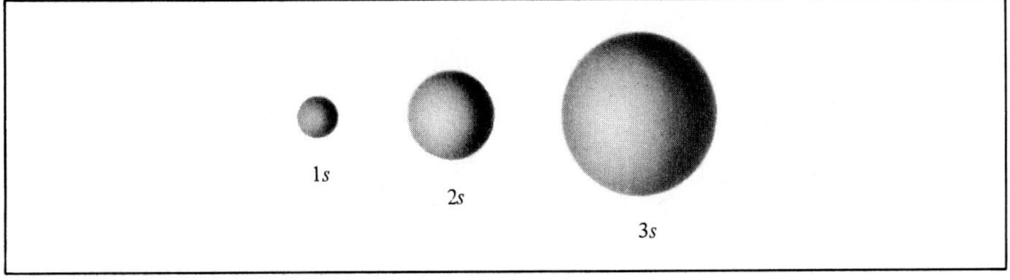

Figure 7.24 *Boundary surface diagrams of the hydrogen 1s, 2s, and 3s orbitals. Each sphere contains about 90 percent of the total electron density. All s orbitals are spherical. Roughly speaking, the size of an orbital is proportional to n^2, where n is the principal quantum number.*

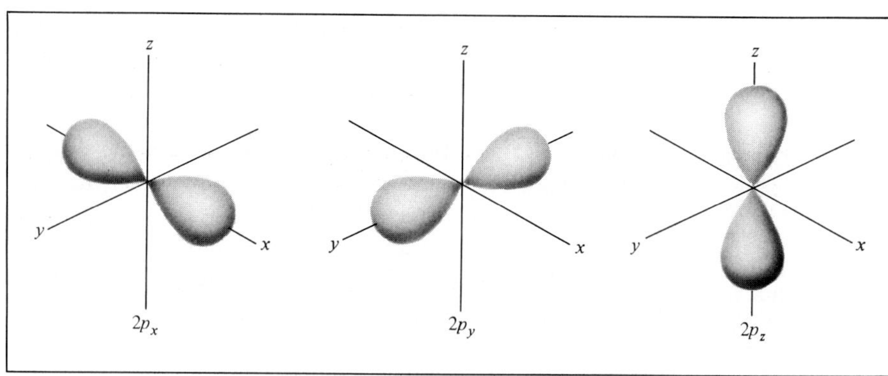

Figure 7.25 *The boundary surface diagrams of the three 2p orbitals. Except for their differing orientations, these orbitals are identical in shape and energy. The p orbitals of higher principal quantum numbers have similar shapes.*

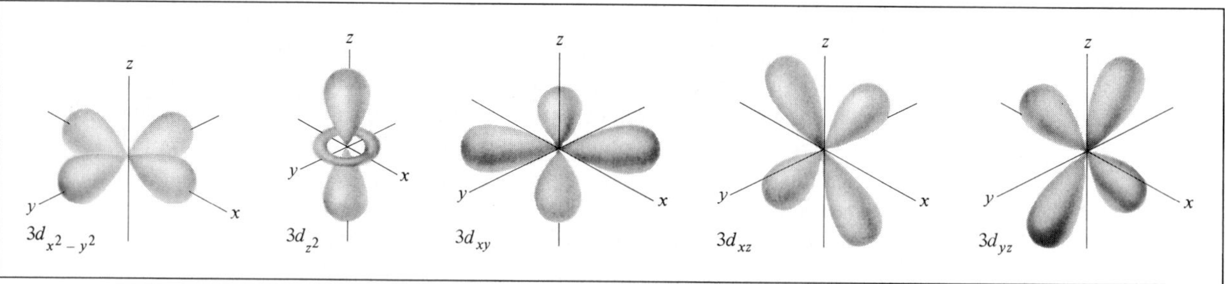

Figure 7.26 *Boundary surface diagrams of the five 3d orbitals. Although the $3d_{z^2}$ orbital looks different, it is equivalent to the other four orbitals in all other respects. The d orbitals of higher principal quantum numbers have similar shapes.*

atomic orbitals. All *s* orbitals are spherical in shape but differ in size, which increases as the principal quantum number increases. Although the details of electron density variation within each boundary surface are lost, there is no serious disadvantage. For us the most important features of atomic orbitals are their shapes and *relative* sizes, which are adequately represented by boundary surface diagrams.

***p* Orbitals.** It should be clear that the *p* orbitals start with the principal quantum number $n = 2$. If $n = 1$, then the angular momentum quantum number ℓ can assume only the value of zero; therefore, there is only a 1*s* orbital. As we saw earlier, when $\ell = 1$, the magnetic quantum number m_ℓ can have values of $-1, 0, 1$. Starting with $n = 2$ and $\ell = 1$, we therefore have three 2*p* orbitals: $2p_x$, $2p_y$, and $2p_z$ (Figure 7.25). The letter subscripts indicate the axes along which the orbitals are oriented. These three *p* orbitals are identical in size, shape, and energy; they differ from one another only in their orientations. Note, however, that there is no simple relation between the values of m_ℓ and the *x*, *y*, and *z* directions. For our purpose, you need only remember that because there are three possible values of m_ℓ, there are three *p* orbitals with different orientations.

The boundary surface diagrams of *p* orbitals in Figure 7.25 show that each *p* orbital can be thought of as two lobes; the nucleus is at the center of the *p* orbital. Like *s* orbitals, *p* orbitals increase in size from 2*p* to 3*p* to 4*p* orbital and so on.

***d* Orbitals and Other Higher-Energy Orbitals.** When $\ell = 2$, there are five values of m_ℓ, which correspond to five *d* orbitals. The lowest value of *n* for a *d* orbital is 3. Because ℓ can never be greater than $n - 1$, when $n = 3$ and $\ell = 2$, we have five 3*d* orbitals ($3d_{xy}$, $3d_{yz}$, $3d_{xz}$, $3d_{x^2-y^2}$, and $3d_{z^2}$), shown in Figure 7.26. As in the case of the *p* orbitals, the different orientations of the *d* orbitals correspond to the different values of m_ℓ, but again there is no direct correspondence between a given orientation and a particular m_ℓ value. All the $3d$ orbitals in an atom are identical in energy. The d orbitals for which n is greater than 3 ($4d$, $5d$, . . .) have similar shapes.

As we saw in Section 7.7, beyond the d orbitals are f, g, . . . orbitals. The f orbitals are important in accounting for the behavior of elements with atomic numbers greater than 57, although their shapes are difficult to represent. We will have no need to concern ourselves with orbitals having ℓ values greater than 3 (the g orbitals and beyond).

The following two examples illustrate the labeling of orbitals with quantum numbers and the calculation of total number of orbitals associated with a given principal quantum number.

EXAMPLE 7.6

List the values of n, ℓ, and m_ℓ for orbitals in the $4d$ subshell.

Answer

As we saw earlier, the number given in the designation of the subshell is the principal quantum number, so in this case $n = 4$. Since we are dealing with d orbitals, $\ell = 2$. The values of m_ℓ can vary from $-\ell$ to ℓ. Therefore, m_ℓ can be $-2, -1, 0, 1, 2$ (which corresponds to the five d orbitals).

Similar problem: 7.60.

EXAMPLE 7.7

What is the total number of orbitals associated with the principal quantum number $n = 3$?

Answer

For $n = 3$, the possible values of ℓ are 0, 1, and 2. Thus there is one $3s$ orbital ($n = 3$, $\ell = 0$, and $m_\ell = 0$); there are three $3p$ orbitals ($n = 3$, $\ell = 1$, and $m_\ell = -1, 0, 1$); there are five $3d$ orbitals ($n = 3$, $\ell = 2$, and $m_\ell = -2, -1, 0, 1, 2$). Therefore the total number of orbitals is $1 + 3 + 5 = 9$.

Similar problem: 7.65.

The Energies of Orbitals

Despite the theoretical difficulties Bohr experienced in extending the usefulness of his model of the atom, his fundamental notion of electronic energy levels has been maintained and incorporated in the quantum mechanical concepts we are reviewing here. Thus along with the shapes and sizes of atomic orbitals, we are now ready to inquire

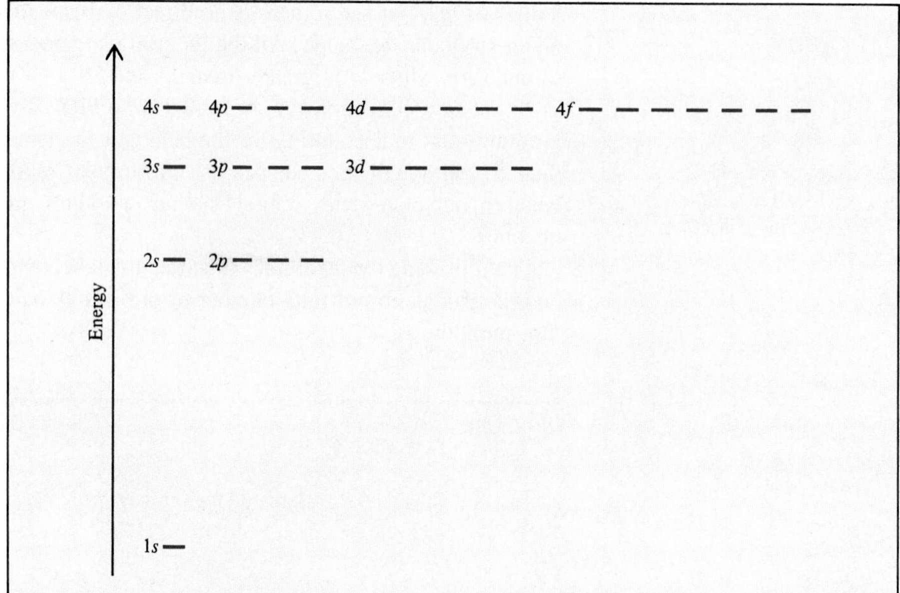

Figure 7.27 *Orbital energy levels in the hydrogen atom. Each short horizontal line here represents one orbital. Orbitals with the same principal quantum number (n) all have the same energy.*

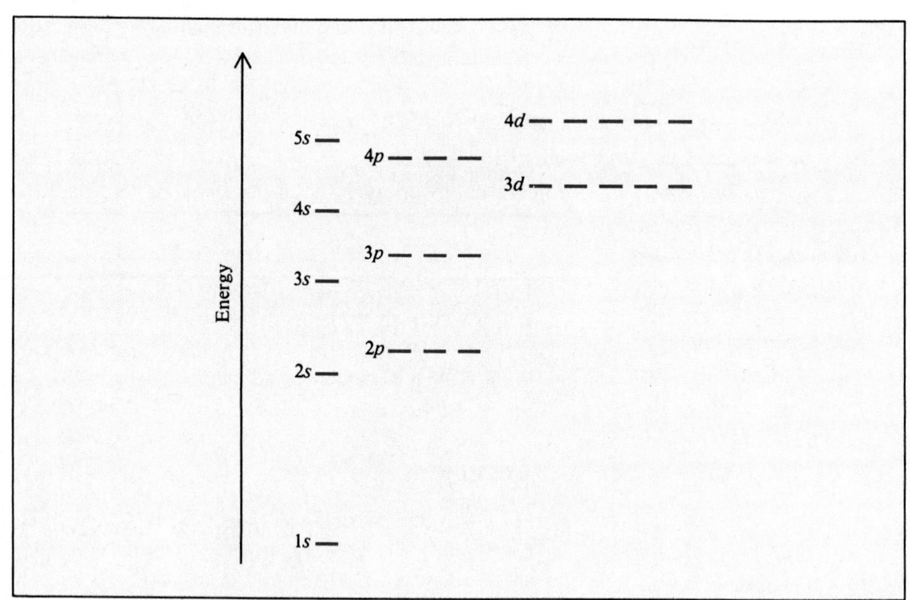

Figure 7.28 *Orbital energy levels in a many-electron atom. Note that the energy level depends on both n and ℓ values.*

into their relative energies and how these energy levels help determine the actual electronic arrangements found in atoms.

According to Equation (7.4), the energy of an electron in a hydrogen atom is determined solely by its principal quantum number. Thus the energies of hydrogen orbitals increase as follows (Figure 7.27):

$$1s < 2s = 2p < 3s = 3p = 3d < 4s = 4p = 4d = 4f < \cdots$$

Although the electron density distributions are different in the 2*s* and 2*p* orbitals, hydrogen's electron has the same energy whether it is in the 2*s* orbital or a 2*p* orbital.

Orbitals that have the same energy are said to be ***degenerate orbitals.*** The 1*s* orbital in a hydrogen atom, as we have said, corresponds to the most stable condition and is called the ground state, and an electron residing in this orbital is most strongly held by the nucleus. An electron in the 2*s*, 2*p*, or higher orbitals in a hydrogen atom is in an excited state.

The arrangement of electrons in a many-electron atom can be understood by examining the atomic orbitals they reside in. As we will see shortly, each orbital can accommodate up to two electrons.

The energy picture is different for many-electron atoms. The energy of an electron in such an atom, unlike that of the hydrogen atom, depends not only on its principal quantum number but also on its angular momentum quantum number (Figure 7.28). For many-electron atoms, the 3*d* energy level is very close to the 4*s* energy level. The total energy of an atom, however, depends not only on the sum of the orbital energies but also on the energy of repulsion between the electrons in these orbitals. It turns out that the total energy of an atom is lower when the 4*s* subshell is filled before a 3*d* subshell. Figure 7.29 is a helpful diagram of the order in which atomic orbitals are filled in a many-electron atom. We will consider specific examples in the next section.

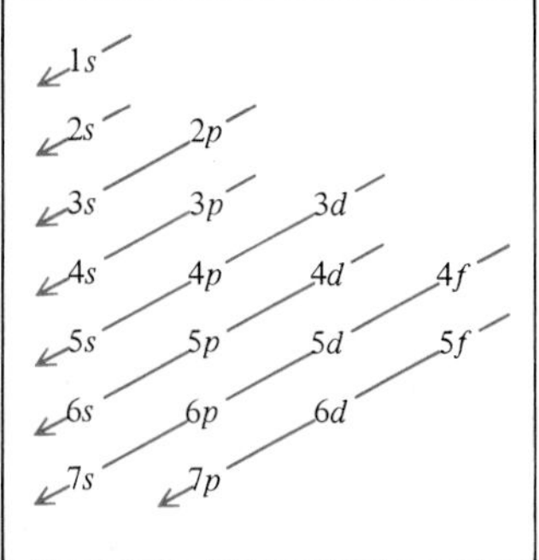

Figure 7.29 *The order in which atomic subshells are filled in a many-electron atom. Start with the 1s orbital and move downward, following the direction of the arrows. Thus the order goes as follows:* *1s* → *2s* → *2p* → *3s* → *3p* → *4s* → *3d* → . . . .

7.9 Electron Configuration

The four quantum numbers n, ℓ, m_ℓ, and m_s enable us to label completely an electron in any orbital in any atom. For example, the four quantum numbers for a 2*s* orbital electron are either $n = 2$, $\ell = 0$, $m_\ell = 0$, and $m_s = +\frac{1}{2}$ or $n = 2$, $\ell = 0$, $m_\ell = 0$, and $m_s = -\frac{1}{2}$. In practice it is inconvenient to write out all the individual quantum numbers, and so we use the simplified notation (n, ℓ, m_ℓ, m_s). For the example above, the quantum numbers are either (2, 0, 0, $+\frac{1}{2}$) or (2, 0, 0, $-\frac{1}{2}$). The value of m_s has no effect on the energy, size, shape, or orientation of an orbital, but it plays a profound role in determining how electrons are arranged in an orbital.

EXAMPLE 7.8

List the different ways to write the four quantum numbers that designate an electron in a 3*p* orbital.

In a sense, we can regard the set of four quantum numbers as the full "address" of an electron in an atom, somewhat in the same way that a postal Zip Code specifies the geographic address of an individual.

Answer

To start with, we know that the principal quantum number n is 3 and the angular momentum quantum number ℓ must be 1 (because we are dealing with a p orbital). For $\ell = 1$, there are three values of m_ℓ given by $-1, 0, 1$. Since the electron spin quantum number m_s can be either $+\frac{1}{2}$ or $-\frac{1}{2}$, we conclude that there are six possible ways to designate the electron:

$$(3, 1, -1, +\tfrac{1}{2}) \qquad (3, 1, -1, -\tfrac{1}{2})$$
$$(3, 1, 0, +\tfrac{1}{2}) \qquad (3, 1, 0, -\tfrac{1}{2})$$
$$(3, 1, 1, +\tfrac{1}{2}) \qquad (3, 1, 1, -\tfrac{1}{2})$$

Similar problems: 7.60, 7.64.

The hydrogen atom is a particularly simple system because it contains only one electron. The electron may reside in the 1*s* orbital (the ground state), or it may be found in some higher orbital (an excited state). The situation is different for many-electron atoms. To understand electronic behavior in a many-electron atom, we must first know the ***electron configuration*** of the atom. The electron configuration of an atom tells us *how the electrons are distributed among the various atomic orbitals*. We will use the first ten elements (hydrogen to neon) to illustrate the basic rules for writing the electron configurations for the *ground states* of the atoms. (Section 7.10 will describe how these rules can be applied to the remainder of the elements in the periodic table.) Recall that the number of electrons in a neutral atom is equal to its atomic number Z.

Hydrogen (Z = 1).

Figure 7.27 indicates that the electron in a ground-state hydrogen atom must be in the $1s$ orbital, so its electron configuration is $1s^1$:

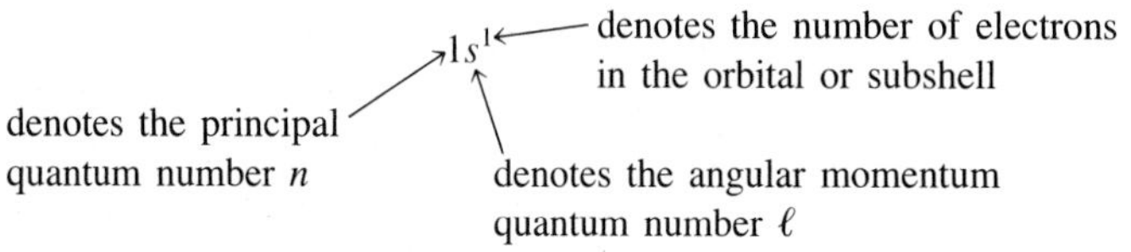

The electron configuration can also be represented by an *orbital diagram* that shows the spin of the electron (see Figure 7.21):

H [↑]
$1s^1$

Remember that the direction of electron spin has no effect on the energy of the electron.

where the upward arrow denotes one of the two possible spinning motions of the electron. (We could have represented the electron with the downward arrow.) The box represents an atomic orbital.

The Pauli Exclusion Principle

The hydrogen atom is a straightforward case because there is only one electron present. How do we deal with the electron configuration of atoms that contain more than one electron? In these cases we use the ***Pauli† exclusion principle*** as our guide. This principle states that *no two electrons in an atom can have the same four quantum numbers*. If two electrons in an atom should have the same n, ℓ, and m_ℓ values (that is, these two electrons are in the *same* atomic orbital), then they must have different values of m_s. Stating the principle in another way: only two electrons may exist in the same atomic orbital, and these electrons must have opposite spins. Consider the helium atom, which has two electrons. The three possible ways of placing two electrons in the $1s$ orbital are as follows:

Helium (Z = 2).

He	[↑↑]	[↓↓]	[↑↓]
	$1s^2$	$1s^2$	$1s^2$
	(a)	(b)	(c)

Diagrams (a) and (b) are ruled out by the Pauli exclusion principle. In (a), both electrons have the same upward spin and would have the quantum numbers $(1, 0, 0, +\frac{1}{2})$; in (b), both electrons have downward spin and would have the quantum numbers (1, 0,

†Wolfgang Pauli (1900–1958). Austrian physicist. One of the founders of quantum mechanics, Pauli was awarded the Nobel Prize in physics in 1945.

0, $-\frac{1}{2}$). Only the configuration in (c) is physically acceptable, because one electron has the quantum numbers (1, 0, 0, $+\frac{1}{2}$) and the other has (1, 0, 0, $-\frac{1}{2}$). Thus the helium atom has the following configuration:

Electrons that have opposite spins are said to be *paired*. In helium, $m_s = +\frac{1}{2}$ for one electron; $m_s = -\frac{1}{2}$ for the other.

He [↑↓]
$1s^2$

Note that $1s^2$ is read "one *s* two," not "one *s* squared."

Diamagnetism and Paramagnetism

The Pauli exclusion principle is one of the fundamental principles of quantum mechanics. It can be tested by a simple observation. If the two electrons in the 1*s* orbital of a helium atom had the same, or parallel, spins (↑↑ or ↓↓), their net magnetic fields would reinforce each other. Such an arrangement would make the helium atom ***paramagnetic*** [Figure 7.30(a)]. Paramagnetic substances are those that are *attracted by a magnet*. On the other hand, if the electron spins are paired, or antiparallel to each other (↑↓ or ↓↑), the magnetic effects cancel out and the atom is ***diamagnetic*** [Figure 7.30(b)]. Diamagnetic substances are those that are slightly *repelled by a magnet*.

By experiment we find that the helium atom is diamagnetic in the ground state, in accord with the Pauli exclusion principle. A useful general rule to keep in mind is that any atom with an odd number of electrons *must be* paramagnetic, because we need an even number of electrons for complete pairing. On the other hand, atoms containing an even number of electrons may be either diamagnetic or paramagnetic. We will see the reason for this behavior shortly.

Lithium ($Z = 3$).

As another example, consider the lithium atom, which has three electrons. The third electron cannot go into the 1*s* orbital because it would inevitably have the same four quantum numbers as one of the first two electrons. Therefore, this electron "enters" the next (energetically) higher orbital, which is the 2*s* orbital (see Figure 7.29). The electron configuration of lithium is $1s^22s^1$, and its orbital diagram is

Experimentally we find that the 2*s* orbital lies at a lower energy level than the 2*p* orbital in a many-electron atom.

Li [↑↓] [↑]
$1s^2$ $2s^1$

The lithium atom contains one unpaired electron and is therefore paramagnetic.

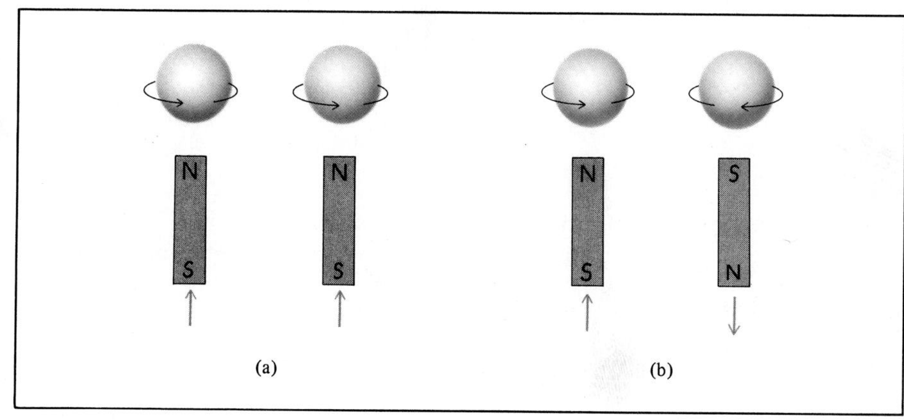

Figure 7.30 *The (a) parallel and (b) antiparallel spins of two electrons. In (a) the two magnetic fields reinforce each other, and the atom is paramagnetic. In (b) the two magnetic fields cancel each other, and the atom is diamagnetic.*

The Shielding Effect in Many-Electron Atoms

But why is the $2s$ orbital lower than the $2p$ orbital on the energy scale in a many-electron atom? In comparing the electron configurations of $1s^2 2s^1$ and $1s^2 2p^1$, we note that, in both cases, the $1s$ orbital is filled with two electrons. Because the $2s$ and $2p$ orbitals are larger than the $1s$ orbital, an electron in either of these orbitals will spend more time away from the nucleus (on the average) than an electron in the $1s$ orbital. Thus, we can speak of a $2s$ or $2p$ electron being partly "shielded" from the attractive force of the nucleus by the $1s$ electrons. The important consequence of the shielding effect is that it *reduces* the electrostatic attraction between protons in the nucleus and the electron in the $2s$ or $2p$ orbital.

The manner in which the electron density varies as we move outward from the nucleus differs for the $2s$ and $2p$ orbitals. The density for the $2s$ electron near the nucleus is greater than that for the $2p$ electron. In other words, a $2s$ electron spends more time near the nucleus than a $2p$ electron does (on the average). For this reason, the $2s$ orbital is said to be more "penetrating" than the $2p$ orbital, and it is less shielded by the $1s$ electrons. In fact, for the same principal quantum number n, the penetrating power decreases as the angular momentum quantum number ℓ increases, or

$$s > p > d > f > \cdots$$

Since the stability of an electron is determined by how strongly the electron is attracted by the nucleus, it follows that a $2s$ electron will be lower in energy than a $2p$ electron. To put it another way, less energy is required to remove a $2p$ electron than a $2s$ electron because a $2p$ electron is not held quite as strongly by the nucleus. The hydrogen atom has only one electron and, therefore, is without such a shielding effect.

Beryllium (Z = 4).

Continuing with our look into atoms of the first ten elements, the electron configuration of beryllium is $1s^2 2s^2$, or

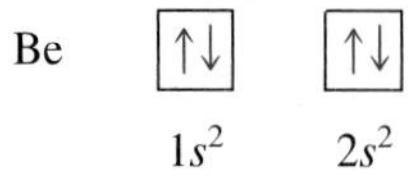

Be	↑↓	↑↓
	$1s^2$	$2s^2$

and beryllium atoms are diamagnetic as we would expect.

Boron (Z = 5).

The electron configuration of boron is $1s^2 2s^2 2p^1$, or

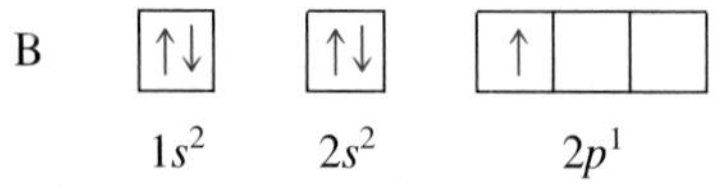

B	↑↓	↑↓	↑		
	$1s^2$	$2s^2$	$2p^1$		

Note that the unpaired electron may be in the $2p_x$, $2p_y$, or $2p_z$ orbital. The choice is completely arbitrary. This follows from the fact that the three p orbitals are equivalent in energy. As the diagram shows, boron atoms are paramagnetic.

Hund's Rule

Carbon (Z = 6).

The electron configuration of carbon is $1s^2 2s^2 2p^2$. The following are different ways for placing two electrons among three p orbitals:

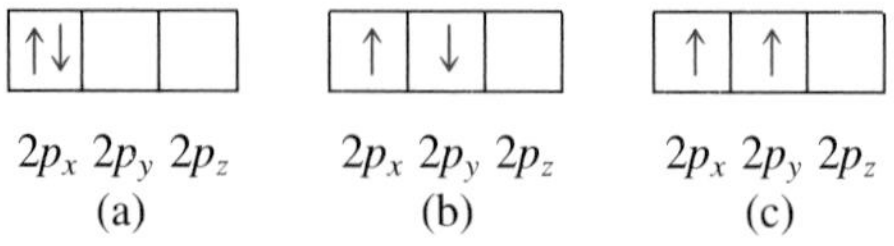

↑↓			↑	↓		↑	↑	
$2p_x$	$2p_y$	$2p_z$	$2p_x$	$2p_y$	$2p_z$	$2p_x$	$2p_y$	$2p_z$
(a)			(b)			(c)		

None of the three arrangements violates the Pauli exclusion principle, so we must determine which one will give the greatest stability. The answer is provided by ***Hund's rule,***† which states that *the most stable arrangement of electrons in subshells is the one with the greatest number of parallel spins*. The arrangement shown in (c) satisfies this condition. In both (a) and (b) the two spins cancel each other. Thus the electron configuration of carbon is $1s^22s^22p^2$, and its orbital diagram is

C	↑↓	↑↓	↑ ↑ ␣
	$1s^2$	$2s^2$	$2p^2$

Qualitatively, we can understand why (c) is preferred to (a). In (a), the two electrons are in the same $2p_x$ orbital, and their proximity results in a greater mutual repulsion than when they occupy two separate orbitals, say $2p_x$ and $2p_y$. The choice of (c) over (b) is more subtle but can be justified on theoretical grounds.

Measurements of magnetic properties provide the most direct evidence supporting specific electron configurations of elements. Advances in instrument design during the last twenty years or so enable us to determine not only whether an atom is paramagnetic but also how many unpaired electrons are present. The fact that carbon atoms are paramagnetic, each containing two unpaired electrons, is in accord with Hund's rule.

Nitrogen (Z = 7). Continuing, the electron configuration of nitrogen is $1s^22s^22p^3$, and the electrons are arranged as follows:

N	↑↓	↑↓	↑ ↑ ↑
	$1s^2$	$2s^2$	$2p^3$

Again, Hund's rule dictates that all three $2p$ electrons have spins parallel to one another; the nitrogen atom is therefore paramagnetic, containing three unpaired electrons.

Oxygen (Z = 8). The electron configuration of oxygen is $1s^22s^22p^4$. An oxygen atom is paramagnetic because it has two unpaired electrons:

O	↑↓	↑↓	↑↓ ↑ ↑
	$1s^2$	$2s^2$	$2p^4$

Fluorine (Z = 9). The electron configuration of fluorine is $1s^22s^22p^5$. The nine electrons are arranged as follows:

F	↑↓	↑↓	↑↓ ↑↓ ↑
	$1s^2$	$2s^2$	$2p^5$

The fluorine atom is thus paramagnetic, having one unpaired electron.

Neon (Z = 10). In neon the $2p$ orbitals are completely filled. The electron configuration of neon is $1s^22s^22p^6$, and *all* the electrons are paired, as follows:

Ne	↑↓	↑↓	↑↓ ↑↓ ↑↓
	$1s^2$	$2s^2$	$2p^6$

† Frederick Hund (1896–). German physicist. Hund's work was mainly in quantum mechanics. He also helped to develop the molecular orbital theory which explains the nature of chemical bonding.

The neon atom is thus predicted to be diamagnetic, which is in accord with experimental observation.

General Rules for Assigning Electrons to Atomic Orbitals

Based on the preceding examples we can formulate some general rules for determining the maximum number of electrons that can be assigned to the various subshells and orbitals for a given value of n:

- Each shell or principal level of quantum number n contains n subshells. For example, if $n = 2$, then there are two subshells (two values of ℓ) of angular momentum quantum numbers 0 and 1.
- Each subshell of quantum number ℓ contains $2\ell + 1$ orbitals. For example, if $\ell = 1$, then there are three p orbitals.
- No more than two electrons can be placed in each orbital. Therefore, the maximum number of electrons is simply twice the number of orbitals that are employed.

The following examples illustrate the procedure for calculating the number of electrons in orbitals and labeling electrons with the four quantum numbers.

EXAMPLE 7.9

What is the maximum number of electrons that can be present in the principal level for which $n = 3$?

Answer

When $n = 3$, then $\ell = 0$, 1, and 2. The number of orbitals for each value of ℓ is given by

Value of ℓ	*Number of orbitals* $(2\ell + 1)$
0	1
1	3
2	5

The total number of orbitals is nine. Since each orbital can accommodate two electrons, the maximum number of electrons that can reside in the orbitals is 2×9, or 18.

This result can be generalized by the formula $2n^2$. Here we have $n = 3$, so $2(3^2) = 18$.

Similar problems: 7.66, 7.67, 7.68.

EXAMPLE 7.10

An oxygen atom has a total of eight electrons. Write the four quantum numbers for each of the eight electrons in the ground state.

Answer

We start with $n = 1$, so $\ell = 0$, a subshell corresponding to the 1*s* orbital. This orbital can accommodate a total of two electrons. Next, $n = 2$, and ℓ may be either 0 or 1. The $\ell = 0$ subshell contains one 2*s* orbital, which can accommodate a total of two electrons. The remaining four electrons are placed in the $\ell = 1$ subshell, which contains three 2*p* orbitals. The orbital diagram is

O [↑↓] [↑↓] [↑↓|↑|↑]

$1s^2$ $2s^2$ $2p^4$

The results are summarized in the following table:

Electron	n	ℓ	m_ℓ	m_s	*Orbital*
1	1	0	0	$+\frac{1}{2}$	1*s*
2	1	0	0	$-\frac{1}{2}$	
3	2	0	0	$+\frac{1}{2}$	2*s*
4	2	0	0	$-\frac{1}{2}$	
5	2	1	−1	$+\frac{1}{2}$	$2p_x$, $2p_y$, $2p_z$
6	2	1	0	$+\frac{1}{2}$	
7	2	1	1	$+\frac{1}{2}$	
8	2	1	1	$-\frac{1}{2}$	

Of course, the placement of the eighth electron in the orbital, labeled $m_\ell = 1$, is completely arbitrary. It would be equally correct to assign it to $m_\ell = 0$ or $m_\ell = -1$.

Similar problem: 7.91.

At this point let's summarize what our examination has revealed about ground-state electron configurations and the properties of electrons in atoms:

- No two electrons in the same atom can have the same four quantum numbers. This is the Pauli exclusion principle.
- Each orbital can be occupied by a maximum of two electrons. They must have opposite spins, or different electron spin quantum numbers.
- The most stable arrangement of electrons in a subshell is the one that has the greatest number of parallel spins. This is Hund's rule.
- Atoms in which one or more electrons are unpaired are paramagnetic. Atoms in which all the electron spins are paired are diamagnetic.
- In a hydrogen atom, the energy of the electron depends only on its principal quantum number n. In a many-electron atom, the energy of an electron depends on both n and its angular momentum quantum number ℓ.
- In a many-electron atom the subshells are filled in the order shown in Figure 7.29.
- For electrons of the same principal quantum number, their penetrating power, or proximity to the nucleus, decreases in the order $s > p > d > f$. This means that, for example, more energy is required to separate a 3*s* electron from an atom than is required for a 3*p* electron.

7.10 The Building-Up Principle

In this final section of the chapter we will extend the rules used in writing electron configurations for the first ten elements to the rest of the elements. This process is based on the Aufbau principle—the German word *Aufbau* means "building up." The ***Aufbau principle*** is based on the fact that *as protons are added one by one to the nucleus to build up the elements, electrons are similarly added to the atomic orbitals.* Through this process we gain a detailed knowledge about the ground-state electron configurations of the elements. As we will see later, a knowledge of the electron configurations helps us understand and predict the properties of the elements; it also explains why the periodic table works so well.

Table 7.3 gives the ground-state electron configurations of all the known elements from H (Z = 1) through Une (Z = 109). We have already discussed the electron configurations from hydrogen through neon in some detail. As you can see, the electron configurations of the elements from sodium (Z = 11) through argon (Z = 18) follow a pattern similar to those of lithium (Z = 3) through neon (Z = 10). (We compare only the highest filled subshells in the outermost shells.)

The "argon core" notation introduced here is an accepted and useful simplification of the full electron configuration notation. In practice, we select the noble gas element that most nearly precedes the element being considered.

As mentioned in Section 7.8, the 4*s* subshell is filled before the 3*d* subshell in a many-electron atom (see Figure 7.29). Thus the electron configuration of potassium is $1s^22s^22p^63s^23p^64s^1$. Since $1s^22s^22p^63s^23p^6$ is also the electron configuration of argon, we can simplify the electron configuration of potassium by writing $[Ar]4s^1$, where [Ar] denotes the "argon core." Similarly, we can write the electron configuration of calcium (Z = 20) as $[Ar]4s^2$. The placement of the outermost electron in the 4*s* orbital (rather than the 3*d* orbital) in potassium is strongly supported by experimental evidence. For example, the chemistry of potassium is very similar to that of lithium and sodium, the first two members of the alkali metals. The outermost electron in both lithium and sodium is in an *s* orbital (there is no ambiguity in assigning their electron configurations); therefore, we expect the last electron in potassium to occupy the 4*s* rather than the 3*d* orbital.

The elements from scandium (Z = 21) to copper (Z = 29) belong to the transition metals. ***Transition metals*** either *have incompletely filled d subshells or readily give rise to cations that have incompletely filled d subshells.* Consider the first transition metal series, from scandium through copper. Along this series the added electrons are placed in the 3*d* orbitals according to Hund's rule. However, there are two irregularities. The electron configuration of chromium (Z = 24) is $[Ar]4s^13d^5$ and not $[Ar]4s^23d^4$, as we might expect. A similar break in the pattern is observed for copper, whose electron configuration is $[Ar]4s^13d^{10}$ rather than $[Ar]4s^23d^9$. The reason for these irregularities is that a slightly greater stability is associated with the half-filled ($3d^5$) and completely filled ($3d^{10}$) subshells. Electrons in the same subshells (in this case, the *d* orbitals) have equal energy but different spatial distributions. Consequently, their shielding of one another is relatively small, and the electrons are more strongly attracted by the nucleus when they have the $3d^5$ configuration. According to Hund's rule, the orbital diagram for Cr is

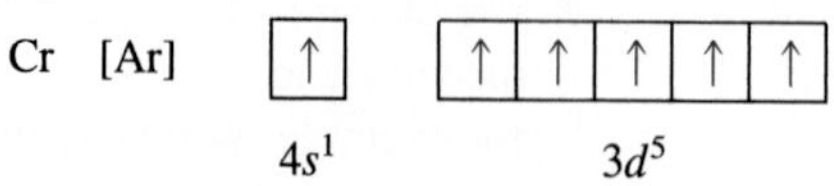

Thus, Cr has a total of six unpaired electrons. The orbital diagram for copper is

Table 7.3 The ground-state electron configurations of the elements*

Atomic number	Symbol	Electron configuration	Atomic number	Symbol	Electron configuration	Atomic number	Symbol	Electron configuration
1	H	$1s^1$	37	Rb	$[Kr]5s^1$	73	Ta	$[Xe]6s^24f^{14}5d^3$
2	He	$1s^2$	38	Sr	$[Kr]5s^2$	74	W	$[Xe]6s^24f^{14}5d^4$
3	Li	$[He]2s^1$	39	Y	$[Kr]5s^24d^1$	75	Re	$[Xe]6s^24f^{14}5d^5$
4	Be	$[He]2s^2$	40	Zr	$[Kr]5s^24d^2$	76	Os	$[Xe]6s^24f^{14}5d^6$
5	B	$[He]2s^22p^1$	41	Nb	$[Kr]5s^14d^4$	77	Ir	$[Xe]6s^24f^{14}5d^7$
6	C	$[He]2s^22p^2$	42	Mo	$[Kr]5s^14d^5$	78	Pt	$[Xe]6s^14f^{14}5d^9$
7	N	$[He]2s^22p^3$	43	Tc	$[Kr]5s^24d^5$	79	Au	$[Xe]6s^14f^{14}5d^{10}$
8	O	$[He]2s^22p^4$	44	Ru	$[Kr]5s^14d^7$	80	Hg	$[Xe]6s^24f^{14}5d^{10}$
9	F	$[He]2s^22p^5$	45	Rh	$[Kr]5s^14d^8$	81	Tl	$[Xe]6s^24f^{14}5d^{10}6p^1$
10	Ne	$[He]2s^22p^6$	46	Pd	$[Kr]4d^{10}$	82	Pb	$[Xe]6s^24f^{14}5d^{10}6p^2$
11	Na	$[Ne]3s^1$	47	Ag	$[Kr]5s^14d^{10}$	83	Bi	$[Xe]6s^24f^{14}5d^{10}6p^3$
12	Mg	$[Ne]3s^2$	48	Cd	$[Kr]5s^24d^{10}$	84	Po	$[Xe]6s^24f^{14}5d^{10}6p^4$
13	Al	$[Ne]3s^23p^1$	49	In	$[Kr]5s^24d^{10}5p^1$	85	At	$[Xe]6s^24f^{14}5d^{10}6p^5$
14	Si	$[Ne]3s^23p^2$	50	Sn	$[Kr]5s^24d^{10}5p^2$	86	Rn	$[Xe]6s^24f^{14}5d^{10}6p^6$
15	P	$[Ne]3s^23p^3$	51	Sb	$[Kr]5s^24d^{10}5p^3$	87	Fr	$[Rn]7s^1$
16	S	$[Ne]3s^23p^4$	52	Te	$[Kr]5s^24d^{10}5p^4$	88	Ra	$[Rn]7s^2$
17	Cl	$[Ne]3s^23p^5$	53	I	$[Kr]5s^24d^{10}5p^5$	89	Ac	$[Rn]7s^26d^1$
18	Ar	$[Ne]3s^23p^6$	54	Xe	$[Kr]5s^24d^{10}5p^6$	90	Th	$[Rn]7s^26d^2$
19	K	$[Ar]4s^1$	55	Cs	$[Xe]6s^1$	91	Pa	$[Rn]7s^25f^26d^1$
20	Ca	$[Ar]4s^2$	56	Ba	$[Xe]6s^2$	92	U	$[Rn]7s^25f^36d^1$
21	Sc	$[Ar]4s^23d^1$	57	La	$[Xe]6s^25d^1$	93	Np	$[Rn]7s^25f^46d^1$
22	Ti	$[Ar]4s^23d^2$	58	Ce	$[Xe]6s^24f^15d^1$	94	Pu	$[Rn]7s^25f^6$
23	V	$[Ar]4s^23d^3$	59	Pr	$[Xe]6s^24f^3$	95	Am	$[Rn]7s^25f^7$
24	Cr	$[Ar]4s^13d^5$	60	Nd	$[Xe]6s^24f^4$	96	Cm	$[Rn]7s^25f^76d^1$
25	Mn	$[Ar]4s^23d^5$	61	Pm	$[Xe]6s^24f^5$	97	Bk	$[Rn]7s^25f^9$
26	Fe	$[Ar]4s^23d^6$	62	Sm	$[Xe]6s^24f^6$	98	Cf	$[Rn]7s^25f^{10}$
27	Co	$[Ar]4s^23d^7$	63	Eu	$[Xe]6s^24f^7$	99	Es	$[Rn]7s^25f^{11}$
28	Ni	$[Ar]4s^23d^8$	64	Gd	$[Xe]6s^24f^75d^1$	100	Fm	$[Rn]7s^25f^{12}$
29	Cu	$[Ar]4s^13d^{10}$	65	Tb	$[Xe]6s^24f^9$	101	Md	$[Rn]7s^25f^{13}$
30	Zn	$[Ar]4s^23d^{10}$	66	Dy	$[Xe]6s^24f^{10}$	102	No	$[Rn]7s^25f^{14}$
31	Ga	$[Ar]4s^23d^{10}4p^1$	67	Ho	$[Xe]6s^24f^{11}$	103	Lr	$[Rn]7s^25f^{14}6d^1$
32	Ge	$[Ar]4s^23d^{10}4p^2$	68	Er	$[Xe]6s^24f^{12}$	104	Unq	$[Rn]7s^25f^{14}6d^2$
33	As	$[Ar]4s^23d^{10}4p^3$	69	Tm	$[Xe]6s^24f^{13}$	105	Unp	$[Rn]7s^25f^{14}6d^3$
34	Se	$[Ar]4s^23d^{10}4p^4$	70	Yb	$[Xe]6s^24f^{14}$	106	Unh	$[Rn]7s^25f^{14}6d^4$
35	Br	$[Ar]4s^23d^{10}4p^5$	71	Lu	$[Xe]6s^24f^{14}5d^1$	107	Uns	$[Rn]7s^25f^{14}6d^5$
36	Kr	$[Ar]4s^23d^{10}4p^6$	72	Hf	$[Xe]6s^24f^{14}5d^2$	108	Uno	$[Rn]7s^25f^{14}6d^6$
						109	Une	$[Rn]7s^25f^{14}6d^7$

*The symbol [He] is called the *helium core* and represents $1s^2$. [Ne] is called the *neon core* and represents $1s^22s^22p^6$. [Ar] is called the *argon core* and represents $[Ne]3s^23p^6$. [Kr] is called the *krypton core* and represents $[Ar]4s^23d^{10}4p^6$. [Xe] is called the *xenon core* and represents $[Kr]5s^24d^{10}5p^6$. [Rn] is called the *radon core* and represents $[Xe]6s^24f^{14}5d^{10}6p^6$.

Cu [Ar] [↑] [↑↓|↑↓|↑↓|↑↓|↑↓]

$4s^1$ $3d^{10}$

Note that extra stability is gained in this case by having the $3d$ orbitals completely filled.

For elements Zn ($Z = 30$) through Kr ($Z = 36$), the $4s$ and $4p$ subshells fill in a straightforward manner. With rubidium ($Z = 37$), electrons begin to enter the $n = 5$ energy level.

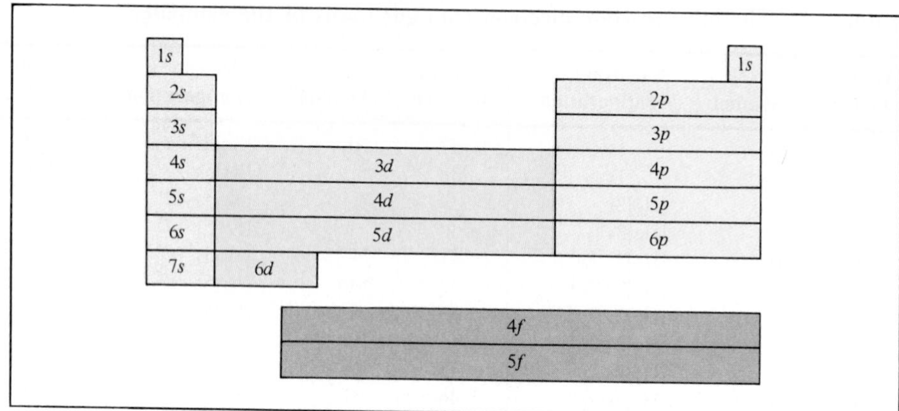

Figure 7.31 *Classification of groups of elements in the periodic table according to the type of subshell being filled with electrons.*

The electron configurations in the second transition metal series [yttrium (Z = 39) to silver (Z = 47)], on the whole, follow a pattern similar to those of the first transition metal series.

The sixth period of the periodic table begins with cesium (Z = 55) and barium (Z = 56), whose electron configurations are $[Xe]6s^1$ and $[Xe]6s^2$, respectively. Next we come to lanthanum (Z = 57). From Figure 7.29 we would expect that after filling the 6*s* orbital we would place the additional electrons in 4*f* orbitals. In reality, the energies of the 5*d* and 4*f* orbitals are very close; in fact, for lanthanum 4*f* is slightly higher in energy than 5*d*. Thus lanthanum's electron configuration is $[Xe]6s^2 5d^1$ and not $[Xe]6s^2 4f^1$. Following lanthanum are the fourteen elements [cerium (Z = 58) to lutetium (Z = 71)] that form the ***lanthanide,*** or ***rare earth, series.*** The rare earth metals *have incompletely filled 4f subshells or readily give rise to cations that have incompletely filled 4f subshells.* In this series, the added electrons are placed in 4*f* orbitals. After the 4*f* subshells are completely filled, the next electron enters the 5*d* subshell, which occurs with lutetium. Note that the electron configuration of gadolinium (Z = 64) is $[Xe]6s^2 4f^7 5d^1$ rather than $[Xe]6s^2 4f^8$. Again, in this instance extra stability is gained by having half-filled subshells ($4f^7$), as was the case for chromium.

Following completion of the lanthanide series, the third transition metal series, including lanthanum and hafnium (Z = 72) through gold (Z = 79), is characterized by the filling of the 5*d* orbitals. The 6*s* and 6*p* subshells are filled next, which takes us to radon (Z = 86).

The last row of elements belongs to the ***actinide series,*** which starts at thorium (Z = 90). *Most of these elements are not found in nature but have been synthesized.*

With few exceptions, you should now be able to write the electron configuration of any element, using Figure 7.29 as a guide. Elements that require particular care are those belonging to the transition metals, the lanthanides, and the actinides. As we noted earlier, at larger values of the principal quantum number *n*, the order of subshell filling may reverse from one element to the next. Figure 7.31 groups the elements according to the type of subshell in which the outermost electrons are placed.

EXAMPLE 7.11

Write the electron configurations for sulfur, mercury, and palladium, which is diamagnetic.

Answer

Sulfur (Z = 16)

1. Sulfur has 16 electrons.
2. It takes 10 electrons to complete the first and second periods ($1s^22s^22p^6$). This leaves us 6 electrons to fill the $3s$ orbital and partially fill the $3p$ orbitals. Thus the electron configuration of S is

$$1s^22s^22p^63s^23p^4$$

or $[Ne]3s^23p^4$.

Mercury (Z = 80)

1. Mercury has 80 electrons.
2. It takes 54 electrons to complete the fifth-period elements. Starting with the sixth period, we need 2 electrons for the $6s$ orbital, 14 electrons for the $4f$ orbitals, and 10 more for the $5d$ orbitals. The total number of electrons adds up to 80; thus the electron configuration of Hg is

$$1s^22s^22p^63s^23p^64s^23d^{10}4p^65s^24d^{10}5p^66s^24f^{14}5d^{10}$$

or simply $[Xe]6s^24f^{14}5d^{10}$.

Palladium (Z = 46)

1. Palladium has 46 electrons.
2. As in the previous cases, we need 36 electrons to complete the fourth period, leaving us with 10 more electrons to distribute among the $5s$ and $4d$ orbitals. The three choices are (a) $4d^{10}$, (b) $4d^95s^1$, and (c) $4d^85s^2$. Since atomic palladium is diamagnetic, its electron configuration must be

$$1s^22s^22p^63s^23p^64s^23d^{10}4p^64d^{10}$$

or simply $[Kr]4d^{10}$. The configurations in (b) and (c) both give paramagnetic Pd atoms.

Similar problems: 7.88, 7.89.

Summary

1. Classical physics, as expressed in Newton's laws of motion, failed to explain the behavior of atoms and molecules. A new theory was needed to account for the behavior of submicroscopic matter.
2. Since 1873, radiation had been known to possess the properties of waves—which we characterize by wavelength, frequency, and amplitude. One failure of classical physics, however, was in explaining the emission of radiation by heated solids. Max Planck solved this mystery in 1900. His theory, known as the quantum theory, revolutionized physics.
3. Planck's quantum theory states that radiant energy is emitted by atoms and molecules in small discrete amounts (quanta), rather than over a continuous range. This behavior is governed by the relationship $E = h\nu$, where E is the energy of the radiation, h is Planck's constant, and ν is the frequency of the radiation. Energy is always emitted in whole-number multiples of $h\nu$ ($1h\nu$, $2h\nu$, $3h\nu$, . . .).

4. By using quantum theory and, like Planck, making a radically new assumption, Albert Einstein in 1905 solved another mystery of physics—the photoelectric effect. Einstein proposed that light can behave like a stream of particles (photons).
5. The line spectrum of hydrogen, yet another mystery to nineteenth-century physicists, was also explained by application of the quantum theory. Niels Bohr developed a model of the hydrogen atom in which the energy of the single electron moving in various circular orbits about the nucleus was quantized—limited to certain energy values determined by an integer, the principal quantum number.
6. An electron in its most stable energy state is said to be in the ground state, and an electron at a higher energy than its most stable state is said to be in an excited state. In the Bohr model, an electron emits a photon when it drops from a higher-energy orbit (an excited state) to a lower-energy orbit (the ground state or another, less excited state). The specific energies released as shown by the lines in the hydrogen emission spectrum can be accounted for in this way.
7. The dual behavior of light in exhibiting both particle and wave properties was extended by de Broglie to all matter in motion. The wavelength of a moving particle of mass m and velocity u is given by the de Broglie equation $\lambda = h/mu$.
8. The Schrödinger equation, formulated by Erwin Schrödinger in 1926, is a fundamental equation of quantum theory that describes the motions and energies of submicroscopic particles. This equation launched quantum mechanics and a new era in physics.
9. From the Schrödinger equation we find the possible energy states of the electron in a hydrogen atom and the probability of its location in a particular region surrounding the nucleus. These results can be applied with reasonable accuracy to many-electron atoms.
10. An atomic orbital (ψ) is a function that defines the distribution of electron density (ψ^2) in space. Orbitals are represented by electron density diagrams or boundary surface diagrams.
11. Four quantum numbers are needed to characterize completely each electron in an atom: the principal quantum number n, which determines the main energy level, or shell, of the orbital; the angular momentum quantum number ℓ, which determines the shape of the orbital; the magnetic quantum number m_ℓ, which determines the orientation of the orbital in space; and the electron spin quantum number m_s, which relates only to the electron's spin on its own axis.
12. The single s orbital in each energy level is spherical and centered on the nucleus. The three p orbitals each have two lobes, and the pairs of lobes are arranged at right angles to one another. There are five d orbitals, of more complex shapes and orientations.
13. The energy of the electron in a hydrogen atom is determined solely by its principal quantum number. In many-electron atoms, both the principal quantum number and the angular momentum quantum number determine the energy of an electron.
14. No two electrons in the same atom can have the same four quantum numbers (the Pauli exclusion principle).
15. The most stable arrangement of electrons in a subshell is the one that has the greatest number of parallel spins (Hund's rule). Atoms with one or more unpaired spins are paramagnetic. Atoms with all electron spins paired are diamagnetic.
16. The periodic table classifies the elements according to their atomic numbers and thus also by the electronic configurations of their atoms.

Key Words

Actinide series, p. 298
Amplitude, p. 263
Atomic orbital, p. 282
Aufbau principle, p. 296
Boundary surface diagram, p. 285
Degenerate orbital, p. 289
Diamagnetic, p. 291
Electromagnetic wave, p. 264
Electron configuration, p. 290
Electron density, p. 282
Emission spectra, p. 269
Excited level, p. 271
Excited state, p. 271
Frequency (ν), p. 262
Ground level, p. 271
Ground state, p. 271
Heisenberg uncertainty principle, p. 280
Hund's rule, p. 293
Lanthanide series, p. 298
Line spectra, p. 269
Many-electron atom, p. 281
Node, p. 277
Orbital, p. 282
Paramagnetic, p. 291
Pauli exclusion principle, p. 290
Photon, p. 267
Quantum, p. 266
Radiation, p. 262
Rare earth series, p. 298
Transition metals, p. 296
Wave, p. 262
Wavelength (λ), p. 262

Exercises

Electromagnetic Radiation and the Quantum Theory

REVIEW QUESTIONS

7.1 Define the following terms: wave, wavelength, frequency, amplitude, electromagnetic radiation, quantum.
7.2 What are the units for wavelength and frequency for electromagnetic waves? What is the speed of light, in meters per second and miles per hour?
7.3 List the electromagnetic radiations, starting with the radiation with the longest wavelength and ending with the radiation with the shortest wavelength.
7.4 Give the high and low wavelength values that define the visible region.
7.5 Briefly explain the Planck quantum theory. What are the units for the Planck constant?
7.6 Give two familiar, everyday examples to illustrate the concept of quantization.

PROBLEMS

7.7 Convert 8.6×10^{13} Hz to nanometers, and 566 nm to hertz.
7.8 The average distance between Mars and Earth is about 1.3×10^{8} miles. How long would it take TV pictures transmitted from the *Viking* space vehicle on Mars' surface to reach Earth? (1 mile = 1.61 km.)
7.9 How long would it take a radio wave to travel from the planet Venus to Earth? (Average distance from Venus to Earth = 28 million miles.)
7.10 (a) What is the frequency of light of wavelength 456 nm? (b) What is the wavelength (in nanometers) of radiation of frequency 2.20×10^{9} Hz?
7.11 Our eyes are sensitive to light in the approximate frequency range of 4.0×10^{14} to 7.5×10^{14} Hz. Calculate the wavelengths in nm that correspond to these two frequencies.
7.12 The basic SI unit of time is the second, which is defined as 9,192,631,770 cycles of radiation associated with a certain emission process in the cesium atom. Calculate the wavelength of this radiation (to three significant figures). In which region of the electromagnetic spectrum is this wavelength found?
7.13 The basic SI unit of length is the meter, which is defined as the length equal to 1,650,763.73 wavelengths of the light emitted by a particular transition in krypton atoms. Calculate the frequency of the light to three significant figures.

The Photoelectric Effect

REVIEW QUESTIONS

7.14 Explain what is meant by the photoelectric effect.
7.15 What are photons? What important role did Einstein's explanation of the photoelectric effect play in the development of the particle-wave interpretation of the nature of electromagnetic radiation?

PROBLEMS

7.16 A photon has a wavelength of 624 nm. Calculate the energy of the photon in joules.
7.17 The blue color of the sky results from the scattering of sunlight by air molecules. The blue light has a frequency of about 7.5×10^{14} Hz. (a) Calculate the wavelength associated with this radiation, and (b) calculate the energy in joules of a single photon associated with this frequency.
7.18 A photon has a frequency of 6.0×10^{4} Hz. (a) Convert this frequency into wavelength (nm). Does this frequency fall in the visible region? (b) Calculate the energy (in joules) of this photon, and (c) calculate the energy (in joules) of 1 mole of photons all with this frequency.

7.19 What is the wavelength in nanometers of radiation that has an energy content of 1.0×10^3 kJ/mol? In which region of the electromagnetic spectrum is this radiation found?

7.20 When copper is bombarded with high-energy electrons, X rays are emitted. Calculate the energy (in joules) associated with the photons if the wavelength of the X rays is 0.154 nm.

7.21 A particular electromagnetic radiation has a frequency of 8.11×10^{14} Hz. (a) What is its wavelength in nanometers? in meters? (b) To what region of the electromagnetic spectrum would you assign it? (c) What is the energy (in joules) of one quantum of this radiation?

7.22 In a photoelectric experiment a student uses a light source whose frequency is greater than that needed to eject electrons from a certain metal. However, after continuously shining the light on the same area of the metal for a long period of time the student notices that the maximum kinetic energy of ejected electrons begins to decrease, even though the frequency of the light is held constant. How would you account for this behavior?

Bohr's Theory of the Hydrogen Atom

REVIEW QUESTIONS

7.23 What are emission spectra? What are line spectra?

7.24 What is meant by the phrase "quantization of energy"?

7.25 What is an energy level? Explain the difference between a ground state and an excited state.

7.26 What does it mean when we say that an atom is excited? Does this atom gain or lose energy?

7.27 Briefly describe Bohr's theory of the hydrogen atom and how it explains the appearance of an emission spectrum. How does Bohr's theory differ from concepts of classical physics?

7.28 Explain the meaning of the negative sign in Equation (7.4).

7.29 Define principal quantum number. What are the lowest and highest values of the principal quantum number?

PROBLEMS

7.30 Explain why elements produce their own characteristic colors when undergoing emission of photons.

7.31 When some compounds containing copper are heated in a flame, green light is emitted. How would you determine whether the light is of one wavelength or a mixture of two or more wavelengths?

7.32 Would it be possible for a fluorescent material to emit radiation in the ultraviolet region after absorbing visible light? Explain your answer.

7.33 Explain how astronomers are able to tell which elements are present in distant stars by analyzing the electromagnetic radiation emitted by the stars.

7.34 Consider the following energy levels of a hypothetical atom:

E_4 ———— -1.0×10^{-19} J
E_3 ———— -5.0×10^{-19} J
E_2 ———— -10×10^{-19} J
E_1 ———— -15×10^{-19} J

(a) What is the wavelength of the photon needed to excite an electron from E_1 to E_4? (b) What is the value (in joules) of the quantum of energy of a photon needed to excite an electron from E_2 to E_3? (c) When an electron drops from the E_3 level to the E_1 level, the atom is said to undergo emission. Calculate the wavelength of the photon emitted in this process.

7.35 Calculate the wavelength of a photon emitted by a hydrogen atom when its electron drops from the $n = 5$ state to the $n = 3$ state.

7.36 The first line of the Balmer series occurs at a wavelength of 656.3 nm. What is the energy difference between the two energy levels involved in the emission that results in this spectral line?

7.37 Calculate the frequency and wavelength of the emitted photon when an electron undergoes a transition from the $n = 4$ to the $n = 2$ level in a hydrogen atom.

7.38 Careful spectral analysis shows that the familiar yellow light of sodium lamps (such as street lamps) is made up of photons of two wavelengths, 589.0 nm and 589.6 nm. What is the difference in energy (in joules) between photons possessing these wavelengths?

7.39 An electron in an orbit of principal quantum number n_i in a hydrogen atom makes a transition to an orbit of principal quantum number 2. The photon emitted has a wavelength of 434 nm. Calculate n_i.

Particle-Wave Duality

REVIEW QUESTIONS

7.40 Explain what is meant by the statement that matter and radiation have a "dual nature."

7.41 How does de Broglie's hypothesis account for the fact that the energies of the electron in a hydrogen atom are quantized?

7.42 Why does Equation (7.7) apply only to submicroscopic particles such as electrons and atoms and not to macroscopic objects?

7.43 Does a baseball in flight possess wave properties? If so, why can we not determine its wave properties?

PROBLEMS

7.44 Calculate the wavelength (in nm) associated with a beam of neutrons moving at 4.00×10^3 cm/s. (Mass of a neutron $= 1.675 \times 10^{-27}$ kg.)

7.45 What is the de Broglie wavelength in cm of a 12.4-g

hummingbird flying at 1.20×10^2 mph? (1 mile = 1.61 km.)

7.46 What is the de Broglie wavelength associated with a 2.5-g Ping-Pong ball traveling at 35 mph?

7.47 What properties of electrons are used in the operation of an electron microscope?

Quantum Mechanical Treatment of the Hydrogen Atom

REVIEW QUESTIONS

7.48 What are the inadequacies of Bohr's theory?

7.49 What is the Heisenberg uncertainty principle? What is the Schrödinger equation?

7.50 What is the physical significance of the wave function?

7.51 How is the concept of electron density used to describe the position of an electron in the quantum mechanical treatment of an atom?

7.52 Define atomic orbital. How does an atomic orbital differ from an orbit?

7.53 Describe the characteristics of an *s* orbital, a *p* orbital, and a *d* orbital.

7.54 Why is a boundary surface diagram useful in representing an atomic orbital?

7.55 Describe the four quantum numbers used to characterize an electron in an atom.

7.56 Which quantum number defines a shell? Which quantum numbers define a subshell?

7.57 Which of the four quantum numbers (n, ℓ, m_ℓ, m_s) determine the energy of an electron in (a) a hydrogen atom and (b) a many-electron atom?

PROBLEMS

7.58 An electron in a certain atom is in the $n = 2$ quantum level. List the possible values of ℓ and m_ℓ that it can have.

7.59 An electron in an atom is in the $n = 3$ quantum level. List the possible values of ℓ and m_ℓ that it can have.

7.60 Give the values of the quantum numbers associated with the following orbitals: (a) 2*p*, (b) 3*s*, (c) 5*d*.

7.61 Discuss the similarities and differences between a 1*s* and a 2*s* orbital.

7.62 What is the difference between a $2p_x$ and a $2p_y$ orbital?

7.63 Draw the shapes (boundary surfaces) of the following orbitals: (a) p_y, (b) d_{z^2}, (c) $d_{x^2-y^2}$. (Show coordinate axes in your sketches.)

7.64 For the following subshells give the values of the quantum numbers (n, ℓ, and m_ℓ) and the number of orbitals in each subshell: (a) 4*p*, (b) 3*d*, (c) 3*s*, (d) 5*f*.

7.65 List all the possible subshells and orbitals associated with the principal quantum number n, if $n = 6$.

7.66 Calculate the total number of electrons that can occupy (a) one *s* orbital, (b) three *p* orbitals, (c) five *d* orbitals, (d) seven *f* orbitals.

7.67 What is the total number of electrons that can be held in all orbitals having the same principal quantum number n?

7.68 What is the maximum number of electrons that can be found in each of the following subshells? 3*s*, 3*d*, 4*p*, 4*f*, 5*f*.

7.69 Indicate the total number of (a) *p* electrons in N (Z = 7); (b) *s* electrons in Si (Z = 14); and (c) 3*d* electrons in S (Z = 16).

7.70 Make a chart of all allowable orbitals in the first four principal energy levels of the hydrogen atom. Designate each by type (for example, *s*, *p*) and indicate how many orbitals of each type there are.

7.71 Why do the 3*s*, 3*p*, and 3*d* orbitals have the same energy in a hydrogen atom but different energies in a many-electron atom?

7.72 For each of the following pairs of hydrogen orbitals, indicate which is higher in energy: (a) 1*s*, 2*s*; (b) 2*p*, 3*p*; (c) $3d_{xy}$, $3d_{yz}$; (d) 3*s*, 3*d*; (e) 5*s*, 4*f*.

7.73 Which orbital in each of the following pairs is lower in energy in a many-electron atom? (a) 2*s*, 2*p*; (b) 3*p*, 3*d*; (c) 3*s*, 4*s*; (d) 4*d*, 5*f*.

Electron Configuration and the Periodic Table

REVIEW QUESTIONS

7.74 Define the following terms: electron configuration, the Pauli exclusion principle, Hund's rule.

7.75 Explain the meaning of diamagnetic and paramagnetic.

7.76 Give an example of an atom that is diamagnetic and one that is paramagnetic.

7.77 Explain the meaning of the symbol $4d^6$.

7.78 What is meant by the term "shielding of electrons" in an atom? Using the Li atom as an example, describe the effect of shielding on the energy of electrons in an atom.

7.79 Define and give an example for each of the following terms: transition metals, lanthanides, actinides.

7.80 Explain why the ground-state electron configurations for Cr and Cu are different from what we might expect.

7.81 Explain what is meant by a noble gas core. Write the electron configuration of a "xenon core."

7.82 Comment on the correctness of the following statement: The probability of finding two electrons with the same four quantum numbers in an atom is zero.

PROBLEMS

7.83 Which of the following sets of quantum numbers are unacceptable? Explain your answers. (a) $(1, 0, \frac{1}{2}, -\frac{1}{2})$, (b) $(3, 0, 0, +\frac{1}{2})$, (c) $(2, 2, 1, +\frac{1}{2})$, (d) $(4, 3, -2, +\frac{1}{2})$, (e) $(3, 2, 1, 1)$.

7.84 Predict the number of spots you would observe on the detecting screen in a Stern-Gerlach experiment for each of the following atoms: He, Li, F, Ne.

7.85 The atomic number of an element is 73. Are the atoms of this element diamagnetic or paramagnetic?

7.86 Indicate the number of unpaired electrons present in each of the following atoms: B, Ne, P, Sc, Mn, Se, Zr, Ru, Cd, I, W, Pb, Ce, Ho.

7.87 Which of the following species has the most unpaired electrons? S^+, S, or S^-. Explain how you arrive at your answer.

7.88 Write the ground-state electron configurations for the following elements: B, V, Ni, As, I, Au.

7.89 Write the ground-state electron configurations for the following elements: Ge, Fe, Zn, Ru, W, Tl.

7.90 The ground-state electron configurations listed here are *incorrect*. Explain what mistakes have been made in each and write the correct electron configurations.

Al $1s^22s^22p^43s^23p^3$
B $1s^22s^22p^6$
F $1s^22s^22p^6$

7.91 The electron configuration of a neutral atom is $1s^22s^22p^63s^2$. Write a complete set of quantum numbers for each of the electrons. Name the element.

7.92 Draw orbital diagrams for atoms with the following electron configurations: (a) $1s^22s^22p^5$, (b) $1s^22s^22p^63s^23p^3$, (c) $1s^22s^22p^63s^23p^64s^23d^7$.

7.93 The electron configurations described in this chapter all refer to atoms in their ground states. An atom may absorb a quantum of energy and promote one of its electrons to a higher-energy orbital. When this happens, we say that the atom is in an excited state. The electron configurations of some excited atoms are given. Identify these atoms and write their ground-state configurations: (a) $1s^22s^22p^23d^1$, (b) $1s^12s^1$, (c) $1s^22s^22p^64s^1$, (d) $[Ar]4s^13d^{10}4p^4$, (e) $[Ne]3s^23p^43d^1$.

Miscellaneous Problems

7.94 When a compound containing cesium ion is heated in a Bunsen burner flame, photons with an energy of 4.30×10^{-19} J are emitted. What color is the cesium flame?

7.95 Distinguish carefully between the following terms: (a) wavelength and frequency, (b) wave properties and particle properties, (c) quantization of energy and continuous variation in energy.

7.96 Identify the following individuals and their contributions to the development of quantum theory: de Broglie, Einstein, Bohr, Planck, Heisenberg, Schrödinger.

7.97 Discuss the current view of the correctness of the following statements. (a) The electron in the hydrogen atom is in an orbit that never brings it closer than 100 pm to the nucleus. (b) Atomic absorption spectra result from transitions of electrons from lower to higher energy levels. (c) A many-electron atom behaves somewhat like a solar system that has a number of planets.

7.98 Spectral lines of the Lyman and Balmer series do not overlap. Verify this statement by calculating the longest wavelength associated with the Lyman series and the shortest wavelength associated with the Balmer series (in nm).

7.99 What is the maximum number of electrons in an atom that can have the following quantum numbers. Label each electron with a particular orbital. (a) $n = 2$, $m_s = +\frac{1}{2}$; (b) $n = 4$, $m_\ell = +1$; (c) $n = 3$, $\ell = 2$; (d) $n = 2$, $\ell = 0$, $m_s = -\frac{1}{2}$; (e) $n = 4$, $\ell = 3$, $m_\ell = -2$.

7.100 Photodissociation of water

$$H_2O(l) + h\nu \longrightarrow H_2(g) + \tfrac{1}{2}O_2(g)$$

has been suggested as a source of hydrogen. The ΔH°_{rxn} for the reaction, calculated from thermochemical data, is 285.8 kJ per mole of water decomposed. Calculate the maximum wavelength (in nm) that would provide the necessary energy. In principle, is it feasible to use sunlight as a source of energy for this process?

7.101 Only a fraction of the electrical energy supplied to a tungsten light bulb is converted to visible light. The rest of the energy shows up as infrared radiation (that is, heat). A 75-W light bulb converts 15.0 percent of the energy supplied to it into visible light (assume the wavelength to be 550 nm). How many photons are emitted by the light bulb per second? (1 W = 1 J/s.)

7.102 Helium was discovered in the sun before it was discovered on Earth. In 1868 the French physicist P. Janssen noticed that there are some dark lines in the emission spectrum of the sun (which is very similar to Figure 7.4). (a) What kind of process will give rise to dark lines in an emission spectrum? (b) How would you confirm that a group of dark lines are associated with the element helium?

7.103 Nolan Ryan's fastballs have been clocked at about 120 mph. (a) Calculate the wavelength of a baseball (in nm) of mass 0.141 kg at this speed. (b) What is the wavelength of a hydrogen atom at the same speed? (1 mile = 1.61 km.)

7.104 A ruby laser produces radiation of wavelength 633 nm in pulses whose durations are 1.00×10^{-9} s. (a) If the laser produces 0.376 J of energy per pulse, how many photons are produced in each pulse? (b) Calculate the power (in watts) delivered by the laser per pulse. (1 W = 1 J/s)

7.105 Considering only the ground-state electron configuration, are there more diamagnetic or paramagnetic atoms of the elements? Explain.

7.106 What does it mean when we say that electrons are paired?

7.107 A 368-g sample of water absorbs infrared radiation at 1.06×10^4 nm from a "carbon dioxide" laser. Suppose all the absorbed radiation is converted to heat. Calculate the number of photons at this wavelength required to raise the temperature of the water by 5.00°C.

7.108 Certain sunglasses have small crystals of silver chloride (AgCl) incorporated in the lenses. When exposed to light of appropriate wavelength, the following reaction occurs:

$$AgCl \longrightarrow Ag + Cl$$

The Ag atoms formed produce a uniform gray color which shades the glare. If ΔH for the reaction is 248 kJ, calculate the maximum wavelength of light that can induce this process.

7.109 If Rutherford and his coworkers had used electrons instead of alpha particles to probe the structure of the nucleus (see Chapter 2), what might have they discovered?

Opposite page: John Dalton's chart of elements, compiled in the early nineteenth century. Today, 109 elements are known.

C H A P T E R 8

PERIODIC RELATIONSHIPS AMONG THE ELEMENTS

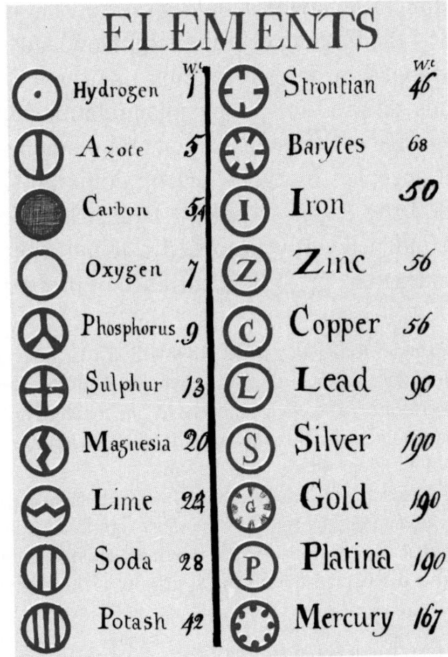

In Chapter 7 we discussed the electronic structure of atoms. As we will continue to see throughout this book, many of the chemical properties of the elements can be understood in terms of their electron configurations. Because electrons fill atomic orbitals in a fairly regular fashion, it is not surprising that elements with similar electron configurations, such as sodium and potassium, behave similarly in many respects and that, in general, the properties of the elements exhibit observable trends. Chemists in the nineteenth century recognized periodic trends in the physical and chemical properties of elements, long before quantum theory came onto the scene. Although these chemists were not aware of the existence of electrons and protons, their efforts to systematize the chemistry of the elements were remarkably successful. Their main sources of information were the atomic masses of the elements and other known physical and chemical properties.

8.1 Development of the Periodic Table

In the last chapter we discussed the building-up, or Aufbau, principle for writing the ground-state electron configuration of the elements. As we will see shortly, elements with similar outer electron configurations behave alike in many ways. This fact helps us group the elements according to the positions they now occupy in the periodic table. But this is *not* the way the original periodic table was constructed in the nineteenth century. The chemists of that time had only a vague idea of atoms and molecules and did not know of the existence of electrons and protons. Instead, they constructed the periodic table using their knowledge of atomic masses. Accurate measurements of the atomic masses of many elements had already been made. Arranging elements according to their atomic masses in a periodic table seemed logical to those chemists, who felt that chemical behavior should somehow be related to atomic mass.

In 1864 the English chemist John Newlands† noticed that when the known elements were arranged in order of atomic mass, every eighth element had similar properties. Newlands referred to this peculiar relationship as the *law of octaves*. However, this "law" turned out to be inadequate for elements beyond calcium, and Newlands's work was not accepted by the scientific community.

In 1869 the Russian chemist Dmitri Mendeleev‡ and the German chemist Lothar Meyer§ independently proposed a much more extensive tabulation of the elements, based on regular, periodic recurrence of properties. Table 8.1 shows an early version of

†John Alexander Reina Newlands (1838–1898). English chemist. Newlands's work was a step in the right direction in the classification of the elements. Unfortunately, because of its shortcomings, he was subjected to much criticism, and even ridicule. At one meeting he was asked if he had ever examined the elements according to the order of their initial letters! Nevertheless, in 1887 Newlands was honored by the Royal Society of London for his contribution.

‡Dmitri Ivanovich Mendeleev (1836–1907). Russian chemist. His work on the periodic classification of elements is regarded by many as the most significant achievement in chemistry in the nineteenth century.

§Julius Lothar Meyer (1830–1895). German chemist. In addition to his contribution to the periodic table, Meyer also discovered the chemical affinity of hemoglobin for oxygen.

Table 8.1 The periodic table as drawn by Mendeleev*

Reihen	Gruppe I — R^2O	Gruppe II — RO	Gruppe III — R^2O^3	Gruppe IV RH^4 RO^2	Gruppe V RH^3 R^2O^5	Gruppe VI RH^2 RO^3	Gruppe VII RH R^2O^7	Gruppe VIII — RO^4
1	H = 1							
2	Li = 7	Be = 9,4	B = 11	C = 12	N = 14	O = 16	F = 19	
3	Na = 23	Mg = 24	Al = 27,3	Si = 28	P = 31	S = 32	Cl = 35,5	
4	K = 39	Ca = 40	– = 44	Ti = 48	V = 51	Cr = 52	Mn = 55	Fe = 56, Co = 59, Ni = 59, Cu = 63.
5	(Cu = 63)	Zn = 65	– = 68	– = 72	As = 75	Se = 78	Br = 80	
6	Rb = 85	Sr = 87	?Yt = 88	Zr = 90	Nb = 94	Mo = 96	– = 100	Ru = 104, Rh = 104, Pd = 106, Ag = 108.
7	(Ag = 108)	Cd = 112	In = 113	Sn = 118	Sb = 122	Te = 125	J = 127	
8	Cs = 133	Ba = 137	?Di = 138	?Ce = 140	–	–	–	— — — —
9	(–)	–	–	–	–	–	–	
10	–	–	?Er = 178	?La = 180	Ta = 182	W = 184	–	Os = 195, Ir = 197, Pt = 198, Au = 199.
11	(Au = 199)	Hg = 200	Ti = 204	Pb = 207	Bi = 208	–	–	
12	–	–	–	Th = 231	–	U = 240	–	— — — —

*Spaces are left for the unknown elements with atomic masses 44, 68, 72, and 100.

Table 8.2 Comparison of the properties for eka-aluminum predicted by Mendeleev and the properties found for gallium*

Eka-Aluminum (Ea)	Gallium (Ga)
Atomic mass about 68.	Atomic mass 69.9.
Metal of specific gravity 5.9; melting point low; nonvolatile; unaffected by air; should decompose steam at red heat; should dissolve slowly in acids and alkalis.	*Metal* of specific gravity 5.94; melting point 30.15°C; nonvolatile at moderate temperatures; does not change in air; action of steam unknown; dissolves slowly in acids and alkalis.
Oxide: formula Ea_2O_3; specific gravity 5.5; should dissolve in acids to form salts of the type EaX_3. The hydroxide should dissolve in acids and alkalis.	*Oxide:* Ga_2O_3; specific gravity unknown; dissolves in acids, forming salts of the type GaX_3. The hydroxide dissolves in acids and alkalis.
Salts should have tendency to form basic salts; the sulfate should form alums; the sulfide should be precipitated by H_2S or $(NH_4)_2S$. The anhydrous chloride should be more volatile than zinc chloride.	*Salts* hydrolyze readily and form basic salts; alums are known; the sulfide is precipitated by H_2S and by $(NH_4)_2S$ under special conditions. The anhydrous chloride is more volatile than zinc chloride.
The element will probably be discovered by spectroscopic analysis.	Gallium was discovered with the aid of the spectroscope.

*From M. E. Weeks, *Discovery of the Elements,* 6th ed., Chemical Education Publishing Company, Easton, Pa., 1956.

Mendeleev's periodic table. Mendeleev's classification was a great improvement over Newlands's in two important ways. First, it grouped the elements together more accurately, according to their properties. Equally important was the fact that it made possible the prediction of the properties of several elements that had not yet been discovered. For example, Mendeleev proposed the existence of an unknown element that he called eka-aluminum. (*Eka* is a Sanskrit word meaning "first"; thus eka-aluminum would be the first element under aluminum in the same group.) When gallium was discovered four years later, the predicted properties of eka-aluminum were found to match the observed properties of gallium very closely (Table 8.2). Figure 8.1 is a chronological chart of the discovery of the elements.

Appendix 1 explains the names and symbols of the elements.

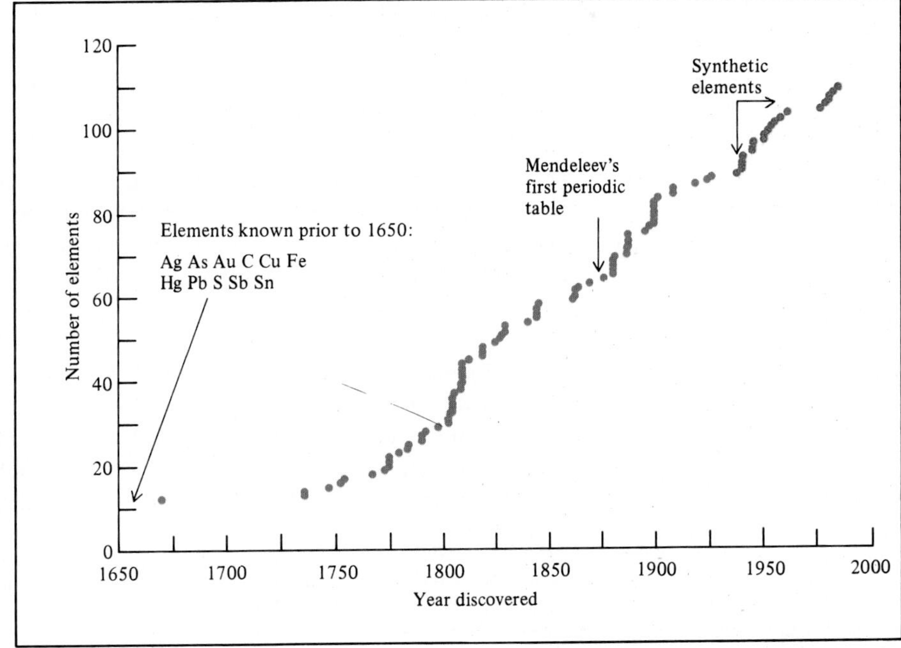

Figure 8.1 *A chronological chart of the discovery of the elements. To date, 109 elements have been identified.*

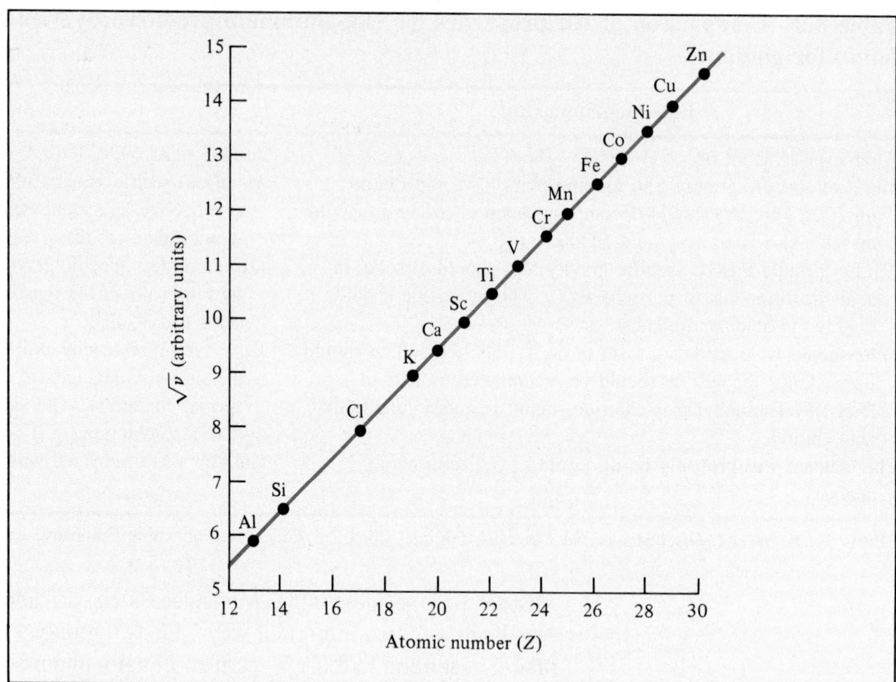

Figure 8.2 *A plot of the square root of the frequency of X rays versus the atomic number of the target element.*

Nevertheless, the early versions of the periodic table had some glaring inconsistencies. For example, the atomic mass of argon (39.95 amu) is greater than that of potassium (39.10 amu). If elements were arranged solely according to increasing atomic mass, argon would appear in the position occupied by potassium in our modern periodic table (see the inside front cover). But of course no chemist would place argon, an inert gas, in the same group as lithium and sodium, two very reactive metals. Such discrepancies suggested that some fundamental property other than atomic mass is the basis of the observed periodicity. This property turned out to be associated with atomic number.

Using data from α-scattering experiments (see Section 2.2), Rutherford was able to estimate the number of positive charges in the nucleus of a few elements, but there was no general procedure for determining the atomic numbers. In 1913 a young English physicist, Henry Moseley,† was able to provide the necessary measurements. When high-energy electrons were focused on a target made of the elements under study, X rays were generated. Moseley noticed that the frequencies of X rays emitted from the elements could be correlated by the equation

$$\sqrt{\nu} = a(Z - b)$$

where ν is the frequency of the emitted X ray and a and b are constants that are the same for all the elements. A plot of $\sqrt{\nu}$ versus Z yields a straight line (Figure 8.2). Thus, from the measured frequency and hence $\sqrt{\nu}$ of emitted X rays, we can determine the atomic number of the element from the plot.

With a few exceptions, Moseley found that the order of increasing atomic number is

†Henry Gwyn-Jeffreys Moseley (1887–1915). English physicist. Moseley discovered the relationship between X-ray spectra and the atomic number of the elements. A lieutenant in the Royal Engineers, he was killed in action at the age of 28 during the British campaign in Gallipoli, Turkey.

the order of increasing atomic mass. For example, calcium is the twentieth element in increasing order of atomic mass, and it has an atomic number of 20. The discrepancies mentioned earlier were now understood. The atomic number of argon is 18 and that of potassium is 19, so potassium should follow argon in the periodic table.

In a modern periodic table the atomic number is usually listed together with the element symbol. As you already know, the atomic number also indicates the number of electrons in the atoms of an element. Electron configurations of elements help explain the recurrence of physical and chemical properties. The importance and usefulness of the periodic table lie in the fact that we can use our understanding of the general properties and trends within a group or a period to predict with considerable accuracy the properties of any element, even though that element may be unfamiliar to us.

8.2 Periodic Classification of the Elements

Figure 8.3 shows the periodic table together with the outermost ground-state electron configurations of the elements. (The electron configurations of all the elements are also given in Table 7.3.) Starting with hydrogen, we see that subshells are filled in the order shown in Figure 7.29. According to the type of subshell being filled, the elements can be divided into categories—the representative elements, the noble gases, the transition elements (or transition metals), the lanthanides, and the actinides. Referring now to Figure 8.3, the ***representative elements*** (also called ***main group elements***) are *the elements in Groups 1A through 7A, all of which have incompletely filled s or p sub-shells of highest principal quantum number*. With the exception of helium, the ***noble gases*** *(the Group 8A elements) all have a completely filled p subshell. (The electron configurations are* $1s^2$ *for helium and* ns^2np^6 *for the other noble gases, where n is the principal quantum number for the outermost shell.)* The transition metals are the elements in Groups 1B and 3B through 8B, which have incompletely filled *d* subshells, or readily produce cations with incompletely filled *d* subshells. (These metals are sometimes referred to as the *d*-block transition elements.) The Group 2B elements are Zn, Cd, and Hg, which are neither representative elements nor transition metals. The lanthanides and actinides are sometimes called *f*-block transition elements because they have incompletely filled *f* subshells. Figure 8.4 distinguishes the groups of elements discussed here.

Group 1A		Group 2A	
Li	$[He]2s^1$	Be	$[He]2s^2$
Na	$[Ne]3s^1$	Mg	$[Ne]3s^2$
K	$[Ar]4s^1$	Ca	$[Ar]4s^2$
Rb	$[Kr]5s^1$	Sr	$[Kr]5s^2$
Cs	$[Xe]6s^1$	Ba	$[Xe]6s^2$
Fr	$[Rn]7s^1$	Ra	$[Rn]7s^2$

A clear pattern emerges when we examine the electron configurations of the elements in a particular group. First we will consider the representative elements. For Groups 1A and 2A elements, the electron configurations are shown in the table on the left. We see that all members of the Group 1A alkali metals have similar outer electron configurations; each has a noble gas core and an ns^1 configuration of the outer electron. Similarly, the Group 2A alkaline earth metals have a noble gas core and an ns^2 configuration of the outer electrons. *The outer electrons of an atom, which are those involved in chemical bonding,* are often called the ***valence electrons.*** The similarity of the outer electron configurations (that is, they have the same number of valence electrons) is what makes the elements in the same group resemble one another in chemical behavior. This observation holds true for the other representative elements. Thus, for instance, the halogens (the Group 7A elements), all with outer electron configurations of ns^2np^5, have very similar properties as a group. We must be careful, however, in predicting properties when elements change from nonmetals through metalloids to metals. For example, the elements in Group 4A all have the same outer electron

For the representative elements, the valence electrons are simply those electrons at the highest principal energy level *n*.

See Figure 1.5 for a classification of elements.

Figure 8.3 *The ground-state electron configurations of the elements. For simplicity, only the configurations of the outer electrons are shown.*

	1A	2A	3B	4B	5B	6B	7B	8B	8B	8B	1B	2B	3A	4A	5A	6A	7A	8A
1	1 **H** $1s^1$																	2 **He** $1s^2$
2	3 **Li** $2s^1$	4 **Be** $2s^2$											5 **B** $2s^22p^1$	6 **C** $2s^22p^2$	7 **N** $2s^22p^3$	8 **O** $2s^22p^4$	9 **F** $2s^22p^5$	10 **Ne** $2s^22p^6$
3	11 **Na** $3s^1$	12 **Mg** $3s^2$											13 **Al** $3s^23p^1$	14 **Si** $3s^23p^2$	15 **P** $3s^23p^3$	16 **S** $3s^23p^4$	17 **Cl** $3s^23p^5$	18 **Ar** $3s^23p^6$
4	19 **K** $4s^1$	20 **Ca** $4s^2$	21 **Sc** $4s^23d^1$	22 **Ti** $4s^23d^2$	23 **V** $4s^23d^3$	24 **Cr** $4s^13d^5$	25 **Mn** $4s^23d^5$	26 **Fe** $4s^23d^6$	27 **Co** $4s^23d^7$	28 **Ni** $4s^23d^8$	29 **Cu** $4s^13d^{10}$	30 **Zn** $4s^23d^{10}$	31 **Ga** $4s^24p^1$	32 **Ge** $4s^24p^2$	33 **As** $4s^24p^3$	34 **Se** $4s^24p^4$	35 **Br** $4s^24p^5$	36 **Kr** $4s^24p^6$
5	37 **Rb** $5s^1$	38 **Sr** $5s^2$	39 **Y** $5s^24d^1$	40 **Zr** $5s^24d^2$	41 **Nb** $5s^14d^4$	42 **Mo** $5s^14d^5$	43 **Tc** $5s^24d^5$	44 **Ru** $5s^14d^7$	45 **Rh** $5s^14d^8$	46 **Pd** $4d^{10}$	47 **Ag** $5s^14d^{10}$	48 **Cd** $5s^24d^{10}$	49 **In** $5s^25p^1$	50 **Sn** $5s^25p^2$	51 **Sb** $5s^25p^3$	52 **Te** $5s^25p^4$	53 **I** $5s^25p^5$	54 **Xe** $5s^25p^6$
6	55 **Cs** $6s^1$	56 **Ba** $6s^2$	57 **La** $6s^25d^1$	72 **Hf** $6s^25d^2$	73 **Ta** $6s^25d^3$	74 **W** $6s^25d^4$	75 **Re** $6s^25d^5$	76 **Os** $6s^25d^6$	77 **Ir** $6s^25d^7$	78 **Pt** $6s^15d^9$	79 **Au** $6s^15d^{10}$	80 **Hg** $6s^25d^{10}$	81 **Tl** $6s^26p^1$	82 **Pb** $6s^26p^2$	83 **Bi** $6s^26p^3$	84 **Po** $6s^26p^4$	85 **At** $6s^26p^5$	86 **Rn** $6s^26p^6$
7	87 **Fr** $7s^1$	88 **Ra** $7s^2$	89 **Ac** $7s^26d^1$	104 **Unq** $7s^26d^2$	105 **Unp** $7s^26d^3$	106 **Unh** $7s^26d^4$	107 **Uns** $7s^26d^5$	108 **Uno** $7s^26d^6$	109 **Une** $7s^26d^7$									

58 **Ce** $6s^24f^15d^1$	59 **Pr** $6s^24f^3$	60 **Nd** $6s^24f^4$	61 **Pm** $6s^24f^5$	62 **Sm** $6s^24f^6$	63 **Eu** $6s^24f^7$	64 **Gd** $6s^24f^75d^1$	65 **Tb** $6s^24f^9$	66 **Dy** $6s^24f^{10}$	67 **Ho** $6s^24f^{11}$	68 **Er** $6s^24f^{12}$	69 **Tm** $6s^24f^{13}$	70 **Yb** $6s^24f^{14}$	71 **Lu** $6s^24f^{14}5d^1$
90 **Th** $7s^26d^2$	91 **Pa** $7s^25f^26d^1$	92 **U** $7s^25f^36d^1$	93 **Np** $7s^25f^46d^1$	94 **Pu** $7s^25f^6$	95 **Am** $7s^25f^7$	96 **Cm** $7s^25f^76d^1$	97 **Bk** $7s^25f^9$	98 **Cf** $7s^25f^{10}$	99 **Es** $7s^25f^{11}$	100 **Fm** $7s^25f^{12}$	101 **Md** $7s^25f^{13}$	102 **No** $7s^25f^{14}$	103 **Lr** $7s^25f^{14}6d^1$

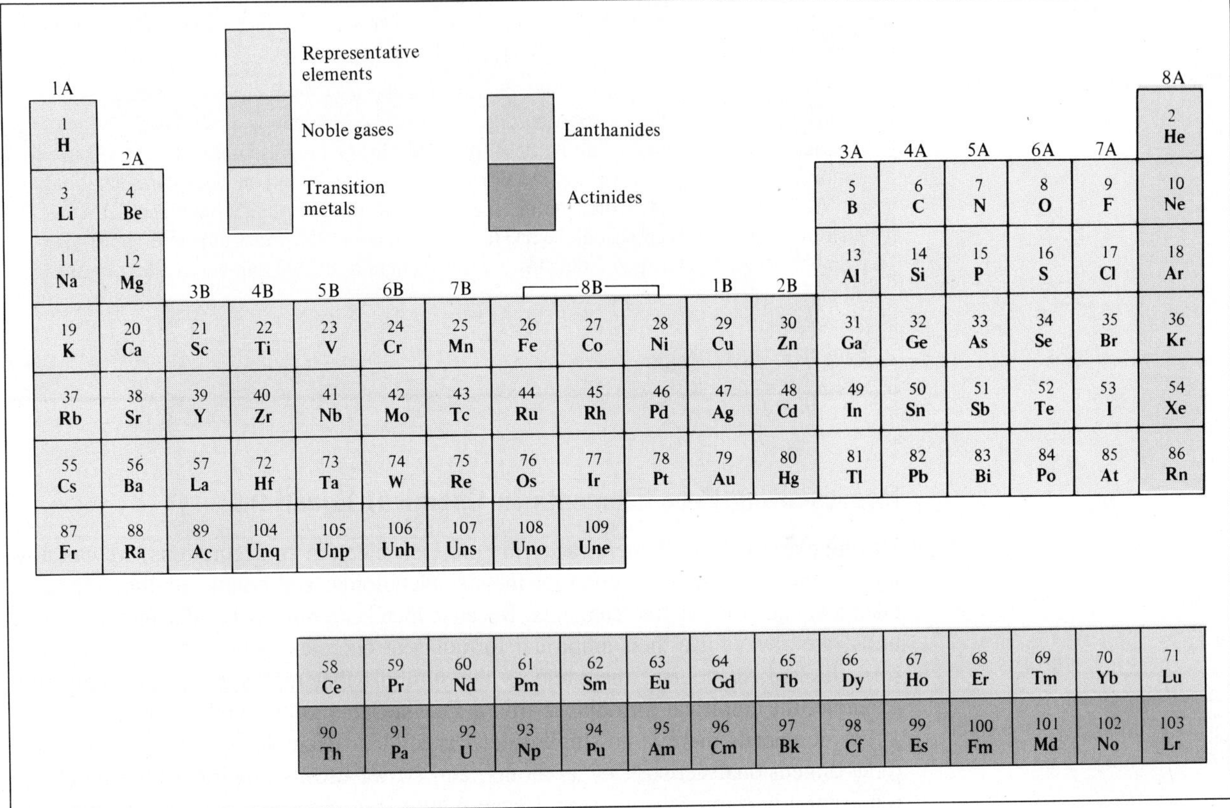

Figure 8.4 *Classification of the elements. Note that the Group 2B elements are often classified as transition metals even though they do not exhibit the characteristics of the transition metals.*

configuration, ns^2np^2, but there is much variation in chemical properties among these elements: Carbon is a nonmetal, silicon and germanium are metalloids, and tin and lead are metals.

As a group, the noble gases behave very similarly. With the exception of krypton and xenon, the rest of these elements are totally inert chemically. The reason is that these elements all have completely filled outer *ns* and *np* subshells, a condition that represents great stability. Although the outer electron configuration of the transition metals is not always the same within a group, and there is no regular pattern in the change of the electron configuration from one metal to the next in the same period, all transition metals share many characteristics that set them apart from other elements. This is because these metals all have an incompletely filled *d* subshell. Likewise, the lanthanide (and the actinide) elements resemble one another within the series because they have incompletely filled *f* subshells.

EXAMPLE 8.1

A neutral atom of a certain element has 15 electrons. Without consulting a periodic table, answer the following questions: (a) What is the electron configuration of the element? (b) How should the element be classified? (c) Are the atoms of this element diamagnetic or paramagnetic?

Answer

(a) Using the building-up principle and knowing the maximum capacity of *s* and *p* subshells, we can write the electron configuration of the element as $1s^22s^22p^63s^23p^3$.
(b) Since the 3*p* subshell is not completely filled, this is a representative element. Based on the information given, we cannot say whether it is a metal, a nonmetal, or a metalloid.
(c) According to Hund's rule, the three electrons in the 3*p* orbitals have parallel spins. Therefore, the atoms of this element are paramagnetic, with three unpaired spins. (Remember, we saw in Chapter 7 that any atom that contains an odd number of electrons *must be* paramagnetic.)

Similar problem: 8.14.

Representing Free Elements in Chemical Equations

Having classified the elements according to their electron configurations, we can now look at the way chemists represent metals, metalloids, and nonmetals that appear in chemical equations as free elements. Because metals do not exist in discrete molecular units, we always use their empirical formulas in chemical equations. The empirical formulas are, of course, the same as the symbols that represent the elements. For example, the empirical formula for iron is Fe, the same as the symbol for the element.

For nonmetals there is no single rule. Carbon, for example, exists as an extensive three-dimensional network of its atoms, and so we use its empirical formula (C) to represent it in chemical equations. Because hydrogen, nitrogen, oxygen, and the halogens exist as diatomic molecules, we use their molecular formulas (H_2, N_2, O_2, F_2, Cl_2, Br_2, I_2). The stable form of phosphorus exists as P_4 molecules, and so we use P_4. Although the stable form of sulfur is S_8, chemists often use its empirical formula in chemical equations. Thus, instead of writing the equation for the combustion of sulfur as

Note that these two equations for the combustion of sulfur have identical stoichiometry. This should not be surprising, since both equations describe the same chemical system. In both cases a number of sulfur atoms react with twice that number of oxygen atoms.

$$S_8(s) + 8O_2(g) \longrightarrow 8SO_2(g)$$

we often write simply

$$S(s) + O_2(g) \longrightarrow SO_2(g)$$

All the noble gases exist as monatomic species; thus we use their symbols: He, Ne, Ar, Kr, Xe, and Rn. The metalloids, like the metals, all exist in complex three-dimensional networks, and we represent them, too, with their empirical formulas, that is, their symbols: B, Si, Ge, and so on.

Electron Configurations of Cations and Anions

Just as for neutral atoms, we use the Pauli exclusion principle and Hund's rule in writing the electron configurations of cations and anions.

We have discussed the electron configurations of elements, but because many ionic compounds are made up of monatomic anions and/or cations, it is helpful to know how to write the electron configurations of these ionic species. The procedure for writing electron configurations of ions requires only a slight extension of the method used for neutral atoms. We will group the ions in two categories for discussion.

Ions Derived from Representative Elements. In the formation of a cation from the neutral atom of a representative element, one or more electrons are removed from the highest occupied *n* shell. Following are the electron configurations of some neutral atoms and their corresponding cations:

Na	[Ne]$3s^1$	Na^+	[Ne]
Ca	[Ar]$4s^2$	Ca^{2+}	[Ar]
Al	[Ne]$3s^23p^1$	Al^{3+}	[Ne]

Note that each ion has a stable noble gas configuration.

In the formation of an anion, one or more electrons are added to the highest partially filled *n* shell. Consider the following examples:

H	$1s^1$	H^-	$1s^2$ or [He]
F	$1s^22s^22p^5$	F^-	$1s^22s^22p^6$ or [Ne]
O	$1s^22s^22p^4$	O^{2-}	$1s^22s^22p^6$ or [Ne]
N	$1s^22s^22p^3$	N^{3-}	$1s^22s^22p^6$ or [Ne]

Again, all the anions have the stable noble gas configurations. Thus a characteristic of most representative elements is that ions derived from their neutral atoms have the noble gas outer electron configuration ns^2np^6. *Ions, or atoms and ions, that have the same number of electrons, and hence the same ground-state electron configuration,* are said to be ***isoelectronic.*** Thus H^- and He are isoelectronic; F^-, Na^+, and Ne are isoelectronic; and so on.

Cations Derived from Transition Metals. Section 7.10 pointed out that in the first-row transition metals (Sc to Cu), the 4*s* orbital is always filled before the 3*d* orbitals. Consider manganese, whose electron configuration is [Ar]$4s^23d^5$. When the Mn^{2+} ion is formed, we might expect the two electrons to be removed from the 3*d* orbitals to yield [Ar]$4s^23d^3$. In fact, the electron configuration of Mn^{2+} is [Ar]$3d^5$! We can understand this rather unexpected sequence if we realize that the electron-electron and electron-nucleus interactions in a neutral atom can be quite different from those in its ion. Thus, whereas the 4*s* orbital is always filled before the 3*d* orbital in Mn, electrons are removed from the 4*s* orbital in forming Mn^{2+} because the 3*d* orbital is more stable than the 4*s* orbital in transition metal ions. Therefore, in forming a cation from an atom of a transition metal, electrons are always removed first from the *ns* orbital and then from the $(n - 1)d$ orbitals.

This rule can be remembered easily if you recognize that electrons are removed from the highest filled subshells first. Bear in mind that the order of electron filling does not determine or predict the order of electron removal for these metals.

Keep in mind that most transition metals can form more than one cation and that frequently the cations are not isoelectronic with the preceding noble gases.

8.3 Periodic Variation in Physical Properties

As we have seen, the electron configurations of the elements show a periodic variation with increasing atomic number. Consequently, the elements also display periodic variations in their physical and chemical behavior. In this section and the next two, we will examine some physical properties of elements in a group and across a period and properties that influence the chemical behavior of elements. Before we do so, however, let's look at the concept of effective nuclear charge, which will help us to understand these topics better.

CHEMISTRY IN ACTION

Today's Periodic Table

The periodic table on the wall of your laboratory or lecture hall has probably become such a familiar part of your day-to-day experience in chemistry that you take it for granted. However, like everything else in science the standard group notation in the periodic table continues to change in order to reflect new discoveries, trends, and developments. Here is another example of chemistry in action.

As mentioned in this chapter, one of the first formal arrangements of elements in a tabular format was completed by Mendeleev in 1869. Apparently he made a card for each of the 63 elements known at that time and recorded the most notable properties of each element on its own card. Placing together the cards for the elements having similar properties, Mendeleev came up with a table having eight vertical groups.

Since then the periodic table has been expanded to over 100 elements, listed by atomic number instead of atomic mass, as in Mendeleev's table. Otherwise, in principle, it is pretty much the same. There are still eight groups, which are designated with the letter A or B. In the United States, the conventional practice has been to use A to designate the representative elements and B to designate transition elements (Figure 8.5). In Europe, the tradition has been to use B for the representative elements (after the alkali and alkaline earth metals) and A for the transition elements. In an attempt to come up with a compromise that would eliminate the

1 — IUPAC convention
1A — European convention
1A — U.S. convention

1	2	3	4	5	6	7	8	9	10	11	12	13	14	15	16	17	18
1A	2A	3A	4A	5A	6A	7A		8A		1B	2B	3B	4B	5B	6B	7B	0
1A	2A	3B	4B	5B	6B	7B		8B		1B	2B	3A	4A	5A	6A	7A	8A
1 H																	2 He
3 Li	4 Be											5 B	6 C	7 N	8 O	9 F	10 Ne
11 Na	12 Mg											13 Al	14 Si	15 P	16 S	17 Cl	18 Ar
19 K	20 Ca	21 Sc	22 Ti	23 V	24 Cr	25 Mn	26 Fe	27 Co	28 Ni	29 Cu	30 Zn	31 Ga	32 Ge	33 As	34 Se	35 Br	36 Kr
37 Rb	38 Sr	39 Y	40 Zr	41 Nb	42 Mo	43 Tc	44 Ru	45 Rh	46 Pd	47 Ag	48 Cd	49 In	50 Sn	51 Sb	52 Te	53 I	54 Xe
55 Cs	56 Ba	57 La	72 Hf	73 Ta	74 W	75 Re	76 Os	77 Ir	78 Pt	79 Au	80 Hg	81 Tl	82 Pb	83 Bi	84 Po	85 At	86 Rn
87 Fr	88 Ra	89 Ac	104 Unq	105 Unp	106 Unh	107 Uns	108 Uno	109 Une									

58 Ce	59 Pr	60 Nd	61 Pm	62 Sm	63 Eu	64 Gd	65 Tb	66 Dy	67 Ho	68 Er	69 Tm	70 Yb	71 Lu
90 Th	91 Pa	92 U	93 Np	94 Pu	95 Am	96 Cm	97 Bk	98 Cf	99 Es	100 Fm	101 Md	102 No	103 Lr

Figure 8.5 *The three different periodic group notations.*

long-standing confusion over the A and B group subdivisions, the International Union of Pure and Applied Chemistry (IUPAC) has recommended adopting a table in which the columns carry Arabic numerals 1 through 18, as shown in Figure 8.5. The proposal has sparked much controversy in the international chemistry community, and its merits and drawbacks will be deliberated for some time to come.

Effective Nuclear Charge

In discussing the properties of many-electron atoms, the concept of effective nuclear charge is quite useful. Consider the helium atom. The nuclear charge of helium is +2, but the full force of this +2 charge is partially offset by the mutual repulsion of the two 1*s* electrons. In other words, as far as each electron is concerned, the nuclear charge *appears* to carry a charge that is less than +2. For this reason we say that each 1*s* electron is *shielded* from the nucleus by the other 1*s* electron, and the *effective nuclear charge,* Z_{eff}, is given by

$$Z_{eff} = Z - \sigma$$

σ is greater than zero but smaller than Z.

where Z is the actual nuclear charge and σ (sigma) is called the *shielding constant* (also called the *screening constant*).

One way to illustrate the effect of electron shielding is to look at the energy required to remove an electron from a many-electron atom. Measurements show that it takes 2373 kJ of energy to remove the first electron from 1 mole of He atoms and 5248 kJ of energy to remove the remaining electron from 1 mole of He^+ ions. The reason it takes less energy for the first step is that the electron-electron repulsion, or electron shielding, results in a reduction of the attraction of the nucleus for each electron. In the He^+ ion there is only one electron present, so there is no shielding and the electron feels the full effect of the +2 nuclear charge. Consequently, it takes considerably more energy to remove the second electron.

Finally, it is useful to remember the following facts about electron shielding: (1) Electrons in a given shell are shielded by electrons in inner shells but not by electrons in outer shells, and (2) inner filled shells shield outer electrons more effectively than electrons in the same subshell shield one another. We will refer to these facts later in our discussions of the physical and chemical properties of the elements.

Atomic Radius

Atomic radius also influences chemical behavior. Here we will concentrate only on the periodic variation of atomic radii.

A number of physical properties, including density, melting point, and boiling point, are related to the sizes of atoms, but atomic size is difficult to define. As we saw in Chapter 7, the electron density in an atom extends far beyond the nucleus. In practice, we normally think of atomic size as the volume containing about 90 percent of the total electron density around the nucleus.

Several techniques allow us to estimate the size of an atom. First consider the metallic elements. The structure of metals is quite varied, but they all share a common characteristic: Their atoms are linked to one another in an extensive three-dimensional network. Thus the ***atomic radius*** of a metal is *one-half the distance between the two nuclei in two adjacent atoms* [Figure 8.6(a)]. *For elements that exist as simple diatomic molecules, the atomic radius is one-half the distance between the nuclei of the two atoms in a particular molecule* [Figure 8.6(b)].

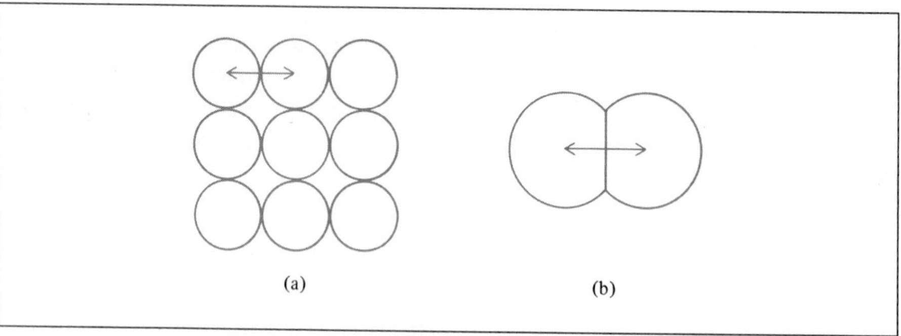

Figure 8.6 *(a) In a metal, the atomic radius is defined as one-half the distance between the centers of two adjacent atoms. (b) For elements such as iodine, which exist as diatomic molecules, the radius of the atom is defined as one-half the distance between the centers of the atoms in the molecule.*

Increasing atomic radius (←)

Increasing atomic radius (↓)

1A	2A	3A	4A	5A	6A	7A	8A
H 32							He 50
Li 152	Be 112	B 98	C 91	N 92	O 73	F 72	Ne 70
Na 186	Mg 160	Al 143	Si 132	P 128	S 127	Cl 99	Ar 98
K 227	Ca 197	Ga 135	Ge 137	As 139	Se 140	Br 114	Kr 112
Rb 248	Sr 215	In 166	Sn 162	Sb 159	Te 160	I 133	Xe 131
Cs 265	Ba 222	Tl 171	Pb 175	Bi 170	Po 164	At 142	Rn 140

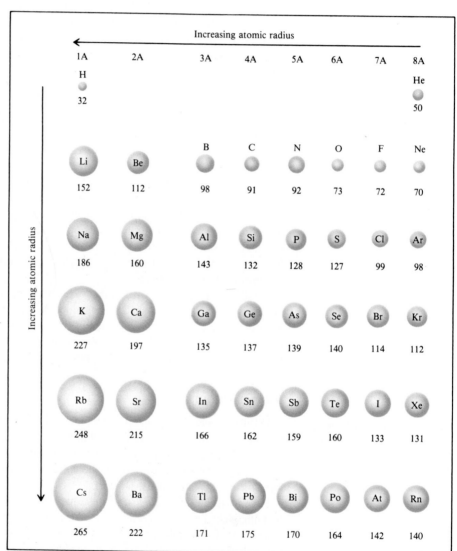

Figure 8.7 *Atomic radii (in picometers) of representative elements according to their positions in the periodic table.*

Figure 8.7 shows the atomic radii of many elements according to their positions in the periodic table, and Figure 8.8 plots the atomic radii of these elements against their

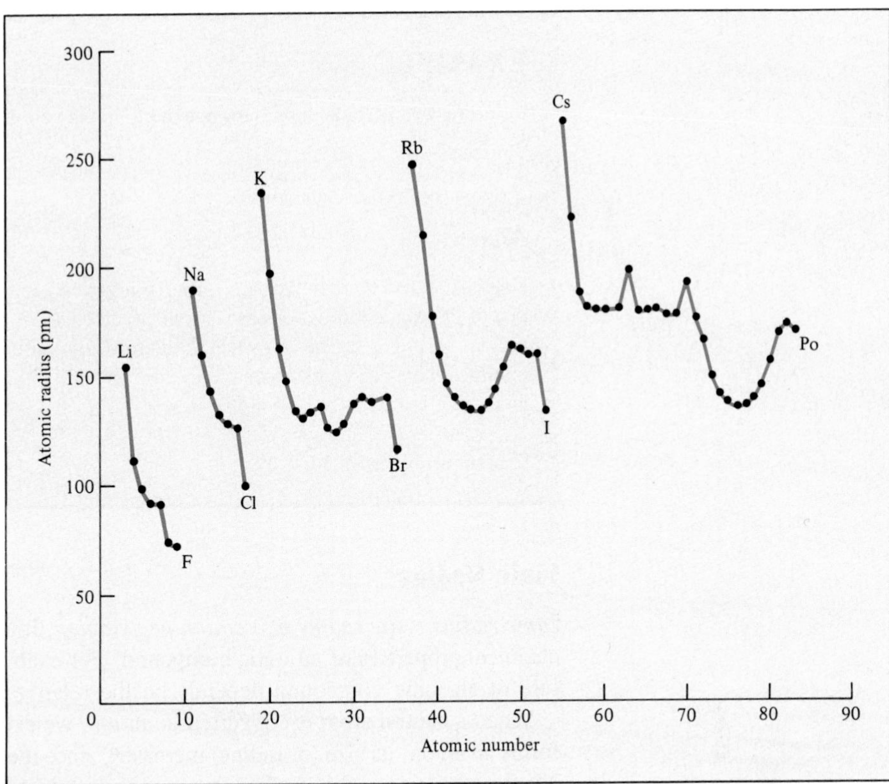

Figure 8.8 *Plot of atomic radii (in picometers) of elements against their atomic numbers.*

atomic numbers. Periodic trends are clearly evident. In studying the trends, bear in mind that the atomic radius is determined to a large extent by how strongly the outer-shell electrons are held by the nucleus. The larger the effective nuclear charge, the more strongly held these electrons are and the smaller the atomic radius is. Consider the second-period elements from Li to F. Moving from left to right, we find that the number of electrons in the inner shell ($1s^2$) remains constant while the nuclear charge increases. The electrons that are added to counterbalance the increasing nuclear charge are ineffective in shielding one another. Consequently, the effective nuclear charge increases steadily while the principal quantum number remains constant ($n = 2$). For example, the outer $2s$ electron on lithium is shielded from the nucleus (which has 3 protons) by the two $1s$ electrons. As an approximation, we assume that the shielding effect of the two $1s$ electrons is to cancel two positive charges in the nucleus. Thus the $2s$ electron only feels the attraction due to one proton in the nucleus, or the effective nuclear charge is $+1$. In beryllium ($1s^2 2s^2$) each of the $2s$ electrons is shielded by the inner two $1s$ electrons, which cancel two of the four positive charges in the nucleus. Because the $2s$ electrons do not shield each other as effectively, the net result is that the effective nuclear charge of each $2s$ electron is greater than $+1$. Thus as the effective nuclear charge increases, the atomic radius decreases steadily from lithium to fluorine.

As we go down a group—say, Group 1A—we find that atomic radius increases with increasing atomic number. For alkali metals the effective nuclear charge of the outermost electron remains constant ($+1$) but n gets larger. Since the orbital size increases with the increasing principal quantum number n, the size of the metal atoms increases from Li to Cs. The same reasoning can be applied to elements in other groups.

EXAMPLE 8.2

Referring to a periodic table, arrange the following atoms in order of increasing radius: P, Si, N.

Answer

Note that N and P are in the same group (Group 5A) so that the radius of N is smaller than that of P (atomic radius increases as we go down a group). Both Si and P are in the third period, and Si is to the left of P. Therefore, the radius of P is smaller than that of Si (atomic radius decreases as we move from left to right across a period). Thus the order of increasing radius is N < P < Si.

Similar problems: 8.34, 8.37.

Ionic Radius

Ionic radius is *the radius of a cation or an anion.* Ionic radius affects the physical and chemical properties of an ionic compound. For example, the three-dimensional structure of an ionic compound depends on the relative sizes of its cations and anions.

When a neutral atom is converted to an ion, we expect a change in size. If the atom forms an anion, its size (or radius) increases, since the nuclear charge remains the same but the repulsion resulting from the additional electron(s) enlarges the domain of the electron cloud. On the other hand, a cation is smaller than the neutral atom, since

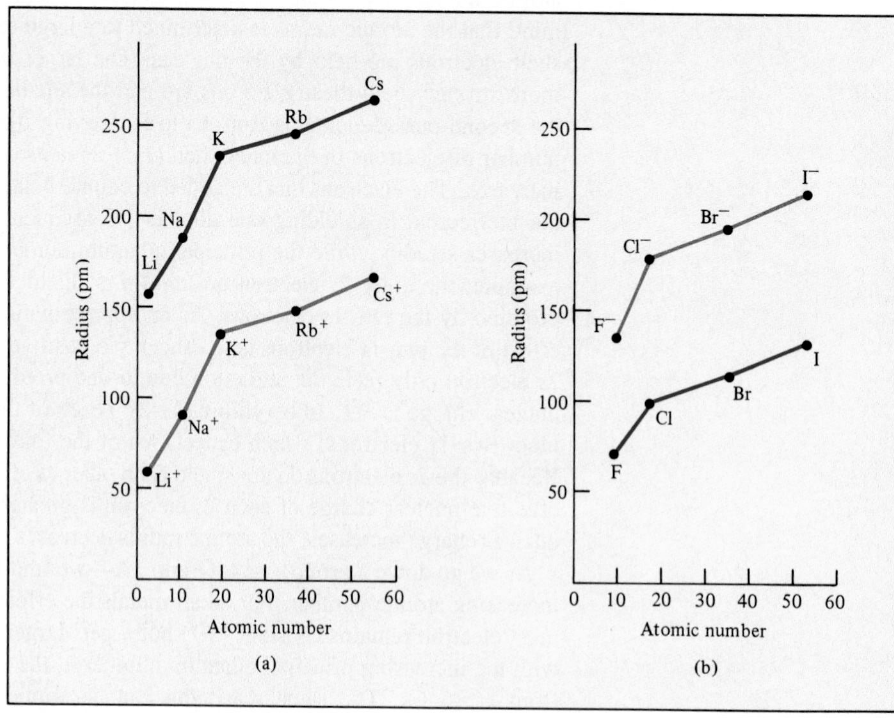

Figure 8.9 *Comparison of atomic radii with ionic radii. (a) Alkali metals and alkali metal cations. (b) Halogens and halide ions.*

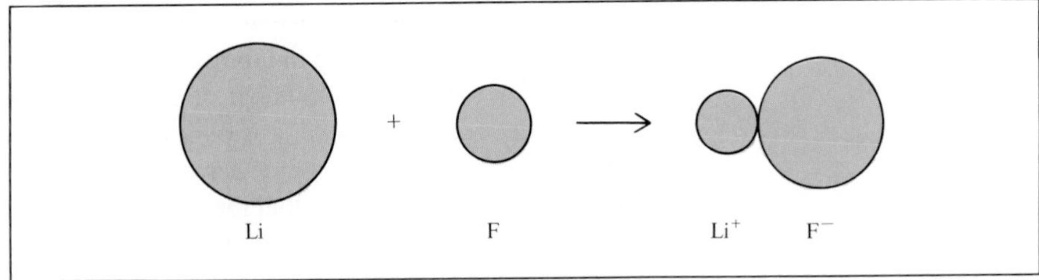

Figure 8.10 *Changes in size when Li reacts with F to form LiF.*

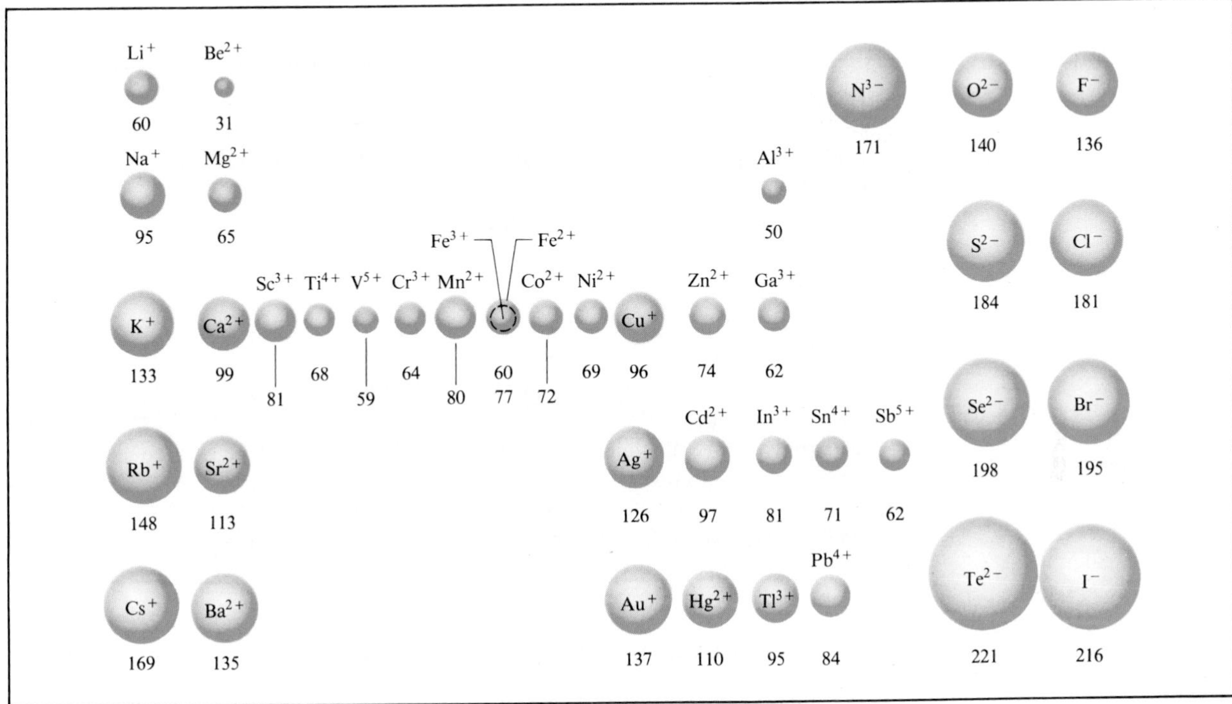

Figure 8.11 *The radii in picometers of some ions of the more familiar elements shown in their positions in the periodic table.*

removing one or more electrons reduces electron-electron repulsion and there will be a shrinkage of the electron cloud. Figure 8.9 shows the changes in size when alkali metals are converted to cations and halogens are converted to anions; Figure 8.10 shows the changes in size when a lithium atom reacts with a fluorine atom to form a LiF unit.

Figure 8.11 shows the radii of ions derived from the more familiar elements, arranged according to their positions in the periodic table. We see that in some places there are parallel trends between atomic radii and ionic radii. For example, from top to bottom of the periodic table both the atomic radius and the ionic radius increase. For ions derived from elements in different groups, the comparison in size is meaningful only if the ions are isoelectronic. If we examine isoelectronic ions, we find that cations are smaller than anions. For example, Na^+ is smaller than F^-. Both ions have the same number of electrons, but Na ($Z = 11$) has more protons than F ($Z = 9$). The larger effective nuclear charge of Na^+ results in a smaller radius.

Focusing on isoelectronic cations, we see that the radii of *tripositive ions* (that is, ions that bear three positive charges) are smaller than those of *dipositive ions* (that is, ions that bear two positive charges), which in turn are smaller than *unipositive ions* (that is, ions that bear one positive charge). This trend is nicely illustrated by the sizes of three isoelectronic ions in the third period: Al^{3+}, Mg^{2+}, and Na^{+} (see Figure 8.11). The Al^{3+} ion has the same number of electrons as Mg^{2+}, but it has one more proton. Thus the electron cloud in Al^{3+} is pulled inward more than that in Mg^{2+}. The smaller radius of Mg^{2+} compared with that of Na^{+} can be similarly explained. Turning to isoelectronic anions, we find that the radius increases as we go from ions with uninegative charge (that is, −) to those with dinegative charge (that is, 2−), and so on. Thus the oxide ion is larger than the fluoride ion because oxygen has one fewer proton than fluorine; the electron cloud is spread out to a greater extent in O^{2-}.

EXAMPLE 8.3

In each of the following pairs, indicate which one of the two species is larger: (a) N^{3-} or F^{-}; (b) Mg^{2+} or Ca^{2+}; (c) Fe^{2+} or Fe^{3+}.

Answer

(a) N^{3-} and F^{-} are isoelectronic anions. Since N^{3-} has three negative charges and F^{-} has one negative charge, N^{3-} is larger.
(b) Both Mg and Ca belong to Group 2A (the alkaline metals). Thus the Ca^{2+} ion is larger than Mg^{2+} because Ca lies below Mg.
(c) Both ions have the same nuclear charge, but Fe^{2+} has one more electron and hence greater electron-electron repulsion. The radius of Fe^{2+} is larger.

Similar problems: 8.40, 8.42, 8.43.

Variation of Physical Properties across a Period

As we move from left to right across a period, there is a transition from metals to metalloids to nonmetals. In this section we will look at the variation in the characteristic physical properties of metallic and nonmetallic elements across a period.

Let's consider the third-period elements from sodium to argon (Figure 8.12). Table

Table 8.3 Variations in some physical properties for the third-period elements

	Na	Mg	Al	Si	P	S	Cl	Ar
Type of element	←	Metal	→	Metalloid	←	Nonmetal		→
Structure	←	Extensive three-dimensional		→	P_4	S_8	Cl_2	Atomic species
Density (g/cm^3)	0.97	1.74	2.70	2.33	1.82	2.07	1.56*	1.40*
Melting point (°C)	98	649	660	1410	44	119	−101	−189
Molar heat of fusion (kJ/mol)	2.6	9.0	10.8	46.4	0.6	1.4	3.2	1.2
Boiling point (°C)	883	1090	2467	2355	280	445	−34.6	−186
Molar heat of vaporization (kJ/mol)	89.0	128.7	293.7	376.7	12.4	9.6	10.2	6.5
Electrical conductivity	Good	Good	Good	Fair	Poor	Poor	Poor	Poor
Thermal conductivity	Good	Good	Good	Fair	Poor	Poor	Poor	Poor

*The densities of Cl_2 and Ar are those of the liquids at their boiling points.

8.3 shows a number of physical properties of the third-period elements. The *molar heats of fusion* and *vaporization* of a substance are the energies (in kJ) needed to melt and vaporize one mole of the substance at its melting point and boiling point, respectively. The terms "electrical conductivity" and "thermal conductivity" are used qualitatively here to indicate an element's ability to conduct electricity and heat. Sodium, the first element in the third period, is a very reactive metal, whereas chlorine, the second-to-last element of that period, is a very reactive nonmetal. In between, the elements show a gradual transition from metallic properties to nonmetallic properties. Note that the melting point, molar heat of fusion, boiling point, and molar heat of vaporization rise from Na to Si and then fall to low values at Ar. Sodium, magnesium, and aluminum all have extensive three-dimensional atomic networks, and these atoms are held together by forces characteristic of the metallic state. Silicon is a metalloid; it has a giant three-dimensional structure in which the Si atoms are held together very

Figure 8.12 *The third-period elements. The photograph of argon, which is a colorless, odorless gas, shows the color emitted by the gas from a discharge tube.*

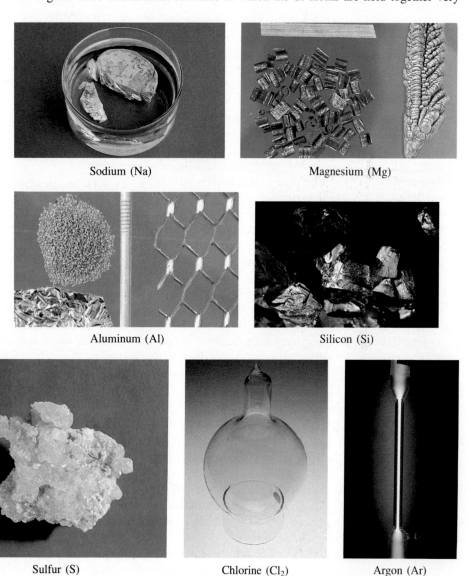

Sodium (Na)

Magnesium (Mg)

Aluminum (Al)

Silicon (Si)

Phosphorus (P)

Sulfur (S)

Chlorine (Cl_2)

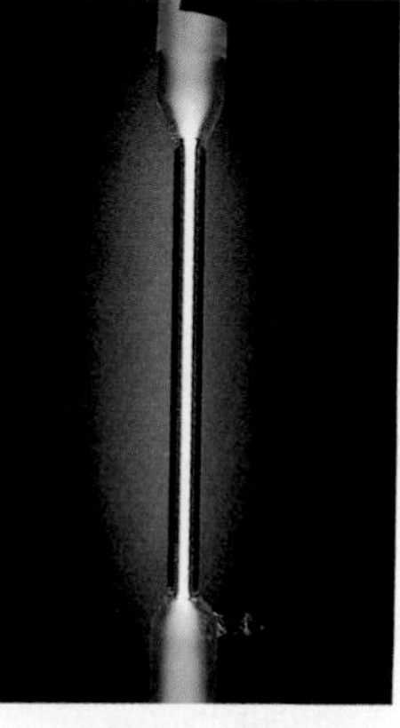

Argon (Ar)

CHEMISTRY IN ACTION

The Third Liquid Element

Of the 109 known elements, 11 are gases under atmospheric conditions. Six of these are the Group 8A elements (the noble gases He, Ne, Ar, Kr, Xe, and Rn), and the other 5 are hydrogen (H_2), nitrogen (N_2), oxygen (O_2), fluorine (F_2), and chlorine (Cl_2). Curiously, only 2 elements are liquids at 25°C: mercury (Hg) and bromine (Br_2).

We do not know the properties of all the known elements because some of them have never been prepared in quantities large enough for investigation. In these cases we must rely on periodic trends to predict their properties. What are the chances, then, of discovering a third liquid element?

Let us look at francium (Fr), the last member of Group 1A, to see if it would qualify as a liquid at 25°C. All of francium's isotopes are radioactive. The most stable isotope is francium-223, which has a half-life of 21 minutes. (*Half-life* is the time it takes for one-half of the atoms in any given amount of a radioactive substance to disintegrate.) This short half-life means that only very small traces of francium could possibly exist on Earth. And although it is feasible to prepare francium in the laboratory, no weighable quantity of the element has been prepared or isolated. Thus we know very little about francium's physical and chemical properties. Yet we can use the group periodic trends to predict some of those properties.

Take francium's melting point as an example. Figure 8.13 shows how the melting points of the alkali metals vary with atomic number. From lithium to sodium, the melting point drops 81.4°; from sodium to potassium, 34.6°; from potassium to rubidium, 24°; from rubidium to cesium, 11°. On the basis of this trend, we can predict that the drop from cesium to francium would be about 5°. If so, the melting point of francium would be 23°C, which would make it a liquid under atmospheric conditions.

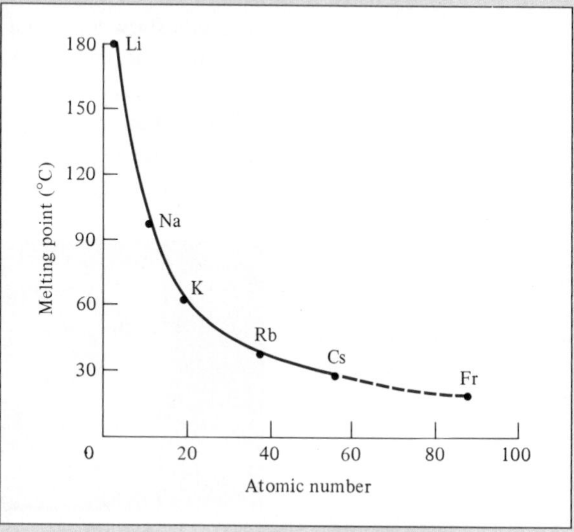

Figure 8.13 *A plot of the melting points of the alkali metals versus their atomic numbers. By extrapolation, francium is predicted to have a melting point of 23°C.*

strongly. That is why the heats of fusion and vaporization reach a maximum at silicon. (The stronger the force of attraction between the atoms, the greater the amount of energy needed to bring about the transition from solid to liquid or from liquid to vapor.) Starting with phosphorus, the elements exist in simple, discrete molecular units (P_4, S_8, Cl_2, and Ar). These molecules or atoms (for argon) are held together by relatively weak forces called *intermolecular forces* that operate among molecules. Therefore, they have lower heats of fusion and vaporization than do the metals and silicon.

The density of an element depends on three quantities: the atomic mass, the size of the atoms, and the way in which the atoms are packed together in the condensed state. (Recall that density is mass/volume.) We see that the density, like some of the other properties, rises as we move to the elements in the middle of the period, and then falls. The electrical conductivity and thermal conductivity are high for the metals and then fall rapidly as we move across the table. This is consistent with the fact that metals are good conductors of heat and electricity, whereas nonmetals are poor conductors in both respects. Note that silicon, a metalloid, has intermediate properties.

Similar trends are observed for representative elements in other periods. As we noted earlier, the transition metals, the lanthanides, and the actinides do not show this kind of periodic variation.

Predicting Physical Properties

One benefit of knowing the periodic trends in physical properties is that we can use this knowledge to predict properties of elements. Suppose we want to predict the boiling point of an element. One approach is to make use of the known boiling points of the element's immediate neighbors in the *same group*. In some cases we find that the boiling point of the element is fairly close to the average of the boiling points of the elements immediately above and below it in the group. The following example illustrates this approach.

EXAMPLE 8.4

The melting points of chlorine (Cl_2) and iodine (I_2) are −101.0°C and 113.5°C, and their boiling points are −34.6°C and 184.4°C, respectively. From these data estimate the melting point and boiling point of bromine (Br_2).

Answer

Cl_2, Br_2, and I_2 belong to Group 7A, the halogens. Since they are all nonmetals, we expect a gradual change in the melting point and boiling point of the elements as we move down the group. Bromine is located between chlorine and iodine; we therefore expect many of its properties to fall between those of chlorine and iodine. Thus we can estimate the melting point and boiling point of bromine by taking the average for those of chlorine and iodine. We set up the following table:

	Melting point (°C)	*Boiling point (°C)*
Cl_2	−101.0	−34.6
Br_2	?	?
I_2	113.5	184.4

$$\text{melting point of } Br_2 = \frac{(-101.0°C) + 113.5°C}{2} = 6.3°C$$

and

$$\text{boiling point of } Br_2 = \frac{(-34.6°C) + 184.4°C}{2} = 74.9°C$$

The actual melting point and boiling point of bromine are −7.2°C and 58.8°C, respectively, and so these estimates can be considered good "ballpark" values.

Similar problem: 8.47.

8.4 Ionization Energy

Before we examine the periodic variations in chemical properties, we need to discuss two concepts that play important roles in determining whether atoms of the elements will preferentially form ionic or molecular compounds: ionization energy and electron affinity (Section 8.5).

Ionization energies are measured in kilojoules per mole (kJ/mol), that is, the amount of energy in kilojoules needed to remove 1 mole of electrons from 1 mole of gaseous atoms (or ions).

Ionization energy is *the minimum energy required to remove an electron from a gaseous atom in its ground state*. The magnitude of ionization energy is a measure of the effort required to force an atom to give up an electron, or of how "tightly" the electron is held in the atom. The higher the ionization energy, the more difficult it is to remove the electron.

Ionization Energies of Many-Electron Atoms

For a many-electron atom, the amount of energy required to remove the first electron from the atom in its ground state,

$$\text{energy} + \text{X}(g) \longrightarrow \text{X}^+(g) + e^- \qquad (8.1)$$

is called the *first ionization energy* (I_1). In Equation (8.1), X represents an atom of any element, e is an electron, and g denotes the gaseous state. Unlike an atom in the condensed phases (liquid and solid), an atom in the gaseous phase is virtually uninfluenced by its neighbors. The *second ionization energy* (I_2) and the *third ionization energy* (I_3) are shown in the following equations:

$$\text{energy} + \text{X}^+(g) \longrightarrow \text{X}^{2+}(g) + e^- \qquad \text{second ionization}$$

$$\text{energy} + \text{X}^{2+}(g) \longrightarrow \text{X}^{3+}(g) + e^- \qquad \text{third ionization}$$

The pattern continues for the removal of subsequent electrons.

After an electron is removed from a neutral atom, the repulsion among the remaining electrons decreases. Since the nuclear charge remains constant, more energy is needed to remove another electron from the positively charged ion. Thus, ionization energies always increase in the following order:

$$I_1 < I_2 < I_3 < \cdots$$

Ionization is always an endothermic process.

Table 8.4 lists the ionization energies of the first 20 elements. By convention, energy absorbed by atoms (or ions) in the ionization process has a positive value. Thus ionization energies are all positive quantities. Figure 8.14 shows the variation of the first ionization energy with atomic number. The plot clearly exhibits the periodicity in the stability of the most loosely held electron. Note that, apart from small irregularities, the ionization energies of elements in a period increase with increasing atomic number. We can explain this trend by referring to the increase in effective nuclear charge from left to right (as in the case of atomic radii variation). A larger effective nuclear charge means a more tightly held outer electron, and hence a higher first ionization energy. A notable feature of Figure 8.14 is the peaks, which correspond to the noble gases. This is consistent with the fact that most noble gases are chemically unreactive because they have such high ionization energies. In fact, helium has the highest first ionization energy of all the elements.

This resistance to chemical change is attributable to the stability of the $1s^2$ configuration of helium and the completely filled outer *s* and *p* subshells of the other noble gases.

The Group 1A elements (the alkali metals) at the bottom of the graph in Figure 8.14 have the lowest ionization energies. Each of these metals has one valence electron (the

Table 8.4 The ionization energies (kJ/mol) of the first 20 elements

Z	Element	First	Second	Third	Fourth	Fifth	Sixth
1	H	1312					
2	He	2373	5248				
3	Li	520	7300	11808			
4	Be	899	1757	14850	20992		
5	B	801	2430	3660	25000	32800	
6	C	1086	2350	4620	6220	38000	47232
7	N	1400	2860	4580	7500	9400	53000
8	O	1314	3390	5300	7470	11000	13000
9	F	1680	3370	6050	8400	11000	15200
10	Ne	2080	3950	6120	9370	12200	15000
11	Na	495.9	4560	6900	9540	13400	16600
12	Mg	738.1	1450	7730	10500	13600	18000
13	Al	577.9	1820	2750	11600	14800	18400
14	Si	786.3	1580	3230	4360	16000	20000
15	P	1012	1904	2910	4960	6240	21000
16	S	999.5	2250	3360	4660	6990	8500
17	Cl	1251	2297	3820	5160	6540	9300
18	Ar	1521	2666	3900	5770	7240	8800
19	K	418.7	3052	4410	5900	8000	9600
20	Ca	589.5	1145	4900	6500	8100	11000

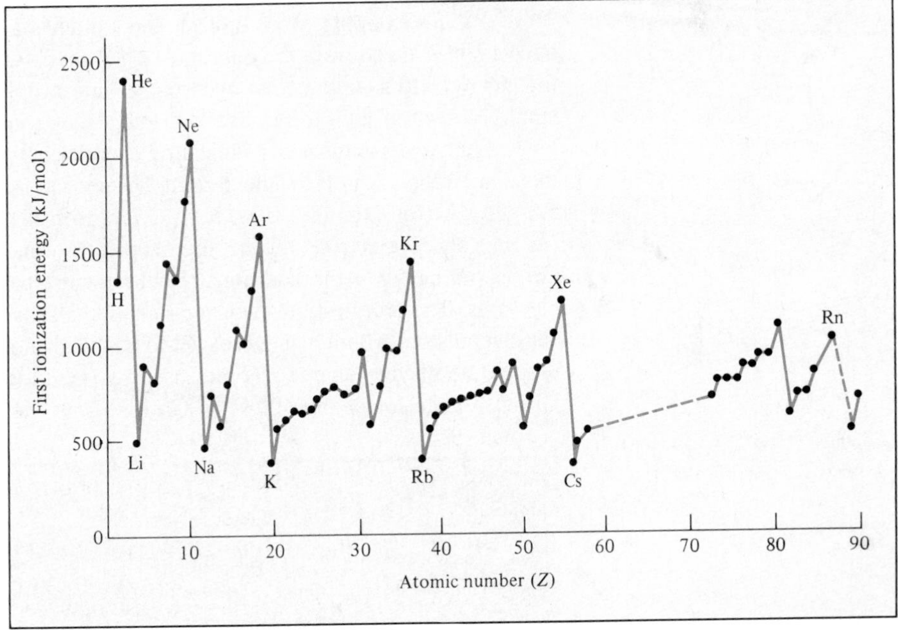

Figure 8.14 *Variation of the first ionization energy with atomic number. Note that the noble gases have high ionization energies, whereas the alkali metals and alkaline earth metals have low ionization energies.*

outermost electron configuration is ns^1) that is effectively shielded by the completely filled inner shells. Consequently, it is energetically easy to remove an electron from the atom of an alkali metal to form a unipositive ion (Li^+, Na^+, K^+, . . .). Significantly, the electron configurations of these ions are isoelectronic with those noble gases just preceding them in the periodic table.

Recall that the electron configurations of noble gases represent stability.

The Group 2A elements (the alkaline earth metals) have higher first ionization energies than the alkali metals do. The alkaline earth metals have two valence electrons (the outermost electron configuration is ns^2). Because these two *s* electrons do not shield each other well, the effective nuclear charge for an alkaline earth metal atom is larger than that for the preceding alkali metal atom. Most alkaline earth compounds contain dipositive ions (Mg^{2+}, Ca^{2+}, Sr^{2+}, Ba^{2+}). The Be^{2+} ion is isoelectronic with Li^+ and with He, Mg^{2+} is isoelectronic with Na^+ and with Ne, and so on.

The only important nonmetallic cation is the ammonium ion, NH_4^+.

From the preceding discussion, we see that metals have relatively low ionization energies, whereas nonmetals possess much higher ionization energies. The ionization energies of the metalloids usually fall between those of metals and nonmetals. The difference in ionization energies suggests why metals always form cations and nonmetals form anions in ionic compounds. For a given group, the ionization energy decreases with increasing atomic number (that is, as we move down the group). Elements in the same group have similar outer electron configurations. However, as the principal quantum number *n* increases, so does the average distance of a valence electron from the nucleus. A greater separation between the electron and the nucleus means a weaker attraction, so that the electron becomes increasingly easier to remove as we go from element to element down a group. Thus the metallic character of the elements within a group increases from top to bottom. This trend is particularly noticeable for elements in Groups 3A to 7A. For example, in Group 4A we note that carbon is a nonmetal, silicon and germanium are metalloids, and tin and lead are metals.

Although the general trend in the periodic table is that of ionization energies increasing from left to right, some irregularities do exist. The first occurs in going from Group 2A to 3A (for example, from Be to B and from Mg to Al). The Group 3A elements all have a single electron in the outermost *p* subshell (ns^2np^1), which is well shielded by the inner electrons and the ns^2 electrons. Therefore, less energy is needed to remove a single *p* electron than to remove a paired *s* electron from the same principal energy level. This explains the *lower* ionization energies in Group 3A elements compared with those in Group 2A in the same period. The second irregularity occurs between Groups 5A and 6A (for example, from N to O and from P to S). In the Group 5A elements (ns^2np^3) the *p* electrons are in three separate orbitals according to Hund's rule. In Group 6A (ns^2np^4) the additional electron must be paired with one of the three *p* electrons. The proximity of two electrons in the same orbital results in greater electrostatic repulsion, which makes it easier to ionize an atom of the Group 6A element, even though the nuclear charge has increased by one unit. Thus the ionization energies in Group 6A are *lower* than those in Group 5A in the same period.

EXAMPLE 8.5

(a) Which atom should have a smaller first ionization energy: oxygen or sulfur? (b) Which atom should have a higher second ionization energy: lithium or beryllium?

Answer

(a) Oxygen and sulfur are members of Group 6A. Since the ionization energy of elements decreases as we move down a periodic group, S should have a smaller first ionization energy.

(b) The electron configurations of Li and Be are $1s^22s^1$ and $1s^22s^2$, respectively. For the second ionization process we write

$$\underset{1s^2}{Li^+(g)} \longrightarrow \underset{1s^1}{Li^{2+}(g)}$$

$$\underset{1s^22s^1}{Be^+(g)} \longrightarrow \underset{1s^2}{Be^{2+}(g)}$$

Since 1*s* electrons shield 2*s* electrons much more effectively than they shield each other, we predict that it should be much easier to remove a 2*s* electron from Be^+ than to remove a 1*s* electron from Li^+. Table 8.4 confirms our prediction.

Similar problem: 8.54.

The importance of ionization energy lies in the close correlation between electron configuration (a microscopic property) and chemical behavior (a macroscopic property). As we will see throughout the book, the chemical properties of any atom are related to the configuration of the atom's valence electrons, and the stability of these electrons is reflected directly in the atom's ionization energies.

8.5 Electron Affinity

Another property of atoms that greatly influences their chemical behavior is their ability to accept one or more electrons. This ability is measured by ***electron affinity,*** which is *the energy change when an electron is accepted by an atom in the gaseous state*. The equation is

$$X(g) + e^- \longrightarrow X^-(g)$$

where X is an atom of an element. In accordance with the convention used in thermochemistry (see Chapter 6), we assign a negative value to the electron affinity when energy is released. The more negative the electron affinity, the greater the tendency of the atom to accept an electron. Table 8.5 shows the electron affinity values of some

Table 8.5 Electron affinities (kJ/mol) of representative elements*

1A	2A	3A	4A	5A	6A	7A	8A
H −77							He (21)
Li −58	Be (241)	B −23	C −123	N 0	O −142	F −333	Ne (29)
Na −53	Mg (230)	Al −44	Si −120	P −74	S −200	Cl −348	Ar (35)
K −48	Ca (154)	Ga (−35)	Ge −118	As −77	Se −195	Br −324	Kr (39)
Rb −47	Sr (120)	In −34	Sn −121	Sb −101	Te −190	I −295	Xe (40)
Cs −45	Ba (52)	Tl −48	Pb −101	Bi −100	Po ?	At ?	Rn ?

*The values in parentheses are estimates.

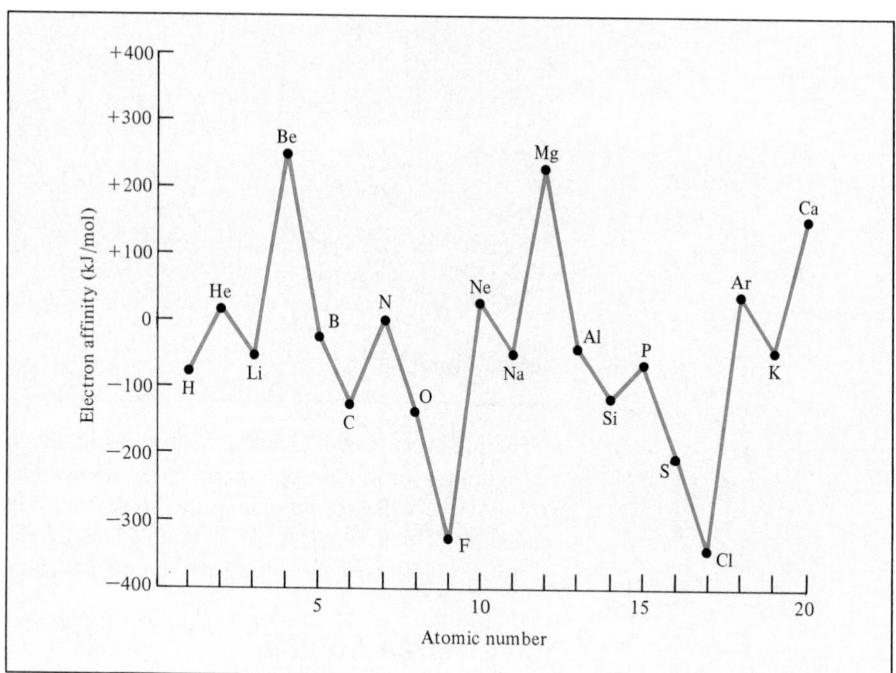

Figure 8.15 *A plot of electron affinity against atomic number for the first 20 elements.*

representative elements and the noble gases, and Figure 8.15 plots the values of the first 20 elements versus atomic number. The overall trend is that the tendency to accept electrons increases (that is, electron affinity values become more negative) as we move from left to right across a period. The electron affinities of metals are generally more positive (or less negative) than those of nonmetals. The values differ little within a given group. The halogens (Group 7A) have the most negative electron affinity values. This is not surprising when we realize that by accepting an electron, each halogen atom assumes the stable electron configuration of the noble gas immediately following it. For example, the electron configuration of F^- is $1s^2 2s^2 2p^6$ or [Ne]; for Cl^- it is $[Ne]3s^2 3p^6$ or [Ar]; and so on. The noble gases, which have filled outer *s* and *p* subshells, have no tendency to accept electrons.

The electron affinity of oxygen has a negative value, which means that the process

$$O(g) + e^- \longrightarrow O^-(g)$$

is favorable. On the other hand, the electron affinity of the O^- ion,

$$O^-(g) + e^- \longrightarrow O^{2-}(g)$$

is positive (780 kJ/mol) even though the O^{2-} ion is isoelectronic with the noble gas Ne. This process is unfavorable in the gas phase because the resulting increase in electron-electron repulsion outweighs the stability gained in achieving a noble gas configuration. However, note that ions such as O^{2-} are common in ionic compounds (for example, Li_2O and MgO); in solids, these ions are stabilized by the neighboring cations.

The following example shows why the alkaline earth metals have no tendency to accept electrons.

EXAMPLE 8.6

Explain why the electron affinities of the alkaline earth metals are all positive.

Answer

The outer-shell electron configuration of the alkaline earth metals is ns^2. For the process

$$M(g) + e^- \longrightarrow M^-(g)$$

where M denotes a member of the Group 2A family, the extra electron must enter the *np* subshell, which is effectively shielded by the two *ns* electrons (the *np* electrons are farther away from the nucleus than the *ns* electrons). Consequently, alkaline earth metals have no tendency to pick up an extra electron.

Similar problem: 8.64.

8.6 Variation in Chemical Properties

On a conceptual level these two fundamental quantities are related in a simple way: Ionization energy indexes the attraction of an atom for its own outer electrons, whereas electron affinity expresses the attraction of an atom for an additional electron from some other source. Together they give us insight into the general attraction of an atom for electrons.

Ionization energy and electron affinity are two characteristics that help chemists understand the types of reactions that elements undergo and the nature of the elements' compounds. Utilizing these concepts, we can survey the chemical behavior of the elements systematically, paying particular attention to the relationship between chemical properties and electron configuration.

We have seen that as we move across a period from left to right, the metallic character of the elements decreases. And as we move down a group in the periodic table, the metallic character of the elements increases. On the basis of these trends and the knowledge that metals usually have low ionization energies while nonmetals usually have high (more negative) electron affinities, we can frequently predict the chemical outcome when atoms of some of these elements react with one another. The following discussion focuses only on the representative elements. We will consider the chemistry of transition metals in Chapter 22.

General Trends in Chemical Properties

Before we study the elements in individual groups, let us look at some overall trends. We have said that elements in the same group resemble one another in chemical behavior because they have similar outer electron configurations. This statement, although correct in the general sense, must be applied with caution. Chemists have long known that the first member of each group (that is, the element in the second period from lithium to fluorine) differs from the rest of the members of the same group. Thus lithium, while exhibiting many of the properties characteristic of the alkali metals, differs in some ways from the rest of the metals in Group 1A. Similarly, beryllium is somewhat atypical of the members of Group 2A, and so on. The reason for the difference is the unusually small size of the first member of each group.

Another trend in chemical behavior of the representative elements is the ***diagonal***

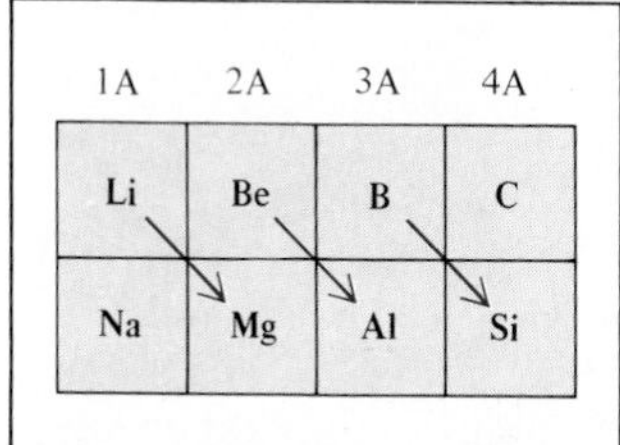

Figure 8.16 *Diagonal relationships in the periodic table. The relationship holds for lithium and magnesium, beryllium and aluminum, and boron and silicon.*

relationship. Diagonal relationship refers to *similarities that exist between pairs of elements in different groups and periods of the periodic table*. Specifically, the first three members of the second period (Li, Be, and B) exhibit many similarities to those elements located diagonally below them in the periodic table (Figure 8.16). The chemistry of lithium resembles that of magnesium in some ways; the same holds for beryllium and aluminum and for boron and silicon. Thus each pair of these elements is said to exhibit a diagonal relationship. We will see a number of examples illustrating this relationship later.

In comparing the properties of elements in the same group, bear in mind that the comparison is most valid if we are dealing with elements of the same type. This guideline applies to the elements in Groups 1A and 2A, which are all metals, and to the elements in Group 7A, which are all nonmetals. In Groups 3A through 6A, where the elements change either from nonmetals to metals or from nonmetals to metalloids, it is natural to expect greater variations in chemical properties even though the members of the same group have similar outer electron configurations.

We will now survey the chemical properties of hydrogen and the elements in Groups 1A through 8A.

Chemical Properties in Individual Groups

Hydrogen ($1s^1$). There is no totally suitable position for hydrogen in the periodic table. It resembles the alkali metals in having a single *s* valence electron and forming the H^+ ion, which is hydrated in solution [similar, for example, to the $Na^+(aq)$ ions]. On the other hand, hydrogen also forms the hydride ion (H^-), which is too reactive to exist in water but does exist in some ionic compounds. In this respect, hydrogen resembles the halogens, since they all form halide ions (F^-, Cl^-, Br^-, and I^-). Traditionally hydrogen is shown in Group 1A in the periodic table, but you should *not* think of it as a member of that group. The most important compound of hydrogen is, of course, water, which is formed when hydrogen burns in air:

$$2H_2(g) + O_2(g) \longrightarrow 2H_2O(l)$$

Li
Na
K
Rb
Cs

Group 1A Elements (ns^1, $n \geq 2$). Figure 8.17 shows the Group 1A elements. The alkali metals all have low ionization energies and therefore a great tendency to lose their single valence electron. In fact, in the vast majority of their compounds they are unipositive ions. These metals are so reactive that they are never found in the free, uncombined state in nature. They react with water to produce hydrogen gas and the corresponding metal hydroxide:

$$2M(s) + 2H_2O(l) \longrightarrow 2MOH(aq) + H_2(g)$$

where M denotes an alkali metal. When exposed to air, they gradually lose their shiny appearance as they combine with oxygen gas to form several different types of oxides or oxygen compounds. Lithium differs from the rest of the alkali metals in that it forms the oxide (containing the O^{2-} ion)

$$4Li(s) + O_2(g) \longrightarrow 2Li_2O(s)$$

The other alkali metals all form peroxides (containing the O_2^{2-} ion) in addition to oxides. For example,

$$2Na(s) + O_2(g) \longrightarrow Na_2O_2(s)$$

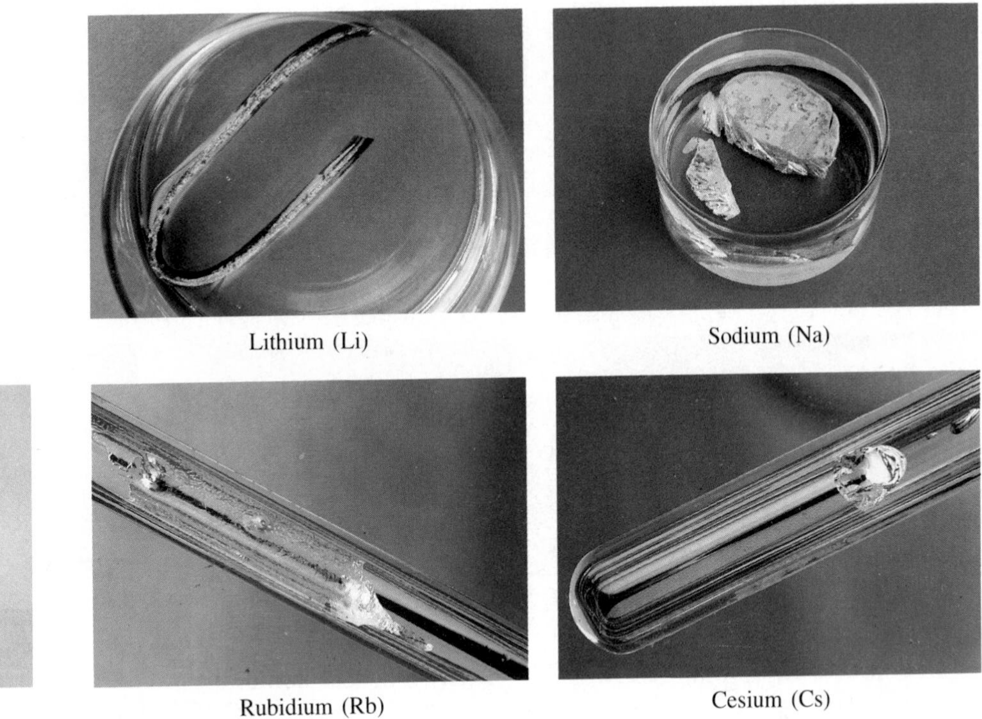

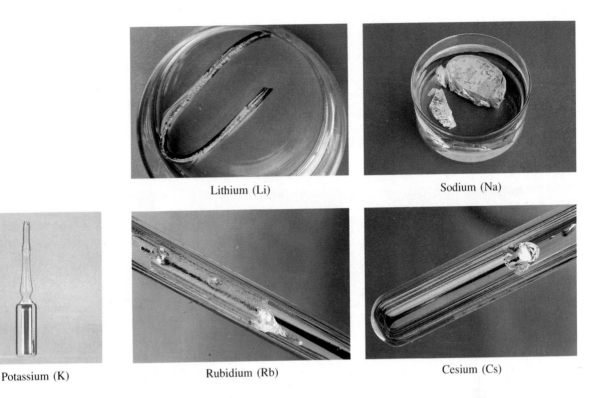

Figure 8.17 *The Group 1A elements: the alkali metals. Francium is a radioactive element.*

Potassium, rubidium, and cesium also form superoxides (containing the O_2^- ion):

$$K(s) + O_2(g) \longrightarrow KO_2(s)$$

The reason that different types of oxides are formed when alkali metals react with oxygen has to do with the stability of the oxides in the solid state. Since these oxides are all ionic compounds, their stability depends on how strongly the cations and anions attract one another (or the lattice energy of the solid). Lithium tends to form predominantly lithium oxide because this compound has a greater stability (larger lattice energy) than lithium peroxide. The formation of other alkali metal oxides can be similarly explained.

Be
Mg
Ca
Sr
Ba

Group 2A Elements (ns^2, $n \geq 2$). Figure 8.18 shows the Group 2A elements. As a group, the alkaline earth metals are also reactive metals, but less so than the alkali metals. Both the first and the second ionization energies decrease from beryllium to barium. Thus the tendency is to form M^{2+} ions (where M denotes an alkaline metal atom), and hence the metallic character increases as we go down the group. Most beryllium compounds are molecular rather than ionic in nature. The reactivities of alkaline earth metals toward water differ quite markedly. Beryllium does not react with water; magnesium reacts slowly with steam; calcium, strontium, and barium are reactive enough to attack cold water:

$$Ba(s) + 2H_2O(l) \longrightarrow Ba(OH)_2(aq) + H_2(g)$$

The reactivities of the alkaline earth metals toward oxygen also increase from Be to Ba. Beryllium and magnesium form oxides (BeO and MgO) only at elevated temperatures, whereas CaO, SrO, and BaO are formed at room temperature.

Figure 8.18 *The Group 2A elements: the alkaline earth metals.*

Magnesium reacts with acids to liberate hydrogen gas:

$$Mg(s) + 2H^+(aq) \longrightarrow Mg^{2+}(aq) + H_2(g)$$

Calcium, strontium, and barium also react with acids to produce hydrogen gas. However, because these metals also attack water, two different reactions will occur simultaneously.

The chemical properties of calcium and strontium provide an interesting example of periodic group similarity. Strontium-90, a radioactive isotope, is a major product of an atomic bomb explosion. If an atomic bomb is exploded in the atmosphere, the strontium-90 formed will eventually settle on land and water alike, and reach our bodies via a relatively short food chain. For example, cows eat contaminated grass and drink contaminated water, and then they pass along strontium-90 in their milk. Because calcium and strontium are chemically similar, Sr^{2+} ions can replace Ca^{2+} ions in our bodies—for example, in bones. Constant exposure of the body to high-energy radiation emitted by the strontium-90 isotopes can lead to anemia, leukemia, and other chronic illnesses.

B
Al
Ga
In
Tl

Group 3A Elements (ns^2np^1, $n \geq 2$). The first member of Group 3A, boron, is a metalloid; the rest are metals (Figure 8.19). Boron does not form binary ionic compounds and is unreactive toward oxygen gas and water. The next element, aluminum, readily forms aluminum oxide when exposed to air:

$$4Al(s) + 3O_2(g) \longrightarrow 2Al_2O_3(s)$$

Aluminum that has a protective layer of aluminum oxide appears less reactive than elemental aluminum. The Group 3A metallic elements form unipositive and tripositive ions. Aluminum, which can form only tripositive ions, reacts with hydrochloric acid as follows:

$$2Al(s) + 6H^+(aq) \longrightarrow 2Al^{3+}(aq) + 3H_2(g)$$

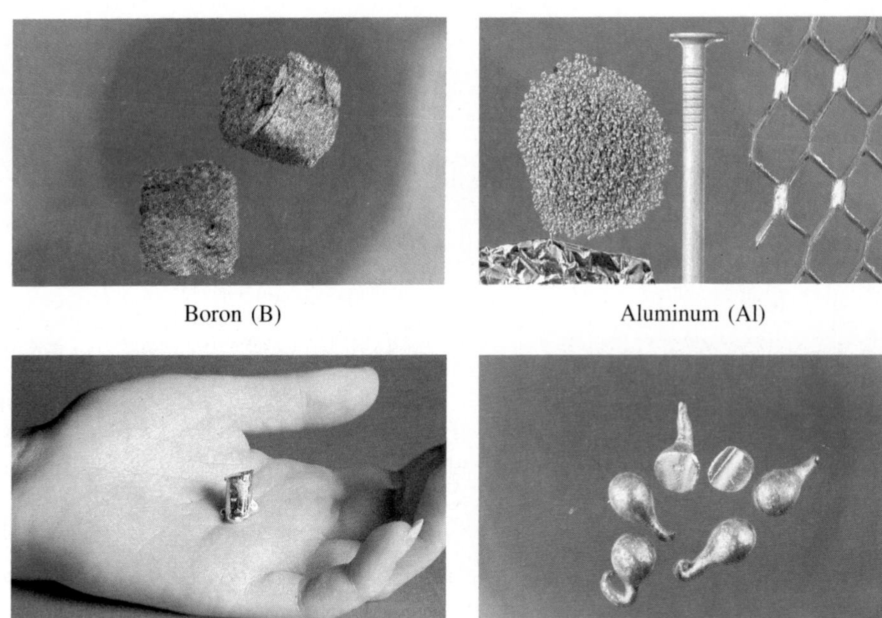

Figure 8.19 *The Group 3A elements.*

Moving down the group, we find that the unipositive ion becomes more stable than the tripositive ion. For example, Tl^{+} is much more stable than Tl^{3+}. The tendency for thallium ($6s^26p^1$) to lose only the $6p$ electron to form Tl^{+} rather than both the $6p$ and the $6s$ electrons to form Tl^{3+} is sometimes called the ***inert-pair effect,*** which refers to the *two relatively stable and unreactive outer s electrons.* The inert-pair effect, which is exhibited by the heavier elements in the group, is also observed for Groups 4A and 5A elements. Note that the metallic elements in Group 3A also form many molecular compounds. Thus, as we move from left to right across the periodic table, we begin to see a gradual shift from metallic character to nonmetallic character for the representative elements.

C
Si
Ge
Sn
Pb

Group 4A Elements (ns^2np^2, $n \geqslant 2$). The first member of Group 4A, carbon, is a nonmetal, and the next two members, silicon and germanium, are metalloids (Figure 8.20). They do not form ionic compounds. The metallic elements of this group, tin and lead, do not react with water but react with acids (hydrochloric acid, for example) to liberate hydrogen gas:

$$Sn(s) + 2H^+(aq) \longrightarrow Sn^{2+}(aq) + H_2(g)$$

$$Pb(s) + 2H^+(aq) \longrightarrow Pb^{2+}(aq) + H_2(g)$$

The Group 4A elements form compounds in both the +2 and +4 oxidation states. For the early members (carbon and silicon), the +4 oxidation state is the more stable one. For example, CO_2 is more stable than CO; and while SiO_2 is a stable compound, SiO does not exist under normal conditions. As we move down the group, however, the trend in stability is reversed. In tin compounds the +4 oxidation is only slightly more stable than the +2 oxidation state. In lead compounds the +2 oxidation state is unquestionably the more stable one. The outer electron configuration of lead is $6s^26p^2$.

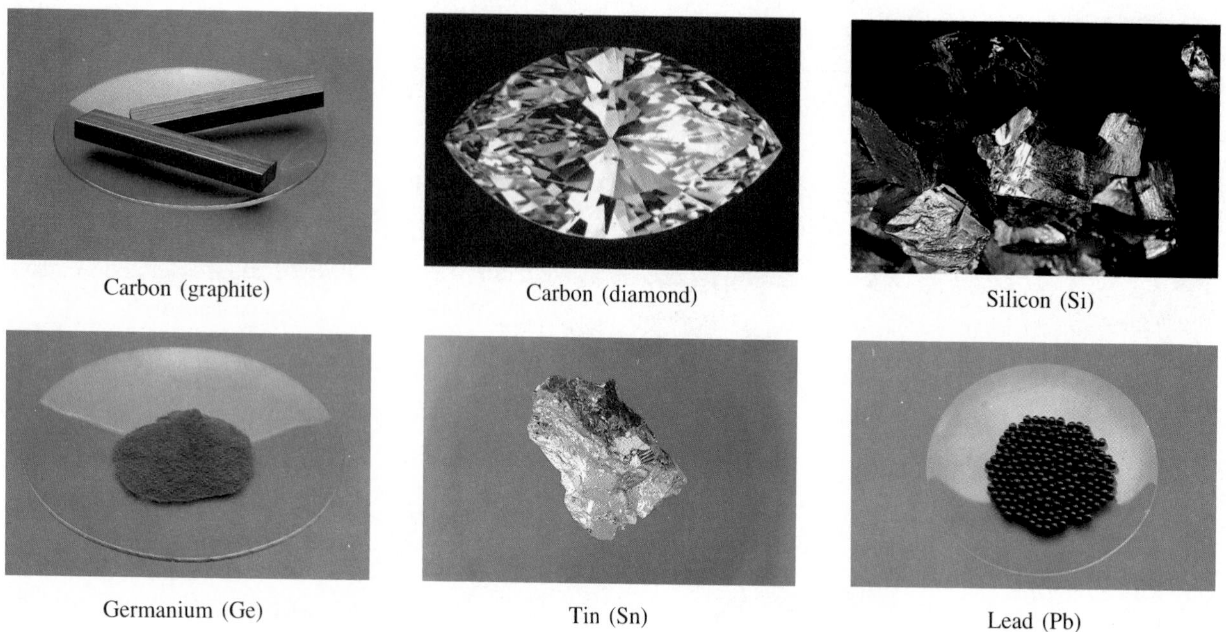

Figure 8.20 *The Group 4A elements.*

The tendency of lead to lose only the 6*p* electrons (to form Pb^{2+}) rather than both the 6*p* and 6*s* electrons (to form Pb^{4+}) is again attributed to the inert pair effect (due to the 6*s* electrons).

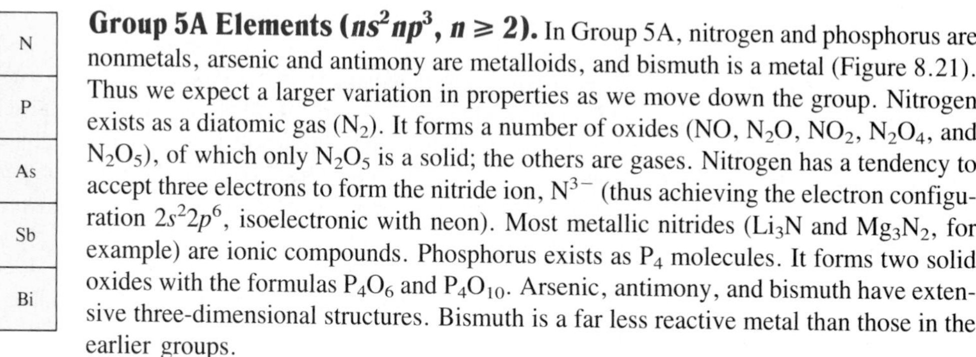

Group 5A Elements (ns^2np^3, $n \geq 2$). In Group 5A, nitrogen and phosphorus are nonmetals, arsenic and antimony are metalloids, and bismuth is a metal (Figure 8.21). Thus we expect a larger variation in properties as we move down the group. Nitrogen exists as a diatomic gas (N_2). It forms a number of oxides (NO, N_2O, NO_2, N_2O_4, and N_2O_5), of which only N_2O_5 is a solid; the others are gases. Nitrogen has a tendency to accept three electrons to form the nitride ion, N^{3-} (thus achieving the electron configuration $2s^22p^6$, isoelectronic with neon). Most metallic nitrides (Li_3N and Mg_3N_2, for example) are ionic compounds. Phosphorus exists as P_4 molecules. It forms two solid oxides with the formulas P_4O_6 and P_4O_{10}. Arsenic, antimony, and bismuth have extensive three-dimensional structures. Bismuth is a far less reactive metal than those in the earlier groups.

O
S
Se
Te
Po

Group 6A Elements (ns^2np^4, $n \geq 2$). The first three members of Group 6A (oxygen, sulfur, and selenium) are nonmetals, and the last two (tellurium and polonium) are metalloids (Figure 8.22). Oxygen is a diatomic gas; sulfur and selenium exist as S_8 and Se_8 units; tellurium and polonium have more extensive three-dimensional structures. (Polonium is a radioactive element that is difficult to study in the laboratory.) Oxygen has a tendency to accept two electrons to form the oxide ion (O^{2-}) in many ionic compounds. (The electron configuration of O^{2-} is $1s^22s^22p^6$, isoelectronic with Ne.) Sulfur, selenium, and tellurium also form anions of the type S^{2-}, Se^{2-}, and Te^{2-}. The elements in this group (especially oxygen) form a large number of molecular compounds with nonmetals.

Group 7A Elements (ns^2np^5, $n \geq 2$). All the halogens are nonmetals with the

Nitrogen (N_2)

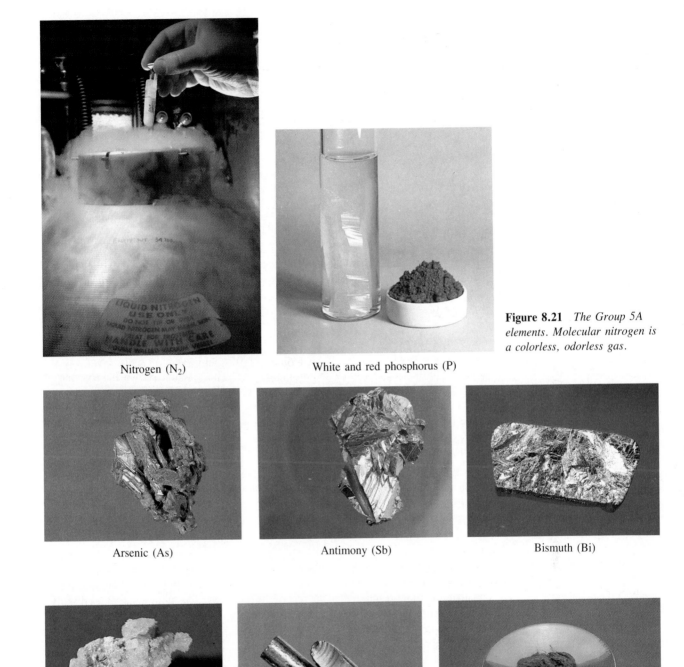

White and red phosphorus (P)

Figure 8.21 *The Group 5A elements. Molecular nitrogen is a colorless, odorless gas.*

Arsenic (As)

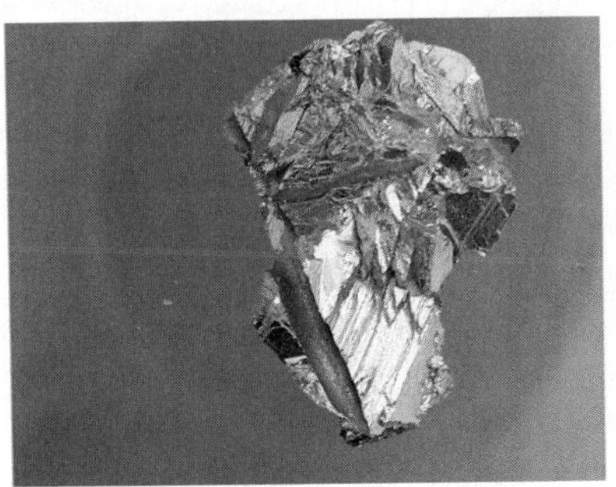

Antimony (Sb)

Bismuth (Bi)

Sulfur (S_8)

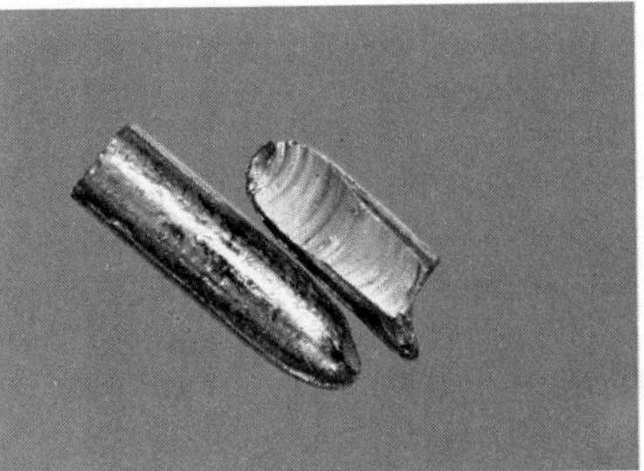

Selenium (Se_8)

Tellurium (Te)

Figure 8.22 *The Group 6A elements sulfur, selenium, and tellurium. Molecular oxygen is a colorless, odorless gas. Polonium is a radioactive element.*

Figure 8.23 *The Group 7A elements chlorine, bromine, and iodine. Fluorine is a greenish-yellow gas that attacks ordinary glassware. Astatine is a radioactive element.*

F
Cl
Br
I
At

general formula X_2, where X denotes a halogen element (Figure 8.23). Because of their great reactivity, the halogens are never found in the elemental form in nature. (The last member of Group 7A is astatine, a radioactive element. Little is known about its properties.) Fluorine is so reactive that it attacks water to generate oxygen:

$$2F_2(g) + 2H_2O(l) \longrightarrow 4HF(aq) + O_2(g)$$

Actually the reaction between molecular fluorine and water is quite complex; the products formed depend on reaction conditions. The reaction shown here is one of several possible changes. The halogens have high ionization energies and large negative electron affinities. These facts suggest that they would preferentially form anions of the type X^-. Anions derived from the halogens (F^-, Cl^-, Br^-, and I^-) are called *halides*. They are isoelectronic with the noble gases. For example, F^- is isoelectronic with Ne, Cl^- with Ar, and so on. The vast majority of the alkali metal and alkaline earth metal halides are ionic compounds. The halogens also form many molecular compounds among themselves (such as ICl and BrF_3) and with nonmetallic elements in other groups (such as PCl_5 and NF_3). The halogens react with hydrogen to form hydrogen halides:

$$H_2(g) + X_2(g) \longrightarrow 2HX(g)$$

When this reaction involves fluorine, it is explosive, but it becomes less and less violent as we substitute chlorine and iodine. This is one indication that fluorine is considerably more reactive than the rest of the halogens. The hydrogen halides dissolve in water to form hydrohalic acids. Of this series, hydrofluoric acid (HF) is a weak acid (that is, it is a weak electrolyte), but the other acids (HCl, HBr, and HI) are strong acids (or strong electrolytes).

He
Ne
Ar
Kr
Xe

Group 8A Elements (ns^2np^6, $n \geqslant 2$). All noble gases exist as monatomic species (Figure 8.24). This fact suggests that they are probably very unreactive and have little or no tendency to combine among themselves or with other elements. That is indeed the case. The electron configurations of the noble gases show that their atoms have completely filled outer *ns* and *np* subshells, indicating great stability. (Helium is $1s^2$.) The Group 8A ionization energies are among the highest of all elements, and they have no tendency to accept extra electrons. For a number of years these elements were

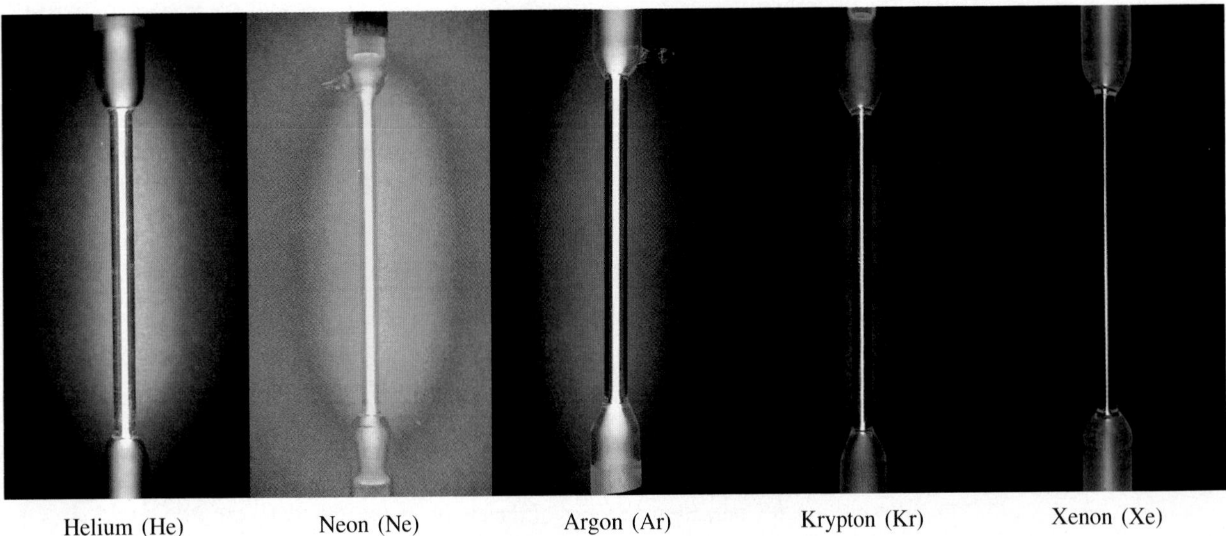

Figure 8.24 *All noble gases are colorless and odorless. These pictures show the colors emitted by the gases from a discharge tube.*

called inert gases due to their observed lack of chemical reactivity. Since 1962, however, a number of krypton and xenon compounds have been synthesized.

Comparison of Group 1A and Group 1B Elements

When we compare the Group 1A (alkali metals) and the Group 1B elements (copper, silver, and gold), we arrive at an interesting conclusion. Although the elements in these two groups have similar outer electron configurations, with one electron in the outermost *s* orbital, their chemical properties are quite different.

The first ionization energies of Cu, Ag, and Au are 745 kJ/mol, 731 kJ/mol, and 890 kJ/mol, respectively. Since these values are considerably larger than those of the alkali metals (see Table 8.4), the Group 1B elements are much less reactive. The higher ionization energies of the Group 1B elements result from incomplete shielding of the nucleus by the inner *d* electrons (compared with the more effective shielding of the completely filled noble gas cores). Consequently the outer *s* electrons of these elements are more strongly attracted by the nucleus. In fact, copper, silver, and gold are so unreactive that they are usually found in the uncombined state in nature. The inertness and rarity of these metals have made them valuable in the manufacture of coins and in jewelry. For this reason, these metals are also called coinage metals. The difference in chemical properties between the Group 2A elements (the alkaline earth metals) and the Group 2B metals (zinc, cadmium, and mercury) can be similarly explained.

Properties of Oxides across a Period

Some elements in the third period (P, S, and Cl) form several types of oxides. We consider only those oxides in which the element has the highest oxidation number.

One way to compare the properties of the representative elements across a period is to examine the properties of a series of similar compounds. Since oxygen combines with almost all elements, we will briefly compare the properties of oxides of the third-period elements to see how metals differ from metalloids and nonmetals. Table 8.6 lists a few general characteristics of the oxides. We observed earlier that oxygen has a tendency to

Table 8.6 Some properties of oxides of the third-period elements

	Na_2O	MgO	Al_2O_3	SiO_2	P_4O_{10}	SO_3	Cl_2O_7
Type of compound	←	Ionic	→	←	Molecular		→
Structure	←	Extensive three-dimensional		→	← Discrete molecular units		→
Melting point (°C)	1275	2800	2045	1610	580	16.8	−91.5
Boiling point (°C)	?	3600	2980	2230	?	44.8	82
Acid-base nature	Basic	Basic	Amphoteric	←	Acidic		→

form the oxide ion. This tendency is greatly favored when oxygen combines with metals that have low ionization energies, namely, those in Groups 1A and 2A and aluminum. Thus Na_2O, MgO, and Al_2O_3 are ionic compounds, as indicated by their high melting points and boiling points. They have extensive three-dimensional structures in which each cation is surrounded by a specific number of anions, and vice versa. As the ionization energies of the elements increase from left to right, so does the molecular nature of the oxides that are formed. Silicon is a metalloid; its oxide (SiO_2) also has a huge three-dimensional network, although no ions are present. The oxides of phosphorus, sulfur, and chlorine are molecular compounds composed of small discrete units. The weak attractions among these molecules result in relatively low melting points and boiling points.

Most oxides can be classified as acidic or basic depending on whether they produce acids or bases when dissolved in water or react as acids or bases in certain reactions. (A special type of oxide that displays both acidic and basic properties is called *amphoteric oxide,* and we will discuss one shortly.) The first two oxides of the third period, Na_2O and MgO, are basic oxides. For example, Na_2O reacts with water to form sodium hydroxide (which is a base):

$$Na_2O(s) + H_2O(l) \longrightarrow 2NaOH(aq)$$

Magnesium oxide is quite insoluble; it does not react with water to any appreciable extent. However, it does react with acids in a manner that resembles an acid-base reaction:

$$MgO(s) + 2HCl(aq) \longrightarrow MgCl_2(aq) + H_2O(l)$$

Note that the products of this reaction are a salt ($MgCl_2$) and water, the usual products from an acid-base neutralization. Aluminum oxide is even less soluble than magnesium oxide; it too does not react with water. However, it shows basic properties by reacting with acids:

$$Al_2O_3(s) + 6HCl(aq) \longrightarrow 2AlCl_3(aq) + 3H_2O(l)$$

Although the sole product is a salt and no water is formed, the reaction between Al_2O_3 and NaOH qualifies as an acid-base reaction.

It also shows acidic properties by reacting with bases:

$$Al_2O_3(s) + 2NaOH(aq) + 3H_2O(l) \longrightarrow 2NaAl(OH)_4(aq)$$

Thus Al_2O_3 is classified as an ***amphoteric oxide*** because it *exhibits both acidic and basic properties.*

Silicon dioxide is insoluble and does not react with water. It has acidic properties, however, because it reacts with very concentrated bases:

$$SiO_2(s) + 2NaOH(aq) \longrightarrow Na_2SiO_3(aq) + H_2O(l)$$

The remaining third-period oxides are acidic, as indicated by their reactions with water to form phosphoric acid (H_3PO_4), sulfuric acid (H_2SO_4), and perchloric acid ($HClO_4$):

$$P_4O_{10}(s) + 6H_2O(l) \longrightarrow 4H_3PO_4(aq)$$

$$SO_3(g) + H_2O(l) \longrightarrow H_2SO_4(aq)$$

$$Cl_2O_7(g) + H_2O(l) \longrightarrow 2HClO_4(aq)$$

This brief examination of oxides of the third-period elements tells us that as the metallic character of the elements decreases from left to right across the period, their oxides change from basic to amphoteric to acidic. Normal metallic oxides are usually basic, and most oxides of nonmetals are acidic. The intermediate properties of the oxides (as shown by the amphoteric oxides) are exhibited by elements whose positions are intermediate within the period. Note also that since the metallic character of the elements increases as we move down a particular group of the representative elements, we would expect oxides of elements with larger atomic numbers to be more basic than the lighter elements. This is indeed the case.

EXAMPLE 8.7

Classify the following oxides as acidic, basic, or amphoteric: (a) Rb_2O, (b) BeO, (c) As_2O_5.

Answer

(a) Since rubidium is an alkali metal, we would expect Rb_2O to be a basic oxide. This is indeed true, as shown by rubidium oxide's reaction with water to form rubidium hydroxide:

$$Rb_2O(s) + H_2O(l) \longrightarrow 2RbOH(aq)$$

(b) Beryllium is an alkaline earth metal. However, because it is the first member of Group 2A, we expect that it may differ somewhat from the other members in the group. Furthermore, beryllium and aluminum exhibit diagonal relationship, so that BeO may resemble Al_2O_3 in properties. It turns out that BeO, like Al_2O_3, is an amphoteric oxide, as shown by its reactions with acids and bases:

These reactions are analogous to those shown above for Al_2O_3.

$$BeO(s) + 2H^+(aq) + 3H_2O(l) \longrightarrow Be(H_2O)_4^{2+}(aq)$$

$$BeO(s) + 2OH^-(aq) + H_2O(l) \longrightarrow Be(OH)_4^{2-}(aq)$$

(c) Since arsenic is a nonmetal, we expect As_2O_5 to be an acidic oxide. This prediction is correct, as shown by the formation of arsenic acid when As_2O_5 reacts with water:

$$As_2O_5(s) + 3H_2O(l) \longrightarrow 2H_3AsO_4(aq)$$

The following Chemistry in Action examines the distribution of elements in Earth's crust and in the human body.

CHEMISTRY IN ACTION

Distribution of Elements in Earth's Crust and in Living Systems

In this chapter we discussed the physical and chemical properties of the elements, the majority of which are naturally occurring. How are these elements distributed in Earth's crust, and which of them are essential to living systems?

By Earth's crust, we mean the layer measured from the surface of Earth to a depth of about 40 km (about 25 miles). Because of technical difficulties, scientists have not been able to study the inner portions of Earth as easily as the crust. It is believed that there is a solid core made up mostly of iron and nickel at the center of Earth. Surrounding the core is a very hot fluid called the *mantle,* which is made up of iron, carbon, silicon, and sulfur (Figure 8.25).

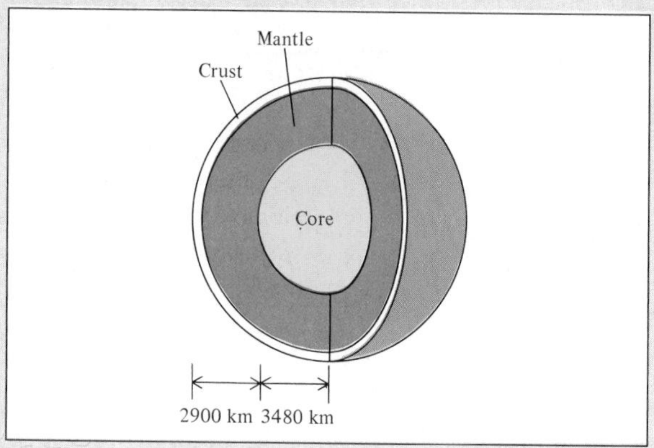

Figure 8.25 *Structure of Earth's interior.*

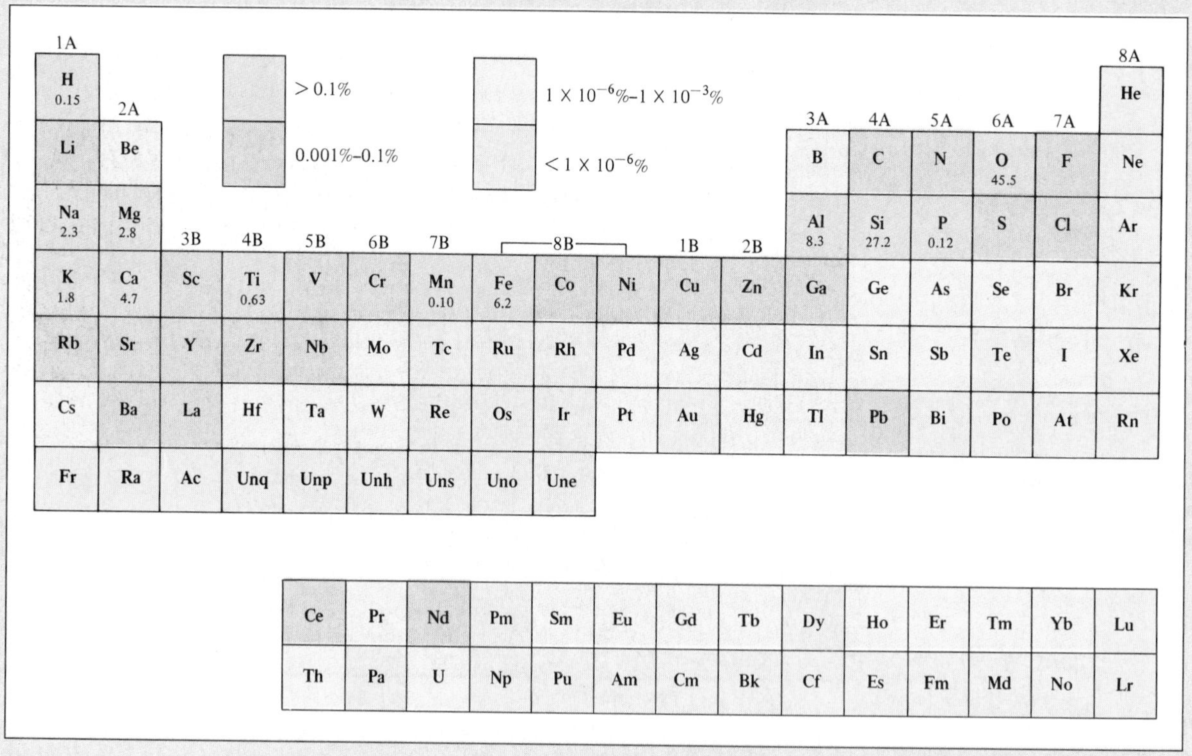

Legend: $> 0.1\%$; $0.001\%–0.1\%$; $1 \times 10^{-6}\%–1 \times 10^{-3}\%$; $< 1 \times 10^{-6}\%$

1A	2A	3B	4B	5B	6B	7B	8B	8B	8B	1B	2B	3A	4A	5A	6A	7A	8A
H 0.15																	He
Li	Be											B	C	N	O 45.5	F	Ne
Na 2.3	Mg 2.8											Al 8.3	Si 27.2	P 0.12	S	Cl	Ar
K 1.8	Ca 4.7	Sc	Ti 0.63	V	Cr	Mn 0.10	Fe 6.2	Co	Ni	Cu	Zn	Ga	Ge	As	Se	Br	Kr
Rb	Sr	Y	Zr	Nb	Mo	Tc	Ru	Rh	Pd	Ag	Cd	In	Sn	Sb	Te	I	Xe
Cs	Ba	La	Hf	Ta	W	Re	Os	Ir	Pt	Au	Hg	Tl	Pb	Bi	Po	At	Rn
Fr	Ra	Ac	Unq	Unp	Unh	Uns	Uno	Une									

Ce	Pr	Nd	Pm	Sm	Eu	Gd	Tb	Dy	Ho	Er	Tm	Yb	Lu
Th	Pa	U	Np	Pu	Am	Cm	Bk	Cf	Es	Fm	Md	No	Lr

Figure 8.26 *Natural abundance of the elements in percent by mass. For example, oxygen's abundance is 45.5 percent. This means that in a 100-g sample of Earth's crust there are, on the average, 45.5 g of the element oxygen.*

Of the 83 elements that are found in nature, 12 of them make up 99.7 percent of Earth's crust by mass. They are, in decreasing order of natural abundance, oxygen (O), silicon (Si), aluminum (Al), iron (Fe), calcium (Ca), magnesium (Mg), sodium (Na), potassium (K), titanium (Ti), hydrogen (H), phosphorus (P), and manganese (Mn) (Figure 8.26). In discussing the natural abundance of the elements, we should keep in mind that (1) the elements are not evenly distributed throughout Earth's crust, and (2) most elements occur in combined forms. These features provide the basis for most methods of obtaining pure elements from their compounds, as we will see in later chapters.

Table 8.7 lists the essential elements in the human body. In addition, the following elements are also found in living systems: aluminum (Al), bromine (Br), chromium (Cr), manganese (Mn), molybdenum (Mo), and silicon (Si).

Table 8.7 Essential elements in the human body

Element	Percent by mass*	Element	Percent by mass*
Oxygen	65	Sodium	0.1
Carbon	18	Magnesium	0.05
Hydrogen	10	Iron	<0.05
Nitrogen	3	Cobalt	<0.05
Calcium	1.5	Copper	<0.05
Phosphorus	1.2	Zinc	<0.05
Potassium	0.2	Iodine	<0.05
Sulfur	0.2	Selenium	<0.01
Chlorine	0.2	Fluorine	<0.01

**Percent by mass* gives the mass of the element present in grams in a 100-g sample.

Summary

1. The periodic table was developed by Newlands, Mendeleev, and Meyer. Nineteenth-century chemists constructed the periodic table by arranging elements in the increasing order of their atomic masses. Some discrepancies in the early version of the periodic table were resolved when Moseley showed that elements should be arranged according to their increasing atomic number.
2. Electron configuration directly influences the properties of the elements. The modern periodic table classifies the elements according to their atomic numbers, and thus also by their electron configurations. The configuration of the outermost electrons (called valence electrons) directly affects the properties of the atoms of the representative elements.
3. As we move across a period, periodic variations are found in physical properties of the elements. These variations can be understood in terms of differences in structure. The metallic character of elements decreases across a period from metals through the metalloids to nonmetals and increases from top to bottom within a particular group of representative elements.
4. The size of an atom, indicated by atomic radius, varies periodically with the arrangement of the elements in the periodic table. It decreases as we move across a period and increases as we move down a group.
5. Ionization energy is a measure of the tendency of an atom to resist the loss of an electron. The higher the ionization energy, the more strongly the nucleus holds the electron. Electron affinity is a measure of the tendency of an atom to gain an electron. The more negative the electron affinity, the greater the tendency for the atom to gain an electron. Metals usually have low ionization energies, and nonmetals usually have high (large negative) electron affinities.
6. Noble gases are very stable because their outer *ns* and *np* subshells are completely filled. The metals of the representative elements (in Groups 1A, 2A, and 3A) tend to lose electrons until the cations of these metals become isoelectronic with the noble gases preceding them in the periodic table. The nonmetals of the representa-

tive elements (in Groups 5A, 6A, and 7A) tend to accept electrons until the anions of these nonmetals become isoelectronic with the noble gases following them in the periodic table.

Key Words

Amphoteric oxide, p. 340
Atomic radius, p. 317
Diagonal relationship, p. 331
Electron affinity, p. 329
Inert-pair effect, p. 335
Ionic radius, p. 320
Ionization energy, p. 326
Isoelectronic, p. 315
Main group elements, p. 311
Noble gases, p. 311
Representative elements, p. 311
Valence electrons, p. 311

Exercises

Development of the Periodic Table

REVIEW QUESTIONS

8.1 Briefly describe the significance of Mendeleev's periodic table.

8.2 What is Moseley's contribution to the modern periodic table?

8.3 Describe the general layout of a modern periodic table.

8.4 What is the most important relationship among elements in the same group in the periodic table?

Periodic Classification of the Elements

REVIEW QUESTIONS

8.5 Give two examples each of a metal, a nonmetal, and a metalloid.

8.6 Which of the following elements are metals, nonmetals, and metalloids? As, Xe, Fe, Li, B, Cl, Ba, P, I, Si.

8.7 Compare the physical and chemical properties of metals and nonmetals.

8.8 Draw a rough sketch of a periodic table (no details are required). Indicate regions where metals, nonmetals, and metalloids are located.

8.9 What is a representative element? Give names and symbols of four representative elements.

8.10 Without referring to a periodic table, write the name and give the symbol for an element in each of the following groups: 1A, 2A, 3A, 4A, 5A, 6A, 7A, 8A, transition metals.

8.11 Indicate whether the following elements exist as atomic species, molecular species, or extensive three-dimensional structures in their most stable states at 25°C and 1 atm and write the molecular or empirical formula for the elements: phosphorus, iodine, magnesium, neon, arsenic, sulfur, boron, selenium, and oxygen.

8.12 What is a noble gas core?

8.13 Define valence electrons. For representative elements, the number of valence electrons of an element is equal to its group number. Show that this is true for the following elements: Al, Sr, K, Br, P, S, C.

8.14 A neutral atom of a certain element has 17 electrons. Without consulting a periodic table, (a) Write the ground-state electron configuration of the element, (b) classify the element, (c) determine whether the atoms of this element are diamagnetic or paramagnetic.

8.15 In the periodic table, the element hydrogen is sometimes grouped with the alkali metals (as in this book) and sometimes with the halogens. Explain why hydrogen can resemble the Group 1A and the Group 7A elements.

8.16 Write the outer electron configurations for (a) the alkali metals, (b) the alkaline earth metals, (c) the halogens, (d) the noble gases.

8.17 Use the first-row transition metals (Sc to Cu) as an example to illustrate the characteristics of the electron configurations of transition metals.

PROBLEMS

8.18 Group the following electron configurations in pairs that would represent similar chemical properties of their atoms:
(a) $1s^22s^22p^63s^23p^5$
(b) $1s^22s^22p^63s^2$
(c) $1s^22s^22p^3$
(d) $1s^22s^22p^63s^23p^64s^23d^{10}4p^6$
(e) $1s^22s^2$
(f) $1s^22s^22p^6$
(g) $1s^22s^22p^63s^23p^3$
(h) $1s^22s^22p^5$

8.19 Without referring to a periodic table, write the electron configuration of elements with the following atomic numbers: (a) 9, (b) 20, (c) 26, (d) 33. Classify the elements.

8.20 Specify in what group of the periodic table each of the following elements is found: (a) [Ne]$3s^1$, (b) [Ne]$3s^23p^3$, (c) [Ne]$3s^23p^6$, (d) [Ar]$4s^23d^8$.

8.21 An ion M^{2+} derived from a metal in the first transition metal series has four and only four electrons in the 3*d* subshell. What element might M be?

Electron Configurations of Ions

REVIEW QUESTIONS

8.22 What is the characteristic of the electron configuration of stable ions derived from representative elements?

8.23 What do we mean when we say that two ions or an atom and an ion are isoelectronic?

8.24 What is wrong with the statement "The atoms of element X are isoelectronic with the atoms of element Y"?

8.25 Give three examples of first-row transition metal (Sc to Cu) ions whose electron configurations are represented by the argon core.

PROBLEMS

8.26 Write ground-state electron configurations for the following ions: (a) Li^+, (b) H^-, (c) N^{3-}, (d) F^-, (e) S^{2-}, (f) Al^{3+}, (g) Se^{2-}, (h) Br^-, (i) Rb^+, (j) Sr^{2+}, (k) Sn^{2+}, (l) Te^{2-}, (m) Ba^{2+}, (n) Pb^{2+}, (o) In^{3+}, (p) Tl^+, (q) Tl^{3+}.

8.27 Write the ground-state electron configurations of the following ions, which play important roles in biochemical processes in our bodies: (a) Na^+, (b) Mg^{2+}, (c) Cl^-, (d) K^+, (e) Ca^{2+}, (f) Fe^{2+}, (g) Cu^{2+}, (h) Zn^{2+}.

8.28 Write ground-state electron configurations of the following transition metal ions: (a) Sc^{3+}, (b) Ti^{4+}, (c) V^{5+}, (d) Cr^{3+}, (e) Mn^{2+}, (f) Fe^{2+}, (g) Fe^{3+}, (h) Co^{2+}, (i) Ni^{2+}, (j) Cu^+, (k) Cu^{2+}, (l) Ag^+, (m) Au^+, (n) Au^{3+}, (o) Pt^{2+}.

8.29 Name the ions with +3 charges that have the following electron configurations: (a) [Ar]$3d^3$, (b) [Ar], (c) [Kr]$4d^6$, (d) [Xe]$4f^{14}5d^6$.

8.30 Which of the following species are isoelectronic with each other? C, Cl^-, Mn^{2+}, B^-, Ar, Zn, Fe^{3+}, Ge^{2+}.

Periodic Variation in Physical Properties

REVIEW QUESTIONS

8.31 Define atomic radius. Does the size of an atom have a precise meaning?

8.32 How does atomic radius change as we move (a) from left to right across the period and (b) from top to bottom in a group?

8.33 Define ionic radius. How does the size change when an atom is converted to (a) an anion and (b) a cation?

PROBLEMS

8.34 On the basis of their positions in the periodic table, select the atom with the larger atomic radius in each of the following pairs: (a) Na, Cs; (b) Be, Ba; (c) N, Sb; (d) F, Br; (e) Ne, Xe.

8.35 Why does the helium atom have a smaller atomic radius than the hydrogen atom?

8.36 Why is the radius of the lithium atom considerably larger than the radius of the hydrogen atom?

8.37 Arrange the following atoms in order of decreasing atomic radius: Na, Al, P, Cl, Mg.

8.38 Which is the smallest atom in Group 7A?

8.39 Use the second period of the periodic table as an example to show that the sizes of atoms decrease as we move from left to right. Explain the trend.

8.40 In each of the following pairs, indicate which one of the two species is smaller: (a) Cl or Cl^-; (b) Na or Na^+; (c) O^{2-} or S^{2-}; (d) Mg^{2+} or Al^{3+}; (e) Au^+ or Au^{3+}.

8.41 List the following ions in order of increasing ionic radius: N^{3-}, Na^+, F^-, Mg^{2+}, O^{2-}.

8.42 Explain which of the following ions is larger, and why: Cu^+ or Cu^{2+}.

8.43 Explain which of the following anions is larger, and why: N^{3-} or P^{3-}.

8.44 Arrange the following anions in order of increasing size: O^{2-}, Te^{2-}, Se^{2-}, S^{2-}. Explain your sequence.

8.45 The H^- ion and the He atom have two 1*s* electrons each. Which of the two species is larger? Explain.

8.46 Give the physical states (gas, liquid, or solid) of the representative elements in the fourth period at 1 atm and 25°C: K, Ca, Ga, Ge, As, Se, Br.

8.47 The boiling points of neon and krypton are −245.9°C and −152.9°C, respectively. Using these data, estimate the boiling point of argon.

Ionization Energy

REVIEW QUESTIONS

8.48 Define ionization energy. Ionization energy measurements are usually carried out with atoms in the gaseous state. Why?

8.49 Sketch an outline of the periodic table and show group and period trends in the first ionization energy of the elements. What types of elements have the highest ionization energies and what types the lowest ionization energies?

8.50 Why is the second ionization energy always greater than the first ionization energy regardless of the element being considered?

PROBLEMS

8.51 In an ionization energy measurement, a beam of energetic electrons is used to ionize the atoms. Atoms can also be ionized by irradiation with light (a process called photoionization). Calculate the frequency and wavelength of light that must be employed to ionize one of the following atoms: (a) Li, (b) Cl, (c) He. Compare this process with the photoelectric effect discussed in Chapter 7.

8.52 Use the third period of the periodic table as an example to illustrate the change in first ionization energies of the elements as we move from left to right. Explain the trend.

8.53 Ionization energy usually increases from left to right across a given period. Aluminum, however, has a lower ionization energy than magnesium. Explain.

8.54 The first and second ionization energies of K are 419 kJ/mol and 3052 kJ/mol, and those of Ca are 590 kJ/mol and 1145 kJ/mol, respectively. Compare their values and comment on the differences.

8.55 Two atoms have the electron configurations $1s^22s^22p^6$ and $1s^22s^22p^63s^1$. The first ionization energy of one is 2080 kJ/mol, and that of the other is 496 kJ/mol. Pair up each ionization energy with one of the given electron configurations. Justify your choice.

8.56 Arrange the following species in isoelectronic pairs: O^+, Ar, S^{2-}, Ne, Zn, Cs^+, N^{3-}, As^{3+}, N, Xe.

8.57 Which one of the following processes requires the greatest input of energy? Why?

$$P(g) \longrightarrow P^+(g) + e^-$$
$$P^+(g) \longrightarrow P^{2+}(g) + e^-$$
$$P^{2+}(g) \longrightarrow P^{3+}(g) + e^-$$

8.58 A hydrogenlike ion is an ion containing only one electron. The energies of the electron in a hydrogenlike ion are given by

$$E_n = -(2.18 \times 10^{-18}\ \text{J})Z^2\left(\frac{1}{n^2}\right)$$

where n is the principal quantum number and Z is the atomic number of the element. Calculate the ionization energy (in kJ/mol) of the He^+ ion.

8.59 Use the equation given in Problem 8.58 to calculate the following ionization energies (in kJ/mol):
(a) $C^{5+}(g) \longrightarrow C^{6+}(g) + e^-$
(b) $Hg^{79+}(g) \longrightarrow Hg^{80+}(g) + e^-$

Electron Affinity

REVIEW QUESTIONS

8.60 Define electron affinity. Electron affinity measurements are usually carried out with atoms in the gaseous state. Why?

8.61 Ionization energy is always a positive quantity, whereas electron affinity may be either positive or negative. Explain.

8.62 Arrange the elements in each of the following groups in increasing order of the most exothermic electron affinity: (a) Li, Na, K, (b) F, Cl, Br, I.

PROBLEMS

8.63 Explain the trends in electron affinity from aluminum to chlorine (see Table 8.5).

8.64 From the electron affinity values for the alkali metals, do you think it is possible for these metals to form an anion like M^-, where M represents an alkali metal?

8.65 Which of the following elements would you expect to have the greatest electron affinity? He, K, Co, S, Cl.

8.66 Explain why alkali metals have a greater affinity for electrons than alkaline earth metals.

Variation in Chemical Properties

REVIEW QUESTIONS

8.67 Explain what is meant by the diagonal relationship. Give two pairs of elements that show this relationship.

8.68 Which elements are more likely to form acidic oxides? basic oxides? amphoteric oxides?

PROBLEMS

8.69 Use the alkali metals and alkaline earth metals as examples to show how we can predict the chemical properties of elements simply from their electron configurations.

8.70 Based on your knowledge of the chemistry of the alkali metals, predict some of the chemical properties of francium, the last member of the group.

8.71 As a group, the noble gases are very stable chemically (only Kr and Xe are known to form some compounds). Why?

8.72 Why are the Group 1B elements more stable than the Group 1A elements even though they seem to have the same outer electron configuration ns^1?

8.73 How do the chemical properties of oxides change as we move across a period from left to right? as we move down a particular group?

8.74 Predict (and give balanced equations for) the reactions between each of the following oxides and water: (a) Li_2O, (b) CaO, (c) SO_2.

8.75 Which oxide is more basic, MgO or BaO? Why?

8.76 Write formulas and give names for the binary hydrogen compounds of the second-period elements (Li to F). Describe the changes in physical and chemical properties of these compounds as we move across the period from left to right.

Miscellaneous Problems

8.77 Determine whether each of the following properties of the representative elements generally increases or decreases upon moving (a) from left to right across a period and (b) from top to bottom in a group: metallic character, atomic size, ionization energy, acidity of oxides.

8.78 With reference to the periodic table, name (a) a halogen element in the fourth period, (b) an element similar to phosphorus in chemical properties, (c) the most reactive metal in the fifth period, (d) an element that has an atomic number smaller than 20 and is similar to strontium.

8.79 Elements that have high ionization energies usually have more negative electron affinities. Why?

8.80 The energies an electron can possess in a hydrogen atom or a hydrogenlike ion are given by $E_n = (-2.18 \times 10^{-18}\ \text{J})Z^2(1/n^2)$. This equation cannot be applied to many-electron atoms. One way to modify it for the more complex atoms is to replace Z with $(Z - \sigma)$, where Z is the atomic number and σ is a positive dimensionless quantity called the shielding constant. Consider the helium atom as an example. The physical significance of σ is that it represents the extent of shielding that the two 1*s* electrons exert on each other. Thus the quantity $(Z - \sigma)$ is appropriately called the "effective nuclear charge." Calculate the value of σ if the first ionization energy of helium is 3.94×10^{-18} J per atom. (Ignore the minus sign in the given equation in your calculation.)

8.81 Match each of the elements on the right with a description on the left:

(a) A dark red liquid	Calcium (Ca)
(b) A colorless gas that burns in oxygen gas	Gold (Au)
	Hydrogen (H_2)
(c) A reactive metal that attacks water	Bromine (Br_2)
	Argon (Ar)
(d) A shiny metal that is used in jewelry	
(e) A totally inert gas	

8.82 Arrange the following isoelectronic species in order of (a) increasing ionic radius and (b) increasing ionization energy: O^{2-}, F^-, Na^+, Mg^{2+}.

8.83 The atomic radius of K is 216 pm and that of K^+ is 133 pm. Calculate the percent decrease in volume that occurs when $K(g)$ is converted to $K^+(g)$. (The volume of a sphere is $\frac{4}{3}\pi r^3$, where r is the radius of the sphere.)

8.84 The atomic radius of F is 72 pm and that of F^- is 136 pm. Calculate the percent increase in volume that occurs when $F(g)$ is converted to $F^-(g)$.

8.85 Write the empirical (or molecular) formulas of compounds that the elements in the third period (sodium to chlorine) are expected to form with (a) molecular oxygen and (b) molecular chlorine. In each case indicate whether you would expect the compound to be ionic or molecular in character.

8.86 Element M is a shiny and highly reactive metal (melting point 63°C), and element X is a highly reactive nonmetal (melting point −7.2°C). They react to form a compound with the empirical formula MX, a colorless, brittle solid that melts at 734°C. When dissolved in water or when in the melt state, the substance conducts electricity. When chlorine gas is bubbled through an aqueous solution containing MX, a reddish-brown liquid appears and Cl^- ions are formed. From these observations, identify M and X. (You may need to consult a handbook of chemistry for the melting-point values.)

8.87 Explain clearly what is meant by periodic relationship.

8.88 The noble gases were among the very last elements to be discovered. However, once the first, argon, was found in 1894, discovery of the remaining members of the group came quickly. (Radon, the last, was discovered in 1900.) Considering how other elements were discovered and the organization of the periodic group, explain why the discovery of the Group 8 elements came so late and why the discovery of one led rapidly to the others.

8.89 A technique called photoelectron spectroscopy is used to measure the ionization energy of atoms. A sample is irradiated with UV light, and electrons are ejected from the valence shell. The kinetic energies of the ejected electrons are measured. Since the energy of the UV photon and the kinetic energy of the ejected electron are known, we can write

$$h\nu = \text{IE} + \tfrac{1}{2}mv^2$$

where ν is the frequency of the UV light, and m and v are the mass and velocity of the electron, respectively. In one experiment the kinetic energy of the ejected electron is found to be 7.00×10^{-19} J using a UV source of wavelength 162 nm. Calculate the ionization energy of potassium. How can you be sure that this ionization energy corresponds to the electron in the valence shell (that is, the most loosely held electron)?

8.90 In which of the following are the species written in decreasing radius? (a) Be, Mg, Ba, (b) N^{3-}, O^{2-}, F^-, (c) Tl^{3+}, Tl^{2+}, Tl^+.

8.91 Which of the following properties show a clear periodic variation? (a) first ionization energy, (b) molar mass of the elements, (c) number of isotopes of an element, (d) atomic radius.

8.92 For each pair of the elements listed below, give three properties that show their chemical similarity: (a) sodium and potassium and (b) chlorine and bromine.

8.93 Explain why the first electron affinity of sulfur is −200 kJ/mol but the second electron affinity is +649 kJ/mol.

8.94 When carbon dioxide is bubbled through a clear calcium hydroxide solution, the solution appears milky. Write an equation for the reaction and explain how this reaction illustrates that CO_2 is an acidic oxide.

8.95 The energy needed for the following process is 1.96×10^4 kJ/mol:

$$\text{Li}(g) \longrightarrow \text{Li}^{3+}(g) + 3e^-$$

If the first ionization of lithium is 520 kJ/mol, calculate the second ionization of lithium, that is, energy required for the process

$$Li^+(g) \longrightarrow Li^{2+}(g) + e^-$$

(*Hint:* You need the equation in Problem 8.58.)

8.96 An element X reacts with hydrogen gas at 200°C to form compound Y. When Y is heated to a higher temperature, it decomposes to the element X and hydrogen gas in the ratio of 559 mL of H_2 (measured at STP) for 1.00 g of X reacted. X also combines with chlorine to form a compound Z which contains 63.89 percent by mass of chlorine. Deduce the identity of X.

8.97 Using the following boiling-point data and following the procedure in Figure 8.13, estimate the boiling point of francium:

metal	Li	Na	K	Rb	Cs
boiling point (°C)	1347	882.9	774	688	678.4

8.98 A student is given samples of three elements X, Y, and Z, which could be an alkali metal, a member of Group 4A, and a member of Group 5A. She makes the following observations: Element X has a metallic luster and conducts electricity. It reacts slowly with hydrochloric acid to produce hydrogen gas. Element Y is a light yellow solid that does not conduct electricity. Element Z has a metallic luster and conducts electricity. When exposed to air, it slowly forms a white powder. A solution of the white powder in water is basic. What can you conclude about the elements from these observations?

8.99 In general, atomic radius and ionization energy have opposite periodic trends. Why?

8.100 Why do noble gases have positive electron affinity values?

8.101 The following are the ionization energies of sodium (in kJ/mol), starting with the first and ending with the eleventh: 495.9, 4560, 6900, 9540, 13400, 16600, 20120, 25490, 28930, 141360, 170000. Plot the log of ionization (*y*-axis) versus the number of ionization (*x*-axis); for example, log 495.9 is plotted versus 1 (labeled I_1, the first ionization energy), log 4560 is plotted versus 2 (labeled I_2, the second ionization energy), and so on. (a) Label I_1 through I_{11} with the electrons in orbitals such as 1*s*, 2*s*, 2*p*, and 3*s*. (b) What can you deduce about electron shells from the breaks in the curve?

8.102 Explain why the electron affinity of nitrogen is approximately zero, while the elements on either side, carbon and oxygen, have substantial electron affinities.

Opposite page: A computer-generated model of the oxygen (O_2) molecule. Electrons play an essential role in holding the atoms in a molecule.

C H A P T E R 9

CHEMICAL BONDING I: BASIC CONCEPTS

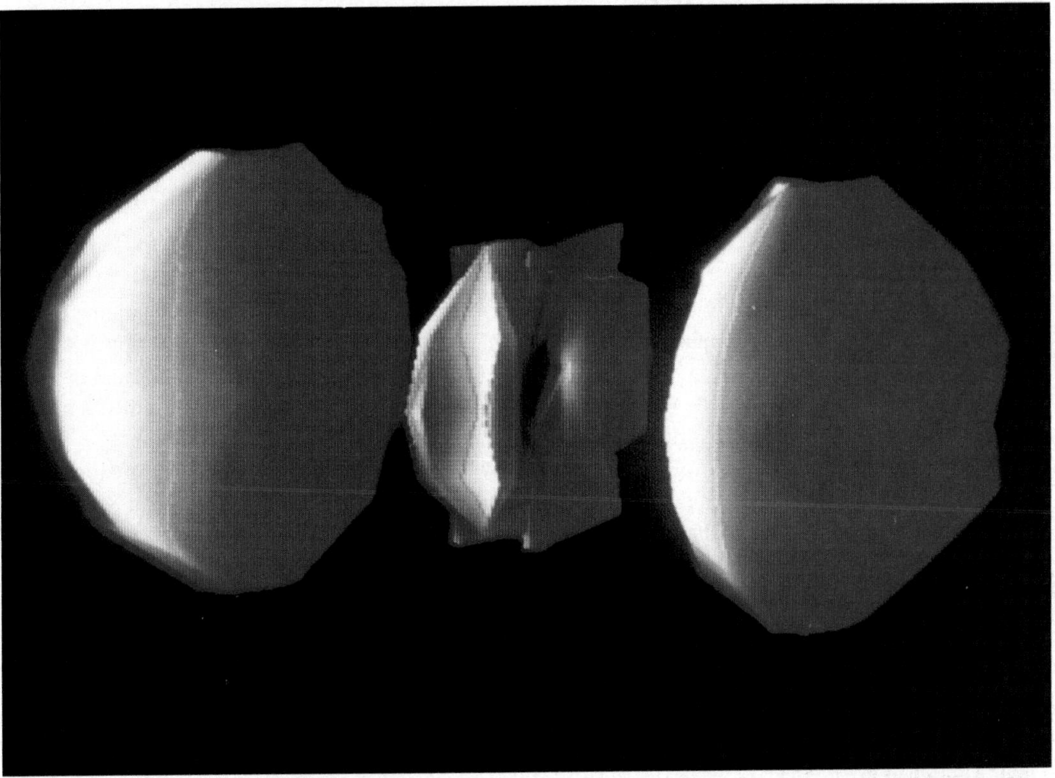

Studying the chemical bond will help us understand the forces that hold atoms together in molecules and ions together in ionic compounds. Using what we know about electron configurations and the periodic table, we will now examine the two major classes of bonds: the ionic bond, in which electrons are transferred from one atom to another, and the covalent bond, in which electrons are shared between atoms. To a great extent, bond formation and the stability of molecules can be predicted by using the octet rule and the concept of resonance, which we will also study in this chapter.

9.1 Lewis Dot Symbols

When atoms interact to form a chemical bond, only their outer regions come into contact. For this reason, as we saw in Chapter 8, elements with similar outer electron configurations behave alike chemically. Therefore, when we study chemical bonding we are usually concerned only with valence electrons.

As we will see shortly, the number of unpaired dots in a Lewis dot symbol corresponds to the number of bonds an atom of the element forms in a compound.

To show the valence electrons of atoms and to keep track of them in a chemical reaction, chemists use Lewis dot symbols. A ***Lewis† dot symbol*** *consists of the symbol of an element and one dot for each valence electron in an atom of the element*. Figure 9.1 shows the Lewis dot symbols of the representative elements and the noble gases. Note that, except for helium, the number of valence electrons in each atom is the same as the group number of the element. For example, Li belongs to Group 1A and has one dot for one valence electron; Be belongs to Group 2A and has two valence electrons (two dots); and so on. Elements in the same group have similar outer electron configurations and hence similar Lewis dot symbols. The transition metals, lanthanides, and actinides all have incompletely filled inner shells, and in general we cannot write simple Lewis dot symbols for them.

9.2 Elements That Form Ionic Compounds

Lithium fluoride.

Learning about Lewis dot symbols is a first step toward understanding the two classes of compounds, ionic compounds and covalent compounds. Section 9.4 deals with the topic of covalent compounds. In this section we study the formation of ionic compounds, beginning with factors that determine whether atoms of different elements will react to form an ionic compound.

Atoms of elements with low ionization energies tend to form cations, while those with high negative electron affinities tend to form anions (Chapter 8). This general rule tells us that the elements most likely to form cations in ionic compounds are the alkali metals and alkaline earth metals, and that the elements most likely to form anions are the halogens and oxygen. Consider the formation of the ionic compound lithium fluoride from lithium and fluorine. The electron configuration of lithium is $1s^22s^1$, and that of fluorine is $1s^22s^22p^5$. When lithium and fluorine atoms come in contact with each other, the outer $2s^1$ valence electron of lithium is transferred to the fluorine atom. Using Lewis dot symbols, we represent the reaction like this:

In the laboratory LiF can be prepared by passing a stream of fluorine gas over heated lithium: $2Li(s) + F_2(g) \rightarrow 2LiF(s)$. Industrially, LiF (like most other ionic compounds) is obtained by recovering and purifying minerals containing the substance.

$$\underset{1s^22s^1}{\cdot\mathrm{Li}} \;+\; \underset{1s^22s^22p^5}{:\!\underset{\cdot\cdot}{\overset{\cdot\cdot}{\mathrm{F}}}\!\cdot} \;\longrightarrow\; \underset{1s^2}{\mathrm{Li}^+}\;\underset{1s^22s^22p^6}{:\!\underset{\cdot\cdot}{\overset{\cdot\cdot}{\mathrm{F}}}\!:^-}\;\text{(or LiF)} \qquad (9.1)$$

For convenience, imagine that this reaction occurs in separate steps: first the ionization of Li

$$\cdot\mathrm{Li} \longrightarrow \mathrm{Li}^+ + e^-$$

and then the acceptance of an electron by F

$$:\!\underset{\cdot\cdot}{\overset{\cdot\cdot}{\mathrm{F}}}\!\cdot + e^- \longrightarrow :\!\underset{\cdot\cdot}{\overset{\cdot\cdot}{\mathrm{F}}}\!:^-$$

†Gilbert Newton Lewis (1875–1946). American chemist. One of the foremost physical chemists of this century, Lewis made many significant contributions in the areas of chemical bonding, thermodynamics, acids and bases, and spectroscopy. Despite the significance of Lewis's work, he was never awarded the Nobel Prize in chemistry.

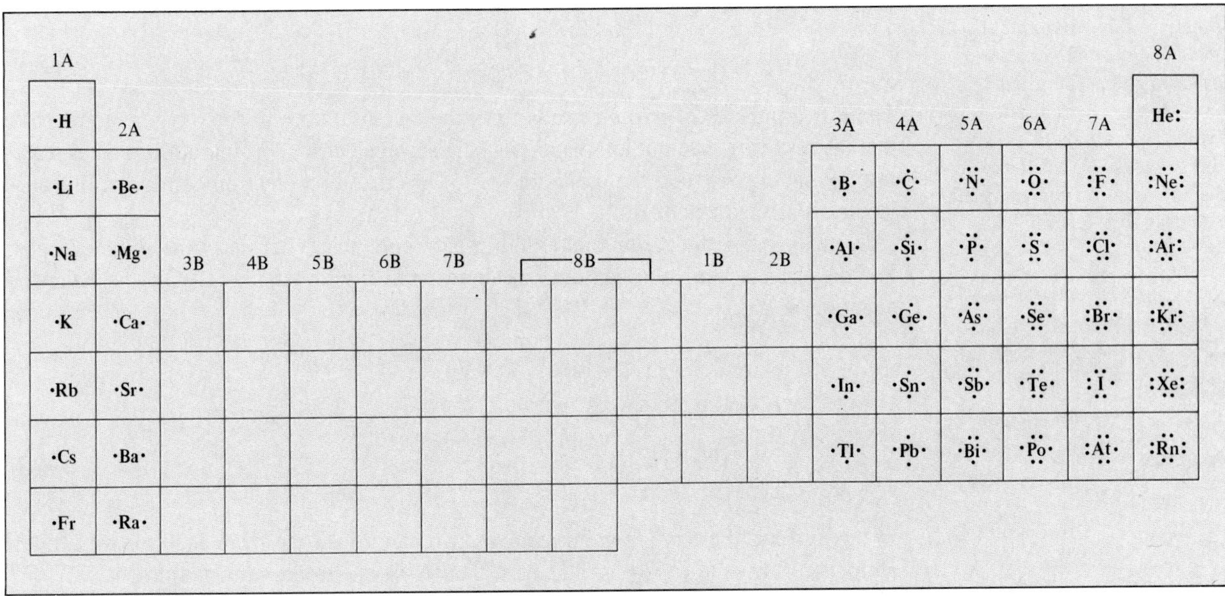

Figure 9.1 *Lewis dot symbols of the representative elements and the noble gases.*

Next, imagine the two separate ions joining to form a LiF unit:

$$Li^+ + :\ddot{\underset{..}{F}}:^- \longrightarrow Li^+:\ddot{\underset{..}{F}}:^-$$

Note that the sum of these three reactions is

$$\cdot Li + :\ddot{\underset{..}{F}}\cdot \longrightarrow Li^+:\ddot{\underset{..}{F}}:^-$$

We normally write the empirical formulas of ionic compounds without showing the charges. The + and − are shown here to emphasize the transfer of electrons.

which is the same as the direct transfer of an electron from lithium to fluorine to form the Li^+F^- unit as represented by Equation (9.1).

The Li^+F^- unit is held together as a result of the electrostatic attraction between the positively charged lithium and the negatively charged fluoride ions. This attraction leads to the formation of an ***ionic bond,*** *the bond that holds ions together in an ionic compound.* According to Coulomb's law, the energy of interaction (E) between these two ions is given by

$$E \propto \frac{Q_{Li^+}Q_{F^-}}{r}$$
$$= k\frac{Q_{Li^+}Q_{F^-}}{r} \tag{9.2}$$

where Q_{Li^+} and Q_{F^-} are the charges on the Li^+ and F^- ions, r is the distance between the centers of the two ions, and k is the proportionality constant. Since Q_{Li^+} is positive and Q_{F^-} is negative, E is a negative quantity, and the formation of an ionic bond from Li^+ and F^- is an exothermic process. A bonded pair of Li^+ and F^- ions is more stable than separate Li^+ and F^- ions.

There are many other common reactions that lead to the formation of ionic compounds. For instance, calcium burns in oxygen to form calcium oxide:

$$2Ca(s) + O_2(g) \longrightarrow 2CaO(s)$$

In Lewis dot symbols this is

To clarify this equation, the Lewis dot symbol equation is based on the assumption that the diatomic O_2 molecule is divided into separate oxygen atoms. Later we will focus in greater detail on the energetics of the O_2 to 2 O step.

$$\cdot\mathrm{Ca}\cdot \;+\; \cdot\ddot{\underset{\cdot\cdot}{\mathrm{O}}}\cdot \;\longrightarrow\; \mathrm{Ca}^{2+}\;:\ddot{\underset{\cdot\cdot}{\mathrm{O}}}:^{2-}$$
$$[\mathrm{Ar}]4s^2 \quad 1s^22s^22p^4 \qquad [\mathrm{Ar}] \quad [\mathrm{Ne}]$$

There is a transfer of two electrons from the calcium atom to the oxygen atom. Note that the resulting calcium ion (Ca^{2+}) has the argon electron configuration (or is isoelectronic with argon), and the oxide ion (O^{2-}) has the neon electron configuration (or is isoelectronic with neon).

In many cases the cation and anion in a compound do not carry the same charges. For instance, when lithium burns in air to form lithium oxide (Li_2O), the balanced equation is

$$4\mathrm{Li}(s) + \mathrm{O_2}(g) \longrightarrow 2\mathrm{Li_2O}(s)$$

Using Lewis dot symbols, we write

$$2\cdot\mathrm{Li} \;+\; \cdot\ddot{\underset{\cdot\cdot}{\mathrm{O}}}\cdot \;\longrightarrow\; 2\mathrm{Li}^{+}\;:\ddot{\underset{\cdot\cdot}{\mathrm{O}}}:^{2-}\;(\text{or } \mathrm{Li_2O})$$
$$1s^22s^1 \quad 1s^22s^22p^4 \qquad [\mathrm{He}] \quad [\mathrm{Ne}]$$

In this process the oxygen atom receives two electrons (one from each of two lithium atoms) to form the oxide ion. The Li^+ ion is isoelectronic with helium.

When magnesium reacts with nitrogen at elevated temperatures, a white solid compound, magnesium nitride (Mg_3N_2), forms:

$$3\mathrm{Mg}(s) + \mathrm{N_2}(g) \longrightarrow \mathrm{Mg_3N_2}(s)$$

or

$$3\cdot\mathrm{Mg}\cdot \;+\; 2\cdot\ddot{\underset{\cdot}{\mathrm{N}}}\cdot \;\longrightarrow\; 3\mathrm{Mg}^{2+}2:\ddot{\underset{\cdot\cdot}{\mathrm{N}}}:^{3-}\;(\text{or } \mathrm{Mg_3N_2})$$
$$[\mathrm{Ne}]3s^2 \quad 1s^22s^22p^3 \qquad [\mathrm{Ne}] \quad [\mathrm{Ne}]$$

The reaction involves the transfer of six electrons (two from each Mg atom) to two nitrogen atoms. The resulting magnesium ion (Mg^{2+}) and the nitride ion (N^{3-}) are both isoelectronic with neon.

In the following example, we apply the Lewis dot symbols to study the formation of an ionic compound.

EXAMPLE 9.1

Use Lewis dot symbols to describe the formation of aluminum oxide (Al_2O_3).

Answer

According to Figure 9.1, the Lewis dot symbols of Al and O are

$$\cdot\underset{\cdot}{\mathrm{Al}}\cdot \qquad \cdot\ddot{\underset{\cdot\cdot}{\mathrm{O}}}\cdot$$

Since aluminum tends to form the cation (Al^{3+}) and oxygen the anion (O^{2-}), the transfer of electrons is from Al to O. There are three valence electrons in each Al atom; each O atom needs two electrons to form the O^{2-} ion, which is isoelectronic with neon. Thus the simplest neutralizing ratio of Al^{3+} to O^{2-} is 2:3; two Al^{3+} ions have a total charge of 6+, and three O^{2-} ions have a total charge of 6−. Thus the empirical formula of aluminum oxide is Al_2O_3 and the reaction is

We use electroneutrality as our guide in writing formulas for ionic compounds.

$$2\cdot\dot{\text{Al}}\cdot \quad + \quad 3\cdot\ddot{\text{O}}\cdot \longrightarrow 2\text{Al}^{3+}3:\ddot{\text{O}}:^{2-} \text{ (or } \text{Al}_2\text{O}_3\text{)}$$

$[Ne]3s^2 3p^1$ $1s^2 2s^2 2p^4$ [Ne] [Ne]

Similar problems: 9.18, 9.19.

Polyatomic Ions

So far we have concentrated on *monatomic ions,* that is, ions containing only one atom. Many ionic compounds contain *polyatomic cations* and/or *polyatomic anions,* ions containing more than one atom. At present, the only polyatomic cations you need to know are the mercury(I) ion (Hg_2^{2+}) and the ammonium ion (NH_4^+), mentioned in Chapter 2. On the other hand, there are many important polyatomic anions with which you do need to become familiar. These anions are listed in Table 2.3.

9.3 Lattice Energy of Ionic Compounds

In this section we will take a closer look at the stability of solid ionic compounds. Section 9.2 mentioned that the formation of ionic compounds depends on the ionization energy and electron affinity of the elements. These important quantities do help predict which elements are most likely to form ionic compounds, but they do not tell the whole story. They are limited because both the ionization process and the acceptance of an electron are defined as taking place in the gas phase. The solid state is a very different environment because each cation in a solid is surrounded by a specific number of anions, and vice versa. Thus the overall stability of a solid ionic compound depends on the interactions of all these ions and not merely on the interaction of a single cation with a single anion. A quantitative measure of the stability of any ionic solid is its lattice energy, defined as the energy required to completely separate one mole of a solid ionic compound into gaseous ions (see Section 6.6).

The Born-Haber Cycle for Determining Lattice Energies

Lattice energy cannot be measured directly. However, if we know the structure and composition of an ionic compound, we can calculate the compound's lattice energy by using Coulomb's law. We can also determine the lattice energy indirectly, by assuming that the formation of an ionic compound takes place in a series of steps known as the ***Born-Haber cycle.*** The Born-Haber cycle *relates lattice energies of ionic compounds to ionization energies, electron affinities, and other atomic and molecular properties.* This approach is based on Hess's law (see Section 6.6).

Developed by Max Born† and Fritz Haber,‡ the Born-Haber cycle defines the vari-

†Max Born (1882–1970). German physicist. Born was one of the founders of modern physics. His work covered a wide range of topics. He received the Nobel Prize in physics in 1954 for his interpretation of the wave function for particles.

‡Fritz Haber (1868–1934). German chemist. Haber's process for synthesizing ammonia from atmospheric nitrogen supplied Germany with nitrates for explosives during World War I. He also did work on gas warfare. Haber received the Nobel Prize in chemistry in 1918.

ous steps that precede the formation of an ionic solid. We will illustrate its use in finding the lattice energy of lithium fluoride.

Consider the reaction between lithium and fluorine gas:

$$\mathrm{Li}(s) + \tfrac{1}{2}\mathrm{F_2}(g) \longrightarrow \mathrm{LiF}(s)$$

The standard enthalpy change for this reaction is -594.1 kJ. (Because this reaction is carried out with its reactants and product in their standard states, the enthalpy change is also the standard enthalpy of formation for LiF.) Applying the Born-Haber cycle, we can trace the formation of LiF through five separate steps, keeping in mind that the sum of enthalpy changes for the steps is equal to the enthalpy change for the overall reaction, that is, -594.1 kJ.

1. Convert solid lithium to lithium vapor (the direct conversion of a solid to a gas is called *sublimation*):

   $$\mathrm{Li}(s) \longrightarrow \mathrm{Li}(g) \qquad \Delta H_1^\circ = 155.2 \text{ kJ}$$

   The energy of sublimation for lithium is 155.2 kJ/mol.
2. Dissociate $\frac{1}{2}$ mole of F_2 gas into separate gaseous F atoms:

   $$\tfrac{1}{2}\mathrm{F_2}(g) \longrightarrow \mathrm{F}(g) \qquad \Delta H_2^\circ = 75.3 \text{ kJ}$$

   The energy needed to break the bonds in 1 mole of F_2 molecules is 150.6 kJ. Here we are breaking the bonds in half a mole of F_2, so the enthalpy change is 150.6/2, or 75.3, kJ.
3. Ionize 1 mole of gaseous Li atoms (see Table 8.4):

   $$\mathrm{Li}(g) \longrightarrow \mathrm{Li^+}(g) + e^- \qquad \Delta H_3^\circ = 520 \text{ kJ}$$

   This process corresponds to the first ionization of lithium.
4. Add 1 mole of electrons to 1 mole of gaseous F atoms (see Table 8.5):

   $$\mathrm{F}(g) + e^- \longrightarrow \mathrm{F^-}(g) \qquad \Delta H_4^\circ = -333 \text{ kJ}$$

   This is the electron affinity of fluorine.
5. Combine 1 mole of gaseous Li^+ and 1 mole of F^- to form 1 mole of solid LiF:

   $$\mathrm{Li^+}(g) + \mathrm{F^-}(g) \longrightarrow \mathrm{LiF}(s) \qquad \Delta H_5^\circ = ?$$

The F atoms in a F_2 molecule are held together by another type of chemical bond called the *covalent bond*. The energy required to break this bond is called the *bond dissociation energy*. More on this in Section 9.10.

The reverse of step 5, as we saw earlier,

$$\text{energy} + \mathrm{LiF}(s) \longrightarrow \mathrm{Li^+}(g) + \mathrm{F^-}(g)$$

defines the lattice energy of LiF. Thus the lattice energy must have the same magnitude as ΔH_5° but an opposite sign. Although we cannot determine ΔH_5° directly, we can calculate its value by the following procedure.

1.	$\mathrm{Li}(s) \longrightarrow \mathrm{Li}(g)$	$\Delta H_1^\circ = 155.2$ kJ
2.	$\tfrac{1}{2}\mathrm{F_2}(g) \longrightarrow \mathrm{F}(g)$	$\Delta H_2^\circ = 75.3$ kJ
3.	$\mathrm{Li}(g) \longrightarrow \mathrm{Li^+}(g) + e^-$	$\Delta H_3^\circ = 520$ kJ
4.	$\mathrm{F}(g) + e^- \longrightarrow \mathrm{F^-}(g)$	$\Delta H_4^\circ = -333$ kJ
5.	$\mathrm{Li^+}(g) + \mathrm{F^-}(g) \longrightarrow \mathrm{LiF}(s)$	$\Delta H_5^\circ = ?$
	$\mathrm{Li}(s) + \tfrac{1}{2}\mathrm{F_2}(g) \longrightarrow \mathrm{LiF}(s)$	$\Delta H_{\text{overall}}^\circ = -594.1$ kJ

The Born-Haber cycle provides a systematic and convenient way to analyze the energetics of ionic compound formation, but note that the actual mechanism by which LiF is formed need not follow these five simplified steps. This pathway allows us, with the assistance of Hess's law, to examine the energies of this class of chemical reactions.

According to Hess's law we can write

$$\Delta H_{\text{overall}}^\circ = \Delta H_1^\circ + \Delta H_2^\circ + \Delta H_3^\circ + \Delta H_4^\circ + \Delta H_5^\circ$$

or $$-594.1 \text{ kJ} = 155.2 \text{ kJ} + 75.3 \text{ kJ} + 520 \text{ kJ} - 333 \text{ kJ} + \Delta H_5^\circ$$

Hence $$\Delta H_5^\circ = -1012 \text{ kJ}$$

and the lattice energy of LiF is +1012 kJ/mol.

Figure 9.2 summarizes the entire Born-Haber cycle for LiF. In this cycle we saw that steps 1, 2, and 3 all require input of energy. On the other hand, steps 4 and 5 release energy. Because ΔH_5° is a large negative quantity, the lattice energy of LiF is a large positive quantity, which accounts for the stability of LiF in the solid state. The greater the lattice energy, the more stable the ionic compound. Keep in mind that lattice energy is *always* a positive quantity because the separation of ions in a solid into ions in the gas phase is, by Coulomb's law, an endothermic process.

Table 9.1 lists the lattice energies and the melting points of several common ionic compounds. There is a rough correlation between lattice energy and melting point. The larger the lattice energy, the more stable the solid and the more tightly held the ions. Thus it would take more energy to melt such a solid, and the solid has a higher melting point than one with a smaller lattice energy. Note the unusually high lattice energies for $MgCl_2$, Na_2O, and MgO. The first one of these ionic compounds involves a doubly

Note that MgO has an unusually high melting point (Table 9.1).

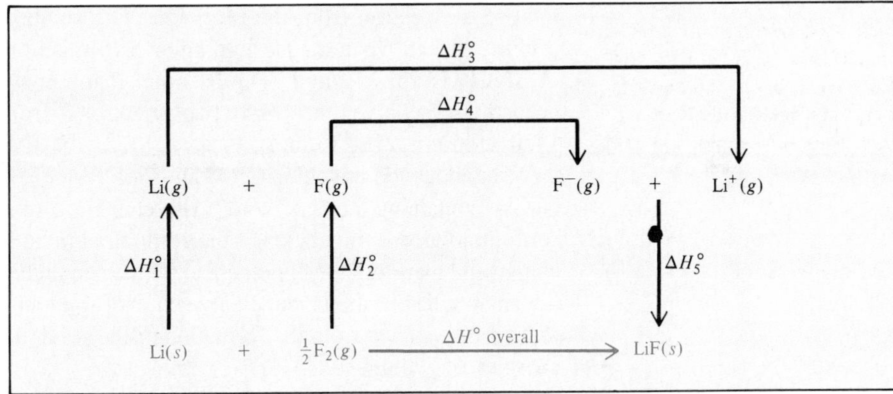

Figure 9.2 *The Born-Haber cycle for the formation of solid LiF.*

Table 9.1 Lattice energies and melting points of some alkali metal and alkaline earth metal halides and oxides

Compound	Lattice energy (kJ/mol)	Melting point (°C)
LiF	1012	845
LiCl	828	610
LiBr	787	550
LiI	732	450
NaCl	788	801
NaBr	736	750
NaI	686	662
KCl	699	772
KBr	689	735
KI	632	680
$MgCl_2$	2527	714
Na_2O	2570	Sub*
MgO	3890	2800

*Na_2O sublimes at 1275°C.

charged cation (Mg^{2+}) and the second a doubly charged anion (O^{2-}); the third involves the interaction of two doubly charged species (Mg^{2+} and O^{2-}). The coulombic attractions between such doubly charged species, or between them and singly charged ions, are much stronger than those between singly charged anions and cations.

Lattice Energy and the Formulas of Ionic Compounds

Because lattice energy is an indication of the stability of ionic compounds, its value can help us rationalize the formulas of these compounds. Consider magnesium chloride as an example. We have seen that the ionization energy of an element increases rapidly as successive electrons are removed from its atom. For example, the first ionization energy of magnesium is 738 kJ/mol, and the second ionization energy is 1450 kJ/mol, almost twice the first. We might ask why, from the standpoint of energy, magnesium does not prefer to form unipositive ions in its compounds. Why doesn't magnesium chloride have the formula MgCl (containing the Mg^+ ion) rather than $MgCl_2$ (containing the Mg^{2+} ion)? Admittedly, the Mg^{2+} ion has the noble gas configuration [Ne], which represents stability because of its completely filled shells. But the stability gained through the filled shells does not, in fact, outweigh the energy input needed to remove an electron from the Mg^+ ion. The solution to this puzzle lies in the extra stability gained when magnesium chloride forms in the solid state. The lattice energy of $MgCl_2$ is 2527 kJ/mol. This is more than enough to compensate for the energy needed to remove the first two electrons from a Mg atom (738 kJ/mol + 1450 kJ/mol = 2188 kJ/mol).

What about sodium chloride? Why is the formula for sodium chloride NaCl and not $NaCl_2$ (containing the Na^{2+} ion)? Although Na^{2+} does not have the noble gas electron configuration, we might expect the compound to be $NaCl_2$ because Na^{2+} has a higher charge and therefore the hypothetical $NaCl_2$ should have a greater lattice energy. Again the answer lies in the balance between energy input (that is, ionization energies) and the stability gained from the formation of the solid. The sum of the first two ionization energies of sodium is

$$496 \text{ kJ/mol} + 4560 \text{ kJ/mol} = 5056 \text{ kJ/mol}$$

Since $NaCl_2$ does not exist, we do not know its lattice energy value. However, if we assume a value roughly the same as that for $MgCl_2$ (2527 kJ/mol), we see that it is far too small an energy yield to compensate for the energy required to produce the Na^{2+} ion.

What has been said about the cations applies also to the anions. In Section 8.5 we observed that the electron affinity of oxygen is a negative value, meaning the following process releases energy (and is therefore favorable):

$$O(g) + e^- \longrightarrow O^-(g)$$

As we would expect, adding another electron to the O^- ion

$$O^-(g) + e^- \longrightarrow O^{2-}(g)$$

would be unfavorable because of the increase in electrostatic repulsion. Indeed, the electron affinity of O^- is positive (+780 kJ/mol). Yet compounds containing the oxide ion (O^{2-}) do exist and are very stable, whereas compounds containing the O^- ion are not known. Again, the high lattice energy realized by the presence of O^{2-} ions in compounds such as Na_2O or MgO far outweighs the energy needed to produce the O^{2-} ion.

CHEMISTRY IN ACTION

Sodium Chloride—A Common and Important Ionic Compound

We are all familiar with sodium chloride as table salt. It is a typical ionic compound, a brittle solid with a high melting point (801°C) that conducts electricity in the molten state and in aqueous solution. The arrangement of Na^+ and Cl^- ions in the solid state is shown in Figure 2.10.

One source of sodium chloride is rock salt, which is found in subterranean deposits often hundreds of meters thick (Figure 9.3). It is also obtained from seawater or brine (a concentrated NaCl solution) by solar evaporation (Figure 9.4). Sodium chloride also occurs in nature as the mineral *halite* (Figure 9.5).

Sodium chloride is used more often than any other material in the manufacture of inorganic chemicals. World consumption of this substance is about 150 million tons per year. As Figure 9.6 shows, the major use of sodium chloride is in the production of other essential inorganic chemicals such as chlorine gas, sodium hydroxide, sodium metal, hydrogen gas, and sodium carbonate. It is also used to melt ice and snow on highways and roads. However, since sodium chloride is harmful to plant life and promotes corrosion of cars, its use for this purpose is of considerable environmental concern.

Figure 9.3 *Underground rock salt mining.*

Figure 9.4 *Solar evaporation process for obtaining sodium chloride.*

Figure 9.5 *The mineral halite (NaCl).*

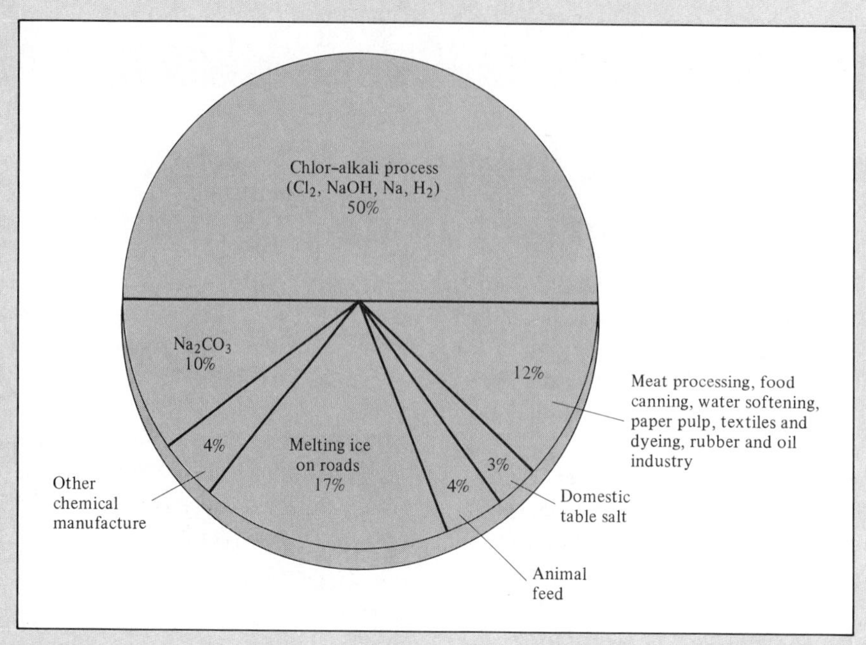

Figure 9.6 *Major uses of sodium chloride.*

9.4 The Covalent Bond

We turn our attention now to covalent bonds, the bonds that hold atoms together in molecules. Although the existence of molecules was hypothesized as early as the seventeenth century, it was not until early in this century that chemists understood how and why molecules form. The first major breakthrough was Gilbert Lewis's suggestion of the role of electrons in chemical bond formation. In particular, Lewis developed the concept that a chemical bond involves two atoms in a molecule sharing a pair of electrons. He depicted the formation of a chemical bond in H_2 as

$$\text{H}\cdot + \cdot\text{H} \longrightarrow \text{H}:\text{H}$$

This type of electron pairing is an example of a ***covalent bond,*** *a bond in which two electrons are shared by two atoms.* For the sake of simplicity, the shared pair of electrons is often represented by a single line. Thus the covalent bond in the hydrogen molecule can be written as H—H. In a covalent bond, each electron in a shared pair is attracted to both of the nuclei involved in the bond. This attraction is responsible for holding the two atoms together in H_2. The same type of attraction is responsible for the formation of covalent bonds in other molecules.

This discussion applies only to representative elements.

When dealing with covalent bonding among many-electron atoms, we need be concerned only with the valence electrons (because these are the ones involved in chemical bonding). Consider the fluorine molecule, F_2. The electron configuration of F is $1s^22s^22p^5$. The $1s$ electrons are low in energy and spend most of the time near the nucleus. For this reason they do not participate in bond formation. Thus each F atom has seven valence electrons (the $2s$ and $2p$ electrons). According to Figure 9.1, there is only one unpaired electron on F, so that the formation of the F_2 molecule is represented as

$$:\ddot{\underset{..}{F}}\cdot + \cdot\ddot{\underset{..}{F}}: \longrightarrow :\ddot{\underset{..}{F}}:\ddot{\underset{..}{F}}: \quad \text{or} \quad :\ddot{\underset{..}{F}}-\ddot{\underset{..}{F}}:$$

Note that some *valence electrons are not involved in covalent bond formation;* these are called ***nonbonding electrons,*** or ***lone pairs.*** Thus each F in F_2 has three lone pairs of electrons:

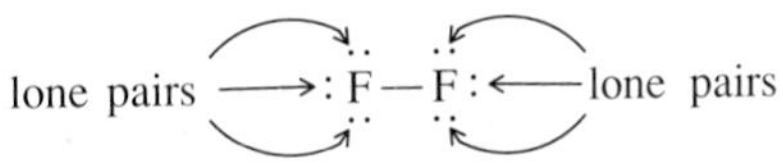

lone pairs ⟶ $:\ddot{\underset{..}{F}}-\ddot{\underset{..}{F}}:$ ⟵ lone pairs

The structures we have shown for H_2 and F_2 are called Lewis structures. A ***Lewis structure*** is *a representation of covalent bonding using Lewis dot symbols in which shared electron pairs are shown either as lines or as pairs of dots between two atoms, and lone pairs are shown as pairs of dots on individual atoms.* Only valence electrons are shown in a Lewis structure.

Next let us consider the Lewis structure of the water molecule. From Figure 9.1 we see that the Lewis dot symbol for oxygen has two unpaired dots or two unpaired electrons, so we expect that O might form two covalent bonds. Since hydrogen has only one electron, it can form only one covalent bond. Thus the Lewis structure for water is

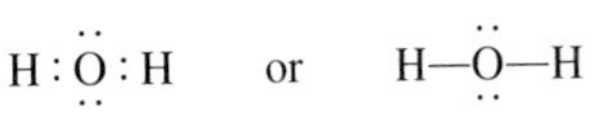

$$H:\ddot{\underset{..}{O}}:H \quad \text{or} \quad H-\ddot{\underset{..}{O}}-H$$

Here we see that the O atom has two lone pairs. The hydrogen atom has no lone pairs because its only electron is used to form a covalent bond.

In the F_2 and H_2O molecules described above, we see that the F and O atoms have obtained the stable noble gas configuration through the sharing of electrons:

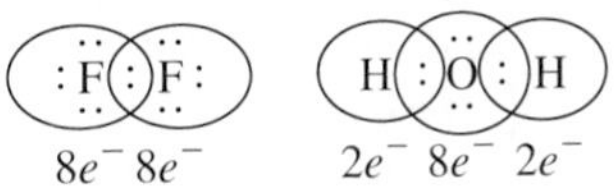

$:\ddot{\underset{..}{F}}:\ddot{\underset{..}{F}}:$ (8e⁻ 8e⁻) $H:\ddot{\underset{..}{O}}:H$ (2e⁻ 8e⁻ 2e⁻)

As we will see later, the octet rule applies mainly to the second-period elements.

This observation can be generalized by the ***octet rule,*** formulated by Lewis: *An atom other than hydrogen tends to form bonds until it is surrounded by eight valence electrons.* Therefore, we can say that a covalent bond forms when there are not enough electrons for each individual atom to have a complete octet. By sharing electrons in a covalent bond, the individual atoms can complete their octets. The requirement for hydrogen is that it obtain the electron configuration of helium, or a total of two electrons.

Two atoms held together by one electron pair are said to be joined by a *single bond.* In many compounds, there are ***multiple bonds;*** that is, bonds formed when *two atoms share two or more pairs of electrons.* If two atoms share two pairs of electrons, the covalent bond is called a *double bond.* Double bonds are found in molecules like carbon dioxide (CO_2):

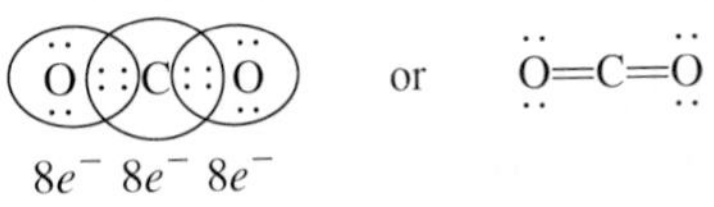

$\ddot{\underset{..}{O}}::C::\ddot{\underset{..}{O}}$ ($8e^-$ $8e^-$ $8e^-$) or $\ddot{\underset{..}{O}}=C=\ddot{\underset{..}{O}}$

Shortly you will be introduced to the rules for writing proper Lewis structures. Here we are simply examining typical Lewis structures to become familiar with the language associated with them.

and ethylene: $H_2C::CH_2$ ($8e^-$ $8e^-$) or $H_2C=CH_2$

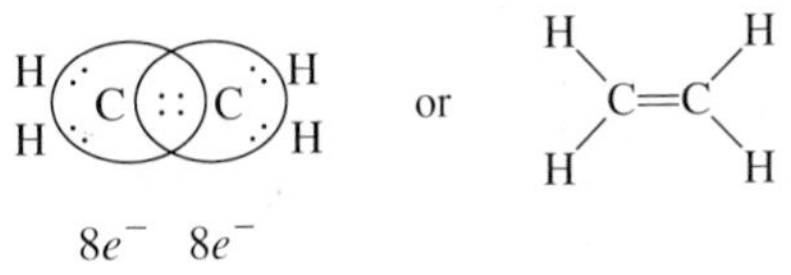

A *triple bond* arises when two atoms share three pairs of electrons, as in the nitrogen molecule (N_2):

$$:N\vdots\vdots N: \quad \text{or} \quad :N{\equiv}N:$$

$8e^-$ $8e^-$

The acetylene molecule (C_2H_2) also contains a triple bond, in this case between two carbon atoms:

$$H:C\vdots\vdots C:H \quad \text{or} \quad H{-}C{\equiv}C{-}H$$

$8e^-$ $8e^-$

Except for carbon monoxide, stable molecules containing carbon do not have lone pairs on the carbon atoms.

Note that in ethylene and acetylene all the valence electrons are used in bonding; there are no lone pairs on the carbon atoms.

Finally we note that *compounds that contain only covalent bonds* are called ***covalent compounds***. There are two types of covalent compounds. One contains discrete molecular units. Such compounds are called molecular covalent compounds. Examples are water, carbon dioxide, and hydrogen chloride. The other type has extensive three-dimensional structures and no discrete molecular units. Solid beryllium chloride ($BeCl_2$) and silicon dioxide (SiO_2) are covalent compounds of this type. Such substances are sometimes called *network covalent compounds,* to distinguish them from molecular covalent compounds. Unless otherwise stated, when we refer to covalent compounds in this chapter, we will mean molecular covalent compounds.

Covalent compounds are the same as molecular compounds, which we discussed in previous chapters.

Comparison of the Properties of Covalent and Ionic Compounds

Ionic and covalent compounds differ markedly in their general physical properties because of the differences in the nature of their bonds. There are two types of attractive forces in covalent compounds. The first type is the force that holds the atoms together in a molecule. A quantitative measure of this attraction is given by bond energy, discussed in Section 9.10. The second type of attractive force operates *between* molecules and is called an *intermolecular force*. Because intermolecular forces are usually quite weak compared with forces holding atoms together within a molecule, molecules

Table 9.2 Comparison of some general properties of an ionic compound and a covalent compound

Property	NaCl	CCl_4
Appearance	White solid	Colorless liquid
Melting point (°C)	801	−23
Molar heat of fusion* (kJ/mol)	30.2	2.5
Boiling point (°C)	1413	76.5
Molar heat of vaporization* (kJ/mol)	600	30
Density (g/cm^3)	2.17	1.59
Solubility in water	High	Very low
Electrical conductivity		
Solid	Poor	Poor
Liquid	Good	Poor

*Molar heat of fusion and molar heat of vaporization are the amounts of heat needed to melt 1 mole of the solid and to vaporize 1 mole of the liquid, respectively.

If intermolecular forces are weak, it is relatively easy to break up aggregates of molecules to form liquids (from solids) and gases (from liquids).

of a covalent compound are not held together tightly. Consequently covalent compounds are usually gases, liquids, or low-melting solids. On the other hand, the electrostatic forces holding the ions together in an ionic compound are usually very strong, so ionic compounds are solids with high melting points. Many ionic compounds are soluble in water, and the resulting aqueous solutions conduct electricity, because the compounds are strong electrolytes. Most covalent compounds are insoluble in water, or if they do dissolve, their aqueous solutions generally do not conduct electricity, because the compounds are nonelectrolytes. Molten ionic compounds conduct electricity because they contain mobile cations and anions; liquid or molten covalent compounds do not conduct electricity because no ions are present. Table 9.2 compares some of the general properties of a typical ionic compound, sodium chloride, with those of a covalent compound, carbon tetrachloride (CCl_4).

9.5 Electronegativity

Before we learn how to write Lewis structures of compounds, we need to become familiar with the concept of electronegativity. Electronegativity not only helps us draw the correct Lewis structures, but also enables us to distinguish between covalent compounds and ionic compounds.

A covalent bond, as we have said, is the sharing of an electron pair between two atoms. In a molecule such as H_2, where the atoms are identical, we expect the electrons to be equally shared; that is, the electrons spend the same amount of time in the vicinity of each atom. The situation is different for the HF molecule. Although the H and F atoms are also joined by a covalent bond, the sharing of the electron pair is no longer equal because H and F are different atoms. Instead, the electrons spend *more* time in the vicinity of one atom than the other. In such a case, the covalent bond is called a *polar covalent bond,* or simply a *polar bond.*

Experimental evidence indicates that in the HF molecule the electrons spend more time near the F atom. We can think of this unequal sharing of electrons in terms of a partial transfer of electrons (or a shift in electron density, as it is more commonly described) from H to F (Figure 9.7). We can represent the electron density shift by adding an arrow (⟶ with a cross at its tail) above the Lewis structure for HF:

$$\overset{+\!\!\longrightarrow}{\text{H—}\ddot{\underset{\cdot\cdot}{\text{F}}}\text{:}}$$

This "unequal sharing" of the bonding electron pair results in a relatively greater electron density near the fluorine atom and a correspondingly lower electron density near the hydrogen atom. The charge separation can be represented as

$$\overset{\delta^+}{\text{H}}\text{—}\overset{\delta^-}{\ddot{\underset{\cdot\cdot}{\text{F}}}}\text{:}$$

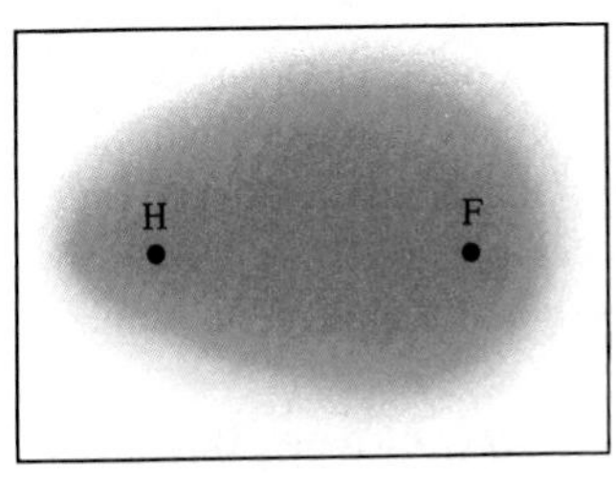

Figure 9.7 *Electron density distribution in the HF molecule. The dots represent the positions of the nuclei.*

where δ (delta) denotes the fractional electron charge separation. The HF bond is a polar bond. A polar bond can be thought of as being intermediate between a (nonpolar) covalent bond, where the sharing of electrons is exactly equal, and an ionic bond, where the transfer of the electron(s) is complete.

A property that helps us distinguish a nonpolar covalent bond from a polar bond is the ***electronegativity*** of the elements, that is, *the ability of an atom to attract toward itself the electrons in a chemical bond.* As we might expect, the electronegativity of an

Increasing electronegativity →

↑ Increasing electronegativity

1A	2A	3B	4B	5B	6B	7B	8B			1B	2B	3A	4A	5A	6A	7A	8A
H 2.1																	
Li 1.0	Be 1.5											B 2.0	C 2.5	N 3.0	O 3.5	F 4.0	
Na 0.9	Mg 1.2											Al 1.5	Si 1.8	P 2.1	S 2.5	Cl 3.0	
K 0.8	Ca 1.0	Sc 1.3	Ti 1.5	V 1.6	Cr 1.6	Mn 1.5	Fe 1.8	Co 1.9	Ni 1.9	Cu 1.9	Zn 1.6	Ga 1.6	Ge 1.8	As 2.0	Se 2.4	Br 2.8	
Rb 0.8	Sr 1.0	Y 1.2	Zr 1.4	Nb 1.6	Mo 1.8	Tc 1.9	Ru 2.2	Rh 2.2	Pd 2.2	Ag 1.9	Cd 1.7	In 1.7	Sn 1.8	Sb 1.9	Te 2.1	I 2.5	
Cs 0.7	Ba 0.9	La–Lu 1.0–1.2	Hf 1.3	Ta 1.5	W 1.7	Re 1.9	Os 2.2	Ir 2.2	Pt 2.2	Au 2.4	Hg 1.9	Tl 1.8	Pb 1.9	Bi 1.9	Po 2.0	At 2.2	
Fr 0.7	Ra 0.9																

Figure 9.8 *The electronegativities of common elements.*

Electronegativity values have no units.

An ionic bond generally involves an atom of a metallic element and an atom of a nonmetallic element, whereas a covalent bond generally involves two atoms of nonmetallic elements.

element is related to its electron affinity and ionization energy. Thus an atom such as fluorine, which has a high electron affinity (tends to pick up electrons easily) and a high ionization energy (does not lose electrons easily), has a high electronegativity. On the other hand, sodium, which has a low electron affinity and a low ionization energy, has a low electronegativity.

Linus Pauling† devised a method for calculating *relative* electronegativities of most elements. These values are shown in Figure 9.8. This chart is important because it shows trends and relationships among electronegativity values of different elements. Figure 9.9 shows the periodic trends in electronegativity. In general, electronegativity increases from left to right across a period in the periodic table, consistent with the decreasing metallic character of the elements. Within each group, electronegativity decreases with increasing atomic number, indicating increasing metallic character (Figure 9.8). Note that the transition metals do not follow these trends. The most electronegative elements—the halogens, oxygen, nitrogen, and sulfur—are found in the upper right-hand corner of the periodic table, and the least electronegative elements (the alkali and alkaline earth metals) are clustered near the lower left-hand corner.

Atoms of elements with widely different electronegativities tend to form ionic bonds with each other since the atom of the less electronegative element gives up its electron(s) to the atom of the more electronegative element. Atoms of elements with more similar electronegativities tend to form covalent bonds or polar bonds with each other because usually only a small shift in electron density takes place. These trends and characteristics are what we would expect from our knowledge of the elements' ionization energies and electron affinities.

Electronegativity and electron affinity are related but different concepts. Both prop-

†Linus Carl Pauling (1901–). American chemist. Regarded by many as the most influential chemist of the twentieth century, Pauling has done research in a remarkably broad range of subjects, from chemical physics to molecular biology. Pauling received the Nobel Prize in chemistry in 1954 for his work on protein structure, and the Nobel Peace Prize in 1962. He is the only person to be the sole recipient of two Nobel Prizes.

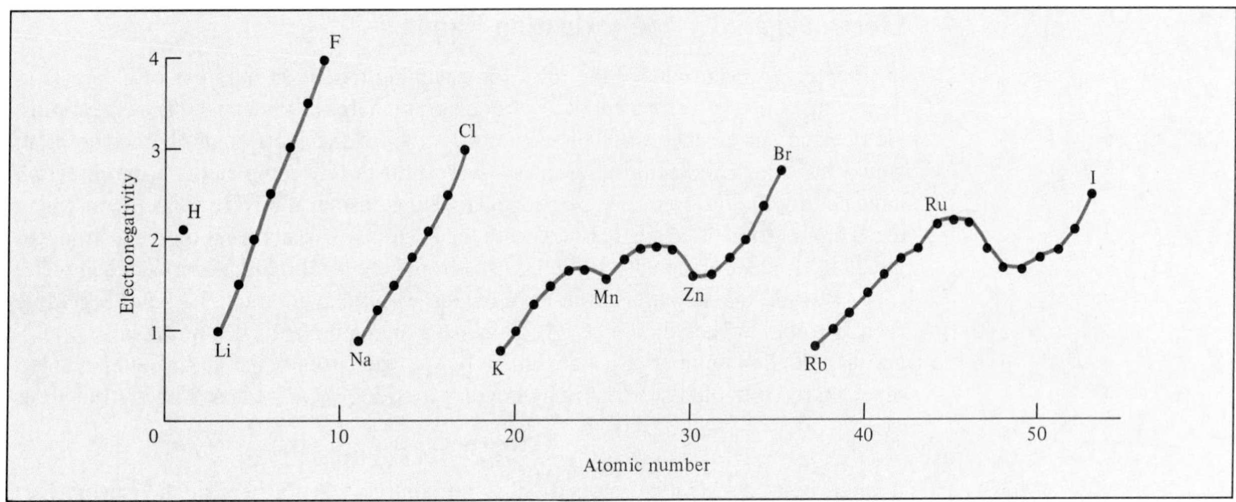

Figure 9.9 *Variation of electronegativity with atomic number. The halogens have the highest electronegativities, and the alkali metals the lowest.*

erties express the tendency of an atom to attract electrons. However, electron affinity refers to an isolated atom's attraction for an additional electron, whereas electronegativity expresses the attraction of an atom in a chemical bond (with another atom) for the shared electrons. Furthermore, electron affinity is an experimentally measurable quantity, whereas electronegativity is a relative number—it is not measurable.

Although there is no sharp distinction between a polar bond and an ionic bond, the following rule is helpful in distinguishing between them. An ionic bond forms when the electronegativity difference between the two bonding atoms is 2.0 or more. This rule applies to most but not all ionic compounds. Sometimes chemists use the quantity *percent ionic character* to describe the nature of a bond. An ionic bond has 100 percent ionic character, whereas a nonpolar or purely covalent bond has 0 percent ionic character.

No known compound has 100 percent ionic character.

The following example shows how a knowledge of electronegativity can help us determine the nature (that is, covalent or ionic) of a chemical bond.

EXAMPLE 9.2

Classify the following bonds as ionic, polar covalent, or covalent: (a) the bond in HCl, (b) the bond in KF, and (c) the CC bond in H_3CCH_3.

Answer

(a) In Figure 9.8 we see that the electronegativity difference between H and Cl is 0.9, which is appreciable but not large enough (by the 2.0 rule) to qualify HCl as an ionic compound. Therefore, the bond between H and Cl is polar covalent.
(b) The electronegativity difference between K and F is 3.2, well above the 2.0 mark; therefore, the bond between K and F is ionic.
(c) The two C atoms are identical in every respect—they are bonded to each other and each is bonded to three other H atoms. Therefore, the bond between them is purely covalent.

Similar problems: 9.37, 9.41.

Electronegativity and Oxidation Number

In Chapter 3 we introduced the rules for assigning oxidation numbers of elements in their compounds. The concept of electronegativity helps us understand how these rules are devised. In essence, oxidation number refers to the number of charges an atom would have in a molecule if electrons were transferred completely in the direction indicated by the difference in electronegativity. Consider the NH_3 molecule in which the N atom forms three single bonds with the H atoms. Because N is more electronegative than H, electron density will be shifted from H to N. If the transfer were complete, each H would donate an electron to N, which would have a total charge of -3 while each H would have a charge of $+1$. Thus we assign an oxidation number of -3 for N and an oxidation number of $+1$ for H in NH_3. Oxygen usually has an oxidation number of -2 in its compounds except in hydrogen peroxide (H_2O_2), whose Lewis structure is

$$\mathrm{H{-}\ddot{\underset{\cdot\cdot}{O}}{-}\ddot{\underset{\cdot\cdot}{O}}{-}H}$$

A bond between identical atoms makes no contribution to the oxidation number of those atoms because the electron pair of that bond is *equally* shared. Since H has an oxidation number of $+1$, each O atom has an oxidation number of -1. The discussion here also explains why fluorine always has an oxidation number of -1. It is the most electronegative element known, and it always forms a single bond in its compounds. Therefore, it would bear a -1 charge assuming complete electron transfer.

9.6 Writing Lewis Structures

Although the octet rule and the Lewis structures do not present a complete picture of covalent bonding, they are very useful in understanding the bonding scheme in many compounds, and chemists frequently refer to them when discussing the properties and reactions of molecules. For this reason, you should practice writing Lewis structures of compounds. The basic steps are as follows:

1. Write the skeletal structure of the compound showing what atoms are bonded to what other atoms. For simple compounds, this task is fairly easy. For more complex compounds, we must either be given the information or make an intelligent guess about it. The following facts are useful in guessing skeletal structures: In general, the least electronegative atom occupies the central position. Hydrogen and fluorine usually occupy the terminal (end) positions in the Lewis structure.
2. Count the total number of valence electrons present, referring, if necessary, to Figure 9.1. For polyatomic anions, add the number of negative charges to that total. (For example, for the CO_3^{2-} ion we add two electrons because the 2− charge indicates that there are two more electrons than are provided by the neutral atoms.) For polyatomic cations, we subtract the number of positive charges from this total. (Thus, for NH_4^+ we subtract one electron because the 1+ charge indicates a loss of one electron from the group of neutral atoms.)
3. Draw a single covalent bond between the central atom and each of the surrounding atoms. Complete the octets of the atoms bonded to the central atom. (Remember that the valence shell of a hydrogen atom is complete with only two electrons.) Electrons belonging to the central or surrounding atoms must be shown as lone pairs if they are not involved in bonding. The total number of electrons to be used is that determined in step 2.

4. If the octet rule is not met for the central atom, then we should try double or triple bonds between the surrounding atoms and the central atom, using the lone pairs from the surrounding atoms.

The following examples illustrate this four-step procedure for writing Lewis structures of compounds.

EXAMPLE 9.3

Write the Lewis structure for nitrogen trifluoride (NF_3), in which all three F atoms are bonded to the N atom.

Answer

Step 1: The skeletal structure of NF_3 is

Note that the central N atom is less electronegative than F.

```
F N F
  F
```

Step 2: The outer-shell electron configurations of N and F are $2s^22p^3$ and $2s^22p^5$, respectively. Thus there are $5 + (3 \times 7)$, or 26, valence electrons to account for in NF_3.

Step 3: We draw a single covalent bond between N and each F, and complete the octets for the F atoms. We place the remaining two electrons on N:

```
 ..  ..  ..
:F—N—F:
 ..  |   ..
    :F:
     ..
```

Since this structure satisfies the octet rule for all the atoms, step 4 is not required. To check, we count the valence electrons in NF_3 (in chemical bonds and in lone pairs). The result is 26, the same as the number of valence electrons on three F atoms and one N atom.

Similar problem: 9.45.

EXAMPLE 9.4

Write the Lewis structure for nitric acid, HNO_3, in which the three O atoms are bonded to the central N atom and the ionizable H atom is bonded to one of the O atoms.

Answer

Step 1: The skeletal structure of HNO_3 is

```
O N O H
  O
```

Step 2: The outer-shell electron configurations of N, O, and H are $2s^22p^3$, $2s^22p^4$, and $1s^1$, respectively. Thus there are $5 + (3 \times 6) + 1$, or 24, valence electrons to account for in HNO_3.

Step 3: We draw a single covalent bond between N and each of the three O atoms and between one O atom and the H atom. Then we fill in electrons to comply with the octet rule for the O atoms:

```
  ..     ..
: O — N — O — H
  ..  |  ..
     :O:
      ..
```

When we have done this, all 24 of the available electrons have been used.

Step 4: We see that this structure satisfies the octet rule for all the O atoms but not for the N atom. Therefore we move a lone pair from one of the end O atoms to form another bond with N. Now the octet rule is also satisfied for the N atom:

```
  ..    ..
  O = N — O — H
  ..  |  ..
     :O:
      ..
```

Similar problem: 9.45.

EXAMPLE 9.5

Write the Lewis structure for the carbonate ion (CO_3^{2-}).

Answer

Step 1: We can deduce the skeletal structure of the carbonate ion by realizing that C is less electronegative than O. Therefore, it is most likely to occupy a central position as follows:

```
   O
O  C  O
```

Another plausible skeletal structure is O C O O. You can show that the octet rule cannot be satisfied for all the atoms in this ion.

Step 2: The outer-shell electron configurations of C and O are $2s^2 2p^2$ and $2s^2 2p^4$, respectively, and the ion itself has two negative charges. Thus the total number of electrons is $4 + (3 \times 6) + 2$, or 24.

Step 3: We draw a single covalent bond between C and each O and comply with the octet rule for the O atoms:

```
       ..
      :O:
  ..   |   ..
: O —  C — O :
  ..       ..
```

This structure shows all 24 electrons.

Step 4: Although the octet rule is satisfied for the O atoms, it is not for the C atom. Therefore we move a lone pair from one of the O atoms to form another bond with C. Now the octet rule is also satisfied for the C atom:

```
[     :O:      ]2−
[  ..  ||  ..  ]
[ : O — C — O :]
[  ..      ..  ]
```

As a final check, we verify that there are 24 valence electrons in the Lewis structure of the carbonate ion.

Similar problem: 9.46.

The octet rule works mainly for elements in the second period of the periodic table. These elements have only 2*s* and 2*p* subshells, which can hold a total of eight electrons. When an atom of one of these elements forms a covalent compound, it can attain the noble gas electron configuration [Ne] by sharing electrons with other atoms in the same compound. However, as we will see later, there are a number of important exceptions to the octet rule that give us further insight into the nature of chemical bonding.

9.7 Formal Charge and Lewis Structure

In drawing the Lewis structure of a molecule, you will often find it useful to compare the number of electrons in an isolated atom with the number that are associated with the same atom in the molecule. This comparison reveals the distribution of electrons in the molecule and, as we will see shortly, also helps us in drawing the most plausible Lewis structures. The electron bookkeeping procedure is as follows: In an isolated atom, the number of electrons associated with the atom is simply the number of valence electrons. (As usual, we need not be concerned with the inner electrons.) In a molecule, electrons associated with the atom are the lone pairs on the atom plus the electrons in the bonding pair(s) between the atom and other atom(s). However, because electrons are shared in a bond, we must divide the electrons in a bonding pair equally between the atoms forming the bond. *The difference between the valence electrons in an isolated atom and the number of electrons assigned to that atom in a Lewis structure* is called that atom's ***formal charge***. The equation for calculating the formal charge on an atom in a molecule is given by

$$\begin{array}{l}\text{formal charge on}\\ \text{an atom in a}\\ \text{Lewis structure}\end{array} = \begin{array}{l}\text{total number of}\\ \text{valence electrons}\\ \text{in the free atom}\end{array} - \begin{array}{l}\text{total number}\\ \text{of nonbonding}\\ \text{electrons}\end{array} - \frac{1}{2}\left(\begin{array}{l}\text{total number}\\ \text{of bonding}\\ \text{electrons}\end{array}\right) \qquad (9.3)$$

Ozone is a toxic, light blue gas (boiling point: −111.3°C) with a pungent odor.

Let us illustrate the concept of formal charge using the ozone molecule (O_3). Experiments show that in O_3 a central O atom is bonded to two end O atoms. Proceeding by steps as we did in Examples 9.3 and 9.4, we draw the skeletal structure of O_3 and then add bonds and electrons to satisfy the octet rule for the two end atoms:

You can see that although all available electrons are used, the octet rule is not satisfied for the central atom. To remedy this, we convert a lone pair on one of the end atoms to a second bond between that end atom and the central atom, as follows:

We can now use Equation (9.3) to calculate the formal charges on the O atoms as follows.

- *The central O atom.* In the preceding Lewis structure the central atom has six valence electrons, one lone pair (or two nonbonding electrons), and three bonds (or six bonding electrons). Substituting in Equation (9.3) we write

$$\text{formal charge} = 6 - 2 - \tfrac{1}{2}(6) = +1$$

- *The end O atom in O=O.* This atom has six valence electrons, two lone pairs (or four nonbonding electrons), and two bonds (or four bonding electrons). Thus we write

$$\text{formal charge} = 6 - 4 - \tfrac{1}{2}(4) = 0$$

- *The end O atom in O—O*. This atom has six valence electrons, three lone pairs (or six nonbonding electrons), and one bond (or two bonding electrons). Thus we write

$$\text{formal charge} = 6 - 6 - \tfrac{1}{2}(2) = -1$$

We can now write the Lewis structure for ozone including the formal charges as

Singly positive and negative charges are normally shown as + and −, respectively, rather than +1 and −1.

$$\ddot{O}^{+} \quad \dot{\dot{O}}{=}\ \ \ {-}\ddot{\underset{\cdot\cdot}{O}}{:}^{-}$$

When you write formal charges, the following rules are helpful:

- For neutral molecules, the sum of the formal charges must add up to zero. (This rule applies, for example, to the O_3 molecule.)
- For cations, the sum of the formal charges must equal the positive charge.
- For anions, the sum of the formal charges must equal the negative charge.

Keep in mind that formal charges do not indicate actual charge separation within the molecule. In the O_3 molecule, for example, there is no evidence that the central atom bears a net +1 charge or that one of the end atoms bears a −1 charge. Writing these charges on the atoms in the Lewis structure merely helps us keep track of the valence electrons in the molecule.

EXAMPLE 9.6

Write formal charges for the carbonate ion.

Answer

The Lewis structure for the carbonate ion was developed in Example 9.5:

$$\left[\ :\ddot{\underset{\cdot\cdot}{O}}\text{—}\overset{\overset{:O:}{\|}}{C}\text{—}\ddot{\underset{\cdot\cdot}{O}}:\ \right]^{2-}$$

The formal charges on the atoms can be calculated as follows:

The C atom:	formal charge = $4 - 0 - \tfrac{1}{2}(8) = 0$
The O atom in C═O:	formal charge = $6 - 4 - \tfrac{1}{2}(4) = 0$
The O atom in C—O:	formal charge = $6 - 6 - \tfrac{1}{2}(2) = -1$

Thus the Lewis formula for CO_3^{2-} with formal charges is

$$^{-}:\ddot{\underset{\cdot\cdot}{O}}\text{—}\overset{\overset{:O:}{\|}}{C}\text{—}\ddot{\underset{\cdot\cdot}{O}}:^{-}$$

Note that the sum of the formal charges is −2, the same as the charge on the carbonate ion.

Formal charges are often useful in selecting a plausible Lewis structure for a given compound. The guidelines we follow are

- For neutral molecules, a Lewis structure in which there are no formal charges is preferable to one in which formal charges are present.
- Lewis structures with large formal charges (+2, +3, and/or −2, −3, and so on) are less plausible than those with small formal charges.
- In choosing among Lewis structures having similar distributions of formal charges, the most plausible Lewis structure is the one in which negative formal charges are placed on the more electronegative atoms.

The following example shows how formal charges can help in selecting the correct Lewis structure for a molecule.

EXAMPLE 9.7

Formaldehyde, a liquid with a disagreeable odor, traditionally has been used as a preservative for dead animals. Its molecular formula is CH_2O. Draw the most likely Lewis structure for the compound.

Answer

The two possible skeletal structures are

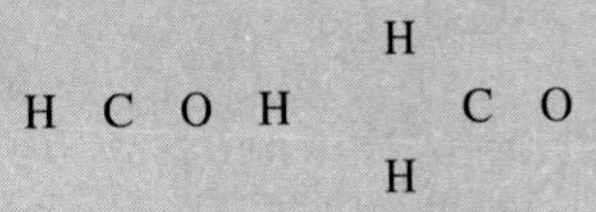

H C O H (a)

H, C O, H (b)

Following the procedures in previous examples we can draw a Lewis structure for each of these possibilities:

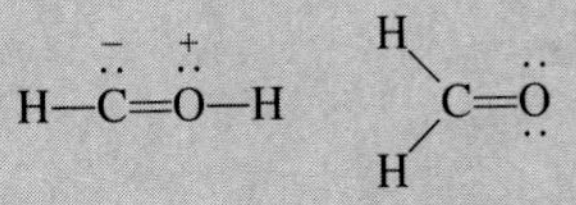

$H—\overset{-}{\ddot{C}}=\overset{+}{\ddot{O}}—H$ (a)

$(H)_2C=\ddot{\underset{..}{O}}$ (b)

Since (b) carries no formal charges, it is the more likely structure. This conclusion is confirmed by experimental evidence.

Can you suggest two other reasons why (a) is less plausible?

9.8 The Concept of Resonance

In drawing a Lewis structure for the ozone, O_3, molecule, we satisfied the octet rule for the central atom by drawing a double bond between it and one of the two end O atoms. Alternatively, we can add the double bond in either of two places, as shown by these two equivalent Lewis structures:

$O=\overset{+}{O}—O^-$ $\quad$ $^-O—\overset{+}{O}=O$

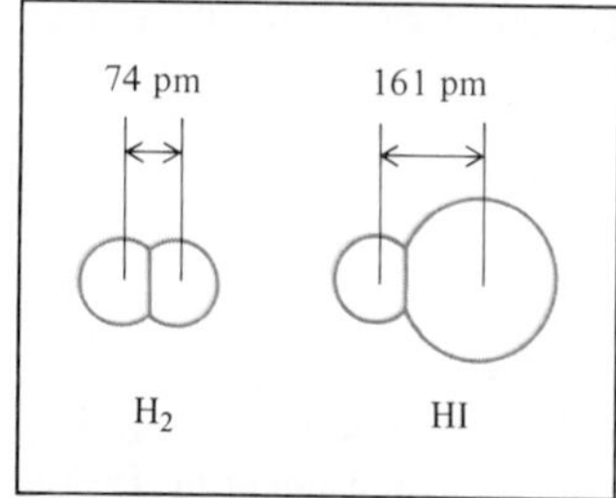

Figure 9.10 *The bond length is the distance between two bonded nuclei. Here the bond lengths (in picometers) for the diatomic molecules H_2 and HI are defined.*

Arbitrarily choosing one of these two Lewis structures to represent ozone presents some problems, however. For one thing, it makes the task of accounting for the known bond lengths in O_3 difficult.

Bond length is *the distance between the nuclei of two bonded atoms in a molecule* (Figure 9.10). Chemists know from experience that bond length depends not only on the nature of the bonded atoms but also on whether the bond joining the atoms is a single bond, a double bond, or a triple bond. Table 9.3 shows some typical bond lengths. As you can see, for a given pair of atoms, triple bonds are shorter than double bonds, which, in turn, are shorter than single bonds. On the basis of these data and the proposed Lewis structure for O_3, we would expect that the O—O bond would be longer than the O═O bond. However, experimental evidence shows that both oxygen-to-oxygen bonds are equal in length (128 pm); therefore, neither of the two Lewis structures shown accurately represents the molecule.

The way we resolve this problem is to use *both* Lewis structures to represent the ozone molecule:

$$\ddot{\mathrm{O}}{=}\overset{+}{\ddot{\mathrm{O}}}{-}\mathrm{O}^{-} \longleftrightarrow {}^{-}\mathrm{O}{-}\overset{+}{\ddot{\mathrm{O}}}{=}\ddot{\mathrm{O}}$$

Table 9.3 Average bond lengths of some common single, double, and triple bonds

Bond type	Bond length (pm)
C—H	107
C—O	143
C═O	121
C—C	154
C═C	133
C≡C	120
C—N	143
C═N	138
C≡N	116
N—O	136
N═O	122
O—H	96

Each of the two Lewis structures is called a ***resonance structure*** or ***resonance form.*** A resonance structure, then, is *one of two or more Lewis structures for a single molecule that cannot be described fully with only one Lewis structure*. The symbol ⟷ indicates that the structures shown are resonance forms. The term ***resonance*** itself means *the use of two or more Lewis structures to represent a particular molecule*. There is an interesting analogy for resonance: A medieval European traveler returned home from Africa and described a rhinoceros as a cross between a griffin and a unicorn, two familiar but imaginary animals. Similarly, we describe ozone, a real molecule, in terms of the two familiar but nonexistent molecules shown.

A common misconception about resonance is the notion that a molecule such as ozone somehow shifts quickly back and forth from one resonance structure to the other. Keep in mind that *neither* resonance structure adequately represents the actual molecule, which has its own unique, stable structure. ''Resonance'' is a human invention, designed to address the limitations in these simple bonding models. To extend the analogy, a rhinoceros is its ''own'' self, not some oscillation between mythical griffin and unicorn!

Other examples of resonance structures are those for the carbonate ion:

$$\begin{array}{c}:\mathrm{O}:\\ \|\\ {}^{-}:\ddot{\mathrm{O}}{-}\mathrm{C}{-}\ddot{\mathrm{O}}:^{-}\end{array} \longleftrightarrow \begin{array}{c}:\ddot{\mathrm{O}}:^{-}\\ |\\ \ddot{\mathrm{O}}{=}\mathrm{C}{-}\ddot{\mathrm{O}}:^{-}\end{array} \longleftrightarrow \begin{array}{c}:\ddot{\mathrm{O}}:^{-}\\ |\\ {}^{-}:\ddot{\mathrm{O}}{-}\mathrm{C}{=}\ddot{\mathrm{O}}\end{array}$$

According to experimental evidence, all carbon-to-oxygen bonds are equivalent. Therefore, the properties of the carbonate ion are best described by considering its resonance structures together, not separately.

The concept of resonance applies equally well to organic systems. A well-known example is the benzene molecule (C_6H_6):

H H
H C H H C H
C C C C
⟷
C C C C
H C H H C H
H H

If benzene actually existed as one of its resonance forms, there would be two different bond lengths between adjacent C atoms, one characteristic of the single bond and the other of the double bond. In fact, the distances between adjacent C atoms in benzene are *all* 140 pm, which is between the length of a C—C bond (154 pm) and that of a C═C bond (133 pm), shown in Table 9.3.

A simpler way of drawing the structure of the benzene molecule or of other compounds containing the "benzene ring" is to show only the skeleton and not the carbon and hydrogen atoms. By this convention the resonance structures are represented by

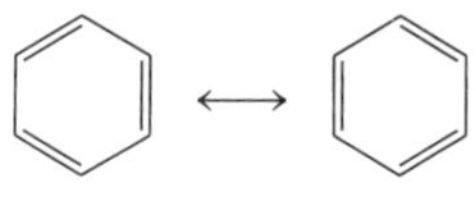

Note that the C atoms at the corners of the hexagon and the H atoms are all omitted, although they are understood to exist. Only the bonds between the C atoms are shown.

Remember this important rule for drawing resonance structures: The positions of electrons, but not those of atoms, can be rearranged in different resonance structures. In other words, the same atoms must be bonded to one another in all the resonance structures for a given species.

The following example illustrates the procedure for drawing resonance structures of a molecule.

EXAMPLE 9.8

Draw resonance structures (including formal charges) for dinitrogen tetroxide (N_2O_4), which has the following skeletal arrangement:

O O
N N
O O

Answer

Since each nitrogen atom has five valence electrons and each oxygen atom has six valence electrons, the total number of valence electrons is $(2 \times 5) + (4 \times 6) = 34$. The following Lewis structure satisfies the octet rule for both the N and O atoms:

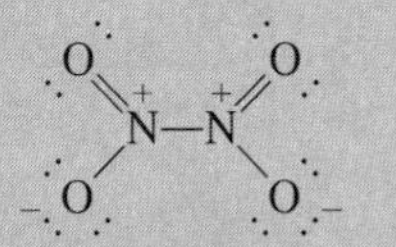

However, since it does not matter where we draw the N—O and N═O bonds, we must include the following resonance structures:

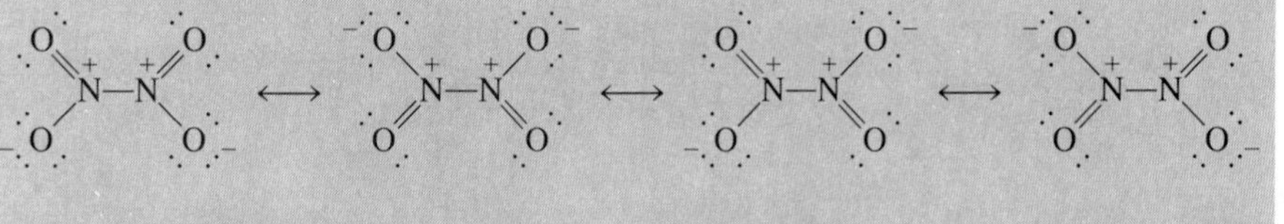

Similar problems: 9.55, 9.58, 9.60.

A final note on resonance: Although including all the resonance structures provides a more accurate description of the properties of a molecule, for simplicity, we will often use only one Lewis structure to represent a molecule.

9.9 Exceptions to the Octet Rule

As mentioned earlier, the octet rule applies mainly to the second-period elements. Following are three types of exceptions to the octet rule.

The Incomplete Octet

Unlike beryllium, which forms mostly covalent compounds, the other Group 2A elements form mostly ionic compounds.

In some compounds the number of electrons surrounding the central atom in a stable molecule is fewer than eight. Consider, for example, beryllium, which is a Group 2A (and a second-period) element. The electron configuration of beryllium is $1s^22s^2$; it has two valence electrons in the 2*s* orbital. In the gas phase, beryllium hydride (BeH_2) exists as discrete molecules. The Lewis structure of BeH_2 is

H—Be—H

As you can see, only four electrons surround the Be atom, and there is no way to satisfy the octet rule for beryllium in this molecule.

Elements in Group 3A, particularly boron and aluminum, also tend to form compounds in which fewer than eight electrons surround each atom. Take boron as an example. Since its electron configuration is $1s^22s^22p^1$, it has a total of three valence electrons. Boron forms with the halogens a class of compounds of the general formula BX_3, where X is a halogen atom. Thus, in boron trifluoride there are only six electrons around the boron atom:

Boron trifluoride is a colorless gas (boiling point: −100°C) with a pungent odor.

```
      ..
     :F:
 ..   |
:F—B
 ..   |
     :F:
      ..
```

A resonance structure with a double bond between B and F can be drawn that satisfies the octet rule for B. However, the properties of BF_3 are more consistent with a Lewis structure in which there are single bonds between B and each F, as shown above.

Although boron trifluoride is stable, it has a tendency to pick up an unshared electron pair from an atom in another compound, as shown by its reaction with ammonia:

```
     ..                      ..
    :F:    H                :F:  H
 ..  |     |             ..  |   |
:F—B  +  :N—H  ⟶  :F—B⁻—N⁺—H
 ..  |     |             ..  |   |
    :F:    H                :F:  H
     ..                      ..
```

This structure satisfies the octet rule for the B, N, and F atoms.

The B—N bond in the above compound is different from the covalent bonds discussed so far in the sense that both electrons are contributed by the N atom. *A covalent bond in which one of the atoms donates both electrons* is called a ***coordinate covalent bond.*** Although the properties of a coordinate covalent bond do not differ from those of a normal covalent bond (because all electrons are alike no matter what their source), the distinction is useful for keeping track of valence electrons and assigning formal charges.

Another name for a coordinate covalent bond is *dative* bond.

Odd-Electron Molecules

Some molecules contain an *odd* number of electrons. Among them are nitric oxide (NO) and nitrogen dioxide (NO_2):

N=O   O=N⁺—O:⁻

Since we need an even number of electrons for complete pairing (to reach eight), the octet rule clearly can *never* be satisfied for all the atoms in any of these molecules.

The Expanded Octet

Atoms of the second-period elements cannot have more than eight valence electrons in a compound.

In a number of compounds there are more than eight valence electrons around an atom. These *expanded octets* occur only around atoms of elements in and beyond the third period of the periodic table. In addition to the 3*s* and 3*p* orbitals, elements in the third period also have 3*d* orbitals that can be used in bonding. One compound in which there is an expanded octet is sulfur hexafluoride, a very stable compound. The electron configuration of sulfur is [Ne]$3s^23p^4$. In SF_6, each of sulfur's six valence electrons forms a covalent bond with a fluorine atom, so there are twelve electrons around the central sulfur atom:

Sulfur hexafluoride is a colorless, odorless gas (boiling point: −64°C).

(Lewis structure of SF_6: S bonded to six F atoms, each F with three lone pairs)

In the next chapter we will see that these twelve electrons, or six bonding pairs, are accommodated in six orbitals that originate from the one 3*s*, the three 3*p*, and two of the five 3*d* orbitals. However, sulfur also forms many compounds in which it does not violate the octet rule. In sulfur dichloride, S has only eight electrons and therefore obeys the octet rule:

:Cl—S—Cl:

Sulfur dichloride is a toxic, foul-smelling cherry red liquid (boiling point: 59°C).

The following examples concern compounds that do not obey the octet rule.

EXAMPLE 9.9

Draw the Lewis structure for aluminum triiodide (AlI_3).

Answer

Like boron, aluminum is a Group 3A element.

The outer-shell electron configuration of Al is $3s^23p^1$. The Al atom forms three covalent bonds with the I atoms as follows:

(Lewis structure: Al bonded to three I atoms, each I with three lone pairs)

Although the octet rule is satisfied for the I atoms, there are only six valence electrons around the Al atom. This molecule is an example of the incomplete octet.

Similar problem: 9.65.

EXAMPLE 9.10

Draw the Lewis structure for phosphorus pentafluoride (PF_5), in which all five F atoms are bonded directly to the P atom.

Answer

Like sulfur, phosphorus is a third-period element.

The outer-shell electron configurations for P and F are $3s^2 3p^3$ and $2s^2 2p^5$, respectively, and so the total number of valence electrons is $5 + (5 \times 7)$, or 40. The Lewis structure of PF_5 is

```
       ..
      :F: ..
   ..  |  F:
  :F—P/   ..
   ..  |\ ..
      :F: F:
       ..  ..
```

Although the octet rule is satisfied for the F atoms, there are ten valence electrons around the P atom, giving it an expanded octet.

Similar problem: 9.69.

EXAMPLE 9.11

Draw a Lewis structure for sulfuric acid (H_2SO_4) in which all four O atoms are bonded to the central S atom and the two ionizable H atoms are terminal.

Answer

Step 1: The skeletal structure of H_2SO_4 is

```
      O
H  O  S  O  H
      O
```

Step 2: Both O and S are Group 6A elements and so have six valence electrons. Therefore, we must account for a total of $6 + (4 \times 6) + 2$, or 32, valence electrons in H_2SO_4.

Step 3: We draw a single covalent bond between all the bonding atoms:

```
        ..
       :O:
    ..  |  ..
  H—O—S—O—H
    ..  |  ..
       :O:
        ..
```

Step 4: Although the above structure satisfies the octet rule and uses all 32 electrons, it is not too plausible because of the separation of formal charges:

```
        ..
       :O:⁻
    ..  |    ..
  H—O—S²⁺—O—H
    ..  |    ..
       :O:₋
        ..
```

Therefore we convert a lone pair from each of two O atoms to form another bond with S:

$$\text{H—}\ddot{\text{O}}\text{—}\overset{:\text{O}:}{\overset{\|}{\underset{\|}{\underset{:\text{O}:}{\text{S}}}}}\text{—}\ddot{\text{O}}\text{—H}$$

There are now 12 electrons around the S atom, and the octet rule is not obeyed. Nevertheless this is a more reasonable Lewis structure because there is no formal charge separation. Sulfur is a third-period element and can therefore form an expanded octet.

Similar problem: 9.70.

9.10 Strength of the Covalent Bond

Bond Dissociation Energy and Bond Energy

The Lewis theory of chemical bonding depicts a covalent bond as the sharing of two electrons between the atoms; it does not indicate the relative strengths of covalent bonds. For example, the bonds in H_2 and Cl_2 are represented by identical single lines:

$$\text{H—H} \qquad :\ddot{\text{Cl}}\text{—}\ddot{\text{Cl}}:$$

yet we know from experiments that it takes more energy to break the bond in H_2 than in Cl_2. In this sense we conclude that H_2 is the more stable molecule of the two. A quantitative measure of the stability of a molecule is ***bond dissociation energy,*** the *enthalpy change required to break a particular bond in 1 mole of gaseous diatomic molecules*. For the hydrogen molecule

We specify the gaseous state here because bond dissociation energies in solids and liquids are affected by neighboring molecules.

$$H_2(g) \longrightarrow H(g) + H(g) \qquad \Delta H° = 436.4 \text{ kJ}$$

This equation tells us that to break the covalent bonds in 1 mole of gaseous H_2 molecules requires 436.4 kJ of energy.

Similarly, for the less stable chlorine molecule,

$$Cl_2(g) \longrightarrow Cl(g) + Cl(g) \qquad \Delta H° = 242.7 \text{ kJ}$$

The term "bond dissociation energy" can also be applied to diatomic molecules containing unlike elements, such as HCl,

$$HCl(g) \longrightarrow H(g) + Cl(g) \qquad \Delta H° = 431.9 \text{ kJ}$$

as well as to molecules containing double and triple bonds:

The Lewis structure of O_2 is $\ddot{\underset{..}{\text{O}}}\text{=}\ddot{\underset{..}{\text{O}}}$.

$$O_2(g) \longrightarrow O(g) + O(g) \qquad \Delta H° = 498.7 \text{ kJ}$$

$$N_2(g) \longrightarrow N(g) + N(g) \qquad \Delta H° = 941.4 \text{ kJ}$$

Measuring the strength of covalent bonds becomes more complicated for polyatomic molecules. For example, measurements show that the energy needed to break the first O—H bond in H_2O is different from that needed to break the second O—H bond:

$$H_2O(g) \longrightarrow H(g) + OH(g) \qquad \Delta H° = 502 \text{ kJ}$$

$$OH(g) \longrightarrow H(g) + O(g) \qquad \Delta H° = 427 \text{ kJ}$$

In each case, an O—H bond is broken, but the first step is more endothermic than the

Table 9.4 Some bond dissociation energies of diatomic molecules* and average bond energies

Bond	Bond energy (kJ/mol)	Bond	Bond energy (kJ/mol)
H—H	436.4	C—S	255
H—N	393	C=S	477
H—O	460	N—N	193
H—S	368	N=N	418
H—P	326	N≡N	941.4
H—F	568.2	N—O	176
H—Cl	431.9	N—P	209
H—Br	366.1	O—O	142
H—I	298.3	O=O	498.7
C—H	414	O—P	502
C—C	347	O=S	469
C=C	620	P—P	197
C≡C	812	P=P	489
C—N	276	S—S	268
C=N	615	S=S	352
C≡N	891	F—F	150.6
C—O	351	Cl—Cl	242.7
C=O†	745	Br—Br	192.5
C—P	263	I—I	151.0

*Bond dissociation energies for diatomic molecules (in color) have more significant figures than bond energies for polyatomic molecules. This is because, for diatomic molecules, the bond dissociation energies are directly measurable quantities and not averaged over many compounds.

†The C=O bond energy in CO_2 is 799 kJ/mol.

second. The difference in the two $\Delta H°$ values suggests that the remaining O—H bond itself undergoes change, because of the changed chemical environment. But the variation is usually not very great, as indicated by the two $\Delta H°$ values for the O—H bond, which are roughly comparable in H_2O and many other molecules.

On the basis of the study of the O—H bond in H_2O, we can understand why the bond energy of the same O—H bond in two different molecules such as methanol (CH_3OH) and water (H_2O) will not be the same because their environments are different. Thus for polyatomic molecules we can only speak of the *average* bond energy of a particular bond. For example, we can measure the energy of the same O—H bond in ten different molecules and obtain the average O—H bond energy by dividing the sum of the bond energies by 10. The situation is different for diatomic molecules because the particular bond (single, double, or triple) only exists for that molecule so it has a unique value which we call bond dissociation energy. Table 9.4 lists the bond dissociation energies of diatomic molecules and average bond energies for a number of covalent bonds in polyatomic molecules.

Use of Bond Energies in Thermochemistry

A comparison of the thermochemical changes of a number of reactions (Chapter 6) reveals a strikingly wide variation in the enthalpies of different reactions. For example, the combustion of hydrogen gas in oxygen gas is fairly exothermic:

$$H_2(g) + \tfrac{1}{2}O_2(g) \longrightarrow H_2O(l) \qquad \Delta H° = -285.8 \text{ kJ}$$

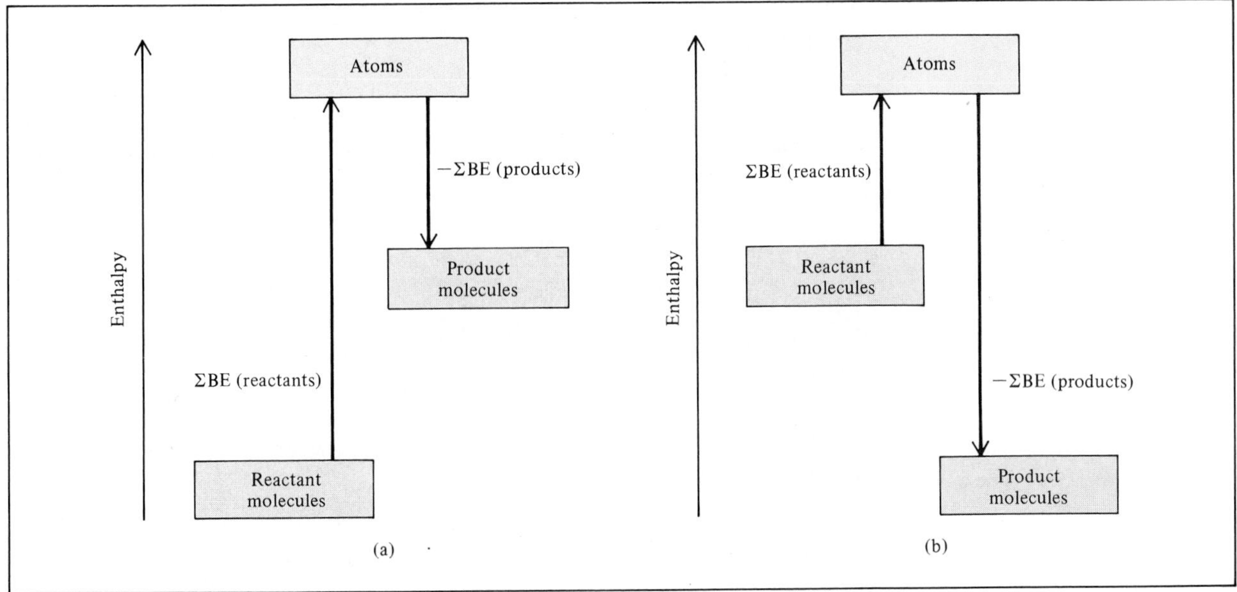

Figure 9.11 *Bond energy changes in (a) an endothermic reaction and (b) an exothermic reaction.*

On the other hand, the formation of glucose ($C_6H_{12}O_6$) from water and carbon dioxide, best achieved by photosynthesis, is highly endothermic:

$$6CO_2(g) + 6H_2O(l) \longrightarrow C_6H_{12}O_6(s) + 6O_2(g) \qquad \Delta H^\circ = 2801 \text{ kJ}$$

To account for such variations, we need to examine the stability of individual reactant and product molecules. After all, most chemical reactions involve the making and breaking of bonds. Therefore, knowing the bond energies and hence the stability of molecules tells us something about the thermochemical nature of reactions that molecules undergo.

In many cases it is possible to predict the approximate enthalpy of reaction for some process by using the average bond energies. Recall that energy is always required to break chemical bonds and that chemical bond formation is always accompanied by a release of energy. To estimate the enthalpy of a reaction, then, all we need to do is count the total number of bonds broken and formed in the reaction and record all the corresponding energy changes. The enthalpy of reaction in the *gas phase* is given by

$$\begin{aligned} \Delta H^\circ &= \Sigma \text{BE(reactants)} - \Sigma \text{BE(products)} \\ &= \text{total energy input} - \text{total energy released} \end{aligned} \qquad (9.4)$$

For diatomic molecules, Equation (9.4) becomes equivalent to Equation (6.9), so that the results obtained from these two equations should be in excellent agreement.

where BE stands for average bond energy and Σ is the summation sign. Equation (9.4) as written takes care of the sign convention for ΔH°. Thus if the total energy input is greater than the total energy released, ΔH° is positive and the reaction is endothermic. On the other hand, if more energy is released than absorbed, ΔH° is negative and the reaction is exothermic (Figure 9.11). Note that if reactants and products are all diatomic molecules, then Equation (9.4) should yield accurate results since the bond dissociation energies of diatomic molecules are accurately known. If some or all of the reactants and products are polyatomic molecules, Equation (9.4) will yield only approximate results because average bond energies will be used in the calculation.

Example 9.12 deals with the calculation of ΔH° of a reaction involving diatomic molecules.

EXAMPLE 9.12

Use Equation (9.4) to calculate the enthalpy of reaction for the process

$$H_2(g) + Cl_2(g) \longrightarrow 2HCl(g)$$

Compare your result with that obtained using Equation (6.9).

Answer

The first step is to count the number of bonds broken and the number of bonds formed. This is best done by using a table:

Refer to Table 9.4 for bond dissociation energies of these diatomic molecules.

Type of bonds broken	*Number of bonds broken*	*Bond energy (kJ/mol)*	*Energy change (kJ)*
H—H (H_2)	1	436.4	436.4
Cl—Cl (Cl_2)	1	242.7	242.7

Type of bonds formed	*Number of bonds formed*	*Bond energy (kJ/mol)*	*Energy change (kJ)*
H—Cl (HCl)	2	431.9	863.8

Next, we obtain the total energy input and total energy released:

$$\text{total energy input} = 436.4 \text{ kJ} + 242.7 \text{ kJ} = 679.1 \text{ kJ}$$

$$\text{total energy released} = 863.8 \text{ kJ}$$

Using Equation (9.4) we write

$$\Delta H° = 679.1 \text{ kJ} - 863.8 \text{ kJ} = -184.7 \text{ kJ}$$

Alternatively, we can use Equation (6.9) and the data in Appendix 3 to calculate the enthalpy of reaction:

$$\begin{aligned}\Delta H° &= 2\Delta H_f°(\text{HCl}) - [\Delta H_f°(\text{H}_2) + \Delta H_f°(\text{Cl}_2)] \\ &= (2 \text{ mol})(-92.3 \text{ kJ/mol}) - 0 - 0 \\ &= -184.6 \text{ kJ}\end{aligned}$$

As stated earlier, the agreement between Equations (9.4) and (6.9) is excellent if the reactants and products are diatomic molecules. If a reaction has one or more polyatomic molecules, either as reactants or as products, then Equation (9.4) can be used only to *estimate* the enthalpy of the reaction, because in that case we have to use average bond energies. The following example shows this approach.

EXAMPLE 9.13

Estimate the enthalpy change for the combustion of hydrogen gas:

$$2H_2(g) + O_2(g) \longrightarrow 2H_2O(g)$$

Answer

As in Example 9.12, we construct a table:

Type of bonds broken	*Number of bonds broken*	*Bond energy (kJ/mol)*	*Energy change (kJ)*
H—H (H_2)	2	436.4	872.8
O=O (O_2)	1	498.7	498.7

Type of bonds formed	*Number of bonds formed*	*Bond energy (kJ/mol)*	*Energy change (kJ)*
O—H (H_2O)	4	460	1840

Next, we obtain the total energy input and total energy released:

$$\text{total energy input} = 872.8 \text{ kJ} + 498.7 \text{ kJ} = 1372 \text{ kJ}$$

$$\text{total energy released} = 1840 \text{ kJ}$$

Using Equation (9.4) we write

$$\Delta H^\circ = 1372 \text{ kJ} - 1840 \text{ kJ} = -468 \text{ kJ}$$

This result is only an estimate because we used the bond energy of O—H, which is an average quantity. Alternatively, we can use Equation (6.9) and the data in Appendix 3 to calculate the enthalpy of reaction:

$$\begin{aligned}\Delta H^\circ &= 2\Delta H^\circ_f(H_2O) - [2\Delta H^\circ_f(H_2) + \Delta H^\circ_f(O_2)]\\ &= (2 \text{ mol})(-241.8 \text{ kJ/mol}) - 0 - 0\\ &= -483.6 \text{ kJ}\end{aligned}$$

Similar problem: 9.78.

Note that the estimated value based on average bond energies is quite close to the value calculated using ΔH°_f data. In general, the bond energy method works best for reactions that are either quite endothermic or quite exothermic, that is, for $\Delta H^\circ >$ 100 kJ or for $\Delta H^\circ < -100$ kJ.

Summary

1. The Lewis dot symbol shows the number of valence electrons possessed by an atom of a given element. Lewis dot symbols are useful mainly for the representative elements.
2. The elements most likely to form ionic compounds are those with low ionization energies (such as the alkali metals and the alkaline earth metals, which form the cations) and those with high negative electron affinities (such as the halogens and oxygen, which form the anions).
3. An ionic bond is the product of the electrostatic forces of attraction between positive and negative ions. An ionic compound is a large collection of ions in which the positive and negative charges are balanced. The structure of a solid ionic compound maximizes the net attractive forces among the ions.
4. Lattice energy is a measure of the stability of an ionic solid. It can be calculated by employing the Born-Haber cycle, which is based on Hess's law.

5. In a covalent bond, two electrons (one pair) are shared between two atoms. In multiple covalent bonds, two or three electron pairs are shared between two atoms. Some bonded atoms possess lone pairs, that is, pairs of valence electrons not involved in bonding. The arrangement of bonding electrons and lone pairs around each atom in a molecule is represented by the Lewis structure.
6. The octet rule predicts that atoms form enough covalent bonds to surround themselves with eight electrons each. When one atom in a covalently bonded pair donates two electrons to the bond, the Lewis structure can include the formal charge on each atom as a means of keeping track of the valence electrons. There are exceptions to the octet rule, particularly for covalent beryllium compounds, elements in Group 3A, and elements in the third period and beyond in the periodic table.
7. Electronegativity is a measure of the ability of an atom to attract electrons in a chemical bond.
8. For some molecules or polyatomic ions, two or more Lewis structures based on the same skeletal structure satisfy the octet rule and appear chemically reasonable. Such resonance structures taken together represent the molecule or ion.
9. The strength of a covalent bond is measured in terms of its bond dissociation energy (for diatomic molecules) or its bond energy (for polyatomic molecules). Bond energies can be used to estimate the enthalpy of reactions.

Applied Chemistry Example: Ethylene Dichloride

Ethylene dichloride ($C_2H_4Cl_2$) is a major intermediate (that is, it is made to be used primarily to make another chemical product) in chemical manufacture. The production of ethylene dichloride in 1988 was 13.7 billion pounds, which ranks it 15th among the top 50 industrial chemicals produced in the United States that year. It is mainly used in the production of vinyl chloride, which in turn is used to manufacture the plastic poly(vinyl chloride) (PVC), and to a lesser extent it is used as a solvent and in the preparation of other chemicals. Ethylene dichloride can be prepared by the oxychlorination process discussed earlier (see p. 209). It is a colorless oily liquid that melts at −35.4°C and boils at 83.5°C. Ethylene dichloride has a pleasant odor and is quite toxic.

1. Write the Lewis structures of ethylene dichloride and vinyl chloride (C_2H_3Cl). Classify the bonds as covalent or polar.

   Answer: We proceed according to the steps outlined in Section 9.6.

   Ethylene dichloride ($C_2H_4Cl_2$)

   Step 1: The skeletal structure of $C_2H_4Cl_2$ is

```
    Cl Cl
H   C  C   H
    H  H
```

   Note that the central atoms (C) are less electronegative than are the end atoms (Cl). As usual, the H atoms occupy the end position.

Step 2: The outer shell electron configurations of C and Cl are $2s^22p^2$ and $3s^23p^5$, respectively. Thus there are $(2 \times 4) + (4 \times 1) + (2 \times 7)$, or 26 valence electrons to account for in $C_2H_4Cl_2$.

Step 3: We draw a single bond between the two C atoms and between each C atom and the two end H atoms and one Cl atom

```
    ..    ..
   :Cl:  :Cl:
    ''    ''
     |     |
  H—C—————C—H
     |     |
     H     H
```

We see that the octet rule is satisfied not only for the end Cl atoms but also for the central C atoms. Therefore, step 4 in writing the Lewis structure procedure is not required. To check, we find that there are 26 valence electrons in $C_2H_4Cl_2$, the same as the number of valence electrons on two C atoms, two Cl atoms, and 4 H atoms.

The C—C bond is covalent because the two bonding atoms are identical. According to Figure 9.8, Cl is more electronegative than C, but the difference in their electronegativities is less than 2. Thus, the C—Cl bonds are polar. The C—H bonds are only slightly polar (because the electronegativities of C and H are nearly equal).

Vinyl chloride (C_2H_3Cl)

Step 1: Vinyl chloride differs from ethylene (C_2H_4) in that one of the H atoms is replaced with a Cl atom. Therefore, its skeletal structure is

```
 H    Cl
   C C
 H    H
```

Step 2: There are $(2 \times 4) + (3 \times 1) + 7$ or 18 valence electrons to account for in C_2H_3Cl.

Step 3: We draw a single covalent bond between C and C, between C and H, and between C and Cl

```
               ..
 H            Cl:
   \  .    . / ..
    C————C
   /       \
 H          H
```

We see that the octet rule is satisfied for the Cl atom. The three lone pairs on Cl and the five single bonds together account for a total of 16 electrons. Since we started with 18 electrons, we place the remaining two electrons on the two C atoms.

Step 4: To satisfy the octet rule for the C atoms, we form an additional bond between them using the two electrons:

```
             ..
 H          Cl:
   \       / ..
    C====C
   /       \
 H          H
```

The C=C bond is covalent, the C—Cl bond is polar, and the C—H bonds are slightly polar.

2. Poly(vinyl chloride) is a polymer, that is, it is a molecule with very high molar mass (on the order of tens of thousands of grams or more). It is formed by joining many vinyl chloride molecules together. The repeating unit in poly(vinyl chloride) is $—CH_2—CHCl—$. Draw a portion of the molecule showing three such repeating units.

   Answer: Following the usual procedure we find that the Lewis structure for the repeating unit is

```
 H  H
 |  |
—C—C—
 |  |
 H :Cl:
     ..
```

   Note that because each single bond is made up of two electrons, the octet rule is satisfied for C. Therefore, the structure of poly(vinyl chloride) with three repeating units is (we omit the lone pairs on Cl for simplicity)

```
 H  H  H  H  H  H
 |  |  |  |  |  |
—C—C—C—C—C—C—
 |  |  |  |  |  |
 H  Cl H  Cl H  Cl
```

3. Calculate the enthalpy change when 1.0×10^3 kg of vinyl chloride react to form poly(vinyl chloride).

   Answer: In the formation of poly(vinyl chloride) from vinyl chloride, for every C=C double bond broken, two C—C single bonds are formed. No other bonds are broken or formed. From Table 9.4 we see that the energy changes for 1 mole of vinylchloride reacted are

$$\text{Total energy input (breaking C=C bonds)} = 620 \text{ kJ}$$

$$\text{Total energy released (forming C—C bonds)} = 2 \times 347 \text{ kJ} = 694 \text{ kJ}$$

   Using Equation (9.4) we write

$$\Delta H^\circ = 620 \text{ kJ} - 694 \text{ kJ} = -74 \text{ kJ}$$

   The negative sign shows that this is an exothermic reaction. To find the total heat released when 1000 kg of vinyl chloride react, we proceed as follows

$$\begin{aligned}\text{Heat released} &= 1000 \times 10^3 \text{ g } C_2H_3Cl \times \frac{1 \text{ mol } C_2H_3Cl}{62.49 \text{ g } C_2H_3Cl} \times \frac{-74 \text{ kJ}}{1 \text{ mol } C_2H_3Cl} \\ &= -1.2 \times 10^6 \text{ kJ}\end{aligned}$$

   In designing a reactor for the formation of poly(vinyl chloride), proper consideration must be paid to the removal of the large amount of heat generated during the reaction.

Key Words

Bond dissociation energy, p. 377
Bond length, p. 372
Born-Haber cycle, p. 355
Coordinate covalent bond, p. 374
Covalent bond, p. 360
Covalent compound, p. 362
Electronegativity, p. 363
Formal charge, p. 369
Ionic bond, p. 353
Lewis dot symbol, p. 352
Lewis structure, p. 361
Lone pair, p. 361
Multiple bond, p. 361
Nonbonding electron, p. 361
Octet rule, p. 361
Resonance, p. 372
Resonance form, p. 372
Resonance structure, p. 372

Exercises

Lewis Dot Symbol

REVIEW QUESTIONS

9.1 What is a Lewis dot symbol? To what elements does the symbol mainly apply?

9.2 Use the second member of each group from Group 1A to Group 7A to show that the number of valence electrons on an atom of the element is the same as its group number.

9.3 Without referring to Figure 9.1, write Lewis dot symbols for atoms of the following elements: (a) Be, (b) K, (c) Ca, (d) Ga, (e) O, (f) Br, (g) N, (h) I, (i) As, (j) F.

9.4 Explain what an ionic bond is.

9.5 Explain how ionization energy and electron affinity determine whether elements will combine to form ionic compounds.

9.6 Name five metals and five nonmetals that are very likely to form ionic compounds. Write formulas for compounds that might result from the combination of these metals and nonmetals. Name these compounds.

9.7 Name one ionic compound that contains only nonmetallic elements.

9.8 Name one ionic compound that contains a polyatomic cation and a polyatomic anion.

9.9 Explain why ions with charges greater than 3 are seldom found in ionic compounds.

9.10 The term "molar mass" was introduced in Chapter 2 to replace "gram molecular weight." What is the advantage in using the term "molar mass" when we discuss ionic compounds?

9.11 Write Lewis dot symbols for the following ions: (a) Li^+, (b) Cl^-, (c) S^{2-}, (d) Mg^{2+}, (e) N^{3-}.

9.12 Write Lewis dot symbols for the following atoms and ions: (a) Br, (b) Br^-, (c) S, (d) S^{2-}, (e) P, (f) P^{3-}, (g) Na, (h) Na^+, (i) Mg, (j) Mg^{2+}, (k) Al, (l) Al^{3+}, (m) Pb, (n) Pb^{2+}.

9.13 List some general characteristics of ionic compounds.

9.14 State in which of the following states NaCl would be electrically conducting: (a) solid, (b) molten (that is, melted), (c) dissolved in water. Explain your answers.

9.15 Beryllium forms a compound with chlorine that has the empirical formula $BeCl_2$. How would you determine whether it is an ionic compound? (The compound is not soluble in water.)

PROBLEMS

9.16 An ionic bond is formed between a cation A^+ and an anion B^-. How would the energy of the ionic bond [see Equation (9.2)] be affected by the following changes? (a) doubling the radius of A^+, (b) tripling the charge on A^+, (c) doubling the charges on A^+ and B^-, (d) decreasing the radii of A^+ and B^- to half their original values.

9.17 Give the empirical formulas and names of the compounds formed from the following pairs of ions: (a) Rb^+ and I^-, (b) Cs^+ and SO_4^{2-}, (c) Sr^{2+} and N^{3-}, (d) Al^{3+} and S^{2-}.

9.18 Use Lewis dot symbols to show electron transfer between the following atoms to form cations and anions: (a) Na and F, (b) K and S, (c) Ba and O, (d) Al and N.

9.19 Write the Lewis dot symbols of the reactants and products in the following reactions. (First balance the equations.)
(a) $Sr + Se \longrightarrow SrSe$
(b) $Ca + H \longrightarrow CaH_2$
(c) $Li + N \longrightarrow Li_3N$
(d) $Al + S \longrightarrow Al_2S_3$

9.20 For each of the following pairs of elements, state whether the binary compound formed is likely to be ionic or molecular. Write the empirical formula and name of the compound: (a) I and Cl, (b) K and Br, (c) Mg and F, (d) Al and F.

Lattice Energy

REVIEW QUESTIONS

9.21 Define lattice energy. What role does it play in the stability of ionic compounds?

9.22 Explain how the lattice energy of an ionic compound such as KCl can be determined by using the Born-Haber cycle. On what law is this procedure based?

9.23 For each of the following pairs of ionic compounds, specify which compound has the higher lattice energy: (a) KCl or MgO, (b) LiF or LiBr, (c) Mg_3N_2 or NaCl. Explain your choice.

9.24 Compare the stability (in the solid state) of the following pairs of compounds: (a) LiF and LiF_2 (containing the Li^{2+} ion), (b) Cs_2O and CsO (containing the O^- ion), (c) $CaBr_2$ and $CaBr_3$ (containing the Ca^{3+} ion).

PROBLEMS

9.25 Use the Born-Haber cycle outlined in Section 9.3 for LiF to calculate the lattice energy of NaCl. [The heat of sublimation of Na is 108 kJ/mol and $\Delta H_f^\circ(\text{NaCl}) = -411$ kJ/mol. Energy needed to dissociate $\frac{1}{2}$ mole of Cl_2 into Cl atoms = 121.4 kJ.]

9.26 Calculate the lattice energy of calcium chloride given that the heat of sublimation of Ca is 121 kJ/mol and $\Delta H_f^\circ(\text{CaCl}_2) = -795$ kJ/mol. (See Tables 8.4 and 8.5 for other data.)

The Covalent Bond

REVIEW QUESTIONS

9.27 What is Lewis's contribution to the formulation of the covalent bond?

9.28 Define the following terms: lone pairs, Lewis structure, molecular covalent compound, and network covalent compound.

9.29 What is the difference between a Lewis dot symbol and a Lewis structure?

9.30 Classify the following substances as ionic compounds, molecular covalent compounds, or network covalent compounds: CH_4, KF, CO, SiO_2, $MgCl_2$.

9.31 How many lone pairs are on the underlined atoms in these compounds? $H\underline{Br}$, $H_2\underline{S}$, $\underline{C}H_4$.

9.32 Distinguish among single, double, and triple bonds in a molecule, and give an example of each.

9.33 Describe the main differences between ionic compounds and molecular covalent compounds.

Electronegativity and Bond Type

REVIEW QUESTIONS

9.34 Define electronegativity, and explain the difference between electronegativity and electron affinity.

9.35 Describe in general how the electronegativities of the elements change according to position in the periodic table.

9.36 What is a polar covalent bond? Give two compounds that contain one or more polar covalent bonds.

PROBLEMS

9.37 List the following bonds in order of increasing ionic character: the lithium-to-fluorine bond in LiF, the potassium-to-oxygen bond in K_2O, the nitrogen-to-nitrogen bond in N_2, the sulfur-to-oxygen bond in SO_2, the chlorine-to-fluorine bond in ClF_3.

9.38 Arrange the following bonds in order of increasing ionic character: carbon to hydrogen, fluorine to hydrogen, bromine to hydrogen, sodium to iodine, potassium to fluorine, lithium to chlorine.

9.39 Four atoms are arbitrarily labeled D, E, F, and G. Their electronegativities are as follows: D = 3.8, E = 3.3, F = 2.8, and G = 1.3. If the atoms of these elements form the molecules DE, DG, EG, and DF, how would you arrange these molecules in order of increasing covalent bond character?

9.40 List the following bonds in order of increasing ionic character: cesium to fluorine, chlorine to chlorine, bromine to chlorine, silicon to carbon.

9.41 Classify the following bonds as ionic, polar covalent, or covalent, and give your reasons: (a) the CC bond in H_3CCH_3, (b) the KI bond in KI, (c) the NB bond in H_3NBCl_3, (d) the ClO bond in ClO_2, (e) the SiSi bond in $Cl_3SiSiCl_3$, (f) the SiCl bond in $Cl_3SiSiCl_3$, (g) the CaF bond in CaF_2.

Lewis Structure and the Octet Rule

REVIEW QUESTIONS

9.42 Summarize the essential features of the Lewis octet rule.

9.43 The octet rule applies mainly to the second-period elements. Explain.

9.44 Explain the concept of formal charge. Do formal charges on a molecule represent actual separation of charges?

PROBLEMS

9.45 Write Lewis structures for the following molecules: (a) ICl, (b) PH_3, (c) CS_2, (d) P_4 (each P is bonded to three other P atoms), (e) H_2S, (f) N_2H_4, (g) $HClO_3$, (h) $COBr_2$ (C is bonded to O and Br atoms).

9.46 Write Lewis structures for the following ions: (a) O_2^{2-}, (b) C_2^{2-}, (c) NO^+, (d) NO_2^-, (e) NH_4^+, (f) NO_3^-.

9.47 The skeletal structure of acetic acid in the following structure is correct, but some of the bonds are wrong. (a) Identify the incorrect bonds and explain what is wrong with them. (b) Write the correct Lewis structure for acetic acid.

```
     H :O:
     |  |
  H═C──C──O──H
     |   ..  ..
     H
```

9.48 The following Lewis structures are incorrect. Explain what is wrong with each one and give a correct Lewis structure for the molecule. (Relative positions of atoms are shown correctly.)

```
(a)      ..  ..
     H—C═N
             ..

(b)  H═C═C═H

(c)  ..        ..
     O—Sn—O
     ..        ..

(d)   ..     ..
     :F     F:
      ..\ ../..
         B
         |
        :F:
         ..

(e)     ..  ..
     H—O═F:
            ..

(f)  H
      \..  ..
       C—F:
      /    ..
    .O.
     .

(g)   ..     ..
     :F     F:
      ..\  /..
         N
         |
        :F:
         ..
```

9.49 Write Lewis structures for the following four isoelectronic species: (a) CO, (b) NO^+, (c) CN^-, (d) N_2. Show formal charges.

Resonance

REVIEW QUESTIONS

9.50 Define bond length, resonance, and resonance structure.

9.51 Is it possible to "trap" a resonance structure of a compound for study? Explain.

9.52 The resonance concept is sometimes described by analogy to a mule, which is a cross between a horse and a donkey. Compare this analogy with that used in this chapter, that is, the description of a rhinoceros as a cross between a griffin and a unicorn. Which description is more appropriate? Why?

9.53 Describe the general rules for drawing plausible resonance structures.

9.54 What are the other two reasons for choosing (b) in Example 9.7?

PROBLEMS

9.55 Write Lewis structures for the following species, including all resonance forms, and showing formal charges: (a) HCO_2^-, (b) $CH_2NO_2^-$. Relative positions of the atoms are as follows:

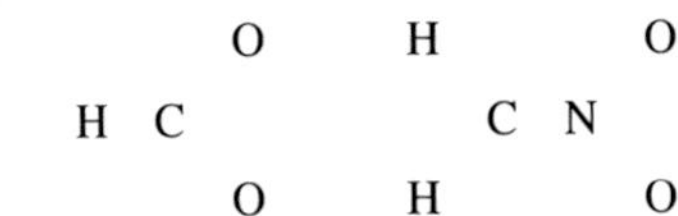

9.56 Draw three resonance structures for the nitrate ion, NO_3^-. Show formal charges.

9.57 Draw three reasonable resonance structures for the OCN^- ion. Show formal charges.

9.58 Write Lewis structures for hydrazoic acid, HN_3, and diazomethane, CH_2N_2. Are there resonance forms of these molecules? If so, draw their Lewis structures. The skeletal structures of the molecules are

H N N N

H
C N N
H

9.59 Some resonance forms of the molecule CO_2 are given here. Explain why some of them are likely to be of little importance in describing the bonding in this molecule.

(a) $\ddot{O}{=}C{=}\ddot{O}$ (c) $:\overset{+}{O}{\equiv}C\ \overset{-}{\ddot{O}}:$

(b) $:\overset{+}{O}{\equiv}C{-}\overset{-}{\ddot{O}}:$ (d) $\overset{-}{:\ddot{O}}{-}\overset{2+}{C}{-}\overset{-}{\ddot{O}}:$

9.60 Draw three resonance structures for the molecule N_2O in which the atoms are arranged in the order NNO. Indicate formal charges.

Exceptions to the Octet Rule

REVIEW QUESTIONS

9.61 Why does the octet rule not hold for many compounds containing elements in the third period of the periodic table and beyond?

9.62 Give three examples of compounds that do not satisfy the octet rule. Write a Lewis structure for each.

9.63 Because fluorine has seven valence electrons ($2s^22p^5$), seven covalent bonds in principle could form around the atom. Such a compound might be FH_7 or FCl_7. These compounds have never been prepared. Why?

9.64 What is a coordinate covalent bond? Is it different from a normal covalent bond?

PROBLEMS

9.65 In the vapor phase, beryllium chloride consists of discrete molecular units $BeCl_2$. Is the octet rule satisfied for Be in this compound? If not, can you form an octet around Be by drawing another resonance structure? How plausible is this structure?

9.66 The AlI_3 molecule (see Example 9.9) has an incomplete octet around Al. Draw three resonance structures of the molecule in which the octet rule is satisfied for both the Al and the I atoms. Show formal charges.

9.67 Of all the noble gases only krypton and xenon are known to form compounds with fluorine and oxygen. These compounds are all covalent in character. Without looking at the formulas of these compounds we can conclude that neither Kr nor Xe obeys the octet rule in any of these compounds. Explain.

9.68 Write Lewis structures for the following molecules: (a) XeF_2, (b) XeF_4, (c) XeF_6, (d) $XeOF_4$, (e) XeO_2F_2. In each case Xe is the central atom.

9.69 Write a Lewis structure for $SbCl_5$. Is the octet rule obeyed in this molecule?

9.70 Write Lewis structures for SeF_4 and SeF_6. Is the octet rule satisfied for Se?

9.71 Write Lewis structures for the reaction

$$AlCl_3 + Cl^- \longrightarrow AlCl_4^-$$

What kind of bond is between Al and Cl in the product?

9.72 Draw a reasonable Lewis structure for nitrous acid (HNO_2).

Bond Energies

REVIEW QUESTIONS

9.73 Define bond energy and bond dissociation energy. Why are bond energies only average values, whereas bond dissociation energies are precisely known values?

9.74 Explain why the bond energy of a molecule is usually defined in terms of a gas phase reaction.

9.75 Why are bond-breaking processes always endothermic and bond-forming processes always exothermic?

PROBLEMS

9.76 From the following data, calculate the average bond energy for the N—H bond:

$$NH_3(g) \longrightarrow NH_2(g) + H(g) \qquad \Delta H° = 435 \text{ kJ}$$
$$NH_2(g) \longrightarrow NH(g) + H(g) \qquad \Delta H° = 381 \text{ kJ}$$
$$NH(g) \longrightarrow N(g) + H(g) \qquad \Delta H° = 360 \text{ kJ}$$

9.77 The bond energy of $F_2(g)$ is 150.6 kJ/mol. Calculate ΔH_f° for $F(g)$.

9.78 For the reaction

$$H_2(g) + C_2H_4(g) \longrightarrow C_2H_6(g)$$

(a) Estimate the enthalpy of reaction, using the bond energy values in Table 9.4. (b) Calculate the enthalpy of reaction, using standard enthalpies of formation. (ΔH_f° for H_2, C_2H_4, and C_2H_6 are 0, 52.3 kJ/mol, and −84.7 kJ/mol, respectively.)

9.79 For the reaction

$$2C_2H_6(g) + 7O_2(g) \longrightarrow 4CO_2(g) + 6H_2O(g)$$

(a) Predict the enthalpy of reaction from the average bond energies in Table 9.4. (b) Calculate the enthalpy of reaction from the standard enthalpies of formation (see Appendix 3) of the reactant and product molecules, and compare the result with your answer for part (a).

Miscellaneous Problems

9.80 Match each of the following energy changes with one of the processes given: ionization energy, electron affinity, bond dissociation energy, standard enthalpy of formation, and lattice energy.
(a) $F(g) + e^- \longrightarrow F^-(g)$
(b) $F_2(g) \longrightarrow 2F(g)$
(c) $Na(g) \longrightarrow Na^+(g) + e^-$
(d) $NaF(s) \longrightarrow Na^+(g) + F^-(g)$
(e) $Na(s) + \frac{1}{2}F_2(g) \longrightarrow NaF(s)$

9.81 The formulas for the fluorides of the third-period elements are NaF, MgF_2, AlF_3, SiF_4, PF_5, SF_6, and ClF_3. Classify these compounds as covalent or ionic.

9.82 Use the ionization energy (see Table 8.4) and electron affinity (see Table 8.5) values to calculate the energy change (in kJ) for the following reactions:
(a) $Li(g) + I(g) \longrightarrow Li^+(g) + I^-(g)$
(b) $Na(g) + F(g) \longrightarrow Na^+(g) + F^-(g)$
(c) $K(g) + Cl(g) \longrightarrow K^+(g) + Cl^-(g)$

9.83 Describe some characteristics of an ionic compound such as KF that would distinguish it from a covalent compound such as benzene (C_6H_6).

9.84 Write Lewis structures for BrF_3, ClF_5, and IF_7. Identify those in which the octet rule is not obeyed.

9.85 Write three reasonable resonance structures of the azide ion N_3^- in which the atoms are arranged as NNN. Show formal charges.

9.86 Give an example of an ion or molecule containing Al that (a) obeys the octet rule, (b) has an expanded octet, and (c) has an incomplete octet.

9.87 The amide group plays an important role in determining the structure of proteins:

```
          :O:
      ..   ||
     —N—C—
      |
      H
```

Draw another resonance structure of this group. Show formal charges.

9.88 Draw four reasonable resonance structures for the PO_3F^{2-} ion. The central P atom is bonded to the three O atoms and to the F atom. Show formal charges.

9.89 Draw reasonable resonance structures for the following ions: (a) HSO_4^-, (b) SO_4^{2-}, (c) HSO_3^-, (d) SO_3^{2-}.

9.90 Attempts to prepare the compounds listed below as stable species under atmospheric conditions have failed. Suggest reasons for the failure.

CF_2 $\quad$ LiO_2 $\quad$ $CsCl_2$ $\quad$ PI_5

9.91 A rule for drawing plausible Lewis structures is that the central atom is invariably less electronegative than the surrounding atoms. Explain why this is so.

9.92 True or false: (a) Formal charges represent actual separation of charges; (b) ΔH_{rxn}° can be estimated from bond energies of reactants and products; (c) all second-period elements obey the octet rule in their compounds; (d) the resonance structures of a molecule can be separated from one another.

9.93 Using the following information

$$C(s) \longrightarrow C(g) \qquad \Delta H_{rxn}^\circ = 716 \text{ kJ}$$
$$2H_2(g) \longrightarrow 4H(g) \qquad \Delta H_{rxn}^\circ = 872.8 \text{ kJ}$$

and the fact that the average C—H bond energy is 414 kJ/mol, estimate the standard enthalpy of formation of methane (CH_4).

9.94 Based on energy considerations, which of the following two reactions will occur more readily?
(a) $Cl(g) + CH_4(g) \longrightarrow CH_3Cl(g) + H(g)$
(b) $Cl(g) + CH_4(g) \longrightarrow CH_3(g) + HCl(g)$
(*Hint:* Refer to Table 9.4, and assume that the average bond energy of the C—Cl bond is 338 kJ/mol.)

9.95 Which of the following molecules has the shortest nitrogen-to-nitrogen bond?

N_2H_4 N_2O N_2 N_2O_4

9.96 Most organic acids can be represented as RCOOH, where COOH is the carboxyl group and R is the rest of the molecule. (For Example, R is CH_3 in acetic acid, CH_3COOH). (a) Draw a Lewis structure of the carboxyl group. (b) Upon ionization, the carboxyl group is converted to the carboxylate group, COO^-. Draw resonance structures of the carboxylate group.

9.97 The following species have been detected in interstellar space: (a) CH, (b) OH, (c) C_2, (d) HNC, (e) HCO. Draw Lewis structures of these species and indicate whether they are diamagnetic or paramagnetic.

9.98 Which of the following molecules or ions are isoelectronic? NH_4^+, C_6H_6, CO, CH_4, N_2, $B_3N_3H_6$.

9.99 The amide ion, NH_2^-, is a Brønsted base. Represent the reaction between the amide ion and water in terms of Lewis structures.

9.100 Calculate the lattice energy of KBr given that the heat of sublimation of K is 90.00 kJ/mol, heat of vaporization of Br_2 is 30.71 kJ/mol, and ΔH_f° (KBr) = −392.2 kJ/mol.

9.101 Draw Lewis structures of the following organic molecules: (a) tetrafluoroethylene (C_2F_4), (b) propane (C_3H_8), (c) butadiene ($CH_2CHCHCH_2$), (d) propyne (CH_3CCH), (e) ethanol (CH_3CH_2OH), (f) benzoic acid (C_6H_5COOH). (This molecule can be drawn by replacing a H atom in benzene with a COOH group.)

Opposite page: Molecular models are used to study complex biochemical reactions such as those between protein and DNA molecules.

C H A P T E R 10

CHEMICAL BONDING II: MOLECULAR GEOMETRY AND MOLECULAR ORBITALS

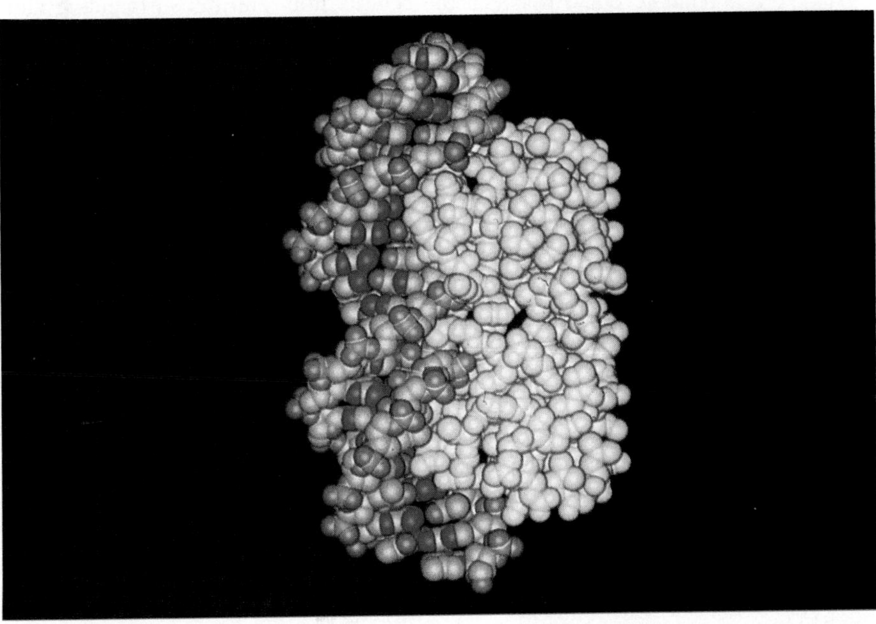

In Chapter 9 we discussed bonding in terms of the Lewis theory. Here we study the shape, or geometry, of molecules. Geometry has an important influence on the physical and chemical properties of molecules, such as melting point, boiling point, and reactivity. We will see that we can predict the shapes of molecules with considerable accuracy using a simple method based on Lewis structures.

The Lewis theory of chemical bonding, although useful and easy to apply, does not tell us how and why bonds form. A proper understanding must come from quantum mechanics. Therefore, in the second part of this chapter we will apply quantum mechanics to the study of the geometry and stability of molecules.

10.1 Molecular Geometry

Molecular geometry refers to the three-dimensional arrangement of atoms in a molecule. Many physical and chemical properties, such as melting point, boiling point, density, and the types of reactions that molecules undergo, are affected by a molecule's geometry. Molecular geometry cannot be predicted from the Lewis structures of molecules. In general, bond lengths and bond angles must be determined by experiment. However, there is a simple procedure that allows us to predict with considerable success the overall geometry of a molecule if we know the number of electrons surrounding a central atom. The basis of this approach is the idea that electron pairs in the valence shell of an atom repel one another. The ***valence shell*** is *the outermost electron-occupied shell of an atom; it holds the electrons that are usually involved in bonding.* In a covalent bond, a pair of electrons (often called the *bonding pair*) is responsible for holding two atoms together. However, in a polyatomic molecule, where there are two or more bonds between the central atom and the surrounding atoms, the repulsion between the electrons in different bonding pairs causes them to remain as far apart as possible. The geometry that the molecule ultimately assumes (as defined by the positions of all the atoms) is such that this repulsion is at a minimum. Therefore this approach to the study of molecular geometry is called the ***valence-shell electron-pair repulsion (VSEPR) model,*** because *it accounts for the geometric arrangements of electron pairs around a central atom in terms of the repulsion between electron pairs.*

VSEPR is pronounced "vesper."

With this model in mind, we can predict the geometry of molecules in a systematic way. To do this, it is convenient to divide molecules into two categories, according to whether or not the central atom has lone pairs. Before we discuss the two categories, make note of the following useful rules in applying VSEPR:

- As far as electron-pair repulsion is concerned, double bonds and triple bonds can be treated as though they were single bonds between neighboring atoms. This approximation is good for qualitative purposes. However, you should realize that in reality multiple bonds are "larger" than single bonds; that is, because there are two or three bonds between two atoms, the electron density occupies more space.
- If two or more resonance structures can be drawn for a molecule, we may apply the VSEPR model to any one of them. Furthermore, formal charges are usually not shown.

Molecules in Which the Central Atom Has No Lone Pairs

For simplicity we will consider molecules that contain atoms of only two elements, A and B, of which A is the central atom. These molecules have the general formula AB_x, where x is an integer 2, 3, ... (If $x = 1$, we have the diatomic molecule AB, which is linear by definition.) In the vast majority of cases, x is between 2 and 6.

Table 10.1 shows five possible arrangements of electron pairs around the central atom A. As a result of mutual repulsion, the electron pairs stay as far from one another as possible. Note that the table shows arrangements of the electron pairs but not the positions of the surrounding atoms. Molecules in this category (that is, the central atom has no lone pairs) have one of these five arrangements of bonding pairs. We will therefore take a close look now at the geometry of molecules with the formulas AB_2, AB_3, AB_4, AB_5, and AB_6.

Table 10.1 Arrangement of electron pairs about a central atom (A) in a molecule

Number of electron pairs	Arrangement of electron pairs*
2	Linear
3	Trigonal planar
4	Tetrahedral
5	Trigonal bipyramidal
6	Octahedral

*The colored lines show the overall geometries.

AB_2: Beryllium Chloride ($BeCl_2$). The Lewis structure of beryllium chloride in the gaseous state is

$$Cl—Be—Cl$$

For simplicity, we omit the lone pairs on the surrounding atoms in considering the geometry of a molecule.

Because the bonding pairs repel each other, they must be at opposite ends of a straight line in order for them to be as far apart as possible. Thus, the ClBeCl angle is predicted to be 180°, and the molecule is linear (see Table 10.1). The "ball-and-stick" model of $BeCl_2$ is

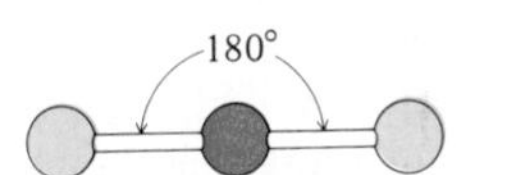

AB$_3$: Boron Trifluoride (BF$_3$). Boron trifluoride contains three covalent bonds, or bonding pairs. In the most stable arrangement, the three BF bonds point to the corners of an equilateral triangle with B in the center of the triangle:

```
    F
    |
    B
   / \
  F   F
```

According to Table 10.1, the geometry of BF_3 is *trigonal planar* because the three end atoms are at the corners of an equilateral triangle which is planar:

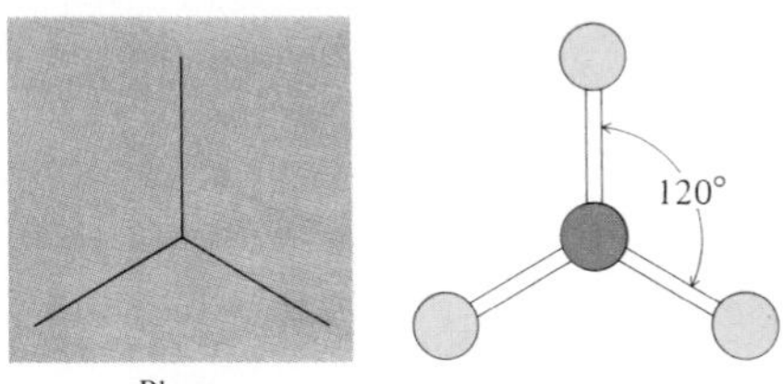

Planar

Thus, each of the three FBF angles is 120°, and all four atoms lie in the same plane.

AB$_4$: Methane (CH$_4$). The Lewis structure of methane is

```
    H
    |
 H—C—H
    |
    H
```

Since there are four bonding pairs, the geometry of CH_4 is tetrahedral (see Table 10.1). A *tetrahedron* has four sides (hence the prefix *tetra*), or four faces, all of which are equilateral triangles. In a tetrahedral molecule, the central atom (C in this case) is located at the center of the tetrahedron and the other four atoms are at the corners. The bond angles are all 109.5°.

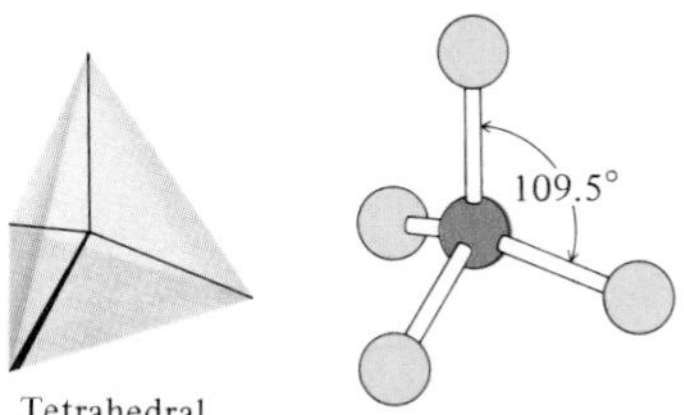

Tetrahedral

AB$_5$: Phosphorus Pentachloride (PCl$_5$). The Lewis structure of phosphorus pentachloride (in the gas phase) is

```
 Cl  Cl
   \ |
     P—Cl
   / |
 Cl  Cl
```

The only way to minimize the repulsive forces among the five bonding pairs is to arrange the PCl bonds in the form of a trigonal bipyramid (see Table 10.1). A trigonal bipyramid can be generated by joining two tetrahedrons along a common base:

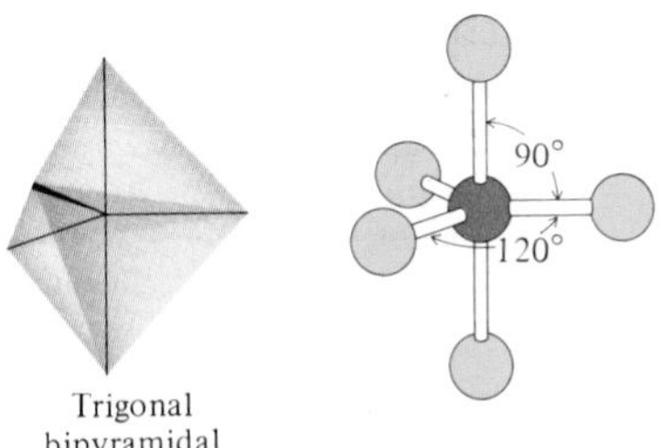

Trigonal bipyramidal

The central atom (P in this case) is located at the center of the common triangle with the other five atoms positioned at the five corners of the trigonal bipyramid. The atoms that are above and below the triangular plane are said to occupy *axial* positions, and those that are in the triangular plane are said to occupy *equatorial* positions. The angle between any two equatorial bonds is 120°; that between an axial bond and an equatorial bond is 90°, and that between the two axial bonds is 180°.

AB_6: Sulfur Hexafluoride (SF_6). The Lewis structure of sulfur hexafluoride is

```
    F
 F\ | /F
    S
 F/ | \F
    F
```

The most stable arrangement of the six SF bonding pairs is that of an octahedron, shown in Table 10.1. An octahedron has eight sides (hence the prefix *octa*). It can be generated by joining two square pyramids on a common base. The central atom (S in this case) is located at the center of the square base and the other six atoms at the six corners. All bond angles are 90° except the one made by the bonds between the central atom and the two atoms diametrically opposite each other. That angle is 180°. Since the six bonds are equivalent in an octahedral molecule, we cannot use the terms "axial" and "equatorial" as in a trigonal bipyramidal molecule.

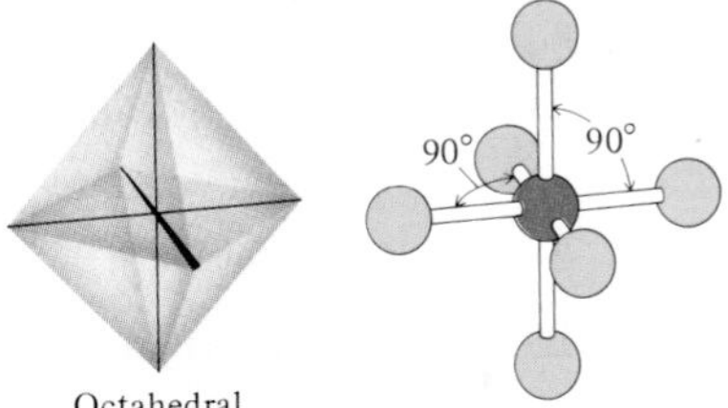

Octahedral

Table 10.2 shows the geometries of some simple molecules as predicted by the VSEPR model. Note that Table 10.2 differs from Table 10.1 in showing the positions of *all* the atoms (and therefore the geometries of the molecules). Table 10.1 shows only the central atom and the arrangement of all the electron pairs. As we will see shortly, Table 10.1 can also be used to study the geometry of molecules in which the central atom possesses one or more lone pairs.

Molecules in Which the Central Atom Has One or More Lone Pairs

Determining the geometry of a molecule is more complicated if the central atom has both lone pairs and bonding pairs. In such molecules there are three types of repulsive

Table 10.2 Geometry of some simple molecules and ions in which the central atom has no lone pairs

Molecule	Geometry*		Examples
AB_2	Linear	B—A—B	$BeCl_2$, $HgCl_2$
AB_3	Trigonal planar		BF_3
AB_4	Tetrahedral		CH_4, NH_4^+
AB_5	Trigonal bipyramidal		PCl_5
AB_6	Octahedral		SF_6

*The colored lines are used only to show the overall shapes; they do not represent bonds.

The forces are electrostatic in origin.

forces—those between bonding pairs, those between lone pairs, and those between a bonding pair and a lone pair. In general, according to VSEPR, the repulsive forces decrease in the following order:

lone-pair vs. lone-pair repulsion > lone-pair vs. bonding-pair repulsion > bonding-pair vs. bonding-pair repulsion

Electrons in a bond are held by the attractive forces exerted by the nuclei of the two bonded atoms. These electrons have less ''spatial distribution''; that is, they take up less space than lone-pair electrons, which are associated with only one particular atom. Consequently, lone-pair electrons in a molecule occupy more space and experience a greater repulsion from neighboring lone pairs and bonding pairs. To keep track of the total number of bonding pairs and lone pairs, we designate molecules with lone pairs as AB_xE_y, where A is the central atom, B is a surrounding atom, and E is a lone pair on A. Both x and y are integers; $x = 2, 3, \ldots$, and $y = 1, 2, \ldots$. Thus the values of x and y indicate the number of surrounding atoms and number of lone pairs on the central atom, respectively. In the scheme here, the simplest molecule would be a triatomic molecule with one lone pair on the central atom; the formula of this molecule is AB_2E.

For molecules in which the central atom has one or more lone pairs, we need to distinguish between the overall arrangement of the electron pairs and the geometry of the molecule. The overall arrangement of the electron pairs refers to the arrangement of *all* electron pairs on the central atom, bonding pairs as well as lone pairs. On the other hand, the geometry of a molecule is described only in terms of the arrangement of its atoms, and hence only the arrangement of bonding pairs is considered. We will see shortly that if lone pairs are present on the central atom, the overall arrangement of the electron pairs is *not* the same as the geometry of the molecule.

This will become clearer as we consider some specific examples.

AB_2E: Sulfur Dioxide (SO_2). The Lewis structure of sulfur dioxide is

Experiments show that the sulfur-to-oxygen bonds are double bonds.

$$\ddot{O}=\ddot{S}=\ddot{O}$$

Because in our simplified scheme we treat double bonds as if they were single (p. 392), the SO_2 molecule can be viewed as consisting of three electron pairs on the central S atom. Of these, two are bonding pairs and one is a lone pair. In Table 10.1 we see that the overall arrangement of three electron pairs is trigonal planar. But because one of the electron pairs is a lone pair, the SO_2 molecule has a "bent" shape.

Since the lone-pair versus bonding-pair repulsion is greater than the bonding-pair versus bonding-pair repulsion, the two sulfur-to-oxygen bonds are pushed together slightly and the OSO angle is less than 120°. By experiment the OSO angle is found to be 119.5°.

AB_3E: Ammonia (NH_3). The ammonia molecule contains three bonding pairs and one lone pair:

$$H-\ddot{N}-H$$
$$|$$
$$H$$

As Table 10.1 shows, the overall arrangement of four electron pairs is tetrahedral. But in NH_3 one of the electron pairs is a lone pair, so the geometry of NH_3 is trigonal pyramidal (so called because it looks like a pyramid, with the N atom at the apex). Because the lone pair repels the bonding pairs more strongly, the three NH bonding pairs are pushed closer together; thus the HNH angle in ammonia is smaller than the ideal tetrahedral angle of 109.5° (Figure 10.1).

AB_2E_2: Water (H_2O). A water molecule contains two bonding and two lone pairs:

$$H-\ddot{\underset{\cdot\cdot}{O}}-H$$

The overall arrangement of the four electron pairs in water is tetrahedral, the same electron-pair arrangement that is found in ammonia. However, unlike ammonia, water has two lone pairs on the central O atom. These lone pairs tend to be as far from each other as possible. Consequently, the two OH bonding pairs are pushed toward each other, so that we can predict an even greater deviation from the tetrahedral angle than in NH_3. As Figure 10.1 shows, the HOH angle is 104.5°. The geometry of H_2O is bent.

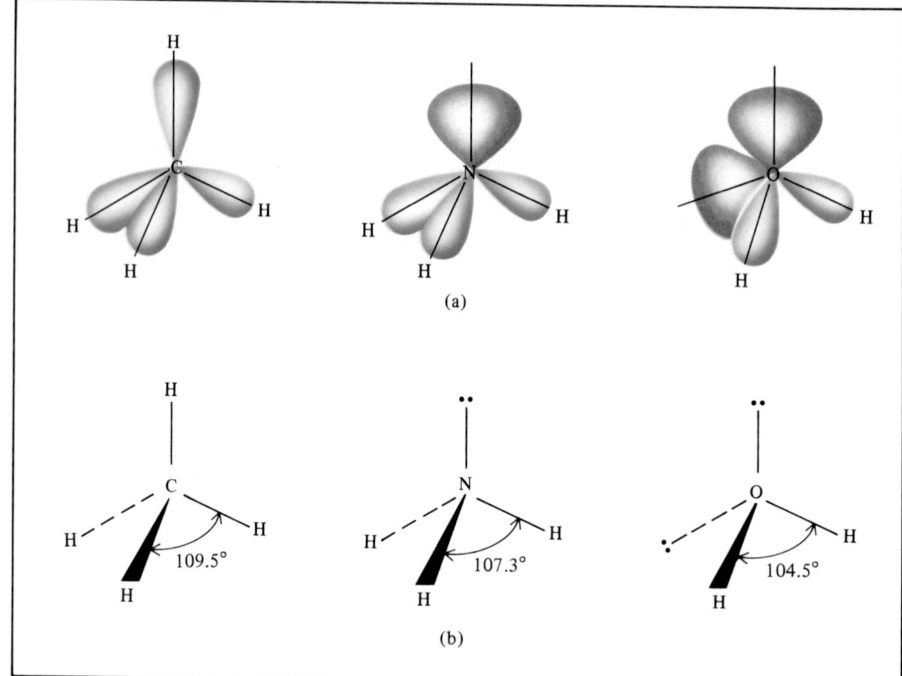

Figure 10.1 *(a) The relative sizes of bonding pair and lone pair in CH_4, NH_3, and H_2O. (b) The bond angles in CH_4, NH_3, and H_2O. In each diagram the dashed line represents a bond axis behind the plane of the paper, the wedged line represents a bond axis in front of the plane of the paper, and the thin solid lines represent bonds in the plane of the paper.*

AB_4E: Sulfur Tetrafluoride (SF_4). The Lewis structure of SF_4 is

F F
S
F F

The central sulfur atom has five electron pairs whose arrangement, according to Table 10.1, is trigonal bipyramidal. In the SF_4 molecule, however, one of the electron pairs is a lone pair, so that the molecule must have one of the following geometries:

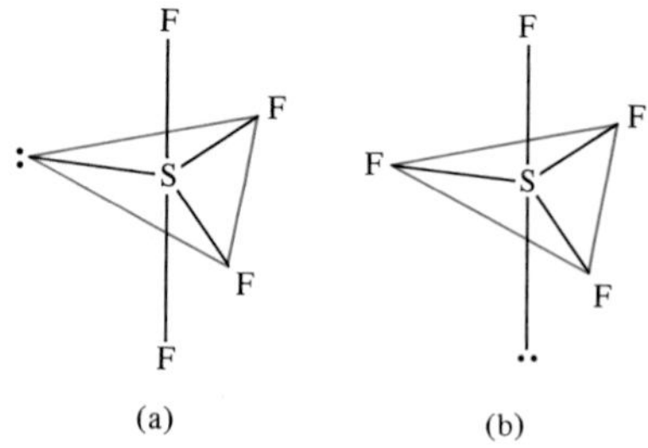

Experimentally, the angle between the axial F atoms and S is found to be 186°, and that between the equatorial F atoms and S is 116°.

In (a) the lone pair occupies an equatorial position, and in (b) it occupies an axial position. The axial position has three neighboring pairs at 90° and one at 180°, while the equatorial position has two neighboring pairs at 90° and two more at 120°. The repulsion is smaller for (a), and indeed (a) is the structure observed experimentally. The shape shown in (a) is sometimes described as a distorted tetrahedron (or a folded square, or seesaw shape).

Table 10.3 shows the geometries of simple molecules in which the central atom has one or more lone pairs, including some that we have not discussed.

Table 10.3 Geometry of simple molecules in which the central atom has one or more lone pairs

Class of molecule	Total number of electron pairs	Number of bonding pairs	Number of lone pairs	Arrangement of electron pairs*	Geometry	Examples
AB_2E	3	2	1	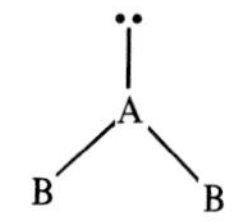 Trigonal planar	Bent	SO_2
AB_3E	4	3	1	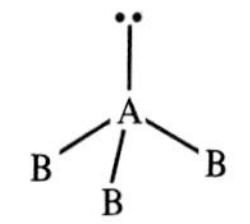 Tetrahedral	Trigonal pyramidal	NH_3
AB_2E_2	4	2	2	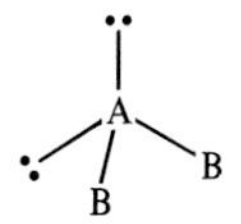 Tetrahedral	Bent	H_2O
AB_4E	5	4	1	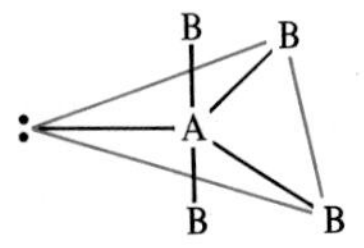 Trigonal bipyramidal	Distorted tetrahedron (or seesaw)	IF_4^+, SF_4, XeO_2F_2
AB_3E_2	5	3	2	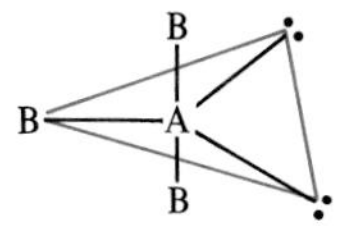 Trigonal bipyramidal	T-shaped	ClF_3
AB_2E_3	5	2	3	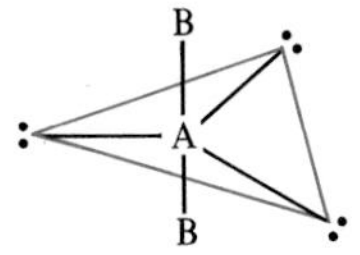 Trigonal bipyramidal	Linear	XeF_2, I_3^-
AB_5E	6	5	1	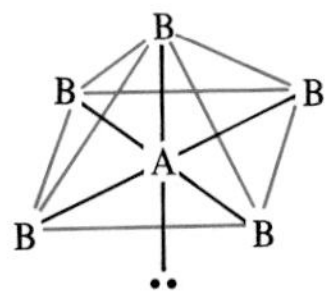 Octahedral	Square pyramidal	BrF_5, $XeOF_4$
AB_4E_2	6	4	2	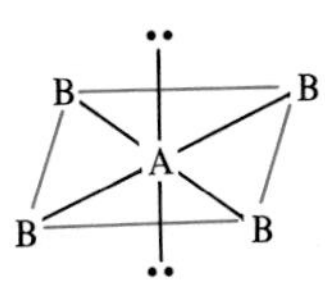 Octahedral	Square planar	XeF_4, ICl_4^-

*The colored lines are used to show the overall shape, not bonds.

Geometry of Molecules with More Than One Central Atom

So far we have discussed the geometry of molecules having only one central atom. (The term ''central atom'' here means an atom that is not a terminal atom in a polyatomic molecule.) The overall geometry of molecules with more than one central atom is difficult to define in most cases. Often we can only describe what the shape is like around each of the central atoms. Consider methanol, CH_3OH, whose Lewis structure is

```
      H
      |   ..
  H—C—O—H
      |   ..
      H
```

The two central (nonterminal) atoms in methanol are C and O. We can say that the three CH and the CO bonding pairs are tetrahedrally arranged about the C atom. The HCH and OCH bond angles are approximately 109°. The O atom here is like the one in water in that it has two lone pairs and two bonding pairs. Therefore, the HOC portion of the molecule is bent, and the angle HOC is approximately equal to 105° (Figure 10.2).

The Chemistry in Action on p. 403 discusses an effective way of studying molecular geometry.

Guidelines for Applying the VSEPR Model

Having studied the geometries of molecules in the two categories (central atoms with and without lone pairs), we summarize some rules that will help you apply the VSEPR model to all types of molecules:

- Write the Lewis structure of the molecule, considering only the electron pairs around the central atom (that is, the atom that is bonded to more than one other atom).
- Count the total number of electron pairs around the central atom (that is, bonding pairs and lone pairs). As a good approximation, treat double and triple bonds as though they were single bonds. Refer to Table 10.1 to predict the overall arrangement of the electron pairs.
- Use Tables 10.2 and 10.3 to predict the geometry of the molecule.

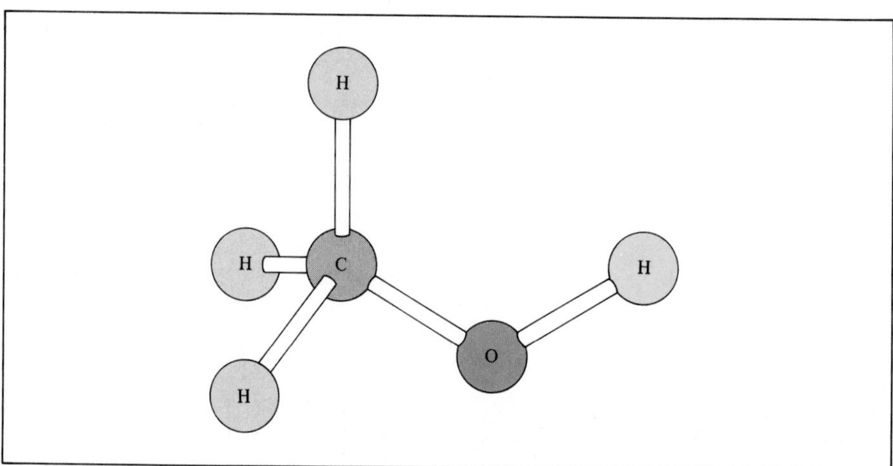

Figure 10.2 *The geometry of CH_3OH.*

- In predicting bond angles, note that a lone pair repels another lone pair or a bonding pair more strongly than a bonding pair repels another bonding pair. Remember that there is no easy way to predict bond angles accurately when the central atom possesses one or more lone pairs.

The VSEPR model, as you have already observed, generates reliable predictions of the geometries of a variety of molecular structures. Chemists accept and use the VSEPR approach because of its simplicity and utility. Although some theoretical concerns have been voiced about whether "electron-pair repulsion" actually determines molecular shapes, it is safe to say that such an assumption leads to useful (and generally reliable) predictions. We need not ask more of any model at this stage in the study of chemistry.

We can now use the VSEPR model to predict the geometry of some simple molecules and polyatomic ions.

EXAMPLE 10.1

Use the VSEPR model to predict the geometry of the following molecules and ions: (a) AsH_3, (b) OF_2, (c) $AlCl_4^-$, (d) I_3^-, (e) C_2H_4.

Answer

(a) The Lewis structure of AsH_3 is

```
      ..
   H—As—H
      |
      H
```

This molecule has three bonding pairs and one lone pair, a combination similar to that in ammonia. Therefore, the geometry of AsH_3 is trigonal pyramidal, like NH_3. We cannot predict the HAsH angle accurately, but we know that it must be less than 109.5°.

(b) The Lewis structure of OF_2 is

```
     ..
   F—O—F
     ..
```

Since there are two lone pairs on the O atom, the OF_2 molecule must have a bent shape, like that of H_2O. Again, all we can say about angle size is that the FOF angle should be less than 109.5°.

(c) The Lewis structure of $AlCl_4^-$ is

```
  [      Cl     ]-
  [      |      ]
  [  Cl—Al—Cl   ]
  [      |      ]
  [      Cl     ]
```

Since the central Al atom has no lone pairs and all four Al—Cl bonds are equivalent, the $AlCl_4^-$ ion should be tetrahedral, and the ClAlCl angles should all be 109.5°.

(d) The Lewis structure of I_3^- is

```
     ...
  [I—I—I]-
     ..
```

The central I atom has two bonding pairs and three lone pairs. From Table 10.3 we see that the three lone pairs all lie in the triangular plane, and the I_3^- ion should be linear, like the XeF_2 molecule.

The I_3^- ion is one of the few structures for which the bond angle (180°) can be predicted accurately even though the central atom contains lone pairs.

(e) The Lewis structure of C_2H_4 is

$$\begin{array}{c} H \quad\quad H \\ \diagdown \quad \diagup \\ C{=}C \\ \diagup \quad \diagdown \\ H \quad\quad H \end{array}$$

The C=C bond is treated as though it were a single bond. Because there are no lone pairs present, the arrangement around each C atom has a trigonal planar shape like BF_3, discussed earlier. Thus the predicted bond angles in C_2H_4 are all 120°.

H 120° H
C=C 120°
H 120° H

Similar problems: 10.8, 10.9, 10.10, 10.11.

In C_2H_4, all six atoms lie in the same plane. The overall planar geometry is not predicted by the VSEPR model, but we will see why the molecule prefers to be planar later. In reality, the angles are close, but not equal, to 120° because the bonds are not all equivalent.

10.2 Dipole Moments

Now that you know how to predict the geometries of simple molecules, you may wonder how the actual geometry of a molecule is determined. A number of spectroscopic techniques and X-ray measurements give us bond lengths and bond angles. But a simpler method—dipole moment measurement—is useful in studying molecular geometry. Although a dipole moment measurement does not give specific bond lengths or bond angles, it can tell us about the overall geometry of a molecule.

In Section 9.5 we learned that in the hydrogen fluoride molecule, there is a shift in electron density from H to F because the F atom is more electronegative than the H atom.

$$\overset{+\!\!\longrightarrow}{\mathrm{H{-}\ddot{\underset{..}{F}}:}}$$

We also noted that there is charge separation in the molecule as follows:

$$\overset{\delta+}{\mathrm{H}}\,{-}\,\overset{\delta-}{\ddot{\underset{..}{\mathrm{F}}}}:$$

The qualitative conclusion we can draw from these observations is that the H—F bond in hydrogen fluoride is a polar bond. But how do we obtain a quantitative measure of the polarity of the bond? The ***dipole moment*** μ, which is *the product of the charge Q and the distance r between charges,* provides a measure of bond polarity:

$$\mu = Q \times r \tag{10.1}$$

In a diatomic molecule like HF, the charge Q is equal in magnitude to $\delta+$ and $\delta-$.

To maintain electrical neutrality, the charges on both ends of an electrically neutral diatomic molecule must be equal in magnitude and opposite in sign. However, the quantity Q in Equation (10.1) refers only to the magnitude of the charge and not its sign, so μ is always positive. *A molecule that possesses a dipole moment* is called a ***polar molecule;*** *a molecule that does not possess a dipole moment* is called a ***nonpolar molecule.*** Thus all ***homonuclear diatomic molecules,*** that is, *diatomic molecules containing atoms of the same element* (for example, H_2, O_2, F_2, and so on), lack a dipole moment because there is no charge separation in these molecules. They are all non-

CHEMISTRY IN ACTION

Molecular Models

We cannot directly observe molecules because of their extremely small size. An effective way to visualize the three-dimensional structure of molecules, which is often difficult to describe with words or diagrams, is by the use of models. Two main types of molecular models are currently in use: ball-and-stick models and space-filling models. In ball-and-stick models the atoms are made of wooden or plastic balls with holes drilled in them. These holes hold the sticks or springs that represent the bonds at angles similar to the actual bond angles in the molecule. Each type of atom is represented by a different color ball, and the balls are all the same size. In space-filling models atoms are represented by truncated balls held together by snap fasteners, so that the bonds are not visible. The atomic radii are accurate to scale in these models.

Ball-and-stick models show the arrangement of bonds clearly, and they are fairly easy to construct. However, treating all atoms as if they were of equal size does not give us a realistic picture of molecules. Space-filling models represent molecules more accurately since different atoms have different sizes. Their drawbacks are that they are time-consuming to construct and they do not show bond orientations clearly. Figure 10.3 shows both types of models for several simple molecules.

Chemists rely heavily on the use of molecular models. In addition to helping us understand the shapes of molecules, an accurately constructed molecular model also provides useful information about the spatial relationship among atoms not bonded to one another and about the geometric requirements in a chemical reaction. In recent years much of our understanding of the structure and reactivity of many biologically important molecules such as proteins and DNA (deoxyribonucleic acid), which contain thousands or even tens of thousands of atoms, has been obtained from the study of models representing these molecules.

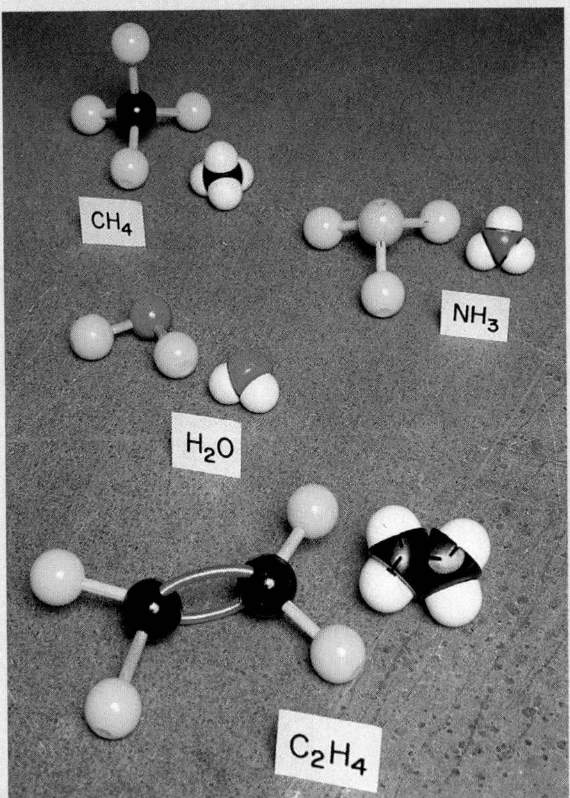

Figure 10.3 *Space-filling models and ball-and-stick models for some simple molecules.*

polar molecules. On the other hand, ***heteronuclear diatomic molecules,*** or *diatomic molecules containing atoms of different elements* (for example, HCl, CO, NO, and so on), generally do possess a dipole moment. They are all polar molecules.

Dipole moments can be measured by studying the behavior of molecules between two oppositely charged plates (Figure 10.4). In the presence of an electric field (due to the + and − charges on the plates), a polar molecule aligns itself with its more negative end oriented toward the positive plate and its more positive end toward the

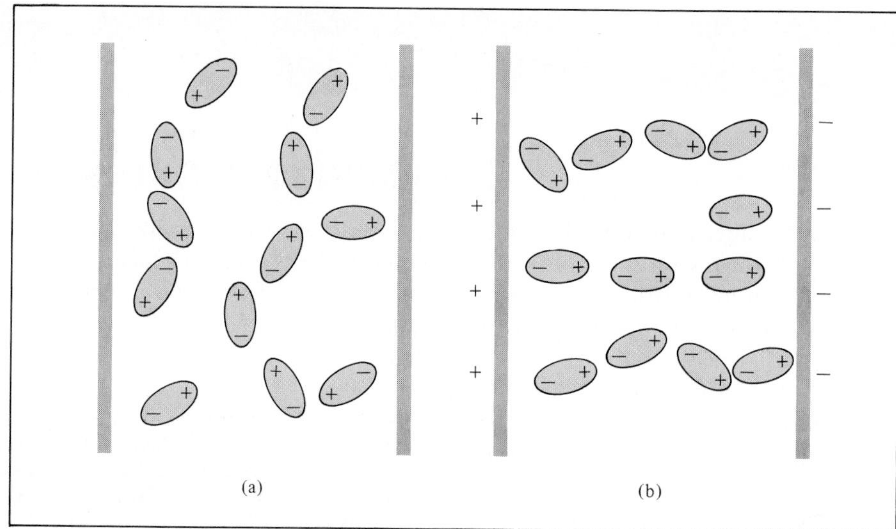

Figure 10.4 *Behavior of polar molecules (a) in the absence and (b) in the presence of an external electric field. Nonpolar molecules are not affected by an electric field.*

negative plate. Nonpolar molecules are unaffected by an electric field. Dipole moments are usually expressed in *debye* units (D), named for Peter Debye.† The conversion factor is

$$1\ \text{D} = 3.33 \times 10^{-30}\ \text{C m}$$

where C is coulomb and m is meter.

The dipole moment of a molecule containing three or more atoms depends on both polarity *and* molecular geometry. For instance, even if polar bonds are present, the molecule itself may not have a dipole moment. Consider the carbon dioxide (CO_2) molecule. Since CO_2 is a triatomic molecule, its geometry is either linear or bent:

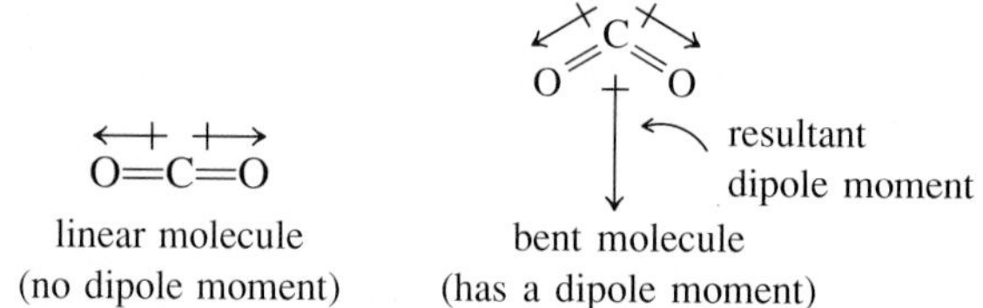

The arrows show the shift of electron density from the less electronegative carbon atom to the more electronegative oxygen atom. In each case, the dipole moment of the entire molecule is made up of two *bond moments,* that is, individual dipole moments in the polar C=O bonds. The measured dipole moment is equal to the sum of these bond moments. The dipole moment, like the bond moment, is a *vector quantity;* that is, it has both magnitude and direction. It is clear that the two bond moments in CO_2 are equal in magnitude. Since they point in opposite directions in a linear CO_2 molecule, the sum or resultant dipole moment would be zero. On the other hand, if the CO_2

†Peter Joseph William Debye (1884–1966). American chemist and physicist of Dutch origin. Debye made many significant contributions in the study of molecular structure, polymer chemistry, X-ray analysis, and electrolyte solution. He was awarded the Nobel Prize in chemistry in 1936.

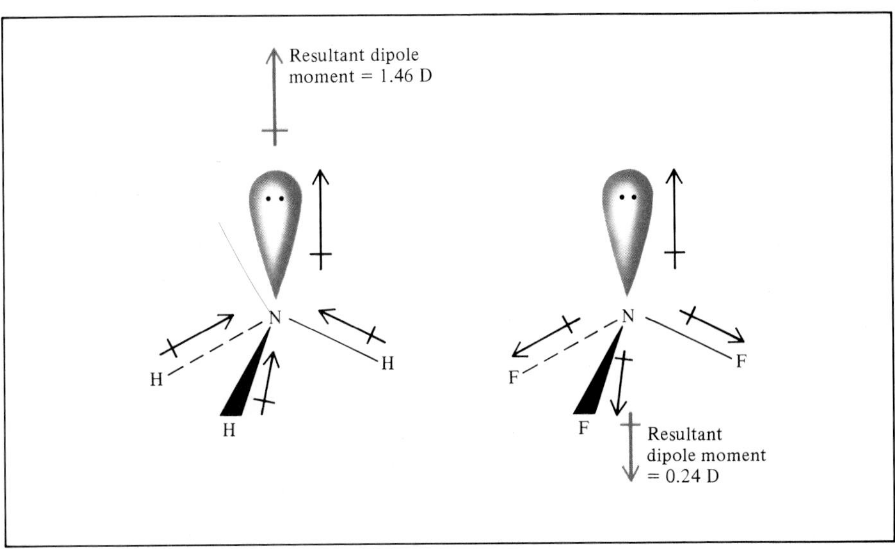

Figure 10.5 *Bond moments and resultant dipole moments in NH_3 and NF_3.*

molecule were bent, the two bond moments would partially reinforce each other, so that the molecule would have a dipole moment. Experimental evidence shows that carbon dioxide has no dipole moment. Therefore we conclude that the carbon dioxide molecule is linear, as is consistent with its nonpolar character. The linear nature of carbon dioxide has, in fact, been confirmed through other experimental measurements.

The VSEPR model predicts that CO_2 is a linear molecule.

Next let us consider the NH_3 and NF_3 molecules shown in Figure 10.5. In both cases the central N atom has a lone pair whose "bond moment" points away from the N atom. From Figure 9.8 we know that N is more electronegative than H, and F is more electronegative than N. Thus the resultant dipole moment in NH_3 is larger than that in NF_3.

Dipole moment measurements can be used to distinguish between molecules. For example, the following two molecules both exist; they have the same molecular formula ($C_2H_2Cl_2$), the same number and type of bonds, but different molecular structures:

resultant dipole moment

Cl Cl
C=C
H H

cis-dichloroethylene
$\mu = 1.89$ D

H Cl
C=C
Cl H

trans-dichloroethylene
$\mu = 0$

Since *cis*-dichloroethylene is a polar molecule but *trans*-dichloroethylene is not, they can readily be distinguished by a dipole moment measurement.

Table 10.4 lists the dipole moments of several polar diatomic and polyatomic molecules.

Example 10.2 shows how we can predict whether a molecule possesses a dipole moment if we know its molecular geometry.

Table 10.4 Dipole moments of some polar molecules

Molecule	Geometry	Dipole moment (D)
HF	Linear	1.92
HCl	Linear	1.08
HBr	Linear	0.78
HI	Linear	0.38
H_2O	Bent	1.87
H_2S	Bent	1.10
NH_3	Pyramidal	1.46
SO_2	Bent	1.60

EXAMPLE 10.2

Predict whether each of the following molecules has a dipole moment: (a) IBr, (b) BF_3 (trigonal planar), (c) CH_2Cl_2 (tetrahedral).

Answer

(a) Since IBr (iodine bromide) is diatomic, it has a linear geometry. Bromine is more electronegative than iodine (see Figure 9.8), so IBr is polar with bromine at the negative end.

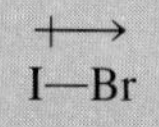

$$\overset{\longmapsto}{\text{I—Br}}$$

Thus the molecule does have a dipole moment.

An analogy is an object that is pulled in the directions shown by the three bond moments. If the forces are equal, the object will not move.

(b) Since fluorine is more electronegative than boron, each B—F bond in BF_3 (boron trifluoride) is polar and the three bond moments are equal. However, the symmetry of a trigonal planar shape means that the three bond moments exactly cancel one another:

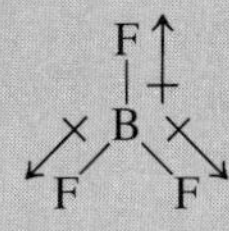

F
|
B
F F

Consequently BF_3 has no dipole moment; it is a nonpolar molecule.

(c) The Lewis structure of CH_2Cl_2 (methylene chloride) is

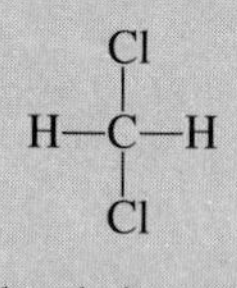

Cl
|
H—C—H
|
Cl

This molecule is similar to CH_4 in that it has an overall tetrahedral shape. However, because not all the bonds are identical, there are three different bond angles: HCH, HCCl, and ClCCl. These bond angles are close to, but not equal to, 109.5°. Since chlorine is more electronegative than carbon, which is more electronegative than hydrogen, the bond moments do not cancel and the molecule possesses a dipole moment:

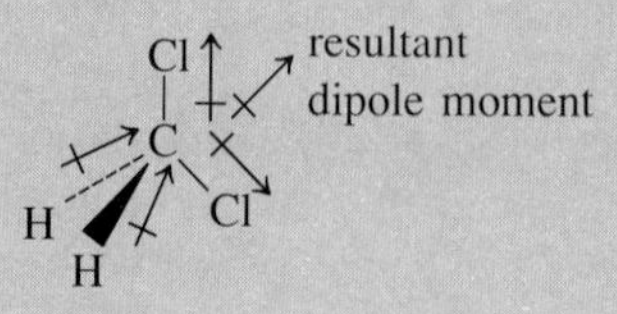

Cl resultant dipole moment
|
C
H Cl
H

Thus CH_2Cl_2 is a polar molecule.

Similar problems: 10.22, 10.23.

10.3 Valence Bond Theory

The VSEPR model, based largely on Lewis structures of molecules, provides a relatively simple and straightforward method for predicting the geometry of molecules. But as we noted earlier, the Lewis theory of chemical bonding does not clearly explain why chemical bonds exist. We need a better understanding of bonding to account for many observed molecular properties.

Lewis's idea of relating the formation of a covalent bond to the pairing of electrons was a step in the right direction, but it did not go far enough. For example, the Lewis theory describes the single bond between the H atoms in H_2 and that between the F atoms in F_2 in essentially the same way—as the pairing of two electrons. Yet these two molecules have quite different bond dissociation energies and bond lengths (436.4 kJ/mol and 74 pm for H_2 and 150.6 kJ/mol and 142 pm for F_2). These and many other facts cannot be explained by the Lewis theory. For a more complete explanation of chemical bond formation we must therefore look to quantum mechanics. In fact, the quantum mechanical study of chemical bonding also provides a means for understanding molecular geometry.

At present, two quantum mechanical theories are used to describe the covalent bond and electronic structure of molecules. *Valence bond (VB) theory* assumes that the electrons in a molecule occupy atomic orbitals of the individual atoms. It permits us to retain a picture of individual atoms taking part in the bond formation. The second theory, called *molecular orbital (MO) theory,* assumes the formation of molecular orbitals from the atomic orbitals. The molecular orbital theory is discussed in Section 10.6; here we focus on the valence bond theory.

Let us start by considering the formation of a H_2 molecule from two H atoms. The Lewis theory describes the H—H bond in terms of the pairing of the two electrons on the H atoms. In the framework of valence bond theory, the covalent H—H bond is formed by the *overlap* of the two 1*s* orbitals in the H atoms. By overlap, we mean that the two orbitals share common region in space.

Chapter 8 discussed electron configuration of isolated atoms. Here we must examine what happens to these atoms as they move toward each other and form a bond. Initially, when the two atoms are far apart, there is no interaction. We say that the potential energy of this system (that is, the two H atoms) is zero. As the atoms approach each other, each electron is attracted by the nucleus in the other atom, and at the same time, the electrons repel each other, as do the nuclei. While the atoms are still separated, attraction is stronger than repulsion, so that the potential energy of the system *decreases* (that is, it becomes negative) as the atoms approach each other (Figure 10.6). This trend continues until the potential energy reaches a minimum value. At this point, where the system has the lowest potential energy, it is most stable. This condition corresponds to substantial overlap of the 1*s* orbitals and formation of a stable H_2 molecule. If the distance between nuclei were to decrease further, the potential energy would rise steeply and finally become positive as a result of the increased electron-electron and nuclear-nuclear repulsions.

Recall that an object has potential energy by virtue of its position.

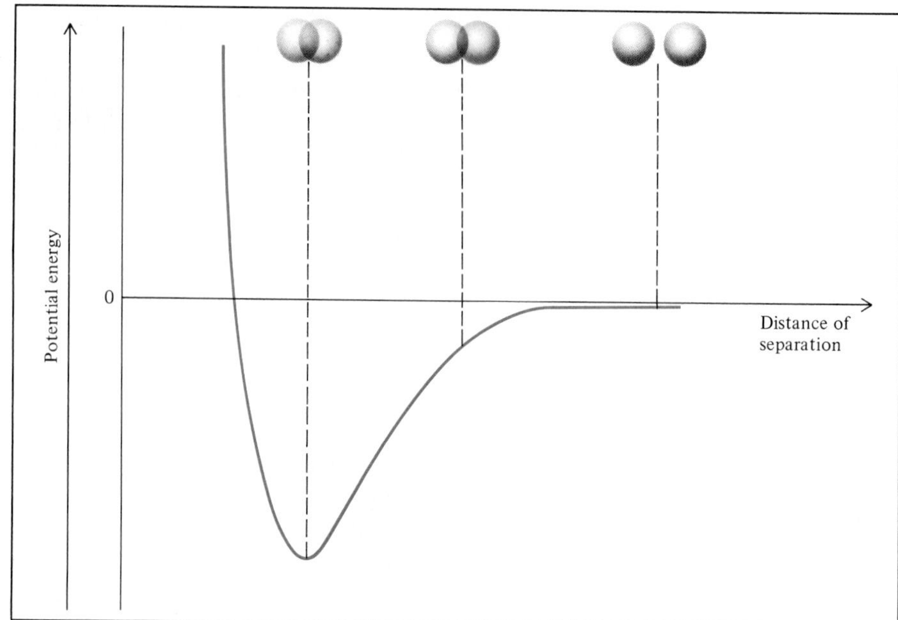

Figure 10.6 *Change in potential energy of two H atoms with their distance of separation. At the point of minimum potential energy, the H_2 molecule is in its most stable state and the bond length is 74 pm.*

In accord with the law of conservation of energy, the decrease in potential energy as a result of H_2 formation must be accompanied by a release of energy. By experiment we find that as a H_2 molecule is formed from two H atoms, heat is given off. The converse is also true. To break a H—H bond, energy must be supplied to the molecule.

Thus we see that valence bond theory gives a clearer picture of chemical bond formation than the Lewis theory does. Valence bond theory states that a stable molecule forms from reacting atoms when the potential energy of the system has decreased to a minimum; the Lewis theory ignores energy changes in chemical bond formation.

The orbital diagram of the F atom is shown on p. 293.

The concept of atomic orbital overlap applies equally well to diatomic molecules other than H_2. Thus a stable F_2 molecule forms when the 2*p* orbitals (containing the unpaired electrons) in the two F atoms overlap to form a covalent bond. Similarly, the formation of the HF molecule can be explained by the overlap of the 1*s* orbital in H with the 2*p* orbital in F. In each case, VB theory accounts for the changes in potential energy as the distance between the reacting atoms changes. Because the orbitals involved are not the same kind in all cases, we can see why the bond energies and bond lengths in H_2, F_2, and HF might be different. As we stated earlier, Lewis theory treats *all* the covalent bonds in the same manner and offers no explanation for why one covalent bond might differ from another.

10.4 Hybridization of Atomic Orbitals

The concept of atomic orbital overlap should apply also to polyatomic molecules. However, a satisfactory bonding scheme for polyatomic molecules must account for molecular geometry. Consider the CH_4 molecule. Focusing only on the valence electrons, we can represent the orbital diagram of C as

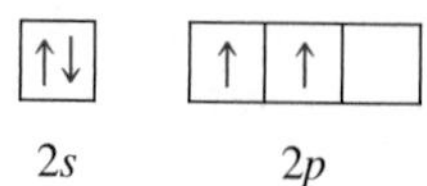

Because the carbon atom has two unpaired electrons (one in each of the two $2p$ orbitals), it can only form two bonds with hydrogen in its ground state. Although the species CH_2 is known, it is very unstable. To account for the four C—H bonds in methane, we might promote (that is, energetically excite) an electron from the $2s$ orbital to the $2p$ orbital:

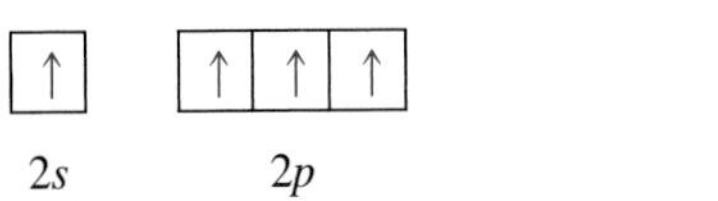

Now there are four unpaired electrons on C, and we can form four C—H bonds. However, the geometry is wrong, because according to this scheme three of the HCH bond angles would be 90° (remember that the three $2p$ orbitals on carbon are mutually perpendicular), and yet both experimental evidence and prediction based on VSEPR indicate that *all* HCH angles are 109.5°.

One way for the VB theory to explain bonding in methane is to introduce the concept of ***hybridization.*** Hybridization is *the mixing of atomic orbitals in an atom (usually a central atom) to generate a set of new atomic orbitals,* called **hybrid orbitals.** Hybrid orbitals, which are *atomic orbitals obtained when two or more nonequivalent orbitals of the same atom combine,* are used to form covalent bonds. We can generate four hybrid orbitals by mixing the $2s$ orbital and the three $2p$ orbitals:

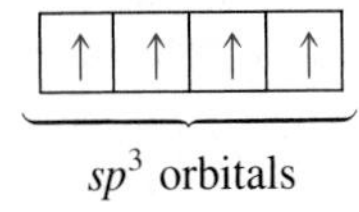

sp^3 is pronounced "*s-p* three."

Because these hybrid orbitals are formed from one s and three p orbitals, they are called sp^3 hybrid orbitals. Figure 10.7 shows the shape and orientations of the sp^3 orbitals. These four equivalent hybrid orbitals are directed toward the four corners of a regular tetrahedron. Figure 10.8 shows how the sp^3 hybrid orbitals of carbon and the $1s$

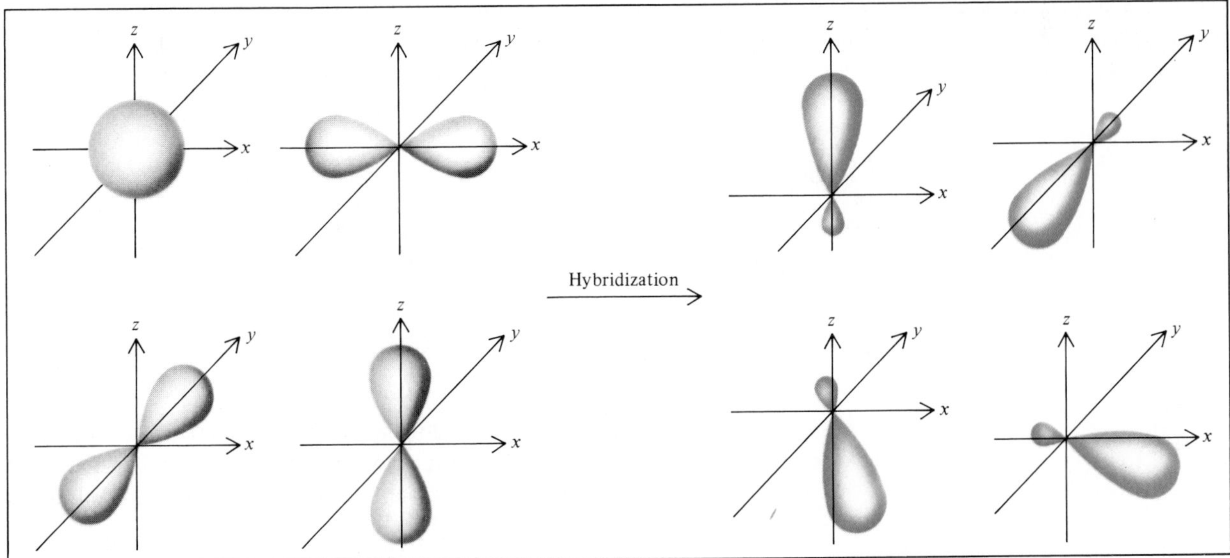

Figure 10.7 *Formation of sp^3 hybrid orbitals.*

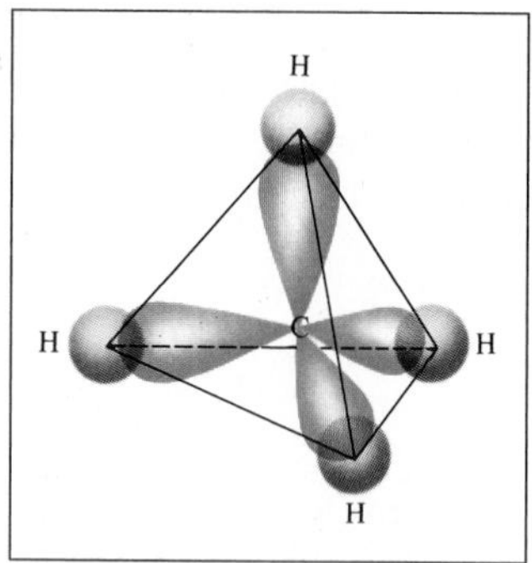

Figure 10.8 *Formation of four bonds between the carbon sp^3 hybrid orbitals and the hydrogen 1s orbitals in CH_4.*

orbitals of hydrogen overlap to form four covalent C—H bonds. Thus CH_4 has a tetrahedral shape, and all the HCH angles are 109.5°. Note that although an input of energy is required to bring about hybridization, this is more than compensated by the energy released upon the formation of C—H bonds. (Bond formations are exothermic processes.)

As another example of sp^3 hybridization, let us consider ammonia (NH_3). Table 10.1 shows that the arrangement of four electron pairs is tetrahedral, so that the bonding in NH_3 can be explained by assuming that N, like C in CH_4, is sp^3-hybridized. The ground-state electron configuration of N is $1s^22s^22p^3$, so that the orbital diagram for the sp^3-hybridized N atom is

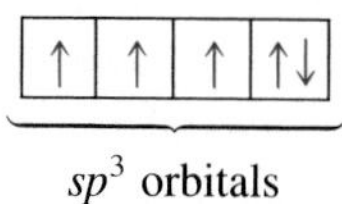

Three of the four hybrid orbitals form covalent N—H bonds, and the fourth hybrid orbital accommodates the lone pair on nitrogen (Figure 10.9). Repulsion between the lone-pair electrons and those in the bonding orbitals decreases the HNH bond angles from 109.5° to 107.3°.

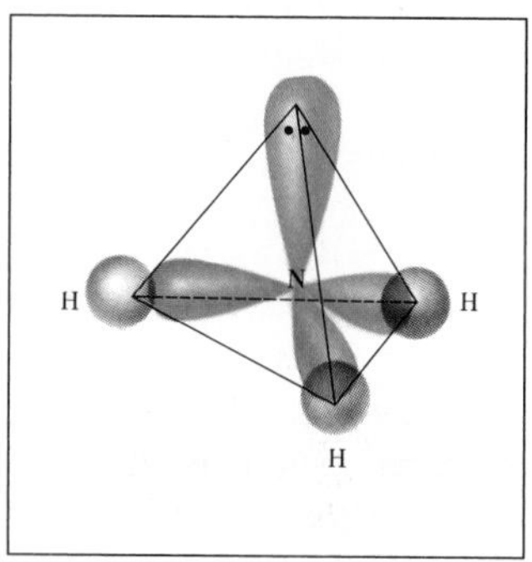

Figure 10.9 *The N atom is sp^3-hybridized in NH_3. It forms three bonds with the H atoms. One of the sp^3 hybrid orbitals is occupied by the lone pair.*

The O atom in H_2O also has four electron pairs. Thus we might expect the O atom to be sp^3-hybridized. However, spectroscopic evidence strongly suggests that the bonding orbitals on O (used to form the O—H bonds) are $2p$ orbitals, rather than sp^3 hybrid orbitals. This means that the O atom in H_2O is *unhybridized*. You might wonder that if this is true, then why isn't the HOH angle 90° since the p orbitals are perpendicular to each other. The reason is that in the H_2O molecule the two O—H bonds need to spread apart to avoid the crowding of the H atoms into the same space (a situation that is energetically unfavorable). Consequently, the bond angle increases from the predicted 90° to 104.5°.

It is important to understand the relationship between hybridization and the VSEPR model. The foregoing discussion shows that we use hybridization to describe the bonding scheme only when the arrangement of electron pairs has been predicted using VSEPR. If the VSEPR model predicts a tetrahedral arrangement of electron pairs, then we assume that one s and three p orbitals are hybridized to form four sp^3 hybrid orbitals. The following are examples of other types of hybridization.

sp Hybridization

Consider the $BeCl_2$ (beryllium chloride) molecule, which is predicted to be linear by VSEPR. The orbital diagram for the valence electrons in Be is

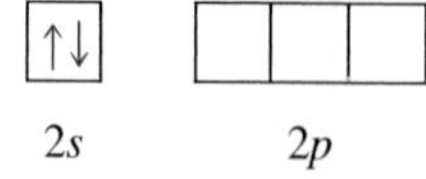

We know that in its ground state Be does not form covalent bonds with Cl because its electrons are paired in the 2s orbital. So we turn to hybridization for an explanation of Be's bonding behavior. First a 2s electron is promoted to a 2p orbital, resulting in

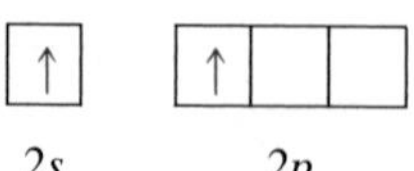

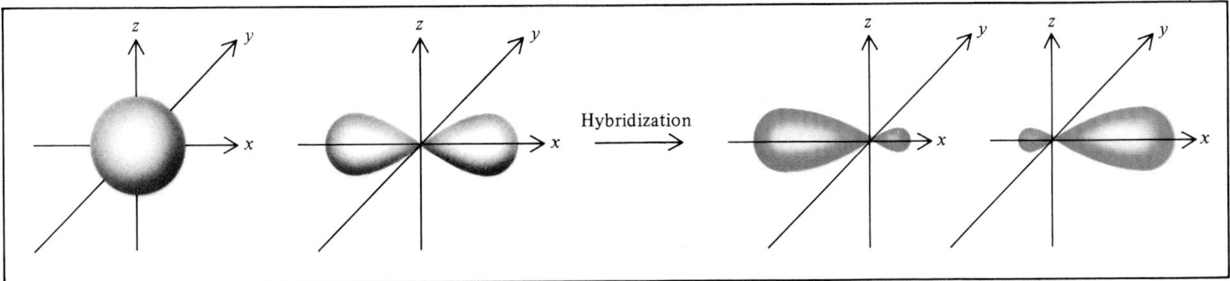

Figure 10.10 *Formation of sp hybrid orbitals.*

Now there are two different Be orbitals available for bonding, the 2*s* and 2*p*. However, we find this is a contradiction of known results from experiment. If two Cl atoms were to combine with Be in this excited state, one Cl atom would share a 2*s* electron and the other Cl would share a 2*p* electron, making two nonequivalent BeCl bonds. In the actual $BeCl_2$ molecule, the two BeCl bonds are identical in every respect. Thus the 2*s* and 2*p* orbitals must be mixed, or hybridized, to form two equivalent *sp* hybrid orbitals:

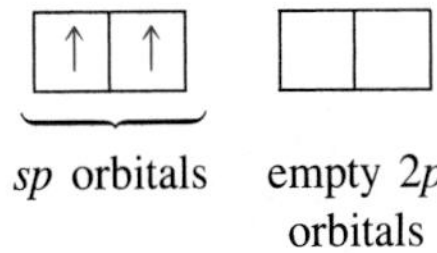

Figure 10.10 shows the shape and orientation of the *sp* orbitals. These two hybrid orbitals lie along the same line, the *x*-axis, so that the angle between them is 180°. Each of the BeCl bonds is then formed by the overlap of a Be *sp* hybrid orbital and a Cl 3*p* orbital, and the resulting $BeCl_2$ molecule has a linear geometry (Figure 10.11).

sp^2 Hybridization

Next we will look at the BF_3 (boron trifluoride) molecule, known to have planar geometry based on VSEPR. Considering only the valence electrons, the orbital diagram of B is

↑↓		↑		
2*s*		2*p*		

First, we promote a 2*s* electron to an empty 2*p* orbital:

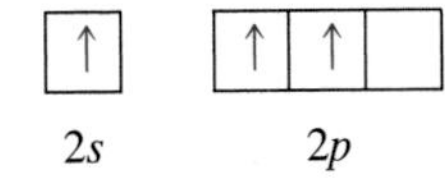

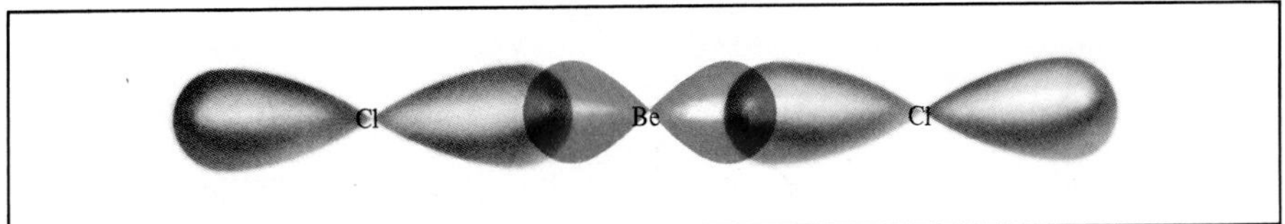

Figure 10.11 *The linear geometry of $BeCl_2$ can be explained by assuming Be to be sp-hybridized. The two sp hybrid orbitals overlap with the two 3p orbitals of chlorine to form two covalent bonds.*

sp^2 is pronounced "s-p two."

Mixing the $2s$ orbital with the two $2p$ orbitals generates three sp^2 hybrid orbitals:

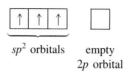

↑	↑	↑		☐
sp^2 orbitals				empty $2p$ orbital

These three sp^2 orbitals lie in a plane, and the angle between any two of them is 120° (Figure 10.12). Each of the BF bonds is formed by the overlap of a boron sp^2 hybrid orbital and a fluorine $2p$ orbital (Figure 10.13). The BF_3 molecule is planar with all the FBF angles equal to 120°. This result conforms to experimental findings and, as mentioned earlier, also to VSEPR predictions.

The main second-period exceptions are Be and B, which sometimes associate themselves in bonding with four and six valence electrons, respectively.

You may have noticed an interesting connection between hybridization and the octet rule. Regardless of the type of hybridization, an atom starting with one s and three p orbitals will still possess four orbitals, enough to accommodate a total of eight electrons in a compound. For elements in the second period of the periodic table, eight is the maximum number of electrons that an atom of any of these elements can accommodate. This is the reason that the octet rule is usually obeyed by the second-period elements.

The situation is different for an atom of a third-period element. If we use only the $3s$ and $3p$ orbitals of the atom to form hybrid orbitals in a molecule, then the octet rule applies. However, in some other molecules the same atom may use one or more $3d$ orbitals in addition to the $3s$ and $3p$ orbitals to form hybrid orbitals. In these cases the octet rule does not hold. We will see specific examples of the participation of the $3d$ orbital in hybridization shortly.

The following points are useful for an understanding of hybridization:

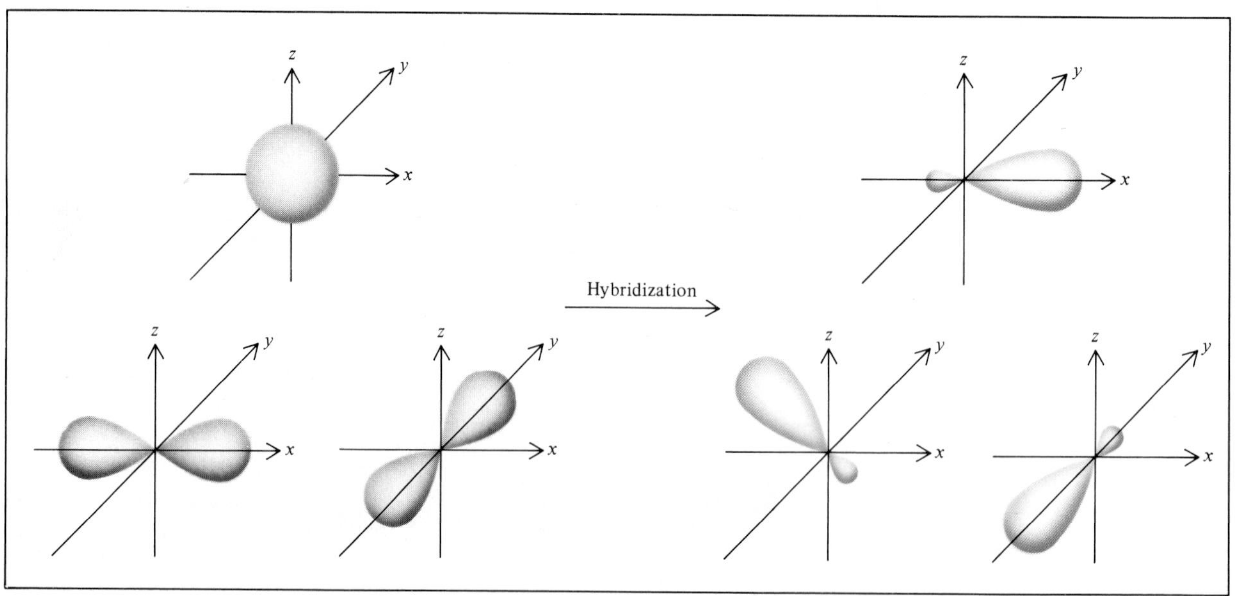

Figure 10.12 *Formation of sp^2 hybrid orbitals.*

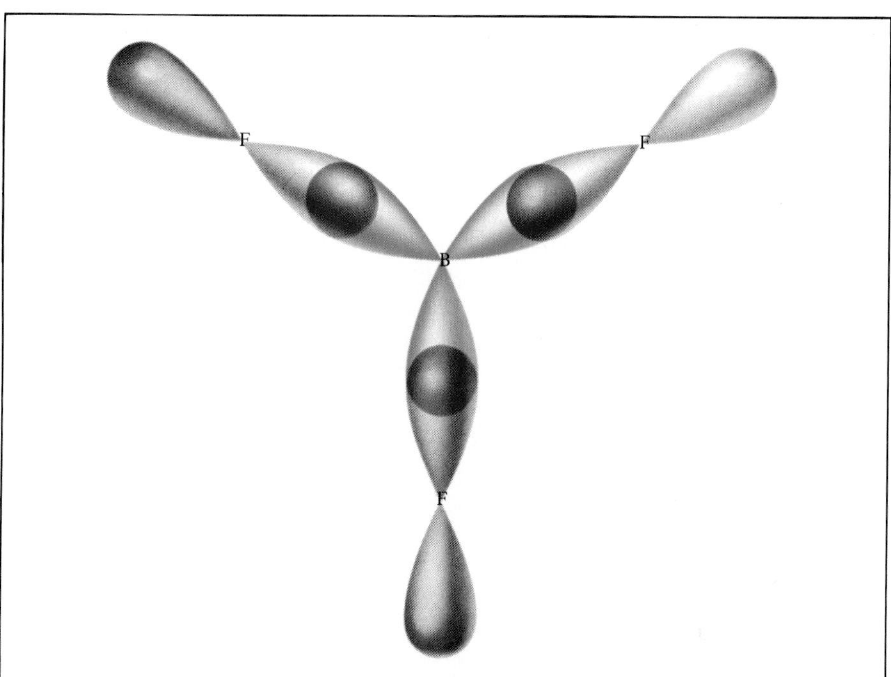

Figure 10.13 *Overlap of the sp^2 hybrid orbitals of the boron atom with the 2p orbitals of the fluorine atoms. The BF_3 molecule is planar, and all the FBF angles are 120°.*

- The concept of hybridization is not applied to isolated atoms. It is used only to explain the bonding scheme in a molecule.
- Hybridization is the mixing of at least two nonequivalent atomic orbitals, for example, *s* and *p* orbitals. Therefore, a hybrid orbital is not a pure (that is, native) atomic orbital. Hybrid orbitals have very different shapes from pure atomic orbitals.
- The number of hybrid orbitals generated is equal to the number of pure atomic orbitals that participate in the hybridization process.
- Hybridization requires an input of energy; however, the system more than recovers this energy during bond formation.
- Covalent bonds in polyatomic molecules are formed by the overlap of hybrid orbitals, or of hybrid orbitals with unhybridized ones. Therefore, the hybridization bonding scheme is still within the framework of valence bond theory; electrons in a molecule are assumed to occupy hybrid orbitals of the individual atoms.

Table 10.5 shows the *sp*, sp^2, and sp^3 hybridizations (as well as other types of hybridizations to be discussed shortly) and the shapes of the hybrid orbitals. In order to assign a suitable state of hybridization of the central atom in a molecule, we must have some idea about the geometry of the molecule. We can start by drawing the Lewis structure of the molecule and predict the overall arrangement of the electron pairs (both bonding pairs and lone pairs) using the VSEPR model (see Table 10.1). We can then deduce the hybridization of the central atom by matching the arrangement of the electron pairs with that of the hybrid orbitals in Table 10.5. The following example illustrates this procedure.

Table 10.5 Important hybrid orbitals and their shapes

Pure atomic orbitals of the central atom	Hybridization of the central atom	Number of hybrid orbitals	Shape of hybrid orbitals	Examples
s, p	*sp*	2	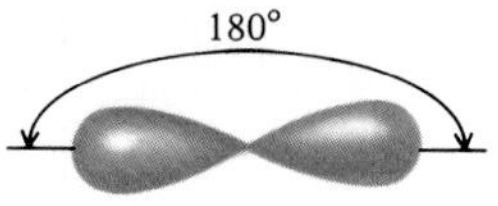 180° Linear	$BeCl_2$
s, p, p	sp^2	3	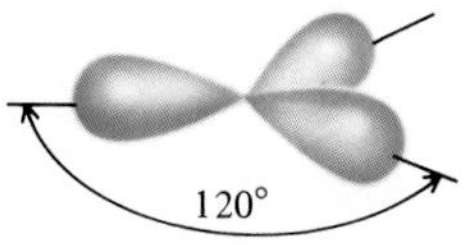 120° Planar	BF_3
s, p, p, p	sp^3	4	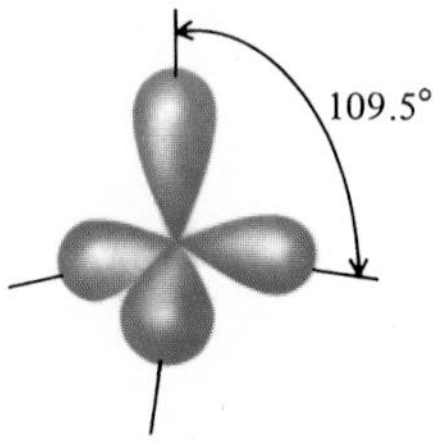 109.5° Tetrahedral	CH_4, NH_4^+
s, p, p, p, d	sp^3d	5	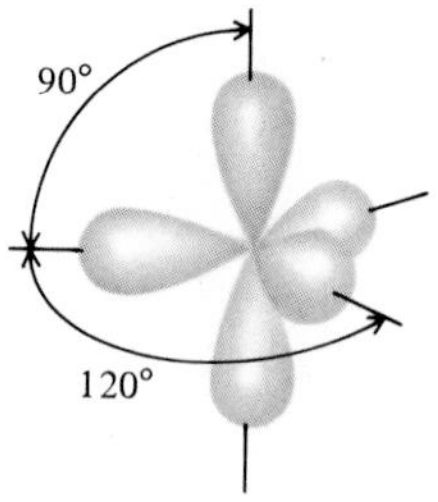 90° 120° Trigonal bipyramidal	PCl_5
s, p, p, p, d, d	sp^3d^2	6	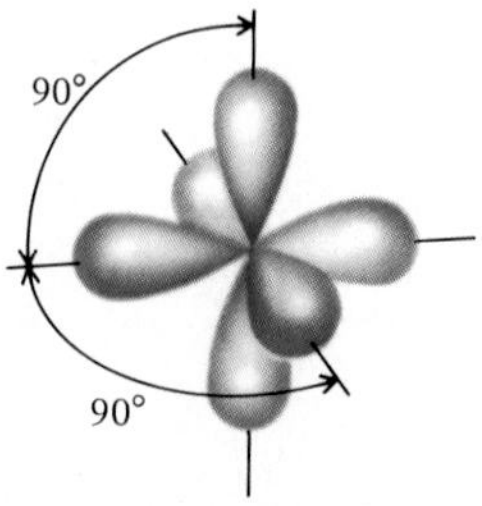 90° 90° Octahedral	SF_6

EXAMPLE 10.3

Determine the hybridization state of the central (underlined) atom in each of the following molecules: (a) $\underline{Hg}Cl_2$, (b) $\underline{Al}I_3$, and (c) $\underline{P}F_3$. Describe the hybridization process and determine the molecular geometry in each case.

Answer

See Table 7.3.

(a) The ground-state electron configuration of Hg is $[Xe]6s^24f^{14}5d^{10}$; therefore, the Hg atom has two valence electrons (the 6*s* electrons). The Lewis structure of $HgCl_2$ is

Cl—Hg—Cl

There are no lone pairs on the Hg atom, so the arrangement of the two electron pairs is linear (see Table 10.1). From Table 10.5 we conclude that Hg is *sp*-hybridized because it has the geometry of the two *sp* hybrid orbitals. The hybridization process can be imagined to take place as follows. First we draw the orbital diagram for the ground state of Hg:

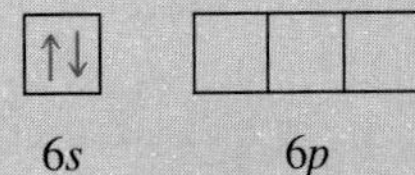

By promoting a 6*s* electron to the 6*p* orbital, we get the excited state:

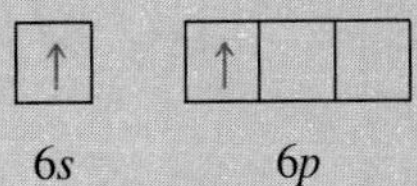

The 6*s* and 6*p* orbitals then mix to form two *sp* hybrid orbitals:

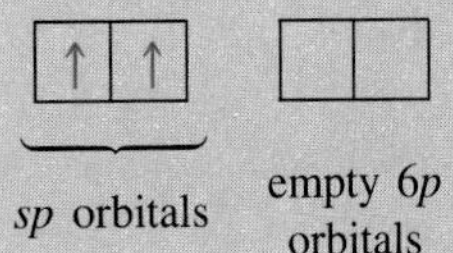

The two Hg—Cl bonds are formed by the overlap of the Hg *sp* hybrid orbitals with the 3*p* orbitals of the Cl atoms. Thus $HgCl_2$ is a linear molecule.

(b) The ground-state electron configuration of Al is $[Ne]3s^23p^1$. Therefore, Al has three valence electrons. The Lewis structure of AlI_3 is

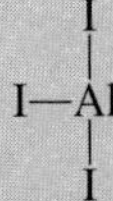

There are three bonding pairs and no lone pair on the Al atom. From Table 10.1 we see that the shape of three electron pairs is trigonal planar, and from Table 10.5 we conclude that Al must be sp^2-hybridized in AlI_3. The orbital diagram of the ground-state Al atom is

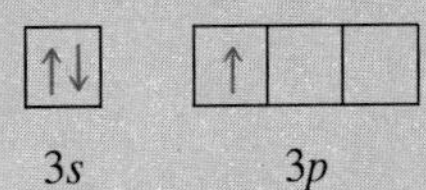

By promoting a 3*s* electron into the 3*p* orbital we obtain the following excited state:

↑ ↑ ↑

3*s* 3*p*

The 3*s* and two 3*p* orbitals then mix to form three sp^2 hybrid orbitals:

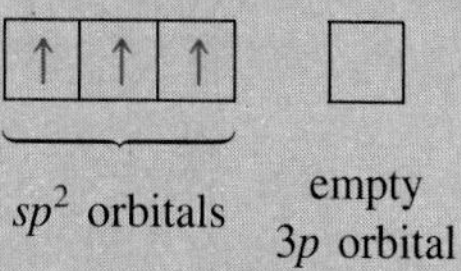

The sp^2 hybrid orbitals overlap with the 5*p* orbitals of I to form three covalent Al—I bonds. We predict the AlI_3 molecule to be planar and all the IAlI angles to be 120°.
(c) The ground-state electron configuration of P is $[Ne]3s^23p^3$. Therefore, the P atom has five valence electrons. The Lewis structure of PF_3 is

```
  ..
F—P—F
  |
  F
```

There are three bonding pairs and one lone pair on the P atom. In Table 10.1 we see that the overall arrangement of four electron pairs is tetrahedral, and from Table 10.5 we conclude that P must be sp^3-hybridized. The orbital diagram of the ground-state P atom is

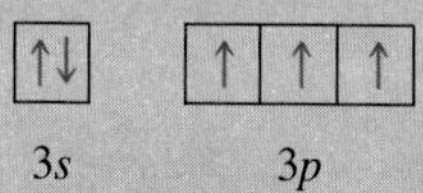

By mixing the 3*s* and 3*p* orbitals, we obtain four sp^3 hybrid orbitals.

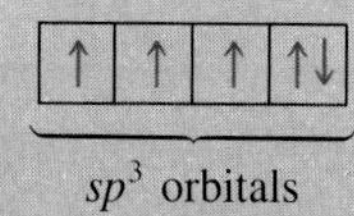

As in the case of NH_3, one of the sp^3 hybrid orbitals is used to accommodate the lone pair on P. The other three sp^3 hybrid orbitals form covalent P—F bonds with the 2*p* orbitals of F. We predict the geometry of the molecule to be pyramidal; the FPF angle should be somewhat less than 109.5°.

Similar problems: 10.34, 10.35.

Hybridization of *s*, *p*, and *d* Orbitals

We have seen that hybridization neatly explains bonding that involves *s* and *p* orbitals. For elements in the third period and beyond, however, we cannot always account for molecular geometry by assuming the hybridization of only *s* and *p* orbitals. To understand the formation of molecules with trigonal bipyramidal and octahedral geometries, for instance, we must include *d* orbitals in the hybridization concept.

Consider the SF_6 molecule as an example. In Section 10.1 we saw that this molecule has an octahedral geometry, which is also the arrangement of the six electron pairs. From Table 10.5 we see that the S atom must be sp^3d^2-hybridized in SF_6. The ground-state electron configuration of S is $[Ne]3s^23p^4$:

sp^3d^2 is pronounced "*s-p* three *d* two."

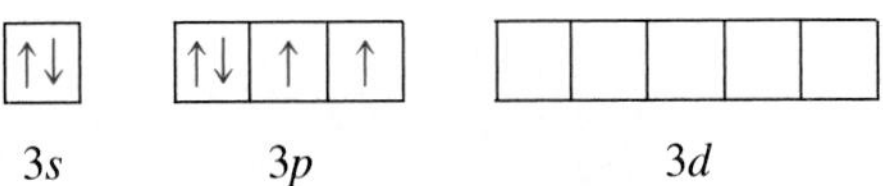

Since the $3d$ level is quite close in energy to the $3s$ and $3p$ levels, we can promote both $3s$ and $3p$ electrons to two of the $3d$ orbitals:

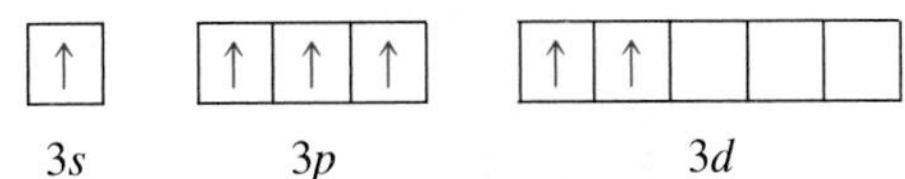

Mixing the $3s$, three $3p$, and two $3d$ orbitals generates six sp^3d^2 hybrid orbitals:

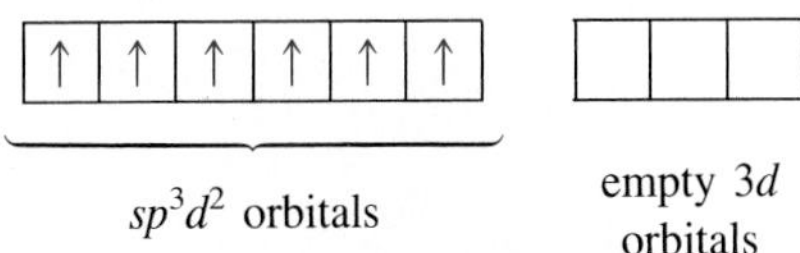

The six S—F bonds are formed by the overlap of the hybrid orbitals of the S atom and the $2p$ orbitals of the F atoms. Since there are 12 electrons around the S atom, the octet rule is exceeded. *The use of d orbitals in addition to s and p orbitals to form covalent bonds* is an example of ***valence-shell expansion,*** which corresponds to the expanded octet case discussed in Section 9.9. The second-period elements, unlike third-period elements, do not have $2d$ levels, so they can never expand their valence shells. Hence atoms of second-period elements can never be surrounded by more than eight electrons in any of their compounds.

Recall that when $n = 2$, $\ell = 0$ and 1. Thus we can only have $2s$ and $2p$ orbitals.

The following example shows the valence-shell expansion of a third-period element in a compound.

EXAMPLE 10.4

Describe the hybridization state of phosphorus in phosphorus pentabromide (PBr_5).

Answer

The Lewis structure of PBr_5 is

```
Br  Br
  \ |
    P—Br
  / |
Br  Br
```

Table 10.1 indicates that the arrangement of five electron pairs is trigonal bipyramidal. Referring to Table 10.5, we find that this is the shape of five sp^3d hybrid orbitals. Thus P must be sp^3d-hybridized in PBr_5. The ground-state electron configuration of P is $[Ne]3s^23p^3$. To describe the hybridization process, we start with the orbital diagram for the ground state of P

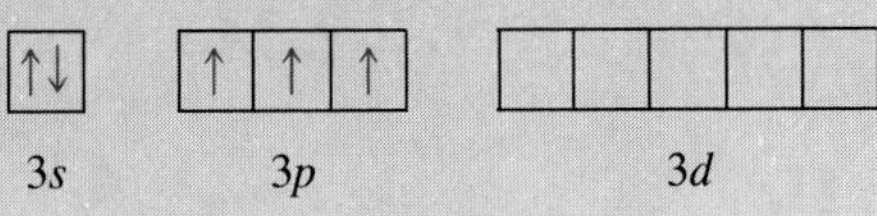

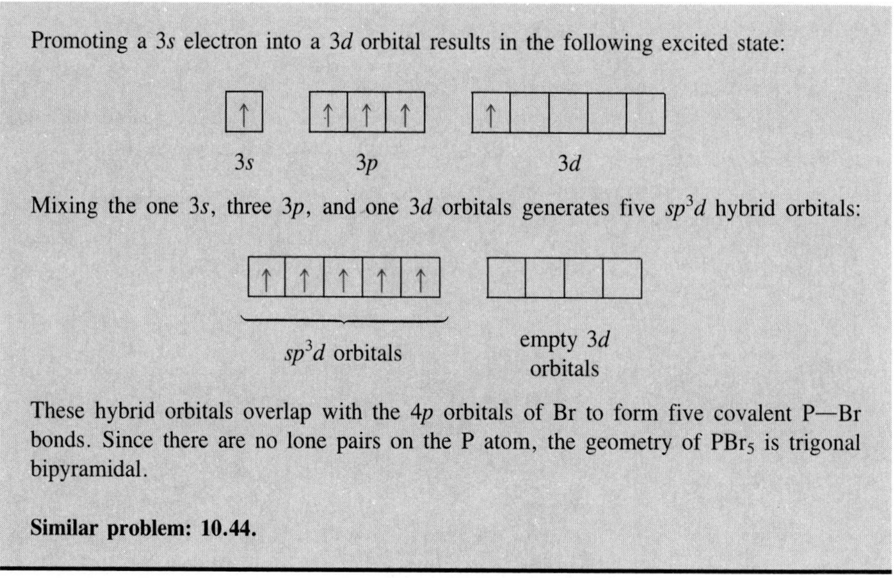
Promoting a $3s$ electron into a $3d$ orbital results in the following excited state:

Mixing the one $3s$, three $3p$, and one $3d$ orbitals generates five sp^3d hybrid orbitals:

These hybrid orbitals overlap with the $4p$ orbitals of Br to form five covalent P—Br bonds. Since there are no lone pairs on the P atom, the geometry of PBr_5 is trigonal bipyramidal.

Similar problem: 10.44.

10.5 Hybridization in Molecules Containing Double and Triple Bonds

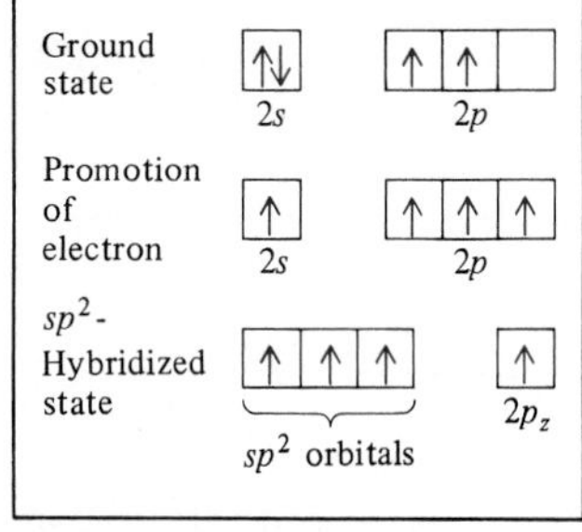

Figure 10.14 *The sp^2 hybridization of a carbon atom. The 2s orbital is mixed with only two 2p orbitals to form three equivalent sp^2 hybrid orbitals. This process leaves an electron in the unhybridized orbital, the $2p_z$ orbital.*

The concept of hybridization is useful also for molecules with double and triple bonds. Consider the ethylene molecule, C_2H_4, as an example. In Example 10.1 we saw that C_2H_4 contains a carbon-carbon double bond and has planar geometry. Both the geometry and the bonding can be understood if we assume that each carbon atom is sp^2-hybridized. Figure 10.14 shows orbital diagrams of this hybridization process. We assume that only the $2p_x$ and $2p_y$ orbitals combine with the $2s$ orbital, and that the $2p_z$ orbital remains unchanged. Figure 10.15 shows that the $2p_z$ orbital is perpendicular to the plane of the hybrid orbitals. Now how do we account for the bonding of the C atoms? As Figure 10.16(a) shows, each carbon atom uses the three sp^2 hybrid orbitals to form two bonds with the two hydrogen $1s$ orbitals and one bond with the sp^2 hybrid orbital of the adjacent C atom. In addition, the two unhybridized $2p_z$ orbitals of the C atoms form another bond by overlapping sideways [Figure 10.16(b)].

A distinction is made between the two types of covalent bonds found in C_2H_4. *In a*

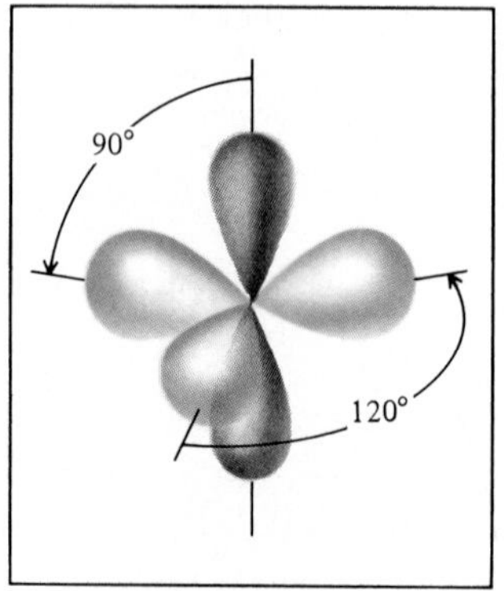

Figure 10.15 *Each carbon atom in the C_2H_4 molecule has three sp^2 hybrid orbitals and one unhybridized $2p_z$ orbital that is perpendicular to the plane of the hybrid orbitals.*

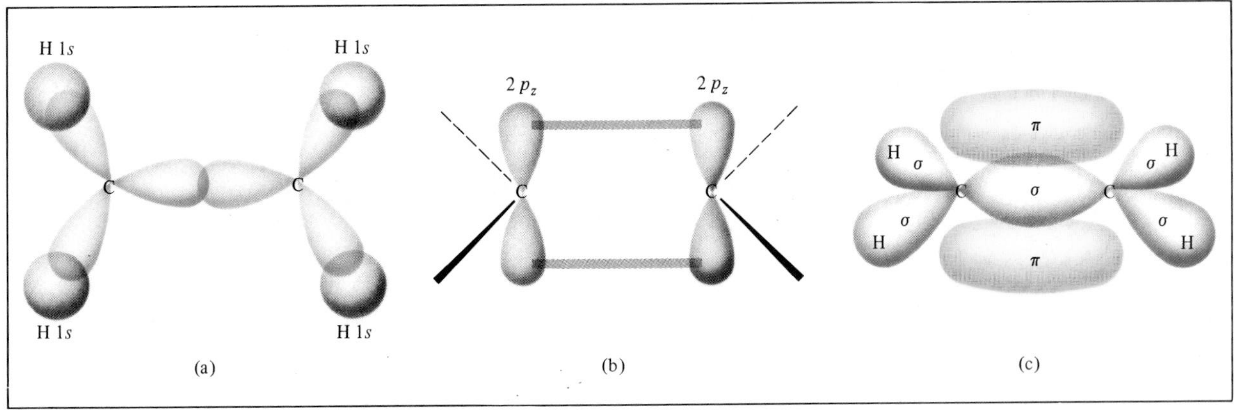

Figure 10.16 *Bonding in ethylene, C_2H_4. (a) Top view of the sigma bonds between carbon atoms and between carbon and hydrogen atoms. All the atoms lie in the same plane, making C_2H_4 a planar molecule. (b) Side view showing how overlap of the two $2p_z$ orbitals on the two carbon atoms takes place, leading to the formation of a pi bond. (c) The sigma bonds and the pi bond in ethylene. Note that the pi bond lies above and below the plane of the molecule.*

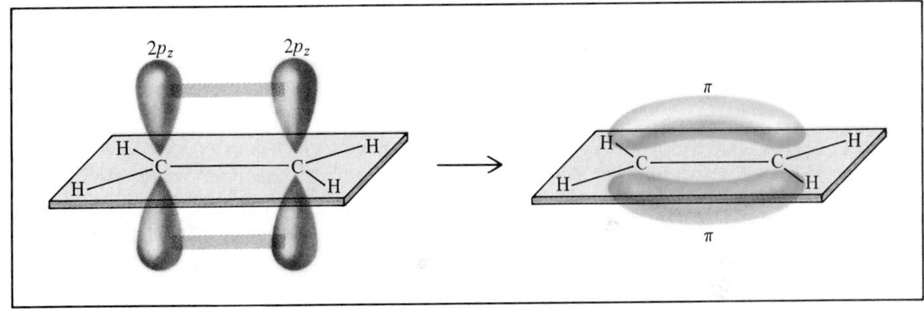

Figure 10.17 *Another view of pi bond formation in the C_2H_4 molecule. Note that all six atoms lie in the same plane. It is the overlap of the $2p_z$ orbitals that causes the molecule to assume a planar structure.*

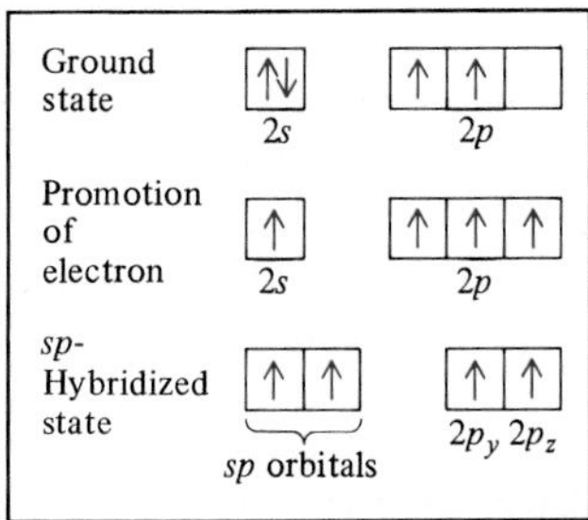

Figure 10.18 *The sp hybridization of a carbon atom. The 2s orbital is mixed with only one 2p orbital to form two sp hybrid orbitals. This process leaves an electron in each of the two unhybridized 2p orbitals, namely, the $2p_y$ and $2p_z$ orbitals.*

covalent bond formed by orbitals overlapping end-to-end, as shown in Figure 10.16(a), *the electron density is concentrated between the nuclei of the bonding atoms.* A bond of this kind is called a ***sigma bond (σ bond).*** The three bonds formed by each C atom in Figure 10.16(a) are all sigma bonds. On the other hand, *a covalent bond formed by sideways overlapping orbitals has electron density concentrated above and below the plane of the nuclei of the bonding atoms* and is called a ***pi bond (π bond).*** The two C atoms form a pi bond as shown in Figure 10.16(b). Figure 10.16(c) shows the orientation of both sigma and pi bonds. Figure 10.17 is yet another way of looking at the planar C_2H_4 molecule and the formation of the pi bond. Although we normally represent the carbon-carbon double bond as C═C (as in a Lewis structure), it is important to keep in mind that the two bonds are different types: One is a sigma bond and the other is a pi bond.

The acetylene molecule (C_2H_2) contains a carbon-carbon triple bond. Since the molecule is linear, we can explain its geometry and bonding by assuming that each C atom is *sp*-hybridized (Figure 10.18). As Figure 10.19 shows, the two *sp* hybrid orbitals of each C atom form one sigma bond with a hydrogen 1*s* orbital and another sigma bond with the other C atom. In addition, two pi bonds are formed by the

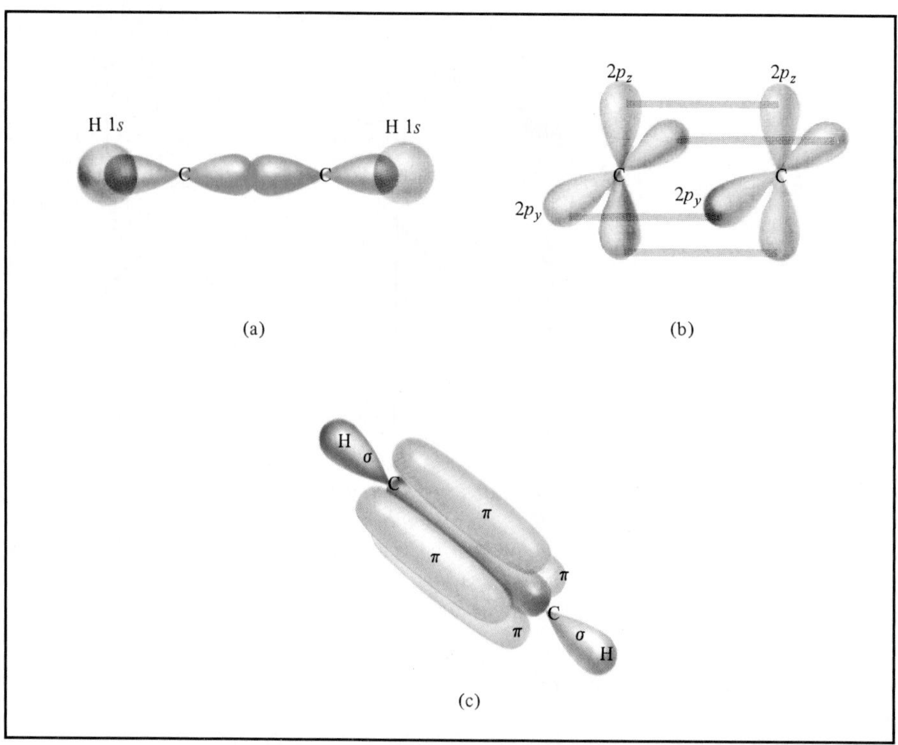

Figure 10.19 *Bonding in acetylene, C_2H_2. (a) Top view showing the sigma bonds between carbon atoms and between carbon and hydrogen atoms. All the atoms lie along a straight line; therefore acetylene is a linear molecule. (b) Side view showing the overlap of the two $2p_y$ orbitals and of the two $2p_z$ orbitals of the two carbon atoms, which leads to the formation of two pi bonds. (c) Diagram showing both the sigma and the pi bonds.*

sideways overlap of the unhybridized $2p_y$ and $2p_z$ orbitals. Thus the C≡C bond is made up of one sigma bond and two pi bonds.

The following rule helps us predict hybridization in molecules containing multiple bonds: If the central atom forms a double bond, it is sp^2-hybridized; if it forms two double bonds or a triple bond, it is *sp*-hybridized. Note that this rule applies only to atoms of the second-period elements. Atoms of third-period elements and beyond that form multiple bonds present a more involved problem and will not be dealt with here.

EXAMPLE 10.5

Describe the bonding in the formaldehyde molecule whose Lewis structure is

```
H
 \     ..
  C=O
 /     ..
H
```

Assume that the O atom is unhybridized.

Answer

We note that the C atom (a second-period element) has a double bond; therefore, it is sp^2-hybridized. The orbital diagram of the O atom is

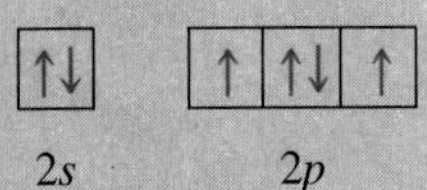

One of its orbitals ($2p_x$) forms a sigma bond with C, and the other orbital ($2p_z$) forms a pi bond with the unhybridized $2p_z$ orbital of C (Figure 10.20). The two lone pairs on the O atom then are the electrons in the $2s$ and $2p_y$ orbitals. The two remaining sp^2 orbitals of the C atom form two sigma bonds with the H atoms, as in ethylene.

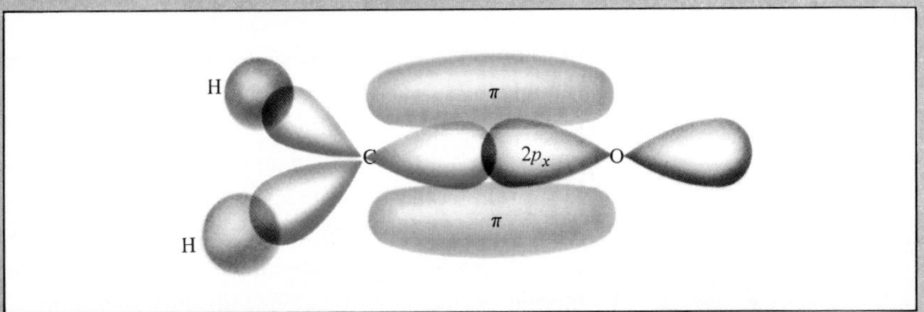

Figure 10.20 *Bonding in the formaldehyde molecule. A sigma bond is formed by the overlap of the sp^2 hybrid orbital of carbon and the $2p_x$ orbital of oxygen; a pi bond is formed by the overlap of the $2p_z$ orbitals of the carbon and oxygen atoms. The two lone pairs on oxygen are placed in the $2s$ and $2p_y$ orbitals (not shown).*

Similar problems: 10.40, 10.45.

10.6 Molecular Orbital Theory

Valence bond theory is one of the two quantum mechanical approaches that explain bonding in molecules. It accounts, at least qualitatively, for the stability of the covalent bond in terms of overlapping atomic orbitals. Using the concept of hybridization, valence bond theory can explain molecular geometries predicted by the VSEPR model. However, the assumption that electrons in a molecule occupy atomic orbitals of the individual atoms can only be an approximation, since each bonding electron in a molecule must be in an orbital that is characteristic of the molecule as a whole.

In some cases, valence bond theory cannot satisfactorily account for observed properties of molecules. Consider the oxygen molecule, whose Lewis structure is

$$\ddot{\underset{\cdot\cdot}{O}}=\ddot{\underset{\cdot\cdot}{O}}$$

According to this description, all the electrons in O_2 are paired and the molecule should therefore be diamagnetic. By experiment we find that the oxygen molecule is paramagnetic, with two unpaired electrons (Figure 10.21). This finding suggests a fundamental deficiency in valence bond theory, one that justifies searching for an alternative bond-

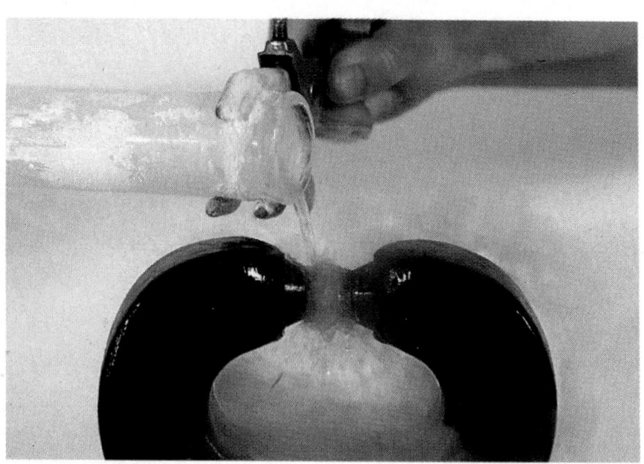

Figure 10.21 *Liquid oxygen is caught between the poles of a magnet because the O_2 molecules are paramagnetic.*

ing approach capable of accounting for the properties of molecules, including commonplace O_2.

The magnetic and other properties of molecules are sometimes better explained by another quantum mechanical approach called *molecular orbital (MO) theory*. Molecular orbital theory describes covalent bonds in terms of ***molecular orbitals,*** which *result from interaction of the atomic orbitals of the bonding atoms and are associated with the entire molecule*. That is how a molecular orbital differs from an atomic orbital—an atomic orbital is associated with only one atom. We must realize that neither theory perfectly explains all aspects of bonding; each has its strengths and weaknesses. We will use both theories, emphasizing one or the other as the situation warrants.

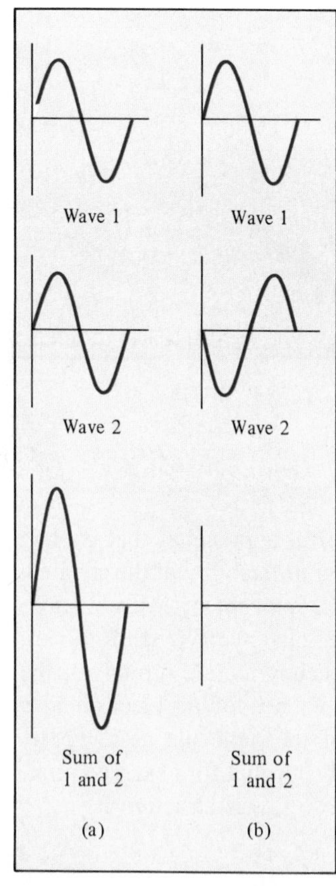

Figure 10.22 *Constructive interference (a) and destructive interference (b) of two waves of the same wavelength and amplitude.*

Bonding and Antibonding Molecular Orbitals

According to MO theory the overlap of the 1*s* orbitals of two hydrogen atoms leads to the formation of two molecular orbitals: one bonding molecular orbital and one antibonding molecular orbital. A ***bonding molecular orbital*** has *lower energy and greater stability than the atomic orbitals from which it was formed*. An ***antibonding molecular orbital*** has *higher energy and lower stability than the atomic orbitals from which it was formed*. As the names ''bonding'' and ''antibonding'' suggest, placing electrons in a bonding molecular orbital yields a stable covalent bond, whereas placing electrons in an antibonding molecular orbital results in an unstable bond.

In the bonding molecular orbital the electron density is greatest between the nuclei of the bonding atoms. In the antibonding molecular orbital, on the other hand, the electron density decreases to zero between the nuclei. We can understand this distinction if we recall that electrons in orbitals have wave characteristics. A property unique to waves allows waves of the same type to interact in such a way that the resultant wave has either an enhanced amplitude or a diminished amplitude. In the former case, we call the interaction *constructive interference;* in the latter case, it is *destructive interference* (Figure 10.22).

The formation of bonding molecular orbitals corresponds to constructive interference (the increase in amplitude is analogous to the buildup of electron density between the two nuclei). The formation of antibonding molecular orbitals corresponds to destructive interference (the decrease in amplitude is analogous to the decrease in electron density between the two nuclei). The constructive and destructive interactions between the two 1*s* orbitals in the H_2 molecule, then, lead to the formation of a sigma bonding molecular orbital (σ_{1s}) and a sigma antibonding molecular orbital ($\sigma^{\star}_{1s}$):

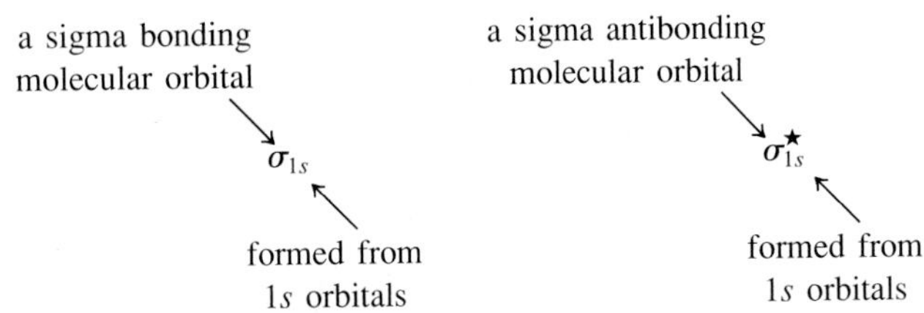

where the star denotes an antibonding molecular orbital.

The two electrons in the sigma molecular orbital are paired. The Pauli exclusion principle applies to molecules as well as to atoms.

In a ***sigma molecular orbital*** (bonding or antibonding) *the electron density is concentrated symmetrically around a line between the two nuclei of the bonding atoms.* Two electrons in a sigma molecular orbital form a sigma bond (see Section 10.5). Remember that a single covalent bond (such as H—H or F—F) is almost always a sigma bond.

Figure 10.23 shows the *molecular orbital energy level diagram*—that is, the relative energy levels of the orbitals produced in the formation of the H_2 molecule, and the constructive and destructive interactions between the two 1*s* orbitals. Notice that in the antibonding molecular orbital there is a *node,* or zero electron density, between the nuclei. The nuclei are repelled by each other's positive charges, rather than held together. Electrons in the bonding molecular orbital have less energy (and hence greater stability) than they would have in the isolated atoms. On the other hand, electrons in the antibonding molecular orbital have higher energy (and less stability) than they would have in the isolated atoms.

So far we have used the hydrogen molecule to illustrate molecular orbital formation, but the concept is equally applicable to other molecules. In the H_2 molecule we consider only the interaction between 1*s* orbitals; with more complex molecules we need to consider additional atomic orbitals as well. Nevertheless, for all *s* orbitals, the process

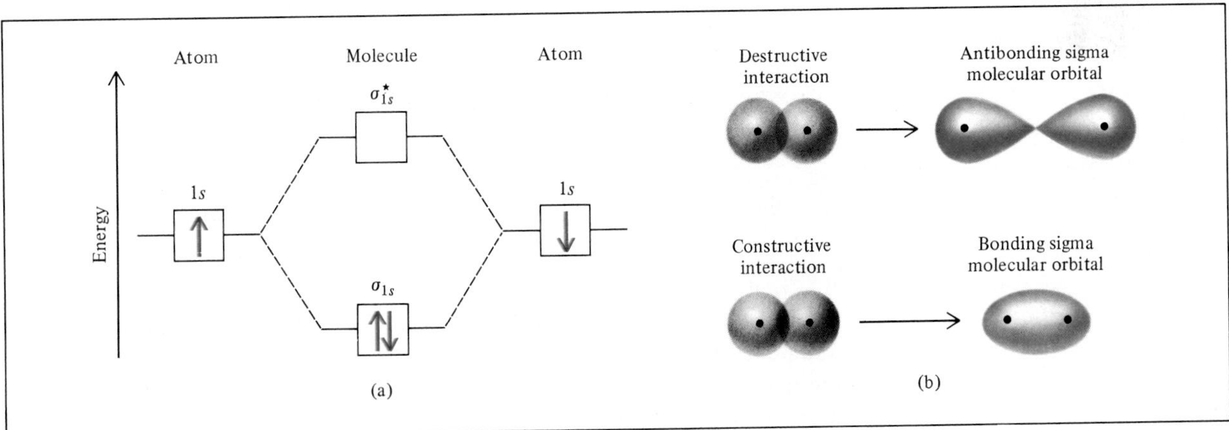

Figure 10.23 *(a) Energy levels of bonding and antibonding molecular orbitals in the H_2 molecule. Note that the two electrons in the σ_{1s} must have opposite spins in accord with the Pauli exclusion principle. Keep in mind that the higher the energy of the molecular orbital, the less stable the electrons in that molecular orbital. (b) Constructive and destructive interactions between the two hydrogen 1s orbitals lead to the formation of a bonding and an antibonding molecular orbital. In the bonding MO, there is a buildup of the electron density between the nuclei, which acts as a negatively charged "glue" holding the positively charged nuclei together.*

is the same as for $1s$ orbitals. Thus, the interaction between two $2s$ or $3s$ orbitals can be understood in terms of the molecular orbital energy level diagram and the formation of bonding and antibonding molecular orbitals shown in Figure 10.23.

For p orbitals the process is more complex because they can interact with each other in two different ways. For example, two $2p$ orbitals can approach each other end-to-end to produce a sigma bonding and a sigma antibonding molecular orbital, as shown in Figure 10.24(a). Alternatively, the two p orbitals can overlap sideways to generate a bonding and an antibonding pi molecular orbital [Figure 10.24(b)].

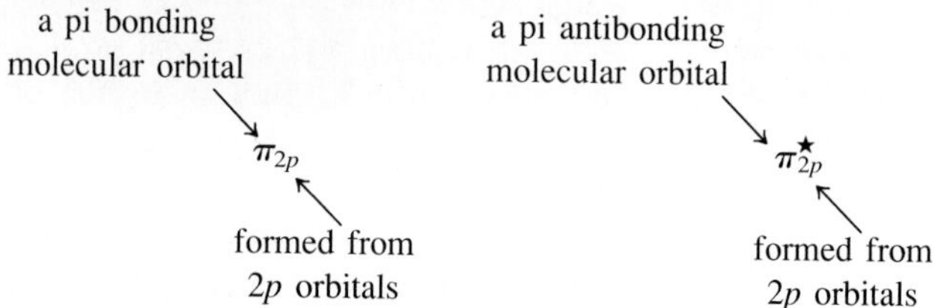

In a ***pi molecular orbital*** (bonding or antibonding), *the electron density is concentrated above and below the line joining the two nuclei of the bonding atoms*. Two electrons in a pi molecular orbital form a pi bond (see Section 10.5). A double bond is almost always composed of a sigma bond and a pi bond; a triple bond is always a sigma bond plus two pi bonds.

10.7 Molecular Orbital Configurations

To understand properties of molecules, we must know their electron configurations, that is, the distribution of electrons among various molecular orbitals. The procedure for determining the distribution is analogous to the one we use to determine the electron configurations of atoms (see Section 7.9).

Rules Governing Molecular Electron Configuration and Stability

In order to write the electron configuration of any molecule, we must first arrange the molecular orbitals that are formed in the molecule in order of increasing energy. Once the order is known, we can use the following rules to guide us in filling these molecular orbitals with electrons. The rules also help us understand the stabilities of the molecular orbitals:

- The number of molecular orbitals formed is always equal to the number of atomic orbitals combined.
- The more stable the bonding molecular orbital, the less stable the corresponding antibonding molecular orbital.
- In a stable molecule, the number of electrons in bonding molecular orbitals is always greater than that in antibonding molecular orbitals.
- Like an atomic orbital, each molecular orbital can accommodate up to two electrons with opposite spins in accordance with the Pauli exclusion principle.
- When electrons are added to molecular orbitals of the same energy, the most stable arrangement is that predicted by Hund's rule; that is, electrons enter these molecular orbitals with parallel spins.

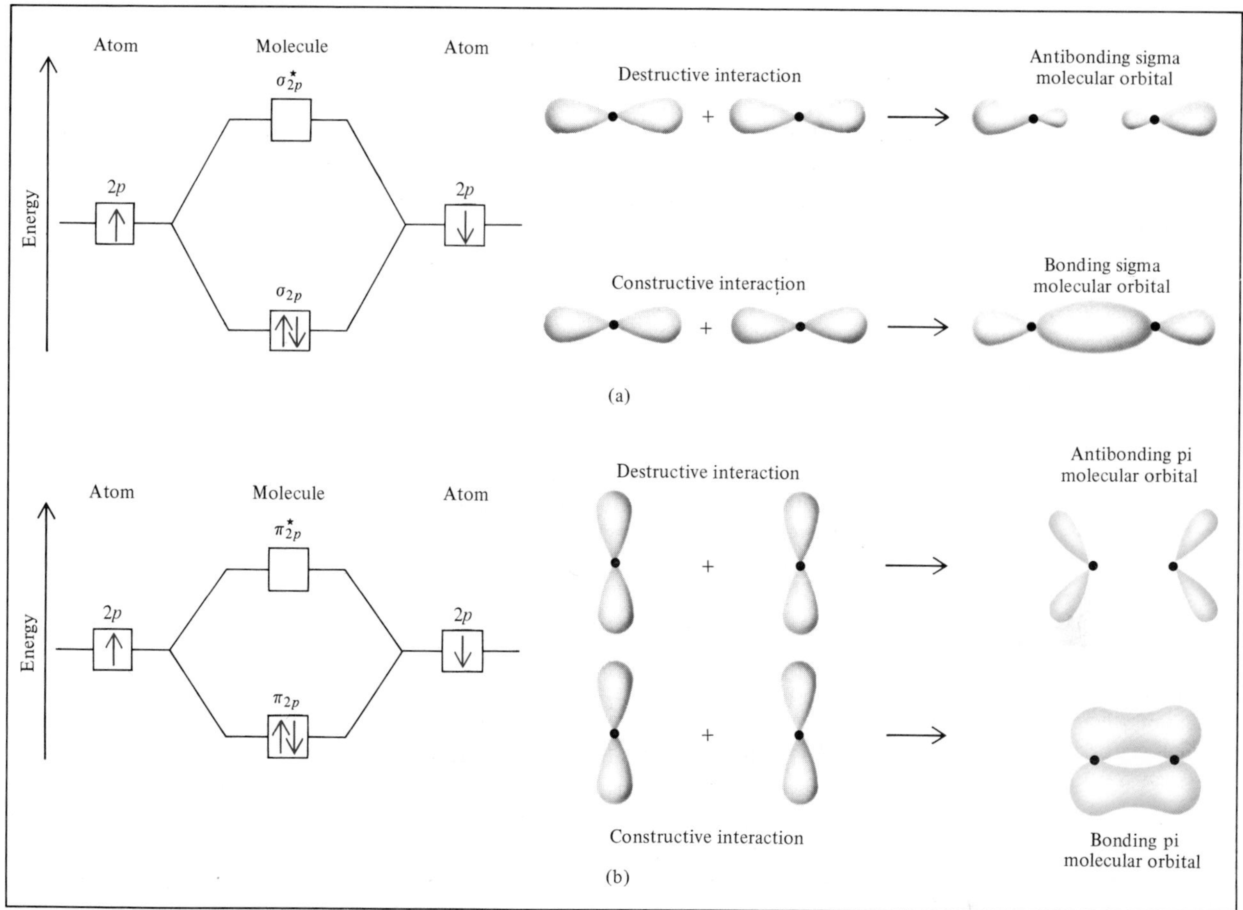

Figure 10.24 *Two possible interactions between two equivalent p orbitals and the corresponding molecular orbitals. (a) When the p orbitals overlap end-to-end, a sigma bonding and a sigma antibonding molecular orbital form. (b) When the p orbitals overlap side-to-side, a pi bonding and a pi antibonding molecular orbital form. Normally, a sigma bonding molecular orbital is more stable than a pi bonding molecular orbital, since side-to-side interaction leads to a smaller overlap of the p orbitals than does end-to-end interaction. We assume here that the $2p_x$ orbitals take part in the sigma molecular orbital formation. The $2p_y$ and $2p_z$ orbitals can interact to form only π molecular orbitals. The behavior shown in (b) represents the interaction between the $2p_y$ orbitals or the $2p_z$ orbitals.*

- The number of electrons in the molecular orbitals is equal to the sum of all the electrons on the bonding atoms.

Hydrogen and Helium Molecules

Later in this section we will study molecules formed by atoms of the second-period elements. Before we do, it will be instructive to predict the relative stabilities of the following simple species H_2^+, H_2, He_2^+, and He_2, using the molecular orbital energy level diagrams shown in Figure 10.25. The σ_{1s} and $\sigma_{1s}^\star$ orbitals can accommodate a maximum of four electrons. The total number of electrons increases from one for H_2^+ to

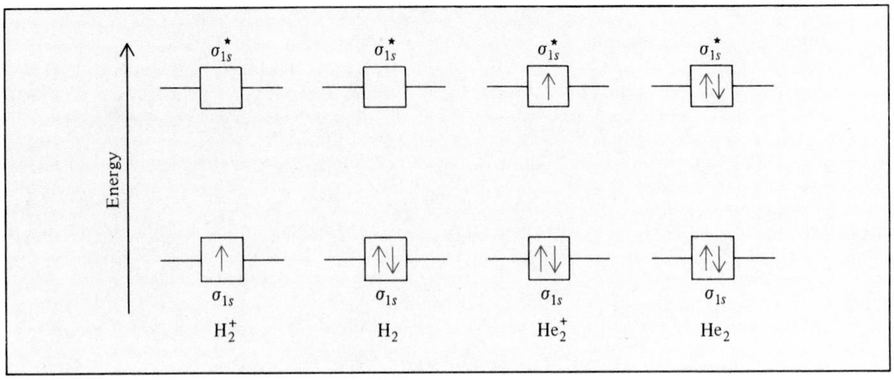

Figure 10.25 *Energy levels of the bonding and antibonding molecular orbitals in H_2^+, H_2, He_2^+, and He_2. In all these species, the molecular orbitals are formed by the interaction of two 1s orbitals.*

four for He_2. Pauli's exclusion principle must be obeyed, so each molecular orbital can accommodate a maximum of two electrons with opposite spins. We are concerned only with the ground-state electron configurations in these cases.

In comparing the stabilities of these species—and, in fact, in general—it is useful to introduce a quantity called ***bond order,*** defined as

$$\text{bond order} = \tfrac{1}{2}\left(\begin{matrix}\text{number of electrons}\\ \text{in bonding MOs}\end{matrix} - \begin{matrix}\text{number of electrons}\\ \text{in antibonding MOs}\end{matrix}\right) \qquad (10.2)$$

The quantitative measure of the strength of a bond is bond dissociation energy, or bond energy (Section 9.10).

The bond order gives us a measure of the strength of the bond. For example, if there are two electrons in the bonding molecular orbital and none in the antibonding molecular orbital, the bond order is one. A bond order of zero (or a negative value) means the bond has no stability, and the molecule cannot exist. It should be understood that bond order can be used only qualitatively for purposes of comparison. For example, two electrons in a bonding sigma molecular orbital or two electrons in a bonding pi molecular orbital would have a bond order of one. Yet, these two bonds must differ in bond strength (and bond length) because of the differences in extent of atomic orbital overlap.

We are ready now to make predictions about the stabilities of H_2^+, H_2, He_2^+, and He_2 (see Figure 10.25). The H_2^+ molecule ion has only one electron, in the σ_{1s} orbital. Since a covalent bond consists of two electrons in a bonding molecular orbital, H_2^+ has only half of one bond, or a bond order of $\frac{1}{2}$. Thus, we predict that the H_2^+ molecule may be a stable species. The electron configuration of H_2^+ is written as $(\sigma_{1s})^1$.

Note that the superscript in $(\sigma_{1s})^1$ indicates that there is one electron in the sigma bonding molecular orbital.

The H_2 molecule has two electrons, both of which are in the σ_{1s} orbital. According to our scheme, two electrons equal one full bond; therefore, the H_2 molecule has a bond order of one, or one full covalent bond. The electron configuration of H_2 is $(\sigma_{1s})^2$.

As for the He_2^+ molecule ion, we find the first two electrons in the σ_{1s} orbital and the third electron in the $\sigma^\star_{1s}$ orbital. Because the antibonding molecular orbital is destabilizing, we expect He_2^+ to be less stable than H_2. Roughly speaking, the instability resulting from the electron in the $\sigma^\star_{1s}$ orbital is canceled by the stability gained by having an electron in the σ_{1s} orbital. Thus, the bond order is $\frac{1}{2}$ and the overall stability of He_2^+ is similar to that of the H_2^+ molecule. The electron configuration of He_2^+ is $(\sigma_{1s})^2(\sigma^\star_{1s})^1$.

In He_2 there would be two electrons in the σ_{1s} orbital and two electrons in the $\sigma^\star_{1s}$ orbital, so the molecule would have no net stability and the bond order is zero. The electron configuration of He_2 would be $(\sigma_{1s})^2(\sigma^\star_{1s})^2$.

To summarize, we can arrange these four species in order of decreasing stability:

$$H_2 > H_2^+, He_2^+ > He_2$$

The hydrogen molecule is, of course, a stable species. Our simple molecular orbital method predicts that H_2^+ and He_2^+ also possess some stability, since both have bond orders of $\frac{1}{2}$. Indeed, their existence has been demonstrated by experiment. It turns out that H_2^+ is somewhat more stable than He_2^+, since there is only one electron in the hydrogen molecule ion and therefore it has less electron-electron repulsion. Furthermore, H_2^+ also has less nuclear repulsion than He_2^+. As for He_2, our prediction that it would have no stability is correct; no one has ever found evidence for the existence of He_2 molecules.

Homonuclear Diatomic Molecules of Second-Period Elements

We are now ready to study the ground-state electron configuration in molecules containing second-period elements. We will consider only the simplest case, that of homonuclear diatomic molecules.

Figure 10.26 shows the molecular orbital energy level diagram for the Li_2 molecule, which involves only *s* orbitals. The situation is more complex when the bonding also involves *p* orbitals. Two *p* orbitals can form either a sigma bond or a pi bond. Since

Figure 10.26 also provides a useful insight regarding antibonding MOs. Note that although the antibonding orbital ($\sigma_{1s}^\star$) has higher energy and is thus less stable than the bonding orbital (σ_{1s}), this antibonding orbital has greater stability than the σ_{2s} bonding orbital. This simply reminds us that the concepts of stability and energy applied here are relative terms rather than absolutes.

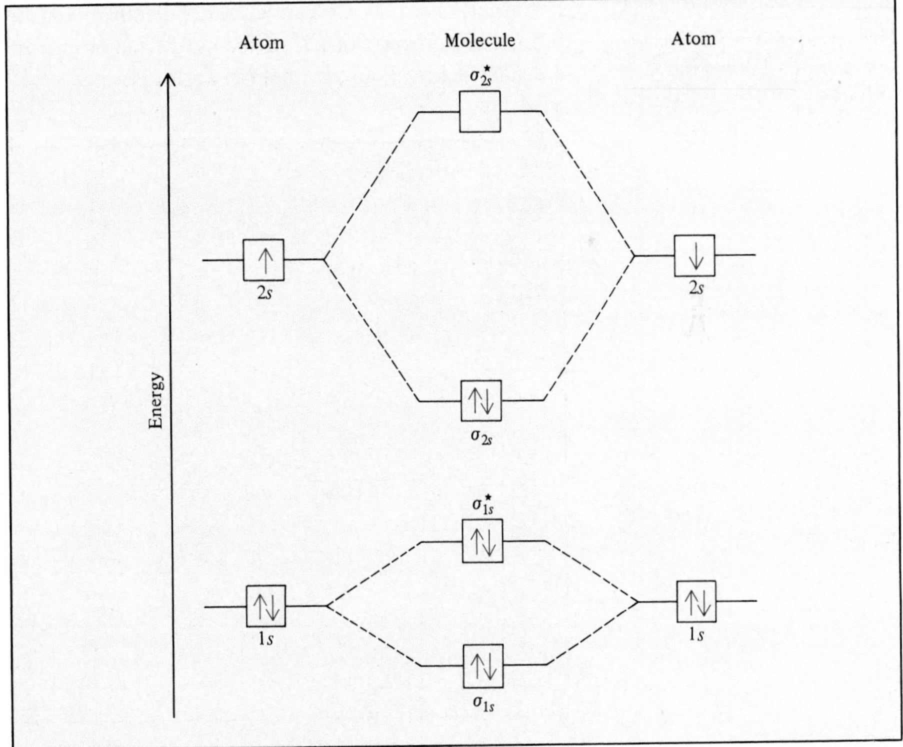

Figure 10.26 *Molecular orbital energy level diagram for the Li_2 molecule. The six electrons in Li_2 (Li is $1s^22s^1$) are in the σ_{1s}, $\sigma_{1s}^\star$, and σ_{2s} orbitals. Since there are two electrons each in σ_{1s} and $\sigma_{1s}^\star$ (just as in He_2), there is no net bonding or antibonding effect. Therefore, the single covalent bond in Li_2 is the result of the two electrons in the bonding molecular orbital σ_{2s}.*

there are three p orbitals for each atom of a second-period element, we know that one sigma and two pi molecular orbitals will result from the constructive interaction. The sigma molecular orbital is formed by the overlap of the $2p_x$ orbitals along the internuclear axis, that is, the x-axis. The $2p_y$ and $2p_z$ orbitals are perpendicular to the x-axis, and they will overlap sideways to give two pi molecular orbitals. The molecular orbitals are called σ_{2p_x}, π_{2p_y}, and π_{2p_z} orbitals, where the subscripts indicate which atomic orbitals take part in forming the molecular orbitals. As Figure 10.24 shows, overlap of the two p orbitals is normally greater in a σ molecular orbital than in a π molecular orbital, so we would expect the former to be lower in energy. However, we must also consider the influence of other filled orbitals. In reality the energies of molecular orbitals actually increase as follows:

$$\sigma_{1s} < \sigma_{1s}^{\star} < \sigma_{2s} < \sigma_{2s}^{\star} < \pi_{2p_y} = \pi_{2p_z} < \sigma_{2p_x} < \pi_{2p_y}^{\star} = \pi_{2p_z}^{\star} < \sigma_{2p_x}^{\star}$$

The reversal of the σ_{2p_x} and the π_{2p_y} and π_{2p_z} energies may be qualitatively understood as follows. The electrons in the σ_{1s}, $\sigma_{1s}^{\star}$, σ_{2s}, and $\sigma_{2s}^{\star}$ orbitals tend to concentrate along the line between the two nuclei. Since the σ_{2p_x} orbital has a geometry similar to that of the σ_s orbitals, its electrons will concentrate in the same region. Consequently, electrons in a σ_{2p_x} orbital will experience greater repulsion as a result of the filled σ_s and $\sigma_s^{\star}$ orbitals than those in the π_{2p_y} and π_{2p_z} orbitals.

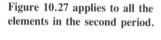

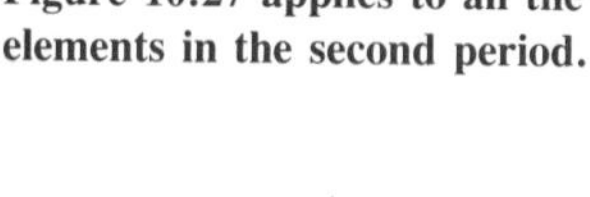
Figure 10.27 applies to all the elements in the second period.

With these concepts and Figure 10.27, which shows the order of increasing molecular orbital energies, we can write the electron configurations and predict the magnetic properties and bond orders of second-period homonuclear diatomic molecules. We will consider a few examples below.

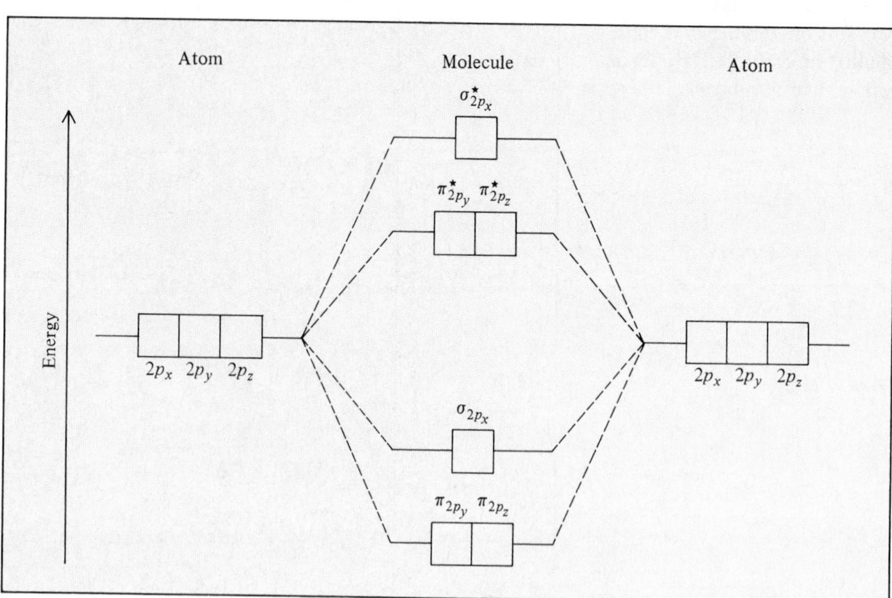

Figure 10.27 *General molecular orbital energy level diagram for the second-period homonuclear diatomic molecules. For simplicity, the σ_{1s} and σ_{2s} orbitals have been omitted. Note that in these molecules the σ_{2p_x} orbital is higher in energy than either the π_{2p_y} or the π_{2p_z} orbitals. This means that electrons in the σ_{2p_x} orbitals are less stable than those in π_{2p_y} and π_{2p_z}. This reversal in relative stability arises as a result of the different interactions between the electrons in the σ_{2p_x} orbital on one hand and π_{2p_y} and π_{2p_z} orbitals on the other hand, with the electrons in the lower σ_s orbitals.*

The Lithium Molecule (Li_2)

The electrons of σ_{1s} and $\sigma^\star_{1s}$ make no net contribution to the bonding.

The electron configuration of Li is $1s^22s^1$, so Li_2 has a total of six electrons. According to Figure 10.26, these electrons are placed (two each) in the σ_{1s}, $\sigma^\star_{1s}$, and σ_{2s} molecular orbitals. Since there are two more electrons in the bonding molecular orbitals than in antibonding orbitals, the bond order is one [see Equation (10.2)]. We conclude that the Li_2 molecule is stable and, since it has no unpaired electron spins, should be diamagnetic. Indeed, diamagnetic Li_2 molecules are known to exist in the vapor phase at high temperatures.

The Carbon Molecule (C_2)

The carbon atom has the electron configuration $1s^22s^22p^2$; thus there are 12 electrons in the C_2 molecule. From the preceding discussion, we see that as we go from Li_2 to C_2 the last four electrons are placed in the π_{2p_y} and π_{2p_z} orbitals. Therefore, C_2 has a bond order of two and is diamagnetic. Again, diamagnetic C_2 molecules have been detected in the vapor state. Note that in the double bond in C_2 both bonds are pi bonds (because of the four electrons in the two pi molecular orbitals). In most other molecules, a double bond is made up of a sigma bond and a pi bond.

The Oxygen Molecule (O_2)

The magnetic properties of O_2 had been known for some time, but the origin of its paramagnetism was not understood.

As we stated earlier, valence bond theory does not account for the magnetic properties of the oxygen molecule. To show the two unpaired electrons on O_2, we need to draw, in addition to the resonance structure on p. 421, another that shows two unpaired electrons:

$$\cdot\ddot{\underset{\cdot\cdot}{O}}-\ddot{\underset{\cdot\cdot}{O}}\cdot$$

This structure is unsatisfactory on at least two counts. First, it implies the presence of a single covalent bond, but experimental evidence strongly suggests that there is a double bond in this molecule. Second, it places seven valence electrons around each oxygen atom, a ''violation'' of the octet rule.

The following example accounts for the paramagnetism of the O_2 molecule.

EXAMPLE 10.6

Write the ground-state electron configuration for O_2 and show that it is paramagnetic.

Answer

The oxygen atom has the electron configuration $1s^22s^22p^4$; thus there are 16 electrons in O_2. Using the order of increasing energies of the molecular orbitals presented on p. 428, we write the ground-state electron configuration of O_2 as

$$(\sigma_{1s})^2(\sigma^\star_{1s})^2(\sigma_{2s})^2(\sigma^\star_{2s})^2(\pi_{2p_y})^2(\pi_{2p_z})^2(\sigma_{2p_x})^2(\pi^\star_{2p_y})^1(\pi^\star_{2p_z})^1$$

According to Hund's rule, the last two electrons enter the $\pi^\star_{2p_y}$ and $\pi^\star_{2p_z}$ orbitals with parallel spins. Therefore, the O_2 molecule has a bond order of two and is paramagnetic with two unpaired spins.

Table 10.6 summarizes the general properties of the stable diatomic molecules of the second period.

Table 10.6 Properties of homonuclear diatomic molecules of the second-period elements*

	Li_2	B_2	C_2	N_2	O_2	F_2
$\sigma^\star_{2p_x}$	☐	☐	☐	☐	☐	☐
$\pi^\star_{2p_y}, \pi^\star_{2p_z}$	☐☐	☐☐	☐☐	☐☐	↑ ↑	↑↓ ↑↓
σ_{2p_x}	☐	☐	☐	↑↓	↑↓	↑↓
π_{2p_y}, π_{2p_z}	☐☐	↑ ↑	↑↓ ↑↓	↑↓ ↑↓	↑↓ ↑↓	↑↓ ↑↓
$\sigma^\star_{2s}$	☐	↑↓	↑↓	↑↓	↑↓	↑↓
σ_{2s}	↑↓	↑↓	↑↓	↑↓	↑↓	↑↓
Bond order	1	1	2	3	2	1
Bond length (pm)	267	159	131	110	121	142
Bond dissociation energy (kJ/mol)	104.6	288.7	627.6	941.4	498.7	150.6
Magnetic properties	Diamagnetic	Paramagnetic	Diamagnetic	Diamagnetic	Paramagnetic	Diamagnetic

*For simplicity the σ_{1s} and $\sigma^\star_{1s}$ orbitals are omitted. These two orbitals hold a total of four electrons.

Example 10.7 shows how MO theory can help predict molecular properties of ions.

EXAMPLE 10.7

The N_2^+ ion can be prepared by bombarding the N_2 molecule with fast-moving electrons. Predict the following properties of N_2^+: (a) electron configuration, (b) bond order, (c) magnetic character, and (d) bond length relative to the bond length of N_2 (is it longer or shorter?).

Answer

(a) Since N_2^+ has one fewer electron than N_2, its electron configuration is (see Table 10.6)

$$(\sigma_{1s})^2(\sigma^\star_{1s})^2(\sigma_{2s})^2(\sigma^\star_{2s})^2(\pi_{2p_y})^2(\pi_{2p_z})^2(\sigma_{2p_x})^1$$

(b) The bond order of N_2^+ is found by using Equation (10.2):

$$\text{bond order} = \tfrac{1}{2}(9 - 4) = 2.5$$

(c) N_2^+ has one unpaired electron, so it is paramagnetic.

(d) Since the electrons in the bonding molecular orbitals are responsible for holding the atoms together, N_2^+ should have a weaker and, therefore, longer bond than N_2. (In fact, the bond length of N_2^+ is 112 pm, compared with 110 pm for N_2.)

Similar problems: 10.59, 10.60.

10.8 Delocalized Molecular Orbitals

We have so far discussed chemical bonding only in terms of electron pairs. However, the properties of a molecule often cannot be explained accurately by a single structure. A case in point is the O_3 molecule, discussed in Section 9.8. There we overcame the dilemma by introducing the concept of resonance. In this section we tackle the problem in another way—by applying the molecular orbital approach. As in Section 9.8, the benzene molecule and the carbonate ion will be used as examples.

The Benzene Molecule

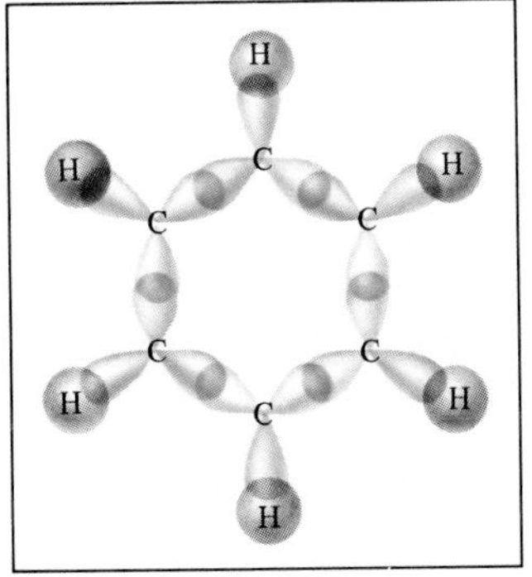

Figure 10.28 *The sigma bond framework in the benzene molecule. Each carbon atom is sp^2-hybridized and forms sigma bonds with two adjacent carbon atoms and another sigma bond with a hydrogen atom.*

Benzene (C_6H_6) is a planar hexagonal molecule with carbon atoms situated at the six corners. All carbon-carbon bonds are equal in length and strength, as are all carbon-hydrogen bonds, and the CCC and HCC angles are all 120°. Therefore, each carbon atom is sp^2-hybridized; it forms three sigma bonds with two adjacent carbon atoms and a hydrogen atom (Figure 10.28). This arrangement leaves an unhybridized $2p_z$ orbital on each carbon atom, perpendicular to the plane of the benzene molecule, or *benzene ring,* as it is often called. So far the description resembles the configuration of ethylene (C_2H_4) as discussed in Section 10.5, except that in this case there are six unhybridized $2p_z$ orbitals in a cyclic arrangement.

Because of their similar shape and orientation, each $2p_z$ orbital overlaps two others, one on each adjacent carbon atom. According to the rules listed on p. 424, the interaction of six $2p_z$ orbitals leads to the formation of six pi molecular orbitals, of which three are bonding and three antibonding. A benzene molecule in the ground state therefore has six electrons in the three pi bonding molecular orbitals, two electrons with paired spins in each orbital (Figure 10.29).

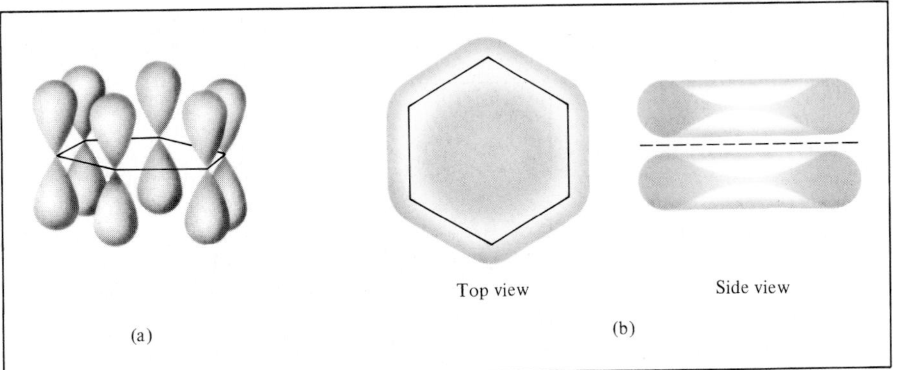

Figure 10.29 *(a) The six $2p_z$ orbitals on the carbon atoms in benzene. (b) The delocalized molecular orbital formed by the overlap of the $2p_z$ orbitals. The delocalized molecular orbital possesses pi symmetry and lies above and below the plane of the benzene ring. Actually, these $2p_z$ orbitals can combine in six different ways to yield three bonding molecular orbitals and three antibonding molecular orbitals. The one shown here is the most stable.*

A characteristic of these pi molecular orbitals is that *they are not confined between two adjacent bonding atoms,* as they are in the case of ethylene, *but actually extend over three or more atoms*. Therefore, electrons residing in any of these orbitals are free to move around the benzene ring. Molecular orbitals of this type are called ***delocalized molecular orbitals.*** For this reason, the structure of benzene is sometimes represented as

in which the circle indicates that the pi bonds between carbon atoms are not confined to individual pairs of atoms; rather, the pi electron densities are evenly distributed throughout the benzene molecule. The carbon and hydrogen atoms are not shown in the simplified diagram.

The resonance structures of benzene are shown on p. 372.

We can now describe each carbon-to-carbon linkage as containing a sigma bond and a "partial" pi bond. The bond order between any two adjacent carbon atoms is therefore somewhere between one and two. Thus molecular orbital theory offers an alternative to the resonance approach, which is based on valence bond theory.

The Carbonate Ion

VSEPR predicts a trigonal planar geometry for the carbonate ion, like that for BF_3.

Cyclic compounds like benzene are not the only ones with delocalized molecular orbitals. Let's look at bonding in the carbonate ion (CO_3^{2-}). The planar structure of the carbonate ion can be understood by assuming the carbon atom to be sp^2-hybridized. The C atom forms sigma bonds with three O atoms. Thus the unhybridized $2p_z$ orbital of the C atom can simultaneously overlap the $2p_z$ orbitals of all three O atoms (Figure 10.30). The result is a delocalized molecular orbital that extends over all four nuclei in such a way that the electron densities (and hence the bond orders) in the carbon-to-oxygen bonds are all the same. Molecular orbital theory therefore provides an acceptable alternative explanation of the properties of the carbonate ion as compared with the resonance structures of the ion shown on p. 372.

We should note that molecules with delocalized molecular orbitals are generally more stable than those containing molecular orbitals extending over only two atoms. For example, the benzene molecule, which contains delocalized molecular orbitals, is

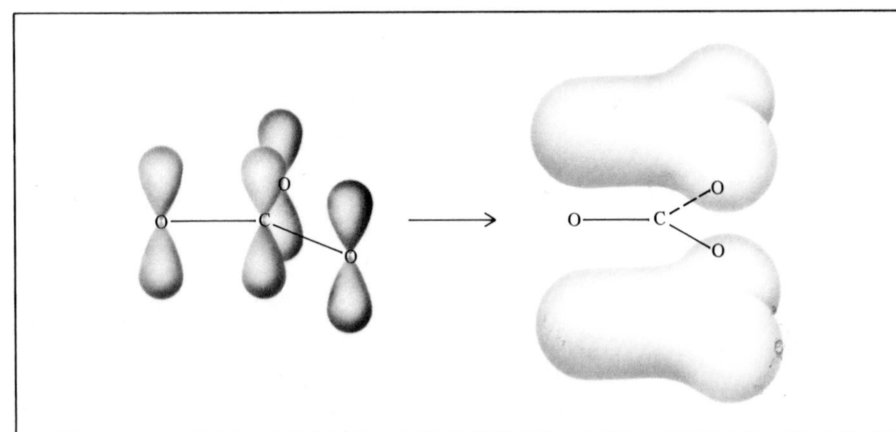

Figure 10.30 *Bonding in the carbonate ion. The carbon atom forms three sigma bonds with the three oxygen atoms. In addition, the $2p_z$ orbitals of the carbon and oxygen atoms overlap to form delocalized molecular orbitals, so that there is also a partial pi bond between the carbon atom and each of the three oxygen atoms.*

chemically less reactive (and hence more stable) than molecules containing "localized" C=C bonds, such as the ethylene molecule.

Summary

1. The VSEPR model for predicting molecular geometry is based on the assumption that valence-shell electron pairs repel one another and tend to stay as far apart as possible.
2. According to the VSEPR model, molecular geometry can be predicted from the number of bonding electron pairs and lone pairs. Lone pairs repel other pairs more strongly than bonding pairs do and thus distort bond angles from those of the ideal geometry.
3. The dipole moment is a measure of the charge separation in molecules containing atoms of different electronegativity. The dipole moment of a molecule is the resultant of whatever bond moments are present in a molecule. Information about molecular geometry can be obtained from dipole moment measurements.
4. In valence bond theory, hybridized atomic orbitals are formed by the combination and rearrangement of (for example, *s* and *p*) orbitals of the same atom. The hybridized orbitals are all of equal energy and electron density, and the number of hybridized orbitals is equal to the number of pure atomic orbitals combined.
5. Valence-shell expansion can be explained by invoking hybridization of *s*, *p*, and *d* orbitals.
6. In *sp* hybridization, the two hybrid orbitals lie in a straight line; in sp^2 hybridization, the three hybrid orbitals are directed toward the corners of a triangle; in sp^3 hybridization, the four hybrid orbitals are directed toward the corners of a tetrahedron; in sp^3d hybridization, the five hybrid orbitals are directed toward the corners of a trigonal bipyramid; in sp^3d^2 hybridization, the six hybrid orbitals are directed toward the corners of an octahedron.
7. In an sp^2-hybridized atom (for example, carbon), the one unhybridized *p* orbital can form a pi bond with another *p* orbital. A carbon-carbon double bond consists of a sigma bond and a pi bond. In an *sp*-hybridized carbon atom, the two unhybridized *p* orbitals can form two pi bonds with two *p* orbitals on another atom (or atoms). A carbon-carbon triple bond consists of one sigma bond and two pi bonds.
8. Molecular orbital theory describes bonding in terms of the combination and rearrangement of atomic orbitals to form orbitals that are associated with the molecule as a whole.
9. Bonding molecular orbitals increase electron density between the nuclei and are lower in energy than individual atomic orbitals. Antibonding molecular orbitals have a region of zero electron density between the nuclei, and an energy level higher than that of the individual atomic orbitals.
10. We write electron configurations for molecular orbitals as we do for atomic orbitals, filling in electrons in the order of increasing energy levels. The number of molecular orbitals always equals the number of atomic orbitals that were combined. In filling the molecular orbitals, the Pauli exclusion principle and Hund's rule are followed.
11. Molecules are stable if the number of electrons in bonding molecular orbitals is greater than that in antibonding molecular orbitals.
12. Delocalized molecular orbitals, in which electrons are free to move around a

whole molecule or group of atoms, are formed by electrons in p orbitals of adjacent atoms. Delocalized molecular orbitals are an alternative to resonance structures in explaining observed molecular properties.

Applied Chemistry Example: Kevlar

The fiber and textile industry has traditionally used many of the natural polymers such as silk, wool, and cotton. Synthetic polymers have been developed that mimic and extend the properties of these natural materials. Since nylon fibers were first synthesized in the 1930s, many synthetic fibers have been reported. Kevlar is one of the newer classes of synthetic fibers. It is heat-resistant and seven times as strong as steel per unit weight. Kevlar is used in belts for automobile tires, ropes, aircraft construction, and lightweight, bulletproof vests. About 45 million pounds of Kevlar were manufactured in the United States in 1988.

Kevlar is prepared as follows:

$$n\ \mathrm{Cl{-}\overset{O}{\overset{\|}{C}}{-}C_6H_4{-}\overset{O}{\overset{\|}{C}}{-}Cl} + n\ \mathrm{H_2N{-}C_6H_4{-}NH_2} \longrightarrow$$

terephthaloyl chloride (I) 1,4-phenylene diamine (II)

$$\left(\mathrm{\overset{O}{\overset{\|}{C}}{-}C_6H_4{-}\overset{O}{\overset{\|}{C}}{-}\underset{H}{\underset{|}{N}}{-}C_6H_4{-}\underset{H}{\underset{|}{N}}}\right)_n + 2n\mathrm{HCl}$$

Kevlar

where n may be in the hundreds or thousands. One feature of Kevlar's structure which is thought to contribute to its unusual strength is the fact that it is quite rigid. The long molecules of Kevlar act as rods and do not easily flex or twist.

1. Describe the hybridization state of the C atom (in the C=O group) in structure I and Kevlar and the N atom in structure II and Kevlar. Write formal charges for the O, C, and N atoms in Kevlar.

   Answer: In I the C atom forms two single bonds and a double bond. Therefore, it is sp^2-hybridized. The same holds for the C atom in Kevlar. In II the N atom forms three single bonds (like N in NH_3) and it is sp^3-hybridized. The same holds for the N atom in Kevlar. From Equation (9.3) we calculate the formal charges on the O, C, and N atoms as follows:

   $$\text{Formal charge on O} = 6 - 4 - \tfrac{1}{2}(4) = 0$$

   $$\text{Formal charge on C} = 4 - 0 - \tfrac{1}{2}(8) = 0$$

   $$\text{Formal charge on N} = 5 - 2 - \tfrac{1}{2}(6) = 0$$

   Thus there are no formal charges on the O, C, and N atoms.

2. Write another resonance structure for the —CO—NH— group (called the amide group) in Kevlar. Determine the state of hybridization and find the formal charges for O, C, and N.

Answer: The other resonance structure is

$$\begin{array}{c} :\ddot{O}:^{-} \\ | \\ -C{=}\overset{+}{N}- \\ \quad\ \ | \\ \quad\ \ H \end{array}$$

In this structure, the C atom has two single bonds and one double bond so it is sp^2-hybridized. The N atom has two single bonds and a double bond so it is also sp^2-hybridized. The formal charges are calculated as follows:

$$\text{Formal charge on O} = 6 - 6 - \tfrac{1}{2}(2) = -1$$

$$\text{Formal charge on C} = 4 - 0 - \tfrac{1}{2}(8) = 0$$

$$\text{Formal charge on N} = 5 - 0 - \tfrac{1}{2}(8) = +1$$

3. In Kevlar the carbon-to-oxygen bond length is 124 pm and the carbon-to-nitrogen bond length is 134 pm. The bond angles around C and N are all approximately 120°. Compare the bond lengths with those listed in Table 9.3 and the bond angles with those predicted based on the hybridizations in 1 and 2.

   Answer: Comparing the bond lengths with those in Table 9.3, we see that the carbon-to-oxygen bond length is close to the C=O bond (121 pm) while the carbon-to-nitrogen bond length is close to the C=N bond (138 pm). Thus we see features that are characteristic of both resonance structures (that is, the C=O bond in one and the C=N bond in the other). The fact that the bond angles are all close to 120° suggests that both C and N are sp^2-hybridized.

4. Describe the bonding in the —CO—NH— group in terms of molecular orbital theory.

   Answer: The discussion in 3 shows that the resonance structure in 2 is an important contribution to the amide group. Since both C and N are sp^2-hybridized, the group is planar, that is, the C, O, N, and H atoms all lie in the same plane (which we define to be the xy plane). We can now imagine that the $2p_z$ orbitals on O, C, and N overlap to form a delocalized molecular orbital as follows:

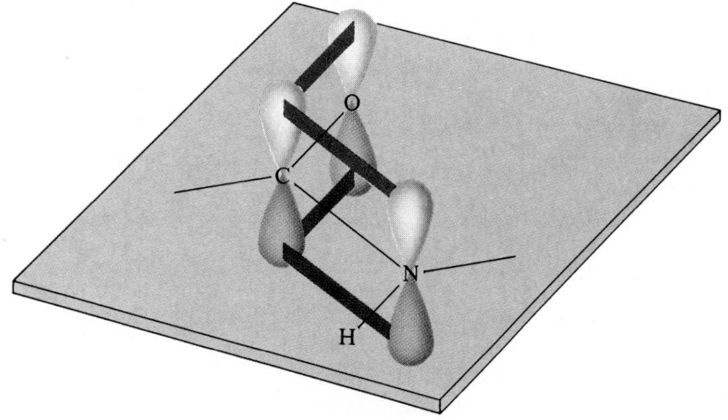

   This scheme shows that both the carbon-to-oxygen and carbon-to-nitrogen bonds have partial pi character.

Key Words

Antibonding molecular orbital, p. 422
Bond order, p. 426
Bonding molecular orbital, p. 422
Delocalized molecular orbital, p. 432
Dipole moment (μ), p. 402
Heteronuclear diatomic molecule, p. 403
Homonuclear diatomic molecule, p. 402
Hybrid orbital, p. 409
Hybridization, p. 409
Molecular orbital, p. 422
Nonpolar molecule, p. 402
Pi bond (π bond), p. 419
Pi molecular orbital, p. 424
Polar molecule, p. 402
Sigma bond (σ bond), p. 419
Sigma molecular orbital, p. 423
Valence shell, p. 392
Valence-shell electron-pair repulsion (VSEPR) model, p. 392
Valence-shell expansion, p. 417

Exercises

Simple Geometric Shapes

REVIEW QUESTIONS

10.1 How is the geometry of a molecule defined? Why is the study of molecular geometry important?

10.2 Sketch the shape of a linear triatomic molecule, a trigonal planar molecule containing four atoms, a tetrahedral molecule, a trigonal bipyramidal molecule, and an octahedral molecule. Give the bond angles in each case.

10.3 How many atoms are directly bonded to the central atom in a tetrahedral molecule, a trigonal bipyramidal molecule, and an octahedral molecule?

VSEPR

REVIEW QUESTIONS

10.4 Discuss the basic features of the VSEPR method.

10.5 Explain why the repulsion decreases in the following order: lone pair–lone pair > lone pair–bonding pair > bonding pair–bonding pair.

10.6 In the trigonal bipyramid arrangement, why does a lone pair occupy an equatorial position rather than an axial position?

10.7 Another possible geometry for CH_4 is square planar, with the four H atoms at the corners of a square and the C atom at the center of the square. Sketch this geometry and compare its stability with that of a tetrahedral CH_4.

PROBLEMS

10.8 Predict the geometries of the following species using the VSEPR method: (a) PCl_3, (b) $CHCl_3$, (c) SiH_4, (d) $TeCl_4$.

10.9 What are the geometries of the following species? (a) $AlCl_3$, (b) $ZnCl_2$, (c) $ZnCl_4^{2-}$, (d) PF_3.

10.10 Predict the geometry of the following molecules using the VSEPR method: (a) $HgBr_2$, (b) N_2O (arrangement of atoms is NNO), (c) SCN^- (arrangement of atoms is SCN).

10.11 What are the geometries of the following ions? (a) NH_4^+, (b) NH_2^-, (c) CO_3^{2-}, (d) ICl_2^-, (e) ICl_4^-, (f) AlH_4^-, (g) $SnCl_5^-$, (h) H_3O^+, (i) BeF_4^{2-}, (j) ClO_2^-.

10.12 Which of the following species are tetrahedral? $SiCl_4$, SeF_4, XeF_4, CI_4, $CdCl_4^{2-}$.

10.13 Describe the geometry around each of the three central atoms in the CH_3COOH molecule.

Dipole Moment and Molecular Geometry

REVIEW QUESTIONS

10.14 Define dipole moment. What are the units and symbol for dipole moment?

10.15 What is the relationship between the dipole moment and bond moment? How is it possible for a molecule to have bond moments yet be nonpolar?

10.16 Explain why an atom cannot have a dipole moment.

10.17 The bonds in beryllium hydride (BeH_2) molecules are polar, yet the dipole moment of the molecule is zero. Explain.

PROBLEMS

10.18 Does the molecule OCS have a higher or lower dipole moment than CS_2?

10.19 Arrange the following molecules in order of increasing dipole moment: H_2O, H_2S, H_2Te, H_2Se. (See Table 10.4.)

10.20 List the following molecules in order of increasing dipole moment: H_2O, CBr_4, H_2S, HF, NH_3, CO_2. (See Table 10.4.)

10.21 The dipole moments of the hydrogen halides decrease from HF to HI (see Table 10.4). Explain this trend.

10.22 Sketch the bond moments and resultant dipole moments in the following molecules: H_2O, PCl_3, XeF_4, PCl_5, SF_6.

10.23 Which of the following molecules has a higher dipole moment?

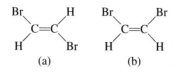

Br H Br Br
C=C C=C
H Br H H
(a) (b)

10.24 Arrange the following compounds in order of increasing dipole moment:

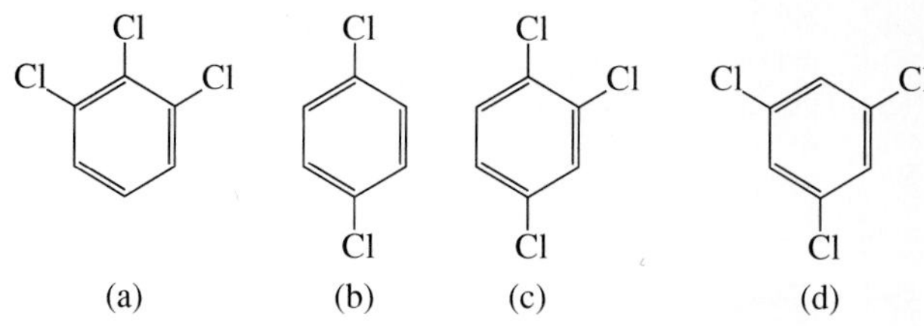

(a) (b) (c) (d)

Hybridization

REVIEW QUESTIONS

10.25 What is valence bond theory? How does it differ from the Lewis concept of chemical bonding?

10.26 Use valence bond theory to explain the bonding in Cl_2 and HCl. Show the overlap of atomic orbitals in the bond formation.

10.27 Draw a potential energy curve for the bond formation in HCl.

10.28 What is the hybridization of atomic orbitals? Why is it impossible for an isolated atom to exist in the hybridized state?

10.29 How does a hybrid orbital differ from a pure atomic orbital?

10.30 Can two $2p$ orbitals of an atom hybridize to give two hybridized orbitals?

10.31 What is the angle between the following two hybrid orbitals on the same atom? (a) sp and sp hybrid orbitals, (b) sp^2 and sp^2 hybrid orbitals, (c) sp^3 and sp^3 hybrid orbitals.

10.32 How would you distinguish between a sigma bond and a pi bond?

10.33 Which of the following pairs of atomic orbitals of adjacent nuclei can overlap to form a sigma bond? Which overlap to form a pi bond? Which cannot overlap (no bond)? Consider the x-axis to be the *internuclear axis,* that is, the line joining the nuclei of the two atoms. (a) $1s$ and $1s$; (b) $1s$ and $2p_x$; (c) $2p_x$ and $2p_y$; (d) $3p_y$ and $3p_y$; (e) $2p_x$ and $2p_x$; (f) $1s$ and $2s$.

PROBLEMS

10.34 Describe the bonding scheme of the AsH_3 molecule in terms of hybridization.

10.35 What is the hybridization of Si in SiH_4 and in $H_3Si—SiH_3$?

10.36 Describe the change in hybridization (if any) of the Al atom in the following reaction:

$$AlCl_3 + Cl^- \longrightarrow AlCl_4^-$$

10.37 Consider the reaction

$$BF_3 + NH_3 \longrightarrow F_3B—NH_3$$

Describe the changes in hybridization (if any) of the B and N atoms as a result of this reaction.

10.38 What hybrid orbitals are used by nitrogen atoms in the following species? (a) NH_3, (b) $H_2N—NH_2$, (c) NO_3^-.

10.39 Draw a diagram representing the formation of a double bond between two carbon atoms in ethylene (C_2H_4).

10.40 What are the hybrid orbitals of the carbon atoms in the following molecules?

(a) $H_3C—CH_3$

(b) $H_3C—CH{=}CH_2$

(c) $CH_3—CH_2—OH$

(d) $CH_3CH{=}O$

(e) CH_3COOH

10.41 What hybrid orbitals are used by carbon atoms in the following species? (a) CO, (b) CO_2, (c) CN^-.

10.42 What is the state of hybridization of the central N atom in the azide ion, N_3^-? (Arrangement of atoms: NNN.)

10.43 The allene molecule $H_2C{=}C{=}CH_2$ is linear (that is, the three C atoms lie on a straight line). What are the hybridization states of the carbon atoms? Draw diagrams to show the formation of sigma bonds and pi bonds in allene.

10.44 Describe the hybridization of phosphorus in PF_5.

10.45 What are the hybridization states of the C, N, and O atoms in this molecule?

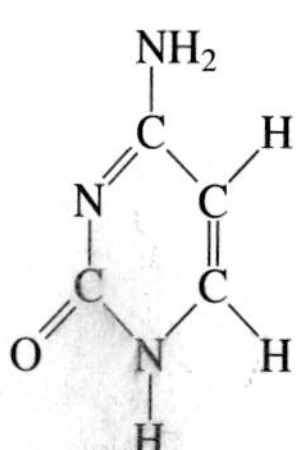

10.46 What is the total number of sigma bonds and pi bonds in each of the following molecules?

H H Cl
Cl—C—Cl C=C
H H H
(a) (b)

```
         H
         |
H3C—C=C—C≡C—H
            |
            H
```

(c)

Molecular Orbital Theory

REVIEW QUESTIONS

10.47 What is molecular orbital theory? How does it differ from valence bond theory?

10.48 Define the following terms: bonding molecular orbital, antibonding molecular orbital, pi molecular orbital, sigma molecular orbital.

10.49 Sketch the shapes of the following molecular orbitals: σ_{1s} and $\sigma^{\star}_{1s}$ and π_{2p} and $\pi^{\star}_{2p}$. How do their energies compare?

10.50 Explain the significance of bond order. Can bond order be used for quantitative comparisons of the strengths of chemical bonds?

PROBLEMS

10.51 Explain in molecular orbital terms the changes in H—H internuclear distance that occur as the molecule H_2 is ionized first to H_2^+ and then to H_2^{2+}.

10.52 Formation of the H_2 molecule from two H atoms is an energetically favorable process. Yet statistically there is less than a 100 percent chance that any two H atoms will undergo the reaction. Apart from energy consideration, how would you account for this observation simply from the electron spins in the two H atoms?

10.53 Draw a molecular orbital energy level diagram for each of the following species: He_2, HHe, He_2^+. Compare their relative stabilities in terms of bond orders. (Treat HHe as a diatomic molecule with three electrons.)

10.54 Arrange the following species in order of increasing stability: Li_2, Li_2^+, Li_2^-. Justify your choice with a molecular orbital energy level diagram.

10.55 Use molecular orbital theory to explain why the Be_2 molecule does not exist.

10.56 Which of these species has a longer bond, B_2 or B_2^+? Explain.

10.57 Acetylene (C_2H_2) has a tendency to lose two protons (H^+) and form the carbide ion (C_2^{2-}), which is present in a number of ionic compounds, such as CaC_2 and MgC_2. Describe the bonding scheme in the C_2^{2-} ion in terms of molecular orbital theory. Compare the bond order in C_2^{2-} with that in C_2.

10.58 Compare the Lewis and molecular orbital theory treatments of the oxygen molecule.

10.59 Explain why the bond order of N_2 is greater than that of N_2^+, but the bond order of O_2 is less than that of O_2^+.

10.60 Compare the relative stability of the following species and indicate their magnetic properties (that is, diamagnetic or paramagnetic): O_2, O_2^+, O_2^- (superoxide ion), O_2^{2-} (peroxide ion).

10.61 Use molecular orbital theory to compare the relative stabilities of F_2 and F_2^+.

10.62 A single bond is almost always a sigma bond, and a double bond is almost always made up of a sigma bond and a pi bond. There are very few exceptions to this rule. Show that the B_2 and C_2 molecules are examples of the exceptions.

Delocalized Molecular Orbitals

REVIEW QUESTIONS

10.63 How does a delocalized molecular orbital differ from a molecular orbital such as that found in H_2 or C_2H_4?

10.64 What do you think are the minimum conditions (for example, number of atoms and types of orbitals) for forming a delocalized molecular orbital?

10.65 In Chapter 9 we saw that the resonance concept is useful for dealing with species such as the benzene molecule and the carbonate ion. How does molecular orbital theory deal with these species?

PROBLEMS

10.66 Both ethylene (C_2H_4) and benzene (C_6H_6) contain the C=C bond. The reactivity of ethylene is greater than that of benzene. For example, ethylene readily reacts with molecular bromine, whereas benzene is normally quite inert toward molecular bromine and many other compounds. Explain this difference in reactivity.

10.67 Explain why the symbol on the left is a better representation for benzene molecules than that on the right:

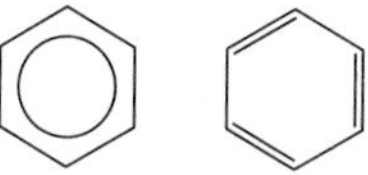

10.68 Determine which of these molecules has a more delocalized orbital and justify your choice:

biphenyl naphthalene

(*Hint:* Both molecules contain two benzene rings. In naphthalene, the two rings are fused together. In biphenyl the two rings are joined by a single bond, around which the two rings can rotate.)

10.69 Nitryl fluoride (FNO_2) is very reactive chemically. The fluorine and oxygen atoms are bonded to the nitrogen atom. (a) Write a Lewis structure for FNO_2. (b) Indicate the hybridization of the nitrogen atom. (c) Describe the bonding in terms of the molecular orbital theory. Where would you expect delocalized molecular orbitals?

10.70 Describe the bonding in the nitrate ion (NO_3^-) in terms of delocalized molecular orbitals.

10.71 What is the state of hybridization of the central O atom in O_3? Describe the bonding in O_3 in terms of delocalized molecular orbitals.

Miscellaneous

10.72 Which of the following species is *not* likely to have a tetrahedral shape? (a) $SiBr_4$, (b) NF_4^+, (c) SF_4, (d) $BeCl_4^{2-}$, (e) BF_4^-, (f) $AlCl_4^-$.

10.73 Draw the Lewis structure of mercury(II) bromide. Is this molecule linear or bent? How would you establish its geometry?

10.74 How many pi bonds and sigma bonds are there in the tetracyanoethylene molecule?

```
N≡C         C≡N
    \      /
     C=C
    /      \
N≡C         C≡N
```

10.75 Although both carbon and silicon are in Group 4A, very few Si=Si bonds are known. Account for the instability of silicon-to-silicon double bonds in general. (*Hint:* The Si atom is larger than the C atom. What effect would the larger size have on pi bond formation?)

10.76 Predict the geometry of sulfur dichloride (SCl_2) and the hybridization of the sulfur atom.

10.77 Describe the bonding in the formate ion ($HCOO^-$) in terms of (a) resonance and (b) molecular orbital theory.

10.78 Antimony pentafluoride, SbF_5, reacts with XeF_4 and XeF_6 to form ionic compounds $XeF_3^+SbF_6^-$ and $XeF_5^+SbF_6^-$. Describe the geometries of the cations and anion in these two compounds.

10.79 Predict the bond angles for the following molecules: (a) $BeCl_2$, (b) BCl_3, (c) CCl_4, (d) CH_3Cl, (e) Hg_2Cl_2 (arrangement of atoms: ClHgHgCl), (f) $SnCl_2$, (g) SCl_2, (h) H_2O_2, (i) SnH_4.

10.80 Write Lewis structures and give any other information requested for the following molecules: (a) BF_3. Shape: planar or nonplanar? (b) ClO_3^-. Shape: planar or nonplanar? (c) H_2O. Show the direction of the resultant dipole moment. (d) OF_2. Polar or nonpolar molecule? (e) NO_2. Estimate the ONO bond angle.

10.81 Briefly compare the VSEPR and hybridization approaches to the study of molecular geometry.

10.82 Describe the state of hybridization of arsenic in arsenic pentafluoride (AsF_5).

10.83 Give Lewis structures and any other information requested for the following: (a) SO_3. Polar or nonpolar molecule? (b) PF_3. Polar or nonpolar molecule? (c) F_3SiH. Show the direction of the resultant dipole moment. (d) SiH_3^-. Planar or pyramidal shape? (e) Br_2CH_2. Polar or nonpolar molecule?

10.84 Which of the following molecules are linear? ICl_2^-, IF_2^+, OF_2, SnI_2, $CdBr_2$

10.85 Draw the Lewis structure for the $BeCl_4^{2-}$ ion. Predict its geometry and describe the state of hybridization of the Be atom.

10.86 The molecule with the formula N_2F_2 can exist in one of the following two forms:

```
        F      F      F
       /        \    /
   N=N           N=N
  /
 F
```

(a) What is the hybridization of N in the molecule? (b) Which of the two forms has a dipole moment?

10.87 Both N and P are members of Group 5A. Elemental nitrogen exists as N_2, while P_2 is unknown. Explain. (*Hint:* Compare the size of N and P.)

10.88 Cyclopropane (C_3H_6) has the shape of a triangle with a C atom bonded to two H atoms and two other C atoms at each corner. Cubane (C_8H_8) has the shape of a cube with a C atom bonded to one H atom and three other C atoms at each corner. (a) Draw structures of these molecules. (b) Compare the CCC angles in these molecules with those predicted for an sp^3-hybridized C atom. (c) Would you expect these molecules to be easy to make?

10.89 The compound 1,2-dichloroethane ($C_2H_4Cl_2$) is nonpolar, while *cis*-dichloroethylene ($C_2H_2Cl_2$) has a dipole moment:

```
     Cl Cl            Cl        Cl
     |  |               \      /
  H—C—C—H                C=C
     |  |               /      \
     H  H              H        H
1,2-dichloroethane   cis-dichloroethylene
```

The reason for the difference is that groups connected by a single bond can rotate with respect to each other, but no rotation occurs when a double bond connects the groups. On the basis of bonding consideration, explain why rotation occurs in 1,2-dichloroethane but not in *cis*-dichloroethylene.

10.90 (a) From the following data:

$$F_2(g) \longrightarrow 2F(g) \qquad \Delta H^\circ_{rxn} = 150.6\ \text{kJ}$$

$$F^-(g) \longrightarrow F(g) + e^- \qquad \Delta H^\circ_{rxn} = 333\ \text{kJ}$$

$$F_2^-(g) \longrightarrow F_2(g) + e^- \qquad \Delta H^\circ_{rxn} = 290\ \text{kJ}$$

calculate the bond energy of the F_2^- ion. (b) Explain the difference between the bond energies of F_2 and F_2^- using the MO theory.

10.91 Referring to Table 10.6, explain why (a) B_2 has a shorter bond length than Li_2 and (b) B_2 has a larger bond dissociation energy than Li_2.

10.92 Which species in each of the following pairs would you expect to have a greater bond dissociation energy? (a) Be_2 and Be_2^+, (b) B_2 and B_2^-, (c) C_2 and C_2^+.

10.93 Does the following molecule have a dipole moment?

$$\text{Cl(H)C}{=}\text{C}{=}\text{C(H)Cl}$$

(*Hint:* See the answer to Problem 10.43.)

10.94 The following molecules (AX_4Y_2) all have octahedral geometry. Group the molecules that are equivalent to each other.

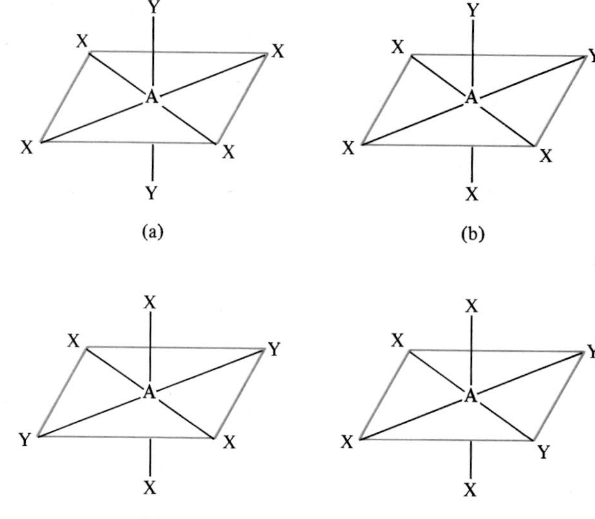

10.95 The bond angle of SO_2 is very close to 120° (see p. 397), even though there is a lone pair on S. Explain.

Opposite page: Because ice is less dense than water, an iceberg floats.

C H A P T E R 1 1

INTERMOLECULAR FORCES AND LIQUIDS AND SOLIDS

Although we live immersed in a mixture of gases that make up Earth's atmosphere, we are more familiar with the behavior of liquids and solids because they are more visible. Every day we use water and other liquids for drinking, bathing, cleaning, and cooking, and we handle, sit upon, and wear solids.

Molecular motion is more restricted in liquids than in gases; and in solids, the atoms and molecules are packed even more tightly together. In fact, in a solid they are held in well-defined positions and are capable of little free motion relative to one another. In this chapter we will examine the structure of liquids and solids and discuss some of their common and important properties.

11.1 The Kinetic Molecular Theory of Liquids and Solids

In Chapter 5 we saw how the kinetic molecular theory could be used to explain the behavior of gases. That explanation was based on the understanding that a gaseous system is a collection of molecules in constant, random motion. In gases, the distances between molecules are so great (compared with their diameters) that at ordinary temperatures and pressures (say, 25°C and 1 atm), there is no appreciable interaction between the molecules. This rather simple description explains several characteristic properties of gases. Because there is a great deal of empty space in a gas, that is, space not occupied by molecules, gases can be readily compressed. The lack of strong forces between molecules allows a gas to expand to the volume of its container. The large amount of empty space also explains why gases have very low densities under normal conditions.

Liquids and solids are quite a different story. The principal difference between the condensed state (liquids or solids) and the gaseous state lies in the distance between molecules. In a liquid the molecules are held close together, so that there is very little empty space. Thus liquids are much more difficult to compress than gases and much denser under normal conditions. Molecules in the liquid are held by one or more types of attractive forces, which will be discussed in the next section. A liquid also has a definite volume, since molecules in a liquid do not break away from the attractive forces. The molecules can, however, move past one another freely, and so a liquid can flow, can be poured, and assumes the shape of its container.

In a solid, molecules are held rigidly in position with virtually no freedom of motion. Many solids are characterized by long-range order; that is, the molecules are arranged in regular configurations in three dimensions. The amount of empty space in a solid is even less than that in a liquid. Thus solids are almost incompressible and possess definite shape and volume. With very few exceptions (water being the most important), the density of the solid is higher than that of the liquid for a given substance. Table 11.1 summarizes some of the characteristic properties of the three states of matter.

11.2 Intermolecular Forces

Attractive forces between molecules, called ***intermolecular forces,*** are responsible for the nonideal behavior of gases that was described in Chapter 5. They are also responsible for the existence of the condensed states of matter—liquids and solids. As the temperature of a gas is lowered, the average kinetic energy of its molecules decreases.

Table 11.1 Characteristic properties of gases, liquids, and solids

State of matter	Volume/shape	Density	Compressibility	Motion of molecules
Gas	Assumes the volume and shape of its container	Low	Very compressible	Very free motion
Liquid	Has a definite volume but assumes the shape of its container	High	Only slightly compressible	Slide past one another freely
Solid	Has a definite volume and shape	High	Virtually incompressible	Vibrate about fixed positions

Eventually, at a sufficiently low temperature, the molecules no longer have enough energy to break away from one another's attraction. At this point, the molecules aggregate to form small drops of liquid. This phenomenon of going from the gaseous to the liquid state is known as *condensation.*

In contrast to intermolecular forces are ***intramolecular forces,*** or *forces that hold atoms together in a molecule.* (Chemical bonding, discussed in Chapters 9 and 10, involves intramolecular forces.) Intramolecular forces are responsible for the stability of individual molecules, whereas intermolecular forces are primarily responsible for the bulk properties of matter (for example, melting point and boiling point).

Generally, intermolecular forces are much weaker than intramolecular forces. Thus it usually requires much less energy to evaporate a liquid than to break the bonds in the molecules of the liquid. For example, it takes about 41 kJ of energy to vaporize 1 mole of water at its boiling point; it takes about 930 kJ of energy to break the two O—H bonds in 1 mole of water molecules. The boiling points of substances often reflect the strength of the intermolecular forces operating among the molecules. At the boiling point, enough energy must be supplied to overcome the attraction among molecules so that they can enter the vapor phase. If it takes more energy to separate molecules of substance A than of substance B because A molecules are held together by stronger intermolecular forces than in substance B, then the boiling point of A is higher than that of B. The same principle applies also to the melting points of the substances. In general the melting points of substances increase with the strength of the intermolecular forces.

To understand the properties of condensed matter, we must understand the different types of intermolecular forces. Some of them can be explained by Coulomb's law. Quantum mechanics provides an explanation for others. Depending on the physical state of a substance (that is, gas, liquid, or solid) and the nature of the molecules, more than one type of interaction may play a role in the total attraction between molecules, as we will see below.

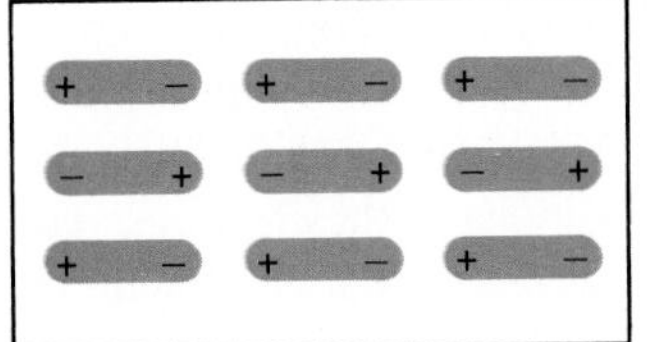

Figure 11.1 *Molecules that have a permanent dipole moment tend to align in the solid state for maximum attractive interaction.*

Dipole-Dipole Forces

Dipole-dipole forces are *forces that act between polar molecules,* that is, between molecules that possess dipole moments (see Section 10.2). Their origin is electrostatic, and they can be understood in terms of Coulomb's law. The larger the dipole moments, the greater the force. Figure 11.1 shows the orientation of polar molecules in a solid. In liquids the molecules are not held as rigidly as in a solid, but they tend to align themselves in such a way that, on the average, the attractive interaction is at a maximum.

Ion-Dipole Forces

Coulomb's law also explains ***ion-dipole forces,*** which *occur between an ion (either a cation or an anion) and a dipole* (Figure 11.2). The strength of this interaction depends on the charge and size of the ion and on the magnitude of the dipole. The charges on cations are generally more concentrated, since cations are usually smaller than anions. Therefore, for equal charges, a cation interacts more strongly with dipoles than does an anion.

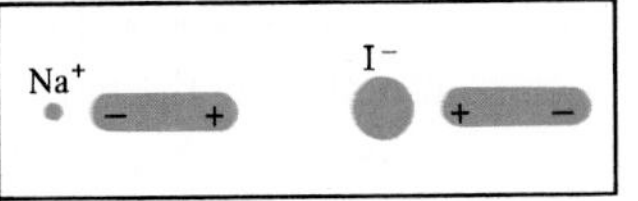

Figure 11.2 *Two types of ion-dipole interaction.*

Hydration, discussed in Section 6.6, is one example of ion-dipole interaction. In an aqueous NaCl solution, the Na^+ and Cl^- ions are surrounded by water molecules, which have a large dipole moment (1.87 D). In this way the water molecules act as an electrical insulator that keeps the ions apart. This process explains what happens when an ionic compound dissolves in water. Carbon tetrachloride (CCl_4), on the other hand, is a nonpolar molecule, and therefore it lacks the ability to participate in ion-dipole interaction. What we find in practice is that carbon tetrachloride is a poor solvent for ionic compounds, as are most nonpolar liquids.

Ionic compounds generally have very low solubilities in nonpolar solvents.

Dispersion Forces

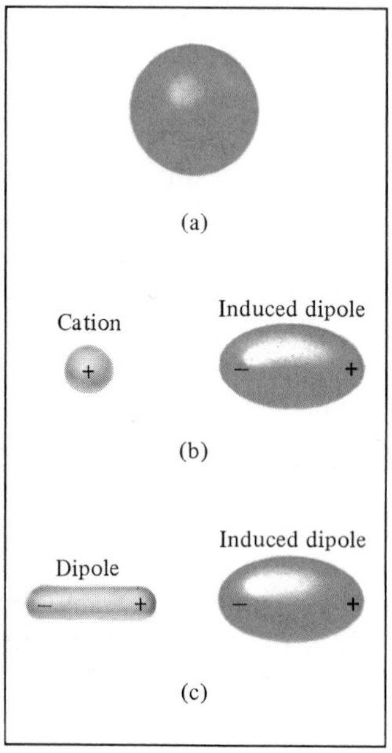

So far only ionic species and polar molecules have been discussed. You may be wondering what kind of attractive interaction exists between nonpolar molecules. To answer this question, consider the arrangement shown in Figure 11.3. If we place an ion or a polar molecule near a neutral atom (or a nonpolar molecule), the electron distribution of the atom (or molecule) is distorted by the force exerted by the ion or the polar molecule. The resulting dipole in the atom (or molecule) is said to be an ***induced dipole*** because *the separation of positive and negative charges in the neutral atom (or a nonpolar molecule) is due to the proximity of an ion or a polar molecule*. The attractive interaction between an ion and the induced dipole is called *ion-induced dipole interaction,* and the attractive interaction between a polar molecule and the induced dipole is called *dipole-induced dipole interaction.*

(a)

Cation

Induced dipole

(b)

Dipole

Induced dipole

(c)

Figure 11.3 *(a) Spherical charge distribution in a helium atom. Distortion caused by (b) the approach of a cation and (c) the approach of a dipole.*

The ease with which a dipole moment can be induced depends not only on the charge on the ion or the strength of the dipole but also on the polarizability of the neutral atom or molecule. ***Polarizability*** is *the ease with which the electron distribution in the neutral atom (or molecule) can be distorted.* Generally, the larger the number of electrons and the more diffuse the electron cloud in the atom or molecule, the greater its polarizability. By *diffuse cloud* we mean an electron cloud that is spread over an appreciable volume, so that the electrons are not held tightly by the nucleus.

Polarizability gives us the clue to why gases containing neutral atoms or nonpolar molecules (for example, He and N_2) are able to condense. In a helium atom the electrons are moving at some distance from the nucleus. At any instant it is likely that the atom has a dipole moment created by the specific positions of the electrons. This dipole moment is called *instantaneous dipole* because it lasts for just a tiny fraction of a second. In the next instant the electrons are in different locations and the atom has a new instantaneous dipole and so on. Averaged over time (that is, the time it takes to make a dipole moment measurement), however, the atom has no dipole moment because the instantaneous dipoles all cancel one another. In a collection of He atoms an instantaneous dipole of one He atom can induce a dipole in each of its nearest neighbors through space (Figure 11.4). At the next moment, a different instantaneous dipole can create temporary dipoles in the surrounding He atoms. The important point is that this kind of interaction results in an attraction between the He atoms. At very low temperatures (and reduced atomic speeds), this attraction is enough to hold the atoms together, causing the helium gas to condense. The attraction between nonpolar molecules can be similarly accounted for.

A quantum mechanical interpretation of temporary dipoles was provided by Fritz London† in 1930. London showed that the magnitude of this attractive interaction is directly proportional to the polarizability of the atom or molecule. As we might expect,

†Fritz London (1900–1954). German physicist. London was a theoretical physicist whose major work was on superconductivity in liquid helium.

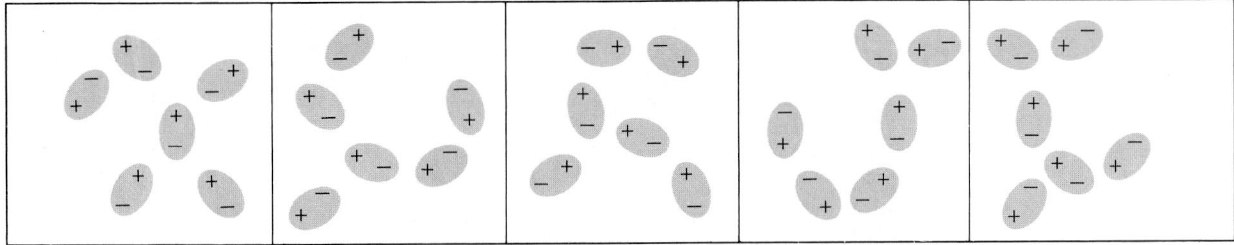

Figure 11.4 *Induced dipoles interacting with each other. Such patterns exist only momentarily; new arrangements are formed in the next instant. This type of interaction is responsible for the condensation of nonpolar gases.*

Strictly speaking, the forces between two nonbonded atoms should be called interatomic forces. However, for simplicity we use the term "intermolecular forces" for both atoms and molecules.

the attractive forces that arise as a result of temporary dipoles induced in the atoms or molecules, called ***dispersion forces,*** may be quite weak. This is certainly true of helium, which has a boiling point of only 4.2 K, or −269°C. (Note that helium has only two electrons that are tightly held in the 1*s* orbital. Therefore, the helium atom has a small polarizability.)

Dispersion forces usually increase with molar mass for the following reason. Molecules with larger molar mass tend to have more electrons, and dispersion forces increase in strength with the number of electrons. Furthermore, larger molar mass often means a bigger atom whose electron distributions are more easily disturbed because the outer electrons are less tightly held by the nuclei. Table 11.2 compares the melting points of some substances that consist of nonpolar molecules. As expected, the melting point increases as the number of electrons in the molecule increases. Since these are all nonpolar molecules, the only attractive intermolecular forces present are the dispersion forces.

Table 11.2 Melting points of similar nonpolar compounds

Compound	Melting point (°C)
CH_4	−182.5
C_2H_6	−183.3
CF_4	−150.0
CCl_4	−23.0
CBr_4	90.0
CI_4	171.0

In many cases dispersion forces can be comparable to or even greater than the dipole-dipole forces between polar molecules. For a dramatic illustration, let us compare the melting points of CH_3F (−141.8°C) and CCl_4 (−23°C). Although CH_3F has a dipole moment of 1.8 D, it melts at a much lower temperature than CCl_4, a nonpolar molecule. CCl_4 melts at a higher temperature simply because it contains more electrons. As a result, the dispersion forces between CCl_4 molecules are stronger than the dispersion forces plus the dipole-dipole forces between CH_3F molecules. (Keep in mind that dispersion forces exist among species of all type, whether they are neutral or bear a net charge, and whether they are polar or nonpolar.)

The following example shows that if we know the kind of species present, we can readily determine the types of intermolecular forces that exist between the species.

EXAMPLE 11.1

What type(s) of intermolecular forces exist between the following pairs: (a) HBr and H_2S, (b) Cl_2 and CBr_4, (c) I_2 and NO_3^-, (d) NH_3 and C_6H_6?

Answer

(a) Both HBr and H_2S are polar molecules, so the forces between them are dipole-dipole forces. There are also dispersion forces between the molecules.

(b) Both Cl_2 and CBr_4 are nonpolar, so there are only dispersion forces between these molecules.
(c) I_2 is nonpolar, so the forces between it and the ion NO_3^- are ion-induced dipole forces and dispersion forces.
(d) NH_3 is polar, and C_6H_6 is nonpolar. The forces are dipole-induced dipole forces and dispersion forces.

Similar problem: 11.13.

van der Waals Forces and van der Waals Radii

The *dipole-dipole, dipole-induced dipole, and dispersion forces* make up what chemists have commonly referred to as ***van der Waals forces,*** after the Dutch physicist Johannes van der Waals (see Section 5.9). These forces are all attractive in nature and play an important role in determining the physical properties of substances, such as melting points, boiling points, and solubilities.

The distance between molecules (or monatomic species) in a solid or liquid is determined by a balance between the van der Waals forces of attraction and the forces of repulsion between electrons and between nuclei. The ***van der Waals radius*** is *one-half the distance between two equivalent nonbonded atoms in their most stable arrangement,* that is, when the net attractive forces are at a maximum.

The space-filling model discussed in Section 9.2 is based on the van der Waals radii.

Table 11.3 Van der Waals radii of some common atoms

Atom	Van der Waals radius (pm)
H	120
C	165
N	150
O	140
F	135
P	190
S	185
Cl	180
Br	195
I	215

The main difference between the atomic radius and ionic radius, on the one hand, and the van der Waals radius, on the other hand, is that the latter applies only to atoms that are not chemically bonded to each other. For instance, two He atoms will not combine to form a stable He_2 molecule, yet He atoms do attract one another through dispersion forces. In liquid helium, we can imagine that the atoms are drawn to one another until the net attractive forces are at their maximum. If any two neighboring atoms are pushed together more closely, repulsive forces between the electrons and between the nuclei cause the system to lose stability. Under the condition of maximum attraction, half the distance between any two adjacent He atoms in liquid helium is the van der Waals radius of He atom.

As another example, consider iodine molecules in the solid state (Figure 11.5). The I_2 molecules are packed in such a way that the total potential energy of the system is at a minimum, which corresponds to maximum net attraction. In this case, half the distance between two adjacent iodine atoms in two *different* I_2 molecules is the van der Waals radius of an iodine atom. Table 11.3 shows the van der Waals radii of selected atoms arranged in increasing atomic number.

The Hydrogen Bond

The ***hydrogen bond*** is *a special type of dipole-dipole interaction between the hydrogen atom in a polar bond, such as O—H or N—H, and electronegative atom O, N, or F.* This interaction is written

Note that these atoms all possess at least one lone pair that can interact with the hydrogen atom in hydrogen bonding.

A—H---B or A—H---A

A and B represent O, N, or F; A—H is one molecule or part of a molecule and B is a part of another molecule; and the dashed line represents the hydrogen bond. The three atoms usually lie along a straight line, but the angle AHB (or AHA) can deviate as much as 30° from linearity.

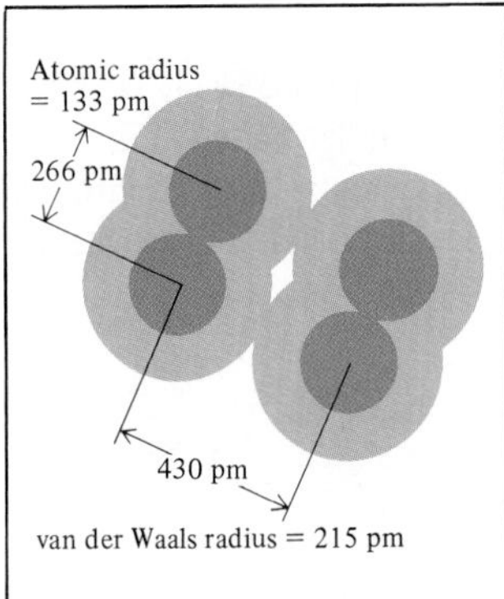

Figure 11.5 *The relationship between van der Waals radius and atomic radius for I_2. In solid I_2 the molecules pack together so that the shortest distance between iodine nuclei in adjacent I_2 molecules is 430 pm. The van der Waals radius of iodine is defined as half this distance, or 215 pm. On the other hand, the atomic radius of iodine is half the distance between two iodine nuclei in the same molecule, which is half of 266 pm, or 133 pm.*

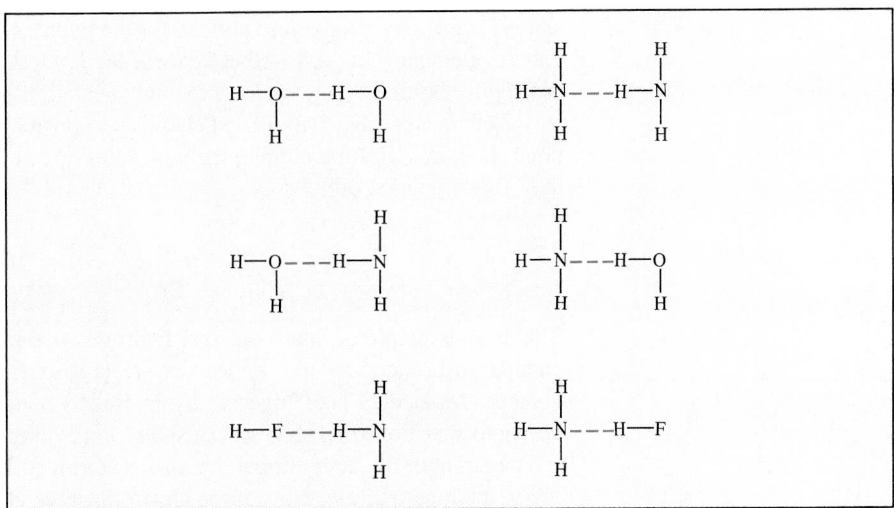

Figure 11.6 *Hydrogen bonding in water, ammonia, and hydrogen fluoride. Solid lines represent covalent bonds, and dashed lines represent hydrogen bonds.*

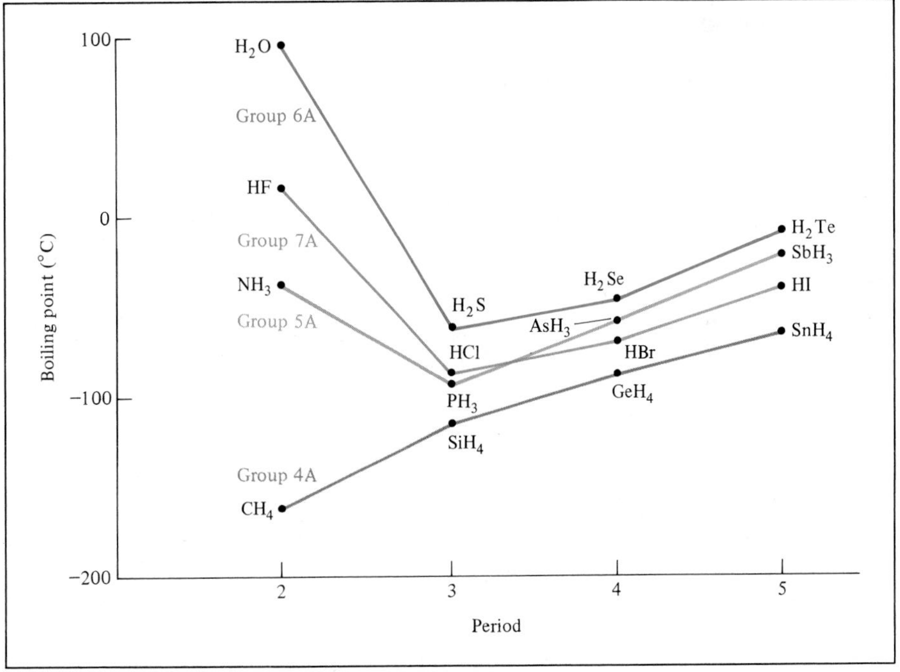

Figure 11.7 *Boiling points of the hydrogen compounds of Groups 4A, 5A, 6A, and 7A elements. Although normally we expect the boiling point to increase as we move down a group, we see that three compounds (NH_3, H_2O, and HF) behave differently. The anomaly can be explained in terms of intermolecular hydrogen bonding.*

The average energy of a hydrogen bond is quite large for a dipole-dipole interaction (up to 40 kJ/mol). Thus, hydrogen bonds are a powerful force in determining the structures and properties of many compounds. Figure 11.6 shows several examples of hydrogen bonding.

Early evidence for hydrogen bonding came from the study of boiling points of compounds. Normally, the boiling points of a series of similar compounds containing elements in the same group increase with increasing molar mass. But, as Figure 11.7

shows, some exceptions were noticed for the hydrogen compounds of Groups 5A, 6A, and 7A elements. In each of these series, the lightest compound (NH_3, H_2O, HF) has the *highest* boiling point, contrary to our expectations based on molar mass. This study and other related observations led chemists to postulate the existence of the hydrogen bond. In solid HF, for example, the molecules do not exist as individual units; instead, they form long zigzag chains:

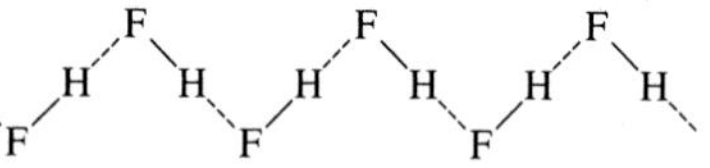

This is an example of intermolecular hydrogen bonding. In the liquid state the zigzag chains are broken, but the molecules are still extensively hydrogen bonded to one another. Molecules held together by hydrogen bonding are more difficult to break apart, so that liquid HF has an unusually high boiling point.

The strength of the hydrogen bond is determined by the coulombic interaction between the lone-pair electrons of the electronegative atom and the hydrogen nucleus. It may seem odd at first that the boiling point of HF is lower than that of water. Fluorine is more electronegative than oxygen, and so we would expect a stronger hydrogen bond to exist in liquid HF than in H_2O. But H_2O is unique in that each molecule takes part in *four* intermolecular hydrogen bonds, and the H_2O molecules are therefore held together more strongly. We will return to this very important property of water in the next section. The following example shows the type of species that can form hydrogen bonds with water.

EXAMPLE 11.2

Which of the following can form hydrogen bonds with water? CH_3OCH_3, CH_4, F^-, HCOOH, Na^+

Answer

There are no electronegative elements (F, O, or N) in either CH_4 or Na^+. Therefore, only CH_3OCH_3, F^-, and HCOOH can form hydrogen bonds with water (Figure 11.8).

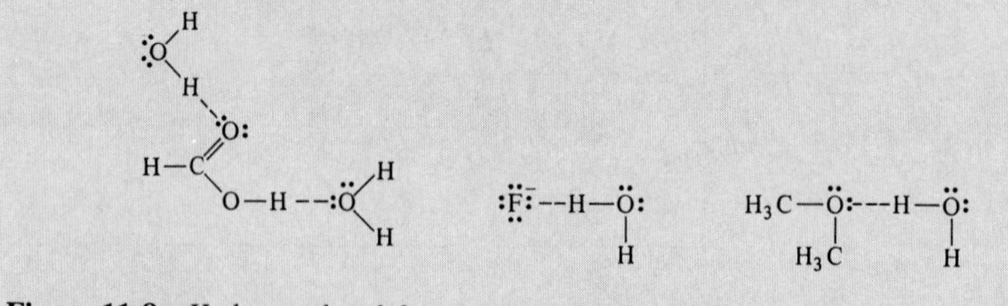

Figure 11.8 *Hydrogen bond formation (dashed lines) between various solutes and water molecules.*

Similar problem: 11.15.

The various types of intermolecular forces discussed so far are all attractive in nature. Keep in mind, though, that molecules also exert repulsive forces on one another. Thus when two molecules are brought into contact with each other, the repulsion between the electrons and between the nuclei in the molecules will come into play. The magnitude of the repulsive force rises very steeply as the distance separating the molecules in a condensed state is shortened. This is the reason that liquids and solids are so hard to compress. In these states, the molecules are already in close contact with one another, so that they greatly resist being further compressed.

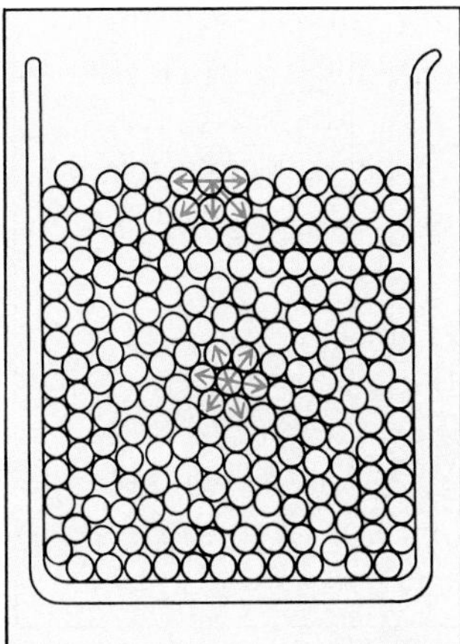

Figure 11.9 *Intermolecular forces acting on a molecule in the surface layer of a liquid and in the interior region of the liquid.*

11.3 The Liquid State

Now that you are familiar with intermolecular forces, we can look at the properties of substances in condensed states, which are largely determined by those forces. We will start with the liquid state. A great many interesting and important chemical reactions occur in water and other liquid solvents, as we will discover in subsequent chapters. In this section we will look at two phenomena associated with liquids: surface tension and viscosity. We will also discuss the structure and properties of water.

Surface Tension

We have seen that one property of liquids is their tendency to assume the shapes of their containers. Why, then, does water bead up on a newly waxed car instead of forming a sheet over it? The answer to this question lies in intermolecular forces.

Molecules within a liquid are pulled in all directions by intermolecular forces; there is no tendency for them to be pulled in any one way. However, molecules at the surface are pulled downward and sideways by other molecules, but not upward away from the surface (Figure 11.9). These intermolecular attractions thus tend to pull the molecules into the liquid and cause the surface to behave as if it were tightened like an elastic film. Since there is little or no attraction between polar water molecules and the wax molecules (which are essentially nonpolar) on a freshly waxed car, a drop of water assumes the shape of a small round bead.

For any given quantity of liquid, the sphere offers the minimum surface area.

A measure of the elasticlike force existing in the surface of a liquid is surface tension. The ***surface tension*** of a liquid is *the amount of energy required to stretch or increase the surface by unit area.* As expected, liquids containing molecules that possess strong intermolecular forces also have high surface tensions. For example, because of hydrogen bonding, water has a considerably greater surface tension than most common liquids.

Another way that surface tension manifests itself is in *capillary action.* Figure 11.10(a) shows water rising spontaneously in a capillary tube. A thin film of water adheres to the wall of the glass tube. The surface tension of water causes this film to contract, and as it does, it pulls the water up the tube. Two types of forces bring about capillary action. One is *the intermolecular attraction between like molecules* (in this case, the water molecules), called ***cohesion.*** The other, which is called ***adhesion,*** is *an attraction between unlike molecules,* such as those in water and in the walls of a glass tube. If adhesion is stronger than cohesion, as it is in Figure 11.10(a), the contents of the tube will be pulled upward along the walls. This process continues until the adhesive force is balanced by the weight of the water in the tube. This action is by no means universal among liquids, as Figure 11.10(b) shows. In mercury, cohesion is greater than the adhesion between mercury and glass, so that a depression in the liquid level actually occurs when a capillary tube is dipped into mercury.

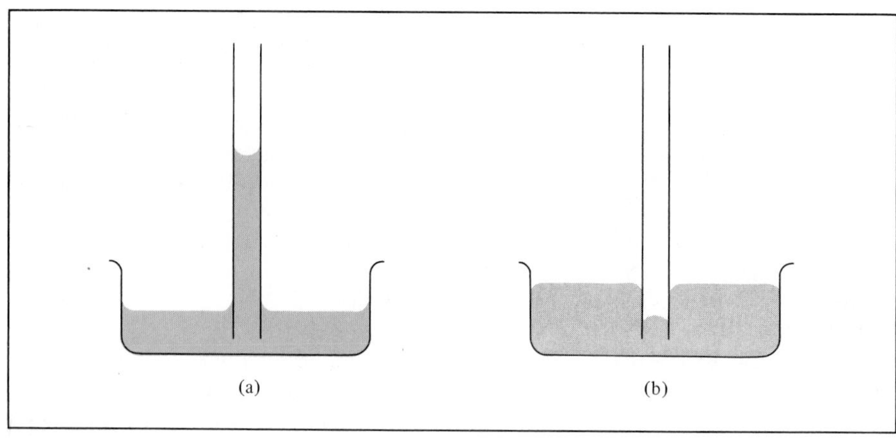

Figure 11.10 *(a) When adhesion is greater than cohesion, the liquid (for example, water) rises in the capillary tube. (b) When cohesion is greater than adhesion, as it is for mercury, a depression of the liquid in the capillary tube results. Note that the meniscus in the tube of water is concave, or rounded downward, whereas that in the tube of mercury is convex, or rounded upward.*

Viscosity

The expression ''slow as molasses in January'' owes its truth to another physical property of liquids called viscosity. ***Viscosity*** is *a measure of a fluid's resistance to flow*. The greater the viscosity, the more slowly the liquid flows. The viscosity of a liquid usually decreases as temperature increases; thus hot molasses flows much faster than cold molasses.

Liquids that have strong intermolecular forces have higher viscosities than those that have weak intermolecular forces (Table 11.4). Water has a higher viscosity than many other liquids because of its ability to form hydrogen bonds. Interestingly, the viscosity of glycerol is significantly higher than that of all the other liquids shown. Glycerol has the structure

Glycerol is a clear, odorless, syrupy liquid used to make explosives, ink, and lubricants, among other things.

$$\begin{array}{l} CH_2\text{—}OH \\ | \\ CH\text{—}OH \\ | \\ CH_2\text{—}OH \end{array}$$

Table 11.4 Viscosity of some common liquids at 20°C

Liquid	Viscosity ($N\ s/m^2$)*
Acetone (C_3H_6O)	3.16×10^{-4}
Benzene (C_6H_6)	6.25×10^{-4}
Carbon tetrachloride (CCl_4)	9.69×10^{-4}
Ethanol (C_2H_5OH)	1.20×10^{-3}
Ethyl ether ($C_2H_5OC_2H_5$)	2.33×10^{-4}
Glycerol ($C_3H_8O_3$)	1.49
Mercury (Hg)	1.55×10^{-3}
Water (H_2O)	1.01×10^{-3}
Blood	4×10^{-3}

*In SI units, viscosities are expressed as newton-second per meter squared.

Like water, glycerol can form hydrogen bonds. We see that each glycerol molecule has three —OH groups that can participate in hydrogen bonding with other glycerol molecules. Furthermore, because of their shape the molecules have a great tendency to become entangled rather than to slip past one another as the molecules in less viscous liquids do. These interactions contribute to its high viscosity.

The Structure and Properties of Water

If water didn't have the ability to form hydrogen bonds, it would be a gas at room temperature.

Water is so common a substance on Earth that we often overlook its unique nature. All life processes involve water. Water is an excellent solvent for many ionic compounds, as well as for other substances capable of forming hydrogen bonds with water. As Table 6.1 shows, water has a high specific heat. The reason for this is that to raise the temperature of water (that is, to increase the average kinetic energy of water molecules), we must first break the many intermolecular hydrogen bonds. Thus, water can absorb a substantial amount of heat while its temperature rises only slightly. The converse is also true: Water can give off much heat with only a slight decrease in its temperature. For this reason, the huge quantities of water that are present in our lakes and oceans can effectively moderate the climate of adjacent land areas by absorbing heat in the summer and giving off heat in the winter, with only small changes in the temperature of the body of water.

The most striking property of water is that its solid form is less dense than its liquid form: an ice cube floats at the surface of water in a glass. This is a virtually unique property. The density of almost all other substances is greater in the solid state than in the liquid state.

To understand why water is different, we have to examine the electronic structure of the H_2O molecule. As we saw in Chapter 9, there are two pairs of nonbonding electrons, or two lone pairs, on the oxygen atom:

```
  ..
  O:
 / \
H   H
```

Although many compounds are capable of forming intermolecular hydrogen bonds, there is a significant difference between H_2O and other polar molecules, such as NH_3 and HF. In water, the number of hydrogen bonds about each oxygen atom is equal to two, the same as the number of lone electron pairs on the oxygen atom. Thus, water molecules are joined together in an extensive three-dimensional network in which each oxygen atom is approximately tetrahedrally bonded to four hydrogen atoms, two by covalent bonds and two by hydrogen bonds. This equality in the number of hydrogen atoms and lone pairs is not characteristic of NH_3 or HF or, for that matter, any other molecule capable of forming hydrogen bonds. Consequently, these other molecules can form rings or chains, but not three-dimensional structures.

The highly ordered three-dimensional structure of ice (Figure 11.11) prevents the molecules from getting too close to one another. But consider what happens when heat is provided and ice melts. There is ample evidence to show that the three-dimensional structure remains largely intact, although the bonds may become somewhat bent and distorted. At the melting point, a relatively small number of water molecules have enough kinetic energy to pull free of the intermolecular hydrogen bonds. These molecules become trapped in the cavities of the three-dimensional structure. As a result there are more molecules per unit volume in liquid water than in ice. Thus, since density = mass/volume, the density of water is greater than that of ice. With further

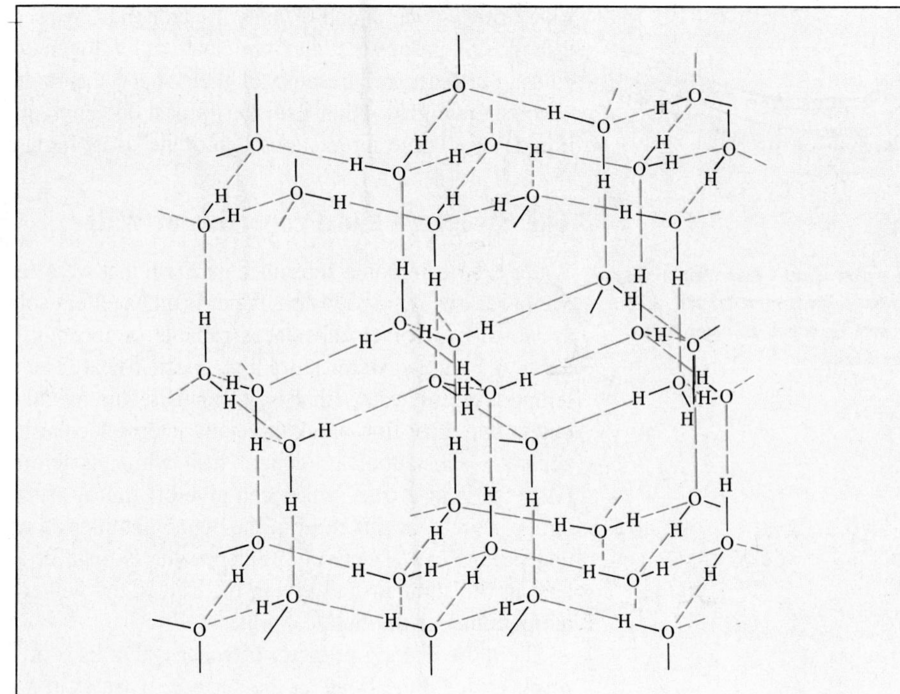

Figure 11.11 *The three-dimensional structure of ice. Each O atom is bonded to four H atoms. The covalent bonds are shown by short solid lines and the weaker hydrogen bonds by long dashed lines between O and H. The empty space in the structure accounts for the low density of ice.*

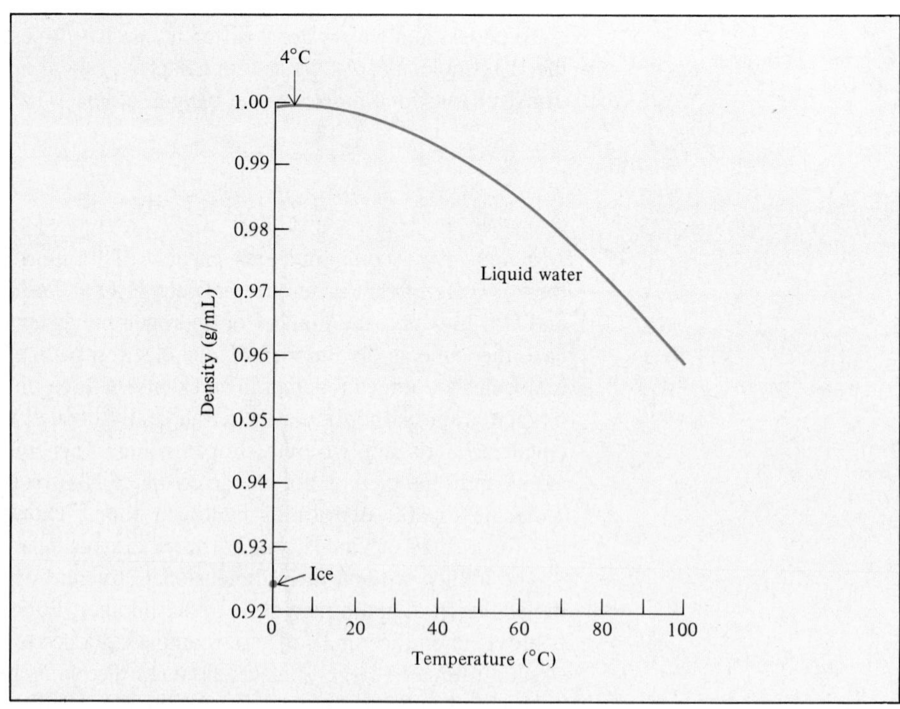

Figure 11.12 *Plot of density versus temperature for liquid water. Note the large difference between the density of ice and that of liquid water at 0°C. The maximum density of water is reached at 4°C.*

heating, more water molecules are released from intermolecular hydrogen bonding, so that just above the melting point water's density tends to rise with temperature. Of course, at the same time, water expands as it is being heated, with the result that its density is decreased. These two processes—the trapping of free water molecules in cavities and thermal expansion—act in opposite directions. From 0°C to 4°C, the trapping prevails and water becomes progressively denser. Beyond 4°C, however, thermal expansion predominates and the density of water decreases with increasing temperature (Figure 11.12).

11.4 Crystal Structure

Solids can be divided into two categories: crystalline and amorphous. A ***crystalline solid,*** such as ice or sodium chloride, *possesses rigid and long-range order; its atoms, molecules, or ions occupy specific positions. The center of each of the positions* is called a ***lattice point,*** and *the geometric order of these lattice points* is called the ***crystal structure.*** The arrangement of atoms, molecules, or ions in a crystalline solid is such

CHEMISTRY IN ACTION

Why Do Lakes Freeze from Top to Bottom?

The fact that ice is less dense than water has a profound ecological significance. Consider, for example, the temperature changes in the fresh water of a lake in a cold climate. As the temperature of the water near the surface is lowered, its density increases. The colder water then sinks toward the bottom, while warmer water, which is less dense, rises to the top. This normal convection motion continues until the temperature throughout the water reaches 4°C. Below this temperature, the density of water begins to decrease with decreasing temperature (see Figure 11.12), so that it no longer sinks. On further cooling, the water begins to freeze at the surface. The ice layer formed does not sink because it is less dense than the liquid; it even acts as a thermal insulator for the water below it. Were ice heavier, it would sink to the bottom of the lake and eventually the water would freeze upward. Most living organisms in the body of water could not survive. Fortunately, this does not happen, and it is this unusual property of water that makes the sport of ice fishing possible (Figure 11.13).

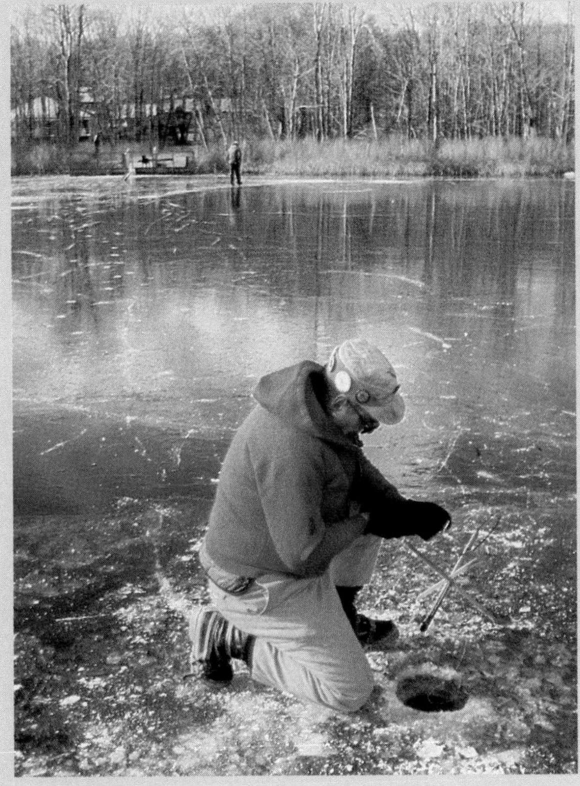

Figure 11.13 *Ice fishing. The ice layer that forms on the surface of a lake insulates the water beneath and maintains a high enough temperature to sustain aquatic life.*

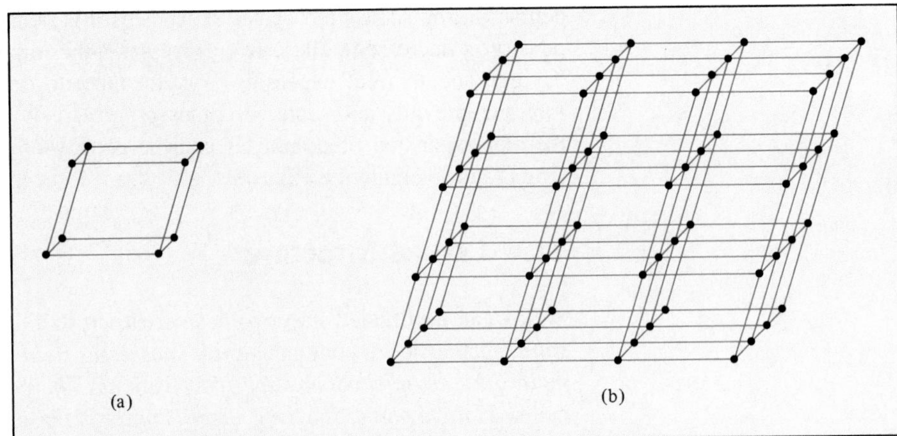

Figure 11.14 *(a) A unit cell and (b) its extension in three dimensions. The black spheres represent either atoms or molecules.*

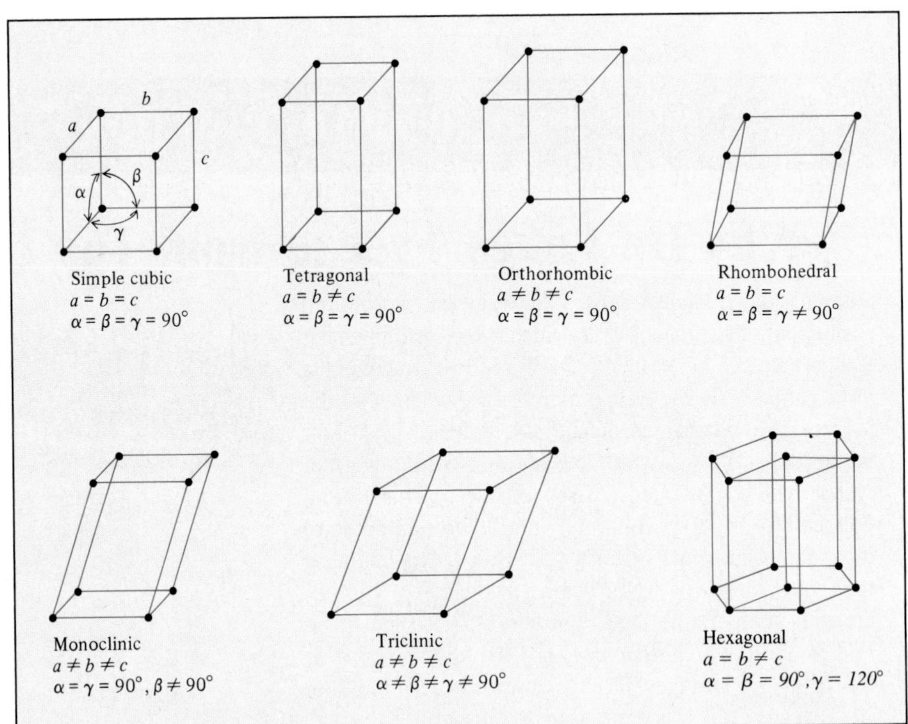

Figure 11.15 *The seven types of unit cells. Angle α is defined by edges a and c, angle β by edges b and c, and angle γ by edges a and b.*

that the net attractive intermolecular forces are at their maximum. The forces responsible for the stability of any crystal can be ionic forces, covalent bonds, van der Waals forces, hydrogen bonds, or a combination of these forces. Amorphous solids such as glass lack well-defined arrangement and long-range molecular order. We will discuss them in Section 11.7. In this section we concentrate on the structure of crystalline solids.

The basic repeating unit of the arrangement of atoms or molecules in a crystalline solid is a ***unit cell.*** Figure 11.14 shows a unit cell and its extension in three dimensions. Every crystalline solid can be described in terms of one of the seven types of unit cells shown in Figure 11.15. Any of these unit cells, when repeated in space, forms the lattice structure characteristic of a crystalline solid. Consider, for example, the cubic unit cell. For simplicity, we can assume that each lattice point is occupied by an atom. The geometry of the cubic unit cell is particularly simple, since all sides and all angles are equal. As Figure 11.16 shows, the location of the atoms determines whether we call a cubic unit cell a *simple cubic cell* (scc), a *body-centered cubic cell* (bcc), or a *face-centered cubic cell* (fcc). Because every unit cell in a crystalline solid is adjacent to other unit cells, most of the atoms are shared by neighboring unit cells. For example, in all types of cubic cells, each corner atom belongs to eight unit cells [Figure 11.17(a)]; a face-centered atom is shared by two unit cells [Figure 11.17(b)].

Packing Spheres

We can understand the general geometric requirements for crystal formation by considering the different ways of packing a number of identical spheres (Ping-Pong balls, for example) to form an ordered three-dimensional structure. In the simplest case, a layer

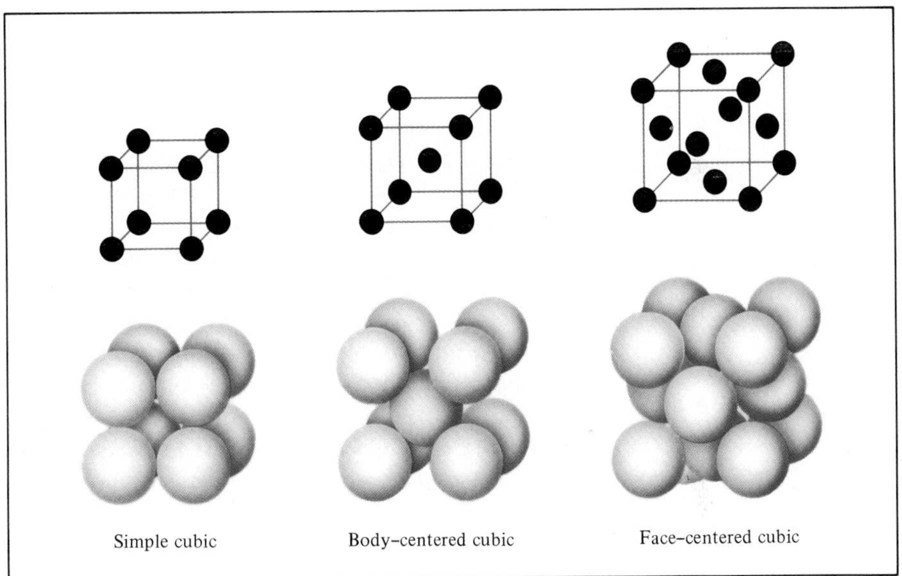

Figure 11.16 *Three types of cubic cells. In reality, the spheres representing atoms, molecules, or ions are in contact with one another in these cubic cells.*

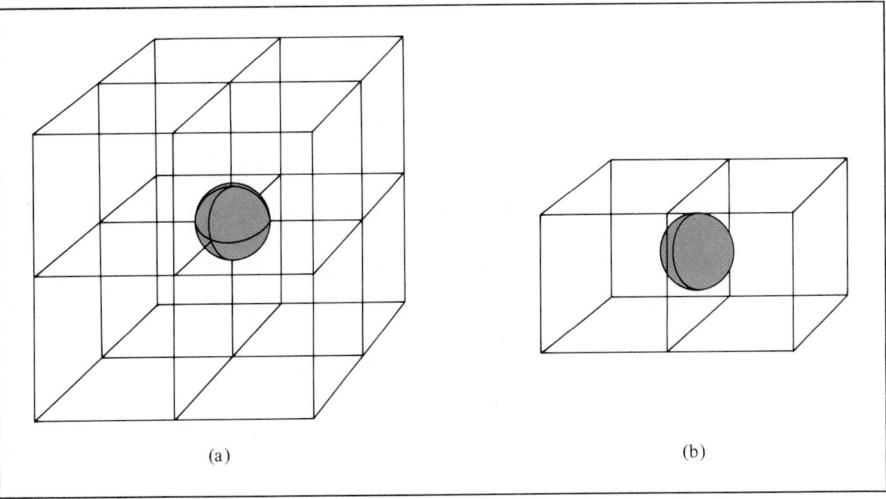

Figure 11.17 *(a) A corner atom in any cubic cell is shared by eight unit cells; (b) a face-centered atom in a cubic cell is shared by two unit cells.*

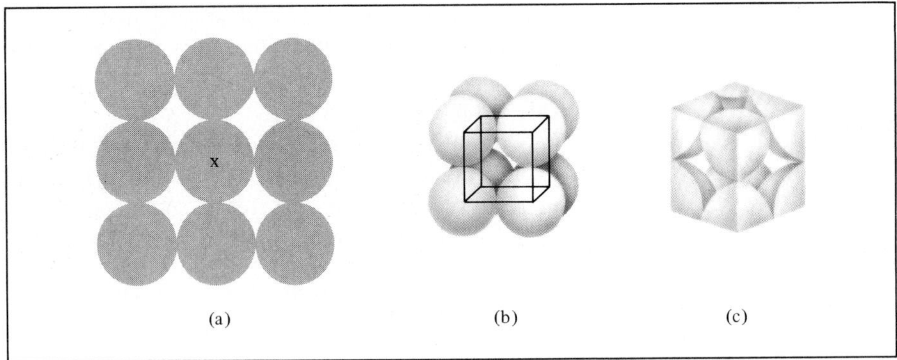

Figure 11.18 *Arrangement of identical spheres in a simple cubic cell. (a) Top view of one layer of spheres. (b) Definition of a simple cubic cell. (c) Since each sphere is shared by eight unit cells and there are eight corners in a cube, there is the equivalent of one complete sphere inside a simple cubic unit cell.*

of spheres can be arranged as shown in Figure 11.18(a). The three-dimensional structure can be generated by placing a layer above and below this layer in such a way that spheres in one layer are directly over the spheres in the layer below it. This procedure can be extended to generate many, many layers, as in the case of a crystal. Focusing on the sphere marked with "x", we see that it is in contact with four spheres in its own layer, one sphere in the layer above, and one sphere in the layer below. Thus each sphere in this arrangement is said to have a ***coordination number*** of 6 because it has six immediate neighbors. The coordination number is therefore *the number of atoms (or ions) surrounding an atom (or ion) in a crystal lattice*. Figure 11.18(b) shows that the basic, repeating unit in this array of spheres is a simple cubic unit cell.

Packing Efficiency. The ***packing efficiency,*** or *percentage of the cell space occupied by the spheres,* is an important crystal property. For one thing, it determines the density of the crystal. For a simple cubic cell, let a be the length of the edge of the cubic unit cell and r the radius of the spheres, so that $a = 2r$ (Figure 11.19). Then

$$\text{volume of one sphere} = \frac{4\pi}{3}r^3 = \frac{4\pi}{3}\left(\frac{a}{2}\right)^3$$

$$\text{volume of the unit cell} = a^3$$

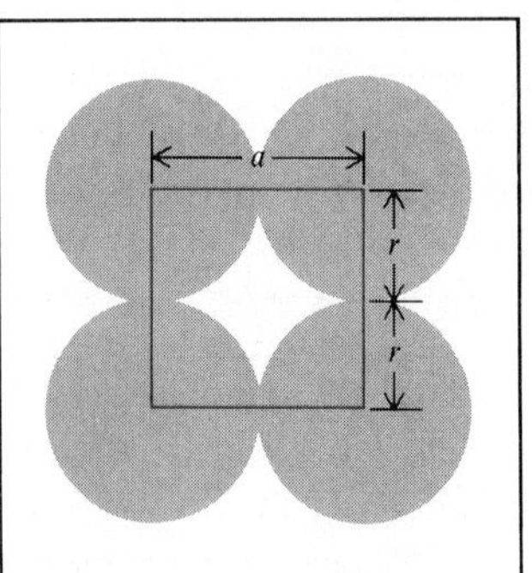

Figure 11.19 *The relation between atomic radius, r, and edge length, a, of a simple cubic cell, a = 2r.*

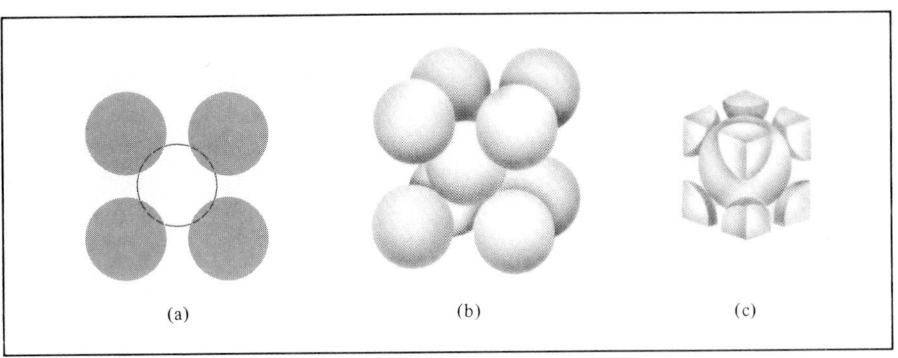

Figure 11.20 *Arrangement of identical spheres in a body-centered cube. (a) Top view. (b) Definition of a body-centered cubic unit cell. (c) There is the equivalent of two complete spheres inside a body-centered cubic unit cell.*

Since each corner sphere is shared by eight unit cells and there are eight corners in a cube, there will be the equivalent of only one complete sphere *inside* a simple cubic unit cell [see Figure 11.18(c)]. We can now write

$$\begin{aligned}\text{packing efficiency} &= \frac{\text{volume of spheres inside the cell}}{\text{volume of the cell}} \times 100\% \\ &= \frac{\frac{4\pi}{3}\left(\frac{a}{2}\right)^3}{a^3} \times 100\% \\ &= \frac{\frac{4\pi a^3}{24}}{a^3} \times 100\% = \frac{\pi}{6} \times 100\% \\ &= 52\%\end{aligned}$$

A body-centered cubic arrangement (Figure 11.20) differs from a simple cube in that the second layer fits into the depressions of the first layer and the third layer into the depressions of the second layer. The coordination number of each sphere in this structure is 8 (each sphere is in contact with four spheres in the layer above and four spheres in the layer below), and the packing efficiency is 68 percent.

Closest Packing

The efficiency of packing can be further increased. We start with the structure shown in Figure 11.21(a), which we call the A layer. Each sphere is in contact with six neighbors in the same layer. In the second layer (which we call the B layer), spheres are packed as closely as possible to the first layer by resting them in the depressions between the spheres in the first layer [Figure 11.21(b)].

There are two ways that a third-layer sphere may cover the second layer to achieve what is called ***closest packing,*** which refers to *the most efficient arrangement for spheres*. The spheres may fit into the depressions so that each third-layer sphere is directly over a first-layer sphere [Figure 11.21(c)]. Since there is no difference between the arrangement of the first and third layers, we call the third layer the A layer. Alternatively, the third-layer spheres may fit into the depressions that lie directly over the depressions in the first layer [Figure 11.21(d)]. In this case we call the third layer the C layer. Figure 11.22 shows the "exploded views" and the resulting structures from these two arrangements. The ABA arrangement generates the *hexagonal close-packed (hcp) structure*, and the ABC arrangement generates the *cubic close-packed (ccp) structure,* which is also called a *face-centered cube*. Note that in the hcp structure

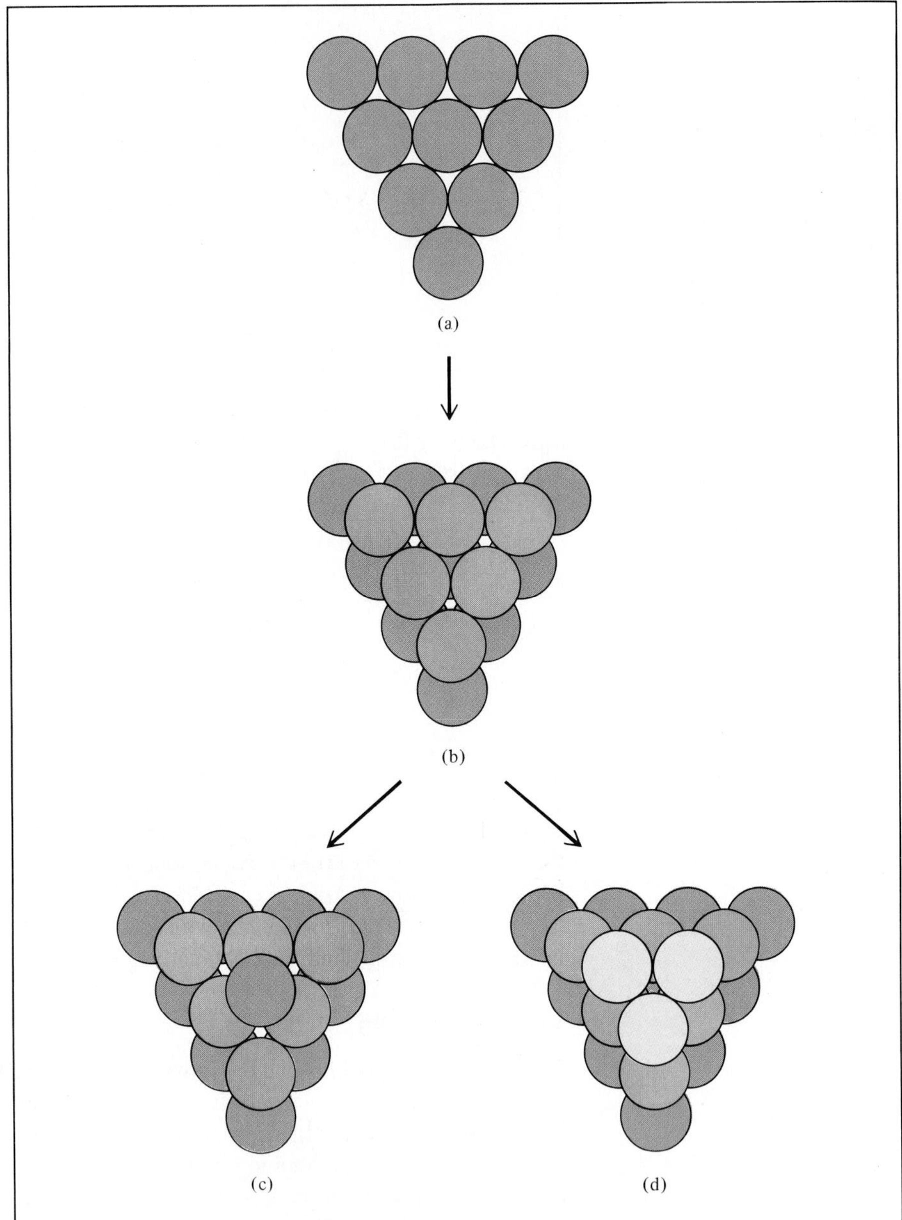

Figure 11.21 *(a) In a close-packed layer, each sphere is in contact with six others. (b) Spheres in the second layer fit into the depressions of the first-layer spheres. (c) In a hexagonal close-packed structure, each third-layer sphere is directly over a first-layer sphere. (d) In the cubic close-packed structure, each third-layer sphere fits into a depression that is directly over a depression in the first layer.*

the spheres in every other layer occupy the same vertical position (ABABAB . . .), while in the ccp structure the spheres in every fourth layer occupy the same vertical position (ABCABCA . . .). In both structures each sphere has a coordination number of 12 (each sphere is in contact with six spheres in its own layer, three spheres in the

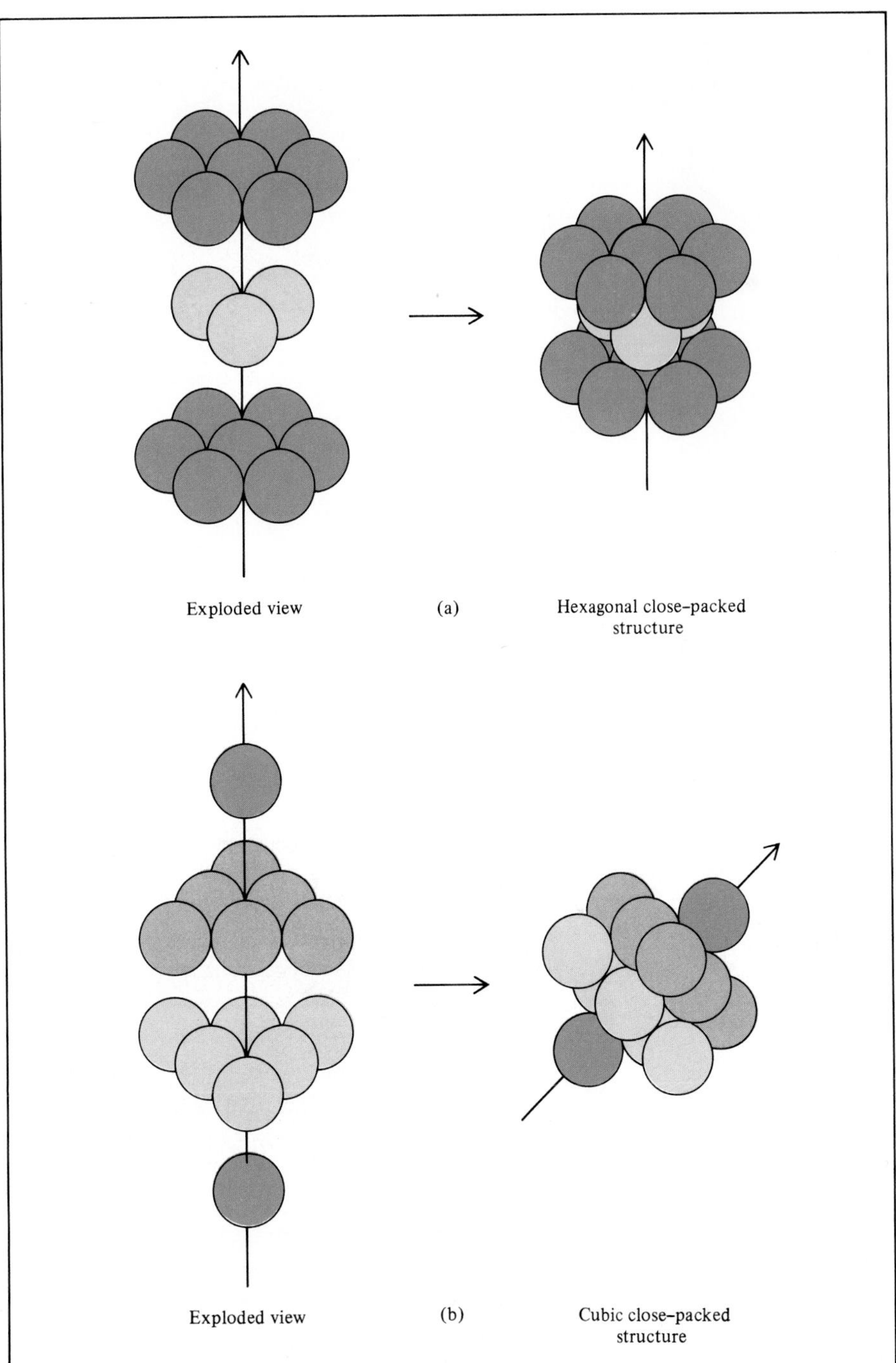

Figure 11.22 *Exploded views of (a) a hexagonal close-packed structure and (b) a cubic close-packed structure. The arrow is tilted to show the face-centered cubic unit cell more clearly.*

layer above, and three spheres in the layer below). The packing efficiency is the same for both: 74 percent.

Figure 11.23 summarizes the relationships between atomic radius, r, and edge length, a, of a simple cubic cell (scc), a body-centered cubic cell (bcc), and a face-centered cubic cell (fcc).

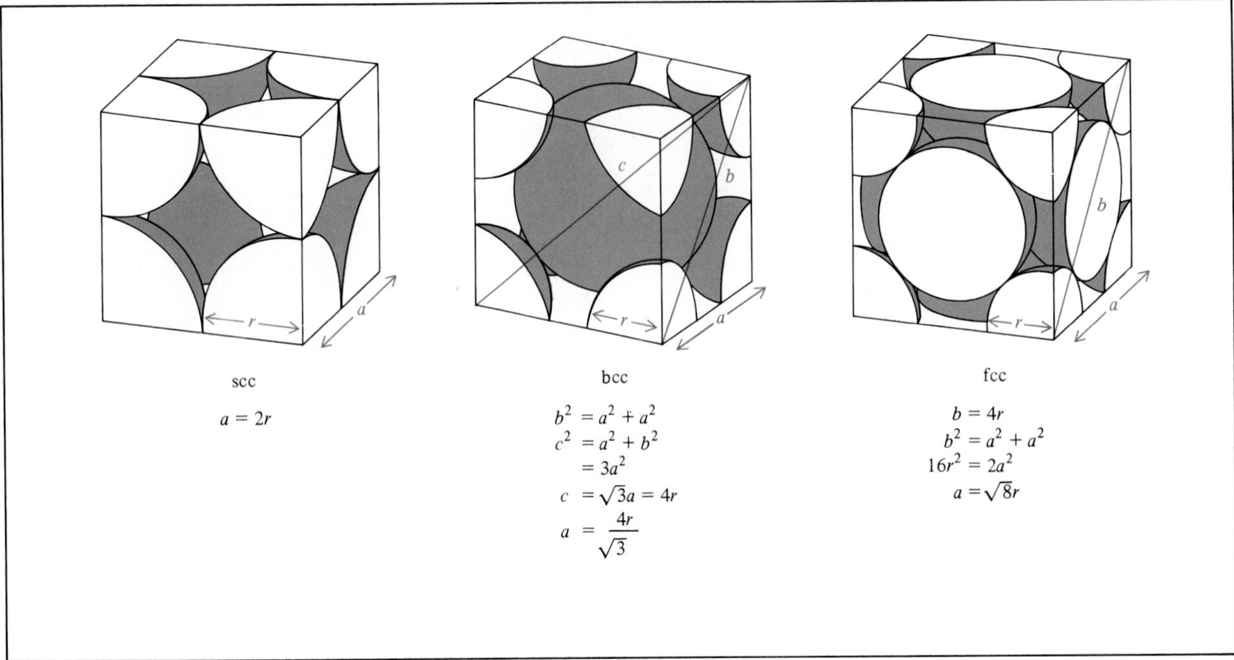

Figure 11.23 *The relationship between the edge length and radius of atoms for simple cubic cell, body-centered cubic cell, and face-centered cubic cell.*

Examples 11.3 and 11.4 show how unit cell dimensions are related to the density of the solid.

EXAMPLE 11.3

Gold crystallizes in a cubic close-packed structure (the face-centered cube). Calculate the density of gold. The atomic radius of gold is 144 pm.

Answer

From Figure 11.23 we see that the relationship between edge length a and atomic radius r is $a = \sqrt{8}r$. Thus

$$a = \sqrt{8}\ (144\ \text{pm}) = 407\ \text{pm}$$

The volume of the unit cell is

$$V = a^3 = (407\ \text{pm})^3 \times \left(\frac{1 \times 10^{-12}\ \text{m}}{1\ \text{pm}}\right)^3 \times \left(\frac{1\ \text{cm}}{1 \times 10^{-2}\ \text{m}}\right)^3$$
$$= 6.74 \times 10^{-23}\ \text{cm}^3$$

Each unit cell has eight corners and six faces. Therefore, the total number of atoms within such a cell is, according to Figure 11.17, $8 \times \frac{1}{8} + 6 \times \frac{1}{2} = 4$. The mass of a unit cell is

$$m = \frac{4\ \text{atoms}}{1\ \text{unit cell}} \times \frac{197.0\ \text{g}}{1\ \text{mol}} \times \frac{1\ \text{mol}}{6.02 \times 10^{23}\ \text{molecules}}$$
$$= 1.31 \times 10^{-21}\ \text{g/unit cell}$$

Finally the density of gold is given by

$$d = \frac{m}{V} = \frac{1.31 \times 10^{-21}\text{ g}}{6.74 \times 10^{-23}\text{ cm}^3} = 19.4\text{ g/cm}^3$$

This value is in excellent agreement with the experimental value of 19.3 g/cm^3.

Similar problem: 11.50.

EXAMPLE 11.4

The metal zinc has a hexagonal close-packed structure with a density of 7.14 g/cm^3. Calculate the radius of a Zn atom in picometers.

Answer

First we calculate the volume of the hcp structure occupied by 1 mole of Zn. Let V_{hcp} be the molar volume, so that we can write

Remember that density is an intensive property, so that it is the same for 1 gram or 1 mole of a substance.

$$V_{hcp} = \frac{\text{molar mass of Zn}}{\text{density of Zn}} = \frac{65.39\text{ g Zn/mol Zn}}{7.14\text{ g/cm}^3} = 9.16\text{ cm}^3\text{/mol Zn}$$

This is molar volume, so it contains 6.02×10^{23} Zn atoms. From the packing efficiency of hcp structure, we know that 74 percent of the volume is occupied by the Zn atoms (the other 26 percent is empty space); that is,

$$\text{volume of Zn atoms} = 0.74 \times 9.16\text{ cm}^3\text{/mol Zn} = 6.78\text{ cm}^3\text{/mol Zn}$$

The volume of a single Zn atom is given by

$$\text{volume of one Zn atom} = \frac{6.78\text{ cm}^3}{1\text{ mol Zn}} \times \frac{1\text{ mol Zn}}{6.02 \times 10^{23}\text{ Zn atoms}} = 1.13 \times 10^{-23}\text{ cm}^3$$

The volume of a sphere of radius r is

$$V = \tfrac{4}{3}\pi r^3$$

so for one spherical Zn atom

$$1.13 \times 10^{-23}\text{ cm}^3 = \tfrac{4}{3}\pi r^3$$
$$r^3 = 2.70 \times 10^{-24}\text{ cm}^3$$
$$r = 1.39 \times 10^{-8}\text{ cm} = 139\text{ pm}$$

This calculated value is in excellent agreement with the experimentally determined value of 138 pm for the radius of a Zn atom.

Similar problem: 11.51.

11.5 X-Ray Diffraction by Crystals

Virtually all we know about crystal structure has been learned from X-ray diffraction studies. ***X-ray diffraction*** refers to *the scattering of X rays by the units of a regular crystalline solid*. The scattering (or diffraction) patterns obtained are used to deduce the arrangement of particles in the solid lattice.

In Section 10.6 we discussed the interference phenomenon associated with waves (see Figure 10.22). Since X rays are one form of electromagnetic radiation, and therefore waves, we would expect them to exhibit such behavior under suitable conditions.

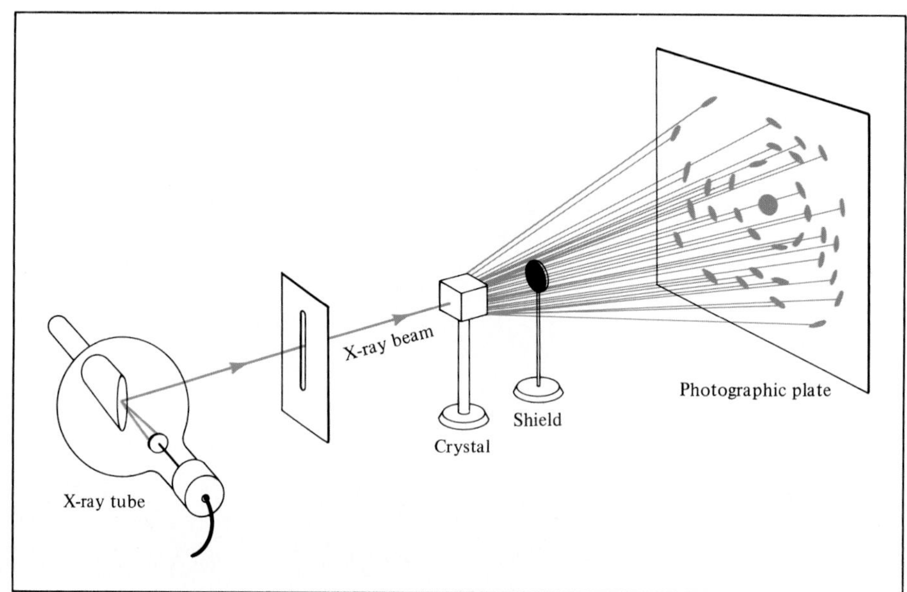

Figure 11.24 *An arrangement for obtaining the X-ray diffraction pattern of a crystal. The shield prevents the strong undiffracted X rays from damaging the photographic plate.*

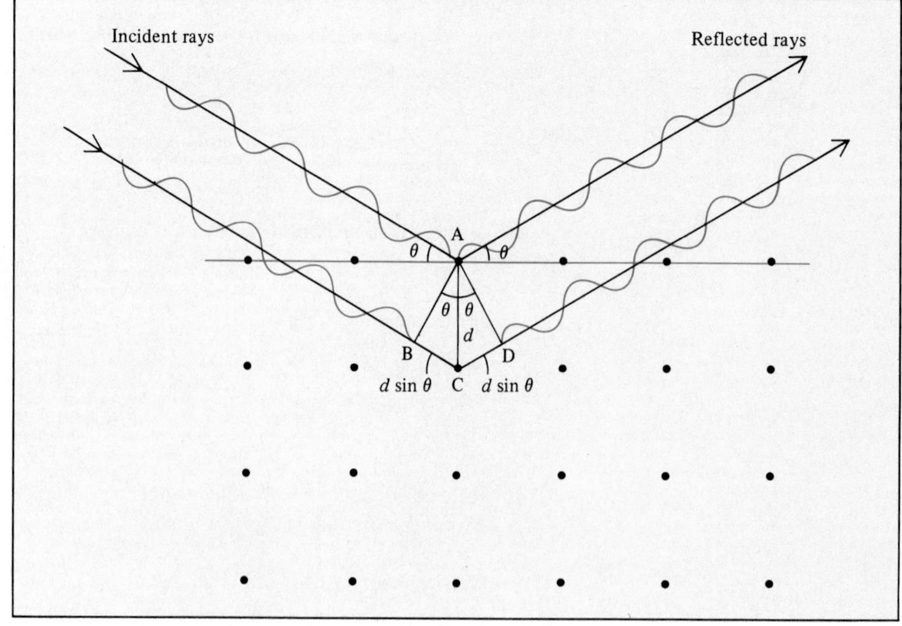

Figure 11.25 *Reflection of X rays from two layers of atoms. The lower wave travels a distance $2d \sin\theta$ longer than the upper wave does. For the two waves to be in phase again after reflection, it must be true that $2d \sin\theta = n\lambda$, where λ is the wavelength of the X ray and $n = 1, 2, 3, \ldots$*

In 1912 the German physicist Max von Laue† correctly suggested that, because the wavelength of X rays is comparable in magnitude to the distances between lattice points in a crystal, the lattice should be able to *diffract* X rays. An X-ray diffraction pattern is a result of interference in the waves associated with X rays.

Figure 11.24 shows a typical arrangement for producing an X-ray diffraction pattern of a crystal. A beam of X rays is directed at a mounted crystal. Atoms in the crystal absorb some of the incoming radiation and then reemit it; the process is called the *scattering of X rays*.

To understand how a diffraction pattern may be generated, consider the scattering of X rays by atoms in two parallel planes (Figure 11.25). Initially, the two incident rays are *in phase* with each other (that is, their maxima and minima occur at the same positions). The upper wave is scattered or reflected by an atom in the first layer, while the lower wave is scattered by an atom in the second layer. In order for these two scattered waves to be in phase again, the extra distance traveled by the lower wave must be an integral multiple of the wavelength (λ) of the X ray; that is,

$$\mathrm{BC} + \mathrm{CD} = 2d \sin \theta = n\lambda \qquad n = 1, 2, 3, \ldots \tag{11.1}$$

where θ is the angle between the X rays and the plane of the crystal and d is the distance between adjacent planes. Equation (11.1) is known as the Bragg equation after William H. Bragg‡ and Sir William L. Bragg.§ The reinforced waves produce a dark spot on a photographic film for each value of θ that satisfies the Bragg equation.

Reinforced waves are waves that have interacted constructively (see Figure 10.22).

The following example illustrates the use of Equation (11.1).

EXAMPLE 11.5

X rays of wavelength 0.154 nm strike an aluminum crystal; the rays are reflected at an angle of 19.3°. Assuming that $n = 1$, calculate the spacing between the planes of aluminum atoms (in pm) that is responsible for this angle of reflection. The conversion factor is obtained from 1 nm = 1000 pm.

Answer

From Equation (11.1)

$$\begin{aligned} d &= \frac{n\lambda}{2 \sin \theta} = \frac{\lambda}{2 \sin \theta} \\ &= \frac{0.154 \text{ nm} \times \dfrac{1000 \text{ pm}}{1 \text{ nm}}}{2 \sin 19.3°} \\ &= 233 \text{ pm} \end{aligned}$$

Similar problems: 11.67, 11.68, 11.69.

†Max Theodor Felix von Laue (1879–1960). German physicist. Von Laue received the Nobel Prize in physics in 1914 for his discovery of X-ray diffraction.

‡William Henry Bragg (1862–1942). English physicist. Bragg's work was mainly in X-ray crystallography. He shared the Nobel Prize in physics with his son Sir William Bragg in 1915.

§Sir William Lawrence Bragg (1890–1972). English physicist. Bragg formulated the fundamental equation for X-ray diffraction and shared the Nobel Prize in physics with his father in 1915.

The X-ray diffraction technique offers the most accurate method for determining bond lengths and bond angles in molecules in the solid state. Because X rays are scattered by electrons, chemists can construct an electron density contour map from the diffraction patterns by using a complex mathematical procedure. Basically, an *electron density contour map* tells us the relative electron densities at various locations in a molecule. The densities reach a maximum near the center of each atom. In this manner, we can determine the positions of the nuclei and hence the geometric parameters of the molecule.

11.6 Types of Crystals

The structures and properties of crystals are determined by the kinds of forces that hold the particles together. We can classify any crystal as one of four types, ionic, covalent, molecular, or metallic (Table 11.5).

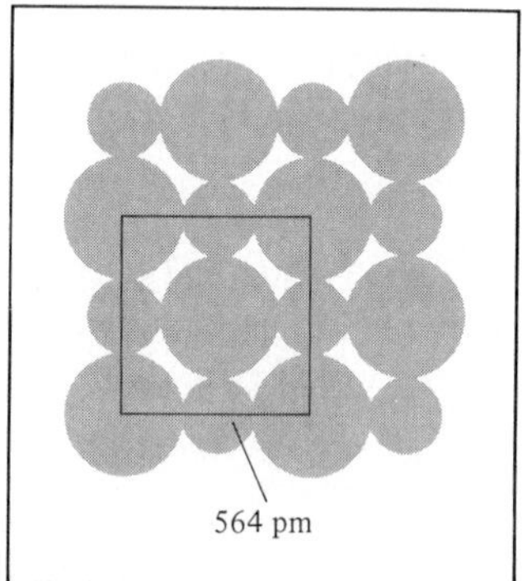

Figure 11.26 *Relation between the radii of Na^+ and Cl^- ions and the unit cell dimension. Here the cell edge length is equal to twice the sum of the two ionic radii.*

Ionic Crystals

Two important characteristics of ionic crystals should be noted: (1) They are composed of charged species, and (2) anions and cations are generally quite different in size. Knowing the radii of the ions is helpful in understanding the structure and stability of these compounds. There is no way to measure the radius of an individual ion, but under favorable circumstances it is possible to estimate the radii of a number of ions. For example, if we know the radius of I^- in KI is about 216 pm, we can proceed to determine the radius of K^+ ion in KI, and from that, the radius of Cl^- in KCl, and so on. Figure 8.11 gives the radii of a number of ions. Keep in mind that these values are averaged over many different compounds. Let us consider the NaCl crystal which has a face-centered cubic lattice (see Figure 2.10). Figure 11.26 shows that the edge length of the unit cell of NaCl is twice the sum of the ionic radius of Na^+ and Cl^-. Using the values in Figure 8.11, we calculate the same edge length to be 2(95 + 181) pm, or

Table 11.5 Types of crystals and general properties

Type of crystal	Units at lattice points	Force(s) holding the units together	General properties	Examples
Ionic	Positive and negative ions	Electrostatic attraction	Hard, brittle, high melting point, poor conductor of heat and electricity	NaCl, LiF, MgO
Covalent	Atoms	Covalent bond	Hard, high melting point, poor conductor of heat and electricity	C (diamond),† SiO_2 (quartz)
Molecular*	Molecules or atoms	Dispersion forces, dipole-dipole forces, hydrogen bonds	Soft, low melting point, poor conductor of heat and electricity	Ar, CO_2, I_2, H_2O, $C_{12}H_{22}O_{11}$ (sucrose)
Metallic	Atoms	Metallic bond	Soft to hard, low to high melting point, good conductor of heat and electricity	All metallic elements; for example, Na, Mg, Fe, Cu

*Included in this category are crystals made up of individual atoms.
†Diamond is a good thermal conductor.

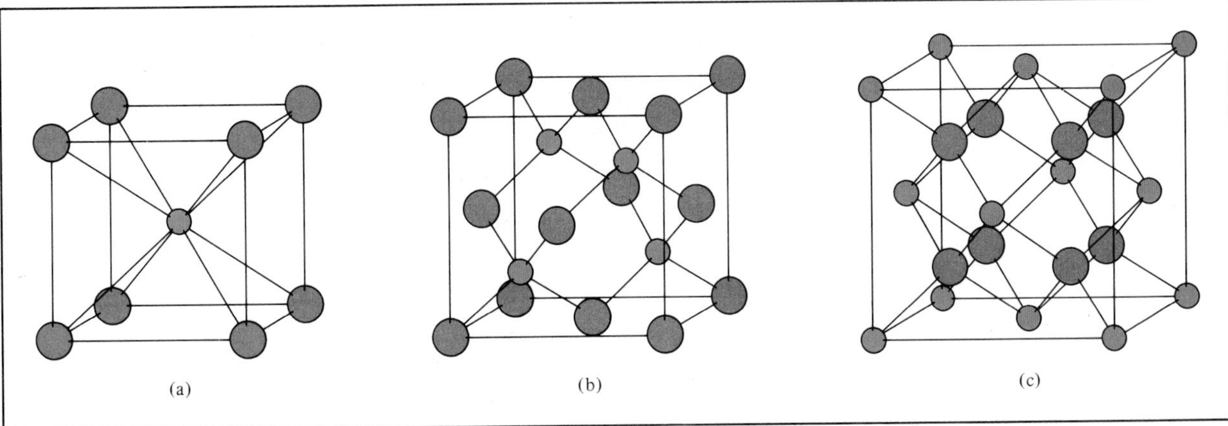

Figure 11.27 *Crystal structures of (a) CsCl, (b) ZnS, and (c) CaF_2. In each case, the cation is the smaller sphere.*

552 pm. But the edge length shown in Figure 11.26 is determined by X-ray diffraction to be 564 pm. The discrepancy between these two values tells us that the radius of an ion actually varies slightly from one compound to another.

Figure 11.27 shows the crystal structures of three ionic compounds: CsCl, ZnS, and CaF_2. Because Cs^+ is considerably larger than Na^+, CsCl has the simple cubic lattice. ZnS has the *zincblende* structure which is based on the face-centered cubic lattice. If the S^{2-} ions are located at the lattice points, the Zn^{2+} ions are located one-fourth of the distance along each body diagonal. Other ionic compounds that have the zincblende structure include CuCl, BeS, CdS, and HgS. CaF_2 has the *fluorite* structure. The Ca^{2+} ions are located at the lattice points, and each F^- ion is tetrahedrally surrounded by four Ca^{2+} ions. Compounds such as SrF_2, BaF_2, $BaCl_2$, and PbF_2 also have the fluorite structure.

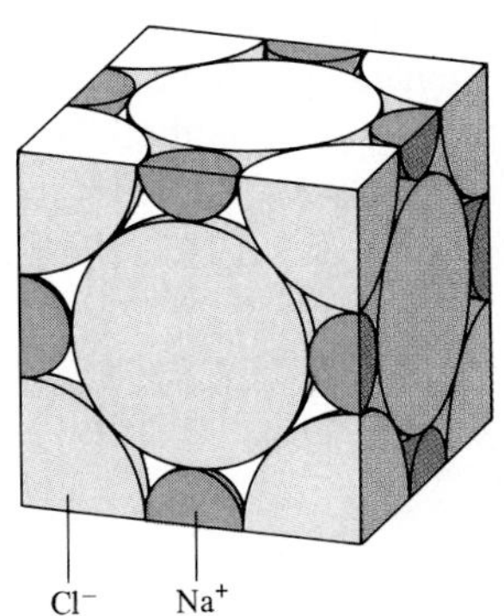

Figure 11.28 *Portions of Na^+ and Cl^- ions within a face-centered cubic unit cell.*

EXAMPLE 11.6

How many Na^+ and Cl^- ions are in each NaCl unit cell?

Answer

NaCl has a structure based on a face-centered cubic lattice. As Figure 2.10 shows, one whole Na^+ ion is at the center of the unit cell and there are twelve Na^+ ions at the edges. Since each edge Na^+ ion is shared by four unit cells, the total number of Na^+ ions is $1 + (12 \times \frac{1}{4}) = 4$. Similarly, there are six Cl^- ions at the face centers and eight Cl^- ions at the corners. Each face-centered ion is shared by two unit cells, and each corner ion is shared by eight unit cells (see Figure 11.17), so the total number of Cl^- ions is $(6 \times \frac{1}{2}) + (8 \times \frac{1}{8}) = 4$. Thus there are four Na^+ ions and four Cl^- ions in each NaCl unit cell. Figure 11.28 shows the portions of the Na^+ and Cl^- ions *within* a unit cell.

EXAMPLE 11.7

The edge length of the NaCl unit cell is 564 pm. What is the density of NaCl in g/cm^3?

Answer

From Example 11.6 we see that there are four Na^+ ions and four Cl^- ions in each unit cell. The total mass (in amu) in a unit cell is therefore

$$\text{mass} = 4(22.99 \text{ amu} + 35.45 \text{ amu}) = 233.8 \text{ amu}$$

The volume of the unit cell is $(564 \text{ pm})^3$. The density of the unit cell is

$$\begin{aligned}\text{density} &= \text{mass/volume}\\ &= \frac{(233.8 \text{ amu})\left(\dfrac{1 \text{ g}}{6.02 \times 10^{23} \text{ amu}}\right)}{(564 \text{ pm})^3\left(\dfrac{1 \text{ cm}}{1 \times 10^{10} \text{ pm}}\right)^3}\\ &= 2.16 \text{ g/cm}^3\end{aligned}$$

Most ionic crystals have high melting points, an indication of the strong cohesive force holding the ions together. A measure of the stability of ionic crystals is the lattice energy (see Section 9.3); the higher the lattice energy, the more stable the compound. Table 9.1 lists the lattice energies of some alkali and alkaline earth metal halides and oxides together with their melting points.

Covalent Crystals

In covalent crystals, atoms are held together entirely by covalent bonds in an extensive three-dimensional network. Well-known examples are the two allotropes of carbon: diamond and graphite (see Figure 8.20). In diamond each carbon atom is bonded to four other atoms (Figure 11.29). The strong covalent bonds in three dimensions contribute to diamond's unusual hardness (it is the hardest material known). In graphite, carbon atoms are arranged in six-membered rings. The atoms are all sp^2-hybridized; each atom is covalently bonded to three other atoms. The remaining unhybridized $2p$ orbital is used in pi bonding. In fact, in each layer of graphite there is the kind of delocalized molecular orbital that is present in benzene (see Section 10.8). Because

Diamond melts at about 3550°C. Since this process involves breaking strong covalent bonds, such a high melting point is not surprising.

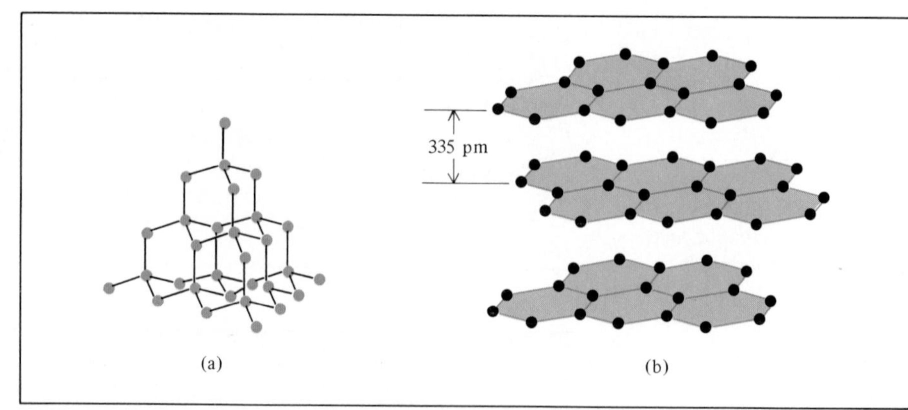

Figure 11.29 *(a) The structure of diamond. Each carbon is tetrahedrally bonded to four other carbon atoms. (b) The structure of graphite. The distance between successive layers is 335 pm.*

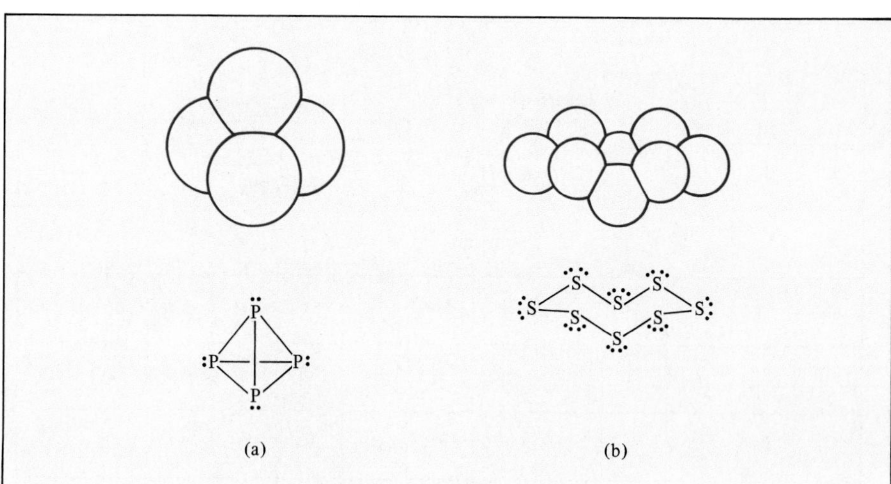

Figure 11.30 *(a) The structure of the P_4 molecule. (b) The structure of the S_8 molecule.*

The central electrode in flashlight batteries is made of graphite.

electrons are free to move around in this extensively delocalized molecular orbital, graphite is a good conductor of electricity in directions along the planes of carbon atoms. The layers are held together by the weak van der Waals forces. The covalent bonds in graphite account for its hardness; however, because the layers can slide over one another, graphite is slippery to the touch and is used as a lubricant. It is also used in pencils and in ribbons used in computer printers and typewriters.

The melting point of quartz is 1610°C.

Another type of covalent crystal is quartz (SiO_2). The arrangement of silicon atoms in quartz is similar to that of carbon in diamond, but in quartz there is an oxygen atom between each pair of Si atoms. Since Si and O have different electronegativities (see Figure 9.8), the Si—O bond is polar. Nevertheless, SiO_2 is similar to diamond in many respects, such as hardness and high melting point.

Molecular Crystals

In a molecular crystal, the lattice points are occupied by molecules, and the attractive forces between them are van der Waals forces and/or hydrogen bonding. An example of a molecular crystal is solid sulfur dioxide (SO_2), in which the predominant attractive force is dipole-dipole interaction. Intermolecular hydrogen bonding is mainly responsible for maintaining the three-dimensional ice lattice (see Figure 11.11). Other examples of molecular crystals are I_2, P_4, and S_8 (Figure 11.30).

In general, except in ice, molecules in molecular crystals are packed together as closely as their size and shape allow. Because van der Waals forces and hydrogen bonding are generally quite weak compared with the covalent and ionic bonds, molecular crystals are more easily broken apart than ionic and covalent crystals. Indeed, most molecular crystals melt below 100°C.

Metallic Crystals

In a sense, the structure of metallic crystals is the simplest to deal with, since every lattice point in a crystal is occupied by an atom of the same metal. Metallic crystals are generally body-centered cubic, face-centered cubic, or hexagonal close-packed (Figure

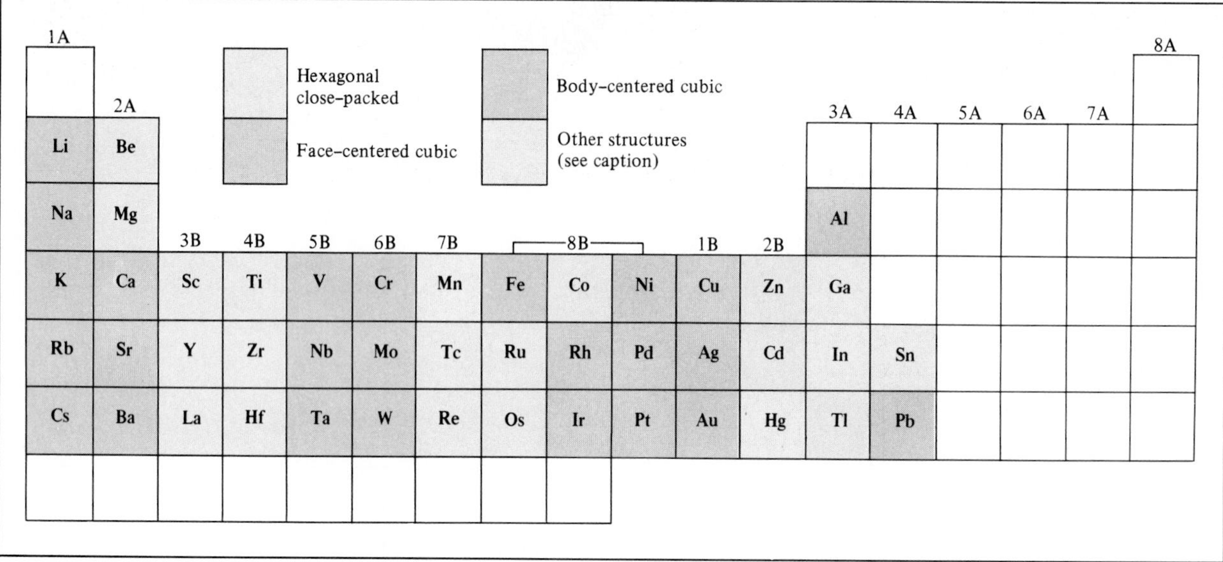

Figure 11.31 *Crystal structures of various metals. The metals are shown in their positions in the periodic table. Mn has a cubic structure, Ga an orthorhombic structure, In and Sn a tetragonal structure, and Hg a rhombohedral structure.*

11.31). As we saw earlier, the packing efficiency of these crystal structures is quite high, so metallic elements are usually very dense.

The bonding in metals is quite different from that in other types of crystals. In a metal the bonding electrons are delocalized over the entire crystal. In fact, metal atoms in a crystal can be imagined as an array of positive ions immersed in a sea of delocalized valence electrons (Figure 11.32). The great cohesive force resulting from delocalization is responsible for the strength of metals. The mobility of the delocalized electrons accounts for metals being good conductors of heat and electricity.

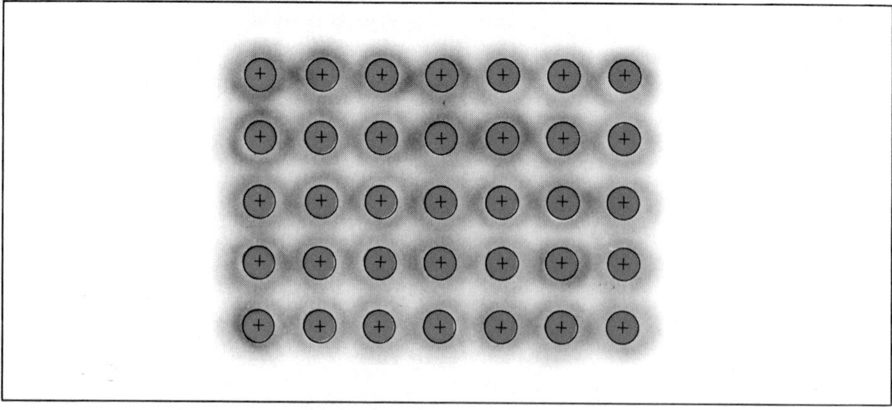

Figure 11.32 *A cross section of the crystal structure of a metal. Each circled positive charge represents the nucleus and inner electrons of a metal atom. The gray area surrounding the positive metal ions indicates the mobile sea of electrons.*

CHEMISTRY IN ACTION

High-Temperature Superconductors

Metals such as copper and aluminum are good conductors of electricity, but they do possess some electrical resistance. In fact, up to about 20 percent of electrical energy may be lost (in the form of heat) when these metals are used as electrical transmission cables. Wouldn't it be marvelous if we had cables that possess no electrical resistance?

Actually it has been known for over seventy years that certain metals and alloys, when cooled to very low temperatures (around the boiling point of liquid helium, or 4 K), lose their resistance totally. However, it is not practical to use these substances, called superconductors, for transmission of electric power because of the prohibitive cost of maintaining electrical cables at such low temperatures. The cost of creating superconductors would far exceed the savings that would result in electricity transmission.

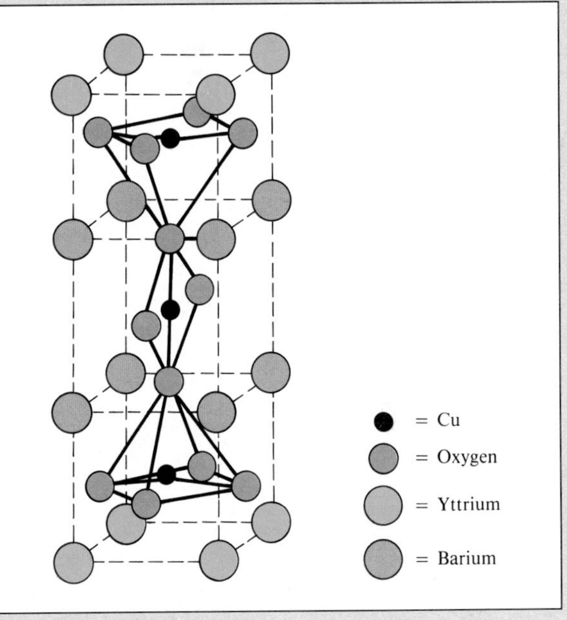

Figure 11.33 *Crystal structure of $YBa_2Cu_3O_x$ ($x = 6$ or 7). Because some of the O atoms are vacant at the sites, the formula is not a constant.*

In 1986 two physicists in Switzerland discovered a new class of materials that were superconducting at around 30 K. Although 30 K is still a very low temperature, the improvement over the 4 K range was so dramatic that their work generated immense interest and triggered a flurry of research activity. Within months, scientists were able to synthesize certain compounds that are superconducting around 95 K, which is well above the boiling point of liquid nitrogen (77 K). The crystal structure of one of these compounds, a mixed oxide of yttrium, barium, and copper with the formula $YBa_2Cu_3O_x$ (where $x = 6$ or 7), is shown in Figure 11.33. Figure 11.34 shows a magnet being levitated above such a superconductor immersed in liquid nitrogen.

Figure 11.34 *The levitation of a magnet above a high-temperature superconductor immersed in liquid nitrogen.*

There is good reason for the tremendous enthusiasm for these materials. Because liquid nitrogen is quite cheap (a liter of liquid nitrogen costs *less* than a liter of milk), we can envision electrical power transmitted over hundreds of miles with no loss of energy. The levitation effect can be used to construct fast as well as quiet and smooth trains that ride above (but not in contact with) the tracks (Figure 11.35). These superconductors can be used to build superfast computers, called supercomputers, whose speeds are limited by how fast electric current flows. The enormous magnetic fields that can be created in superconductors will result in more powerful particle accelerators, efficient devices for nuclear fusion, and more accurate magnetic resonance image techniques for medical use.

The outlook is not entirely bright, however. Recently (1989) scientists have discovered an obstacle called a *magnetic flux lattice*. The unexpected weakness of this lattice in high-temperature superconductors means that they may not be able to carry enough current or hold a high enough magnetic field to perform many of the tasks once envisioned for them (such as those mentioned above). The future of high-temperature superconductors, that is, whether they will truly revolutionize science and technology or only have limited applications, will become more clear in the next few years.

Figure 11.35 *An experimental levitation train that operates on superconducting material at liquid helium temperature.*

11.7 Amorphous Solids

Solids that are not crystalline in structure are said to be *amorphous*. ***Amorphous solids,*** such as glass, *lack a regular three-dimensional arrangement of atoms*. In this section, we will discuss briefly some interesting and important properties of glass.

Glass is one of civilization's most valuable and versatile materials. It is also one of the oldest—glass articles date as far back as 1000 B.C. The term ***glass*** is commonly used to mean *an optically transparent fusion product of inorganic materials that has*

Table 11.6 Composition and properties of three types of glass

Name	Composition	Properties and uses
Pure quartz glass	100% SiO_2	Low thermal expansion, transparent to wide range of wavelengths. Used in optical research.
Pyrex glass	SiO_2, 60–80% B_2O_3, 10–25% Al_2O_3, small amount	Low thermal expansion; transparent to visible and infrared, but not to UV radiation. Used mainly in laboratory and household cooking glassware.
Soda-lime glass	SiO_2, 75% Na_2O, 15% CaO, 10%	Easily attacked by chemicals and sensitive to thermal shocks. Transmits visible light, but absorbs UV radiation. Used mainly in windows and bottles.

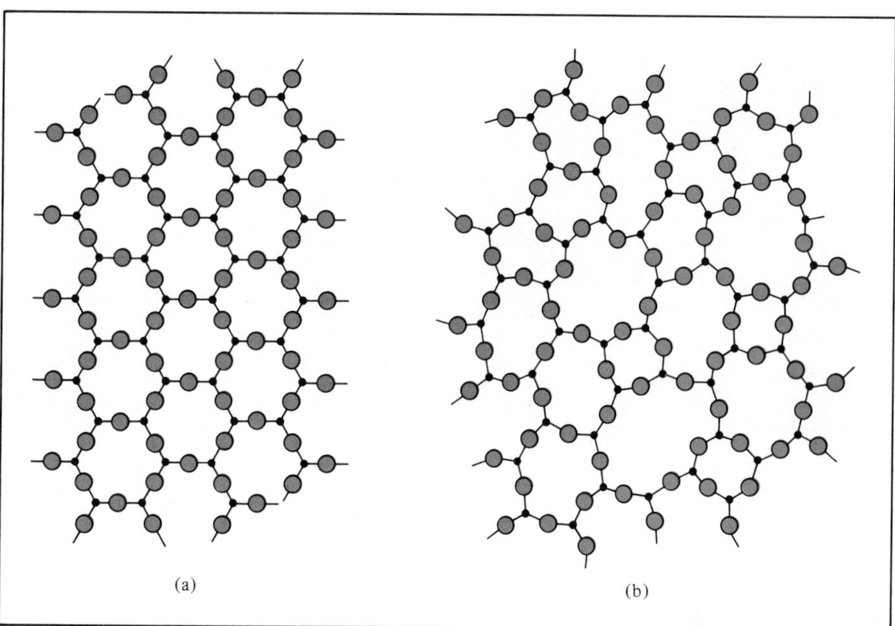

Figure 11.36 *Two-dimensional representation of (a) crystalline quartz and (b) noncrystalline quartz glass. The small spheres represent silicon.*

cooled to a rigid state without crystallizing. In some respects glass behaves more like a liquid than a solid. X-ray diffraction studies show that glass lacks long-range periodic order.

There are about 800 different types of glass in common use today. Table 11.6 shows the composition and properties of three of them. Figure 11.36 shows two-dimensional schematic representations of crystalline quartz and quartz glass.

Note that most of the ions are derived from the transition metals.

The color of glass is due largely to the presence of metal ions (as oxides). For example, green glass contains iron(III) oxide, Fe_2O_3, or copper(II) oxide, CuO; yellow glass contains uranium(IV) oxide, UO_2; blue glass contains cobalt(II) and copper(II) oxides, CoO and CuO; and red glass contains small particles of gold and copper.

CHEMISTRY IN ACTION

Optical Fibers

It has been known for years that glass can be drawn into fibers thinner than human hair and still possess great mechanical strength. However, scientists and engineers had to wait for two recent technological advances—development of the laser and the perfection of high-purity glass—to make use of this property in communication and medicine. The material used in these areas is called *optical fiber*.

Figure 11.37 shows a glass fiber that acts as a medium for a beam of laser light whose source can be smaller than the period at the end of this sentence. Note that the glass fiber is able to "bend" the light by internal reflection with very little leakage through the glass. By varying the amplitude of the light wave, such a light beam can be made to carry messages just as a radio wave does. Theoretically, one light beam could accom-

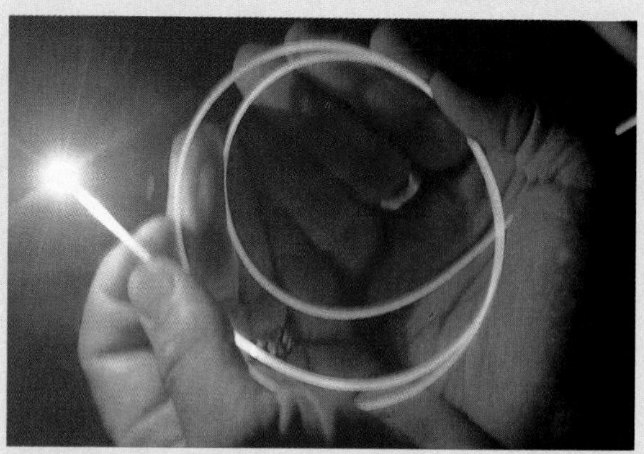

Figure 11.37 *Laser light traveling along an optical fiber.*

Figure 11.38 *The thin optical fiber can transmit the same number of messages as the copper wires shown with it.*

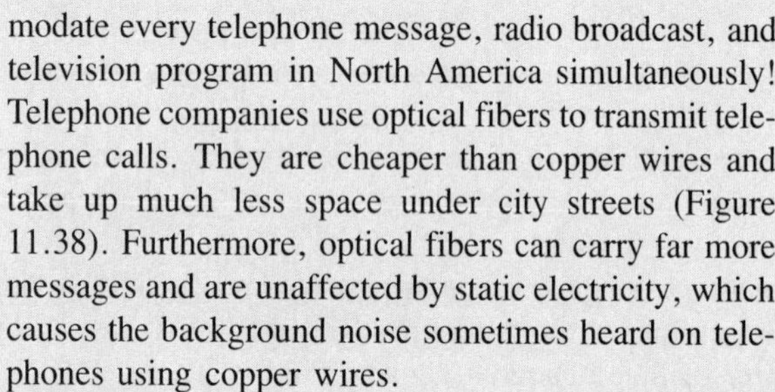

modate every telephone message, radio broadcast, and television program in North America simultaneously! Telephone companies use optical fibers to transmit telephone calls. They are cheaper than copper wires and take up much less space under city streets (Figure 11.38). Furthermore, optical fibers can carry far more messages and are unaffected by static electricity, which causes the background noise sometimes heard on telephones using copper wires.

Optical fibers are made of the same materials that go into windowpanes: silica (SiO_2), soda (Na_2CO_3), and lime (CaO). But unlike ordinary glass, which stops a light beam in less than a meter, optical fibers for communication transmit light over kilometers without any appreciable decrease in light intensity.

Since light stays within a fiber even when the fiber is bent, doctors sometimes use optical fibers to see inside parts of the human body. This procedure avoids expensive and painful exploratory surgery and involves less risk. Because these fibers are so fine, doctors can even insert a fiber through a vein to photograph the inside of a patient's heart!

11.8 Phase Changes

The discussions in Chapter 5 and in this chapter have given us an overview of the properties of the three states of matter. A logical next question is how substances behave as they change from one state to another. As temperature changes, they undergo what is known as ***phase changes,*** or *transformation from one phase to another*. Molecules in the solid state have the most order; those in the gas phase have the greatest randomness. We know from experience that energy (usually in the form of heat) is required to bring about solid-to-liquid-to-gas changes, which are accompanied by increases in randomness or disorder. As we study phase transformations, keep in mind the relationship between energy change and the increase or decrease in molecular order. These parameters will help us understand the nature of these physical changes.

Liquid-Vapor Equilibrium

Vapor Pressure. Molecules in a liquid are not in a rigid lattice. Although they lack the total freedom of gaseous molecules, they are in constant motion. Because liquids

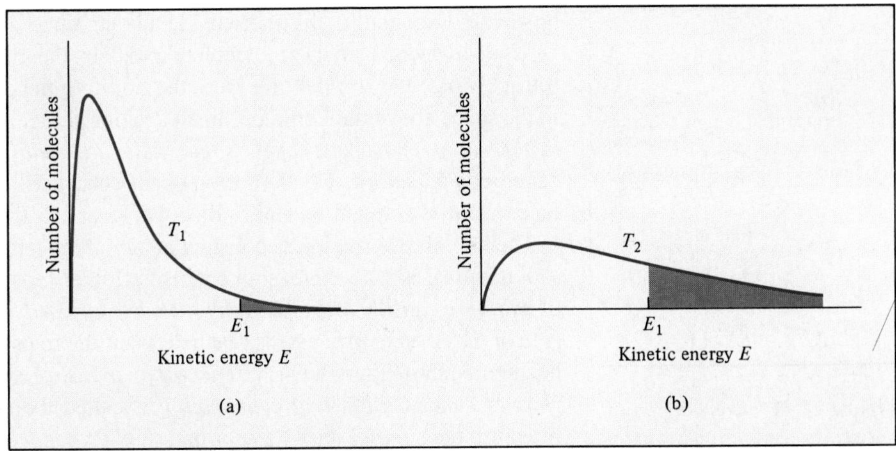

Figure 11.39 *Kinetic energy distribution curves for molecules in a liquid (a) at a temperature T_1 and (b) at a higher temperature T_2. Note that at the higher temperature the curve flattens out. The shaded areas represent the number of molecules possessing kinetic energy equal to or greater than a certain kinetic energy E_1. The higher the temperature, the greater the number of molecules with high kinetic energy.*

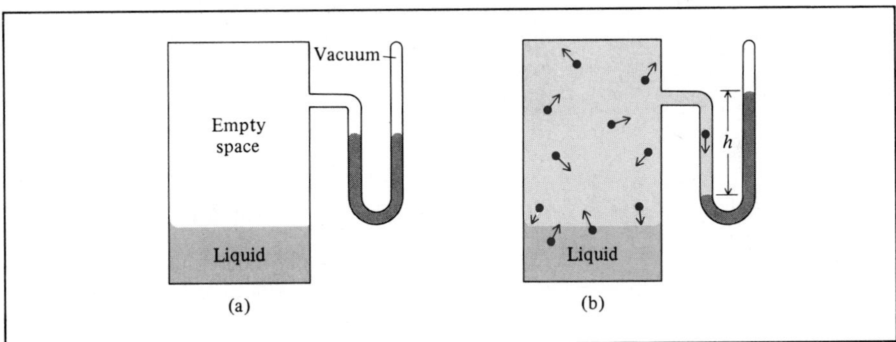

Figure 11.40 *Apparatus for the measurement of the vapor pressure of a liquid (a) before the evaporation begins and (b) at equilibrium. In (b) the number of molecules leaving the liquid is equal to the number of molecules returning to the liquid. The difference in the mercury levels (h) gives the equilibrium vapor pressure of the liquid at the specified temperature.*

are denser than gases, the collision rate among molecules is much higher in the liquid phase than in the gas phase.

Figure 11.39 shows the kinetic energy distribution of molecules in a liquid. *At any given temperature, a certain number of the molecules in a liquid possess sufficient kinetic energy to escape from the surface*. This process is called ***evaporation,*** or ***vaporization.***

The difference between a gas and a vapor is explained in Section 5.1.

When a liquid evaporates, its molecules in the gas phase exert a vapor pressure. Consider the apparatus shown in Figure 11.40. Before the evaporation process starts, the mercury levels in the manometer are equal. As soon as a few molecules leave the liquid, a vapor phase is established. The vapor pressure is measurable only when a fair amount of vapor is present. The process of evaporation does not continue indefinitely, however. Eventually, the mercury levels stabilize and no further changes are seen.

What happened at the molecular level? In the beginning, there is only one-way traffic: Molecules are moving from the liquid to the empty space. Soon the molecules in the space above the liquid establish a vapor phase. *As the concentration of molecules in the vapor phase increases, some molecules return to the liquid phase,* a process called ***condensation.*** Condensation occurs because a molecule striking the liquid surface becomes trapped by intermolecular forces in the liquid.

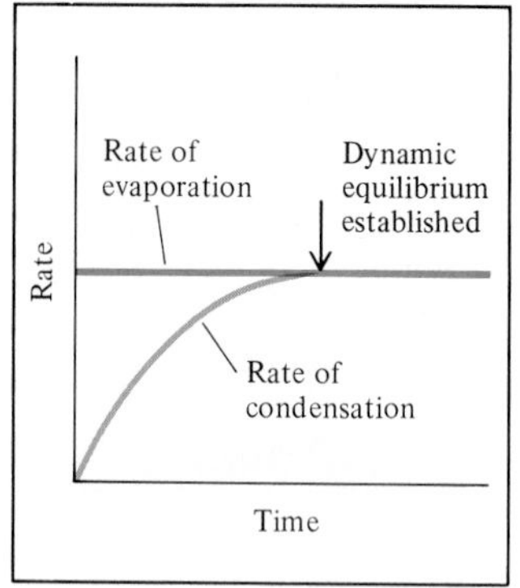

Figure 11.41 *Comparison of the rates of evaporation and condensation at constant temperature.*

Equilibrium vapor pressure is independent of the amount of liquid as long as there is some liquid present.

The rate of evaporation is constant at any given temperature, and the rate of condensation increases with increasing concentration of molecules in the vapor phase. A state of ***dynamic equilibrium*** *(when the rate of a forward process is exactly balanced by the rate of the reverse process)* is reached when the rates of condensation and evaporation become equal (Figure 11.41). *The vapor pressure measured under dynamic equilibrium of condensation and evaporation* is called the ***equilibrium vapor pressure.*** We often use the simpler term "vapor pressure" when we talk about the equilibrium vapor pressure of a liquid. This is acceptable as long as we know the meaning of the abbreviated term.

It is important to note that the equilibrium vapor pressure is the *maximum* vapor pressure a liquid exerts at a given temperature and that it is a constant at constant temperature. Vapor pressure does, however, change with temperature. Plots of vapor pressure versus temperature for three different liquids are shown in Figure 11.42. From the distribution curve in Figure 11.39 we know that the number of molecules with higher kinetic energies is greater at the higher temperature, and that therefore so is the evaporation rate. For this reason, the vapor pressure of a liquid always increases with temperature. For example, the vapor pressure of water is 17.5 mmHg at 20°C, but it rises to 760 mmHg at 100°C (see Figure 5.15).

Heat of Vaporization and Boiling Point

A measure of how strongly molecules are held in a liquid is its ***molar heat of vaporization*** **(ΔH_{vap})**, defined as *the energy* (usually in kilojoules) *required to vaporize one mole of a liquid.* The molar heat of vaporization is directly related to the strength of intermolecular forces that exist in the liquid. If intermolecular attraction is strong, then molecules in a liquid cannot easily escape into the vapor phase. Consequently, the

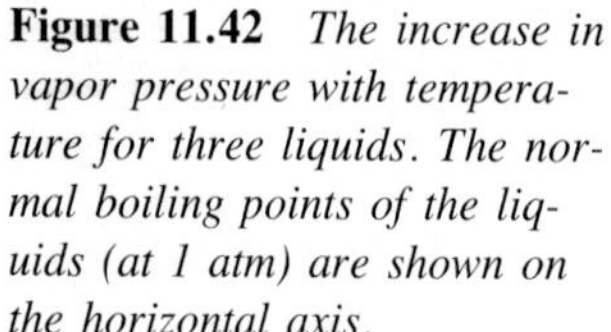

Figure 11.42 *The increase in vapor pressure with temperature for three liquids. The normal boiling points of the liquids (at 1 atm) are shown on the horizontal axis.*

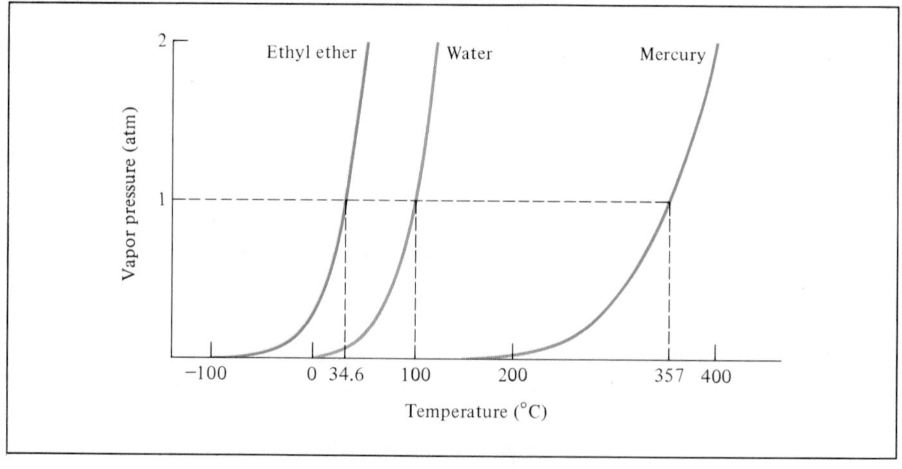

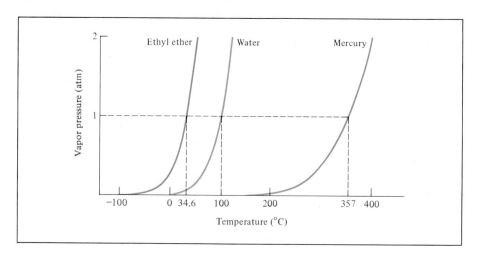

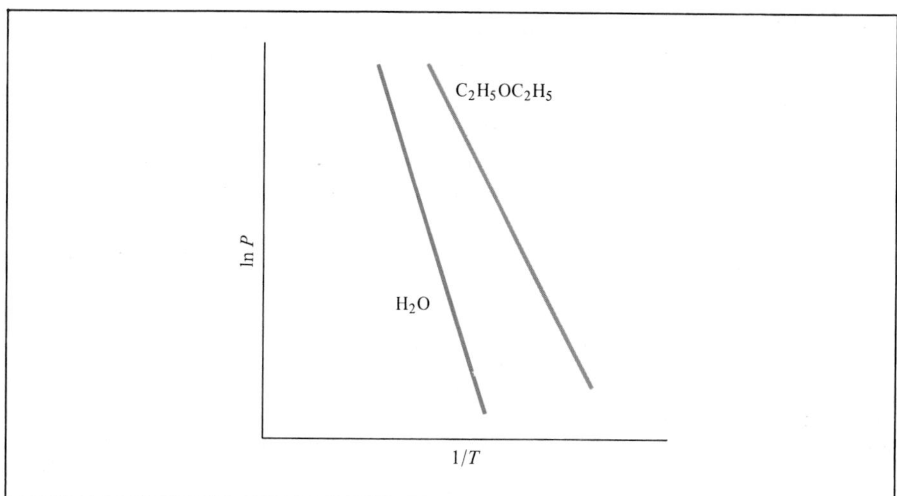

Figure 11.43 *Plots of ln P versus 1/T for water and ethyl ether.*

liquid has a relatively low vapor pressure and a high molar heat of vaporization because much energy is needed to increase the kinetic motion of individual molecules to free them from intermolecular attraction.

The quantity ΔH_{vap} can be determined experimentally. As Figure 11.42 shows, vapor pressure increases with increasing temperature. The quantitative relationship between the vapor pressure P of a liquid and the absolute temperature T is given by the Clausius†-Clapeyron‡ equation

$$\ln P = -\frac{\Delta H_{vap}}{RT} + C \tag{11.2}$$

where ln is the natural logarithm, R is the gas constant (8.314 J/K · mol), and C is a constant. In terms of common, or base 10, logarithms, Equation (11.2) can be written as follows:

$$\log P = -\frac{\Delta H_{vap}}{2.303RT} + C$$

If you have a scientific calculator, you will probably find it easier to work with natural logarithms.

The factor 2.303 converts common logs to natural logs (see Appendix 4). The Clausius-Clapeyron equation has the form of the linear equation $y = mx + b$:

$$\begin{array}{ccccc} \ln P = & \left(-\dfrac{\Delta H_{vap}}{R}\right) & \left(\dfrac{1}{T}\right) & + & C \\ \updownarrow & \updownarrow & \updownarrow & & \updownarrow \\ y = & m & x & + & b \end{array}$$

Figure 11.43 shows that by plotting ln P versus $1/T$, we obtain a straight line with slope (which is negative) equal to $-\Delta H_{vap}/R$. Table 11.7 lists the ΔH_{vap} values of a number of common liquids.

The Clausius-Clapeyron equation can be used in a different way. Suppose we know

†Rudolf Julius Emanuel Clausius (1822–1888). German physicist. Clausius's work was mainly in electricity, kinetic theory of gases, and thermodynamics.

‡Benoit Paul Emile Clapeyron (1799–1864). French engineer. Clapeyron made contributions to the thermodynamic aspects of steam engines.

Table 11.7 Molar heats of vaporization for selected liquids

Substance	Boiling point* (°C)	ΔH_{vap} (kJ/mol)
Argon (Ar)	−186	6.3
Methane (CH_4)	−164	9.2
Ethyl ether ($C_2H_5OC_2H_5$)	34.6	26.0
Ethanol (C_2H_5OH)	78.3	39.3
Benzene (C_6H_6)	80.1	31.0
Water (H_2O)	100	40.79
Mercury (Hg)	357	59.0

*Measured at 1 atm.

ΔH_{vap} is independent of temperature.

the values of ΔH_{vap} and P of a liquid at one temperature. We can calculate the vapor pressure of the liquid at a different temperature as follows. At temperatures T_1 and T_2 the vapor pressures are P_1 and P_2. From Equation (11.2) we can write

$$\ln P_1 = -\frac{\Delta H_{vap}}{RT_1} + C \tag{11.3}$$

$$\ln P_2 = -\frac{\Delta H_{vap}}{RT_2} + C \tag{11.4}$$

Subtracting Equation (11.4) from Equation (11.3) we obtain

$$\begin{aligned}\ln P_1 - \ln P_2 &= -\frac{\Delta H_{vap}}{RT_1} - \left(-\frac{\Delta H_{vap}}{T_2}\right)\\ &= \frac{\Delta H_{vap}}{R}\left(\frac{1}{T_2} - \frac{1}{T_1}\right)\end{aligned}$$

Hence

$$\begin{aligned}\ln \frac{P_1}{P_2} &= \frac{\Delta H_{vap}}{R}\left(\frac{1}{T_2} - \frac{1}{T_1}\right)\\ &= \frac{\Delta H_{vap}}{R}\left(\frac{T_1 - T_2}{T_1T_2}\right)\end{aligned} \tag{11.5}$$

The following example illustrates the use of Equation (11.5).

EXAMPLE 11.8

The vapor pressure of ethyl ether is 401 mmHg at 18.0°C. Calculate its vapor pressure at 32.0°C.

Answer

From Table 11.7, ΔH_{vap} = 26.0 kJ/mol = 26,000 J/mol. The data are

P_1 = 401 mmHg $\quad$ P_2 = ?

T_1 = 18.0°C = 291 K $\quad$ T_2 = 32.0°C = 305 K

From Equation (11.5) we have

$$\ln \frac{401}{P_2} = \frac{26{,}000 \text{ J/mol}}{8.314 \text{ J/K} \cdot \text{mol}} \left[\frac{291 \text{ K} - 305 \text{ K}}{(291 \text{ K})(305 \text{ K})} \right]$$

Taking the antilog of both sides (see Appendix 4), we obtain

$$\frac{401}{P_2} = 0.6106$$

Hence

$$P_2 = 657 \text{ mmHg}$$

Similar problem: 11.107.

A practical way to demonstrate the molar heat of vaporization is by rubbing alcohol on your hands. The heat from your hands increases the kinetic energy of the alcohol molecules. The alcohol evaporates rapidly, thus extracting heat from your hands and cooling them. The process is similar to perspiration, which is one of the effective means by which the human body maintains a constant temperature. Because of the strong intermolecular hydrogen bonding that exists in water, a considerable amount of energy is needed to vaporize the water of perspiration from the body's surface. This energy is supplied by the heat generated in various metabolic processes.

You have already seen that the vapor pressure of a liquid increases with temperature. For every liquid there exists a temperature at which it begins to boil. The ***boiling point*** is *the temperature at which the vapor pressure of a liquid is equal to the external pressure*. The *normal* boiling point of a liquid is the boiling point when the external pressure is 1 atm.

At the boiling point, bubbles form within the liquid. When a bubble forms, the liquid originally occupying that space is pushed aside, and the level of the liquid in the container is forced to rise. The pressure exerted *on* the bubble is largely atmospheric pressure, plus some *hydrostatic pressure* (that is, pressure due to the presence of liquid). The pressure *inside* the bubble is due solely to the vapor pressure of the liquid. When the vapor pressure becomes equal to the external pressure, the bubble rises to the surface of the liquid and bursts. If the vapor pressure in the bubble were lower than the external pressure, the bubble would collapse before it could rise. We can thus conclude that the boiling point of a liquid depends on the external pressure. (We usually ignore the small contribution due to the hydrostatic pressure.) For example, at 1 atm water boils at 100°C, but if the pressure is reduced to 0.5 atm, water boils at only 82°C.

Since the boiling point is defined in terms of the vapor pressure of the liquid, we expect the boiling point to be related to the molar heat of vaporization: The higher the ΔH_{vap}, the higher the boiling point. The data in Table 11.7 roughly confirm our prediction. Ultimately, both the boiling point and ΔH_{vap} are determined by the strength of intermolecular forces. For example, argon (Ar) and methane (CH_4), which have weak dispersion forces, have low boiling points and small molar heats of vaporization. Ethyl ether ($C_2H_5OC_2H_5$) has a dipole moment, and the dipole-dipole forces account for its moderately high boiling point and ΔH_{vap}. Both ethanol (C_2H_5OH) and water have strong hydrogen bonding, which accounts for their high boiling points and large ΔH_{vap} values. Strong metallic bonding causes mercury to have the highest boiling point and ΔH_{vap} of this group of liquids. Interestingly, the boiling point of benzene, which is

CHEMISTRY IN ACTION

Supercritical Fluid Extraction

The critical phenomenon not only is of scientific interest, but has recently found many interesting commercial and industrial applications.

As we have seen in this section, above its critical temperature a substance cannot be liquefied. In this state, the substance exists as a supercritical fluid (SCF). Like a gas, a SCF expands to fill its container. But like a liquid, a SCF can act as a solvent for solids and liquids. By carefully controlling the temperature and pressure of a SCF, we can vary its density and hence alter its solvent properties so that it may selectively dissolve a particular component in a mixture of substances. To date the most successful solvent for many applications is SCF CO_2. As Table 11.8 shows, CO_2 becomes a SCF at 31°C, and under appropriate conditions it can dissolve a variety of substances, including caffeine. Some coffee producers now treat raw coffee beans with SCF CO_2, which dissolves and extracts the caffeine, leaving largely undissolved chemicals that give coffee its characteristic flavor and smell (Figure 11.44). When the SCF caffeine solution is pumped away, the coffee beans are left free of both caffeine and solvent residue. Since residual CO_2 is a gas at room temperature, it disperses immediately. This technique has considerable advantage over the use of the conventional solvent methylene chloride (CH_2Cl_2) to extract caffeine. Carbon dioxide is nontoxic and unreactive toward food constituents, whereas methylene chloride in high dose has been shown to act as a carcinogen in animals. It is difficult to remove traces of methylene chloride (boiling point = 40°C) from coffee beans completely. Other food industry applications under development involve the use of SCF CO_2 to remove (dissolve) the chemicals responsible for the unpleasant smell of fish and vegetable oils and to extract cooking oil from potato chips to produce crisp, oil-free chips.

SCFs are being used in chemical analysis. In a technique called chromatography, a mixture of substances to be analyzed is dispersed in a gas or liquid and passed through a tube packed with an absorbent material. The larger molecules pass through the tube more slowly than the smaller ones, and this difference in speed spreads out the various components so that they can be studied separately. The use of SCFs such as CO_2 as a carrier medium has been shown to be effective for substances that are unstable at high temperatures (for example, biological molecules such as antibiotics and hormones).

Figure 11.44 *An apparatus for supercritical fluid extraction.*

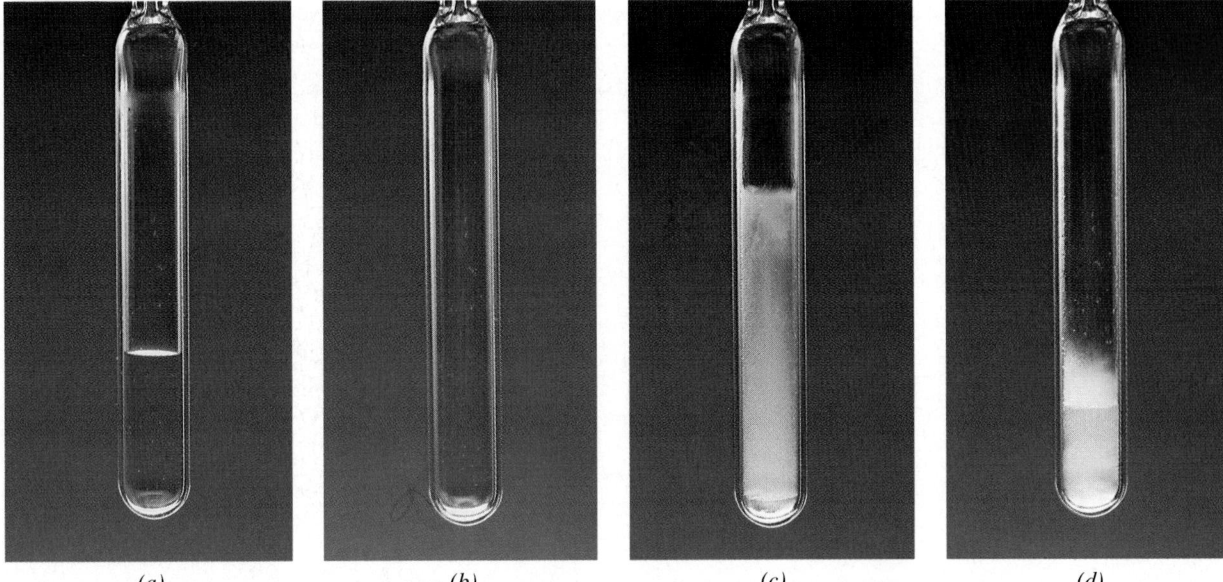

(a) (b) (c) (d)

Figure 11.45 *The critical phenomenon of sulfur hexafluoride. (a) Below the critical temperature. (b) Above the critical temperature. Note that the liquid phase disappears. (c) The substance is cooled just below its critical temperature. The fog represents the condensation of vapor. (d) Finally the appearance of the liquid phase.*

Putting it another way, above the critical temperature there is no fundamental distinction between a liquid and a gas—we simply have a fluid.

Keep in mind that intermolecular forces are independent of temperature, whereas the kinetic energy of molecules increases with temperature.

nonpolar, is comparable to that of ethanol. Benzene has a large polarizability due to its electrons in the delocalized pi molecular orbitals, and the dispersion forces among benzene molecules can become as strong as or even stronger than dipole-dipole forces and/or hydrogen bonds.

Critical Temperature and Pressure. The opposite of evaporation is condensation. In principle, a gas can be made to liquefy by either one of two techniques. By cooling a sample of gas we decrease the kinetic energy of its molecules, and eventually molecules aggregate to form small drops of liquid. Alternatively, we may apply pressure to the gas. Under compression, the average distance between molecules is reduced so that they are, in this way too, held together by mutual attraction. Industrial liquefaction processes use a combination of these two methods.

Every substance has *a temperature,* called the ***critical temperature*** (T_c), *above which its gas form cannot be made to liquefy, no matter how great the applied pressure.* This is also *the highest temperature at which a substance can exist as a liquid. The minimum pressure that must be applied to bring about liquefaction at the critical temperature* is called the ***critical pressure*** (P_c). The existence of the critical temperature can be qualitatively explained as follows. The intermolecular attraction is a finite quantity for any given substance. Below T_c, this force is sufficiently strong to hold the molecules together (under some appropriate pressure) in a liquid. Above T_c, molecular motion becomes so energetic that the molecules can always break away from this attraction. Figure 11.45 shows what happens when sulfur hexafluoride is heated above its critical temperature (45.5°C) and then cooled down to below 45.5°C.

Table 11.8 lists the critical temperatures and critical pressures of a number of common substances. The critical temperature of a substance is determined by the strength of its intermolecular force. Substances such as benzene, ethanol, mercury, and water, which have strong intermolecular forces, also have high critical temperatures compared with the other substances listed in the table.

The Chemistry in Action on the opposite page describes a recently developed application of supercritical fluids.

Table 11.8 Critical temperatures and critical pressures of selected substances

Substance	T_c (°C)	P_c (atm)
Ammonia (NH_3)	132.4	111.5
Argon (Ar)	−186	6.3
Benzene (C_6H_6)	288.9	47.9
Carbon dioxide (CO_2)	31.0	73.0
Ethanol (C_2H_5OH)	243	63.0
Ethyl ether ($C_2H_5OC_2H_5$)	192.6	35.6
Mercury (Hg)	1462	1036
Methane (CH_4)	−83.0	45.6
Molecular hydrogen (H_2)	−239.9	12.8
Molecular nitrogen (N_2)	−147.1	33.5
Molecular oxygen (O_2)	−118.8	49.7
Sulfur hexafluoride (SF_6)	45.5	37.6
Water (H_2O)	374.4	219.5

Liquid-Solid Equilibrium

"Fusion" refers to the process of melting. Thus a "fuse" breaks an electrical circuit when a metallic strip melts due to the heat generated by excessively high electrical current.

The transformation of liquid to solid is called *freezing*, and the reverse process is called *melting* or *fusion*. The **melting point** of a solid (or the freezing point of a liquid) is *the temperature at which solid and liquid phases coexist in equilibrium*. The *normal* melting point (or the *normal* freezing point) of a substance is the melting point (or freezing point) measured at 1 atm pressure. We generally omit the word "normal" in referring to the melting point of a substance at 1 atm.

The most familiar liquid-solid equilibrium occurs between water and ice. At 0°C and 1 atm, the dynamic equilibrium is represented by

$$\text{ice} \rightleftharpoons \text{water}$$

A practical illustration of this dynamic equilibrium is provided by a glass of ice water. As the ice cubes melt to form water, some of the water between ice cubes may freeze, thus joining the cubes together. This is not a true dynamic equilibrium; since the glass is not kept at 0°C, all the ice cubes will eventually melt away.

Because molecules are more strongly held in the solid state than in the liquid state, heating is required to bring about the solid-liquid phase transition. *The energy* (usually in kilojoules) *required to melt one mole of a solid* is called the **molar heat of fusion (ΔH_{fus})**. Table 11.9 shows the molar heats of fusion for the substances listed in Table

Table 11.9 Molar heats of fusion for selected substances

Substance	Melting point* (°C)	ΔH_{fus} (kJ/mol)
Argon (Ar)	−190	1.3
Methane (CH_4)	−183	0.84
Ethyl ether ($C_2H_5OC_2H_5$)	−116.2	6.90
Ethanol (C_2H_5OH)	−117.3	7.61
Benzene (C_6H_6)	5.5	10.9
Water (H_2O)	0	6.01
Mercury (Hg)	−39	23.4

*Measured at 1 atm.

11.7. A comparison of the data in the two tables shows that ΔH_{fus} is smaller than ΔH_{vap} for the same substance. This is consistent with the fact that molecules in a liquid are still fairly closely packed together, so that some energy is needed to bring about the rearrangement from solid to liquid. On the other hand, when a liquid evaporates, its molecules become completely separated from one another and considerably more energy is required to overcome the attractive force.

Heating and Cooling Curves. *Heating curves* such as the one shown in Figure 11.46 are helpful in the study of phase transitions. When a solid sample is heated, its temperature increases gradually until point A is reached. At this point, the solid begins to melt. During the melting period (A $\rightarrow$ B), the first flat portion of the curve in Figure 11.46, heat is being absorbed by the system, yet its temperature remains constant. The heat helps the molecules overcome the attractive forces in the solid. Once the sample has melted completely (point B), the heat absorbed increases the average kinetic energy of the liquid molecules, and the liquid temperature rises (B $\rightarrow$ C). The vaporization process (C $\rightarrow$ D) can be explained similarly. The temperature remains constant during the period when the increased kinetic energy is used to overcome the cohesive forces in the liquid. When all molecules are in the gas phase, the temperature rises again.

As we would expect, the *cooling curve* of a substance is the reverse of its heating curve. If we remove heat from a gas sample at a steady rate, its temperature decreases. As the liquid is being formed, heat is given off by the system, since its potential energy is decreasing. For this reason, the temperature of the system remains constant over the condensation period (D $\rightarrow$ C). After all the vapor has condensed, the temperature of the liquid begins to drop. Continued cooling of the liquid finally leads to freezing (B $\rightarrow$ A).

A liquid can be temporarily cooled to below its freezing point. This phenomenon, called ***supercooling,*** may occur when heat is removed from a liquid so rapidly that the molecules literally have no time to assume the ordered structure of a solid. A super-

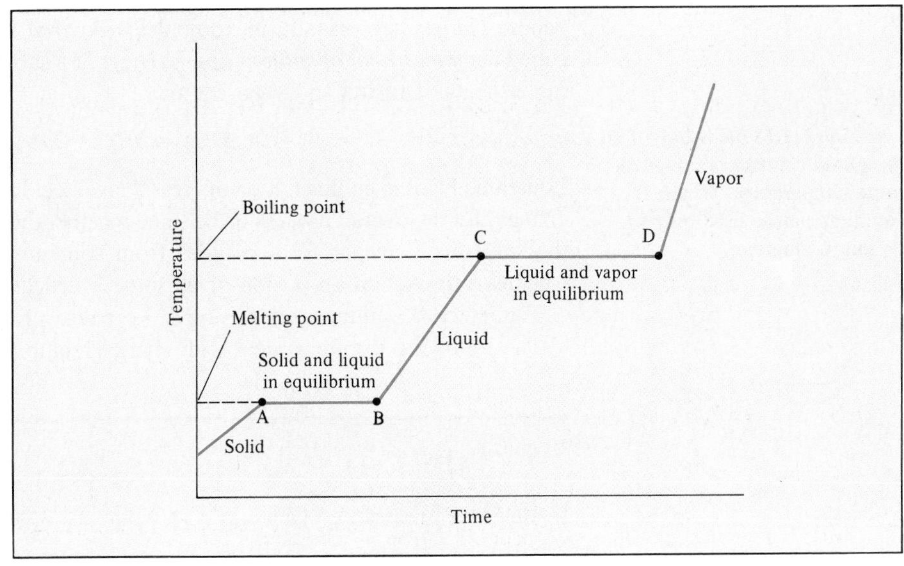

Figure 11.46 *A typical heating curve, from the solid phase through the liquid phase to the gas phase of a substance. Because ΔH_{fus} is smaller than ΔH_{vap}, a substance melts in less time than it takes to boil. This explains why AB is shorter than CD. The steepness of the solid, liquid, and vapor heating lines is determined by the specific heat of the substance in each state.*

Figure 11.47 *Solid iodine in equilibrium with its vapor at room temperature.*

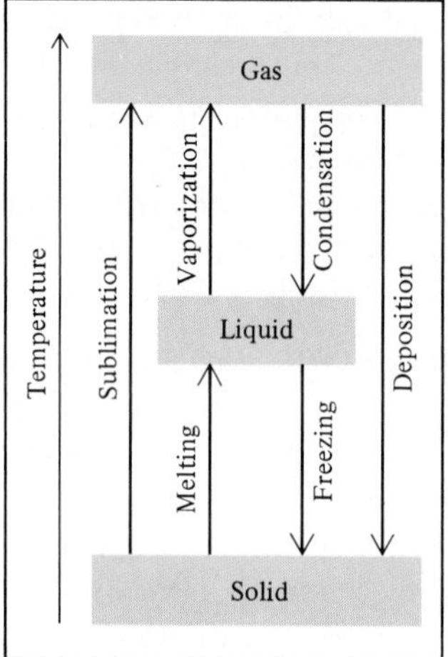

Figure 11.48 *The various phase changes that a substance can undergo.*

cooled liquid is unstable; gentle stirring or the addition to it of a small "seed" crystal of the same substance will cause it to solidify quickly.

Solid-Vapor Equilibrium

Solids, too, undergo evaporation and, therefore, possess a vapor pressure. Consider the following dynamic equilibrium:

$$\text{solid} \rightleftharpoons \text{vapor}$$

The process in which molecules go directly from the solid into the vapor phase is called ***sublimation,*** and the reverse process (that is, *from vapor directly to solid*) is called ***deposition.*** Naphthalene (the substance used to make moth balls) has a fairly high (equilibrium) vapor pressure for a solid (1 mmHg at 53°C); thus its pungent vapor quickly permeates an enclosed space. Because molecules are more tightly held in a solid, its vapor pressure is generally much less than that of the corresponding liquid. Figure 11.47 shows another volatile solid, molecular iodine, in equilibrium with its vapor. *The energy* (usually in kilojoules) *required to sublime one mole of a solid,* called the ***molar heat of sublimation (*** ΔH_{sub} ***)*** of a substance, is given by the sum of the molar heats of fusion and vaporization:

Equation (11.6) holds only if all the phase changes occur at the same temperature. If not, the equation can be used only as an approximation.

$$\Delta H_{\text{sub}} = \Delta H_{\text{fus}} + \Delta H_{\text{vap}} \qquad (11.6)$$

Equation (11.6) is an illustration of Hess's law (see Section 6.5). The enthalpy, or heat change, for the overall process is the same whether the substance changes directly from the solid to the vapor form or goes from solid to liquid and then to vapor. (The Chemistry in Action on p. 486 gives some practical examples of sublimation.)

Figure 11.48 summarizes the types of phase changes discussed in this section. The following examples deal with energy relationships in phase changes.

EXAMPLE 11.9

How much heat (in kilojoules) is required to melt 86.0 g of ice at 0°C?

Answer

We need to convert 86.0 g into moles. Then, with the molar heat of fusion from Table 11.9, we can calculate the amount of heat required for the melting process:

$$\text{heat required} = 86.0 \text{ g } H_2O \times \frac{1 \text{ mol } H_2O}{18.02 \text{ g } H_2O} \times \frac{6.01 \text{ kJ}}{1 \text{ mol } H_2O}$$
$$= 28.7 \text{ kJ}$$

Similar problem: 11.95.

EXAMPLE 11.10

Calculate the amount of energy (in kilojoules) needed to heat 346 g of liquid water from 0°C to 182°C. Assume that the specific heat of water is 4.184 J/g · °C over the entire liquid range and the specific heat of steam is 1.99 J/g · °C.

Answer

The calculation can be broken down into three steps.

Step 1: Heating water from 0°C to 100°C

Using Equation (6.3) we write

$$\begin{aligned} q_1 &= ms\,\Delta t \\ &= (346 \text{ g})(4.184 \text{ J/g} \cdot {}^\circ\text{C})(100^\circ\text{C} - 0^\circ\text{C}) \\ &= 1.45 \times 10^5 \text{ J} \\ &= 145 \text{ kJ} \end{aligned}$$

Step 2: Evaporating 346 g of water at 100°C

In Table 11.7 we see $\Delta H_{vap} = 40.79$ kJ/mol for water, so

$$q_2 = 346 \text{ g } H_2O \times \frac{1 \text{ mol } H_2O}{18.02 \text{ g } H_2O} \times \frac{40.79 \text{ kJ}}{1 \text{ mol } H_2O}$$
$$= 783 \text{ kJ}$$

Step 3: Heating steam from 100°C to 182°C

$$\begin{aligned} q_3 &= ms\,\Delta t \\ &= (346 \text{ g})(1.99 \text{ J/g} \cdot {}^\circ\text{C})(182^\circ\text{C} - 100^\circ\text{C}) \\ &= 5.65 \times 10^4 \text{ J} \\ &= 56.5 \text{ kJ} \end{aligned}$$

The overall energy required is given by

$$\begin{aligned} q_{overall} &= q_1 + q_2 + q_3 \\ &= 145 \text{ kJ} + 783 \text{ kJ} + 56.5 \text{ kJ} \\ &= 985 \text{ kJ} \end{aligned}$$

Similar problem: 11.96.

CHEMISTRY IN ACTION

Freeze-Dried Coffee and Cloud Seeding

Of the three types of phase changes (melting, vaporization, and sublimation) discussed in Section 11.8, sublimation is the least familiar to the nonscientist. Yet the sublimation process plays many important roles in our daily lives. Two examples are discussed here.

Freeze-Dried Coffee

Better-quality instant coffee is often made by a "freeze-drying" procedure. A batch of freshly brewed coffee is frozen and then placed in a container from which air is removed by a vacuum pump. The vacuum causes the ice component to sublime. When most of the ice has been removed, the freeze-dried coffee is ready for packaging. Freeze drying has an important advantage over other methods of instant coffee processing—the molecules responsible for the flavor are not destroyed, as they are when coffee is dried by prolonged heating. Furthermore, freeze-dried coffee has a long shelf life and weighs less than other forms of coffee.

Cloud Seeding

For centuries, people have talked about the weather but could do little to control it. To be sure, global weather patterns are very complex, and our knowledge is still limited, so that we cannot predict the weather accurately, let alone control it. Nevertheless, in recent years there have been a number of attempts to change weather in small, localized regions. We cannot convert gloomy, rainy days into bright, sunlit ones, but it is sometimes possible to induce a downpour from an overcast sky. Clouds consist of fine water droplets. In order for these droplets to aggregate and form raindrops, nuclei—small particles onto which water molecules can cluster—must be present in the clouds. In principle, ice crystals, which can act as nuclei for precipitation, should form at 0°C. However, because of supercooling, they seldom develop unless the temperature drops below −10°C. To promote ice crystal formation, clouds are sometimes seeded by dispersing granulated dry ice (a trademark for solid carbon dioxide), which is dispersed into the clouds from an airplane. When solid carbon dioxide sublimes, it absorbs heat from the surrounding clouds and reduces the temperature to and below that required for ice crystal formation. Consequently, the presence of dry ice can often induce a small-scale rainfall (Figure 11.49).

Figure 11.49 *Effects produced by seeding a cloud deck with dry ice. Within an hour a hole (showing the disappearance of cloud) developed in the seeded area.*

11.9 Phase Diagrams

The overall relationships among the solid, liquid, and vapor phases are best represented in a single graph known as a phase diagram. A ***phase diagram*** summarizes *the conditions at which a substance exists as a solid, liquid, or gas*. In this section we will briefly discuss the phase diagrams of water and carbon dioxide.

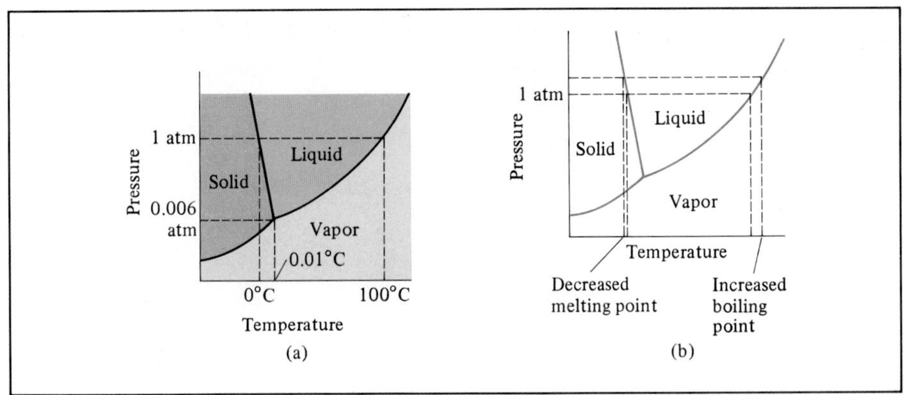

Figure 11.50 *(a) The phase diagram of water. Each solid line between two phases specifies the conditions of pressure and temperature under which the two phases can exist in equilibrium. The point at which all three phases can exist in equilibrium (0.006 atm and 0.01°C) is called the triple point. Note that the liquid-vapor line extends only to the critical temperature of water. (b) This phase diagram tells us that increasing the pressure on ice lowers its melting point and that increasing the pressure of liquid water raises its boiling point.*

Water

Figure 11.50(a) shows the phase diagram of water. The graph is divided into three regions, each of which represents a pure phase. The line separating any two regions indicates conditions under which these two phases can exist in equilibrium. For example, the curve between the liquid and vapor phases shows the variation of vapor pressure with temperature. (Compare this curve with Figure 11.42.) The other two curves similarly indicate conditions for equilibrium between ice and liquid water and between ice and water vapor. The point at which all three curves meet is called the ***triple point.*** For water, this point is at 0.01°C and 0.006 atm. This is the only *condition under which all three phases can be in equilibrium with one another.*

Phase diagrams enable us to predict changes in the melting point and boiling point of a substance as a result of changes in the external pressure; we can also anticipate directions of phase transitions brought about by changes in temperature and pressure. The normal melting point and boiling point of water, measured at 1 atm, are 0°C and 100°C, respectively. What would happen if the melting and boiling were carried out at some other pressure? Figure 11.50(b) shows clearly that increasing the pressure above 1 atm will raise the boiling point and lower the melting point. A decrease in pressure will lower the boiling point and raise the melting point.

Carbon Dioxide

The phase diagram of carbon dioxide (Figure 11.51) is generally similar to that of

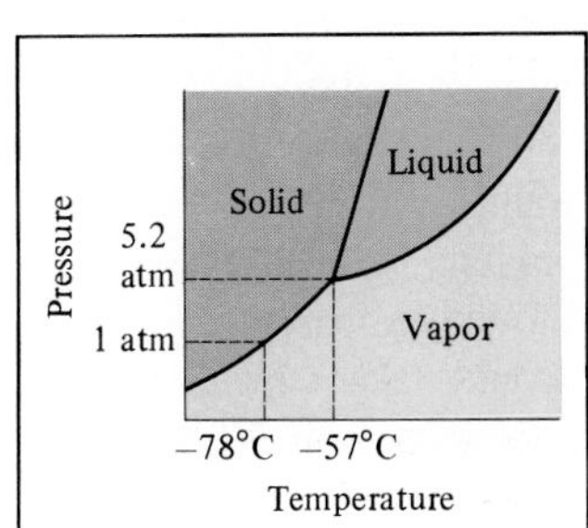

Figure 11.51 *The phase diagram of carbon dioxide. This diagram differs from Figure 11.50(a) in that the solid-liquid boundary line has a positive slope. The triple point is at 5.2 atm and −57°C. The liquid-vapor line does not extend beyond the critical temperature of CO_2. As you can see, the liquid phase is not stable below 5.2 atm, so that only the solid and vapor phases of carbon dioxide can exist under atmospheric conditions.*

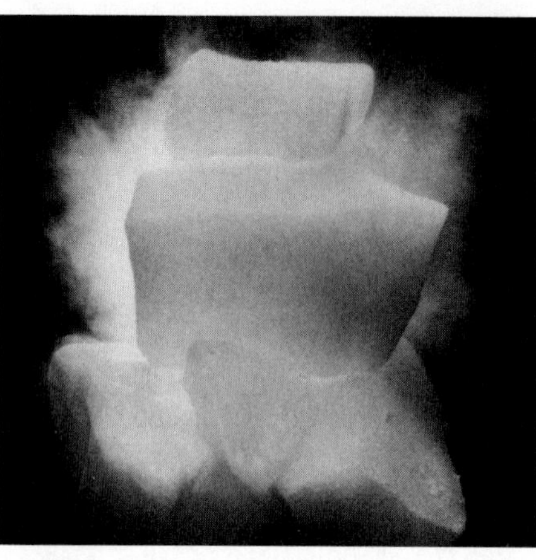

Figure 11.52 *Under atmospheric conditions, dry ice does not melt; it can only sublime.*

water, with one important exception—the slope of the curve between solid and liquid is positive. In fact, this holds true for almost all other substances. Water behaves differently because ice is *less* dense than liquid water. The triple point of carbon dioxide is at 5.2 atm and −57°C.

An interesting observation can be made about the phase diagram in Figure 11.51. As you can see, the entire liquid phase lies well above atmospheric pressure; therefore, it is impossible for solid carbon dioxide to melt at 1 atm. Instead, when solid CO_2 is heated to −78°C at 1 atm, it sublimes. In fact, solid carbon dioxide is called dry ice because it looks like ice and *does not melt* (Figure 11.52).

CHEMISTRY IN ACTION

Hard-Boiling an Egg on a Mountaintop, Pressure Cookers, and Ice Skating

Phase equilibria are affected by external pressure. Depending on atmospheric conditions, the boiling point and freezing point of water may deviate appreciably from 100°C and 0°C, respectively, as we see below.

Hard-Boiling an Egg on a Mountaintop

Suppose you have just scaled Pike's Peak in Colorado. To help regain your strength following the strenuous work, you decide to hard-boil an egg and eat it. To your surprise, water seems to boil more quickly than usual but, after ten minutes in boiling water, the egg is still not cooked. A little knowledge of phase equilibria could have saved you the disappointment of cracking open an uncooked egg (especially if it is the only egg you brought with you). The summit of Pike's Peak is 14,000 ft above sea level. At this altitude, the atmospheric pressure is only about 0.6 atm. From Figure 11.50(b), we see that the boiling point of a liquid decreases with decreasing pressure, so at the lower pressure water will boil at about 86°C. However, it is not the boiling action but the amount of heat delivered to the egg that does the actual cooking, and the amount of heat delivered is proportional to the temperature of the water. For this reason, it would take considerably longer, perhaps 30 minutes, to hard-cook your egg.

Pressure Cookers

The effect of pressure on boiling point also explains why pressure cookers save time in the kitchen. A pressure cooker is a sealed container which allows steam to escape only when it exceeds a certain pressure. The pressure above the water in the cooker is the sum of the atmospheric pressure and the pressure of the steam. Consequently, the water in the pressure cooker will boil above 100°C and the food in it will be hotter and cook faster.

Ice Skating

Let us now turn to the ice-water equilibrium. The negative slope of the solid-liquid curve means that the melting point of ice *decreases* with increasing external pressure, as shown in Figure 11.50(b). This phenomenon makes ice skating possible. Because skates have very thin runners, a 130-lb person can exert a pressure equivalent to 500 atm on the ice. (Remember that pressure is defined as force per unit area.) Consequently, at a temperature lower than 0°C the ice under the skates melts and the film of water formed under the runner facilitates the movement of the skater over ice (Figure 11.53). Calculations show that the melting point of ice decreases by 7.4×10^{-3} °C when the pressure increases by 1 atm. Thus, when the pressure exerted on the ice by the skater is 500 atm, the melting point falls to $-(500 \times 7.4 \times 10^{-3})$, or -3.7°C. Outdoor skating is possible, however, even when the temperature drops below -20°C because friction between the blades and the ice helps to melt the ice.

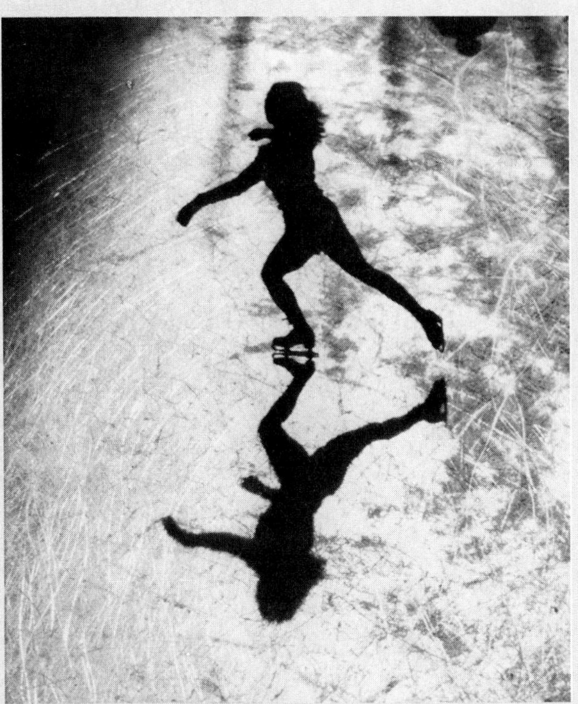

Figure 11.53 *The pressure exerted by the skater on ice lowers its melting point, and the film of water formed under the blades acts as a lubricant between the skate and the ice.*

Summary

1. All substances exist in one of three states: gas, liquid, or solid. The major difference between the condensed state and the gaseous state is the distance of separation between molecules.
2. Intermolecular forces act between molecules or between molecules and ions. Generally, these forces are much weaker than bonding forces.
3. Dipole-dipole forces and ion-dipole forces are involved in the attraction of molecules with dipole moments.
4. Dispersion forces are the result of temporary dipole moments induced in ordinarily nonpolar molecules. The extent to which a dipole moment can be induced in a molecule is called its polarizability. The term "van der Waals forces" refers to the total effect of dipole-dipole, dipole-induced dipole, and dispersion forces. The van der Waals radius is one-half the distance at which these net attractive forces are at their maximum between nonbonded atoms.
5. Hydrogen bonding is a relatively strong dipole-dipole force that acts between a

polar bond containing a hydrogen atom and the bonded electronegative atoms, O, N, or F. Hydrogen bonds between water molecules are particularly strong.

6. Liquids tend to assume a geometry that ensures the minimum surface area. Surface tension is the energy needed to expand a liquid surface area; strong intermolecular forces lead to greater surface tension.
7. Viscosity is a measure of the resistance of a liquid to flow; it decreases with increasing temperature.
8. Water molecules in both the solid and liquid states form a three-dimensional network in which each oxygen atom is covalently bonded to two hydrogen atoms and is hydrogen-bonded to two hydrogen atoms. This unique structure accounts for the fact that ice is less dense than liquid water, a property that allows life to survive under the ice in ponds and lakes in cold climates.
9. Water is also ideally suited for its ecological role by its high specific heat, another property imparted by its strong hydrogen bonding. Large bodies of water are able to moderate the climate by giving off and absorbing substantial amounts of heat with only small changes in the water temperature.
10. All solids are either crystalline (with a regular structure of atoms, ions, or molecules) or amorphous (without this regular structure). Glass is an example of an amorphous solid.
11. Any crystal structure can be classified as one of seven basic types, built up of unit cells that are repeated throughout the crystal. X-ray diffraction has provided much of our knowledge about crystal structure.
12. The four types of crystals and the forces that hold their particles together are ionic crystals, held together by ionic bonding; covalent crystals, covalent bonding; molecular crystals, van der Waals forces and/or hydrogen bonding; and metallic crystals, metallic bonding.
13. A liquid in a closed vessel eventually establishes a dynamic equilibrium between evaporation and condensation. The vapor pressure over the liquid under these conditions is the equilibrium vapor pressure, which is often referred to simply as "vapor pressure."
14. At the boiling point, the vapor pressure of a liquid equals the external pressure. The molar heat of vaporization of a liquid is the energy required to vaporize 1 mole of the liquid. It can be obtained by measuring the vapor pressure of the liquid as a function of temperature and using the Clausius-Clapeyron equation [Equation (11.2)]. The molar heat of fusion of a solid is the energy required to melt 1 mole of the solid.
15. For every substance there is a temperature, called the critical temperature, above which its gas form cannot be made to liquefy.
16. The relationships among the phases of a single substance are represented by a phase diagram, in which each region represents a pure phase and the boundaries between the regions show the temperatures and pressures at which the two phases are in equilibrium. At the triple point, all three phases are in equilibrium.

Applied Chemistry Example: Silicon

Silicon is the second most abundant element (after oxygen) in Earth's crust. Its applications range from bulk commodities such as alloys, concretes, and ceramics through more chemically modified systems like glasses and glazes to the recent industries based on silicone polymers. Silicon-based photovoltaic cells can convert solar energy to electricity with efficiency as high as 15 percent. However, its most profound impact on our society is in semiconductor technology. For example, millions of electronic components may be placed on a single chip no larger than the eraser of a pencil—the so-called one-chip computer used in devices such as the pocket calculator.

Ultrapure silicon is needed for the electronic industry. The production of transistors requires the routine preparation of silicon crystals with impurity levels below 10^{-9} (that is, fewer than one impurity atom for every 10^9, or billion, Si atoms). Purity levels approaching 10^{-12} can be attained in special cases! Silicon (99 percent purity) is first made by the reduction of quartz (SiO_2: melting point = 1602°C, boiling point = 2230°C) with coke in an electric furnace at about 2000°C:

$$SiO_2(s) + 2C(s) \longrightarrow Si(l) + 2CO(g)$$

Silicon (melting point = 1413°C, boiling point = 2355°C) produced this way, called metallurgical grade (MG) silicon, is used in the metallurgical industry to manufacture corrosion-resistant iron and steel. MG silicon can be further purified by treatment with hydrogen chloride at 350°C to form trichlorosilane ($SiCl_3H$)

$$Si(s) + 3HCl(g) \longrightarrow SiCl_3H(g) + H_2(g)$$

Trichlorosilane is distilled to separate it from impurities. Pure silicon is obtained by reversing the above reaction at 1000°C

$$SiCl_3H(g) + H_2(g) \longrightarrow Si(s) + 3HCl(g)$$

Silicon prepared in this manner is called electronic grade (EG) silicon and is suitable for use in electronic components.

1. Trichlorosilane has a vapor pressure of 0.258 atm at −2°C. At what temperature will it boil at 1 atm (the normal boiling point)? Is trichlorosilane's boiling point consistent with the type of intermolecular forces that exist among its molecules? (The molar heat of vaporization of trichlorosilane is 28.8 kJ/mol.)

   Answer: To calculate the boiling point of trichlorosilane, we rearrange Equation (11.5) to get

$$\frac{1}{T_2} = \frac{R}{\Delta H_{vap}} \ln \frac{P_1}{P_2} + \frac{1}{T_1}$$

   where T_2 is the normal boiling point of trichlorosilane. Setting $P_1 = 0.258$ atm, $T_1 = (-2 + 273)$ K = 271 K, $P_2 = 1.00$ atm, we write

$$\frac{1}{T_2} = \frac{8.314 \text{ J/K}\cdot\text{mol}}{28.8 \times 10^3 \text{ J/mol}} \ln \frac{0.258}{1.00} + \frac{1}{271 \text{ K}}$$

$$T_2 = 303 \text{ K or } 30°\text{C}$$

   The Si atom is sp^3-hybridized and the $SiCl_3H$ molecule has a tetrahedral geometry and a dipole moment [like CH_2Cl_2 in Example 10.2(c)]. Thus trichlorosilane is polar and the predominant forces among its molecules are dipole-dipole forces. Since dipole-dipole forces are normally fairly weak, we expect trichlorosilane to have a low boiling point which is consistent with the calculated value of 30°C.

2. What types of crystals do Si and SiO_2 form?

   Answer: From Section 11.6 we see that SiO_2 forms a covalent crystal. Silicon, like carbon in Group 4A, also forms a covalent crystal. The strong covalent bonds between Si atoms (in silicon) and between Si and O atoms (in quartz) account for their high melting points and boiling points.

3. Silicon has a diamond crystal structure (see Figure 11.29). Each cubic unit cell (edge length a = 543 pm) contains eight Si atoms. If there are 1.0×10^{13} impurity boron atoms per cubic centimeter in a sample of EG silicon, how many Si atoms are there for every B atom in the sample? Does this sample satisfy the 10^{-9} purity requirement?

Answer: To test the 10^{-9} purity requirement, we need to calculate the number of Si atoms in 1 cm^3. We can arrive at the answer by carrying out the following three steps: (*a*) Determine the volume of an Si unit cell in cubic centimeters, (*b*) determine the number of Si unit cells in 1 cm^3, and (*c*) multiply the number of unit cells in 1 cm^3 by 8, the number of Si atoms in a unit cell.

Step a: The volume of the unit cell V is

$$\begin{aligned} V &= a^3 \\ &= (543\ \text{pm})^3 \times \left(\frac{1 \times 10^{-12}\ \text{m}}{1\ \text{pm}} \times \frac{1\ \text{cm}}{1 \times 10^{-2}\ \text{m}}\right)^3 \\ &= 1.60 \times 10^{-22}\ \text{cm}^3 \end{aligned}$$

Step b: The number of cells per cubic centimeter is given by

$$\text{number of unit cells} = \frac{1\ \text{cm}^3}{1.60 \times 10^{-22}\ \text{cm}^3} = 6.25 \times 10^{21}$$

Step c: Since there are 8 Si atom per unit cell, the total number of Si atoms is

$$\begin{aligned} \text{number of Si atoms} &= (8\ \text{Si atoms/unit cell})(6.25 \times 10^{21}\ \text{unit cells}) \\ &= 5.00 \times 10^{22}\ \text{Si atoms} \end{aligned}$$

Finally, to calculate the purity of the Si crystal, we write

$$\begin{aligned} \frac{\text{B atoms}}{\text{Si atoms}} &= \frac{1.0 \times 10^{13}\ \text{B atoms}}{5.00 \times 10^{22}\ \text{Si atoms}} \\ &= 2.0 \times 10^{-10} \end{aligned}$$

Since this number is smaller than 10^{-9}, the purity requirement is satisfied.

Key Words

Packing efficiency, p. 458
Phase changes, p. 474
Phase diagram, p. 486
Polarizability, p. 446
Sublimation, p. 484
Supercooling, p. 483
Surface tension, p. 451
Triple point, p. 487
Unit cell, p. 456
van der Waals forces, p. 448
van der Waals radius, p. 448
Vaporization, p. 475
Viscosity, p. 452
X-ray diffraction, p. 464

Exercises

Intermolecular Forces

REVIEW QUESTIONS

11.1 Define the following terms and give an example for each category: (a) dipole-dipole interaction, (b) dipole-induced dipole interaction, (c) ion-dipole interaction, (d) dispersion forces, (e) van der Waals forces.

11.2 Explain the term "polarizability." What kind of molecules tend to have high polarizabilities? What is the relationship between polarizability and intermolecular forces?

11.3 How does the van der Waals radius differ from atomic and ionic radius? Describe one practical use of this radius.

11.4 Explain the difference between the temporary dipole moment induced in an atom or a molecule and the permanent dipole moment in a polar molecule.

11.5 Give some evidence that all nonbonded atoms and molecules exert attractive forces on one another.

11.6 What type of physical properties would you need to consider in comparing the strength of intermolecular forces in solids and in liquids?

11.7 Which elements are able to take part in hydrogen bonding?

11.8 Why is ice less dense than water?

11.9 Outdoor water pipes have to be drained or insulated in winter in a cold climate. Why?

PROBLEMS

11.10 The compounds Br_2 and ICl have the same number of electrons and almost the same molar masses, yet Br_2 melts at $-7.2°C$, whereas ICl melts at $27.2°C$. Explain.

11.11 If you lived in Alaska, state which of the following natural gases you would keep in an outdoor storage tank in winter and explain why: methane (CH_4), propane (C_3H_8), or butane (C_4H_{10}).

11.12 The binary hydrogen compounds of the Group 4A elements are CH_4 ($-162°C$), SiH_4 ($-112°C$), GeH_4 ($-88°C$), and SnH_4 ($-52°C$). The temperatures in parentheses are the corresponding boiling points. Explain the increase in boiling points from CH_4 to SnH_4.

11.13 List the types of intermolecular forces that exist between molecules (or basic units) in each of the following species: (a) benzene (C_6H_6), (b) CH_3Cl, (c) PF_3, (d) NaCl, (e) CS_2.

11.14 Ammonia is both a donor and an acceptor of hydrogen in hydrogen bond formation. Draw a diagram to show the hydrogen bonding of an ammonia molecule with two other ammonia molecules.

11.15 Which of the following species are capable of hydrogen bonding among themselves? (a) C_2H_6, (b) HI, (c) KF, (d) BeH_2, (e) CH_3COOH.

11.16 Arrange the following in order of increasing boiling point: RbF, CO_2, CH_3OH, CH_3Br. Explain your choice.

11.17 Diethyl ether has a boiling point of $34.5°C$, and 1-butanol has a boiling point of $117°C$:

```
   H  H     H  H             H  H  H  H
   |  |     |  |             |  |  |  |
H—C—C—O—C—C—H         H—C—C—C—C—OH
   |  |     |  |             |  |  |  |
   H  H     H  H             H  H  H  H
    diethyl ether               1-butanol
```

Both of these compounds have the same numbers and types of atoms. Explain the difference in their boiling points.

11.18 People living in the northeastern United States who have swimming pools usually drain the water before winter every year. Why?

11.19 Which member of each of the following pairs of substances would you expect to have a higher boiling point? (a) O_2 and N_2, (b) SO_2 and CO_2, (c) HF and HI.

11.20 Explain in terms of intermolecular forces why (a) NH_3 has a higher boiling point than CH_4, (b) KCl has a higher melting point than I_2, and (c) naphthalene ($C_{10}H_8$) is more soluble in benzene than is LiBr.

11.21 Which of the following statements are false? (a) Dipole-dipole interactions between molecules are greatest if the molecules possess only temporary dipole moments. (b) All compounds containing hydrogen atoms can participate in hydrogen bond formation. (c) Dispersion forces exist between all atoms, molecules, and ions. (d) The extent of ion-induced dipole interaction depends only on the charge on the ion.

11.22 What kind of attractive forces must be overcome in order to (a) melt ice, (b) boil molecular bromine, (c) melt solid iodine, and (d) dissociate F_2 into F atoms?

11.23 State which substance in each of the following pairs you would expect to have the higher boiling point and explain why: (a) Ne or Xe, (b) CO_2 or CS_2, (c) CH_4 or Cl_2, (d) F_2 or LiF, (e) NH_3 or PH_3.

11.24 The following compounds have the same number and type of atoms. Which one would you expect to have a higher boiling point?

$$CH_3—CH_2—CH_2—CH_3 \qquad CH_3—\underset{}{\overset{CH_3}{\overset{|}{C}}}H—CH_3$$

(*Hint:* Molecules that can be stacked together more easily have greater intermolecular attraction.)

11.25 Explain the difference in the melting points of the following compounds:

NO_2, OH (ortho-nitrophenol) m.p. 45°C NO_2, OH (para-nitrophenol) m.p. 115°C

(*Hint:* Only one of the two can form intramolecular hydrogen bonds.)

The Liquid State

REVIEW QUESTIONS

11.26 Explain why liquids, unlike gases, are virtually incompressible.

11.27 Define surface tension. What is the relationship between the intermolecular forces that exist in a liquid and its surface tension?

11.28 Despite the fact that stainless steel is much denser than water, a razor blade can be made to float on water. Why?

11.29 Use water and mercury as examples to explain adhesion and cohesion.

11.30 A glass can be filled slightly above the rim with water. Explain why the water does not overflow.

11.31 Draw diagrams showing the capillary action of (a) water and (b) mercury in three tubes of different radii.

11.32 What is viscosity? What is the relationship between the intermolecular forces that exist in a liquid and its viscosity?

11.33 Why does the viscosity of a liquid decrease with increasing temperature?

PROBLEMS

11.34 Predict which liquid has the greater surface tension, ethanol (C_2H_5OH) or dimethyl ether (CH_3OCH_3).

11.35 Predict the viscosity of ethylene glycol

$$\begin{array}{l} CH_2—OH \\ | \\ CH_2—OH \end{array}$$

relative to that of ethanol and glycerol (see Table 11.4).

11.36 Which liquid would you expect to have a greater viscosity, water or diethyl ether? The structure of diethyl ether is shown in Problem 11.17.

Crystalline Solids

REVIEW QUESTIONS

11.37 Define the following terms: crystalline solid, lattice point, crystalline structure, unit cell, coordination number, closest packing, packing efficiency.

11.38 Describe the geometries of the following cubic cells: simple cubic cell, body-centered cubic cell, face-centered cubic cell. Which of these cells would give the highest density for the same type of atoms?

11.39 Describe, and give examples of, the following types of crystals: (a) ionic crystals, (b) covalent crystals, (c) molecular crystals, (d) metallic crystals.

11.40 A solid is hard, brittle, and electrically nonconducting. Its melt (the liquid form of the substance) and an aqueous solution containing the substance do conduct electricity. Classify the solid.

11.41 A solid is soft and has a low melting point (below 100°C). The solid, its melt, and a solution containing the substance are all nonconductors of electricity. Classify the solid.

11.42 A solid is very hard and has a high melting point. Neither the solid nor its melt conducts electricity. Classify the solid.

11.43 Why are metals good conductors of heat and electricity?

11.44 Why does the ability of a metal to conduct electricity decrease with increasing temperature?

11.45 Classify the solid states of the elements in the second period of the periodic table according to crystal type.

11.46 The melting points of the oxides of the third-period elements are given in parentheses: Na_2O (1275°C), MgO (2800°C), Al_2O_3 (2045°C), SiO_2 (1610°C), P_4O_{10} (580°C), SO_3 (16.8°C), Cl_2O_7 (−91.5°C). Classify these solids.

11.47 Which of the following are molecular solids and which are covalent solids? Se_8, HBr, Si, CO_2, C, P_4O_6, B, SiH_4.

PROBLEMS

11.48 What is the coordination number of each sphere in (a) a simple cubic lattice, (b) a body-centered cubic lattice, and (c) a face-centered cubic lattice? Assume the spheres to be of equal size.

11.49 Calculate the number of spheres in the simple cubic, body-centered cubic, and face-centered cubic cells. Assume that the spheres are of equal size and that they are only at the lattice points. Also calculate the packing efficiency for each type of cell.

11.50 Silver crystallizes in a cubic close-packed structure (the face-centered cube). The unit cell edge length is 408.7 pm. Calculate the density of silver.

11.51 The metal polonium crystallizes in a primitive cubic lattice (the Po atoms occupy only the lattice points). If the unit cell length is 336 pm, what is the atomic radius of Po?

11.52 Tungsten crystallizes in a body-centered cubic lattice (the W atoms occupy only the lattice points). How many W atoms are present in a unit cell?

11.53 Metallic iron crystallizes in a cubic lattice. The unit cell edge length is 287 pm. The density of iron is 7.87 g/cm^3. How many iron atoms are there within a unit cell?

11.54 Barium metal crystallizes in a body-centered cubic lattice (the Ba atoms are at the lattice points only). The unit cell edge length is 502 pm, and the density of the metal is 3.50 g/cm^3. Using this information, calculate Avogadro's number. (*Hint:* First calculate the volume occupied by 1 mole of Ba atoms in the unit cells. Next calculate the volume occupied by one Ba atom in the unit cell.)

11.55 Vanadium crystallizes in a body-centered cubic lattice (the V atoms occupy only the lattice points). How many V atoms are present in a unit cell?

11.56 Europium crystallizes in a body-centered cubic lattice (the Eu atoms occupy only the lattice points). The density of Eu is 5.26 g/cm^3. Calculate the unit cell edge length in pm.

11.57 Copper crystallizes in a face-centered cubic lattice (the Cu atoms are at the lattice points only). If the density of metallic copper is 8.96 g/cm^3, what is the unit cell edge length in pm?

11.58 Crystalline silicon has a cubic structure. The unit cell edge length is 543 pm. The density of the solid is 2.33 g/cm^3. Calculate the number of Si atoms in one unit cell.

11.59 A face-centered cubic cell contains 8 X atoms at the corners of the cell and 6 Y atoms at the faces. What is the empirical formula of the solid?

11.60 Describe the general properties of (a) covalent crystals, (b) molecular crystals, (c) ionic crystals, (d) metallic crystals. Give two examples of each type of crystal.

11.61 Classify the crystalline state of the following substances as ionic crystals, covalent crystals, molecular crystals, or metallic crystals: (a) CO_2, (b) B, (c) S_8, (d) KBr, (e) Mg, (f) SiO_2, (g) LiCl, (h) Cr.

11.62 Explain why diamond is harder than graphite.

11.63 Carbon and silicon belong to Group 4A of the periodic table and have the same valence electron configuration (ns^2np^2). Why does silicon dioxide (SiO_2) have a much higher melting point than carbon dioxide (CO_2)?

11.64 The interionic distances of several alkali halides are:

NaCl	NaBr	NaI	KCl	KBr	KI
282 pm	299 pm	324 pm	315 pm	330 pm	353 pm

(a) Plot lattice energy versus interionic distance for these compounds. (b) Plot lattice energy versus the reciprocal of the interionic distance. Which graph is more linear? Why? (For lattice energies see Table 9.1.)

X-Ray Diffraction

REVIEW QUESTIONS

11.65 Define X-ray diffraction. What are the typical wavelengths (in nanometers) of X rays?

11.66 Write the Bragg equation. Define every term and describe how this equation can be used to measure interatomic distances.

PROBLEMS

11.67 When X rays of wavelength 0.090 nm are diffracted by a metallic crystal, the angle of first-order diffraction ($n = 1$) is measured to be 15.2°. What is the distance (in pm) between the layers of atoms responsible for the diffraction?

11.68 The distance between layers in a NaCl crystal is 282 pm. X rays are diffracted from these layers at an angle of 23.0°. Assuming that $n = 1$, calculate the wavelength of the X rays in nm.

11.69 X rays of wavelength 0.065 nm are diffracted from a crystal at an angle of 46°. Assuming that $n = 1$, calculate the distance (in pm) between layers in the crystal.

Amorphous Solids

REVIEW QUESTIONS

11.70 Define amorphous solid. How does it differ from crystalline solid?

11.71 Define glass. What is the chief component of glass? Name three types of glass.

11.72 Which has a greater density, crystalline SiO_2 or amorphous SiO_2? Explain.

Phase Changes

REVIEW QUESTIONS

11.73 Define phase change. Name all possible changes that can occur among the vapor, liquid, and solid states of a substance.

11.74 What is the equilibrium vapor pressure of a liquid? How does it change with temperature?

11.75 Use any one of the phase changes to explain what is meant by dynamic equilibrium.

11.76 Define the following terms: (a) molar heat of vaporization, (b) molar heat of fusion, (c) molar heat of sublimation. What are their units?

11.77 How is the molar heat of sublimation related to the molar heats of vaporization and fusion? On what law is this relation based?

11.78 What can we learn about the strength of intermolecular forces in a liquid from its molar heat of vaporization?

11.79 The greater the molar heat of vaporization of a liquid, the greater its vapor pressure. True or false?

11.80 Write the Clausius-Clapeyron equation. Define all the terms and indicate their units. Explain how you can use the equation to determine experimentally the molar heat of vaporization of a liquid.

11.81 Define boiling point. How does the boiling point of a liquid depend on external pressure? Referring to Table 5.3, what is the boiling point of water when the external pressure is 187.5 mmHg?

11.82 As a liquid is heated at constant pressure, its temperature rises. This trend continues until the boiling point of the liquid is reached. No further rise in temperature of the liquid can be induced by heating. Explain.

11.83 Define critical temperature. What is the significance of critical temperature in liquefaction of gases?

11.84 What is the relationship between intermolecular forces in a liquid and the liquid's boiling point and critical temperature?

11.85 Why is the critical temperature of water greater than that of most other substances?

11.86 What is a supercooled liquid? How can it be produced?

11.87 How do the boiling points and melting points of water and carbon tetrachloride vary with pressure? Explain any difference in behavior of these two substances.

11.88 Why is solid carbon dioxide called dry ice?

11.89 The vapor pressure of a liquid in a closed container depends on which of the following? (a) the volume above the liquid, (b) the amount of liquid present, (c) temperature.

11.90 Referring to Figure 11.42, estimate the boiling points of ethyl ether, water, and mercury at 0.5 atm.

11.91 Wet clothes dry more quickly on a hot, dry day than on a hot, humid day. Explain.

11.92 Explain why the molar heat of vaporization of a substance is always greater than its molar heat of fusion.

11.93 Which of the following phase transitions gives off more heat? (a) 1 mole of steam to 1 mole of water at 100°C, or (b) 1 mole of water to 1 mole of ice at 0°C.

11.94 A beaker of water is heated to boiling by a Bunsen burner. Would adding another burner raise the boiling point of water? Explain.

PROBLEMS

11.95 Calculate the amount of heat (in kJ) required to convert 74.6 g of water to steam at 100°C.

11.96 How much heat (in kJ) is needed to convert 866 g of ice at −10°C to steam at 126°C? (The specific heats of ice and steam are 2.03 J/g · °C and 1.99 J/g · °C, respectively.)

11.97 The molar heats of fusion and sublimation of molecular iodine are 15.27 kJ/mol and 62.30 kJ/mol, respectively. Use Equation (11.6) to estimate the molar heat of vaporization of liquid iodine.

11.98 How is the rate of evaporation of a liquid affected by (a) temperature, (b) the surface area of a liquid exposed to air, (c) intermolecular forces?

11.99 The following compounds are liquid at −10°C; their boiling points are given: butane, −0.5°C; ethanol, 78.3°C; toluene, 110.6°C. At −10°C, which of these liquids would you expect to have the highest vapor pressure? Which the lowest? Explain.

11.100 Freeze-dried coffee is prepared by freezing a sample of brewed coffee and then removing the ice component by vacuum-pumping the sample. Describe the phase changes taking place during these processes.

11.101 A student hangs wet clothes outdoors on a winter day when the temperature is −15°C. After a few hours, the clothes are found to be fairly dry. Describe the phase changes in this drying process.

11.102 Steam at 100°C causes more serious burns than water at 100°C. Why?

11.103 A pressure cooker is a sealed container that allows steam to escape when it exceeds a predetermined pressure. How does this device reduce the time needed for cooking?

11.104 The blades of ice skates are quite thin, so that the pressure exerted on ice by a skater can be substantial. Explain how this fact enables a person to skate on ice.

11.105 A length of wire is placed on top of a block of ice. The ends of the wire extend over the edges of the ice, and a heavy weight is attached to each end. It is found that the ice under the wire gradually melts, so that the wire slowly moves through the ice block. At the same time, the water above the wire refreezes. Explain the phase changes that accompany this phenomenon.

11.106 From the following data, determine graphically the molar heat of vaporization for mercury:

t (°C)	200	250	300	320	340
P (mmHg)	17.3	74.4	246.8	376.3	557.9

11.107 The vapor pressure of benzene, C_6H_6, is 40.1 mmHg at 7.6°C. What is its vapor pressure at 60.6°C? The molar heat of vaporization of benzene is 31.0 kJ/mol.

Phase Diagrams

REVIEW QUESTION

11.108 What is a phase diagram? What useful information can be obtained from study of a phase diagram?

PROBLEMS

11.109 The boiling point and freezing point of sulfur dioxide are $-10°C$ and $-72.7°C$ (at 1 atm), respectively. The triple point is $-75.5°C$ and 1.65×10^{-3} atm, and its critical point is at 157°C and 78 atm. On the basis of this information, draw a rough sketch of the phase diagram of SO_2.

11.110 Consider the phase diagram of water shown at the end of this problem. Label the regions. Predict what would happen if we did the following: (a) Starting at A, we raise the temperature at constant pressure. (b) Starting at C, we lower the temperature at constant pressure. (c) Starting at B, we lower the pressure at constant temperature.

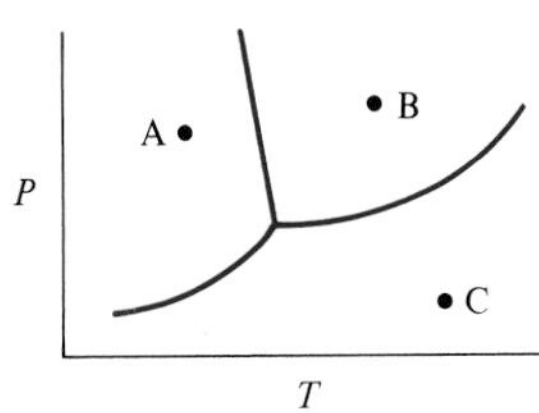

Miscellaneous Problems

11.111 The properties of gases, liquids, and solids differ in a number of respects. How would you use the kinetic molecular theory (see Section 5.7) to explain the following observations? (a) Ease of compressibility decreases from gas to liquid to solid. (b) Solids retain a definite shape, but gases and liquids do not. (c) For most substances, the volume of a given amount of material increases as it changes from solid to liquid to gas.

11.112 Name the kinds of attractive forces that must be overcome in order to (a) boil liquid ammonia, (b) melt solid phosphorus (P_4), (c) dissolve CsI in liquid HF, (d) melt potassium metal.

11.113 Select the substance in each pair that should have the higher boiling point. In each case identify the principal intermolecular forces involved and account briefly for your choice. (a) K_2S or $(CH_3)_3N$, (b) Br_2 or $CH_3CH_2CH_2CH_3$.

11.114 At $-35°C$, liquid HI has a higher vapor pressure than liquid HF. Explain.

11.115 Which of the following indicates very strong intermolecular forces in a liquid? (a) a very low surface tension, (b) a very low critical temperature, (c) a very low boiling point, (d) a very low vapor pressure.

11.116 Under the same conditions of temperature and density, which of the following gases would you expect to behave *less* ideally: CH_4, SO_2? Explain.

11.117 Explain the critical temperature phenomenon in terms of the kinetic molecular theory.

11.118 A small drop of oil in water assumes a spherical shape. Explain. (*Hint:* Oil is made up of nonpolar molecules, which tend to avoid contact with water.)

11.119 The standard enthalpy of formation of gaseous molecular bromine is 30.7 kJ/mol. From these data calculate the molar heat of vaporization of molecular bromine at 25°C. (*Hint:* The standard enthalpy of formation of liquid molecular bromine is zero.)

11.120 A solid contains X, Y, and Z atoms in a cubic lattice with X atoms occupying the corners, Y atoms in the body-centered positions, and Z atoms on the faces of the cell. What is the empirical formula of the compound?

11.121 From the following properties of elemental boron, classify it as one of the crystalline solids discussed in Section 11.6: high melting point (2300°C), poor conductor of heat and electricity, insoluble in water, very hard substance.

11.122 Referring to Figure 11.51, determine the stable phase of CO_2 at (a) 4 atm and $-60°C$ and (b) 0.5 atm and $-20°C$.

11.123 A sample of water of mass 1.20 g is injected into an evacuated 5.00-L flask at 65°C. What percentage of the water will be vapor when the system reaches equilibrium? Assume ideal behavior of water vapor and that the volume of liquid water is negligible. The vapor pressure of water at 65°C is 187.5 mmHg.

11.124 The melting points of the fluorides of the second-period elements are LiF (845°C), BeF_2 (800°C), BF_3 ($-126.7°C$), CF_4($-184°C$), NF_3 ($-206.6°C$), OF_2 ($-223.8°C$), F_2 ($-219.6°C$). Classify the type(s) of intermolecular forces present in each compound.

11.125 The distance between Li^+ and Cl^- is 257 pm in solid LiCl and 203 pm in a LiCl molecule in the gas phase. Explain the difference in the bond lengths.

11.126 Heats of hydration (see Section 6.6) of ionic compounds are largely due to ion-dipole interactions. The values for the alkali metal ions are Li^+ (-520 kJ/mol), Na^+ (-405 kJ/mol), K^+ (-321 kJ/mol). Account for the trend in these values.

11.127 What is the vapor pressure of mercury at its normal boiling point (357°C)?

11.128 A student is given four samples of solids W, X, Y, and Z, all of which have a metallic luster. She is told that the solids could be gold, lead sulfide, mica (which is quartz, or SiO_2), and iodine. The results of her investigations are (a) W is a good electrical conductor; X, Y, and Z are poor electrical conductors; (b) when the solids are hit with a hammer, W flattens out, X shatters into many pieces, Y is smashed into a powder, and Z is not affected; (c) when the solids are heated with a Bunsen burner, Y melts with some sublimation, but X, W, and Z do not melt; (d) in treatment with 6 *M* HNO_3, X dissolves; there is no effect on W or Z. On the basis of her studies, identify the solids.

11.129 In March of 1989 some scientists claimed to have observed nuclear fusion of deuterium atoms in palladium metal at room temperature. While the validity of this "cold fusion" remains highly controversial, it has been known for many years that hydrogen (and deuterium) gas dissolves in the metal as atoms. As Figure 11.31 shows, palladium forms a face-centered lattice. (a) Calculate the atomic radius of Pd (in pm). (b) Calculate the maximum radius of an atom that could fit in between Pd atoms in the lattice and compare this value with 52.9 pm (the atomic radius of H, which is very close to that of D). (c) How many D atoms can fit into a unit cell of Pd? The density of Pd is 12.02 g/cm^3.

11.130 Classify the unit cell of molecular iodine:

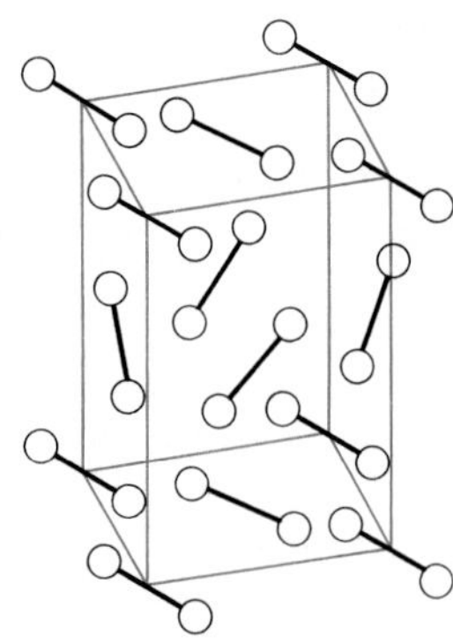

(*Hint:* See Figure 11.15.)

11.131 A CO_2 fire extinguisher is located on the outside of a building in Massachusetts. During the winter months, one can hear a sloshing sound when the extinguisher is gently shaken. In the summertime the sound is often absent. Explain. Assume that the extinguisher has no leaks and that it has not been used.

Opposite page: One application of the freezing-point depression phenomenon is the making of ice cream. Adding salt to ice and water produces a mixture with a freezing point below 0°C.

CHAPTER 12

PHYSICAL PROPERTIES OF SOLUTIONS

Most chemical reactions take place, not between pure solids, liquids, or gases, but among ions and molecules dissolved in water or other solvents. In Chapters 5 and 11 we looked at the properties of gases, liquids, and solids. But what happens when substances are dissolved in solutions? In this chapter we examine the properties of solutions, concentrating mainly on the role of intermolecular forces in solubility and other physical properties of solutions.

12.1 Types of Solutions

We have noted (Section 3.2) that a solution is a homogeneous mixture of two or more substances. Since this definition places no restriction on the nature of the substances involved, we can consider six basic types of solutions, depending on the original states (solid, liquid, or gas) of the solution components. Table 12.1 gives examples of each of these types.

Our focus here will be on solutions involving at least one liquid component—that is, gas-liquid, liquid-liquid, and solid-liquid solutions. And, perhaps not too surprisingly, the liquid we will most often consider as the solvent in our study of solutions is water. But let's start this discussion of solutions at the molecular level, where all "dissolving" ultimately must take place.

12.2 A Molecular View of the Solution Process

In Section 6.6 we discussed the solution process from a macroscopic point of view. Here we will look at the same process at the molecular level. In liquid and solid states molecules are held together by intermolecular attractions. These forces also play a central role in the formation of solutions. The saying "like dissolves like" is based on the action of intermolecular forces. When one substance (the solute) dissolves in another (the solvent)—for example, a solid dissolving in a liquid—particles of the solute disperse uniformly throughout the solvent. The solute particles occupy positions that are normally taken by solvent molecules. The ease with which a solute particle may replace a solvent molecule depends on the relative strengths of three types of interactions:

- solvent-solvent interaction
- solute-solute interaction
- solvent-solute interaction

We can imagine the solution process as taking place in three separate steps (Figure 12.1). Step 1 involves the separation of solvent molecules, and step 2 involves the separation of solute molecules. These steps require energy input to break up attractive intermolecular forces; therefore, they are endothermic. In step 3 the solvent and solute molecules mix. The mixing may be exothermic or endothermic. The heat of solution ΔH_{soln} (see Section 6.6) is given by

Table 12.1 Types of solutions

Component 1	Component 2	State of resulting solution	Examples
Gas	Gas	Gas	Air
Gas	Liquid	Liquid	Soda water (CO_2 in water)
Gas	Solid	Solid	H_2 gas in palladium
Liquid	Liquid	Liquid	Ethanol in water
Solid	Liquid	Liquid	NaCl in water
Solid	Solid	Solid	Brass (Cu/Zn), solder (Sn/Pb)

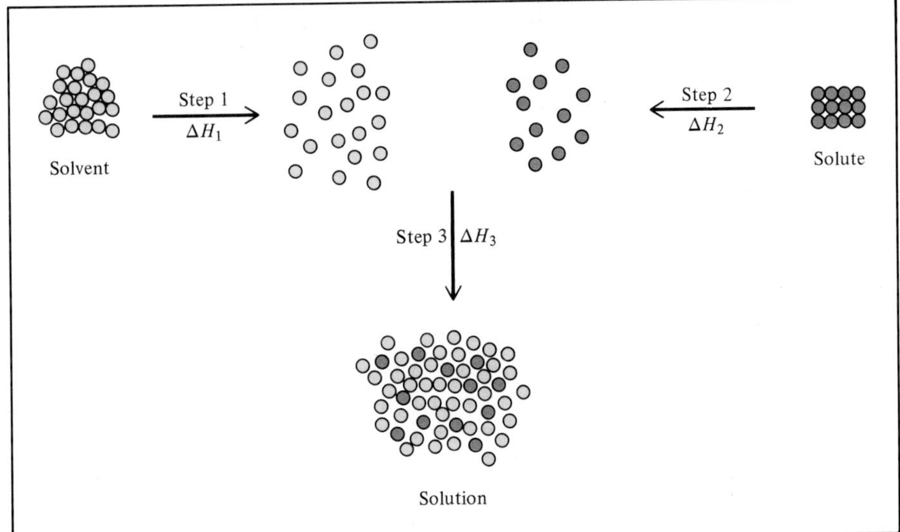

Figure 12.1 *A molecular view of the solution process portrayed as taking place in three steps: First the solvent and solute molecules are separated (steps 1 and 2). Then the solvent and solute molecules mix (step 3).*

This equation is an application of Hess's law.

$$\Delta H_{soln} = \Delta H_1 + \Delta H_2 + \Delta H_3$$

If the solute-solvent attraction is stronger than the solvent-solvent attraction and solute-solute attraction, the solution process will be favorable; that is, the solute will dissolve in the solvent. Such a solution process is exothermic ($\Delta H_{soln} < 0$). If the solute-solvent interaction is weaker than the solvent-solvent and solute-solute interactions, then the solution process is endothermic ($\Delta H_{soln} > 0$).

You may wonder why a solute dissolves in a solvent at all if the attraction among its own molecules is stronger than that between its molecules and the solvent molecules. The solution process, like all physical and chemical processes, is governed by two factors. One is the energy factor, which here determines whether a solution process is exothermic or endothermic. Only exothermic solution processes are related to favorable solute-solvent interactions.

The other factor that needs to be considered is the disorder or randomness that results when solute and solvent molecules mix to form a solution. In the pure state, the solvent and solute possess a fair degree of order. By order we mean the more or less regular arrangement of atoms, molecules, or ions in three-dimensional space. Much of this order is destroyed when the solute dissolves in the solvent (see Figure 12.1). Therefore, the solution process is *always* accompanied by an increase in disorder or randomness. It is the increase in disorder of the system that favors the solubility of any substance. This explains why a substance can be soluble in a solvent even though the solution process is endothermic.

This gaseous solution—assuming a certain proportion of oxygen gas and nitrogen gas are mixed with small amounts of carbon dioxide and argon—is known, of course, as "air"!

A good illustration of the importance of the increase in disorder as a driving force in the solution process is provided by the mixing of two nonreacting gases, such as N_2 and O_2. Under atmospheric conditions these gases behave almost like ideal gases; the intermolecular interaction between them is negligible. Consequently, the mixing process is neither exothermic nor endothermic. Yet we know that these gases easily and spontaneously mix in any proportion, which means that the process is highly favorable. Indeed, it is the disorder created when these two pure gases mix to form a mixture (or a solution) of gases that is responsible for the process taking place so readily.

12.3 Solutions of Liquids in Liquids

Both carbon tetrachloride [CCl_4, b.p. (boiling point) = 76.5°C] and benzene (C_6H_6, b.p. = 80.1°C) are nonpolar liquids. The only intermolecular forces present in these substances are dispersion forces (see Section 11.2). When these two liquids are mixed, they readily dissolve in each other. This is because the attraction between CCl_4 and C_6H_6 molecules is comparable in magnitude to that between CCl_4 molecules and between C_6H_6 molecules. When *two liquids are completely soluble in each other in all proportions,* as in this case, they are said to be ***miscible.*** Ethanol (C_2H_5OH) and water also are miscible liquids. In a solution of ethanol and water the solute-solvent interaction takes the form of hydrogen bonding that is comparable in magnitude to the hydrogen bonding between water molecules and between ethanol molecules.

In either gas-gas or liquid-liquid solutions, distinctions between solutes and solvents become largely a matter of definition. In these cases the solvent is usually assumed to be the component present in the greatest amount.

What would happen if we tried to dissolve carbon tetrachloride in water? To form a solution, the CCl_4 molecules would have to replace some of the H_2O molecules. However, the attractive forces between CCl_4 and H_2O molecules are dipole-induced dipole and dispersion forces, which in this case are much weaker than the hydrogen bonds in water (and the dispersion forces in CCl_4). Consequently, the two liquids do not mix; they are said to be *immiscible*. What we obtain, then, are two distinct liquid phases: one consisting of water with a very small amount of CCl_4 in it, and one consisting of CCl_4 with a very small amount of water in it (Figure 12.2).

Figure 12.2 *A two-phase mixture of water and carbon tetrachloride. The more dense CCl_4 forms the bottom layer.*

These examples illustrate that "like dissolves like": Two nonpolar liquids are often miscible and so are two polar liquids, but a nonpolar one usually dissolves only slightly in a polar solvent. Other examples are sugar and ethylene glycol (a substance used as an antifreeze) dissolving in water (sugar and ethylene glycol molecules can form hydrogen bonds with water) and elemental sulfur (containing the nonpolar S_8 molecules) dissolving in the nonpolar carbon disulfide (CS_2) solvent. The forces of attraction between the S_8 molecules and the CS_2 molecules are solely dispersion forces.

The "like dissolves like" saying, while useful, should be applied with caution. For example, acetic acid (CH_3COOH) is totally miscible with water. We can understand this fact because CH_3COOH molecules can form hydrogen bonds with H_2O molecules. Yet acetic acid is also soluble in nonpolar solvents, such as CCl_4 and C_6H_6! Experimental evidence shows that in nonpolar solvents acetic acid forms ***dimers,*** which are *molecules made up of two identical molecules*. The acetic acid dimer has the structure

```
          O---H—O
        //       \
  H3C—C            C—CH3
        \        //
          O—H---O
```

in which two *(di-)* molecules of the acid are bonded together. As the formula shows, the dimer is held together by two hydrogen bonds. Although acetic acid is a polar molecule, the dimeric species is much less polar because the symmetric structure created by hydrogen bonding reduces the polarity of the molecules. The attractive forces between the dimers and CCl_4 or C_6H_6 molecules are mainly dispersion forces.

12.4 Solutions of Solids in Liquids

The solubility of a solid in a liquid does not depend on whether it is in crystalline or noncrystalline form.

In discussing the solution process of solids in liquids, we will divide solids into four categories, as outlined in Section 11.6: ionic, covalent, molecular, and metallic.

Ionic Crystals

Rules were given in Section 3.3 to help us predict the solubility of a particular ionic compound in water. Section 6.6 described the solution process of ionic compounds in some detail. Using NaCl as an example, we saw that ions are stabilized in solution by hydration, which involves ion-dipole interaction. We also analyzed the heat of solution in terms of the lattice energy of the compound and the heat of hydration. In general, we predict that ionic compounds should be much more soluble in polar solvents, such as water, liquid ammonia, and liquid hydrogen fluoride, than in nonpolar solvents, such as benzene and carbon tetrachloride. Since the molecules of nonpolar solvents lack a dipole moment, they cannot effectively solvate the Na^+ and Cl^- ions. The predominant intermolecular interaction is ion-induced dipole interaction, which is much weaker than ion-dipole interaction. Consequently, ionic compounds usually have extremely low solubility in nonpolar solvents.

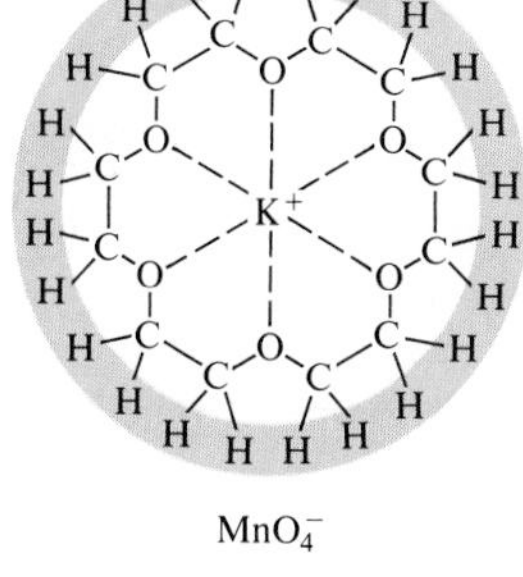

The solubility of ionic compounds in nonpolar solvents can be drastically increased by using a class of compounds called *crown ethers*. The compound shown in Figure 12.3 is particularly effective in binding alkali metal ions such as K^+. For example, potassium permanganate, which is practically insoluble in benzene, can be made to dissolve to a certain extent in that solvent if a suitable crown ether is added to it (Figure 12.4). The cation is situated in the center of the ring and is stabilized by ion-dipole interaction with the surrounding O atoms (see diagram in margin). Because the exterior of the crown ether is largely nonpolar (colored portion), the crown ether–K^+ complex can readily dissolve in nonpolar solvents. The resulting solution is useful because it allows oxidation with potassium permanganate to be carried out in nonpolar organic solvents.

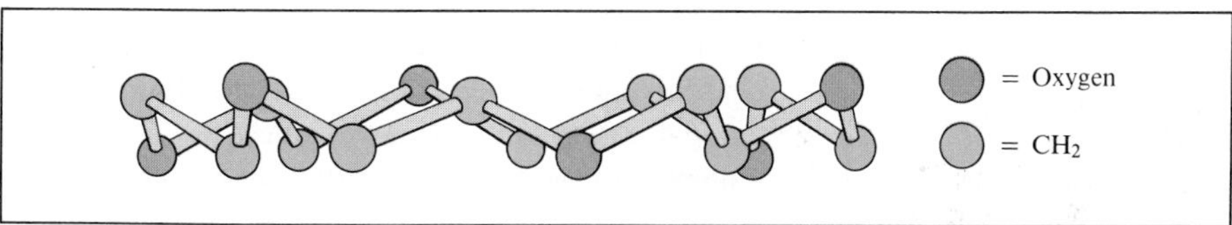

Figure 12.3 *The structure of 18-crown-6, where the number 18 refers to the total number of atoms in the ring and 6 to the number of O atoms in the ring. As you can see, the three-dimensional structure of the molecule resembles a crown, and hence the name of this class of compounds.*

Figure 12.4 *Left: Potassium permanganate (bottom of flask) is completely insoluble in benzene. Right: Upon the addition of 18-crown-6, the purple color appears, showing that some potassium permanganate has dissolved.*

Figure 12.5 *Solubilities of iodine in water (upper layer) and in carbon tetrachloride (lower layer).*

Covalent Crystals

Covalent crystals, such as graphite and quartz (SiO_2), generally do not dissolve in any solvent, polar or nonpolar.

Molecular Crystals

The attractive forces between molecules of a molecular crystal are relatively weak dipole-dipole, dispersion type, and/or hydrogen bonding. Consider naphthalene ($C_{10}H_8$) as an example. In a naphthalene crystal the $C_{10}H_8$ molecules are held together solely by dispersion forces. Thus we can predict that naphthalene should dissolve readily in nonpolar solvents, such as benzene, but only slightly in water. This is indeed the case. On the other hand, we find that urea (H_2NCONH_2)

$$H_2N-\underset{\underset{O}{\|}}{C}-NH_2$$

is much more soluble in water and ethanol than in CCl_4 or C_6H_6 because it has the capacity to form hydrogen bonds with water. (The H atoms and the O and N atoms in urea can form hydrogen bonds with H_2O and C_2H_5OH molecules.) Figure 12.5 compares the solubilities of molecular iodine, a nonpolar substance, in water and in carbon tetrachloride. The darker color of the lower layer shows that iodine is much more soluble in carbon tetrachloride than in water.

Metallic Crystals

In general, metals are not soluble in any solvent, polar or nonpolar. A number of metals instead react chemically with the solvent. For example, the alkali metals and some alkaline earth metals (Ca, Sr, and Ba) react with water to produce hydrogen gas and the corresponding metal hydroxide. The alkali metals also "dissolve" in liquid ammonia to produce a beautiful blue solution containing solvated electrons:

Although we use the term "dissolve," this is really a chemical rather than a physical process.

$$Na(s) \xrightarrow{NH_3} Na^+ + e^-$$

The Na^+ ions and the electrons are stabilized by ion-dipole interaction with the polar NH_3 molecules.

These general principles allow us to predict solubilities in various solvents.

EXAMPLE 12.1

Recall that μ represents dipole moment.

Predict the relative solubilities in the following cases: (a) Br_2 in benzene (C_6H_6) (μ = 0 D) and in water (μ = 1.87 D), (b) KCl in carbon tetrachloride (μ = 0 D) and in liquid ammonia (μ = 1.46 D).

Answer

(a) Br_2 is a nonpolar molecule and therefore should be more soluble in C_6H_6, which is also nonpolar, than in water. The only intermolecular forces between Br_2 and C_6H_6 are dispersion forces.

(b) KCl is an ionic compound. For it to dissolve, the individual K^+ and Cl^- ions must be stabilized by ion-dipole interaction. Since carbon tetrachloride has no dipole moment, potassium chloride should be more soluble in liquid ammonia, a polar molecule with a large dipole moment.

Similar problems: 12.10, 12.12.

12.5 Concentration Units

Quantitative study of a solution requires that we know its *concentration,* that is, the amount of solute present in a given amount of solution. Chemists use several different concentration units, each of which has advantages as well as limitations. The choice of concentration unit is generally based on the kind of measurement made of the solution, to be discussed later in this section. First, though, let us examine the four most common units of concentration: percent by mass, mole fraction, molarity, and molality.

Types of Concentration Units

Percent by Mass. The ***percent by mass*** (also called the *percent by weight* or the *weight percent*) is defined as

$$\text{percent by mass of solute} = \frac{\text{mass of solute}}{\text{mass of solute} + \text{mass of solvent}} \times 100\%$$

$$= \frac{\text{mass of solute}}{\text{mass of soln}} \times 100\%$$

The percent by mass has no units because it is a ratio of two similar quantities.

EXAMPLE 12.2

A sample of 0.892 g of potassium chloride (KCl) is dissolved in 54.6 g of water. What is the percent by mass of KCl in this solution?

Answer

$$\text{percent by mass of KCl} = \frac{\text{mass of solute}}{\text{mass of soln}} \times 100\%$$

$$= \frac{0.892\text{ g}}{0.892\text{ g} + 54.6\text{ g}} \times 100\%$$

$$= 1.61\%$$

Similar problem: 12.14.

Mole Fraction (X). The concept of mole fraction was first introduced in Section 5.6. The mole fraction of a component of a solution, say, component A, is written X_A and is defined as

$$\text{mole fraction of component A} = X_A = \frac{\text{moles of A}}{\text{sum of moles of all components}}$$

The mole fraction has no units, since it too is a ratio of two similar quantities.

EXAMPLE 12.3

A chemist prepared a solution by adding 200.4 g of pure ethanol (C_2H_5OH) to 143.9 g of water. Calculate the mole fractions of these two components. The molar masses of ethanol and water are 46.02 g and 18.02 g, respectively.

Answer

The number of moles of C_2H_5OH and H_2O present are

$$\text{moles of } C_2H_5OH = 200.4 \text{ g } C_2H_5OH \times \frac{1 \text{ mol } C_2H_5OH}{46.02 \text{ g } C_2H_5OH}$$
$$= 4.355 \text{ mol } C_2H_5OH$$

$$\text{moles of } H_2O = 143.9 \text{ g } H_2O \times \frac{1 \text{ mol } H_2O}{18.02 \text{ g } H_2O}$$
$$= 7.986 \text{ mol } H_2O$$

In a two-component system made up of A and B molecules, the mole fraction of A is given by

$$X_A = \frac{\text{moles of A}}{\text{sum of moles of A and B}}$$

Using this equation we can write the mole fractions of ethanol and water as

$$X_{C_2H_5OH} = \frac{4.355 \text{ mol}}{(4.355 + 7.986) \text{ mol}} = 0.3529$$

$$X_{H_2O} = \frac{7.986 \text{ mol}}{(4.355 + 7.986) \text{ mol}} = 0.6471$$

By definition, the sum of the mole fractions of all components in a solution must be 1. Thus

$$X_{C_2H_5OH} + X_{H_2O} = 0.3529 + 0.6471 = 1.0000$$

Similar problem: 12.16.

For calculations involving molarities, see Example 4.7 and 4.8.

Molarity (*M*). The molarity unit was first introduced in Section 4.4; it is defined as the number of moles of solute in 1 liter of solution; that is,

$$\text{molarity} = \frac{\text{moles of solute}}{\text{liters of soln}}$$

Thus, molarity has the units of mol/L.

Molality (m). ***Molality*** is *the number of moles of solute dissolved in 1 kg (1000 g) of solvent*—that is,

$$\text{molality} = \frac{\text{moles of solute}}{\text{mass of solvent (kg)}}$$

For example, to prepare a 1 *molal,* or 1 m, sodium sulfate (Na_2SO_4) aqueous solution, we need to dissolve 1 mole (142.0 g) of the substance in 1000 g (1 kg) of water. Depending on the nature of the solute-solvent interaction, the final volume of the solution will be either greater or less than 1000 mL. It is also possible, though very unlikely, that the final volume could be equal to 1000 mL.

EXAMPLE 12.4

Calculate the molality of a sulfuric acid solution containing 24.4 g of sulfuric acid in 198 g of water. The molar mass of sulfuric acid is 98.08 g.

Answer

From the known molar mass of sulfuric acid, we can calculate the molality in two steps. First we need to find the number of grams of sulfuric acid dissolved in 1000 g (1 kg) of water. Next we must convert the number of grams into the number of moles. Combining these two steps we write

$$\begin{aligned}\text{molality} &= \frac{\text{moles of solute}}{\text{mass of solvent (kg)}}\\ &= \frac{24.4 \text{ g } H_2SO_4}{198 \text{ g } H_2O} \times \frac{1000 \text{ g } H_2O}{1 \text{ kg } H_2O} \times \frac{1 \text{ mol } H_2SO_4}{98.08 \text{ g } H_2SO_4}\\ &= 1.26 \text{ mol } H_2SO_4/\text{kg } H_2O\\ &= 1.26\ m\end{aligned}$$

Similar problem: 12.17.

EXAMPLE 12.5

In how many grams of water should 18.7 g of ammonium nitrate (NH_4NO_3) be dissolved to prepare a 0.542 m solution?

Answer

The number of moles of NH_4NO_3 in 18.7 g of NH_4NO_3 is

$$\text{moles} = 18.7 \text{ g } NH_4NO_3 \times \frac{1 \text{ mol } NH_4NO_3}{80.06 \text{ g } NH_4NO_3} = 0.2336 \text{ mol } NH_4NO_3$$

Since

$$\text{molality} = \frac{\text{moles of solute}}{\text{mass of solvent (kg)}}$$

we write

$$\text{mass of water} = \frac{\text{mol } NH_4NO_3}{\text{molality}}$$
$$= 0.2336 \text{ mol } NH_4NO_3 \times \frac{1 \text{ kg } H_2O}{0.542 \text{ mol } NH_4NO_3}$$
$$= 0.431 \text{ kg } H_2O$$
$$= 431 \text{ g } H_2O$$

Comparison of Concentration Units

As mentioned, the four concentration terms have various uses. For instance, the mole fraction unit is not generally used to express the concentrations of solutions for titrations and gravimetric analyses, but it is useful for calculating partial pressures of gases (see Section 5.6) and for dealing with vapor pressures of solutions (to be discussed later in this chapter).

We saw in Section 4.4 that the advantage of molarity is that it is generally easier to measure the volume of a solution, using precisely calibrated volumetric flasks, than to weigh the solvent. For this reason, molarity is often preferred over molality. On the other hand, molality is independent of temperature, since the concentration is expressed in number of moles of solute and mass of solvent, whereas the volume of a solution generally increases with increasing temperature. A solution that is 1.0 *M* at 25°C may become 0.97 *M* at 45°C because of the increase in volume. This concentration dependence on temperature can significantly affect the accuracy of an experiment.

Percent by mass is similar to molality in that it too is independent of temperature since it is defined in terms of ratio of mass of solute to mass of solution. Furthermore, we do not need to know the molar mass of the solute in order to calculate the percent by mass.

The following examples show that these ways of expressing concentration are interconvertible when the density of the solution is known.

EXAMPLE 12.6

Calculate the molarity of a 0.396 molal glucose ($C_6H_{12}O_6$) solution. The molar mass of glucose is 180.2 g, and the density of the solution is 1.16 g/mL.

Answer

The mass of the solution must be converted to volume in working problems of this type. The density of the solution is used as a conversion factor. Since a 0.396 *m* glucose solution contains 0.396 mole of glucose in 1 kg of water, the total mass of the solution is

$$\left(0.396 \text{ mol } C_6H_{12}O_6 \times \frac{180.2 \text{ g}}{1 \text{ mol } C_6H_{12}O_6}\right) + 1000 \text{ g } H_2O \text{ solution} = 1071 \text{ g}$$

From the known density of the solution (1.16 g/mL), we can calculate the molarity as follows:

$$\text{molarity} = \frac{0.396 \text{ mol } C_6H_{12}O_6}{1071 \text{ g soln}} \times \frac{1.16 \text{ g soln}}{1 \text{ mL soln}} \times \frac{1000 \text{ mL soln}}{1 \text{ L soln}}$$

$$= \frac{0.429 \text{ mol } C_6H_{12}O_6}{1 \text{ L soln}}$$

$$= 0.429\ M$$

EXAMPLE 12.7

The density of a 2.45 M aqueous methanol (CH_3OH) solution is 0.976 g/mL. What is the molality of the solution? The molar mass of methanol is 32.04 g.

Answer

The total mass of 1 L of a 2.45 M solution of methanol is

$$1 \text{ L soln} \times \frac{1000 \text{ mL soln}}{1 \text{ L soln}} \times \frac{0.976 \text{ g soln}}{1 \text{ mL soln}} = 976 \text{ g soln}$$

Since this solution contains 2.45 moles of methanol, the amount of water in the solution is

$$976 \text{ g soln} - \left(2.45 \text{ mol } CH_3OH \times \frac{32.04 \text{ g } CH_3OH}{1 \text{ mol } CH_3OH}\right) = 898 \text{ g } H_2O$$

Now the molality of the solution can be calculated:

$$\text{molality} = \frac{2.45 \text{ mol } CH_3OH}{898 \text{ g } H_2O} \times \frac{1000 \text{ g } H_2O}{1 \text{ kg } H_2O}$$

$$= \frac{2.73 \text{ mol } CH_3OH}{1 \text{ kg } H_2O}$$

$$= 2.73\ m$$

Similar problems: 12.19, 12.20.

EXAMPLE 12.8

Calculate the molality of a 35.4 percent (by mass) aqueous solution of phosphoric acid (H_3PO_4). The molar mass of phosphoric acid is 98.00 g.

Answer

100% − 35.4% = 64.6%, so the ratio of phosphoric acid to water in this solution is 35.4 g H_3PO_4 to 64.6 g H_2O. From the known molar mass of phosphoric acid, we can calculate the molality in two steps. First we need to find the number of grams of phosphoric acid dissolved in 1000 g (1 kg) of water. Next, we must convert the number of grams into the number of moles. Combining these two steps we write

$$\text{molality} = \frac{35.4 \text{ g } H_3PO_4}{64.6 \text{ g } H_2O} \times \frac{1000 \text{ g } H_2O}{1 \text{ kg } H_2O} \times \frac{1 \text{ mol } H_3PO_4}{98.00 \text{ g } H_3PO_4}$$
$$= 5.59 \text{ mol } H_3PO_4/\text{kg } H_2O$$
$$= 5.59\ m$$

Similar problem: 12.24.

12.6 Effect of Temperature on Solubility

Solid Solubility and Temperature

At a given temperature, when the maximum amount of a substance has dissolved in a solvent, we have a ***saturated solution.*** Before this point is reached, *when a solution contains less solute than it has the capacity to dissolve,* we have an ***unsaturated solution.*** A dynamic equilibrium process exists in a saturated solution. Take a saturated LiBr solution as an example. The solution can be prepared by adding an *excess* of solid LiBr to a beaker of water at 25°C. At every instant some of the Li^+ and Br^- ions in the solid dissolve in water, while an *equal* number of Li^+ and Br^- ions in solution aggregate to re-form solid LiBr. *The process in which dissolved solute comes out of solution and forms crystals* is called ***crystallization.***

For most substances, solution temperature affects solubility. Figure 12.6 shows the temperature dependence of the solubility in water of some ionic compounds. In most but certainly not all cases the solubility of a solid substance increases with temperature. However, there is no clear correlation between the sign of ΔH_{soln} and the variation of

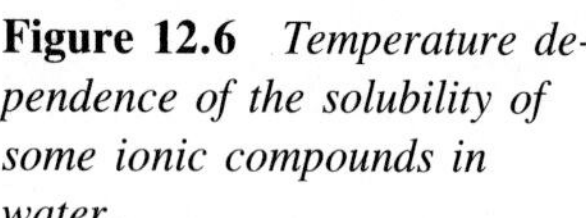

Figure 12.6 *Temperature dependence of the solubility of some ionic compounds in water.*

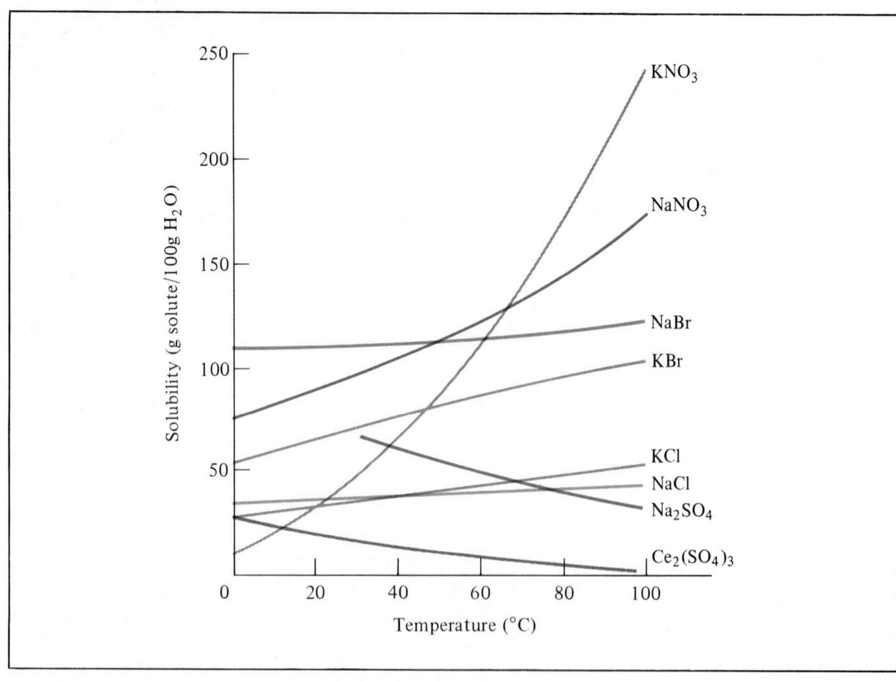

Figure 12.7 *In a supersaturated sodium acetate solution (left), sodium acetate crystals rapidly form after a small seed crystal is added.*

solubility with temperature. For example, the solution process of $CaCl_2$ is exothermic, and that of NH_4NO_3 is endothermic. But the solubility of both compounds increases with increasing temperature. In general the dependence of solubility on temperature is best determined by experiment.

Many substances can also form ***supersaturated solutions,*** *solutions that contain more of the solute than is present in saturated solutions*. For instance, a supersaturated solution of sodium acetate (CH_3COONa) can be prepared by cooling a saturated solution of it. If the process is carried out slowly and carefully, the excess sodium acetate will not crystallize out as the temperature decreases. However, supersaturated solutions are unstable. Once the supersaturated solution has been prepared, addition of a very small amount of solid sodium acetate (called a *seed crystal*) will cause the excess sodium acetate to crystallize out of the solution immediately (Figure 12.7).

Note that both precipitation and crystallization describe the separation of excess solid substance from a supersaturated solution. However, the solids formed in the two processes differ in appearance. We normally think of precipitates as being made up of small particles, whereas crystals may be large and well formed.

Fractional Crystallization

As Figure 12.6 shows, the dependence of the solubility of a substance on temperature varies considerably. The solubility of sodium nitrate, for example, increases sharply with temperature, while that of sodium bromide changes hardly at all. ***Fractional crystallization,*** or *the separation of a mixture of substances into pure components on the basis of their differing solubilities,* makes use of this fact. Suppose we have a sample of 90 g of KNO_3 that is contaminated with 10 g of NaCl. To purify the KNO_3 sample, we dissolve the mixture in 100 mL of water at 60°C and then gradually cool the solution to 0°C. At this temperature the solubilities of KNO_3 and NaCl are 12.1 g/100 g H_2O and 34.2 g/100 g H_2O, respectively. Thus (90 − 12) g, or 78 g, of KNO_3 will separate from the solution, but all of the NaCl will remain dissolved (Figure 12.8). In this manner, about 90 percent of the original KNO_3 can be obtained in pure form. The KNO_3 crystals can be separated from the solution by filtration.

Many of the commercially available solid inorganic and organic compounds that are used in the laboratory are purified by fractional crystallization. Generally the method works best if the compound to be purified has a steep solubility curve, that is, if it is considerably more soluble at high temperatures than at low temperatures. Otherwise, much of it will remain dissolved as the solution is cooled. Fractional crystallization also works well if the amount of impurity in the solution is relatively small.

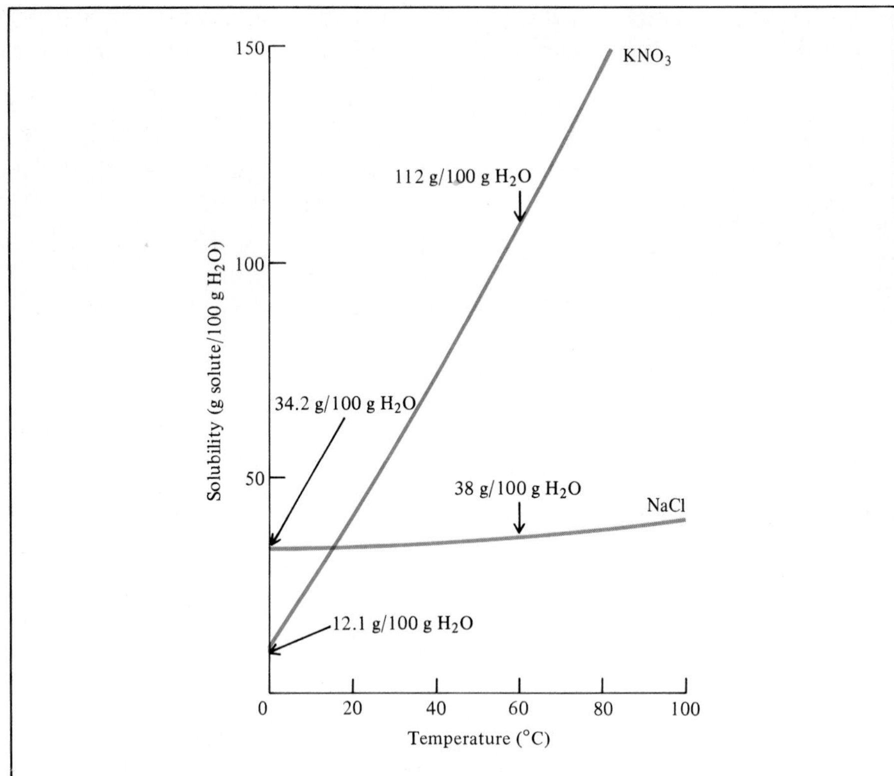

Figure 12.8 *The solubilities of KNO_3 and NaCl at 0°C and 60°C. The difference in temperature dependence enables us to isolate one of these salts from a solution containing both of them, through fractional crystallization.*

Gas Solubility and Temperature

In contrast to the solubility of solids, the solubility of gases in water usually decreases with increasing temperature (Figure 12.9). The next time you heat water in a beaker, notice the bubbles of air forming on the side of the glass before the water boils. As the temperature rises, the dissolved air molecules begin to ''boil out'' of the solution long before the water itself boils.

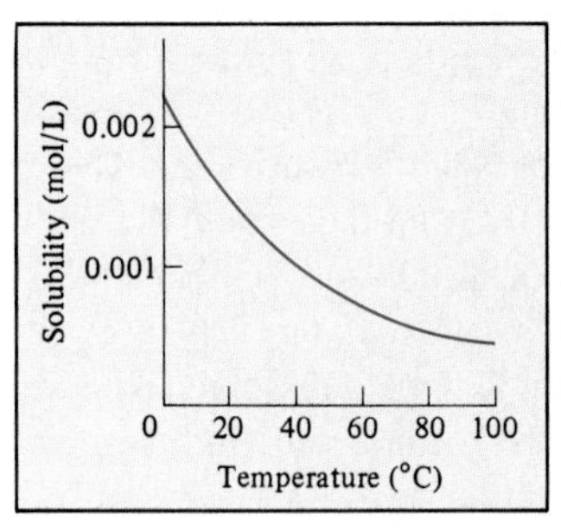

Figure 12.9 *Dependence on temperature of the solubility of O_2 gas in water. Note that the solubility decreases as temperature increases. The pressure of the gas over the solution is 1 atm.*

The reduced solubility of molecular oxygen in hot water has a direct bearing on ***thermal pollution,*** that is, *the heating of the environment*—usually waterways—*to temperatures that are harmful to its living inhabitants*. It is estimated that every year in the United States some 100,000 billion gallons of water are used for industrial cooling, mostly in electric power and nuclear power production. This process heats up the water, which is then returned to the rivers and lakes from which it was taken. Ecologists have become increasingly concerned about the effect of thermal pollution on aquatic life. Fish, like all other cold-blooded animals, have much more difficulty coping with rapid temperature fluctuation in the environment than humans do. An increase in water temperature accelerates their rate of metabolism, which generally doubles with each 10°C rise. The speedup of metabolism increases the fishes' need for oxygen at the same time that the supply of oxygen decreases because of its lower solubility in heated water. Effective ways to cool power plants while doing only minimal damage to the biological environment are being sought.

On the lighter side, knowledge of the variation of gas solubility with temperature can improve one's performance in a popular recreational sport—fishing. On a hot summer day, an experienced fisherman usually picks a deep spot in the river or lake to cast the bait. Because the oxygen content is greater in the deeper, cooler region, most fish will be found there.

12.7 Effect of Pressure on the Solubility of Gases

For all practical purposes, external pressure has no influence on the solubilities of liquids and solids, but it does greatly affect the solubility of gases. The quantitative relationship between gas solubility and pressure is given by ***Henry's*† *law,*** which states that *the solubility of a gas in a liquid is proportional to the pressure of the gas over the solution:*

$$c \propto P$$

$$c = kP \tag{12.1}$$

If several gases are present, *P* becomes the partial pressure.

Here c is the molar concentration (mol/L) of the dissolved gas; P is the pressure (in atm) of the gas over the solution; and, for a given gas, k is a constant that depends only on temperature. The constant k has the units mol/L · atm. You can see that when the pressure of the gas is 1 atm, c is *numerically* equal to k.

Henry's law can be understood qualitatively in terms of the kinetic molecular theory. The amount of gas that will dissolve in a solvent depends on how frequently the molecules in the gas phase collide with the liquid surface and become trapped by the condensed phase. Suppose we have a gas in dynamic equilibrium with a solution [Figure 12.10(a)]. At every instant, the number of gas molecules entering the solution is equal to the number of dissolved molecules moving into the gas phase. When the partial pressure is increased, more molecules dissolve in the liquid because more molecules are striking the surface of the liquid. This process continues until the concentration of the solution is again such that the number of molecules leaving the solution per second equals the number entering the solution [Figure 12.10(b)]. Because of the

† William Henry (1775–1836). English chemist. Henry's major contribution to science is his discovery of the law describing the solubility of gases that now bears his name.

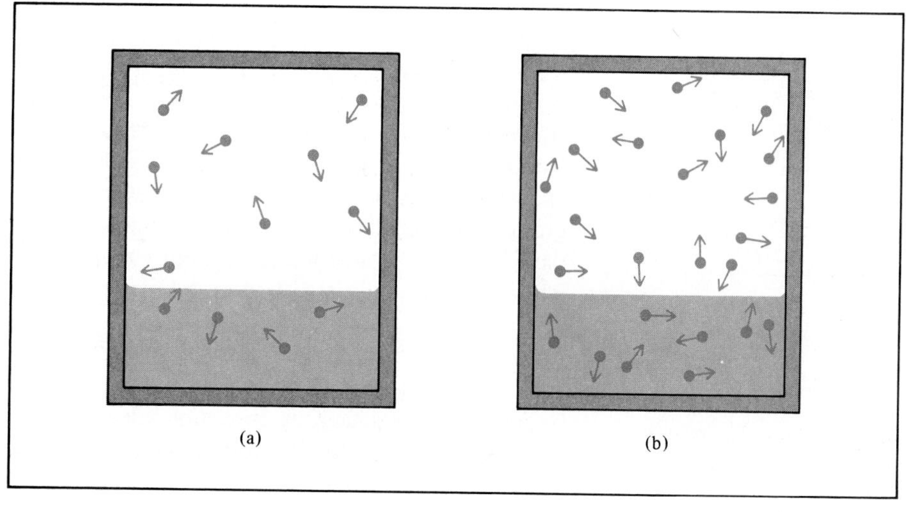

Figure 12.10 *A molecular interpretation of Henry's law. When the partial pressure of the gas over the solution increases from (a) to (b), the concentration of the dissolved gas also increases according to Equation (12.1).*

increased concentration of molecules in both the gas and solution phases, this number is greater in (b) than in (a) when the partial pressure was lower.

Example 12.9 applies Henry's law to nitrogen gas.

EXAMPLE 12.9

The solubility of pure nitrogen gas at 25°C and 1 atm is 6.8×10^{-4} mol/L. What is the concentration of nitrogen dissolved in water under atmospheric conditions? The partial pressure of nitrogen gas in the atmosphere is 0.78 atm.

Answer

The first step is to calculate the quantity k in Equation (12.1):

$$c = kP$$

$$6.8 \times 10^{-4}\ \text{mol/L} = k(1\ \text{atm})$$

$$k = 6.8 \times 10^{-4}\ \text{mol/L} \cdot \text{atm}$$

Therefore, the solubility of nitrogen gas in water is

$$\begin{aligned} c &= (6.8 \times 10^{-4}\ \text{mol/L} \cdot \text{atm})(0.78\ \text{atm}) \\ &= 5.3 \times 10^{-4}\ \text{mol/L} \\ &= 5.3 \times 10^{-4}\ M \end{aligned}$$

The decrease in solubility is the result of lowering the pressure from 1 atm to 0.78 atm.

Similar problem: 12.40.

Most gases obey Henry's law, but there are some important exceptions. For example, if the dissolved gas *reacts* with water, higher solubilities can result. The solubility of ammonia is much higher than expected because of the reaction

$$NH_3 + H_2O \rightleftharpoons NH_4^+ + OH^-$$

Carbon dioxide also reacts with water, as follows:

$$CO_2 + H_2O \rightleftharpoons H_2CO_3$$

Another interesting example concerns the dissolution of molecular oxygen in blood. Normally, oxygen gas is only sparingly soluble in water (2.9×10^{-4} mol/L at 24°C and partial pressure of 0.2 atm). However, its solubility in blood is dramatically greater because of the high content of hemoglobin (Hb) molecules in blood. As an oxygen carrier, each hemoglobin molecule can bind up to four oxygen molecules, which are eventually delivered to the tissues for use in metabolism:

$$Hb + 4O_2 \rightleftharpoons Hb(O_2)_4$$

It is this process that accounts for the high solubility of molecular oxygen in blood.

The following Chemistry in Action relates gas solubility to a commonly observed phenomenon and to a dangerous situation sometimes encountered by scuba divers.

CHEMISTRY IN ACTION

The Soft-Drink Bottle, "The Bends," and Gas Solubility

As we saw in Section 12.7, the solubility of a gas can be increased by increasing pressure. The following two examples show what happens to gas solubilities when external conditions change suddenly.

Every time we open a bottle of soda, beer, or champagne, we observe the formation of gas bubbles (Figure 12.11). Before the beverage bottle is sealed, it is pressurized with a mixture of air and CO_2 saturated with water vapor. Because of its high partial pressure in the pressurizing gas mixture, the amount of CO_2 dissolved in the soft drink is many times the amount that would dissolve under normal atmospheric conditions. When the cap is removed, the pressurized gases escape, the pressure in the bottle falls to atmospheric pressure, and the amount of CO_2 remaining in the beverage is determined only by the normal atmospheric partial pressure of CO_2, 0.0003 atm. The excess dissolved CO_2 comes out of solution, causing the effervescence.

We have already seen in Chapter 5 some of the applications of the gas laws to scuba diving. A knowledge of Henry's law can protect divers from a sometimes fatal condition known as "the bends." When a diver at a depth of more than 15 m breathes in compressed air from the supply tank, more nitrogen dissolves in the blood and other body fluids than would dissolve at the surface because the pressure at that depth is far greater than surface atmospheric pressure. If the diver ascends too quickly, much of this nitrogen will boil off rapidly as its partial pressure decreases, forming bubbles in the bloodstream. These bubbles restrict blood flow, affect the transmission of nerve impulses, and can cause death. As a precautionary step, most professional divers today use a helium-oxygen mixture instead of a nitrogen-oxygen mixture in their compressed gas tanks, because of the lower solubility of helium in blood. The excess oxygen dissolved in the blood is used in metabolism, so it does not afflict divers with the bends.

Figure 12.11 *The effervescence of a soft drink. The bottle was shaken slightly before opening to dramatize the escape of CO_2.*

12.8 Colligative Properties of Nonelectrolyte Solutions

Several important properties of solutions depend on the number of solute particles in solution and not on the nature of the solute particles. These properties are called ***colligative properties*** (or collective properties) because they are bound together by a common origin; that is, they all depend on the number of solute particles present, whether these particles are atoms, ions, or molecules. The colligative properties are vapor-pressure lowering, boiling-point elevation, freezing-point depression, and osmotic pressure. For our discussion of colligative properties of nonelectrolyte solutions it is important to keep in mind that we are talking about relatively dilute solutions, that is, solutions whose concentrations are $\lesssim 0.2\ M$.

Vapor-Pressure Lowering

To review the concept of equilibrium vapor pressure as it applies to pure liquids, see Section 11.8.

If a solute is ***nonvolatile*** (that is, it *does not have a measurable vapor pressure*), the vapor pressure of its solution is always less than that of the pure solvent. Thus the relationship between solution vapor pressure and solvent vapor pressure depends on the concentration of the solute in the solution. This relationship is given by ***Raoult's† law***, which states that *the partial pressure of a solvent over a solution, P_1, is given by the vapor pressure of the pure solvent, P_1°, times the mole fraction of the solvent in the solution, X_1:*

$$P_1 = X_1 P_1^\circ \qquad (12.2)$$

In a solution containing only one solute, $X_1 = 1 - X_2$, where X_2 is the mole fraction of the solute. Equation (12.2) can therefore be rewritten as

$$P_1 = (1 - X_2)P_1^\circ$$
$$P_1^\circ - P_1 = \Delta P = X_2 P_1^\circ \qquad (12.3)$$

We see that the *decrease* in vapor pressure, ΔP, is directly proportional to the concentration (measured in mole fraction) of the solute present.

The following example applies Equation (12.3).

† François Marie Raoult (1830–1901). French chemist. Raoult's work was mainly in solution properties and electrochemistry.

EXAMPLE 12.10

At 25°C, the vapor pressure of pure water is 23.76 mmHg and that of a dilute aqueous urea solution is 22.98 mmHg. Estimate the molality of the solution.

Answer

From Equation (12.3) we write

$$\Delta P = (23.76 - 22.98)\ \text{mmHg} = X_2(23.76\ \text{mmHg})$$
$$X_2 = 0.033$$

By definition

$$X_2 = \frac{n_2}{n_1 + n_2}$$

where n_1 and n_2 are the numbers of moles of solvent and solute, respectively. Since the solution is dilute, we can assume that n_1 is much larger than n_2, and we can write

$$X_2 = \frac{n_2}{n_1 + n_2} \simeq \frac{n_2}{n_1} \qquad (n_1 \gg n_2)$$
$$n_2 = n_1 X_2$$

The number of moles of water in 1 kg of water is

$$1000 \text{ g } H_2O \times \frac{1 \text{ mol } H_2O}{18.02 \text{ g } H_2O} = 55.49 \text{ mol } H_2O$$

and the number of moles of urea present in 1 kg of water is

$$\begin{aligned} n_2 = n_1X_2 &= (55.49 \text{ mol})(0.033) \\ &= 1.8 \text{ mol} \end{aligned}$$

Thus, the concentration of the urea solution is 1.8 *m*.

Similar problems: 12.60, 12.61.

Why is the vapor pressure of a solution less than that of its pure solvent? As was mentioned in Section 12.2, one driving force in physical and chemical processes is the increase in disorder—the greater the disorder created, the more favorable the process. Vaporization increases the disorder of a system because molecules in a vapor are not as closely packed and therefore have less order than those in a liquid. Because a solution is more disordered than a pure solvent, the difference in disorder between a solution and a vapor is less than that between a pure solvent and a vapor. Thus solvent molecules have less of a tendency to leave a solution than to leave the pure solvent to become vapor, and the vapor pressure of a solution is less than that of the solvent.

If both components of a solution are ***volatile*** (that is, *have measurable vapor pressure*), the vapor pressure of the solution is the sum of the individual partial pressures. Raoult's law holds equally well in this case:

$$P_A = X_A P_A^\circ$$
$$P_B = X_B P_B^\circ$$

where P_A and P_B are the partial pressures over the solution for components A and B; P_A° and P_B° are the vapor pressures of the pure substances; and X_A and X_B are their mole fractions. The total pressure is given by Dalton's law of partial pressure [Equation (5.11)]:

$$P_T = P_A + P_B$$

Benzene and toluene have similar structures and therefore similar intermolecular forces:

CH3

benzene toluene

In a solution of benzene and toluene, the vapor pressure of each component obeys Raoult's law. Figure 12.12 shows the dependence of the total vapor pressure (P_T) in a benzene-toluene solution on the composition of the solution. Note that we need only express the composition of the solution in terms of the mole fraction of one component. For every value of X_{benzene}, the mole fraction of toluene, X_{toluene}, is given by $(1 - X_{\text{benzene}})$. The benzene-toluene solution is one of the few examples of an ***ideal solu-***

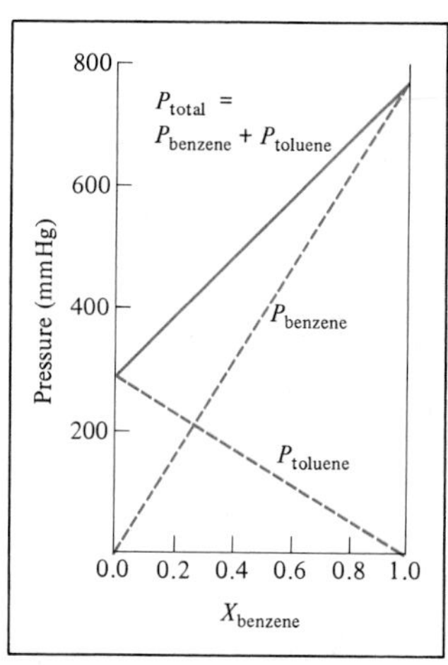

Figure 12.12 *The dependence of the partial pressures of benzene and toluene on their mole fraction in a benzene-toluene solution ($X_{toluene} = 1 - X_{benzene}$) at 80°C. This solution is said to be ideal because the vapor pressures obey Raoult's law [Equation (12.2)].*

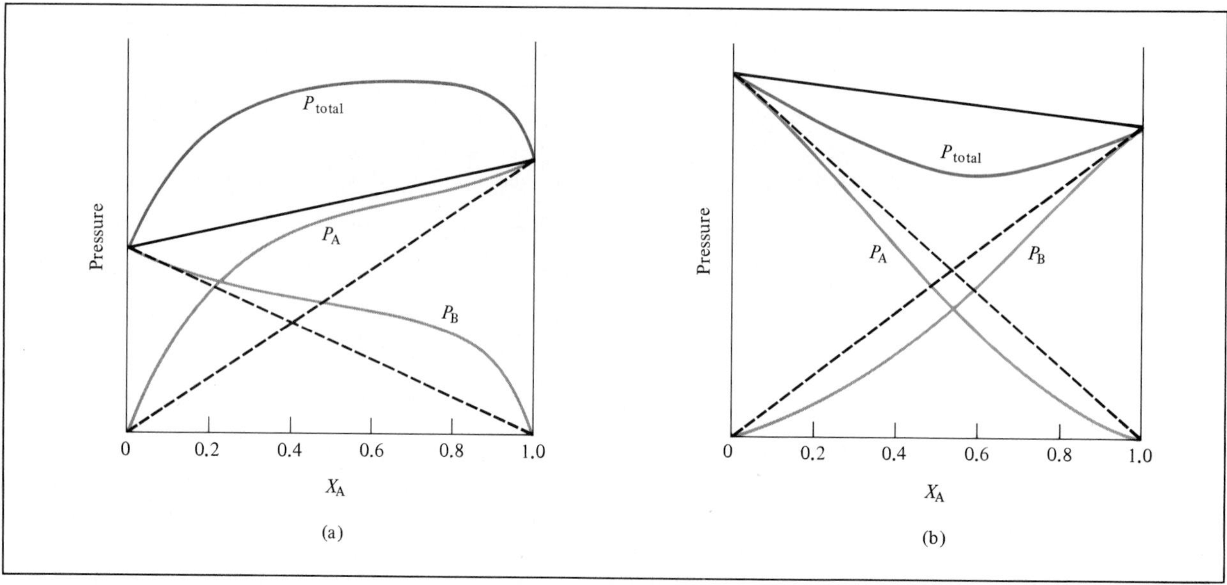

Figure 12.13 *Nonideal solutions. (a) Positive deviation occurs when P_{total} is greater than that predicted by Raoult's law (the solid black line). (b) Negative deviation. Here, P_{total} is less than that predicted by Raoult's law (the solid black line).*

tion, which is *any solution that obeys Raoult's law*. One characteristic of an ideal solution is that the heat of solution, ΔH_{soln}, is always zero.

Most solutions do not behave ideally in this respect. Designating two volatile substances as A and B, we can consider the following two cases:

Case 1: If the intermolecular forces between A and B molecules are weaker than those between A molecules and between B molecules, then there is a greater tendency for these molecules to leave the solution than in the case of an ideal solution. Consequently, the vapor pressure of the solution is greater than the sum of the vapor pressures as predicted by Raoult's law at the same concentration. This behavior gives rise to the *positive deviation* [Figure 12.13(a)]. The heat of solution is positive (that is, mixing is an endothermic process).

Case 2: If A molecules attract B molecules more strongly than they do their own kind, the vapor pressure of the solution is less than the sum of the vapor pressures as predicted by Raoult's law at the same concentration. Here we have a *negative deviation* [Figure 12.13(b)]. The heat of solution is negative (that is, mixing is an exothermic process).

Fractional Distillation. The discussion of solution vapor pressure has a direct bearing on ***fractional distillation,*** *a procedure for separating liquid components of a solution that is based on their different boiling points*. Fractional distillation is somewhat analogous to fractional crystallization. Suppose we want to separate a *binary system* (a system with two components), say, benzene-toluene. Both benzene and toluene are relatively volatile, yet their boiling points are appreciably different (80.1°C and 110.6°C, respectively). When we boil a solution containing these two substances, the vapor formed is somewhat richer in the more volatile component, benzene. If the vapor is condensed in a separate container and that liquid is boiled again, a still higher concentration of benzene will be obtained in the vapor phase. By repeating this process many times, it is possible to separate benzene completely from toluene.

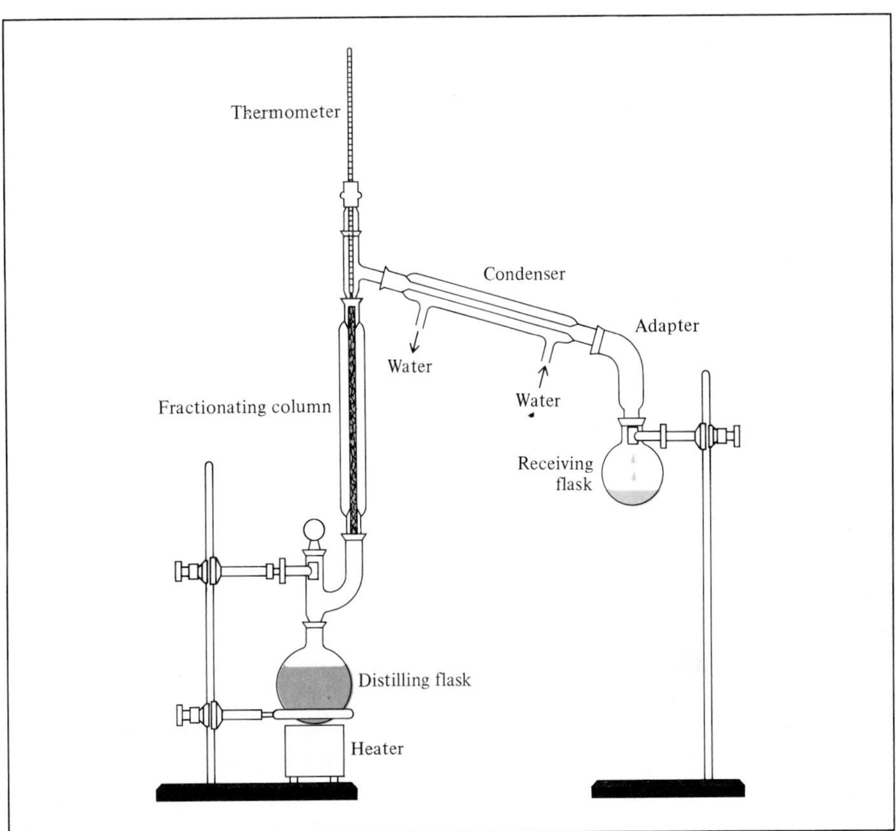

Figure 12.14 *An apparatus for small-scale fractional distillation. The fractionating column is packed with tiny glass beads.*

The longer the fractionating column, the more complete the separation of the volatile liquids.

In practice, chemists use an apparatus like that shown in Figure 12.14 to separate volatile liquids. The round-bottomed flask containing the benzene-toluene solution is fitted with a long column packed with small glass beads. When the solution boils, the vapor condenses on the beads in the lower portion of the column, and the liquid falls back into the distilling flask. As time goes on, the beads gradually become heated, allowing the vapor to move upward slowly. In essence, the packing material causes the benzene-toluene mixture to be subjected continuously to numerous vaporization-condensation steps. At each step the composition of the vapor in the column will be richer in the more volatile, or lower boiling-point, component (in this case, benzene). Finally, essentially pure benzene vapor rises to the top of the column and is then condensed and collected in a receiving flask.

Fractional distillation is as important in industry as it is in the laboratory. The petroleum industry employs fractional distillation on a large scale to separate the various components of crude oil. More will be said of this process in Chapter 24.

Boiling-Point Elevation

Because the presence of a nonvolatile solute lowers the vapor pressure of a solution, it must also affect the boiling point of the solution. The boiling point of a solution is the temperature at which its vapor pressure equals the external atmospheric pressure (see Section 11.8). Figure 12.15 shows the phase diagram of water and the changes that occur in an aqueous solution. Because at any temperature the vapor pressure of the

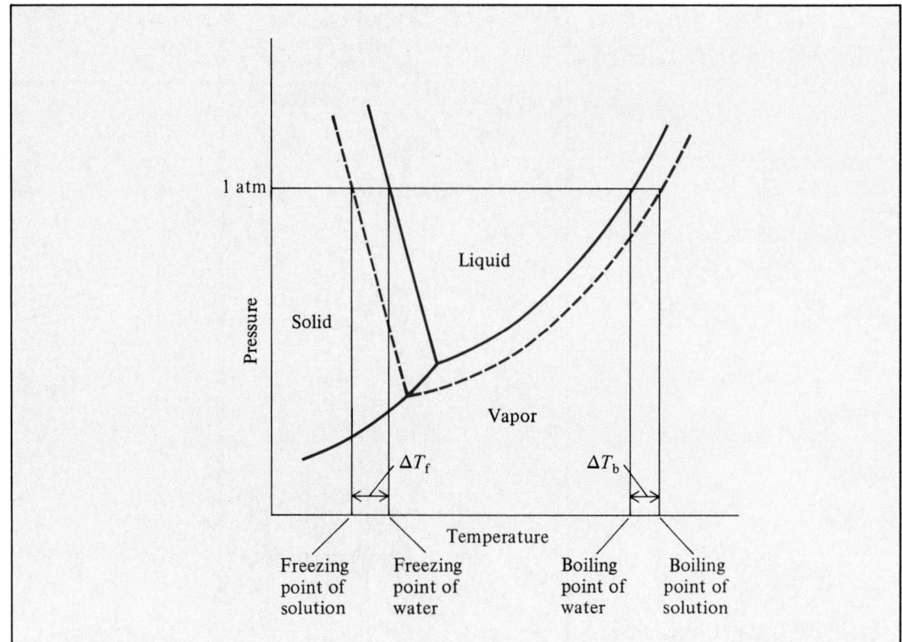

Figure 12.15 *Phase diagram illustrating the boiling-point elevation and freezing-point depression of aqueous solutions. The dashed curves pertain to the solution, and the solid curves to the pure solvent. As you can see, the boiling point of the solution is higher than that of water, and the freezing point of the solution is lower than that of water.*

solution is lower than that of the pure solvent, the liquid-vapor curve (the dotted line) for the solution lies below that for the pure solvent (the solid line). Consequently, the dotted line intersects the horizontal line that marks $P = 1$ atm at a *higher* temperature than the normal boiling point of the pure solvent. This graphical anlaysis shows that the boiling point of the solution is higher than that of water. The boiling-point elevation, ΔT_b, is defined as

Since $T_b > T_b^\circ$, ΔT_b is a positive quantity.

$$\Delta T_b = T_b - T_b^\circ$$

where T_b is the boiling point of the solution and T_b° the boiling point of the pure solvent. Because ΔT_b is proportional to the vapor-pressure lowering, it is also proportional to the concentration (molality) of the solution. That is,

$$\Delta T_b \propto m$$

$$\Delta T_b = K_b m \qquad (12.4)$$

where m is the molal concentration of the solute and K_b is the proportionality constant. The latter is called the *molal boiling-point elevation constant;* it has the units of °C/m.

It is important to understand the choice of concentration unit here. We are dealing with a system (the solution) whose temperature is *not* kept constant, so we cannot express the concentration units in molarity because molarity changes with temperature.

Table 12.2 lists the value of K_b for several common solvents. Using the boiling-point elevation constant for water and Equation (12.4), you can see that if the molality m of an aqueous solution is 1.00, the boiling point will be 100.52°C.

Freezing-Point Depression

A nonscientist may remain forever unaware of the boiling-point elevation phenomenon, but a careful observer living in a cold climate is familiar with freezing-point depression. Ice on frozen roads and sidewalks melts when sprinkled with salts such as

Table 12.2 Molal boiling-point elevation and freezing-point depression constants of several common liquids

Solvent	Normal freezing point (°C)*	K_f (°C/m)	Normal boiling point (°C)*	K_b (°C/m)
Water	0	1.86	100	0.52
Benzene	5.5	5.12	80.1	2.53
Ethanol	−117.3	1.99	78.4	1.22
Acetic acid	16.6	3.90	117.9	2.93
Cyclohexane	6.6	20.0	80.7	2.79

*Measured at 1 atm.

NaCl or $CaCl_2$. This method of thawing succeeds because it depresses the freezing point of water.

Figure 12.15 shows that lowering the vapor pressure of the solution shifts the solid-liquid curve to the left (see the dashed line). Consequently, this dashed line intersects the horizontal line at a temperature *lower* than the freezing point of water. The depression in freezing point, ΔT_f, is defined as

Since $T_f^\circ > T_f$, ΔT_f is a positive quantity.

$$\Delta T_f = T_f^\circ - T_f$$

where T_f° is the freezing point of the pure solvent and T_f the freezing point of the solution. Again, ΔT_f is proportional to the concentration of the solution:

$$\Delta T_f \propto m$$

$$\Delta T_f = K_f m \qquad (12.5)$$

When an aqueous solution freezes, the solid first formed is almost always ice.

where m is the concentration of the solute in molality units, and K_f is the *molal freezing-point depression constant* (see Table 12.2). Like K_b, K_f has the units °C/m.

Note that we require that the solute be nonvolatile in the case of boiling-point elevation, but no such restriction applies to freezing-point depression. For example, methanol (CH_3OH), a fairly volatile liquid that boils at only 65°C, has sometimes been used as an antifreeze in automobile radiators.

A practical application of the freezing-point depression is shown in the following example.

EXAMPLE 12.11

Ethylene glycol (EG), $CH_2(OH)CH_2(OH)$, is a common automobile antifreeze. It is water soluble and fairly nonvolatile (b.p. 197°C). Calculate the freezing point of a solution containing 651 g of this substance in 2505 g of water. Would you keep this substance in your car radiator during the summer? The molar mass of ethylene glycol is 62.01 g.

Answer

The number of moles of ethylene glycol in 1000 g of water is

$$651 \text{ g EG} \times \frac{1 \text{ mol EG}}{62.01 \text{ g EG}} \times \frac{1000 \text{ g}}{2505 \text{ g}} = 4.19 \text{ mol EG}$$

Thus, the molality of the solution is 4.19 *m*. From Equation (12.5) and Table 12.2 we have

$$\Delta T_f = (1.86°C/m)(4.19\ m)$$
$$= 7.79°C$$

Since pure water freezes at 0°C, the solution will freeze at −7.79°C. We can calculate boiling-point elevation in the same way as follows:

$$\Delta T_b = (0.52°C/m)(4.19\ m)$$
$$= 2.2°C$$

Because the solution will boil at 102.2°C, it would be preferable to leave the antifreeze in your car radiator in summer to prevent the solution from boiling.

Similar problems: 12.68, 12.70.

Osmotic Pressure

The phenomenon of osmotic pressure is illustrated in Figure 12.16. The left compartment of the apparatus contains pure solvent; the right compartment contains a solution. The two compartments are separated by a ***semipermeable membrane,*** which *allows solvent molecules to pass through but blocks the passage of solute molecules*. At the start, the water levels in the two tubes are equal [see Figure 12.16(a)]. After some time, the level in the right tube begins to rise; this continues until equilibrium is reached. *The net movement of solvent molecules through a semipermeable membrane from a pure solvent or from a dilute solution to a more concentrated solution* is called ***osmosis.*** The ***osmotic pressure*** (π) of a solution is *the pressure required to stop osmosis from pure solvent into the solution*. As shown in Figure 12.16(b), this pressure can be measured directly from the difference in the final fluid levels.

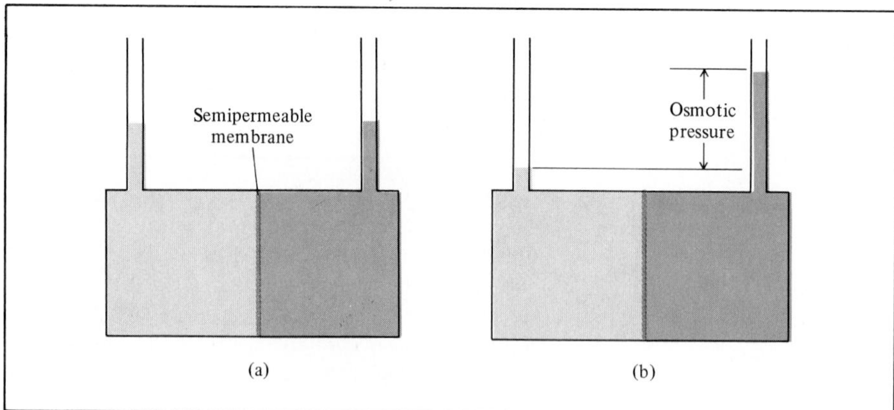

Figure 12.16 *Osmotic pressure. (a) The levels of the pure solvent (left) and of the solution (right) are equal at the start. (b) During osmosis, the level on the solution side rises as a result of the net flow of solvent from left to right. The osmotic pressure is equal to the hydrostatic pressure exerted by the column of fluid in the right tube at equilibrium. Basically the same effect is observed when the pure solvent is replaced by a more dilute solution than that on the right.*

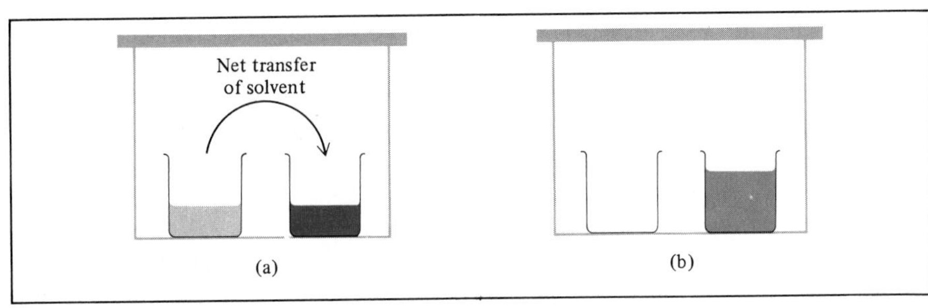

Figure 12.17 *Unequal vapor pressures inside the container lead to a net transfer of water from the beaker in (a) to the solution in the beaker in (b), shown at equilibrium. This driving force for solvent transfer is analogous to the osmotic phenomenon shown in Figure 12.16.*

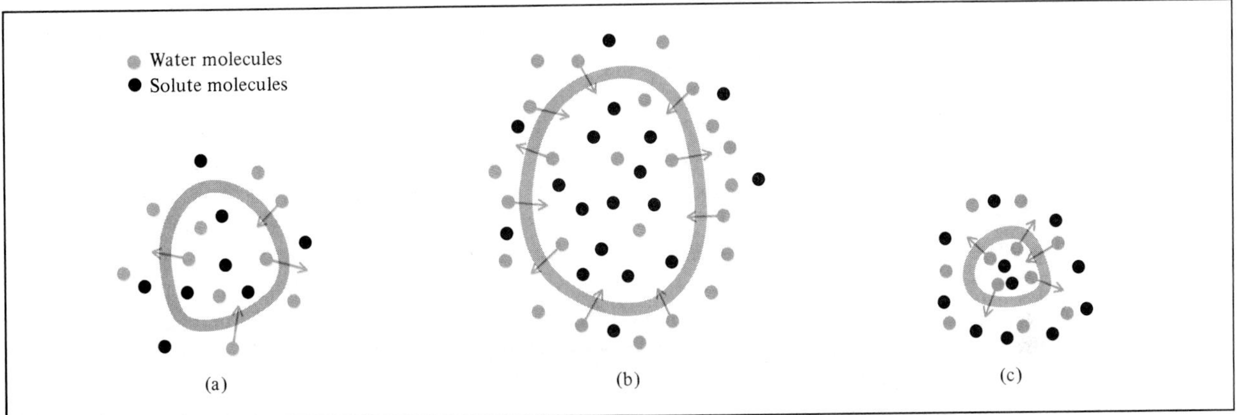

Figure 12.18 *A cell in (a) an isotonic solution, (b) a hypotonic solution, and (c) a hypertonic solution. The cell remains unchanged in (a), swells in (b), and shrinks in (c).*

What causes water to move spontaneously from left to right in this case? Compare the vapor pressure of water and that of a solution (Figure 12.17). Because the vapor pressure of pure water is higher, there is a tendency for net transfer of water from the left beaker to the right one. Given enough time, the transfer will continue to completion. A similar force causes water to move into the solution during osmosis.

Although osmosis is a common and well-studied phenomenon, relatively little is known as to how the semipermeable membrane stops some molecules yet allows others to pass. In some cases, it is simply a matter of size. A semipermeable membrane may have pores small enough to let only the solvent molecules through. In other cases, a different mechanism may be responsible for the membrane's selectivity—for example, the solvent's greater "solubility" in the membrane.

The osmotic pressure of a solution is given by

$$\pi = MRT \tag{12.6}$$

where M is the molarity of solution, R the gas constant (0.0821 L · atm/K · mol), and T the absolute temperature. The osmotic pressure, π, is expressed in atm. Since osmotic pressure measurements are carried out at constant temperature, we express the concentration here in terms of the more convenient units of molarity rather than molality.

As in the boiling-point elevation and freezing-point depression equations, osmotic pressure too is directly proportional to the concentration of solution. This is what we would expect, bearing in mind that all colligative properties depend only on the number of solute particles in solution. If two solutions are of equal concentration and, hence, of the same osmotic pressure, they are said to be *isotonic*. If two solutions are of unequal osmotic pressures, the more concentrated solution is said to be *hypertonic* and the more dilute solution is described as *hypotonic* (Figure 12.18).

The osmotic pressure phenomenon manifests itself in many interesting applications. To study the contents of red blood cells, which are protected from the external environment by a semipermeable membrane, biochemists use a technique called *hemolysis*. The red blood cells are placed in a hypotonic solution. Because the hypotonic solution is less concentrated than the interior of the cell, water moves into the cells, as shown in Figure 12.18(b). The cells swell and eventually burst, releasing hemoglobin and other molecules.

Home preserving of jam and jelly has recently been revived as a popular (and money-saving) hobby in the United States. Use of a large quantity of sugar is actually essential to the preservation process, since the sugar is partly responsible for killing bacteria that may cause botulism. As Figure 12.18(c) shows, when a bacterial cell is in a hypertonic (high concentration) sugar solution, the intracellular water tends to move out of the bacterial cell to the more concentrated solution by osmosis. This process, known as *crenation,* causes the cell to shrink and, eventually, to cease to function. The acidic medium due to the presence of citric acid and other acids in fruits also helps kill the bacteria.

The capillary action discussed in Section 11.3 is responsible for the rise of water only up to a few centimeters.

Osmotic pressure is the major mechanism transporting water upward in plants. Because the leaves of trees constantly lose water to the air, in a process called *transpiration,* the solute concentrations in leaf fluids increase. Water is pushed up through the trunk, branches, and stems by osmotic pressure, which may have to be as high as 10 to 15 atm in order to reach leaves at the tops of California's redwoods, which reach about 120 m in height.

The following example shows that an osmotic pressure measurement can give us the concentration of the solution.

EXAMPLE 12.12

When we say that seawater has an osmotic pressure of 30 atm, we mean that when seawater is placed in the apparatus shown in Figure 12.16, we obtain a measurement of 30 atm.

The average osmotic pressure of seawater is about 30.0 atm at 25°C. Calculate the molar concentration of an aqueous solution of urea (NH_2CONH_2) that is isotonic with seawater.

Answer

A solution of urea that is isotonic with seawater must have the same osmotic pressure, 30.0 atm. Using Equation (12.6),

$$\pi = MRT$$

$$M = \frac{\pi}{RT} = \frac{30.0 \text{ atm}}{(0.0821 \text{ L} \cdot \text{atm/K} \cdot \text{mol})(298 \text{ K})}$$

$$= 1.23 \text{ mol/L}$$

$$= 1.23\ M$$

Similar problem: 12.77.

Using Colligative Properties to Determine Molar Mass

The colligative properties of nonelectrolyte solutions provide us with a means of determining the molar mass of a solute. Theoretically, any of the four colligative properties can be used for this purpose. In practice, however, only freezing-point depression and osmotic pressure are used because those cases show the most pronounced changes.

The following examples demonstrate the determination of molar masses from freezing-point depression and osmotic pressure measurements.

EXAMPLE 12.13

A 7.85 g sample of a compound with empirical formula C_5H_4 is dissolved in 301 g of benzene. The freezing point of the solution is 1.05°C below that of pure benzene. What are the molar mass and molecular formula of this compound?

Answer

From Equation (12.5) and Table 12.2 we can write

$$\text{molality} = \frac{\Delta T_f}{K_f} = \frac{1.05°\text{C}}{5.12°\text{C}/m} = 0.205\ m$$

Thus, the solution contains 0.205 mole of solute per kilogram of benzene. Since the solution was prepared by dissolving 7.85 g of solute in 301 g of benzene, the number of grams of solute in a kilogram of solvent is

$$\frac{\text{g solute}}{\text{kg benzene}} = \frac{7.85\text{ g solute}}{301\text{ g benzene}} \times \frac{1000\text{ g benzene}}{1\text{ kg benzene}}$$
$$= \frac{26.1\text{ g solute}}{1\text{ kg benzene}}$$

Now that we know 0.205 mole of solute is 26.1 g, the molar mass of solute must be

$$\text{molar mass} = \frac{26.1\text{ g}}{0.205\text{ mol}} = 127\text{ g/mol}$$

Since the formula mass of C_5H_4 is 64 g and the molar mass is found to be 127 g, the molecular formula of the compound is $C_{10}H_8$.

Similar problems: 12.69, 12.71.

EXAMPLE 12.14

A solution is prepared by dissolving 35.0 g of hemoglobin (Hb) in enough water to make up 1 L in volume. If the osmotic pressure of the solution is found to be 10.0 mmHg at 25°C, calculate the molar mass of hemoglobin.

Answer

First we calculate the concentration of the solution:

$$\pi = MRT$$

$$M = \frac{\pi}{RT}$$

$$= \frac{10.0 \text{ mmHg} \times \frac{1 \text{ atm}}{760 \text{ mmHg}}}{(0.0821 \text{ L} \cdot \text{atm/K} \cdot \text{mol})(298 \text{ K})}$$

$$= 5.38 \times 10^{-4}\, M$$

The volume of the solution is 1 L, so it must contain 5.38×10^{-4} mol of Hb. We use this quantity to calculate the molar mass:

$$\text{moles of Hb} = \frac{\text{mass of Hb}}{\text{molar mass of Hb}}$$

$$\text{molar mass of Hb} = \frac{\text{mass of Hb}}{\text{moles of Hb}}$$

$$= \frac{35.0 \text{ g}}{5.38 \times 10^{-4} \text{ mol}}$$

$$= 6.51 \times 10^{4} \text{ g/mol}$$

Similar problems: 12.78, 12.79.

A pressure such as 10.0 mmHg in the preceding example can be measured easily and accurately. For this reason, osmotic pressure measurements are among the most useful techniques for determining the molar masses of large molecules such as proteins. To see how much more useful the osmotic pressure technique is than freezing-point depression would be, let us estimate the change in freezing point of the same hemoglobin solution. If a solution is quite dilute, we can assume that molarity is roughly equal to molality. (Molarity would be equal to molality if the density of the solution were 1 g/mL.) Hence, from Equation (12.5) we write

$$\Delta T_f = (1.86°\text{C}/m)(5.38 \times 10^{-4}\, m)$$

$$= 1.00 \times 10^{-3}\ °\text{C}$$

Because of low sensitivity and experimental difficulties, vapor-pressure lowering and boiling-point elevation are not normally used for molar mass determination.

The thousandth-of-a-degree freezing-point depression is too small a temperature change to measure accurately. For this reason, the freezing-point depression technique is more suitable for measuring the molar mass of smaller and more soluble molecules, those having molar masses of 500 g or less, since the freezing-point depressions of their solutions are much greater.

12.9 Colligative Properties of Electrolyte Solutions

The colligative properties of nonelectrolyte solutions are better understood than those of electrolyte solutions. The study of the "structure" of electrolyte solutions, that is, how each ion is arranged relative to its neighbors and how it interacts with the solvent molecules, has occupied some of the most prominent chemists of the past hundred years. Yet no completely satisfactory theory of electrolyte solution has been formulated.

Qualitatively, at least, we know the following facts. In a dilute solution, say,

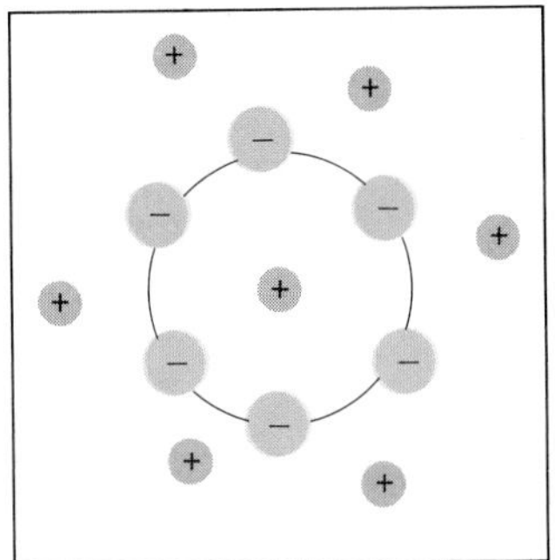

Figure 12.19 *An ionic atmosphere surrounding a cation in solution. If the center ion is an anion, then the ionic atmosphere would be made up mainly of cations.*

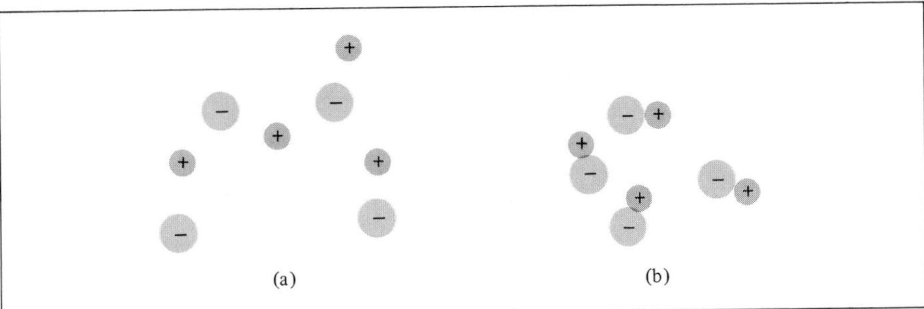

Figure 12.20 *(a) Free ions and (b) ion pairs in solution. Such an ion pair bears no net charge and therefore cannot conduct electricity in solution.*

0.01 *M* or less, each ion is surrounded by a number of water molecules. The degree of hydration depends on the nature of the ion: It is greater for small ions with high charges. The size of this "hydration sphere" affects the movement of cations and anions toward the negative and positive electrodes in a solution (see Figure 3.3).

The interionic attraction in an electrolyte solution also affects its properties. It is reasonable to assume that each ion is surrounded, on the average, by ions bearing the opposite charge. Thus, we can think of each ion as being the center of an ionic atmosphere (Figure 12.19). Indeed, experimental evidence strongly supports this notion.

Ion pairs effect an apparent decrease in the number of independently moving particles.

A fully hydrated (or solvated) ion is called a *free ion*. At higher concentrations, cations and anions have less than complete hydration spheres and so tend to associate with each other to form ***ion pairs***. An ion pair consists of *a cation and an anion held closely together by attractive forces with few or no water molecules between them* (Figure 12.20). The presence of ion pairs in a solution decreases the electrical conductivity. Since the cation and anion in a neutral ion pair cannot move freely as individual units, there can be no net migration in solution. Electrolytes containing multicharged ions such as Mg^{2+}, Al^{3+}, SO_4^{2-}, CO_3^{2-}, and PO_4^{3-} have a greater tendency to form ion pairs than electrolytes such as NaCl or KNO_3.

Dissociation of electrolytes into ions has a direct bearing on the colligative properties of solutions, which are determined only by the number of particles present. For example, the depression in freezing point of a 0.1 *m* NaCl solution is about twice as great as that of a nonelectrolyte solution of 0.1 *m* containing cane sugar or urea as solute since every mole of NaCl produces 2 moles of particles in solution. The same holds true for boiling-point elevation and osmotic pressure. Thus Equations (12.4), (12.5), and (12.6) should be modified as follows:

$$\Delta T_b = iK_b m \tag{12.7}$$

$$\Delta T_f = iK_f m \tag{12.8}$$

$$\pi = iMRT \tag{12.9}$$

where *i* is called the *van't Hoff† factor* and is defined as follows:

$$i = \frac{\text{actual number of particles in solution after dissociation}}{\text{number of formula units initially dissolved in solution}}$$

†Jacobus Hendricus van't Hoff (1852–1911). Dutch chemist. One of the most prominent chemists of his time, van't Hoff did significant work in thermodynamics, molecular structure and optical activity, and solution chemistry. In 1901 he received the first Nobel Prize in chemistry.

Table 12.3 The van't Hoff factor of 0.05 *M* electrolyte solutions at 25°C

Electrolyte	i (measured)	i (calculated)
Sucrose*	1.0	1.0
HCl	1.9	2.0
NaCl	1.9	2.0
$MgSO_4$	1.3	2.0
$MgCl_2$	2.7	3.0
$FeCl_3$	3.4	4.0

*Sucrose is a nonelectrolyte. It is listed here for comparison purposes only.

Every unit of NaCl or KNO_3 that dissociates gives two ions ($i = 2$); every unit of Na_2SO_4 or $MgCl_2$ that dissociates produces three ions ($i = 3$).

Thus i should be 1 for all nonelectrolytes. For strong electrolytes such as NaCl and KNO_3, i should be 2, and for strong electrolytes such as Na_2SO_4 and $MgCl_2$, i should be 3.

Table 12.3 shows the van't Hoff factor for several strong electrolytes at a concentration of 0.05 M. For electrolytes that form only singly charged ions, such as HCl and NaCl, the agreement between measured and calculated i values is quite good. Significant deviations are observed for electrolytes containing Mg^{2+}, Fe^{3+}, and SO_4^{2-} ions, suggesting the formation of some ion pairs.

As the following example shows, the van't Hoff factor can be determined from osmotic pressure measurements.

EXAMPLE 12.15

The osmotic pressures of 0.010 M solutions of potassium iodide (KI) and of sucrose at 25°C are 0.465 atm and 0.245 atm, respectively. Calculate the van't Hoff factor for KI.

Answer

This comparison works if the concentrations of the original solutions are the same.

Since osmotic pressure is directly proportional to the number of solute particles present in solution and since there are more particles in the KI solution than in the sucrose solution, we can express the van't Hoff factor for KI as follows:

$$i = \frac{\text{number of particles in KI soln}}{\text{number of particles in sucrose soln}}$$

$$= \frac{\text{osmotic pressure of KI soln}}{\text{osmotic pressure of sucrose soln}}$$

$$= \frac{0.465 \text{ atm}}{0.245 \text{ atm}}$$

$$= 1.90$$

Similar problem: 12.92.

The following Chemistry in Action describes three physical techniques for obtaining the pure solvent (water) from a solution (seawater).

CHEMISTRY IN ACTION

Desalination

Over the centuries, scientists have sought ways of *removing salts from seawater,* a process called ***desalination,*** to augment the supply of fresh water. The ocean is an enormous and extremely complex aqueous solution. There are about 1.5×10^{21} L of seawater in the ocean, of which 3.5 percent (by mass) is dissolved material. Table 12.4 lists the concentrations of seven substances that, together, comprise more than 99 percent of the dissolved constituents of ocean water. In an age when astronauts have landed on the moon and spectacular advances in science and medicine have been made, desalination may seem a simple enough objective. However, although desalination technology exists, it remains very costly. It is an interesting paradox that in our technological society, accomplishing something simple like desalination at a socially acceptable cost is often as difficult as achieving something complex like sending an astronaut to the moon.

Table 12.4 Composition of seawater

Ions	g/kg of seawater
Chloride (Cl^-)	19.35
Sodium (Na^+)	10.76
Sulfate (SO_4^{2-})	2.71
Magnesium (Mg^{2+})	1.29
Calcium (Ca^{2+})	0.41
Potassium (K^+)	0.39
Bicarbonate (HCO_3^-)	0.14

Distillation

The oldest method of desalination, distillation, accounts for more than 90 percent of the approximately 500 million gallons per day capacity of the desalination systems currently in operation worldwide. The process involves vaporizing seawater and then condensing the pure water vapor. Most distillation systems use heat energy to do this. To reduce the cost of distillation, attempts have been made to utilize solar radiation, as shown in Figure 12.21. This approach is attractive because sunshine is normally more intense in arid lands, where the need for water is also greatest. However, despite intensive research and development efforts, several engineering problems persist, and "solar stills" do not yet operate on a large scale.

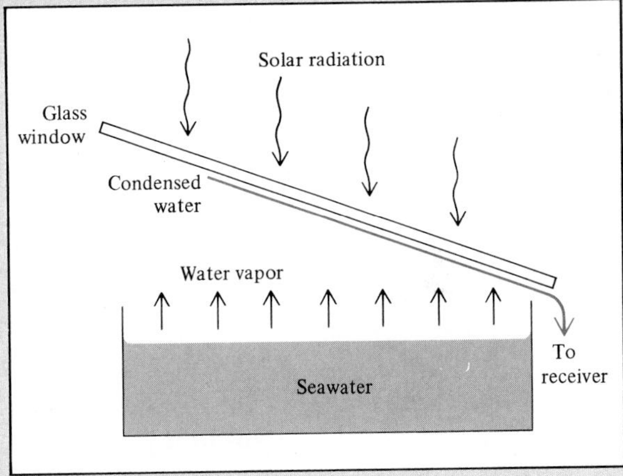

Figure 12.21 *A solar still for desalinating seawater.*

Freezing

Desalination by freezing has also been under development for a number of years, but it has not yet become commercially feasible. This method is based on the fact that when an aqueous solution (in this case, seawater) freezes, the solid that separates from solution is almost pure water. Thus, ice crystals from frozen seawater at desalination plants could be rinsed off and thawed to provide usable water. The main advantage of freezing is its low energy consumption, as compared with distillation. The heat of vaporization of water is 40.79 kJ/mol, whereas that of fusion is only 6.01 kJ/mol. Some scientists have suggested that a partial solution to the water shortage in California would be to tow icebergs from the Arctic down to the West Coast. The major disadvantages of freezing are associated with the slow growth of ice crystals and with washing the salt deposits off the crystals.

Reverse Osmosis

Both distillation and freezing involve phase changes that require considerable energy. On the other hand, desalination by reverse osmosis does not involve a phase change and is economically more desirable. ***Reverse osmosis** uses high pressure to force water from a more concentrated solution to a less concentrated one through a semipermeable membrane* (Figure 12.22). The osmotic pressure of seawater is about 30 atm—this is the pressure that must be applied to the saline solution in order to stop the flow of water from left to right. If the pressure on the salt solution were increased beyond 30 atm, the osmotic flow would be reversed, and fresh water would actually pass from the solution through the membrane into the left compartment. Desalination by reverse osmosis is considerably cheaper than by distillation and avoids the technical difficulties associated with freezing. The main obstacle to this method is the development of a membrane that is permeable to water but not to other dissolved substances and that can be used on a large scale for prolonged periods under high-pressure conditions. Once this problem has been solved, and present signs are encouraging, reverse osmosis could become a major desalination technique.

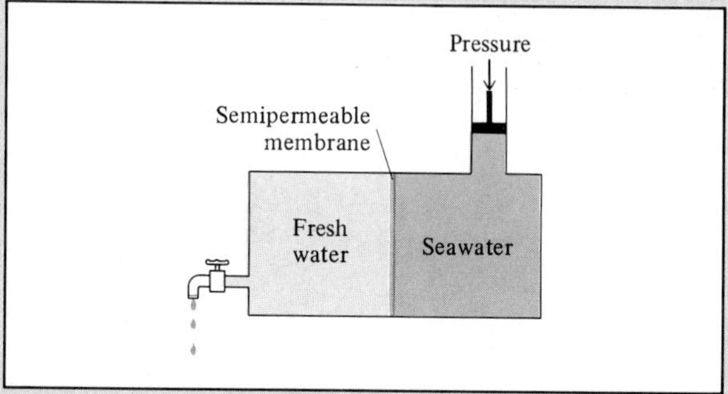

Figure 12.22 *Reverse osmosis. By applying enough pressure on the solution side, fresh water can be made to flow from right to left. The semipermeable membrane allows the passage of water molecules but not of dissolved ions.*

Summary

1. Solutions are homogeneous mixtures of two or more substances, which may be solids, liquids, or gases.
2. The ease of dissolution of a solute in a solvent is governed by intermolecular forces. In addition to energy consideration, the other driving force for the solution process is the disorder resulting when molecules of the solute and solvent mix to form a solution.
3. The concentration of a solution can be expressed as percent by mass, mole fraction, molarity, and molality. These differing units are useful in differing circumstances.
4. Increasing temperature usually increases the solubility of solid and liquid substances and always decreases the solubility of gases.
5. According to Henry's law, the solubility of a gas in a liquid is directly proportional to the partial pressure of the gas over the solution.
6. Raoult's law states that the partial pressure of a substance A over a solution is related to the mole fraction (X_A) of A and to the vapor pressure (P_A°) of pure A as follows: $P_A = X_A P_A^\circ$. An ideal solution obeys Raoult's law over the entire range of concentration. In practice, very few solutions behave ideally in this manner.
7. Vapor-pressure lowering, boiling-point elevation, freezing-point depression, and osmotic pressure are colligative properties of solutions; that is, they are properties that depend only on the number of solute particles that are present and not on their nature.
8. In electrolyte solutions, the interaction between ions leads to the formation of ion pairs. The van't Hoff factor provides a measure of the extent of dissociation in solution.

Applied Chemistry Example: Benzene

Benzene (C_6H_6) is one of the most important organic industrial chemicals. About 90 percent of benzene is used in the polymer industry. It is also used as a solvent and in the synthesis of compounds. Approximately 12 billion pounds of benzene were produced in the United States in 1988, which ranked it 16th among the top 50 industrial chemicals.

Benzene is a toxic substance and is considered to be a carcinogen. Because of these undesirable properties there is a need to limit its use. Toluene (C_7H_8) is a suitable substitute for benzene as a solvent. However, as a reagent in polymer synthesis, there is no substitute for benzene.

Until about the late 1940s the main source of benzene was coal. In recent years benzene is almost exclusively obtained from petroleum in a process called *catalytic reforming*. This process converts a mixture of hydrocarbons obtained from the distillation of petroleum at 450°C, in the presence of a catalyst, to new hydrocarbon products such as benzene and toluene. (Benzene and toluene are called *aromatic hydrocarbons* because they contain the "benzene ring." Their structures are shown on p. 519.) The formation of benzene is as follows:

$$\underset{\text{hexane}}{C_6H_{14}(g)} \longrightarrow \underset{\text{cyclohexane}}{C_6H_{12}(g)} + H_2(g)$$

$$\underset{\text{cyclohexane}}{C_6H_{12}(g)} \longrightarrow \underset{\text{benzene}}{C_6H_6(g)} + 3H_2(g)$$

The reaction mixture is cooled and the condensed liquid is first distilled to remove the volatile components. Further distillation and condensation of the vapor produces a liquid fraction containing about 25 percent benzene by mass.

In the *Udex process* this liquid fraction is treated with the solvent diethylene glycol ($C_4H_{10}O_3$):

$$HO{-}CH_2{-}CH_2{-}O{-}CH_2{-}CH_2{-}OH$$

Benzene and toluene are readily soluble in diethylene glycol, but nonaromatic hydrocarbons (such as hexane and cyclohexane) are not. Finally, benzene is separated from other aromatic compounds by distillation of the diethylene glycol solution.

1. Explain the high solubility of benzene in diethylene glycol in terms of intermolecular forces.

   Answer: Benzene has a regular hexagonal structure and is a nonpolar molecule while diethylene glycol is polar. Benzene has a large polarizability as a result of the delocalized molecular orbitals (see Section 10.9). The strong dipole-induced dipole forces between benzene and diethylene glycol account for the high solubility of benzene. Hexane and cyclohexane are also nonpolar. However, because they have much smaller polarizabilities (they do not possess delocalized molecular orbitals), they are much less soluble in diethylene glycol.

2. A certain diethylene glycol solution contains 7.5 percent benzene by mass. Calculate (*a*) the mole fractions and (*b*) the molality of the solution.

   Answer: (*a*) We need the number of moles of benzene and diethylene glycol to calculate the mole fractions. In 100 g of the solution, there are 7.5 g benzene and 92.5 g diethylene glycol. The mole fraction of benzene is

$$\begin{aligned} X_{C_6H_6} &= \frac{\text{mol } C_6H_6}{\text{mol } C_6H_6 + \text{mol } C_4H_{10}O_3} \\ &= \frac{7.5 \text{ g/(78.11 g/mol)}}{7.5 \text{ g/(78.11 g/mol)} + 92.5 \text{ g/(106.1 g/mol)}} \\ &= 9.9 \times 10^{-2} \end{aligned}$$

By definition, the sum of the mole fractions is 1:

$$X_{C_6H_6} + X_{C_4H_{10}O_3} = 1$$

Hence

$$\begin{aligned} X_{C_4H_{10}O_3} &= 1 - X_{C_6H_6} \\ &= 1 - 9.9 \times 10^{-2} \\ &= 0.90 \end{aligned}$$

This procedure is similar to that shown in Example 12.3.

(*b*) The definition of molality is

$$m = \frac{\text{moles of solute}}{\text{kg solvent}}$$

From (*a*) we see that there is 7.5 g/(78.11 g/mol), or 0.096 mole, of benzene in 92.5 g, or 0.0925 kg, of diethylene glycol. Therefore, the molality of the solution is

$$m = \frac{0.096 \text{ mol}}{0.0925 \text{ kg}} = 1.0 \ m$$

3. Benzene is recovered from diethylene glycol by distillation at 65°C. Toluene is the major impurity. If the vapor pressures of pure benzene, toluene, and diethylene glycol are 369 mmHg, 149 mmHg, and 0.170 mmHg, respectively, at 65°C, calculate the percent by mass of toluene in the liquid condensed from the vapor at this temperature when the total vapor pressure of the liquid is 368 mmHg.

Answer: The total pressure of the vapor, P_T, is given by Dalton's law of partial pressures [Equation (5.11)]

$$P_T = P_{C_6H_6} + P_{C_7H_8} + P_{C_4H_{10}O_3} \qquad (1)$$

To calculate individual pressures, we must turn to Raoult's law [Equation (12.2)]:

$$P_i = X_i P_i^\circ$$

Equation (1) now becomes

$$P_T = X_{C_6H_6}P^\circ_{C_6H_6} + X_{C_7H_8}P^\circ_{C_7H_8} + X_{C_4H_{10}O_3}P^\circ_{C_4H_{10}O_3}$$

However, since the vapor pressure of diethylene glycol is much lower than that of benzene and toluene, only a tiny amount of it was originally present in the vapor phase during distillation. Consequently, its mole fraction in the condensed liquid phase will be quite small, so we can ignore the last term in the above equation and write

$$P_T = X_{C_6H_6}P^\circ_{C_6H_6} + X_{C_7H_8}P^\circ_{C_7H_8} \qquad (2)$$

In a two-component system we have

$$X_{C_6H_6} + X_{C_7H_8} = 1$$

or

$$X_{C_7H_8} = 1 - X_{C_6H_6} \tag{3}$$

Substituting (3) into (2) we get

$$P_T = X_{C_6H_6}P^\circ_{C_6H_6} + (1 - X_{C_6H_6})P^\circ_{C_7H_8} \tag{4}$$

Rearrangement of Equation (4) gives

$$\begin{aligned} X_{C_6H_6} &= \frac{P_T - P^\circ_{C_7H_8}}{P^\circ_{C_6H_6} - P^\circ_{C_7H_8}} \\ &= \frac{368 \text{ mmHg} - 149 \text{ mmHg}}{369 \text{ mmHg} - 149 \text{ mmHg}} \\ &= 0.995 \end{aligned}$$

The mole fraction of toluene can now be calculated as

$$\begin{aligned} X_{C_7H_8} &= 1 - X_{C_6H_6} \\ &= 1 - 0.995 \\ &= 0.005 \end{aligned}$$

To calculate the percent by mass of toluene we proceed as follows. From the mole fractions we see that in one mole of the mixture there are 0.995 mole of benzene and 0.005 mole of toluene so that

$$\begin{aligned} \text{percent by mass of } C_7H_8 &= \frac{0.005 \text{ mol}(92.13 \text{ g/mol})}{0.005 \text{ mol}(92.13 \text{ g/mol}) + 0.995 \text{ mol}(78.11 \text{ g/mol})} \times 100\% \\ &= 0.6\% \end{aligned}$$

Thus the benzene obtained by the Udex process has a purity of (100% − 0.6%), or 99.4%, by mass.

Key Words

Colligative properties, p. 517
Crystallization, p. 512
Desalination, p. 531
Dimer, p. 504
Fractional crystallization, p. 513
Fractional distillation, p. 520
Henry's law, p. 515
Ideal solution, p. 519
Ion pairs, p. 529
Miscible, p. 504
Molality, p. 509
Nonvolatile, p. 518
Osmosis, p. 524
Osmotic pressure, p. 524
Percent by mass, p. 507
Raoult's law, p. 518
Reverse osmosis, p. 532
Saturated solution, p. 512
Semipermeable membrane, p. 524
Supersaturated solution, p. 513
Thermal pollution, p. 514
Unsaturated solution, p. 512
Volatile, p. 519

Exercises

The Solution Process

REVIEW QUESTIONS

12.1 Briefly describe the solution process at the molecular level. Use the dissolution of a solid in a liquid as an example.

12.2 Based on intermolecular force consideration, explain what ''like dissolves like'' means.

12.3 What is solvation? What are the factors that influence the extent to which solvation occurs? Give two examples of solvation, one involving ion-dipole interaction and the other dispersion forces.

12.4 As you know, some solution processes are endothermic and others are exothermic. Provide a molecular interpretation for the difference.

12.5 Explain why the solution process invariably leads to an increase in disorder.

12.6 Describe the factors that affect the solubility of a solid in a liquid.

12.7 What does it mean to say that two liquids are miscible?

PROBLEMS

12.8 Why is naphthalene ($C_{10}H_8$) more soluble than CsF in benzene?

12.9 Explain why ethanol (C_2H_5OH) is not soluble in cyclohexane (C_6H_{12}).

12.10 Arrange the following compounds in order of increasing solubility in water: O_2, LiCl, Br_2, methanol (CH_3OH).

12.11 Explain the variations in solubility in water of the alcohols listed below:

Compound	Solubility in water g/100 g, 20°C
CH_3OH	∞
CH_3CH_2OH	∞
$CH_3CH_2CH_2OH$	∞
$CH_3CH_2CH_2CH_2OH$	9
$CH_3CH_2CH_2CH_2CH_2OH$	2.7

(*Note:* ∞ means the alcohol and water are completely miscible in all proportions.)

12.12 State which of the alcohols listed in Problem 12.11 you would expect to be the best solvent for each of the following, and explain why: (a) I_2, (b) KBr, (c) $CH_3CH_2CH_2CH_2CH_3$.

Concentration Units

REVIEW QUESTION

12.13 Define the following concentration terms and give their units: percent by mass, mole fraction, molarity, molality. Compare their advantages and disadvantages.

PROBLEMS

12.14 Calculate the percent by mass of the solute in each of the following aqueous solutions: (a) 5.50 g of NaBr in 78.2 g of solution, (b) 31.0 g of KCl in 152 g of water, (c) 4.5 g of toluene in 29 g of benzene.

12.15 Calculate the amount of water (in grams) that must be added to (a) 5.00 g of urea (H_2NCONH_2) in the preparation of a 16.2 percent by mass solution and (b) 26.2 g of $MgCl_2$ in the preparation of a 1.5 percent by mass solution.

12.16 A solution is prepared by mixing 62.6 mL of benzene (C_6H_6) with 80.3 mL of toluene (C_7H_8). Calculate the mole fractions of these two components. The densities are: benzene, 0.879 g/cm^3; and toluene, 0.867 g/cm^3.

12.17 Calculate the molality of each of the following solutions: (a) 14.3 g of sucrose ($C_{12}H_{22}O_{11}$) in 676 g of water, (b) 7.20 mole of ethylene glycol ($C_2H_6O_2$) in 3546 g of water.

12.18 For dilute solutions, in which the density of the solution is roughly equal to that of the pure solvent, the molarity of the solution is equal to its molality. Show that this statement is correct for a 0.010 *M* aqueous urea (H_2NCONH_2) solution.

12.19 Calculate the molality of each of the following aqueous solutions: (a) 2.50 *M* NaCl solution (density of solution = 1.08 g/mL), (b) 5.86 *M* ethanol solution (density of solution = 0.927 g/mL), (c) 48.2 percent by mass KBr solution.

12.20 Calculate the molalities of the following aqueous solutions: (a) 1.22 *M* sugar ($C_{12}H_{22}O_{11}$) solution (density of solution = 1.12 g/mL), (b) 0.87 *M* NaOH solution (density of solution = 1.04 g/mL), (c) 5.24 *M* $NaHCO_3$ solution (density of solution = 1.19 g/mL).

12.21 The alcohol content of hard liquor is normally given in terms of the "proof," which is defined as twice the percentage by volume of ethanol (C_2H_5OH) present. Calculate the number of grams of alcohol present in 1.00 L of 75 proof gin. The density of ethanol is 0.798 g/mL.

12.22 The concentrated sulfuric acid we use in the laboratory is 98.0 percent H_2SO_4 by mass. Calculate the molality and molarity of the acid solution. The density of the solution is 1.83 g/mL.

12.23 Calculate the molarity, molality, and the mole fraction of NH_3 for a solution of 30.0 g of NH_3 in 70.0 g of water. The density of the solution is 0.982 g/mL.

12.24 The density of an aqueous solution containing 10.0 percent of ethanol (C_2H_5OH) by mass is 0.984 g/mL. (a) Calculate the molality of this solution. (b) Calculate its molarity. (c) What volume of the solution would contain 0.125 mole of ethanol? (d) Calculate the mole fraction of water in the solution.

12.25 It is estimated that 1.0 mL of seawater contains about 4.0×10^{-12} g of gold. The total volume of ocean water is 1.5×10^{21} L. Calculate the total amount of gold present in seawater. With so much gold out there, why hasn't someone become rich by mining gold from the ocean?

Solubility and Fractional Crystallization

REVIEW QUESTIONS

12.26 Define the following terms: saturated solution, unsaturated solution, supersaturated solution, crystallization, fractional crystallization.

12.27 How do the solubilities of most ionic compounds in water change with temperature?

12.28 How do a crystal and a precipitate differ?

12.29 What is the effect of pressure on the solubility of a liquid in liquid and of a solid in liquid?

12.30 What is the practical application of fractional crystallization?

PROBLEMS

12.31 A 3.20-g sample of a salt dissolves in 9.10 g of water to give a saturated solution at 25°C. What is the solubility (in g salt/100 g of H_2O) of the salt?

12.32 The solubility of KNO_3 is 155 g per 100 g of water at 75°C and 38.0 g at 25°C. What mass (in grams) of KNO_3 will crystallize out of solution if exactly 100 g of its saturated solution at 75°C are cooled to 25°C?

12.33 A 50-g sample of impure $KClO_3$ (solubility = 7.1 g per 100 g H_2O at 20°C) is contaminated with 10 percent of KCl (solubility = 25.5 g per 100 g of H_2O at 20°C). Calculate the minimum quantity of 20°C water needed to dissolve all the KCl from the sample. How much $KClO_3$ will be left after this treatment? (Assume that the solubilities are unaffected by the presence of the other compound.)

Gas Solubility

REVIEW QUESTIONS

12.34 Discuss the factors that influence the solubility of a gas in a liquid. Explain why the solubility of a gas in a liquid always decreases with increasing temperature.

12.35 What is thermal pollution? Why is it harmful to aquatic life?

12.36 What is Henry's law? Define each term in the equation, and give its units. How would you account for the law in terms of the kinetic molecular theory of gases?

12.37 Give two exceptions to Henry's law.

PROBLEMS

12.38 A student is observing two beakers of water. One beaker is heated to 30°C, and the other is heated to 100°C. In each case, bubbles form in the water. Are these bubbles of the same origin? Explain.

12.39 A man bought a goldfish in a pet shop. Upon returning home, he put the goldfish in a bowl of recently boiled water that had been cooled quickly. A few minutes later the fish was found dead. Explain what happened to the fish.

12.40 The solubility of CO_2 in water at 25°C and 1 atm is 0.034 mol/L. What is its solubility under atmospheric conditions? (The partial pressure of CO_2 in air is 0.030 atm.) Assume that CO_2 obeys Henry's law.

12.41 A beaker of water is initially saturated with dissolved air. Explain what happens when He gas at 1 atm is bubbled through the solution for a long time.

12.42 Neither HCl nor NH_3 gas obeys Henry's law. Explain.

12.43 A miner working 260 m below sea level opened a carbonated soft drink during a lunch break. To his surprise, the soft drink tasted rather "flat." Shortly afterward, the miner took an elevator to the surface. During the trip up, he could not stop belching. Why?

12.44 The solubility of N_2 in blood at 37°C and at a partial pressure of 0.80 atm is 5.6×10^{-4} mol/L. A deep-sea diver breathes compressed air with the partial pressure of N_2 equal to 4.0 atm. Assume that the total volume of blood in the body is 5.0 L. Calculate the amount of N_2 gas released (in liters) when the diver returns to the surface of the water, where the partial pressure of N_2 is 0.80 atm.

Colligative Properties of Nonelectrolytes

REVIEW QUESTIONS

12.45 What are colligative properties? What is the meaning of the word "colligative" in this context?

12.46 Give two examples of a volatile liquid and two examples of a nonvolatile liquid.

12.47 Define Raoult's law. Define each term in the equation representing Raoult's law, and give its units.

12.48 What is an ideal solution?

12.49 Give a molecular interpretation of positive deviation and negative deviation.

12.50 Explain fractional distillation. What is its practical application?

12.51 Define boiling-point elevation and freezing-point depression. Write the equations relating boiling-point elevation and freezing-point depression to the concentration of the solution. Define all the terms, and give their units.

12.52 How is the lowering in vapor pressure related to a rise in the boiling point of a solution?

12.53 Use a phase diagram to show the difference in freezing point and boiling point between an aqueous urea solution and pure water.

12.54 What is osmosis? What is a semipermeable membrane?

12.55 Write the equation relating osmotic pressure to the concentration of a solution. Define all the terms and give their units.

12.56 What does it mean when we say that the osmotic pressure of a sample of seawater is 25 atm at a certain temperature?

12.57 Explain why molality is used for boiling-point elevation and freezing-point depression calculations and molarity is used in osmotic pressure calculations.

12.58 Describe how you would use the freezing-point depression and osmotic pressure measurements to determine the molar mass of a compound. Why is the boiling-point elevation phenomenon normally not used for this purpose?

12.59 Explain why it is essential that fluids used in intravenous injections have approximately the same osmotic pressure as blood.

PROBLEMS

12.60 A solution is prepared by dissolving 396 g of sucrose ($C_{12}H_{22}O_{11}$) in 624 g of water. What is the vapor pressure of this solution at 30°C? (The vapor pressure of water is 31.8 mmHg at 30°C.)

12.61 How many grams of sucrose ($C_{12}H_{22}O_{11}$) must be added to 552 g of water to give a solution with a vapor pressure 2.0 mmHg less than that of pure water at 20°C? (The vapor pressure of water at 20°C is 17.5 mmHg.)

12.62 The vapor pressure of benzene is 100.0 mmHg at 26.1°C. Calculate the vapor pressure of a solution containing 24.6 g of camphor ($C_{10}H_{16}O$) dissolved in 98.5 g of benzene. (Camphor is a low-volatility solid.)

12.63 The vapor pressures of ethanol (C_2H_5OH) and 1-propanol (C_3H_7OH) at 35°C are 100 mmHg and 37.6 mmHg, respectively. Assume ideal behavior and calculate the partial pressures of ethanol and 1-propanol at 35°C over a solution of ethanol in 1-propanol, in which the mole fraction of ethanol is 0.300.

12.64 The vapor pressure of ethanol (C_2H_5OH) at 20°C is 44 mmHg, and the vapor pressure of methanol (CH_3OH) at the same temperature is 94 mmHg. A mixture of 30.0 g of methanol and 45.0 g of ethanol is prepared (and may be assumed to behave as an ideal solution). (a) Calculate the vapor pressure of methanol and ethanol above this solution at 20°C. (b) Calculate the mole fraction of methanol and ethanol in the vapor above this solution at 20°C. (c) Suggest a method for separating the two components of the solution.

12.65 Two beakers, one containing an 8.0 *M* glucose aqueous solution and the other containing pure water, are placed under a tightly sealed bell jar at room temperature. After a few months one beaker is completely dry and the other has increased in water by an amount equal to that originally present in the other beaker. Account for this phenomenon.

12.66 How many grams of urea (H_2NCONH_2) must be added to 450 g of water to give a solution with a vapor pressure 2.50 mmHg less than that of pure water at 30°C? (The vapor pressure of water at 30°C is 31.8 mmHg.)

12.67 What are the boiling point and freezing point of a 2.47 *m* solution of naphthalene in benzene? (The boiling point and freezing point of benzene are 80.1°C and 5.5°C, respectively.)

12.68 An aqueous solution contains the amino acid glycine (NH_2CH_2COOH). Assuming no ionization of the acid, calculate the molality of the solution if it freezes at −1.1°C.

12.69 Pheromones are compounds secreted by the females of many insect species to attract males. One of these compounds contains 80.78 percent C, 13.56 percent H, and 5.66 percent O. A solution of 1.00 g of this pheromone in 8.50 g of benzene freezes at 3.37°C. What are the molecular formula and molar mass of the compound? (The normal freezing point of pure benzene is 5.50°C.)

12.70 How many liters of the antifreeze ethylene glycol [$CH_2(OH)CH_2(OH)$] would you add to a car radiator containing 6.50 L of water if the coldest winter temperature in your area is −20°C? Calculate the boiling point of this water-ethylene glycol mixture. The density of ethylene glycol is 1.11 g/mL.

12.71 A solution of 0.85 g of the organic compound mesitol in 100.0 g of benzene is observed to have a freezing point of 5.16°C. What are the molality of the mesitol solution and the molar mass of mesitol?

12.72 A solution of 2.50 g of a compound of empirical formula C_6H_5P in 25.0 g of benzene is observed to freeze at 4.3°C. Calculate the molar mass of the solute and its molecular formula.

12.73 A solution is prepared by condensing 4.00 L of a gas, measured at 27°C and 748 mmHg pressure, into 58.0 g of benzene. Calculate the freezing point of this solution.

12.74 The elemental analysis of an organic solid extracted from gum arabic showed that it contained 40.0 percent C, 6.7 percent H, and 53.3 percent O. A solution of 0.650 g of the solid in 27.8 g of the solvent diphenyl gave a freezing-point depression of 1.56°C. Calculate the molar mass and molecular formula of the solid. (K_f for diphenyl is 8.00°C/*m*.)

12.75 The molar mass of benzoic acid (C_6H_5COOH) determined by measuring the freezing-point depression in benzene is twice that expected for the molecular formula, $C_7H_6O_2$. Explain this apparent anomaly.

12.76 A quantity of 7.480 g of an organic compound is dissolved in water to make 300.0 mL of solution. The solution has an osmotic pressure of 1.43 atm at 27°C. The analysis of this compound shows it to contain 41.8 percent C, 4.7 percent H, 37.3 percent O, and 16.3 percent N. Calculate the molecular formula of the compound.

12.77 What is the osmotic pressure (in atm) of a 1.36 *M* aqueous urea solution at 22.0°C?

12.78 A solution containing 0.8330 g of a protein of unknown structure in 170.0 mL of aqueous solution was found to have an osmotic pressure of 5.20 mmHg at 25°C. Determine the molar mass of the protein.

12.79 A solution of 6.85 g of a carbohydrate in 100.0 g of water has a density of 1.024 g/mL and an osmotic pressure of 4.61 atm at 20.0°C. Calculate the molar mass of the carbohydrate.

Colligative Properties of Electrolyte Solutions

REVIEW QUESTIONS

12.80 Why is the discussion of the colligative properties of electrolyte solutions more involved than that of nonelectrolyte solutions?

12.81 Define ion pairs. What effect does ion-pair formation have on the colligative properties of a solution?

12.82 How does the ease of ion-pair formation depend on (a) charges on the ions, (b) size of the ions, (c) nature of the solvent (polar versus nonpolar)?

12.83 In each case, indicate which of the following pairs of compounds is more likely to form ion pairs in water: (a) NaCl or Na_2SO_4, (b) $MgCl_2$ or $MgSO_4$, (c) LiBr or KBr.

12.84 Define the van't Hoff factor. What information does this quantity provide?

PROBLEMS

12.85 Which of the following two aqueous solutions has (a) the higher boiling point, (b) the higher freezing point, and (c) the lower vapor pressure: 0.35 *m* $CaCl_2$ or 0.90 *m* urea? State your reasons.

12.86 Consider two aqueous solutions, one of sucrose ($C_{12}H_{22}O_{11}$) and the other of nitric acid (HNO_3), both of which freeze at $-1.5°C$. What other properties do these solutions have in common?

12.87 Arrange the following solutions in order of decreasing freezing point: 0.10 *m* Na_3PO_4, 0.35 *m* NaCl, 0.20 *m* $MgCl_2$, 0.15 *m* $C_6H_{12}O_6$, 0.15 *m* CH_3COOH.

12.88 Both NaCl and $CaCl_2$ are used to melt ice on roads in winter. What advantages do these substances have over sucrose or urea in lowering the freezing point of water?

12.89 Arrange the following aqueous solutions in order of decreasing freezing point and explain your reasons: 0.50 *m* HCl, 0.50 *m* glucose, 0.50 *m* acetic acid.

12.90 What are the normal freezing points and boiling points of the following solutions? (a) 21.2 g NaCl in 135 mL of water, (b) 15.4 g of urea in 66.7 mL of water.

12.91 At 25°C the vapor pressure of pure water is 23.76 mmHg and that of seawater is 22.98 mmHg. Assuming that seawater contains only NaCl, estimate its concentration in molality units.

12.92 The osmotic pressure of 0.010 *M* solutions of $CaCl_2$ and urea at 25°C are 0.605 atm and 0.245 atm, respectively. Calculate the van't Hoff factor for the $CaCl_2$ solution.

12.93 A 0.86 percent by mass solution of NaCl is called "physiological saline" because its osmotic pressure is equal to that of the solution in blood cells. Calculate the osmotic pressure of this solution at normal body temperature (37°C). Note that the density of the saline solution is 1.005 g/mL.

12.94 Consider the following initial arrangement:

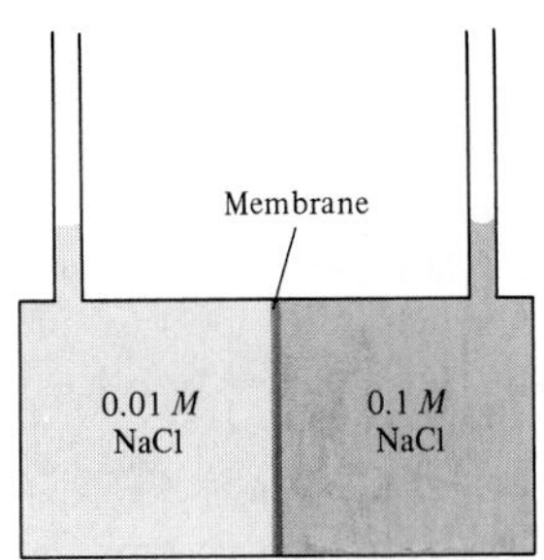

What will happen if the membrane is (a) permeable to both water and the Na^+ and Cl^- ions, (b) permeable to water and Na^+ ions but not to Cl^- ions, (c) permeable to water but not to Na^+ and Cl^- ions?

MISCELLANEOUS PROBLEMS

12.95 A solution is made up by mixing two volatile liquids A and B. Complete the following table where the symbol $\longleftrightarrow$ indicates intermolecular attractive forces.

Attractive forces	Deviation from Raoult's law	ΔH_{soln}
$A \longleftrightarrow A$, $B \longleftrightarrow B > A \longleftrightarrow B$		
	Negative	
		Zero

12.96 Explain each of the following statements: (a) The boiling point of seawater is greater than that of pure water. (b) Carbon dioxide escapes from the solution when the cap is removed from a soft-drink bottle. (c) Concentrations of dilute solutions expressed in molalities are approximately equal to those expressed in molarities. (d) In discussing the colligative properties of a solution (other than osmotic pressure), it is preferable to express the concentration units in molalities rather than in molarities. (e) Methanol (b.p. 65°C) is useful as an antifreeze, but it should be removed from the car radiator during the summer season.

12.97 Lysozyme is an enzyme that cleaves bacterial cell walls. A sample of lysozyme extracted from chicken egg white has a molar mass of 13,930 g. A quantity of 0.100 g of this enzyme is dissolved in 150 g of water at 25°C. Calculate the vapor-pressure lowering, the depression in freezing point, the elevation in boiling point, and the osmotic pressure of this solution. (The vapor pressure of water at 25°C = 23.76 mmHg.)

12.98 A solution of 1.00 g of anhydrous aluminum chloride, $AlCl_3$, in 50.0 g of water is found to have a freezing point of $-1.11°C$. Explain this observation.

12.99 A cucumber placed in concentrated brine (salt water) shrivels into a pickle. Explain.

12.100 Solutions A and B have osmotic pressures of 2.4 atm and 4.6 atm, respectively, at a certain temperature. What is the osmotic pressure of a solution prepared by mixing equal volumes of A and B at the same temperature?

12.101 Define reverse osmosis. Explain why reverse osmosis is (theoretically) more desirable as a method for desalination than distillation or freezing.

12.102 What minimum pressure must be applied to seawater at 25°C in order to carry out the reverse osmosis process? (Treat seawater as a 0.70 *M* NaCl solution.)

12.103 A protein has been isolated as a salt with the formula $Na_{20}P$ (this means that there are 20 Na^+ ions associated with a negatively charged protein P^{20-}). The osmotic pressure of a 10.0-mL solution containing 0.225 g of the protein is 0.257 atm at 25.0°C. (a) Calculate the molar mass of the protein from these data. (b) Calculate the actual molar mass of the protein.

12.104 What masses of sodium chloride, magnesium chloride, sodium sulfate, calcium chloride, potassium chloride, and sodium bicarbonate are needed to yield 1 L of artificial seawater for an aquarium? The required ionic concentrations are $[Na^+] = 2.56\ M$, $[K^+] = 0.0090\ M$, $[Mg^{2+}] = 0.054\ M$, $[Ca^{2+}] = 0.010\ M$, $[HCO_3^-] = 0.0020\ M$, $[Cl^-] = 2.60\ M$, $[SO_4^{2-}] = 0.051\ M$.

12.105 Suggest two ways to separate a mixture of iodine and sodium chloride.

12.106 Two liquids A and B have vapor pressures of 76 mmHg and 132 mmHg, respectively, at 25°C. What is the total vapor pressure of the ideal solution made up of (a) 1.00 mole of A and 1.00 mole of B and (b) 2.00 moles of A and 5.00 moles of B?

12.107 Calculate the van't Hoff factor of Na_3PO_4 in a 0.40 *m* solution whose boiling point is 100.78°C.

12.108 A 262-mL sample of a sugar solution containing 1.22 g of the sugar has an osmotic pressure of 30.3 mmHg at 35°C. What is the molar mass of the sugar?

12.109 A nonvolatile organic compound Z was used to make up two solutions. In solution A, 5.00 g of Z was dissolved in 100 g of water; and in solution B, 2.31 g of Z was dissolved in 100 g of benzene. Solution A had a vapor pressure of 754.5 mmHg at the normal boiling point of water, and solution B had the same vapor pressure at the normal boiling point of benzene. Calculate the molar mass of Z in solutions A and B and account for the difference.

12.110 Consider the following three mercury manometers. One of them has 1 mL of water placed on top of the mercury, another has 1 mL of a 1 *m* urea solution placed on top of the mercury, and the third one has 1 mL of a 1 *m* NaCl solution placed on top of the mercury. Identify X, Y, and Z with these solutions.

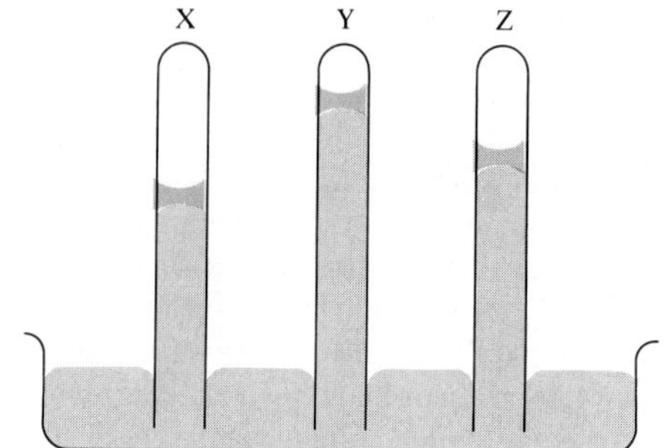

Opposite page: Changes in cloud condensation nuclei often affect Earth's climate. Kinetic models are used to understand how the increase in concentration of sulfur dioxide, which is oxidized to sulfate aerosols, changes the climate.

CHAPTER 13

CHEMICAL KINETICS

By now we have studied basic definitions in chemistry, and we have examined the properties of gases, liquids, solids, and solutions. We have discussed some molecular properties and looked at several types of reactions in some detail. In this chapter and in the four that follow, we look more closely at the relationships and the laws that govern chemical reactions.

How can we predict whether or not a reaction will take place? Once started, how fast does the reaction proceed? How far will the reaction go before it is finished? The laws of thermodynamics (to be discussed in Chapter 18) help us answer the first question. Chemical kinetics, the subject of this chapter, provides answers to the second question, how fast a reaction proceeds. The last question is one of many answered by the study of equilibrium in chemical reactions, which we will consider in Chapters 14, 16, 17, and 18.

13.1 The Rate of a Reaction

The area of chemistry concerned with the speeds, or rates, at which a chemical reaction occurs is called ***chemical kinetics.*** The word "kinetic" suggests motion. (Recall that in Chapter 5 we studied the kinetic molecular theory of gases, which describes the random motion of gas molecules.) Here kinetics refers to the rate of a reaction, or the ***reaction rate,*** which is the *change of the concentration of reactant or product with time*. (For simplicity, we will generally speak of "the rate" rather than "the reaction rate.")

We know that any reaction can be represented by the general equation

$$\text{reactants} \longrightarrow \text{products}$$

This equation tells us that, during the course of a reaction, reactant molecules are consumed while product molecules are formed. As a result, we can follow the progress of a reaction by monitoring either the decrease in concentration of the reactants or the increase in concentration of the products.

Figure 13.1 shows the progress of a simple reaction in which reactant molecules A are converted to product molecules B:

$$\text{A} \longrightarrow \text{B}$$

The decrease in the number of A molecules and the increase in the number of B molecules with time are shown in Figure 13.2.

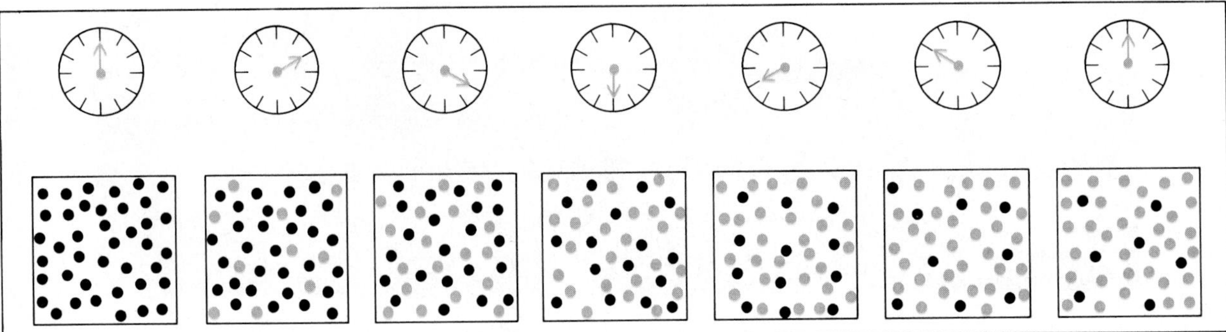

Figure 13.1 *The progress of reaction A → B at 10-s intervals over a period of 60 s. Initially, only A molecules (black dots) are present. As time progresses, B molecules (colored dots) are formed.*

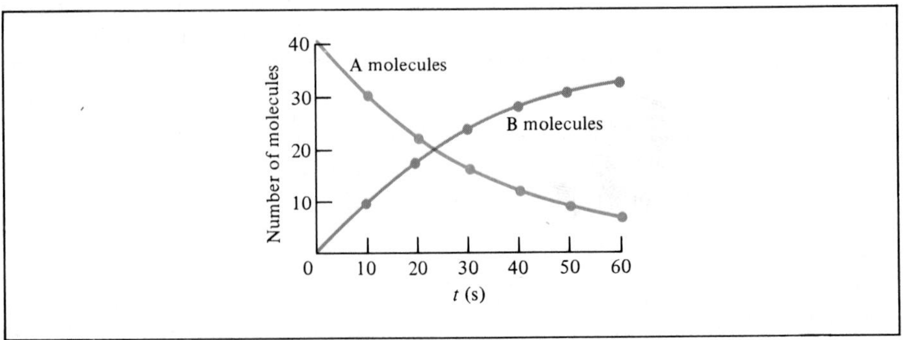

Figure 13.2 *The rate of reaction A → B, represented as the decrease of A molecules with time and as the increase of B molecules with time.*

To see how the rate of a chemical reaction is actually measured, we will consider a specific example: the reaction of molecular bromine with formic acid. After we have seen how rates are measured experimentally, we will learn to write rate expressions using the experimental data.

Reaction of Molecular Bromine and Formic Acid. In aqueous solutions, molecular bromine reacts with formic acid (HCOOH) as follows:

$$Br_2(aq) + HCOOH(aq) \longrightarrow 2Br^-(aq) + 2H^+(aq) + CO_2(g)$$

Molecular bromine has a characteristic red brown color. As the reaction progresses, the concentration of Br_2 steadily decreases (Figure 13.3). This change can be monitored easily by measuring the fading of molecular bromine's color with a spectrometer. Table 13.1 shows the molar concentration of Br_2 at different times; these data are plotted in Figure 13.4. Note that time zero is just after the mixing of the reactants.

The average rate of the reaction can be defined as the change in the reactant concentration over a certain time interval. That is,

Since [ ] is mol/L or M, and time is in seconds, s, the units of rate are M/s.

$$\text{average rate} = -\frac{[Br_2]_{\text{final}} - [Br_2]_{\text{initial}}}{t_{\text{final}} - t_{\text{initial}}} = -\frac{\Delta[Br_2]}{\Delta t}$$

where $\Delta[Br_2] = [Br_2]_{\text{final}} - [Br_2]_{\text{initial}}$ and $\Delta t = t_{\text{final}} - t_{\text{initial}}$. Because the concentra-

Figure 13.3 *From left to right. The decrease in bromine concentration as time elapses shows up as a loss in color.*

Table 13.1 Bromine concentration in the reaction between molecular bromine and formic acid at 25°C

Time (s)	$[Br_2]$ (M)	Average rate (M/s)
0.0	0.0120	
		3.80×10^{-5}
50.0	0.0101	
		3.28×10^{-5}
100.0	0.00846	
		2.72×10^{-5}
150.0	0.00710	
		2.28×10^{-5}
200.0	0.00596	
		1.92×10^{-5}
250.0	0.00500	
		1.60×10^{-5}
300.0	0.00420	
		1.34×10^{-5}
350.0	0.00353	
		1.14×10^{-5}
400.0	0.00296	

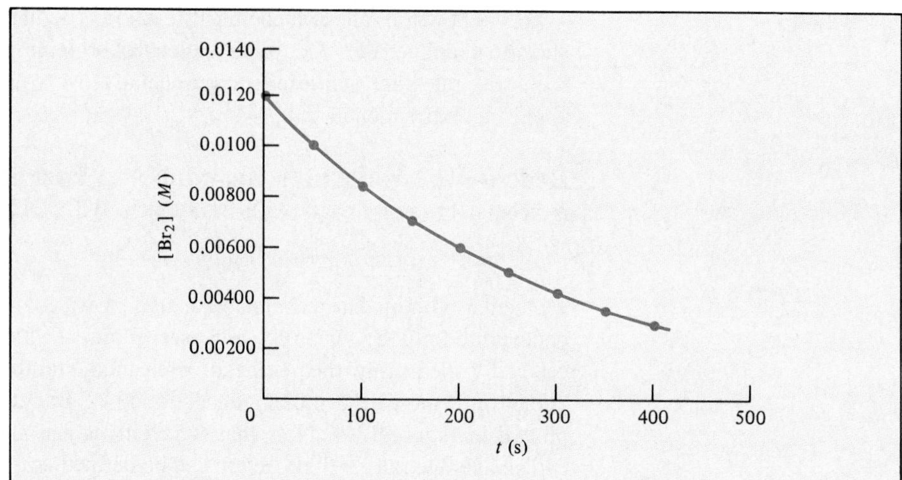

Figure 13.4 *The change in molecular bromine concentration as a function of time in a reaction between molecular bromine and formic acid.*

tion of Br_2 *decreases* during the time interval, $\Delta[Br_2]$ is a negative quantity. Since the rate of a reaction is a positive quantity, so a minus sign is needed in the rate expression to make the rate positive.

From Table 13.1 we can calculate the average rate over the first 50-s time interval as follows:

$$\begin{aligned}\text{average rate} &= -\frac{\Delta[Br_2]}{\Delta t}\\ &= -\frac{(0.0101 - 0.0120)\ M}{50.0\ \text{s}}\\ &= 3.80 \times 10^{-5}\ M/\text{s}\end{aligned}$$

The average rate during the next 50-s time interval (from t = 50.0 s to t = 100.0 s) is similarly given by

$$\text{average rate} = -\frac{(0.00846 - 0.0101)\ M}{50.0\ \text{s}} = 3.28 \times 10^{-5}\ M/\text{s}$$

These calculations demonstrate that the average rate of the reaction is not a constant but changes with the concentration of reactant molecules present. Initially, when the concentration of Br_2 is high, the rate is correspondingly high. The rate gradually decreases, and it eventually becomes zero when all the molecular bromine has reacted.

The rate of the reaction also depends on the concentration of formic acid. However, by using a large excess of formic acid in the reaction mixture we can ensure that the concentration of formic acid remains virtually constant throughout the course of the reaction. Under this condition the change in the amount of formic acid present in solution has no effect on the rate measured.

Using average rates to describe the progress of a reaction is unsatisfactory because the rate of 3.80×10^{-5} M/s is not associated with any particular instant of time; rather, it represents the average value of the rates from time zero to 50.0 s. If we had chosen the first 100 s as the time interval, the rate would have been 3.54×10^{-5} M/s, and so on. Thus average rate does not tell us what the rate is at any specific instant.

There is much advantage in speaking of the rate of a reaction at a specific time. For one thing, it makes the comparison of two reaction rates more meaningful. We have seen that the value of an average rate depends on the time interval we choose. However, we can gradually eliminate this arbitrary choice of time interval by calculating the rate over a smaller and smaller time interval. In fact, to find the rate at a particular time instant we draw a line tangent to the concentration versus time curve at the point corresponding to that time. The slope of the tangent then gives the desired rate (Figure 13.5). To distinguish this rate from the average rate, we call it the *instantaneous rate*. Note that the instantaneous rate determined in this manner will have the same value for the same concentrations of reactants, as long as the temperature is kept constant. In the following discussion, we will frequently refer to the instantaneous rate merely as the rate.

Table 13.2 lists the rates of the bromine–formic acid reaction at specific times. These rates are obtained from the curve shown in Figure 13.5. The rate at $t = 0$ is called the *initial rate*, the rate of the reaction immediately after the mixing of the reactants.

Figure 13.6 is a plot of the rates versus Br_2 concentration. The fact that this gives a

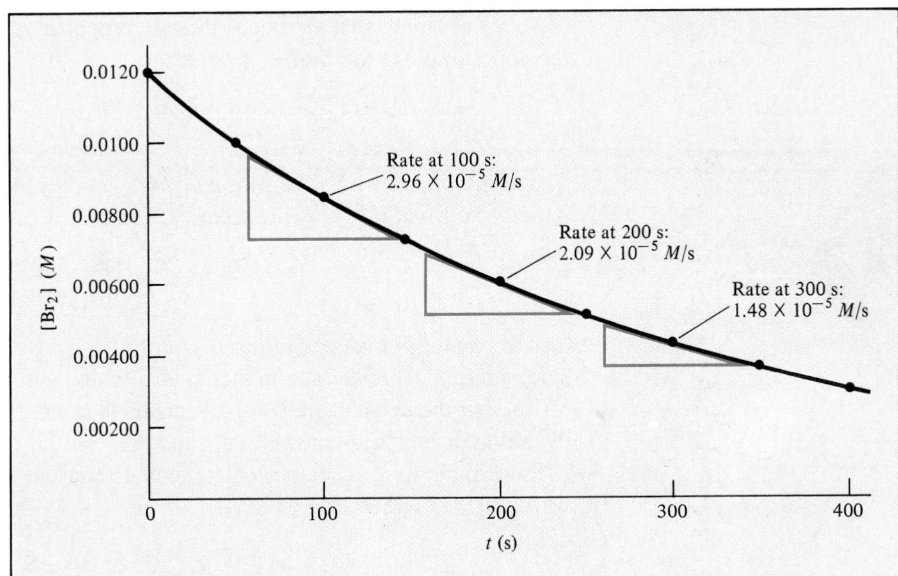

Figure 13.5 *The instantaneous rates of the reaction between molecular bromine and formic acid at t = 100 s, 200 s, and 300 s are given by the slopes of the tangents at these times.*

Table 13.2 Rates of the reaction between molecular bromine and formic acid at 25°C

Time (s)	$[Br_2]$ (M)	Rate (M/s)	$k = \dfrac{\text{rate}}{[Br_2]}$ (s^{-1})
0.0	0.0120	4.20×10^{-5}	3.50×10^{-3}
50.0	0.0101	3.52×10^{-5}	3.49×10^{-3}
100.0	0.00846	2.96×10^{-5}	3.50×10^{-3}
150.0	0.00710	2.49×10^{-5}	3.51×10^{-3}
200.0	0.00596	2.09×10^{-5}	3.51×10^{-3}
250.0	0.00500	1.75×10^{-5}	3.50×10^{-3}
300.0	0.00420	1.48×10^{-5}	3.52×10^{-3}
350.0	0.00353	1.23×10^{-5}	3.48×10^{-3}
400.0	0.00296	1.04×10^{-5}	3.51×10^{-3}

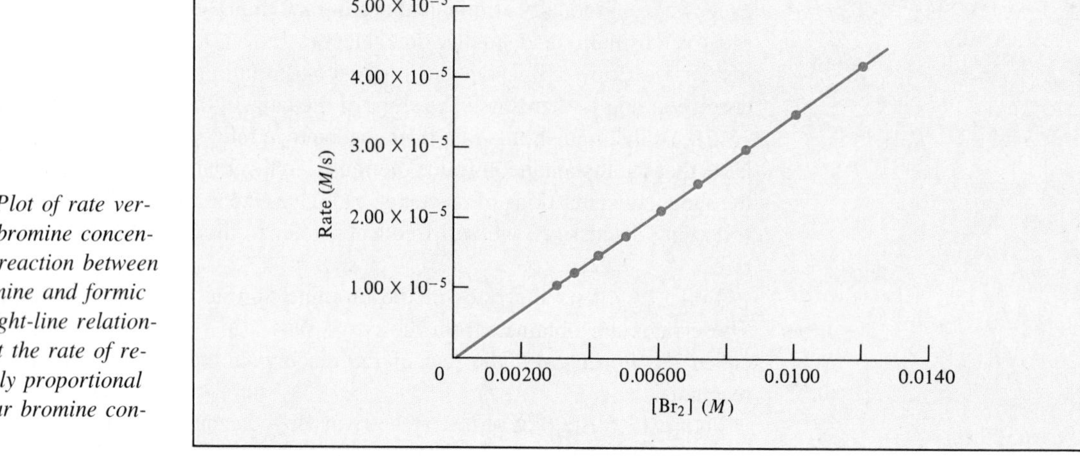

Figure 13.6 *Plot of rate versus molecular bromine concentration, in the reaction between molecular bromine and formic acid. The straight-line relationship shows that the rate of reaction is directly proportional to the molecular bromine concentration.*

straight line shows that the rate is directly proportional to the concentration; the higher the concentration, the higher the rate:

$$\begin{aligned} \text{rate} &\propto [\mathrm{Br_2}] \\ &= k[\mathrm{Br_2}] \end{aligned}$$

The term k is known as the ***rate constant,*** *a constant of proportionality between the reaction rate and the concentrations of reactants*. Rearrangement of the equation gives

$$k = \frac{\text{rate}}{[\mathrm{Br_2}]}$$

It is important to understand that k is *not* affected by the concentration of Br_2. To be sure, the *rate* is greater at a higher concentration and smaller at a lower concentration of Br_2, but the *ratio*—rate/[Br_2]—remains the same. As we will see in Section 13.3, the value of any rate constant depends only on the temperature.

From Table 13.2 we can calculate the rate constant for the reaction. Taking the data for $t = 0$, we write

$$\begin{aligned} k &= \frac{\text{rate}}{[\mathrm{Br_2}]} \\ &= \frac{4.20 \times 10^{-5}\ M/\mathrm{s}}{0.0120\ M} \\ &= 3.50 \times 10^{-3}\ \mathrm{s}^{-1} \end{aligned}$$

Similarly, we can use the data for $t = 50$ s to calculate k again, and so on. The slight variations in the values of k listed in the last column of Table 13.2 are due to experimental deviations in rate measurements.

Reaction Rates and Stoichiometry

In the example of the bromine–formic acid reaction, the rate is measured by monitoring the change in concentration of the reactant. Because bromine has a characteristic red color, it is easy to measure the concentration change by following the color change. For many other reactions, we may find it more convenient to determine the rate of a reaction by measuring the change in product concentration. For stoichiometrically simple reactions such as $A \rightarrow B$, we can express the rate in terms of the change of either the concentration of the reactant or that of the product:

$$\text{rate} = -\frac{\Delta[\text{A}]}{\Delta t} \qquad \text{or} \qquad \text{rate} = \frac{\Delta[\text{B}]}{\Delta t}$$

The rate of product formation does not require the minus sign because $\Delta[\text{B}]$ is a positive quantity. For more complex reactions, we must be careful in writing the rate expressions. Consider, for example, the reaction

$$2\text{A} \longrightarrow \text{B}$$

Two moles of A disappear for each mole of B that forms—that is, the rate of disappearance of A is twice as fast as the rate of appearance of B. We write the rate as either

$$\text{rate} = -\frac{1}{2}\frac{\Delta[\text{A}]}{\Delta t} \qquad \text{or} \qquad \text{rate} = \frac{\Delta[\text{B}]}{\Delta t}$$

In general, for the reaction

$$a\text{A} + b\text{B} \longrightarrow c\text{C} + d\text{D}$$

the rate is given by

$$\text{rate} = -\frac{1}{a}\frac{\Delta[\text{A}]}{\Delta t} = -\frac{1}{b}\frac{\Delta[\text{B}]}{\Delta t} = \frac{1}{c}\frac{\Delta[\text{C}]}{\Delta t} = \frac{1}{d}\frac{\Delta[\text{D}]}{\Delta t}$$

The following example requires writing the reaction rate in terms of both reactants and products.

EXAMPLE 13.1

Write the rate expressions for the following reactions in terms of the disappearance of the reactants and the appearance of the products:

(a) $I^-(aq) + OCl^-(aq) \longrightarrow Cl^-(aq) + OI^-(aq)$
(b) $3O_2(g) \longrightarrow 2O_3(g)$
(c) $4NH_3(g) + 5O_2(g) \longrightarrow 4NO(g) + 6H_2O(g)$

Answer

(a) Since each of the stoichiometric coefficients equals 1,

$$\text{rate} = -\frac{\Delta[\text{I}^-]}{\Delta t} = -\frac{\Delta[\text{OCl}^-]}{\Delta t} = \frac{\Delta[\text{Cl}^-]}{\Delta t} = \frac{\Delta[\text{OI}^-]}{\Delta t}$$

(b) Here the coefficients are 3 and 2, so

$$\text{rate} = -\frac{1}{3}\frac{\Delta[\text{O}_2]}{\Delta t} = \frac{1}{2}\frac{\Delta[\text{O}_3]}{\Delta t}$$

(c) In this reaction

$$\text{rate} = -\frac{1}{4}\frac{\Delta[\text{NH}_3]}{\Delta t} = -\frac{1}{5}\frac{\Delta[\text{O}_2]}{\Delta t} = \frac{1}{4}\frac{\Delta[\text{NO}]}{\Delta t} = \frac{1}{6}\frac{\Delta[\text{H}_2\text{O}]}{\Delta t}$$

Similar problem: 13.6.

13.2 The Rate Laws

The rate of a reaction is related to the concentrations of the reacting species. One way to study the effect of concentration on reaction rate is to determine how the rate at the beginning of a reaction, called the *initial rate,* depends on the starting concentrations. In general it is preferable to measure the initial rates because as the reaction proceeds, the concentrations of the reactants decrease and it may become difficult to measure the changes accurately. Also, there may be a reverse reaction of the type

$$\text{products} \longrightarrow \text{reactants}$$

which would introduce error in the rate measurement. Both of these complications are virtually absent during the early stages of the reaction.

Let us consider the following reaction:

$$F_2(g) + 2ClO_2(g) \longrightarrow 2FClO_2(g)$$

Table 13.3 shows three rate measurements of this reaction. Looking at entries 1 and 3, we see that as we double $[F_2]$ while holding $[ClO_2]$ constant, the rate doubles. Thus the rate is directly proportional to $[F_2]$. Similarly, the data in 1 and 2 show that as we quadruple $[ClO_2]$ at constant $[F_2]$, the rate increases by four times, so that the rate is also directly proportional to $[ClO_2]$. We can summarize the observations by writing

$$\begin{aligned}\text{rate} &\propto [F_2][ClO_2] \\ &= k[F_2][ClO_2]\end{aligned}$$

where k is the rate constant. The above equation is known as the ***rate law,*** *an expression relating the rate of a reaction to the rate constant and the concentrations of the reactants*. For a general reaction of the type

$$aA + bB \longrightarrow cC + dD$$

the rate law takes the form

$$\text{rate} = k[A]^x\,[B]^y \tag{13.1}$$

Rate laws are always determined experimentally.

The usefulness of the rate law is that if we know the values of k, x, and y, we can always calculate the rate of the reaction, given the concentrations of A and B. As we will see shortly, x and y, like k, must be determined experimentally. *The sum of the powers to which all reactant concentrations appearing in the rate law are raised* is called the overall ***reaction order.*** In the rate law expression shown above, the overall reaction order is given by $x + y$. For the reaction involving F_2 and ClO_2, the overall order is 1 + 1, or 2. We say that the reaction is first order in F_2 and first order in ClO_2, or second order overall.

Reaction order is always defined in terms of reactant (not product) concentrations.

Having defined reaction order, we can now better appreciate the dependence of rate on reactant concentrations as follows. Suppose, for example, that for a certain reaction $x = 1$ and $y = 2$. The rate law for the reaction is

Table 13.3 Rate data for the reaction between F_2 and ClO_2

$[F_2](M)$	$[ClO_2](M)$	Initial rate (M/s)
1. 0.10	0.010	1.2×10^{-3}
2. 0.10	0.040	4.8×10^{-3}
3. 0.20	0.010	2.4×10^{-3}

This reaction is first order in A, second order in B, and third order overall (1 + 2 = 3).

$$\text{rate} = k[\text{A}][\text{B}]^2$$

Let us assume that initially [A] = 1.0 *M* and [B] = 1.0 *M*. The equation tells us that if we double the concentration of A from 1.0 *M* to 2.0 *M* at constant [B], we also double the rate:

$$\text{for } [\text{A}] = 1.0\ M \qquad \text{rate}_1 = k(1.0\ M)(1.0\ M)^2 = k(1.0\ M^3)$$

$$\text{for } [\text{A}] = 2.0\ M \qquad \text{rate}_2 = k(2.0\ M)(1.0\ M)^2 = k(2.0\ M^3)$$

Hence $\text{rate}_2 = 2(\text{rate}_1)$

On the other hand, if we double the concentration of B from 1.0 *M* to 2.0 *M* at constant [A], the rate will increase by a factor of 4 because of the power 2 in the exponent:

$$\text{for } [\text{B}] = 1.0\ M \qquad \text{rate}_1 = k(1.0\ M)(1.0\ M)^2 = k(1.0\ M^3)$$

$$\text{for } [\text{B}] = 2.0\ M \qquad \text{rate}_2 = k(1.0\ M)(2.0\ M)^2 = k(4.0\ M^3)$$

Hence $\text{rate}_2 = 4(\text{rate}_1)$

If, for a certain reaction, $x = 0$ and $y = 1$, then the rate law is

$$\text{rate} = k[\text{A}]^0[\text{B}] = k[\text{B}]$$

This reaction is zero order in A, first order in B, and first order overall. Thus the rate of this reaction is *independent* of the concentration of A.

Experimental Determination of Rate Law

The first step in the determination of the rate law of a reaction involves the measurement of the rate. For reactions in solution, the concentration of a species can usually be followed by spectrophotometric means, as in the case of bromine reacting with formic acid discussed earlier. If ions are involved, the change in concentration may also be monitored by an electrical conductance measurement. Reactions involving gases are most conveniently followed by pressure measurements.

If a reaction involves only one reactant, the rate law can be readily determined by measuring the rate of the reaction as a function of the reactant's concentration. For example, if the rate doubles when the concentration of the reactant doubles, then the reaction is first order in the reactant. If the rate quadruples when the concentration doubles, the reaction is second order in the reactant.

For a reaction involving more than one reactant, we can find the rate law by measuring the dependence of the reaction rate on the concentration of each reactant, one at a time. We fix the concentrations of all but one reactant and record the rate of the reaction as a function of the concentration of that reactant. Any changes in the rate must be due only to changes in that substance. The dependence thus observed gives us the order in that particular reactant. The same procedure is then applied to the next reactant, and so on. This method is known as the *isolation method*.

The following example illustrates the procedure for determining the rate law of a reaction.

EXAMPLE 13.2

The rate of the reaction A + 2B ⟶ C has been observed at 25°C. From the following data, determine the rate law for the reaction and calculate the rate constant.

Experiment	Initial [A] (*M*)	Initial [B] (*M*)	Initial rate (*M*/s)
1	0.100	0.100	5.50×10^{-6}
2	0.200	0.100	2.20×10^{-5}
3	0.400	0.100	8.80×10^{-5}
4	0.100	0.300	1.65×10^{-5}
5	0.100	0.600	3.30×10^{-5}

Answer

We assume that the rate law takes the form

$$\text{rate} = k[\text{A}]^x[\text{B}]^y$$

Experiments 1 and 2 show that when we double the concentration of A at constant [B], the rate increases by fourfold. Thus the reaction is second order in A. Experiments 4 and 5 indicate that doubling [B] at constant [A] doubles the rate; the reaction is first order in B.

For more complex reactions involving fractional orders, we must determine the order in a formal way, that is, by taking ratios of the rates.

We can arrive at the same conclusion by taking the ratios of rates as follows. Using the data from experiments 1 and 2, we can write the ratio of the rates as

$$\frac{\text{rate}_1}{\text{rate}_2} = \frac{5.50 \times 10^{-6}\ M/\text{s}}{2.20 \times 10^{-5}\ M/\text{s}} = \frac{1}{4}$$

However, the ratio of the rates can also be expressed in terms of the rate law, as follows:

$$\frac{\text{rate}_1}{\text{rate}_2} = \frac{1}{4} = \frac{\cancel{k}(0.100\ M)^x\cancel{(0.100\ M)^y}}{\cancel{k}(0.200\ M)^x\cancel{(0.100\ M)^y}} = \left(\frac{0.100\ M}{0.200\ M}\right)^x = \left(\frac{1}{2}\right)^x$$

Thus $x = 2$, and the reaction is second order in A.

Similarly, using the data for experiments 4 and 5, we have

$$\frac{\text{rate}_4}{\text{rate}_5} = \frac{1.65 \times 10^{-5}\ M/\text{s}}{3.30 \times 10^{-5}\ M/\text{s}} = \frac{1}{2}$$

From the rate law we write

$$\frac{\text{rate}_4}{\text{rate}_5} = \frac{1}{2} = \frac{\cancel{k(0.100\ M)^x}(0.300\ M)^y}{\cancel{k(0.100\ M)^x}(0.600\ M)^y} = \left(\frac{0.300\ M}{0.600\ M}\right)^y = \left(\frac{1}{2}\right)^y$$

Therefore, $y = 1$, so the reaction is first order in B. The overall order is (2 + 1), or 3 (third order), and the rate law is

$$\text{rate} = k[\text{A}]^2[\text{B}]$$

The rate constant k can be calculated using the data from any one of the experiments.

Since

$$k = \frac{\text{rate}}{[A]^2[B]}$$

data from experiment 1 give us

$$k = \frac{5.50 \times 10^{-6}\ M/s}{(0.100\ M)^2(0.100\ M)}$$
$$= 5.50 \times 10^{-3}/M^2s$$

Similar problem: 13.18.

Example 13.2 brings out an important point about rate law expressions; that is, in general there is no relationship between the exponents x and y and the stoichiometric coefficients in the balanced equation. Consider the following examples.

- For the thermal decomposition of N_2O_5

  $$2N_2O_5(g) \longrightarrow 4NO_2(g) + O_2(g)$$

  the rate law is

  $$\text{rate} = k[N_2O_5]$$

  and *not* rate $= k[N_2O_5]^2$, as we might have guessed from the balanced equation.
- The rate law for the thermal decomposition of acetaldehyde (CH_3CHO)

  $$CH_3CHO(g) \longrightarrow CH_4(g) + CO(g)$$

  has been determined experimentally to be

  $$\text{rate} = k[CH_3CHO]^{3/2}$$

  and *not* rate $= k[CH_3CHO]$.
- The reaction of hydrogen peroxide with iodide ion in an acidic medium is

  $$H_2O_2(aq) + 3I^-(aq) + 2H^+(aq) \longrightarrow I_3^-(aq) + 2H_2O(l)$$

  and the rate law is

  $$\text{rate} = k[H_2O_2][I^-]$$

  Thus the reaction is first order in H_2O_2, first order in I^-, and zero order in H^+. The concentration of H^+ ions has *no* effect on the rate of the reaction. Therefore the overall reaction order is 2.

Sometimes we find that the reaction orders are the same as the stoichiometric coefficients. For example, the reaction

$$H_2(g) + I_2(g) \longrightarrow 2HI(g)$$

has the rate law

$$\text{rate} = k[H_2][I_2]$$

which is first order in H_2 and first order in I_2. This situation is really the exception rather than the rule, however. In general, *the order of a reaction must be determined by experiment; it cannot be deduced from the overall balanced equation.*

13.3 Relation between Reactant Concentrations and Time

Rate law expressions enable us to calculate the rate of a reaction from the rate constant and reactant concentrations. They can also be converted into equations that allow us to determine the concentrations of reactants at any time during the course of a reaction. We will illustrate this application by considering two of the simplest kinds of rate laws—those applying to reactions that are first order overall and those applying to reactions that are second order overall.

First-Order Reactions

A ***first-order reaction*** is a *reaction whose rate depends on the reactant concentration raised to the first power.* In a first-order reaction of the type

$$\text{A} \longrightarrow \text{product}$$

the rate is

$$\text{rate} = -\frac{\Delta[\text{A}]}{\Delta t}$$

Also, from the rate law we know that

For a first-order reaction, doubling the concentration of the reactant doubles the rate.

$$\text{rate} = k[\text{A}]$$

Thus

$$-\frac{\Delta[\text{A}]}{\Delta t} = k[\text{A}] \qquad (13.2)$$

We can determine the units of the first-order rate constant k by transposing:

$$k = -\frac{\Delta[\text{A}]}{[\text{A}]}\frac{1}{\Delta t} = \frac{M}{M\ \text{s}} = \frac{1}{\text{s}} = \text{s}^{-1}$$

(The minus sign does not enter into the evaluation of units.) Using calculus, we can show from Equation (13.2) that

$$\ln \frac{[\text{A}]_0}{[\text{A}]} = kt \qquad (13.3)$$

See Appendix 4 for a discussion of natural logarithms and of logarithms and significant figures.

If you have a scientific calculator, you will probably find it easier to work with natural logarithms.

where ln is the natural logarithm, and $[\text{A}]_0$ and $[\text{A}]$ are the concentrations of A at times $t = 0$ and $t = t$, respectively. In terms of common or base 10 logarithms, Equation (13.3) can be written as

$$\log \frac{[\text{A}]_0}{[\text{A}]} = \frac{kt}{2.303}$$

It should be understood that $t = 0$ need not correspond to the beginning of the experiment; it can be any arbitrarily chosen time when we start to monitor the change in the concentration of A.

Equation (13.3) can be rearranged as follows:

$$\ln\,[\text{A}]_0 - \ln\,[\text{A}] = kt$$

or

$$\ln\,[\text{A}] = -kt + \ln\,[\text{A}]_0 \qquad (13.4)$$

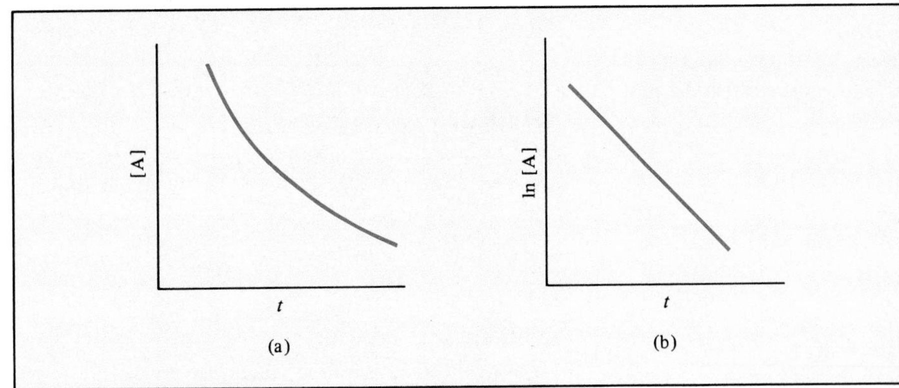

Figure 13.7 *First-order reaction characteristics: (a) decrease of reactant concentration with time; (b) plot of the straight-line relationship to obtain the rate constant. The slope of the line is equal to* $-k$.

The linear plot of ln [A] versus t is a characteristic of first-order reactions.

Equation (13.4) has the form of the linear equation $y = mx + b$, in which m is the slope of the line that is the graph of the equation:

$$\begin{array}{ccccc} \ln [A] & = & (-k)(t) & + & \ln [A]_0 \\ \updownarrow & & \updownarrow \ \updownarrow & & \updownarrow \\ y & = & m \ \ x & + & b \end{array}$$

Thus a plot of ln [A] versus t (or y versus x) gives a straight line with a slope of $-k$ (or m). This allows us to calculate the rate constant k. Figure 13.7 shows the characteristics of a first-order reaction.

There are many known first-order reactions. The decomposition of N_2O_5, which we discussed earlier, is first order in N_2O_5. Another example is the conversion of *cis*-2-butene to *trans*-2-butene:

$$\underset{cis\text{-2-butene}}{(H_3C)(H)C{=}C(CH_3)(H)} \longrightarrow \underset{trans\text{-2-butene}}{(H_3C)(H)C{=}C(H)(CH_3)}$$

In the following example we apply Equation (13.3) to a first-order reaction.

N_2O_5 decomposes to give NO_2.

EXAMPLE 13.3

The decomposition of dinitrogen pentoxide is a first-order reaction with a rate constant of $5.1 \times 10^{-4}\ s^{-1}$ at 45°C:

$$2N_2O_5(g) \longrightarrow 4NO_2(g) + O_2(g)$$

(a) If the initial concentration of N_2O_5 was 0.25 M, what is the concentration after 3.2 min? (b) How long will it take for the concentration of N_2O_5 to decrease from 0.25 M to 0.15 M? (c) How long will it take to convert 62 percent of the starting material?

Answer

(a) Applying Equation (13.3),

$$\ln \frac{[A]_0}{[A]} = kt$$

Because k is given in units of s^{-1}, we must convert 3.2 min to seconds.

$$\ln \frac{0.25\ M}{[\mathrm{A}]} = (5.1 \times 10^{-4}\ \mathrm{s}^{-1})\left(3.2\ \mathrm{min} \times \frac{60\ \mathrm{s}}{1\ \mathrm{min}}\right)$$

Solving the equation, we obtain

$$\ln \frac{0.25\ M}{[\mathrm{A}]} = 0.098$$

$$\frac{0.25\ M}{[\mathrm{A}]} = e^{0.098} = 1.1$$

$$[\mathrm{A}] = 0.23\ M$$

(b) Again using Equation (13.3), we have

$$\ln \frac{0.25\ M}{0.15\ M} = (5.1 \times 10^{-4}\ \mathrm{s}^{-1})t$$

$$t = 1.0 \times 10^{3}\ \mathrm{s} = 17\ \mathrm{min}$$

(c) In a calculation of this type, we do not need to know the actual concentration of the starting material. If 62 percent of the starting material has reacted, then the amount left after time t is (100% − 62%), or 38%. Thus $[\mathrm{A}]/[\mathrm{A}]_0$ = 38%/100%, or 0.38. From Equation (13.3), we write

$$t = \frac{1}{k} \ln \frac{[\mathrm{A}]_0}{[\mathrm{A}]} = \frac{1}{5.1 \times 10^{-4}\ \mathrm{s}^{-1}} \ln \frac{1.0}{0.38} = 1.9 \times 10^{3}\ \mathrm{s} = 32\ \mathrm{min}$$

Similar problems: 13.20, 13.21.

The ***half-life*** of a reaction, $t_{\frac{1}{2}}$, is the *time required for the concentration of a reactant to decrease to half of its initial concentration*. We can obtain an expression for $t_{\frac{1}{2}}$ for a first-order reaction as follows. From Equation (13.3) we write

$$t = \frac{1}{k} \ln \frac{[\mathrm{A}]_0}{[\mathrm{A}]}$$

By the definition of half-life, when $t = t_{\frac{1}{2}}$, $[\mathrm{A}] = [\mathrm{A}]_0/2$, so

$$t_{\frac{1}{2}} = \frac{1}{k} \ln \frac{[\mathrm{A}]_0}{[\mathrm{A}]_0/2}$$

or

$$t_{\frac{1}{2}} = \frac{1}{k} \ln 2 = \frac{0.693}{k} \tag{13.5}$$

Equation (13.5) tells us that the half-life of a first-order reaction is independent of the initial concentration of the reactant. Thus, it takes the same time for the concentration of the reactant to decrease from 1.0 M to 0.50 M, say, as it does for a decrease in concentration from 0.10 M to 0.050 M (Figure 13.8). Measuring the half-life of a

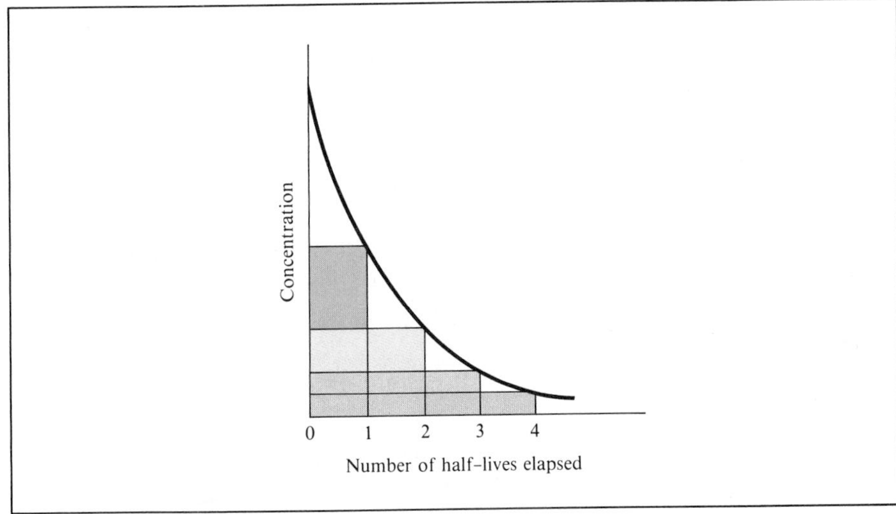

Figure 13.8 *Change in concentration of a reactant with number of half-lives for a first-order reaction.*

reaction is one way to determine the rate constant of a first-order reaction.

Example 13.4 calculates the half-life of a first-order reaction.

EXAMPLE 13.4

The conversion of cyclopropane to propene in the gas phase

$$\underset{\text{cyclopropane}}{\overset{CH_2}{CH_2-CH_2}} \longrightarrow \underset{\text{propene}}{CH_3-CH=CH_2}$$

is a first-order reaction with a rate constant of $6.7 \times 10^{-4}\ s^{-1}$ at 500°C. Calculate the half-life of the reaction.

Answer

(a) Applying Equation (13.5),

$$\begin{aligned} t_{\frac{1}{2}} &= \frac{0.693}{k} \\ &= \frac{0.693}{6.7 \times 10^{-4}\ s^{-1}} \\ &= 1.0 \times 10^3\ s \end{aligned}$$

Thus it will take 1.0×10^3 s, or about 17 min, for *any* initial concentration of cyclopropane to decrease to half its value.

Similar problem: 13.31.

Equations (13.3) and (13.4) can help us evaluate a first-order rate constant. From experimental data for concentration versus time, we can determine k graphically (as shown in Figure 13.7), or we can use the half-life method, as Example 13.5 will show.

For gas-phase reactions we can replace the concentration terms in Equation (13.3) with the pressures of the gaseous reactant. Consider the first-order reaction

$$A(g) \longrightarrow \text{product}$$

Using the ideal gas equation [Equation (5.7)] we write

$$PV = n_A RT$$

or

$$\frac{n_A}{V} = [A] = \frac{P}{RT}$$

Substituting $[A] = P/RT$ in Equation (13.3) we get

$$\ln\frac{[A]_0}{[A]} = \ln\frac{P_0/RT}{P/RT} = \ln\frac{P_0}{P} = kt$$

The following example shows the use of pressure measurements to study the kinetics of a reaction.

EXAMPLE 13.5

The rate of decomposition of azomethane is studied by monitoring the partial pressure of the reactant as a function of time:

$$CH_3{-}N{=}N{-}CH_3(g) \longrightarrow N_2(g) + C_2H_6(g)$$

The data obtained at 300°C are shown in the following table:

Time (s)	Partial pressure of azomethane (mmHg)
0	284
100	220
150	193
200	170
250	150
300	132

(a) Are these values consistent with first-order kinetics? If so, determine the rate constant (b) by plotting the data as shown in Figure 13.7 and (c) by using the half-life method.

Answer

(a) The partial pressure of azomethane at any time is directly proportional to its concentration (in mol/L). Therefore Equation (13.4) can be written in terms of partial pressures as

$$\ln P = -kt + \ln P_0$$

where P_0 and P are the partial pressures of azomethane at time $t = 0$ and $t = t$. This equation has the form of the linear equation $y = mx + b$. Figure 13.9(a), which is based on the data given in the table below, shows that a plot of $\ln P$ versus t yields a straight line, so the reaction is indeed first order.

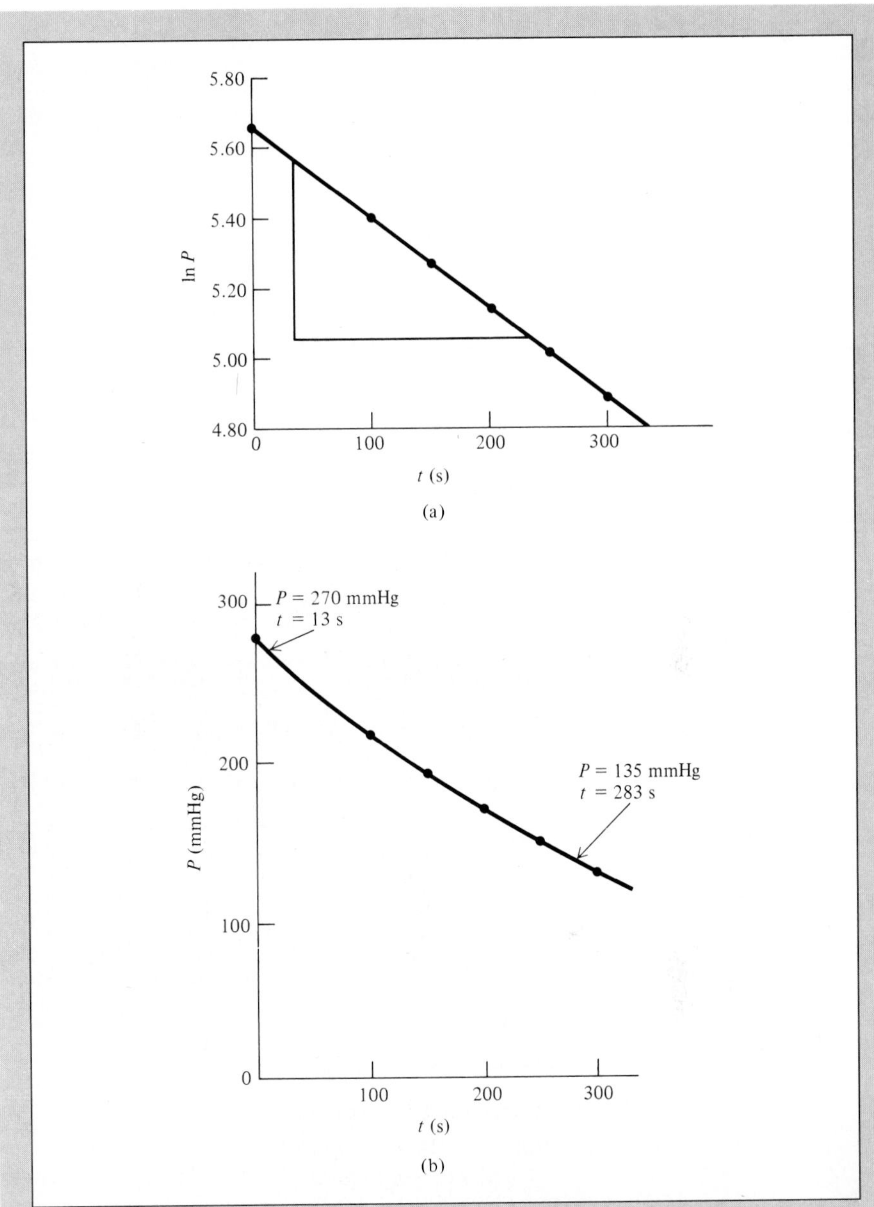

Figure 13.9 *(a) Plot of ln P versus time. The slope of the line is calculated from two pairs of coordinates:*

$$slope = \frac{5.05 - 5.56}{(233 - 33)\ s} = -2.55 \times 10^{-3}\ s^{-1}$$

According to Equation (13.4), the slope is equal to $-k$. (b) The half-life method for determining the rate constant of the first-order reaction. The time it takes for the pressure to drop from 270 mmHg (an arbitrary choice) to 135 mmHg is the half-life of the reaction. Here $t_{\frac{1}{2}}$ = 283 s − 13 s, or 270 s.

t (s)	ln P
0	5.649
100	5.394
150	5.263
200	5.136
250	5.011
300	4.883

(b) From Equation (13.4) we know that the slope of the line for a first-order reaction is equal to $-k$. In Figure 13.9(a) the slope is -2.55×10^{-3} s^{-1}. Therefore

$$-k = \text{slope}$$
$$= -2.55 \times 10^{-3}\ \text{s}^{-1}$$
$$k = 2.55 \times 10^{-3}\ \text{s}^{-1}$$

(c) To calculate the rate constant by the half-life method we need to plot the partial pressure versus time, as shown in Figure 13.9(b), because the given data do not show the time it takes for the partial pressure to decrease to half of its value. From the curve we find that $t_{\frac{1}{2}} = 270$ s, so the rate constant can be calculated by using Equation (13.5):

$$t_{\frac{1}{2}} = \frac{0.693}{k}$$

$$k = \frac{0.693}{270\ \text{s}}$$
$$= 2.57 \times 10^{-3}\ \text{s}^{-1}$$

The small difference between the two rate constants calculated in (b) and (c) for the same reaction should not be surprising because they are arrived at by different ways of analyzing the data.

Similar problems: 13.23, 13.24, 13.25.

Second-Order Reactions

A ***second-order reaction*** is a *reaction whose rate depends on reactant concentration raised to the second power or on the concentrations of two different reactants, each raised to the first power*. The simpler type involves only one kind of reactant molecule:

$$\text{A} \longrightarrow \text{product}$$

where

$$\text{rate} = -\frac{\Delta[\text{A}]}{\Delta t}$$

From the rate law

$$\text{rate} = k[\text{A}]^2$$

Thus

$$-\frac{\Delta[\text{A}]}{\Delta t} = k[\text{A}]^2$$

We can determine the units of the second-order rate constant k by solving for k:

$$k = -\frac{\Delta[\text{A}]}{[\text{A}]^2}\frac{1}{\Delta t} = \frac{M}{M^2\ \text{s}} = \frac{1}{M\ \text{s}}$$

(Remember, we do not use the minus sign for units.)

Another type of second-order reaction is represented as

$$A + B \longrightarrow \text{product}$$

From the stoichiometry of the reaction we see that the rate of decrease in A is the same as that in B; that is,

$$\text{rate} = -\frac{\Delta[A]}{\Delta t} = -\frac{\Delta[B]}{\Delta t}$$

According to the rate law

$$\text{rate} = k[A][B]$$

Thus

$$-\frac{\Delta[A]}{\Delta t} = -\frac{\Delta[B]}{\Delta t} = k[A][B]$$

The reaction is first order in A and first order in B, so it has an overall order of 2. Again, the units of k are $1/M \cdot s$.

Using calculus we can obtain the following expressions for "A → product" reactions:

$$\frac{1}{[A]} = \frac{1}{[A]_0} + kt \qquad (13.6)$$

(The corresponding equation for "A + B → product" reactions is too complex to discuss here.) We can obtain an equation for the half-life of a second-order reaction by setting $[A] = [A]_0/2$ in Equation (13.6).

$$\frac{1}{[A]_0/2} = \frac{1}{[A]_0} + kt_{\frac{1}{2}}$$

Solving for $t_{\frac{1}{2}}$ we obtain

$$t_{\frac{1}{2}} = \frac{1}{k[A]_0} \qquad (13.7)$$

Note that the half-life of a second-order reaction depends on the initial concentration, unlike the half-life of a first-order reaction [see Equation (13.5)]. This is one way to distinguish between first-order and second-order reactions.

The kinetic analysis of a second-order reaction is shown in the following example.

EXAMPLE 13.6

The recombination of iodine atoms to form molecular iodine in the gas phase

$$I(g) + I(g) \longrightarrow I_2(g)$$

follows second-order kinetics and has the high rate constant $7.0 \times 10^9/M \cdot s$ at 23°C. (a) If the initial concentration of I was 0.086 M, calculate the concentration after 2.0 min. (b) Calculate the half-life of the reaction if the initial concentration of I is 0.60 M and if it is 0.42 M.

Answer

(a) We begin with Equation (13.6):

$$\frac{1}{[A]} = \frac{1}{[A]_0} + kt$$

$$\frac{1}{[A]} = \frac{1}{0.086\ M} + (7.0 \times 10^9/M \cdot s)\left(2.0\ \text{min} \times \frac{60\ s}{1\ \text{min}}\right)$$

where [A] is the desired concentration at $t = 2.0$ min. Solving the equation, we get

$$[A] = 1.2 \times 10^{-12}\ M$$

This is such a low concentration that it is virtually undetectable. The very large rate constant for the reaction means that practically all the I atoms combine after only 2.0 min of reaction time.

(b) We need Equation (13.7) for this part.

For $[I]_0 = 0.60\ M$

$$\begin{aligned} t_{\frac{1}{2}} &= \frac{1}{k[A]_0} \\ &= \frac{1}{(7.0 \times 10^9/M \cdot s)(0.60\ M)} \\ &= 2.4 \times 10^{-10}\ s \end{aligned}$$

For $[I]_0 = 0.42\ M$

$$\begin{aligned} t_{\frac{1}{2}} &= \frac{1}{(7.0 \times 10^9/M \cdot s)(0.42\ M)} \\ &= 3.4 \times 10^{-10}\ s \end{aligned}$$

These results confirm that the half-life of a second-order reaction is not a constant but depends on the initial concentration of the reactant(s).

Similar problem: 13.33.

First- and second-order reactions are the most common reaction types. Reactions whose order is zero are rare. Mathematically speaking, zero-order reactions are easy to deal with. The rate law is

$$\begin{aligned} \text{rate} &= k[A]^0 \\ &= k \end{aligned}$$

Thus the rate of a zero-order reaction is a *constant,* independent of its reactant concentrations. Third-order and higher-order reactions are quite complex; they will not be studied in this book. Table 13.4 summarizes the kinetics for first-order and second-order reactions of the type A → product.

Table 13.4 Summary of the kinetics of first-order and second-order reactions

Order	Rate law	Concentration-time equation	Half-life
1	Rate $= k[A]$	$\ln \frac{[A]_0}{[A]} = kt$	$\frac{0.693}{k}$
2	Rate $= k[A]^2$	$\frac{1}{[A]} = \frac{1}{[A]_0} + kt$	$\frac{1}{k[A]_0}$

The following Chemistry in Action describes an interesting application of chemical kinetics: estimating the age of objects.

CHEMISTRY IN ACTION

Determining the Age of the Shroud of Turin

How do scientists determine the ages of artifacts from archaeological excavations? If someone tried to sell you a manuscript supposedly dating from 1000 B.C., how could you be certain of its authenticity? Is a mummy found in an Egyptian pyramid *really* three thousand years old? The answers to these and other similar questions can usually be obtained by applying chemical kinetics and the *radiocarbon dating technique*.

Earth's atmosphere is constantly being bombarded by cosmic rays of extremely high penetrating power. These rays, which originate in outer space, consist of electrons, neutrons, and atomic nuclei. One of the important reactions between the atmosphere and cosmic rays is the capture of neutrons by atmospheric nitrogen (nitrogen-14 isotope) to produce the radioactive carbon-14 isotope and hydrogen. The unstable carbon atoms eventually form $^{14}CO_2$, which mixes with the ordinary carbon dioxide ($^{12}CO_2$) in the air. The carbon-14 isotope decays with the emission of β particles (electrons). The rate of decay (as measured by the number of electrons emitted per second) obeys first-order kinetics. It is customary in the study of radioactive decays to write the rate law as

$$\text{rate} = kN$$

where k is the first-order rate constant and N the number of ^{14}C nuclei present. The half-life of the decay, $t_{\frac{1}{2}}$, is 5.73×10^3 yr, so that from Equation (13.5) we write

$$k = \frac{0.693}{5.73 \times 10^3 \text{ yr}} = 1.21 \times 10^{-4} \text{ yr}^{-1}$$

The carbon-14 isotopes enter the biosphere when carbon dioxide is taken up in plant photosynthesis. Plants are eaten by animals, which exhale carbon-14 in CO_2. Eventually, carbon-14 participates in many aspects of the carbon cycle. The ^{14}C lost by radioactive decay is constantly replenished by the production of new isotopes in the atmosphere. In this decay-replenishment process, a dynamic equilibrium is established whereby the ratio of ^{14}C to ^{12}C remains constant in living matter. But when an individual plant or an animal dies, the carbon-14 isotope in it is no longer replenished, so the ratio decreases as ^{14}C decays. This same change occurs when carbon atoms are trapped in coal, petroleum, or wood preserved underground, and, of course, in Egyptian mummies. After a number of years, there are proportionately fewer ^{14}C nuclei in, say, a mummy than in a living person.

In 1955, Willard F. Libby† suggested that this fact could be used to estimate the length of time the carbon-14 isotope of a particular specimen has been decaying without replenishment. Using Equation (13.3), we can write

$$\ln \frac{N_0}{N} = kt$$

where N_0 and N are the number of ^{14}C nuclei present at $t = 0$ and $t = t$, respectively. Since the rate of decay is directly proportional to the number of ^{14}C nuclei present, the above equation can be rewritten as

$$\begin{aligned} t &= \frac{1}{k} \ln \frac{N_0}{N} \\ &= \frac{1}{1.21 \times 10^{-4} \text{ yr}^{-1}} \ln \frac{\text{decay rate at } t = 0}{\text{decay rate at } t = t} \\ &= \frac{1}{1.21 \times 10^{-4} \text{ yr}^{-1}} \ln \frac{\text{decay rate of fresh sample}}{\text{decay rate of old sample}} \end{aligned}$$

Knowing k and the decay rates of the fresh sample and the old sample, we can calculate t, which is the age of the old sample. This ingenious technique is based on a

†Willard Frank Libby (1908–1980). American chemist. Libby received the Nobel Prize in chemistry in 1960 for his work on radiocarbon dating.

remarkably simple idea; its success depends on how accurately we can measure the rate of decay. In fresh samples, the ratio $^{14}C/^{12}C$ is about $1/10^{12}$, so the equipment used to monitor the radioactive decay must be very sensitive. Precision is more difficult with older samples because they contain even fewer ^{14}C nuclei. Nevertheless, radiocarbon dating has become an extremely valuable tool for estimating the age of archaeological artifacts, paintings, and other objects dating back 1000 to 50,000 years.

The latest major application of radiocarbon dating is the determination of the age of the Shroud of Turin (Figure 13.10). For generations there was controversy about whether the shroud, a piece of linen bearing the yellowish image of a man, was the burial cloth of Jesus Christ. Then in 1988 three laboratories in Europe and the United States, working on less than 50-mg samples of the shroud, independently showed by carbon-14 dating that the shroud dates from between A.D. 1260 and A.D. 1390. Thus the shroud could not have been the burial cloth of Christ.

Figure 13.10 *The Shroud of Turin. The age of the shroud has been determined by radiocarbon dating.*

13.4 Activation Energy and Temperature Dependence of Rate Constants

The Collision Theory of Chemical Kinetics

With very few exceptions, reaction rates increase with increasing temperature. For example, the time required to hard-boil an egg is much shorter if the "reaction" is carried out at 100°C (about 10 min) than at 80°C (about 30 min). Conversely, an effective way to preserve foods is to store them at subzero temperatures, thereby slowing the rate of bacterial decay. Figure 13.11 shows a typical dependence of the rate constant of a reaction on temperature. In order to explain this behavior, we must ask how reactions get started in the first place.

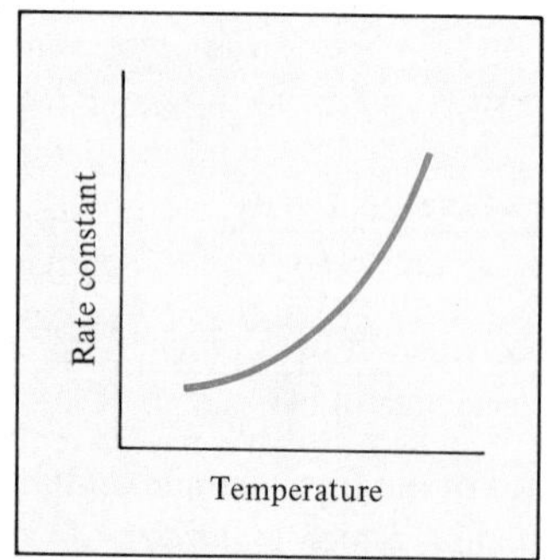

Figure 13.11 *Dependence of rate constant on temperature. The rate constants of most reactions increase with increasing temperature.*

It seems logical to assume—and it is generally true—that chemical reactions occur as a result of collisions between reacting molecules. In terms of the *collision theory* of chemical kinetics, then, we expect the rate of a reaction to be directly proportional to the number of molecular collisions per second, or to the frequency of molecular collisions:

$$\text{rate} \propto \frac{\text{number of collisions}}{\text{s}}$$

This simple relationship explains the dependence of reaction rate on concentration.

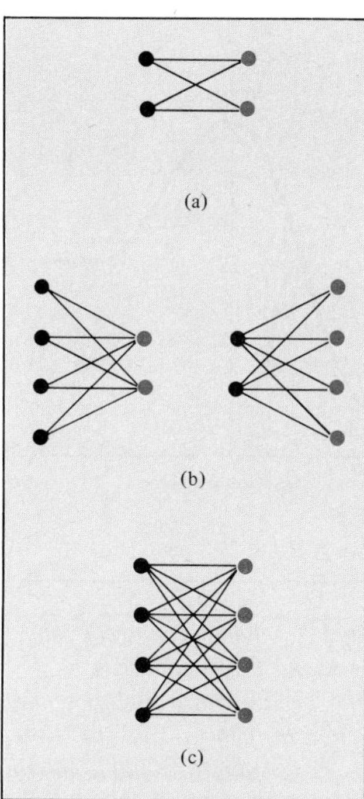

Figure 13.12 *Dependence of number of collisions on concentration. We consider here only A-B collisions, which can lead to formation of products. (a) There are 4 possible collisions among two A and two B molecules. (b) Doubling the number of either type of molecule (but not both) increases the number of collisions to 8. (c) Doubling both the A and B molecules increases the number of collisions to 16.*

Consider the reaction of A molecules with B molecules to form some product. Suppose that each product molecule is formed by the direct combination of an A molecule and a B molecule. If we doubled the concentration of A, say, then the number of A-B collisions would also double, because, in any given volume, there would be twice as many A molecules that could collide with B molecules (Figure 13.12). Consequently, the rate would increase by a factor of 2. Similarly, doubling the concentration of B molecules would increase the rate twofold. Thus, we can express the rate law as

$$\text{rate} = k[\text{A}][\text{B}]$$

The reaction is first order in both A and B and obeys second-order kinetics.

The collision theory is intuitively appealing, but the relationship between rate and molecular collision is more complicated than you might expect. The implication of the collision theory is that a reaction always occurs when an A and a B molecule collide. However, not all collisions lead to reactions. Calculations based on the kinetic molecular theory show that, at ordinary pressures (say, 1 atm) and temperatures (say, 298 K), there are about 1×10^{27} binary collisions (collisions between two molecules) in 1 mL of volume every second, in the gas phase. Even more collisions per second occur in liquids. If every binary collision led to a product, then most reactions would be complete almost instantaneously. In practice, we find that the rates of reactions differ greatly. This means that, in many cases, collisions alone do not guarantee that a reaction will take place.

Any molecule in motion possesses kinetic energy; the faster it moves, the greater the kinetic energy. But a fast-moving molecule will not break up into fragments on its own. To react, it must collide with another molecule. To use a simple analogy, a car traveling at 80 km/h (50 mph) will not begin to disintegrate on its own (assuming that it is in good running condition), but if it meets another car in a head-on collision, then quite a few pieces of both cars will fly apart. When molecules collide, part of their kinetic energy is converted to vibrational energy. If the initial kinetic energies are large, then the colliding molecules will vibrate so strongly as to break some of the chemical bonds. This bond fracture is the first step toward product formation. If the initial kinetic energies are small, the molecules will merely bounce off each other intact. (Note that this is an important difference between car collisions and molecular collisions.) Energetically speaking, there is some minimum collision energy below which no reaction occurs.

We postulate that, in order to react, the colliding molecules must have a total kinetic energy equal to or greater than the ***activation energy*** (E_a), which is the *minimum amount of energy required to initiate a chemical reaction*. Lacking this energy, the molecules remain intact, and no change results from the collision. *The species temporarily formed by the reactant molecules as a result of the collision before they form the product* is called the ***activated complex.***

Figure 13.13 shows two different potential energy profiles for the reaction

$$\text{A} + \text{B} \longrightarrow \text{C} + \text{D}$$

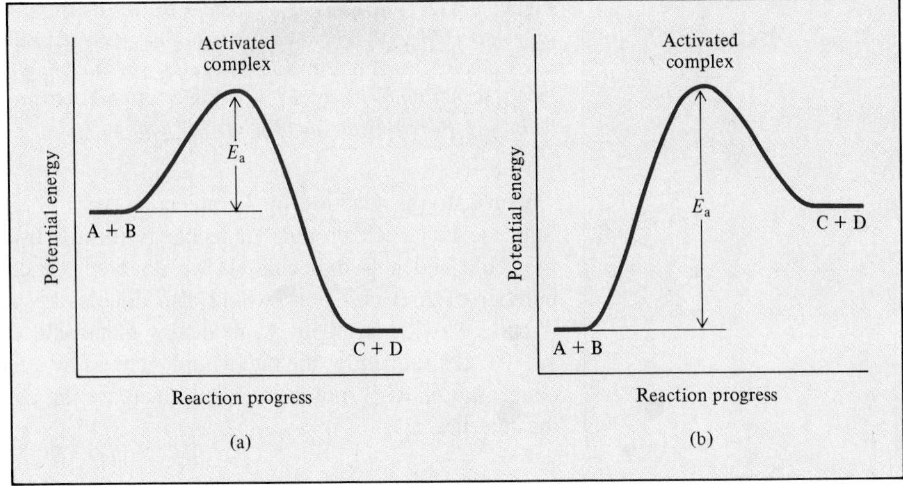

Figure 13.13 *Potential energy profiles for (a) exothermic and (b) endothermic reactions. These plots show the change in potential energy as reactants A and B are converted to products C and D. The activated complex is a highly unstable species with a high potential energy. The activation energy is defined for the forward reaction in both (a) and (b). Note that the products C and D are more stable than the reactants in (a), and less stable in (b).*

If the products are more stable than the reactants, then the reaction will be accompanied by a release of heat; that is, the reaction is exothermic [Figure 13.13(a)]. On the other hand, if the products are less stable than the reactants, then heat will be absorbed by the reacting mixture from the surroundings and we have an endothermic reaction [Figure 13.13(b)]. In both cases we plot the potential energy of the reacting system versus the progress of the reaction. Qualitatively, these plots show the potential energy changes as reactants are converted to products.

We can think of activation energy as a barrier that prevents less energetic molecules from reacting. Because the number of reactant molecules in an ordinary reaction is very large, the speeds, and hence also the kinetic energies of the molecules, vary greatly. Normally, only a small fraction of the colliding molecules—the fastest-moving ones—have enough kinetic energy to exceed the activation energy. These molecules can therefore take part in the reaction. The increase in the rate (or the rate constant) with temperature can now be explained: The speeds of the molecules obey the Maxwell distributions shown in Figure 5.17. Compare the speed distributions at two different temperatures. Since more high-energy molecules are present at the higher temperature, the rate of product formation is also greater at the higher temperature.

The Arrhenius Equation

In 1889 Svante Arrhenius showed that the dependence of the rate constant of a reaction on temperature can be expressed by the following equation, now known as the *Arrhenius equation:*

$$k = Ae^{-E_a/RT} \tag{13.8}$$

where E_a is the activation energy of the reaction (in kJ/mol), R the gas constant (8.314 J/K · mol), T the absolute temperature, and e the base of the natural logarithm scale (see Appendix 4). The quantity A represents the collision frequency and is called the *frequency factor*. It can be treated as a constant for a given reacting system over a fairly wide temperature range. Equation (13.8) shows that the rate constant is directly proportional to A and, therefore, to the collision frequency. Further, because of the minus sign associated with the exponent E_a/RT, the rate constant decreases with increasing activation energy and increases with increasing temperature. This equation can be expressed in a more useful form by taking the natural logarithm of both sides:

$$\ln k = \ln Ae^{-E_a/RT}$$

$$= \ln A - \frac{E_a}{RT} \tag{13.9}$$

Equation (13.9) can take the form of a linear equation:

$$\begin{array}{ccccccc} \ln k & = & \left(-\frac{E_a}{R}\right) & \left(\frac{1}{T}\right) & + & \ln A \\ \updownarrow & & \updownarrow & \updownarrow & & \updownarrow \\ y & = & m & x & + & b \end{array}$$

Thus, a plot of $\ln k$ versus $1/T$ gives a straight line whose slope m is equal to $-E_a/R$ and whose intercept b with the ordinate (the y-axis) is $\ln A$.

The following example demonstrates a graphical method for determining the activation energy of a reaction.

EXAMPLE 13.7

The rate constants for the decomposition of acetaldehyde

$$CH_3CHO(g) \longrightarrow CH_4(g) + CO(g)$$

were measured at five different temperatures. The data are shown below. Plot $\ln k$ versus $1/T$, and determine the activation energy (in kJ/mol) for the reaction.

k ($1/M^{\frac{1}{2}}s$)	T (K)
0.011	700
0.035	730
0.105	760
0.343	790
0.789	810

Answer

We need to plot $\ln k$ on the y-axis versus $1/T$ on the x-axis. From the given data we obtain

$\ln k$	$1/T$ (K^{-1})
−4.51	1.43×10^{-3}
−3.35	1.37×10^{-3}
−2.254	1.32×10^{-3}
−1.070	1.27×10^{-3}
−0.237	1.23×10^{-3}

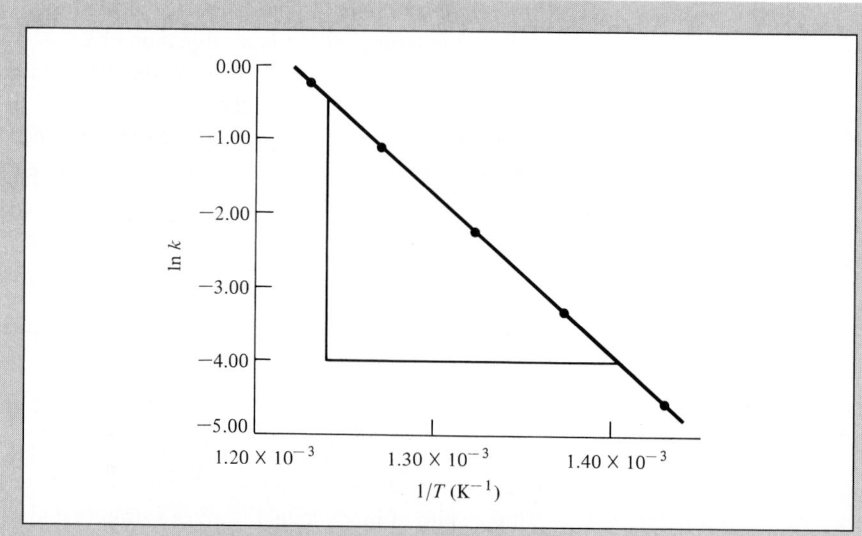

Figure 13.14 *Plot of ln k versus 1/T. The slope of the line is calculated from two pair of coordinates:*

$$slope = \frac{-4.00 - (-0.45)}{(1.41 - 1.24) \times 10^{-3}\ K^{-1}} = -2.09 \times 10^4\ K$$

The slope is equal to $-E_a/RT$. (The intercept of the straight line on the ln k axis is not shown here because it would require that the horizontal axis be too long.)

These data, when plotted, yield the graph shown in Figure 13.14. The slope of the straight line is -2.09×10^4 K. From the linear form of Equation (13.9)

$$\text{slope} = -\frac{E_a}{R} = -2.09 \times 10^4\ \text{K}$$

$$\begin{aligned} E_a &= (8.314\ \text{J/K} \cdot \text{mol})(2.09 \times 10^4\ \text{K}) \\ &= 1.74 \times 10^5\ \text{J/mol} \\ &= 1.74 \times 10^2\ \text{kJ/mol} \end{aligned}$$

It is important to note that although the rate constant itself has the units $1/M^{\frac{1}{2}}$s, the quantity ln k has no units (we cannot take the logarithm of any unit).

Similar problems: 13.42, 13.43.

An equation relating the rate constants k_1 and k_2 at temperatures T_1 and T_2 can be used to calculate the activation energy or to find the rate constant at another temperature if the activation energy is known. To derive such an equation we start with Equation (13.9):

$$\ln k_1 = \ln A - \frac{E_a}{RT_1}$$

$$\ln k_2 = \ln A - \frac{E_a}{RT_2}$$

Subtracting $\ln k_2$ from $\ln k_1$ gives

$$\ln k_1 - \ln k_2 = \frac{E_a}{R}\left(\frac{1}{T_2} - \frac{1}{T_1}\right)$$

$$\ln \frac{k_1}{k_2} = \frac{E_a}{R}\left(\frac{1}{T_2} - \frac{1}{T_1}\right)$$

$$= \frac{E_a}{R}\left(\frac{T_1 - T_2}{T_1 T_2}\right) \tag{13.10}$$

The following example illustrates the use of the equation we have just derived.

EXAMPLE 13.8

The rate constant of a first-order reaction is $3.46 \times 10^{-2}\ s^{-1}$ at 298 K. What is the rate constant at 350 K if the activation energy for the reaction is 50.2 kJ/mol?

Answer

The data are

$$k_1 = 3.46 \times 10^{-2}\ s^{-1} \qquad k_2 = ?$$

$$T_1 = 298\ K \qquad T_2 = 350\ K$$

Substituting in Equation (13.10),

$$\ln \frac{3.46 \times 10^{-2}}{k_2} = \frac{50.2 \times 10^3\ J/mol}{8.314\ J/K \cdot mol}\left[\frac{298\ K - 350\ K}{(298\ K)(350\ K)}\right]$$

Solving the equation gives

$$\ln \frac{3.46 \times 10^{-2}}{k_2} = -3.01$$

$$\frac{3.46 \times 10^{-2}}{k_2} = 0.0493$$

$$k_2 = 0.702\ s^{-1}$$

Similar problem: 13.47.

The Arrhenius equation is useful in studying reactions involving simple species (atoms or diatomic molecules). But for more complex reactions the frequency factor A in Equation (13.8) does not depend solely on the collision frequency. To understand why, let us consider the reaction between carbon monoxide and nitrogen dioxide that produces carbon dioxide and nitric oxide.

$$CO(g) + NO_2(g) \longrightarrow CO_2(g) + NO(g)$$

The reaction will take place if the reactants collide as shown in Figure 13.15(a). On the other hand, the collision shown in Figure 13.15(b) will not yield products, even though the reacting species may possess enough kinetic energy to exceed the activation en-

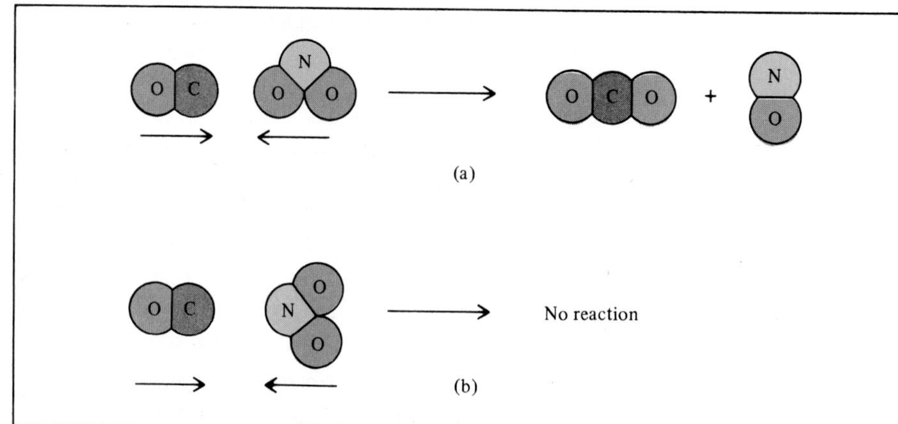

Figure 13.15 *Relative orientation of reacting molecules. (a) An effective collision and (b) an ineffective collision.*

ergy. Thus the "orientation" of reactant molecules when they collide is an important factor in reactions involving complex molecules. This orientation factor can be accounted for in a more detailed treatment of the Arrhenius equation.

13.5 Reaction Mechanisms

To use a travel metaphor, an overall chemical equation specifies the origin and destination, but not the actual route followed during a trip. The reaction mechanism is comparable to the route.

As we mentioned earlier, an overall balanced chemical equation does not tell us much about how a reaction actually takes place. In many cases, it merely represents the sum of *a series of simple reactions* that are often called the ***elementary steps*** (or ***elementary reactions***) because they *represent the progress of the overall reaction at the molecular level. The sequence of elementary steps that leads to product formation* is called the ***reaction mechanism.*** To illustrate what we mean by the study of reaction mechanism, let us consider the reaction between nitric oxide and oxygen:

$$2NO(g) + O_2(g) \longrightarrow 2NO_2(g)$$

We know that the products are not formed directly from the collision of a NO molecule with an O_2 molecule because a species, N_2O_2, is detected during the course of the reaction. A plausible mechanism is to assume that the reaction actually takes place via two elementary steps as follows:

$$2NO(g) \longrightarrow N_2O_2(g)$$
$$N_2O_2(g) + O_2(g) \longrightarrow 2NO_2(g)$$

In the first elementary step, two NO molecules collide to form a N_2O_2 molecule. This is followed by the reaction between N_2O_2 and O_2 to give two molecules of NO_2. The net chemical equation, which represents the overall change, is given by the sum of the elementary steps:

The sum of the elementary steps must give the overall balanced equation.

$$\begin{array}{rcll} NO + NO & \longrightarrow & N_2O_2 & \text{elementary step} \\ N_2O_2 + O_2 & \longrightarrow & 2NO_2 & \text{elementary step} \\ \hline 2NO + \cancel{N_2O_2} + O_2 & \longrightarrow & \cancel{N_2O_2} + 2NO_2 & \text{net reaction} \end{array}$$

Species such as N_2O_2 are called ***intermediates*** because they *appear in the mechanism of the reaction (that is, the elementary steps) but not in the overall balanced equation.*

Keep in mind that an intermediate is always formed in an earlier elementary step and is consumed in a later elementary step.

The term that describes *the number of molecules reacting in an elementary step* is called the ***molecularity of a reaction.*** Each of the elementary steps discussed above is called a ***bimolecular reaction,*** *an elementary step that involves two molecules.* An example of a ***unimolecular reaction,*** *an elementary step in which only one reacting molecule participates,* is the conversion of cyclopropane to propene shown in Example 13.4. Very few ***termolecular reactions,*** *reactions that involve the participation of three molecules in one elementary step,* are known. The reason is that in a termolecular reaction the product forms as a result of the simultaneous encounter of three molecules, which is a far less likely event than a bimolecular collision.

Rate Laws and Elementary Steps

Knowing the elementary steps of a reaction enables us to deduce the rate law. Suppose we have the following unimolecular elementary reaction:

$$\text{A} \longrightarrow \text{products}$$

Because this *is* the process that occurs at the molecular level, the larger the number of A molecules present, the faster the rate of formation of the products. Thus the rate of a unimolecular reaction will be first order in A:

$$\text{rate} = k[\text{A}]$$

Recalling the collision theory of chemical kinetics, we see that for a bimolecular elementary reaction involving A and B molecules

$$\text{A} + \text{B} \longrightarrow \text{product}$$

The rate of product formation depends on how frequently A and B collide, which in turn depends on the concentrations of A and B. Thus we can express the rate as

$$\text{rate} = k[\text{A}][\text{B}]$$

Similarly, for a bimolecular elementary reaction of the type

$$\text{A} + \text{A} \longrightarrow \text{products}$$

or

$$2\text{A} \longrightarrow \text{products}$$

the rate becomes

$$\text{rate} = k[\text{A}]^2$$

The above examples show that the reaction order for each reactant in an elementary reaction is equal to its stoichiometric coefficient in the chemical equation for that step. In general we cannot tell by merely looking at a balanced equation whether the reaction occurs as shown or whether the reaction occurs in a series of steps. The following two examples further illustrate the elucidation of reaction mechanism by experimental studies.

Hydrogen Peroxide Decomposition. The decomposition of hydrogen peroxide is facilitated by iodide ions. The overall reaction is

$$2H_2O_2(aq) \longrightarrow 2H_2O(l) + O_2(g)$$

In studying reaction mechanisms, the first step invariably is to determine the rate law of the reaction.

By experiment, the rate law is found to be

$$\text{rate} = k[H_2O_2][I^-]$$

Thus the reaction is first order with respect to both H_2O_2 and I^-. You can see that decomposition does not occur in a single elementary step corresponding to the overall balanced equation. If it did, the reaction would be second order in H_2O_2 (as a result of the collision of two H_2O_2 molecules). What's more, the I^- ion, which is not even in the overall equation, appears in the rate law expression. How can we reconcile these facts?

We can account for the observed rate law by assuming that the reaction takes place in two separate elementary steps, each of which is bimolecular:

Step 1: $$H_2O_2 + I^- \xrightarrow{k_1} H_2O + OI^-$$

Step 2: $$H_2O_2 + OI^- \xrightarrow{k_2} H_2O + O_2 + I^-$$

An analogy for the rate-determining step is the flow of traffic along a narrow road. Assuming the cars cannot pass one another on the road, the rate at which the cars travel is governed by the slowest-moving car.

If we further assume that the second step is much faster than the first step, then the overall rate is controlled by the rate of the first step (1), which is aptly called the ***rate-determining step.*** The rate-determining step is *the slowest step in the sequence of steps leading to the formation of products*. In our example, after the OI^- ion is formed in the first step it is immediately consumed in the second step. So the rate of the reaction can be determined from the first step alone:

$$\text{rate} = k_1[H_2O_2][I^-]$$

Note that the OI^- ion is an intermediate because it does not appear in the overall balanced equation. Although the I^- ion also does not appear in the overall equation, I^- differs from OI^- in that the former is present at the start of the reaction and at its completion. The function of I^- is to speed up the reaction; that is, it is a catalyst. We will discuss catalysis in the next section.

The Hydrogen Iodide Reaction A common reaction mechanism is one that involves at least two elementary steps, the first of which is very rapid in both the forward and reverse directions compared with the second step. An example is the reaction between molecular hydrogen and molecular iodine to produce hydrogen iodide:

$$H_2(g) + I_2(g) \longrightarrow 2HI(g)$$

Experimentally, the rate law is found to be

$$\text{rate} = k[H_2][I_2]$$

For many years it was thought that the reaction occurred just as written; that is, it was thought to be simply a bimolecular reaction involving a hydrogen molecule and an iodine molecule. However, in the 1960s chemists found that the actual mechanism is more complicated than that shown above. A two-step mechanism was proposed:

Step 1: $$I_2 \underset{k_{-1}}{\overset{k_1}{\rightleftharpoons}} 2I$$

Step 2: $$H_2 + 2I \xrightarrow{k_2} 2HI$$

where k_1, k_{-1}, and k_2 are the rate constants for the reactions. The I atoms are the intermediate in this reaction.

Chemical equilibrium will be discussed in the next chapter.

When the reaction first begins, there are very few I atoms present. But as I_2 dissociates, the concentration of I_2 decreases while that of I increases. Therefore, the forward rate of step 1 decreases and the reverse rate increases. Soon the two rates become equal, and a dynamic equilibrium is established. Because the elementary reactions in step 1 are much faster than the second step, equilibrium is reached before any significant reaction with hydrogen occurs and it persists throughout the reaction.

In the equilibrium condition of step 1 the forward rate is equal to the reverse rate; that is,

$$k_1[I_2] = k_{-1}[I]^2$$

or

$$[I]^2 = \frac{k_1}{k_{-1}}[I_2]$$

The rate of the reaction is given by the slow, rate-determining step, which is step 2:

$$\text{rate} = k_2[H_2][I]^2$$

Substituting the expression for $[I]^2$ into this rate law we obtain

$$\begin{aligned}\text{rate} &= \frac{k_1k_2}{k_{-1}}[H_2][I_2] \\ &= k[H_2][I_2]\end{aligned}$$

where $k = k_1k_2/k_{-1}$. As you can see, this two-step mechanism also gives the correct rate law for the reaction. This agreement along with the observation of intermediate I atoms provides strong evidence for believing the mechanism to be correct.

Let us summarize the sequence involved in studying reaction mechanisms: First, we collect data (measurement of rates). Second, we analyze these data to determine the rate constant and order of the reaction, and we write the rate law. Third, we suggest a plausible mechanism for the reaction in terms of elementary steps (Figure 13.16). The elementary steps must satisfy two requirements:

- The sum of the elementary steps must give the overall balanced equation for the reaction.
- The rate-determining step should predict the same rate law as is determined experimentally.

Keep in mind that for any proposed reaction scheme, we must be able to detect the presence of any intermediate(s) formed in one or more elementary steps.

The following example concerns the mechanistic study of a relatively simple reaction.

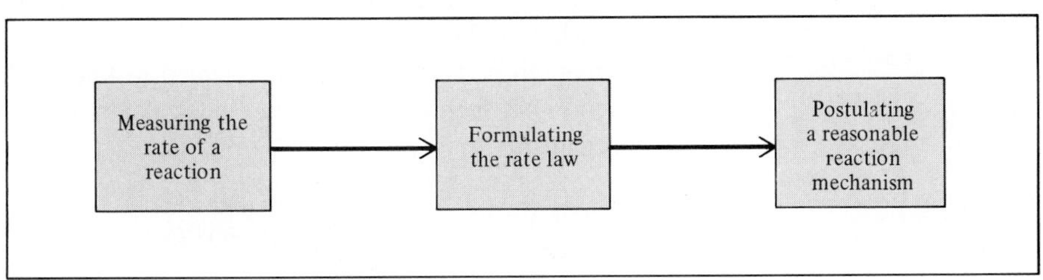

Figure 13.16 *Sequence of steps in the study of a reaction mechanism.*

EXAMPLE 13.9

The gas-phase decomposition of nitrous oxide (N_2O) is believed to occur via two elementary steps:

Step 1: $$N_2O \xrightarrow{k_1} N_2 + O$$

Step 2: $$N_2O + O \xrightarrow{k_2} N_2 + O_2$$

Experimentally the rate law is found to be rate $= k[N_2O]$. (a) Write the equation for the overall reaction. (b) Which of the species are intermediates? (c) What can you say about the relative rates of steps 1 and 2?

Answer

(a) Adding the equations for steps 1 and 2 gives the overall reaction

$$2N_2O \longrightarrow 2N_2 + O_2$$

(b) Since the O atom is produced in the first elementary step and it does not appear in the overall balanced equation, it is an intermediate.
(c) If we assume that step 1 is the rate-determining step (that is, if $k_2 \gg k_1$), then the rate of the overall reaction is given by

$$\text{rate} = k_1[N_2O]$$

and $k = k_1$.

Similar problem: 13.58.

Finally we note that not all reactions have a rate-determining step. For example, a reaction may have two or more comparably slow steps. The kinetic analysis of such reactions is generally more involved.

Experimental Support for Reaction Mechanisms

How can we find out whether the proposed mechanism for a particular reaction is correct? In the case of hydrogen peroxide decomposition we might try to detect the presence of the OI^- ions by spectroscopic means. Evidence of their existence would support the reaction scheme. Similarly, for the hydrogen iodide reaction, detection of iodine atoms would lend support to the two-step mechanism. For example, it is known that I_2 dissociates into atoms when it is irradiated with visible light. Thus we might predict that the formation of HI from H_2 and I_2 would speed up as the intensity of light is increased because that should increase the concentration of I atoms. Indeed, this is just what is observed.

In another instance, chemists wanted to know which C—O bond is broken in the reaction between methyl acetate and water in order to better understand the reaction mechanism.

$$\underset{\text{methyl acetate}}{CH_3{-}\overset{\overset{\displaystyle O}{\|}}{C}{-}O{-}CH_3} + H_2O \longrightarrow \underset{\text{acetic acid}}{CH_3{-}\overset{\overset{\displaystyle O}{\|}}{C}{-}OH} + \underset{\text{methanol}}{CH_3OH}$$

The two possibilities are

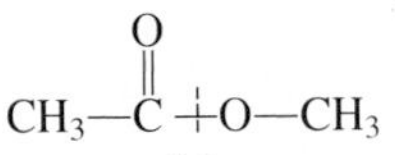

$$\underset{\text{(a)}}{CH_3{-}\overset{\overset{\displaystyle O}{\|}}{C}{+}O{-}CH_3} \qquad \underset{\text{(b)}}{CH_3{-}\overset{\overset{\displaystyle O}{\|}}{C}{-}O{+}CH_3}$$

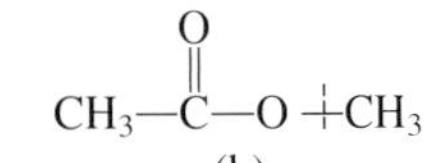

A mass spectrometer can be used to distinguish compounds containing different isotopes.

To distinguish between schemes (a) and (b), chemists used water containing the oxygen-18 isotope instead of ordinary water (which contains the oxygen-16 isotope). When the oxygen-18 water was used, only the acetic acid formed contained the oxygen-18 isotope:

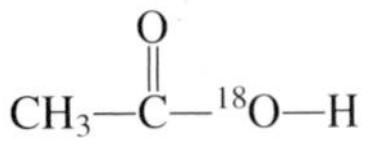

$$CH_3{-}\overset{\overset{\displaystyle O}{\|}}{C}{-}{}^{18}O{-}H$$

Thus, the reaction must have occurred via bond-breaking scheme (a), because the product formed via scheme (b) would retain both of its original oxygen atoms. These examples give some idea of how inventive chemists must be in studying reaction mechanisms, but you must realize that for complex reactions it is impossible to prove the uniqueness of any particular mechanism.

13.6 Catalysis

A rise in temperature also increases the rate of a reaction. However, at high temperatures the products formed may undergo other reactions, thereby reducing the yield.

We saw in studying the decomposition of hydrogen peroxide that the reaction rate depends on the concentration of iodide ions even though I^- does not appear in the overall equation. We noted there that I^- acts as a catalyst for that reaction. A ***catalyst*** is *a substance that increases the rate of a chemical reaction without itself being consumed.* The catalyst may react to form an intermediate, but it is regenerated in a subsequent step of the reaction. In the laboratory preparation of molecular oxygen, a sample of potassium chlorate is heated, as shown in Figure 3.11; the reaction is

$$2KClO_3(s) \longrightarrow 2KCl(s) + 3O_2(g)$$

However, this thermal decomposition is very slow in the absence of a catalyst. The rate of decomposition can be increased dramatically by adding a small amount of the catalyst manganese dioxide (MnO_2), a black powdery substance. All the MnO_2 can be recovered at the end of the reaction, just as all the I^- ions remain following H_2O_2 decomposition.

To continue the traffic analogy, adding a catalyst can be compared with building a tunnel through a mountain to connect two towns that were previously linked by a winding road over the mountain.

Regardless of its nature, a catalyst speeds up a reaction by providing a set of elementary steps with more favorable kinetics than those that exist in its absence. From Equation (13.9) we know that the rate constant k (and hence the rate) of a reaction depends on the frequency factor A and the activation energy E_a—the larger the A or the smaller the E_a, the greater the rate. In many cases, a catalyst increases the rate by lowering the activation energy for the reaction.

Let us assume that the following reaction has a certain rate constant k and an activation energy E_a.

$$A + B \xrightarrow{k} C + D$$

In the presence of a catalyst, however, the rate constant is k_c (called the *catalytic rate constant*):

$$A + B \xrightarrow{k_c} C + D$$

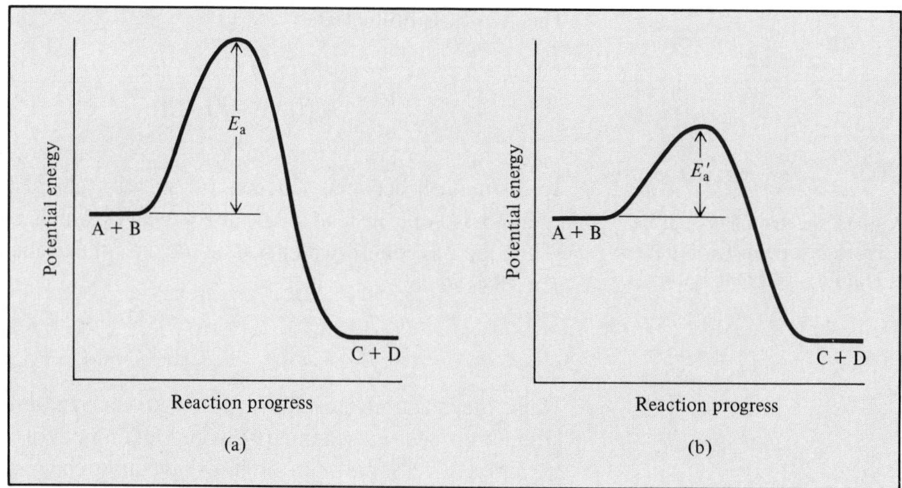

Figure 13.17 *Comparison of the activation energy barriers of an uncatalyzed reaction and the same reaction with a catalyst. Note that the catalyst lowers the energy barrier but does not affect the actual energies of the reactants or products. Although the reactants and products are the same in both cases, the reaction mechanisms and rate laws are different in (a) and (b).*

By the definition of a catalyst,

$$\text{rate}_{\text{catalyzed}} > \text{rate}_{\text{uncatalyzed}}$$

A catalyst lowers the activation energy for both the forward and reverse reactions.

Figure 13.17 shows the potential energy profiles for both reactions. Note that the total energies of the reactants (A and B) and those of the products (C and D) are unaffected by the catalyst; the only difference between the two is a lowering of activation energy from E_a to E'_a. Because the activation energy for the reverse reaction is also lowered, a catalyst enhances the rate of the reverse reaction to the same extent as it does the forward reaction.

There are three general types of catalysis, depending on the nature of the rate-increasing substance: heterogeneous catalysis, homogeneous catalysis, and enzyme catalysis.

Heterogeneous Catalysis

In *heterogeneous catalysis,* reactants and catalyst are in different phases. Usually the catalyst is a solid and the reactants are either gases or liquids. Heterogeneous catalysis is by far the most important type of catalysis in industrial processes. It plays an important role in the synthesis of many key chemicals and other chemical processes. Here we describe five specific examples of heterogeneous catalysis that account for millions of tons of chemicals produced annually on an industrial scale.

The Haber Synthesis of Ammonia. Ammonia is an extremely valuable inorganic substance used as raw material in the fertilizer industry, the manufacture of explosives, and many other areas. Around the turn of the century, many chemists strove to synthesize ammonia from nitrogen and hydrogen. The supply of atmospheric nitrogen is virtually inexhaustible, and hydrogen gas can be produced readily by passing steam over heated coal:

$$H_2O(g) + C(s) \longrightarrow CO(g) + H_2(g)$$

Hydrogen is also obtained as a by-product of petroleum refining.

Predictions based on the laws of thermodynamics showed that the reaction

$$N_2(g) + 3H_2(g) \longrightarrow 2NH_3(g) \qquad \Delta H^\circ = -92.6 \text{ kJ}$$

should go a long way toward completion at room temperature, but the extremely slow rate of the reaction makes any large-scale operation virtually impossible. To make the reaction feasible two requirements must be met simultaneously: an appreciable rate and a high yield of the product. Raising the temperature does accelerate the reaction, but at the same time it promotes the decomposition of NH_3 molecules into N_2 and H_2, thus lowering the yield of NH_3.

In 1905, after testing literally thousands of compounds at various temperatures and pressures, Fritz Haber discovered that iron plus a few percent of oxides of potassium and aluminum catalyze the reaction of hydrogen with nitrogen to yield ammonia at about 500°C. This procedure is known as the *Haber process*.

In heterogeneous catalysis, the surface of the solid catalyst is usually the site of the reaction. The initial step in the Haber process involves the dissociation of N_2 and H_2 on the metal surface (Figure 13.18). Although the dissociated species are not truly free atoms because they are bonded to the metal surface, they are highly reactive. The two reactant molecules behave very differently on the catalyst surface. Studies show that H_2 dissociates into atomic hydrogen at temperatures as low as −196°C (the boiling point of liquid nitrogen). Nitrogen molecules, on the other hand, dissociate at about 500°C. The highly reactive N and H atoms combine rapidly at high temperatures to produce the desired NH_3 molecules:

$$N + 3H \longrightarrow NH_3$$

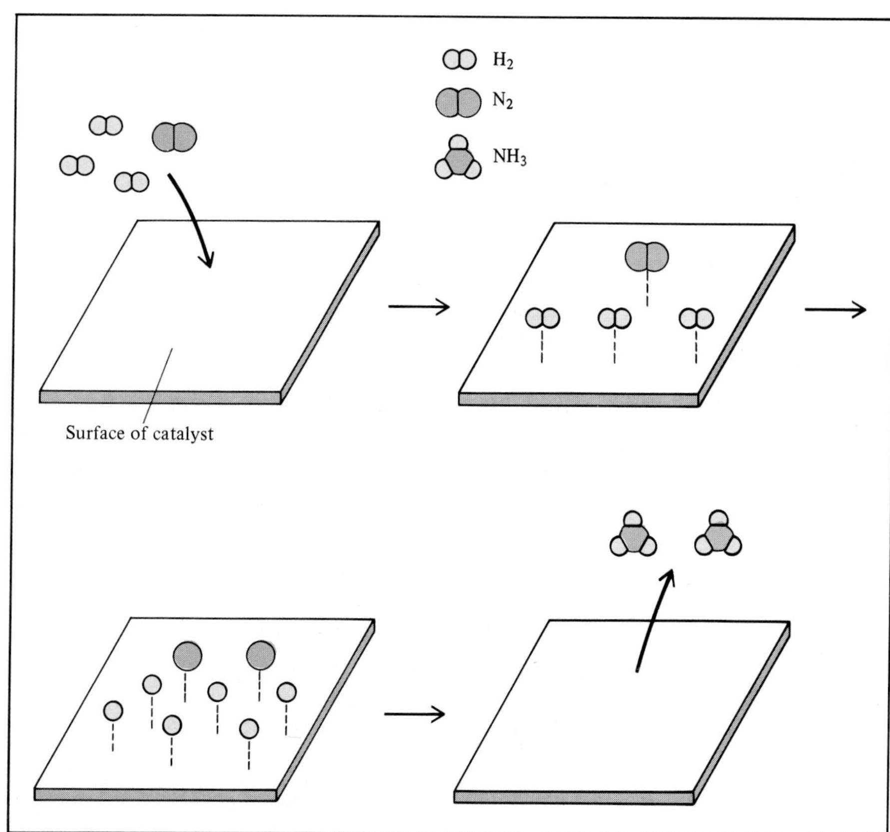

Figure 13.18 *The catalytic action in the synthesis of ammonia. First the H_2 and N_2 molecules bind to the surface of the catalyst. This interaction weakens the covalent bonds within the molecules and eventually causes the molecules to dissociate. The highly reactive H and N atoms combine to form NH_3 molecules, which then leave the surface.*

Figure 13.19 *Vanadium(V) oxide on alumina (Al_2O_3)—used to catalyze the conversion of sulfur dioxide to sulfur trioxide.*

Figure 13.20 *Platinum-rhodium catalyst used in the Ostwald process.*

The Manufacture of Sulfuric Acid. Sulfuric acid is prepared industrially by first burning sulfur in air:

$$S(s) + O_2(g) \longrightarrow SO_2(g)$$

Next is the key step of converting sulfur dioxide to sulfur trioxide:

$$2SO_2(g) + O_2(g) \longrightarrow 2SO_3(g)$$

Finally sulfur trioxide is treated with water to produce sulfuric acid:

$$SO_3(g) + H_2O(l) \longrightarrow H_2SO_4(aq)$$

Vanadium(V) oxide (V_2O_5) is the catalyst used for the second step (Figure 13.19). Because the sulfur dioxide and oxygen molecules react in contact with the surface of solid V_2O_5, the process is referred to as the *contact process*.

The Manufacture of Nitric Acid. Like sulfuric acid, nitric acid is one of the most important inorganic acids. It is used in the production of fertilizers, dyes, drugs, and explosives. The major industrial method of producing nitric acid is the *Ostwald*† *process*. The starting materials, ammonia and molecular oxygen, are heated in the presence of a platinum-rhodium catalyst (Figure 13.20) to about 800°C:

$$4NH_3(g) + 5O_2(g) \longrightarrow 4NO(g) + 6H_2O(g)$$

The nitric oxide formed readily oxidizes (without catalysis) to nitrogen dioxide:

$$2NO(g) + O_2(g) \longrightarrow 2NO_2(g)$$

When dissolved in water, NO_2 forms both nitrous acid and nitric acid:

† Wilhelm Ostwald (1853–1932). German chemist. Ostwald made important contributions to chemical kinetics, thermodynamics, and electrolyte solutions. He developed the industrial process for preparing nitric acid that now bears his name. He received the Nobel Prize in chemistry in 1909.

$$2NO_2(g) + H_2O(l) \longrightarrow HNO_2(aq) + HNO_3(aq)$$

On heating, nitrous acid is converted to nitric acid as follows:

Note that this is a disproportionation reaction (see Section 3.5).

$$3HNO_2(aq) \longrightarrow HNO_3(aq) + H_2O(l) + 2NO(g)$$

The NO generated can be recycled to produce NO_2 in the second step.

Hydrogenation. ***Hydrogenation*** is *the addition of hydrogen to compounds containing multiple bonds, especially C═C and C≡C bonds*. A simple hydrogenation reaction is the conversion of ethylene to ethane:

$$H_2 + \underset{\text{ethylene}}{H_2C{=}CH_2} \longrightarrow \underset{\text{ethane}}{H_3C{-}CH_3}$$

Figure 13.21 *Platinum catalyst on alumina (Al_2O_3)—used in hydrogenations.*

This reaction is quite slow under normal conditions, but the reaction rate can be greatly increased by the presence of a catalyst such as nickel or platinum (Figure 13.21). As in the Haber synthesis of ammonia, the main function of the catalyst is to weaken the H—H bond and facilitate the reaction.

Hydrogenation is an important process in the food industry. Vegetable oils have considerable nutritional value, but some oils must be hydrogenated before we can use them because of their unsavory flavor and their inappropriate molecular structures (that is, there are too many C═C bonds present). Upon exposure to air, these *polyunsaturated* molecules (that is, molecules with many C═C bonds) undergo oxidation to yield unpleasant-tasting products (oil that has oxidized is said to be rancid). In the hydrogenation process, a small amount of nickel (about 0.1 percent by mass) is added to the oil and the mixture is exposed to hydrogen gas at high temperature and pressure. Afterward, the nickel is removed by filtration. Hydrogenation reduces the number of double bonds in the molecule but does not completely eliminate them. If all the double bonds are eliminated, the oil becomes hard and brittle. Under controlled conditions, suitable cooking oils and margarine may be prepared by hydrogenation from vegetable oils extracted from cottonseed, corn, and soybeans.

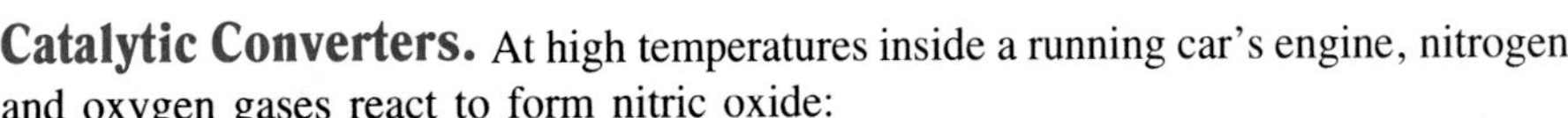

Catalytic Converters. At high temperatures inside a running car's engine, nitrogen and oxygen gases react to form nitric oxide:

$$N_2(g) + O_2(g) \rightleftharpoons 2NO(g)$$

When released to the atmosphere, NO rapidly combines with O_2 to form NO_2, a poisonous dark brown gas with a choking odor:

$$2NO(g) + O_2(g) \longrightarrow 2NO_2(g)$$

This nitrogen dioxide and other undesirable gases emitted by an automobile such as carbon monoxide (CO) and various unburned hydrocarbons make automobile exhaust a major source of air pollution.

A *hydrocarbon* is a compound containing only carbon and hydrogen atoms. Hydrocarbons are the main constituents of natural gas and gasoline.

Most autos are now manufactured with catalytic converters (Figure 13.22). An efficient catalytic converter serves two purposes: It oxidizes CO and unburned hydrocarbons to CO_2 and H_2O, and it reduces NO and NO_2 to N_2 and O_2. Hot exhaust gases into which air has been injected are passed through the first chamber of one converter

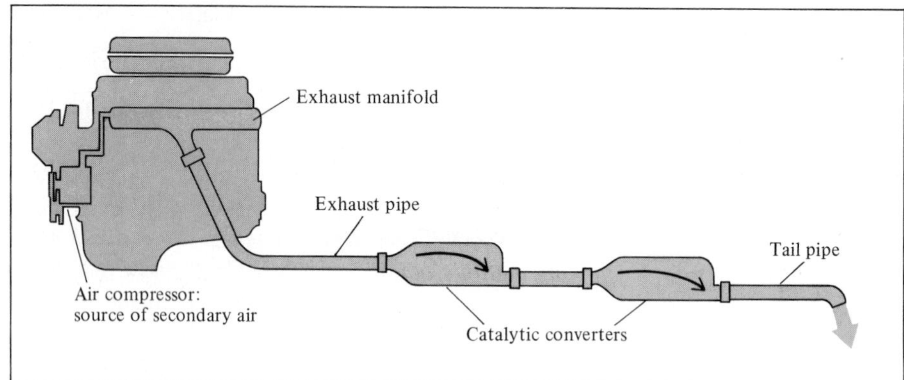

Figure 13.22 *A two-stage catalytic converter for an automobile.*

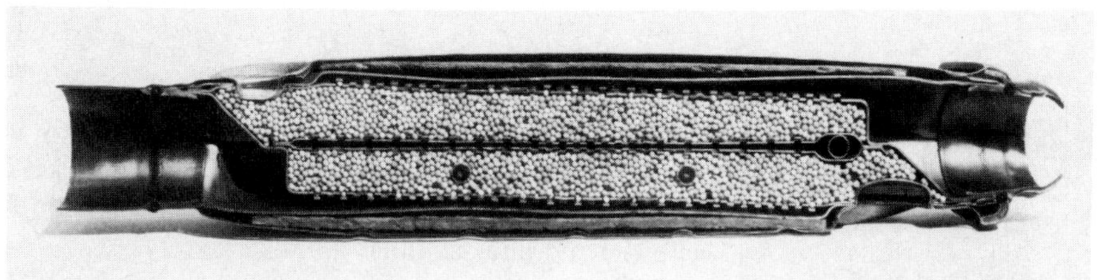

Figure 13.23 *A cross-sectional view of a catalytic converter. The beads contain platinum, palladium, and rhodium, which catalyze the combustion of CO and hydrocarbons.*

to accelerate the complete burning of hydrocarbons and to decrease CO emission. (A cross section of the catalytic converter, containing Pt or Pd, or a transition metal oxide such as CuO or Cr_2O_3, is shown in Figure 13.23.) However, since high temperatures increase NO production, a second chamber containing a different catalyst (again a transition metal or a transition metal oxide) operating at a lower temperature is required to dissociate NO into N_2 and O_2 before the exhaust is discharged through the tailpipe.

Homogeneous Catalysis

In *homogeneous catalysis* the reactants, products, and catalyst are all dispersed in a single phase, usually liquid. Acid and base catalyses are the most important types of homogeneous catalysis in liquid solution. For example, the reaction of ethyl acetate with water to form acetic acid and ethanol normally occurs too slowly to be measured.

$$\underset{\text{ethyl acetate}}{CH_3{-}\overset{\overset{\displaystyle O}{\|}}{C}{-}O{-}C_2H_5} + H_2O \longrightarrow \underset{\text{acetic acid}}{CH_3{-}\overset{\overset{\displaystyle O}{\|}}{C}{-}OH} + \underset{\text{ethanol}}{C_2H_5OH}$$

In the absence of the catalyst, the rate law is given by

$$\text{rate} = k[CH_3COOC_2H_5]$$

We assume that $k_c \gg k$, so that the rate is determined solely by the catalyzed portion of the reaction.

However, the reaction can be catalyzed by acids. In the presence of hydrochloric acid, the rate is given by

$$\text{rate} = k_c[CH_3COOC_2H_5][H^+]$$

Homogeneous catalysis can also take place in the gas phase. A well-known example of catalyzed gas-phase reactions is the lead chamber process, which for many years was the major process for the manufacture of sulfuric acid. Starting with sulfur, we would expect the production of sulfuric acid to occur in the following steps:

$$S(s) + O_2(g) \longrightarrow SO_2(g)$$
$$2SO_2(g) + O_2(g) \longrightarrow 2SO_3(g)$$
$$H_2O(l) + SO_3(g) \longrightarrow H_2SO_4(aq)$$

In reality, however, sulfur dioxide is not converted directly to sulfur trioxide; rather, the oxidation is more efficiently carried out in the presence of the catalyst nitrogen dioxide:

$$2SO_2(g) + 2NO_2(g) \longrightarrow 2SO_3(g) + 2NO(g)$$
$$2NO(g) + O_2(g) \longrightarrow 2NO_2(g)$$

Overall reaction: $$2SO_2(g) + O_2(g) \longrightarrow 2SO_3(g)$$

Note that there is no net loss of NO_2 in the overall reaction, so that NO_2 meets the criteria for a catalyst. (See the Chemistry in Action on p. 583 for an example of homogeneous catalysis in a reaction in Earth's atmosphere.)

In recent years chemists have devoted much effort to developing a class of metallic compounds to serve as homogeneous catalysts. These compounds are soluble in various organic solvents and therefore can catalyze reactions in the same phase as the dissolved reactants. Many of the processes they catalyze are organic. For example, a red violet compound of rhodium, $[(C_6H_5)_3P]_3RhCl$, catalyzes the hydrogenation of a carbon-carbon double bond as follows:

$$\mathrm{>C{=}C<} + H_2 \longrightarrow \mathrm{-\overset{|}{\underset{H}{C}}-\overset{|}{\underset{H}{C}}-}$$

Some advantages of homogeneous catalysis over heterogeneous catalysis are: (1) the reactions can often be carried out under atmospheric conditions, thus reducing costs of production and minimizing the decomposition of products at high temperatures; (2) homogeneous catalysts can often be designed to function selectively for a particular type of reaction; and (3) homogeneous catalysts cost less than the precious metals (for example, platinum and gold) used in heterogeneous catalysis.

Enzyme Catalysis

Enzymes are made of protein molecules that have molar masses ranging from thousands to millions of grams.

Of all the intricate processes that have evolved in living systems, none is more striking or more essential than enzyme catalysis. ***Enzymes*** are *biological catalysts*. The amazing fact about enzymes is not only that they can increase the rate of biochemical reactions by factors ranging from 10^6 to 10^{12}, but that they are so highly specific. They act only on certain molecules, called *substrates* (that is, reactants), while leaving the rest of the system unaffected. It has been estimated that an average living cell may contain some 3000 different enzymes, each of them catalyzing a specific reaction in which substrates are converted into the appropriate products. Enzyme-catalyzed reactions are examples of homogeneous catalysis because the substrate, enzyme, and product are all present in the aqueous solution.

But enzymes do not operate independently. There is, so to speak, a built-in control that turns on certain enzymes for action when the need arises and then turns them off when enough of the product has been formed. One theory for the specificity of enzymes—the so-called ''lock-and-key'' theory—was developed by the German chemist

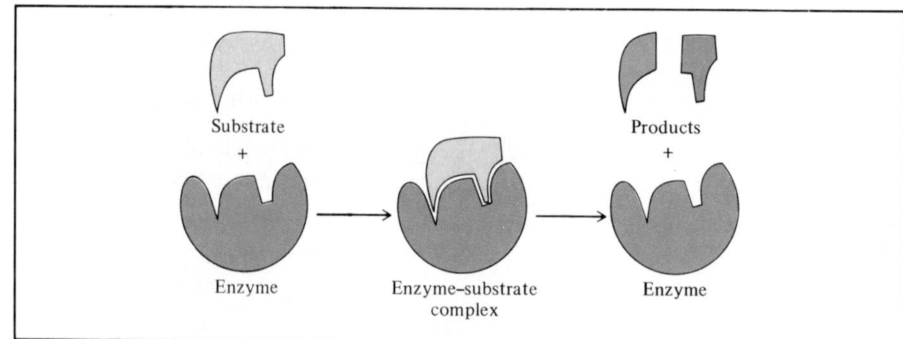

Figure 13.24 *The lock-and-key model of an enzyme's specificity for substrate molecules.*

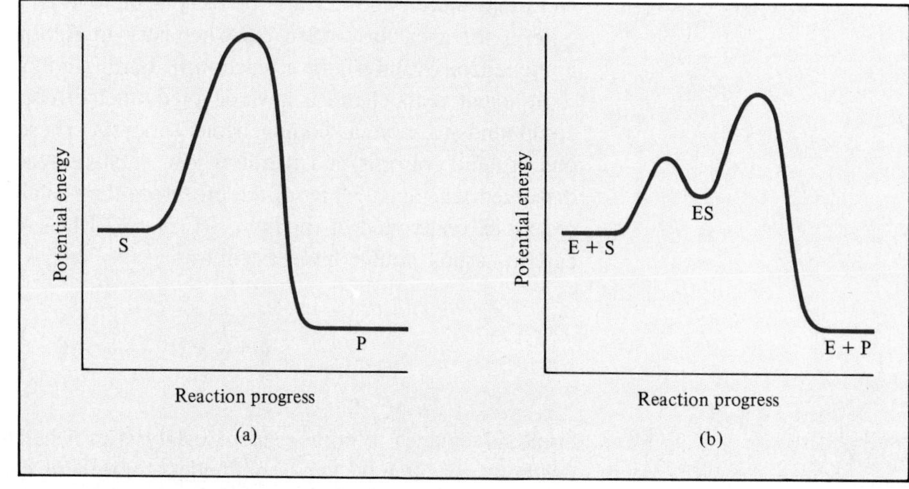

Figure 13.25 *Comparison of (a) an uncatalyzed reaction and (b) the same reaction catalyzed by an enzyme. The plot in (b) assumes that the catalyzed reaction has a two-step mechanism, in which the second step (ES → E + P) is rate determining.*

Emil Fischer† in 1894. An enzyme is ordinarily a very large protein molecule that contains one or more *active sites* where reactions with substrates take place; the rest of the molecule maintains the three-dimensional integrity of the network. According to Fischer's hypothesis, the active site has a rigid structure, similar to a lock. A substrate molecule has the complementary structure that causes it to fit and function like a key (Figure 13.24). Although appealing in many respects, this theory has some glaring discrepancies. For instance, substrates of quite different sizes and shapes are known to bind to the same type of enzyme. Therefore the enzyme's specificity cannot be explained in terms of structural rigidity.

Chemists now know that an enzyme molecule (or at least its active site) actually has a fair amount of flexibility. During a reaction it changes to accommodate the substrate molecule. Just how these changes are brought about is not entirely clear.

The mathematical treatment of enzyme kinetics is quite complex, even when we know the basic steps involved in the reaction. A simplified scheme is as follows:

$$\mathrm{E + S \rightleftharpoons ES}$$

$$\mathrm{ES \xrightarrow{k} P + E}$$

†Emil Fischer (1852–1919). German chemist. Regarded by many as the greatest organic chemist of the nineteenth century, Fischer made many important contributions in the synthesis of sugars and other important molecules. He was awarded the Nobel Prize in chemistry in 1902.

where E, S, and P represent enzyme, substrate, and product, and ES is the enzyme-substrate intermediate. Figure 13.25 shows the potential energy profile for the reaction. It is often assumed that the formation of ES and its decomposition back to enzyme and substrate molecules occur rapidly and that the rate-determining step is the formation of product. In general, the rate of such an enzyme-catalyzed reaction is given by the equation

$$\begin{aligned} \text{rate} &= \frac{\Delta[\text{P}]}{\Delta t} \\ &= k[\text{ES}] \end{aligned}$$

The concentration of the ES intermediate is itself proportional to the amount of the substrate present, and a plot of the rate versus the concentration of substrate typically yields a curve such as that shown in Figure 13.26. Initially the rate rises rapidly with increasing substrate concentration. However, above a certain concentration all the active sites are occupied, and the reaction becomes zero order in the substrate. That is, the rate remains the same even though the substrate concentration increases. At and beyond this point, the rate of formation of product depends only on how fast the ES intermediate breaks down, not on the number of substrate molecules present.

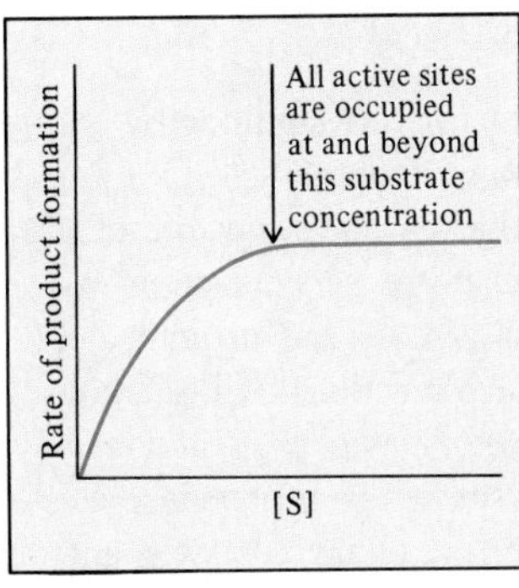

Figure 13.26 *Plot of the rate of product formation versus substrate concentration in an enzyme-catalyzed reaction. When all the enzymes are tied up by the substrate molecules to form the ES intermediate, the rate in the substrate concentration becomes zero order; it no longer changes with increasing substrate concentration.*

CHEMISTRY IN ACTION

Depletion of Ozone in the Stratosphere

It is generally believed that three or four billion years ago, Earth's atmosphere consisted mainly of ammonia, methane, and water. There was little, if any, free O_2 present. Over the years, the concentration of O_2 gas began to increase, largely as a result of photosynthesis and photochemical decomposition of water vapor. Ozone molecules (O_3) in the atmosphere are formed by a process involving O_2 and solar radiation at wavelengths shorter than 260 nm:

$$O_2(g) + h\nu \longrightarrow O(g) + O(g)$$
$$O(g) + O_2(g) \longrightarrow O_3(g)$$

where $h\nu$ is the energy of a photon. Most atmospheric ozone is found in the stratosphere (Figure 13.27), where its concentration is about 10 ppm (parts per million) by volume. Ozone has the important photochemical property of absorbing solar radiation between the wavelengths of 200 nm and 300 nm:

$$O_3(g) + h\nu \longrightarrow O(g) + O_2(g)$$

The reactive atomic oxygen formed when this takes place recombines with molecular oxygen to form ozone, thereby completing the ozone cycle. By absorbing short-wavelength ultraviolet (UV) solar radiation, the ozone layer in the stratosphere acts as our protective shield. The damaging effects of UV radiation are well documented. It can induce skin cancer and cause mutations, for example. Without ozone, life on Earth would gradually disappear.

In recent years scientists have become alarmed about the effects of certain chlorofluorocarbons (CFCs) on

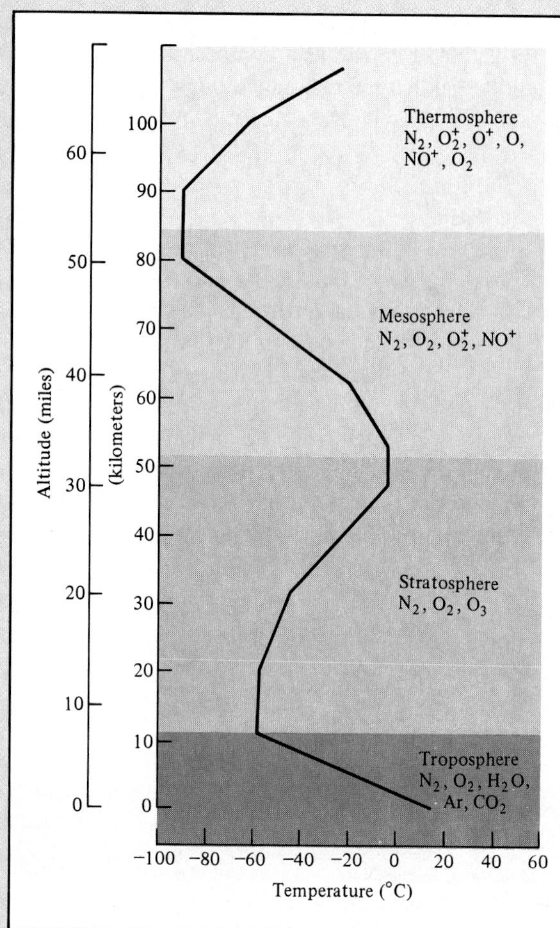

Figure 13.27 *The different regions of Earth's atmosphere. Temperature variation with altitude is represented by the black curve. Ozone concentrates in the stratosphere.*

the ozone layer in the atmosphere. These compounds, known commercially as Freons, have the following structures: $CFCl_3$ (Freon 11), CF_2Cl_2 (Freon 12), $C_2F_3Cl_3$ (Freon 113), and $C_2F_4Cl_2$ (Freon 114). Because they are readily liquefied, relatively inert, noncombustible, and volatile, the Freons are used as aerosol propellants in spray cans, as a coolant in air-conditioning refrigeration, as solvents to clean electronic circuit boards, and in the manufacture of plastic foams. Once released into the atmosphere, they slowly diffuse up to the stratosphere, where UV radiation of wavelengths between 175 nm and 220 nm causes them to decompose:

$$CFCl_3 \longrightarrow CFCl_2 + Cl$$

$$CF_2Cl_2 \longrightarrow CF_2Cl + Cl$$

The reactive chlorine atoms formed then undergo the following reactions:

$$Cl + O_3 \longrightarrow ClO + O_2 \qquad (1)$$

$$ClO + O \longrightarrow Cl + O_2 \qquad (2)$$

The overall result is a net removal of O_3 molecules from the stratosphere:

$$O_3 + O \longrightarrow 2O_2$$

The oxygen atoms in this reaction are supplied by the photochemical decomposition of ozone ($O_3 \rightarrow O + O_2$ as described earlier). Note that Cl plays the role of a catalyst in the reaction mechanism scheme given by steps (1) and (2) because it is not used up and can therefore take part in many such reactions. (Cl is acting as a homogeneous catalyst in the gas-phase reactions.) On the other hand, the ClO species is an intermediate because it is produced in the first elementary step and consumed in the second step. The above mechanism for the destruction of ozone has been supported by the detection of ClO in the stratosphere in recent years.

How bad is the stratospheric ozone depletion? Because of the complexities of the reactions and the difficulties in measuring the concentrations of Freons and reaction intermediates, there was considerable debate about whether the CFCs were really harmful to the ozone layer. However, the growing scientific evidence of the last few years has shown that significant depletion of ozone has already occurred in the stratosphere. Perhaps the most dramatic case was the discovery in 1986 of an "ozone hole" in the South Pole (Figure 13.28). In the winter, temperatures in the Antarctic stratosphere can drop to below −80°C, cold enough that water condenses even in the extremely dry stratosphere to form ice crystals. The ice crystals act as a heterogeneous catalyst that converts man-made hydrogen chloride (HCl) and chlorine nitrate ($ClONO_2$) into molecular chlorine and hypochlorous acid (HOCl), which are released into the gas phase (the HNO_3 formed remains in the condensed, that is, ice, phase):

$$HCl + ClONO_2 \longrightarrow HNO_3 + Cl_2$$

$$H_2O + ClONO_2 \longrightarrow HNO_3 + HOCl$$

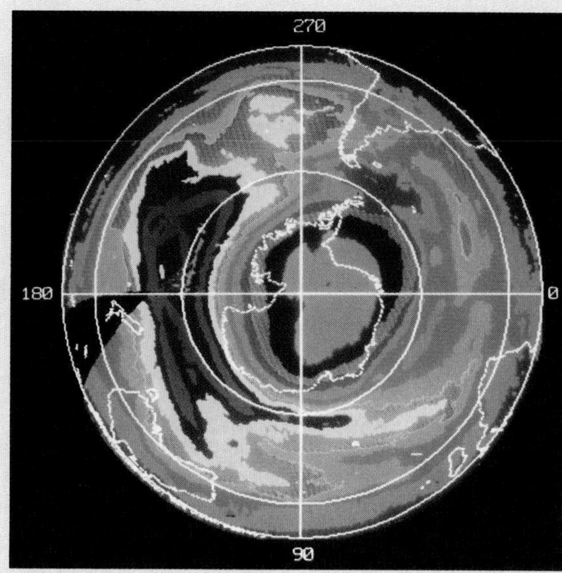

Figure 13.28 *In recent years, scientists have found that the ozone layer in the stratosphere over the South Pole has become thinner. This photo, taken in October 1986, shows the depletion of ozone in an area (purple color) roughly the size of the United States.*

Both Cl_2 and HOCl can be easily photolyzed to Cl atoms that can participate in the ozone destruction described earlier. This is believed to be a major cause of the observed ozone hole. Because of different weather patterns, the polar stratospheric clouds that are key to the Antarctic ozone hole occur to a much less extent in the Arctic (Figure 13.29). Recently (1989), it has been found that the thinning of the ozone layer has spread from the Antarctic to regions over southern Australia. The studies so far show that Freons, as well as other halogen-containing compounds, can lead to ozone destruction by a number of mechanisms in different regions.

The U.S. government banned the use of Freons in aerosol sprays several years ago, but other uses continue. Furthermore, the ozone depletion is a global issue, so that international accord must be reached. Some progress has been made in this respect, and it will take great determination and considerable economic sacrifices on the part of many nations to halt production of CFCs and search for less harmful substitutes. In view of the seriousness of the situation, there is no other choice.

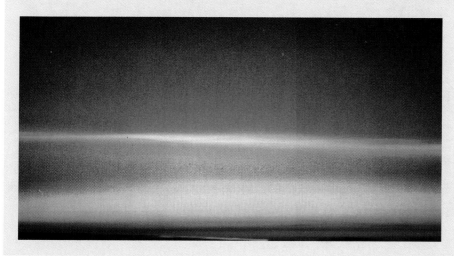

Figure 13.29 *Polar stratospheric clouds containing ice particles can catalyze the formation of Cl atoms and lead to the destruction of ozone.*

Summary

1. The rate of a chemical reaction is shown by the change in the concentration of reactant or product molecules over time. The rate is not constant, but varies continuously as concentration changes.
2. The rate law is an expression relating the rate of a reaction to the rate constant and

the concentrations of the reactants raised to appropriate powers. The rate constant k for a given reaction changes only with temperature.

3. Reaction order is the sum of the powers to which reactant concentrations are raised in the rate law. The rate law and the reaction order cannot be determined from the stoichiometry of the overall equation for a reaction; they must be determined by experiments.
4. The half-life of a reaction (the time it takes for the concentration of a reactant to decrease by one-half) can be used to determine the rate constant of a first-order reaction.
5. According to collision theory, a reaction occurs when molecules collide with sufficient energy, called the activation energy, to break the bonds and initiate the reaction.
6. The rate constant and the activation energy are related by the Arrhenius equation, $k = Ae^{-E_a/RT}$.
7. The overall balanced equation for a reaction may be the sum of a series of simple reactions, called elementary steps. The complete series of elementary steps for a reaction is the reaction mechanism.
8. If one step in a reaction mechanism is much slower than all other steps, it is called the rate-determining step.
9. A catalyst speeds up a reaction usually by lowering the value of E_a. A catalyst can be recovered unchanged at the end of a reaction.
10. In heterogeneous catalysis, which is of great industrial importance, the catalyst is a solid and the reactants are gases or liquids. In homogeneous catalysis, the catalyst and the reactants are in the same phase. Enzymes are catalysts in living systems.

Applied Chemistry Example: Polyethylene

Kinetics usually plays an important role in industrial processes, since the products must be manufactured in the minimum time under the most economical conditions. Consider the synthesis of polyethylene, which is used in many items in every day life such as piping, bottles, electrical insulation, toys, and mailer envelopes. Polyethylene is a polymer which is a molecule of very high molar mass (thousands to hundreds of thousands of grams). It is made by joining many units of ethylene molecules (called monomers, M) together in a process called polymerization. The mechanism of this reaction involves first heating an initiator molecule R_2

$$R_2 \xrightarrow{k_i} 2R\cdot \qquad \text{initiation}$$

where k_i is the rate constant for the initiation and $R\cdot$ is a radical. A radical has an unpaired electron and is highly reactive. In the next elementary step, the radical reacts with a monomer molecule M to generate a new radical M_1.

$$R\cdot + M \longrightarrow M_1\cdot$$

Since the monomer is ethylene, the reaction is

$$R\cdot + H_2C{=}CH_2 \longrightarrow R{-}\overset{\displaystyle H}{\underset{\displaystyle H}{\overset{|}{\underset{|}{C}}}}{-}\overset{\displaystyle H}{\underset{\displaystyle H}{\overset{|}{\underset{|}{C}}}}\cdot$$

Reaction of another monomer with $M_1\cdot$ leads to the growth (or propagation) of the polymer molecule

$$M_1\cdot + M \xrightarrow{k_p} M_2\cdot \qquad \text{propagation}$$

where k_p is the rate constant for the propagation. This step can be repeated to hundreds or thousands of monomer units. The propagation terminates when the radical end containing the unpaired electron forms a covalent bond with the unpaired electron on another radical

$$M'\cdot + M''\cdot \xrightarrow{k_t} M'\text{—}M'' \qquad \text{termination}$$

where k_t is the rate constant for the termination and $M'\cdot$ and $M''\cdot$ represent radicals of different lengths.

1. The initiator frequently used in the polymerization process is benzoyl peroxide $[(C_6H_5COO)_2]$:

$$\underset{R_2}{(C_6H_5COO)_2} \longrightarrow \underset{2R\cdot}{2C_6H_5COO\cdot}$$

This is a first-order reaction and the half-life of benzoyl peroxide at 100°C is 19.8 min. Calculate the rate constant (in min^{-1}) of the reaction.

Answer: The relationship between half-life and rate constant is that given in Equation (13.5):

$$\begin{aligned} k &= \frac{0.693}{t_{\frac{1}{2}}} \\ &= \frac{0.693}{19.8\ \text{min}} \\ &= 0.0350\ \text{min}^{-1} \end{aligned}$$

[This part is similar to Example 13.5(c)].

2. If the half-life of benzoyl peroxide is 7.30 h or 438 min at 70°C, what is the activation energy (in kJ/mol) for the decomposition of benzoyl peroxide?

Answer: Following the procedure in 1 we find the rate constant at 70°C to be $1.58 \times 10^{-3}\ \text{min}^{-1}$. We now have two values of rate constants (k_1 and k_2) at two temperatures (T_1 and T_2). This information allows us to calculate the activation energy E_a using Equation (13.10) as follows:

$$\ln \frac{k_1}{k_2} = \frac{E_a}{R}\left(\frac{T_1 - T_2}{T_1 T_2}\right)$$

Rearranging, we get

$$\begin{aligned} E_a &= R\left(\frac{T_1 T_2}{T_1 - T_2}\right)\ln \frac{k_1}{k_2} \\ &= (8.314\ \text{J/K}\cdot\text{mol})\left(\frac{343\ \text{K} \times 373\ \text{K}}{373\ \text{K} - 343\ \text{K}}\right)\ln\left(\frac{0.0350\ \text{min}^{-1}}{1.58 \times 10^{-3}\ \text{min}^{-1}}\right) \\ &= 1.10 \times 10^5\ \text{J/mol} \\ &= 110\ \text{kJ/mol} \end{aligned}$$

3. Write the rate laws for the elementary steps in the above polymerization process and identify the reactant, product, and intermediates.

Answer: Since all the above steps are elementary steps, we can deduce the rate law simply from the equations representing the steps. The rate laws are

Initiation: rate $= k_i[R_2]$

Propagation: rate $= k_p[M][M_1 \cdot]$

Termination: rate $= k_t[M' \cdot][M'' \cdot]$

The reactant molecules are the ethylene monomers and the product is polyethylene. Recalling that intermediates are species that are formed in an early elementary step and consumed in a later step, we see that they are the radicals $M' \cdot$, $M'' \cdot$, and so on. (The $R \cdot$ species also qualify as an intermediate.)

4. What condition would favor the growth of long high-molar-mass polyethylenes?

Answer: The growth of long polymers would be favored by a high rate of propagation and a low rate of termination. Since the rate law of propagation depends on the concentration of monomer, an increase in the concentration of ethylene would increase the propagation (growth) rate. From the rate law for termination we see that a low concentration of the radical fragment $M' \cdot$ or $M'' \cdot$ would lead to a slower rate of termination. This can be accomplished by using a low concentration of the initiator R_2.

Key Words

Activated complex, p. 565
Activation energy, p. 565
Bimolecular reaction, p. 571
Catalyst, p. 575
Chemical kinetics, p. 544
Elementary reaction, p. 570
Elementary step, p. 570
Enzyme, p. 581
First-order reaction, p. 554
Half-life, p. 556
Hydrogenation, p. 579
Intermediate, p. 570
Molecularity of a reaction, p. 571
Rate constant, p. 548
Rate-determining step, p. 572
Rate law, p. 550
Reaction mechanism, p. 570
Reaction order, p. 550
Reaction rate, p. 544
Second-order reaction, p. 560
Termolecular reaction, p. 571
Unimolecular reaction, p. 571

Exercises

Reaction Rate

REVIEW QUESTIONS

13.1 What is meant by the rate of a chemical reaction?

13.2 What are the units of the rate of a reaction?

13.3 Distinguish between average rate and instantaneous rate. Which of the two rates gives us an unambiguous measurement of reaction rate? Why?

13.4 What are the advantages of measuring the initial rate of a reaction?

13.5 Can you suggest two reactions that are very slow (take days or longer to complete) and two reactions that are very fast (reactions that are over in minutes or seconds)?

PROBLEMS

13.6 Write the reaction rate expressions for the following reactions in terms of the disappearance of the reactants and the appearance of products:
(a) $H_2(g) + I_2(g) \longrightarrow 2HI(g)$
(b) $2H_2(g) + O_2(g) \longrightarrow 2H_2O(g)$
(c) $5Br^-(aq) + BrO_3^-(aq) + 6H^+(aq) \longrightarrow 3Br_2(aq) + 3H_2O(l)$

13.7 Consider the reaction

$$N_2(g) + 3H_2(g) \longrightarrow 2NH_3(g)$$

Suppose that at a particular moment during the reaction

molecular hydrogen is reacting at the rate of 0.074 M/s. (a) What is the rate at which ammonia is being formed? (b) What is the rate at which molecular nitrogen is reacting?

13.8 Consider the reaction

$$C_2H_5I(aq) + H_2O(l) \longrightarrow C_2H_5OH(aq) + H^+(aq) + I^-(aq)$$

How could you follow the progress of the reaction by measuring the electrical conductance of the solution?

Rate Laws

REVIEW QUESTIONS

13.9 Explain what is meant by the rate law of a reaction.

13.10 What is meant by the order of a reaction?

13.11 What are the units for the rate constants of first-order and second-order reactions?

13.12 Write an equation relating the concentration of a reactant A at $t = 0$ to that at $t = t$ for a first-order reaction. Define all the terms and give their units. Do the same for a second-order reaction.

13.13 Consider the zero-order reaction: A $\rightarrow$ product. (a) Write the rate law for the reaction. (b) What are the units for the rate constant? (c) Plot the rate of the reaction versus [A].

13.14 The rate constant of a first-order reaction is 66 s^{-1}. What is the rate constant in units of minutes?

13.15 On which of the following quantities does the rate constant of a reaction depend? (a) concentrations of reactants, (b) nature of reactants, (c) temperature.

13.16 For each of the following pairs of reaction conditions, indicate which has the faster rate of formation of hydrogen gas: (a) sodium or potassium with water; (b) magnesium or iron with 1.0 M HCl; (c) magnesium rod or magnesium powder with 1.0 M HCl; (d) magnesium with 0.10 M HCl or magnesium with 1.0 M HCl.

PROBLEMS

13.17 The rate law for the reaction

$$NH_4^+(aq) + NO_2^-(aq) \longrightarrow N_2(g) + 2H_2O(l)$$

is given by rate = $k[NH_4^+][NO_2^-]$. At 25°C, the rate constant is $3.0 \times 10^{-4}/M \cdot s$. Calculate the rate of the reaction at this temperature if $[NH_4^+] = 0.26\ M$ and $[NO_2^-] = 0.080\ M$.

13.18 Consider the reaction

$$A + B \longrightarrow \text{products}$$

From the following data obtained at a certain temperature, determine the order of the reaction and calculate the rate constant:

[A] (M)	[B] (M)	Rate (M/s)
1.50	1.50	3.20×10^{-1}
1.50	2.50	3.20×10^{-1}
3.00	1.50	6.40×10^{-1}

13.19 Determine the overall orders of the reactions to which the following rate laws apply: (a) rate = $k[NO_2]^2$; (b) rate = k; (c) rate = $k[H_2][Br_2]^{\frac{1}{2}}$; (d) rate = $k[NO]^2[O_2]$.

13.20 A certain first-order reaction is 35.5 percent complete in 4.90 min at 25°C. What is its rate constant?

13.21 Consider the reaction

$$A \longrightarrow B$$

The rate of the reaction is 1.6×10^{-2} M/s when the concentration of A is 0.35 M. Calculate the rate constant if the reaction is (a) first order in A and (b) second order in A.

13.22 Starting with the data in Table 13.3, (a) deduce the rate law for the reaction, (b) calculate the rate constant, and (c) calculate the rate of the reaction at the time when $[F_2] = 0.010\ M$ and $[ClO_2] = 0.020\ M$.

13.23 At 500 K, butadiene gas converts to cyclobutene gas:

$$\underset{\text{butadiene }(g)}{CH_2{=}CH{-}CH{=}CH_2} \longrightarrow \underset{\text{cyclobutene }(g)}{\begin{matrix} CH_2{-}CH \\ | \quad\ \ \| \\ CH_2{-}CH \end{matrix}}$$

From the following rate data determine graphically the order of the reaction in butadiene and the rate constant.

Time from start (s)	Concentration of butadiene (M)
195	0.0162
604	0.0147
1246	0.0129
2180	0.0110
4140	0.0084
8135	0.0057

13.24 The following gas-phase reaction was studied at 290°C by observing the change in pressure as a function of time in a constant-volume vessel.

$$ClCO_2CCl_3(g) \longrightarrow 2COCl_2(g)$$

Time (s)	P (mmHg)
0	15.76
181	18.88
513	22.79
1164	27.08

Is the reaction first or second order in [$ClCO_2CCl_3$]?

13.25 Azomethane, $C_2H_6N_2$, decomposes at 297°C in the gas phase as follows:

$$C_2H_6N_2(g) \longrightarrow C_2H_6(g) + N_2(g)$$

From the data in the following table determine the order of the reaction in [$C_2H_6N_2$] and the rate constant.

Time (min)	[$C_2H_6N_2$] (*M*)
0	0.36
15	0.30
30	0.25
48	0.19
75	0.13

13.26 Consider the reaction

$$X + Y \longrightarrow Z$$

The following data are obtained at 360 K:

Initial rate of disappearance of X (*M*/s)	[X] (*M*)	[Y] (*M*)
0.147	0.10	0.50
0.127	0.20	0.30
4.064	0.40	0.60
1.016	0.20	0.60
0.508	0.40	0.30

(a) Determine the order of the reaction. (b) Determine the initial rate of disappearance of X when the concentration of X is 0.20 *M* and that of Y is 0.30 *M*.

13.27 The rate constant for the second-order reaction

$$2NOBr(g) \longrightarrow 2NO(g) + Br_2(g)$$

is 0.80/*M* · s at 10°C. Starting with a concentration of 0.086 *M* of NOBr, calculate its concentration after 22 s.

Half-Life

REVIEW QUESTIONS

13.28 Define the half-life of a reaction.

13.29 Write the equation relating the half-life of a first-order reaction to the rate constant. How does it differ from that of a second-order reaction?

13.30 For a first-order reaction, how long will it take for the concentration of reactant to fall to one-eighth its original value? Express your answer in terms of the half-life ($t_{\frac{1}{2}}$) and in terms of the rate constant *k*.

PROBLEMS

13.31 What is the half-life of a compound if 75 percent of a given sample of the compound decomposes in 60 min? Assume first-order kinetics.

13.32 The thermal decomposition of phosphine (PH_3) into phosphorus and molecular hydrogen is a first-order reaction:

$$4PH_3(g) \longrightarrow P_4(g) + 6H_2(g)$$

The half-life of the reaction is 35.0 s at 680°C. Calculate (a) the first-order rate constant for the reaction and (b) the time required for 95 percent of the phosphine to decompose.

13.33 The following reaction is second order in A:

$$A \longrightarrow \text{product}$$

At a certain temperature the second-order rate constant is 1.46/*M* · s. Calculate the half-life of the reaction if the initial concentration of A is 0.86 *M*.

13.34 The thermal decomposition of N_2O_5 obeys first-order kinetics. At 45°C, a plot of ln[N_2O_5] versus *t* gives a slope of -6.18×10^{-4} min^{-1}. What is the half-life of the reaction?

Activation Energy

REVIEW QUESTIONS

13.35 Define activation energy. What role does activation energy play in chemical kinetics?

13.36 Write the Arrhenius equation and define all terms.

13.37 Use the Arrhenius equation to show why the rate constant of a reaction (a) decreases with increasing activation energy and (b) increases with increasing temperature.

13.38 As we know, methane burns readily in oxygen in a highly exothermic reaction. Yet a mixture of methane and oxygen gas can be kept indefinitely without any apparent change. Explain.

13.39 Sketch a potential energy versus reaction progress plot for the following reactions:

(a) $S(s) + O_2(g) \longrightarrow SO_2(g)$ $\Delta H° = -296.06$ kJ

(b) $Cl_2(g) \longrightarrow Cl(g) + Cl(g)$ $\Delta H° = 242.7$ kJ

13.40 The reaction $H + H_2 \longrightarrow H_2 + H$ has been studied for many years. Sketch a potential energy versus reaction progress diagram for this reaction.

13.41 The reaction of G_2 with E_2 to form 2EG is exothermic, and the reaction of G_2 with X_2 to form 2XG is endothermic. The activation energy of the exothermic reaction is greater than that of the endothermic reaction. Sketch the potential energy profile diagrams for these two reactions on the same graph.

PROBLEMS

13.42 Variation of the rate constant with temperature for the first-order reaction

$$2N_2O_5(g) \longrightarrow 2N_2O_4(g) + O_2(g)$$

is given in the following table. Determine graphically the activation energy for the reaction.

T(K)	k(s^{-1})
273	7.87×10^3
298	3.46×10^5
318	4.98×10^6
338	4.87×10^7

13.43 The first-order rate constant for the reaction A → B has been measured at a series of temperatures:

k(s^{-1})	t(°C)
10.6	10
47.4	20
162	30
577	40

Determine graphically the activation energy for the reaction.

13.44 Given the same concentrations, the reaction

$$CO(g) + Cl_2(g) \longrightarrow COCl_2(g)$$

at 250°C is 1.50×10^3 times as fast as the same reaction at 150°C. Calculate the energy of activation for this reaction. Assume the frequency factor to remain constant.

13.45 For the reaction

$$NO(g) + O_3(g) \longrightarrow NO_2(g) + O_2(g)$$

the frequency factor A is 8.7×10^{12} s^{-1} and the activation energy is 63 kJ/mol. What is the rate constant for the reaction at 75°C?

13.46 The rate constant of a first-order reaction is 4.60×10^{-4} s^{-1} at 350°C. If the activation energy is 104 kJ/mol, calculate the temperature at which its rate constant is 8.80×10^{-4} s^{-1}.

13.47 The rate constant for the decomposition of the AB molecule at 25°C is 5.8×10^{-2} s^{-1}. Calculate the rate constant at 125°C if the activation energy is 52 kJ/mol.

13.48 The rate constants of some reactions double with every 10-degree rise in temperature. Assume a reaction takes place at 295 K and 305 K. What must the activation energy be for the rate constant to double as described?

13.49 The rate at which tree crickets chirp is 2.0×10^2 per minute at 27°C but only 39.6 per minute at 5°C. From these data, calculate the "energy of activation" for the chirping process. (*Hint:* The ratio of rates is equal to the ratio of rate constants.)

Reaction Mechanism

REVIEW QUESTIONS

13.50 What do we mean by the mechanism of a reaction?

13.51 What is an elementary step?

13.52 What is the molecularity of a reaction?

13.53 Reactions can be classified as unimolecular, bimolecular, and so on. Why are there no zero-molecular reactions?

13.54 Explain why termolecular reactions are rare.

13.55 What is the rate-determining step of a reaction? Give an everyday analogy to illustrate the meaning of the term "rate determining."

13.56 The equation for the combustion of ethane (C_2H_6) is

$$2C_2H_6 + 7O_2 \longrightarrow 4CO_2 + 6H_2O$$

Explain why it is unlikely that this equation also represents the elementary step for the reaction.

13.57 Which of the following species cannot be isolated in a reaction? activated complex, product, intermediate

PROBLEMS

13.58 The rate law for the reaction

$$2NO(g) + Cl_2(g) \longrightarrow 2NOCl(g)$$

is given by rate = $k[NO][Cl_2]$. (a) What is the order of the reaction? (b) A mechanism involving the following steps has been proposed for the reaction

$$NO(g) + Cl_2(g) \longrightarrow NOCl_2(g)$$

$$NOCl_2(g) + NO(g) \longrightarrow 2NOCl(g)$$

If this mechanism is correct, what does it imply about the relative rates of these two steps?

13.59 For the reaction $X_2 + Y + Z \rightarrow XY + XZ$ it is found that doubling the concentration of X_2 doubles the reaction rate, tripling the concentration of Y triples the rate, and doubling the concentration of Z has no effect. (a) What is the rate law for this reaction? (b) Why is it that the change in the concentration of Z has no effect on the rate? (c) Suggest a mechanism for the reaction that is consistent with the rate law.

13.60 Consider the following elementary step:

$$X + 2Y \longrightarrow XY_2$$

(a) Write a rate law for this reaction. (b) If the initial rate of formation of XY_2 is 3.8×10^{-3} M/s and the initial concentrations of X and Y are 0.26 M and 0.88 M, what is the rate constant of the reaction?

13.61 The rate of the reaction between H_2 and I_2 to form HI (discussed on p. 572) increases with the intensity of visible light. (a) Explain why this fact supports the two-step mechanism given. (The color of I_2 vapor is shown in Figure 11.47.) (b) Explain why the visible light has no effect on the formation of H atoms.

13.62 The rate law for the decomposition of ozone to molecular oxygen

$$2O_3(g) \longrightarrow 3O_2(g)$$

is

$$\text{rate} = k\frac{[O_3]^2}{[O_2]}$$

The mechanism proposed for this process is

$$O_3 \underset{k_{-1}}{\overset{k_1}{\rightleftharpoons}} O + O_2$$

$$O + O_3 \xrightarrow{k_2} 2O_2$$

Derive the rate law from these elementary steps. Clearly state the assumptions you use in the derivation. Explain why the rate decreases with increasing O_2 concentration.

13.63 The rate law for the reaction

$$2H_2(g) + 2NO(g) \longrightarrow N_2(g) + 2H_2O(g)$$

is rate $= k[H_2][NO]^2$. Which of the following mechanisms can be ruled out on the basis of the observed rate expression?

Mechanism I

$$H_2 + NO \longrightarrow H_2O + N \quad \text{(slow)}$$
$$N + NO \longrightarrow N_2 + O \quad \text{(fast)}$$
$$O + H_2 \longrightarrow H_2O \quad \text{(fast)}$$

Mechanism II

$$H_2 + 2NO \longrightarrow N_2O + H_2O \quad \text{(slow)}$$
$$N_2O + H_2 \longrightarrow N_2 + H_2O \quad \text{(fast)}$$

Mechanism III

$$2NO \rightleftharpoons N_2O_2 \quad \text{(fast equilibrium)}$$
$$N_2O_2 + H_2 \longrightarrow N_2O + H_2O \quad \text{(slow)}$$
$$N_2O + H_2 \longrightarrow N_2 + H_2O \quad \text{(fast)}$$

Catalysis

REVIEW QUESTIONS

13.64 How does a catalyst increase the rate of a reaction?

13.65 What are the characteristics of a catalyst?

13.66 A certain reaction is known to proceed slowly at room temperature. Is it possible to make the reaction proceed at a faster rate without changing the temperature?

13.67 Distinguish between homogeneous catalysis and heterogeneous catalysis. Describe three important industrial processes that utilize heterogeneous catalysis.

13.68 Are enzyme-catalyzed reactions examples of homogeneous or heterogeneous catalysis? Explain.

13.69 The concentrations of enzymes in cells are usually quite small. What is the biological significance of this fact?

PROBLEMS

13.70 Most reactions, including enzyme-catalyzed reactions, proceed faster at higher temperatures. However, for a given enzyme, the rate drops off abruptly at a certain temperature. Account for this behavior. (*Hint:* The three-dimensional structure of an enzyme molecule is maintained by relatively weak intermolecular forces within the molecule.)

13.71 Consider the following mechanism for the enzyme-catalyzed reaction:

$$E + S \underset{k_{-1}}{\overset{k_1}{\rightleftharpoons}} ES \quad \text{(fast equilibrium)}$$

$$ES \xrightarrow{k_2} E + P \quad \text{(slow)}$$

Derive an expression for the rate law of the reaction in terms of the concentrations of E and S.

Miscellaneous Problems

13.72 Suggest experimental means by which the rates of the following reactions could be followed:
(a) $CaCO_3(s) \longrightarrow CaO(s) + CO_2(g)$
(b) $Cl_2(g) + 2Br^-(aq) \longrightarrow Br_2(aq) + 2Cl^-(aq)$
(c) $C_2H_6(g) \longrightarrow C_2H_4(g) + H_2(g)$

13.73 List four factors that influence the rate of a reaction.

13.74 "The rate constant for the reaction

$$NO_2(g) + CO(g) \longrightarrow NO(g) + CO_2(g)$$

is $1.64 \times 10^{-6}/M \cdot s$." What is incomplete about this statement?

13.75 The bromination of acetone is acid-catalyzed:

$$CH_3COCH_3 + Br_2 \xrightarrow[\text{catalyst}]{H^+} CH_3COCH_2Br + H^+ + Br^-$$

The rate of disappearance of bromine was measured for several different concentrations of acetone, bromine, and H^+ ions at a certain temperature:

$[CH_3COCH_3]$	$[Br_2]$	$[H^+]$	Rate of disappearance of Br_2 (M/s)
(a) 0.30	0.050	0.050	5.7×10^{-5}
(b) 0.30	0.10	0.050	5.7×10^{-5}
(c) 0.30	0.050	0.10	1.2×10^{-4}
(d) 0.40	0.050	0.20	3.1×10^{-4}
(e) 0.40	0.050	0.050	7.6×10^{-5}

(a) What is the rate law for the reaction? (b) Determine the rate constant.

13.76 When methyl phosphate is heated in acid solution, it reacts with water:

$$CH_3OPO_3H_2 + H_2O \longrightarrow CH_3OH + H_3PO_4$$

If the reaction is carried out in water enriched with ^{18}O, the oxygen-18 isotope is found in the phosphoric acid product but not in the methanol. What does this tell us about the mechanism of the reaction?

13.77 The rate of the reaction

$$CH_3COOC_2H_5(aq) + H_2O(l) \longrightarrow CH_3COOH(aq) + C_2H_5OH(aq)$$

shows first-order characteristics—that is, rate $= k[CH_3COOC_2H_5]$—even though this is a second-order reaction (first order in $CH_3COOC_2H_5$ and first order in H_2O). Explain.

13.78 Explain why most metals used in catalysis are transition metals.

13.79 In a certain industrial process using a heterogeneous catalyst, the volume of the catalyst (in the shape of a sphere) is 10.0 cm^3. Calculate the surface area of the catalyst. If the sphere is broken down into eight spheres, each of which has a volume of 1.25 cm^3, what is the total surface area of the spheres? Which of the two geometric configurations of the catalyst is more effective? Explain. (The surface area of a sphere is $4\pi r^2$, where r is the radius of the sphere.)

13.80 The reaction $2A + 3B \rightarrow C$ is first order with respect to A and B. When the initial concentrations are $[A] = 1.6 \times 10^{-2}\ M$ and $[B] = 2.4 \times 10^{-3}\ M$, the rate is $4.1 \times 10^{-4}\ M/s$. Calculate the rate constant of the reaction.

13.81 The decomposition of N_2O to N_2 and O_2 is a first-order reaction. At 730°C the half-life of the reaction is 3.58×10^3 min. If the initial pressure of N_2O is 2.10 atm at 730°C, calculate the total gas pressure after one half-life. Assume that the volume remains constant.

13.82 The reaction $S_2O_8^{2-} + 2I^- \rightarrow 2SO_4^{2-} + I_2$ proceeds slowly in aqueous solution, but it can be catalyzed by the Fe^{3+} ion. Given that Fe^{3+} can oxidize I^- and Fe^{2+} can reduce $S_2O_8^{2-}$, write a plausible two-step mechanism for this reaction. Explain why the uncatalyzed reaction is slow.

13.83 The carbon-14 decay rate of a sample obtained from a young tree is 0.260 disintegration per second per gram of the sample. Another wood sample prepared from an object recovered at an archaeological excavation gives a decay rate of 0.186 disintegration per second per gram of the sample. What is the age of the object?

13.84 In a certain experiment, the rate of hydrogen peroxide decomposition, $2H_2O_2 \rightarrow 2H_2O + O_2$, is followed by titration against a potassium permanganate solution. At regular time intervals, equal volume of H_2O_2 is withdrawn to given the following data:

Time(min)	0	10.0	20.0	30.0
Volume of $KMnO_4$ used (mL)	22.8	13.8	8.25	5.00

Determine the order of the reaction and calculate the rate constant.

13.85 What are the units of the rate constant for a third-order reaction?

13.86 Consider the zero-order reaction $A \rightarrow B$. Sketch the following plots: (a) rate versus [A] and (b) [A] versus t.

13.87 The usefulness of radiocarbon dating is limited to objects no older than 50,000 years. What percent of the carbon-14, originally present in the sample, remains after this period of time?

13.88 Consider the two competing first-order reactions: (1) $A \rightarrow B$ (rate constant k_1) and (2) $A \rightarrow C$ (rate constant k_2). The activation energies of reactions (1) and (2) are 48.6 kJ/mol and 61.3 kJ/mol, respectively. If $k_1 = k_2$ at 300 K, at what temperature is the ratio k_1/k_2 equal to 2?

13.89 A flask contains a mixture of compounds A and B. Both compounds decompose by first-order kinetics. The half-lives are 50.0 min for A and 18.0 min for B. If the concentrations of A and B are equal initially, how long will it take for the concentration of A to be four times that of B?

13.90 The rate law for the reaction $2NO_2(g) \rightarrow N_2O_4(g)$ is rate $= k[NO_2]^2$. Which of the following changes will change the value of k? (a) The pressure of NO_2 is doubled. (b) The reaction is run in an organic solvent. (c) The volume of the container is doubled. (d) The temperature is decreased. (e) A catalyst is added to the container.

13.91 Referring to Example 13.5, explain how you would measure experimentally the partial pressure of azomethane as a function of time.

Opposite page: Equilibrium is a balancing act, much like the balance of forces holding up the leaning tower of Pisa shown here. In chemical equilibrium, the net concentrations of components of a reaction do not change with time because the forward and reverse rates are equal.

C H A P T E R 14

CHEMICAL EQUILIBRIUM

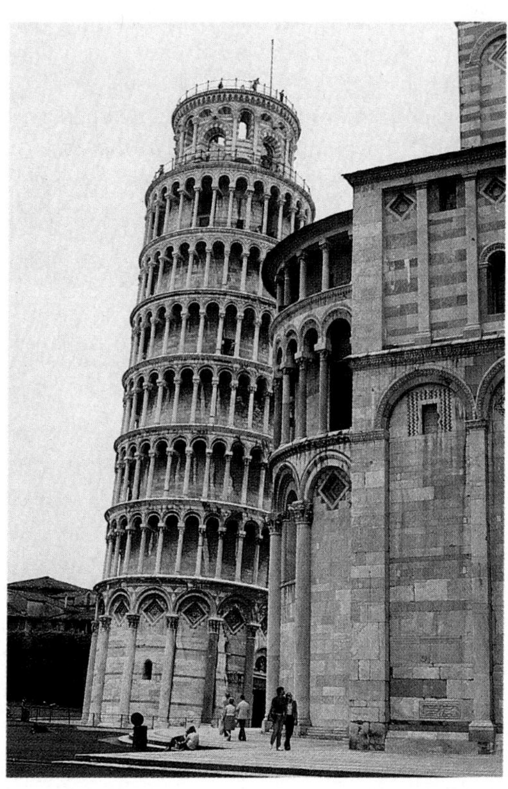

Few chemical reactions proceed in only one direction; most are, at least to some extent, reversible. At the start of a reversible reaction, the reaction proceeds toward the formation of products. As soon as some product molecules are formed, the reverse process—that is, the formation of reactant molecules from product molecules—begins to take place. When the rates of the forward and reverse processes are equal, a state of chemical equilibrium exists and the concentrations of reactants and products no longer change with time. We can obtain much useful information about a reaction after it has reached equilibrium.

This chapter discusses several different types of equilibrium, the meaning of the equilibrium constant, and factors that affect the attainment of equilibrium.

14.1 The Concept of Equilibrium

Equilibrium is *a state in which there are no observable changes as time goes by*. For example, when a chemical reaction has reached the equilibrium state, the concentrations of reactants or products remain unchanged with time. However, there is still much activity at the molecular level because constant conversion between the reactant molecules and product molecules continues to occur. Equilibrium is a dynamic situation. The concept of dynamic equilibrium is analogous to the movement of skiers on a busy day at a ski resort if the number of skiers carried up on the chair lift and the number coming down the slopes are the same. Thus while there is a constant transfer of skiers, the number of people at the top and the number at the bottom of the slope remain unchanged.

The vaporization of water in a closed container at a given temperature is another example of dynamic equilibrium. In this instance, the numbers of H_2O molecules leaving and returning to the liquid phase per second are equal:

$$H_2O(l) \rightleftharpoons H_2O(g)$$

where the double arrow ($\rightleftharpoons$) indicates that the process is reversible. The pressure of the water vapor measured at this temperature is called the equilibrium vapor pressure (see Section 11.8).

14.2 Chemical Equilibrium

Equilibrium between two phases of the same substance, such as the system containing liquid water and water vapor, is called ***physical equilibrium*** because *the changes that occur are physical processes*. The study of physical equilibrium leads to useful information such as the equilibrium vapor pressure. To chemists *the dynamic equilibrium that takes the form of a chemical reaction* is of particular interest. This kind of equilibrium is called *chemical equilibrium*.

Let us begin by considering the reversible reaction involving nitrogen dioxide (NO_2) and dinitrogen tetroxide (N_2O_4). The progress of the reaction

$$N_2O_4(g) \rightleftharpoons 2NO_2(g)$$

can be monitored easily because N_2O_4 is a colorless gas, whereas NO_2 has a dark brown color that makes it sometimes visible in polluted air. Suppose that a known amount of N_2O_4 is injected into an evacuated flask. Some brown color appears immediately, indicating the formation of NO_2 molecules. The color intensifies as the dissociation of N_2O_4 continues until eventually equilibrium is reached. Beyond that point, no further change in color is observed. By experiment we find that we can also reach the equilibrium state by starting with pure NO_2 or with a mixture of NO_2 and N_2O_4. In each case, we observe an initial change in color, due either to the formation of NO_2 (if the color intensifies) or to the depletion of NO_2 (if the color fades), and then the final state in which the color of NO_2 no longer changes. Depending on the temperature of the reacting system and on the initial amounts of NO_2 and N_2O_4, the concentrations of NO_2 and N_2O_4 at equilibrium differ from system to system (Figure 14.1).

Table 14.1 shows some experimental data for this reaction at 25°C. The gas concentrations are expressed in molarity, which can be calculated from the number of moles of the gases present initially and at equilibrium and the volume of the flask in liters.

Case 3: If K is neither very large nor very small compared with 1, then the quantities of reactants and products present at equilibrium will be comparable. Consider the following equilibrium system. At 830°C

$$CO(g) + H_2O(g) \rightleftharpoons H_2(g) + CO_2(g) \qquad K = \frac{[H_2][CO_2]}{[CO][H_2O]} = 5.10$$

If $[CO] = 0.200\ M$, $[H_2O] = 0.400\ M$, and $[H_2] = 0.300\ M$ at equilibrium, then

$$[CO_2] = \frac{(0.200)(0.400)}{(0.300)}(5.10) = 1.36\ M$$

14.3 Ways of Expressing Equilibrium Constants

The concept of equilibrium constants is extremely important in chemistry. As you will soon see, they are the key to solving a wide variety of stoichiometry problems involving equilibrium systems. To use equilibrium constants, we must know how to express them in terms of the reactant and product concentrations. Our only guidance is the law of mass action [Equation (14.2)]. However, because the concentrations of the reactants and products can be expressed in several types of units and because the reacting species are not always in the same phase, there may be more than one way to express the equilibrium constant for the *same* reaction. To begin with, we will consider reactions in which the reactants and products are in the same phase. Then we will look at another type of equilibrium reaction.

Homogeneous Equilibria

The term ***homogeneous equilibrium*** applies to reactions in which *all reacting species are in the same phase*. An example of homogeneous gas-phase equilibrium is the dissociation of N_2O_4. The equilibrium constant, as given in Equation (14.1), is

$$K_c = \frac{[NO_2]^2}{[N_2O_4]}$$

Note that the subscript we have added in K_c denotes that in this form of the equilibrium constant the concentrations of the reacting species are expressed in moles per liter. The concentrations of reactants and products in gaseous reactions can also be expressed in terms of their partial pressures. From Equation (5.7) we see that at constant temperature the pressure P of a gas is directly related to the concentration in mol/L of the gas; that is, $P = (n/V)RT$. Thus, for the equilibrium process

$$N_2O_4(g) \rightleftharpoons 2NO_2(g)$$

we can write

$$K_P = \frac{P_{NO_2}^2}{P_{N_2O_4}} \tag{14.3}$$

where P_{NO_2} and $P_{N_2O_4}$ are the equilibrium partial pressures (in atm) of NO_2 and N_2O_4, respectively. The subscript in K_P tells us that equilibrium concentrations are expressed in terms of pressure.

In general, K_c is not equal to K_P, since the partial pressures of reactants and products are not equal to their concentrations expressed in moles per liter. A simple relationship between K_P and K_c can be derived as follows. Let us consider the following equilibrium in the gas phase:

$$aA(g) \rightleftharpoons bB(g)$$

where a and b are stoichiometric coefficients. The equilibrium constant K_c is given by

$$K_c = \frac{[B]^b}{[A]^a}$$

and the expression for K_P is

$$K_P = \frac{P_B^b}{P_A^a}$$

where P_A and P_B are the partial pressures of A and B. Assuming ideal gas behavior,

$$P_A V = n_A RT$$

$$P_A = \frac{n_A RT}{V}$$

where V is the volume of the container in liters. Also

$$P_B V = n_B RT$$

$$P_B = \frac{n_B RT}{V}$$

Substituting these relations into the expression for K_P, we obtain

$$K_P = \frac{\left(\frac{n_B RT}{V}\right)^b}{\left(\frac{n_A RT}{V}\right)^a} = \frac{\left(\frac{n_B}{V}\right)^b}{\left(\frac{n_A}{V}\right)^a}(RT)^{b-a}$$

Now both n_A/V and n_B/V have the units of mol/L and can be replaced by [A] and [B], so that

$$K_P = \frac{[B]^b}{[A]^a}(RT)^{\Delta n}$$
$$= K_c(RT)^{\Delta n} \quad (14.4)$$

where $\Delta n = b - a$
$= \text{moles of gaseous products} - \text{moles of gaseous reactants}$

Since pressures are usually expressed in atm, the gas constant R is given by 0.0821 L · atm/K · mol (see Appendix 2), and we can write the relationship between K_P and K_c as

To use this equation, the pressures in K_P must be in atm.

$$K_P = K_c(0.0821\ T)^{\Delta n} \quad (14.5)$$

In general $K_P \neq K_c$ except in the special case when $\Delta n = 0$. In that case Equation (14.5) can be written as

Any number raised to the zero power is equal to 1.

$$K_P = K_c(0.0821\ T)^0$$
$$= K_c$$

As another example of homogeneous equilibrium, let's consider the ionization of acetic acid (CH_3COOH) in water:

$$CH_3COOH(aq) + H_2O(l) \rightleftharpoons CH_3COO^-(aq) + H_3O^+(aq)$$

The equilibrium constant is

$$K_c' = \frac{[CH_3COO^-][H_3O^+]}{[CH_3COOH][H_2O]}$$

In 1 L, or 1000 g, of water there are 1000 g/(18.02 g/mol), or 55.5 moles, of water. Therefore, the "concentration" of water, or $[H_2O]$, is 55.5 mol/L, or 55.5 *M*. This value can be used for water in most aqueous solutions.

However, when the acid concentration is low ($\lesssim 1\ M$) and/or the equilibrium constant is small ($\lesssim 1$), as it is in this case, the amount of water consumed in this process is negligible compared with the total amount of water present. Thus we may treat $[H_2O]$ as a constant and rewrite the equilibrium constant as

$$K_c = \frac{[CH_3COO^-][H_3O^+]}{[CH_3COOH]}$$

where

$$K_c = K_c'[H_2O]$$

Note that it is general practice not to include units for the equilibrium constant. In thermodynamics, *K* is defined to have no units since every concentration (molarity) or pressure (atm) term is actually a ratio to a standard value, which is 1 *M* or 1 atm. This procedure eliminates all units but does not alter the numerical parts of the concentration or pressure. Consequently, *K* has no units. We will extend this practice to acid-base equilibria (to be discussed in Chapter 16) and to solubility equilibria (to be discussed in Chapter 17).

The following examples illustrate the procedure for writing equilibrium constant expressions and calculating equilibrium constants and equilibrium concentrations.

EXAMPLE 14.1

Write expressions for K_c, and K_P if applicable, for the following reversible reactions at equilibrium:

(a) $HF(aq) + H_2O(l) \rightleftharpoons H_3O^+(aq) + F^-(aq)$
(b) $2NO(g) + O_2(g) \rightleftharpoons 2NO_2(g)$
(c) $2H_2S(g) + 3O_2(g) \rightleftharpoons 2H_2O(g) + 2SO_2(g)$
(d) $CH_3COOH(aq) + C_2H_5OH(aq) \rightleftharpoons CH_3COOC_2H_5(aq) + H_2O(l)$

Answer

(a) Since there are no gases present, K_P does not apply and we have only K_c':

$$K_c' = \frac{[H_3O^+][F^-]}{[HF][H_2O]}$$

HF is a weak acid, so that the amount of water consumed in acid ionizations is negligible compared with the total amount of water present as solvent. Thus we can rewrite the equilibrium constant as

$$K_c = \frac{[H_3O^+][F^-]}{[HF]}$$

(b) $$K_c = \frac{[NO_2]^2}{[NO]^2[O_2]} \qquad K_P = \frac{P_{NO_2}^2}{P_{NO}^2 P_{O_2}}$$

(c) $$K_c = \frac{[H_2O]^2[SO_2]^2}{[H_2S]^2[O_2]^3} \qquad K_P = \frac{P_{H_2O}^2 P_{SO_2}^2}{P_{H_2S}^2 P_{O_2}^3}$$

(d) The equilibrium constant K'_c is given by

$$K'_c = \frac{[CH_3COOC_2H_5][H_2O]}{[CH_3COOH][C_2H_5OH]}$$

Because the water produced in the reaction is negligible compared with the water solvent, the concentration of water does not change. Thus we can write the new equilibrium constant as

$$K_c = \frac{[CH_3COOC_2H_5]}{[CH_3COOH][C_2H_5OH]}$$

Similar problems: 14.7, 14.9.

EXAMPLE 14.2

The following equilibrium process has been studied at 230°C:

$$2NO(g) + O_2(g) \rightleftharpoons 2NO_2(g)$$

In one experiment the concentrations of the reacting species at equilibrium are found to be $[NO] = 0.0542\ M$, $[O_2] = 0.127\ M$, and $[NO_2] = 15.5\ M$. Calculate the equilibrium constant (K_c) of the reaction at this temperature.

Answer

The equilibrium constant is given by

$$K_c = \frac{[NO_2]^2}{[NO]^2[O_2]}$$

Substituting the concentrations, we find that

$$K_c = \frac{(15.5)^2}{(0.0542)^2(0.127)} = 6.44 \times 10^5$$

Note that K_c is given without units.

Similar problem: 14.13.

EXAMPLE 14.3

The equilibrium constant K_P for the reaction

$$PCl_5(g) \rightleftharpoons PCl_3(g) + Cl_2(g)$$

is found to be 1.05 at 250°C. If the equilibrium partial pressures of PCl_5 and PCl_3 are 0.875 atm and 0.463 atm, respectively, what is the equilibrium partial pressure of Cl_2 at 250°C?

Answer

First, we write K_P in terms of the partial pressures of the reacting species.

$$K_P = \frac{P_{PCl_3}P_{Cl_2}}{P_{PCl_5}}$$

Knowing the partial pressures, we write

$$1.05 = \frac{(0.463)(P_{Cl_2})}{(0.875)}$$

or

$$P_{Cl_2} = \frac{(1.05)(0.875)}{(0.463)} = 1.98 \text{ atm}$$

Note that we have added atm as the unit for P_{Cl_2}.

Similar problem: 14.16.

EXAMPLE 14.4

The equilibrium constant (K_c) for the reaction

$$N_2O_4(g) \rightleftharpoons 2NO_2(g)$$

is 4.63×10^{-3} at 25°C. What is the value of K_P at this temperature?

Answer

From Equation (14.5) we write

$$K_P = K_c(0.0821\ T)^{\Delta n}$$

Since $T = 298$ K and $\Delta n = 2 - 1 = 1$, we have

$$K_P = (4.63 \times 10^{-3})(0.0821 \times 298)$$
$$= 0.113$$

Note that K_P, like K_c, is treated as a dimensionless quantity. This example shows that we can get quite a different value for the equilibrium constant for the same reaction, depending on whether we express the concentrations in moles per liter or in atmospheres.

Similar problems: 14.14, 14.17.

Heterogeneous Equilibria

A reversible reaction involving reactants and products that are in different phases leads to a ***heterogeneous equilibrium.*** For example, when calcium carbonate is heated in a closed vessel, the following equilibrium is attained:

$$CaCO_3(s) \rightleftharpoons CaO(s) + CO_2(g)$$

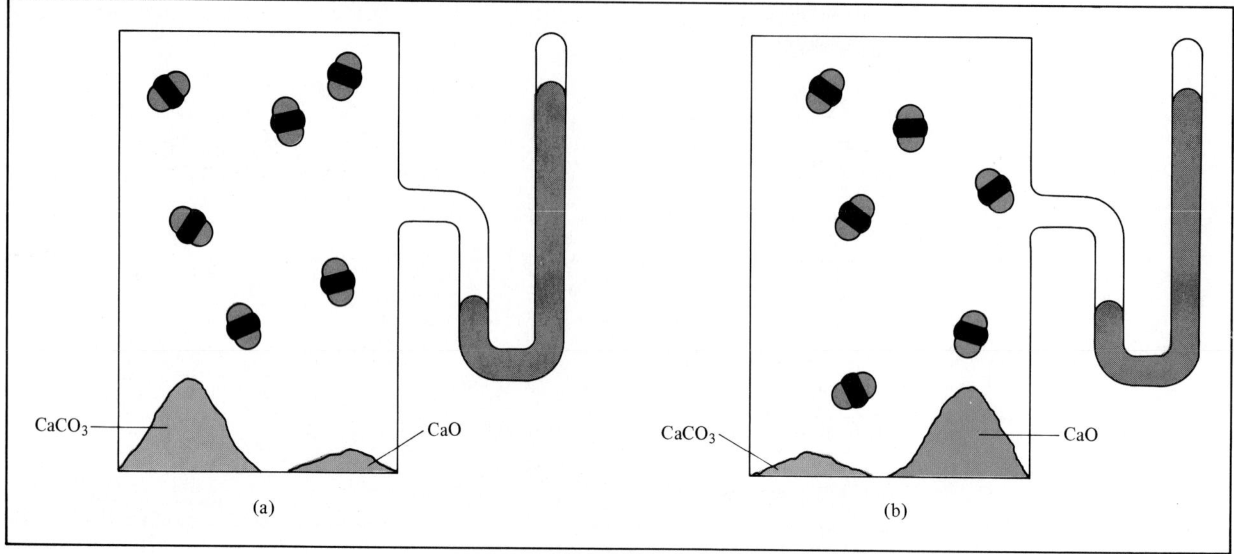

Figure 14.2 *Regardless of the amount of $CaCO_3$ and CaO present, the equilibrium pressure of CO_2 is the same in (a) and (b) at the same temperature.*

The two solids and one gas constitute three separate phases. At equilibrium, we might expect the equilibrium constant to be

$$K_c' = \frac{[CaO][CO_2]}{[CaCO_3]} \tag{14.6}$$

But how do we express the concentration of a solid substance? Because both $CaCO_3$ and CaO are pure solids, their "concentrations" do not change as the reaction proceeds. This follows from the fact that the molar concentration of a pure solid (or a pure liquid) is a constant at a given temperature; it does not depend on the quantity of the substance present. For example, the molar concentration at 20°C of copper (density: 8.96 g/cm^3) is the same, whether we have 1 gram or 1 ton of the metal:

$$\frac{8.96 \text{ g}}{\text{cm}^3} \times \frac{1 \text{ mol}}{63.55 \text{ g}} = 0.141 \text{ mol/cm}^3 = 141 \text{ mol/L}$$

From the above discussion we can express the equilibrium constant for the decomposition of $CaCO_3$ in a different way. Rearranging Equation (14.6), we obtain

$$\frac{[CaCO_3]}{[CaO]}K_c' = [CO_2] \tag{14.7}$$

Since both $[CaCO_3]$ and [CaO] are constants and K_c' is an equilibrium constant, all the terms on the left-hand side of the equation are constants. We can simplify the equation by writing

$$\frac{[CaCO_3]}{[CaO]}K_c' = K_c = [CO_2]$$

where K_c, the "new" equilibrium constant, is now conveniently expressed in terms of a single concentration, that of CO_2. Keep in mind that the value of K_c does not depend on how much $CaCO_3$ and CaO are present, as long as some of each is present at equilibrium (Figure 14.2).

Alternatively, we can express the equilibrium constant as

$$K_P = P_{CO_2} \tag{14.8}$$

The equilibrium constant in this case is numerically equal to the pressure of the CO_2 gas, an easily measurable quantity.

What has been said about pure solids also applies to liquids. Thus if a pure liquid is a reactant or a product in a reaction, we can treat its concentration as constant and omit it from the equilibrium constant expression.

Reactions involving heterogeneous equilibria are shown in the following examples.

EXAMPLE 14.5

Write the equilibrium constant expression K_c, and K_P if applicable, for each of the following heterogeneous systems:

(a) $(NH_4)_2Se(s) \rightleftharpoons 2NH_3(g) + H_2Se(g)$
(b) $AgCl(s) \rightleftharpoons Ag^+(aq) + Cl^-(aq)$
(c) $P_4(s) + 6Cl_2(g) \rightleftharpoons 4PCl_3(l)$

Answer

(a) The equilibrium constant is given by

$$K_c' = \frac{[NH_3]^2[H_2Se]}{[(NH_4)_2Se]}$$

However, since $(NH_4)_2Se$ is a solid, we write the new equilibrium constant as

$$K_c = [NH_3]^2[H_2Se]$$

where $K_c = K_c'[(NH_4)_2Se]$. Alternatively, we can express the equilibrium constant K_P in terms of the partial pressures of NH_3 and H_2Se:

$$K_P = P_{NH_3}^2 P_{H_2Se}$$

(b)

$$K_c' = \frac{[Ag^+][Cl^-]}{[AgCl]}$$

$$K_c = [Ag^+][Cl^-]$$

Again, we have incorporated [AgCl] into K_c because AgCl is a solid.

(c) The equilibrium constant is

$$K_c' = \frac{[PCl_3]^4}{[P_4][Cl_2]^6}$$

Since pure solids and pure liquids do not appear in the equilibrium constant expression, we write

$$K_c = \frac{1}{[Cl_2]^6}$$

Alternatively, we can express the equilibrium constant in terms of the pressure of Cl_2:

$$K_P = \frac{1}{P_{Cl_2}^6}$$

Similar problems: 14.8, 14.9(d), (f), (g).

EXAMPLE 14.6

Consider the following heterogeneous equilibrium:

$$CaCO_3(s) \rightleftharpoons CaO(s) + CO_2(g)$$

At 800°C, the pressure of CO_2 is 0.236 atm. Calculate (a) K_P and (b) K_c for the reaction at this temperature.

Answer

(a) Using Equation (14.8), we write

$$K_P = P_{CO_2}$$
$$= 0.236$$

(b) From Equation (14.5), we know

$$K_P = K_c(0.0821\ T)^{\Delta n}$$

$T = 800 + 273 = 1073\ K$ and $\Delta n = 1$, so we substitute these in the equation and obtain

$$0.236 = K_c(0.0821 \times 1073)$$
$$K_c = 2.68 \times 10^{-3}$$

Similar problem: 14.18.

Multiple Equilibria

The reactions we have considered so far are all relatively simple. Let us now consider a more complicated situation in which the product molecules in one equilibrium system are involved in a second equilibrium process:

$$A + B \rightleftharpoons C + D$$
$$C + D \rightleftharpoons E + F$$

The products formed in the first reaction, C and D, react further to form products E and F. At equilibrium we can write two separate equilibrium constants:

$$K_c' = \frac{[C][D]}{[A][B]}$$

and

$$K_c'' = \frac{[E][F]}{[C][D]}$$

The overall reaction is given by the sum of the two reactions

$$\begin{array}{lll} & A + B \rightleftharpoons C + D & K_c' \\ & C + D \rightleftharpoons E + F & K_c'' \\ \hline \text{Overall reaction:} & A + B \rightleftharpoons E + F & K_c \end{array}$$

and the equilibrium constant K_c for the overall reaction is

$$K_c = \frac{[E][F]}{[A][B]}$$

We obtain this same expression if we take the product of the expressions for K'_c and K''_c:

$$K'_cK''_c = \frac{[C][D]}{[A][B]} \times \frac{[E][F]}{[C][D]} = \frac{[E][F]}{[A][B]}$$

Therefore we arrive at

$$K_c = K'_cK''_c \tag{14.9}$$

We can now make an important statement about multiple equilibria: *If a reaction can be expressed as the sum of two or more reactions, the equilibrium constant for the overall reaction is given by the product of the equilibrium constants of the individual reactions.*

Among the many known examples of multiple equilibria is the ionization of diprotic acids in aqueous solution. The following equilibrium constants have been determined for carbonic acid (H_2CO_3) at 25°C:

$$H_2CO_3(aq) \rightleftharpoons H^+(aq) + HCO_3^-(aq) \qquad K'_c = \frac{[H^+][HCO_3^-]}{[H_2CO_3]} = 4.2 \times 10^{-7}$$

$$HCO_3^-(aq) \rightleftharpoons H^+(aq) + CO_3^{2-}(aq) \qquad K''_c = \frac{[H^+][CO_3^{2-}]}{[HCO_3^-]} = 4.8 \times 10^{-11}$$

The overall reaction is the sum of these two reactions

$$H_2CO_3(aq) \rightleftharpoons 2H^+(aq) + CO_3^{2-}(aq)$$

and the corresponding equilibrium constant is given by

$$K_c = \frac{[H^+]^2[CO_3^{2-}]}{[H_2CO_3]}$$

Using Equation (14.9) we arrive at

$$\begin{aligned} K_c &= K'_cK''_c \\ &= (4.2 \times 10^{-7})(4.8 \times 10^{-11}) \\ &= 2.0 \times 10^{-17} \end{aligned}$$

The Form of *K* and the Equilibrium Equation

Before closing this section, we should note the two following important rules about writing equilibrium constants:

The reciprocal of x is $1/x$.

- When the equation for a reversible reaction is written in the opposite direction, the equilibrium constant becomes the reciprocal of the original equilibrium constant. Thus if we write the NO_2–N_2O_4 equilibrium as

  $$N_2O_4(g) \rightleftharpoons 2NO_2(g)$$

  then

  $$K_c = \frac{[NO_2]^2}{[N_2O_4]} = 4.63 \times 10^{-3}$$

  However, we can represent the equilibrium equally well as

  $$2NO_2(g) \rightleftharpoons N_2O_4(g)$$

and the equilibrium constant is now given by

$$K_c' = \frac{[N_2O_4]}{[NO_2]^2} = \frac{1}{K_c} = \frac{1}{4.63 \times 10^{-3}} = 216$$

You can see that $K_c = 1/K_c'$ or $K_cK_c' = 1.00$. Either K_c or K_c' is a valid equilibrium constant, but it is meaningless to say that the equilibrium constant for the NO_2–N_2O_4 system is 4.63×10^{-3}, or 216, unless we also specify how the equilibrium equation is written.

- The value of K also depends on how the equilibrium equation is balanced. Consider the following two ways of describing the same equilibrium:

$$\tfrac{1}{2}N_2O_4(g) \rightleftharpoons NO_2(g) \qquad K_c' = \frac{[NO_2]}{[N_2O_4]^{\frac{1}{2}}}$$

$$N_2O_4(g) \rightleftharpoons 2NO_2(g) \qquad K_c = \frac{[NO_2]^2}{[N_2O_4]}$$

Looking at the exponents we see that $K_c' = \sqrt{K_c}$. In Table 14.1 we find $K_c = 4.63 \times 10^{-3}$; therefore $K_c' = 0.0680$.

According to the law of mass action, each concentration term in the equilibrium constant expression is raised to a power equal to its stoichiometric coefficient. Thus if you double a chemical equation throughout, the corresponding equilibrium constant will be the square of the original value; if you triple the equation, the equilibrium constant will be the cube of the original value, and so on. The NO_2–N_2O_4 example illustrates once again the need to write the particular chemical equation when quoting the numerical value of an equilibrium constant.

The following example deals with the relationship between the equilibrium constants for differently balanced equations describing the same reaction.

EXAMPLE 14.7

The reaction for the production of ammonia can be written in a number of ways:

(a) $N_2(g) + 3H_2(g) \rightleftharpoons 2NH_3(g)$
(b) $\tfrac{1}{2}N_2(g) + \tfrac{3}{2}H_2(g) \rightleftharpoons NH_3(g)$
(c) $\tfrac{1}{3}N_2(g) + H_2(g) \rightleftharpoons \tfrac{2}{3}NH_3(g)$

Write the equilibrium constant expression for each formulation. (Express the concentrations of the reacting species in mol/L.)

(d) How are the equilibrium constants related to one another?

Answer

(a)
$$K_a = \frac{[NH_3]^2}{[N_2][H_2]^3}$$

(b)
$$K_b = \frac{[NH_3]}{[N_2]^{\frac{1}{2}}[H_2]^{\frac{3}{2}}}$$

(c)

$$K_c = \frac{[NH_3]^{\frac{2}{3}}}{[N_2]^{\frac{1}{3}}[H_2]}$$

(d)

$$K_a = K_b^2$$

$$K_a = K_c^3$$

$$K_b^2 = K_c^3 \quad \text{or} \quad K_b = K_c^{\frac{3}{2}}$$

Similar problems: 14.15, 14.22.

Summary of the Rules for Writing Equilibrium Constant Expressions

- The concentrations of the reacting species in the condensed phase are expressed in mol/L; in the gaseous phase, the concentrations can be expressed in mol/L or in atm. K_c is related to K_P by a simple equation [Equation (14.5)].
- The concentrations of pure solids, pure liquids (in heterogeneous equilibria), and solvents (in homogeneous equilibria) do not appear in the equilibrium constant expressions.
- The equilibrium constant (K_c or K_P) is treated as a dimensionless quantity.
- In quoting a value for the equilibrium constant, we must specify the balanced equation and the temperature.
- If a reaction can be expressed as the sum of two or more reactions, the equilibrium constant for the overall reaction is given by the product of the equilibrium constants of the individual reactions.

14.4 Relationship between Chemical Kinetics and Chemical Equilibrium

We have seen that the quantity K, defined in Equation (14.2), is always constant at a given temperature regardless of the variations in individual equilibrium concentrations (review Table 14.1). It is appropriate to ask why this is so. The answer and insight into the equilibrium process can be gained by considering the kinetics of chemical reactions.

Let us suppose that the following reversible reaction occurs via a mechanism consisting of a single *elementary step* (see p. 570) in both the forward and reverse directions:

$$A + 2B \underset{k_r}{\overset{k_f}{\rightleftharpoons}} AB_2$$

The forward rate is given by

$$\text{rate}_f = k_f[A][B]^2$$

and the reverse rate is given by

$$\text{rate}_r = k_r[AB_2]$$

where k_f and k_r are the rate constants for the forward and reverse directions. At equilibrium, when no net changes occur, the two rates must be equal:

$$\text{rate}_f = \text{rate}_r$$

or

$$k_f[A][B]^2 = k_r[AB_2]$$

$$\frac{k_f}{k_r} = \frac{[AB_2]}{[A][B]^2}$$

Since both k_f and k_r are constants at a given temperature, their ratio is also a constant, and equal to the equilibrium constant K_c.

$$\frac{k_f}{k_r} = K_c = \frac{[AB_2]}{[A][B]^2}$$

We can now understand why K is always a constant regardless of the equilibrium concentrations of the reacting species: It is always equal to k_f/k_r, the quotient of two quantities that are themselves constant at a given temperature. Because rate constants are temperature-dependent [see Equation (13.8)], it follows that the equilibrium constant must also change with temperature.

Now suppose the same reaction has a mechanism with more than one elementary step. Suppose it occurs via a two-step mechanism as follows:

Step 1: $$2B \underset{k'_r}{\overset{k'_f}{\rightleftharpoons}} B_2$$

Step 2: $$A + B_2 \underset{k''_r}{\overset{k''_f}{\rightleftharpoons}} AB_2$$

Overall reaction: $$A + 2B \rightleftharpoons AB_2$$

This is an example of multiple equilibria, discussed in Section 14.3. We write the expressions for the equilibrium constants:

$$K' = \frac{k'_f}{k'_r} = \frac{[B_2]}{[B]^2} \tag{14.10}$$

$$K'' = \frac{k''_f}{k''_r} = \frac{[AB_2]}{[A][B_2]} \tag{14.11}$$

Multiplying Equation (14.10) by Equation (14.11), we get

$$K'K'' = \frac{[B_2][AB_2]}{[B]^2[A][B_2]} = \frac{[AB_2]}{[A][B]^2}$$

$$K_c = \frac{[AB_2]}{[A][B]^2}$$

Since both K' and K'' are constants, K_c is also a constant. This result lets us generalize our treatment of the reaction

$$aA + bB \rightleftharpoons cC + dD$$

Regardless of whether this reaction occurs via a single-step or a multistep mechanism, we can write the equilibrium constant expression according to the law of mass action shown in Equation (14.2):

$$K = \frac{[C]^c[D]^d}{[A]^a[B]^b}$$

In summary, we see that from the chemical kinetics viewpoint, the equilibrium constant of a reaction can be expressed as a ratio of the rate constants of the forward and reverse reactions. This analysis explains why the equilibrium constant is a constant and why its value changes with temperature.

14.5 What Does the Equilibrium Constant Tell Us?

We have seen that the equilibrium constant for a given reaction can be calculated from known equilibrium concentrations. Once we know the value of the equilibrium constant, we can use Equation (14.2) to calculate unknown equilibrium concentrations—remembering, of course, that the equilibrium constant is constant only if the temperature does not change. In general, the equilibrium constant helps us to predict the direction in which a reaction mixture will proceed to achieve equilibrium and to calculate the concentrations of reactants and products once equilibrium has been reached. These uses of the equilibrium constant will be explored in this section.

Predicting the Direction of a Reaction

The equilibrium constant K_c for the reaction

$$H_2(g) + I_2(g) \rightleftharpoons 2HI(g)$$

is 54.3 at 430°C. Suppose that in a certain experiment we place 0.243 mole of H_2, 0.146 mole of I_2, and 1.98 moles of HI all in a 1-L container at 430°C. Will there be a net reaction to form more H_2 and I_2 or more HI? Inserting the starting concentrations in the equilibrium constant expression, we write

$$\frac{[HI]_0^2}{[H_2]_0[I_2]_0} = \frac{(1.98)^2}{(0.243)(0.146)} = 111$$

where the subscript 0 indicates initial concentrations. Because the quotient $[HI]_0^2/[H_2]_0[I_2]_0$ is greater than K_c, this system is not at equilibrium. Consequently, some of the HI will react to form more H_2 and I_2 (decreasing the value of the quotient). Thus the net reaction proceeds from right to left to reach equilibrium.

Substituting the initial concentrations into the equilibrium constant expression gives us a quantity called the ***reaction quotient (Q_c)***. To determine in which direction a reaction will proceed to achieve equilibrium, we compare the values of Q_c and K_c. The three possible cases are as follows:

- $Q_c > K_c$ The ratio of initial concentrations of products to reactants is too large. To reach equilibrium, products must be converted to reactants. The system proceeds from right to left (consuming products, forming reactants) to reach equilibrium.
- $Q_c = K_c$ The initial concentrations are equilibrium concentrations. The system is at equilibrium.
- $Q_c < K_c$ The ratio of initial concentrations of products to reactants is too small. To reach equilibrium, reactants must be converted to products. The system proceeds from left to right (consuming reactants, forming products) to reach equilibrium.

A comparison of K_c with Q_c for a reaction is shown in the following example:

EXAMPLE 14.8

At the start of a reaction, there are 0.249 mol N_2, 3.21×10^{-2} mol H_2, and 6.42×10^{-4} mol NH_3 in a 3.50-L reaction vessel at 200°C. If the equilibrium constant (K_c) for the reaction

$$N_2(g) + 3H_2(g) \rightleftharpoons 2NH_3(g)$$

is 0.65 at this temperature, decide whether the system is at equilibrium. If it is not, predict which way the net reaction will proceed.

Answer

The initial concentrations of the reacting species are:

$$[N_2]_0 = \frac{0.249 \text{ mol}}{3.50 \text{ L}} = 0.0711\ M$$

$$[H_2]_0 = \frac{3.21 \times 10^{-2} \text{ mol}}{3.50 \text{ L}} = 9.17 \times 10^{-3}\ M$$

$$[NH_3]_0 = \frac{6.42 \times 10^{-4} \text{ mol}}{3.50 \text{ L}} = 1.83 \times 10^{-4}\ M$$

Next we write

$$\frac{[NH_3]_0^2}{[N_2]_0[H_2]_0^3} = \frac{(1.83 \times 10^{-4})^2}{(0.0711)(9.17 \times 10^{-3})^3} = 0.611 = Q_c$$

Since Q_c is smaller than K_c (0.65), the system is not at equilibrium. The net result will be an increase in the concentration of NH_3 and a decrease in the concentrations of N_2 and H_2. That is, the net reaction will proceed from left to right until equilibrium is reached.

Similar problems: 14.38, 14.40.

Calculating Equilibrium Concentrations

If we know the equilibrium constant for a particular reaction, we can calculate the concentrations in the equilibrium mixture from a knowledge of the initial concentrations. Depending on the information given, the calculation may be straightforward or complex. In the most common situation, only the initial reactant concentrations are given. Let us consider the following system, which has an equilibrium constant (K_c) of 24.0 at a certain temperature:

$$A \rightleftharpoons B$$

Suppose that A is initially present at a concentration of 0.850 mol/L. How do we calculate the concentrations of A and B at equilibrium? From the stoichiometry of the reaction we see that for every mole of A converted, 1 mole of B is formed. Let x be the equilibrium concentration of B in mol/L; therefore, the equilibrium concentration of A must be $(0.850 - x)$ mol/L. It is useful to summarize the changes in concentration as follows:

	A	$\rightleftharpoons$	B
Initial:	0.850 M		0 M
Change:	$-x$ M		$+x$ M
Equilibrium:	$(0.850 - x)$ M		x M

A positive (+) change represents an increase and a negative (−) change indicates a decrease in concentration at equilibrium. Next we set up the equilibrium constant expression

$$K_c = \frac{[B]}{[A]}$$

$$24.0 = \frac{x}{0.850 - x}$$

$$x = 0.816\ M$$

Having solved for x, we calculate the equilibrium concentrations of A and B as follows:

$$[A] = (0.850 - 0.816)\ M = 0.034\ M$$
$$[B] = 0.816\ M$$

We summarize our approach to solving equilibrium constant problems as follows:

1. Express the equilibrium concentrations of all species in terms of the initial concentrations and a single unknown x, which represents the change in concentration.
2. Write the equilibrium constant expression in terms of the equilibrium concentrations. Knowing the value of the equilibrium constant, solve for x.
3. Having solved for x, calculate the equilibrium concentrations of all species.

The following examples illustrate the application of the three-step approach.

EXAMPLE 14.9

A mixture of 0.500 mol H_2 and 0.500 mol I_2 was placed in a 1.00-L stainless-steel flask at 430°C. Calculate the concentrations of H_2, I_2, and HI at equilibrium. The equilibrium constant K_c for the reaction $H_2(g) + I_2(g) \rightleftharpoons 2HI(g)$ is 54.3 at this temperature.

Answer

Step 1: The stoichiometry of the reaction is 1 mol H_2 reacting with 1 mol I_2 to yield 2 mol HI. Let x be the depletion in concentration (mol/L) of either H_2 or I_2 at equilibrium. It follows that the equilibrium concentration of HI must be $2x$. We summarize the changes in concentrations as follows:

	H_2	+	I_2	$\rightleftharpoons$	2HI
Initial:	0.500 M		0.500 M		0.000 M
Change:	$-x$ M		$-x$ M		$+2x$ M
Equilibrium:	$(0.500 - x)$ M		$(0.500 - x)$ M		$2x$ M

Step 2: The equilibrium constant is given by

$$K_c = \frac{[HI]^2}{[H_2][I_2]}$$

Substituting $$54.3 = \frac{(2x)^2}{(0.500 - x)(0.500 - x)}$$

Taking the square root of both sides, we get

$$7.37 = \frac{2x}{0.500 - x}$$

$$x = 0.393\ M$$

You can check your answers by calculating K_c using the equilibrium concentrations.

Step 3: At equilibrium, the concentrations are

$$[H_2] = (0.500 - 0.393)\ M = 0.107\ M$$

$$[I_2] = (0.500 - 0.393)\ M = 0.107\ M$$

$$[HI] = 2 \times 0.393\ M = 0.786\ M$$

Similar problem: 14.49.

EXAMPLE 14.10

Following the procedure in Example 14.8, convince yourself that the initial concentrations do not correspond to the equilibrium concentrations.

For the same reaction and temperature as in Example 14.9, suppose that the initial concentrations of H_2, I_2, and HI are 0.00623 *M*, 0.00414 *M*, and 0.0224 *M*, respectively. Calculate the concentrations of these species at equilibrium.

Answer

Step 1: Let x be the depletion in concentration (mol/L) for H_2 and I_2 at equilibrium. From the stoichiometry of the reaction it follows that the increase in concentration for HI must be $2x$. Next we write

	H_2	+	I_2	$\rightleftharpoons$	2HI
Initial:	0.00623 *M*		0.00414 *M*		0.0224 *M*
Change:	$-x$ *M*		$-x$ *M*		$+2x$ *M*
Equilibrium:	$(0.00623 - x)$ *M*		$(0.00414 - x)$ *M*		$(0.0224 + 2x)$ *M*

Step 2: The equilibrium constant is

$$K_c = \frac{[HI]^2}{[H_2][I_2]}$$

Substituting, we get $$54.3 = \frac{(0.0224 + 2x)^2}{(0.00623 - x)(0.00414 - x)}$$

It is not possible to solve this equation by the square root shortcut, as the starting concentrations $[H_2]$ and $[I_2]$ are unequal. Instead, we must first carry out the multiplications

$$54.3(2.58 \times 10^{-5} - 0.0104x + x^2) = 5.02 \times 10^{-4} + 0.0896x + 4x^2$$

Collecting terms, we get

$$50.3x^2 - 0.654x + 8.98 \times 10^{-4} = 0$$

This is a quadratic equation of the form $ax^2 + bx + c = 0$. The solution for a quadratic equation (see Appendix 4) is

$$x = \frac{-b \pm \sqrt{b^2 - 4ac}}{2a}$$

Here we have $a = 50.3$, $b = -0.654$, and $c = 8.98 \times 10^{-4}$, so that

$$x = \frac{0.654 \pm \sqrt{(-0.654)^2 - 4(50.3)(8.98 \times 10^{-4})}}{2 \times 50.3}$$

$$x = 0.0114\ M \qquad \text{or} \qquad x = 0.00156\ M$$

The first solution is physically impossible since the amounts of H_2 and I_2 reacted would be more than those originally present. The second solution gives the correct answer. Note that in solving quadratic equations of this type, one answer is always physically impossible, so the choice of which value to use for x is easy to make.

Using the initial concentrations you can calculate Q_c and predict which species will increase and which species will decrease in concentration at equilibrium.

Step 3: At equilibrium, the concentrations are

$$[H_2] = (0.00623 - 0.00156)\ M = 0.00467\ M$$

$$[I_2] = (0.00414 - 0.00156)\ M = 0.00258\ M$$

$$[HI] = (0.0224 + 2 \times 0.00156)\ M = 0.0255\ M$$

Similar problem: 14.50.

In some cases simplifications are possible that greatly reduce the mathematical difficulties in equilibrium constant calculations. Example 14.11 shows the conditions under which valid approximations can be made.

EXAMPLE 14.11

The equilibrium constant (K_p) for the following reaction is found to be 4.31×10^{-4} at 200°C.

$$N_2(g) + 3H_2(g) \rightleftharpoons 2NH_3(g)$$

In a certain experiment a student starts with 0.862 atm of N_2 and 0.373 atm of H_2 in a constant-volume vessel at 200°C. Calculate the pressures of all species when equilibrium is reached.

Answer

Step 1: Since initially there is no NH_3 present, we conclude that the reaction must proceed from left to right to establish equilibrium. In Examples 14.9 and 14.10 we used molarities in calculations. However, it is equally valid to use pressures for a gas-phase reaction at constant temperature and volume. The ideal gas equation [Equation (5.7)] shows that the pressure of a gas is directly proportional to the number of moles per liter of volume:

$$P = \left(\frac{n}{V}\right)RT$$

Thus we can represent the change needed to establish equilibrium in terms of pressures. We summarize the changes in pressures as follows:

	N_2	+	$3H_2$	$\rightleftharpoons$	$2NH_3$
Initial:	0.862 atm		0.373 atm		0 atm
Change:	$-x$ atm		$-3x$ atm		$+2x$ atm
Equilibrium:	$(0.862 - x)$ atm		$(0.373 - 3x)$ atm		$2x$ atm

Step 2: The equilibrium constant is given by

$$K_P = \frac{P^2_{NH_3}}{P_{N_2}P^3_{H_2}}$$

Substituting, we get

$$4.31 \times 10^{-4} = \frac{(2x)^2}{(0.862 - x)(0.373 - 3x)^3}$$

Multiplying and collecting terms would give an equation with terms containing x^4, x^3, x^2, and x, which would be very difficult to solve. Fortunately, we can greatly simplify the above equation by realizing that, because K_P is a small number, the amount of NH_3 formed at equilibrium must also be very small. Thus we make the following approximations: $0.862 - x \simeq 0.862$ and $0.373 - 3x \simeq 0.373$. The equilibrium constant expression can now be written as

$$4.31 \times 10^{-4} = \frac{(2x)^2}{(0.862)(0.373)^3}$$

Solving for x gives $$x = 2.20 \times 10^{-3} \text{ atm}$$

Step 3: The equilibrium pressures are

$$P_{N_2} = (0.862 - 2.20 \times 10^{-3}) \text{ atm} = 0.860 \text{ atm}$$

$$P_{H_2} = (0.373 - 3 \times 2.20 \times 10^{-3}) \text{ atm} = 0.366 \text{ atm}$$

$$P_{NH_3} = 2(2.20 \times 10^{-3}) \text{ atm} = 4.40 \times 10^{-3} \text{ atm}$$

Finally we note that as a rule of thumb, the approximation is considered valid if the quantity ignored is equal to or less than 5 percent of the quantity from which it is subtracted (or to which it is added). To check our approximations we write

$$\frac{2.20 \times 10^{-3}}{0.862} \times 100\% = 0.255\% \qquad \frac{3 \times 2.20 \times 10^{-3}}{0.373} \times 100\% = 1.77\%$$

Thus our approximations are valid.

Similar problem: 14.79.

Examples 14.9–14.11 show that we can calculate the concentrations of all the reacting species at equilibrium if we know the equilibrium constant and the initial concentrations. This information is of great value for estimating the yield of a reaction. For example, if the reaction between H_2 and I_2 to form HI were to go to completion, the number of moles of HI formed in Example 14.9 would be 2×0.500 mol, or 1.00 mol. However, because of the equilibrium process, the actual amount of HI formed can be no more than 2×0.393 mol, or 0.786 mol, a 78.6 percent yield.

14.6 Factors That Affect Chemical Equilibrium

When we say that an equilibrium position shifts to the right, for example, we mean that the net reaction is now from left to right.

Chemical equilibrium represents a balance between forward and reverse reactions. In most cases, this balance is quite delicate. Changes in experimental conditions may disturb the balance and shift the equilibrium position so that more or less of the desired product is formed. At our disposal are the following experimentally controllable variables: concentration, pressure, volume, and temperature. Here we will examine how each of these variables affects a reacting system at equilibrium. In addition, we will examine the effect of a catalyst on equilibrium.

Le Chatelier's Principle

There is a general rule that help us to predict the direction in which an equilibrium reaction will move when a change in concentration, pressure, volume, or temperature occurs. The rule, known as ***Le Chatelier's† principle,*** states that *if an external stress is applied to a system at equilibrium, the system adjusts itself in such a way that the stress is partially offset*. The word "stress" here means a change in concentration, pressure, volume, or temperature that removes a system from the equilibrium state. We will use Le Chatelier's principle to assess the effects of such changes.

Changes in Concentrations

Iron(III) thiocyanate [$Fe(SCN)_3$] dissolves readily in water to give a solution of red color due to the presence of $FeSCN^{2+}$ ion. The equilibrium between undissociated $FeSCN^{2+}$ and the Fe^{3+} and SCN^- ions is given by

$$\underset{\text{red}}{FeSCN^{2+}(aq)} \rightleftharpoons \underset{\text{pale yellow}}{Fe^{3+}(aq)} + \underset{\text{colorless}}{SCN^-(aq)}$$

What would happen if we were to add some sodium thiocyanate (NaSCN) to this solution? In this case, the stress applied to the equilibrium system is an increase in the concentration of SCN^- (from the dissociation of NaSCN). To offset this stress, some Fe^{3+} ions react with the added SCN^- ions, and the equilibrium shifts from right to left:

$$FeSCN^{2+}(aq) \longleftarrow Fe^{3+}(aq) + SCN^-(aq)$$

Both Na^+ and NO_3^- are colorless spectator ions.

Consequently, the red color of the solution deepens (Figure 14.3). Similarly, if we added iron(III) nitrate [$Fe(NO_3)_3$] to the original solution, the red color would also deepen because the additional Fe^{3+} ions [from $Fe(NO_3)_3$] would shift the equilibrium from right to left.

Oxalic acid is sometimes used to remove bathtub rings that consist of rust, or Fe_2O_3.

Now suppose we add some oxalic acid ($H_2C_2O_4$) to the original solution. Oxalic acid ionizes in water to form the oxalate ion, $C_2O_4^{2-}$, which binds strongly to the Fe^{3+} ions. The formation of the stable yellow ion $Fe(C_2O_4)_3^{3-}$ removes free Fe^{3+} ions in solution.

†Henry Louis Le Chatelier (1850–1936). French chemist. Le Chatelier did work on metallurgy, cements, glasses, fuels, and explosives. Although a scientist, he was also noted for his skills in industrial management.

(a) (b) (c) (d)

Figure 14.3 *Effect of concentration change on the position of equilibrium. (a) An aqueous $Fe(SCN)_3$ solution. The color of the solution is due to both the red $FeSCN^{2+}$ and the yellow Fe^{3+} species. (b) After the addition of some NaSCN to the solution in (a), the equilibrium shifts to the left. (c) After the addition of some $Fe(NO_3)_3$ to the solution in (a), the equilibrium shifts to the left. (d) After the addition of some $H_2C_2O_4$ to the solution in (a), the equilibrium shifts to the right. The yellow color is due to the $Fe(C_2O_4)_3^{3-}$ ions.*

Consequently, more $FeSCN^{2+}$ units dissociate and the equilibrium shifts from left to right:

$$FeSCN^{2+}(aq) \longrightarrow Fe^{3+}(aq) + SCN^-(aq)$$

The red solution will turn yellow due to the formation of $Fe(C_2O_4)_3^{3-}$ ions.

This experiment demonstrates that at equilibrium all reactants and products are present in the reacting system. Second, increasing the concentrations of the products (Fe^{3+} or SCN^-) shifts the equilibrium to the left, and decreasing the concentration of the product Fe^{3+} shifts the equilibrium to the right. These results are just as predicted by Le Chatelier's principle.

Since Le Chatelier's principle simply summarizes the observed behavior of equilibrium systems, it is incorrect to say that a given equilibrium shift occurs "because of" Le Chatelier's principle.

The effect of change in concentration on the equilibrium position is shown in the following example.

EXAMPLE 14.12

At 350°C, the equilibrium constant K_c for the reaction

$$N_2(g) + 3H_2(g) \rightleftharpoons 2NH_3(g)$$

is 2.37×10^{-3}. In a certain experiment, the equilibrium concentrations are $[N_2] = 0.683\ M$, $[H_2] = 8.80\ M$, and $[NH_3] = 1.05\ M$. Suppose some of the NH_3 is added to the mixture so that its concentration is increased to $3.65\ M$. (a) Use Le Chatelier's principle to predict the direction that the net reaction will shift to reach a new equilibrium. (b) Confirm your prediction by calculating the reaction quotient Q_c and comparing its value with K_c.

Answer

(a) The stress applied to the system is the addition of NH_3. To offset this stress, some NH_3 reacts to produce N_2 and H_2 until a new equilibrium is established. The net reaction therefore shifts from right to left; that is,

$$N_2(g) + 3H_2(g) \longleftarrow 2NH_3(g)$$

(b) At the instant when some of the NH_3 is added, the system is no longer at equilibrium. The reaction quotient is given by

$$Q_c = \frac{[NH_3]_0^2}{[N_2]_0[H_2]_0^3}$$

$$= \frac{(3.65)^2}{(0.683)(8.80)^3}$$

$$= 2.86 \times 10^{-2}$$

Since this value is greater than 2.37×10^{-3}, the net reaction shifts from right to left until Q_c equals K_c.

Figure 14.4 shows qualitatively the changes in concentrations of the reacting species.

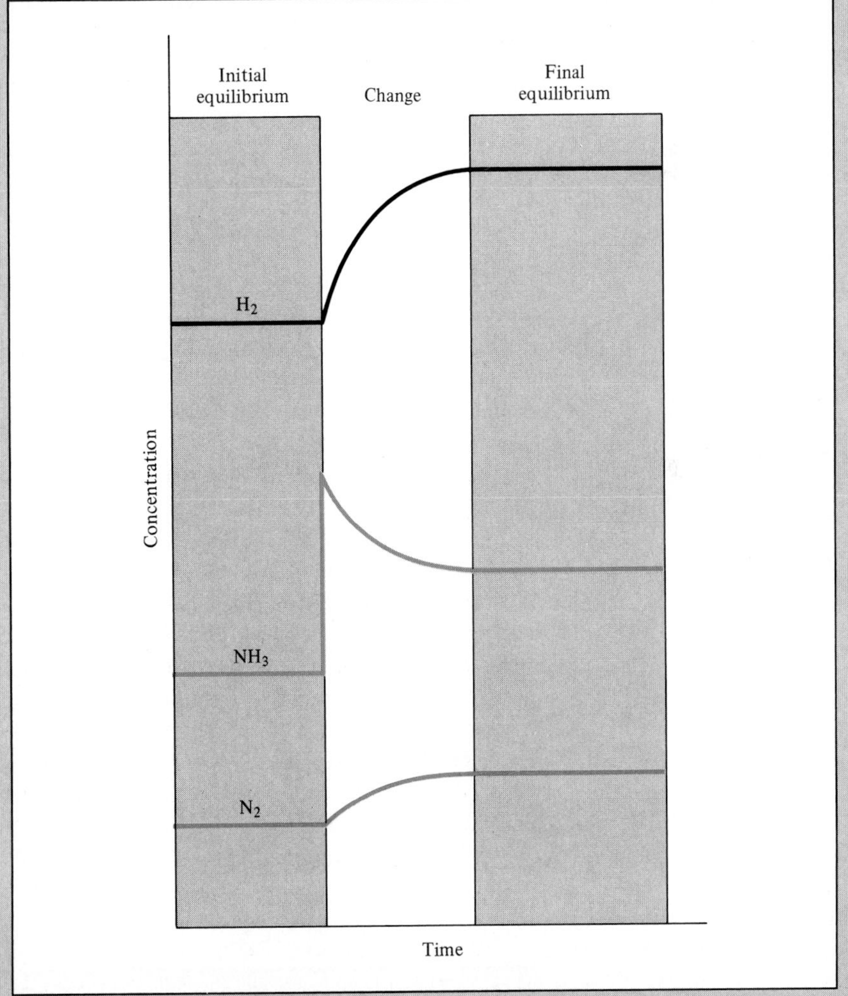

Figure 14.4 *Changes in concentration of H_2, N_2, and NH_3 after the addition of NH_3 to the equilibrium mixture.*

Changes in Volume and Pressure

Changes in pressure ordinarily do not affect the concentrations of reacting species in condensed phases (say, in an aqueous solution) because liquids and solids are virtually incompressible. On the other hand, concentrations of gases are greatly affected by changes in pressure. Let us look again at Equation (5.7):

$$PV = nRT$$

$$P = \left(\frac{n}{V}\right)RT$$

Thus P and V are related to each other inversely: The greater the pressure, the smaller the volume, and vice versa. Note, too, that the term (n/V) is the concentration of the gas in mol/L.

Suppose that the equilibrium system

$$N_2O_4(g) \rightleftharpoons 2NO_2(g)$$

is in a cylinder fitted with a movable piston. What would happen if we increase the pressure on the gases by pushing down on the piston at constant temperature? Since the volume decreases, the concentration (n/V) of both NO_2 and N_2O_4 increases. Because the concentration of NO_2 is squared, the increase in pressure will increase the numerator more than the denominator. The system is no longer at equilibrium, so we write

$$Q_c = \frac{[NO_2]_0^2}{[N_2O_4]_0}$$

Thus $Q_c > K_c$, and the net reaction will shift to the left until $Q_c = K_c$. Conversely, a decrease in pressure (increase in volume) would result in $Q_c < K_c$; the net reaction would shift to the right until $Q_c = K_c$.

In general, an increase in pressure (decrease in volume) favors the net reaction that decreases the total number of moles of gases (the reverse reaction, in this case), and a decrease in pressure (increase in volume) favors the net reaction that increases the total number of moles of gases (here, the forward reaction). For reactions in which there is no change in the number of moles of gases, a pressure (or volume) change has no effect on the position of equilibrium.

Note that it is possible to change the pressure of a system without changing its volume. Suppose the NO_2–N_2O_4 system is contained in a stainless steel vessel whose volume is constant. We can increase the total pressure in the vessel by adding an inert gas (helium, for example) to the equilibrium system. Adding helium to the equilibrium mixture at constant volume increases the total gas pressure and decreases the mole fractions of both NO_2 and N_2O_4; but the partial pressure of each gas, given by the product of its mole fraction and total pressure (see Section 5.6), does not change. Thus the presence of an inert gas in such a case does not affect the equilibrium.

The following example illustrates the effect of change in pressure on the equilibrium position.

EXAMPLE 14.13

Consider the following equilibrium systems:

(a) $2PbS(s) + 3O_2(g) \rightleftharpoons 2PbO(s) + 2SO_2(g)$
(b) $PCl_5(g) \rightleftharpoons PCl_3(g) + Cl_2(g)$
(c) $H_2(g) + CO_2(g) \rightleftharpoons H_2O(g) + CO(g)$

Predict the direction of the net reaction in each case as a result of increasing the pressure (decreasing the volume) on the system at constant temperature.

Answer

(a) Consider only the gaseous molecules. In the balanced equation there are 3 moles of gaseous reactants and 2 moles of gaseous products. Therefore the net reaction will shift toward products (to the right) when the pressure is increased.
(b) The number of moles of products is 2 and that of reactants is 1; therefore the net reaction will shift to the left, toward reactants.
(c) The number of moles of products is equal to the number of moles of reactants, so a change in pressure has no effect on the equilibrium.

Similar problem: 14.59.

Changes in Temperature

A change in concentration, pressure, or volume may alter the equilibrium position, but it does not change the value of the equilibrium constant. Only a change in temperature can alter the equilibrium constant.

The formation of NO_2 from N_2O_4 is an endothermic process:

$$N_2O_4(g) \longrightarrow 2NO_2(g) \qquad \Delta H° = 58.0 \text{ kJ}$$

and the reverse reaction is exothermic:

$$2NO_2(g) \longrightarrow N_2O_4(g) \qquad \Delta H° = -58.0 \text{ kJ}$$

At equilibrium the net heat effect is zero because there is no net reaction. What happens if the following equilibrium system

$$N_2O_4(g) \rightleftharpoons 2NO_2(g)$$

is heated at constant volume? Since endothermic processes absorb heat from the surroundings, heating favors the dissociation of N_2O_4 into NO_2 molecules. Consequently, the equilibrium constant, given by

$$K_c = \frac{[NO_2]^2}{[N_2O_4]}$$

increases with temperature (Figure 14.5).

As another example, let us consider the equilibrium between $CoCl_4^{2-}$ and $Co(H_2O)_6^{2+}$ ions:

$$\underset{\text{blue}}{CoCl_4^{2-}} + 6H_2O \rightleftharpoons \underset{\text{pink}}{Co(H_2O)_6^{2+}} + 4Cl^-$$

The formation of $CoCl_4^{2-}$ is endothermic. On heating, the equilibrium shifts to the left

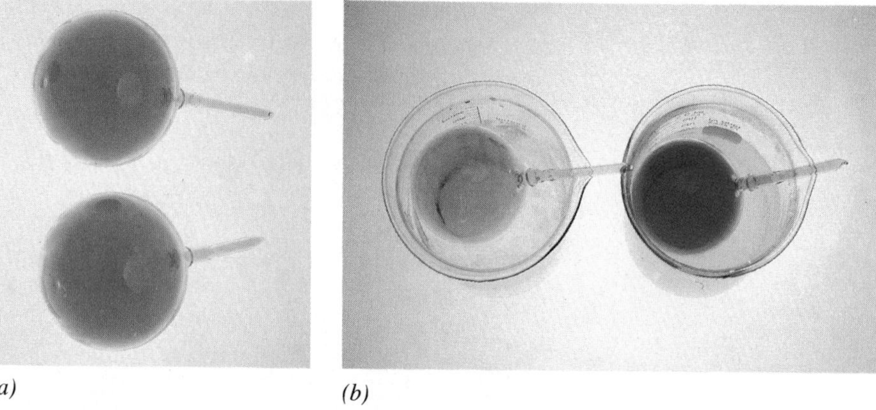

(a) (b)

Figure 14.5 *(a) Two bulbs containing a mixture of NO_2 and N_2O_4 gases at equilibrium. (b) When the bulb on the left is immersed in ice water, its color becomes lighter, indicating the formation of the colorless gas N_2O_4. When the bulb on the right is immersed in hot water, its color darkens, indicating the formation of NO_2.*

Figure 14.6 *Left: Heating favors the formation of the blue $CoCl_4^{2-}$ ion. Right: Cooling favors the formation of the $Co(H_2O)_6^{2+}$ ion.*

and the solution turns blue. Cooling favors the exothermic reaction [the formation of $Co(H_2O)_6^{2+}$] and the solution turns pink (Figure 14.6).

We summarize the results with the statement: *A temperature increase favors an endothermic reaction, and a temperature decrease favors an exothermic reaction.*

The Effect of a Catalyst

We know that a catalyst enhances the rate of a reaction by lowering the reaction's activation energy (Chapter 13). However, as Figure 13.17 shows, a catalyst lowers the activation energy of the forward reaction to the same extent as it lowers that for the reverse reaction. This means that both the forward and the reverse rates are affected to exactly the same degree. We can therefore conclude that the presence of a catalyst does not alter the equilibrium constant nor does it shift the position of an equilibrium system. Adding a catalyst to a reaction mixture that is not at equilibrium will speed up both the forward and reverse rates to achieve an equilibrium mixture much faster. The same equilibrium mixture could be obtained without the catalyst, but we might have to wait much longer for it to happen.

Summary of Factors That May Affect the Equilibrium Position

We have considered four ways to affect a reacting system at equilibrium. It is important to remember that of the four *only a change in temperature changes the value of the equilibrium constant*. Changes in concentration, pressure, and volume can alter the equilibrium concentrations of the reacting mixture, but they cannot change the equilibrium constant as long as the temperature does not change. A catalyst can help establish equilibrium faster, but it too has no effect on the equilibrium constant, nor on the equilibrium concentrations of the reacting species.

The factors affecting equilibrium are illustrated in an industrial and a physiological process in the two Chemistry in Action essays on pp. 624 and 625.

The effects of heating, concentration, and pressure changes and the addition of an inert gas on an equilibrium system are treated in the following example.

EXAMPLE 14.14

Consider the following equilibrium process:

$$N_2F_4(g) \rightleftharpoons 2NF_2(g) \qquad \Delta H° = 38.5 \text{ kJ}$$

Predict the changes in the equilibrium if (a) the reacting mixture is heated at constant volume; (b) NF_2 gas is removed from the reacting mixture at constant temperature and volume; (c) the pressure on the reacting mixture is decreased at constant temperature; and (d) an inert gas, such as helium, is added to the reacting mixture at constant volume and temperature.

Answer

(a) Since the forward reaction is endothermic, an increase in temperature favors the formation of NF_2. The equilibrium constant

$$K_c = \frac{[NF_2]^2}{[N_2F_4]}$$

will therefore increase with increasing temperature.

(b) The stress here is the removal of NF_2 gas. To offset it, more N_2F_4 will decompose to form NF_2. The equilibrium constant K_c remains unchanged, however.

(c) A decrease in pressure (which is accompanied by an increase in gas volume) favors the formation of more gas molecules, that is, the forward reaction. Thus, more NF_2 gas will be formed. The equilibrium constant will remain unchanged.

(d) Adding helium to the equilibrium mixture at constant volume will not shift the equilibrium.

Similar problems: 14.60, 14.61, 14.64.

CHEMISTRY IN ACTION

The Haber Process

Having discussed the factors that affect chemical equilibrium, we can now apply what we have learned to a system of great practical importance—the synthesis of ammonia. We know from Chapter 13 that the Haber process of ammonia synthesis is a gas-phase reaction in which a heterogeneous catalyst is used to promote the speed of the reaction. Here we will concentrate on factors other than the reaction rate.

Suppose, as a prominent industrial chemist at the turn of the century, you are asked to design an efficient procedure for synthesizing ammonia from hydrogen and nitrogen. Your main objective is to obtain a high yield of the product while keeping the production costs down. Your first step is to take a careful look at the balanced equation for the production of ammonia:

$$N_2(g) + 3H_2(g) \rightleftharpoons 2NH_3(g) \qquad \Delta H° = -92.6\ \text{kJ}$$

Two ideas strike you: *First,* since 1 mole of N_2 reacts with 3 moles of H_2 to produce 2 moles of NH_3, a higher yield of NH_3 can be obtained at equilibrium if the reaction is carried out at high pressures. This is indeed the case, as shown by the plot of mole percent of NH_3 versus the total pressure of the reacting system (Figure 14.7). *Second,* the exothermic nature of the forward reaction tells you that the equilibrium constant for the reaction will decrease with increasing temperature (see Table 14.2). Thus, for maximum yield of NH_3, the reaction should be run at the lowest possible temperature. The plot in Figure 14.8 shows that the yield of ammonia increases with decreasing temperature. A low-temperature operation (say, 220 K or −53°C) is desirable in other respects too. The boiling point of NH_3 is −33.5°C, so as it formed it would quickly condense to a liquid, which could be conveniently removed from the reacting system. (Both H_2 and N_2 are still gases at this temperature.) Consequently, the net reaction would shift from left to right, just as desired.

Table 14.2 Variation with temperature of the equilibrium constant for the synthesis of ammonia

t(°C)	K_c
25	6.0×10^5
200	0.65
300	0.011
400	6.2×10^{-4}
500	7.4×10^{-5}

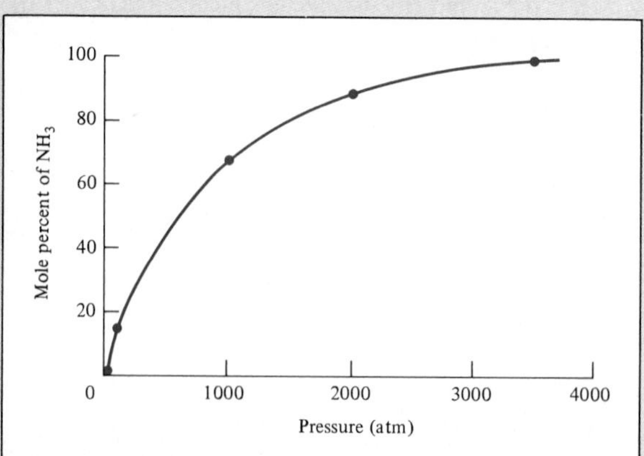

Figure 14.7 *Mole percent of NH_3 as a function of total pressures of the gases at 425°C.*

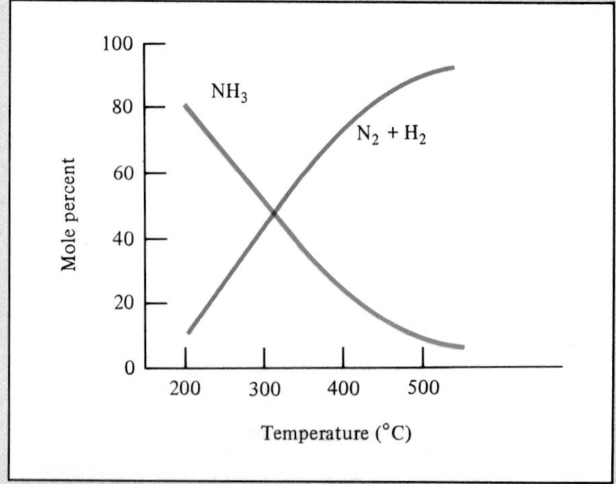

Figure 14.8 *The composition (mole percent) of $H_2 + N_2$ and NH_3 at equilibrium (for a certain starting mixture) as a function of temperature.*

On paper, then, these are your conclusions. Let us compare your recommendations with the actual conditions employed in industrial plants. Typically, the operating pressures are between 500 atm and 1000 atm, so you are right to advocate high pressure. Furthermore, in the industrial process the NH_3 never reaches its equilibrium value but is constantly removed from the reaction mixture in a continuous process operation. This design makes sense, too, as you had anticipated. The only discrepancy is that the operation is usually carried out at about 500°C. This high-temperature operation is costly and the yield of NH_3 is low. The justification for this choice is that the *rate* of NH_3 production increases with increasing temperature. Commercially, a faster production of NH_3 is desirable even if it means a lower yield and a higher cost of operation. But as we know, raising the temperature alone is not sufficient; the proper catalysts must be employed to speed up the process, as discussed in Chapter 13. Figure 14.9 is a schematic diagram of the industrial synthesis of NH_3 from N_2 and H_2, by the Haber process.

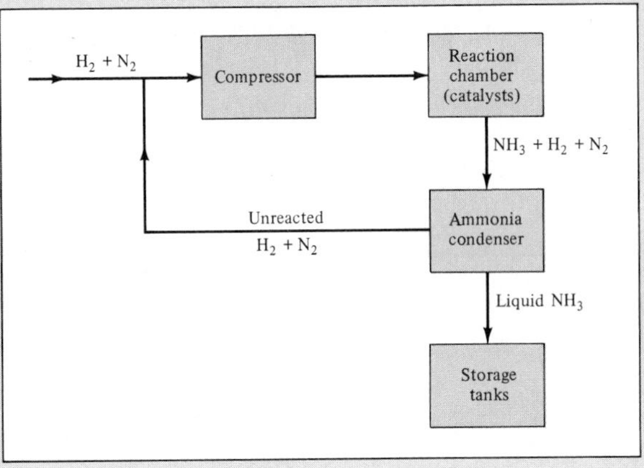

Figure 14.9 *Schematic diagram of the Haber process for ammonia synthesis.*

CHEMISTRY IN ACTION

Life at High Altitude and Hemoglobin Production

Physiology is affected by environmental conditions. The consequences of a sudden change in altitude dramatize this fact. Flying from San Francisco, which is at sea level, to Mexico City, where the elevation is 2.3 km (1.4 mi), or scaling a 3-km mountain in two days can cause headache, nausea, unusual fatigue, and other discomforts (Figure 14.10). These are all symptoms of hypoxia, a deficiency in the amount of oxygen reaching body tissues. In serious cases, the victim may slip into a coma and die if not treated quickly. And yet a person living at a high altitude for weeks or months gradually recovers from altitude sickness and becomes acclimatized to the low oxygen content in the atmosphere, able to function normally.

The combination of oxygen with the hemoglobin (Hb) molecule, which carries oxygen through the

Figure 14.10 *Mountaineers need weeks or even months to become acclimatized before scaling summits such as Mount Everest.*

blood, is a complex reaction, but for our purposes here it can be represented by a simplified equation:

$$Hb(aq) + O_2(aq) \rightleftharpoons HbO_2(aq)$$

where HbO_2 is oxyhemoglobin, the hemoglobin-oxygen complex that actually transports oxygen to tissue. The equilibrium constant is

$$K_c = \frac{[HbO_2]}{[Hb][O_2]}$$

At an altitude of 3 km the partial pressure of oxygen is only about 0.14 atm, compared with 0.2 atm at sea level. According to Le Chatelier's principle, a decrease in oxygen concentration will shift the equilibrium shown in the equation above from right to left. This change depletes oxyhemoglobin, causing hypoxia. Given enough time, the body can cope with this adversity by producing more hemoglobin molecules. The equilibrium will then gradually shift again from left to right, favoring the formation of oxyhemoglobin. The increase in hemoglobin production proceeds slowly, requiring two or three weeks to develop. Full capacity may require several years to develop. Studies show that long-time residents of high-altitude areas have high hemoglobin levels in their blood—sometimes as much as 50 percent more than individuals living at sea level!

Summary

1. Chemists are concerned with dynamic equilibria between phases (physical equilibria) and between reacting substances (chemical equilibria).
2. For the general chemical reaction

$$aA + bB \rightleftharpoons cC + dD$$

the concentrations of reactants and products at equilibrium (in moles per liter) are related by the equilibrium constant expression

$$K_c = \frac{[C]^c[D]^d}{[A]^a[B]^b}$$

3. The equilibrium constant may also be expressed in terms of the equilibrium partial pressures (in atm) of gases, as K_P.
4. An equilibrium in which all reactants and products are in the same phase is a homogeneous equilibrium. If the reactants and products are not all in the same phase, the equilibrium is called heterogeneous. The concentrations of pure solids, pure liquids, and solvents are constant and do not appear in the equilibrium constant expression of a reaction.
5. If a reaction can be expressed as the sum of two or more reactions, the equilibrium constant for the overall reaction is given by the product of the equilibrium constants of the individual reactions.
6. The value of K depends on how the chemical equation is balanced, and the equilibrium constant for the reverse of a particular reaction is the reciprocal of the equilibrium constant of that reaction.
7. The equilibrium constant is the ratio of the rate constant for the forward reaction to that for the reverse reaction.
8. The reaction quotient Q has the same form as the equilibrium constant expression, but it applies to a reaction that may not be at equilibrium. If $Q > K$, the reaction will proceed from right to left to achieve equilibrium. If $Q < K$, the reaction will proceed from left to right to achieve equilibrium.
9. Le Chatelier's principle states that if an external stress is applied to a system at chemical equilibrium, the system will adjust to partially offset the stress.

10. Only a change in temperature changes the value of the equilibrium constant for a particular reaction. Changes in concentration, pressure, or volume may change the equilibrium concentrations of reactants and products. The addition of a catalyst hastens the attainment of equilibrium but does not affect the equilibrium concentrations of reactants and products.

Applied Chemistry Example: Hydrogen

A notable absence among the top 50 industrial chemicals produced in the United States is hydrogen, which is used in the synthesis of ammonia, in the food industry (to prepare saturated oils), and in the production of many other chemicals. The reason is that hydrogen is often used captively, that is, it is used at the site where it is prepared. Therefore, it is difficult to know exactly how much is produced annually. About 75 percent of hydrogen is produced by the *steam-reforming* process. In this process, a mixture of methane and water are reacted at high temperatures to form carbon monoxide and hydrogen.

The steam-reforming reaction is carried out in two stages, called primary and secondary reforming. In the primary stage, a mixture of steam and methane at about 30 atm is heated over a nickel catalyst at 800°C to give hydrogen and carbon monoxide:

$$CH_4(g) + H_2O(g) \longrightarrow CO(g) + 3H_2(g) \qquad \Delta H^\circ = 206 \text{ kJ}$$

The secondary stage is carried out at about 1000°C, in the presence of air, to convert the remaining methane to hydrogen:

$$CH_4(g) + \tfrac{1}{2}O_2(g) \longrightarrow CO(g) + 2H_2(g) \qquad \Delta H^\circ = 35.7 \text{ kJ}$$

1. What conditions of temperature and pressure would favor the formation of products in both the primary and the secondary stage?

   Answer: Since both reactions are endothermic (ΔH° is positive), according to Le Chatelier's principle the products would be favored at high temperatures. Indeed, the steam-reforming process is carried out at very high temperatures (between 800°C and 1000°C). It is interesting to note that in a plant that uses natural gas (methane) both for hydrogen generation and heating, about one-third of the gas is burned to maintain the high temperatures.

   In each reaction there are more moles of products than reactants; therefore, we expect products to be favored at low pressures. In reality, the reactions are carried out at high pressures. The reason is that when the hydrogen gas produced is used captively (usually in the synthesis of ammonia), high pressure leads to higher yields of ammonia.

2. The equilibrium constant (K_c) for the primary stage is about 18 at 800°C. (*a*) Calculate K_P for the reaction. (*b*) If the partial pressures of methane and steam were both 15 atm at the start, what are the pressures of all the species at equilibrium?

   Answer: (*a*) The relation between K_c and K_P is given by Equation (14.5):

$$K_P = K_c(0.0821T)^{\Delta n}$$

   Since $\Delta n = 4 - 2 = 2$, we write

$$\begin{aligned} K_P &= (18)(0.0821 \times 1073)^2 \\ &= 1.4 \times 10^5 \end{aligned}$$

(*b*) Following the procedure outlined in Section 14.5, we proceed as follows.

Step 1: Let x be the amount of CH_4 and H_2O (in atm) reacted. We write

	CH_4	+	H_2O	$\rightleftharpoons$	CO	+	$3H_2$
Initial:	15 atm		15 atm		0 atm		0 atm
Change:	$-x$ atm		$-x$ atm		$+x$ atm		$+3x$ atm
Equilibrium:	$(15 - x)$ atm		$(15 - x)$ atm		x atm		$3x$ atm

Step 2: The equilibrium constant is given by

$$K_P = \frac{P_{CO}P_{H_2}^3}{P_{CH_4}P_{H_2O}}$$

$$1.4 \times 10^5 = \frac{(x)(3x)^3}{(15 - x)(15 - x)} = \frac{27x^4}{(15 - x)^2}$$

Taking the square roots of both sides, we obtain

$$3.7 \times 10^2 = \frac{5.2x^2}{15 - x}$$

which can be expressed as

$$5.2x^2 + 3.7 \times 10^2\, x - 5.6 \times 10^3 = 0$$

Solving the quadratic equation, we get

$$x = 13 \text{ atm}$$

(The other solution for x is negative and is physically impossible.)

Step 3: At equilibrium, the pressures are

$$P_{CH_4} = (15 - 13) \text{ atm} = 2 \text{ atm}$$

$$P_{H_2O} = (15 - 13) \text{ atm} = 2 \text{ atm}$$

$$P_{CO} = 13 \text{ atm}$$

$$P_{H_2} = 3(13 \text{ atm}) = 39 \text{ atm}$$

Key Words

Equilibrium, p. 596
Equilibrium constant, p. 598
Heterogeneous equilibrium, p. 603
Homogeneous equilibrium, p. 599
Le Chatelier's principle, p. 617
Physical equilibrium, p. 596
Reaction quotient (Q_c), p. 611

Exercises

Concept of Equilibrium

REVIEW QUESTIONS

14.1 Define equilibrium. Give two examples of a dynamic equilibrium.

14.2 Explain the difference between physical equilibrium and chemical equilibrium. Give two examples of each.

14.3 Briefly describe the importance of equilibrium in the study of chemical reactions.

14.4 Consider the equilibrium system 3A $\rightleftharpoons$ B. Sketch the change in concentrations of A and B with time for the following situations: (a) initially only A is present; (b) initially only B is present; (c) initially both A and B are present (with A in higher concentration). In each case, assume that the concentration of B is higher than that of A at equilibrium.

Equilibrium Constant Expressions

REVIEW QUESTIONS

14.5 Define homogeneous equilibrium and heterogeneous equilibrium. Give two examples of each.

14.6 What do the symbols K_c and K_P represent?

14.7 Write the expression for the equilibrium constants K_c and K_P, if applicable, of each of the following reactions:
(a) $H_2O(l) \rightleftharpoons H_2O(g)$
(b) $H_2O(g) + CO(g) \rightleftharpoons H_2(g) + CO_2(g)$
(c) $2Mg(s) + O_2(g) \rightleftharpoons 2MgO(s)$
(d) $PCl_5(g) \rightleftharpoons PCl_3(g) + Cl_2(g)$

14.8 Write the expressions for the equilibrium constants K_P of the following thermal decompositions:
(a) $2NaHCO_3(s) \rightleftharpoons Na_2CO_3(s) + CO_2(g) + H_2O(g)$
(b) $2CaSO_4(s) \rightleftharpoons 2CaO(s) + 2SO_2(g) + O_2(g)$

14.9 Write equilibrium constant expressions for K_c and for K_P, if applicable, for the following processes:
(a) $2CO_2(g) \rightleftharpoons 2CO(g) + O_2(g)$
(b) $3O_2(g) \rightleftharpoons 2O_3(g)$
(c) $CO(g) + Cl_2(g) \rightleftharpoons COCl_2(g)$
(d) $H_2O(g) + C(s) \rightleftharpoons CO(g) + H_2(g)$
(e) $HCOOH(aq) \rightleftharpoons H^+(aq) + HCOO^-(aq)$
(f) $2HgO(s) \rightleftharpoons 2Hg(l) + O_2(g)$
(g) $Ni(s) + 4CO(g) \rightleftharpoons Ni(CO)_4(g)$

14.10 Write the equilibrium constant expressions for K_c and K_P, if applicable, for the following reactions:
(a) $2NO_2(g) + 7H_2(g) \rightleftharpoons 2NH_3(g) + 4H_2O(l)$
(b) $2ZnS(s) + 3O_2(g) \rightleftharpoons 2ZnO(s) + 2SO_2(g)$
(c) $C(s) + CO_2(g) \rightleftharpoons 2CO(g)$
(d) $N_2O_5(g) \rightleftharpoons 2NO_2(g) + \frac{1}{2}O_2(g)$
(e) $C_6H_5COOH(aq) \rightleftharpoons C_6H_5COO^-(aq) + H^+(aq)$

Calculating Equilibrium Constants

REVIEW QUESTION

14.11 Write the equation relating K_c to K_P and define all the terms.

PROBLEMS

14.12 The equilibrium constant for the reaction

$$2HCl(g) \rightleftharpoons H_2(g) + Cl_2(g)$$

is 4.17×10^{-34} at 25°C. What is the equilibrium constant for the reaction

$$H_2(g) + Cl_2(g) \rightleftharpoons 2HCl(g)$$

at the same temperature?

14.13 Consider the following equilibrium process at 700°C:

$$2H_2(g) + S_2(g) \rightleftharpoons 2H_2S(g)$$

Analysis shows that there are 2.50 moles of H_2, 1.35×10^{-5} mole of S_2, and 8.70 moles of H_2S present in a 12.0 L-flask. Calculate the equilibrium constant K_c for the reaction.

14.14 What is the K_P at 1273°C for the reaction

$$2CO(g) + O_2(g) \rightleftharpoons 2CO_2(g)$$

if K_c is 2.24×10^{22} at the same temperature?

14.15 A reaction vessel contains NH_3, N_2, and H_2 at equilibrium at a certain temperature. The equilibrium concentrations are $[NH_3] = 0.25\ M$, $[N_2] = 0.11\ M$, and $[H_2] = 1.91\ M$. Calculate the equilibrium constant K_c for the synthesis of ammonia if the reaction is represented as
(a) $N_2(g) + 3H_2(g) \rightleftharpoons 2NH_3(g)$
(b) $\frac{1}{2}N_2(g) + \frac{3}{2}H_2(g) \rightleftharpoons NH_3(g)$

14.16 Consider the following reaction:

$$N_2(g) + O_2(g) \rightleftharpoons 2NO(g)$$

If the equilibrium partial pressures of N_2, O_2, and NO are 0.15 atm, 0.33 atm, and 0.050 atm, respectively, at 2200°C, what is K_P?

14.17 The equilibrium constant K_c for the reaction

$$I_2(g) \rightleftharpoons 2I(g)$$

is 3.8×10^{-5} at 727°C. Calculate K_c and K_P for the equilibrium

$$2I(g) \rightleftharpoons I_2(g)$$

at the same temperature.

14.18 The pressure of the reacting mixture at equilibrium

$$CaCO_3(s) \rightleftharpoons CaO(s) + CO_2(g)$$

is 0.105 atm at 350°C. Calculate K_P and K_c for this reaction.

14.19 A 2.50-mole quantity of NOCl was initially in a 1.50-L reaction chamber at 400°C. After equilibrium was established, it was found that 28.0 percent of the NOCl had dissociated:

$$2NOCl(g) \rightleftharpoons 2NO(g) + Cl_2(g)$$

Calculate the equilibrium constant K_c for the reaction.

14.20 The equilibrium constant K_P for the reaction

$$PCl_5(g) \rightleftharpoons PCl_3(g) + Cl_2(g)$$

is 1.05 at 250°C. The reaction starts with a mixture of PCl_5, PCl_3, and Cl_2 at pressures 0.177 atm, 0.223 atm, and 0.111 atm, respectively, at 250°C. When the mixture comes to equilibrium at that temperature, which pressures will have decreased and which will have increased? Explain why.

14.21 Ammonium carbamate, $NH_4CO_2NH_2$, decomposes as follows:

$$NH_4CO_2NH_2(s) \rightleftharpoons 2NH_3(g) + CO_2(g)$$

Starting with only the solid, it is found that at 40°C the total gas pressure (NH_3 and CO_2) is 0.363 atm. Calculate the equilibrium constant, K_P.

14.22 The equilibrium constant K_c for the following reaction is 0.65 at 200°C.

$$N_2(g) + 3H_2(g) \rightleftharpoons 2NH_3(g)$$

(a) What is the value of K_P for this reaction?
(b) What is the value of the equilibrium constant K_c for $2NH_3(g) \rightleftharpoons N_2(g) + 3H_2(g)$?
(c) What is K_c for $\frac{1}{2}N_2(g) + \frac{3}{2}H_2(g) \rightleftharpoons NH_3(g)$?
(d) What are the values of K_P for the reactions described in (b) and (c)?

14.23 At 20°C, the vapor pressure of water is 0.0231 atm. Calculate K_P and K_c for the "reaction"

$$H_2O(l) \rightleftharpoons H_2O(g)$$

14.24 Consider the following reaction at 1600°C:

$$Br_2(g) \rightleftharpoons 2Br(g)$$

When 1.05 moles of Br_2 are put in a 0.980-L flask, 1.20 percent of the Br_2 undergoes dissociation. Calculate the equilibrium constant K_c for the reaction.

14.25 Consider the equilibrium

$$2NOBr(g) \rightleftharpoons 2NO(g) + Br_2(g)$$

If nitrosyl bromide, NOBr, is 34 percent dissociated at 25°C and the total pressure is 0.25 atm, calculate K_P and K_c for the dissociation at this temperature.

14.26 Pure phosgene gas ($COCl_2$), 3.00×10^{-2} mol, was placed in a 1.50-L container. It was heated to 800 K, and at equilibrium the pressure of CO was found to be 0.497 atm. Calculate the equilibrium constant K_P for the reaction

$$CO(g) + Cl_2(g) \rightleftharpoons COCl_2(g)$$

Multiple Equilibria

REVIEW QUESTION

14.27 What is the rule for writing the equilibrium constant for the overall reaction involving two or more reactions?

PROBLEMS

14.28 The following equilibrium constants have been determined for hydrosulfuric acid at 25°C:

$$H_2S(aq) \rightleftharpoons H^+(aq) + HS^-(aq) \qquad K_c' = 9.5 \times 10^{-8}$$

$$HS^-(aq) \rightleftharpoons H^+(aq) + S^{2-}(aq) \qquad K_c'' = 1.0 \times 10^{-19}$$

Calculate the equilibrium constant for the following reaction at the same temperature:

$$H_2S(aq) \rightleftharpoons 2H^+(aq) + S^{2-}(aq)$$

14.29 Consider the following equilibria:

$$2SO_2(g) + O_2(g) \rightleftharpoons 2SO_3(g)$$
$$2NO_2(g) \rightleftharpoons 2NO(g) + O_2(g)$$

Express the equilibrium constant K_c for the process

$$SO_2(g) + NO_2(g) \rightleftharpoons SO_3(g) + NO(g)$$

in terms of the equilibrium constants for the first two processes.

14.30 The following equilibrium constants were determined at 1123 K:

$$C(s) + CO_2(g) \rightleftharpoons 2CO(g) \qquad K_P' = 1.3 \times 10^{14}$$
$$CO(g) + Cl_2(g) \rightleftharpoons COCl_2(g) \qquad K_P'' = 6.0 \times 10^{-3}$$

Write the equilibrium constant expression K_P, and calculate the equilibrium constant at 1123 K for

$$C(s) + CO_2(g) + 2Cl_2(g) \rightleftharpoons 2COCl_2(g)$$

14.31 At a certain temperature the following reactions have the constants shown:

$$S(s) + O_2(g) \rightleftharpoons SO_2(g) \qquad K_c' = 4.2 \times 10^{52}$$
$$2S(s) + 3O_2(g) \rightleftharpoons 2SO_3(g) \qquad K_c'' = 9.8 \times 10^{128}$$

Calculate the equilibrium constant K_c for the following reaction at that temperature:

$$2SO_2(g) + O_2(g) \rightleftharpoons 2SO_3(g)$$

Relationship between Chemical Kinetics and Chemical Equilibrium

REVIEW QUESTION

14.32 Based on rate constant considerations, explain why the equilibrium constant depends on temperature.

PROBLEMS

14.33 Water is a very weak electrolyte that undergoes the following ionization (called self-ionization):

$$H_2O(l) \underset{k_{-1}}{\overset{k_1}{\rightleftharpoons}} H^+(aq) + OH^-(aq)$$

(a) If $k_1 = 2.4 \times 10^{-5}\ s^{-1}$ and $k_{-1} = 1.3 \times 10^{11}/M \cdot s$, calculate the equilibrium constant K where $K = [H^+][OH^-]/[H_2O]$. (b) Calculate the product $[H^+][OH^-]$ and hence $[H^+]$ and $[OH^-]$.

14.34 Consider the following reaction, which takes place in a single elementary step:

$$2A + B \underset{k_r}{\overset{k_f}{\rightleftharpoons}} A_2B$$

If the equilibrium constant K_c is 12.6 at a certain temperature and if $k_r = 5.1 \times 10^{-2}\ s^{-1}$, calculate the value of k_f.

Calculating Equilibrium Concentrations

REVIEW QUESTIONS

14.35 Define reaction quotient. How does it differ from equilibrium constant?

14.36 Outline the steps for calculating the concentrations of reacting species in an equilibrium reaction.

PROBLEMS

14.37 Consider the following equilibrium reaction

$$2NO(g) + Cl_2(g) \rightleftharpoons 2NOCl(g)$$

The equilibrium constant is 6.5×10^4 at 35°C. In a certain experiment 2.0×10^{-2} mole of NO, 8.3×10^{-3} mole of Cl_2, and 6.8 moles of NOCl are mixed in a 2.0-L flask. In which direction will the system proceed to reach equilibrium?

14.38 The equilibrium constant K_P for the reaction

$$2SO_2(g) + O_2(g) \rightleftharpoons 2SO_3(g)$$

is 5.60×10^4 at 350°C. SO_2 and O_2 are mixed initially at 0.350 atm and 0.762 atm, respectively, at 350°C. When the mixture equilibrates, is the total pressure less than or greater than the sum of the initial pressures, 1.112 atm?

14.39 The equilibrium constant K_P for the reaction

$$2NO_2(g) \rightleftharpoons 2NO(g) + O_2(g)$$

is 158 at 1000 K. Calculate P_{O_2} if $P_{NO_2} = 0.400$ atm and $P_{NO} = 0.270$ atm.

14.40 For the synthesis of ammonia

$$N_2(g) + 3H_2(g) \rightleftharpoons 2NH_3(g)$$

the equilibrium constant K_c at 200°C is 0.65. Starting with $[H_2]_0 = 0.76\ M$, $[N_2]_0 = 0.60\ M$, and $[NH_3]_0 = 0.48\ M$, when this mixture comes to equilibrium, which gases will have increased in concentration and which will have decreased in concentration?

14.41 For the reaction

$$H_2(g) + CO_2(g) \rightleftharpoons H_2O(g) + CO(g)$$

at 700°C, $K_c = 0.534$. Calculate the number of moles of H_2 formed at equilibrium if a mixture of 0.300 mole of CO and 0.300 mole of H_2O is heated to 700°C in a 10.0-L container.

14.42 A sample of pure NO_2 gas heated to 1000 K decomposes:

$$2NO_2(g) \rightleftharpoons 2NO(g) + O_2(g)$$

The equilibrium constant K_P is 158. Analysis shows that the partial pressure of O_2 is 0.25 atm at equilibrium. Calculate the pressure of NO and NO_2 in the mixture.

14.43 The equilibrium constant K_c for the reaction

$$H_2(g) + Br_2(g) \rightleftharpoons 2HBr(g)$$

is 2.18×10^6 at 730°C. Starting with 3.20 mol HBr in a 12.0-L reaction vessel, calculate the concentrations of H_2, Br_2, and HBr at equilibrium.

14.44 The dissociation of molecular iodine into iodine atoms is represented as

$$I_2(g) \rightleftharpoons 2I(g)$$

At 1000 K, the equilibrium constant K_c for the reaction is 3.80×10^{-5}. Suppose you start with 0.0456 mole of I_2 in a 2.30-L flask at 1000 K. What are the concentrations of the gases at equilibrium?

14.45 The equilibrium constant K_c for the decomposition of phosgene, $COCl_2$, is 4.63×10^{-3} at 527°C:

$$COCl_2(g) \rightleftharpoons CO(g) + Cl_2(g)$$

Calculate the equilibrium partial pressure of all the components, starting with pure phosgene at 0.760 atm.

14.46 Consider the following equilibrium process at 686°C:

$$CO_2(g) + H_2(g) \rightleftharpoons CO(g) + H_2O(g)$$

The equilibrium concentrations of the reacting species are $[CO] = 0.050\ M$, $[H_2] = 0.045\ M$, $[CO_2] = 0.086\ M$, and $[H_2O] = 0.040\ M$. (a) Calculate K_c for the reaction at 686°C. (b) If the concentration of CO_2 were raised to 0.50 mol/L by the addition of CO_2, what would be the concentrations of all the gases when equilibrium is reestablished.

14.47 For the dissociation of N_2O_4

$$N_2O_4(g) \rightleftharpoons 2NO_2(g)$$

the equilibrium constant K_c is 4.63×10^{-3} at 25°C. Starting with 0.100 mole of N_2O_4 in a 5.00-L reaction vessel, calculate the concentrations of N_2O_4 and NO_2 at equilibrium.

14.48 Consider the heterogeneous equilibrium process:

$$C(s) + CO_2(g) \rightleftharpoons 2CO(g)$$

At 700°C, the total pressure of the system is found to be 4.50 atm. If the equilibrium constant K_P is 1.52, calculate the equilibrium partial pressures of CO_2 and CO.

14.49 The equilibrium constant K_c for the reaction

$$H_2(g) + CO_2(g) \rightleftharpoons H_2O(g) + CO(g)$$

is 4.2 at 1650°C. Initially 0.80 mol H_2 and 0.80 mol CO_2 are injected into a 5.0-L flask. Calculate the concentration of each species at equilibrium.

14.50 The equilibrium constant K_c for the reaction

$$H_2(g) + I_2(g) \rightleftharpoons 2HI(g)$$

is 54.3 at 430°C. At the start of the reaction there are 0.714 mole of H_2, 0.984 mole of I_2, and 0.886 mole of HI in a 2.40-L reaction chamber. Calculate the concentrations of the gases at equilibrium.

Le Chatelier's Principle

REVIEW QUESTIONS

14.51 Explain Le Chatelier's principle. How can this principle help us maximize the yields of reactions?

14.52 Use Le Chatelier's principle to explain why the equilibrium vapor pressure of a liquid increases with increasing temperature.

14.53 What is meant by "the position of an equilibrium"?

14.54 List four factors that can shift the position of an equilibrium. Only one of these factors can alter the value of the equilibrium constant. Which one is it?

14.55 Does the addition of a catalyst have any effects on the position of an equilibrium? Explain.

PROBLEMS

14.56 Consider the following equilibrium system:

$$SO_2(g) + Cl_2(g) \rightleftharpoons SO_2Cl_2(g)$$

Predict how the equilibrium position would change if (a) Cl_2 gas were added to the system; (b) SO_2Cl_2 were removed from the system; (c) SO_2 were removed from the system. The temperature remains constant.

14.57 Heating solid sodium bicarbonate in a closed vessel establishes the following equilibrium:

$$2NaHCO_3(s) \rightleftharpoons Na_2CO_3(s) + H_2O(g) + CO_2(g)$$

What would happen to the equilibrium position if (a) some of the CO_2 were removed from the system; (b) some solid Na_2CO_3 were added to the system; (c) some of the solid $NaHCO_3$ were removed from the system? The temperature remains constant.

14.58 Consider the following equilibrium systems:

(a) $A \rightleftharpoons 2B$ $\quad \Delta H° = 20.0$ kJ

(b) $A + B \rightleftharpoons C$ $\quad \Delta H° = -5.4$ kJ

(c) $A \rightleftharpoons B$ $\quad \Delta H° = 0.0$ kJ

Predict the change in the equilibrium constant K_c that would occur in each case if the temperature of the reacting system were raised.

14.59 What effect does an increase in pressure have on each of the following systems at equilibrium?

(a) $A(s) \rightleftharpoons 2B(s)$

(b) $2A(l) \rightleftharpoons B(l)$

(c) $A(s) \rightleftharpoons B(g)$

(d) $A(g) \rightleftharpoons B(g)$

(e) $A(g) \rightleftharpoons 2B(g)$

The temperature is kept constant. In each case, the reactants are in a cylinder fitted with a movable piston.

14.60 Consider the equilibrium

$$3O_2(g) \rightleftharpoons 2O_3(g) \qquad \Delta H° = 284 \text{ kJ}$$

What would be the effect on the position of equilibrium of (a) increasing the total pressure on the system by decreasing its volume; (b) adding O_2 to the reaction mixture; and (c) decreasing the temperature?

14.61 Consider the following equilibrium process:

$$PCl_5(g) \rightleftharpoons PCl_3(g) + Cl_2(g) \qquad \Delta H° = 92.5 \text{ kJ}$$

Predict the direction of the shift in equilibrium when (a) the temperature is raised; (b) more chlorine gas is added to the reaction mixture; (c) some PCl_3 is removed from the mixture; (d) the pressure on the gases is increased; (e) a catalyst is added to the reaction mixture.

14.62 Consider the reaction

$$2SO_2(g) + O_2(g) \rightleftharpoons 2SO_3(g) \qquad \Delta H° = -198.2 \text{ kJ}$$

Comment on the changes in the concentrations of SO_2, O_2, and SO_3 at equilibrium if we were to (a) increase the temperature; (b) increase the pressure; (c) increase SO_2; (d) add a catalyst; (e) add helium at constant volume.

14.63 In the uncatalyzed reaction at 100°C

$$N_2O_4(g) \rightleftharpoons 2NO_2(g)$$

the pressure of the gases at equilibrium are $P_{N_2O_4} =$ 0.377 atm and $P_{NO_2} = 1.56$ atm. What would happen to these pressures if a catalyst were present?

14.64 Consider the gas-phase reaction

$$2CO(g) + O_2(g) \rightleftharpoons 2CO_2(g)$$

Predict the shift in the equilibrium position when helium gas is added to the equilibrium mixture (a) at constant pressure and (b) at constant volume.

14.65 Consider the following reaction at equilibrium in a closed container:

$$CaCO_3(s) \rightleftharpoons CaO(s) + CO_2(g)$$

What would happen if (a) the volume is increased; (b) some CaO is added to the mixture; (c) some $CaCO_3$ is removed; (d) some CO_2 is added to the mixture; (e) a few drops of a NaOH solution are added to the mixture; (f) a few drops of a HCl solution are added to the mixture (ignore the reaction between CO_2 and water); (g) temperature is increased?

14.66 Consider the following reacting system:

$$2NO(g) + Cl_2(g) \rightleftharpoons 2NOCl(g)$$

What combination of temperature and pressure would maximize the yield of NOCl? (*Hint:* $\Delta H_f°(NOCl) =$ 51.7 kJ/mol. You will also need to consult Appendix 3.)

Miscellaneous Problems

14.67 Consider the statement: "The equilibrium constant of a reacting mixture of solid NH_4Cl and gaseous NH_3 and HCl is 0.316." List three important pieces of information that are missing from this statement.

14.68 Initially pure NOCl gas was heated to 240°C in a 1.00-L container. At equilibrium it was found that the total pressure was 1.00 atm and the NOCl pressure was 0.64 atm.

$$2NOCl(g) \rightleftharpoons 2NO(g) + Cl_2(g)$$

(a) Calculate the partial pressures of NO and Cl_2 in the system. (b) Calculate the equilibrium constant K_P.

14.69 Consider the following reaction:

$$N_2(g) + O_2(g) \rightleftharpoons 2NO(g)$$

The equilibrium constant K_P for the reaction is 1.0×10^{-15} at 25°C and 0.050 at 2200°C. Is the formation of nitric oxide endothermic or exothermic? Explain your answer.

14.70 Baking soda (sodium bicarbonate) undergoes thermal decomposition as follows:

$$2NaHCO_3(s) \rightleftharpoons Na_2CO_3(s) + CO_2(g) + H_2O(g)$$

Would we obtain more CO_2 and H_2O by adding extra baking soda to the reaction mixture in (a) a closed vessel or (b) an open vessel?

14.71 Consider the following reaction at equilibrium:

$$A(g) \rightleftharpoons 2B(g)$$

From the following data, calculate the equilibrium constant (both K_P and K_c) at each temperature. Is the reaction endothermic or exothermic?

Temperature (°C)	[A]	[B]
200	0.0125	0.843
300	0.171	0.764
400	0.250	0.724

14.72 The equilibrium constant K_P for the reaction

$$2H_2O(g) \rightleftharpoons 2H_2(g) + O_2(g)$$

is found to be 2×10^{-42} at 25°C. (a) What is K_c for the reaction at the same temperature? (b) The very small value of K_P (and K_c) indicates that the reaction overwhelmingly favors the formation of water molecules. Explain why, despite this fact, a mixture of hydrogen and oxygen gases can be kept at room temperature without any change.

14.73 At a certain temperature and a total pressure of 1.2 atm, the partial pressures of an equilibrium mixture

$$2A(g) \rightleftharpoons B(g)$$

are $P_A = 0.60$ atm and $P_B = 0.60$ atm. (a) Calculate the K_P for the reaction at this temperature. (b) If the total pressure were increased to 1.5 atm, what would be the partial pressures of A and B at equilibrium?

14.74 The decomposition of ammonium hydrogen sulfide

$$NH_4HS(s) \rightleftharpoons NH_3(g) + H_2S(g)$$

is an endothermic process. A 6.1589-g sample of the solid is placed in an evacuated 4.000-L vessel at exactly 24°C. After equilibrium has been established, the total pressure inside is 0.709 atm. Some solid NH_4HS remains in the vessel. (a) What is the K_P for the reaction? (b) What percentage of the solid has decomposed? (c) If the volume of the vessel were doubled at constant temperature, what would happen to the amount of solid in the vessel?

14.75 Consider the reaction

$$2NO(g) + O_2(g) \rightleftharpoons 2NO_2(g)$$

At 430°C, an equilibrium mixture consists of 0.020 mole of O_2, 0.040 mole of NO, and 0.96 mole of NO_2. Calculate K_P for the reaction, given that the total pressure is 0.20 atm.

14.76 When heated, ammonium carbamate decomposes as follows:

$$NH_4CO_2NH_2(s) \rightleftharpoons 2NH_3(g) + CO_2(g)$$

At a certain temperature the equilibrium pressure of the system is 0.318 atm. Calculate K_P for the reaction.

14.77 A mixture of 0.47 mole of H_2 and 3.59 mole of HCl is heated to 2800°C. Calculate the equilibrium partial pressures of H_2, Cl_2, and HCl if the total pressure is 2.00 atm. The K_P for the reaction $H_2(g) + Cl_2(g) \rightleftharpoons 2HCl(g)$ is 193 at 2800°C.

14.78 Consider the reaction in a closed container:

$$N_2O_4(g) \rightleftharpoons 2NO_2(g)$$

Initially there is 1 mole of N_2O_4 present. At equilibrium, α mole of N_2O_4 has dissociated to form NO_2. (a) Derive an expression for K_P in terms of α and P, the total pressure. (b) How does the expression in (a) help you predict the shift in equilibrium due to an increase in P? Does your prediction agree with Le Chatelier's principle?

14.79 One mole of N_2 and three moles of H_2 are placed in a flask at 200°C. Calculate the total pressure of the system at equilibrium if the mole fraction of NH_3 is found to be 0.21. The K_P for the reaction is 4.31×10^{-4}.

14.80 At 1130°C the equilibrium constant (K_c) for the reaction

$$2H_2S(g) \rightleftharpoons 2H_2(g) + S_2(g)$$

is 2.25×10^{-4}. If $[H_2S] = 4.84 \times 10^{-3}$ M and $[H_2] = 1.50 \times 10^{-3}$ M, calculate $[S_2]$.

14.81 A quantity of 6.75 g of SO_2Cl_2 was placed in a 2.00-L flask. At 648 K, there is 0.0345 mole of SO_2 present.

Calculate K_c for the reaction

$$SO_2Cl_2(g) \rightleftharpoons SO_2(g) + Cl_2(g)$$

14.82 The formation of SO_3 from SO_2 and O_2 is an intermediate step in the manufacture of sulfuric acid, and it is also responsible for the acid rain phenomenon. The equilibrium constant (K_P) for the reaction

$$2SO_2(g) + O_2(g) \rightleftharpoons 2SO_3(g)$$

is 0.13 at 830°C. In one experiment 2.00 mol SO_2 and 2.00 mol O_2 were initially present in a flask. What must the total pressure at equilibrium be in order to have an 80.0 percent yield of SO_3?

14.83 Consider the dissociation of iodine:

$$I_2(g) \rightleftharpoons 2I(g)$$

A 1.00-g sample of I_2 is heated at 1200°C in a 500-mL flask. At equilibrium the total pressure is 1.51 atm. Calculate K_P for the reaction. [*Hint:* Use the result in 14.78(a). The degree of dissociation α is given by the ratio of observed pressure over calculated pressure assuming no dissociation.]

14.84 In this chapter we learned that a catalyst has no effect on the position of an equilibrium because it speeds up both the forward and reverse rates to the same extent. To test this statement, consider a situation in which an equilibrium of the type

$$2A(g) \rightleftharpoons B(g)$$

is established inside a cylinder fitted with a weightless piston. The piston is attached to the cover of a box containing a catalyst with a string. When the piston moves upward (expanding against atmospheric pressure), the cover is lifted and the catalyst is exposed to the gases. When the piston moves downward, the box is closed. Assume that the catalyst only speeds up the forward reaction (2A → B) but does not affect the reverse process (B → 2A). Supposing the catalyst is suddenly exposed to the equilibrium system as shown below, describe what would happen subsequently. How does this ''thought'' experiment convince you that no such catalyst can exist?

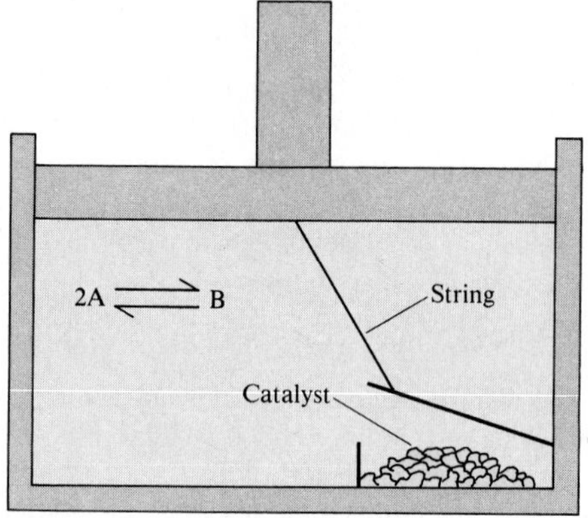

Opposite page: Some commonly encountered acids and bases in food, medicine, and cleaning solution.

CHAPTER 15

ACIDS AND BASES: GENERAL PROPERTIES

Acid-base reactions in aqueous solution are some of the most important processes in chemical and biological systems. In this first of two chapters dealing with the properties of acids and bases, we study some important definitions of acids and bases, the pH scale, and the relationship between acid strengths and molecular structure. We also look at some typical acid-base reactions and at certain oxides that can act as acids and bases.

15.1 Brønsted Acids and Bases

Acids and bases were first defined in Section 3.4. It was pointed out that the Brønsted definitions are generally most suitable to discuss the properties and reactions of acids and bases. According to Brønsted, an acid is a substance capable of donating a proton, and a base is a substance capable of accepting a proton.

Conjugate Acid-Base Pair

Conjugate means "joined together."

An extension of the Brønsted definition of acids and bases is the concept of the ***conjugate acid-base pair,*** which can be defined as *an acid and its conjugate base or a base and its conjugate acid.* The conjugate base of a Brønsted acid is the species that remains when a proton has been removed from the acid. Conversely, a conjugate acid results from the addition of a proton to a Brønsted base. Every Brønsted acid has a conjugate base, and every Brønsted base has a conjugate acid. For example, the chloride ion (Cl^-) is the conjugate base formed from the acid HCl, and H_2O is the conjugate base of the acid H_3O^+. Similarly, the ionization of acetic acid can be represented as

$$\mathrm{H{-}\underset{H}{\overset{H}{C}}{-}\overset{:O:}{\overset{\|}{C}}{-}\ddot{\underset{..}{O}}{-}H + H{-}\underset{H}{\ddot{O}}: \rightleftharpoons H{-}\underset{H}{\overset{H}{C}}{-}\overset{:O:}{\overset{\|}{C}}{-}\ddot{\underset{..}{O}}:^- + \left[H{-}\underset{H}{\ddot{O}}{-}H\right]^+}$$

$$\underset{\text{acid}_1}{CH_3COOH(aq)} + \underset{\text{base}_2}{H_2O(l)} \rightleftharpoons \underset{\text{base}_1}{CH_3COO^-(aq)} + \underset{\text{acid}_2}{H_3O^+(aq)}$$

The formula of a conjugate base always has one fewer hydrogen atom and one more negative charge (or one fewer positive charge) than the formula of the corresponding acid.

The subscripts 1 and 2 designate the two conjugate acid-base pairs. Thus the acetate ion (CH_3COO^-) is the conjugate base of CH_3COOH. Both the ionization of HCl (shown earlier) and the ionization of CH_3COOH are examples of Brønsted acid-base reactions. Unlike Arrhenius acid-base reactions, Brønsted acid-base reactions do not produce a salt and water.

Strictly speaking, NaOH is not a Brønsted base since it cannot accept a proton. However, NaOH is a strong electrolyte that ionizes completely in solution. The hydroxide ion (OH^-) formed by that ionization *is* a Brønsted base because it can accept a proton:

$$H_3O^+(aq) + OH^-(aq) \longrightarrow 2H_2O(l)$$

Thus, when we call NaOH or any other metal hydroxide a base, we are actually referring to the OH^- species derived from the hydroxide.

The Brønsted definition also allows us to classify ammonia as a base because of its ability to accept a proton:

$$\mathrm{H{-}\underset{H}{\ddot{N}}{-}H + H{-}\underset{H}{\ddot{O}}: \rightleftharpoons \left[H{-}\underset{H}{\overset{H}{N}}{-}H\right]^+ + H{-}\ddot{\underset{..}{O}}:^-}$$

$$\underset{\text{base}_1}{NH_3(aq)} + \underset{\text{acid}_2}{H_2O(l)} \rightleftharpoons \underset{\text{acid}_1}{NH_4^+(aq)} + \underset{\text{base}_2}{OH^-(aq)}$$

In this case, NH_4^+ is the conjugate acid of the base NH_3, and OH^- is the conjugate base of the acid H_2O. Table 15.1 lists a number of common acids and their conjugate bases.

Table 15.1 Some common acids and their conjugate bases

Acid		Conjugate base	
Name	**Formula**	**Name**	**Formula**
Hydrochloric acid	HCl	Chloride ion	Cl^-
Nitric acid	HNO_3	Nitrate ion	NO_3^-
Hydrocyanic acid	HCN	Cyanide ion	CN^-
Perchloric acid	$HClO_4$	Perchlorate ion	ClO_4^-
Sulfuric acid	H_2SO_4	Hydrogen sulfate ion	HSO_4^-
Hydrogen sulfate ion	HSO_4^-	Sulfate ion	SO_4^{2-}
Carbonic acid	H_2CO_3	Hydrogen carbonate ion	HCO_3^-
Hydrogen carbonate ion	HCO_3^-	Carbonate ion	CO_3^{2-}
Ammonium ion	NH_4^+	Ammonia	NH_3

Example 15.1 shows the conjugate pairs in an acid-base reaction.

EXAMPLE 15.1

Identify the conjugate acid-base pairs in the following reaction:

$$NH_3(aq) + HF(aq) \rightleftharpoons NH_4^+(aq) + F^-(aq)$$

Answer

The conjugate acid-base pairs are (1) HF (acid) and F^- (base) and (2) NH_4^+ (acid) and NH_3 (base).

Similar problem: 15.9.

The Hydrated Proton

The H^+ ion is hydrated in aqueous solution. The reaction between H^+ and water is usually given as

$$H^+(aq) + H_2O(l) \longrightarrow H_3O^+(aq)$$

However, evidence suggests that the proton may be associated with more than one water molecule (Figure 15.1). In practice, a quantitative treatment of acid ionization is unaffected by whether we write H^+, H_3O^+, $H_7O_3^+$, or $H_9O_4^+$ for the proton in water. In this book we will write the formula of a proton in aqueous solution as either H^+ or H_3O^+. The formula H^+ is less cumbersome in calculations involving hydrogen ion concentrations (see next section) and in calculations involving equilibrium constants (see Chapter 16), whereas the formula H_3O^+ is more useful when discussing Brønsted acid-base properties.

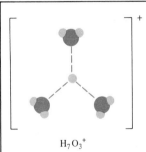

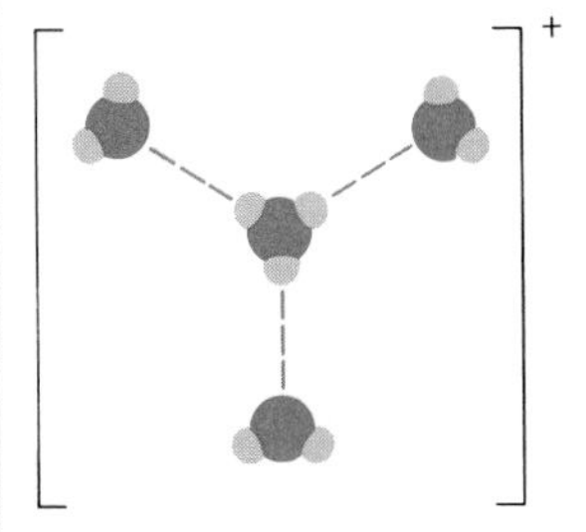

Figure 15.1 *Probable structures for the hydrated H^+ ion in water: $H_7O_3^+$ and $H_9O_4^+$. The dashes represent hydrogen bonds.*

15.2 Autoionization of Water and the pH Scale

As we know from preceding chapters, water is a unique solvent. One of its special properties is its ability to act both as an acid and as a base. Water functions as a base in reactions with acids such as HCl and CH_3COOH, and it functions as an acid in reactions with bases such as NH_3. In fact, water itself undergoes ionization to a small extent:

$$H_2O(l) \rightleftharpoons H^+(aq) + OH^-(aq) \tag{15.1}$$

This reaction is sometimes called the *autoionization* of water. To describe the acid-base properties of water in the Brønsted framework, it is preferable to express the autoionization as

$$\mathrm{H{-}\ddot{\underset{\underset{H}{|}}{O}}{:}} + \mathrm{H{-}\ddot{\underset{\underset{H}{|}}{O}}{:}} \rightleftharpoons \left[\mathrm{H{-}\ddot{\underset{\underset{H}{|}}{O}}{-}H}\right]^+ + \mathrm{H{-}\ddot{\underset{\cdot\cdot}{O}}{:}^-}$$

or

$$\underset{\text{acid}_1}{H_2O} + \underset{\text{base}_2}{H_2O} \rightleftharpoons \underset{\text{acid}_2}{H_3O^+} + \underset{\text{base}_1}{OH^-}$$

The acid-base conjugate pairs are (1) acid_1 and base_1 (H_2O and OH^-) and (2) acid_2 and base_2 (H_3O^+ and H_2O).

The Ion Product of Water

In the study of acid-base reactions in aqueous solutions, the important quantity is the hydrogen ion concentration. (As we will see later, the hydroxide ion concentration is related to the hydrogen ion concentration, so we need not consider it here.) Although water undergoes autoionization, it is a very weak electrolyte and, therefore, a poor electrical conductor.

Tap water and water from underground sources do conduct electricity because they contain many dissolved ions.

Expressing the hydrated proton as H^+ rather than H_3O^+, we can write the equilibrium constant for the autoionization of water, Equation (15.1), as

$$K_c = \frac{[H^+][OH^-]}{[H_2O]}$$

Recall that in pure water, $[H_2O] = 55.5\ M$ (see p. 601).

Since a very small fraction of water molecules are ionized, the concentration of water, that is, $[H_2O]$, remains virtually unchanged. Therefore

$$K_c[H_2O] = K_w = [H^+][OH^-] \tag{15.2}$$

The "new" equilibrium constant, K_w, is called the ***ion-product constant,*** which is the *product of the molar concentrations of H^+ and OH^- ions at a particular temperature.* In pure water at 25°C, the concentrations of H^+ and OH^- ions are equal and found to be $[H^+] = 1.0 \times 10^{-7}\ M$ and $[OH^-] = 1.0 \times 10^{-7}\ M$. Thus, from Equation (15.2)

To dramatize these concentrations, imagine that you could randomly draw out and examine ten particles (H_2O, H^+, or OH^-) per second from a beaker of water. On average, it would take you nearly two years, working nonstop, to find one H^+ ion!

$$K_w = (1.0 \times 10^{-7})(1.0 \times 10^{-7}) = 1.0 \times 10^{-14} \quad \text{at } 25°\text{C}$$

Note that whether we have pure water or a solution of dissolved species, the following relation *always* holds at 25°C:

$$K_w = [H^+][OH^-] = 1.0 \times 10^{-14}$$

Whenever $[H^+] = [OH^-]$, the aqueous solution is said to be neutral. In an acidic solution there is an excess of H^+ ions, or $[H^+] > [OH^-]$. In a basic solution there is an excess of hydroxide ions, or $[H^+] < [OH^-]$. In practice we can adjust the concentration of either H^+ or OH^- ions in solution, but we cannot vary both of them independently. If we adjust the solution so that $[H^+] = 1.0 \times 10^{-6}\ M$, the OH^- concentration *must* change to

$$[OH^-] = \frac{K_w}{[H^+]} = \frac{1.0 \times 10^{-14}}{1.0 \times 10^{-6}} = 1.0 \times 10^{-8}\ M$$

K_W increases with temperature because the ionization of water is endothermic.

It is important to remember that, because K_w is an equilibrium constant, its value changes with temperature. Thus at 40°C, $K_w = 3.8 \times 10^{-14}$ and Equation (15.2) becomes

$$3.8 \times 10^{-14} = [H^+][OH^-] = K_w$$

Since the solution is still neutral, $[H^+] = [OH^-]$. Therefore, at this temperature the concentrations are

$$[H^+] = \sqrt{3.8 \times 10^{-14}} = 1.9 \times 10^{-7}\ M$$

and

$$[OH^-] = \sqrt{3.8 \times 10^{-14}} = 1.9 \times 10^{-7}\ M$$

A neutral solution at 40°C will have a higher H^+ ion concentration ($1.9 \times 10^{-7}\ M$) than a neutral solution at 25°C ($1.0 \times 10^{-7}\ M$). Unless otherwise stated, we will assume in all subsequent discussions that the temperature is 25°C.

An application of Equation (15.2) is given in Example 15.2.

EXAMPLE 15.2

The concentration of OH^- ions in a certain household ammonia cleaning solution is 0.0025 *M*. Calculate the concentration of the H^+ ions.

Answer

Rearranging Equation (15.2), we write

$$[H^+] = \frac{K_w}{[OH^-]} = \frac{1.0 \times 10^{-14}}{0.0025} = 4.0 \times 10^{-12}\ M$$

Since $[H^+] < [OH^-]$, the solution is basic, as we would expect from the earlier discussion of the reaction of ammonia with water.

Similar problem: 15.22.

pH—A Measure of Acidity

Because the concentrations of H^+ and OH^- ions are frequently very small numbers and therefore inconvenient to work with, Soren Sorensen† in 1909 proposed a more practical measure called pH. The ***pH*** of a solution is defined as

$$pH = -\log [H^+] \tag{15.3}$$

† Soren Peer Lauritz Sorensen (1868–1939). Danish biochemist. Sorensen originally wrote the symbol as p_H and called p the "hydrogen ion exponent" (*Wasserstoffionexponent*); it is the initial letter of *Potenz* (German), *puissance* (French), and *power* (English). It is now customary to write the symbol as pH.

Table 15.2 The pHs of some common fluids

Sample	pH value
Gastric juice in the stomach	1.0–2.0
Lemon juice	2.4
Vinegar	3.0
Grapefruit juice	3.2
Orange juice	3.5
Urine	4.8–7.5
Water exposed to air*	5.5
Saliva	6.4–6.9
Milk	6.5
Pure water	7.0
Blood	7.35–7.45
Tears	7.4
Milk of magnesia	10.6
Household ammonia	11.5

*Water exposed to air for a long period of time absorbs atmospheric CO_2 to form carbonic acid, H_2CO_3.

Thus we see that the *pH of a solution is given by the negative logarithm of the hydrogen ion concentration (in mol/L).*

Keep in mind that Equation (15.3) is simply a definition designed to give us convenient numbers to work with. Furthermore, the term $[H^+]$ in Equation (15.3) pertains only to the *numerical part* of the expression for hydrogen ion concentration, for we cannot take the logarithm of units. Thus, like the equilibrium constant, the pH of a solution is a dimensionless quantity.

Since pH is simply a way to express hydrogen ion concentration, acidic and basic solutions at 25°C can be identified by their pH values, as follows:

Acidic solutions: $[H^+] > 1.0 \times 10^{-7}\ M$, pH < 7.00

Basic solutions: $[H^+] < 1.0 \times 10^{-7}\ M$, pH > 7.00

Neutral solutions: $[H^+] = 1.0 \times 10^{-7}\ M$, pH $= 7.00$

In the laboratory, the pH of a solution is measured with a pH meter (Figure 15.2). Electrodes attached to the meter are dipped into the test solution, and the pH is read directly on the scale. Table 15.2 lists the pHs for a number of common fluids.

A scale analogous to the pH scale can be devised using the negative logarithm of the hydroxide ion concentration. Thus we define pOH as

$$\text{pOH} = -\log\,[OH^-] \qquad (15.4)$$

Now consider again the ion-product constant for water:

$$[H^+][OH^-] = K_w = 1.0 \times 10^{-14}$$

Taking the negative logarithm of both sides, we obtain

$$-(\log\,[H^+] + \log\,[OH^-]) = -\log\,(1.0 \times 10^{-14})$$

$$-\log\,[H^+] - \log\,[OH^-] = 14.00$$

From the definitions of pH and pOH we obtain

$$\text{pH} + \text{pOH} = 14.00 \qquad (15.5)$$

Equation (15.5) provides us with another way to express the relationship between the H^+ ion concentration and the OH^- ion concentration.

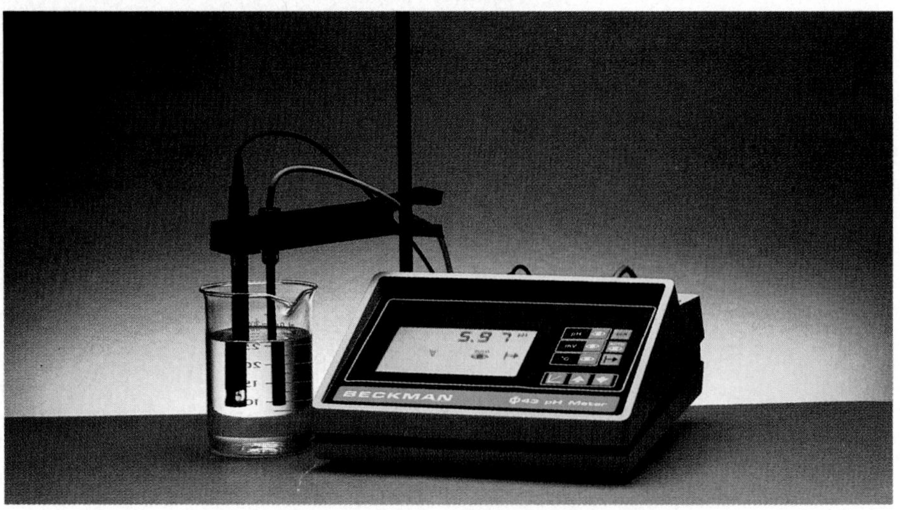

Figure 15.2 *A pH meter is commonly used in the laboratory to determine the pH of a solution. Although many pH meters have scales marked with values from 1 to 14, pH values can, in fact, be less than 1 and greater than 14.*

The following examples illustrate the calculation of pH and related quantities.

EXAMPLE 15.3

The concentration of H^+ ions in a bottle of table wine was 3.2×10^{-4} *M* right after the cork was removed. Only half of the wine was consumed. The other half, after it had been standing open to the air for a month, was found to have a hydrogen ion concentration equal to 1.0×10^{-3} *M*. Calculate the pH of the wine on these two occasions.

Answer

When the bottle was first opened, $[H^+] = 3.2 \times 10^{-4}$ *M*, which we substitute in Equation (15.3).

$$\begin{aligned} pH &= -\log [H^+] \\ &= -\log (3.2 \times 10^{-4}) = 3.49 \end{aligned}$$

In each case the pH has only two significant figures. The three in 3.49 serves only to locate the decimal point in the number 3.2×10^{-4}.

On the second occasion, $[H^+] = 1.0 \times 10^{-3}$ *M*, so that

$$pH = -\log (1.0 \times 10^{-3}) = 3.00$$

The increase in hydrogen ion concentration (or decrease in pH) is largely the result of the conversion of some of the alcohol (ethanol) to acetic acid, a reaction that takes place in the presence of molecular oxygen.

Similar problems: 15.23, 15.24, 15.25.

EXAMPLE 15.4

The pH of rainwater collected in a certain region of the U.S. Northeast on a particular day was 4.82. Calculate the H^+ ion concentration of the rainwater.

Answer

From Equation (15.3)

$$4.82 = -\log [H^+]$$

Taking the antilog of both sides of this equation (see Appendix 4) gives

$$1.5 \times 10^{-5}\ M = [H^+]$$

Similar problems: 15.23, 15.24, 15.25.

EXAMPLE 15.5

In a NaOH solution $[OH^-]$ is 2.9×10^{-4} *M*. Calculate the pH of the solution.

Answer

We use Equation (15.4):

$$\begin{aligned} pOH &= -\log [OH^-] \\ &= -\log (2.9 \times 10^{-4}) \\ &= 3.54 \end{aligned}$$

Now we use Equation (15.5):

$$\begin{aligned} pH + pOH &= 14.00 \\ pH &= 14.00 - pOH \\ &= 14.00 - 3.54 = 10.46 \end{aligned}$$

Similar problems: 15.23, 15.24, 15.25.

15.3 Strengths of Acids and Bases

Depending on the nature of an acid, some or all of its molecules may ionize when the acid is dissolved in water. The *strength* of an acid can be measured by the fraction of molecules that undergo ionization. Quantitatively, acid strength may be expressed in terms of ***percent ionization,*** which is defined as

Percent ionization refers only to the equilibrium condition.

$$\text{percent ionization} = \frac{\text{ionized acid concentration at equilibrium}}{\text{initial concentration of acid}} \times 100\%$$

Suppose we have two aqueous solutions containing acids HA and HB at the *same* concentration. Using the Brønsted definition, let's say that HA is a stronger acid than HB because it can transfer a proton to water (which acts as a Brønsted base in this case) more readily than HB can. Consequently the percent ionization is greater for HA than for HB.

$$HA(aq) + H_2O(l) \rightleftharpoons H_3O^+(aq) + A^-(aq)$$

$$HB(aq) + H_2O(l) \rightleftharpoons H_3O^+(aq) + B^-(aq)$$

At equilibrium, the solution containing HA will have a higher concentration of H^+ ions and a lower pH than the one containing HB.

Strong acids—hydrochloric acid (HCl), nitric acid (HNO_3), perchloric acid ($HClO_4$), and sulfuric acid (H_2SO_4), for example—are all strong electrolytes. For most practical purposes, they may be assumed to be 100 percent ionized in water:

$$HCl(aq) + H_2O(l) \longrightarrow H_3O^+(aq) + Cl^-(aq)$$

$$HNO_3(aq) + H_2O(l) \longrightarrow H_3O^+(aq) + NO_3^-(aq)$$

$$HClO_4(aq) + H_2O(l) \longrightarrow H_3O^+(aq) + ClO_4^-(aq)$$

$$H_2SO_4(aq) + H_2O(l) \longrightarrow H_3O^+(aq) + HSO_4^-(aq)$$

Note that H_2SO_4 is a diprotic acid; we consider only the first stage of ionization here.

Most acids, however, are classified as weak acids. Within this category the strength of acids can actually vary greatly. For example, both hydrofluoric acid (HF) and hydrocyanic acid (HCN) are classified as weak acids, yet in 0.10 *M* solutions the percent ionizations at 25°C of HF and HCN at equilibrium are 8.4 percent and 0.0070 percent, respectively. Thus HF is a much stronger acid than HCN. The limited ionization of weak acids is related to the equilibrium constant for ionization, a relationship we will study in the next chapter.

A reference acid is an acid chosen to compare the strengths of several bases.

Much of what we have said so far about acids also applies to bases. The strength of a base refers to its ability to accept a proton from a reference acid, which is normally the solvent. Hydroxides of alkali metals and alkaline earth metals, such as NaOH, KOH, and $Ba(OH)_2$, are strong bases. These substances are all strong electrolytes that ionize completely in solution:

All alkali metal hydroxides are soluble. Of the alkaline earth metal hydroxides, $Be(OH)_2$, $Mg(OH)_2$, and $Ca(OH)_2$ are insoluble; $Sr(OH)_2$ is slightly soluble; and $Ba(OH)_2$ is soluble.

$$NaOH(aq) \longrightarrow Na^+(aq) + OH^-(aq)$$

$$KOH(aq) \longrightarrow K^+(aq) + OH^-(aq)$$

$$Ba(OH)_2(aq) \longrightarrow Ba^{2+}(aq) + 2OH^-(aq)$$

On the other hand, ammonia is a weak base. Measurements show that the percent ionization of a 0.10 *M* NH_3 solution at 25°C is only 1.3 percent:

Note that NH_3 does not ionize like an acid because it does not split up to form ions the way, say, HCl does.

$$NH_3(aq) + H_2O(l) \rightleftharpoons NH_4^+(aq) + OH^-(aq)$$

Table 15.3 lists some important conjugate acid-base pairs, in order of their relative strengths. We can make several observations in examining this table:

- If an acid is strong, its conjugate base has no measurable strength. Within a series of weak acids, the stronger the acid, the weaker its conjugate base, and vice versa. For example, HNO_2 is a stronger acid than CH_3COOH; therefore, NO_2^- is a weaker base than CH_3COO^-.
- H_3O^+ is the strongest acid that can exist in aqueous solution. Acids stronger than H_3O^+ react with water to produce H_3O^+ and their conjugate bases. Thus HCl, which is a stronger acid than H_3O^+, reacts with water completely to form H_3O^+ and Cl^-:

A bottle labeled "0.10 *M* HCl" actually contains 0.10 *M* H_3O^+, and no molecular HCl at all!

$$HCl(aq) + H_2O(l) \longrightarrow H_3O^+(aq) + Cl^-(aq)$$

Table 15.3 Relative strengths of conjugate acid-base pairs

	Acid		Conjugate base	
↑ Acid strength increases	$HClO_4$	Strong acids. Assumed to be 100 percent ionized in aqueous solution.	ClO_4^-	Base strength increases ↓
	HI		I^-	
	HBr		Br^-	
	HCl		Cl^-	
	H_2SO_4		HSO_4^-	
	HNO_3		NO_3^-	
	H_3O^+		H_2O	
	HSO_4^-	Weak acids. At equilibrium, there is a mixture of nonionized acid molecules, as well as the H^+ ions and the conjugate base.	SO_4^{2-}	
	HF		F^-	
	HNO_2		NO_2^-	
	$HCOOH$		$HCOO^-$	
	CH_3COOH		CH_3COO^-	
	NH_4^+		NH_3	
	HCN		CN^-	
	H_2O		OH^-	
	NH_3		NH_2^-	

Acids weaker than H_3O^+ react with water to a much smaller extent, producing H_3O^+ and their conjugate bases. For example, the following equilibrium lies primarily to the left:

$$HF(aq) + H_2O(l) \rightleftharpoons H_3O^+(aq) + F^-(aq)$$

- The OH^- ion is the strongest base that can exist in aqueous solution. Bases stronger than OH^- react with water to produce OH^- and their conjugate acids. For example, the amide ion (NH_2^-) is a stronger base than OH^-, so it reacts with water completely as follows:

$$NH_2^-(aq) + H_2O(l) \longrightarrow NH_3(aq) + OH^-(aq)$$

The amide ion *does* exist in liquid ammonia.

For this reason the amide ion does not exist in aqueous solutions.

The following example shows calculations of pH for solutions containing a strong acid and a strong base.

EXAMPLE 15.6

Calculate the pH of (a) a 1.0×10^{-3} *M* HCl solution and (b) a 0.020 *M* $Ba(OH)_2$ solution.

Answer

(a) Since HCl is a strong acid, it is completely ionized in solution:

$$HCl(aq) \longrightarrow H^+(aq) + Cl^-(aq)$$

The concentrations of all the species (HCl, H^+, and Cl^-) before and after ionization can be represented as follows:

	$HCl(aq)$	$\longrightarrow$	$H^+(aq)$	+	$Cl^-(aq)$
Initial:	1.0×10^{-3} *M*		0.0 *M*		0.0 *M*
Change:	-1.0×10^{-3} *M*		$+1.0 \times 10^{-3}$ *M*		$+1.0 \times 10^{-3}$ *M*
Final:	0.0 *M*		1.0×10^{-3} *M*		1.0×10^{-3} *M*

A positive (+) change represents an increase and a negative (−) change indicates a decrease in concentration. Thus

$$[H^+] = 1.0 \times 10^{-3}\ M$$
$$pH = -\log(1.0 \times 10^{-3})$$
$$= 3.00$$

(b) $Ba(OH)_2$ is a strong base; each $Ba(OH)_2$ produces two OH^- ions:

$$Ba(OH)_2(aq) \longrightarrow Ba^{2+}(aq) + 2OH^-(aq)$$

The changes in the concentrations of all the species can be represented as follows:

	$Ba(OH)_2(aq)$	$\longrightarrow$	$Ba^{2+}(aq)$	+	$2OH^-(aq)$
Initial:	0.020 *M*		0.00 *M*		0.00 *M*
Change:	−0.020 *M*		+0.020 *M*		+2(0.020) *M*
Final:	0.00 *M*		0.020 *M*		0.040 *M*

Thus

$$[OH^-] = 0.040\ M$$

$$pOH = -\log 0.040 = 1.40$$

Therefore

$$\begin{aligned} pH &= 14.00 - pOH \\ &= 14.00 - 1.40 \\ &= 12.60 \end{aligned}$$

Note that in both (a) and (b) we have neglected the contribution of the autoionization of water to $[H^+]$ and $[OH^-]$ because $1.0 \times 10^{-7}\ M$ is so small compared with $1.0 \times 10^{-3}\ M$ and $0.040\ M$.

Similar problem: 15.25.

The equilibrium position between two weak acids is illustrated in Example 15.7.

EXAMPLE 15.7

Predict the direction of the following reaction in aqueous solution:

$$HNO_2(aq) + CN^-(aq) \rightleftharpoons HCN(aq) + NO_2^-(aq)$$

Answer

From Table 15.3 we see that HNO_2 is a stronger acid than HCN. Thus CN^- is a stronger base than NO_2^-. The net reaction will proceed from left to right as written because HNO_2 is a better proton donor than HCN (and CN^- is a better proton acceptor than NO_2^-).

Similar problem: 15.43.

The Leveling Effect

We have seen that one way to compare the strengths of acids is to choose a reference base (usually the solvent) and measure the extent to which the acid donates a proton to the base. However, since the reaction between any acid stronger than H_3O^+ and water goes completely to the right, strong acids such as $HClO_4$, HCl, HNO_3, and H_2SO_4 will appear to be of equal strength in aqueous solution. Therefore we cannot compare the relative strengths of all acids on the basis of their readiness to lose a proton in aqueous solution. *The inability of a solvent to differentiate among the relative strengths of all acids stronger than the solvent's conjugate acid* is known as the ***leveling effect*** because the solvent is said to level the strengths of these acids, making them seem identical.

A simple analogy to the leveling effect is the following. Suppose an average adult and an Olympic weight lifter want to compare their strengths. First they attempt to lift a 15-kg weight. However, since they both can do this with ease, the result is inconclusive. But when they try to lift a 50-kg weight, the Olympic weight lifter lifts it readily but the average adult must struggle to do it and may even fail. The weights here correspond to the solvents in which we study the ionization of acids. The lighter weight corresponds to water, which exerts the leveling effect that renders comparison of acid strengths inconclusive.

Pure acetic acid is also called "glacial acetic acid."

To compensate for the leveling effect of water we can use a more weakly basic solvent like acetic acid. (This procedure corresponds to the use of a heavier weight to compare the relative strengths of the average adult and the Olympic weight lifter.) Acetic acid can function as a base by accepting a proton:

$$CH_3COOH + H^+ \rightleftharpoons CH_3COOH_2^+$$

The structure of a protonated acetic acid molecule is

This is one of two equivalent resonance structures.

$$CH_3-C\begin{matrix} \diagup\!\!\diagup \overset{+}{\ddot{O}}-H \\ \diagdown \ddot{\underset{..}{O}}-H \end{matrix}$$

Since acetic acid is a much weaker base than water, it is not as easily protonated. Thus there are appreciable differences in the extent to which the following reactions proceed from left to right:

$$HNO_3 + CH_3COOH \rightleftharpoons CH_3COOH_2^+ + NO_3^-$$

$$HCl + CH_3COOH \rightleftharpoons CH_3COOH_2^+ + Cl^-$$

$$HClO_4 + CH_3COOH \rightleftharpoons CH_3COOH_2^+ + ClO_4^-$$

$$H_2SO_4 + CH_3COOH \rightleftharpoons CH_3COOH_2^+ + HSO_4^-$$

Measuring the extent of ionization of these and other acids in acetic acid solvent shows that their relative strengths increase as follows:

$$HNO_3 < H_2SO_4 < HCl < HBr < HI < HClO_4$$

15.4 Molecular Structure and the Strength of Acids

What makes one acid stronger than another? How can we predict the relative strengths of acids in a series of similar compounds? To begin with, we must realize that the strength of an acid depends on a number of factors, such as the properties of the solvent, the temperature, and, of course, the molecular structure of the acid. When we compare the strengths of two acids, we can eliminate some variables by considering their properties in the same solvent and at the same temperature. Then we can focus on the structures of the acids.

Let us consider a certain acid HX. The factors that determine the strength of the acid are the polarity and bond energy of the H—X bond. The more polar the bond, the more readily the acid ionizes into H^+ and X^-. On the other hand, strong bonds (that is, bonds with high bond energies) are less easily ionized than weaker ones.

In this section we concentrate on two types of acids: ***binary acids,*** *acids that contain only two different elements,* and ***ternary acids,*** *acids that contain three different elements.*

Binary Acids

All binary acids contain hydrogen plus another element. The halogens form a particularly important series of binary acids called the *hydrohalic acids*. As we saw earlier,

Table 15.4 Bond dissociation energies for hydrogen halides and acid strengths for hydrohalic acids

Bond	Bond dissociation energy (kJ/mol)	Acid strength (in water)
H—F	568.2	weak
H—Cl	431.9	strong
H—Br	366.1	strong
H—I	298.3	strong

The ≪ sign means "much less than."

the strengths of the hydrohalic acids increase in the following order:

$$\text{HF} \ll \text{HCl} < \text{HBr} < \text{HI}$$

Because of the leveling effect of water, the relative strengths of the strong acids HCl, HBr, and HI must be studied in some other solvent, such as acetic acid. As Table 15.4 shows, HF has the highest bond dissociation energy of the four hydrogen halides. Since more energy is required to break the H—F bond, HF is a weak acid. At the other extreme in the series, HI has the lowest bond energy, so HI is the strongest acid of the group. In this series of acids the polarity of the bond actually decreases from HF to HI because F is the most electronegative of the halogens (see Figure 9.8). This property should enhance the acidity of HF relative to the other hydrohalic acids, but its magnitude is not great enough to break the trend in bond dissociation energies.

Any molecule containing a H atom is potentially a Brønsted acid.

The relative importance of bond energy versus bond polarity in determining acid strength is reversed for binary compounds in a particular period of the periodic table. Consider the binary hydrogen compounds of the second-period elements: CH_4, NH_3, H_2O, and HF. Methane (CH_4) has no measurable acidic properties whatsoever, whereas hydrofluoric acid (HF) is an acid of measurable strength in water. The strengths of the compounds as acids increase as follows:

$$CH_4 < NH_3 < H_2O < HF$$

This trend is the reverse of what we would expect on the basis of bond energy considerations. (From Table 9.4 we see that the HF molecule has the highest bond energy of the four molecules.) We can explain the trend by noting the increase in electronegativity as we go from C to F. As electronegativity increases, the H—X bond becomes more polar (X denotes the C, N, O, or F atom), so the compound has a greater tendency to ionize into H^+ and the corresponding anion.

Ternary Acids

To review the nomenclature of inorganic acids see Section 2.9.

Most ternary (three-element) acids are oxoacids. Oxoacids, as we learned in Chapter 2, contain hydrogen, oxygen, and one other element Z, which occupies a central position. Figure 15.3 shows the Lewis structures of several common oxoacids. As you can see, these acids are characterized by the presence of one or more O—H bonds. Let us consider a OH group bonded to some other atom Z which might in turn have other groups attached to it:

$$-\!\!\!\!\!\!\!\!\!\!\;\;\diagdown\!\!\!\!\!\!\diagup\,\text{Z—O—H}$$

At one extreme, Z might be an electropositive metal such as an alkali or alkaline earth metal. In that case the Z—O bond is so polar that the compound is actually ionic:

$$\overset{\delta+}{\text{Z}}\text{—}\overset{\delta-}{\text{O}}\text{—H} \longrightarrow \text{Z}^+ + \text{OH}^-$$

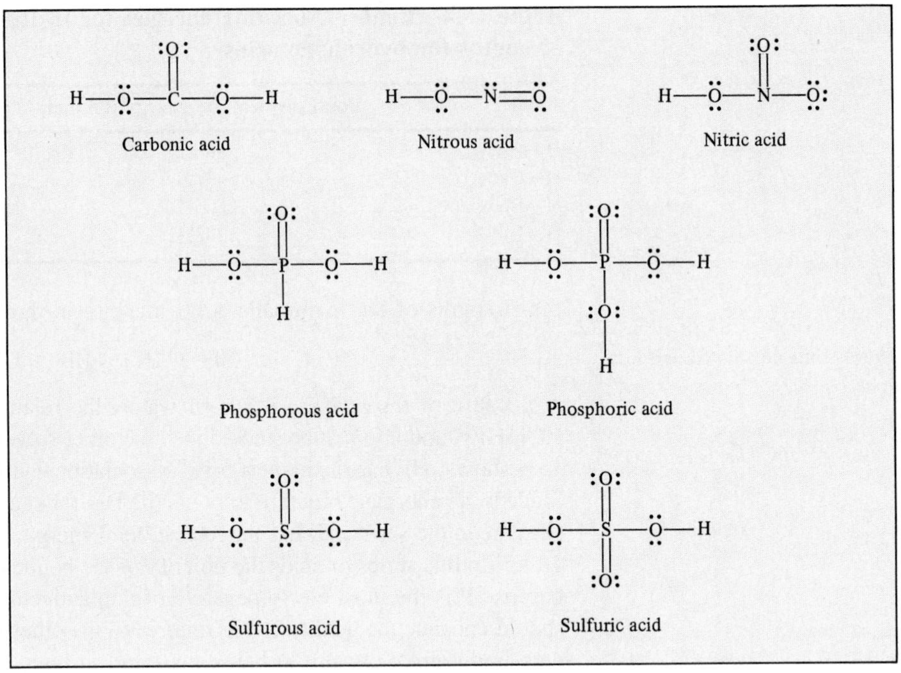

Carbonic acid   Nitrous acid   Nitric acid

Phosphorous acid   Phosphoric acid

Sulfurous acid   Sulfuric acid

Figure 15.3 *Lewis structures of some common oxoacids. For simplicity, the formal charges have been omitted.*

As the oxidation number of an atom becomes larger, its ability to draw electrons in a bond toward itself increases.

This is the reason that alkali and alkaline earth metal hydroxides are all bases.

If Z is an electronegative element, or is in a high oxidation state, it will attract electrons, thus making the Z—O bond more covalent (less polar) and the O—H bond more polar. Consequently, the tendency for the hydrogen to be donated as a H^+ ion increases:

$$\text{—Z—}\overset{\delta-}{\text{O}}\text{—}\overset{\delta+}{\text{H}} \longrightarrow \text{—Z—O}^- + \text{H}^+$$

In this case the compound behaves like an oxoacid.

To compare the strengths of oxoacids, it is convenient to divide the oxoacids into two groups.

1. *Oxoacids that have different central atoms which are from the same group of the periodic table and which have the same oxidation number.* Within this group, the strengths of the acids increase with increasing electronegativity of the central atom, as the following example illustrates:

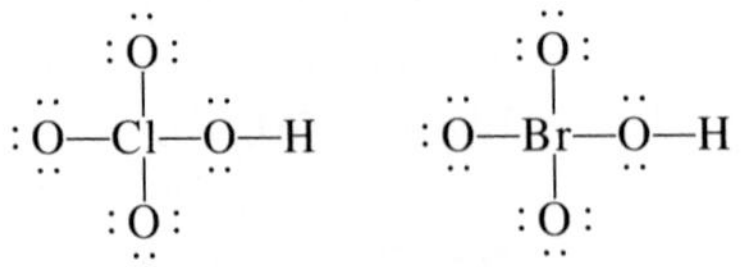

$$\text{O—Cl(O)(O)—O—H} \qquad \text{O—Br(O)(O)—O—H}$$

Cl and Br have the same oxidation number, +7. However, since Cl is more electronegative than Br (see Figure 9.8), it attracts the electron pair it shares with the oxygen (in the Cl—O—H group) to a greater extent than Br does. Consequently, the O—H bond is more polar in perchloric acid and ionizes more readily. Thus the relative acid strengths are

$$HClO_4 > HBrO_4$$

2. *Oxoacids that have the same central atom but different numbers of attached*

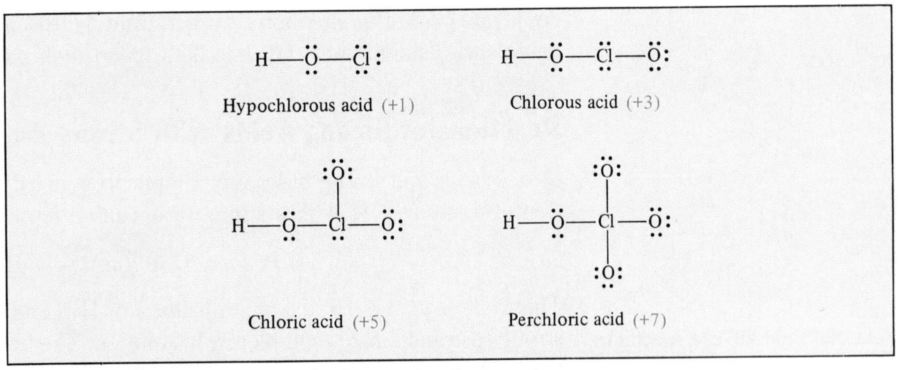

H—O—Cl: Hypochlorous acid (+1)

H—O—Cl—O: Chlorous acid (+3)

H—O—Cl(—O)—O: Chloric acid (+5)

H—O—Cl(—O)(—O)—O: Perchloric acid (+7)

Figure 15.4 *Lewis structures of the oxoacids of chlorine. The oxidation number of the Cl atom is shown in parentheses. For simplicity, the formal charges have been omitted.*

groups. Within this group the acid strength increases as the oxidation number of the central atom increases. Consider the oxoacids of chlorine shown in Figure 15.4. In this series the ability of chlorine to withdraw electrons away from the OH group (thus making the O—H bond more polar) increases with the number of electronegative O atoms attached to Cl. Thus $HClO_4$ is the strongest acid because it has the largest number of O atoms attached to Cl, and the acid strength decreases as follows:

$$HClO_4 > HClO_3 > HClO_2 > HOCl$$

The following example compares the strengths of acids based on their molecular structures.

EXAMPLE 15.8

Predict the relative strengths of the oxoacids in each of the following groups: (a) HOCl, HOBr, and HOI; (b) HNO_3 and HNO_2.

Answer

(a) These acids all have the same structure, and the halogens all have the same oxidation number (+1). Since the electronegativity decreases from Cl to I, the polarity of the X—O bond (where X denotes a halogen atom) increases from HOCl to HOI, and the polarity of the O—H bond decreases from HOCl to HOI. Thus the acid strength decreases as follows:

$$HOCl > HOBr > HOI$$

(b) The structures of HNO_3 and HNO_2 are shown in Figure 15.3. Since the oxidation number of N is +5 in HNO_3 and +3 in HNO_2, HNO_3 is a stronger acid than HNO_2.

Similar problems: 15.46, 15.47.

15.5 Some Typical Acid-Base Reactions

Having discussed the strengths of acids and bases, we look now at some typical acid-base reactions in aqueous solution. These types of reactions were first described in Section 3.4. Here we will study them within the framework of Brønsted's acid-base conjugate pairs. For simplicity we will limit our discussion to monoprotic acids and to bases such as ammonia and the alkali metal hydroxides.

Reactions of Strong Acids with Strong Bases

Strong acids and strong bases are completely ionized in solution. Therefore, the reaction of a strong acid with a strong base can be represented by the net ionic equation

$$H^+(aq) + OH^-(aq) \longrightarrow H_2O(l)$$

Both Na^+ and Cl^- are spectator ions.

This is the equation for the neutralization of HCl and NaOH, for example. NaOH is a strong base and ionizes completely in solution. The hydrated Na^+ ion, $Na^+(aq)$, has no tendency to accept or donate protons; therefore, it has no effect on the pH of the solution. HCl is a strong acid and ionizes completely. The Cl^- ion is an extremely weak Brønsted base (it is the conjugate base of a strong acid). Therefore, the Cl^- ion has no tendency to accept a H^+ ion from H_2O. Consequently, the Cl^- ions also do not affect the pH of the solution.

A solution prepared by mixing equimolar amounts of a strong monoprotic acid and an alkali metal hydroxide is neutral, with a pH of 7. [The Chemistry in Action on pp. 656–657 discusses an interesting strong acid–strong base reaction, that between HCl and $Mg(OH)_2$.]

Reactions of Weak Acids with Strong Bases

Weak acids are largely nonionized in solution, so the ionic equation representing the reaction between a weak acid such as CH_3COOH and a strong base such as NaOH is

$$CH_3COOH(aq) + OH^-(aq) \longrightarrow CH_3COO^-(aq) + H_2O(l)$$

The quantitative relationship between weak acids and their conjugate bases is given in the next chapter.

where the OH^- ions are supplied by the base. We noted earlier that conjugate bases of strong acids have no measurable base strength. The conjugate bases of weak acids do have measurable strength, however, and in general they behave as weak bases. The acetate ion (CH_3COO^-) produced in the above reaction is the conjugate base of a weak acid; therefore, it reacts to a certain extent with water:

$$\underset{\text{base}_1}{CH_3COO^-(aq)} + \underset{\text{acid}_2}{H_2O(l)} \rightleftharpoons \underset{\text{acid}_1}{CH_3COOH(aq)} + \underset{\text{base}_2}{OH^-(aq)}$$

Thus the resulting solution is basic as a result of the surplus OH^- ions. The hydrated Na^+ ions are present as spectator ions.

Reactions of Strong Acids with Weak Bases

A typical reaction between a strong acid and a weak base is that of nitric acid (HNO_3) and ammonia (NH_3). The ionic equation for this reaction is

$$H^+(aq) + NH_3(aq) \longrightarrow NH_4^+(aq)$$

where the H^+ ions are supplied by the acid. Since NH_4^+ is the weak conjugate acid of the weak base NH_3, it reacts with water to a certain extent:

$$\underset{\text{acid}_1}{NH_4^+(aq)} + \underset{\text{base}_2}{H_2O(l)} \rightleftharpoons \underset{\text{base}_1}{NH_3(aq)} + \underset{\text{acid}_2}{H_3O^+(aq)}$$

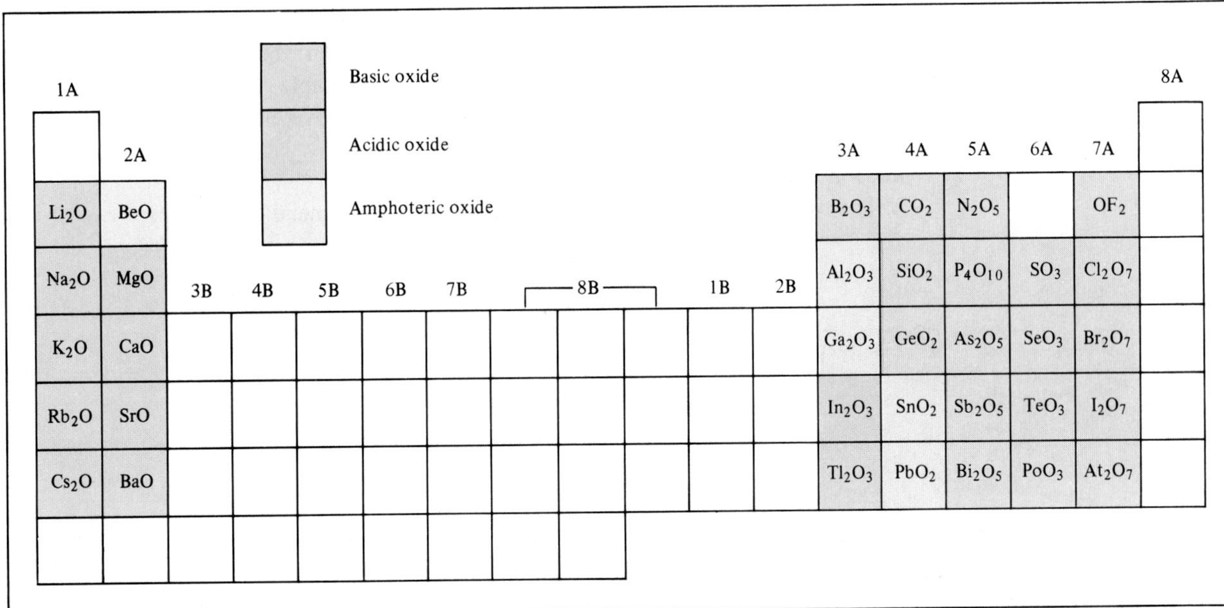

Figure 15.5 *The formulas of a number of oxides of the representative elements in their highest oxidation states.*

Thus, a solution prepared by mixing equimolar amounts of HNO_3 and NH_3 is acidic. (Note that this reaction produces an excess of H^+ ions.) The NO_3^- ion is an extremely weak conjugate base of the strong acid HNO_3 and therefore has no tendency to accept a H^+ ion from H_2O.

Reactions of Weak Acids with Weak Bases

Neither weak acids nor weak bases ionize appreciably in solution. The equation representing the reaction between a weak acid such as CH_3COOH and a weak base such as NH_3 must therefore be written in molecular form:

$$CH_3COOH(aq) + NH_3(aq) \longrightarrow CH_3COO^-(aq) + NH_4^+(aq)$$

In this case, both the anion and the cation react with water:

$$CH_3COO^-(aq) + H_2O(l) \rightleftharpoons CH_3COOH(aq) + OH^-(aq)$$

$$NH_4^+(aq) + H_2O(l) \rightleftharpoons NH_3(aq) + H_3O^+(aq)$$

Whether the resulting solution is acidic, basic, or neutral depends on the relative extent to which the CH_3COO^- and NH_4^+ ions react with water. We will present a quantitative treatment of the reactions between the cation and anion of a salt (CH_3COONH_4 in this case) with water in the next chapter.

Acidic, Basic, and Amphoteric Oxides

As we saw in Chapter 8, oxides can be classified as acidic, basic, or amphoteric. Therefore, a discussion of acid-base reactions would be incomplete without examining the properties of these compounds.

Figure 15.5 shows the formulas of a number of oxides of the representative elements in their highest oxidation states. Note that all alkali metal oxides and all alkaline earth metal oxides except BeO are basic. Beryllium oxide and several metallic oxides in Groups 3A and 4A are amphoteric. Nonmetallic oxides in which the oxidation number of the representative element is high are acidic (for example, N_2O_5, SO_3, and Cl_2O_7). Those in which the oxidation number of the representative element is low (for example, CO and NO) show no measurable acidic properties. No nonmetallic oxides are known to have basic properties.

The basic metallic oxides react with water to form metal hydroxides:

$$Na_2O(s) + H_2O(l) \longrightarrow 2NaOH(aq)$$

$$BaO(s) + H_2O(l) \longrightarrow Ba(OH)_2(aq)$$

The reactions between acidic oxides and water are as follows:

$$CO_2(g) + H_2O(l) \rightleftharpoons H_2CO_3(aq)$$

$$SO_3(g) + H_2O(l) \rightleftharpoons H_2SO_4(aq)$$

$$N_2O_5(g) + H_2O(l) \rightleftharpoons 2HNO_3(aq)$$

$$P_4O_{10}(s) + 6H_2O(l) \rightleftharpoons 4H_3PO_4(aq)$$

$$Cl_2O_7(g) + H_2O(l) \rightleftharpoons 2HClO_4(aq)$$

The reaction between CO_2 and H_2O explains why when pure water is exposed to air (which contains CO_2) it gradually reaches a pH of about 5.5 (Figure 15.6). The reaction between SO_3 and H_2O is partly responsible for the acid rain phenomenon (see the Chemistry in Action on p. 659).

Reactions between acidic oxides and bases and those between basic oxides and acids resemble normal acid-base reactions in that the products are a salt and water:

$$\underset{\text{acidic oxide}}{CO_2(g)} + \underset{\text{base}}{2NaOH(aq)} \longrightarrow \underset{\text{salt}}{Na_2CO_3(aq)} + \underset{\text{water}}{H_2O(l)}$$

$$\underset{\text{basic oxide}}{BaO(s)} + \underset{\text{acid}}{2HNO_3(aq)} \longrightarrow \underset{\text{salt}}{Ba(NO_3)_2(aq)} + \underset{\text{water}}{H_2O(l)}$$

Figure 15.6 *Left: A beaker of water to which a few drops of bromothymol blue indicator have been added. Right: As the dry ice is added to the water, the CO_2 reacts to form carbonic acid, which turns the solution acidic and the color from blue to yellow.*

As Figure 15.5 shows, aluminum oxide (Al_2O_3) is an amphoteric oxide. Depending on the reaction conditions, it can behave either as an acidic oxide or as a basic oxide. For example, Al_2O_3 acts as a base with hydrochloric acid to produce a salt ($AlCl_3$) and water:

$$Al_2O_3(s) + 6HCl(aq) \longrightarrow 2AlCl_3(aq) + 3H_2O(l)$$

and acts as an acid with sodium hydroxide:

$$Al_2O_3(s) + 2NaOH(aq) + 3H_2O(l) \longrightarrow 2NaAl(OH)_4(aq)$$

Note that only a salt, $NaAl(OH)_4$ [containing the Na^+ and $Al(OH)_4^-$ ions], is formed in the latter reaction; no water is produced. Nevertheless, this reaction can still be classified as an acid-base reaction because Al_2O_3 neutralizes NaOH.

The higher the oxidation number of the metal, the more covalent the compound. The lower the oxidation number of the metal, the more ionic the compound.

Some transition metal oxides in which the metal has a high oxidation number act as acidic oxides. Two examples are manganese(VII) oxide (Mn_2O_7) and chromium(VI) oxide (CrO_3), both of which react with water to produce acids:

$$Mn_2O_7(l) + H_2O(l) \longrightarrow \underset{\text{permanganic acid}}{2HMnO_4(aq)}$$

$$CrO_3(s) + H_2O(l) \longrightarrow \underset{\text{chromic acid}}{H_2CrO_4(aq)}$$

Basic and Amphoteric Hydroxides

All amphoteric hydroxides are insoluble.

We have seen that the alkali and alkaline earth metal hydroxides [except $Be(OH)_2$] are basic in properties. The following hydroxides are amphoteric: $Be(OH)_2$, $Al(OH)_3$, $Sn(OH)_2$, $Pb(OH)_2$, $Cr(OH)_3$, $Ni(OH)_2$, $Cu(OH)_2$, $Zn(OH)_2$, and $Cd(OH)_2$. For example, aluminum hydroxide reacts with both acids and bases:

$$Al(OH)_3(s) + 3H^+(aq) \longrightarrow Al^{3+}(aq) + 3H_2O(l)$$

$$Al(OH)_3(s) + OH^-(aq) \rightleftharpoons Al(OH)_4^-(aq)$$

It is interesting to note that beryllium hydroxide, like aluminum hydroxide, exhibits amphoterism:

$$Be(OH)_2(s) + 2H^+(aq) \longrightarrow Be^{2+}(aq) + 2H_2O(l)$$

$$Be(OH)_2(s) + 2OH^-(aq) \rightleftharpoons Be(OH)_4^{2-}(aq)$$

This is another example of the diagonal relationship between beryllium and aluminum.

15.6 Lewis Acids and Bases

Acid-base properties so far have been discussed in terms of the Brønsted theory. To behave as a Brønsted base, for example, a substance must be able to accept protons. By this definition both the hydroxide ion and ammonia are bases:

$$H^+ + {}^-:\ddot{\underset{..}{O}}—H \longrightarrow H—\ddot{\underset{..}{O}}—H$$

$$H^+ + :\overset{\displaystyle H \atop |}{\underset{| \atop \displaystyle H}{N}}—H \longrightarrow \left[H—\overset{\displaystyle H \atop |}{\underset{| \atop \displaystyle H}{N}}—H \right]^+$$

CHEMISTRY IN ACTION

Antacids and the pH Balance in Your Stomach

An average adult produces between 2 and 3 L of gastric juice daily. Gastric juice is a thin, acidic digestive fluid secreted by glands in the mucous membrane lining the stomach. It contains, among other substances, hydrochloric acid. The pH of the gastric juice is about 1.5, which corresponds to a hydrochloric acid concentration of 0.03 *M*—a concentration strong enough to dissolve zinc metal! What is the purpose of this highly acidic medium? Where do the H^+ ions come from? What happens when there is an excess of H^+ ions present in the stomach?

Figure 15.7 is a simplified diagram of the stomach. The inside lining is made up of parietal cells, which are fused together to form tight junctions. The interiors of the cells are protected from the surroundings by cell membranes. These membranes permit passage of water and neutral molecules, but usually block the movement of ions such as H^+, Na^+, K^+, and Cl^- ions. The H^+ ions come from the carbonic acid (H_2CO_3) formed as a result of the hydration of CO_2, an end product of metabolism:

$$CO_2(g) + H_2O(l) \rightleftharpoons H_2CO_3(aq)$$

$$H_2CO_3(aq) \rightleftharpoons H^+(aq) + HCO_3^-(aq)$$

These reactions take place in the blood plasma bathing the cells in the mucosa. By a process known as *active transport,* H^+ ions move across the membrane into the stomach interior. (Active transport processes are known to be carried out with the aid of enzymes, but the details are not clearly understood at present.) To maintain electrical balance, an equal number of Cl^- ions also move from the blood plasma into the stomach. Once in the stomach, most of these ions are prevented from diffusing back into the blood plasma by cell membranes.

The purpose of the highly acidic medium within the stomach is to digest food and to activate certain digestive enzymes. Eating stimulates H^+ ion secretion. A small fraction of these ions are reabsorbed by the mucosa, and many tiny hemorrhages result, a normal process. About half a million cells are shed every minute, and a healthy stomach is completely relined every three days or so. However, if the acid content is excessively high, the constant influx of the H^+ ions through the membrane back to the blood plasma can cause muscular contraction, pain, swelling, inflammation, and bleeding.

One way to temporarily reduce the H^+ ion concentration in the stomach is to take an antacid. The major function of antacids is to neutralize excess HCl in gastric juice. Table 15.5 lists the active ingredients of some popular antacids. The reactions by which these antacids neutralize stomach acid are as follows:

$$NaHCO_3(aq) + HCl(aq) \longrightarrow NaCl(aq) + H_2O(l) + CO_2(g)$$

$$CaCO_3(s) + 2HCl(aq) \longrightarrow CaCl_2(aq) + H_2O(l) + CO_2(g)$$

$$MgCO_3(s) + 2HCl(aq) \longrightarrow MgCl_2(aq) + H_2O(l) + CO_2(g)$$

$$Mg(OH)_2(s) + 2HCl(aq) \longrightarrow MgCl_2(aq) + 2H_2O(l)$$

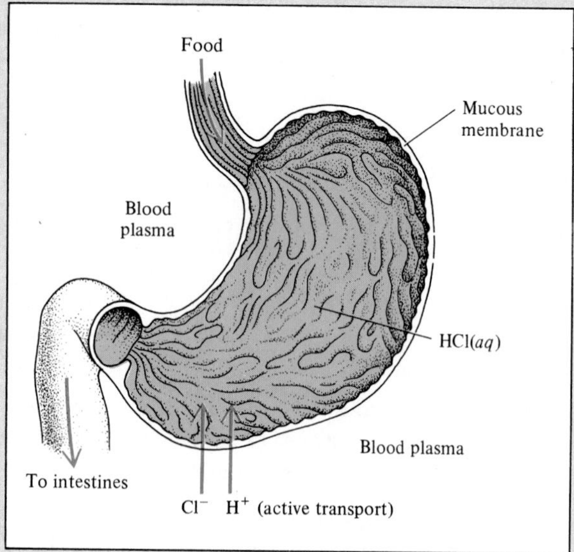

Figure 15.7 *A simplified diagram of the human stomach.*

Table 15.5 Some common commercial antacid preparations

Commercial name	Active ingredients
Alka-2	Calcium carbonate
Alka-Seltzer	Aspirin, sodium bicarbonate, citric acid
Bufferin	Aspirin, magnesium carbonate, aluminum glycinate
Buffered aspirin	Aspirin, magnesium carbonate, aluminum hydroxide-glycine
Milk of magnesia	Magnesium hydroxide
Rolaids	Dihydroxy aluminum sodium carbonate
Tums	Calcium carbonate

$$Al(OH)_2NaCO_3(s) + 4HCl(aq) \longrightarrow AlCl_3(aq) + NaCl(aq) + 3H_2O(l) + CO_2(g)$$

The CO_2 released by most of these reactions increases the gas pressure in the stomach, causing the person to belch. The fizzing that takes place when an Alka-Seltzer tablet dissolves in water is caused by carbon dioxide, which is released by the reaction between citric acid and sodium bicarbonate:

$$\underset{\text{citric acid}}{C_5H_7O_5(COOH)(aq)} + NaHCO_3(aq) \longrightarrow \underset{\text{sodium citrate}}{C_5H_7O_5COONa(aq)} + H_2O(l) + CO_2(g)$$

The mucosa is also damaged by the action of aspirin. Aspirin, or acetylsalicylic acid, is itself a moderately weak acid:

$$\underset{\text{acetylsalicylic acid}}{C_6H_4(OCOCH_3)(COOH)} \rightleftharpoons \underset{\text{acetylsalicylate ion}}{C_6H_4(OCOCH_3)(COO^-)} + H^+$$

In the presence of the high concentration of H^+ ions in the stomach, this acid remains largely un-ionized. Acetylsalicylic acid is a relatively nonpolar molecule and, as such, has the ability to penetrate membrane barriers that are also made up of nonpolar molecules. However, inside the membrane are many small water pockets. Once an acetylsalicylic acid molecule enters such a pocket, it ionizes into H^+ and acetylsalicylate ions. These ionic species then become trapped in the interior regions of the membrane. The continued buildup of ions in this fashion weakens the structure of the membrane, and eventually causes bleeding. Approximately 2 mL of blood is usually lost for every aspirin tablet taken, an amount not generally considered harmful. However, the action of aspirin can result in severe bleeding for some individuals. It is interesting to note that the presence of alcohol makes acetylsalicylic acid even more soluble in the membrane, and so further promotes the bleeding.

In each case, the atom to which the proton becomes attached possesses at least one unshared pair of electrons. This characteristic property of the OH^- ion, of NH_3, and of other Brønsted bases suggests a more general definition of acids and bases.

Lewis presented his theory of acids and bases in 1932, the same year that Brønsted introduced his.

The American chemist G. N. Lewis formulated such a definition. According to Lewis's definition, a base is *a substance that can donate a pair of electrons,* and an acid is *a substance that can accept a pair of electrons*. For example, in the protonation of ammonia, NH_3 acts as a ***Lewis base*** because it donates a pair of electrons to the proton H^+, which acts as a ***Lewis acid*** by accepting the pair of electrons. A Lewis acid-base reaction, therefore, is one that involves the donation of a pair of electrons from one species to another. As we will see below, such a reaction does not produce a salt and water.

All Brønsted bases are Lewis bases.

The significance of the Lewis concept is that it is much more general than other definitions; it includes as acid-base reactions many reactions that do not involve Brønsted acids. Consider, for example, the reaction between boron trifluoride (BF_3) and ammonia (Figure 15.8):

A coordinate covalent bond (see p. 374) is always formed in a Lewis acid-base reaction.

$$\begin{array}{c} \text{F} \\ | \\ \text{F—B} \\ | \\ \text{F} \\ \text{acid} \end{array} + \begin{array}{c} \text{H} \\ | \\ \text{:N—H} \\ | \\ \text{H} \\ \text{base} \end{array} \longrightarrow \begin{array}{c} \text{F}\;\;\text{H} \\ |\;\;\;\;| \\ \text{F—B—N—H} \\ |\;\;\;\;| \\ \text{F}\;\;\text{H} \end{array}$$

Figure 15.8 *A Lewis acid-base reaction involving BF_3 and NH_3.*

In Section 10.4 we saw that the B atom in BF_3 is sp^2-hybridized. The vacant, unhybridized $2p$ orbital accepts the pair of electrons from NH_3. So we see that BF_3 functions as an acid according to the Lewis definition, even though it does not contain an ionizable proton.

Another Lewis acid containing boron is boric acid. Boric acid (a weak acid used in eyewash) is an oxoacid with the following structure:

$$\begin{array}{c} \text{H} \\ | \\ \text{:}\ddot{\text{O}}\text{:} \\ | \\ \text{H—}\ddot{\underset{\cdot\cdot}{\text{O}}}\text{—B—}\ddot{\underset{\cdot\cdot}{\text{O}}}\text{—H} \end{array}$$

boric acid

However, unlike the oxoacids shown in Figure 15.3, boric acid does not ionize in water to produce a H^+ ion. Instead, its reaction with water is

$$B(OH)_3(aq) + H_2O(l) \rightleftharpoons B(OH)_4^-(aq) + H^+(aq)$$

In this Lewis acid-base reaction, boric acid accepts a pair of electrons from the hydroxide ion that is derived from the H_2O molecule.

The hydration of carbon dioxide to produce carbonic acid

$$CO_2(g) + H_2O(l) \rightleftharpoons H_2CO_3(aq)$$

Carbon is a second-period element. It cannot expand its valence shell and can only accommodate eight valence electrons.

can be understood in the Lewis framework as follows: The first step involves donation of a lone pair on the oxygen atom in H_2O to the carbon atom in CO_2. An orbital is vacated on the C atom to accommodate the lone pair by removal of the electron pair in the C—O pi bond. These shifts of electrons are indicated by the curved arrows.

$$\begin{array}{c} \text{H} \\ \diagdown \\ \ddot{\text{O}}\text{:} \\ \diagup \\ \text{H} \end{array} \begin{array}{c} \ddot{\text{O}}\text{:} \\ \| \\ \text{C} \\ \| \\ \text{:}\ddot{\text{O}}\text{:} \end{array} \longrightarrow \begin{array}{c} \text{:}\ddot{\text{O}}\text{:}^- \\ | \\ \text{H—}\overset{\cdot\cdot +}{\text{O}}\text{—C} \\ |\;\;\;\;\| \\ \text{H}\;\;\text{:}\ddot{\text{O}}\text{:} \end{array}$$

Therefore, H_2O is a Lewis base and CO_2 is a Lewis acid. Next, a proton is transferred onto the O atom bearing a negative charge to form H_2CO_3.

$$\begin{array}{c} \text{:}\ddot{\text{O}}\text{:}^- \\ | \\ \text{H—}\overset{\cdot\cdot +}{\text{O}}\text{—C} \\ |\;\;\;\;\| \\ \text{H}\;\;\text{:}\ddot{\text{O}}\text{:} \end{array} \longrightarrow \begin{array}{c} \text{H—}\ddot{\text{O}}\text{:} \\ | \\ \text{:}\ddot{\text{O}}\text{—C} \\ |\;\;\;\;\| \\ \text{H}\;\;\text{:}\ddot{\text{O}}\text{:} \end{array}$$

Other examples of Lewis acid-base reactions are

$$\underset{\text{acid}}{Ag^+(aq)} + \underset{\text{base}}{2NH_3(aq)} \rightleftharpoons Ag(NH_3)_2^+(aq)$$

$$\underset{\text{acid}}{Cd^{2+}(aq)} + \underset{\text{base}}{4I^-(aq)} \rightleftharpoons CdI_4^{2-}(aq)$$

$$\underset{\text{acid}}{Ni(s)} + \underset{\text{base}}{4CO(g)} \rightleftharpoons Ni(CO)_4(g)$$

It is important to note that the hydration of metal ions in solution is in itself a Lewis acid-base reaction. Thus, when copper(II) sulfate ($CuSO_4$) dissolves in water, each Cu^{2+} ion is associated with six water molecules as $Cu(H_2O)_6^{2+}$. In this case, the Cu^{2+} ion acts as the acid and the H_2O molecules act as the base.

Although the Lewis definition of acids and bases is of greater significance because of its generality, we normally speak of "an acid" and "a base" in terms of the Brønsted definition. The term "Lewis acid" usually is reserved for substances that can accept a pair of electrons but do not contain ionizable hydrogen atoms.

Classification of Lewis acids and bases is demonstrated in the following example.

EXAMPLE 15.9

Identify the Lewis acid and Lewis base in each of the following reactions:

(a) $SnCl_4(s) + 2Cl^-(aq) \rightleftharpoons SnCl_6^{2-}(aq)$
(b) $Hg^{2+}(aq) + 4CN^-(aq) \rightleftharpoons Hg(CN)_4^{2-}(aq)$
(c) $Co^{3+}(aq) + 6NH_3(aq) \rightleftharpoons Co(NH_3)_6^{3+}(aq)$

Answer

(a) In this reaction, $SnCl_4$ accepts two pairs of electrons from the Cl^- ions. Therefore $SnCl_4$ is the Lewis acid and Cl^- is the Lewis base.
(b) Here the Hg^{2+} ion accepts four pairs of electrons from the CN^- ions. Therefore Hg^{2+} is the Lewis acid and CN^- is the Lewis base.
(c) In this reaction, the Co^{3+} ion accepts six pairs of electrons from the NH_3 molecules. Therefore, Co^{3+} is the Lewis acid and NH_3 is the Lewis base.

Similar problem: 15.55.

CHEMISTRY IN ACTION

Acid Rain

Both sulfur dioxide (SO_2) and sulfur trioxide (SO_3) are highly toxic substances. Acidic oxides, they react with water to give the corresponding acids sulfurous acid (H_2SO_3) and sulfuric acid (H_2SO_4). The increasing concentration of these molecules in the atmosphere has posed a serious environmental problem in recent years—acid rain. Precipitation in the northeastern United States, for example, has an average pH of about 4. Since atmospheric CO_2 in equilibrium with rainwater would not be expected to result in a pH less than 5.5, SO_2 and to a lesser extent nitrogen oxides from auto emissions are responsible for the high acidity. The problem is not unique to the United States; any region with a high density of industrial installations also has precipitation with a pH of 4 or lower.

There are several sources of SO_2. Nature itself is

Figure 15.9 *Sulfur dioxide and other air pollutants being released into the atmosphere from a coal-burning power plant.*

responsible for much SO_2 emission in the form of volcanic eruptions. Also, many metals exist in the combined form with sulfur in nature. The first step in refining these ores is often *smelting* or *roasting*, the process of heating the metal sulfide in air to form the metal oxide:

$$2ZnS(s) + 3O_2(g) \longrightarrow 2ZnO(s) + 2SO_2(g)$$

The oxide can be more easily reduced (by a more reactive metal or in some cases by carbon) to the free metal. Although this is an important source of SO_2, burning fossil fuels (natural gas, oil, and coal) in industry, in power plants of electrical generating stations, and in homes accounts for most of the SO_2 emitted to the atmosphere (Figure 15.9). The sulfur content of coal ranges from 0.5 to 5 percent by mass, depending on the source of the coal. The sulfur content of other fossil fuels is similarly variable. Oil from the Middle East is low in sulfur, and that from Venezuela high. All in all, some 50 to 60 million tons of SO_2 are released to the atmosphere each year!

Some of the SO_2 in air is oxidized to SO_3 by any of several pathways. For example, it may react with ozone as follows:

$$SO_2(g) + O_3(g) \longrightarrow SO_3(g) + O_2(g)$$

A SO_2 molecule may be photoexcited by sunlight

$$SO_2(g) + h\nu \longrightarrow SO_2^*(g)$$

and then undergo the reactions

$$SO_2^*(g) + O_2(g) \longrightarrow SO_4(g)$$

$$SO_4(g) + SO_2(g) \longrightarrow 2SO_3(g)$$

where the SO_4 species is a reactive intermediate. In another reaction sequence, dust or other solid particles present in the atmosphere can act as heterogeneous catalysts for the reaction

$$2SO_2(g) + O_2(g) \longrightarrow 2SO_3(g)$$

Both SO_2 and SO_3 are converted to their acids by rainwater. The resulting acids can corrode buildings made of limestone and marble ($CaCO_3$). A typical reaction is

$$CaCO_3(s) + H_2SO_4(aq) \longrightarrow CaSO_4(s) + H_2O(l) + CO_2(g)$$

Sulfur dioxide can also attack calcium carbonate directly:

$$2CaCO_3(s) + 2SO_2(g) + O_2(g) \longrightarrow 2CaSO_4(s) + 2CO_2(g)$$

Every year, acid rain causes hundreds of millions of dollars' worth of damage to buildings and statues. The term "stone leprosy" is used by some environmental chemists to describe the corrosion of stone caused by acid rain (Figure 15.10). Acid rain is also extremely harmful to vegetation and aquatic life. Many well-documented cases show dramatically how acid rain has destroyed agricultural and forest lands and killed fish in lakes (Figure 15.11).

There are two ways to minimize the effects of SO_2 pollution. The most direct approach is to remove sulfur from fossil fuels before combustion, but this is technologically difficult to accomplish. A cheaper (although less efficient) way is to remove SO_2 as it is formed. For example, in one process powdered limestone is injected into the power plant boiler or furnace along with coal (Figure 15.12). At high temperatures the following decomposition occurs:

Figure 15.10 *Photos of a statue, taken about sixty years apart (in 1908 and 1969), show the damaging effects of air pollutants such as sulfur dioxide.*

Figure 15.11 *A forest damaged by acid rain.*

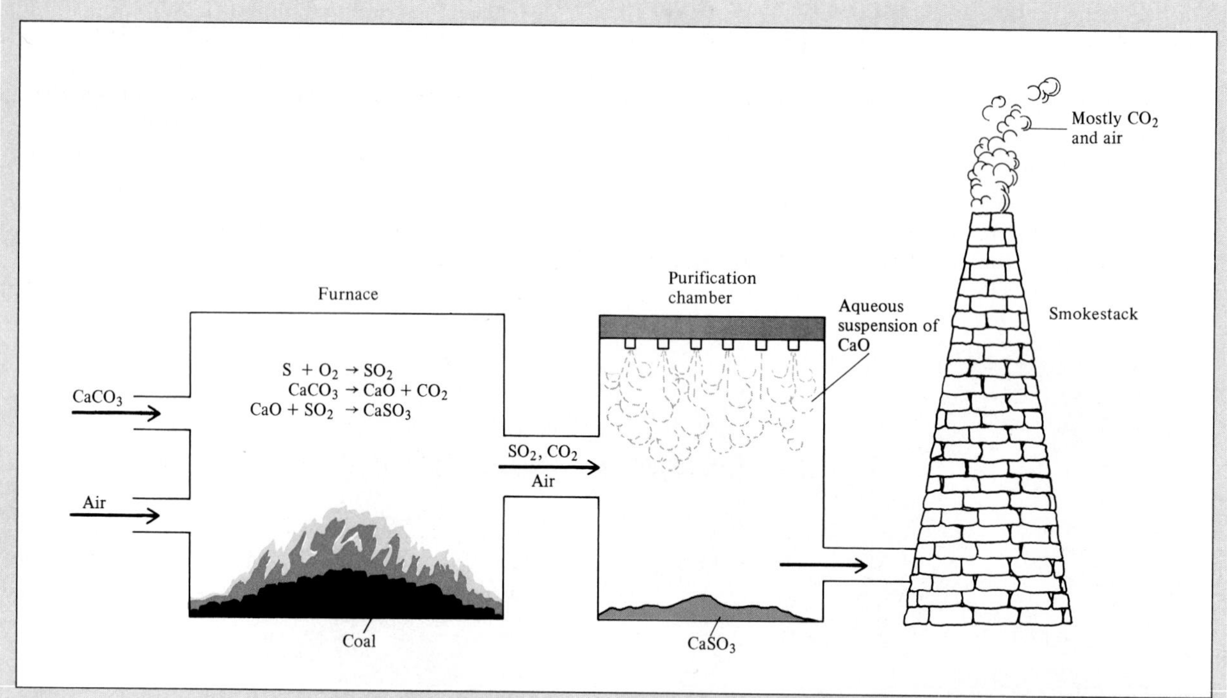

Figure 15.12 *Common procedure for removing sulfur dioxide from burning fossil fuel. Powdered limestone decomposes into CaO, which reacts with SO_2 to form $CaSO_3$. The remaining SO_2 then enters a chamber where it reacts with an aqueous suspension of CaO to form $CaSO_3$.*

$$\underset{\text{limestone}}{CaCO_3(s)} \longrightarrow \underset{\text{quicklime}}{CaO(s)} + CO_2(g)$$

The quicklime reacts with SO_2 produced by combustion to form calcium sulfite and some calcium sulfate:

$$CaO(s) + SO_2(g) \longrightarrow CaSO_3(s)$$

$$2CaO(s) + 2SO_2(g) + O_2(g) \longrightarrow 2CaSO_4(s)$$

To remove any remaining unreacted SO_2, an aqueous suspension of quicklime is injected into a purification chamber prior to the gases' escape through the smokestack.

Finally we note that some metal ore refining sites have installed a sulfuric acid plant nearby. The SO_2 produced in the process of roasting the metal sulfides is captured for use in the first step of sulfuric acid synthesis. This is a very sensible way to turn what is a pollutant in one process into a useful starting material for another process!

Summary

1. Brønsted acids donate protons, and Brønsted bases accept protons. This is the definition that normally underlies the use of the terms "acid" and "base."
2. The acidity of an aqueous solution is expressed as its pH, which is defined as the negative logarithm of the hydrogen ion concentration (in mol/L).
3. At 25°C, an acidic solution has pH < 7, a basic solution has pH > 7, and a neutral solution has pH $= 7$.
4. The strength of an acid or a base is measured by the extent of its ionization in solution.
5. The solvent plays an important role in determining the strength of an acid. The leveling effect is the inability of a solvent to differentiate among the relative

strengths of all acids stronger than its conjugate acid.

6. In aqueous solution, the following are classified as strong acids: $HClO_4$, HI, HBr, HCl, H_2SO_4 (first stage of ionization), and HNO_3. The following are classified as strong bases in aqueous solution: hydroxides of alkali metals and of alkaline earth metals (except beryllium).
7. The relative strengths of acids can be explained qualitatively in terms of their molecular structures.
8. Most oxides can be classified as acidic, basic, or amphoteric. Metal hydroxides are either basic or amphoteric.
9. Lewis acids accept pairs of electrons and Lewis bases donate pairs of electrons. The term "Lewis acid" is generally reserved for substances that can accept electron pairs but do not contain ionizable hydrogen atoms.

Applied Chemistry Example: Hydrochloric Acid

The United States produced 5.87 billion pounds of hydrochloric acid in 1988, which ranked it 26th among the top 50 industrial chemicals manufactured that year. Most hydrochloric acid is obtained as a by-product in the preparation of chlorinated hydrocarbons; for example,

$$CH_4(g) + Cl_2(g) \longrightarrow CH_3Cl(g) + HCl(g)$$

The HCl gas is separated from the mixture by reacting it with water to form hydrochloric acid (the other gases are all insoluble in water). Very pure hydrochloric acid can be prepared by burning hydrogen in chlorine:

$$H_2(g) + Cl_2(g) \longrightarrow 2HCl(g)$$

About half the hydrochloric acid produced is used in metal pickling. This process involves the removal of metal oxide layers from metal surfaces to prepare them for finishing with coatings (for example, chrome or paint). Pickling steel is the largest single use of hydrochloric acid and accounts for about 600,000 tons of 30 percent HCl solution (by mass) produced anually. Hydrochloric acid is also used in the manufacture of chemicals, food processing, and oil recovery.

1. Write the overall and net ionic equations for the reaction between iron(III) oxide which represents the rust layer on iron and hydrochloric acid. Identify the Brønsted acid and base.

   Answer: The overall equation is

   $$Fe_2O_3(s) + 6HCl(aq) \longrightarrow 2FeCl_3(aq) + 3H_2O(aq)$$

   and the net ionic equation is

   $$Fe_2O_3(s) + 6H^+(aq) \longrightarrow 2Fe^{3+}(aq) + 3H_2O(l)$$

   Since HCl donates the H^+ ion, it is the Brønsted acid. Each Fe_2O_3 unit accepts six H^+ ions; therefore, it is the Brønsted base.

2. Hydrochloric acid is also used to remove scales (which are mostly $CaCO_3$) in water pipes (see p. 102). Hydrochloric acid reacts with calcium carbonate in two stages; the first stage forms the bicarbonate ion which then reacts further to form carbon dioxide. Write equations for these two stages and the overall equation.

Answer: The first stage is

$$CaCO_3(s) + HCl(aq) \longrightarrow Ca^{2+}(aq) + HCO_3^-(aq) + Cl^-(aq)$$

and the second stage is

$$HCl(aq) + HCO_3^-(aq) \longrightarrow CO_2(g) + Cl^-(aq) + H_2O(l)$$

The overall equation is

$$CaCO_3(s) + 2HCl(aq) \longrightarrow CaCl_2(aq) + H_2O(l) + CO_2(g)$$

The $CaCl_2$ formed is soluble in water.

3. Hydrochloric acid is used to recover oil in oil wells by dissolving rocks (often $CaCO_3$) so that the oil may flow more easily. In one process, a 15 percent (by mass) HCl solution is injected into an oil well to dissolve the rocks. If the density of the acid solution is 1.073 g/mL, what is the pH of the solution?

Answer: We need to find the concentration of the HCl solution in order to determine its pH. The mass of 1000-mL solution is

$$1000 \text{ mL} \times \frac{1.073 \text{ g}}{1 \text{ mL}} = 1073 \text{ g}$$

The number of moles of HCl in a 15 percent solution is

$$0.15 \times 1073 \text{ g} \times \frac{1 \text{ mol HCl}}{36.46 \text{ g HCl}} = 4.4 \text{ mol HCl}$$

Thus there are 4.4 moles of HCl in one liter of solution and the concentration is 4.4 *M*. The pH of the solution is

$$\begin{aligned} pH &= -\log(4.4) \\ &= -0.64 \end{aligned}$$

This is a highly acidic solution (note that the pH is negative) which is needed to dissolve large quantities of rocks in the oil recovery process.

Key Words

Binary acid, p. 648
Conjugate acid-base pair, p. 638
Ion-product constant, p. 640
Leveling effect, p. 647
Lewis acid, p. 657
Lewis base, p. 657
Percent ionization, p. 644
pH, p. 641
Ternary acid, p. 648

Exercises†

General Properties of Acids and Bases

REVIEW QUESTIONS

15.1 List some general properties of acids.
15.2 List some general properties of bases.

15.3 Name the following acids, and identify them as inorganic or organic: (a) HBr, (b) HNO_2, (c) CH_3COOH, (d) H_3PO_4, (e) H_2CO_3, (f) HCOOH, (g) $HClO_4$, (h) $HClO_3$, (i) HF. (The nomenclature of inorganic acids is discussed in Section 2.9.)

† Unless otherwise stated, the temperature is assumed to be 25°C for all problems.

Brønsted Acids and Bases

REVIEW QUESTIONS

15.4 Define Brønsted acids and bases. How do the Brønsted definitions differ from Arrhenius's definitions of acids and bases?

15.5 In order for a species to act as a Brønsted base, an atom in the species must possess a lone pair. Explain why this is so.

15.6 Give an example of a conjugate pair in an acid-base reaction.

PROBLEMS

15.7 Classify each of the following species as a Brønsted acid or base, or both: (a) H_2O, (b) OH^-, (c) H_3O^+, (d) NH_3, (e) NH_4^+, (f) NH_2^-, (g) NO_3^-, (h) CO_3^{2-}, (i) HBr, (j) HCN.

15.8 What are the names and formulas of the conjugate bases of the following acids? (a) HNO_2, (b) H_2SO_4, (c) H_2S, (d) HCN, (e) HCOOH (formic acid).

15.9 Identify the acid-base conjugate pairs in each of the following reactions:

(a) $CH_3COO^- + HCN \rightleftharpoons CH_3COOH + CN^-$
(b) $NH_2^- + NH_3 \rightleftharpoons NH_3 + NH_2^-$
(c) $HF + NH_3 \rightleftharpoons NH_4^+ + F^-$
(d) $HCO_3^- + HCO_3^- \rightleftharpoons H_2CO_3 + CO_3^{2-}$
(e) $H_2PO_4^- + NH_3 \rightleftharpoons HPO_4^{2-} + NH_4^+$
(f) $HOCl + CH_3NH_2 \rightleftharpoons CH_3NH_3^+ + ClO^-$
(g) $CO_3^{2-} + H_2O \rightleftharpoons HCO_3^- + OH^-$
(h) $CH_3COO^- + H_2O \rightleftharpoons CH_3COOH + OH^-$

15.10 Give the conjugate acid of each of the following bases: (a) HS^-, (b) HCO_3^-, (c) CO_3^{2-}, (d) $H_2PO_4^-$, (e) HPO_4^{2-}, (f) PO_4^{3-}, (g) HSO_4^-, (h) SO_4^{2-}, (i) HSO_3^-, (j) SO_3^{2-}.

15.11 Give the conjugate base of each of the following acids: (a) $CH_2ClCOOH$, (b) HIO_4, (c) H_3PO_4, (d) $H_2PO_4^-$, (e) HPO_4^{2-}, (f) H_2SO_4, (g) HSO_4^-, (h) H_2SO_3, (i) HSO_3^-, (j) NH_4^+, (k) H_2S, (l) HS^-, (m) HOCl.

15.12 Oxalic acid ($C_2H_2O_4$) has the following structure:

```
O=C—O—H
  |
O=C—O—H
```

An oxalic acid solution contains the following species in varying concentrations: $C_2H_2O_4$, $C_2HO_4^-$, $C_2O_4^{2-}$, and H^+. (a) Draw Lewis structures of $C_2HO_4^-$ and $C_2O_4^{2-}$. (b) Which of the above four species can act only as acids, which can act only as bases, and which can act as both acids and bases?

pH and pOH Calculations

REVIEW QUESTIONS

15.13 What is the ion-product constant for water?

15.14 Write an equation relating $[H^+]$ and $[OH^-]$ in solution at 25°C.

15.15 The ion-product constant for water is 1.0×10^{-14} at 25°C and 3.8×10^{-14} at 40°C. Is the process

$$H_2O(l) \longrightarrow H^+(aq) + OH^-(aq)$$

endothermic or exothermic?

15.16 Define pH. Why do chemists normally choose to discuss the acidity of a solution in terms of pH rather than hydrogen ion concentration, $[H^+]$?

15.17 Complete the following table for a solution:

pH	$[H^+]$	Solution is
<7		
	$<1.0 \times 10^{-7}$ *M*	
		Neutral

15.18 The pH of a solution is 6.7. From this statement alone, can you conclude that the solution is acidic? If not, what additional information would you need?

15.19 Define pOH. Write an equation relating pH and pOH.

15.20 Calculate the pH of a 0.26 *M* HNO_3 solution.

15.21 Can the pH of a solution be zero or negative? If so, give examples to illustrate these values.

PROBLEMS

15.22 Indicate whether the following solutions are acidic, basic, or neutral: (a) 0.62 *M* NaOH, (b) 1.4×10^{-3} *M* HCl, (c) 2.5×10^{-11} *M* H^+, (d) 3.3×10^{-10} *M* OH^-.

15.23 Calculate the hydrogen ion concentration for solutions with the following pH values: (a) 2.42, (b) 11.21, (c) 6.96, (d) 15.00.

15.24 Calculate the hydrogen ion concentration in mol/L for each of the following solutions: (a) a solution whose pH is 5.20, (b) a solution whose pH is 16.00, (c) a solution whose hydroxide concentration is 3.7×10^{-9} *M*.

15.25 Calculate the pH of each of the following solutions: (a) 0.0010 *M* HCl, (b) 0.76 *M* KOH, (c) 2.8×10^{-4} *M* $Ba(OH)_2$, (d) 5.2×10^{-4} *M* HNO_3.

15.26 Calculate the pH of water at 40°C, given that K_W is 3.8×10^{-14} at this temperature.

15.27 Fill in the word *acidic, basic,* or *neutral* for the following solutions:

(a) pOH > 7; solution is ________
(b) pOH = 7; solution is ________
(c) pOH < 7; solution is ________

15.28 The pOH of a solution is 9.40. Calculate the hydrogen ion concentration of the solution.

15.29 A solution is made by dissolving 18.4 g of HCl in 662 mL of water. Calculate the pH of the solution. (Assume that the volume of the solution is also 662 mL.)

15.30 How much NaOH (in grams) is needed to prepare 546 mL of solution with a pH of 10.00?

15.31 Calculate the number of moles of KOH in 5.50 mL of a 0.360 *M* KOH solution. What is the pOH of the solution?

Strengths of Acids

REVIEW QUESTIONS

15.32 Explain what is meant by the strength of an acid.

15.33 Without referring to the text, write the formulas of four strong acids and four weak acids.

15.34 What are the strongest acid and strongest base that can exist in water?

15.35 H_2SO_4 is a strong acid, but HSO_4^- is a weak acid. Account for the difference in strength of these two related species.

15.36 Explain the leveling effect and its influence on the study of the strength of acids.

15.37 Give two examples of a strong oxoacid and two examples of a weak oxoacid.

PROBLEMS

15.38 Classify each of the following species as a weak or strong acid: (a) HNO_3, (b) HF, (c) H_2SO_4, (d) HSO_4^-, (e) H_2CO_3, (f) HCO_3^-, (g) HCl, (h) HCN, (i) HNO_2.

15.39 Classify each of the following species as a weak or strong base: (a) LiOH, (b) CN^-, (c) H_2O, (d) ClO_4^-, (e) NH_2^-.

15.40 You are given two aqueous solutions containing a strong acid (HA) and a weak acid (HB), respectively. Describe how you would compare the strengths of these two acids by (a) pH measurement, (b) electrical conductance measurement, (c) studying the rate of hydrogen gas evolution when these solutions are reacted with an active metal such as Mg or Zn.

15.41 Which of the following statements is/are true regarding a 0.10 *M* solution of a weak acid HA?
(a) The pH is 1.00.
(b) $[H^+] \gg [A^-]$
(c) $[H^+] = [A^-]$
(d) The pH is less than 1.

15.42 Which of the following statements is/are true regarding a 1.0 *M* solution of a strong acid HA?
(a) $[A^-] > [H^+]$
(b) The pH is 0.00.
(c) $[H^+] = 1.0\ M$
(d) $[HA] = 1.0\ M$

15.43 Predict the direction that predominates in this reaction:

$$F^-(aq) + H_2O(l) \rightleftharpoons HF(aq) + OH^-(aq)$$

15.44 What is the strongest acid that can exist in (a) water, (b) glacial acetic acid, (c) liquid ammonia? What is the strongest base that can exist in (a) water, (b) liquid ammonia?

15.45 Predict whether the following reaction will proceed from left to right to any measurable extent:

$$CH_3COOH(aq) + Cl^-(aq) \longrightarrow$$

15.46 Compare the strengths of the following pairs of oxoacids: (a) H_2SO_4 and H_2SeO_4, (b) H_2SO_3 and H_2SeO_3, (c) H_3PO_4 and H_3AsO_4, (d) $HBrO_4$ and HIO_4.

15.47 Predict the acid strengths of the following compounds: H_2O, H_2S, and H_2Se.

15.48 Which of the following acids is the stronger: CH_3COOH or $CH_2ClCOOH$? Explain your choice.

15.49 Consider the following compounds:

[benzene ring]—O—H CH_3—O—H
phenol   methanol

Experimentally phenol is found to be a stronger acid than methanol. Explain this difference in terms of the structures of the conjugate bases. (*Hint:* A more stable conjugate base favors ionization. Only one of the conjugate bases can be stabilized by resonance.)

Acid-Base Reactions

REVIEW QUESTIONS

15.50 Write balanced ionic equations for the following: (a) NaOH solution reacts with NH_4Cl to form H_2O and the conjugate base of NH_4^+. (b) HCl solution reacts with CH_3COONa to form the conjugate acid of CH_3COO^-.

15.51 Write balanced net ionic equations for the reactions when solutions of the following substances are mixed: (a) perchloric acid and ammonia; (b) carbonic acid and potassium hydroxide; (c) barium hydroxide and acetic acid; (d) nitric acid and sodium carbonate.

Lewis Acids and Bases

REVIEW QUESTIONS

15.52 What are the Lewis definitions of an acid and a base? In what way are they more general than the Brønsted definitions?

15.53 In terms of orbitals and electron arrangements, what must be present for a molecule or an ion to act as a Lewis acid (use H^+ and BF_3 as examples)? What must be present for a molecule or ion to act as a Lewis base (use OH^- and NH_3 as examples)?

PROBLEMS

15.54 Classify each of the following species as a Lewis acid or a Lewis base: (a) CO_2, (b) H_2O, (c) I^-, (d) SO_2, (e) NH_3, (f) OH^-, (g) H^+, (h) BCl_3.

15.55 Describe the following reaction according to the Lewis theory of acids and bases:

$$AlCl_3(s) + Cl^-(aq) \longrightarrow AlCl_4^-(aq)$$

15.56 Which would be considered a stronger Lewis acid: (a) BF_3 or BCl_3, (b) Fe^{2+} or Fe^{3+}? Explain.

15.57 Describe the hydration of SO_2 as a Lewis acid-base reaction. (*Hint:* Follow the procedure for the hydration of CO_2 on p. 658.)

15.58 All Brønsted acids are Lewis acids, but the reverse is not true. Give two examples of Lewis acids that are not Brønsted acids.

Miscellaneous Problems

15.59 Write balanced molecular and ionic equations for the following neutralization reactions: (a) $KOH + H_3PO_4$, (b) $NH_3 + H_2SO_4$, (c) $Mg(OH)_2 + HClO_4$, (d) $NaOH + H_2CO_3$.

15.60 Use the ionization of HCN in water as an example to illustrate the meaning of dynamic equilibrium.

15.61 Give an example of (a) a weak acid that contains oxygen atoms, (b) a weak acid that does not contain oxygen atoms, (c) a neutral molecule that acts as a Lewis acid, (d) a neutral molecule that acts as a Lewis base, (e) a weak acid that contains two ionizable H atoms, (f) a conjugate acid-base pair, both of which react with HCl to give carbon dioxide gas.

15.62 A typical reaction between an antacid and the hydrochloric acid in gastric juice is

$$NaHCO_3(aq) + HCl(aq) \longrightarrow NaCl(aq) + H_2O(l) + CO_2(g)$$

Calculate the volume (in L) of CO_2 generated from 0.350 g of $NaHCO_3$ and excess gastric juice at 1.00 atm and 37.0°C.

15.63 Explain why metal oxides tend to be basic if the oxidation number of the metal is low and acidic if the oxidation number of the metal is high. (*Hint:* Metallic compounds with low oxidation numbers of the metals are more ionic than those in which the oxidation numbers of the metals are high.)

15.64 Arrange the oxides in each of the following groups in order of increasing basicity: (a) K_2O, Al_2O_3, BaO, (b) CrO_3, CrO, Cr_2O_3.

15.65 $Zn(OH)_2$ is an amphoteric hydroxide. Write balanced ionic equations to show its reaction with (a) HCl and (b) NaOH [the product is $Zn(OH)_4^{2-}$].

15.66 $Al(OH)_3$ is an insoluble compound. It dissolves in excess NaOH in solution. Write a balanced ionic equation for this reaction. What type of reaction is this?

15.67 Classify the following oxides as acidic, basic, amphoteric, or neutral: (a) CO_2, (b) K_2O, (c) CaO, (d) N_2O_5, (e) CO, (f) NO, (g) SnO_2, (h) SO_3, (i) Al_2O_3, (j) BaO.

15.68 Which of the following is the stronger base: NF_3 or NH_3? (*Hint:* F is more electronegative than H.)

15.69 Which of the following is the stronger base: NH_3 or PH_3? (*Hint:* The N—H bond is stronger than the P—H bond.)

15.70 Use VSEPR to predict the geometry of the hydronium ion, H_3O^+.

15.71 The ion product of D_2O is 1.35×10^{-15} at 25°C. (a) Calculate pD of pure D_2O where $pD = -\log [D^+]$. (b) For what values of pD will a solution be acidic in D_2O? (c) Derive a relation between pD and pOD.

15.72 What is the pH of 250.0 mL of an aqueous solution containing 0.616 g of the strong acid trifluoromethane sulfonic acid (CF_3SO_3H)?

15.73 To which of the following would addition of an equal volume of 0.60 *M* NaOH lead to a solution having a lower pH? (a) Water, (b) 0.30 *M* HCl, (c) 0.70 *M* KOH, (d) 0.40 *M* $NaNO_3$.

15.74 The pH of a 0.0642 *M* solution of a monoprotic acid is 3.86. Is this a strong acid?

15.75 When chlorine reacts with water, the resulting solution is weakly acidic and reacts with $AgNO_3$ to give a white precipitate. Write balanced equations to represent these reactions. Explain why manufacturers of household bleaches add bases such as NaOH to their products to increase their effectiveness.

15.76 HF is a weak acid, but its strength increases with concentration. Explain. (*Hint:* F^- reacts with HF to form HF_2^-. The equilibrium constant for this reaction is 5.2 at 25°C.)

15.77 When the concentration of a strong acid is not substantially higher than 1.0×10^{-7} *M*, the ionization of water must be taken into account in the calculation of the pH of the solution. (a) Derive an expression for the pH of a strong acid solution, including the contribution to $[H^+]$ from H_2O. (b) Calculate the pH of a 1.0×10^{-7} *M* HCl solution.

Opposite page: The pH of the cells of gastric glands, monitored by using a fluorescent dye. Although local changes in pH occur all the time in the human body, the overall pH in various regions is maintained by one or more buffer systems.

C H A P T E R 1 6

ACID-BASE EQUILIBRIA

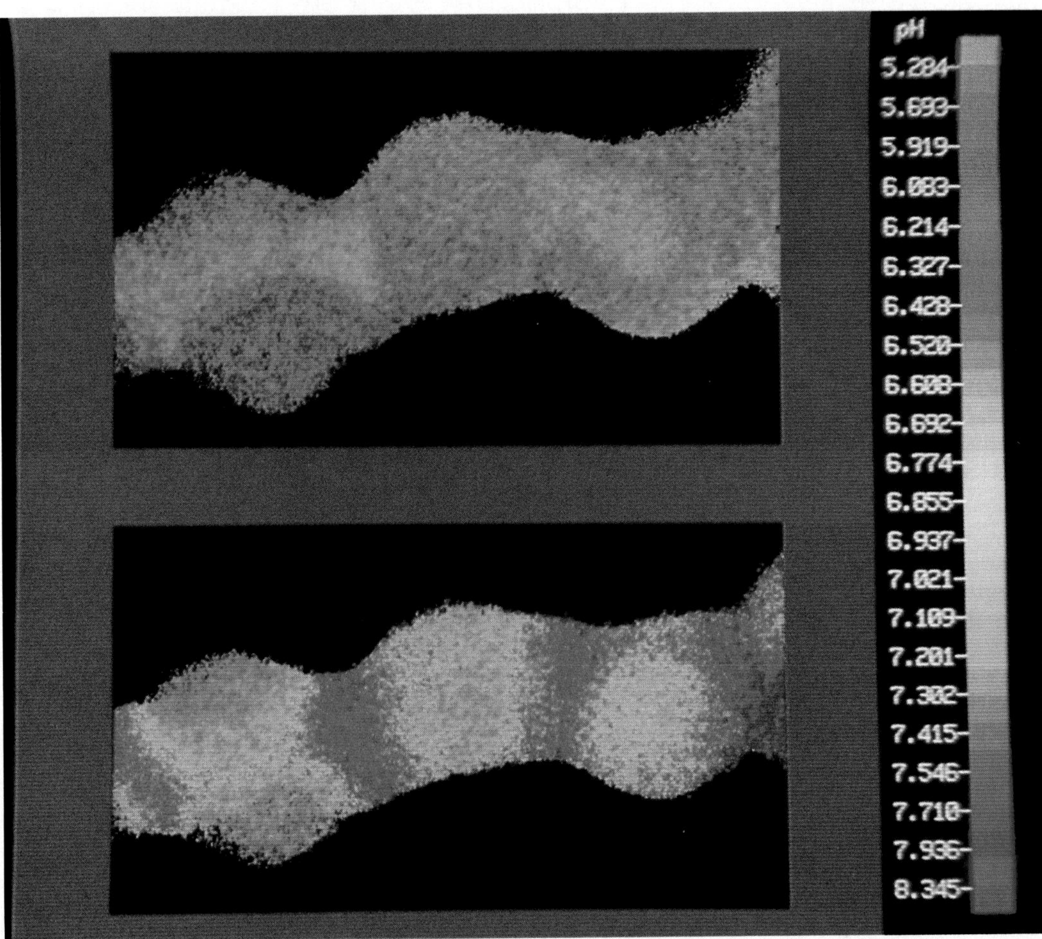

Chapter 15 described the general properties of acids and bases. This chapter deals quantitatively with acid and base ionization in water. Our discussion here is based on a single concept—chemical equilibria in solution.

16.1 Weak Acids and Acid Ionization Constants

Consider a weak monoprotic acid HA. Its ionization in water is represented by

$$HA(aq) + H_2O(l) \rightleftharpoons H_3O^+(aq) + A^-(aq)$$

or simply

$$HA(aq) \rightleftharpoons H^+(aq) + A^-(aq)$$

The equilibrium constant for this *acid ionization,* which we call the ***acid ionization constant, K_a,*** is given by

All concentrations in this equation are equilibrium concentrations.

$$K_a = \frac{[H^+][A^-]}{[HA]}$$

At a given temperature, the strength of the acid HA is measured quantitatively by the magnitude of K_a. The larger K_a, the stronger the acid—that is, the greater the concentration of H^+ ions at equilibrium due to its ionization.

Because the ionization of weak acids is never complete, all species (the nonionized

Table 16.1 Ionization constants of some weak acids at 25°C

Name of acid	Formula	Structure	K_a	Conjugate base	K_b
Hydrofluoric acid	HF	H—F	7.1×10^{-4}	F^-	1.4×10^{-11}
Nitrous acid	HNO_2	O=N—O—H	4.5×10^{-4}	NO_2^-	2.2×10^{-11}
Acetylsalicylic acid (aspirin)	$C_9H_8O_4$	(benzene ring)—C(=O)—O—H; —O—C(=O)—CH_3	3.0×10^{-4}	$C_9H_7O_4^-$	3.3×10^{-11}
Formic acid	HCOOH	H—C(=O)—O—H	1.7×10^{-4}	$HCOO^-$	5.9×10^{-11}
Ascorbic acid*	$C_6H_8O_6$	H—O, OH, H, C=C, C, C=O, CHOH, O, CH_2OH	8.0×10^{-5}	$C_6H_7O_6^-$	1.3×10^{-10}
Benzoic acid	C_6H_5COOH	(benzene ring)—C(=O)—O—H	6.5×10^{-5}	$C_6H_5COO^-$	1.5×10^{-10}
Acetic acid	CH_3COOH	CH_3—C(=O)—O—H	1.8×10^{-5}	CH_3COO^-	5.6×10^{-10}
Hydrocyanic acid	HCN	H—C≡N	4.9×10^{-10}	CN^-	2.0×10^{-5}
Phenol	C_6H_5OH	(benzene ring)—O—H	1.3×10^{-10}	$C_6H_5O^-$	7.7×10^{-5}

*For ascorbic acid it is the upper left hydroxyl group that is associated with this ionization constant.

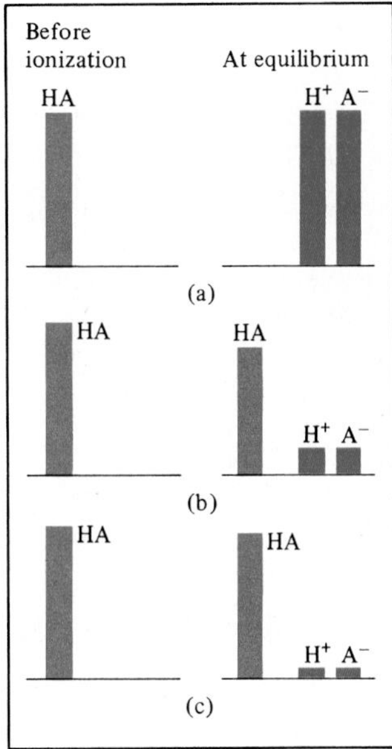

Figure 16.1 *The extent of ionization of (a) a strong acid that undergoes 100 percent ionization, (b) a weak acid, and (c) a very weak acid.*

acid, the H^+ ions, and the A^- ions) are present at equilibrium (Figure 16.1). Table 16.1 lists a number of weak acids and their K_a values in order of decreasing acid strength. Note that at equilibrium the predominant species in solution (other than the solvent) is the nonionized acid. Although the compounds listed in Table 16.1 are all weak acids, within the group there is great variation in acid strength. For example, K_a for HF (7.1×10^{-4}) is about 1.5 million times that for HCN (4.9×10^{-10}).

We can calculate K_a from the initial concentration of the acid and the pH of the solution, and we can use K_a and the initial concentration of the acid to calculate equilibrium concentrations of all the species and pH of solution. In calculating the equilibrium concentrations in a weak acid, we follow essentially the same procedure outlined in Section 14.5; the systems may be different, but the calculations are based on the same principle, the law of mass action [Equation (14.2)]. The three basic steps are:

1. Express the equilibrium concentrations of all species in terms of the initial concentrations and a single unknown x, which represents the change in concentration.
2. Write the acid ionization constant in terms of the equilibrium concentrations. Knowing the value of K_a, we can solve for x.
3. Having solved for x, calculate the equilibrium concentrations of all species and/or the pH of the solution.

Unless otherwise stated, we will assume that the temperature is 25°C for all such calculations.

The following three examples illustrate the calculations using the above procedure.

EXAMPLE 16.1

Calculate the concentrations of the nonionized acid and of the ions of a 0.100 M formic acid (HCOOH) solution at equilibrium.

Answer

Step 1: Table 16.1 shows that HCOOH is a weak acid. Since it is monoprotic, one HCOOH molecule ionizes to give one H^+ ion and one $HCOO^-$ ion. Let x be the equilibrium concentration of H^+ and $HCOO^-$ ions in mol/L. Then the equilibrium concentration of HCOOH must be $(0.100 - x)$ mol/L or $(0.100 - x)\,M$. We can now summarize the changes in concentrations as follows:

	$HCOOH(aq)$	$\rightleftharpoons$	$H^+(aq)$	+ $HCOO^-(aq)$
Initial:	$0.100\,M$		$0.000\,M$	$0.000\,M$
Change:	$-x\,M$		$+x\,M$	$+x\,M$
Equilibrium:	$(0.100 - x)\,M$		$x\,M$	$x\,M$

Step 2: According to Table 16.1

$$K_a = \frac{[H^+][HCOO^-]}{[HCOOH]} = 1.7 \times 10^{-4}$$

$$\frac{x^2}{0.100 - x} = 1.7 \times 10^{-4}$$

See Appendix 4 for a discussion of the quadratic equation.

The sign $\simeq$ means "approximately equal to."

This equation can be rewritten as

$$x^2 + 1.7 \times 10^{-4}x - 1.7 \times 10^{-5} = 0$$

which fits the quadratic equation $ax^2 + bx + c = 0$.

To avoid solving a quadratic equation we can often apply a simplifying approximation to this type of problem. Since HCOOH is a weak acid, the extent of its ionization must be small. Therefore x is small compared with 0.100. As a general rule, if the quantity x that is subtracted from the original concentration of the acid (0.100 M in this case) is equal to or less than 5 percent of the original concentration, we can assume $0.100 - x \simeq 0.100$. This removes x from the denominator of the equilibrium constant expression and we avoid a quadratic equation. If x is more than 5 percent of the original value, then we must solve the quadratic equation. Normally we can apply the approximation in cases where K_a is small (equal to or less than 1×10^{-4}) and the initial concentration of the acid is high (equal to or greater than 0.1 M). In case of doubt, we can always solve for x by the approximate method and then check the validity of our approximation. Assuming that $0.100 - x \simeq 0.100$, then

$$\frac{x^2}{0.100 - x} \simeq \frac{x^2}{0.100} = 1.7 \times 10^{-4}$$

$$x^2 = 1.7 \times 10^{-5}$$

Taking the square root of both sides, we obtain

$$x = 4.1 \times 10^{-3}\ M$$

Step 3: At equilibrium, therefore,

$$[H^+] = 4.1 \times 10^{-3}\ M$$

$$[HCOO^-] = 4.1 \times 10^{-3}\ M$$

$$[HCOOH] = (0.100 - 0.0041)\ M = 0.096\ M$$

To check the validity of our approximation

$$\frac{0.0041\ M}{0.100\ M} \times 100\% = 4.1\%$$

This shows that the quantity x is less than 5 percent of the original concentration of the acid. Thus our approximation was justified.

Note that we omitted the contribution to $[H^+]$ by water. Except in very dilute acid solutions, this omission is always valid, because the hydrogen ion concentration due to water is negligibly small compared with that due to the acid.

Similar problem: 16.6.

EXAMPLE 16.2

Calculate the pH of a 0.050 M nitrous acid (HNO_2) solution.

Answer

Step 1: From Table 16.1 we see that HNO_2 is a weak acid. Letting x be the equilibrium concentration of H^+ and NO_2^- ions in mol/L, we summarize:

	$HNO_2(aq)$	$\rightleftharpoons$	$H^+(aq)$	+	$NO_2^-(aq)$
Initial:	$0.050\ M$		$0.00\ M$		$0.00\ M$
Change:	$-x\ M$		$+x\ M$		$+x\ M$
Equilibrium:	$(0.050 - x)\ M$		$x\ M$		$x\ M$

Step 2: From Table 16.1,

$$K_a = \frac{[H^+][NO_2^-]}{[HNO_2]} = 4.5 \times 10^{-4}$$

$$\frac{x^2}{0.050 - x} = 4.5 \times 10^{-4}$$

Applying the approximation $0.050 - x \simeq 0.050$, we obtain

$$\frac{x^2}{0.050 - x} \simeq \frac{x^2}{0.050} = 4.5 \times 10^{-4}$$

$$x^2 = 2.3 \times 10^{-5}$$

Taking the square root of both sides gives

$$x = 4.8 \times 10^{-3}\ M$$

To test the approximation $$\frac{0.0048\ M}{0.050\ M} \times 100\% = 9.6\%$$

This shows that x is more than 5 percent of the original concentration. Thus our approximation is *not* valid, and we must solve the quadratic equation in step 2, as follows:

$$x^2 + 4.5 \times 10^{-4}x - 2.3 \times 10^{-5} = 0$$

Using the quadratic formula,

$$x = \frac{-b \pm \sqrt{b^2 - 4ac}}{2a}$$

$$= \frac{-4.5 \times 10^{-4} \pm \sqrt{(4.5 \times 10^{-4})^2 - 4(1)(-2.3 \times 10^{-5})}}{2(1)}$$

Thus $x = 4.6 \times 10^{-3}\ M$ or $x = -5.0 \times 10^{-3}\ M$

For this kind of problem, one of the solutions is always physically impossible.

The second solution is physically impossible, since the concentrations of ions produced as a result of ionization cannot be negative. Therefore, the solution to the quadratic equation is given by the positive root—that is, $x = 4.6 \times 10^{-3}\ M$.

Step 3: At equilibrium

$$[H^+] = 4.6 \times 10^{-3}\ M$$

and $$\text{pH} = -\log (4.6 \times 10^{-3}) = 2.34$$

Similar problem: 16.6.

EXAMPLE 16.3

The pH of a 0.100 *M* solution of a weak monoprotic acid HA is 2.85. What is the K_a of the acid?

Answer

In this case we are given the pH, from which we can obtain the equilibrium concentrations, and we are asked to calculate the acid ionization constant. We can follow the same basic steps.

Step 1: First we need to calculate the hydrogen ion concentration from the pH value.

$$pH = -\log [H^+]$$

$$2.85 = -\log [H^+]$$

Taking the antilog of both sides, we get

$$[H^+] = 1.4 \times 10^{-3}\ M$$

Next we summarize the changes:

	$HA(aq)$	$\rightleftharpoons$	$H^+(aq)$	+	$A^-(aq)$
Initial:	0.100 *M*		0.000 *M*		0.000 *M*
Change:	−0.0014 *M*		+0.0014 *M*		+0.0014 *M*
Equilibrium:	(0.100 − 0.0014) *M*		0.0014 *M*		0.0014 *M*

Step 2: The acid ionization constant is given by

$$K_a = \frac{[H^+][A^-]}{[HA]} = \frac{(0.0014)(0.0014)}{(0.100 - 0.0014)}$$
$$= 2.0 \times 10^{-5}$$

Similar problems: 16.7, 16.10.

When Can the Ionization of Water Be Ignored?

In all three above examples we ignored the contribution to $[H^+]$ due to the ionization of water. The assumption is valid because in each case the H^+ ion concentration from the ionization of the acid is at least 10,000 times greater than that from pure water (1×10^{-7} *M*). The general rule of thumb is that the ionization of water can be ignored when the H^+ ion concentration from water is less than 5 percent of the total H^+ concentration.

We can calculate what this total H^+ ion concentration is by the following procedure. In a solution of a weak monoprotic acid we have

$$HA(aq) \rightleftharpoons H^+(aq) + A^-(aq)$$

Simultaneously, we have the autoionization of water:

$$H_2O(l) \rightleftharpoons H^+(aq) + OH^-(aq)$$

The total H^+ ion concentration is given by

$$[H^+]_T = [H^+]_{HA} + [H^+]_W$$

where the subscripts T, HA, and W denote total, acid, and water, respectively. From the stoichiometry of water ionization we write

$$[H^+]_W = [OH^-]_W$$

However, in an acid solution we no longer have $[H^+]_W = 1.0 \times 10^{-7}$ *M* because, according to Le Chatelier's principle, the presence of H^+ ions from the acid suppresses the ionization of water. This means that $[H^+]_W$ is less than 1.0×10^{-7} *M*. The ion product of water in the acid solution is given by

The ion product is an equilibrium constant. Its value is the same in pure water or in an acid solution at the same temperature.

$$[H^+]_T[OH^-]_W = 1.0 \times 10^{-14}$$

To apply the 5 percent approximation we set $[H^+]_W = 0.05[H^+]_T$. Since $[H^+]_W = [OH^-]_W$, the ion-product expression can be rewritten as

$$[H^+]_T(0.05[H^+]_T) = 1.0 \times 10^{-14}$$

Solving the above equation, we get $[H^+]_T = 4.5 \times 10^{-7}$ *M*. Thus we arrive at the conclusion that as long as the total H^+ ion concentration is equal to or greater than 4.5×10^{-7} *M*, we can ignore the contribution from water in our calculations.

Percent Ionization

In Chapter 15 we saw that the percent ionization can also be used to compare the strengths of acids. For monoprotic acids, the concentration of acid that undergoes ionization is equal to the concentration of the H^+ ions and the concentration of the conjugate base at equilibrium. Therefore, we can write the percent ionization of a monoprotic acid as

$$\% \text{ ionization} = \frac{\text{hydrogen ion concentration at equilibrium}}{\text{initial concentration of acid}} \times 100\%$$

$$= \frac{\text{equilibrium concentration of conjugate base}}{\text{initial concentration of acid}} \times 100\%$$

Example 16.4 compares the percent ionization of two weak monoprotic acids.

EXAMPLE 16.4

Calculate the percent ionization of (a) a 0.60 *M* hydrofluoric acid (HF) solution and (b) a 0.60 *M* hydrocyanic acid (HCN) solution.

Answer

(a) HF

Step 1: Let *x* be the concentrations of H^+ and F^- ions at equilibrium in mol/L. We summarize:

	$HF(aq)$	$\rightleftharpoons$	$H^+(aq)$	+	$F^-(aq)$
Initial:	0.60 *M*		0.00 *M*		0.00 *M*
Change:	$-x$ *M*		$+x$ *M*		$+x$ *M*
Equilibrium:	$(0.60 - x)$ *M*		x *M*		x *M*

Step 2: From Table 16.1

$$K_a = \frac{[H^+][F^-]}{[HF]} = 7.1 \times 10^{-4}$$

$$\frac{x^2}{0.60 - x} = 7.1 \times 10^{-4}$$

Assuming that $0.60 - x \simeq 0.60$, then

$$\frac{x^2}{0.60 - x} \simeq \frac{x^2}{0.60} = 7.1 \times 10^{-4}$$

$$x^2 = 4.3 \times 10^{-4}$$

$$x = 0.021\ M$$

Because HF is monoprotic, the concentration of hydrogen ions produced by the acid at equilibrium is equal to that of the F^- ions. Thus

$$\%\ \text{ionization} = \frac{0.021\ M}{0.60\ M} \times 100\%$$
$$= 3.5\%$$

(b) HCN

Step 1: Let x be the concentrations of H^+ and CN^- ions at equilibrium in mol/L. We summarize:

	$HCN(aq)$	$\rightleftharpoons$	$H^+(aq)$	+ $CN^-(aq)$
Initial:	$0.60\ M$		$0.00\ M$	$0.00\ M$
Change:	$-x\ M$		$+x\ M$	$+x\ M$
Equilibrium:	$(0.60 - x)\ M$		$x\ M$	$x\ M$

Step 2: Referring to Table 16.1, we obtain

$$K_a = \frac{[H^+][CN^-]}{[HCN]} = 4.9 \times 10^{-10}$$

$$\frac{x^2}{0.60 - x} = 4.9 \times 10^{-10}$$

We assume that $0.60 - x \simeq 0.60$:

$$\frac{x^2}{0.60 - x} \simeq \frac{x^2}{0.60} = 4.9 \times 10^{-10}$$

$$x^2 = 2.9 \times 10^{-10}$$

$$x = 1.7 \times 10^{-5}\ M$$

Like HF, HCN is monoprotic. Therefore

$$\%\ \text{ionization} = \frac{1.7 \times 10^{-5}\ M}{0.60\ M} \times 100\%$$
$$= 0.0028\%$$

Thus, at the same concentration, HF ionizes to a much greater extent than does HCN. Note that the form of the percent ionization calculation is exactly the same as the calculation used to check the approximation that avoids a quadratic equation solution. Since the simplifying assumption is regarded as valid for acids whose percent ionization is less than 5 percent, both calculations in Example 16.4 qualify.

Similar problems: 16.11, 16.13.

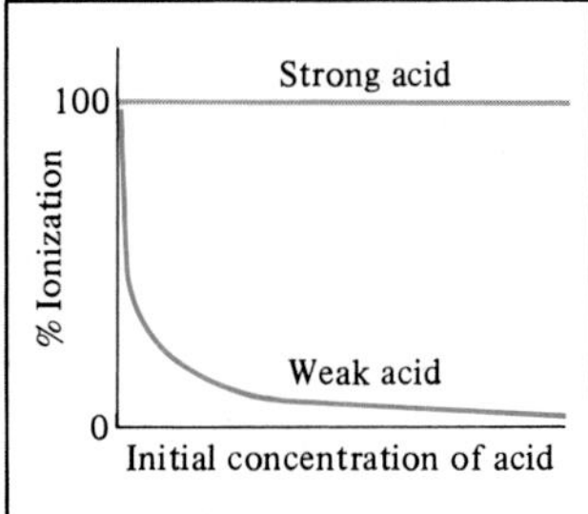

Figure 16.2 *Dependence of percent ionization on initial concentration of acid. Note that at very low concentrations all acids (weak and strong) are almost completely ionized.*

The extent to which a weak acid ionizes depends on the initial concentration of the acid. The more dilute the solution, the greater the percent ionization (Figure 16.2). In qualitative terms, when an acid is diluted, initially the number of particles (nonionized acid molecules plus ions) per unit volume is reduced. According to Le Chatelier's principle (see Section 14.6), to counteract this "stress" (that is, the dilution), the equilibrium shifts from un-ionized acid to H^+ and its conjugate base to produce more particles (ions).

The following example shows that the percent ionization, unlike the equilibrium constant, depends on the initial acid concentration.

EXAMPLE 16.5

Compare the percent ionization of HF at 0.60 M and at 0.00060 M.

Answer

In Example 16.4 we found the percent ionization at 0.60 M HF to be 3.5 percent. In order to compare we must find the percent ionization of 0.00060 M HF.

Step 1: Let x be the concentration in mol/L of H^+ and F^- at equilibrium. We summarize the changes:

	$HF(aq)$	$\rightleftharpoons$	$H^+(aq)$ +	$F^-(aq)$
Initial:	0.00060 M		0.00 M	0.00 M
Change:	$-x$ M		$+x$ M	$+x$ M
Equilibrium:	$(0.00060 - x)$ M		x M	x M

Step 2:

$$\frac{x^2}{0.00060 - x} = 7.1 \times 10^{-4}$$

Since the concentration of the acid is very low and the ionization constant is fairly large, the approximation method is not applicable. We express the equation in quadratic form and then substitute in the quadratic formula:

$$x^2 + 7.1 \times 10^{-4}x - 4.3 \times 10^{-7} = 0$$

$$x = \frac{-7.1 \times 10^{-4} \pm \sqrt{(7.1 \times 10^{-4})^2 - 4(1)(-4.3 \times 10^{-7})}}{2(1)}$$

$$x = 3.9 \times 10^{-4}\ M$$

Therefore

$$\%\ \text{ionization} = \frac{3.9 \times 10^{-4}\ M}{0.00060\ M} \times 100\%$$
$$= 65\%$$

The large percent ionization here shows clearly that the approximation $0.00060 - x \simeq 0.00060$ would not be valid. It also confirms our prediction that the extent of acid ionization increases with dilution.

Similar problem: 16.12.

16.2 Weak Bases and Base Ionization Constants

Strong bases such as the hydroxides of alkali metals and of the alkaline earth metals other than beryllium are completely ionized in water:

$$NaOH(aq) \longrightarrow Na^+(aq) + OH^-(aq)$$
$$KOH(aq) \longrightarrow K^+(aq) + OH^-(aq)$$
$$Ba(OH)_2(aq) \longrightarrow Ba^{2+}(aq) + 2OH^-(aq)$$

Table 16.2 Ionization constants of some common weak bases at 25°C

Name of base	Formula	Structure	K_b*	Conjugate acid	K_a
Ethylamine	$C_2H_5NH_2$	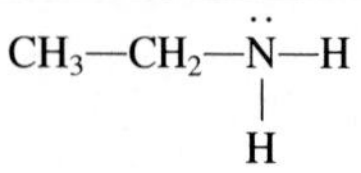 $CH_3—CH_2—\ddot{N}H—H$ (N bonded to H below)	5.6×10^{-4}	$C_2H_5\overset{+}{N}H_3$	1.8×10^{-11}
Methylamine	CH_3NH_2	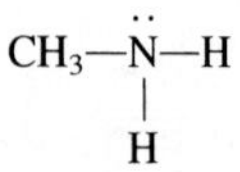 $CH_3—\ddot{N}—H$ (N bonded to H below)	4.4×10^{-4}	$CH_3\overset{+}{N}H_3$	2.3×10^{-11}
Caffeine	$C_8H_{10}N_4O_2$	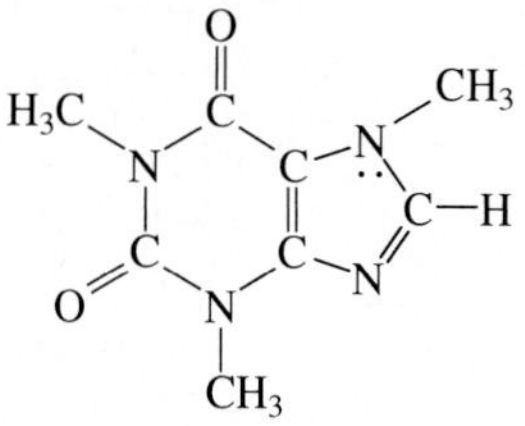 H$_3$C, O, C, N, C, O, N, CH$_3$, C, C, N̈, CH$_3$, C—H, N (fused ring structure)	4.1×10^{-4}	$\overset{+}{C_8H_{11}N_4O_2}$	2.4×10^{-11}
Ammonia	NH_3	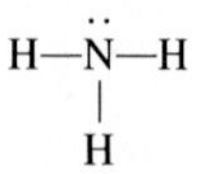 $H—\ddot{N}—H$ (N bonded to H below)	1.8×10^{-5}	NH_4^+	5.6×10^{-10}
Pyridine	C_5H_5N	 (ring) N:	1.7×10^{-9}	$C_5H_5\overset{+}{N}H$	5.9×10^{-6}
Aniline	$C_6H_5NH_2$	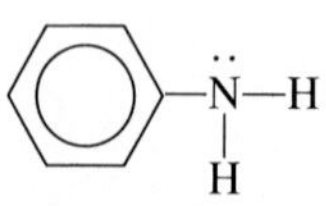 (ring)$—\ddot{N}—H$ (N bonded to H below)	3.8×10^{-10}	$C_6H_5\overset{+}{N}H_3$	2.6×10^{-5}
Urea	N_2H_4CO	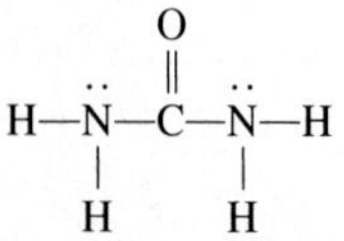 $H—\ddot{N}—\overset{O}{\overset{\Vert}{C}}—\ddot{N}—H$ (each N bonded to H below)	1.5×10^{-14}	$H_2NCON\overset{+}{H}_3$	0.67

*The nitrogen atom with the lone pair accounts for each compound's basicity. In the case of urea, K_b can be associated with either nitrogen atom.

Recall that OH^- is the strongest base that can exist in aqueous solutions.

Weak bases are treated like weak acids. When ammonia dissolves in water, it undergoes the reaction

$$NH_3(aq) + H_2O(l) \rightleftharpoons NH_4^+(aq) + OH^-(aq)$$

The production of hydroxide ions in this *base ionization* reaction means that, in this solution at 25°C, $[OH^-] > [H^+]$, and therefore pH > 7.

Because, compared with the total concentration of water, very few water molecules are consumed by the reaction, we can treat $[H_2O]$ as a constant. Thus we can write the equilibrium constant as

$$K[H_2O] = K_b = \frac{[NH_4^+][OH^-]}{[NH_3]}$$
$$= 1.8 \times 10^{-5}$$

The ability of the lone pair to accept a H^+ ion makes these substances Brønsted bases.

where K_b, *the equilibrium constant for base ionization,* is called the ***base ionization constant.*** Table 16.2 lists a number of common weak bases and their ionization constants. Note that the basicity of all these compounds is attributable to the lone pair of electrons on the nitrogen atom.

The ionization of a weak base is treated in the following example.

EXAMPLE 16.6

What is the pH of a 0.400 *M* ammonia solution?

Answer

The procedure is essentially the same as the one we use for weak acids.

Step 1: Let x be the concentration in mol/L of NH_4^+ and OH^- ions at equilibrium. Next we summarize:

	$NH_3(aq)$	$+ H_2O(l) \rightleftharpoons$	$NH_4^+(aq)$	$+ OH^-(aq)$
Initial:	0.400 *M*		0.000 *M*	0.000 *M*
Change:	$-x$ *M*		$+x$ *M*	$+x$ *M*
Equilibrium:	$(0.400 - x)$ *M*		x *M*	x *M*

Step 2: Using the base ionization constant listed in Table 16.2, we write

$$K_b = \frac{[NH_4^+][OH^-]}{[NH_3]} = 1.8 \times 10^{-5}$$

$$\frac{x^2}{0.400 - x} = 1.8 \times 10^{-5}$$

You should confirm the validity of this approximation.

Applying the approximation $0.400 - x \simeq 0.400$ gives

$$\frac{x^2}{0.400 - x} \simeq \frac{x^2}{0.400} = 1.8 \times 10^{-5}$$

$$x^2 = 7.2 \times 10^{-6}$$

$$x = 2.7 \times 10^{-3}\ M$$

Step 3: At equilibrium, $[OH^-] = 2.7 \times 10^{-3}\ M$. Thus

$$\begin{aligned} pOH &= -\log(2.7 \times 10^{-3}) \\ &= 2.57 \end{aligned}$$

$$\begin{aligned} pH &= 14.00 - 2.57 \\ &= 11.43 \end{aligned}$$

Note that we omitted the contribution to $[OH^-]$ from water.

The condition is similar to that for ignoring the contribution to $[H^+]$ from water (see p. 674).

Similar problems: 16.18, 16.19.

16.3 The Relationship between Conjugate Acid-Base Ionization Constants

An important relationship between the acid ionization constant and the ionization constant of its conjugate base can be derived as follows, using acetic acid as an example:

$$CH_3COOH(aq) \rightleftharpoons H^+(aq) + CH_3COO^-(aq)$$

$$K_a = \frac{[H^+][CH_3COO^-]}{[CH_3COOH]}$$

The conjugate base, CH_3COO^-, reacts with water according to the equation

$$CH_3COO^-(aq) + H_2O(l) \rightleftharpoons CH_3COOH(aq) + OH^-(aq)$$

and we can write the base ionization constant as

$$K_b = \frac{[CH_3COOH][OH^-]}{[CH_3COO^-]}$$

The product of these two ionization constants is given by

$$\begin{aligned} K_aK_b &= \frac{[H^+][CH_3COO^-]}{[CH_3COOH]} \times \frac{[CH_3COOH][OH^-]}{[CH_3COO^-]} \\ &= [H^+][OH^-] \\ &= K_w \end{aligned}$$

This result may seem strange at first, but it can be understood by realizing that the sum of reactions (1) and (2) below is simply the autoionization of water.

$$\begin{array}{llcll} (1) & CH_3COOH(aq) & \rightleftharpoons & H^+(aq) + CH_3COO^-(aq) & K_a \\ (2) & CH_3COO^-(aq) + H_2O(l) & \rightleftharpoons & CH_3COOH(aq) + OH^-(aq) & K_b \\ \hline (3) & H_2O(l) & \rightleftharpoons & H^+(aq) + OH^-(aq) & K_w \end{array}$$

This example illustrates one of the rules we learned about chemical equilibria: When two reactions are added to give a third reaction, the equilibrium constant for the third reaction is the product of the equilibrium constants for the two added reactions (see Section 14.3). Thus, for any conjugate acid-base pair it is always true that

$$K_aK_b = K_w \qquad (16.1)$$

Expressing Equation (16.1) in the following ways

$$K_a = \frac{K_w}{K_b} \qquad K_b = \frac{K_w}{K_a}$$

enables us to draw an important conclusion: The stronger the acid (the larger K_a), the weaker its conjugate base (the smaller K_b), and vice versa (see Tables 16.1 and 16.2).

We can use Equation (16.1) to calculate the K_b of the conjugate base (CH_3COO^-) of CH_3COOH as follows. We find the K_a value of CH_3COOH in Table 16.1 and write

$$\begin{aligned} K_b &= \frac{K_w}{K_a} \\ &= \frac{1.0 \times 10^{-14}}{1.8 \times 10^{-5}} \\ &= 5.6 \times 10^{-10} \end{aligned}$$

16.4 Diprotic and Polyprotic Acids

The treatment of diprotic and polyprotic acids is more involved than that of monoprotic acids because these substances may yield more than one hydrogen ion per molecule. These acids ionize in a stepwise manner; that is, they lose one proton at a time. An ionization constant expression can be written for each step of ionization. Consequently, two or more equilibrium constant expressions must often be used to calculate the concentrations of species in the acid solution. For example, for H_2CO_3 we write

$$H_2CO_3(aq) \rightleftharpoons H^+(aq) + HCO_3^-(aq) \qquad K_{a_1} = \frac{[H^+][HCO_3^-]}{[H_2CO_3]}$$

$$HCO_3^-(aq) \rightleftharpoons H^+(aq) + CO_3^{2-}(aq) \qquad K_{a_2} = \frac{[H^+][CO_3^{2-}]}{[HCO_3^-]}$$

Note that the conjugate base in the first step of ionization becomes the acid in the second step of ionization.

Table 16.3 shows the ionization constants of several diprotic acids and a polyprotic acid. For a given acid, the first ionization constant is much larger than the second ionization constant, and so on. This trend seems logical when we realize that it is easier to remove a H^+ ion from a neutral molecule than to remove another H^+ from a negatively charged ion derived from the molecule.

As the following examples show, calculation involving ionization constants of diprotic or polyprotic acids is more complex than the calculation for a monoprotic acid because it involves more than one stage of ionization.

EXAMPLE 16.7

Calculate the pH of a 0.010 *M* H_2SO_4 solution.

Answer

We note that H_2SO_4 is a strong acid for the first stage of ionization and that HSO_4^- is a weak acid. We summarize the changes in the first stage of ionization:

Table 16.3 Ionization constants of some common diprotic and polyprotic acids in water at 25°C

Name of acid	Formula	Structure	K_a	Conjugate base	K_b
Sulfuric acid	H_2SO_4	O ‖ H—O—S—O—H ‖ O	very large	HSO_4^-	very small
Hydrogen sulfate ion	HSO_4^-	O ‖ H—O—S—O⁻ ‖ O	1.3×10^{-2}	SO_4^{2-}	7.7×10^{-13}
Oxalic acid	$C_2H_2O_4$	O O ‖ ‖ H—O—C—C—O—H	6.5×10^{-2}	$C_2HO_4^-$	1.5×10^{-13}
Hydrogen oxalate ion	$C_2HO_4^-$	O O ‖ ‖ H—O—C—C—O⁻	6.1×10^{-5}	$C_2O_4^{2-}$	1.6×10^{-10}
Sulfurous acid*	H_2SO_3	O ‖ H—O—S—O—H	1.3×10^{-2}	HSO_3^-	7.7×10^{-13}
Hydrogen sulfite ion	HSO_3^-	O ‖ H—O—S—O⁻	6.3×10^{-8}	SO_3^{2-}	1.6×10^{-7}
Carbonic acid	H_2CO_3	O ‖ H—O—C—O—H	4.2×10^{-7}	HCO_3^-	2.4×10^{-8}
Hydrogen carbonate ion	HCO_3^-	O ‖ H—O—C—O⁻	4.8×10^{-11}	CO_3^{2-}	2.1×10^{-4}
Hydrosulfuric acid	H_2S	H—S—H	9.5×10^{-8}	HS^-	1.1×10^{-7}
Hydrogen sulfide ion†	HS^-	H—S⁻	1×10^{-19}	S^{2-}	1×10^{5}
Phosphoric acid	H_3PO_4	O ‖ H—O—P—O—H | O | H	7.5×10^{-3}	$H_2PO_4^-$	1.3×10^{-12}
Dihydrogen phosphate ion	$H_2PO_4^-$	O ‖ H—O—P—O⁻ | O | H	6.2×10^{-8}	HPO_4^{2-}	1.6×10^{-7}
Hydrogen phosphate ion	HPO_4^{2-}	O ‖ H—O—P—O⁻ ‖ O⁻	4.8×10^{-13}	PO_4^{3-}	2.1×10^{-2}

* H_2SO_3 has never been isolated and exists in only minute concentration in aqueous solution of SO_2. The K_a value here refers to the process $SO_2(g) + H_2O(l) \rightleftharpoons H^+(aq) + HSO_3^-(aq)$.

† The ionization constant of HS^- is very low and difficult to measure. The value listed here is only an estimate.

	$H_2SO_4(aq)$ $\longrightarrow$	$H^+(aq)$	$+ HSO_4^-(aq)$
Initial:	0.010 M	0.00 M	0.00 M
Change:	−0.010 M	+0.010 M	+0.010 M
Final:	0.00 M	0.010 M	0.010 M

For the second stage of ionization, we proceed as for a weak monoprotic acid.

Step 1: Let x be the concentration in mol/L of H^+ and SO_4^{2-} produced by the ionization of HSO_4^-. The total concentration of the H^+ ions at equilibrium must be the sum of H^+ ion concentrations due to both stages of ionization, that is, $(0.010 + x)\ M$.

	$HSO_4^-(aq)$ $\rightleftharpoons$	$H^+(aq)$	$+ SO_4^{2-}(aq)$
Initial:	0.010 M	0.010 M	0.00 M
Change:	$-x\ M$	$+x\ M$	$+x\ M$
Equilibrium:	$(0.010 - x)\ M$	$(0.010 + x)\ M$	$x\ M$

Step 2: From Table 16.3 we write

$$K_{a_2} = \frac{[H^+][SO_4^{2-}]}{[HSO_4^-]} = 1.3 \times 10^{-2}$$

$$\frac{(0.010 + x)x}{0.010 - x} = 1.3 \times 10^{-2}$$

Since the K_{a_2} of HSO_4^- is quite large, we must solve the quadratic equation, which simplifies to

$$x^2 + 0.023x - 1.3 \times 10^{-4} = 0$$

$$x = \frac{-0.023 \pm \sqrt{(0.023)^2 - 4(1)(-1.3 \times 10^{-4})}}{2(1)}$$

$$= 4.7 \times 10^{-3}\ M$$

The total H^+ ion concentration at equilibrium is the sum of the concentrations due to both stages of ionization $(0.010 + 4.7 \times 10^{-3})\ M$, or 0.015 M. Finally

$$\begin{aligned} pH &= -\log [H^+] \\ &= -\log 0.015 \\ &= 1.82 \end{aligned}$$

Similar problem: 16.24.

EXAMPLE 16.8

Oxalic acid is a poisonous substance used chiefly as a bleaching and cleansing agent (for example, to remove bathtub rings). Calculate the concentrations of all the species present at equilibrium in a solution of concentration 0.10 M.

Answer

Oxalic acid is a diprotic acid (see Table 16.3). We begin with the first stage of ionization.

Step 1:

	$C_2H_2O_4(aq)$	$\rightleftharpoons$	$H^+(aq)$	$+ \; C_2HO_4^-(aq)$
Initial:	$0.10\ M$		$0.00\ M$	$0.00\ M$
Change:	$-x\ M$		$+x\ M$	$+x\ M$
Equilibrium:	$(0.10 - x)\ M$		$x\ M$	$x\ M$

Step 2: With the ionization constant from Table 16.3, we write

$$K_{a_1} = \frac{[H^+][C_2HO_4^-]}{[C_2H_2O_4]} = 6.5 \times 10^{-2}$$

$$\frac{x^2}{0.10 - x} = 6.5 \times 10^{-2}$$

Since the ionization constant is fairly large, we cannot apply the approximation $0.10 - x \simeq 0.10$. Instead, we must solve the quadratic equation

$$x^2 + 6.5 \times 10^{-2}x - 6.5 \times 10^{-3} = 0$$

The solution gives $x = 0.054\ M$.

Step 3: When the equilibrium for the first stage of ionization is reached, the concentrations are

$$[H^+] = 0.054\ M$$

$$[C_2HO_4^-] = 0.054\ M$$

$$[C_2H_2O_4] = (0.10 - 0.054)\ M = 0.046\ M$$

Next we consider the second stage of ionization.

Step 1: Let y be the equilibrium concentration of $C_2O_4^{2-}$ in mol/L. Thus the equilibrium concentration of $C_2HO_4^-$ must be $(0.054 - y)\ M$. We have

	$C_2HO_4^-(aq)$	$\rightleftharpoons$	$H^+(aq)$	$+ \; C_2O_4^{2-}(aq)$
Initial:	$0.054\ M$		$0.054\ M$	$0.00\ M$
Change:	$-y\ M$		$+y\ M$	$+y\ M$
Equilibrium:	$(0.054 - y)\ M$		$(0.054 + y)\ M$	$y\ M$

Step 2: Using the ionization constant in Table 16.3, we write

$$K_{a_2} = \frac{[H^+][C_2O_4^{2-}]}{[C_2HO_4^-]} = 6.1 \times 10^{-5}$$

$$\frac{(0.054 + y)y}{(0.054 - y)} = 6.1 \times 10^{-5}$$

Because the ionization constant is small, we can make the following approximations:

$$0.054 + y \simeq 0.054$$

$$0.054 - y \simeq 0.054$$

Next we substitute them in the equation, which reduces to

$$y = 6.1 \times 10^{-5}\ M$$

Step 3: At equilibrium, therefore

$$[C_2H_2O_4] = 0.046\ M$$

$$[C_2HO_4^-] = (0.054 - 6.1 \times 10^{-5})\ M \simeq 0.054\ M$$

$$[H^+] = (0.054 + 6.1 \times 10^{-5})\ M \simeq 0.054\ M$$

$$[C_2O_4^{2-}] = 6.1 \times 10^{-5}\ M$$

$$[OH^-] = 1.0 \times 10^{-14}/0.054 = 1.9 \times 10^{-13}\ M$$

Similar problems: 16.25, 16.26.

Example 16.8 shows that for diprotic acids if $K_{a_1} \gg K_{a_2}$, then the concentration of the H^+ ions at equilibrium may be assumed to result only from the first stage of ionization.

Phosphoric acid is responsible for much of the "tangy" flavor of popular cola drinks.

Phosphoric acid (H_3PO_4) is an important polyprotic acid with three ionizable hydrogen atoms:

$$H_3PO_4(aq) \rightleftharpoons H^+(aq) + H_2PO_4^-(aq) \qquad K_{a_1} = \frac{[H^+][H_2PO_4^-]}{[H_3PO_4]} = 7.5 \times 10^{-3}$$

$$H_2PO_4^-(aq) \rightleftharpoons H^+(aq) + HPO_4^{2-}(aq) \qquad K_{a_2} = \frac{[H^+][HPO_4^{2-}]}{[H_2PO_4^-]} = 6.2 \times 10^{-8}$$

$$HPO_4^{2-}(aq) \rightleftharpoons H^+(aq) + PO_4^{3-}(aq) \qquad K_{a_3} = \frac{[H^+][PO_4^{3-}]}{[HPO_4^{2-}]} = 4.8 \times 10^{-13}$$

We see that phosphoric acid is a weak polyprotic acid and that its ionization constants decrease rapidly. Thus we can predict that, in a solution containing phosphoric acid, the concentration of the nonionized acid is the highest and the only other species present in significant concentrations are H^+ and $H_2PO_4^-$ ions.

16.5 Acid-Base Properties of Salts

The word "hydrolysis" is derived from the Greek words *hydro*, meaning "water," and *lysis*, meaning "to split apart."

As defined earlier, a salt is an ionic compound formed by the reaction between an acid and a base. Salts are strong electrolytes that completely dissociate in water. The term ***salt hydrolysis*** *describes the reaction of an anion or a cation of a salt, or both, with water.* Salt hydrolysis usually affects the pH of a solution. In this section we discuss various types of salt hydrolysis.

Salts That Produce Neutral Solutions

When $NaNO_3$, formed by the reaction between NaOH and HNO_3, dissolves in water, it dissociates as follows:

$$NaNO_3(aq) \xrightarrow{H_2O} Na^+(aq) + NO_3^-(aq)$$

As we saw in Section 15.5, the hydrated Na^+ ion neither donates nor accepts H^+ ions. The NO_3^- ion is the conjugate base of the strong acid HNO_3, and it has no affinity for H^+ ions. Consequently, a solution containing Na^+ and NO_3^- is neutral, with a pH of 7. It is generally true that salts containing an alkali metal ion or alkaline earth metal ion (except Be^{2+}) and the conjugate base of a strong acid (for example, Cl^-, Br^-, and NO_3^-) do not undergo hydrolysis, and their solutions are also neutral.

Salts That Produce Basic Solutions

The dissociation of sodium acetate (CH_3COONa) in water is given by

$$CH_3COONa(s) \xrightarrow{H_2O} Na^+(aq) + CH_3COO^-(aq)$$

The hydrated Na^+ ion has no acidic or basic properties. The acetate ion CH_3COO^-, however, is the conjugate base of the weak acid CH_3COOH and therefore has an affinity for H^+ ions. The hydrolysis reaction is given by

$$CH_3COO^-(aq) + H_2O(l) \rightleftharpoons CH_3COOH(aq) + OH^-(aq)$$

Because this reaction produces OH^- ions, the sodium acetate solution will be basic. The equilibrium constant for this hydrolysis reaction is identical to the base ionization constant expression for CH_3COO^-, so we write

$$K_b = \frac{[CH_3COOH][OH^-]}{[CH_3COO^-]}$$

The value of K_b can be obtained from Equation (16.1):

$$K_b = \frac{K_w}{K_a} = \frac{1.0 \times 10^{-14}}{1.8 \times 10^{-5}} = 5.6 \times 10^{-10}$$

where K_a is the acid ionization of CH_3COOH.

Since each CH_3COO^- ion that hydrolyzes produces one OH^- ion, the concentration of OH^- at equilibrium is the same as the concentration of CH_3COO^- that hydrolyzed. We can define the *percent hydrolysis* as

$$\begin{aligned}\% \text{ hydrolysis} &= \frac{[CH_3COO^-]_{\text{hydrolyzed}}}{[CH_3COO^-]_{\text{initial}}} \times 100\% \\ &= \frac{[OH^-]_{\text{equilibrium}}}{[CH_3COO^-]_{\text{initial}}} \times 100\%\end{aligned}$$

The CH_3COONa example shows that the solution of a salt derived from a strong base and a weak acid is basic.

A calculation based on the hydrolysis of CH_3COONa is illustrated in Example 16.9.

EXAMPLE 16.9

Calculate the pH of a 0.15 *M* solution of sodium acetate (CH_3COONa). What is the percent hydrolysis?

Answer

We note that CH_3COONa is the salt formed from a weak acid (CH_3COOH) and a strong base (NaOH). Consequently, only the anion (CH_3COO^-) will hydrolyze. The initial dissociation of the salt is

$$\begin{array}{lccc} CH_3COONa(s) \xrightarrow{H_2O} & CH_3COO^-(aq) & + & Na^+(aq) \\ & 0.15\ M & & 0.15\ M \end{array}$$

We can now treat the hydrolysis of CH_3COO^-. The acetate ion acts as a weak Brønsted base.

Step 1: Let x be the equilibrium concentration of CH_3COOH and OH^- ions in mol/L. We summarize the changes:

	$CH_3COO^-(aq)$ + $H_2O(l)$ ⇌	$CH_3COOH(aq)$ +	$OH^-(aq)$
Initial:	$0.15\ M$	$0.00\ M$	$0.00\ M$
Change:	$-x\ M$	$+x\ M$	$+x\ M$
Equilibrium:	$(0.15 - x)\ M$	$x\ M$	$x\ M$

Step 2:

$$K_b = \frac{[CH_3COOH][OH^-]}{[CH_3COO^-]} = 5.6 \times 10^{-10}$$

$$\frac{x^2}{0.15 - x} = 5.6 \times 10^{-10}$$

Since K_b is very small and the initial concentration of the base is large, we can apply the approximation $0.15 - x \simeq 0.15$:

$$\frac{x^2}{0.15 - x} \simeq \frac{x^2}{0.15} = 5.6 \times 10^{-10}$$

$$x = 9.2 \times 10^{-6}\ M$$

Step 3: At equilibrium

$$[OH^-] = 9.2 \times 10^{-6}\ M$$

$$\begin{aligned} pOH &= -\log(9.2 \times 10^{-6}) \\ &= 5.04 \end{aligned}$$

$$\begin{aligned} pH &= 14.00 - 5.04 \\ &= 8.96 \end{aligned}$$

Thus the solution is basic, as we would expect. The percent hydrolysis is given by

$$\begin{aligned} \%\ \text{hydrolysis} &= \frac{9.2 \times 10^{-6}\ M}{0.15\ M} \times 100\% \\ &= 0.0061\% \end{aligned}$$

The result shows that only a very small portion of the anion undergoes hydrolysis. The small percent hydrolysis also justifies the approximation that $0.15 - x \simeq 0.15$.

Similar problem: 16.36.

Salts That Produce Acidic Solutions

When a salt derived from a strong acid and a weak base dissolves in water, the solution becomes acidic. For example, consider the process

$$NH_4Cl(s) \xrightarrow{H_2O} NH_4^+(aq) + Cl^-(aq)$$

The Cl^- ion has no affinity for H^+. The ammonium ion NH_4^+ is the weak conjugate acid of the weak base NH_3 and ionizes as follows:

$$NH_4^+(aq) + H_2O(l) \rightleftharpoons NH_3(aq) + H_3O^+(aq)$$

or simply

$$NH_4^+(aq) \rightleftharpoons NH_3(aq) + H^+(aq)$$

Since this reaction produces H^+ ions, the pH of the solution decreases. As you can see, the hydrolysis of the NH_4^+ ion is the same as the ionization of the NH_4^+ acid. The equilibrium constant (or ionization constant) for this process is given by

By coincidence, K_a of NH_4^+ has the same numerical value as K_b of CH_3COO^-.

$$K_a = \frac{[NH_3][H^+]}{[NH_4^+]} = \frac{K_w}{K_b} = \frac{1.0 \times 10^{-14}}{1.8 \times 10^{-5}} = 5.6 \times 10^{-10}$$

The following example deals with the pH change and percent hydrolysis of an NH_4Cl solution.

EXAMPLE 16.10

What are the pH and percent hydrolysis of a 0.10 *M* NH_4Cl solution?

Answer

We note that NH_4Cl is the salt formed from a strong acid (HCl) and a weak base (NH_3). Consequently, only the cation (NH_4^+) of the salt will hydrolyze. The initial dissociation of the salt is

$$\begin{array}{lcll} NH_4Cl(s) & \xrightarrow{H_2O} & NH_4^+(aq) & + \ Cl^-(aq) \\ & & 0.10\ M & \quad 0.10\ M \end{array}$$

The Cl^- ion is an extremely weak Brønsted base.

We can now treat the hydrolysis of the cation as the ionization of the acid.

Step 1: We represent the hydrolysis of the cation NH_4^+, and let x be the equilibrium concentration of NH_3 and H^+ ions in mol/L:

$$\begin{array}{lccc} & NH_4^+(aq) & \rightleftharpoons NH_3(aq) & + \ H^+(aq) \\ \text{Initial:} & 0.10\ M & 0.00\ M & 0.00\ M \\ \text{Change:} & -x\ M & +x\ M & +x\ M \\ \hline \text{Equilibrium:} & (0.10 - x)\ M & x\ M & x\ M \end{array}$$

Step 2: From Table 16.2 we obtain the K_a for NH_4^+:

$$K_a = \frac{[NH_3][H^+]}{[NH_4^+]} = 5.6 \times 10^{-10}$$

$$\frac{x^2}{0.10 - x} = 5.6 \times 10^{-10}$$

Applying the approximation $0.10 - x \simeq 0.10$, we get

$$\frac{x^2}{0.10 - x} \simeq \frac{x^2}{0.10} = 5.6 \times 10^{-10}$$

$$x = 7.5 \times 10^{-6}\ M$$

Thus the pH is given by

$$\begin{aligned} pH &= -\log(7.5 \times 10^{-6}) \\ &= 5.12 \end{aligned}$$

The percent hydrolysis is

$$\% \text{ hydrolysis} = \frac{7.5 \times 10^{-6}\, M}{0.10\, M} \times 100\%$$
$$= 0.0075\%$$

We see that the extent of hydrolysis is very small in a 0.10 M ammonium chloride solution and that the approximation $0.10 - x \approx 0.10$ was justified.

Similar problem: 16.37.

Salts that contain small, highly charged metal cations (for example, Al^{3+}, Cr^{3+}, Fe^{3+}, Bi^{3+}, and Be^{2+}) and the conjugate bases of strong acids also produce an acidic solution. For example, when aluminum chloride ($AlCl_3$) dissolves in water, the Al^{3+} ions take the hydrated form $Al(H_2O)_6^{3+}$ (Figure 16.3). Let's consider one bond between the metal ion and the oxygen atom of one of the six water molecules in $Al(H_2O)_6^{3+}$:

$$\text{Al} \Leftarrow\!\!\!\!- \text{O} \begin{smallmatrix} \nearrow \text{H} \\ \searrow \text{H} \end{smallmatrix}$$

The positively charged Al^{3+} ion draws electron density toward itself, making the O—H bond more polar. Consequently, the H atoms have a greater tendency to ionize than those in water molecules not involved in hydration. The resulting ionization process can be written as

Note that the hydrated Al^{3+} ion qualifies as a proton donor and thus a Brønsted acid in this reaction.

$$Al(H_2O)_6^{3+}(aq) + H_2O(l) \rightleftharpoons Al(OH)(H_2O)_5^{2+}(aq) + H_3O^+(aq)$$

or simply

$$Al(H_2O)_6^{3+}(aq) \rightleftharpoons Al(OH)(H_2O)_5^{2+}(aq) + H^+(aq)$$

The equilibrium constant for the metal cation hydrolysis is given by

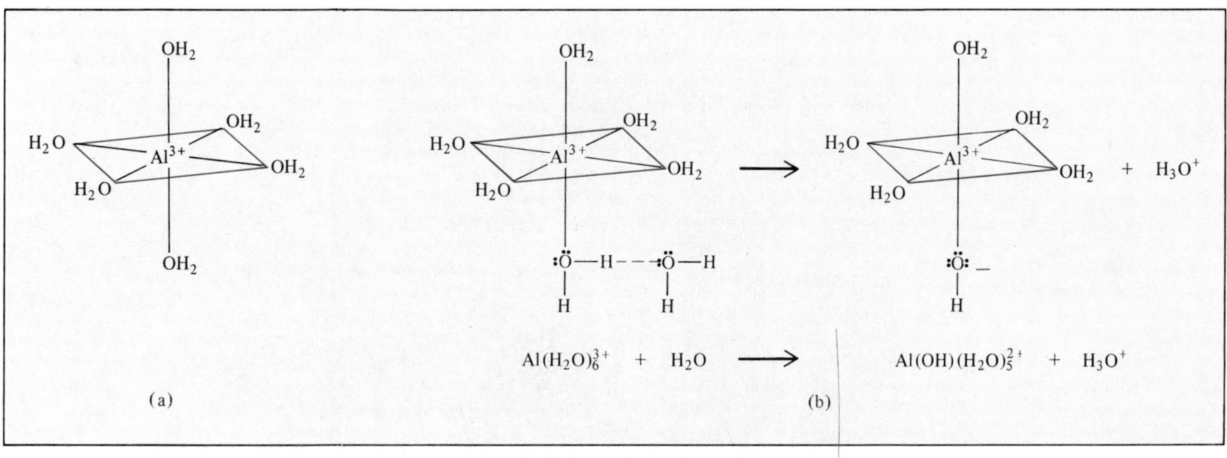

Figure 16.3 *(a) The $Al(H_2O)_6^{3+}$ ion. The six H_2O molecules surround the Al^{3+} ion octahedrally. (b) The attraction of the small Al^{3+} ion for the lone pairs on the oxygen atoms is so great that the O—H bonds in a H_2O molecule attached to the metal cation are weakened, allowing the loss of a proton (H^+) to an incoming H_2O molecule. This metal cation hydrolysis makes the solution acidic.*

Note that $Al(H_2O)_6^{3+}$ is roughly as strong an acid as CH_3COOH.

$$K_a = \frac{[Al(OH)(H_2O)_5^{2+}][H^+]}{[Al(H_2O)_6^{3+}]} = 1.3 \times 10^{-5}$$

Note that the $Al(OH)(H_2O)_5^{2+}$ species can undergo further ionization as

$$Al(OH)(H_2O)_5^{2+}(aq) \rightleftharpoons Al(OH)_2(H_2O)_4^{+}(aq) + H^+(aq)$$

and so on. However, generally it is sufficient to consider only the first step of hydrolysis.

The extent of hydrolysis is greatest for the smallest and most highly charged ions because a "compact" highly charged ion is more effective in polarizing the O—H bond and facilitating ionization. This is the reason that relatively large ions of low charge such as Na^+ and K^+ do not undergo hydrolysis.

The following example shows the calculation of the pH of an $AlCl_3$ solution.

EXAMPLE 16.11

Calculate the pH of a 0.020 *M* $AlCl_3$ solution.

Answer

Since $AlCl_3$ is a strong electrolyte, the initial dissociation is

$$\underset{}{AlCl_3(s)} \xrightarrow{H_2O} \underset{0.020\,M}{Al^{3+}(aq)} + \underset{0.060\,M}{3Cl^-(aq)}$$

Only the Al^{3+} ion will hydrolyze. As in the case of NH_4^+, we can treat the hydrolysis of Al^{3+} as the ionization of its hydrated ion.

Step 1: Let x be the equilibrium concentration of $Al(OH)(H_2O)_5^{2+}$ and of H^+ in mol/L:

	$Al(H_2O)_6^{3+}(aq)$	$\rightleftharpoons$ $Al(OH)(H_2O)_5^{2+}(aq)$	+ $H^+(aq)$
Initial:	0.020 *M*	0.00 *M*	0.00 *M*
Change:	$-x$ *M*	$+x$ *M*	$+x$ *M*
Equilibrium:	$(0.020 - x)$ *M*	x *M*	x *M*

Step 2: The equilibrium constant for the ionization is

$$K_a = \frac{[Al(OH)(H_2O)_5^{2+}][H^+]}{[Al(H_2O)_6^{3+}]} = 1.3 \times 10^{-5}$$

$$\frac{x^2}{0.020 - x} = 1.3 \times 10^{-5}$$

Applying the approximation $0.020 - x \simeq 0.020$, we have

$$\frac{x^2}{0.020 - x} \simeq \frac{x^2}{0.020} = 1.3 \times 10^{-5}$$

$$x = 5.1 \times 10^{-4}\ M$$

Thus the pH of the solution is

$$pH = -\log(5.1 \times 10^{-4}) = 3.29$$

Similar problem: 16.39.

Salts in Which Both the Cation and the Anion Hydrolyze

So far we have considered salts in which only one ion undergoes hydrolysis. For salts derived from a weak acid and a weak base, both the cation and the anion hydrolyze. However, whether a solution containing such a salt is acidic, basic, or neutral depends on the relative strengths of the weak acid and the weak base. Since the mathematics associated with this type of system is rather involved, we limit ourselves to making qualitative predictions about these solutions. We consider three situations:

- $K_b > K_a$. If K_b for the anion is greater than K_a for the cation, then the solution must be basic because the anion will hydrolyze to a greater extent than the cation. At equilibrium, there will be more OH^- ions than H^+ ions.
- $K_b < K_a$. Conversely, if K_b of the anion is smaller than K_a of the cation, the solution will be acidic because cation hydrolysis will be more extensive than anion hydrolysis.
- $K_a \simeq K_b$. If K_a is approximately equal to K_b, the solution will be nearly neutral.

Table 16.4 summarizes the behavior in aqueous solution of the salts discussed in this section.

16.6 The Common Ion Effect

Our discussion of acid-base ionization and salt hydrolysis so far has been limited to solutions containing a single solute. What happens when two different compounds are dissolved? If both sodium acetate and acetic acid are in solution, they both dissociate and ionize to produce CH_3COO^- ions:

$$CH_3COONa(s) \xrightarrow{H_2O} CH_3COO^-(aq) + Na^+(aq)$$

$$CH_3COOH(aq) \rightleftharpoons CH_3COO^-(aq) + H^+(aq)$$

CH_3COONa is a strong electrolyte, so it dissociates completely in solution, but CH_3COOH, a weak acid, ionizes only slightly. According to Le Chatelier's principle, the addition of CH_3COO^- ions from CH_3COONa to a solution of CH_3COOH will suppress the ionization of CH_3COOH (that is, shift the equilibrium from right to left), thereby decreasing the hydrogen ion concentration. Thus a solution containing both CH_3COOH and CH_3COONa will be *less* acidic than a solution containing only CH_3COOH at the same concentration. The shift in equilibrium of the acetic acid ionization is caused by the additional acetate ions from the salt. The CH_3COO^- ion is

Table 16.4 Acid-base properties of salts

Type of salt	Examples	Ions that undergo hydrolysis	pH of solution
Cation from strong base; anion from strong acid	$NaCl$, KI, KNO_3, $RbBr$, $BaCl_2$	None	$\simeq 7$
Cation from strong base; anion from weak acid	CH_3COONa, KNO_2	Anion	>7
Cation from weak base; anion from strong acid	NH_4Cl, NH_4NO_3	Cation	<7
Cation from weak base; anion from weak acid	NH_4NO_2, CH_3COONH_4, NH_4CN	Anion and cation	<7 if $K_b < K_a$ $\simeq 7$ if $K_b \simeq K_a$ >7 if $K_b > K_a$
Small, highly charged cation; anion from strong acid	$AlCl_3$, $Fe(NO_3)_3$	Hydrated cation	<7

called the *common ion* because it is supplied by both CH_3COOH and CH_3COONa.

The shift in equilibrium caused by the addition of a compound having an ion in common with the dissolved substance is called the ***common ion effect.*** The common ion effect plays an important role in determining the pH of a solution and the solubility of a slightly soluble salt. We will deal with the latter in Chapter 17. Here we will study the common ion effect as it relates to the pH of a solution. Keep in mind that despite its distinctive name, the common ion effect is simply a special case of Le Chatelier's principle.

Let us consider the pH of a solution containing a weak acid HA and a soluble salt of the weak acid, such as NaA. We start by writing

$$HA(aq) + H_2O(l) \rightleftharpoons H_3O^+(aq) + A^-(aq)$$

or simply

$$HA(aq) \rightleftharpoons H^+(aq) + A^-(aq)$$

The ionization constant K_a is given by

$$K_a = \frac{[H^+][A^-]}{[HA]} \tag{16.2}$$

Rearranging Equation (16.2) gives

$$[H^+] = \frac{K_a[HA]}{[A^-]}$$

Taking the negative logarithm of both sides, we obtain

$$-\log\,[H^+] = -\log K_a - \log\frac{[HA]}{[A^-]}$$

or

$$-\log\,[H^+] = -\log K_a + \log\frac{[A^-]}{[HA]}$$

So

$$\mathrm{pH} = \mathrm{p}K_a + \log\frac{[A^-]}{[HA]} \tag{16.3}$$

pK_a is related to K_a as pH is related to $[H^+]$. Remember that the stronger the acid (that is, the larger the K_a), the smaller the pK_a.

where

$$\mathrm{p}K_a = -\log K_a \tag{16.4}$$

Equation (16.3) is called the *Henderson-Hasselbalch equation*. In a more general form it can be expressed as

$$\mathrm{pH} = \mathrm{p}K_a + \log\frac{[\text{conjugate base}]}{[\text{acid}]} \tag{16.5}$$

In our example, HA is the acid and A^- is the conjugate base. Thus, if we know K_a and the concentrations of the acid and the salt of the acid, we can calculate the pH of the solution.

It is important to remember that the Henderson-Hasselbalch equation is derived from the equilibrium constant expression. It is valid regardless of the source of the conjugate base (that is, whether it comes from the acid alone or is supplied by both the acid and its salt).

In solving problems that involve the common ion effect, we are usually given the starting concentrations of a weak acid HA and of its salt, such as NaA. As long as the concentrations of these species are reasonably high ($\geq 0.1\ M$), we can neglect the ionization of the acid and the hydrolysis of the salt. Thus we can use the starting concentrations as the equilibrium concentrations in Equation (16.2) or Equation (16.5).

The following example deals with calculation of the pH of a solution containing a common ion.

EXAMPLE 16.12

(a) Calculate the pH of a solution containing 0.20 M CH_3COOH and 0.30 M CH_3COONa. (b) What would be the pH of a 0.20 M CH_3COOH solution if no salt were present?

Answer

(a) Sodium acetate is a strong electrolyte, so it dissociates completely in solution:

$$\underset{}{CH_3COONa(s)} \xrightarrow{H_2O} \underset{0.30\,M}{CH_3COO^-(aq)} + \underset{0.30\,M}{Na^+(aq)}$$

The equilibrium concentrations of both the acid and the conjugate base are assumed to be the same as the starting concentrations; that is,

$$[CH_3COOH] = 0.20\,M \quad \text{and} \quad [CH_3COO^-] = 0.30\,M$$

This is a valid assumption because (1) CH_3COOH is a weak acid and the extent of hydrolysis of the CH_3COO^- ion is very small (see Example 16.9); and (2) the presence of CH_3COO^- ions further suppresses the ionization of CH_3COOH, and the presence of CH_3COOH further suppresses the hydrolysis of the CH_3COO^- ions. From Equation (16.2) we have

$$[H^+] = \frac{K_a[HA]}{[A^-]} = \frac{(1.8 \times 10^{-5})(0.20)}{0.30} = 1.2 \times 10^{-5}\,M$$

Thus

$$pH = -\log [H^+] = -\log (1.2 \times 10^{-5}) = 4.92$$

Alternatively, we can calculate the pH of the solution by using the Henderson-Hasselbalch equation. In this case we need to calculate pK_a of the acid first [see Equation (16.4)]:

$$pK_a = -\log K_a = -\log (1.8 \times 10^{-5}) = 4.74$$

We can calculate the pH of the solution by substituting the value of pK_a and the concentrations of the acid and its conjugate base in Equation (16.5):

$$pH = pK_a + \log \frac{[CH_3COO^-]}{[CH_3COOH]} = 4.74 + \log \frac{0.30\,M}{0.20\,M} = 4.92$$

(b) Following the procedure in Example 16.1, we find that the pH of a 0.20 *M* CH_3COOH solution is:

$$[H^+] = 1.9 \times 10^{-3}\ M$$

$$\begin{aligned} pH &= -\log(1.9 \times 10^{-3}) \\ &= 2.72 \end{aligned}$$

Thus, without the common ion effect, the pH of a 0.20 *M* CH_3COOH solution is 2.72, considerably lower than 4.92, the pH in the presence of CH_3COONa, as calculated in (a). The presence of the common ion CH_3COO^- clearly suppresses the ionization of the acid CH_3COOH.

Similar problem: 16.44.

The common ion effect also operates in a solution containing a weak base (for example, NH_3) and a salt of the base (for example, NH_4Cl). At equilibrium

$$NH_4^+(aq) \rightleftharpoons NH_3(aq) + H^+(aq)$$

$$K_a = \frac{[NH_3][H^+]}{[NH_4^+]}$$

We can derive the Henderson-Hasselbalch equation for this system as follows. Rearranging the above equation we obtain

$$[H^+] = \frac{K_a[NH_4^+]}{[NH_3]}$$

Taking the negative logarithm of both sides gives

$$-\log[H^+] = -\log K_a - \log\frac{[NH_4^+]}{[NH_3]}$$

$$-\log[H^+] = -\log K_a + \log\frac{[NH_3]}{[NH_4^+]}$$

or

$$pH = pK_a + \log\frac{[NH_3]}{[NH_4^+]}$$

A solution containing both NH_3 and its salt NH_4Cl is *less* basic than a solution containing only NH_3 at the same concentration. The common ion NH_4^+ suppresses the ionization of NH_3 in the solution containing both the base and the salt.

16.7 Buffer Solutions

A ***buffer solution*** is a *solution of (1) a weak acid or base and (2) its salt; both components must be present. The solution has the ability to resist changes in pH upon the addition of small amounts of either acid or base.* Buffers are very important to chemical and biological systems. The pH in the human body varies greatly from one fluid to another; for example, the pH of blood is about 7.4, whereas the gastric juice in our stomachs has a pH of about 1.5. These pH values, which are crucial for the proper functioning of enzymes and the balance of osmotic pressure, are maintained by buffers in most cases.

A buffer solution must contain an acid to react with any OH^- ions that may be added to it and a base to react with any added H^+ ions. Furthermore, the acid and the base components of the buffer must not consume each other in a neutralization reaction. These requirements are satisfied by an acid-base conjugate pair (a weak acid and its conjugate base or a weak base and its conjugate acid).

One of the simplest buffers is the acetic acid-sodium acetate system. A solution containing these two substances has the ability to neutralize either added acid or added base as follows. If a base is added to the buffer system, the OH^- ions will be neutralized by the acid in the buffer:

$$CH_3COOH(aq) + OH^-(aq) \longrightarrow CH_3COO^-(aq) + H_2O(l)$$

If an acid is added, the H^+ ions will be consumed by the conjugate base in the buffer, according to the equation

$$CH_3COO^-(aq) + H^+(aq) \longrightarrow CH_3COOH(aq)$$

Here the acetate ions are provided by the dissociation of the sodium acetate:

$$CH_3COONa(s) \xrightarrow{H_2O} CH_3COO^-(aq) + Na^+(aq)$$

As you can see, the two reactions that characterize this buffer system are identical to those for the common ion effect described in Example 16.12. In general, a buffer system can be represented as salt/acid or conjugate base/acid. Thus the sodium acetate–acetic acid buffer system can be written as CH_3COONa/CH_3COOH or CH_3COO^-/CH_3COOH. Figure 16.4 shows the buffer system in action.

(a) (b) (c) (d)

Figure 16.4 *The acid-base indicator bromophenol blue (added to all solutions) is used to illustrate buffer action. The indicator's color is blue-purple above pH 4.6 and changes to yellow below pH 3.0. (a) A buffer solution made up of 50 mL of 0.1 M CH_3COOH and 50 mL of 0.1 M CH_3COONa. The solution has a pH of 4.7 and turns the indicator blue-purple. (b) After the addition of 40 mL of 0.1 M HCl solution to the solution in (a), the color remains blue-purple. (c) A 100-mL CH_3COOH solution whose pH is 4.7. (d) After the addition of 6 drops (about 0.3 mL) of 0.1 M HCl solution, the color turns yellow. Without buffer action, the pH of the solution decreases rapidly to below 3.0 upon the addition of 0.1 M HCl.*

Example 16.13 distinguishes buffer systems from acid-salt combinations that do not function as buffers.

EXAMPLE 16.13

Which of the following solutions are buffer systems? (a) KH_2PO_4/H_3PO_4, (b) $NaClO_4/HClO_4$, (c) C_5H_5N/C_5H_5NHCl (C_5H_5N is pyridine; its K_b is given in Table 16.2). Explain your answer.

Answer

(a) H_3PO_4 is a weak acid and its conjugate base, $H_2PO_4^{2-}$, is a weak base (see Table 16.3). Therefore, this is a buffer system.
(b) Because $HClO_4$ is a strong acid, its conjugate base, ClO_4^-, is an extremely weak base. This means that the ClO_4^- ion will not combine with a H^+ ion in solution to form $HClO_4$. Thus the system cannot act as a buffer system.
(c) As Table 16.2 shows, C_5H_5N is a weak base and its conjugate acid, $C_5H_5\overset{+}{N}H$ (the cation of the salt C_5H_5NHCl), is a weak acid. Therefore, this is a buffer system.

Similar problem: 16.53.

The action of a buffer solution is demonstrated by the following example.

EXAMPLE 16.14

(a) Calculate the pH of a buffer system containing 1.0 M CH_3COOH and 1.0 M CH_3COONa. (b) What is the pH of the buffer system after the addition of 0.10 mole of gaseous HCl to 1 L of the solution? Assume that the volume of the solution does not change when the HCl is added.

Answer

(a) The pH of the buffer system before the addition of HCl can be calculated according to the procedure described in Example 16.12. Assuming negligible ionization of the acetic acid and hydrolysis of the acetate ions, we have, at equilibrium,

$$[CH_3COOH] = 1.0\ M \qquad \text{and} \qquad [CH_3COO^-] = 1.0\ M$$

$$K_a = \frac{[H^+][CH_3COO^-]}{[CH_3COOH]} = 1.8 \times 10^{-5}$$

$$\begin{aligned}[H^+] &= \frac{K_a[CH_3COOH]}{[CH_3COO^-]} \\ &= \frac{(1.8 \times 10^{-5})(1.0)}{(1.0)} \\ &= 1.8 \times 10^{-5}\ M\end{aligned}$$

$$\text{pH} = -\log (1.8 \times 10^{-5}) = 4.74$$

Thus when the concentrations of the acid and the conjugate base are the same, the pH of the buffer is equal to the pK_a of the acid.

(b) After the addition of HCl, complete ionization of HCl acid occurs:

$$\underset{0.10 \text{ mol}}{\text{HCl}(aq)} \longrightarrow \underset{0.10 \text{ mol}}{\text{H}^+(aq)} + \underset{0.10 \text{ mol}}{\text{Cl}^-(aq)}$$

Originally, there were 1.0 mol CH_3COOH and 1.0 mol CH_3COO^- present in 1 L of the solution. After neutralization of the HCl acid by CH_3COO^-, which we write as

$$\underset{0.10 \text{ mol}}{\text{CH}_3\text{COO}^-(aq)} + \underset{0.10 \text{ mol}}{\text{H}^+(aq)} \longrightarrow \underset{0.10 \text{ mol}}{\text{CH}_3\text{COOH}(aq)}$$

the number of moles of acetic acid and acetate ions present are

CH_3COOH: $(1.0 + 0.1) \text{ mol} = 1.1 \text{ mol}$

CH_3COO^-: $(1.0 - 0.1) \text{ mol} = 0.90 \text{ mol}$

Next we calculate the hydrogen ion concentration:

$$[\text{H}^+] = \frac{K_a[\text{CH}_3\text{COOH}]}{[\text{CH}_3\text{COO}^-]} = \frac{(1.8 \times 10^{-5})(1.1)}{0.90} = 2.2 \times 10^{-5}\, M$$

The pH of the solution becomes

$$\text{pH} = -\log (2.2 \times 10^{-5}) = 4.66$$

Note that since the volume of the solution is the same for both species, we replaced the ratio of their molar concentrations by the ratio of the number of moles present, that is, (1.1 mol/L)/(0.90 mol/L) = (1.1 mol/0.90 mol).

Similar problem: 16.60.

We see that in this buffer solution there is a decrease in pH of 0.08 (becoming more acidic) as a result of the addition of HCl. We can also compare the change in H^+ ion concentrations as follows:

Before addition of HCl: $[\text{H}^+] = 1.8 \times 10^{-5}\, M$

After addition of HCl: $[\text{H}^+] = 2.2 \times 10^{-5}\, M$

Thus the H^+ ion concentration increases by a factor of

$$\frac{2.2 \times 10^{-5}\, M}{1.8 \times 10^{-5}\, M} = 1.2$$

To appreciate the effectiveness of the CH_3COONa/CH_3COOH buffer, let us find out what would happen if 0.10 mol HCl were added to 1 L of water, and compare it with the result in the preceding example. In this case

Before addition of HCl: $[H^+] = 1.0 \times 10^{-7}\ M$

After addition of HCl: $[H^+] = 0.10\ M$

Thus, as a result of the addition of HCl, the H^+ ion concentration increases by a factor of

$$\frac{0.10\ M}{1.0 \times 10^{-7}\ M} = 1.0 \times 10^6$$

amounting to a millionfold increase! This comparison shows that a properly chosen buffer solution can maintain a fairly constant H^+ ion concentration, or pH.

Distribution Curves

The relationship between pH and the amount of either the acid or the conjugate base present is best understood by studying the *distribution curve* shown in Figure 16.5. In this case, the distribution curve gives the fraction of acetic acid and of acetate ion present in solution as a function of pH. At low pH the concentration of CH_3COOH is much greater than the concentration of CH_3COO^- ion, because the presence of H^+ ions drives the equilibrium from right to left (Le Chatelier's principle):

$$CH_3COOH(aq) \rightleftharpoons H^+(aq) + CH_3COO^-(aq)$$

Thus, the species present are mostly nonionized acetic acid molecules. The opposite effect occurs at high pH. Here, the hydroxide ions deplete the concentration of the acid, so the concentration of the acetate ion increases:

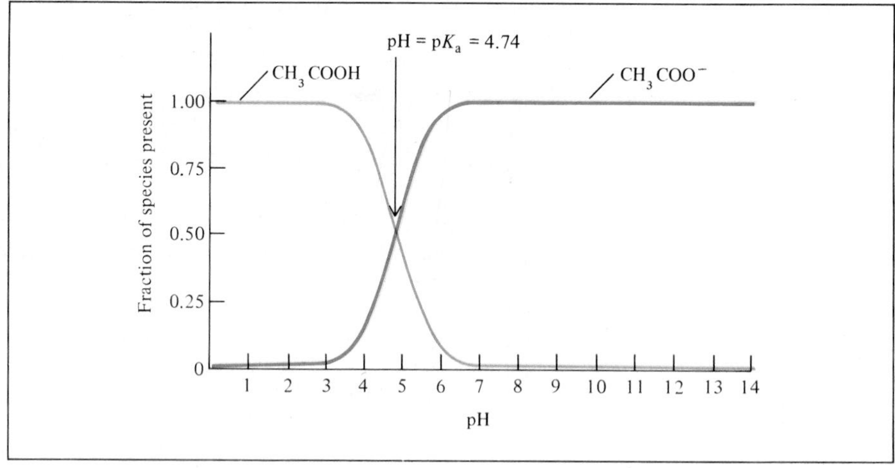

Figure 16.5 *Distribution curves for acetic acid and acetate ion as a function of pH. The fraction of species present is given by the ratio of the concentration of either CH_3COOH or CH_3COO^- to the total concentration of CH_3COOH plus CH_3COO^- in solution. Thus at very low pH (very acidic media), most of the species present in solution are CH_3COOH molecules, and the concentration of CH_3COO^- is very low. The reverse holds true for very high pH (very basic media). The pH range over which the CH_3COO^-/CH_3COOH buffer system functions most effectively is between 3.74 and 5.74. When $[CH_3COO^-] = [CH_3COOH]$, or when the fraction is 0.5, the pH of the solution is numerically equal to the pK_a of the acid (4.74).*

$$CH_3COOH(aq) + OH^-(aq) \rightleftharpoons CH_3COO^-(aq) + H_2O(l)$$

Now there are many more CH_3COO^- ions present than CH_3COOH molecules. In order for the buffer to function properly, it must contain comparable amounts of the acid and its conjugate base. There is a rather limited range over which the buffer is most effective. The range is often called the ***buffer range,*** that is, *the pH range in which a buffer is effective,* and is defined by the following expression:

$$\text{pH range} = pK_a \pm 1.00 \tag{16.6}$$

Since pK_a for acetic acid is 4.74, the buffer range for the CH_3COO^-/CH_3COOH system is given by pH = 3.74 to 5.74.

Note that the buffer functions best at pH = 4.74, that is, when $[CH_3COOH] = [CH_3COO^-]$. An additional requirement for a buffer system to function effectively is that the concentrations of the weak acid and its conjugate base should be high enough that the buffer can neutralize an appreciable amount of added H^+ or OH^- ions without significantly changing pH.

The CH_3COO^-/CH_3COOH buffer system has no physiological importance. On the other hand, the bicarbonate–carbonic acid (HCO_3^-/H_2CO_3) buffer plays an important role in many biological systems. Because H_2CO_3 is diprotic, there are actually two different buffer systems, that is, HCO_3^-/H_2CO_3 and CO_3^{2-}/HCO_3^-. Bicarbonate ion HCO_3^- acts as the conjugate base of the first system and the acid of the second. From Table 16.3 we can calculate the buffer ranges as follows:

HCO_3^-/H_2CO_3 $pH = pK_{a_1} \pm 1.00 = 6.38 \pm 1.00$
Buffer range: pH = 5.38 to 7.38

CO_3^{2-}/HCO_3^- $pH = pK_{a_2} \pm 1.00 = 10.32 \pm 1.00$
Buffer range: pH = 9.32 to 11.32

Figure 16.6 shows the distribution curves for these two buffers. They clearly function over quite different pH regions.

Preparing a Buffer Solution with a Specific pH

Now a question arises: How do we prepare a buffer solution with a specific pH? Equation (16.5) indicates that if the molar concentrations of the acid and its conjugate base are approximately equal, then

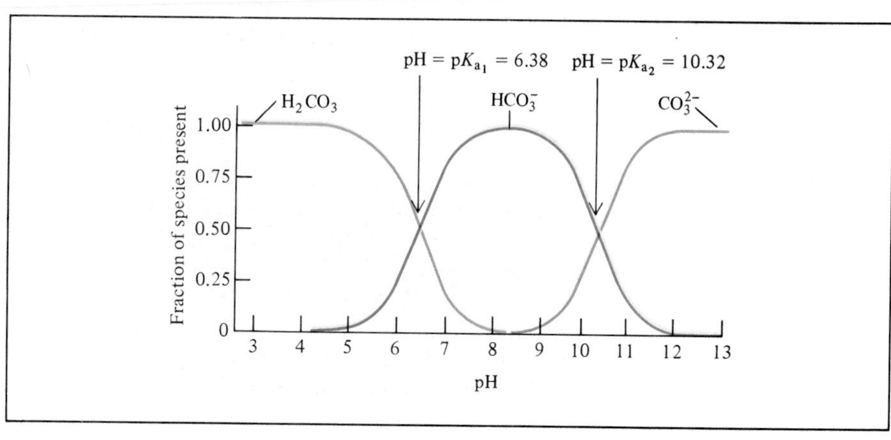

Figure 16.6 *Distribution curves for carbonic acid and bicarbonate ion as a function of pH. At any pH, there are only two predominant species (H_2CO_3 and HCO_3^-, or HCO_3^- and CO_3^{2-}) present in solution. When $[H_2CO_3] = [HCO_3^-]$, we find that $pH = pK_{a_1} = 6.38$. When $[HCO_3^-] = [CO_3^{2-}]$, we find that $pH = pK_{a_2} = 10.32$.*

$$\log \frac{[\text{conjugate base}]}{[\text{acid}]} \simeq 0$$

or

$$\text{pH} \simeq \text{p}K_a$$

Thus, to prepare a buffer solution, we choose a weak acid whose pK_a is equal to or close to the desired pH. This choice not only gives the correct pH value of the buffer system, but also ensures that we have comparable amounts of the acid and its conjugate base present; both are prerequisites for the buffer system to function effectively. The following example shows this approach.

EXAMPLE 16.15

Describe how you would prepare a "phosphate buffer" at a pH of about 7.40.

Answer

We write three stages of ionization of phosphoric acid as follows. The K_a values are obtained from Table 16.3, and the pK_a values are found by applying Equation (16.4).

$$H_3PO_4(aq) \rightleftharpoons H^+(aq) + H_2PO_4^-(aq) \qquad K_{a_1} = 7.5 \times 10^{-3};\ \text{p}K_{a_1} = 2.12$$

$$H_2PO_4^-(aq) \rightleftharpoons H^+(aq) + HPO_4^{2-}(aq) \qquad K_{a_2} = 6.2 \times 10^{-8};\ \text{p}K_{a_2} = 7.21$$

$$HPO_4^{2-}(aq) \rightleftharpoons H^+(aq) + PO_4^{3-}(aq) \qquad K_{a_3} = 4.8 \times 10^{-13};\ \text{p}K_{a_3} = 12.32$$

The most suitable of the three buffer systems is $HPO_4^{2-}/H_2PO_4^-$, because the pK_a of the acid $H_2PO_4^-$ is closest to the desired pH. From the Henderson-Hasselbalch equation we write

$$\text{pH} = \text{p}K_a + \log \frac{[\text{conjugate base}]}{[\text{acid}]}$$

$$7.40 = 7.21 + \log \frac{[HPO_4^{2-}]}{[H_2PO_4^-]}$$

$$\log \frac{[HPO_4^{2-}]}{[H_2PO_4^-]} = 0.19$$

Taking the antilog, we obtain

$$\frac{[HPO_4^{2-}]}{[H_2PO_4^-]} = 1.5$$

Thus, one way to prepare a phosphate buffer with a pH of 7.40 is to dissolve disodium hydrogen phosphate (Na_2HPO_4) and sodium dihydrogen phosphate (NaH_2PO_4) in molar ratio of 1.5:1.0 in water. For example, we could dissolve 1.5 moles of Na_2HPO_4 and 1.0 mole of NaH_2PO_4 in enough water to make up a 1-L solution.

Similar problem: 16.62.

The following Chemistry in Action illustrates the importance of buffer systems in the human body.

CHEMISTRY IN ACTION

Maintaining the pH of Blood

All higher animals need a circulatory system with which to fuel their life processes and remove wastes. In the human body this vital exchange takes place in the versatile fluid known as blood, of which there are about 5 L (10.6 pints) in an average adult. Blood circulating deep in the tissues carries oxygen and nutrients to keep cells alive, and removes carbon dioxide and other waste materials. Using several buffer systems, nature has provided an extremely efficient method for the delivery of oxygen and the removal of carbon dioxide.

Blood is an enormously complex system, but for our purposes we need look at only two essential components: blood plasma and red blood cells, or *erythrocytes*. Blood plasma contains many compounds, including proteins, metal ions, and inorganic phosphates. The erythrocytes contain hemoglobin molecules, as well as the enzyme *carbonic anhydrase,* which catalyzes both the formation of carbonic acid (H_2CO_3) and its decomposition:

$$CO_2(aq) + H_2O(l) \rightleftharpoons H_2CO_3(aq)$$

The substances inside the erythrocyte are protected from extracellular fluid (blood plasma) by a cell membrane that allows only certain molecules to diffuse through it.

The pH of blood plasma is maintained at about 7.40 by several buffer systems, the most important of which is the HCO_3^-/H_2CO_3 system. In the erythrocyte, where the pH is 7.25, the principal buffer systems are HCO_3^-/H_2CO_3 and hemoglobin. The hemoglobin molecule is a complex protein molecule (molar mass 65,000 g) that contains a number of ionizable protons. As a very rough approximation, we can treat it as a monoprotic acid of the form HHb:

$$HHb(aq) \rightleftharpoons H^+(aq) + Hb^-(aq)$$

where HHb represents the hemoglobin molecule and Hb^- the conjugate base of HHb. Oxyhemoglobin ($HHbO_2$), formed by the combination of oxygen with hemoglobin, is a stronger acid than HHb:

$$HHbO_2(aq) \rightleftharpoons H^+(aq) + HbO_2^-(aq)$$

As Figure 16.7(a) shows, carbon dioxide produced by metabolic processes diffuses into the erythrocyte, where it is rapidly converted to H_2CO_3 by carbonic anhydrase:

$$CO_2(aq) + H_2O(l) \rightleftharpoons H_2CO_3(aq)$$

The ionization of the carbonic acid

$$H_2CO_3(aq) \rightleftharpoons H^+(aq) + HCO_3^-(aq)$$

has two important consequences. First, the bicarbonate ion diffuses out of the erythrocyte and is carried by the blood plasma to the lungs. This is the major mechanism for removing carbon dioxide. Second, the H^+ ions produced now shift the equilibrium in favor of the nonionized oxyhemoglobin molecule:

$$H^+(aq) + HbO_2^-(aq) \longrightarrow HHbO_2(aq)$$

Since $HHbO_2$ releases oxygen more readily than does its conjugate base (HbO_2^-), the formation of the acid promotes the following reaction from left to right:

$$HHbO_2(aq) \rightleftharpoons HHb(aq) + O_2(aq)$$

The O_2 molecules diffuse out of the erythrocyte and are taken up by other cells to carry out metabolism.

When the venous blood returns to the lungs, the above processes are reversed [see Figure 16.7(b)]. The bicarbonate ions now diffuse into the erythrocyte, where they react with hemoglobin to form carbonic acid:

$$HHb(aq) + HCO_3^-(aq) \rightleftharpoons Hb^-(aq) + H_2CO_3(aq)$$

Most of the acid is then converted to CO_2 by carbonic anhydrase:

$$H_2CO_3(aq) \rightleftharpoons H_2O(l) + CO_2(aq)$$

The carbon dioxide diffuses to the lungs and is eventually exhaled. The formation of the Hb^- ions (due to the reaction between HHb and HCO_3^- shown above) also favors the uptake of oxygen at the lungs

$$Hb^-(aq) + O_2(aq) \rightleftharpoons HbO_2^-(aq)$$

because Hb^- has a greater affinity for oxygen than does HHb.

When the arterial blood flows back to the body tissues, the entire cycle is repeated.

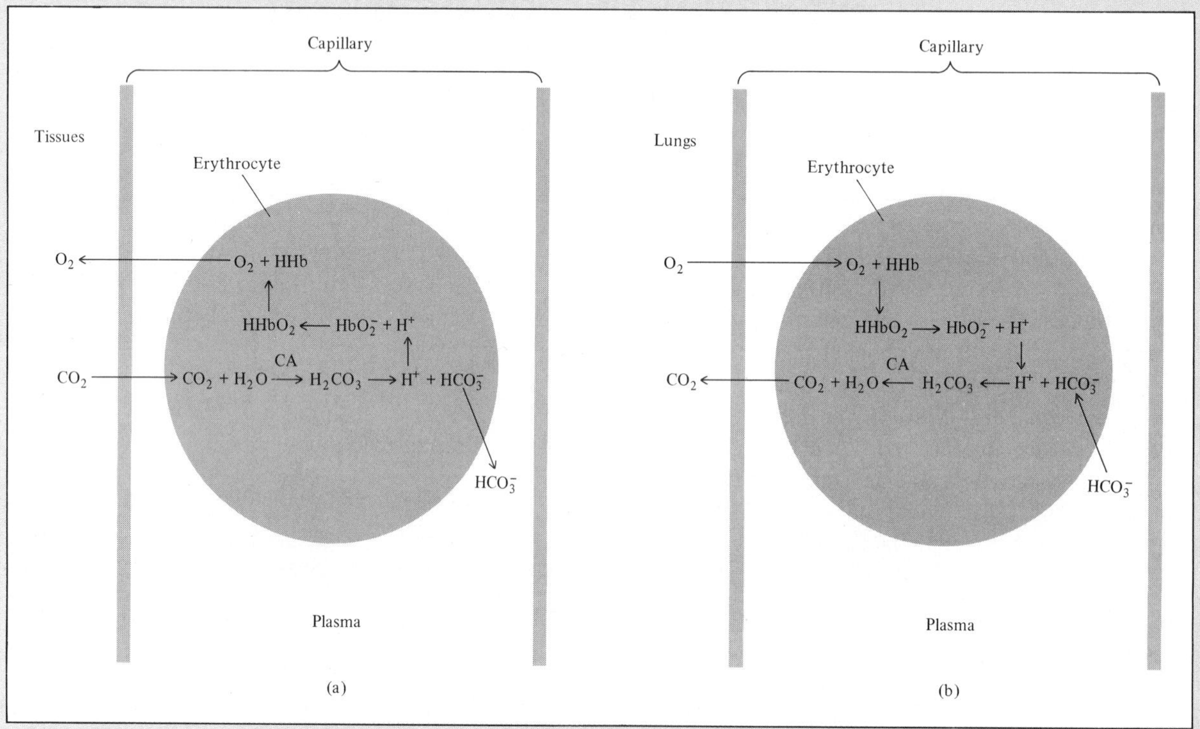

Figure 16.7 *Oxygen–carbon dioxide transport and release by blood. (a) The partial pressure of CO_2 is higher in the metabolizing tissues than in the plasma. Thus, it diffuses into the blood capillaries and then into erythrocytes. There it is converted to carbonic acid by the enzyme carbonic anhydrase (CA). The protons provided by the carbonic acid then combine with the HbO_2^- anions to form $HHbO_2$, which eventually dissociates into HHb and O_2. Because the partial pressure of O_2 is higher in the erythrocytes than in the tissues, oxygen molecules diffuse out of the erythrocytes and then into the tissues. The bicarbonate ions also diffuse out of the erythrocytes and are carried by the plasma to the lungs. (b) In the lungs, the processes are exactly reversed. Oxygen molecules diffuse from the lungs, where they have a higher partial pressure, into the erythrocytes. There they combine with HHb to form $HHbO_2$. The protons provided by $HHbO_2$ combine with the bicarbonate ions diffused into the erythrocytes from the plasma to form carbonic acid. In the presence of carbonic anhydrase, carbonic acid is converted to H_2O and CO_2. The CO_2 then diffuses out of the erythrocytes and into the lungs, where it is exhaled.*

16.8 A Closer Look at Acid-Base Titrations

Having discussed salt hydrolysis and buffer solutions, we can now look in more detail at the quantitative aspects of acid-base reactions as reflected in titrations (see Section

4.6). A question of particular importance is: How does the pH of the solution change during the course of a titration? We will consider three types of reactions: (1) titrations involving a strong acid and a strong base, (2) titrations involving a weak acid and a strong base, and (3) titrations involving a strong acid and a weak base. Titrations involving a weak acid and a weak base are more complicated because of the hydrolysis of both the cation and the anion of the salt formed. These titrations will not be dealt with here. The reactions we will study are similar to those discussed in Section 15.5.

Titrations Involving a Strong Acid and a Strong Base

The reaction between a strong acid (say, HCl) and a strong base (say, NaOH) is represented by

$$HCl(aq) + NaOH(aq) \longrightarrow NaCl(aq) + H_2O(l)$$

or in terms of a net ionic equation by

$$H^+(aq) + OH^-(aq) \longrightarrow H_2O(l)$$

For convenience we will use only three significant figures for titration calculations and two significant figures (two decimal places) for pH values.

Consider the addition of a 0.100 *M* NaOH solution (from a buret) to an Erlenmeyer flask containing 25.0 mL of 0.100 *M* HCl. Figure 16.8 shows the pH profile of the titration (also known as the *titration curve*). Before the addition of NaOH, the pH of the acid is given by $-\log(0.100)$, or 1.00. When NaOH is added, the pH of the solution increases slowly at first. Near the equivalence point the pH begins to rise steeply, and at the equivalence point (that is, the point at which equimolar amounts of acid and base have reacted) the curve rises almost vertically. In a strong acid–strong base titration both the hydrogen ion and hydroxide ion concentrations are very small at the equivalence point; consequently, the addition of a single drop of the base can cause a large increase in $[OH^-]$ and hence the pH of the solution. Beyond the equivalence point, the pH again increases slowly with the addition of NaOH.

It is possible to calculate the pH of the solution at every stage of titration. Here are three sample calculations:

1. After the addition of 10.0 mL of 0.100 *M* NaOH to 25.0 mL of 0.100 *M* HCl.

   The total volume of the solution is 35.0 mL. The number of moles of NaOH in 10.0 mL is

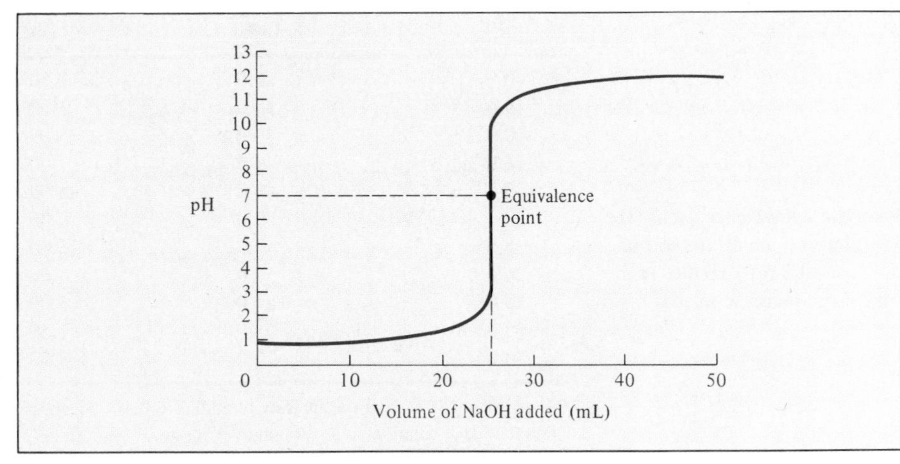

Figure 16.8 *pH profile of a strong acid–strong base titration. A 0.100 M NaOH solution is added from a buret to 25.0 mL of a 0.100 M HCl solution in an Erlenmeyer flask (see Figure 4.7). This curve is sometimes referred to as a titration curve.*

$$10.0 \text{ mL} \times \frac{0.100 \text{ mol NaOH}}{1 \text{ L NaOH}} \times \frac{1 \text{ L}}{1000 \text{ mL}} = 1.00 \times 10^{-3} \text{ mol}$$

The number of moles of HCl originally present in 25.0 mL of solution is

$$25.0 \text{ mL} \times \frac{0.100 \text{ mol HCl}}{1 \text{ L HCl}} \times \frac{1 \text{ L}}{1000 \text{ mL}} = 2.50 \times 10^{-3} \text{ mol}$$

Keep in mind that 1 mol NaOH≎1 mol HCl.

Thus, the amount of HCl left after partial neutralization is $(2.50 \times 10^{-3}) - (1.00 \times 10^{-3})$, or 1.50×10^{-3} mol. Next, the concentration of H^+ ions in 35.0 mL of solution is found as follows:

$$\frac{1.50 \times 10^{-3} \text{ mol HCl}}{35.0 \text{ mL}} \times \frac{1000 \text{ mL}}{1 \text{ L}} = 0.0429 \text{ mol HCl/L}$$
$$= 0.0429\ M \text{ HCl}$$

Thus $[H^+] = 0.0429\ M$, and the pH of the solution is

$$\text{pH} = -\log 0.0429 = 1.37$$

Neither Na^+ nor Cl^- undergoes hydrolysis.

2. After the addition of 25.0 mL of 0.100 *M* NaOH to 25.0 mL of 0.100 *M* HCl.
   This is a simple calculation, because it involves a complete neutralization reaction. At the equivalence point, $[H^+] = [OH^-]$ and the pH of the solution is 7.00.
3. After the addition of 35.0 mL of 0.100 *M* NaOH to 25.0 mL of 0.100 *M* HCl.
   The total volume of the solution is 60.0 mL. The number of moles of NaOH added is

$$35.0 \text{ mL} \times \frac{0.100 \text{ mol NaOH}}{1 \text{ L NaOH}} \times \frac{1 \text{ L}}{1000 \text{ mL}} = 3.50 \times 10^{-3} \text{ mol}$$

The number of moles of HCl in 25.0 mL solution is 2.50×10^{-3}. After complete neutralization of HCl, the number of moles of NaOH left is $(3.50 \times 10^{-3}) - (2.50 \times 10^{-3})$, or 1.00×10^{-3} mol. The concentration of NaOH in 60.0 mL of solution is

Table 16.5 Titration data for HCl versus NaOH*

Volume of NaOH added (mL)	Total volume (mL)	Mole excess of acid or base	pH of solution
0.0	25.0	$2.50 \times 10^{-3}\ H^+$	1.00
10.0	35.0	1.50×10^{-3}	1.37
20.0	45.0	5.00×10^{-4}	1.95
24.0	49.0	1.00×10^{-4}	2.69
24.5	49.5	5.00×10^{-5}	3.00
25.0	50.0	0.00	7.00
25.5	50.5	$5.00 \times 10^{-5}\ OH^-$	11.00
26.0	51.0	1.00×10^{-4}	11.29
30.0	55.0	5.00×10^{-4}	11.96
40.0	65.0	1.50×10^{-3}	12.36
50.0	75.0	2.50×10^{-3}	12.52

Near the equivalence point, the addition of a small amount of the base (0.5 mL) results in a very large increase in pH.

*The titration is carried out by slowly adding a 0.100 *M* NaOH solution from a buret to 25.0 mL of a 0.100 *M* HCl solution in an Erlenmeyer flask.

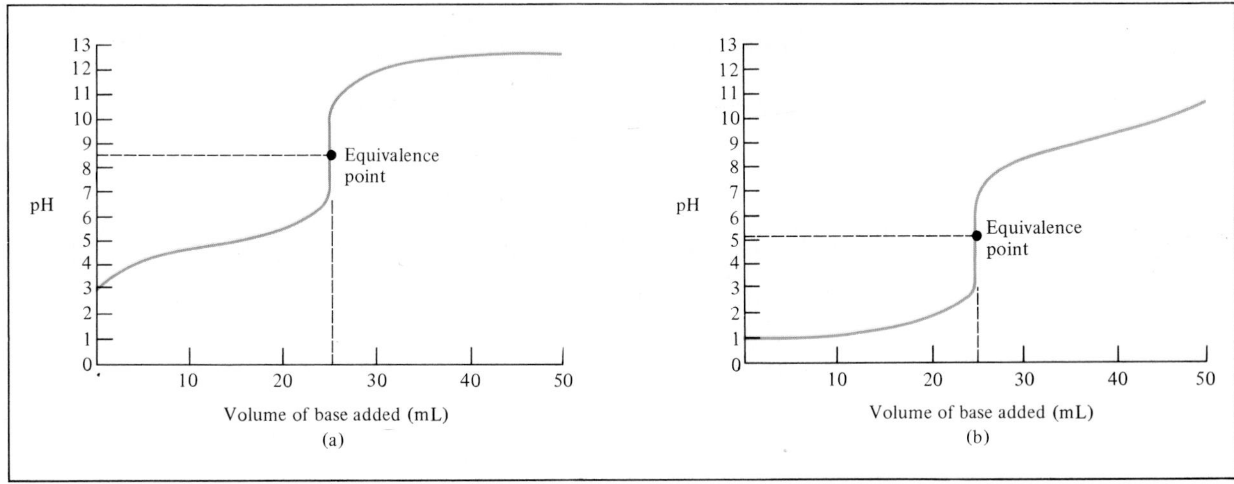

Figure 16.9 *Titration curves. (a) Weak acid versus strong base. (b) Strong acid versus weak base. In each case, a 0.100 M base solution is added from a buret to 25.0 mL of a 0.100 M acid solution in an Erlenmeyer flask. As a result of salt hydrolysis, the pH at the equivalence point is greater than 7 for (a) and less than 7 for (b).*

$$\frac{1.00 \times 10^{-3} \text{ mol NaOH}}{60.0 \text{ mL}} \times \frac{1000 \text{ mL}}{1 \text{ L}} = 0.0167 \text{ mol NaOH/L}$$
$$= 0.0167\ M \text{ NaOH}$$

Thus $[OH^-] = 0.0167\ M$ and $pOH = -\log 0.0167 = 1.78$. Hence, the pH of the solution is

$$\begin{aligned} pH &= 14.00 - pOH \\ &= 14.00 - 1.78 \\ &= 12.22 \end{aligned}$$

Table 16.5 shows a more complete set of data for the titration.

Titrations Involving a Weak Acid and a Strong Base

Consider the neutralization reaction between acetic acid (a weak acid) and sodium hydroxide (a strong base):

$$CH_3COOH(aq) + NaOH(aq) \longrightarrow CH_3COONa(aq) + H_2O(l)$$

This equation can be simplified to

$$CH_3COOH(aq) + OH^-(aq) \longrightarrow CH_3COO^-(aq) + H_2O(l)$$

The acetate ion undergoes hydrolysis as follows:

$$CH_3COO^-(aq) + H_2O(l) \rightleftharpoons CH_3COOH(aq) + OH^-(aq)$$

Therefore, at the equivalence point the pH will be *greater than 7* as a result of the excess OH^- ions formed [Figure 16.9(a)].

EXAMPLE 16.16

Calculate the pH in the titration of 25.0 mL of 0.100 *M* acetic acid by sodium hydroxide after the addition to the acid solution of (a) 10.0 mL of 0.100 *M* NaOH, (b) 25.0 mL of 0.100 *M* NaOH, (c) 35.0 mL of 0.100 *M* NaOH.

Answer

The neutralization reaction is

$$CH_3COOH(aq) + NaOH(aq) \longrightarrow CH_3COONa(aq) + H_2O(l)$$

For each of the three stages of the titration we first calculate the number of moles of NaOH added to the acetic acid solution. Next, we calculate the number of moles of the acid (or the base) left over after neutralization. Then we determine the pH of the solution.

(a) The number of moles of NaOH in 10.0 mL is

$$10.0 \text{ mL} \times \frac{0.100 \text{ mol NaOH}}{1 \text{ L NaOH soln}} \times \frac{1 \text{ L}}{1000 \text{ mL}} = 1.00 \times 10^{-3} \text{ mol}$$

The number of moles of CH_3COOH originally present in 25.0 mL of solution is

$$25.0 \text{ mL} \times \frac{0.100 \text{ mol } CH_3COOH}{1 \text{ L } CH_3COOH \text{ soln}} \times \frac{1 \text{ L}}{1000 \text{ mL}} = 2.50 \times 10^{-3} \text{ mol}$$

Thus, the amount of CH_3COOH left after all added base has been neutralized is

$$(2.50 \times 10^{-3} - 1.00 \times 10^{-3}) \text{ mol} = 1.50 \times 10^{-3} \text{ mol}$$

The amount of CH_3COONa formed is 1.00×10^{-3} mole:

$$\begin{array}{ccccc} CH_3COOH(aq) & + \; NaOH(aq) & \longrightarrow & CH_3COONa(aq) & + \; H_2O(l) \\ 1.00 \times 10^{-3} \text{ mol} & 1.00 \times 10^{-3} \text{ mol} & & 1.00 \times 10^{-3} \text{ mol} & \end{array}$$

At this stage we have a buffer system made up of CH_3COONa and CH_3COOH. To calculate the pH of this solution we write

Since the volume of the solution is the same for CH_3COOH and CH_3COO^-, the ratio of the number of moles present is equal to the ratio of their molar concentrations.

$$\begin{aligned} [H^+] &= \frac{[CH_3COOH]K_a}{[CH_3COO^-]} \\ &= \frac{(1.50 \times 10^{-3})(1.8 \times 10^{-5})}{1.00 \times 10^{-3}} \\ &= 2.7 \times 10^{-5} \, M \end{aligned}$$

$$\begin{aligned} \text{pH} &= -\log (2.7 \times 10^{-5}) \\ &= 4.57 \end{aligned}$$

(b) These quantities (that is, 25.0 mL of 0.100 M NaOH reacting with 25.0 mL of 0.100 M CH_3COOH) correspond to the equivalence point. The number of moles of both NaOH and CH_3COOH in 25.0 mL is 2.50×10^{-3} mol, so the number of moles of the salt formed is

$$\begin{array}{ccccc} CH_3COOH(aq) & + \; NaOH(aq) & \longrightarrow & CH_3COONa(aq) & + \; H_2O(l) \\ 2.50 \times 10^{-3} \text{ mol} & 2.50 \times 10^{-3} \text{ mol} & & 2.50 \times 10^{-3} \text{ mol} & \end{array}$$

The total volume is 50.0 mL, so the concentration of the salt is

$$\begin{aligned} [CH_3COONa] &= \frac{2.50 \times 10^{-3} \text{ mol}}{50.0 \text{ mL}} \times \frac{1000 \text{ mL}}{1 \text{ L}} \\ &= 0.0500 \text{ mol/L} = 0.0500 \, M \end{aligned}$$

The next step is to calculate the pH of the solution that results from the hydrolysis of the CH_3COO^- ions. Following the procedure described in Example 16.9, we find that the pH of the solution at the equivalence point is 8.72.

(c) After the addition of 35.0 mL of NaOH the solution is well past the equivalence point. At this stage we have two species that are responsible for making the solution basic:

CH_3COO^- and OH^-. However, since OH^- is a much stronger base than CH_3COO^-, we can safely neglect the CH_3COO^- ions and calculate the pH of the solution using only the concentration of the OH^- ions. Only 25.0 mL of NaOH are needed for complete neutralization, so the number of moles of NaOH left after neutralization is

$$(35.0 - 25.0)\text{ mL} \times \frac{0.100\text{ mol NaOH}}{1\text{ L NaOH soln}} \times \frac{1\text{ L}}{1000\text{ mL}} = 1.00 \times 10^{-3}\text{ mol}$$

The total volume of the combined solutions is now 60.0 mL, so we calculate OH^- concentration as follows:

$$\begin{aligned}[OH^-] &= \frac{1.00 \times 10^{-3}\text{ mol}}{60.0\text{ mL}} \times \frac{1000\text{ mL}}{1\text{ L}} \\ &= 0.0167\text{ mol/L} = 0.0167\ M\end{aligned}$$

$$\begin{aligned}\text{pOH} &= -\log 0.0167 \\ &= 1.78\end{aligned}$$

$$\begin{aligned}\text{pH} &= 14.00 - \text{pOH} \\ &= 14.00 - 1.78 \\ &= 12.22\end{aligned}$$

Similar problem: 16.68.

Titrations Involving a Strong Acid and a Weak Base

Consider the titration of hydrochloric acid (a strong acid) and ammonia (a weak base):

$$HCl(aq) + NH_3(aq) \longrightarrow NH_4Cl(aq)$$

or simply

$$H^+(aq) + NH_3(aq) \longrightarrow NH_4^+(aq)$$

The pH at the equivalence point is *less than 7* as a result of salt hydrolysis [Figure 16.9(b)]:

$$NH_4^+(aq) + H_2O(l) \longrightarrow NH_3(aq) + H_3O^+(aq)$$

or simply

$$NH_4^+(aq) \longrightarrow NH_3(aq) + H^+(aq)$$

EXAMPLE 16.17

Calculate the pH in the titration of hydrochloric acid by ammonia after adding (a) 10.0 mL of 0.100 M NH_3, (b) 25.0 mL of 0.100 M NH_3, and (c) 35.0 mL of 0.100 M NH_3 to 25.0 mL of 0.100 M HCl.

Answer

The neutralization reaction is

$$HCl(aq) + NH_3(aq) \longrightarrow NH_4Cl(aq)$$

At each stage of the titration we first calculate the number of moles of NH_3 added to the hydrochloric acid solution. Next, we calculate the number of moles of the acid (or the base) left over after neutralization. Then we determine the pH of the solution.

(a) The number of moles of NH_3 in 10.0 mL of 0.100 *M* NH_3 is

$$10.0 \text{ mL} \times \frac{0.100 \text{ mol NH}_3}{1 \text{ L NH}_3 \text{ soln}} \times \frac{1 \text{ L}}{1000 \text{ mL}} = 1.00 \times 10^{-3} \text{ mol}$$

The number of moles of HCl originally present in 25.0 mL of solution is

$$25.0 \text{ mL} \times \frac{0.100 \text{ mol HCl}}{1 \text{ L HCl soln}} \times \frac{1 \text{ L}}{1000 \text{ mL}} = 2.50 \times 10^{-3} \text{ mol}$$

Thus, the amount of HCl left after all added base has been neutralized is

$$(2.50 \times 10^{-3} - 1.00 \times 10^{-3}) \text{ mol} = 1.50 \times 10^{-3} \text{ mol}$$

At this stage, two types of species in solution act to decrease the pH of the solution: The NH_4^+ ions hydrolyze to produce H^+ ions, and the HCl ionizes to yield H^+ ions. Since the percent hydrolysis of NH_4^+ is small (see Example 16.10) and HCl is a strong acid, the pH of the solution is largely determined by the remaining acid, whose concentration is

$$[\text{HCl}] = \frac{1.50 \times 10^{-3} \text{ mol}}{35.0 \text{ mL}} \times \frac{1000 \text{ mL}}{1 \text{ L}}$$
$$= 0.0429 \text{ mol/L} = 0.0429\ M$$

Thus the pH of the solution is

$$\text{pH} = -\log 0.0429$$
$$= 1.37$$

(b) These quantities (that is, 25.0 mL of 0.100 *M* NH_3 reacting with 25.0 mL of 0.100 *M* HCl) correspond to the equivalence point. The amount of the salt formed is

$$\begin{array}{ccccc} \text{HCl}(aq) & + & \text{NH}_3(aq) & \longrightarrow & \text{NH}_4\text{Cl}(aq) \\ 2.50 \times 10^{-3} \text{ mol} & & 2.50 \times 10^{-3} \text{ mol} & & 2.50 \times 10^{-3} \text{ mol} \end{array}$$

The total volume is 50.0 mL, so the concentration of NH_4Cl is

$$[\text{NH}_4\text{Cl}] = \frac{2.50 \times 10^{-3} \text{ mol}}{50.0 \text{ mL}} \times \frac{1000 \text{ mL}}{1 \text{ L}}$$
$$= 0.0500 \text{ mol/L} = 0.0500\ M$$

Following the procedure described in Example 16.10, we find that the pH of the solution due to the hydrolysis of NH_4^+ ions is 5.28.

(c) After 35.0 mL of NH_3 are added, all the HCl will be neutralized and the solution will now contain NH_3 and NH_4^+, which are the ingredients for a buffer system. Since 25.0 mL of NH_3 are consumed in reacting with HCl, the amount of NH_3 left over is that present in 10.0 mL of the solution, which is 1.00×10^{-3} mol [see part (a)]. From part (b) we find that the number of moles of NH_4^+ present is 2.50×10^{-3}. Therefore the H^+ concentration is

Since the volume of the solution is the same for NH_4^+ and NH_3, the ratio of the number of moles present is equal to the ratio of their molar concentrations.

$$[\text{H}^+] = \frac{[\text{NH}_4^+]K_a}{[\text{NH}_3]}$$
$$= \frac{(2.50 \times 10^{-3})(5.6 \times 10^{-10})}{1.00 \times 10^{-3}}$$
$$= 1.4 \times 10^{-9} \text{ mol/L} = 1.4 \times 10^{-9}\ M$$

Thus the pH of the solution is

$$\text{pH} = -\log (1.4 \times 10^{-9}) \\ = 8.85$$

Similar problem: 16.69.

16.9 Acid-Base Indicators

The end point or equivalence point in an acid-base titration is often signaled by a change in the color of an acid-base indicator. You will recall from Chapter 4 that an indicator is usually a weak organic acid or base that has distinctly different colors in its nonionized and ionized forms. These two forms are related to the pH of the solution in which the indicator is dissolved, as we will soon see. A change in the color of an indicator can be used to follow the progress of a titration.

Indicators do not all change color at the same pH, so the choice of indicator for a particular titration depends on the nature of acid and base used (that is, whether they are strong or weak). Let us consider a weak monoprotic acid, HIn, which acts as an indicator. In solution

We assume that HIn and In^- have different colors.

$$\text{HIn}(aq) \rightleftharpoons \text{H}^+(aq) + \text{In}^-(aq)$$

If the indicator is in a sufficiently acidic medium, the equilibrium, according to Le Chatelier's principle, shifts to the left and the predominant color of the indicator is that of the nonionized form (HIn). On the other hand, in a basic medium the equilibrium shifts to the right and the color of the conjugate base (In^-) predominates. Roughly speaking, we can use the following ratios of concentrations to predict the perceived color of the indicator:

$$\frac{[\text{HIn}]}{[\text{In}^-]} \geq 10 \qquad \text{color of acid (HIn) predominates}$$

$$\frac{[\text{In}^-]}{[\text{HIn}]} \geq 10 \qquad \text{color of conjugate base (In}^-\text{) predominates}$$

If $[\text{HIn}] \simeq [\text{In}^-]$, then the indicator color is a combination of the colors of HIn and In^-.

In Section 4.6 we mentioned that phenolphthalein is a suitable indicator for the titration of NaOH and HCl. Phenolphthalein is colorless in acidic and neutral solutions, but reddish pink in basic solutions. Measurements show that at $\text{pH} < 8.3$ the indicator is colorless but that it begins to turn reddish pink when the pH exceeds 8.3. Because of the steepness of the pH curve, near the equivalence point in Figure 16.8, the addition of a very small quantity of NaOH (say, 0.05 mL, which is about the volume of a drop from the buret) brings about a large increase in the pH of the solution. What is important, however, is the fact that the steep portion of the pH profile includes the range over which phenolphthalein changes from colorless to reddish pink. Whenever such a matching occurs, the indicator can be used to locate the equivalence point of the titration.

Many acid-base indicators are plant pigments. For example, by boiling chopped red cabbage in water we can extract pigments that exhibit many different colors at various pHs (Figure 16.10).

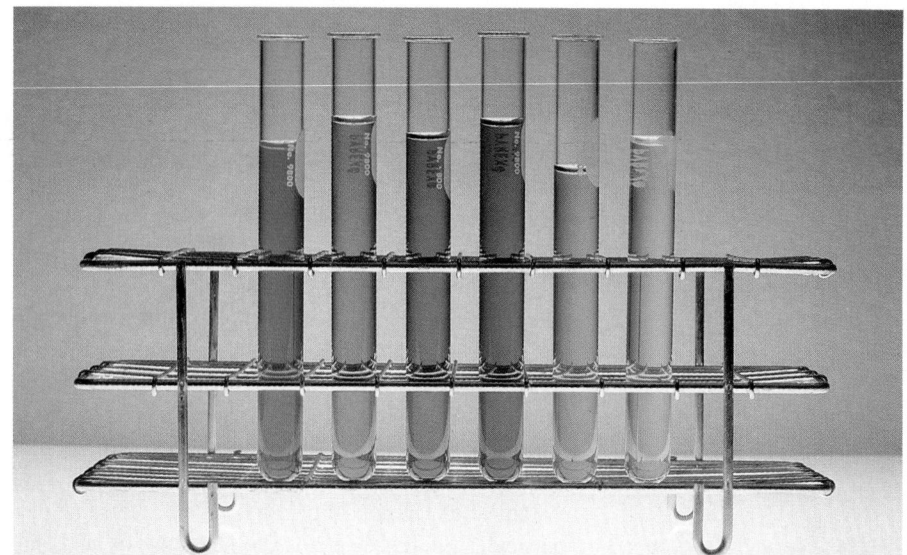

Figure 16.10 *Solutions containing extracts of red cabbage (obtained by boiling the cabbage in water) develop different colors when treated with an acid and a base. The pH of the solutions increases from left to right.*

Table 16.6 lists a number of indicators commonly used in acid-base titrations. The choice of indicator depends on the strengths of the acid and base in a particular titration, as discussed in Section 16.8. The criterion for choosing an appropriate indicator for a given titration, therefore, is whether the pH range over which the indicator changes color corresponds with the steep portion of the titration curves shown in Figures 16.8 and 16.9. If this criterion is not met, then the indicator will not accurately identify the equivalence point.

Table 16.6 Some common acid-base indicators

	Color		
Indicator	**In acid**	**In base**	**pH range***
Thymol blue	Red	Yellow	1.2–2.8
Bromophenol blue	Yellow	Bluish purple	3.0–4.6
Methyl orange	Orange	Yellow	3.1–4.4
Methyl red	Red	Yellow	4.2–6.3
Chlorophenol blue	Yellow	Red	4.8–6.4
Bromothymol blue	Yellow	Blue	6.0–7.6
Cresol red	Yellow	Red	7.2–8.8
Phenolphthalein	Colorless	Reddish pink	8.3–10.0

*The pH range is defined as the range over which the indicator changes from the acid color to the base color.

The choice of indicators for acid-base titrations is discussed in Example 16.18.

EXAMPLE 16.18

Which indicator or indicators listed in Table 16.6 would you use for the acid-base titrations shown in (a) Figure 16.8, (b) Figure 16.9(a), and (c) Figure 16.9(b)?

Answer

(a) Near the equivalence point, the pH of the solution changes abruptly from 4 to 10. Therefore all the indicators except thymol blue, bromophenol blue, and methyl orange are suitable for use in the titration.
(b) Here the steep portion covers the pH range between 7 and 10; therefore, the suitable indicators are cresol red and phenolphthalein.
(c) Here the steep portion of the pH curve covers the pH range between 3 and 7; therefore, the suitable indicators are bromophenol blue, methyl orange, methyl red, and chlorophenol blue.

Similar problem: 16.77.

Summary

1. The acid ionization constant K_a is larger for stronger acids and smaller for weaker acids. K_b similarly expresses the strengths of bases.
2. Percent ionization is another measure of the strength of acids. The more dilute a solution of a weak acid, the greater the percent ionization of the acid.
3. The product of the ionization constant of an acid and the ionization constant of its conjugate base is equal to the ion-product constant of water; that is, $K_aK_b = K_w$.
4. Most salts are strong electrolytes and dissociate completely into ions in solution. The reaction of these ions with water, called salt hydrolysis, can lead to acidic or basic solutions. In salt hydrolysis, the conjugate bases of weak acids yield basic solutions, and the conjugate acids of weak bases yield acidic solutions.
5. Small, highly charged metal ions such as Al^{3+} and Fe^{3+} hydrolyze to yield acidic solutions.
6. The common ion effect tends to suppress the ionization of a weak acid or a weak base. Its action can be explained by Le Chatelier's principle.
7. A buffer solution is a combination of a weak acid and its weak conjugate base; the solution reacts with small amounts of added acid or base in such a way that the pH of the solution remains nearly constant. Buffer systems play an important role in maintaining the pH of body fluids.
8. The pH at the equivalence point of an acid-base titration depends on the hydrolysis of the salt formed in the neutralization reaction. For strong acid–strong base titrations, the pH at the equivalence point is 7; for weak acid–strong base titrations, the pH at the equivalence point is greater than 7; for strong acid–weak base titrations, the pH at the equivalence point is less than 7.
9. Acid-base indicators are weak organic acids or bases that can be used to detect the equivalence point in an acid-base neutralization reaction.

Applied Chemistry Example: Citric Acid

Citric acid ($C_6H_8O_7$, abbreviated as H_3Ct), present in ''citric fruits'' such as orange, lemon, and grapefruit, is an important chemical used in the food industry. The addition of citric acid to beverages and fruit juices imparts a sour taste, often imitating the flavor of the fruit for which the beverage is named. The acidic medium also prevents the growth of bacteria and mold. It has the following structure:

$$\begin{array}{c} \text{COOH} \\ | \\ \text{CH}_2 \\ | \\ \text{HO—C—CH}_2\text{—COOH} \\ | \\ \text{COOH} \end{array}$$

In the pharmaceutical industry the acid, which is a solid, is used in effervescent tablets such as Alka-Seltzer (see p. 657.) Citric acid is a triprotic acid

$$H_3Ct(aq) \rightleftharpoons H^+(aq) + H_2Ct^-(aq) \qquad K_{a_1} = 7.1 \times 10^{-4}$$

$$H_2Ct^-(aq) \rightleftharpoons H^+(aq) + HCt^{2-}(aq) \qquad K_{a_2} = 1.7 \times 10^{-5}$$

$$HCt^{2-}(aq) \rightleftharpoons H^+(aq) + Ct^{3-}(aq) \qquad K_{a_3} = 6.4 \times 10^{-6}$$

1. A certain fruit juice contains 0.26 oz of citric acid per gallon. Use only the first two stages of ionization to calculate the pH of the juice. (1 oz = 28.4 g, 1 gal = 3.785 L, and the molar mass of citric acid is 180.1 g/mol.)

   Answer: First we need to calculate the citric acid concentration of the juice in molarity. We write

$$\text{molarity} = 0.26 \text{ oz} \times \frac{28.4 \text{ g}}{1 \text{ oz}} \times \frac{1 \text{ mol}}{180.1 \text{ g}} \times \frac{1}{3.785 \text{ L}}$$

$$= 0.011\ M$$

   To calculate the pH of the juice, we proceed with the first stage of ionization as follows:

   Step 1:

	$H_3Ct(aq)$	$\rightleftharpoons$ $H^+(aq)$	+ $H_2Ct^-(aq)$
Initial:	$0.011\ M$	$0.00\ M$	$0.00\ M$
Change:	$-x\ M$	$+x\ M$	$+x\ M$
Equilibrium:	$(0.011 - x)\ M$	$x\ M$	$x\ M$

   Step 2:

$$K_{a_1} = \frac{[H^+][H_2Ct^-]}{[H_3Ct]} = 7.1 \times 10^{-4}$$

$$\frac{x^2}{0.011 - x} = 7.1 \times 10^{-4}$$

   Solving, we get $x = 2.5 \times 10^{-3}\ M$.

   Step 3: When the equilibrium for the first stage is reached, the concentrations are

$$[H^+] = 0.0025\ M$$

$$[H_2Ct^-] = 0.0025\ M$$

$$[H_3Ct] = 0.011\ M - 0.0025\ M = 0.0085\ M$$

   For the second stage of ionization, we have

Step 1:

	$H_2Ct^-(aq)$	$\rightleftharpoons$	$H^+(aq)$	$+\ HCt^{2-}(aq)$
Initial:	$0.0025\ M$		$0.0025\ M$	$0.00\ M$
Change:	$-y\ M$		$+y\ M$	$+y\ M$
Equilibrium:	$(0.0025 - y)\ M$		$(0.0025 + y)\ M$	$y\ M$

Step 2:

$$K_{a_2} = \frac{[H^+][HCt^{2-}]}{[H_2Ct^-]} = 1.7 \times 10^{-5}$$

$$= \frac{(0.0025 + y)y}{0.0025 - y} = 1.7 \times 10^{-5}$$

Assuming that $0.0025 + y \simeq 0.0025$ and $0.0025 - y \simeq 0.0025$, we get $y = 1.7 \times 10^{-5}\ M$.

Step 3: At equilibrium

$$[H^+] = (0.0025 + 1.7 \times 10^{-5})\ M \simeq 0.0025\ M$$

$$pH = 2.60$$

The pH of fruit juices is typically between 2.5 and 4.5

2. "Citrate buffers" are commonly used in foods and in biochemical research. Describe how you would prepare a citrate buffer at pH of about 5.40.

Answer: From the three K_a values shown above we can write their pK_a values as

$$K_{a_1} = 7.1 \times 10^{-4} \qquad pK_{a_1} = 3.15$$
$$K_{a_2} = 1.7 \times 10^{-5} \qquad pK_{a_2} = 4.77$$
$$K_{a_3} = 6.4 \times 10^{-6} \qquad pK_{a_3} = 5.19$$

Since pK_{a_3} is closest to the desired pH, the most effective buffer system will be Ct^{3-}/HCt^{2-}. From Equation (16.5)

$$pH = pK_a + \log \frac{[\text{conjugate base}]}{[\text{acid}]}$$

we write

$$5.40 = 5.19 + \log \frac{[Ct^{3-}]}{[HCt^{2-}]}$$

or

$$\frac{[Ct^{3-}]}{[HCt^{2-}]} = 1.62$$

Thus one way to prepare the buffer is to dissolve 1.62 moles of Na_3Ct and 1 mole of Na_2HCt in enough water to make up 1 liter of solution. This procedure is similar to that shown in Example 16.15.

Key Words

Acid ionization constant (K_a), p. 670
Base ionization constant (K_b), p. 679
Buffer range, p. 699
Buffer solution, p. 694
Common ion effect, p. 692
Salt hydrolysis, p. 685

Exercises†

Weak Acid Ionization Constants

REVIEW QUESTIONS

16.1 What does the ionization constant tell us about the strength of an acid?

16.2 List the factors on which the K_a of a weak acid depends.

16.3 Why do we normally not quote K_a values for strong acids such as HCl and HNO_3?

16.4 Why is it necessary to specify temperature when giving K_a values?

16.5 Which of the following solutions has the highest pH? (a) 0.40 *M* HCOOH, (b) 0.40 *M* $HClO_4$, (c) 0.40 *M* CH_3COOH.

PROBLEMS

16.6 The K_a for benzoic acid is 6.5×10^{-5}. Calculate the concentrations of all the species (C_6H_5COOH, $C_6H_5COO^-$, H^+, and OH^-) in a 0.10 *M* benzoic acid solution.

16.7 The pH of a 0.060 *M* weak monoprotic acid is 3.44. Calculate the K_a of the acid.

16.8 The pH of a HF solution is 6.20. Calculate the ratio [conjugate base]/[acid] for this acid at this pH.

16.9 A 0.0560-g quantity of acetic acid is dissolved in enough water to make 50.0 mL of solution. Calculate the concentrations of H^+, CH_3COO^-, and CH_3COOH at equilibrium. (K_a for acetic acid = 1.8×10^{-5}.)

16.10 What is the original molarity of a solution of formic acid (HCOOH) whose pH is 3.26 at equilibrium?

16.11 Calculate the percent ionization of benzoic acid at the following concentrations: (a) 0.20 *M*, (b) 0.00020 *M*.

16.12 Calculate the percent ionization of hydrofluoric acid at the following concentrations: (a) 1.00 *M*, (b) 0.60 *M*, (c) 0.080 *M*, (d) 0.0046 *M*, (e) 0.00028 *M*. Comment on the trends.

16.13 A 0.040 *M* solution of a monoprotic acid in water is 14 percent ionized. Calculate the ionization constant of the acid.

16.14 (a) Calculate the percent ionization of a 0.20 *M* solution of the monoprotic acetylsalicylic acid (aspirin). (K_a = 3.0×10^{-4}.) (b) The pH of gastric juice in the stomach of a certain individual is 1.00. After a few aspirin tablets have been swallowed, the concentration of acetylsalicylic acid in the stomach is 0.20 *M*. Calculate the percent ionization of the acid under these conditions. What effect does the nonionized acid have on the membranes lining the stomach? (*Hint:* See the Chemistry in Action on pp. 656.)

†Unless otherwise stated, the temperature is assumed to be 25°C for all problems.

Weak Base Ionization Constants; K_a–K_b Relationship

REVIEW QUESTIONS

16.15 Use NH_3 to illustrate what we mean by the strength of a base.

16.16 Write the equation relating K_a for a weak acid and K_b for its conjugate base.

16.17 Use NH_3 and its conjugate acid NH_4^+ to derive the relationship between K_a and K_b.

PROBLEMS

16.18 Calculate the pH for each of the following solutions: (a) 0.10 *M* NH_3, (b) 0.050 *M* pyridine, (c) 0.260 *M* methylamine.

16.19 The pH of a 0.30 *M* solution of a weak base is 10.66. What is the K_b of the base?

16.20 What is the original molarity of a solution of ammonia whose pH is 11.22?

16.21 In a 0.080 *M* NH_3 solution, what percent of the NH_3 is present as NH_4^+?

Diprotic and Polyprotic Acids

REVIEW QUESTIONS

16.22 Malonic acid is a diprotic acid. Explain what that means.

```
O=C—OH
  |
  CH2
  |
O=C—OH
```

16.23 Write all the species (except water) that are present in a phosphoric acid solution. Indicate which species can act as a Brønsted acid, which as a Brønsted base, and which as both a Brønsted acid and a Brønsted base.

PROBLEMS

16.24 What are the concentrations of HSO_4^-, SO_4^{2-}, and H^+ in a 0.20 *M* $KHSO_4$ solution? (*Hint:* H_2SO_4 is a strong acid; K_a for HSO_4^- = 1.3×10^{-2}.)

16.25 Oxalic acid ($C_2H_2O_4$) is a diprotic acid (see Table 16.3). Calculate the concentrations of $C_2H_2O_4$, $C_2HO_4^-$, $C_2O_4^{2-}$, and H^+ in a 0.20 *M* oxalic acid solution.

16.26 Calculate the concentrations of H^+, HCO_3^-, and CO_3^{2-} in a 0.025 *M* H_2CO_3 solution.

16.27 Calculate the concentrations of all species in a 0.100 *M* H_3PO_4 solution.

Acid-Base Properties of Salt Solutions

REVIEW QUESTIONS

16.28 Define salt hydrolysis. Categorize salts according to how they affect the pH of a solution.

16.29 Explain why small, highly charged metal ions are able to undergo hydrolysis.

16.30 Al^{3+} is not a Brønsted acid but $Al(H_2O)_6^{3+}$ is. Explain.

16.31 Specify which of the following salts will undergo hydrolysis: KF, $NaNO_3$, NH_4NO_2, $MgSO_4$, KCN, C_6H_5COONa, RbI, Na_2CO_3, $CaCl_2$, HCOOK.

16.32 Predict the pH (>7, <7, or ≃7) of the aqueous solutions containing the following salts: (a) KBr, (b) $Al(NO_3)_3$, (c) $BaCl_2$, (d) $Bi(NO_3)_3$.

16.33 Which ion of the alkaline earth metals is most likely to undergo hydrolysis?

PROBLEMS

16.34 A certain salt, MX (containing the M^+ and X^- ions), is dissolved in water, and the pH of the resulting solution is 7.0. Can you say anything about the strengths of the acid and the base from which the salt is derived?

16.35 In a certain experiment a student finds the pHs of 0.10 *M* solutions of three salts KX, KY, and KZ are 7.0, 9.0, and 11.0, respectively. Arrange the acids HX, HY, and HZ in the order of increasing acid strength.

16.36 Calculate the pH of a 0.36 *M* CH_3COONa solution.

16.37 Calculate the pH of a 0.42 *M* NH_4Cl solution.

16.38 Calculate the pH of a 0.20 *M* ammonium acetate (CH_3COONH_4) solution.

16.39 What is the pH of a 0.050 *M* $AlCl_3$ solution?

16.40 How many grams of NaCN would you need to dissolve in enough water to make up an exactly 250-mL solution whose pH is 10.00?

16.41 Predict whether a solution containing the salt K_2HPO_4 will be acidic, neutral, or basic. (*Hint:* You need to consider both the ionization and hydrolysis of HPO_4^{2-}.)

The Common Ion Effect

REVIEW QUESTIONS

16.42 Define the term ''common ion effect.'' Explain the common ion effect in terms of Le Chatelier's principle.

16.43 Describe the change in pH (increase, decrease, or no change) that results from each of the following additions: (a) potassium acetate to an acetic acid solution; (b) ammonium nitrate to an ammonia solution; (c) sodium formate (HCOONa) to a formic acid (HCOOH) solution; (d) potassium chloride to a hydrochloric acid solution; (e) barium iodide to a hydroiodic acid solution.

PROBLEMS

16.44 Determine the pH of (a) a 0.40 *M* CH_3COOH solution and (b) a solution that is 0.40 *M* in CH_3COOH and 0.20 *M* in CH_3COONa.

16.45 Determine the pH of (a) a 0.20 *M* NH_3 solution and (b) a solution that is 0.20 *M* in NH_3 and 0.30 *M* in NH_4Cl.

16.46 Determine (a) the hydrogen ion and acetate ion concentrations in a 0.100 *M* CH_3COOH solution and (b) the concentrations of the same ions in a solution that is 0.100 *M* in both CH_3COOH and HCl.

Buffer Solutions

REVIEW QUESTIONS

16.47 Define buffer solution. What constitutes a buffer solution?

16.48 Define buffer range.

16.49 Define pK_a for a weak acid. What is the relationship between the value of the pK_a and the strength of the acid? Do the same for a weak base.

16.50 The pK_as of two monoprotic acids HA and HB are 5.9 and 8.1, respectively. Which of the two is the stronger acid?

16.51 The pK_bs for the bases X^-, Y^-, and Z^- are 2.72, 8.66, and 4.57, respectively. Arrange the following acids in order of increasing strength: HX, HY, HZ.

16.52 Calculate the pH of the following two buffer solutions: (a) 2.0 *M* CH_3COONa/2.0 *M* CH_3COOH and (b) 0.20 *M* CH_3COONa/0.20 *M* CH_3COOH. Which is the more effective buffer? Why?

PROBLEMS

16.53 Specify which of the following systems can be classified as a buffer system: (a) KCl/HCl, (b) NH_3/NH_4NO_3, (c) Na_2HPO_4/NaH_2PO_4, (d) KNO_2/HNO_2, (e) $KHSO_4/H_2SO_4$, (f) HCOOK/HCOOH.

16.54 Calculate the pH of the buffer system 0.15 *M* NH_3/0.35 *M* NH_4Cl.

16.55 The pH of a sodium acetate/acetic acid buffer is 4.50. Calculate the ratio $[CH_3COO^-]/[CH_3COOH]$.

16.56 The pH of blood plasma is 7.40. Assuming the principal buffer system is HCO_3^-/H_2CO_3, calculate the ratio $[HCO_3^-]/[H_2CO_3]$. Is this buffer more effective against an added acid or an added base?

16.57 The pH of a bicarbonate–carbonic acid buffer is 8.00. Calculate the ratio of the concentration of carbonic acid to that of the bicarbonate ion.

16.58 What is the pH of the buffer 0.10 *M* Na_2HPO_4/0.15 *M* KH_2PO_4?

16.59 Calculate the pH of a buffer solution prepared by adding 20.5 g of CH_3COOH and 17.8 g of CH_3COONa to enough water to make 5.00×10^2 mL of solution.

16.60 Calculate the pH of 1.00 L of the buffer 1.00 *M* CH_3COONa/1.00 *M* CH_3COOH before and after the addition of (a) 0.080 mol NaOH and (b) 0.12 mol HCl. (Assume that there is no change in volume.)

16.61 Calculate the pH of the 0.20 *M* NH_3/0.20 *M* NH_4Cl buffer. What is the pH of the buffer after the addition of 10.0 mL of 0.10 *M* HCl to 65.0 mL of the buffer?

16.62 A diprotic acid, H_2A, has the following ionization constants: $K_{a_1} = 1.1 \times 10^{-3}$ and $K_{a_2} = 2.5 \times 10^{-6}$. In order to make up a buffer solution of pH 5.80, which combination would you choose? $NaHA/H_2A$ or $Na_2A/NaHA$.

16.63 A student wishes to prepare a buffer solution at pH = 8.60. Which of the following weak acids should she choose and why? HA ($K_a = 2.7 \times 10^{-3}$), HB ($K_a = 4.4 \times 10^{-6}$), or HC ($K_a = 2.6 \times 10^{-9}$).

Acid-Base Titrations

PROBLEMS

16.64 A 12.5-mL volume of 0.500 *M* H_2SO_4 neutralizes 50.0 mL of NaOH. What is the concentration of the NaOH solution?

16.65 A 0.2688-g sample of a monoprotic acid neutralizes 16.4 mL of 0.08133 *M* KOH solution. Calculate the molar mass of the acid.

16.66 A 5.00-g quantity of a diprotic acid is dissolved in water and made up to exactly 250 mL. Calculate the molar mass of the acid if 25.0 mL of this solution required 11.1 mL of 1.00 *M* KOH for neutralization. Assume that both protons of the acid are titrated.

16.67 Calculate the pH at the equivalence point for the following titrations: (a) 0.10 *M* HCl versus 0.10 *M* NH_3, (b) 0.10 *M* CH_3COOH versus 0.10 *M* NaOH.

16.68 Exactly 100 mL of 0.100 *M* nitrous acid solution are titrated against a 0.100 *M* NaOH solution. Compute the pH for (a) the initial solution, (b) the point when 80.0 mL of the base have been added, (c) the equivalence point, (d) the point at which 105 mL of the base have been added.

16.69 Calculate the pH at the following points in the titration of 50.0 mL of 0.100 *M* methylamine (CH_3NH_2), $K_b = 4.4 \times 10^{-4}$, with a 0.200 *M* HCl solution: (a) before any acid has been added, (b) after adding 15.0 mL of HCl solution, (c) at the equivalence point.

16.70 A sample of 0.1276 g of an unknown monoprotic acid was dissolved in 25.0 mL of water and titrated with 0.0633 *M* NaOH solution. The volume of base required to reach the equivalence point was 18.4 mL. (a) Calculate the molar mass of the acid. (b) After 10.0 mL of base had been added in the titration, the pH was determined to be 5.87. What is the K_a of the unknown acid?

16.71 A solution is made by mixing exactly 500 mL of 0.167 *M* NaOH with exactly 500 mL 0.100 *M* CH_3COOH. Calculate the equilibrium concentrations of H^+, CH_3COOH, CH_3COO^-, OH^-, and Na^+.

16.72 Sketch titration curves for the following acid-base titrations: (a) HCl versus NaOH, (b) HCl versus NH_3, (c) CH_3COOH versus NaOH. In each case, the acid is added to the base in an Erlenmeyer flask. Your graphs should show pH as the *y*-axis and volume of acid added as the *x*-axis.

Acid-Base Indicators

REVIEW QUESTIONS

16.73 Explain how an acid-base indicator works in a titration.

16.74 What are the criteria for choosing an indicator for a particular acid-base titration?

16.75 The amount of indicator used in an acid-base titration must be small. Why?

16.76 A student carried out an acid-base titration by adding NaOH solution from a buret to an Erlenmeyer flask containing HCl solution and using phenolphthalein as indicator. At the equivalence point, she observed a faint reddish-pink color. However, after a few minutes, the solution gradually turned colorless. What do you suppose happened?

PROBLEMS

16.77 Referring to Table 16.6, specify which indicator or indicators you would use for the following titrations: (a) HCOOH versus NaOH, (b) HCl versus KOH, (c) HNO_3 versus NH_3.

16.78 The ionization constant K_a of an indicator HIn is 1.0×10^{-6}. The color of the nonionized form is red and that of the ionized form is yellow. What is the color of this indicator in a solution whose pH is 4.00? (*Hint:* The color of an indicator can be estimated by considering the ratio $[HIn]/[In^-]$. If the ratio is equal to or greater than 10, the color will be that of the nonionized form. If the ratio is equal to or smaller than 0.1, the color will be that of the ionized form.)

16.79 The K_a of a certain indicator is 2.0×10^{-6}. The color of HIn is green, and that of In^- is red. A few drops of the indicator are added to a HCl solution, which is then titrated against a NaOH solution. At what pH will the indicator change color?

Miscellaneous Problems

16.80 Compare the pH of a 0.040 *M* HCl solution with that of a 0.040 *M* H_2SO_4 solution.

16.81 The first and second ionization constants of a diprotic acid H_2A are K_{a_1} and K_{a_2} at a certain temperature. Under what conditions will $[A^{2-}] = K_{a_2}$?

16.82 The K_a of formic acid is 1.7×10^{-4} at 25°C. Will the acid become stronger or weaker at 40°C? Explain. (*Hint:* Consider the dependence of K_a on temperature.)

16.83 Use the data in Table 16.1 to calculate the equilibrium constant for the following reaction:

$$CH_3COOH(aq) + NO_2^-(aq) \rightleftharpoons CH_3COO^-(aq) + HNO_2(aq)$$

16.84 Sketch a distribution curve for oxalic acid ($C_2H_2O_4$) such as those shown in Figures 16.5 and 16.6.

16.85 Cacodylic acid is $(CH_3)_2AsO_2H$. Its ionization constant is 6.4×10^{-7}. (a) Calculate the pH of 50.0 mL of a 0.10 *M* solution of the acid. (b) Calculate the pH of 25.0 mL of 0.15 *M* $(CH_3)_2AsO_2Na$. (c) Mix the solutions in part (a) and part (b). Calculate the pH of the resulting solution.

16.86 A quantity of 0.560 g of KOH is added to 25.0 mL of 1.00 *M* HCl. Excess Na_2CO_3 is then added to the solution. What mass (in grams) of CO_2 is formed?

16.87 Calculate x, the number of molecules of water in oxalic acid hydrate $H_2C_2O_4 \cdot xH_2O$, from the following data: 5.00 g of the compound is made up to exactly 250 mL solution and 25.0 mL of this solution requires 15.9 mL of 0.500 *M* sodium hydroxide solution for neutralization.

16.88 A volume of 25.0 mL of 0.100 *M* HCl is titrated against a 0.100 *M* NH_3 solution added to it from a buret. Calculate the pH values of the solution (a) after 10.0 mL of NH_3 solution has been added, (b) after 25.0 mL of NH_3 solution has been added, (c) after 35.0 mL of NH_3 solution has been added.

16.89 The pK_a of butyric acid (HBut) is 4.7. Calculate K_b for the butyrate ion (But^-).

16.90 A solution of formic acid (HCOOH) has a pH of 2.53. How many grams of formic acid are there in 100.0 mL of the solution?

16.91 Calculate the concentrations of all the species in a 0.100 *M* Na_2CO_3 solution.

16.92 What is the equilibrium constant for the following reaction:

$$CH_3COOH(aq) + OH^-(aq) \rightleftharpoons CH_3COO^-(aq) + H_2O(l)$$

16.93 Henry's law constant for CO_2 at 38°C is 2.28×10^{-3} mol/L · atm. Calculate the pH of a solution of CO_2 at 38°C in equilibrium with the gas at a partial pressure of 3.20 atm.

16.94 Phenolphthalein is the common indicator for the titration of a strong acid with a strong base. (a) If the pK_a of phenolphthalein is 9.10, what is the ratio of the un-ionized form of the indicator (colorless) to the ionized form (reddish pink) at pH 8.00? (b) If 2 drops of 0.060 *M* phenolphthalein are used in a titration involving a 50.0-mL volume, what is the concentration of the ionized form at pH 8.00? (Assume that 1 drop = 0.050 mL.)

16.95 Calculate the pH of a 1-L solution containing 0.150 mole of CH_3COOH and 0.100 mole of HCl.

16.96 Describe how you would prepare a 1-L 0.20 *M* CH_3COONa/0.20 *M* CH_3COOH buffer system by (a) mixing a solution of CH_3COOH with a solution of CH_3COONa, (b) reacting a solution of CH_3COOH with a solution of NaOH, and (c) reacting a solution of CH_3COONa with a solution of HCl.

16.97 The buffer range is defined by the equation $pH = pK_a \pm 1$. Calculate the range of the ratio [conjugate base]/[acid] that corresponds to this equation.

Opposite page: Giant subtidal stromatolites in a tidal channel in the Bahamas. Stromatolites are the oldest macrofossils of the earliest forms of life. They are composed of pelletal carbonate sand, microbes, and carbonate cement. Calcium and magnesium carbonates are insoluble in water, accounting for their precipitation as components of the stromatolites.

C H A P T E R 1 7

SOLUBILITY EQUILIBRIA

The acid-base equilibria discussed in Chapter 16 are homogeneous equilibria, that is, equilibria of reactions that occur in a single phase. Another important type of solution equilibria involves the dissolution and precipitation of slightly soluble substances. These processes are examples of heterogeneous equilibria; that is, they pertain to reactions that occur in more than one phase.

17.1 Solubility and Solubility Product

The general rules for predicting the solubility of ionic compounds in water were introduced in Section 3.3. These solubility rules, although useful, do not allow us to make quantitative predictions about how much of a given ionic compound will dissolve in water. To develop a quantitative approach, all we need is the realization that solubility can be thought of as a special case of chemical equilibrium.

Solubility Product

Unless otherwise stated, the solvent in all discussions of solubility is understood to be water at a temperature of 25°C.

Consider a saturated solution of silver chloride that is in contact with solid silver chloride. The solubility equilibrium can be represented as

$$AgCl(s) \rightleftharpoons Ag^+(aq) + Cl^-(aq)$$

Because salts such as AgCl are treated as strong electrolytes, all the AgCl that dissolves in water is assumed to dissociate completely into Ag^+ and Cl^- ions. We know from Chapter 14 that for heterogeneous reactions the concentration of the solid is a constant. Thus we can write the equilibrium constant for the dissolution of AgCl (see Example 14.5) as

$$K_{sp} = [Ag^+][Cl^-]$$

where K_{sp} is called the solubility product constant or simply the solubility product. In general, the ***solubility product*** of a compound is *the product of the molar concentrations of the constituent ions, each raised to the power of its stoichiometric coefficient in the equilibrium equation*. The value of K_{sp} indicates the solubility of a compound—the smaller the K_{sp}, the less soluble the compound.

Because each AgCl unit contains only one Ag^+ ion and one Cl^- ion, its solubility product expression is particularly simple to write. The following three cases are more complex.

- MgF_2

$$MgF_2(s) \rightleftharpoons Mg^{2+}(aq) + 2F^-(aq) \qquad K_{sp} = [Mg^{2+}][F^-]^2$$

- Ag_2CO_3

$$Ag_2CO_3(s) \rightleftharpoons 2Ag^+(aq) + CO_3^{2-}(aq) \qquad K_{sp} = [Ag^+]^2[CO_3^{2-}]$$

- $Ca_3(PO_4)_2$

$$Ca_3(PO_4)_2(s) \rightleftharpoons 3Ca^{2+}(aq) + 2PO_4^{3-}(aq) \qquad K_{sp} = [Ca^{2+}]^3[PO_4^{3-}]^2$$

Not listing K_{sp} values for soluble salts is somewhat analogous to not listing K_a values for strong acids in Table 16.1.

Table 17.1 lists the solubility products for a number of salts of low solubility. Soluble salts such as NaCl and KNO_3, which have very large K_{sp} values, are not listed in the table.

Consider an aqueous solution containing Ag^+ and Cl^- ions at 25°C. Any one of the following conditions may exist: (1) the solution is unsaturated, (2) the solution is saturated, or (3) the solution is supersaturated. Following the procedure in Section 14.4 we use Q, called the *ion product,* to represent the product of the molar concentrations of the ions raised to the power of their stoichiometric coefficients:

$$Q = [Ag^+]_0[Cl^-]_0$$

The subscript 0 reminds us that these are initial concentrations and do not necessarily correspond to those at equilibrium. The possible relationships between Q and K_{sp} are

Table 17.1 Solubility products of some slightly soluble ionic compounds at 25°C

Name	Formula	K_{sp}
Aluminum hydroxide	$Al(OH)_3$	1.8×10^{-33}
Barium carbonate	$BaCO_3$	8.1×10^{-9}
Barium fluoride	BaF_2	1.7×10^{-6}
Barium sulfate	$BaSO_4$	1.1×10^{-10}
Bismuth sulfide	Bi_2S_3	1.6×10^{-72}
Cadmium sulfide	CdS	8.0×10^{-28}
Calcium carbonate	$CaCO_3$	8.7×10^{-9}
Calcium fluoride	CaF_2	4.0×10^{-11}
Calcium hydroxide	$Ca(OH)_2$	8.0×10^{-6}
Calcium phosphate	$Ca_3(PO_4)_2$	1.2×10^{-26}
Chromium(III) hydroxide	$Cr(OH)_3$	3.0×10^{-29}
Cobalt(II) sulfide	CoS	4.0×10^{-21}
Copper(I) bromide	$CuBr$	4.2×10^{-8}
Copper(I) iodide	CuI	5.1×10^{-12}
Copper(II) hydroxide	$Cu(OH)_2$	2.2×10^{-20}
Copper(II) sulfide	CuS	6.0×10^{-37}
Iron(II) hydroxide	$Fe(OH)_2$	1.6×10^{-14}
Iron(III) hydroxide	$Fe(OH)_3$	1.1×10^{-36}
Iron(II) sulfide	FeS	6.0×10^{-19}
Lead(II) carbonate	$PbCO_3$	3.3×10^{-14}
Lead(II) chloride	$PbCl_2$	2.4×10^{-4}
Lead(II) chromate	$PbCrO_4$	2.0×10^{-14}
Lead(II) fluoride	PbF_2	4.1×10^{-8}
Lead(II) iodide	PbI_2	1.4×10^{-8}
Lead(II) sulfide	PbS	3.4×10^{-28}
Magnesium carbonate	$MgCO_3$	4.0×10^{-5}
Magnesium hydroxide	$Mg(OH)_2$	1.2×10^{-11}
Manganese(II) sulfide	MnS	3.0×10^{-14}
Mercury(I) chloride	Hg_2Cl_2	3.5×10^{-18}
Mercury(II) sulfide	HgS	4.0×10^{-54}
Nickel(II) sulfide	NiS	1.4×10^{-24}
Silver bromide	$AgBr$	7.7×10^{-13}
Silver carbonate	Ag_2CO_3	8.1×10^{-12}
Silver chloride	$AgCl$	1.6×10^{-10}
Silver iodide	AgI	8.3×10^{-17}
Silver sulfate	Ag_2SO_4	1.4×10^{-5}
Silver sulfide	Ag_2S	6.0×10^{-51}
Strontium carbonate	$SrCO_3$	1.6×10^{-9}
Strontium sulfate	$SrSO_4$	3.8×10^{-7}
Tin(II) sulfide	SnS	1.0×10^{-26}
Zinc hydroxide	$Zn(OH)_2$	1.8×10^{-14}
Zinc sulfide	ZnS	3.0×10^{-23}

$Q < K_{sp}$ $[Ag^+]_0[Cl^-]_0 < 1.6 \times 10^{-10}$	Unsaturated solution
$Q = K_{sp}$ $[Ag^+][Cl^-] = 1.6 \times 10^{-10}$	Saturated solution
$Q > K_{sp}$ $[Ag^+]_0[Cl^-]_0 > 1.6 \times 10^{-10}$	Supersaturated solution; AgCl will precipitate out until the product of the ionic concentrations is equal to 1.6×10^{-10}.

Solubility Equilibria for Sulfides

From the ongoing discussion we might expect the solubility equilibrium for cadmium sulfide (CdS) to be written as

$$CdS(s) \rightleftharpoons Cd^{2+}(aq) + S^{2-}(aq)$$

where

$$K_{sp} = [Cd^{2+}][S^{2-}]$$

Table 16.3 shows that K_a for $HS^- \rightleftharpoons H^+ + S^{2-}$ is very small (1×10^{-19}).

However, studies show that the S^{2-} ion is a strong Brønsted base and, like the O^{2-} ion, is extensively hydrolyzed in solution. For this reason the solubility equilibrium for CdS is correctly written as

$$CdS(s) \rightleftharpoons Cd^{2+}(aq) + S^{2-}(aq)$$
$$S^{2-}(aq) + H_2O(l) \rightleftharpoons HS^-(aq) + OH^-(aq)$$

$$\text{Overall:} \quad CdS(s) + \cancel{S^{2-}(aq)} + H_2O(l) \rightleftharpoons Cd^{2+}(aq) + \cancel{S^{2-}(aq)} + HS^-(aq) + OH^-(aq)$$

or simply

$$CdS(s) + H_2O(l) \rightleftharpoons Cd^{2+}(aq) + HS^-(aq) + OH^-(aq)$$

Thus the solubility product expression for CdS, which is the equilibrium constant for the overall reaction, is given by

$$K_{sp} = [Cd^{2+}][HS^-][OH^-] = 8.0 \times 10^{-28}$$

The O^{2-} and S^{2-} ions exist in the solid lattice, but not in aqueous solution.

Note that this is the value quoted for the K_{sp} of CdS (see Table 17.1), and we should not interpret it as $K_{sp} = [Cd^{2+}][S^{2-}] = 8.0 \times 10^{-28}$. The important point is that the S^{2-} ion concentration in solution is usually very low because of hydrolysis.

Anions such as F^-, CO_3^{2-}, and PO_4^{3-} also hydrolyze in water but to a much lesser extent than S^{2-}. Thus we write the solubility product expressions for salts containing these anions [for example, CaF_2, Ag_2CO_3, $Mg_3(PO_4)_2$] in the manner shown on p. 720.

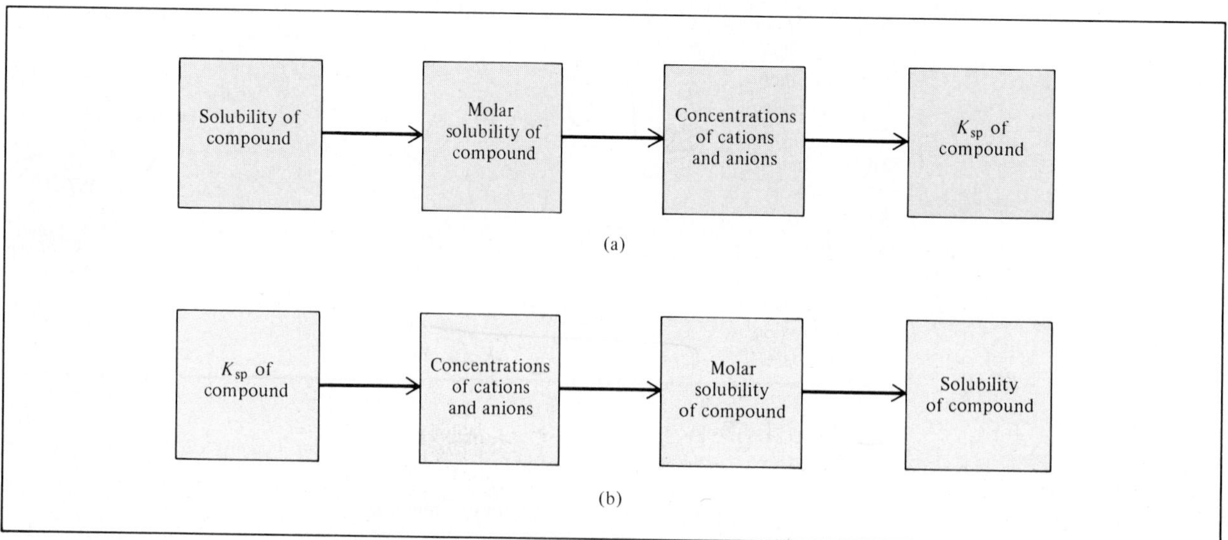

Figure 17.1 *Sequence of steps (a) for calculating K_{sp} from solubility data and (b) for calculating solubility from K_{sp} data.*

Molar Solubility and Solubility

In using K_{sp} values to compare solubilities, you should choose compounds that have similar formulas such as AgCl and ZnS or CaF_2 and $Fe(OH)_2$.

As we have seen, the value of K_{sp} indicates how soluble an ionic compound is in water; the larger the K_{sp}, the more soluble the compound. There are two other quantities that express a substance's solubility: ***molar solubility,*** that is, *the number of moles of solute in one liter of a saturated solution (mol/L),* and ***solubility,*** that is, *the number of grams of solute in one liter of a saturated solution (g/L).* Note that all these expressions refer to the concentration of saturated solutions at some given temperature (usually 25°C). Both molar solubility and solubility are convenient quantities to use in the laboratory.

In studying solubility equilibria, we need to become familiar with the conversions among K_{sp}, molar solubility, and solubility. In Examples 17.1 and 17.2 we will see how to calculate K_{sp} from molar solubility and solubility data. The steps involved in solving such problems are shown in Figure 17.1(a).

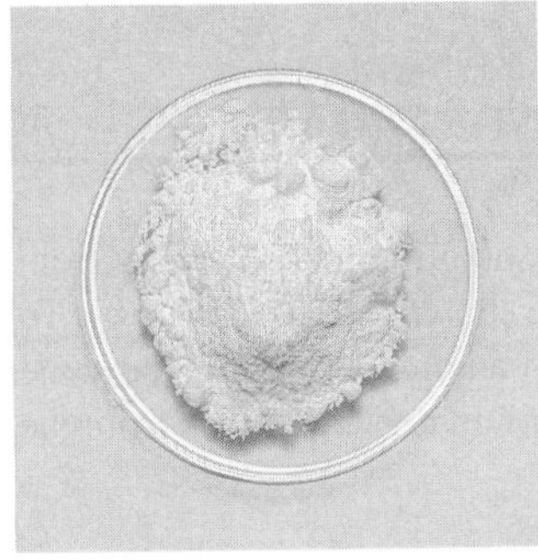

Silver sulfate.

EXAMPLE 17.1

The molar solubility of silver sulfate is 1.5×10^{-2} mol/L. Calculate the solubility product of the salt.

Answer

First we write the solubility equilibrium equation:

$$Ag_2SO_4(s) \rightleftharpoons 2Ag^+(aq) + SO_4^{2-}(aq)$$

From the stoichiometry we see that 1 mole of Ag_2SO_4 produces 2 moles of Ag^+ and 1 mole of SO_4^{2-} in solution. Therefore, when 1.5×10^{-2} mol Ag_2SO_4 is dissolved in 1 L of solution, the concentrations are

$$[Ag^+] = 2(1.5 \times 10^{-2}\ M) = 3.0 \times 10^{-2}\ M$$

$$[SO_4^{2-}] = 1.5 \times 10^{-2}\ M$$

Now we can calculate the solubility product:

$$\begin{aligned} K_{sp} &= [Ag^+]^2[SO_4^{2-}] \\ &= (3.0 \times 10^{-2})^2(1.5 \times 10^{-2}) \\ &= 1.4 \times 10^{-5} \end{aligned}$$

Similar problem: 17.8.

Calcium sulfate.

EXAMPLE 17.2

It is found experimentally that the solubility of calcium sulfate is 0.67 g/L. Calculate the value of K_{sp} for calcium sulfate.

Answer

First we calculate the number of moles of $CaSO_4$ dissolved in 1 L of solution:

$$\frac{0.67 \text{ g CaSO}_4}{1 \text{ L soln}} \times \frac{1 \text{ mol CaSO}_4}{136.2 \text{ g CaSO}_4} = 4.9 \times 10^{-3} \text{ mol/L}$$

In the solubility equilibrium

$$CaSO_4(s) \rightleftharpoons Ca^{2+}(aq) + SO_4^{2-}(aq)$$

we see that for every mole of $CaSO_4$ that dissolves, 1 mole of Ca^{2+} and 1 mole of SO_4^{2-} are produced. Thus, at equilibrium

$$[Ca^{2+}] = 4.9 \times 10^{-3}\, M \qquad \text{and} \qquad [SO_4^{2-}] = 4.9 \times 10^{-3}\, M$$

Now we can calculate K_{sp}:

$$\begin{aligned} K_{sp} &= [Ca^{2+}][SO_4^{2-}] \\ &= (4.9 \times 10^{-3})(4.9 \times 10^{-3}) \\ &= 2.4 \times 10^{-5} \end{aligned}$$

Similar problems: 17.7, 17.9, 17.10.

Examples 17.3 and 17.4 show how to calculate molar solubility and solubility from K_{sp} data. The steps involved in this type of calculation are shown in Figure 17.1(b). To convert the K_{sp} value to the concentrations of cations and anions, we follow the same procedure outlined in Section 14.5 for calculating equilibrium concentrations from the equilibrium constant.

Copper(I) iodide.

EXAMPLE 17.3

Using the data in Table 17.1, calculate the molar solubility of copper(I) iodide; that is, calculate the number of moles of CuI dissolved in 1 L of solution.

Answer

Step 1: Let s be the molar solubility (in mol/L) of CuI. Since one unit of CuI yields one Cu^+ ion and one I^- ion, at equilibrium both $[Cu^+]$ and $[I^-]$ are equal to s. We summarize the changes in concentrations as follows:

	$CuI(s) \rightleftharpoons$	$Cu^+(aq)$ +	$I^-(aq)$
Initial:		$0.00\, M$	$0.00\, M$
Change:		$+s\, M$	$+s\, M$
Equilibrium:		$s\, M$	$s\, M$

Step 2:

$$\begin{aligned} K_{sp} &= [Cu^+][I^-] \\ 5.1 \times 10^{-12} &= (s)(s) \\ s &= (5.1 \times 10^{-12})^{\frac{1}{2}} = 2.3 \times 10^{-6}\, M \end{aligned}$$

Step 3: Therefore, at equilibrium

$$[Cu^+] = 2.3 \times 10^{-6}\, M$$

$$[I^-] = 2.3 \times 10^{-6}\, M$$

Because 1 mole of CuI yields 1 mole of Cu^+ and 1 mole of I^-, the concentration of Cu^+ or I^- ions is equal to that of CuI dissociated. Thus the molar solubility of CuI also is $2.3 \times 10^{-6}\ M$.

Similar problem: 17.11.

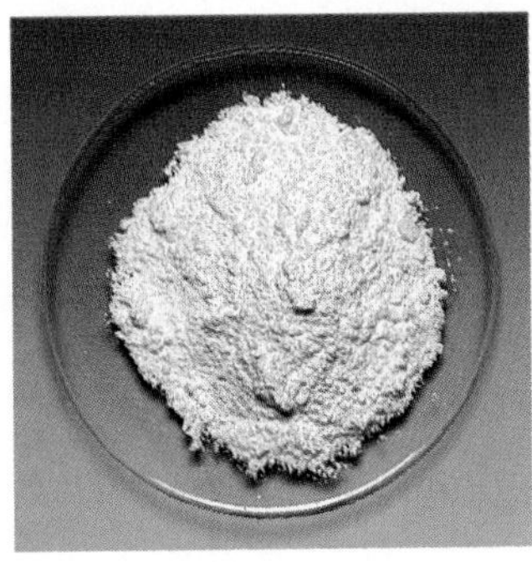

Calcium phosphate.

EXAMPLE 17.4

Using the data in Table 17.1, calculate the solubility of calcium phosphate, $Ca_3(PO_4)_2$, in g/L.

Answer

Step 1: Let s be the molar solubility of $Ca_3(PO_4)_2$. Since one unit of $Ca_3(PO_4)_2$ yields three Ca^{2+} ions and two PO_4^{3-} ions, at equilibrium $[Ca^{2+}]$ is $3s$ and $[PO_4^{3-}]$ is $2s$. We summarize the changes in concentrations as follows:

	$Ca_3(PO_4)_2(s) \rightleftharpoons$	$3Ca^{2+}(aq)$ +	$2PO_4^{3-}(aq)$
Initial:		$0.00\ M$	$0.00\ M$
Change:		$+3s\ M$	$+2s\ M$
Equilibrium:		$3s\ M$	$2s\ M$

Step 2:

$$K_{sp} = [Ca^{2+}]^3[PO_4^{3-}]^2$$

$$1.2 \times 10^{-26} = (3s)^3(2s)^2$$

$$s^5 = \frac{1.2 \times 10^{-26}}{108} = 1.1 \times 10^{-28}$$

Solving for s, we get

$$s = 2.6 \times 10^{-6}\ M$$

Knowing that the molar mass of $Ca_3(PO_4)_2$ is 310.2 g and knowing its molar solubility, we can calculate its solubility in g/L as follows:

$$\text{solubility of } Ca_3(PO_4)_2 = \frac{2.6 \times 10^{-6} \text{ mol } Ca_3(PO_4)_2}{1 \text{ L soln}} \times \frac{310.2 \text{ g } Ca_3(PO_4)_2}{1 \text{ mol } Ca_3(PO_4)_2}$$
$$= 8.1 \times 10^{-4} \text{ g/L}$$

Similar problem: 17.12.

As Examples 17.1–17.4 show, solubility and solubility product are related. If we know one, we can calculate the other, but each quantity provides different information. Table 17.2 shows the relationship between molar solubility and solubility product for a number of ionic compounds.

When carrying out solubility and/or solubility product calculations, keep the following important points in mind:

Table 17.2 Relationship between K_{sp} and molar solubility (s)

Compound	K_{sp} expression	Equilibrium concentration (M)		Relation between K_{sp} and s
		Cation	Anion	
$AgCl$	$[Ag^+][Cl^-]$	s	s	$K_{sp} = s^2;\ s = (K_{sp})^{\frac{1}{2}}$
$BaSO_4$	$[Ba^{2+}][SO_4^{2-}]$	s	s	$K_{sp} = s^2;\ s = (K_{sp})^{\frac{1}{2}}$
Ag_2CO_3	$[Ag^+]^2[CO_3^{2-}]$	$2s$	s	$K_{sp} = 4s^3;\ s = \left(\frac{K_{sp}}{4}\right)^{\frac{1}{3}}$
PbF_2	$[Pb^{2+}][F^-]^2$	s	$2s$	$K_{sp} = 4s^3;\ s = \left(\frac{K_{sp}}{4}\right)^{\frac{1}{3}}$
$Al(OH)_3$	$[Al^{3+}][OH^-]^3$	s	$3s$	$K_{sp} = 27s^4;\ s = \left(\frac{K_{sp}}{27}\right)^{\frac{1}{4}}$
$Ca_3(PO_4)_2$	$[Ca^{2+}]^3[PO_4^{3-}]^2$	$3s$	$2s$	$K_{sp} = 108s^5;\ s = \left(\frac{K_{sp}}{108}\right)^{\frac{1}{5}}$

You should be able to derive the relationships. Do not memorize them.

- The solubility is the quantity of substance that dissolves in a certain quantity of water. In solubility equilibria calculations, it is usually expressed as *grams* of solute per liter of solution. Molar solubility is the number of *moles* of solute per liter of solution.
- The solubility product is an equilibrium constant.
- Molar solubility, solubility, and solubility product all refer to a *saturated solution*.

Predicting Precipitation Reactions

From a knowledge of the solubility rules (see Section 3.3) and the solubility products listed in Table 17.1, we can predict whether a precipitate will form when we mix two solutions. Example 17.5 illustrates the steps involved in the calculation.

EXAMPLE 17.5

Exactly 200 mL of 0.0040 M $BaCl_2$ are added to exactly 600 mL of 0.0080 M K_2SO_4. Will a precipitate form?

Answer

According to solubility rule 3 on p. 100 the only precipitate that might form is $BaSO_4$:

$$Ba^{2+}(aq) + SO_4^{2-}(aq) \longrightarrow BaSO_4(s)$$

The number of moles of Ba^{2+} present in the original 200 mL of solution is

$$200\text{ mL} \times \frac{0.0040\text{ mol Ba}^{2+}}{1\text{ L soln}} \times \frac{1\text{ L}}{1000\text{ mL}} = 8.0 \times 10^{-4}\text{ mol Ba}^{2+}$$

We assume that the volumes are additive.

The total volume after combining the two solutions is 800 mL. The concentration of Ba^{2+} in the 800 mL volume is

$$[Ba^{2+}] = \frac{8.0 \times 10^{-4}\ \text{mol}}{800\ \text{mL}} \times \frac{1000\ \text{mL}}{1\ \text{L soln}}$$
$$= 1.0 \times 10^{-3}\ M$$

The number of moles of SO_4^{2-} in the original 600 mL solution is

$$600\ \text{mL} \times \frac{0.0080\ \text{mol}\ SO_4^{2-}}{1\ \text{L soln}} \times \frac{1\ \text{L}}{1000\ \text{mL}} = 4.8 \times 10^{-3}\ \text{mol}\ SO_4^{2-}$$

The concentration of SO_4^{2-} in the 800 mL of the combined solution is

$$[SO_4^{2-}] = \frac{4.8 \times 10^{-3}\ \text{mol}}{800\ \text{mL}} \times \frac{1000\ \text{mL}}{1\ \text{L soln}}$$
$$= 6.0 \times 10^{-3}\ M$$

Now we must compare Q and K_{sp}. From Table 17.1, the K_{sp} for $BaSO_4$ is 1.1×10^{-10}. As for Q,

$$Q = [Ba^{2+}]_0[SO_4^{2-}]_0 = (1.0 \times 10^{-3})(6.0 \times 10^{-3})$$
$$= 6.0 \times 10^{-6}$$

Compare Q and K_{sp}:

$$Q = 6.0 \times 10^{-6}$$
$$K_{sp} = 1.1 \times 10^{-10}$$
$$\therefore Q > K_{sp}$$

The solution is supersaturated because the value of Q indicates that the concentrations of the ions are too large. Thus some of the $BaSO_4$ will precipitate out of solution until

$$[Ba^{2+}][SO_4^{2-}] = 1.1 \times 10^{-10}$$

Similar problems: 17.13, 17.14.

17.2 Separation of Ions by Fractional Precipitation

Sometimes it is desirable to remove one type of ion from solution by precipitation while leaving other ions in solution. For instance, the addition of sulfate ions to a solution containing both potassium and barium ions causes $BaSO_4$ to precipitate out, thereby removing most of the Ba^{2+} ions from the solution. The other "product," K_2SO_4, is soluble and will remain in solution. The $BaSO_4$ precipitate can be separated from the solution by filtration.

Even when *both* products are insoluble, we can still achieve some degree of separation by choosing the proper reagent to bring about precipitation. Consider a solution that contains Cl^-, Br^-, and I^- ions. One way to separate these ions is to convert them into insoluble silver halides. As the K_{sp} values in the margin show, the solubility of the halides decreases from AgCl to AgI. Thus, when a soluble compound such as silver nitrate is slowly added to this solution, AgI begins to precipitate first, followed by AgBr and then AgCl.

Compound	K_{sp}
AgCl	1.6×10^{-10}
AgBr	7.7×10^{-13}
AgI	8.3×10^{-17}

The following example describes the separation of only two ions (Cl^- and Br^-), but the procedure can be applied to a solution containing more than two different types of ions if precipitates of differing solubility can be formed.

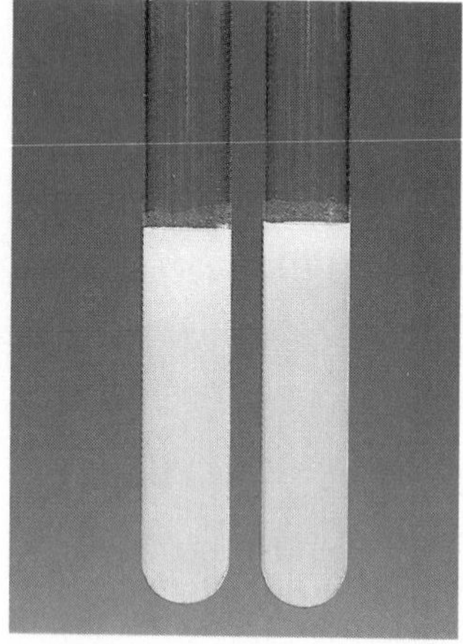

AgCl *AgBr*

EXAMPLE 17.6

Silver nitrate is slowly added to a solution that is 0.020 *M* in Cl^- ions and 0.020 *M* in Br^- ions. Calculate the concentration of Ag^+ ions (in mol/L) required to initiate (a) the precipitation of AgBr and (b) the precipitation of AgCl.

Answer

(a) From the K_{sp} values, we know that AgBr will precipitate before AgCl. So for AgBr we write

$$K_{sp} = [Ag^+][Br^-]$$

Since $[Br^-] = 0.020\ M$, the concentration of Ag^+ that must be exceeded to initiate the precipitation of AgBr is

$$[Ag^+] = \frac{K_{sp}}{[Br^-]} = \frac{7.7 \times 10^{-13}}{0.020}$$
$$= 3.9 \times 10^{-11}\ M$$

Thus $[Ag^+] > 3.9 \times 10^{-11}\ M$ is required to start the precipitation of AgBr.

(b) For AgCl

$$K_{sp} = [Ag^+][Cl^-]$$

$$[Ag^+] = \frac{K_{sp}}{[Cl^-]} = \frac{1.6 \times 10^{-10}}{0.020}$$
$$= 8.0 \times 10^{-9}\ M$$

Therefore $[Ag^+] > 8.0 \times 10^{-9}\ M$ is needed to initiate the precipitation of AgCl.

Similar problems: 17.18, 17.19.

Example 17.6, suggests a question. What is the concentration of Br^- ions remaining in solution just before AgCl begins to precipitate? To answer this question we let $[Ag^+] = 8.0 \times 10^{-9}\ M$. Then

$$[Br^-] = \frac{K_{sp}}{[Ag^+]}$$
$$= \frac{7.7 \times 10^{-13}}{8.0 \times 10^{-9}}$$
$$= 9.6 \times 10^{-5}\ M$$

The percent of Br^- remaining in solution (the *unprecipitated* Br^-) at the critical concentration of Ag^+ is

$$\%\ Br^- = \frac{[Br^-]_{unppt'd}}{[Br^-]_{original}} \times 100\%$$
$$= \frac{9.6 \times 10^{-5}\ M}{0.020\ M} \times 100\%$$
$$= 0.48\%\ \text{unprecipitated}$$

Thus, (100 − 0.48) percent, or 99.52 percent, of Br^- will have precipitated just before AgCl begins to precipitate. By this procedure, the Br^- ions can be quantitatively separated from the Cl^- ions by fractional precipitation.

17.3 The Common Ion Effect and Solubility

We saw in Chapter 16 the effect that the presence of a common ion can have on acid and base ionizations. Here we will examine the relationship between the common ion effect and solubility.

As we have noted, the solubility product is an equilibrium constant; precipitation of ionic compound from the solution occurs whenever the ion product exceeds K_{sp} for that substance. In a saturated solution of AgCl, for example, the ion product $[Ag^+][Cl^-]$ is, of course, equal to K_{sp}. Furthermore, simple stoichiometry tells us that $[Ag^+] = [Cl^-]$. But this equality does not hold in all situations.

Suppose we study a solution containing two dissolved substances that share a common ion, say, AgCl and $AgNO_3$. In addition to the dissociation of AgCl, the following process also contributes to the total concentration of the common silver ions in solution:

$$AgNO_3(s) \xrightarrow{H_2O} Ag^+(aq) + NO_3^-(aq)$$

If $AgNO_3$ is added to a saturated AgCl solution, the increase in $[Ag^+]$ will make the ion product greater than the solubility product:

$$Q = [Ag^+]_0[Cl^-]_0 > K_{sp}$$

To reestablish equilibrium, some AgCl will precipitate out of the solution, in accordance with Le Chatelier's principle, until the ion product is once again equal to K_{sp}. The effect of adding a common ion, then, is a *decrease* in the solubility of the salt (AgCl) in solution. Note that in this case $[Ag^+]$ is no longer equal to $[Cl^-]$ at equilibrium; rather, $[Ag^+] > [Cl^-]$.

EXAMPLE 17.7

Calculate the solubility of silver chloride (in g/L) in a 6.5×10^{-3} *M* silver nitrate solution.

Answer

Step 1: Since $AgNO_3$ is a soluble strong electrolyte, it dissociates completely:

$$\begin{array}{cccc} AgNO_3(s) \xrightarrow{H_2O} & Ag^+(aq) & + & NO_3^-(aq) \\ & 6.5 \times 10^{-3}\,M & & 6.5 \times 10^{-3}\,M \end{array}$$

Let *s* be the molar solubility of AgCl in $AgNO_3$ solution. We summarize the changes in concentrations as follows:

	$AgCl(s) \rightleftharpoons$	$Ag^+(aq)$	$+ Cl^-(aq)$
Initial:		$6.5 \times 10^{-3}\,M$	$0.0\,M$
Change:		$+s\,M$	$+s\,M$
Equilibrium:		$(6.5 \times 10^{-3} + s)\,M$	$s\,M$

Step 2:

$$K_{sp} = [Ag^+][Cl^-]$$

$$1.6 \times 10^{-10} = (6.5 \times 10^{-3} + s)(s)$$

Since AgCl is quite insoluble and the presence of Ag^+ ions from $AgNO_3$ further lowers the solubility of AgCl, s must be very small compared with 6.5×10^{-3}. Therefore, applying the approximation $6.5 \times 10^{-3} + s \simeq 6.5 \times 10^{-3}$, we obtain

$$1.6 \times 10^{-10} = 6.5 \times 10^{-3}s$$

$$s = 2.5 \times 10^{-8}\ M$$

Step 3: At equilibrium

$$[Ag^+] = (6.5 \times 10^{-3} + 2.5 \times 10^{-8})\ M \simeq 6.5 \times 10^{-3}\ M$$

$$[Cl^-] = 2.5 \times 10^{-8}\ M$$

and so our approximation $6.5 \times 10^{-3} + 2.5 \times 10^{-8} \simeq 6.5 \times 10^{-3}$ was justified in step 2. Since all the Cl^- ions must come from AgCl, the amount of AgCl dissolved in $AgNO_3$ solution also is $2.5 \times 10^{-8}\ M$. Then, knowing the molar mass of AgCl (143.4 g), we can calculate the solubility of the AgCl as follows:

$$\text{solubility of AgCl in AgNO}_3\text{ solution} = \frac{2.5 \times 10^{-8}\text{ mol AgCl}}{1\text{ L soln}} \times \frac{143.4\text{ g AgCl}}{1\text{ mol AgCl}}$$
$$= 3.6 \times 10^{-6}\text{ g/L}$$

Similar problem: 17.25.

For comparison with the result obtained in Example 17.7, let us find the solubility of AgCl in pure water. We start with

$$K_{sp} = [Ag^+][Cl^-] = 1.6 \times 10^{-10}$$

Hence

$$[Ag^+] = [Cl^-] = 1.3 \times 10^{-5}\ M$$

At a given temperature, only the solubility of a compound is altered (decreased) by the common ion effect. Its solubility product, which is an equilibrium constant, remains the same whether or not other substances are present in the solution.

The solubility of AgCl in g/L therefore is

$$\text{solubility of AgCl in water} = \frac{1.3 \times 10^{-5}\text{ mol AgCl}}{1\text{ L soln}} \times \frac{143.4\text{ g AgCl}}{1\text{ mol AgCl}}$$
$$= 1.9 \times 10^{-3}\text{ g/L}$$

As we might have expected, the solubility of AgCl is lower by a considerable amount in the presence of the common ion Ag^+ in $AgNO_3$ solution than it is in pure water.

EXAMPLE 17.8

What is the molar solubility of lead(II) iodide in a 0.050 M solution of sodium iodide?

Answer

Step 1: Sodium iodide is a soluble strong electrolyte, so

$$\underset{}{\text{NaI}(s)} \xrightarrow{H_2O} \underset{0.050\ M}{\text{Na}^+(aq)} + \underset{0.050\ M}{\text{I}^-(aq)}$$

Let s be the molar solubility of PbI_2 in mol/L. We summarize the changes in concentrations as follows:

	$PbI_2(s) \rightleftharpoons$	$Pb^{2+}(aq)$ +	$2I^-(aq)$
Initial:		$0.00\ M$	$0.050\ M$
Change:		$+s\ M$	$+2s\ M$
Equilibrium:		$s\ M$	$(0.050 + 2s)\ M$

Step 2:

$$K_{sp} = [Pb^{2+}][I^-]^2$$

$$1.4 \times 10^{-8} = (s)(0.050 + 2s)^2$$

The low solubility of PbI_2 and the effect of the common ion I^- in decreasing the molar solubility, s, of PbI_2 allow us to make the approximation $0.050 + 2s \simeq 0.050$ so that

$$1.4 \times 10^{-8} = (0.050)^2 s$$

$$s = 5.6 \times 10^{-6}\ M$$

Step 3: At equilibrium

$$[Pb^{2+}] = 5.6 \times 10^{-6}\ M$$

$$[I^-] = (0.050\ M) + 2(5.6 \times 10^{-6}\ M) \simeq 0.050\ M$$

showing that our approximation was appropriate.

Since all the Pb^{2+} ions must come from PbI_2, the molar solubility of PbI_2 in the NaI solution also is $5.6 \times 10^{-6}\ M$.

As an exercise, show that the molar solubility of PbI_2 in pure water is $1.5 \times 10^{-3}\ M$. You will note that the presence of the common ion I^- greatly decreases the solubility of PbI_2.

Similar problem: 17.26.

17.4 pH and Solubility

In addition to depending on the presence of a common ion, the solubility of many substances also depends on the pH of solution. Consider the solubility equilibrium of magnesium hydroxide:

$$Mg(OH)_2(s) \rightleftharpoons Mg^{2+}(aq) + 2OH^-(aq)$$

This is why milk of magnesia, $Mg(OH)_2$, dissolves in the acidic gastric juice in a person's stomach (see p. 656).

Adding OH^- ions (increasing the pH) shifts the equilibrium from right to left, thereby decreasing the solubility of $Mg(OH)_2$. (This is another example of the common ion effect.) On the other hand, adding H^+ ions (decreasing the pH) shifts the equilibrium from left to right, and the solubility of $Mg(OH)_2$ increases. Thus, insoluble bases tend to dissolve in acidic solutions. Similar logic shows why insoluble acids dissolve in basic solutions.

To see the quantitative effect of pH on the solubility of $Mg(OH)_2$, let us first calculate the pH of a saturated $Mg(OH)_2$ solution. We write

$$K_{sp} = [Mg^{2+}][OH^-]^2 = 1.2 \times 10^{-11}$$

Let s be the molar solubility of $Mg(OH)_2$. Proceeding as in Example 17.3,

$$K_{sp} = (s)(2s)^2 = 4s^3$$
$$4s^3 = 1.2 \times 10^{-11}$$
$$s^3 = 3.0 \times 10^{-12}$$
$$s = 1.4 \times 10^{-4}\ M$$

At equilibrium, therefore,

$$[OH^-] = 2 \times 1.4 \times 10^{-4}\ M = 2.8 \times 10^{-4}\ M$$
$$pOH = -\log (2.8 \times 10^{-4}) = 3.55$$
$$pH = 14.00 - 3.55 = 10.45$$

In a medium whose pH is less than 10.45, the solubility of $Mg(OH)_2$ would increase. This follows from the fact that a lower pH indicates a higher $[H^+]$ and thus a lower $[OH^-]$, as we would expect from $K_w = [H^+][OH^-]$. Consequently, $[Mg^{2+}]$ rises to maintain the equilibrium condition, and more $Mg(OH)_2$ dissolves. The dissolution process and the effect of extra H^+ ions can be summarized as follows:

$$Mg(OH)_2(s) \rightleftharpoons Mg^{2+}(aq) + 2OH^-(aq)$$
$$2H^+(aq) + 2OH^-(aq) \rightleftharpoons 2H_2O(l)$$
$$\text{Overall:}\quad Mg(OH)_2(s) + 2H^+(aq) \rightleftharpoons Mg^{2+}(aq) + 2H_2O(l)$$

If the pH of the medium were higher than 10.45, $[OH^-]$ would be higher and the solubility of $Mg(OH)_2$ would decrease because of the common ion (OH^-) effect.

The pH also influences the solubility of salts that contain a basic anion. For example, the solubility equilibrium for BaF_2 is

$$BaF_2(s) \rightleftharpoons Ba^{2+}(aq) + 2F^-(aq)$$

and

$$K_{sp} = [Ba^{2+}][F^-]^2$$

In an acidic medium, the high $[H^+]$ will shift the following equilibrium from left to right:

HF is a weak acid.

$$H^+(aq) + F^-(aq) \rightleftharpoons HF(aq)$$

As $[F^-]$ decreases, $[Ba^{2+}]$ must increase to maintain the equilibrium condition. Thus more BaF_2 dissolves. The dissolution process and the effect of pH on the solubility of BaF_2 can be summarized as follows:

$$BaF_2(s) \rightleftharpoons Ba^{2+}(aq) + 2F^-(aq)$$
$$2H^+(aq) + 2F^-(aq) \rightleftharpoons 2HF(aq)$$
$$\text{Overall:}\quad BaF_2(s) + 2H^+(aq) \rightleftharpoons Ba^{2+}(aq) + 2HF(aq)$$

The solubilities of salts containing anions that do not hydrolyze (for example, Cl^-, Br^-, and I^-) are unaffected by pH.

Examples 17.9 and 17.10 deal with the effect of pH on the solubility of a slightly soluble substance and a precipitation reaction.

EXAMPLE 17.9

At 25°C the molar solubility of $Mg(OH)_2$ in pure water is $1.4 \times 10^{-4}\ M$. Calculate its molar solubility in a buffer medium whose pH is (a) 12.00 and (b) 9.00.

Answer

(a) We write

$$Mg(OH)_2(s) \rightleftharpoons Mg^{2+}(aq) + 2OH^-(aq)$$

$$pH = 12.00$$

$$pOH = 14.00 - 12.00 = 2.00$$

$$[OH^-] = 1.0 \times 10^{-2}\ M$$

Therefore

$$K_{sp} = [Mg^{2+}][OH^-]^2 = 1.2 \times 10^{-11}$$

$$[Mg^{2+}] = \frac{1.2 \times 10^{-11}}{(1.0 \times 10^{-2})^2} = 1.2 \times 10^{-7}\ M$$

Since 1 mol $Mg(OH)_2$ ≎ 1 mol Mg^{2+}, the molar solubility of $Mg(OH)_2$ is 1.2×10^{-7} *M*. Due to the common ion (OH^-) effect, the molar solubility of $Mg(OH)_2$ is markedly lower than its solubility in pure water (1.4×10^{-4} *M*).

(b) In this case we write

$$pH = 9.00$$

$$pOH = 14.00 - 9.00 = 5.00$$

$$[OH^-] = 1.0 \times 10^{-5}\ M$$

$$[Mg^{2+}] = \frac{1.2 \times 10^{-11}}{(1.0 \times 10^{-5})^2} = 0.12\ M$$

Therefore, the molar solubility of $Mg(OH)_2$ is 0.12 *M*. The increase in molar solubility from (a) to (b) is the result of the partial removal of OH^- ions by the extra H^+ ions present.

Similar problem: 17.34.

EXAMPLE 17.10

Calculate the concentration of aqueous ammonia necessary to initiate the precipitation of iron(II) hydroxide from a 0.0030 *M* solution of $FeCl_2$.

Answer

The equilibria of interest are

$$NH_3(aq) + H_2O(l) \rightleftharpoons NH_4^+(aq) + OH^-(aq)$$

$$Fe^{2+}(aq) + 2OH^-(aq) \rightleftharpoons Fe(OH)_2(s)$$

First we find the OH^- concentration above which $Fe(OH)_2$ begins to precipitate. We write

$$K_{sp} = [Fe^{2+}][OH^-]^2 = 1.6 \times 10^{-14}$$

Since $FeCl_2$ is a strong electrolyte, $[Fe^{2+}] = 0.0030\ M$ and

$$[OH^-]^2 = \frac{1.6 \times 10^{-14}}{0.0030} = 5.3 \times 10^{-12}$$

$$[OH^-] = 2.3 \times 10^{-6}\ M$$

Next, we calculate the concentration of NH_3 that will supply $2.3 \times 10^{-6}\ M$ OH^- ions. Let x be the initial concentration of NH_3 in mol/L. We summarize the changes in concentrations resulting from the ionization of NH_3 as follows:

	$NH_3(aq) + H_2O(l) \rightleftharpoons$	$NH_4^+(aq)$ +	$OH^-(aq)$
Initial:	$x\ M$	$0.00\ M$	$0.00\ M$
Change:	$-2.3 \times 10^{-6}\ M$	$+2.3 \times 10^{-6}\ M$	$+2.3 \times 10^{-6}\ M$
Equilibrium:	$(x - 2.3 \times 10^{-6})\ M$	$2.3 \times 10^{-6}\ M$	$2.3 \times 10^{-6}\ M$

Substituting the equilibrium concentrations in the expression for the ionization constant,

$$K_b = \frac{[NH_4^+][OH^-]}{[NH_3]} = 1.8 \times 10^{-5}$$

$$\frac{(2.3 \times 10^{-6})(2.3 \times 10^{-6})}{(x - 2.3 \times 10^{-6})} = 1.8 \times 10^{-5}$$

Solving for x, we obtain

$$x = 2.6 \times 10^{-6}\ M$$

Therefore the concentration of NH_3 must be slightly greater than $2.6 \times 10^{-6}\ M$ to initiate the precipitation of $Fe(OH)_2$.

Similar problems: 17.35, 17.36.

CHEMISTRY IN ACTION

pH, Solubility, and Tooth Decay

Tooth decay has plagued humans for centuries. Although its cause is fairly well understood, total prevention of tooth decay is still not possible.

Teeth are protected by a hard enamel layer about 2 mm thick that is composed of a mineral called hydroxyapatite, $Ca_5(PO_4)_3OH$. When it dissolves (a process called *demineralization*), the ions go into solution in the saliva:

$$Ca_5(PO_4)_3OH(s) \longrightarrow 5Ca^{2+}(aq) + 3PO_4^{3-}(aq) + OH^-(aq)$$

Because phosphates of alkaline earth metals such as calcium are insoluble, this reaction proceeds only to a small extent. The reverse process, called *remineralization,* is the body's natural defense against tooth decay:

$$5Ca^{2+}(aq) + 3PO_4^{3-}(aq) + OH^-(aq) \longrightarrow Ca_5(PO_4)_3OH(s)$$

In children, the growth of the enamel layer (mineralization) occurs faster than demineralization; in adults, demineralization and remineralization take place at roughly the same rate.

After a meal, bacteria in the mouth break down some food to produce organic acids such as acetic acid and lactic acid, $CH_3CH(OH)COOH$. Acid production is greatest from food with high sugar content, such as candies, ice cream, and sugary beverages. The decrease in pH results in the removal of OH^- ions

$$H^+(aq) + OH^-(aq) \longrightarrow H_2O(l)$$

promoting demineralization. Once the protective enamel layer is weakened, tooth decay starts.

The best way to fight tooth decay is to eat a diet low in sugar and always brush one's teeth immediately after eating. Most toothpastes contain a fluoride compound such as NaF or SnF_2 that also helps to reduce tooth decay (Figure 17.2). The F^- ions from these compounds replace some of the OH^- ions during the remineralization process

$$5Ca^{2+}(aq) + 3PO_4^{3-}(aq) + F^-(aq) \longrightarrow Ca_5(PO_4)_3F(s)$$

Because F^- is a weaker base than OH^-, the modified enamel, called fluorapatite, is more resistant to acid.

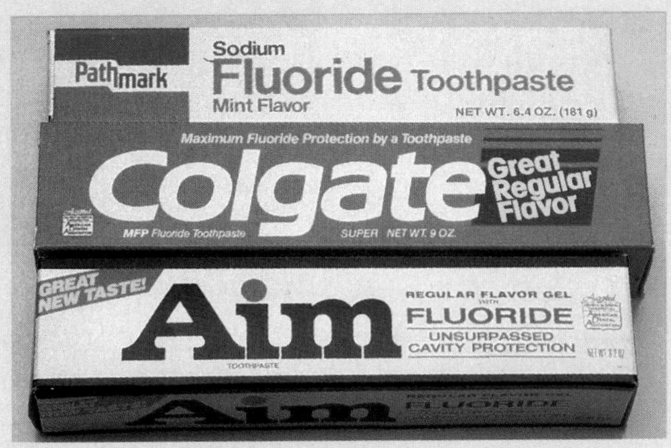

Figure 17.2 *Commercial toothpastes contain fluoride compounds to fight tooth decay.*

17.5 Complex Ion Equilibria and Solubility

Some metal ions, especially those of transition metals, form complex ions in solution. A ***complex ion*** may be defined as *an ion containing a central metal cation bonded to one or more molecules or ions*. Cobalt forms such complexes. A solution of cobalt(II) chloride is pink because of the presence of the $Co(H_2O)_6^{2+}$ ions (Figure 17.3). When HCl acid is added, the solution turns blue as a result of the formation of the complex ion $CoCl_4^{2-}$:

According to our definition, $Co(H_2O)_6^{2+}$ itself is a complex ion. When we write $Co^{2+}(aq)$, we mean the hydrated Co^{2+} ion.

$$Co^{2+}(aq) + 4Cl^-(aq) \rightleftharpoons CoCl_4^{2-}(aq)$$

Complex ion formations are Lewis acid-base reactions in which the metal ion acts as the acid and the anions or molecules as the base (see Section 15.6). Complex ions play an important role in many chemical and biological processes. We will consider the effect of complex ion formation on solubility.

Copper(II) sulfate ($CuSO_4$) dissolves in water to produce a blue solution. The hy-

Figure 17.3 *Left: An aqueous cobalt(II) chloride solution. The pink color is due to the $Co(H_2O)_6^{2+}$ ions. Right: After the addition of HCl solution, the solution turns blue because of the formation of the complex $CoCl_4^{2-}$ ions.*

Figure 17.4 *Left: A beaker containing an aqueous solution of copper(II) sulfate. Center: After the addition of a few drops of a concentrated aqueous ammonia solution, a light blue precipitate of $Cu(OH)_2$ is formed. Right: When an excess of concentrated aqueous ammonia solution is added, the $Cu(OH)_2$ precipitate dissolves to form the dark blue complex ions $Cu(NH_3)_4^{2+}$.*

drated copper(II) ions are responsible for this color; many other sulfates (Na_2SO_4, for example) are colorless. Adding a *few drops* of concentrated ammonia solution to a $CuSO_4$ solution causes a light blue precipitate, copper(II) hydroxide, to form:

$$Cu^{2+}(aq) + 2OH^-(aq) \longrightarrow Cu(OH)_2(s)$$

where the OH^- ions are supplied by the ammonia solution. If an *excess* of NH_3 is then added, the blue precipitate redissolves to produce a beautiful dark blue solution, this time due to the formation of the complex ion $Cu(NH_3)_4^{2+}$ (Figure 17.4):

$$Cu(OH)_2(s) + 4NH_3(aq) \rightleftharpoons Cu(NH_3)_4^{2+}(aq) + 2OH^-(aq)$$

Thus the formation of the complex ion $Cu(NH_3)_4^{2+}$ increases the solubility of $Cu(OH)_2$.

A measure of the tendency of a metal ion to form a particular complex ion is given by the ***formation constant*** K_f (also called the ***stability constant***), which is *the equilibrium constant for the complex ion formation*. The larger the K_f, the more stable the complex ion. Table 17.3 lists the formation constants of a number of complex ions. The formation of the $Cu(NH_3)_4^{2+}$ ion can be expressed as

$$Cu^{2+}(aq) + 4NH_3(aq) \rightleftharpoons Cu(NH_3)_4^{2+}(aq)$$

Table 17.3 Formation constants of selected complex ions in water at 25°C

Complex ion	Equilibrium expression	Formation constant (K_f)
$Ag(NH_3)_2^+$	$Ag^+ + 2NH_3 \rightleftharpoons Ag(NH_3)_2^+$	1.5×10^7
$Ag(CN)_2^-$	$Ag^+ + 2CN^- \rightleftharpoons Ag(CN)_2^-$	1.0×10^{21}
$Cu(CN)_4^{2-}$	$Cu^{2+} + 4CN^- \rightleftharpoons Cu(CN)_4^{2-}$	1.0×10^{25}
$Cu(NH_3)_4^{2+}$	$Cu^{2+} + 4NH_3 \rightleftharpoons Cu(NH_3)_4^{2+}$	5.0×10^{13}
$Cd(CN)_4^{2-}$	$Cd^{2+} + 4CN^- \rightleftharpoons Cd(CN)_4^{2-}$	7.1×10^{16}
CdI_4^{2-}	$Cd^{2+} + 4I^- \rightleftharpoons CdI_4^{2-}$	2.0×10^6
$HgCl_4^{2-}$	$Hg^{2+} + 4Cl^- \rightleftharpoons HgCl_4^{2-}$	1.7×10^{16}
HgI_4^{2-}	$Hg^{2+} + 4I^- \rightleftharpoons HgI_4^{2-}$	2.0×10^{30}
$Hg(CN)_4^{2-}$	$Hg^{2+} + 4CN^- \rightleftharpoons Hg(CN)_4^{2-}$	2.5×10^{41}
$Co(NH_3)_6^{3+}$	$Co^{3+} + 6NH_3 \rightleftharpoons Co(NH_3)_6^{3+}$	5.0×10^{31}
$Zn(NH_3)_4^{2+}$	$Zn^{2+} + 4NH_3 \rightleftharpoons Zn(NH_3)_4^{2+}$	2.9×10^9

for which the formation constant is

$$K_f = \frac{[Cu(NH_3)_4^{2+}]}{[Cu^{2+}][NH_3]^4} = 5.0 \times 10^{13}$$

The very large value of K_f in this case indicates the great stability of the complex ion in solution and accounts for the very low concentration of copper(II) ions at equilibrium.

EXAMPLE 17.11

A 0.20 mole quantity of $CuSO_4$ is added to a liter of 1.20 M NH_3 solution. What is the concentration of Cu^{2+} ions at equilibrium?

Answer

The addition of $CuSO_4$ to the NH_3 solution results in the reaction

$$Cu^{2+}(aq) + 4NH_3(aq) \rightleftharpoons Cu(NH_3)_4^{2+}(aq)$$

Since K_f is very large (5.0×10^{13}), the reaction lies mostly to the right. As a good approximation, we can assume that essentially all the dissolved Cu^{2+} ions end up as $Cu(NH_3)_4^{2+}$ ions. Thus the amount of NH_3 consumed in forming the complex ions is 4×0.20 mol, or 0.80 mol. (Note that 0.20 mol Cu^{2+} is initially present in solution and four NH_3 molecules are needed to "complex" one Cu^{2+} ion.) The concentration of NH_3 at equilibrium therefore is (1.20 − 0.80) M, or 0.40 M, and that of $Cu(NH_3)_4^{2+}$ is 0.20 M, the same as the initial concentration of Cu^{2+}. Since $Cu(NH_3)_4^{2+}$ does dissociate to a slight extent, we call the concentration of Cu^{2+} ions at equilibrium x and write

$$K_f = \frac{[Cu(NH_3)_4^{2+}]}{[Cu^{2+}][NH_3]^4} = 5.0 \times 10^{13}$$

$$\frac{0.20}{x(0.40)^4} = 5.0 \times 10^{13}$$

Solving for x, we obtain

$$x = 1.6 \times 10^{-13}\ M = [Cu^{2+}]$$

The small value of $[Cu^{2+}]$ at equilibrium, compared with 0.20 M, certainly justifies our approximation.

Similar problem: 17.41.

The effect of complex ion formation generally is to *increase* the solubility of a substance, as the following example shows.

EXAMPLE 17.12

Calculate the molar solubility of silver chloride in a 1.0 M NH_3 solution.

Answer

Step 1: The equilibria reactions are

$$AgCl(s) \rightleftharpoons Ag^+(aq) + Cl^-(aq)$$
$$K_{sp} = [Ag^+][Cl^-] = 1.6 \times 10^{-10}$$
$$Ag^+(aq) + 2NH_3(aq) \rightleftharpoons Ag(NH_3)_2^+(aq)$$
$$K_f = \frac{[Ag(NH_3)_2^+]}{[Ag^+][NH_3]^2} = 1.5 \times 10^7$$

Overall: $AgCl(s) + 2NH_3(aq) \rightleftharpoons Ag(NH_3)_2^+(aq) + Cl^-(aq)$

The equilibrium constant K for the overall reaction is the product of the equilibrium constants of the individual reactions (see Section 14.3):

$$K = K_{sp}K_f = \frac{[Ag(NH_3)_2^+][Cl^-]}{[NH_3]^2}$$
$$= (1.6 \times 10^{-10})(1.5 \times 10^7)$$
$$= 2.4 \times 10^{-3}$$

Let s be the molar solubility of AgCl (mol/L). We summarize the changes in concentrations that result from formation of the complex ion as follows:

	$AgCl(s) +$	$2NH_3(aq)$	$\rightleftharpoons$	$Ag(NH_3)_2^+(aq)$	$+ Cl^-(aq)$
Initial:		$1.0\ M$		$0.0\ M$	$0.0\ M$
Change:		$-2s\ M$		$+s\ M$	$+s\ M$
Equilibrium:		$(1.0 - 2s)\ M$		$s\ M$	$s\ M$

The formation constant for $Ag(NH_3)_2^+$ is quite large, so most of the silver ions exist in the complexed form. In the absence of ammonia we have, at equilibrium, $[Ag^+] = [Cl^-]$. As a result of complex ion formation, however, we can write $[Ag(NH_3)_2^+] = [Cl^-]$.

Step 2:

$$K = \frac{(s)(s)}{(1.0 - 2s)^2}$$

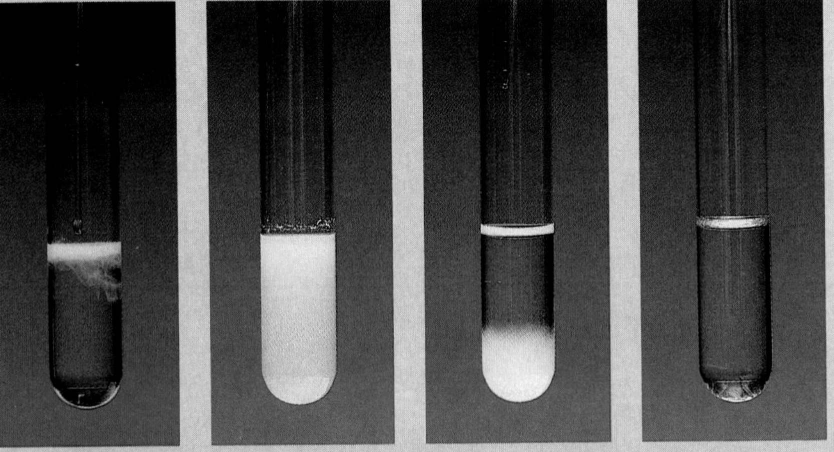

Figure 17.5 *From left to right: Formation of AgCl precipitate when $AgNO_3$ solution is added to NaCl solution. Upon the addition of NH_3 solution, the AgCl precipitate dissolves as the soluble $Ag(NH_3)_2^+$ forms.*

$$2.4 \times 10^{-3} = \frac{s^2}{(1.0 - 2s)^2}$$

Taking the square root of both sides, we obtain

$$0.049 = \frac{s}{1.0 - 2s}$$

$$s = 0.045\ M$$

Step 3: At equilibrium, 0.045 mole of AgCl dissolves in 1 L of 1.0 M NH_3 solution.

We saw earlier (p. 730) that the molar solubility of AgCl in pure water is $1.3 \times 10^{-5}\ M$. Thus, the formation of the complex ion $Ag(NH_3)_2^+$ enhances the solubility of AgCl. Figure 17.5 shows the formation of AgCl precipitate and its subsequent dissolution as a result of complex ion formation.

Similar problem: 17.45.

According to Le Chatelier's principle, the removal of Ag^+ ions from the solution to form $Ag(NH_3)_2^+$ ions will cause more AgCl to dissolve.

17.6 Application of the Solubility Product Principle to Qualitative Analysis

In Section 4.5, we discussed the principle of gravimetric analysis, by which we measure the amount of an ion in an unknown sample. Here we will briefly discuss qualitative analysis, *the determination of the types of ions present in a solution*. We will focus on the cations.

Do not confuse the groups in Table 17.4, which are based on solubility products, with those in the periodic table, which are based on the electron configurations of the elements.

There are some twenty common cations that can be analyzed readily in aqueous solution. These cations can be divided into five groups according to the solubility products of their insoluble salts (Table 17.4). Since an unknown solution may contain any one or all twenty ions, any analysis must be carried out systematically from group 1 through group 5. We will not give a comprehensive discussion of how each cation can be identified; instead, we will outline the procedure for separating these twenty ions by adding precipitating reagents to an unknown solution.

- *Group 1 cations*. When dilute HCl is added to the unknown solution, only the Ag^+, Hg_2^{2+}, and Pb^{2+} ions precipitate as insoluble chlorides. All other ions form soluble chlorides and remain in solution.
- *Group 2 cations*. After the chloride precipitates have been removed by filtration, hydrogen sulfide is reacted with the unknown acidic solution. As mentioned earlier (p. 722), the concentration of the S^{2-} ion in solution is negligible. Therefore the precipitation of metal sulfides is best represented as

$$M^{2+}(aq) + H_2S(aq) \rightleftharpoons MS(s) + 2H^+(aq)$$

  Adding acid to the solution shifts this equilibrium to the left so that only the least soluble metal sulfides, that is, those with the smallest K_{sp} values, will precipitate out of solution. These are Bi_2S_3, CdS, CuS, and SnS (see Table 17.4).
- *Group 3 cations*. At this stage, sodium hydroxide is added to the solution to make it basic. In a basic solution, the above equilibrium shifts to the right. Therefore, the more soluble sulfides (CoS, FeS, MnS, NiS, ZnS) now precipitate out of solution. Note that the Al^{3+} and Cr^{3+} ions actually precipitate as the hydroxides $Al(OH)_3$ and $Cr(OH)_3$, rather than as the sulfides, because the hydroxides are less

Table 17.4 Separation of cations into groups according to their precipitation reactions with various reagents

Group	Cation	Precipitating reagents	Insoluble compound	K_{sp}
1	Ag^+	HCl	AgCl	1.6×10^{-10}
	Hg_2^{2+}	↓	Hg_2Cl_2	3.5×10^{-18}
	Pb^{2+}	↓	$PbCl_2$	2.4×10^{-4}
2	Bi^{3+}	H_2S	Bi_2S_3	1.6×10^{-72}
	Cd^{2+}	in acidic	CdS	8.0×10^{-28}
	Cu^{2+}	solutions	CuS	6.0×10^{-37}
	Sn^{2+}	↓	SnS	1.0×10^{-26}
3	Al^{3+}	H_2S	$Al(OH)_3$	1.8×10^{-33}
	Co^{2+}	in basic	CoS	4.0×10^{-21}
	Cr^{3+}	solutions	$Cr(OH)_3$	3.0×10^{-29}
	Fe^{2+}	│	FeS	6.0×10^{-19}
	Mn^{2+}	│	MnS	3.0×10^{-14}
	Ni^{2+}	│	NiS	1.4×10^{-24}
	Zn^{2+}	↓	ZnS	3.0×10^{-23}
4	Ba^{2+}	Na_2CO_3	$BaCO_3$	8.1×10^{-9}
	Ca^{2+}	│	$CaCO_3$	8.7×10^{-9}
	Sr^{2+}	↓	$SrCO_3$	1.6×10^{-9}
5	K^+	No precipitating	None	
	Na^+	reagent	None	
	NH_4^+		None	

soluble. The solution is then filtered to remove the insoluble sulfides and hydroxides.

- *Group 4 cations*. After all the group 1, 2, and 3 cations have been removed from solution, sodium carbonate is added to the basic solution to precipitate Ba^{2+}, Ca^{2+}, and Sr^{2+} ions as $BaCO_3$, $CaCO_3$, and $SrCO_3$. These precipitates too are removed from solution by filtration.
- *Group 5 cations*. At this stage, the only cations possibly remaining in solution are Na^+, K^+, and NH_4^+. The presence of NH_4^+ can be determined by adding sodium hydroxide:

$$NaOH(aq) + NH_4^+(aq) \longrightarrow Na^+(aq) + H_2O(l) + NH_3(g)$$

The ammonia gas is detected either by noting its characteristic odor or by observing a piece of wet red litmus paper turning blue when placed above (not in contact with) the solution. To confirm the presence of Na^+ and K^+ ions, we usually use a flame test, as follows: A piece of platinum wire (chosen because platinum is inert) is moistened with the solution and is then held over a Bunsen burner flame. Each type of metal ion gives a characteristic color when heated in this manner. For example, the color emitted by Na^+ ions is yellow, that of K^+ ions is violet, and that of Cu^{2+} ions is green (Figure 17.6).

Because NaOH is added in group 3 and Na_2CO_3 is added in group 4, the flame test for Na^+ ions is carried out using the original solution.

The flow chart shown in Figure 17.7 summarizes this scheme for separating metal ions.

Two points regarding qualitative analysis must be mentioned. First, the separation of the cations into groups is made as selective as possible; that is, the anions that are added as reagents must be such that they will precipitate the fewest types of cations.

Figure 17.6 *Left to right: Flame colors of lithium, sodium, potassium, and copper.*

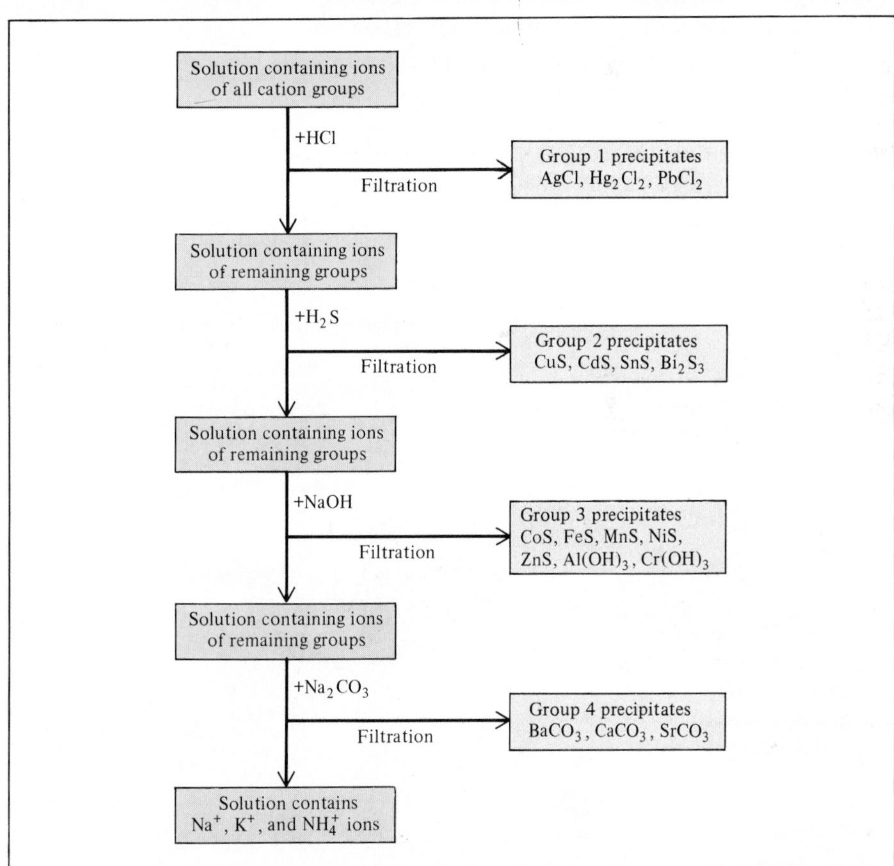

Figure 17.7 *A flow chart for the separation of cations in qualitative analysis.*

For example, all the cations in group 1 form insoluble sulfides. Thus, if H_2S were reacted with the solution at the start, as many as seven different sulfides might precipitate out of solution (group 1 *and* group 2 sulfides), an undesirable outcome. Second, the separation of cations at each step must be carried out as completely as possible. For example, if we do not add enough HCl to the unknown solution to remove all the group 1 cations, they will precipitate with the group 2 cations as insoluble sulfides; this too would interfere with further chemical analysis and lead us to draw erroneous conclusions.

CHEMISTRY IN ACTION

Solubility Equilibria and the Formation of Sinkholes, Stalagmites, and Stalactites

If you have visited the Carlsbad Caverns in New Mexico or other limestone caverns, you must have been impressed by the fantastic rock formations, called *stalactites,* that hang icicle-like from the cavern ceiling, and *stalagmites,* the columns that rise from the cavern floor. How are these objects formed?

The principal nonsilicate mineral in rocks is limestone, $CaCO_3$. The solubility product for the dissolving of limestone

$$CaCO_3(s) \rightleftharpoons Ca^{2+}(aq) + CO_3^{2-}(aq)$$

is

$$K_{sp} = [Ca^{2+}][CO_3^{2-}] = 8.7 \times 10^{-9}$$

Thus $CaCO_3$ is rather insoluble. However, $CaCO_3$ dissolves readily in acid solutions as a result of the following reaction:

$$CaCO_3(s) + 2HCl(aq) \longrightarrow Ca^{2+}(aq) + 2Cl^-(aq) + H_2O(l) + CO_2(g)$$

Since soil moisture commonly contains humic acids derived from the decay of vegetation, most groundwater can dissolve limestone. Furthermore, water contain-

Figure 17.8 *This landscape in Kentucky is dotted with sinkholes, which appear as depressions.*

ing dissolved CO_2 from the atmosphere is acidic and reacts with $CaCO_3$ as follows:

$$CaCO_3(s) + CO_2(aq) + H_2O(l) \rightleftharpoons Ca^{2+}(aq) + 2HCO_3^-(aq)$$

The extent to which this reaction takes place depends on the partial pressure of CO_2. At high partial pressures of CO_2 the amount of dissolved CO_2 is high and the above equilibrium shifts to the right; at low partial pressures, the equilibrium shifts to the left.

When acidic surface waters seep underground, they slowly dissolve away limestone deposits. If the limestone deposits are close to the surface of the earth, their dissolution leads to the collapse of the thin layer of earth that lies above them. The resulting depressions in the landscape are called *sinkholes* (Figure 17.8). The dissolution of deep deposits of limestone produces underground caves. When a solution containing Ca^{2+} and HCO_3^- ions drips through the cracks of a cave's ceiling into the cave, where the partial pressure of carbon dioxide is lower than it is in the soil, the solution becomes supersaturated in carbon dioxide. The escape of the CO_2 gas from the solution

$$CO_2(aq) \longrightarrow CO_2(g)$$

shifts the above equilibrium to the left:

$$CaCO_3(s) + CO_2(aq) + H_2O(l) \longleftarrow Ca^{2+}(aq) + 2HCO_3^-(aq)$$

Consequently, a precipitate of $CaCO_3$ forms and stalactites and stalagmites develop (Figure 17.9). In time, stalactites and stalagmites may join to form columns that reach from the floor to the ceiling of the cave.

Figure 17.9 *Downward-growing, icicle-like stalactites and upward-growing, columnar stalagmites. It may take hundreds of years for these structures to form.*

Summary

1. The solubility product K_{sp} expresses the equilibrium between a solid and its ions in solution. Solubility can be found from K_{sp} and vice versa.
2. The presence of a common ion decreases the solubility of a salt.
3. The solubility of slightly soluble salts containing basic anions increases as the hydrogen ion concentration increases. The solubility of salts with anions derived from strong acids is unaffected by pH.
4. Complex ions are formed in solution by the combination of a metal cation with molecules or anions. The formation constant K_f measures the tendency toward the formation of a specific complex ion. Complex ion formation can increase the solubility of an insoluble substance.
5. Qualitative analysis is the identification of cations and anions in solution.

Applied Chemistry Example: Water Softening

Water containing Ca^{2+} and Mg^{2+} ions (from limestone $CaCO_3$ and dolomite $CaCO_3 \cdot MgCO_3$) is called *hard water* and water that is mostly free of these ions is called *soft water* (see p. 102). Hard water is unsuitable for some household and industrial use because the Ca^{2+} and Mg^{2+} ions react with soap to form insoluble salts or curds. Hard water also containing bicarbonate ions (HCO_3^-) is called temporary hard water because on heating the bicarbonates are converted to carbonates ions

$$2HCO_3^-(aq) \longrightarrow CO_3^{2-}(aq) + H_2O(l) + CO_2(g)$$

which then precipitate out the Ca^{2+} and Mg^{2+} ions as $CaCO_3$ and $MgCO_3$.

One industrial procedure for "water softening," that is, the removal of the undesirable cations in a solution containing bicarbonate ions, uses slaked lime [$Ca(OH)_2$]. It may seem strange that a calcium compound is used to remove calcium ions. However, the following equation shows the chemical reasoning for this choice:

$$Ca(HCO_3)_2(aq) + Ca(OH)_2(aq) \longrightarrow 2CaCO_3(s) + 2H_2O(l)$$

For every mole of $Ca(OH)_2$ added, two moles of Ca^{2+} ions are removed as $CaCO_3$ (K_{sp} for $CaCO_3$ is 8.7×10^{-9}). The equations for the removal of Mg^{2+} ions from solution are

$$Mg(HCO_3)_2(aq) + Ca(OH)_2(aq) \longrightarrow MgCO_3(aq) + CaCO_3(s) + 2H_2O(l)$$

$$MgCO_3(aq) + Ca(OH)_2(aq) \longrightarrow Mg(OH)_2(s) + CaCO_3(s)$$

and the equation for the overall process is

$$Mg(HCO_3)_2(aq) + 2Ca(OH)_2(aq) \longrightarrow 2CaCO_3(s) + Mg(OH)_2(s) + 2H_2O(l)$$

We see that two moles of $Ca(OH)_2$ are needed to remove one mole of $Mg(HCO_3)_2$ because $MgCO_3$ ($K_{sp} = 4.0 \times 10^{-5}$) is much more soluble than $Mg(OH)_2$ ($K_{sp} = 1.2 \times 10^{-11}$). The precipitates formed are removed from water by filtration. Hard water not containing bicarbonate ions is called permanent hard water. In this case the Ca^{2+} ions can be removed as $CaCO_3$ by treatment with washing soda ($Na_2CO_3 \cdot 10H_2O$) which supplies the carbonate ions and the Mg^{2+} ions can be removed as $Mg(OH)_2$ by treatment with $Ca(OH)_2$.

1. The molar solubility of calcium carbonate is $9.3 \times 10^{-5}\ M$. What is its molar solubility in a $0.050\ M$ Na_2CO_3 solution?

   Answer: This is a common ion (CO_3^{2-}) problem. Following the procedure outlined in Section 17.3, we write

   Step 1: The dissociation of Na_2CO_3 is

$$\underset{}{Na_2CO_3(s)} \xrightarrow{H_2O} \underset{2 \times 0.050\ M}{2Na^+(aq)} + \underset{0.050\ M}{CO_3^{2-}(aq)}$$

   Let s be the molar solubility of $CaCO_3$ in Na_2CO_3 solution. We summarize the changes as

	$CaCO_3(s) \rightleftharpoons$	$Ca^{2+}(aq)$ +	$CO_3^{2-}(aq)$
Initial:		$0.00\ M$	$0.050\ M$
Change:		$+s\ M$	$+s\ M$
Equilibrium:		$s\ M$	$(0.050 + s)\ M$

Step 2:

$$K_{sp} = [Ca^{2+}][CO_3^{2-}]$$
$$8.7 \times 10^{-9} = s(0.050 + s)$$

Since s is very small, we assume that $0.050 + s \simeq 0.050$ and obtain $s = 1.7 \times 10^{-7}\ M$. Thus the addition of washing soda to permanent hard water removes most of the Ca^{2+} ions as a result of the common ion effect.

2. Slaked lime is added to permanent hard water to produce a saturated solution with a pH of 12.41. Calculate the concentration of Mg^{2+} ions in this solution.

Answer: First we calculate the concentration of OH^- ions. Since

$$\begin{aligned} pOH &= 14.00 - pH \\ &= 14.00 - 12.41 \\ &= 1.59 \end{aligned}$$

Thus

$$[OH^-] = 0.026\ M$$

At this rather high concentration of OH^- ions, most of the Mg^{2+} ions will be removed as $Mg(OH)_2$. The small amount of Mg^{2+} remaining in solution is due to the equilibrium

$$Mg(OH)_2(s) \rightleftharpoons Mg^{2+}(aq) + 2OH^-(aq)$$
$$K_{sp} = [Mg^{2+}][OH^-]^2$$

Rearranging, we get

$$\begin{aligned} [Mg^{2+}] &= \frac{K_{sp}}{[OH^-]^2} \\ &= \frac{1.2 \times 10^{-11}}{(0.026)^2} \\ &= 1.8 \times 10^{-8}\ M \end{aligned}$$

The calculations in 1 and 2 show that after treating permanent hard water with washing soda and slaked lime, the concentrations of Ca^{2+} and Mg^{2+} ions are so low that they will have a negligible effect on the action of soap.

Key Words

Complex ion, p. 735
Formation constant (K_f), p. 736
Molar solubility, p. 723
Solubility, p. 723
Solubility product, p. 720
Stability constant, p. 736

Exercises†

Solubility and Solubility Product

REVIEW QUESTIONS

17.1 Define solubility, molar solubility, and solubility product. Explain the difference between solubility and the solubility product of a slightly soluble substance such as $BaSO_4$.

17.2 Why do we usually not quote the K_{sp} values for soluble ionic compounds?

17.3 Write balanced equations and solubility product expressions for the solubility equilibria of the following compounds: (a) CuBr, (b) ZnC_2O_4, (c) Ag_2CrO_4, (d) Hg_2Cl_2, (e) $AuCl_3$, (f) $Mn_3(PO_4)_2$.

17.4 Write the solubility product expression for the ionic compound A_xB_y.

17.5 How can we predict whether a precipitate will form when two solutions are mixed?

† The temperature is assumed to be 25°C for all the problems.

PROBLEMS

17.6 Calculate the concentration of ions in the following saturated solutions:
(a) $[I^-]$ in AgI solution with $[Ag^+] = 9.1 \times 10^{-9}\ M$
(b) $[Al^{3+}]$ in $Al(OH)_3$ with $[OH^-] = 2.9 \times 10^{-9}\ M$

17.7 From the solubility data given, calculate the solubility products for the following compounds:
(a) SrF_2, 7.3×10^{-2} g/L
(b) $PbCrO_4$, 4.5×10^{-5} g/L
(c) Ag_3PO_4, 6.7×10^{-3} g/L

17.8 The molar solubility of $MnCO_3$ is $4.2 \times 10^{-6}\ M$. What is K_{sp} for this compound?

17.9 The solubility of an ionic compound MX (molar mass = 346 g) is 4.63×10^{-3} g/L. What is K_{sp} for the compound?

17.10 The solubility of an ionic compound M_2X_3 (molar mass = 288 g) is 3.6×10^{-17} g/L. What is K_{sp} for the compound?

17.11 Using data from Table 17.1, calculate the molar solubility of the following compounds: (a) $PbCO_3$, (b) CaF_2.

17.12 Calculate the solubility (in g/L) of Ag_2CO_3.

17.13 A sample of 20.0 mL of 0.10 M $Ba(NO_3)_2$ is added to 50.0 mL of 0.10 M Na_2CO_3. Will $BaCO_3$ precipitate?

17.14 If 2.00 mL of 0.200 M NaOH are added to 1.00 L of 0.100 M $CaCl_2$, will precipitation occur? [K_{sp} of $Ca(OH)_2 = 8.0 \times 10^{-6}$.]

17.15 What is the pH of a saturated zinc hydroxide solution?

17.16 The pH of a saturated solution of a metal hydroxide MOH is 9.68. Calculate the K_{sp} for the compound.

17.17 A volume of 75 mL of 0.060 M NaF is mixed with 25 mL of 0.15 M $Sr(NO_3)_2$. Calculate the concentrations in the final solution of NO_3^-, Na^+, Sr^{2+}, and F^-. (K_{sp} for $SrF_2 = 2.0 \times 10^{-10}$.)

Fractional Precipitation

PROBLEMS

17.18 Solid NaI is slowly added to a solution that is 0.010 M in Cu^+ and 0.010 M in Ag^+. (a) Which compound will begin to precipitate first? (b) Calculate $[Ag^+]$ when CuI just begins to precipitate. (c) What percent of Ag^+ remains in solution at this point?

17.19 The solubility products of AgCl and Ag_3PO_4 are 1.6×10^{-10} and 1.8×10^{-18}, respectively. (a) If Ag^+ is added (without changing the volume) to 1.00 L of solution containing 0.10 mol Cl^- and 0.10 mol PO_4^{3-}, which compound will precipitate first? (b) What is the concentration of the first anion just before the salt of the second anion begins to precipitate?

17.20 Find the approximate pH range suitable for the separation of Fe^{3+} and Zn^{2+} by precipitation of $Fe(OH)_3$ from a solution that is initially 0.010 M in Fe^{3+} and Zn^{2+}.

The Common Ion Effect

REVIEW QUESTIONS

17.21 How does the common ion effect for solubility equilibria differ from that in acid-base equilibria discussed in Chapter 16?

17.22 Use Le Chatelier's principle to explain the decrease in solubility of $CaCO_3$ in a Na_2CO_3 solution.

17.23 Use $CaCO_3$ and $Ca(NO_3)_2$ to show how the presence of a common ion can reduce the solubility of an insoluble salt.

17.24 The molar solubility of AgCl in $6.5 \times 10^{-3}\ M$ $AgNO_3$ is $2.5 \times 10^{-8}\ M$. In deriving K_{sp} from these data, which of the following assumptions are reasonable?
(a) K_{sp} is the same as solubility.
(b) K_{sp} of AgCl is the same in $6.5 \times 10^{-3}\ M$ $AgNO_3$ as in pure water.
(c) Solubility of AgCl is independent of the concentration of $AgNO_3$.
(d) $[Ag^+]$ in solution does not change significantly upon the addition of AgCl to $6.5 \times 10^{-3}\ M$ $AgNO_3$.
(e) $[Ag^+]$ in solution after the addition of AgCl to $6.5 \times 10^{-3}\ M$ $AgNO_3$ is the same as it would be in pure water.

PROBLEMS

17.25 Calculate the solubility in g/L of AgBr (a) in pure water and (b) in 0.0010 M NaBr.

17.26 The solubility product of $PbBr_2$ is 8.9×10^{-6}. Determine the molar solubility (a) in pure water, (b) in 0.20 M KBr solution, (c) in 0.20 M $Pb(NO_3)_2$ solution.

17.27 How many grams of $CaCO_3$ will dissolve in 3.0×10^2 mL of 0.050 M $Ca(NO_3)_2$?

17.28 Calculate the molar solubility of AgCl in a solution made by dissolving 10.0 g of $CaCl_2$ in 1.00 L of solution.

17.29 Calculate the molar solubility of $BaSO_4$ (a) in water and (b) in a solution containing 1.0 M SO_4^{2-} ions.

17.30 The molar solubility of $Pb(IO_3)_2$ in a 0.10 M $NaIO_3$ solution is 2.4×10^{-11} mol/L. What is K_{sp} for $Pb(IO_3)_2$?

pH and Solubility

PROBLEMS

17.31 Which of the following ionic compounds will be more soluble in acid solution than in water? (a) $BaSO_4$, (b) $PbCl_2$, (c) $Fe(OH)_3$, (d) $CaCO_3$

17.32 Which of the following substances will be more soluble in acid solution than in pure water? (a) CuI, (b) Ag_2SO_4, (c) $Zn(OH)_2$, (d) BaC_2O_4, (e) $Ca_3(PO_4)_2$

17.33 What is the pH of a saturated solution of aluminum hydroxide?

17.34 Calculate the molar solubility of $Fe(OH)_2$ at (a) pH 8.00 and (b) pH 10.00.

17.35 The solubility product of $Mg(OH)_2$ is 1.2×10^{-11}. What minimum OH^- concentration must be attained (for example, by adding NaOH) to make the Mg^{2+} concentration in a solution of $Mg(NO_3)_2$ less than 1.0×10^{-10} *M*?

17.36 Calculate whether or not a precipitate will form if 2.00 mL of 0.60 *M* NH_3 are added to 1.0 L of 1.0×10^{-3} *M* of $FeSO_4$.

Complex Ions

REVIEW QUESTIONS

17.37 Explain the formation of complexes in Table 17.4 in terms of Lewis acid-base theory.

17.38 Give an example to illustrate the general effect of complex ion formation on solubility.

PROBLEMS

17.39 Write the formation constant expressions for the following complex ions: (a) $Zn(OH)_4^{2-}$, (b) $Co(NH_3)_6^{3+}$, (c) HgI_4^{2-}.

17.40 Explain, with balanced ionic equations, why (a) CuI_2 dissolves in ammonia solution, (b) AgBr dissolves in NaCN solution, (c) $HgCl_2$ dissolves in KCl solution.

17.41 If 2.50 g of $CuSO_4$ are dissolved in 9.0×10^2 mL of 0.30 *M* NH_3, what are the concentrations of Cu^{2+}, $Cu(NH_3)_4^{2+}$, and NH_3 at equilibrium?

17.42 Calculate the concentrations of Cd^{2+}, $Cd(CN)_4^{2-}$, and CN^- at equilibrium when 0.50 g of $Cd(NO_3)_2$ dissolves in 5.0×10^2 mL of 0.50 *M* NaCN.

17.43 If NaOH is added to 0.010 *M* Al^{3+}, which will be the predominant species at equilibrium: $Al(OH)_3$ or $Al(OH)_4^-$? The pH of the solution is 14.00. [K_f for $Al(OH)_4^- = 2.0 \times 10^{33}$.]

17.44 Both Ag^+ and Zn^{2+} form complex ions with NH_3. Write balanced equations for the reactions. However, $Zn(OH)_2$ is soluble in 6 *M* NaOH, and AgOH is not. Explain.

17.45 Calculate the molar solubility of AgBr in a 1.0 *M* NH_3 solution.

Qualitative Analysis

REVIEW QUESTIONS

17.46 Outline the general principle of qualitative analysis.

17.47 Give two examples of metal ions in each group (1 through 5) in the qualitative analysis scheme.

PROBLEMS

17.48 In a group 1 analysis, a student obtained a precipitate containing both AgCl and $PbCl_2$. Suggest one reagent that would allow her to separate AgCl(*s*) from $PbCl_2$(*s*).

17.49 In a group 1 analysis, a student adds hydrochloric acid to the unknown solution to make $[Cl^-] = 0.15$ *M*. Some $PbCl_2$ precipitates. Calculate the concentration of Pb^{2+} remaining in solution.

17.50 The precipitation of metal sulfides in groups 2 and 3 can be represented by

$$M^{2+} + H_2S \rightleftharpoons MS + 2H^+$$

where the equilibrium constant K is given by $[H^+]^2/[M^{2+}][H_2S]$. (a) Show that K is equal to $K_wK_{a_1}/K_{sp}$, where K_w is the ion product of water, K_{a_1} is the first ionization constant of H_2S, and K_{sp} is the solubility product of MS. (*Hint:* To derive this expression, you need to multiply the numerator and denominator in K by $[OH^-][HS^-]$.) (b) Calculate the value of K for Sn^{2+} and Mn^{2+}.

17.51 Both KCl and NH_4Cl are white solids. Suggest one reagent that would allow you to distinguish between these two compounds.

Miscellaneous Problems

17.52 For which of the following reactions is the equilibrium constant called a solubility product:
(a) $Zn(OH)_2(s) + 2OH^-(aq) \rightleftharpoons Zn(OH)_4^{2-}(aq)$
(b) $3Ca^{2+}(aq) + 2PO_4^{3-}(aq) \rightleftharpoons Ca_3(PO_4)_2(s)$
(c) $CaCO_3(s) + 2H^+(aq) \rightleftharpoons Ca^{2+}(aq) + H_2O(l) + CO_2(g)$
(d) $PbI_2(s) \rightleftharpoons Pb^{2+}(aq) + 2I^-(aq)$

17.53 Describe a simple test that would allow you to distinguish between $AgNO_3(s)$ and $Cu(NO_3)_2(s)$.

17.54 Write the electron configuration for Al^{3+} and predict the geometry of the $Al(OH)_4^-$ ion.

17.55 Suggest a reagent that would allow you to separate Al^{3+} from Fe^{3+}.

17.56 All metal hydroxides react with acids. However, a number of them also react with bases. Such compounds are said to be amphoteric. Identify the amphoteric hydroxides among the following compounds: NaOH, $Al(OH)_3$, $Cu(OH)_2$, $Mg(OH)_2$, $Zn(OH)_2$, $Fe(OH)_3$, $Ni(OH)_2$. Write ionic equations for the complex ion formation between the amphoteric hydroxide and the OH^- ions. (*Hint:* See Section 15.5.)

17.57 The Hg_2^{2+} ion is highly unstable with respect to disproportionation in liquid ammonia: $Hg_2^{2+} \rightarrow Hg^{2+} + Hg$, but is relatively stable in water. Explain.

17.58 A solution contains 0.070 *M* Ca^{2+} and 0.070 *M* Mg^{2+}. Is it possible to precipitate 99.9 percent of the Ca^{2+} ions as $CaCO_3$ without precipitating $MgCO_3$?

Opposite page: The efficiency of converting heat to work, as in the case of a steam locomotive, is governed by the second law of thermodynamics.

C H A P T E R 18

ENTROPY, FREE ENERGY, AND EQUILIBRIUM

Thermodynamics is an extensive and far-reaching scientific discipline that deals with the interconversion of heat and other forms of energy. Thermodynamics enables us to use information gained from experiments on a system to draw conclusions about other aspects of the same system without further experimentation. For example, we saw in Chapter 6 that it is possible to calculate the heat of reaction from the standard enthalpies of formation of the reactant and product molecules. This chapter introduces the second law of thermodynamics and the Gibbs free-energy function. It also discusses the relationship between Gibbs free energy and chemical equilibrium.

18.1 Spontaneous Processes and Entropy

Spontaneous Processes

One of the main objectives in studying thermodynamics, as far as chemists are concerned, is to be able to predict whether or not a reaction will occur when reactants are brought together under a special set of conditions (for example, at a certain temperature, pressure, and concentration). A reaction that *does* occur under the given set of conditions is called a *spontaneous reaction*. If it does not occur under that set of conditions, it is said to be nonspontaneous. To look into the idea of spontaneity, let us consider the following physical and chemical processes:

- A waterfall runs downhill, but never up, spontaneously.
- A lump of sugar spontaneously dissolves in a cup of coffee, but dissolved sugar does not spontaneously reappear in its original form.
- Water freezes spontaneously below 0°C, and ice melts spontaneously above 0°C (at 1 atm).
- Heat flows from a hotter object to a colder one, but the reverse never happens spontaneously.
- The expansion of a gas in an evacuated bulb is a spontaneous process [Figure 18.1(a)]. The reverse process, that is, the gathering of all the molecules into one bulb, is not spontaneous [Figure 18.1(b)].
- A piece of sodium metal reacts violently with water to form sodium hydroxide and hydrogen gas. However, hydrogen gas does not react with sodium hydroxide to form water and sodium.
- Iron exposed to water and oxygen forms rust, but rust does not spontaneously change back to iron.

These examples and many others show that processes that occur spontaneously in one direction cannot also take place spontaneously in the opposite direction. However, an important question remains: Is there a thermodynamic quantity that can help us predict whether a process will occur spontaneously?

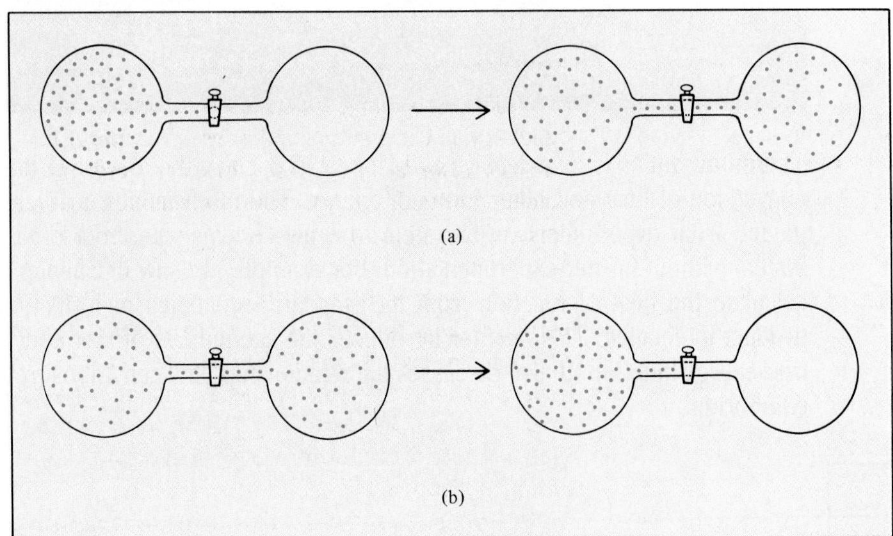

Figure 18.1 *(a) A spontaneous process. (b) A nonspontaneous process.*

It seems logical to assume that spontaneous processes occur so as to decrease the energy of a system. This assumption helps explain why things fall downward and why springs in a clock unwind. In chemical reactions, too, we find that a large number of exothermic reactions are spontaneous. An example is the combustion of methane:

$$CH_4(g) + 2O_2(g) \longrightarrow CO_2(g) + 2H_2O(l) \qquad \Delta H° = -890.4 \text{ kJ}$$

Another example is the acid-base neutralization reaction:

$$H^+(aq) + OH^-(aq) \longrightarrow H_2O(l) \qquad \Delta H° = -56.2 \text{ kJ}$$

But the assumption that spontaneous processes always decrease a system's energy fails in a number of cases. Consider a solid-to-liquid phase transition such as

$$H_2O(s) \longrightarrow H_2O(l) \qquad \Delta H° = 6.01 \text{ kJ}$$

Experience tells us that ice melts spontaneously above 0°C even though the process is endothermic. As another example, consider the cooling that results when ammonium nitrate dissolves in water:

$$NH_4NO_3(s) \xrightarrow{H_2O} NH_4^+(aq) + NO_3^-(aq) \qquad \Delta H° = 25 \text{ kJ}$$

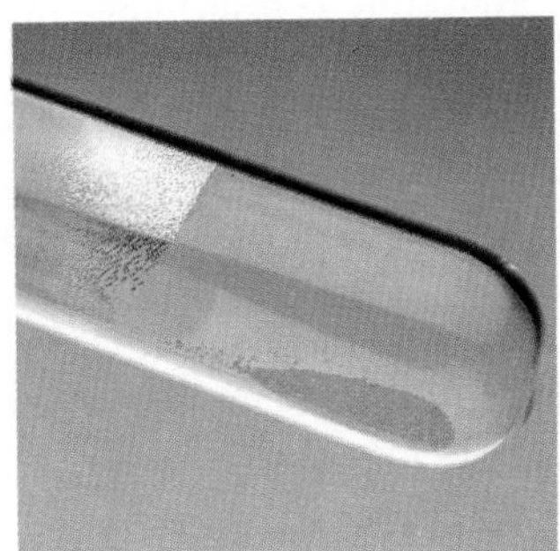

On heating, HgO decomposes to give Hg and O_2.

The dissolution process is spontaneous, yet it is also endothermic. The decomposition of mercury(II) oxide is an endothermic reaction that is nonspontaneous at room temperature but becomes spontaneous when the temperature is raised:

$$2HgO(s) \longrightarrow 2Hg(l) + O_2(g) \qquad \Delta H° = 90.7 \text{ kJ}$$

From a study of the examples mentioned and many more cases, we come to the following conclusion: Exothermicity favors the spontaneity of a reaction but does not guarantee it. It is possible for an exothermic reaction to be nonspontaneous, just as it is possible for an endothermic reaction to be spontaneous. Consideration of energy changes alone is not enough to predict spontaneity or nonspontaneity. Thus, it becomes necessary to look for another thermodynamic quantity (in addition to enthalpy) to help predict the direction of chemical reactions. This quantity turns out to be *entropy,* as we will see shortly.

Entropy

For most processes, especially those in condensed phase, ΔE and ΔH are practically the same.

In order to predict the spontaneity of a process, we need to know the change in both the enthalpy and the entropy of the system. ***Entropy* (*S*)** is *a direct measure of the randomness or disorder of a system.* In other words, entropy describes the extent to which atoms, molecules, or ions are distributed in a disorderly fashion in a given region in space. The greater the disorder of a system, the greater its entropy. Conversely, the more ordered a system, the smaller its entropy. At the macroscopic level, we can think of order and disorder in terms of a deck of playing cards. A new deck of cards is arranged in an ordered fashion (the cards are from ace to king and the suits are in the order of spades to hearts to diamonds to clubs), but once the deck is shuffled, the cards appear disordered both in numerical and in suit sequence. It is possible, but extremely unlikely, that after reshuffling that we restore the original order. There are many ways for the cards to be out of sequence but only one way for them to be ordered according to our definition.

The notion of order and disorder is related to probability. A probable event is one that can happen in many ways, and an improbable event is one that can happen in only one or a few ways. The idea of probability can be applied to the situation shown in Figure 18.1. Referring to Figure 18.1(a), suppose that initially we have only one molecule in the left flask. Since the two flasks have equal volume, the probability of finding the molecule in either flask after the stopcock is opened is 1/2. If the number of molecules is increased to two, the probability of finding both molecules in the *same* flask becomes 1/2 × 1/2. or 1/4. Now 1/4 is an appreciable quantity, so that it would not be surprising to find them in the same flask after opening the stopcock. However, it is not difficult to see that as the number of molecules increases, the probability (P) of finding all the molecules in the *same* flask becomes progressively smaller:

$$P = (\tfrac{1}{2}) \times (\tfrac{1}{2}) \times (\tfrac{1}{2}) \cdots$$
$$= (\tfrac{1}{2})^N$$

where N is the total number of molecules present. If $N = 100$, we have

$$P = (\tfrac{1}{2})^{100} = 8 \times 10^{-31}$$

If N is of the order of 6×10^{23}, the number of molecules in one mole of a gas, the probability becomes $(\frac{1}{2})^{6 \times 10^{23}}$, which is such a small number that, for all practical purposes, it can be regarded as zero. On the basis of probability consideration, we would expect the gas to fill both flasks spontaneously. By the same token, we now understand why the situation depicted in Figure 18.1(b) is not spontaneous, because it represents a highly improbable event. To summarize our discussion: An ordered state has a low probability of occurring and a small entropy, while a disordered state has a high probability of occurring and a large entropy.

It is possible to determine the *absolute* entropy of a substance, something we cannot do for energy or enthalpy, but the details of entropy measurements need not concern us here. A few general points about entropy serve our needs in this chapter:

- Listed entropy values are for substances at 1 atm and 25°C; these are called *standard* entropies ($S°$).
- Entropies of both elements and compounds are positive (that is, $S° > 0$). By contrast, the standard enthalpy of formation ($\Delta H°$) for elements in their stable forms is zero, but for compounds it may be positive or negative.
- The units of entropy are J/K or J/K · mol for 1 mole of the substance. (Because entropy values are usually quite small, we use joules rather than kilojoules.)

Consider the entropies of the following substances at 25°C:

Appendix 3 lists the standard entropy values of a number of elements and compounds.

Substance	$S°$ (J/K · mol)
$H_2O(l)$	69.9
$H_2O(g)$	188.7
$Br_2(l)$	152.3
$Br_2(g)$	245.3
$I_2(s)$	116.7
$I_2(g)$	260.6
C(diamond)	2.44
C(graphite)	5.69
He(g)	126.1
Ne(g)	146.2

For any substance, the particles in the solid are more ordered than those in the liquid state, which in turn are more ordered than those in the gaseous state. Water vapor has a greater entropy than liquid water, because a mole of water has a much smaller volume in the liquid state than it does in the gaseous state. Consequently, the water molecules are more ordered in the liquid state because there are fewer positions for them to occupy. Similarly, bromine vapor has a greater entropy than liquid bromine, and iodine vapor has a greater entropy than solid iodine. Thus for any substance, the entropy always increases in the following order:

The comparison is valid only for the same molar amount of the substance.

$$S_{\text{solid}} < S_{\text{liquid}} \ll S_{\text{gas}}$$

Both diamond and graphite are solids, but diamond has a more ordered structure (see Figure 11.29). Therefore, diamond has a smaller entropy than graphite. Neon has a greater entropy than helium because atoms that have more electrons have larger entropies than atoms with fewer electrons.

Like energy and enthalpy, entropy is a state function. Consider a certain process in which a system changes from some initial state to some final state. The entropy change for the process, ΔS, is

$$\Delta S = S_{\text{f}} - S_{\text{i}}$$

where S_{f} and S_{i} are the entropies of the system in the final and initial states, respectively. If the change results in an increase in randomness, or disorder, then $S_{\text{f}} > S_{\text{i}}$, or $\Delta S > 0$. Figure 18.2 shows several processes that lead to an increase in entropy. In

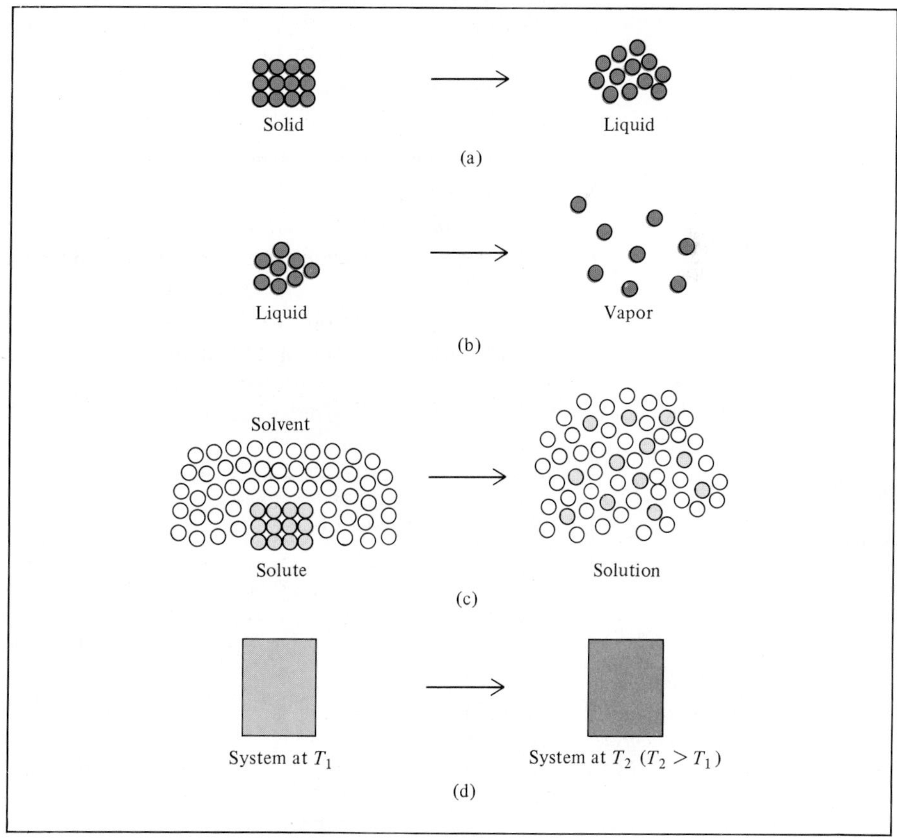

Figure 18.2 *Processes that lead to an increase in entropy of the system: (a) melting: $S_{liquid} > S_{solid}$; (b) vaporization: $S_{vapor} > S_{liquid}$; (c) dissolving: $S_{soln} > (S_{solvent} + S_{solute})$; (d) heating: $S_{T_2} > S_{T_1}$.*

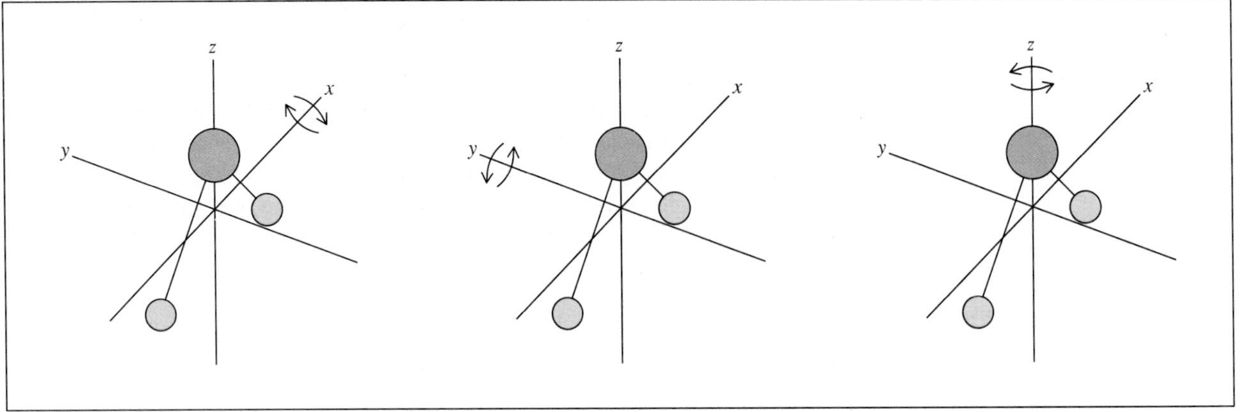

Figure 18.3 *Rotational motion of a molecule. A water molecule can rotate in three different ways about the x-, y-, and z-axes.*

each case you can see that the system changes from a more ordered state to a less ordered one. It is clear that both melting and vaporization processes have $\Delta S > 0$. When a solid solute dissolves in water, the highly ordered crystalline structure of the solid and part of the ordered structure of water are destroyed. Consequently, the solution possesses greater disorder than the pure solute and pure solvent. Heating also increases the entropy of a system. In addition to translational motion (that is, motion from one point to another), molecules can also execute rotational motion (Figure 18.3) and vibrational motion (Figure 18.4). The higher the temperature, the more energetic the various types of molecular motion. This means an increase in randomness at the molecular level, which corresponds to an increase in entropy.

The following example deals with the entropy changes of a system as a result of physical changes.

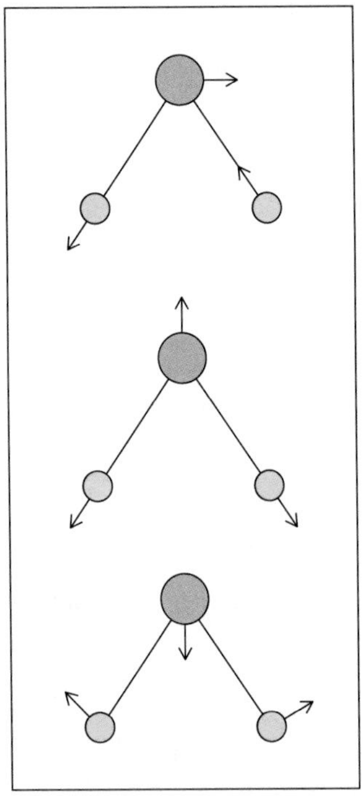

Figure 18.4 *The three different modes of vibration of a water molecule. Each mode of vibration can be imagined by moving the atoms along the arrows and then reversing their directions.*

EXAMPLE 18.1

Predict whether the entropy change is greater than or less than zero for each of the following processes: (a) freezing liquid bromine, (b) evaporating a beaker of ethanol at room temperature, (c) dissolving sucrose in water, (d) cooling nitrogen gas from 80°C to 20°C.

Answer

(a) This is a liquid-to-solid phase transition. The system becomes more ordered, so that $\Delta S < 0$.
(b) This is a liquid-to-vapor phase transition. The system becomes more disordered, and $\Delta S > 0$.
(c) A solution is invariably more disordered than its components (the solute and solvent). Therefore, $\Delta S > 0$.
(d) Cooling decreases molecular motion; therefore, $\Delta S < 0$.

Similar problem: 18.6.

18.2 The Second Law of Thermodynamics

The connection between entropy and spontaneity of a reaction is the subject of the ***second law of thermodynamics:*** *The entropy of the universe increases in a spontaneous process and remains unchanged in an equilibrium process.* Since the universe is made up of the system and the surroundings, the entropy change in the universe (ΔS_{univ}) for any process is the *sum* of the entropy changes in the system (ΔS_{sys}) and in the surroundings (ΔS_{surr}). Mathematically, we can express the second law of thermodynamics as follows:

$$\text{A spontaneous process:} \quad \Delta S_{univ} = \Delta S_{sys} + \Delta S_{surr} > 0 \tag{18.1}$$

$$\text{An equilibrium process:} \quad \Delta S_{univ} = \Delta S_{sys} + \Delta S_{surr} = 0 \tag{18.2}$$

For a spontaneous process this law says that ΔS_{univ} must be greater than zero, but it does not place a restriction on either ΔS_{sys} or ΔS_{surr}. Thus it is possible for either ΔS_{sys} or ΔS_{surr} to be negative, as long as the sum of these two quantities is greater than zero. For an equilibrium process, ΔS_{univ} is zero. In this case ΔS_{sys} and ΔS_{surr} must be equal in magnitude, but opposite in sign. What if for some process we find that S_{univ} is negative? What this means is that the process is not spontaneous in the direction described. Rather, the process is spontaneous in the *opposite* direction.

Entropy Changes in the System

To calculate S_{univ}, we need to know both ΔS_{sys} and ΔS_{surr}. Here we will focus on ΔS_{sys}. Let us suppose that the system is represented by the following reaction:

$$a\text{A} + b\text{B} \longrightarrow c\text{C} + d\text{D}$$

As in the case of calculating the enthalpy of a reaction [see Equation (6.8)], the standard entropy change $\Delta S°$ is given by

The units of $S°$ in this equation are J/K · mol.

$$\Delta S°_{rxn} = [cS°(\text{C}) + dS°(\text{D})] - [aS°(\text{A}) + bS°(\text{B})] \tag{18.3}$$

or, in general, using Σ to represent summation and m and n for the stoichiometric coefficients in the reaction,

$$\Delta S°_{rxn} = \Sigma nS°(\text{products}) - \Sigma mS°(\text{reactants}) \tag{18.4}$$

The standard entropy values of a large number of compounds have been measured; therefore, to calculate $\Delta S°_{rxn}$ (which is ΔS_{sys}), we look up their values in Appendix 3 and proceed according to the following example.

EXAMPLE 18.2

From the absolute entropy values in Appendix 3, calculate the standard entropy changes for the following reactions at 25°C.

(a) $CaCO_3(s) \longrightarrow CaO(s) + CO_2(g)$
(b) $N_2(g) + 3H_2(g) \longrightarrow 2NH_3(g)$
(c) $H_2(g) + Cl_2(g) \longrightarrow 2HCl(g)$

Answer

We can calculate $\Delta S°$ by using Equation (18.4).

(a) $\Delta S°_{rxn} = [S°(CaO) + S°(CO_2)] - [S°(CaCO_3)]$
$= (1 \text{ mol})(39.8 \text{ J/K} \cdot \text{mol}) + (1 \text{ mol})(213.6 \text{ J/K} \cdot \text{mol}) - (1 \text{ mol})(92.9 \text{ J/K} \cdot \text{mol})$
$= 160.5 \text{ J/K}$

Thus, when 1 mole of $CaCO_3$ decomposes to form 1 mole of CaO and 1 mole of gaseous CO_2, there is an increase in entropy equal to 160.5 J/K.

(b) $\Delta S°_{rxn} = [2S°(NH_3)] - [S°(N_2) + 3S°(H_2)]$
$= (2 \text{ mol})(193 \text{ J/K} \cdot \text{mol}) - [(1 \text{ mol})(192 \text{ J/K} \cdot \text{mol}) + (3 \text{ mol})(131 \text{ J/K} \cdot \text{mol})]$
$= -199 \text{ J/K}$

This result shows that when 1 mole of gaseous nitrogen reacts with 3 moles of gaseous hydrogen to form 2 moles of gaseous ammonia, there is a decrease in entropy equal to -199 J/K.

(c) $\Delta S°_{rxn} = [2S°(HCl)] - [S°(H_2) + S°(Cl_2)]$
$= (2 \text{ mol})(187 \text{ J/K} \cdot \text{mol}) - [(1 \text{ mol})(131 \text{ J/K} \cdot \text{mol}) + (1 \text{ mol})(223 \text{ J/K} \cdot \text{mol})]$
$= 20 \text{ J/K}$

Thus the formation of 2 moles of gaseous HCl from 1 mole of gaseous H_2 and 1 mole of gaseous Cl_2 results in a small increase in entropy equal to 20 J/K.

Similar problem: 18.10.

The results of Example 18.2 are consistent with many other reactions, which allow us to state the following general rules:

- If a reaction produces more gas molecules than it consumes [Example 18.2(a)], $\Delta S°$ is positive. If the total number of gas molecules diminishes [Example 18.2(b)], $\Delta S°$ is negative.
- If there is no net change in the total number of gas molecules [Example 18.2(c)], then $\Delta S°$ may be positive or negative, but will be relatively small numerically.

These conclusions are based on the fact that gases invariably have greater entropy than liquids and solids. For reactions involving only liquids and solids prediction is more difficult, but the general rule still holds in many cases that an increase in the total number of molecules and/or ions is accompanied by an increase in entropy.

The following example shows how knowing the nature of reactants and products makes it possible to predict entropy changes.

EXAMPLE 18.3

Predict whether the entropy change of the system in each of the following reactions is positive or negative.

(a) $Ag^+(aq) + Cl^-(aq) \longrightarrow AgCl(s)$
(b) $NH_4Cl(s) \longrightarrow NH_3(g) + HCl(g)$
(c) $H_2(g) + Br_2(g) \longrightarrow 2HBr(g)$

Answer

(a) The Ag^+ and Cl^- ions are free to move in solution, whereas AgCl is a solid. Furthermore, the number of particles decreases from left to right. Therefore, ΔS is negative.
(b) Since the solid is converted to two gaseous products, ΔS is positive.
(c) We see that the same number of moles of gas is involved in the reactants as in the product. Therefore we cannot predict the sign of ΔS but we know the change must be quite small.

Similar problems: 18.12, 18.13.

Entropy Changes in the Surroundings

Next we see how the ΔS_{surr} is calculated. When an exothermic process takes place in the system, the heat transferred to the surroundings will be used to enhance various types of motions of the molecules. Consequently, there is an increase in disorder at the molecular level and the entropy of the surroundings increases. Conversely, an endothermic process in the system absorbs heat from the surroundings and this will result in a decrease in the entropy of the surroundings because molecular motions become less energetic (Figure 18.5). For constant-pressure processes the heat change is equal to the enthalpy change of the system, ΔH_{sys}. The entropy change of the surroundings, ΔS_{surr}, is therefore proportional to ΔH_{sys}:

$$\Delta S_{surr} \propto -\Delta H_{sys}$$

The minus sign is needed here because if the process is exothermic, ΔH_{sys} is negative and ΔS_{surr} is a positive quantity which corresponds to an increase in entropy. On the other hand, for an endothermic process ΔH_{sys} is positive and the negative sign ensures that the entropy of the surroundings decreases.

The change in entropy for a given amount of heat also depends on the temperature. If the surroundings are at a high temperature, the various types of molecular motions are already sufficiently energetic. Therefore, the absorption of heat from an exothermic process in the system will have relatively little impact on molecular motions and the resulting increase in entropy will be small. However, if the temperature of the sur-

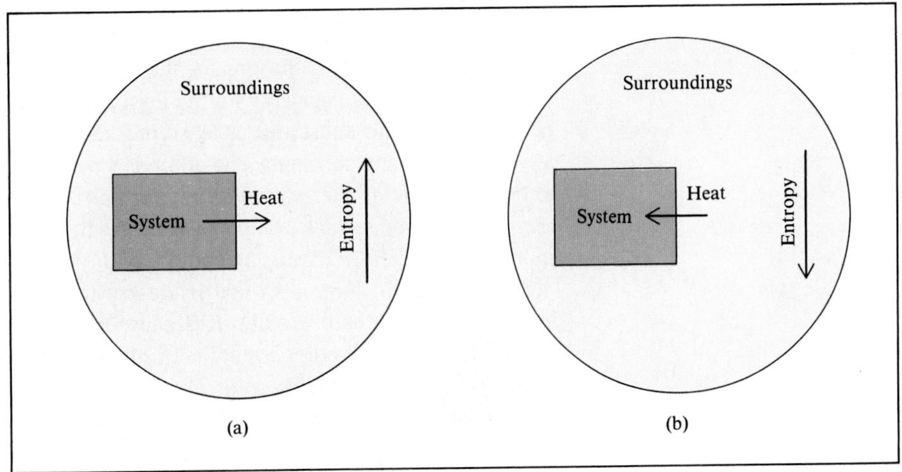

Figure 18.5 *(a) An exothermic process transfers heat from the system to the surroundings and results in an increase in the entropy of the surroundings. (b) An endothermic process absorbs heat from the surroundings and thereby decreases the entropy of the surroundings.*

roundings is low, then the addition of the same amount of heat will cause a more drastic increase in molecular motions and hence a larger increase in entropy. An analogy of the influence of temperature is someone coughing in a crowded restaurant, which will not disturb too many people, and someone coughing in a library, which definitely will. From the inverse relationship between ΔS_{surr} and temperature T (in kelvins), that is, the higher the temperature, the smaller the ΔS_{surr} and vice versa, we can rewrite the above relationship as

This equation, which can be derived from the laws of thermodynamics, assumes that both the system and the surroundings are at temperature *T*.

$$\Delta S_{surr} = \frac{-\Delta H_{sys}}{T} \tag{18.5}$$

Let us now apply the procedure for calculating ΔS_{sys} and ΔS_{surr} to the synthesis of ammonia and ask whether the reaction is spontaneous at 25°C:

$$N_2(g) + 3H_2(g) \longrightarrow 2NH_3(g) \qquad \Delta H^\circ = -92.6 \text{ kJ}$$

From Example 18.2(b) we have $\Delta S_{sys} = -199$ J/K, and substituting ΔH_{sys} (-92.6 kJ) in Equation (18.5), we obtain

$$\Delta S_{surr} = \frac{-(-92.6 \times 1000) \text{ J}}{298 \text{ K}} = 311 \text{ J/K}$$

The change in entropy of the universe is

$$\begin{aligned}\Delta S_{univ} &= \Delta S_{sys} + \Delta S_{surr}\\ &= -199 \text{ J/K} + 311 \text{ J/K}\\ &= 112 \text{ J/K}\end{aligned}$$

Because ΔS_{univ} is positive, we predict that the reaction is spontaneous at 25°C. It is important to keep in mind that just because a reaction is spontaneous does not mean that it will occur at an observable rate. The synthesis of ammonia is, in fact, extremely slow at room temperature. Thermodynamics can tell us whether a reaction will occur, but it does not say how fast it will occur. Reaction rates are the subject of chemical kinetics (see Chapter 13).

The Third Law of Thermodynamics and Absolute Entropy

Finally it is appropriate to mention briefly the *third law of thermodynamics,* which is related to the determination of entropy values. So far we have related entropy to molecular disorder—the greater the disorder or freedom of motion of the atoms or molecules in a system, the greater the entropy of the system. The most ordered arrangement of any substance with the least freedom of atomic or molecular motion is a perfect crystalline substance at absolute zero (0 K). It follows, therefore, that the lowest entropy any substance can attain is a perfect crystal at absolute zero. According to the ***third law of thermodynamics,*** *the entropy of a perfect crystalline substance is zero at the absolute zero of temperature*. As the temperature increases, the freedom of motion also increases. Thus the entropy of any substance at a temperature above 0 K is greater than zero. Note also that if the crystal is impure or if it has defects, then its entropy is greater than zero at 0 K because we do not have a perfectly ordered arrangement. The important point about the third law of thermodynamics is that it allows us to determine the *absolute* entropies of substances. Starting with the knowledge that the entropy of a pure crystalline substance is zero at absolute zero, we can measure the increase in entropy of the substance when it is heated to, say, 25°C. The change in entropy, ΔS, is given by

$$\Delta S = S_f - S_i$$
$$= S_f$$

since S_i is zero. The entropy of the substance at 25°C, then, is given by ΔS or S_f, which is called the absolute entropy because this is the *true* value and not a value derived using some arbitrary reference. Thus the entropy values quoted so far are all absolute entropies. In contrast, we cannot have the absolute energy or enthalpy of a substance because the zero of energy or enthalpy is undefined.

18.3 Gibbs Free Energy

The second law of thermodynamics tells us that to be spontaneous a reaction must lead to an increase in the entropy of the universe, that is, $\Delta S_{univ} > 0$. This is the criterion for spontaneity we were looking for. In order to determine the sign of ΔS_{univ} for a reaction, however, we would need to calculate both ΔS_{sys} and ΔS_{surr}. Therefore, because we are usually concerned only with what happens in a particular system rather than with what happens in the entire universe, and because the calculation of ΔS_{surr} often can be quite difficult, it is desirable to have another thermodynamic function that will help us determine whether a reaction will occur spontaneously even though we consider only the system itself.

We know that for a spontaneous process, we have

$$\Delta S_{univ} = \Delta S_{sys} + \Delta S_{surr} > 0$$

Substituting $-\Delta H_{sys}/T$ for ΔS_{surr}, we write

$$\Delta S_{univ} = \Delta S_{sys} - \frac{\Delta H_{sys}}{T} > 0$$

Multiplying both sides of the equation by T gives

$$T\Delta S_{univ} = -\Delta H_{sys} + T\Delta S_{sys} > 0$$

$6 > 0;\ -6 < 0.$

If we multiply the equation throughout by -1, the "$>$" sign is replaced by "$<$":

$$-T\Delta S_{univ} = \Delta H_{sys} - T\Delta S_{sys} < 0$$

Now we have a criterion for a spontaneous reaction that is expressed only in terms of the properties of the system and we no longer need be concerned with the surroundings.

We introduce a new thermodynamic function, called the ***Gibbs*† *free energy,*** or simply ***free energy,*** as follows:

$$G = H - TS \tag{18.6}$$

Remember that *all* the quantities in Equation (18.6) pertain to the system, and T is the temperature of the system. You can see that G has the units of energy (both H and TS have the units of energy) and, like H and S, it is a state function.

†Josiah Willard Gibbs (1839–1903). American physicist. One of the founders of thermodynamics, Gibbs is considered by many to be the most brilliant of the native-born American scientists (Benjamin Franklin comes second in this ranking). Gibbs was a modest and private individual, who spent almost all of his professional life at Yale University. Because he published most of his works in obscure journals, Gibbs never gained the eminence that his contemporary and admirer James Maxwell did in Europe. Even today, very few people outside of chemistry and physics have ever heard of Gibbs.

The change in free energy (ΔG) of a system for a process *at constant temperature* is given by

$$\Delta G = \Delta H - T\Delta S \tag{18.7}$$

Up to this point we have done nothing except reorganize the expression for the entropy change of the universe. Specifically, we have identified the change in free energy of the system with the $-T\Delta S_{univ}$ term. We can now summarize the conditions for spontaneity and equilibrium at constant temperature and pressure in terms of ΔG as follows:

$\Delta G < 0$ A spontaneous reaction.

$\Delta G > 0$ A nonspontaneous reaction. The reaction is spontaneous in the reverse direction.

$\Delta G = 0$ The system is at equilibrium. There is no net change.

The word "free" here does not mean without cost.

One way to appreciate the significance of free energy is to realize that the adjective "free" describes the usable, or available, energy—energy in forms that could be used to do work. Thus, if a particular reaction is accompanied by a release of usable energy (that is, if ΔG is negative), this fact alone guarantees that it must be spontaneous, and there is no need to worry about what happens to the rest of the universe.

Calculating Free-Energy Changes

Because free-energy change under conditions of constant pressure and temperature can be used to predict the outcome of a process, we will now look at ways to calculate its value. Consider the following reaction:

$$a\text{A} + b\text{B} \longrightarrow c\text{C} + d\text{D}$$

If the reaction is carried out under standard-state conditions, that is, reactants in their standard states are converted to products in their standard states, the free-energy change is called the standard free-energy change, ΔG°. Table 18.1 summarizes the conventions used by chemists to define the standard states of pure substances as well as solutions. For the above reaction the standard free-energy change is given by

$$\Delta G^\circ_{rxn} = [c\Delta G^\circ_f(\text{C}) + d\Delta G^\circ_f(\text{D})] - [a\Delta G^\circ_f(\text{A}) + b\Delta G^\circ_f(\text{B})]$$

or, in general,

$$\Delta G^\circ_{rxn} = \Sigma n\Delta G^\circ_f(\text{products}) - \Sigma m\Delta G^\circ_f(\text{reactants}) \tag{18.8}$$

where m and n are stoichiometric coefficients. The standard free energy of formation of a compound (ΔG°_f) is the free-energy change that occurs when 1 mole of the compound is synthesized from its elements in their standard states. For the combustion of graphite

$$\text{C(graphite)} + O_2(g) \longrightarrow CO_2(g)$$

the standard free-energy change [from Equation (18.8)] is

$$\Delta G^\circ_{rxn} = \Delta G^\circ_f(CO_2) - [\Delta G^\circ_f(\text{C}) + \Delta G^\circ_f(O_2)]$$

As in the case of the standard enthalpy of formation (p. 234), we define the standard free energy of formation of any element in its stable form as zero. Thus

$$\Delta G^\circ_f(\text{C}) = 0 \quad \text{and} \quad \Delta G^\circ_f(O_2) = 0$$

Table 18.1 Conventions for standard states

State of matter	Standard state
Gas	1 atm pressure
Liquid	Pure liquid
Solid	Pure solid
Elements*	ΔG°_f(element) $= 0$
Solution	1 molar concentration

*The most stable allotropic form at 25°C and 1 atm.

Note that ΔG° is in kJ, but ΔG_f° is in kJ/mol. The equation holds because the coefficient in front of ΔG_f° (1 in this case) has the unit "mol."

The standard free-energy change for the reaction is numerically equal to the standard free energy of formation of CO_2:

$$\Delta G^\circ_{rxn} = \Delta G_f^\circ(CO_2)$$

Appendix 3 lists the values of ΔG_f° for a number of compounds.

Calculations of standard free-energy changes are shown in the following example.

EXAMPLE 18.4

Calculate the standard free-energy changes for the following reactions at 25°C.

(a) $CH_4(g) + 2O_2(g) \longrightarrow CO_2(g) + 2H_2O(l)$
(b) $2MgO(s) \longrightarrow 2Mg(s) + O_2(g)$

Answer

(a) According to Equation (18.8), we write

$$\Delta G^\circ_{rxn} = [\Delta G_f^\circ(CO_2) + 2\Delta G_f^\circ(H_2O)] - [\Delta G_f^\circ(CH_4) + 2\Delta G_f^\circ(O_2)]$$

We insert the values from Appendix 3:

$$\begin{aligned}\Delta G^\circ_{rxn} &= [(1\text{ mol})(-394.4\text{ kJ/mol}) + (2\text{ mol})(-237.2\text{ kJ/mol})] \\ &\quad - [(1\text{ mol})(-50.8\text{ kJ/mol}) + (2\text{ mol})(0\text{ kJ/mol})] \\ &= -818.0\text{ kJ}\end{aligned}$$

(b) The equation is

$$\Delta G^\circ_{rxn} = [2\Delta G_f^\circ(Mg) + \Delta G_f^\circ(O_2)] - [2\Delta G_f^\circ(MgO)]$$

From data in Appendix 3 we write

$$\begin{aligned}\Delta G^\circ_{rxn} &= [(2\text{ mol})(0\text{ kJ/mol}) + (1\text{ mol})(0\text{ kJ/mol})] - [(2\text{ mol})(-569.6\text{ kJ/mol})] \\ &= 1139\text{ kJ}\end{aligned}$$

Similar problems: 18.18, 18.19.

In the preceding example the large negative value of ΔG° for the combustion of methane (a) means that the reaction is a highly spontaneous process under standard-state conditions, whereas the decomposition of MgO in (b) is nonspontaneous because ΔG° is a large, positive quantity. Remember, however, that a large, negative ΔG° does not tell us anything about the actual *rate* of the spontaneous process; a mixture of CH_4 and O_2 at 25°C could sit unchanged for quite some time in the absence of a spark or flame.

Applications of Equation (18.7)

Free energy is useful because it allows us to predict whether a given process is spontaneous. In order to predict the sign of ΔG, we need to determine both ΔH and ΔS, as shown by Equation (18.7). The factors that tend to make ΔG negative are a negative ΔH (an exothermic reaction) and a positive ΔS (a reaction that results in an increase in

Table 18.2 Factors affecting the sign of ΔG in the relationship $\Delta G = \Delta H - T\Delta S$

ΔH	ΔS	ΔG	Example
+	+	Reaction proceeds spontaneously at high temperatures. At low temperatures, reaction is spontaneous in the reverse direction.	$H_2(g) + I_2(g) \longrightarrow 2HI(g)$
+	−	ΔG is always positive. Reaction is spontaneous in the reverse direction at all temperatures.	$3O_2(g) \longrightarrow 2O_3(g)$
−	+	ΔG is always negative. Reaction proceeds spontaneously at all temperatures.	$2H_2O_2(l) \longrightarrow 2H_2O(l) + O_2(g)$
−	−	Reaction proceeds spontaneously at low temperatures. At high temperatures, the reverse reaction becomes spontaneous.	$NH_3(g) + HCl(g) \longrightarrow NH_4Cl(s)$

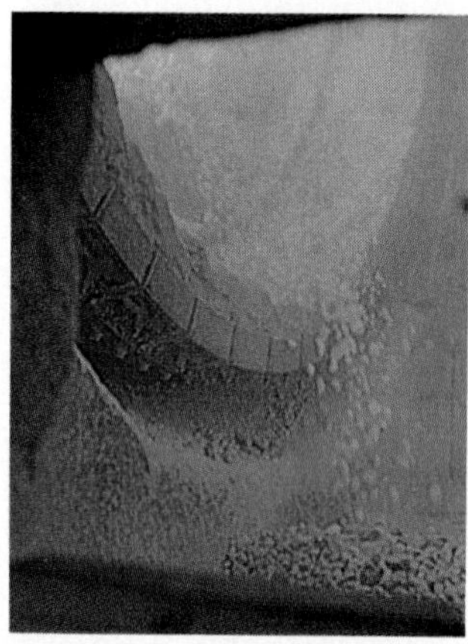

Figure 18.6 *The production of CaO in a rotary kiln.*

disorder of the system). In addition, temperature may also influence the direction of a spontaneous reaction. We discuss the four possibilities below.

- If both ΔH and ΔS are positive, then ΔG will be negative only when the $T\Delta S$ term is greater in magnitude than ΔH. This condition is met when T is large.
- If ΔH is positive and ΔS is negative, ΔG will always be positive, regardless of temperature.
- If ΔH is negative and ΔS is positive, then ΔG will always be negative regardless of temperature.
- If ΔH is negative and ΔS is negative, then ΔG will be negative only when $T\Delta S$ is smaller in magnitude than ΔH. This condition is met when T is small.

What temperatures will cause ΔG to be negative for the first and last cases depends on the actual values of ΔH and ΔS of the system. Table 18.2 summarizes the effects of the four possibilities just described.

We will now consider two specific applications of Equation (18.7).

Temperature and Chemical Reactions. Calcium oxide (CaO), also called quicklime, is an extremely valuable inorganic substance used in steelmaking, production of calcium metal, the paper industry, water treatment, and pollution control. It is prepared by decomposing limestone ($CaCO_3$) in a kiln at high temperatures (Figure 18.6):

$$CaCO_3(s) \rightleftharpoons CaO(s) + CO_2(g)$$

Le Chatelier's principle predicts that the forward, endothermic reaction is favored by heating.

The reaction is reversible and CaO readily combines with CO_2 to form $CaCO_3$. The pressure of CO_2 in equilibrium with $CaCO_3$ and CaO increases with temperature. In actual production the system is never maintained at equilibrium; rather, CO_2 is constantly removed from the kiln to shift the equilibrium from left to right, promoting the formation of calcium oxide.

The important information for the practical chemist is the temperature at which the decomposition of $CaCO_3$ becomes appreciable (that is, at what temperature does the reaction become spontaneous). We can make a reliable estimate of that temperature as follows. First we calculate $\Delta H°$ [using Equation (6.9)] and $\Delta S°$ [using Equation (18.4)] for the reaction at 25°C, using the data in Appendix 3.

$$\begin{aligned}\Delta H° &= [\Delta H_f°(CaO) + \Delta H_f°(CO_2)] - [\Delta H_f°(CaCO_3)] \\ &= [(1\ \text{mol})(-635.6\ \text{kJ/mol}) + (1\ \text{mol})(-393.5\ \text{kJ/mol})] - [(1\ \text{mol})(-1206.9\ \text{kJ/mol})] \\ &= 177.8\ \text{kJ}\end{aligned}$$

$$\begin{aligned}\Delta S^\circ &= [S^\circ(\text{CaO}) + S^\circ(\text{CO}_2)] - [S^\circ(\text{CaCO}_3)]\\ &= [(1\text{ mol})(39.8\text{ J/K}\cdot\text{mol}) + (1\text{ mol})(213.6\text{ J/K}\cdot\text{mol})] - [(1\text{ mol})(92.9\text{ J/K}\cdot\text{mol})]\\ &= 160.5\text{ J/K}\end{aligned}$$

For reactions carried out under standard-state conditions, Equation (18.7) takes the form

$$\Delta G^\circ = \Delta H^\circ - T\Delta S^\circ$$

so we obtain

$$\begin{aligned}\Delta G^\circ &= 177.8\text{ kJ} - (298\text{ K})(160.5\text{ J/K})\left(\frac{1\text{ kJ}}{1000\text{ J}}\right)\\ &= 130.0\text{ kJ}\end{aligned}$$

Since ΔG° is a large positive quantity, we conclude that the reaction is not favored at 25°C (or 298 K). In order to make ΔG° negative, we first have to find the temperature at which ΔG° is zero; that is,

$$0 = \Delta H^\circ - T\Delta S^\circ$$

or

$$\begin{aligned}T &= \frac{\Delta H^\circ}{\Delta S^\circ}\\ &= \frac{(177.8\text{ kJ})(1000\text{ J/1 kJ})}{160.5\text{ J/K}}\\ &= 1108\text{ K}\\ &= 835^\circ\text{C}\end{aligned}$$

At a temperature higher than 835°C, ΔG° becomes negative, indicating that the decomposition is spontaneous. For example, at 840°C or 1113 K

$$\begin{aligned}\Delta G^\circ &= \Delta H^\circ - T\Delta S^\circ\\ &= 177.8\text{ kJ} - (1113\text{ K})(160.5\text{ J/K})\left(\frac{1\text{ kJ}}{1000\text{ J}}\right)\\ &= -0.8\text{ kJ}\end{aligned}$$

Two points are worth making about such a calculation. First, we used the ΔH° and ΔS° values at 25°C to calculate changes that occur at a much higher temperature. Since both ΔH° and ΔS° change with temperature, this approach will not give us an accurate value of ΔG°, but it is good enough for "ball park" estimates. Second, we should not be misled into thinking that nothing happens below 835°C and that at 835°C $CaCO_3$ suddenly begins to decompose. Far from it. The fact that ΔG° is a positive value at some temperature below 835°C does not mean that no CO_2 is produced. It only says that the pressure of the CO_2 gas formed at that temperature will be below 1 atm (its standard state; see Table 18.1). As Figure 18.7 shows, the pressure of CO_2 at first increases very slowly with temperature; it becomes easily measurable above 700°C. The significance of 835°C is that this is the temperature at which the equilibrium pressure of CO_2 reaches 1 atm. Above 835°C, the equilibrium pressure of CO_2 exceeds 1 atm. (In Section 18.4 we will see how ΔG° is related to the equilibrium constant of a given reaction.)

The equilibrium constant of this reaction is $K_P = P_{CO_2}$.

Phase Transitions. Equation (18.7) can also be applied to phase transitions. At the transition temperature (that is, at the melting point or the boiling point) the system is at equilibrium ($\Delta G = 0$), so Equation (18.7) becomes

$$0 = \Delta H - T\Delta S$$

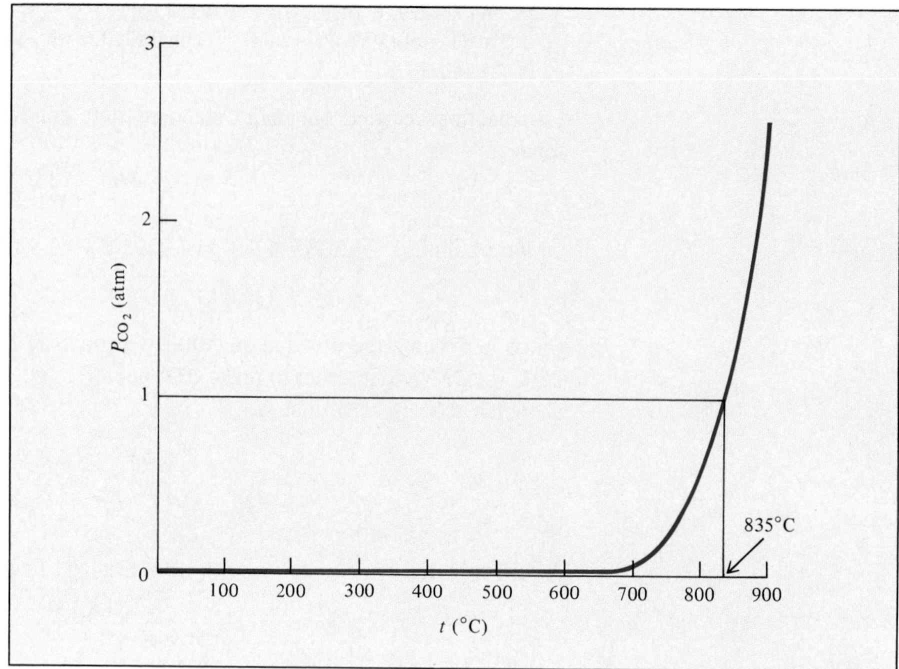

Figure 18.7 *Equilibrium pressure of CO_2 from the decomposition of $CaCO_3$, as a function of temperature. This curve is calculated by assuming that $\Delta H°$ and $\Delta S°$ of the reaction do not change with temperature.*

$$\Delta S = \frac{\Delta H}{T}$$

The melting of ice is an endothermic process (ΔH is positive), and the freezing of water is exothermic (ΔH is negative).

Let us first consider the ice-water equilibrium. For the ice → water transition, ΔH is the molar heat of fusion (see Table 11.9) and T is the melting point. The entropy change is therefore

$$\Delta S_{\text{ice}\rightarrow\text{water}} = \frac{6010 \text{ J/mol}}{273 \text{ K}}$$
$$= 22.0 \text{ J/K} \cdot \text{mol}$$

Thus when 1 mole of ice melts at 0°C, there is an increase in entropy of 22.0 J/K. The increase in entropy is consistent with the increase in disorder from solid to liquid. Conversely, for the water → ice transition, the decrease in entropy is given by

$$\Delta S_{\text{water}\rightarrow\text{ice}} = \frac{-6010 \text{ J/mol}}{273 \text{ K}}$$
$$= -22.0 \text{ J/K} \cdot \text{mol}$$

In the laboratory we normally carry out unidirectional changes, that is, either ice to water or water to ice. We can calculate entropy changes using the equation $\Delta S = \Delta H/T$ as long as the temperature remains 0°C. The same procedure can be applied to the water-steam transition. In this case ΔH is the heat of vaporization and T is the boiling point of water. We make these calculations for benzene in the following example.

EXAMPLE 18.5

The molar heats of fusion and vaporization of benzene are 10.9 kJ/mol and 31.0 kJ/mol, respectively. Calculate the entropy changes for the solid → liquid and liquid → vapor transitions for benzene. At 1 atm pressure, benzene melts at 5.5°C and boils at 80.1°C.

Answer

At the melting point, the system is at equilibrium. Therefore $\Delta G = 0$ and the entropy of fusion is given by

$$0 = \Delta H_{\text{fus}} - T_{\text{m.p.}}\,\Delta S_{\text{fus}}$$

$$\Delta S_{\text{fus}} = \frac{\Delta H_{\text{fus}}}{T_{\text{m.p.}}} = \frac{(10.9\ \text{kJ/mol})(1000\ \text{J}/1\ \text{kJ})}{(5.5 + 273)\ \text{K}} = 39.1\ \text{J/K}\cdot\text{mol}$$

Similarly, at the boiling point, $\Delta G = 0$ and we have

$$\Delta S_{\text{vap}} = \frac{\Delta H_{\text{vap}}}{T_{\text{b.p.}}} = \frac{(31.0\ \text{kJ/mol})(1000\ \text{J}/1\ \text{kJ})}{(80.1 + 273)\ \text{K}} = 87.8\ \text{J/K}\cdot\text{mol}$$

Since vaporization creates more disorder than the melting process, $\Delta S_{\text{vap}} > \Delta S_{\text{fus}}$.

Similar problem: 18.22.

18.4 Free Energy and Chemical Equilibrium

In this section we take a closer look at the difference between ΔG and $\Delta G°$ and at the relationship between free-energy changes and equilibrium constants.

The equations for free-energy change and standard free-energy change are, respectively,

$$\Delta G = \Delta H - T\Delta S$$

$$\Delta G° = \Delta H° - T\Delta S°$$

It is important that we understand the conditions under which these equations should be used and what kind of information can be obtained from ΔG and $\Delta G°$. Let us illustrate by considering the following reaction:

$$\text{reactants} \longrightarrow \text{products}$$

The standard free-energy change for the reaction is given by

$$\Delta G° = G°(\text{products}) - G°(\text{reactants})$$

The quantity $\Delta G°$ represents the change when the reactants in their standard states are converted to products in their standard states (see Table 18.1 for definitions of standard states). We have already seen in Example 18.4 how $\Delta G°$ for a reaction is calculated from the known standard free energies of formation of reactants and products. Suppose we start the reaction in solution with all the reactants in their standard states (that is, all at 1 *M* concentration). As soon as the reaction starts, the standard-state condition no longer exists for the reactants or the products since neither remains at 1 *M* concentration. Under conditions that are *not* standard state, we must use ΔG rather than $\Delta G°$ to predict the direction of the reaction. The relationship between ΔG and $\Delta G°$ is

$$\Delta G = \Delta G° + RT \ln Q \qquad (18.9)$$

where R is the gas constant (8.314 J/K · mol), T the absolute temperature of the reaction, and Q the reaction quotient defined on p. 611. We see that ΔG depends on two quantities: $\Delta G°$ and $RT \ln Q$. For a given reaction at temperature T the value of $\Delta G°$ is fixed but that of $RT \ln Q$ is not, because Q varies according to the composition of the reacting mixture. Let us consider the two cases:

Case 1: If $\Delta G°$ is a large negative value, the $RT \ln Q$ term will not become positive enough to match the $\Delta G°$ term until significant product formation has occurred.

Case 2: If $\Delta G°$ is a large positive value, the $RT \ln Q$ term will be more negative than $\Delta G°$ is positive only as long as very little product formation has occurred and the concentration of the reactant is high relative to that of the product.

Sooner or later a reversible reaction will reach equilibrium.

At equilibrium, by definition, $\Delta G = 0$ and $Q = K$, where K is the equilibrium constant. Thus

$$0 = \Delta G° + RT \ln K$$

In this equation K_P is used for gases and K_c for reactions in solution. Note that the larger the K, the more negative the $\Delta G°$.

or

$$\Delta G° = -RT \ln K \qquad (18.10)$$

For chemists, Equation (18.10) is one of the most important equations in thermodynamics because it relates the equilibrium constant of a reaction to the change in standard free energy. Thus, from a knowledge of K we can calculate $\Delta G°$, and vice versa.

For reactions having very large or very small equilibrium constants, it is generally very difficult, if not impossible, to measure the K values by monitoring the concentrations of all the reacting species. Consider, for example, the formation of nitric oxide from molecular nitrogen and molecular oxygen:

$$N_2(g) + O_2(g) \rightleftharpoons 2NO(g)$$

At 25°C, the equilibrium constant K_c is

$$K_c = \frac{[NO]^2}{[N_2][O_2]} = 4.0 \times 10^{-31}$$

The very small value of K_c means that the concentration of NO at equilibrium will be exceedingly low. In such a case the equilibrium constant is more conveniently obtained from the $\Delta G°$ of the reaction. (As we have seen, $\Delta G°$ can be calculated from the $\Delta H°$ and $\Delta S°$ values.) On the other hand, the equilibrium constant for the formation of hydrogen iodide from molecular hydrogen and molecular iodine is near unity at room temperature:

$$H_2(g) + I_2(g) \rightleftharpoons 2HI(g)$$

This makes it easier to measure K for this reaction and then calculate $\Delta G°$ using Equation (18.10) than to measure $\Delta H°$ and $\Delta S°$ of the reaction.

Figure 18.8 shows plots of free energy of a reacting system versus the extent of the reaction for two types of reactions. As you can see, if $\Delta G° < 0$, the products are

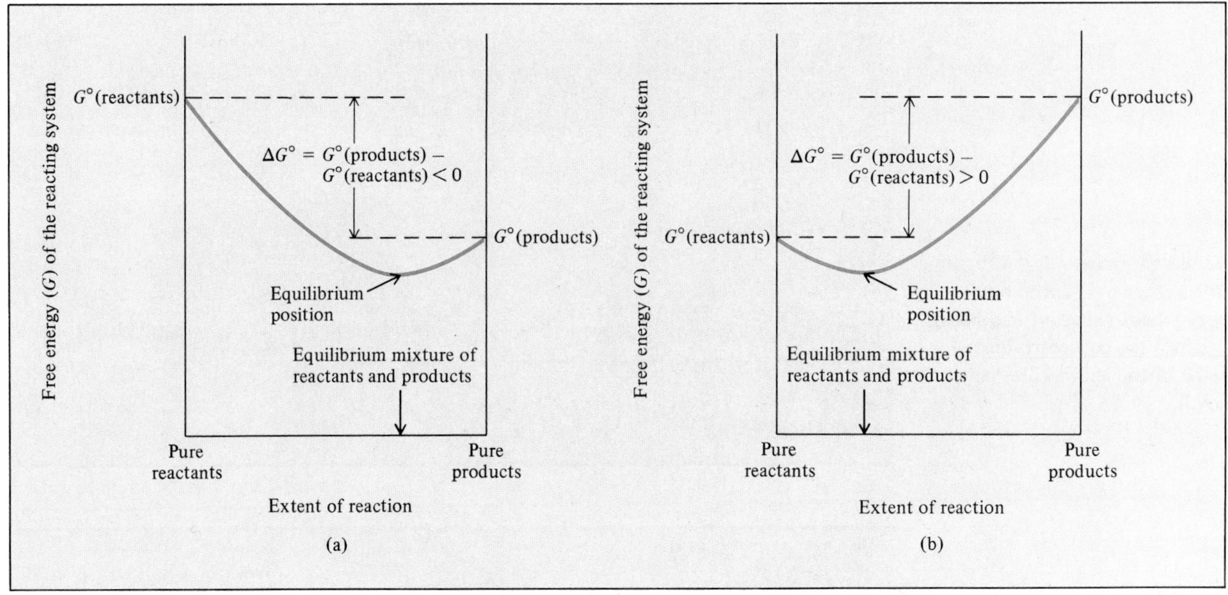

Figure 18.8 *(a)* $G^\circ_{products} < G^\circ_{reactants}$, *so* $\Delta G^\circ < 0$. *At equilibrium, there is a significant conversion of reactants to products. (b)* $G^\circ_{products} > G^\circ_{reactants}$, *so* $\Delta G^\circ > 0$. *At equilibrium, reactants are favored over products.*

Table 18.3 Relation between ΔG° and K as predicted by the equation $\Delta G^\circ = -RT \ln K$

K	$\ln K$	ΔG°	Comments
>1	Positive	Negative	Products are favored over reactants at equilibrium.
$=1$	0	0	Products and reactants are equally favored at equilibrium.
<1	Negative	Positive	Reactants are favored over products at equilibrium.

favored over reactants at equilibrium. Conversely, if $\Delta G^\circ > 0$, there will be more reactants than products at equilibrium.

Table 18.3 summarizes the three possible relations between ΔG° and K, as predicted by Equation (18.10).

The following examples illustrate the use of Equations (18.9) and (18.10).

EXAMPLE 18.6

Using data listed in Appendix 3, calculate the equilibrium constant for the following reaction at 25°C:

$$2H_2O(l) \rightleftharpoons 2H_2(g) + O_2(g)$$

Answer

According to Equation (18.8)

$$\begin{aligned}\Delta G^\circ_{rxn} &= [2\Delta G^\circ_f(H_2) + \Delta G^\circ_f(O_2)] - [2\Delta G^\circ_f(H_2O)] \\ &= [(2\text{ mol})(0\text{ kJ/mol}) + (1\text{ mol})(0\text{ kJ/mol})] - [(2\text{ mol})(-237.2\text{ kJ/mol})] \\ &= 474.4\text{ kJ}\end{aligned}$$

Using Equation (18.8) $\Delta G^\circ_{rxn} = -RT \ln K$

$$474.4 \text{ kJ} \times \frac{1000 \text{ J}}{1 \text{ kJ}} = -(8.314 \text{ J/K} \cdot \text{mol})(298 \text{ K}) \ln K$$

$$\ln K = -191.5$$

$$K = e^{-191.5}$$

$$= 7 \times 10^{-84}$$

As stated earlier, it is difficult to measure very large (and very small) values of K directly because the concentrations of some of the species are very small.

This extremely small equilibrium constant is consistent with the fact that water does not decompose into hydrogen and oxygen gases at 25°C. Thus a large positive ΔG° favors reactants over products at equilibrium.

Similar problems: 18.26, 18.29.

EXAMPLE 18.7

In Chapter 17 we discussed the solubility product of slightly soluble substances. Using the solubility product of silver chloride at 25°C (1.6×10^{-10}), calculate ΔG° for the process

$$AgCl(s) \rightleftharpoons Ag^+(aq) + Cl^-(aq)$$

Answer

Because this is a heterogeneous equilibrium, the solubility product is the equilibrium constant.

$$K_{sp} = [Ag^+][Cl^-] = 1.6 \times 10^{-10}$$

Using Equation (18.10) we obtain

$$\Delta G^\circ = -(8.314 \text{ J/K} \cdot \text{mol})(298 \text{ K}) \ln 1.6 \times 10^{-10}$$
$$= 5.6 \times 10^4 \text{ J}$$
$$= 56 \text{ kJ}$$

The large, positive ΔG° is consistent with the fact that AgCl is slightly soluble and that the equilibrium lies mostly to the left.

Similar problem: 18.28.

EXAMPLE 18.8

The standard free-energy change for the reaction

$$N_2(g) + 3H_2(g) \rightleftharpoons 2NH_3(g)$$

is −33.2 kJ and the equilibrium constant K_p is 6.59×10^5 at 25°C. In a certain experiment, the initial pressures are $P_{H_2} = 0.250$ atm, $P_{N_2} = 0.870$ atm, and $P_{NH_3} = 12.9$ atm. Calculate ΔG for the reaction at these pressures, and predict the direction of reaction.

Answer

Equation (18.9) can be written as

$$\begin{aligned}\Delta G &= \Delta G^\circ + RT \ln Q_P \\ &= \Delta G^\circ + RT \ln \frac{P_{NH_3}^2}{P_{H_2}^3 P_{N_2}} \\ &= -33.2 \times 1000\text{ J} + (8.314\text{ J/K}\cdot\text{mol})(298\text{ K}) \times \ln \frac{(12.9)^2}{(0.250)^3(0.870)} \\ &= -33.2 \times 10^3\text{ J} + 23.3 \times 10^3\text{ J} \\ &= -9.9 \times 10^3\text{ J} = -9.9\text{ kJ}\end{aligned}$$

Since ΔG is negative, the net reaction proceeds from left to right. As an exercise, confirm this prediction by calculating the reaction quotient Q_P and comparing it with the equilibrium constant K_P.

Similar problems: 18.33, 18.34.

CHEMISTRY IN ACTION

The Thermodynamics of a Rubber Band

We all know how useful a rubber band can be. But not everyone is aware that a rubber band has some very interesting thermodynamic properties based on its structure.

You can easily perform the following experiments. Obtain a rubber band at least 0.5 cm wide. Quickly stretch the rubber band and then press it against your lips. You will feel a slight warming effect. You can also carry out the reverse process. First, stretch a rubber band and hold it in position for a few seconds. Then quickly release the tension and press the rubber band against your lips. This time you will feel a slight cooling effect. A thermodynamic analysis of these two experiments can tell us something about the molecular structure of rubber (Figure 18.9).

Rearrangement of Equation (18.7) ($\Delta G = \Delta H - T\Delta S$) gives

$$T\Delta S = \Delta H - \Delta G$$

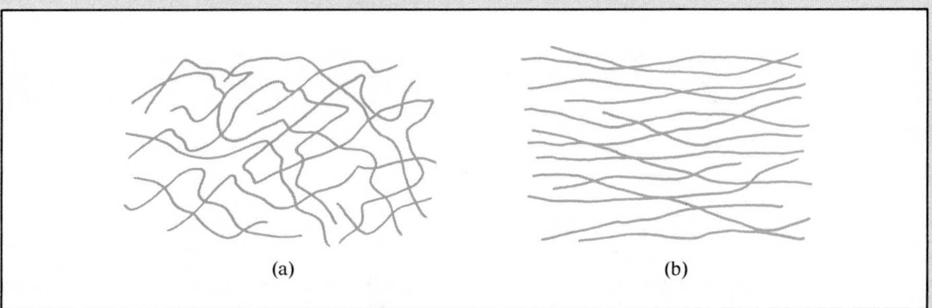

Figure 18.9 *(a) Rubber molecules in their normal state. Note the high degree of disorder (high entropy). (b) Under tension, the molecules line up and the arrangement becomes much more ordered (low entropy).*

The warming effect (an exothermic process) due to stretching means that $\Delta H < 0$, and since stretching is nonspontaneous (that is, $\Delta G > 0$ and $-\Delta G < 0$), $T\Delta S$ must be negative. Since T, the absolute temperature, is always positive, we conclude that ΔS due to stretching must be negative. This observation tells us that rubber in its natural state is more disordered than when it is under tension.

When the tension is removed, the stretched rubber band spontaneously snaps back to its original shape; that is, ΔG is negative and $-\Delta G$ is positive. The cooling effect means that it is an endothermic process ($\Delta H > 0$) so that $T\Delta S$ is positive. Thus the entropy of the rubber band increases when it goes from the stretched state to the natural state.

Another way to study the thermodynamic properties of a rubber band is as follows. Figure 18.10(a) shows a weight attached to a rubber band. When the rubber band is warmed by a hair dryer, it contracts [Figure 18.10(b)]! Since heat is absorbed by the rubber band, ΔH is positive, and the contraction occurs spontaneously so that ΔG is negative and $-\Delta G$ is positive. Thus ΔS is positive for contraction, which is consistent with what we now know about the structure of rubber.

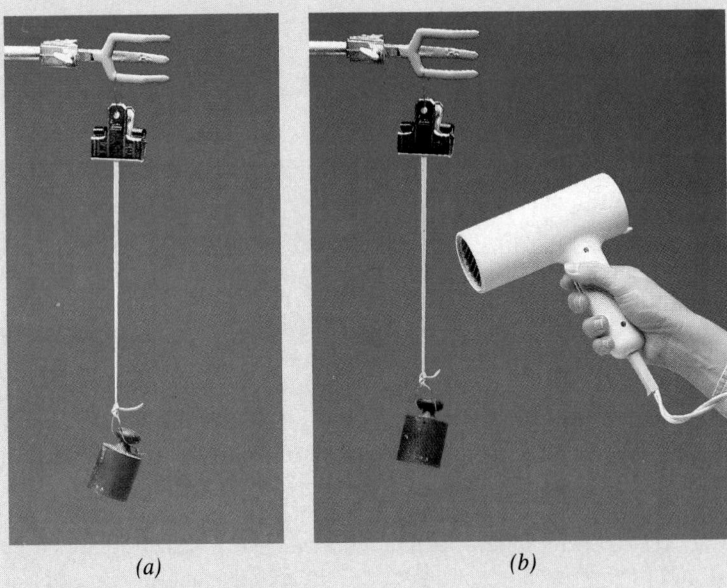

(a) (b)

Figure 18.10 *(a) A stretched piece of rubber band. (b) When heated, the rubber band contracts spontaneously.*

CHEMISTRY IN ACTION

The Efficiency of Heat Engines

An engine is a machine that converts energy to work; a *heat engine* is a machine that converts *thermal energy* to work. Heat engines play an essential role in our technological society; they range from automobile engines to the giant steam turbines that run generators to produce electricity. Regardless of the type of heat engine, its efficiency level is of great importance; that is, for a given amount of heat input, how much useful work can

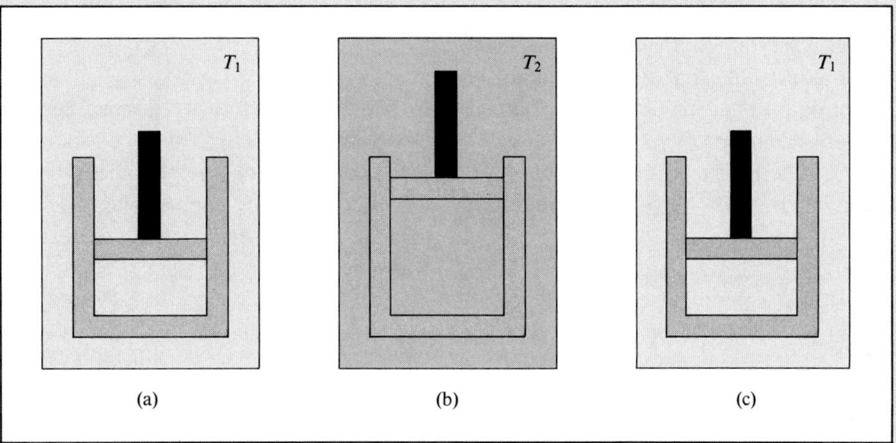

Figure 18.11 *A simple heat engine. (a) The engine is initially at T_1. (b) When heated to T_2, the piston is pushed up due to gas expansion. (c) When cooled to T_1, the piston returns to the original position.*

we get out of the engine? The second law of thermodynamics helps us answer this question.

Figure 18.11 shows a simple form of a heat engine. A cylinder fitted with a weightless piston is initially at temperature T_1. Next, the cylinder is heated to a higher temperature T_2. The gas in the cylinder expands, so the piston is pushed up. Finally the cylinder is cooled down to T_1 and the apparatus returns to its original state. By repeating this cycle, the up-and-down movement of the piston can be made to do mechanical work.

A unique feature of heat engines is that some heat must be given off to the surroundings when they do work. With the piston in the up position [Figure 18.11(b)], no further work can be done if we do not cool the cylinder back to T_1. The cooling process removes some of the thermal energy that could otherwise be converted to work and thereby places a limit on the efficiency of heat engines.

Figure 18.12 shows the heat transfer processes in a heat engine. Initially, a certain amount of heat flows from the heat reservoir (at temperature T_h) into the engine. As the engine does work, some of the heat is given off to the surroundings or the heat sink (at temperature T_c). By definition, the efficiency of a heat engine is

$$\text{efficiency} = \frac{\text{useful work output}}{\text{energy input}} \times 100\%$$

Analysis based on the second law shows that this efficiency can also be expressed as

$$\text{efficiency} = \left(1 - \frac{T_\mathrm{c}}{T_\mathrm{h}}\right) \times 100\%$$
$$= \frac{T_\mathrm{h} - T_\mathrm{c}}{T_\mathrm{h}} \times 100\%$$

Thus the efficiency of a heat engine is given by the difference in temperatures between the heat reservoir

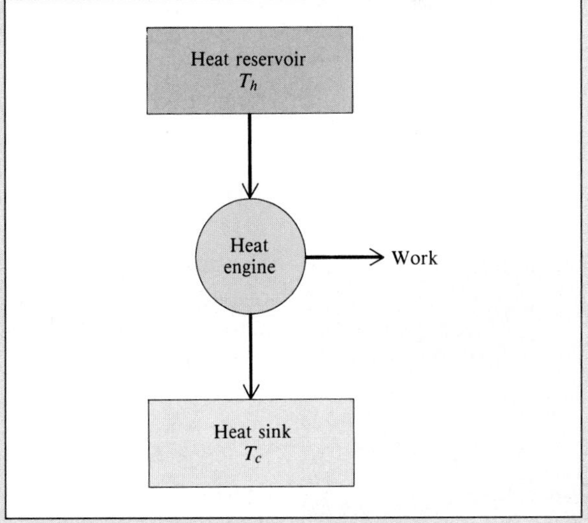

Figure 18.12 *Heat transfers during the operation of a heat engine.*

and the heat sink (both in kelvins), divided by the temperature of the heat reservoir. In practice we can make $(T_h - T_c)$ as large as possible, but since T_c cannot be zero and T_h cannot be infinite, the efficiency of a heat engine can therefore never be 100 percent.

At a power plant, superheated steam (at about 560°C or 833 K) is used to drive the turbine for electricity generation. The temperature of the heat sink is about 38°C (or 311 K). The efficiency is given by

$$\text{efficiency} = \frac{833\text{ K} - 311\text{ K}}{833\text{ K}} \times 100\%$$
$$= 63\%$$

In practice, because of friction, heat loss, and other complications, the maximum efficiency of a steam turbine is only about 40 percent. This means that for every ton of coal used at a power plant, 0.40 ton of it is used to generate electricity while the rest of it ends up in warming the surroundings!

Summary

1. Entropy is a measure of the disorder of a system. Any spontaneous process must lead to a net increase in entropy in the universe (second law of thermodynamics) or, stated mathematically, $\Delta S_{univ} = \Delta S_{sys} + \Delta S_{surr} > 0$.
2. The standard entropy change of a chemical reaction can be calculated from the absolute entropies of reactants and products.
3. The third law of thermodynamics states that the entropy of a perfect crystalline substance is zero at the absolute zero of temperature. This law enables us to measure the absolute entropies of substances.
4. Under conditions of constant temperature and pressure, the free-energy change ΔG is less than zero for a spontaneous process and greater than zero for a nonspontaneous process. For an equilibrium process, $\Delta G = 0$.
5. For a chemical or physical process at constant temperature and pressure, $\Delta G = \Delta H - T\Delta S$. This equation can be used to predict the spontaneity of a process.
6. The standard free-energy change for a reaction, $\Delta G°$, can be found from the standard free energies of formation of reactants and products.
7. The equilibrium constant of a reaction and the standard free-energy change of the reaction are related by the equation $\Delta G° = -RT \ln K$.

Applied Chemistry Example: The Shift Reaction

The steam reforming process discussed on p. 627 produces H_2 and CO. If the hydrogen is used to synthesize ammonia, then carbon monoxide must be removed since it will poison the catalysts. The removal can be carried out by reacting carbon monoxide with steam at about 300°C in the presence of a copper-zinc catalyst:

$$CO(g) + H_2O(g) \rightleftharpoons CO_2(g) + H_2(g)$$

Note that this process, which is called the *shift reaction,* produces more hydrogen for ammonia synthesis. The carbon dioxide formed can be removed by treating the gas mixture with an aqueous solution of potassium carbonate:

$$H_2O(l) + CO_2(g) + K_2CO_3(aq) \longrightarrow 2KHCO_3(aq)$$

1. Use the thermodynamic data in Appendix 3 to calculate the equilibrium constants

for the shift reaction at 25°C and 300°C. Assume $\Delta H°$ and $\Delta S°$ to be independent of temperature.

Answer: To calculate K at 25°C and 300°C we need the corresponding $\Delta G°$ values. We can calculate $\Delta G°$ if we know both $\Delta H°$ and $\Delta S°$ of the reaction. From Equation (6.8) we write

$$\begin{aligned}\Delta H°_{\text{rxn}} &= [(1\text{ mol})\Delta H_f°(H_2) + (1\text{ mol})\Delta H_f°(CO_2)] - [(1\text{ mol})\Delta H_f°(CO) + (1\text{ mol})\Delta H_f°(H_2O)]\\ &= [(1\text{ mol})(0) + (1\text{ mol})(-393.5\text{ kJ/mol})] - [(1\text{ mol})(-110.5\text{ kJ/mol}) + \\ &\qquad (1\text{ mol})(-241.8\text{ kJ/mol})]\\ &= -41.2\text{ kJ}\end{aligned}$$

and from Equation (18.3) we have

$$\begin{aligned}\Delta S°_{\text{rxn}} &= [(1\text{ mol})S°(H_2) + (1\text{ mol})S°(CO_2)] - [(1\text{ mol})S°(CO) + (1\text{ mol})S°(H_2O)]\\ &= [(1\text{ mol})(131.0\text{ J/K}\cdot\text{mol}) + (1\text{ mol})(213.6\text{ J/K}\cdot\text{mol})] - \\ &\quad [(1\text{ mol})(197.9\text{ J/K}\cdot\text{mol}) + (1\text{ mol})(188.7\text{ J/K}\cdot\text{mol})]\\ &= -42.0\text{ J/K}\end{aligned}$$

Next, we calculate $\Delta G°$. From Equation (18.7) and at 25°C (298 K),

$$\begin{aligned}\Delta G° &= \Delta H° - T\Delta S°\\ &= -41.2 \times 1000\text{ J} - (298\text{ K})(-42.0\text{ J/K})\\ &= -2.87 \times 10^4\text{ J}\end{aligned}$$

Applying Equation (18.10), we write

$$\Delta G° = -RT\ln K_P$$

$$-2.87 \times 10^4\text{ J} = -(8.314\text{ J/K}\cdot\text{mol})(298\text{ K})\ln K_P$$

$$\ln K_P = 11.6$$

$$K_P = 1.1 \times 10^5$$

Similarly for the process at 300°C (573 K) we have

$$\begin{aligned}\Delta G° &= -41.2 \times 1000\text{ J} - (573\text{ K})(-42.0\text{ J/K})\\ &= -1.71 \times 10^4\text{ J}\end{aligned}$$

Hence

$$-1.71 \times 10^4\text{ J} = -(8.314\text{ J/K}\cdot\text{mol})(573\text{ K})\ln K_P$$

$$\ln K_P = 3.59$$

$$K_P = 36$$

Note that because the reaction is exothermic ($\Delta H°_{\text{rxn}} = -41.2$ kJ), we expect it to be favored at a lower temperature. This is indeed true since the equilibrium constant at 25°C is greater than that at 300°C. However, in practice the reaction is run at much higher temperatures than 25°C because the rate is too slow at room temperature.

2. Above what temperature will the shift reaction begin to favor the reactants (CO and H_2O)?

Answer: A reaction that favors the reactants has a positive $\Delta G°$ value. The results in 1 show that $\Delta G°$ increases (becoming more positive) with increasing temperature. Therefore, we expect that $\Delta G°$ will be positive (or K_P will become smaller than 1) at a temperature above 300°C. However, first we need to find the temperature at which $\Delta G°$ becomes zero. This is the temperature at which reactants and products are equally favored. Setting $\Delta G° = 0$ in Equation (18.7) we get

$$0 = \Delta H° - T\Delta S°$$

$$T = \frac{\Delta H°}{\Delta S°}$$

$$= \frac{-41.2 \times 10^3 \text{ J}}{-42.0 \text{ J/K}}$$

$$= 981 \text{ K}$$

$$= 708°\text{C}$$

This calculation shows that at 708°C, $\Delta G° = 0$ and the equilibrium constant K_P is 1. Above 708°C, $\Delta G°$ is positive and K_P will be smaller than 1, meaning that the reactants will be favored over products. Note that the temperature 708°C is only an estimate, as we have assumed that both $\Delta H°$ and $\Delta S°$ are independent of temperature.

Key Words

Entropy(S), p. 751
Free energy, p. 759
Gibbs free energy, p. 759
Second law of thermodynamics, p. 755
Third law of thermodynamics, p. 758

Exercises

Spontaneous Processes

REVIEW QUESTIONS

18.1 Explain what is meant by a spontaneous process. Give two examples each of spontaneous and nonspontaneous processes.

18.2 Which of the following processes are spontaneous and which are nonspontaneous? (a) dissolving table salt (NaCl) in hot soup; (b) climbing Mt. Everest; (c) spreading fragrance in a room by removing the cap from a perfume bottle; (d) separating $^{235}UF_6$ from $^{238}UF_6$ by gaseous effusion (see Section 5.8)

18.3 Which of the following processes are spontaneous and which are nonspontaneous at a given temperature?

(a) $NaNO_3(s) \xrightarrow{H_2O} NaNO_3(aq)$ saturated soln

(b) $NaNO_3(s) \xrightarrow{H_2O} NaNO_3(aq)$ unsaturated soln

(c) $NaNO_3(s) \xrightarrow{H_2O} NaNO_3(aq)$ supersaturated soln

PROBLEM

18.4 Referring to Figure 18.1(a), calculate the probability of all the molecules in the same flask if the number is (a) 6, (b) 60, (c) 600.

Entropy; Second Law of Thermodynamics

REVIEW QUESTIONS

18.5 Define entropy. What are the units of entropy?

18.6 How does the entropy of a system change for each of the following processes?
(a) A solid melts.
(b) A liquid freezes.
(c) A liquid boils.
(d) A vapor is converted to a solid.
(e) A vapor condenses to a liquid.
(f) A solid sublimes.

18.7 State the second law of thermodynamics in words and express it mathematically.

PROBLEMS

18.8 In each pair of substances listed here, choose the one having the larger standard entropy at 25°C. Explain the basis for your choice. (a) Li(s) or Li(l); (b) $C_2H_5OH(l)$ or $CH_3OCH_3(l)$ (*Hint:* Which molecule can hydrogen bond?); (c) Ar(g) or Xe(g); (d) CO(g) or $CO_2(g)$; (e) graphite or diamond; (f) $NO_2(g)$ or $N_2O_4(g)$. The same molar amount is used in the comparison.

18.9 Arrange the following substances (1 mole each) in the

order of increasing entropy at 25°C: (a) Ne(g), (b) $SO_2(g)$, (c) Na(s), (d) NaCl(s), (e) $NH_3(g)$. Give the reasons for your arrangement.

18.10 Using the data in Appendix 3, calculate the standard entropy change for the following reactions at 25°C:
(a) $S(s) + O_2(g) \longrightarrow SO_2(g)$
(b) $MgCO_3(s) \longrightarrow MgO(s) + CO_2(g)$
(c) $H_2(g) + CuO(s) \longrightarrow Cu(s) + H_2O(g)$
(d) $2Al(s) + 3ZnO(s) \longrightarrow Al_2O_3(s) + 3Zn(s)$
(e) $CH_4(g) + 2O_2(g) \longrightarrow CO_2(g) + 2H_2O(l)$

18.11 Without consulting Appendix 3, predict whether the entropy change is positive or negative for each of the following reactions. Give reasons for your predictions.
(a) $2KClO_4(s) \longrightarrow 2KClO_3(s) + O_2(g)$
(b) $H_2O(g) \longrightarrow H_2O(l)$
(c) $S(s) + O_2(g) \longrightarrow SO_2(g)$
(d) $2Na(s) + 2H_2O(l) \longrightarrow 2NaOH(aq) + H_2(g)$
(e) $CH_4(g) + 2O_2(g) \longrightarrow CO_2(g) + 2H_2O(l)$
(f) $N_2(g) \longrightarrow 2N(g)$
(g) $2LiOH(aq) + CO_2(g) \longrightarrow Li_2CO_3(aq) + H_2O(l)$
(h) $NH_4Cl(s) \longrightarrow NH_3(g) + HCl(g)$

18.12 Discuss qualitatively the sign of the entropy change expected for each of the following processes.
(a) $PCl_3(l) + Cl_2(g) \longrightarrow PCl_5(s)$
(b) $2HgO(s) \longrightarrow 2Hg(l) + O_2(g)$
(c) $H_2(g) \longrightarrow 2H(g)$
(d) $H_2(g) + Cl_2(g) \longrightarrow 2HCl(g)$
(e) $U(s) + 3F_2(g) \longrightarrow UF_6(s)$

18.13 Predict whether the entropy change is positive or negative for each of the following reactions:
(a) $Zn(s) + 2HCl(aq) \longrightarrow ZnCl_2(aq) + H_2(g)$
(b) $O(g) + O(g) \longrightarrow O_2(g)$
(c) $NH_4NO_3(s) \longrightarrow N_2O(g) + 2H_2O(g)$
(d) $Ba^{2+}(aq) + SO_4^{2-}(aq) \longrightarrow BaSO_4(s)$
(e) $2H_2O_2(l) \longrightarrow 2H_2O(l) + O_2(g)$

18.14 The reaction $NH_3(g) + HCl(g) \longrightarrow NH_4Cl(s)$ proceeds spontaneously at 25°C even though there is a decrease in disorder in the system (gases are converted to a solid). Explain.

Free Energy

REVIEW QUESTIONS

18.15 Define free energy. What are its units?

18.16 Why is it more convenient to predict the direction of a reaction in terms of ΔG_{sys} rather than ΔS_{univ}?

18.17 (a) Do the changes in Equation (18.7) refer to the system or the surroundings? (b) Under what conditions can this equation be used to predict the spontaneity of a reaction?

PROBLEMS

18.18 Calculate $\Delta G°$ for the following reactions at 25°C:
(a) $N_2(g) + O_2(g) \longrightarrow 2NO(g)$
(b) $H_2O(l) \longrightarrow H_2O(g)$
(c) $2C_2H_2(g) + 5O_2(g) \longrightarrow 4CO_2(g) + 2H_2O(l)$
(d) $2Mg(s) + O_2(g) \longrightarrow 2MgO(s)$
(e) $2SO_2(g) + O_2(g) \longrightarrow 2SO_3(g)$
(*Hint:* Look up the standard free energies of formation of the reactants and products in Appendix 3.)

18.19 Calculate $\Delta G°$ for the combustion of ethane:

$$2C_2H_6(g) + 7O_2(g) \longrightarrow 4CO_2(g) + 6H_2O(l)$$

See Appendix 3 for thermodynamic data.

18.20 From the following ΔH and ΔS values, predict whether each of the reactions would be spontaneous at 25°C. If not, at what temperature might the reaction become spontaneous? Reaction A: $\Delta H = 10.5$ kJ, $\Delta S = 30$ J/K; Reaction B: $\Delta H = 1.8$ kJ, $\Delta S = -113$ J/K; Reaction C: $\Delta H = -126$ kJ, $\Delta S = 84$ J/K; Reaction D: $\Delta H = -11.7$ kJ, $\Delta S = -105$ J/K.

18.21 The molar heat of vaporization of ethanol is 39.3 kJ/mol and the boiling point of ethanol is 78.3°C. Calculate ΔS for the vaporization of 0.50 mol ethanol.

18.22 Use the following data to determine the normal boiling point, in K, of mercury. What assumptions must you make in order to do the calculation?

Hg(l) $\Delta H_f^\circ = 0$ (by definition)
$S° = 77.4$ J/K · mol

Hg(g) $\Delta H_f^\circ = 60.78$ kJ/mol
$S° = 174.7$ J/K · mol

18.23 A certain reaction is known to have a $\Delta G°$ value of -122 kJ. Will the reaction necessarily occur if the reagents are mixed together?

Free Energy and Chemical Equilibrium

REVIEW QUESTIONS

18.24 Explain why Equation (18.10) is of great importance in chemistry.

18.25 Explain clearly the difference between ΔG and $\Delta G°$.

PROBLEMS

18.26 For the reaction

$$H_2(g) + I_2(g) \rightleftharpoons 2HI(g)$$

$\Delta G° = 2.60$ kJ. Calculate K_P for the reaction at 25°C.

18.27 K_w for the autoionization of water

$$H_2O(l) \rightleftharpoons H^+(aq) + OH^-(aq)$$

is 1.0×10^{-14}. What is $\Delta G°$ for the process?

18.28 Consider the following reaction at 25°C:

$$Fe(OH)_2(s) \rightleftharpoons Fe^{2+}(aq) + 2OH^-(aq)$$

Calculate $\Delta G°$ for the reaction. K_{sp} for $Fe(OH)_2$ is 1.6×10^{-14}.

18.29 Calculate $\Delta G°$ and K_P for the equilibrium reaction

$$2H_2O(g) \rightleftharpoons 2H_2(g) + O_2(g)$$

at 25°C.

18.30 Calculate K_P for the following reaction at 25°C:

$$N_2(g) + O_2(g) \rightleftharpoons 2NO(g)$$

From your result, what can you conclude about the stability of a mixture of O_2 and N_2 gases in the atmosphere?

18.31 Calculate $\Delta G°$ and K_P for the following processes at 25°C:

(a) $H_2(g) + Br_2(l) \rightleftharpoons 2HBr(g)$
(b) $\frac{1}{2}H_2(g) + \frac{1}{2}Br_2(l) \rightleftharpoons HBr(g)$

Account for the differences in $\Delta G°$ and K_P obtained for (a) and (b).

18.32 Consider the decomposition of calcium carbonate:

$$CaCO_3(s) \rightleftharpoons CaO(s) + CO_2(g)$$

Calculate the pressure in atm of CO_2 in an equilibrium process (a) at 25°C and (b) at 800°C. Assume that $\Delta H° = 177.8$ kJ and $\Delta S° = 160.5$ J/K.

18.33 (a) Calculate $\Delta G°$ and K_P for the following equilibrium reaction at 25°C. The $\Delta G_f°$ values are, for $Cl_2(g)$, 0; for $PCl_3(g)$, −286 kJ/mol; and for $PCl_5(g)$, −325 kJ/mol.

$$PCl_5(g) \rightleftharpoons PCl_3(g) + Cl_2(g)$$

(b) Calculate ΔG for the reaction if the partial pressures of the initial mixture are $P_{PCl_5} = 0.0029$ atm, $P_{PCl_3} = 0.27$ atm, and $P_{Cl_2} = 0.40$ atm.

18.34 The equilibrium constant (K_P) for the reaction

$$H_2(g) + CO_2(g) \rightleftharpoons H_2O(g) + CO(g)$$

is 4.40 at 2000 K. (a) Calculate $\Delta G°$ for the reaction. (b) Calculate ΔG for the reaction when the partial pressures are $P_{H_2} = 0.25$ atm, $P_{CO_2} = 0.78$ atm, $P_{H_2O} = 0.66$ atm, and $P_{CO} = 1.20$ atm.

18.35 The equilibrium constant K_P for the reaction

$$CO(g) + Cl_2(g) \rightleftharpoons COCl_2(g)$$

is 5.62×10^{35} at 25°C. Calculate $\Delta G_f°$ for $COCl_2$ at 25°C.

18.36 At 25°C, $\Delta G°$ for the process

$$H_2O(l) \rightleftharpoons H_2O(g)$$

is 8.6 kJ. Calculate the "equilibrium constant" for the process.

18.37 Calculate $\Delta G°$ for the process

$$C(\text{diamond}) \longrightarrow C(\text{graphite})$$

Is the reaction spontaneous at 25°C? If so, why is it that diamonds do not become graphite on standing?

18.38 Calculate the pressure of O_2 (in atm) over a sample of NiO at 25°C if $\Delta G° = 212$ kJ for the reaction

$$NiO(s) \rightleftharpoons Ni(s) + \tfrac{1}{2}O_2(g)$$

Miscellaneous Problems

18.39 Explain the following nursery rhyme in terms of the second law of thermodynamics.

Humpty Dumpty sat on a wall;
Humpty Dumpty had a great fall.
All the King's horses and all the King's men
Couldn't put Humpty together again.

18.40 Calculate ΔG for the reaction

$$H_2O(l) \rightleftharpoons H^+(aq) + OH^-(aq)$$

at 25°C for the following conditions:
(a) $[H^+] = 1.0 \times 10^{-7}\ M$, $[OH^-] = 1.0 \times 10^{-7}\ M$
(b) $[H^+] = 1.0 \times 10^{-3}\ M$, $[OH^-] = 1.0 \times 10^{-4}\ M$
(c) $[H^+] = 1.0 \times 10^{-12}\ M$, $[OH^-] = 2.0 \times 10^{-8}\ M$
(d) $[H^+] = 3.5\ M$, $[OH^-] = 4.8 \times 10^{-4}\ M$

18.41 Which of the following thermodynamic functions are associated only with the first law of thermodynamics: S, E, G, and H?

18.42 A student placed 1 g of each of three compounds A, B, and C in a container and found that after one week no change had occurred. Offer some possible explanations for the fact that no reactions took place. Assume that A, B, and C are totally miscible liquids.

18.43 Predict the signs of ΔH, ΔS, and ΔG of the system for the following processes at 1 atm: (a) ammonia melts at −60°C, (b) ammonia melts at −77.7°C, (c) ammonia melts at −100°C. (The normal melting point of ammonia is −77.7°C.)

18.44 Give a detailed example of each of the following, with an explanation: (a) a thermodynamically spontaneous process; (b) a process that would violate the first law of thermodynamics; (c) a process that would violate the second law of thermodynamics; (d) an irreversible process; (e) an equilibrium process.

18.45 Gaseous diffusion is a spontaneous process. In most cases, the enthalpy change is quite small. Use Equation (18.7) to predict the entropy change of the system (that is, whether $\Delta S > 0$ or $\Delta S < 0$).

18.46 Ammonium nitrate dissolves spontaneously and endothermically in water. What can you deduce about the sign of ΔS for the solution process?

18.47 A carbon monoxide (CO) crystal is found to have an entropy greater than zero at the temperature absolute zero. Give two possible reasons that can account for this observation.

18.48 Consider the following facts: Water freezes spontaneously at −5°C and 1 atm, and ice has a more ordered structure than liquid water. Explain how a spontaneous process can lead to a decrease in entropy.

18.49 Calculate the equilibrium pressure of CO_2 due to the decomposition of barium carbonate ($BaCO_3$) at 25°C.

18.50 For reactions carried out under standard-state conditions, Equation (18.7) takes the form $\Delta G° = \Delta H° - T\Delta S°$. (a) Assuming $\Delta H°$ and $\Delta S°$ are independent of temperature, derive the following equation:

$$\ln \frac{K_2}{K_1} = \frac{\Delta H°}{R}\left(\frac{T_2 - T_1}{T_1 T_2}\right)$$

where K_1 and K_2 are the equilibrium constants at T_1 and T_2, respectively. (b) Given that the K_c for the reaction

$$N_2O_4(g) \rightleftharpoons 2NO_2(g) \qquad \Delta H° = 58.0 \text{ kJ}$$

is 4.63×10^{-3} at 25°C, calculate the equilibrium constant at 65°C.

18.51 (a) Trouton's rule states that the ratio of the molar heat of vaporization of a liquid (ΔH_{vap}) to its boiling point in kelvins is approximately 90 J/K · mol. Show that this is the case with the following data:

	$t_{b.p.}$(°C)	ΔH_{vap}(kJ/mol)
Benzene	80.1	31.0
Hexane	68.7	30.8
Mercury	357	59.0
Toluene	110.6	35.2

and explain why this rule holds true. (b) Use the values in Table 11.7 to calculate the same ratio for ethanol and water. Explain why Trouton's rule does not apply to these two substances as well as it does to other liquids.

18.52 Referring to the above problem, explain why the ratio is considerably smaller than 90 J/K · mol for the HF liquid.

18.53 Carbon monoxide (CO) and nitric oxide (NO) are undesirable gases emitted in an automobile's exhaust. Under suitable conditions, these gases can be made to react to form nitrogen (N_2) and the less harmful carbon dioxide (CO_2). (a) Write an equation for this reaction. (b) Identify the oxidizing and reducing agents. (c) Calculate the K_P for the reaction at 25°C. (d) Under normal atmospheric conditions, the partial pressures are $P_{N_2} = 0.80$ atm, $P_{CO_2} = 3.0 \times 10^{-4}$ atm, $P_{CO} = 5.0 \times 10^{-5}$ atm, and $P_{NO} = 5.0 \times 10^{-7}$ atm. Calculate Q_P and predict the direction toward which the reaction will proceed. (e) Will raising the temperature favor the formation of N_2 and CO_2?

18.54 Some copper(II) sulfate pentahydrate is placed in a closed container, and the following equilibrium is established at 25°C:

$$CuSO_4 \cdot 5H_2O(s) \rightleftharpoons CuSO_4(s) + 5H_2O(g)$$

Is the pressure of the water vapor measurable? ($\Delta G_f°$ for $CuSO_4 \cdot 5H_2O$ is −1879.7 kJ/mol.)

18.55 The K_{sp} of AgCl is given in Table 17.1. What is its value at 60°C? (*Hint:* You need the result in Problem 18.50 and the data in Appendix 3 to calculate $\Delta H°$.)

18.56 In a coal-fired power plant, the hot "reservoir" is a supply of high-pressure steam at a temperature of about 500°C. The cold reservoir will typically be a river at a temperature of about 15°C. Calculate the thermodynamic efficiency of the power plant. In reality, the efficiency of such a power plant is always less than the thermodynamic efficiency. Explain why this is so.

18.57 (a) Over the years there have been numerous claims about "perpetual motion machines," machines that will produce useful work with no input of energy. Explain why the first law of thermodynamics prohibits the possibility of such a machine existing. (b) Another kind of machine, sometimes called a "perpetual motion of the second kind," operates as follows. Suppose an ocean liner sails by scooping up water from the ocean and then extracting heat from the water, converting the heat to electric power to run the ship, and dumping the water back into the ocean. This process does not violate the first law of thermodynamics, for no energy is created—energy from the ocean is just converted to electrical energy. Show that the second law of thermodynamics prohibits the existence of such a machine.

18.58 Water gas is a fuel made by reacting steam with red-hot coke (coke is the product of coal distillation):

$$H_2O(g) + C(s) \rightarrow CO(g) + H_2(g)$$

From the data in Appendix 3, estimate the temperature at which the reaction becomes favorable.

Opposite page: Corrosion of metals by salts in aqueous solution is an electrochemical process which creates enormous economic problems. Shown here are corroded coins together with a sheet of metal alloy that has been treated to be corrosion resistant.

CHAPTER 19

ELECTROCHEMISTRY

One of the forms of energy of greatest practical significance to us on a day-to-day basis is electrical energy. Imagine, for example, a day without electrical power from either the power company or batteries. The area of chemistry that deals with the interconversion of electrical energy and chemical energy is electrochemistry. Because all electrochemical reactions involve the transfer of electrons from one substance to another, their action can be understood in terms of redox reactions, which we discussed in Chapter 3.

There are two types of cells in which electrochemical processes are carried out. A *galvanic cell* uses the energy released from a spontaneous reaction to generate electricity. An *electrolytic cell,* on the other hand, uses electrical energy to cause a nonspontaneous chemical reaction to occur; this process is called electrolysis.

This chapter discusses the fundamental principles and applications of electrochemical cells, the thermodynamics of electrochemical reactions, and the cause and prevention of corrosion by electrochemical means. Some simple electrolytic processes and the quantitative aspects of electrolysis are also discussed.

19.1 Review of Redox Reactions

Redox, or oxidation-reduction, reactions were discussed in Chapter 3. Because electrochemistry is based on redox processes, it is helpful to review some of the basic redox concepts that will come up later in this chapter.

The reaction between magnesium metal and hydrochloric acid is an example of a redox reaction:

$$\overset{0}{Mg}(s) + 2\overset{+1}{H}Cl(aq) \longrightarrow \overset{+2}{Mg}Cl_2(aq) + \overset{0}{H_2}(g)$$

Recall that the numbers written above the elements are the oxidation numbers of the elements. The loss of electrons that occurs during oxidation is marked by an increase in the oxidation number of an element in a reaction. In reduction, there is a gain of electrons, indicated by a decrease in the oxidation number of an element in a reaction. In the above reaction the metal Mg is oxidized and H^+ ions are reduced.

Oxidation and reduction occur simultaneously. We cannot have one without the other; every electron lost by one species is gained by another. In studying a redox reaction, we often think of it as being made up of two separate reactions, called half-reactions, one representing the oxidation process and the other the reduction process. In Section 3.5 we saw that when a piece of zinc metal is placed in a $CuSO_4$ solution, Zn is oxidized to Zn^{2+} ions while Cu^{2+} ions are reduced to metallic copper (see Figure 3.14). The net ionic equation for this process is

$$Zn(s) + Cu^{2+}(aq) \longrightarrow Zn^{2+}(aq) + Cu(s)$$

Breaking this equation into two half-reactions, we have

$$Zn(s) \longrightarrow Zn^{2+}(aq) + 2e^-$$

$$Cu^{2+}(aq) + 2e^- \longrightarrow Cu(s)$$

As a Zn atom in the metal is oxidized, it loses two electrons to a Cu^{2+} ion, which is reduced. The sum of these two half-reactions gives us the overall reaction.

19.2 Galvanic Cells

Under normal conditions a redox reaction occurs when the oxidizing agent is in contact with the reducing agent, as in the reaction between Zn and Cu^{2+} described in the last section. The electrons are transferred directly from the reducing agent to the oxidizing agent in solution. If we could physically separate the oxidizing agent from the reducing agent, it would be possible for the transfer of electrons to take place via an external conducting medium, rather than directly in solution. Thus, as the reaction progressed, it would set up a constant flow of electrons and hence generate electricity (that is, it would produce electrical work).

The experimental apparatus for generating electricity through the use of a redox reaction is called an *electrochemical cell*. Figure 19.1 shows the essential components of a *galvanic*† *cell*, also called a *voltaic cell*, in which electricity is produced by means

†Luigi Galvani (1737–1798). Italian physiologist. Galvani pioneered in the field of electrophysiology, that is, the study of the relationship between electricity and living organisms. He discovered that the legs of a frog touched with a scalpel during dissection began to twitch whenever an electrical discharge occurred nearby.

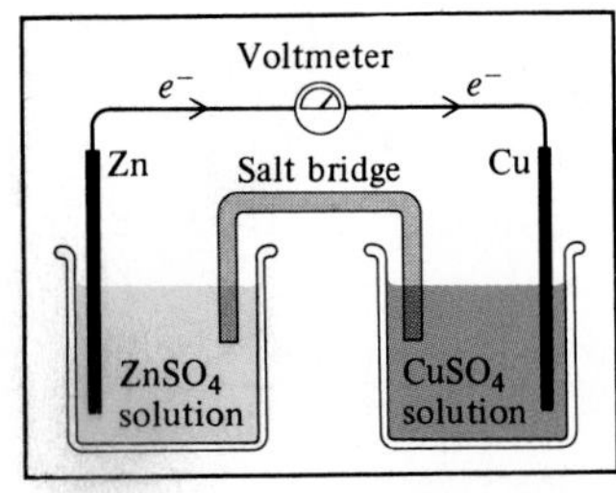

Figure 19.1 *A galvanic cell. The salt bridge (an inverted U tube) containing a KCl solution provides an electrically conducting medium between two solutions. The openings of the U tube are loosely plugged with cotton balls (not shown) to prevent the KCl solution from running into the containers while allowing the anions and cations to move across. Electrons flow externally from the Zn electrode (anode) to the Cu electrode (cathode).*

of a spontaneous redox reaction. A bar of zinc metal is dipped into a $ZnSO_4$ solution and a bar of copper metal is dipped into a $CuSO_4$ solution. The galvanic cell operates on the principle that the oxidation of Zn to Zn^{2+} and the reduction of Cu^{2+} to Cu can be made to take place simultaneously in separate locations with the transfer of electrons through the external wire. The zinc bar and the copper bar are called *electrodes*. By definition, *the electrode at which oxidation occurs* is called the ***anode; the electrode at which reduction occurs*** is the ***cathode.***

Half-cell reactions are similar to the half-reactions discussed above.

For this system, *the oxidation and reduction reactions at the electrodes,* which are called ***half-cell reactions,*** are

$$\text{Zn electrode (anode):} \quad Zn(s) \longrightarrow Zn^{2+}(aq) + 2e^-$$

$$\text{Cu electrode (cathode):} \quad Cu^{2+}(aq) + 2e^- \longrightarrow Cu(s)$$

Note that unless the two solutions are separated from each other, the Cu^{2+} ions will react directly with the zinc bar:

$$Cu^{2+}(aq) + Zn(s) \longrightarrow Cu(s) + Zn^{2+}(aq)$$

and no useful electrical work will be obtained.

To complete the electric circuit, the solutions must be connected by a conducting medium through which the cations and anions can move. This requirement is satisfied by a *salt bridge,* which, in its simplest form, is an inverted U tube containing an inert electrolyte such as KCl or NH_4NO_3, whose ions will not react with other ions in solution or with the electrodes (Figure 19.1). During the course of the overall redox reaction, electrons flow externally from the anode (Zn electrode) through the wire and voltmeter to the cathode (Cu electrode). In the solution, the cations (Zn^{2+}, Cu^{2+}, and K^+) move toward the cathode, while the anions (SO_4^{2-} and Cl^-) move in the opposite direction, toward the anode.

The fact that electrons flow from one electrode to the other indicates that there is a voltage difference between the two electrodes. This *voltage difference between the electrodes,* called the ***electromotive force,*** or ***emf*** ($\mathscr{E}$), can be measured by connecting a voltmeter between the two electrodes. The emf of a galvanic cell is usually measured in volts; it is also referred to as *cell voltage* or *cell potential*. We will see later that the emf of a cell depends not only on the nature of the electrodes and the ions, but also on the concentrations of the ions and the temperature at which the cell is operated.

The conventional notation for representing galvanic cells is the *cell diagram*. For the galvanic cell just described, using KCl as the electrolyte in the salt bridge and assuming 1 *M* concentrations, the cell diagram is

$$Zn(s)|Zn^{2+}(aq, 1\ M)|KCl(\text{sat'd})|Cu^{2+}(aq, 1\ M)|Cu(s)$$

where the vertical lines represent phase boundaries. For example, the zinc electrode is a solid and the Zn^{2+} ions (from $ZnSO_4$) are in solution. Thus we draw a line between Zn and Zn^{2+} to show the phase boundaries. Note that there is also a line between the $ZnSO_4$ solution and the KCl solution in the salt bridge because these two solutions are not mixed and therefore constitute two separate phases. By convention, the anode is written first at the left and the other components appear in the order in which we would encounter them in moving from the anode to the cathode.

19.3 Standard Electrode Potentials

Emf is an intensive property and therefore is independent of the volume of the solution and the size of the electrodes.

This method is analogous to choosing the surface of the ocean as the reference for altitude, arbitrarily assigning it to be zero meters, and then referring to any terrestrial altitude as being a certain number of meters above or below sea level.

Standard states are defined in Table 18.3.

When the concentrations of the Cu^{2+} and Zn^{2+} ions are both 1.0 M, we find that the emf of the cell shown in Figure 19.1 is 1.10 V at 25°C. This voltage must be related directly to the redox reactions, but how? Just as the overall cell reaction can be thought of as the sum of two half-cell reactions, the measured emf of the cell can be treated as the sum of the electrical potentials at the Zn and Cu electrodes. Knowing one of these electrode potentials, we could obtain the other by subtraction (from 1.10 V). It is impossible to measure the potential of just a single electrode, but if we arbitrarily set the potential value of a particular electrode at zero, we can use it to determine the relative potentials of other electrodes. The hydrogen electrode, shown in Figure 19.2, serves as the standard for this purpose. Hydrogen gas is bubbled into a hydrochloric acid solution at 25°C. The platinum electrode has two functions. First, it provides a surface on which the dissociation of hydrogen molecules can take place:

$$H_2 \longrightarrow 2H^+ + 2e^-$$

Second, it serves as an electrical conductor to the external circuit. Under standard-state conditions (that is, when the pressure of H_2 is 1 atm and the concentration of the HCl solution is 1 M) and at 25°C, the potential for the following reduction is defined to be *exactly* zero:

$$2H^+(aq, 1\ M) + 2e^- \longrightarrow H_2(g, 1\text{ atm}) \qquad \mathscr{E}° = 0\text{ V}$$

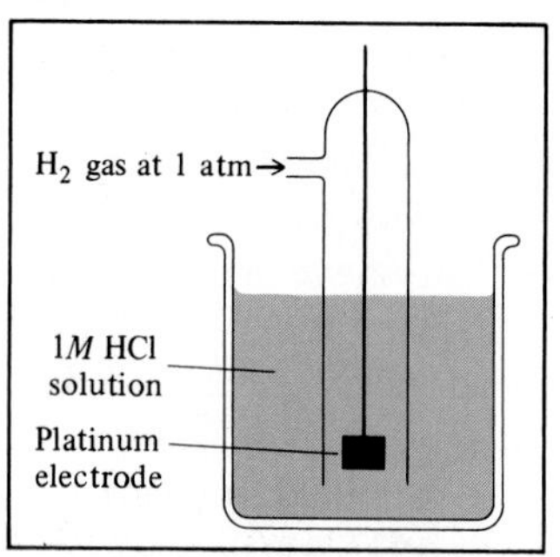

Figure 19.2 *A hydrogen electrode operating under standard-state conditions. Hydrogen gas at 1 atm is bubbled through a 1 M HCl solution. The platinum electrode is part of the hydrogen electrode.*

As usual, the superscript o denotes standard-state conditions. *For a reduction reaction at an electrode when all solutes are 1 M and all gases are at 1 atm, the voltage* is called the ***standard reduction potential.*** Thus, the standard reduction potential of the hydrogen electrode is defined as zero. The electrode itself is known as the *standard hydrogen electrode (SHE)*.

We can use the SHE to measure the potentials of other kinds of electrodes. Figure 19.3(a) shows a galvanic cell with a zinc electrode and a SHE electrode. Experiments show that the zinc electrode is the anode and the SHE the cathode. We deduce this because the mass of the zinc electrode decreases during the operation of the cell, a fact consistent with the loss of zinc to the solution by the oxidation reaction:

$$Zn(s) \longrightarrow Zn^{2+}(aq) + 2e^-$$

The cell diagram is

$$Zn(s)|Zn^{2+}(aq, 1\ M)|KCl(\text{sat'd})|H^+(aq, 1\ M)|H_2(g, 1\text{ atm})|Pt(s)$$

When all the reactants are in their standard states (that is, H_2 at 1 atm, H^+ and Zn^{2+} ions at 1 M), the emf of the cell is 0.76 V. We can write the half-cell reactions as follows:

$$\begin{array}{llll}
\text{Anode:} & Zn(s) & \longrightarrow Zn^{2+}(aq, 1\ M) + 2e^- & \mathscr{E}°_{Zn/Zn^{2+}} \\
\text{Cathode:} & 2H^+(aq, 1\ M) + 2e^- & \longrightarrow H_2(g, 1\text{ atm}) & \mathscr{E}°_{H^+/H_2} \\
\hline
\text{Overall:} & Zn(s) + 2H^+(aq, 1\ M) & \longrightarrow Zn^{2+}(aq, 1\ M) + H_2(g, 1\text{ atm}) & \mathscr{E}°_{cell}
\end{array}$$

Zn/Zn^{2+} means $Zn \rightarrow Zn^{2+} + 2e^-$.

where $\mathscr{E}°_{Zn/Zn^{2+}}$ is the ***standard oxidation potential*** for the oxidation half-reaction:

$$Zn(s) \longrightarrow Zn^{2+}(aq, 1\ M) + 2e^-$$

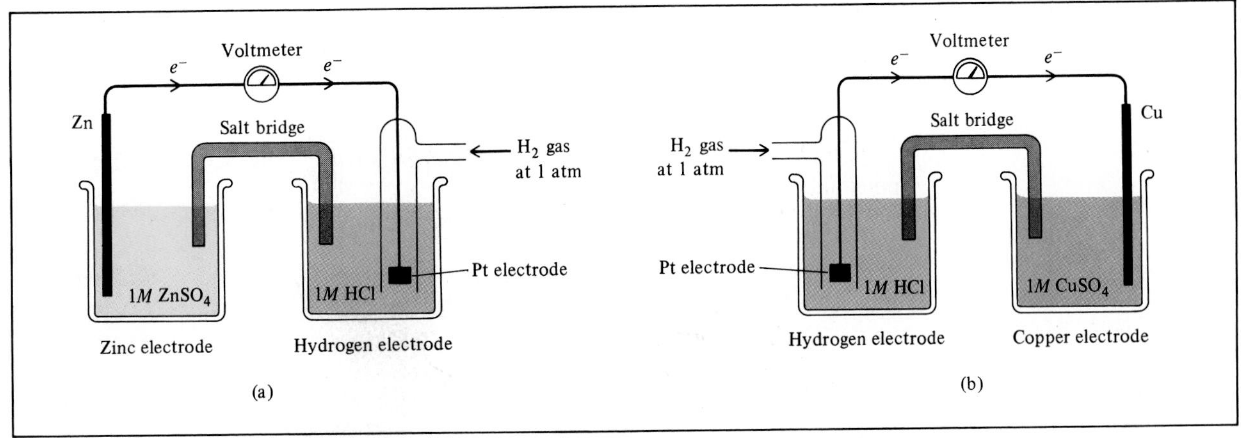

Figure 19.3 *(a) A cell consisting of a zinc electrode and a hydrogen electrode. (b) A cell consisting of a copper electrode and a hydrogen electrode. Both cells are operating under standard-state conditions. Note that in (a) the SHE acts as the cathode, but in (b) the SHE is the anode.*

H^+/H_2 means $2H^+ + 2e^- \rightarrow H_2$.

and $\mathscr{E}^\circ_{H^+/H_2}$ the standard reduction potential previously defined for the SHE. The ***standard emf*** of the cell, $\mathscr{E}^\circ_{cell}$, is *the sum of the standard oxidation potential and the standard reduction potential:*

$$\mathscr{E}^\circ_{cell} = \mathscr{E}^\circ_{ox} + \mathscr{E}^\circ_{red} \tag{19.1}$$

where the subscripts ''ox'' and ''red'' indicate oxidation and reduction, respectively. For our galvanic cell,

$$\mathscr{E}^\circ_{cell} = \mathscr{E}^\circ_{Zn/Zn^{2+}} + \mathscr{E}^\circ_{H^+/H_2}$$

$$0.76\text{ V} = \mathscr{E}^\circ_{Zn/Zn^{2+}} + 0$$

Thus the standard oxidation potential of zinc is 0.76 V.

We can evaluate the standard reduction potential of zinc, $\mathscr{E}^\circ_{Zn^{2+}/Zn}$, by reversing the oxidation half-reaction:

$$Zn^{2+}(aq,\ 1\ M) + 2e^- \longrightarrow Zn(s) \qquad \mathscr{E}^\circ_{Zn^{2+}/Zn} = -0.76\text{ V}$$

This result fits a general pattern: Whenever we reverse a half-cell reaction, $\mathscr{E}^\circ$ changes sign.

The standard electrode potentials of copper can similarly be obtained using a galvanic cell with a copper electrode and a SHE [Figure 19.3(b)]. Here we know that the copper electrode is the cathode because its mass increases during the operation of the cell, consistent with the reduction reaction:

$$Cu^{2+}(aq) + 2e^- \longrightarrow Cu(s)$$

The cell diagram is

$$Pt(s)|H_2(g,\ 1\text{ atm})|H^+(aq,\ 1\ M)|KCl(\text{sat'd})|Cu^{2+}(aq,\ 1\ M)|Cu(s)$$

and the half-cell reactions are

Anode:	$H_2(g,\ 1\text{ atm}) \longrightarrow 2H^+(aq,\ 1\ M) + 2e^-$	$\mathscr{E}^\circ_{H_2/H^+}$
Cathode:	$Cu^{2+}(aq,\ 1\ M) + 2e^- \longrightarrow Cu(s)$	$\mathscr{E}^\circ_{Cu^{2+}/Cu}$
Overall:	$H_2(g,\ 1\text{ atm}) + Cu^{2+}(aq,\ 1\ M) \longrightarrow 2H^+(aq,\ 1\ M) + Cu(s)$	$\mathscr{E}^\circ_{cell}$

Under standard-state conditions and at 25°C, the emf of the cell is 0.34 V, so we write

$$\mathscr{E}^\circ_{\text{cell}} = \mathscr{E}^\circ_{H_2/H^+} + \mathscr{E}^\circ_{Cu^{2+}/Cu}$$

$$0.34\text{ V} = 0 + \mathscr{E}^\circ_{Cu^{2+}/Cu}$$

Thus the standard reduction potential of copper is 0.34 V and its standard oxidation potential, $\mathscr{E}^\circ_{Cu/Cu^{2+}}$, is −0.34 V.

For the cell shown in Figure 19.1, we can now write

Anode:	$Zn(s) \longrightarrow Zn^{2+}(aq, 1\,M) + 2e^-$		$\mathscr{E}^\circ_{Zn/Zn^{2+}}$
Cathode:	$Cu^{2+}(aq, 1\,M) + 2e^- \longrightarrow Cu(s)$		$\mathscr{E}^\circ_{Cu^{2+}/Cu}$
Overall:	$Zn(s) + Cu^{2+}(aq, 1\,M) \longrightarrow Zn^{2+}(aq, 1\,M) + Cu(s)$		$\mathscr{E}^\circ_{\text{cell}}$

and the emf of the cell is

$$\begin{aligned}\mathscr{E}^\circ_{\text{cell}} &= \mathscr{E}^\circ_{Zn/Zn^{2+}} + \mathscr{E}^\circ_{Cu^{2+}/Cu}\\ &= 0.76\text{ V} + 0.34\text{ V}\\ &= 1.10\text{ V}\end{aligned}$$

This example illustrates how we can use the *sign* of the emf of the cell to predict the spontaneity of a redox reaction. Under standard-state conditions for reactants and products, the redox reaction is spontaneous in the forward direction if the standard emf of the cell is positive. If it is negative, the reaction is spontaneous in the opposite direction. It is important to keep in mind that a negative $\mathscr{E}^\circ_{\text{cell}}$ does *not* mean that a reaction will not occur if the reactants are mixed at 1 *M* concentrations. It merely means that the equilibrium of a redox reaction, when reached, will lie to the left. We will examine the relationships among $\mathscr{E}^\circ_{\text{cell}}$, ΔG°, and K later in this chapter.

The activity series in Figure 3.15 is based on data given in Table 19.1.

By convention, we list only standard reduction potentials of half-reactions (Table 19.1). By definition the SHE has an $\mathscr{E}^\circ$ value of 0.00 V. The negative standard reduction potentials increase above it, and the positive standard reduction potentials increase below it. It is important to understand the following points about the table:

- The $\mathscr{E}^\circ$ values apply to the half-cell reactions as read in the forward (left to right) direction.
- The more positive $\mathscr{E}^\circ$, the greater the tendency for the substance to be reduced. For example, the half-cell reaction

  $$F_2(g, 1\text{ atm}) + 2e^- \longrightarrow 2F^-(aq, 1\,M) \qquad \mathscr{E}^\circ = 2.87\text{ V}$$

  has the highest $\mathscr{E}^\circ$ value among all the half-cell reactions. Thus F_2 is the *strongest* oxidizing agent because it has the greatest tendency to be reduced. At the other extreme is the reaction

  $$Li^+(aq, 1\,M) + e^- \longrightarrow Li(s) \qquad \mathscr{E}^\circ = -3.05\text{ V}$$

  which has the most negative $\mathscr{E}^\circ$ value. Thus the Li^+ ion is the *weakest* oxidizing agent because it is the most difficult species to reduce. Conversely, we say that the F^- ion is the weakest reducing agent and that the Li metal is the strongest reducing agent. Under standard-state conditions, the oxidizing agents (the species on the left-hand side of the half-reactions in Table 19.1) increase in strength from top to bottom and the reducing agents (the species on the right-hand side of the half-reactions) increase in strength from bottom to top.
- The half-cell reactions are reversible. Depending on the conditions, any electrode can act either as an anode or as a cathode. Earlier we saw that the SHE is the cathode (H^+ is reduced to H_2) when coupled with zinc in a cell and that it becomes the anode (H_2 is oxidized to H^+) when in a cell with copper.

Table 19.1 Standard reduction potentials at 25°C*

(↓ Increasing strength as oxidizing agent — left side; ↑ Increasing strength as reducing agent — right side)

Half-Reaction	$\mathscr{E}°$(V)
$Li^+(aq) + e^- \longrightarrow Li(s)$	−3.05
$K^+(aq) + e^- \longrightarrow K(s)$	−2.93
$Ba^{2+}(aq) + 2e^- \longrightarrow Ba(s)$	−2.90
$Sr^{2+}(aq) + 2e^- \longrightarrow Sr(s)$	−2.89
$Ca^{2+}(aq) + 2e^- \longrightarrow Ca(s)$	−2.87
$Na^+(aq) + e^- \longrightarrow Na(s)$	−2.71
$Mg^{2+}(aq) + 2e^- \longrightarrow Mg(s)$	−2.37
$Be^{2+}(aq) + 2e^- \longrightarrow Be(s)$	−1.85
$Al^{3+}(aq) + 3e^- \longrightarrow Al(s)$	−1.66
$Mn^{2+}(aq) + 2e^- \longrightarrow Mn(s)$	−1.18
$2H_2O + 2e^- \longrightarrow H_2(g) + 2OH^-(aq)$	−0.83
$Zn^{2+}(aq) + 2e^- \longrightarrow Zn(s)$	−0.76
$Cr^{3+}(aq) + 3e^- \longrightarrow Cr(s)$	−0.74
$Fe^{2+}(aq) + 2e^- \longrightarrow Fe(s)$	−0.44
$Cd^{2+}(aq) + 2e^- \longrightarrow Cd(s)$	−0.40
$PbSO_4(s) + 2e^- \longrightarrow Pb(s) + SO_4^{2-}(aq)$	−0.31
$Co^{2+}(aq) + 2e^- \longrightarrow Co(s)$	−0.28
$Ni^{2+}(aq) + 2e^- \longrightarrow Ni(s)$	−0.25
$Sn^{2+}(aq) + 2e^- \longrightarrow Sn(s)$	−0.14
$Pb^{2+}(aq) + 2e^- \longrightarrow Pb(s)$	−0.13
$2H^+(aq) + 2e^- \longrightarrow H_2(g)$	0.00
$Sn^{4+}(aq) + 2e^- \longrightarrow Sn^{2+}(aq)$	+0.13
$Cu^{2+}(aq) + e^- \longrightarrow Cu^+(aq)$	+0.15
$SO_4^{2-}(aq) + 4H^+(aq) + 2e^- \longrightarrow SO_2(g) + 2H_2O$	+0.20
$AgCl(s) + e^- \longrightarrow Ag(s) + Cl^-(aq)$	+0.22
$Cu^{2+}(aq) + 2e^- \longrightarrow Cu(s)$	+0.34
$O_2(g) + 2H_2O + 4e^- \longrightarrow 4OH^-(aq)$	+0.40
$I_2(s) + 2e^- \longrightarrow 2I^-(aq)$	+0.53
$MnO_4^-(aq) + 2H_2O + 3e^- \longrightarrow MnO_2(s) + 4OH^-(aq)$	+0.59
$O_2(g) + 2H^+(aq) + 2e^- \longrightarrow H_2O_2(aq)$	+0.68
$Fe^{3+}(aq) + e^- \longrightarrow Fe^{2+}(aq)$	+0.77
$Ag^+(aq) + e^- \longrightarrow Ag(s)$	+0.80
$Hg_2^{2+}(aq) + 2e^- \longrightarrow 2Hg(l)$	+0.85
$2Hg^{2+}(aq) + 2e^- \longrightarrow Hg_2^{2+}(aq)$	+0.92
$NO_3^-(aq) + 4H^+(aq) + 3e^- \longrightarrow NO(g) + 2H_2O$	+0.96
$Br_2(l) + 2e^- \longrightarrow 2Br^-(aq)$	+1.07
$O_2(g) + 4H^+(aq) + 4e^- \longrightarrow 2H_2O$	+1.23
$MnO_2(s) + 4H^+(aq) + 2e^- \longrightarrow Mn^{2+}(aq) + 2H_2O$	+1.23
$Cr_2O_7^{2-}(aq) + 14H^+(aq) + 6e^- \longrightarrow 2Cr^{3+}(aq) + 7H_2O$	+1.33
$Cl_2(g) + 2e^- \longrightarrow 2Cl^-(aq)$	+1.36
$Au^{3+}(aq) + 3e^- \longrightarrow Au(s)$	+1.50
$MnO_4^-(aq) + 8H^+(aq) + 5e^- \longrightarrow Mn^{2+}(aq) + 4H_2O$	+1.51
$Ce^{4+}(aq) + e^- \longrightarrow Ce^{3+}(aq)$	+1.61
$PbO_2(s) + 4H^+(aq) + SO_4^{2-}(aq) + 2e^- \longrightarrow PbSO_4(s) + 2H_2O$	+1.70
$H_2O_2(aq) + 2H^+(aq) + 2e^- \longrightarrow 2H_2O$	+1.77
$Co^{3+}(aq) + e^- \longrightarrow Co^{2+}(aq)$	+1.82
$O_3(g) + 2H^+(aq) + 2e^- \longrightarrow O_2(g) + H_2O(l)$	+2.07
$F_2(g) + 2e^- \longrightarrow 2F^-(aq)$	+2.87

*For all half-reactions the concentration is 1 *M* for dissolved species and the pressure is 1 atm for gases.

- Under standard-state conditions, any species on the left of a given half-cell reaction will react spontaneously with a species that appears on the right of any half-cell reaction located *above it* in Table 19.1. This is sometimes called the *diagonal rule*. In the case of the Cu/Zn galvanic cell

$$Zn^{2+}(aq,\ 1\ M) + 2e^- \longrightarrow Zn(s) \qquad \mathscr{E}^\circ = -0.76\ V$$

$$Cu^{2+}(aq,\ 1\ M) + 2e^- \longrightarrow Cu(s) \qquad \mathscr{E}^\circ = +0.34\ V$$

We see that the substance on the left of the second half-cell reaction is Cu^{2+} and the substance on the right in the first half-cell reaction is Zn. Therefore, as we saw earlier, Zn spontaneously reduces Cu^{2+} to form Zn^{2+} and Cu.

- Changing the stoichiometric coefficients of a half-cell reaction *does not* affect the value of $\mathscr{E}^\circ$ because electrode potentials are intensive properties. For example, from Table 19.1

$$I_2(s) + 2e^- \longrightarrow 2I^-(aq,\ 1\ M) \qquad \mathscr{E}^\circ = 0.53\ V$$

but $\mathscr{E}^\circ$ does not change if we write the reaction

$$2I_2(s) + 4e^- \longrightarrow 4I^-(aq,\ 1\ M) \qquad \mathscr{E}^\circ = 0.53\ V$$

- As noted earlier, $\mathscr{E}^\circ$ changes sign whenever we reverse a half-cell reaction.

Of the following examples, one deals with redox reactions in which the oxidizing agent and the reducing agent are in the same medium; in another they are linked in a galvanic cell; and one example applies equally to either situation. Using Table 19.1 and the diagonal rule, we can compare the strengths of oxidizing (or reducing) agents (Example 19.1) and the spontaneity of a redox process (Example 19.2). In calculating the emf of a cell (Example 19.3), we first have to identify the anode and the cathode (again using the diagonal rule). Next we reverse one of the reduction half-reactions in Table 19.1 to write the half-reactions of the cell and to calculate $\mathscr{E}^\circ_{cell}$.

EXAMPLE 19.1

Arrange the following species in order of increasing strength as oxidizing agents: MnO_4^- (in acidic solution), Sn^{2+}, Al^{3+}, Co^{3+}, and Ag^+. Assume all species are in their standard states.

Answer

Consulting Table 19.1, we write the half-reactions in the order in which they appear there:

$$Al^{3+}(aq,\ 1\ M) + 3e^- \longrightarrow Al(s) \qquad \mathscr{E}^\circ = -1.66\ V$$

$$Sn^{2+}(aq,\ 1\ M) + 2e^- \longrightarrow Sn(s) \qquad \mathscr{E}^\circ = -0.14\ V$$

$$Ag^+(aq,\ 1\ M) + e^- \longrightarrow Ag(s) \qquad \mathscr{E}^\circ = 0.80\ V$$

$$MnO_4^-(aq,\ 1\ M) + 8H^+(aq,\ 1\ M) + 5e^- \longrightarrow Mn^{2+}(aq,\ 1\ M) + 4H_2O(l) \qquad \mathscr{E}^\circ = 1.51\ V$$

$$Co^{3+}(aq,\ 1\ M) + e^- \longrightarrow Co^{2+}(aq,\ 1\ M) \qquad \mathscr{E}^\circ = 1.82\ V$$

Remember that the more positive the standard reduction potential, the greater the tendency of the species to be reduced, or the stronger the species as an oxidizing agent. We find that the oxidizing agents increase in strength as follows:

$$Al^{3+} < Sn^{2+} < Ag^{+} < MnO_4^{-} < Co^{3+}$$

Similar problem: 19.13.

EXAMPLE 19.2

Predict what will happen if molecular bromine (Br_2) is added to a solution containing NaCl and NaI at 25°C. Assume all species are in their standard states.

Answer

To predict what redox reaction(s) will take place, we need to compare the standard reduction potentials for the following half-reactions:

$$I_2(s) + 2e^- \longrightarrow 2I^-(aq,\ 1\ M) \qquad \mathscr{E}° = 0.53\ \text{V}$$

$$Br_2(l) + 2e^- \longrightarrow 2Br^-(aq,\ 1\ M) \qquad \mathscr{E}° = 1.07\ \text{V}$$

$$Cl_2(g,\ 1\ \text{atm}) + 2e^- \longrightarrow 2Cl^-(aq,\ 1\ M) \qquad \mathscr{E}° = 1.36\ \text{V}$$

Applying the diagonal rule we see that Br_2 will oxidize I^- but will not oxidize Cl^-. Therefore, the only redox reaction that will occur appreciably under standard-state conditions is

$$\begin{array}{ll} \text{Oxidation:} & 2I^-(aq,\ 1\ M) \longrightarrow I_2(s) + 2e^- \\ \text{Reduction:} & Br_2(l) + 2e^- \longrightarrow 2Br^-(aq,\ 1\ M) \\ \hline \text{Overall:} & 2I^-(aq,\ 1\ M) + Br_2(l) \longrightarrow I_2(s) + 2Br^-(aq,\ 1\ M) \end{array}$$

Note that the Na^+ ions are inert and do not enter into the redox reaction.

Similar problems: 19.11, 19.14.

EXAMPLE 19.3

A galvanic cell consists of a Mg electrode in a 1.0 *M* $Mg(NO_3)_2$ solution and a Ag electrode in a 1.0 *M* $AgNO_3$ solution. Calculate the standard emf of this electrochemical cell at 25°C.

Answer

Table 19.1 gives the standard reduction potentials of the two electrodes:

$$Mg^{2+}(aq,\ 1\ M) + 2e^- \longrightarrow Mg(s) \qquad \mathscr{E}° = -2.37\ \text{V}$$

$$Ag^+(aq,\ 1\ M) + e^- \longrightarrow Ag(s) \qquad \mathscr{E}° = 0.80\ \text{V}$$

Applying the diagonal rule, we see that Ag^+ will oxidize Mg:

$$\begin{array}{lrl} \text{Anode:} & Mg(s) & \longrightarrow Mg^{2+}(aq,\ 1\ M) + 2e^- \\ \text{Cathode:} & 2Ag^+(aq,\ 1\ M) + 2e^- & \longrightarrow 2Ag(s) \\ \hline \text{Overall:} & Mg(s) + 2Ag^+(aq,\ 1\ M) & \longrightarrow Mg^{2+}(aq,\ 1\ M) + 2Ag(s) \end{array}$$

Note that in order to balance the overall equation we multiplied the reduction of Ag^+ by 2. We can do so because as an intensive property $\mathscr{E}°$ is not affected by this procedure. We find the emf of the cell by using Equation (19.1) and Table 19.1:

$$\begin{aligned} \mathscr{E}°_{cell} &= \mathscr{E}°_{Mg/Mg^{2+}} + \mathscr{E}°_{Ag^+/Ag} \\ &= 2.37\ V + 0.80\ V \\ &= 3.17\ V \end{aligned}$$

The positive value of $\mathscr{E}°$ shows that the overall reaction is spontaneous.

Similar problems: 19.7, 19.8, 19.9.

19.4 Spontaneity of Redox Reactions

Table 19.1 enables us to predict the outcome of redox reactions under standard-state conditions, whether they take place in an electrochemical cell, where the reducing agent and oxidizing agent are physically separated from each other, or in a beaker, where the reactants are all mixed together. Our next step is to see how $\mathscr{E}°_{cell}$ is related to other thermodynamic quantities.

In Chapter 18 we saw that the free-energy change (decrease) in a spontaneous process is the energy available to do work (p. 760). In fact, for a process carried out at constant temperature and pressure

$$\Delta G = w_{max}$$

where w_{max} is the maximum amount of work that can be done.

In a galvanic cell, chemical energy is converted into electrical energy. Electrical energy in this case is the product of the emf of the cell and the total electrical charge (in coulombs) that passes through the cell:

$$\begin{aligned} \text{electrical energy} &= \text{volts} \times \text{coulombs} \\ &= \text{joules} \end{aligned}$$

***n* is the number of electrons exchanged between the reducing agent and the oxidizing agent in the overall redox equation.**

The total charge is determined by the number of moles of electrons (n) that pass through the circuit. By definition

$$\text{total charge} = nF$$

where F, the Faraday† constant, is the electrical charge contained in 1 mole of electrons. Experimentally, 1 ***faraday*** has been found to be *equivalent to 96,487 coulombs,* or 96,500 coulombs, rounded off to three significant figures. Thus

$$1\ F = 96{,}500\ \text{C/mol}$$

† Michael Faraday (1791–1867). English chemist and physicist. Faraday is regarded by many as the greatest experimental scientist of the nineteenth century. He started as an apprentice to a bookbinder at the age of 13, but became interested in science after reading a book on chemistry. Faraday invented the electric motor and was the first person to demonstrate the principle governing electrical generators. Besides making notable contributions to the fields of electricity and magnetism, Faraday also worked on optical activity, and discovered and named benzene.

Since $$1 \text{ J} = 1 \text{ C} \times 1 \text{ V}$$

we can also express the units of faraday as

$$1\ F = 96{,}500 \text{ J/V} \cdot \text{mol}$$

A common device for measuring the cell's emf is the *potentiometer,* which can precisely match the voltage of the cell without actually draining any current from the cell. Thus the measured emf is the *maximum* voltage that the cell can achieve. This value is used to calculate the maximum amount of electrical energy that can be obtained from the chemical reaction. This energy is used to do electrical work (w_{ele}), so

$$\begin{aligned} w_{max} &= w_{ele} \\ &= -nF\mathscr{E}_{cell} \end{aligned}$$

The sign convention for electrical work is the same as that for *P-V* work discussed in Section 6.7.

The negative sign on the right-hand side indicates that the electrical work is done by the system on the surroundings. Now, since

$$\Delta G = w_{max}$$

we obtain $$\Delta G = -nF\mathscr{E}_{cell} \tag{19.2}$$

Both n and F are positive quantities and ΔG is negative for a spontaneous process, so $\mathscr{E}_{cell}$ is positive. For reactions in which reactants and products are in their standard states, Equation (19.2) becomes

$$\Delta G^\circ = -nF\mathscr{E}^\circ_{cell} \tag{19.3}$$

Here again, $\mathscr{E}^\circ_{cell}$ is positive for a spontaneous process.

In Section 18.4 we saw that the standard free-energy change ΔG° for a reaction is related to its equilibrium constant as follows [see Equation (18.10)]:

$$\Delta G^\circ = -RT \ln K$$

Therefore, from Equations (19.3) and (18.10) we obtain

$$-nF\mathscr{E}^\circ_{cell} = -RT \ln K$$

Solving for $\mathscr{E}^\circ_{cell}$ $$\mathscr{E}^\circ_{cell} = \frac{RT}{nF} \ln K \tag{19.4}$$

When $T = 298$ K, Equation (19.4) can be simplified by substituting for R and F and converting the natural logarithm to the common (base-10) logarithm:

$$\begin{aligned} \mathscr{E}^\circ_{cell} &= \frac{2.303(8.314 \text{ J/K} \cdot \text{mol})(298 \text{ K})}{n(96{,}500 \text{ J/V} \cdot \text{mol})} \log K \\ &= \frac{0.0591 \text{ V}}{n} \log K \end{aligned} \tag{19.5}$$

Thus, if any one of the three quantities ΔG°, K, or $\mathscr{E}^\circ_{cell}$ is known, the other two can be calculated using Equation (18.10), Equation (19.3), or Equation (19.4). We can summarize the relationships among ΔG°, K, and $\mathscr{E}^\circ_{cell}$ and characterize the spontaneity of a redox reaction as follows:

ΔG°	K	$\mathscr{E}^\circ_{cell}$	Reaction under standard-state conditions
Negative	> 1	Positive	Spontaneous
0	$= 1$	0	At equilibrium
Positive	< 1	Negative	Nonspontaneous. Reaction is spontaneous in the reverse direction.

The following examples deal with the spontaneity of redox reactions.

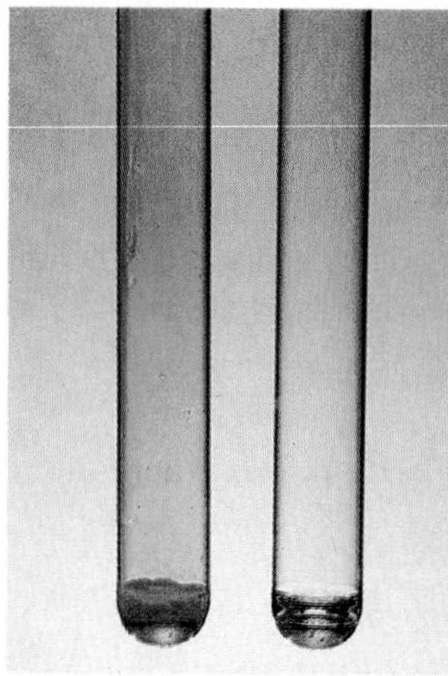

HNO_3 (left) can oxidize Hg, but HCl (right) cannot.

EXAMPLE 19.4

Using data in Table 19.1, calculate $\mathscr{E}°$ for the reactions of mercury with (a) 1 *M* HCl and (b) 1 *M* HNO_3. Which acid will oxidize Hg to Hg_2^{2+} under standard-state conditions?

Answer

(a) HCl: First we write the half-reactions:

$$\begin{array}{lrcl} \text{Oxidation:} & 2Hg(l) & \longrightarrow & Hg_2^{2+}(aq, 1\,M) + 2e^- \\ \text{Reduction:} & 2H^+(aq, 1\,M) + 2e^- & \longrightarrow & H_2(g, 1\text{ atm}) \\ \hline \text{Overall:} & 2Hg(l) + 2H^+(aq, 1\,M) & \longrightarrow & Hg_2^{2+}(aq, 1\,M) + H_2(g, 1\text{ atm}) \end{array}$$

The standard emf, $\mathscr{E}°$, is given by

$$\begin{aligned} \mathscr{E}° &= \mathscr{E}°_{Hg/Hg_2^{2+}} + \mathscr{E}°_{H^+/H_2} \\ &= -0.85\text{ V} + 0 \\ &= -0.85\text{ V} \end{aligned}$$

(We omit the subscript "cell" because this reaction is not carried out in an electrochemical cell.) Since $\mathscr{E}°$ is negative, we conclude that mercury is not oxidized by hydrochloric acid under standard-state conditions.

(b) HNO_3: The reactions are

$$\begin{array}{ll} \text{Oxidation:} & 3[2Hg(l) \longrightarrow Hg_2^{2+}(aq, 1\,M) + 2e^-] \\ \text{Reduction:} & 2[NO_3^-(aq, 1\,M) + 4H^+(aq, 1\,M) + 3e^- \longrightarrow NO(g, 1\text{ atm}) + 2H_2O(l)] \\ \hline \text{Overall:} & 6Hg(l) + 2NO_3^-(aq, 1\,M) + 8H^+(aq, 1\,M) \longrightarrow \\ & \qquad 3Hg_2^{2+}(aq, 1\,M) + 2NO(g, 1\text{ atm}) + 4H_2O(l) \end{array}$$

Thus

$$\begin{aligned} \mathscr{E}° &= \mathscr{E}°_{Hg/Hg_2^{2+}} + \mathscr{E}°_{NO_3^-/NO} \\ &= -0.85\text{ V} + 0.96\text{ V} \\ &= 0.11\text{ V} \end{aligned}$$

Since $\mathscr{E}°$ is positive, the reaction is spontaneous under standard-state conditions.

Similar problem: 19.17.

EXAMPLE 19.5

Calculate the equilibrium constant for the following reaction at 25°C:

$$Sn(s) + 2Cu^{2+}(aq) \rightleftharpoons Sn^{2+}(aq) + 2Cu^+(aq)$$

Answer

The two half-reactions for the overall process are

$$\text{Oxidation:} \qquad \text{Sn}(s) \longrightarrow \text{Sn}^{2+}(aq) + 2e^-$$

$$\text{Reduction:} \qquad 2\text{Cu}^{2+}(aq) + 2e^- \longrightarrow 2\text{Cu}^+(aq)$$

In Table 19.1 we find that $\mathscr{E}^\circ_{\text{Sn}^{2+}/\text{Sn}} = -0.14$ V and $\mathscr{E}^\circ_{\text{Cu}^{2+}/\text{Cu}^+} = 0.15$ V. Thus

$$\begin{aligned} \mathscr{E}^\circ &= \mathscr{E}^\circ_{\text{Sn/Sn}^{2+}} + \mathscr{E}^\circ_{\text{Cu}^{2+}/\text{Cu}^+} \\ &= 0.14 \text{ V} + 0.15 \text{ V} \\ &= 0.29 \text{ V} \end{aligned}$$

Equation (19.5) can be written

$$\log K = \frac{n\mathscr{E}^\circ}{0.0591 \text{ V}}$$

In the overall reaction we find $n = 2$. Therefore

$$\begin{aligned} \log K &= \frac{(2)(0.29 \text{ V})}{0.0591 \text{ V}} = 9.8 \\ K &= 6 \times 10^9 \end{aligned}$$

Similar problems: 19.21, 19.22.

EXAMPLE 19.6

Calculate the standard free-energy change for the following reaction at 25°C:

$$2\text{Au}(s) + 3\text{Ca}^{2+}(aq, 1\,M) \longrightarrow 2\text{Au}^{3+}(aq, 1\,M) + 3\text{Ca}(s)$$

Answer

First we break up the overall reaction into half-reactions:

$$\text{Oxidation:} \qquad 2\text{Au}(s) \longrightarrow 2\text{Au}^{3+}(aq, 1\,M) + 6e^-$$

$$\text{Reduction:} \qquad 3\text{Ca}^{2+}(aq, 1\,M) + 6e^- \longrightarrow 3\text{Ca}(s)$$

In Table 19.1 we find that $\mathscr{E}^\circ_{\text{Au}^{3+}/\text{Au}} = 1.50$ V and $\mathscr{E}^\circ_{\text{Ca}^{2+}/\text{Ca}} = -2.87$ V. Therefore

$$\begin{aligned} \mathscr{E}^\circ &= \mathscr{E}^\circ_{\text{Au/Au}^{3+}} + \mathscr{E}^\circ_{\text{Ca}^{2+}/\text{Ca}} \\ &= -1.50 \text{ V} - 2.87 \text{ V} \\ &= -4.37 \text{ V} \end{aligned}$$

Now we use Equation (19.3): $\Delta G^\circ = -nF\mathscr{E}^\circ$

The overall reaction shows that $n = 6$, so

$$\begin{aligned} \Delta G^\circ &= -(6 \text{ mol})(96{,}500 \text{ J/V} \cdot \text{mol})(-4.37 \text{ V}) \\ &= 2.53 \times 10^6 \text{ J} \\ &= 2.53 \times 10^3 \text{ kJ} \end{aligned}$$

The large positive value of ΔG° tells us that the reaction is not spontaneous under standard-state conditions at 25°C.

Similar problem: 19.23.

The Chemistry in Action on p. 810 shows that dental filling discomfort is an electrochemical phenomenon.

19.5 Effect of Concentration on Cell EMF

The Nernst Equation

So far we have concentrated on redox reactions in which reactants and products are in their standard states. However, standard-state conditions are often difficult and sometimes impossible to maintain. The relationship between the emf of a cell and the concentrations of reactants and products in a redox reaction of the type

$$a\text{A} + b\text{B} \longrightarrow c\text{C} + d\text{D}$$

can be derived as follows. From Equation (18.9) we write

$$\Delta G = \Delta G^\circ + RT \ln Q$$

We omit the subscript "cell" in $\mathscr{E}$ and $\mathscr{E}^\circ$ for convenience.

Since $\Delta G = -nF\mathscr{E}$ and $\Delta G^\circ = -nF\mathscr{E}^\circ$, the equation can be expressed as

$$-nF\mathscr{E} = -nF\mathscr{E}^\circ + RT \ln Q$$

Dividing the equation through by $-nF$ and converting to common logarithms, we get

$$\mathscr{E} = \mathscr{E}^\circ - \frac{2.303RT}{nF} \log Q \qquad (19.6)$$

Equation (19.6) is known as the Nernst† equation in which Q is the reaction quotient (see Section 14.5). At 298 K, Equation (19.6) can be rewritten as

$$\mathscr{E} = \mathscr{E}^\circ - \frac{0.0591 \text{ V}}{n} \log Q \qquad (19.7)$$

At equilibrium, there is no net transfer of electrons, so $\mathscr{E} = 0$ and $Q = K$, where K is the equilibrium constant. In this case, Equation (19.6) takes the form

$$0 = \mathscr{E}^\circ - \frac{2.303RT}{nF} \log K$$

or

$$\mathscr{E}^\circ = \frac{2.303RT}{nF} \log K$$

which is the same as Equation (19.4) except for the conversion to common logs.

The Nernst equation enables us to calculate $\mathscr{E}$ as a function of reactant and product concentrations in a redox reaction. Referring to the galvanic cell in Figure 19.1,

$$\text{Zn}(s) + \text{Cu}^{2+}(aq) \longrightarrow \text{Zn}^{2+}(aq) + \text{Cu}(s)$$

The Nernst equation for this cell at 25°C can be written as

Remember that concentrations of pure solids (and pure liquids) do not appear in the expression for Q.

$$\mathscr{E} = 1.10 \text{ V} - \frac{0.0591 \text{ V}}{2} \log \frac{[\text{Zn}^{2+}]}{[\text{Cu}^{2+}]}$$

† Walter Hermann Nernst (1864–1941). German chemist and physicist. Nernst's work was mainly on electrolyte solution and thermodynamics. He also invented an electric piano. Nernst was awarded the Nobel Prize in chemistry in 1920 for his contribution to thermodynamics.

If the ratio $[Zn^{2+}]/[Cu^{2+}]$ is less than 1, $\log [Zn^{2+}]/[Cu^{2+}]$ is a negative number so that the second term on the right-hand side of the above equation is positive. Under this condition $\mathscr{E}$ is greater than the standard emf $\mathscr{E}°$. If the ratio is greater than 1, $\mathscr{E}$ is smaller than $\mathscr{E}°$.

The following example illustrates the use of the Nernst equation.

EXAMPLE 19.7

Predict whether the following reaction would proceed spontaneously as written at 298 K:

$$Co(s) + Fe^{2+}(aq) \longrightarrow Co^{2+}(aq) + Fe(s)$$

given that $[Co^{2+}] = 0.15\ M$ and $[Fe^{2+}] = 0.68\ M$.

Answer

The half-reactions are

$$\text{Oxidation:} \quad Co(s) \longrightarrow Co^{2+}(aq) + 2e^-$$

$$\text{Reduction:} \quad Fe^{2+}(aq) + 2e^- \longrightarrow Fe(s)$$

In Table 19.1 we find that $\mathscr{E}°_{Co^{2+}/Co} = -0.28$ V and $\mathscr{E}°_{Fe^{2+}/Fe} = -0.44$ V. Therefore, the standard emf is

$$\begin{aligned}\mathscr{E}° &= \mathscr{E}°_{Co/Co^{2+}} + \mathscr{E}°_{Fe^{2+}/Fe} \\ &= 0.28\ V + (-0.44\ V) \\ &= -0.16\ V\end{aligned}$$

From Equation (19.7) we write

$$\begin{aligned}\mathscr{E} &= \mathscr{E}° - \frac{0.0591\ V}{n} \log \frac{[Co^{2+}]}{[Fe^{2+}]} \\ &= -0.16\ V - \frac{0.0591\ V}{2} \log \frac{0.15}{0.68} \\ &= -0.16\ V + 0.019\ V \\ &= -0.14\ V\end{aligned}$$

Since $\mathscr{E}$ is negative (or ΔG is positive), the reaction is *not* spontaneous in the direction written.

Similar problems: 19.28, 19.31.

It is interesting to determine at what ratio of $[Co^{2+}]$ to $[Fe^{2+}]$ the reaction in Example 19.7 will become spontaneous. We start with the Nernst equation as rewritten for 298 K [Equation (19.7)]:

$$\mathscr{E} = \mathscr{E}° - \frac{0.0591\ V}{n} \log Q$$

When $\mathscr{E} = 0$, $Q = K$.

We set $\mathscr{E}$ equal to zero since this corresponds to the equilibrium situation.

$$0 = -0.16\ V - \frac{0.0591\ V}{2} \log \frac{[Co^{2+}]}{[Fe^{2+}]}$$

$$\log \frac{[Co^{2+}]}{[Fe^{2+}]} = -5.4$$

Taking the antilog of both sides, we obtain

$$\frac{[Co^{2+}]}{[Fe^{2+}]} = 4 \times 10^{-6} = K$$

Thus for the reaction to be spontaneous, the ratio $[Co^{2+}]/[Fe^{2+}]$ must be smaller than 4×10^{-6}.

As the following example shows, if gases are involved in the cell reaction, their concentrations should be expressed in atm.

EXAMPLE 19.8

Consider the galvanic cell shown in Figure 19.3(a). In a certain experiment, the emf ($\mathscr{E}$) of the cell is found to be 0.54 V at 25°C. Suppose that $[Zn^{2+}] = 1.0\ M$ and $P_{H_2} =$ 1.0 atm. Calculate the molar concentration of H^+.

Answer

The overall cell reaction is

$$Zn(s) + 2H^+(aq,\ ?\ M) \longrightarrow Zn^{2+}(aq,\ 1\ M) + H_2(g,\ 1\ \text{atm})$$

As we saw earlier (p. 783), the standard emf for the cell is 0.76 V. From Equation (19.7), we write

$$\mathscr{E} = \mathscr{E}° - \frac{0.0591\ \text{V}}{n} \log \frac{[Zn^{2+}]P_{H_2}}{[H^+]^2}$$

$$0.54\ \text{V} = 0.76\ \text{V} - \frac{0.0591\ \text{V}}{2} \log \frac{(1.0)(1.0)}{[H^+]^2}$$

$$-0.22\ \text{V} = -\frac{0.0591\ \text{V}}{2} \log \frac{1}{[H^+]^2}$$

$$7.4 = \log \frac{1}{[H^+]^2}$$

$$7.4 = -2 \log [H^+]$$

$$\log [H^+] = -3.7$$

$$[H^+] = 2 \times 10^{-4}\ M$$

Similar problems: 19.29, 19.30.

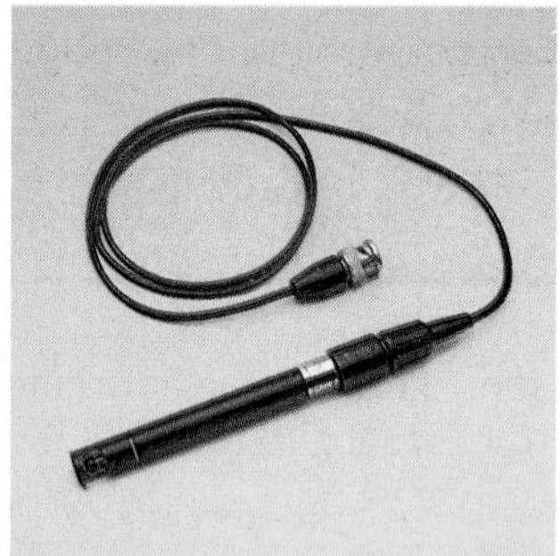

Figure 19.4 *A glass electrode that is used in conjunction with a reference electrode in a pH meter.*

The preceding example shows that a galvanic cell whose cell reaction involves H^+ ions can be used to measure $[H^+]$ or pH. The pH meter described in Section 15.2 is based on this principle, but for practical reasons the electrodes used in a pH meter are quite different from the SHE and zinc electrode in the galvanic cell (Figure 19.4).

Concentration Cells

We have seen that an electrode potential depends on the concentrations of the ions used. In practice, a cell may be constructed from two half-cells composed of the *same* material but differing in ion concentrations. Such a cell is called a *concentration cell.*

Consider a situation in which zinc electrodes are put into two aqueous solutions of zinc sulfate at 0.10 *M* and 1.0 *M* concentrations. The two solutions are connected by a salt bridge, and the electrodes are joined by a piece of wire in an arrangement like that shown in Figure 19.1. According to Le Chatelier's principle, the tendency for the reduction

$$Zn^{2+}(aq) + 2e^- \longrightarrow Zn(s)$$

increases with increasing concentration of Zn^{2+} ions. Therefore, reduction should occur in the more concentrated compartment and oxidation should take place on the more dilute side. The cell diagram is

$$Zn(s)|Zn^{2+}(aq,\ 0.10\ M)|KCl(sat'd)|Zn^{2+}(aq,\ 1.0\ M)|Zn(s)$$

and the half-reactions are

$$\begin{array}{lrcl} \text{Oxidation:} & Zn(s) & \longrightarrow & Zn^{2+}(aq,\ 0.10\ M) + 2e^- \\ \text{Reduction:} & Zn^{2+}(aq,\ 1.0\ M) + 2e^- & \longrightarrow & Zn(s) \\ \hline \text{Overall:} & Zn^{2+}(aq,\ 1.0\ M) & \longrightarrow & Zn^{2+}(aq,\ 0.10\ M) \end{array}$$

The emf of the cell is

$$\mathscr{E} = \mathscr{E}° - \frac{0.0591\ \text{V}}{n} \log \frac{[\text{Zn}]_{\text{dil}}}{[\text{Zn}]_{\text{conc}}}$$

where the subscripts ''dil'' and ''conc'' refer to the 0.10 *M* and 1.0 *M* concentrations, respectively. $\mathscr{E}°$ for this cell is zero (the *same* electrode and the same type of ions are involved), so

$$\begin{aligned} \mathscr{E} &= -\frac{0.0591\ \text{V}}{2} \log \frac{0.10}{1.0} \\ &= 0.0296\ \text{V} \end{aligned}$$

The emf of concentration cells is usually small and decreases continually during the operation of the cell as the concentrations in the two compartments approach each other. When the concentrations of the ions in the two compartments are the same, $\mathscr{E}$ becomes zero.

19.6 Batteries

A ***battery*** is an electrochemical cell, or often, *several electrochemical cells connected in series, that can be used as a source of direct electric current at a constant voltage.* Although the operation of a battery is similar in principle to that of the galvanic cells described in Section 19.2, a battery has the advantage of being completely self-contained and requiring no auxiliary components such as salt bridges. We will describe several types of batteries that are in widespread use.

The Dry Cell Battery

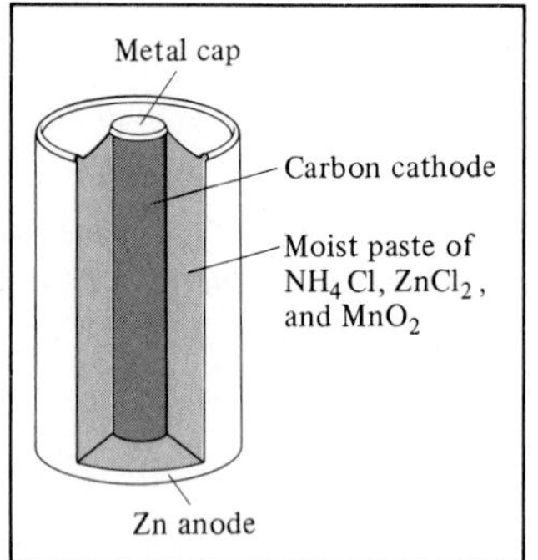

Figure 19.5 *Interior section of a dry cell of the kind used in flashlights and transistor radios. Actually, the cell is not completely dry, as it contains a moist electrolyte paste.*

The most common dry cell, that is, a cell without fluid component, is the *Leclanché cell,* used in flashlights and transistor radios. The anode of the cell consists of a zinc can or container that is in contact with manganese dioxide (MnO_2) and an electrolyte. The electrolyte consists of ammonium chloride and zinc chloride in water, to which starch is added to thicken the solution to a pastelike consistency so that it is less likely to leak (Figure 19.5). A carbon rod serves as the cathode, which is immersed in the electrolyte in the center of the cell. The cell reactions are

$$\begin{array}{ll} \text{Anode:} & Zn(s) \longrightarrow Zn^{2+}(aq) + 2e^- \\ \text{Cathode:} & 2NH_4^+(aq) + 2MnO_2(s) + 2e^- \longrightarrow Mn_2O_3(s) + 2NH_3(aq) + H_2O(l) \\ \hline \text{Overall:} & Zn(s) + 2NH_4^+(aq) + 2MnO_2(s) \longrightarrow \\ & \quad Zn^{2+}(aq) + 2NH_3(aq) + H_2O(l) + Mn_2O_3(s) \end{array}$$

Actually, this equation is an oversimplification, because the reactions that occur in the cell are quite complex. The voltage produced by a dry cell is about 1.5 V.

The Mercury Battery

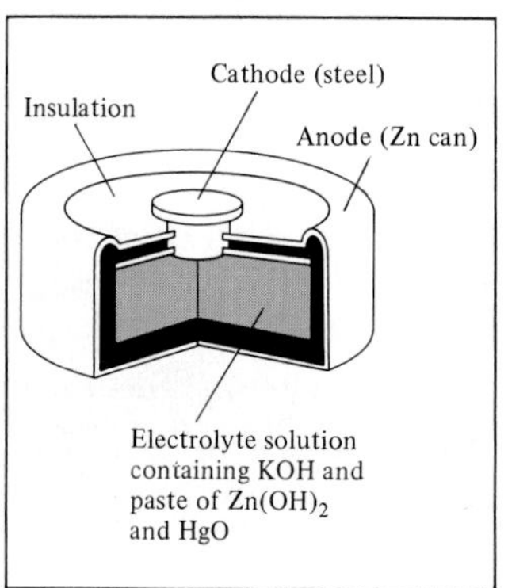

Figure 19.6 *Interior section of a mercury battery.*

The mercury battery is used extensively in medicine and electronic industries and is more expensive than the common dry cell. Contained in a stainless steel cylinder, the mercury battery consists of a zinc anode (amalgamated with mercury) in contact with a strongly alkaline electrolyte containing zinc oxide and mercury(II) oxide (Figure 19.6). The cell reactions are

$$\begin{array}{ll} \text{Anode:} & Zn(Hg) + 2OH^-(aq) \longrightarrow ZnO(s) + H_2O(l) + 2e^- \\ \text{Cathode:} & HgO(s) + H_2O(l) + 2e^- \longrightarrow Hg(l) + 2OH^-(aq) \\ \hline \text{Overall:} & Zn(Hg) + HgO(s) \longrightarrow ZnO(s) + Hg(l) \end{array}$$

Because there is no change in electrolyte composition during operation—the overall cell reaction involves only solid substances—the mercury battery provides a more constant voltage (1.35 V) than the Leclanché cell. It also has a considerably higher capacity and longer life. These qualities make the mercury battery ideal for use in pacemakers, hearing aids, electric watches, and light meters.

The Lead Storage Battery

The lead storage battery commonly used in automobiles consists of six identical cells joined together in series. Each cell has a lead anode and a cathode made of lead dioxide (PbO_2) packed on a metal plate (Figure 19.7). Both the cathode and the anode are immersed in an aqueous solution of sulfuric acid, which acts as the electrolyte. The cell reactions are

$$\begin{array}{ll} \text{Anode:} & Pb(s) + SO_4^{2-}(aq) \longrightarrow PbSO_4(s) + 2e^- \\ \text{Cathode:} & PbO_2(s) + 4H^+(aq) + SO_4^{2-}(aq) + 2e^- \longrightarrow PbSO_4(s) + 2H_2O(l) \\ \hline \text{Overall:} & Pb(s) + PbO_2(s) + 4H^+(aq) + 2SO_4^{2-}(aq) \longrightarrow 2PbSO_4(s) + 2H_2O(l) \end{array}$$

Under normal operating conditions, each cell produces 2 V; a total of 12 V from the six cells is used to power the ignition circuit of the automobile and its other electrical systems. The lead storage battery can deliver large amounts of current for a short time, such as the time it takes to start up the engine.

Unlike the Leclanché cell and the mercury battery, the lead storage battery is re-

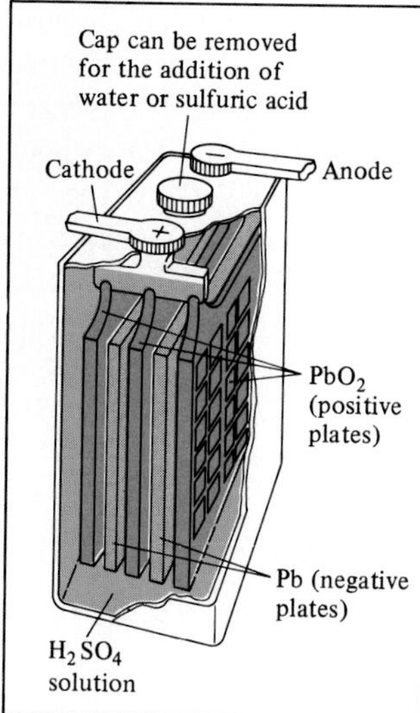

Figure 19.7 *Interior section of a lead storage battery. Under normal operating conditions, the concentration of the sulfuric acid solution is about 38 percent by mass.*

chargeable. Recharging the battery means reversing the normal electrochemical reaction by applying an external voltage at the cathode and the anode. (This kind of process is called *electrolysis,* which we will discuss in Section 19.8.) The reactions that replenish the original materials are

$$\begin{aligned}
&\text{Anode:} && PbSO_4(s) + 2e^- \longrightarrow Pb(s) + SO_4^{2-}(aq) \\
&\text{Cathode:} && PbSO_4(s) + 2H_2O(l) \longrightarrow PbO_2(s) + SO_4^{2-}(aq) + 4H^+(aq) + 2e^- \\
&\text{Overall:} && 2PbSO_4(s) + 2H_2O(l) \longrightarrow Pb(s) + PbO_2(s) + 4H^+(aq) + 2SO_4^{2-}(aq)
\end{aligned}$$

The overall reaction is exactly the opposite of the normal cell reaction.

Two aspects of the operation of a lead storage battery are worth noting. First, because the electrochemical reaction consumes sulfuric acid, the degree to which the battery has been discharged can be checked by measuring the density of the electrolyte with a hydrometer, as is usually done at gas stations. The density of the fluid in a "healthy," fully charged battery should be equal to or greater than 1.2 g/cm^3. Second, people living in cold climates sometimes have trouble starting their cars because the battery has "gone dead." Thermodynamic calculations show that the emf of many electrochemical cells decreases with decreasing temperature. However, for a lead storage battery, the temperature coefficient is about 1.5×10^{-4} V/°C; that is, there is a decrease in voltage of 1.5×10^{-4} V for every degree drop in temperature. Thus, even allowing for a 40°C change in temperature, the decrease in voltage amounts to only 6×10^{-3} V, which is about

$$\frac{6 \times 10^{-3}\ \text{V}}{12\ \text{V}} \times 100\% = 0.05\%$$

of the operating voltage, an insignificant change. The real cause of a battery's apparent breakdown is an increase in the viscosity of the electrolyte as the temperature decreases. For the battery to function properly, the electrolyte must be fully conducting. However, the ions move much more slowly in a viscous medium, so the resistance of the fluid increases, leading to a decrease in the power output of the battery. If an apparently "dead battery" is warmed to near room temperature, it recovers its ability to deliver normal power.

Solid-State Lithium Batteries

Unlike the batteries discussed so far, a solid-state battery employs a solid (rather than an aqueous solution or a water-based paste) as the electrolyte connecting the electrodes. Figure 19.8 shows the schematic of a solid-state lithium battery. There is much

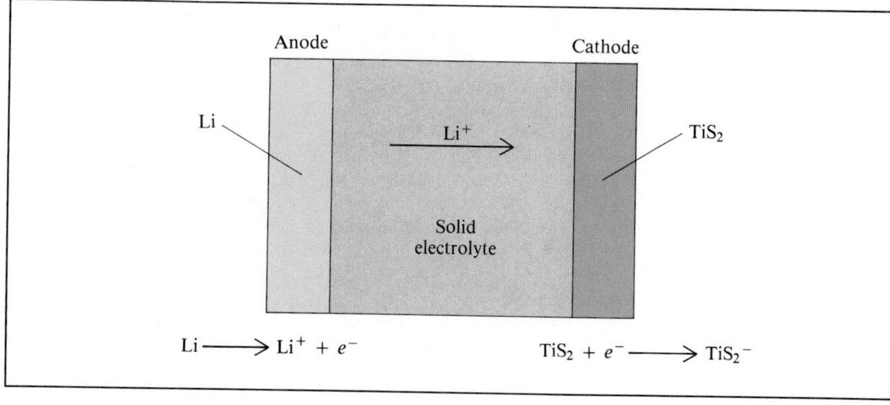

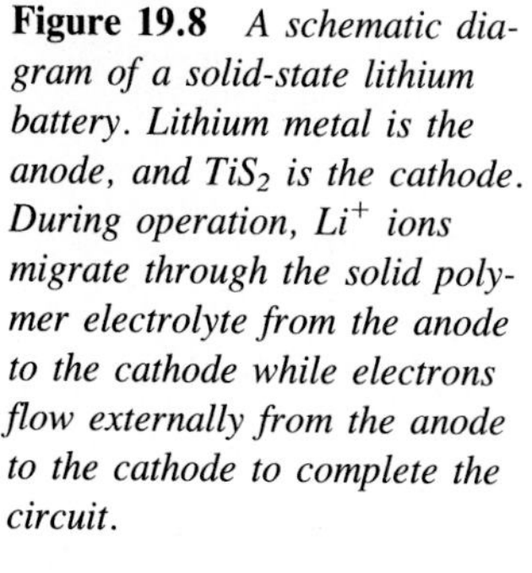
Figure 19.8 *A schematic diagram of a solid-state lithium battery. Lithium metal is the anode, and TiS_2 is the cathode. During operation, Li^+ ions migrate through the solid polymer electrolyte from the anode to the cathode while electrons flow externally from the anode to the cathode to complete the circuit.*

advantage in choosing lithium as the anode since, as Table 19.1 shows, it has the most negative $\mathscr{E}°$ value. Furthermore, lithium is a lightweight metal, so that only 6.941 g of Li (its molar mass) are needed to produce 1 mole of electrons. The electrolyte is made of a polymer material that permits the passage of ions but not of electrons. The cathode is made of a substance called an *insertion compound* (either TiS_2 or V_6O_{13}). The cell voltage of a solid-state lithium battery can be as high as 3 V, and it can be recharged in the same way as a lead storage battery. Although much work is still needed to develop these kinds of batteries with high reliability and long lifetimes, they are already being hailed as the batteries of the future.

Fuel Cells

Fossil fuels are a major source of energy, but conversion of fossil fuel into electrical energy is a highly inefficient process. Consider the combustion of methane:

$$CH_4(g) + 2O_2(g) \longrightarrow CO_2(g) + 2H_2O(l) + \text{energy}$$

To generate electricity, heat produced by the reaction is first used to convert water to steam, which then drives a turbine that drives a generator. An appreciable fraction of the energy released in the form of heat is lost to the surroundings at each step; the most efficient power plant now in existence converts only about 40 percent of the original chemical energy into electricity. Because combustion reactions are redox reactions, it would be more desirable to carry them out directly by electrochemical means, thereby greatly increasing the efficiency of power production. This objective has been accomplished by devices known as fuel cells.

In its simplest form, a hydrogen-oxygen fuel cell consists of an electrolyte solution, such as potassium hydroxide solution, and two inert electrodes. Hydrogen and oxygen gases are bubbled through the anode and cathode compartments (Figure 19.9), where the following reactions take place:

$$\begin{array}{ll} \text{Anode:} & 2H_2(g) + 4OH^-(aq) \longrightarrow 4H_2O(l) + 4e^- \\ \text{Cathode:} & O_2(g) + 2H_2O(l) + 4e^- \longrightarrow 4OH^-(aq) \\ \hline \text{Overall:} & 2H_2(g) + O_2(g) \longrightarrow 2H_2O(l) \end{array}$$

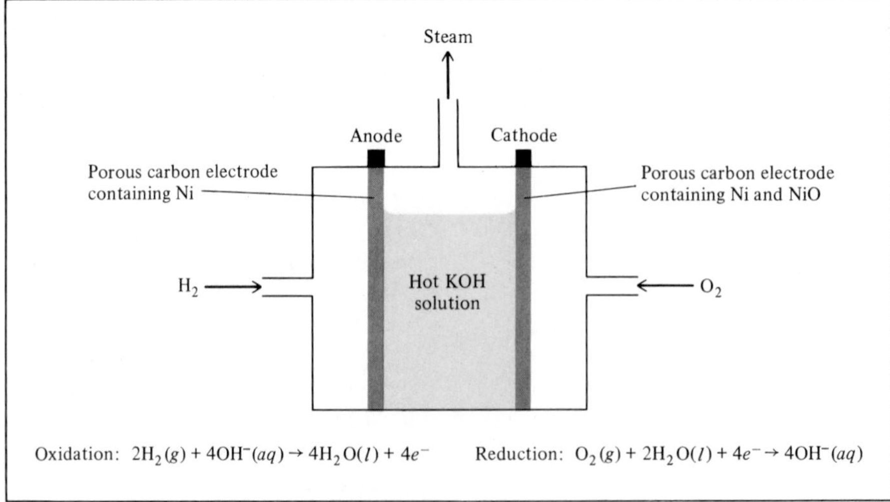

Figure 19.9 *A hydrogen-oxygen fuel cell. The Ni and NiO embedded in the porous carbon electrodes are electrocatalysts. When the cell is in operation, the electrodes are connected to an electrical appliance by conducting wires.*

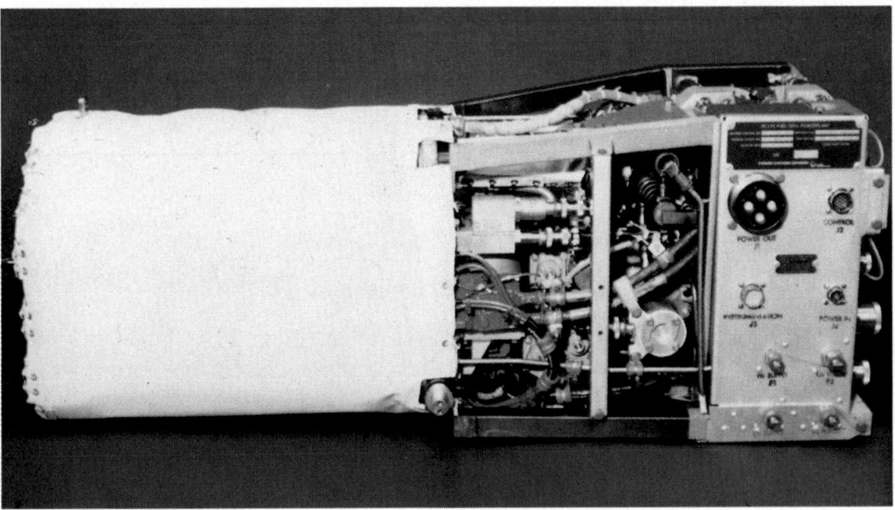

Figure 19.10 *A hydrogen-oxygen fuel cell used in the space program. The pure water produced by the cell is used by the astronauts.*

The standard emf of the cell can be calculated as follows, with data from Table 19.1:

$$\begin{aligned}\mathscr{E}^\circ &= \mathscr{E}^\circ_{\text{ox}} + \mathscr{E}^\circ_{\text{red}} \\ &= 0.83\text{ V} + 0.40\text{ V} \\ &= 1.23\text{ V}\end{aligned}$$

Thus the cell reaction is spontaneous under standard-state conditions. Note that the reaction is the same as the hydrogen combustion reaction, but the oxidation and reduction are carried out separately at the anode and the cathode. Like platinum in the standard hydrogen electrode, the electrodes have a twofold function. They serve as electrical conductors, and they provide the necessary surfaces for the initial decomposition of the molecules into atomic species, prior to electron transfer. They are *electrocatalysts*. Metals such as platinum, nickel, and rhodium are good electrocatalysts.

In addition to the H_2-O_2 system, a number of other fuel cells have been developed. Among these is the propane-oxygen fuel cell. The half-cell reactions are

$$\begin{array}{lrcl}\text{Anode:} & C_3H_8(g) + 6H_2O(l) & \longrightarrow & 3CO_2(g) + 20H^+(aq) + 20e^- \\ \text{Cathode:} & 5O_2(g) + 20H^+(aq) + 20e^- & \longrightarrow & 10H_2O(l) \\ \hline \text{Overall:} & C_3H_8(g) + 5O_2(g) & \longrightarrow & 3CO_2(g) + 4H_2O(l)\end{array}$$

The overall reaction is identical to the burning of propane in oxygen.

Unlike batteries, fuel cells do not store chemical energy. Reactants must be constantly resupplied, and products must be constantly removed from a fuel cell. In this respect, a fuel cell resembles an engine more than it does a battery. However, the fuel cell does not operate like a heat engine and therefore is not subject to the same kind of thermodynamic limitations in energy conversion (see Chemistry in Action on p. 770).

Properly designed fuel cells may be as much as 70 percent efficient, about twice as efficient as an internal combustion engine. In addition, fuel-cell generators are free of the noise, vibration, heat transfer, thermal pollution, and other problems normally associated with conventional power plants. Despite all these features, fuel cells are not yet in large-scale operation. A major problem lies in the choice and availability of suitable electrocatalysts able to function efficiently for long periods of time without contamination. The most successful application of fuel cells to date has been in space vehicles (Figure 19.10).

The Aluminum-Air Battery

The aluminum-air battery is really a combination of a battery and a fuel cell; it is one of the more promising new developments in electrical power generation. Aluminum is a

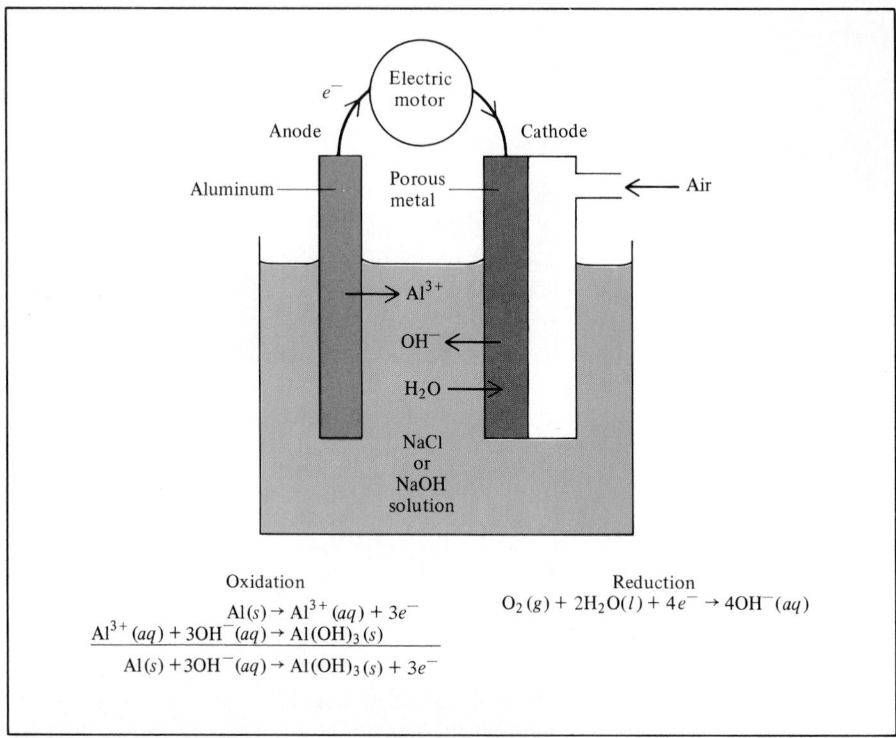

Figure 19.11 *The half-cell reactions in an aluminum-air battery.*

Figure 19.12 *An aluminum-air battery in operation.*

high-energy density fuel in the sense that it is capable of releasing large amounts of electrical energy per kilogram. This energy may be released by electrochemical conversion. Figure 19.11 shows a schematic diagram of the battery. Very high purity aluminum ($\gtrsim$99.99 percent) is used as the anode, and a constant supply of air is bubbled through the solution at the cathode, which is made of an inert, porous metal. The electrolyte can be an aqueous solution of either NaCl or NaOH. As you can see, at the anode the aluminum metal is converted to Al^{3+} and then to the insoluble $Al(OH)_3$, while O_2 is reduced to OH^- at the cathode. The overall reaction is

$$4Al(s) + 3O_2(g) + 6H_2O(l) \longrightarrow 4Al(OH)_3(s)$$

Since only Al and water are consumed in this reaction, recharging the battery involves only adding water, replacing the Al electrode (which acts as a solid fuel), and removing the $Al(OH)_3$ precipitate. Under maximum operating conditions, the cell emf is about 2.7 V. These features make the aluminum-air battery ideally suited for use in electric cars and other similar devices. Figure 19.12 shows such a battery in operation.

19.7 Corrosion

Corrosion is the term usually applied to *the deterioration of metals by an electrochemical process*. We see many examples of corrosion around us—iron rust, silver tarnish, the green patina formed on copper and brass (Figure 19.13). Corrosion causes enormous damage to buildings, bridges, ships, and cars. One estimate put the cost of metallic corrosion to the U.S. economy in 1984 at about 80 billion dollars, or 3 percent of the gross national product for that year! This section discusses some of the fundamental processes that occur in corrosion and methods used to protect metals against it.

By far the most familiar example of corrosion is the formation of rust on iron. Oxygen gas and water must be present for iron to rust. Although the reactions involved are quite complex and not completely understood, the main steps are believed to be as follows. A region of the metal's surface serves as the anode, where oxidation occurs:

$$Fe(s) \longrightarrow Fe^{2+}(aq) + 2e^-$$

The electrons given up by iron reduce atmospheric oxygen to water at the cathode, which is another region of the same metal's surface:

$$O_2(g) + 4H^+(aq) + 4e^- \longrightarrow 2H_2O(l)$$

The overall redox reaction is

$$2Fe(s) + O_2(g) + 4H^+(aq) \longrightarrow 2Fe^{2+}(aq) + 2H_2O(l)$$

With data from Table 19.1, we find the standard emf for this process:

$$\begin{aligned}\mathscr{E}^\circ &= \mathscr{E}^\circ_{ox} + \mathscr{E}^\circ_{red}\\ &= 0.44\ V + 1.23\ V\\ &= 1.67\ V\end{aligned}$$

Note that this reaction occurs in an acidic medium; the H^+ ions are supplied in part by the reaction of atmospheric carbon dioxide with water to form H_2CO_3.

The Fe^{2+} ions formed at the anode are further oxidized by oxygen:

$$4Fe^{2+}(aq) + O_2(g) + (4 + 2x)H_2O(l) \longrightarrow 2Fe_2O_3 \cdot xH_2O(s) + 8H^+(aq)$$

(a)

(b)

(c)

Figure 19.13 *Examples of corrosion: (a) a rusted ship, (b) a half-tarnished silver dish, and (c) the Statue of Liberty coated with patina before its restoration in 1986.*

This hydrated form of iron(III) oxide is known as rust. The amount of water associated with the iron oxide varies, so we represent the formula as $Fe_2O_3 \cdot xH_2O$.

Figure 19.14 shows the mechanism of rust formation. The electric circuit is completed by the migration of electrons and ions; this is the reason that rusting occurs so rapidly in salt water. In cold climates, salts (NaCl or $CaCl_2$) spread on roadways to

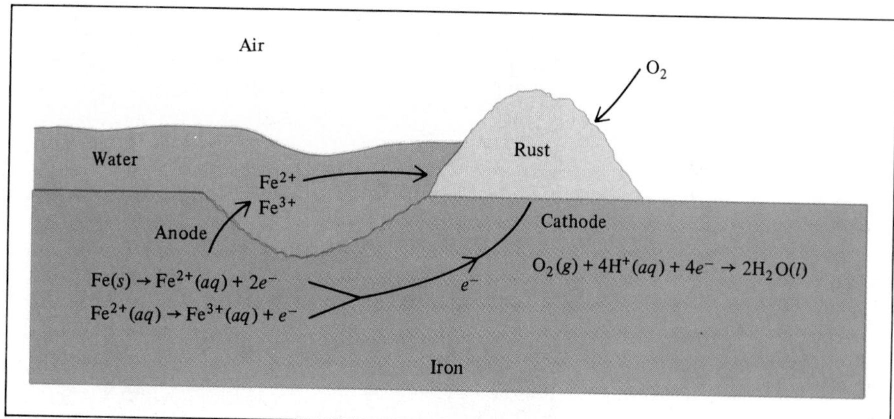

Figure 19.14 *The electrochemical process involved in rust formation. The H^+ ions are supplied by H_2CO_3, which forms when CO_2 dissolves in water.*

melt ice and snow are a major cause of rust formation on automobiles.

Metallic corrosion is not limited to iron. Consider aluminum, a metal used to make many useful things, including airplanes and beverage cans. Aluminum has a much greater tendency to oxidize than does iron; in Table 19.1 we see that Al has a more negative standard reduction potential than Fe. Based on this fact alone, we might expect to see airplanes slowly corrode away in rainstorms, and soda cans transformed into piles of corroded aluminum. These processes do not occur because the layer of insoluble aluminum oxide (Al_2O_3) that forms on its surface when the metal is exposed to air serves to protect the aluminum underneath from further corrosion. The rust that forms on the surface of iron, however, is too porous to protect the underlying metal.

Coinage metals such as copper and silver also corrode, but much more slowly.

$$Cu(s) \longrightarrow Cu^{2+}(aq) + 2e^-$$

$$Ag(s) \longrightarrow Ag^+(aq) + e^-$$

In normal atmospheric exposure, copper forms a layer of copper carbonate ($CuCO_3$), a green substance also called *patina,* that protects the metal underneath from further corrosion. Likewise, silverware that comes into contact with foodstuffs develops a layer of silver sulfide (Ag_2S).

A number of methods have been devised to protect metals from corrosion. Most of these methods are aimed at preventing rust formation. The most obvious approach is to coat the metal surface with paint. However, if the paint is scratched, pitted, or dented to expose even the smallest area of bare metal, rust will form under the paint layer. The surface of iron metal can be made inactive by a process called *passivation.* A thin oxide layer is formed when the metal is treated with a strong oxidizing agent such as concentrated nitric acid. A solution of sodium chromate is often added to cooling systems and radiators to prevent rust formation.

The tendency for iron to oxidize is greatly reduced by alloying with certain other metals. For example, when iron is alloyed with chromium and nickel to become stainless steel, the layer of chromium oxide that is formed protects the iron from corrosion.

An iron container can be covered with a layer of another metal such as tin or zinc. A "tin" can is made by applying a thin layer of tin over iron. Rust formation is prevented as long as the tin layer remains intact. However, once the surface has been scratched, rusting occurs rapidly. If we look up the standard reduction potentials, we find that iron acts as the anode and tin as the cathode in the corrosion process:

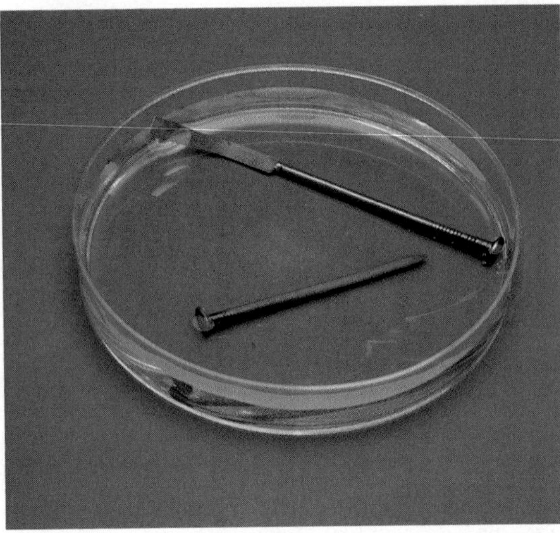

Figure 19.15 *An iron nail that is cathodically protected by a piece of zinc strip does not rust in water, while an iron nail without such protection rusts readily.*

$$Fe(s) \longrightarrow Fe^{2+} + 2e^- \qquad \mathscr{E}^\circ = +0.44\ V$$

$$Sn^{2+}(aq) + 2e^- \longrightarrow Sn(s) \qquad \mathscr{E}^\circ = -0.14\ V$$

The protective process is different for zinc-plated, or *galvanized,* iron. Zinc is more easily oxidized than iron (see Table 19.1):

$$Zn(s) \longrightarrow Zn^{2+}(aq) + 2e^- \qquad \mathscr{E}^\circ = +0.76\ V$$

So even if a scratch exposes the iron, the zinc is still attacked. In this case, the zinc metal serves as the anode and the iron is the cathode.

Cathodic protection is a process in which the metal that is to be protected from corrosion is made the cathode in what amounts to an electrochemical cell. Figure 19.15 shows how an iron nail can be protected from rusting by connecting the nail to a piece of zinc. Without such protection, an iron nail quickly rusts in water. Rusting of underground iron pipes and iron storage tanks can be prevented or greatly reduced by connecting them to metals such as zinc and magnesium, which oxidize more readily than iron (Figure 19.16).

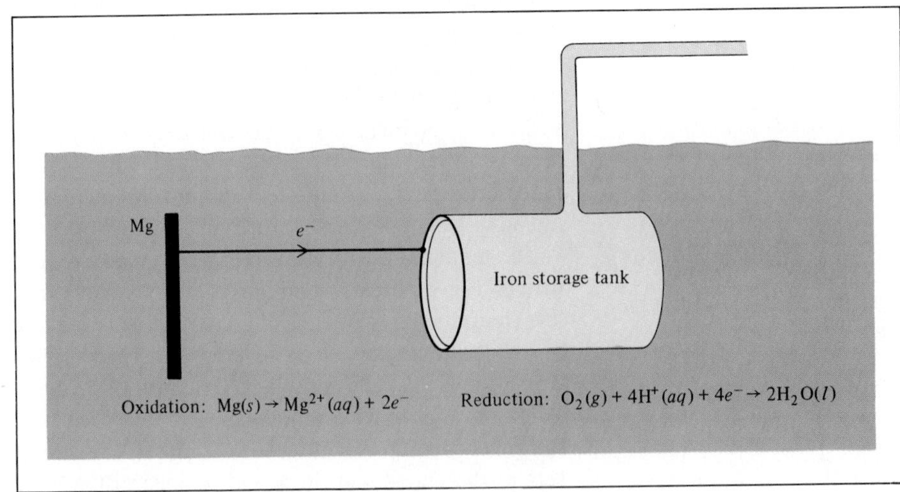

Figure 19.16 *Cathodic protection of an iron storage tank (cathode) by magnesium, a more electropositive metal (anode). Since only the magnesium is depleted in the electrochemical process, it is sometimes called the sacrificial anode.*

19.8 Electrolysis

In contrast to spontaneous redox reactions, which result in the conversion of chemical energy into electrical energy, ***electrolysis*** is the process in which *electrical energy is used to cause a nonspontaneous chemical reaction to occur*. Here we will discuss three examples of electrolysis, all of which show that electrolysis is based on the same principles as those that apply to electrochemical processes carried out in galvanic cells. Then we will look at the quantitative aspects of electrolysis.

Electrolysis of Molten Sodium Chloride

In its molten state, sodium chloride, an ionic compound, can be electrolyzed to form sodium metal and chlorine. Figure 19.17(a) is a diagram of a *Downs cell,* which is used for large-scale electrolysis of NaCl. In molten NaCl, the cations and anions are the Na^+ and Cl^- ions, respectively. Figure 19.17(b) shows the reactions that occur at the electrodes. The *electrolytic cell* contains a pair of electrodes connected to the battery. The battery serves as an "electron pump," driving electrons to the cathode, where reduction occurs, and withdrawing electrons from the anode, where oxidation occurs. The reactions at the electrodes are

These are nonspontaneous reactions, which are driven by the battery.

$$\begin{array}{lrcl} \text{Anode (oxidation):} & 2Cl^- & \longrightarrow & Cl_2(g) + 2e^- \\ \text{Cathode (reduction):} & 2Na^+ + 2e^- & \longrightarrow & 2Na(l) \\ \hline \text{Overall:} & 2Na^+ + 2Cl^- & \longrightarrow & 2Na(l) + Cl_2(g) \end{array}$$

This process is a major source of pure sodium metal and chlorine gas.

Electrolysis of Water

Water in a beaker under atmospheric conditions (1 atm and 25°C) will not spontane-

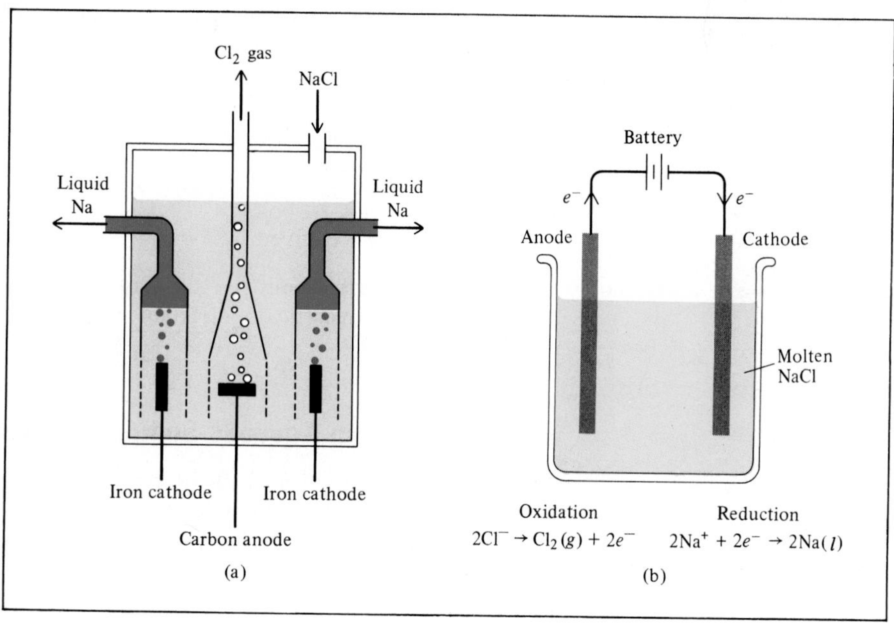

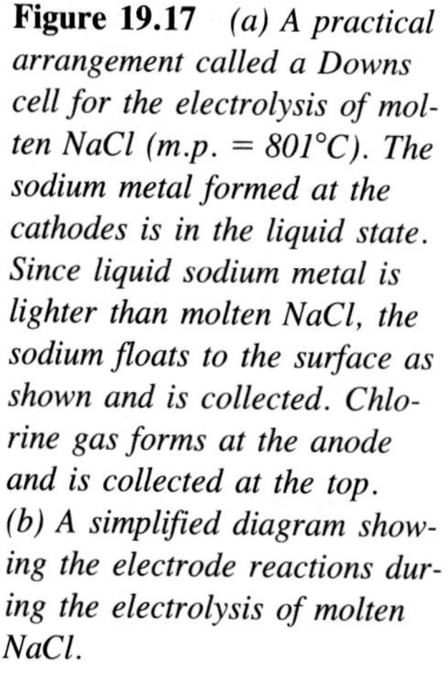
Figure 19.17 *(a) A practical arrangement called a Downs cell for the electrolysis of molten NaCl (m.p. = 801°C). The sodium metal formed at the cathodes is in the liquid state. Since liquid sodium metal is lighter than molten NaCl, the sodium floats to the surface as shown and is collected. Chlorine gas forms at the anode and is collected at the top. (b) A simplified diagram showing the electrode reactions during the electrolysis of molten NaCl.*

Figure 19.18 *Apparatus for small-scale electrolysis of water. The volume of hydrogen gas generated (left column) is twice that of oxygen gas (right column).*

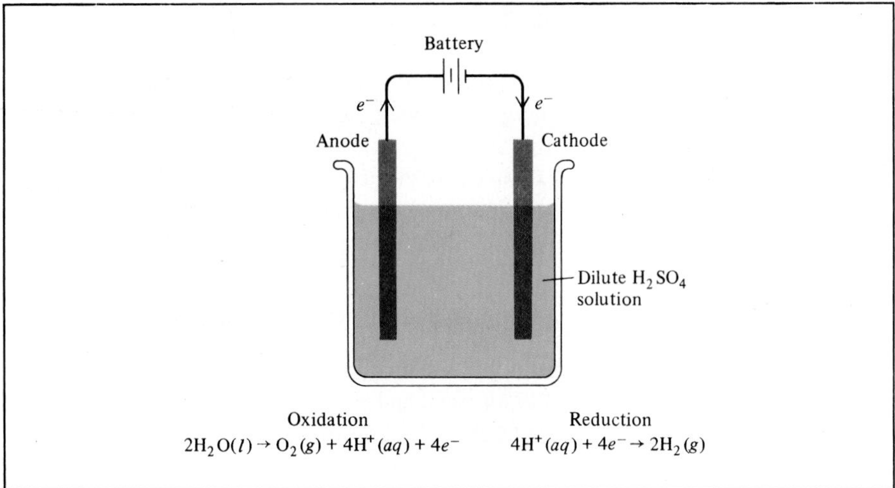

Figure 19.19 *A diagram showing the electrode reactions during the electrolysis of water.*

ously decompose to form hydrogen and oxygen gas because the standard free-energy change for the reaction is a large positive quantity:

$$2H_2O(l) \longrightarrow 2H_2(g) + O_2(g) \qquad \Delta G^\circ = 474.4 \text{ kJ}$$

However, this reaction can be induced by electrolyzing water in a cell like the one shown in Figure 19.18. This electrolytic cell consists of a pair of electrodes made of a nonreactive metal such as platinum immersed in water. When the electrodes are connected to the battery, nothing happens because there are not enough ions in pure water to carry much of an electric current. (Remember that at 25°C, pure water has only 1×10^{-7} M H^+ ions and 1×10^{-7} M OH^- ions.)

On the other hand, the reaction occurs readily in a 0.1 M H_2SO_4 solution because there are a sufficient number of ions to conduct electricity. Immediately, gas bubbles begin to appear at both electrodes. The process at the anode is (Figure 19.19)

$$2H_2O(l) \longrightarrow O_2(g) + 4H^+(aq) + 4e^-$$

while at the cathode we have

$$H^+(aq) + e^- \longrightarrow \tfrac{1}{2}H_2(g)$$

The overall reaction is given by

Note that no net H_2SO_4 is consumed.

$$\begin{array}{lrl} \text{Anode (oxidation):} & 2H_2O(l) & \longrightarrow O_2(g) + 4H^+(aq) + 4e^- \\ \text{Cathode (reduction):} & 4[H^+(aq) + e^- & \longrightarrow \tfrac{1}{2}H_2(g)] \\ \hline \text{Overall:} & 2H_2O(l) & \longrightarrow 2H_2(g) + O_2(g) \end{array}$$

Electrolysis of an Aqueous Sodium Chloride Solution

This is the most complicated of the three examples of electrolysis considered here because aqueous sodium chloride solution contains several species that could possibly be oxidized and reduced. The oxidation reactions that might occur at the anode are

$$(1) \quad 2H_2O(l) \longrightarrow O_2(g) + 4H^+(aq) + 4e^-$$

$$(2) \quad 2Cl^-(aq) \longrightarrow Cl_2(g) + 2e^-$$

Referring to Table 19.1, we find

$$O_2(g) + 4H^+(aq) + 4e^- \longrightarrow 2H_2O(l) \qquad \mathscr{E}° = 1.23\text{ V}$$

$$Cl_2(g) + 2e^- \longrightarrow 2Cl^-(aq) \qquad \mathscr{E}° = 1.36\text{ V}$$

Because Cl_2 is more easily reduced than O_2, it follows that it would be more difficult to oxidize Cl^- than H_2O at the anode.

The standard reduction potentials of (1) and (2) are not very different, but the values do suggest that H_2O should be preferentially oxidized at the anode. However, by experiment we find that the gas liberated at the anode is Cl_2, not O_2! In studying electrolytic processes we sometimes find that the voltage required for a reaction is considerably higher than the electrode potential indicates. *The additional voltage required to cause electrolysis* is called the ***overvoltage.*** The overvoltage for O_2 formation is quite high. Therefore, under normal operating conditions Cl_2 gas is actually formed at the anode instead of O_2.

The possible reductions that can occur at the cathode are

$$(3) \quad Na^+(aq) + e^- \longrightarrow Na(s) \qquad \mathscr{E}° = -2.71\text{ V}$$

$$(4) \quad 2H_2O(l) + 2e^- \longrightarrow H_2(g) + 2OH^-(aq) \qquad \mathscr{E}° = -0.83\text{ V}$$

$$(5) \quad 2H^+(aq) + 2e^- \longrightarrow H_2(g) \qquad \mathscr{E}° = 0.00\text{ V}$$

Reaction (3) is ruled out because it has a very negative standard reduction potential. Reaction (5) is preferred over (4) under standard-state conditions. At a pH of 7 (as is the case for a NaCl solution), however, they are equally probable. We generally use (4) to describe the cathode reaction because the concentration of H^+ ions is too low (about 1×10^{-7} *M*) to make (5) a reasonable choice.

Thus, the reactions in the electrolysis of aqueous sodium chloride are

Anode (oxidation):	$2Cl^-(aq) \longrightarrow Cl_2(g) + 2e^-$
Cathode (reduction):	$2H_2O(l) + 2e^- \longrightarrow H_2(g) + 2OH^-(aq)$
Overall:	$2H_2O(l) + 2Cl^-(aq) \longrightarrow H_2(g) + Cl_2(g) + 2OH^-(aq)$

As the overall reaction shows, the concentration of the Cl^- ions decreases during electrolysis and that of the OH^- ions increases. Therefore, in addition to H_2 and Cl_2, the useful by-product NaOH can be obtained by evaporating the aqueous solution at the end of the electrolysis.

The following example deals with the electrolysis of another aqueous salt solution.

EXAMPLE 19.9

An aqueous Na_2SO_4 solution is electrolyzed, using the apparatus shown in Figure 19.18. If the products formed at the anode and cathode are oxygen gas and hydrogen gas, respectively, describe the electrolysis in terms of the reactions at the electrodes.

Answer

The SO_4^{2-} ion is the conjugate base of the weak acid HSO_4^- ($K_a = 1.3 \times 10^{-2}$). However, the extent to which SO_4^{2-} hydrolyzes is negligible.

Before we look at the electrode reactions, we should consider the following facts: (1) Since Na_2SO_4 does not hydrolyze in water, the pH of the solution is close to 7. (2) The Na^+ ions are not reduced at the cathode, and the SO_4^{2-} ions are not oxidized at the anode. These conclusions are drawn from the electrolysis of water in the presence of sulfuric acid and in aqueous sodium chloride solution. Thus the electrode reactions are

$$\text{Anode:} \quad 2H_2O(l) \longrightarrow O_2(g) + 4H^+(aq) + 4e^-$$

$$\text{Cathode:} \quad 2H_2O(l) + 2e^- \longrightarrow H_2(g) + 2OH^-$$

The overall reaction, obtained by doubling the cathode reaction coefficients and adding the result to the anode reaction, is

$$6H_2O(l) \longrightarrow 2H_2(g) + O_2(g) + 4H^+(aq) + 4OH^-(aq)$$

If the H^+ and OH^- ions are allowed to mix, then

$$4H^+(aq) + 4OH^-(aq) \longrightarrow 4H_2O(l)$$

and the overall reaction becomes

$$2H_2O(l) \longrightarrow 2H_2(g) + O_2(g)$$

Similar problem: 19.48.

Electrolysis has many important applications in industry, mainly in the extraction and purification of metals. We will discuss some of these applications in Chapter 20.

Quantitative Aspects of Electrolysis

The quantitative treatment of electrolysis was developed primarily by Faraday. He observed that the mass of product formed (or reactant consumed) at an electrode is proportional to both the amount of electricity transferred at the electrode and the molar mass of the substance in question. For example, in the electrolysis of molten NaCl, the cathode reaction tells us that one Na atom is produced when one Na^+ ion accepts an electron from the electrode. To reduce 1 mole of Na^+ ions, we must supply Avogadro's number (6.02×10^{23}) of electrons to the cathode. On the other hand, the stoichiometry of the anode reaction shows that oxidation of two Cl^- ions yields one chlorine molecule. Therefore, the formation of 1 mole of Cl_2 results in the transfer of 2 moles of electrons from the Cl^- ions to the anode. Similarly, it takes 2 moles of electrons to reduce 1 mole of Mg^{2+} ions and 3 moles of electrons to reduce 1 mole of Al^{3+} ions:

$$Mg^{2+} + 2e^- \longrightarrow Mg$$

$$Al^{3+} + 3e^- \longrightarrow Al$$

Therefore

$$2\,F \simeq 1 \text{ mol } Mg^{2+}$$

$$3\,F \simeq 1 \text{ mol } Al^{3+}$$

where F is the faraday.

In an electrolysis experiment, we generally measure the current (in amperes, A) that passes through an electrolytic cell in a given period of time. The relationship between charge (in coulombs, C) and current is

$$1 \text{ C} = 1 \text{ A} \times 1 \text{ s}$$

that is, a coulomb is the quantity of electrical charge passing any point in the circuit in 1 second when the current is 1 ampere.

Figure 19.20 shows the steps involved in calculating the quantities of substances produced in electrolysis. These steps are illustrated in the following example.

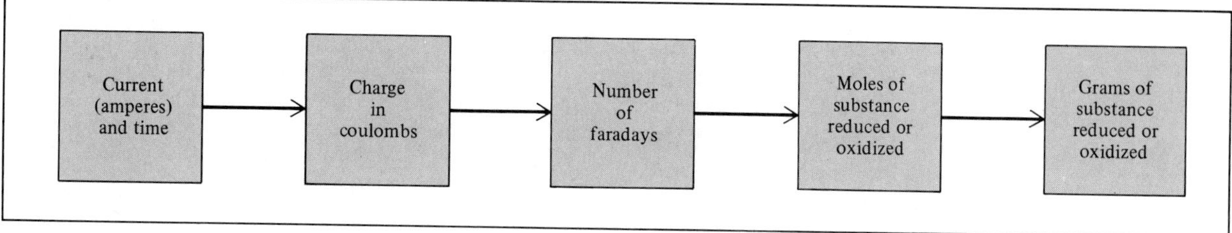

Figure 19.20 *Steps involved in calculating amounts of substances oxidized or reduced in electrolysis.*

EXAMPLE 19.10

A current of 0.452 A is passed through an electrolytic cell containing molten $CaCl_2$ for 1.50 hours. Write the electrode reactions and calculate the quantity of products (in grams) formed at the electrodes.

Answer

Since the only ions present in molten $CaCl_2$ are Ca^{2+} and Cl^-, the reactions are

$$\begin{array}{lrcl} \text{Anode:} & 2Cl^- & \longrightarrow & Cl_2(g) + 2e^- \\ \text{Cathode:} & Ca^{2+} + 2e^- & \longrightarrow & Ca \\ \hline \text{Overall:} & Ca^{2+} + 2Cl^- & \longrightarrow & Ca + Cl_2(g) \end{array}$$

The quantities of Ca metal and chlorine gas formed depend on the number of electrons that pass through the electrolytic cell, which in turn depends on the current and time, or charge:

$$? \text{ C} = 0.452 \text{ A} \times 1.50 \text{ h} \times \frac{3600 \text{ s}}{1 \text{ h}} \times \frac{1 \text{ C}}{1 \text{ A} \cdot \text{s}} = 2.44 \times 10^3 \text{ C}$$

Since 1 F = 96,500 C and 2 F are required to reduce 1 mole of Ca^{2+} ions, the mass of Ca metal formed at the cathode is calculated as follows:

$$? \text{ g Ca} = 2.44 \times 10^3 \text{ C} \times \frac{1\,F}{96{,}500 \text{ C}} \times \frac{1 \text{ mol Ca}}{2\,F} \times \frac{40.08 \text{ g Ca}}{1 \text{ mol Ca}} = 0.507 \text{ g Ca}$$

The anode reaction indicates that 1 mole of chlorine is produced per 2 F of electricity. Hence the mass of chlorine gas formed is

$$? \text{ g } Cl_2 = 2.44 \times 10^3 \text{ C} \times \frac{1\,F}{96{,}500 \text{ C}} \times \frac{1 \text{ mol } Cl_2}{2\,F} \times \frac{70.90 \text{ g } Cl_2}{1 \text{ mol } Cl_2} = 0.896 \text{ g } Cl_2$$

Similar problems: 19.45, 19.46.

Quantitative electrolysis also enables us to determine the charge of an ion. The charge is calculated by electrolyzing a solution containing the ion and relating the quantity of the element produced (either oxidized or reduced) to the number of coulombs of charge passing through the cell. The following example illustrates this approach.

EXAMPLE 19.11

An aqueous solution of a platinum salt is electrolyzed by passing a current of 2.50 A for 2.00 h. As a result, 9.09 g of metallic platinum are formed at the cathode. Calculate the charge on the platinum ions in this solution.

Answer

Our first step is to calculate the quantity of charge in coulombs passing through the cell:

$$? \text{ C} = 2.00 \text{ h} \times \frac{3600 \text{ s}}{1 \text{ h}} \times \frac{2.50 \text{ C}}{1 \text{ s}} = 1.80 \times 10^4 \text{ C}$$

The number of faradays corresponding to this quantity of charge is

$$? F = 1.80 \times 10^4 \text{ C} \times \frac{1 F}{96{,}500 \text{ C}} = 0.187 F$$

The half-cell reaction for the reduction is

$$\begin{array}{ccccc} Pt^{n+} & + & ne^- & \longrightarrow & Pt \\ 1 \text{ mol} & & n(6.02 \times 10^{23}) & & 1 \text{ mol} \\ 195.1 \text{ g} & & n(96{,}500 \text{ C}) & & 195.1 \text{ g} \\ 195.1 \text{ g} & & nF & & 195.1 \text{ g} \end{array}$$

where n is the number of charges on the platinum ion. The mass of Pt produced is

$$? \text{ g Pt} = 0.187 F \times \frac{1 \text{ mol Pt}}{nF} \times \frac{195.1 \text{ g Pt}}{1 \text{ mol Pt}} = 9.09 \text{ g Pt}$$

Solving for n, we get

$$n = \frac{36.5}{9.09} = 4$$

Thus the platinum ion is Pt^{4+}.

Similar problem: 19.58.

CHEMISTRY IN ACTION

Dental Filling Discomfort

In modern dentistry the material most commonly used for filling decaying teeth is a composition known as *dental amalgam*. (An amalgam is a substance made by combining mercury with another metal or metals.) It actually consists of three solid phases having stoichiometries approximately corresponding to Ag_2Hg_3, Ag_3Sn, and Sn_8Hg. The standard reduction potentials for these phases are Hg_2^{2+}/Ag_2Hg_3, 0.85 V; Sn^{2+}/Ag_3Sn, −0.05 V; Sn^{2+}/Sn_8Hg, −0.13 V.

Anyone who bites a piece of aluminum foil (such as

that used for wrapping candies) in such a way that the foil presses against a dental filling will probably experience a momentary sharp pain. In effect, a galvanic cell has been created in the mouth, with aluminum ($\mathscr{E}° = -1.66$ V) as the anode, the filling as the cathode, and saliva as the electrolyte. Contact between the aluminum foil and the filling short-circuits the cell, causing a weak current to flow between the electrodes. This current is detected by the sensitive nerve of the tooth as an unpleasant sensation.

Another type of discomfort results when a less electropositive metal touches a dental filling. For example, if a filling makes contact with a gold inlay in a nearby tooth, corrosion of the filling will occur (Figure 19.21). In this case, the dental filling acts as the anode and the gold inlay as the cathode. Referring to the $\mathscr{E}°$ values given for the three phases, we see that the Sn_8Hg phase is most likely to corrode. When that happens, release of Sn(II) ions in the mouth accounts for the unpleasant metallic taste. Prolonged corrosion will eventually result in another visit to the dentist for a "refilling" job.

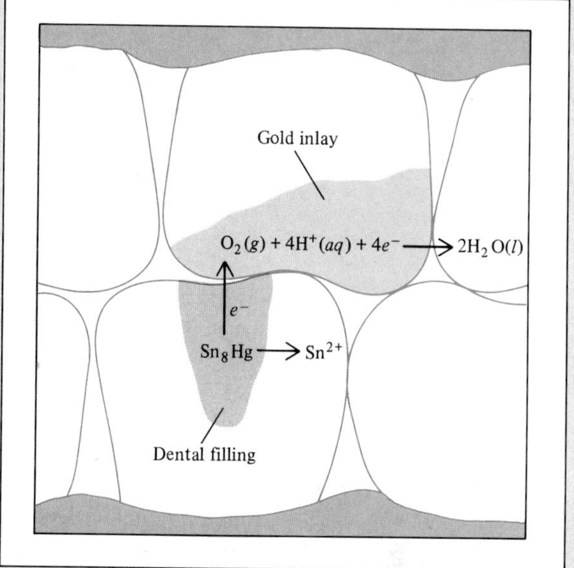

Figure 19.21 *Corrosion of dental filling when it is in contact with gold inlay.*

Summary

1. All electrochemical reactions involve the transfer of electrons and are therefore redox reactions.
2. In a galvanic cell, electricity is produced by a spontaneous chemical reaction. The oxidation at the anode and the reduction at the cathode take place separately, and the electrons flow through an external circuit.
3. The two parts of a galvanic cell are the half-cells, and the reactions at the electrodes are the half-cell reactions. A salt bridge allows ions to flow between the half-cells.
4. The electromotive force (emf) of a cell is the voltage difference between the two electrodes. In the external circuit, electrons flow from the anode to the cathode in a galvanic cell. In solution, the anions move toward the anode and the cations move toward the cathode.
5. The quantity of electricity carried by 1 mole of electrons is called a faraday, which is equal to 96,500 coulombs.
6. Standard reduction potentials show the relative likelihood of half-cell reduction reactions and can be used to predict the products, direction, and spontaneity of redox reactions between various substances.
7. The decrease in free energy of the system in a spontaneous redox reaction is equal to the electrical work done by the system on the surroundings, or $\Delta G = -nF\mathscr{E}$.
8. The equilibrium constant for a redox reaction can be found from the standard electromotive force of a cell.
9. The Nernst equation gives the relationship between the cell emf and the concentra-

tions of the reactants and products under conditions other than the standard state.

10. Batteries, which consist of one or more electrochemical cells, are used widely as self-contained power sources. Some of the better-known batteries are the dry cell, such as the Leclanché cell, the mercury battery, the nickel-cadmium battery, and the lead storage battery used in automobiles. Fuel cells produce electrical energy from a continuous supply of reactants. The aluminum-air battery is a combination of a battery and a fuel cell.
11. The corrosion of metals, the most familiar example of which is the rusting of iron, is an electrochemical phenomenon.
12. Electric current from an external source is used to drive a nonspontaneous chemical reaction in an electrolytic cell. The amount of product formed or reactant consumed depends on the quantity of electricity transferred at the electrode.

Applied Chemistry Example: Magnesium

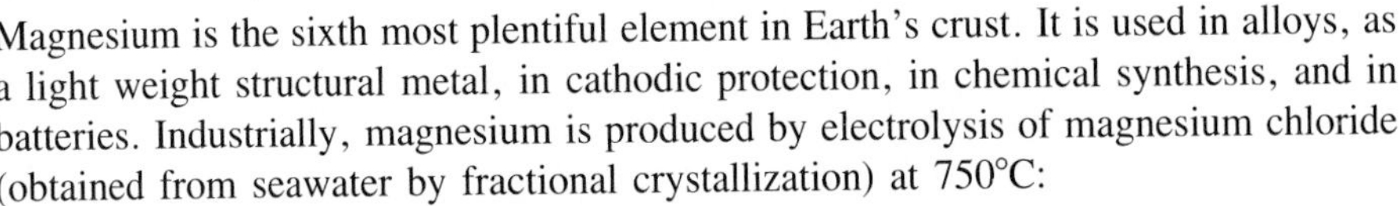

Magnesium is the sixth most plentiful element in Earth's crust. It is used in alloys, as a light weight structural metal, in cathodic protection, in chemical synthesis, and in batteries. Industrially, magnesium is produced by electrolysis of magnesium chloride (obtained from seawater by fractional crystallization) at 750°C:

$$MgCl_2(l) \longrightarrow Mg(l) + Cl_2(g)$$

The magnesium metal produced at the cathode floats to the top of the melt and is skimmed off. The chlorine gas generated at the anode is a valuable by-product.

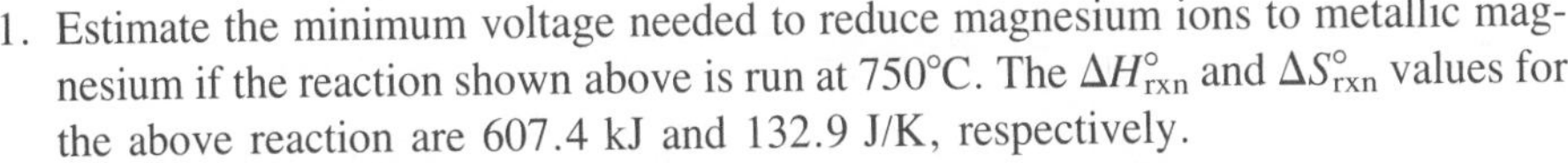

1. Estimate the minimum voltage needed to reduce magnesium ions to metallic magnesium if the reaction shown above is run at 750°C. The ΔH°_{rxn} and ΔS°_{rxn} values for the above reaction are 607.4 kJ and 132.9 J/K, respectively.

Answer: This calculation requires two steps. First we calculate the ΔG°_{rxn} for the reaction and then the emf needed for the reduction of Mg^{2+} ions. From Equation (18.7) we write

$$\begin{aligned}\Delta G^\circ_{rxn} &= \Delta H^\circ_{rxn} - T\Delta S^\circ_{rxn}\\ &= 607.4 \times 10^3\ \text{J} - (1023\ \text{K})(132.9\ \text{J/K})\\ &= 4.71 \times 10^5\ \text{J}\end{aligned}$$

The reduction of Mg^{2+} is

$$Mg^{2+} + 2e^- \longrightarrow Mg$$

We use Equation (19.3) to calculate the desired emf

$$\begin{aligned}\mathscr{E} &= \frac{-\Delta G^\circ_{rxn}}{nF}\\ &= \frac{-4.71 \times 10^5\ \text{J}}{(2\ \text{mol})(96{,}500\ \text{J/V}\cdot\text{mol})}\\ &= -2.44\ \text{V}\end{aligned}$$

The value −2.44 V means that we must supply a minimum of +2.44 V to drive the nonspontaneous reaction. In practice, the applied voltage is much higher (about 6 V) to make the process more efficient.

2. A current of 2.0×10^5 A is passed through the electrolytic cell for 18 hours. If the process is 90 percent efficient, calculate the magnesium metal produced in grams. How many grams of chlorine are generated at the anode?

Answer: We recall that 1 C = 1 A × 1 s. The charge (in coulombs) corresponding to the current passed through the cell is

$$? \text{ C} = 2.0 \times 10^5 \text{ A} \times 18 \text{ h} \times \frac{3600 \text{ s}}{1 \text{ h}} \times \frac{1 \text{ C}}{1 \text{ A} \cdot \text{s}} = 1.3 \times 10^{10} \text{ C}$$

Since 1 F = 96,500 C and 2 F are required to reduce 1 mole of Mg^{2+} ions, the mass of Mg metal formed at the cathode is

$$? \text{ g Mg} = 1.3 \times 10^{10} \text{ C} \times \frac{1\ F}{96{,}500} \times \frac{1 \text{ mol Mg}}{2\ F} \times \frac{24.31 \text{ g Mg}}{1 \text{ mol Mg}} \times 0.90$$
$$= 1.5 \times 10^6 \text{ g Mg}$$

Note that the factor 0.90 accounts for the 90 percent efficiency.

The anode reaction is

$$2Cl^- \longrightarrow Cl_2 + 2e^-$$

which indicates that 1 mole of chlorine is produced per 2 faraday of electricity. Hence the mass of chlorine gas formed is

$$? \text{ g } Cl_2 = 1.3 \times 10^{10} \text{ C} \times \frac{1\ F}{96{,}500 \text{ C}} \times \frac{1 \text{ mol } Cl_2}{2\ F} \times \frac{70.90 \text{ g } Cl_2}{1 \text{ mol } Cl_2} \times 0.90$$
$$= 4.3 \times 10^6 \text{ g } Cl_2$$

(This part is similar to Example 19.10.)

Key Words

Exercises

Galvanic Cells

REVIEW QUESTIONS

19.1 Discuss the role of redox reactions in electrochemical processes.

19.2 Define the following terms: anode, cathode, electromotive force, standard oxidation potential, standard reduction potential.

19.3 Describe the basic features of a galvanic cell. Why are the two components in a galvanic cell separated from each other?

19.4 What is the function of a salt bridge in a galvanic cell?

19.5 What is the difference between the half-reactions discussed in redox processes in Chapter 3 and the half-cell reactions discussed in Section 19.2?

19.6 After operating a galvanic cell like the one shown in Figure 19.1 for a few minutes, a student notices that the cell emf begins to drop. Why?

PROBLEMS

19.7 Calculate the standard emf of a cell that uses the Mg/Mg^{2+} and Cu/Cu^{2+} half-cell reactions at 25°C. Write the equation for the cell reaction occurring under standard-state conditions.

19.8 Consider a galvanic cell consisting of a magnesium electrode in contact with 1.0 M $Mg(NO_3)_2$ and a cadmium electrode in contact with 1.0 M $Cd(NO_3)_2$. Calculate $\mathscr{E}°$ for the cell and draw a diagram showing the cathode, the anode, and the direction of electron flow.

19.9 Calculate the standard emf of a cell that uses Ag/Ag^+ and Al/Al^{3+} half-cell reactions. Write the cell reaction occurring under standard-state conditions.

Spontaneity of Redox Reactions

REVIEW QUESTION

19.10 Referring to Table 19.1, explain the diagonal rule. How can it be used to predict the spontaneity of a redox reaction?

PROBLEMS

19.11 Predict whether Fe^{3+} can oxidize I^- to I_2 under standard-state conditions.

19.12 Predict whether the following reactions would occur spontaneously in aqueous solution at 25°C. Assume that the initial concentrations of dissolved species are all 1.0 M.
(a) $Ca(s) + Cd^{2+}(aq) \longrightarrow Ca^{2+}(aq) + Cd(s)$
(b) $2Br^-(aq) + Sn^{2+}(aq) \longrightarrow Br_2(l) + Sn(s)$
(c) $2Ag(s) + Ni^{2+}(aq) \longrightarrow 2Ag^+(aq) + Ni(s)$
(d) $Cu^+(aq) + Fe^{3+}(aq) \longrightarrow Cu^{2+}(aq) + Fe^{2+}(aq)$

19.13 Which of the following reagents can oxidize H_2O to $O_2(g)$ under standard-state conditions? $H^+(aq)$, $Cl^-(aq)$, $Cl_2(g)$, $Cu^{2+}(aq)$, $Pb^{2+}(aq)$, $MnO_4^-(aq)$ (in acid).

19.14 Predict whether a spontaneous reaction will occur (a) when a piece of silver wire is dipped in a $ZnSO_4$ solution; (b) when iodine is added to a NaBr solution; (c) when a piece of zinc metal is dipped in a $NiSO_4$ solution; (d) when chlorine gas is bubbled through a KI solution. Assume all species to be in their standard states.

19.15 For each of the following redox reactions (i) write the half-reactions; (ii) write a balanced equation for the whole reaction; (iii) determine in which direction the reaction will proceed spontaneously under standard-state conditions.
(a) $H_2(g) + Ni^{2+}(aq) \longrightarrow H^+(aq) + Ni(s)$
(b) $Ni^{2+}(aq) + Cd(s) \longrightarrow Ni(s) + Cd^{2+}(aq)$
(c) $MnO_4^-(aq) + Cl^-(aq) \longrightarrow Mn^{2+}(aq) + Cl_2(g)$ (in acid solution)
(d) $Ce^{3+}(aq) + H^+(aq) \longrightarrow Ce^{4+}(aq) + H_2(g)$
(e) $Cr(s) + Zn^{2+}(aq) \longrightarrow Cr^{3+}(aq) + Zn(s)$

19.16 Consider the half-cell reactions involving the following oxidized and reduced forms of a number of elements: $Sn^{2+}(aq)$ and $Sn(s)$; $Cl_2(g)$ and $Cl^-(aq)$; $Ca^{2+}(aq)$ and $Ca(s)$; $Fe^{2+}(aq)$ and $Fe(s)$; $Ag^+(aq)$ and $Ag(s)$. (a) Of the species listed, which is the strongest oxidizing agent under standard-state conditions? (b) Which is the strongest reducing agent under standard-state conditions? (c) What are the standard cell potential and the overall spontaneous cell reaction for the galvanic cell constructed from the $Ag(s)$, $Ag^+(aq)$ and the $Fe(s)$, $Fe^{2+}(aq)$ half-cells? (d) Draw a diagram of this cell as you would construct it in order to measure its potential, and indicate the direction of electron flow.

19.17 Consider the following half-reactions:

$$MnO_4^-(aq) + 8H^+(aq) + 5e^- \longrightarrow Mn^{2+}(aq) + 4H_2O(l)$$

$$NO_3^-(aq) + 4H^+(aq) + 3e^- \longrightarrow NO(g) + 2H_2O(l)$$

Predict whether NO_3^- ions will oxidize Mn^{2+} to MnO_4^- under standard-state conditions.

19.18 Which species in each pair is a better oxidizing agent under standard-state conditions? (a) Br_2 or Au^{3+}, (b) H_2 or Ag^+, (c) Cd^{2+} or Cr^{3+}, (d) O_2 in acidic media or O_2 in basic media.

19.19 Which species in each pair is a better reducing agent under standard-state conditions? (a) Na or Li, (b) H_2 or I_2, (c) Fe^{2+} or Ag, (d) Br^- or Co^{2+}.

Relationship among $\mathscr{E}°$, $\Delta G°$, and K

REVIEW QUESTION

19.20 Write the equations relating $\Delta G°$ and K to the standard emf of a cell. Define all the terms.

PROBLEMS

19.21 Use the standard reduction potentials to find the equilibrium constant for each of the following reactions at 25°C:
(a) $Br_2(l) + 2I^-(aq) \rightleftharpoons 2Br^-(aq) + I_2(s)$
(b) $2Ce^{4+}(aq) + 2Cl^-(aq) \rightleftharpoons Cl_2(g) + 2Ce^{3+}(aq)$
(c) $5Fe^{2+}(aq) + MnO_4^-(aq) + 8H^+(aq) \rightleftharpoons Mn^{2+}(aq) + 4H_2O + 5Fe^{3+}(aq)$

19.22 The equilibrium constant for the reaction

$$Sr(s) + Mg^{2+}(aq) \rightleftharpoons Sr^{2+}(aq) + Mg(s)$$

is 2.69×10^{12} at 25°C. Calculate $\mathscr{E}°$ for a cell made up of the Sr/Sr^{2+} and Mg/Mg^{2+} half-cells.

19.23 Calculate $\Delta G°$ and K_c for the following reactions at 25°C:
(a) $Mg(s) + Pb^{2+}(aq) \rightleftharpoons Mg^{2+}(aq) + Pb(s)$
(b) $Br_2(l) + 2I^-(aq) \rightleftharpoons 2Br^-(aq) + I_2(s)$
(c) $O_2(g) + 4H^+(aq) + 4Fe^{2+}(aq) \rightleftharpoons 2H_2O(l) + 4Fe^{3+}(aq)$
(d) $2Al(s) + 3I_2(s) \longrightarrow 2Al^{3+}(aq) + 6I^-(aq)$

19.24 What is the equilibrium constant for the following reaction at 25°C?

$$Mg(s) + Zn^{2+}(aq) \rightleftharpoons Mg^{2+}(aq) + Zn(s)$$

19.25 What spontaneous reaction will occur in aqueous solution, under standard-state conditions, among the ions Ce^{4+}, Ce^{3+}, Fe^{3+}, and Fe^{2+}? Calculate $\Delta G°$ and K_c for the reaction.

19.26 Given that $\mathscr{E}° = 0.52$ V for the reduction $Cu^+(aq) + e^- \longrightarrow Cu(s)$, calculate $\mathscr{E}°$, $\Delta G°$, and K for the following reaction at 25°C:

$$2Cu^+(aq) \longrightarrow Cu^{2+}(aq) + Cu(s)$$

The Nernst Equation

REVIEW QUESTION

19.27 What is the Nernst equation? Write the Nernst equation for the following processes at some temperature T K:
(a) $Mg(s) + Sn^{2+}(aq) \longrightarrow Mg^{2+}(aq) + Sn(s)$
(b) $2Cr(s) + 3Pb^{2+}(aq) \longrightarrow 2Cr^{3+}(aq) + 3Pb(s)$
Explain all terms.

PROBLEMS

19.28 What is the potential of a cell made up of Zn/Zn^{2+} and Cu/Cu^{2+} half-cells at 25°C if $[Zn^{2+}] = 0.25\ M$ and $[Cu^{2+}] = 0.15\ M$?

19.29 Calculate the standard potential of the cell consisting of the Zn/Zn^{2+} half-cell and the SHE. What will be the emf of the cell if $[Zn^{2+}] = 0.45\ M$, $P_{H_2} = 2.0$ atm, and $[H^+] = 1.8\ M$?

19.30 What is the emf of a cell consisting of a Pb/Pb^{2+} half-cell and a $Pt/H_2/H^+$ half-cell if $[Pb^{2+}] = 0.10\ M$, $[H^+] = 0.050\ M$, and $P_{H_2} = 1.0$ atm?

19.31 Calculate $\mathscr{E}°$, $\mathscr{E}$, and ΔG from the following cell reactions.
(a) $Mg(s) + Sn^{2+}(aq) \longrightarrow Mg^{2+}(aq) + Sn(s)$
$[Mg^{2+}] = 0.045\ M$, $[Sn^{2+}] = 0.035\ M$
(b) $3Zn(s) + 2Cr^{3+}(aq) \longrightarrow 3Zn^{2+}(aq) + 2Cr(s)$
$[Cr^{3+}] = 0.010\ M$, $[Zn^{2+}] = 0.0085\ M$

19.32 Referring to Figure 19.1, calculate the $[Cu^{2+}]/[Zn^{2+}]$ ratio at which the following reaction will become spontaneous at 25°C:

$$Cu(s) + Zn^{2+}(aq) \longrightarrow Cu^{2+}(aq) + Zn(s)$$

19.33 Calculate the emf of the following concentration cell:

$$Mg(s)|Mg^{2+}(aq,\ 0.24\ M)|KCl(\text{sat'd})|Mg^{2+}(aq,\ 0.53\ M)|Mg(s)$$

Batteries and Fuel Cells

REVIEW QUESTIONS

19.34 Explain clearly the differences between a primary galvanic cell—one that is not rechargeable—and a storage cell (for example, the lead storage battery), which is rechargeable.

19.35 Discuss the advantages and disadvantages of fuel cells over conventional power plants in producing electricity.

19.36 What are the advantages of the aluminum-air battery over a fuel cell using only gases, such as the hydrogen-oxygen fuel cell?

PROBLEMS

19.37 The hydrogen-oxygen fuel cell is described in Section 19.6. (a) What volume of $H_2(g)$, stored at 25°C at a pressure of 155 atm, would be needed to run an electric motor drawing a current of 8.5 A for 3.0 h? (b) What volume (L) of air at 25°C and 1.00 atm will have to pass into the cell per minute to run the motor? Assume that air is 20 percent O_2 by volume, and that all the O_2 is consumed in the cell. The other components of air do not affect the fuel-cell reactions. Assume ideal gas behavior.

19.38 Calculate the standard emf of the propane fuel cell discussed on p. 798 at 25°C, given that $\Delta G_f°$ for propane is −23.5 kJ/mol.

Corrosion

REVIEW QUESTIONS

19.39 Steel hardware, including nuts and bolts, is often coated with a thin cadmium plating. Explain the function of the cadmium layer.

19.40 "Galvanized iron" is steel sheet coated with zinc; "tin" cans are made of steel sheet coated with tin. Discuss the functions of these coatings and the electrochemistry of the corrosion reactions that occur if an electrolyte contacts the scratched surface of a galvanized iron sheet or a tin can.

19.41 Tarnished silver contains Ag_2S. The tarnish can be removed from silverware by placing the silver in an aluminum pan containing an inert electrolyte solution, such as NaCl. Explain the electrochemical principle for this procedure. [The standard reduction potential for the half-cell reaction

$$Ag_2S(s) + 2e^- \longrightarrow 2Ag(s) + S^{2-}(aq)$$

is −0.71 V.]

19.42 How does the tendency of iron to rust depend on the pH of solution?

Electrolysis

REVIEW QUESTIONS

19.43 What is the difference between an electrochemical cell (such as a galvanic cell) and an electrolytic cell?

19.44 What is Faraday's contribution to quantitative electrolysis?

PROBLEMS

19.45 The half-reaction at an electrode is

$$Mg^{2+}(\text{molten}) + 2e^- \longrightarrow Mg(s)$$

Calculate the number of grams of magnesium that can be produced by passing 1.00 F through the electrode.

19.46 Consider the electrolysis of molten calcium chloride, $CaCl_2$. (a) Write the electrode reactions. (b) How many grams of calcium metal can be produced by passing 0.50 A for 30 min?

19.47 In the electrolysis of water one of the half-reactions is

$$2H_2O(l) \longrightarrow O_2(g) + 4H^+(aq) + 4e^-$$

Suppose 0.076 L of O_2 is collected at 25°C and 755 mmHg. How many faradays of electricity must have passed through the solution?

19.48 Explain why different products are obtained in the electrolysis of molten $ZnCl_2$ and in the electrolysis of an aqueous solution of $ZnCl_2$.

19.49 Considering only the cost of electricity, would it be cheaper to produce a ton of sodium or a ton of aluminum by electrolysis?

19.50 How many faradays of electricity are required to produce (a) 0.84 L of O_2 at exactly 1 atm and 25°C from aqueous H_2SO_4 solution; (b) 1.50 L of Cl_2 at 750 mmHg and 20°C from molten NaCl; (c) 6.0 g of Sn from molten $SnCl_2$?

19.51 Calculate the amounts of Cu and Br_2 produced at inert electrodes by passing a current of 4.50 A through a solution of $CuBr_2$ for 1.0 h.

19.52 In the electrolysis of an aqueous $AgNO_3$ solution, 0.67 g of Ag is deposited after a certain period of time. (a) Write the half-reaction for the reduction of Ag^+. (b) What is the probable oxidation half-reaction? (c) Calculate the quantity of electricity used, in coulombs.

19.53 A steady current was passed through molten $CoSO_4$ until 2.35 g of metallic cobalt was produced. Calculate the number of coulombs of electricity used.

19.54 A constant electric current flows for 3.75 h through two electrolytic cells connected in series. One contains a solution of $AgNO_3$ and the second a solution of $CuCl_2$. During this time 2.00 g of silver are deposited in the first cell. (a) How many grams of copper are deposited in the second cell? (b) What is the current flowing, in amperes?

19.55 What is the hourly production rate of chlorine gas (in kg) from an electrolytic cell using aqueous NaCl electrolyte and carrying a current of 1.500×10^3 A? The anode efficiency for the oxidation of Cl^- is 93.0 percent.

19.56 Chromium plating is applied by electrolysis to objects suspended in a dichromate solution, according to the following (unbalanced) half-reaction:

$$Cr_2O_7^{2-}(aq) + e^- + H^+(aq) \longrightarrow Cr(s) + H_2O(l)$$

How long (in hours) would it take to apply a chromium plating of thickness 1.0×10^{-2} mm to a car bumper of surface area 0.25 m^2 in an electrolysis cell carrying a current of 25.0 A? (The density of chromium is 7.19 g/cm^3.)

19.57 The passage of a current of 0.750 A for 25.0 min deposited 0.369 g of copper from a $CuSO_4$ solution. From this information, calculate the molar mass of copper.

19.58 A quantity of 0.300 g of copper was deposited from a $CuSO_4$ solution by passing a current of 3.00 A for 304 s. Calculate the value of the faraday constant.

19.59 In a certain electrolysis experiment, 1.44 g of Ag were deposited in one cell (containing an aqueous $AgNO_3$ solution), while 0.120 g of an unknown metal X was deposited in another cell (containing an aqueous XCl_3 solution) in series. Calculate the molar mass of X.

19.60 If the cost of electricity to produce magnesium by the electrolysis of molten magnesium chloride is $155 per ton of metal, what is the cost (in dollars) of electricity to produce (a) 10.0 tons of aluminum, (b) 30.0 tons of sodium, (c) 50.0 tons of calcium?

19.61 In the electrolysis of water one of the half-reactions is

$$4H^+(aq) + 4e^- \longrightarrow 2H_2(g)$$

Suppose 0.845 L of H_2 is collected at 25°C and 782 mmHg. How many faradays of electricity must have passed through the solution?

19.62 A current of 6.00 A passes for 3.40 h through an electrolytic cell containing dilute sulfuric acid. If the volume of O_2 gas generated at the anode is 4.26 L (at STP), calculate the charge (in coulombs) on an electron.

Miscellaneous Problems

19.63 Complete the following table.

$\mathscr{E}$	ΔG	Cell reaction
>0		
	>0	
= 0		

State whether the cell reaction is spontaneous, nonspontaneous, or at equilibrium.

19.64 Distinguish between processes carried out in an electrolytic cell and those carried out in a galvanic cell.

19.65 Most useful galvanic cells give voltages of no more than 1.5 to 2.5 V. Explain why this is so. What are the prospects for developing practical galvanic cells with voltages of 5 V or more?

19.66 Consider the cell composed of the SHE and a half-cell using the reaction $Ag^+(aq) + e^- \rightarrow Ag(s)$. (a) Calculate the standard cell potential. (b) What is the spontaneous cell reaction under standard-state conditions? (c) Calculate the cell potential when $[H^+]$ in the hydrogen electrode is changed to (i) 1.0×10^{-2} M, and (ii) 1.0×10^{-5} M, all other reagents being held at standard-state conditions. (d) Based on this cell arrangement, suggest a design for a pH meter.

19.67 From the following information, calculate the solubility product of AgBr.

$$AgBr(s) + e^- \longrightarrow Ag(s) + Br^-(aq) \qquad \mathscr{E}° = 0.07 \text{ V}$$

$$Ag^+(aq) + e^- \longrightarrow Ag(s) \qquad \mathscr{E}° = 0.80 \text{ V}$$

19.68 A silver rod and a SHE are dipped into a saturated aqueous solution of silver oxalate, $Ag_2C_2O_4$, at 25°C. The measured potential difference between the rod and the SHE is 0.589 V, the rod being positive. Calculate the solubility product constant for silver oxalate.

19.69 A galvanic cell consists of a silver electrode in contact with 346 mL of 0.100 *M* $AgNO_3$ solution and a magnesium electrode in contact with 288 mL of 0.100 *M* $Mg(NO_3)_2$ solution. (a) Calculate $\mathscr{E}$ for the cell at 25°C. (b) A current is drawn from the cell until 1.20 g of silver have been deposited at the silver electrode. Calculate $\mathscr{E}$ for the cell at this stage of operation.

19.70 Use data in Table 19.1 to explain why nitric acid oxidizes Cu to Cu^{2+} but hydrochloric acid does not.

19.71 Consider a galvanic cell for which the anode reaction is

$$Pb(s) \longrightarrow Pb^{2+}(aq, 0.10\ M) + 2e^-$$

and the cathode reaction is

$$VO^{2+}(aq, 0.10\ M) + 2H^+(aq, 0.10\ M) + e^- \longrightarrow V^{3+}(aq, 1.0 \times 10^{-5}\ M) + H_2O(l)$$

At 25°C the measured emf of the cell is 0.67 V. Calculate (a) the standard reduction potential for the VO^{2+}/V^{3+} couple; (b) the equilibrium constant for the reaction

$$Pb(s) + 2VO^{2+}(aq) + 4H^+(aq) \rightleftharpoons Pb^{2+}(aq) + 2V^{3+}(aq) + 2H_2O(l)$$

19.72 Explain why chlorine gas can be prepared by electrolyzing an aqueous solution of NaCl but fluorine gas cannot be prepared by electrolyzing an aqueous solution of NaF.

19.73 Suppose you are asked to verify experimentally the electrode reactions shown in Example 19.9. In addition to the apparatus and the solution, you are also given two pieces of litmus paper, one blue and the other red. Describe what steps you would take in this experiment.

19.74 Calculate the emf of the following concentration cell at 25°C:

$$Cu(s)|Cu^{2+}(aq, 0.080\ M)|KCl(\text{sat'd})|Cu^{2+}(aq, 1.2\ M)|Cu(s)$$

19.75 The cathode reaction in the Leclanché cell is given by

$$2MnO_2(s) + Zn^{2+} + 2e^- \longrightarrow ZnMn_2O_4(s)$$

If a Leclanché cell produces a current of 0.0050 A, calculate how many hours this current supply can last if there is initially 4.0 g of MnO_2 present in the cell. Assume that there is an excess of Zn^{2+} ions.

19.76 For a number of years it was not clear whether mercury(I) ions existed in solution as Hg^+ or as Hg_2^{2+}. To distinguish between these two possibilities, we could set up the following system:

$$Hg(l)|\text{soln A}|KCl(\text{sat'd})|\text{soln B}|Hg(l)$$

where soln A contained 0.263 g mercury(I) nitrate per liter and soln B contained 2.63 g mercury(I) nitrate per liter. If the measured emf of such a cell is 0.0289 V at 18°C, what can you deduce about the nature of the mercury ions?

19.77 An aqueous KI solution to which a few drops of phenolphthalein have been added is electrolyzed using an apparatus similar to that shown in Figure 19.18. Describe what you would observe at the anode and the cathode. (*Hint:* Molecular iodine is only slightly soluble in water, but in the presence of I^- ions, it forms the brown color of I_3^- ions according to the equation: $I^- + I_2 \rightarrow I_3^-$.)

19.78 A piece of magnesium metal weighing 1.56 g is placed in 100.0 mL of 0.100 *M* $AgNO_3$ at 25°C. Calculate $[Mg^{2+}]$ and $[Ag^+]$ in solution at equilibrium. What is the mass of the magnesium left? Assume that the volume remains constant.

19.79 An acidified solution was electrolyzed using copper electrodes. After passing a constant current of 1.18 A for 1.52×10^3 s, the anode was found to have lost 0.584 g. (a) What is the gas produced at the cathode and what is its volume at STP? (b) Given that the charge of an electron is 1.6022×10^{-19} C, calculate Avogadro's number. Assume that copper is oxidized to Cu^{2+} ions.

19.80 Describe an experiment that would enable you to determine which is the cathode and which is the anode in a galvanic cell using copper and zinc electrodes.

19.81 A solution containing H^+ and D^+ ions is in equilibrium with a mixture of H_2 and D_2 gases at 25°C. If the partial pressures of the gases are both 1.0 atm, calculate the ratio $[D^+]/[H^+]$. ($\mathscr{E}°_{D^+/D_2} = -0.003$ V.)

19.82 Use the data in Table 19.1 to determine if hydrogen peroxide is stable to disproportionation in an acid medium: $2H_2O_2 \rightarrow 2H_2O + O_2$.

19.83 A galvanic cell is constructed as follows. A half-cell consists of a platinum wire immersed in a solution containing 1.0 *M* Sn^{2+} and 1.0 *M* Sn^{4+}, and another half-cell has a thallium rod immersed in a solution of 1.0 *M* Tl^+. (a) Write the half-cell reactions and the overall reaction. (b) What is the equilibrium constant at 25°C? (c) What is the cell voltage if the Tl^+ concentration is increased by tenfold? ($\mathscr{E}°_{Tl^+/Tl} = -0.34$ V.)

19.84 Zinc is an amphoteric metal; that is, it reacts with both acids and bases. If the standard oxidation potential for the reaction $Zn(s) + 4OH^-(aq) \rightarrow Zn(OH)_4^{2-}(aq) + 2e^-$ is 1.36 V, calculate the formation constant (K_f) for the reaction $Zn^{2+}(aq) + 4OH^-(aq) \rightleftharpoons Zn(OH)_4^{2-}(aq)$.

19.85 In a certain electrolysis experiment involving Al^{3+} ions, 60.2 g of Al is recovered when a current of 0.352 A is used. How long in minutes did the electrolysis take?

19.86 Consider the oxidation of ammonia:

$$4NH_3(g) + 3O_2(g) \longrightarrow 2N_2(g) + 6H_2O(l)$$

(a) Calculate the $\Delta G°$ for the reaction. (b) If this reaction were used in a fuel cell, what would be the standard cell potential?

19.87 The magnitudes (but *not* the signs) of the standard electrode potentials of two metals X and Y are

$$X^{2+} + 2e^- \longrightarrow X \qquad |\mathscr{E}°| = 0.25 \text{ V}$$

$$Y^{2+} + 2e^- \longrightarrow Y \qquad |\mathscr{E}°| = 0.34 \text{ V}$$

where the | | notation denotes that only the magnitude (but not the sign) of the $\mathscr{E}°$ value is shown. When the half-cells of X and Y are connected, electrons flow from X to Y. When X is connected to a SHE, electrons flow from X to SHE. (a) What are the signs of the $\mathscr{E}°$ values of the half-reactions? (b) What is the standard emf of a cell made up of X and Y?

19.88 A galvanic cell is constructed by immersing a piece of copper wire in 25.0 mL of a 0.20 *M* $CuSO_4$ solution and a zinc strip in 25.0 mL of a 0.20 *M* $ZnSO_4$ solution. (a) Calculate the emf of the cell at 25°C and predict what would happen if a small amount of concentrated NH_3 solution is added to (i) the $CuSO_4$ solution and (ii) the $ZnSO_4$ solution. Assume the volume in each compartment to remain at 25.00 mL. (b) In a separate experiment, 25.0 mL of 3.00 *M* NH_3 are added to the $CuSO_4$ solution. If the emf of cell is 0.68 V, calculate the formation constant (K_f) of $Cu(NH_3)_4^{2+}$.

Opposite page: Three gold tablets with Etruscan and Phoenician writings discovered in 1964 demonstrate one use of metals in early history. Since metals are malleable, they could be pounded, shaped, and inscribed. This property makes metals useful for making art objects, jewelry, and tools.

C H A P T E R 20

METALLURGY AND THE CHEMISTRY OF METALS

Up to this point we have concentrated mainly on fundamental principles: theories of chemical bonding, intermolecular forces, rates and mechanisms of chemical reactions, equilibrium, the laws of thermodynamics, and electrochemistry. Familiarity with these topics is necessary for an understanding of the properties of the elements and their compounds.

We have seen that elements can be classified as metals, nonmetals, or metalloids. In this chapter we will study the methods for extracting, refining, and purifying metals and the properties of metals that belong to the representative elements. We will emphasize (1) the occurrence and preparation of metals, (2) the physical and chemical properties of some of their compounds, and (3) their uses in modern society and their roles in biological systems.

20.1 Occurrence of Metals

Most metals occur in nature in a chemically combined state as minerals. A ***mineral*** is *a naturally occurring substance with a characteristic range of chemical composition. A mineral deposit concentrated enough to allow economical recovery of a desired metal* is known as ***ore***. Table 20.1 lists the principal types of minerals, and Figure 20.1 shows a classification of metals according to their minerals. In addition to the minerals

Table 20.1 Principal types of minerals

Type	Minerals
Native metals	Ag, Au, Bi, Cu, Pd, Pt
Carbonates	$BaCO_3$ (witherite), $CaCO_3$ (calcite, limestone), $MgCO_3$ (magnesite), $CaCO_3 \cdot MgCO_3$ (dolomite), $PbCO_3$ (cerussite), $ZnCO_3$ (smithsonite)
Halides	CaF_2 (fluorite), NaCl (halite), KCl (sylvite), Na_3AlF_6 (cryolite)
Oxides	$Al_2O_3 \cdot 2H_2O$ (bauxite), Al_2O_3 (corundum), Fe_2O_3 (hematite), Fe_3O_4 (magnetite), Cu_2O (cuprite), MnO_2 (pyrolusite), SnO_2 (cassiterite), TiO_2 (rutile), ZnO (zincite)
Phosphates	$Ca_3(PO_4)_2$ (phosphate rock), $Ca_5(PO_4)_3OH$ (hydroxyapatite)
Silicates	$Be_3Al_2Si_6O_{18}$ (beryl), $ZrSiO_4$ (zircon), $NaAlSi_3O_8$ (albite), $Mg_3(Si_4O_{10})(OH)_2$ (talc)
Sulfides	Ag_2S (argentite), CdS (greenockite), Cu_2S (chalcocite), FeS_2 (pyrite), HgS (cinnabar), PbS (galena), ZnS (sphalerite)
Sulfates	$BaSO_4$ (barite), $CaSO_4$ (anhydrite), $PbSO_4$ (anglesite), $SrSO_4$ (celestite), $MgSO_4 \cdot 7H_2O$ (epsomite)

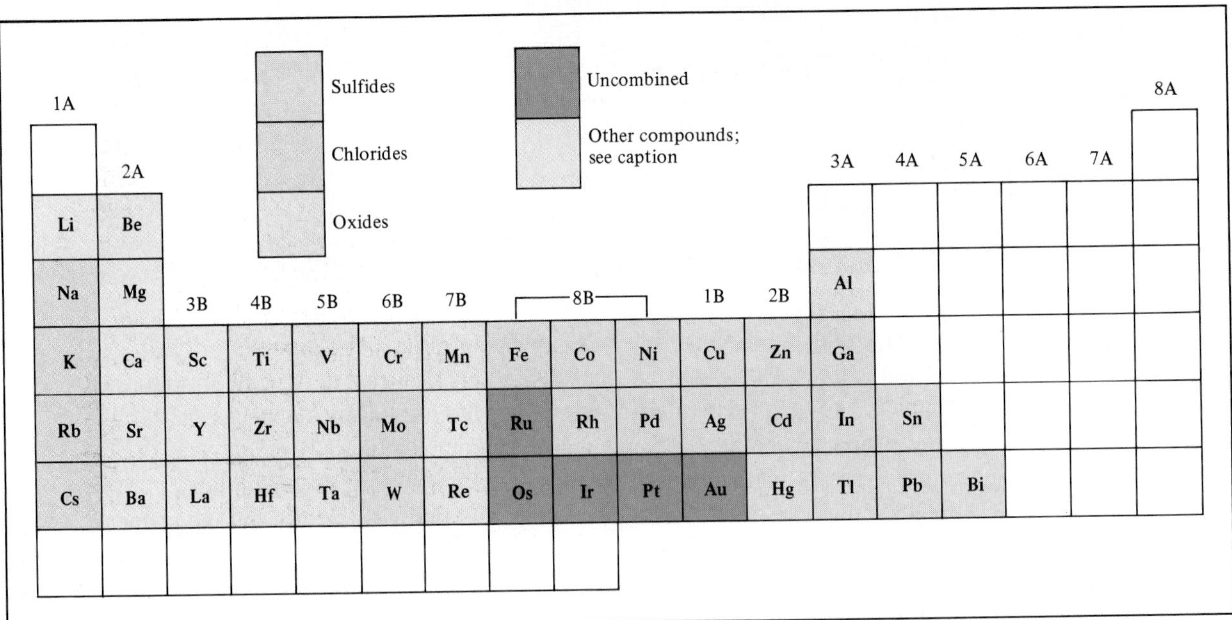

Figure 20.1 *Some of the best-known minerals of the metals. The lithium-containing mineral is spodumene ($LiAlSi_2O_6$); the beryllium-containing mineral is beryl (see Table 20.1). The minerals containing the rest of the alkaline earth metals are the carbonates and sulfates. The minerals for Sc, Y, and La are the phosphates. Some metals have more than one type of important mineral. For example, in addition to the sulfide, iron is found as the oxides hematite (Fe_2O_3) and magnetite (Fe_3O_4); and aluminum, in addition to the oxide, is found in beryl ($Be_3Al_2Si_6O_{18}$). Technetium (Tc) is a synthetic element.*

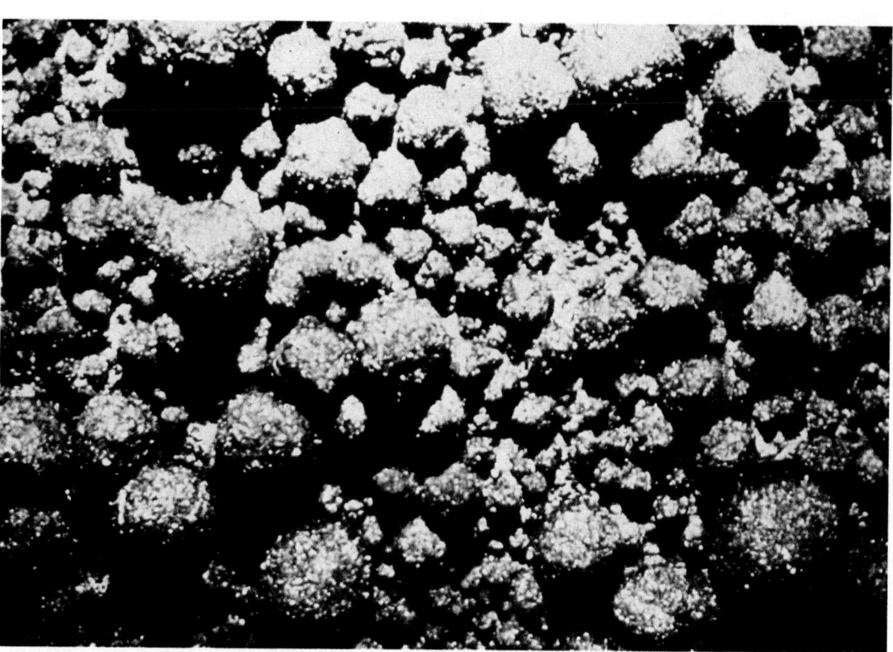

Figure 20.2 *Manganese nodules on the ocean floor.*

found in Earth's crust, seawater is a rich source of some metal ions, such as Mg^{2+} and Ca^{2+}. Furthermore, vast areas of the ocean floor are covered with *manganese nodules,* which are made up mostly of manganese, along with iron, nickel, copper, and cobalt (Figure 20.2).

20.2 Metallurgical Processes

Metallurgy is *the science and technology of separating metals from their ores and of compounding alloys.* (An ***alloy*** is *a solid solution composed of two or more metals, or of a metal or metals with one or more nonmetals.*) The three principal steps in the recovery of a metal from its ore are (1) preparation of the ore, (2) production of the metal, and (3) purification of the metal.

Preparation of the Ore

In the preliminary treatment of an ore, the desired mineral is separated from waste materials—usually clay and silicate minerals—which are collectively called the *gangue*. One very useful method for carrying out such a separation is called *flotation*. In this process the ore is finely ground and added to water containing oil and detergent. The liquid mixture is then beaten or blown to form a froth. The oil preferentially wets the mineral particles, which are then carried to the top in the froth, while the gangue settles to the bottom. The froth is skimmed off, allowed to collapse, and dried to recover the mineral particles.

Another physical separation process makes use of the magnetic properties of certain minerals. The mineral magnetite (Fe_3O_4), in particular, can be separated from the gangue by using a strong electromagnet. Substances like iron and cobalt that are strongly attracted to magnets are called ***ferromagnetic.***

Mercury forms amalgams with a number of metals. (An ***amalgam*** is *an alloy of* *mercury with another metal or metals.*) It can therefore be used to extract metal from ore. Mercury dissolves the silver and gold in an ore to form a liquid amalgam, which is easily separated from the remaining ore. The gold or silver is then recovered by distilling off the mercury.

Production of Metals

Because metals in their combined forms always have positive oxidation numbers, the production of a free metal is always a reduction process. Preliminary operations may be necessary to convert the ore to a chemical state more suitable for reduction. For example, an ore may be *roasted* to drive off volatile impurities and at the same time to convert the carbonates and sulfides to the corresponding oxides, which can be reduced more conveniently to yield the pure metals:

$$CaCO_3(s) \longrightarrow CaO(s) + CO_2(g)$$

$$2PbS(s) + 3O_2(g) \longrightarrow 2PbO(s) + 2SO_2(g)$$

This last equation points up the fact that the conversion of sulfides to oxides is a major source of sulfur dioxide, a notorious air pollutant (p. 659). The development in recent years of processes for converting this by-product to sulfuric acid instead of releasing it into the air has helped to reduce SO_2 emissions in some parts of the country.

How a pure metal is obtained by reduction from its combined form depends on the standard reduction potential of the metal (see Table 19.1). Table 20.2 outlines the reduction processes for several metals. Currently the major *metallurgical processes* are *carried out at high temperatures* in a procedure known as ***pyrometallurgy.*** The reduction in these procedures may be accomplished either chemically or electrolytically.

A more electropositive metal has a more negative standard reduction potential (see Table 19.1).

Chemical Reduction. Theoretically we can use a more electropositive metal as a reducing agent to separate a less electropositive metal from its compound at high temperatures:

$$V_2O_5(s) + 5Ca(l) \longrightarrow 2V(l) + 5CaO(s)$$

$$TiCl_4(g) + 2Mg(l) \longrightarrow Ti(s) + 2MgCl_2(l)$$

$$Cr_2O_3(s) + 2Al(s) \longrightarrow 2Cr(l) + Al_2O_3(s)$$

$$3Mn_3O_4(s) + 8Al(s) \longrightarrow 9Mn(l) + 4Al_2O_3(s)$$

In some cases even molecular hydrogen can be used as a reducing agent, as in the preparation of tungsten (for making filaments in light bulbs) from tungsten(VI) oxide:

Table 20.2 Reduction processes for some common metals

Decreasing activity of metals ↓	Metal	Reduction process
	Lithium, sodium, magnesium, calcium	Electrolytic reduction of the molten chloride
	Aluminum	Electrolytic reduction of anhydrous oxide (in molten cryolite)
	Chromium, manganese, titanium, vanadium, iron, zinc	Reduction of the metal oxide with a more electropositive metal, or reduction with coke and carbon monoxide
	Mercury, silver, platinum, copper, gold	These metals occur in the free (uncombined) state or can be obtained by roasting their sulfides

$$WO_3(s) + 3H_2(g) \longrightarrow W(s) + 3H_2O(g)$$

Electrolytic Reduction. Electrolytic reduction is suitable for very electropositive metals, such as sodium, magnesium, and aluminum. The process is usually carried out on the anhydrous molten oxide or halide of the metal:

$$2MO(l) \longrightarrow 2M \text{ (at cathode)} + O_2 \text{ (at anode)}$$

$$2MCl(l) \longrightarrow 2M \text{ (at cathode)} + Cl_2 \text{ (at anode)}$$

We will describe the specific procedures later in this chapter.

The Metallurgy of Iron

Use of FeS_2 leads to SO_2 production and acid rain.

Iron exists in Earth's crust in many different minerals, such as iron pyrite (FeS_2), siderite ($FeCO_3$), hematite (Fe_2O_3), and magnetite (Fe_3O_4, often represented as $FeO \cdot Fe_2O_3$). Of these, hematite and magnetite are particularly suitable for the extraction of iron. The metallurgical processing of iron involves the chemical reduction of the minerals by carbon (in the form of coke) in a blast furnace (Figure 20.3). The concentrated iron ore, limestone ($CaCO_3$), and coke are introduced into the furnace from the top. A

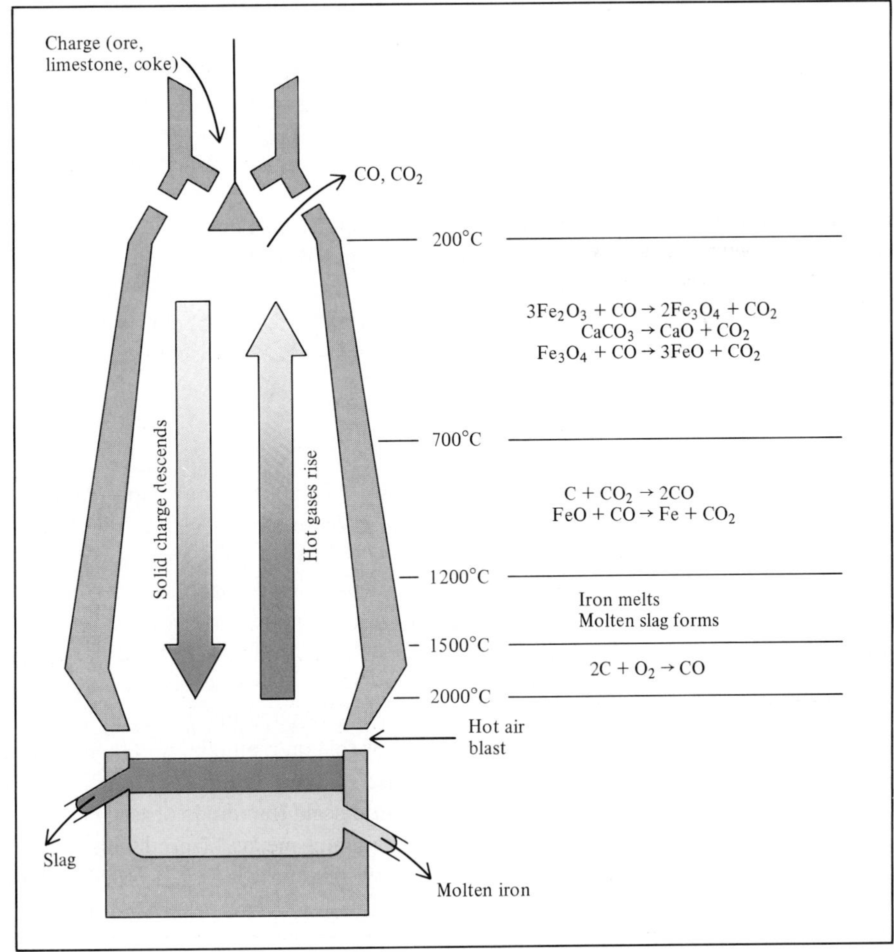

Figure 20.3 *A blast furnace. Iron ore, limestone, and coke are introduced at the top of the furnace. Iron is obtained from the ore by reduction with carbon.*

blast of preheated air is forced up the furnace from the bottom—hence the name *blast furnace*. The oxygen gas reacts with the carbon in the coke to form mostly carbon monoxide and some carbon dioxide. These reactions are highly exothermic, and, as the hot CO and CO_2 gases rise, they react with the iron oxides in different temperature zones as shown in Figure 20.3. The key steps in the extraction of iron are

$$3Fe_2O_3(s) + CO(g) \longrightarrow 2Fe_3O_4(s) + CO_2(g)$$

$$Fe_3O_4(s) + CO(g) \longrightarrow 3FeO(s) + CO_2(g)$$

$$FeO(s) + CO(g) \longrightarrow Fe(l) + CO_2(g)$$

Compounds such as $CaCO_3$ which are used to form a molten mixture with the impurities in the ore for easy removal are called *flux*.

The limestone decomposes in the furnace as follows:

$$CaCO_3(s) \longrightarrow CaO(s) + CO_2(g)$$

The calcium oxide then reacts with the impurities in the iron, which are mostly sand (SiO_2) and aluminum oxide (Al_2O_3):

$$CaO(s) + SiO_2(s) \longrightarrow CaSiO_3(l)$$

$$CaO(s) + Al_2O_3(s) \longrightarrow Ca(AlO_2)_2(l)$$

The mixture of calcium silicate and calcium aluminate that remains molten at the furnace temperature is known as *slag*.

By the time the ore works its way down to the bottom of the furnace, most of it has already been reduced to iron. The temperature of the lower part of the furnace is above the melting point of impure iron, and so the molten iron at the lower level can be run off to a receiver. The slag, because it is less dense, forms the top layer above the molten iron and can be run off at that level, as shown in Figure 20.3.

Iron extracted in this way contains many impurities and is called *pig iron;* it may contain up to 5 percent carbon and some silicon, phosphorus, manganese, and sulfur. Some of the impurities stem from the silicate and phosphate minerals (for example, manganese, silicon, and phosphorus), while carbon and sulfur come from coke. Pig iron is granular and brittle. It has a relatively low melting point (about 1180°C), so it can be cast; for this reason it is also called *cast iron*.

Steelmaking

Steel manufacturing is one of the most important metal industries. In the United States, the annual consumption of steel is well above 100 million tons. Steel is an iron alloy that contains from 0.03 to 1.4 percent carbon plus various amounts of other elements. The wide range of useful mechanical properties associated with steel is primarily a function of chemical composition and heat treatment of a particular type of steel.

Whereas the production of iron is basically a reduction process (converting iron oxides to metallic iron), the conversion of iron to steel is essentially an oxidation process in which the unwanted impurities are removed from the iron by reaction with oxygen gas. One of several methods used in steelmaking is the *basic oxygen process*. Because of its ease of operation and the relatively short time (about 20 minutes) required for each large-scale (hundreds of tons) conversion, the basic oxygen process is by far the most common means of producing steel today.

Figure 20.4 shows the basic oxygen process. Molten iron from the blast furnace (described in the previous section) is poured into a cylindrical vessel positioned origi-

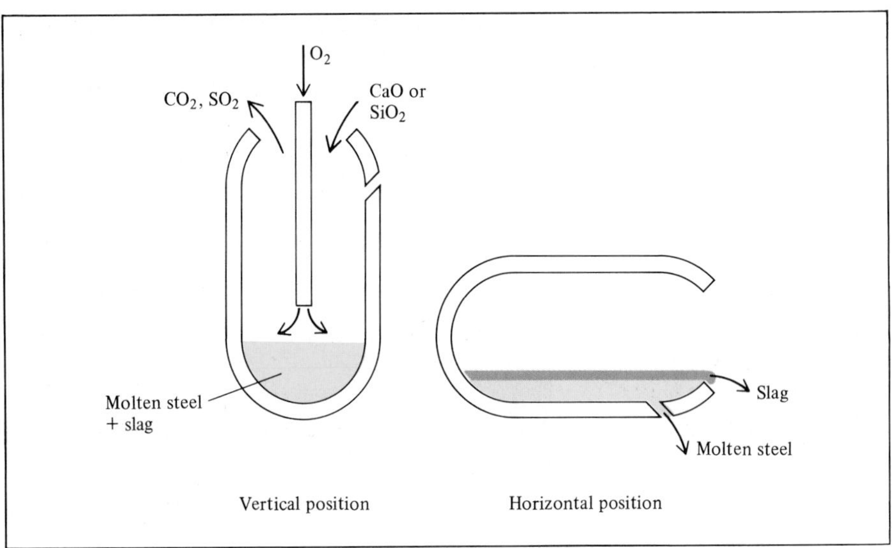

Figure 20.4 *Arrangement for steelmaking. The capacity of a typical vessel is 100 tons of cast iron.*

nally in the vertical position. Pressurized oxygen gas is then introduced via a water-cooled tube above the molten metal. Under these conditions, manganese, phosphorus, and silicon, as well as the excess carbon, react with oxygen to form the oxides. These oxides are then reacted with the appropriate fluxes (for example, CaO or SiO_2) to form slag. The type of flux chosen depends on the composition of the iron. If the main impurities are silicon and phosphorus, a basic flux such as CaO is added to the iron:

$$SiO_2(s) + CaO(s) \longrightarrow CaSiO_3(l)$$

$$P_4O_{10}(l) + 6CaO(s) \longrightarrow 2Ca_3(PO_4)_2(l)$$

On the other hand, if manganese is the main impurity, then an acidic flux such as SiO_2 is needed to form the slag:

$$MnO(s) + SiO_2(s) \longrightarrow MnSiO_3(l)$$

The molten steel is sampled at intervals. When the desired blend of carbon and other impurities has been reached, the vessel is rotated into a horizontal position so that the molten steel can be tapped off (Figure 20.5).

The properties of steel depend not only on its chemical composition but also on the heat treatment. At high temperatures, iron and carbon in steel combine to form iron carbide, Fe_3C, called *cementite:*

$$3Fe(s) + C(s) \rightleftharpoons Fe_3C(s)$$

The forward reaction is endothermic so that the formation of cementite is favored at high temperatures. When steel containing cementite is cooled slowly, the above equilibrium shifts to the left and the carbon separates as small particles of graphite, which give the steel a gray color. (Very slow decomposition of cementite also takes place at room temperature.) If the steel is cooled rapidly, equilibrium is not attained and the carbon remains largely in the form of cementite, Fe_3C. Steel containing cementite is light in color, and it is harder and more brittle than that containing graphite.

It is apparent, then, that the mechanical properties of steel can be altered by "tempering," that is, by heating the steel to some appropriate temperature for a short time

Figure 20.5 *Steelmaking.*

and then cooling it rapidly. In this way, the ratio of carbon present as graphite and as cementite can be varied within rather wide limits. Table 20.3 shows the composition, properties, and uses of various types of steel.

Purification of Metals

Metals prepared by reduction usually need further treatment to remove various impurities. The extent of purification, of course, depends on the use to be made of the metal. Below we describe three common purification procedures.

Distillation. Metals that have low boiling points, such as mercury, magnesium, and zinc, may be separated from other metals by fractional distillation. One well-known method of fractional distillation is the *Mond† process* for the purification of nickel. Carbon monoxide gas is passed over the impure nickel metal at about 70°C to form the volatile tetracarbonylnickel (b.p. 43°C), a highly toxic substance, which is separated from the less volatile impurities by distillation:

$$Ni(s) + 4CO(g) \longrightarrow Ni(CO)_4(g)$$

Pure metallic nickel is then recovered from $Ni(CO)_4$ by heating the gas at 200°C:

$$Ni(CO)_4(g) \longrightarrow Ni(s) + 4CO(g)$$

The carbon monoxide that is released is recycled back into the process.

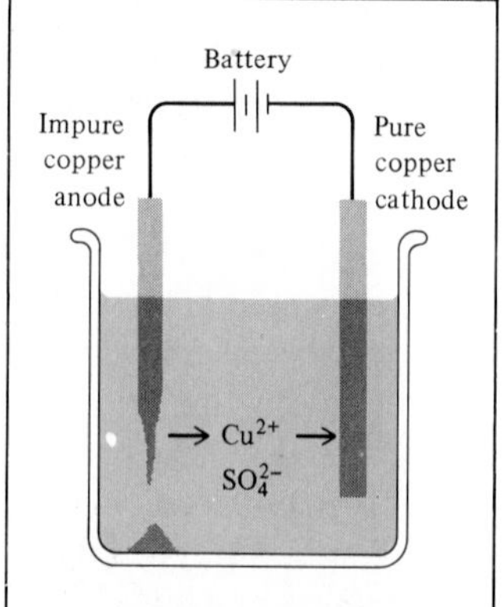

Figure 20.6 *Electrolytic purification of copper.*

†Ludwig Mond (1839–1909). British chemist of German origin. Mond made many important contributions to industrial chemistry. His method for purifying nickel by converting it to the volatile $Ni(CO)_4$ compound has been described as having given "wings" to the metal.

Table 20.3 Types of steel

Type	Composition (percent by mass)*								Uses
	C	Mn	P	S	Si	Ni	Cr	Others	
Plain	1.35	1.65	0.04	0.05	0.06	—	—	Cu (0.2–0.6)	Sheet products, tools
High-strength	0.25	1.65	0.04	0.05	0.15–0.9	0.4–1.0	0.3–1.3	Cu (0.01–0.08)	Construction, steam turbines
Stainless	0.03–1.2	1.0–10	0.04–0.06	0.03	1–3	1–22	4.0–27	—	Kitchen utensils, razor blades

*A single number indicates the maximum amount of the substance present.

Electrolysis. Electrolysis is another important purification technique. The copper metal obtained by roasting copper sulfide usually contains a number of impurities such as zinc, iron, silver, and gold. The more electropositive metals are removed by an electrolysis process in which the impure copper acts as the anode and *pure* copper acts as the cathode in a sulfuric acid solution containing Cu^{2+} ions (Figure 20.6). The reactions are

$$\text{Anode (oxidation):} \quad Cu(s) \longrightarrow Cu^{2+}(aq) + 2e^-$$

$$\text{Cathode (reduction):} \quad Cu^{2+}(aq) + 2e^- \longrightarrow Cu(s)$$

The metal impurities separated from the copper anode are valuable by-products that can often pay for the electricity used.

Iron and zinc, like the other more reactive metals in the copper anode, are also oxidized at the anode and enter the solution as Fe^{2+} and Zn^{2+}. They are not reduced at the cathode, however. The less electropositive metals, such as gold and silver, are not oxidized at the anode. Eventually, as the copper anode dissolves, these metals fall to the bottom of the cell. Thus, the net result of this electrolysis process is the transfer of copper from the anode to the cathode. Copper prepared this way has a purity greater than 99.5 percent (Figure 20.7).

Figure 20.7 *Copper cathodes used in the electrorefining process.*

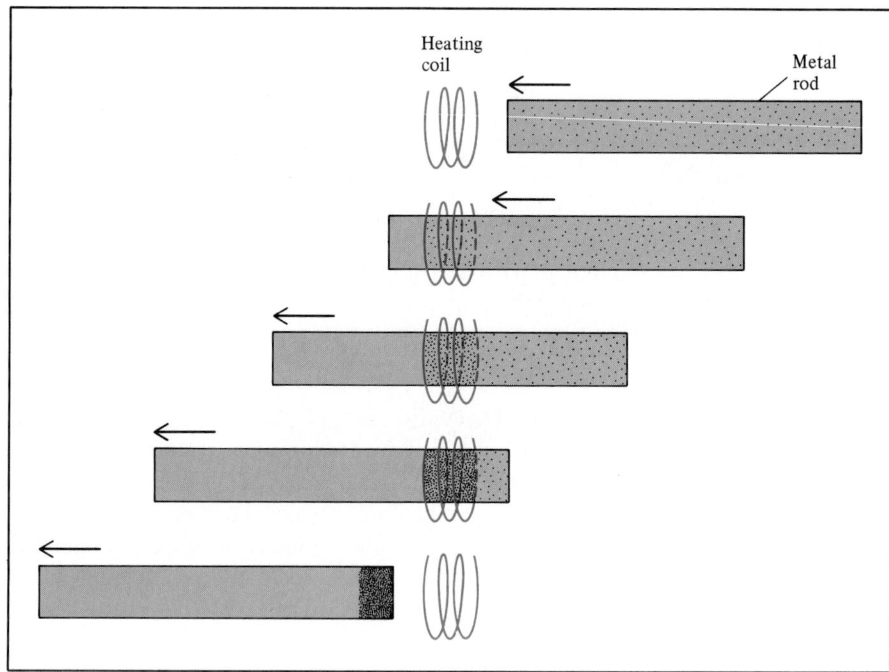

Figure 20.8 *Zone-refining technique for purifying metals. Top to bottom: An impure metal rod is moved slowly through a heating coil. As the metal rod moves forward, the impurities dissolve in the molten portion of the metal while pure metal crystallizes out in front of the molten zone. Eventually the end portion of the rod, which contains most of the impurities, is allowed to cool and is cut off from the rest of the rod.*

Zone Refining. Another often-used method of obtaining extremely pure metals is zone refining. In this process a metal rod containing a few impurities is drawn through an electrical heating coil that melts the metal (Figure 20.8). Most impurities dissolve in the molten metal. As the metal rod emerges from the heating coil, it cools and the pure metal crystallizes, leaving the impurities in the molten metal portion that is still in the heating coil. (This is analogous to the freezing of seawater, in which the solid that separates is mostly pure solvent—water. In zone refining the liquid metal acts as the solvent and the impurities as the solutes.) When the molten zone carrying the impurities, now at increased concentration, reaches the end of the rod, it is allowed to cool and is then cut off. Repeating this procedure a number of times results in metal with a purity greater than 99.99 percent.

20.3 Bonding in Metals and Semiconducting Elements

Having briefly acquainted ourselves with the metallic state in Section 11.6, we can now take a closer look at metals, particularly with respect to bonding. Quantitative treatment of metallic bonding, which is beyond our needs here, requires a thorough knowledge of quantum mechanics. However, we can gain a qualitative understanding of the bonding in metals through a discussion of the *band theory*. We will also apply the band theory to a class of elements called the semiconducting elements.

Conductors

Metals are characterized by their high electrical conductivity. Consider, for example, magnesium metal. The electron configuration of Mg is [Ne]$3s^2$, so each atom has two valence electrons in the 3*s* orbital. In a metal the atoms are packed closely together, so the energy levels of each magnesium atom are affected by the immediate neighbors of the atom as a result of orbital overlaps. In Chapter 10 we saw that the interaction between two atomic orbitals leads to the formation of a bonding and an antibonding molecular orbital. Since the number of atoms in even a small piece of magnesium metal is enormously large (on the order of 10^{20} atoms), the corresponding number of molecular orbitals formed is also very large. These molecular orbitals are so closely spaced on the energy scale that they are more appropriately described as forming a "band" (Figure 20.9). This set of closely spaced *filled* levels is called the *valence band,* as shown in Figure 20.10. The upper half of the energy levels corresponds to the empty, delocalized molecular orbitals formed by the overlap of the 3*p* orbitals. This set of closely spaced *empty* levels is called the *conduction band.*

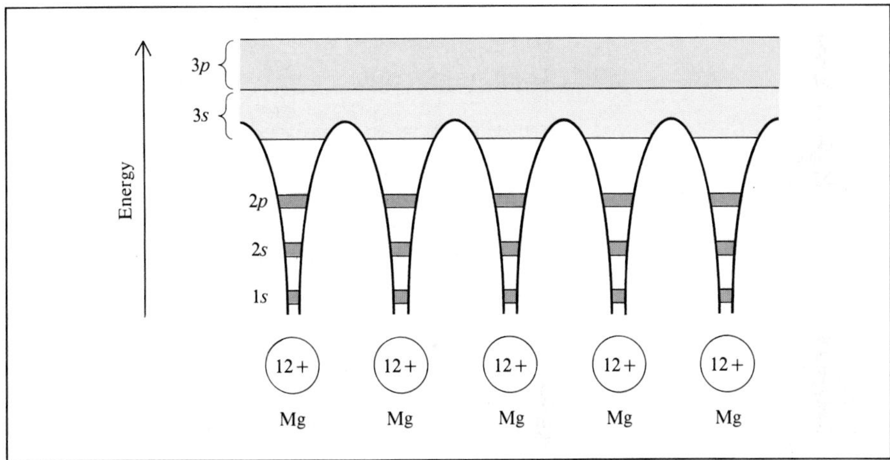

Figure 20.9 *Formation of conduction bands in magnesium metal. The electrons in the 1s, 2s, and 2p orbitals are localized on each Mg atom. However, the 3s and 3p orbitals overlap to form delocalized molecular orbitals. Electrons in these orbitals can travel throughout the metal, and this accounts for the electrical conductivity of the metal.*

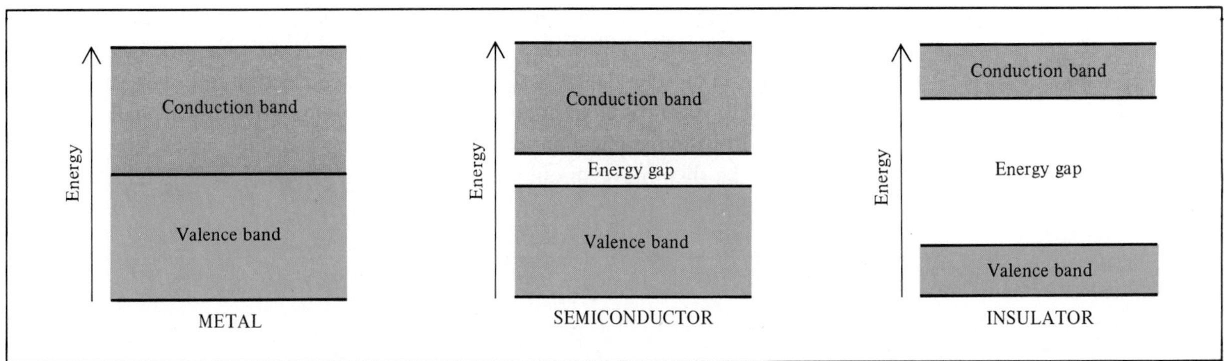

Figure 20.10 *Comparison of the energy gaps between valence band and conduction band in a metal, a semiconductor, and an insulator. In a metal the energy gap is virtually nonexistent; in a semiconductor the energy gap is small; and in an insulator the energy gap is very large, thus making the promotion of an electron from the valence band to the conduction band difficult.*

We can imagine magnesium metal as an array of positive ions immersed in a sea of delocalized valence electrons (see Figure 11.32). The great cohesive force resulting from the delocalization is partly responsible for the strength noted in most metals. Because the valence band and the conduction band are adjacent to each other, only a negligible amount of energy is needed to promote a valence electron to the conduction band, where it is free to move through the entire metal, since the conduction band is void of electrons. This freedom of movement accounts for the fact that metals are *capable of conducting electric current;* that is, they are good ***conductors.***

Why don't substances like wood and glass conduct electricity, as metals do? Figure 20.10 provides an answer to this question. Basically, the electrical conductivity of a solid depends on the spacing and the state of occupancy of the energy bands. Other metals resemble magnesium in that their valence bands are adjacent to their conduction bands, and, therefore, these metals readily act as conductors. In an ***insulator*** *the gap between the valence band and the conduction band is considerably greater than that in a metal;* consequently, much more energy is needed to excite an electron into the conduction band. Lacking this energy, free motion of the electrons is not possible. Typical insulators are glass, wood, and rubber.

Semiconductors

A number of elements, especially Si and Ge in Group 4A, *have properties that are intermediate between those of metals and nonmetals* and are therefore called ***semiconducting elements.*** The energy gap between the filled and empty bands for these solids is much smaller than that for insulators (see Figure 20.10). If the energy needed to excite electrons from the valence band into the conduction band is provided, the solid becomes a conductor. Note that this behavior is opposite that of the metals. A metal's ability to conduct electricity *decreases* with increasing temperature, because the enhanced vibration of atoms at higher temperatures tends to disrupt the flow of electrons.

The ability of a semiconductor to conduct electricity can also be enhanced by adding small amounts of certain impurities to the element (a process called *doping*). Let us consider what happens when a trace amount of boron or phosphorus is added to solid silicon. (In this doping, about five out of every million Si atoms are replaced by B or P atoms.) The structure of solid silicon is similar to that of diamond; each Si atom is covalently bonded to four Si atoms. Phosphorus ([Ne]$3s^23p^3$) has one more valence electron than silicon ([Ne]$3s^23p^2$), so there is a valence electron left over after four of them are used to form covalent bonds with silicon (Figure 20.11). This extra electron can be removed from the phosphorus atom by applying a voltage across the solid. The free electron can then move through the structure and function as a conduction electron. Impurities of this type are known as ***donor impurities,*** since they *provide conduc-*

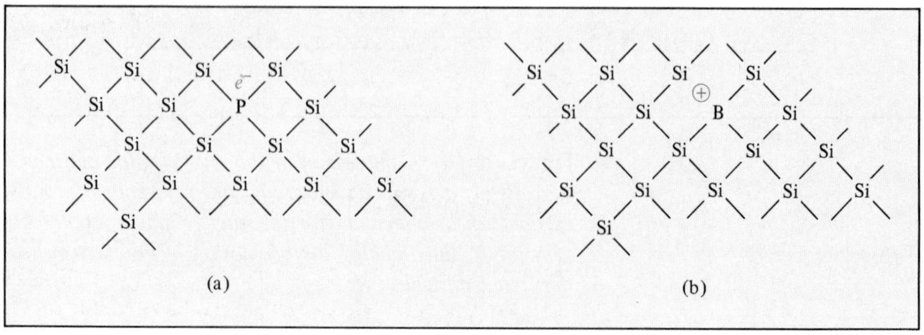

Figure 20.11 *(a) Silicon crystal doped with phosphorus. (b) Silicon crystal doped with boron. Note the formation of a negative center in (a) and of a positive center in (b).*

tion electrons. Solids containing donor impurities are called ***n*-type semiconductors,** where *n* stands for negative (the charge of the ''extra'' electron).

The opposite effect occurs if boron is added to silicon. A boron atom has three valence electrons ($1s^22s^22p^1$). Thus, for every boron atom in the silicon crystal there is a single *vacancy* in a bonding orbital. It is possible to excite a valence electron from a nearby Si into this vacant orbital. A vacancy created at that Si atom can then be filled by an electron from a neighboring Si atom, and so on. In this manner, electrons can move through the crystal in one direction while the vacancies, or ''positive holes,'' move in the opposite direction, and the solid becomes an electrical conductor. Semiconductors that *contain acceptor impurities* are called ***p*-type semiconductors,** where *p* stands for positive. *Impurities that are electron deficient* are called ***acceptor impurities.***

In both the *p*-type and *n*-type semiconductors the energy gap between the valence band and the conduction band is effectively reduced, so that only a small amount of energy is needed to excite the electrons. Typically, the conductivity of a semiconductor is increased by a factor of 100,000 or so by the presence of impurity atoms.

The growth of the semiconductor industry since the early 1960s has been truly remarkable. Today semiconductors are essential components of nearly all electronic equipment, ranging from radio and television sets to pocket calculators and computer facilities. One of the main advantages of solid-state devices over vacuum tube electronics is that the former can be made on a single ''chip'' of silicon no larger than the cross section of a pencil eraser. In this manner, much more equipment can be packed into a small volume—a point of particular importance in space travel, as well as in hand-held calculators and microprocessors (computers-on-a-chip).

20.4 Periodic Trends in Metallic Properties

Before we study specific metals, let us look at the general properties of metallic elements. Metals are lustrous in appearance, solid at room temperature (with the exception of mercury), good conductors of heat and electricity, malleable (can be hammered flat), and ductile (can be drawn into wire).

Figure 20.12 shows the positions of the representative metals and the group 2B metals in the periodic table. (The transition metals are discussed in Chapter 22.) As we saw in Chapter 9, the electronegativity of elements increases from left to right across a period and from bottom to top in a group (see Figure 9.8). The metallic character of metals increases in just the opposite directions, that is, from right to left across a period and from top to bottom in a group. Because metals generally have low electronegativities, they tend to form cations and almost always have positive oxidation numbers in their compounds. However, beryllium and magnesium in Group 2A and metals in Group 3A and beyond also form covalent compounds.

In the next five sections we will study the chemistry of the following metals: Group 1A (the alkali metals), Group 2A (the alkaline earth metals), Group 3A (aluminum), Group 4A (tin and lead), and Group 2B (zinc, cadmium, and mercury).

20.5 The Alkali Metals

As a group, the alkali metals (the Group 1A elements) are the most electropositive (or the least electronegative) elements known. They exhibit many similar properties. Table

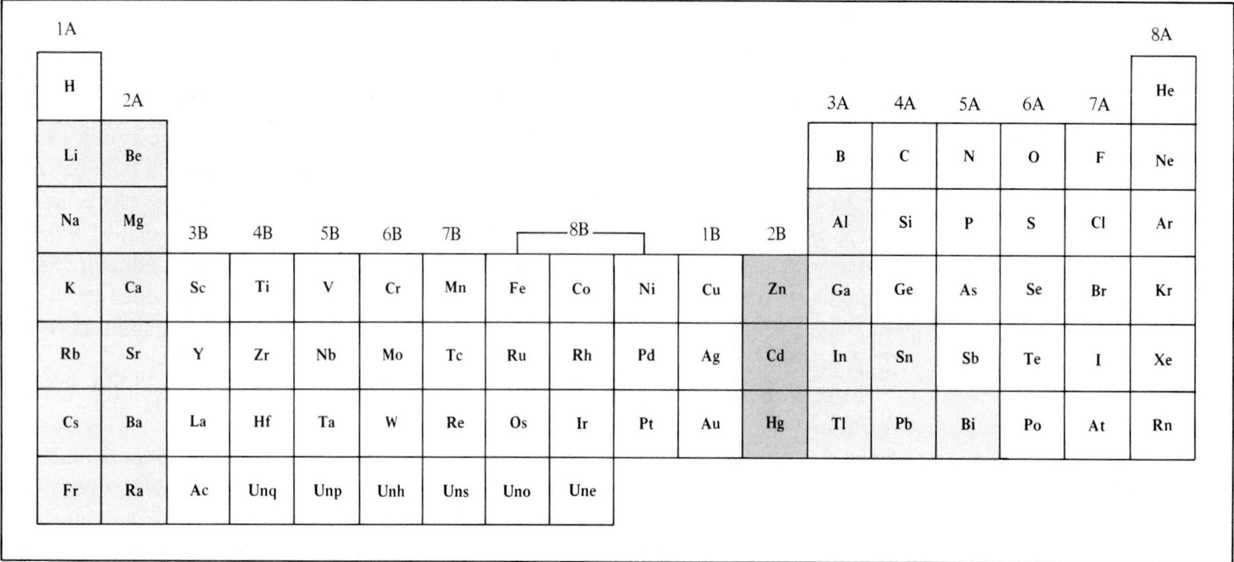

Figure 20.12 *Representative metals and Group 2B metals according to their positions in the periodic table.*

20.4 lists some common properties of the alkali metals. From their electron configurations we expect the oxidation number of these elements in their compounds to be +1 since the cations would be isoelectronic with the noble gases. This is indeed the case.

The alkali metals have low melting points and are soft enough to be sliced with a knife (see Figure 8.17). These metals all possess a body-centered crystal structure (see Figure 11.31) with low packing efficiency. This accounts for their low densities. In fact, lithium is the lightest metal known.

Because of their great chemical reactivity, the alkali metals never occur naturally in elemental form; they are found combined with halide, sulfate, carbonate, and silicate ions. In this section we will describe the chemistry of the first three members of Group 1A: lithium, sodium, and potassium. The chemistry of rubidium and cesium is less important; all isotopes of francium, the last member of the group, are radioactive.

Table 20.4 Properties of alkali metals

	Li	Na	K	Rb	Cs
Valence electron configuration	$2s^1$	$3s^1$	$4s^1$	$5s^1$	$6s^1$
Density (g/cm^3)	0.534	0.97	0.86	1.53	1.87
Melting point (°C)	179	97.6	63	39	28
Boiling point (°C)	1317	892	770	688	678
Atomic radius (pm)	155	190	235	248	267
Ionic radius (pm)*	60	95	133	148	169
Ionization energy (kJ/mol)	520	496	419	403	375
Electronegativity	1.0	0.9	0.8	0.8	0.7
Standard reduction potential (V)†	−3.05	−2.71	−2.93	−2.93	−2.92

*Refers to the cation M^+, where M denotes an alkali metal atom.
†The half-reaction is $M^+(aq) + e^- \longrightarrow M(s)$.

Lithium

Earth's crust is about 0.006 percent lithium by mass. The element is also present in seawater to the extent of about 0.1 ppm by mass. The most important lithium-containing mineral is spodumene ($LiAlSi_2O_6$), shown in Figure 20.13. Lithium metal is obtained by the electrolysis of molten LiCl to which some inert salts have been added to lower the melting point to about 500°C.

Like all alkali metals, lithium reacts with cold water to produce hydrogen gas:

$$2Li(s) + 2H_2O(l) \longrightarrow 2LiOH(aq) + H_2(g)$$

But, as we mentioned in Chapter 8 (p. 331), the chemistry of the first member of a group differs in some ways from the rest of the members of the same group. The following examples illustrate the differences between lithium and the rest of the alkali metals:

These properties parallel those of magnesium closely; an example of the diagonal relationship.

- On combustion, lithium forms the oxide (containing the O^{2-} ion):

  $$4Li(s) + O_2(g) \longrightarrow 2Li_2O(s)$$

  while sodium forms the peroxide (containing the O_2^{2-} ion) and potassium forms both the peroxide and the superoxide (containing the O_2^- ion).
- Lithium nitride is formed from the direct combination of the metal and molecular nitrogen. The nitrides of other alkali metals are formed more indirectly.
- Lithium carbonate and lithium phosphate are much less soluble than the carbonates and phosphates of other alkali metals.

Like the other alkali metal oxides, Li_2O is basic (see Figure 15.5) and reacts with water to yield the corresponding hydroxide:

$$Li_2O(s) + H_2O(l) \longrightarrow 2LiOH(aq)$$

When this is written as a net ionic equation, we see that this reaction is the hydrolysis of the oxide ion, which is a strong Brønsted base:

$$O^{2-}(aq) + H_2O(l) \longrightarrow 2OH^-(aq)$$

NaOH and KOH react with CO_2 in a similar way. However, because LiOH is lighter, it is more suitable for space vehicles.

The ability of lithium hydroxide to react with carbon dioxide to form lithium carbonate makes it a useful air purifier in space vehicles and submarines:

$$2LiOH(aq) + CO_2(g) \longrightarrow Li_2CO_3(aq) + H_2O(l)$$

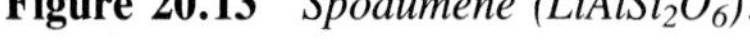

Figure 20.13 *Spodumene ($LiAlSi_2O_6$).*

Lithium combines with molecular hydrogen at high temperatures to yield lithium hydride:

$$2Li(s) + H_2(g) \longrightarrow 2LiH(s)$$

Lithium hydride reacts readily with water as follows:

$$2LiH(s) + 2H_2O(l) \longrightarrow 2LiOH(aq) + 2H_2(g)$$

This property makes LiH useful for drying organic solvents. Lithium aluminum hydride, $LiAlH_4$, is a powerful reducing agent that has many uses in organic synthesis. It can be prepared by reacting lithium hydride with aluminum chloride:

$$4LiH(s) + AlCl_3(s) \longrightarrow LiAlH_4(s) + 3LiCl(s)$$

The stability of alkali halides can be studied by the Born-Haber cycle (see Section 9.3).

Lithium chloride and bromide are highly ***hygroscopic;*** that is, they *have a great tendency to absorb water*. For this reason, they are sometimes used in dehumidifiers and air conditioners. Certain lithium salts, particularly lithium carbonate, are valuable drugs in the treatment of manic-depressive patients.

Sodium and Potassium

Figure 20.14 *Halite (NaCl).*

Sodium and potassium are about equally abundant in nature. They occur in silicate minerals such as albite ($NaAlSi_3O_8$) and orthoclase ($KAlSi_3O_8$). Over long periods of time (on a geologic scale), silicate minerals are slowly decomposed by wind and rain, and their sodium and potassium ions are converted to more soluble compounds. Eventually rain leaches these compounds out of the soil and carries them to the sea. Yet when we look at the composition of seawater, we find that the concentration ratio of sodium to potassium is about 28 to 1. The reason for this uneven distribution is that potassium is essential to plant growth, while sodium is not. Plants take up many of the potassium ions along the way, while sodium ions are free to move on to the sea. Other minerals that contain sodium or potassium are halite (NaCl), shown in Figure 20.14, Chile saltpeter ($NaNO_3$), and sylvite (KCl). Sodium chloride is also obtained from rock salt (see Figure 9.3).

Remember that Ca^{2+} is harder to reduce than Na^+.

Metallic sodium is most conveniently obtained from molten sodium chloride by electrolysis in the Downs cell (see Section 19.8). The melting point of sodium chloride is rather high (801°C), and much energy is needed to keep large amounts of the substance molten. Adding a suitable substance, such as $CaCl_2$, lowers the melting point to about 600°C—a more convenient temperature for the electrolysis process.

Metallic potassium cannot be easily prepared by the electrolysis of molten KCl because it is too soluble in the molten KCl to float on top of the cell for collection. Moreover, it vaporizes readily at the operating temperatures, creating hazardous conditions. Instead, it is usually obtained by the distillation of molten KCl in the presence of sodium vapor at 892°C. The reaction that takes place at this temperature is

Note that this is a chemical rather than electrolytic reduction.

$$Na(g) + KCl(l) \rightleftharpoons NaCl(l) + K(g)$$

This reaction may seem strange because in Table 20.4 we have seen that potassium is a stronger reducing agent than sodium. However, potassium has a lower boiling point (770°C) than sodium (892°C), so it is more volatile at 892°C and distills off more easily. According to Le Chatelier's principle, constant removal of potassium vapor shifts the above equilibrium from left to right, assuring recovery of metallic potassium.

Sodium and potassium are both extremely reactive, but potassium is the more reactive of the two. Both react with water to form the corresponding hydroxides. In a limited supply of oxygen, sodium burns to form sodium oxide (Na_2O). However, in the presence of excess oxygen, sodium forms the pale yellow peroxide:

$$2Na(s) + O_2(g) \longrightarrow Na_2O_2(s)$$

Sodium peroxide reacts with water to give an alkaline solution and hydrogen peroxide:

$$Na_2O_2(s) + 2H_2O(l) \longrightarrow 2NaOH(aq) + H_2O_2(aq)$$

Like sodium, potassium forms the peroxide. In addition, potassium also forms the superoxide when it burns in air:

$$K(s) + O_2(g) \longrightarrow KO_2(s)$$

When potassium superoxide reacts with water, oxygen gas is evolved:

$$2KO_2(s) + 2H_2O(l) \longrightarrow 2KOH(aq) + O_2(g) + H_2O_2(aq)$$

This reaction is utilized in breathing equipment (Figure 20.15). Exhaled air contains both moisture and carbon dioxide. The moisture reacts with KO_2 in the apparatus to generate oxygen gas as shown above. Furthermore, KO_2 also reacts with the CO_2 in the exhaled air, which produces more oxygen gas:

$$4KO_2(s) + 2CO_2(g) \longrightarrow 2K_2CO_3(s) + 3O_2(g)$$

In this manner, a user can continue to breathe in oxygen generated internally in the apparatus without being exposed to the toxic fumes outside.

Sodium and potassium metals dissolve in liquid ammonia to produce a beautiful blue solution:

$$Na \longrightarrow Na^+ + e^-$$

$$K \longrightarrow K^+ + e^-$$

Both the cation and the electron exist in the solvated form; the solvated electrons are responsible for the characteristic blue color of such solutions. Metal-ammonia solutions are powerful reducing agents (because they contain free electrons); they are useful in synthesizing both organic and inorganic compounds. It was discovered recently that the hitherto unknown alkali metal anions, M^-, are also formed in such solutions. This means that an ammonia solution of an alkali metal contains ion pairs

Figure 20.15 *Self-contained breathing apparatus.*

Figure 20.16 *Crystals of a salt composed of sodium anions and a complex of sodium cations.*

such as Na^+Na^- and K^+K^-! (Keep in mind that in each case the metal cation exists as a complex ion.) In fact, these "salts" are so stable that they can be isolated in crystalline form (Figure 20.16). This finding is of considerable theoretical interest, for it shows clearly that the alkali metals can have an oxidation number of -1, although -1 is not found in ordinary compounds.

Sodium and potassium are essential elements of living matter. Sodium ions and potassium ions are present in intracellular and extracellular fluids, and they are essential for osmotic balance and enzyme functions. We now describe the preparations and uses of several of the important compounds of sodium and potassium.

Sodium Chloride. The source, properties, and uses of sodium chloride were discussed in Chapter 9 (see p. 359).

Sodium Carbonate. Sodium carbonate is an important compound used in all kinds of industrial processes, including water treatment and the manufacture of soaps, detergents, medicines, and food additives. Today about half of all Na_2CO_3 produced is used in the glass industry (in soda-lime glass; see Section 11.7). Sodium carbonate ranks eleventh among the chemicals produced in the United States (9.5 million tons in 1988). For many years Na_2CO_3 was produced by the Solvay† process in which ammonia is first dissolved in a saturated solution of sodium chloride. Bubbling carbon dioxide into the solution results in the precipitation of sodium bicarbonate as follows:

$$NH_3(aq) + NaCl(aq) + H_2CO_3(aq) \longrightarrow NaHCO_3(s) + NH_4Cl(aq)$$

Sodium bicarbonate is then separated from the solution and heated to give sodium carbonate:

$$2NaHCO_3(s) \longrightarrow Na_2CO_3(s) + CO_2(g) + H_2O(g)$$

The last plant using the Solvay process in the United States closed in 1986.

However, the rising cost of ammonia and the pollution problem resulting from byproducts have prompted chemists to look for other sources of sodium carbonate. In recent years, discoveries of large deposits of the mineral *trona* $[Na_5(CO_3)_2(HCO_3) \cdot 2H_2O]$ have been made in Wyoming. When trona is crushed and heated, the hydrogen carbonate ion decomposes as follows:

†Ernest Solvay (1838–1922). Belgian chemist. Solvay's major contribution to industrial chemistry was the development of the process for the production of sodium carbonate that now bears his name.

$$2HCO_3^- \longrightarrow CO_3^{2-} + H_2O + CO_2$$

Removal of gaseous H_2O and CO_2 drives the reaction from left to right. The overall reaction is

$$2Na_5(CO_3)_2(HCO_3) \cdot 2H_2O(s) \longrightarrow 5Na_2CO_3(s) + CO_2(g) + 3H_2O(g)$$

The sodium carbonate obtained this way is dissolved in water, the solution is filtered to remove the insoluble impurities, and the sodium carbonate is crystallized as $Na_2CO_3 \cdot 10H_2O$. Finally, the hydrate is heated to give pure, anhydrous sodium carbonate.

Sodium Hydroxide and Potassium Hydroxide. The properties of sodium hydroxide and potassium hydroxide are very similar. Both hydroxides are prepared by the electrolysis of aqueous NaCl and KCl solutions (see Section 19.8); both hydroxides are strong bases and very soluble in water. Sodium hydroxide is used in the manufacture of soap and many organic and inorganic compounds. Potassium hydroxide is used as an electrolyte in some storage batteries, and aqueous potassium hydroxide is used to remove carbon dioxide and sulfur dioxide from air.

Sodium Nitrate and Potassium Nitrate. Large deposits of sodium nitrate (*saltpeter*) are found in Chile. It decomposes with the evolution of oxygen at about 500°C:

$$2NaNO_3(s) \longrightarrow 2NaNO_2(s) + O_2(g)$$

Potassium nitrate is prepared beginning with the ''reaction''

$$KCl(aq) + NaNO_3(aq) \longrightarrow KNO_3(aq) + NaCl(aq)$$

This process is carried out just below 100°C. Because KNO_3 is the least soluble salt at room temperature, it is separated from the solution by fractional crystallization. Like $NaNO_3$, KNO_3 decomposes when heated.

Gunpowder consists of potassium nitrate, wood charcoal, and sulfur in the approximate proportions of 6:1:1 by mass. When gunpowder is heated, the reaction is

$$2KNO_3(s) + S(s) + 3C(s) \longrightarrow K_2S(s) + N_2(g) + 3CO_2(g)$$

The sudden formation of hot expanding gases causes an explosion.

20.6 The Alkaline Earth Metals

All isotopes of radium are radioactive. Therefore, it is difficult and expensive to study the chemistry of radium.

The alkaline earth metals are somewhat less electropositive and less reactive than the alkali metals. Except for the first member of the family, beryllium, which resembles aluminum (a Group 3A metal) in some respects, the alkaline earth metals have similar chemical properties. Because their M^{2+} ions attain the stable noble gas electron configuration, the oxidation number of alkaline earth metals in the combined form is almost always +2. Table 20.5 lists some common properties of these metals.

Beryllium

Beryllium (see Figure 8.18) is relatively rare, ranking 32 in order of elemental abundance in the Earth's crust. The only important beryllium ore is beryl (beryllium aluminum silicate, $Be_3Al_2Si_6O_{18}$) (Figure 20.17). Pure beryllium is obtained by first con-

Table 20.5 Properties of alkaline earth metals

	Be	Mg	Ca	Sr	Ba
Valence electron configuration	$2s^2$	$3s^2$	$4s^2$	$5s^2$	$6s^2$
Density (g/cm^3)	1.86	1.74	1.55	2.6	3.5
Melting point (°C)	1280	650	838	770	714
Boiling point (°C)	2770	1107	1484	1380	1640
Atomic radius (pm)	112	160	197	215	222
Ionic radius (pm)*	31	65	99	113	135
First and second ionization energies (kJ/mol)	899 1757	738 1450	590 1145	548 1058	502 958
Electronegativity	1.5	1.2	1.0	1.0	0.9
Standard reduction potential (V)†	−1.85	−2.37	−2.87	−2.89	−2.90

*Refers to the cation M^{2+}, where M denotes an alkaline earth metal atom.
†The half-reaction is $M^{2+}(aq) + 2e^- \rightarrow M(s)$.

Figure 20.17 *Beryllium ore, beryl ($Be_3Al_2Si_6O_{18}$).*

verting the ore to the oxide (BeO). The oxide is then converted to the chloride or fluoride. Beryllium fluoride is heated with magnesium in a furnace at about 1000°C to produce metallic beryllium:

$$BeF_2(s) + Mg(l) \longrightarrow Be(s) + MgF_2(s)$$

Beryllium is highly toxic, especially if inhaled as a dust. Its toxicity is the result of the ability of Be^{2+} ion to compete with Mg^{2+} at many enzyme sites. Beryllium is used mainly in alloys with copper to enhance the hardness of copper. Beryllium atoms do not absorb high-energy radiation readily, so beryllium is used also as a "window" for X-ray tubes. It is also used as a moderator for neutrons produced in nuclear reactions.

Because only two electrons (the 1*s* electrons) shield the outer 2*s* electrons, the first and second ionization energies of Be are higher than those of other alkaline earth metals. On the other hand, it takes less energy to promote one of the 2*s* electrons to the 2*p* orbital, followed by *sp* hybridization (see p. 410). For this reason beryllium forms a number of simple covalent gaseous compounds such as BeH_2, $BeCl_2$, and $BeBr_2$. In many of its compounds beryllium forms four tetrahedral bonds, with Be at the center, for example, in complexes like BeF_4^{2-}, $BeCl_4^{2-}$, and $BeBr_4^{2-}$.

Beryllium being the first member of the Group 2A, its chemistry differs in a number of respects from that of the other alkaline earth metals. For example, there are no known compounds containing the uncomplexed Be^{2+} ion. Beryllium oxide and beryllium halides are essentially covalent, whereas other alkaline earth oxides, chlorides, and fluorides are ionic.

As an example of the diagonal relationship (see Section 8.6), beryllium resembles aluminum in a number of respects.

Magnesium

Magnesium (see Figure 8.18) is the sixth most plentiful element in Earth's crust (about 2.5 percent by mass). Among the principal magnesium ores are brucite, $Mg(OH)_2$; dolomite, $CaCO_3 \cdot MgCO_3$ (Figure 20.18); and epsomite, $MgSO_4 \cdot 7H_2O$. Seawater is a good source of magnesium; there are about 1.3 g of magnesium in each kilogram of

Figure 20.18 *Dolomite ($CaCO_3 \cdot MgCO_3$).*

seawater. As is the case with most alkali and alkaline earth metals, metallic magnesium is obtained by the electrolysis of its molten chloride, $MgCl_2$ (obtained from seawater).

The chemistry of magnesium is intermediate between that of beryllium and the heavier Group 2A elements. Magnesium does not react with cold water but does react slowly with steam:

$$Mg(s) + H_2O(g) \longrightarrow MgO(s) + H_2(g)$$

It burns brilliantly in air to produce magnesium oxide and magnesium nitride (see Figure 3.10):

$$2Mg(s) + O_2(g) \longrightarrow 2MgO(s)$$

$$3Mg(s) + N_2(g) \longrightarrow Mg_3N_2(s)$$

This property makes magnesium (in the form of thin ribbons or fibers) useful in flash photography and flares.

Magnesium oxide reacts very slowly with water to form magnesium hydroxide, a white solid suspension called *milk of magnesia* (Figure 20.19) which is used to treat acid indigestion:

$$MgO(s) + H_2O(l) \longrightarrow Mg(OH)_2(s)$$

Magnesium is a typical alkaline earth metal in that its hydroxide is a strong base. [The only alkaline earth hydroxide that is not a strong base is $Be(OH)_2$, which is amphoteric.] As Table 20.6 shows, all alkaline earth hydroxides except $Ba(OH)_2$ have low solubilities.

The major uses of magnesium are in alloys, as a lightweight structural metal, for cathodic protection (see Section 19.7), in organic synthesis, and in batteries. Magnesium is essential to plant and animal life, and Mg^{2+} ions are not toxic. It is estimated

Table 20.6 Solubilities of alkaline earth hydroxides at 25°C

	Solubility product (K_{sp})	Molar solubility (M)
$Be(OH)_2$	1×10^{-19}	3×10^{-7}
$Mg(OH)_2$	1.2×10^{-11}	1.4×10^{-4}
$Ca(OH)_2$	5.4×10^{-6}	0.011
$Sr(OH)_2$	3.1×10^{-4}	0.043
$Ba(OH)_2$	5.0×10^{-3}	0.11

Figure 20.19 *A commercially available milk of magnesia.*

Figure 20.20 *Fluorite (CaF_2).*

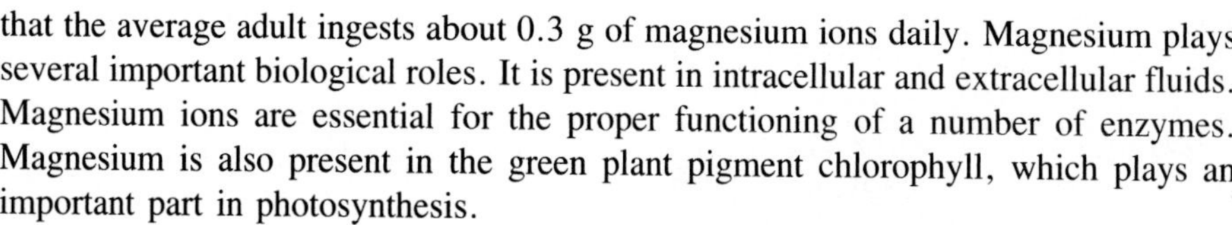

that the average adult ingests about 0.3 g of magnesium ions daily. Magnesium plays several important biological roles. It is present in intracellular and extracellular fluids. Magnesium ions are essential for the proper functioning of a number of enzymes. Magnesium is also present in the green plant pigment chlorophyll, which plays an important part in photosynthesis.

Calcium

Earth's crust contains about 3.4 percent calcium (see Figure 8.18) by mass. Calcium occurs in limestone, calcite, chalk, and marble as $CaCO_3$; in dolomite as $CaCO_3 \cdot MgCO_3$ (see Figure 20.18); in gypsum as $CaSO_4 \cdot 2H_2O$; and in fluorite as CaF_2 (Figure 20.20). Metallic calcium is best prepared by the electrolysis of molten calcium chloride ($CaCl_2$).

As we read down Group 2A from beryllium to barium, we note an increase in metallic properties. Unlike beryllium and magnesium, calcium (like strontium and barium) reacts with cold water to yield the corresponding hydroxide, although the rate of reaction is much slower than those involving the alkali metals (see Figure 3.12):

$$Ca(s) + 2H_2O(l) \longrightarrow Ca(OH)_2(aq) + H_2(g)$$

The term "lime" does not refer to a single substance. Rather, it includes quicklime, or calcium oxide (CaO), and slaked lime, or hydrated lime, which is calcium hydroxide [$Ca(OH)_2$]. Lime is one of the oldest materials known to mankind. Quicklime is produced by the thermal decomposition of calcium carbonate (see Section 18.3):

$$CaCO_3(s) \longrightarrow CaO(s) + CO_2(g)$$

while slaked lime is produced by the reaction between quicklime and water:

$$CaO(s) + H_2O(l) \longrightarrow Ca(OH)_2(s)$$

Quicklime is used in metallurgy (see Section 20.2) and the removal of SO_2 when fossil fuel is burned (see p. 659). Slaked lime is used in water treatment (see p. 744).

For many years, farmers have used lime to lower the acidity of soil for their crops (a process called *liming*), and nowadays liming is also applied to lakes affected by acid rain.

Metallic calcium has rather limited uses. It is used mainly as an alloying agent for metals like aluminum and copper and in the preparation of beryllium metal from its compounds. It is also used as a dehydrating agent for organic solvents.

Calcium is an essential element in living matter. It is the major component of bones and teeth; the calcium ion is present in a complex phosphate salt, hydroxyapatite, $Ca_5(PO_4)_3OH$ (see p. 734). A characteristic function of Ca^{2+} ions in living systems is the activation of a variety of metabolic processes. Calcium plays an important role in heart action, blood clotting, muscle contraction, and nerve transmission.

Figure 20.21 *Celestite ($SrSO_4$).*

Strontium and Barium

Strontium and barium have properties similar to those of calcium except that, in general, they are more reactive. Strontium occurs as the carbonate $SrCO_3$ (strontianite) and as the sulfate $SrSO_4$ (celestite) (Figure 20.21). There is no large-scale use of strontium. It is extracted by the electrolysis of its fused chloride ($SrCl_2$).

Figure 20.22 *Red fireworks contain strontium compounds.*

Barium occurs as the carbonate $BaCO_3$ (witherite) and as the sulfate $BaSO_4$ (barite). Metallic barium can be prepared by the electrolysis of the fused chloride ($BaCl_2$) or by reduction of its oxide with aluminum:

$$3BaO(s) + 2Al(s) \longrightarrow 3Ba(s) + Al_2O_3(s)$$

The similarity between calcium and the radioactive strontium-90 was discussed on p. 334. Strontium salts such as $Sr(NO_3)_2$ and $SrCO_3$ are used in red fireworks (Figure 20.22) as well as in the familiar red warning flares on highways.

Unlike that of strontium, barium's toxicity to the human body is chemical rather than radioactive in nature. If taken into the body as a soluble salt (for example, $BaCl_2$), barium causes serious deterioration of the heart's function—a phenomenon known as *ventricular fibrillation*. Interestingly, the insoluble barium sulfate, $BaSO_4$, is nontoxic to humans, presumably because of the low Ba^{2+} ion concentration in solution. This dense barium salt absorbs X rays and acts as an opaque barrier to them. Thus, X-ray examination of a patient who has swallowed a solution containing finely divided $BaSO_4$ particles allows the radiologist to diagnose an ailment of the patient's digestive tract (Figure 20.23). Barium sulfate is quite insoluble ($K_{sp} = 1.1 \times 10^{-10}$), so it passes through the digestive system and no appreciable amounts of Ba^{2+} ions are taken up by the body.

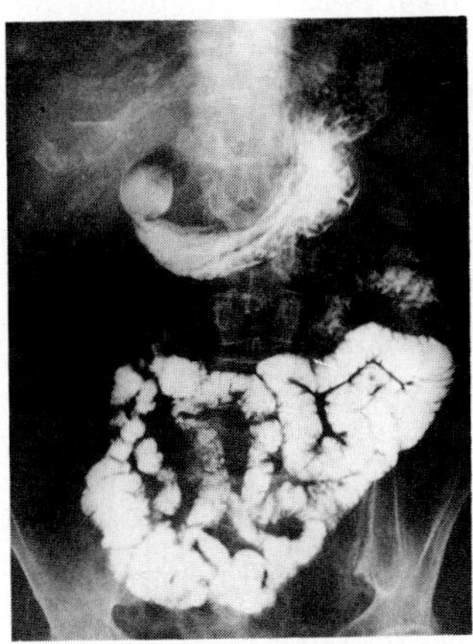
Figure 20.23 *Barium sulfate is opaque to X rays. An aqueous suspension of the substance is used to examine digestive tracts.*

20.7 Aluminum

Aluminum (see Figure 8.19) is the most abundant metal and the third most plentiful element in Earth's crust (7.5 percent by mass). The elemental form does not occur in nature; its principal ore is bauxite ($Al_2O_3 \cdot 2H_2O$). Other minerals containing aluminum are orthoclase ($KAlSi_3O_8$), beryl ($Be_3Al_2Si_6O_{18}$) (see Figure 20.17), cryolite (Na_3AlF_6), and corundum (Al_2O_3) (Figure 20.24).

Aluminum is usually prepared from bauxite, which is frequently contaminated with silica (SiO_2), iron oxides, and titanium(IV) oxide. The ore is first heated in sodium hydroxide solution to convert the silica into soluble silicates:

$$SiO_2(s) + 2OH^-(aq) \longrightarrow SiO_3^{2-}(aq) + H_2O(l)$$

At the same time, aluminum oxide is converted to the aluminate ion (AlO_2^-):

$$Al_2O_3(s) + 2OH^-(aq) \longrightarrow 2AlO_2^-(aq) + H_2O(l)$$

Iron oxide and titanium oxide are unaffected by this treatment and are filtered off. Next, the solution is treated with acid to precipitate the insoluble aluminum hydroxide:

Figure 20.24 *Corundum (Al_2O_3).*

$$AlO_2^-(aq) + H_3O^+(aq) \longrightarrow Al(OH)_3(s)$$

After filtration, the aluminum hydroxide is heated to obtain aluminum oxide:

$$2Al(OH)_3(s) \longrightarrow Al_2O_3(s) + 3H_2O(g)$$

Anhydrous aluminum oxide, or *corundum,* is reduced to aluminum by the *Hall† process.* Figure 20.25 shows a Hall electrolytic cell, which contains a series of carbon anodes. The cathode is also made of carbon and constitutes the lining inside the cell. The key to the Hall process is the use of cryolite, or Na_3AlF_6 (m.p. 1000°C), as solvent for aluminum oxide (m.p. 2045°C). The mixture is electrolyzed to produce aluminum and oxygen gas:

Molten cryolite provides a good conducting medium for electrolysis.

$$\begin{array}{lrl} \text{Anode (oxidation):} & 3[2O^{2-} & \longrightarrow O_2(g) + 4e^-] \\ \text{Cathode (reduction):} & 4[Al^{3+} + 3e^- & \longrightarrow Al(l)] \\ \hline \text{Overall:} & 2Al_2O_3 & \longrightarrow 4Al(l) + 3O_2(g) \end{array}$$

Oxygen gas reacts with the carbon anodes (at elevated temperatures) to form carbon monoxide, which escapes as a gas. The liquid aluminum metal (m.p. 660.2°C) sinks to the bottom of the vessel, from which it can be drained from time to time during the procedure.

Aluminum is one of the most versatile metals known. It has a low density (2.7 g/cm^3) and high tensile strength (that is, it can be stretched or drawn out). Aluminum is malleable, it can be rolled into thin foils, and it is an excellent electrical conductor. Its conductivity is about 65 percent that of copper. However, because aluminum is cheaper and lighter than copper, it is widely used in high-voltage transmission lines. Although aluminum's chief use is in aircraft construction, the pure metal itself is too soft and weak to withstand much strain. Its mechanical properties are greatly improved by alloying it with small amounts of metals such as copper, magnesium, and manganese, as well as silicon. Aluminum is not involved in living systems and is generally considered to be nontoxic.

As we read across the periodic table from left to right in a given period, we note a gradual decrease in metallic properties. Thus, although aluminum is considered an active metal, it does not react with water as do sodium and magnesium. Aluminum reacts with hydrochloric acid and with strong bases as follows:

$$2Al(s) + 6HCl(aq) \longrightarrow 2AlCl_3(aq) + 3H_2(g)$$

$$2Al(s) + 2NaOH(aq) + 2H_2O(l) \longrightarrow 2NaAlO_2(aq) + 3H_2(g)$$

Aluminum readily forms the oxide Al_2O_3 when exposed to air:

$$4Al(s) + 3O_2(g) \longrightarrow 2Al_2O_3(s)$$

A tenacious film of this oxide protects metallic aluminum from further corrosion and accounts for some of the unexpected inertness of aluminum.

Aluminum oxide has a very large exothermic enthalpy of formation ($\Delta H_f^\circ = -1670$ kJ/mol). This property makes aluminum suitable for use in solid propellants for

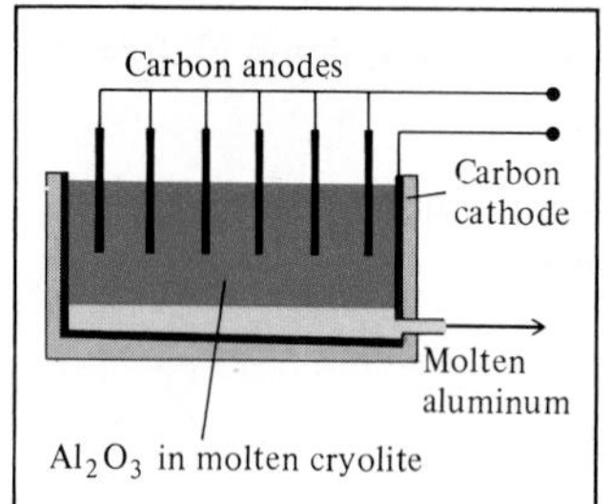

Figure 20.25 *Electrolytic production of aluminum based on the Hall process.*

†Charles Martin Hall (1863–1914). American inventor. While Hall was an undergraduate at Oberlin College, he became interested in finding an inexpensive way to extract aluminum. Shortly after graduation, when he was only 22 years old, Hall succeeded in obtaining aluminum from aluminum oxide in a backyard woodshed. Amazingly, the same discovery was made at almost the same moment in France, by Paul Héroult, another 22-year-old inventor working in a similar makeshift laboratory.

Figure 20.26 *A thermite reaction.*

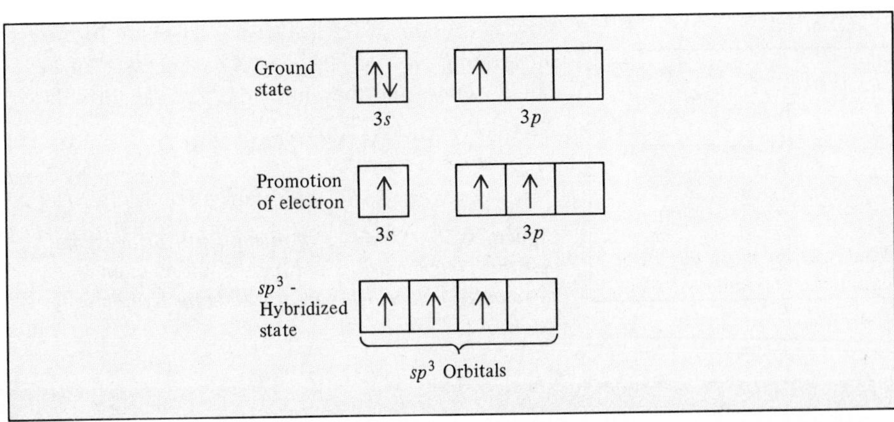

Figure 20.27 *The sp^3 hybridization of an Al atom in Al_2Cl_6. Each Al atom has one vacant sp^3 hybrid orbital that can accept a lone pair from the bridging Cl atom.*

rockets such as those used for the space shuttle *Columbia*. When a mixture of aluminum and ammonium perchlorate (NH_4ClO_4) is ignited, aluminum is oxidized to Al_2O_3, and the heat liberated in the reaction causes the gases that are formed to expand with great force. This action is responsible for lifting the rocket. The great affinity of aluminum for oxygen is illustrated nicely by the reaction of aluminum powder with a variety of metal oxides, particularly the transition metal oxides, to produce the corresponding metals. A typical reaction is (Figure 20.26)

$$2Al(s) + Fe_2O_3(s) \longrightarrow Al_2O_3(l) + 2Fe(l) \qquad \Delta H^\circ = -852 \text{ kJ}$$

which can result in temperatures approaching 3000°C. This reaction is called the *thermite reaction,* which is used in the welding of steel and iron. Aluminum chloride exists as a dimer:

```
Cl    Cl     Cl
  \  /  \   /
   Al    Al
  /  \  /  \
Cl    Cl     Cl
```

Each of the bridging chlorine atoms forms a normal covalent bond and a coordinate covalent bond (indicated by →) with two aluminum atoms. Each aluminum atom is assumed to be sp^3-hybridized, so the vacant sp^3 hybrid orbital can accept a lone pair from the chlorine atom (Figure 20.27). Aluminum chloride undergoes hydrolysis as follows:

$$AlCl_3(s) + 3H_2O(l) \longrightarrow Al(OH)_3(s) + 3HCl(aq)$$

Aluminum hydroxide is an amphoteric hydroxide:

$$Al(OH)_3(s) + 3H^+(aq) \longrightarrow Al^{3+}(aq) + 3H_2O(l)$$

$$Al(OH)_3(s) + OH^-(aq) \longrightarrow Al(OH)_4^-(aq)$$

A polymer is a large molecule containing hundreds or thousands of atoms that are covalently bonded together.

In contrast to the boron hydrides, which are a well-defined series of compounds, aluminum hydride is a polymeric substance in which each aluminum atom is surrounded octahedrally by bridging hydrogen atoms (Figure 20.28).

When an aqueous mixture of aluminum sulfate and potassium sulfate is evaporated slowly, crystals of $KAl(SO_4)_2 \cdot 12H_2O$ are formed. Similar crystals can be formed by substituting Na^+ or NH_4^+ for K^+, and Cr^{3+} or Fe^{3+} for Al^{3+}. These compounds are called *alums*, and they have the general formula

$$M^+M^{3+}(SO_4)_2 \cdot 12H_2O \qquad \begin{array}{l} M^+: \; K^+, Na^+, NH_4^+ \\ M^{3+}: \; Al^{3+}, Cr^{3+}, Fe^{3+} \end{array}$$

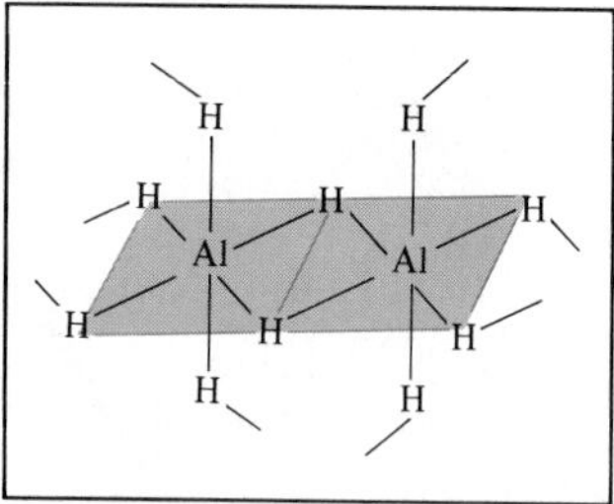

Figure 20.28 *Structure of aluminum hydride. Note that this compound is a polymer. Each Al atom is surrounded octahedrally by six bridging H atoms.*

CHEMISTRY IN ACTION

Recycling Aluminum

Aluminum beverage cans were virtually unknown in 1960, yet by the early 1970s over 1.3 billion pounds of aluminum had been used for these containers. What properties of aluminum account for its popularity in the beverage industry? The metal is nontoxic, odorless, tasteless, and lightweight. Furthermore, it is thermally conducting, so the fluid inside the container can be chilled rapidly. Such a tremendous increase in the use of aluminum does have definite drawbacks, however. More than 3 billion pounds of the metal cans and foils are discarded annually. The immense litter and landfills of our throwaway society stand as a testimony to a way of life we know to be both wrong and costly. The best solution to this environmental problem is recycling.

Let us compare the energy consumed in the production of aluminum from bauxite with that consumed when aluminum is recycled. The overall reaction for the Hall process can be represented as

$$Al_2O_3 \text{ (in molten cryolite)} + 3C(s) \longrightarrow 2Al(l) + 3CO(g)$$

for which $\Delta H° = 1340$ kJ and $\Delta S° = 586$ J/K. At 1000°C, which is the temperature of the process, the standard free-energy change for the reaction is given by

$$\begin{aligned} \Delta G° &= \Delta H° - T\Delta S° \\ &= 1340 \text{ kJ} - (1273 \text{ K})\left(\frac{586 \text{ J}}{\text{K}}\right)\left(\frac{1 \text{ kJ}}{1000 \text{ J}}\right) \\ &= 594 \text{ kJ} \end{aligned}$$

Equation (19.3) states that $\Delta G° = -nF\mathscr{E}°$; therefore, the amount of electrical energy needed to produce 1 mole of Al from bauxite is 594 kJ/2, or 297 kJ.

When aluminum metal is recycled, the only energy required is the energy to heat the can to its melting point (660°C) plus the heat of fusion (10.7 kJ/mol). The heat change for heating 1 mole of the metal from 25°C to 660°C is

$$\begin{aligned} \text{heat input} &= \mathcal{M}\rho\Delta t \\ &= (27.0 \text{ g})(0.900 \text{ J/g} \cdot °\text{C})(660 - 25)°\text{C} \\ &= 15.4 \text{ kJ} \end{aligned}$$

where $\mathcal{M}$ is the molar mass and ρ the specific heat of Al and Δt is the temperature change. Thus, the total energy needed to recycle 1 mole of Al is given by

$$\begin{aligned} \text{total energy} &= 15.4 \text{ kJ} + 10.7 \text{ kJ} \\ &= 26.1 \text{ kJ} \end{aligned}$$

To compare the energy requirements of the two methods we write

$$\frac{\text{energy needed to recycle 1 mol Al}}{\text{energy needed to produce 1 mol Al by electrolysis}} = \frac{26.1 \text{ kJ}}{297 \text{ kJ}} \times 100\% = 8.8\%$$

Thus, by recycling aluminum cans we can save about 91 percent of the energy required to make new metal from bauxite. By recycling most of the aluminum cans thrown away each year, the saving would amount to 20 billion kilowatt-hours of electricity—about 1 percent of the electric power used in the United States

(a)

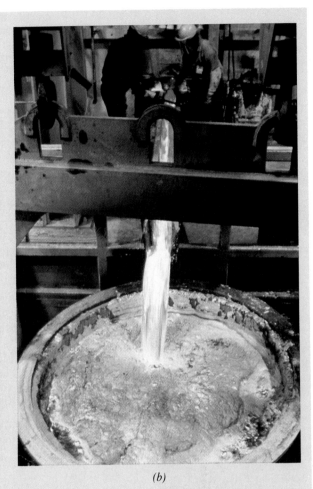

(b)

annually. (Watt is the unit for power; 1 watt = 1 joule per second.) Figure 20.29 shows aluminum cans being recycled.

Figure 20.29 *Recycling aluminum: (a) collecting aluminum cans; (b) melting and purifying aluminum.*

20.8 Tin and Lead

Tin

Tin and lead (see Figure 8.20) are the only metallic elements in Group 4A (Table 20.7). They were among the first metals used by humans. The principal ore of tin is Sn(IV) oxide, or cassiterite (Figure 20.30). Metallic tin is prepared by reducing SnO_2 with carbon at elevated temperatures:

$$SnO_2(s) + 2C(s) \longrightarrow Sn(l) + 2CO(g)$$

There are three allotropic forms of tin with the following transition temperatures:

$$\text{gray tin} \xrightarrow{13^\circ\text{C}} \text{white tin} \xrightarrow{161^\circ\text{C}} \gamma\text{-tin}$$

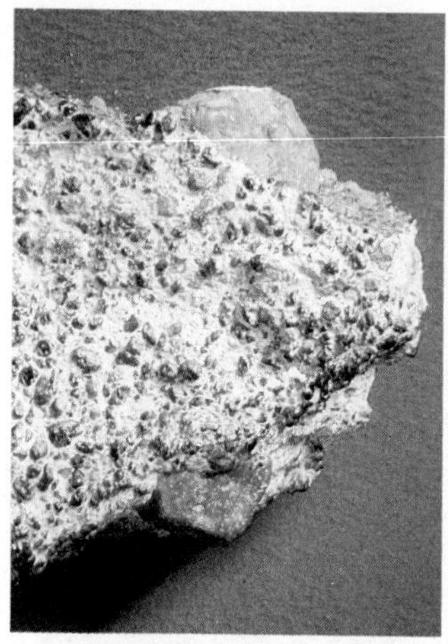

Figure 20.30 *The mineral cassiterite (SnO_2).*

Table 20.7 Properties of tin and lead

	Sn	Pb
Valence electron configuration	$5s^25p^2$	$6s^26p^2$
Density (g/cm^3)	7.28	11.4
Melting point (°C)	231.9	327.5
Boiling point (°C)	2270	1740
Atomic radius (pm)	162	175
Ionic radius (pm)*	112	120
First and second ionization energies (kJ/mol)	709; 1413	716; 1451
Electronegativity	1.8	1.9
Standard reduction potential (V)†	−0.14	−0.13

*Refers to the cation M^{2+}, where M denotes Sn or Pb.
†The half-reaction is $M^{2+}(aq) + 2e^- \rightarrow M(s)$.

Metallic tin (that is, white tin) is soft and malleable. Its valence electron configuration is $5s^25p^2$; the metal forms compounds with oxidation numbers of +2 and +4. Tin(II) compounds (called *stannous compounds*) are generally more ionic and reducing, while tin(IV) compounds (called *stannic compounds*) are more covalent and oxidizing. Tin reacts with hydrochloric acid to give tin(II) chloride:

$$Sn(s) + 2HCl(aq) \longrightarrow SnCl_2(aq) + H_2(g)$$

and with oxidizing acids such as nitric acid to give Sn(IV) compounds:

$$Sn(s) + 4HNO_3(aq) \longrightarrow SnO_2(s) + 4NO_2(g) + 2H_2O(l)$$

Tin also reacts with hot concentrated aqueous solutions of sodium hydroxide and potassium hydroxide to form the stannate ion (SnO_3^{2-}):

$$Sn(s) + 2OH^-(aq) + H_2O(l) \longrightarrow SnO_3^{2-}(aq) + 2H_2(g)$$

Thus, tin exhibits amphoteric properties. It combines directly with chlorine to form tin(IV) chloride, which is a liquid:

$$Sn(s) + 2Cl_2(g) \longrightarrow SnCl_4(l)$$

Unlike carbon and silicon, tin forms only two hydrides: SnH_4 and Sn_2H_6, neither of which is very stable. In both compounds the tin atom is sp^3-hybridized.

Figure 20.31 *Solder is used extensively in the construction of electronic circuits.*

Tin is used mainly to form alloys. For example, bronze is 20 percent tin and 80 percent copper; soft solder is 33 percent tin and 67 percent lead (Figure 20.31); and pewter is about 85 percent tin, 6.8 percent copper, 6 percent bismuth, and 1.7 percent antimony. Of course, tin is also used in the manufacture of "tin" cans (see Section 19.7).

Lead

The chief ore of lead is galena, PbS (Figure 20.32). Metallic lead is obtained by first roasting the sulfide in air:

$$2PbS(s) + 3O_2(g) \longrightarrow 2PbO(s) + 2SO_2(g)$$

The oxide is then reduced in a blast furnace with coke:

$$PbO(s) + C(s) \longrightarrow Pb(l) + CO(g)$$

Figure 20.32 *Galena (PbS).*

$$PbO(s) + CO(g) \longrightarrow Pb(l) + CO_2(g)$$

Because both tin and lead are much less electropositive than the alkali and alkaline earth metals, they can be prepared more readily by chemical reduction than by the expensive process of electrolysis.

As mentioned in Chapter 8, as a result of the inert-pair effect, lead(II) compounds are more stable than lead(IV) compounds. Practically all common lead compounds contain lead in the +2 oxidation state (PbO, $PbCl_2$, PbS, $PbSO_4$, $PbCO_3$, and $PbCrO_4$). Except for lead acetate and lead nitrate, most lead compounds are insoluble. Lead(II) oxide, known as *litharge,* was used to glaze ceramic vessels. For a number of years, white lead, $Pb_3(OH)_2(CO_3)_2$, was used as a pigment for paints, but because of its toxocity, its use in the United States has been banned. Lead(IV) oxide is a covalent compound and a strong oxidizing agent. It can oxidize hydrochloric acid to molecular chlorine:

$$PbO_2(s) + 4HCl(aq) \longrightarrow PbCl_2(s) + Cl_2(g) + 2H_2O(l)$$

The major use of lead is in lead storage batteries (see Section 19.6). Because lead is relatively impenetrable to high-energy radiation, it is also used in protective shields for nuclear chemists, X-ray operators, and radiologists.

CHEMISTRY IN ACTION

The Toxicity of Lead

Lead has no known beneficial function in human metabolism. On the contrary, its toxicity has been known for over 2000 years. Although we know lead is a poison, it is still widely used in our society. At present, lead poisoning is one of the most serious environmental concerns. Tetraethyllead [$(C_2H_5)_4Pb$] has long been used as an antiknocking agent for improving the performance of gasoline. Some "leaded" gasolines contained as much as 2 to 4 g of the substance per gallon of gasoline. Figure 20.33 shows the annual consumption of lead-containing antiknock additives in the United States since 1925. In cities with heavy automobile traffic, lead concentrations in the atmosphere may be as high as 10 micrograms per cubic meter of air, which endangers the health of countless people. In recent years the phasing out of leaded gasoline has helped to reduce the lead content in the environment. Others exposed to lead include young children living in dilapidated houses who nibble sweet-tasting chips of leaded paint, whiskey drinkers who consume quantities of lead-contaminated moonshine, and people who eat or drink from improperly lead-glazed earthenware. Soldered joints (which contain lead) in plumbing systems for drinking water are also a major source of lead.

Lead is extremely toxic; its effect on humans is cumulative. It enters the body either as inorganic lead

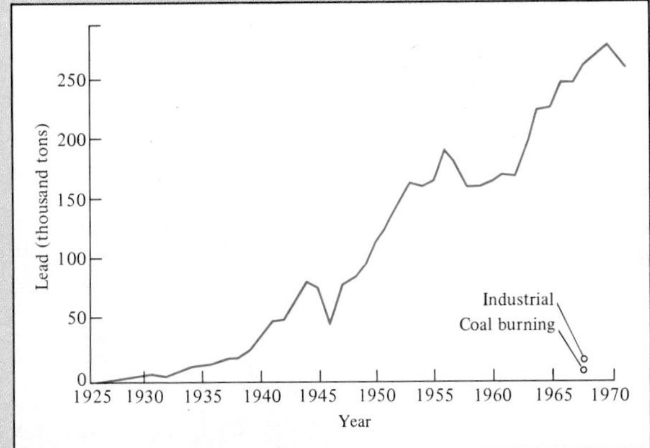

Figure 20.33 *Annual consumption of gasoline antiknock additives in the United States since 1925. Note the magnitude of lead consumption in gasoline compared with that in industry and coal burning for the year 1967.*

(Pb^{2+}) ion or as tetraethyllead. Inhaled or ingested lead concentrates in the blood, tissues, and bones. It is known that lead ions inhibit enzymes that catalyze the reactions for biosynthesis of hemoglobin. Thus, one symptom of lead poisoning is anemia.

Tetraethyllead is even more poisonous than Pb^{2+}. In the liver it is converted to the $(C_2H_5)_3Pb^+$ ion. Because the hydrocarbon group is nonpolar, this ion can move across membrane layers more easily than uncomplexed lead ions, and can therefore attack enzymes in various locations, such as in the brain. Indeed, brain damage is the most common symptom of those—especially children—afflicted with acute lead poisoning. Lead also affects the central nervous system and impairs kidney functions.

Lead poisoning is usually treated with chelating agents—substances that can form stable complex ions with Pb^{2+}. Two particularly effective compounds for removing lead ions from blood and tissues are ethylenediaminetetraacetate ion (EDTA) and 2,3-dimercaptopropanol, which is more commonly called BAL (British anti-lewisite):

$$(^{-}OOCCH_2)_2N\text{—}CH_2\text{—}CH_2\text{—}N(CH_2COO^{-})_2 \qquad CH_2(SH)\text{—}CH(SH)\text{—}CH_2OH$$

EDTA BAL

BAL was developed during World War II as an antidote for *lewisite*, a poison gas containing arsenic. In ionized form, both compounds yield very stable complexes with lead (Figure 20.34) that are eventually excreted through the kidney.

Figure 20.34 *(a) EDTA complex of lead. The complex bears a net charge of −2, since each O donor atom has one negative charge and the lead ion carries two positive charges. Note the octahedral geometry around the Pb^{2+} ion. (b) BAL complex of lead. This complex also bears a net charge of −2, since each S atom has one negative charge. The Pb^{2+} ion is tetrahedrally bonded to the four S atoms.*

20.9 Zinc, Cadmium, and Mercury

The Group 2B elements—zinc, cadmium, and mercury (Figure 20.35)—resemble the alkaline earth metals in that they all have the same valence electron configuration, ns^2. Yet their chemical properties are quite different (Table 20.8). This difference is a result of the difference in the number of electrons in the next-to-outermost shell; Ca, Sr, and

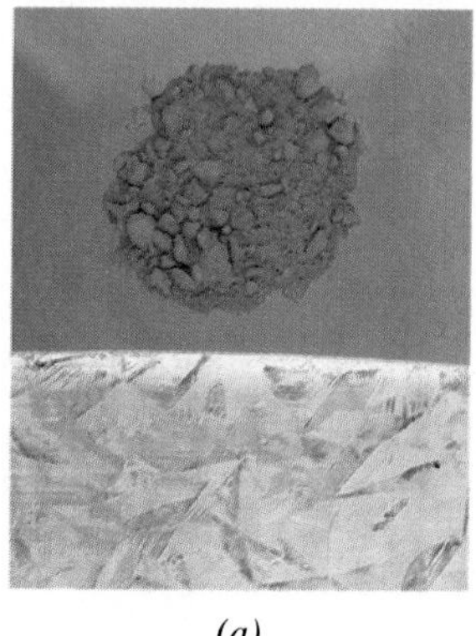
(a)

(b)

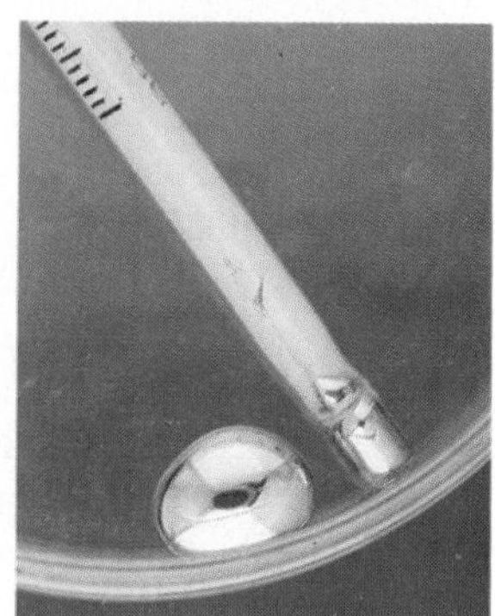
(c)

Figure 20.35 *(a) Zinc, (b) cadmium, and (c) mercury.*

Table 20.8 Properties of the Group 2B elements

	Zn	Cd	Hg
Valence electron configuration	$4s^23d^{10}$	$5s^24d^{10}$	$6s^25d^{10}$
Density (g/cm^3)	7.14	8.64	13.59
Melting point (°C)	419.5	320.9	−38.9
Boiling point (°C)	907	767	357
Atomic radius (pm)	138	154	157
Ionic radius (pm)*	74	97	110
First and second ionization energies (kJ/mol)	906; 1733	867; 1631	1006; 1809
Electronegativity	1.6	1.7	1.9
Standard reduction potential (V)†	−0.76	−0.40	0.85

*Refers to the cation M^{2+}, where M denotes Zn, Cd, or Hg.

†The half-reaction is $M^{2+}(aq) + 2e^- \rightarrow M(s)$ for Zn and Cd. For Hg, it is $M_2^{2+}(aq) + 2e^- \rightarrow 2M(l)$.

Ba have 8 electrons (s^2p^6), whereas Zn, Cd, and Hg have 18 electrons ($s^2p^6d^{10}$). Because the d orbitals do not shield outer electrons effectively from the nucleus, the increase in nuclear charge from calcium to zinc results in an increase in the ionization energies of the valence electrons. Further, we expect the zinc atom to be smaller than the calcium atom, and this is indeed the case (Ca: 197 pm; Zn: 138 pm). For similar reasons, cadmium and mercury lose electrons less readily than strontium and barium. Thus, the Group 2B elements are less reactive and less electropositive and have a greater tendency to form covalent compounds than the alkaline earth metals.

Zinc and Cadmium

Figure 20.36 *Sphalerite (ZnS).*

The properties of zinc and cadmium are so similar that we will describe their chemistries together. Not an abundant element (about 0.007 percent of Earth's crust by mass), zinc occurs principally as the mineral sphalerite, ZnS (Figure 20.36), also called zincblende. Metallic zinc is obtained by roasting the sulfide in air to convert it to the oxide and then reducing the oxide with finely divided carbon:

$$2ZnS(s) + 3O_2(g) \longrightarrow 2ZnO(s) + 2SO_2(g)$$

$$ZnO(s) + C(s) \longrightarrow Zn(s) + CO(g)$$

Cadmium is only about one-thousandth as abundant as zinc and is present in small amounts in most zinc ores as CdS. Cadmium is obtained from flue dust emitted during the purification of zinc by distillation. Because cadmium is more volatile (b.p. 767°C) than zinc (b.p. 907°C), it evaporates first and concentrates in the first distillates.

Both zinc and cadmium are silvery metals in the pure state. Zinc is hard and brittle, but cadmium is soft enough to be cut with a knife.

The only known oxidation number of these metals is +2. Both metals react with strong acids to produce hydrogen gas:

$$Zn(s) + 2H^+(aq) \longrightarrow Zn^{2+}(aq) + H_2(g)$$

$$Cd(s) + 2H^+(aq) \longrightarrow Cd^{2+}(aq) + H_2(g)$$

Zinc and cadmium oxides are formed through direct combination of these metals with oxygen. Zinc is amphoteric, but cadmium is not. In addition to its reaction with strong acids, zinc also reacts with strong bases:

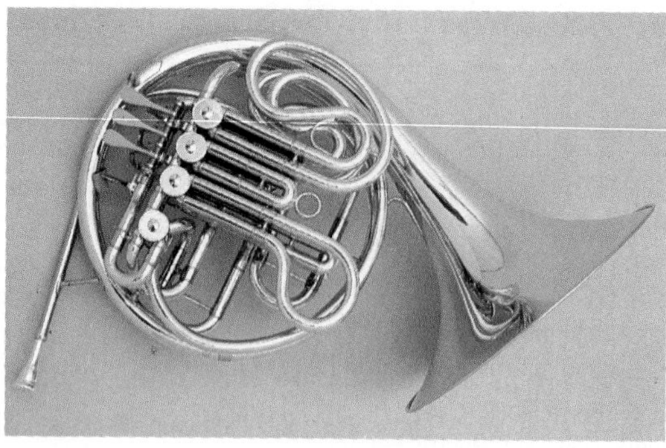

Figure 20.37 *Brass is used to make many wind instruments, such as the French horn.*

$$Zn(s) + 2OH^-(aq) + 2H_2O(l) \longrightarrow Zn(OH)_4^{2-}(aq) + H_2(g)$$

The inertness of cadmium toward basic solutions makes it a useful metal for plating other metals used in alkaline solutions.

Metallic zinc is used to form alloys. For example, brass is an alloy containing about 20 percent zinc and 80 percent copper by mass (Figure 20.37). Because zinc is oxidized easily, it is used to provide cathodic protection (see Section 19.7) for less electropositive metals. Iron protected this way, called *galvanized iron,* is prepared by dipping iron into molten zinc or by plating zinc on iron by electrolysis (using iron as the cathode in a solution containing Zn^{2+} ions).

Zinc sulfide is used in the white pigment lithopone, which contains a roughly equimolar mixture of ZnS and $BaSO_4$. Unlike the white lead discussed earlier, this substance is nontoxic. Zinc sulfide emits light when struck by X rays or an electron beam. Therefore it is used in screens of television sets, oscilloscopes, and X-ray fluoroscopes.

Zinc oxide, an important zinc compound, is prepared by burning the metal in air,

$$2Zn(s) + O_2(g) \longrightarrow 2ZnO(s)$$

and by the thermal decomposition of zinc nitrate or zinc carbonate,

$$2Zn(NO_3)_2(s) \longrightarrow 2ZnO(s) + 4NO_2(g) + O_2(g)$$

$$ZnCO_3(s) \longrightarrow ZnO(s) + CO_2(g)$$

Zinc oxide is amphoteric:

$$ZnO(s) + 2HCl(aq) \longrightarrow ZnCl_2(aq) + H_2O(l)$$

$$ZnO(s) + 2NaOH(aq) \longrightarrow Na_2ZnO_2(aq) + H_2O(l)$$

The compound is used in the treatment of rubber. It is also used as a white pigment in paint, as an ointment base, and in cement.

Zinc is an essential element in living matter; its presence in the human body is necessary for the activity of a number of enzymes such as carbonic anhydrase, which catalyzes the hydration and dehydration of carbon dioxide.

The major use of cadmium is to plate other metals. In the electroplating of steel, for example, the anode (Cd) and cathode (steel) are dipped into a solution containing $Cd(CN)_4^{2-}$ ions. Normally, a very thin film of metallic cadmium (about 0.0005 cm thick) is sufficient for protecting the metal.

Although cadmium has no biological significance, there is growing concern over its

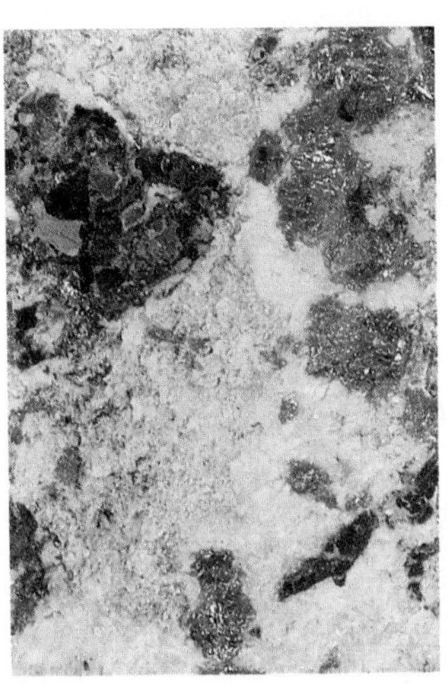

Figure 20.38 *Cinnabar (HgS).*

effect as an environmental pollutant. Like compounds of most heavy metals, cadmium compounds are exceedingly toxic. The common symptoms of cadmium poisoning are hypertension (high blood pressure), anemia, and kidney failure. The action of cadmium on our bodies is not as well understood as that of lead; presumably the Cd^{2+} ions also react with proteins. Electroplating industries are the main source of cadmium pollution, by discharging waste solutions into lakes and rivers, but there may be enough cadmium in the air alone to cause health problems.

As mentioned earlier, zinc oxide is used to treat rubber. Because zinc and cadmium have similar properties, zinc oxide is always contaminated with a small amount of cadmium oxide. As an automobile tire wears, both ZnO and CdO are liberated to the atmosphere as fine dust, which may not settle for quite some time. Tobacco, too, contains a small amount of cadmium, so smokers probably have a higher body content of cadmium than nonsmokers.

Mercury

Mercury has been known since ancient times. It is the only metal that exists as a liquid at room temperature. (Two other metals—cesium, m.p. 28.7°C, and gallium, m.p. 29.8°C—come quite close to being liquids at room temperature.) Although mercury is rarer than gold and platinum (it constitutes about 8×10^{-6} percent of Earth's crust by mass), its sources are so much more concentrated that the metal can be obtained readily. The principal ore of mercury is mercury(II) sulfide, HgS, called *cinnabar* (Figure 20.38). The ore is first concentrated by flotation and then roasted in air to yield mercury(II) oxide,

$$2HgS(s) + 3O_2(g) \longrightarrow 2HgO(s) + 2SO_2(g)$$

which decomposes to yield mercury vapor,

$$2HgO(s) \longrightarrow 2Hg(g) + O_2(g)$$

Mercury does not dissolve iron; for this reason mercury has been stored in iron containers since ancient times.

Metallic mercury is a bright, silvery, dense liquid that freezes at −38.9°C and boils at 357°C. Liquid mercury dissolves many metals, such as copper, silver, gold, and the alkali metals, to form amalgams that may be either solids or liquids. The reactivity of a metal is generally lowered when it is amalgamated with mercury. For example, the sodium-mercury amalgam is so much less reactive than sodium that it decomposes water at a considerably slower rate. Such amalgams are useful as mild reducing agents in organic synthesis.

Unlike cadmium and zinc, mercury has both +1 (*mercurous*) and +2 (*mercuric*) oxidation numbers in its compounds. Because the standard reduction potential of mercury is positive, the metal does not react with water and does not decompose nonoxidizing acids like HCl. Many of the mercury(II) salts (HgS and HgI_2, for example) are insoluble. This is particularly significant, for if HgS were soluble, rain and weathering might have distributed mercury throughout the planet in sufficient concentration to poison the environment. Life as we know it might have been impossible. (On the other hand, mercury might have become an essential element to other kinds of life systems.)

Mercury(II) oxide is formed by heating the metal in air slowly:

$$2Hg(l) + O_2(g) \longrightarrow 2HgO(s)$$

At higher temperatures, the oxide decomposes to yield elemental mercury and molecular oxygen again:

$$2HgO(s) \longrightarrow 2Hg(l) + O_2(g)$$

In fact, this was the famous experiment conducted by Joseph Priestley† in 1774 when he discovered oxygen.

Mercury with an oxidation number of +2 has a greater tendency to form covalent bonds than do zinc and cadmium. Mercury(II) chloride has a linear structure, as predicted by the VSEPR model:

$$:\ddot{\underset{\cdot\cdot}{Cl}}—Hg—\ddot{\underset{\cdot\cdot}{Cl}}:$$

Mercury(II) fluoride, HgF_2, on the other hand, is essentially ionic. The Hg^{2+} ion also has a tendency to form very stable complex ions, such as $HgCl_4^{2-}$, $Hg(NH_3)_4^{2+}$, and $Hg(CN)_4^{2-}$.

For a number of years the true identity of mercurous ion was not clear; the two possibilities are Hg^+ and Hg_2^{2+}. If the mercury(I) ion existed as Hg^+ (electron configuration $6s^1$), it would be paramagnetic, but there is no experimental evidence to support this structure. Instead, the ion consists of two mercury atoms held together by a sigma bond:

$$[Hg—Hg]^{2+}$$

This was one of the first metal-metal bonds known in a compound.

The most common mercury(I) compound is probably mercury(I) chloride, Hg_2Cl_2, an insoluble white solid called *calomel*. At one time, calomel was used in medicine as a purgative. However, it has a tendency to disproportionate to Hg and $HgCl_2$, which is quite poisonous, so calomel is no longer used in internal medicine.

It has been estimated that mercury and its compounds have over 2000 uses. Only the more familiar and important applications will be mentioned here. Mercury has a large and uniform coefficient of volume expansion (with temperature) and is therefore suitable for thermometers. The liquid has a very high density (13.6 g/cm^3) and is employed in barometers. Although the conductivity of mercury is only about 2 percent that of copper, the advantages of its fluidity are so great that the metal is used in making electrical contacts in household light switches and thermostats. Mercury is also used as the cathode in the electrolytic production of many elements, for example, chlorine.

†Joseph Priestley (1733–1804). English chemist. In addition to conducting the famous oxygen experiment, Priestley discovered HCl gas and showed that O_2 is produced by green plants.

CHEMISTRY IN ACTION

Mad as a Hatter

Mercury is a heavy metal poison whose influence is cumulative and whose effects on neurological behavior are notorious. However, the toxicity of mercury very much depends on the physical and chemical states of the element. Pure metallic mercury is not particularly poisonous; in fact, ingestion of a very small amount of mercury (as from the dental amalgam mentioned on p. 810) produces no noticeable ill effects. The metal apparently passes through the body without undergoing chemical change. Mercury vapor, on the other hand, is very dangerous because it causes irritation and destruction of lung tissues. Although the liquid is not very volatile (vapor pressure is only 2.5×10^{-6} atm at 25°C), prolonged exposure to mercury vapor should be avoided. Because the substance is so widely used in households and laboratories, spillage of mercury occurs frequently. Simple calculations show that as little as 3 mL of mercury could saturate a large (and poorly ventilated) room with its vapor within a week, making it unsafe to work in. Spilled mercury is almost impossible to recover completely because the liquid enters the cracks in the floor. The usual remedy is to cover the mercury with yellow sulfur powder, which retards the evaporation rate by reducing its surface area and also slowly converts the mercury to the less harmful compound HgS.

The toxicity of inorganic mercury compounds depends on their solubilities. For example, the insoluble mercurous chloride is not considered very toxic and, in fact, has been used in medicine as a purgative and a drug to kill intestinal worms. Such treatment may be termed selective poisoning. The substance is harmful to the human body; but, by a judicious choice of the quantity administered, the poison kills the worms before it endangers the patient. Because mercuric salts are generally more soluble, they are considerably more toxic. The Hg^{2+} ions concentrate chiefly in the liver and kidneys. The harmful effects are usually slow to develop. Some symptoms are sore gums and loose teeth. Mercury poisoning can cause brain damage to unborn infants and can lead to a condition called *erethism,* which is characterized by jerking, irritability, and mental and emotional disturbances.

The mad hatter of *Alice in Wonderland* was no whimsical invention. Solutions of mercuric chloride and mercuric nitrate were once used in the production of felt hats, and mercury poisoning was a serious occupational hazard for workers in that industry. The most toxic of all the mercury compounds are believed to be dimethylmercury, $(CH_3)_2Hg$, and the methylmercury ion, CH_3Hg^+. As in the case of alkylated lead compounds, the presence of the methyl groups makes these substances much more soluble in the membranes separating the bloodstream from the brain and in those membranes in the placenta through which nutrients must pass to a developing fetus.

Because of the enormously damaging effects of mercury on the biological environment, an intense effort is under way at present to trace the sources of all mercury pollution in order to eliminate them. Mercuric salts such as $HgCl_2$ have been used for a number of years as disinfectants for seeds and to control diseases of tubers and bulbs (including potatoes). Methylated mercury compounds are used as fungicides, pesticides, and mildew killers. Another organomercury compound is phenylmercury acetate, $C_6H_5HgOOCCH_3$, used in the paper industry for in-process slime control.

It was once believed that elemental mercury was sufficiently inert that it would stay at the bottom of a lake or river and be slowly converted to the rather harmless HgS. This, unfortunately, is not the case. We now know that certain bacteria and microorganisms can metabolize mercury first to Hg^{2+} ions and, eventually, to CH_3Hg^+ and $(CH_3)_2Hg$ (Figure 20.39). Fish take in bacteria, and methylmercury compounds slowly con-

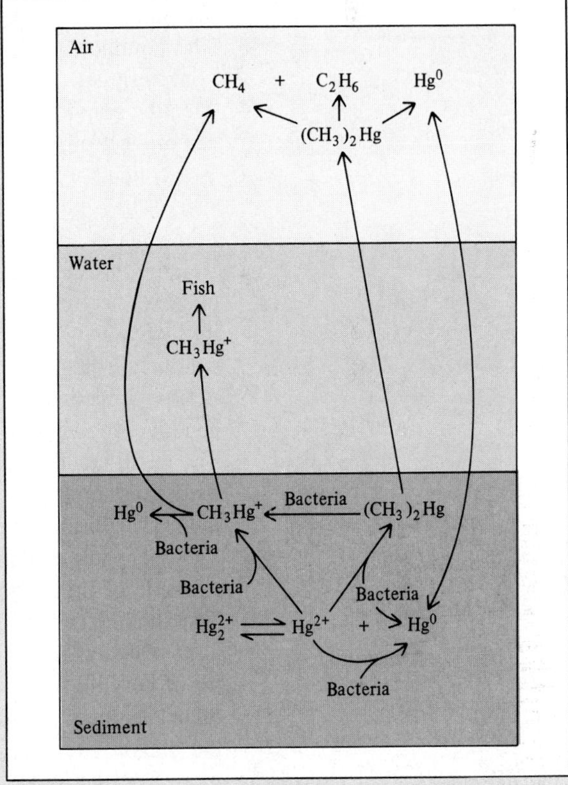

Figure 20.39 *The biological cycle of mercury. Through the action of bacteria, the relatively harmless metallic mercury can be converted into the highly toxic substances* CH_3Hg^+ *and* $(CH_3)_2Hg$.

centrate in the fatty tissues of their bodies. Small fish are eaten by large fish until, finally, at the end of this food chain, we find humans. Because of the cumulative effect, by the time we eat a walleye caught from a mercury-contaminated lake, the amount of the metal ion in its body may be 50,000 times greater than the mercury concentration in the lake water! *The buildup of any poison along a food chain* is sometimes called ***bioamplification.*** As in the case of lead, the best treatment for mercury poisoning is a dosage of a chelating agent such as the calcium salts of EDTA.

The impact of mercury pollution has been so strong that many mercury-containing compounds have been banned in industry and agriculture. Specific guidelines have been set up to determine the toxicity level. But the mercury problem will remain with us for quite some time. The discharge of mercury into the environment is by no means totally halted. It will take years to clean up lakes, rivers, and other waterways that already contain large deposits of mercury compounds.

Summary

1. Depending on their reactivities, metals exist in nature in either the free or combined state.
2. To recover a metal from its ore, the ore must first be prepared. The metal is then separated (usually by a reduction process) and purified.
3. The common procedures for purifying metals are distillation, electrolysis, and zone refining.
4. Metallic bonding can be pictured as the force between positive ions immersed in a sea of electrons. The atomic orbitals merge to form energy bands, and a substance is a conductor when electrons can be readily promoted to the conduction band, where they are free to move through the substance.
5. In insulators, the energy gap between the valence band and the conduction band is so large that electrons cannot be promoted into the conduction band. In semiconductors, electrons can cross the energy gap at higher temperatures, and conductivity increases with increasing temperatures as more electrons are able to reach the conduction band.
6. *n*-Type semiconductors contain donor impurities and extra electrons. *p*-Type semiconductors contain acceptor impurities and positive holes.
7. The alkali metals are the most reactive of all the metallic elements. They have an oxidation number of +1 in their compounds. Under special conditions, some of them also form uninegative anions.
8. The chemistry of lithium resembles that of magnesium in some ways. This is an example of the diagonal relationship.
9. The alkaline earth metals are somewhat less reactive than the alkali metals. They almost always have an oxidation number of +2 in their compounds. The chemistry of beryllium, the first member of Group 2A, differs appreciably from that of other members. It resembles aluminum in some ways. This is another example of the diagonal relationship. The properties of the alkaline earth elements become more metallic as we go down their periodic table group.
10. Aluminum does not react with water due to the formation of a protective oxide, and its oxide is amphoteric.
11. Tin and lead both have oxidation numbers of +2 and +4 in their compounds.
12. Zinc, cadmium, and mercury are less reactive and less electropositive and have a greater tendency to form covalent compounds than the alkaline earth metals.

13. Zinc and cadmium have very similar properties, although zinc is amphoteric and cadmium is not. Both of these elements always have an oxidation number of +2 in their compounds, and both form a variety of complex ions in solution.
14. Mercury forms Hg_2^{2+} and Hg^{2+} ions, and also forms a number of stable complex ions in solution. Mercury(II) compounds tend to be covalent.
15. Lead and its compounds, cadmium and its compounds, mercury vapor, soluble inorganic mercury, and organic mercury compounds are all very toxic substances.

Key Words

Acceptor impurity, p. 833
Alloy, p. 823
Amalgam, p. 823
Bioamplification, p. 856
Conductor, p. 832
Donor impurity, p. 832
Ferromagnetic, p. 823
Hygroscopic, p. 836
Insulator, p. 832
Metallurgy, p. 823
Mineral, p. 822
n-Type semiconductor, p. 833
Ore, p. 822
p-Type semiconductor, p. 833
Pyrometallurgy, p. 824
Semiconducting elements, p. 832

Exercises

Minerals

REVIEW QUESTIONS

20.1 Define mineral, ore, and metallurgy.

20.2 List three metals that are usually found in an uncombined state in nature and three metals that are always found in a combined state in nature.

20.3 What are manganese nodules?

20.4 Write chemical formulas for the following minerals: (a) calcite, (b) dolomite, (c) fluorite, (d) halite, (e) corundum, (f) magnetite, (g) beryl, (h) galena, (i) epsomite, (j) anhydrite.

20.5 Name the following minerals: (a) $MgCO_3$, (b) Na_3AlF_6, (c) Al_2O_3, (d) Ag_2S, (e) HgS, (f) ZnS, (g) $SrSO_4$, (h) $PbCO_3$, (i) MnO_2, (j) TiO_2.

Metallurgical Processes

REVIEW QUESTIONS

20.6 Describe the main steps involved in the preparation of an ore.

20.7 What does roasting mean in metallurgy? Why is it a main source of air pollution and acid rain?

20.8 What is pyrometallurgy?

20.9 Describe with examples the chemical and electrolytic reduction processes used in the production of metals.

20.10 Describe the main steps used to purify metals.

20.11 Describe the extraction of iron in a blast furnace.

20.12 Briefly discuss the steelmaking process.

PROBLEMS

20.13 In the Mond process for the purification of nickel, CO is passed over metallic nickel to give $Ni(CO)_4$:

$$Ni(s) + 4CO(g) \rightleftharpoons Ni(CO)_4(g)$$

Given that the standard free energies of formation of CO(*g*) and $Ni(CO)_4(g)$ are −137.3 kJ/mol and −587.4 kJ/mol, respectively, calculate the equilibrium constant of the reaction at 80°C. (Assume ΔG_f° to be independent of temperature.)

20.14 Copper is purified by electrolysis (see Figure 20.6). A 5.00-kg anode is used in a cell where the current is 37.8 A. How long (in hours) must the current run to dissolve this anode and electroplate it on the cathode?

20.15 Consider the electrolytic procedure for purifying copper described in Figure 20.6. Suppose that a sample of copper contains the following impurities: Fe, Ag, Zn, Au, Co, Pt, and Pb. Which of the metals will be oxidized and dissolved in solution and which will be unaffected and simply form the sludge that accumulates at the bottom of the cell?

20.16 How would you obtain zinc from sphalerite (ZnS)?

20.17 Starting with rutile (TiO_2), explain how you would obtain pure titanium metal. (*Hint:* First convert TiO_2 to $TiCl_4$. Next, reduce $TiCl_4$ with Mg. Look up physical properties of $TiCl_4$, Mg, and $MgCl_2$ in a chemistry handbook.)

20.18 A certain mine produces 2.0×10^8 kg of copper from chalcopyrite ($CuFeS_2$) each year. The ore contains only 0.80 percent Cu by mass. (a) If the density of the ore is 2.8 g/cm^3, calculate the volume (in cm^3) of ore removed each year. (b) Calculate the mass (in kg) of SO_2 produced by roasting (assume chalcopyrite to be the only source of sulfur).

20.19 Which of the following compounds would require electrolysis to yield the free metals? Ag_2S, $CaCl_2$, NaCl, Fe_2O_3, Al_2O_3, $TiCl_4$.

20.20 Although iron is only about two-thirds as abundant as aluminum in Earth's crust, mass for mass it costs only about one-quarter as much to produce. Why?

20.21 In steelmaking, nonmetallic impurities such as P, S, and Si are removed as the corresponding oxides. The inside of the furnace is usually lined with $CaCO_3$ and $MgCO_3$, which decompose at high temperatures to yield CaO and MgO. How do CaO and MgO help in the removal of the nonmetallic oxides?

Conductors, Semiconductors, and Insulators

REVIEW QUESTIONS

20.22 Define the following terms: conductor, insulator, semiconducting elements, donor impurities, acceptor impurities, *n*-type semiconductors, *p*-type semiconductors.

20.23 Briefly discuss the nature of bonding in metals, insulators, and semiconducting elements.

20.24 Describe the general characteristics of *n*-type and *p*-type semiconductors.

20.25 Which of the following elements would form *n*-type or *p*-type semiconductors with silicon? Ga, Sb, Al, As.

20.26 An early view of metallic bonding assumed that bonding in metals consisted of localized, shared electron-pair bonds between metal atoms. List the evidence that would help you to argue against this viewpoint.

Alkali Metals

PROBLEMS

20.27 How are lithium and sodium prepared commercially?

20.28 Why is potassium usually not prepared electrolytically from one of its salts?

20.29 Describe the uses of the following compounds: LiH, $LiAlH_4$, Li_2CO_3, NaCl, Na_2CO_3, NaOH, KOH, KO_2.

20.30 Under what conditions do sodium and potassium form Na^- and K^- ions?

20.31 How does lithium differ from the other alkali metals?

20.32 Complete and balance the following equations:
(a) $K(s) + H_2O(l) \longrightarrow$
(b) $Li(s) + N_2(g) \longrightarrow$
(c) $CsH(s) + H_2O(l) \longrightarrow$
(d) $Li(s) + O_2(g) \longrightarrow$
(e) $Na(s) + O_2(g) \longrightarrow$
(f) $K(s) + O_2(g) \longrightarrow$
(g) $Rb(s) + O_2(g) \longrightarrow$

20.33 Write a balanced equation for each of the following reactions: (a) potassium reacts with water; (b) an aqueous solution of NaOH reacts with CO_2; (c) solid Na_2CO_3 reacts with a HCl solution; (d) solid $NaHCO_3$ reacts with a HCl solution; (e) solid $NaHCO_3$ is heated; (f) solid Na_2CO_3 is heated.

20.34 Calculate the volume of CO_2 at 10.0°C and 746 mmHg pressure that is obtained by treating 25.0 g of Na_2CO_3 with an excess of hydrochloric acid.

Alkaline Earth Metals

PROBLEMS

20.35 How are beryllium, magnesium, calcium, strontium, and barium obtained commercially?

20.36 How does beryllium differ from the rest of the alkaline earth metals?

20.37 Write balanced equations for the reactions between metallic beryllium and (a) an acid, (b) a base.

20.38 From the thermodynamic data in Appendix 3, calculate the $\Delta H°$ values for the following decompositions:
(a) $MgCO_3(s) \longrightarrow MgO(s) + CO_2(g)$
(b) $CaCO_3(s) \longrightarrow CaO(s) + CO_2(g)$
Which of the two compounds is more easily decomposed by heat?

20.39 Starting with magnesium and concentrated nitric acid, describe how you would prepare magnesium oxide. [*Hint:* First convert Mg to $MgNO_3$. Next, MgO can be obtained by heating $Mg(NO_3)_2$.]

20.40 Describe two ways of preparing magnesium chloride.

20.41 The second ionization energy of magnesium is only about twice as great as the first, but the third ionization energy is 10 times as great. Why does it take so much more energy to remove the third electron?

20.42 Helium contains the same number of electrons in its outer shell as do the alkaline earth metals. Explain why helium is inert whereas the Group 2A metals are not.

20.43 List the sulfates of the Group 2A metals in order of increasing solubility in water. Explain the trend. (*Hint:* You need to consult a chemistry handbook.)

20.44 When exposed to air, calcium first forms calcium oxide, which is then converted to calcium hydroxide, and finally to calcium carbonate. Write a balanced equation for each step.

20.45 Write chemical formulas for (a) quicklime, (b) slaked lime, (c) lime water.

20.46 How would you prepare the following compounds? (a) BaO, (b) $Ba(OH)_2$, (c) $BaCO_3$, (d) $BaSO_4$.

20.47 Barium ions are toxic. Which of the following barium salts is more harmful if ingested? $BaCl_2$ or $BaSO_4$. Explain.

Aluminum

PROBLEMS

20.48 Describe the Hall process for preparing aluminum.

20.49 How long (in hours) will it take to deposit 664 g of Al in the Hall process with a current of 32.6 A?

20.50 Before Hall invented his electrolytic process, aluminum was produced by reduction of its chloride with an active metal. Which metals would you use for the production of aluminum in that way?

20.51 The overall reaction for the electrolytic production of aluminum in the Hall process may be represented as

$$Al_2O_3(s) + 3C(s) \longrightarrow 2Al(l) + 3CO(g)$$

At 1000°C, the standard free-energy change for this process is 594 kJ. (a) Calculate the minimum voltage required to produce 1 mole of aluminum at this temperature. (b) If the actual voltage applied is exactly three times the ideal value, calculate the energy required to produce 1.00 kg of the metal.

20.52 Aluminum does not rust as iron does. Why?

20.53 Aluminum forms the complex ions $AlCl_4^-$ and AlF_6^{3-}. Describe the shapes of these ions. $AlCl_6^{3-}$ does not form. Why? (*Hint:* Consider the relative sizes of Al^{3+}, F^-, and Cl^- ions.)

20.54 In basic solution aluminum metal is a strong reducing agent, being oxidized to AlO_2^-. Give balanced equations for the reaction of Al in basic solution with the following: (a) $NaNO_3$, to give ammonia; (b) water, to give hydrogen; (c) Na_2SnO_3, to give metallic tin.

20.55 Write a balanced equation for the thermal decomposition of aluminum nitrate to form aluminum oxide, nitrogen dioxide, and oxygen gas.

20.56 Describe some important properties of aluminum that make it one of the most versatile metals known.

20.57 Starting with aluminum, describe with balanced equations how you would prepare (a) Al_2Cl_6, (b) Al_2O_3, (c) $Al_2(SO_4)_3$, (d) $NH_4Al(SO_4)_2 \cdot 12H_2O$.

20.58 The pressure of gaseous Al_2Cl_6 increases more rapidly with temperature than that predicted by the ideal gas equation ($PV = nRT$) even though Al_2Cl_6 behaves like an ideal gas. Explain.

20.59 Explain the change in bonding when Al_2Cl_6 dissociates to form $AlCl_3$ in the gas phase.

Tin and Lead

PROBLEMS

20.60 How are tin and lead prepared commercially?

20.61 Briefly describe the change in chemical properties as we move down in Group 4A from carbon to lead.

20.62 Use the compounds of tin and lead to illustrate what is meant by the inert-pair effect.

20.63 Define the following alloys: solder, bronze, pewter.

20.64 Tin(II) chloride is a reducing agent. Write balanced ionic equations for the reaction between Sn^{2+} and (a) acidified permanganate solution, (b) acidified dichromate solution, (c) Fe^{3+} ions, (d) Hg^{2+} ions, (e) Hg_2^{2+} ions.

20.65 A 0.753-g sample of bronze dissolves in nitric acid and gives 0.171 g of SnO_2. Calculate the percent by mass of tin in the bronze.

20.66 Near room temperature tin exists in two allotropic forms called white tin and gray tin. White tin has a close-packed structure and is stable above 13°C; and gray tin, which has a diamond structure, is stable below this temperature. (a) Predict which of the two allotropic forms has a greater density. (b) Predict which of the two allotropic forms is a better conductor of electricity. (c) The standard entropies of gray tin and white tin at 13°C are 44.4 J/K · mol and 53.6 J/K · mol, respectively. Calculate the enthalpy change for the reaction Sn(gray) → Sn(white) at this temperature.

20.67 Explain why Pb(II) compounds are usually ionic and Pb(IV) compounds are covalent.

20.68 Oil paintings containing lead(II) compounds as constituents of their pigments darken over the years. Suggest a chemical reason for the color change.

20.69 Despite the known hazards of lead, lead pipes are still in use in many places. In such places it is safer to drink water that contains carbonate ions than water without carbonate ions. Why?

Zinc, Cadmium, and Mercury

PROBLEMS

20.70 How are zinc, cadmium, and mercury obtained commercially?

20.71 Although zinc is not a toxic substance, water running through galvanized pipes often picks up substantial amounts of a toxic metal ion. Suggest what this metal ion might be.

20.72 Starting with zinc, describe how you would prepare (a) ZnO, (b) $Zn(OH)_2$, (c) $ZnCO_3$, (d) ZnS, (e) $ZnCl_2$.

20.73 Describe a method, giving equations, for the removal of mercury from a solution of $Cd(NO_3)_2$ and $Hg_2(NO_3)_2$.

20.74 Look up the melting points of the following compounds in a handbook: HgF_2, $HgCl_2$, $HgBr_2$, and HgI_2. What can you deduce about the bonding in these compounds?

20.75 The ingestion of a very small quantity of mercury is not considered too harmful. Would this statement still hold if the gastric juice in your stomach were mostly nitric acid instead of hydrochloric acid?

20.76 Given that

$$2Hg^{2+}(aq) + 2e^- \longrightarrow Hg_2^{2+}(aq) \qquad \mathscr{E}° = 0.92\ V$$
$$Hg_2^{2+}(aq) + 2e^- \longrightarrow 2Hg(l) \qquad \mathscr{E}° = 0.85\ V$$

calculate $\Delta G°$ and K for the disproportionation at 25°C:

$$Hg_2^{2+}(aq) \longrightarrow Hg^{2+}(aq) + Hg(l)$$

Miscellaneous Problems

20.77 The quantity of 1.164 g of a certain metal sulfide was roasted in air. As a result, 0.972 g of the metal oxide was formed. If the oxidation number of the metal is +2, calculate the molar mass of the metal.

20.78 Referring to Figure 20.6, would you expect H_2O and H^+ to be reduced at the cathode and H_2O oxidized at the anode?

20.79 A 0.450-g sample of steel contains manganese as an impurity. The sample is dissolved in acidic solution and the manganese is oxidized to the permanganate ion MnO_4^-. The MnO_4^- ion is reduced to Mn^{2+} by reacting with 50.0 mL of 0.0800 *M* $FeSO_4$ solution. The excess Fe^{2+} ions are then oxidized to Fe^{3+} by 22.4 mL of 0.0100 *M* $K_2Cr_2O_7$. Calculate the percent by mass of manganese in the sample.

20.80 Given that $\Delta G_f^\circ(Fe_2O_3) = -741.0$ kJ/mol and that $\Delta G_f^\circ(HgS) = -48.8$ kJ/mol, calculate ΔG° for the following reactions at 25°C:
(a) $HgS(s) \longrightarrow Hg(l) + S(s)$
(b) $2Fe_2O_3(s) \longrightarrow 4Fe(s) + 3O_2(g)$
(c) $2Al_2O_3(s) \longrightarrow 4Al(s) + 3O_2(g)$

20.81 When an inert atmosphere is needed in a metallurgical process, nitrogen is frequently used. However, in the reduction of $TiCl_4$ by magnesium, helium is used. Explain why nitrogen is not suitable for this process.

20.82 Describe, with examples, the diagonal relationship between lithium and magnesium and between beryllium and aluminum.

20.83 Explain what is meant by amphoterism. Use compounds of beryllium and aluminum as examples.

20.84 Explain each of the following statements: (a) An aqueous solution of $AlCl_3$ is acidic. (b) $Al(OH)_3$ is soluble in NaOH solutions but not in NH_3 solution. (c) The melting point of Sn is lower than that of C. (d) Pb^{2+} is a weaker reducing agent than Sn^{2+}.

20.85 It has been shown that Na_2 species are formed in the vapor phase. Describe the formation of the "disodium molecule" in terms of a molecular orbital energy-level diagram. Would you expect the alkaline earth metals to exhibit a similar property?

20.86 Write balanced equations for the following reactions: (a) the heating of aluminum carbonate; (b) the hydrolysis of the $Be(H_2O)_4^{2+}$ ion (the complex ion formed is $[Be(H_2O)_3(OH)]^+$); (c) the reaction between lead(II) oxide and molecular hydrogen; (d) the reaction between $AlCl_3$ and K; (e) the reaction between Sn and hot concentrated H_2SO_4; (f) the reaction between Na_2CO_3 and $Ca(OH)_2$.

20.87 What reagents would you employ to separate the following pairs of ions in solution? (a) Na^+ and Ba^{2+}, (b) Mg^{2+} and Pb^{2+}, (c) Zn^{2+} and Hg^{2+}.

20.88 Write balanced equations for the following reactions: (a) calcium oxide and dilute HCl solution; (b) a solution containing ammonium ions with $Ba(OH)_2$; (c) a solution containing Pb^{2+} ions and H_2S gas.

20.89 Explain the following statements: (a) An aqueous solution of aluminum chloride has a pH less than 7. (b) Molten beryllium chloride is a poor conductor of electricity. (c) Mercury(II) iodide, an insoluble compound, is soluble in a solution containing iodide ions.

20.90 Compare the properties (such as bonding, structure, and acid-base character) of the following Group 4A oxides: CO_2, SiO_2, SnO_2, PbO_2.

Opposite page: Carbon fibers being prepared here are often used to reinforce fabrics.

C H A P T E R 2 1

NONMETALLIC ELEMENTS AND THEIR COMPOUNDS

In this chapter we continue our survey of the elements by concentrating on the nonmetals. Again the emphasis will be on important chemical properties and the roles of nonmetals and their compounds in industrial, chemical, and biological processes.

21.1 General Properties of Nonmetallic Elements

Properties of nonmetals are more varied than those of metals. A number of nonmetals are gases in the elemental state: hydrogen, oxygen, nitrogen, fluorine, chlorine, and the noble gases. Only one, bromine, is a liquid. All the remaining nonmetals are solids at room temperature. Unlike metals, nonmetallic elements are poor conductors of heat and electricity, and many of their compounds exhibit both positive and negative oxidation numbers.

A small group of elements have properties characteristic of both metals and nonmetals. These elements are called metalloids (see Figure 1.5). As we saw in Section 20.3, some of the metalloids, such as boron, silicon, germanium, and arsenic, are semiconducting elements.

Nonmetals are more electronegative than metals. The electronegativity of elements increases from left to right across any period and from bottom to top in any group in the periodic table (see Figure 9.8). The locations of the nonmetals in the periodic table are consistent with these trends. With the exception of hydrogen, the nonmetals are concentrated in the upper right-hand corner of the periodic table (Figure 21.1). Compounds formed between metals and nonmetals often tend to be ionic, having a metallic cation and a nonmetallic anion.

In this chapter we discuss the chemistry of a number of common and important nonmetallic elements: hydrogen, Group 3A (boron), Group 4A (carbon and silicon), Group 5A (nitrogen and phosphorus), Group 6A (oxygen and sulfur), Group 7A (fluorine, chlorine, bromine, and iodine), and Group 8A (helium, neon, argon, krypton, xenon, and radon).

21.2 Hydrogen

Hydrogen is the simplest element known—its most common atomic form contains only one proton and one electron. In the atomic state hydrogen tends to combine in an exothermic reaction that results in molecular hydrogen:

$$H(g) + H(g) \longrightarrow H_2(g) \qquad \Delta H^\circ = -436.4 \text{ kJ}$$

Molecular hydrogen is a colorless, odorless, and nonpoisonous gas that boils at −252.9°C (20.3 K).

Hydrogen is the most abundant element in the universe, accounting for about 70 percent of the universe's total mass. It is the third most abundant element in Earth's crust. Unlike Jupiter and Saturn, Earth does not have a strong enough gravitational pull to retain the lightweight H_2 molecules, so hydrogen is not found in our atmosphere.

We saw in Chapter 8 that there is no totally suitable position for hydrogen in the periodic table. The electron configuration of H is $1s^1$. It resembles the alkali metals in one respect—it can be oxidized to the H^+ ion, which exists in aqueous solutions in the hydrated form. On the other hand, hydrogen can be reduced to the hydride ion, H^-, which is isoelectronic with helium ($1s^2$). Hydrogen resembles the halogens in that it forms the uninegative ion. Hydrogen forms a large number of covalent compounds. It also takes part in hydrogen bond formation (see Section 11.2). We will discuss the properties of some of hydrogen's compounds shortly.

Hydrogen gas plays an important role in industrial processes. About 95 percent of

1A	2A	3B	4B	5B	6B	7B	8B	8B	8B	1B	2B	3A	4A	5A	6A	7A	8A
1 H																	2 He
3 Li	4 Be											5 B	6 C	7 N	8 O	9 F	10 Ne
11 Na	12 Mg											13 Al	14 Si	15 P	16 S	17 Cl	18 Ar
19 K	20 Ca	21 Sc	22 Ti	23 V	24 Cr	25 Mn	26 Fe	27 Co	28 Ni	29 Cu	30 Zn	31 Ga	32 Ge	33 As	34 Se	35 Br	36 Kr
37 Rb	38 Sr	39 Y	40 Zr	41 Nb	42 Mo	43 Tc	44 Ru	45 Rh	46 Pd	47 Ag	48 Cd	49 In	50 Sn	51 Sb	52 Te	53 I	54 Xe
55 Cs	56 Ba	57 La	72 Hf	73 Ta	74 W	75 Re	76 Os	77 Ir	78 Pt	79 Au	80 Hg	81 Tl	82 Pb	83 Bi	84 Po	85 At	86 Rn
87 Fr	88 Ra	89 Ac	104 Unq	105 Unp	106 Unh	107 Uns	108 Uno	109 Une									

Figure 21.1 *Representative nonmetallic elements according to their positions in the periodic table. The metalloids are shown in yellow.*

the hydrogen produced is used captively; that is, it is produced at or near the plant where it is used for industrial processes such as the synthesis of ammonia or hydrogenation. The most important large-scale industrial preparation is the reaction between propane (from natural gas and also as a product from oil refinery) and steam in the presence of a catalyst at 900°C:

$$C_3H_8(g) + 3H_2O(g) \longrightarrow 3CO(g) + 7H_2(g)$$

In another process, steam is passed over a bed of red-hot coke:

$$C(s) + H_2O(g) \longrightarrow CO(g) + H_2(g)$$

The mixture of carbon monoxide and hydrogen gas produced in this reaction is commonly known as *water gas*. Because both CO and H_2 burn in air, water gas was used as a fuel for many years. But since CO is poisonous, water gas has been replaced by natural gases, such as methane and propane.

Small quantities of hydrogen gas can be prepared conveniently in the laboratory by the reaction of zinc with dilute hydrochloric acid (Figure 21.2):

$$Zn(s) + 2HCl(aq) \longrightarrow ZnCl_2(aq) + H_2(g)$$

Hydrogen gas is also produced by the reaction between an alkali metal or an alkaline earth metal (Ca or Ba) and water (see Section 3.5), but the reactions are too violent to be suitable for the laboratory preparation of hydrogen gas. Very pure hydrogen gas can be obtained by the electrolysis of water (see Section 19.8).

Binary Hydrides

Binary hydrides are compounds containing hydrogen and another element, either a metal or a nonmetal. Depending on structure and properties, these hydrides can be broadly divided into three types: (1) ionic hydrides, (2) covalent hydrides, and (3) interstitial hydrides.

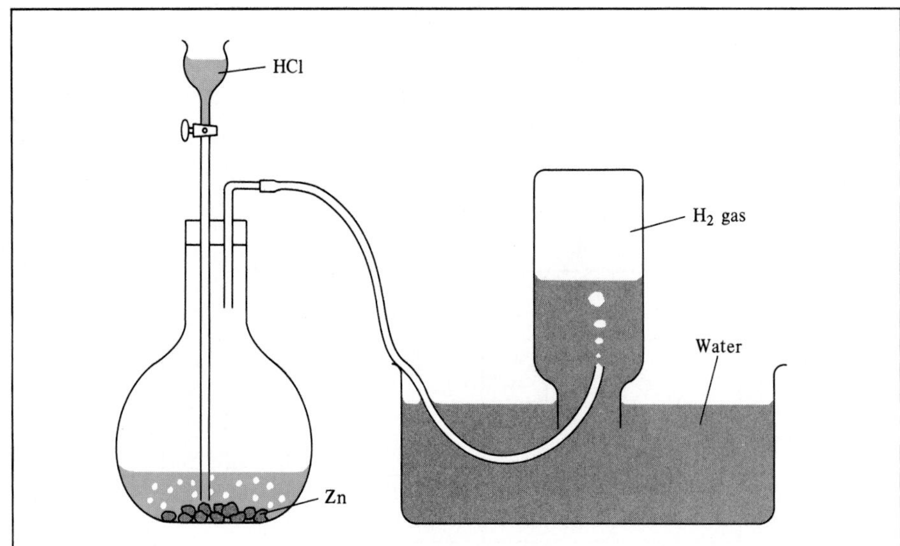

Figure 21.2 *Apparatus for the laboratory preparation of hydrogen gas. The gas is collected over water as in the case of oxygen gas (see Figure 5.14).*

Ionic Hydrides. *Ionic hydrides* are formed when molecular hydrogen combines directly with any alkali metal or with some of the alkaline earth metals (Ca, Sr, and Ba):

$$2Li(s) + H_2(g) \longrightarrow 2LiH(s)$$

$$Ca(s) + H_2(g) \longrightarrow CaH_2(s)$$

All ionic hydrides are solids that have the high melting points characteristic of ionic compounds. The anion in these compounds is the hydride ion, H^-, which is a very strong Brønsted base. It readily accepts a proton from a proton donor such as water:

$$H^-(aq) + H_2O(l) \longrightarrow OH^-(aq) + H_2(g)$$

As a result of their high reactivity with water, ionic hydrides are frequently used to remove traces of water from organic solvents.

Covalent Hydrides. In *covalent hydrides* the hydrogen atom is covalently bonded to the atom of another element. There are two types of covalent hydrides—those containing discrete molecular units, such as CH_4 and NH_3, and those having complex polymeric structure, such as $(BeH_2)_x$ and $(AlH_3)_x$, where x is a very large number.

This is an example of the diagonal relationship between Be and Al (see p. 331).

Figure 21.3 shows the binary ionic and covalent hydrides of the representative elements. The physical and chemical properties of these compounds change from ionic to covalent across a given period. Consider, for example, the hydrides of the second-period elements:

$$LiH \quad BeH_2 \quad B_2H_6 \quad CH_4 \quad NH_3 \quad H_2O \quad HF$$

LiH is an ionic compound with a high melting point (680°C). The structure of BeH_2 (in the solid state) is that of a polymer; it is a covalent compound. The molecules B_2H_6 and CH_4 are nonpolar. In contrast, NH_3, H_2O, and HF are all polar molecules in which the hydrogen atom is the *positive* end of the polar bond. Of this group of hydrides (NH_3, H_2O, and HF) only HF is acidic in water.

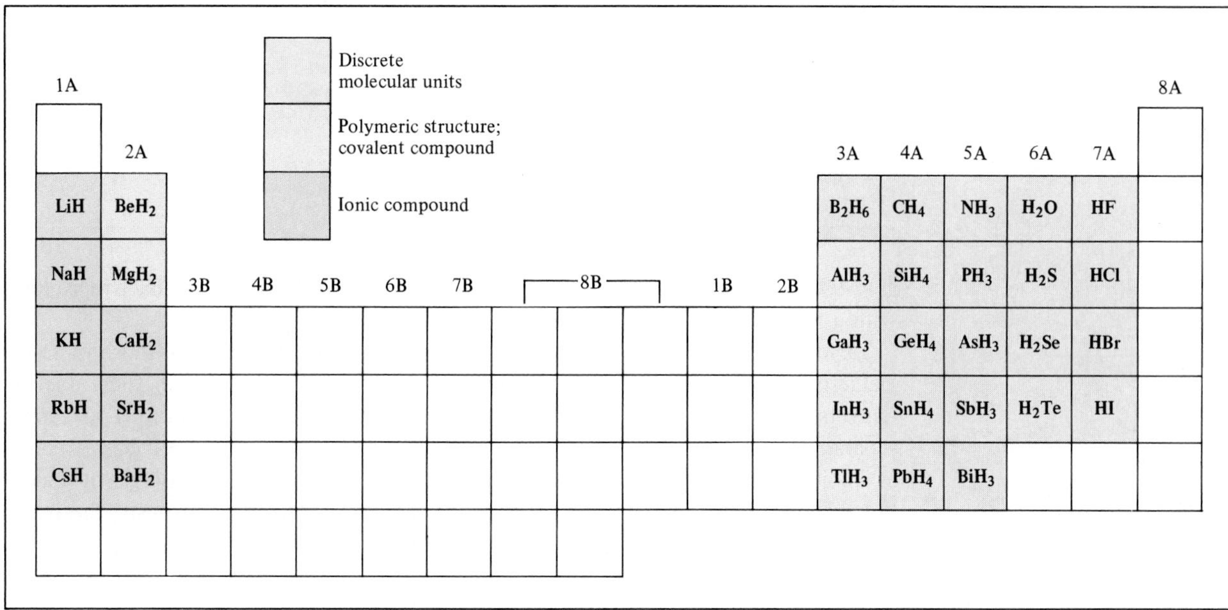

Figure 21.3 *Binary hydrides of the representative elements. In cases in which hydrogen forms more than one compound with the same element, only the formula of the simplest hydride is shown. The properties of many of the transition metal hydrides are not well characterized.*

As we move down any group in Figure 21.3, for example, Group 2A, the compounds change from covalent (BeH_2 and MgH_2) to ionic (CaH_2, SrH_2, and BaH_2), as might be expected from the increase in metallic character as we go down any group in the periodic table.

These compounds are sometimes called *nonstoichiometric* compounds. Note that they do not obey the law of definite proportion (see Section 2.7).

Interstitial Hydrides. Molecular hydrogen forms a number of hydrides with transition metals. In some of these compounds, the ratio of hydrogen atoms to metal atoms is *not* a constant. Such compounds are called *interstitial hydrides*. For example, depending on conditions, the formula for titanium hydride can vary between $TiH_{1.8}$ and TiH_2.

Many of the interstitial hydrides retain metallic properties such as electrical conductivity. Yet it is known that hydrogen is definitely bonded to the metal in these compounds, although the exact nature of bonding is often not clear.

A particularly interesting reaction is that between H_2 and palladium (Pd). Hydrogen gas is readily adsorbed onto the surface of the palladium metal, which it dissociates into atomic hydrogen. The H atoms then ''dissolve'' into the metal. On heating and under the pressure of H_2 gas on one side of the metal, these atoms diffuse through the metal and recombine to form molecular hydrogen, which emerges as the gas from the other side. Because no other gas behaves in this way with palladium, this process has been used to separate hydrogen gas from other gases.

The ^1_1H isotope is also called *protium*. Hydrogen is the only element whose isotopes are given different names.

Isotopes of Hydrogen

Hydrogen has three isotopes: ^1_1H (hydrogen), ^2_1H (deuterium, symbol D), and ^3_1H (tritium, symbol T). The natural abundances of the stable hydrogen isotopes are: hydro-

Table 21.1 Properties of H_2O and D_2O

Property	H_2O	D_2O
Molar mass (g/mol)	18.0	20.0
Melting point (°C)	0	3.8
Boiling point (°C)	100	101.4
Density at 4°C (g/cm^3)	1.000	1.108

gen, 99.985 percent; and deuterium, 0.015 percent. Tritium is a radioactive isotope with a half-life of about 12.5 years.

Table 21.1 compares some of the common properties of H_2O with those of D_2O. Deuterium oxide, or heavy water as it is commonly called, is used in some nuclear reactors as a coolant and a moderator of nuclear reactions (see Chapter 23). D_2O can be separated from H_2O by fractional distillation because H_2O boils at a lower temperature, as Table 21.1 shows. Another technique for its separation is electrolysis of water. Since H_2 gas is formed about eight times as fast as D_2 during electrolysis, the water remaining in the electrolytic cell becomes progressively enriched with D_2O. Interestingly, the Dead Sea, which for thousands of years has entrapped water that has no outlets other than through evaporation, has a higher $[D_2O]/[H_2O]$ ratio than water found elsewhere.

Although D_2O chemically resembles H_2O in most respects, it is a toxic substance. This is so because deuterium is heavier than hydrogen; thus, its compounds often react more slowly than those of the lighter isotope. Regular drinking of D_2O instead of H_2O could prove fatal because of the slower rate of transfer of D^+ compared with that of H^+ in the acid-base reactions involved in enzyme catalysis. This *kinetic isotope effect* is also manifest in acid ionization constants. For example, the ionization constant of acetic acid

$$CH_3COOH(aq) \rightleftharpoons CH_3COO^-(aq) + H^+(aq) \qquad K_a = 1.8 \times 10^{-5}$$

is about three times as large as that of deuterated acetic acid:

$$CH_3COOD(aq) \rightleftharpoons CH_3COO^-(aq) + D^+(aq) \qquad K_a = 6 \times 10^{-6}$$

Hydrogenation

Hydrogenation is the addition of hydrogen to compounds containing multiple bonds, especially C=C and C≡C bonds. In Section 13.6 we saw that hydrogenation is an important process in food industry.

The Hydrogen Economy

The world's fossil fuel reserve is being depleted at an alarmingly fast rate. Faced with this dilemma, scientists have made intensive efforts in recent years to develop a method of obtaining hydrogen gas as an alternate energy source. Hydrogen gas could replace gasoline to power automobiles (after considerable modification of the engine, of course) or be used with oxygen gas in fuel cells to generate electricity (see p. 798). One major advantage of using hydrogen gas in these ways is that the reactions are essentially free of pollutants; the end product formed in a hydrogen-powered engine or in a fuel cell would be water, just as in the burning of hydrogen gas in air:

$$2H_2(g) + O_2(g) \longrightarrow 2H_2O(l)$$

The total volume of ocean water is about 1×10^{21} L. Thus, the ocean contains an almost inexhaustible supply of hydrogen.

Of course, the potential success of a so-called hydrogen economy would depend on how cheaply we could produce hydrogen gas and how easily we could store it.

As you know, hydrogen gas can be obtained from water by electrolysis, but this method consumes too much energy. A more attractive approach, which is currently in the early stages of development, is to use solar energy to "split" water molecules. In this scheme a *catalyst* (a complex molecule containing one or more transition metal atoms, such as ruthenium) absorbs a photon from solar radiation and becomes energetically excited. In its excited state the catalyst is capable of reducing water to molecular hydrogen.

Some of the interstitial hydrides we have discussed are suitable candidates for storing hydrogen gas. The reactions that form these hydrides are usually reversible, so hydrogen gas can be obtained simply by reducing the pressure of the hydrogen gas above the metal. The advantages of using interstitial hydrides are as follows: (1) many metals have a high capacity to take up hydrogen gas—sometimes up to three times as many hydrogen atoms as there are metal atoms; and (2) because these hydrides are solids, they can be stored and transported more easily than gases or liquids.

21.3 Boron

Since boron has semimetallic properties, it is classified as a metalloid. It is a rare element, constituting less than 0.0003 percent of Earth's crust by mass. Pure boron does not occur in nature; it exists in the form of oxygen compounds. Some of these compounds are borax, $Na_2B_4O_7 \cdot 10H_2O$ (Figure 21.4); orthoboric acid, H_3BO_3; and kernite, $Na_2B_4O_7 \cdot 4H_2O$. Elemental boron, a transparent, crystalline substance that is almost as hard as diamond (see Figure 8.19), can be prepared by reducing boron oxide with magnesium powder:

Figure 21.4 *Borax ($Na_2B_4O_7 \cdot 10H_2O$).*

$$B_2O_3(s) + 3Mg(s) \longrightarrow 2B(s) + 3MgO(s)$$

A binary compound formed by boron and hydrogen is called a ***borane.*** There are a number of boranes, the simplest and most important of which is diborane, B_2H_6. There has been a great deal of interest in the bonding scheme of this molecule. Because the electron configuration of boron is $1s^22s^22p^1$, there is a total of twelve valence electrons in diborane (three for each boron atom and six for the hydrogen atoms). This means that it is impossible to represent diborane as

```
    H   H
    |   |
H — B — B — H
    |   |
    H   H
```

because this structure would require fourteen valence electrons (each of the seven covalent bonds would be made up of two electrons). Measurements have shown that there are actually two types of H atoms in diborane, in the ratio of 4:2. In the simple molecular orbital description, each B atom is assumed to be sp^3-hybridized. Four of the six H atoms are bonded to borons by sigma bonds. The other two are held together by *three-center molecular orbitals* (Figure 21.5). Each of the three-center bonds extends over three atoms: the two boron atoms and a bridging hydrogen atom.

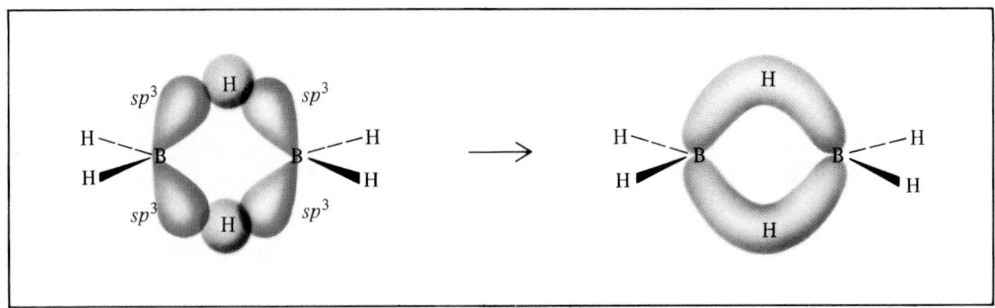

Figure 21.5 *Three-center molecular orbitals in diborane. Each molecular orbital is formed by the overlap of the sp^3 hybrid orbitals of boron and the 1s orbital of hydrogen. No more than two electrons can be placed in each molecular orbital.*

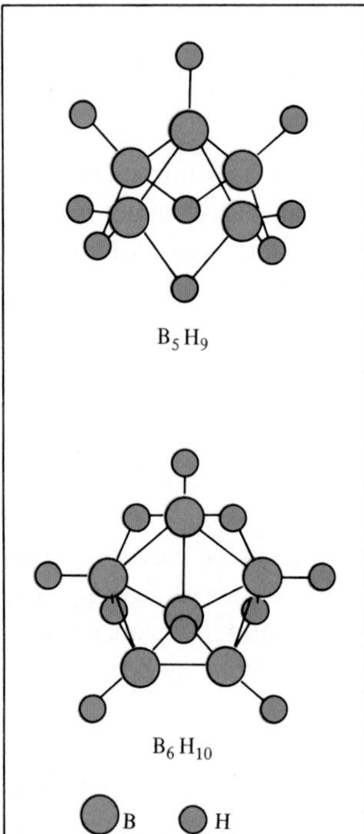

Figure 21.6 *Structures of B_5H_9 and B_6H_{10}.*

The three-center bond is not restricted to diborane. Bridging hydrogen atoms are present also in higher boranes, such as B_5H_9 and B_6H_{10} (Figure 21.6). Because of their high heats of combustion, the boranes were once considered for use as high-energy fuels for jet engines and rockets. However, these compounds are expensive to prepare and difficult to store, so the idea was abandoned.

The boron halides (BF_3, BCl_3, BBr_3, and BI_3) are another series of interesting compounds. Since they have planar structures (see Section 10.4), the boron atom in each is assumed to be sp^2-hybridized (see Figure 10.13). The unhybridized $2p_z$ orbital is vacant and can accept a pair of electrons from another molecule to form an addition compound:

$$BF_3 + NH_3 \longrightarrow F_3B{-}NH_3$$

As a result, the boron atom is sp^3-hybridized and the octet rule is satisfied for boron. This is a Lewis-type acid-base reaction as discussed in Section 15.6.

The only important oxide of boron is boron oxide, B_2O_3. Prepared by heating boric acid, boron oxide is used in the production of boron, in heat-resistant glassware, in electronic components, and in herbicides. Like most other nonmetallic oxides, it is acidic (see Figure 15.5). When dissolved in water, it yields boric acid:

$$B_2O_3(s) + 3H_2O(l) \longrightarrow 2B(OH)_3(aq)$$

which is moderately soluble in water. Boric acid, which is often written as H_3BO_3, is a very weak, monoprotic acid (Figure 21.7). It behaves as a Lewis acid by accepting a pair of electrons from the hydroxide ion, which originates from the H_2O molecule:

$$B(OH)_3(aq) + H_2O(l) \rightleftharpoons B(OH)_4^-(aq) + H^+(aq) \qquad K_a = 7.3 \times 10^{-10}$$

Its weak acid strength and antiseptic properties make boric acid a suitable eyewash. Industrially, boric acid is used to make heat-resistant glass, called *borosilicate glass*.

Pyrex glass is made of borosilicate glass.

Finally, we note that boron exemplifies the diagonal relationship between elements in the periodic table. The chemistry of boron resembles that of silicon in several ways:

- The oxides B_2O_3 and SiO_2 are both acidic; they form the oxoacids H_3BO_3 and H_2SiO_3.
- Both boron and silicon form a series of complex hydrides.
- They form metallic borides and silicides, which react with hydrochloric acid to give boron hydrides and silicon hydrides, respectively.
- Boron and silicon form similar glassy materials.

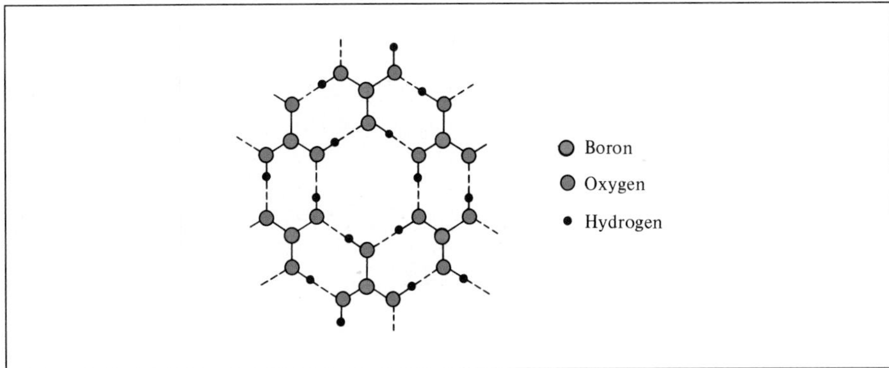

Figure 21.7 *Layer structure of boric acid. Dotted lines denote the hydrogen bonds.*

21.4 Carbon and Silicon

Carbon

Although it constitutes only about 0.09 percent by mass of Earth's crust, carbon is an essential element of living matter. It is found free in the form of diamond and graphite (see Figure 8.20) and also occurs in natural gas, petroleum, and coal. (*Coal* is a natural dark brown to black solid used as a fuel; it is formed from fossilized plants and consists of amorphous carbon with various organic and some inorganic compounds.) Carbon combines with oxygen to form carbon dioxide in the atmosphere and occurs as carbonate in limestone and chalk.

Diamond and graphite are allotropes of carbon (see Chapter 2). Figure 21.8 shows the phase diagram of carbon. Although graphite is the stable form of carbon at 1 atm and 25°C, owners of diamond jewelry need not be alarmed, for the rate of the spontaneous process

$$\text{C(diamond)} \longrightarrow \text{C(graphite)} \qquad \Delta G^\circ = -2.87 \text{ kJ}$$

is extremely slow. Millions of years may pass before a diamond turns to graphite.

Synthetic diamond can be prepared from graphite by applying very high pressures and temperatures. Figure 21.9 shows a synthetic diamond and its starting material,

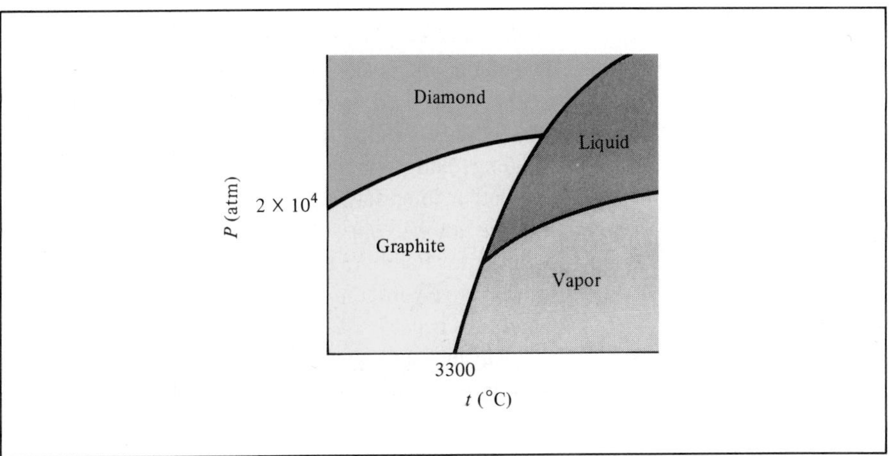

Figure 21.8 *Phase diagram of carbon. Note that under atmospheric conditions, graphite is the stable form of carbon.*

Figure 21.9 *A synthetic diamond and the starting material—graphite.*

graphite. Synthetic diamonds generally lack the optical properties of natural diamonds. They are useful, however, as abrasives and in cutting concrete and many other hard substances, including metals and alloys. The uses of graphite are described on p. 469.

Carbon combines with hydrogen to form a very large number of compounds called *hydrocarbons*. The chemistry of hydrocarbons is discussed in Chapter 24.

Carbides and Cyanides. *Carbon combines with metals to form a number of ionic compounds* called ***carbides,*** in which carbon is in the form of C_2^{2-} or C^{4-} ions. Two examples are CaC_2 and Be_2C. These ions are strong Brønsted bases and react with water as follows:

$$C_2^{2-}(aq) + 2H_2O(l) \longrightarrow 2OH^-(aq) + C_2H_2(g)$$

$$C^{4-}(aq) + 4H_2O(l) \longrightarrow 4OH^-(aq) + CH_4(g)$$

Carbon also forms a covalent compound with silicon. Silicon carbide, SiC, is called *carborundum* and is prepared as follows:

$$SiO_2(s) + 3C(s) \longrightarrow SiC(s) + 2CO(g)$$

The compound is also formed by heating silicon with carbon at 1500°C. Carborundum is almost as hard as diamond and has the diamond structure; each carbon atom is tetrahedrally bonded to four Si atoms, and vice versa. It is used mainly for cutting, grinding, and polishing metals and glasses.

Another important class of carbon compounds consists of the ***cyanides,*** which *contain the anion group* $:C{\equiv}N:^-$. Cyanide ions are extremely toxic because they bind almost irreversibly to the Fe(III) ion in cytochrome oxidase, a key enzyme in metabolic processes. Hydrogen cyanide, which has the aroma of bitter almonds, is even more dangerous because of its volatility (b.p. 26°C). A few tenths of 1 percent by volume of HCN in air can cause death within minutes. Hydrogen cyanide can be prepared by treating sodium cyanide or potassium cyanide with acid:

HCN is the gas used in gas execution chambers.

$$NaCN(s) + HCl(aq) \longrightarrow NaCl(aq) + HCN(aq)$$

Because HCN (in solution, called hydrocyanic acid) is a very weak acid ($K_a = 4.9 \times$

10^{-10}), most of the HCN produced in this reaction is in the nonionized form and leaves the solution as hydrogen cyanide gas. For this reason acids should never be mixed with metal cyanides in the laboratory without proper ventilation.

Cyanide ions are used to extract gold and silver from their ores. These metals are usually found in the uncombined state in nature. In other metal ores silver and gold may be present in relatively small concentrations and are more difficult to extract. In a typical process, the crushed ore is treated with an aqueous cyanide solution in the presence of air to dissolve the gold by forming the soluble complex ion $[Au(CN)_2]^-$:

$$4Au(s) + 8CN^-(aq) + O_2(g) + 2H_2O(l) \longrightarrow 4[Au(CN)_2]^-(aq) + 4OH^-(aq)$$

The complex ion $[Au(CN)_2]^-$ is separated from other insoluble materials by filtration (accompanied by a suitable cation such as Na^+) and treated with an electropositive metal such as zinc to recover the gold:

$$Zn(s) + 2[Au(CN)_2]^-(aq) \longrightarrow [Zn(CN)_4]^{2-}(aq) + 2Au(s)$$

Figure 21.10 shows an aerial view of a "cyanide pond" used for the extraction of gold.

Oxides of Carbon. Of the several oxides of carbon, the most important are carbon monoxide, CO, and carbon dioxide, CO_2. Carbon monoxide is a colorless, odorless gas formed by the incomplete combustion of carbon or carbon-containing compounds:

$$2C(s) + O_2(g) \longrightarrow 2CO(g)$$

Industrially, CO is prepared by passing steam over heated coke. Carbon monoxide burns readily in oxygen to form carbon dioxide:

$$2CO(g) + O_2(g) \longrightarrow 2CO_2(g) \qquad \Delta H^\circ = -566 \text{ kJ}$$

Carbon monoxide is not an acidic oxide (it differs from carbon dioxide in that regard) and is only slightly soluble in water.

Although CO is relatively unreactive, it is a very poisonous gas because it has the

Figure 21.10 *A cyanide pond for extracting gold.*

unusual ability to bind very strongly to hemoglobin, the oxygen carrier in blood. Both molecular oxygen and carbon monoxide bind to the Fe(II) ion in hemoglobin. Unfortunately, the affinity of hemoglobin for CO is about 200 times greater than that for O_2. Hemoglobin molecules with tightly bound CO (called carboxyhemoglobin) cannot carry the oxygen needed for metabolic processes. A small amount of carbon monoxide intake can cause drowsiness and headache; death may result when about half the hemoglobin molecules are complexed with CO. The best first-aid response to carbon monoxide poisoning is to remove the victim immediately to an atmosphere with a plentiful oxygen supply or to give mouth-to-mouth resuscitation.

The abbreviation *ppm* means "parts per million."

Carbon monoxide is a major air pollutant. In cities with heavy automobile traffic, the concentration of CO in air is about 40 ppm by volume, approximately 95 percent of which is due to automobile exhaust. Note also that the concentration of carboxyhemoglobin in the blood of people who smoke is two to five times higher than that in nonsmokers.

Carbon dioxide is produced when any form of carbon and carbon-containing compounds are burned in an excess of oxygen. Many carbonates give off CO_2 when heated, and all give off CO_2 when treated with acid:

$$CaCO_3(s) \longrightarrow CaO(s) + CO_2(g)$$

$$CaCO_3(s) + 2HCl(aq) \longrightarrow CaCl_2(aq) + H_2O(l) + CO_2(g)$$

Carbon dioxide is also a by-product of the fermentation of sugar:

$$\underset{\text{glucose}}{C_6H_{12}O_6(aq)} \xrightarrow{\text{yeast}} \underset{\text{ethanol}}{2C_2H_5OH(aq)} + 2CO_2(g)$$

and an end product of metabolism in animals:

$$C_6H_{12}O_6(aq) + 6O_2(g) \longrightarrow 6CO_2(g) + 6H_2O(l)$$

Carbon dioxide is a colorless and odorless gas. Unlike carbon monoxide, CO_2 is nontoxic. It is an acidic oxide (see p. 654).

Carbon dioxide is used in beverages, in fire extinguishers, and in the manufacture of baking soda, $NaHCO_3$, and soda ash, Na_2CO_3. Solid carbon dioxide, called *dry ice,* is used as a refrigerant (see Figure 11.52). Dry ice is also used in cloud seeding to induce rain (see p. 486).

The Carbon Cycle. Figure 21.11 shows the carbon cycle in our global ecosystem. The transfer of carbon dioxide to and from the atmosphere is an essential part of the carbon cycle, which begins with the process of photosynthesis conducted by plants and certain microorganisms:

This reaction is not spontaneous and requires radiant energy (visible light).

$$6CO_2 + 6H_2O \longrightarrow C_6H_{12}O_6 + 6O_2 \qquad \Delta G° = 2862 \text{ kJ}$$

Carbohydrates and other complex carbon-containing molecules are consumed by animals, which respire and release CO_2. After plants and animals die, they are decomposed by microorganisms in the soil; the carbon in their tissues is oxidized to CO_2 and returns to the atmosphere. In addition, there is a dynamic equilibrium between atmospheric CO_2 and carbonates in the oceans and lakes. Two other major sources of CO_2 production are volcanoes and combustion in homes and factories.

The following Chemistry in Action sections deal respectively with the use of coal as a fuel source and the role of carbon dioxide in controlling our climate.

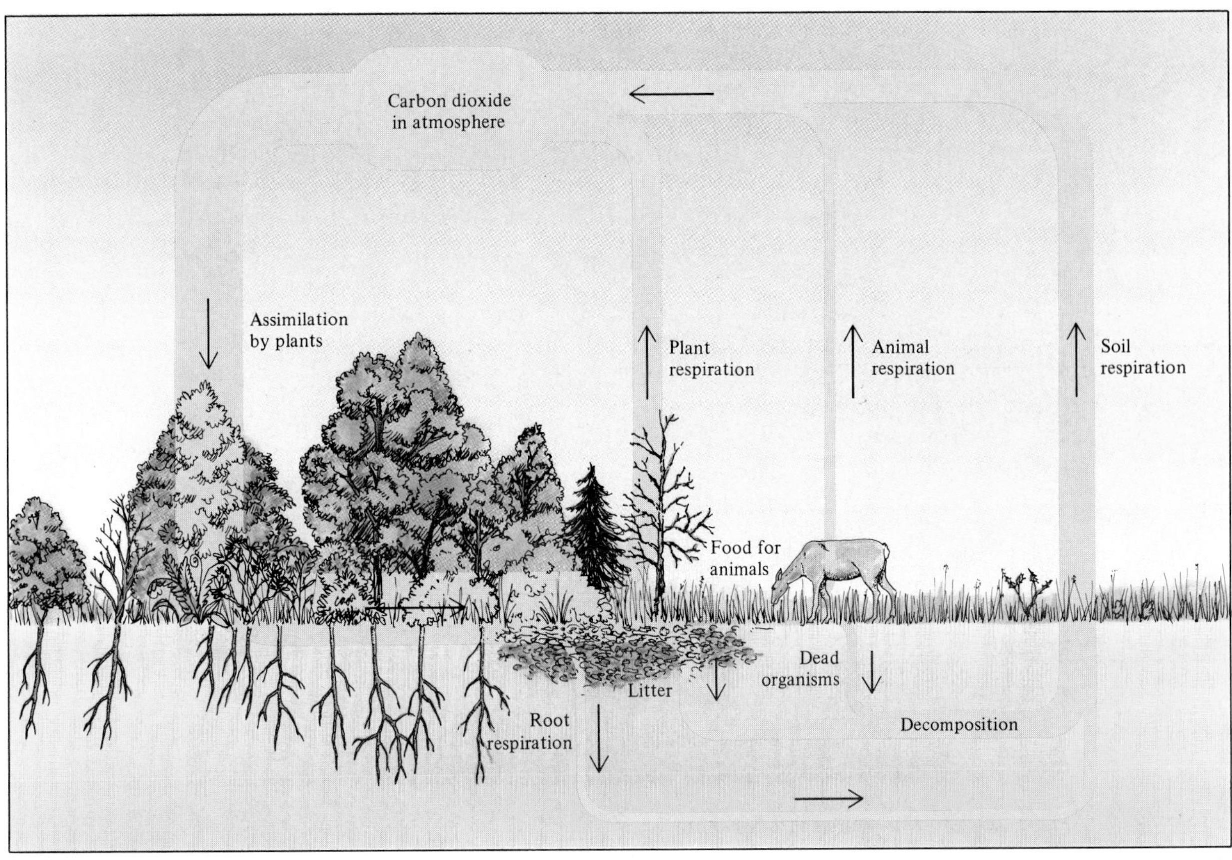

Figure 21.11 *The carbon cycle begins with the fixation of atmospheric carbon dioxide by the process of photosynthesis, conducted by plants and microorganisms. In this process carbon dioxide and water react to form carbohydrates; the reaction simultaneously releases molecular oxygen. Some of the carbohydrates are consumed directly by the plant for energy; the carbon dioxide so generated is released either through the plant's leaves or through its roots. Part of the carbon fixed by plants is consumed by animals, which respire carbon dioxide. Plants and animals die and ultimately are decomposed by microorganisms in the soil; the carbon in their tissues is oxidized to carbon dioxide and returns to the atmosphere. The widths of the pathways shown are roughly proportional to the quantities involved. A similar carbon cycle takes place within the sea.*

CHEMISTRY IN ACTION

Synthetic Gas from Coal

The very existence of our technological society depends on an abundant supply of energy. Although the United States has only 5 percent of the world's population, we consume about 20 percent of the world's energy! At present, the two major sources of energy are fossil fuels and nuclear fission (discussed in Chapters 23 and 24, respectively). Coal, oil (which is also known as petroleum), and natural gas (mostly methane) are collectively called fossil fuels because they are the end result of the decomposition of plants and animals

Figure 21.12 *Underground coal mining.*

over tens or hundreds of millions of years. Oil and natural gas are cleaner-burning and more efficient fuels than coal, so they are preferred for most purposes. However, supplies of oil and natural gas are being depleted at an alarming rate, and research is under way to devise ways of making coal a more versatile source of energy.

Coal consists of many high molar mass carbon compounds that also contain oxygen, hydrogen, and, to a lesser extent, nitrogen and sulfur. Coal constitutes about 90 percent of the world's fossil fuel reserves. For centuries coal has been used as a fuel both in homes and in industry. However, underground coal mining (Figure 21.12) is expensive and dangerous, and strip mining (that is, mining in an open pit after removal of the earth and rock covering coal) is tremendously harmful to the environment. Another problem, this one associated with the burning of coal, is the formation of sulfur dioxide (SO_2) from the sulfur-containing compounds. This process leads to the formation of "acid rain," discussed on p. 659.

One of the most promising methods for making coal a more efficient and cleaner fuel involves the conver-

Figure 21.13 *A coal gasification plant for power generation.*

sion of coal to a gaseous form, called *syngas* for "synthetic gas." This process is called *coal gasification*. In the presence of very hot steam and air, coal decomposes and reacts according to the following simplified scheme:

$$C(s) + H_2O(g) \longrightarrow CO(g) + H_2(g)$$

$$C(s) + 2H_2(g) \longrightarrow CH_4(g)$$

The main component of syngas is methane. In addition, the reactions yield hydrogen and carbon monoxide gases and other useful by-products. Under suitable conditions, CO and H_2 combine to form methanol:

$$CO(g) + 2H_2(g) \longrightarrow CH_3OH(l)$$

Methanol has many uses—as a solvent, a starting material for plastics, and so on. Syngas is easier than coal to store and transport in pipelines. What's more, it is not a major source of air pollution because sulfur is removed in the process. Figure 21.13 shows a coal gasification plant.

CHEMISTRY IN ACTION

Carbon Dioxide and Climate

Although carbon dioxide is only a trace gas in Earth's atmosphere, with a concentration of about 0.03 percent by volume, it plays a critical role in controlling our climate. The solar radiant energy received by Earth is distributed over a band of wavelengths. Energy from the sun is emitted primarily at wavelengths shorter than 4000 nm, with much of the energy concentrated in the visible region of the spectrum. By contrast, the thermal radiation emitted by Earth's surface is confined to wavelengths longer than 4000 nm because of the much lower average surface temperature of 280 K (Figure 21.14).

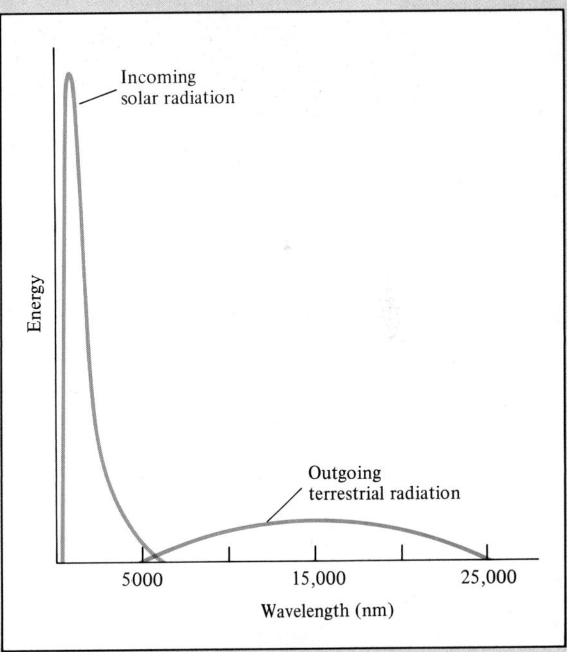

Figure 21.14 *The incoming radiation from the sun and the outgoing radiation from Earth's surface.*

Both water vapor and carbon dioxide in the atmosphere can absorb much of the outgoing radiation and, therefore, affect the overall thermal balance of Earth with its surroundings. Upon receiving a photon of the appropriate wavelength in the infrared region, a H_2O or a CO_2 molecule is promoted to a higher energy level:

$$H_2O + h\nu \longrightarrow H_2O^*$$

$$CO_2 + h\nu \longrightarrow CO_2^*$$

where $h\nu$ represents the appropriate radiation energy and the asterisk denotes a molecule in the excited state. Energetically excited molecules are unstable, and they quickly lose their excess energy either by collision with other molecules or by spontaneous emission of radiation. Part of this radiation is emitted to outer space and part of it returns to Earth's surface.

The concentration of water vapor is highest near Earth's surface (about 30,000 ppm by volume), but it decreases rapidly with altitude because of the decreasing temperature. Because water molecules absorb infrared radiation very strongly, the presence of water vapor affects the atmospheric temperature at night, when Earth is emitting radiation into space and is not receiving energy from the sun. The term "radiational cooling" is often used to account for the cold morning that follows a cloudless night.

Whereas the total amount of water vapor in our atmosphere has not altered noticeably over the years, the concentration of carbon dioxide has been steadily rising since the turn of the century as a result of the burning of fossil fuels. Figure 21.15 shows the variation of carbon dioxide concentration over a period of years, as measured in Hawaii. The seasonal oscillations are caused by removal of carbon dioxide by photosynthesis during the growing season in the Northern Hemisphere and its buildup during the fall and winter months. Clearly, the trend points to an increase of CO_2. The current rate is about 1 ppm per year, equivalent to 9×10^9 tons of CO_2! Scientists have estimated that by the year 2000 the CO_2 concentration will exceed preindustrial levels by about 25 percent. It is predicted by some meteorologists that this increase will raise Earth's average temperature by about 3–5°C.

Carbon dioxide's influence on Earth's temperature is often called the *greenhouse effect*. The glass roof of a greenhouse transmits visible sunlight and absorbs some of the outgoing infrared radiation, thereby trapping the heat. Carbon dioxide acts somewhat like a glass roof, except that the temperature rise in the greenhouse is due mainly to the restricted air circulation inside. While carbon dioxide is the main culprit in warming Earth's atmosphere, other "greenhouse" gases such as methane (from natural gas, sewage treatment, and cattle as they digest), chlorofluorocarbons (see p. 583), and nitrogen oxides (from car exhausts) also contribute to the heating of Earth.

Although a temperature increase of 3–5°C may seem insignificant, it is actually large enough to affect the delicate thermal balance on Earth and could cause glaciers and icecaps to melt. This, in turn, would raise the sea level, resulting in flooding of coastal areas. Current measurements show that Earth's temperature is indeed rising, and much more work is needed to understand how the greenhouse effect will affect Earth's climate. It is clear that the greenhouse effect, like acid rain and the depletion of ozone in the stratosphere, is one of the most pressing environmental issues facing the world today.

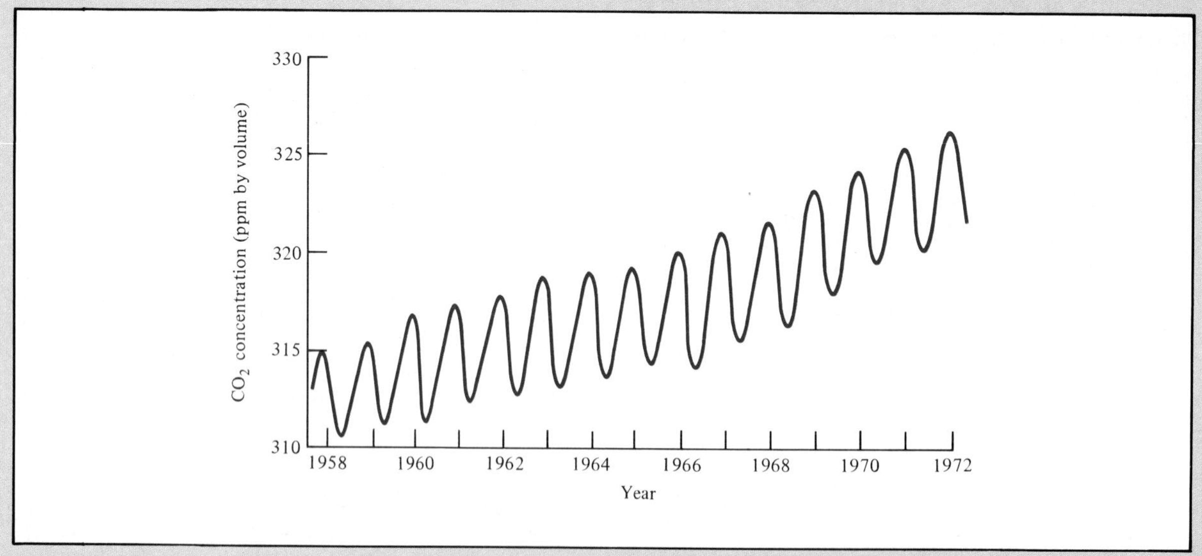

Figure 21.15 *Yearly variation of carbon dioxide concentration at Mauna Loa, Hawaii. The general trend clearly points to an increase of carbon dioxide in the atmosphere. The concentration of CO_2 had reached 340 ppm by 1988.*

Silicon

Although both carbon and silicon (a metalloid) belong to Group 4A of the periodic table and therefore have similar outer electron configurations—ns^2np^2—they have quite different chemical properties. The differences between the chemistry of carbon and that of silicon not only are interesting and useful, but also provide us with some insight into the evolution of life.

Silicon is the second most abundant element, constituting about 26 percent by mass of Earth's crust. It does not occur in the free form and is most commonly combined with oxygen in silicates. Some silicate minerals are zircon, $ZrSiO_4$, and beryl, $Be_3Al_2Si_6O_{18}$. Silica, SiO_2, constitutes most beach sands (Figure 21.16).

Silicon can be prepared by heating a mixture of silica sand and coke to about 3000°C:

$$SiO_2(s) + 2C(s) \longrightarrow Si(l) + 2CO(g)$$

Pure silicon has a diamond structure (see Figure 11.29); each Si atom is tetrahedrally bonded to four other Si atoms. No silicon analog of graphite is known. Silicon combines with carbon to form silicon carbide, or carborundum, which was discussed earlier. Ultrapure silicon, prepared by the zone-refining technique (Figure 21.17), is used in solid-state electronics. For example, the so-called one-chip computer used in pocket calculators is made of silicon and other components (Figure 21.18).

Figure 21.16 *Sand is mostly silica (SiO_2).*

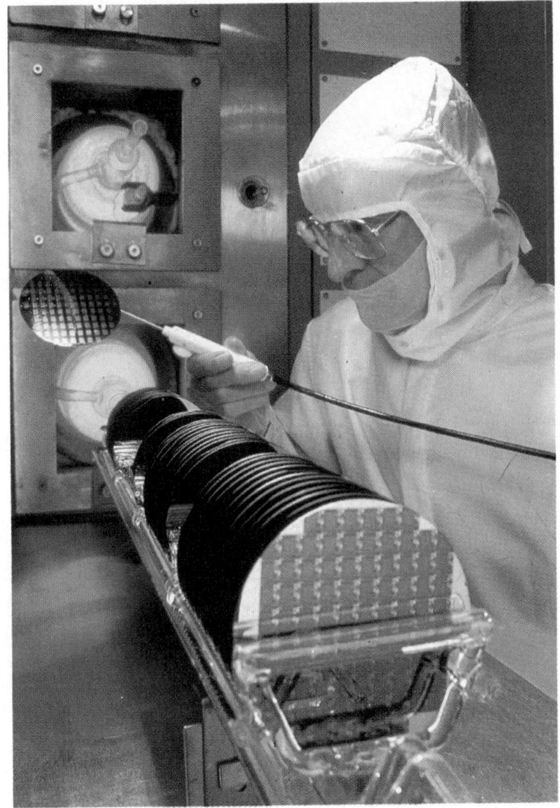

Figure 21.17 *Production of ultrapure silicon wafers by the zone-refining process.*

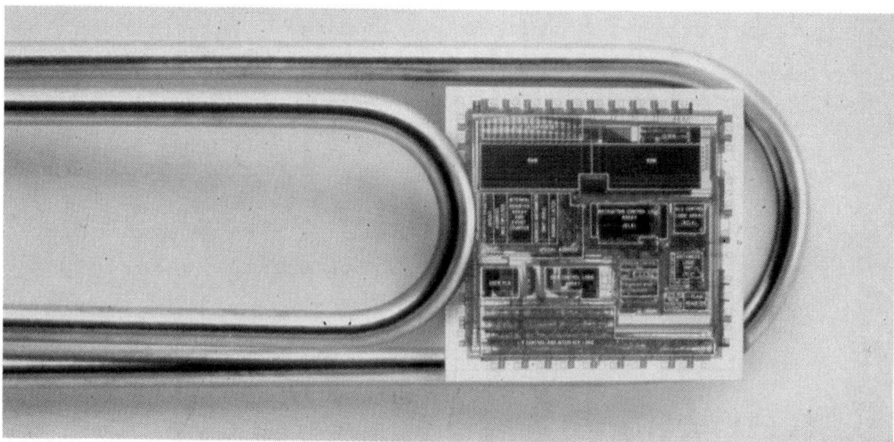

Figure 21.18 *Silicon in solid-state electronics—the one-chip computer.*

Silanes. *Silicon forms a series of covalent hydrides* called ***silanes,*** which have the general formula Si_nH_{2n+2} (SiH_4, Si_2H_6, Si_3H_8, Si_4H_{10}, and so on). Like the hydrocarbons, the lower silanes are gases at room temperature. Chemically, they are much more reactive than their carbon counterparts. For example, when exposed to air, monosilane, SiH_4, ignites spontaneously:

$$SiH_4(g) + 2O_2(g) \longrightarrow SiO_2(s) + 2H_2O(l)$$

In many respects, the silanes behave more like metal hydrides than like hydrocarbons. In the presence of an alkaline catalyst, they will reduce water to hydrogen:

$$SiH_4(aq) + 2H_2O(l) \longrightarrow SiO_2(s) + 4H_2(g)$$

Methane, CH_4, the carbon counterpart of SiH_4, does not undergo such reactions.

No silicon analogs of ethylene and acetylene are known, and only a very few compounds containing Si=Si and Si=C bonds have been prepared. The atomic radius of Si (132 pm) is considerably larger than that of C (91 pm); consequently, the 3*p* orbital on a Si atom cannot overlap effectively with a *p* orbital on a neighboring atom (such as Si, C, or O) to form a pi bond. The larger size of Si also explains the difference between CO_2 and SiO_2. Carbon dioxide, as you know, is a discrete molecular species whose Lewis structure is

$$\ddot{\underset{..}{O}}=C=\ddot{\underset{..}{O}}$$

A similar structure for SiO_2

$$\ddot{\underset{..}{O}}=Si=\ddot{\underset{..}{O}}$$

would be unstable because the small overlap between the 3*p* orbital of Si and the 2*p* orbital of O results in a weak pi bond. Instead, the structure of silica is a giant three-dimensional network consisting of Si atoms that are tetrahedrally bonded to four O atoms (see Figure 11.36). As we saw in Section 11.7, silica is a chief component of glass. However, because silica dissolves slowly in a strong alkaline solution

The Si atom is sp^3-hybridized.

$$SiO_2(s) + 2OH^-(aq) \longrightarrow SiO_3^{2-}(aq) + H_2O(l)$$

concentrated alkaline solutions such as NaOH and KOH must be stored in polyethylene containers rather than in glass bottles.

About 90 percent of Earth's crust consists of ***silicates,*** which are *anion groups*

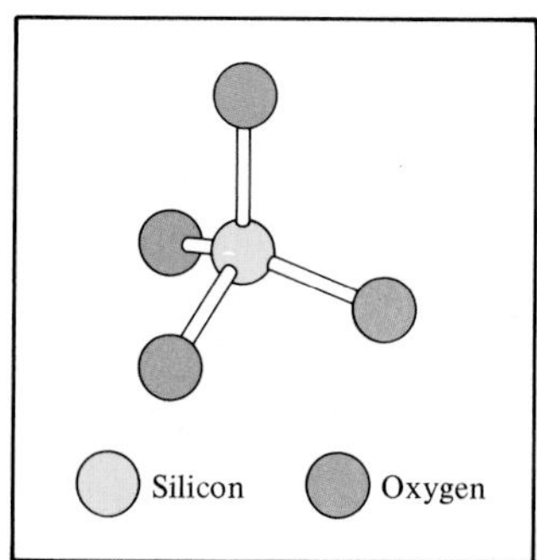

Figure 21.19 *The tetrahedral arrangement of the SiO_4^{2-} ion. This is the basic unit for all the higher silicates.*

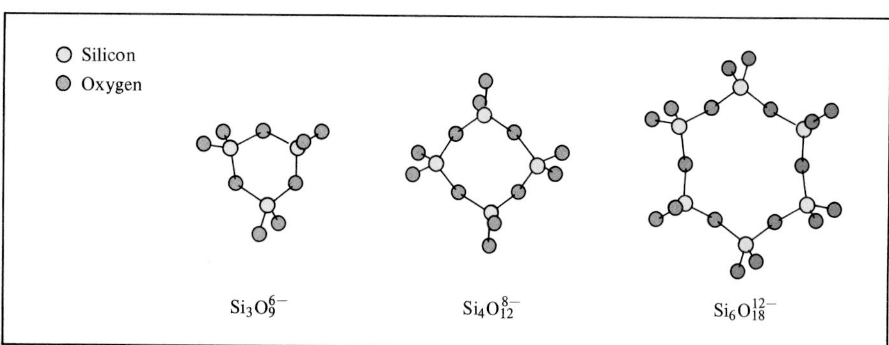

Figure 21.20 *The structures of several silicates.*

consisting of Si and O atoms. Although silicates vary in the ratio of Si to O atoms, the basic unit in all silicates consists of a Si atom tetrahedrally bonded to four O atoms (Figure 21.19). In higher silicates, these units are joined together by sharing an O atom between two Si atoms (Figure 21.20).

The fibrous silicate mineral known as *asbestos* has been used for thousands of years. Actually, asbestos is not a discrete chemical compound or even a single group of minerals. The term is a commercial one that refers to several fibrous inorganic materials that are used in industry for their mechanical strength and resistance to heat. Figure 21.21 shows one type of asbestos fiber, the mineral cummingtonite-grunerite, which has the formula $(Mg, Fe^{2+})_7Si_8O_{22}(OH)_2$. Until recently, asbestos was used as a thermal insulator in buildings. It is now well established that prolonged exposure to airborne suspensions of asbestos fiber dust can be very dangerous to the lung tissues and

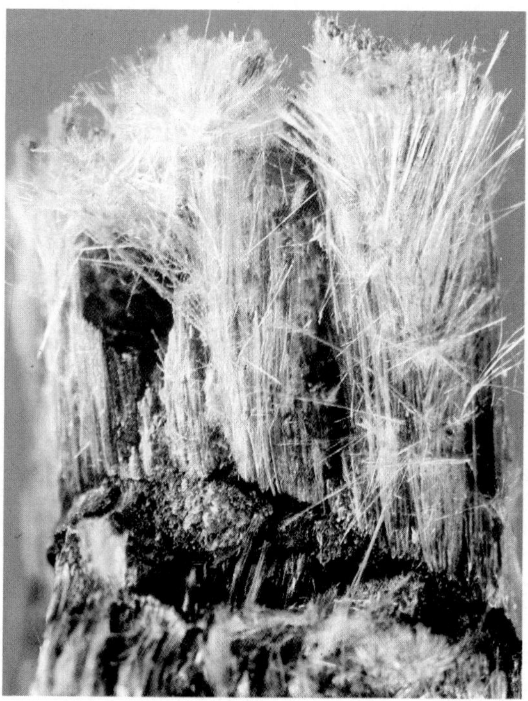

Figure 21.21 *Asbestos fibers.*

the digestive tract, and often results in cancer. The symptoms, however, may take years to develop.

Silicon has no known role in the human body, but *diatoms,* which are microscopic unicellular algae, make their skeletons from SiO_2 and have an active metabolism involving Si. It has also been shown that silicon is an essential element in baby chicks, but its precise function is not known. In view of the fact that silicon is 146 times more plentiful than carbon and exhibits some of the same properties, scientists have wondered why it does not play a more significant part in directing the chemistry of living forms.

There are at least two reasons that favor the role of carbon as the basis for life on Earth. First, carbon dioxide is a monomeric (that is, a single molecular unit), stable molecule that is readily soluble in water. In contrast, silica comprises a giant Si—O network that forms a macromolecule of silicon dioxide. Further, since silicon dioxide is not soluble in water, it cannot take part in acid-base reactions. The second reason concerns ***catenation,*** which is *the linking of like atoms*. Carbon has the unique ability to form long chains (consisting of more than 50 C atoms) and stable rings with five or six members. This versatility is responsible for the millions of organic compounds found on Earth. Silicon atoms, on the other hand, can form only relatively short chains. Therefore, the number of different silicon compounds is severely limited.

The low stability of many silicon compounds is the result of the weak Si—Si bond, about 226 kJ/mol, as compared with the comparatively strong C—C bond of about 347 kJ/mol.

21.5 Nitrogen and Phosphorus

Nitrogen

About 78 percent of air by volume is nitrogen. The most important mineral sources of nitrogen are saltpeter (KNO_3) and Chile saltpeter ($NaNO_3$). Nitrogen is an essential element of life; it is a component of proteins and nucleic acids.

Molecular nitrogen will boil off before molecular oxygen does during the fractional distillation of liquid air.

Molecular nitrogen is obtained by fractional distillation of air (the boiling points of liquid nitrogen and liquid oxygen are −196°C and −183°C, respectively). In the laboratory, very pure nitrogen gas can be prepared by the thermal decomposition of ammonium nitrite:

$$NH_4NO_2(s) \longrightarrow 2H_2O(g) + N_2(g)$$

The N_2 molecule contains a triple bond and is very stable with respect to dissociation into atomic species. However, nitrogen forms a large number of compounds with hydrogen and oxygen in which the oxidation number of nitrogen varies from −3 to +5 (Table 21.2). Most of the nitrogen compounds are covalent; however, when heated with certain metals, nitrogen forms ionic nitrides containing N^{3-} ions:

$$6Li(s) + N_2(g) \longrightarrow 2Li_3N(s)$$

The nitride ion is a strong Brønsted base and reacts with water to produce ammonia and hydroxide ion:

$$N^{3-}(aq) + 3H_2O(l) \longrightarrow NH_3(g) + 3OH^-(aq)$$

Ammonia. Ammonia is one of the best-known nitrogen compounds. It is prepared

Table 21.2 Common compounds of nitrogen

Oxidation number	Compound	Formula	Structure
−3	Ammonia	NH_3	H—N̈—H | H
−2	Hydrazine	N_2H_4	H—N̈—N̈—H | | H H
−1	Hydroxylamine	NH_2OH	H—N̈—Ö—H | H
0	Nitrogen* (dinitrogen)	N_2	:N≡N:
+1	Nitrous oxide (dinitrogen monoxide)	N_2O	:N≡N—Ö:
+2	Nitric oxide (nitrogen monoxide)	NO	:Ṅ=Ö
+3	Nitrous acid	HNO_2	Ö=N̈—Ö—H
+4	Nitrogen dioxide	NO_2	:Ö—Ṅ=Ö
+5	Nitric acid	HNO_3	Ö=N—Ö—H | :Ö:

*We list the element here as a reference.

industrially from nitrogen and hydrogen by the Haber process (see Section 13.6 and p. 624). It can be prepared in the laboratory by treating ammonium chloride with sodium hydroxide:

$$NH_4Cl(aq) + NaOH(aq) \longrightarrow NaCl(aq) + H_2O(l) + NH_3(g)$$

Ammonia is a colorless gas (b.p. −33.4°C) with an irritating odor. About three-quarters of the ammonia produced annually in the United States (17 million tons in 1988) is used in fertilizers.

Liquid ammonia resembles water in that it undergoes autoionization:

$$2NH_3(l) \rightleftharpoons NH_4^+ + NH_2^-$$

or simply

$$NH_3(l) \rightleftharpoons H^+ + NH_2^-$$

The amide ion is a strong Brønsted base and does not exist in water.

where NH_2^- is called the *amide ion*. Note that both H^+ and NH_2^- are solvated with the NH_3 molecules. (Here is an example of ion-dipole interaction.) At −50°C, the ion product $[H^+][NH_2^-]$ is about 1×10^{-33}, considerably smaller than 1×10^{-14} for water at 25°C. Nevertheless, liquid ammonia is a suitable solvent for many electrolytes, especially when a more basic medium is required or if the solutes react with water. The ability of liquid ammonia to dissolve alkali metals was discussed in Section 20.5.

Hydrazine. Another important hydride of nitrogen is hydrazine:

$$\begin{array}{c} H \quad\quad H \\ \diagdown\ \ddot{N}-\ddot{N}\ \diagup \\ \diagup \quad\quad \diagdown \\ H \quad\quad H \end{array}$$

Each N atom is sp^3-hybridized. Hydrazine is a colorless liquid that smells like ammonia. It melts at 2°C and boils at 114°C.

Hydrazine is a base that can be protonated to give the $N_2H_5^+$ and $N_2H_6^{2+}$ ions. A reducing agent, it can reduce Fe^{3+} to Fe^{2+}, MnO_4^- to Mn^{2+}, and I_2 to I^-. Its reaction with oxygen is highly exothermic:

$$N_2H_4(l) + O_2(g) \longrightarrow N_2(g) + 2H_2O(l) \qquad \Delta H^\circ = -666.6 \text{ kJ}$$

Hydrazine is used as a rocket fuel in the space program (for example, the Apollo missions to the moon). Hydrazine and its derivative methylhydrazine, $N_2H_3(CH_3)$, together with the oxidizer dinitrogen tetroxide (N_2O_4), are used as rocket fuels. Hydrazine also plays a role in the polymer industry and in the manufacture of pesticides.

Oxides and Oxoacids of Nitrogen. Many nitrogen oxides exist, but we will discuss only three important ones: nitrous oxide, nitric oxide, and nitrogen dioxide:

Nitrous oxide, N_2O, is a colorless gas with a pleasing odor and sweet taste. It is prepared by heating ammonium nitrate to about 270°C:

$$NH_4NO_3(s) \longrightarrow N_2O(g) + 2H_2O(g)$$

Nitrous oxide resembles molecular oxygen in that it supports combustion. It does so because it decomposes when heated to form molecular nitrogen and molecular oxygen:

$$2N_2O(g) \longrightarrow 2N_2(g) + O_2(g)$$

It is chiefly used as an anesthetic in dental and other minor surgery. Nitrous oxide is also called ''laughing gas'' because a person inhaling the gas becomes somewhat giddy. No satisfactory explanation has yet been proposed for this unusual physiological response.

Nitric oxide, NO, is a colorless gas. The reaction of N_2 and O_2 in the atmosphere

According to Le Chatelier's principle, the forward endothermic reaction is favored by heating.

$$N_2(g) + O_2(g) \rightleftharpoons 2NO(g) \qquad \Delta G^\circ = 173.4 \text{ kJ}$$

might be considered a way of fixing nitrogen. ***Nitrogen fixation*** is *the conversion of molecular nitrogen into nitrogen compounds*. The equilibrium constant for the above reaction is very small at room temperature: K_P is only 4.0×10^{-31} at 25°C, so very little NO will form at that temperature. However, the equilibrium constant increases rapidly with temperature, for example, in a running auto engine. An appreciable amount of nitric oxide is formed in the atmosphere by the action of lightning. In the laboratory the gas can be prepared by the reduction of dilute nitric acid with copper:

$$3Cu(s) + 8HNO_3(aq) \longrightarrow 3Cu(NO_3)_2(aq) + 4H_2O(l) + 2NO(g)$$

The nitric oxide molecule is paramagnetic, containing one unpaired electron. It can be represented by the following resonance structures:

$$\dot{N}=\ddot{O} \longleftrightarrow {}^-\ddot{N}=\dot{O}^+$$

As we noted in Chapter 9, this molecule does not obey the octet rule. Nitric oxide

Figure 21.22 *The production of NO_2 gas when copper reacts with concentrated nitric acid.*

forms brown fumes of nitrogen dioxide instantaneously in the presence of air:

$$2NO(g) + O_2(g) \longrightarrow 2NO_2(g)$$

Consequently, we do not know what nitric oxide would smell like. It is only slightly soluble in water, and the resulting solution is not acidic.

Unlike nitrous oxide and nitric oxide, nitrogen dioxide is a highly toxic yellow-brown gas with a choking odor. In the laboratory nitrogen dioxide is prepared by the action of concentrated nitric acid on copper (Figure 21.22):

$$Cu(s) + 4HNO_3(aq) \longrightarrow Cu(NO_3)_2(aq) + 2H_2O(l) + 2NO_2(g)$$

Nitrogen dioxide is paramagnetic. It has a strong tendency to dimerize to dinitrogen tetroxide, which is a diamagnetic molecule:

$$2NO_2 \rightleftharpoons N_2O_4$$

This reaction occurs in both the gas phase and the liquid phase.

Nitrogen dioxide is an acidic oxide; it reacts rapidly with cold water to form both nitrous acid, HNO_2, and nitric acid:

Neither N_2O nor NO reacts with water.

$$2NO_2(g) + H_2O(l) \longrightarrow HNO_2(aq) + HNO_3(aq)$$

This is a disproportionation reaction (see p. 116) in which the oxidation number of nitrogen changes from +4 (in NO_2) to +3 (in HNO_2) and +5 (in HNO_3). Note that this reaction is quite different from that between CO_2 and H_2O, in which only one acid (carbonic acid) is formed.

Nitric acid is one of the most important inorganic acids. The major industrial method of producing nitric acid is the Ostwald process, discussed in Section 13.6.

Nitric acid is a liquid that boils at 82.6°C. It does not exist as a pure liquid because it decomposes spontaneously to some extent as follows:

On standing, a concentrated nitric acid solution turns slightly yellow as a result of NO_2 formation.

$$4HNO_3(l) \longrightarrow 4NO_2(g) + 2H_2O(l) + O_2(g)$$

The concentrated nitric acid used in the laboratory is 68 percent HNO_3 by mass (density 1.42 g/cm^3), which corresponds to 15.7 *M*.

Nitric acid is a powerful oxidizing agent. The oxidation number of N in HNO_3 is +5. The most common reduction products of nitric acid are NO_2 (N = +4), NO (N = +2), and NH_4^+ (N = −3). Nitric acid can oxidize metals both below and above hydrogen in the activity series (see Figure 3.15). For example, copper is oxidized by concentrated nitric acid as discussed earlier.

In the presence of a strong reducing agent, such as zinc metal, nitric acid can be reduced all the way to the ammonium ion:

$$4Zn(s) + 10H^+(aq) + NO_3^-(aq) \longrightarrow 4Zn^{2+}(aq) + NH_4^+(aq) + 3H_2O(l)$$

Concentrated nitric acid does not oxidize gold. However, when the acid is added to concentrated hydrochloric acid in a 1:3 ratio by volume (one part HNO_3 to three parts HCl), the resulting solution, called *aqua regia,* can oxidize gold, as follows:

The oxidation of Au is promoted by the complexing ability of the Cl^- ion (to form the $AuCl_4^-$ ion).

$$Au(s) + 3HNO_3(aq) + 4HCl(aq) \longrightarrow HAuCl_4(aq) + 3H_2O(l) + 3NO_2(g)$$

Concentrated nitric acid also oxidizes a number of nonmetals to their corresponding oxoacids:

$$P_4(s) + 20HNO_3(aq) \longrightarrow 4H_3PO_4(aq) + 20NO_2(g) + 4H_2O(l)$$

$$S(s) + 6HNO_3(aq) \longrightarrow H_2SO_4(aq) + 6NO_2(g) + 2H_2O(l)$$

Nitric acid is used in the manufacture of fertilizers, dyes, drugs, and explosives.

The Nitrogen Cycle. Figure 21.23 summarizes the major processes involved in the cycle of nitrogen in nature. As stated earlier, molecular nitrogen is very stable, but through *biological* and *industrial* fixation, atmospheric nitrogen gas is converted into compounds suitable for assimilation (for example, nitrates) by higher plants. As with nitric oxide, a major mechanism for producing nitrates from nitrogen gas is lightning—about 30 million tons of HNO_3 are produced this way annually. The nitric acid is converted to nitrate salts in the soil. These nutrients are taken up by plants that, in turn, are ingested by animals to make proteins and other essential biomolecules.

The term "denitrification" applies to processes that reverse nitrogen fixation. For example, certain anaerobic organisms decompose animal wastes as well as dead plants and animals to produce free molecular nitrogen from nitrates, thereby completing the cycle.

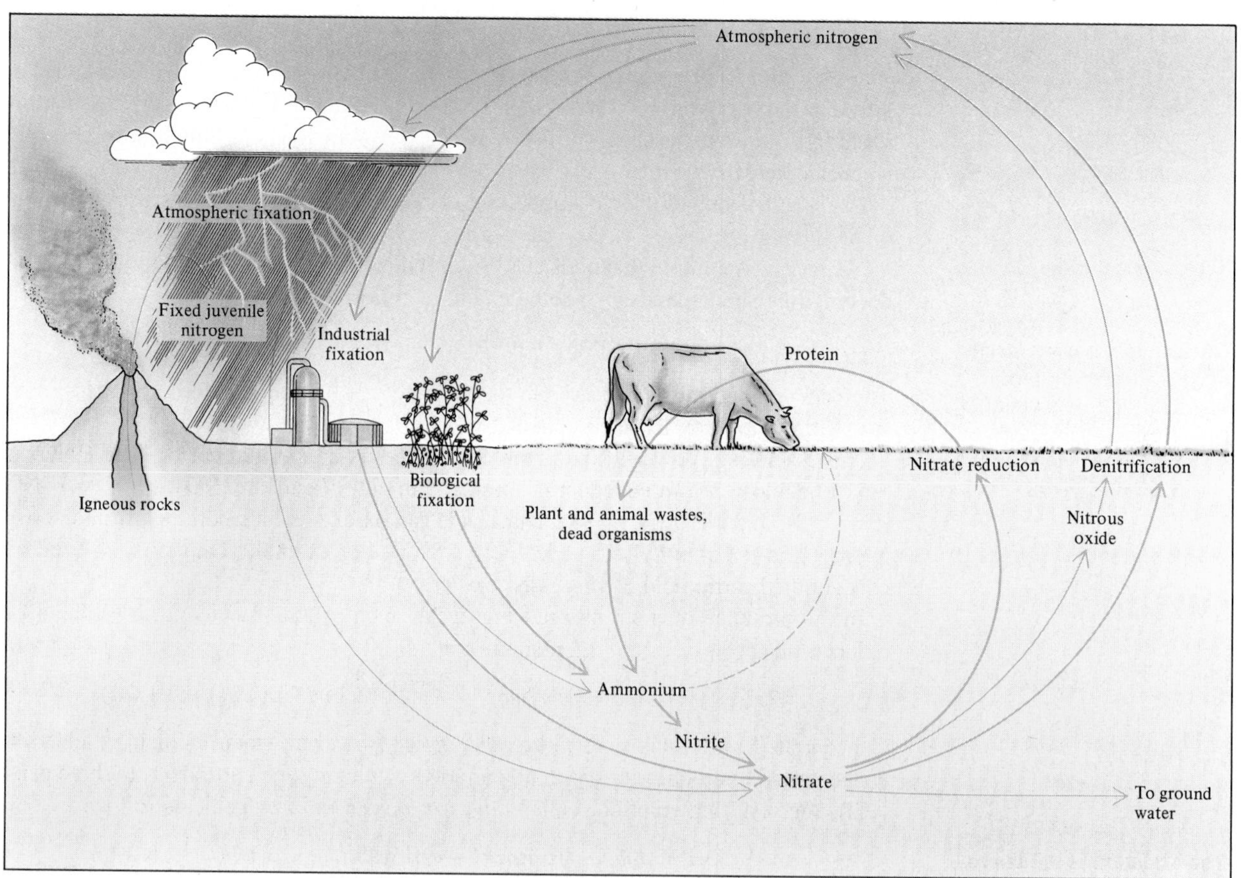

Figure 21.23 *The nitrogen cycle. Although the supply of nitrogen in the atmosphere is virtually inexhaustible, it must be combined with hydrogen or oxygen before it can be assimilated by higher plants, which in turn are consumed by animals. Juvenile nitrogen is nitrogen that has not previously participated in the nitrogen cycle.*

CHEMISTRY IN ACTION

Photochemical Smog

In recent years there has been a great deal of interest in the chemistry of NO and NO_2 because of their leading role in the formation of smog. The word ''smog'' was originally coined to describe the combination of smoke and fog occurring in London during the 1950s, much of which was due to the presence of sulfur dioxide in the atmosphere. *Photochemical smog* is formed by the photochemical reactions of automobile exhaust in the presence of sunlight.

Smog begins with primary pollutants, substances that may be relatively harmless and unreactive by themselves. Secondary pollutants, formed photochemically from the primary pollutants, are responsible for the buildup of smog. Primary pollutants consist mainly of nitric oxide, formed by the reaction between atmospheric nitrogen and oxygen at high temperatures inside the automobile engine:

$$N_2(g) + O_2(g) \longrightarrow 2NO(g)$$

as well as carbon monoxide and various unburned hydrocarbons. Once released into the atmosphere, nitric oxide is quickly oxidized to the toxic nitrogen dioxide:

$$2NO(g) + O_2(g) \longrightarrow 2NO_2(g)$$

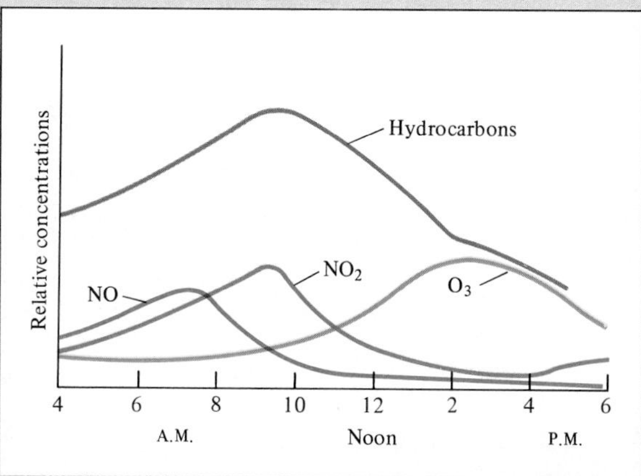

Figure 21.24 *Typical variations with time in concentration of air pollutants on a smoggy day.*

Sunlight causes the photochemical decomposition of NO_2 (at a wavelength of about 400 nm) into NO and O:

$$NO_2(g) + h\nu \longrightarrow NO(g) + O(g)$$

Atomic oxygen is a highly reactive species that can initiate a number of important reactions, one of which is the formation of ozone:

$$O(g) + O_2(g) + M \longrightarrow O_3(g) + M$$

where M is some inert substance such as N_2. The role of M is to absorb some of the excess energy of the ozone formed by collision; otherwise, the O_3 molecule may spontaneously decompose to form O and O_2. Ozone attacks the C=C linkage in rubber:

$$R_2C{=}CR_2 + O_3 \longrightarrow R_2C\!\!\begin{array}{c}O\\ \\ O{-}O\end{array}\!\!CR_2 \xrightarrow{H_2O} R_2C{=}O + O{=}CR_2 + H_2O_2$$

This reaction causes cracks in automobile tires and may lead to such damage in ''smoggy'' areas. Similar reactions are also damaging to lung tissues and other biological substances.

Ozone can be produced also by a series of very complex reactions involving unburned hydrocarbons, aldehydes, nitrogen oxides, and oxygen. One of the products of these reactions is peroxyacetyl nitrate, PAN:

$$H_3C{-}\overset{\displaystyle O}{\overset{\|}{C}}{-}O{-}O{-}NO_2$$

PAN is a powerful lachrymator, or tear producer, and causes breathing difficulties.

Figure 21.24 shows typical variations with time of primary and secondary pollutants. Initially, the concentration of NO_2 is quite low. As soon as solar radiation penetrates the atmosphere, more NO_2 is formed from NO and O_2. Note that the concentration of ozone remains fairly constant at a low level in the early morning hours. As the concentration of unburned hydrocarbons and aldehydes increases in the air, the concentrations of NO_2 and O_3 also rise rapidly. The actual values, of course, depend on the location, traffic, and weather conditions, as well as the time of day. Figure 21.25 shows the visible effect of smog.

As the mechanism of photochemical smog formation has become better understood, major efforts have been made to reduce the buildup of primary pollutants. Most automobiles now are equipped with catalytic converters whose functions are to oxidize CO and unburnt hydrocarbons to CO_2 and H_2O and to reduce NO and NO_2 to N_2 and O_2 (see Section 13.6).

Figure 21.25 *A smoggy day in New York City.*

Phosphorus

Like nitrogen, phosphorus is a member of the Group 5A family; in some respects the chemistry of phosphorus resembles that of nitrogen. Phosphorus occurs most commonly in nature as *phosphate rocks,* which are mostly calcium phosphate, $Ca_3(PO_4)_2$, and fluoroapatite, $Ca_5(PO_4)_3F$ (Figure 21.26). Free phosphorus can be obtained by heating calcium phosphate with coke and silica sand:

$$2Ca_3(PO_4)_2(s) + 10C(s) + 6SiO_2(s) \longrightarrow 6CaSiO_3(s) + 10CO(g) + P_4(s)$$

There are several allotropic forms of phosphorus, but only two—white phosphorus

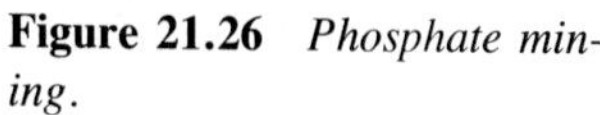

Figure 21.26 *Phosphate mining.*

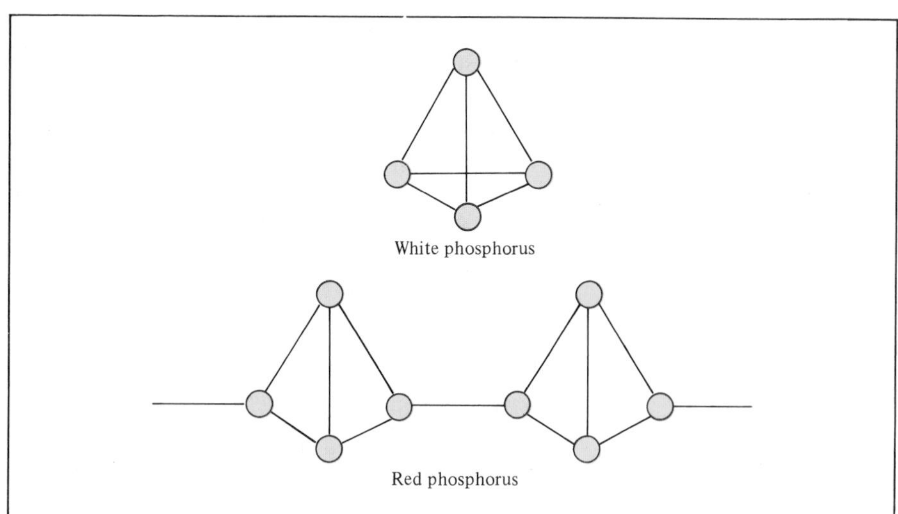

Figure 21.27 *The structures of white and red phosphorus. Red phosphorus is believed to consist of a chain structure, as shown.*

and red phosphorus (see Figure 8.21)—are of importance. White phosphorus consists of discrete tetrahedral P_4 molecules (Figure 21.27). A white solid (m.p. 44.2°C), white phosphorus is insoluble in water but quite soluble in carbon disulfide (CS_2) and in organic solvents such as chloroform ($CHCl_3$). White phosphorus is a highly toxic substance. It bursts into flames spontaneously when exposed to air; hence it is used in incendiary bombs and grenades:

$$P_4(s) + 5O_2(g) \longrightarrow P_4O_{10}(s)$$

Unlike N_2, elemental phosphorus contains no P=P or P≡P bonds because P is considerably larger than N (see Figure 8.7).

The high reactivity of white phosphorus is believed to be the result of the strain in the P_4 molecule: The P—P bonds are compressed in the tetrahedral structure. White phosphorus was once used in the manufacture of matches, but because of its toxicity it has been replaced by tetraphosphorus trisulfide, P_4S_3.

When heated in the absence of air, white phosphorus is slowly converted to red phosphorus at about 300°C:

$$nP_4 \text{ (white phosphorus)} \longrightarrow (P_4)_n \text{ (red phosphorus)}$$

Red phosphorus has a polymeric structure (see Figure 21.27) and is more stable and less volatile than white phosphorus.

Hydride of Phosphorus. The most important hydride of phosphorus is phosphine, PH_3, a colorless, very poisonous gas formed by heating white phosphorus in concentrated sodium hydroxide:

$$P_4(s) + 3NaOH(aq) + 3H_2O(l) \longrightarrow \underset{\text{sodium hypophosphite}}{3NaH_2PO_2(aq)} + \underset{\text{phosphine}}{PH_3(g)}$$

Phosphine is moderately soluble in water and more soluble in carbon disulfide and organic solvents. Unlike those of ammonia, its aqueous solutions are neutral. In liquid ammonia, phosphine dissolves to give $NH_4^+PH_2^-$. Phosphine is a strong reducing agent; it reduces many metal salts to the corresponding metal. The gas burns in air:

$$PH_3(g) + 2O_2(g) \longrightarrow H_3PO_4(s)$$

Halides of Phosphorus. Phosphorus forms binary compounds with halogens: the trihalides, PX_3, and the pentahalides, PX_5, where X denotes a halogen atom. In contrast, nitrogen can form only trihalides (NX_3). Unlike nitrogen, phosphorus has a 3*d* subshell that can be used for valence-shell expansion. We can explain the bonding in PCl_5 by assuming a hybridization process involving 3*s*, 3*p*, and 3*d* orbitals of phosphorus, that is, an sp^3d hybridization (see Example 10.4). The five sp^3d hybrid orbitals also explain satisfactorily the trigonal bipyramid geometry of the PCl_5 molecule (see Table 10.5).

Phosphorus trichloride is prepared by heating white phosphorus in an atmosphere of chlorine:

$$P_4(l) + 6Cl_2(g) \longrightarrow 4PCl_3(g)$$

A colorless liquid (b.p. 76°C), it is hydrolyzed according to the equation:

$$PCl_3(l) + 3H_2O(l) \longrightarrow \underset{\text{phosphorous acid}}{H_3PO_3(aq)} + 3HCl(g)$$

In the presence of an excess of chlorine gas, PCl_3 is converted to phosphorus pentachloride, which is a light yellow solid:

$$PCl_3(l) + Cl_2(g) \longrightarrow PCl_5(s)$$

X-ray studies have shown that solid phosphorus pentachloride exists as $[PCl_4^+][PCl_6^-]$, in which the PCl_4^+ ion has the tetrahedral geometry and the PCl_6^- ion has the octahedral geometry. In the gas phase, PCl_5 (which has the trigonal bipyramidal geometry) is in equilibrium with PCl_3 and Cl_2:

$$PCl_5(g) \rightleftharpoons PCl_3(g) + Cl_2(g)$$

Phosphorus pentachloride reacts with water as follows:

$$PCl_5(s) + 4H_2O(l) \longrightarrow H_3PO_4(aq) + 5HCl(aq)$$

Phosphorus burning in oxygen.

Oxides and Oxoacids of Phosphorus. The two important oxides of phosphorus are tetraphosphorus hexaoxide, P_4O_6, and tetraphosphorus decaoxide, P_4O_{10}. The oxides are obtained by burning white phosphorus in limited and excess amounts of oxygen gas, respectively:

$$P_4(s) + 3O_2(g) \longrightarrow P_4O_6(s)$$

$$P_4(s) + 5O_2(g) \longrightarrow P_4O_{10}(s)$$

Figure 21.28 shows their structures. Both oxides are acidic; that is, they are converted to acids in water. The compound P_4O_{10} is a white flocculent powder (m.p. 420°C) that has a great affinity for water:

$$P_4O_{10}(s) + 6H_2O(l) \longrightarrow 4H_3PO_4(aq)$$

For this reason it is often used for drying gases and for removing water from solvents.

Phosphoric acid is the most important phosphorus-containing oxoacid.

There are many oxoacids containing phosphorus; some examples are phosphorous acid, H_3PO_3; phosphoric acid, H_3PO_4; hypophosphorous acid, H_3PO_2; and triphosphoric acid, $H_5P_3O_{10}$ (Figure 21.29). Phosphoric acid is also called orthophosphoric acid; it is a weak triprotic acid (see Section 16.4). It is prepared industrially by the reaction of calcium phosphate with sulfuric acid:

$$Ca_3(PO_4)_2(s) + 3H_2SO_4(aq) \longrightarrow 2H_3PO_4(aq) + 3CaSO_4(aq)$$

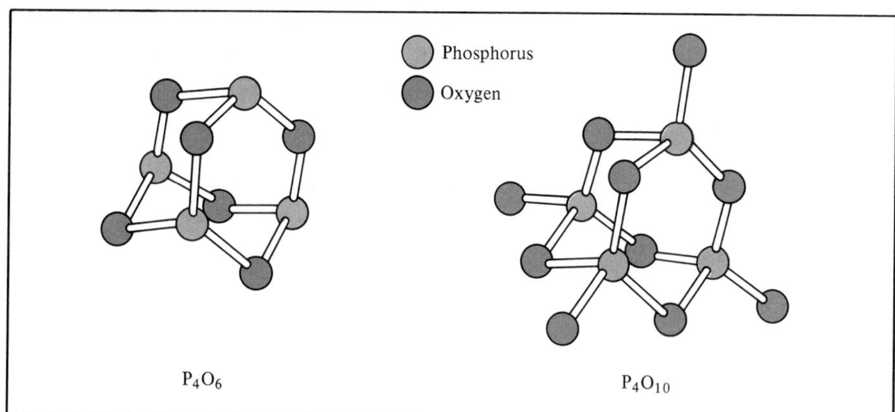

Figure 21.28 *The structures of P_4O_6 and P_4O_{10}. Note the tetrahedral arrangement of the P atoms in P_4O_{10}.*

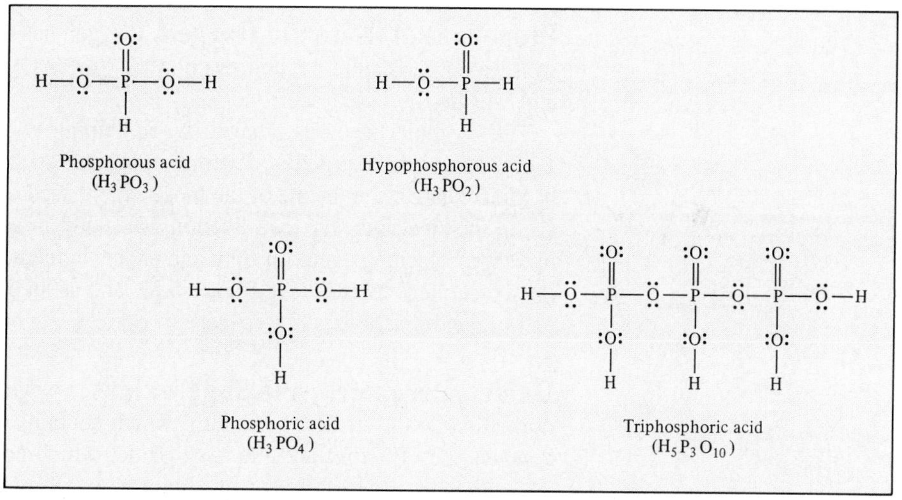

Phosphorous acid (H_3PO_3)

Hypophosphorous acid (H_3PO_2)

Phosphoric acid (H_3PO_4)

Triphosphoric acid ($H_5P_3O_{10}$)

Figure 21.29 *Structures of some common phosphorus-containing oxoacids.*

In the pure form phosphoric acid is a colorless solid (m.p. 42.2°C). The phosphoric acid we use in the laboratory is usually an 82 percent H_3PO_4 solution (by mass). Phosphoric acid and phosphates have many commercial applications in detergents, fertilizers, flame retardants, and toothpastes, and as buffers in carbonated beverages to maintain a constant pH.

Like nitrogen, phosphorus is an element that is essential to life. It constitutes only about 1 percent by mass of the human body, but it is a very important 1 percent. About 23 percent of the human skeleton is mineral matter. The phosphorus content of this mineral matter, calcium phosphate, $Ca_3(PO_4)_2$, is 20 percent. Our teeth are basically $Ca_3(PO_4)_2$ and $Ca_5(PO_4)_3OH$. The phosphates are important components of the genetic materials deoxyribonucleic acid (DNA) and ribonucleic acid (RNA).

21.6 Oxygen and Sulfur

Oxygen

Oxygen is by far the most abundant element in Earth's crust, constituting about 46 percent of its mass. The atmosphere contains about 21 percent of molecular oxygen by volume (23 percent by mass). Like nitrogen, oxygen occurs in the free state as a diatomic molecule (O_2). In the laboratory, oxygen gas can be obtained by heating potassium chlorate (see Figure 5.14):

$$2KClO_3(s) \longrightarrow 2KCl(s) + 3O_2(g)$$

The reaction is usually catalyzed by manganese(IV) dioxide, MnO_2. Pure oxygen gas can be prepared by electrolyzing water (p. 805). Industrially, oxygen gas is prepared by the fractional distillation of liquefied air (p. 882). Oxygen gas is colorless and odorless.

Oxygen is a fundamental building block of practically all vital biomolecules, accounting for about a fourth of the atoms in living matter. Molecular oxygen is the essential oxidant in the metabolic breakdown of food molecules. Without it, a human being cannot survive for more than a few minutes.

Properties of Diatomic Oxygen. Oxygen has two allotropes, O_2 and O_3. When we speak of molecular oxygen, we normally mean O_2. The O_3 molecule is called *ozone* and will be discussed shortly.

The O_2 molecule is paramagnetic, containing two unpaired electrons (see Section 10.7). Molecular oxygen is a strong oxidizing agent.

Molecular oxygen is one of the most widely used industrial chemicals. Its main uses are in the steel industry (see Section 20.2) and in sewage treatment. Oxygen is also used as a bleaching agent for pulp and paper, in medicine to ease breathing difficulties, in oxyacetylene torches (see Figure 6.8), and as an oxidizing agent in many inorganic and organic reactions.

Oxides, Peroxides, and Superoxides. Oxygen forms three types of oxides: the normal oxide (or simply the oxide), which contains the O^{2-} ion; the peroxide, which contains the O_2^{2-} ion; and the superoxide, which contains the O_2^- ion:

$\ddot{\underset{..}{O}}::\ddot{\underset{..}{O}}$	$:\ddot{\underset{..}{O}}:^{2-}$	$:\ddot{\underset{..}{O}}:\ddot{\underset{..}{O}}:^{2-}$	$:\ddot{\underset{..}{O}}:\ddot{\underset{..}{O}}:^{-}$
oxygen	oxide	peroxide	superoxide

These ions are all strong Brønsted bases and react with water as follows:

Oxide:	$O^{2-}(aq) + H_2O(l) \longrightarrow 2OH^-(aq)$
Peroxide:	$2O_2^{2-}(aq) + 2H_2O(l) \longrightarrow O_2(g) + 4OH^-(aq)$
Superoxide:	$4O_2^-(aq) + 2H_2O(l) \longrightarrow 3O_2(g) + 4OH^-(aq)$

Note that the reaction of O^{2-} with water is a hydrolysis reaction, but those involving O_2^{2-} and O_2^- are redox processes.

The nature of bonding in oxides changes across any period in the periodic table (see Figure 15.5). Oxides of elements on the left side of the periodic table, such as those of the alkali metals and alkaline earth metals, are generally ionic solids and have high melting points. Oxides of the metalloids and of the metallic elements toward the middle of the periodic table are also solids, but they have much less ionic character. Oxides of nonmetals are covalent compounds that generally exist as liquids or gases at room temperature. The acidic character of the oxides increases from left to right. Consider the oxides of the third-period elements:

Na_2O MgO	Al_2O_3	SiO_2 P_4O_{10} SO_3 Cl_2O_7
basic	amphoteric	acidic

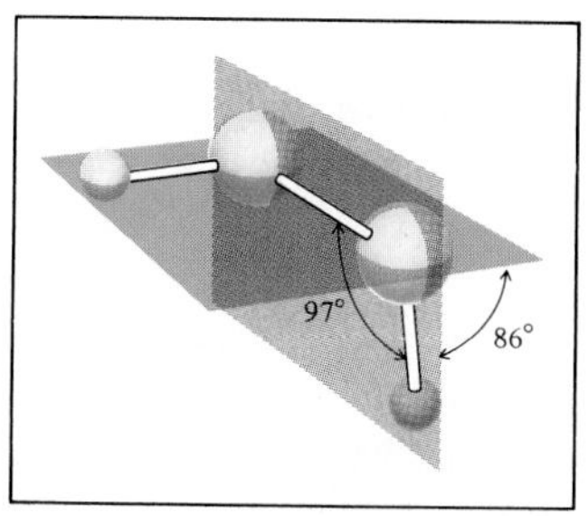

Figure 21.30 *The structure of H_2O_2.*

The basic character of the oxides increases as we move down a particular group. MgO does not react with water but reacts with acid as follows:

$$MgO(s) + 2H^+(aq) \longrightarrow Mg^{2+}(aq) + H_2O(l)$$

On the other hand, BaO, which is more basic, undergoes the hydrolysis to yield the corresponding hydroxide:

$$BaO(s) + H_2O(l) \longrightarrow Ba(OH)_2(aq)$$

The best-known peroxide is hydrogen peroxide (H_2O_2). It is a colorless, syrupy liquid (m.p. −0.9°C), prepared in the laboratory by the action of cold dilute sulfuric acid on barium peroxide octahydrate:

$$BaO_2 \cdot 8H_2O(s) + H_2SO_4(aq) \longrightarrow BaSO_4(s) + H_2O_2(aq) + 8H_2O(l)$$

The structure of hydrogen peroxide is shown in Figure 21.30. Using the VSEPR method we see that the H—O and O—O bonds are bent around each oxygen atom, similar to the structure of water. The lone-pair–bonding-pair repulsion is greater in H_2O_2 than in H_2O, so that the HOO angle is only 97° (compared with 104.5° for HOH in H_2O). Hydrogen peroxide is a polar molecule (μ = 2.16 D).

Hydrogen peroxide readily decomposes on heating or even in the presence of dust particles or certain metals, including iron and copper:

$$2H_2O_2(l) \longrightarrow 2H_2O(l) + O_2(g) \qquad \Delta H° = -196.4 \text{ kJ}$$

Note that this is a disproportionation reaction. The oxidation number of oxygen changes from −1 to −2 and 0.

Hydrogen peroxide is miscible with water in all proportions, due to its ability to hydrogen-bond with water. Dilute hydrogen peroxide solutions (3 percent by mass), available in drugstores, are used as mild antiseptics; more concentrated H_2O_2 solutions are employed as bleaching agents for textiles, fur, and hair. The high heat of decomposition of hydrogen peroxide also makes it a suitable component in rocket fuel.

Hydrogen peroxide is a strong oxidizing agent; it can oxidize Fe^{2+} ions to Fe^{3+} ions in an acidic solution:

$$H_2O_2(aq) + 2Fe^{2+}(aq) + 2H^+(aq) \longrightarrow 2Fe^{3+}(aq) + 2H_2O(l)$$

It also oxidizes SO_3^{2-} ions to SO_4^{2-} ions:

$$H_2O_2(aq) + SO_3^{2-}(aq) \longrightarrow SO_4^{2-}(aq) + H_2O(l)$$

Hydrogen peroxide can act also as a reducing agent toward substances that are stronger oxidizing agents than itself. For example, hydrogen peroxide reduces silver oxide to metallic silver:

$$H_2O_2(aq) + Ag_2O(s) \longrightarrow 2Ag(s) + H_2O(l) + O_2(g)$$

and permanganate, MnO_4^-, to manganese(II) in an acidic solution:

$$5H_2O_2(aq) + 2MnO_4^-(aq) + 6H^+(aq) \longrightarrow 2Mn^{2+}(aq) + 5O_2(g) + 8H_2O(l)$$

If we want to determine hydrogen peroxide concentration, this reaction can be carried out as a redox titration, using a standard permanganate solution.

There are relatively few known superoxides, that is, compounds containing the O_2^- ion. In general, only the most reactive alkali metals (K, Rb, and Cs) form superoxides (see Section 20.5).

We should take note of the fact that both the peroxide ion and superoxide ion are by-products of metabolism. Because these ions are highly reactive, they can inflict great damage on various cellular components. Fortunately, our bodies have been equipped with the enzymes needed to convert these toxic substances to water and molecular oxygen.

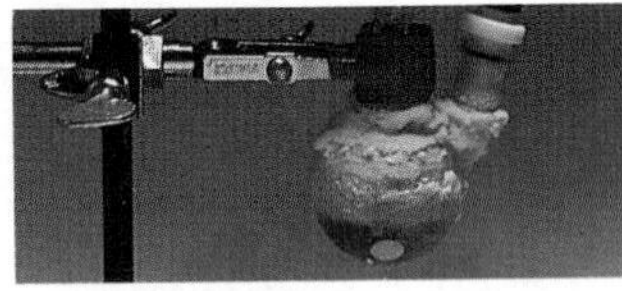

Liquid ozone.

Ozone. Ozone is a rather toxic, light blue gas (b.p. $-111.3°C$) with a pungent odor. The odor of ozone is evident wherever significant electrical discharges are occurring (for example, near a subway train). Ozone can be prepared from molecular oxygen, either photochemically or by subjecting O_2 to an electrical discharge (Figure 21.31):

$$3O_2(g) \longrightarrow 2O_3(g) \qquad \Delta G° = 326.8 \text{ kJ}$$

Since the standard free energy of formation of ozone is a large positive quantity—$\Delta G_f^\circ = (326.8/2)$ kJ/mol or 163.4 kJ/mol—ozone is less stable than molecular oxygen. The ozone molecule has a bent structure in which the bond angle is 116.5°:

$$\ddot{O}{:}\!-\!\ddot{O}\!=\!\ddot{O} \longleftrightarrow \ddot{O}\!=\!\ddot{O}\!-\!\ddot{O}{:}$$

Ozone is mainly used to purify drinking water, to deodorize air and sewage gases, and to bleach waxes, oils, and textiles.

Ozone is a very powerful oxidizing agent—its oxidizing power is exceeded only by that of molecular fluorine (see Table 19.1). For example, ozone can oxidize sulfides of many metals to the corresponding sulfates:

$$4O_3(g) + PbS(s) \longrightarrow PbSO_4(s) + 4O_2(g)$$

Ozone oxidizes all the common metals except gold and platinum. In fact, a convenient test for ozone is based on its action on mercury. When exposed to ozone, the metal

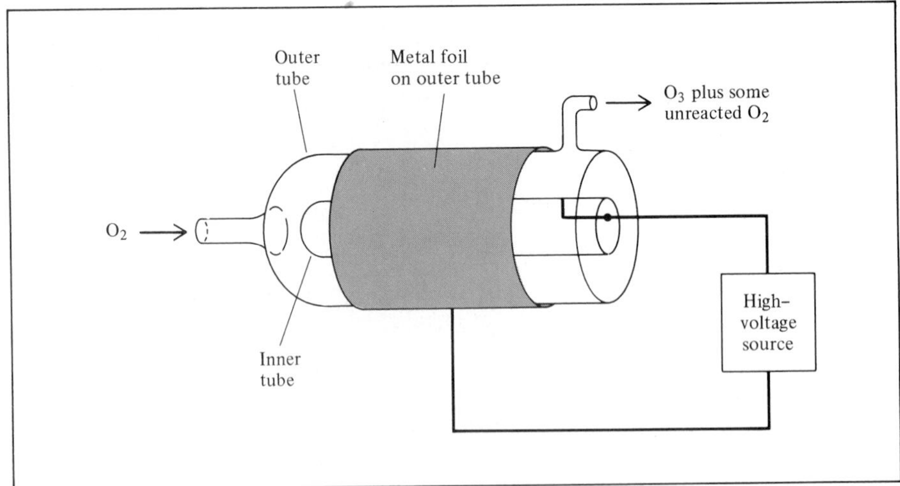

Figure 21.31 *The preparation of O_3 from O_2 by electrical discharge. The outside of the outer tube and the inside of the inner tube are coated with metal foils that are connected to a high-voltage source. (The metal foil on the inside of the inner tube is not shown.) During the electrical discharge, O_2 gas is passed through the tube. The O_3 gas formed plus some unreacted O_2 gas exits at the upper right-hand tube.*

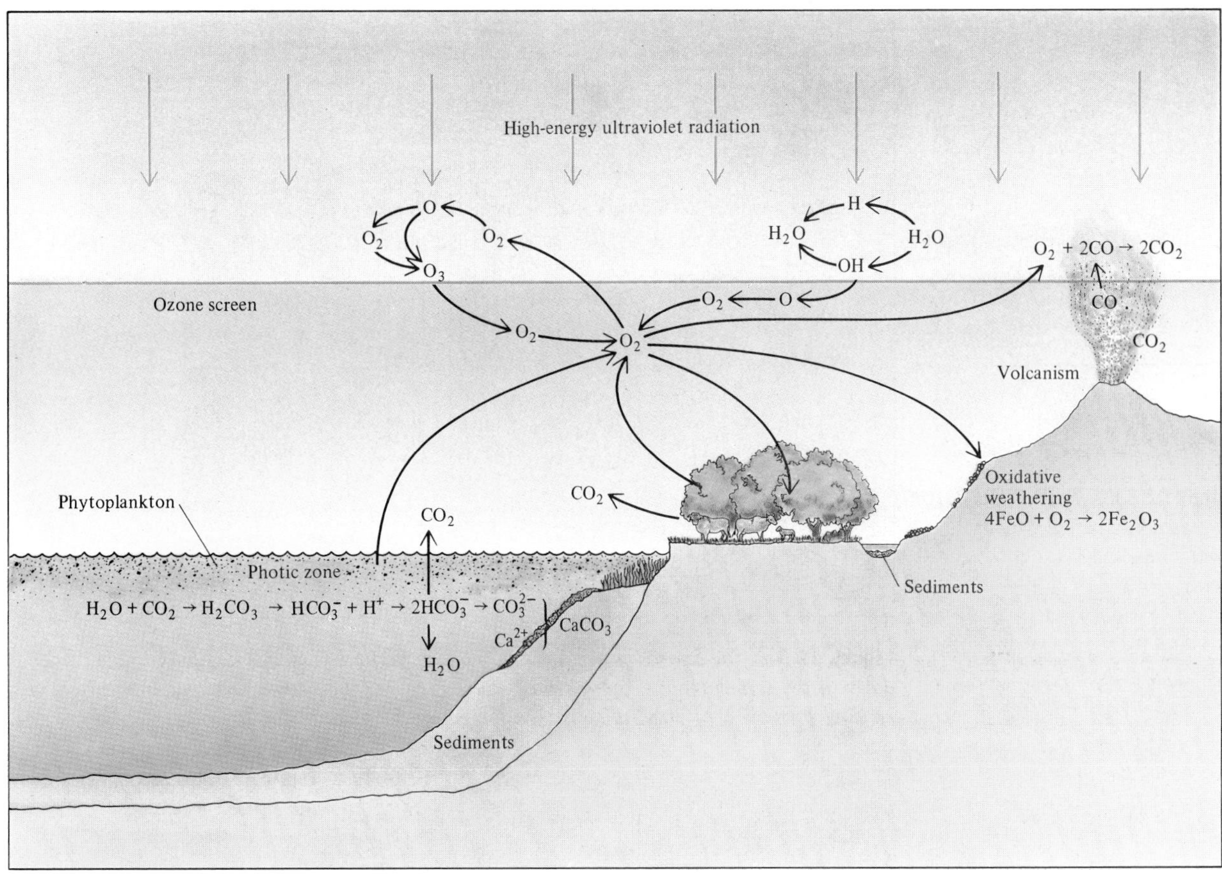

Figure 21.32 *The oxygen cycle. The cycle is complicated because oxygen appears in so many chemical forms and combinations, primarily as molecular oxygen, in water, and in organic and inorganic compounds.*

becomes dull looking and sticks to glass tubing (instead of flowing freely through it). This behavior is attributed to the change in surface tension caused by the formation of mercury(II) oxide:

$$O_3(g) + 3Hg(l) \longrightarrow 3HgO(s)$$

The beneficial effect of ozone in the stratosphere and its undesirable action in smog formation were discussed on pp. 583 and 887, respectively.

The Oxygen Cycle. Figure 21.32 shows the main processes in the global oxygen cycle. The cycle is complicated by the fact that oxygen appears in so many different chemical forms. Atmospheric oxygen is removed through respiration and various industrial processes. The depletion of oxygen is intimately related to the production of carbon dioxide and water. Photosynthesis is the major mechanism by which molecular oxygen is regenerated from carbon dioxide. As in the carbon and nitrogen cycles, the overall balance of oxygen on Earth is ultimately tied to the energy supplied by the sun.

Sulfur

Although sulfur is not a very abundant element (it constitutes only about 0.06 percent of Earth's crust by mass), it is readily available because it occurs commonly in nature in the elemental form. The largest known reserves of sulfur are found in sedimentary

Figure 21.33 *Pyrite (FeS_2). It is commonly called "fool's gold" because it looks like gold.*

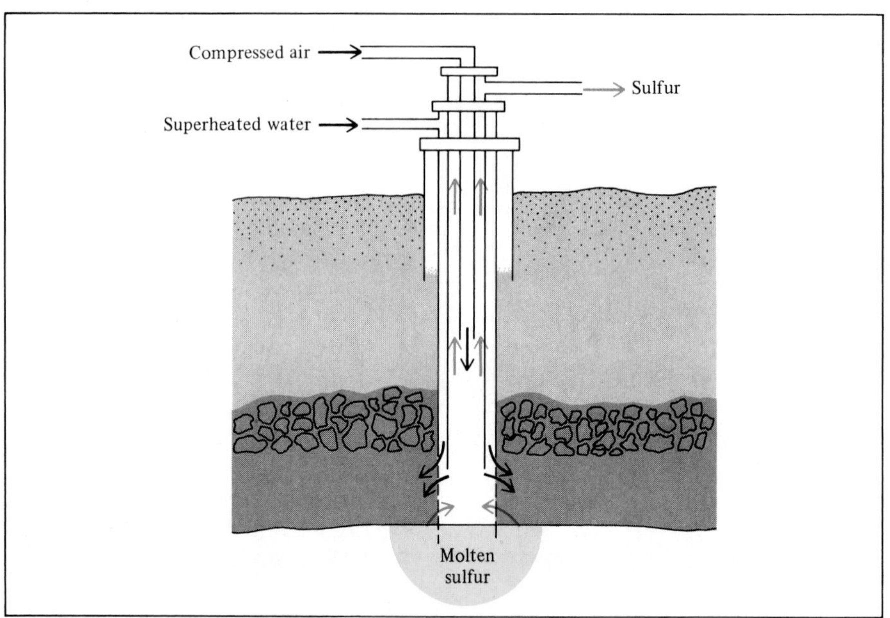

Figure 21.34 *The Frasch process. Three concentric pipes are inserted into a hole drilled down to the sulfur deposit. Superheated water is forced down the outer pipe into the sulfur, causing it to melt. Molten sulfur is then forced up the middle pipe by compressed air.*

Figure 21.35 *Steam injection in the Frasch process.*

deposits. In addition, sulfur occurs widely in gypsum ($CaSO_4 \cdot 2H_2O$) and various sulfide minerals such as pyrite (FeS_2) (Figure 21.33). Sulfur is also present in natural gas as H_2S, SO_2, and other sulfur-containing compounds.

Sulfur is extracted from underground deposits by the *Frasch† process,* shown in Figure 21.34. In this process, superheated water (liquid water heated to about 160°C under high pressure to prevent it from boiling) is pumped down the outermost pipe to melt the sulfur. Next, compressed air is forced down the innermost pipe. Liquid sulfur mixed with air forms an emulsion that is less dense than water and therefore rises to the surface as it is forced up the middle pipe (Figure 21.35). Sulfur produced in this

†Herman Frasch (1851–1914). German chemical engineer. Besides inventing the process for obtaining pure sulfur, for which he is most famous, Frasch also developed methods for refining petroleum.

manner, which amounts to about 10 million tons per year, has a purity of about 99.5 percent.

There are several allotropic forms of sulfur in existence, the most important being the rhombic and monoclinic forms. Rhombic sulfur is thermodynamically the most stable form; it has a puckered S_8 ring structure:

Because S is considerably larger than O (see Figure 8.7), it does not form S═S bonds in the elemental form.

It is a yellow, tasteless, and odorless solid (m.p. 112°C) (see Figure 8.22) that is insoluble in water but soluble in carbon disulfide. On heating it is slowly converted to monoclinic sulfur (m.p. 119°C), which also consists of the S_8 units. When liquid sulfur is heated above 150°C, the rings begin to break up and the entangling of the sulfur chains results in a sharp increase in the liquid's viscosity. Further heating tends to rupture the chains, and the viscosity begins to decrease.

Like nitrogen, sulfur shows a wide variation of oxidation numbers in its compounds (Table 21.3). The best-known hydrogen compound of sulfur is hydrogen sulfide, which is prepared by the action of an acid on a sulfide; for example

$$FeS(s) + H_2SO_4(aq) \longrightarrow FeSO_4(aq) + H_2S(g)$$

Table 21.3 Common compounds of sulfur

Oxidation number	Compound	Formula	Structure
−2	Hydrogen sulfide	H_2S	
0	Sulfur*	S_8	
+1	Disulfur dichloride	S_2Cl_2	
+2	Sulfur dichloride	SCl_2	
+4	Sulfur dioxide	SO_2	
+6	Sulfur trioxide	SO_3	

*We list the element here as a reference.

Nowadays, hydrogen sulfide used in qualitative analysis (see Section 17.6) is prepared by the hydrolysis of thioacetamide:

$$\underset{\text{thioacetamide}}{H_3C{-}C({=}S){-}NH_2} + 2H_2O + H^+ \longrightarrow \underset{\text{acetic acid}}{H_3C{-}C({=}O){-}O{-}H} + H_2S + NH_4^+$$

Hydrogen sulfide is a colorless gas (b.p. $-60.2°C$) with an offensive odor resembling that of rotten eggs. (The smell of rotten eggs actually does come from hydrogen sulfide, which is formed by the bacterial decomposition of sulfur-containing proteins.) Hydrogen sulfide is a highly toxic substance that, like hydrogen cyanide, attacks respiratory enzymes. It is a very weak diprotic acid (see Table 16.3). In basic solution, H_2S is a stronger reducing agent. For example, it is oxidized by permanganate to elemental sulfur:

$$3H_2S(aq) + 2MnO_4^-(aq) \longrightarrow 3S(s) + 2MnO_2(s) + 2H_2O(l) + 2OH^-(aq)$$

Oxides of Sulfur. Sulfur has two important oxides: sulfur dioxide, SO_2; and sulfur trioxide, SO_3. Sulfur dioxide is formed when sulfur burns in air:

$$S(s) + O_2(g) \longrightarrow SO_2(g)$$

In the laboratory, it can be prepared by the action of an acid on a sulfite; for example,

$$2HCl(aq) + Na_2SO_3(aq) \longrightarrow 2NaCl(aq) + H_2O(l) + SO_2(g)$$

or by the action of concentrated sulfuric acid on copper:

$$Cu(s) + 2H_2SO_4(aq) \longrightarrow CuSO_4(aq) + 2H_2O(l) + SO_2(g)$$

There is no evidence for the formation of sulfurous acid, H_2SO_3, in water.

Sulfur dioxide (b.p. $-10°C$) is a colorless gas with a pungent odor, and it is quite toxic. An acidic oxide, it reacts with water as follows:

$$SO_2(g) + H_2O(l) \rightleftharpoons H^+(aq) + HSO_3^-(aq)$$

Sulfur dioxide is slowly oxidized to sulfur trioxide, but the reaction rate can be greatly enhanced by a platinum or vanadium oxide catalyst (see Section 13.6):

$$2SO_2(g) + O_2(g) \longrightarrow 2SO_3(g)$$

Sulfur trioxide dissolves in water to form sulfuric acid:

$$SO_3(g) + H_2O(l) \longrightarrow H_2SO_4(aq)$$

The role of sulfur dioxide in acid rain is discussed on p. 659.

Sulfuric Acid. Sulfuric acid is the world's most important industrial chemical. Its preparation, properties, and uses have been discussed previously (see Section 13.6).

Sulfuric acid is a diprotic acid (see Table 16.3). It is a colorless, viscous liquid (m.p. 10.4°C). The concentrated sulfuric acid we use in the laboratory is 98 percent H_2SO_4 by mass (density: 1.84 g/cm^3), which corresponds to a concentration of 18 *M*. The oxidizing strength of sulfuric acid depends on its temperature and concentration. A cold, dilute sulfuric acid solution reacts with metals above hydrogen in the activity series (see Figure 3.15), thereby liberating molecular hydrogen in a displacement reaction:

$$Mg(s) + H_2SO_4(aq) \longrightarrow MgSO_4(aq) + H_2(g)$$

This is a typical reaction of an active metal with an acid. The strength of sulfuric acid as an oxidizing agent is greatly enhanced when it is both hot and concentrated. In such a solution, the oxidizing agent is actually the sulfate ion rather than the hydrated proton, $H^+(aq)$. Thus, copper reacts with concentrated sulfuric acid as follows:

$$Cu(s) + 2H_2SO_4(aq) \longrightarrow CuSO_4(aq) + SO_2(g) + 2H_2O(l)$$

Depending on the nature of the reducing agents, the sulfate ion may be further reduced to elemental sulfur or the sulfide ion. For example, reduction of H_2SO_4 by HI yields H_2S and I_2:

$$8HI(aq) + H_2SO_4(aq) \longrightarrow H_2S(aq) + 4I_2(s) + 4H_2O(l)$$

Concentrated sulfuric acid oxidizes nonmetals. For example, it oxidizes carbon to carbon dioxide and sulfur to sulfur dioxide:

$$C(s) + 2H_2SO_4(aq) \longrightarrow CO_2(g) + 2SO_2(g) + 2H_2O(l)$$

$$S(s) + 2H_2SO_4(aq) \longrightarrow 3SO_2(g) + 2H_2O(l)$$

Other Compounds of Sulfur. Carbon disulfide, a colorless, flammable liquid (b.p. 46°C), is formed by heating carbon and sulfur at a high temperature:

$$C(s) + 2S(l) \longrightarrow CS_2(l)$$

It is only slightly soluble in water. Carbon disulfide is a good solvent for sulfur, phosphorus, iodine, and other nonpolar substances, such as waxes and rubber.

Another interesting compound of sulfur is sulfur hexafluoride, SF_6, which is prepared by heating sulfur in an atmosphere of fluorine:

$$S(l) + 3F_2(g) \longrightarrow SF_6(g)$$

Sulfur hexafluoride is a nontoxic, colorless gas (b.p. −63.8°C). It is the most inert of all sulfur compounds; it resists attack even by molten KOH. The structure and bonding of SF_6 were discussed in Chapters 9 and 10 and its critical phenomenon illustrated in Chapter 11 (see Figure 11.45).

CHEMISTRY IN ACTION

Volcanoes

Volcanic eruptions, Earth's most spectacular displays of energy, are instrumental in forming large parts of Earth's crust. Earth's upper mantle, under the outer shell, is nearly molten. A slight increase in heat, such as that generated by the movement of one plate under another, will melt the rock. The molten rock, called *magma,* rises to the surface and is responsible for some types of volcanic eruptions.

Volcanoes are also a major source of atmospheric pollutants. An active volcano emits gases, liquids, and solids (Figure 21.36). The gases are mainly N_2, CO_2, HCl, HF, H_2S, and water vapor. It is the sulfur-containing compounds that are of interest here.

It is estimated that volcanoes are the source of about two-thirds of the sulfur emitted into the atmosphere. On the slopes of Mount St. Helens, which last erupted

(a) (b)

Figure 21.36 *(a) A volcanic eruption on the island of Hawaii. (b) The eruption of Mount St. Helens in 1980. An active volcano emits gases, liquids, and solid particles.*

in 1980, deposits of elemental sulfur are visible near the eruption site. At high temperatures the hydrogen sulfide gas emitted by a volcano is oxidized by air:

$$2H_2S(g) + 3O_2(g) \longrightarrow 2SO_2(g) + 2H_2O(g)$$

The SO_2 is reduced by more H_2S from the volcano to produce elemental sulfur:

$$2H_2S(g) + SO_2(g) \longrightarrow 3S(s) + 2H_2O(l)$$

Volcanic eruptions contribute to the acid rain phenomenon (see p. 659), and they also have a long-term effect on weather conditions. The tremendous force of the eruption carries a sizable amount of gas into the stratosphere (see Figure 13.27), where SO_2 is oxidized to SO_3. When the SO_3 reacts with water vapor, it forms sulfuric acid *aerosols*. (An aerosol is a gaseous suspension of fine solid or liquid particles.) Because the stratosphere is above the atmospheric weather patterns, it is very stable, and aerosol clouds can often persist for more than a year. Sulfuric acid aerosols absorb part of solar radiation, and the net effect is a drop in temperature at Earth's surface. However, this cooling effect is local rather than global, since it depends on the location and frequency of volcanic eruptions.

21.7 The Halogens

The halogens—fluorine, chlorine, bromine, and iodine—are reactive nonmetals (see Figure 8.23). Table 21.4 lists some of the properties of these elements. The element astatine also belongs to the Group 7A family. However, all isotopes of astatine are radioactive; its longest-lived isotope is astatine-210, which has a half-life of 8.3 hours. Therefore it is both difficult and expensive to study astatine in the laboratory.

The halogens form a very large number of compounds. In the elemental state they form diatomic molecules, X_2. In nature, however, because of their high reactivity, halogens are always found combined with other elements. Chlorine, bromine, and iodine occur as halides in seawater, and fluorine occurs in the minerals fluorite (CaF_2) (see Figure 20.20) and cryolite (Na_3AlF_6).

Preparations and General Properties of the Halogens

Because fluorine and chlorine are strong oxidizing agents, they must be prepared by

Table 21.4 Properties of the halogens

Property	F	Cl	Br	I
Valence electron configuration	$2s^22p^5$	$3s^23p^5$	$4s^24p^5$	$5s^25p^5$
Melting point (°C)*	−223	−102	−7	114
Boiling point (°C)*	−187	−35	59	183
Appearance*	Pale yellow gas	Yellow-green gas	Red-brown liquid	Dark violet vapor Dark metallic-looking solid
Atomic radius (pm)	72	99	114	133
Ionic radius (pm)†	136	181	195	216
Ionization energy (kJ/mol)	1680	1251	1139	1003
Electronegativity	4.0	3.0	2.8	2.5
Standard reduction potential (V)*	2.87	1.36	1.07	0.53
Bond energy (kJ/mol)*	150.6	242.7	192.5	151.0

*These values and descriptions apply to the diatomic species X_2, where X represents a halogen atom. The half-reaction is: $X_2(g) + 2e^- \rightarrow 2X^-(aq)$.

†Refers to the anion X^-.

electrolytic rather than chemical oxidation of the fluoride and chloride ions. Electrolysis does not work for aqueous solutions of fluorides, however, because fluorine is a stronger oxidizing agent than oxygen. From Table 19.1 we find that

$$O_2(g) + 4H^+(aq) + 4e^- \longrightarrow 2H_2O(l) \qquad \mathscr{E}^\circ = 1.23\text{ V}$$

$$F_2(g) + 2e^- \longrightarrow 2F^-(aq) \qquad \mathscr{E}^\circ = 2.87\text{ V}$$

If F_2 were formed by the electrolysis of an aqueous fluoride solution, it would immediately oxidize water to oxygen. For this reason, fluorine is prepared by electrolyzing liquid hydrogen fluoride containing potassium fluoride to increase its conductivity, at about 70°C (Figure 21.37):

$$\begin{array}{ll} \text{Anode:} & 2F^- \longrightarrow F_2(g) + 2e^- \\ \text{Cathode:} & 2H^+ + 2e^- \longrightarrow H_2(g) \\ \hline \text{Overall reaction:} & 2HF(l) \longrightarrow H_2(g) + F_2(g) \end{array}$$

Chlorine gas, Cl_2, is prepared industrially by the electrolysis of molten NaCl (see Section 19.8) or by *the electrolysis of a concentrated aqueous NaCl solution* (called brine). The latter process is called the ***chlor-alkali process*** (where *chlor* denotes chlorine and *alkali* denotes an alkali metal, such as sodium). Two of the common cells used in the chlor-alkali process are the mercury cell and the diaphragm cell. In both cells the overall reaction is

$$2NaCl(aq) + 2H_2O(l) \xrightarrow{\text{electrolysis}} 2NaOH(aq) + H_2(g) + Cl_2(g)$$

As you can see, this reaction produces two useful by-products, NaOH and H_2. The cells are designed to separate the molecular chlorine from the sodium hydroxide solution and the molecular hydrogen that are generated. Otherwise, side reactions such as

$$2NaOH(aq) + Cl_2(g) \longrightarrow NaOCl(aq) + NaCl(aq) + H_2O(l)$$

$$H_2(g) + Cl_2(g) \longrightarrow 2HCl(g)$$

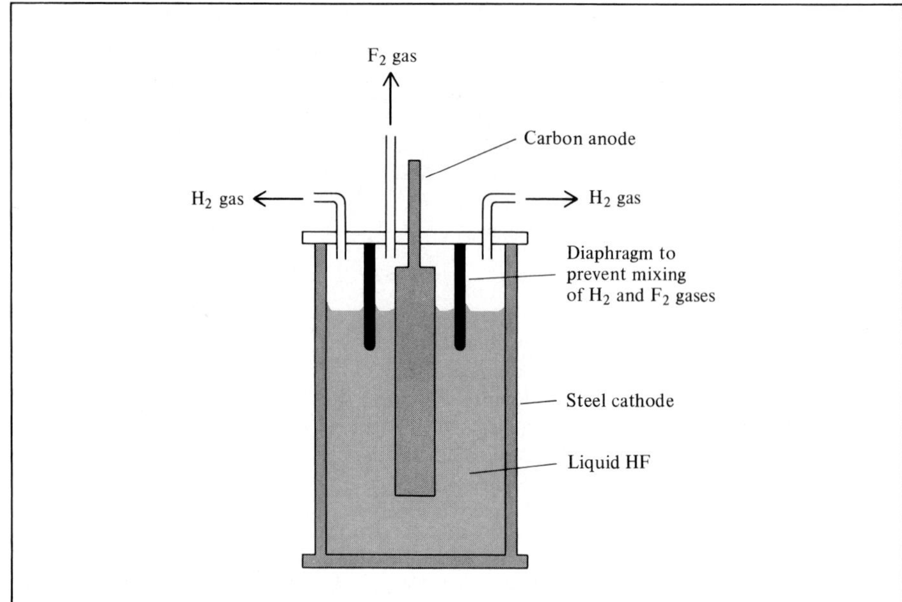

Figure 21.37 *Electrolytic cell for the preparation of fluorine gas. Note that because H_2 and F_2 form an explosive mixture, these gases must be separated from each other.*

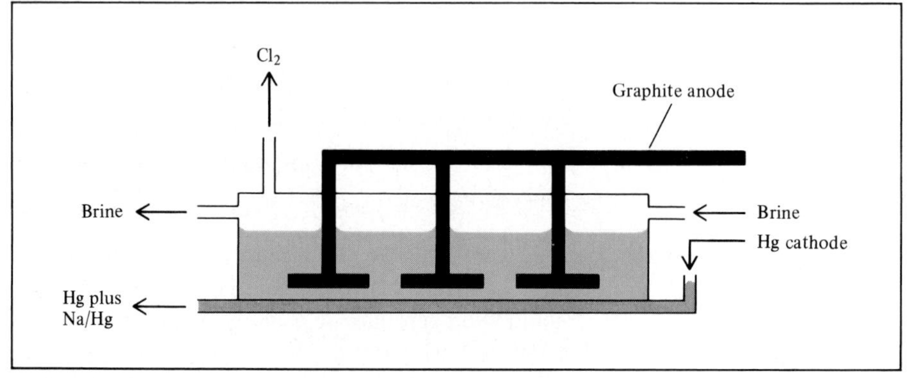

Figure 21.38 *Mercury cell used in the chlor-alkali process. The cathode contains mercury. The sodium-mercury amalgam is treated with water outside the cell to produce sodium hydroxide and hydrogen gas.*

would occur. These reactions consume the desired products and can be dangerous because a mixture of H_2 and Cl_2 is explosive.

Figure 21.38 shows the mercury cell used in the chlor-alkali process. The cathode is a liquid mercury pool at the bottom of the cell, and the anode is made of either graphite or titanium coated with platinum. Brine is continuously passed through the cell as shown. The electrode reactions are

$$\begin{array}{lrcl} \text{Anode:} & 2\text{Cl}^-(aq) & \longrightarrow & \text{Cl}_2(g) + 2e^- \\ \text{Cathode:} & 2\text{Na}^+(aq) + 2e^- & \xrightarrow{\text{Hg}(l)} & 2\text{Na/Hg} \\ \hline \text{Overall reaction:} & 2\text{NaCl}(aq) & \longrightarrow & 2\text{Na/Hg} + \text{Cl}_2(g) \end{array}$$

where Na/Hg denotes the formation of sodium amalgam. (An *amalgam* is a substance made by combining mercury with another metal or metals.) The chlorine gas generated this way is very pure. The sodium amalgam does not react with the brine solution but decomposes as follows when treated with pure water outside the cell:

$$2\text{Na/Hg} + 2\text{H}_2\text{O}(l) \longrightarrow 2\text{NaOH}(aq) + \text{H}_2(g) + 2\text{Hg}(l)$$

Again, the by-products of the chlor-alkali process are sodium hydroxide and hydrogen gas. Although the mercury is cycled back into the cell for reuse, some of it is always discharged with waste solutions into the environment, resulting in mercury pollution (see p. 854). This is a major drawback of the mercury cell. Figure 21.39 shows the industrial manufacture of chlorine gas.

The half-cell reactions in a diaphragm cell are shown in Figure 21.40. The asbestos diaphragm is permeable to the ions but not to the hydrogen and chlorine gases and so prevents the gases from mixing. During electrolysis a positive pressure is applied on the side of the anode compartment to prevent the migration of the OH^- ions from the cathode compartment. Periodically fresh brine solution is added to the cell and the sodium hydroxide solution is run off as shown. The diaphragm cell presents no pollution problems. Its main disadvantage is that the sodium hydroxide solution is contaminated with unreacted sodium chloride.

From Table 19.1 we see that the oxidizing strength decreases from Cl_2 to Br_2 to I_2.

The preparation of molecular bromine and iodine from seawater by oxidation with chlorine was discussed in Section 3.5. In the laboratory, chlorine, bromine, and iodine can be prepared by heating the alkali halides (NaCl, KBr, or KI) in concentrated sulfuric acid in the presence of manganese(IV) oxide. A representative reaction is

$$MnO_2(s) + 2H_2SO_4(aq) + 2NaCl(aq) \longrightarrow MnSO_4(aq) + Na_2SO_4(aq) + 2H_2O(l) + Cl_2(g)$$

Figure 21.39 *The industrial manufacture of chlorine gas.*

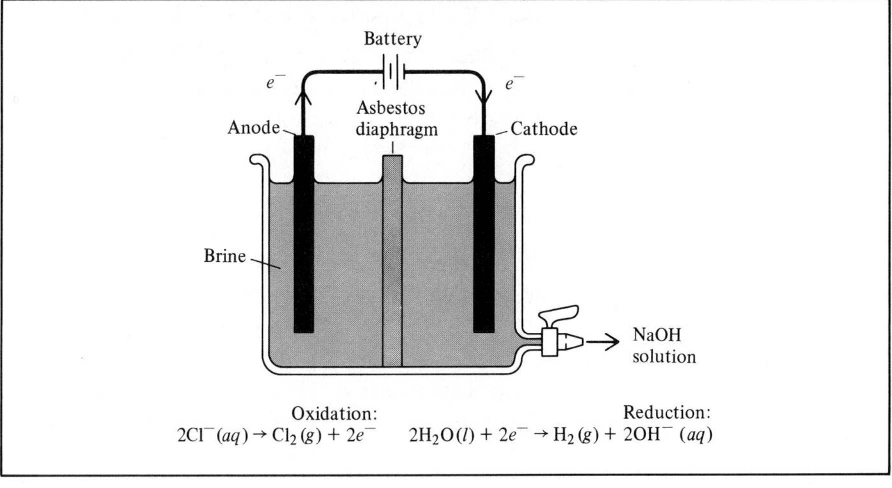

Figure 21.40 *Diaphragm cell used in the chlor-alkali process.*

Although all halogens are highly reactive and toxic, the magnitude of reactivity and toxicity generally decreases from fluorine to iodine. Most of the halides can be classified into two categories. The fluorides and chlorides of many metallic elements, especially those belonging to the alkali metal and alkaline earth metal (except beryllium) families, are ionic compounds. Most of the halides of nonmetals such as sulfur and phosphorus are covalent compounds. As Figure 3.9 shows, the oxidation numbers of the halogens can vary from -1 to $+7$. The only exception is fluorine. Because it is the most electronegative element, it can have only two oxidation numbers, 0 (as in F_2) and -1, in its compounds.

Recall that the first member of a group usually differs in properties from the rest of the members of the group (see p. 331).

The chemistry of fluorine differs in some ways from that of the rest of the halogens. The following are some of the differences:

- Fluorine is the most reactive of all the halogens. The difference in reactivity between fluorine and chlorine is greater than that between chlorine and bromine. Table 21.4 shows that the F—F bond is considerably weaker than the Cl—Cl bond. The weak bond in F_2 can be explained in terms of the lone pairs on the F atoms:

$$:\ddot{\underset{..}{F}}-\ddot{\underset{..}{F}}:$$

The small size of the F atoms (see Table 21.4) allows a close approach of the three lone pairs on each of the F atoms, resulting in a greater repulsion than that found in Cl_2, which consists of larger atoms.
- Hydrogen fluoride, HF, has a high boiling point (19.5°C) as a result of strong intermolecular hydrogen bonding, whereas all other hydrogen halides have much lower boiling points (see Figure 11.7).
- Hydrofluoric acid is a weak acid, whereas all other hydrohalic acids (HCl, HBr, and HI) are strong acids.
- Fluorine reacts with cold sodium hydroxide solution to produce oxygen difluoride as follows:

$$2F_2(g) + 2NaOH(aq) \longrightarrow 2NaF(aq) + H_2O(l) + OF_2(g)$$

The same reaction with chlorine or bromine, on the other hand, produces a halide and a hypohalite:

$$X_2(g) + 2NaOH(aq) \longrightarrow NaX(aq) + NaOX(aq) + H_2O(l)$$

where X stands for Cl or Br. Iodine does not react under the same conditions.
- Silver fluoride, AgF, is soluble. All other silver halides (AgCl, AgBr, and AgI) are insoluble (see solubility rules in Section 3.3).

The Hydrogen Halides

The hydrogen halides, an important class of halogen compounds, can be formed by the direct combination of the elements:

$$H_2(g) + X_2(g) \rightleftharpoons 2HX(g)$$

where X denotes a halogen atom. These reactions (especially those involving F_2 and Cl_2) can occur with explosive violence. Industrially hydrogen chloride is prepared as a by-product in the manufacture of chlorinated hydrocarbons:

$$C_2H_6(g) + Cl_2(g) \longrightarrow C_2H_5Cl(g) + HCl(g)$$

In the laboratory, hydrogen fluoride and hydrogen chloride can be prepared by reacting the metal halides with concentrated sulfuric acid:

$$CaF_2(s) + H_2SO_4(aq) \longrightarrow 2HF(g) + CaSO_4(s)$$

$$2NaCl(s) + H_2SO_4(aq) \longrightarrow 2HCl(g) + Na_2SO_4(aq)$$

Hydrogen bromide and hydrogen iodide cannot be prepared this way because they are oxidized to elemental bromine and iodine. For example, the reaction between NaBr and H_2SO_4 is

$$2NaBr(s) + 2H_2SO_4(aq) \longrightarrow Br_2(l) + SO_2(g) + Na_2SO_4(aq) + 2H_2O(l)$$

Instead, hydrogen bromide is prepared by first reacting bromine with phosphorus to form phosphorus tribromide:

$$P_4(s) + 6Br_2(l) \longrightarrow 4PBr_3(l)$$

Next, PBr_3 is treated with water to yield HBr:

$$PBr_3(l) + 3H_2O(l) \longrightarrow 3HBr(g) + H_3PO_3(aq)$$

Hydrogen iodide can be prepared in the same manner.

The high reactivity of HF is demonstrated by the fact that it attacks silica and silicates:

$$6HF(aq) + SiO_2(g) \longrightarrow H_2SiF_6(aq) + 2H_2O(l)$$

This property makes HF suitable for etching glass and is the reason that hydrogen fluoride must be kept in plastic or inert metal (for example, Pt) containers. Hydrogen fluoride is used in the manufacture of Freons (see Chapter 13), for example:

$$CCl_4(l) + HF(g) \longrightarrow CFCl_3(g) + HCl(g)$$

$$CFCl_3(g) + HF(g) \longrightarrow CF_2Cl_2(g) + HCl(g)$$

and in the production of aluminum (see Section 20.7). Hydrogen chloride is used in the preparation of hydrochloric acid, inorganic chlorides, and various metallurgical processes. Hydrogen bromide and hydrogen iodide do not have any major uses.

Aqueous solutions of hydrogen halides are acidic. As mentioned earlier, the strength of the acids increases as follows:

$$HF \ll HCl < HBr < HI$$

Oxoacids of the Halogens

The halogens also form a series of oxoacids with the following general formulas:

HOX	HXO_2	HXO_3	HXO_4
hypohalous acid	halous acid	halic acid	perhalic acid

Chlorous acid, $HClO_2$, is the only known halous acid. All the halogens except fluorine form halic and perhalic acids. The Lewis structures of the chlorine oxoacids are

```
                                               ..
                                              :O:
        .. ..          .. .. ..       .. .. ..     .. .. ..
  H:O:Cl:        H:O:Cl:O:      H:O:Cl:O:   H:O:Cl:O:
    .. ..          .. .. ..       .. .. ..     .. .. ..
                                    :O:          :O:
                                     ..           ..
```

hypochlorous acid	chlorous acid	chloric acid	perchloric acid

Table 21.5 Common compounds of halogens*

Compound	F	Cl	Br	I
Hydrogen halide	HF (−1)	HCl (−1)	HBr (−1)	HI (−1)
Oxides	OF_2 (−1)	Cl_2O (+1)	Br_2O (+1)	I_2O_5 (+5)
		ClO_2 (+4)	BrO_2 (+4)	
		Cl_2O_7 (+7)		
Oxoacids	HOF (−1)	HOCl (+1)	HOBr (+1)	HOI (+1)
		$HClO_2$ (+3)		
		$HClO_3$ (+5)	$HBrO_3$ (+5)	HIO_3 (+5)
		$HClO_4$ (+7)	$HBrO_4$ (+7)	H_5IO_6 (+7)

*The number in parentheses indicates the oxidation number of the halogen.

For a given halogen, the acid strength decreases from perhalic acid to hypohalous acid; the origin of this trend is discussed in Section 15.4.

Table 21.5 lists some of the halogen compounds. Note that periodic acid, HIO_4, does not appear. This is because it cannot be isolated in the pure form. Instead, a substance with the formula H_5IO_6 is often used to represent periodic acid.

Interhalogen Compounds

Because the halogens are very reactive elements, it is not surprising that they also form binary compounds among themselves. ***Interhalogen compounds,*** or *compounds formed between two different halogen elements,* have the formulas

$$XX' \qquad XX'_3 \qquad XX'_5 \qquad XX'_7$$

where X and X′ are two different halogens and X is the larger atom of the two. Many of these compounds can be prepared by direct combination:

$$Cl_2(g) + F_2(g) \longrightarrow 2ClF(g)$$

$$Cl_2(g) + 3F_2(g) \longrightarrow 2ClF_3(g)$$

$$Br_2(l) + 3F_2(g) \longrightarrow 2BrF_3(l)$$

Others require more indirect routes:

$$KCl(s) + 3F_2(g) \longrightarrow KF(s) + ClF_5(g)$$

$$KI(s) + 4F_2(g) \longrightarrow KF(s) + IF_7(g)$$

Many of the interhalogen compounds are unstable and react violently with water. Note that all the interhalogen molecules with the exception of the XX′ type violate the octet rule. The geometry of these compounds can be predicted by using the VSEPR model (Figure 21.41).

Uses of the Halogens

Fluorine. The halogens and their compounds find many applications in industry, health care, and other areas. One is fluoridation, the practice of adding small quantities of fluorides (about 1 ppm by mass) such as NaF to drinking water to reduce dental caries (see p. 734).

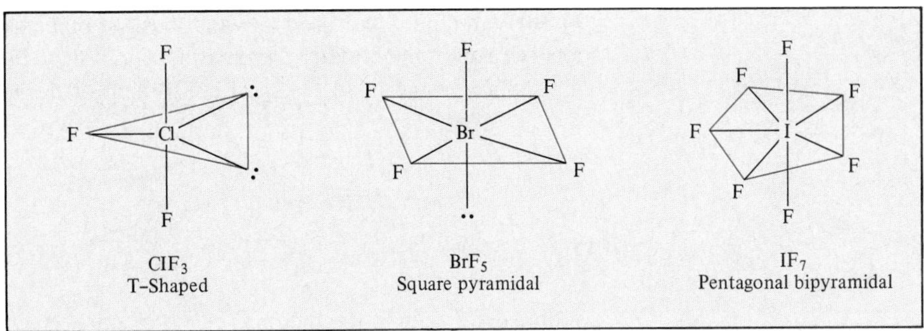

Figure 21.41 *Structures of some interhalogen compounds.*

One of the most important inorganic fluorides is uranium hexafluoride, UF_6, which is essential to the gaseous effusion process for separating isotopes of uranium (U-235 and U-238). Industrially, fluorine is used to produce polytetrafluoroethylene, a polymer better known as Teflon:

$$-\!\!(CF_2-CF_2)\!\!-_n$$

where n is a large number. Teflon is used in electrical insulators, high-temperature plastics, cooking utensils, and so on.

Chlorine. Chlorine plays an important biological role in the human body, because the chloride ion is the principal anion in intracellular and extracellular fluids. Chlorine is widely used as an industrial bleaching agent for paper and textiles. Ordinary household laundry bleach contains the active ingredient sodium hypochlorite (about 5 percent by mass), which is prepared by reacting chlorine gas with a cold solution of sodium hydroxide:

$$Cl_2(g) + 2NaOH(aq) \longrightarrow NaCl(aq) + NaOCl(aq) + H_2O(l)$$

Chlorine is also used to purify water and disinfect swimming pools. When chlorine dissolves in water, it undergoes the following reaction:

$$Cl_2(g) + H_2O(l) \longrightarrow HCl(aq) + HOCl(aq)$$

It is thought that the OCl^- ions destroy bacteria by oxidizing life-sustaining compounds within them.

Chlorinated methanes, such as carbon tetrachloride and chloroform, are useful organic solvents. Large quantities of chlorine are used to produce insecticides, such as DDT. However, in view of the damage they inflict on the environment, the use of many of these compounds is either totally banned or greatly restricted in the United States. Chlorine is also used to produce polymers such as poly(vinyl chloride).

Bromine. So far as we know, bromine compounds occur naturally only in some marine organisms. Seawater is about 1×10^{-3} M Br^-; therefore, it is the main source of bromine. Bromine is used to prepare ethylene dibromide ($BrCH_2CH_2Br$), which is used as an insecticide and as a scavenger for lead (that is, to combine with lead) in gasoline to keep lead from depositing in engines. Recent studies have shown that ethylene dibromide is a very potent carcinogen.

Bromine combines directly with silver to form silver bromide (AgBr), which is used in photographic films (see p. 120).

Iodine. Iodine is not used as widely as the other halogens. A 50 percent (by mass) alcohol solution of iodine, known as *tincture of iodine,* is used medicinally as an antiseptic. Iodine is an essential constituent of the thyroid hormone thyroxine:

$$\text{HO}-\text{C}_6\text{H}_2\text{I}_2-\text{O}-\text{C}_6\text{H}_2\text{I}_2-\text{CH}_2-\underset{\text{NH}_2}{\overset{\text{H}}{\text{C}}}-\text{C}\overset{\text{O}}{\underset{\text{OH}}{}}$$

Iodine deficiency in the diet may result in enlargement of the thyroid gland (known as goiter). Iodized table salt sold in the United States usually contains 0.01 percent KI or NaI, which is more than sufficient to satisfy the 1 mg of iodine per week required for the formation of thyroxine in the human body.

A compound of iodine that deserves mention is silver iodide, AgI. It is a pale yellow solid that darkens when exposed to light. In this respect it is similar to silver bromide. In Chapter 11 we saw that dry ice is used in cloud seeding (see p. 486). Silver iodide, too, can be used in this process. The advantage of using silver iodide is that enormous numbers of nuclei (that is, small particles on which ice crystals can form) become available. About 10^{15} nuclei are produced from 1 g of AgI by vaporizing an acetone solution of silver iodide in a hot flame. The nuclei are then dispersed into the clouds from an airplane (Figure 21.42).

21.8 The Noble Gases

We come now to the Group 8A elements (sometimes known as the Group 0 elements): helium, neon, argon, krypton, xenon, and radon (see Figure 8.24). Because their outer *s* and *p* subshells are completely filled, these elements are chemically unreactive. Unlike the other nonmetallic gases, they occur naturally in the atomic state. As we will see later, not all these elements are totally inert. Table 21.6 lists some of their common properties.

Figure 21.42 *Cloud seeding using AgI particles.*

Table 21.6 Properties of noble gases

Property	He	Ne	Ar	Kr	Xe
Valence electron configuration	$1s^2$	$2s^22p^6$	$3s^23p^6$	$4s^24p^6$	$5s^25p^6$
Atmospheric abundance (% by volume)	5×10^{-4}	0.0012	0.94	1.1×10^{-4}	9×10^{-6}
Melting point (°C)	−272.3	−248.7	−189.2	−157.1	−111.9
Boiling point (°C)	−268.9	−245.9	−185.7	−152.9	−106.9
First ionization energy (kJ/mol)	2373	2080	1521	1241	1167
Examples of compounds	None	None	None	KrF_2	XeF_4, XeO_4

Helium

Helium was detected in the sun by spectroscopic measurements of the emission spectrum before it was definitely identified on Earth. In the interior of the sun, helium is produced by the fusion of hydrogen atoms at temperatures exceeding 15 million degrees Celsius. Terrestrial helium results from alpha particles emitted by the radioactive decay of elements such as uranium and thorium in Earth's crust (such nuclear reactions are discussed in Chapter 23):

$$^{238}_{92}\text{U} \longrightarrow {}^{234}_{90}\text{Th} + {}^{4}_{2}\text{He}$$

$$^{230}_{90}\text{Th} \longrightarrow {}^{226}_{88}\text{Ra} + {}^{4}_{2}\text{He}$$

Helium nuclei (He^{2+}) readily pick up two electrons from the surroundings and become neutral helium atoms. The helium gas so formed mixes with natural gas and is found in concentrations ranging from 0.5 to 2.4 percent by volume. It can be separated from other gases by applying enough pressure at low temperature to liquefy all other gases, such as nitrogen, oxygen, and hydrocarbons. In Table 21.6 we see that helium has a very low boiling point of −268.9°C (4.2 K). In fact, it is the most difficult gas to condense because the dispersion forces between the helium atoms are so weak.

No stable compounds of helium have ever been prepared; hence there is no chemistry of helium to speak of. Helium itself, however, is a very interesting and important substance. The earliest practical use of helium was to replace hydrogen as a lifting gas in balloons. It has about 90 percent of the buoyancy of hydrogen, but it is not flammable and is nontoxic.

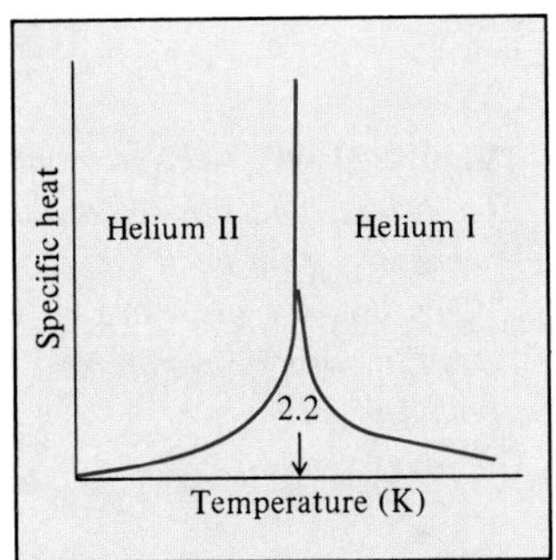

Figure 21.43 *The lambda point of liquid helium. The choice of the name is based on the similarity between the curves and the Greek letter lambda (λ).*

Most helium produced today is used in the space industry to pressurize rocket fuels, as a coolant, and as a substance to dilute oxygen gas in spacecraft atmospheres. In this role it has two advantages over nitrogen: It is lighter, and because it is less soluble in blood, it does not induce any unusual physiological effects in humans. The latter property makes helium suitable to be added to oxygen as a breathing mixture for divers.

Liquid helium has some unique thermodynamic properties. Figure 21.43 shows the specific heat of liquid helium as a function of temperature. When the liquid is cooled to −271.0°C (2.2 K), there is a very sharp transition, and the normal liquid helium (called helium I) becomes a superfluid (called helium II). As a superfluid, helium II is almost a perfect heat conductor. It also has nearly zero viscosity. This property is dramatically demonstrated by partly immersing a precooled beaker in helium II and observing the spontaneous climb of the liquid (as a thin layer) up the outer surface of the beaker and down into the beaker, until the levels inside and outside are equal.

Neon and Argon

Both neon and argon have important practical uses. Neon is used in luminescent electric lighting, or "neon lights." Liquid neon is used in low-temperature research. Argon is used mainly in electric light bulbs to provide an inert atmosphere. Because of its inert nature, argon is also employed to flush dissolved oxygen gas from molten metals in metallurgical processes.

Krypton and Xenon

The Group 8A elements were called the inert gases until 1963, and rightly so. No one had ever prepared a compound containing any of these elements. But an experiment carried out by Neil Bartlett† in that year shattered chemists' long-held views of these elements. Bartlett at that time was investigating the great oxidizing power of platinum hexafluoride, PtF_6. He noticed that when this compound came into contact with oxygen, it changed color and yielded an ionic substance having the formula $O_2^+PtF_6^-$. Since the ionization energy of O_2 (1176 kJ/mol) is very close to that of xenon (1167 kJ/mol), Bartlett reasoned that xenon should also be oxidized by PtF_6. He was correct. The reaction is rapid and visibly dramatic (Figure 21.44):

It has since been found that the product of this reaction is more complex than shown. However, this does not diminish Bartlett's achievement of making the first noble gas compound.

$$\underset{\text{}}{Xe(g)} + \underset{\text{red}}{PtF_6(g)} \longrightarrow \underset{\text{yellow orange}}{Xe^+PtF_6^-(s)}$$

The report of this reaction triggered tremendous interest and excitement. Only a few months later, scientists at the Argonne National Laboratories succeeded in preparing binary xenon fluorides. By heating xenon and fluorine (at a 1:5 ratio) in a nickel container at 400°C and 6 atm for 1 hour, they obtained xenon tetrafluoride, XeF_4 (Figure 21.45). Two other fluorides, XeF_2 and XeF_6, were also synthesized. A number

†Neil Bartlett (1932–). English chemist. Bartlett's work is mainly in the preparation and study of compounds of unusual oxidation numbers and in solid-state chemistry.

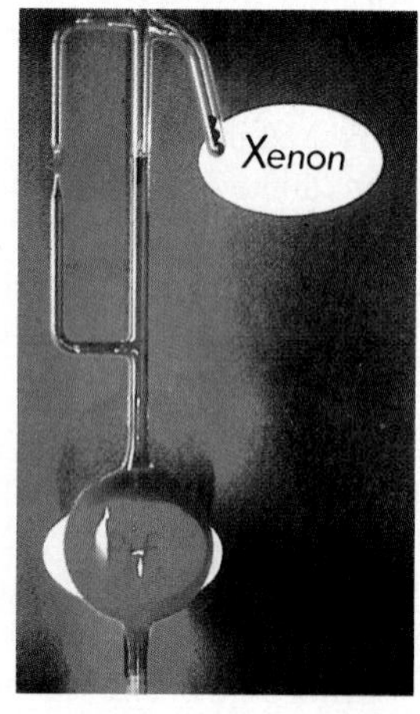

(a)

(b)

Figure 21.44 *(a) Xenon gas (colorless) and PtF_6 (red gas) separated from each other. (b) When the two gases are mixed, a yellow orange solid compound is formed.*

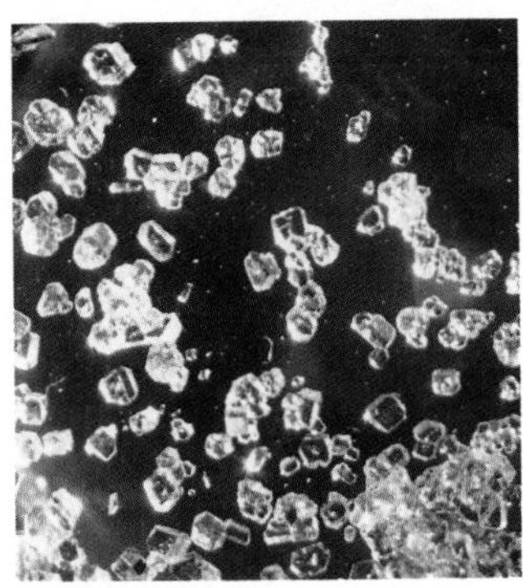

Figure 21.45 *Crystals of xenon tetrafluoride.*

of xenon-oxygen compounds (XeO_3, XeO_4) and ternary compounds containing Xe, F, and O ($XeOF_4$, XeO_3F_2) have also been prepared. The success with xenon suggested that other noble gases might undergo similar reactions. Indeed, a few compounds of krypton (KrF_2, for example) have since been prepared, but helium, argon, and neon remain "confirmed bachelors" to date. That this is so should not be surprising to us. As Table 21.6 shows, the ionization energies of the noble gases decrease from helium to xenon; therefore, the two elements on the far right, Kr and Xe, can be oxidized more easily than the others.

The structure and properties of noble gas compounds have been thoroughly investigated by many physical and chemical techniques. Today we know more about the chemistry of xenon than about some of the more abundant elements in spite of the fact that its compounds do not appear to have any commercial applications, and they are certainly *not* involved in natural biological processes.

At first it was thought that a unique type of bonding might be involved in the noble gas compounds. This turned out to be incorrect. For the most part the geometry of these compounds can be accounted for satisfactorily by the VSEPR method outlined in Chapter 10. Figure 21.46 shows the structure of some Xe compounds as predicted by the VSEPR method. Unlike most other hexafluorides, such as SF_6, PtF_6, and UF_6, the XeF_6 molecule does not have an octahedral structure. In xenon hexafluoride, the Xe atom has 14 valence electrons (8 from xenon plus 1 from each of the F atoms). Thus, the structure of XeF_6 must be based on the preferred arrangement of a total of seven electron pairs, of which one is a nonbonding pair. Xenon has limited applications in medicine as an anesthetic.

Radon

All isotopes of radon are radioactive; the longest-lived isotope is radon-222, with a half-life of 3.8 days. Therefore, it is rather difficult to study radon in the laboratory.

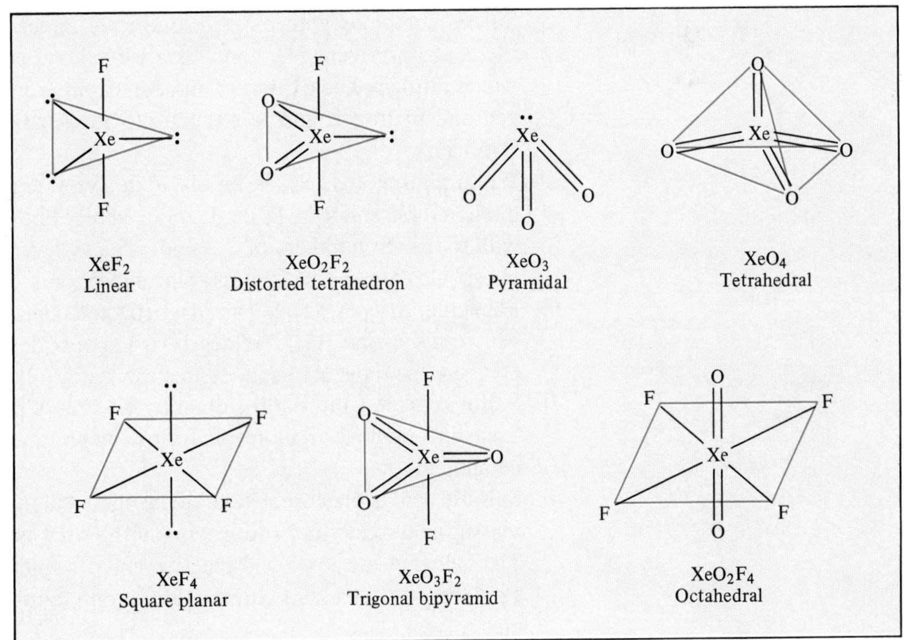

Figure 21.46 *The structures of xenon compounds predicted by the VSEPR method. The exact structure of XeF_6 (not shown here) has not been determined, although it is thought to be a distorted octahedron. Note that none of the compounds obeys the octet rule.*

There is some evidence that radon forms a difluoride (RnF_2) and some complexes.

The uranium decay series will be discussed in Chapter 23.

Radon is very much in the news these days because it has been identified as an indoor air pollutant. Radon is formed as a product in the uranium-238 decay series. Being an alpha particle emitter, radon produces radioactive polonium, which in turn decays to other radioactive isotopes. It was once thought that radon was only released over mining sites (particularly uranium and phosphate mines, which have high concentrations of uranium). However, we now know that radon is released from the soil in various amounts all over the country. Because it is largely unreactive, once formed radon is free to diffuse upward and enter buildings through the cracks of the foundation. Radon in the air is breathed in and exhaled. But its decay products are solids, and a fraction of these can deposit on the airways' surfaces. The long-term, cumulative effect of radiation can lead to lung cancer. It is estimated that about 10,000 deaths annually may be attributed to radon. The only solution to the ''radon pollution'' is to regularly ventillate the indoor air.

Summary

1. Hydrogen atoms contain one proton and one electron. They are the simplest atoms. Hydrogen combines with many metals and nonmetals to form hydrides; some hydrides are ionic and some are covalent.
2. There are three isotopes of hydrogen: ^1_1H, ^2_1H (deuterium), and ^3_1H (tritium). Heavy water contains deuterium.
3. Boron, a metalloid, forms unusual three-center bonds in its binary compounds with hydrogen, the boranes.
4. The important inorganic compounds of carbon are the carbides; the cyanides, most of which are extremely toxic; carbon monoxide, also toxic and a major air pollutant; the carbonates and bicarbonates; and carbon dioxide, an end product of metabolism and a component in the global carbon cycle.
5. Silicon combines with oxygen in silicate minerals and in silica (SiO_2).
6. Elemental nitrogen, N_2, contains a triple bond and is very stable. Compounds in which nitrogen has oxidation numbers from -3 to $+5$ are formed between nitrogen and hydrogen and/or oxygen atoms. Ammonia, NH_3, is widely used as a fertilizer.
7. White phosphorus, P_4, is highly toxic, very reactive, and flammable; the polymeric red phosphorus, $(P_4)_n$, is more stable. Phosphorus forms oxides and halides with oxidation numbers of $+3$ and $+5$ and several oxygen-containing acids. The phosphates are the most important phosphorus compounds.
8. Elemental oxygen, O_2, is paramagnetic and contains two unpaired electrons. Oxygen forms ozone (O_3), oxides (O^{2-}), peroxides (O_2^{2-}), and superoxides (O_2^-). Oxygen, the most abundant element in Earth's crust, is essential for life on Earth.
9. Sulfur is taken from Earth's crust by the Frasch process as a molten liquid. Sulfur exists in a number of allotropic forms and has a variety of oxidation numbers in its compounds.
10. Sulfuric acid is the cornerstone of the chemical industry. It is produced from sulfur via sulfur dioxide and sulfur trioxide in what is called the contact process.
11. The halogens are toxic and reactive and are found only in compounds with other elements. Fluorine and chlorine are strong oxidizing agents and are prepared by electrolysis.

12. The reactivity, toxicity, and oxidizing ability of the halogens decrease from fluorine to iodine. The halogens all form binary acids (HX) and series of oxoacids.
13. The noble gases are quite unreactive and are present in the atmosphere in atomic form. Compounds of xenon and krypton have, quite unexpectedly, been prepared in recent years.

Key Words

Borane, p. 869
Carbide, p. 872
Catenation, p. 882
Chlor-alkali process, p. 901
Cyanide, p. 872
Interhalogen compounds, p. 906
Nitrogen fixation, p. 884
Silane, p. 880
Silicate, p. 880

Exercises

General Properties of Nonmetals

REVIEW QUESTIONS

21.1 Without referring to Figure 21.1, determine for each of the following elements whether they are metals, metalloids, or nonmetals: (a) Cs, (b) Ge, (c) I, (d) Kr, (e) W, (f) Ga, (g) Te, (h) Bi.
21.2 List two chemical and two physical properties that distinguish a metal from a nonmetal.
21.3 Make a list of physical and chemical properties of chlorine (Cl_2) and magnesium. Comment on their differences with reference to the fact that one is a metal and the other is a nonmetal.
21.4 Carbon is usually classified as a nonmetal. However, the graphite used in "lead" pencils conducts electricity. Look at a pencil and list two nonmetallic properties of graphite.

Hydrogen

PROBLEMS

21.5 Explain why hydrogen has no unique position in the periodic table.
21.6 Describe two laboratory and two industrial preparations for hydrogen.
21.7 Hydrogen exhibits three types of bonding in its compounds. Describe each type of bonding with an example.
21.8 Elements number 17 and 20 form compounds with hydrogen. Look up the formulas for these two compounds and compare their chemical behavior in water.
21.9 Give an example of hydrogen as (a) an oxidizing agent and (b) a reducing agent.
21.10 Give the name of (a) an ionic hydride and (b) a covalent hydride. In each case describe the preparation and give the structure of the compound.
21.11 Compare the physical and chemical properties of the hydrides of each of the following elements: Na, Ca, C, N, O, Cl.
21.12 What are interstitial hydrides?
21.13 Suggest a physical method other than effusion that would allow you to separate hydrogen gas from neon gas.
21.14 Write a balanced equation to show the reaction between CaH_2 and H_2O. How many grams of CaH_2 are needed to produce 26.4 L of H_2 gas at 20°C and 746 mmHg?
21.15 How many kilograms of water must be processed to obtain 2.0 L of D_2 at 25°C and 0.90 atm pressure? Assume that deuterium abundance is 0.015 percent and that recovery is 80 percent.
21.16 Predict the outcome of the following reactions:
(a) $CuO(s) + H_2(g) \longrightarrow$
(b) $Na_2O(s) + H_2(g) \longrightarrow$
21.17 Lubricants used for watches usually consist of long-chain hydrocarbons. Oxidation by air forms solid polymers that eventually destroy the effectiveness of the lubricants. It is believed that one of the initial steps in the oxidation is removal of a hydrogen atom (hydrogen abstraction). By replacing the hydrogen atoms at reactive sites with deuterium atoms, it is possible to substantially slow down the overall oxidation rate. Why? (*Hint:* Consider the kinetic isotope effect.)
21.18 Describe what is meant by the "hydrogen economy."

Boron

PROBLEMS

21.19 Describe an industrial preparation of boron.
21.20 The standard enthalpy of formation of $B_2H_6(g)$ is 31.4 kJ/mol. Find in Appendix 3 the ΔH_f° of C_2H_6 and C_2H_4. How do these compounds compare in thermody-

namic stability?

21.21 Equal volumes of NH_3 gas and BF_3 gas at the same temperature and pressure react to form a solid product. What is its structure?

21.22 A gaseous compound contains only B and H. The density of the gas is 1.24 g/L at STP. Calculate the molar mass of the compound and suggest a molecular formula for it.

21.23 A solid compound of boron and hydrogen has the empirical formula B_5H_7. A solution consisting of 0.103 g of this solid dissolved in 5.00 g of benzene freezes 0.87°C below the freezing point of pure benzene. Determine the molecular formula of the compound of boron and hydrogen. The molal freezing-point depression constant for benzene is 5.12°C/*m*.

21.24 Write a balanced equation for the oxidation of boron to boric acid by nitric acid.

21.25 Consider the borohydride ion BH_4^-. (a) Is the ion isoelectronic with CH_4? (b) Describe the bonding in BH_4^- in terms of hybridization.

Carbon and Silicon

PROBLEMS

21.26 Give an example of a carbide and a cyanide.

21.27 How are cyanide ions used in metallurgy?

21.28 Briefly discuss the preparation and properties of carbon monoxide and carbon dioxide.

21.29 What is coal?

21.30 Explain what is meant by coal gasification.

21.31 How does carbon dioxide influence Earth's climate?

21.32 Describe the reaction between CO_2 and OH^- in terms of a Lewis acid-base reaction such as that shown on p. 658.

21.33 Draw a Lewis structure for the C_2^{2-} ion.

21.34 Balance the following equations:
(a) $Be_2C(s) + H_2O(l) \longrightarrow$
(b) $CaC_2(s) + H_2O(l) \longrightarrow$

21.35 Unlike $CaCO_3$, Na_2CO_3 does not yield CO_2 on heating. On the other hand, $NaHCO_3$ undergoes thermal decomposition to produce CO_2 and Na_2CO_3. (a) Write a balanced equation for the reaction. (b) How would you test for the CO_2 evolved? [*Hint:* Treat the gas with limewater, that is, an aqueous solution of $Ca(OH)_2$.]

21.36 Two solutions are labeled A and B. Solution A contains Na_2CO_3 and solution B contains $NaHCO_3$. Describe how you would distinguish between the two solutions if you were provided with a $MgCl_2$ solution. (*Hint:* You need to know the solubilities of $MgCO_3$ and $MgHCO_3$.)

21.37 Magnesium chloride is dissolved in a solution containing sodium bicarbonate. On heating, a white precipitate is formed. Explain what causes the precipitation.

21.38 A few drops of concentrated ammonia solution added to a calcium bicarbonate solution causes a white precipitate to form. Write a balanced equation for the reaction.

21.39 Sodium hydroxide is hygroscopic; that is, it absorbs moisture when exposed to the atmosphere. A student placed a pellet of NaOH on a watch glass. A few days later, she noticed that the pellet was covered with a white solid. What is the identity of this solid? (*Hint:* Air contains CO_2.)

21.40 A piece of red-hot magnesium ribbon will continue to burn in an atmosphere of CO_2 even though CO_2 does not support combustion. Explain.

21.41 Is carbon monoxide isoelectronic with nitrogen (N_2)?

21.42 A concentration of 8.00×10^2 ppm by volume of CO is considered lethal to humans. Calculate the minimum mass of CO in grams that would become a lethal concentration in a closed room 17.6 m long, 8.80 m wide, and 2.64 m high. The temperature and pressure are 20.0°C and 756 mmHg, respectively.

21.43 Describe one way of preparing silicon.

21.44 What is the major component of sand?

21.45 What is the basic unit in silicates?

21.46 When heated strongly a silane decomposes into solid silicon and H_2. A sample yielded 0.465 g of Si and 0.0220 mole of H_2. Determine the empirical formula of the compound.

21.47 Why is silicon unable to form double bonds as readily as does carbon?

21.48 Explain why Si=Si bonds are less stable then C=C bonds.

21.49 Under certain conditions the reaction of Cl_2 with Mg_2Si forms a compound that has the empirical formula $SiCl_3$. Its molar mass is 269 g. What is the structure of the compound?

21.50 Suggest some reasons why carbon is more suitable as an essential element of terrestrial life than silicon would be.

21.51 Explain why $SiCl_4$ undergoes hydrolysis, whereas CCl_4 does not. (*Hint:* Silicon has a low-lying 3*d* orbital, so it can undergo valence-shell expansion.)

21.52 Glass containers are not suitable for storing very concentrated alkaline solutions. Why?

Nitrogen and Phosphorus

PROBLEMS

21.53 Describe a laboratory and an industrial preparation of nitrogen gas.

21.54 Nitrogen can be prepared by (a) passing ammonia over red-hot copper(II) oxide and (b) heating ammonium dichromate [one of the products is Cr(III) oxide]. Write a balanced equation for each of the preparations.

21.55 Write balanced equations for the preparation of sodium nitrite by (a) heating sodium nitrate and (b) heating sodium nitrate with carbon.

21.56 Sodium amide ($NaNH_2$) reacts with water to produce sodium hydroxide and ammonia. Describe this reaction as a Brønsted acid-base reaction.

21.57 Write a balanced equation for the formation of urea, $(NH_2)_2CO$, from carbon dioxide and ammonia. Should the reaction be run at a high or low pressure to maximize the yield?

21.58 Some farmers feel that lightning helps produce a better crop. What is the scientific basis for this belief?

21.59 At 620 K the vapor density of ammonium chloride relative to hydrogen (H_2) under the same conditions of temperature and pressure is 14.5, although, according to its formula mass, it should have a vapor density of 26.8. How would you account for this discrepancy?

21.60 Describe the Ostwald synthesis of nitric acid.

21.61 Describe the Haber synthesis of ammonia.

21.62 What is meant by nitrogen fixation? Describe a process for fixation of nitrogen on an industrial scale.

21.63 Explain why nitric acid can be reduced but not oxidized.

21.64 Explain, giving one example in each case, why nitrous acid can act both as a reducing agent and as an oxidizing agent.

21.65 Write a balanced equation for each of the following processes: (a) On heating, ammonium nitrate produces nitric oxide. (b) On heating, potassium nitrate produces potassium nitrite and oxygen gas. (c) On heating, lead nitrate produces lead(II) oxide, nitrogen dioxide (NO_2), and oxygen gas.

21.66 Explain why, under normal conditions, the reaction of zinc with nitric acid does not produce hydrogen.

21.67 Potassium nitrite may be produced by heating a mixture of potassium nitrate and carbon. Write a balanced equation for this reaction. Calculate the theoretical yield of KNO_2 produced by heating 57.0 g of KNO_3 with an excess of carbon.

21.68 Consider the reaction

$$N_2(g) + O_2(g) \rightleftharpoons 2NO(g)$$

Given that the $\Delta G°$ for the reaction at 298 K is 173.4 kJ, calculate (a) the standard free energy of formation of NO, (b) K_P for the reaction, and (c) K_c for the reaction.

21.69 Predict the geometry of nitrous oxide, N_2O, by the VSEPR method and draw resonance structures for the molecule. (*Hint:* The atoms are arranged as NNO.)

21.70 From the data in Appendix 3, calculate $\Delta H°$ for the synthesis of NO at 25°C:

$$4NH_3(g) + 5O_2(g) \longrightarrow 4NO(g) + 6H_2O(l)$$

21.71 Describe an industrial preparation of phosphorus.

21.72 Explain why two N atoms can form a double bond or a triple bond, whereas two P atoms can form only a single bond.

21.73 Why is the P_4 molecule unstable?

21.74 Starting with elemental phosphorus, P_4, show how you would prepare phosphoric acid.

21.75 When 1.645 g of white phosphorus is dissolved in 75.5 g of CS_2, the solution boils at 46.709°C, whereas pure CS_2 boils at 46.300°C. The molal boiling-point elevation constant for CS_2 is 2.34°C/*m*. Calculate the molar mass of white phosphorus and give the molecular formula.

21.76 Dinitrogen pentoxide can be prepared by the reaction between P_4O_{10} and HNO_3. Write a balanced equation for this reaction. Calculate the theoretical yield of N_2O_5 if 79.4 g of P_4O_{10} are used in the reaction and HNO_3 is the excess reagent. (*Hint:* One of the products is HPO_3.)

21.77 What is the hybridization of phosphorus in the phosphonium ion, PH_4^+?

21.78 Explain why (a) NH_3 is more basic than PH_3, (b) NH_3 has a higher boiling point than PH_3, (c) PCl_5 exists but NCl_5 does not, (d) N_2 is more inert than P_4.

21.79 Solid PCl_5 exists as $[PCl_4^+][PCl_6^-]$. Draw Lewis structures for these ions. Describe the state of hybridization of the P atoms.

Oxygen and Sulfur

PROBLEMS

21.80 Describe one industrial and one laboratory preparation of O_2.

21.81 Give an account of the various kinds of oxides that exist and illustrate each type by two examples.

21.82 Compare the structures of the following pairs of oxides: (a) CO_2 and SiO_2, (b) H_2O and H_2O_2, (c) MgO and NO.

21.83 Describe some unusual properties (both physical and chemical) of water.

21.84 How do hydrolysis and hydration differ?

21.85 Draw molecular orbital energy level diagrams for O_2, O_2^-, and O_2^{2-}.

21.86 Hydrogen peroxide can be prepared by treating barium peroxide with an aqueous solution of sulfuric acid. Write a balanced equation for this reaction.

21.87 One of the steps involved in the depletion of ozone in the stratosphere by nitric oxide may be represented as

$$NO(g) + O_3(g) \rightleftharpoons NO_2(g) + O_2(g)$$

From the data in Appendix 3 calculate $\Delta G°$, K_P, and K_c for the reaction at 25°C.

21.88 Hydrogen peroxide is unstable and decomposes readily:

$$2H_2O_2(aq) \longrightarrow 2H_2O(l) + O_2(g)$$

This reaction is accelerated by light, heat, or a catalyst. (a) Explain why hydrogen peroxide sold in drugstores is usually kept in dark bottles. (b) The concentrations of aqueous hydrogen peroxide solutions are normally expressed as percent by mass. In the decomposition of hydrogen peroxide, how many liters of oxygen gas can be produced at STP from 15.0 g of a 7.50 percent hydrogen peroxide solution?

21.89 What are the oxidation numbers of O and F in HOF?

21.90 Oxygen forms double bonds in O_2, but sulfur forms single bonds in S_8. Explain.

21.91 Describe the contact process (see p. 578) for the production of sulfuric acid.

21.92 The total production of sulfuric acid in the United States in 1981 was 41 million tons. Calculate the amount of sulfur (in grams and moles) used in the production.

21.93 Sulfuric acid is a dehydrating agent. Write balanced equations for the reactions between sulfuric acid and the following substances: (a) HCOOH, (b) H_3PO_4, (c) HNO_3, (d) $HClO_3$. (*Hint:* Sulfuric acid is not decomposed by the dehydrating action.)

21.94 Describe two reactions in which sulfuric acid acts as an oxidizing agent.

21.95 SF_6 exists but OF_6 does not. Explain.

21.96 Explain why SCl_6, SBr_6, and SI_6 cannot be prepared.

21.97 Compare the physical and chemical properties of H_2O and H_2S.

21.98 Calculate the amount of $CaCO_3$ (in grams) that would be required to react with 50.6 g of SO_2 emitted by a power plant.

21.99 The bad smell in water caused by hydrogen sulfide can be removed by the action of chlorine. The reaction is

$$H_2S(aq) + Cl_2(aq) \longrightarrow 2HCl(aq) + S(s)$$

If the hydrogen sulfide content of contaminated water is 22 ppm by mass, calculate the amount of Cl_2 (in grams) required to remove all the H_2S from 1.5×10^3 gallons of water, which is the average amount of water used daily per person in the United States (1 gallon = 3.785 L).

21.100 Concentrated sulfuric acid reacts with sodium iodide to produce molecular iodine, hydrogen sulfide, and sodium hydrogen sulfate. Write a balanced equation for the reaction.

The Halogens

PROBLEMS

21.101 Describe an industrial preparation for each of the halogens.

21.102 Metal chlorides can be prepared in a number of ways: (a) direct combination of metal and molecular chlorine, (b) reaction between metal and hydrochloric acid, (c) acid-base neutralization, (d) metal carbonate treated with hydrochloric acid, (e) precipitation reaction. Give an example for each type of preparation.

21.103 Show that chlorine, bromine, and iodine are very much alike by giving an account of their behavior (a) with hydrogen, (b) in producing silver salts, (c) as oxidizing agents, and (d) with sodium hydroxide. (e) In what respects is fluorine not a typical halogen element?

21.104 Aqueous copper(II) sulfate solution is blue in color. When aqueous potassium fluoride is added to the $CuSO_4$ solution, a green precipitate is formed. If aqueous potassium chloride is added instead, a bright green solution is formed. Explain what happens in each case.

21.105 Draw structures for (a) $(HF)_2$ and (b) HF_2^-.

21.106 Sulfuric acid is a weaker acid than hydrochloric acid. Yet hydrogen chloride is evolved when concentrated sulfuric acid is added to sodium chloride. Explain.

21.107 A 375-gallon tank is filled with water containing 167 g of bromine in the form of Br^- ions. How many liters of Cl_2 gas at 1.00 atm and 20°C will be required to oxidize all the bromide to molecular bromine?

21.108 Hydrogen fluoride can be prepared by the action of sulfuric acid on sodium fluoride. Explain why hydrogen bromide cannot be prepared by the action of the same acid on sodium bromide.

21.109 What volume of bromine (Br_2) measured at 100°C and 700 mmHg pressure would be obtained if 2.00 L of dry chlorine (Cl_2), measured at 15.0°C and 760 mmHg, were absorbed by a potassium bromide solution?

21.110 Use the VSEPR method to predict the geometries of the following species: (a) IF_7, (b) I_3^-, (c) $SiCl_4$, (d) PF_5, (e) SF_4, (f) ClO_2^-.

21.111 Iodine pentoxide, I_2O_5, is sometimes used to remove carbon monoxide from the air by forming carbon dioxide and iodine. Write a balanced equation for this reaction and identify species that are oxidized and reduced.

The Noble Gases

PROBLEMS

21.112 As a group, the noble gases are more stable than other elements. Why?

21.113 Helium gas is used to dilute oxygen gas for deep-sea divers. If a diver has to work at a depth where the pressure is 4.0 atm, what must be the content of He (percent by volume) in a He/O_2 mixture so that the partial pressure of O_2 is 0.20 atm?

21.114 Why do all noble gas compounds violate the octet rule?

21.115 Balance the following equations:
(a) $XeF_4 + H_2O \longrightarrow XeO_3 + Xe + HF$
(b) $XeF_6 + H_2O \longrightarrow XeO_3 + HF$

21.116 Heated xenon(VI) fluoride undergoes decomposition:

$$XeF_6(s) \longrightarrow XeF_4(g) + F_2(g)$$

Starting with 8.96 g of XeF_6, calculate the combined volume of the product gases at 107°C and 374 mmHg. Assume complete decomposition.

21.117 Give oxidation numbers of Xe in the following compounds: XeF_2, XeF_4, XeF_6, $XeOF_4$, XeO_3, XeO_2F_2, XeO_4.

Miscellaneous Problems

21.118 Write a balanced equation for each of the following reactions: (a) Heating phosphorus acid yields phosphoric acid and phosphine (PH_3). (b) Lithium carbide reacts with hydrochloric acid to give lithium chloride and methane. (c) Bubbling HI gas through an aqueous solution of HNO_2 yields molecular iodine and nitric oxide. (d) Hydrogen sulfide is oxidized by chlorine to give HCl and SCl_2.

21.119 (a) Which of the following compounds has the greatest ionic character? PCl_5, $SiCl_4$, CCl_4, BCl_3 (b) Which of the following ions has the smallest ionic radius? F^-, C^{4-}, N^{3-}, O^{2-}. (c) Which of the following atoms has the highest ionization energy? F, Cl, Br, I. (d) Which of the following oxides is most acidic? H_2O, SiO_2, CO_2.

21.120 Starting with deuterium oxide, describe how you would prepare (a) ND_3 and (b) C_2D_2.

21.121 Both N_2O and O_2 support combustion. Suggest one physical and one chemical test to distinguish between the two gases.

21.122 What is the change in oxidation number for the following reaction?

$$3O_2 \longrightarrow 2O_3$$

21.123 Describe the bonding in the C_2^{2-} ion in terms of the molecular orbital theory.

21.124 Write an overall equation for photosynthesis. How does photosynthesis influence the greenhouse effect?

21.125 Briefly describe the role of nitrogen oxides and ozone in smog formation.

21.126 The yolks of eggs boiled too long often turn green. What causes this color? (*Hint:* The proteins in the egg white contain sulfur, and some hydrogen sulfide is evolved when the proteins decompose on heating. The egg yolks contain Fe^{2+} ions.)

Opposite page: Copper ions implanted in Al_2O_3 emit visible radiation (luminescence) when excited by UV light. The color of the light emitted depends on the environment of the copper ion and can, therefore, be changed by adding other elements in small amounts. The color of light absorbed and emitted by transition metal complexes likewise depends on the bonding environment.

C H A P T E R 2 2

TRANSITION METAL CHEMISTRY AND COORDINATION COMPOUNDS

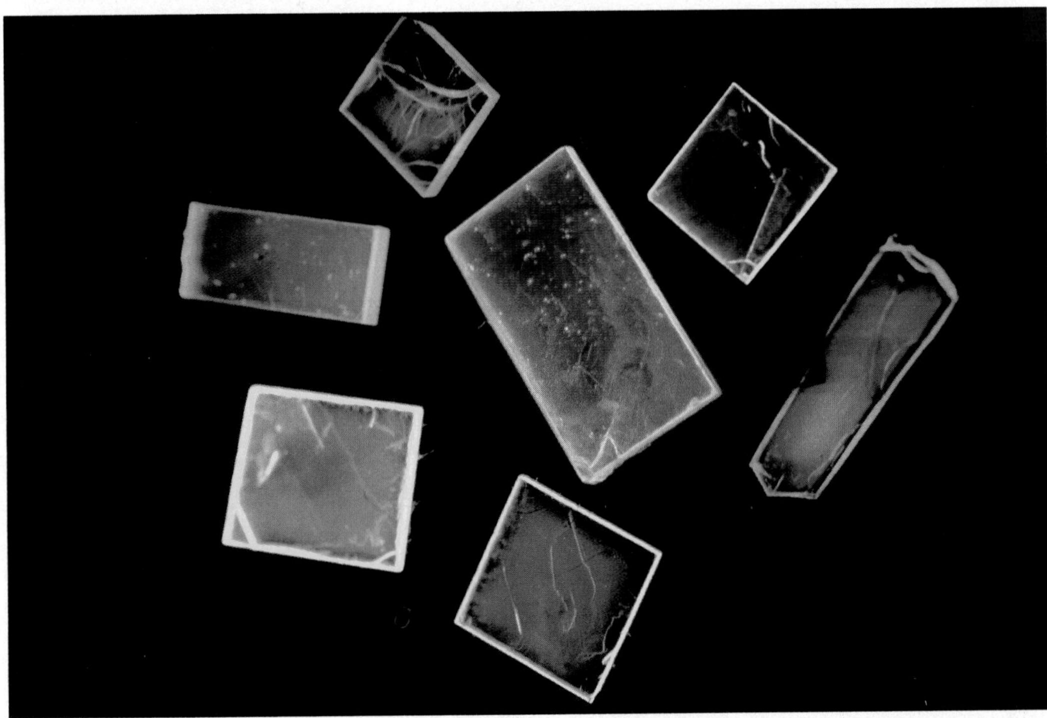

The series of elements in the periodic table in which the *d* and *f* orbitals are gradually filled are called the transition elements. There are about 50 transition elements, and they have widely varying and fascinating properties. To present even one interesting feature of each transition element, however, is beyond the scope of this book. We therefore limit ourselves to discussion of the transition elements that have incompletely filled *d* orbitals and their most commonly encountered property—the tendency to form complex ions.

22.1 Properties of the Transition Metals

Transition metals characteristically have incompletely filled *d* subshells or readily give rise to ions with incompletely filled *d* subshells (Figure 22.1). In this chapter we concentrate on the first-row transition metals from scandium to copper; they are the most common of all the transition metals. Table 22.1 lists some of their properties.

As we read across any period from left to right, atomic numbers increase, electrons are being added to the outer shell, and the nuclear charge increases by the addition of protons. In the third-period elements—sodium to argon—the outer electrons shield one another weakly from the extra nuclear charge. Consequently, atomic radii decrease rapidly from sodium to argon and the electronegativities and ionization energies increase steadily (see Figures 8.7, 8.14, and 9.8).

For the transition metals, these tendencies are different. As we read across from scandium to copper, the nuclear charge, of course, increases, but electrons are being added to the inner 3*d* subshell. These 3*d* electrons shield the 4*s* electrons from the increasing nuclear charge somewhat more effectively than outer-shell electrons can shield one another, so the atomic radii decrease less rapidly. For the same reason, electronegativities and ionization energies increase only slightly from scandium across to copper compared with the increases from sodium to argon.

Although the transition metals are less electropositive (or more electronegative) than the alkali and alkaline earth metals, their standard reduction potentials seem to indicate that all of them except copper should react with strong acids such as hydrochloric acid to produce hydrogen gas. However, most transition metals are inert toward acids or react slowly because of a protective layer of oxide. A case in point is chromium: Despite its rather negative standard reduction potential (see Table 19.1), it is quite inert

1A	2A	3B	4B	5B	6B	7B	8B	8B	8B	1B	2B	3A	4A	5A	6A	7A	8A
1 H																	2 He
3 Li	4 Be											5 B	6 C	7 N	8 O	9 F	10 Ne
11 Na	12 Mg											13 Al	14 Si	15 P	16 S	17 Cl	18 Ar
19 K	20 Ca	21 Sc	22 Ti	23 V	24 Cr	25 Mn	26 Fe	27 Co	28 Ni	29 Cu	30 Zn	31 Ga	32 Ge	33 As	34 Se	35 Br	36 Kr
37 Rb	38 Sr	39 Y	40 Zr	41 Nb	42 Mo	43 Tc	44 Ru	45 Rh	46 Pd	47 Ag	48 Cd	49 In	50 Sn	51 Sb	52 Te	53 I	54 Xe
55 Cs	56 Ba	57 La	72 Hf	73 Ta	74 W	75 Re	76 Os	77 Ir	78 Pt	79 Au	80 Hg	81 Tl	82 Pb	83 Bi	84 Po	85 At	86 Rn
87 Fr	88 Ra	89 Ac	104 Unq	105 Unp	106 Unh	107 Uns	108 Uno	109 Une									

Figure 22.1 *The transition metals. Note that although the Group 2B elements (Zn, Cd, Hg) are described as transition metals by some chemists, neither the metals nor their ions possess incompletely filled d subshells.*

Table 22.1 Electron configuration and other properties of the first-row transition metals

	Sc	Ti	V	Cr	Mn	Fe	Co	Ni	Cu
Electron configuration									
M	$4s^23d^1$	$4s^23d^2$	$4s^23d^3$	$4s^13d^5$	$4s^23d^5$	$4s^23d^6$	$4s^23d^7$	$4s^23d^8$	$4s^13d^{10}$
M^{2+}	—	$3d^2$	$3d^3$	$3d^4$	$3d^5$	$3d^6$	$3d^7$	$3d^8$	$3d^9$
M^{3+}	[Ar]	$3d^1$	$3d^2$	$3d^3$	$3d^4$	$3d^5$	$3d^6$	$3d^7$	$3d^8$
Electronegativity	1.3	1.5	1.6	1.6	1.5	1.8	1.9	1.9	1.9
Ionization energy (kJ/mol)									
First	631	658	650	652	717	759	760	736	745
Second	1235	1309	1413	1591	1509	1561	1645	1751	1958
Third	2389	2650	2828	2986	3250	2956	3231	3393	3578
Radius (pm)									
M	162	147	134	130	135	126	125	124	128
M^{2+}	—	90	88	85	80	77	75	69	72
M^{3+}	81	77	74	64	66	60	64	—	—
Standard reduction potential (V)*	−2.08	−1.63	−1.2	−0.74	−1.18	−0.44	−0.28	−0.25	0.34

*The half-reaction is $M^{2+}(aq) + 2e^- \longrightarrow M(s)$ (except for Sc and Cr, where the ions are Sc^{3+} and Cr^{3+}, respectively).

Table 22.2 Physical properties of elements K to Zn

	1A	2A	Transition metals									2B
	K	Ca	Sc	Ti	V	Cr	Mn	Fe	Co	Ni	Cu	Zn
Atomic radius (pm)	235	197	162	147	134	130	135	126	125	124	128	138
Melting point (°C)	63.7	838	1539	1668	1900	1875	1245	1536	1495	1453	1083	419.5
Boiling point (°C)	760	1440	2730	3260	3450	2665	2150	3000	2900	2730	2595	906
Density (g/cm^3)	0.86	1.54	3.0	4.51	6.1	7.19	7.43	7.86	8.9	8.9	8.96	7.14

chemically because of the formation on its surfaces of chromium(III) oxide, Cr_2O_3. Consequently, chromium is commonly used as a protective and noncorrosive plating on other metals.

The following characteristics of the transition metals are both interesting to consider and important to understand.

General Physical Properties

Most of the transition metals have a close-packed structure (see Figure 11.31) in which each atom has a coordination number of 12. Furthermore, these elements have relatively small atomic radii. The combined effect of closest packing and small atomic size results in strong metallic bonds. Therefore, transition metals have higher densities, higher melting points and boiling points, and higher heats of fusion and vaporization than the Group 1A and 2A metals, as well as the Group 2B metals (Table 22.2).

Electron Configurations

The electron configurations of the first-row transition metals were discussed in Section 7.10. Calcium has the electron configuration $[Ar]4s^2$. From scandium across to copper, electrons are added to the $3d$ orbitals. Thus, the outer electron configuration of scandium is $4s^23d^1$, that of titanium is $4s^23d^2$, and so on. The two exceptions are chromium and copper, whose outer electron configurations are $4s^13d^5$ and $4s^13d^{10}$, respectively. These irregularities are the result of the extra stability associated with half-filled and completely filled $3d$ subshells.

When the first-row transition metals form cations, we know that electrons are removed first from the $4s$ orbitals and then from the $3d$ orbitals. (This is the opposite of the order in which orbitals are filled in neutral atoms.) For example, the outer electron configuration of Fe^{2+} is $3d^6$, not $4s^23d^4$.

Oxidation States

The transition metals exhibit variable oxidation states in their compounds by losing one or more electrons. Figure 22.2 shows the oxidation states from scandium to copper. Note that the common oxidation states for each element include +2, +3, or both. The +3 oxidation states are more stable at the beginning of the series, whereas toward the end the +2 oxidation states are more stable. The reason for this trend can be understood by examining the ionization energy plots in Figure 22.3. In general, the ionization energies increase gradually from left to right. However, the third ionization energy (when an electron is removed from the $3d$ orbital) increases more rapidly than the first and second ionization energy. Because it takes more energy to remove the third electron from the metals near the end of the row than from those near the beginning, the metals near the end tend to form M^{2+} ions rather than M^{3+} ions.

The relative stability of the +2 and +3 oxidation states also correlates with the standard reduction potential $\mathscr{E}°$ for the half-cell reaction

Sc	Ti	V	Cr	Mn	Fe	Co	Ni	Cu
				+7				
			+6	+6	+6			
		+5	+5	+5	+5			
	+4	+4	+4	+4	+4	+4		
+3	+3	+3	+3	+3	+3	+3	+3	+3
	+2	+2	+2	+2	+2	+2	+2	+2
								+1

Figure 22.2 *Oxidation states of the first-row transition metals. The most stable oxidation numbers are shown in color. The zero oxidation state is encountered in some compounds, such as $Ni(CO)_4$ and $Fe(CO)_5$.*

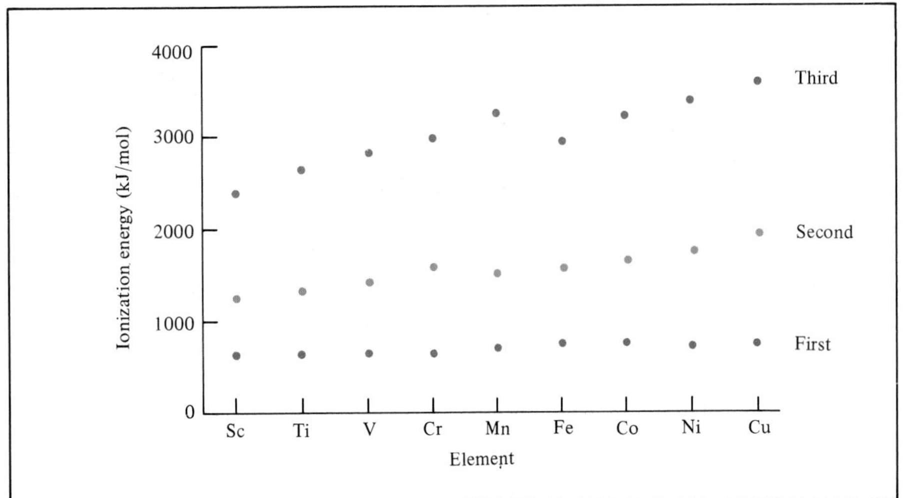

Figure 22.3 *Variation of the first, second, and third ionization energies for the first-row transition metals.*

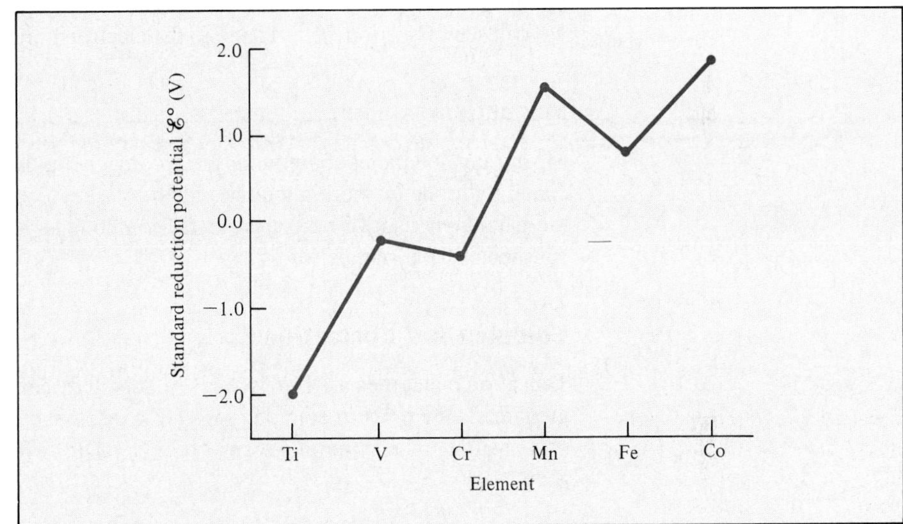

Figure 22.4 *Standard reduction potentials for the half-reaction $M^{3+} + e^- \rightarrow M^{2+}$ for several transition metals.*

$$M^{3+}(aq) + e^- \longrightarrow M^{2+}(aq)$$

in which M stands for a transition metal. The more positive the $\mathscr{E}°$ value, the more likely the reduction will proceed to the right and the more stable the +2 oxidation state is compared with the +3 state. Figure 22.4 shows how $\mathscr{E}°$ varies for several metals. Titanium, vanadium, and chromium all have negative $\mathscr{E}°$ values, indicating that their M^{3+} ions are more stable than the M^{2+} ions. Manganese has a large positive $\mathscr{E}°$, which means that Mn^{2+} is more stable than Mn^{3+}. The difference in stability between Fe^{2+} and Fe^{3+} ions is not nearly as marked as that between Mn^{2+} and Mn^{3+}. In fact, both Fe^{2+} and Fe^{3+} ions are stable species in solution. The plot also shows that Co^{2+} is more stable than Co^{3+}.

Recall that oxides in which the metal has a high oxidation number are covalent and acidic, whereas those in which the metal has a low oxidation number are ionic and basic.

The highest oxidation state is +7, for manganese ($4s^23d^5$). For elements to its right (Fe to Cu), oxidation numbers are lower. Transition metals usually exhibit their highest oxidation states in compounds with very electronegative elements such as oxygen and fluorine—for example, V_2O_5, CrO_3, and Mn_2O_7.

Figure 22.5 *Colors of some of the first-row transition metal ions in solution. From left to right:* Ti^{3+}, Cr^{3+}, Mn^{2+}, Fe^{3+}, Co^{2+}, Ni^{2+}, Cu^{2+}. *The* Sc^{3+} *and* V^{5+} *ions are colorless.*

Color

Many transition metal ions and complex ions and anions containing transition metals are distinctively colored (Figure 22.5; also see Figure 4.8.) The origin of the color, as we will see in Section 22.5, is due to the electronic transition involving the *d* electrons.

Magnetism

Most transition metal compounds are paramagnetic. Recall that paramagnetism is associated with one or more unpaired electrons (see Section 7.9). As we will see shortly, the paramagnetism of transition metal compounds is due to the incompletely filled *d* subshells of the metals.

Complex Ion Formation

Transition metal ions are Lewis acids (that is, they can accept electron pairs) and have a great tendency to form complex ions (first discussed in Section 17.5). Some examples of transition metal complexes are $[Cu(NH_3)_4]^{2+}$, $[Co(NH_3)_6]^{3+}$, $[Ni(CN)_4]^{2-}$, and $[Fe(CN)_6]^{3-}$.

Catalytic Properties

Many of the transition metals and their compounds are good catalysts both for inorganic and organic reactions and for electrochemical processes. We have already described the use of iron as a catalyst in the Haber synthesis of ammonia (p. 576); the use of vanadium oxide, V_2O_5, in the production of sulfuric acid (p. 578); the use of platinum as a catalyst in hydrogenation reactions (p. 579); and the use of nickel and nickel oxide as electrocatalysts in fuel cells (Section 19.6). Metals such as copper, ruthenium, rhodium, and gold are used as catalysts in many reactions.

22.2 Chemistry of the First-Row Transition Metals

In this section we will briefly survey the chemistry of the first-row transition metals, paying particular attention to their occurrence, preparation, uses, and important compounds. These metals are pictured in Figure 22.6.

Scandium

Scandium, a rare element (2.5×10^{-3} percent of Earth's crust by mass), is difficult to obtain in its pure form. Prepared by the electrolysis of molten scandium chloride, $ScCl_3$, scandium is silvery white and develops a yellowish cast when exposed to air. The oxidation state of scandium is $+3$ in most of its compounds. Because the Sc^{3+} ion does not possess any *d* electrons, scandium compounds are colorless and diamagnetic. Scandium and its compounds have no important uses.

Titanium

Titanium is the most abundant transition metal after iron (0.63 percent of Earth's crust by mass). It occurs as *rutile,* TiO_2, and as *ilmenite,* $FeTiO_3$. The metal is prepared by heating TiO_2 with coke (C) and chlorine gas at about 900°C:

$$TiO_2(s) + 2C(s) + 2Cl_2(g) \longrightarrow TiCl_4(l) + 2CO_2(g)$$

Figure 22.6 *The first-row transition metals.*

Reduction of $TiCl_4$ with molten magnesium in a steel crucible between 950°C and 1150°C produces titanium metal:

$$TiCl_4(g) + 2Mg(l) \longrightarrow Ti(s) + 2MgCl_2(l)$$

Titanium is a strong, lightweight, corrosion-resistant metal that is used in the construction of rockets, aircrafts, and jet engines. Because it does not corrode easily and is resistant to attack by acids and chlorine, titanium also has applications in the chemical industry. Titanium dissolves in concentrated sulfuric acid as follows:

$$2Ti(s) + 6H_2SO_4(aq) \longrightarrow Ti_2(SO_4)_3(aq) + 6H_2O(l) + 3SO_2(g)$$

The most important oxidation states of titanium are +3 and +4. The Ti(IV) compounds are colorless and diamagnetic. In contrast, the Ti(III) compounds are colored and paramagnetic because Ti^{3+} contains one *d* electron. Titanium(IV) oxide, TiO_2, is very stable, nontoxic, opaque, and brilliant white. These properties make it suitable as a pigment in white paint. The annual production of TiO_2 in the United States is about 800,000 tons. Titanium(IV) chloride is a colorless liquid that hydrolyzes as follows:

$$TiCl_4(l) + 2H_2O(l) \longrightarrow TiO_2(s) + 4HCl(g)$$

Titanium(IV) chloride, together with trimethylaluminum, $(CH_3)_3Al$, is an important catalyst in the polymer industry (see Section 25.2).

Vanadium

Vanadium constitutes 0.014 percent of Earth's crust by mass. It occurs as *vanadinite,* $Pb_5(VO_4)_3Cl$, and *patronite,* VS_4. The first step in extracting vanadium is to obtain vanadium(V) oxide from the sulfide ore or, in a series of complex reactions, from vanadinite. Next, V_2O_5 is reduced by molten calcium to give the metal:

$$V_2O_5(s) + 5Ca(l) \longrightarrow 5CaO(s) + 2V(l)$$

Pure vanadium is a bright, shiny metal that is fairly soft and ductile and quite resistant to corrosion. Because of these properties it is used in steel alloys. It combines with the carbon present in steel to form V_4C_3, which disperses to produce a fine-grained steel that is more resistant to wear and tear and performs better at high temperatures than ordinary steel. Vanadium forms compounds in the +2, +3, +4, and +5 oxidation states, of which the +4 and +5 are the most important (Figure 22.7). Vanadium(IV) oxide is a dark blue solid; in solution the blue color is due to the VO^{2+} ion. Vanadium(IV) chloride, VCl_4, is a reddish-brown liquid; it is a molecular compound. Vanadium(V) oxide is a yellow-orange compound that is best prepared by the thermal decomposition of ammonium metavanadate, NH_4VO_3:

$$2NH_4VO_3(s) \longrightarrow V_2O_5(s) + 2NH_3(g) + H_2O(g)$$

Vanadium(V) oxide is used as a catalyst in the contact process for the manufacture of sulfuric acid (see Figure 13.19).

Chromium

Chromium is relatively rare (0.0122 percent of Earth's crust by mass). Its most important ore is *chromite,* $FeO \cdot Cr_2O_3$; it is also found in *crocoite,* $PbCrO_4$ (Figure 22.8). The metal is extracted by the reaction with aluminum at high temperature:

Figure 22.7 *Left to right: Colors of aqueous solutions of compounds containing vanadium in four oxidation states (V, VI, III, and II).*

$$2Al(s) + Cr_2O_3(s) \longrightarrow Al_2O_3(s) + 2Cr(l)$$

Chromium is a hard, brittle metal that forms a protective oxide, Cr_2O_3, which accounts for the metal's resistance to corrosion. Its chief use is in the manufacture of stainless steel. Another use of chromium is in the ''chromium-plating'' process. Chromium(III) oxide, Cr_2O_3, is dissolved in sulfuric acid and the Cr^{3+} ions produced are electrolytically reduced onto the surface of a metal that acts as the cathode. Car bumpers protected this way are both resistant to corrosion and decorative.

The common oxidation states of chromium are +2, +3, and +6, of which the +3 state is the most stable. The green Cr^{3+} ion (see Figure 4.8) forms many complex ions. Like Al^{3+}, it hydrolyzes to give an acidic solution. Chromium(VI) oxide, CrO_3, exists as dark red crystals; it is a strong oxidizing agent. Chromium(III) oxide, Cr_2O_3, a green powder, is prepared by the thermal decomposition of ammonium dichromate (Figure 22.9). Chromium(III) oxide has amphoteric properties and is very stable. It is

Figure 22.8 *The chromium ore crocoite, $PbCrO_4$.*

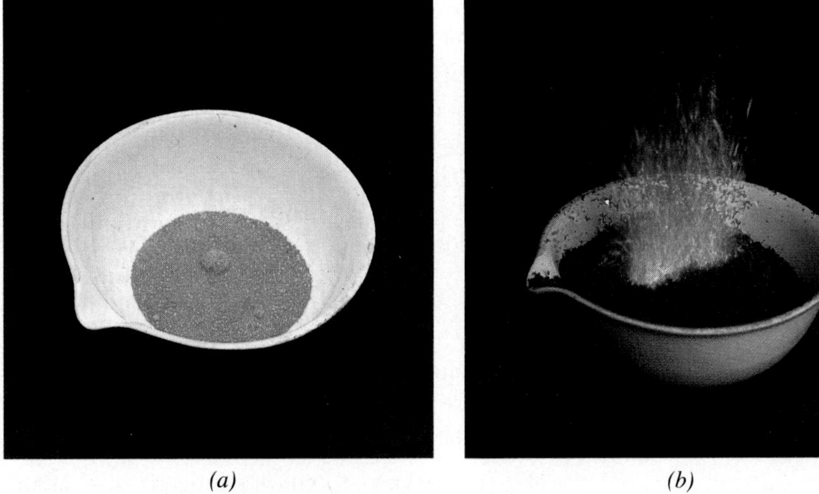

(a) (b)

Figure 22.9 *Thermal decomposition of ammonium dichromate. (a) Before heating. (b) The solid product is chromium(III) oxide.*

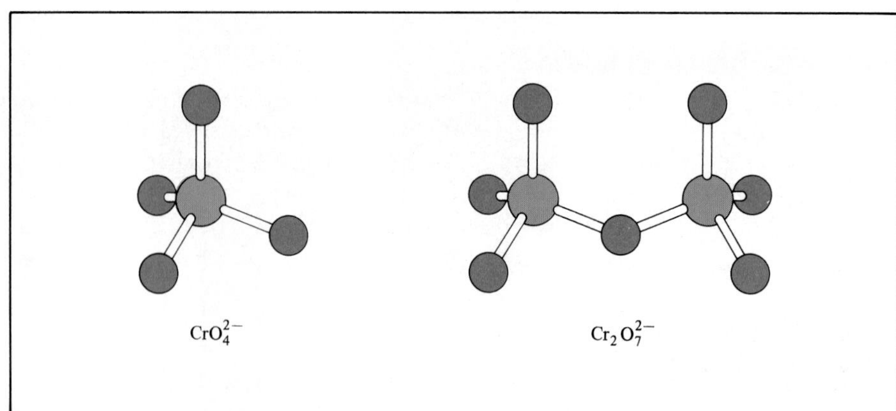

Figure 22.10 *Geometry of the CrO_4^{2-} and $Cr_2O_7^{2-}$ ions. In both cases the Cr atom is tetrahedrally bonded to four O atoms.*

used as a green pigment in paints. Chromium(IV) oxide, CrO_2, is a brownish-black solid. Its high electrical conductivity and magnetic properties make it suitable in the manufacture of magnetic recording tapes which perform better than those made of iron oxides. Figure 22.10 shows the structures of two important chromium-containing anions, chromate, CrO_4^{2-}, and dichromate, $Cr_2O_7^{2-}$. The oxidizing properties of the $Cr_2O_7^{2-}$ ions are discussed in Sections 4.7.

Manganese

Manganese is a relatively abundant element (0.11 percent of Earth's crust by mass). The major source of manganese is *pyrolusite,* MnO_2, but it is also found in manganese nodules (see Section 20.1). The metal is prepared by the thermite process. First, MnO_2 is converted to Mn_3O_4 by heating:

$$3MnO_2(s) \longrightarrow Mn_3O_4(s) + O_2(g)$$

Then Mn_3O_4 is treated with aluminum to yield liquid manganese:

$$3Mn_3O_4(s) + 8Al(s) \longrightarrow 4Al_2O_3(s) + 9Mn(l)$$

(The reaction between MnO_2 and Al is too violent.) The metal is then purified by distillation. In pure form, manganese is a brittle, hard metal that has a silvery white appearance. Its main use is in making steel. Steel containing 12 percent manganese is very hard and is used for railroad tracks and similar metal products that have to stand up under severe wear and tear.

The known oxidation states of manganese range from +2 to +7. The +2 state is particularly stable (it has a half-filled 3*d* subshell). The Mn^{2+} ions have a light pink color in solution (see Figure 4.8). The Mn^{3+} ion is unstable in solution. Manganese(IV) oxide, normally available as a dark powder, is used in the laboratory preparation of oxygen (see Figure 5.14), in glassmaking, and in the LeClanché cell (see Figure 19.5). Potassium permanganate is a dark purple compound. It is prepared as follows:

$$2MnO_2(s) + 2KOH(l) + KClO_3(l) \longrightarrow 2KMnO_4(l) + H_2O(g) + 2KCl(l)$$

The oxidizing properties of MnO_4^- are discussed in Section 4.7.

Iron

Figure 22.11 *The iron ore magnetite, Fe_3O_4.*

After aluminum, iron is the most abundant metal in Earth's crust (6.2 percent by mass). It is found in many ores; some of the important ones are *hematite,* Fe_2O_3; *siderite,* $FeCO_3$; and *magnetite,* Fe_3O_4 (Figure 22.11). Figure 22.12 shows an aerial view of open-pit mining. The preparation of iron in a blast furnace and steelmaking were discussed in Section 20.2. Pure iron is a gray metal and is not particularly hard. It is an essential element in living systems (to be discussed later).

Iron reacts with hydrochloric acid to give hydrogen gas:

$$Fe(s) + 2H^+(aq) \longrightarrow Fe^{2+}(aq) + H_2(g)$$

Concentrated sulfuric acid oxidizes the metal to Fe^{3+}, but concentrated nitric acid renders the metal "passive" by forming a thin layer of Fe_3O_4 over the metal. One of the best-known reactions of iron is rust formation (see Section 19.7). The two oxidation states of iron are +2 and +3. Iron(II) compounds include FeO (black powder), $FeSO_4 \cdot 7H_2O$ (green), $FeCl_2$ (yellow), and FeS (black). In the presence of oxygen, Fe^{2+} ions in solution are readily oxidized to Fe^{3+} ions. Iron(III) oxide is reddish brown, and iron(III) chloride is brownish black.

Cobalt

Cobalt is a rare element (3×10^{-3} percent of Earth's crust by mass). It is found in association with iron, nickel, and silver. Its major ores are *smaltite,* $CoAs_2$, and *cobalitite,* CoAsS. The metal is prepared by roasting CoAsS in air and then reducing the metal oxide with carbon. Cobalt is a bright, bluish-white metal. It is used in alloys such as stellite, an alloy containing cobalt, chromium, and tungsten. Stellite is used in making surgical instruments.

The oxidation states of cobalt are +2 and +3; the +2 is the more stable of the two.

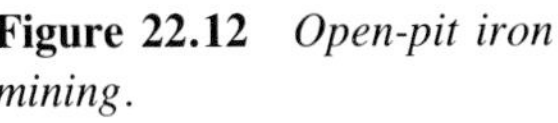

Figure 22.12 *Open-pit iron mining.*

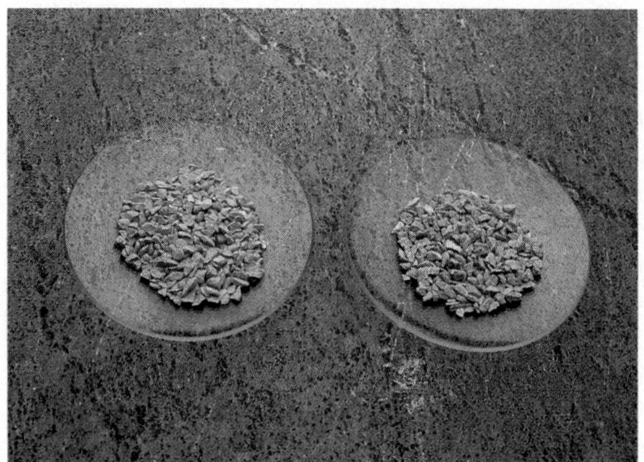

Figure 22.13 *Left: In the anhydrous form, $CoCl_2$ is blue. Right: In the presence of moist air, its color gradually turns pink due to the formation of $CoCl_2 \cdot 6H_2O$.*

Cobalt reacts with hot, dilute sulfuric acid and hot hydrochloric acid to form Co(II) salts. Cobalt(II) oxide, CoO, is a greenish powdery substance. Cobalt(II) chloride, $CoCl_2$, is blue in the anhydrous state but the hydrated form, $CoCl_2 \cdot 6H_2O$, is pink. This property makes the compound useful as an indicator of moisture (Figure 22.13).

Nickel

Nickel is a rare element (1.0×10^{-2} percent of Earth's crust by mass). The ore *millerite,* NiS, is associated with *pyrite,* FeS_2, and *chalcopyrite,* $CuFeS_2$. The metal is obtained by first roasting the ore in air and then reducing it with hydrogen gas. Nickel is purified by the Mond process, described in Section 20.2. Nickel is a silvery metal having high electrical and thermal conductivities. It is mainly used in making alloys. The coin we call a "nickel" is made of copper and nickel. Nickel is also used as a catalyst in hydrogenation reactions (see Section 13.6) and as electrodes in batteries and fuel cells (see Section 19.6).

Nickel is attacked only by nitric acid. Its stable oxidation state is +2. Hydrated Ni^{2+} ions have an emerald green color. Some important nickel compounds are NiO (green), $NiCl_2$ (yellow), and NiS (black).

Copper

Copper, a rare element (6.8×10^{-3} percent of Earth's crust by mass), is found in nature in the uncombined state as well as in ores such as chalcopyrite, $CuFeS_2$ (Figure 22.14). The metal is obtained by roasting the ore to give Cu_2S and then metallic copper:

Figure 22.14 *Chalcopyrite, $CuFeS_2$.*

$$2CuFeS_2(s) + 4O_2(g) \longrightarrow Cu_2S(s) + 2FeO(s) + 3SO_2(g)$$

$$Cu_2S(s) + O_2(g) \longrightarrow 2Cu(l) + SO_2(g)$$

Impure copper can be purified by the electrolytic technique (see Section 20.2). Copper is a reddish-brown metal. Aside from silver, which is too expensive for large-scale use, copper has the highest electrical conductivity. It is also a good thermal conductor. Copper is used in alloys, electrical cables, plumbing (pipes), and coins (Figure 22.15).

Copper reacts only with hot concentrated sulfuric acid and nitric acid (see Figure 21.22). Its two important oxidation states are +1 and +2. The +1 state is less stable and disproportionates in solution:

$$2Cu^{+}(aq) \longrightarrow Cu(s) + Cu^{2+}(aq)$$

All compounds of Cu(I) are diamagnetic and colorless except for Cu_2O, which is red. The Cu(II) compounds are all paramagnetic and colored. The hydrated Cu^{2+} ion is blue. Some important Cu(II) compounds are CuO (black), $CuSO_4 \cdot 5H_2O$ (blue), and CuS (black).

22.3 Coordination Compounds

Recall that a complex ion contains a central metal ion bonded to one or more ions or molecules.

We have noted that transition metals have a distinct tendency to form complex ions. *A neutral species containing one or more complex ions* is called a ***coordination compound.*** Coordination compounds usually have rather complicated formulas in which we enclose the complex ion in brackets. Thus, $[Ni(NH_3)_6]^{2+}$ is a complex ion, and $[Ni(NH_3)_6]Cl_2$ is a coordination compound. Most, but not all, of the metals in coordination compounds are transition metals.

The molecules or ions that surround the metal in a complex ion are called ***ligands.*** The interactions between a metal atom and the ligands can be thought of as Lewis acid-base reactions. As we saw in Section 15.6, a Lewis base is a substance capable of donating one or more electron pairs. Every ligand has at least one unshared pair of valence electrons, as these examples show:

$H_2\ddot{O}$: $\quad$ $\ddot{N}H_3$ $\quad$ $:\ddot{Cl}:^-$ $\quad$ $:C{\equiv}O:$

Therefore, ligands play the role of Lewis bases. On the other hand, a transition metal atom (in either its neutral or positively charged state) acts as a Lewis acid, accepting (and sharing) pairs of electrons from the Lewis bases. Thus the metal-ligand bonds are usually coordinate covalent bonds (see Section 9.9).

The atom in a ligand that is bound directly to the metal atom is known as the ***donor***

Figure 22.15 *Copper sheets awaiting stamping into pennies.*

Table 22.3 Some common ligands

Name	Structure
	Monodentate ligands
Ammonia	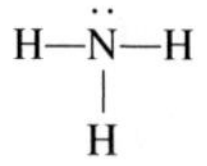 $\mathrm{H-\ddot{N}(-H)-H}$
Pyridine	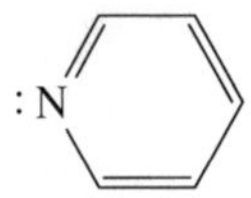 $:\mathrm{N}$ (ring)
Carbon monoxide	$:\mathrm{C{\equiv}O}:$
Chloride ion	$:\ddot{\underset{..}{\mathrm{Cl}}}:^{-}$
Cyanide ion	$[:\mathrm{C{\equiv}N}:]^{-}$
Thiocyanate ion	$[:\ddot{\underset{..}{\mathrm{S}}}-\mathrm{C{\equiv}N}:]^{-}$
Triphenylphosphine	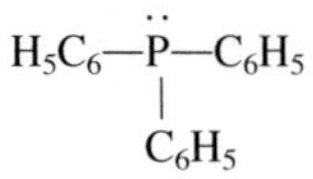 $\mathrm{H_5C_6-\ddot{P}(-C_6H_5)-C_6H_5}$
	Bidentate ligands
Ethylenediamine	$\mathrm{H_2\ddot{N}-CH_2-CH_2-\ddot{N}H_2}$
Acetylacetonate ion	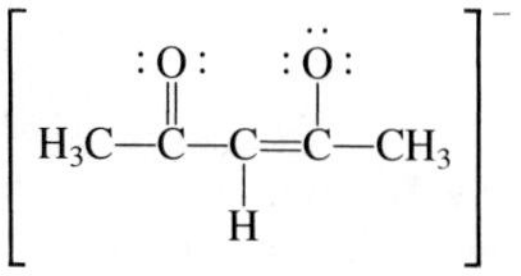 $\left[\mathrm{H_3C-C({=}\ddot{O}{:})-C(H){=}C(-\ddot{O}{:})-CH_3}\right]^{-}$
Oxalate ion	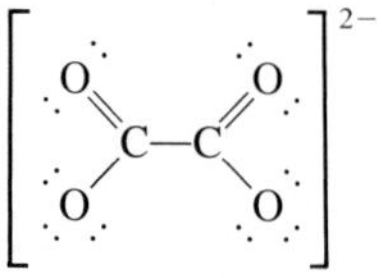 $\left[\mathrm{(\ddot{O}{=})(\ddot{O}{-})C-C(=\ddot{O})(-\ddot{O})}\right]^{2-}$
	Polydendate ligands
Ethylenediaminetetraacetate ion (EDTA)	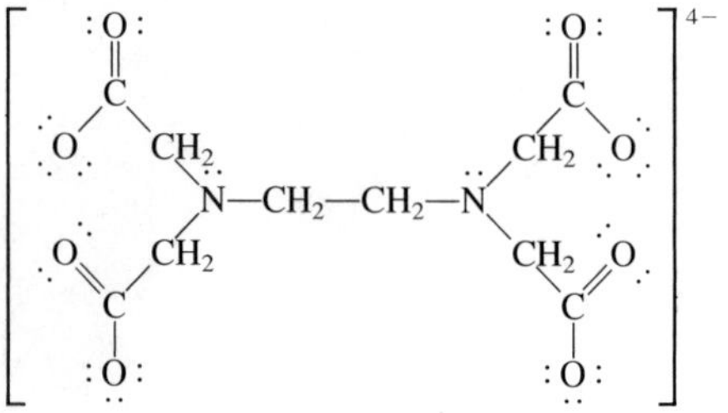 $\left[\mathrm{(^{-}OOC-CH_2)_2\ddot{N}-CH_2-CH_2-\ddot{N}(CH_2-COO^{-})_2}\right]^{4-}$
Tripolyphosphate ion	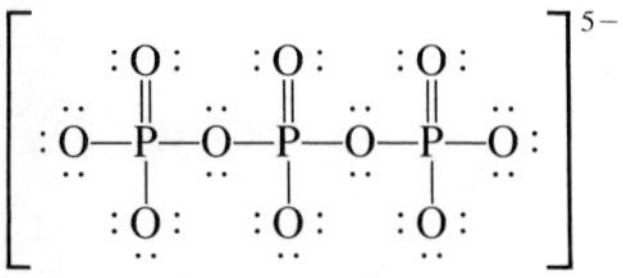 $\left[\mathrm{:\ddot{O}-P(=O)(-O)-\ddot{O}-P(=O)(-O)-\ddot{O}-P(=O)(-O)-\ddot{O}:}\right]^{5-}$

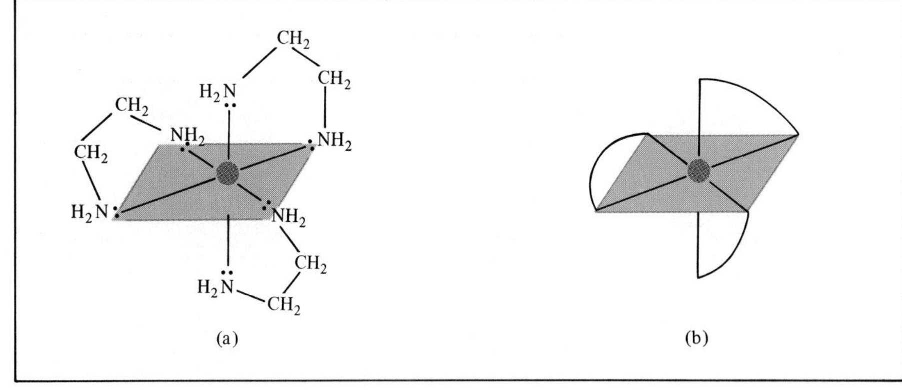

Figure 22.16 *(a) Structure of metal-ethylenediamine complex. Each ethylenediamine molecule provides two N donor atoms and is therefore a bidentate ligand. (b) Simplified structure of the same complex.*

In a crystal lattice the *coordination number* of an atom (or ion) is defined as the number of atoms (or ions) surrounding the atom (or ion).

atom. For example, nitrogen is the donor atom in the $[Cu(NH_3)_4]^{2+}$ complex ion. The ***coordination number*** in coordination compounds is defined as *the number of donor atoms surrounding the central metal atom in a complex ion.* For example, the coordination number of Ag^+ in $[Ag(NH_3)_2]^+$ is 2, that of Cu^{2+} in $[Cu(NH_3)_4]^{2+}$ is 4, and that of Fe^{3+} in $[Fe(CN)_6]^{3-}$ is 6. The most common coordination numbers are 4 and 6, but coordination numbers such as 2 and 5 are also known.

Depending on the number of donor atoms present in the molecule or ion, ligands can be classified as *monodentate, bidentate,* or *polydentate.* Ligands such as H_2O and NH_3 are monodentate because there is only one donor atom per ligand. A common bidentate ligand is ethylenediamine:

Ethylenediamine is sometimes abbreviated *en.*

$$H_2\ddot{N}—CH_2—CH_2—\ddot{N}H_2$$

The two nitrogen atoms can coordinate with a metal atom as shown in Figure 22.16.

The use of EDTA to treat metal poisoning is discussed in Chapter 20.

Ethylenediaminetetraacetate ion (EDTA) is a polydentate ligand (Table 22.3) containing six donor atoms—two nitrogen atoms and four oxygen atoms. The four oxygen atoms are in the four $—COO^-$ groups that are single-bonded to the carbon atoms. *Bi- and polydentate ligands* are also called ***chelating agents*** because of *their ability to hold the metal atom like a claw* (from the Greek *chele,* meaning "claw").

Oxidation Number of Metals in Coordination Compounds

Another important property of coordination compounds is the oxidation number of the central metal atom. The net charge of a complex ion is the sum of the charges on the central metal atom and its surrounding ligands. In the $[PtCl_6]^{2-}$ ion, for example, each chloride ion has an oxidation number of -1, so the oxidation number of Pt must be $+4$. If the ligands do not bear net charges, the oxidation number of the metal is equal to the charge of the complex ion. Thus, in $[Cu(NH_3)_4]^{2+}$ each NH_3 is neutral, so the oxidation number of Cu is $+2$.

Example 22.1 deals with oxidation number of metals in coordination compounds.

EXAMPLE 22.1

Find the oxidation number of the central metal atom in each of the following compounds: (a) $[Ru(NH_3)_5(H_2O)]Cl_2$, (b) $[Cr(NH_3)_6](NO_3)_3$, (c) $[Fe(CO)_5]$, (d) $K_4[Fe(CN)_6]$.

Answer

(a) Both NH_3 and H_2O are neutral species. Since each chloride ion carries a -1 charge, and there are two Cl^- ions, the oxidation number of Ru must be $+2$.
(b) Each nitrate ion has a charge of -1; therefore, the cation must be $[Cr(NH_3)_6]^{3+}$. NH_3 is neutral, so the oxidation number of Cr is $+3$.
(c) Since the CO species are neutral, the oxidation number of Fe is zero.
(d) Each potassium ion has a charge of $+1$; therefore, the anion is $[Fe(CN)_6]^{4-}$. Next, we know that each cyanide group bears a charge of -1, so Fe must have an oxidation number of $+2$.

Similar problem: 22.16.

Naming of Coordination Compounds

Now that we have discussed the various types of ligands and the oxidation number of metals, our next step is to learn what to call these coordination compounds. The rules for naming coordination compounds are as follows:

- The cation is named before the anion, as is the case for other ionic compounds. The rule holds regardless of whether the complex ion bears a net positive or a negative charge. For example, in $K_3[Fe(CN)_6]$ and $[Co(NH_3)_4Cl_2]Cl$, we name the K^+ and $[Co(NH_3)_4Cl_2]^+$ cations first, respectively.
- Within a complex ion the ligands are named first, in alphabetical order, and the metal ion is named last.
- The names of anionic ligands end with the letter *o*, whereas a neutral ligand is usually called by the name of the molecule. The exceptions are H_2O (aquo), CO (carbonyl), and NH_3 (ammine). Table 22.4 lists some common ligands.
- When several ligands of a particular kind are present, we use the Greek prefixes *di-*, *tri-*, *tetra-*, *penta-*, and *hexa-* to name them. Thus the ligands in $[Co(NH_3)_4Cl_2]^+$ are "tetraamminedichloro." (Note that prefixes play no role in the alphabetical order of ligands.) If the ligand itself contains a Greek prefix, we

Table 22.4 Names of common ligands in coordination compounds

Ligand	Name of ligand in coordination compound
Bromide, Br^-	Bromo
Chloride, Cl^-	Chloro
Cyanide, CN^-	Cyano
Hydroxide, OH^-	Hydroxo
Oxide, O^{2-}	Oxo
Carbonate, CO_3^{2-}	Carbonato
Nitrite, NO_2^-	Nitro
Oxalate, $C_2O_4^{2-}$	Oxalato
Ammonia, NH_3	Ammine
Carbon monoxide, CO	Carbonyl
Water, H_2O	Aquo
Ethylenediamine	Ethylenediamine
Ethylenediaminetetraacetate	Ethylenediaminetetraacetato

Table 22.5 Names of anions containing metal atoms

Metal	Name of metal in anionic complex
Aluminum	Aluminate
Chromium	Chromate
Cobalt	Cobaltate
Copper	Cuprate
Gold	Aurate
Iron	Ferrate
Lead	Plumbate
Manganese	Manganate
Molybdenum	Molybdate
Nickel	Nickelate
Silver	Argentate
Tin	Stannate
Tungsten	Tungstate
Zinc	Zincate

use the prefixes *bis (2), tris (3),* and *tetrakis (4)* to indicate the number of ligands present. For example, the ligand ethylenediamine already contains *di;* therefore, if two such ligands are present the name is *bis(ethylenediamine).*

- The oxidation number of the metal is written in Roman numerals following the name of the metal. For example, the Roman numeral III is used to indicate the +3 oxidation state of chromium in $[Cr(NH_3)_4Cl_2]^+$, which is called tetraamminedichlorochromium(III) ion.
- If the complex is an anion, its name ends in *-ate*. For example, in $K_4[Fe(CN)_6]$ the anion $[Fe(CN)_6]^{4-}$ is called hexacyanoferrate(II) ion. Note that the Roman numeral II indicates the oxidation state of iron. Table 22.5 gives the names of anions containing metal atoms.

The following examples deal with the nomenclature of coordination compounds.

EXAMPLE 22.2

Give the systematic names of the following compounds: (a) $Ni(CO)_4$, (b) $[Co(NH_3)_4Cl_2]Cl$, (c) $K_3[Fe(CN)_6]$, (d) $[Cr(en)_3]Cl_3$.

Answer

(a) The CO ligands are neutral species and the nickel atom bears no net charge, so the compound is called *tetracarbonylnickel(0),* or, more commonly, *nickel tetracarbonyl.*
(b) Starting with the cation, each of the two chloride ligands bears a negative charge and the ammonia molecules are neutral. Thus, the cobalt atom must have an oxidation number of +3 (to balance the chloride anion). The compound is called *tetraamminedichlorocobalt(III) chloride.*
(c) The complex ion is the anion and it bears three negative charges. Thus, the iron atom must have an oxidation number of +3. The compound is *potassium hexacyanoferrate(III).* This compound is commonly called *potassium ferricyanide.*
(d) As we saw earlier, *en* is the abbreviation for the ligand ethylenediamine. Since there are three *en* groups present and the name of the ligand already contains *di,* the name of the compound is *tris(ethylenediamine)chromium(III) chloride.*

Similar problem: 22.17.

EXAMPLE 22.3

Write the formulas for the following compounds: (a) pentaamminechlorocobalt(III) chloride, (b) dichlorobis(ethylenediamine)platinum(IV) nitrate, (c) sodium hexanitrocobaltate(III).

Answer

(a) The complex cation contains five NH_3 groups, a chloride ion, and a cobalt ion with a

+3 oxidation number. The net charge on the cation must be 2+. Therefore, the formula for the compound is $[Co(NH_3)_5Cl]Cl_2$.
(b) There are two chloride ions, two ethylenediamine groups, and a platinum ion with an oxidation number of +4 in the complex cation. Therefore, the formula for the compound is $[Pt(en)_2Cl_2](NO_3)_2$.
(c) The complex anion contains six nitro groups and a cobalt ion with an oxidation number of +3. Therefore, the formula for the compound is $Na_3[Co(NO_2)_6]$.

Similar problem: 22.18.

22.4 Stereochemistry of Coordination Compounds

In studying the geometry of coordination compounds, we often find that there is more than one way to arrange ligands around the central atom. Compounds rearranged in this fashion have distinctly different physical and chemical properties. Figure 22.17 shows four different geometric arrangements for metal atoms with monodentate ligands. In these diagrams we see that structure and coordination number of the metal atom relate to each other as follows:

Coordination number	Structure
2	Linear
4	Tetrahedral or square planar
6	Octahedral

The term ***stereoisomerism*** describes *the occurrence of two or more compounds with the same types and numbers of atoms and the same chemical bonds but different spatial arrangements*. ***Stereoisomers*** are *compounds possessing the same formula and bonding arrangement but different spatial arrangements of atoms*. There are two types of stereoisomers: geometric isomers and optical isomers. We will look into the importance of these stereoisomers in coordination compounds.

Geometric Isomers

Geometric isomers are *compounds with the same type and number of atoms and the same chemical bonds but different spatial arrangements; such isomers cannot be inter-*

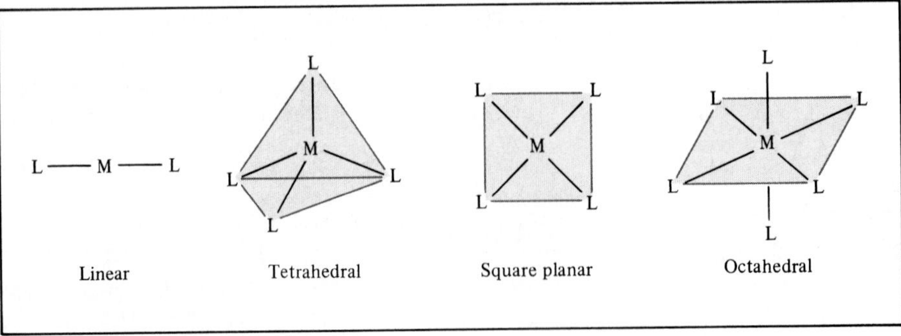

Figure 22.17 *Common geometries of complex ions. In each case M is a metal and L is a monodentate ligand.*

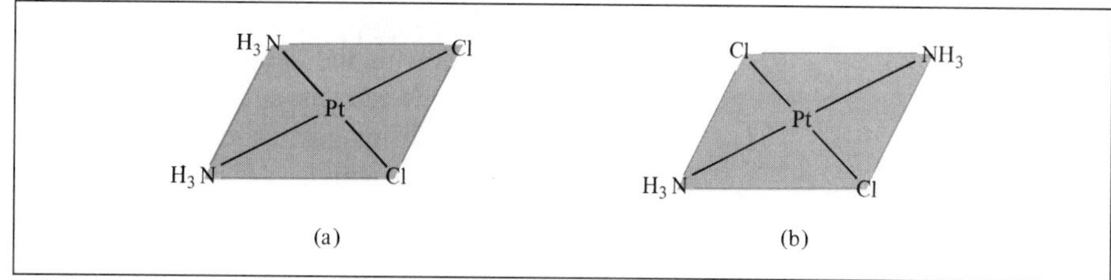

Figure 22.18 *The (a) cis and (b) trans isomers of diamminedichloroplatinum(II). Note that the two Cl atoms (and the two NH_3 molecules) are adjacent to each other in the cis isomer and diagonally across from each other in the trans isomer.*

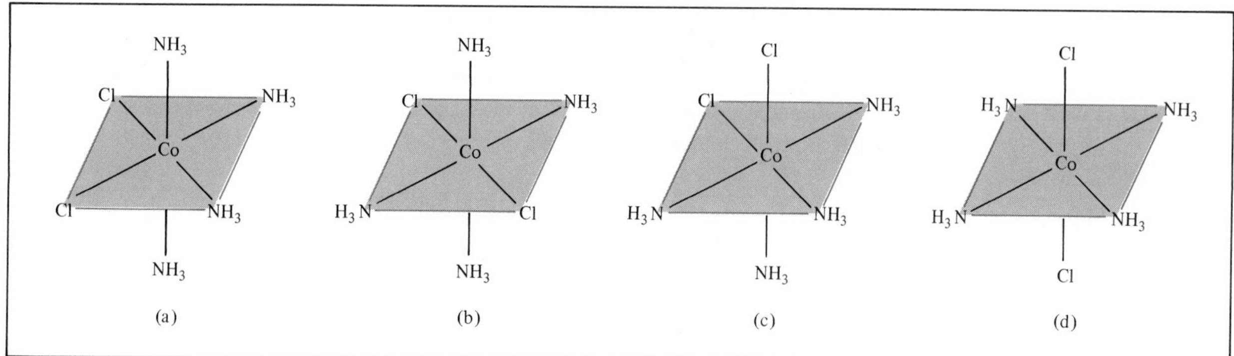

Figure 22.19 *The (a) cis and (b) trans isomers of tetraamminedichlorocobalt(III) ion. The structure shown in (c) can be generated by rotating that in (a), and the structure shown in (d) can be generated by rotating that in (b). Therefore, the ion has only two geometric isomers, (a) and (b) or (c) and (d).*

converted without breaking a chemical bond. We use the terms ''cis'' and ''trans'' to distinguish one geometric isomer from another of the same compound: Cis means that two particular atoms (or groups of atoms) are adjacent to each other, and trans means that the atoms (or groups of atoms) are on opposite sides in the structural formula. The cis and trans isomers of coordination compounds generally have quite different colors, melting points, dipole moments, and chemical reactivities. Figure 22.18 shows the cis and trans isomers of diamminedichloroplatinum(II). Note that although the types of bonds are the same in both isomers (two Pt—N and two Pt—Cl bonds), the spatial arrangements are different. Another example is tetraamminedichlorocobalt(III) ion, shown in Figure 22.19.

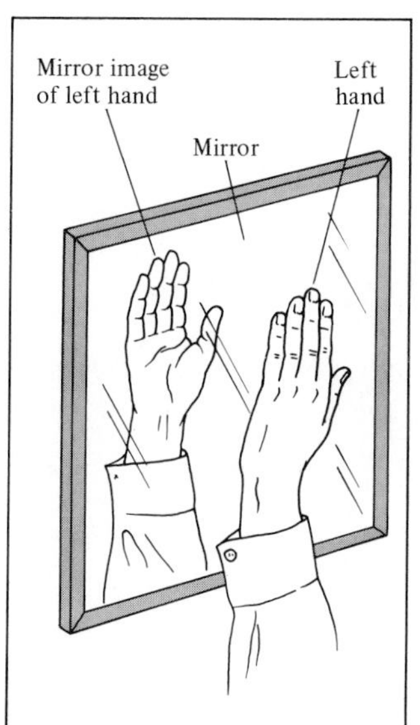

Figure 22.20 *A left hand and its mirror image, which looks the same as the right hand.*

Optical Isomers

Optical isomers are *compounds that are nonsuperimposable mirror images.* (*Superimposable* means that if one ion or molecule is placed over the other, the positions of all the atoms will match.) Like geometric isomers, optical isomers of a compound have the same type and number of atoms and the same chemical bonds. Optical isomers differ from geometric isomers, however, in that the former have *identical* physical and chemical properties such as melting point, boiling point, and chemical reactivity toward molecules that are not optical isomers themselves. The structural relationship between two optical isomers is analogous to the relationship between your left and right hands. If you place your left hand in front of a mirror, you will see that the image of your left hand in the mirror looks the same as your right hand (Figure 22.20)—we say that your left and right hands are mirror images of each other. However, they are nonsuperimposable, because when you place your left hand over your right hand (with both palms facing down), the hands do not match.

Figure 22.21 shows the cis and trans isomers of dichlorobis(ethylenediamine)-cobalt(III) ion and their images. Careful examination shows that the trans isomer and its mirror image are superimposable, but the same does not hold for the cis isomer and its mirror image. Therefore, the cis isomer and its mirror image are optical isomers. *Compounds or ions that are not superimposable with their mirror images* are called ***chiral*** and are said to be optically active; if they are superimposable with their mirror images, they are said to be achiral, or optically inactive. Thus the cis isomer shown in Figure 22.21 is chiral but the trans isomer is achiral. *Optical isomers, that is, compounds and their nonsuperimposable mirror images,* are also called ***enantiomers.***

Enantiomers differ with respect to their interaction with ***plane-polarized light,*** that is, *light in which the electric field and magnetic field components are confined to*

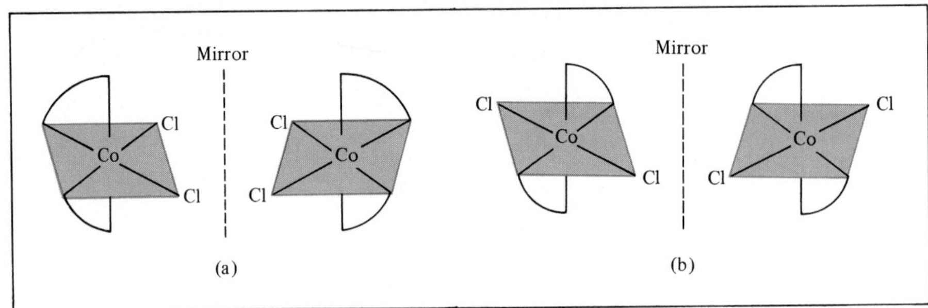

Figure 22.21 *The (a) cis and (b) trans isomers of dichlorobis(ethylenediamine)cobalt(III) ion and their mirror images. If you rotate the mirror image of the trans isomer 90° clockwise about the vertical position and place the ion over the trans isomer, you will find that the two are superimposable. The same does not hold for the cis isomer and its mirror image.*

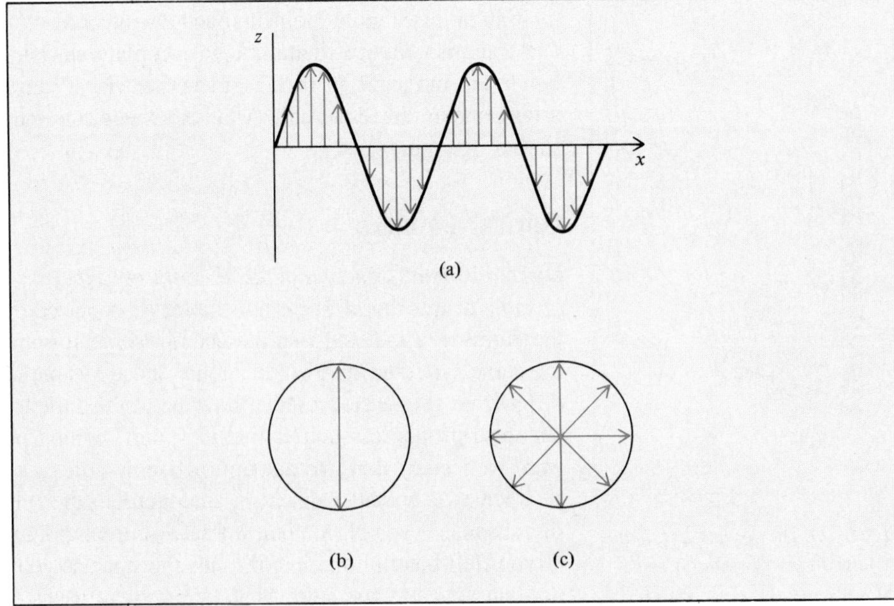

Figure 22.22 *(a) Propagation of an electromagnetic wave along the x-axis. Only the electric field component is shown. (b) End view (looking along the x-axis toward the source of radiation) of the vibration of the electric field component in one plane. (c) End view of the electric field component vibrating in many planes. The two-headed arrows represent the maximum of the amplitude of the wave in the positive and negative directions.*

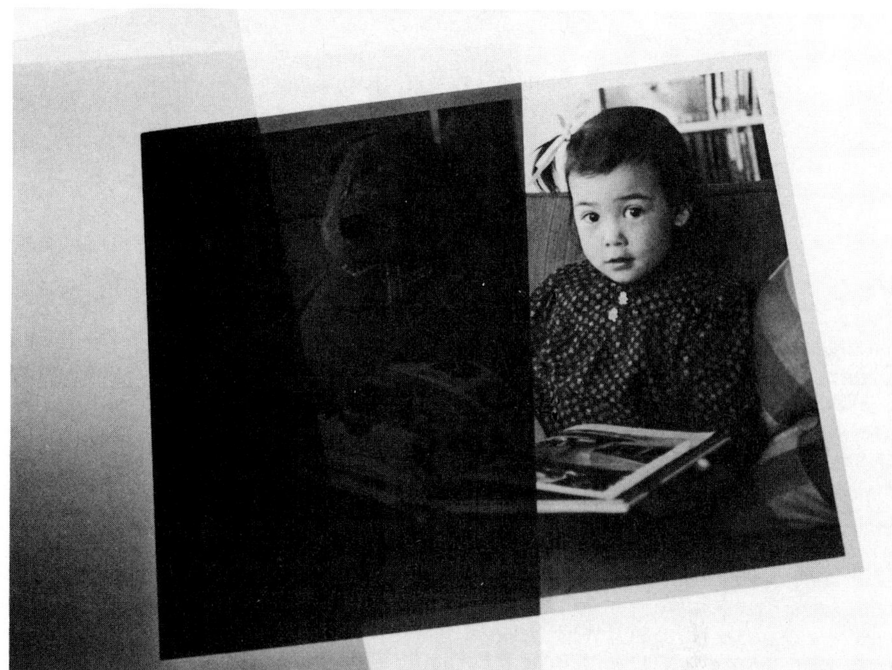

Figure 22.23 *With one Polaroid sheet over a picture, light passes through. With a second sheet of Polaroid placed over the first so that the axes of polarization of the sheets are perpendicular, little or no light passes through. If the axes of polarization of the two sheets were parallel, light would pass through.*

specific planes. In Section 7.1 we saw that the properties of light can be understood in terms of an oscillating electric field and a magnetic field component. Only the former is important in our discussion here, so let us examine the variation of this component along a given direction, as shown in Figure 22.22. In ordinary or unpolarized light, the waves are not confined to one plane as in Figure 22.22(b); rather, they vibrate in all planes, as represented by Figure 22.22(c). A simple way to produce polarized light is to pass a beam of ordinary light through a Polaroid sheet, which transmits light vibrating only in a particular plane (Figure 22.23).

Polaroid sheets are used to make Polaroid sunglasses.

A chiral molecule differs from an achiral molecule only in that the former can rotate the plane of polarization of plane-polarized light when the light is passed through the substance. If this plane is rotated to the right, the substance is *dextrorotatory (d);* to the left, *levorotatory (l)*. (The *d*- and *l*-isomers of the same compound are the two enantiomers.) The two enantiomers of the same chiral substance always rotate the light by the same amount, but in opposite directions. Thus, in *an equimolar mixture of the two optical isomers,* called a ***racemic mixture,*** the net rotation is zero.

The instrument for studying interaction between plane-polarized light and chiral molecules is known as a ***polarimeter*** (Figure 22.24). Initially, a beam of unpolarized light is passed through a Polaroid sheet, called the *polarizer,* and then through a sample tube. By rotating the *analyzer,* another Polaroid sheet, minimal light transmission can be achieved, as is shown in Figure 22.23. Next, the sample tube is filled with a solution containing an optically active compound. As the plane-polarized light passes through the sample tube this time, its plane of polarization is rotated either to the right or to the left by a certain amount, depending on whether the optical isomer is in the *d*- or *l*-form. This rotation can be measured readily by turning the analyzer in the appropriate direction until minimal light transmission is again achieved. The angle of rotation in degrees, α, depends not only on the nature of the molecules, but also on the concentration of the solution and the length of the sample tube.

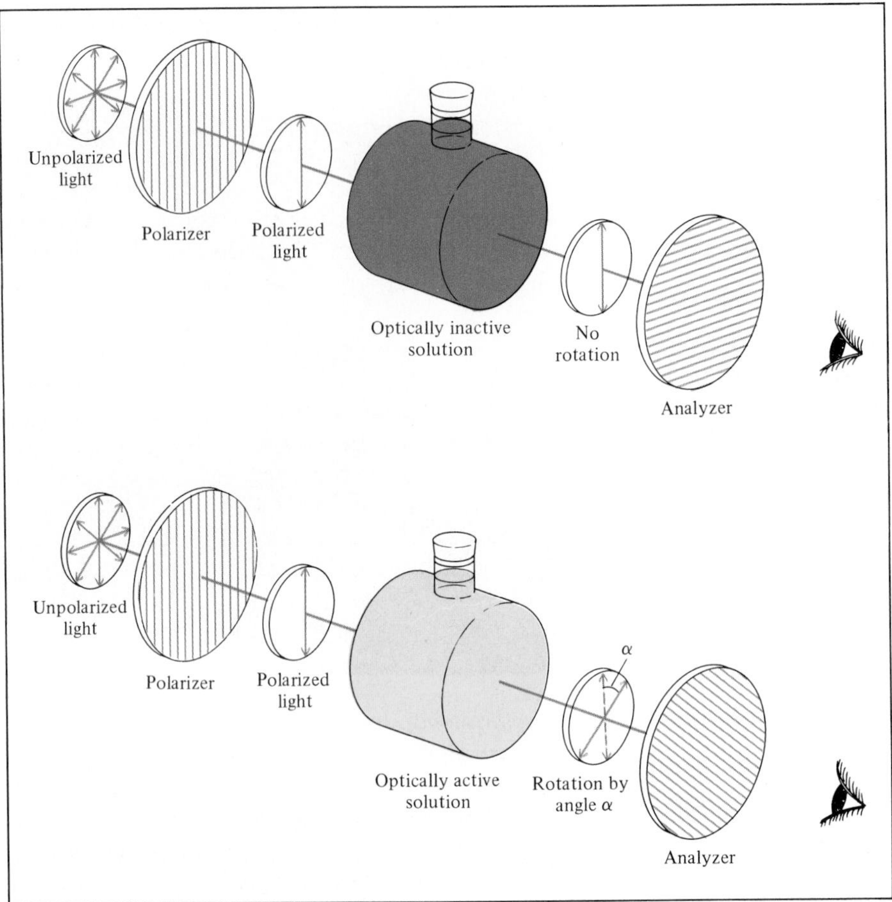

Figure 22.24 *An optical rotation experiment. Top: Two Polaroid sheets aligned for minimum light transmission. The achiral solution does not rotate the plane of the polarized light. Bottom: In a chiral medium, the plane of the polarized light is rotated by a certain angle α, so some light does pass through. To measure this angle, the analyzer is rotated by the same amount (but in the opposite direction) to again achieve minimum light transmission. A compound that rotates the plane of polarized light to the right as one looks toward the light source is called dextrorotatory (d); to the left, levorotatory (l).*

22.5 Bonding in Coordination Compounds

Our understanding of the nature of coordination compounds is the result of the classic work of Alfred Werner.† In 1893, at the age of 26, he proposed what is now commonly referred to as *Werner's coordination theory*.

The chemists of that day had been puzzled by a certain class of reactions. For example, it was known that the valences of the elements in cobalt(III) chloride and in ammonia seem to be completely satisfied. Yet these two substances react to form a stable compound having the formula $CoCl_3 \cdot 6NH_3$. To explain this behavior, Werner postulated that most elements exhibit two types of valence: *primary valence* and *secondary valence*, which, in modern terminology, correspond to the oxidation number and coordination number of the element. In $CoCl_3 \cdot 6NH_3$, then, cobalt has a primary valence of 3 and a secondary valence of 6.

Nowadays we write the formula of the compound as $[Co(NH_3)_6]Cl_3$ to indicate that the ammonia molecules and the cobalt atom form a complex ion; the chloride ions are not part of the complex, but are held to it by ionic forces.

† Alfred Werner (1866–1919). Swiss chemist. Werner started as an organic chemist, but became interested in coordination chemistry. He prepared and characterized hundreds of coordination compounds. For his major contribution, Werner was awarded the Nobel Prize in chemistry in 1913.

Werner also laid the basis for the stereochemistry of coordination compounds. During his relatively short scientific career, Werner synthesized hundreds of coordination compounds whose structures and properties provided strong support for his theory.

A satisfactory theory of bonding for coordination compounds must account for properties such as color and magnetism, as well as stereochemistry and bond strengths. No single theory as yet does all of this for us. Rather, several different approaches have been applied to transition metal complexes. We will consider only one of them here—the crystal field theory—because it is straightforward and it accounts satisfactorily for the colors and magnetic properties of many coordination compounds.

Crystal Field Theory

Crystal field theory considers the bonding in complex ions purely in terms of electrostatic interactions between the metal atom and the ligands. We will begin by considering octahedral complex ions.

The first question we address is: What effect will the surrounding ligands have on the energies of the metal atom's five *d* orbitals? As we observed in Chapter 7, *d* orbitals have various orientations, but in the absence of external disturbance they all have the same energy (see Figure 7.26). When such a metal ion is in the center of an octahedron surrounded by six lone pairs of electrons (on the six ligands), two types of electrostatic interaction come into play. First, there is the attraction between the positive metal ion and the negatively charged ligand or the negative end of a polar ligand. This is the force that holds the ligands to the metal in the complex. In addition, there is the electrostatic repulsion between the lone pairs on the ligands and the electrons in the *d* orbitals of the metal. However, the magnitude of this repulsion depends on the particular *d* orbital that is involved. Take the $d_{x^2-y^2}$ orbital as an example. We see that the lobes of this orbital point along the *x*- and *y*-axes, where the lone-pair electrons are positioned (Figure 22.25). Thus, an electron residing in this orbital would experience a

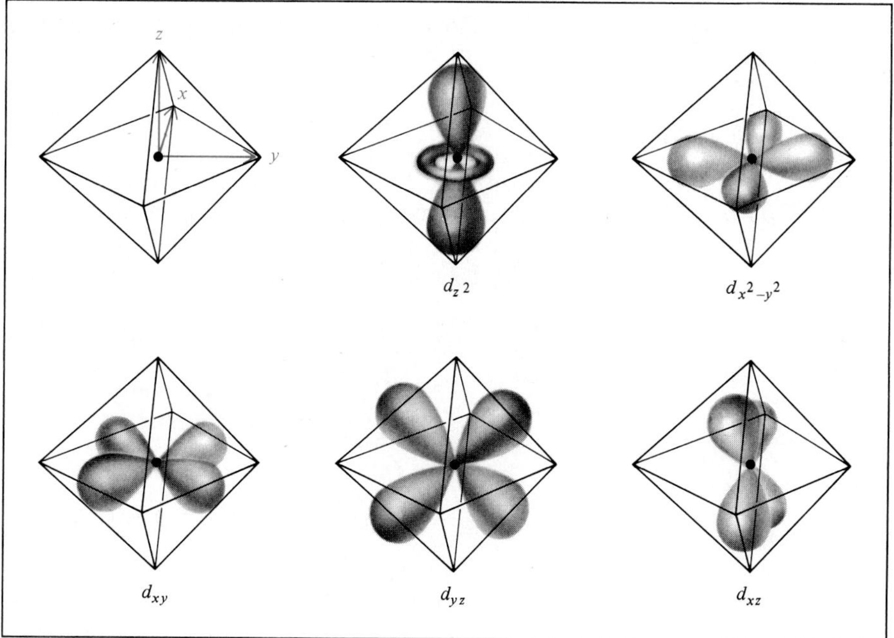

Figure 22.25 *The five d orbitals in an octahedral environment. The metal atom (or ion) is at the center of the octahedron, and the six lone pairs on the donor atoms of the ligands are at the corners.*

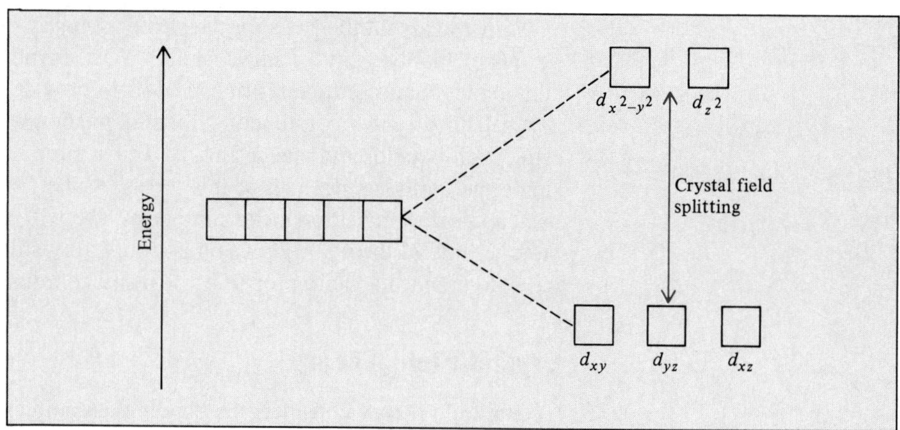

Figure 22.26 *Crystal field splitting between d orbitals in an octahedral complex.*

greater repulsion from the ligands than an electron would in, say, the d_{xy} orbital. For this reason the energy of the $d_{x^2-y^2}$ orbital is increased relative to the d_{xy}, d_{yz}, and d_{xz} orbitals. The d_{z^2} orbital energy is also increased, because its lobes are pointed at the ligands along the z-axis.

As a result of these metal-ligand interactions, the equality (in energy) of the five d orbitals is nullified to give two high-lying equal-energy levels ($d_{x^2-y^2}$ and d_{z^2}) and three low-lying equal-energy levels (d_{xy}, d_{yz}, and d_{xz}), shown in Figure 22.26. *The energy difference between these two sets of d orbitals* is called the ***crystal field splitting*** (Δ); its magnitude depends on the metal and the nature of the ligands. With this basic energy level scheme in mind, we are ready to discuss two characteristics of octahedral transition metal complex ions: color and magnetic properties.

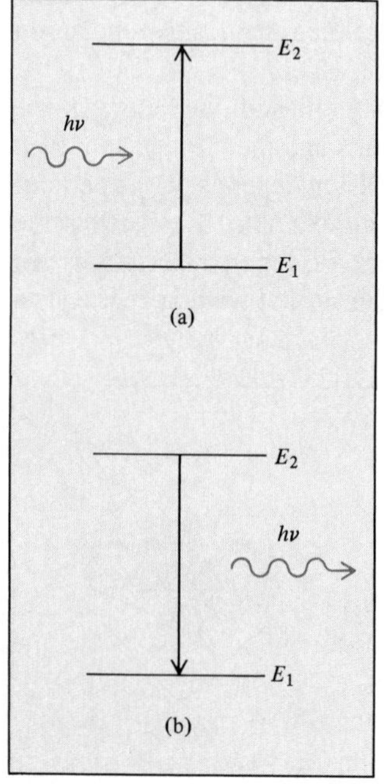

Figure 22.27 *(a) A molecule absorbs a photon with energy $h\nu$. (b) In the emission process, a molecule gives off a photon with energy $h\nu$. The frequency—and hence the wavelength—of the photon involved in either case depends on the difference between the two energy levels.*

Color. A substance appears colored because it absorbs light at one or more wavelengths in the visible part of the electromagnetic spectrum (400 to 700 nm) (see Figure 7.4) and reflects or transmits the others. Each wavelength of light in this region appears as a different color. A combination of all colors appears white (as in sunlight); an absence of lightwaves appears black. Table 22.6 shows the relationship of absorbed wavelength to the observed color. From this table, we can infer, for example, that the blue hydrated cupric ion $Cu(H_2O)_6^{2+}$ absorbs light in the orange region.

Figure 22.27 shows the quantum-mechanical description of the absorption and emission of light. For the sake of simplicity, we consider two particular electron energy levels involved in a transition. Absorption of light may occur when the frequency of the incoming photon, multiplied by the Planck constant, is equal to the difference in energy between these two levels; that is (see Section 7.3),

Table 22.6 Relationship of wavelength to color

Wavelength absorbed (nm)	Color observed
400 (violet)	Greenish yellow
450 (blue)	Orange yellow
490 (blue green)	Red
570 (yellow green)	Purple
580 (yellow)	Dark blue
600 (orange)	Blue
650 (red)	Green

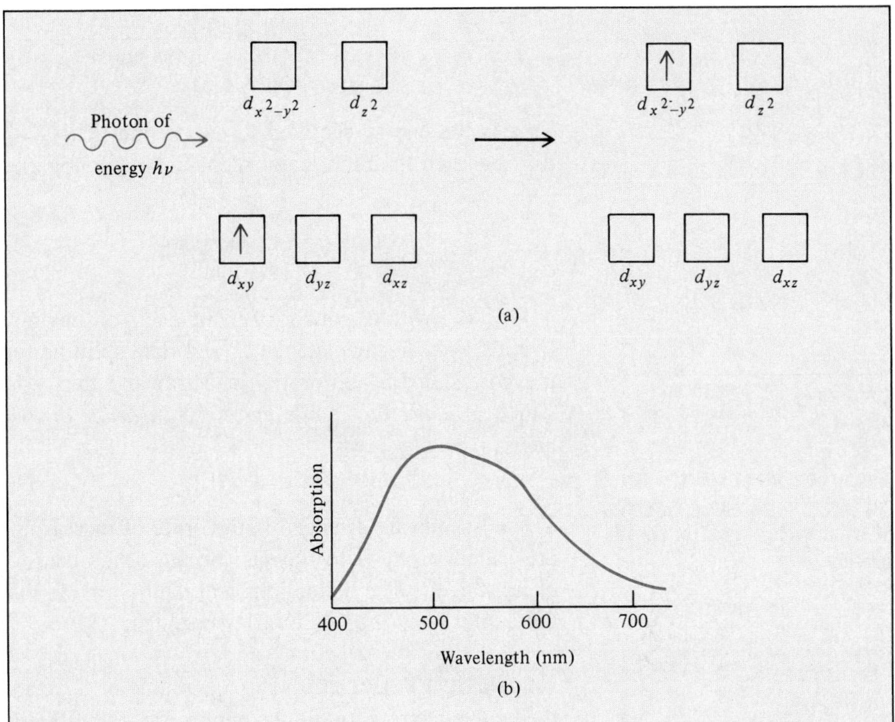

Figure 22.28 *(a) The absorptive process of a photon and (b) a graph of the absorption spectrum of $[Ti(H_2O)_6]^{3+}$. The energy of the incoming photon is equal to the crystal field splitting. The maximum absorption peak in the visible region occurs at 498 nm.*

$$\Delta E = h\nu$$

When we say that the hydrated cupric ion is blue, we mean that each ion absorbs a photon whose frequency is about 5×10^{14} Hz, which corresponds to a wavelength of about 600 nm (orange light). When this component of light is removed, the transmitted light no longer looks white but appears blue to our eyes. With this knowledge, we can calculate the energy change involved in the electron transition that occurs in the cupric ion:

$$\begin{aligned} \Delta E &= (6.63 \times 10^{-34} \text{ J s})(5 \times 10^{14}/\text{s}) \\ &= 3 \times 10^{-19} \text{ J} \end{aligned}$$

Figure 22.29 *A titanium(III) chloride solution.*

If the wavelength of the photon absorbed by a molecule lies outside the visible region, then the transmitted light looks the same (to us) as the incident light—white—and the substance made up of these molecules appears colorless.

Spectroscopic techiques offer the best means of measuring crystal field splitting. The $[Ti(H_2O)_6]^{3+}$ ion provides a particularly simple example, because Ti^{3+} has only one $3d$ electron (Figure 22.28). Figure 22.29 shows an aqueous solution of titanium(III) chloride. The $[Ti(H_2O)_6]^{3+}$ ion absorbs light in the visible region; the wavelength corresponding to maximum absorption is 498 nm [Figure 22.28(b)]. This information enables us to calculate the crystal field splitting as follows. We start by writing

$$\Delta = h\nu$$

Also

$$\nu = \frac{c}{\lambda}$$

where c is the velocity of light and λ is the wavelength. Therefore

$$\Delta = \frac{hc}{\lambda} = \frac{(6.63 \times 10^{-34}\text{ J s})(3.00 \times 10^{8}\text{ m/s})}{(498\text{ nm})(1 \times 10^{-9}\text{ m/1 nm})}$$
$$= 3.99 \times 10^{-19}\text{ J}$$

This is the energy required to excite one $[Ti(H_2O)_6]^{3+}$ ion. To express this energy difference in the more convenient units of kilojoules per mole, we write

$$\Delta = (3.99 \times 10^{-19}\text{ J/ion})(6.02 \times 10^{23}\text{ ions/mol})$$
$$= 240{,}000\text{ J/mol}$$
$$= 240\text{ kJ/mol}$$

Working with a number of complexes, all having the same metal ion but different ligands, we can calculate the crystal field splitting for each ligand (aided by data from the corresponding absorption spectra) and thus establish a ***spectrochemical series,*** which is *a list of ligands arranged in order of their abilities to split the d orbital energies:*

The order in the spectrochemical series is the same regardless of what metal atom (or ion) is present.

$$I^- < Br^- < Cl^- < OH^- < F^- < H_2O < NH_3 < \text{en} < CN^- < CO$$

These ligands are arranged in the order of increasing value of Δ. Thus, CO and CN^- are called strong-field ligands, because they cause a large splitting of the *d* orbital energy levels. The halide ions and the hydroxide ion are weak-field ligands, because they split the *d* orbitals to a lesser extent.

Magnetic Properties. The magnitude of the crystal field splitting also determines the magnetic properties of a complex ion. For $[Ti(H_2O)_6]^{3+}$, the single *d* electron must be in one of the three lower orbitals, and the ion is always paramagnetic. However, in an ion where several *d* electrons are present, as in Fe^{3+} complexes, the situation becomes more involved. Consider the octahedral complexes $[FeF_6]^{3-}$ and $[Fe(CN)_6]^{3-}$ (Figure 22.30). The electron configuration of Fe^{3+} is $3d^5$, and there are two possibilities for placing the five *d* electrons in the five *d* orbitals. According to Hund's rule (see Section 7.9), maximum stability is reached when the electrons enter five separate orbitals with parallel spins. But this arrangement can be achieved only at a cost; two of the five electrons must be energetically promoted to the high-lying $d_{x^2-y^2}$ and d_{z^2} orbitals. No such energy investment is needed if all five electrons enter the d_{xy}, d_{xz}, and d_{yz} orbitals. According to Pauli's exclusion principle (p. 290), there will only be one unpaired electron present in this case.

The actual arrangement of the electrons is determined by the amount of stability gained by having maximum parallel spins versus the investment in energy required to

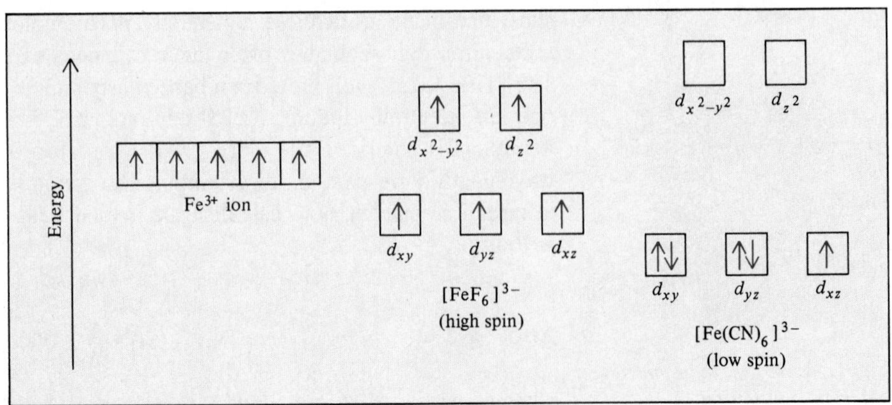

Figure 22.30 *Energy level diagrams for the Fe^{3+} ion and $[FeF_6]^{3-}$ and $[Fe(CN)_6]^{3-}$ complex ions.*

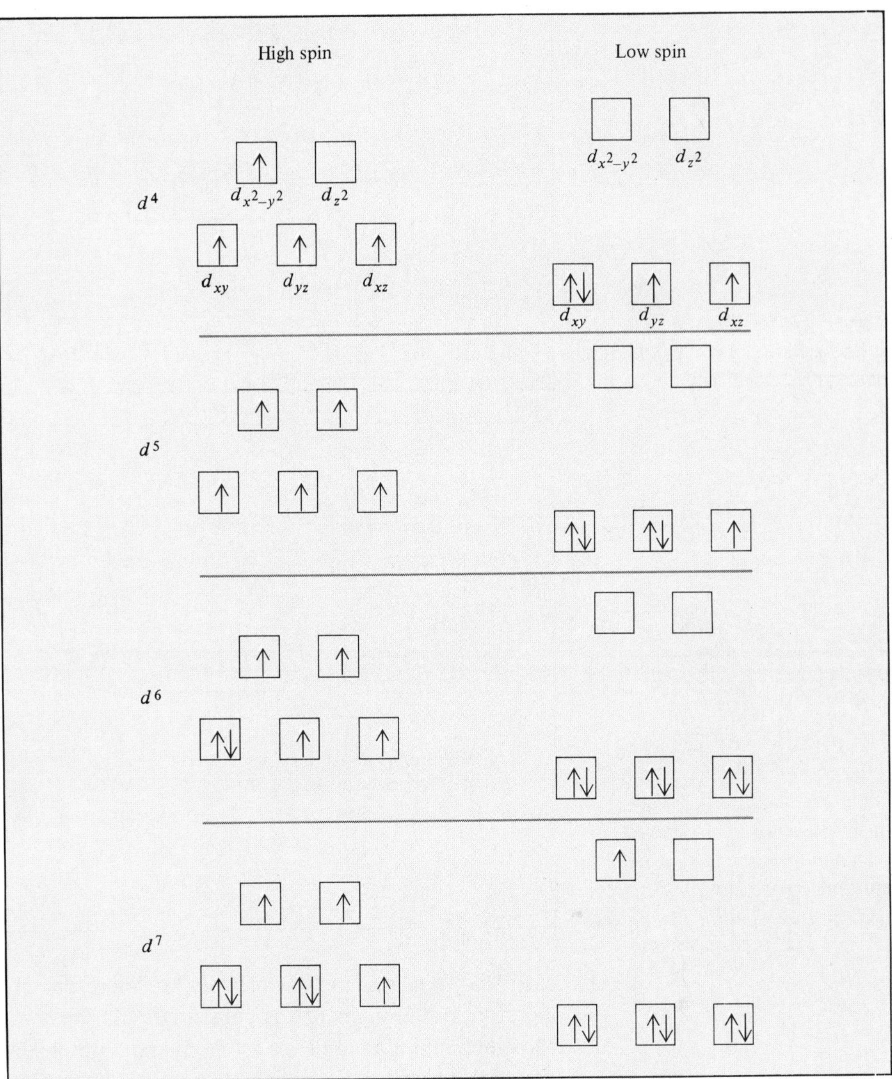

Figure 22.31 *Orbital diagrams for the high-spin and low-spin octahedral complexes corresponding to the electron configurations d^4, d^5, d^6, and d^7. No such distinctions can be made for d^1, d^2, d^3, d^8, d^9, and d^{10}.*

promote electrons to higher *d* orbitals. Because F^- is a weak-field ligand, the five *d* electrons enter five separate *d* orbitals with parallel spins to create a *high-spin* complex. On the other hand, the cyanide ion is a strong-field ligand, so it is energetically preferable for all five electrons to be in the lower orbitals and so a *low-spin* complex is formed.

The magnetic properties of a complex ion depend on the number of unpaired electrons present. High-spin complexes are more paramagnetic than low-spin complexes.

The actual number of unpaired electrons (or spins) in a complex ion can be found by magnetic measurements, and the general agreement between theory and experiment supports the usefulness of the crystal field theory. A distinction between low- and high-spin complexes can be made only if the metal ion contains more than three and fewer than eight *d* electrons. Figure 22.31 shows the distribution of electrons among *d* orbitals that results in low- and high-spin complexes.

So far we have concentrated on octahedral complexes. The splitting of the *d* orbital energy levels in two other types of complexes—tetrahedral and square-planar—can also be accounted for satisfactorily by the crystal field theory. In fact, the splitting pattern for a tetrahedral ion is just the reverse of that for octahedral complexes. In this

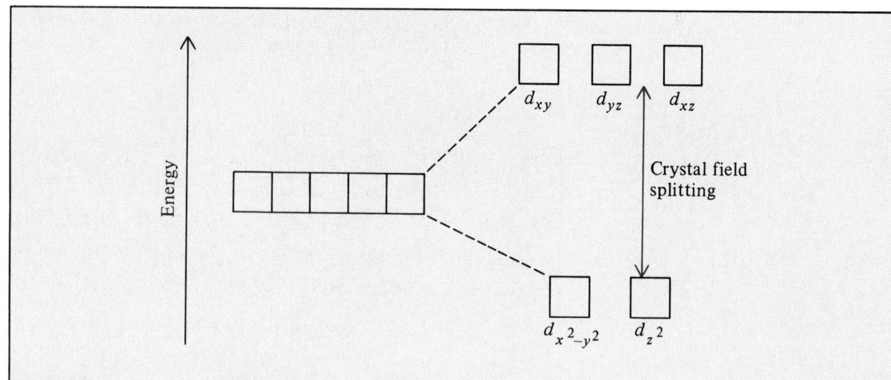

Figure 22.32 *Crystal field splitting between d orbitals in a tetrahedral complex.*

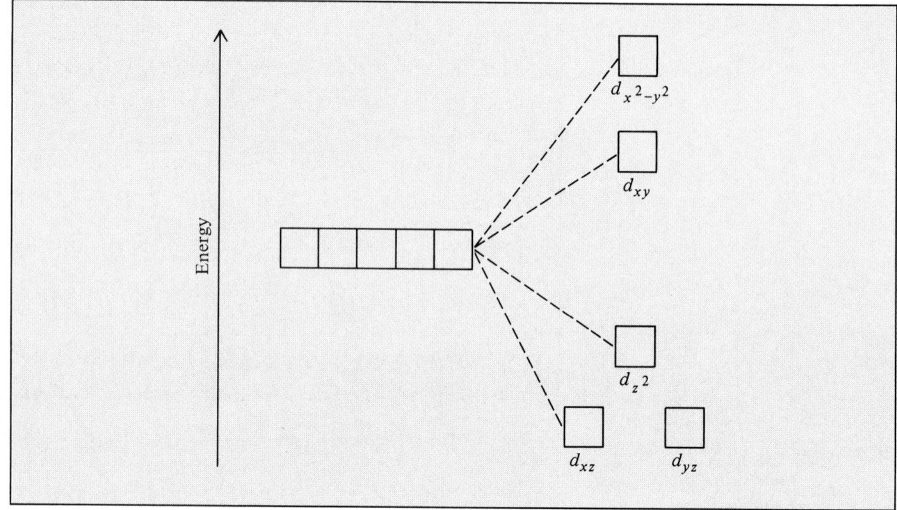

Figure 22.33 *Energy level diagram for a square-planar complex. Because there are more than two energy levels, we cannot define crystal field splitting as we can for octahedral and tetrahedral complexes.*

case, the d_{xy}, d_{xz}, and d_{yz} orbitals are more closely directed at the ligands and are therefore raised in energy as compared with the $d_{x^2-y^2}$ and d_{z^2} orbitals (Figure 22.32). Most tetrahedral complexes are high-spin complexes. Presumably, the tetrahedral arrangement reduces the magnitude of metal-ligand interactions, resulting in a smaller Δ value. This is a reasonable assumption since the number of ligands is smaller in a tetrahedral complex.

As Figure 22.33 shows, the splitting pattern for square-planar complexes is the most complicated. Clearly, the $d_{x^2-y^2}$ orbital possesses the highest energy (as in the octahedral case), and the d_{xy} orbital the next highest. However, the relative placement of the d_{z^2} and the d_{xz} and d_{yz} orbitals cannot be determined simply by inspection and must be calculated.

22.6 Reactions of Coordination Compounds

Complex ions undergo ligand exchange (or substitution) reactions in solution. The rates of these reactions vary widely, depending on the nature of the metal ion and the ligands. In studying ligand exchange reactions, it is often useful to distinguish between the stability of a complex ion (as measured by its formation constant; see p. 736) and its tendency to react, which we call *kinetic lability*. For example, when we say that tetracyanonickelate(II) complex ion is stable, we are talking about its large formation constant (about 1×10^{30}):

$$Ni^{2+} + 4CN^- \rightleftharpoons [Ni(CN)_4]^{2-}$$

Therefore, the stability of a complex ion refers to the thermodynamic property of the species because it is measured in terms of the magnitude of the equilibrium constant. By using cyanide ions labeled with the radioactive isotope carbon-14, it has been found that the complex ion undergoes ligand exchange very rapidly in solution (the following equilibrium is established almost as soon as the species are mixed):

$$[Ni(CN)_4]^{2-} + 4{*CN^-} \rightleftharpoons [Ni(*CN)_4]^{2-} + 4CN^-$$

where the asterisk denotes a ^{14}C atom. *Complex ions* such as $[Ni(CN)_4]^{2-}$ *that undergo rapid ligand exchange reactions* are called ***labile complexes.*** Thus a thermodynamically stable species (that is, one that has a large formation constant) is not necessarily unreactive. The above exchange reaction is fast because it has a small energy of activation. (In Section 13.4 we saw that the smaller the activation energy, the larger the rate constant, and hence the greater the rate.)

An ***inert complex*** is *a complex ion that undergoes very slow ligand exchange reactions* (on the order of hours or even days). An example is

$$[Cr(CN)_6]^{3-} + 6{*CN^-} \rightleftharpoons [Cr(*CN)_6]^{3-} + 6CN^-$$

This reaction takes days to complete. Here the slow rate is due to a high activation energy.

An inert complex that is thermodynamically unstable in acidic solution is $[Co(NH_3)_6]^{3+}$. The equilibrium constant for the following reaction is about 1×10^{20}:

$$[Co(NH_3)_6]^{3+} + 6H^+ + 6H_2O \rightleftharpoons [Co(H_2O)_6]^{3+} + 6NH_4^+$$

When equilibrium is reached, the concentration of the $[Co(NH_3)_6]^{3+}$ ion is very low. However, this reaction requires several days to complete because of the inertness of the $[Co(NH_3)_6]^{3+}$ ion. This example shows that a thermodynamically unstable species is not necessarily chemically reactive. The rate of reaction is determined by the energy of activation, which is high in this case.

Most complex ions containing Co^{3+}, Cr^{3+}, and Pt^{2+} are kinetically inert. Because they exchange ligands so slowly, they are easy to study in solution. As a result, our knowledge of the bonding, structure, and isomerism of coordination compounds has come largely from studies of these compounds.

22.7 Applications of Coordination Compounds

Coordination compounds find many uses in households, industry, and medicine. We describe a few representative examples here.

Metallurgy

The extraction of silver and gold by the formation of cyanide complexes (p. 873) and the purification of nickel (p. 828) by converting the metal to the gaseous compound $Ni(CO)_4$ are typical examples of the use of coordination compounds in metallurgical processes.

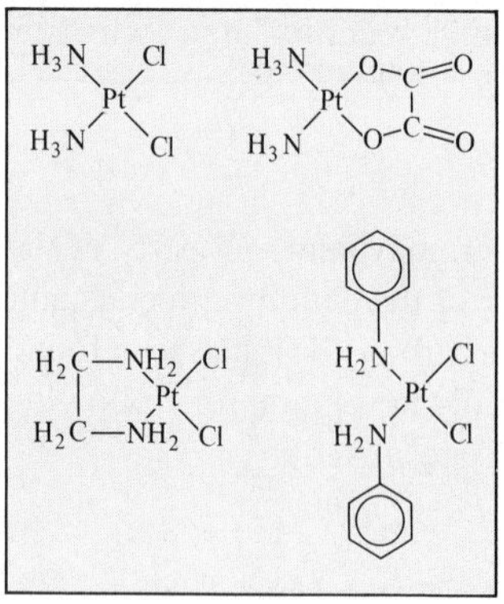

Figure 22.34 *Structures of some platinum complexes used in cancer treatment. Only the cis isomers are effective antitumor chelating agents.*

Therapeutic Chelating Agents

In Sections 20.8 and 20.9 we mentioned the use of chelating agents such as EDTA and BAL (2,3-dimercaptopropanol) in the treatment of lead and mercury poisoning. Recent studies have shown that some of the coordination complexes of platinum effectively inhibit the growth of cancerous cells. It has long been suspected that chelation processes are associated both with the development of cancer as well as with antitumor activity in the living cell. In fact, a compound that has antitumor activity in one environment generally functions as a carcinogen in another. Figure 22.34 shows some of the platinum(II) complexes that exhibit antitumor activity. The striking feature of all these complexes is the cis arrangement of identical ligands. *trans*-Diamminedichloroplatinum(II)

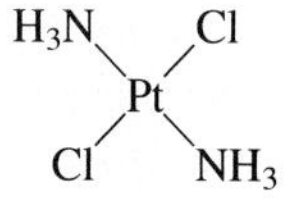

and other trans isomers are not active as antitumor agents. Apparently, the antitumor activity is associated in some way with reactions involving chelation. In *cis*-diamminedichloroplatinum(II), the two chlorine atoms are more easily removed by another chelating agent than the ammonia ligands. Replacement of the chlorine atoms in the trans isomer by a chelating agent does not occur readily. The mechanism by which the platinum complexes act is not entirely clear.

Chemical Analysis

Although EDTA has a great affinity for a large number of metal ions (especially 2+ and 3+ ions), other chelates are more selective in binding. For example, dimethylglyoxime,

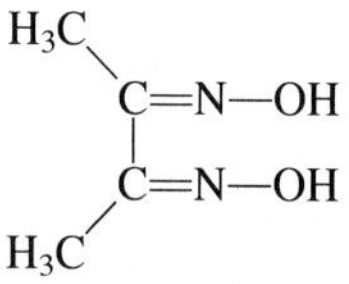

forms an insoluble brick red solid with Ni^{2+} (Figure 22.35) and an insoluble bright yellow solid with Pd^{2+}. These characteristic colors are used in qualitative analysis to identify nickel and palladium. Further, the quantities of ions present can be determined by gravimetric analysis (see Section 4.5), as follows: To a solution containing Ni^{2+} ions, say, we add an excess of dimethylglyoxime reagent, and a brick red precipitate

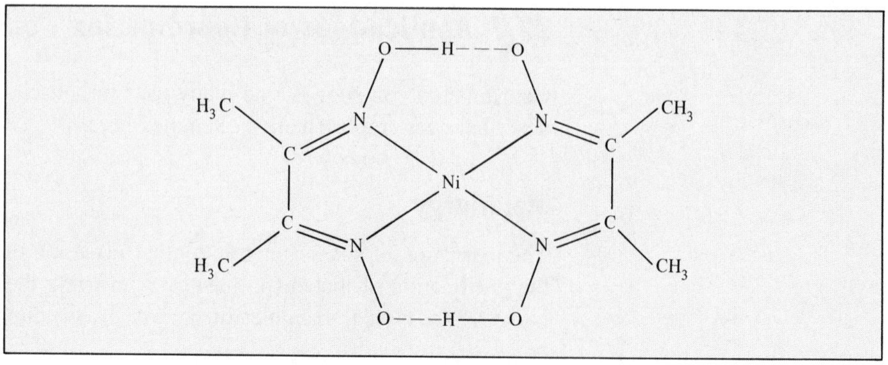

Figure 22.35 *Structure of nickel dimethylglyoxime. Note that the overall structure is stabilized by hydrogen bonds.*

forms. The precipitate is then filtered, dried, and weighed. Knowing the formula of the complex (see Figure 22.35), we can readily calculate the amount of nickel present in the original solution.

Plant Growth

Plants need various nutrients for healthy growth. Essential nutrients include a number of metals such as iron, zinc, copper, manganese, and molybdenum. Iron in the +3 state in the soil is mostly hydrolyzed to form insoluble iron hydroxides such as $Fe(OH)_3$ which cannot be taken up by plants. Plants deficient in iron are likely to develop a disorder known as *iron chlorosis,* evidenced by yellowing leaves. Iron chlorosis particularly affects the yield of fruit from citrus trees. The standard treatment, frequently used in citrus groves, supplies the trees with Fe(III)–EDTA complex. This complex is soluble in water and readily enters the roots of trees, where it is eventually converted into a utilizable form. Certain strains of soybeans growing in alkaline soil, which favors the formation of iron hydroxides, generate and secrete into the soil chelating agents that solubilize the iron needed for plant growth.

Detergents

We saw in Section 20.6 that the cleansing action of soap in hard water is hampered by the reaction of the Ca^{2+} ions in the water with the soap molecules to form insoluble salts or curds. In the late 1940s the detergent industry began to introduce a "builder," usually sodium tripolyphosphate, to circumvent this problem. The tripolyphosphate ion (see Table 22.3) is an effective chelating agent that forms stable, soluble complexes with Ca^{2+} ions. Sodium tripolyphosphate was introduced in the synthetic detergent Tide in 1947, and its use has revolutionized the industry. However, because phosphates are plant nutrients, waste waters containing phosphates discharged into lakes cause algae to grow, resulting in oxygen depletion. Under these conditions most or all aquatic life eventually succumbs. This process is called *eutrophication*. Since the 1970s, the use of phosphates in detergents has been banned in some states such as Minnesota and Wisconsin.

CHEMISTRY IN ACTION

Coordination Compounds in Living Systems

Coordination compounds play many important roles in animals and plants. They are essential in the storage and transport of oxygen, as electron transfer agents, as catalysts, and in photosynthesis. Here we focus on coordination compounds containing iron and magnesium.

Because of its central role as an oxygen carrier for metabolic processes, hemoglobin is probably the most studied of all the proteins. The molecule contains four folded long chains called *subunits*. The main function of hemoglobin is to carry oxygen in the blood from the lungs to the tissues, where it delivers the oxygen molecules to myoglobin. Myoglobin, which is made up of only one subunit, stores oxygen for metabolic processes in muscle.

Figure 22.36 shows the structure of the porphine molecule, which forms an important part of the hemo-

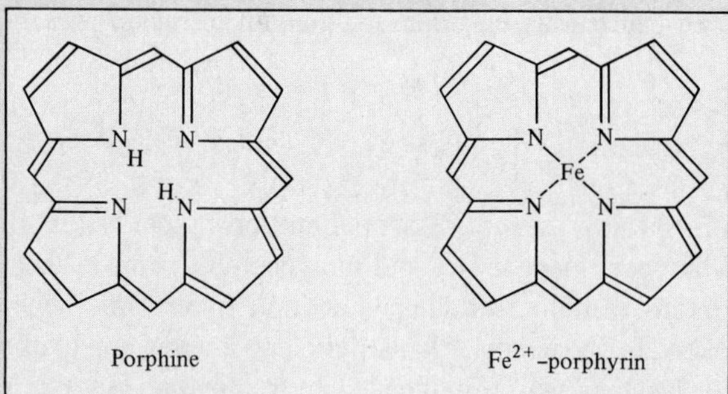

Figure 22.36 *Structures of the porphine molecule and the Fe^{2+}–porphyrin complex. The dotted lines indicate coordinate covalent bonds.*

globin structure. Upon coordination to a metal, the two H^+ ions shown bonded to nitrogen atoms are displaced. Complexes derived from porphine are called *porphyrins,* and the iron-porphyrin combination is called the *heme* group. The iron in the heme group has the oxidation number +2; it is coordinated to the four nitrogen atoms in the porphine group and also to a nitrogen donor atom in a ligand group which is attached to the protein (Figure 22.37). In the absence of oxygen, the sixth ligand is a water molecule, which binds to the Fe^{2+} ion on the other side of the ring to complete the octahedral complex. Under these conditions, the hemoglobin molecule is called *deoxyhemoglobin* and has a blue color characteristic of venous blood. The water ligand can be replaced readily by molecular oxygen to form the red-colored oxyhemoglobin present in arterial blood. Each subunit contains a heme group, so each hemoglobin molecule can bind up to four O_2 molecules.

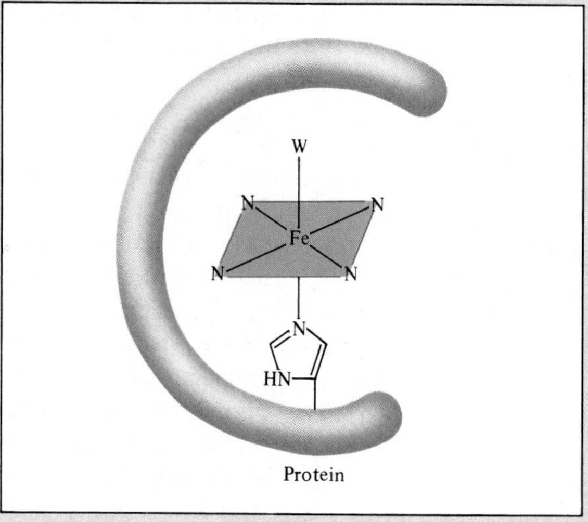

Figure 22.37 *The heme group in hemoglobin. The Fe^{2+} ion is coordinated with the nitrogen atoms of the heme group. The ligand below the porphyrin is the histidine group that is attached to the protein. The sixth ligand is a water molecule. The water molecule is denoted by W because the exact geometry of the bonding of H_2O to Fe^{2+} is not known.*

For a number of years, the exact arrangement of the oxygen molecule relative to the porphyrin group was not clear. Figure 22.38 shows three possible arrange-

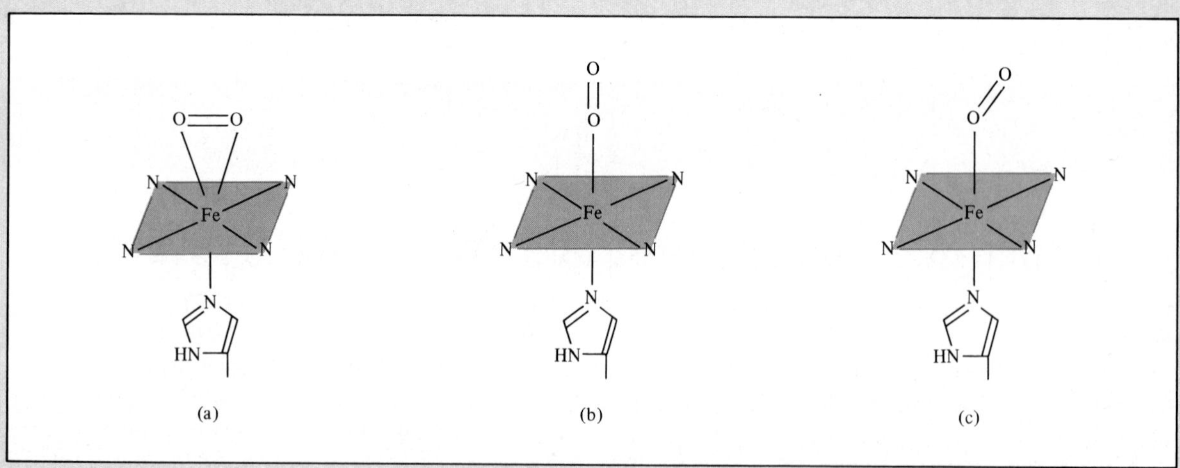

Figure 22.38 *Three possible ways for molecular oxygen to bind to the heme group in hemoglobin. The structure shown in (c) is the most likely arrangement.*

ments in oxyhemoglobin. Figure 22.38(a) would necessitate a coordination number of 7, which is considered unlikely for Fe(II) complexes. Of the remaining two structures, the end-on arrangement shown in Figure 22.38(b) may seem more reasonable; however, evidence points to the structure in Figure 22.38(c) as the most plausible.

The porphyrin group is a very effective chelating agent, and, not surprisingly, we find it in a number of biological systems. The iron-heme complex is present in another class of proteins, called the *cytochromes*. Figure 22.39 shows the structure of cytochrome *c*, which is the most studied member of this class of compounds. The iron forms an octahedral complex, but because both the histidine and the methionine groups are firmly bound to the metal ion, these ligands cannot be displaced by oxygen or other ligands. Instead, the cytochromes act as electron carriers, which also play an essential part in metabolic processes. In cytochromes iron undergoes rapid reversible redox reactions:

$$Fe^{3+} + e^- \rightleftharpoons Fe^{2+}$$

which are coupled to the oxidation of organic molecules such as the carbohydrates.

The chlorophyll molecule, which plays an essential role in photosynthesis, also contains the porphyrin ring, but the metal ion there is Mg^{2+} rather than Fe^{2+} (Figure 22.40). Figure 22.41 shows the natural green color of chlorophyll (in a solution) at the bottom of a

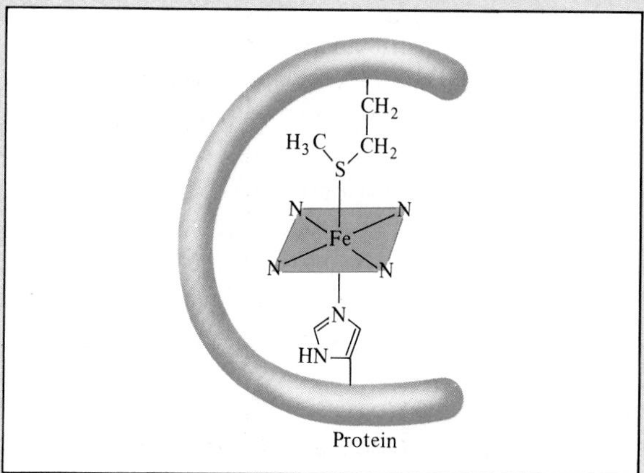

Figure 22.39 *The heme group in cytochrome c. The ligands above and below the porphyrin are the methionine group and histidine group of the protein, respectively.*

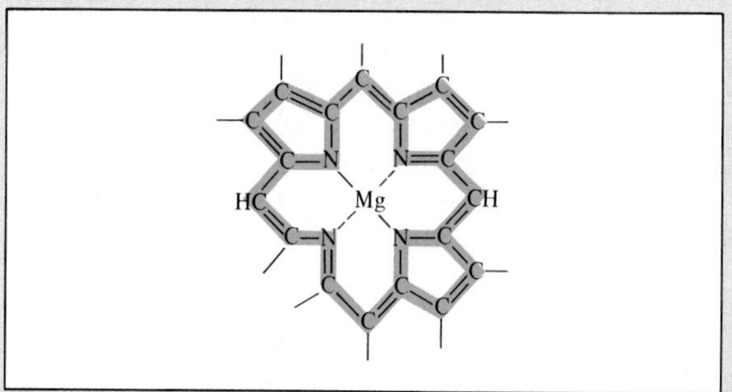

Figure 22.40 *The porphyrin structure in chlorophyll. The dotted lines indicate the coordinate covalent bonds. The electron delocalized portion of the molecule is shown in color.*

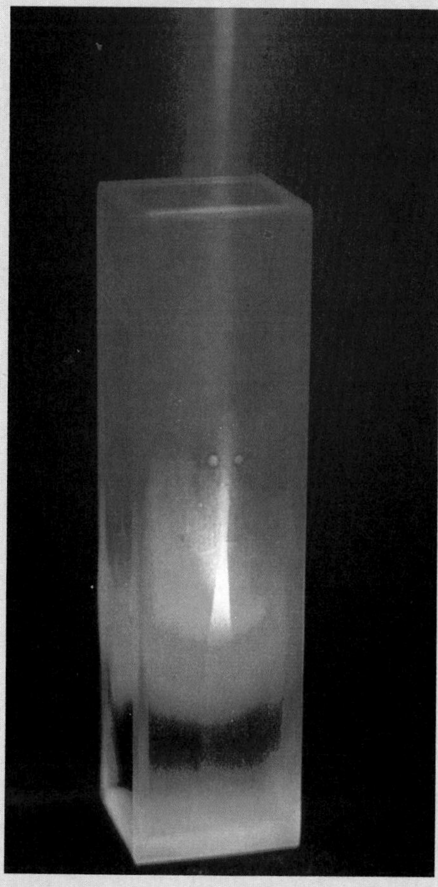

Figure 22.41 *Fluorescence (red color) of the plant pigment chlorophyll in an organic solvent is stimulated by illumination with blue light. The natural green color of chlorophyll can be seen at the bottom of the container.*

container. The beam of light enters at the top and is absorbed by the chlorophyll molecules. Some of the energy of the absorbed light is dissipated as heat; the rest is emitted as fluorescence in the red region of the visible spectrum. ***Fluorescence*** *is the emission of electromagnetic radiation from an atom or a molecule, particularly in the visible region, after an initial absorption of a photon.*

Summary

1. Transition metals usually have incompletely filled *d* orbitals and have a pronounced tendency to form complexes. Compounds that contain complex ions are called coordination compounds.
2. The first-row transition metals (scandium to copper) are the most common of all the transition metals; their chemistry is characteristic, in many ways, of the entire group.
3. The donor atoms in the ligands each contribute an electron pair to the central metal ion in a complex.
4. Coordination compounds may display geometric and/or optical isomerism. All octahedral complexes containing three bidentate ligands are optically active.
5. Crystal field theory explains bonding in complexes in terms of electrostatic interactions. According to crystal field theory, the *d* orbitals are split into two higher-energy and three lower-energy orbitals in an octahedral complex. The energy difference between these two sets of *d* orbitals is the crystal field splitting.
6. Strong-field ligands cause a large splitting, and weak-field ligands cause a small splitting. Electron spins tend to be parallel with weak-field ligands and paired with strong-field ligands, where a greater investment of energy is required to promote electrons into the high-lying *d* orbitals.
7. Complex ions undergo ligand exchange reactions in solution.
8. Coordination compounds find application in many different areas—for example, in water treatment, as antidotes for metal poisoning, and as therapeutic agents.

Key Words

Chelating agent, p. 933
Chiral, p. 938
Coordination compound, p. 931
Coordination number, p. 933
Crystal field splitting (Δ), p. 942
Donor atom, p. 931
Enantiomers, p. 938
Fluorescence, p. 952
Geometric isomers, p. 936
Inert complex, p. 947
Labile complex, p. 947
Ligand, p. 931
Optical isomers, p. 937
Plane-polarized light, p. 938
Polarimeter, p. 939
Racemic mixture, p. 939
Spectrochemical series, p. 944
Stereoisomerism, p. 936
Stereoisomers, p. 936

Exercises

Properties of Transition Metals

REVIEW QUESTIONS

22.1 What distinguishes a transition metal from a representative metal?

22.2 Why is zinc not considered a transition metal?

22.3 Explain why atomic radii decrease very gradually from scandium to copper.

22.4 Without referring to the text, write the ground-state electron configurations of the first-row transition metals.

Explain any irregularities.

22.5 Write the electron configuration of the following ions: V^{5+}, Cr^{3+}, Mn^{2+}, Fe^{3+}, Cu^{2+}, Sc^{3+}, Ti^{4+}.

22.6 Why do transition metals have more oxidation states than other elements?

22.7 Give the highest oxidation states for scandium to copper.

22.8 As we read across the first-row transition metals from left to right, the +2 oxidation state becomes more stable in comparison with the +3 state. Why is this so?

22.9 Why does chromium appear less reactive than its standard reduction potential indicates?

22.10 Which is a stronger oxidizing agent, Mn^{3+} or Cr^{3+}? Explain.

22.11 What are the oxidation states of Fe and Ti in the ore ilmentile, $FeTiO_3$? (*Hint:* Look up the ionization energies of Fe and Ti in Table 22.1; the fourth ionization energy of Ti is 4180 kJ/mol.)

Coordination Compounds: Nomenclature; Oxidation Number

REVIEW QUESTIONS

22.12 Define the following terms: coordination compound, ligand, donor atom, coordination number, chelating agent.

22.13 Describe the interaction between a donor atom and a metal atom in terms of a Lewis acid-base reaction.

PROBLEMS

22.14 Complete the following statements for the complex ion $[Co(en)_2(H_2O)CN]^{2+}$. (a) en is the abbreviation for ______. (b) The oxidation number of Co is ______. (c) The coordination number of Co is ______. (d) ______ is a bidentate ligand.

22.15 Complete the following statements for the complex ion $[Cr(C_2O_4)_2(H_2O)_2]^-$. (a) The oxidation number of Cr is ______. (b) The coordination number of Cr is ______. (c) ______ is a bidentate ligand.

22.16 Give the oxidation numbers of the metals in the following species:
(a) $K_3[Fe(CN)_6]$
(b) $K_3[Cr(C_2O_4)_3]$
(c) $[Ni(CN)_4]^{2-}$
(d) Na_2MoO_4
(e) $MgWO_4$
(f) $K[Au(OH)_4]$

22.17 What are the systematic names for the following ions and compounds?
(a) $[Co(NH_3)_4Cl_2]^+$
(b) $Cr(NH_3)_3Cl_3$
(c) $[Co(en)_2Br_2]^+$
(d) $Fe(CO)_5$
(e) *trans*-$Pt(NH_3)_2Cl_2$
(f) [*cis*-$Co(en)_2Cl_2]^+$
(g) $[Pt(NH_3)_5Cl]Cl_3$
(h) $[Co(NH_3)_6]Cl_3$
(i) $[Co(NH_3)_5Cl]Cl_2$
(j) $[Cr(H_2O)_4Cl_2]Cl$

22.18 Write the formulas for each of the following ions and compounds: (a) tetrahydroxozincate(II), (b) chloropentaaquochromium(III) chloride, (c) tetrabromocuprate(II), (d) ethylenediaminetetraacetatoferrate(II), (e) bis(ethylenediamine)dichlorochromium(III), (f) pentacarbonyliron(0), (g) potassium tetracyanocuprate(II), (h) tetraammineaquochlorocobalt(III) chloride, (i) tris(ethylenediamine)cobalt(III) sulfate.

22.19 What is the systematic name for the compound $[Fe(en)_3][Fe(CO)_4]$? (*Hint:* The oxidation number of Fe in the complex cation is +2.)

Structure of Coordination Compounds

REVIEW QUESTIONS

22.20 Define the following terms: stereoisomerism, stereoisomers, geometric isomers, optical isomers, plane-polarized light.

22.21 Specify which of the following structures can exhibit geometric isomerism: (a) linear, (b) square-planar, (c) tetrahedral, (d) octahedral.

22.22 What determines whether a molecule is chiral?

22.23 Explain the following terms: (a) enantiomers, (b) racemic mixtures.

22.24 How does a polarimeter work?

PROBLEMS

22.25 How many geometric isomers are in the following species? (a) $[Co(NH_3)_2Cl_4]^-$, (b) $[Co(NH_3)_3Cl_3]$.

22.26 Draw structures of all the geometric and optical isomers of each of the following cobalt complexes:
(a) $[Co(NH_3)_6]^{3+}$
(b) $[Co(NH_3)_5Cl]^{2+}$
(c) $[Co(NH_3)_4Cl_2]^+$
(d) $[Co(en)_3]^{3+}$
(e) $[Co(C_2O_4)_3]^{3-}$

22.27 The complex ion $[Ni(CN)_2Br_2]^{2-}$ has a square-planar geometry. Draw the structures of the geometric isomers of this complex.

22.28 A student has prepared a cobalt complex which has one of the following three structures: $[Co(NH_3)_6]Cl_3$, $[Co(NH_3)_5Cl]Cl_2$, or $[Co(NH_3)_4Cl_2]Cl$. Explain how the student would distinguish among these possibilities by an electrolytic conductance experiment. At the student's disposal are three strong electrolytes: NaCl, $MgCl_2$, and $FeCl_3$, which may be used for comparison purposes.

Bonding, Color, Magnetism

REVIEW QUESTIONS

22.29 Briefly describe the crystal field theory.

22.30 Define the following terms: crystal field splitting, high-spin complex, low-spin complex, spectrochemical series.

22.31 What is the origin of color in a compound?

22.32 Compounds containing the Sc^{3+} ion are colorless, whereas those containing the Ti^{3+} ion are colored. Explain.

22.33 What factors determine whether a given complex will be diamagnetic or paramagnetic?

PROBLEMS

22.34 For the same type of ligands, explain why the crystal field splitting for an octahedral complex is always greater than that for a tetrahedral complex.

22.35 Transition metal complexes containing CN^- ligands are often yellow in color, whereas those containing H_2O ligands are often green or blue. Explain.

22.36 The $[Ni(CN)_4]^{2-}$ ion, which has a square-planar geometry, is diamagnetic, whereas the $[NiCl_4]^{2-}$ ion, which has a tetrahedral geometry, is paramagnetic. Show the crystal field splitting diagrams for those two complexes.

22.37 Predict the number of unpaired electrons in the following complex ions: (a) $[Cr(CN)_6]^{4-}$, (b) $[Cr(H_2O)_6]^{2+}$.

22.38 The absorption maximum for the complex ion $[Co(NH_3)_6]^{3+}$ occurs at 470 nm. (a) Predict the color of the complex and (b) calculate the crystal field splitting in kJ/mol.

22.39 In each of the following pairs of complexes, choose the one that absorbs light at a longer wavelength: (a) $[Co(NH_3)_6]^{2+}$, $[Co(H_2O)_6]^{2+}$; (b) $[FeF_6]^{3-}$, $[Fe(CN)_6]^{3-}$; (c) $[Cu(NH_3)_4]^{2+}$, $[CuCl_4]^{2-}$.

22.40 A solution made by dissolving 0.875 g of $Co(NH_3)_4Cl_3$ in 25.0 g of water freezes 0.56°C below the freezing point of pure water. Calculate the number of moles of ions produced when 1 mole of $Co(NH_3)_4Cl_3$ is dissolved in water, and suggest a structure for the complex ion present in this compound.

22.41 A student in 1895 prepared three coordination compounds containing chromium, with the following properties:

Formula	Color	Cl^- ions in solution per formula unit
(a) $CrCl_3 \cdot 6H_2O$	Violet	3
(b) $CrCl_3 \cdot 6H_2O$	Light green	2
(c) $CrCl_3 \cdot 6H_2O$	Dark green	1

Write modern formulas for these compounds and suggest a method for confirming the number of Cl^- ions present in solution in each case. (*Hint:* Some of the compounds may exist as hydrates.)

Reactions of Coordination Compounds

REVIEW QUESTIONS

22.42 Define the following terms: (a) labile complex, (b) inert complex.

22.43 Explain why it is that a thermodynamically stable species may be chemically reactive and a thermodynamically unstable species may be unreactive.

PROBLEMS

22.44 Oxalic acid, $H_2C_2O_4$, is sometimes used to clean rust stains from sinks and bathtubs. Explain the chemistry involved in this cleaning action.

22.45 The $[Fe(CN)_6]^{3-}$ complex is more labile than the $[Fe(CN)_6]^{4-}$ complex. Suggest an experiment that would prove the labile nature of the $[Fe(CN)_6]^{3-}$ complex.

22.46 Aqueous copper(II) sulfate solution is blue in color. When aqueous potassium fluoride is added, a green precipitate is formed. When aqueous potassium chloride is added instead, a bright green solution is formed. Explain what is happening in these two cases.

22.47 When aqueous potassium cyanide is added to a solution of copper(II) sulfate, a white precipitate, soluble in an excess of potassium cyanide, is formed. No precipitate is formed when hydrogen sulfide is bubbled through the solution at this point. Explain.

22.48 A concentrated aqueous copper(II) chloride solution is bright green in color. On dilution with water, the solution becomes light blue. Explain.

22.49 In dilute nitric acid solution, Fe^{3+} reacts with thiocyanate ion (SCN^-) to form a dark red complex:

$$[Fe(H_2O)_6]^{3+} + SCN^- \rightleftharpoons H_2O + [Fe(H_2O)_5NCS]^{2+}$$

The equilibrium concentration of $[Fe(H_2O)_5NCS]^{2+}$ may be determined by how darkly colored the solution is (measured by a spectrometer). In one such experiment, 1.0 mL of 0.20 *M* $Fe(NO_3)_3$ was mixed with 1.0 mL of 1.0×10^{-3} *M* KSCN and 8.0 mL of dilute HNO_3. The color of the solution indicated that the $[Fe(H_2O)_5NCS]^{2+}$ concentration was 7.3×10^{-5} *M*. Calculate the formation constant for $[Fe(H_2O)_5NCS]^{2+}$.

Miscellaneous Problems

22.50 Chemical analysis shows that hemoglobin contains 0.34 percent of Fe by mass. What is the minimum possible molar mass of hemoglobin? The actual molar mass of hemoglobin is about 65,000 g. How do you account for the discrepancy between your minimum value and the actual value?

22.51 Explain the following facts: (a) Copper and iron have several oxidation states, whereas zinc exists in only one. (b) Copper and iron form colored ions, whereas zinc does not.

22.52 The formation constant for the reaction $Ag^+ + 2NH_3 \rightleftharpoons [Ag(NH_3)_2]^+$ is 1.5×10^7 and that for the reaction $Ag^+ + 2CN^- \rightleftharpoons [Ag(CN)_2]^-$ is 1.0×10^{21} at 25°C (see Table 17.3). Calculate the equilibrium constant and $\Delta G°$ at 25°C for the reaction

$$[Ag(NH_3)_2]^+ + 2CN^- \rightleftharpoons [Ag(CN)_2]^- + 2NH_3$$

22.53 From the standard reduction potentials listed in Table 19.1 for Zn/Zn^{2+} and Cu^{+}/Cu^{2+}, calculate $\Delta G°$ and the equilibrium constant for the reaction

$$Zn(s) + 2Cu^{2+}(aq) \longrightarrow Zn^{2+}(aq) + 2Cu^{+}(aq)$$

22.54 Using the standard reduction potentials listed in Table 19.1 and the *Handbook of Chemistry and Physics*, show that the following reaction is favorable under standard-state conditions:

$$2Ag(s) + Pt^{2+}(aq) \longrightarrow 2Ag^{+}(aq) + Pt(s)$$

What is the equilibrium constant of this reaction at 25°C?

22.55 Give three examples of coordination compounds in biological systems.

22.56 Discuss, giving suitable examples, the role of chelating agents in medicine.

22.57 The Co^{2+}–porphyrin complex is more stable than the Fe^{2+}–porphyrin complex. Why, then, is iron the metal ion in hemoglobin (and other heme-containing proteins)?

22.58 What are the differences between geometric isomers and optical isomers?

22.59 Oxyhemoglobin is bright red, whereas deoxyhemoglobin is purple. Show that the difference in color can be accounted for qualitatively on the basis of high-spin and low-spin complexes. (*Hint:* O_2 is a strong-field ligand.)

22.60 Hydrated Mn^{2+} ions are practically colorless (see Figure 22.5) even though they possess five $3d$ electrons. Explain. (*Hint:* Electronic transitions in which there is a change in the number of unpaired electrons do not occur readily.)

Opposite page: The Great Nebula in the constellation Orion shown here is proposed to be a site where Li, Be, and B are synthesized in nuclear reactions of hydrogen and other elements.

CHAPTER 23

NUCLEAR CHEMISTRY

Nuclear chemistry is the study of reactions involving changes in atomic nuclei. This branch of chemistry began with the discovery of natural radioactivity by Becquerel and grew as a result of subsequent investigations by Pierre and Marie Curie and many others.

Nuclear chemistry is very much in the news today. In addition to applications in the manufacture of atomic bombs, hydrogen bombs, and neutron bombs, even the peaceful use of nuclear energy has become controversial, as evidenced by the accidents at the Three Mile Island and Chernobyl nuclear plants. In this chapter we consider nuclear reactions, the stability of the atomic nucleus, radioactivity, and the effects of radiation on biological systems.

23.1 The Nature of Nuclear Reactions

With the exception of hydrogen, ${}^1_1H$, all nuclei contain two kinds of fundamental particles, called *protons* and *neutrons*. Some nuclei are unstable; they emit particles and/or electromagnetic radiation spontaneously (see Section 2.2). This phenomenon is known as *radioactivity*. *Nuclei can also undergo change as a result of bombardment by neutrons, electrons, or other nuclei;* nuclear changes that are induced in this way are known as ***nuclear transmutation***. Both radioactive decay and nuclear transmutation are examples of *nuclear reactions*. As Table 23.1 shows, nuclear reactions differ from ordinary chemical reactions in several important ways.

Before we discuss specific nuclear reactions, let us see how to write and balance the equations. Writing an equation for a nuclear reaction is more involved than writing an equation for an ordinary chemical reaction. In addition to writing the symbols for various chemical elements, we must also explicitly indicate protons, neutrons, and electrons. In fact, we must show the numbers of protons and neutrons present in *every* species in such an equation.

The symbols for elementary particles are as follows:

${}^1_1p$ or ${}^1_1H$	${}^1_0n$	${}^{0}_{-1}e$ or ${}^{0}_{-1}\beta$	${}^{0}_{+1}e$ or ${}^{0}_{+1}\beta$	${}^4_2He$ or ${}^4_2\alpha$
proton	neutron	electron	positron	α particle

In accordance with the notation used in Section 2.3, the superscript in each case denotes the mass number and the subscript is the atomic number. Thus, the "atomic number" of a proton is 1, because there is one proton present, and the "mass number" is also 1, because there is one proton but no neutrons present. On the other hand, the "mass number" of a neutron is 1, but its "atomic number" is zero, because there are no protons present. For the electron, the "mass number" is zero (there are neither protons nor neutrons present), but the "atomic number" is -1, because the electron possesses a unit negative charge.

The symbol ${}^{0}_{-1}e$ represents an electron in or from an atomic orbital. The symbol ${}^{0}_{-1}\beta$ represents an electron that, although physically identical to any other electron, comes from a nucleus and not from an atomic orbital.

The positron has the same mass as the electron, but bears a $+1$ charge. The α particle has two protons and two neutrons, so its atomic number is 2 and its mass number is 4.

Table 23.1 Comparison of chemical reactions and nuclear reactions

Chemical reactions	Nuclear reactions
1. Atoms are rearranged by the breaking and forming of chemical bonds.	1. Elements (or isotopes of the same elements) are converted from one to another.
2. Only electrons in atomic orbitals are involved in the breaking and forming of bonds.	2. Protons, neutrons, electrons, and other elementary particles may be involved.
3. Reactions are accompanied by absorption or release of relatively small amounts of energy.	3. Reactions are accompanied by absorption or release of tremendous amounts of energy.
4. Rates of reaction are influenced by temperature, pressure, concentration, and catalysts.	4. Rates of reaction normally are not affected by temperature, pressure, and catalysts.

In balancing any nuclear equation, we observe the following rules:

- The total number of protons plus neutrons in the products and in the reactants must be the same (conservation of mass number).
- The total number of nuclear charges in the products and in the reactants must be the same (conservation of atomic number).

If the atomic numbers and mass numbers of all the species but one in a nuclear equation are known, the unknown species can be identified by applying these rules, as shown in the following example, which illustrates balancing nuclear decay equations.

EXAMPLE 23.1

Balance the following nuclear equations (that is, identify the product X):

(a) $^{212}_{84}\text{Po} \longrightarrow {}^{208}_{82}\text{Pb} + \text{X}$

(b) $^{137}_{55}\text{Cs} \longrightarrow {}^{137}_{56}\text{Ba} + \text{X}$

(c) $^{20}_{11}\text{Na} \longrightarrow {}^{20}_{10}\text{Ne} + \text{X}$

Answer

(a) The mass number and atomic number are 212 and 84, respectively, on the left-hand side and 208 and 82, respectively, on the right-hand side. Thus, X must have a mass number of 4 and an atomic number of 2, which means that it is an α particle. The balanced equation is

$$^{212}_{84}\text{Po} \longrightarrow {}^{208}_{82}\text{Pb} + {}^{4}_{2}\text{He}$$

(b) In this case the mass number is the same on both sides of the equation, but the atomic number of the product is 1 more than that of the reactant. The only way this change can come about is to have a neutron in the Cs nucleus transformed into a proton and an electron; that is, $^{1}_{0}\text{n} \rightarrow {}^{1}_{1}\text{p} + {}^{\ 0}_{-1}\beta$ (note that this process does not alter the mass number). Thus, the balanced equation is

We use the $^{-1}_{0}\beta$ notation here because the electron came from the nucleus.

$$^{137}_{55}\text{Cs} \longrightarrow {}^{137}_{56}\text{Ba} + {}^{\ 0}_{-1}\beta$$

(c) As in (b), the mass number is the same on both sides of the equation, but the atomic number of the product is *smaller* by 1 than that of the reactant. The only way this change can come about is to have one of the protons in the Na nucleus transformed into a neutron and a positron; that is, $^{1}_{1}\text{p} \rightarrow {}^{1}_{0}\text{n} + {}^{\ 0}_{+1}\beta$ (note that this process does not alter the mass number). Thus, the balanced equation is

$$^{20}_{11}\text{Na} \longrightarrow {}^{20}_{10}\text{Ne} + {}^{\ 0}_{+1}\beta$$

Similar problem: 23.6.

23.2 Nuclear Stability

The nucleus occupies a very small portion of the total volume of an atom, but it contains most of the atom's mass because both the protons and the neutrons reside there. In studying the stability of the atomic nucleus, it is helpful to know something about its density, because it tells us how tightly the particles are packed together. As a sample calculation, let us assume that a nucleus has a radius of 5×10^{-3} pm and a mass of 1×10^{-22} g. These figures correspond roughly to a nucleus containing 30 protons and 30 neutrons. Density is mass/volume, and we can calculate the volume from the known radius (the volume of a sphere is $\frac{4}{3}\pi r^3$, where r is the radius of the sphere). First we convert the pm units to cm. Then we calculate the density in g/cm^3:

$$r = 5 \times 10^{-3} \text{ pm} \times \frac{1 \times 10^{-12} \text{ m}}{1 \text{ pm}} \times \frac{100 \text{ cm}}{1 \text{ m}} = 5 \times 10^{-13} \text{ cm}$$

$$\text{density} = \frac{\text{mass}}{\text{volume}} = \frac{1 \times 10^{-22} \text{ g}}{\frac{4}{3}\pi r^3} = \frac{1 \times 10^{-22} \text{ g}}{\frac{4}{3}\pi (5 \times 10^{-13} \text{ cm})^3}$$
$$= 2 \times 10^{14} \text{ g/cm}^3$$

To dramatize the almost incomprehensibly high density, it has been suggested that it is equivalent to packing the mass of all the world's automobiles into one thimble.

This is an exceedingly high density. The highest density known for an element is only 22.6 g/cm^3, for osmium (Os). Thus the average atomic nucleus is roughly 9×10^{12} (or 9 trillion) times more dense than the densest element known!

The enormously high density of the nucleus prompts us to wonder what holds the particles together so tightly. From *Coulomb's law* we know that like charges repel and unlike charges attract one another. We would thus expect the protons to repel one another strongly, particularly when we consider how close they must be to each other. This indeed is so. However, in addition to the repulsion, there are also short-range attractions between proton and proton, proton and neutron, and neutron and neutron. The study of these interactions is beyond the scope of this book. What you should understand is that the stability of any nucleus is determined by the difference between coulombic repulsion and this attraction. If the repulsion outweighs attraction, the nucleus disintegrates, emitting particles and/or radiation. This is the phenomenon of radioactivity we discussed in Chapter 2. If the attraction prevails, the nucleus will be stable.

The principal factor for determining whether a nucleus is stable is the *neutron-to-proton ratio* ($n{:}p$). For stable atoms of elements of low atomic number, the n:p value is close to 1. As the atomic number increases, the neutron-to-proton ratios of the stable nuclei become greater than 1. This deviation at higher atomic numbers arises because a larger number of neutrons is needed to stabilize the nucleus by counteracting the strong repulsion among the protons. The following rules are useful in predicting nuclear stability:

- Nuclei that contain 2, 8, 20, 50, 82, or 126 protons or neutrons are generally more stable than nuclei that do not possess these numbers. For example, there are ten stable isotopes of tin (Sn) with the atomic number 50 and only two stable isotopes of antimony (Sb) with the atomic number 51. The numbers 2, 8, 20, 50, 82, and 126 are called *magic numbers*. These numbers have a significance in nuclear stability similar to the numbers of electrons associated with the very stable noble gases (that is, 2, 10, 18, 36, 54, and 86 electrons).
- Nuclei with even numbers of both protons and neutrons are generally more stable than those with odd numbers of these particles (Table 23.2).
- All isotopes of the elements starting with polonium (Po, $Z = 84$) are radioactive. All isotopes of technetium (Tc, $Z = 43$) and promethium (Pm, $Z = 61$) are also radioactive.

Figure 23.1 shows a plot of the number of neutrons versus the number of protons in various isotopes. The stable nuclei are located in an area of the graph known as the *belt*

Table 23.2 Number of stable isotopes with even and odd numbers of protons and neutrons

Protons	Neutrons	Number of stable isotopes
Odd	Odd	8
Odd	Even	50
Even	Odd	52
Even	Even	157

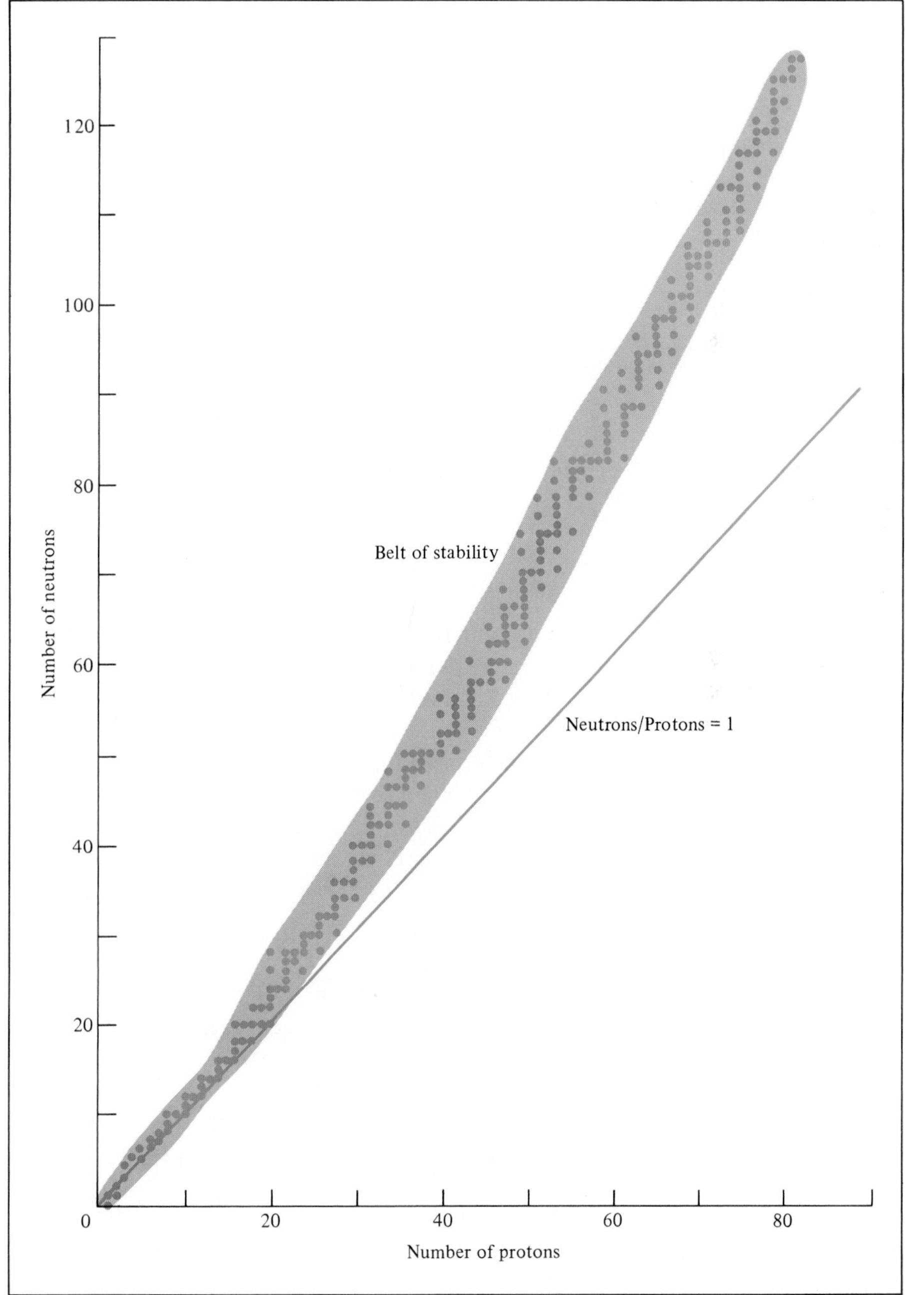

Figure 23.1 *Plot of neutrons versus protons for various stable isotopes, represented by dots. The straight line represents the points at which the neutron-to-proton ratio equals 1. The shaded area represents the belt of stability.*

of stability. Most of the radioactive nuclei lie outside this belt. Above the stability belt, the nuclei have higher neutron-to-proton ratios than those within the belt (for the same number of protons). To lower this ratio (and hence move down toward the belt of stability), these nuclei undergo the following process, called *β-particle emission:*

The proton stays in the nucleus.

$$^{1}_{0}n \longrightarrow ^{1}_{1}p + ^{0}_{-1}\beta$$

Beta-particle emission leads to an increase in the number of protons and a simultaneous decrease in the number of neutrons. Some examples are

$$^{14}_{6}C \longrightarrow ^{14}_{7}N + ^{0}_{-1}\beta$$

$$^{40}_{19}K \longrightarrow ^{40}_{20}Ca + ^{0}_{-1}\beta$$

$$^{97}_{40}Zr \longrightarrow ^{97}_{41}Nb + ^{0}_{-1}\beta$$

Below the stability belt the nuclei have lower neutron-to-proton ratios than those in the belt (for the same number of protons). To increase this ratio (and hence move up toward the belt of stability), these nuclei either emit a positron

$$^{1}_{1}p \longrightarrow ^{1}_{0}n + ^{0}_{+1}\beta$$

or undergo electron capture. An example of positron emission is

$$^{38}_{19}K \longrightarrow ^{38}_{18}Ar + ^{0}_{+1}\beta$$

Electron capture is the capture of an electron by the nucleus in an atom, usually a 1*s* electron. This process has the same net effect as positron emission:

We use $^{-1}_{0}e$ rather than $^{-1}_{0}\beta$ here because the electron came from an atomic orbital and not from the nucleus.

$$^{37}_{18}Ar + ^{0}_{-1}e \longrightarrow ^{37}_{17}Cl$$

$$^{55}_{26}Fe + ^{0}_{-1}e \longrightarrow ^{55}_{25}Mn$$

Nuclear Binding Energy

A quantitative measure of nuclear stability is the ***nuclear binding energy,*** which is *the energy required to break up a nucleus into its component protons and neutrons*. It was noticed in the study of nuclear properties that the masses of nuclei are always less than the sums of the masses of the *nucleons*. (The term ''nucleon'' refers to both protons and neutrons in a nucleus.) For example, the $^{19}_{9}F$ isotope has an atomic mass of 18.9984 amu. The nucleus has 9 protons and 10 neutrons and therefore a total of 19 nucleons. Using the known mass of the $^{1}_{1}H$ atom (1.007825 amu) and the neutron (1.008665 amu), we can carry out the following analysis. The mass of 9 $^{1}_{1}H$ atoms (that is, the mass of 9 protons and 9 electrons) is

$$9 \times 1.007825 \text{ amu} = 9.070425 \text{ amu}$$

and the mass of 10 neutrons is

$$10 \times 1.008665 \text{ amu} = 10.08665 \text{ amu}$$

Therefore, the atomic mass of a $^{19}_{9}F$ atom calculated from the known numbers of electrons, protons, and neutrons is

$$9.070425 \text{ amu} + 10.08665 \text{ amu} = 19.15708 \text{ amu}$$

There is no change in the electron's mass since it is not a nucleon.

which is larger than 18.9984 amu (the measured mass of $^{19}_{9}F$) by 0.1587 amu. *This difference between the mass of the atom and the sum of the masses of its protons, neutrons, and electrons* is called the ***mass defect.*** Where did the mass go?

Einstein's relativity theory tells us that the loss in mass shows up as energy (heat) given off to the surroundings. Thus the formation of $^{19}_{9}F$ is exothermic. According to *Einstein's mass-energy equivalence relationship* ($E = mc^2$, where E is energy, m is mass, and c is the velocity of light), we can calculate the energy released as follows. We start by writing

$$\Delta E = (\Delta m)c^2$$

where ΔE and Δm are defined as follows:

$$\Delta E = \text{energy of product} - \text{energy of reactants}$$

$$\Delta m = \text{mass of product} - \text{mass of reactants}$$

Thus we have

$$\begin{aligned}\Delta m &= 18.9984 \text{ amu} - 19.15708 \text{ amu} \\ &= -0.1587 \text{ amu}\end{aligned}$$

Because $^{19}_{9}F$ has a mass that is less than the mass calculated from the number of electrons and nucleons present, Δm is a negative quantity. Consequently, ΔE is also a negative quantity; that is, energy is released to the surroundings as a result of the formation of the fluorine-19 nucleus. So we calculate ΔE as follows:

$$\begin{aligned}\Delta E &= (-0.1587 \text{ amu})(3.00 \times 10^8 \text{ m/s})^2 \\ &= -1.43 \times 10^{16} \text{ amu m}^2/\text{s}^2\end{aligned}$$

With the conversion factors

$$1 \text{ kg} = 6.022 \times 10^{26} \text{ amu}$$

$$1 \text{ J} = 1 \text{ kg m}^2/\text{s}^2$$

we obtain

$$\begin{aligned}\Delta E &= \left(-1.43 \times 10^{16} \frac{\text{amu m}^2}{\text{s}^2}\right) \times \left(\frac{1.00 \text{ kg}}{6.022 \times 10^{26} \text{ amu}}\right) \times \left(\frac{1 \text{ J}}{1 \text{ kg m}^2/\text{s}^2}\right) \\ &= -2.37 \times 10^{-11} \text{ J}\end{aligned}$$

This is the energy released (note the negative sign, which indicates that this is an exothermic process) when one fluorine-19 nucleus is formed from 9 protons and 10 neutrons. The nuclear binding energy of the nucleus has the value of 2.37×10^{-11} J, which is the amount of energy needed to decompose the nucleus into separate protons and neutrons. In general, the larger the mass defect, the greater the nuclear binding energy and the more stable the nucleus. In the formation of 1 mole of fluorine nuclei, for instance, the energy released is

$$\begin{aligned}\Delta E &= (-2.37 \times 10^{-11} \text{ J})(6.022 \times 10^{23}/\text{mol}) \\ &= -1.43 \times 10^{13} \text{ J/mol} \\ &= -1.43 \times 10^{10} \text{ kJ/mol}\end{aligned}$$

The nuclear binding energy, therefore, is 1.43×10^{10} kJ for 1 mole of fluorine-19 nuclei, which is a tremendously large quantity when we consider that the enthalpies of ordinary chemical reactions are of the order of only 200 kJ. The procedure we have followed can be used to calculate the nuclear binding energy of any nucleus.

As we have noted, nuclear binding energy is an indication of the stability of a nucleus. However, in comparing the stability of any two nuclei we must account for the fact that they have different numbers of nucleons. For this reason it is more meaningful to use the *nuclear binding energy per nucleon,* defined as

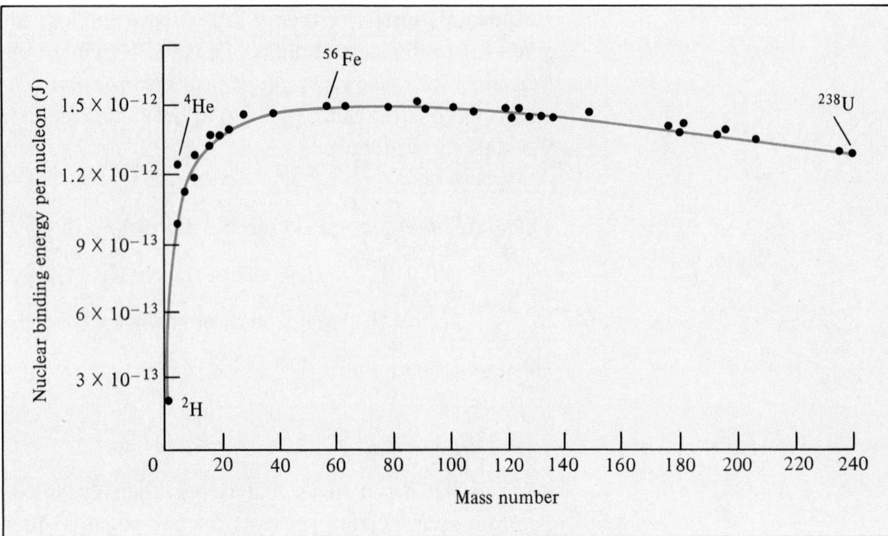

Figure 23.2 *Plot of nuclear binding energy per nucleon versus mass number.*

$$\text{nuclear binding energy per nucleon} = \frac{\text{nuclear binding energy}}{\text{number of nucleons}}$$

For the fluorine-19 nucleus,

$$\begin{aligned}\text{nuclear binding energy per nucleon} &= \frac{2.37 \times 10^{-11}\ \text{J}}{19\ \text{nucleons}} \\ &= 1.25 \times 10^{-12}\ \text{J/nucleon}\end{aligned}$$

In this manner we can compare the stability of all nuclei on a common basis. Figure 23.2 shows the variation of nuclear binding energy per nucleon plotted against mass number. As you can see, the curve rises rather steeply; the highest binding energies per nucleon belong to elements with intermediate mass numbers—between 40 and 100—and are greatest for elements in the iron, cobalt, and nickel region (the Group 8B elements) of the periodic table. This result shows that the *net* attractive forces among the particles (protons and neutrons) are greatest for the nuclei of these elements.

Nuclear binding energy and nuclear binding energy per nucleon are calculated for an iodine nucleus in the following example.

EXAMPLE 23.2

The atomic mass of $^{127}_{53}\text{I}$ is 126.9004 amu. Calculate the nuclear binding energy of this nucleus and the corresponding nuclear binding energy per nucleon.

Answer

There are 53 protons and 74 neutrons in the nucleus. The mass of 53 $^{1}_{1}\text{H}$ atoms is

$$53 \times 1.007825\ \text{amu} = 53.41473\ \text{amu}$$

and the mass of 74 neutrons is

$$74 \times 1.008665 \text{ amu} = 74.64121 \text{ amu}$$

Therefore, the predicted mass for $^{127}_{53}I$ is 53.41473 + 74.64121 = 128.05594 amu, and the mass defect is

$$\begin{aligned} \Delta m &= 126.9004 \text{ amu} - 128.05594 \text{ amu} \\ &= -1.1555 \text{ amu} \end{aligned}$$

The energy released is

$$\begin{aligned} \Delta E &= (\Delta m)c^2 \\ &= (-1.1555 \text{ amu})(3.00 \times 10^8 \text{ m/s})^2 \\ &= -1.04 \times 10^{17} \text{ amu m}^2\text{/s}^2 \\ &= \left(-1.04 \times 10^{17} \frac{\text{amu m}^2}{\text{s}^2}\right) \times \left(\frac{1.00 \text{ kg}}{6.022 \times 10^{26} \text{ amu}}\right) \times \left(\frac{1 \text{ J}}{1 \text{ kg m}^2\text{/s}^2}\right) \\ &= -1.73 \times 10^{-10} \text{ J} \end{aligned}$$

The neutron-to-proton ratio is 1.4, which places iodine-127 in the belt of stability.

Thus the nuclear binding energy is 1.73×10^{-10} J. The nuclear binding energy per nucleon is obtained as follows:

$$\frac{1.73 \times 10^{-10} \text{ J}}{127 \text{ nucleons}} = 1.36 \times 10^{-12} \text{ J/nucleon}$$

Similar problem: 23.23.

23.3 Natural Radioactivity

Nuclei outside the belt of stability, as well as nuclei with more than 83 protons, tend to be unstable. Radioactivity is the spontaneous emission by unstable nuclei of particles or electromagnetic radiation, or both. The main types of radiation are α particles (or doubly charged helium nuclei, He^{2+}), β particles (or electrons), γ rays, which are very-short-wavelength (0.1 nm to 10^{-4} nm) electromagnetic waves, positron emission, and electron capture.

When a radioactive nucleus disintegrates, the products formed may also be unstable and therefore will undergo further disintegration. This process is repeated until a stable product finally is formed. Starting with the original radioactive nucleus, the sequence of disintegration steps is called a *decay series*. Table 23.3 shows the decay series of naturally occurring uranium-238, which involves 14 steps. This decay scheme is known as the *uranium decay series*.

It is important to be able to balance the nuclear reaction for each of the steps. For example, the first step is the decay of uranium-238 to thorium-234, with the emission of an α particle. Hence, the reaction is

$$^{238}_{92}U \longrightarrow \, ^{234}_{90}Th + \, ^{4}_{2}He$$

The next step is represented by

$$^{234}_{90}Th \longrightarrow \, ^{234}_{91}Pa + \, ^{0}_{-1}\beta$$

and so on. In a discussion of radioactive decay steps, the beginning radioactive isotope is called the *parent* and the product, the *daughter*.

Table 23.3 The uranium decay series

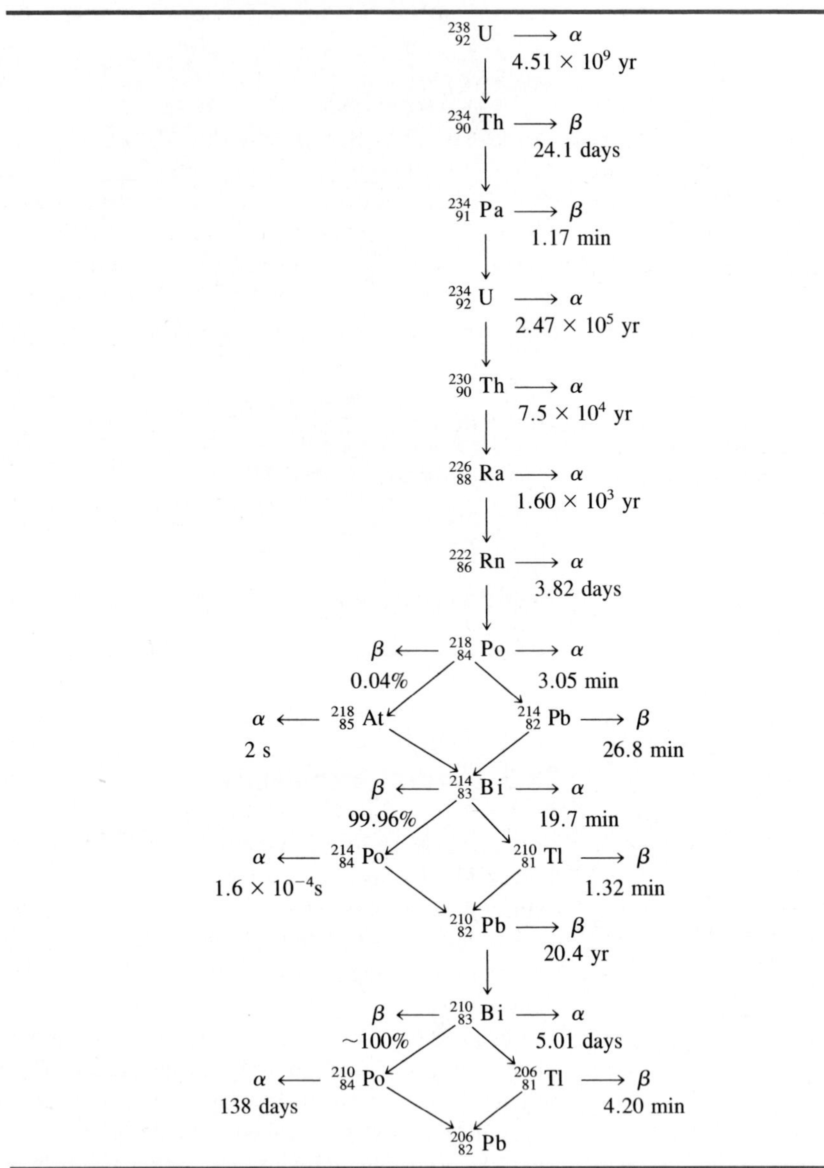

Kinetics of Radioactive Decay

All radioactive decays obey first-order kinetics. This means that the rate of radioactive decay at any time t is given by

$$\text{rate of decay at time } t = kN$$

where k is the first-order rate constant and N is the number of radioactive nuclei present at time t. The half-life for such a reaction is given by [see Equation (13.5)]

$$t_{\frac{1}{2}} = \frac{0.693}{k}$$

The half-lives (hence the rate constants) of radioactive decay vary greatly from nucleus to nucleus. For example, looking at Table 23.3, we find two extreme cases:

We do not have to wait 4.51 × 10^9 yr to make a half-life measurement of uranium-238. Its value can be calculated from the rate constant using Equation (13.5).

$$^{238}_{92}\text{U} \longrightarrow ^{234}_{90}\text{Th} + ^{4}_{2}\text{He} \qquad t_{\frac{1}{2}} = 4.51 \times 10^9 \text{ yr}$$

$$^{214}_{84}\text{Po} \longrightarrow ^{210}_{82}\text{Pb} + ^{4}_{2}\text{He} \qquad t_{\frac{1}{2}} = 1.6 \times 10^{-4} \text{ s}$$

The ratio of these two rate constants after conversion to the same time unit is about 1×10^{21}, an enormously large number. Furthermore, the rate constants are unaffected by changes in environmental conditions such as temperature and pressure. These highly unusual features are not seen in ordinary chemical reactions (see Table 23.1).

Dating Based on Radioactive Decay

The half-lives of radioactive decay have been used as ''atomic clocks'' to determine the ages of certain objects. Some examples of dating by radioactive decay measurements will be described here.

Radiocarbon Dating. This technique is discussed on p. 563. The carbon-14 isotope is produced when atmospheric nitrogen is bombarded by cosmic rays:

$$^{14}_{7}\text{N} + ^{1}_{0}\text{n} \longrightarrow ^{14}_{6}\text{C} + ^{1}_{1}\text{H}$$

The radioactive carbon-14 isotope decays according to the equation

$$^{14}_{6}\text{C} \longrightarrow ^{14}_{7}\text{N} + ^{0}_{-1}\beta$$

Dating Using Uranium-238 Isotopes. Because some members of the uranium series have very long half-lives (see Table 23.3), it is particularly suitable for estimating the age of rocks in Earth and of extraterrestrial objects. The half-life for the first step ($^{238}_{92}$U to $^{234}_{90}$Th) is 4.51×10^9 yr. This is about 20,000 times the second largest value (that is, 2.47×10^5 yr), which is the half-life for $^{234}_{92}$U to $^{230}_{90}$Th. Therefore, as a good approximation we can assume that the half-life for the overall process (that is, from $^{238}_{92}$U to $^{206}_{82}$Pb) is governed solely by the first step:

We can think of the first step as the rate-determining step in the overall process.

$$^{238}_{92}\text{U} \longrightarrow ^{206}_{82}\text{Pb} + 8\,^{4}_{2}\text{He} + 6\,^{0}_{-1}\beta \qquad t_{\frac{1}{2}} = 4.51 \times 10^9 \text{ yr}$$

In naturally occurring uranium minerals we should and do find some lead-206 isotopes formed by radioactive decay. Assuming that no lead was present when the mineral was formed and that the mineral has not undergone chemical changes that would allow the lead-206 isotope to be separated from the parent uranium-238, it is possible to estimate the age of the rocks from the mass ratio of $^{206}_{82}$Pb to $^{238}_{92}$U. The preceding equation tells us that for every mole, or 238 g, of uranium that undergoes complete decay, 1 mole, or 206 g, of lead is formed. If only half a mole of uranium-238 has undergone decay, the mass ratio Pb-206/U-238 becomes

$$\frac{206 \text{ g}/2}{238 \text{ g}/2} = 0.866$$

and the process would have taken a half-life of 4.51×10^9 yr to complete (Figure

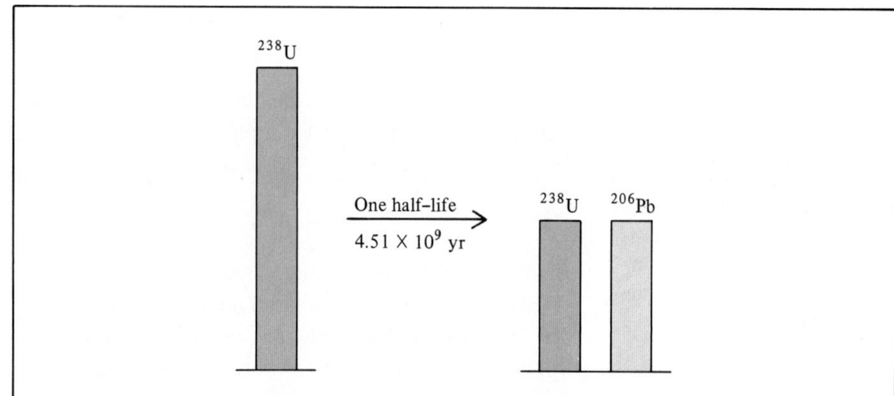

Figure 23.3 *After one half-life, nearly half of the original uranium-238 is converted to lead-206.*

23.3). Ratios lower than 0.866 mean that the rocks are less than 4.51×10^9 yr old, and higher ratios suggest a greater age. Interestingly, studies based on the uranium series as well as other decay series put the age of the oldest rocks and, therefore, probably the age of Earth itself at 4.5×10^9, or 4.5 billion, years.

Dating Using Potassium-40 Isotopes. This is one of the most important techniques in geochemistry. The radioactive potassium-40 isotope decays by several different modes, but the relevant one as far as dating is concerned is that of electron capture:

$$^{40}_{19}\text{K} + ^{0}_{-1}e \longrightarrow ^{40}_{18}\text{Ar} \qquad t_{\frac{1}{2}} = 1.2 \times 10^9 \text{ yr}$$

The accumulation of gaseous argon-40 is used to gauge the age of a specimen. When a potassium-40 atom in a mineral decays, the argon-40 formed is trapped in the lattice of the mineral and can escape only if the material is melted. Melting, therefore, is the procedure for analyzing a mineral sample in the laboratory. The amount of argon-40 present can be conveniently measured with a mass spectrometer. Knowing the ratio of argon-40 to potassium-40 in the mineral and the half-life of decay makes it possible to establish the ages of rocks ranging from thousands to billions of years old.

23.4 Artificial Radioactivity

Nuclear Transmutation

The scope of nuclear chemistry would have been rather narrow if study had been limited to natural radioactive elements. An experiment performed by Rutherford in 1919, however, suggested the possibility of observing artificial radioactivity. When he bombarded a sample of nitrogen with α particles, the following reaction took place:

$$^{14}_{7}\text{N} + ^{4}_{2}\text{He} \longrightarrow ^{17}_{8}\text{O} + ^{1}_{1}\text{p}$$

An oxygen-17 isotope was produced with the emission of a proton. This reaction demonstrated for the first time the feasibility of converting one element into another—that is, the possibility of *nuclear transmutation*.

The above reaction can be abbreviated as $^{14}_{7}\text{N}(\alpha,\text{p})^{17}_{8}\text{O}$. Note that in the parentheses the bombarding particle is written first, followed by the ejected particle. Nuclear transmutations differ from radioactive decay in that the latter is a spontaneous process;

consequently, in decay equations only *one* reactant appears on the left side of the equation.

EXAMPLE 23.3

Write the balanced equation for the nuclear reaction $^{56}_{26}\text{Fe}(\text{d},\alpha)^{54}_{25}\text{Mn}$, where d represents the deuterium nucleus (that is, $^{2}_{1}\text{H}$).

Answer

The abbreviation tells us that when iron-56 is bombarded with a deuterium nucleus, it produces the manganese-54 nucleus plus an α particle, $^{4}_{2}\text{He}$. Thus, the equation for this reaction is

$$^{56}_{26}\text{Fe} + {}^{2}_{1}\text{H} \longrightarrow {}^{4}_{2}\text{He} + {}^{54}_{25}\text{Mn}$$

Similar problem: 23.7.

Although light elements are generally not radioactive, they can be made so by bombarding their nuclei with appropriate particles. As we saw earlier, the radioactive carbon-14 isotope can be prepared by bombarding nitrogen-14 with neutrons. Tritium, $^{3}_{1}\text{H}$, is prepared according to the following bombardment:

$$^{6}_{3}\text{Li} + {}^{1}_{0}\text{n} \longrightarrow {}^{3}_{1}\text{H} + {}^{4}_{2}\text{He}$$

Tritium decays with the emission of β particles:

$$^{3}_{1}\text{H} \longrightarrow {}^{3}_{2}\text{He} + {}^{0}_{-1}\beta \qquad t_{\frac{1}{2}} = 12.5\ \text{yr}$$

Many synthetic isotopes are prepared by using neutrons as projectiles. This is particularly convenient because neutrons carry no charges and therefore are not repelled by the targets—the nuclei. The situation is different when the projectiles are positively charged particles—for example, when protons or α particles are used, as in

$$^{27}_{13}\text{Al} + {}^{4}_{2}\text{He} \longrightarrow {}^{30}_{15}\text{P} + {}^{1}_{0}\text{n}$$

To react with the aluminum nuclei, the α particles must have considerable kinetic energy in order to overcome the electrostatic repulsion between themselves and the target atoms.

In a *particle accelerator,* electric and magnetic fields are used to increase the kinetic energy of charged species to the point that a reaction will occur. Alternating the polarity (that is, + and −) on specially constructed plates causes the particles to accelerate along a spiral path. When they have the energy necessary to carry out the desired nuclear reaction, they are guided out of the accelerator into a collision with a target substance (Figure 23.4).

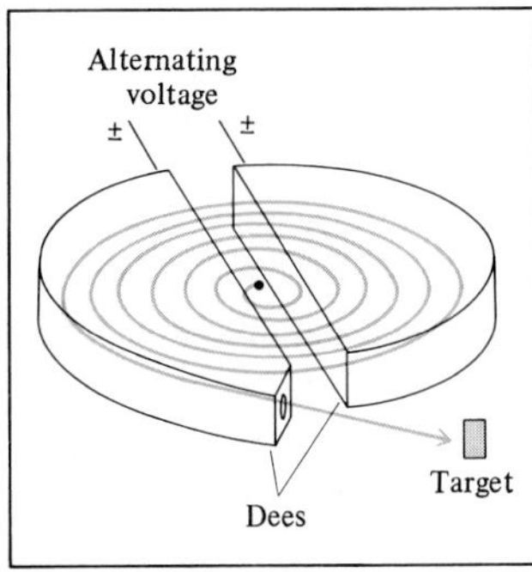

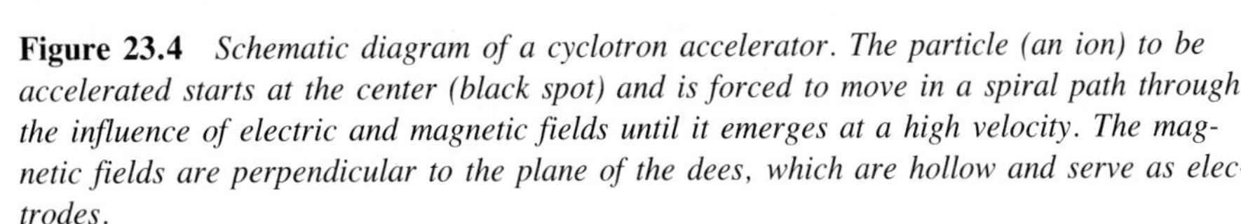

Figure 23.4 *Schematic diagram of a cyclotron accelerator. The particle (an ion) to be accelerated starts at the center (black spot) and is forced to move in a spiral path through the influence of electric and magnetic fields until it emerges at a high velocity. The magnetic fields are perpendicular to the plane of the dees, which are hollow and serve as electrodes.*

Figure 23.5 *A section of a particle accelerator.*

Various designs have been developed for particle accelerators, one of which accelerates particles along a linear path of about 3 km (2 miles). It is now possible to accelerate particles to a speed well above 90 percent of the speed of light. (According to Einstein's theory of relativity, it is impossible for a particle to move *at* the speed of light. The only exception is the photon, which has a zero rest mass.) The extremely energetic particles produced in accelerators are employed by physicists to smash atomic nuclei to fragments. Studying the debris from such disintegrations provides valuable information about nuclear structure and binding forces (Figure 23.5).

The Transuranium Elements

The advent of particle accelerators has also made possible the synthesis of *elements with atomic numbers greater than 92,* called ***transuranium elements.*** Since neptunium ($Z = 93$) was first prepared in 1940, 16 other transuranium elements have been synthesized. All isotopes of these elements are radioactive. Table 23.4 lists the transuranium elements and the reactions through which they are formed.

Naming of Elements 104 and Beyond. Until recently, the honor of naming a newly synthesized element went to its discoverers. However, considerable controversy has arisen over the naming of some of the transuranium elements that have been independently synthesized by different groups of workers at about the same time. As a result, some of the elements are identified only by their atomic numbers (such as elements 104 through 109) and not by the names coined by their discoverers (see Table 23.4). To systematize the naming of these elements, the International Union of Pure and Applied Chemistry (IUPAC) has recommended the use of three-letter symbols for elements having atomic numbers greater than 103. All the names end in *-ium,* whether the element is a metal or a nonmetal. The following roots are used to derive the names of the elements:

0 = nil	2 = bi
1 = un	3 = tri

Table 23.4 The transuranium elements

Atomic number	Name	Symbol	Preparation
93	Neptunium	Np	$^{238}_{92}U + ^{1}_{0}n \longrightarrow ^{239}_{93}Np + ^{0}_{-1}\beta$
94	Plutonium	Pu	$^{239}_{93}Np \longrightarrow ^{239}_{94}Pu + ^{0}_{-1}\beta$
95	Americium	Am	$^{239}_{94}Pu + ^{1}_{0}n \longrightarrow ^{240}_{95}Am + ^{0}_{-1}\beta$
96	Curium	Cm	$^{239}_{94}Pu + ^{4}_{2}He \longrightarrow ^{242}_{96}Cm + ^{1}_{0}n$
97	Berkelium	Bk	$^{241}_{95}Am + ^{4}_{2}He \longrightarrow ^{243}_{97}Bk + 2^{1}_{0}n$
98	Californium	Cf	$^{242}_{96}Cm + ^{4}_{2}He \longrightarrow ^{245}_{98}Cf + ^{1}_{0}n$
99	Einsteinium	Es	$^{238}_{92}U + 15^{1}_{0}n \longrightarrow ^{253}_{99}Es + 7^{0}_{-1}\beta$
100	Fermium	Fm	$^{238}_{92}U + 17^{1}_{0}n \longrightarrow ^{255}_{100}Fm + 8^{0}_{-1}\beta$
101	Mendelevium	Md	$^{253}_{99}Es + ^{4}_{2}He \longrightarrow ^{256}_{101}Md + ^{1}_{0}n$
102	Nobelium	No	$^{246}_{96}Cm + ^{12}_{6}C \longrightarrow ^{254}_{102}No + 4^{1}_{0}n$
103	Lawrencium	Lr	$^{252}_{98}Cf + ^{10}_{5}B \longrightarrow ^{257}_{103}Lr + 5^{1}_{0}n$
104	Unnilquadium	Unq	$^{249}_{98}Cf + ^{12}_{6}C \longrightarrow ^{257}_{104}Unq + 4^{1}_{0}n$
105	Unnilpentium	Unp	$^{249}_{98}Cf + ^{15}_{7}N \longrightarrow ^{260}_{105}Unp + 4^{1}_{0}n$
106	Unnilhexium	Unh	$^{249}_{98}Cf + ^{18}_{8}O \longrightarrow ^{263}_{106}Unh + 4^{1}_{0}n$
107	Unnilseptium	Uns	$^{209}_{83}Bi + ^{54}_{24}Cr \longrightarrow ^{262}_{107}Uns + ^{1}_{0}n$
108	Unniloctium	Uno	$^{208}_{82}Pb + ^{58}_{26}Fe \longrightarrow ^{265}_{108}Uno + ^{1}_{0}n$
109	Unnilennium	Une	$^{209}_{83}Bi + ^{58}_{26}Fe \longrightarrow ^{266}_{109}Une + ^{1}_{0}n$

4 = quad	7 = sept
5 = pent	8 = oct
6 = hex	9 = enn

The name of an element is formed by joining the roots in the order of the digits that make up the atomic number and ending with *-ium*. For example, element 104 is unnilquadium (Unq), element 116 would be ununhexium (Uuh), element 202 would be binilbium (Bnb), and so on. (The *i* in *bi* is omitted when the root precedes *ium*.)

23.5 Nuclear Fission

In ***nuclear fission*** *a heavy nucleus (mass number > 200) divides to form smaller nuclei of intermediate mass and one or more neutrons.* Because the heavy nucleus is less stable than its products (see Figure 23.2), this process releases a large amount of energy.

The first nuclear fission reaction studied was that of uranium-235 bombarded with slow neutrons, whose speed is comparable to that of air molecules at room temperature. Uranium-235 in such bombardment undergoes fission, as shown in Figure 23.6.

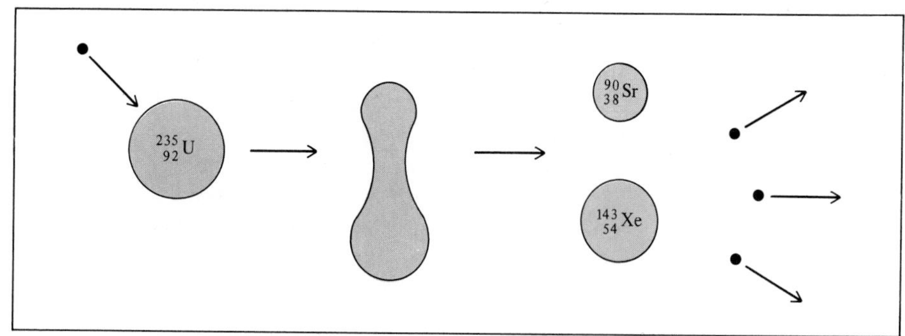

Figure 23.6 *Nuclear fission of U-235. When a U-235 nucleus captures a neutron (black sphere), it undergoes fission to yield two smaller nuclei. On the average, three neutrons are emitted for every U-235 nucleus that divides.*

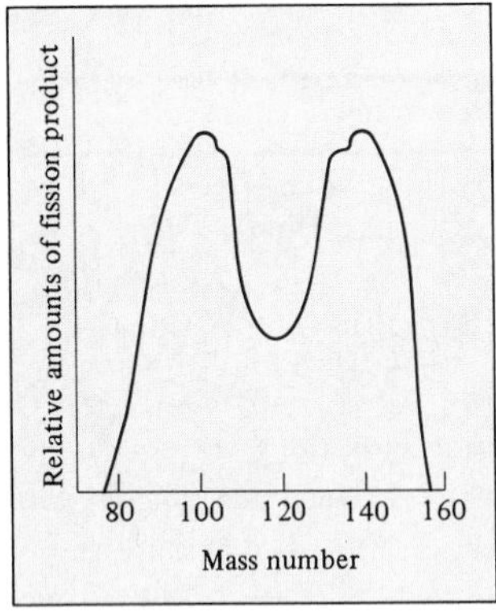

Figure 23.7 *Relative yields of the products resulting from the fission of U-235, as a function of mass number.*

Actually, this reaction is very complex: More than 30 different elements have been found among the fission products (Figure 23.7). A representative reaction is

$$^{235}_{92}\text{U} + ^{1}_{0}\text{n} \longrightarrow ^{90}_{38}\text{Sr} + ^{143}_{54}\text{Xe} + 3^{1}_{0}\text{n}$$

Although many heavy nuclei can be made to undergo fission, only the fission of naturally occurring uranium-235 and of the artificial plutonium-239 has any practical importance. As Figure 23.2 shows, the binding energy per nucleon for uranium-235 is less than the sum of the binding energies for strontium-90 and xenon-143. Therefore, when a uranium-235 nucleus is split into two smaller nuclei, a certain amount of energy is released. Let us estimate the magnitude of this energy, using the preceding reaction as an example. Table 23.5 shows the nuclear binding energies of uranium-235 and its fission products. The difference between the binding energies of the reactants and products is $(1.23 \times 10^{-10} + 1.92 \times 10^{-10})$ J − (2.82×10^{-10}) J, or 3.3×10^{-11} J per uranium-235 nucleus. For 1 mole of uranium-235, the energy released would be $(3.3 \times 10^{-11})(6.02 \times 10^{23})$, or 2.0×10^{13} J. This is an extremely exothermic reaction, considering that the heat of combustion of 1 ton of coal is only about 8×10^{7} J.

Table 23.5 Nuclear binding energies of ^{235}U and its fission products

	Nuclear binding energy
^{235}U	2.82×10^{-10} J
^{90}Sr	1.23×10^{-10} J
^{143}Xe	1.92×10^{-10} J

The significant feature of uranium-235 fission is not just the enormous amount of energy released, but the fact that more neutrons are produced than are originally captured in the process. This property makes possible a ***nuclear chain reaction,*** which is *a self-sustaining sequence of nuclear fission reactions.* The neutrons generated during the initial stages of fission can induce fission in other uranium-235 nuclei, which in turn produce more neutrons, and so on. In less than a second, the reaction can become uncontrollable, liberating a tremendous amount of heat to the surroundings.

Figure 23.8 shows two types of fission reactions. For a chain reaction to occur, enough uranium-235 must be present in the sample to capture the neutrons. Otherwise, many of the neutrons will escape from the sample and the chain reaction will not occur [Figure 23.8(a)]. In this situation the mass of the sample is said to be *subcritical.* Figure 23.8(b) shows what happens when the amount of the fissionable material is equal to or greater than the ***critical mass,*** *the minimum mass of fissionable material required to generate a self-sustaining nuclear chain reaction.* In this case most of the neutrons will be captured by uranium-235 and a chain reaction will occur.

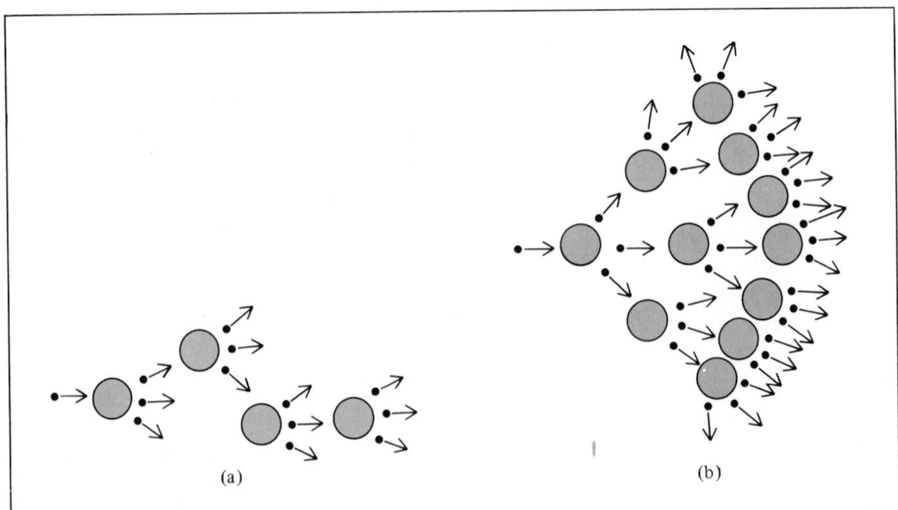

Figure 23.8 *Two types of nuclear fission. (a) If the mass of U-235 is subcritical, no chain reaction will result. Many of the neutrons produced will escape to the surroundings. (b) If the mass is critical, most of the neutrons will be captured by U-235 nuclei and an uncontrollable chain reaction will occur.*

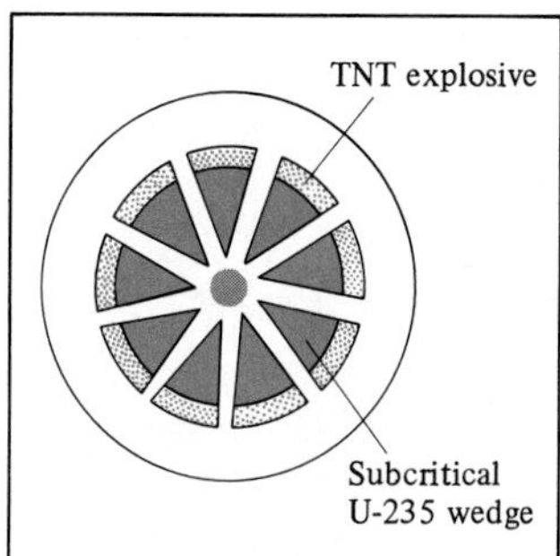

Figure 23.9 *Schematic cross section of an atomic bomb. The TNT explosives are set off first. The explosion forces the sections of fissionable material together to form an amount considerably larger than the critical mass.*

TNT stands for trinitrotoluene; it is a powerful explosive.

The Atomic Bomb

The first application of nuclear fission was in the development of the atomic bomb. How is such a bomb made and detonated? The crucial factor in the bomb's design is the determination of the critical mass for the bomb. A small atomic bomb is equivalent to 20,000 tons of TNT. Since 1 ton of TNT releases about 4×10^9 J of energy, 20,000 tons would produce 8×10^{13} J. Earlier we saw that 1 mole, or 235 g, of uranium-235 liberates 2.0×10^{13} J of energy when it undergoes fission. Thus the mass of the isotope present in a small bomb must be at least

$$235 \text{ g} \times \frac{8 \times 10^{13} \text{ J}}{2.0 \times 10^{13} \text{ J}} \simeq 1 \times 10^3 \text{ g} = 1 \text{ kg}$$

For obvious reasons, an atomic bomb is never assembled with the critical mass already present. Instead, the critical mass is formed by using a conventional explosive (for example, TNT) to force the fissionable sections together, as shown in Figure 23.9. Neutrons from a source at the center of the device trigger the nuclear chain reaction. Uranium-235 was the fissionable material in the bomb dropped on Hiroshima, Japan, on August 6, 1945. Plutonium-239 was used in the bomb exploded over Nagasaki three days later. The fission reactions were similar in these two cases, as was the extent of the destruction.

Nuclear Reactors

In Europe, nuclear reactors provide about 40 percent of the electrical energy consumed.

A peaceful but controversial application of nuclear fission is the generation of electricity using heat from a controlled chain reaction in a nuclear reactor. Currently, nuclear reactors provide about 20 percent of the electrical energy in the United States. This is a small but by no means negligible contribution to the nation's energy production. Several different types of nuclear reactors are in operation; we will briefly discuss the main features of three of them: light water reactors, heavy water reactors, and breeder reactors.

Light Water Reactors. Most of the nuclear reactors in the United States are *light water reactors*. Figure 23.10 is a schematic diagram of such a reactor, and Figure 23.11 shows the refueling process in the core of a nuclear reactor.

An important aspect of the fission process is the speed of the neutrons. Slow neutrons split uranium-235 nuclei more efficiently than do fast ones. Because fission reactions are so exothermic, the neutrons produced usually move at high velocities. For greater efficiency they must be slowed down before they can be used to induce nuclear disintegration. To accomplish this goal, scientists use ***moderators,*** which are *substances that can reduce the kinetic energy of neutrons*. A good moderator must satisfy several requirements: It must be a fluid so it can be used also as a coolant; it should possess a high specific heat; it should be nontoxic and inexpensive (as very large quantities of it are necessary); and it should resist conversion into a radioactive substance by neutron bombardment. No substance fits all these requirements, although water comes closer than many others that have been considered. Nuclear reactors using

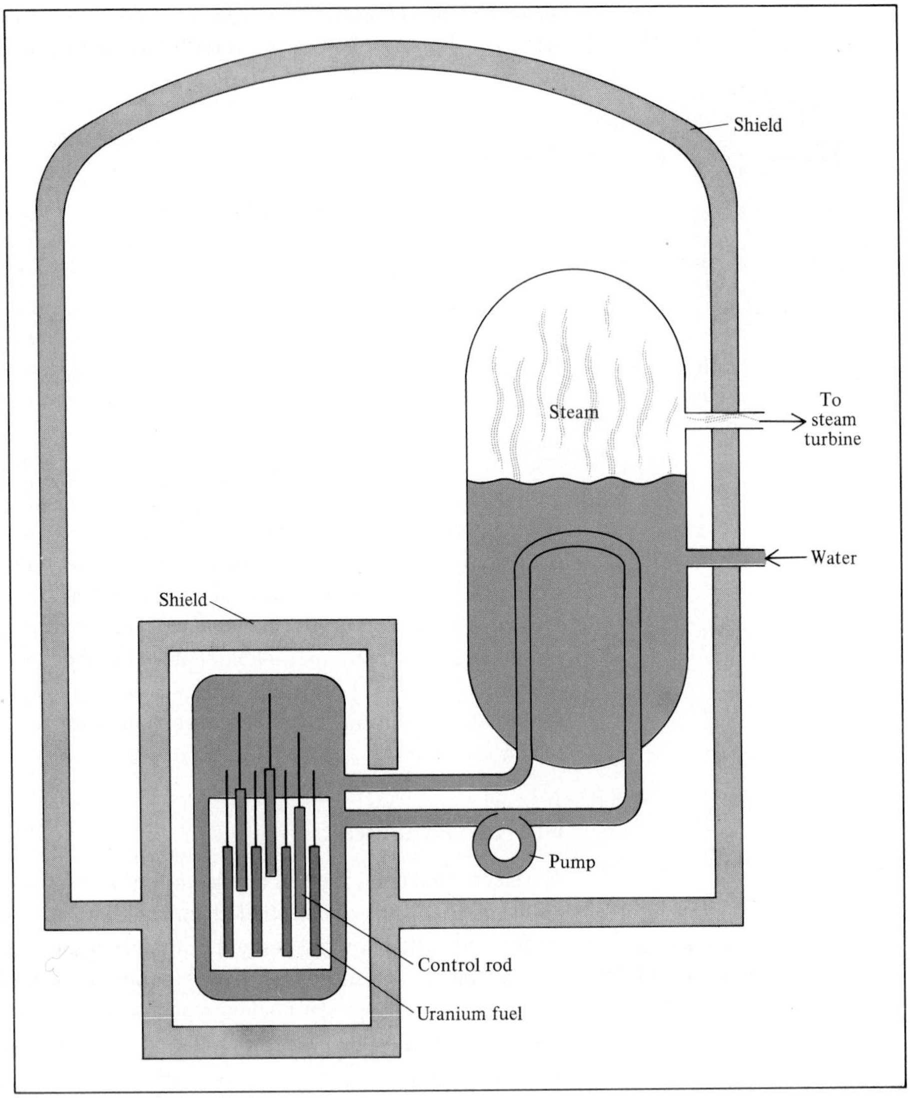

Figure 23.10 *Schematic diagram of a nuclear fission reactor. The fission process is controlled by cadmium or boron rods. The heat generated by the process is used to produce steam for the generation of electricity via a heat exchange system.*

Figure 23.11 *Refueling the core of a nuclear reactor.*

light water as a moderator are called light water reactors because H is the lightest isotope of the element hydrogen.

The nuclear fuel consists of uranium, usually in the form of its oxide, U_3O_8 (Figure 23.12). Naturally occurring uranium contains about 0.7 percent of the uranium-235 isotope, which is too low a concentration to sustain a small-scale chain reaction. For effective operation of a light water reactor, uranium-235 must be enriched to a concentration of 3 or 4 percent. In principle, the main difference between an atomic bomb and a nuclear reactor is that the chain reaction that takes place in a nuclear reactor is kept under control at all times. The factor limiting the rate of the reaction is the number of

Figure 23.12 *Uranium oxide, U_3O_8.*

neutrons present. This can be controlled by lowering cadmium or boron rods between the fuel elements. These rods capture neutrons according to the equations

$$^{113}_{48}\text{Cd} + ^{1}_{0}\text{n} \longrightarrow ^{114}_{48}\text{Cd} + \gamma$$

$$^{10}_{5}\text{B} + ^{1}_{0}\text{n} \longrightarrow ^{7}_{3}\text{Li} + ^{4}_{2}\text{He}$$

where γ denotes gamma rays. Without the control rods the heat generated would melt down the reactor core, releasing radioactive materials into the environment.

Nuclear reactors have rather elaborate cooling systems that absorb the heat generated by the nuclear reaction and transfer it outside the reactor core, where it is used to produce enough steam to drive an electric generator. In this respect a nuclear power plant is similar to a conventional power plant that burns fossil fuel. In both cases large quantities of cooling water are needed to condense steam for reuse. Thus, most nuclear power plants are built near a river or a lake. Unfortunately this method of cooling causes thermal pollution (see Section 12.6).

Figure 23.13 *The red glow of the radioactive plutonium-239 isotope. The orange color is due to the presence of its oxide.*

Heavy Water Reactors. Another type of nuclear reactor uses D_2O, or heavy water, as the moderator, rather than H_2O. Heavy water slows down the neutrons emerging from the fission reaction less efficiently than does light water. Consequently, there is no need to use enriched uranium for fission in a *heavy water reactor*. The faster-moving neutrons travel greater distances, and so the probability that they will strike the proper targets—uranium-235 isotopes—is correspondingly high. Eventually most of the uranium-235 isotopes will take part in the fission process.

The main advantage of a heavy water reactor is that it eliminates the need for building expensive uranium enrichment facilities. However, D_2O must be prepared by either fractional distillation or electrolysis of ordinary water, which can be very expensive considering the quantity of water used in a nuclear reactor. In countries such as Canada and Norway, where hydroelectric power is abundant, the cost of producing D_2O by electrolysis can be reasonably low. At present, Canada is the only nation successfully using heavy water nuclear reactors. The fact that no enriched uranium is required in a heavy water reactor allows a country to enjoy the benefits of nuclear power without undertaking work that is closely associated with weapons technology.

Breeder Reactors. A ***breeder reactor*** uses uranium fuel, but unlike a conventional nuclear reactor, it *produces more fissionable materials than it uses.*

We know that when uranium-238 is bombarded with fast neutrons the following reactions take place:

$$^{238}_{92}\text{U} + ^{1}_{0}\text{n} \longrightarrow ^{239}_{92}\text{U}$$

$$^{239}_{92}\text{U} \longrightarrow ^{239}_{93}\text{Np} + ^{0}_{-1}\beta \qquad t_{\frac{1}{2}} = 23.4 \text{ min}$$

$$^{239}_{93}\text{Np} \longrightarrow ^{239}_{94}\text{Pu} + ^{0}_{-1}\beta \qquad t_{\frac{1}{2}} = 2.35 \text{ days}$$

The plutonium produced forms the oxide and can be readily separated from uranium.

In this manner the nonfissionable uranium-238 is transmuted into the fissionable isotope plutonium-239, which has a half-life of 24,400 yr (Figure 23.13).

In a typical breeder reactor, nuclear fuel containing uranium-235 or plutonium-239 is mixed with uranium-238 so that breeding takes place within the core. For every uranium-235 (or plutonium-239) nucleus undergoing fission, more than one neutron is captured by uranium-238 to generate plutonium-239. Thus, the stockpile of fissionable material can be steadily increased as the starting nuclear fuels are consumed. An important factor is the *doubling time,* the time required to produce as much net additional nuclear fuel as was originally present in the reactor. It takes about 7 to 10 years to regenerate the sizable amount of material needed to refuel the original reactor and to fuel another reactor of comparable size.

Another fertile isotope is $^{232}_{90}Th$. Upon capturing slow neutrons, thorium is transmuted to uranium-233, which, like uranium-235, is a fissionable isotope:

$$^{232}_{90}\text{Th} + ^{1}_{0}\text{n} \longrightarrow ^{233}_{90}\text{Th}$$

$$^{233}_{90}\text{Th} \longrightarrow ^{233}_{91}\text{Pa} + ^{0}_{-1}\beta \qquad t_{\frac{1}{2}} = 22 \text{ min}$$

$$^{233}_{91}\text{Pa} \longrightarrow ^{233}_{92}\text{U} + ^{0}_{-1}\beta \qquad t_{\frac{1}{2}} = 27.4 \text{ days}$$

Uranium-233 is stable enough for long-term storage. The amounts of uranium-238 and thorium-232 in Earth's crust are relatively plentiful (4 ppm and 12 ppm by mass, respectively).

Despite the promising prospects, the development of breeder reactors has been very slow. To date, the United States does not have a single operating breeder reactor, and only a few have been built in other countries, such as France and Russia. One problem is economics; breeder reactors are more expensive to build than conventional reactors. There are also more technical difficulties associated with the construction of such reactors. As a result, the future of breeder reactors, in the United States at least, is rather uncertain.

Hazards of Nuclear Energy. There are many people, including environmentalists, who regard nuclear fission as a highly undesirable method of energy production. Many fission products such as strontium-90 are dangerous radioactive isotopes with long half-lives. Plutonium-239, used as a nuclear fuel and produced in breeder reactors, is one of the most toxic substances known. It is an α emitter with a half-life of 24,400 yr.

Accidents, too, present many dangers. The accident at the Three Mile Island reactor in Pennsylvania in 1979 first brought the potential hazards of nuclear plants to public attention. Only a few years later, the disaster at the Chernobyl nuclear plant in the Soviet Union, on April 26, 1986, was a tragic reminder of just how catastrophic a runaway nuclear reaction can be. On that day, a reactor at the plant surged out of control. The fire and explosion that followed released much radioactive material into the environment. People working near the plant died within weeks as a result of the exposure to the intense radiation. The long-term effect of the radioactive fallout in the western Soviet Union and Europe has not yet been clearly assessed, although agriculture and dairy farming have already been affected by the fallout. The number of potential cancer deaths attributable to the radiation contamination is estimated to be between a few thousand and more than 100,000.

Figure 23.14 *Molten glass is poured over nuclear wastes before burial.*

Furthermore, the problem of radioactive waste disposal has not been satisfactorily resolved even for safely operated nuclear plants. Many suggestions have been made as to where to store or dispose of nuclear waste, including burial underground (Figures 23.14 and 23.15), burial beneath the ocean floor, and storage in deep geologic formations. But none of these sites has proved absolutely safe in the long run. Leakage of radioactive wastes into underground water, for example, can endanger nearby communities. The ideal disposal site would seem to be the sun, where a bit more radiation would make little difference, but this kind of operation requires 100 percent reliability in space technology.

Because of the hazards, the future of nuclear reactors is clouded. What was once

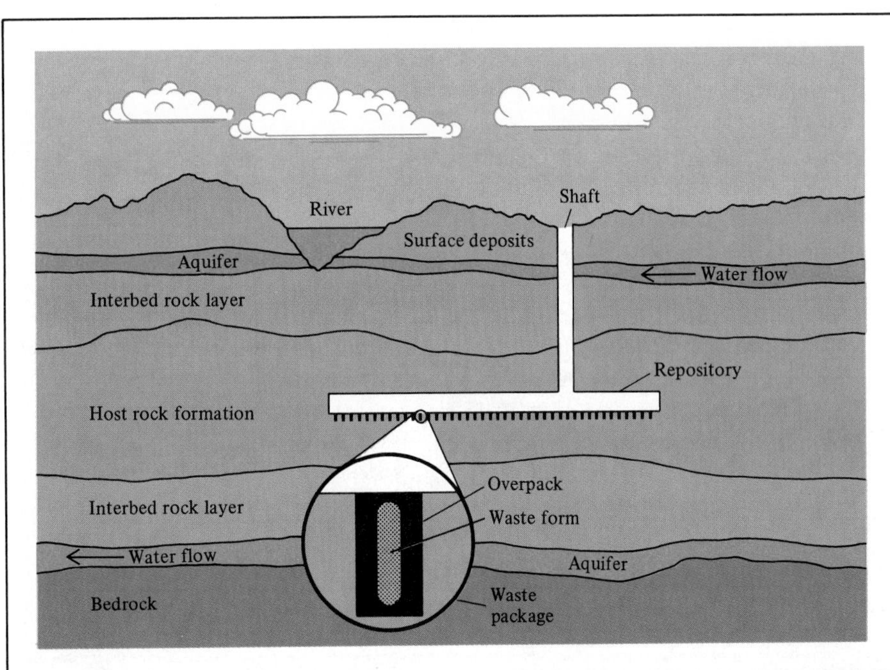

Figure 23.15 *Schematic diagram for the deep underground burial of high-level radioactive waste.*

hailed as the ultimate solution to our energy needs in the twenty-first century is now being debated and questioned by both the scientific community and laypeople. It seems likely that the controversy will continue for some time.

CHEMISTRY IN ACTION

Nature's Own Fission Reactor

It all started with a routine analysis in May 1972 at the nuclear fuel processing plant in Pierrelatte, France. A staff member was checking the isotope ratio of U-235 to U-238 in a uranium ore and obtained a puzzling result. It had long been known that the relative natural occurrence of U-235 and U-238 is 0.7202 percent and 99.2798 percent, respectively. In this case, however, the amount of U-235 present was only 0.7171 percent. This may seem like a very small deviation, but the measurements were so precise that this difference was considered highly significant. The ore had come from the Oklo mine in the Gabon Republic, a small country on the west coast of Africa. Subsequent analyses of other samples showed that some contained even less U-235, in some cases as little as 0.44 percent.

The logical explanation for the low percentages of U-235 was that a nuclear fission reaction at the site of ore extraction must have consumed some of the U-235 isotopes. But how did this happen? There are several conditions under which such a nuclear fission reaction could take place. In the presence of heavy water, for example, a chain reaction is possible with unenriched uranium. Without heavy water, such a fission reaction could still occur if the uranium ore and the moderator were arranged according to some specific geometric constraints at the site of the reaction. Both of the possibilities seem rather farfetched. The most plausible explanation is that the uranium ore originally present in the mine was enriched with U-235 and that a nuclear fission reaction took place with light water, as in a conventional nuclear reactor.

As mentioned earlier, the natural abundance of U-235 is 0.7202 percent, but it has not always been that low. The half-lives of U-235 and U-238 are 700 million and 4.51 billion years, respectively. This means that U-235 must have been *more* abundant in the past, because it has a shorter half-life. In fact, at the time Earth was formed, the natural abundance of U-235 was as high as 17 percent! Since the lowest concentration of U-235 required for the operation of a fission reactor is 1 percent, a nuclear chain reaction could have taken place as recently as 400 million years ago. By analyzing the amounts of radioactive fission products left in the ore, scientists concluded that the Gabon "reactor" operated about 2 billion years ago.

Having an enriched uranium sample is only one of the requirements for starting a controlled chain reaction. There must also have been a sufficient amount of the ore and an appropriate moderator present. It appears that as a result of a geological transformation, uranium ore was continually being washed into the Oklo region to yield concentrated deposits. The moderator needed for the fission process was largely water, present as water of crystallization in the sedimentary ore. Figure 23.16 shows a section of the reaction zone on the floor of the pit at Oklo.

Thus, in a series of extraordinary events, a natural nuclear fission reactor operated at the time when the first sign of living systems began to appear on Earth. As is often the case in scientific endeavors, humans are not necessarily the innovators but merely the imitators of nature.

Figure 23.16 *Photo showing the natural nuclear reactor site (lower right-hand corner) at Oklo, Gabon Republic.*

23.6 Nuclear Fusion

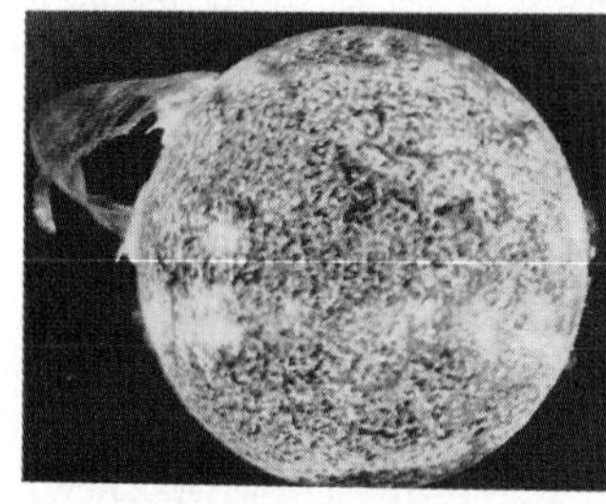

As a result of nuclear fusion, the temperature in the interior of the sun is about 15 million °C.

In contrast to the nuclear fission process, ***nuclear fusion,*** *the combining of small nuclei into larger ones,* is largely exempt from the waste disposal problem.

Figure 23.2 showed that for the lightest elements, nuclear stability increases with increasing mass number. This behavior suggests that if two light nuclei combine or fuse together to form a larger, more stable nucleus, an appreciable amount of energy will be released in the process. This is the basis for ongoing research into the harnessing of nuclear fusion for the production of energy.

Nuclear fusion occurs constantly in the sun. The sun is made up mostly of hydrogen and helium. In its interior, where temperatures reach about 15 million degrees Celsius, the following fusion reactions are believed to take place:

$$^{1}_{1}\text{H} + ^{2}_{1}\text{H} \longrightarrow ^{3}_{2}\text{He}$$

$$^{3}_{2}\text{He} + ^{3}_{2}\text{He} \longrightarrow ^{4}_{2}\text{He} + 2^{1}_{1}\text{H}$$

$$^{1}_{1}\text{H} + ^{1}_{1}\text{H} \longrightarrow ^{2}_{1}\text{H} + ^{0}_{+1}\beta$$

Because *fusion reactions take place only at very high temperatures,* they are often called ***thermonuclear reactions.***

Fusion Reactors

A major concern in choosing the proper nuclear fusion process for energy production is the temperature necessary to carry out the process. Some of the more promising reactions are

Reaction	Energy released
$^{2}_{1}\text{H} + ^{2}_{1}\text{H} \longrightarrow ^{3}_{1}\text{H} + ^{1}_{1}\text{H}$	6.3×10^{-13} J
$^{2}_{1}\text{H} + ^{3}_{1}\text{H} \longrightarrow ^{4}_{2}\text{He} + ^{1}_{0}\text{n}$	2.8×10^{-12} J
$^{6}_{3}\text{Li} + ^{2}_{1}\text{H} \longrightarrow 2^{4}_{2}\text{He}$	3.6×10^{-12} J

These reactions take place at extremely high temperatures, of the order of 100 million degrees Celsius, to overcome the repulsive forces between the nuclei. The first reaction is particularly attractive because the world's supply of deuterium is virtually inexhaustible. The total volume of water on Earth is about 1.5×10^{21} L. Since the natural abundance of deuterium is 1.5×10^{-2} percent, the total amount of deuterium present is roughly 4.5×10^{21} g, or 5.0×10^{15} tons. The cost of preparing deuterium is minimal compared with the value of the energy released by the reaction.

Compared with the fission process, nuclear fusion looks like a very promising energy source, at least "on paper." Its advantages are that (1) the fuels are cheap and almost inexhaustible; (2) the process is "clean"; that is, except for thermal pollution, it produces little radioactive waste; and (3) it is a safe process. If a fusion machine were turned off, it would shut down completely and instantly; there would be no possibility of a meltdown.

If nuclear fusion is so great, why isn't there even one fusion reactor producing energy? Although we command the scientific knowledge to design such a reactor, the technical difficulties have not yet been solved. The basic problem is finding a way to hold the nuclei together long enough, and at the appropriate temperature, for fusion to occur. At temperatures of about 100 million degrees Celsius, molecules cannot exist, and most or all of the atoms are stripped of their electrons. This *state of matter, in which a gaseous system consists of positive ions and electrons,* is called ***plasma.*** The problem of containing this plasma is a formidable one. What solid container can exist at such temperatures? None, unless the amount of plasma is small; but then the solid surface would immediately cool the sample and quench the fusion reaction. One approach to solving this problem is to use *magnetic confinement*. Since a plasma consists of charged particles moving at high speeds, a magnetic field would exert force on it. Figure 23.17 shows a recent magnetic confinement design, called *tokamak*. The plasma moves through this doughnut-shaped tunnel, confined by a complex magnetic field. Thus the plasma never comes in contact with the walls of the container.

Another promising development employs high-power lasers to initiate the fusion reaction. In test runs a number of laser beams transfer energy to a small fuel pellet, heating it and causing it to *implode,* that is, to collapse inward from all sides and compress into a small volume (Figure 23.18). Consequently, fusion occurs. Like the magnetic confinement approach, laser fusion presents a number of technical difficulties that still need to be overcome before it can be put to practical use on a large scale.

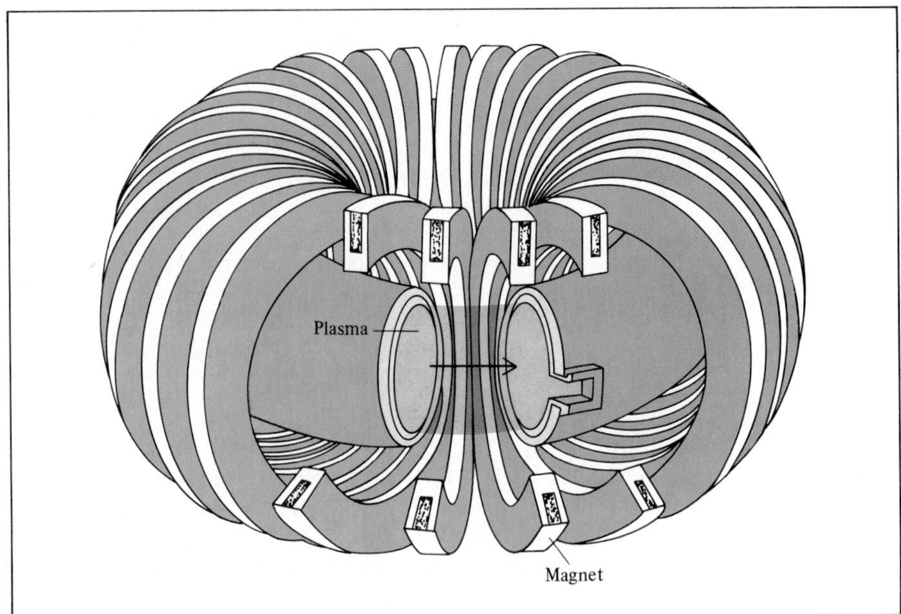

Figure 23.17 *Magnetic confinement of plasma.*

Figure 23.18 *This small-scale fusion reaction was created in 1986 at the Lawrence Livermore National Laboratory using the world's most powerful laser, Nova.*

Figure 23.19 *Explosion of a thermonuclear bomb.*

The Hydrogen Bomb

The technical problems inherent in the design of a nuclear fusion reactor do not affect the production of a hydrogen bomb, also called a thermonuclear bomb. In this case the objective is all power and no control. Hydrogen bombs do not contain gaseous hydrogen or gaseous deuterium; they contain solid lithium deuteride (LiD), which can be packed very tightly. The detonation of a hydrogen bomb occurs in two stages—first a fission reaction and then a fusion reaction. The required temperature for fusion is derived from an atomic bomb. Immediately after the atomic bomb explodes, the following fusions occur, releasing vast amounts of energy (Figure 23.19):

$$^{6}_{3}Li + ^{2}_{1}H \longrightarrow 2^{4}_{2}He$$

$$^{2}_{1}H + ^{2}_{1}H \longrightarrow ^{3}_{1}H + ^{1}_{1}H$$

There is no critical mass in a fusion bomb, and the force of explosion is limited only by the quantity of reactants present. Thermonuclear bombs are described as being "cleaner" than atomic bombs because they do not produce radioactive isotopes except for tritium, which is a weak β-particle emitter ($t_{\frac{1}{2}}$ = 12.5 yr), and the products from the fission starter. Their damaging effects on the environment can be aggravated, however, by incorporating in the construction some nonfissionable material such as cobalt. Upon bombardment by neutrons, cobalt-59 is converted to cobalt-60, which is a very strong γ-ray emitter with a half-life of 5.2 yr. The presence of these radioactive cobalt isotopes in the debris or fallout from a thermonuclear explosion would cause the deaths of those who survived the initial blast.

CHEMISTRY IN ACTION

The Confusion about Cold Fusion

The nuclear fusion reactor described in this section represents the brute-force way of bringing about nuclear fusion, using extremely high temperature and strong magnetic fields. Despite decades of hard work and billions of dollars spent on research, physicists have not yet reached the break-even point, that is, the point at which energy produced is equal to the energy input. It is therefore not surprising that the scientific community and the general public were tremendously excited when they heard the announcement in March of 1989 by two chemists, B. Stanley Pons of the University of Utah and Martin Fleischmann of the University of Southampton, England, that they had achieved nuclear fusion at room temperature.

The amazing fact about Pons and Fleischmann's claim is that the fusion process occurs in a simple electrolytic cell (Figure 23.20), one that could be used in an introductory chemistry experiment! Pons and Fleischmann passed a current between a platinum anode and a palladium cathode in a 0.1 *M* LiOD solution in D_2O. They found extremely high heat output for the energy input into the cell. Pons and Fleischmann

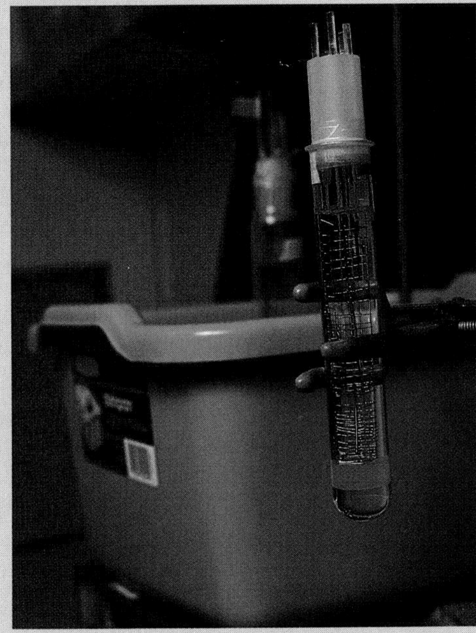

Figure 23.20 *An electrolytic cell used by Pons and Fleischmann in their cold fusion experiment.*

were unable to explain such a large excess of heat by normal chemical reactions, so they concluded that a nuclear fusion process must be taking place. These scientists suggest that deuterium formed at the cathode diffuses into the palladium rod. Eventually, it fills up the vacant sites in the palladium lattice. Invoking quantum mechanical effects, they propose that the crowding together of deuteriums leads to their fusing. It is known that deuterium-deuterium fusion follows two different paths, with about equal probability. One path leads to helium-3 and a neutron:

$$^{2}_{1}H + ^{2}_{1}H \longrightarrow ^{3}_{2}He + ^{1}_{0}n$$

and the other yields tritium and a proton:

$$^{2}_{1}H + ^{2}_{1}H \longrightarrow ^{3}_{1}H + ^{1}_{1}H$$

Pons and Fleischmann claim to have detected both tritium and neutrons during electrolysis, but the levels are much lower than one would expect on the basis of the extra heat generated. Thus they suggest that it must be due to a hitherto unknown nuclear process.

If Pons and Fleischmann's experiment is correct, then their discovery would rank among the most significant of this century. It would mean that we would have an almost inexhaustible supply of energy without polluting the environment. The scientific community has been galvanized into feverish activity by Pons and Fleischmann's work. In the months following their announcement, there were countless reports of experiments confirming some of their observations, and others showing no effect at all. At present there is considerable skepticism about Pons and Fleischmann's claim, while some scientists still believe that an unusual nuclear process may be taking place. It is hoped that the current confusion about cold fusion will be resolved in the next year or so.

23.7 Applications of Isotopes

Radioactive and stable isotopes alike have many applications in science and medicine. We have previously described the use of isotopes in the study of reaction mechanisms (see Section 13.4) and in dating (p. 563 and Section 23.3). In this section we discuss a few more examples.

Structural Determination

The formula of the thiosulfate ion is $S_2O_3^{2-}$. For some years chemists were uncertain as to whether the two sulfur atoms occupied equivalent positions in the ion. The thiosulfate ion is prepared by treatment of the sulfite ion with elemental sulfur:

$$SO_3^{2-}(aq) + S(s) \longrightarrow S_2O_3^{2-}(aq) \qquad (23.1)$$

When thiosulfate is treated with dilute acid, the sulfite ion is re-formed and elemental sulfur precipitates:

$$S_2O_3^{2-}(aq) \xrightarrow{H^+} SO_3^{2-}(aq) + S(s) \qquad (23.2)$$

If this sequence is started with the elemental sulfur enriched in the radioactive sulfur-35 isotope, the isotope acts as a "label" for S atoms. All the labels are found in the sulfur precipitate in Equation (23.2); none of them appears in the final sulfite ions. Clearly, then, the two atoms of sulfur in $S_2O_3^{2-}$ are not structurally equivalent, as would be the case if the structure were

$$\left[:\ddot{\underset{..}{O}}-\ddot{\underset{..}{S}}-\ddot{\underset{..}{O}}-\ddot{\underset{..}{S}}-\ddot{\underset{..}{O}}: \right]^{2-}$$

Otherwise, the radioactive isotope would be present in both the elemental sulfur precipitate and the sulfite ion. Based on other studies, we know that the structure of the thiosulfate ion is

$$\left[\begin{array}{c} :\ddot{O}: \\ | \\ :\ddot{O}-S-\ddot{S}: \\ | \\ :\ddot{O}: \end{array}\right]^{2-}$$

Study of Photosynthesis

The study of photosynthesis is also rich with isotope applications. The overall photosynthesis reaction can be represented as

$$6CO_2 + 6H_2O \longrightarrow C_6H_{12}O_6 + 6O_2$$

A question that arose early in studies of photosynthesis was whether the molecular oxygen was derived from water, from carbon dioxide, or from both. By the use of $H_2{}^{18}O$, it was demonstrated that the evolved oxygen came from water, and none came from carbon dioxide, because the O_2 formed contained only the ^{18}O isotopes. This result established the mechanism in which water molecules are "split" by light:

$$2H_2O + h\nu \longrightarrow O_2 + 4H^+ + 4e^-$$

The protons and electrons can then be used to drive energetically unfavorable reactions which are necessary for a plant's growth and function.

The radioactive carbon-14 isotope helped to determine the path of carbon in photosynthesis. Starting with $^{14}CO_2$, it was possible to isolate the intermediate products during photosynthesis and measure the amount of radioactivity of each carbon-containing compound. In this manner the path from CO_2 through various intermediate compounds to carbohydrate could be clearly charted. *Isotopes, especially radioactive isotopes that are used to trace the path of the atoms of an element in a chemical or biological process,* are called ***tracers.***

Isotopes in Medicine

Tracers are used also for diagnosis in medicine. Sodium-24 (a β emitter with a half-life of 14.8 h) injected into the bloodstream as a salt solution can be monitored to trace the flow of blood and detect possible constrictions or obstructions in the circulatory system. Iodine-131 (a β emitter with a half-life of 8 days) has been used to test the activity of the thyroid gland. A malfunctioning thyroid can be detected by giving the patient a drink of a solution containing a known amount of $Na^{131}I$ and measuring the radioactivity just above the thyroid to see if the iodine is absorbed at the normal rate. Of course, the amounts of radioisotope used in the human body must always be kept small; otherwise, the patient might suffer permanent damage from the high-energy radiation. Another radioactive isotope of iodine, iodine-123 (a γ-ray emitter), is used to image the brain (Figure 23.21).

Technetium is the first artificially prepared element.

Technetium is one of the most useful elements in nuclear medicine. Although technetium is a transition metal, all its isotopes are radioactive. Therefore, technetium does not occur naturally on Earth. In the laboratory it is prepared by the nuclear reactions

$$^{98}_{42}Mo + ^{1}_{0}n \longrightarrow ^{99}_{42}Mo$$

$$^{99}_{42}Mo \longrightarrow ^{99m}_{43}Tc + ^{0}_{-1}\beta$$

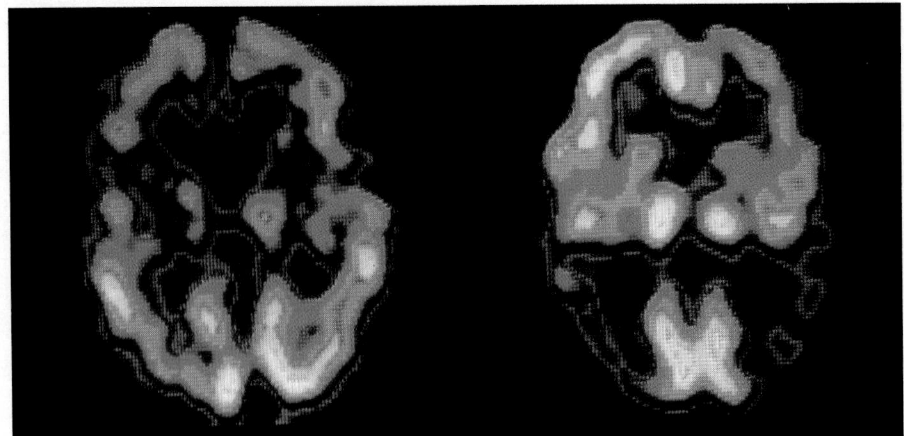

Figure 23.21 *A compound labeled with iodine-123 is used to image the brain. Left: A normal brain. Right: The brain of a patient with Alzheimer's disease.*

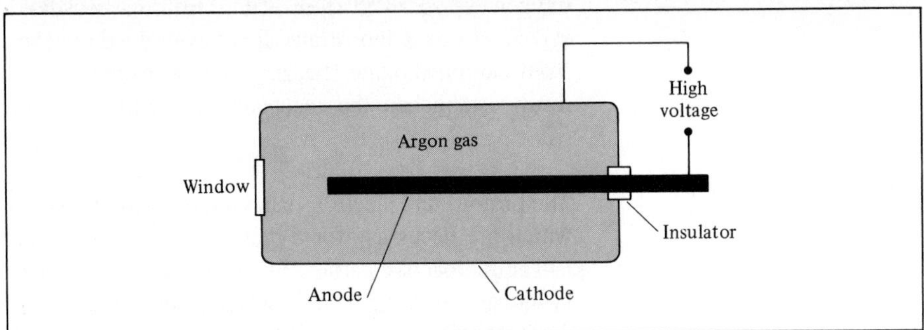

Figure 23.22 *Schematic diagram of a Geiger counter. Radiation (α, β, or γ rays) entering through the window causes ionization of the argon gas, which allows a small current to flow between the electrodes. This current is amplified and is used to flash a light or operate a counter with a clicking sound (not shown).*

where the superscript m denotes that the technetium-99 isotope is produced in its excited nuclear state. This isotope has a half-life of about 6 hours, decaying by γ radiation to technetium-99 in its nuclear ground state. Thus it is a valuable diagnostic tool. The patient either drinks or is injected with a solution containing ^{99m}Tc. By detecting the γ rays emitted by ^{99m}Tc, doctors can obtain images of organs such as the heart, liver, and lungs.

A major advantage of using radioactive isotopes as tracers is that they are easy to detect. Their presence even in very small amounts can be detected by photographic techniques or by devices known as counters. Figure 23.22 is a diagram of a Geiger counter, an instrument widely used in scientific work and medical laboratories to detect radiation.

23.8 Biological Effects of Radiation

In this section we will examine briefly the effects of radiation on biological systems. But first let us define quantitative measures of radiation. The fundamental unit of radioactivity is the *curie* (Ci); 1 Ci corresponds to exactly 3.70×10^{10} nuclear disintegrations per second. This decay rate is equivalent to that of 1 g of radium. A *millicurie* (mCi) is one-thousandth of a curie. Thus, 10 mCi of a carbon-14 sample is the quantity that undergoes

$$(10 \times 10^{-3})(3.70 \times 10^{10}) = 3.70 \times 10^{8}$$

disintegrations per second. The intensity of radiation depends on the number of disintegrations as well as on the energy and type of radiation emitted. One common unit for the absorbed dose of radiation is the *rad* (*r*adiation *a*bsorbed *d*ose), which is the amount of radiation that results in the absorption of 1×10^{-5} J per gram of irradiated material. The biological effect of radiation depends on the part of the body irradiated and the type of radiation. For this reason the rad is often multiplied by a factor called *RBE* (*r*elative *b*iological *e*ffectiveness). The product is called a *rem* (*r*oentgen *e*quivalent for *m*an):

$$1 \text{ rem} = 1 \text{ rad} \times 1 \text{ RBE}$$

Of the three types of nuclear radiation, α particles usually have the least penetrating power. Beta particles are more penetrating than α particles, but less so than γ rays. Gamma rays have very short wavelengths and high energies. Furthermore, since they carry no charge, they cannot be stopped by shielding materials as easily as α and β particles. However, if α or β emitters are ingested, their damaging effects are greatly aggravated because the organs will be constantly subject to damaging radiation at close range. For example, strontium-90, a β emitter, can replace calcium in bones, where it does most of the damage. Table 23.6 lists the average amounts of radiation an American receives every year. It should be pointed out that for short-term exposures to radiation, a dosage of 50–200 rem will cause a decrease in white blood cell counts and other complications while a dosage of 500 rem or greater will result in death within weeks.

The chemical basis of radiation damage is that of ionizing radiation. Radiation of either particles or γ rays can remove electrons from atoms and molecules in its path, leading to the formation of ions and radicals. ***Radicals*** are *molecular fragments having one or more unpaired electrons; they are usually short-lived and highly reactive*. For example, when water is irradiated with γ rays, the following reactions take place:

$$H_2O \xrightarrow{\text{radiation}} H_2O^+ + e^-$$

$$H_2O^+ + H_2O \longrightarrow H_3O^+ + \cdot OH$$

Table 23.6 Average yearly radiation doses for Americans

Source	Dose (mrem/yr)*
Cosmic rays	20–50
Ground and surroundings	25
Human body†	26
Medical and dental X rays	50–75
Air travel	5
Fallout from weapons tests	5
Nuclear waste	2
Total	133–188

*1 mrem = 1 millirem = 1×10^{-3} rem.
†The radioactivity in the body comes from food and air.

Current safety standards permit nuclear workers to be exposed to no more than 5 rem per year and specify that members of the general public should not be exposed to more than 0.5 rem of man-made radiation per year.

The electron (in the hydrated form) can subsequently react with water or with a hydrogen ion to form atomic hydrogen, and with oxygen to produce the superoxide ion, O_2^-:

$$e^- + H_2O \longrightarrow H\cdot + OH^-$$

$$e^- + H^+ \longrightarrow H\cdot$$

$$e^- + O_2 \longrightarrow \cdot O_2^-$$

In the tissues the superoxide ions and other free radicals attack cell membranes and a host of organic compounds, such as enzymes and DNA molecules. Organic compounds can themselves be directly ionized and destroyed by high-energy radiation.

It has long been known that exposure to high-energy radiation can induce cancer in humans and other animals. Cancer is characterized by uncontrolled cellular growth. On the other hand, it is also well established that cancer cells can be destroyed by proper radiation treatment. In radiation therapy, a compromise is sought. The radiation to which the patient is exposed must be sufficient to destroy cancer cells without killing too many normal cells and, it is hoped, without inducing another form of cancer.

Radiation damage to living systems is generally classified as *somatic* or *genetic*. Somatic injuries are those that affect the organism during its own lifetime. Sunburn, skin rash, cancer, and cataracts are examples of somatic damage. Genetic damage means inheritable changes or gene mutations. For example, a person whose chromosomes have been damaged or altered by radiation may have deformed offspring.

Chromosomes are parts of the cell structure that contain the genetic material (DNA).

CHEMISTRY IN ACTION

Tobacco Radioactivity

"SURGEON GENERAL'S WARNING: Smoking Is Hazardous to Your Health." Warning labels such as this appear on every package of cigarettes sold in the United States. The link between cigarette smoke and cancer has long been established. There is, however, another cancer-causing mechanism in smokers. The culprit in this case is a radioactive environmental pollutant present in the tobacco leaves from which cigarettes are made.

The soil in which tobacco is grown is heavily treated with phosphate fertilizers, which are rich in uranium and its decay products. Consider a particularly important step in the uranium-238 decay series:

$$^{226}_{88}Ra \longrightarrow ^{222}_{86}Rn + \alpha$$

The product formed, radon-222, is an unreactive gas (radon is the only gaseous product in the uranium-238 decay series). Radon-222 emanates from radium-226 and is present at high concentrations in soil gas and in the surface air layer under the vegetation canopy provided by the field of growing tobacco (Figure 23.23). In this layer some of the daughters of radon-222 such as polonium-218 and lead-214 become firmly attached to the surface and interior of tobacco leaves. As Table 23.3 shows, the next few decay reactions leading to the formation of lead-210 proceed rapidly. Gradually, the concentration of radioactive lead-210 can build to quite a high level.

During the combustion of a cigarette, small insoluble smoke particles are inhaled and deposited in the respiratory tract of the smoker and are eventually transported and stored at sites in the liver, spleen, and bone marrow. Measurements have shown that there is a high lead-210 content in these particles. (Note that the lead-210 content is not high enough to be hazardous *chemically,* but because it is radioactive it is hazardous.)

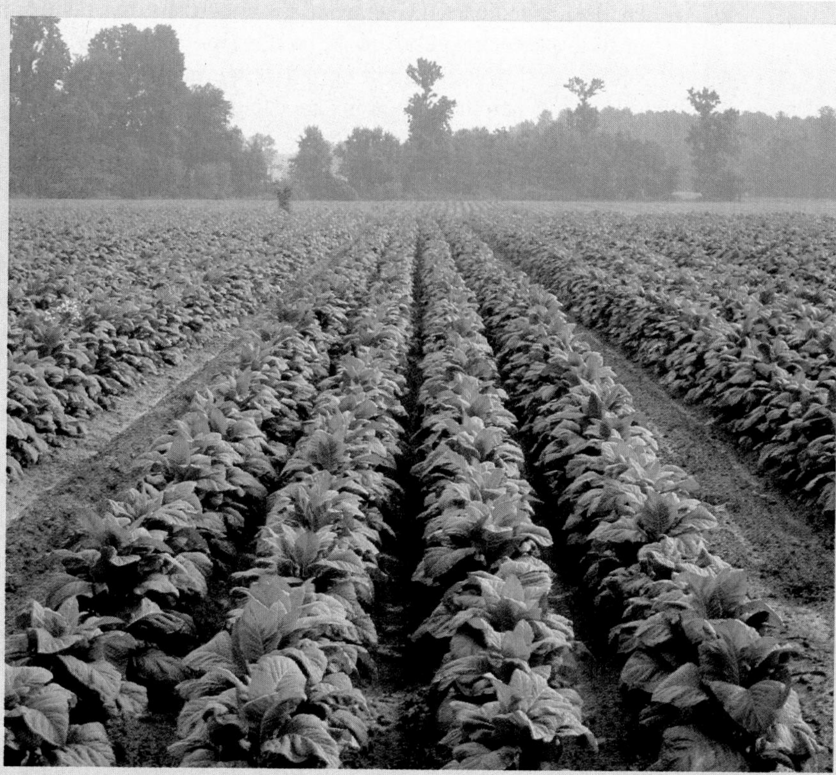

Figure 23.23 *A field of tobacco.*

Because of its long half-life (20.4 yr), lead-210 and its radioactive daughters—bismuth-210 and polonium-210—can continue to build up in the body throughout the period of smoking. Constant exposure of organs and bone marrow to α- and β-particle radiation increases the probability of cancer development in the smoker. The overall damaging effect on a person is quite similar to that caused by radon gas indoors (see Section 21.8).

Summary

1. For stable nuclei of low atomic number, the neutron-to-proton ratio is close to 1. For heavier stable nuclei, the ratio becomes greater than 1. All nuclei with 84 or more protons are unstable and radioactive. Nuclei with even atomic numbers are more stable than those with odd atomic numbers.
2. A quantitative measure of nuclear stability is the nuclear binding energy. Nuclear binding energy can be calculated from a knowledge of the mass defect of the nucleus.
3. Radioactive nuclei emit α particles, β particles, positrons, or γ rays. The equation for a nuclear reaction includes the particles emitted, and both the mass numbers and the atomic numbers must balance.
4. Uranium-238 is the parent of a natural radioactive decay series which can be used to determine the ages of rocks.

5. Artificially radioactive elements are created by the bombardment of other elements by accelerated neutrons, protons, or α particles.
6. Nuclear fission is the splitting of a large nucleus into two smaller nuclei, plus neutrons. When these neutrons are captured efficiently by other nuclei, an uncontrollable chain reaction can occur.
7. Nuclear reactors use the heat from a controlled nuclear fission reaction to produce power. The three important types of reactors are light water reactors, heavy water reactors, and breeder reactors.
8. Nuclear fusion, the type of reaction that occurs in the sun, is the combination of two light nuclei to form one heavy nucleus. Fusion takes place only at very high temperatures—so high that controlled large-scale nuclear fusion has so far not been achieved.
9. Radioactive isotopes are easy to detect and thus make excellent tracers in chemical reactions and in medical practice.
10. High-energy radiation damages living systems by causing ionization and the formation of free radicals.

Applied Chemistry Example: Nuclear Waste Disposal

As was pointed out in the chapter, a major problem facing the nuclear industry is the disposal of nuclear wastes. The variety of highly radioactive isotopes arises from the fission "daughters," reaction of neutrons with the building materials (for example, iron and cobalt in the steel used in construction), and formation of transuranium elements from the bombardment of ^{238}U by neutrons and α particles. Among the radioactive isotopes generated are ^{55}Fe, ^{60}Co, ^{85}Kr, ^{90}Sr, ^{137}Cs, ^{129}I, and ^{106}Ru. Obviously isotopes with longer half-lives present a greater problem as they will remain radioactive for years or even hundreds of years. Nuclear wastes are usually classified as high or low level on the basis of the concentration and toxicity of the radioactive constituents present. The total amount of radioactivity present in nuclear waste in the United States today is estimated to be about 1.5 billion curies. In choosing a safe and stable environment for these nuclear wastes, especially high-level wastes, consideration must be given to the heat released due to decay. As an illustrative example, consider the β decay of ^{90}Sr (89.907738 amu):

$$^{90}_{38}\text{Sr} \longrightarrow\ ^{90}_{39}\text{Y} +\ ^{0}_{-1}\beta \qquad t_{\frac{1}{2}} = 28.1 \text{ yr}$$

The ^{90}Y (89.907152 amu) formed further decays as follows:

$$^{90}_{39}\text{Y} \longrightarrow\ ^{90}_{40}\text{Zr} +\ ^{0}_{-1}\beta \qquad t_{\frac{1}{2}} = 64 \text{ h}$$

^{90}Zr (89.904703 amu) is a stable isotope. The question of interest is: How much heat is generated in a year starting with one mole of ^{90}Sr? We arrive at the answer by calculating the heat released in each of the above two steps and the fraction of ^{90}Sr that undergoes decay in one year, as shown below.

1. Use the mass defect to calculate the energy released (in joules) in each of the above two decays. Given that the mass of an electron is 5.4857×10^{-4} amu.

   Answer: Considering the ^{90}Sr decay, the mass defect is

   $$\Delta m = \text{mass}(^{90}\text{Y}) + \text{mass}(e^{-}) - \text{mass}(^{90}\text{Sr})$$

$$= (89.907152 \text{ amu} + 5.4857 \times 10^{-4} \text{ amu} - 89.907738 \text{ amu}) \times \frac{1 \text{ kg}}{6.022 \times 10^{26} \text{ amu}}$$
$$= -6.216 \times 10^{-32} \text{ kg}$$

The energy change is given by

$$\begin{aligned} \Delta E &= (\Delta m)c^2 \\ &= (-6.216 \times 10^{-32} \text{ kg})(3.00 \times 10^8 \text{ m/s})^2 \\ &= -5.59 \times 10^{-15} \text{ kg m}^2/\text{s}^2 \\ &= -5.59 \times 10^{-15} \text{ J} \end{aligned}$$

Similarly, for the ^{90}Y decay, we have

$$\Delta m = \text{mass}(^{90}\text{Zr}) + \text{mass}(e^-) - \text{mass}(^{90}\text{Y})$$
$$= (89.904703 \text{ amu} + 5.4857 \times 10^{-4} \text{ amu} - 89.907152 \text{ amu}) \times \frac{1 \text{ kg}}{6.022 \times 10^{26} \text{ amu}}$$
$$= -3.156 \times 10^{-30} \text{ kg}$$

and the energy change is

$$\begin{aligned} \Delta E &= (-3.156 \times 10^{-30} \text{ kg})(3.00 \times 10^8 \text{ m/s})^2 \\ &= -2.84 \times 10^{-13} \text{ J} \end{aligned}$$

Thus the total amount of energy released is $(-5.59 \times 10^{-15} \text{ J} - 2.84 \times 10^{-13} \text{ J})$ or -2.90×10^{-13} J.

2. Starting with one mole of ^{90}Sr, calculate the number of moles of ^{90}Sr that will decay in a year.

   Answer: This calculation requires that we know the rate constant for the decay. From the half-life and Equation (13.5) we write

$$\begin{aligned} k &= \frac{0.693}{t_{\frac{1}{2}}} \\ &= \frac{0.693}{28.1 \text{ yr}} \\ &= 0.0247 \text{ yr}^{-1} \end{aligned}$$

   To calculate the number of moles of ^{90}Sr decayed in a year, we apply Equation (13.3).

$$\ln \frac{[\text{A}]_0}{[\text{A}]} = kt$$

$$\ln \frac{1.00}{x} = (0.0247 \text{ yr}^{-1})(1.00 \text{ yr})$$

   where x is the number of moles of ^{90}Sr nuclei left over. Solving, we obtain

$$x = 0.976 \text{ mole}$$

   Thus the number of moles of nuclei which decay in a year is 1 − 0.976, or 0.024 mole. This is a reasonable number since it takes 28.1 years for 0.5 mole of ^{90}Sr to decay.

3. Calculate the amount of heat released (in kilojoules) corresponding to the number of moles of ^{90}Sr decayed to ^{90}Zr in 2.

Answer: Since the half-life of ^{90}Y is much shorter than that of ^{90}Sr, we can safely assume that *all* the ^{90}Y formed from ^{90}Sr will be converted to ^{90}Zr. The energy changes calculated in 1 refer to the decay of individual nuclei. In 0.024 mole, the number of nuclei that have decayed is $0.024 \times 6.022 \times 10^{23}$. Thus the heat released from 1 mole of ^{90}Sr waste in a year is given by

$$\begin{aligned}\text{heat released} &= (-2.90 \times 10^{-13}\ \text{J})(0.024 \times 6.022 \times 10^{23})\\ &= -4.19 \times 10^{9}\ \text{J}\\ &= -4.19 \times 10^{6}\ \text{kJ}\end{aligned}$$

It is roughly quivalent to the heat generated by burning 50 tons of coal! Although the heat is slowly released during the course of a year, effective ways must be devised for cooling to prevent damage to the storage containers and subsequent leakage of radioactive materials to the surroundings.

Key Words

Breeder reactor, p. 975
Critical mass, p. 972
Mass defect, p. 962
Moderators, p. 973
Nuclear binding energy, p. 962
Nuclear chain reaction, p. 972
Nuclear fission, p. 971
Nuclear fusion, p. 978
Nuclear transmutation, p. 958
Plasma, p. 979
Radical, p. 985
Thermonuclear reaction, p. 979
Tracer, p. 983
Transuranium elements, p. 970

Exercises

Nuclear Reactions

REVIEW QUESTIONS

23.1 How do nuclear reactions differ from ordinary chemical reactions?

23.2 What are the steps in balancing nuclear equations?

23.3 What is the difference between $_{-1}^{0}e$ and $_{-1}^{0}\beta$?

23.4 What is the difference between an electron and a positron?

PROBLEMS

23.5 Write complete nuclear equations for the following processes: (a) tritium, $_{1}^{3}H$, undergoes β decay; (b) ^{242}Pu undergoes α-particle emission; (c) ^{131}I undergoes β decay; (d) ^{251}Cf emits an α particle.

23.6 Complete the following nuclear equations and identify X in each case:

(a) $_{12}^{26}Mg + _{1}^{1}p \longrightarrow _{2}^{4}He + X$

(b) $_{27}^{59}Co + _{1}^{2}H \longrightarrow _{27}^{60}Co + X$

(c) $_{92}^{235}U + _{0}^{1}n \longrightarrow _{36}^{94}Kr + _{56}^{139}Ba + 3X$

(d) $_{92}^{235}U + _{0}^{1}n \longrightarrow _{40}^{99}Sr + _{52}^{135}Te + 2X$

(e) $_{24}^{53}Cr + _{2}^{4}He \longrightarrow _{0}^{1}n + X$

(f) $_{8}^{20}O \longrightarrow _{9}^{20}F + X$

(g) $_{53}^{135}I \longrightarrow _{54}^{135}Xe + X$

(h) $_{19}^{40}K \longrightarrow _{-1}^{0}\beta + X$

(i) $_{27}^{59}Co + _{0}^{1}n \longrightarrow _{25}^{56}Mn + X$

(j) $_{33}^{78}As \longrightarrow _{-1}^{0}\beta + X$

23.7 Write balanced nuclear equations for the following reactions and identify X:

(a) $_{7}^{15}N(p,\alpha)_{6}^{12}C$

(b) $_{13}^{27}Al(d,\alpha)_{12}^{25}Mg$

(c) $_{25}^{55}Mn(n,\gamma)X$

(d) $_{34}^{80}Se(d,p)X$

(e) $_{4}^{9}Be(d,2p)_{3}^{9}Li$

(f) $_{46}^{106}Pd(\alpha,p)_{47}^{109}Ag$

23.8 Describe how you would prepare astatine-211, starting with bismuth-209.

23.9 A long-cherished dream of alchemists was to produce gold from cheaper and more abundant elements. This dream was finally realized when $_{80}^{198}Hg$ was converted into gold by neutron bombardment. Write a balanced equation for this reaction.

Nuclear Stability

REVIEW QUESTIONS

23.10 State the general rules for predicting nuclear stability.
23.11 What is the belt of stability?
23.12 Why is it impossible for the isotope $^{2}_{2}He$ to exist?
23.13 Define nuclear binding energy, mass defect, and nucleon.
23.14 How does Einstein's equation, $E = mc^2$, allow us to calculate nuclear binding energy?
23.15 Why is it preferable to compare the stability of nuclei in terms of nuclear binding energy per nucleon?

PROBLEMS

23.16 The radius of a uranium-235 nucleus is about 7.0×10^{-3} pm. Calculate the density of the nucleus in g/cm^3. (Assume the atomic mass to be 235 amu.)
23.17 For each pair of isotopes listed, predict which one is less stable: (a) $^{6}_{3}Li$ or $^{9}_{3}Li$, (b) $^{22}_{11}Na$ or $^{25}_{11}Na$, (c) $^{48}_{20}Ca$ or $^{48}_{21}Sc$.
23.18 For each pair of elements listed, predict which one has more stable isotopes: (a) Co or Ni, (b) F or Se, (c) Ag or Cd.
23.19 The nucleus of nitrogen-18 lies above the stability belt. Write an equation for a nuclear reaction by which nitrogen-18 can achieve stability.
23.20 In each pair of isotopes shown, indicate which one you would expect to be radioactive: (a) $^{20}_{10}Ne$ and $^{17}_{10}Ne$, (b) $^{40}_{20}Ca$ and $^{45}_{20}Ca$, (c) $^{95}_{42}Mo$ and $^{92}_{43}Tc$, (d) $^{195}_{80}Hg$ and $^{196}_{80}Hg$, (e) $^{209}_{83}Bi$ and $^{242}_{96}Cm$.
23.21 Given that

$$H(g) + H(g) \longrightarrow H_2(g) \qquad \Delta H^\circ = -436.4 \text{ kJ}$$

Calculate the change in mass (in kg) per mole of H_2 formed.
23.22 Calculate the nuclear binding energy (in J) and the binding energy per nucleon of the following isotopes: (a) $^{7}_{3}Li$ (7.01600 amu), (b) $^{35}_{17}Cl$ (34.95952 amu), (c) $^{4}_{2}He$ (4.0026 amu), (d) $^{209}_{83}Bi$ (208.9804 amu).
23.23 Estimates show that the total energy output of the sun is 5×10^{26} J/s. What is the corresponding mass loss in kg/s of the sun?

Radioactive Decay; Dating

PROBLEMS

23.24 Fill in the blanks in the following radioactive decay series:

(a) $^{232}Th \xrightarrow{\alpha}$ ____ $\xrightarrow{\beta}$ ____ $\xrightarrow{\beta}$ ^{228}Th

(b) $^{235}U \xrightarrow{\alpha}$ ____ $\xrightarrow{\beta}$ ____ $\xrightarrow{\alpha}$ ^{227}Ac

(c) ____ $\xrightarrow{\alpha}$ $^{233}Pa \xrightarrow{\beta}$ ____ $\xrightarrow{\alpha}$ ____

23.25 A radioactive substance decays as follows:

Time (days)	Mass (g)
0	500
1	389
2	303
3	236
4	184
5	143
6	112

Calculate the first-order decay constant and the half-life of the reaction.
23.26 The radioactive decay of Tl-206 to Pb-206 has a half-life of 4.20 min. Starting with 5.00×10^{22} atoms of Tl-206, calculate the number of such atoms left after 42.0 min.
23.27 A freshly isolated sample of ^{90}Y was found to have an activity of 9.8×10^5 disintegrations per minute at 1:00 P.M. on December 3, 1982. At 2:15 P.M. on December 17, 1982, its activity was redetermined and found to be 2.6×10^4 disintegrations per minute. Calculate the half-life of ^{90}Y.
23.28 In the thorium decay series, thorium-232 loses a total of 6 α particles and 4 β particles in a 10-stage process. What is the final isotope produced?
23.29 Why do radioactive decay series obey first-order kinetics?
23.30 Strontium-90 is one of the products of the fission of uranium-235. This isotope of strontium is radioactive, with a half-life of 28.1 yr. Calculate how long (in yr) it will take for 1.00 g of the isotope to be reduced to 0.200 g by decay.
23.31 Consider the decay series

$$A \longrightarrow B \longrightarrow C \longrightarrow D$$

where A, B, and C are radioactive with half-lives of 4.50 s, 15.0 days, and 1.00 s, respectively, and D is nonradioactive. Starting with 1.00 mole of A, and none of B, C, or D, calculate the number of moles of A, B, C, and D left after 30 days.
23.32 The radioactive potassium-40 isotope decays to argon-40 with a half-life of 1.2×10^9 yr. (a) Write a balanced equation for the reaction. (b) A sample of moon rock is found to contain 18 percent potassium-40 and 82 percent argon by mass. Calculate the age of the rock in years.

Nuclear Fission; Nuclear Reactors

REVIEW QUESTIONS

23.33 Define nuclear fission, nuclear chain reaction, and critical mass.
23.34 Which isotopes can undergo nuclear fission?
23.35 Explain how an atomic bomb works.

23.36 Explain the functions of a moderator and a control rod in a nuclear reactor.

23.37 Discuss the differences between a light water and a heavy water nuclear fission reactor.

23.38 What are the advantages of a breeder reactor over a conventional nuclear fission reactor?

23.39 No form of energy production is without risk. Make a list of the risks to society involved in fueling and operating a conventional coal-fired electric power plant, and compare them with the risks of fueling and operating a nuclear fission-powered electric plant.

Nuclear Fusion

REVIEW QUESTIONS

23.40 Define nuclear fusion, thermonuclear reaction, and plasma.

23.41 Why do heavy elements such as uranium undergo fission while light elements such as hydrogen and lithium undergo fusion?

23.42 How does a hydrogen bomb work?

23.43 What are the advantages of a fusion reactor over a fission reactor? What are the practical difficulties in operating a large-scale fusion reactor?

Applications of Isotopes

REVIEW QUESTION

23.44 Define a tracer and describe three applications of isotopes in chemistry and medicine.

PROBLEMS

23.45 Describe how you would use a radioactive iodine isotope to demonstrate that the following process is in dynamic equilibrium:

$$PbI_2(s) \rightleftharpoons Pb^{2+}(aq) + 2I^-(aq)$$

23.46 Consider the following redox reaction:

$$IO_4^-(aq) + 2I^-(aq) + H_2O(l) \longrightarrow I_2(s) + IO_3^-(aq) + 2OH^-(aq)$$

When KIO_4 is added to a solution containing iodide ions labeled with radioactive iodine-128, all the radioactivity appears in I_2 and none in the IO_3^- ion. What can you deduce about the mechanism for the redox process?

23.47 Explain how you might use a radioactive tracer to show that ions are not completely motionless in crystals.

23.48 Each hemoglobin molecule, the oxygen carrier in blood, contains four Fe atoms. Explain how you would use the radioactive $^{59}_{26}Fe$ ($t_{\frac{1}{2}}$ = 46 days) to show that the iron in a certain diet is converted into hemoglobin.

Miscellaneous Problems

23.49 How does a Geiger counter work?

23.50 Chlorine-39 is radioactive and undergoes β decay. What differences would you expect in the chemical behavior of HCl molecules containing chlorine-39, compared with those containing chlorine-35? Cl-35 is nonradioactive.

23.51 Name the elements with atomic numbers (a) 110, (b) 116, (c) 125, (d) 218.

23.52 Write atomic numbers of the following elements: (a) ununbium, (b) untrioctium, (c) unsepthexium, (d) binilquadium.

23.53 Nuclei with an even number of protons and an even number of neutrons are more stable than those with an odd number of protons and/or an odd number of neutrons. What is the significance of the even number of protons and of neutrons in this case?

23.54 Tritium, ^3_1H, is radioactive and decays by electron emission. Its half-life is 12.5 yr. In ordinary water the ratio of ^1H to ^3H atoms is 1.0×10^{17} to 1. (a) Write a balanced nuclear equation for tritium decay. (b) How many disintegrations will be observed per minute in a 1.00-kg sample of water?

23.55 Balance the following equations, which are for nuclear reactions that are known to occur in the explosion of an atomic bomb:

(a) $^{235}_{92}U + ^1_0n \longrightarrow ^{140}_{56}Ba + 3^1_0n + X$

(b) $^{235}_{92}U + ^1_0n \longrightarrow ^{144}_{55}Cs + ^{90}_{37}Rb + 2X$

(c) $^{235}_{92}U + ^1_0n \longrightarrow ^{87}_{35}Br + 3^1_0n + X$

(d) $^{235}_{92}U + ^1_0n \longrightarrow ^{160}_{62}Sm + ^{72}_{30}Zn + 4X$

23.56 (a) What is the activity, in millicuries, of a 0.500-g sample of $^{237}_{93}Np$? (This isotope decays by α-particle emission and has a half-life of 2.20×10^6 yr.) (b) Write a balanced nuclear equation for the decay of $^{237}_{93}Np$.

23.57 Calculate the nuclear binding energies, in J/nucleon, for the following species: (a) ^{10}B (10.0129 amu), (b) ^{11}B (11.00931 amu), (c) ^{14}N (14.00307 amu), (d) ^{56}Fe (55.9349 amu), (e) ^{184}W (183.9510 amu).

23.58 Why is strontium-90 a particularly dangerous isotope for humans?

23.59 How are scientists able to tell the age of a fossil?

23.60 After the Chernobyl accident, people living close to the nuclear reactor site were urged to take large amounts of potassium iodide as a safety precaution. What is the chemical basis for this action?

23.61 Astatine, the last member of Group 7A, can be prepared by bombarding bismuth-209 with α particles. (a) Write an equation for the reaction. (b) Represent the equation in the abbreviated form as discussed in Section 23.4.

23.62 To detect bombs that may be smuggled onto airplanes, the

Federal Aviation Administration (FAA) will soon require all major airports in the United States to install thermal neutron analyzers. A thermal neutron analyzer bombards baggages with low-energy neutrons, converting some of the nitrogen-14 nuclei to nitrogen-15, with simultaneous emission of γ rays. Because nitrogen content is usually high in explosives, detection of a high dosage of γ rays will suggest that a bomb may be present. (a) Write an equation for the nuclear process. (b) Compare this technique with the conventional X-ray detection method.

Opposite page: An organic research laboratory.

C H A P T E R 24

ORGANIC CHEMISTRY

Organic chemistry deals with virtually all carbon compounds. (Traditionally, compounds such as CO, CO_2, CS_2, and various bicarbonates, carbonates, and cyanides are considered to be inorganic compounds.) The word "organic" was originally used by eighteenth-century chemists to describe substances obtained from living sources—from plants and animals. They believed that nature possessed a certain vital force and that it alone could produce organic compounds. This romantic notion was disproved by experiments like those first carried out in 1828 by Friedrich Wohler, who prepared urea, an organic compound, from the reaction between two inorganic compounds, lead cyanate and aqueous ammonia:

$$Pb(OCN)_2 + 2NH_3 + 2H_2O \longrightarrow \underset{\text{urea}}{2(NH_2)_2CO} + Pb(OH)_2$$

Today, well over 6 million synthetic and natural organic compounds are known. This number is significantly greater than the 100,000 or so inorganic compounds known to exist.

24.1 Hydrocarbons

Recall that the linking of like atoms is called *catenation*. The ability of carbon to catenate is discussed in Section 21.4.

Carbon can form more compounds than any other element because carbon atoms are able to link up with each other in straight chains and branched chains. Although the number of known organic compounds is enormous, the study of ***organic chemistry,*** that is, *the branch of chemistry that deals with carbon compounds,* is not as difficult as it may seem. Most organic compounds can be divided into relatively few classes according to the *functional groups* they contain. A ***functional group*** is *the part of a molecule having a special arrangement of atoms that is largely responsible for the chemical behavior of the parent molecule.* Different molecules containing the same kind of functional group or groups react similarly. Thus, by learning the characteristic properties of a few functional groups, we can study and understand the properties of many organic compounds.

Before discussing functional groups we will examine a class of compounds that forms the framework of all organic compounds—the hydrocarbons. ***Hydrocarbons*** are *made up of only two elements, hydrogen and carbon.* On the basis of structure, hydrocarbons are divided into two main classes—aliphatic and aromatic. ***Aliphatic hydrocarbons*** *do not contain the benzene group, or the benzene ring,* whereas ***aromatic hydrocarbons*** *contain one or more benzene rings.* Aliphatic hydrocarbons are further divided into alkanes, alkenes, and alkynes (Figure 24.1).

Alkanes

For a given number of carbon atoms, the saturated hydrocarbon contains the largest number of hydrogen atoms.

Alkanes *have the general formula C_nH_{2n+2}, where $n = 1, 2, \ldots$.* The essential characteristic of alkane hydrocarbon molecules is that *only single covalent bonds are present.* In these compounds the bonds are said to be saturated. Thus the alkanes are known as ***saturated hydrocarbons.***

Anaerobic organisms do not require oxygen to function.

The simplest member of the family (that is, with $n = 1$) is CH_4, which occurs naturally by the anaerobic bacterial decomposition of vegetable matter under water. Because it was first collected in marshes, methane became known as "marsh gas." A rather improbable but proven source of methane is termites. When these voracious

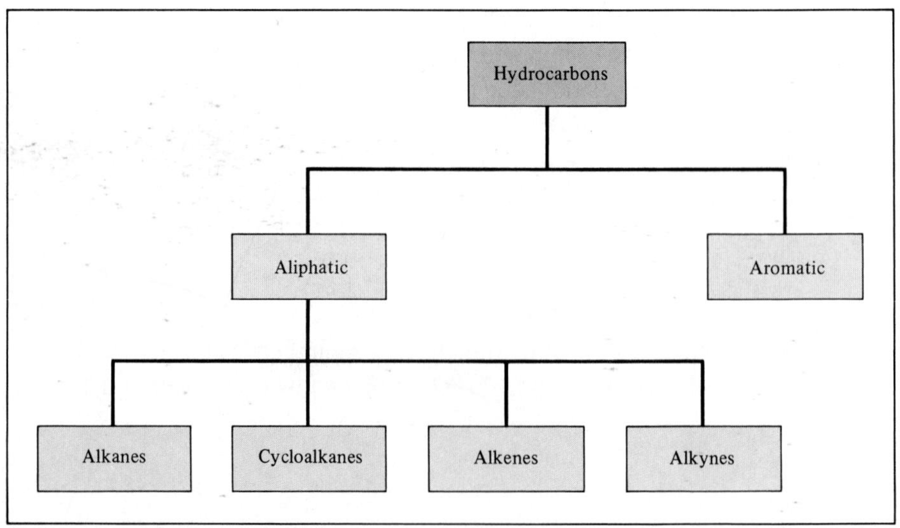

Figure 24.1 *Classification of hydrocarbons.*

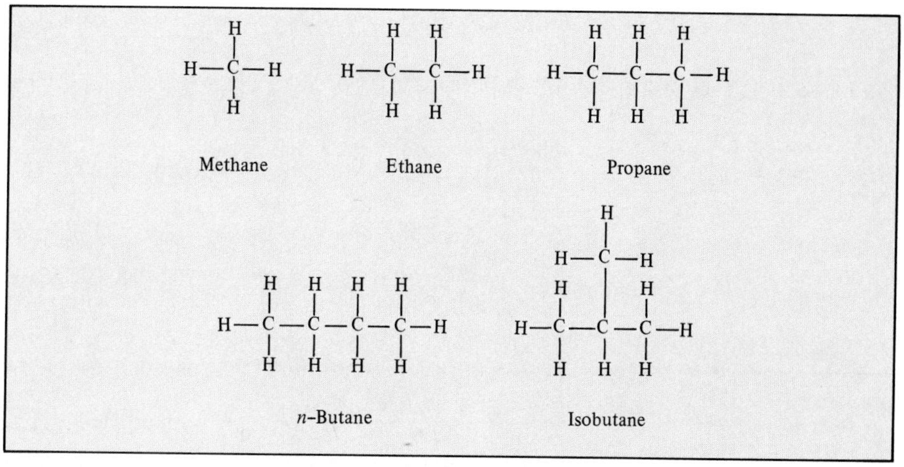

```
        H               H   H               H   H   H
        |               |   |               |   |   |
    H—C—H         H—C—C—H           H—C—C—C—H
        |               |   |               |   |   |
        H               H   H               H   H   H

    Methane           Ethane              Propane

                                              H
                                              |
                                          H—C—H
      H   H   H   H                       H   |   H
      |   |   |   |                       |   |   |
  H—C—C—C—C—H                 H—C—C—C—H
      |   |   |   |                       |   |   |
      H   H   H   H                       H   H   H

        n-Butane                        Isobutane
```

Figure 24.2 *Structures of the first four alkanes. Note that butane can exist in two structurally different forms, called structural isomers.*

insects consume wood, cellulose (the major component of wood) is broken down into methane, carbon dioxide, and other compounds by microorganisms that inhabit their digestive system. An estimated 170 million tons of methane are produced annually by termites! Commercially, methane is obtained from natural gas. It is also produced in some sewage treatment processes.

Figure 24.2 shows the structures of the first four alkanes (that is, from $n = 1$ to $n = 4$). Natural gas is a mixture of methane, ethane, and a small amount of propane. We discussed the bonding scheme of methane in Chapter 10. Indeed the carbon atoms in all the alkanes can be assumed to be sp^3-hybridized. The structures of ethane and propane are straightforward, for there is only one way to join the carbon atoms in these molecules. Butane, however, has two possible bonding schemes and can exist as structural isomers called *n*-butane (*n* stands for normal) and isobutane. ***Structural isomers*** are *molecules that have the same molecular formula, but different structures.*

In the alkane series, as the number of carbon atoms increases, the number of structural isomers increases rapidly. For example, butane has two isomers; decane, $C_{10}H_{22}$, has 75 isomers; and the alkane $C_{30}H_{62}$ has over 400 million, or 4×10^8, isomers! Obviously, most of these isomers do not exist in nature and have not been synthesized. Nevertheless, the numbers help to explain why carbon is found in so many more compounds than is any other element.

The following example deals with the number of structural isomers of an alkane.

EXAMPLE 24.1

How many structural isomers can be identified for pentane, C_5H_{12}?

Answer

For small hydrocarbon molecules (eight or fewer carbon atoms), it is relatively easy to determine the number of structural isomers by trial and error. The first step is to write down the straight-chain structure:

$$\begin{array}{ccccccccccc} & & H & & H & & H & & H & & H & & \\ & & | & & | & & | & & | & & | & & \\ H & - & C & - & C & - & C & - & C & - & C & - & H \\ & & | & & | & & | & & | & & | & & \\ & & H & & H & & H & & H & & H & & \end{array}$$

n-pentane
(b.p. 36.1°C)

The second structure, by necessity, must be a branched chain:

$$\begin{array}{ccccccccccc} & & H & & CH_3 & & H & & H & & \\ & & | & & | & & | & & | & & \\ H & - & C & - & C & - & C & - & C & - & H \\ & & | & & | & & | & & | & & \\ & & H & & H & & H & & H & & \end{array}$$

2-methylbutane
(b.p. 27.9°C)

Yet another branched-chain structure is possible:

$$\begin{array}{ccccccccc} & & H & & CH_3 & & H & & \\ & & | & & | & & | & & \\ H & - & C & - & C & - & C & - & H \\ & & | & & | & & | & & \\ & & H & & CH_3 & & H & & \end{array}$$

2,2-dimethylpropane
(b.p. 9.5°C)

We can draw no other structure for an alkane having the molecular formula C_5H_{12}. Thus pentane has three structural isomers, in which the numbers of carbon and hydrogen atoms remain unchanged despite the differences in structure.

Similar problem: 24.9.

Table 24.1 shows the melting and boiling points of the straight-chain isomers of the first 10 alkanes. The first four are gases at room temperature; and pentane through decane are liquids. As molecular size increases, so does the boiling point, because of the increasing dispersion forces (see Section 11.2).

Table 24.1 The first ten straight-chain alkanes*

Name of hydrocarbon	Molecular formula	Number of carbon atoms	Melting point (°C)	Boiling point (°C)
Methane	CH_4	1	−182.5	−161.6
Ethane	$CH_3—CH_3$	2	−183.3	−88.6
Propane	$CH_3—CH_2—CH_3$	3	−189.7	−42.1
Butane	$CH_3—(CH_2)_2—CH_3$	4	−138.3	−0.5
Pentane	$CH_3—(CH_2)_3—CH_3$	5	−129.8	36.1
Hexane	$CH_3—(CH_2)_4—CH_3$	6	−95.3	68.7
Heptane	$CH_3—(CH_2)_5—CH_3$	7	−90.6	98.4
Octane	$CH_3—(CH_2)_6—CH_3$	8	−56.8	125.7
Nonane	$CH_3—(CH_2)_7—CH_3$	9	−53.5	150.8
Decane	$CH_3—(CH_2)_8—CH_3$	10	−29.7	174.0

*By "straight chain," we mean that the carbon atoms are joined along one line. This does not mean, however, that these molecules are linear. Each carbon is sp^3- (not sp-) hybridized.

The system of naming organic compounds is based on the recommendations of the International Union of Pure and Applied Chemistry (IUPAC).

Table 24.2 Common alkyl groups

Name	Formula
Methyl	$—CH_3$
Ethyl	$—CH_2—CH_3$
n-Propyl	$—CH_2—CH_2—CH_3$
n-Butyl	$—CH_2—CH_2—CH_2—CH_3$
Isopropyl	$—C(CH_3)(CH_3)—H$
t-Butyl*	$—C(CH_3)_2—CH_3$

*The letter *t* stands for tertiary.

Table 24.3 Common functional groups

Functional group	Name
$—NH_2$	Amino
—F	Fluoro
—Cl	Chloro
—Br	Bromo
—I	Iodo
$—NO_2$	Nitro
$—CH{=}CH_2$	Vinyl
$—C_6H_5$ (benzene ring)	Phenyl

Alkane Nomenclature. The names of the straight-chain alkanes are fairly easy to remember. As Table 24.1 shows, except for the first four members, the number of carbon atoms in each alkane is reflected in the Greek prefix (see Table 2.4). When one or more hydrogen atoms are replaced by other groups, the name of the compound must indicate the locations of carbon atoms where replacements are made. Let us first consider groups that are themselves derived from alkanes. For example, when a hydrogen atom is removed from methane, we are left with the CH_3 fragment, which is called a *methyl* group. Similarly, removing a hydrogen atom from the ethane molecule gives an *ethyl* group, or C_2H_5. Because they are derived from alkanes, they are also called ***alkyl groups.*** Table 24.2 lists the names of several common alkyl groups.

With the aid of Tables 24.1 and 24.2 and the rule that the locations of other groups of carbon atoms are denoted by numbering the atoms along the longest carbon chain, we are now ready to name some substituted alkanes. Note that we always start numbering from the end that is closer to the carbon atom bearing the substituted group. Consider the following example:

$$\overset{1}{H_3C}—\overset{2}{C}(CH_3)(H)—\overset{3}{CH_2}—\overset{4}{CH_3}$$

Since the longest carbon chain contains four C atoms, the parent name of the compound is butane. We note that a methyl group is attached to carbon number 2; therefore, the name of the compound is 2-methylbutane. Here is another example:

$$\overset{1}{H_3C}—\overset{2}{C}(CH_3)_2—\overset{3}{CH_2}—\overset{4}{CH}(CH_3)—\overset{5}{CH_2}—\overset{6}{CH_3}$$

The parent name of this compound is hexane. Noting that there are two methyl groups attached to carbon number 2 and one methyl group attached to carbon number 4, we call the compound 2,2,4-trimethylhexane.

Of course, we can have many different types of substituents on alkanes. Table 24.3 lists the names of some common functional groups. Thus the compound

$$\overset{1}{H_3C}—\overset{2}{CH}(Br)—\overset{3}{CH}(NO_2)—\overset{4}{CH_3}$$

is called 2-bromo-3-nitrobutane. If the Br is attached to the first carbon atom

$$Br—\overset{1}{CH_2}—\overset{2}{CH_2}—\overset{3}{CH}(NO_2)—\overset{4}{CH_3}$$

the compound becomes 1-bromo-3-nitrobutane. Now consider the structure

$$H_3C—CH(C_6H_5)—CH_3$$

We call this compound 2-phenylpropane.

Reactions of Alkanes. Alkanes are generally not considered to be very reactive substances. However, under suitable conditions they do undergo several kinds of reactions—including combustion. The burning of natural gas, gasoline, and fuel oil involves the combustion of alkanes. These reactions are all highly exothermic:

$$CH_4(g) + 2O_2(g) \longrightarrow CO_2(g) + 2H_2O(l) \qquad \Delta H° = -890.4 \text{ kJ}$$

$$2C_2H_6(g) + 7O_2(g) \longrightarrow 4CO_2(g) + 6H_2O(l) \qquad \Delta H° = -3119 \text{ kJ}$$

These, and similar reactions, have long been utilized in industrial processes and in domestic heating and cooking.

The *halogenation* of alkanes—that is, the replacement of one or more hydrogen atoms by halogen atoms—is another well-studied kind of reaction. When a mixture of methane and chlorine is heated above 100°C or irradiated with light of a suitable wavelength, methyl chloride is produced:

The systematic names of methyl chloride, methylene chloride, and chloroform are monochloromethane, dichloromethane, and trichloromethane, respectively.

$$CH_4(g) + Cl_2(g) \longrightarrow \underset{\text{methyl chloride}}{CH_3Cl(g)} + HCl(g)$$

If chlorine gas is present in a sufficient amount, the reaction can proceed further:

$$CH_3Cl(g) + Cl_2(g) \longrightarrow \underset{\text{methylene chloride}}{CH_2Cl_2(l)} + HCl(g)$$

$$CH_2Cl_2(g) + Cl_2(g) \longrightarrow \underset{\text{chloroform}}{CHCl_3(l)} + HCl(g)$$

$$CHCl_3(l) + Cl_2(g) \longrightarrow \underset{\text{carbon tetrachloride}}{CCl_4(l)} + HCl(g)$$

A great deal of experimental evidence suggests that the initial step of the first halogenation reaction occurs as follows:

$$Cl_2 + \text{energy} \longrightarrow Cl\cdot + Cl\cdot$$

Thus the covalent bond in Cl_2 breaks and two chlorine atoms form. We know it is the Cl—Cl bond that breaks when the mixture is heated or irradiated because the bond energy of Cl_2 is 242.7 kJ/mol, whereas about 414 kJ/mol is needed to break C—H bonds in CH_4.

A free chlorine atom contains an unpaired electron, shown by a single dot. These atoms are highly reactive and attack methane molecules according to the equation

$$CH_4 + Cl\cdot \longrightarrow \cdot CH_3 + HCl$$

This reaction produces hydrogen chloride and the methyl radical $\cdot CH_3$. The methyl radical is another reactive species; it combines with molecular chlorine to give methyl chloride and a chlorine atom:

$$\cdot CH_3 + Cl_2 \longrightarrow CH_3Cl + Cl\cdot$$

The production of methylene chloride from methyl chloride and further reactions can be explained in the same way. The actual mechanism is more complex than the scheme we have shown because "side reactions" that do not lead to the desired products often take place, such as

$$Cl\cdot + Cl\cdot \longrightarrow Cl_2$$

$$CH_3 + \cdot CH_3 \longrightarrow C_2H_6$$

Alkanes in which one or more hydrogen atoms have been replaced by a halogen atom are called *alkyl halides*. Among the large number of alkyl halides, the best known are chloroform ($CHCl_3$), carbon tetrachloride (CCl_4), methylene chloride (CH_2Cl_2), and the chlorofluorohydrocarbons.

Chloroform is a volatile, sweet-tasting liquid used for many years as an anesthetic. However, because of its toxicity—it can severely damage the liver, kidneys, and heart—it has been replaced by other compounds. Carbon tetrachloride, also a toxic substance, serves as a cleaning liquid, for it removes grease stains from clothing. Methylene chloride is used as a solvent to decaffeinate coffee and as a paint remover.

The preparation of chlorofluorocarbons and the effect of these compounds on ozone in the stratosphere were discussed in Chapter 13 and Section 21.7.

Optical isomerism was first discussed in Section 22.4.

Optical Isomerism of Substituted Alkanes. Optical isomers are compounds that are nonsuperimposable mirror images. Consider the following substituted methanes: CH_2ClBr and CHFClBr. Figure 24.3 shows perspective drawings of these two molecules and their mirror images. The two mirror images of (a) are superimposable, but those of (b) are not, no matter how we rotate the molecules. Thus the CHFClBr molecule is chiral. Careful observation shows that most simple chiral molecules contain at least one asymmetric carbon atom—that is, a carbon atom bonded to four different atoms or groups of atoms.

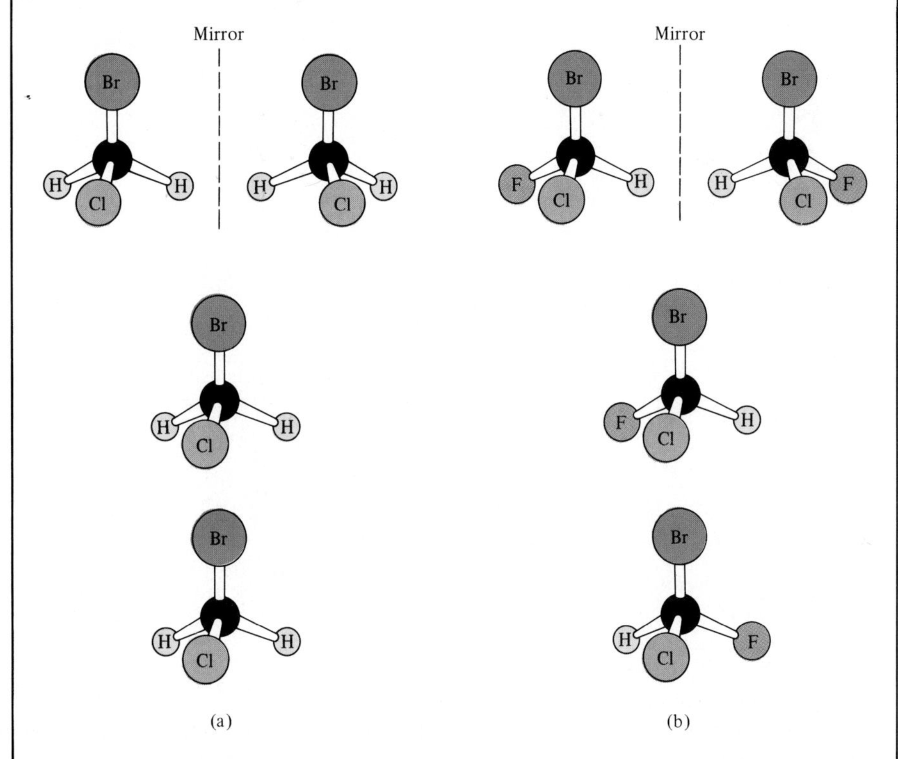

Figure 24.3 *(a) The CH_2ClBr molecule and its mirror image. Since the molecule and its mirror image are superimposable, the molecule is said to be achiral. (b) The CHFClBr molecule and its mirror image. Since the molecule and its mirror image are not superimposable, no matter how we rotate one with respect to the other, the molecule is said to be chiral.*

EXAMPLE 24.2

Is the following molecule chiral?

$$H-\underset{\displaystyle CH_3}{\overset{\displaystyle Cl}{\underset{|}{\overset{|}{C}}}}-CH_2-CH_3$$

Answer

We note that the central carbon atom is bonded to a hydrogen atom, a chlorine atom, a $-CH_3$ group, and a $-CH_2-CH_3$ group. Therefore the central carbon atom is asymmetric and the molecule is chiral.

Similar problem: 24.19.

Cycloalkanes. *Alkanes whose carbon atoms are joined in rings* are known as ***cycloalkanes***. They have the general formula C_nH_{2n}, where $n = 3, 4, \ldots$. The structures of some simple cycloalkanes are given in Figure 24.4. Theoretical analysis shows that cyclohexane can assume two different geometries that are relatively free of strain

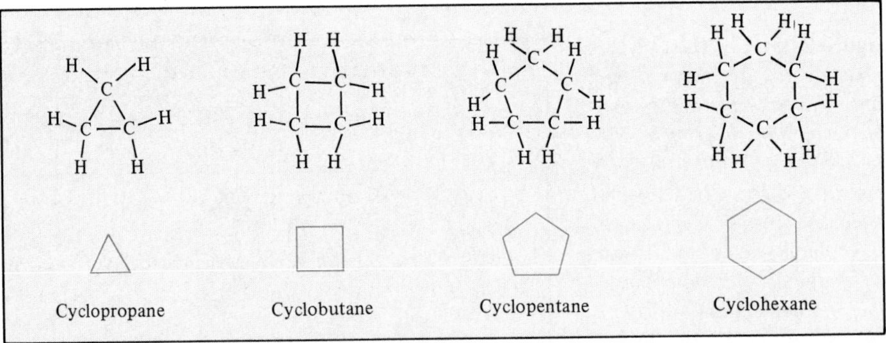

Figure 24.4 *Structures of the first four cyclic alkanes and their simplified forms.*

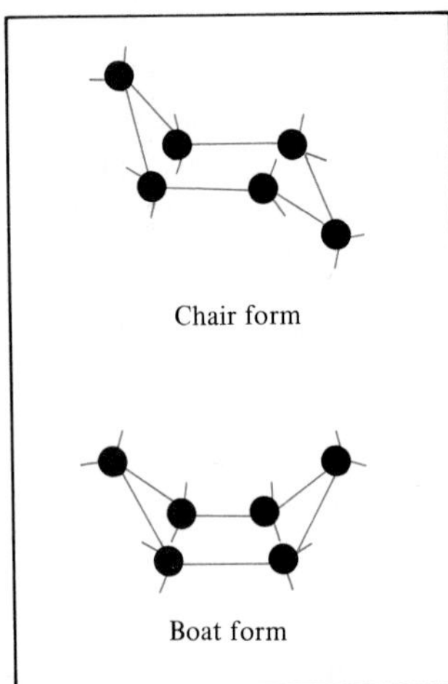

Figure 24.5 *The cyclohexane molecule can exist in various shapes. The most stable shape is the chair form. The boat form is less stable than the chair form. Hydrogen atoms are not shown in these diagrams.*

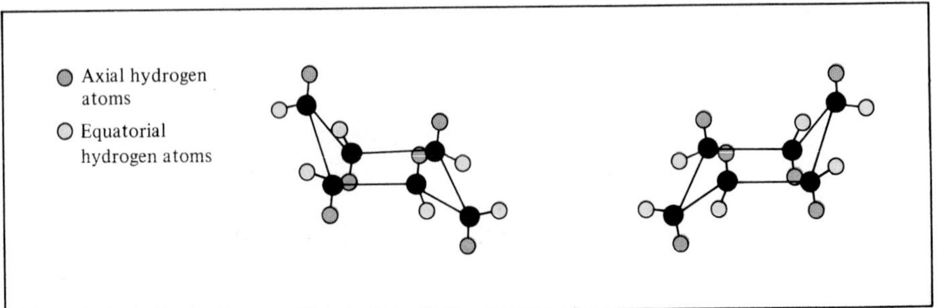

Figure 24.6 *Hydrogen atoms in the chair form of cyclohexane can be classified as axial or equatorial. When the chair flips from one form to its mirror image, axial hydrogen atoms become equatorial, and vice versa.*

(Figure 24.5). By "strain" we mean that bonds are compressed, stretched, or twisted out of their normal geometric shapes as predicted by hybridization. The most stable geometry is the *chair form*.

As Figure 24.6 shows, there are two kinds of hydrogen atoms in the chair form of cyclohexane. Six of these are called *equatorial* hydrogens and the other six are called *axial* hydrogens. The equatorial C—H bonds are in the plane of the "seat" of the chair, whereas the axial C—H bonds are perpendicular to this plane. These atoms differ with respect to their immediate environment in the same molecule.

Study of the various shapes and stability of the cyclohexane ring is important because many biologically significant substances contain one or more such ring systems.

Alkenes

There are two types of ***unsaturated hydrocarbons:*** *hydrocarbons that contain carbon-carbon double bonds and hydrocarbons with carbon-carbon triple bonds*. The ***alkenes*** (also called *olefins*) *contain at least one carbon-carbon double bond*. (Hydrocarbons containing carbon-carbon triple bonds will be discussed shortly.) *Alkenes have the general formula* C_nH_{2n}, *where* $n = 2, 3, \ldots$. The simplest alkene is C_2H_4, ethylene, in which both carbon atoms are sp^2-hybridized and the double bond is made up of a sigma bond and a pi bond (see Chapter 10).

Alkene Nomenclature. In naming alkenes it is necessary to indicate the positions of the carbon-carbon double bonds. The names of compounds containing C=C bonds end with *-ene*. As with the alkanes, the name of the parent compound is determined by the number of carbon atoms in the longest chain (see Table 24.1), as shown here:

$$\underset{\text{1-butene}}{CH_2{=}CH{-}CH_2{-}CH_3} \qquad \underset{\text{2-butene}}{H_3C{-}CH{=}CH{-}CH_3}$$

The numbers in the names for 1-butene and 2-butene indicate the lowest numbered carbon atom in the chain that is part of the C=C bond of the alkene. The name "butene" means that this is a compound with four carbon atoms in the longest chain. We must also consider the possibilities of geometric isomers, such as

In the cis isomer, the two H atoms are on the same side of the C=C bond; in the trans isomer, the two H atoms are across from each other.

```
                CH3
 1             4|     5      6
H3C             CH—CH2—CH3
   \  2  3    /
    C=C
   /        \
 H           H
  4-methyl-cis-2-hexene

  1
H3C        H
   \ 2  3 /
    C=C
   /      \ 4    5      6
 H         CH—CH2—CH3
           |
           CH3
  4-methyl-trans-2-hexene
```

4-methyl-*cis*-2-hexene 4-methyl-*trans*-2-hexene

Properties and Reactions of Alkenes. Ethylene is an extremely important substance because it is used in large quantities for the manufacture of organic polymers (to be discussed in the next chapter) and in the preparation of many other organic chemicals. Ethylene is prepared industrially by the *cracking* process, that is, the thermal decomposition of a large hydrocarbon into smaller molecules. When ethane is heated to about 800°C, it undergoes the following reaction:

$$C_2H_6(g) \xrightarrow{\text{Pt catalyst}} CH_2{=}CH_2(g) + H_2(g)$$

Other alkenes can be prepared by cracking the higher members of the alkane family.

In Section 13.6 we discussed the hydrogenation of ethylene to produce ethane. Hydrogenation is sometimes described as an "addition reaction." Other addition reactions to the C=C bond include

In these reactions it is the weaker pi bond of C=C that is broken; the sigma bond of C—C remains intact.

$$C_2H_4(g) + HX(g) \longrightarrow CH_3\text{—}CH_2X(g)$$

$$C_2H_4(g) + X_2(g) \longrightarrow CH_2X\text{—}CH_2X(g)$$

where X represents a halogen (Cl, Br, or I). The addition of a hydrogen halide to an unsymmetrical alkene such as propylene is more complicated because two products are possible:

$$\underset{\text{propylene}}{(H_3C)HC{=}CH_2} + HBr \longrightarrow \underset{\text{1-bromopropane}}{H_3C\text{—}CH_2\text{—}CH_2Br} \quad \text{and/or} \quad \underset{\text{2-bromopropane}}{H_3C\text{—}CHBr\text{—}CH_3}$$

In practice, however, only one product is formed, and that product is 2-bromopropane. This phenomenon was observed in all reactions between unsymmetrical reagents and alkenes. In 1871, Vladimir Markovnikov† postulated a generalization that enables us to predict the outcome of such an addition reaction. This generalization, now known as *Markovnikov's rule,* states that in the addition of unsymmetrical (that is, polar) reagents to alkenes, the positive portion of the reagent (usually hydrogen) adds to the carbon atom that already has the most hydrogen atoms.

Geometric Isomers of Alkenes. In a compound such as ethane, C_2H_6, the rotation of the two methyl groups about the carbon-carbon single bond (which is a sigma bond) is quite free. The situation is different for molecules that contain carbon-carbon double bonds, such as ethylene, C_2H_4. In addition to the sigma bond, there is a pi bond between the two carbon atoms. Rotation about the carbon-carbon linkage does not affect the sigma bond, but it does move the two $2p_z$ orbitals out of alignment for overlap and, hence, partially or totally destroys the pi bond (see Figure 10.16). This process requires an input of energy on the order of 270 kJ/mol. For this reason, the rotation of a carbon-carbon double bond is considerably restricted, but not impossible. Consequently, molecules containing carbon-carbon double bonds (that is, the alkenes) may have geometric isomers, which are compounds that are related to each other by a restricted rotation around a bond.

The molecule dichloroethylene, ClHC=CHCl, can exist as one of the two geometric isomers called *cis*-dichloroethylene and *trans*-dichloroethylene:

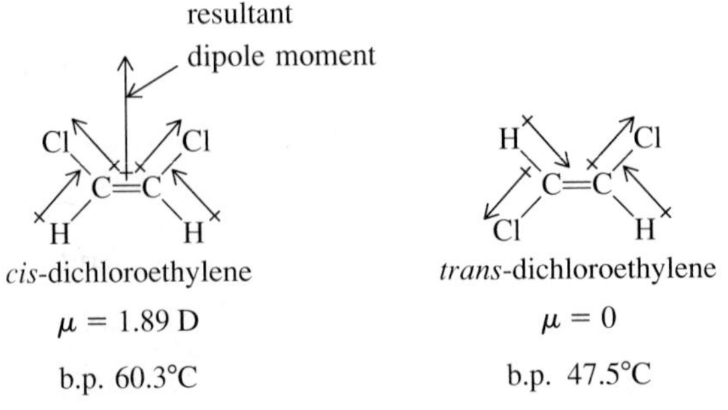

cis-dichloroethylene $\mu = 1.89$ D b.p. 60.3°C

trans-dichloroethylene $\mu = 0$ b.p. 47.5°C

† Vladimir W. Markovnikov (1838–1904). Russian chemist. Markovnikov's observations of the addition reactions to alkenes were published a year after his death.

where the term *cis* means that two particular atoms (or groups of atoms) are adjacent to each other, and *trans* means that the two atoms (or groups of atoms) are across from each other. Generally, cis and trans isomers have distinctly different physical and chemical properties. Heat or irradiation with light is commonly used to bring about the conversion of one geometric isomer to another, a process called cis-trans isomerization, or geometric isomerization. As the above data show, dipole moment measurements can be used to distinguish between geometric isomers. In general, cis isomers possess a dipole moment, whereas trans isomers do not.

Alkynes

***Alkynes** contain at least one carbon-carbon triple bond.* They have the *general formula C_nH_{2n-2}, where $n = 2, 3, \ldots$.*

Alkyne Nomenclature. Names of compounds containing C≡C bonds end with *-yne*. Again the name of the parent compound is determined by the number of carbon atoms in the longest chain (see Table 24.1 for names of alkane counterparts). As in the case of alkenes, the names of alkynes indicate the position of the carbon-carbon triple bond, as, for example, in

$$\underset{\text{1-butyne}}{HC{\equiv}C{-}CH_2{-}CH_3} \qquad \underset{\text{2-butyne}}{H_3C{-}C{\equiv}C{-}CH_3}$$

Properties and Reactions of Alkynes. The simplest alkyne is acetylene, C_2H_2, whose structure and bonding were discussed in Section 10.5. Acetylene is a colorless gas (b.p. −84°C) prepared by the reaction between calcium carbide and water:

$$CaC_2(s) + 2H_2O(l) \longrightarrow C_2H_2(g) + Ca(OH)_2(aq)$$

The reaction of calcium carbide with water produces acetylene, a flammable gas.

Acetylene has many important uses in industry. Because of its high heat of combustion

$$2C_2H_2(g) + 5O_2(g) \longrightarrow 4CO_2(g) + 2H_2O(l) \qquad \Delta H^\circ = -2599.2 \text{ kJ}$$

acetylene burned in an ''oxyacetylene torch'' gives an extremely hot flame (about 3000°C). Thus, oxyacetylene torches are used to weld metals (see Figure 6.8).

The standard free energy of formation of acetylene is positive ($\Delta G_f^\circ = 209.2$ kJ/mol), unlike that of the alkanes. This means that the molecule is unstable (relative to its elements) and has a tendency to decompose:

$$C_2H_2(g) \longrightarrow 2C(s) + H_2(g)$$

In the presence of a suitable catalyst or when the gas is kept under pressure, this reaction can occur with explosive violence. For the gas to be transported safely, it must be dissolved in an inert organic solvent such as acetone at moderate pressure. In the liquid state, acetylene is very sensitive to shock and is highly explosive.

Acetylene can be hydrogenated to yield ethylene:

$$C_2H_2(g) + H_2(g) \longrightarrow C_2H_4(g)$$

It undergoes the following addition reactions with hydrogen halides and halogens:

$$C_2H_2(g) + HX(g) \longrightarrow CH_2{=}CHX(g)$$

$$C_2H_2(g) + X_2(g) \longrightarrow CHX{=}CHX(g)$$

$$C_2H_2(g) + 2X_2(g) \longrightarrow CHX_2{-}CHX_2(g)$$

Ladenburg's formula | Dewar's formula | Armstrong's formula | Kekulé's formula

Figure 24.7 *Proposed structures of benzene. Molecules corresponding to Ladenburg's formula and Dewar's formula have been synthesized. These compounds bear no resemblance to benzene in their physical or chemical properties.*

Methylacetylene (propyne), $CH_3—C≡C—H$, is the next member in the alkyne family. It undergoes reactions similar to those of acetylene. The addition reactions of propyne also obey Markovnikov's rule, as in the case of alkenes:

$$H_3C—C≡C—H + HBr \longrightarrow (H_3C)(Br)C=C(H)(H)$$

propyne → 2-bromopropene

Aromatic Hydrocarbons

Benzene, the parent compound of this large family of organic substances, was discovered by Michael Faraday in 1826. Over the next 40 years, chemists were preoccupied with determining its molecular structure. Despite the small number of atoms in the molecule, there are quite a few ways to represent the structure of benzene without violating the tetravalency of carbon (Figure 24.7). However, most proposed structures were rejected because they could not explain the known properties of benzene. Finally, in 1865, August Kekulé† deduced that the benzene molecule could be best represented by a ring structure—a cyclic compound consisting of six carbon atoms:

An electron micrograph of benzene molecules which shows clearly the ring structure.

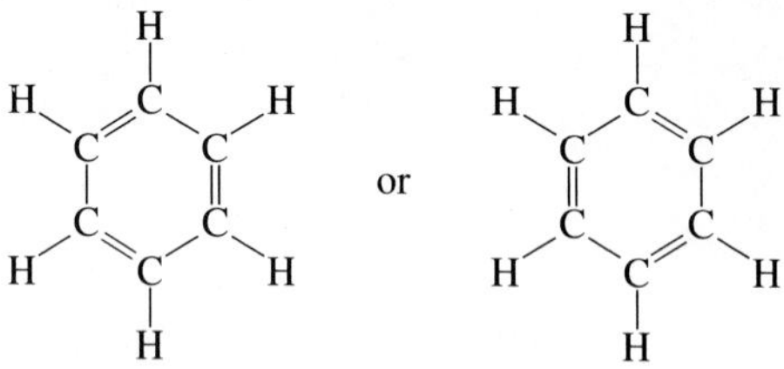

As we saw in Section 9.8, the properties of benzene are best represented by both of the above structures.

†August Kekulé (1829–1896). German chemist. Kekulé was a student of architecture before he became interested in chemistry. He supposedly solved the riddle of the structure of the benzene molecule after having a dream in which dancing snakes bit their own tails. Kekulé's work is regarded by many as the crowning achievement of theoretical organic chemistry of the nineteenth century.

Nomenclature of Aromatic Compounds. The naming of monosubstituted benzenes is quite straightforward, as shown below:

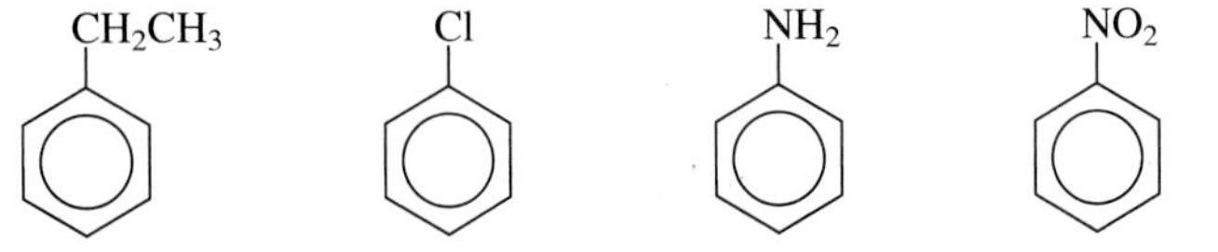

CH_2CH_3 Cl NH_2 NO_2

ethylbenzene chlorobenzene aminobenzene (aniline) nitrobenzene

If more than one substituent is present, we must indicate the location of the second group relative to the first. The systematic way to accomplish this is to number the carbon atoms as follows:

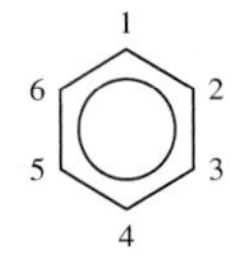

1 2 3 4 5 6

Three different dibromobenzenes are possible:

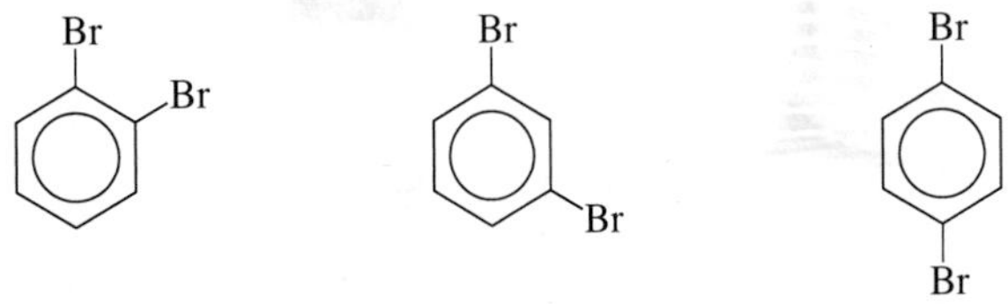

Br Br Br Br Br Br

1,2-dibromobenzene (*o*-dibromobenzene) 1,3-dibromobenzene (*m*-dibromobenzene) 1,4-dibromobenzene (*p*-dibromobenzene)

Frequently, the prefix *o*- (*ortho*-), *m*- (*meta*-), or *p*- (*para*-) is used to denote the relative positions of the two substituted groups. For the dibromobenzenes, the names are as indicated. If the two groups are different, as they are in

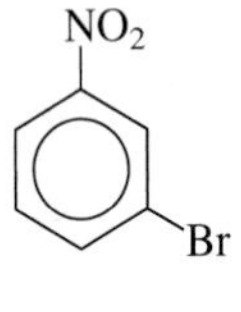

NO_2 Br

the systematic name is 3-bromonitrobenzene and the common name is *m*-bromonitrobenzene.

Properties and Reactions of Aromatic Compounds. Benzene is a colorless, flammable liquid obtained chiefly from petroleum and to a lesser extent coal tar. Perhaps the most remarkable chemical property of benzene is its relative inertness. Although it has the same empirical formula as acetylene (CH) and a high degree of unsaturation, it is much less reactive than either ethylene or acetylene. We now understand that the stability of benzene is the result of electron delocalization. In fact, benzene can be hydrogenated, but only with difficulty. The following reaction is carried out at significantly higher temperatures and pressures than are similar reactions for the alkenes:

$$C_6H_6 + 3H_2 \xrightarrow[\text{catalyst}]{\text{Pt}} C_6H_{12}$$

cyclohexane

We saw earlier that alkenes react readily with halogens to form addition products, because the pi bond in C=C can be broken easily. On the other hand, the most common reaction of halogens with benzene is one of *substitution:*

A substitution reaction is the replacement of an atom or a group of atoms in a compound by another atom or group of atoms.

$$C_6H_6 + Br_2 \xrightarrow[\text{catalyst}]{FeBr_3} C_6H_5Br + HBr$$

bromobenzene

Note that if the reaction were an addition reaction, electron delocalization would be destroyed in the product

$$C_6H_6Br_2$$

and the molecule would not have the aromatic characteristic of chemical unreactivity.

Bromobenzene, C_6H_5Br, contains a benzene ring that has lost one hydrogen atom. The C_6H_5 group is called the *phenyl group. Any group that contains one or more fused benzene rings* is called an ***aryl group.*** Thus, the phenyl group is the simplest aryl group. Naphthalene, $C_{10}H_8$, less one hydrogen atom ($C_{10}H_7$) is the next simplest aryl group.

Alkyl groups can be introduced into the ring system by allowing benzene to react with an alkyl halide using $AlCl_3$ as the catalyst:

$$C_6H_6 + CH_3CH_2Cl \xrightarrow[\text{catalyst}]{AlCl_3} C_6H_5CH_2CH_3 + HCl$$

ethyl chloride ethylbenzene

An enormously large number of compounds can be generated from substances in which benzene rings are fused together. Some of these *polycyclic* aromatic hydrocarbons are shown in Figure 24.8. The best known of these compounds is naphthalene, an important component of mothballs. These and many other similar compounds are present in coal tar. Some of the compounds with larger numbers of rings are powerful carcinogens—that is, they can cause cancer in humans and other animals.

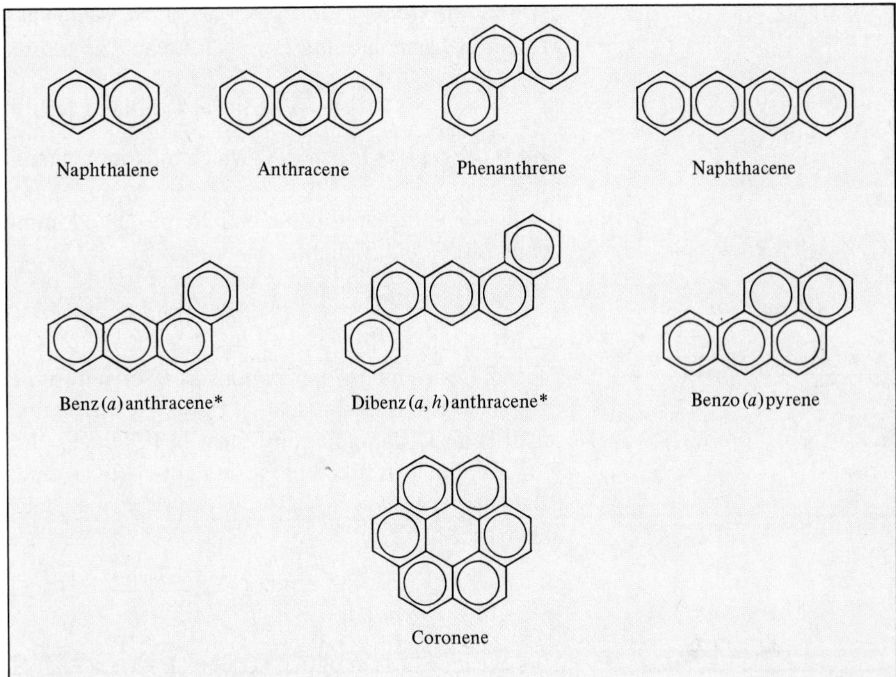

Figure 24.8 *Some polycyclic aromatic hydrocarbons. Compounds denoted by * are potent carcinogens. An enormous number of such compounds exist in nature.*

24.2 Functional Groups

We now examine in greater depth some organic functional groups—groups that are responsible for most of the reactions of the parent compounds. In particular, we focus on oxygen-containing and nitrogen-containing compounds.

Alcohols

The functional group in alcohols is the hydroxyl group, —OH.

All ***alcohols*** *contain the hydroxyl group, —OH.* Some common alcohols are shown in Figure 24.9. Ethyl alcohol, or ethanol, is by far the best known. It is produced biologi-

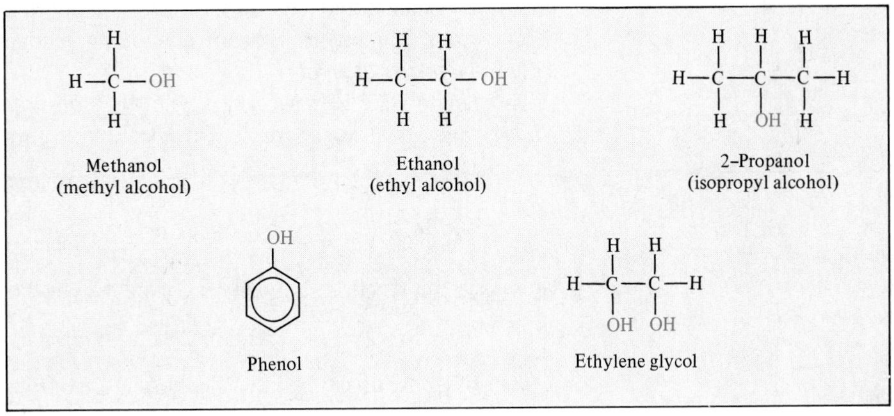

Methanol (methyl alcohol)

Ethanol (ethyl alcohol)

2-Propanol (isopropyl alcohol)

Phenol

Ethylene glycol

Figure 24.9 *Common alcohols. Note that all the compounds contain the OH group. The properties of phenol are quite different from those of the aliphatic alcohols.*

cally by the fermentation of sugar or starch. In the absence of oxygen the enzymes present in bacterial cultures or yeast catalyze the reaction

$$C_6H_{12}O_6(aq) \xrightarrow{\text{enzyme}} 2CH_3CH_2OH(aq) + 2CO_2(g)$$

This process gives off energy, which microorganisms, in turn, use for growth and other functions.

Ethanol is prepared commercially by the addition reaction of water to ethylene at about 280°C and 300 atm:

$$CH_2{=}CH_2(g) + H_2O(g) \xrightarrow[\text{catalyst}]{H_3PO_4} CH_3CH_2OH(g)$$

Ethanol has countless applications as a solvent for organic chemicals and as a starting compound for the manufacture of dyes, synthetic drugs, cosmetics, explosives, and so on. It is also a familiar constituent of beverages. Ethanol is the only nontoxic (more properly, the least toxic) of the straight-chain alcohols; our bodies produce an enzyme, called *alcohol dehydrogenase,* which helps metabolize ethanol by oxidizing it to acetaldehyde:

$$CH_3CH_2OH \xrightarrow{\text{enzyme}} \underset{\text{acetaldehyde}}{CH_3CHO} + H_2$$

This equation is a simplified version of what actually takes place; the H atoms are taken up by other molecules, so that no H_2 gas is evolved.

Ethanol can be oxidized also by inorganic oxidizing agents, such as acidified dichromate, to acetaldehyde and acetic acid:

$$CH_3CH_2OH \xrightarrow[H^+]{Cr_2O_7^{2-}} CH_3CHO \xrightarrow[H^+]{Cr_2O_7^{2-}} CH_3COOH$$

Aliphatic alcohols are derived from alkanes, which are all aliphatic hydrocarbons.

The simplest aliphatic alcohol is methanol, CH_3OH. Called *wood alcohol,* it was prepared at one time by the dry distillation of wood. It is now synthesized industrially by the reaction of carbon monoxide and molecular hydrogen at high temperatures and pressures:

$$CO(g) + 2H_2(g) \xrightarrow[\text{catalyst}]{Fe_2O_3} CH_3OH(g)$$

Methanol is highly toxic. Ingestion of only a few milliliters can cause nausea and blindness. Ethanol intended for industrial use is often mixed with methanol to prevent people from drinking it. Ethanol containing methanol or other toxic substances is called *denatured alcohol.*

The alcohols are very weakly acidic; they do not react with strong bases, such as NaOH. The alkali metals react with alcohols to produce hydrogen:

$$2CH_3OH + 2Na \longrightarrow \underset{\substack{\text{sodium}\\\text{methoxide}}}{2NaOCH_3} + H_2$$

However, the reaction is much less violent than that between Na and water:

$$2H_2O + 2Na \longrightarrow 2NaOH + H_2$$

Two other familiar aliphatic alcohols are isopropyl alcohol, commonly known as rubbing alcohol, and ethylene glycol, which is used as an antifreeze. Note that ethylene glycol has two —OH groups and so can form hydrogen bonds with water molecules more effectively than compounds that have only one —OH group (see Figure 24.9). Most alcohols—especially those with low molar masses—are highly flammable.

Ethers

Ethers** contain the R—O—R′ linkage, where R and R′ are either an alkyl group or an aryl group*. They are formed by the condensation reaction of alcohols. A ***condensation reaction is characterized by *the joining of two molecules and the elimination of a small molecule, usually water:*

$$\underset{}{CH_3OH} + HOCH_3 \xrightarrow[\text{catalyst}]{H_2SO_4} \underset{\text{dimethyl ether}}{CH_3OCH_3} + H_2O$$

Like alcohols, ethers are extremely flammable. When left standing in air, they have a tendency to slowly form explosive peroxides:

$$\underset{\text{diethyl ether}}{C_2H_5OC_2H_5} + O_2 \longrightarrow C_2H_5O{-}\underset{H}{\overset{CH_3}{\overset{|}{\underset{|}{C}}}}{-}O{-}O{-}H$$

Peroxides contain the —O—O— linkage; the simplest peroxide is hydrogen peroxide, H_2O_2. Diethyl ether, commonly known as "ether," was used as an anesthetic for many years. It produces unconsciousness by depressing the activity of the central nervous system. The major disadvantages of diethyl ether are its irritating effects on the respiratory system and the occurrence of postanesthetic nausea and vomiting. "Neothyl," or methyl propyl ether, $CH_3OCH_2CH_2CH_3$, is currently favored as an anesthetic because it is relatively free of side effects.

Aldehydes and Ketones

Under mild oxidation conditions alcohols can be converted to aldehydes and ketones:

$$CH_3OH + \tfrac{1}{2}O_2 \longrightarrow \underset{\text{formaldehyde}}{H_2C{=}O} + H_2O$$

$$C_2H_5OH + \tfrac{1}{2}O_2 \longrightarrow \underset{\text{acetaldehyde}}{(H_3C)(H)C{=}O} + H_2O$$

$$CH_3{-}\underset{OH}{\overset{H}{\overset{|}{\underset{|}{C}}}}{-}CH_3 + \tfrac{1}{2}O_2 \longrightarrow \underset{\text{acetone}}{(H_3C)_2C{=}O} + H_2O$$

The functional group in these compounds is the carbonyl group, $\rangle C{=}O$. The difference between ***aldehydes*** and ***ketones*** is that *in aldehydes at least one hydrogen atom is bonded to the carbon atom in the carbonyl group, whereas in ketones no hydrogen atoms are bonded to this carbon atom.*

Formaldehyde ($H_2C{=}O$) has a tendency to *polymerize;* that is, the individual molecules join together to form a compound of high molar mass. This action gives off much heat and is often explosive, so formaldehyde is usually prepared and stored in aqueous solution (to reduce the concentration). This rather disagreeable-smelling liquid is used as a starting material in the polymer industry (see Chapter 25) and in the laboratory as a preservative for dead animals. Interestingly, the higher molar mass aldehydes, such as cinnamic aldehyde

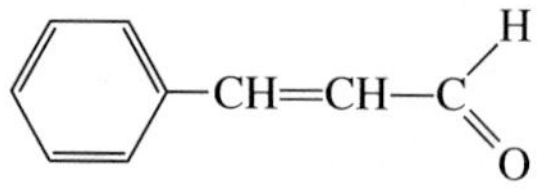

$$C_6H_5-CH{=}CH-C\begin{smallmatrix}\diagup H\\ \diagdown\!\!\diagdown O\end{smallmatrix}$$

have a pleasant odor and are used in the manufacture of perfumes.

Ketones generally are less reactive than aldehydes. The simplest ketone is acetone, a pleasant-smelling liquid that is used mainly as a solvent for organic compounds and nail polish remover. Acetone is much more difficult to oxidize than is acetaldehyde.

Carboxylic Acids

Under appropriate conditions both alcohols and aldehydes can be oxidized to ***carboxylic acids,*** *acids that contain the carboxyl group, —COOH:*

$$CH_3CH_2OH + O_2 \longrightarrow CH_3COOH + H_2O$$

$$CH_3CHO + \tfrac{1}{2}O_2 \longrightarrow CH_3COOH$$

The oxidization of ethanol to acetic acid in wine is catalyzed by enzymes.

These reactions occur so readily, in fact, that wine must be protected from atmospheric oxygen while in storage. Otherwise, it would soon taste like vinegar (due to the formation of acetic acid). Figure 24.10 shows the structure of some of the common carboxylic acids.

Carboxylic acids are widely distributed in nature; they are found in both the plant and animal kingdoms. All protein molecules are made of amino acids, a special kind of carboxylic acid, in which an amino functional group and a carboxylic acid group are present in the same molecule.

Unlike the inorganic acids HCl, HNO_3, and H_2SO_4, carboxylic acids are usually weak. They react with alcohols to form pleasant-smelling esters:

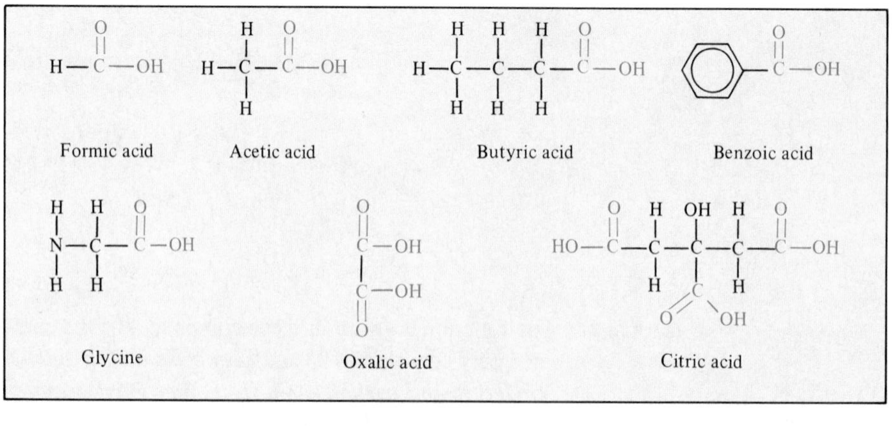

Figure 24.10 *Some common carboxylic acids. Note that they all contain the COOH group. (Glycine is one of the amino acids found in proteins.)*

$$CH_3COOH + HOCH_2CH_3 \longrightarrow \underset{\text{ethyl acetate}}{CH_3{-}\overset{\overset{\displaystyle O}{\|}}{C}{-}O{-}CH_2CH_3} + H_2O$$

Other common reactions of carboxylic acids are neutralization

$$CH_3COOH + NaOH \longrightarrow CH_3COONa + H_2O$$

and formation of acid halides, such as acetyl chloride

$$CH_3COOH + PCl_5 \longrightarrow \underset{\text{acetyl chloride}}{CH_3COCl} + HCl + \underset{\text{phosphoryl chloride}}{POCl_3}$$

Acid halides are reactive compounds used as intermediates in the preparation of many other organic compounds. They hydrolyze in much the same way as many nonmetallic halides, such as $SiCl_4$:

$$CH_3COCl(l) + H_2O(l) \longrightarrow CH_3COOH(aq) + HCl(g)$$

$$SiCl_4(l) + 3H_2O(l) \longrightarrow \underset{\text{silicic acid}}{H_2SiO_3(s)} + 4HCl(g)$$

Esters

Esters *have the general formula R′COOR, where R′ can be H or an alkyl or aryl group and R is an alkyl or aryl group*. Esters are used in the manufacture of perfumes and as flavoring agents in the confectionery and soft-drink industries. Many fruits owe their characteristic smell and flavor to the presence of small quantities of esters.

The functional group in esters is the —COOR group. In the presence of an acid catalyst, such as HCl, esters undergo hydrolysis to yield a carboxylic acid and an alcohol. For example, in acid solution, ethyl acetate hydrolyzes as follows:

$$\underset{\text{ethyl acetate}}{CH_3COOC_2H_5} + H_2O \rightleftharpoons \underset{\text{acetic acid}}{CH_3COOH} + \underset{\text{ethanol}}{C_2H_5OH}$$

However, this reaction does not go to completion because the reverse reaction, that is, the formation of an ester from an alcohol and an acid, also occurs to an appreciable extent. On the other hand, when NaOH solution is used in hydrolysis the sodium acetate does not react with ethanol, so this reaction does go to completion from left to right:

$$\underset{\text{ethyl acetate}}{CH_3COOC_2H_5} + NaOH \longrightarrow \underset{\text{sodium acetate}}{CH_3COO^-Na^+} + \underset{\text{ethanol}}{C_2H_5OH}$$

For this reason, ester hydrolysis is usually carried out in basic solutions. Note that NaOH does not act as a catalyst; rather, it is consumed by the reaction. The term ***saponification*** (meaning *soapmaking*) was originally used to describe the alkaline hydrolysis of fatty acid esters to yield soap molecules (sodium stearate):

$$\underset{\text{ethyl stearate}}{C_{17}H_{35}COOC_2H_5} + NaOH \longrightarrow \underset{\text{sodium stearate}}{C_{17}H_{35}COO^-Na^+} + C_2H_5OH$$

Saponification has now become *a general term for alkaline hydrolysis of any type of ester*.

Amines

Amines are *organic bases. They have the general formula R_3N, where R may be H, an alkyl group, or an aryl group.* As with ammonia, the reaction of amines with water is

$$RNH_2 + H_2O \rightleftharpoons RNH_3^+ + OH^-$$

where R represents either an alkyl or an aryl group. Like all bases, the amines form salts when allowed to react with acids:

$$\underset{\text{ethylamine}}{CH_3CH_2NH_2} + HCl \longrightarrow \underset{\text{ethylammonium chloride}}{CH_3CH_2NH_3^+Cl^-}$$

These salts are usually colorless, odorless solids.

Aromatic amines are used mainly in the manufacture of dyes. Aniline, the simplest aromatic amine, itself is a toxic compound; a number of other aromatic amines such as 2-naphthylamine and benzidine are potent carcinogens:

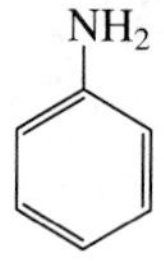

NH_2 NH_2 H_2N NH_2

aniline 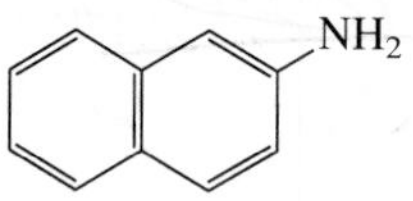 2-naphthylamine 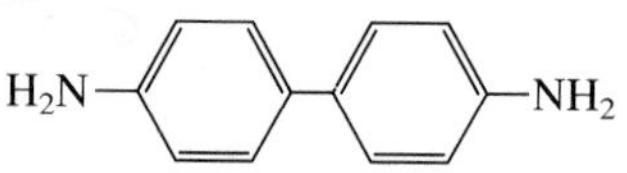 benzidine

Summary of Functional Groups

Table 24.4 summarizes the common functional groups. A compound can and often does contain more than one functional group. Generally, the reactivity of a compound is determined by the number and types of functional groups in its makeup.

Example 24.3 shows how we can use the functional groups to predict reactions.

EXAMPLE 24.3

Cholesterol is a major component of gallstones. It has received much attention because of the suspected correlation between the cholesterol level in the blood and certain types of heart disease, for example, atherosclerosis. From the following structure of the compound, predict its reaction with (a) Br_2, (b) H_2 (in the presence of Pt catalyst), (c) CH_3COOH:

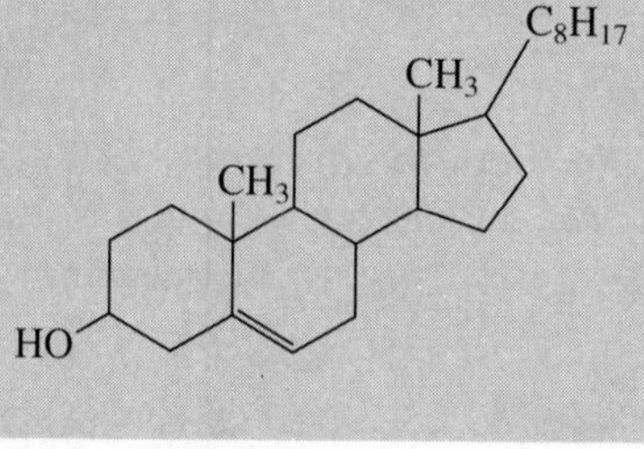

C_8H_{17} CH_3 CH_3 HO

Answer

The first step is to identify the functional groups in cholesterol. There are two: the hydroxyl group and the carbon-carbon double bond.

(a) The reaction with bromine results in the addition of bromine to the double-bonded carbons, which become single-bonded.
(b) This is a hydrogenation reaction. Again, the carbon-carbon double bond is converted to a carbon-carbon single bond.
(c) The acid reacts with the hydroxyl group to form an ester and water. Figure 24.11 shows the products of these reactions.

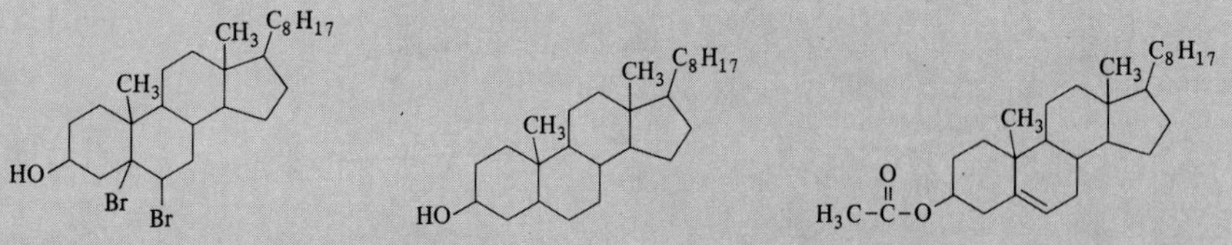

Figure 24.11 *The products formed by the reaction of cholesterol with (a) molecular bromine, (b) molecular hydrogen, and (c) acetic acid.*

Similar problem: 24.34.

Table 24.4 Important functional groups and their reactions

Functional group	Name	Typical reactions
$>C=C<$	Carbon-carbon double bond	Addition reactions with halogens, hydrogen halides, and water; hydrogenation to yield alkanes
$-C\equiv C-$	Carbon-carbon triple bond	Addition reactions with halogens, hydrogen halides; hydrogenation to yield alkenes and alkanes
$-\ddot{X}:$ (X = F, Cl, Br, I)	Halogen	Exchange reactions: $CH_3CH_2Br + KI \longrightarrow CH_3CH_2I + KBr$
$-\ddot{O}-H$	Hydroxyl	Esterification (formation of an ester) with carboxylic acids; oxidation to aldehydes, ketones, and carboxylic acids
$>C=\ddot{O}$	Carbonyl	Reduction to yield alcohols; oxidation of aldehydes to yield carboxylic acids
$-C(=O)-\ddot{O}-H$	Carboxyl	Esterification with alcohols; reaction with phosphorus pentachloride to yield acid chlorides
$-C(=O)-\ddot{O}-R$ (R = alkyl or aryl)	Ester	Hydrolysis to yield acids and alcohols
$-\ddot{N}R_2$ (R = H, alkyl, or aryl)	Amine	Formation of ammonium salts with acids

CHEMISTRY IN ACTION

The Petroleum Industry

It is estimated that in 1988 about 42 percent of the energy needs of the United States were supplied by oil or petroleum. The rest was provided by (approximately) natural gas (23 percent), coal (24 percent), hydroelectric power (3 percent), nuclear power (7 percent), and others (1 percent). In addition to energy, petroleum also supplies numerous organic chemicals used to manufacture drugs, clothing, and many other products.

Petroleum is a complex mixture of alkanes, alkenes, cycloalkanes, and aromatic compounds. Before refinement, petroleum is often called crude oil, which is a viscous, dark brown liquid (Figure 24.12). It is formed in Earth's crust over the course of millions of years by the decomposing action of anaerobic bacterial organisms on animal and vegetable matter.

Petroleum deposits are widely distributed throughout the world, mainly in North America, Mexico, western U.S.S.R., China, Venezuela, and, of course, the Middle East. The actual composition of petroleum varies with location. In the United States, for example, Pennsylvania crude oils are mostly aliphatic hydrocarbons, whereas the major components of western crude oil are aromatic in nature.

Although petroleum contains literally thousands of hydrocarbon compounds, we can classify its components according to the range of their boiling points (Table 24.5). These hydrocarbons can be separated on the basis of molar mass by fractional distillation, as shown in Figure 24.13. Heating crude oil to about 400°C converts the viscous oil into hot vapor and fluid. In this form it enters the fractionating tower. The vapor rises and condenses on various collecting trays according to the temperatures at which the various components of the vapor liquefy. Some gases are drawn off at the top of the column, and the unvaporized residual oil is collected at the bottom.

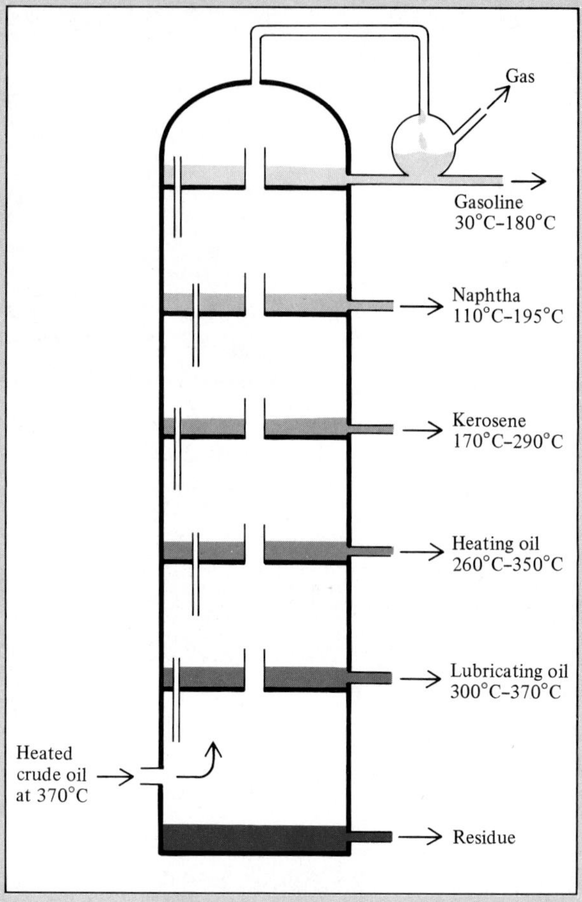

Figure 24.13 *Fractional distillation column for separating the components of petroleum crude oil. As the hot vapor of the crude oil moves upward, it condenses and the various components of the crude oil are separated according to their boiling points and are drawn off as shown.*

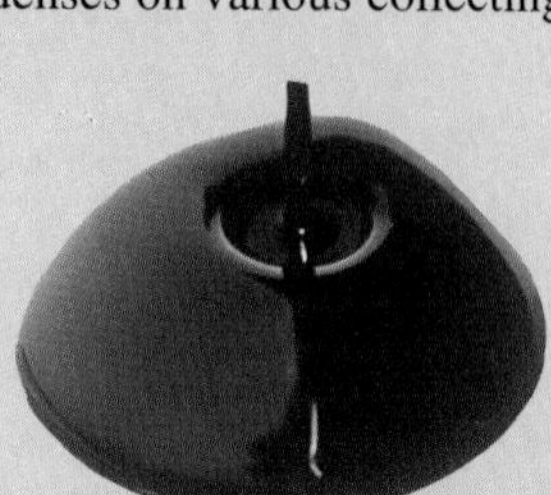

Figure 24.12 *Crude oil.*

Gasoline is probably the best-known component of petroleum. Actually, gasoline is itself a mixture of volatile hydrocarbons. It contains mostly alkanes, cyclo-

Table 24.5 Major fractions of petroleum

Fraction	Carbon atoms*	Boiling point range (°C)	Uses
Natural gas	C_1–C_4	−161 to 20	Fuel and cooking gas
Petroleum ether	C_5–C_6	30–60	Solvent for organic compounds
Ligroin	C_7	20–135	Solvent for organic compounds
Gasoline	C_6–C_{12}	30–180	Automobile fuels
Kerosene	C_{11}–C_{16}	170–290	Rocket and jet engine fuels, domestic heating
Heating fuel oil	C_{14}–C_{18}	260–350	Domestic heating and fuel for electricity production
Lubricating oil	C_{15}–C_{24}	300–370	Lubricants for automobiles and machines

*The entries in this column indicate the numbers of carbon atoms in the compounds involved. For example, C_1–C_4 tells us that in natural gas the compounds contain 1 to 4 carbon atoms, and so on.

alkanes, and a few aromatic hydrocarbons. Some of these compounds are far more suitable for fueling an automobile engine than others, and herein lies the problem of the further treatment and refinement of gasoline.

Figure 24.14 shows a schematic diagram of the four-stroke operation of the *Otto cycle* engine that drives most automobiles. A major engineering concern is to control the burning of the gasoline-air mixture inside each cylinder to obtain a smooth expansion of the gas mixture. If the mixture burns too rapidly, the piston receives a hard jerk rather than a smooth, strong push. This action produces a "knocking" or "pinging" sound, as well as a decrease in efficiency in the conversion of combustion energy to mechanical energy. It turns out that straight-chain hydrocarbons have the greatest tendency to produce knocking, whereas the branched-chain and aromatic hydrocarbons give the desired smooth push. For this reason, gasolines are usually rated according to the ***octane number,*** *a measure of their tendency to cause knocking.* On this scale, a branched C_8 compound (2,2,4-trimethylpentane, or isooctane) has been arbitrarily assigned an octane number of 100, and that of *n*-heptane is zero (Table 24.6). The higher the octane number of the hydrocarbon, the

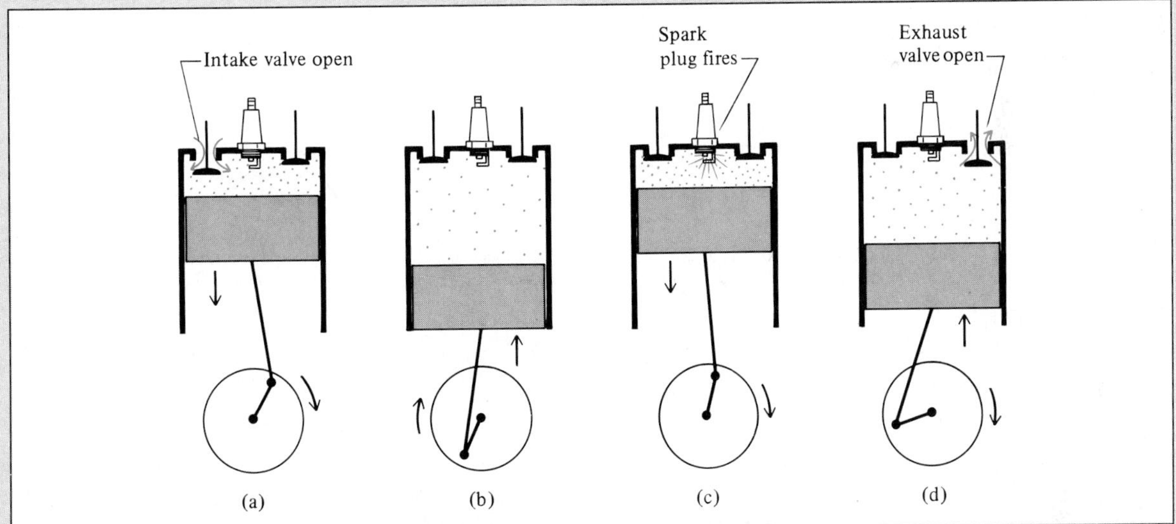

Figure 24.14 *The four stages of operation of an internal combustion engine. This is the type of engine used in practically all automobiles and is described technically as a four-stroke Otto cycle engine. (a) Intake valve opens to let in a gasoline-air mixture. (b) During the compression stage the two valves are closed. (c) The spark plug fires and the piston is pushed outward. (d) Finally, as the piston is pushed downward, the exhaust valve opens to let out the exhaust gas.*

Table 24.6 Octane ratings of some hydrocarbons

Hydrocarbon	Molecular structure	Type of structure	Octane rating
n-Heptane	$CH_3-CH_2-CH_2-CH_2-CH_2-CH_2-CH_3$	Straight chain	0*
n-Hexane	$CH_3-CH_2-CH_2-CH_2-CH_2-CH_3$	Straight chain	25
2-Methylhexane	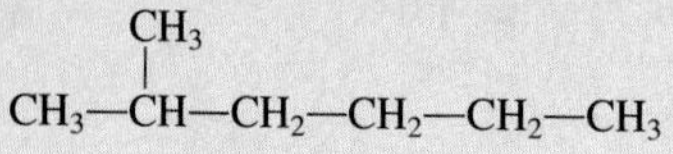 CH_3 $CH_3-CH-CH_2-CH_2-CH_2-CH_3$	Branched chain	42
2-Methylbutane	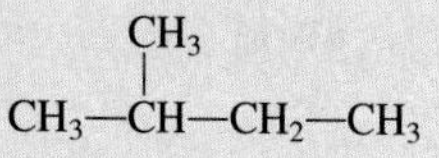 CH_3 $CH_3-CH-CH_2-CH_3$	Branched chain	93
2,2,4-Trimethylpentane (isooctane)	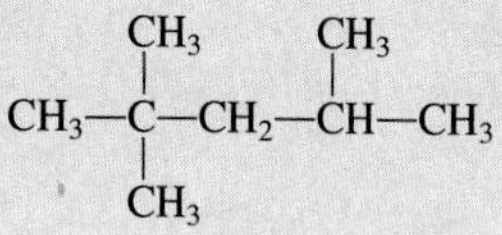 CH_3 CH_3 $CH_3-C-CH_2-CH-CH_3$ CH_3	Branched chain	100*
Toluene	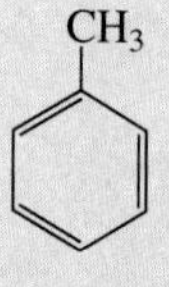 CH_3	Aromatic	120
Benzene	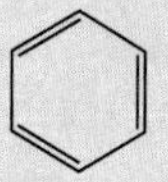	Aromatic	106
2,2,3-Trimethylbutane	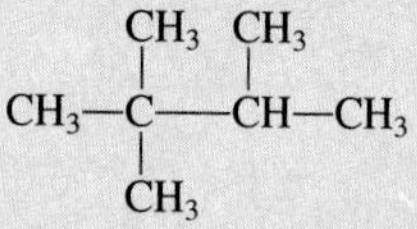 CH_3 CH_3 $CH_3-C-CH-CH_3$ CH_3	Branched chain	125

*Reference points for octane rating.

better its performance in the internal combustion engine.

The octane rating of hydrocarbons can be improved by the addition of small quantities of compounds called *antiknocking agents*. Among the most widely used substances are the following:

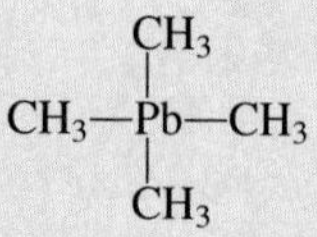

CH_3
$CH_3-Pb-CH_3$
CH_3

tetramethyllead

CH_3
CH_2
$CH_3-CH_2-Pb-CH_2-CH_3$
CH_2
CH_3

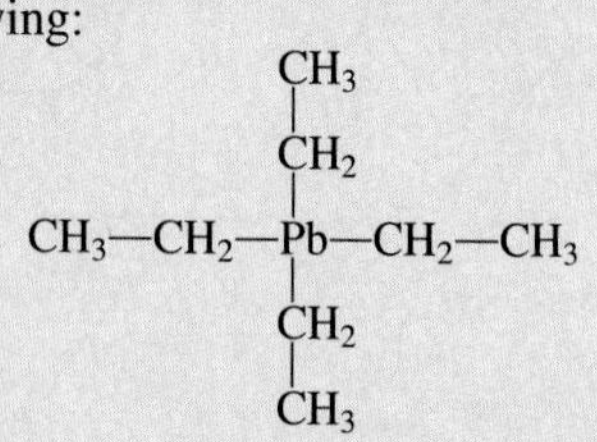

tetraethyllead

The addition of 2 to 4 g of either of these compounds to a gallon of gasoline increases the octane rating by 10 or more. However, lead is a highly toxic metal, and the constant discharge of automobile exhaust into the atmosphere has become a serious environmental problem. Federal regulations require that all automobiles made after 1974 use "unleaded" gasolines. The catalytic converters with which late-model automobiles are equipped can be "poisoned" by lead, another reason for its exclusion from gasoline. To minimize knocking, unleaded gasolines contain a higher proportion of branched-chain hydrocarbons.

Summary

1. Because carbon atoms can link up with other carbon atoms in straight and branched chains, carbon can form more compounds than any other element.
2. Methane, CH_4, is the simplest of the alkanes, a family of hydrocarbons with the general formula C_nH_{2n+2}. Cyclopropane, C_3H_6, is the simplest of the cycloalkanes, a family of alkanes whose carbon atoms are joined in a ring. Alkanes and cycloalkanes are saturated hydrocarbons.
3. Ethylene, $CH_2{=}CH_2$, is the simplest of the olefins, or alkenes, a class of hydrocarbons containing carbon-carbon double bonds and having the general formula C_nH_{2n}. Acetylene, $CH{\equiv}CH$, is the simplest of the alkynes, which are compounds that have the general formula C_nH_{2n-2} and contain carbon-carbon triple bonds.
4. Compounds that contain one or more benzene rings are called aromatic hydrocarbons. These compounds undergo substitution by halogens and alkyl groups.
5. Functional groups impart specific types of chemical reactivity to molecules. Classes of compounds characterized by their functional groups include alcohols, ethers, aldehydes and ketones, carboxylic acids and esters, and amines.

Applied Chemistry Example: Isopropanol

Isopropanol (C_3H_8O) is reported to be the first *petrochemical,* that is, a chemical derived from petroleum. About 1.42 billion pounds of this compound were produced in 1988, which ranked it 48th among the top 50 industrial chemicals produced in the United States.

Isopropanol is a colorless liquid with an odor characteristic of alcohols. It melts at −89.5°C and boils at 82.5°C. Isopropanol is soluble in water, ethanol, and ether. The major uses of isopropanol are as a component in the manufacture of acetone, as a solvent to dissolve the components of shellac and other resinous finishes, as a disinfectant, and as rubbing alcohol.

1. Isopropanol is prepared by reacting propylene ($CH_3CH{=}CH_2$) with sulfuric acid, followed by treatment with water. Show the sequence of steps leading to the product. What is the role of sulfuric acid?

Answer: Sulfuric acid ionizes as follows:

$$H_2SO_4(aq) \longrightarrow H^+(aq) + HSO_4^-(aq)$$

The cation (H^+) and anion (HSO_4^-) add to the double bond in propylene according to Markovnikov's rule:

$$CH_3{-}CH{=}CH_2 + H^+ + HSO_4^- \longrightarrow CH_3{-}\underset{\displaystyle H}{\overset{\displaystyle OSO_3H}{\underset{|}{\overset{|}{C}}}}{-}CH_3$$

Reaction of the intermediate with water then yields isopropanol:

$$CH_3{-}\underset{\displaystyle H}{\overset{\displaystyle OSO_3H}{\underset{|}{\overset{|}{C}}}}{-}CH_3 + H_2O \longrightarrow CH_3{-}\underset{\displaystyle H}{\overset{\displaystyle OH}{\underset{|}{\overset{|}{C}}}}{-}CH_3 + H_2SO_4$$

Since sulfuric acid is regenerated, it plays the role of a catalyst.

2. Draw the structure of the alcohol that is an isomer of isopropanol.

   Answer: The other structure containing the —OH group is

$$CH_3{-}CH_2{-}CH_2{-}OH$$

propanol

3. Is isopropanol a chiral molecule?

   Answer: From the structure of isopropanol shown above we see that the molecule does not have an asymmetric carbon atom. Therefore, isopropanol is achiral.

4. What property of isopropanol makes it useful as a rubbing alcohol?

   Answer: Isopropanol is fairly volatile (b.p. 82.5°C) and the —OH group allows it to form hydrogen bond with water molecules. Thus as it evaporates, it produces a cooling and soothing effect on the skin.

Key Words

Alcohol, p. 1009
Aldehyde, p. 1011
Aliphatic hydrocarbon, p. 996
Alkane, p. 996
Alkene, p. 1003
Alkyl group, p. 999
Alkyne, p. 1005
Amine, p. 1014
Aromatic hydrocarbon, p. 996
Aryl group, p. 1008
Carboxylic acid, p. 1012
Condensation reaction, p. 1011
Cycloalkane, p. 1002
Ester, p. 1013
Ether, p. 1011
Functional group, p. 996
Hydrocarbon, p. 996
Ketone, p. 1011
Octane number, p. 1017
Organic chemistry, p. 996
Saponification, p. 1013
Saturated hydrocarbon, p. 996
Structural isomer, p. 997
Unsaturated hydrocarbon, p. 1003

Exercises

Hydrocarbons

REVIEW QUESTIONS

24.1 Define the following terms: hydrocarbon, aliphatic hydrocarbon, aromatic hydrocarbon, alkane, cycloalkane, alkene, alkyne, radical.

24.2 Give examples of a saturated hydrocarbon and an unsaturated hydrocarbon.

24.3 Give examples of a chiral substituted alkane and an achiral substituted alkane.

24.4 Alkenes exhibit geometric isomerism because rotation about the C=C bond is restricted. Explain what this means.

24.5 Why is it that alkanes and alkynes, unlike alkenes, have no geometric isomers?

24.6 What is Markovnikov's rule?

24.7 Both benzene and ethylene contain the C=C bond. Why is benzene more stable than ethylene?

24.8 Explain why carbon is able to form so many more compounds than any other element.

PROBLEMS

24.9 Draw all possible structural isomers for the following alkanes: (a) C_4H_{10}, (b) C_6H_{14}, (c) C_7H_{16}.

24.10 Draw all possible isomers for the molecule C_4H_8.

24.11 Discuss how you can determine which of the following compounds might be alkanes, cycloalkanes, alkenes, or alkynes, without drawing their formulas: (a) C_6H_{12}, (b) C_4H_6, (c) C_5H_{12}, (d) C_7H_{14}, (e) C_3H_4.

24.12 The benzene and cyclohexane molecules both contain six-membered rings. Benzene is a planar molecule and cyclohexane is nonplanar. Explain. Would you expect cyclobutadiene to be a stable molecule? Explain.

```
H       H
 \     /
  C—C
  ‖  ‖
  C—C
 /     \
H       H
```

(*Hint:* Consider the angles between carbon atoms.)

24.13 Suggest two chemical tests that would help you distinguish between these two compounds:

$$CH_3CH_2CH_2CH_2CH_3 \qquad CH_3CH_2CH_2CH{=}CH_2$$

24.14 Draw the structures of *cis*-2-butene and *trans*-2-butene. Which of the two compounds would have the higher heat of hydrogenation? Explain.

24.15 Draw the structures of the geometric isomers of the molecule $CH_3{-}CH{=}CH{-}CH{=}CH{-}CH_3$.

24.16 When a mixture of methane and bromine is exposed to light, the following reaction occurs slowly:

$$CH_4(g) + Br_2(g) \longrightarrow CH_3Br(g) + HBr(g)$$

Suggest a reasonable mechanism for this reaction. (*Hint:* Bromine vapor is deep red; methane is colorless.)

24.17 Sulfuric acid, H_2SO_4, adds across the double bond of alkenes as H^+ and $^-OSO_3H$. Draw structures for the products of addition of sulfuric acid to (a) ethylene, (b) propylene, (c) 1-butene, and (d) 2-butene.

24.18 Draw all possible isomers for the molecule C_3H_5Br.

24.19 Indicate the asymmetric carbon atoms in the following compounds:

$$CH_3{-}CH_2{-}\underset{}{\overset{CH_3}{\overset{|}{C}H}}{-}\underset{\underset{NH_2}{|}}{C}H{-}\overset{\overset{O}{\|}}{C}{-}NH_2$$

H, Br (top carbon); H, H (left carbon); H, Br (right carbon) — cyclopropane ring

24.20 State which member of each of the following pairs of compounds is the more reactive and explain why: (a) propane and cyclopropane, (b) ethylene and methane, (c) cyclohexane and benzene

24.21 How many distinct chloropentanes, $C_5H_{11}Cl$, could be produced in the direct chlorination of *n*-pentane, $CH_3(CH_2)_3CH_3$? Draw the structure of each and identify any that are chiral.

24.22 The structural isomers of pentane, C_5H_{12}, have substantially different boiling points (see Example 24.1). Explain the observed variation in boiling point, using structure as a reason.

24.23 Under certain conditions acetylene can trimerize to benzene:

$$3C_2H_2(g) \longrightarrow C_6H_6(l)$$

Calculate the standard enthalpy change in kJ for this reaction at 25°C.

24.24 Geometric isomers are not restricted to compounds containing the C═C bond. For example, certain disubstituted cycloalkanes can exist in the cis and the trans forms. Name the following compounds:

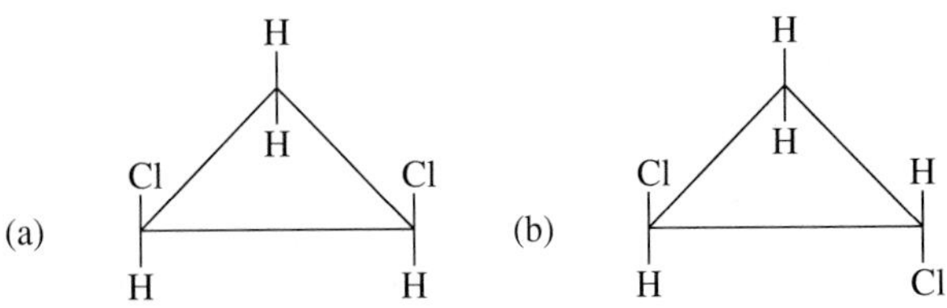

Functional Groups

REVIEW QUESTIONS

24.25 Define "functional group." Why is it logical to classify organic compounds according to their functional groups?

24.26 Draw the Lewis structure for each of the following functional groups: alcohol, ether, aldehyde, ketone, carboxylic acid, ester, amine.

PROBLEMS

24.27 Classify each of the following molecules as alcohol, aldehyde, ketone, carboxylic acid, amine, or ether:

(a) $H_3C{-}O{-}CH_2{-}CH_3$ (b) $H_3C{-}CH_2{-}NH_2$

(c) $H_3C{-}CH_2{-}C({=}O){-}H$ (d) $H_3C{-}\underset{\underset{O}{\|}}{C}{-}CH_2{-}CH_3$

(e) $H{-}\overset{\overset{O}{\|}}{C}{-}OH$ (f) $H_3C{-}CH_2CH_2{-}OH$

(g) $C_6H_5{-}CH_2{-}\overset{\overset{NH_2}{|}}{\underset{\underset{H}{|}}{C}}{-}\overset{\overset{O}{\|}}{C}{-}OH$

24.28 Ethanol reacts with (a) gaseous HCl, (b) sodium metal. Predict the products formed in each reaction and write balanced equations for each.

24.29 Complete the following equation and identify the products:

$$CH_3{-}CH_2{-}COOH + CH_3OH \longrightarrow$$

24.30 Generally aldehydes are more susceptible to air oxidation than are ketones. Use acetaldehyde and acetone as examples and show why ketones are more stable than aldehydes in this respect.

24.31 Draw structures for molecules with the following formulas: (a) CH_4O, (b) C_2H_6O, (c) $C_3H_6O_2$, (d) C_3H_8O.

24.32 A compound having the molecular formula $C_4H_{10}O$ does not react with sodium metal. In the presence of light, the compound reacts with Cl_2 to form three compounds having the formula C_4H_9OCl. Draw a structure for the original compound that is consistent with this information.

24.33 Indicate the classes of compounds having the formulas

(a) $H—C{\equiv}C—CH_3$
(b) C_4H_9OH
(c) $CH_3OC_2H_5$
(d) C_2H_5CHO
(e) C_6H_5COOH
(f) CH_3NH_2

24.34 Predict the product or products of each of the following reactions:

(a) $C_2H_5OH + HCOOH \longrightarrow$
(b) $H—C{\equiv}C—CH_3 + H_2 \longrightarrow$
(c) $(C_2H_5)(H)C{=}C(H)(H) + HBr \longrightarrow$

24.35 Identify the functional groups in each of the following molecules:

(a) $CH_3CH_2COCH_2CH_2CH_3$
(b) $CH_3COOC_2H_5$
(c) $CH_3CH_2OCH_2CH_2CH_2CH_3$

24.36 Beginning with 3-methyl-1-butyne, show how you would prepare the following compounds:

(a) $CH_2{=}C(Br)—CH(CH_3)—CH_3$
(b) $CH_2Br—CBr_2—CH(CH_3)—CH_3$
(c) $CH_3—CH(Br)—CH(CH_3)—CH_3$

24.37 An organic compound has the empirical formula $C_5H_{12}O$. Upon controlled oxidation, it is converted into a compound of empirical formula $C_5H_{10}O$, which behaves as a ketone. Draw structures of the original compound and the final compound.

Nomenclature

PROBLEMS

24.38 Name the following compounds:

(a) $CH_3—CH(CH_3)—CH_2—CH_2—CH_3$
(b) $CH_3—CH(C_2H_5)—CH(CH_3)—CH(CH_3)—CH_3$
(c) $CH_3—CH_2—CH(CH_2—CH_2—CH_3)—CH_2—CH_3$
(d) $CH_2{=}CH—CH(CH_3)—CH{=}CH_2$
(e) $CH_3—C{\equiv}C—CH_2—CH_3$
(f) $CH_3—CH_2—CH(C_6H_5)—CH{=}CH_2$

24.39 Name the following compounds:

(a) benzene ring with Cl, Cl, CH_3 substituents
(b) benzene ring with NO_2, NO_2, CH_2CH_3 substituents
(c) benzene ring with CH_3, CH_3, H_3C, CH_3 substituents

24.40 Write the structural formulas for the following organic compounds: (a) 3-methylhexane, (b) 1,3,5-trichlorocyclohexane, (c) 2,3-dimethylpentane, (d) 2-phenyl-4-bromopentane, (e) 3,4,5-trimethyloctane.

24.41 Write the structural formulas for the following compounds: (a) *trans*-2-pentene, (b) 2-ethyl-1-butene, (c) 4-ethyl-*trans*-2-heptene, (d) 3-phenylbutyne, (e) 2-nitro-4-bromotoluene.

24.42 Draw structures for the following compounds: (a) cyclopentane, (b) *cis*-2-butene, (c) 2-hexanol, (d) 1,4-dibromobenzene, (e) 2-butyne.

Miscellaneous Problems

24.43 Draw all the possible structural isomers for the molecule having the formula C_7H_7Cl. The molecule contains one benzene ring.

24.44 Given these data

$$C_2H_4(g) + 3O_2(g) \longrightarrow 2CO_2(g) + 2H_2O(l) \qquad \Delta H^\circ = -1411 \text{ kJ}$$

$$2C_2H_2(g) + 5O_2(g) \longrightarrow 4CO_2(g) + 2H_2O(l) \qquad \Delta H^\circ = -2599 \text{ kJ}$$

$$H_2(g) + \tfrac{1}{2}O_2(g) \longrightarrow H_2O(l) \qquad \Delta H^\circ = -285.8 \text{ kJ}$$

calculate the heat of hydrogenation for acetylene:

$$C_2H_2(g) + H_2(g) \longrightarrow C_2H_4(g)$$

24.45 An organic compound is found to contain 37.5 percent carbon, 3.2 percent hydrogen, and 59.3 percent fluorine by mass. The following pressure and volume data were obtained for 1.00 g of this substance at 90°C:

P (atm)	*V* (L)
2.00	0.332
1.50	0.409
1.00	0.564
0.50	1.028

The molecule is known to have no dipole moment. (a) What is the empirical formula of this substance? (b) Does this substance behave as an ideal gas? (c) What is its molecular formula? (d) Draw the Lewis structure of this molecule and describe its geometry. (e) What is the systematic name of this compound?

24.46 Which of the following compounds can form hydrogen bonds with water molecules? (a) carboxylic acids, (b) alkenes, (c) ethers, (d) aldehydes, (e) alkanes, (f) amines.

24.47 Name at least one commercial use for each of the following compounds: (a) 2-propanol (isopropyl alcohol), (b) acetic acid, (c) naphthalene, (d) methanol, (e) ethanol, (f) ethylene glycol, (g) methane, (h) ethylene.

24.48 How many liters of air (78 percent N_2, 22 percent O_2 by volume) at 20°C and 1.00 atm are needed for the complete combustion of 1.0 L of octane, C_8H_{18}, a typical gasoline component, of density 0.70 g/mL?

24.49 How many carbon-carbon sigma bonds are present in each of the following molecules? (a) benzene, (b) cyclobutane, (c) 2-methyl-3-ethylpentane, (d) 2-butyne, (e) anthracene (See Figure 24.8.)

24.50 Draw all the structural isomers of the formula $C_4H_8Cl_2$. Indicate those that are chiral and give them systematic names.

24.51 Combustion of 20.63 mg of compound Y, which contains only C, H, and O, with excess oxygen gave 57.94 mg of CO_2 and 11.85 mg of H_2O. (a) Calculate how many milligrams of C, H, and O were present in the original sample of Y. (b) Derive the empirical formula of Y. (c) Suggest a plausible structure for Y if the empirical formula is the same as the molecular formula.

24.52 Combustion of 3.795 mg of liquid B, which contains only C, H, and O, with excess oxygen gave 9.708 mg of CO_2 and 3.969 mg of H_2O. In a molar mass determination, 0.205 g of B vaporized at 1.00 atm and 200.0°C occupied a volume of 89.8 mL. Derive the empirical formula, molar mass, and molecular formula of B and draw three plausible structures.

24.53 Discuss what is meant by the term "geometric isomerism," using N_2F_2 (difluorodiazine) and $C_2H_2Cl_2$ (1,2-dichloroethylene) as examples.

24.54 Suppose that benzene contained three distinct single bonds and three distinct double bonds. How many different isomers would there be for dichlorobenzene ($C_6H_4Cl_2$)? Draw all your proposed structures.

24.55 Ethanol, C_2H_5OH, and dimethyl ether, CH_3OCH_3, are structural isomers. Compare their melting points, boiling points, and solubilities in water.

24.56 The compound $CH_3—C≡C—CH_3$ is hydrogenated to an alkene using platinum as the catalyst. Predict whether the product is the pure trans isomer, the pure cis isomer, or a mixture of cis and trans isomers. Based on your prediction, comment on the mechanism of the heterogeneous catalysis.

24.57 How many asymmetric carbon atoms are present in each of the following compounds?

```
        H  H  H
        |  |  |
(a) H—C—C—C—Cl
        |  |  |
        H  Cl H

         OH   CH3
         |    |
(b) H3C—C————C—CH2OH
         |    |
         H    H

          CH2OH
          |
          C———O
     H  /  |    \   OH
     | /   H     \  |
(c)  C     OH  H   C
     | \   |   | / |
    HO   C———C     H
         |   |
         H   OH
```

Opposite page: Mechanical parts, connectors, and integrated-circuit boards made of plastics. Polymeric materials have created a revolution in packaging and engineering.

C H A P T E R 25

ORGANIC POLYMERS: SYNTHETIC AND NATURAL

Polymers are very large molecules containing hundreds or thousands of atoms. People have been using polymers since prehistoric time, and chemists have been synthesizing them for the past century. Natural polymers are the basis of all life processes, and our technological society is largely dependent on synthetic polymers.

This chapter discusses some of the preparation and properties of important synthetic organic polymers in addition to two naturally occurring polymers in living systems—proteins and nucleic acids.

25.1 Properties of Polymers

A ***polymer*** is *a chemical species distinguished by a high molar mass, ranging into thousands and millions of grams*. Polymers are often called *macromolecules*. The physical properties of polymers differ greatly from those of small, ordinary molecules, and special techniques are required to study these giant molecules.

Polymers are divided into two classes: natural and synthetic. Examples of natural polymers are proteins, nucleic acids, cellulose (polysaccharides), and rubber (polyisoprene). Most synthetic polymers are organic compounds. Familiar examples are nylon, poly(hexamethylene adipamide); Dacron, poly(ethylene terephthalate); and Lucite or Plexiglas, poly(methyl methacrylate).

The development of polymer chemistry began in the 1920s. Chemists were making great progress in clarifying the chemical structures of various substances, but they were generally puzzled by the behavior of certain materials, including wood, gelatin, cotton, and rubber. For example, when rubber, with the known empirical formula of C_5H_8, was dissolved in an organic solvent, the solution displayed several unusual properties—high viscosity, low osmotic pressure, and negligible depression in the freezing point of the solvent.

These observations strongly suggested the presence of very high molar mass solutes, but chemists were not ready at that time to accept the idea that such giant molecules could exist. Instead, they postulated that materials such as rubber consist of aggregates of small molecular units, like C_5H_8 or $C_{10}H_{16}$, held together by intermolecular forces. This misconception persisted for a number of years, until Hermann Staudinger† clearly showed that these so-called aggregates are, in fact, enormously large molecules, each of which contains many thousands of atoms held together by covalent bonds.

Once the properties of these macromolecules were understood, the way was open for manufacturing polymers, which play such an important role in almost every aspect of our daily lives. About 90 percent of today's chemists, including biochemists, work with polymers.

25.2 Synthetic Organic Polymers

Because of their size, polymers may seem inexplicably complex. We might expect molecules containing thousands of carbon and hydrogen atoms to form an enormous number of structural and geometric isomers (if C═C bonds are present). However, these molecules all contain rather *simple repeating units,* called ***monomers,*** which severely restrict the number of isomers that can exist. Synthetic polymers are made by joining monomers together, one at a time. A great deal is known about the mechanism of polymerization. We discuss two important types of polymerization reaction here: addition reactions and condensation reactions.

Addition Reactions

In an addition reaction the polymerization process is initiated by a radical, a cation, or an anion. In Chapter 13 we discussed the kinetics of radical-initiated polymerization of ethylene to form polyethylene (see p. 586). Polyethylene (Figure 25.1) is a very stable

† Hermann Staudinger (1881–1963). German chemist. One of the pioneers in polymer chemistry, Staudinger was awarded the Nobel Prize in chemistry in 1953.

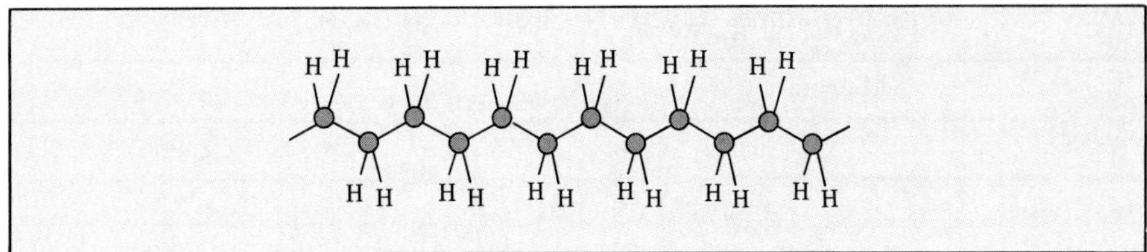

Figure 25.1 *Structure of polyethylene. Each carbon atom is* sp^3*-hybridized.*

and quite inert substance. The individual chains pack together well and so account for the crystallike properties of the material. Polyethylene is mainly used in making films in frozen food packaging and other product wrappings. A specially treated type of polyethylene called Tyvek is used as an insulating layer in houses.

Polymers that are made from only one type of monomer, such as polyethylene, are called ***homopolymers.*** Other important homopolymers synthesized by the radical mechanism are Teflon, polytetrafluoroethylene (Figure 25.2):

$$\text{-(}CF_2\text{—}CF_2\text{)}_n\text{-}$$

and poly(vinyl chloride):

$$\text{-(}CH_2\text{—}\underset{\displaystyle Cl}{\underset{|}{C}}H\text{)}_n\text{-}$$

Table 25.1 lists a number of monomers and their synthetic polymers prepared by the addition reaction.

The chemistry becomes more complex if the starting units are asymmetric; for example,

$$\begin{matrix} H_3C & & H \\ & C{=}C & \\ H & & H \end{matrix}$$

propylene

Figure 25.2 *A cooking utensil made of Silverstone, which contains polytetrafluoroethylene.*

Table 25.1 Some monomers and their common synthetic polymers

Monomer		Polymer	
Formula	**Name**	**Name and formula**	**Uses**
$H_2C{=}CH_2$	Ethylene	Polyethylene $\text{-(}CH_2{-}CH_2\text{)-}_n$	Plastic piping, bottles, electrical insulation, toys
$H_2C{=}CH{-}CH_3$ (H and CH_3 on C)	Propylene	Polypropylene $\text{-(}CH(CH_3){-}CH_2{-}CH(CH_3){-}CH_2\text{)-}_n$	Packaging film, carpets, crates for soft-drink bottles, lab wares, toys
$H_2C{=}CHCl$ (H and Cl on C)	Vinyl chloride	Poly(vinyl chloride) (PVC) $\text{-(}CH_2{-}CH(Cl)\text{)-}_n$	Piping, siding, gutters, floor tile, clothing, toys
$H_2C{=}CH{-}CN$ (H and CN on C)	Acrylonitrile	Polyacrylonitrile (PAN) $\text{-(}CH_2{-}CH(CN)\text{)-}_n$	Carpets, knitwear
$F_2C{=}CF_2$	Tetrafluoroethylene	Polytetrafluoroethylene (Teflon) $\text{-(}CF_2{-}CF_2\text{)-}_n$	Cooking utensils, electrical insulation, bearings
$H_2C{=}C(COOCH_3)(CH_3)$	Methyl methacrylate	Poly(methyl methacrylate) (Plexiglas) $\text{-(}CH_2{-}C(COOCH_3)(CH_3)\text{)-}_n$	Optical equipment, home furnishing
$H_2C{=}CH{-}C_6H_5$ (H and benzene ring on C)	Styrene	Polystyrene $\text{-(}CH_2{-}CH(C_6H_5)\text{)-}_n$	Containers, thermal insulation (ice buckets, water coolers), toys
$H_2C{=}CH{-}CH{=}CH_2$	Butadiene	Polybutadiene $\text{-(}CH_2CH{=}CHCH_2\text{)-}_n$	Tire tread, coating resin
See above structures	Butadiene and stryrene	Stryrene-butadiene rubber (SBR) $\text{-(}CH(C_6H_5){-}CH_2{-}CH_2{-}CH{=}CH{-}CH_2\text{)-}_n$	Synthetic rubber

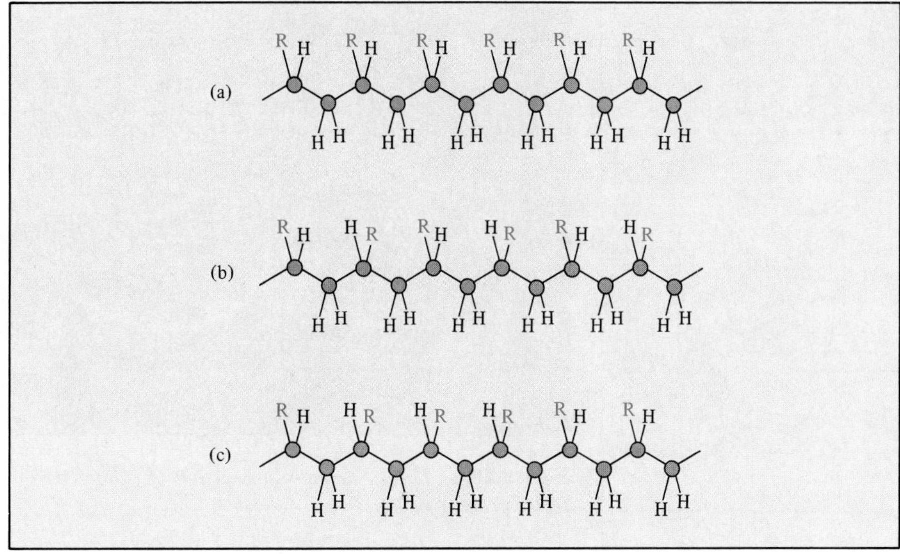

Figure 25.3 *Stereoisomers of polymers. When the R group is CH_3, the polymer is polypropylene. (a) When the R groups are all on one side of the chain, the polymer is said to be isotactic. (b) When R groups alternate from side to side, the polymer is said to be syndiotactic. (c) When R groups are disposed at random, the polymer is atactic.*

Several structurally different polymers can result from an addition reaction of asymmetric propylenes. If the additions occur randomly, we obtain the *atactic* polypropylenes, which do not pack together well (Figure 25.3). The result is a rubbery, amorphous polymer of relatively low strength. Two other possibilities yield *isotactic* and *syndiotactic* polypropylenes, also shown in Figure 25.3. Of these, the former has a higher melting point and greater crystallinity and is endowed with superior mechanical properties.

One of the major problems that the polymer industry faced was how to selectively synthesize either the isotactic or syndiotactic polymer without having it contaminated by other products. The answer was provided by the work of Giulio Natta† and Karl Ziegler,‡ who demonstrated that the product with the desired structure can be obtained by using certain "stereo catalysts" such as triethylaluminum, $Al(C_2H_5)_3$, and titanium trichloride, $TiCl_3$, which promote the formation only of specific isomers. Their work opened up almost limitless possibilities for chemists to design and synthesize polymers to suit any purpose.

Rubber is the only true hydrocarbon polymer found in nature; it is probably the best known organic polymer. It is formed by the radical addition of the monomer isoprene:

$$CH_2{=}\overset{\overset{\displaystyle CH_3}{|}}{C}{-}CH{=}CH_2$$

Actually, polymerization can result in either poly-*cis*-isoprene or poly-*trans*-isoprene, or a mixture of both, depending on reaction conditions:

$$\left(\begin{array}{c} CH_3 \quad\quad H \\ \diagdown\quad\diagup \\ C{=}C \\ \diagup\quad\diagdown \\ {-}CH_2 \quad\quad CH_2{-} \end{array}\right)_n \qquad \left(\begin{array}{c} {-}CH_2 \quad\quad H \\ \diagdown\quad\diagup \\ C{=}C \\ \diagup\quad\diagdown \\ CH_3 \quad\quad CH_2{-} \end{array}\right)_n$$

poly-*cis*-isoprene   poly-*trans*-isoprene

† Giulio Natta (1903–1979). Italian chemist. Natta received the Nobel Prize in chemistry in 1963 for discovering stereospecific catalysts for polymer synthesis.

‡ Karl Ziegler (1898–1976). German chemist. Ziegler shared the Nobel Prize in chemistry in 1963 with Natta for his work in polymer synthesis.

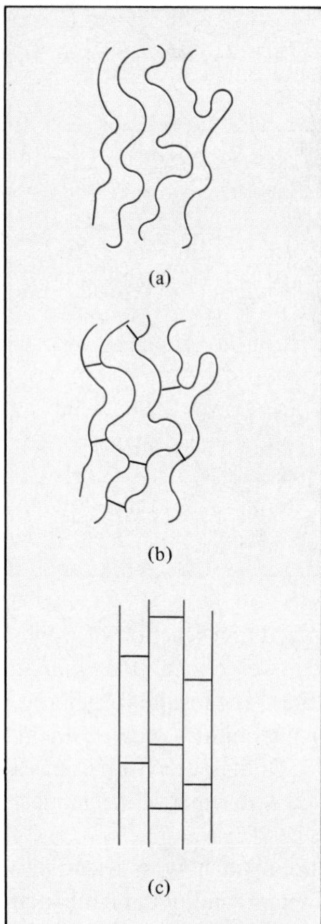

Figure 25.5 *Rubber molecules ordinarily are bent and convoluted. (a) and (b) represent the long chains before and after vulcanization, respectively; (c) shows the alignment of molecules when stretched. Without vulcanization these molecules would slip past one another.*

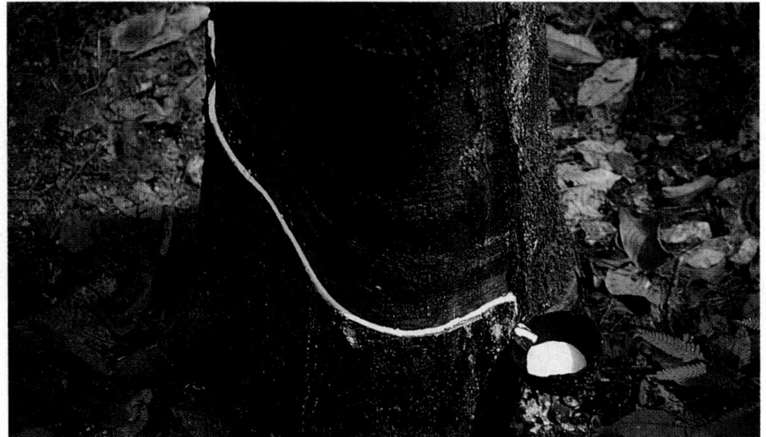

Figure 25.4 *Latex (aqueous suspension of rubber particles) being collected from a rubber tree.*

Note that in the cis isomer the two CH_2 groups are on the same side of the C=C bond, whereas the same groups are across from each other in the trans isomer. Natural rubber is poly-*cis*-isoprene, which is extracted from the tree *Hevea brasiliensis* (Figure 25.4). An unusual and very useful property of rubber is its elasticity. For example, rubber can be extended up to 10 times its length and return to its original size. In contrast, a piece of copper wire can be stretched only a small percentage of its length and still return to its original size. Unstretched rubber has no regular X-ray diffraction pattern and is therefore amorphous. Stretched rubber, however, possesses a fair amount of crystallinity and order.

The elastic property of rubber is due to the flexibility of the long-chain molecules. In the bulk state, however, rubber is a tangle of polymeric chains; and if the external force is strong enough, individual chains slip past one another, thereby causing the rubber to lose most of its elasticity. In 1839, Charles Goodyear† discovered that natural rubber could be crosslinked with sulfur (using zinc oxide as the catalyst) to prevent chain slippage (Figure 25.5). His process is known as *vulcanization*. This discovery paved the way for many practical and commercial uses of rubber, such as in automobile tires.

During World War II a shortage of natural rubber in the United States prompted an intensive program to produce synthetic rubber. Most synthetic rubbers (called *elastomers*) are formed from products of the petroleum industry such as ethylene, propylene, and butadiene. For example, chloroprene molecules polymerize readily to form polychloroprene, commonly known as *neoprene,* which has properties that are comparable or even superior to those of natural rubber:

$$\underset{\text{chloroprene}}{H_2C{=}CH{-}CCl{=}CH_2} \qquad \underset{\text{polychloroprene}}{\left(\begin{array}{ccc} -CH_2 & & H \\ & C{=}C & \\ Cl & & CH_2- \end{array} \right)_n}$$

Another important synthetic rubber is formed by the addition of butadiene to styrene

†Charles Goodyear (1800–1860). American chemist. Goodyear was the first person to realize the potential of natural rubber. His vulcanization process made rubber usable in countless ways and opened the way for the development of the automobile industry.

in a 3:1 ratio to give the styrene-butadiene rubber (SBR) (see Table 25.1). Because two different types of monomer are needed to synthesize SBR, this process is an example of copolymerization. ***Copolymerization*** is *the formation of a polymer containing two or more different monomers*.

Condensation Reactions

The monomers used in a condensation reaction are bifunctional; that is, they have a chemically reactive group on each end of their molecules. To link these monomers together it is necessary to remove a small molecule, usually water.

One of the best-known polycondensation processes is the reaction between hexamethylenediamine and adipic acid, shown in Figure 25.6. The final product, called nylon 66, was first made by Wallace Carothers† at Du Pont in 1931. (The word *nylon* was coined for synthetic polyamides. The nylons are described by a numbering system that indicates the number of carbon atoms in the monomer units. As there are six carbon atoms each in hexamethylenediamine and adipic acid, it was logical to name the product nylon 66.) The versatility of nylons is so great that the annual production of nylons and related substances now amounts to several billion pounds. Figure 25.7 shows how nylon 66 is prepared in the laboratory.

Condensation reactions are also used to produce polymers such as Dacron (polyester)

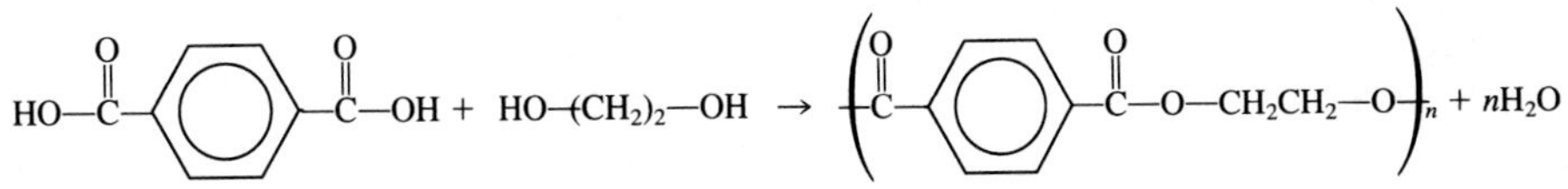

$$HO-\overset{O}{\overset{\|}{C}}-C_6H_4-\overset{O}{\overset{\|}{C}}-OH + HO{-}(CH_2)_2{-}OH \rightarrow \left(\overset{O}{\overset{\|}{C}}-C_6H_4-\overset{O}{\overset{\|}{C}}-O-CH_2CH_2-O\right)_n + nH_2O$$

Terephthalic acid 1,2-Ethylene glycol

Figure 25.6 *The formation of nylon by the condensation reaction between hexamethylenediamine and adipic acid.*

$$H_2N-(CH_2)_6-NH_2 + HOOC-(CH_2)_4-COOH$$

Hexamethylenediamine Adipic acid

↓ Condensation

$$H_2N-(CH_2)_6-\underset{H}{\underset{|}{N}}-\overset{O}{\overset{\|}{C}}-(CH_2)_4-COOH + H_2O$$

↓ Further condensation reactions

$$-(CH_2)_4-\overset{O}{\overset{\|}{C}}-\underset{H}{\underset{|}{N}}-(CH_2)_6-\underset{H}{\underset{|}{N}}-\overset{O}{\overset{\|}{C}}-(CH_2)_4-\overset{O}{\overset{\|}{C}}-\underset{H}{\underset{|}{N}}-(CH_2)_6-$$

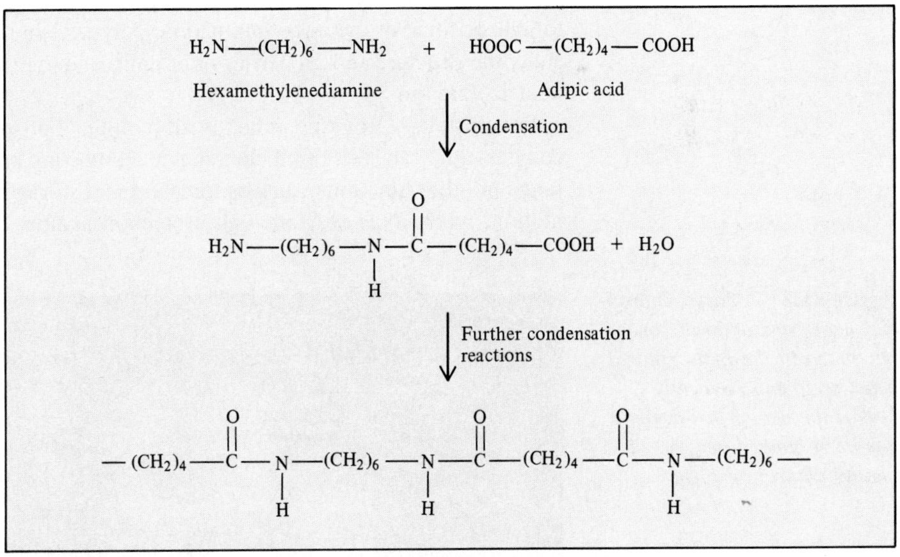

†Wallace H. Carothers (1896–1937). American chemist. Besides its enormous commercial success, Carothers' work on nylon is ranked with that of Staudinger in clearly elucidating macromolecular structure and properties. Depressed by the death of his sister and convinced that his life's work was a failure, Carothers committed suicide at the age of 41.

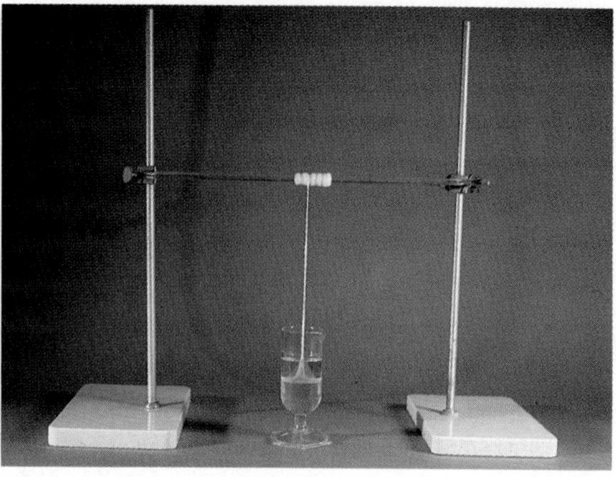

Figure 25.7 *The nylon rope trick. Adding a solution of adipoyl chloride (an adipic acid derivative in which the OH groups have been replaced by Cl groups) in cyclohexane to an aqueous solution of hexamethylenediamine causes nylon to form at the interfacial layer. It can then be drawn off.*

and polyurethane

$$\underset{\text{1,2-ethylene diisocyanate}}{OCN{-}(CH_2)_2{-}NCO} + HO{-}(CH_2)_2{-}OH \longrightarrow \underset{\text{1,2-ethylene glycol}}{\left(\overset{\overset{\displaystyle O}{\|}}{C}{-}NH{-}(CH_2)_2{-}NH{-}\overset{\overset{\displaystyle O}{\|}}{C}{-}O{-}(CH_2)_2{-}O\right)_n}$$

The first artificial heart (known as the Jarvik-7) implanted in a human (1982) was made of polyurethane.

Polyesters are used in fibers, films, and plastic bottles. Polyurethane is used in coatings, insulators, and adhesives. Figure 25.8 shows an artificial heart and artificial artery sections made of polyurethane supported on an aluminum base.

25.3 Proteins

Having discussed the synthetic organic polymers and natural rubber, we are now ready to look at some of the important natural polymers in living systems. In this section we study the structure and properties of proteins and then go on to discuss the properties of nucleic acids in Section 25.4.

Proteins play a key role in nearly all biological processes. As we saw in Chapter 13, enzymes, the catalysts of biochemical reactions, are proteins. Proteins facilitate a wide range of other functions, such as transport and storage of vital substances, coordinated motion, mechanical support, and protection against diseases.

Figure 25.8 *A Jarvik-7 artificial heart similar to the ones implanted in the first artificial heart recipients. Recently (1990) the use of this design has been banned because of quality control problems.*

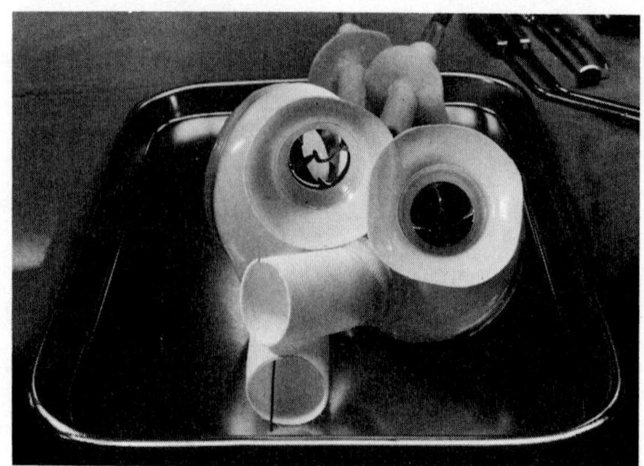

The basic structural units of proteins are *amino acids*. An amino acid is a compound that contains at least one amino group ($-NH_2$) and at least one carboxyl group ($-COOH$):

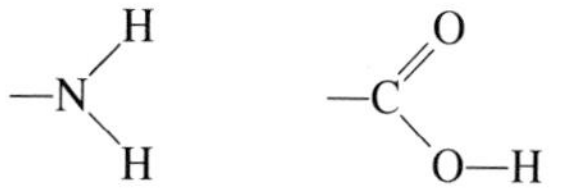

amino group carboxyl group

The simplest amino acid is glycine

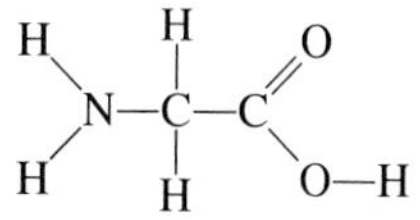

which is usually written as NH_2CH_2COOH. Twenty different amino acids are the building blocks for the tens of thousands of different proteins in the human body (Table 25.2).

Proteins have high molar masses, ranging from about 5000 g to 1×10^7 g. The mass composition of proteins by elements is remarkably constant: carbon, 50 to 55 percent; hydrogen, 7 percent; oxygen, 23 percent; nitrogen, 16 percent; and sulfur, 1 percent.

Proteins can be divided into two classes: simple proteins and conjugated proteins. ***Simple proteins*** *contain only amino acids.* ***Conjugated proteins,*** on the other hand, *may contain a prosthetic group in addition to the amino acids.* (A ***prosthetic group*** is *part of the protein structure that is not made up of amino acid components;* an example is the heme group in hemoglobin, discussed in Section 22.7.) Conjugated proteins may also be complexed with carbohydrates (to give *glycoproteins*), with nucleic acids (to give *nucleoproteins*), and with lipids (to give *lipoproteins*). In this section we concentrate on the simple proteins.

It is interesting to compare this reaction with that shown in Figure 25.6.

The first step in the synthesis of a protein molecule is a condensation reaction between two amino acids, as shown in Figure 25.9. The molecule formed from two amino acids is called a *dipeptide,* and the $-CO-NH-$ group is called the *amide* group. Either end of the dipeptide can engage in a condensation reaction with another

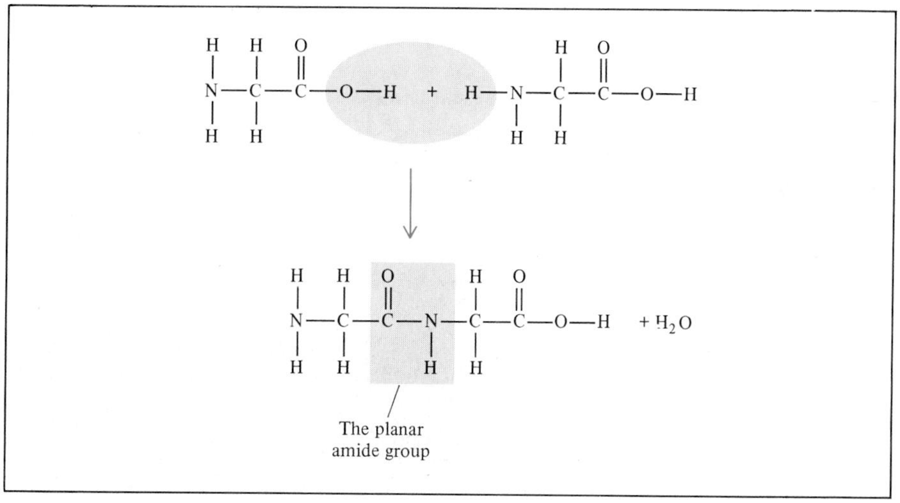

$+ H_2O$

The planar amide group

Figure 25.9 *Condensation reaction between two amino acid molecules (glycine). An amino acid contains at least one amino group (NH_2) and one carboxyl group ($COOH$). The reaction joins the two molecules together, with the elimination of one water molecule. The resultant bond is called a peptide bond.*

Table 25.2 The 20 amino acids essential to living organisms

Name	Abbreviation	Structure
Alanine	Ala	$H_3C-C(H)(NH_2)-COOH$
Arginine	Arg	$H_2N-C(=NH)-N(H)-CH_2-CH_2-CH_2-C(H)(NH_2)-COOH$
Asparagine	Asn	$H_2N-C(=O)-CH_2-C(H)(NH_2)-COOH$
Aspartic acid	Asp	$HOOC-CH_2-C(H)(NH_2)-COOH$
Cysteine	Cys	$HS-CH_2-C(H)(NH_2)-COOH$
Glutamic acid	Glu	$HOOC-CH_2-CH_2-C(H)(NH_2)-COOH$
Glutamine	Gln	$H_2N-C(=O)-CH_2-CH_2-C(H)(NH_2)-COOH$
Glycine	Gly	$H-C(H)(NH_2)-COOH$
Histidine	His	$HC{=}C-CH_2-C(H)(NH_2)-COOH$ (ring: $HC{=}C$, N, NH, $C(H)$; $N{=}C(H)-NH$)
Isoleucine	Ile	$H_3C-CH_2-C(CH_3)(H)-C(H)(NH_2)-COOH$

Table 25.2 The 20 amino acids essential to living organisms (*continued*)

Name	Abbreviation	Structure
Leucine	Leu	$(H_3C)_2CH{-}CH_2{-}CH(NH_2){-}COOH$
Lysine	Lys	$H_2N{-}CH_2{-}CH_2{-}CH_2{-}CH_2{-}CH(NH_2){-}COOH$
Methionine	Met	$H_3C{-}S{-}CH_2{-}CH_2{-}CH(NH_2){-}COOH$
Phenylalanine	Phe	$C_6H_5{-}CH_2{-}CH(NH_2){-}COOH$
Proline	Pro	HN—C(H)—COOH ring closed through H_2C, CH_2, CH_2
Serine	Ser	$HO{-}CH_2{-}CH(NH_2){-}COOH$
Threonine	Thr	$H_3C{-}CH(OH){-}CH(NH_2){-}COOH$
Tryptophan	Trp	indole (N—H) C—$CH_2{-}CH(NH_2){-}COOH$, with =CH
Tyrosine	Tyr	$HO{-}C_6H_4{-}CH_2{-}CH(NH_2){-}COOH$
Valine	Val	$(H_3C)_2CH{-}CH(NH_2){-}COOH$

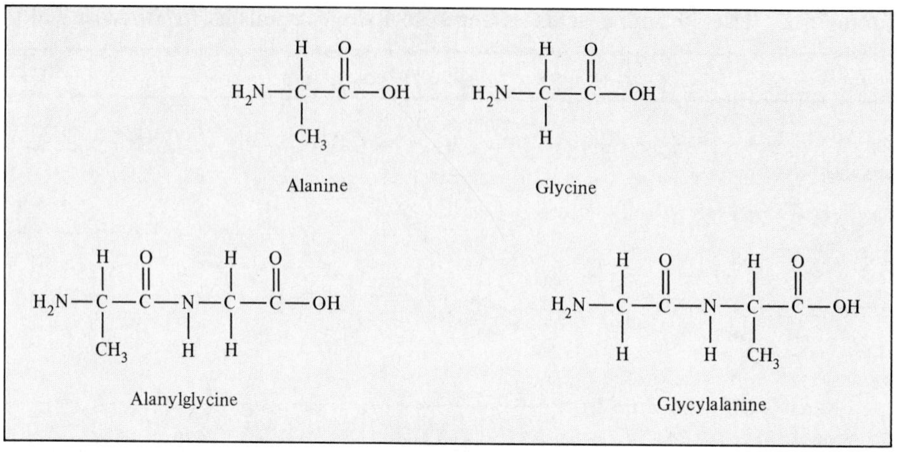

Alanine

Glycine

Alanylglycine

Glycylalanine

Figure 25.10 *The formation of two dipeptides from two different amino acids. Alanylglycine is different from glycylalanine in that in alanylglycine the amino and methyl groups are bonded to the same carbon atom.*

amino acid to form a *tripeptide,* and so on. The final product, the protein molecule, is a *polypeptide;* it can also be thought of as a polymer of amino acids.

An amino acid unit in a polypeptide chain is called a *residue*. Typically, a polypeptide chain contains 100 or more amino acid residues. The sequence of amino acids in a polypeptide chain is written conventionally from left to right, starting with the amino-terminal residue and ending with the carboxyl-terminal residue. Let us consider a dipeptide formed from glycine and alanine. Figure 25.10 shows that alanylglycine and glycylalanine are different molecules. With 20 different amino acids to choose from, 20^2, or 400, different dipeptides can be generated. Even for a very small protein such as insulin, which contains only 50 amino acid residues, the number of chemically different structures that is possible is of the order of 20^{50} or 10^{65}! This is an incredibly large number when you consider that the total number of atoms in our galaxy is about 10^{68}.

It is estimated that the human body contains about 100,000 different kinds of protein molecules, a large number by ordinary standards, but very small compared with 10^{65}. With so many possibilities for protein synthesis, it is remarkable that generation after generation of cells can produce identical proteins for specific physiological functions.

To understand the properties of a protein molecule, we must know its structure. That leads us to ask two questions: What is the amino acid sequence in the polypeptide chain? What is the overall shape of the molecule?

Under suitable conditions, proteins can be hydrolyzed to yield their amino acid components. The type and number of amino acids present in the original protein can be determined readily by a number of techniques. But the task of determining the sequence or order in which these amino acids are joined together is more involved. Advances in instrumentation have resulted in an amino acid ''sequenator'' capable of cleaving amino acid residues from a polypeptide chain one at a time for analysis. Consequently, the sequence of amino acids in the polypeptide chain can be determined much more easily than when biochemistry was a young science.

Once the amino acid sequence is known, we can move on to the question about the shape of the protein molecule. Is the polypeptide chain stretched out like a raw noodle, or is it folded in a particular pattern? In the 1930s Linus Pauling and his coworkers conducted a systematic investigation of protein structure. First they studied the geometry of the basic repeating group, that is, the amide group. The structure and bonding of the amide group was discussed in Chapter 10 (see p. 434). Because it is more difficult

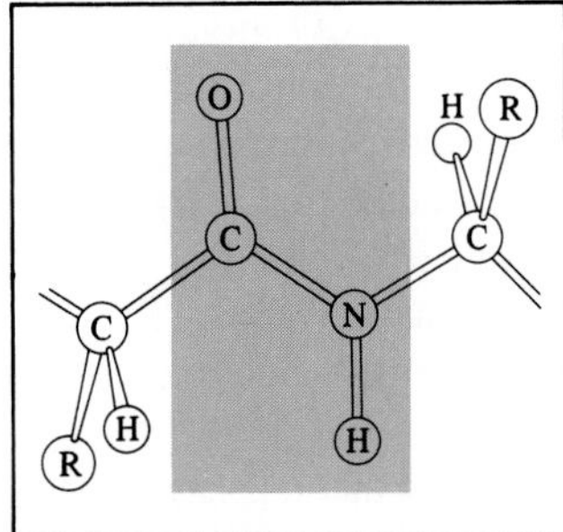

Figure 25.11 *The planar amide group in protein. Rotation about the C—N bond in the amide group is hindered.*

(that is, it would take more energy) to twist a double bond than a single bond, the four atoms in the amide group become locked in the same plane (Figure 25.11). Figure 25.12 depicts the repeating amide group in a polypeptide chain.

On the basis of models and X-ray diffraction data, Pauling suggested that there are two common structures for protein molecules, called *α helix* and *β-pleated sheet*. The α-helical structure of a polypeptide chain is shown in Figure 25.13. The helix is

```
   H   O       H   O       H   O       H   O
   |   ||      |   ||      |   ||      |   ||
 —C——C——N——C——C——N——C——C——N——C——C——N—
   |       |   |       |   |       |   |       |
   R       H   R       H   R       H   H       H
```

Figure 25.12 *A polypeptide chain. Note the repeating units of the amide group. The symbol R represents part of the structure characteristic of the individual amino acids. For glycine, R is simply a H atom.*

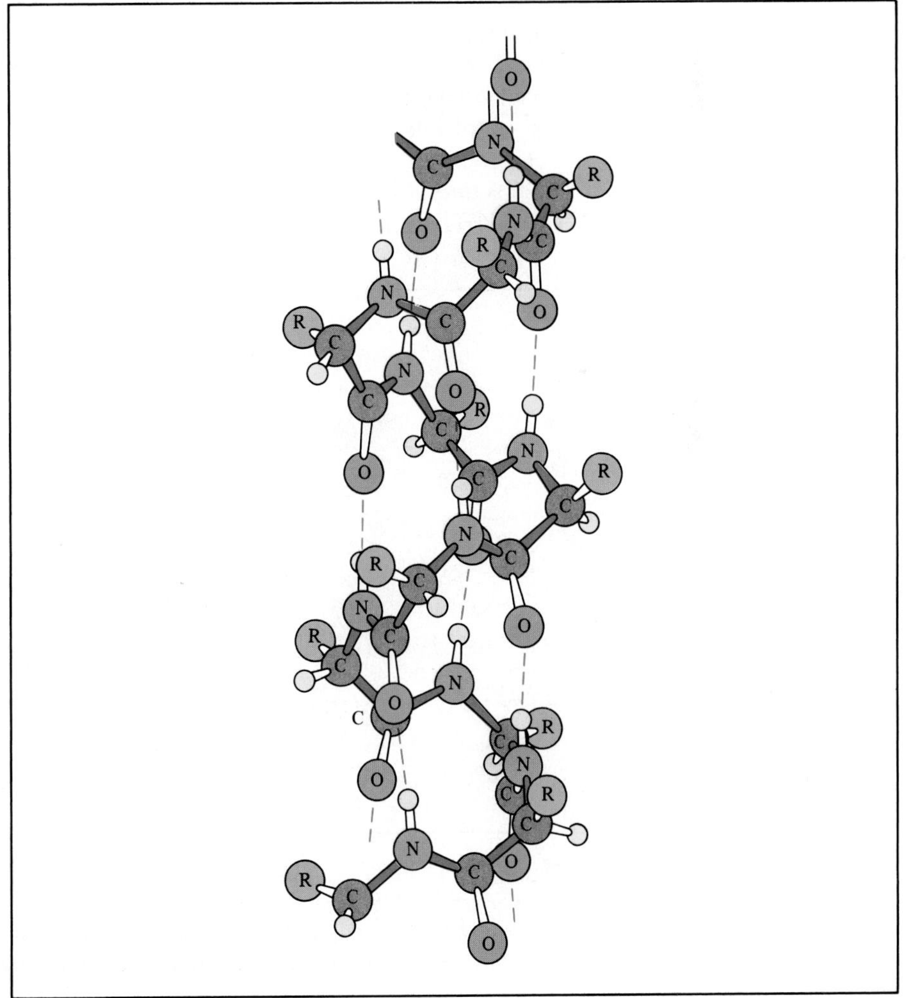

Figure 25.13 *The α-helical structure of a polypeptide chain. The yellow spheres are hydrogen atoms. The structure is held in position by intramolecular hydrogen bonds, shown as dashed lines.*

stabilized by *intramolecular* hydrogen bonds between the NH and CO groups of the main chain, giving rise to an overall rodlike shape. The CO group of each amino acid is hydrogen-bonded to the NH group of the amino acid that is four residues away in the sequence. In this manner all the main-chain CO and NH groups take part in hydrogen bonding. X-ray studies have shown that the structure of a number of proteins, including myoglobin and hemoglobin, is to a great extent α-helical in nature (Figure 25.14). Other proteins, such as the digestive enzyme chymotrypsin and the electron carrier cytochrome *c*, are almost devoid of α-helical structure.

The β-pleated structure is markedly different from the α helix in that it is like a sheet rather than a rod. The polypeptide chain is almost fully extended, and each chain forms many *intermolecular* hydrogen bonds with adjacent chains. Figure 25.15 shows the two different types of β-pleated structures, called *parallel* and *antiparallel*. Silk molecules possess the β structure. Because its polypeptide chains are already in extended form, silk lacks elasticity and extensibility, but it is quite strong due to the intermolecular hydrogen bonds.

The structure of a protein is of great importance, for it is ultimately responsible for the protein's activities: catalytic action, binding of oxygen and other molecules, and so on. Consider the enzyme chymotrypsin. Chymotrypsin belongs to a class of enzymes called *digestive enzymes*. Without these enzymes it would take about 50 years to digest a meal. We owe our lives to their efficiency. If the polypeptide chain of chymotrypsin were to fold in a different way, the enzyme would not be able to catalyze the degradation of food molecules.

The work of Pauling's group was a great triumph in protein chemistry. It showed for the first time how to predict a protein structure purely from a knowledge of the geometry of its fundamental building blocks—amino acids. However, there are many pro-

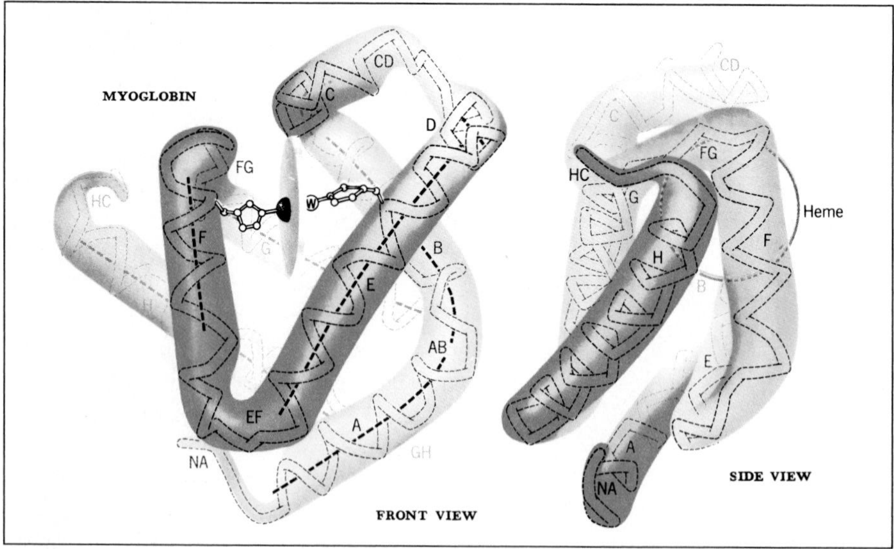

Figure 25.14 *Front and side views of myoglobin. Each myoglobin molecule contains a heme group at whose center is an Fe atom (the black half-sphere in the left diagram). The Fe atom binds to an O_2 molecule or, in its absence, to a water molecule (denoted by a W). The α-helical structure in myoglobin is quite evident. Other letters denote regions of the helical structure.*

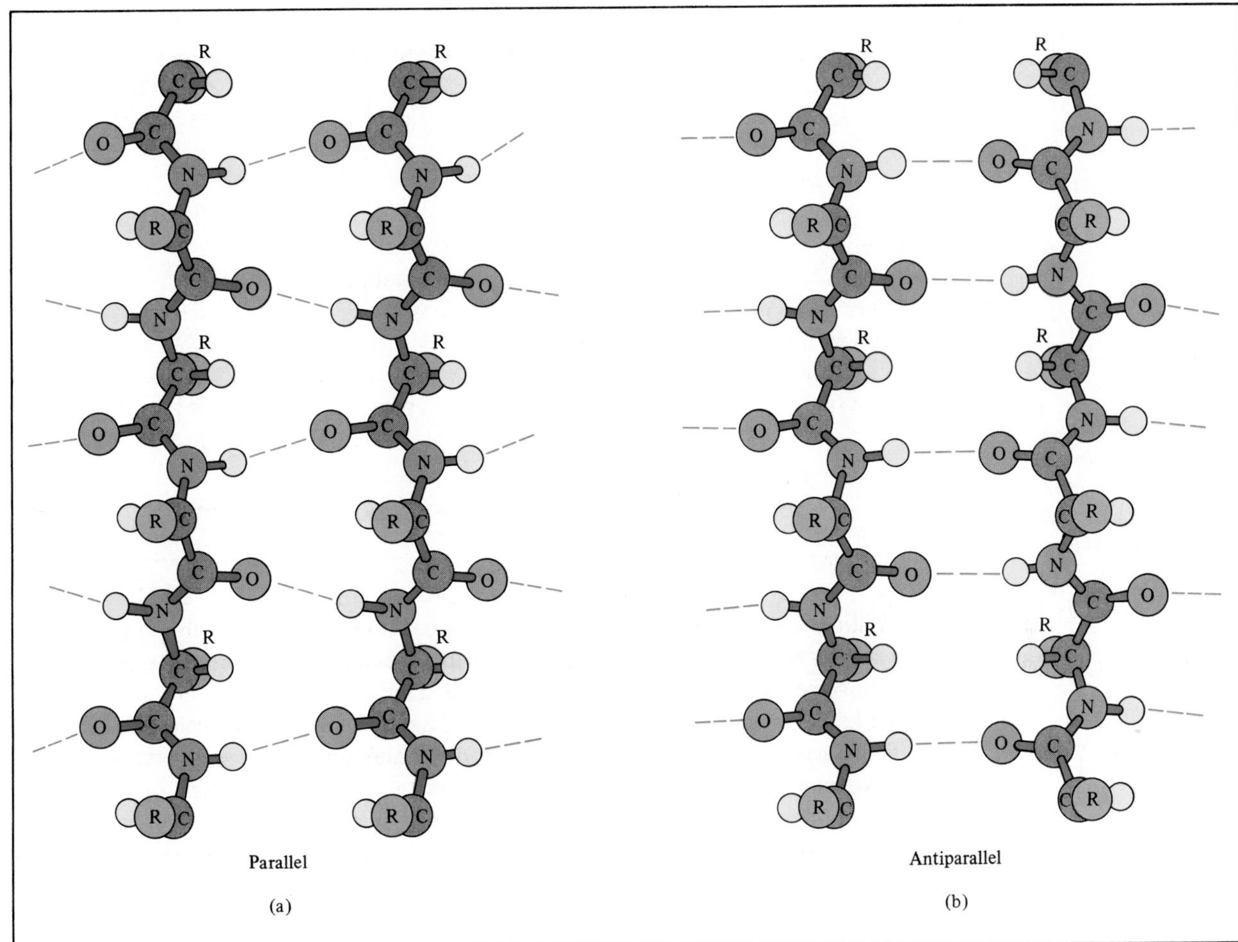

Figure 25.15 *Hydrogen bonds (a) in parallel β-pleated sheet structure, in which all the polypeptide chains run in the same direction; and (b) in antiparallel β-pleated sheet, in which adjacent polypeptide chains run in opposite directions.*

teins whose structures do not correspond to the α-helical or β structure. Chemists now realize that the three-dimensional structures of these biopolymers are maintained by several types of intermolecular forces in addition to hydrogen bonding. These forces include van der Waals forces (see Chapter 11) and other intermolecular forces (Figure 25.16). The delicate balance of the various interactions can be appreciated by considering an example: When glutamic acid, one of the amino acid residues in two of the four polypeptide chains in hemoglobin, is replaced by valine, another amino acid, the protein molecules aggregate to form insoluble polymers, causing the disease known as sickle cell anemia (see the Chemistry in Action on p. 1041).

It is customary to divide protein structure into four levels of organization. The *primary structure* refers to the unique amino acid sequence of the polypeptide chain. The *secondary structure* refers to those parts of the polypeptide chain that are stabilized by a regular pattern of hydrogen bonds involving the CO and NH groups of the backbone, for example, the α helix. The term *tertiary structure* applies to the three-dimensional structure stabilized by dispersion forces, hydrogen bonding, and other intermolecular forces. It differs from secondary structure in that the amino acids taking part in these interactions may be far apart in the linear sequence.

A protein molecule may be made up of more than one polypeptide chain. Thus, in

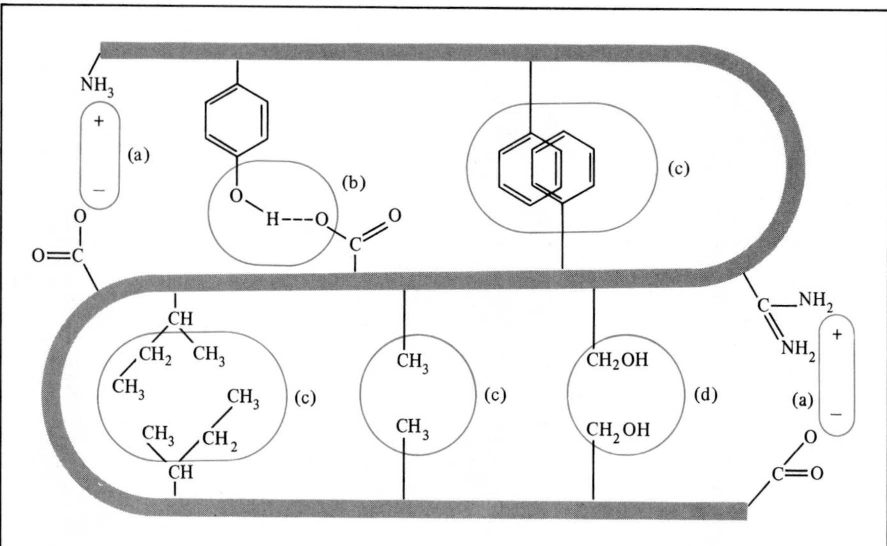

Figure 25.16 *Intermolecular forces in a protein molecule: (a) ionic forces, (b) hydrogen bonding, (c) dispersion forces, and (d) dipole–dipole forces.*

addition to the various interactions *within* a chain that give rise to the secondary and tertiary structures, we must also consider the interaction *between* chains. The overall arrangement of the polypeptide chains is called the *quaternary structure*. For example, the hemoglobin molecule consists of four separate polypeptide chains, or *subunits*. These subunits are held together by van der Waals forces and ionic forces.

When we consider protein structure, we must not think of a protein as a rigid entity. Most proteins have a certain amount of flexibility. Enzymes, for example, are flexible enough to change their geometry to fit substrates of various sizes and shapes. Another interesting example of protein flexibility is found in the binding of hemoglobin to oxygen. Each of the four polypeptide chains in hemoglobin contains a heme group that can bind to an oxygen molecule (see Section 22.7). In deoxyhemoglobin, the affinity of each of the heme groups for oxygen is about the same. However, as soon as one of the heme groups becomes oxygenated, the affinity of the other three hemes for oxygen is greatly enhanced, and so on. This phenomenon, called *cooperativity,* makes hemoglobin a particularly suitable substance for the uptake of oxygen in the lungs. By the same token, once a fully oxygenated hemoglobin releases an oxygen molecule (to myoglobin in the tissues), the other three oxygen molecules will depart with increasing ease. The cooperative nature of the binding is such that information about the presence (or absence) of oxygen molecules is transmitted from one subunit to another along the polypeptide chains, a process made possible by the flexibility of the three-dimensional structure (Figure 25.17). It is believed that the Fe^{2+} ion has too large a radius to fit into the porphyrin ring of deoxyhemoglobin. When O_2 binds to Fe^{2+}, however, the ion shrinks somewhat so that it can fit into the plane of the ring. As the ion slips into the ring, it pulls the histidine residue toward the ring and thereby sets off a sequence of structural changes from one subunit to another. Although the details of the changes are not clear, biochemists believe that this is how the bonding of an oxygen molecule to one heme group affects another heme group. The structural changes drastically affect the affinity of the remaining heme groups for oxygen molecules.

When proteins are heated above body temperature or when they are subjected to unusual acid or base conditions or treated with special reagents called *denaturants,*

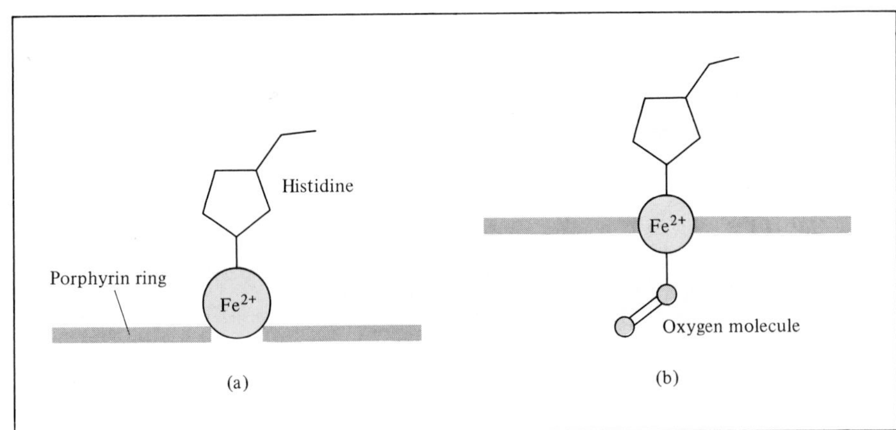

Figure 25.17 *The structural changes that occur when the heme group in hemoglobin binds an oxygen molecule. (a) The heme group in deoxyhemoglobin. (b) Oxyhemoglobin.*

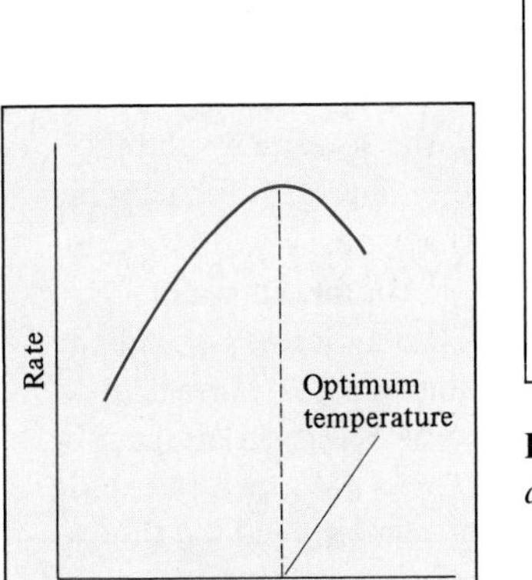

Figure 25.18 *Dependence of the rate of an enzyme-catalyzed reaction on temperature. An enzyme has an optimum temperature at which it is most effective. Above this temperature its activity drops off.*

they lose some or all of their tertiary and secondary structure. *Proteins in this state no longer exhibit normal biological activities* and are said to be ***denatured proteins.*** Figure 25.18 shows the variation of rate with temperature for a typical enzyme-catalyzed reaction. Initially, the rate increases with increasing temperature, as we would expect. Beyond the optimum temperature, however, the enzyme begins to denature and the rate falls. If a protein is denatured under mild conditions, its original structure can often be regenerated by removing the denaturant or by restoring the temperature to normal conditions. This process is called reversible denaturation.

CHEMISTRY IN ACTION

Sickle Cell Anemia—A Molecular Disease

Sickle cell anemia is caused by a hereditary defect in hemoglobin, the oxygen-carrying protein in red blood cells. The hemoglobin molecule is large (molar mass about 65,000 g). In normal human hemoglobin, or HbA, there are two α chains consisting of 141 amino acids each and two β chains consisting of 146 amino acids each (Figure 25.19). These four polypeptide chains, or subunits, are held together by ionic and van der Waals forces.

The functions of a protein depend on its three-dimensional structure, which, in turn, depends on the various amino acids that make up the polypeptide chain. When a polypeptide chain folds, amino acids that were originally far apart along the chain may be brought into proximity with one another. The final overall shape of the protein molecule must be influenced by the organization of these amino acids relative to one another and, hence, by the makeup of the chain.

There are many mutant hemoglobin molecules—molecules with an amino acid sequence that differs somewhat from the sequence in HbA. Most of the mutant hemoglobins are harmless, but a few are known to cause serious diseases. Sickle cell hemoglobin, or HbS, belongs to the latter category.

HbS differs from HbA in only one very small detail. A valine molecule replaces a glutamic acid molecule on

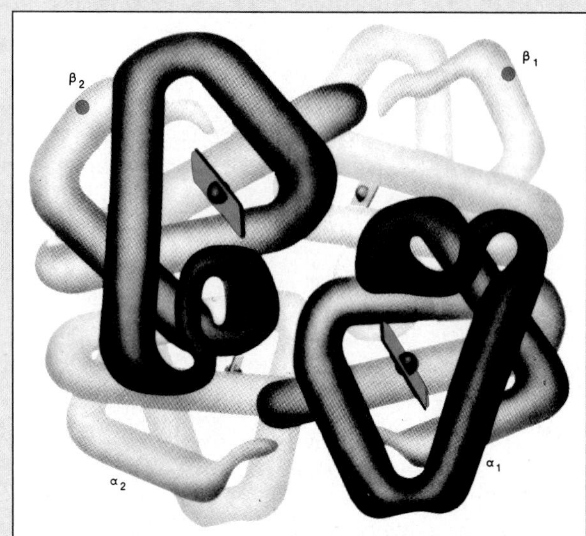

Figure 25.19 *The overall structure of hemoglobin. Each hemoglobin molecule contains two α chains and two β chains. Each of the four chains is similar to a myoglobin molecule in structure, and each also contains a heme group for binding oxygen. In sickle cell hemoglobin, the defective regions (the valine groups) are located near the ends of the β chains, as indicated by the dots.*

each of the two β chains:

$$\mathrm{HOOC{-}CH_2{-}CH_2{-}\underset{\underset{\displaystyle NH_2}{|}}{\overset{\overset{\displaystyle H}{|}}{C}}{-}COOH}$$

glutamic acid

$$\mathrm{(H_3C)_2CH{-}\underset{\underset{\displaystyle NH_2}{|}}{\overset{\overset{\displaystyle H}{|}}{C}}{-}COOH}$$

valine

Yet this small change (one amino acid out of 146) has a profound effect on the stability of HbS in solution. The valine groups are located at the bottom outside of the molecule to form a protruding "key" on each of the β chains. The nonpolar portion of valine

$$\mathrm{(H_3C)_2CH{-}}$$

can attract another nonpolar group in the α chain of an adjacent HbS molecule through dispersion forces. Biochemists often refer to *this kind of attraction between nonpolar groups* as ***hydrophobic*** (*"water-fearing"*) *interaction.* (***Hydrophilic*** *interaction occurs between water and other polar substances.*) Gradually, enough HbS molecules will aggregate to form a "superpolymer."

A general rule about the solubility of a substance is that the larger its molecules, the lower its solubility, because the solvation process becomes unfavorable with increasing molecular surface area. For this reason, proteins generally are not very soluble in water.

Eventually the aggregated HbS molecules precipitate out of solution. The precipitate causes normal disk-shaped red blood cells to assume a warped crescent or sickle shape (Figure 25.20). These deformed cells clog the narrow capillaries, thereby restricting blood flow to organs of the body. The usual symptoms of sickle cell anemia are swelling, severe pain, and other complications. Sickle cell anemia has been termed a molecular disease by Linus Pauling, who did some of the early important chemical research on the nature of the affliction, because the destructive action occurs at the molecular level and the disease is, in effect, due to a molecular defect.

Some substances, such as urea and the cyanate ion,

$$\mathrm{H_2N{-}\underset{\underset{\displaystyle O}{\|}}{C}{-}NH_2} \qquad \mathrm{O{=}C{=}N^-}$$

urea — cyanate ion

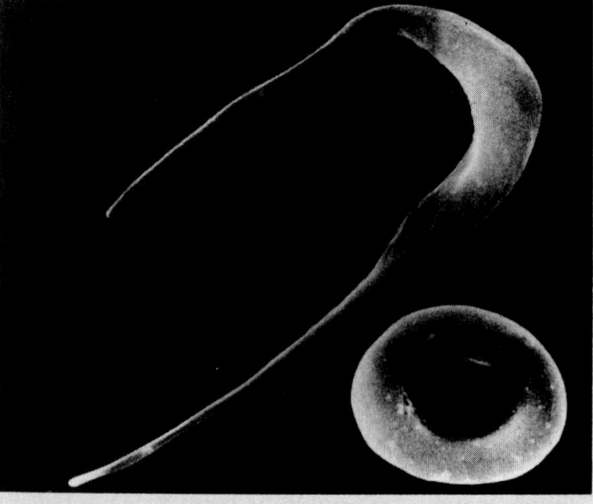

Figure 25.20 *Electron micrograph showing a normal red blood cell and a sickled red blood cell from the same person.*

can break up the hydrophobic interaction between HbS molecules and have been applied with some success to reverse the "sickling" of red blood cells. Nevertheless, despite intensive research efforts, no cure for sickle cell anemia is yet known.

25.4 Nucleic Acids

Nucleic acids are high molar mass polymers that play an essential role in protein synthesis. There are two types of nucleic acid: deoxyribonucleic acids (DNA) and ribonucleic acid (RNA). In this section we discuss their composition and structure.

DNA molecules are among the largest molecules known; they have molar masses of up to tens of billions of grams. On the other hand, RNA molecules vary greatly in size, some having a molar mass of about 25,000 g. Compared with proteins, which are made of up to 20 different amino acids, the composition of nucleic acids is considerably simpler. A DNA or RNA molecule contains only four types of building blocks: purines, pyrimidines, furanose sugars, and phosphate groups (Figure 25.21). Each purine or pyrimidine is called a *base*.

Despite their comparatively simple composition, the structures of nucleic acids were not known for many years. In the 1940s, Erwin Chargaff† studied DNA molecules obtained from various sources and noticed the following regularities in base composition:

1. The amount of adenine is equal to that of thymine; that is, A = T, or A/T = 1.
2. The amount of cytosine is equal to that of guanine; that is, C = G, or C/G = 1.
3. The total number of purine bases is equal to the total number of pyrimidine bases; that is, A + G = C + T.

An electron micrograph of a DNA molecule obtained in 1989. The double helical structure is evident.

These observations are now known as *Chargaff's rules*. Based on chemical analyses and information obtained from X-ray diffraction measurements, James Watson‡ and Francis Crick§ formulated the double-helical structure for the DNA molecule in 1953. Watson and Crick determined that the DNA molecule is made up of two helical strands. *The repeating unit in each strand* is called a ***nucleotide,*** *which consists of a base-deoxyribose-phosphate linkage* (Figure 25.22).

The key to the double-helical structure is the formation of hydrogen bonds between bases in the two strands. Although hydrogen bonds can form between any two bases, called *base pairs,* Watson and Crick found that the most favorable couplings are between adenine and thymine and between cytosine and guanine (Figure 25.23). Note that this scheme is consistent with Chargaff's rules, because every purine base is hydrogen-bonded to a pyrimidine base, and vice versa (A + G = C + T). Other attractive forces such as dipole-dipole and van der Waals forces between the base pairs also help to stabilize the double helix.

†Erwin Chargaff (1905–). American biochemist. Chargaff was the first to show that different biological species contain different DNA molecules.

‡James Dewey Watson (1928–). American biologist. Watson shared the 1962 Nobel Prize in medicine and physiology with Crick and Wilkins for their work on the DNA structure, which is considered by many to be the most significant development in biology of the twentieth century.

§Francis Harry Compton Crick (1916–). British biologist. Crick started as a physicist but became interested in biology after reading the book *What Is Life?* by Erwin Schrödinger (see Chapter 7). In addition to elucidating the structure of DNA, for which he was a corecipient of the Nobel Prize in medicine and physiology in 1962, Crick has made many significant contributions to molecular biology.

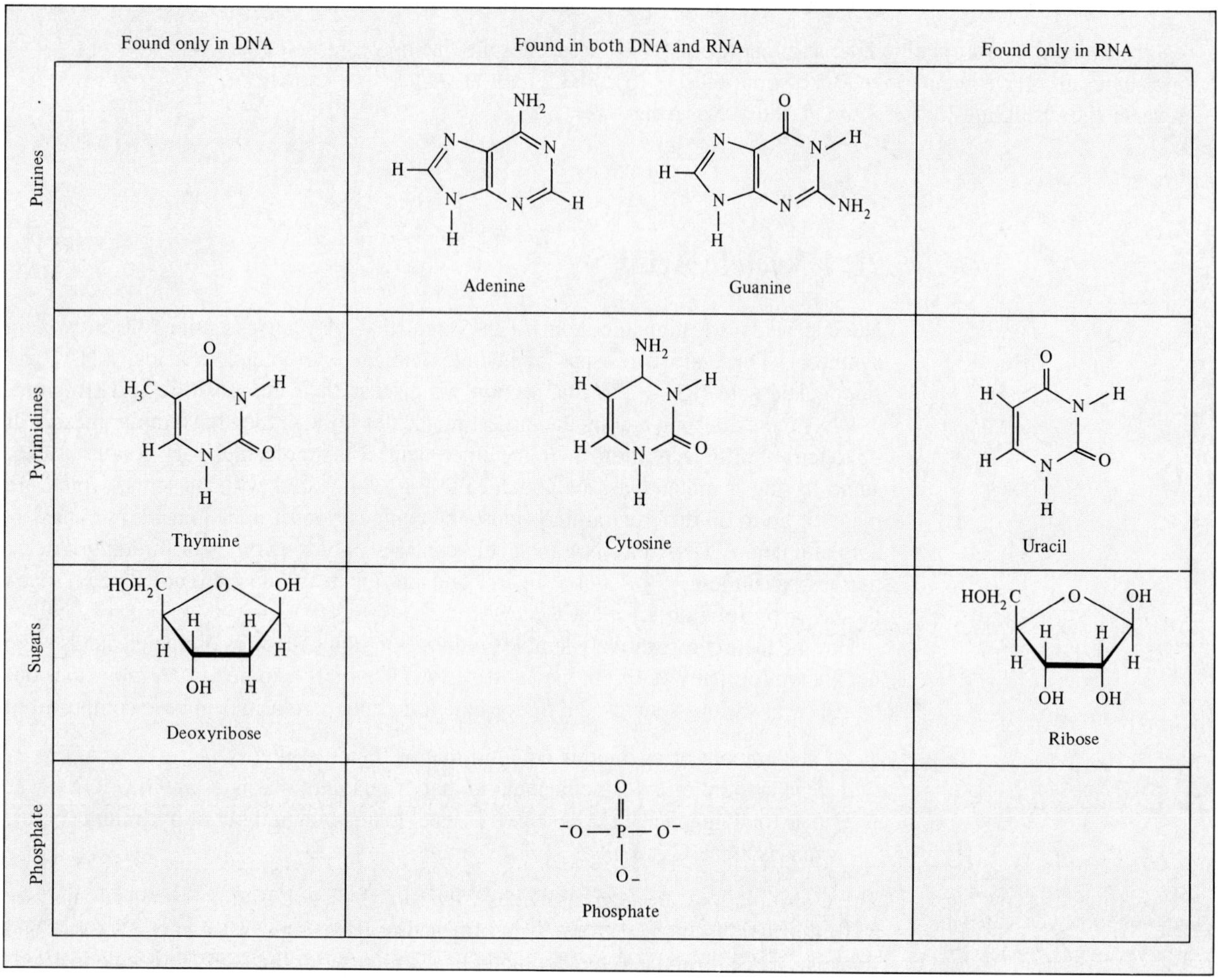

Figure 25.21 *The components of the nucleic acids DNA and RNA.*

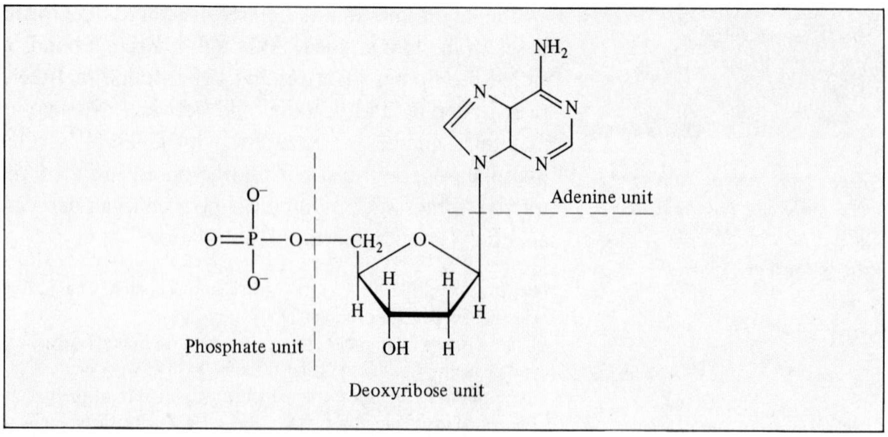

Figure 25.22 *Structure of a nucleotide, one of the repeating units in DNA.*

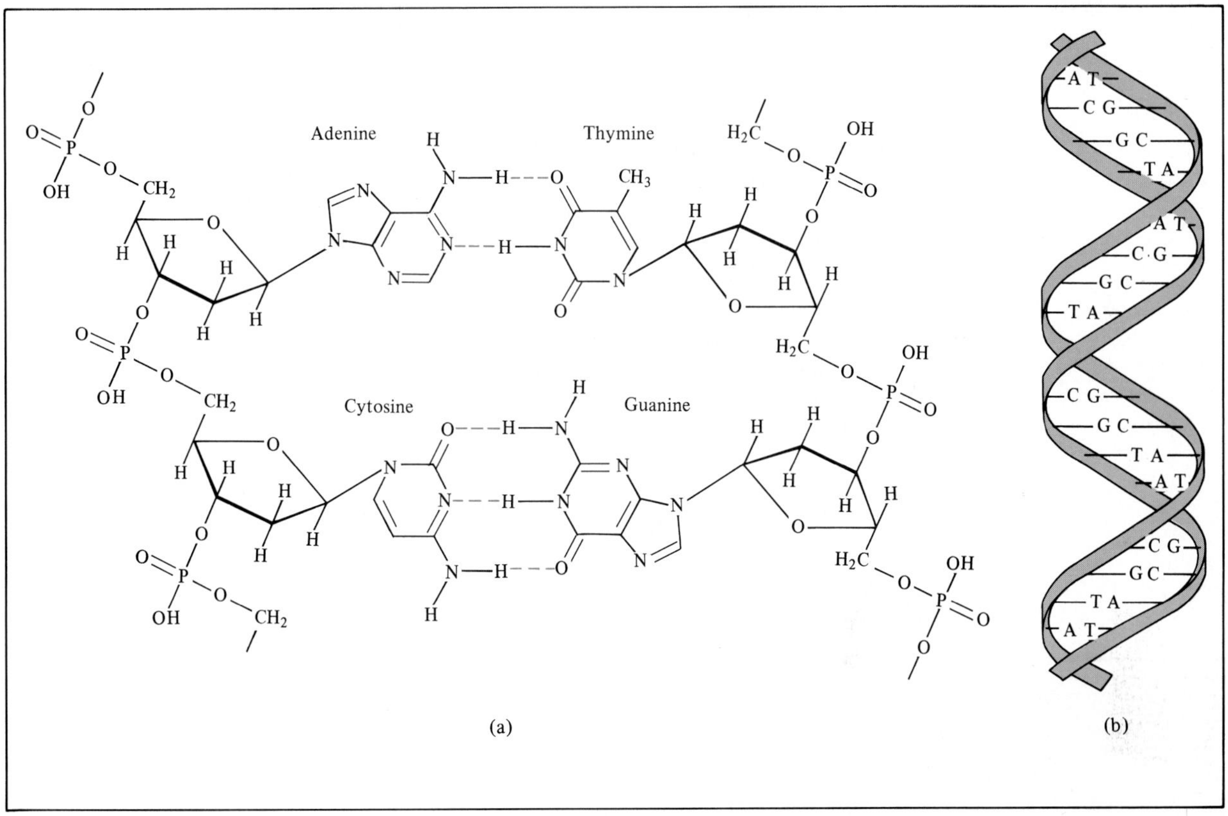

Figure 25.23 *(a) Base-pair formation by adenine and thymine and by cytosine and guanine. (b) The double-helical strand of a DNA molecule held together by hydrogen bonds (and other intermolecular forces) between base pairs A-T and C-G.*

The structure of RNA differs from that of DNA in several respects. First, as shown in Figure 25.21, the four bases found in RNA molecules are adenine, cytosine, guanine, and uracil. Second, RNA contains the sugar ribose rather than the 2-deoxyribose of DNA. Third, chemical analysis shows that the composition of RNA does not obey Chargaff's rules. In other words, the purine-to-pyrimidine ratio is not equal to 1 as in the case of DNA. This and other evidence rule out a double-helical structure. In fact, the RNA molecule exists as a single-strand polynucleotide. There are actually three types of RNA molecules, called *messenger* RNA (*m*RNA), *ribosomal* RNA (*r*RNA), and *transfer* RNA (*t*RNA). These RNAs have similar nucleotides but differ from one another in molar mass, overall structure, and biological functions.

DNA and RNA molecules play a key role in the synthesis of proteins in the cell, a subject that is beyond the scope of this book. Interested students should consult introductory texts in biochemistry and molecular biology.

Summary

1. Polymers are large molecules made up of small, repeating units called monomers.
2. Proteins, nucleic acids, cellulose, and rubber are natural polymers. Nylon, Dacron, and Lucite are examples of synthetic polymers.
3. Organic polymers can be synthesized via addition reactions or condensation reactions.

4. By varying the structure of a polymer from atactic to isotactic to syndiotactic, a wide variety of properties can be imparted to the polymer.
5. Synthetic rubbers include polychloroprene and styrene-butadiene rubber, which is made by the copolymerization of styrene and butadiene.
6. Simple proteins contain only amino acids; conjugated proteins contain prosthetic groups such as the heme group in hemoglobin. All enzymes are proteins.
7. Structure is very closely related to the function and properties of proteins. Hydrogen bonding and other intermolecular forces determine the structure of proteins to a great extent.
8. The term *primary protein structure* refers to the sequence of amino acids. The α helix is an example of what is referred to as secondary structure. We also speak of tertiary and quaternary structures, which determine the three-dimensional arrangement of proteins.
9. Nucleotides are the building blocks of DNA and RNA. The DNA nucleotides each contain a purine or pyrimidine base, a deoxyribose molecule, and a phosphate group. The nucleotides of RNA are similar but contain different bases and ribose instead of deoxyribose.

Applied Chemistry Example: Nylon and Silk

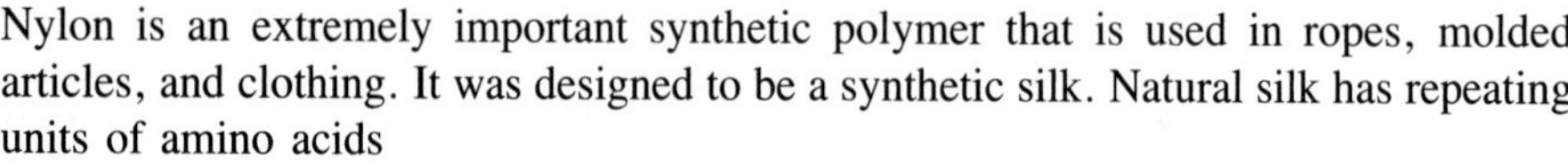

Nylon is an extremely important synthetic polymer that is used in ropes, molded articles, and clothing. It was designed to be a synthetic silk. Natural silk has repeating units of amino acids

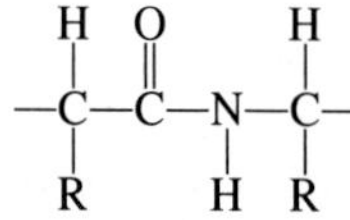

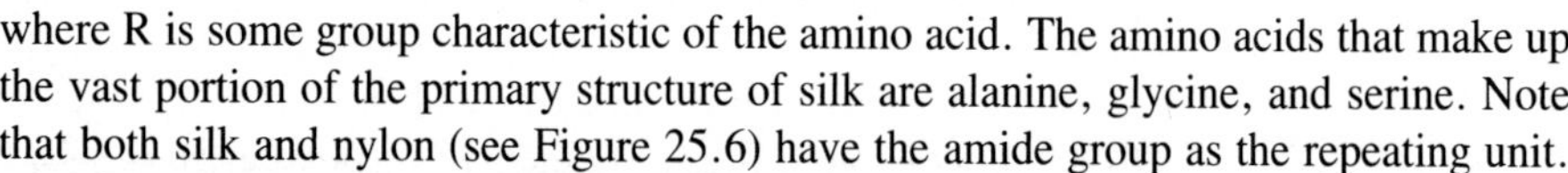

where R is some group characteristic of the amino acid. The amino acids that make up the vast portion of the primary structure of silk are alanine, glycine, and serine. Note that both silk and nylon (see Figure 25.6) have the amide group as the repeating unit.

The production of nylon 66 from adipic acid and hexamethylenediamine is carried out at about 300°C under vacuum. (The vacuum condition helps to remove water and drives the reaction from left to right, as shown on p. 1031.) The hot polymer formed in this reaction is squeezed out into a ribbon which is cooled and broken up into chips before being converted into fibers, which is the main use of nylon. Fibers of nylon are obtained by ''melt-spinning.'' In this process, the chips of nylon are melted and then forced through small holes to produce fibers. The molar mass of nylon originally produced is critical since if it is too low, fibers will not form and if it is too high, it will require higher temperature and pressure, resulting in a greater cost.

1. The average molar mass of a batch of nylon 66 is 12,000 g/mol. How many monomer units are there in this sample?

   Answer: The repeating unit in nylon 66 is (see Figure 25.6)

$$-(CH_2)_4-\overset{O}{\overset{\|}{C}}-\underset{H}{\underset{|}{N}}-(CH_2)_6-\underset{H}{\underset{|}{N}}-\overset{O}{\overset{\|}{C}}-$$

   and the molar mass of the unit is 226.3 g/mol. Therefore, the number of repeating units (n) is $$n = \frac{12000 \text{ g/mol}}{226.3 \text{ g/mol}} = 53$$

2. Which part of nylon's structure is similar to that in polypeptides?

   Answer: The most obvious feature is the presence of the amide group in the repeating unit (see p. 1031). Another important and related feature that makes the two types of polymers similar is the ability of the molecules to form intramolecular hydrogen bonds.

3. How many different tripeptides (made up of three amino acids) can be formed from the amino acids alanine (Ala), glycine (Gly), and Serine (Ser) which account for most of the amino acids in silk?

Answer: We approach this question systematically. First, there are three tripeptides made up of only one type of amino acid:

Ala-Ala-Ala Gly-Gly-Gly Ser-Ser-Ser

Next, there are eighteen tripeptides made up of two types of amino acids:

Ala-Ala-Ser	Ser-Ser-Ala	Ala-Ala-Gly	Gly-Gly-Ala
Ala-Ser-Ala	Ala-Ser-Ser	Ala-Gly-Ala	Ala-Gly-Gly
Ser-Ala-Ala	Ser-Ala-Ser	Gly-Ala-Ala	Gly-Ala-Gly
Gly-Gly-Ser	Ser-Ser-Gly	Gly-Ser-Gly	Gly-Ser-Ser
Ser-Gly-Gly	Ser-Gly-Ser		

Finally, there are six different tripeptides from three different amino acids

Ala-Gly-Ser	Ser-Ala-Gly	Ala-Ser-Gly
Ser-Gly-Ala	Gly-Ala-Ser	Gly-Ser-Ala

Thus there is a total of twenty-seven ways to synthesize a tripeptide from three amino acids. In silk, a basic six-residue unit repeats for long distances in the chain:

—Gly-Ser-Gly-Ala-Gly-Ala—

The ability of living organisms to reproduce the correct sequence is truly remarkable. It is also interesting to note that we can emulate the properties of silk with such a simple structure as nylon.

Key Words

Conjugated protein, p. 1033
Copolymerization, p. 1031
Denatured protein, p. 1041
Homopolymer, p. 1027
Hydrophilic, p. 1042
Hydrophobic, p. 1042
Monomer, p. 1026
Nucleotide, p. 1043
Polymer, p. 1026
Prosthetic group, p. 1033
Simple protein, p. 1033

Exercises

Synthetic Organic Polymers

REVIEW QUESTIONS

25.1 Define the following terms: copolymerization, homopolymer, monomer, polymer.

25.2 Name 10 objects that are either totally or partly made of synthetic organic polymers.

25.3 Calculate the molar mass of a particular polyethylene sample, $\text{-}(CH_2\text{—}CH_2)_n\text{-}$, where $n = 4600$.

25.4 Describe the two major mechanisms in organic polymer

synthesis.

25.5 What are Natta-Ziegler catalysts? What are their roles in polymer synthesis?

25.6 In Chapter 12 you learned about the colligative properties of solutions. Which of the colligative properties is suitable for determining the molar mass of a polymer? Why?

PROBLEMS

25.7 Teflon is formed by radical addition reaction involving the monomer tetrafluoroethylene. Show the mechanism for this reaction.

25.8 Vinyl chloride, $H_2C{=}CHCl$, undergoes copolymerization with 1,1-dichloroethylene, $H_2C{=}CCl_2$, to form a polymer commercially known as Saran. Draw the structure of the polymer, showing the repeating monomer units.

25.9 Kevlar is formed by the copolymerization between the following two monomers:

$H_2N—C_6H_4—NH_2$ $\quad$ $HO—\overset{\overset{\displaystyle O}{\|}}{C}—C_6H_4—\overset{\overset{\displaystyle O}{\|}}{C}—OH$

Sketch a portion of the polymer chain showing several monomer units. Write an overall equation for the condensation reaction.

25.10 Describe the formation of polystyrene.

25.11 Deduce plausible monomers for polymers with the following repeating units:

(a) $-\!\!(CH_2—CF_2)\!\!-_n$

(b) $-\!\!\left(CO—C_6H_4—CONH—C_6H_4—NH\right)\!\!-_n$

(c) $-\!\!(CH_2—CH{=}CH—CH_2)\!\!-_n$

(d) $-\!\!(CO—(CH_2)_6NH)\!\!-_n$

25.12 Nylon can be destroyed easily by strong acids. Explain the chemical basis for the destruction.

Proteins

REVIEW QUESTIONS

25.13 Define the following terms: simple protein, conjugated protein, prosthetic group, denatured protein.

25.14 Discuss the characteristics of an amide group and its importance in protein structure.

25.15 In what respect does a nylon resemble a polypeptide?

25.16 Briefly describe the following terms used in the study of protein structure: primary structure, secondary structure, tertiary structure, quaternary structure.

25.17 What is the α-helical structure in proteins?

25.18 Describe the β-pleated structure present in some proteins.

25.19 Discuss some of the main functions of proteins in living systems.

25.20 List the similarities of myoglobin and hemoglobin, and describe their differences.

25.21 Why is sickle cell anemia sometimes called a molecular disease?

25.22 Briefly explain the phenomenon of cooperativity exhibited by the hemoglobin molecule in binding oxygen.

PROBLEMS

25.23 Chemical analysis shows that hemoglobin contains 0.34 percent of Fe by mass. What is the minimum possible molar mass of hemoglobin? The actual molar mass of hemoglobin is four times this minimum value. What conclusion can you draw from these data?

25.24 Draw structures of the dipeptides that can be formed from the following pairs of amino acids: (a) glycine and alanine, (b) glycine and lysine (see Table 25.2 for structures of amino acids).

25.25 Draw structures of all the tripeptides that can possibly be formed from one molecule of each of the following three amino acids: glycine, serine, and phenylalanine.

25.26 The amino acid glycine can be condensed to form a polymer called polyglycine. Draw the structure of this polymer.

25.27 How many different tripeptides can be formed by lysine and alanine?

25.28 When a nonapeptide (containing nine amino acid residues) isolated from rat brains was hydrolyzed, it gave the following smaller peptides as identifiable products: Gly-Ala-Phe, Ala-Leu-Val, Gly-Ala-Leu, Phe-Glu-His, and His-Gly-Ala-Leu.
Reconstruct the amino acid sequence in the nonapeptide, giving your reasons. (Remember the convention used in writing peptides.)

25.29 The folding of a polypeptide chain depends not only on its amino acid sequence but also on the nature of the solvent. Discuss the types of interactions that might occur between water molecules and the amino acid residues of the polypeptide chain. Which groups (hydrophilic or hydrophobic) would be exposed on the exterior of the protein in contact with water and which groups would be buried in the interior of the protein?

25.30 The following are data obtained on the rate of product formation of an enzyme-catalyzed reaction:

Temperature (°C)	**Rate of product formation (*M*/s)**
10	0.0025
20	0.0048
30	0.0090
35	0.0086
45	0.0012

Comment on the dependence of rate on temperature. (No calculations are required.)

25.31 Referring to Figure 25.17, explain why the radius of the Fe^{2+} ion is smaller when it is complexed with molecular oxygen. (*Hint:* Iron in deoxyhemoglobin is in the high-spin state; in oxyhemoglobin it is in the low-spin state.)

Nucleic Acids

REVIEW QUESTIONS

25.32 Describe the structure of a nucleotide.

25.33 What is the difference between ribose and deoxyribose?

25.34 What are Chargaff's rules?

25.35 Draw structures of the nucleotides containing the following components: (a) deoxyribose and cytosine, (b) deoxyribose and thymine, (c) ribose and uracil, (d) ribose and cytosine.

Miscellaneous Problems

25.36 Discuss the importance of hydrogen bonding in biological systems. Use proteins and nucleic acids as examples.

25.37 Proteins vary widely in structure, whereas nucleic acids have rather uniform structures. How do you account for this major difference?

25.38 If untreated, fevers of 104°F or higher may lead to brain damage. Why?

Opposite page: The chemical industry is one of the most vital and essential industries of our modern society. The photo shows an oil refinery.

CHAPTER 26

INDUSTRIAL CHEMISTRY

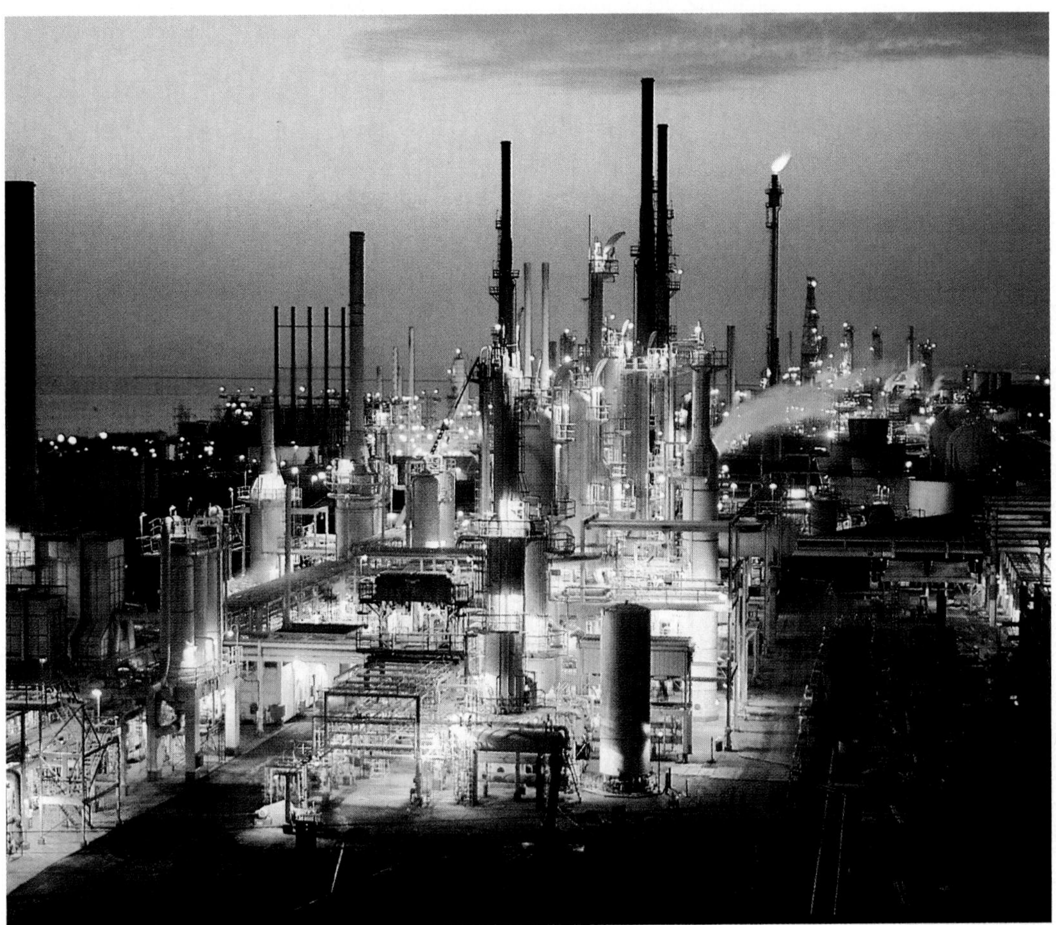

In this final chapter we will take a brief look at a very important branch of chemistry that deals with industrial processes. The chemical industry is essential for our modern society, as it provides us with fuels, fertilizers, detergents, pharmaceuticals, fabrics, and thousands of other items. Many of the key inorganic and organic industrial chemicals have been discussed in this book. We will also examine the impact of the chemical industry on the environment.

26.1 The Nature and Scope of the Chemical Industry

When we speak of the chemical industry, we normally think of the manufacture of chemicals such as ammonia and sulfuric acid. Actually the chemical industry is a set of related industries with many diverse functions and products. The basic activities of the chemical industry are to isolate raw materials and to prepare key intermediates and finished products. The following list gives the products that are directly or indirectly related to the chemical industry:

Industrial inorganic chemicals	Paper
Industrial organic chemicals	Textile mill products
Petroleum and coal products	Pesticides
Fertilizers	Pharmaceuticals
Metals	Food
Glass	Photographic products
Polymers	Electrical and electronic equipment
Soaps and detergents	Products of the nuclear industry
Paints	Personal care products

The chemical industry is one of the most important industries, and it accounts for about 25 percent of the U.S. gross national product. Next to agricultural commodities, it has the largest surplus in international trade.

Laboratory Scale versus Industrial Scale

Running a reaction at a chemical plant is very different from doing an experiment in a chemical laboratory. The reason is that on the laboratory scale we normally use grams of starting materials, whereas tons of the same substances may be used in an industrial process. The large quantity of chemicals used presents many problems not encountered in the laboratory. The major ones are: (1) Special designs of apparatus are required to allow for proper heat and mass transfer and to prevent corrosion. Industrial processes are often carried out at very high temperatures and pressures and in a *continuous flow;* that is, reactants are added and products removed continuously. A laboratory preparation is usually run at much lower temperatures and pressures, and products are isolated at the end of the experiment. (2) Because huge quantities of chemicals are involved, the cost is always a primary consideration. In addition to varying conditions to maximize the yield, catalysts play a central role in many industrial processes because they shorten the time needed for conversion to products and often permit the process to be carried out under milder conditions. Electrical heaters are usually employed in the laboratory; in industry, oil and natural gas are used as fuels. To reduce cost, heat generated from some intermediate exothermic reaction is used to promote other steps. (3) By-products are recycled or sold in industry. In the laboratory they are usually discarded. (4) Worker safety and disposal of hazardous materials are of major concern for the chemical industry. These problems are more easily handled in the laboratory because of the small quantity of chemicals involved.

How is an industrial process developed? First, a reaction is tested in the laboratory to get an estimate of the time required and the yield. If the reaction looks promising, the next step is to build a small-scale model, or pilot plant, using quantities intermediate between laboratory scale and industrial scale. This is the critical stage of finding out all the practical problems of running the reaction on a large scale. Engineering designs are needed to allow chemists to monitor the process and check the yield and quality of the intermediates formed. The problems of recycling unreacted starting materials and using heat generated in some intermediate step must also be solved in a pilot plant. If all the requirements are satisfactorily met, a much larger manufacturing plant will be built to carry out the desired process. It should be kept in mind that it is virtually impossible to anticipate all the problems of running a reaction on an industrial scale, and the many chemical engineers and chemists employed in the industry are constantly seeking ways to improve the efficiency of the process and to minimize pollution to the environment.

26.2 Raw Materials for the Chemical Industry

All the chemicals used in industry have to be manufactured from raw materials available in nature. If these materials are scarce, the price of the product will be high. For this reason, major chemical industries have developed around the most plentiful raw materials. The most common raw materials are air, seawater, rock salt, minerals, coal, natural gas, petroleum, and vegetation, discussed below.

Air

Table 26.1 shows the composition of dry air at sea level; it is the source of six industrial gases (N_2, O_2, Ne, Ar, Kr, and Xe). The mass of Earth's atmosphere is approximately 5×10^{15} tons; therefore, the supply of the gases is virtually unlimited. The gases are separated from one another by fractional distillation of liquid air. The uses of oxygen and nitrogen were discussed in Chapter 21. Note that only nitrogen and oxygen are produced on a large scale. The noble gases are used mostly in carrying out chemical research and in providing an inert atmosphere in the recovery of metals and in the electronics industry.

Seawater and Rock Salt

The ocean is an enormous and extremely complex aqueous solution. There are about 1.5×10^{21} L of seawater in the ocean, of which 3.5 percent by mass is dissolved material. The composition of seawater is shown on p. 531. Seawater is a good source of sodium chloride, magnesium, and bromine (the concentration of bromine in seawater is about 65 mg Br/kg seawater). Other elements present in seawater can be more conveniently prepared from their minerals. Sodium chloride is obtained from seawater by evaporation (see p. 359). Another major source of sodium chloride is rock salt.

Table 26.1 Composition of dry air (% by volume) at sea level

Gas	Boiling point (°C)*
N_2 (78.03)	−196.0
O_2 (20.99)	−183.1
CO_2 (0.033)	—
Ar (0.94)	−185.7
Ne (0.0015)	−245.9
Kr (0.00014)	−152.9
Xe (0.000006)	−106.9

*Measured at 1 atm. Carbon dioxide sublimes at −78.5°C.

Minerals

The vast majority of the elements are obtained from Earth's crust. Reactive metals and nonmetals are obtained by chemical or electrolytic reduction and oxidation. Less reactive elements such as platinum, gold, and sulfur occur in the uncombined state. Cheap and abundant minerals such as limestone ($CaCO_3$) and dolomite ($CaCO_3 \cdot MgCO_3$) are used in the steel industry, water treatment, paper and pulp industries, agriculture, and many other areas.

Coal

Coal, which is formed by the gradual decomposition of plant vegetation over millions of years, is distributed throughout the world. There are several forms of coal that differ in composition and texture; for example, *anthracite coal* is the hardest form of coal, while *bituminous coal* is much softer. Coal has served as the main source of solid fuel for thousands of years. Today about 25 percent of the energy consumption in the United States is supplied by coal. The coal reserve in this country is so large that we can continue using it at the present level for well over another thousand years!

Destructive Distillation of Coal. Besides being a valuable fuel, coal is also a source of many chemicals essential to the chemical industry. When coal is heated to a temperature between 600°C and 1000°C in the absence of air, it decomposes to yield a number of solid, liquid, and gaseous products; this process is called the *destructive distillation of coal:*

$$\text{coal} \longrightarrow \text{coke} + \text{coal tar} + \text{coal gas}$$

The solid product, coke, contains most of the nonvolatile substances originally present in coal. Coke is used in the manufacture of water gas (see Section 21.2) and in the extraction of iron (see Section 20.2) and less reactive metals like tin. The liquid fraction contains *coal tar,* a viscous substance from which more than a hundred other products (mostly aromatic compounds such as benzene, toluene, and more complex molecules) can be obtained by distillation. After the removal of coal tar and ammoniacal liquor, a black, very viscous substance called *pitch* is left over. Pitch is used mainly for road construction. The gaseous product, called *coal gas,* contains hydrogen, methane, ethane, carbon monoxide, and to a lesser extent ammonia and hydrogen sulfide.

Section 21.4 describes the gasification of coal. In the presence of appropriate catalysts and hydrogen, coal can be converted to gasoline and other liquid products. This process is called *coal liquefaction*. Besides yielding valuable chemicals for fuel and as starting materials for other processes, this procedure also removes much of the sulfur in coal as hydrogen sulfide.

Natural Gas and Petroleum

Natural gas is a mixture of gaseous hydrocarbons (methane, ethane, and to a lesser extent propane and butane). It occurs in reservoirs of porous rock (sand or sandstone) capped by impervious strata. Natural gas is often associated with petroleum (see Chapter 24), with which it has a common origin in the decomposition of organic matter in sedimentary deposits. Together petroleum and natural gas supply about 70 percent of the energy needs in the United States. In addition, they are also sources of valuable chemicals for the industry. Figure 26.1 summarizes some of the important uses of compounds obtained from petroleum. Note that in addition to the aliphatic and aromatic hydrocarbons, petroleum contains a sizable amount of inorganic materials, most of which are compounds of nitrogen and sulfur. Natural gas is also a source of sulfur (from hydrogen sulfide present in the natural gas) and helium.

Vegetation

Many vegetable products are used in the chemical industry. For example, natural oils such as castor oil, palm oil, and olive oil are used in the manufactur of soap and

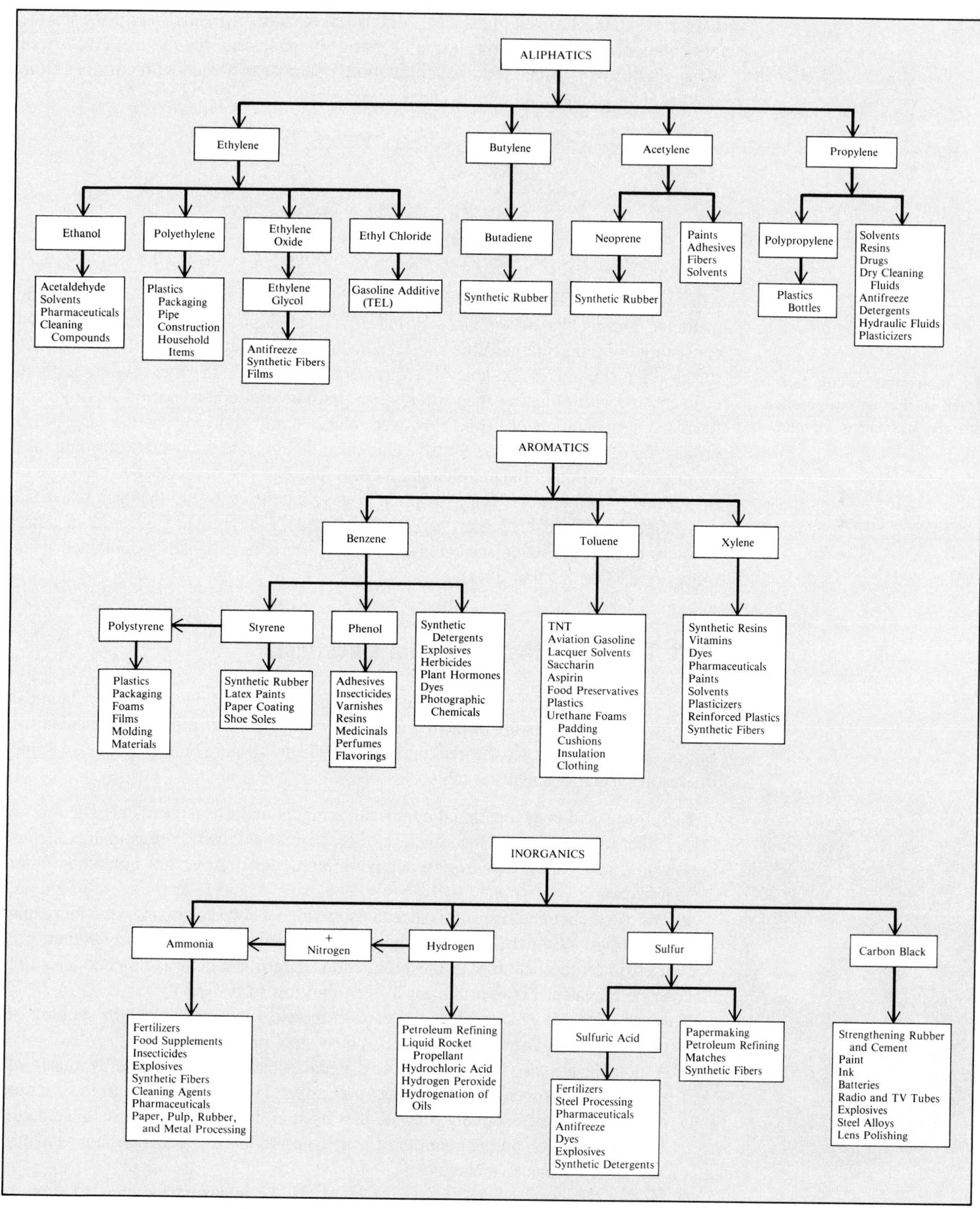

Figure 26.1 *The components of petroleum and their uses.*

margarine. In the 1930s, almost all the industrial ethanol (C_2H_5OH) was made via the fermentation of sugars. During the processing of sugar cane for the manufacture of sugar, waste syrup (molasses) containing mostly sucrose is treated with enzymes from yeast to give ethanol and carbon dioxide:

$$\underset{\text{sucrose}}{C_{12}H_{22}O_{11}} + H_2O \xrightarrow{\text{invertase}} 2C_6H_{12}O_6$$

$$\underset{\text{glucose and fructose}}{C_6H_{12}O_6} \xrightarrow{\text{zymase}} 2C_2H_5OH + 2CO_2$$

After World War II the manufacture of ethanol was replaced by more efficient synthetic methods. During the oil embargo in the 1970s, attempts were made to use a fuel consisting of 90 percent gasoline and 10 percent ethanol (called *gasohol*) to substitute for gasoline (Figure 26.2). This idea was attractive because the fuel mixture is more efficient and less polluting than the regular gasoline and ethanol could be produced from the fermentation of corn. However, closer study showed that the large-scale production of ethanol was more costly than once thought and the operation would take a substantial portion of the nation's grain crop, thus causing an increase in the price of many foods. Furthermore, since the price of gasoline never rose to the once feared \$2 per gallon level, the idea was largely abandoned. Today ethanol is used in some gasoline to increase its octane number. Finally, we note that latex is obtained from rubber trees (see Section 25.2).

It is interesting to note that Brazil is phasing out gasoline entirely, in favor of ethanol.

26.3 The Top Fifty Industrial Chemicals

Every year *Chemical and Engineering News*, a publication of the American Chemical Society, provides a ranking of the top fifty industrial chemicals produced in the United States. Table 26.2 shows the rankings from 1982 to 1988. The following are some interesting features about the table.

- Sulfuric acid is the perennial number-one-ranked industrial chemical. The reason is that it is used in the manufacture of fertilizers, polymers, drugs, paints, detergents, and paper, and in petroleum refining, metallurgy, and numerous other processes.
- The fifty chemicals are almost evenly divided between organic and inorganic categories. While the inorganic chemicals dominate the top ten (many of them are used in large quantities in the production of fertilizers), organic chemicals are more prevalent between the top fifteen and top forty-five.
- Much of the oxygen produced is used in the steel industry. Thus the ranking of oxygen can be correlated with the annual steel production.
- A notable absence in the list is hydrogen, which is used in the synthesis of ammonia, hydrogenation, and other processes. The reason is that in most cases hydrogen is used captively; that is, it is used at the plant where it is generated. Thus information on the quantity of hydrogen produced is generally not as available as it is for other chemicals.
- Methyl *tert*-butyl ether (MTBE), $(CH_3)_3COCH_3$, appeared for the first time in 1984 on the top fifty list. The reason is that MTBE has replaced tetraethyl lead as an antiknocking agent in gasoline (see Chapter 24). Both tetraethyl lead and

Figure 26.2 *Gasohol is available at some of the gas stations in the United States.*

Table 26.2 The top fifty industrial chemicals*

Chemical	1988	1987	1986	1985	1984	1983	1982
Sulfuric acid	1(85.6)	1(77.5)	1(73.6)	1(79.3)	1(83.6)	1(73.2)	1(65.4)
Nitrogen	2(52.1)	2(47.4)	2(48.6)	2(47.5)	2(47.9)	2(40.6)	2(35.1)
Oxygen	3(37.1)	5(32.3)	3(33.0)	4(32.5)	5(31.9)	5(28.4)	4(28.1)
Ethylene	4(36.6)	3(35.0)	4(32.8)	6(29.8)	6(31.4)	4(28.7)	6(24.5)
Ammonia	5(33.9)	4(32.3)	6(28.0)	3(34.6)	3(33.4)	6(28.1)	3(31.6)
Lime	6(32.3)	6(30.4)	5(30.3)	5(31.6)	4(32.0)	3(29.7)	5(28.2)
Sodium hydroxide	7(24.0)	7(23.0)	7(22.0)	7(21.8)	8(21.8)	7(20.5)	7(18.8)
Phosphoric acid	8(23.4)	9(21.0)	9(18.4)	8(21.0)	7(22.7)	9(19.5)	9(16.5)
Chlorine	9(22.7)	8(22.0)	8(21.0)	9(20.8)	9(21.4)	8(19.9)	8(18.4)
Propylene	10(20.0)	10(18.5)	10(17.3)	11(14.9)	11(15.6)	11(14.0)	14(12.5)
Sodium carbonate	11(19.1)	11(17.8)	11(17.2)	10(17.2)	10(17.0)	10(16.9)	10(15.6)
Nitric acid	12(15.8)	13(14.2)	13(13.1)	12(14.7)	12(15.4)	12(13.9)	11(14.8)
Urea	13(15.8)	12(14.9)	14(12.1)	14(13.4)	13(14.9)	14(12.3)	13(13.0)
Ammonium nitrate	14(14.4)	15(12.8)	15(11.1)	13(13.6)	14(14.3)	13(12.5)	12(14.2)
Ethylene dichloride	15(13.7)	14(13.8)	12(14.5)	15(12.1)	15(10.7)	15(11.5)	16(7.62)
Benzene	16(11.8)	16(11.7)	16(10.2)	17(9.39)	16(9.70)	16(9.08)	15(7.75)
Ethylbenzene	17(9.94)	18(9.41)	17(8.92)	20(7.39)	20(7.61)	18(7.88)	19(6.66)
Terephthalic acid	18(9.60)	20(8.11)	21(7.68)	21(6.49)	23(5.91)	23(5.63)	25(4.84)
Carbon dioxide	19(9.38)	17(9.86)	18(8.50)	18(9.25)	17(9.13)	19(7.76)	18(7.36)
Vinyl chloride	20(9.06)	19(8.23)	19(8.42)	16(9.46)	22(9.10)	20(6.88)	24(4.90)
Styrene	21(8.59)	21(8.09)	20(7.84)	19(7.62)	19(7.71)	21(6.80)	20(5.94)
Methanol	22(7.34)	22(7.29)	22(7.33)	27(5.00)	18(8.19)	17(7.98)	17(7.56)
Formaldehyde	23(6.73)	24(6.08)	25(5.89)	22(5.61)	24(5.82)	25(5.47)	26(4.82)
Toluene	24(6.47)	23(6.73)	26(5.82)	26(5.07)	27(5.28)	22(5.66)	21(5.18)
Xylene	24(6.47)	23(6.73)	27(5.55)	25(5.31)	21(6.15)	26(5.23)	27(4.74)
Hydrochloric acid	26(5.87)	28(4.99)	23(5.97)	22(5.61)	26(5.46)	27(4.84)	23(4.92)
p-Xylene	27(5.60)	27(5.15)	29(4.62)	28(4.78)	29(4.26)	29(4.11)	31(3.21)
Ethylene oxide	28(5.37)	26(5.62)	24(5.94)	24(5.43)	25(5.70)	24(5.53)	22(4.99)
Ethylene glycol	29(4.90)	29(4.52)	28(4.76)	30(4.18)	28(4.82)	28(4.43)	28(4.31)
Cumene	30(4.80)	31(4.15)	31(3.70)	31(3.35)	31(3.75)	31(3.35)	33(2.74)
Methyl *tert*-butyl ether	31(4.68)	32(3.37)	41(2.24)	41(1.89)	49(1.38)	—	—
Ammonium sulfate	32(4.67)	30(4.37)	30(4.17)	29(4.19)	30(4.13)	30(3.91)	30(3.54)
Phenol	33(3.53)	33(3.24)	33(2.92)	34(2.78)	34(2.89)	34(2.64)	37(2.02)
Potash	34(3.35)	36(2.78)	36(2.58)	33(2.84)	32(3.45)	32(3.15)	29(3.93)
Butadiene	35(3.19)	35(3.00)	34(2.59)	39(2.34)	36(2.45)	36(2.35)	38(1.92)
Acetic acid	36(3.16)	34(3.23)	32(2.93)	32(2.90)	35(2.62)	33(2.81)	32(2.75)
Propylene oxide	37(3.11)	38(2.61)	38(2.48)	37(2.40)	43(1.78)	41(1.80)	43(1.48)
Carbon black	38(2.92)	37(2.72)	35(2.59)	35(2.57)	33(2.89)	35(2.50)	35(2.30)
Aluminum sulfate	39(2.59)	41(2.45)	37(2.52)	36(2.54)	37(2.26)	37(2.21)	34(2.37)
Acrylonitrile	40(2.58)	39(2.55)	39(2.31)	28(2.35)	38(2.22)	38(2.15)	36(2.04)
Vinyl acetate	41(2.56)	40(2.51)	40(2.25)	40(2.11)	40(2.02)	40(1.96)	39(1.88)
Cyclohexane	42(2.32)	42(2.25)	42(2.07)	44(1.66)	41(1.99)	44(1.66)	48(1.28)
Acetone	43(2.29)	43(2.08)	43(1.94)	42(1.79)	42(1.86)	40(1.86)	42(1.69)
Titanium dioxide	44(2.04)	44(1.90)	44(1.83)	43(1.72)	45(1.67)	45(1.51)	46(1.31)
Sodium sulfate	45(1.69)	46(1.61)	47(1.55)	45(1.65)	44(1.74)	43(1.71)	41(1.73)
Sodium silicate	46(1.64)	44(1.90)	45(1.57)	48(1.44)	46(1.50)	46(1.46)	45(1.33)
Adipic acid	47(1.59)	47(1.59)	48(1.52)	47(1.45)	48(1.39)	48(1.28)	49(1.20)
Isopropyl alcohol	48(1.42)	49(1.31)	49(1.28)	50(1.24)	47(1.39)	49(1.21)	44(1.38)
Calcium chloride	49(1.31)	48(1.55)	46(1.56)	46(1.88)	39(2.10)	42(1.71)	40(1.75)
Caprolactam	50(1.16)	50(0.85)	—	—	—	—	—

*The first number (the number to the left of the parentheses) denotes ranking, and the number in parentheses denotes the production in billions of pounds.

MTBE effectively increase the octane number; however, tetraethyl lead poses an environmental hazard since lead is a toxic metal. As a result of the federal regulation requiring all automobiles to use "unleaded" gasolines, the vast demand for MTBE has resulted in its phenomenal growth in recent years.

The Seven Basic Organic Chemicals

The number in parentheses denotes the 1988 ranking.

Figure 26.3 shows the structure of the seven basic organic chemicals from which all other important industrial organic chemicals are derived. Methane is obtained from natural gas. Ethylene (4) is prepared by the cracking process (see Section 24.1). Similarly, propylene (10) is prepared by the cracking of propane at 850°C:

$$\underset{\text{propane}}{2C_3H_8(g)} \longrightarrow \underset{\text{propylene}}{CH_3CH{=}CH_2(g)} + C_2H_4(g) + CH_4(g) + H_2(g)$$

The major uses of ethylene and propylene are in the manufacture of polymers and other chemicals.

Dehydrogenation means the removal of hydrogen atoms.

Butadiene (35) is prepared by the catalytic dehydrogenation of butane at around 650°C:

$$\underset{\text{butane}}{CH_3CH_2CH_2CH_3(g)} \xrightarrow{Fe_2O_3} \underset{\text{butadiene}}{CH_2{=}CH{-}CH{=}CH_2(g)} + 2H_2(g)$$

The main use of butadiene is in the manufacture of a synthetic rubber called stryrene-butadiene rubber (SBR) (see Section 25.2). It is also used as a starting material in the production of other polymers.

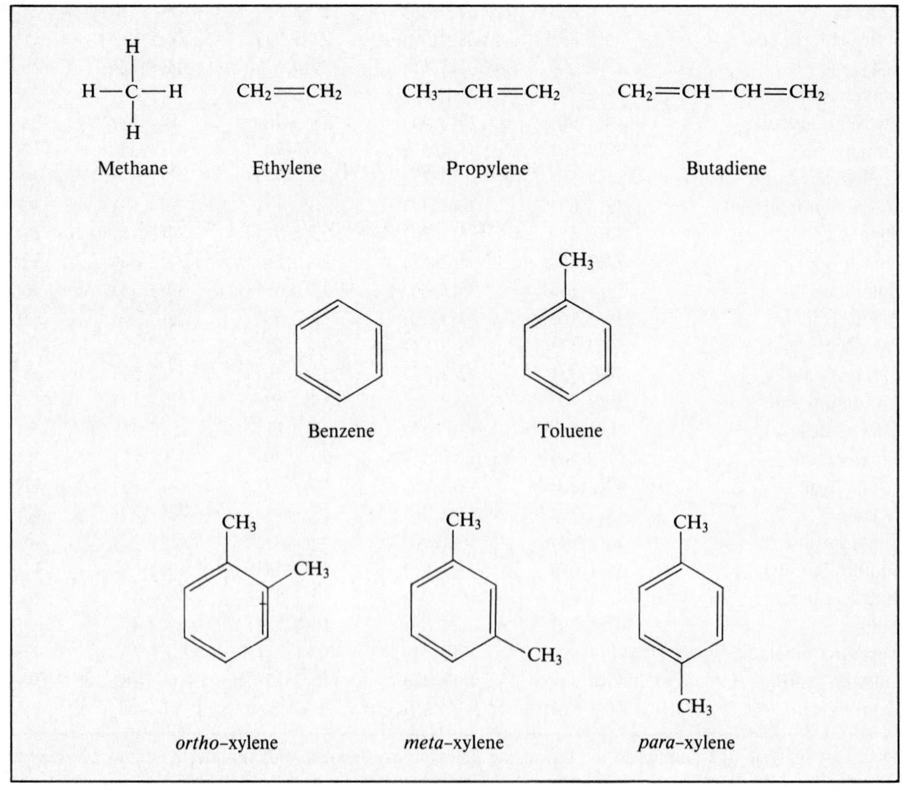

Figure 26.3 *The structure of the seven basic organic chemicals. The three xylenes are treated as one type of compound.*

Benzene (16) is obtained chiefly from hydrocarbons derived from petroleum (see Chapter 24) and to a lesser extent from coal tar. First, the hydrocarbons are converted to cycloalkanes. Next, hydrogen atoms are displaced to give the desired product. For example, hexane from petroleum is first converted to cyclohexane:

$$\underset{\text{hexane}}{C_6H_{14}(g)} \longrightarrow \underset{\text{cyclohexane}}{C_6H_{12}(g)} + H_2(g)$$

This is followed by a dehydrogenation reaction:

$$C_6H_{12}(g) \longrightarrow C_6H_6 \text{ (benzene ring)} + 3H_2(g)$$

Benzene is used in the production of polymers and other chemicals. Toluene (24) and xylenes (24) are prepared in a similar manner. About 50 percent of toluene is used to prepare benzene; it is also used in gasoline (to increase octane number) and as a solvent. The xylenes are used as solvents and in gasoline to upgrade octane number.

Inorganic Industrial Chemicals

The preparation of many of the top fifty inorganic chemicals has been discussed in the text. Here we will briefly show how some of these chemicals are obtained from the raw materials.

Air. Nitrogen (2), obtained from air by fractional distillation (see Section 21.5), is converted to ammonia (5) by the Haber process (see Section 13.6). Ammonia is used to prepare nitric acid (12) by the Ostwald process (see Section 13.6) and ammonium nitrate (14) by reacting with nitric acid:

$$NH_3(g) + HNO_3(aq) \longrightarrow NH_4NO_3(aq)$$

Fractional distillation of air also produces oxygen (3).

Seawater, Rock Salt, and Brines. Electrolysis of fused sodium chloride obtained from seawater and rock salt in a Downs cell (see Section 19.8) produces sodium metal and chlorine (9). Electrolysis of an aqueous solution of sodium chloride, which is the basis of the chlor-alkali industry (see Section 21.7), produces sodium hydroxide (7), chlorine (9), and hydrogen. Brines, which are concentrated aqueous solutions of mixed salts found in lakes, are good sources of sodium sulfate (45) and calcium chloride (49). These salts are separated from other dissolved components by fractional crystallization.

Potash is any potassium-containing mineral such as KCl and K_2SO_4 (see p. 79).

Minerals. Several of the top fifty chemicals are obtained directly from minerals. They are: sodium carbonate (11) from the mineral trona (see Section 20.5), potash (34), and sodium silicate (46). Lime (6) is prepared by the thermal decomposition of the mineral calcite ($CaCO_3$). Sulfur, which exists in nature in the uncombined state, is used to manufacture sulfuric acid (1) (see Section 21.6). Sulfuric acid plays a key role in the preparation of the following top fifty inorganic chemicals: phosphoric acid (8), ammonium sulfate (32), aluminum sulfate (39), titanium dioxide (44), and sodium sulfate (45).

Table 26.3 lists the major uses of the top inorganic industrial chemicals.

Table 26.3 Major uses of inorganic industrial chemicals*

Chemical	Uses
Sulfuric acid (H_2SO_4)	Fertilizers, metallurgical processes, chemical manufacture, petroleum refining, drugs, dyes, paint, paper, explosives
Nitrogen (N_2)	Synthesis of ammonia, oil recovery, metallurgical processes, blanketing atmosphere, aerospace industry
Oxygen (O_2)	Steelmaking, chemical manufacture, sewage treatment, rocket fuel, paper, medicine, metal fabrication
Ammonia (NH_3)	Fertilizers, fibers and plastics, explosives
Lime [CaO, $Ca(OH)_2$]	Metallurgy, pollution control, water treatment, chemical manufacture
Sodium hydroxide ($NaOH$)	Chemical manufacture, paper industry, water treatment, aluminum processing, petroleum refining
Phosphoric acid (H_3PO_4)	Fertilizers, detergents
Chlorine (Cl_2)	Chemical manufacture, bleaching, water treatment
Sodium carbonate (Na_2CO_3)	Glass, chemical manufacture, soaps and detergents
Nitric acid (HNO_3)	Fertilizers, explosives, chemical manufacture
Ammonium nitrate (NH_4NO_3)	Fertilizers, explosives
Carbon dioxide (CO_2)	Refrigerant, beverage carbonation, oil recovery, chemical syntheses, metal fabrication
Hydrochloric acid (HCl)	Metal pickling, manufacture of chemicals, food processing, oil recovery
Ammonium sulfate [$(NH_4)_2SO_4$]	Fertilizers, water treatment, fermentation processes, leather tanning
Potash (KCl, K_2SO_4)	Fertilizers, manufacture of chemicals, food, medicine
Carbon black†	Automobile tires, industrial rubber parts, inks and pigments
Aluminum sulfate [$Al_2(SO_4)_3$]	Paper production, water treatment
Titanium dioxide (TiO_2)	Paint, paper, plastics, ceramics, floor coverings
Sodium sulfate (Na_2SO_4)	Detergents, paper, glass
Sodium silicate ($Na_2Si_4O_9$)‡	Soaps and detergents, silica gel, pigments, adhesives
Calcium chloride ($CaCl_2$)	Road de-icing, dust control, industrial refrigeration, concrete, tire weighting

*The chemicals are arranged according to their 1988 ranking (see Table 26.2).

†Carbon black contains 90 to 99% carbon by weight, plus oxygen, hydrogen, and sulfur.

‡Sodium silicate refers to a series of some forty compounds of which sodium tetrasilicate ($Na_2Si_4O_9$) is the most common example.

26.4 The Chemical Industry and the Environment

In this final section we will briefly survey the impact of the chemical industry on the environment. Nowadays it is difficult to read any newspaper without seeing some account of how one or another type of pollution is plaguing our nation. Many of the environmental problems are directly or indirectly related to the activities of the chemical industry. There are many areas of the chemical industry that must be carefully monitored and controlled to avoid damaging effects on the environment. It is helpful to consider the following categories which contribute to various forms of pollution.

Sources of Energy

The chemical industry is energy intensive. Most of the energy is supplied by the combustion of oil and natural gas, and to a lesser extent, coal. Combustion produces carbon dioxide (which contributes to the greenhouse effect), sulfur dioxide (which is responsible for acid rain), and various nitrogen oxides and unreacted hydrocarbons (which result in acid rain and smog formation). It should be pointed out that other industries, automobiles, and domestic use all contribute heavily to various types of air pollution.

Isolation of Raw Materials

Mining, especially strip mining such as that used for coal and iron (Figure 26.4), is damaging to the environment, and it may take years for the land to recover and vegetation to grow. Oil drilling on land or beneath the sea is also potentially harmful to the environment. Large-scale oil spills during transportation such as that which occurred in Alaska in 1989 (Figure 26.5) are a major concern, and this type of accident cannot be totally prevented.

Figure 26.4 *Strip mining for coal.*

Figure 26.5 *On March 24, 1989, the oil tanker* Valdez *of Exxon spilled 10.1 million gallons of crude oil in Alaska's Prince William Sound.*

Chemical Processes

During the operation at a chemical plant or during a metallurgical process, harmful gases are often emitted into the atmosphere. For example, roasting of sulfide ores produces sulfur dioxide. In addition, various particulate matters such as soot, ashes which contain heavy metal salts, and dust are also generated by the plant. For many years the chlor-alkali industry used mercury as the cathode (see Section 21.7). Although mercury does not participate in the overall reaction, some of it is inevitably discharged with waste solutions into the environment.

Disposal of Solid Wastes

Most industrial processes result in one or more unwanted by-products which must be discarded. The problem of solid waste disposal (both industrial and domestic) is one of the most serious problems facing our society today (Figure 26.6). Waste chemicals

Figure 26.6 *At the rate we are generating solid wastes, there will not be much room left to hold them. Here we see an unwanted garbage barge afloat off New York.*

Figure 26.7 *The difficult task of cleaning up chemicals at a waste site.*

must be stored in a safe site and constantly monitored. One of the most serious cases of improper disposal of toxic waste was discovered in 1978 when an old chemical dump in Niagara Falls, New York, near Love Canal, began leaking into the environment. As a result, the neighborhood had to be evacuated. Much effort is needed to clean up many of the similarly affected landfills in the nation (Figure 26.7). A related problem is the disposal of nuclear wastes, which does not yet have a totally satisfactory solution (see Section 23.5).

Potential Hazards of Chemicals

A herbicide is a pesticide that kills weeds.

Asbestos had been used as a thermal insulator for many years in buildings although it was known that asbestos dust, if inhaled, can cause lung cancer. Dioxin is a common contaminant of certain herbicides including Agent Orange, which was used as a defoliant during the Vietnam War. Actually dioxin is a synonym for a group of similar compounds of which the following is the most studied:

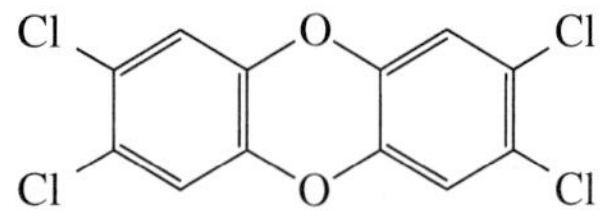

2,3,7,8-tetrachloro-dibenzo-*p*-dioxin

These compounds are extremely toxic. In the early 1970s, dirt roads in Times Beach, Missouri, were sprayed with waste automobile oil to control dust. Several years later, prompted by reports of mysterious animal and human illness, tests on the oil sprayed on the road revealed that it was contaminated with dioxins. In fact, the soil in the Times Beach suburb was so contaminated that in 1983 the Environmental Protection Agency (EPA) had to buy out the entire town and relocate some 2200 people!

Like dioxins, the polychlorinated biphenyls (PCBs) are a group of about 70 different but closely related chlorinated hydrocarbons. A typical compound in the PCB category is

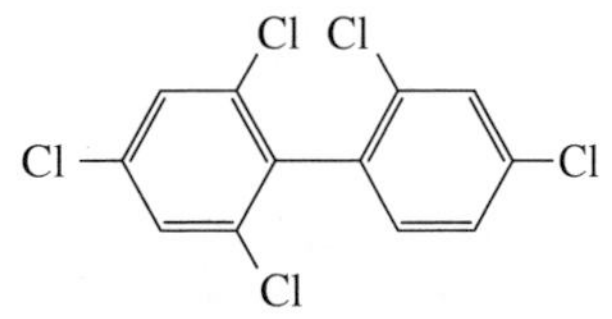

2,4,6,2′,4′-pentachlorobiphenyl

PCBs have been widely used as insulating and cooling fluids in electrical transformers and capacitors. They were also used in the production of plastics, paints, rubber, adhesives, and carbonless copy paper and for dust control on roads. In 1968 it was reported that some people in Japan suffered from liver and kidney damage after they had eaten food contaminated with PCBs. In 1976 the U.S. Congress banned the further manufacture and use of PCBs, except in existing electrical transformers. However, the EPA estimates that some 75,000 tons of PCBs had entered the environment because of indiscriminate dumping at landfills.

Even in cases when a chemical is known to be totally harmless to humans and vegetation, its long-term effect on the environment cannot be easily assessed. A case in point is the group of compounds called chlorofluorocarbons (CFCs; see Section 13.6). When these compounds were first synthesized in the early 1930s as refrigerants, they were considered ideal substances because they are nontoxic and noncombustible. It took nearly thirty years before chemists began to realize the potential hazard of CFCs to the ozone in the stratosphere.

Despite the various environmental problems facing us today, the outlook is not entirely bleak. For the last twenty years or so the chemical industry has become much more aware of pollution control than it once was. Innovative procedures have been introduced to minimize the damage to the environment. As mentioned earlier, sulfur dioxide generated in metallurgical processes is now used as starting material in the production of sulfuric acid at some plants. Diaphragm cells are replacing mercury cells in the chlor-alkali industry. An intense effort is under way to find replacements for the CFCs which will not deplete ozone in the stratosphere. Slow but promising progress is being made to utilize nonpolluting sources of energy such as solar energy at an economically acceptable level. Scientists are developing bacteria to attack a wide range of toxic wastes. A large-scale experiment is currently (1989) in progress to use bacteria to break down the crude oil spilled by the oil tanker *Valdez* in Alaska. In the area of solid waste treatment, recycling offers one of the best solutions. Consider aluminum as an example. Each year Americans throw away enough aluminum to rebuild the entire U.S. commercial airline fleet every three months! Aluminum recycling (see p. 846) is economically sound and avoids further damage to the environment from mining. Literally thousands of household articles are made of seemingly indestructible plastic, and they are a main contributor to solid waste and litter on the road. An effective way to

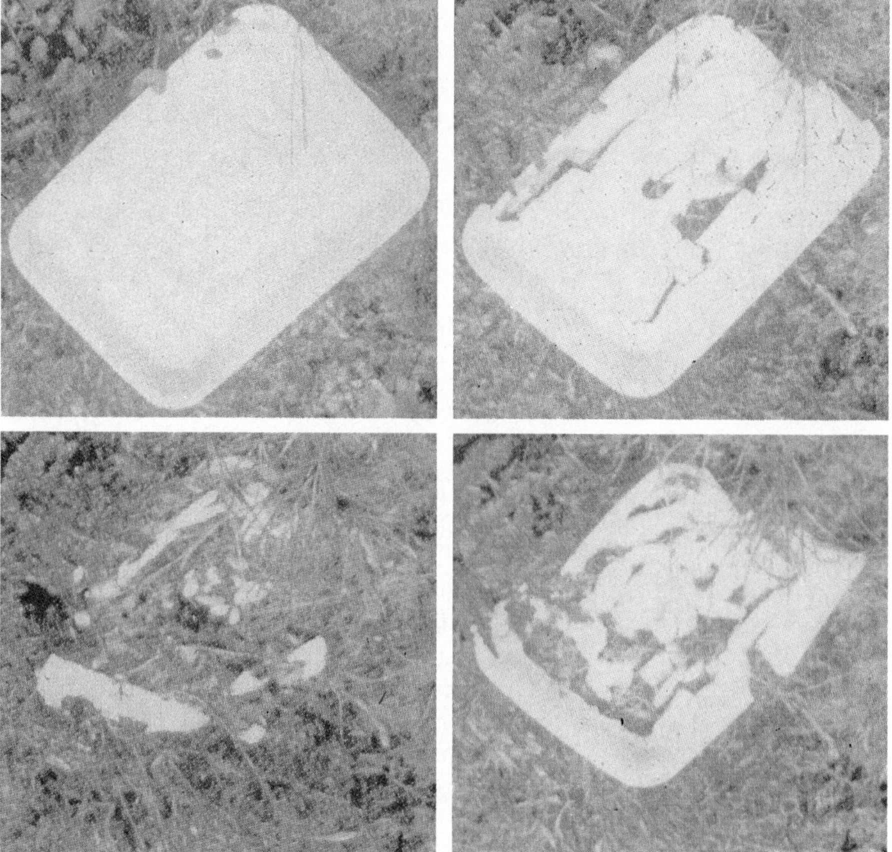

Figure 26.8 *Photodegradable plastics at work. Clockwise starting from the upper left photo: a plastic food tray left outdoors after 30 days; after 40 days; after 45 days; after 55 days.*

reduce the "plastic" pollution is the manufacture of photodegradable plastics. Light-sensitive compounds are incorporated into the polymer chains that make up the plastic. When exposed to the sun's ultraviolet rays, these compounds become photoexcited and they can cause the polymers to break down into smaller units (Figure 26.8).

In recent years the U.S. Congress has enacted a number of important environmental and resource protection laws. Among these are the National Energy Acts of 1978 and 1980; the Toxic Substances Control Act of 1976; the Clean Air Acts of 1965, 1970, and 1977; and the Clean Water Acts of 1977 and 1987. The Congress also passed the Superfund program (1980, 1986) to deal with the problems of illegal hazardous waste disposal. The EPA estimates that the total cost of the cleanup may be as high as $250 billion—an average of $1000 per person! To be sure, it will take a tremendous effort (both time and money) to clean up our land, water, and air, but this is a small price to pay considering that the future of planet Earth is at stake.

Summary

1. The chemical industry is one of the most important industries. Its activities include the isolation of raw materials as well as the preparation of intermediates and finished products.
2. The raw materials for the chemical industry are derived from air, seawater, minerals, coal, petroleum, natural gas, and vegetation.
3. Organic industrial chemistry is based on several basic organic chemicals: methane, ethylene, propylene, butadiene, benzene, toluene, and xylenes. Sulfuric acid is the most important industrial chemical.
4. The activities of the chemical industry have a profound impact on the environment.

Exercises

PROBLEMS

26.1 Referring to Table 26.1, explain why carbon dioxide, which is an important industrial gas, is not obtained from air.

26.2 Describe how you can obtain the following substances from seawater: sodium, magnesium, chlorine, bromine, and sodium hydroxide.

26.3 Suggest a reason why methane is not listed among the top fifty chemicals.

26.4 How does an experiment carried out in the laboratory differ from a similar process carried out on an industrial scale?

26.5 Briefly describe the preparation of the following top fifty inorganic chemicals: (a) sulfuric acid, (b) nitrogen, (c) oxygen, (d) ammonia, (e) lime, (f) sodium hydroxide, (g) phosphoric acid, (h) chlorine, (i) sodium carbonate, (j) nitric acid, (k) ammonium nitrate, (l) carbon dioxide, (m) hydrochloric acid, (n) ammonium sulfate, (o) aluminum sulfate, (p) sodium sulfate, (q) calcium chloride.

26.6 Titanium dioxide is a white, opaque, nontoxic solid used mainly as a white pigment in paint, paper, linoleum, nylon fibers, and cosmetics. It is prepared by treating the ore ilmenite ($FeTiO_3$) with sulfuric acid. Write an equation for this process.

26.7 The following environmental pollutions have been discussed in previous chapters: depletion of ozone in the stratosphere, acid rain, the greenhouse effect, smog, mercury. Show that each type of pollution can be directly or indirectly tied to the activity of the chemical industry.

26.8 Suggest ways to deal with the solid waste crisis in our society.

26.9 List five major energy sources and discuss their relative merits. (*Hint:* See Chemistry in Action: The Petroleum Industry in Chapter 24.)

26.10 Name a polymer that is derived from each of the following monomers, which are all top fifty industrial chemicals: ethylene, propylene, vinyl chloride, butadiene, acrylonitrile, styrene. (*Hint:* See Table 25.1.)

APPENDIX 1

THE ELEMENTS AND THE DERIVATION OF THEIR NAMES AND SYMBOLS*

Element	Symbol	Atomic No.	Atomic Mass†	Date of Discovery	Discoverer and Nationality‡	Derivation
Actinium	Ac	89	(227)	1899	A. Debierne (Fr.)	Gr. *aktis,* beam or ray
Aluminum	Al	13	26.98	1827	F. Woehler (Ge.)	Alum, the aluminum compound in which it was discovered; derived from L. *alumen,* astringent taste
Americium	Am	95	(243)	1944	A. Ghiorso (USA) R. A. James (USA) G. T. Seaborg (USA) S. G. Thompson (USA)	The Americas
Antimony	Sb	51	121.8	Ancient		L. *antimonium* (*anti,* opposite of; *monium,* isolated condition), so named because it is a tangible (metallic) substance which combines readily; symbol, L. *stibium,* mark
Argon	Ar	18	39.95	1894	Lord Raleigh (GB) Sir William Ramsay (GB)	Gr. *argos,* inactive
Arsenic	As	33	74.92	1250	Albertus Magnus (Ge.)	Gr. *aksenikon,* yellow pigment; L. *arsenicum,* orpiment; the Greeks once used arsenic trisulfide as a pigment
Astatine	At	85	(210)	1940	D. R. Corson (USA) K. R. MacKenzie (USA) E. Segre (USA)	Gr. *astatos,* unstable

(Continued)

Source: Reprinted with permission from "The Elements and Derivation of Their Names and Symbols," G. P. Dinga, *Chemistry* **41** (2), 20–22 (1968). Copyright by the American Chemical Society.

*At the time this table was drawn up, only 103 elements were known to exist.

†The atomic masses given here correspond to the 1961 values of the Commission on Atomic Weights. Masses in parentheses are those of the most stable or most common isotopes.

‡The abbreviations are (Ar.) Arabic; (Au.) Austrian; (Du.) Dutch; (Fr.) French; (Ge.) German; (GB) British; (Gr.) Greek; (H.) Hungarian; (I.) Italian; (L.) Latin; (P.) Polish; (R.) Russian; (Sp.) Spanish; (Swe.) Swedish; (USA) American.

Element	Symbol	Atomic No.	Atomic Mass	Date of Discovery	Discoverer and Nationality	Derivation
Barium	Ba	56	137.3	1808	Sir Humphry Davy (GB)	barite, a heavy spar, derived from Gr. *barys,* heavy
Berkelium	Bk	97	(247)	1950	G. T. Seaborg (USA) S. G. Thompson (USA) A. Ghiorso (USA)	Berkeley, Calif.
Beryllium	Be	4	9.012	1828	F. Woehler (Ge.) A. A. B. Bussy (Fr.)	Fr. L. *beryl,* sweet
Bismuth	Bi	83	209.0	1753	Claude Geoffroy (Fr.)	Ge. *bismuth,* probably a distortion of *weisse masse* (white mass) in which it was found
Boron	B	5	10.81	1808	Sir Humphry Davy (GB) J. L. Gay-Lussac (Fr.) L. J. Thenard (Fr.)	The compound borax, derived from Ar. *buraq,* white
Bromine	Br	35	79.90	1826	A. J. Balard (Fr.)	Gr. *bromos,* stench
Cadmium	Cd	48	112.4	1817	Fr. Stromeyer (Ge.)	Gr. *kadmia,* earth; L. *cadmia,* calamine (because it is found along with calamine)
Calcium	Ca	20	40.08	1808	Sir Humphry Davy (GB)	L. *calx,* lime
Californium	Cf	98	(249)	1950	G. T. Seaborg (USA) S. G. Thompson (USA) A. Ghiorso (USA) K. Street, Jr. (USA)	California
Carbon	C	6	12.01	Ancient		L. *carbo,* charcoal
Cerium	Ce	58	140.1	1803	J. J. Berzelius (Swe.) William Hisinger (Swe.) M. H. Klaproth (Ge.)	Asteroid Ceres
Cesium	Cs	55	132.9	1860	R. Bunsen (Ge.) G. R. Kirchhoff (Ge.)	L. *caesium,* blue (cesium was discovered by its spectral lines, which are blue)
Chlorine	Cl	17	35.45	1774	K. W. Scheele (Swe.)	Gr. *chloros,* light green
Chromium	Cr	24	52.00	1797	L. N. Vauquelin (Fr.)	Gr. *chroma,* color (because it is used in pigments)
Cobalt	Co	27	58.93	1735	G. Brandt (Ge.)	Ge. *Kobold,* goblin (because the ore yielded cobalt instead of the expected metal, copper, it was attributed to goblins)
Copper	Cu	29	63.55	Ancient		L. *cuprum,* copper, derived from *cyprium,* Island of Cyprus, the main source of ancient copper
Curium	Cm	96	(247)	1944	G. T. Seaborg (USA) R. A. James (USA) A. Ghiorso (USA)	Pierre and Marie Curie
Dysprosium	Dy	66	162.5	1886	Lecoq de Boisbaudran (Fr.)	Gr. *dysprositos,* hard to get at
Einsteinium	Es	99	(254)	1952	A. Ghiorso (USA)	Albert Einstein
Erbium	Er	68	167.3	1843	C. G. Mosander (Swe.)	Ytterby, Sweden, where many rare earths were discovered

Element	Symbol	Atomic No.	Atomic Mass	Date of Discovery	Discoverer and Nationality	Derivation
Europium	Eu	63	152.0	1896	E. Demarcay (Fr.)	Europe
Fermium	Fm	100	(253)	1953	A. Ghiorso (USA)	Enrico Fermi
Fluorine	F	9	19.00	1886	H. Moissan (Fr.)	Mineral fluorspar, from L. *fluere,* flow (because fluorspar was used as a flux)
Francium	Fr	87	(223)	1939	Marguerite Perey (Fr.)	France
Gadolinium	Gd	64	157.3	1880	J. C. Marignac (Fr.)	Johan Gadolin, Finnish rare earth chemist
Gallium	Ga	31	69.72	1875	Lecoq de Boisbaudran (Fr.)	L. *Gallia,* France
Germanium	Ge	32	72.59	1886	Clemens Winkler (Ge.)	L. *Germania,* Germany
Gold	Au	79	197.0	Ancient		L. *aurum,* shining dawn
Hafnium	Hf	72	178.5	1923	D. Coster (Du.) G. von Hevesey (H.)	L. *Hafnia,* Copenhagen
Helium	He	2	4.003	1868	P. Janssen (spectr) (Fr.) Sir William Ramsay (isolated) (GB)	Gr. *helios,* sun (because it was first discovered in the sun's spectrum)
Holmium	Ho	67	164.9	1879	P. T. Cleve (Swe.)	L. *Holmia,* Stockholm
Hydrogen	H	1	1.008	1766	Sir Henry Cavendish (GB)	Gr. *hydro,* water; *genes,* forming (because it produces water when burned with oxygen)
Indium	In	49	114.8	1863	F. Reich (Ge.) T. Richter (Ge.)	Indigo, because of its indigo blue lines in the spectrum
Iodine	I	53	126.9	1811	B. Courtois (Fr.)	Gr. *iodes,* violet
Iridium	Ir	77	192.2	1803	S. Tennant (GB)	L. *iris,* rainbow
Iron	Fe	26	55.85	Ancient		L. *ferrum,* iron
Krypton	Kr	36	83.80	1898	Sir William Ramsay (GB) M. W. Travers (GB)	Gr. *kryptos,* hidden
Lanthanum	La	57	138.9	1839	C. G. Mosander (Swe.)	Gr. *lanthanein,* concealed
Lawrencium	Lr	103	(257)	1961	A. Ghiorso (USA) T. Sikkeland (USA) A. E. Larsh (USA) R. M. Latimer (USA)	E. O. Lawrence (USA), inventor of the cyclotron
Lead	Pb	82	207.2	Ancient		Symbol, L. *plumbum,* lead, meaning heavy
Lithium	Li	3	6.941	1817	A. Arfvedson (Swe.)	Gr. *lithos,* rock (because it occurs in rocks)
Lutetium	Lu	71	175.0	1907	G. Urbain (Fr.) C. A. von Welsbach (Au.)	*Lutetia,* ancient name for Paris
Magnesium	Mg	12	24.31	1808	Sir Humphry Davy (GB)	*Magnesia,* a district in Thessaly; possibly derived from L. *magnesia*
Manganese	Mn	25	54.94	1774	J. G. Gahn (Swe.)	L. *magnes,* magnet
Mendelevium	Md	101	(256)	1955	A. Ghiorso (USA) G. R. Choppin (USA) G. T. Seaborg (USA) B. G. Harvey (USA) S. G. Thompson (USA)	Mendeleev, Russian chemist who prepared the periodic chart and predicted properties of undiscovered elements
Mercury	Hg	80	200.6	Ancient		Symbol, L. *hydrargyrum,* liquid silver

(Continued)

Element	Symbol	Atomic No.	Atomic Mass	Date of Discovery	Discoverer and Nationality	Derivation
Molybdenum	Mo	42	95.94	1778	G. W. Scheele (Swe.)	Gr. *molybdos,* lead
Neodymium	Nd	60	144.2	1885	C. A. von Welsbach (Au.)	Gr. *neos,* new; *didymos,* twin
Neon	Ne	10	20.183	1898	Sir William Ramsay (GB) M. W. Travers (GB)	Gr. *neos,* new
Neptunium	Np	93	(237)	1940	E. M. McMillan (USA) P. H. Abelson (USA)	Planet Neptune
Nickel	Ni	28	58.69	1751	A. F. Cronstedt (Swe.)	Swe. *kopparnickel,* false copper; also Ge. *nickel,* referring to the devil that prevented copper from being extracted from nickel ores
Niobium	Nb	41	92.91	1801	Charles Hatchett (GB)	Gr. *Niobe,* daughter of Tantalus (niobium was considered identical to tantalum, named after *Tantalus,* until 1884; originally called columbium, with symbol Cb)
Nitrogen	N	7	14.01	1772	Daniel Rutherford (GB)	Fr. *nitrogene,* derived from L. *nitrum,* native soda, or Gr. *nitron,* native soda, and Gr. *genes,* forming
Nobelium	No	102	(253)	1958	A. Ghiorso (USA) T. Sikkeland (USA) J. R. Walton (USA) G. T. Seaborg (USA)	Alfred Nobel
Osmium	Os	76	190.2	1803	S. Tennant (GB)	Gr. *osme,* odor
Oxygen	O	8	16.00	1774	Joseph Priestley (GB) C. W. Scheele (Swe.)	Fr. *oxygene,* generator of acid, derived from Gr. *oxys,* acid, and L. *genes,* forming (because it was once thought to be a part of all acids)
Palladium	Pd	46	106.4	1803	W. H. Wollaston (GB)	Asteroid Pallas
Phosphorus	P	15	30.97	1669	H. Brandt (Ge.)	Gr. *phosphoros,* light bearing
Platinum	Pt	78	195.1	1735 1741	A. de Ulloa (Sp.) Charles Wood (GB)	Sp. *platina,* silver
Plutonium	Pu	94	(242)	1940	G. T. Seaborg (USA) E. M. McMillan (USA) J. W. Kennedy (USA) A. C. Wahl (USA)	Planet Pluto
Polonium	Po	84	(210)	1898	Marie Curie (P.)	Poland
Potassium	K	19	39.10	1807	Sir Humphry Davy (GB)	Symbol, L. *kalium,* potash
Praseodymium	Pr	59	140.9	1885	C. A. von Welsbach (Au.)	Gr. *prasios,* green; *didymos,* twin
Promethium	Pm	61	(147)	1945	J. A. Marinsky (USA) L. E. Glendenin (USA)	Gr. mythology, *Prometheus,* the Greek

Element	Symbol	Atomic No.	Atomic Mass	Date of Discovery	Discoverer and Nationality	Derivation
					C. D. Coryell (USA)	Titan who stole fire from heaven
Protactinium	Pa	91	(231)	1917	O. Hahn (Ge.) L. Meitner (Au.)	Gr. *protos,* first; *actinium* (because it disintegrates into actinium)
Radium	Ra	88	(226)	1898	Pierre and Marie Curie (Fr.; P.)	L. *radius,* ray
Radon	Rn	86	(222)	1900	F. E. Dorn (Ge.)	Derived from radium with suffix "on" common to inert gases (once called nitron, meaning shining, with symbol Nt)
Rhenium	Re	75	186.2	1925	W. Noddack (Ge.) I. Tacke (Ge.) Otto Berg (Ge.)	L. *Rhenus,* Rhine
Rhodium	Rh	45	102.9	1804	W. H. Wollaston, (GB)	Gr. *rhodon,* rose (because some of its salts are rose-colored)
Rubidium	Rb	37	85.47	1861	R. W. Bunsen (Ge.) G. Kirchhoff (Ge.)	L. *rubidius,* dark red (discovered with the spectroscope, its spectrum shows red lines)
Ruthenium	Ru	44	101.1	1844	K. K. Klaus (R.)	L. *Ruthenia,* Russia
Samarium	Sm	62	150.4	1879	Lecoq de Boisbaudran (Fr.)	Samarskite, after Samarski, Samarski, a Russian engineer
Scandium	Sc	21	44.96	1879	L. F. Nilson (Swe.)	Scandinavia
Selenium	Se	34	78.96	1817	J. J. Berzelius (Swe.)	Gr. *selene,* moon (because it resembles tellurium, named for the earth)
Silicon	Si	14	28.09	1824	J. J. Berzelius (Swe.)	L. *silex, silicis,* flint
Silver	Ag	47	107.9	Ancient		Symbol, L. *argentum,* silver
Sodium	Na	11	22.99	1807	Sir Humphry Davy (GB)	L. *sodanum,* headache remedy; symbol, L. *natrium,* soda
Strontium	Sr	38	87.62	1808	Sir Humphry Davy (GB)	Strontian, Scotland, derived from mineral strontionite
Sulfur	S	16	32.07	Ancient		L. *sulphurium* (Sanskrit, *sulvere*)
Tantalum	Ta	73	180.9	1802	A. G. Ekeberg (Swe.)	Gr. mythology, *Tantalus,* because of difficulty in isolating it (Tantalus, son of Zeus, was punished by being forced to stand up to his chin in water which receded whenever he tried to drink)
Technetium	Tc	43	(99)	1937	C. Perrier (I.)	Gr. *technetos,* artificial (because it was the first artificial element)
Tellurium	Te	52	127.6	1782	F. J. Müller (Au.)	L. *tellus,* earth

(Continued)

Element	Symbol	Atomic No.	Atomic Mass	Date of Discovery	Discoverer and Nationality	Derivation
Terbium	Tb	65	158.9	1843	C. G. Mosander (Swe.)	Ytterby, Sweden
Thallium	Tl	81	204.4	1861	Sir William Crookes (GB)	Gr. *thallos,* a budding twig (because its spectrum shows a bright green line)
Thorium	Th	90	232.0	1828	J. J. Berzelius (Swe.)	Mineral thorite, derived from *Thor,* Norse god of war
Thulium	Tm	69	168.9	1879	P. T. Cleve (Swe.)	*Thule,* early name for Scandinavia
Tin	Sn	50	118.7	Ancient		Symbol, L. *stannum,* tin
Titanium	Ti	22	47.88	1791	W. Gregor (GB)	Gr. giants, the Titans, and L. *titans,* giant deities
Tungsten	W	74	183.9	1783	J. J. and F. de Elhuyar (Sp.)	Swe. *tung sten,* heavy stone; symbol, wolframite, a mineral
Uranium	U	92	238.0	1789 1841	M. H. Klaproth (Ge.) E. M. Peligot (Fr.)	Planet Uranus
Vanadium	V	23	50.94	1801 1830	A. M. del Rio (Sp.) N. G. Sefstrom (Swe.)	*Vanadis,* Norse goddess of love and beauty
Xenon	Xe	54	131.3	1898	Sir William Ramsay (GB) M. W. Travers (GB)	Gr. *xenos,* stranger
Ytterbium	Yb	70	173.0	1907	G. Urbain (Fr.)	Ytterby, Sweden
Yttrium	Y	39	88.91	1843	C. G. Mosander (Swe.)	Ytterby, Sweden
Zinc	Zn	30	65.39	1746	A. S. Marggraf (Ge.)	Ge. *zink,* of obscure origin
Zirconium	Zr	40	91.22	1789	M. H. Klaproth (Ge.)	Zircon, in which it was found, derived from *Ar. zargum,* gold color

A P P E N D I X 2

UNITS FOR THE GAS CONSTANT

In this appendix we will see how the gas constant R can be expressed in units J/K · mol. Our first step is to derive a relationship between atm and pascal. We start with

$$\text{pressure} = \frac{\text{force}}{\text{area}}$$

$$= \frac{\text{mass} \times \text{acceleration}}{\text{area}}$$

$$= \frac{\text{volume} \times \text{density} \times \text{acceleration}}{\text{area}}$$

$$= \text{length} \times \text{density} \times \text{acceleration}$$

By definition, the standard atmosphere is the pressure exerted by a column of mercury exactly 76 cm high of density 13.5951 g/cm^3, in a place where acceleration due to gravity is 980.665 cm/s^2. However, to express pressure in N/m^2 it is necessary to write

$$\text{density of mercury} = 1.35951 \times 10^4 \text{ kg/m}^3$$

$$\text{acceleration due to gravity} = 9.80665 \text{ m/s}^2$$

The standard atmosphere is given by

$$\begin{aligned} 1 \text{ atm} &= (0.76 \text{ m Hg})(1.35951 \times 10^4 \text{ kg/m}^3)(9.80665 \text{ m/s}^2) \\ &= 101325 \text{ kg m/m}^2 \cdot \text{s}^2 \\ &= 101325 \text{ N/m}^2 \\ &= 101325 \text{ Pa} \end{aligned}$$

From Section 5.5 we see that the gas constant R is given by 0.082057 L · atm/K · mol. Using the conversion factors

$$1 \text{ L} = 1 \times 10^{-3} \text{ m}^3$$

$$1 \text{ atm} = 101325 \text{ N/m}^2$$

we write

$$\begin{aligned} R &= \left(0.082057 \frac{\text{L atm}}{\text{K mol}}\right)\left(\frac{1 \times 10^{-3} \text{ m}^3}{1 \text{ L}}\right)\left(\frac{101325 \text{ N/m}^2}{1 \text{ atm}}\right) \\ &= 8.314 \frac{\text{N m}}{\text{K mol}} \\ &= 8.314 \frac{\text{J}}{\text{K mol}} \end{aligned}$$

and

$$\begin{aligned} 1 \text{ L} \cdot \text{atm} &= (1 \times 10^{-3} \text{ m}^3)(101325 \text{ N/m}^2) \\ &= 101.3 \text{ N m} \\ &= 101.3 \text{ J} \end{aligned}$$

A P P E N D I X 3

SELECTED THERMODYNAMIC DATA AT 1 atm AND 25°C

Inorganic Substances

Substance	ΔH_f° (kJ/mol)	ΔG_f° (kJ/mol)	S° (J/K · mol)
$Ag(s)$	0	0	42.7
$Ag^+(aq)$	105.9	77.1	73.9
$AgCl(s)$	−127.0	−109.7	96.1
$AgBr(s)$	−99.5	−95.9	107.1
$AgI(s)$	−62.4	−66.3	114.2
$AgNO_3(s)$	−123.1	−32.2	140.9
$Al(s)$	0	0	28.3
$Al^{3+}(aq)$	−524.7	−481.2	−313.38
$Al_2O_3(s)$	−1669.8	−1576.4	50.99
$As(s)$	0	0	35.15
$AsO_4^{3-}(aq)$	−870.3	−635.97	−144.77
$AsH_3(g)$	171.5		
$H_3AsO_4(s)$	−900.4		
$Au(s)$	0	0	47.7
$Au_2O_3(s)$	80.8	163.2	125.5
$AuCl(s)$	−35.2		
$AuCl_3(s)$	−118.4		
$B(s)$	0	0	6.5
$B_2O_3(s)$	−1263.6	−1184.1	54.0
$H_3BO_3(s)$	−1087.9	−963.16	89.58
$H_3BO_3(aq)$	−1067.8	−963.3	159.8
$Ba(s)$	0	0	66.9
$Ba^{2+}(aq)$	−538.4	−560.66	12.55
$BaO(s)$	−558.2	−528.4	70.3
$BaCl_2(s)$	−860.1	−810.86	125.5
$BaSO_4(s)$	−1464.4	−1353.1	132.2
$BaCO_3(s)$	−1218.8	−1138.9	112.1
$Be(s)$	0	0	9.5
$BeO(s)$	−610.9	−581.58	14.1

Substance	ΔH_f° (kJ/mol)	ΔG_f° (kJ/mol)	S° (J/K · mol)
$Br_2(l)$	0	0	152.3
$Br^-(aq)$	−120.9	102.8	80.7
$HBr(g)$	−36.2	−53.2	198.48
C(graphite)	0	0	5.69
C(diamond)	1.90	2.87	2.4
$CO(g)$	−110.5	−137.3	197.9
$CO_2(g)$	−393.5	−394.4	213.6
$CO_2(aq)$	−412.9	−386.2	121.3
$CO_3^{2-}(aq)$	−676.3	−528.1	−53.1
$HCO_3^-(aq)$	−691.1	−587.1	94.98
$H_2CO_3(aq)$	−699.7	−623.2	187.4
$CS_2(g)$	115.3	65.1	237.8
$CS_2(l)$	87.9	63.6	151.0
$HCN(aq)$	105.4	112.1	128.9
$CN^-(aq)$	151.0	165.69	117.99
$CO(NH_2)_2(s)$	−333.19	−197.15	104.6
$CO(NH_2)_2(aq)$	−319.2	−203.84	173.85
$Ca(s)$	0	0	41.6
$Ca^{2+}(aq)$	−542.96	−553.0	−55.2
$CaO(s)$	−635.6	−604.2	39.8
$Ca(OH)_2(s)$	−986.6	−896.8	76.2
$CaF_2(s)$	−1214.6	−1161.9	68.87
$CaCl_2(s)$	−794.96	−750.19	113.8
$CaSO_4(s)$	−1432.69	−1320.3	106.69
$CaCO_3(s)$	−1206.9	−1128.8	92.9
$Cd(s)$	0	0	51.46
$Cd^{2+}(aq)$	−72.38	−77.7	−61.09
$CdO(s)$	−254.6	−225.06	54.8
$CdCl_2(s)$	−389.1	−342.59	118.4
$CdSO_4(s)$	−926.17	−820.2	137.2
$Cl_2(g)$	0	0	223.0
$Cl^-(aq)$	−167.2	−131.2	56.5
$HCl(g)$	−92.3	−95.27	187.0
$Co(s)$	0	0	28.45
$Co^{2+}(aq)$	−67.36	−51.46	155.2
$CoO(s)$	−239.3	−213.38	43.9
$Cr(s)$	0	0	23.77
$Cr^{2+}(aq)$	−138.9		
$Cr_2O_3(s)$	−1128.4	−1046.8	81.17
$CrO_4^{2-}(aq)$	−863.16	−706.26	38.49
$Cr_2O_7^{2-}(aq)$	−1460.6	−1257.29	213.8
$Cs(s)$	0	0	82.8
$Cs^+(aq)$	−247.69	−282.0	133.05
$Cu(s)$	0	0	33.3
$Cu^+(aq)$	51.88	50.2	−26.36

(Continued)

Substance	ΔH_f° (kJ/mol)	ΔG_f° (kJ/mol)	S° (J/K · mol)
$Cu^{2+}(aq)$	64.39	64.98	98.7
$CuO(s)$	−155.2	−127.2	43.5
$Cu_2O(s)$	−166.69	−146.36	100.8
$CuCl(s)$	−134.7	−118.8	91.6
$CuCl_2(s)$	−205.85		
$CuS(s)$	−48.5	−49.0	66.5
$CuSO_4(s)$	−769.86	−661.9	113.39
$F_2(g)$	0	0	203.34
$F^-(aq)$	−329.1	−276.48	−9.6
$HF(g)$	−268.6	−270.7	173.5
$Fe(s)$	0	0	27.2
$Fe^{2+}(aq)$	−87.86	−84.9	−113.39
$Fe^{3+}(aq)$	−47.7	−10.5	−293.3
$Fe_2O_3(s)$	−822.2	−741.0	90.0
$Fe(OH)_2(s)$	−568.19	−483.55	79.5
$Fe(OH)_3(s)$	−824.25		
$H(g)$	218.2	203.2	114.6
$H_2(g)$	0	0	131.0
$H^+(aq)$	0	0	0
$OH^-(aq)$	−229.94	−157.30	−10.5
$H_2O(g)$	−241.8	−228.6	188.7
$H_2O(l)$	−285.8	−237.2	69.9
$H_2O_2(l)$	−187.6	−118.1	?
$Hg(l)$	0	0	77.4
$Hg^{2+}(aq)$		−164.38	
$HgO(s)$	−90.7	−58.5	72.0
$HgCl_2(s)$	−230.1		
$Hg_2Cl_2(s)$	−264.9	−210.66	196.2
$HgS(s)$	−58.16	−48.8	77.8
$HgSO_4(s)$	−704.17		
$Hg_2SO_4(s)$	−741.99	−623.92	200.75
$I_2(s)$	0	0	116.7
$I^-(aq)$	55.9	51.67	109.37
$HI(g)$	25.9	1.30	206.3
$K(s)$	0	0	63.6
$K^+(aq)$	−251.2	−282.28	102.5
$KOH(s)$	−425.85		
$KCl(s)$	−435.87	−408.3	82.68
$KClO_3(s)$	−391.20	−289.9	142.97
$KClO_4(s)$	−433.46	−304.18	151.0
$KBr(s)$	−392.17	−379.2	96.4
$KI(s)$	−327.65	−322.29	104.35
$KNO_3(s)$	−492.7	−393.1	132.9
$Li(s)$	0	0	28.0
$Li^+(aq)$	−278.46	−293.8	14.2

Substance	ΔH_f° (kJ/mol)	ΔG_f° (kJ/mol)	S° (J/K · mol)
$Li_2O(s)$	−595.8	?	?
$LiOH(s)$	−487.2	−443.9	50.2
$Mg(s)$	0	0	32.5
$Mg^{2+}(aq)$	−461.96	−456.0	−117.99
$MgO(s)$	−601.8	−569.6	26.78
$Mg(OH)_2(s)$	−924.66	−833.75	63.1
$MgCl_2(s)$	−641.8	−592.3	89.5
$MgSO_4(s)$	−1278.2	−1173.6	91.6
$MgCO_3(s)$	−1112.9	−1029.3	65.69
$Mn(s)$	0	0	31.76
$Mn^{2+}(aq)$	−218.8	−223.4	−83.68
$MnO_2(s)$	−520.9	−466.1	53.1
$N_2(g)$	0	0	191.5
$N_3^-(aq)$	245.18	?	?
$NH_3(g)$	−46.3	−16.6	193.0
$NH_4^+(aq)$	−132.80	−79.5	112.8
$NH_4Cl(s)$	−315.39	−203.89	94.56
$NH_3(aq)$	−366.1	−263.76	181.17
$N_2H_4(l)$	50.4		
$NO(g)$	90.4	86.7	210.6
$NO_2(g)$	33.85	51.8	240.46
$N_2O_4(g)$	9.66	98.29	304.3
$N_2O(g)$	81.56	103.6	219.99
$HNO_2(aq)$	−118.8	−53.6	
$HNO_3(l)$	−173.2	−79.9	155.6
$NO_3^-(aq)$	−206.57	−110.5	146.4
$Na(s)$	0	0	51.05
$Na^+(aq)$	−239.66	−261.87	60.25
$Na_2O(s)$	−415.89	−376.56	72.8
$NaCl(s)$	−411.0	−384.0	72.38
$NaI(s)$	−288.0		
$Na_2SO_4(s)$	−1384.49	−1266.8	149.49
$NaNO_3(s)$	−466.68	−365.89	116.3
$Na_2CO_3(s)$	−1130.9	−1047.67	135.98
$NaHCO_3(s)$	−947.68	−851.86	102.09
$Ni(s)$	0	0	30.1
$Ni^{2+}(aq)$	−64.0	−46.4	159.4
$NiO(s)$	−244.35	−216.3	38.58
$Ni(OH)_2(s)$	−538.06	−453.1	79.5
$O(g)$	249.4	230.1	160.95
$O_2(g)$	0	0	205.0
$O_3(aq)$	−12.09	16.3	110.88
$O_3(g)$	142.2	163.4	237.6
P(white)	0	0	44.0
P(red)	−18.4	13.8	29.3
$PO_4^{3-}(aq)$	−1284.07	−1025.59	−217.57
$P_4O_{10}(s)$	−3012.48		

(Continued)

Substance	ΔH_f° (kJ/mol)	ΔG_f° (kJ/mol)	S° (J/K · mol)
$PH_3(g)$	9.25	18.2	210.0
$HPO_4^{2-}(aq)$	−1298.7	−1094.1	−35.98
$H_2PO_4^-(aq)$	−1302.48	−1135.1	89.1
$Pb(s)$	0	0	64.89
$Pb^{2+}(aq)$	1.6	24.3	21.3
$PbO(s)$	−217.86	−188.49	69.45
$PbO_2(s)$	−276.65	−218.99	76.57
$PbCl_2(s)$	−359.2	−313.97	136.4
$PbS(s)$	−94.3	−92.68	91.2
$PbSO_4(s)$	−918.4	−811.2	147.28
$Pt(s)$	0	0	41.84
$PtCl_4^{2-}(aq)$	−516.3	−384.5	175.7
$Rb(s)$	0	0	69.45
$Rb^+(aq)$	−246.4	−282.2	124.27
S(rhombic)	0	0	31.88
S(monoclinic)	0.30	0.10	32.55
$SO_2(g)$	−296.1	−300.4	248.5
$SO_3(g)$	−395.2	−370.4	256.2
$SO_3^{2-}(aq)$	−624.25	−497.06	43.5
$SO_4^{2-}(aq)$	−907.5	−741.99	17.15
$H_2S(g)$	−20.15	−33.0	205.64
$HSO_3^-(aq)$	−627.98	−527.3	132.38
$HSO_4^-(aq)$	−885.75	−752.87	126.86
$H_2SO_4(l)$	−811.3	?	?
$SF_6(g)$	−1096.2	?	?
$Se(s)$	0	0	42.44
$SeO_2(s)$	−225.35		
$H_2Se(g)$	29.7	15.90	218.9
$Si(s)$	0	0	18.70
$SiO_2(s)$	−859.3	−805.0	41.84
$Sr(s)$	0	0	54.39
$Sr^{2+}(aq)$	−545.5	−557.3	39.33
$SrCl_2(s)$	−828.4	−781.15	117.15
$SrSO_4(s)$	−1444.74	−1334.28	121.75
$SrCO_3(s)$	−1218.38	−1137.6	97.07
$W(s)$	0	0	33.47
$WO_3(s)$	−840.3	−763.45	83.26
$WO_4^-(aq)$	−1115.45		
$Zn(s)$	0	0	41.6
$Zn^{2+}(aq)$	−152.4	−147.2	106.48
$ZnO(s)$	−348.0	−318.2	43.9
$ZnCl_2(s)$	−415.89	−369.26	108.37
$ZnS(s)$	−202.9	−198.3	57.7
$ZnSO_4(s)$	−978.6	−871.6	124.7

Organic Substances

Substance	Formula	ΔH_f° (kJ/mol)	ΔG_f° (kJ/mol)	S° (J/K · mol)
Acetic acid(*l*)	CH_3COOH	−484.2	−389.45	159.83
Acetaldehyde(*g*)	CH_3CHO	−166.35	−139.08	264.2
Acetone(*l*)	CH_3COCH_3	−246.8	−153.55	198.74
Acetylene(*g*)	C_2H_2	226.6	209.2	200.8
Benzene(*l*)	C_6H_6	49.04	124.5	124.5
Ethanol(*l*)	C_2H_5OH	−276.98	−174.18	161.04
Ethane(*g*)	C_2H_6	−84.7	−32.89	229.49
Ethylene(*g*)	C_2H_4	52.3	68.1	219.45
Formic acid(*l*)	$HCOOH$	−409.2	−346.0	128.95
Glucose(*s*)	$C_6H_{12}O_6$	−1274.5	−910.56	212.1
Methane(*g*)	CH_4	−74.85	−50.8	186.19
Methanol(*l*)	CH_3OH	−238.7	−166.3	126.78
Sucrose(*s*)	$C_{12}H_{22}O_{11}$	−2221.7	−1544.3	360.24

APPENDIX 4

MATHEMATICAL OPERATIONS

Logarithms

Common Logarithms. The concept of the logarithm is an extension of the concept of exponents, which is discussed in Chapter 1. The *common,* or base-10, logarithm of any number is the power to which 10 must be raised to equal the number. The following examples illustrate this relationship:

Logarithm	Exponent
$\log 1 = 0$	$10^0 = 1$
$\log 10 = 1$	$10^1 = 10$
$\log 100 = 2$	$10^2 = 100$
$\log 10^{-1} = -1$	$10^{-1} = 0.1$
$\log 10^{-2} = -2$	$10^{-2} = 0.01$

In each case the logarithm of the number can be obtained by inspection.

Since the logarithms of numbers are exponents, they have the same properties as exponents. Thus, we have

Logarithm	Exponent
$\log AB = \log A + \log B$	$10^A \times 10^B = 10^{A+B}$
$\log \frac{A}{B} = \log A - \log B$	$\frac{10^A}{10^B} = 10^{A-B}$

Furthermore, $\log A^n = n \log A$.

Now suppose we want to find the common logarithm of 6.7×10^{-4}. On most electronic calculators, the number is entered first and then the log key is punched. This operation gives us

$$\log 6.7 \times 10^{-4} = -3.17$$

Note that there are as many digits *after* the decimal point as there are significant figures in the original number. The original number has two significant figures and the ''17'' in −3.17 tells us that the log has two significant figures. Other examples are

Number	Common Logarithm
62	1.79
0.872	−0.0595
1.0×10^{-7}	−7.00

Sometimes (as in the case of pH calculations) it is necessary to obtain the number whose logarithm is known. This procedure is known as taking the antilogarithm; it is simply the reverse of taking the logarithm of a number. Suppose in a certain calculation we have pH = 1.46 and are asked to calculate $[H^+]$. From the definition of pH ($pH = -\log [H^+]$) we can write

$$[H^+] = 10^{-1.46}$$

Many calculators have a key labeled $\log^{-1}$ or INV log to obtain antilogs. Other calculators have a 10^x or y^x key (where x corresponds to −1.46 in our example and y is 10 for base-10 logarithm). Therefore, we find that $[H^+] = 0.035\ M$.

Natural Logarithms. Logarithms taken to the base e instead of 10 are known as natural logarithms (denoted by ln or $\log_e$); e is equal to 2.7183. The relationship between common logarithms and natural logarithms is as follows:

$$\log 10 = 1 \qquad 10^1 = 10$$

$$\ln 10 = 2.303 \qquad e^{2.303} = 10$$

Thus

$$\ln x = 2.303 \log x$$

To find the natural logarithm of 2.27, say, we first enter the number on the electronic calculator and then punch the ln key to get

$$\ln 2.27 = 0.820$$

If no ln key is provided, we can proceed as follows:

$$2.303 \log 2.27 = 2.303 \times 0.356$$
$$= 0.820$$

Sometimes we may be given the natural logarithm and asked to find the number it represents. For example

$$\ln x = 59.7$$

On many calculators, we simply enter the number and punch the e key:

$$e^{59.7} = 8.46 \times 10^{25}$$

The Quadratic Equation

A quadratic equation takes the form

$$ax^2 + bx + c = 0$$

If coefficients a, b, and c are known, then x is given by

$$x = \frac{-b \pm \sqrt{b^2 - 4ac}}{2a}$$

Suppose we have the following quadratic equation:

$$2x^2 + 5x - 12 = 0$$

Solving for x, we write

$$x = \frac{-5 \pm \sqrt{(5)^2 - 4(2)(-12)}}{2(2)}$$

$$= \frac{-5 \pm \sqrt{25 + 96}}{4}$$

Therefore

$$x = \frac{-5 + 11}{4} = \frac{3}{2}$$

and

$$x = \frac{-5 - 11}{4} = -4$$

GLOSSARY*

Absolute temperature scale. A temperature scale that uses the absolute zero of temperature as the lowest temperature. (5.3)

Absolute zero. Theoretically the lowest attainable temperature. (5.3)

Acceptor impurities. Impurities that can accept electrons from semiconductors. (20.3)

Accuracy. The closeness of a measurement to the true value of the quantity that is measured. (1.8)

Acid. A substance that yields hydrogen ions (H^+) when dissolved in water. (2.9)

Acid ionization constant. The equilibrium constant for the acid ionization. (16.1)

Actinide series. Elements that have incompletely filled 5*f* subshells or readily give rise to cations that have incompletely filled 5*f* subshells. (7.10)

Activated complex. The species temporarily formed by the reactant molecules as a result of the collision before they form the product. (13.4)

Activation energy. The minimum amount of energy required to initiate a chemical reaction. (13.4)

Activity series. A summary of the results of many possible displacement reactions. (3.5)

Actual yield. The amount of product actually obtained in a reaction. (4.3)

Adiabatic process. A process in which no heat exchange occurs between the system and its surroundings. (6.7)

Alcohol. An organic compound containing the hydroxyl group —OH. (24.2)

Aldehydes. Compounds with a carbonyl functional group and the general formula RCHO, where R is an H atom, an alkyl, or an aryl group. (24.2)

Aliphatic hydrocarbons. Hydrocarbons that do not contain the benzene group or the benzene ring. (24.1)

Alkali metals. The Group 1A elements (Li, Na, K, Rb, Cs, and Fr). (1.4)

Alkaline earth metals. The Group 2A elements (Be, Mg, Ca, Sr, Ba, and Ra). (1.4)

Alkanes. Hydrocarbons having the general formula C_nH_{2n+2}, where $n = 1, 2, \ldots$. (24.1)

Alkenes. Hydrocarbons that contain one or more carbon-carbon double bonds. They have the general formula C_nH_{2n}, where $n = 2, 3, \ldots$. (24.1)

Alkynes. Hydrocarbons that contain one or more carbon-carbon triple bonds. They have the general formula C_nH_{2n-2} where $n = 2, 3, \ldots$. (24.1)

Allotropes. Two or more forms of the same element that differ significantly in chemical and physical properties. (2.4)

Alloy. A solid solution composed of two or more metals, or of a metal or metals with one or more nonmetals. (20.2)

Alpha particles. See alpha rays.

Alpha (α) rays. Helium ions with a positive charge of +2. (2.2)

Amalgam. An alloy of mercury with another metal or metals. (20.2)

Amines. Organic bases that have the functional group $—NR_2$, where R may be H, an alkyl group, or an aryl group. (24.2)

Amorphous solid. A solid that lacks a regular three-dimensional arrangement of atoms or molecules. (11.7)

Amphoteric oxide. An oxide that exhibits both acidic and basic properties. (8.6)

Amplitude. The vertical distance from the middle of a wave to the peak or trough. (7.1)

Anion. An ion with a net negative charge. (2.5)

Anode. The electrode at which oxidation occurs. (19.2)

Antibonding molecular orbital. A molecular orbital that is of higher energy and lower stability than the atomic orbitals from which it was formed. (10.6)

Aromatic hydrocarbon. Hydrocarbons that contain one or more benzene rings. (24.1)

Aryl group. A group of atoms equivalent to a benzene ring or a set of fused benzene rings, less one hydrogen atom. (24.1)

Atmospheric pressure. The pressure exerted by Earth's atmosphere. (1.7)

Atom. The basic unit of an element that can enter into chemical combination. (2.2)

Atomic mass. The mass of an atom in atomic mass units. (2.3)

Atomic mass unit. A mass exactly equal to $\frac{1}{12}$th the mass of one carbon-12 atom. (2.3)

Atomic number (Z). The number of protons in the nucleus of an atom. (2.3)

Atomic orbital. The probability function that defines the distribution of electron

*The number in parentheses is the number of the section in which the term first appears.

density in space around the atomic nucleus. (7.6)

Atomic radius. One-half the distance between the two nuclei in two adjacent atoms of the same element in a metal. For elements that exist as diatomic units, the atomic radius is one-half the distance between the nuclei of the two atoms in a particular molecule. (8.3)

Aufbau principle. As protons are added one by one to the nucleus to build up the elements, electrons similarly are added to the atomic orbitals. (7.10)

Avogadro's law. At constant pressure and temperature, the volume of a gas is directly proportional to the number of moles of the gas present. (5.3)

Avogadro's number. 6.022×10^{23}; the number of particles in a mole. (2.3)

Barometer. An instrument that measures atmospheric pressure. (5.2)

Base. A substance that yields hydroxide ions (OH^-) when dissolved in water. (2.9)

Base ionization constant. The equilibrium constant for the base ionization. (16.2)

Battery. An electrochemical cell, or often several electrochemical cells connected in series, that can be used as a source of direct electric current at a constant voltage. (19.6)

Beta (β) rays. Electrons. (2.2)

Bimolecular reaction. An elementary step that involves two molecules. (13.5)

Binary acid. An acid that contains only two elements. (15.4)

Binary compounds. A compound formed from just two elements. (2.9)

Bioamplification. The buildup of any poison along a food chain. (20.9)

Boiling point. The temperature at which the vapor pressure of a liquid is equal to the external atmospheric pressure. (11.8)

Bond dissociation energy. The enthalpy change required to break a particular bond in a mole of gaseous diatomic molecules. (9.10)

Bond length. The distance between the centers of two bonded atoms in a molecule. (9.8)

Bond order. The difference between the numbers of electrons in bonding molecular orbitals and antibonding molecular orbitals, divided by two. (10.7)

Bonding molecular orbital. A molecular orbital that is of lower energy and greater stability than the atomic orbitals from which it was formed. (10.6)

Boranes. Compounds containing only boron and hydrogen. (21.3)

Born-Haber cycle. The cycle that relates lattice energies of ionic compounds to ionization energies, electron affinities, heats of sublimation and formation, and bond dissociation energies. (9.3)

Boundary surface diagram. Diagram of the region containing a substantial amount of the electron density (about 90 percent) in an orbital. (7.8)

Boyle's law. The volume of a fixed amount of gas maintained at constant temperature is inversely proportional to the gas pressure. (5.3)

Breeder reactor. A nuclear reactor that produces more fissionable materials than it uses. (23.5)

Brønsted acid. A substance capable of donating a proton. (3.4)

Brønsted base. A substance capable of accepting a proton. (3.4)

Buffer range. The pH range over which a buffer solution is most effective. (16.7)

Buffer solution. A solution of (a) a weak acid or base and (b) its salt; both components must be present. The solution has the ability to resist changes in pH upon the addition of small amounts of either acid or base. (16.7)

Calorimetry. The measurement of heat changes. (6.4)

Carbides. Ionic compounds containing the C_2^{2-} or C^{4-} ion. (21.4)

Carboxylic acids. Acids that contain the carboxyl group —COOH. (24.2)

Catalyst. A substance that increases the rate of a chemical reaction without itself being consumed. (13.6)

Catenation. The ability of the atoms of an element to form bonds with one another. (21.4)

Cathode. The electrode at which reduction occurs. (19.2)

Cation. An ion with a net positive charge. (2.5)

Charles and Gay-Lussac's law. See Charles' law.

Charles' law. The volume of a fixed amount of gas maintained at constant pressure is directly proportional to the absolute temperature of the gas. (5.3)

Chelating agent. A substance that forms complex ions with metal ions in solution. (22.3)

Chemical energy. Energy stored within the structural units of chemical substances. (6.1)

Chemical equilibrium. A chemical state in which no net change can be observed. (3.2)

Chemical formula. An expression showing the chemical composition of a compound in terms of the symbols for the atoms of the elements involved. (2.4)

Chemical kinetics. The area of chemistry concerned with the speeds, or rates, at which chemical reactions occur. (13.1)

Chemical property. Any property of a substance that cannot be studied without converting the substance into some other substance. (1.3)

Chemistry. The science that studies the properties of substances and how substances react with one another. (1.1)

Chiral. Compounds or ions that are not superimposable with mirror images. (22.4)

Chlo-alkali process. The production of chlorine gas by the electrolysis of aqueous NaCl solution. (21.7)

Closed system. A system that allows the exchange of energy (usually in the form of heat) but not mass with its surroundings. (6.1)

Closest packing. The most efficient arrangements for packing atoms, molecules, or ions in a crystal. (11.4)

Cohesion. The intermolecular attraction between like molecules. (11.3)

Colligative properties. Properties of solutions that depend on the number of solute particles in solution and not on the nature of the solute particles. (12.8)

Common ion effect. The shift in equilibrium caused by the addition of a compound having an ion in common with the dissolved substances. (16.6)

Complex ion. An ion containing a central metal cation bonded to one or more molecules or ions. (17.5)

Compound. A substance composed of atoms of two or more elements chemically united in fixed proportions. (1.3)

Concentration of a solution. The amount of solute present in a given quantity of solution. (4.4)

Condensation. The phenomenon of going from the gaseous state to the liquid state. (11.8)

Condensation reaction. A reaction in which two smaller molecules combine to form a larger molecule. Water is invariably one of the products of such a reaction. (24.2)

Conductor. Substance capable of conducting electric current. (20.3)

Conjugate acid-base pair. An acid and its conjugate base or a base and its conjugate acid. (15.1)

Coordinate covalent bond. A bond in which the pair of electrons is supplied by one of the two bonded atoms; also called a dative bond. (9.9)

Coordination compound. A neutral species containing a complex ion. (22.3)

Coordination number. In a crystal lattice it is defined as the number of atoms (or ions) surrounding an atom (or ion). In coordination compounds it is defined as the number of donor atoms surrounding the central metal atom in a complex. (11.4, 22.3)

Copolymerization. The formation of a polymer that contains two or more different monomers. (25.2)

Corrosion. The deterioration of metals by an electrochemical process. (19.7)

Covalent bond. A bond in which two electrons are shared by two atoms. (9.4)

Covalent compounds. Compounds containing only covalent bonds. (9.4)

Critical mass. The minimum mass of fissionable material required to generate a self-sustaining nuclear chain reaction. (23.5)

Critical pressure. The minimum pressure that must be applied to bring about liquefaction at the critical temperature. (11.8)

Critical temperature. The temperature above which a gas will not liquefy. (11.8)

Crystal field splitting. The energy difference between two sets of *d* orbitals of a metal atom in the presence of ligands. (22.5)

Crystal structure. The geometrical order of the lattice points. (11.4)

Crystalline solid. A solid that possesses rigid and long-range order; its atoms, molecules, or ions occupy specific positions. (11.4)

Crystallization. The process in which dissolved solute comes out of solution and forms crystals. (12.6)

Cyanides. Compounds containing the CN^- ion. (21.4)

Dalton's law of partial pressures. The total pressure of a mixture of gases is just the sum of the pressures that each gas would exert if it were present alone. (5.6)

Degenerate orbitals. Orbitals that have the same energy. (7.8)

Delocalized molecular orbitals. Molecular orbitals that are not confined between two adjacent bonding atoms but actually extend over three or more atoms. (10.8)

Denatured protein. Protein that does not exhibit normal biological activities. (25.3)

Density. The mass of a substance divided by its volume. (1.7)

Deposition. The process in which the molecules go directly from the vapor into the solid phase. (11.8)

Desalination. Purification of sea water by the removal of dissolved salts. (12.9)

Diagonal relationship. Similarities between pairs of elements in different groups and periods of the periodic table. (8.6)

Diamagnetic. Repelled by a magnet; a diamagnetic substance contains only paired electrons. (7.9)

Diatomic molecule. A molecule that consists of two atoms. (2.4)

Diffusion. The gradual mixing of molecules of one gas with the molecules of another by virtue of their kinetic properties. (5.8)

Dilution. A procedure for preparing a less concentrated solution from a more concentrated solution. (4.4)

Dimer. A molecule made up of two identical molecules. (12.3)

Dipole moment. The product of charge and the distance between the charges in a molecule. (10.2)

Dipole-dipole forces. Forces that act between polar molecules. (11.2)

Diprotic acid. Each unit of the acid yields two hydrogen ions. (3.4)

Dispersion forces. The attractive forces that arise as a result of temporary dipoles induced in the atoms or molecules; also called London forces. (11.2)

Displacement reaction. An atom or an ion in a compound is replaced by an atom of another element. (3.5)

Disproportionation reaction. A reaction in which an element in one oxidation state is both oxidized and reduced. (3.5)

Donor atom. The atom in a ligand that is bonded directly to the metal atom. (22.3)

Donor impurities. Impurities that provide conduction electrons to semiconductors. (20.3)

Dynamic equilibrium. The condition in which the rate of a forward process is exactly balanced by the rate of a reverse process. (11.8)

Effusion. The process by which a gas under pressure escapes from one compartment of a container to another by passing through a small opening. (5.8)

Electrolysis. A process in which electrical energy is used to cause a nonspontaneous chemical reaction to occur. (19.8)

Electrolyte. A substance that, when dissolved in water, results in a solution that can conduct electricity. (3.2)

Electromagnetic wave. A wave that has an electric field component and a magnetic field component. (7.1)

Electromotive force (emf). The voltage difference between electrodes. (19.2)

Electromotive series. See activity series.

Electron. A subatomic particle that has a very low mass and carries a single negative electric charge. (2.2)

Electron affinity. The energy change when an electron is accepted by an atom (or an ion) in the gaseous state. (8.5)

Electron configuration. The distribution of electrons among the various orbitals in an atom or molecule. (7.9)

Electron density. The probability that an electron will be found at a particular region in an atomic orbital. (7.6)

Electronegativity. The ability of an atom to attract electrons toward itself in a chemical bond. (9.5)

Element. A substance that cannot be separated into simpler substances by chemical means. (1.3)

Elementary steps. A series of simple reactions that represent the progress of the overall reaction at the molecular level. (13.5)

Emission spectra. Continuous or line spectra emitted by substances. (7.3)

Empirical formula. An expression showing the types of elements present and the ratios of the different kinds of atoms. (2.4)

Enantiomers. Optical isomers, that is, compounds and their nonsuperimposable mirror images. (22.4)

Endothermic processes. Processes that absorb heat from the surroundings. (6.2)

Energy. The capacity to do work or to produce change. (1.7)

Enthalpy. A thermodynamic quantity used to describe heat changes taking place at constant pressure. (6.3)

Enthalpy of reaction. The difference between the enthalpies of the products and the enthalpies of the reactants. (6.3)

Enthalpy of solution. The heat generated or absorbed when a certain amount of solute is dissolved in a certain amount of solvent. (6.6)

Entropy. A direct measure of the randomness or disorder of a system. (18.1)

Enzyme. A biological catalyst. (13.6)

Equilibrium. A state in which there are no observable changes as time goes by. (14.1)

Equilibrium constant. A number equal to the ratio of the equilibrium concentrations of products to the equilibrium concentrations of reactants, each raised to the power of its stoichiometric coefficient. (14.2)

Equilibrium vapor pressure. The vapor pressure measured under dynamic equilibrium of condensation and evaporation. (11.8)

Equivalence point. The point at which the acid is completely reacted with or neutralized by the base. (4.6)

Esters. Compounds that have the general formula R′COOR, where R′ can be H or an alkyl group or an aryl group and R is an alkyl group or an aryl group. (24.2)

Ether. An organic compound containing the R—O—R′ linkage, where R and R′ are alkyl and/or aryl groups. (24.2)

Evaporation. The escape of molecules from the surface of a liquid; also called vaporization. (11.8)

Excess reagents. One or more reactants present in quantities greater than those needed to react with the quantity of the limiting reagent. (4.2)

Excited state. A state that has higher energy than the ground state. (7.3)

Exothermic processes. Processes that give off heat to the surroundings. (6.2)

Extensive property. A property that depends on how much matter is being considered. (1.3)

Family. The elements in a vertical column of the periodic table. (1.4)

Faraday. Charge contained in 1 mole of electrons, equivalent to 96,487 coulombs. (19.4)

First law of thermodynamics. Energy can be converted from one form to another, but cannot be created or destroyed. (6.7)

First-order reaction. A reaction whose rate depends on reactant concentration raised to the first power. (13.2)

Fluorescence. The emission of electromagnetic radiation from an atom or a molecule, particularly in the visible region, after an initial absorption of a photon. (22.7)

Formal charge. The difference between the valence electrons in an isolated atom and the number of electrons assigned to that atom in a Lewis structure. (9.7)

Formation constant. The equilibrium constant for the complex ion formation. (17.5)

Fractional crystallization. The separation of a mixture of substances into pure components on the basis of their differing solubilities. (12.6)

Fractional distillation. A procedure for separating liquid components of a solution that is based on their different boiling points. (12.8)

Free energy. The energy available to do useful work. (18.3)

Frequency. The number of waves that pass through a particular point per unit time. (7.1)

Functional group. That part of a molecule characterized by a special arrangement of atoms that is largely responsible for the chemical behavior of the parent molecule. (24.1)

Gamma (γ) rays. High-energy radiation. (2.2)

Gas constant (*R*). The constant that appears in the ideal gas equation ($PV = nRT$). It is usually expressed as 0.08206 L · atm/K · mol, or 8.314 J/K · mol. (5.4)

Geometric isomers. Compounds with the same type and number of atoms and the same chemical bonds but different spatial arrangements; such isomers cannot be interconverted without breaking a chemical bond. (22.4)

Gibbs free energy. See free energy.

Glass. The optically transparent fusion product of inorganic materials that has cooled to a rigid state without crystallizing. (11.7)

Graham's law of diffusion. Under the same conditions of temperature and pressure, rates of diffusion for gaseous substances are inversely proportional to the square roots of their molar masses. (5.8)

Gravimetric analysis. An analytical procedure that involves the measurement of mass. (4.5)

Ground state. The lowest energy state of

a system. (7.3)

Group. The elements in a vertical column of the periodic table. (1.4)

Half-cell reactions. Oxidation and reduction reactions at the electrodes. (19.2)

Half-life. The time required for the concentration of a reactant to decrease to half of its initial concentration. (13.3)

Half reaction. A reaction that explicitly shows electrons involved in either oxidation or reduction. (3.5)

Halogens. The nonmetallic elements in Group 7A (F, Cl, Br, I, and At). (1.4)

Hard water. Water that contains Ca^{2+} and Mg^{2+} ions. (3.3)

Heat. Transfer of energy (usually thermal energy) between two bodies that are at different temperatures. (6.1)

Heat capacity. The amount of heat required to raise the temperature of a given quantity of the substance by one degree Celsius. (6.4)

Heat content. See enthalpy.

Heat of dilution. The heat change associated with the dilution process.

Heat of hydration. The heat change (absorbed or released) associated with the hydration process. (6.6)

Heat of solution. See enthalpy of solution.

Heisenberg uncertainty principle. It is impossible to know simultaneously both the momentum and the position of a particle with certainty. (7.5)

Henry's law. The solubility of a gas in a liquid is proportional to the pressure of the gas over the solution. (12.7)

Hess's law. When reactants are converted to products, the change in enthalpy is the same whether the reaction takes place in one step or in a series of steps. (6.5)

Heterogeneous equilibrium. An equilibrium state in which the reacting species are not all in the same phase. (14.3)

Heterogeneous mixture. The individual components of a mixture remain physically separated and can be seen as separate components. (1.3)

Heteronuclear diatomic molecule. A diatomic molecule containing atoms of different elements. (10.2)

Homogeneous equilibrium. An equilibrium state in which all reacting species are in the same phase. (14.3)

Homogeneous mixture. The composition of the mixture, after sufficient stirring, is the same throughout the solution. (1.3)

Homonuclear diatomic molecule. A diatomic molecule containing atoms of the same element. (10.2)

Homopolymer. Polymer that is made from only one type of monomer. (25.2)

Hund's rule. The most stable arrangement of electrons in subshells is the one with the greatest number of parallel spins. (7.9)

Hybrid orbitals. Atomic orbitals obtained when two or more nonequivalent orbitals of the same atom combine. (10.5)

Hybridization. The process of mixing the atomic orbitals in an atom (usually the central atom) to generate a set of new atomic orbitals. (10.4)

Hydrates. Compounds that have a specific number of water molecules attached to them. (2.9)

Hydration. A process in which an ion or a molecule is surrounded by water molecules arranged in a specific manner. (3.2)

Hydrocarbons. Compounds made up only of carbon and hydrogen. (24.1)

Hydrogen bond. A special type of dipole-dipole interaction between the hydrogen atom bonded to an atom of a very electronegative element (F, N, O) and another atom of one of the three electronegative elements. (11.2)

Hydrogenation. The addition of hydrogen, especially to compounds with double and triple carbon-carbon bonds. (13.6)

Hydrolysis. The reaction of a substance with water that usually changes the pH of the solution. (16.5)

Hydrophilic. Water-liking. (25.3)

Hydrophobic. Water-fearing. (25.3)

Hygroscopic. Having a tendency to absorb water. (20.5)

Hypothesis. A tentative explanation for a set of observations. (1.2)

Ideal gas. A hypothetical gas whose pressure-volume-temperature behavior can be completely accounted for by the ideal gas equation. (5.4)

Ideal gas equation. An equation expressing the relationships among pressure, volume, temperature, and amount of gas ($PV = nRT$, where R is the gas constant). (5.4)

Ideal solution. Any solution that obeys Raoult's law. (12.8)

Indicators. Substances that have distinctly different colors in acidic and basic media. (4.6)

Induced dipole. The separation of positive and negative charges in a neutral atom (or a nonpolar molecule) caused by the proximity of an ion or a polar molecule. (11.2)

Inert complex. A complex ion that undergoes very slow ligand exchange reactions. (22.6)

Inert pair effect. The two relatively stable and unreactive outer *s* electrons. (8.6)

Inorganic compounds. Compounds other than organic compounds. (2.9)

Insulator. A substance incapable of conducting electricity. (20.3)

Intensive property. Properties that do not depend on how much matter is being considered. (1.3)

Interhalogen compounds. Compounds formed between two halogen elements. (21.7)

Intermediate. A species that appears in the mechanism of the reaction (that is, the elementary steps) but not in the overall balanced equation. (13.5)

Intermolecular forces. Attractive forces that exist among molecules. (11.2)

Intramolecular forces. Forces that hold atoms together in a molecule. (11.2)

Ion. A charged particle formed when a neutral atom or group of atoms gain or lose one or more electrons. (2.5)

Ionic bond. The electrostatic force that holds ions together in an ionic compound. (9.2)

Ionic compound. Any neutral compound containing cations and anions. (2.5)

Ionic equation. An equation that shows

dissolved ionic compounds in terms of their free ions. (3.3)

Ionic radius. The radius of a cation or an anion as measured in an ionic compound. (8.3)

Ionization energy. The minimum energy required to remove an electron from an isolated atom (or an ion) in its ground state. (8.4)

Ion pair. A species made up of at least one cation and at least one anion held together by electrostatic forces. (12.9)

Ion-dipole forces. Forces that operate between an ion and a dipole. (11.2)

Ion-product constant. Product of hydrogen ion concentration and hydroxide ion concentration (both in molarity) at a particular temperature. (15.2)

Isoelectronic. Ions, or atoms and ions, that possess the same number of electrons, and hence the same ground-state electron configuration, are said to be isoelectronic. (8.2)

Isolated system. A system that does not allow the transfer of either mass or energy to or from its surroundings. (6.1)

Isotopes. Atoms having the same atomic number but different mass numbers. (2.3)

Joule. Unit of energy given by newtons × meters. (1.7)

Kelvin temperature scale. See absolute temperature scale.

Ketones. Compounds with a carbonyl functional group and the general formula RR′CO, where R and R′ are alkyl and/or aryl groups. (24.2)

Kinetic energy. Energy available because of the motion of an object. (6.1)

Lanthanide series. Elements that have incompletely filled $4f$ subshells or readily give rise to cations that have incompletely filled $4f$ subshells. (7.10)

Lattice energy. The energy required to completely separate one mole of a solid ionic compound into gaseous ions. (6.6)

Lattice points. The positions occupied by atoms, molecules, or ions that define the geometry of a unit cell. (11.4)

Law. A concise verbal or mathematical statement of a relationship between phenomena that is always the same under the same conditions. (1.2)

Law of conservation of energy. The total quantity of energy in the universe is constant. (6.1)

Law of conservation of mass. Matter can be neither created nor destroyed. (2.1)

Law of definite proportions. Different samples of the same compound always contain its constituent elements in the same proportions by mass. (2.7)

Law of multiple proportions. If two elements can combine to form more than one type of compound, the masses of one element that combine with a fixed mass of the other element are in ratios of small whole numbers. (2.7)

Le Chatelier's principle. If an external stress is applied to a system at equilibrium, the system will adjust itself in such a way as to partially offset the stress. (14.6)

Leveling effect. The inability of a solvent to differentiate among the relative strengths of all acids stronger than the solvent's conjugate acid. (15.3)

Lewis acid. A substance that can accept a pair of electrons. (15.6)

Lewis base. A substance that can donate a pair of electrons. (15.6)

Lewis dot symbol. The symbol of an element with one or more dots that represent the number of valence electrons in an atom of the element. (9.1)

Lewis structure. A representation of covalent bonding using Lewis symbols. Shared electron pairs are shown either as lines or as pairs of dots between two atoms, and lone pairs are shown as pairs of dots on individual atoms. (9.4)

Ligand. A molecule or an ion that is bonded to the metal ion in a complex ion. (22.3)

Limiting reagent. The reactant used up first in a reaction. (4.2)

Line spectra. Spectra produced when radiation is absorbed or emitted by substances only at some wavelengths. (7.3)

Liter. The volume occupied by one cubic decimeter. (1.7)

Lone pairs. Valence electrons that are not involved in covalent bond formation. (9.4)

Macroscopic properties. Properties that can be measured directly. (1.6)

Many-electron atoms. Atoms that contain two or more electrons. (7.6)

Mass. A measure of the quantity of matter contained in an object. (1.3)

Mass defect. The difference between the mass of an atom and the sum of the masses of its protons, neutrons, and electrons. (23.2)

Mass number (*A*). The total number of neutrons and protons present in the nucleus of an atom. (2.3)

Matter. Anything that occupies space and possesses mass. (1.3)

Mean free path. The average distance traveled by a molecule between successive collisions. (5.7)

Melting point. The temperature at which solid and liquid phases coexist in equilibrium. (11.8)

Metals. Elements that are good conductors of heat and electricity and have the tendency to form positive ions in ionic compounds. (1.4)

Metalloid. An element with properties intermediate between those of metals and nonmetals. (1.4)

Metallurgy. The science and technology of separating metals from their ores and of compounding alloys. (20.2)

Microscopic properties. Properties that cannot be measured directly without the aid of a microscope or other special instrument. (1.6)

Mineral. A naturally occurring substance with a characteristic range of chemical composition. (20.1)

Miscible. Two liquids that are completely soluble in each other in all proportions are said to be miscible. (12.3)

Mixture. A combination of two or more substances in which the substances retain their identity. (1.3)

Moderator. A substance that can reduce the kinetic energy of neutrons. (23.5)

Molality. The number of moles of solute dissolved in one kilogram of solvent. (12.5)

Molar heat of fusion. The energy (in kilojoules) required to melt one mole of a solid. (11.8)

Molar heat of sublimation. The energy (in kilojoules) required to sublime one mole of a solid. (11.8)

Molar heat of vaporization. The energy (in kilojoules) required to vaporize one mole of a liquid. (11.8)

Molar concentration. See molarity.

Molar mass. The mass (in grams or kilograms) of one mole of atoms, molecules, or other particles. (2.3)

Molar mass of a compound. The mass (in grams or kilograms) of one mole of the compound. (2.4)

Molar solubility. The number of moles of solute in one liter of a saturated solution (mol/L). (17.1)

Molarity. The number of moles of solute in one liter of solution. (4.4)

Mole. The amount of substance that contains as many elementary entities (atoms, molecules, or other particles) as there are atoms in exactly 12 grams (or 0.012 kilograms) of the carbon-12 isotope. (2.3)

Mole fraction. Ratio of the number of moles of one component of a mixture to the total number of moles of all components in the mixture. (5.6)

Mole method. An approach for determining the amount of product formed in a reaction. (4.1)

Molecular equations. Equations in which the formulas of the compounds are written as though all species existed as molecules or whole units. (3.3)

Molecular formula. An expression showing the exact numbers of atoms of each element in a molecule. (2.4)

Molecular mass. The sum of the atomic masses (in amu) present in the molecule. (2.4)

Molecular orbital. An orbital that results from the interaction of the atomic orbitals of the bonding atoms. (10.6)

Molecularity of a reaction. The number of molecules reacting in an elementary step. (13.5)

Molecule. An aggregate of at least two atoms in a definite arrangement held together by special forces. (2.4)

Monatomic ion. An ion that contains only one atom. (2.5)

Monomer. The single repeating unit of a polymer. (25.2)

Monoprotic acid. Each unit of the acid yields one hydrogen ion. (3.4)

Multiple bonds. Bonds formed when two atoms share two or more pairs of electrons. (9.4)

Net ionic equation. An equation that indicates only the ionic species that actually take part in the reaction. (3.3)

Neutralization reaction. A reaction between an acid and a base. (3.3)

Neutron. A subatomic particle that bears no net electric charge. Its mass is slightly greater than that of the proton. (2.2)

Newton. The SI unit for force. (1.6)

Noble gases. The nonmetallic elements in Group 8A (He, Ne, Ar, Kr, Xe, and Rn). With the exception of helium, these elements all have completely filled *p* subshells. (The electron configurations are $1s^2$ for helium and ns^2np^6 for the other noble gases, where *n* is the principal quantum number for the outermost shell.) (8.2)

Nonbonding electrons. Valence electrons that are not involved in covalent bond formation. (9.4)

Nonelectrolyte. A substance that, when dissolved in water, gives a solution that is not electrically conducting. (3.2)

Nonmetals. Elements that are usually poor conductors of heat and electricity. (1.4)

Nonpolar molecule. A molecule that does not possess a dipole moment. (10.2)

Nonvolatile. Does not have a measurable vapor pressure. (12.8)

***n*-Type semiconductors.** Semiconductors that contain donor impurities. (20.3)

Nuclear binding energy. The energy required to break up a nucleus into its component protons and neutrons. (23.2)

Nuclear chain reaction. A self-sustaining sequence of nuclear fission reactions. (23.5)

Nuclear fission. A heavy nucleus (mass number > 200) divides to form smaller nuclei of intermediate mass and one or more neutrons. (23.5)

Nuclear fusion. The combining of small nuclei into larger ones. (23.6)

Nuclear transmutation. The change undergone by a nucleus as a result of bombardment by neutrons or other particles. (23.1)

Nucleotide. The repeating unit in each strand of a DNA molecule which consists of a base-deoxyribose-phosphate linkage. (25.4)

Nucleus. The central core of an atom. (2.2)

Octane number. A measure of gasoline's tendency to cause knocking. (24.2)

Octet rule. An atom other than hydrogen tends to form bonds until it is surrounded by eight valence electrons. (9.4)

Open system. A system that can exchange mass and energy (usually in the form of heat) with its surroundings. (6.1)

Optical isomers. Compounds that are nonsuperimposable mirror images. (22.4)

Orbital. See atomic orbital and molecular orbital.

Ore. The material of a mineral deposit in a sufficiently concentrated form to allow economical recovery of a desired metal. (20.1)

Organic chemistry. The branch of chemistry that deals with carbon compounds. (24.1)

Organic compounds. Compounds that contain carbon, usually in combination with elements such as hydrogen, oxygen, nitrogen, and sulfur. (2.9)

Osmosis. The net movement of solvent molecules through a semipermeable membrane from a pure solvent or from a dilute solution to a more concentrated solution. (12.8)

Osmotic pressure. The pressure required to stop osmosis from pure solvent into the solution. (12.8)

Overvoltage. The additional voltage required to cause electrolysis. (19.8)

Oxidation number. The number of

charges an atom would have in a molecule if electrons were transferred completely in the direction indicated by the difference in electronegativity. (3.5)

Oxidation reaction. The half-reaction that involves the loss of electrons. (3.5)

Oxidation state. See oxidation number.

Oxidizing agent. A substance that can accept electrons from another substance or increase the oxidation numbers in another substance. (3.5)

Oxoacid. Acids containing hydrogen, oxygen, and another element (the central element). (2.9)

Oxoanion. Anions derived from oxoacids. (2.9)

Packing efficiency. Percentage of the unit cell space occupied by the spheres. (11.4)

Paramagnetic. Attracted by a magnet. A paramagnetic substance contains one or more unpaired electrons. (7.9)

Partial pressure. Pressure of one component in a mixture of gases. (5.6)

Pascal. A pressure of one newton per square meter ($1\ N/m^2$). (1.7)

Pauli exclusion principle. No two electrons in an atom can have the same four quantum numbers. (7.9)

Percent composition by mass. The percent by mass of each element in a compound. (2.6)

Percent composition by weight. See percent composition by mass.

Percent ionization. Ratio of ionized acid concentration at equilibrium to the initial concentration of acid. (15.3)

Percent yield. The ratio of actual yield to theoretical yield, multiplied by 100%. (4.3)

Period. A horizontal row of the periodic table. (1.4)

Periodic table. A tabular arrangement of the elements. (1.4)

pH. The negative logarithm of the hydrogen ion concentration. (15.2)

Phase change. Transformation from one phase to another. (11.8)

Phase diagram. A diagram showing the conditions at which a substance exists as a solid, liquid, or vapor. (11.9)

Photon. A particle of light. (7.2)

Physical equilibrium. An equilibrium in which only physical properties change. (14.2)

Physical property. Any property of a substance that can be observed without transforming the substance into some other substance. (1.3)

Pi bond. A covalent bond formed by sideways overlapping orbitals; its electron density is concentrated above and below the plane of the nuclei of the bonding atoms. (10.5)

Pi molecular orbital. A molecular orbital in which the electron density is concentrated above and below the line joining the two nuclei of the bonding atoms. (10.7)

Plane-polarized light. Light in which the electric field and magnetic field components are confined to specific planes. (22.4)

Plasma. A state of matter in which a gaseous system consists of positive ions and electrons. (23.6)

Polar molecule. A molecule that possesses a dipole moment. (10.2)

Polarimeter. The instrument for studying interaction between plane-polarized light and chiral molecules. (22.4)

Polarizability. The ease with which the electron density in a neutral atom (or molecule) can be distorted. (11.2)

Polyatomic ion. An ion that contains more than one atom. (2.5)

Polyatomic molecule. A molecule that consists of more than two atoms. (2.4)

Polymer. A chemical species distinguished by a high molar mass, ranging into thousands and millions of grams. (25.1)

Potential energy. Energy available by virtue of an object's position. (6.1)

Precipitate. An insoluble solid that separates from the solution. (3.3)

Precision. The closeness of agreement of two or more measurements of the same quantity. (1.8)

Pressure. Force applied per unit area. (1.7)

Product. The substance formed as a result of a chemical reaction. (3.1)

Proton. A subatomic particle having a single positive electric charge. The mass of a proton is about 1840 times that of an electron. (2.2)

***p*-Type semiconductors.** Semiconductors that contain acceptor impurities. (20.3)

Pyrometallurgy. Metallurgical processes that are carried out at high temperatures. (20.2)

Qualitative. Consisting of general observations about the system. (1.2)

Qualitative analysis. The determination of the types of ions present in a solution. (17.6)

Quantitative. Comprising numbers obtained by various measurements of the system. (1.2)

Quantitative analysis. The determination of the amount of substances present in a sample. (4.4)

Quantum. The smallest quantity of energy that can be emitted (or absorbed) in the form of electromagnetic radiation. (7.1)

Racemic mixture. An equimolar mixture of the two enantiomers. (22.4)

Radiation. The emission and transmission of energy through space in the form of waves. (2.2)

Radical. Any neutral fragment of a molecule containing an unpaired electron. (23.8)

Radioactivity. The spontaneous breakdown of an atom by emission of particles and/or radiation. (2.2)

Raoult's law. The partial pressure of the solvent over a solution is given by the product of the vapor pressure of the pure solvent and the mole fraction of the solvent in the solution. (12.8)

Rare earth series. See lanthanide series.

Rate constant. Constant of proportionality between the reaction rate and the concentrations of reactants. (13.1)

Rate law. An expression relating the rate of a reaction to the rate constant and the concentrations of the reactants. (13.2)

Rate-determining step. The slowest step in the sequence of steps leading to the formation of products. (13.5)

Reactants. The starting substances in a chemical reaction. (3.1)

Reaction mechanism. The sequence of

elementary steps that leads to product formation. (13.5)

Reaction order. The sum of the powers to which all reactant concentrations appearing in the rate law are raised. (13.2)

Reaction quotient. A number equal to the ratio of product concentrations to reactant concentrations, each raised to the power of its stoichiometric coefficient at some point other than equilibrium. (14.5)

Reaction rate. The change in the concentration of reactant or product with time. (13.1)

Redox reaction. A reaction in which there is either a transfer of electrons or a change in the oxidation numbers of the substances taking part in the reaction. (3.5)

Reducing agent. A substance that can donate electrons to another substance or decrease the oxidation numbers in another substance. (3.5)

Reduction reaction. The half-reaction that involves the gain of electrons. (3.5)

Representative elements. Elements in Groups 1A through 7A, all of which have incompletely filled *s* or *p* subshell of highest principal quantum number. (8.2)

Resonance. The use of two or more Lewis structures to represent a particular molecule. (9.8)

Resonance form. See resonance structure. (9.8)

Resonance structure. One of two or more alternative Lewis structures for a single molecule that cannot be described fully with a single Lewis structure. (9.8)

Reverse osmosis. A method of desalination using high pressure to force water through a semipermeable membrane from a more concentrated solution to a less concentrated solution. (12.9)

Reversible reaction. A reaction that can occur in both directions. (3.2)

Salt. An ionic compound made up of a cation other than H^+ and an anion other than OH^- or O^{2-}. (3.4)

Salt hydrolysis. The reaction of the anion or cation, or both, of a salt with water. (16.5)

Saponification. Soapmaking. (24.2)

Saturated hydrocarbons. Hydrocarbons that contain only single covalent bonds. (24.1)

Saturated solution. At a given temperature, the solution that results when the maximum amount of a substance has dissolved in a solvent. (12.6)

Scientific method. A systematic approach to research. (1.2)

Second law of thermodynamics. The entropy of the universe increases in a spontaneous process and remains unchanged in an equilibrium process. (18.2)

Second-order reaction. A reaction whose rate depends on reactant concentration raised to the second power or on the concentrations of two different reactants, each raised to the first power. (13.3)

Semiconducting elements. Elements that normally cannot conduct electricity, but can have their conductivity greatly enhanced either by raising the temperature or by adding certain impurities. (20.3)

Semipermeable membrane. A membrane that allows solvent molecules to pass through, but blocks the movement of solute molecules. (12.8)

Sigma bond. A covalent bond formed by orbitals overlapping end-to-end; it has its electron density concentrated between the nuclei of the bonding atoms. (10.5)

Sigma molecular orbital. A molecular orbital in which the electron density is concentrated around a line between the two nuclei of the bonding atoms. (10.7)

Significant figures. The number of meaningful digits in a measured or calculated quantity. (1.8)

Silanes. Binary compounds containing silicon and hydrogen. (21.4)

Silicates. Compounds whose anions consist of Si and O atoms. (21.4)

Simple protein. Protein that contains only amino acids. (25.3)

Soft water. Water that is mostly free of Ca^{2+} and Mg^{2+} ions. (3.3)

Solubility. The maximum amount of solute that can be dissolved in a given quantity of solvent at a specific temperature. (12.4)

Solubility product. The product of the molar concentrations of the constituent ions, each raised to the power of its stoichiometric coefficient in the equilibrium equation. (17.1)

Solute. The substance present in smaller amount in a solution. (3.2)

Solution. A homogeneous mixture of two or more substances. (3.2)

Solvent. The substance present in larger amount in a solution. (3.2)

Specific heat. The amount of heat energy required to raise the temperature of one gram of the substance by one degree Celsius. (6.4)

Spectator ions. Ions that are not involved in the overall reaction. (3.3)

Spectrochemical series. A list of ligands arranged in order of their abilities to split the *d*-orbital energies. (22.5)

Stability constant. See formation constant.

Standard emf. The sum of the standard reduction potential of the substance that undergoes reduction and the oxidation potential of the substance that undergoes oxidation. (19.3)

Standard enthalpy of formation. The heat change that results when one mole of a compound is formed from its elements in their standard states. (6.5)

Standard enthalpy of reaction. The enthalpy change when the reaction is carried out under standard state conditions. (6.5)

Standard oxidation potential. The voltage measured as oxidation occurs at the electrode when all solutes are 1 *M* and all gases are at 1 atm. (19.3)

Standard reduction potential. The voltage measured as a reduction reaction occurs at the electrode when all solutes are 1 *M* and all gases are at 1 atm. (19.3)

Standard solution. A solution of accurately known concentration. (4.6)

Standard state. The condition of 1 atm of pressure. (6.5)

Standard temperature and pressure (STP). 0°C and 1 atm. (5.4)

State function. A property that is determined by the state of the system. (6.7)

State of a system. The values of all pertinent macroscopic variables (for example, composition, volume, pressure, and temperature) of a system. (6.7)

Stereoisomerism. The occurrence of two or more compounds with the same types and numbers of atoms and the same chemical bonds but different spatial arrangements. (22.4)

Stereoisomers. Compounds possessing the same formula and bonding arrangement but different spatial arrangements of atoms. (22.4)

Stoichiometric amounts. The exact molar amounts of reactants and products that appear in the balanced chemical equation. (4.1)

Stoichiometry. The mass relationships among reactants and products in chemical reactions. (4.1)

Structural isomers. Molecules that have the same molecular formula but different structures. (24.1)

Sublimation. The process in which molecules go directly from the solid into the vapor phase. (11.8)

Substance. A form of matter that has a definite or constant composition (the number and type of basic units present) and distinct properties. (1.3)

Supercooling. Cooling of a liquid below its freezing point without forming the solid. (11.8)

Supersaturated solution. A solution that contains more of the solute than is present in a saturated solution. (12.6)

Surface tension. The amount of energy required to stretch or increase the surface of a liquid by a unit area. (11.3)

Surroundings. The rest of the universe outside the system. (6.1)

System. Any specific part of the universe that is of interest to us. (6.1)

Termolecular reaction. An elementary step that involves three molecules. (13.5)

Ternary acid. An acid that contains three different elements. (15.4)

Ternary compounds. Compounds consisting of three elements. (2.9)

Theoretical yield. The amount of product predicted by the balanced equation when all of the limiting reagent has reacted. (4.3)

Theory. A unifying principle that explains a body of facts and the laws that are based on them. (1.2)

Thermal energy. Energy associated with the random motion of atoms and molecules. (6.1)

Thermal pollution. The heating of the environment to temperatures that are harmful to its living inhabitants. (12.6)

Thermochemical equation. An equation that shows both the mass and enthalpy relations. (6.3)

Thermochemistry. The study of heat changes in chemical reactions. (6.2)

Thermodynamics. The scientific study of the interconversion of heat and other forms of energy. (6.7)

Thermonuclear reactions. Nuclear fusion reactions that occur at very high temperatures. (23.6)

Third law of thermodynamics. The entropy of a perfect crystalline substance is zero at the absolute zero of temperature. (18.2)

Titration. The gradual addition of a solution of accurately known concentration to another solution of unknown concentration until the chemical reaction between the two solutions is complete. (4.6)

Tracers. Isotopes, especially radioactive isotopes, that are used to trace the path of the atoms of an element in a chemical or biological process. (23.7)

Transition metals. Elements that have incompletely filled *d* subshells or readily give rise to cations that have incompletely filled *d* subshells. (7.10)

Transuranium elements. Elements with atomic numbers greater than 92. (23.4)

Triple point. The point at which the vapor, liquid, and solid states of a substance are in equilibrium. (11.9)

Unimolecular reaction. An elementary step in which only one reacting molecule participates. (13.5)

Unit cell. The basic repeating unit of the arrangement of atoms, molecules, or ions in a crystalline solid. (11.4)

Unsaturated hydrocarbons. Hydrocarbons that contain carbon-carbon double bonds or carbon-carbon triple bonds. (24.1)

Unsaturated solution. A solution that contains less solute than it has the capacity to dissolve. (12.6)

Valence electrons. The outer electrons of an atom, which are those involved in chemical bonding. (8.2)

Valence shell. The outermost electron-occupied shell of an atom, which holds the electrons that are usually involved in bonding. (10.1)

Valence-shell electron-pair repulsion (VSEPR) model. A model that accounts for the geometrical arrangements of shared and unshared electron pairs around a central atom in terms of the repulsions between electron pairs. (10.1)

Valence-shell expansion. The use of *d* orbitals in addition to *s* and *p* orbitals to form covalent bonds. (10.4)

Van der Waals forces. The dipole-dipole, dipole-induced dipole, and dispersion forces. (11.2)

Van der Waals radius. One-half the distance between two equivalent nonbonded atoms in their most stable arrangement. (11.2)

Vaporization. The escape of molecules from the surface of a liquid; also called evaporation. (11.8)

Viscosity. A measure of a fluid's resistance to flow. (11.3)

Volatile. Has a measurable vapor pressure. (12.8)

Wave. A vibrating disturbance by which energy is transmitted. (7.1)

Wavelength. The distance between identical points on successive waves. (7.1)

Weight. The force that gravity exerts on an object. (1.3)

X-ray diffraction. The scattering of X rays by the units of a regular crystalline solid. (11.5)

Yield of the reaction. The quantity of product obtained from the reaction. (4.3)

ANSWERS TO EVEN-NUMBERED NUMERICAL PROBLEMS

Chapter 1

1.30 1.30×10^3 g. **1.32** 2.3×10^3 cm^3. **1.34** (a) 41°C, (b) 11.3°F, (c) 1.1×10^4°F. **1.38** (a) 0.0152, (b) 0.0000000778. **1.42** (a) 1, (b) 3, (c) 3, (d) 4, (e) 2 or 3, (f) 1, (g) 1 or 2. **1.44** 226 dm. **1.46** 1.10×10^8 mg. **1.48** \$11.50/g. **1.50** 8.3 min. **1.52** (a) 1.8 m; 71.7 kg, (b) 88 km/h, (c) 6.7×10^8 mph, (d) 6.9×10^{-3} g. **1.54** (a) 8.35×10^{12} mi, (b) 2.96×10^3 cm, (c) 9.8×10^8 ft/s, (d) 8.6°C, (e) −459.67°F, (f) 7.12×10^{-5} m^3, (g) 7.2×10^3 L. **1.56** 6.25×10^{-4} g/cm^3. **1.66** (a) 8.08×10^4 g, (b) 1.4×10^{-6} g, (c) 39.9 g. **1.68** 31.35 cm^3.

Chapter 2

2.10 0.62 mi. **2.28** 9.648×10^4 C; 5.486×10^{-4} g. **2.30** 35.45 amu. **2.32** 6.022×10^{23}. **2.34** 5.1×10^{24} amu. **2.36** 9.96×10^{-15} mol. **2.38** 3.01×10^3 g. **2.40** 3.44×10^{-10} g. **2.44** 65.4 g (Zn). **2.56** (a) 16.04 amu, (b) 18.02 amu, (c) 34.02 amu, (d) 78.11 amu, (e) 208.22 amu. **2.58** 893.5 g. **2.62** 8.56×10^{22} molecules. **2.64** H: 6.02×10^{22} atoms; C and O: 3.01×10^{22} atoms. **2.66** 2.1×10^9 molecules. **2.72** 9.24×10^{18} KCN units. **2.78** 78.77% Sn; 21.23% O. **2.80** 10.06% C; 0.8442% H; 89.07% Cl. **2.82** (a) 80.56% C; 7.51% H; 11.93% O, (b) 2.11×10^{21} molecules. **2.84** 0.50 mol Cl_2. **2.86** 39.3 g. **2.88** 5.97 g. **2.92** $C_8H_{10}N_4O_2$. **2.100** 2 peaks. **2.124** 51.9 g/mol. **2.128** 1.6×10^4 g Mb/mol Fe. **2.130** (a) K^+, Br^-: 4.24×10^{22} ions, (b) Na^+: 4.58×10^{22} ions; SO_4^{2-}: 2.29×10^{22} ions, (c) Ca^{2+}: 4.34×10^{22} ions; PO_4^{3-}: 2.89×10^{22} ions. **2.136** (c) ±0.030 amu.

Chapter 4

4.6 1.01 mol. **4.8** (b) 78.3 g. **4.10** 0.324 L. **4.12** 0.294 mol. **4.14** 2.88 mol. **4.16** (b) 2.0×10 g. **4.20** 0.886 mol. **4.22** 0.709 g; 6.9×10^{-3} mol. **4.24** 23.4 g. **4.28** 92.9%. **4.30** 87.2%. **4.34** 0.131 *M*. **4.36** 2.04 *M*. **4.38** (a) 1.16 *M*, (b) 0.608 *M*, (c) 1.78 *M*. **4.40** (a) 1.37 *M*, (b) 0.426 *M*, (c) 0.716 *M*. **4.42** (a) 6.50 g, (b) 2.45 g, (c) 2.65 g, (d) 7.36 g, (e) 3.95 g. **4.48** 0.0433 *M*. **4.50** 126 mL. **4.52** 1.09 *M* **4.56** 35.72%. **4.58** 2.31×10^{-4} *M*. **4.62** (a) 42.78 mL, (b) 158.5 mL, (c) 79.23 mL. **4.64** (a) 6.0 mL, (b) 8.0 mL, (c) 23 mL. **4.66** 0.156 *M*. **4.68** 45.1%. **4.70** 0. **4.72** 0.09216 *M*. **4.74** 0.493 g. **4.76** 0.231 mg/mL blood. **4.80** 89.6%. **4.82** Mg_3N_2. **4.84** 1.85×10^5 kg. **4.88** (b) 9.09×10^2 kg. **4.90** 43.97% NaCl; 56.03% KCl. **4.92** 0.773 L. **4.94** 1145 g/mol. **4.96** 1.33 g. **4.98** 0.2525%. **4.100** 37.72%. **4.102** 4.99 grain. **4.104** 86.49%. **4.106** 3.48×10^3 g.

Chapter 5

5.12 74.9 kPa; 15 mmHg. **5.20** 1.30×10^3 mL. **5.22** 457 mmHg. **5.24** 1.41. **5.26** 44.1 L. **5.38** 6.2 atm. **5.40** (b) 0.0820 L atm/K · mol. **5.42** 8.4×10^2 L. **5.44** 9.0 L. **5.46** 71 mL. **5.48** 6.17 L. **5.50** 5.90×10^2 cm^3. **5.52** 2.9×10^{19} molecules. **5.54** 3.3×10^{13} molecules. **5.56** 2.98 g/L. **5.58** 32.0 g/mol. **5.62** 3.88 L. **5.68** 94.7%. **5.70** (b) 1.71×10^3 L. **5.76** (a) $P_{total} = 0.89$ atm, (b) 1.4 L. **5.78** 349 mmHg. **5.80** 19.8 g. **5.82** $P_{N_2} = 217$ mmHg; $P_{H_2} = 650$ mmHg. **5.92** N_2: 471.7 m/s; O_2: 441.4 m/s; O_3: 360.4 m/s. **5.94** 4.2×10^5 m^2/s^2. **5.100** 13 g/mol. **5.102** 2.74. **5.104** 4. **5.110** 18.5 atm. **5.118** (b) 0.273 g. **5.120** $P_{Ne} = 2.09$ atm; $P_{SF_6} = 1.24$ atm. **5.122** 54.4% CO, 45.6% CO_2. **5.126** 33.4%. **5.128** 716 **5.130** (a) 61.3 m/s, (b) 4.57×10^{-4} s, (c) 366.1 m/s.

Chapter 6

6.16 (a) −2905.6 kJ, (b) 1452.8 kJ, (c) −1276.8 kJ. **6.22** −3.31 kJ. **6.26** 24.8 kJ/g Mg; 603 kJ/mol Mg. **6.28** 25.03°C. **6.38** 0.30 kJ. **6.40** −238.7 kJ/mol. **6.42** 86 kJ/mol. **6.46** 177.8 kJ. **6.48** (a) −571.6 kJ, (b) −2599 kJ, (c) −1411 kJ, (d) −1124 kJ. **6.50** -3.43×10^4 kJ. **6.52** −3924 kJ. **6.54** 2.70×10^2 kJ. **6.70** −3.1 kJ. **6.72** −44.35 kJ/mol. **6.74** 9.9 kJ; −175.3 kJ. **6.78** 0. **6.80** −17 J. **6.82** −367 kJ. **6.84** −238.7 kJ/mol. **6.86** (a) 1.34 kJ/°C, (b) −2799 kJ/mol; $\Delta H_f^\circ = -1277$ kJ/mol, (c) −1347.2 kJ/mol; $\Delta H_f^\circ = -690.7$ kJ/mol, (d) $\Delta H_{rxn}^\circ = -104.4$ kJ. **6.90** 1.09×10^4 L. **6.92** 4.10 L.

Chapter 7

7.8 7.0 × 10^2 s. **7.10** (a) 6.58 × 10^{14} s^{-1}, (b) 1.36 × 10^8 nm. **7.12** 3.26 × 10^7 nm; microwave region. **7.16** 3.19 × 10^{-19} J. **7.18** (a) 5.0 × 10^{12} nm, (b) 4.0 × 10^{-29} J, (c) 2.4 × 10^{-5} J. **7.20** 1.29 × 10^{-15} J. **7.34** (a) 1.4 × 10^2 nm, (b) 5 × 10^{-19} J, (c) 2.0 × 10^2 nm. **7.36** 3.027 × 10^{-19} J. **7.38** 2 × 10^2 J/mol. **7.98** 121 nm; 365 nm. **7.100** 419 nm. **7.104** (a) 1.20 × 10^{18} photons, (b) 3.76 × 10^8 W. **7.108** 483 nm.

Chapter 8

8.58 5.25 × 10^3 kJ/mol. **8.80** 0.65. **8.84** 574%.

Chapter 9

9.26 2197 kJ/mol. **9.76** 392 kJ/mol. **9.78** (a) −119 kJ, (b) −137.0 kJ. **9.82** (a) 225 kJ, (b) 163 kJ, (c) 71 kJ. **9.100** 689 kJ/mol.

Chapter 10

10.90 (a) 108 kJ.

Chapter 11

11.50 10.50 g/cm^3. **11.52** 2. **11.54** 6.21 × 10^{23} atom/mol. **11.56** 458 pm. **11.58** 8. **11.68** 0.220 nm. **11.96** 2670 kJ. **11.106** 60.1 kJ/mol.

Chapter 12

12.14 (a) 7.03%, (b) 16.9%, (c) 13%. **12.16** $X_{benzene}$ = 0.482, $X_{toluene}$ = 0.518. **12.20** (a) 1.74 *m*, (b) 0.87 *m*, (c) 6.99 *m*. **12.22** 5.0 × 10^2 *m*; 18.3 *M*. **12.24** (a) 2.41 *m*, (b) 2.13 *M*, (c) 0.0587 L, (d) 0.958. **12.32** 45.9 g. **12.40** 1.0 × 10^{-3} *M*. **12.44** 0.28 L. **12.60** 30.8 mmHg. **12.62** 88.6 mmHg. **12.64** (a) $P_{methanol}$ = 46 mmHg; $P_{ethanol}$ = 22 mmHg, (b) $X_{methanol}$ = 0.68; $X_{ethanol}$ = 0.32. **12.66** 128 g. **12.68** 0.59 *m*. **12.70** 3.93 L; 105.6°C. **12.72** 108.1 g/mol. **12.74** 120 g/mol. **12.78** 1.75 × 10^4 g/mol. **12.90** (a) −10.0°C; 102.8°C, (b) −7.15°C; 102.0°C. **12.92** 2.47. **12.100** 3.5 atm. **12.102** 34 atm. **12.104** NaCl: 143.8 g; $MgCl_2$: 5.14 g; Na_2SO_4: 7.25 g; $CaCl_2$: 1.11 g; KCl: 0.67 g; $NaHCO_3$: 0.17 g. **12.106** (a) P_A = 38 mmHg, P_B = 66 mmHg, (b) P_A = 22 mmHg, P_B = 94 mmHg. **12.108** 2.95 × 10^3 g/mol.

Chapter 13

13.18 0.213 s^{-1}. **13.20** 0.0896 min^{-1}. **13.22** (b) 1.2/*M* · s, (c) 2.4 × 10^{-4} *M*/s. **13.26** (b) 0.13 *M*/s. **13.32** (a) 0.0198 s^{-1}, (b) 151 s. **13.34** 1.12 × 10^3 min. **13.42** E_a = 104 kJ/mol. **13.44** 135 kJ/mol. **13.46** 644 K. **13.48** 51.9 kJ/mol. **13.60** (b) 1.9 × 10^{-2}/M^2 · s. **13.80** 10.7/*M* · s. **13.84** 0.0504 min^{-1}. **13.88** 265 K.

Chapter 14

14.12 2.40 × 10^{33}. **14.14** 1.76 × 10^{20}. **14.16** 0.051. **14.18** K_P: 0.105; K_c: 2.05 × 10^{-3}. **14.22** (a) 4.3 × 10^{-4}, (b) 1.5, (c) 0.81, (d) 2.3 × 10^3 for (b) and 0.021 for (c). **14.24** 6.2 × 10^{-4}. **14.26** 3.3. **14.28** 9.5 × 10^{-27}. **14.30** 4.7 × 10^9. **14.34** 0.64/M^2 · s. **14.42** P_{NO_2}: 0.020 atm; P_{NO}: 0.50 atm. **14.44** [I] = 8.58 × 10^{-4} *M*; [I_2] = 0.0194 *M*. **14.46** (a) K_c = 0.52, (b) [CO_2] = 0.48 *M*; [H_2] = 0.020 *M*; [CO] = 0.075 *M*; [H_2O] = 0.065 *M*. **14.48** P_{CO} = 1.96 atm; P_{CO_2} = 2.54 atm. **14.50** [H_2] = 0.070 *M*; [I_2] = 0.182 *M*; [HI] = 0.825 *M*. **14.68** (a) P_{Cl_2} = 0.12 atm; P_{NO} = 0.24 atm, (b) 1.7 × 10^{-2}. **14.72** 8 × 10^{44}. **14.74** (a) 0.126, (b) 48.3%, (c) 0.0041 mol. **14.76** 4.76 × 10^{-3}. **14.80** 2.34 × 10^{-3} *M*. **14.82** 328 atm.

Chapter 15

15.24 (a) 6.3 × 10^{-6} *M*, (b) 1.0 × 10^{-16} *M*, (c) 2.7 × 10^{-6} *M*. **15.26** 6.72. **15.28** 2.5 × 10^{-5} *M*. **15.30** 2.2 × 10^{-3} g. **15.62** 0.106 L. **15.72** 1.79.

Chapter 16

16.6 [$C_6H_5COO^-$] = 2.5 × 10^{-3} *M*; [OH^-] = 4.0 × 10^{-12} *M*; [C_6H_5COOH] = 0.10 *M*. **16.8** 1.1 × 10^3. **16.10** 2.3 × 10^{-3} *M*. **16.12** (a) 2.7%, (b) 3.5%, (c) 9.0%, (d) 33%, (e) 79%. **16.14** 0.30%. **16.18** (a) 11.11, (b) 8.96, (c) 12.03. **16.20** 0.15 *M*. **16.24** [H^+] = [SO_4^{2-}] = 0.045 *M*; [HSO_4^-] = 0.16 *M*. **16.26** [H^+] = 1.0 × 10^{-4} *M*; [HCO_3^-] = 1.0 × 10^{-4} *M*; [CO_3^{2-}] = 4.8 × 10^{-11} *M*. **16.36** 9.15. **16.38** 7.00. **16.40** 7.4 × 10^{-3} g. **16.44** (a) 2.57, (b) 4.44. **16.46** (a) [H^+] = 1.3 × 10^{-3} *M*; [CH_3COO^-] = 1.3 × 10^{-3} *M*, (b) [H^+] = 0.100 *M*; [CH_3COO^-] = 1.8 × 10^{-5} *M*. **16.54** 8.88. **16.56** 10. **16.58** 7.03. **16.60** (a) 4.81, (b) 4.64. **16.64** 0.25 *M*. **16.66** 89 g/mol. **16.68** (a) 2.19, (b) 3.95, (c) 8.00, (d) 11.39. **16.70** (a) 110 g/mol, (b) 1.6 × 10^{-6}. **16.80** 1.40; 1.31. **16.86** 0.330 g. **16.88** (a) 1.37, (b) 5.28, (c) 8.85. **16.90** 0.25 g. **16.92** 1.8 × 10^9. **16.94** (a) 12.6, (b) 8.8 × 10^{-6} *M*.

Chapter 17

17.6 (a) [I^-] = 9.1 × 10^{-9} *M*, (b) [Al^{3+}] = 7.4 × 10^{-8} *M*. **17.8** 1.8 × 10^{-11}. **17.10** 4.0 × 10^{-93}. **17.12** 0.036 g/L. **17.16** 2.3 × 10^{-9}. **17.18** (b) 1.6 × 10^{-7} *M*, (c) 1.6 × 10^{-3}%. **17.26** (a) 0.013 *M*, (b) 2.2 × 10^{-4} *M*, (c) 3.4 × 10^{-3} *M*. **17.28** 8.9 × 10^{-10} mol/L. **17.30** 2.4 × 10^{-13}. **17.34** (a) 0.016 *M*, (b) 1.6 × 10^{-6} *M*. **17.42** [Cd^{2+}] = 1.1 × 10^{-18} *M*, [CN^-] = 0.48 *M*, [$Cd(CN)_4^{2-}$] = 4.2 × 10^{-3} *M*. **17.50** (b) 3.2 × 10^{-8}.

Chapter 18

18.4 (a) 0.016, (b) 8.7 × 10^{-19}, (c) ≈0. **18.10** (a) 11.6 J/K, (b) 174.7 J/K, (c) 47.5 J/K, (d) −12.5 J/K, (e) −242.8 J/K. **18.18** (a) 173.4 kJ, (b) 8.6 kJ, (c) −2470 kJ,

(d) −1139 kJ, (e) −140 kJ. **18.20** A: 350 K; B: no temperature; C: all temperatures; D: 111 K. **18.22** 352°C. **18.26** 0.350. **18.28** 79 kJ. **18.30** 4.9×10^{-31}. **18.32** (a) 1.6×10^{-23} atm, (b) 0.535 atm. **18.34** (a) −24.6 kJ, (b) −1.3 kJ. **18.36** 3.1×10^{-2} atm. **18.38** 4.8×10^{-75} atm. **18.40** (a) 8.0×10^4 J, (b) 4.0×10^4 J, (c) -3.2×10^4 J, (d) 6.4×10^4 J. **18.50** 0.074. **18.54** 1.5 mmHg. **18.56** 62.7%.

Chapter 19

19.8 1.97 V. **19.22** 0.367 V. **19.24** 3×10^{54}. **19.26** 0.37 V; −36 kJ; 2×10^6. **19.28** 1.09 V. **19.30** 0.083 V. **19.32** $<6.0 \times 10^{-38}$. **19.38** 1.09 V. **19.46** (b) 0.19 g. **19.50** (a) 0.14 *F*, (b) 0.123 *F*, (c) 0.10 *F*. **19.52** (c) 6.0×10^2 C. **19.54** (a) 0.589 g Cu, (b) 0.133 A. **19.56** 2.2 hr. **19.58** 9.66×10^4 C. **19.60** (a) $\$2.09 \times 10^3$, (b) $\$2.46 \times 10^3$, (c) $\$4.70 \times 10^3$. **19.62** 1.60×10^{-19} C/e^-. **19.66** (a) 0.80 V, i) 0.92 V, ii) 1.10 V. **19.68** 9.8×10^{-12}. **19.74** 0.035 V. **19.78** $[Mg^{2+}] = 0.0500$ *M*; $[Ag^+] = 2 \times 10^{-55}$ *M*; 1.44 g Mg. **19.84** 2×10^{20}. **19.86** (a) −1356.8 kJ, (b) 1.17 V. **19.88** (a) 1.10 V, (b) 5.7×10^{13}.

Chapter 20

20.14 112 h. **20.18** (a) 8.9×10^{12} cm^3, (b) 4.0×10^8 kg. **20.34** 5.59 L. **20.38** (a) 117.6 kJ, (b) 177.8 kJ; $MgCO_3$. **20.66** (c) 2.6 kJ/mol. **20.76** 14 kJ, 4×10^{-3}. **20.80** (a) 48.8 kJ, (b) 1482 kJ, (c) 3152.8 kJ.

Chapter 21

21.14 22.7 g CaH_2. **21.22** 27.8 g/mol. **21.42** 379 g. **21.68** (a) 86.7 kJ/mol, (b) 4×10^{-31}, (c) 4×10^{-31}. **21.70** −1168 kJ. **21.76** 60.4 g. **21.88** (b) 0.371 L. **21.92** 1.2×10^{13} g; 3.7×10^{11} mol. **21.98** 79.1 g. **21.116** 4.63 L.

Chapter 22

22.38 (b) 255 kJ/mol. **22.40** 2. **22.50** 1.6×10^4 g/mol. **22.52** 6.7×10^{13}, −79 kJ. **22.54** 3×10^{13}.

Chapter 23

23.16 2.8×10^{14} g/cm^3. **23.22** (a) 6.30×10^{-12} J, 9.00×10^{-13} J/nucleon, (b) 4.92×10^{-11} J, 1.41×10^{-12} J/nucleon, (c) 4.54×10^{-12} J, 1.14×10^{-12} J/nucleon, (d) 2.63×10^{-10} J, 1.26×10^{-12} J/nucleon. **23.26** 4.89×10^{19} atoms. **23.30** 65.2 yr. **23.32** 3.0×10^9 yr. **23.54** (b) 70.5 disintegrations/min. **23.56** (a) 0.343 millicurie.

Chapter 24

24.44 −174 kJ. **24.48** 8.4×10^3 L. **24.52** 88.7 g/mol.

INDEX

PHOTO CREDITS

Page i: McGraw-Hill photo by Ken Karp.

Chapter 1

Opener: Courtesy of Stanley Smith, University of Illinois, Urbana-Champaign, from the program ''Exploring Chemistry'' by Loretta Jones and Stanley Smith, Falcon Software, Wenthworth, NH. Figure 1.2: Mark Antman/The Image Works. Figure 1.3: McGraw-Hill photo by Ken Karp. Figure 1.7: Fritz Goro, Life Magazine © Time, Inc. Figure 1.11: Courtesy of Mettler. Figure 1.12: Courtesy of The American Museum of Natural History. Page 19: E. R. Degginger.

Chapter 2

Opener: J-L Charmet/Science Photo Library Photo Researchers. Figures 2.3, 2.7 and 2.16: McGraw-Hill photos by Ken Karp. Figure 2.17: a, McGraw-Hill photo by Ken Karp; b, Courtesy of the Diamond Information Center. Page 46: Andrew Popper/Picture Group. Page 49: Both, E. R. Degginger. Page 50: L. V. Bergman & Associates. Page 59: Ward's Natural Science Establishment.

Chapter 3

Opener: Dr. E. R. Degginger. Figures 3.4 and 3.5: McGraw-Hill photos by Ken Karp. Figure 3.6: Courtesy of the Permutit Company, Inc. Figure 3.7: McGraw-Hill photo by Ken Karp. Figure 3.8: Courtesy of IBM. Figure 3.10: a, b, c, McGraw-Hill photos by Ken Karp; d, E. R. Degginger; e, McGraw-Hill photo by Ken Karp. Figures 3.11, 3.12 and 3.13: All, McGraw-Hill photos by Ken Karp. Figure 3.14: Joel Gordon. Figure 3.16: McGraw-Hill photo by Ken Karp. Figure 3.17: Mula & Haramaty/Phototake. Figure 3.18: Joel Gordon. Page 90: Joel Gordon. Pages 99 bottom and 101: McGraw-Hill photos by Ken Karp.

Chapter 4

Opener: Courtesy of Gary E. Thomas, Laboratory of Atmospheric & Space Physics, University of Colorado at Boulder. Photo by Pekka Parviainen. Figure 4.3: Grant Heilman/Grant Heilman Photography. Figures 4.6, 4.7 and 4.8: McGraw-Hill photos by Ken Karp. Figure 4.9: National Draeger, Inc. Pages 133, 137, 138, 140, 149, 153 bottom and 156: McGraw-Hill photos by Ken Karp.

Chapter 5

Opener: Courtesy of General Motors Research Laboratory. Figure 5.12: McGraw-Hill Photo by Ken Karp. Figure 5.16: Jeff Rotman. Figures 5.19 and 5.22: McGraw-Hill photos by Ken Karp. Figure 5.23: James L. Ruhle & Associates, Fullerton, CA. Page 169: McGraw-Hill photo by Ken Karp. Page 174: Courtesy National Scientific Balloon Facility, Palestine, Texas. Page 188: Courtesy of General Motors.

Chapter 6

Opener: Courtesy of Sandia National Laboratories, photo by Cary Chin. Figure 6.2: UPI/Bettmann Newsphotos. Figure 6.6: All, McGraw-Hill photos by Ken Karp. Figure 6.8: E. R. Degginer. Figure 6.12: Courtesy of T. Eisner and Daniel aneshansley, Cornell University. Figure 6.14: Courtesy of Physicians & Nurses Manufacturing Corporation, Larchmont, N.Y. Figure 6.17: Grafton M. Smith/The Image Bank. Page 235: Top, McGraw-Hill photo by Ken Karp; bottom, Courtesy of Diamond Information Center. Page 248: McGraw-Hill photo by Ken Karp.

Chapter 7

Opener: Terry E. Schmidt/TIARA Observatory. Figure 7.7: Joel Gordon. Figure 7.12: Helium, neon, and sodium, McGraw-Hill photos by Ken Karp; mercury, Joel Gordon. Figure 7.15: Professor Ahmed H. Zewail of CAL TECH, Chem. Dept. Figure 7.18: By permission of Sargent-Welch Scientific Company, Skokie, Ill. Figure 7.19: San-il Park, Stanford University, Applied Physics.

Chapter 8

Opener: Science Museum, London. Figure 8.12: Sodium, McGraw-Hill photo by Ken Karp; magnesium and aluminum, L. V. Bergman & Associates; silicon, Frank Wing/Stock, Boston; phosphorus, Life Science Library/MATTER, photo Albert Fenn © 1963 Time-Life Books, Inc.; sulfur, L. V. Bergman & Associates; chlorine and argon, McGraw-Hill photos by Ken Karp. Figure 8.17: Lithium and sodium, McGraw-Hill photos by Ken Karp; potassium, Life Science Library/MATTER, photo Albert Fenn © 1963 Time-Life Books, Inc.; rubidium and cesium, L. V. Bergman & Associates. Figure 8.18: Beryllium, magnesium, calcium, strontium and barium, L. V. Bergman & Associates; radium, Life Science Library/MATTER, photo by Phil Brodatz © Time-Life Books, Inc. Figure 8.19: All, L. V. Bergman & associates. Figure 8.20: Graphite, McGraw-Hill photo by Ken Karp; diamond, Courtesy of Diamond Information Center; silicon, Frank Wing/Stock, Boston; germanium, McGraw-Hill photo by Ken Karp; tin, L. V. Bergman & Associates; lead, McGraw-Hill photo by Ken Karp. Figure 8.21: Nitrogen, Joe McNally/Wheeler Pictures; phosphorus, Life Science Library/MATTER, photo Albert Fenn © 1963 Time-Life Books, Inc.; arsenic, antimony and bismuth, L. V. Bergman & Associates. Figure 8.22: Sulfur and selenium, L. V. Bergman & Associates; tellurium, McGraw-Hill photo by Ken Karp. Figure 8.23: Joel Gordon. Figure 8.24: All, McGraw-Hill photos by Ken Karp.

Chapter 9

Opener: Butterworth and Company, Ltd. 1986; first published in Journal of Molecular Graphics. Vol. 4, No. 3, Sept. 1986. Figure 9.3: Courtesy of International Salt Company. Figure 9.4: Liane Enkelis/Stock, Boston. Figure 9.5: Ward's Natural Science Establishment, Inc. Page 352: McGraw-Hill Photo by Ken Karp.

Chapter 10

Opener: Courtesy Aneel Aggarwal, Ph.D. and Stephen C. Harrison, Department of Biochemistry and Molecular Biology, Harvard University. Figure 10.3: Joel Gordon. Figure 10.21: Donald Clegg.

Chapter 11

Opener: John Eastcott/Yva Momatiuk/Woodfin Camp & Associates. Figure 11.13: Alan Carey/The Image Works. Figure 11.34: Courtesy of Edmund Catalogs. Figure 11.35: Courtesy Railway Technical Research Institute, Tokyo, Japan. Figure 11.37: Hank Morgan/Rainbow. Figure 11.38: Courtesy of Corning Glass Works. Figure 11.44: All, McGraw-Hill photos by Ken Karp. Figure 11.45: Cornell University Photograph Charles Harrington. Figure 11.47: McGraw-Hill photo by Ken Karp. Figure 11.49: General Electric. Figure 11.52: Ken Karp. Figure 11.53: Tim Eagan/Woodfin Camp & Associates.

Chapter 12

Opener: McGraw-Hill photo by Ken Karp. Figures 12.2, 12.4, 12.5 and 12.7: McGraw-Hill photos by Ken Karp. Figure 12.11: Joel Gordon.

Chapter 13

Opener: Henry Lansford/Photo Researchers. Figure 13.3: McGraw-Hill photo by Ken Karp. Figure 13.10: Francois Lochon/Gamma Liaison. Figure 13.19: McGraw-Hill photo by Ken Karp. Figure 13.20: Courtesy of Johnson Matthey. Figure 13.21: McGraw-Hill photo by Ken Karp. Figure 13.23: Courtesy of General Motors Corporation. Figure 13.28: NASA/Goddard Space Flight Center. Figure 13.29: Courtesy of NASA. Page 555: McGraw-Hill photo by Ken Karp.

Chapter 14

Opener: Starlene Frontino/West Light. Figures 14.3, 14.5 and 14.6: McGraw-Hill photos by Ken Karp. Figures 14.1: J. M. Boivin/Gamma Liaison.

Chapter 15

Opener: McGraw-Hill photo by Ken Karp. Figures 15.2, 15.5 left and right and 15.8: McGraw-Hill photos by Ken Karp. Figure 15.9: Owen Franken. Figure 15.10: Landschafts-Verband Westfalen-Lippe, Munich. Figure 15.11: Michael Melford/Wheeler Pictures.

Chapter 16

Opener: Dr. R. Y. Tsier and Dr. T. E. Maeher. Figures 16.4 and 16.10: McGraw-Hill photos by Ken Karp.

Chapter 17

Opener: Courtesy of R. F. Dill. Figures 17.2, 17.3, 17.4 and 17.5: McGraw-Hill photos by Ken Karp. Figure 17.6: Fujimoto/Kodansha. Figure 17.8: John S. Shelton from Geology Illustrated, 1966, W. H. Freeman. Figure 17.9: C. Allen Morgan/Peter Arnold. Pages 723 top and bottom, 724, 725 and 728: McGraw-Hill photos by Ken Karp.

Chapter 18

Opener: Farrell Grehan/Photo Researchers. Figure 18.6: Courtery National Lime Association. Figure 18.10: Both, McGraw-Hill photos by Ken Karp. Page 751: McGraw-Hill photo by Ken Karp.

Chapter 19

Opener: Courtesy of Jessop Steel Company and Van Dine Humphrey Adversiting. Figure 19.4: McGraw-Hill photo by Ken Karp. Figure 19.10: NASA. Figure 19.12: Courtesy of Alupower. Figure 19.13: a, Dr. E. R. Degginger; b, McGraw-Hill photo by Ken Karp; c, Donald Dietz/Stock, Boston. Figures 19.15 and 19.18: McGraw-Hill photos by Ken Karp. Page 790: McGraw-Hill photo by Ken Karp.

Chapter 20

Opener: National Geographic Society, photo by O. Louis Mazzatenta, June 1988, p. 726 top. Figure 20.2: Dr. Bruce Heezen, Lamong-Doherty. Figure 20.5: Jeff Smith. Figure 20.7: Don Green. Kennecott. Figure 20.13: Richard Megna/Fundamental Photographers. Figure 20.14: Ward's Natural Science Establishment. Figure 20.15: Aronson Photo/Stock, Boston. Figure 20.16: James L. Dye. Figures 20.17 and 20.18: Ward's Natural Science Establishment. Figure 20.19: McGraw-Hill photo by Ken Karp. Figures 20.20 and 20.21: Ward's Natural Science Establishment. Figure 20.22: Stuart Cohen/Stock, Boston. Figure 20.23: L. V. Bergman & Associates. Figure 20.24: Ward's Natural Science Establishment. Figure 20.26: E. R. Degginger. Figure 20.29: Both, courtesy of Aluminum Company of America. Figure 20.30: Ward's Natural Science Establishment. Figure 20.31: McGraw-Hill photo by Ken Karp. Figure 20.32: Ward's Natural Science Establishment. Figure 20.35: All, L. V. Bergman & Associates. Figure 20.36: Ward's Natural Science Establishment. Figure 20.37: Courtesy of Yamaha International Corporation. Figure 20.38: Ward's Natural Science Establishment.

Chapter 21

Opener: Courtesy of Courtaulds, London. Figure 21.4: Ward's Natural Science Establishment. Figure 21.9: Courtesy of General Electric Company and Development Center. Figure 21.10: Paul Logsdon. Figure 21.12: Jeff Smith. Figure 21.13: Courtesy of US Department of Energy. Figure 21.16: Bill Belknap. Figure 21.17: Jeff Smith. Figure 21.18: Courtesy of Bell Labs. Figure 21.21: Courtesy of Avondale Research Center, US Bureau of Mines. Figure 21.22: McGraw-Hill photo by Ken Karp. Figure 21.25: Stan Ries/The Picture Cube. Figure 21.26: Courtesy of Amax, Inc. Figure 21.33: L. V. Bergman & Associates. Figure 21.35: C. B. Jones/Taurus Photos. Figure 21.36: a, Dr. E. R. Degginger, FPSA; b, Roger Werth/Longview Daily News/Woodfin Camp & Associates. Figure 21-39: Vulcan Materials Company, photo by Charles Beck. Figure 21.42: Jim Brandenberg. Figure 21.44: Both, Neil Bartlett. Figure 21.45: Courtesy of Argonne National Laboratory. Page 890: Dr. E. R. Degginger, FPSA. Page 894: McGraw-Hill photo by Ken Karp.

Chapter 22

Opener: J. D. Barrie & C. H. Barrie, Jr. Figure 22.5: McGraw-Hill photo by Ken Karp. Figure 22.6: Scandium, titanium, vanadium, chromium, manganese, iron, cobalt, and nickel, McGraw-Hill photos by Ken Karp; copper, L. V. Bergman & Associates. Figure 22.7: All, McGraw-Hill photos by Ken Karp. Figure 22.8: Ward's Natural Science Establishment. Figure 22.9: Both, McGraw-Hill photos by Ken Karp. Figure 22.11: Ward's Natural Science Establishment. Figure 22.12: Grant Heilman. Figure 22.13: McGraw-Hill photo by Ken Karp. Figure 22.14: Ward's Natural Science Establishment. Figure 22.15: Andrew Popper/Picture Cube. Figure 22.23: Joel Gorden. Figure 22.29: McGraw-Hill photo by Ken Karp. Figure 22.41: Fritz Goro Al Lamme'.

Chapter 23

Opener: Courtesy of National Optical Astronomy Observatories, Kitt Peak National Observatory. Figure 23.5: Fermi National Accelerator Laboratory. Figure 23.11: Pierre Kopp/West Light. Figure 23.12: M. Lazarus/Photo Researchers. Figure 23.13: Los Alamos National Laboratory. Figure 23.14: U.S. Department of Energy/Science Photo Library/Photo Researchers. Figure 23.16: W. J. Maeck. Figure 23.18: Lawrence Livermore National Laboratory. Figure 23.19: Department of Defense, Still Media Records Center, U.S. Navy Photo. Figure 23.20: Courtesy of the University of Utah. Figure 23.21: Courtesy B. Leonard Holman, M.D., Brigham & Women's Hospital, Harvard Medical Center. Figure 23.23: Grant Heilman Photography. Page 978: NASA.

Chapter 24

Opener: James Scherer, Boston, MA. Courtesy of Prof. G. M. Whitesides, Harvard University. Figure 24.12: Courtesy of American Petroleum Institute. Page 1005: Dr. E. R. Degginger, FPSA. Page 1006: IBM Photo-Almaden Research Center.

Chapter 25

Opener: Courtesy of General Electric. Figure 25.2: McGraw-Hill photo by Ken Karp. Figure 25.4: Charles Weckler/The Image Bank. Figure 25.7: Donald Klegg. Figure 25.8: William Strode/Humana, Inc./Black Star. Figure 25.20: Courtesy of Dr. Bruce F. Cameron and Dr. Robert Zucker, Miami Comprehensive Sickle Cell Centers. Page 1043: Lawrence Berkeley Laboratory.

Chapter 26

Opener: Jed Share/The Stock Market. Figure 26.2: Herman K. Kokojan/Black Star. Figure 26.4: Hank Morgan/Photo Researchers. Figure 26.5: J. L. Atlan/Sygma. Figure 26.6: Dennis Capolongo/Black Star. Figure 26.7: Alon Reininger/Contact Press Images. Figure 26-8: Courtesy of Eco Corporation, from Ecoplastics Limited, Willowdale, Ontario.

Applied Chemistry Examples Photo: courtesy American Petroleum Institute.

Fundamental Constants

Avogadro's number	6.022×10^{23}
Electron charge (e)	1.6022×10^{-19} C
Electron mass	9.1095×10^{-28} g
Faraday constant (F)	96,487 C/mol electron
Gas constant (R)	8.314 J/K · mol (0.08206 L · atm/K · mol)
Planck's constant (h)	6.6256×10^{-34} J s
Proton mass	1.67252×10^{-24} g
Neutron mass	1.67495×10^{-24} g
Speed of light in vacuum	2.99792458×10^{8} m/s

Useful Conversion Factors and Relationships

1 lb = 453.6 g
1 in = 2.54 cm (exactly)
1 mi = 1.609 km
1 km = 0.6215 mi
1 pm = 1×10^{-12} m = 1×10^{-10} cm
1 atm = 760 mmHg = 760 torr = 101,325 N/m^2 = 101,325 Pa
1 cal = 4.184 J (exactly)
1 L atm = 101.325 J
1 J = 1 C × 1 V

$$?^\circ C = (^\circ F - 32^\circ F) \times \frac{5^\circ C}{9^\circ F}$$

$$?^\circ F = \frac{9^\circ F}{5^\circ C} \times (^\circ C) + 32^\circ F$$

$$K = (^\circ C + 273.15^\circ C)\left(\frac{1\ K}{1^\circ C}\right)$$

Some Prefixes Used with SI Units

Tera (T)	10^{12}
Giga (G)	10^{9}
Mega (M)	10^{6}
Kilo (K)	10^{3}
Deci (d)	10^{-1}
Centi (c)	10^{-2}
Milli (m)	10^{-3}
Micro (μ)	10^{-6}
Nano (n)	10^{-9}
Pico (p)	10^{-12}